Introduction to Modern Planar Transmission Lines

Introduction to Modern Planar Transmission Lines

Physical, Analytical, and Circuit Models Approach

Anand K. Verma
Adjunct Professor, School of Engineering
Macquarie University, Sydney, Australia
Formerly Professor, Department of Electronic Science
South Campus, Delhi University
New Delhi, India

IEEE PRESS

WILEY

This edition first published 2021

Registered Office(s)
John Wiley & Sons, Inc., 111 River Street, Hoboken, NJ 07030, USA
John Wiley & Sons Ltd, The Atrium, Southern Gate, Chichester, West Sussex, PO19 8SQ, UK

Editorial Office
The Atrium, Southern Gate, Chichester, West Sussex, PO19 8SQ, UK

For details of our global editorial offices, customer services, and more information about Wiley products visit us at www.wiley.com.

Wiley also publishes its books in a variety of electronic formats and by print-on-demand. Some content that appears in standard print versions of this book may not be available in other formats.

Library of Congress Cataloging-in-Publication Data:

Names: Verma, Anand K., 1948– author.
Title: Introduction to modern planar transmission lines : physical, analytical, and circuit models approach / Anand K. Verma.
Description: Hoboken, New Jersey : Wiley-IEEE Press, [2021] | Includes bibliographical references and index.
Identifiers: LCCN 2020050198 (print) | LCCN 2020050199 (ebook) | ISBN 9781119632276 (cloth) | ISBN 9781119632450 (adobe pdf) | ISBN 9781119632474 (epub)
Subjects: LCSH: Electric lines.
Classification: LCC TK3221 .V47 2021 (print) | LCC TK3221 (ebook) | DDC 621.3815–dc23
LC record available at https://lccn.loc.gov/2020050198
LC ebook record available at https://lccn.loc.gov/2020050199

Cover Design: Wiley
Cover Image: © DamienGeso/Getty Images

Set in 10/12.5pt STIXTwoText by Straive, Pondicherry, India

To my loving and caring wife Kamini Verma and ever-smiling grandchildren Naina, Tinu, and Nupur.

Contents

Preface

The planar transmission lines form the core of modern high-frequency communication, computer, and other related technology. The subject has come up to the present level of maturity over the past three to four decades. The planar transmission lines are used not only as interconnects on the PCB board and IC chips; these are directly needed for the development of microwave and mm-wave components in the form of microwave integrated circuits (MIC). The types of planar transmission lines, i.e. their physical structures and material medium, have been changing with the growth of technology in many other disciplines. Such efforts during recent years propelled the MIC to move in many exotic directions – MMIC, MEMS, LTCC, use of ferroelectrics, and high-temperature superconductors, optically controlled microwave devices, nonlinear planar transmission lines, DGS, EBG, metamaterials, etc. The researchers with varying backgrounds have contributed much to research activities. Already the divergent planar technology has contributed significantly to the advancement of high-frequency electronics and in the near future, more contribution will be made by it. The exotic planar transmission lines are not covered comprehensively in a single book. The present book is an attempt in this direction. *The proposed book aims to provide a comprehensive discussion of planar transmission lines and their applications. It focuses on physical understanding, analytical approach, and circuit models for planar transmission lines and resonators in the complex environment.*

The present book has evolved from the lecture notes, workshop, seminar presentation, and invited lectures delivered by the author at many universities and R&D centers. Some chapters were also initially written for the Ph.D. students to help them to understand the topics. Finally, it has evolved from notes prepared by the author as a scheme for the self-study. The author started his academic career after 17 years of professional experience in the field of electrical engineering, broadcast transmitters, and satellite communication.

At present, the planar transmission lines are taught as part of the course on RF and microwave packaging, advanced electromagnetic field theory, and microwave design. It is also taught as an independent paper. However, a teacher has to consult divergent sources to prepare the lecture notes, as no single source at the teaching-level is available. Moreover, the classroom teaching of the planar transmission lines is not as systematic as the classical metallic waveguide structures. It is due to the very nature of the subject itself. *The available books are usually not classroom oriented. Usually, they can be grouped into two categories – 1. Design-oriented books, 2. Monograph kind of books.* Once we use the first category of books in the classroom, we end up writing only closed-form expressions without any systematic derivation of the expressions. The systematic approach is important in the classroom environment. The second category of books is suitable for an experienced researcher or specialist. It is difficult to use them in the classroom. Thus, a teacher of this subject has to struggle between these two extremes to balance teaching throughout the semester. Finally, a teacher has to depend on personal experience and lecture notes.

There is a need to present, in one cover, the divergent topics of the planar transmission lines in a student-friendly format. The researchers with varying backgrounds in physics, chemistry, engineering, and other fields have joined activities in the expanding area of the planar technology. The early researchers, R&D professionals in the industry, teachers, and students need a text that could be useful in faster acquisition of the physical modeling process and theoretical formulations used in the classical planar transmission lines. Similar treatment is also needed for the modern engineered EBG and metamaterial lines and surfaces.

Therefore, the real motivation for writing this *intermediate-level* book is to fill the gap for a *textbook* on the planar transmission line that caters to the need for classroom teaching, early researchers with divergent backgrounds, and designers working in the microwave industry. The book is intended to help students both at the undergraduate and postgraduate levels. It also serves the purpose of a resource book for self-study. *The detailed derivations of results and physical modeling of the planar transmission lines are two basic concepts followed through the book*. The present book is neither a design-oriented book nor an advanced monograph. The book correlates the physical process with mathematical treatment. The advanced mathematical methods such as the *conformal mapping method, variational method, and spectral domain method* applied to planar lines are worked out in adequate details. The book further covers modern topics such as the *DGS/EBG, metamaterial-based planar transmission lines,* and *surfaces*. The approach used in writing the book is perhaps less formal than most available texts. This approach is helpful for classroom teaching. It also assists the reader to follow the contemporary developments in planar technology.

Acknowledgments

The author is thankful to Prof. Karu P. Esselle and Prof. Graham Town, *School of Engineering, Macquarie University, Sydney,* for supporting Adjunct Professorship at Macquarie University. The author is also thankful to Prof. Enakshi K. Sharma, *Department of Electronic Science, South Campus, Delhi University,* for continuous discussions on topics related to EM-Theory, Wave propagation, etc. The author sincerely appreciates the help and guidance provided by Prof. Kai Chang, *Department of Electrical Engineering, Texas A&M University* during the review process of the book. The author also appreciates the active interest taken by Mr. Brett Kurzman and his team of *Wiley Publishing* for the review and friendly administrative support.

The author is particularly grateful to Dr. I.J. Bahl, *Editor-in-Chief, Int. J. of RF & Microwave Computer-Aided Engineering, John Wiley, USA*; Prof. Zhongxiang Shen, *School of Electrical and Electronic Engineering, Nanyang Technological University, Singapore*; and Prof. Ladislau Matekovits, *Department of Electronics and Telecommunications, Politecnico di Torino, Italy,* for reading book chapters and providing valuable suggestions to improve the book.

The author has benefitted from the comments, suggestions, and corrections of many colleagues, teachers, and students. The author would like to thank the following people for their useful contributions toward corrections and useful discussions: Dr. Koteshwar Rao, Dr. Harsupreet Kaur, Dr. Kamlesh Patel, Mr. Amit Birwal, Dr. Ashwani Kumar, Dr. Paramjeet Singh, Dr. Y.K. Awasthi, and Mr. Prashant Chaudhary from the *Department of Electronic Science, Delhi University*; Dr. Raheel Hashmi, Dr. Sudipta Chakraborty, and Dr. Rajas Prakash Khokle of *School of Engineering, Macquarie University, Sydney*; Dr. Nasimuddin of *Institute for Infocomm Research, Singapore*; Prof. Asoke De, Dr. Priyanka Jain, Ms. Priyanka Garg of *Department of Electrical Engineering, DTU, Delhi*; Mr. Shailendra Singh of *Product Development and Innovation Center, Bharat Electronics Ltd., Bengaluru, India*; Dr. Rajesh Singh, *Microwave Radiation Laboratory, University of Pisa, Italy*; Dr. Archna Rajput, *IIT Jammu, India;* Dr. Ravi Kumar Arya, *Dept. of ECE, NIT, Delhi.* The author is especially thankful to his students, Mr. Shailendra Singh and Mr. Prashant Chaudhary for their continuous help in correction of all chapters.

The author expresses his unbounded love and regards to his parents – Late Sh. A.P. Verma and Late Tara Devi, Uncle Late Sh. Hira Prasad, to grandmother Late Radhika Devi, and Uncle and Life-guide Sh. Girija Pd. Srivastava. The author also is grateful to his teacher Prof. M.K.P. Mishra for excellent teaching of Circuit Theory in unique style, and for providing support in many ways. The author is thankful to his family members for their encouragement and support. Finally, I wish to express my heartfelt thanks and deepest appreciation to my wife, Kamini. The smiling faces of my grandchildren, Naina, Tinu, and Nupur have always kept me going on with the tiring work of book writing.

Anand K. Verma
New Delhi, Sydney

Author Biography

Anand K. Verma, PhD, is an adjunct professor in the School of Engineering, Macquarie University, Sydney, Australia. Formerly, he was professor and head, Department of Electronic Science, South Campus, University of Delhi, New Delhi, India. He has been a visiting professor at Otto van Guericke University, Magdeburg, Germany (2002–2003), and a Tan Chin Tuan Scholar (2001) at Nanyang Technological University, Singapore. He holds a German patent on a microstrip antenna. He has introduced *the concept and design method of surface-mounted compact horn antenna* used for high gain, wideband, and ultra-wideband quasi-planar antenna applicable to both linear and circular polarization. He has organized and attended many international symposia and workshops. He has conducted short-term courses and delivered invited lectures at the research institutes in India and several countries. He was also chairman of the TPC, APMC-2004, New Delhi, India. Professor Verma has published over 250 papers in international journals and the proceedings of international and national symposia. He has introduced the *concept of single layer reduction (SLR) formulation* for the CAD-oriented modeling of multilayer planar lines.

1

Overview of Transmission Lines

(Historical Perspective, Overview of Present Book)

Introduction

The transmission line is at the core of the communication technology system. It forms a medium for signal transmission, and also helps to develop high-frequency passive components and circuit blocks. Historically, both experimental investigations and analytical theories have played significant roles in the growth of transmission line technology. Each type of distinct line structure is responsible for the development of distinct communication technology. The *single-wire transmission line* with the Earth as a return conductor is responsible for the operation of Telegraphy. It evolved into the *coaxial cable* that made the *Transatlantic Telegraphy* possible. The *two-wire open line* became a medium for the *Telephonic transmission*. These two line structures are behind the development of the monopole and dipole antenna that made possible the growth of the high-frequency communication using the medium wave (MW), short wave (SW), very high-frequency (VHF), and ultra-high frequency (UHF) bands. The microwave and mm-wave transmission systems are developed mostly around the metallic waveguides, and subsequently also using the nonmetallic dielectric waveguides. Finally, it has resulted in modern optical fiber technology. The planar transmission lines are behind the modern advanced microwave communication components and systems.

The present chapter provides a very brief historical overview of the classical and modern planar transmission lines. The chapter presents a historical survey of the development of the electromagnetic (EM) theory also. Next, a brief overview of the organization of the book is discussed.

Objectives

- *To present a survey of the developments of the classical EM-theory.*
- *To present brief historical notes on the classical transmission lines and development of transmission line theory.*
- *To present brief historical notes on planar transmission lines.*
- *To present an overview of the contents of the book.*

1.1 Overview of the Classical Transmission Lines

The classical transmission lines such as a single-wire line with the earth as a return conductor, coaxial cable, two-wire line, multi-conductor lines, and waveguides are reviewed very briefly in this section. The historical development of the *Telegrapher's Equations* is also presented. The developments of the theoretical concepts of EM-theory are reviewed below. The data related to the review of the EM-theory and transmission lines are collected from the published books [B.1–B.7] and journal articles referred at the end of the chapter.

1.1.1 Telegraph Line

The telegraph is the first coded point-to-point electrical communication system. As early as 1747, William Watson showed the possibility of transmitting an electrical current on a wire using the earth as a return conductor. Thus, *overhead single-wire* with the earth as a return conductor is the first transmission line. It is to be noted that even the *voltaic pile*, i.e. the chemical battery of Volta was nonexistent at that time. The *Leiden jar*, a capacitor to store the static electrical charges, was invented just two years before in 1745. However, much earlier in the year 1663, Otto Von Guericke studied the phenomenon of *static electricity* and designed a machine to produce it. Thus, the charged Leiden jar became a source of

Introduction to Modern Planar Transmission Lines: Physical, Analytical, and Circuit Models Approach, First Edition. Anand K. Verma.

electricity, and single-wire transmission was the communication medium. These were two important ingredients to establish the telegraph link. The third ingredient of telegraphy, the *electroscope*, invented by William Gilbert around 1600, acted as the receiver for the coded signals. It is interesting to note that the telegraph was conceived without any theoretical investigations on electricity. Even Coulomb's law was discovered later in 1785. However, the proper telegraphy could be developed only after the invention of *voltaic pile, i.e. a chemical battery* by Volta in 1799. Further development of the telegraph has an involved history. In 1837, Morse patented his telegraph in the United States, and on January 6, 1838, the first telegram was sent over 3 km distance. Cooke commercialized the telegraph in England and established the 21 km link on April 9, 1839. Thus, an era of electrical communication heralded. In 1844, long-distance Morse's telegraph between Washington, DC, and Baltimore, Maryland was established. Before further reviewing the growth of the classical transmission lines, it is useful to have a quick look at the theoretical developments related to electricity [B.5].

1.1.2 Development of Theoretical Concepts in EM-Theory

The ancients unfolded our story of Electromagnetism through careful observations of the phenomena of static electricity and natural magnetism. The developments of basic concepts, analytical modeling, and theoretical formulations used in the EM-Theory are emphasized in the review. The theoretical concepts of the electric and magnetic fields followed the mathematical models developed for the gravitational force by Newton in 1687, and subsequently refined by other investigators. The development of the theoretical models of transmission lines inherited the modeling process, and mathematical method of Fourier developed for the transmission of heat in a rod.

Electrostatics and Scalar Potential

Newton published his theory of gravitation in his monograph *Philosophiae Naturalis Principia Mathematica*. Newton viewed the gravitational interaction between two masses through *force*. The effect of static electricity was known for a long time, at least since 600 BC. However, only in 1600, Gilbert carried out systematic studies of both magnetism and static electricity. The *static electricity* was generated by the rubbing of two specific objects. He suggested the word *electricus* for electricity, and the English word "*electricity*" was suggested by Thomas Browne in 1646. Gilbert also suggested that the electrical effect is due to the flow of a small stream of weightless particles called *effluvium*. This concept helped the formulation of *one- and two-fluid model* of electricity. He also invented the first electrical measuring instrument, the *electroscope*, which helped further experimental investigations on electricity.

In 1733, Fay proposed that electricity comes in two forms – *vitreous* and *resinous,* and on combination, they cancel each other. The flow of the two forms of electricity was explained by the *two-fluid model*. During this time interval, around up to 1745, the electrical attraction and repulsion were explained using the flow of Gilbert's particle effluvium. In 1750, Benjamin Franklin proposed the *one-fluid model* of electricity. The matter containing a very small quantity of electric fluid was treated as negatively charged, and the matter with excess electric fluid was treated positively charged. Thus, the *negative charge* was resinous electricity, and the *positive charge* was vitreous electricity. Now, the stage was ready for further theoretical and experimental investigations on electricity.

In the year 1773, Lagrange introduced the concept of the *gravitational field*, now called *the scalar potential field*, created by a mass. The gravitational force of Newton was conceived as working through the *gravitational field*. The scalar potential field has appeared as a mechanism to explain the gravitational force interaction between two masses. Thus, a mass located in the potential field, described by a function called the *potential function*, experiences the gravitational force. In 1777, Lagrange also introduced the *divergence theorem* for the gravitational field. The nomenclature *potential field* was introduced by Green in 1828. Subsequently, Gauss in 1840 called it "*potential.*" Laplace in 1782 showed that the potential function ϕ (x,y,z) satisfies the equation $\nabla^2\phi = 0$. Now the equation is called *Laplace's equation*.

Following the Law of Gravitation, Coulomb postulated similar inverse square law, now called *Coulomb's law*, for the electrically charged bodies. He experimentally demonstrated the inverse square law for the charged bodies in 1785. Thus, the mathematical foundation of the EM-theory was laid by Coulomb. The law was also applicable to magnetic objects. The interaction between charged bodies was described by the electric force. In the year 1812, Poisson extended the concept of potential function from the gravitation to electrostatics. Incorporating the charge distribution function ρ, he obtained the modified Laplace's equation, written in modern terminology, $\nabla^2\phi = -\rho/\varepsilon_0$. This equation now called *Poisson's equation* is the key equation to describe the potential field due to the charge distribution. In the same year, Gauss rediscovered the divergence theorem originally discovered by Lagrange for the gravitational field.

In the year 1828, Green coined the nomenclature – the *potential function*, for the function of Lagrange and modern concept of the *scalar potential field* came into existence. Green also showed an important relation between the surface and volume integrals, now known as *Green's Theorem*. Green applied his method to the static magnetic field also. Green also introduced a method to solve the 3D inhomogeneous Poisson partial differential equation where the considered source is a point charge. The point charge is described by the *Dirac's delta function*. The solution of the Poisson's partial differential equation, using Dirac's delta function, is now called *Green's function*. Neumann (1832–1925) extended the *Green's function* method to solve the 2D potential problem and obtained the eigenfunction expansion of 2D Green's function [J.1–J.5, B.2, B.7].

Magnetic Effect of Current

So far, we have paid attention to the electrostatics. At this stage, *Leiden jar* was the only source of static electricity. A source for the continuous electric current was not available. Volta in 1799 invented the *voltaic pile, i.e. a chemical battery*, and the first time a continuous source of electric current came into existence. On April 21, 1820, it led to the *discovery of the magnetic effect of current* flowing in a wire. The electric current became the source of the magnetic field, encircling the current-carrying wire. The magnetic effect of current was discovered by Orsted (Oersted). In the same year, Ampere showed that the co-directional parallel currents flowing in two wires attract each other, and the counter-currents repel each other. It was a very significant discovery, i.e. creation of the attractive and repulsive magnetic forces without any physical magnet. It firmly established the relation between electricity and magnetism. Ampere further developed an equation, presently called *Ampere's Circuital Law*, to connect the current flowing in a wire to the magnetic field around it and developed the right-hand rule. He called the new field of electricity *Electrodynamics* and Maxwell recognized him as the *Father of Electrodynamics*. Ampere further modeled the natural magnetic materials as the materials composed of perpetual tiny circulating electric currents. He demonstrated the validity of his concept using the current-carrying conductor in the helical form called a *solenoid*. The solenoid worked like an artificial bar magnet. In the year 1820 itself, Biot–Savart obtained the equation using the line integral to compute the magnetic field at a position in the space due to the current flowing in a wire [J.2, B.6, B.7].

Ohm's Law

The voltaic pile helped the discovery of the magnetic effect of current; however, surprisingly the relation between the current flowing in resistance and voltage across it, known as the *Ohm's law*, remained undiscovered. The primary reason was the unstable voltage supplied by the voltaic pile. The discovery of *thermoelectricity* by Seebeck in 1822 provided a constant voltage source to supply continuous electric current. Using the thermo-piles in the year 1826, Ohm obtained a simple but powerful relation among voltage, current, and resistance. It was the beginning of the *Electric Circuit Theory*. However, only in 1850 Kirchhoff published his two *circuital laws* and opened the path for the development of the *Network Theory*. Kirchhoff also showed that *Ohm's electroscopic force (voltage)* and classical potential of Lagrange, Laplace, and Poisson are identical. Interestingly, Ohm's law could be viewed as a symbol of the International Scientific unity relating to Italy (Volta), Germany (Ohm), and France (Ampere). Based on the magnetic effect of current, in the same year, Johann Christian Poggendorff invented the *galvanometer* to detect the current in a wire. Kelvin improved its sensitivity by designing the *mirror galvanometer* in 1858 [B.1, B.6, B.7].

Electric Effect of the Time-Varying Magnetic Field

On knowing the magnetic effect created by an electric current, Faraday argued that the magnetic field can also produce the electric effect. After some attempts, he realized that such an effect can't be produced by the stationary magnet. In 1831, he could generate the electric potential (*electromotive force*) and electric current by the time-varying magnetic field of a moving magnet. The phenomenon is called the *induction effect*. The voltage induction effect demonstrated that electricity could be generated by a purely mechanical process, converting the mechanical energy into electrical energy via the medium of the moving magnetic field. The first *DC generator* was demonstrated by Faraday himself, and next year French instrument maker Hippolyte Pixii built the first *A.C. generator* inaugurating the Electrical Age. Now, the electricity was ready to accelerate the growth of human civilization at an unprecedented rate [B.6, B.7].

Concept of the Magnetic Vector Potential

In the process of discovery of induction, Faraday introduced the *concept of fields*, and also suggested that the electric energy resides in the field around the charged body and the magnetic energy resides in the field around the magnetized body. Thus, he viewed that the electric and magnetic energies reside in the space around the charged or magnetized body, not in the charge or magnet.

The *field concept* has greatly influenced the further development of EM-theory. The field provided a mechanism of interaction between charged bodies. Using Ampere-Biot–Savart law of magnetic forces, and electromagnetic induction of Faraday, Neumann in 1845 introduced the concept of the *magnetic vector potential* $\vec{A}$ to describe the magnetic field. Subsequently, Maxwell showed that the time derivative of $\vec{A}$ computes the induced electric field $\vec{E}$. Kelvin in 1847 further extended the concept of the *magnetic vector potential* $\vec{A}$ to compute the magnetic field using the relation $\vec{B} = \nabla \times \vec{A}$. This relation comes as a solution of the Gauss divergence equation $\nabla \cdot \vec{B} = 0$ due to the closed-loop of the magnetic field, showing the nonexistence of a magnetic charge. Kelvin further elaborated on the mathematical theory of magnetism in 1851. It is interesting to note that at any location in the space once time-dependent magnetic vector potential function is known, both the magnetic and electric fields could be computed as,

$$\begin{aligned}
&\vec{B} = \nabla \times \vec{A} && \text{(a)} \\
&\nabla \times \vec{E} = -\partial \vec{B}/\partial t = \nabla \times \left(-\partial \vec{A}/\partial t\right) && (1.1.1) \\
&\Rightarrow \vec{E} = -\partial \vec{A}/\partial t && \text{(b).}
\end{aligned}$$

Maxwell shared the views of Neumann and Kelvin. However, *time-retardation* was not incorporated in the scalar and vector potentials. In 1867, Lorentz introduced the *concept of retardation* in both the scalar and vector potentials to develop the EM-theory of light, independent of Maxwell. The time-retardation only in the scalar potential was first suggested by Riemann in 1858, but his work was published posthumously in 1867 [J.1, J.2, B.6, B.7].

Maxwell's Dynamic Electromagnetic Theory

At this stage of developments in the EM-theory, the electric field was described in terms of the scalar electric potential, and the magnetic field was described by the magnetic vector potential. Several laws were in existence, such as Faraday's law, Ampere's law, Gauss's law, and Ohm's law. Now Maxwell, Newton of the EM-theory, arrived at the scene to combine all the laws in one harmonious concept, i.e. in the *Dynamic Electromagnetic Theory*. He introduced the brilliant concept of the *displacement current,* created not by any new kind of charge but simply by the time-dependent electric field. Unlike the usual electric current supported by a conductor, this new current was predominantly supported by the dielectric medium. However, both currents were in a position to generate the magnetic fields. Thus, Maxwell modified Ampere's circuital law by incorporating the displacement current in it. The outcome was dramatic; the *electromagnetic wave equation*. Despite such success, the concept and physical existence of displacement current created a controversy that continues even in our time, and its measurement is a controversial issue [J.6–J.8].

In the year 1856, Maxwell formulated the *Faraday's law of induction* mathematically, and modified Ampere's circuital law in 1861 by adding the displacement current to it. Finally in 1865 after a time lag of nearly 10 years, Maxwell could consolidate all available knowledge of the electric and magnetic phenomena in *a set of 20 equations* with 20 unknowns. However, he could solve the equations to get the wave equations for the EM-wave with velocity same as the velocity of light. Now, the light became simply an EM-wave. In the year 1884, *Heaviside reformulated the Maxwell equations* in a modern set of four vector differential equation. The new formulation of Maxwell equations was in terms of the electric and magnetic field quantities and completely removed the concept of potentials, considering them unnecessary and unphysical. Hertz has independently rewritten the Maxwell equation in the scalar form using 12 equations without potential function. Hertz worked out these equations only after Heaviside. In 1884, Poynting computed the power transported by the EM-waves. Recognizing the contributions of both Heaviside and Hertz in reformulating Maxwell's set of equations, Lorentz called the EM-fields equations *Maxwell–Heaviside–Hertz equations*. However, in due course of time, the other two names were dropped and the four-vector differential equations are now popularly known as "*Maxwell's Equations*" [J.1, J.6, J.9, J.10, B.5–B.7].

Generation and Transmission of Electromagnetic Waves

Maxwell's EM-theory was a controversial theory, and physicists such as Kelvin never accepted it. Hertz finally generated, transmitted, and detected the EM-waves in 1887 at wavelengths of 5 m and 50 cm. In the process, he invented the *loaded dipole* as the transmitting antenna, rectangular *wire-loop receiving antenna*, and *spark-gap* both as transmitter and detector to detect the propagated EM-waves. Thus, he experimentally confirmed the validity of Maxwell equations and opened the magnificent gateway of wireless communication.

In the year 1895, Marconi transmitted and received a coded telegraphic message at a distance of 1.75 miles. Marconi continued his works and finally on December 12, 1901, he succeeded in establishing the 1700 miles long-distance wireless communication link between England and Canada. The transmission took place using the *Hertzian spark-gap transmitter* operating at the wavelength of 366m. In the year 1895 itself, J.C. Bose generated, transmitted, and detected the 6 mm EM-wave. He used circular waveguide and horn antenna in his system. In 1897, Bose reported his microwave and mm-wave researches in the wavelengths ranging from 2.5 cm to 5 mm at Royal Institution, London. Of course, the Hertzian spark-gap transmitter was at the core of his communication system. Bose was much ahead of his time as the commercial communication system grew around the low frequency, and the microwave phase of communication was yet to come in the future. In 1902, Max Abraham introduced the concept of the *radiation resistance of* an antenna [J.11–J.13, B.1–B6].

Further Information on Potentials

Hertz is known for his outstanding experimental works. However, as a student of Helmholtz, he was a high ranking theoretical physicist. Although, he considered, like Heaviside, electric and magnetic fields as the real physical quantities, still he used the vector potentials, now called *Hertzian potentials* $\vec{\pi}^{e}$ and $\vec{\pi}^{m}$, to solve Maxwell's wave equation for the radiation problem. These potentials are closely related to the electric scalar potential ϕ and magnetic vector potential $\vec{A}$. Stratton further used Hertzian potentials in elaborating the EM-theory [B.8]. Collin continued the use of Hertzian potentials for the analysis of the guided waves. He also used the $\vec{A}$ and ϕ potentials in the radiation problems [B.9, B.10]. The use of Hertzian potentials gradually declined. However, its usefulness in problem-solving has been highlighted [J.1, J.11, J.12].

Gradually, the magnetic vector potential became the problem-solving tool if not the physical reality. Further, by using the *retarded scalar and vector potentials* and *Lorentz gauge condition* $\nabla\cdot\vec{A} = -\mu\varepsilon\partial\phi/\partial t$ connecting both the vector and scalar potentials, Lorentz formulated the EM-theory of Maxwell in terms of the magnetic vector potential. In his formulation, a current is the source of the magnetic vector potential $\vec{A}$. So, Lorentz considered the propagation of both the magnetic vector and electric scalar potential with a finite velocity that resulted in the retarded time at the field point. However, Maxwell's scalar potential was nonpropagating. Maxwell did not write a wave equation for the scalar potential, as his use of *Coulomb gauge* $\nabla\cdot\vec{A} = 0$ was inconsistent with it. Later on, even *electric vector potential* $\vec{F}$ was introduced in the formulation of EM-theory. The nonphysical magnetic current, introduced in Maxwell's equations by Heaviside, is the source of potential $\vec{F}$. The use of vector potentials simplified the computation of the fields due to radiation from wire antenna and aperture antenna. A component of the magnetic/electric vector potential is a scalar quantity. It has further helped the reformulation of EM-field theory in terms of the *electric scalar* and *magnetic scalar potentials* [B.9, B.10]. Such formulations are used in the guided-waves analysis. In recent years, it has been pointed out that the *Lorentz gauge condition* and *retarded potentials* were formulated by Lorenz in 1867, much before the formulation of famous H. A. Lorentz [J.14, J.15]. However, most of the textbooks refer to the name of Lorentz.

Both Heaviside and Hertz considered only the electric and magnetic fields as real physical quantities, and magnetic vector and electric scalar potentials as merely auxiliary nonphysical mathematical concepts to solve the EM-field problem. Possibly, this was not the attitude of Kelvin and Maxwell. They identified the electrical potential with energy, and magnetic vector potential with momentum. The magnetic vector potential $\vec{A}$ could be considered as the *potential momentum per unit charge*, just as the electric scalar potential ϕ is the *potential energy per unit charge* [J.16]. The potential momentum $\vec{P}$ is obtained as follows:

$$\vec{F} = \frac{d\vec{P}}{dt} = q\vec{E} = -\frac{\partial\left(q\vec{A}\right)}{\partial t}, \Rightarrow \vec{P} = q\vec{A}. \tag{1.1.2}$$

In the above equation, $\vec{F}$ is the force acting on the charge q, and $\vec{E}$ is given by equation (1.1.1b).

Lebedev in 1900 experimentally demonstrated the radiation pressure, demonstrating momentum carried by the EM-wave. The energy and momentum carried by the EM-wave indicate that the light radiation could be viewed as some kind of particle, not a wave phenomenon. A particle is characterized by energy and momentum. Such a dual nature of light is a quantum mechanical duality phenomenon. Einstein introduced the concept of the light particle, called "photon" to explain the interaction of light with matter, i.e. the *photoelectric effect*. However, Lorentz retained the classical wave model to explain the interaction between radiation and matter via *polarization of dipoles* in a material creating its frequency-dependent permittivity.

It is to be noted that at a location in the space, even for zero $\vec{B}$ and $\vec{E}$ fields, the potentials $\vec{A}$ and ϕ could exist [B.11]. Aharonov–Bohm predicted that the potential fields $\vec{A}$ and ϕ, in the absence of $\vec{B}$ and $\vec{E}$, could influence a charged particle. Tonomura and collaborators experimentally confirmed the validity of Aharonov–Bohm prediction. The *Aharonov–Bohm effect* demonstrates that $\vec{B}$ and $\vec{E}$ fields only partly describe the EM-fields in quantum mechanics. The vector potential also has to be retained for a complete description of the EM-field quantum mechanically [J.3, J.16, J.17]. However, to solve the classical electromagnetic problems, such as guided wave propagation and radiation from antenna, Heaviside formulation of Maxwell equations and potential functions as additional tools is adequate.

EM-Modeling of Medium

The above brief review omitted developments in the electromagnetic properties of the material medium. A few important developments could be summarized. In 1837, Faraday introduced the concept of the *dielectric constant* of a material. In 1838, he introduced the concept of *electric polarization* $\vec{P}$ in dielectrics under the influence of the external electric field. Soon after the discovery of the electron in 1897 by J.J. Thomson, the models around electrons were developed to describe the electromagnetic properties of a material. Around 1898, John Gaston Leathem obtained an important relation $\vec{D} = \vec{E} + \vec{P}$, connecting the displacement of charges in a material with polarization. Kelvin, in the year 1850, developed the concept of *magnetic permeability* and *susceptibility* with separate concepts of $\vec{B}$, $\vec{M}$, and $\vec{H}$ to characterize a magnetic material. In 1900, Drude developed the electrical conduction model, now known as the *Drude model*, after electron theory. Subsequently, the model was extended to the dielectric medium by Lorentz in 1905. The model called the *Drude-Lorentz model* explains the dispersive property of dielectrics. In the year 1912, Debye developed the concept of *dipole moment* and obtained equations relating it to the dielectric constant. These models laid the foundation to study of the electric and magnetic properties of natural and engineered materials under the influence of external fields [B.4, B.6, B.7, B.12]

1.1.3 Development of the Transmission Line Equations

Kelvin's Cable Theory

During the period 1840–1850, several persons conceived the idea of telegraph across the Atlantic Ocean. Finally, in the year 1850, the first *under-sea telegraphy*, between Dover (*Kent, England*) and Calais (*France*), was made operational. However, no cable theory was available at that time to understand the electrical behavior of signal transmission over the undersea cable.

In 1854, Kelvin modeled the under-sea cable as a *coaxial cable* with an inner conductor of wire surrounded by an insulating dielectric layer, followed by the saline seawater acting as the outer conductor [J.18, B.1]. The coaxial cable was modeled by him as a *distributed RC circuit* with the series resistance R *per unit length (p.u.l.)* and shunt capacitance C *p.u.l.* It was the time of the *fluid model* of electricity. Kelvin further conceived the flow of electricity similar to the flow of heat in a conductor. Fourier analysis of 1D heat flow was in existence since 1822. Following the analogy of heat equation of Fourier, Kelvin obtained the *diffusion type equation* for the transmitted voltage signal over the under-sea coaxial cable:

$$\frac{\partial^2 v}{\partial x^2} = RC\frac{\partial v}{\partial t}. \tag{1.1.3}$$

This is the first *Cable Theory*; Kelvin called the above equation the equation of electric excitation in a submarine telegraph wire. Kelvin's model did not account for the inductance L *p.u.l.* and the conductance G *p.u.l.* of the cable. The cable inductance L is due to the magnetic effect of current, and G is due to the leakage current between the inner and outer conductors. However, cable theory was a great success. Following the method of Fourier, he solved the equation for both the voltage and current signals. At any distance x on the cable, a definite time-interval was needed to get the maximum current of the received signal. The galvanometer was used to detect the received current. This time-interval called the *retardation time* of the received current signal also depends on the square of the distance. Moreover, the telegraph signals constituted of several waves of different frequencies, and their propagation velocities were frequency-dependent. It limited the speed of signal transmission for long-distance telegraphy. The conclusions of Kelvin's analysis were ignored, and 1858 transatlantic cable worked only for three weeks. It failed due to the application of 2000 V potential pulse on the cable. The speed of transmission was just 0.1 words per minute. Finally, following Kelvin's advice and using a very sensitive mirror galvanometer invented by him, the *transatlantic telegraph* was successfully completed in 1865 with eight words per minute transmission speed [B.1–B.3].

Heaviside Transmission Line Equation

The limitation of the speed of telegraph signals was not understood at that time. The RC model of the cable, leading to the diffusion equation, and use of the time-domain pulse could not explain it. Moreover, it became obvious that the *RC model* couldn't be used to understand the problems related to voice transmission over telephonic channels. The telephony was coming into existence. The modern telephone system is an outcome of the efforts of several innovators. However, Graham Bell got the *first patent of a telephone* in the year 1874. The transmitted telephonic voice signal was distorted. Therefore, an analytical model was urgently needed to improve the quality of telephonic transmission. Heaviside in 1876 introduced the *line inductance* L *p.u.l.* and reformulated the cable theory of Kelvin using Kirchhoff circuital laws [B1, B.3]. The formulation resulted in the *wave equation* for both the voltage (V) and current (I) waves on the line:

$$\frac{\partial^2 T}{\partial x^2} = RC\frac{\partial T}{\partial t} + LC\frac{\partial^2 T}{\partial t^2}, \quad (\text{where } T = V, I). \qquad (1.1.4)$$

In the case of line inductance L = 0, the above equation is reduced to the diffusion type cable equation (1.1.3) of Kelvin. Using the Fourier method, Heaviside solved the aforementioned time-domain equation. Only in 1887, he could introduce the *line conductance* G *p.u.l* in his formulation to account for the leakage current in an imperfect insulating layer between two conductors. Finally, Heaviside obtained a set of coupled transmission line equations using all four line constants R, L, C, and G. Subsequently, the coupled transmission line equations were called the *Telegrapher's equations*. At the end, Heaviside obtained the following modified wave equation:

For lossy line:

$$\frac{\partial^2 T}{\partial x^2} = (RC + LG)\frac{\partial T}{\partial t} + LC\frac{\partial^2 T}{\partial t^2} + RG\,T \qquad (a)$$

For lossless line:

$$\frac{\partial^2 T}{\partial x^2} = LC\frac{\partial^2 V}{\partial t^2}, (\text{where } T = V, I) \qquad (b). \qquad (1.1.5)$$

To solve the above time-domain equation, Heaviside developed his own intuitive *operational method approach* by defining the operator $\partial/\partial t \rightarrow p$. The use of the operator reduced the above partial differential equation to the ordinary second-order differential equation. Finally, he solved the equation under initial and final conditions at the ends of a finite length line. In the process, he obtained the expressions for the *characteristic impedance* and *propagation constant* in terms of line parameters. Heaviside could obtain results for the line under different conditions. For a lossless line, R = G = 0, the equation (1.1.5b) is obtained. Conceptually, the characteristic impedance provided a mechanism to explain the phenomenon of wave propagation on an infinite line. At each section of the line, it behaved like a secondary Huygens's source providing the forward-moving wave motion. Heaviside also obtained the condition for the *dispersionless transmission* on a real lossy line, and suggested the *inductive loading* of a line to reduce the distortion in both the telegraph and telephone lines. Afterward, his intuitive *operational method approach* developed into the formal *Laplace transform method*, widely used to solve the differential equations [J.19, J.20, B.1–B.3, B.13].

The method of Heaviside was further extended by Pupin in 1899 and 1900. Pupin introduced the *harmonic excitation* in the wave equation as a real part of the source V_0e^{jpt} [J.21, J.22]. This was an indication of the use of the modern phasor solution of the wave equation. Similar analytical works, and also practical inductive loading of the line was done by Campbell at Bell Laboratory. He published the results in 1903 [J.23]. In July 1893, Steinmetz introduced the *concept of phasor* to solve the AC networks of RLC circuits. In 1893, Kennelly also published the use of complex notation in Ohm's law for the AC circuits [J.24]. Carson in 1921 applied the method to solve Maxwell's equations for the wave propagation on closely spaced lines, and also analyzed for the mutual impedances. Carson in 1927 developed the electromagnetic theory of the Electric Circuits, and paved the way for the modeling of the wave phenomena using the circuit models [J.25, J.26].

Peijel in 1918, and Levin in 1927 analyzed the wave propagation on the parallel lines. Levin extended the telegrapher's equations to the *multiconductor transmission lines* using Maxwell's equations [J.27]. In 1931 Bewley presented a set of wave equations on the coupled multiconductor lines. Subsequently, Pipes introduced the *matrix method* to formulate the wave propagation problem on the multiconductor lines [J.28, J.29]. Thus, the theoretical foundation was laid to deal with the complex technical problems related to transmission lines. Starting with Marconi wireless in 1895, several improvements took place in the long-range wireless telegraphy. Also, the audio broadcasting was developed between 1905 and 1906 and commercially, around 1920–1923, in the long-wave, medium-wave, and short-wave RF frequency bands [B.5]. Now, the time was ripe for microwave and mm-wave communication.

The above discussion shows that the Telegrapher's equations have come in existence due to the contributions of both Kelvin and Heaviside. To recognize their contributions, we call in this book the Telegrapher's equations as the *Kelvin-Heaviside transmission line equations*. Also, as the characteristic impedance behaves as the secondary Huygens's source, so it can also be viewed as the *Huygens's load*. Such Huygens's load distributed over a surface forms the modern *Huygens's metasurface*, discussed in the chapter 22 of this book.

1.1.4 Waveguides as Propagation Medium

Heaviside reformulated Maxwell equation in 1884. He rejected the idea of EM-wave propagation in a hollow metallic cylinder. In his opinion, two conductors, alternatively one conductor and the earth as a ground conductor are essential for the EM-wave propagation. However, in 1893 J.J. Thomson expressed the possibility of the EM-wave propagation in a hollow cylinder [B.12]. Next year, Oliver Lodge verified it experimentally. In the year 1895, J.C. Bose used the waveguide and horn antenna for the mm-wave transmission and reception. In 1897 he reported the work at Royal Institution in London [B.5]. However, it was Rayleigh who carried out a detailed solution of boundary-value problems. He obtained the *normal mode* solution, showing wave propagation in the form of the distinct discrete modes, i.e. the normal modes. He obtained his solutions for both the TE and TM modes, and introduced the concept of the cutoff frequency for modes. He further examined the EM-wave propagation on a dielectric waveguide [J.30]. In 1920 Rayleigh, Sommerfeld and Debye continued the researches in this direction.

However, only in 1930 proper experimental investigations of the wave propagation in the waveguides were undertaken by G. C. Southworth at Bell Labs, and W.L. Barrow at MIT. In 1934, *microwave commercial link* was established, and in 1936, Southworth and Barrow discovered the possibility of using the waveguide as a transmission medium. However, they published their works only in 1936 [J.31, J.32, B.5]. During the same time-period, Brillouin also investigated the wave propagation in a tube [J.33]. Serious analytical work on waveguides was further undertaken by J.R Carson, S.P. Mead, and S.A. Schelkunoff around 1933 [J.34]. Almost forgotten analytical works of Rayleigh was reinvented. Chu and Barrow further investigated the EM-waves propagation in the elliptical and rectangular hollow metallic pipes [J.35]. During 1934, Schelkunoff extended the *concept of impedance* to the EM-wave propagation in the coaxial line, and obtained the transmission line equations using the electromagnetic theory [J.36]. In 1937, he further extended the theory to the TE and TM mode guided wave propagations, and obtained the circuit models of mode supporting waveguides. Finally, Schelkunoff generalized the standard telegrapher's equation, using Maxwell's EM-theory to represent an infinite set of uncoupled and coupled modes of a waveguide by the system of uncoupled and coupled transmission line equations [J.37–J.39]. Subsequently, his method has been extended to planar lines in an inhomogeneous medium supporting the hybrid modes [B.13].

During the World War-II period, important theoretical and practical works were done in the field of waveguide technology for the development of the waveguide-based components and systems. The development of Radar provided the impetus for such research activities.

1.2 Planar Transmission Lines

A brief review of the development of planar transmission lines, influencing modern microwave technology, is presented below. A review is also given for the analytical methods as applied to the planar line parameters.

1.2.1 Development of Planar Transmission Lines

The waveguide is a low-loss transmission medium capable of handling high power transmission. However, it is a bulky structure with limited bandwidth. The fabrication of waveguide-based components is a complex and expensive machining process. The limitations of the waveguide provided an impetus for the growth of planar lines, and technology based on the planar lines. H.A Wheeler, in 1936, developed a low-loss *coplanar stripline*, and in 1942 created parallel plate strip transmission line on a high permittivity substrate. The line structure was compact and suitable at low RF frequency from 150 MHz to 1500 MHz. However, properly documented *stripline* was reported by R.M. Barrett only in 1951. Just next year, i.e. in 1952, Grieg and Engelmann reported *microstrip* line. Both structures competed with each other. Initially, the stripline in the homogeneous medium was a preferred line, as it is a dispersionless line with a larger bandwidth. It supports the TEM mode propagation. As it is a shielded line, so it also has a higher Q-factor. Whereas, the microstrip in the inhomogeneous medium is a dispersive line as it supports the dominant hybrid mode. It has a smaller bandwidth and lower Q-factor. During 1960, solid-state components started appearing, and microstrip became the preferred line structure for the MIC environment. The microstrip is an open structure that provided easier access for the interconnections. It led to the development of miniaturized *microstrip integrated circuit (MIC)* technology. Gradually, the discrete active devices were combined with the planar passive microwave components, and the *hybrid MIC (HMIC)* came into existence. The sixties were a very creative period for the planar line technology. In 1968, Cohn reported the *slot line* followed by the *coplanar waveguide (CPW)* that became the medium of MMIC. C.P. Wen in 1969 developed the CPW. It is an interesting and unusual coincidence that the abbreviation of both the line name and inventor's name is CPW. The integration of the slot line with waveguide took place in 1972 when Meier reported the quasi-planar *fin line* [J.40–J.44].

Further compactness in the microwave circuits and systems took place through the development of the *monolithic MIC (MMIC)* circuit concept in the year 1964. At this stage, the MMIC was based on *silicon technology*. Unfortunately, the program was not successful due to the very lossy Si-substrate. The semi-insulating Si-substrate deteriorated in the process of the formation of active devices, such as bipolar junction transistors (BJTs) on a Si-substrate. The next phase of MMIC development took place for the GaAs substrate-based technology in 1968. It required nearly 10–12 years for its more meaningful development. The span of 1980–1986 was a period of rapid growth for *MMIC technology*. In 1990s, SiGe based technology was developed that permitted operation of high-efficiency circuits at higher frequencies. The MMIC technology achieved its maturity for the MMIC based on the silicon and indium–phosphide (InP) substrates apart from the GaAs substrate. At the core of the development were the *multilayer planar lines* and new varieties of active devices [J.45].

Another kind of Si-based technology, namely the *micro-electro-mechanical system (MEMS)* gradually came to the fields of RF and microwave. Petersen's reported the MEMS membrane-based switches in July 1979. However, after a long gap, Yao and Chang developed the surface MEMS switch for DC-4 GHz operation and high-quality MEMS inductor chip could be realized in 1997. Subsequent years witnessed a reduction in operating voltage of MEMS switches. The operation of MEMS in the microwave and mm-wave ranges expanded their applications in the field of the antenna and other microwave systems [J.46–J.48].

The robust and compact multilayer ceramic tape-based microwave technology, called the *low-temperature co-fired ceramics (LTCC)* gradually acquired significance for the development of the hybrid integrated circuits. It started in 1950–1960 to develop more robust capacitors. The several layers of different materials are used in a single multilayer laminated package to design multi-functionality circuit-blocks. The planar lines in the LTCC are used in the multilayer and multilevel formats as a medium to develop the components and interconnect [J.49, J.50].

Further innovations in the planar microwave technology were added by incorporating the periodic reactive loading of planar lines and planar surfaces resulting in the *electronic band-gap (EBG) lines and EBG surfaces* for a wide range of applications. Long ago, the theoretical basis for the analysis of the periodic structures was summarized by L. Brillouin [B.14]. The theoretical concept of the *metamaterial* as a double negative (negative permittivity and negative permeability) material medium, and its radical impact on behaviors of the electromagnetic phenomena were worked out by Victor Veselago in 1967. The practical development of the metamaterials is an outcome of a long history of *artificial dielectrics* and *mixture medium*. However, only in 1996–1999, Pendry and co-workers suggested, and further experimentally demonstrated, the artificial negative permittivity below the controlled plasma frequency. It was realized by using the periodic arrangement of thin conducting wires. Further, in 1999 Pendry and co-workers suggested and experimentally produced resonance type magnetic behavior in the split coaxial conducting cylinders. However, only Smith and co-workers worked out the simultaneous negative permittivity and negative permeability in 2000, and experimentally verified it in 2001. Gradually, the concept of metamaterials was added to the planar lines and surfaces resulting in the realization of *metalines and metasurfaces*. These artificial structures have significantly influenced the design and development of unique antenna, components, and circuits with new characteristics and multifunctionality. Present researches in these fields are in progress in many directions [J.51–J.55].

1.2.2 Analytical Methods Applied to Planar Transmission Lines

Assaudourion and Rimai considered the *microstrip* in the quasi-static limit. They assumed the TEM mode propagation on it. They applied, in 1952, the well-established *conformal mapping method* to compute the characteristic impedance, dielectric, and conductor losses. Between the years 1954 and 1955, Cohn also used the conformal mapping method to get the design-oriented results for the characteristic impedance, dielectric and conductor losses of the *stripline*. He further used the conformal mapping method to get the odd-even mode impedances of the *edge-coupled strip lines* in 1955 itself. He further obtained these results for the *broadside-coupled strip lines* in 1960. Following the conformal mapping method, in 1964 and 1965 Wheeler produced more accurate and design-oriented expressions for the computation of characteristic impedance of microstrip line. He extended his analysis to get further results in 1977 and 1978 [J.56–J.64].

In 1969 Cohn suggested another planar line, i.e. the *slot line*. It is a complementary structure of the microstrip line. He also presented the *equivalent waveguide model* of the slot line, and obtained the frequency-dependent propagation parameters of the slot line. Next in the group of the planar lines is the *coplanar waveguide (CPW)* proposed by C.P. Wen. He obtained the initial quasi-static line parameters of CPW using the conformal mapping method. Subsequently, the conformal mapping method was applied to analyze several variants of the planar lines [J.65–J.67].

Other quasi-analytical and numerical methods were also used for the analysis of microstrip lines. For instance, in 1968 Yamashita and Mitra introduced the quasi-analytical *variational method in the Fourier domain* to obtain the quasi-static line parameters of the microstrip line. It was the prelude to the quasi-analytical *dynamic spectral domain analysis* (SDA) of microstrip and other planar lines. The dynamic SDA is a full-wave analysis method that considers the hybrid mode nature of planar lines. After a gap of nearly 10 years, Itoh used the concept of the *discrete Fourier transform* and *Galerkin's method to* get the static line parameters of suspended coupled microstrip lines, and also extended the method to suspended multiconductor microstrip structures. The Fourier domain method was significantly extended by many investigators to other planar structures such as the CPW [J.68–J.71, B.15, B.16].

In 1973 and 1974, Itoh and Mitra introduced the *dynamic SDA* to obtain dispersion characteristics of the slot line, and also microstrip line. Jansen extended the dynamic SDA to analyze the higher order modes in the microstrip. The method is very powerful and analytically elegant. It has been used and improved by other researchers in the field of planar resonators, antenna, and line structures. Other powerful methods, such as *the method of moments, finite elements, finite-difference time-domain method,* and so on have also been developed to analyze the 2D and 3D complex planar structures. The contemporary *EM-Simulators* are based on these numerical methods. The *closed-form models* for faster computation of the static and frequency-dependent line parameters of planar lines have also been developed by several investigators. The closed-form models of lines, discontinuities, and so on helped the development of the *Circuit Simulators* [J.72–J.75, B.15, B.16].

1.3 Overview of Present Book

The book presents a seamless treatment of the *classical planar transmission lines* and *modern engineered planar lines* using the concept of the engineered electromagnetic bandgap (EBG) structures and metamaterials. The modern EBG and metamaterials based planar lines are the outcome of the classical researches in the artificial dielectrics and concept of *homogenization of mixing of inclusions in the host medium*. Gradually, the modern microwave planar transmissions became a complex medium of wave propagations on the 1D lines and 2D surfaces. It demanded serious considerations of wave–matter interactions, especially in the engineered materials by the microwaves researchers and engineers. It demanded a physical understanding of various electromagnetic phenomena taking place in the artificially engineered complex medium. It also required the analytical and circuit modeling of the planar transmission lines under the complex environment. The present book: **Introduction to Modern Planar Transmission Lines** *(Physical, Analytical, and Circuit Models Approach)* addresses these problems from the very basics, making it suitable for the early comers to the fields. However, the detailed treatment of topics could be also useful to more experienced professionals and engineers. The numerical methods used in the analysis of the planar structures and basis of the EM-simulators are more specialized topics beyond the scope and line of thought followed in the present book.

The key concept used throughout the book is the modeling, *physical, analytical,* and *circuit,* of the planar structures. However, what is the meaning of modeling itself? *Scientific modeling is a process of understanding the unknown with the help of known*. The reverse is not possible. The *method of analogy* is a great tool in such a modeling process. The growth of electromagnetic field theories at different stages has evolved from the previously known results of the gravitational field. Likewise, the gradual development of the transmission line theory has used the analogy of heat flow. These are two important illustrative examples discussed in the previous section. The experimental observation and the experimental verification of the theoretically predicted results are further contributors to the modeling process. The scientific modeling process has been examined in depth by the modern educationists [B.17]. The reader can observe such a modeling process in the development of models for the complex planar medium exhibiting unique properties.

1.3.1 The Organization of Chapters in This Book

The chapters of this book are organized into four distinct groups as follows:

i) Introductory transmission line and EM wave theory.
ii) Basic planar lines and Resonators: Microstrip, CPW, Slot lines, Coupled lines, and Resonators.
iii) Analytical Methods: Conformal mapping method, Variational method, Full- wave SDA, and SLR formulation.
iv) Contemporary engineered planar structures: Periodic planar lines and surfaces, Metamaterials – Bulk, 1D metalines, 2D metasurfaces.

The group i reviews the *transmission line and the EM-theory* to assist the reader to follow the rest of the chapters with ease. The groups ii and iii form the *classical transmission lines*, and the group iv is the *modern transmission lines and surfaces*. The book presents a seamless treatment of the classical planar transmission lines and the modern engineered planar lines and surfaces using the concept of EBG and metamaterials. The modern EBG and metamaterials based planar lines are the outcome of the classical researches in the artificial dielectrics and concept of homogenization of mixing of inclusions in the host medium. The topics of the chapters are selected to provide comprehensive coverage of the needed background to understand the functioning of both the classical and modern lines and surfaces. Each chapter follows a uniform style. The topics within a chapter start with simple concepts and move to a higher complexity level. Likewise, the chapters are also arranged from the simpler to complex.

The distribution of the chapters among the groups is discussed below. The key features of the chapters are also summarized.

Introductory Transmission Line and EM-Wave Theory

The six chapters, chapters 2–7, on the transmission lines and various aspects of the EM-theory are introduced in the book before even commencing with the microstrip in chapter 8. These topics provide the essential background to follow smoothly the topics covered in this book. It could be useful in understanding the analysis and modeling of the planar line structures, the EBG based lines, and surfaces, and also the metamaterials and metasurfaces. Moreover, the topics discussed may also help to understand the modern publications in these fields. The usual undergraduate textbooks on the EM-theory do not cover all the topics. However, the reader's familiarity with the transmission line and EM-theory is assumed. The reader interested in a more detailed study of these topics can follow the references given at the end of the chapters. Some contents of the chapters are highlighted below.

The chapters 2 and 3 on transmission lines are written as a review. However, it goes beyond a regular review, although it starts with the familiar notion of oscillation and wave propagation on lines. Usually, the available textbooks present the transmission line equations and wave equations for the uniform lines only, without any source. The present book covers the transmission line equations and wave equations with a source, and the analysis of the multisection transmission lines is also introduced. Such formulation is used in the chapters 14 and 16 to obtain the Green's functions of planar transmission lines used with the variational method and full-wave spectral-domain analysis (SDA) method. The chapter on the transmission line adequately covers the concept of dispersion in the wave supporting medium. Also, the impact of the reactive loading of the line on the nature of wave propagation is discussed. Such treatment prepares a reader for the periodically loaded engineered lines and surfaces, both as the bandgap medium and homogenized metamaterial medium. These topics are discussed in chapters 19–22. Chapter 3 covers various parameters used for the characterization of a line section. Understanding of this topic is essential for understanding the microwave components design, the results obtained from EM-simulation, and to develop the circuit models.

Chapters 4 and 5 cover the wave's propagation in the material medium. Again, primarily it is a review of the

EM-theory. However, its perspective is very broadly applicable to the topics discussed in other chapters. The chapter 4 commences with a basic review of the electromagnetics and elementary electrical properties of the material medium, such as linearity, nonlinearity, homogeneity, inhomogeneity, anisotropy, and losses. It further covers the topic of the circuit modeling of a medium. The known topics of Maxwell's equations in the differential form, as well as in the vector-algebraic form are presented. The wave propagation is discussed not only in the isotropic and conducting media but also in the *anisotropic, uniaxial, gyroelectric*, and *biaxial* media. The complex media are encountered by the wave propagating in the metamaterials. While reviewing the wave polarizations, the *Jones matrix* description of polarization states is also discussed. It is needed to follow the contemporary developments in the metasurfaces. Chapter 4 ends with the concepts of *isofrequency contours* and *isofrequency surfaces*, and the dispersion relations in the uniaxial medium.

Chapter 5 reviews both the normal and polarization-dependent oblique incidence of the waves at the interface of two media. It also presents the equivalent transmission line model of the wave's incidence at the interface of two media. The model can be extended to more number of layers. The formulation has many applications. The model is used for instance in chapter 20 on the EBG surface. The chapter 5 also presents the basic electrodynamics of the engineered metamaterials and formulate the basic characteristics of the wave propagation in the metamaterials. It also discusses some important directions for applications of metamaterials. However, the realization of the bulk metamaterials, metalines, and metasurfaces are followed up in the chapters 21 and 22.

Chapter 6 covers a review of the electrical properties of the natural and artificial dielectric media. It also presents various static and frequency-dependent models of the mixture media. The artificial dielectric medium finds its application in modeling the metamaterials. The Lorentz, Drude, and Debye models applicable to the frequency-dependent permittivity are discussed. Chapter 6 further discusses the interfacial polarization and its circuit model. This important topic is usually not discussed in popular textbooks. The modeling of the substrates, using the single term Debye and Lorentz models, as well as the multi-term and wideband Debye models are elaborated. These models help to get the causal effective permittivity of the planar lines of the substrates, useful in the time-domain analysis of pulse propagation on the planar lines. Finally, the chapter ends with a novel concept of artificial metasubstrate.

Chapter 7 comprehensively treats basic waveguide structures. It begins with the classification of the modal EM-fields, and the sources of their generation. The waveguides are analyzed using the scalar electric and magnetic potentials. The spectral domain analysis (SDA) method discussed in chapter 16 is based on these scalar potentials. The concept of the *perfect electric conductor (PEC)* and the *perfect magnetic conductor (PMC)* with boundary conditions are introduced. The analysis of the rectangular geometry of the waveguides formed with these surfaces is presented. Thus, apart from the usual all metallic walls, i.e. the PEC based waveguides, all PMC and two PEC and two PMC walls waveguides are also discussed. The dielectric slab waveguides and surface-waveguides are also presented. The concept of the odd/even mode analysis is introduced. These concepts are used in the book for the analysis of symmetrically coupled planar lines in chapters 11 and 12. The simple and powerful *transverse resonance method (TRM)* is introduced to get the propagation characteristics of the dielectric-loaded waveguides and the multilayer surface-waveguides. Finally, chapter 7 ends with the contemporary *substrate integrated waveguide (SIW)* developed in the environment of the planar technology.

Basic Planar Lines and Resonators

The planar line structures – microstrip, CPW, and slot line are discussed in chapters 8–10, respectively. The chapters 11 and 12 cover the theory of the coupled transmission lines and their realization and analysis in the planar technology environment. The theory of resonating structures and planar lines version of the resonators are discussed in chapters 17 and 18, respectively. The fabrication technologies – MIC, MMIC, MEMS, and LTCC used in the planar lines and components are reviewed in chapter 13.

The microstrip is the most commonly used planar line in planar technology. It is in the inhomogeneous medium supporting the *hybrid-mode* that is approximated as the dispersive *quasi-TEM mode*. However, at the lower frequency, it is treated in the nondispersive static condition. Chapter 8 introduces the concept of medium transformation from the inhomogeneous medium to the homogeneous medium using *Wheeler's transformation* for the lossy microstrip medium. The results on the static microstrip line parameters are summarized. The *dispersion law* is discussed to get the dispersion model of microstrip. Some other dispersion models are also summarized. The losses and their computation are presented in detail. Finally, chapter 8 ends with the circuit model of the microstrip line giving the complex frequency-dependent characteristic impedance and propagation constant. The circuit model explains the behavior of the low-frequency dispersion due to the finite conductivity of the conductors. Several topics are covered for the first time in a book form. The derivations of some frequently used expressions are provided.

The *coplanar waveguides (CPW)* and the *coplanar stripline structures (CPS)* and their variations are discussed in chapter 9. The approach used in this chapter is based on the detailed derivation of the results using the *conformal mapping method*. Usually, the available books only summarize the results of the conformal mapping method. However, chapter 9 briefly presents the conformal mapping method as applied to the CPW and CPS. The characteristics of the modes, dispersion, and losses are presented in detail. The results are also presented for the synthesis of the CPW and CPS line structures. Finally, the circuit models of the lossy CPW and CPS are given to get the frequency-dependent complex characteristics impedance and propagation constant.

The modeling of the third important planar line, i.e. the *slot line* is presented in chapter 10. The modeling process is based on the *unique waveguide model* of Cohn. The model provides the frequency-dependent characteristic impedance and propagation constant of the slot line, supporting the hybrid mode. The waveguide model of the slot line treats the hybrid-mode as a linear combination of the TE and TM modes. The equivalent waveguide model is further extended to the multilayer and shielded slot line structures. The chapter ends with the *closed-form integrated model* of the slot line to compute the dispersion and loss parameters.

The next two chapters 11 and 12 cover the basic characteristic of the coupled lines theory and implementation of theory in the planar technology. Chapter 11 discusses the coupling mechanism and the analysis of symmetrical and asymmetrical coupled lines. The wave equation of the coupled transmission is obtained and solved in some cases. Chapter 12 summarizes the design expressions for the edge coupled and broadside coupled microstrip lines. Similar expressions are also summarized for the coupled CPW line structures. The network parameters for both symmetrical and asymmetrical coupled line sections are discussed in detail. Such an analysis is useful for the design of the filters.

At this stage, the further discussion of the planar line structure is discontinued and the fabrication technologies suitable for the planar lines and components are introduced in chapter 13. The chapter 13 discusses, in brief, the four kinds of fabrication technologies – the *hybrid microwave integrated circuit (HMIC)* suitable for the PCB board medium, the semiconductor based *monolithic MIC (MMIC)* technology, the silicon-based *micro-electro-mechanical systems (MEMS)* technology, and the ceramic tape-based *low temperature co-fired ceramic (LTCC)* technology. The typical details of the material and conductor parameters used in these technologies are also summarized. The familiarity of the fabrication process could be useful to the researchers and designers developing planar lines models and circuits.

The basic discussion and basic analysis of the resonator circuits, lumped and distributed line type, are presented in chapter 17. The implementation of the theory of transmission line resonators, and also the patch resonators, is the subject matter of chapter 18. The emphasis is placed on the circuit modeling of the resonating structures. Chapter 18 also discusses the fractal resonators and the dual-mode resonators with the illustrative examples of their applications. The resonating structures are important components for the development of the planar EBG and metamaterials.

Analytical Methods

Most of the discussions on the planar transmission lines and resonators are centered on the physical models and closed-form expression using circuit modeling. However, the analytical and quasi-analytical methods have been developed in the literature for more versatile and accurate modeling of the planar line structures in the multilayered medium. The *conformal mapping method*, as applied to the CPW and CPS line structures, is presented in chapter 9. The chapter 14 presents the *static variational methods*, both in the space-domain and Fourier domain, for the analysis of the microstrip and coupled microstrip. Using the *transverse transmission line (TTL) technique*, the variational method is extended to the multilayer microstrip lines. The method is extended to the boxed microstrip line and CPW using the Galerkin's method. However, the planar lines are both lossy and dispersive medium. The closed-form models are normally not available for the multilayer lossy planar lines. Chapter 15 presents the scheme of the *single-layer reduction (SLR) formulation* that utilizes the variational method and available single layer closed-form expressions to compute the dispersion and losses in the multilayer planar lines. The SLR models could be incorporated in the microwave CAD packages for the synthesis of the planar microwave components in the multilayer environment. Finally, the semi-analytical full-wave method, i.e. the *dynamic SDA* is elaborated in chapter 16. The treatment is at the introductory level with the detailed derivation of expressions. The method also incorporates the multilayered planar lines.

Contemporary Engineered Planar Structures

These are the topics of current interest and have wide-ranging applications in the development of novel devices and antenna structures with controlled functionalities. Chapter 19 is dedicated to the periodically loaded transmission lines and their circuit models. The method of the circuit models based analysis of the *electronic band-gap (EBG) line* is elaborated in detail and applied to the loaded microstrip and CPW line structures. The method is extended to the planar EBG surfaces in chapter 20. The accurate circuit model of the EBG surfaces is presented, and also the process of the EM-simulation and interpretation of the results are discussed.

The analyses of the engineered artificial media called the bulk metamaterials, metalines, and metasurfaces are presented in the chapters 21 and 22. The modeling and parameter extraction of the engineered media are also discussed. The metasurfaces are interesting structures with several possible applications both in the development of the circuits and planar antenna with multi-dimensional functionalities. The unique *generalized Snell's laws* are discussed in chapter 22 that helps to control the wave propagation. The polarization control and conversion of the state of polarization of the incident waves are also achieved with metasurfaces. Several applications of the metasurface are summarized.

1.3.2 Key Features, Intended Audience, and Some Suggestions

Key Features

Possibly, it is the first comprehensive book dedicated entirely to the planar transmission lines. The book covers the microstrip lines, CPW, and other classical resonating structures from basic to advanced form in one cover. The book further covers the EBG and metamaterial-based planar transmission lines and surfaces. These are the topics of current interest. The sequence of the chapter is logical and of increasing complexity.

The emphasis of the book is on the modeling of the planar lines and engineered surfaces using the physical concepts, circuit-models, closed-form expressions, and the derivation of a large number of expressions. It provides an in-depth review of the classical transmission line theory, electromagnetics, modeling of the material medium, and waveguide structures compactly without sacrificing the clarity of presentation.

The advanced mathematical treatment of the topics, such as the variational method, conformal mapping method, and SDA for several kinds of planar line structures is carried out in detail to help the new readers unfamiliar with these topics. A large number of illustrative examples from the published literature are given to clarify the theory and physical principles involved in the contemporary topics of the EBG lines and surfaces, metamaterials, metalines, and metasurfaces.

The multilayer structures useful for the MMIC, MEMS, and LTCC technology are covered. The closed-form modeling of dispersion, frequency-dependent characteristic impedance, and losses in the multilayer planar transmission lines is covered for the first time in a book-form. Also, the basic fabrication technology of the planar transmission lines in the MMIC, MEMS, and LTCC technology is described to help the modelers of the EM-phenomenon and microwave circuit designers.

Thus, the book prepares the reader to follow the modern designs of the planar circuits and also to undertake independent researches in the field of planar microwave technology.

Intended Audience

The book is intended to help undergraduate students of third/fourth year and also postgraduate students. It is useful to the teachers of microwave engineering in preparing lectures, assignments, and projects. The new researchers in the field of microwave engineering will find the book useful to improve their skills in the modeling of the planar structures. It is also suitable for the self-study of the RF/Microwave professionals in the industries. The selected chapters could be used in classroom teaching. It could be the main text for conducting elective courses at the university level. Further, the book can serve as a reference book even to more experienced users in the industry.

Some Suggestions

The students and new researchers can have a fast review of the transmission line theory, EM-theory, and so on from chapters 2–7. Then it will be much easier to follow the remaining text. It will be also helpful in following other advanced books and published current literature. However, experienced readers can read any chapter or topic. The concept for the *forward and backward referencing* of chapters, sections, and subsections are provided to help the reader to skip the chapters or to go back to an intended topic inside the book.

A large number of closed-form expressions are given for the modeling purpose. A reader is encouraged to write the Matlab codes or codes in any other language familiar to him/her. The teacher can assign to students the code development as homework. Such activities will enhance the learning process and skill development. More efforts will be needed to write the codes for the variational method, Galerkin's method, and the SDA. Of course, the effort is rewarding.

The researchers must look into the journals to follow newer investigations and decide the direction of his/her research activities. Good luck!

References

Books

B.1 Nahin Paul, J.: *Oliver Heaviside, Sage in Solitude: The Life, Work, and Times of an Electrical Genius of the Victorian Age*, IEEE Press, New York, 1988.

B.2 Mahon, B.: *Oliver Heaviside: Maverick Mastermind of Electricity*, IET, London, 2009.

B.3 Yavetz, I.: *From Obscurity to Enigma: The Works of Oliver Heaviside 1872–1889*, Birkhäuser, Basel, Switzerland, 2011.

B.4 Whittaker, E.: *A History of Theories of Aether and Electricity (Two Volumes)*, Dover Publications, Dallas, TX, USA, 1989.

B.5 Sarkar, T.K.; Mailloux, R.J.; Oliner, A.A.; Salazar-Palma, M.; Sengupta, D.L.: *History of Wireless*, John Wiley & Sons, Inc., Hoboken, New Jersey, USA, 2006.

B.6 Elliott, R.S.: *Electromagnetics: History, Theory, and Applications*, IEEE Press, Piscataway, NJ, 1993.

B.7 Lorentz, H.A.: *The Theory of Electrons and Its Applications to the Phenomena of Light and Radiant Heat*, 2nd Edition, B.G. Teubner, Leipzig, 1916.

B.8 Stratton, J.A.: *Electromagnetic Theory*, McGraw-Hill, New York, 1941.

B.9 Collin, R.E.: *Field Theory of Guided Waves*, 2nd Edition, IEEE Press, New York, 1991.

B.10 Harrington, R.F.: *Time-Harmonic Electromagnetic Fields*, McGraw-Hill, New York, 1961.

B.11 Feynman, R.P.; Leighton, R.B.; Sands, M.: *The Feynman Lectures on Physic, Vol. II, Electromagnetic Field*, pp. 15-8, 15-14, Addison – Wesley Pub. Co. Inc., Boston, Indian Reprint, B.I. Publications, New Delhi, 1969.

B.12 Thomson, J.J.: *Recent Researches in Electricity and Magnetism*, Clarendon Press, Oxford, 1893.

B.13 Fricke, H.M.: Two rival programme in 19th-century classical electrodynamics action – at- distance versus field theory, Ph. D. Thesis, London School of Economics & Political Science, 1982, London.

B.14 Fache, N.; Olyslager, F.; Zutter, D.D.: *Electromagnetic and Circuit Modeling of Multiconductor Transmission Lines*, Clarendon Press, Oxford, 1993.

B.15 Brillouin, L.: *Wave Propagation in Periodic Structures - Electric Filters and Crystal Lattice*, 2nd Edition, Dover Publications, New York, 1953.

B.16 Itoh, T. (Editor): *Planar Transmission Lines*, IEEE Press, Piscataway, NJ, 1987.

B.17 Khine, M.S.; Saleh, I.M. (Editors): *Models and Modeling in Science Education*, Vol. 6, Springer Dordrecht Heidelberg, London, New York, 2011.

Journals

J.1 Elliott, R.S.: *The history of electromagnetics as Hertz would have known it, IEEE Antennas Propag. Soc. Newsl.*, Vol. 30, No. 3, pp. 5–18, June 1988

J.2 Grattan-Guinness, I.: *Why did George Green write his essay of 1828 on electricity and magnetism?, Am. Math. Mon.*, Vol. 102, No. 5, pp. 387–397, DOI: 10.1080/00029890.1995.12004591http://about.jstor.org/terms.

J.3 Wu, A.C.T.; Yang, C.N.: *Evolution of the concept of the vector potential in the description of potential interactions, Int. J. Mod. Phys. A*, Vol. 21, No. 16, pp. 3235–3277, 2006.

J.4 Eliseo Fernández, E.: *Concepts and instruments: the potential from Green to Kelvin*, The Midwest Junto for the History of Science, March 25–27, 1994.

J.5 Challis, L.; Sheard, F.: *The green of Green's function, Phys. Today*, Vol. 56, No. 12, pp. 41–46, 2003.

J.6 John Roche, R.: *The present status of Maxwell's displacement current, Eur. J. Phys.*, Vol. 19, pp. 155–16n, 1998.

J.7 French, A.P.; Tessman, J.R.: *Displacement currents and magnetic fields, Am. J. Phys.*, Vol. 31, pp. 201–204, 1963.

J.8 Scheler, G.; Paulus, G.G.: *Measurement of Maxwell's displacement current, Eur. J. Phys.*, Vol. 36, No. 055048, pp. 1–9, 2015.

J.9 Selvan, K.T.: *Presentation of Maxwell's equations in historical perspective and the likely desirable outcomes, IEEE Antennas Propag. Mag.*, Vol. 49, No. 5, pp. 155–160, Oct. 2007.

J.10 Selvan, K.T.: *A revisiting of scientific and philosophical perspectives on Maxwell's displacement current, IEEE Antennas Propag. Mag.*, Vol. 51, No.3, pp. 36–46, June 2009.

J.11 Essex, E.A.: *Hertz vector potentials of electromagnetic theory, Am. J. Phys.*, Vol. 45, pp. 1099–1101, 1977.

J.12 Kraus, J.D.: *Heinrich Hertz-theorist and experimenter, IEEE Trans. Microwave Theory Tech.*, Vol. 36, No. 5, pp. 824–829, May 1988.

J.13 Susskind, C.: *Heinrich Hertz: A short life, IEEE Trans. Microwave Theory Tech.*, Vol. 36, No. 5, pp. 802–805, May 1988.

J.14 Bladel, J.V.: *Lorenz or Lorentz?, IEEE Antennas Propag. Mag.*, Vol. 33, No. 2, pp. 69, April 1991.

J.15 Jackson, J.D.; Okun, L.B.: *Historical roots of gauge invariance, Rev. Mod. Phys.*, Vol. 73, pp. 663–680, July 2001.

J.16 Griffiths, D.J.: *Resource letter EM-1: Electromagnetic momentum, Am. J. Phys.*, Vol. 80, No. 1, pp. 7–18, Jan. 2012.

J.17 Aharonov, Y.; Bohm, D.: *Significance of electromagnetic potentials in quantum theory, Phys. Rev.*, Vol. 115, pp. 485–491, 1959.

J.18 Thomson, W. (Kelvin): *On the theory of the electric telegraph, Proc. R. Soc. London*, Vol. 4, No. 11, pp. 382–399, May 1855.

J.19 Searle, G.F.C., et al. *The Heaviside Centenary Volume*, The Institution of Electrical Engineers, London, 1950.

J.20 Whittaker, E.T.: *Oliver Heaviside*, In *Electromagnetic Theory* Vol. 1, Oliver Heaviside, Reprint, Chelsea Publishing Company, New York, 1971.

J.21 Pupin, M.I.: *Wave propagation over non-uniform cables and long-distance airlines, Trans. Am. Inst. Electr. Eng.*, Vol. 17, pp. 445–507, (discussion on pp. 508–512), May 1900.

J.22 Pupin, M.I.: *Propagation of long electrical waves, Trans. Am. Inst. Electr. Eng.*, Vol. xv, No. 144, pp. 93–142, March 1899.

J.23 Campbell, G.A.: *On loaded lines in telephonic transmission, London, Edinburg Dublin Philos. Mag. J. Sci.*, Vol. 5, No. 27, pp. 313–330, 1903.

J.24 Kennelly, A.E.: *Impedance*, 76th Meeting of American Institute of Electrical Engineers, pp. 175–232, April 1893.

J.25 Carson, J.R.: *Wave propagation over parallel wires: The proximity effect, London, Edinburgh Dublin Philos. Mag. J. Sci.*, Vol. 41, No. 244, pp. 607–633, 1921

J.26 Carson, J.R.: *Electromagnetic theory and the foundations of electrical circuit theory, Bell Syst. Tech. J.*, Vol. 6, pp. 1–17, Jan. 1927.

J.27 Levin, A.: *Electromagnetic waves guided by parallel wires with particular reference to the effect of the earth, Trans. Am. Inst. Electr. Eng.*, Vol. 46, pp. 983–989, June 1927.

J.28 Bewley, L.V.: *Traveling waves on transmission systems, Trans. Am. Inst. Electr. Eng.*, Vol. 50, No. 2, pp. 532–550, June 1931.

J.29 Pipes, L.A.: *Matrix theory of multiconductor transmission lines, London, Edinburgh Dublin Philos. Mag. J. Sci.*, Vol. 24, No. 159, pp. 97–113, July 1937.

J.30 Rayleigh, J.W.S.: *On the passage of electric waves through tubes, or vibrations of dielectric cylinders, Philos. Mag.*, Vol. 43, pp. 125–132, Feb. 1897.

J.31 Southworth, G.C.: *Hyper – frequency waveguides – General consideration and experimental results, Bell Sys. Tech. J.*, Vol. 15, pp. 284–309, 1936.

J.32 Barrow, W.L.: *Transmission of electromagnetic waves in hollow tubes of metal, Proc. IRE*, Vol. 24, pp. 1298–1328, Oct. 1936.

J.33 Brillouin, L.: *Propagation of electromagnetic waves in a tube, Rev. GMn. de l'Elec.*, Vol. 40, pp. 227–239, Aug. 1936.

J.34 Carson, J.R.; Mead, S.P.; Schelkunoff, S.A.: *Hyper-frequency wave guides – mathematical theory, Bell Sys. Tech. J.*, Vol. 15, pp. 310–333, 1936.

J.35 Chu, L.I.; Barrow, W.L.: *Electromagnetic waves in hollow metal tubes of rectangular cross- section, Proc. IRE*, Vol. 26, 1520, 1938.

J.36 Schelkunoff, S.A.: *The electromagnetic theory of coaxial transmission lines and cylindrical shields, Bell Sys. Tech. J.*, Vol. 13, No. 4, pp. 532–579, Oct. 1934.

J.37 Schelkunoff, S.A.: *Transmission theory of plane electromagnetic waves, Proc. IRE*, Vol. 25, p. 1437, 1937.

J.38 Schelkunoff, S.A.: *Generalized telegraphist's equations for waveguides, Bell Sys. Tech. J.*, Vol. 31, No. 4, pp. 784–801, July 1952.

J.39 Schelkunoff, S.A.: *Conversion of Maxwell's equations into generalized telegraphist's equations, Bell Sys. Tech. J.*, Vol. 34, No. 5, pp. 995–1043, Sept. 1955.

J.40 Niehenke, E.C.; Pucel, R.A.; Bahl, I.J.: *Microwave and millimeter-wave integrated circuits, IEEE Trans. Microwave Theory Tech.*, Vol. 50, No. 3, pp. 846–857, March 2002.

J.41 Greig D.D.; Engelmann, H.: *Microstrips – A new transmission technique for the kilomegacycle range, Proc. IRE*, Vol. 40, pp. 1644–1650, Dec. 1952.

J.42 Barrett, R.M.: *Microwave printed circuits-A historical survey, IEEE Trans. Microwave Theory Tech.*, Vol. 3, No. 2, pp. 1–9, Mar. 1955.

J.43 Barrett, R.M.: *Microwave printed circuits - The early years, IEEE Trans. Microwave Theory Tech.*, Vol. MTT-32, No. 9, pp. 983–990, Sept. 1984.

J.44 Cohn, S.B.: *Characteristic impedance of the shielded-strip transmission line, IRE Trans. Microwave Theory Tech.*, Vol. MTT-2, pp. 52–57, July 1954.

J.45 McQuiddy, D.N.; Wassel, J.W.; LaGrange, J.B.; Wisseman, W.R.: *Monolithic microwave integrated circuits: An historical perspective, IEEE Trans. Microwave Theory Tech.*, Vol. 32, No. 9, pp. 997–1008, 1984.

J.46 Petersen, K.E.: *Micro-mechanical membrane switches on Silicon, IBM J. Res. Dev.*, Vol. 23, No. 4, pp. 376–385, July 1979.

J.47 Yao, J.; Chang, M.F.: *A surface micro-machined miniature switch for telecommunication application* *with single freq. from DC up to 4 GHz*, Proc. Transducer'95, pp. 384–387, June 1995.

J.48 Goldsmith, C.; Lin, T.H.; Powers, B.; Wu, W.R.; Norvell, B.: *Micro-mechanical membrane switches for microwave application*, IEEE Microwave Theory Tech. Symp., MTT-S Digest, pp. 91–94, May 1995.

J.49 Rodriguez, A.R.; Wallace, A.B.: *Ceramic capacitor and method of making it*, Patent US 3004197, issued 10/10/1961.

J.50 Hajian, A.; Müftüoglu, D.; Konegger, T.; Schneider, M.; Schmid, U.: *On the porosification of LTCC substrates with sodium hydroxide, Compos. Part B: Eng.*, Vol. 157, pp. 14–23, 2019.

J.51 Veselago, V.: *The electrodynamics of substances with simultaneously negative values of* ε *and* μ, *Soviet Physics Uspekhi*, Vol. 10, No. 4, pp. 509–514, Jan., Feb. 1968.

J.52 Pendry, J.B.; Holden, A.J.; Stewart, W.J.; Youngs, I.: *Extremely low-frequency plasmons in metallic mesostructures, Phys. Rev. Lett.*, Vol. 76, No. 25, pp. 4773–4776, June 1996.

J.53 Pendry, J.B.; Holden, A.J.; Robbins, D.J.; Stewart, W.J.: *Magnetism from conductors and enhanced nonlinear phenomena, IEEE Trans. Microwave Theory Tech.*, Vol. 47, No. 11, pp. 2075–2084, Nov. 1999.

J.54 Shelby, R.A.; Smith, D.R.; Schultz, S.: *Experimental verification of a negative index of refraction, Science*, Vol. 292, No. 5514, pp. 77–79, April 2001.

J.55 Smith, D.R.; Padilla, W.J.; Vier, D.C.; Nemat-Nasser, S. C.; Schultz, S.: *Composite medium with simultaneously negative permeability and permittivity, Phys. Rev. Lett.*, Vol. 84, No. 18, pp. 4184–4187, May 2000.

J.56 Assaudourion, F.; Rimai, E.: *Simplified theory of microstrip transmission systems, Proc. IRE*, Vol. 40, pp. 1651–1657, Dec. 1952.

J.57 Cohn, S.B.: *Characteristic impedance of the shielded-strip transmission line, IRE Trans. Microwave Theory Tech.*, Vol. MTT-2, pp. 52–57, July 1954.

J.58 Cohn, S.B.: *Problems in strip transmission lines, IEEE Trans. Microwave Theory Tech.*, Vol. MTT-3, No. 2, pp. 119–126, March 1955.

J.59 Cohn, S.B.: *Shielded coupled-strip transmission line, IRE Trans. Microwave Theory Tech.*, Vol. MTT-3, pp. 29–38, Oct. 1955.

J.60 Cohn, S.B.: *Characteristic impedances of broadside-coupled strip transmission lines, IRE Trans. Microwave Theory Tech.*, Vol. MTT-8, pp. 633–637, Nov. 1960.

J.61 Wheeler, H.A.: *Transmission-line properties of parallel wide strips by a conformal-mapping approximation, IEEE Trans. Microwave Theory Tech.*, Vol. MTT-12, pp. 280–289, May 1964.

J.62 Wheeler, H.A.: *Transmission-line properties of parallel strips separated by a dielectric sheet, IEEE Trans. Microwave Theory Tech.*, Vol. MTT-13, pp. 172–185, March 1965.

J.63 Wheeler, H.A.: *Transmission-line properties of a strip on a dielectric sheet on a plane, IEEE Trans. Microwave Theory Tech.*, Vol. MTT-25, pp. 631–647, Aug. 1977.

J.64 Wheeler, H.A.: *Transmission line properties of a stripline between parallel planes, IEEE Trans. Microwave Theory Tech.*, Vol. MTT-26, pp. 866–876, Nov. 1978.

J.65 Cohn, S.B.: *Slotline on a dielectric substrate, IEEE Trans. Microwave Theory Tech.*, Vol. MTT-17, pp. 768–778, Oct. 1969.

J.66 Wen, C.P.: *Coplanar waveguide: a surface strip transmission line suitable for nonreciprocal gyromagnetic device application, IEEE Trans. Microwave Theory Tech.*, Vol. MTT-17, pp. 1087–1090, Dec. 1969.

J.67 Meier, P.J.: *Two new integrated-circuit media with special advantages at millimeter-wave lengths*, IEEE MTT-S Int. Microwave Symp. Dig., pp. 221–223, 1972.

J.68 Yamashita, E.; Mitra, R.: *Variational method for analysis of microstrip lines, IEEE Trans. Microwave Theory Tech.*, Vol. MTT-16, No. 4, pp. 251–256, April 1968.

J.69 Yamashita, E.: *Variational method for analysis of microstrip – like transmission lines, IEEE Trans. Microwave Theory Tech.*, Vol. MTT-16, No. 8, pp. 529–535, Aug. 1968.

J.70 Itoh, T.; Herbert, A.S.: *A generalized spectral domain analysis for coupled suspended microstrip lines with tuning septum, IEEE Trans. Microwave Theory Tech.*, Vol. MTT-26, No. 10, pp. 820–826, Oct. 1978.

J.71 Itoh, T.: *A generalized spectral domain analysis for multiconductor printed lines and its application to tunable suspended microstrips, IEEE Trans. Microwave Theory Tech.*, Vol. MTT-26, No. 12, pp. 983–987, Dec. 1978.

J.72 Itoh, T.; Mittra, R.: *Spectral-domain approach for calculating the dispersion characteristics of microstrip lines, IEEE Trans. Microwave Theory Tech.*, Vol. MTT-21, pp. 496–499, July 1973.

J.73 Itoh, T.; Mittra, R.: *Dispersion characteristics of slot lines, Electron. Lett.*, Vol. 7, pp. 364–365, 1971.

J.74 Itoh, T.: *Spectral-domain immittance approach for dispersion characteristics of generalized printed transmission lines, IEEE Trans. Microwave Theory Tech.*, Vol. MTT-28, pp. 733–736, 1980.

J.75 Jansen, R.H.: *High-speed computation of single and coupled microstrip parameters including dispersion, high-order modes, loss and finite strip thickness, IEEE Trans. Microwave Theory Tech.*, Vol. MTT-26, pp. 75–82, 1978.

2

Waves on Transmission Lines – I

(Basic Equations, Multisection Transmission Lines)

Introduction

The two-wire transmission line is a useful medium for the propagation of the voltage and current waves. It is also useful in the modeling of planar transmission lines. The EM-wave propagating on multilayered planar transmission lines could also be analyzed with the help of the multisection transmission lines. The primary purpose of this chapter is to review in detail the wave propagation on a transmission line without and with sources. Several other important topics such as the characterization of a line section, nature of wave velocities, dispersion, and reactively loaded lines are further discussed in chapter 3.

Objectives

- *To formulate Kelvin–Heaviside transmission line equations.*
- *To obtain the solution of the wave equation.*
- *To compute the power flow on a transmission line.*
- *To use Thevenin's theorem on a transmission line section to obtain its transfer function.*
- *To consider the wave propagation on a multi-section transmission line with the voltage and current sources.*
- *To understand the nature of the wave propagation on a nonuniform transmission line.*

2.1 Uniform Transmission Lines

This section presents the basic understanding of wave propagation on a uniform transmission line. The *Kelvin–Heaviside transmission line equations* are formulated using the *lumped circuit elements model* of the uniform transmission line.

The voltage and current wave equations are obtained and solved for a terminated line section. The phenomenon of the standing wave is discussed. The *Thevenin theorem* of the transmission line network is discussed and the transfer function of a line section is obtained.

2.1.1 Wave Motion

The wave motion could be treated as the transfer of oscillation from one location to another location. A harmonic oscillation is described either by a sine or by a cosine function. A periodic oscillation has a fixed period. An oscillation repeats itself after the periodic time (T). In general, an oscillation, i.e. an oscillatory motion can have any shape such as square, triangular, and so on. However, with the help of the Fourier series, such periodic oscillations can be decomposed to the harmonic functions. Likewise, wave motion can also acquire an arbitrary shape. The arbitrary periodic shape of a wave can be decomposed into the harmonic waves.

Figure (2.1) shows the instantaneous amplitude v(t) of the wave generated by an oscillating quantity, such as an oscillating ball (particle) at the location A. It has an angular frequency of oscillation ω radian/sec and maximum amplitude $V_{max.}$ The ball is attached to a spring and immersed in a water tank. It generates a wave motion at the surface of the water. The water wave travels with velocity v_p and causes another *delayed oscillation,* at the location B, in a similar ball attached to a spring. *Through the mechanism of wave motion, the oscillation is transferred from the location A to the location B separated by distance x.* Likewise, the oscillating quantity could be a charge, electric field, magnetic field, or voltage obtained from an oscillator connected to a cable.

The equation of the harmonic oscillation at location A is described by the cosine function,

$$v_a(t) = V_{max} \cos \omega t. \tag{2.1.1}$$

The equation of harmonic oscillation that appears at location B after a delayed time t' is

$$v_b(t) = V_{max} \cos \omega(t - t'). \tag{2.1.2}$$

The wave motion, created by the oscillation at the location A, travels with velocity v_p. The time delay in

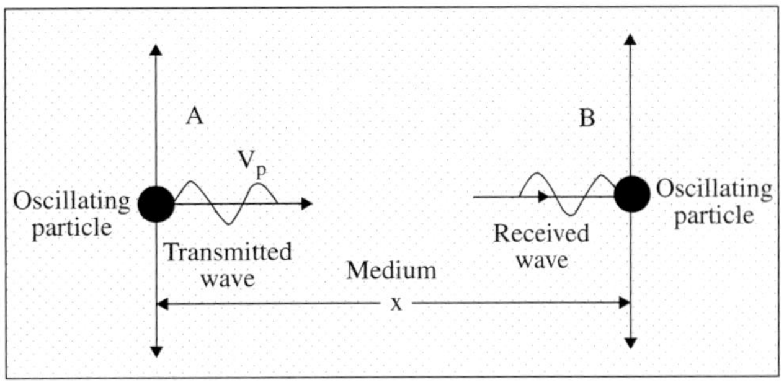

Figure 2.1 Delayed oscillation as a wave motion-initial oscillation v(t) at location A, and delayed oscillation $v(t - x/v_p)$ at location B. The waves propagate from the locations A to B.

setting up the oscillation at the location B is $t' = x/v_p$. So the equation of oscillation at the location B is

$$v_b(t) = V_{max} \cos \omega (t - x/v_p). \tag{2.1.3}$$

Equation (2.1.3) describes the propagating wave through the medium between locations A and B. It is called the *wave equation*. The *phase constant* β of the propagating wave is defined as $\beta = \omega/v_p$. On dropping the subscript "b," equation (2.1.3) is rewritten in the usual form,

$$v(t, x) = V_{max} \cos(\omega t - \beta x). \tag{2.1.4}$$

In equation (2.1.4), the *lagging phase* $\phi = \beta x$ is caused by the delay in oscillation at the receiving end. The medium supporting the wave motion is assumed to be lossless. The wave variable v(t, x) is a doubly periodic function of both time and space coordinates as shown in Fig (2.2a and b), respectively.

The *temporal period*, i.e. time-period T is related to the angular frequency ω as follows:

$$T = \frac{2\pi}{\omega} = \frac{1}{f}, \tag{2.1.5}$$

where f is the frequency in Hz (Hertz), i.e. the number of cycles per sec. The wavelength (λ) describes the *spatial period*, i.e. space periodicity. It is related to the *phase-shift constant (or phase constant)*, i.e. the propagation constant β, as follows:

$$\lambda = \frac{2\pi}{\beta}. \tag{2.1.6}$$

The wave motion or velocity of a *monochromatic wave* (the wave of a single frequency) is the motion of its *wavefront*. For the 3D wave motion in an isotropic space, the wavefront is a surface of the constant phase. In the case of the 1D wave motion, the wavefront is a line, whereas, for the 2D wave motion on an isotropic surface, the wavefront is a circle.

Figure (2.3) shows the peak point P at the wavefront of the ID wave. The motion of a constant phase surface is known as its *phase velocity* v_p. Thus, the peak point P at

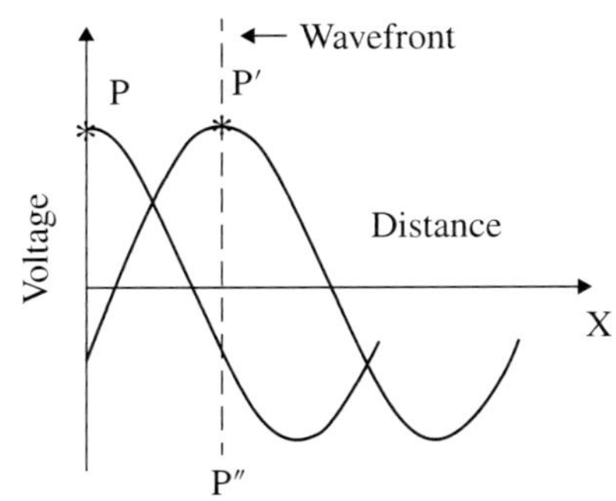

Figure 2.3 Wave motion as a motion of constant phase surface. It is shown as the moving point P.

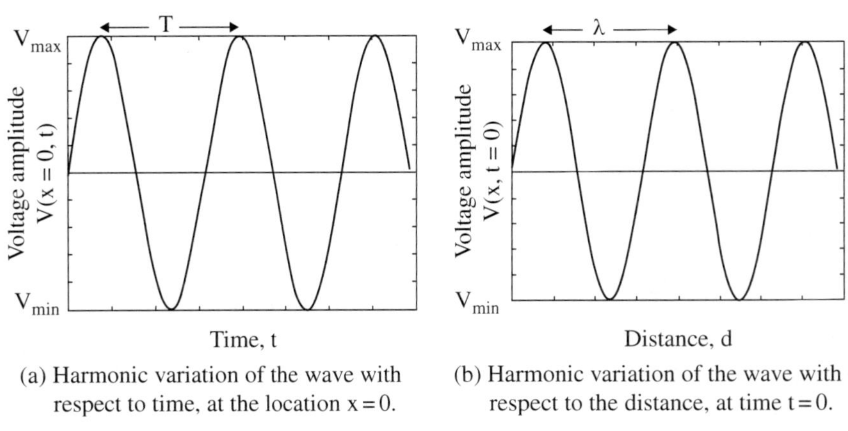

Figure 2.2 Double periodic variations of wave motion.

the wavefront has moved to a new location P′ in such a way that phase of the propagating wave remains constant, i.e. $\omega t - \beta x = \text{constant}$. On differentiating the constant phase with respect to time t, the expression for the phase velocity is obtained:

$$v_p = \frac{dx}{dt} = \frac{\omega}{\beta}. \tag{2.1.7}$$

In a *nondispersive* medium, the phase velocity of a wave is independent of frequency, i.e. the waves of all frequencies travel at the same velocity. Figure (2.4) shows the nondispersive wave motion on the $(\omega - \beta)$ diagram. It is a *linear graph*. The slope of the point Q on the dispersion diagram subtends an angle θ at the origin that corresponds to the *phase velocity* of a propagating wave,

$$v_p = \tan\theta = \frac{\omega}{\beta}. \tag{2.1.8}$$

Thus, for a frequency-independent nondispersive wave, the phase constant is a linear function of angular frequency.

$$\beta = \text{Constant} \times \omega. \tag{2.1.9}$$

The dispersive nature of the 1D wave motion is further discussed in sections (3.3) and (3.4) of the chapter 3. The phase velocity in a dispersive medium is usually frequency-dependent. It is known as the *temporal dispersion*. In some cases, the phase velocity could also depend on wavevector (β, or k). It is known as *spatial dispersion*. The spatial dispersion is discussed in the subsection (21.1.1) of the chapter 21. The dispersion diagrams of the wave propagation in the isotropic and anisotropic media are further discussed in the section (4.7) of the chapter 4, and also in the section (5.2) of chapter 5.

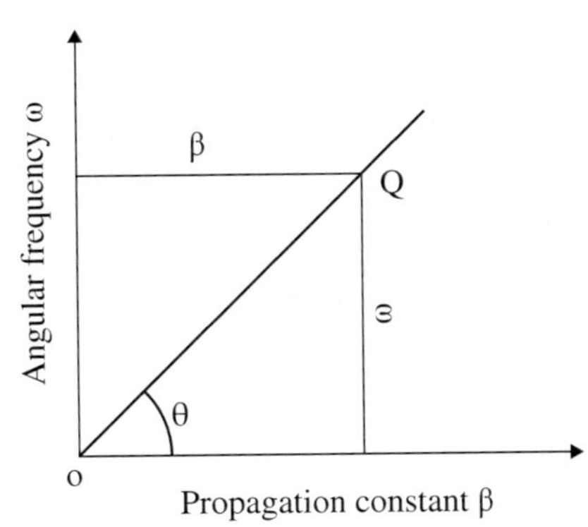

Figure 2.4 The ω-β dispersion diagram of nondispersive wave.

The one-dimensional (1D) wave travels both in the forward and in the backward directions. It can have any arbitrary shape. In general, the 1D wave propagation can be described by the following *wave function*, known as the *general wave equation*:

$$\psi(t, x) = \psi\left(t \mp \frac{x}{v_p}\right). \tag{2.1.10}$$

On taking the second-order partial derivative of the wave function $\psi(t, x)$ with respect to space-time coordinate x and t, the 1D second-order partial differential equation (PDE) of the wave equation is obtained. Similarly, wave functions $\psi(t, x, y)$ and $\psi(t, x, y, z)$ are solutions of two-dimensional (2D) and three-dimensional (3D) wave equations supported by the surface and free space medium, respectively. These PDEs are summarized below:

$$\frac{\partial^2\psi}{\partial x^2} = \frac{1}{v_p^2}\frac{\partial^2\psi}{\partial t^2} \quad (a), \qquad \frac{\partial^2\psi}{\partial x^2} + \frac{\partial^2\psi}{\partial y^2} = \frac{1}{v_p^2}\frac{\partial^2\psi}{\partial t^2} \quad (b)$$
$$\frac{\partial^2\psi}{\partial x^2} + \frac{\partial^2\psi}{\partial y^2} + \frac{\partial^2\psi}{\partial z^2} = \frac{1}{v_p^2}\frac{\partial^2\psi}{\partial t^2} \quad (c), \Rightarrow \nabla^2\psi = \frac{1}{v_p^2}\frac{\partial^2\psi}{\partial t^2} \quad (d). \tag{2.1.11}$$

The dispersion diagram of the 2D wave propagation over the (x, y) surface is obtained by revolving the slant line of Fig (2.4) around the ω-axis with propagation constants β_x, β_y in the *x*- and *y*-direction. It is discussed in subsection (4.7.4) of chapter 4.

2.1.2 Circuit Model of Transmission Line

A physical transmission line, supporting the voltage/current wave, can be modeled by the lumped R, L, C, G components, i.e. the resistance, inductance, capacitance, and conductance per unit length (*p.u.l.*), respectively. The two-conductor transmission line can acquire many physical forms. A few of these forms are shown in Fig (2.5). The lines as shown in Fig (2.5a–c) support the wave propagation in the **t**ransverse **e**lectro**m**agnetic mode, i.e. in the *TEM-mode; while Fig (2.5c) shows the quasi-TEM mode-supporting microstrip*. For TEM mode wave propagation, the electric field and magnetic field are normal to each other and also normal to the direction of wave propagation. For the TEM mode, there is no field component along the direction of propagation. However, the quasi-TEM mode also has component of weak fields along the longitudinal direction of wave propagation. The quasi-TEM mode is a *hybrid mode* discussed in the subsection (7.1.4) of chapter 7.

All TEM mode supporting transmission lines can be represented by a parallel two-wire transmission line

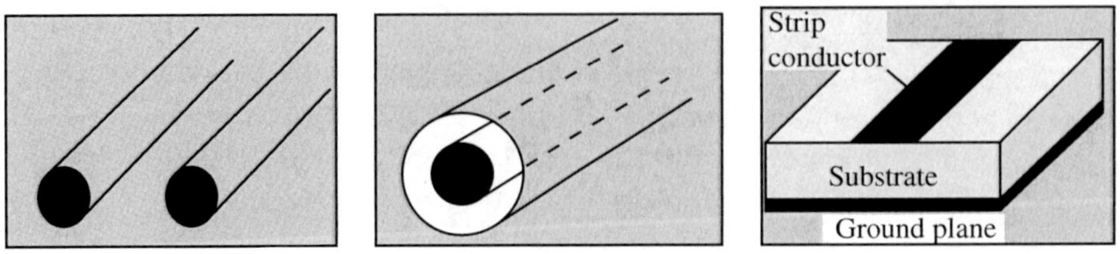

(a) Two-wire line. (b) Co-axial line. (c) Microstrip line.

Figure 2.5 Cross-section of a few two-conductor transmission lines.

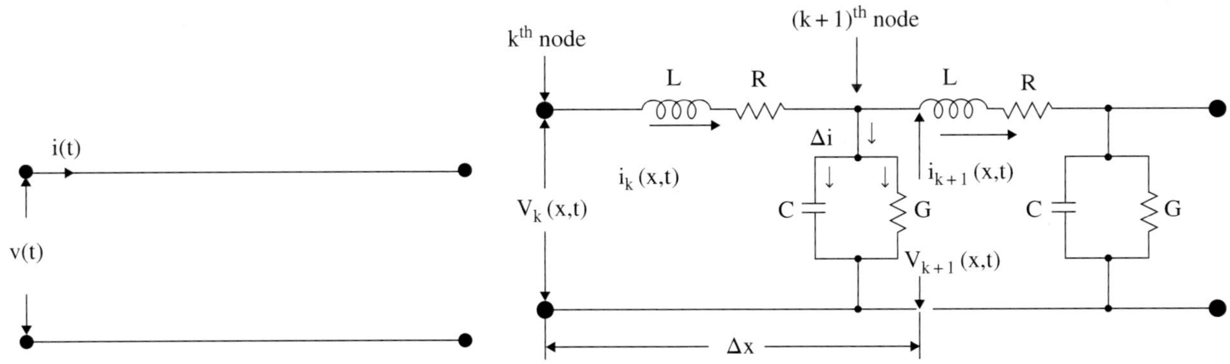

(a) Physical two-wire transmission line. (b) Equivalent circuit of the two-wire transmission line.

Figure 2.6 RLCG lumped circuit model of a transmission line.

shown in Fig (2.6a). A transmission line is a 1D wave supporting structure. Its cross-sectional dimension is much less than $\lambda/4$; otherwise, its TEM nature is changed. The longitudinal dimension can have any value, from a fraction of a wavelength to several wavelengths. The mode theory of the electromagnetic (EM) wave propagation is further discussed in chapter 7.

A two-conductor transmission line, or any other line supporting the TEM mode, is modeled as a chain of discrete passive RLCG components. As a matter of fact, by cascading several sections of discrete L-network of the series L and shunt C elements, or even discrete L-network of the series C and shunt L elements, an *artificial transmission line* can also be constructed. The artificial transmission line is discussed in section (3.4) of chapter 3, and also in the chapters 19 and 22. It plays a very important role in modern microwave planar technology. The behavior of a transmission line is determined in terms of the resistor (R), inductor (L), capacitor (C), and conductance (G); all line elements are in per unit length, i.e. *p.u.l.* Kelvin introduced the modeling of the telegraph cable laid in the ocean using the *RC circuit model*. Heaviside further introduced L and G components in the circuit model to improve the modeling of the lossy transmission line [B.1, B.2, J.1, J.2]. Kelvin RC-circuit model of the transmission line leads to the diffusion equation, not to the wave equation. Whereas using the RLGC circuit model, Heaviside finally obtained the wave equation for the voltage/current on a transmission line. Using the RLCG circuit model, shown in Fig (2.6b), the voltage, and current equations are obtained for the transmission line. The set of the coupled voltage and current equations are normally called the *telegrapher's equations;* as it was originally developed for the telegraph cables. However, the set of coupled transmission line equations can be called the *Kelvin–Heaviside transmission line equations* to recognize their contributions.

The Resistance of a Line

The electrical loss in a transmission line, known as the *conductor loss*, is due to the finite conductivity of the line. It is modeled as the resistance R *p.u.l.* It is also influenced by the skin effect phenomena at a higher frequency. The instantaneous current $i(t)$ flowing through a lumped resistance R^{lum} is related to the instantaneous voltage drop $v(t)$ by Ohm's law:

$$v(t) = R^{lum} i(t). \tag{2.1.12}$$

The Inductance of a Line

The current flowing in a conductor creates the magnetic field around itself. So the magnetic energy stored in the space around the transmission line, i.e. the time-varying current supporting line section, is modeled by a series inductor L *p.u.l.* The inductance of a line is not lumped at one point, i.e. it is not confined at one point. It is distributed over the whole length of a line. The instantaneous voltage across a lumped inductor L^{lum} is related to the current flowing through it by

$$v(t) = L^{lum}\frac{di(t)}{dt}. \tag{2.1.13}$$

The Capacitance of a Line

In a two-conductor transmission line, the conductors separated by a dielectric medium form a distributed system of capacitance. The electric field energy stored in a line is modeled by the shunt capacitance C *p.u.l.* The instantaneous shunt current through a lumped capacitor C^{lum} is related to the instantaneous voltage across it by

$$i_{sh}(t) = C^{lum}\frac{dv(t)}{dt}. \tag{2.1.14}$$

The Conductance of a Line

If the medium between two conductors of the transmission line is not a perfect dielectric, i.e. if it has finite conductivity, then a part of the line current shunts through the medium causing the *dielectric loss*. The dielectric loss of the line is modeled by shunt conductance G *p.u.l.* The instantaneous shunt current is related to the instantaneous voltage across a lumped conductance by

$$i_{sh}(t) = G^{lum}v(t). \tag{2.1.15}$$

Figure (2.6a) shows a physical transmission line that supports the TEM mode wave propagation. Figure (2.6b) shows that this line could be modeled as a chain of the lumped RLCG structure. More numbers of RLCG sections per wavelength are needed to model a transmission line. The RLCG model is the modeling of a transmission line by the lossy series inductor and lossy shunt capacitor. The transmission line supports the wave propagation from power AC to RF and above. Likewise, the corresponding lumped components model of a transmission line also supports such waves. The transmission line structure behaves like a low-pass filter.

2.1.3 Kelvin–Heaviside Transmission Line Equations in Time Domain

Figure (2.6b) shows the *lumped element equivalent circuit model* of a section Δx of the transmission line. The primary line constants R, L, C, G on the per unit length (*p.u.l.*) basis are related to the lumped circuit elements as $R^{lum} = R\Delta x$, $L^{lum} = L\Delta x$, $C^{lum} = C\Delta x$, and $G^{lum} = G\Delta x$. Figure (2.6b) shows the *voltage loop equation* and the *current node equation* for a small line section Δx. These are written as follows:

The Loop Equation

Differential voltage change across line length Δx = Voltage drop across inductor + Voltage drop across resistance

$$v_k - v_{k+1} = \Delta v = -(L\Delta x)\frac{\partial i}{\partial t} - (R\Delta x)i. \tag{2.1.16}$$

The instantaneous line voltage v(t) and the instantaneous line current i(t) are functions of the space–time variables x and t. In the limiting case of $\Delta x \to 0$, equation (2.1.16) is written as

$$\frac{\partial v}{\partial x} = -\left(Ri + L\frac{\partial i}{dt}\right). \tag{2.1.17}$$

The Node Equation

Differential shunt current at the node = Current through conductance + Current through capacitor

$$i_k - i_{k+1} = \Delta i = -(G\Delta x)v - (C\Delta x)\frac{\partial v}{\partial t}. \tag{2.1.18}$$

In the limiting case, $\Delta x \to 0$; and the above equation is written as

$$\frac{\partial i}{\partial x} = -\left(Gv + C\frac{\partial v}{\partial t}\right). \tag{2.1.19}$$

The pair of coupled voltage and current transmission line equations in the time-domain summarized below is known as "*the time domain telegrapher's equations*":

$$\frac{\partial v}{\partial x} = -\left(Ri + L\frac{\partial i}{\partial t}\right) \quad (a), \quad \frac{\partial i}{\partial x} = -\left(Gv + C\frac{\partial v}{\partial t}\right) \quad (b). \tag{2.1.20}$$

The above pair of equations can also be called the *Kelvin–Heaviside transmission line equations.* The coupled Kelvin–Heaviside transmission line equations relate the voltage and current on a transmission line through the line parameters, RLCG. These parameters are known as the *primary constants* of a line. The RLCG parameters depend on the physical configuration of a line, i.e. on its physical shapes, dimensions, and the electrical properties of the medium. They could be frequency-dependent parameters also. For simple transmission lines such as a pair of wires and coaxial lines, the closed-form formulas are available to compute them. However, these parameters could also be obtained through measurements. The empirical expressions for the RLCG parameters of a *microstrip line* are also available [J.3]. The transmission line equations are the complementary parts of Maxwell's field equations that relate the time-varying electric and magnetic fields in a physical medium through the primary constants of a medium-conductivity (σ), permittivity (ε) and permeability (μ). It is discussed in chapter 4. The transmission line equations can be obtained from Maxwell's equations. It is discussed in section (7.3) of chapter 7.

The voltage and current coupled variables of equation (2.1.20) can be separated. The separation of the variables leads to the wave equations for the voltage and current waves on a transmission line. On differentiating equation (2.1.20a) with respect to the variable x, the *voltage wave equation* is obtained:

$$\frac{\partial^2 v}{\partial x^2} = -\left[R\frac{\partial i}{\partial x} + L\frac{\partial}{\partial x}\left(\frac{\partial i}{\partial t}\right)\right] = -\left[R\frac{\partial i}{\partial x} + L\frac{\partial}{\partial t}\left(\frac{\partial i}{\partial x}\right)\right]. \tag{2.1.21}$$

On substituting $\frac{\partial i}{\partial x}$ from equation (2.1.20b) in equation (2.1.21), the voltage wave equation is

$$\frac{\partial^2 v}{\partial x^2} = \left[RGv + (RC + LG)\frac{\partial v}{\partial t} + LC\frac{\partial^2 v}{\partial t^2}\right]. \tag{2.1.22}$$

The above partial differential equation describes the time-domain voltage wave on a lossy transmission line. Likewise, an equation could be written to describe the current wave on a transmission line:

$$\frac{\partial^2 i}{\partial x^2} = \left[RGi + (RC + LG)\frac{\partial i}{\partial t} + LC\frac{\partial^2 i}{\partial t^2}\right]. \tag{2.1.23}$$

A lossless transmission line has, $R = G = 0$. The voltage and current waves on a lossless line are given by the following 1D PDEs:

$$\frac{\partial^2 v}{\partial x^2} = LC\frac{\partial^2 v}{\partial t^2}. \tag{2.1.24}$$

$$\frac{\partial^2 i}{\partial x^2} = LC\frac{\partial^2 i}{\partial t^2}. \tag{2.1.25}$$

On comparing the above equations with equation (2.1.11a), the velocity of propagation, for both the current and voltage waves, is

$$v_p = \frac{1}{\sqrt{LC}}. \tag{2.1.26}$$

It is like the velocity of propagation of an electromagnetic wave in a dielectric medium obtained from Maxwell's equations, where the primary constant of the line L and C are replaced by the medium constants permeability μ and permittivity ε. The EM-wave is discussed in chapter 4.

2.1.4 Kelvin–Heaviside Transmission Line Equations in Frequency-Domain

The time-harmonic instantaneous voltage in the frequency domain, i.e. in the phasor form, is written as

$$v(t) = V_{max}\cos(\omega t + \phi) = \text{Re}\left(\tilde{V}e^{j\omega t}\right). \tag{2.1.27}$$

where "Re" stands for the real part of the voltage phasor $\tilde{V}$. The voltage phasor $\tilde{V}$ is given by the following expression:

$$\tilde{V} = \tilde{V}_{max}e^{j\phi}. \tag{2.1.28}$$

The phasor is nothing but a polar form of a complex quantity. Likewise, the instantaneous current in the phasor form is

$$i(t) = I_{max}\cos(\omega t + \varphi) = \text{Re}\left(\tilde{I}e^{j\omega t}\right), \tag{2.1.29}$$

where current phasor is

$$\tilde{I} = \tilde{I}_{max}e^{j\varphi}. \tag{2.1.30}$$

The phasor is either a constant or a function of only the space variable. It is not a function of time *t*. The phasor is shown with a tilde (~) sign in this chapter. However, in the subsequent chapters, the tilde (~) sign is dropped. The phasor is used at a single frequency. Using the phasor notation, the voltage across R, L, and the current through C, G; given by equations (2.1.12)–(2.1.15) in the time domain, can be rewritten in the frequency-domain:

$$\begin{aligned} &\tilde{V} = R\tilde{I} \quad (a), \quad \tilde{V} = j\omega L\tilde{I} \quad (b) \\ &\tilde{I} = j\omega C\tilde{V} \quad (c), \quad \tilde{I} = G\tilde{V} \quad (d). \end{aligned} \tag{2.1.31}$$

In the above equations, the time derivative $\partial/\partial t$ is replaced by $j\omega$ converting the expression from the time-domain to frequency-domain. Following the conversion process, the time-domain coupled voltage-current transmission line equation (2.1.20), is rewritten in the frequency-domain:

$$\frac{d\widetilde{V}}{dx} = -(R + j\omega L)\widetilde{I} \quad \text{(a)}$$
$$\frac{d\widetilde{I}}{dx} = -(G + j\omega C)\widetilde{V} \quad \text{(b)}. \qquad (2.1.32)$$

On separation of the voltage and current variables, the following voltage and current wave equations are obtained in the frequency-domain:

$$\frac{d^2\widetilde{V}}{dx^2} = (R + j\omega L)(G + j\omega C)\widetilde{V} \quad \text{(a)}$$
$$\frac{d^2\widetilde{I}}{dx^2} = (R + j\omega L)(G + j\omega C)\widetilde{I} \quad \text{(b)}. \qquad (2.1.33)$$

It is noted that the second-order partial differential wave equations in the time-domain are converted to the second-order ordinary differential equations in the frequency-domain. The factors at the right-hand side of the above equations help to define a *secondary parameter* γ, known as the *complex propagation constant* of a transmission line:

$$\gamma = [(R + j\omega L)(G + j\omega C)]^{\frac{1}{2}}, \qquad (2.1.34)$$

where $\gamma = \alpha + j\beta$. The real part α(Np/m) of the complex propagation constant γ is called the *attenuation constant* and the imaginary part β (rad/sec) is the *propagation constant* of a lossy transmission line. The parameter β is also known as the *phase-shift constant or phase constant*. On separating the real and imaginary parts of the above equation, the following expressions are obtained:

$$\alpha = \text{Re}[\gamma] = \left\{\frac{RG - \omega^2 LC + \sqrt{(RG - \omega^2 LC)^2 + \omega^2(LG + RC)^2}}{2}\right\}^{\frac{1}{2}}. \qquad (2.1.35)$$

$$\beta = \text{Im}[\gamma] = \left\{\frac{\omega^2 LC - RG + \sqrt{(RG - \omega^2 LC)^2 + \omega^2(LG + RC)^2}}{2}\right\}^{\frac{1}{2}}. \qquad (2.1.36)$$

The attenuation constant α, and propagation constant β are given in terms of the primary line constants, R, L, C, G. Normally α and β are frequency-dependent. Thus, the phase velocity of both the current and voltage waves, given by $v_p = \omega/\beta$ is frequency-dependent. This kind of transmission line is known as the *dispersive transmission line*. A complex wave traveling on a lossy dispersive line gets distorted as each component of the complex wave travels with different phase velocity.

Using the complex propagation constant, the wave equation (2.1.33a and b) are rewritten as

$$\frac{d^2\widetilde{V}}{dx^2} - \gamma^2\widetilde{V} = 0 \quad \text{(a)}$$
$$\frac{d^2\widetilde{I}}{dx^2} - \gamma^2\widetilde{I} = 0 \quad \text{(b)}. \qquad (2.1.37)$$

The above homogeneous wave equations on a transmission line could be treated as a *boundary value problem* to get the voltage and current at any location on the line. If a transmission line is infinitely long and excited from one end, then the voltage and current waves on the line always move in the forward direction without any reflection. At any location on the line, the voltage and current are related by another secondary parameter called "*the characteristic impedance*" of a transmission line:

$$\frac{\widetilde{V}}{\widetilde{I}} = Z_0. \qquad (2.1.38)$$

The characteristic impedance of a transmission line could be viewed as a mechanism that explains the wave propagation on a line. It recasts the *Huygens' Principle* of the secondary wave formation in terms of the characteristic impedance. The characteristic impedance could be called the *Huygens' load*. It is an unusual load impedance with a special property. It absorbs power from the source and itself becomes a secondary source for the further transmission of power in the form of wave motion. In this manner, the wave on a transmission line moves; as the characteristic impedance, i.e. Huygens' load, acts both as a load and also as a source of the wave motion. The process is similar to the Huygens' secondary source for the wavefront propagation. The concept of the Huygens' load is further extended to engineer the *Huygens' metasurface* with unique characteristics to control the reflected and transmitted (refracted) EM-waves. It is discussed in subsection (22.5.2) of chapter 22.

The characteristic impedance, i.e. *Huygens'* load of a lossy line is a complex quantity. Its real part does not dissipate energy like the real part of a normal complex load. The imaginary part of Huygens's load indicates the presence of losses in a transmission line, whereas in the case of a normal complex load its imaginary part shows the storage of the reactive energy. For a lossless transmission line, Huygens' load is a real quantity $\sqrt{L/C}$ that is non-dissipative. Huygens adopted the secondary source model to explain the propagation of the light wave in the space [B.3, B.4]. The expression for the characteristic impedance of a line is obtained from equation (2.1.32).

$$Z_0^2 = \frac{\Delta \tilde{V}}{\tilde{I}} \times \frac{\tilde{V}}{\Delta \tilde{I}} = \frac{R + j\omega L}{G + j\omega C}, \Rightarrow Z_0 = \sqrt{\frac{R + j\omega L}{G + j\omega C}}. \tag{2.1.39}$$

In general, the characteristic impedance of a lossy transmission line is a complex quantity. However, for a lossless line, the lossy elements are zero, i.e. $R = G = 0$. It leads to the following expressions:

$$\alpha = 0 \quad (a), \quad \beta = \omega\sqrt{LC} \quad (b), \quad Z_0 = \sqrt{\frac{L}{C}} \quad (c). \tag{2.1.40}$$

There is no attenuation in the propagating wave on a lossless line. If the line inductance L and the line capacitance C are frequency-independent, the transmission line is nondispersive. The characteristic impedance is a real quantity. The line parameters such as the attenuation constant (α), propagation constant (β), and characteristic impedance (Z_0) are known as the *secondary parameters* of a transmission line. These secondary parameters are finally expressed in terms of the physical geometry and the electrical characteristics of material medium of a line. A microwave circuit designer is more interested in these secondary parameters as compared to the primary line constants, RLCG, of a transmission line. The secondary parameters are more conveniently measured and numerically computed for a large class of the transmission line structures. For any practical transmission line, the losses are always present on a line.

2.1.5 Characteristic of Lossy Transmission Line

A transmission line, such as a coaxial cable, a flat cable, used in the computer, or a feeder to TV receiver, is embedded in a lossy dielectric medium. The loss in a dielectric medium is known as the *dielectric loss* of a transmission line. Of course, when the line is in the air medium, the dielectric loss could be neglected. Likewise, there is another kind of loss, known as the *conductor loss*, on a transmission line. It is due to the finite conductivity of the conducting wires or the strip conductors forming a line. All open transmission lines tend to radiate some power, leading to *radiation loss*. In the case of a planar transmission line, there are also other mechanisms of losses. However, this section considers only the conductor and the dielectric losses of a line and their effect on the propagation characteristics of the line.

Characteristic Impedance

The characteristic impedance Z_0 of a uniform lossy transmission line is given by equation (2.1.39). In the limiting case, $\omega \to 0$, i.e. at a lower frequency, it is reduced to a real quantity that is dominated by the lossy elements of a line:

$$Z_0 = \sqrt{\frac{R}{G}}. \tag{2.1.41}$$

The characteristic impedance Z_0 at very high frequency, i.e. for $\omega \to \infty$, is also reduced to a real quantity. However, now it is dominated by the lossless reactive elements:

$$Z_0 = \sqrt{\frac{L}{C}}. \tag{2.1.42}$$

At higher frequency, we have $\omega L >> R$ and $\omega C >> G$. Therefore, the R and G are normally ignored for the computation of characteristic impedance of a low-loss microwave transmission line. In the intermediate frequency range, the characteristic impedance of a line is a complex quantity. Its *imaginary part indicates the presence of the loss* in a line. Equation (2.1.42) is also applicable to a lossless line.

The characteristic impedance of transmission line in the lossless dielectric medium, or a moderately lossy medium where G could be neglected, is obtained in equation (2.1.43). However, the conductor loss is present on the line:

$$Z_0 = \left[\frac{R + j\omega L}{G + j\omega C}\right]^{\frac{1}{2}} = \left[\frac{L}{C}\left(1 + \frac{R}{j\omega L}\right)\right]^{\frac{1}{2}} \approx \sqrt{\frac{L}{C}}\left(1 - j\frac{R}{2\omega L}\right). \tag{2.1.43}$$

The measured or computed complex characteristic impedance of a line, over a certain frequency range, with a negative imaginary part, indicates that the loss in the line is mainly due to the conductor loss [J.4].

The alternative case of a lossy line, with $G \neq 0$, $R = 0$, could be also considered. In this case, the conductor loss is ignored; however, the dielectric loss is dominant. The characteristic impedance of such line is approximated as follows:

$$Z_0 = \left[\frac{j\omega L}{G + j\omega C}\right]^{\frac{1}{2}} \approx \sqrt{\frac{L}{C}}\left(1 + j\frac{G}{2\omega C}\right). \tag{2.1.44}$$

If the imaginary part of the characteristic impedance of a line is positive over some frequency range, then the dielectric loss dominates the loss in the line. However, if both R and G are moderately present, with $\omega L >> R$ and $\omega C >> G$, the real and imaginary parts of the characteristic impedance could be approximated by using the binomial expansion as follows:

$$(1 \pm x)^{\frac{1}{2}} \approx 1 \pm \frac{x}{2} - \frac{x^2}{8}. \tag{2.1.45}$$

$$Z_0 = \sqrt{\frac{L}{C}}\left[\left(1 - j\frac{R}{\omega L}\right)\left(1 + j\frac{G}{\omega C}\right)\right]^{\frac{1}{2}}. \tag{2.1.46}$$

$$\Rightarrow Z_0 = \sqrt{\frac{L}{C}}\left[\left(1 + \frac{R^2}{8\omega^2 L^2} + \frac{G^2}{8\omega^2 C^2} + \frac{RG}{4\omega^2 LC}\right)\right.$$
$$\left. + j\left(\frac{G}{2\omega C} - \frac{R}{2\omega L}\right)\right]. \tag{2.1.47}$$

In the above expression, ω^3 and ω^4 terms are neglected. If we neglect ω^2 terms and also take $G = 0$ or $R = 0$, equation (2.1.47) reduces to either equation (2.1.43) or equation (2.1.44), respectively. It is also possible that with the change of frequency, the imaginary part of characteristic impedance changes from a positive to a negative value indicating that the dominant loss can change from the dielectric loss to the conductor loss. For such cases, R and G are usually frequency-dependent [J.4]. Over a band of frequencies, the imaginary part of the characteristic impedance could be zero leading to

$$\frac{R}{L} = \frac{G}{C}. \tag{2.1.48}$$

It is well known as *Heaviside's condition*. On meeting it, a lossy line becomes dispersion-less and the propagation constant β becomes a linear function of frequency, while the attenuation constant becomes frequency-independent. Following the above equation (2.1.48), a lossy line could be made dispersionless by the inductive loading [B.5, B.6].

Propagation Constant

The propagation constant γ of a uniform lossy transmission line is given by equation (2.1.34). It could be approximated under the low-loss condition. Its real and imaginary parts are separated to get the frequently used approximate expressions for the attenuation and phase constants of a line:

$$\gamma = j\omega\sqrt{LC}\left[\left(1 - j\frac{R}{\omega L}\right)\left(1 - j\frac{G}{\omega C}\right)\right]^{\frac{1}{2}}$$
$$\Rightarrow \gamma \approx j\omega\sqrt{LC}\left[\left(1 - j\frac{R}{2\omega L} + \frac{R^2}{8\omega^2 L^2}\right)\right.$$
$$\left.\left(1 - j\frac{G}{2\omega C} + \frac{G^2}{8\omega^2 C^2}\right)\right]. \tag{2.1.49}$$

On neglecting ω^2, ω^3, and ω^4 terms, the real part of the propagation constant γ provides the attenuation constant, whereas the imaginary part gives the propagation constant:

$$\alpha \approx \frac{R}{2\sqrt{L/C}} + \frac{G\sqrt{L/C}}{2} = \frac{R}{2Z_0} + \frac{GZ_0}{2} \quad \text{(a)}$$
$$\beta \approx \omega\sqrt{LC} \quad \text{(b)}. \tag{2.1.50}$$

The *first term* of the above equation (2.1.50a) shows the conductor loss of a line, while the *second term* shows its dielectric loss. If R and G are frequency-independent, the attenuation in a line would be frequency-independent under $\omega L >> R$ and $\omega C >> G$ conditions. However, usually, R is frequency-dependent due to the skin effect. In some cases, G could also be frequency-dependent [B.7].

The dispersive phase constant β is obtained from the imaginary part of equation (2.1.49):

$$\beta \approx \omega\sqrt{LC}\left[\left(1 - \frac{RG}{4\omega^2 LC} + \frac{R^2}{8\omega^2 L^2} + \frac{G^2}{8\omega^2 C^2}\right)\right]. \tag{2.1.51}$$

On neglecting the second-order term, β becomes a linear function of frequency and the line is dispersionless. In that case, its phase velocity is also independent of frequency. *A lossy line is dispersive. However, it also becomes dispersionless under the Heaviside's condition – (2.1.48).* A transmission line, such as a microstrip in the inhomogeneous medium, can have dispersion even without losses.

2.1.6 Wave Equation with Source

In the above discussion, the development of the voltage and current wave equations has ignored the voltage or current source. However, a voltage or current source is always required to launch the voltage and current waves on a line. Therefore, it is appropriate to develop the transmission line equation with a source [B.8]. The consideration of a voltage/current source is important to solve the electromagnetic field problems of the layered medium planar lines, discussed in chapters 14 and 16.

Shunt Current Source

Figure (2.7) shows the lumped element model of a transmission line section of length Δx with a shunt current source $\tilde{I}_s \Delta x$ located at $x = x_0$. It is expressed through Dirac's delta function as $\tilde{I}_s \Delta x \delta(x - x_0)$. The lumped line constants R, L, C, G are given for *p.u.l.*

The loop and node equations are written below to develop the Kelvin–Heaviside transmission line equations with a current source:

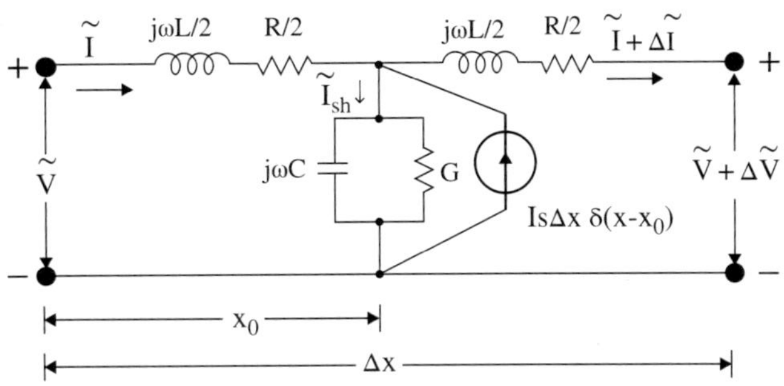

Figure 2.7 Equivalent lumped circuit of a transmission line with a shunt current source.

The Loop Equation

$$\tilde{V} - [(R + j\omega L)\Delta x/2]\tilde{I} - [(R + j\omega L)\Delta x/2]\left(\tilde{I} + \Delta\tilde{I}\right)$$
$$- \left(\tilde{V} + \Delta\tilde{V}\right) = 0$$

$$\Rightarrow -\Delta\tilde{V} - [(R + j\omega L)\Delta x]\tilde{I} - \left[(R + j\omega L)\Delta x\Delta\tilde{I}/2\right] = 0,$$

By taking $\Delta x\Delta\tilde{I}/2 \approx 0$

$$\Rightarrow \frac{\Delta\tilde{V}}{\Delta x} = -[(R + j\omega L)]\tilde{I}$$

The Node Equation

$$\tilde{I} + \tilde{I}_s\Delta x\delta(x - x_0) - \left(\tilde{I} + \Delta\tilde{I}\right) - \tilde{I}_{sh} = 0$$

$$\Rightarrow \tilde{I} + \tilde{I}_s\Delta x\delta(x - x_0) - \left(\tilde{I} + \Delta\tilde{I}\right) - [(G + j\omega C)\Delta x]$$
$$\left[\tilde{V} - \tilde{I}(R + j\omega L)\Delta x/2\right] = 0$$

$$\Rightarrow \frac{\Delta\tilde{I}}{\Delta x} = -(G + j\omega C)\left[\tilde{V} - \tilde{I}(R + j\omega L)\Delta x/2\right] + \tilde{I}_s\delta(x - x_0)$$

For $\Delta x \rightarrow 0$, the above equations are reduced to

$$\frac{d\tilde{V}}{dx} = -(R + j\omega L)\tilde{I} \quad \text{(a)}$$
$$\frac{d\tilde{I}}{dx} = -(G + j\omega C)\tilde{V} + \tilde{I}_s\delta(x - x_0) \quad \text{(b)}. \tag{2.1.52}$$

The above equations are rewritten below in term of the characteristic impedance (Z_0) and propagation constant (γ) of a transmission line:

$$\frac{d\tilde{V}}{dx} = -\gamma Z_0\tilde{I} \quad \text{(a)}, \qquad \frac{d\tilde{I}}{dx} = -\frac{\gamma}{Z_0}\tilde{V} + \tilde{I}_s\delta(x - x_0) \quad \text{(b)}. \tag{2.1.53}$$

On solving the above equations for the voltage, the following inhomogeneous voltage wave equation, with a current source, is obtained:

$$\frac{d^2\tilde{V}}{dx^2} - \gamma^2\tilde{V} = -\gamma Z_0\tilde{I}_s\delta(x - x_0). \tag{2.1.54}$$

Away from the location of the current source, i.e. for $x \neq x_0$, equation (2.1.54) reduces to the homogeneous equation (2.1.37a). The wave equation for the current wave, with a shunt current source, could also be rewritten.

2.1.7 Solution of Voltage and Current-Wave Equation

The voltage and current wave equations in the phasor form are given in equation (2.1.37). The solution of a wave equation is written either in terms of the hyperbolic functions or in terms of the exponential functions. The first form is suitable for a line terminated in an arbitrary load. A section of the line transforms the load impedance into the input impedance at any location on the line. The impedance transformation takes place due to the standing wave formation. The *hyperbolic form* of the solution also provides the voltage and current distributions along the line. The *exponential form* of the solution demonstrates the traveling waves on a line, both in the forward and in the backward directions. A combination of the forward-moving and the backward-moving waves produces the *standing wave* on a transmission line.

The Hyperbolic Form of a Solution

Figure (2.8a) shows a section of the transmission line having a length ℓ. It is fed by a voltage source, $\tilde{V}_g$ with Z_g internal impedance. The general solutions for the line voltage $\tilde{V}(x)$ and line current $\tilde{I}(x)$ of the wave equation (2.1.37) are

$$\tilde{V}(x) = A_1\cosh(\gamma x) + B_1\sinh(\gamma x) \quad \text{(a)}$$
$$\tilde{I}(x) = A_2\cosh(\gamma x) + B_2\sinh(\gamma x) \quad \text{(b)}. \tag{2.1.55}$$

At any section on the line, its characteristic impedance Z_0 relates the line voltage $\tilde{V}(x)$ and line current $\tilde{I}(x)$. So the constants A_2, B_2 are related to the constants A_1 and B_1. In Fig (2.8a), the point P on the line is located at a *distance x from the source end*, i.e. at a *distance* $d = (\ell - x)$ *from the load end*. The load is located at $d = 0$, and the source is located at $d = -\ell$. The $\tilde{V}_s$, and $\tilde{I}_s$ are the input voltage and the input current at the port-aa, i.e at $x = 0$. At $x = \ell$, i.e. at the port-bb, $\tilde{V}_R$ and $\tilde{I}_R$ are the load end voltage and current, respectively.

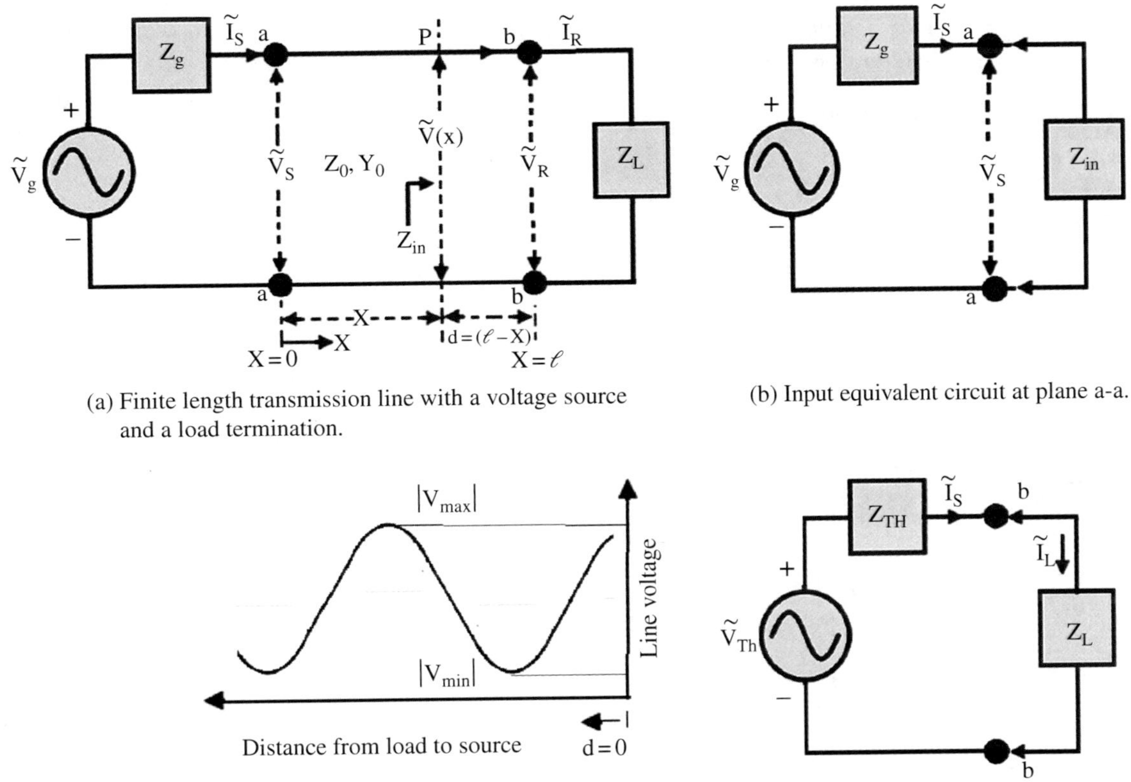

(a) Finite length transmission line with a voltage source and a load termination.

(b) Input equivalent circuit at plane a-a.

(c) A standing wave on the line.

(d) Output Thevenin equivalent circuit at plane b-b.

Figure 2.8 Transmission line circuit. The distance x is measured from the source end; whereas the distance d is measured from the load.

The ideal voltage generator $\tilde{V}_g$ has the internal impedance, $Z_g = 0$, i.e. $\tilde{V}_S = \tilde{V}_g$. The phasor form of the line current, from equations (2.1.32b) and (2.1.55a), is written below:

$$\tilde{I}(x) = -(G + j\omega C)\int [A_1 \cosh(\gamma x) + B_1 \sinh(\gamma x)]dx$$
$$\Rightarrow \tilde{I}(x) = -\frac{1}{Z_0}[A_1 \sinh(\gamma x) + B_1 \cosh(\gamma x)]. \tag{2.1.56}$$

On comparing the coefficients of sinh(γx) and cosh (γx), of equations (2.1.55b) and (2.1.56), two constants A_2 and B_2 are determined:

$$B_2 = -\frac{A_1}{Z_0}, \quad A_2 = -\frac{B_1}{Z_0}. \tag{2.1.57}$$

The phasor line voltage and line current are written as follows:

$$\tilde{V}(x) = A_1 \cosh(\gamma x) + B_1 \sinh(\gamma x) \qquad (a)$$
$$\tilde{I}(x) = -\frac{1}{Z_0}[A_1 \sinh(\gamma x) + B_1 \cosh(\gamma x)] \qquad (b). \tag{2.1.58}$$

The constants A_1 and B_1 are determined by using the boundary conditions at input $x = 0$ and output $x = \ell$.

- At $x = 0$, the line input voltage is $\tilde{V}_S$, giving the value of A_1:

$$\tilde{V}(x = 0) = A_1 = \tilde{V}_S. \tag{2.1.59}$$

At the receiving end, $x = \ell$, the load end voltage and current are

$$\tilde{V}(x = \ell) = \tilde{V}_R = \tilde{V}_S \cosh(\gamma\ell) + B_1 \sinh(\gamma\ell) \qquad (a)$$
$$\tilde{I}(x = \ell) = \tilde{I}_R = -\frac{1}{Z_0}\left[\tilde{V}_S \sinh(\gamma\ell) + B_1 \cosh(\gamma\ell)\right] \qquad (b). \tag{2.1.60}$$

- At $x = \ell$, i.e. at the receiving end, the voltage across load Z_L is

$$\tilde{V}_R = Z_L \tilde{I}_R. \tag{2.1.61}$$

The constant B_1 is evaluated on substituting $\tilde{V}_R$ and $\tilde{I}_R$, from equation (2.1.60), in the above equation:

$$B_1 = -\frac{\tilde{V}_S[Z_0 \cosh(\gamma\ell) + Z_L \sinh(\gamma\ell)]}{Z_L \cosh(\gamma\ell) + Z_0 \sinh(\gamma\ell)}. \tag{2.1.62}$$

On substituting constants A_1 and B_1 in equation (2.1.58a), the expression for the line voltage at location P, from the source or load end, is

$$\tilde{V}(x) = V_S\left[\frac{Z_L\cosh[\gamma(\ell - x)] + Z_0\sinh[\gamma(\ell - x)]}{Z_L\cosh(\gamma\ell) + Z_0\sinh(\gamma\ell)}\right] \quad (a)$$
$$\Rightarrow \tilde{V}(d) = V_S\left[\frac{Z_L\cosh(\gamma d) + Z_0\sinh(\gamma d)}{Z_L\cosh(\gamma\ell) + Z_0\sinh(\gamma\ell)}\right] \quad (b). \tag{2.1.63}$$

Similarly, the line current at the location P is obtained as follows:

$$\tilde{I}(x) = \frac{V_S}{Z_0}\left[\frac{Z_L\sinh[\gamma(\ell - x)] + Z_0\cosh[\gamma(\ell - x)]}{Z_L\cosh(\gamma\ell) + Z_0\sinh(\gamma\ell)}\right] \quad (a)$$
$$\Rightarrow \tilde{I}(d) = \frac{V_S}{Z_0}\left[\frac{Z_L\sinh(\gamma d) + Z_0\cosh(\gamma d)}{Z_L\cosh(\gamma\ell) + Z_0\sinh(\gamma\ell)}\right] \quad (b). \tag{2.1.64}$$

At any location P on the line, the load impedance is transformed as input impedance by the line length $d = (\ell - x)$:

$$Z_{in}(x) = \frac{\tilde{V}(x)}{\tilde{I}(x)} = Z_0\left[\frac{Z_L + Z_0\tanh[\gamma(\ell - x)]}{Z_0 + Z_L\tanh[\gamma(\ell - x)]}\right] \quad (a)$$
$$\Rightarrow Z_{in}(d) = \frac{\tilde{V}(d)}{\tilde{I}(d)} = Z_0\left[\frac{Z_L + Z_0\tanh(\gamma d)}{Z_0 + Z_L\tanh(\gamma d)}\right] \quad (b)$$
$$\text{for } \alpha = 0,\ Z_{in}(d) = Z_0\left[\frac{Z_L + jZ_0\tan(\beta d)}{Z_0 + jZ_L\tan(\beta d)}\right] \quad (c). \tag{2.1.65}$$

Equations (2.1.65a,b) take care of the losses in a line through the complex propagation constant, $\gamma = \alpha + j\beta$. However, for a lossless line $\alpha = 0$, $\gamma = j\beta$ and the hyperbolic functions are replaced by the trigonometric functions shown in equation (2.1.65c). It shows the impedance transformation characteristics of $d = \lambda/4$ transmission line section.

Equations (2.1.63) and (2.1.64) could be further written in terms of the generator voltage $\tilde{V}_g$ for the case, $Z_g \neq 0$. Figure (2.8b) shows that at the source end $x = 0$, the load appears as the input impedance Z_{in}. The sending end voltage is obtained as follows:

$\tilde{V}_s = \tilde{V}_g - Z_g\tilde{I}_s$ where $\tilde{I}_s = \frac{V_g}{Z_g + Z_{in}}$, and

$$Z_{in}(x = 0) = Z_0\left[\frac{Z_L + Z_0\tanh\gamma\ell}{Z_0 + Z_L\tanh\gamma\ell}\right].$$

$$\tilde{V}_s = \tilde{V}_g - \tilde{V}_g\frac{Z_g}{Z_{in} + Z_g} = \tilde{V}_g\frac{Z_{in}}{Z_{in} + Z_g}$$

$$\Rightarrow \tilde{V}_s = \tilde{V}_g\left[\frac{Z_0Z_L + Z_0^2\tanh\gamma\ell}{(Z_0Z_L + Z_0Z_g) + (Z_0^2 + Z_LZ_g)\tanh\gamma\ell}\right]. \tag{2.1.66}$$

The line voltage, in terms of $\tilde{V}_g$, Z_g, and Z_L, is obtained on substituting equation (2.1.66) in equation (2.1.63):

$$\tilde{V}(x) = V_g\left[\frac{Z_0[Z_L\cosh\gamma(\ell - x) + Z_0\sinh\gamma(\ell - x)]}{(Z_0Z_L + Z_0Z_g)\cosh\gamma\ell + (Z_0^2 + Z_LZ_g)\sinh\gamma\ell}\right]. \tag{2.1.67}$$

Likewise, from equations (2.1.64) and (2.1.66), an expression for the line current is obtained:

$$\tilde{I}(x) = V_g\left[\frac{[Z_L\sinh\gamma(\ell - x) + Z_0\cosh\gamma(\ell - x)]}{(Z_0Z_L + Z_0Z_g)\cosh\gamma\ell + (Z_0^2 + Z_LZ_g)\sinh\gamma\ell}\right]. \tag{2.1.68}$$

The above equations could be reduced to the following equations for a lossless line, i.e. for $\alpha = 0$, $\gamma = j\beta$, $\cosh(j\beta) = \cos\beta$ and $\sinh(j\beta) = j\sin\beta$:

$$\tilde{V}(x) = V_g\left[\frac{Z_0[Z_L\cos\beta(\ell - x) + jZ_0\sin\beta(\ell - x)]}{(Z_0Z_L + Z_0Z_g)\cos\beta\ell + j(Z_0^2 + Z_LZ_g)\sin\beta\ell}\right]. \tag{2.1.69}$$

$$\tilde{I}(x) = V_g\left[\frac{[Z_0\cos\beta(\ell - x) + jZ_L\sin\beta(\ell - x)]}{(Z_0Z_L + Z_0Z_g)\cos\beta\ell + j(Z_0^2 + Z_LZ_g)\sin\beta\ell}\right]. \tag{2.1.70}$$

Equation (2.1.65c), for the input impedance, could be obtained from the above two equations. The sending end voltage and current are obtained at the input port – aa, $x = 0$:

$$\tilde{V}_s = \tilde{V}(x = 0)$$
$$= V_g\left[\frac{Z_0[Z_L\cos\beta\ell + jZ_0\sin\beta\ell]}{(Z_0Z_L + Z_0Z_g)\cos\beta\ell + j(Z_0^2 + Z_LZ_g)\sin\beta\ell}\right]. \tag{2.1.71}$$

$$\tilde{I}_s = \tilde{I}(x = 0)$$
$$= V_g\left[\frac{[Z_0\cos\beta\ell + jZ_L\sin\beta\ell]}{(Z_0Z_L + Z_0Z_g)\cos\beta\ell + j(Z_0^2 + Z_LZ_g)\sin\beta\ell}\right]. \tag{2.1.72}$$

Likewise, the expressions for the voltage and current at the output port – bb, i.e. at the receiving end for $x = \ell$, are obtained:

$$\tilde{V}_R = \tilde{V}(x = \ell)$$
$$= \tilde{V}_g\left[\frac{Z_0Z_L}{(Z_0Z_L + Z_0Z_g)\cos\beta\ell + j(Z_0^2 + Z_LZ_g)\sin\beta\ell}\right]. \tag{2.1.73}$$

$$\tilde{I}_R = \tilde{I}(x = \ell) = \tilde{V}_g\left[\frac{Z_0}{(Z_0 Z_L + Z_0 Z_g)\cos\beta\ell + j(Z_0^2 + Z_L Z_g)\sin\beta\ell}\right]. \quad (2.1.74)$$

Two special cases of the load termination, i.e. the short-circuited load and the open-circuited load, are discussed below. The voltage and current distributions on a transmission line for both the cases are also obtained.

Short-Circuited Receiving End

For the short-circuited load $Z_L = 0$, the line voltage at the load end is zero $\tilde{V}(x = \ell) = 0$. However, the voltage on the line is not zero. Equations (2.1.63) and (2.1.64) provide the voltage and current distributions on a short-circuited line:

$$\tilde{V}(x) = \tilde{V}_S \frac{\sinh\gamma(\ell - x)}{\sinh\gamma\ell} \quad (a)$$

$$\tilde{I}(x) = \frac{\tilde{V}_S}{Z_0}\frac{\cosh\gamma(\ell - x)}{\sinh\gamma\ell} \quad (b). \quad (2.1.75)$$

The input impedance at any distance $(\ell - x)$ from the source end is

$$Z_{in}(x) = Z_0 \tanh\gamma(\ell - x) \quad (a)$$

$$Z_{in}(d) = Z_0 \tanh\gamma d \quad (b)$$

$$\text{for } \alpha = 0, Z_{in}(d) = jZ_0 \tan\beta d \quad (c). \quad (2.1.76)$$

At the load end, the voltage is zero. However, the line current is not infinite like the lumped element circuit with a short-circuited termination at output. A short-circuited transmission line draws only a finite current from the source. *The $\ell < \lambda/4$ short-circuited line section behaves as an inductive element.* The electrical nature of the line section can be controlled by changing its electrical length [B.9–B.15].

Open-Circuited Receiving End

The load impedance is $Z_L \to \infty$ for an open-circuited transmission line and the load current $\tilde{I}(x = \ell) = 0$. Again, equations (2.1.63) and (2.1.64) provide the voltage and current distributions on an open-circuited transmission line. The voltage and current waves and the input impedance at location P from the load end can be computed for the open-circuited load as follows:

$$\tilde{V}(x) = \tilde{V}_S \frac{\cosh\gamma(\ell - x)}{\cosh\gamma\ell} \quad (a)$$

$$\tilde{I}(x) = \frac{\tilde{V}_S}{Z_0}\frac{\sinh\gamma(\ell - x)}{\cosh\gamma\ell} \quad (b). \quad (2.1.77)$$

$$Z_{in}(x) = Z_0 \coth\gamma(\ell - x) \quad (a)$$

$$\Rightarrow Z_{in}(d) = Z_0 \coth\gamma d \quad (b)$$

$$\text{for } \alpha = 0, Z_{in}(d) = -jZ_0 \cot\beta d \quad (c). \quad (2.1.78)$$

The $\ell < \lambda/4$ open-circuited line section behaves as a capacitive element. The electrical nature of the line section can be controlled by changing its electrical length.

Matched and Mismatched Termination

The input impedance at any location on a line is $Z_{in}(x) = Z_0$ if it is matched terminated in its characteristic impedance, i.e. $Z_L = Z_0$. Normally, the characteristics impedance of a microwave line is a real quantity. The line terminated in Z_0 does not create any reflected wave on a transmission line. However, for the mismatched termination $Z_L \neq Z_0$, there is a reflected wave on a transmission line, traveling from the load end to the source end.

Exponential Form of Solution

The wave nature of the line voltage and line current becomes more obvious from the exponential form of solutions of the wave equations. The solution of wave equation (2.1.37), for the phasor line voltage and line current, can also be written in the exponential form:

$$\tilde{V}(x) = V^+ e^{-\gamma x} + V^- e^{\gamma x} \quad (a)$$

$$\tilde{I}(x) = \frac{1}{Z_0}(V^+ e^{-\gamma x} - V^- e^{\gamma x}) \quad (b). \quad (2.1.79)$$

The distance x is measured from the source end. The time-dependent harmonic form of the voltage wave is $v(x,t) = \text{Re}\left[\tilde{V}(x)\exp(j\omega t)\right]$. Finally, it is written as follows:

$$v(x,t) = \underbrace{V^+ e^{-\alpha x}\cos(\omega t - \beta x)}_{\text{Forward traveling wave}} + \underbrace{V^- e^{\alpha x}\cos(\omega t + \beta x)}_{\text{Backward traveling wave}}. \quad (2.1.80)$$

For an outgoing wave on a lossy line, the wave amplitude decays and its phase lags; whether the distance is measured from source end or load end. It is accounted for by the proper sign of distance x. The amplitude of a wave is exponentially decaying due to the line losses. It is expressed by the attenuation constant α (Np/m).

The expression of the traveling current wave on a line could be written as follows:

$$i(x,t) = \frac{1}{Z_0}[V^+ e^{-\alpha x}\cos(\omega t - \beta x) - V^- e^{\alpha x}\cos(\omega t + \beta x)].$$

↓ Forward traveling wave ↓ Backward traveling wave

(2.1.81)

If the source at $x = 0$ is connected to a line of infinite extent, there is no reflection from the load end, as the wave will never reach to the load end to get reflected. Therefore, for the forward traveling voltage and current waves on an infinite line, $V^- = 0$ and the above solutions of the wave equations are written as follows:

$$\tilde{V}(x) = V^+ e^{-\gamma x} \quad (a)$$

$$\tilde{I}(x) = \frac{V^+ e^{-\gamma x}}{Z_0} \quad (b)$$

$$v(x,t) = V^+ e^{-\alpha x}\cos(\omega t - \beta x) \quad (c)$$

$$i(x,t) = \frac{V^+}{Z_0} e^{-\alpha x}\cos(\omega t - \beta x) \quad (d). \qquad (2.1.82)$$

The input impedance of an infinite line at any location x is $Z_{in}(x) = Z_0$. Thus, *a finite line terminated in its characteristic impedance, i.e. $Z_L = Z_0$ behaves as an infinite extent transmission line without any reflection*. The characteristic impedance shows a specific feature of a line that is determined by the geometry and physical medium of the line. Once a finite extent line is terminated in any other load impedance, i.e. $Z_L \neq Z_0$, the voltage and current waves are reflected from the load. The wave reflection is expressed by *the reflection coefficient*, $\Gamma(x)$. The reflection coefficient is a complex quantity and its phase changes over the length of a line. However, its magnitude remains constant over the length of a lossless line. The interference of the forward-moving and backward-moving reflected waves produces the *standing wave* with maxima and minima of the voltage and current on a line. Figure (2.8c) shows the voltage standing wave. The position and magnitude of the voltage maxima V_{max} and voltage minima V_{min} are measurable quantities. Their positions are measured from the load end. The *reflection coefficient*, and also the *voltage standing wave ratio*, VSWR, i.e. S, is defined using the V_{max} and V_{min}. The VSWR could be measured with a VSWR meter. Therefore, the reflection on a transmission line is expressed by both the reflection coefficient and the VSWR. The reflection at the input terminal of a line is also expressed as the *return-loss*, $RL = -20\log_{10}|\Gamma_{in}|$. The wave behavior in terms of the reflection parameter is an important quantity in the design of circuits and components in the microwave and RF engineering.

At the load end, the total line voltage is a sum of the forward and reflected waves, and it is equal to the voltage drop V_R across the load impedance Z_L. It is shown in Fig (2.8a). To simplify expressions for the line voltage and current, the origin is shifted from the source end to the load end. In that case, expressions for both the line voltage and line current, given by equation (2.1.79), have to be modified for the new distance variable $x < 0$. Now the source and load are located at $x = -\ell$ and at $x = 0$, respectively. For the origin at the load end, the reflected wave $V^- e^{(\omega t + \gamma x)}$ appears as the "forward-traveling wave" from load to the source, whereas the wave incident at the load $V^+ e^{(\omega t - \gamma x)}$ appears as the "backward-traveling" wave. The voltage drop across the load is

$$\tilde{V}_R = Z_L \tilde{I}_R. \qquad (2.1.83)$$

Keeping in view that the origin of the distance ($x = 0$) is at the load, the line voltage and current at the load end are written, from equation (2.1.79), as follows:

$$\tilde{V}_R = \tilde{V}(x=0) = V^+ + V^- \quad (a)$$

$$\tilde{I}_R = \tilde{I}(x=0) = \frac{1}{Z_0}(V^+ - V^-) \quad (b). \qquad (2.1.84)$$

The *voltage reflection coefficient* at the load end is defined as follows:

$$\Gamma_L = \frac{\text{Amplitude of reflected voltage}}{\text{Amplitude of forward(incident)voltage}} = \frac{V^-}{V^+}. \qquad (2.1.85)$$

The expression to compute the reflection coefficient at the load end is obtained from the above equations (2.1.83) – (2.1.85): equations:

$$1 + \Gamma_L = \frac{Z_L}{Z_0}(1 - \Gamma_L) \quad (a)$$

$$\Rightarrow \Gamma_L = \frac{Z_L - Z_0}{Z_L + Z_0} \quad (b). \qquad (2.1.86)$$

The mismatch of a load impedance Z_L with the characteristic impedance Z_0 of a line is the cause of the reflection at the load end. For the condition $Z_L = Z_0$, the matched load terminated line avoids the reflection on a line, as $\Gamma_L = 0$. At any distance x, the reflection coefficient is a complex quantity with both the magnitude and phase expressed as follows:

$$\Gamma(x) = |\Gamma(x)|e^{j\phi} = \Gamma_r(x) + j\Gamma_i(x) \quad (a)$$

$$|\Gamma(x)| = \sqrt{(\Gamma_r(x))^2 + (\Gamma_i(x))^2}$$

$$\phi = \tan^{-1}(\Gamma_i(x)/\Gamma_r(x)) \quad (b). \qquad (2.1.87)$$

A lossless line has $|\Gamma(x)| = |\Gamma_L|$, i.e. on a lossless line magnitude of the reflection is the same at all locations on a line. However, the lagging phase ϕ changes with distance. It has 180° periodicity, i.e. for an inductive load, the range of phase is $0 < \phi < \pi$, and a capacitive load has the phase in the range $-\pi < \phi < 0$. Using equation (2.1.79), the line voltage and line current are written in term of the load reflection coefficient:

$$\tilde{V}(x) = V^+ e^{-\gamma x}\left[1 + \Gamma_L e^{2\gamma x}\right] \quad (a)$$

$$\tilde{I}(x) = \frac{V^+ e^{-\gamma x}}{Z_0}\left[1 - \Gamma_L e^{2\gamma x}\right] \quad (b). \quad (2.1.88)$$

For a lossless transmission line, the above equations are written as follows:

$$\tilde{V}(x) = |V^+|\left[1 + |\Gamma_L| e^{j(\phi + 2\beta x)}\right] \quad (a)$$

$$\tilde{I}(x) = \frac{|V^+|}{Z_0}\left[1 - |\Gamma_L| e^{j(\phi + 2\beta x)}\right] \quad (b). \quad (2.1.89)$$

In the above equations, the origin is at the load end, i.e. $x < 0$. The maxima and minima of the voltage and current waves along the line occur due to the phase variation along the line. The voltage maximum occurs at $e^{j(\phi + 2\beta x)} = +1$. In this case, both the forward and reflected waves are in-phase. The voltage minimum occurs at $e^{j(\phi + 2\beta x)} = -1$. In this case, both the forward and reflected waves are out of phase. Finally, the maxima and minima of the voltage on a line are given as follows:

$$\tilde{V}_{max}(x) = |V^+|[1 + |\Gamma_L|] \quad (a)$$

$$\tilde{V}_{min}(x) = |V^+|[1 - |\Gamma_L|] \quad (b). \quad (2.1.90)$$

The reflection coefficient $\Gamma(x)$ at any location x from the load end is related to the reflection coefficient at the load Γ_L by

$$\text{For a lossy line}: \quad \Gamma(x) = \Gamma_L e^{2\gamma x} \quad (a)$$

$$\text{for a lossless line}: \quad \Gamma(x) = \Gamma_L e^{j2\beta x} \quad (b). \quad (2.1.91)$$

The measurable quantity *voltage standing wave ratio* (VSWR) is defined as follows:

$$S = VSWR = \frac{\tilde{V}_{max}(x)}{\tilde{V}_{min}(x)} = \frac{1 + |\Gamma(x)|}{1 - |\Gamma(x)|}. \quad (2.1.92)$$

For a lossless line, the VSWR is constant along the length of a line. Likewise, the current standing wave ratio is also defined.

The wave reflection also takes place at the sending end when the source impedance Z_g is not matched to the characteristic impedance of a line. The reflection coefficient at $x = -\ell$, i.e. at the generator (source) end is defined as $\Gamma(x = -\ell) = \Gamma_g$. The voltage and current at the generator end are obtained from equation (2.1.88),

$$\tilde{V}_s(x) = \tilde{V}(x = -\ell) = V^+ e^{+\gamma l}\left[1 + \Gamma_L e^{-2\gamma\ell}\right] \quad (a)$$

$$\tilde{I}_s(x) = \tilde{I}(x = -\ell) = \frac{V^+ e^{+\gamma l}}{Z_0}\left[1 - \Gamma_L e^{-2\gamma\ell}\right] \quad (b). \quad (2.1.93)$$

The amplitude factor V^+ is determined by the reflections at both the source and load ends.

Figure (2.8b) shows that the port voltage $\tilde{V}_s$ and the line current $\tilde{I}_s$, at the input port – aa, are related to the source voltage $\tilde{V}_g$ and its internal impedance Z_g by

$$\tilde{I}_s = \frac{\tilde{V}_g - \tilde{V}_s}{Z_g}. \quad (2.1.94)$$

On substitution of equation (2.1.93) in (2.1.94), the voltage wave amplitude V^+ is obtained as follows:

$$V^+\left\{(Z_g + Z_0)e^{+\gamma l} - (Z_g - Z_0)e^{-\gamma l}\Gamma_L\right\} = Z_0\tilde{V}_g. \quad (2.1.95)$$

However, the reflection coefficient at the source end is

$$\Gamma_g = \frac{Z_g - Z_0}{Z_g + Z_0}. \quad (2.1.96)$$

Therefore, the amplitude of the voltage wave launched by the source is

$$V^+ = \tilde{V}_g \frac{Z_0 e^{-\gamma\ell}}{(Z_g + Z_0)\left[1 - \Gamma_g\Gamma_L e^{-2\gamma\ell}\right]}. \quad (2.1.97)$$

Equation (2.1.88a and b) give the voltage and current waves on a transmission line with the amplitude factor V^+. The amplitude factor V^+ is given by equation (2.1.97).

2.1.8 Application of Thevenin's Theorem to Transmission Line

Thevenin's theorem is a very popular concept used in the analysis of the low-frequency lumped element circuits. It is equally applicable to a transmission line network. At the output end of the line, the input source voltage $\tilde{V}_g$ and the line section are replaced by the equivalent Thevenin's voltage, $\tilde{V}_{TH}(x = 0)$ with internal impedance, i.e. the Thevenin's impedance Z_{TH} [B.12]. Figure (2.8d) shows it. The distance is measured from the load end. *Thevenin's voltage* is an open-circuit voltage at the load end. In the case of the open-circuited load, $Z_L \to \infty$, equation (2.1.86) provides a reflection coefficient $\Gamma_L = 1$. Thevenin's voltage is obtained from equations (2.1.88a) and (2.1.97):

$$\tilde{V}_{TH}(x=0) = \frac{2\tilde{V}_g Z_o e^{-\gamma\ell}}{(Z_g + Z_o)\left[1-\Gamma_g e^{-2\gamma\ell}\right]}. \quad (2.1.98)$$

On replacing Γ_g from equation (2.1.96), Thevenin's voltage is

$$\tilde{V}_{TH}(x=0) = \frac{\tilde{V}_g Z_o}{\left[Z_o \cosh(\gamma\ell) + Z_g \sinh(\gamma\ell)\right]}. \quad (2.1.99)$$

Thevenin's impedance Z_{TH} is obtained from equation (2.1.88b) by computing *Norton current*, i.e. the short-circuit current at $x = 0$. Under the short-circuited load condition at $x = 0$, $\Gamma_L = -1$, and the Norton current is

$$\tilde{I}_N(x=0) = \frac{2\tilde{V}_g e^{-\gamma\ell}}{(Z_g + Z_o)\left[1+\Gamma_g e^{-2\gamma\ell}\right]}. \quad (2.1.100)$$

Thevenin's impedance is obtained as follows:

$$Z_{TH} = \frac{\tilde{V}_{TH}}{\tilde{I}_N}, \Rightarrow Z_{TH} = Z_0 \frac{1+\Gamma_g e^{-2\gamma\ell}}{1-\Gamma_g e^{-2\gamma\ell}}. \quad (2.1.101)$$

Transfer Function

The transmission line section could be treated as a circuit element. Its transfer function is obtained either with respect to the source voltage V_g or with respect to the input voltage V_s at the port- aa, as shown in Fig (2.8a). The load current is obtained from Fig (2.8d):

$$\tilde{I}_L = \frac{\tilde{V}_{TH}}{(Z_{TH} + Z_L)}. \quad (2.1.102)$$

The voltage across the load is

$$\tilde{V}_R = Z_L \tilde{I}_L = \frac{2Z_0 Z_L \tilde{V}_g}{(Z_{TH} + Z_L)(Z_g + Z_0)}\left[\frac{e^{-\gamma\ell}}{1-\Gamma_g e^{-2\gamma\ell}}\right]. \quad (2.1.103)$$

The transfer function of a transmission line with respect to the source voltage $\tilde{V}_g$ is

$$H(\omega) = \frac{\tilde{V}_R}{\tilde{V}_g} = \frac{2Z_0 Z_L}{(Z_{TH} + Z_L)(Z_g + Z_0)}\left[\frac{e^{-\gamma\ell}}{1-\Gamma_g e^{-2\gamma\ell}}\right]. \quad (2.1.104)$$

For a lossless transmission line connected to a matched source and a matched load, i.e. $\gamma\ell = j\beta\ell$, $Z_g = Z_0$, $Z_{TH} = Z_0$, $Z_L = Z_0$, $\Gamma_g = 0$ the transfer function is

$$H(\omega) = \frac{e^{-\gamma\ell}}{2}. \quad (2.1.105)$$

However, if the transfer function is defined by the ratio of the input voltage $\tilde{V}_s$ at the port – aa to the output voltage $\tilde{V}_R$, $H(\omega) = e^{-\gamma\ell}$. It is obtained from equation (2.1.104) for $Z_g = 0$.

2.1.9 Power Relation on Transmission Line

The average power over a time-period T in any time-harmonic periodic signal is [J.5, B.10]

$$P_{av} = \frac{1}{T}\int_0^T v(t)\cdot i(t)dt. \quad (2.1.106)$$

where the time-harmonic instantaneous voltage and current waveforms are

$$v(t) = V_0 \cos(\omega t + \phi) \quad (a)$$
$$i(t) = I_0 \cos(\omega t + \phi') \quad (b). \quad (2.1.107)$$

The voltage and current in the phasor form are written as follows:

$$\tilde{V} = V_0 e^{j\phi} \quad (a), \quad \tilde{I} = I_0 e^{j\phi'} \quad (b). \quad (2.1.108)$$

A complex number $X = a + jb$ has its complex conjugate, $X^* = a - jb$. Thus, the real (Re) and imaginary (Im) parts of a complex number are written as follows:

$$\mathrm{Re}(X) = \frac{1}{2}(X + X^*) \quad (a)$$
$$\mathrm{Im}(X) = \frac{1}{2}(X - X^*) \quad (b). \quad (2.1.109)$$

On using the above property, the instantaneous voltage and current are written as follows:

$$v(t) = \frac{1}{2}\left(\tilde{V}e^{j\omega t} + \tilde{V}^* e^{-j\omega t}\right) \quad (a)$$
$$i(t) = \frac{1}{2}\left(\tilde{I}e^{j\omega t} + \tilde{I}^* e^{-j\omega t}\right) \quad (b). \quad (2.1.110)$$

The average power in phasor form is obtained from equations (2.1.106) and (2.1.110),

$$P_{av} = \frac{1}{2}\mathrm{Re}\left[\widetilde{V}\widetilde{I}^*\right]. \quad (2.1.111)$$

It can be expressed in the usual AC form,

$$P_{av} = \frac{1}{2}\mathrm{Re}\left[V_0 e^{j\phi}\cdot I_0^* e^{-j\phi'}\right] = \frac{1}{2}V_0 I_0^* \cos(\phi - \phi'). \quad (2.1.112)$$

Available Power from Generator

Figure (2.9a) shows that the maximum available power from a source is computed by directly connecting the load to it. The average power supplied to the load is

$$P_{av} = \frac{1}{2}\,\text{Re}\,[V_L \cdot I^*] = \frac{1}{2}\,\text{Re}\left[|I|^2 Z_L\right] = \frac{1}{2}|I|^2 R_L. \quad (2.1.113)$$

The load current is

$$I = \frac{\tilde{V}_g}{Z_g + Z_L} = \frac{\tilde{V}_g}{(R_g + R_L) + j(X_g + X_L)}$$

$$\Rightarrow |I|^2 = \frac{|\tilde{V}_g|^2}{(R_g + R_L)^2 + (X_g + X_L)^2}.$$

Therefore, the average power supplied to the load is

$$P_{av} = \frac{1}{2}\left[\frac{|\tilde{V}_g|^2 R_L}{(R_g + R_L)^2 + (X_g + X_L)^2}\right]. \quad (2.1.114)$$

Under the *conjugate matching*, $X_L = -X_g$ and $R_L = R_g$, the average power supplied to load is maximum:

$$P_{av} = \frac{1}{8}\frac{|\tilde{V}_g|^2}{R_g}. \quad (2.1.115)$$

This is the maximum power available from a generator under the matching condition and delivered to a load R_L. At this stage, the maximum power delivered to a load is computed in the absence of the transmission line. For a matched terminated lossless line, the maximum available power from the source is delivered to the load. It is examined below.

The voltage and current waves on a line under no reflection case are

$$\tilde{V}(x) = \tilde{V}_S e^{-j\beta x} \quad (a), \quad \tilde{I}(x) = \frac{\tilde{V}_S e^{-j\beta x}}{Z_0} \quad (b). \quad (2.1.116)$$

The average power on the line is

$$P_{av} = \frac{1}{2}\,\text{Re}\left[\tilde{V}(x)\tilde{I}^*(x)\right] = \frac{1}{2}\,\text{Re}\left[\tilde{V}_S e^{-j\beta x}\frac{\tilde{V}_S^*}{Z_0}e^{j\beta x}\right] = \frac{|\tilde{V}_S|^2}{2Z_0}. \quad (2.1.117)$$

On a lossless line, the average power is independent of the distance *x* from a source. Physically it makes a sense, as the same amount of power flows at any location on the line. Under the matched load termination, $Z_L = Z_0$, the input impedance at the source end is Z_0 itself. It is shown in Fig (2.9b). The sending end voltage at the input port – aa of a transmission line is

$$|\tilde{V}_S| = \left|\frac{\tilde{V}_g Z_0}{Z_g + Z_0}\right|. \quad (2.1.118)$$

The maximum power is transferred from a generator to the transmission line under the matching condition, $R_g = Z_0$. The maximum available power from the generator to feed the line is given by equation (2.1.115). It is identical to the power determined from equation (2.1.117), as $\tilde{V}_s = \tilde{V}_g/2$.

If the line is not terminated in its characteristic impedance, then a reflection takes place at the load end. The reflected wave travels from the load toward the generator given by

$$\tilde{V}(x) = V^- e^{+j\beta x} \quad (a), \quad \tilde{I}(x) = -\frac{V^- e^{+j\beta x}}{Z_0} \quad (b). \quad (2.1.119)$$

The average power in the reflected wave is

$$P_{av} = \frac{1}{2}\,\text{Re}\left[\tilde{V}(x)\tilde{I}^*(x)\right] = -\frac{|V^-|^2}{2Z_0}. \quad (2.1.120)$$

However, at the load end amplitude of the reflected voltage wave is $V^- = \Gamma_L V^+$; where $V^+ = \tilde{V}_S$. Therefore, the average reflected power on the line is

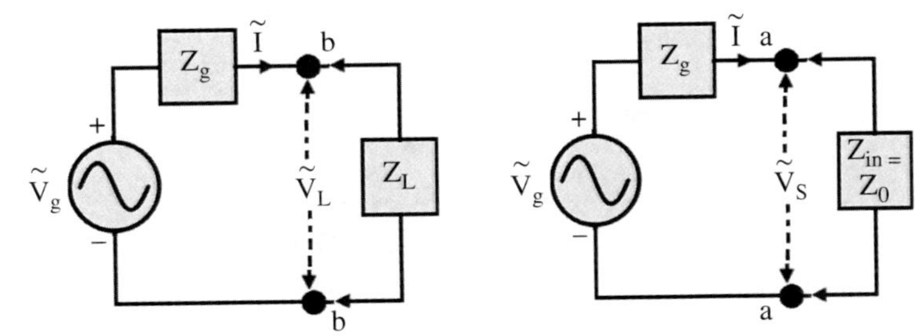

(a) Load directly connected to a source. (b) Line terminated in a matched load.

Figure 2.9 Load connections to a source.

$$P_{av} = -\frac{|\Gamma_L|^2 |V^+|^2}{2Z_0} = -\frac{|\Gamma_L|^2 \tilde{V}_s^2}{2Z_0}. \qquad (2.1.121)$$

The negative sign (−) shows that the reflected power travels from the load toward the source. Finally, under the mismatched load, the power delivered to the load is obtained from equations (2.1.117) and (2.1.121)

$$P_L = \frac{\tilde{V}_S^2}{2Z_0}\left(1 - |\Gamma_L|^2\right). \qquad (2.1.122)$$

For a lossless line, the power balance is written as follows:

$$P_{in} = \frac{1}{2Z_0}\left|\tilde{V}_s\right|^2\left[1 - |\Gamma_L|^2\right] = P_{out}. \qquad (2.1.123)$$

↓ Incident power ↓ Reflected power

In equation (2.1.123), input power P_{in} at the input port – aa of the line enters into the line. It is supplied by a source. The output power P_{out} is the power supplied to the load.

2.2 Multisection Transmission Lines and Source Excitation

This section extends the solution of the voltage wave equation to the multisection transmission line [B.8, B.16]. Next, the voltage responses are obtained for the shunt connected current source, and also the series-connected voltage source, at any location on a line. This treatment is used in chapters 14 and 16 for the *spectral domain analysis* of the multilayer planar transmission lines.

2.2.1 Multisection Transmission Lines

Figure (2.10a) shows a multisection transmission line, consisting of the N number of line sections. Each line section has different lengths (d_1, d_2,...,d_N), different

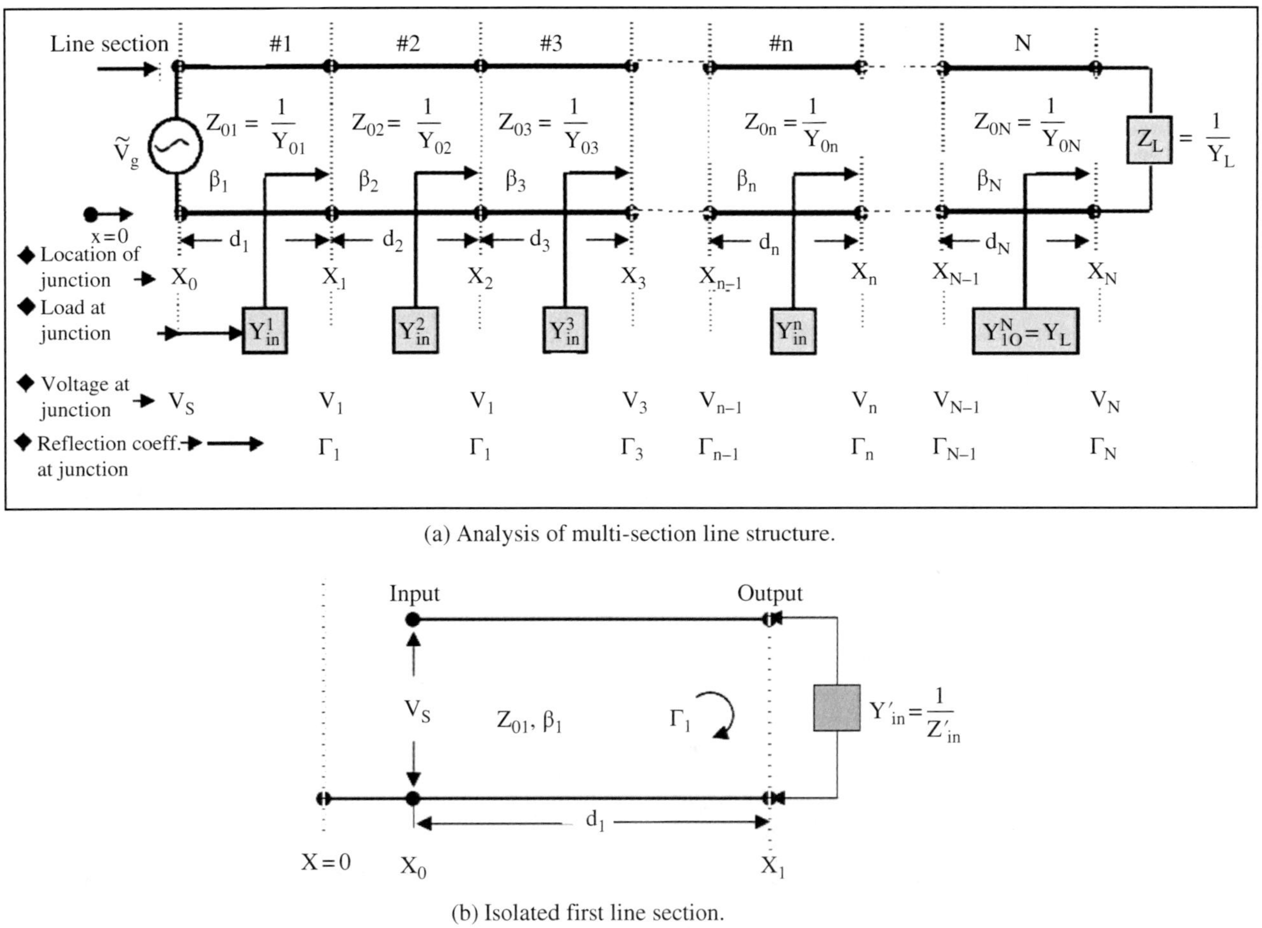

Figure 2.10 The multisection transmission line.

characteristic impedances ($Z_{01}, Z_{02},\ldots,Z_{0N}$), or different characteristic admittances ($Y_{01}, Y_{02},\ldots,Y_{0N}$) and different propagation constants ($\beta_1, \beta_2, \ldots, \beta_N$). At each junction of two dissimilar lines, the voltage wave reflection occurs with the reflection coefficient $\Gamma_1, \Gamma_2, \ldots, \Gamma_N$. At each junction, the input admittance of all succeeding line sections appears as a load. The input admittances at junctions ($x_1, x_2, \ldots, x_N$) are $\left(Y_{in}^1, Y_{in}^2, \ldots, Y_{in}^N\right)$. The distances of the junctions are measured from the origin that is located on the left-hand side. The input terminals of line sections 1,2,..., $N-1$ are located at the junctions ($x_0, x_1, x_2,\ldots,x_N$). The voltage source V_s is located at the input of the first line section. The last line section is terminated in the load $Z_L = 1/Y_L$.

The objective is to find the voltage at each junction of the multisection line. Further, the voltage distribution on each line section is determined due to the input voltage $\tilde{V}_s$.

The solutions for the voltage and current wave equations involve four constants. The constants of the current wave are related to two constants of the voltage wave through the characteristic impedance of a line. Out of two constants of the voltage wave, one is expressed in terms of the reflection coefficient at the load end; that itself is expressed by the characteristic impedance and the terminated load impedance. The reflection coefficient can also be expressed by the characteristic admittance and the terminated load admittance. The second constant is evaluated by the source condition at the input end. Figure (2.10b) shows the first isolated line section. The voltage and current waves, with respect to the origin at the load end x_1, on the line section ($x_0 \le x \le x_1$) are written from equation (2.1.88):

$$\tilde{V}(x) = V^+\left[e^{-j\beta_1(x-x_1)} + \Gamma_1 e^{j\beta_1(x-x_1)}\right] \quad \text{(a)}$$

$$\tilde{I}(x) = \frac{V^+}{Z_{01}}\left[e^{-j\beta_1(x-x_1)} - \Gamma_1 e^{j(x-x_1)}\right], (x_0 \le x \le x_1) \quad \text{(b)}. \tag{2.2.1}$$

The reflection coefficient Γ_1 at the load end, i.e. at $x = x_1$ is given by

$$\Gamma_1 = \frac{Z_{in}^1 - Z_{01}}{Z_{in}^1 + Z_{01}} = \frac{Y_{01} - Y_{in}^1}{Y_{01} + Y_{in}^1}. \tag{2.2.2}$$

The load at the $x = x_1$ end is formed by the cascaded line sections after location $x = x_1$. The voltage amplitude V^+ is evaluated by the boundary condition at the input, $x = x_0$, of the first line section. At $x = x_0$, shown in Fig (2.10b), the source voltage $\tilde{V}(x_0)$ is $\tilde{V}_s$ and V^+ is evaluated as follows:

$$\tilde{V}(x_0) = \tilde{V}_S = V^+\left[e^{-j\beta_1(x_0-x_1)} + \Gamma_1 e^{j\beta_1(x_0-x_1)}\right]$$

$$\Rightarrow V^+ = \frac{\tilde{V}_S}{e^{-j\beta_1(x_0-x_1)} + \Gamma_1 e^{j\beta_1(x_0-x_1)}}. \tag{2.2.3}$$

The voltage wave on the transmission line section #1 is

$$\tilde{V}_1(x) = \frac{\tilde{V}_S\left[e^{-j\beta_1(x-x_1)} + \Gamma_1 e^{j\beta_1(x-x_1)}\right]}{e^{-j\beta_1(x_0-x_1)} + \Gamma_1 e^{j\beta_1(x_0-x_1)}}. \tag{2.2.4}$$

The above expression is valid over the range $x_0 \le x \le x_1$. The voltage at the output of the line section #1 ($x = x_1$), that is at the junction of line #1 and line #2, is

$$\tilde{V}_1(x_1) = \frac{\tilde{V}_S[1+\Gamma_1]}{e^{-j\beta_1(x_0-x_1)} + \Gamma_1 e^{j\beta_1(x_0-x_1)}} = \frac{\tilde{V}_S[1+\Gamma_1]}{e^{+j\beta_1 d_1} + \Gamma_1 e^{-j\beta_1 d_1}}, \tag{2.2.5}$$

where $d_1 = x_1 - x_0$ is the length of the line section #1. The above voltage is input to the line section #2. Equations (2.2.4) and (2.2.5) apply to any line section and at any junction. The voltage $\tilde{V}_{n-1}(x_{n-1})$ is treated as the input voltage of the nth line section. It is the same as the output voltage of the $(n-1)^{th}$ line section. The line length d_1 and reflection coefficient Γ_1 are replaced by d_n and Γ_n, respectively. The cascaded line sections to the right-hand side of any junction act as a load at the junction and the reflection coefficient at the junction is

At load end of the line section#1 :

$$\Gamma_1 = \frac{Y_{01} - Y_{in}^1}{Y_{01} + Y_{in}^1} \quad \text{(a)}$$

At load end of the line section#n :

$$\Gamma_n = \frac{Y_{0n} - Y_{in}^n}{Y_{0n} + Y_{in}^n} \quad \text{(b)}$$

Length of the line section#n :

$$d_n = x_n - x_{n-1}, \quad (n = 1, 2, \ldots N) \quad \text{(c)}. \tag{2.2.6}$$

Equation (2.2.4) is applied to Fig (2.10a) to compute the voltage distribution on any line section. The voltage on line section #2 is

$$\tilde{V}_2(x) = \frac{\tilde{V}_1\left[e^{-j\beta_2(x-x_2)} + \Gamma_2 e^{j\beta_2(x-x_2)}\right]}{e^{-j\beta_2(x_1-x_2)} + \Gamma_2 e^{j\beta_2(x_1-x_2)}}, x_1 \le x \le x_2. \tag{2.2.7}$$

The voltage at the output of the line section #2, i.e. the junction voltage of the line sections #2 and # 3 at $x = x_2$, is obtained from the above equation:

$$\tilde{V}_2(x_2) = \frac{\tilde{V}_1[1+\Gamma_2]}{e^{+j\beta_2 d_2} + \Gamma_2 e^{-j\beta_2 d_2}}. \tag{2.2.8}$$

Using equation (2.2.5) and above equations, the voltage distribution on the line section #2, and also the junction voltage at $x = x_2$, are obtained:

$$\tilde{V}_2(x) = \frac{\tilde{V}_S[1+\Gamma_1]}{e^{+j\beta_1 d_1} + \Gamma_1 e^{-j\beta_1 d_1}} \left\{ \frac{e^{-j\beta_2(x-x_2)} + \Gamma_2 e^{j\beta_2(x-x_2)}}{e^{+j\beta_2 d_2} + \Gamma_2 e^{-j\beta_2 d_2}} \right\}. \tag{2.2.9}$$

$$\tilde{V}_2(x_2) = \frac{\tilde{V}_S[1+\Gamma_1]}{e^{+j\beta_1 d_1} + \Gamma_1 e^{-j\beta_1 d_1}} \left\{ \frac{1+\Gamma_2}{e^{+j\beta_2 d_2} + \Gamma_2 e^{-j\beta_2 d_2}} \right\}. \tag{2.2.10}$$

Finally, the voltage distribution on the nth line section and the voltage at the nth line junction can be written as follows:

$$\tilde{V}_n(x) = \left[\prod_{n=1}^{N} \left\{ \frac{V_S(1+\Gamma_{n-1})}{e^{j\beta_{n-1} d_{n-1}} + \Gamma_{n-1} e^{-j\beta_{n-1} d_{n-1}}} \right\} \right] \left\{ \frac{e^{-j\beta_n(x-x_n)} + \Gamma_n e^{j\beta_n(x-x_n)}}{e^{+j\beta_n d_n} + \Gamma_n e^{-j\beta_n d_n}} \right\} \tag{2.2.11}$$

$$\tilde{V}_n(x_n) = \left[\prod_{n=1}^{N} \left\{ \frac{V_S(1+\Gamma_n)}{e^{jd_n\beta_n} + \Gamma_n e^{-jd_n\beta_n}} \right\} \right]. \tag{2.2.12}$$

2.2.2 Location of Sources

The shunt voltage $\tilde{V}_s$ could be located at any junction and the voltage distribution is computed on any line section due to it. However, it is also interesting to consider a *shunt current source* and a *series voltage source* located anywhere on a multisection transmission line. Both kinds of sources create the voltage wave on a line.

Current Source at the Junction of Finite Length Line and Infinite Length Line

Figure (2.11a) shows a transmission line circuit with a current source I_S located at $x = 0$ that is the junction of two lines of different electrical characteristics. The open-circuited line #1, with length $x = -d_1$, is located at the left-hand side of the current source. Its characteristics impedance/admittance is (Z_{01}/Y_{01}) and its propagation constant is β_1. The infinite length line #2, with

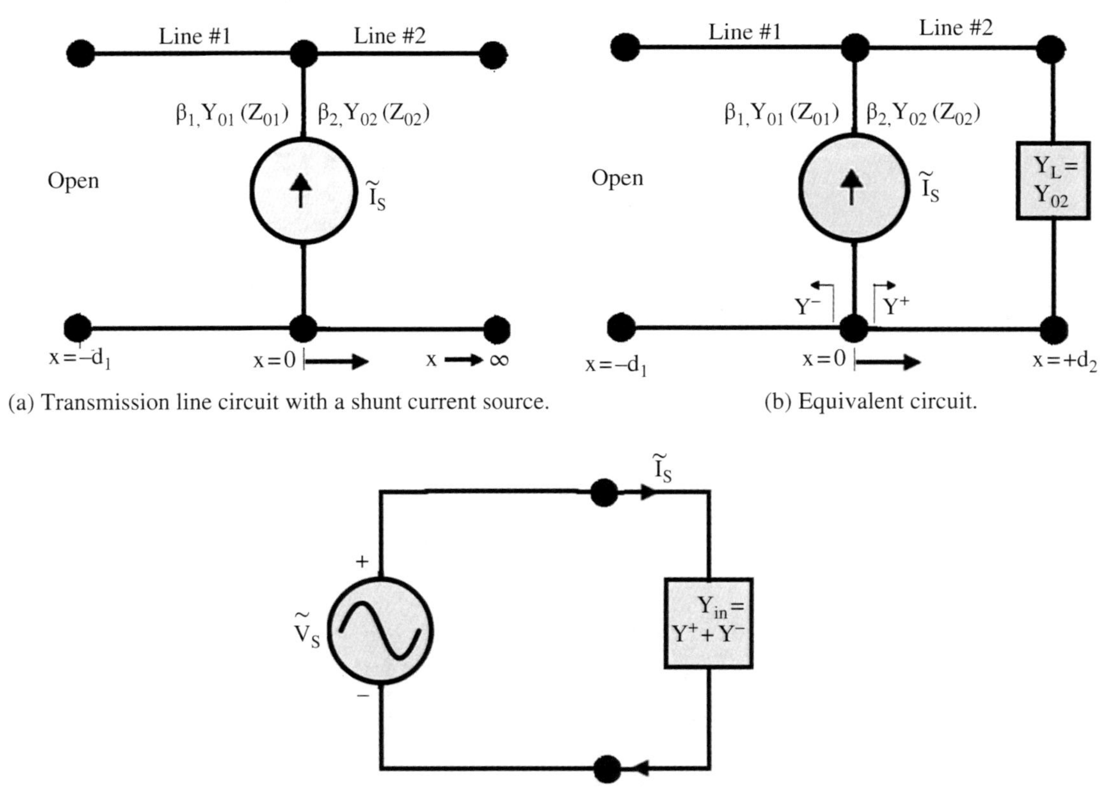

(a) Transmission line circuit with a shunt current source.

(b) Equivalent circuit.

(c) Replacement of shunt current source by the equivalent voltage source.

Figure 2.11 A shunt current source at the junction of two-line sections.

characteristics impedance/admittance (Z_{02}/Y_{02}) and the propagation constant β_2, is located at the right-hand side of the current source. It can be replaced by a load admittance $Y_L = Y_{02}$ at a distance $x = d_2$, shown in Fig (2.11b). The objective is to find out the voltage waves on both the lines as excited by the current source.

The current source I_S can be replaced by an equivalent voltage source V_s, shown in Fig (2.11c), at $x = 0$:

$$\tilde{V}_S = \frac{I_S}{Y_{in}} = \frac{I_S}{Y^- + Y^+}, \tag{2.2.13}$$

where Y_{in} is the total load admittance at the plane containing the current source I_S. Y^- and Y^+ are left-hand and right-hand side admittances at $x = 0$ given by

$$\begin{aligned} Y^- &= Y_{01} \tanh(\gamma_1 d_1) \\ &= jY_{01} \tan \beta_1 d_1 (\text{For a losslessline}), \quad Y^+ = Y_{02}. \end{aligned} \tag{2.2.14}$$

The general solution of a voltage wave is given by equation (2.1.79a). The constants V^+ and V^- are evaluated for the left-hand side of a lossless transmission line. At $x = 0$, $V(x = 0) = V_s$. On using this boundary condition in equation (2.1.79a): $V_S = V^+ + V^-$. At $x = -d_1$ the line is open-circuited with $I(x = -d_1) = 0$. On using this boundary condition in equation (2.1.79b): $V^+ e^{j\beta_1 d_1} - V^- e^{-j\beta_1 d_1} = 0$. The constants V^+ and V^-, from these two equations, are obtained as

$$V^+ = \frac{\tilde{V}_S}{1 + e^{j2\beta_1 d_1}} \quad (a), \quad V^- = V^+ e^{j2\beta_1 d_1} \quad (b). \tag{2.2.15}$$

The voltage wave on the left-hand line #1 is obtained by substituting equation (2.2.15) in equation (2.1.79a):

$$\tilde{V}(x) = \frac{\tilde{V}_S}{\cos(\beta_1 d_1)} \cos[\beta_1 (x + d_1)], \quad -d_1 \le x \le 0. \tag{2.2.16}$$

The line at the right-hand side of the current source is an infinite length line that supports a traveling wave without any reflection. Therefore, at $x = 0$, $V^- = 0$ and $V^+ = V_S$. The voltage wave on line #2 at the right-hand side is

$$\tilde{V}(x) = \tilde{V}_S e^{-j\beta_2 x}, \quad x \ge 0. \tag{2.2.17}$$

The method can be easily extended to a multi-section line structure. For this purpose, the left-hand and right-hand side admittances Y^- and Y^+ are determined at the plane containing the current source.

Series Voltage Source

Figure (2.12a) shows the series-connected voltage source V_S at $x = 0$. The location $x = 0$ is a junction of two transmission lines – line #1 open-circuited finite-length line and line #2 infinite length line. The lines at the left-hand and right-hand sides of the voltage source can be replaced by the equivalent impedances Z^- and Z^+, respectively. It is shown in the equivalent circuit, Fig (2.12b). Again, the voltage waves on both lines, excited by a series voltage source, could be determined.

The voltages across loads Z^- (Z_1) and Z^+ (Z_2), shown in Fig (2.12b), are obtained as follows:

$$\begin{aligned} &\text{Line current}: \ \tilde{I} = \frac{\tilde{V}_S}{Z^- + Z^+} && (a) \\ &\text{Voltage across } Z_1: \ \tilde{V}(x = 0^-) = -\frac{\tilde{V}_S Z_1}{Z_1 + Z_2} && (b) \\ &\text{Voltage across } Z_2: \ \tilde{V}(x = 0^+) = \frac{\tilde{V}_S Z_2}{Z_1 + Z_2} && (c). \end{aligned} \tag{2.2.18}$$

Line #1 is open-circuited and line #2 is of infinite extent. Therefore, their input impedances at $x = 0^-$ and $x = 0^+$ are

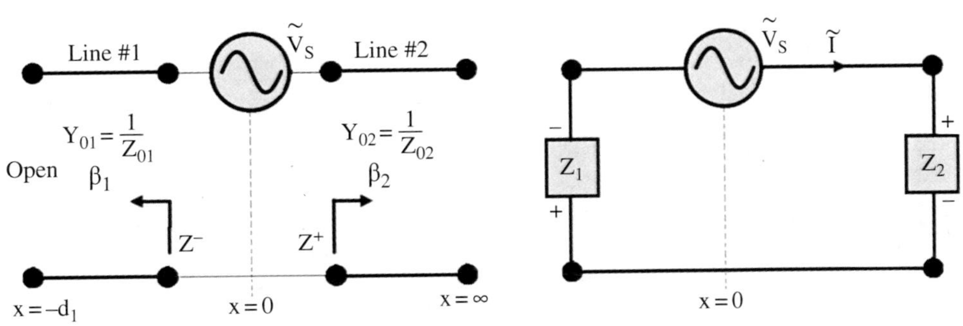

(a) Transmission line circuit with a series voltage source.

(b) Equivalent circuit.

Figure 2.12 A series voltage source at the junction of two-line sections.

$$Z_1 = -jZ_{01}\cot(\beta_1 d_1) \quad (a), \quad Z_2 = Z_{02} \quad (b). \tag{2.2.19}$$

The voltage at $x = 0^+$ from equations (2.2.18c) and (2.2.19) is

$$\widetilde{V}(x=0^+) = \widetilde{V}_S \frac{1}{Y_{02}} \times \frac{1}{\left[\dfrac{1}{jY_{01}\tan(\beta_1 d_1)} + \dfrac{1}{Y_{02}}\right]}$$

$$= \widetilde{V}_S \frac{jY_{01}\tan(\beta_1 d_1)}{Y_{02} + jY_{01}\tan(\beta_1 d_1)}. \tag{2.2.20}$$

For a lossy transmission line, the above equation could be written as follows:

$$\widetilde{V}(x=0^+) = \widetilde{V}_S \frac{Y_{01}\tanh(\gamma_1 d_1)}{Y_{02} + Y_{01}\tanh(\gamma_1 d_1)}. \tag{2.1.21}$$

The voltage wave on the infinite extent lossy line #2 is

$$\widetilde{V}(x) = \widetilde{V}(x=0^+)e^{-\gamma_2 x}$$

$$= \widetilde{V}_S \frac{Y_{01}\tanh(\gamma_1 d_1)}{Y_{02} + Y_{01}\tanh(\gamma_1 d_1)} e^{-\gamma_2 x}, \quad x \geq 0. \tag{2.2.22}$$

The voltage at $x = 0^-$ on the lossless line #1 is

$$\widetilde{V}(x=0^-) = -\widetilde{V}_S \frac{Y_{02}}{Y_{02} + jY_{01}\tan(\beta_1 d_1)}. \tag{2.2.23}$$

However, the voltage at $x = 0^-$ on a lossy line #1 is

$$\widetilde{V}(x=0^-) = -\widetilde{V}_S \frac{Y_{02}}{Y_{02} + Y_{01}\tanh(\gamma_1 d_1)}. \tag{2.2.24}$$

The voltage wave on the open-circuited lossless line #1 is obtained from equations (2.2.16) and (2.2.23):

$$\widetilde{V}(x) = \frac{V(x=0^-)}{\cos(\beta_1 d_1)}\cos[\beta_1(x+d_1)]$$

$$\Rightarrow \widetilde{V}(x) = -\widetilde{V}_S \frac{Y_{02}}{Y_{02} + jY_{01}\tan(\beta_1 d_1)} \times \frac{\cos[\beta_1(x+d_1)]}{\cos(\beta_1 d_1)}, \quad -d_1 \leq x \leq 0. \tag{2.2.25}$$

2.3 Nonuniform Transmission Lines

The previous sections have presented the voltage and current waves on the uniform transmission line that has no change in the geometry along the direction of propagation. For a uniform line, the relative permittivity and relative permeability also do not change along the line. For such a uniform lossless or low-loss transmission line, the voltage and current waves travel at a definite velocity from low frequency to high frequency. The uniform transmission line behaves as a low-pass filter (LPF) section. The propagation constant for such a uniform transmission line is the same at any section of the line. The line also has a unique characteristic impedance that is independent of the location on a line. However, if the geometry of a transmission line or an electrical property of the medium of a line changes in the direction of propagation, such a line is no longer a uniform transmission line. It is a *nonuniform transmission line*. Its electrical properties, such as the RLCG primary constants, propagation constant, phase velocity, and characteristic impedance, become a function of the space coordinate along the direction of propagation. The characteristic impedance of a nonuniform transmission line changes from one end to another end; therefore, it finds application in the broadband impedance matching [B.9, B.10, B.12]. It is also used for the design of delay equalizers, filters, wave-shaping circuits, etc. [J.6–J.9]. It is an essential section of the on-chip measurement system [J.10]. *Unlike a uniform transmission line, it shows a cut-off phenomenon, i.e. the wave propagates on the line only above the cut-off frequency*. Below the cut-off frequency, the wave only attenuates with distance. Thus, the nonuniform transmission line behaves like a high-pass filter (HPF) section [J.11–J.13, B.17].

The present section obtains the wave equations for a nonuniform transmission line. However, like the uniform transmission lines, the nonuniform transmission lines do not have closed-form solutions for the voltage and current waves. The numerical methods have been used to determine the response of an arbitrarily shaped nonuniform transmission line [J.10, J.11]. However, this section discusses only the *exponential nonuniform transmission line* to understand its characteristics.

2.3.1 Wave Equation for Nonuniform Transmission Line

Figure (2.13) shows a nonuniform transmission line. The line parameters (primary line constants) $R(x)$, $L(x)$, $C(x)$, $G(x)$ are distance-dependent. It results in the distance-dependent characteristic impedance, $Z_0(x)$, and propagation constant, $\gamma(x)$. Using equation (2.1.20), the voltage and current equations for a nonuniform transmission line are written as follows:

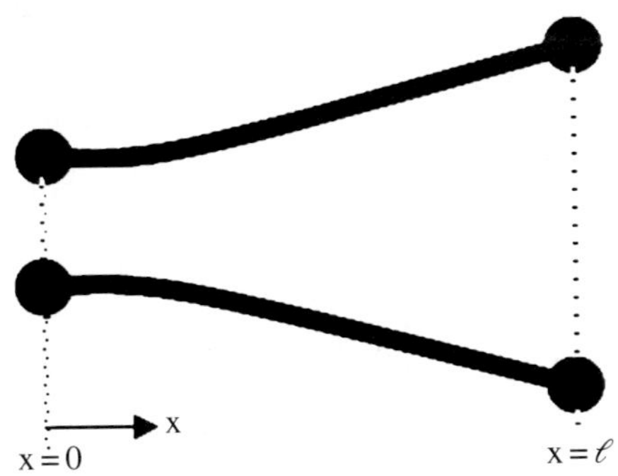

Figure 2.13 Nonuniform transmission line.

$$\frac{\partial v}{\partial x} = -\left(R(x)i + L(x)\frac{\partial i}{\partial t}\right) \quad (a)$$
$$\frac{\partial i}{\partial x} = -\left(G(x)v + C(x)\frac{\partial v}{\partial t}\right) \quad (b). \quad (2.3.1)$$

The following expressions are obtained on differentiating equation (2.3.1a) with x and equation (2.3.1b) with t:

$$\frac{\partial^2 v}{\partial x^2} = -\left(R(x)\frac{\partial i}{\partial x} + i\frac{\partial R(x)}{\partial x} + L(x)\frac{\partial^2 i}{\partial x \partial t} + \frac{\partial i}{\partial t}\frac{\partial L(x)}{\partial x}\right). \quad (2.3.2)$$

$$\frac{\partial^2 i}{\partial t \partial x} = -\left(G(x)\frac{\partial v}{\partial t} + C(x)\frac{\partial^2 v}{\partial t^2}\right). \quad (2.3.3)$$

On substituting equations (2.3.1b) and (2.3.3) in equation (2.3.2):

$$-\frac{\partial^2 v}{\partial x^2} = -R(x)\left(G(x)v + C(x)\frac{\partial v}{\partial t}\right) + i\frac{\partial R(x)}{\partial x} - L(x)\left(G(x)\frac{\partial v}{\partial t} + C(x)\frac{\partial^2 v}{\partial t^2}\right) + \frac{\partial i}{\partial t}\frac{\partial L(x)}{\partial x}. \quad (2.3.4)$$

This equation has both the voltage and current variables v(x, t) and i(x, t). However, most of the transmission lines are low-loss lines. Thus, using $R(x) \to 0$ $G(x) \to 0$ in equations (2.3.1a) and (2.3.4), the following *voltage wave equation* is obtained for a lossless nonuniform transmission line:

$$\frac{\partial^2 v}{\partial x^2} = L(x)C(x)\frac{\partial^2 v}{\partial t^2} + \left(\frac{1}{L(x)}\frac{\partial L(x)}{\partial x}\right)\frac{\partial v}{\partial x}. \quad (2.3.5)$$

Likewise, the *current wave equation* is obtained as,

$$\frac{\partial^2 i}{\partial x^2} = L(x)C(x)\frac{\partial^2 i}{\partial t^2} + \left(\frac{1}{C(x)}\frac{\partial C(x)}{\partial x}\right)\frac{\partial i}{\partial x}. \quad (2.3.6)$$

If L(x) and C(x) are not a function of x, then equations (2.3.5) and (2.3.6) reduce to the familiar wave equations (2.1.24) and (2.1.25) on a uniform transmission line.

For a lossy nonuniform transmission line, it is not possible to get separate voltage and current wave equations in the time domain. However, separate voltage and current wave equations can be obtained in the frequency domain by using the phasor form of voltage and current. The transmission line equations in the phasor form are

$$\frac{d\tilde{V}(x)}{dx} = -Z(x)\tilde{I}(x) \quad (a)$$
$$\frac{d\tilde{I}(x)}{dx} = -Y(x)\tilde{V}(x) \quad (b), \quad (2.3.7)$$

where the line series impedance and shunt admittance *p.u.l.* are given by

$$Z(x) = R(x) + j\omega L(x) \quad (a)$$
$$Y(x) = G(x) + j\omega C(x) \quad (b). \quad (2.3.8)$$

The following wave equations for the nonuniform transmission line are obtained:

$$\frac{d^2\tilde{V}(x)}{dx^2} - \left(\frac{1}{Z(x)}\frac{dZ(x)}{dx}\right)\frac{d\tilde{V}(x)}{dx} - Z(x)Y(x)\tilde{V}(x) = 0. \quad (2.3.9)$$

$$\frac{d^2\tilde{I}(x)}{dx^2} - \left(\frac{1}{Y(x)}\frac{dY(x)}{dx}\right)\frac{d\tilde{I}(x)}{dx} - Z(x)Y(x)\tilde{I}(x) = 0. \quad (2.3.10)$$

If Z(x) and Y(x) are not a function of x, the above wave equations reduce to wave equation (2.1.37a and b) for a uniform transmission line. For a lossless nonuniform line, the series impedance and shunt admittance per unit length are $Z(x) = j\omega L(x)$, $Y(x) = j\omega C(x)$. The voltage wave equation (2.3.9) could be written as

$$\frac{d^2\tilde{V}(x)}{dx^2} - \left[\frac{1}{L(x)}\frac{dL(x)}{dx}\right]\frac{d\tilde{V}(x)}{dx} + \left(\frac{\omega}{v_p(x)}\right)^2\tilde{V}(x) = 0, \quad (2.3.11)$$

where position-dependent *nominal* phase velocity of a nonuniform transmission line is given by

$$v_p = \frac{1}{\sqrt{L(x)C(x)}}. \quad (2.3.12)$$

It is difficult to get a general solution for the above wave equations. However, under the case of no reflection on a line, and the line with a small fractional change in L(x) and C(x) over a wavelength, Lewis and Wells, and Wohler [B.17, J.11] have given the following solution of wave equation (2.3.11):

$$\tilde{V}(x) = \tilde{V}(0)\left[\frac{Z_0(x)}{Z_0(0)}\right]^{\frac{1}{2}} \exp\left[-j\omega\int_0^x \frac{dx}{v_p(x)}\right]. \tag{2.3.13}$$

In this expression $Z_0(x)$ is the *nominal* characteristic impedance at any location x on the nonuniform transmission line, whereas characteristic impedance $Z_0(0)$ is the nominal characteristic impedance at $x = 0$. For a uniform line, the phase velocity $v_p(x)$ is constant and $Z_0(x) = Z_0(0)$, $\frac{dL(x)}{dx} = 0$. The wave equation (2.3.11) is reduced to the wave equation of a uniform transmission line. The solution (2.3.13) is also reduced to the standard solution, $\tilde{V}(x) = \tilde{V}(0)e^{-j\beta x}$.

Equation (2.3.13) shows that for increasing characteristic impedance $Z_0(x)$ along the line length, the voltage amplitude also increases as the square root of nominal characteristic impedance. Lewis and Wells [B.17] have also given an expression for the reflection coefficient of the nonuniform transmission line terminated in the load Z_L at $x = \ell$:

$$\Gamma(x) = \frac{\left[1 - \frac{1}{j2\omega}\frac{v_p(x=\ell)}{Z_0(x=\ell)}\left(\frac{dZ_0(x)}{dx}\right)_{x=\ell}\right]Z_L - Z_0(x=\ell)}{\left[1 + \frac{1}{j2\omega}\frac{v_p(x=\ell)}{Z_0(x=\ell)}\left(\frac{dZ_0(x)}{dx}\right)_{x=\ell}\right]Z_L + Z_0(x=\ell)}. \tag{2.3.14}$$

For a uniform transmission line $Z_0(x = \ell) = Z_0$, and equation (2.3.14) is reduced to the nominal reflection coefficient,

$$\Gamma_{nom}(x) = \frac{Z_L - Z_0(x=\ell)}{Z_L + Z_0(x=\ell)} = \frac{Z_L - Z_0}{Z_L + Z_0}. \tag{2.3.15}$$

At higher operating frequency ω, the reflection coefficient for any termination, given by equation (2.3.14), is also reduced to equation (2.3.15). However, reflection occurs at a lower frequency ω on a nonuniform transmission line, even if the nominal reflection coefficient $\Gamma_{nom}(x = \ell)$ zero, i.e. even if the line is matched at the load end. This behavior is different from that of a uniform transmission line.

2.3.2 Lossless Exponential Transmission Line

The general solution of the wave equation for a nonuniform transmission line is not available. However, the closed-form solution is obtained for an exponential transmission line [J.11, J.13]. This case demonstrates the properties of a nonuniform line. The following exponential variation is assumed for the line inductance and capacitance of a nonuniform transmission line:

$$L(x) = L_0e^{2px}, \quad C(x) = C_0e^{-2px}, \tag{2.3.16}$$

where L_0 and C_0 are primary line constants at $x = 0$ and p is a *parameter controlling the propagation characteristics*. The above choice of line inductance and capacitance maintains a constant phase velocity that is independent of the location along the line length. The characteristic impedance of a lossless exponential transmission line changes exponentially along the line length. Its propagation constant is also frequency-dependent. Therefore, *a lossless nonuniform line is dispersive*. The *nominal* characteristic impedance at any location x on the line is

$$Z_0(x) = \sqrt{\frac{L(x)}{C(x)}} = \sqrt{\frac{L_0}{C_0}}e^{2px} = Z_0(x=0)e^{2px}. \tag{2.3.17}$$

The parameter p, defined below, could be determined from the characteristic impedance at the input and output ends of the line:

$$p = \frac{1}{2\ell}\log_e\frac{Z_0(\ell)}{Z_0(x=0)}. \tag{2.3.18}$$

If the impedances at both ends of a line are fixed, changing the line length, ℓ, can change the parameter p. The parameter p also determines the propagation characteristics of a nonuniform transmission line. The series impedance and shunt admittance *p.u.l.* of the exponential line can be written as follows:

$$Z(x) = j\omega L_0e^{2px} \quad (a), \qquad Y(x) = j\omega C_0e^{-2px} \quad (b). \tag{2.3.19}$$

In case of an exponential line, the voltage and current wave equations (2.3.9) and (2.3.10) reduce to

$$\frac{d^2\tilde{V}(x)}{dx^2} - 2p\frac{d\tilde{V}}{dx} + \omega^2L_0C_0\tilde{V}(x) = 0. \tag{2.3.20}$$

$$\frac{d^2\tilde{I}(x)}{dx^2} + 2p\frac{d\tilde{I}}{dx} + \omega^2L_0C_0\tilde{I}(x) = 0. \tag{2.3.21}$$

Let us assume the following exponential form of the solution for the above wave equations with separate propagation constants for the voltage and current waves:

$$\tilde{V}(x) = \tilde{V}_0 e^{-\gamma_1 x} \quad (a), \quad \tilde{I}(x) = \tilde{I}_0 e^{-\gamma_2 x} \quad (b). \tag{2.3.22}$$

The above differential equations provide the following characteristic equations:

$$\gamma_1^2 + 2p\gamma_1 + \omega^2 L_0 C_0 = 0 \quad (a)$$
$$\gamma_2^2 - 2p\gamma_2 + \omega^2 L_0 C_0 = 0 \quad (b). \tag{2.3.23}$$

On solving the above equations, the following expressions are obtained for the complex propagation constants:

$$\gamma_1 = \alpha_1 \pm j\beta_1 = -p \pm \sqrt{p^2 - \omega^2 L_0 C_0} \quad (a)$$
$$\gamma_2 = \alpha_2 \pm j\beta_2 = p \pm \sqrt{p^2 - \omega^2 L_0 C_0} \quad (b). \tag{2.3.24}$$

In the case of a uniform transmission line ($p = 0$), the propagation constants for the voltage and current waves are identical. The parameter p determines the attenuation constant, i.e. α of a nonuniform line. It is positive for the condition $Z_0(x = \ell) > Z_0(x = 0)$. Thus, there is an attenuation factor even for a lossless nonuniform line. The factor under the radical sign provides the propagation constant, i.e. the phase-shift constant β. At the cut-off frequency, $\omega = \omega_c$, β is zero. The cut-off frequency is given by

$$\omega_c = \frac{p}{\sqrt{L_0 C_0}}, \tag{2.3.25}$$

where phase velocity of the voltage and current waves on the line at $x = 0$ is

$$V_p(x = 0) = \frac{1}{\sqrt{L_0 C_0}}. \tag{2.3.26}$$

The complex propagation constants can be rewritten as follows:

$$\gamma_1 = \alpha_1 \pm j\beta_1 = -p \pm jp\sqrt{\left(\frac{\omega}{\omega_c}\right)^2 - 1} \quad (a)$$
$$\gamma_2 = \alpha_2 \pm j\beta_2 = p \pm jp\sqrt{\left(\frac{\omega}{\omega_c}\right)^2 - 1} \quad (b). \tag{2.3.27}$$

The propagation constants β_1 and β_2 are imaginary quantities for the signal below the cut-off frequency $\omega < \omega_c$. Under such conditions, no wave propagates on the nonuniform line. The initial signal only gets attenuated. It is called the *evanescent mode*. The wave propagation takes place only for $\omega > \omega_c$. Therefore, a nonuniform transmission line behaves like a high-pass filter (HPF).

However, real parts of the complex propagation constants γ_1 and γ_2 are nonzero. For $p > 0$, the voltage wave gets attenuated while the current wave is increased in the positive direction of wave propagation. In the backward direction, the reflected voltage and current waves have opposite behavior. The attenuation in the signal is not due to any ohmic loss of a line. It is due to the continuous reflection of the wave as it progresses on the line. The opposite behavior of the voltage and current waves maintains the constant flow of power (P) at any location on a line:

$$P(x) = \frac{1}{2}\tilde{V}(x)\tilde{I}^*(x) = \frac{1}{2}V_0 I_0 e^{-(\gamma_1 + \gamma_2^*)x}, \tag{2.3.28}$$

where $\gamma_1 + \gamma_2^* = 0$. Unlike a lossless uniform transmission line, the phase velocity of the voltage and current waves on a lossless nonuniform transmission line is dispersive as given below:

$$v_p = \frac{\omega}{\beta} = \frac{\omega}{p\sqrt{\left(\frac{\omega}{\omega_c}\right)^2 - 1}}. \tag{2.3.29}$$

The phase velocity shows singularity at the cut-off frequency. After the cut-off frequency, i.e. for $\omega >> \omega_c$, it decreases, with an increase in frequency, to a value given by expression (2.3.26).

References

Books

B.1 Nahin Paul, J.: *Oliver Heaviside, Sage in Solitude: The Life, Work, and Times of an Electrical Genius of the Victorian Age*, IEEE Press, New York, 1988.

B.2 MacCluer, C.R.: *Boundary Value Problems and Fourier Expansions*, Dover Publications, Mineola, NY, 2004.

B.3 Sears, F.W.; Zemansky, M.W.: *University Physics*, Addition-Wesley, Boston, MA, 1973.

B.4 Huygens, C.: *Treatise on Light*, Macmillan, London, 1912.

B.5 Karakash, J.J.: *Transmission Lines and Filter Networks*, Macmillan, New York, 1950.

B.6 Johnson, W.C. *Transmission Lines and Networks*, McGraw-Hill, Inc., New York, 1950.

B.7 Mattick, R.E.: *Transmission Lines for Digital and Communication Networks*, IEEE Press, New York, 1995.

B.8 Weeks, W.L.: *Electromagnetic Theory for Engineering Applications*, John Wiley & Sons, New York, 1964.

B.9 Rizzi, P.A.: *Microwave Engineering- Passive Circuits*, Prentice-Hall International Edition, Englewood Cliff, NJ, 1988.

B.10 Pozar, D.M.: *Microwave Engineering*, 2nd Edition, John Wiley & Sons, Singapore, 1999.

B.11 Ramo, S.; Whinnery, J.R.; Van Duzer, T.: *Fields, and Waves in Communication Electronics*, 3rd Edition, John Wiley & Sons, Singapore, 1994.

B.12 Collin, R.E.: *Foundations for Microwave Engineering*, 2nd Edition, McGraw-Hill, Inc., New York, 1992.

B.13 Rao, N.N.: *Elements of Engineering Electromagnetics*, 3rd Edition, Prentice-Hall, Englewood Cliff, NJ, 1991.

B.14 Sadiku, M.N.O.: *Elements of Electromagnetics*, 3rd Edition, Oxford University Press, New York, 2001.

B.15 Cheng, D.K.: *Fields and Wave Electromagnetics*, 2nd Edition, Pearson Education, Singapore, 1089.

B.16 Bhattacharyya, A.K.: *Electromagnetic Fields in Multilayered Structures*, Artech House, Norwood, MA, 1994.

B.17 Lewis, I.A.D.; Wells, F.H.: *Millimicrosecond Pulse Techniques*, 2nd Edition, Pergamon Press, London, 1939.

Journals

J.1 Searle, G.F.C., et al. *The Heaviside Centenary Volume*, The Institution of Electrical Engineers, London, 1950.

J.2 Whittaker, E.T.: *Oliver Heaviside*, In *Electromagnetic Theory* Vol. 1, Oliver Heaviside, Reprint, Chelsea Publishing Company, New York, 1971.

J.3 Verma, A.K.; Nasimuddin: *Quasistatic RLCG parameters of lossy microstrip line for CAD application*, *Microwave Opt. Tech. Lett.*, Vol. 28, No. 3, pp. 209–212, Feb. 2001

J.4 Hasegawa, H.; Furukawa, M.; Yanai, H.: *Properties of microstrip line on Si-SiO2 system*, *IEEE Trans. Microwave Theory Tech.*, Vol. MTT-19, pp. 869–881, 1971.

J.5 Kurokawa, K.: *Power waves and the scattering matrix*, *IEEE Trans. Microwave Theory Tech.*, Vol. 13, No. 2, pp. 607–610, 1965.

J.6 Tang, C.C.H.: Delay equalization by tapered cutoff waveguides, *IEEE Trans. Microwave Theory Tech.*, Vol. 12, No. 6, pp. 608–615, Nov. 1964.

J.7 Roberts, P.P.; Town, G.E.: *Design of microwave filters by inverse scattering*, *IEEE Trans. Microwave Theory Tech.*, Vol. 7, pp. 39–743, April 1995.

J.8 Burkhart, S.C.; Wilcox R.B.: *Arbitrary pulse shape synthesis via nonuniform transmission lines*, *IEEE Trans. Microwave Theory Tech.*, Vol. 38, No. 10, pp.1514–1518, Oct. 1990.

J.9 Hayden, L.A.; Tripathi, V.K.: *Nonuniform coupled microstrip transversal filters for analog signal processing*, *IEEE Trans. Microwave Theory Tech.*, Vol. 39, No. 1, pp. 47–53, Jan. 1991.

J.10 Young, P.R.; McPherson, D.S.; Chrisostomidis, C.; Elgaid, K.; Thayne, I.G.; Lucyszyn, S.; Robertson I.D.: *Accurate non-uniform transmission line model and its application to the de-embedding of on-wafer measurements*, *IEEE Proc. Microwave Antennas Propag.*, Vol. 148, No. 1, pp. 153–156, June 2001.

J.11 Wohlers, M.R.: *Approximate analysis of lossless tapered transmission lines with arbitrary terminations*, *Proc. IRE*, Vol. 52, No. 11, 1365, Dec. 1964.

J.12 Khalaj-Amirhosseini, M.: *Analysis of periodic and aperiodic coupled nonuniform transmission lines using the Fourier series expansion*, *Prog. Electromagn. Res., PIER*, Vol. 65, pp. 15–26, 2006.

J.13 Ghose, R.N.: *Exponential transmission lines as resonators and transformers*, *IRE Trans. Microwave Theory Tech.*, Vol. 5, No. 3, pp. 213–217, July 1957.

3

Waves on Transmission Lines – II

(Network Parameters, Wave Velocities, Loaded Lines)

Introduction

The transmission line sections are used to develop various passive components. These are characterized by several kinds of matrix parameters. This chapter discusses the matrix parameters and their conversion among themselves. It also discusses various kinds of dispersion and wave propagation encountered on transmission lines. The transmission lines could be loaded by the reactive elements and resonating circuits to modify the nature of the wave propagation on the lines. Such loaded lines are important in modern planar microwave technology. Such loaded lines are introduced in this chapter. The primary purpose of this chapter is to review in detail the matrix description of lines and wave propagations on the dispersive transmission line that supports various kinds of wave phenomena.

Objectives

- *To review the matrix representations of the two-port networks using the [Z], [Y], and [ABCD] parameters.*
- *To discuss the basic properties and use of the scattering [S] parameters.*
- *To understand the process of de-embedding of true [S] parameters of a device.*
- *To understand the process of extraction of the propagation constant from the [S] parameters.*
- *To understand the phase and group velocities in a dispersive medium.*
- *To discuss the circuit modeling of the reactively loaded line supporting both the forward and backward waves.*

3.1 Matrix Description of Microwave Network

At low frequency, the circuit is described in terms of several kinds of matrices that relate the port voltages to the port currents. These matrices could be the *impedance matrix* [Z], *admittance matrix* [Y], and *hybrid matrix* [H]. The *transmission matrix* is defined as the [ABCD] matrix. It is useful in cascading of two or more networks or transmission line sections. At low radio frequency, the voltage and current are measurable parameters. Therefore, the matrix elements of a network and device could be experimentally determined.

Normally, the microwave passive components, circuits, and networks are constructed around the transmission lines supporting the TEM or the *quasi*-TEM mode. Sometimes, the lumped elements are also used. The waveguide sections supporting non-TEM mode are also used to develop components and circuits. As a matter of fact, the voltage and current can be uniquely defined only for the TEM mode supporting structures. However, for non-TEM line structures, only the *equivalent* voltage and current, based on the power equivalence principle, is defined [B.1, B.2].

The abovementioned parameters are discussed in this section, as these are important for the analysis of the line networks and the networks involving both the line sections and lumped circuit elements. The results of the analysis and measurement are also presented using these parameters. The reader can study these parameters in detail from any of the excellent textbooks [B.1, B.3–B.7]. One basic difference could be seen between the lumped elements based low-frequency circuits and the transmission line sections based on high-frequency circuits. The low-frequency circuits are the *oscillation type circuits*, whereas the high-frequency microwave circuits are the *wave type circuits*. In the case of the low-frequency oscillation type circuits, the port voltage and port current are described by a single voltage or current. In general at any port, for the high-frequency wave-type circuits, the port voltage is described by a sum of the incident and reflected voltages, also the port current is a sum of the incident and reflected currents. It is illustrated in the discussions on the evaluation of the parameters.

Introduction to Modern Planar Transmission Lines: Physical, Analytical, and Circuit Models Approach, First Edition. Anand K. Verma.

How to characterize the components, circuits, and network made of the transmission line sections and waveguide sections? At the microwave frequency, port voltage and port current are not the measurable quantities. However, from the analysis point of view, the networks can be characterized by the [Z], [Y], and [ABCD] parameters. But these are not the measurable parameters at microwave frequencies. A different kind of matrix parameter, called the *scattering or [S]-parameters*, is used for the practical characterization of the microwave network and the transmission line structures [J.1]. The [S] parameter is a measurable quantity. A *Scalar Network Analyzer* is used to measure the magnitude |S| of S-parameter of any microwave circuit and network. For the measurement of the complex [S] parameters, i.e. both the magnitude and phase response of a network, a *Vector Network Analyzer* (VNA) is used. The *Circuit Simulators* and the *EM-Simulators* (Electromagnetic field simulators) are also used to get the frequency-dependent [S] parameters response of the microwave circuits.

3.1.1 [Z] Parameters

The [Z] matrix defines the impedance parameter of a two-port or a multiport network. The matrix elements are evaluated by *open circuiting* the ports. Therefore, the [Z] parameters or the impedance parameters are also called the *open-circuit parameters*. The *port voltage* (V_N) and the *port current* (I_N) are the sums of the reflected and incident voltages and currents, respectively. The port current is an independent variable, whereas the port voltage is the dependent variable. Therefore, the port currents are the *excitation* sources creating the port voltages as the *response*. The response voltage is proportional to the excitation current and the proportionality constant has the dimension of impedance.

The impedance matrix could be obtained for a general linear two-port network, shown in Fig (3.1). The wave entering the port is the *incident voltage* (V_n^+) or the *incident current* (I_n^+) wave. The *reflected voltage* (V_n^-) and *reflected current* (I_n^-) waves are also present at the ports. The total port voltage (V_n) or current (I_n) is the sum of the incident and reflected voltage or current:

$$V_n = V_n^+ + V_n^- \text{ (a)}, I_n = I_n^+ - I_n^- \text{ (b)}; n = 1, 2. \tag{3.1.1}$$

In equation (3.1.1), n = 1, 2 is the port number, i.e. port-1, port-2. The power entering the network is taken as a positive quantity for the incident wave, so the power

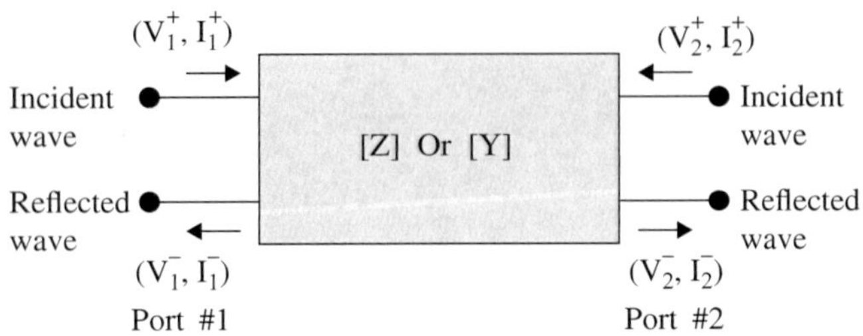

Figure 3.1 Two-port network to determine [Z] and [Y] parameters.

coming out of the network, i.e. the power of the reflected wave, is taken as a negative quantity. The reflected voltage (V_n^-) is positive. To maintain the negative direction of power flow, the reflected current wave (I_n^-) is taken as negative in the above equation. At each port, the current entering the port from outside is positive, whereas the current leaving the port is negative. For the linear networks, the voltage at any port is a combined response of the currents applied to all ports. On using the superposition of the voltage responses, the following set of equations is written:

$$\begin{aligned} V_1 &= Z_{11}I_1 + Z_{12}I_2 \\ V_2 &= Z_{21}I_1 + Z_{22}I_2. \end{aligned} \tag{3.1.2}$$

Equation (3.1.2) is written in a more compact matrix form:

$$[V] = [Z]\,[I], \tag{3.1.3}$$

where [V] and [I] are the column matrices. The two-port impedance matrix is

$$[Z] = \begin{bmatrix} Z_{11} & Z_{12} \\ Z_{21} & Z_{22} \end{bmatrix}. \tag{3.1.4}$$

The [Z] parameter can be easily extended to the N-port networks [B.1, B.3–B.5]. The Z-parameters are the *open-circuited parameters*. The coefficient of the matrix can be defined in terms of the open circuit condition at the ports:

$$Z_{ii} = \left.\frac{V_i}{I_i}\right|_{I_k = 0}, \quad i = 1, 2..., \text{and } k \neq i. \tag{3.1.5}$$

All the ports are open-circuited, except the ith port at which the matrix element Z_{ii} is defined. For instance, in the case of a two-port network, Z_{11} is obtained when current I_1 is applied to port-1 and the voltage response is also obtained at the port-1, while keeping the port-2 open-circuited, i.e. $I_2 = 0$. The coefficient, Z_{11}, is known as the *self-impedance* of the network. These are the diagonal elements of a [Z] matrix. The off-diagonal elements of a [Z] matrix are defined as follows:

$$Z_{ij} = \left.\frac{V_i}{I_j}\right|_{I_k = 0}, \quad k \neq j. \tag{3.1.6}$$

In this case, the current excitation is applied at the port-j and the voltage response is obtained at the port-i. All other ports are kept open-circuited allowing $I_k = 0$, except at the port-j. For instance, in the case of a two-port network to evaluate Z_{12}, the current source is applied at the port-2, and the voltage response is obtained at the port-1, while keeping the port-1 open-circuited. The coefficient Z_{12} is the *mutual impedance* that describes the coupling of port-2 with the port-1. A network can have $Z_{11} = Z_{22}$, i.e. both of the ports are electrically identical. Such a network is known as the *symmetrical* network. Furthermore, the voltage response of a network at the port-1 due to the current at the port-2 can be identical to the voltage response at the port-2, due to the current at the port-1. This kind of network is a *reciprocal* network. It has a $Z_{12} = Z_{21}$. If $Z_{12} = Z_{21} = 0$, the ports are *isolated* one.

Example 3.1

Figure (3.2) shows lumped elements T-network. Determine the [Z] parameter of the network.

Solution

For the port-2 open-circuited, $I_2 = 0$. The voltage at the port-1 is

$$V_1 = (Z_A + Z_C)I_1, \quad Z_{11} = \left.\frac{V_1}{I_1}\right|_{I_2 = 0} = Z_A + Z_C.$$

$$V_2 = I_1 Z_C, \quad Z_{21} = \left.\frac{V_2}{I_1}\right|_{I_2 = 0} = Z_C.$$

Likewise, for the port-1 open-circuited, $I_1 = 0$, and the parameters are $Z_{22} = Z_B + Z_C$, $Z_{12} = Z_C$. The [Z] matrix description of a T-network is

$$[Z] = \begin{bmatrix} Z_A + Z_C & Z_C \\ Z_C & Z_B + Z_C \end{bmatrix}. \tag{3.1.7}$$

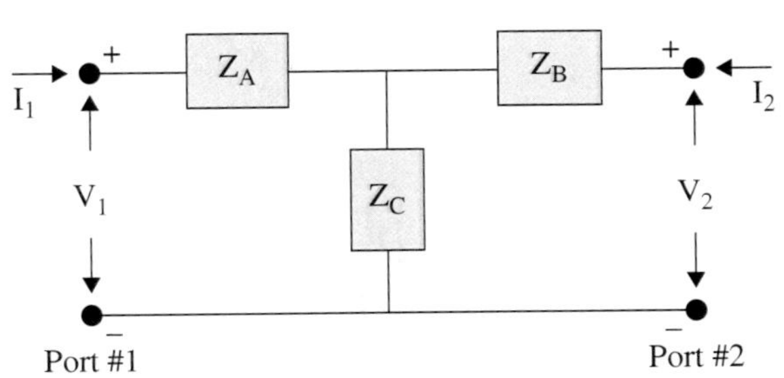

Figure 3.2 Lumped T-network.

The given circuit is asymmetrical. However, it is a reciprocal circuit. It becomes symmetrical for $Z_A = Z_B$.

Example 3.2

Determine the [Z]-parameter of a section of the transmission line of length ℓ shown in Fig (3.3).

Solution

Let the port-2 be open-circuited and an incident voltage $V^{inc} = V_1^+$ is applied at the port-1. The voltage wave reaches to the port-2 and reflects from there. It reaches the port-1 as the voltage wave $V_1^+ \, e^{-2\gamma\ell}$. The maximum of the voltage wave occurs at the port-2. The total port voltages are given below:

At the port-1: $V_1 = V_{inc} + V_{ref} = V_1^+ + V_1^+ e^{-2\gamma\ell}$.

At the port-2: $V_2 = V_{inc} + V_{ref} = 2V_1^+ e^{-\gamma\ell}$.

At the port-1: $I_1 = I_{inc} - I_{ref} = \frac{V_1^+}{Z_0} - \frac{V_1^+}{Z_0} e^{-2\gamma\ell}$.

At the port-2: $I_2 = 0$.

Thus, the [Z] parameters are obtained as follows:

$$Z_{11} = \left.\frac{V_1}{I_1}\right|_{I_2 = 0} = Z_0\left(\frac{1 + e^{-2\gamma\ell}}{1 - e^{-2\gamma\ell}}\right) = Z_0 \coth(\gamma\ell).$$

$$Z_{21} = \left.\frac{V_2}{I_1}\right|_{I_2 = 0} = Z_0\left(\frac{2\,e^{-\gamma\ell}}{1 - e^{-2\gamma\ell}}\right) = Z_0 \operatorname{cos\,ech}(\gamma\ell).$$

The following [Z] matrix of a line section is obtained by keeping in view that the uniform transmission line is a symmetrical and reciprocal network:

$$[Z] = \begin{bmatrix} Z_0 \coth(\gamma\ell) & Z_o \operatorname{cosech}(\gamma\ell) \\ Z_0 \operatorname{cosech}(\gamma\ell) & Z_0 \coth(\gamma\ell) \end{bmatrix} \quad \text{(a)}.$$

For the lossless transmission line, $\gamma = j\beta$, $\alpha = 0$ and [Z] is

$$[Z] = -\begin{bmatrix} j Z_0 \cot(\beta\ell) & j Z_o \operatorname{cosec}(\beta\ell) \\ j Z_0 \operatorname{cosec}(\beta\ell) & j Z_0 \cot(\beta\ell) \end{bmatrix} \quad \text{(b)}.$$

(3.1.8)

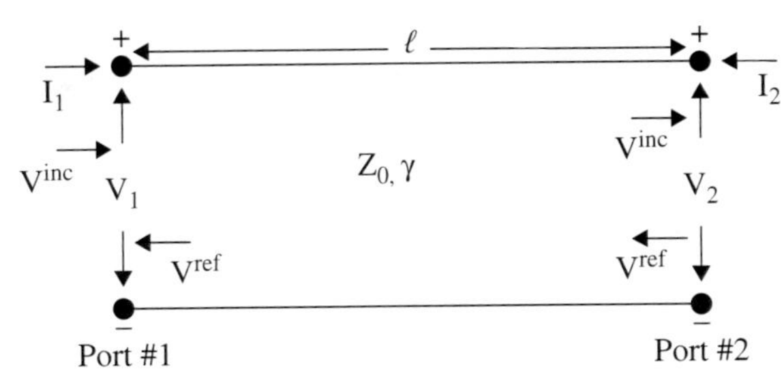

Figure 3.3 A transmission line section.

3.1.2 Admittance Matrix

To define the [Y] parameters, the voltage is taken as an independent variable and current as of the dependent one for a two-port network shown in Fig (3.1). In this case, the voltage is a source of *excitation*, and current at the port is the *response*. Thus, for a linear network, the total port current is a superposition of currents due to the voltages applied at both the ports:

$$\begin{matrix} I_1 = Y_{11}V_1 + Y_{12}V_2 \\ I_2 = Y_{21}V_1 + Y_{22}V_2 \end{matrix} \text{(a)}, \Rightarrow \quad [I] = [Y]\,[V]\ \text{(b)}, \tag{3.1.9}$$

where [V] and [I] are the voltage and current column matrices. The admittance matrix of the two-port network is

$$[Y] = \begin{bmatrix} Y_{11} & Y_{12} \\ Y_{21} & Y_{22} \end{bmatrix}. \tag{3.1.10}$$

The Y-parameters are defined as the *short-circuited parameters*. For the short-circuited port-2, $V_2 = 0$, and Y_{11} and Y_{21} are defined from equation (3.1.9):

$$Y_{11} = \left.\frac{I_1}{V_1}\right|_{V_2=0}, \quad Y_{21} = \left.\frac{I_2}{V_1}\right|_{V_2=0}. \tag{3.1.11}$$

Likewise, for the short-circuited port-1, the Y-parameters are

$$Y_{22} = \left.\frac{I_2}{V_2}\right|_{V_1=0}, \quad Y_{12} = \left.\frac{I_1}{V_2}\right|_{V_1=0}. \tag{3.1.12}$$

The [Y] parameters are extended to a multiport network by defining its matrix elements as follows:

$$Y_{ii} = \left.\frac{I_i}{V_i}\right|_{V_k=0}, \quad i = 1, 2\ldots, \text{and } k \neq i. \tag{3.1.13}$$

Equation (3.1.13) shows that to get Y_{ii}, i.e. the diagonal elements of the [Y] matrix, all the ports are short-circuited, except the ith port. The current is evaluated at the ith port for the voltage applied at the ith port itself. To get Y_{ij}, i.e. the off-diagonal elements of the [Y] matrix, the voltage is applied at the jth port. Y_{ij} is the mutual admittance describing the coupling between the jth port and the ith port. The current at the ith port is evaluated or measured, while all other ports are short-circuited. The admittance element Y_{ij} is evaluated as

$$Y_{ij} = \left.\frac{I_i}{V_j}\right|_{V_k=0}, \quad i = 1, 2\ldots, \text{and } k \neq j. \tag{3.1.14}$$

Example 3.3

Fig (3.2) shows the T-network. Determine the [Y] parameter of the network.

Solution

The loop equations for the circuit are written as

$$V_1 = (Z_A + Z_C)\,I_1 + Z_C\,I_2, V_2 = Z_C\,I_1 + (Z_B + Z_C)\,I_2.$$

For the short-circuited port-2, $V_2 = 0$: $I_2 = -\frac{Z_C}{(Z_B + Z_c)} I_1$.

From the above equations:

$$Y_{11} = \frac{I_1}{V_1} = \frac{Z_B + Z_C}{(Z_A + Z_C)(Z_B + Z_c) - Z_C^2}.$$
$$Y_{21} = \frac{I_2}{V_1} = -\frac{Z_C}{(Z_A + Z_C)(Z_B + Z_c) - Z_C^2}.$$

Likewise, the expressions for Y_{22} and Y_{12} could be computed by short-circuiting the port-1, $V_1 = 0$. Final [Y] matrix of the T-network is

$$[Y] = \frac{1}{\Delta Z}\begin{bmatrix} Z_B + Z_C & -Z_C \\ -Z_C & Z_A + Z_C \end{bmatrix}, \quad \text{where} \tag{3.1.15}$$
$$\Delta Z = (Z_A + Z_C)(Z_B + Z_C) - Z_C^2.$$

The above matrix is a reciprocal of the [Z] matrix, given in of length ℓ equation (3.1.7).

Example 3.4

Determine the [Y] parameter of a section of the transmission line of length ℓ shown in Fig (3.3).

Solution

The incident voltage $V^{inc} = V_1^+$ excites the port-1, and it reaches the port-2 as $V_1^+ e^{-\gamma\ell}$. The port-2 is short-circuited to determine the [Y] parameter. Under the short-circuit condition, the reflected voltage at the port-2 is $-V_1^+ e^{-\gamma\ell}$ such that the total voltage at the port-2 is zero. The reflected voltage at the port-1 is $-V_1^+ e^{-2\gamma\ell}$. The total voltage and the total current at the port-1 are

$$V_1 = V_{inc} + V_{ref}, \quad \Rightarrow V_1 = V_1^+ - V_1^+ e^{-2\gamma\ell}.$$
$$I_1 = I_{inc} - I_{ref}, \quad \Rightarrow I_1 = \frac{V_1^+}{Z_0} + \frac{V_1^+}{Z_0} e^{-2\gamma\ell}.$$

At the port-1, the incident current I_{inc} enters the port, so it is positive, whereas at the port-1, the reflected current I_{ref} leaves the port, so it is negative. At the port-2, the incident current I_{inc} enters the port-2 from the port-1 side and leaves the port-2, so it is negative, whereas at the port-2, the reflected current I_{ref} from the terminated load, enters the port-2, so it is positive. The total voltage and the total current at the port-2 are

$$V_2 = V_1^+ e^{-\gamma\ell} + \left(-V_1^+ e^{-\gamma\ell}\right) = 0, I_2 = -I_{inc} + I_{ref}$$

$$\Rightarrow I_2 = -\frac{V_1^+}{Z_0} e^{-\gamma\ell} + \left(-\frac{V_1^+}{Z_0} e^{-\gamma\ell}\right) = -\frac{2V_1^+}{Z_0} e^{-\gamma\ell}.$$

$$Y_{11} = \left.\frac{I_1}{V_1}\right|_{V_2=0} = \frac{1}{Z_0}\frac{1+e^{-2\gamma\ell}}{1-e^{-2\gamma\ell}} = Y_0 \coth(\gamma\ell).$$

$$Y_{21} = \left.\frac{I_2}{V_1}\right|_{V_2=0} = -\frac{2e^{-\gamma\ell}}{Z_0(1-e^{-2\gamma\ell})} = -Y_0 \operatorname{cosech}(\gamma\ell).$$

The line section is symmetrical and reciprocal giving the [Y] parameter:

for the lossy line section:

$$[Y] = \begin{bmatrix} Y_0 \coth(\gamma\ell) & -Y_o \cos\text{ech}(\gamma\ell) \\ -Y_0 \cos\text{ech}(\gamma\ell) & Y_0 \coth(\gamma\ell) \end{bmatrix} \quad \text{(a)}$$

and for the lossless line section:

$$[Y] = \begin{bmatrix} -jY_0 \cot(\beta\ell) & jY_o \cos\text{ec}(\beta\ell) \\ jY_0 \cos\text{ec}(\beta\ell) & -jY_0 \cot(\beta\ell) \end{bmatrix} \quad \text{(b)}. \tag{3.1.16}$$

3.1.3 Transmission [ABCD] Parameter

On many occasions, two or more circuit elements or circuit blocks are interconnected in such a way that the output voltage and current of the first circuit block become the input to the next circuit block. To facilitate such combination or cascading, the circuit elements and blocks are characterized using the *transmission parameters*, i.e. the [ABCD] matrix, instead of [Z] or [Y] matrix. The great strength of the transmission parameter, i.e. the [ABCD] parameter, is due to its ability to provide [ABCD] matrix of the complete cascaded network, as a multiplication of the [ABCD] matrices of the individual circuit element or circuit block. The [ABCD] parameter, different from the T-matrix, is applicable to a two-port network only.

To obtain the transmission matrix description of a two-port network, the output voltage and current are treated as the independent variables. The following expressions relate to the input and output voltage and current of the two-port network shown in Fig (3.4):

$$\begin{aligned} V_1 &= AV_2 + BI_2 \\ I_1 &= CV_2 + DI_2. \end{aligned} \tag{3.1.17}$$

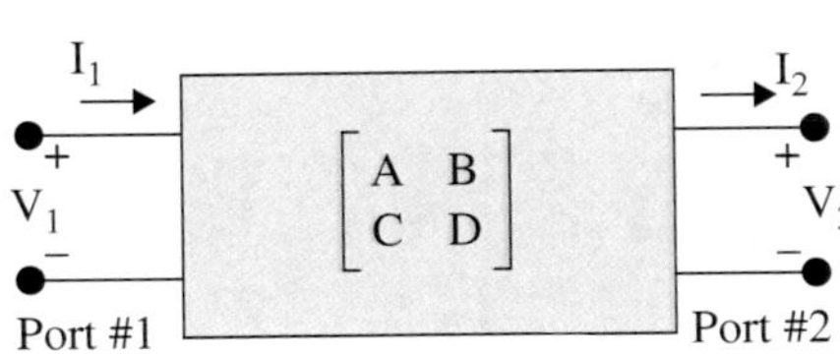

Figure 3.4 Two-port network for transmission parameter.

These expressions can be written in the matrix form,

$$\begin{bmatrix} V_1 \\ I_1 \end{bmatrix} = \begin{bmatrix} A & B \\ C & D \end{bmatrix} \begin{bmatrix} V_2 \\ I_2 \end{bmatrix}. \tag{3.1.18}$$

In the case of the [Z] and [Y] parameters, the positive current I_2 enters the port, while in the above network defining the [ABCD] parameter in Fig (3.4), *the output current I_2 leaving the port is taken as positive* [B.1, B.3]. It is an input to the next circuit block, as shown in Fig (3.5). However, like defining the [Z] and [Y] parameters, to define the [ABCD] parameter current, I_2 could be taken as the current entering the output port. In this case, I_2 in equation (3.1.18) is replaced by $(-I_2)$ [B.1, B.4].

The matrix elements A, B, C, D can be determined from the open and short circuit conditions at the output port. When the output is open-circuited, $I_2 = 0$. Equation (3.1.17) provides the parameter-A and C:

$$A = \left.\frac{V_1}{V_2}\right|_{I_2=0} \quad \text{(a)}, \quad C = \left.\frac{I_1}{V_2}\right|_{I_2=0} \quad \text{(b)}. \tag{3.1.19}$$

The parameter A is the *voltage ratio* that is a reciprocal of the voltage gain. The parameter C is the *trans-admittance* of a network. It relates the output voltage of a network to its input current source.

When the output is short-circuited, $V_2 = 0$. Equation (3.1.17) again provides the parameters-B and D:

$$B = \left.\frac{V_1}{I_2}\right|_{V_2=0}, \quad D = \left.\frac{I_1}{I_2}\right|_{V_2=0}. \tag{3.1.20}$$

The parameter B is the *trans-impedance* of a network. It provides the output current when the input of a network is excited by the voltage source. The parameter D is the *current ratio* giving a reciprocal of the current gain of a network.

Fig (3.5) demonstrates the usefulness of the transmission parameters to obtain an equivalent [ABCD] parameter of the cascaded networks. The [ABCD] parameters for the first and the second network are written as

$$\begin{bmatrix} V_1 \\ I_1 \end{bmatrix} = \begin{bmatrix} A_1 & B_1 \\ C_1 & D_1 \end{bmatrix} \begin{bmatrix} V_2 \\ I_2 \end{bmatrix}, \begin{bmatrix} V_3 \\ I_3 \end{bmatrix} = \begin{bmatrix} A_2 & B_2 \\ C_2 & D_2 \end{bmatrix} \begin{bmatrix} V_4 \\ I_4 \end{bmatrix}.$$

At the junction of two networks, $I_2 = I_3$ and $V_2 = V_3$. Therefore, from the above equations, the following expression is obtained:

$$\begin{bmatrix} V_1 \\ I_1 \end{bmatrix} = \begin{bmatrix} A_1 & B_1 \\ C_1 & D_1 \end{bmatrix} \begin{bmatrix} A_2 & B_2 \\ C_2 & D_2 \end{bmatrix} \begin{bmatrix} V_4 \\ I_4 \end{bmatrix}. \tag{3.1.21}$$

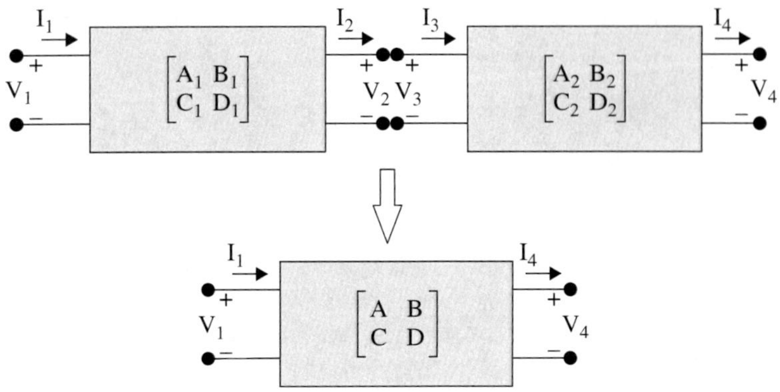

Figure 3.5 Cascading of two networks to get one equivalent network.

Finally, two cascaded networks can be replaced by one equivalent 2-port network having equivalent [ABCD] parameter. It is given by the following expression:

$$\begin{bmatrix} A & B \\ C & D \end{bmatrix} = \begin{bmatrix} A_1 & B_1 \\ C_1 & D_1 \end{bmatrix} \begin{bmatrix} A_2 & B_2 \\ C_2 & D_2 \end{bmatrix}. \tag{3.1.22}$$

Expression (3.1.22) can be extended to the cascading of N-networks by multiplying the individual matrix of each network.

Example 3.5

Determine the [ABCD] parameters of the series impedance as shown in Fig (3.6).

Solution

The output port is open-circuited, $I_2 = 0$. Therefore, equation (3.1.17) provides $V_1 = AV_2$ and $I_1 = CV_2$. For the port 2 of Fig (3.6) open-circuited, $I_2 = 0$, $V_1 = V_2$ and $I_2 = I_1 = 0$. On comparing these equations, the computed parameters are $A = 1$ and $C = 0$.

For the output port is short-circuited, $V_2 = 0$. Therefore, equation (3.1.17) helps to get, $V_1 = BI_2$ and $I_1 = DI_2$. Using Fig (3.6) shows, $V_2 = 0$, $V_1 = ZI_2$ and $I_1 = I_2$. The comparison of these equations provide $B = Z$ and $D = 1$.

Thus, the [ABCD] matrix of series impedance is written as

$$\begin{bmatrix} A & B \\ C & D \end{bmatrix} = \begin{bmatrix} 1 & Z \\ 0 & 1 \end{bmatrix}. \tag{3.1.23}$$

Example 3.6

Determine the [ABCD] parameters of a shunt admittance shown in Fig (3.7).

Solution

The output port-2 is open-circuited, $I_2 = 0$. Therefore, from matrix equation (3.1.17): $V_1 = AV_2$ and $I_1 = CV_2$.At the open-circuited output port 2: $I_2 = 0$, $V_1 = V_2$ and $I_1 = YV_2$. On comparing these equations: $A = 1$ and $C = Y$. At the short-circuited output port 2: $V_2 = 0, V_1 = BI_2$ and $I_1 = DI_2$.Using Fig (3.7), for $V_2 = 0$, $V_1 = 0$ and $I_1 = I_2$. On comparing these equations: $B = 0$ and $D = 1$. Finally, the [ABCD] matrix of shunt admittance can be written as

$$\begin{bmatrix} A & B \\ C & D \end{bmatrix} = \begin{bmatrix} 1 & 0 \\ Y & 1 \end{bmatrix}. \tag{3.1.24}$$

The [ABCD] matrix could be easily evaluated for the L, T, and π networks, shown in Fig (3.8). The [ABCD] matrix of each element is known and the complete circuit is a cascading of the elements.

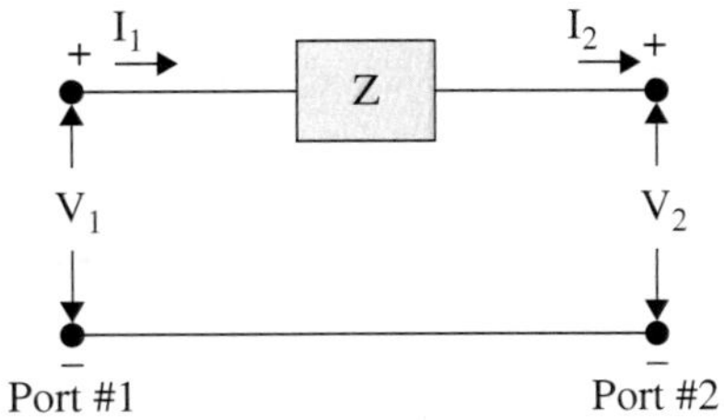

Figure 3.6 Series impedance.

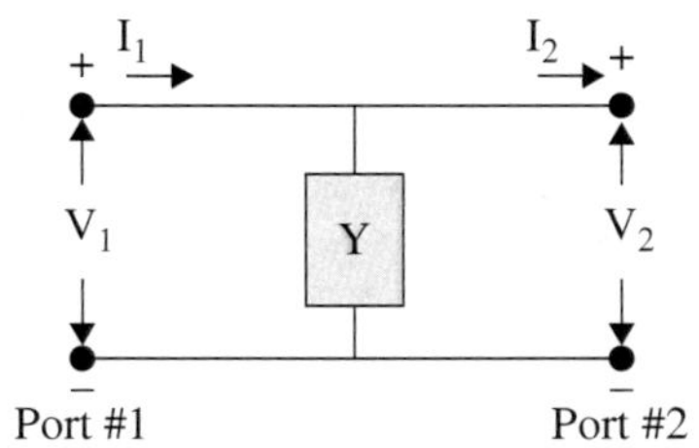

Figure 3.7 Shunt admittance.

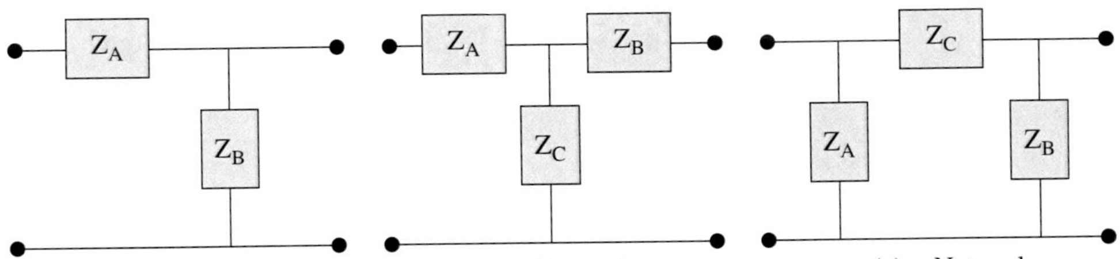

(a) L-Network. (b) T-Network. (c) π-Network.

Figure 3.8 Basic networks.

Example 3.7

Determine the [ABCD] parameters of a section of transmission line shown in Fig (3.3).

Solution

Equations (2.1.79) of chapter 2 provide the voltage and current waves on a transmission line:

$$V(x) = V^+ e^{-\gamma x} + V^- e^{\gamma x}, I(x) = \frac{V^+}{Z_0} e^{-\gamma x} - \frac{V^-}{Z_0} e^{\gamma x}.$$

The V^+ and V^- are the amplitudes of the forward and reflected waves, respectively. For convenience, the distance x is measured from the port-2. The voltage and current at the port-2 are

$$V(x=0) = V_2 = V^+ + V^-, I(x=0) = Z_0 I_2 = V^+ - V^-.$$

The amplitudes of the forward and reflected voltages in terms of the port voltage and port current are

$$V^+ = \frac{V_2 + Z_0 I_2}{2}, \quad V^- = \frac{V_2 - Z_0 I_2}{2}.$$

The voltage and current on a transmission line can be written as

$$V(x) = \left[\frac{e^{\gamma x} + e^{-\gamma x}}{2}\right] V_2 + \left[\frac{e^{-\gamma x} - e^{\gamma x}}{2}\right] Z_0 I_2.$$
$$I(x) = \frac{1}{Z_0}\left[\frac{e^{-\gamma x} - e^{\gamma x}}{2}\right] V_2 + \left[\frac{e^{-\gamma x} + e^{\gamma x}}{2}\right] I_2.$$

The voltage and current at the input port-1 are obtained for $x = -\ell$:

$$V(x = -\ell) = V_1 = \cosh(\gamma\ell) V_2 + Z_0 \sinh(\gamma\ell) I_2$$
$$I(x = -\ell) = I_1 = \frac{\sinh(\gamma\ell)}{Z_0} V_2 + \cosh(\gamma\ell) I_2.$$

Above equations can be written in the matrix form:

$$\begin{bmatrix} V_1 \\ I_1 \end{bmatrix} = \begin{bmatrix} \cosh(\gamma\ell) & Z_o \sinh(\gamma\ell) \\ Y_0 \sinh(\gamma\ell) & \cosh(\gamma\ell) \end{bmatrix} \begin{bmatrix} V_2 \\ I_2 \end{bmatrix}.$$

The [ABCD] parameters of the lossy and lossless transmission line sections are given by equation (3.1.25a) and equation (3.1.25b), respectively:

$$\begin{bmatrix} A & B \\ C & D \end{bmatrix} = \begin{bmatrix} \cosh(\gamma\ell) & Z_o \sinh(\gamma\ell) \\ Y_0 \sinh(\gamma\ell) & \cosh(\gamma\ell) \end{bmatrix} \quad \text{(a)}$$
$$\begin{bmatrix} A & B \\ C & D \end{bmatrix} = \begin{bmatrix} \cos(\beta\ell) & j Z_o \sin(\beta\ell) \\ j Y_0 \sin(\beta\ell) & \cos(\beta\ell) \end{bmatrix}; \ \alpha = 0, \gamma = j\beta \quad \text{(b)}. \tag{3.1.25}$$

The above example can be further extended to a network of several cascaded transmission line sections having different ℓ, Z_0, and γ. The overall [ABCD] parameter of the multisection transmission line can be obtained by a multiplication of the [ABCD] matrix of each line section. The line sections can be attached to the series and the shunt lumped elements. Even in such cases, one can find the overall [ABCD] parameter of a complete network. The input impedance, output impedance, Thevenin and Norton equivalent circuits, and power transfer relation, etc. of a complete circuit can be written in terms of the [ABCD] matrix. However, a detailed discussion of these aspects is out of the scope of this book. The reader can follow many available texts for this purpose [B.1, B.2–B.5, B.7, B.8].

3.1.4 Scattering [S] Parameters

The [Z], [Y], and [ABCD] matrix descriptions of any microwave network or component are based on the port voltage and port current relations. The evaluation of these parameters requires the short-circuiting and open-circuiting of the ports. At the microwave frequency, usually, it is difficult to measure the voltage and current. Similarly, the short-circuiting and open-circuiting of the ports may not be always possible at the microwave frequency. Thus, these parameters are normally not measurable quantities. However, these parameters are useful for the analysis of the microwave circuits built around the lumped and distributed circuit elements. At this stage, another kind of measurable parameters is needed to characterize the microwave circuits. At the microwave frequency, the power could be measured, and also the forward and reflected power waves could be obtained. The frequency and phase of a microwave signal are also measurable quantities. The *scattering parameters*, also called the *S-parameters*, are defined for any two-port network, or even the multiport microwave network, in terms of the measurable incident and reflected power waves [J.1].

Basic Concept

A commonly used two-port network is suitable to develop the concept of the S-parameter. Even the S-parameters of a multiport network are measured as the two-port parameters, while other ports are terminated in the matched loads. Figure (3.9) shows the two-port network. It is to be characterized by the S-parameters. The port-1 and port-2 are terminated with the line sections of characteristic impedance Z_{01} and Z_{02}, respectively. However, most of the two-port networks have $Z_{01} = Z_{02} = Z_0$, i.e. the identical transmission line sections at both the ports. The *reference* impedance Z_0 is normally 50Ω. The incident voltage waves at both the ports-V_1^+ and V_2^+, enter the ports and the reflected voltage waves at both the ports-V_1^- and V_2^-, come out of the ports.

The forward power, i.e. the incident power entering the port-1, is

$$P_1^+ = \frac{V_1^{+2}}{Z_0} \ (a), \qquad P_1^+ = \frac{V_{1\,max}^{+2}}{2Z_0} \ (b). \tag{3.1.26}$$

In equation (3.1.26b), $V_1^+ = V_{1\,max}^+/\sqrt{2}$ is the RMS voltage of the voltage wave. In general for the two-port or N-port network, the forward power entering the ith port is written as

$$P_i^+ = \frac{V_i^{+2}}{Z_{0i}}, \quad i = 1, 2, \ldots, N. \tag{3.1.27}$$

The *incident power variable* a_i at the ith port is defined in a way that the power entering the port is given by the square of the *power variable*:

$$P_i^+ = a_i^2. \tag{3.1.28}$$

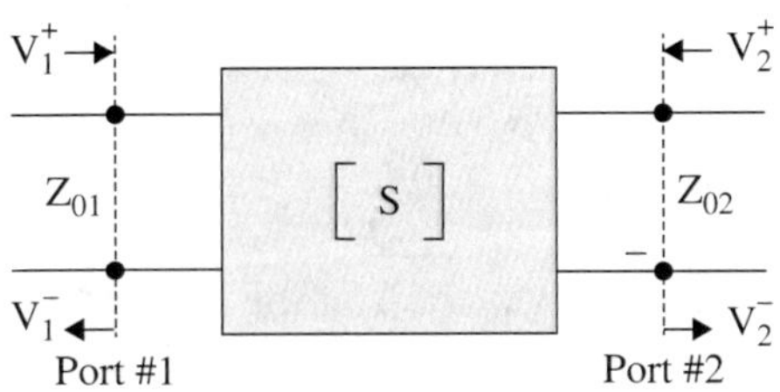

Figure 3.9 Two-port network for evaluation of S-parameter.

Using equations (3.1.27) and (3.1.28), the power variable a_i is written in terms of the forward RMS voltage V_i^+ at the ith port

$$a_i = \frac{V_i^+}{\sqrt{Z_{0i}}}, \quad i = 1, 2, \ldots, N. \tag{3.1.29}$$

The forward voltage can also be written in term of the power variable as

$$V_i^+ = \sqrt{Z_{0i}}\,a_i. \tag{3.1.30}$$

The power variable a_i is simply a normalized forward voltage wave, incident on the ith port. The normalization is done with respect to the square root of the characteristic impedance at the port. The forward power variable can also be viewed as the incident normalized current. The power entering the ith port, in terms of the incident RMS current I_1^+, is given below:

$$P_i^+ = I_i^{+2} Z_{0i} = a_i^2 \Rightarrow a_i = I_i^+ \sqrt{Z_{0i}}, \quad i = 1, 2, \ldots, N. \tag{3.1.31}$$

The forward port current in terms of the forward power variable is

$$I_i^+ = \frac{a_i}{\sqrt{Z_{0i}}}, \quad i = 1, 2, \ldots, N. \tag{3.1.32}$$

The multiplication of the voltage and current of equations (3.1.30) and (3.1.32), again provides the forward power, $P_i^+ = V_i^+ I_i^+ = a_i^2$. Thus, the definitions of the power variable both as the *normalized voltage wave* and as *the normalized current wave* are consistent. However, one must be careful about the presence of the square root of the characteristic impedance in the numerator and denominator for two definitions.

Consider the reflected power wave at the ith port with characteristic impedance Z_{0i}. Figure (3.10) shows that the ith port is connected to a source with impedance Z_0. For the sake of clarity, the port is taken out of the

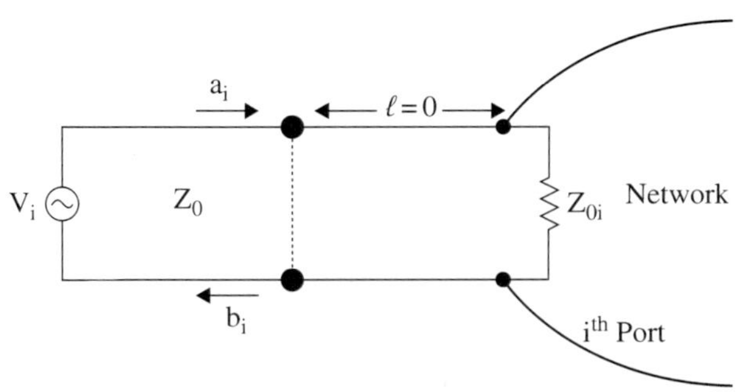

Figure 3.10 A section of the multiport network. Port is shown extended with length ℓ = 0.

network using an interconnect line of characteristic impedance Z_0 with zero length, $\ell = 0$. The total power available from the source does not enter the network. A part of it gets reflected. The reflected power in terms of the *reflected power variable* b_i is

$$P_i^- = b_i^2 = \frac{{V_i^-}^2}{Z_{0i}}, \quad i = 1, 2, \ldots, N. \tag{3.1.33}$$

The reflected power variable is related to the reflected port voltage and the reflected port current as follows:

$$b_i = \frac{V_i^-}{\sqrt{Z_{0i}}}, \quad \Rightarrow \quad V_i^- = \sqrt{Z_{0i}}\, b_i. \tag{3.1.34}$$

$$b_i = \sqrt{Z_{0i}}\, I_i^-, \quad \Rightarrow \quad I_i^- = \frac{b_i}{\sqrt{Z_{0i}}}. \tag{3.1.35}$$

The power entering the ith port is

$$P_{i,in} = P_i^+ - P_i^- = a_i^2 - b_i^2 = a_i\left(1 - |\Gamma_i|^2\right), \tag{3.1.36}$$

where the *reflection coefficient* at the ith port is $\Gamma_i = b_i/a_i$. The *total port voltage* and the *total port current* in term of the power variables can be written as

$$\begin{aligned} V_i &= V_i^+ + V_i^- = \sqrt{Z_{0i}}(a_i + b_i) \quad \text{(a)} \\ I_i &= I_i^+ - I_i^- = \frac{1}{\sqrt{Z_{0i}}}(a_i - b_i) \quad \text{(b)}. \end{aligned} \tag{3.1.37}$$

The reflected port current I_i^- is negative, such that the reflected power $P_i^- = V_i^- I_i^-$ travels in the opposite direction, i.e. it travels away from the port. Using equation (3.1.37), the power variables can be written in terms of the total port voltage and the total port current:

$$\begin{aligned} a_i &= \frac{1}{2}\left(\frac{V_i}{\sqrt{Z_{0i}}} + I_i\sqrt{Z_{0i}}\right) \quad \text{(a)} \\ b_i &= \frac{1}{2}\left(\frac{V_i}{\sqrt{Z_{0i}}} - I_i\sqrt{Z_{0i}}\right) \quad \text{(b)}. \end{aligned} \tag{3.1.38}$$

The definition of the power wave variables given by the equations (3.1.30), (3.1.32), (3.1.34), and (3.1.35) are valid for the special cases of the forward wave and reflected waves. The definition, given in equation (3.1.38), is valid for the general case. It is applicable at any port for any kind of termination. *The power-variables a_i and b_i are complex quantities.* The incident and reflected power are

$$P_i^+ = |a_i|^2 = a_i\, a_i^*, \quad P_i^- = |b_i|^2 = b_i\, b_i^*. \tag{3.1.39}$$

Scattering [S] Matrix

Figure (3.11) shows the N-port network. The power *entering* the ith port is given in terms of the forward voltage wave V_i^+ or the forward power variable a_i and the power *leaving* the ith port is given in terms of the reflected (backward) voltage wave V_i^- or the reflected (backward) power variable b_i. A part of the microwave power is reflected at the ith port itself and remaining power entering the network comes out of all other ports as $(V_1^-, V_2^-, \ldots, V_N^-)$ or as $(b_1^-, b_2^-, \ldots, b_N^-)$. The outcoming power from any port is a linear combination of the transmitted power from all other ports. Using either the voltage variables or the power variables, the incident power and reflected power at various ports are correlated as follows:

$$\underbrace{\begin{bmatrix} V_1^- \\ V_2^- \\ \cdot\cdot \\ \cdot\cdot \\ V_N^- \end{bmatrix}}_{\text{Response}} = \underbrace{\begin{bmatrix} S_{11} & S_{12} & \cdot\cdot & S_{1N} \\ S_{21} & S_{22} & \cdot\cdot & S_{2N} \\ \cdot\cdot & \cdot\cdot & \cdot\cdot & \cdot\cdot \\ \cdot\cdot & \cdot\cdot & \cdot\cdot & \cdot\cdot \\ S_{N1} & S_{N2} & \cdot\cdot & S_{NN} \end{bmatrix}}_{\text{Network}} \underbrace{\begin{bmatrix} V_1^+ \\ V_2^+ \\ \cdot\cdot \\ \cdot\cdot \\ V_N^+ \end{bmatrix}}_{\text{Excitation}} \text{(a)}, \quad \underbrace{\begin{bmatrix} b_1 \\ b_2 \\ \cdot\cdot \\ \cdot\cdot \\ b_N \end{bmatrix}}_{\text{Response}} = \underbrace{\begin{bmatrix} S_{11} & S_{12} & \cdot\cdot & S_{1N} \\ S_{21} & S_{22} & \cdot\cdot & S_{2N} \\ \cdot\cdot & \cdot\cdot & \cdot\cdot & \cdot\cdot \\ \cdot\cdot & \cdot\cdot & \cdot\cdot & \cdot\cdot \\ S_{N1} & S_{N2} & \cdot\cdot & S_{NN} \end{bmatrix}}_{\text{Network}} \underbrace{\begin{bmatrix} a_1 \\ a_2 \\ \cdot\cdot \\ \cdot\cdot \\ a_N \end{bmatrix}}_{\text{Excitation}} \text{(b)}. \tag{3.1.40}$$

The incident power at the port is treated as the *excitation*, and reflected/transmitted power at the port is considered as the *response*. The network is characterized by the S-parameters.

Therefore, the matrix elements S_{ij} relating the excitation (V_i^+ or a_i) to the response (V_i^- or b_i) are described as follows:

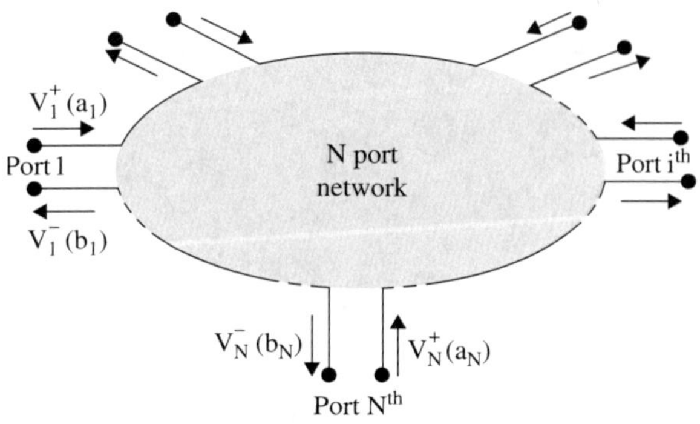

Figure 3.11 N-port network showing power variables (a_i, b_i) in terms of voltage variables (V_i^+, V_i^-).

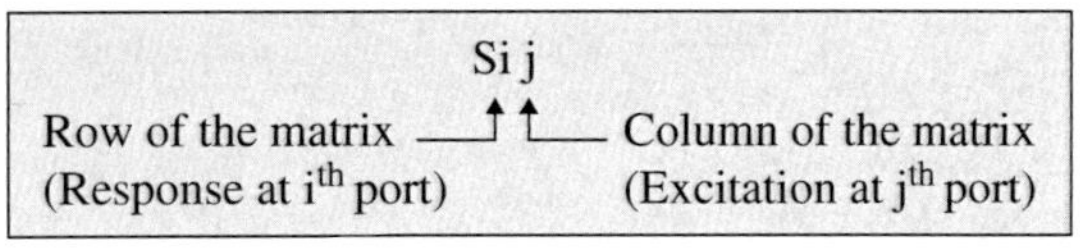

The S_{ij} is defined with the help of the matched termination. The matched termination also helps to measure the matrix elements S_{ij}.

Reflection Coefficient S_{ii}

Figure (3.12) shows that the ith port of a multiport network is terminated in a load equal to the characteristic impedance of the port. The power wave b_i coming out of the port is incident on the load Z_L, whereas the incident power wave a_i is the reflected wave from the load. If a port is terminated in its characteristic impedance, i.e. $Z_L = Z_0$, then the reflection from the load at the port is zero, i.e. $a_i = 0$. Thus, for the excitation a_j applied at the jth port, while all other ports are terminated in their characteristic impedances, the ports have $V_1^+(a_1) = V_2^+(a_2) = \cdots = V_i^+(a_i) = 0, (i \neq j)$. However, the power b_j is reflected from the jth port.

The reflection coefficient at the jth port is obtained from equation (3.1.40):

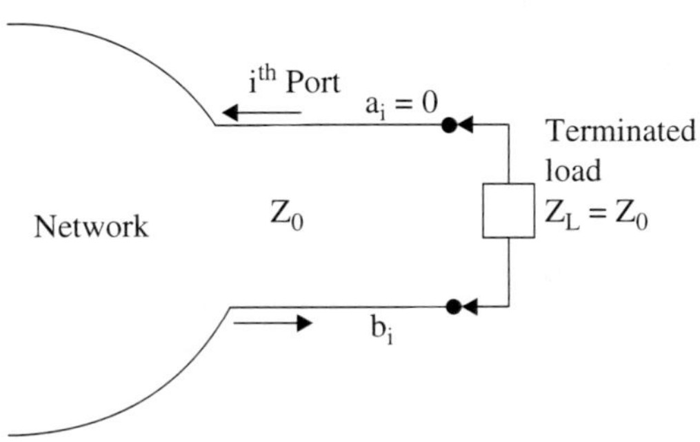

Figure 3.12 At the ith port, the load is terminated in port characteristic impedance.

$$S_{jj} = \left.\frac{b_j}{a_j}\right|_{a_i=0} = \left.\frac{V_j^-}{V_j^+}\right|_{V_i^+=0}, \quad i \neq j. \tag{3.1.41}$$

Therefore, S_{jj} is the reflection coefficient (Γ_j) at the jth port, provided all other ports are terminated in their characteristic impedances. However, if other ports are not terminated in their characteristic impedances, then S_{jj} is not a measure of the true reflection coefficient of the network or a device at the jth port. The true reflection coefficient at the jth port, under the unmatched load condition, is more than S_{jj} that is defined under the matched load condition.

Transmission Coefficient S_{ij}

If the excitation source is connected only to the jth port and the response is seen at the ith port, while all other ports are terminated in their characteristic impedances, it leads to $V_1^+ = V_2^+ = \cdots = V_k^+ = 0$, i.e. $a_1 = a_2 = \cdots = a_k = 0, (k \neq j)$. It shows that the excitation is zero at all ports, except at the jth port. The transmitted power, i.e. the scattered power, from the jth port is available at all ports, i = 1, 2, ..., k. However, at the jth port, a part of the incident power appears as the reflected power. The *transmission coefficient*, S_{ij} for the power transfer from the jth port to the ith port is defined as

$$S_{ij} = \left.\frac{b_i}{a_j}\right|_{a_k=0,k\neq j} = \left.\frac{V_i^-}{V_j^+}\sqrt{\frac{Z_{0j}}{Z_{0i}}}\right|_{V_k^+=0,k\neq j} = \left.\frac{I_i^-}{I_j^+}\sqrt{\frac{Z_{0i}}{Z_{0j}}}\right|_{V_k^+=0,k\neq j}. \tag{3.1.42}$$

Normally, the network has identical port impedances and equal to the system impedance, i.e. $Z_{0i} = Z_{0j} = Z_0$. Equation (3.1.40) is written in compact form as

$$[V^-] = [S][V^+], \quad \Rightarrow [b] = [S][a]. \tag{3.1.43}$$

The elements of the [S] matrix are determined using equations (3.1.41) and (3.1.42).

Properties of [S] Matrix

Some important properties of the [S] matrix description of the network are summarized below, without going for the formal proof of these statements. Usually, elements of the [S] matrix are complex quantities. The detailed discussion is available in the well-known textbooks [B.1–B.5, B.7].

Reciprocity Property

The [S] matrix of a *reciprocal* network is a symmetric matrix, i.e. the transpose $[S]^T$ of the [S] matrix is equal to the [S] matrix itself:

$$[S] = [S]^T. \tag{3.1.44}$$

Unitary Property

The [S] matrix of a *lossless* network is a unitary one. However, if the network is not lossless, then it is not unitary. The definition of the unitary matrix provides the following relation for the given [S] matrix:

$$[S]^T\,[S]^* = [I], \tag{3.1.45}$$

where $[S]^T$ is the transpose of the [S] matrix, $[S]^*$ is a complex conjugate of the complex [S] matrix and [I] is the identity matrix. Thus, for a given 2-port [S] matrix, we have

$$[S] = \begin{bmatrix} S_{11} & S_{12} \\ S_{21} & S_{22} \end{bmatrix},\quad [S]^* = \begin{bmatrix} S_{11}^* & S_{12}^* \\ S_{21}^* & S_{22}^* \end{bmatrix},\quad [S]^T = \begin{bmatrix} S_{11} & S_{21} \\ S_{12} & S_{22} \end{bmatrix}.$$

On substituting these expressions in the unitary relation (3.1.45), the following result is obtained:

$$\begin{bmatrix} S_{11} & S_{21} \\ S_{12} & S_{22} \end{bmatrix}\begin{bmatrix} S_{11}^* & S_{12}^* \\ S_{21}^* & S_{22}^* \end{bmatrix} = \begin{bmatrix} 1 & 0 \\ 0 & 1 \end{bmatrix}$$

$$\Rightarrow \begin{bmatrix} S_{11}S_{11}^* + S_{21}S_{21}^* & S_{11}S_{12}^* + S_{21}S_{22}^* \\ S_{12}S_{11}^* + S_{22}S_{21}^* & S_{12}S_{12}^* + S_{22}S_{22}^* \end{bmatrix} = \begin{bmatrix} 1 & 0 \\ 0 & 1 \end{bmatrix}. \tag{3.1.46}$$

On equating each element of matrix equation (3.1.46), the following relations are obtained:

$$\begin{matrix} S_{11}S_{11}^* + S_{21}\,S_{21}^* = 1 \\ S_{12}S_{12}^* + S_{22}\,S_{22}^* = 1 \end{matrix}\ (a),\quad \begin{matrix} S_{11}S_{12}^* + S_{21}\,S_{22}^* = 0 \\ S_{12}S_{11}^* + S_{22}\,S_{21}^* = 0 \end{matrix}\ (b). \tag{3.1.47}$$

Equations (3.1.47) are generalized for the N-port network:

$$\sum_{i=1}^{N}\left(S_{ik}S_{ik}^*\right) = 1,\ \ k = 1, 2, \ldots, N. \tag{3.1.48}$$

$$\sum_{i=1}^{N}\left(S_{ik}S_{ij}^*\right) = 0,\ \ j \neq k\quad j, k = 1, 2, \ldots, N. \tag{3.1.49}$$

Equation (3.1.48) shows that both elements have identical columns, whereas in equation (3.1.49) column are not identical. The [S] matrix is formed by the column vector as follows:

$$[S] = \left[\vec{p}\,,\,\vec{q}\right],\text{where } \left[\vec{p}\right] = \begin{bmatrix} S_{11} \\ S_{21} \end{bmatrix} \text{ and } \left[\vec{q}\right] = \begin{bmatrix} S_{12} \\ S_{22} \end{bmatrix}. \tag{3.1.50}$$

Therefore, in the usual vector notation we have

$$\begin{aligned} \vec{p} &= \hat{i}\,S_{11} + \hat{j}\,S_{21} \\ \vec{q} &= \hat{i}\,S_{12} + \hat{j}\,S_{22}. \end{aligned} \tag{3.1.51}$$

Hence, for a lossless network the following statements, based on equations (3.1.48) and (3.1.49) are made:

- The dot product of any column vector with its complex conjugate is unity, $\vec{p}\cdot\vec{p}^{\,*} = 1$.
- The dot product of any column vector with the complex conjugate of any other column vector is zero, $\vec{p}\,.\,\vec{q}^{\,*} = 0$.
- The [S] matrix forms an orthogonal set of the vectors.

The following expressions are written from equation (3.1.47):

$$|S_{11}|^2 + |S_{21}|^2 = 1\ (a),\ \ |S_{12}|^2 + |S_{22}|^2 = 1\ (b). \tag{3.1.52}$$

Equation (3.1.52) is the power balance equations for the lossless two-port networks. The unit input power fed to the port-1 is a sum of the reflected power ($|S_{11}|^2$) at the port-1 and the transmitted power $|S_{21}|^2$ to the port-2. In the case $|S_{11}|^2 + |S_{21}|^2$ is less than unity, some power is lost in the network through the mechanism of conductor, dielectric, and radiation losses. The lost power, i.e. the power dissipation in the network, is

$$P_d = 1 - \left(|S_{11}|^2 + |S_{21}|^2\right). \tag{3.1.53}$$

Phase Shift Property

The [S] parameter is a complex quantity. It has both magnitude and phase. Thus, the [S]-parameter is always defined with respect to a *reference plane*. In Fig (3.13) [S]-parameter of the N-port network is known at the location x = 0. It is determined at the new location, $x = -\ell_n$. Alternatively, once the [S] parameters are known at $x = -\ell_n$, these are determined at x = 0, i.e. at the port of the network. The location ℓ_n shows the length of the line connected to each port of an N-port network. Normally, it is the point of measurement of the [S] parameters of the network or device. The interconnecting transmission line is lossless and has propagation constant β_n. Thus, the electrical length of the connecting line is $\theta_n = \beta_n\,\ell_n$.

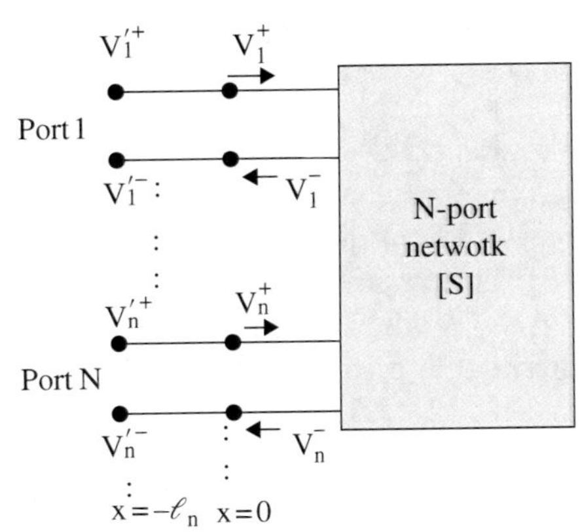

Figure 3.13 N-port network showing phase-shifting property.

For an N-port network, the incident wave at the nth port, $x = -\ell_n$, after reflection from the port at $x = 0$, returns to $x = -\ell_n$. In the process, it travels the electrical length $2\theta_n$. Similarly, if the wave is incident at port #1, located at $x = -\ell_1$ and arrives at the port-2, located at $x = -\ell_2$; the electrical length traveled by the wave is $\theta_1 + \theta_2 = \beta_1 \ell_1 + \beta_2 \ell_2$, or $2\theta_1$, on the assumption that $\beta_1 = \beta_2$, and $\ell_1 = \ell_2$, i.e. the transmission lines connected at both the ports are identical. The measured or simulated scattering matrix $[S']$ at the location $x = -\ell_n$ is related to the [S] parameters of the network by the following expression

$$[S'] = [S]\left[e^{-j\,2\theta_n}\right]. \tag{3.1.54}$$

The [S]-parameter of the network is extracted from equation (3.1.54), as

$$[S] = [S']\left[e^{+j\,2\theta_n}\right]. \tag{3.1.55}$$

For reducing the cascaded network to a single equivalent network, the [S] parameters cannot be cascaded like the [ABCD] parameters. The [ABCD] matrix is suitable for this purpose. However, it is not defined in terms of the power variables. Therefore, another suitable transmission matrix, called [T] *matrix* has been defined in terms of the power variables to cascade the microwave networks. The [S] matrix is easily converted to the [T] parameters [B.1, B.2–B.5, B.7, B.9].

The concept of the [S] matrix is used below to some simple, but useful circuits. These examples would help to appreciate the applications of the [S] parameters.

Example 3.8

Determine the S-parameters and return loss of a 2-port network with arbitrary termination shown in Fig (3.14).

Solution

The 2-port network (device) is connected to a source at the port-1 and a load Z_L at the port-2. The source has voltage V_g with internal impedance Z_g. The network scattering parameters-[S] are computed under the matched condition. The characteristic impedance of the connecting line between the port-1 and the source is Z_{01}, whereas the characteristic impedance of the connecting line between the port-2 and the load is Z_{02}. The lengths of the connecting lines are zero. The reflection and transmission coefficients are to be determined at the input and output terminals. This is a practical problem for the measurement and simulation of the 2-port network:

$$\left.\begin{aligned} b_1 &= S_{11}a_1 + S_{12}a_2 && (a)\\ b_2 &= S_{21}a_1 + S_{22}a_2 && (b).\end{aligned}\right\} \quad (i)$$

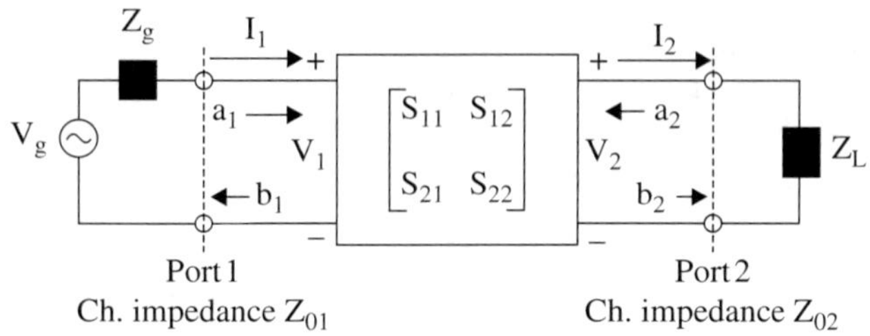

Figure 3.14 A two-port network with arbitrary termination.

Figure (3.14) shows that the power variable b_2 is the incident wave at the load Z_L and the power variable a_2 is the reflected wave from the load. Thus, the reflection coefficient at the load is

$$\Gamma_L = \frac{Z_L - Z_{02}}{Z_L + Z_{02}} = \frac{a_2}{b_2}. \tag{ii}$$

From above equations (i) and (ii):

$$\left.\begin{aligned} b_1 &= S_{11}a_1 + S_{12}\Gamma_L b_2 && (a)\\ b_2 &= S_{21}a_1 + S_{22}\Gamma_L b_2 &&\\ \Rightarrow b_2 &= \frac{S_{21}}{1 - S_{22}\Gamma_L}a_1 && (b).\end{aligned}\right\} \quad (iii)$$

On substituting b_2 from equation (b) in equation (a):

$$b_1 = \left[S_{11} + \frac{S_{12}S_{21}\Gamma_L}{1 - S_{22}\Gamma_L}\right]a_1. \tag{iv}$$

The input reflection coefficient at the port-1 is

$$\Gamma_{in} = \Gamma_1 = \frac{\text{Reflected wave from port} - 1}{\text{Incident wave at port} - 1} = \frac{b_1}{a_1}$$

$$\Rightarrow \quad \Gamma_1 = \left[S_{11} + \frac{S_{12}S_{21}\Gamma_L}{1 - S_{22}\Gamma_L}\right]. \tag{3.1.56}$$

The reflection coefficient Γ_1 is more than S_{11} of the network. The mismatch at the load degrades the return loss (RL) of the network. It is given by

$$RL = -20\log_{10}|\Gamma_1|. \tag{3.1.57}$$

For the port 2 open-circuited ($Z_L \to \infty$), the waves get reflected in-phase, i.e. $\Gamma_L = 1$, and for a short-circuited load ($Z_L = 0$) the total reflection is out of phase, i.e. $\Gamma_L = -1$. If the network is terminated in a matched load ($Z_L = Z_{02}$), the incident waves are absorbed with $\Gamma_L = 0$ and $\Gamma_1 = S_{11}$. Likewise, the source reflection coefficient Γ_g could be defined at the input port-1. Figure (3.14) again shows that b_1 is the incident wave on the internal impedance of the source Z_g and a_1 is the reflected wave from Z_g. Thus,

$$\Gamma_g = \frac{Z_g - Z_{01}}{Z_g + Z_{01}} = \frac{a_1}{b_1}. \tag{v}$$

The output reflection coefficient Γ_2 at the port-2 is obtained from equations (i) and (v):

$$\Gamma_2 = \frac{b_2}{a_2} = \left[S_{22} + \frac{S_{21}S_{12}\Gamma_g}{1 - S_{11}\Gamma_g}\right]. \qquad (3.1.58)$$

Again under the matched condition ($Z_g = Z_{01}$) at the input port, $\Gamma_g = 0$. For most of the applications, 50 Ω system impedance is used, i.e. $Z_{01} = Z_{02} = Z_0 = 50\ \Omega$. For a 2-port lossless network, we have the following expressions:

$$S_{11}S_{11}^* + S_{21}S_{21}^* = 1\ (a), \quad S_{12}S_{12}^* + S_{22}S_{22}^* = 1\ (b).$$

However, for a reciprocal network $S_{12} = S_{21}$. Thus, the above equations provide

$$|S_{11}| = |S_{22}| \quad (a), \quad |S_{12}| = \sqrt{1 - |S_{11}|^2} \quad (b). \qquad (3.1.59)$$

The network also follows $S_{11}S_{12}^* + S_{21}S_{22}^* = 0$. The S-parameters are complex quantities. The S-parameters are written in the phasor form: $S_{11} = |S_{11}|\,e^{j\theta_1}$, $S_{22} = |S_{22}|\,e^{j\theta_2}$ and $S_{12} = |S_{12}|\,e^{j\phi}$. From the above equation, the phase relation is obtained:

$$|S_{11}|\sqrt{1 - |S_{11}|^2}\left[e^{j(\theta_1 - \phi)} + e^{j(\phi - \theta_2)}\right] = 0 \quad (a)$$

$$\Rightarrow \phi = \frac{\theta_1 + \theta_2}{2} + \frac{\pi}{2} \mp n\pi \quad (b). \qquad (3.1.60)$$

Therefore, once the complex S_{11} and S_{22} are measured, both the magnitude and phase of the S_{21} are determined. However, usually, both S_{11} and S_{21} are obtained from a VNA and also from the circuit simulator or EM-simulator. The magnitude of S_{21} provides the *insertion-loss* of the network and ϕ is the phase shift at the output of the network.

Example 3.9
Determine the S-parameters of the series impedance shown in Fig (3.15). Also, compute the attenuation and the phase shift offered by the series impedance.

Solution

To compute S_{11} that is the reflection coefficient of a network under the matched condition, the port-2 is terminated in Z_0. Thus, $Z_{in} = Z + Z_0$ and the reflection coefficient at port-1 is

$$S_{11} = \frac{Z_{in} - Z_0}{Z_{in} + Z_0} = \frac{Z}{Z + 2Z_0}. \qquad (3.1.61)$$

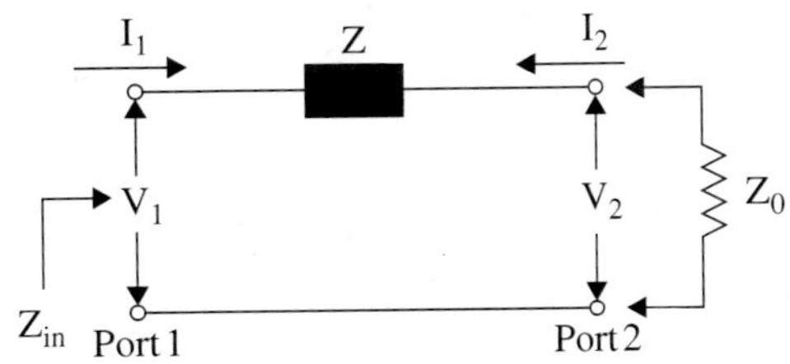

Figure 3.15 Network of series impedance.

Likewise, to compute S_{22}, the port-1 is terminated in Z_0. It gives $Z_{out} = Z + Z_0$ at the port-2. The S_{22} is

$$S_{22} = \frac{Z_{out} - Z_0}{Z_{out} + Z_0} = \frac{Z}{Z + 2Z_0}. \qquad (3.1.62)$$

The total port voltage at the port-1 is a sum of the forward and reflected voltages:

$$V_1 = V_1^+ + V_1^- = (S_{11} + 1)V_1^+ \quad (i) \quad \left\{\text{Note: } S_{11} = \frac{V_1^-}{V_1^+}\right\}.$$

To compute S_{21}, i.e. the transmission coefficient from the port-1 to the port-2 under the matched termination, at first, the total port voltage at the port-2 is obtained:

$$V_2 = V_2^+ + V_2^- = V_2^- \qquad (ii)$$

$$\{\text{Note: } V_2^+ = 0 \text{ due to the matched termination.}\}$$

Therefore, from equations (i) and (ii):

$$S_{21} = \left.\frac{V_2^-}{V_1^+}\right|_{V_2^+ = 0} = \frac{V_2}{V_1}(S_{11} + 1) \qquad (iii)$$

However, the port voltage V_2 computed from the port current is

$$I_1 = \frac{V_1}{Z + Z_0} \text{ and } V_2 = Z_0 I_1 = \frac{Z_0 V_1}{Z + Z_0} \qquad (iv)$$

Finally, S_{21} is obtained from equations (iii), (iv) and (3.1.61):

$$S_{21} = \frac{V_2}{V_1}(S_{11} + 1) = \frac{Z_0}{Z + Z_0}\left(1 + \frac{Z}{Z + 2Z_0}\right) = \frac{2Z_0}{Z + 2Z_0}. \qquad (3.1.63)$$

Equations (3.1.61) and (3.1.63) provide the following relation:

$$S_{11} + S_{21} = 1. \qquad (3.1.64)$$

The [S] matrix of the series impedance is

$$[S] = \begin{bmatrix} S_{11} & 1 - S_{11} \\ 1 - S_{11} & S_{11} \end{bmatrix}. \qquad (3.1.65)$$

The attenuation and phase shift of a signal, applied at the input port-1 of series impedance $Z = R + jX$, are computed below.

Using S_{21} from equation (3.1.63), the attenuation offered by the series impedance is

$$\alpha(dB) = 20\log_{10}\left|\frac{1}{S_{21}}\right| = 20\log_{10}\left|1 + \frac{R + jX}{2Z_0}\right|$$
$$\Rightarrow \alpha(dB) = 10\log\left[\left(1 + \frac{R}{2Z_0}\right)^2 + \left(\frac{X}{2Z_0}\right)^2\right]. \tag{3.1.66}$$

The lagging phase shift of the signal at the output port-2, due to the series element, is

$$\phi = -\tan^{-1}\left(\frac{X}{R + 2Z_0}\right). \tag{3.1.67}$$

Example 3.10
Determine the S-parameter of a shunt admittance shown in Fig (3.16). Also, compute the attenuation and the phase shift offered by the shunt admittance.

Solution

The shunt admittance is $Y = G + jB$. To compute S_{11}, the port-2 is terminated in Z_0 $(=1/Y_0)$ giving $Y_{in} = Y + Y_0$. The reflection coefficient of the shunt admittance under matched termination is

$$S_{11} = \frac{Y_0 - Y_{in}}{Y_o + Y_{in}} = \frac{-Y}{2Y_0 + Y}. \tag{3.1.68}$$

Likewise, to compute S_{22} of the shunt admittance, the port-1 is terminated in Z_0:

$$S_{22} = \frac{-Y}{2Y_0 + Y}. \tag{3.1.69}$$

Following the previous case of the series impedance, the S_{21} is computed:
$S_{21} = \frac{V_2}{V_1}(S_{11} + 1)$. Fig (3.16) shows $V_1 = V_2$; therefore,

$$S_{21} = 1 + S_{11} = \frac{2}{2 + YZ_0}. \tag{3.1.70}$$

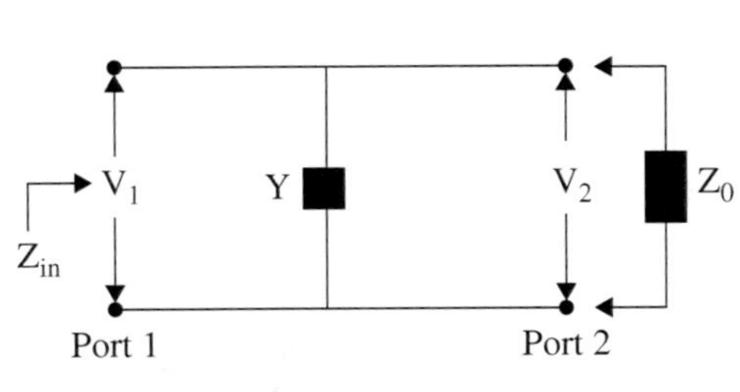

Figure 3.16 Network of shunt admittance.

The [S] matrix of the shunt admittance is

$$[S] = \begin{bmatrix} \frac{-YZ_0}{2 + YZ_0} & \frac{2}{2 + YZ_0} \\ \frac{2}{2 + YZ_0} & \frac{-YZ_0}{2 + YZ_0} \end{bmatrix}. \tag{3.1.71}$$

The attenuation of the input signal due to the shunt admittance is

$$\alpha(dB) = 20\log_{10}\left|\frac{1}{S_{21}}\right| = 20\log_{10}\left|1 + \frac{Y}{2Y_0}\right|$$
$$\Rightarrow \alpha(dB) = 10\log\left[\left(1 + \frac{G}{2Y_0}\right)^2 + \left(\frac{B}{2Y_0}\right)^2\right]. \tag{3.1.72}$$

The lagging phase shift of the signal at the output port-2, due to the shunt admittance, is

$$\phi = -\tan^{-1}\left(\frac{B}{G + 2Y_0}\right). \tag{3.1.73}$$

Example 3.11
Determine the S-parameters of a transmission line section, shown in Fig (3.17), with an arbitrary characteristic impedance.

Solution

The line has an arbitrary characteristic impedance nZ_0 and propagation constant β. The Z_0 is taken as the reference impedance to define the S-parameter. The reflection coefficient at the load end is

$$\Gamma_L(x = 0) = \frac{Z_0 - nZ_0}{Z_0 + nZ_0} = \frac{1 - n}{1 + n}. \tag{3.1.74}$$

Using equation (2.1.88) of chapter 2, the input impedance at the port-1 of the transmission line having characteristic impedance nZ_0 is

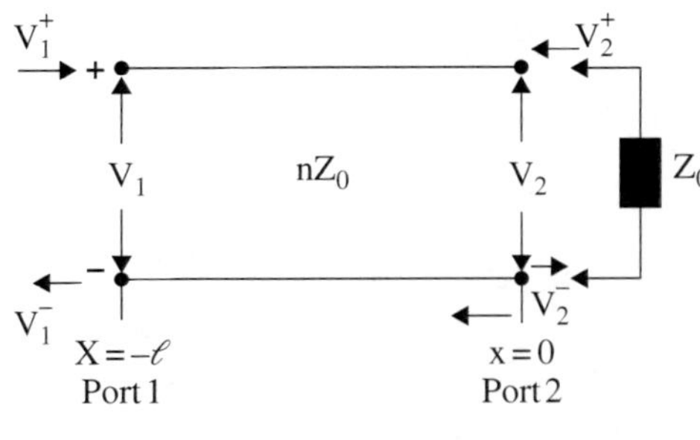

Figure 3.17 A transmission line circuit with an arbitrary characteristic impedance.

$$Z_{in} = nZ_0 \left[\frac{1 + \Gamma_L(x=0)\, e^{-j\,2\beta\ell}}{1 - \Gamma_L(x=0)\, e^{-j\,2\beta\ell}}\right]. \text{On substituting } \Gamma_L,$$

$$Z_{in} = nZ_0 \left[\frac{(1+n) + (1-n)\, e^{-j\,2\beta\ell}}{(1+n) - (1-n)\, e^{-j\,2\beta\ell}}\right]. \tag{3.1.75}$$

Thus, the reflection coefficient at the port-1 is

$$S_{11} = \frac{Z_{in} - Z_0}{Z_{in} + Z_0} = \frac{(n^2-1)(1-e^{-j\,2\beta\ell})}{(n+1)^2 + (n-1)^2 e^{-j\,2\beta\ell}}. \tag{3.1.76}$$

The transmission parameter S_{21} is computed in terms of S_{11}. If the amplitude of the forward traveling voltage wave on the transmission line is V_1^+, the total voltage on the transmission line is given by

$$V(x) = V_1^+ \left[e^{-j\beta x} + \Gamma_L(x=0)e^{j\beta x}\right], \tag{3.1.77}$$

where x is measured from the load end, as shown in Fig (3.17). The input port-1 is located at $x = -\ell$. The voltage at the port-1 is

$$V_1 = V_1^+ \left[e^{j\beta\ell} + \Gamma_L(x=0)e^{-j\beta\ell}\right]. \tag{3.1.78}$$

The port voltage V_1 is obtained as a sum of the incident and reflected voltages at the port-1:

$$V_1 = V_1^+ + V_1^- = V_1^+[1 + S_{11}], \{S_{11} = V_1^-/V_1^+\}. \tag{3.1.79}$$

At the port-2, under the matched termination, $Z_L = nZ_0$ giving $V_2^+ = 0$ and $V_2 = V_2^-$. Equation (3.1.77) shows that the voltage at the port-2, i.e. at $x = 0$ is

$$V_2 = V_1^+[1 + \Gamma_L(x=0)]. \tag{3.1.80}$$

Using equation (3.1.79), the transmission coefficient, S_{21} of the circuit shown in Fig (3.17) is obtained as

$$S_{21} = \frac{V_2^-}{V_1^+} = \frac{V_2}{V_1}(S_{11}+1). \tag{3.1.81}$$

On substituting V_1 from equation (3.1.78) and V_2 from equation (3.1.80) in the above equation S_{21} is obtained:

$$\begin{aligned} S_{21} &= \frac{[1 + \Gamma_L(x=0)]\,(1+S_{11})}{e^{j\beta\ell} + \Gamma_L(x=0)e^{-j\beta\ell}} \\ &= \frac{2(1+S_{11})}{(1+n)e^{j\beta\ell} + (1-n)e^{-j\beta\ell}}. \end{aligned} \tag{3.1.82}$$

The present line network is symmetrical and reciprocal. It has $S_{11} = S_{22}$ and $S_{21} = S_{12}$. The above expressions are checked for $n = 1$, i.e. for a transmission line of characteristic impedance Z_0. For this case, $S_{11} = S_{22} = 0$ and $S_{21} = S_{12} = e^{-j\beta\ell}$. These are expressions of the S-parameters for a line having characteristic impedance Z_0.

3.2 Conversion and Extraction of Parameters

Sometimes, the conversion of one kind of network parameter to another kind is needed for the analysis of a circuit. For instance, if several circuit blocks comprising of the lumped elements and the transmission line sections are cascaded, each circuit block could be expressed by its [ABCD] matrix. It helps to get an overall [ABCD] matrix of the cascaded network. However, the final [ABCD] matrix, describing the cascaded network is further converted to the [S] matrix. Similarly, the [S] matrix of each building block of the cascaded network has to be converted to the [ABCD] matrix to get the overall [ABCD] matrix of the cascaded network. Finally, the overall [ABCD] matrix is converted to the [S] matrix of the cascaded network. The S-parameters are measurable quantities. The performance of a network is measured in the [S] matrix using a *VNA*.

On several occasions, the S-parameters of a line section or a network are known either from the simulations or from the measurements. The S-parameters are used to get the characteristic impedance and the propagation constant of a line, or a network. However, the *true S-parameters* of a network are needed for this purpose. The true S-parameters are normally *embedded* in the measured or the simulated S-parameters at the ports of measurement, or the ports of simulation. The true S-parameters of a line or a network are extracted, i.e. de-embedded, from the measured, or simulated, S-parameters at the ports. This is known as the *de-embedding* process [B.10]. The EM-Simulators have provision to de-embed the true S-parameters from the S-parameters obtained at the measurement or simulation ports.

This section presents the conversion of matrix parameters, de-embedding of the S-parameters, and extraction of the propagation characteristics.

3.2.1 Relation Between Matrix Parameters

[Z] and [ABCD] Parameters

Figure (3.18) shows a network with its known [Z] parameters. It requires conversion to the [ABCD] parameters. The [ABCD] and [Z] parameters of the network are summarized below:

$$\left.\begin{aligned} V_1 &= A\,V_2 + B\,I_2 \quad &(a) \\ I_1 &= C\,V_2 + D\,I_2 \quad &(b). \end{aligned}\right\} \tag{3.2.1}$$

$$\left.\begin{aligned} V_1 &= Z_{11}I_1 - Z_{12}\,I_2 \quad &(a) \\ V_2 &= Z_{21}I_1 - Z_{22}\,I_2 \quad &(b). \end{aligned}\right\} \tag{3.2.2}$$

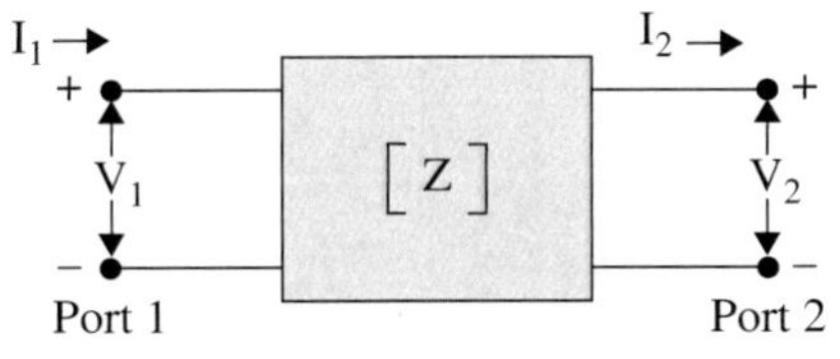

Figure 3.18 Network for Z-parameter.

The current (I_2) entering the port-2 is taken positively. However, the Z-parameter is defined in Fig (3.18) for the output current leaving the port. In this case, I_2 is negative. Equation (3.2.2) is rearranged to get the port voltage and current V_1 and I_1 at the port-1 in terms of the V_2 and I_2 at the port-2:

$$V_1 = \frac{Z_{11}}{Z_{21}} V_2 + \left[\frac{Z_{11} Z_{22}}{Z_{21}} - Z_{12}\right] I_2 \quad (a)$$
$$I_1 = \frac{1}{Z_{21}} V_2 + \frac{Z_{22}}{Z_{21}} I_2 \quad (b). \qquad (3.2.3)$$

On comparing equations (3.2.1) and (3.2.3), the following conversion expressions are obtained:

$$A = \frac{Z_{11}}{Z_{21}}, \quad B = \left[\frac{Z_{11} Z_{22}}{Z_{21}} - Z_{12}\right], \quad C = \frac{1}{Z_{21}}, \quad D = \frac{Z_{22}}{Z_{21}}. \qquad (3.2.4)$$

Likewise, the relations between [Y] and [ABCD] parameters are obtained:

$$A = \frac{-Y_{22}}{Y_{21}}, \qquad B = \frac{-1}{Y_{21}}$$
$$C = -\left[\frac{Y_{11} Y_{22}}{Y_{21}} - Y_{12}\right], \quad D = \frac{-Y_{11}}{Y_{21}}. \qquad (3.2.5)$$

A complete set of the conversion table of parameters is available in textbooks [B.1, B.5, B.7].

[S] and [Z] Parameters

The N-port network, having normalized reference port impedance, $Z_{0n} = 1$, is considered. The port voltage and port current in terms of the incident and reflected voltage can be written as

$$V_n = V_n^+ + V_n^- \quad (a), \quad I_n = I_n^+ - I_n^- = V_n^+ - V_n^- \quad (b). \qquad (3.2.6)$$

The above equations are written in the column matrix form:

$$[V] = [V^+] + [V^-] \quad (a), \quad [I] = [V^+] - [V^-] \quad (b). \qquad (3.2.7)$$

The port voltage is related to the port current through the [Z]-matrix:

$$[V] = [Z]\,[I] \Rightarrow [V^+] + [V^-] = [Z]\,\{[V^+] - [V^-]\}$$
$$\Rightarrow [V^-] = \left\{\frac{[Z] - [I]}{[Z] + [I]}\right\}[V^+], \qquad (3.2.8)$$

where [I] is a unit or identity matrix. Keeping in view the definition of the [S] matrix, the following relations, between the [S] matrix and [Z] matrix, are obtained:

$$[S] = \frac{[Z] - [I]}{[Z] + [I]} \quad (a), \quad \Rightarrow [Z] = \frac{[I] + [S]}{[I] - [S]} \quad (b). \qquad (3.2.9)$$

Similarly, the following expressions, relating [S] and [Y]-parameters are obtained:

$$[S] = \frac{[I] - [Y]}{[I] + [Y]} \quad (a), \quad \Rightarrow [Y] = \frac{[I] - [S]}{[I] + [S]} \quad (b). \qquad (3.2.10)$$

[ABCD] and [S] Parameters

Figure (3.19) shows a 2-port network. The known [ABCD] parameters of the network are to be converted to the [S] parameters. The voltage pair (V_1^+, V_2^+), and (V_1^-, V_2^-) are the incident voltage and reflected voltage at both port-1 and port-2. Figure (3.19) also shows the total port voltage (V_1, V_2) and total port current (I_1, I_2). The port currents, at both the ports, enter into the network. Therefore, the current I_2 is negative. The input port voltage and port current are related to the output port voltage and port current through the [ABCD] parameters as follows:

$$V_1 = A\,V_2 + B\,(-I_2) \quad (a)$$
$$I_1 = C\,V_2 + D\,(-I_2) \quad (b). \qquad (3.2.11)$$

The port voltage and current are a linear combination of the incident and reflected voltages and currents:

$$V_n = V_n^+ + V_n^-; \quad n = 1, 2 \quad (a)$$
$$I_n = \frac{1}{Z_0}(V_n^+ - V_n^-) \quad (b). \qquad (3.2.12)$$

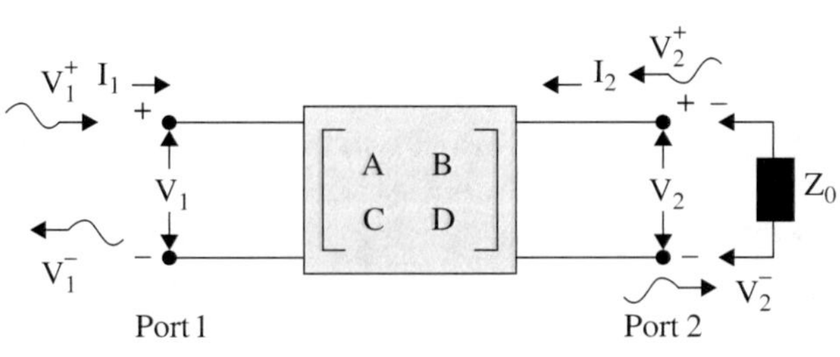

Figure 3.19 Network for [ABCD] parameter.

On substituting equation (3.2.11) in equation (3.2.12):

$$V_1 = V_1^+ + V_1^- = A\,(V_2^+ + V_2^-) - B\frac{(V_2^+ - V_2^-)}{Z_0} \quad \text{(a)}$$
$$I_1 = \frac{1}{Z_0}(V_1^+ - V_1^-) = C\,(V_2^+ + V_2^-) - D\frac{(V_2^+ - V_2^-)}{Z_0} \quad \text{(b)}. \qquad (3.2.13)$$

To define the [S] parameters, port-2 is terminated in the reference impedance Z_0 giving $V_2^+ = 0$, i.e. the reflection from the matched terminated load is zero. The voltage V_2^- is the incident wave on the load ($Z_L = Z_0$), whereas the voltage V_2^+ is the reflected wave from the load. The above equations are reduced to the following expressions:

$$V_1^+ + V_1^- = \left(A + \frac{B}{Z_0}\right)V_2^- \quad \text{(a)}$$
$$V_1^+ - V_1^- = (C\,Z_0 + D)\,V_2^- \quad \text{(b)}. \qquad (3.2.14)$$

On adding the above equations, the following expression is obtained:

$$2\,V_1^+ = (A + B/Z_0 + CZ_0 + D)\,V_2^-. \qquad (3.2.15)$$

Equation (3.2.15) provides the transmission coefficient S_{21}, defined as follows:

$$S_{21} = \left.\frac{\text{Response at port-2}}{\text{Excitation at port-1}}\right|_{V_2^+ = 0} = \left.\frac{V_2^-}{V_1^+}\right|_{V_2^+ = 0}$$
$$\Rightarrow S_{21} = \frac{2}{A + B/Z_0 + CZ_0 + D}. \qquad (3.2.16)$$

The following expression is obtained from equation (3.2.14):

$$\frac{V_1^+ + V_1^-}{V_1^+ - V_1^-} = \frac{A + B/Z_0}{CZ_0 + D}. \qquad (3.2.17)$$

Equation (3.2.17) gives the following reflection coefficient S_{11} at the port-1, while port-2 is terminated in the matched load Z_0:

$$S_{11} = \left.\frac{\text{Response at port-1}}{\text{Excitation at port-1}}\right|_{V_2^+ = 0} = \left.\frac{V_1^-}{V_1^+}\right|_{V_2^+ = 0}$$
$$\Rightarrow S_{11} = \frac{A + B/Z_0 - CZ_0 - D}{A + B/Z_0 + CZ_0 + D}. \qquad (3.2.18)$$

Similarly, the following expressions are obtained from equation (3.2.13), where the port-1 is terminated in the matched load Z_0, i.e. $V_1^+ = 0$:

$$V_1^- = (A - B/Z_0)V_2^+ + (A + B/Z_0)V_2^- \quad \text{(a)}$$
$$-V_1^- = (CZ_0 - D)V_2^+ + (CZ_0 + D)V_2^- \quad \text{(b)}. \qquad (3.2.19)$$

On eliminating V_1^- the S_{22} is obtained, whereas S_{12} is obtained eliminating V_2^-:

$$S_{22} = \left.\frac{V_2^-}{V_2^+}\right|_{V_1^+ = 0} = \frac{-A + B/Z_0 - CZ_0 + D}{A + B/Z_0 + CZ_0 + D} \quad \text{(a)}$$
$$S_{12} = \left.\frac{V_1^-}{V_2^+}\right|_{V_1^+ = 0} = \frac{2(AD - BC)}{A + B/Z_0 + CZ_0 + D} \quad \text{(b)}. \qquad (3.2.20)$$

If the network is reciprocal, $AD - BC = 1$, i.e. $S_{12} = S_{21}$. For the symmetrical network, $S_{11} = S_{22}$ leading to $A = D$. The known [S] parameters can also be converted to the [A, B, C, D] parameters. Similarly, the [Z], [Y], [ABCD] and [S] parameters are also converted among themselves [B.1, B.3, B.5].

3.2.2 De-Embedding of True S-Parameters

A transmission line section could be treated as a device and its performance can be evaluated by using a VNA. The device is connected to the VNA through connecting cables and connectors. The S-parameters of a device is measured at the external circuit ports that include the effect of the cables and connectors on the S-parameters of the device. Thus, the true S-parameter of a device is *embedded* in the measured S-parameters of the device. However, it is desired to obtain the true S-parameters of the device under test (DUT). The line section could be the DUT. The process of extracting the S-parameters of the device at the *internal device ports* (1_{in}, 2_{in}), from the measured S-parameters, at the *external circuit ports* (1_{ex}, 2_{ex}) is known as the *de-embedding* process. It is achieved through a *calibration process* in which the S-parameters of two *error boxes* are quantified. The error box represents errors in the S-parameters due to cables and connectors connecting the device to the external circuit ports [B.1]. The S or [ABCD] parameter representation of the device at internal ports (1_{in}, 2_{in}) along with the error boxes is shown in Fig (3.20a). The location of the measurement ports, i.e. the external ports (1_{ex}, 2_{ex}) and the device internal ports (1_{in}, 2_{in}), are further shown in Fig (3.20b).

Once the error boxes are characterized through their S-parameters, it can be converted to the [A^e B^e C^e D^e] parameters. Similarly, the measured S-parameters of the device and the error box combined are available at the external ports. These can be converted to the measured [A^m B^m C^m D^m] parameters. The device [A^d B^d C^d D^d] parameters are related to the other two parameters by the following equation:

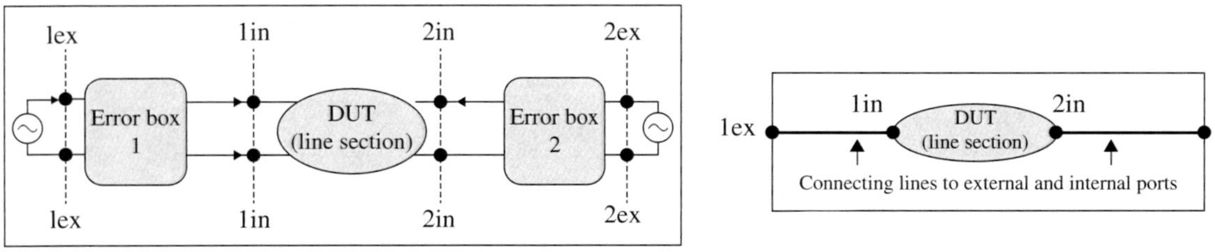

(a) Representation of measurement process with external measurement ports and internal device ports.

(b) Location of ports.

Figure 3.20 Calibration process in the measurement of S-parameters of a device.

$$\begin{bmatrix} A^m & B^m \\ C^m & D^m \end{bmatrix} = \begin{bmatrix} A^e & B^e \\ C^e & D^e \end{bmatrix} \begin{bmatrix} A^d & B^d \\ C^d & D^d \end{bmatrix} \begin{bmatrix} A^e & B^e \\ C^e & D^e \end{bmatrix}^{-1}. \tag{3.2.21}$$

The error box 2 is the mirror image of the error box 1 with respect to the DUT. So in the above-given matrix sequence, the third matrix is inverse of the first matrix [B.11]. At the internal device ports, the device $[A^d\ B^d\ C^d\ D^d]$ parameters are de-embedding as follows:

$$\begin{bmatrix} A^d & B^d \\ C^d & D^d \end{bmatrix} = \begin{bmatrix} A^e & B^e \\ C^e & D^e \end{bmatrix}^{-1} \begin{bmatrix} A^m & B^m \\ C^m & D^m \end{bmatrix} \begin{bmatrix} A^e & B^e \\ C^e & D^e \end{bmatrix}. \tag{3.2.22}$$

The de-embedded device $[A^d\ B^d\ C^d\ D^d]$ parameters are converted to the de-embedded S-parameters of the device. The de-embedded S-parameters could be further converted to the Z and Y-parameters. Thus, any two-port device can be characterized through measurements using suitable parameters- S, Z, or Y. In the case of a transmission line section, the de-embedded S-parameters can be converted to the propagation parameters and the characteristic impedance of the line.

The above-mentioned concept of de-embedding of the device S-parameters at the internal port of a device is equally applicable to the EM-Simulators – both 2.5D and 3D simulators [B.10]. In EM-simulators, the *delta-gap voltage source* could be used to launch the wave on a line section or a device. It also generates the non-propagating evanescent modes at the ports. They cause a discontinuity at the external circuit ports, i.e. at the ports of simulation. The port discontinuity affects the S-parameters of the device that is removed by the process of de-embedding [J.2]. The EM-simulators could be used to extract the propagation parameters and the characteristic impedance of a line.

3.2.3 Extraction of Propagation Characteristics

The true S-parameters of the line section, i.e. its de-embedded S-parameters over a range of frequencies are known either through the measurement or through the EM-simulation. This information can be converted to the [ABCD] parameters of a line section as follows [B.1]:

$$\begin{bmatrix} A & B \\ C & D \end{bmatrix} = \begin{bmatrix} \dfrac{(1+S_{11})(1-S_{22})+S_{12}S_{21}}{2S_{21}} & Z_o\dfrac{(1+S_{11})(1+S_{22})-S_{12}S_{21}}{2S_{21}} \\ \dfrac{1}{Z_o}\dfrac{(1-S_{11})(1-S_{22})-S_{12}S_{21}}{2S_{21}} & \dfrac{(1-S_{11})(1+S_{22})+S_{12}S_{21}}{2S_{21}} \end{bmatrix}. \tag{3.2.23}$$

Usually, $S_{11} = S_{22}$ and $S_{12} = S_{21}$ because a section of the transmission line is treated as the symmetrical and reciprocal network. The frequency-dependent [ABCD] parameters of a lossy line of length ℓ are used to compute the characteristic impedance Z_0 and propagation constant $\gamma = \alpha + j\beta$ over a range of frequencies [B.1]:

$$Z_{01} = \sqrt{\frac{B}{C}} \quad (a), \quad \gamma = \alpha + j\beta = \frac{1}{\ell}\cosh^{-1}(A) = \frac{1}{\ell}\ln\left(A \pm \sqrt{A^2-1}\right) \quad (b). \tag{3.2.24}$$

The frequency-dependent secondary line parameters are extracted to get the dispersion characteristics of a line. Once the secondary line parameters of a line section are known, their frequency-dependent primary constants are computed as follows:

$$R = \mathrm{Re}(\gamma Z_0) \quad (a), \quad L = \frac{\mathrm{Im}(\gamma Z_0)}{\omega} \quad (b)$$
$$G = \mathrm{Re}\left(\frac{\gamma}{Z_0}\right) \quad (c), \quad C = \frac{\mathrm{Im}\,(\gamma/Z_0)}{\omega} \quad (d). \tag{3.2.25}$$

3.3 Wave Velocity on Transmission Line

In a communication network, several kinds of electrical signals propagate on a transmission line. The signal could be a modulated or unmodulated carrier wave, the baseband analog signal, or the digital pulses. The TEM mode transmission lines, and also various kinds of non-TEM waveguide structures support wave propagation. The parameters defining these transmitting media could be either frequency-independent or frequency-dependent. The property of the medium has a significant impact on the nature of wave propagation through a medium. The wave velocity has no simple or unique meaning, like the meaning of the velocity of a particle. There are several kinds of wave velocities – *phase velocity, group velocity, energy velocity, signal velocity,* etc., applied to wave propagation. The significance of several types of wave velocities is inherent both in the complexity of a signal and also in the complexity of the wave supporting medium. This section focuses attention on the meaning of the phase and group velocities only. Section (3.4) demonstrates these two wave velocities as applied to several kinds of the *artificial linear dispersive transmission lines.*

3.3.1 Phase Velocity

The concept of phase velocity is applicable to a single frequency wave, i.e. to a *monochromatic* wave discussed in Section (2.1) of chapter 2. The phase velocity is just the movement of the *wavefront*. The wavefront is a surface of constant phase, like maximum, minimum, or zero-level points shown in Fig (2.3). It is given by equation (2.1.8) of chapter 2 and reproduced below:

$$v_p = \frac{\omega}{\beta}. \tag{3.3.1}$$

The propagation constant β is influenced by the wave-supporting medium. For a lossless TEM transmission line and lossless unbounded space, β is given by

$$\beta = \omega\sqrt{LC} \quad (a), \quad \beta = \omega\sqrt{\mu\varepsilon} \quad (b), \tag{3.3.2}$$

where ε and μ are permittivity and permeability of a medium. Thus, pairs (L, C) and (ε, μ) are the parameters that characterize the electrical property of the wave supporting-media. The unbounded medium supports the plane wave propagation. If these parameters are not frequency-dependent, the medium is known as *nondispersive*. In such a medium, the phase velocity remains constant at every frequency. However, if any of these parameters are frequency-dependent, the propagation constant β is frequency-dependent and consequently, the phase velocity is frequency-dependent. The medium that supports the frequency-dependent phase velocity is known as the *dispersive* medium. Normally, the characteristic impedance or intrinsic impedance of a dispersive medium is also frequency-dependent. The parameters (L, C) and (ε, μ) are usually independent of signal strength. Such a medium is called a *linear* medium, whereas the signal strength dependent medium is a *nonlinear* medium. The characteristics of the medium are discussed in Section (4.2) of chapter 4. The present discussion is only about the linear and dispersive transmission lines.

Why a medium becomes dispersive? One reason for dispersion is the loss associated with a medium. The geometry of a wave supporting inhomogeneous structures, commonly encountered in the planar technology, is another source of the dispersion. In the case of a transmission line, the parameters R and G are associated with losses and they make propagation constant β frequency-dependent. Likewise, losses make permittivity ε and permeability μ of material medium frequency-dependent complex quantities. However, a low-loss dielectric medium can be nondispersive in the useful frequency band. For such cases, the attenuation and propagation constants are given by

$$\alpha = \frac{\omega\varepsilon_r''\sqrt{\mu_r/\varepsilon_r'}}{2c} \quad (a), \quad \beta = \frac{\omega}{c}\sqrt{\mu_r\varepsilon_r'}\left[1 + \frac{1}{8}\left(\frac{\varepsilon_r''}{\varepsilon_r'}\right)^2\right] \quad (b). \tag{3.3.3}$$

In equation (3.3.3), c is the velocity of EM-wave in free space. The above expressions are obtained from equations (4.5.16) and (4.5.19) of chapter 4. The complex permittivity is given by $\varepsilon = \varepsilon' - j\varepsilon''$. The imaginary part, showing a loss in a medium, is related to the conductivity of a medium through relation $\sigma = \omega\varepsilon''$. If ε' and ε'' are not frequency-dependent, the propagation constant β is not frequency-dependent and the phase velocity is also not frequency-dependent. However, if they are frequency dependent, the phase velocity is also frequency-dependent. The phase velocity in the low-loss dielectric medium is

$$v_p = \frac{\omega}{\beta} = \frac{1}{\sqrt{\mu\varepsilon'}}\left[1 - \frac{1}{8}\left(\frac{\varepsilon''}{\varepsilon'}\right)^2\right]. \qquad (3.3.4)$$

Therefore, the presence of loss decreases the phase velocity of EM-wave. This kind of wave is known as the *slow-wave*. The slow-wave can be dispersive or non-dispersive. However, it is associated with a loss. This aspect is further illustrated through the EM-wave propagation in a high conductivity medium. The conducting medium is discussed in subsection (4.5.5) of chapter 4. The attenuation (α), phase constant (β), and phase velocity ($v_{p,con}$) of a highly conducting medium are given by equation (4.5.35b) of chapter 4 [B.3]:

$$\alpha = \beta = \sqrt{\pi f \mu \sigma}. \qquad (3.3.5)$$

$$v_{p,con} = \frac{\omega}{\beta} = v_p\sqrt{\frac{2\omega\varepsilon'}{\sigma}}. \qquad (3.3.6)$$

The above expressions are obtained for a highly conducting medium, $\sigma/\omega\varepsilon >> 1$. It does not apply to the lossless medium with $\sigma = 0$. The wave propagation is associated with significant loss (α) given by equation (3.3.5). Moreover, both the attenuation constant and phase velocity are frequency dependent, so a conducting medium is highly dispersive. However, the periodic structures and other mechanisms give slow-wave structures with a small loss. Such structures are useful for the development of compact microwave components and devices [B.1, B.3–B.5, B7, B.12, B.13]. The slow-wave periodic transmission line structures are discussed in chapter 19.

Some EM-wave supporting media have *cut-off* property. They support the wave propagation only above the certain characteristic frequency of a medium or a structure. These media and structures are also dispersive. For instance, the *nonmagnetic plasma* medium has such cut-off property [B.4, B.14]. The plasma medium is discussed in the subsection (6.5.2) of chapter 6. The permittivity of a plasma medium is given by equation (6.5.16):

$$\varepsilon_p = \varepsilon_0\left[1 - \left(\frac{f_p}{f}\right)^2\right] \quad (a), \qquad f_p = \frac{1}{2\pi}\sqrt{\frac{Ne^2}{m_e\varepsilon_0}} \quad (b). \qquad (3.3.7)$$

In the above expression, f_p is the plasma frequency that is a characteristic cut-off frequency of the plasma medium [B.4, B.14]. The permeability of nonmagnetized plasma is $\mu = \mu_0$. Other parameters are as follows-ε_0: permittivity of free space, N: electron density, e: electron charge, and m_e: electron mass. The propagation constant, phase velocity, and plasma wavelength λ_{plasma} of the EM-wave wave in a plasma medium are given below:

$$\beta = \omega\sqrt{\mu\varepsilon_0}\left[1 - \left(\frac{f_p}{f}\right)^2\right]^{\frac{1}{2}} \qquad (a)$$

$$v_p = \frac{1}{\sqrt{\mu\varepsilon_0}}\left[1 - \left(\frac{f_p}{f}\right)^2\right]^{-\frac{1}{2}} = v\left[1 - \left(\frac{f_p}{f}\right)^2\right]^{-\frac{1}{2}} \qquad (b)$$

$$\lambda_{plasma} = \frac{2\pi}{\beta} = \frac{v}{f}\left[1 - \left(\frac{f_p}{f}\right)^2\right]^{-\frac{1}{2}} \qquad (c)$$

$$v_p = \left(\frac{\lambda_{plasma}}{\lambda}\right)v \qquad (d). \qquad (3.3.8)$$

In equation (3.3.8), $v = 1/\sqrt{\mu\varepsilon_0}$ is the velocity of EM-wave in the homogeneous medium with parameters ε_0 and μ. The wavelength in the homogeneous medium is $\lambda = v/f$. However, the nonmagnetized plasma medium has the parameters ε_0, μ_o supporting the wavelength $\lambda_0 = c/f$. The nonmagnetized plasma medium behaves as free space.

The phase velocity of the EM-waves in a plasma medium is frequency-dependent. Therefore, it is a dispersive medium that supports a *fast-wave*. It is fast in the sense that the phase velocity is higher than the phase velocity of the EM-wave in free space given by $c = 1/\sqrt{\mu_0\varepsilon_0}$. The plasma medium exhibits the *cut-off phenomenon,* similar to the cut-off behavior of the waveguide medium. The waveguide medium is discussed in the section (7.4) of chapter 7. There is no wave propagation at the plasma frequency $f = f_p$. The plasma frequency f_p behaves like the cut-off frequency f_c of a waveguide. *Thus, the waveguide can be used to simulate the electrical behavior of plasma*. For $f < f_p$, no wave propagation takes place, as the propagation constant β becomes an imaginary quantity. Such a wave is known as an *evanescent wave*. It is an exponentially decaying nonpropagating wave ($E = E_0\,e^{-\alpha z}$). The standard metallic waveguide also supports the cut-off phenomenon and has a frequency-dependent phase velocity [B.1, B.5, B.7, B.8, B.15–B.17].

The dispersion is a property of the wave-supporting medium. The phase velocity of a wave in a dispersive medium can either decrease or increase with the increase in frequency. Thus, all dispersive media could be put into two groups – (i) *normal dispersive medium* and (ii) *abnormal or anomalous dispersive medium.*

Figure 3.21 Nature of normal (positive) dispersion.

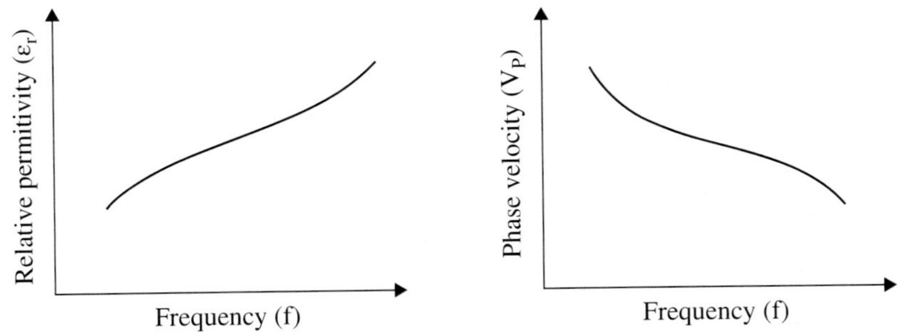

(a) Frequency-dependent relative permittivity $d\varepsilon_r / df > 0$.

(b) Frequency-dependent phase velocity $dv_p / df < 0$.

Figure (3.21a and b) show the general behavior of a medium having *normal* dispersion. The relative permittivity of such a medium increases with frequency, i.e. $d\varepsilon_r/df$ is positive, and the phase velocity decreases with frequency, i.e. dv_p/df is negative. A microstrip line provides such a medium for the normal dispersion. The effective relative permittivity of a microstrip line increases with frequency leading to a decrease in the phase velocity with an increase in frequency. The microstrip is discussed in chapter 8.

Figure (3.22a and b) show the general behavior of an *anomalous dispersive* medium. The relative permittivity of such a medium decreases with an increase in frequency, i.e. $d\varepsilon_r/df < 0$ (negative). It leads to an increase in the phase velocity with an increase in frequency, i.e. $dv_p/df > 0$ (positive). A microstrip line on a semiconductor substrate having the *Metal, Insulator, Semiconductor* (MIS) or the Schottky structure, in the transition region, is an anomalous dispersive medium [J.3, J.4].

It is emphasized that there is nothing abnormal with the anomalous dispersion. Both kinds of dispersions exist in reality. The normal dispersion is also called the *positive dispersion* as the gradient of ε_r with frequency is positive, i.e. $d\varepsilon_r/df > 0$. Similarly, the anomalous dispersion is called the *negative dispersion* with $d\varepsilon_r/df < 0$. The relative permittivity of material undergoes both kinds of dispersion depending upon the physical cause of dispersion. The dispersion is caused by several kinds of material polarizations – dipolar, ionic, electronic, and interfacial polarization. Once the frequency is varied from low-frequency to the optical frequency, the material medium undergoes these polarization changes, and the propagating wave experiences both the normal and anomalous dispersion at different frequencies [B.17, B.18]. It is discussed in chapter 6.

The concept of phase velocity applies to a single frequency signal. Now the question is to apply it to a complex baseband signal and a modulated signal. It is possible to use the phase velocity concept to such waveforms through the Fourier series of a periodic signal and using the Fourier integral for a nonperiodic signal. Any signal, periodic or nonperiodic, is composed of a large number of sinusoidal signals. They have a definite amplitude and phase relationship with the fundamental frequency of the signal. A combination of all

Figure 3.22 Nature of anomalous (negative) dispersion.

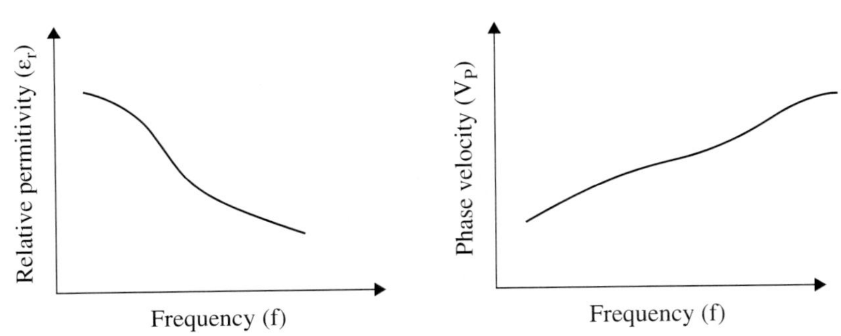

(a) Frequency-dependent relative permittivity $d\varepsilon_r / df < 0$.

(b) Frequency-dependent phase velocity $dv_p / df > 0$.

sinusoidal components gives a complex signal of definite wave-shape. If the complex waveshape travels through a dispersive medium having frequency-dependent attenuation constant $\alpha(f)$, the amplitude of each signal component changes differently. Similarly, in a dispersive medium having a frequency-dependent propagation constant $\beta(f)$, each signal component travels with a different velocity. It results in different phase-change for each frequency component of the complex wave; so the shape of the wave changes while traveling on a line or through the medium. The numerical inverse Fourier transform provides the wave-shape of a signal in the time-domain at any location in the medium. Thus, the Fourier method helps to apply the concept of phase velocity to complex waveform propagation [J.5, J.6]. Such investigations are important to maintain the signal integrity on the IC and MMIC chips.

3.3.2 Group Velocity

A complex signal composed of two or more frequency components forms a *wave-packet*. However, the frequency components should not be much different from each other like an amplitude modulated signal. Figure (3.23) shows the wave-packet formed by a group of a narrowband complex signal. The wave-packet has a central or a *carrier* frequency of higher value, superimposed with a low-frequency *envelope*. The carrier wave travels with the phase velocity v_p, whereas the envelope, i.e. the wave shape travels with the *group velocity* v_g. In a nondispersive medium, the carrier wave and envelope both travel with the same velocity without a change in the waveshape. However, in the case of a dispersive medium, velocities of the carrier and envelope are different. Depending on the nature of dispersion, the wave could be the *forward wave* or the *backward wave*. If the medium has *normal (positive) dispersion*, the phase velocity, i.e. the velocity of the carrier wave, and the velocity of the envelope, i.e. the group velocity, are in the same direction, as shown in Fig (3.23). The wave is known as the forward wave. However, in the case of a highly *anomalous (negative) dispersive* medium, under a certain condition, the phase and group velocities are in the opposite directions, forming the backward wave.

The carrier and envelope are combined to form a unified wave structure called the wave-packet. In the case of normal dispersion, the group velocity is the *energy velocity* of a signal and the information travels with the group velocity [B.1, B.4, B.5, B.7, B.14, B.16]. However, in the case of anomalous dispersion, the energy velocity and group velocity are different. In this case, group velocity is not velocity of information. Moreover, the concept of group velocity applies only to a narrow-band wave-packet, not to the wideband signal. The controversy exists at present on the travel of information with a velocity more than the velocity of light [J.7].

Formation of Two-Frequency Wave-Packet

A wave-packet is formed by a linear combination of two signals of equal magnitude with a small difference in angular frequency and phase constant. It is shown in Fig (3.24). The composite voltage wave is given by

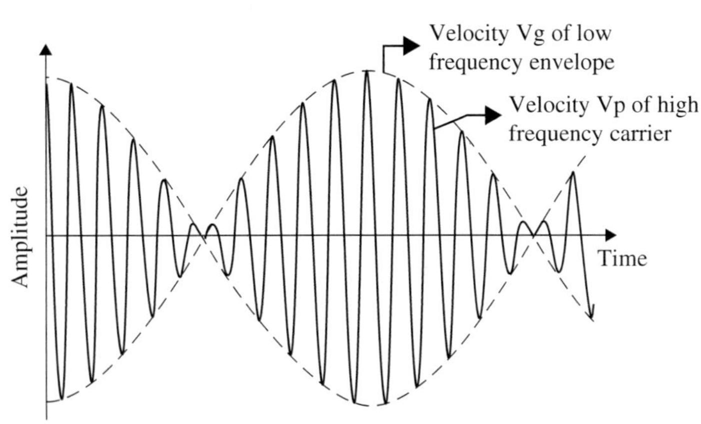

Figure 3.23 Description of phase and group velocities of a forward-moving modulated wave.

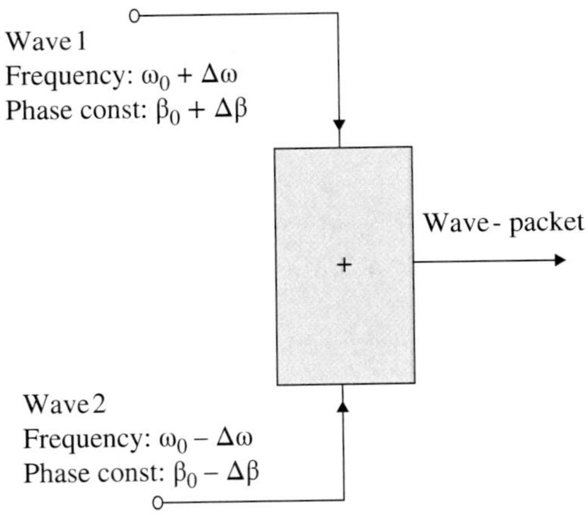

Figure 3.24 Formation of a wave-packet.

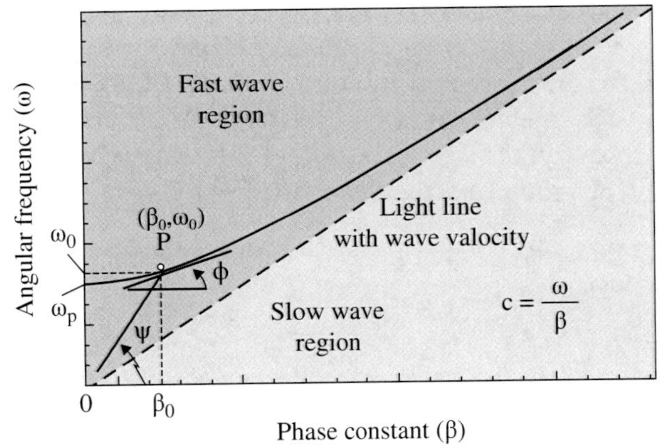

Figure 3.25 $\omega - \beta$ diagram to get phase and group velocities.

$$v(x,t) = \underbrace{V_0 \cos\{(\omega_0 + \Delta\omega)t - (\beta_0 + \Delta\beta)x\}}_{\text{Wave\#1}} + \underbrace{V_0 \cos\{(\omega_0 - \Delta\omega)t - (\beta_0 - \Delta\beta)x\}}_{\text{Wave\#2}} \quad (3.3.9)$$

$$\Rightarrow v(x,t) = \underbrace{\{2V_0 \cos(\Delta\omega t - \Delta\beta x)\}}_{\text{Envelope}} \underbrace{\cos(\omega_0 t - \beta_0 x)}_{\text{Carrier wave}}.$$

The carrier wave has frequency ω_0 and propagation constant β_0. The above expression applies to a narrow-band signal, $\Delta\omega << \omega_0$. The envelope frequency is $\Delta\omega$ and its propagation constant is $\Delta\beta$. The carrier wave inside the envelope, shown in Fig (3.23), moves with phase velocity:

$$\omega_0 t - \beta_0 x = \text{Constant}, \qquad \Rightarrow v_p = \frac{dx}{dt} = \frac{\omega_0}{\beta_0}. \quad (3.3.10)$$

The velocity of the envelope, i.e. the group velocity, is obtained from the constant phase point on the envelope

$$\Delta\omega t - \Delta\beta x = \text{Const.}, \Rightarrow \qquad v_g = \frac{\Delta x}{\Delta t} = \frac{\Delta\omega}{\Delta\beta}$$

$$\text{For} \quad \Delta\omega \to 0, \qquad v_g = \frac{1}{d\beta/d\omega}. \quad (3.3.11)$$

Both the phase velocity and group velocity are represented over the $(\omega - \beta)$ dispersion diagram of the wave-supporting medium. The $(\omega - \beta)$ diagram plays a very significant role in the wave propagation in periodic and metamaterial media. These are discussed in the chapters 18–22. Figure (3.25) shows the dispersion relation given by equation (3.3.8a). The point P on the dispersion diagram locates (β_0, ω_0). The slope ψ of a point P on the dispersion diagram, with respect to the origin, gives the phase velocity of a carrier wave, whereas the *local slope* ϕ at $\omega = \omega_0$, i.e. a local tangent at point P, provides the group velocity:

$$v_p = \tan^{-1}\psi = \frac{\omega_0}{\beta_0} \quad (a)$$

$$v_g = \tan^{-1}\phi = \left.\frac{d\omega}{d\beta}\right|_{\omega=\omega_0} = \left.\frac{1}{d\beta/d\omega}\right|_{\omega=\omega_0} \quad (b). \quad (3.3.12)$$

At the cut-off frequency ω_p, $\beta \to 0$, results in an infinite extent of the phase velocity, while the group velocity is zero, $v_g = 0$. However, away from the cut-off frequency, the phase velocity decreases to a limiting value $c = 1/\sqrt{\mu_0\varepsilon_0}$. It is the slope of the *light–line* as shown in Fig (3.25). The light-line divides the $(\omega - \beta)$ diagram into two regions – within the *light-cone* it is the fast-wave region, and outside the light-cone, it is the slow-wave region. At much higher frequency, both velocities become equal. For the plasma medium, equations (3.3.8a) and (3.3.8b) provide the propagation constant and phase velocity.

As the propagation constant increases from $\beta = 0$ at ω_p to $\omega\sqrt{\mu\varepsilon_0}$ for $\omega \to \infty$, and the phase velocity decreases from $v_p \approx \infty$ at ω_p to $1/\sqrt{\mu\varepsilon_0}$ for $\omega \to \infty$, the plasma forms a normal, i.e. the positive dispersive medium. However, the phase velocity above the cut-off frequency in a plasma medium is always greater than the velocity of EM-wave in the free space, $v_p \geq c$. This wave is a *fast-wave*. The group velocity is computed as follows:

$$\frac{d\beta}{d\omega} = \frac{1}{c\sqrt{1-(\omega_p/\omega)^2}} \quad \text{(a)}$$

$$v_g = \frac{1}{d\beta/d\omega} = c\left[1-\left(\frac{\omega_p}{\omega}\right)^2\right]^{\frac{1}{2}} \quad \text{(b)}$$

$$\text{also,} \quad v_g = \left(\frac{\lambda_0}{\lambda_{plasma}}\right)c \quad \text{(c).} \tag{3.3.13}$$

The group velocity is less than the velocity of the EM-wave in the free space, $v_g \leq c$. In summary, the plasma forms a *fast-wave normal dispersive* medium that supports the *forward wave* having the same direction for both the phase and group velocities. Equations (3.3.8d) and (3.3.13c) for the wave propagation in a nonmagnetized plasma with $\mu = \mu_0$, ε_0 and $v = c$ show that the phase and group velocities are related through the following expression:

$$v_g v_p = c^2 \quad \text{(a), or in general} \quad v_g v_p = v^2 \quad \text{(b).} \tag{3.3.14}$$

In equation (3.3.14), the general isotropic medium supports EM-wave propagation with velocity $v = c/\sqrt{\varepsilon_r\mu_r}$.

A more general relation between the phase and group velocities could be obtained. In the dispersive medium, phase velocity (v_p) is a function of frequency; therefore,

$$\frac{d\beta}{d\omega} = \frac{d}{d\omega}(\omega/v_p) = \frac{1}{v_p}\left[1-\frac{\omega}{v_p}\frac{dv_p}{d\omega}\right]$$

$$\Rightarrow \quad v_g = \frac{v_p}{1-\dfrac{\omega}{v_p}\dfrac{dv_p}{d\omega}}. \tag{3.3.15}$$

Figure (3.26a and b) show the dispersive behaviors of the phase velocity (v_p) and propagation constant (β). The expression for the group velocity in a dispersive medium, i.e. a medium with frequency-dependent refractive index $n(\omega) = \sqrt{\varepsilon_r(\omega)}$, can be rewritten as follows:

$$n = \frac{c}{v_p} = c\frac{\beta}{\omega} \quad \text{(a),} \quad \Rightarrow \frac{dn}{d\omega} = \frac{c}{\omega}\left[\frac{d\beta}{d\omega} - \frac{\beta}{\omega}\right]$$

$$v_g = \frac{d\omega}{d\beta} = \frac{c}{n+\omega\dfrac{dn}{d\omega}} \quad \text{(b).} \tag{3.3.16}$$

In equation (3.3.16), c is the velocity of EM-wave in free space. The cases of dispersion are considered below:

Case-I: $\dfrac{dn}{d\omega} = 0$, i.e. $\dfrac{dv_p}{d\omega} = 0$. It is the no dispersion case. In this case, the above equation provides $v_g = v_p$. Figure (3.26a) shows that the phase velocity in a non-dispersive medium remains unchanged with angular frequency. Figure (3.26b) shows that the propagation constant (β) is a linear function of angular frequency (ω). The free space is one such medium.

Case-II: $\dfrac{dn}{d\omega} > 0$, i.e. $\dfrac{dv_p}{d\omega} < 0$. It is the case of normal, i.e. the positive dispersion. Figure (3.26b) indicates that the slope of the propagation constant β in the normal dispersive medium increases nonlinearly with

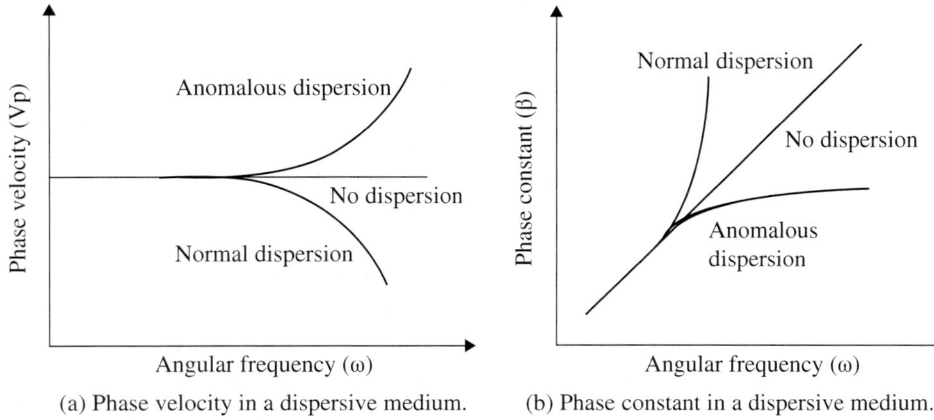

Figure 3.26 Nature of dispersion on ($\omega - \beta$) the diagram.

angular frequency ω. Therefore, the phase velocity decreases with angular frequency, i.e. $dv_p/d\omega$ is negative. It is shown in Fig (3.26a). For this case, equations (3.3.15) and (3.3.16) show that $v_g < v_p$. It is the case applicable to a dispersive microstrip line.

Case-III: $\frac{dn}{d\omega} < 0$, i.e. $\frac{dv_p}{d\omega} > 0$. It is the case of the anomalous (abnormal), i.e. the negative dispersion. The propagation constant β of such an anomalous dispersive medium increases nonlinearly with angular frequency ω. However, on the (ω – β) diagram, shown in Fig (3.26b), its value is below the dispersionless medium of the case-I. The slope of the propagation constant β, in the nonlinear region, decreases with an increase in frequency ω, i.e. $dn/d\omega < 0$. Figure (3.26a) shows that the phase velocity increases with angular frequency, i.e. $dv_p/d\omega$ is positive. Equations (3.3.15) and (3.3.16) show that $v_g > v_p$. However, the group velocity in an anomalous dispersive medium is not the velocity of energy transportation. This case is applicable to the dispersive MIS or Schottky microstrip lines in the transition region [J.3, J.4]. The equations (3.3.15) and (3.3.16) further indicate a possibility of *backward wave propagation* (v_g negative) in an anomalous dispersive medium, if $dn/d\omega$ is significantly negative, i.e. the medium has a very large negative dispersion. This is the special condition for the existence of backward wave in the anomalous dispersive medium. However, such a medium is very lossy. Lorentz model, discussed in subsection (6.5.1) of chapter 6, explains the phenomenon.

The relations between two velocities are obtained below by using the propagation constant (β) and wavelength (λ), instead of the angular frequency (ω):

$$v_g = v_p + \beta\,\frac{dv_p}{d\beta}\ \text{(a)},\quad \Rightarrow v_g = v_p - \lambda\,\frac{dv_p}{d\lambda}\ \text{(b)}. \tag{3.3.17}$$

The condition for the normal dispersion is $dv_p/d\beta < 0$, $dv_p/d\lambda > 0$ and for the anomalous dispersion, it is $dv_p/d\beta > 0$, $dv_p/d\lambda < 0$.

3.4 Linear Dispersive Transmission Lines

The velocity of EM-wave propagation in a dispersive medium has been discussed in the previous section. The transmission line is a 1D wave-supporting medium. The lossy line is a dispersive medium, whereas a lossless line, modeled through the line constants L and C, is a nondispersive medium. It acts as a low-pass filter (LPF). This section shows that *a reactively loaded lossless line could be a dispersive medium*. A variety of transmission line structures with interesting properties can be developed by using several additional combinations of C and L. Such line structures can be realized with the lumped elements and also by the modification of planar transmission lines. The transmission medium with *negative relative permittivity* and *negative relative permeability* has been synthesized with the help of the modified, i.e. reactively loaded line structures. These wave supporting media form a new class of materials known as the *metamaterials*. They do not exist in nature. However, these novel media have been developed with the *defects* in the transmission line and using the embedded resonators in the transmission line [B.19, J.8]. The present section considers only an infinitesimal section of the transmission line, modeled as a lumped circuit elements network. The modeled network is reactively loaded to get the loaded line. However, in practice, a finite length of the line is periodically loaded with reactance. Such reactively loaded lines offer novel and improved designs of microwave components and circuits. The periodically loaded line, creating the 1D – EBG and metalines are discussed in the chapters 19 and 22.

3.4.1 Wave Equation of Dispersive Transmission Lines

Figure (3.27a) shows the *shunt inductor loaded* line structure. An infinitesimally small Δx line length is considered. The cascading of a large number line sections, under the condition $\Delta x \to 0$, forms a continuous line. The standard Δx long *host* transmission line section is modeled by the distributed line constants L and C p.u.l. The host line section is loaded with the *inclusion*, i.e. with additional shunt inductance L_{sh}. The loading element, i.e. inclusion is shown in the gray box, so the periodic embedding of reactive inclusion in the host line medium creates an artificial 1D mixture medium. The concept of the mixture material is further discussed in subsection (6.3.1) of chapter 6.

The inclusion lumped shunt inductance L_{sh} distributed over the length Δx is $(L_{sh}/\Delta x)$ *p.u.l.* The total series inductance, $(L\,\Delta x)$, gives the total series impedance $Z = j\,\omega L\,\Delta x$ and a combination of shunt capacitance $(C\,\Delta x)$ and shunt inductance $(L_{sh}/\Delta x)$ gives the total shunt admittance, $Y = j\,(\omega C\Delta x - \Delta x/(\omega L_{sh}))$. The transmission line equations for the shunt inductor loaded line are obtained as follows:

∗Drop in the line voltage

= Voltage drop across the series inductor

$$\Delta v = -(L\Delta x)\frac{\Delta i}{\Delta t} \qquad (a)$$

∗Drop in the line current =

Currents through the shunt capacitor and shunt inductor

$$\Delta i = -(C\Delta x)\frac{\Delta v}{\Delta t} - i_2 \qquad (b). \qquad (3.4.1)$$

For the limiting case, $\Delta x \to 0$, $\Delta t \to 0$, the following expressions are obtained:

$$\frac{\partial v}{\partial x} = -L\frac{\partial i}{\partial t} \qquad (a)$$
$$\frac{\partial i}{\partial x} = -C\frac{\partial v}{\partial t} - \frac{i_2}{dx} \qquad (b). \qquad (3.4.2)$$

The voltage v across the shunt inductor is

$$v = \left(\frac{L_{sh}}{dx}\right)\frac{\partial i_2}{\partial t}. \qquad (3.4.3)$$

From equation (3.4.2a,b), the following voltage wave equation is obtained:

$$\frac{\partial^2 v}{\partial x^2} = -L\frac{\partial^2 i}{\partial x \partial t} \Rightarrow \frac{\partial^2 v}{\partial x^2} = L\frac{\partial}{\partial t}\left[C\frac{\partial v}{\partial t} + \frac{i_2}{dx}\right]$$
$$\Rightarrow \frac{\partial^2 v}{\partial x^2} = LC\frac{\partial^2 v}{\partial t^2} + \frac{L}{dx}\left(\frac{\partial i_2}{\partial t}\right). \qquad (3.4.4)$$

On substituting $\partial i_2/\partial t$ from equation (3.4.3) in the above equation, the above wave equation is reduced to

$$\frac{\partial^2 v}{\partial x^2} = LC\frac{\partial^2 v}{\partial t^2} + \frac{L}{L_{sh}}v, \Rightarrow \frac{\partial^2 v}{\partial t^2} - \frac{1}{LC}\frac{\partial^2 v}{\partial x^2} + \frac{1}{CL_{sh}}v = 0, \qquad (3.4.5)$$

where $v_0 = 1/\sqrt{LC}$ is the phase velocity of the standard nondispersive line without shunt loaded inductance, i.e. for the case $L_{sh} \to \infty$. For a line in the air medium, the phase velocity is the velocity of the EM-wave in a vacuum, i.e. $v_0 = c$. The cut-off frequency is defined as $\omega_c = 1/\sqrt{L_{sh}C}$. Finally, the above voltage wave equation of the dispersive transmission line is rewritten as

$$\frac{\partial^2 v}{\partial t^2} - v_0^2\frac{\partial^2 v}{\partial x^2} + \omega_c^2 v = 0. \qquad (3.4.6)$$

Equation (3.4.6) is known as the *Klein–Gordon equation* [B.20]. For a line without cut-off frequency, i.e. for $\omega_c = 0$, the above equation is reduced to the standard dispersionless transmission line equation (2.1.24) of chapter 2. On assuming the harmonic solution for the forward-moving voltage, the voltage on the loaded line is

$$v(x,t) = V_{max}\cos(\omega t - \beta x)$$
$$\Rightarrow v(x,t) = V_{max}\,\mathrm{Re}\left[e^{j(\omega t - \beta x)}\right]. \qquad (3.4.7)$$

On substituting equation (3.4.7) in equation (3.4.6), the following *dispersion relation* is obtained:

$$\omega^2 = \omega_c^2 + v_0^2\beta^2, \quad \Rightarrow \quad \beta = \frac{\omega}{v_0}\sqrt{1 - \left(\frac{\omega_c}{\omega}\right)^2}. \qquad (3.4.8)$$

Equation (3.4.8) is identical to equation (3.3.8a) for the plasma medium. The cut-off frequency ω_c of the loaded line corresponds to the plasma frequency ω_p. *Thus, a plasma medium can be modeled by the shunt inductor loaded transmission line.* Further, Fig (3.27a) shows that below the cut-off frequency, i.e. for $\omega < \omega_c$ the circuit is reduced to the L-L circuit presenting inductors in both the series and shunt arm. Such a medium is called the *epsilon negative (ENG) medium*, discussed in section (5.5) of chapter 5.

For frequency $\omega < \omega_c$, the propagation constant β is the imaginary quantity and the voltage wave cannot propagate on the shunt inductor loaded line. The voltage wave propagates only for $\omega > \omega_c$. Thus, the shunt inductor loaded line behaves like a high-pass filter (HPF). It is like a metallic waveguide discussed in subsection (7.4.1) of chapter 7. For $\omega < \omega_c$, the loaded line supports the non-propagating attenuated wave, called the *evanescent wave*. Its attenuation constant α is obtained by the following expression:

$$\alpha = \frac{\omega}{v_0}\sqrt{\left(\frac{\omega_c}{\omega}\right)^2 - 1}. \qquad (3.4.9)$$

The phase and group velocities of the wave on the loaded dispersive line are

$$v_p = \frac{\omega}{\beta} = \frac{v_0}{\sqrt{1 - \left(\frac{\omega_c}{\omega}\right)^2}} \quad (a), \quad v_g = v_0\sqrt{1 - \left(\frac{\omega_c}{\omega}\right)^2} \quad (b). \qquad (3.4.10)$$

Figure (3.27b) shows the dispersion behavior, i.e. the function $\omega = f(\beta)$, on the $(\omega - \beta)$ diagram for the shunt inductor loaded line. It shows the frequency-dependent behavior of the propagation constant β and attenuation constant α. The dashed line is the *light line*. Its slope ψ at the origin is the phase velocity of the EM-wave in free space, i.e. $v_p = v_0 = c$. The propagation constant β exists

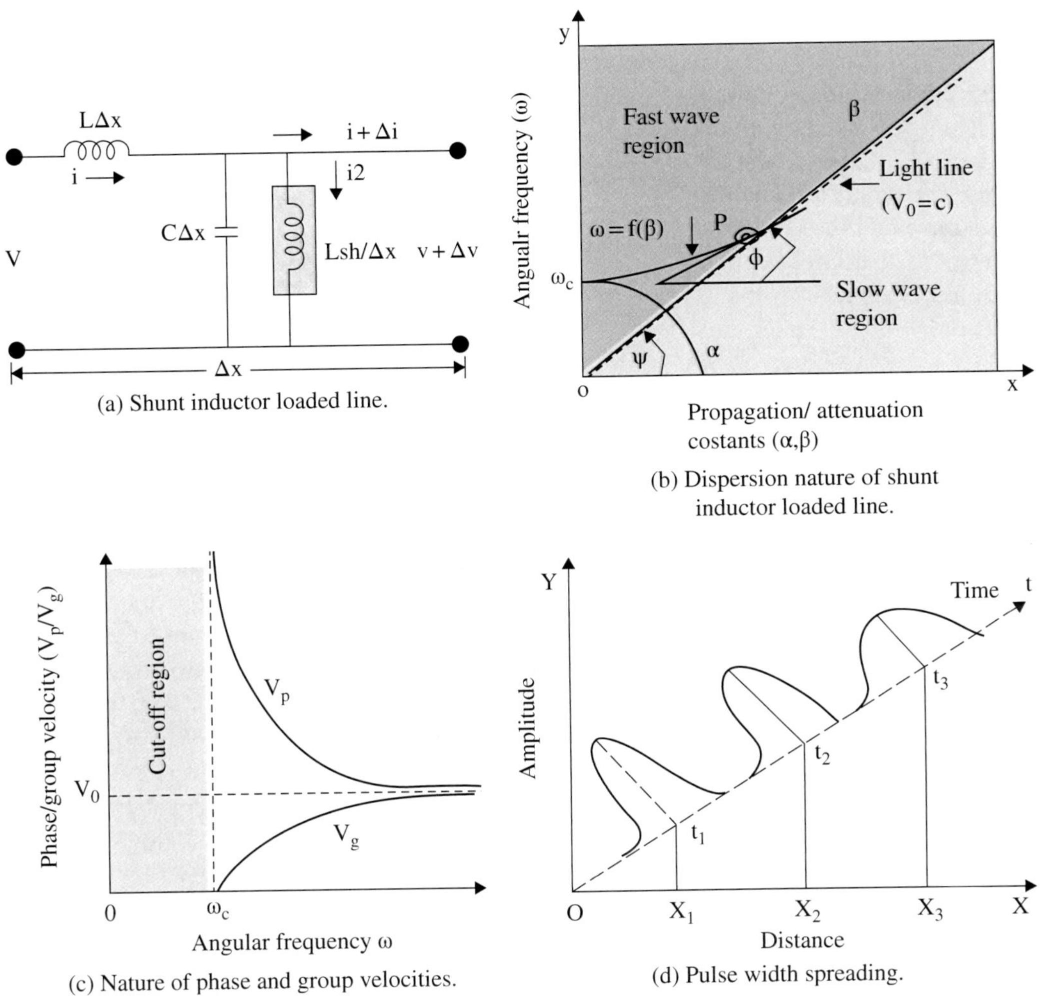

Figure 3.27 Shunt inductor loaded line and its characteristics.

above the light line only for $\omega > \omega_c$ and wave attenuation occurs for $\omega < \omega_c$. At location P, the local slope ϕ provides the group velocity v_g, whereas the slope ψ of the point P at the origin O provides the phase velocity v_p of the propagating EM-wave on the loaded transmission line.

Figure (3.27c) shows the variations in the phase velocity and the group velocity with frequency. For $\omega \to \omega_c$, $v_p \to \infty$, $v_g \to 0$, and for $\omega \to \infty$, both wave velocities move toward the light line, i.e. toward the phase velocity (v_0) of the unloaded host transmission line. From equation (3.4.10a) $dv_p/d\omega$ is negative, i.e. the phase velocity decreases with frequency, whereas from the equation (3.4.10b), $dv_g/d\omega$ is positive, i.e. the group velocity increases with frequency. However, both velocities tend toward v_0, i.e. phase velocity in the unloaded host medium. Figure (3.27b) also shows that the propagation constant β increases with frequency. Thus, the shunt inductor loaded line has *normal dispersion*. The wave considered is a forward-moving wave with both the phase and group velocities in the same direction, as both anti-clockwise gradients ψ and ϕ defining the phase and group velocities respectively are positive. Above the cut-off frequency, i.e. above the light line in the fast-wave region, the phase velocity is higher than the phase velocity of the unloaded L-C type transmission line. The shunt inductor loaded line supports the *fast-wave*. The wave is fast as compared to the wave velocity, shown as the slope of the light line, on an unloaded transmission line.

The group velocity v_g is also frequency-dependent, causing dispersion in the envelope of a wave-packet. Therefore, if a voltage wave-packet, say a Gaussian wave-packet, propagates on a shunt inductor loaded line, it disperses while moving forward. Figure (3.27d) shows that its amplitude decreases and the pulse width increases. This is known as the *group dispersion*. The time delay of the envelope over the distance d is

$$\tau_d = \frac{d}{v_g} = d\frac{d\beta}{d\omega}. \tag{3.4.11}$$

In the present subsection, the propagation constant of the loaded line is obtained by solving the wave equation. However, the dispersion relation and also the characteristics impedance of the loaded line could be obtained from the circuit analysis. This simple method is applicable to several interesting cases of loaded lines forming the basis for the modern *Electromagnetic bandgap (EBG)* materials and metamaterials.

3.4.2 Circuit Models of Dispersive Transmission Lines

The above discussion demonstrates that the reactive loading of a line modifies the electrical characteristics of an unloaded host line. This section considers a few such modifications.

Shunt Inductor Loaded Line

The total series impedance and total shunt admittance of the shunt inductor loaded dispersive line, shown in Fig (3.27a), are given by

$$Z = j\,\omega L\,\Delta x \quad (a), \quad Y = j\left(\omega C - \frac{1}{\omega L_{sh}}\right)\Delta x \quad (b). \tag{3.4.12}$$

The series impedance and shunt admittance per unit length (*p.u.l.*) are

$$z = j\,\omega L \qquad (a), \quad y = j\left(\omega C - \frac{1}{\omega L_{sh}}\right) \qquad (b). \tag{3.4.13}$$

The complex propagation constant of the wave on the shunt inductor loaded line is

$$\gamma = \alpha + j\,\beta = \sqrt{zy} = \sqrt{-\omega L\left(\omega C - \frac{1}{\omega L_{sh}}\right)}$$
$$= j\frac{\omega}{v_0}\left[1 - \left(\frac{\omega_c}{\omega}\right)^2\right]^{\frac{1}{2}}. \tag{3.4.14}$$

The propagation constant β for $\omega > \omega_c$ is identical to equation (3.4.8). The attenuation constant α is obtained for $\omega < \omega_c$. The expressions for the phase velocity and group velocity follow from the expression of β as discussed previously. The characteristic impedance of the loaded dispersive transmission line $Z_{0d}(\omega)$ is given by

$$Z_{0d}(\omega) = \sqrt{\frac{z}{y}} = \left[\frac{\omega L}{\omega C - 1/\omega L_{sh}}\right]^{\frac{1}{2}}$$
$$= \left[\frac{L}{C\{1 - (\omega_c/\omega)^2\}}\right]^{\frac{1}{2}} = \frac{Z_0}{\sqrt{1 - (\omega_c/\omega)^2}}. \tag{3.4.15}$$

For the case, $\omega < \omega_c$, the characteristic impedance Z_{0d} is an imaginary quantity that stops the signal transmission through the line. The line behaves like a high-pass filter (HPF). For the case $\omega > \omega_c$, the characteristic impedance Z_{0d} is a real quantity that allows the signal propagation on the line. However, the characteristic impedance Z_{0d} in the pass-band is frequency-dependent.

Backward Wave Supporting Line

Figure (3.28a) shows the circuit model of the standard low-pass filter (LPF) type *LC transmission line* that can also be realized by cascading several lumped elements LC unit cells. However, the distance between two units cells should be a fraction of wavelength, i.e. number of LC unit cells should 10 or more per wavelength. Figure (3.28b) shows the dual of the LC line realized again by cascading several lumped elements CL unit cells. It is called the *CL transmission line.* It is a high-pass filter (HPF) type transmission line. Its propagation characteristic is obtained using the circuit analysis:

$$\gamma = j\beta = \pm\sqrt{zy} = \pm\left[\left(\frac{1}{j\,\omega C}\right)\left(\frac{1}{j\,\omega L}\right)\right]^{\frac{1}{2}}$$
$$\Rightarrow \beta = \mp\frac{1}{\omega\sqrt{LC}}. \tag{3.4.16}$$

Equation (3.4.16) gives the rectangular hyperbolic relationship for ω and β in the first and second quadrants of the $(\omega - \beta)$ diagram shown in Fig (3.28c). The phase velocity is

$$v_p = \frac{\omega}{\beta} = \mp\omega^2\sqrt{LC}. \tag{3.4.17}$$

The group velocity is obtained separately for both the negative and positive propagation constants:

$$\frac{d\beta}{d\omega} = \pm\frac{1}{\omega^2}\frac{1}{\sqrt{LC}} \quad \text{leading to}$$
$$v_g = \frac{1}{d\beta/d\omega} = +\omega^2\sqrt{LC} \quad \text{for } \beta - \text{tive} \quad (a)$$
$$v_g = -\omega^2\sqrt{LC} \qquad \text{for } \beta + \text{tive} \quad (b). \tag{3.4.18}$$

The phase velocity and group velocity are in opposite directions to each other, i.e. $v_g > 0$, $v_p < 0$ or $v_g < 0$, $v_p > 0$. This kind of wave is called the *backward wave.*

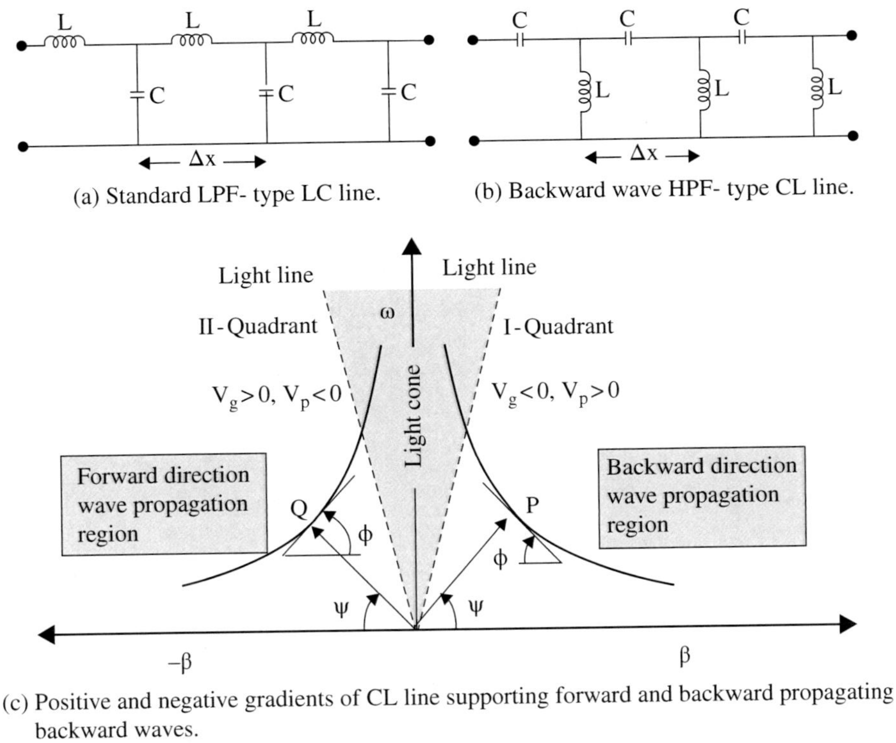

Figure 3.28 Lumped elements models of short transmission line sections and dispersion diagram of CL line.

If the power is moving from the source to load in the positive direction, the group velocity is in the positive direction from the source to the load, i.e. from the left side to the right side of Fig (3.28b). *The power always flows from the source to load. Therefore, the direction of the power flow decides the direction of the EM-wave propagation.* Normally, the direction of the group velocity is in the direction of the power flow. However, the phase velocity appears to be from load to the source. *It means at the end of a CL line section the leading phase shift is obtained. It is unlike the lagging phase shift for an LC line section.*

Figure (3.28c) shows the above results on the $(\omega - \beta)$ diagram. In both the quadrants, the slope ψ at the origin determines the phase velocity, and the local slopes ϕ at the points P and Q determine the group velocities. Both these slopes rapidly increase with an increase in frequency, showing the fast increase in both velocities as given by equations (3.4.17) and (3.4.18). *Thus, the lossless CL-line is a highly dispersive medium.* The slope at the origin and the local slope have opposite sense, i.e. one is clockwise and another is anti-clockwise, showing the antiparallel nature of the phase and group velocities. Such waves are the *backward waves*. The second quadrant with $v_g > 0$, $v_p < 0$ shows the positive direction of backward wave propagation, while the first quadrant with $v_g < 0$, $v_p > 0$ shows the negative direction of backward wave propagation, i.e. the reflected wave on the CL-line. The wave in the shaded light-cone region is fast-wave, whereas outside the light-cone it is the slow-wave.

Equations (3.4.16) and (3.4.17) show that the propagation constant β decreases with an increase in frequency, whereas the phase velocity increases with an increase in frequency. Therefore, the CL-line has an anomalous dispersion. It is a highly dispersive transmission line. In the case of the material medium, the backward wave with the *anomalous dispersion* is associated with the strong absorption region. It is discussed in the subsection (6.5.1) of chapter-6 using the Lorentz model. The standard LPF type transmission line is known as the *right hand, i.e. the RH-transmission line* and the HPF type transmission line supporting the backward wave is known as *the left hand, i.e. the LH-transmission line*. The RH-line corresponds to the *double-positive* (*DPS*), i.e. both μ, ε positive medium, whereas the LH-line corresponds to the *double negative* (*DNG*), i.e. both μ, ε negative medium. These media are discussed in section (5.5) of chapter 5. The composite *RH-LH (CRLH)-transmission line,* giving some unique transmission characteristics, models the so-called *metamaterials* with negative permittivity and negative

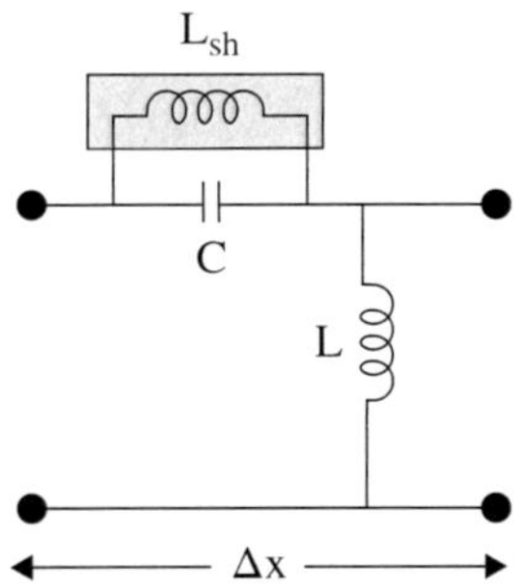

Figure 3.29 Inductor loaded CL-line.

permeability [B.19, J.8]. This section considers only one isolated LC unit. The cascading of several such units forms the backward wave supporting LH-transmission line known as the *metalines*. It is further discussed in chapter 22.

Series Connected Parallel L_{sh}-C Type Line

The backward wave supporting CL-line, discussed above, has no cut-off frequency. Figure (3.29) shows the modified CL- line by adding a shunt inductor L_{sh}, inside the gray box, across the series-connected capacitor C. It is an HPF type CL-line that supports the backward wave with a cut-off frequency. The propagation characteristics of this line are obtained from the series impedance and shunt admittance *p.u.l.*:

$$z = \frac{1}{-(j/\omega L_{sh}) + j\,\omega C} = \frac{\omega L_{sh}}{j(\omega^2 L_{sh} C - 1)}, \quad y = \frac{1}{j\,\omega L}. \tag{3.4.19}$$

The propagation constant of the line is

$$\gamma = j\beta = \pm\sqrt{zy} = \pm\left[\frac{\omega L_{sh}}{j(\omega^2 C L_{sh} - 1)}\frac{1}{j\omega L}\right]^{\frac{1}{2}}$$
$$= \mp j\left[\frac{L_{sh}}{L(\omega^2 C L_{sh} - 1)}\right]^{\frac{1}{2}}$$
$$\Rightarrow \beta = \mp\sqrt{\frac{L_{sh}}{L}}\frac{1}{\sqrt{(\omega/\omega_c)^2 - 1}}. \tag{3.4.20}$$

The cut-off frequency is $\omega_c = 1/\sqrt{CL_{sh}}$. The phase velocity of the usual LC-line is $v_0 = 1/\sqrt{LC}$. Therefore, the propagation constant of the shunt inductor L_{sh} in the series arm loaded CL-line is

$$\beta = \pm\sqrt{\frac{L_{sh}}{L}}\frac{\omega_c}{\omega}\frac{1}{\sqrt{1-(\omega_c/\omega)^2}}$$
$$= \pm\sqrt{\frac{L_{sh}}{L}}\frac{1}{\sqrt{L_{sh}C}}\frac{1}{\omega}\frac{1}{\sqrt{1-(\omega_c/\omega)^2}}$$
$$\Rightarrow \beta = \pm\frac{v_0}{\omega}\frac{1}{\sqrt{1-(\omega_c/\omega)^2}}. \tag{3.4.21}$$

The inductor loaded CL-line behaves like a high-pass filter. The wave propagates for $\omega > \omega_c$. For the frequency below cut-off, i.e. for $\omega < \omega_c$, the wave is in the evanescent mode. The $(\omega - \beta)$ diagram of the inductor loaded CL-line is similar to the $(\omega - \beta)$ diagram of Fig (3.28c). However, the cut-off frequency is not shown in Fig (3.28c). A reader can easily add the cut-off frequency ω_c in the dispersion diagram of Fig (3.28c). Unlike the unloaded CL, the present loaded CL line shows the cut-off frequency behavior. The present HPF type loaded CL line also supports the dispersive backward wave with phase velocity and group velocity opposite to each other. The propagation constant β decreases with frequency, whereas the phase velocity increases with frequency. It shows that the loaded CL-line has anomalous dispersion. The phase and group velocities of the backward wave are

$$v_p = \frac{\omega}{\beta} = \pm\frac{\omega^2}{v_0}\sqrt{1-\left(\frac{\omega_c}{\omega}\right)^2} = \pm v_p^{CL}\sqrt{1-\left(\frac{\omega_c}{\omega}\right)^2} \quad \text{(a)}$$
$$v_g = \frac{1}{(d\beta/d\omega)} = \mp\frac{\omega^2}{v_0}\left[1-\left(\frac{\omega_c}{\omega}\right)^2\right]^{\frac{3}{2}} = \mp v_p^{CL}\left[1-\left(\frac{\omega_c}{\omega}\right)^2\right]^{\frac{3}{2}} \quad \text{(b)}. \tag{3.4.22}$$

In summary, the inductor loaded CL-line shown in Fig (3.29) supports the *backward wave*. Above the cut-off frequency, i.e. for $\omega > \omega_c$, and for the limiting case $\omega \to \infty$, equations (3.4.22a) and (3.4.22b) reduce to equations (3.4.17) and (3.4.18), respectively. The phase velocity of the unloaded CL-line is $v_p^{CL} = \omega^2\sqrt{LC} = \omega^2/v_0$. For $\omega_C = 0$ also, the above equations are reduced to equations (3.4.17) and (3.1.18). Therefore, above the cut-off frequency, i.e. for $\omega > \omega_c$, the inductive loading of the transmission line shown in Fig (3.29) supports the backward *slow-wave* outside the light cone and the backward *fast-wave* within the light cone. Its dual structure, shown in Fig (3.27a), always supports the forward fast-wave.

Series Capacitor Loaded LC-Line

The normal LC-line can also be loaded with a series capacitor C_s in the series arm. Figure (3.30) shows the

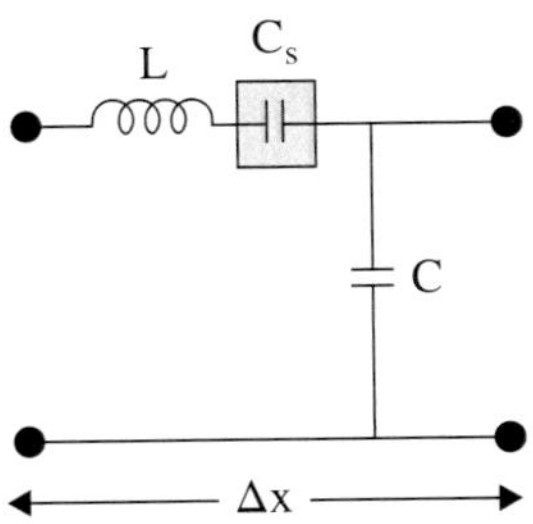

Figure 3.30 Series capacitor loaded LC-line.

series capacitor loaded LC line. The propagation parameters of the loaded line are computed using the circuit analysis. The series arm impedance and the shunt arm admittance *p.u.l.* are given below:

$$z = j\left(\omega L - \frac{1}{\omega C_s}\right) \quad (a), \quad y = j\,\omega C \qquad (b). \tag{3.4.23}$$

The propagation constant of the capacitor loaded LC-line is

$$\gamma = j\beta = \pm\sqrt{zy} = \pm\left[j\omega C \times j\omega L\left(1 - \frac{1}{\omega^2 LC_s}\right)\right]^{\frac{1}{2}}$$
$$= \pm j\omega\sqrt{LC}\left[1 - \left(\frac{\omega_c}{\omega}\right)^2\right]^{\frac{1}{2}} \Rightarrow \beta = \pm\frac{\omega}{v_0}\left[1 - \left(\frac{\omega_c}{\omega}\right)^2\right]^{\frac{1}{2}}, \tag{3.4.24}$$

where the phase velocity of the unloaded LC-line is $v_0 = 1/\sqrt{LC}$. The cut-off frequency of the loaded line is $\omega_c = 1/\sqrt{LC_s}$. For $\omega > \omega_c$, the line behaves as the HPF. In absence of the series capacitance C_s loading, the line behaves like the LPF. The series-arm impedance is capacitive at a frequency below cut-off, i.e. for $\omega < \omega_c$. The circuit of Fig (3.30) is reduced to the C-C line, i.e. a line with capacitive elements in both the series and shunt arms. It corresponds to the *mu-negative (MNG) medium* discussed in the section (5.5) of chapter 5. The C-C line blocks the low-frequency signal. Therefore, for the frequency $\omega < \omega_c$, the propagation is in the evanescent mode with high attenuation. The phase and group velocities of the propagating waves are obtained as

$$v_p = v_0\left[1 - \left(\frac{\omega_c}{\omega}\right)^2\right]^{-\frac{1}{2}} \quad (a), \quad v_g = v_0\left[1 - \left(\frac{\omega_c}{\omega}\right)^2\right]^{\frac{1}{2}} \quad (b). \tag{3.4.25}$$

Equations (3.4.24) and (3.4.25) for β, v_p, and v_g are identical to equations (3.4.10a) and (3.4.10b) for the transmission line shown in Fig (3.27a). Thus, Fig (3.27b and c) also show the behavior of the present series capacitor loaded LC-line. This line supports the forward wave and it is a *normal dispersive* transmission line. It supports the *fast-wave* above the cut-off frequency. This line is the dual structure of the line shown in Fig (3.29) that supports the backward wave propagation.

Additional numbers of configurations for the loaded transmission lines could be obtained. For instance, the L-C section of a line, supporting the forward wave, could be cascaded with the C-L section of a line, supporting the backward wave. The composite line forms an interesting kind of the transmission line structure [B.19, J.8]. Both the series and parallel reactive loading of the lines can be done. Such loaded line structures have been realized in the planar technology to obtain novel properties useful for the development of novel microwave devices. They form the so-called *metamaterials.* The concept of metamaterials has been introduced in chapter 5 and elaborated in chapter 21. Chapter 22 considers the planar *1D-metalines* and *2D-planar metasurfaces,* and chapter 19 discusses the planar *periodic transmission lines.*

References

Books

B.1 Pozar, D.M.: *Microwave Engineering*, 2nd Edition, John Wiley & Sons, Singapore, 1999.

B.2 Fache, N.; Olyslager, F.; De Zutter, D.: *Electromagnetic and Circuit Modeling of Multiconductor Transmission Lines*, Clarendon Press, Oxford, NY, 1993.

B.3 Rizzi, P.A.: *Microwave Engineering-Passive Circuits*, Prentice-Hall International Edition, Englewood Cliff, NJ, 1988.

B.4 Ramo Simon, W.J.R.: *Van Duzer Theodore, Fields, and Waves in Communication Electronics*, 3rd Edition, John Wiley & Sons, Singapore, 1994.

B.5 Collin, R.E.: *Foundations for Microwave Engineering*, 2nd Edition, McGraw-Hill, Inc., New York, 1992

B.6 Carson, R.S.: *High-Frequency Amplifiers*, 2nd Edition, John Wiley & Sons, New York, 1982.

B.7 Elliott, R.S.: *An Introduction to Guided-Waves and Microwave Circuits*, Prentice-Hall, Englewood Cliff, NJ, 1993.

B.8 Gardial, F.E.: *Lossy Transmission Lines*, Artech House, Boston, MA, 1987.

B.9 Freeman, J.C.: *Fundamentals of Microwave Transmission Lines*, John Wiley, New York., 1996.

B.10 Swanson, D.G.; Hoefer, W.J.R.: *Microwave Circuit Modeling Using Electromagnetic Field Simulation*, Artech House, Boston, MA, 2003.

B.11 Weber, R.J.: *Introduction to Microwave Circuits, Radio Frequency and Design Applications*, IEEE Press, New York, 2001.

B.12 Collin, R.E: *Field Theory of Guided Waves*, IEEE Press, New York, 1991.

B.13 Orfanidis, S.J.: *Electromagnetic Waves and Antenna*, Free Book on Web.

B.14 Staelin, D.H.; Morgenthaler, A.W.; Kong, J.A.: *Electromagnetic Waves*, Prentice-Hall, Englewood Cliff, NJ, 1994.

B.15 Sadiku, M.N.O.: *Elements of Electromagnetics*, 3rd Edition, Oxford University Press, New York, 2001.

B.16 Cheng, D.K.: *Fields and Wave Electromagnetics*, 2nd Edition, Pearson Education, Singapore, 1089.

B.17 Balanis, C.A.: *Advanced Engineering Electromagnetics*, John Wiley & Sons, New York, 1989.

B.18 Mattick, R.E.: *Transmission Lines for Digital and Communication Networks*. IEEE Press, New York, 1995.

B.19 Engheta, N.; Ziolkowski, R.W.: *Metamaterials: Physics and Engineering Explorations*, John Wiley & Sons, Inc., New York, 2006.

B.20 Remoissenet, M.: *Waves Called Solitons: Concepts and Experiments*, Springer, New York, 1996.

Journals

J.1 Kurokawa, K.: *Power waves and the scattering matrix*, *IEEE Trans. Microwave Theory Tech.* Vol. 13. No. 2, pp. 607–610, 1965.

J.2 Lei, Z.; Wu, K.: *Short-open calibration technique for field theory-based parameter extraction of lumped elements of planar integrated circuits*, *IEEE Trans. Microwave Theory Tech.*, Vol. 50, No. 8, pp. 1861–1869, Aug. 2002.

J.3 Hasegawa, H.; Furukawa, M.; Yanai, H.: *Properties of microstrip line on Si-SiO2 system*, *IEEE Trans. Microwave Theory Tech.*, Vol. MTT-19, pp. 869–881, 1971.

J.4 Jager, D.; Rabus, W.: *Bias-dependent phase delay of Schottky contact microstrip line*, *Electron. Lett.*, Vol. 9, pp. 201–202, 1973.

J.5 Veghte, R.L.; Balanis, C.A.: *Dispersion of transient signal in microstrip transmission lines*, *IEEE Trans. Microwave Theory Tech.* Vol. 34, pp. 1427–1436, Dec. 1986.

J.6 Verma, A.K.; Kumar, R.: *Distortion in gaussian pulse on microstrip-like transmission lines*, *Microw. Opt. Technol. Lett.*, Vol. 17, No. 4, pp. 253–255, March 1998.

J.7 Hua, C.; Dogariu, A.; Wang, L.J.: *Negative group delay and pulse compression in superluminal pulse propagation*, *IEEE J. Sel. Top. Quantum Electron.*, Vol. 9, No. 1, pp. 52–58, Jan–Feb 2003.

J.8 Lai, A.; Caloz, C.; Itoh, T.: *Transmission line based metamaterials and their microwave applications*, *Microwave Mag.*, Vol. 5, No. 3, pp. 34–50, Sept. 2004.

4

Waves in Material Medium – I

(Waves in Isotropic and Anisotropic Media, Polarization of Waves)

Introduction

The characteristics of EM-wave propagating on a planar line are strongly dependent on the nature of the materials used in planar technology. The familiarity with the characteristics of the medium and EM-wave propagation in the unbounded medium is important to understand the working of the planar transmission lines. These topics are extensively covered in several books [B.1–B.15].

Broadly speaking, the present chapter covers basic electrical characteristics of the material media and the EM-waves propagation in the unbounded dielectric media – both isotropic and anisotropic. In the first part of the present chapter, and also in chapter 6, attention is paid to the physical processes and the circuit models to understand the electrical properties of the material medium. The electrical and magnetic properties of the materials appear as the *responses* to the electric and magnetic *excitations*. Such excitations could be in the form of the *circuit sources,* such as the voltage and a current source. It could also be in the form of the *field sources*, such as the electric field intensity (E) and magnetic field intensity (H). The excitation could be any of three forms, namely (i) time-independent, i.e. the static or DC type; (ii) frequency-dependent, i.e. the time-harmonic dependent, or AC (phasor) type; and (iii) arbitrary time-dependent, i.e. the transient type. The discussion is limited to the static and time-harmonic type of responses of the materials, i.e. the material response and behavior in the frequency-domain.

Objectives

- *To review the EM-field quantities and medium parameters.*
- *To review the basic electrical properties of media.*
- *To obtain elementary circuit models of media.*
- *To review Maxwell's equations.*
- *To present the wave equation in the unbounded lossless and lossy isotropic dielectric medium.*
- *To review wave polarizations.*
- *Jones matrix description of polarization states.*
- *To present the wave equation in the unbounded lossless anisotropic dielectric medium.*

4.1 Basic Electrical Quantities and Parameters

The electrical charge and the electric current are the primary electrical sources for the creation of the electric field and the magnetic field, respectively. The charge, also current (*displacement current*), is described by the *flux field,* i.e. the *flux density* ($\vec{D}$, $\vec{B}$). Two electrically charged bodies or two current-carrying conductors interact through the *force fields*, i.e. the *field intensity* ($\vec{E}$, $\vec{H}$). The static charge creates the static electric field around itself, whereas the electric current creates the magnetic field around itself. The magnetic charge does not exist in nature. Sometimes, we talk about the magnetic charge, only as a mathematical source for the magnetic field. It is a hypothetical creation to maintain the symmetry of the field equations. The charge and current, i.e. *flux fields* are not determined by a medium, whereas the electric and magnetic interactions, i.e. *force fields*, between two separate bodies in a medium, are influenced by the electromagnetic parameters of the medium.

4.1.1 Flux Field and Force Field

The electric and magnetic fields are visualized through *the line of flux*. Thus, the electric charge (Q_e) is described by the electric flux (Ψ_e). The total charge (Q_T) on a physical body and the corresponding electric flux are related by *Gauss's law*:

Introduction to Modern Planar Transmission Lines: Physical, Analytical, and Circuit Models Approach, First Edition. Anand K. Verma.

$$Q_T = \Psi_e. \tag{4.1.1}$$

If the charge is distributed throughout the volume of a body, in the form of volume charge density ρ_e, the elemental charge in the volume element dv is $\rho_e$dv. The flux is further expressed as the *electric flux density* (D), i.e. the flux per unit area. The flux through the elemental surface area is $d\Psi_e = \vec{D}.d\vec{s}$. It is shown in Fig. (4.1). The flux density is a vector quantity. It is also called the *electric displacement vector*.

Gauss's Law for Electric Flux

Total electric flux coming out of a closed surface = Total charge enclosed inside the volume of a closed surface, i.e.

$$\oiint_{sur} \vec{D}.d\vec{s} = \iiint_{vol} \rho_e dv. \tag{4.1.2}$$

The above expression is the *integral form* of *Gauss's law*. It can be converted to the *differential form* by using Gauss's vector integral identity,

$$\iiint_{vol} \vec{\nabla}.\vec{D}\, dv = \iiint_{vol} \rho_e\, dv, \quad \Rightarrow \quad \vec{\nabla}.\vec{D} = \rho_e. \tag{4.1.3}$$

Gauss's Law for Magnetic Flux

Similar to the electric charge distribution, the magnetic charge distribution can be assumed in a volume of the body. The *magnetic charge density* is expressed as ρ_m. The magnetic charge creates a magnetic flux Ψ_m. Similar to the case of the electric charge, the elemental magnetic charge in the volume dv is $\rho_m$dv, and the elemental magnetic flux coming out of the surface is $d\Psi_m = \vec{B}.d\vec{s}$, where $\vec{B}$ is the magnetic flux density as ds is the elemental surface of the enclosed volume v. It is also called the *magnetic displacement vector*. The Gauss's law for the magnetic charge and magnetic flux can be written in the integral form as follows:

$$\oiint_{sur} \vec{B}.d\vec{s} = \iiint_{vol} \rho_m dv. \tag{4.1.4}$$

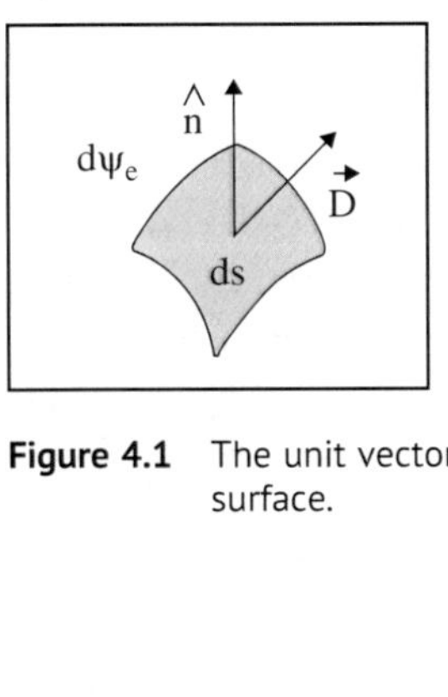

Figure 4.1 The unit vector n̂ is in the direction normal to the surface.

Again, by using Gauss's vector integral identity, the above expression is written below in the differential form:

$$\vec{\nabla}.\vec{B} = \rho_m. \tag{4.1.5}$$

However, the magnetic charges are not found in nature, i.e. $\rho_m = 0$. Therefore,

$$\vec{\nabla}.\vec{B} = 0. \tag{4.1.6}$$

The amount of charge, or current, is an absolute quantity. It does not dependent on the material medium. Thus, the corresponding flux or the flux density is also not dependent on the surrounding medium. *In brief, the charge and current create the electric and magnetic flux field, i.e. the flux densities* $(\vec{D}, \vec{B})$*; and these are not influenced by a material medium.*

Experiments demonstrate that the electrically charged body, or a current-carrying conductor, interacts with other charged body, or another current-carrying conductor. Such *interaction*, i.e. the *mutual force*, is influenced by the medium surrounding these bodies. Therefore, medium-independent flux densities $(\vec{D}, \vec{B})$ cannot explain the interaction between two charged bodies or current-carrying conductors. The interactions between the charges and current-carrying conductors take place through the *force fields*, expressed by *the electric field intensity* $\vec{E}$ and *magnetic field intensity* $\vec{H}$. The field intensities $(\vec{E}, \vec{H})$ are also responsible for the electromagnetic (EM)-power transportation through a medium. However, the field intensities are influenced by the electrical and magnetic properties of a medium.

4.1.2 Constitutive Relations

It is noted above that the charge and current create the electric and magnetic *flux fields*, described by the flux densities $(\vec{D}, \vec{B})$. They also create the *force field* described by the electric and magnetic field intensities $(\vec{E}, \vec{H})$. The *flux density* parameters $(\vec{D}, \vec{B})$ are related to the *force field intensity* parameters $(\vec{E}, \vec{H})$ by the following *constitutive relations*:

$$\begin{aligned} \vec{D} &= \varepsilon_0\varepsilon_r\vec{E} = \varepsilon\vec{E} && \text{(a)} \\ \vec{B} &= \mu_0\mu_r\vec{H} = \mu\vec{H} && \text{(b)}, \end{aligned} \tag{4.1.7}$$

where ε_0 and μ_0 are the permittivity and permeability of the free space. The permittivity of any medium is its ability to store electric energy, such as a capacitor.

Therefore, it is identified as the capacitance of the free space. Any dielectric material medium can store more electric energy through the mechanism of *electric polarization*. The electric dipole is created during the process of polarization and the total induced charge is shown as electric flux density $\vec{D}$. Chapter 6 presents a detailed discussion of material polarization. The electric polarization of material under the influence of an external electric field gives a higher value of permittivity as compared to the permittivity of the free space. This is known as the *relative permittivity* ε_r of a medium. It is also known as the *dielectric constant* of a medium. The permittivity of a medium is $\varepsilon = \varepsilon_0\varepsilon_r$. For an isotropic dielectric medium, ε_r is a *scalar* quantity. However, for an anisotropic medium, it is a *tensor* quantity.

Likewise, the free space has also an ability to store magnetic energy. It is expressed as its *permeability*. Magnetic material is magnetized by the process of *magnetization* under the influence of an external magnetic field. Thus, magnetic material stores more magnetic energy as compared to the free space. The ability of a magnetic material to store magnetic energy is expressed through its *relative permeability*. The permeability of the medium is $\mu = \mu_0\mu_r$. The permittivity ε_0 and permeability μ_0 of the free space are the *primary physical constants* $\varepsilon_0 = 8.854 \times 10^{-12}$ F/m, $\mu_0 = 4\pi \times 10^{-7}$ H/m. Again, for the isotropic magnetic medium, the relative permeability μ_r is a *scalar* quantity, and for an anisotropic medium, it is a *tensor* quantity.

It is interesting to note that the velocity of EM-wave and the characteristic (*intrinsic*) impedance of the free space are given in terms of these primary constants,

$$c = \frac{1}{\sqrt{\mu_0\varepsilon_0}} \approx 3 \times 10^8 \text{ m/sec}, \quad \eta = \sqrt{\frac{\mu_0}{\varepsilon_0}} = 377\ \Omega. \tag{4.1.8}$$

A material medium with a finite conductivity *dissipates* energy in the form of heat. The finite conductivity of a medium is due to the presence of *free charge carriers*. The free space is considered as a lossless medium because it has no free charge carrier. The *conduction current* flows through a medium under the influence of an external electric field. The conduction current density ($\vec{J}_c$) in a medium is related to the electric field intensity ($\vec{E}$) by *Ohm's law*:

$$\vec{J}_c = \sigma\vec{E}, \tag{4.1.9}$$

where σ is the conductivity of a conducting medium. The conductivity (σ) of the isotropic conducting medium is a scalar and for an anisotropic conducting medium it is a tensor. As free space has no conductivity, there is nothing like the relative conductivity of a medium. However, sometimes the conductivity of a medium is expressed in terms of the conductivity of copper.

In summary, the electrical properties of a material are described by the relative *permittivity* (ε_r), *relative permeability* (μ_r), and *conductivity* (σ). A material can have all three properties at a time, or it can have one predominant property at a time. Assuming the case of one predominant property at a time, all materials are classified into three basic categories.

4.1.3 Category of Materials

Dielectric Materials

The dielectric materials support electric polarization of bound charges and electric displacement current through it. At the *micro-level,* the electric polarization creates dipole moments that appear as the permittivity of the dielectric material at the *macro-level*. The permittivity is frequency-dependent and lossy. So, permittivity is a complex quantity showing the lossy nature of dielectric materials. This is discussed in chapter 6. The relative permittivity of natural dielectric material is always positive and more than unity. However, engineered *artificial dielectrics* can have relative permittivity $0 < \varepsilon_r < 1$. Such materials are known as the *epsilon near zero* (ENZ) materials. Under certain conditions, it can also acquire negative permittivity, creating the *epsilon negative* (ENG) medium [B.16]. It is discussed in section (5.5) of chapter 5.

The *electric displacement current* flowing through a dielectric medium does not involve the flow of current through free charges. It is associated with time-dependent electric fields. The electric *displacement current density* is a vector quantity. It expressed as follows:

$$\vec{J}_d = \frac{\partial\vec{D}}{\partial t} = \varepsilon\frac{\partial\vec{E}}{\partial t}. \tag{4.1.10}$$

Magnetic Materials

The magnetic materials support *magnetization,* i.e. magnetic polarization by the magnetic field. It also supports *magnetic displacement current* due to the time-dependent magnetic field:

$$\vec{M}_d = \frac{\partial\vec{B}}{\partial t} = \mu\frac{\partial\vec{H}}{\partial t}. \tag{4.1.11}$$

A *diamagnetic material* has permeability $\mu \leq \mu_0$ and a *paramagnetic material* has permeability $\mu > \mu_0$. The diamagnetic behavior is caused by induced magnetic dipole moments, creating a magnetic field that opposes the externally applied magnetic field. However, the paramagnetic property comes into existence due to the alignment of already randomly existing magnetic dipole moments in the material. The conducting ferromagnetic materials have much larger permeability due to domain formation. However, the *engineered artificial magnetic materials* can have relative permeability less than unity, $0 < \mu_r < 1$. These are known as the *mu near-zero (MNZ) materials* [B.16]. Again, the engineered materials can also have mu negative, giving the *mu negative (MNG) materials*. It is discussed in section (5.5) of chapter 5.

Conductors

It supports the flow of the free charge carrier, i.e. electrons. The free charge carriers are responsible for the conduction current density. It is given by equation (4.1.9). The free electrons confined within a neutral conductor form the *plasma medium*. The artificial dielectrics, including the *wire-medium* (i.e. rodded medium), are also modeled as the plasma medium. The ionosphere also supports the plasma medium [B.2]. The plasma medium is modeled using the Drude model, discussed in subsection (6.5.2) of chapter 6. The plasma medium provides a means to engineer materials with negative permittivity. These materials are called the epsilon negative materials, i.e. the ENG materials.

4.2 Electrical Property of Medium

Any material can be electrically characterized by the relative permittivity (ε_r), relative permeability (μ_r) and conductivity (σ), or resistivity (ρ). However, these electrical parameters are not *constants* for any given material. For instance, these are both temperature and frequency-dependent. Also, these may not be uniform throughout the volume of a material. Further, the characterizing parameters may depend on the field intensity, the direction of the applied field, operating frequency, working temperature, and pressure. The parameters can also depend on the history of a medium. However, the static value of these parameters, at room temperature, is treated as constant. A special kind of material, called *chiral* material, requires another parameter called *chirality*, i.e. the handedness of materials for its characterization [B.13, B.17]. Several properties of the medium are described briefly in this section.

4.2.1 Linear and Nonlinear Medium

The relative permittivity (ε_r) and the relative permeability (μ_r) of a *linear* material do not depend on the magnitude of the electric or magnetic field intensity, respectively. However, for *nonlinear* materials, the relative permittivity and relative permeability are functions of the electric and magnetic field intensity, respectively, and expressed as $\varepsilon_r(E)$ and $\mu_r(H)$. Thus, for an isotropic nonlinear medium, equation (4.1.7a) is written as

$$\vec{D} = \varepsilon_0 \varepsilon_r(E)\,\vec{E}, \tag{4.2.1}$$

where $\varepsilon_r(E) = \varepsilon_{r1} + \varepsilon_{r2}E + \varepsilon_{r3}E^2 + \cdots$ and so forth. The coefficients ε_{r2}, ε_{r3}, ... indicate the order of nonlinearity in the nonlinear relative permittivity. In the case of a time-harmonic electric field, i.e. $E = E_0 e^{j\omega t}$, equation (4.2.1) is written as,

$$\vec{D} = \varepsilon_0\,\varepsilon_{r1} E_0 e^{j\omega t} + \varepsilon_0 \varepsilon_{r2} E_0^2 e^{j2\omega t} + \varepsilon_0 \varepsilon_{r3} E_0^3 e^{j3\omega t} + \cdots. \tag{4.2.2}$$

It is obvious that while the input signal has only one frequency ω, shown in Fig. (4.2), the output of a nonlinear medium has several harmonically related frequency components, ω, 2ω, 3ω, ... and so forth. Thus, a sinusoidal input signal gets distorted, once it passes through a nonlinear medium. Such distortion also occurs in an amplifier in the nonlinear region.

Similarly, the relative permeability of a nonlinear magnetic medium is a function of the amplitude of the magnetic field. The constitutive relation, given by equation (4.1.7b), is written as

$$\vec{B} = \mu_0\,\mu_r\,(H)\,\vec{H}. \tag{4.2.3}$$

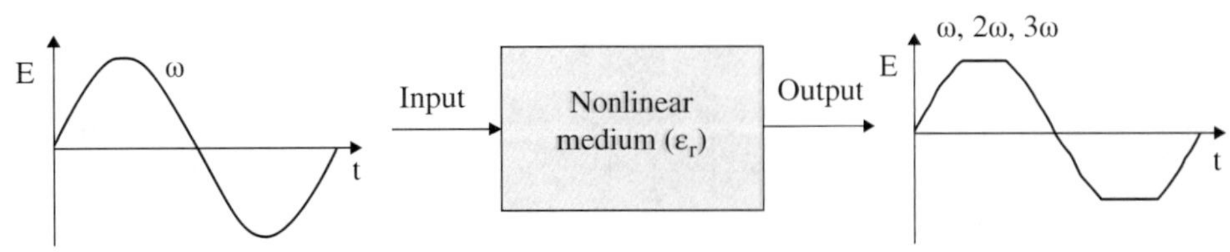

Figure 4.2 Response of nonlinear medium showing generation of harmonics.

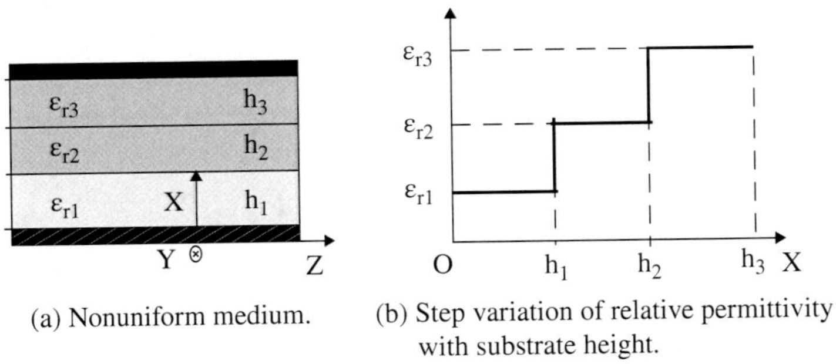

Figure 4.3 Inhomogeneous medium showing a step variation of relative permittivity with substrate height.

where $\mu_r(H) = \mu_{r1} + \mu_{r2}H + \mu_{r3}H^2 + \cdots$ and so forth. The coefficients μ_{r2}, μ_{r3}, ... indicate the order of nonlinearity in the nonlinear relative permeability of a magnetic medium.

4.2.2 Homogeneous and Nonhomogeneous Medium

The relative permittivity (ε_r) and the permeability (μ_r) are not necessarily uniform throughout the volume of a medium. These parameters could also be position-dependent. The variation in ε_r and μ_r could be in discrete steps, or they could be continuous functions of the position. Likewise, the conductivity of a medium can also be a function of position. If the parameters ε_r, μ_r, and σ are uniform throughout the medium, the medium is *called homogeneous*; otherwise, it is a *nonhomogeneous* medium or an *inhomogeneous* medium. A multilayer dielectric medium, forming a parallel capacitor, as shown in Fig. (4.3), is a nonhomogeneous medium, where the relative permittivity $\varepsilon_r(x)$ is a function of position x in discrete steps. The conductivity of a doped Si substrate is a function of the depth of penetration of the charged carrier, forming a continuously variable nonhomogeneous medium.

4.2.3 Isotropic and Anisotropic Medium

Inside the *isotropic dielectric medium*, the electric displacement vector $\vec{D}$ and the electric field intensity $\vec{E}$ are parallel to each other, i.e. the applied electric field views the same relative permittivity of a medium in all directions. Likewise, the magnetic displacement vector $\vec{B}$ is parallel to the magnetic field intensity $\vec{H}$ within the isotropic medium. These properties are expressed through constitutive relations (4.1.7a) and (4.1.7b). For the isotropic media, permittivity and permeability are scalar quantities.

However, there are dielectrics, such as quartz, sapphire, alumina, MgO, and so forth, where $\vec{D}$ and $\vec{E}$ are not parallel to each other, i.e. they are not in the same direction. Such dielectrics form the *anisotropic* medium. In such a medium, the relative permittivity viewed by the applied electric field is direction-dependent. For instance, Fig. (4.3a) forms a composite anisotropic medium as the effective permittivity along the x-axis is different from the effective permittivity along the z-axis. Similarly, magnetic materials such as ferrite, garnet, and so forth are also anisotropic because $\vec{B}$ and $\vec{H}$ vectors are not in the same direction. Several authors have treated the properties of anisotropic medium and EM-wave propagation through such media in detail [B.1–B.4, B.9, B.11, B.13–B.15, B.17–B.23]. This subsection reviews basic concepts related to anisotropic media.

The relative permittivity and relative permeability of these anisotropic media are not scalar quantities. They are tensor quantity, $\bar{\bar{\varepsilon}}_r, \bar{\bar{\mu}}_r$ described by 3×3 matrices. The constitutive relations of such electric and magnetic media are written as follows:

$$\vec{D} = \varepsilon_0 \bar{\bar{\varepsilon}}_r . \vec{E}, \quad \Rightarrow \begin{bmatrix} D_x \\ D_y \\ D_z \end{bmatrix} = \varepsilon_0 \begin{bmatrix} \varepsilon_{r,xx} & \varepsilon_{r,xy} & \varepsilon_{r,xz} \\ \varepsilon_{r,yx} & \varepsilon_{r,yy} & \varepsilon_{r,yz} \\ \varepsilon_{r,zx} & \varepsilon_{r,zy} & \varepsilon_{r,zz} \end{bmatrix} \begin{bmatrix} E_x \\ E_y \\ E_z \end{bmatrix} \quad \text{(a)}$$

$$\vec{B} = \mu_0 \bar{\bar{\mu}}_r . \vec{H}, \quad \Rightarrow \begin{bmatrix} B_x \\ B_y \\ B_z \end{bmatrix} = \mu_0 \begin{bmatrix} \mu_{r,xx} & \mu_{r,xy} & \mu_{r,xz} \\ \mu_{r,yx} & \mu_{r,yy} & \mu_{r,yz} \\ \mu_{r,zx} & \mu_{r,zy} & \mu_{r,zz} \end{bmatrix} \begin{bmatrix} H_x \\ H_y \\ H_z \end{bmatrix} \quad \text{(b).}$$

(4.2.4)

In general, elements of permittivity and permeability matrices could be complex quantities and also frequency-dependent, accounting for the losses and dispersion in the material medium. Equation (4.2.4a) shows that for the anisotropic dielectric medium, the electric flux density $\vec{D}$ is not parallel to electric field intensity $\vec{E}$. Likewise, equation (4.2.4b) shows that the vector $\vec{B}$ is not parallel to the vector $\vec{H}$. For instance, if the incident field on an anisotropic dielectric medium has only E_x component, i.e. x-polarized incident E-field, it generates all three components of electric flux density – D_x, D_y, and D_z. The same applies to the anisotropic magnetic medium.

The above equations can be written in a more compact form as

$$[\vec{D}] = \varepsilon_0 [\varepsilon_r].[\vec{E}] \quad \text{(a)}$$
$$[\vec{B}] = \mu_0 [\mu_r].[\vec{H}] \quad \text{(b).} \qquad (4.2.5)$$

The above permittivity and permeability matrices could be either symmetric or anti-symmetric. Thus, the anisotropic materials could be divided into two broad groups: (i) *symmetric anisotropic materials* and (ii) *anti-symmetric anisotropic materials*. The symmetric anisotropic materials support linearly polarized EM-waves propagating as the *normal modes* of the homogeneous unbounded medium. However, circularly polarized EM-waves are the normal modes of the anti-symmetric anisotropic medium. The normal modes of media travel without any change in polarization.

Symmetric Anisotropic Materials

The complex permittivity matrix $[\varepsilon_r^*]$, showing the permittivity tensor, of symmetric anisotropic dielectric material is a Hermitian symmetric matrix, i.e. the following relation holds:

$$[\varepsilon_r^*] = [\bar{\varepsilon}_r^*]^T. \qquad (4.2.6)$$

In the above equation, the matrix elements $\bar{\varepsilon}_r^*$ are the complex conjugate of the matrix elements ε_r^*. The superscript T shows the transpose of the permittivity matrix.

The above relation also holds for a real permittivity tensor of symmetric anisotropic dielectric material. In a dielectric material case, the off-diagonal elements of the matrix are symmetrical, i.e. $\varepsilon_{r,\,xy} = \varepsilon_{r,\,yx}$, and so forth. Similar expression can also be obtained for the symmetric anisotropic magnetic material. However, some media do not follow this symmetry rule.

To describe an anisotropic medium, two sets of the coordinate systems are used: one set is for the crystal structure axes of the anisotropic medium; and another set is used for the physical axes of the line structure. Figure (4.4) shows the *crystal axes* (ξ, η, ς), i.e. (*xi, eta, zeta*), and the *physical axes* (x, y, z) of a microstrip line on an anisotropic substrate.

The crystal axes (ξ, η, ς) are rotated with respect to the physical axes (x, y, z) by the angles θ_1, θ_2, and θ_3. The off-diagonal elements of $[\varepsilon_r]$ in equation (4.2.4) are present due to the nonalignment of two coordinate systems. However, if they are aligned, i.e. $\theta_1 = \theta_2 = \theta_3 = 0$, then the off-diagonal elements of $[\varepsilon_r]$ are zero; and the constitutive relation (4.2.4a) reduces to

$$\begin{bmatrix} D_x \\ D_y \\ D_z \end{bmatrix} = \varepsilon_0 \begin{bmatrix} \varepsilon_{r\xi} & 0 & 0 \\ 0 & \varepsilon_{r\eta} & 0 \\ 0 & 0 & \varepsilon_{r\zeta} \end{bmatrix} \begin{bmatrix} E_x \\ E_y \\ E_z \end{bmatrix}. \qquad (4.2.7)$$

The crystal axes (ξ, η, ς) are also known as the principal coordinate system of a material medium; and the diagonal relative permittivity components $\varepsilon_{r\xi}$, $\varepsilon_{r\eta}$ and $\varepsilon_{r\zeta}$ are known as the *principal relative permittivity components* [B.1, B.3, B.10, B.12–B.14, B.24]. Normally, all relative permittivity components are positive quantities. However, it is possible to get one component as a

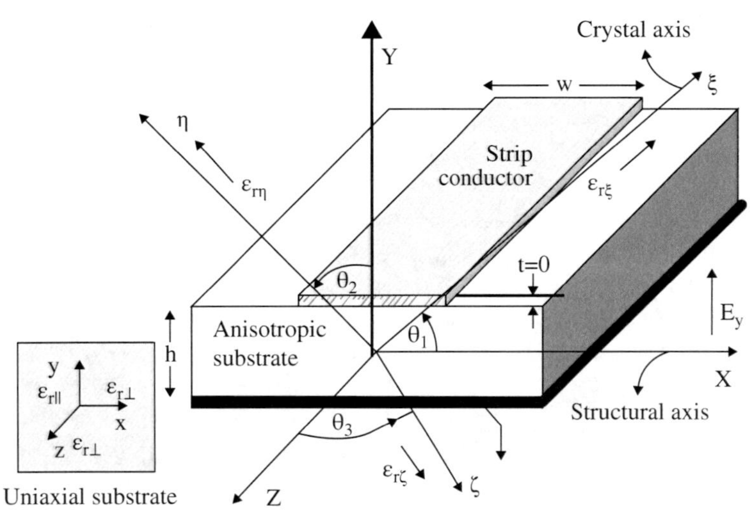

Figure 4.4 The crystal axes (ξ, η, ς) and the physical axes (x, y, z) of a planar anisotropic sheet.

negative quantity in the *engineered composites* that provide unique EM-wave characteristics [J.1, J.2, B.16]. Along the principal axes, the components of the vector $\vec{D}$ are parallel to vector $\vec{E}$.

The dielectric materials are further classified into three categories:

Type I: Isotropic materials. The relative permittivity components of these materials are identical, i.e. $\varepsilon_{r\zeta} = \varepsilon_{r\eta} = \varepsilon_{r\zeta}$. Thus, the relative permittivity of isotropic material is a scalar quantity.

Type II: Uniaxial materials. These are anisotropic materials with relative permittivity components $\varepsilon_{r\zeta} = \varepsilon_{r\xi} = \varepsilon_{r\perp}$, and $\varepsilon_{r\eta} = \varepsilon_{r||}$. It is shown within a box in Fig. (4.4). In the case of alignment of crystal axes along the physical axes, permittivity components are expressed as $\varepsilon_{xx} = \varepsilon_{zz} = \varepsilon_{r\perp}$, $\varepsilon_{yy} = \varepsilon_{r||}$. Thus, the permittivity tensor of the uniaxial anisotropic substrate is expressed as follows

$$\bar{\bar{\varepsilon}}_r = \varepsilon_0 \begin{bmatrix} \varepsilon_{r,xx} & \varepsilon_{r,xy} \\ \varepsilon_{r,yx} & \varepsilon_{r,yy} \end{bmatrix}. \tag{4.2.8}$$

For a uniaxial substrate, shown in Fig. (4.4), the applied external electric field E_y faces the relative permittivity component $\varepsilon_{r||}$. The permittivity component $\varepsilon_{r||}$ is parallel (||) to the normal (y-axis) of an anisotropic substrate surface located in the (x-z) plane. In the (x-z) plane, the remaining two relative permittivity components have identical values $\varepsilon_{r\perp}$. The relative permittivity components $\varepsilon_{r\perp}$ are in the plane normal ($\perp$) to the y-axis, and also normal to the external electric field E_y. The y-axis is the main axis. It is also known as the *C-axis,* or the *optic-axis,* or the *extraordinary axis*. In the direction of the optic axis, the permittivity is different. The other two x- and z-axes are known as the *ordinary axes*. The x- and z-polarized EM-waves, known as the *ordinary waves*, in the (x-z)-plane travel with the same velocity, whereas y-polarized EM-wave, known as the *extraordinary wave,* travels with another velocity. The nonhomogeneous medium as shown in Fig. (4.3) is also a uniaxial medium with the x-axis as the optic axis, supporting the extraordinary wave propagation. The y and z-axes are the ordinary axes, supporting the ordinary wave propagation.

Figure (4.4) shows a microstrip line of width w on an anisotropic substrate of thickness h. It forms a parallel-plate capacitor placed in the (x-z)-plane. It views $\varepsilon_{r||}$ component of the uniaxial relative permittivity. Whereas, if the parallel plates of the capacitor are placed either in the (x-y)-plane or the (y-z)-plane, it will view the $\varepsilon_{r\perp}$ component of a uniaxial substrate medium. Thus, a uniaxial dielectric medium offers two different values of capacitance, depending on the placement of the parallel plates, connected to a voltage source. Normally, manufacturers provide data for $\varepsilon_{r||}$ and $\varepsilon_{r\perp}$ of the uniaxial substrates. The constitutive relation for the uniaxial medium aligned to the physical axes is

$$\begin{bmatrix} D_x \\ D_y \\ D_z \end{bmatrix} = \varepsilon_0 \begin{bmatrix} \varepsilon_{r\perp} & 0 & 0 \\ 0 & \varepsilon_{r||} & 0 \\ 0 & 0 & \varepsilon_{r\perp} \end{bmatrix} \begin{bmatrix} E_x \\ E_y \\ E_z \end{bmatrix}. \tag{4.2.9}$$

In the case of the crystal axes ξ, η of an anisotropic substrate, shown in Fig. (4.4), have an angle $\theta = \theta_1 = \theta_2$ with respect to the physical axes x, y; the relative permittivity components could be computed by the following expressions [B.24]:

$$\begin{aligned} \varepsilon_{rxx} &= \varepsilon_{r\xi}\cos^2\theta + \varepsilon_{r\eta}\sin^2\theta \\ \varepsilon_{ryy} &= \varepsilon_{r\xi}\sin^2\theta + \varepsilon_{r\eta}\cos^2\theta \\ \varepsilon_{rxy} &= \varepsilon_{ryx} = \left(\varepsilon_{r\xi} - \varepsilon_{r\eta}\right)\sin\theta\cos\theta. \end{aligned} \tag{4.2.10}$$

Type III: Biaxial materials. For such an anisotropic medium, all three principal relative permittivity components are different, i.e. $\varepsilon_{r\xi} \neq \varepsilon_{r\eta} \neq \varepsilon_{r\zeta}$. Such a medium is known as the biaxial medium [B.3].

The above nomenclatures are also applicable to the permeability of a magnetic material.

Anti-symmetric Anisotropic Materials

The plasma medium, i.e. the electron gas model of metal or ionosphere, is treated as an isotropic medium. However, in the presence of a static biasing magnetic field, it becomes a uniaxial dielectric material with off-diagonal elements for the permittivity matrix. This permittivity matrix is anti-symmetric. It does not support linearly polarized characteristic waves as the normal modes. It forms a *gyroelectric medium, i.e. electrically gyrotropic medium*. Similarly, in the presence of a static biased magnetic field in the z-direction, ferrite medium becomes a *gyromagnetic medium,* i.e. *magnetically gyrotropic medium*. Both plasma and ferrite media, under the magnetic biasing, support propagation of circularly polarized characteristic waves as the normal modes [B.2–B.4, B.21, B.22]. The permittivity and permeability tensors in matrix form for these media are summarized below.

Gyroelectric Medium

The relative permittivity matrix for gyroelectric medium, under the applied magnetic field in the z-direction, is expressed as follows:

$$\bar{\bar{\varepsilon}}_r = [\varepsilon_r] = \begin{bmatrix} \varepsilon_r & j\kappa & 0 \\ -j\kappa & \varepsilon_r & 0 \\ 0 & 0 & \varepsilon_{r,zz} \end{bmatrix} \quad \text{(a)}$$

$$\text{where, } \varepsilon_r = 1 + \frac{\omega_p^2}{\omega_c^2 - \omega^2}, \quad \kappa = \frac{\omega_p^2 \omega_c}{(\omega_c^2 - \omega^2)\,\omega}, \quad \varepsilon_{r,zz} = 1 - \frac{\omega_p^2}{\omega^2} \quad \text{(b)}$$

$$\omega_p = \sqrt{\frac{Ne^2}{m\varepsilon_0}}, \qquad \omega_c = \frac{eB_0}{m} \quad \text{(c).} \qquad (4.2.11)$$

In the above expressions, ω_p and ω_c are the plasma frequency and *cyclotron frequency*. The cyclotron frequency is also called the gyrofrequency. The $j\kappa$ is a *cross-coupling factor* responsible for the gyration property of the relative permittivity of a medium. Other parameters are N: electron density, e: electron charge, m: electron mass, and B_0: DC magnetic field. In the case of the extremely high magnetic field, $\omega_c \to \infty$ and $\varepsilon_r = 1$, $\kappa = 0$. Also, in the absence of magnetic field, $B_0 = 0$, $\kappa = 0$, and the gyrotropic medium is reduced to a uniaxial dielectric medium. In some cases, there is no plasma medium in the z-direction and $\varepsilon_{r,\,zz} = 1$.

Gyromagnetic Medium

The relative permeability matrix of a gyromagnetic ferrite medium is given below:

$$\bar{\bar{\mu}}_r = [\mu_r] = \begin{bmatrix} \mu_r & j\kappa & 0 \\ -j\kappa & \mu_r & 0 \\ 0 & 0 & 1 \end{bmatrix} \quad \text{(a)}$$

$$\text{where, } \mu_r = 1 + \frac{\omega_0 \omega_M}{\omega_0^2 - \omega^2}, \quad \kappa = \frac{\omega \omega_M}{(\omega_0^2 - \omega^2)} \quad \text{(b)}$$

$$\text{and } \omega_M = \mu_0 \gamma M_s, \quad \omega_0 = \mu_0 \gamma H_0 \quad \text{(c).} \qquad (4.2.12)$$

In the above equations, H_0 is the biasing magnetic field in the z-direction, M_s is the saturation magnetization, and the parameter γ is the *gyromagnetic ratio*. Again the *cross-coupling factor* $j\kappa$ is responsible for the gyration property of the relative permeability of a medium. The permeability function shows a singularity at the frequency $\omega = \omega_0$. It is suppressed in the presence of losses. In the case of unbiased and demagnetized ferrite, $H_0 = 0$ and $M_s = 0$ leading to $k = 0$ and $\mu_r = 1$. The ferrite, in this case, acts as a dielectric medium with permeability $\mu = \mu_0$.

Magneto-dielectric Composite Materials

The magneto-dielectric (MD) materials have both electric and magnetic properties. These materials are characterized by the simultaneous presence of permittivity and permeability tensors. These tensors for uniaxial MD materials are expressed as follows:

$$\bar{\bar{\varepsilon}}_r = \begin{bmatrix} \varepsilon_{r,0} & 0 & 0 \\ 0 & \varepsilon_{r,0} & 0 \\ 0 & 0 & \varepsilon_{r,e} \end{bmatrix} \quad \text{(a)}$$

$$\bar{\bar{\mu}}_r = \begin{bmatrix} \mu_{r,0} & 0 & 0 \\ 0 & \mu_{r,0} & 0 \\ 0 & 0 & \mu_{r,e} \end{bmatrix} \quad \text{(b).} \qquad (4.2.13)$$

The above-given expressions for the uncoupled independent relative permittivity and permeability tensors show that there is no coupling between the electric and magnetic fields. Ferrites have such property. However, ferrites are very lossy at microwave frequency. The low-loss MD materials are synthesized by mixing ferrites/hexaferrite and their composites with a polymer as a host medium [J.3]. The periodic structures are embedded in the host medium to engineer MD materials for antenna applications in VHF and UHF bands [J.4]. The *metamaterial composites* are also MD materials with simultaneously negative permittivity and permeability in a certain frequency band.

Magnetoelectric Materials

The magnetoelectric materials are general *bianisotropic* electromagnetic materials with *cross-coupling of electric and magnetic fields*. These materials have anisotropy for both the permittivity and permeability with additional cross-coupling of the electric and magnetic field. In such materials, the electric flux density vector $\vec{D}$ and also magnetic flux density $\vec{B}$ depends on both applied $\vec{E}$ and $\vec{H}$. It shows that in the bianisotropic materials, the $\vec{E}$ fields not only generate an electric polarization but also create the magnetic polarization, i.e. magnetization. Similarly, $\vec{H}$ fields applied to such materials create both magnetization and electric polarization. The constitutive relation relating four flux and field vectors for a linear magneto-electric medium is expressed as follows [B.21–B.23]:

$$\vec{D} = \bar{\bar{\varepsilon}}.\vec{E} + \bar{\bar{\xi}}.\vec{H} \quad \text{(a)}$$

$$\vec{B} = \bar{\bar{\zeta}}.\vec{E} + \bar{\bar{\mu}}.\vec{H} \quad \text{(b).} \qquad (4.2.14)$$

The above given four material parameter tensors $\bar{\bar{\varepsilon}}$, $\bar{\bar{\mu}}$, $\bar{\bar{\xi}}$, and $\bar{\bar{\zeta}}$ describe the bianisotropic magnetoelectric materials. The tensors $\bar{\bar{\varepsilon}}$ and $\bar{\bar{\mu}}$ are usual permittivity and permeability tensors, whereas $\bar{\bar{\xi}}$ and $\bar{\bar{\zeta}}$ are *magneto-electric coupling tensors*. In general, 36 complex material parameters are required to characterize bianisotropic materials. However, the material parameter matrix could be diagonalized. So, for the *uniaxial bianisotropic*, i.e. the *magneto-electric*, materials, constitutive relations are written as follows [B.15, B.21]:

$$\vec{D} = \begin{bmatrix} \varepsilon & 0 & 0 \\ 0 & \varepsilon & 0 \\ 0 & 0 & \varepsilon_z \end{bmatrix} . \vec{E} + \begin{bmatrix} \xi & 0 & 0 \\ 0 & \xi & 0 \\ 0 & 0 & \xi_z \end{bmatrix} . \vec{H} \quad \text{(a)}$$

$$\vec{B} = \begin{bmatrix} \zeta & 0 & 0 \\ 0 & \zeta & 0 \\ 0 & 0 & \zeta_z \end{bmatrix} . \vec{E} + \begin{bmatrix} \mu & 0 & 0 \\ 0 & \mu & 0 \\ 0 & 0 & \mu_z \end{bmatrix} . \vec{H} \quad \text{(b).} \qquad (4.2.15)$$

In the absence of cross-coupling, i.e. for $\bar{\bar{\xi}} = \bar{\bar{\zeta}} = 0$, the bianisotropic medium is reduced to an MD medium. Figure (4.5a) shows groups of *general bianisotropic medium- isotropic, bi-isotropic, biaxial anisotropic*, and *bianisotropic media* [B.25]. Two cases have already been discussed.

The *bi-isotropic* materials are isotropic materials also showing cross-coupling of electric and magnetic fields. However, Fig. (4.5b) shows that the *general bi-isotropic medium* has special forms – *isotropic, Pasture, Tellegen*, and *bi-isotropic*. For general bi-isotropic medium, the medium tensors are reduced to scalars, and the constitutive relations given by equation (4.2.14) are reduced to the following simpler form [B.23]:

$$\vec{D} = \varepsilon\vec{E} + \xi\vec{H} \quad \text{(a)}, \quad \vec{B} = \zeta\vec{E} + \mu\vec{H} \quad \text{(b)}$$
$$\text{where, } \xi = (\chi + j\kappa)\sqrt{\mu_0\varepsilon_0} \quad \text{(c)}, \quad \zeta = (\chi - j\kappa)\sqrt{\mu_0\varepsilon_0} \quad \text{(d)}. \qquad (4.2.16)$$

where $\xi^2/\mu\varepsilon$ is nearly unity. The magnetoelectric coupling parameters ξ and ζ have two components: the *chirality parameter* κ *(kappa)* and the *cross-coupling parameter* χ *(chi)*. The chirality parameter κ measures the degree of the handedness of the medium. The parameter χ is due to the cross-coupling of fields. It decides the reciprocity ($\chi = 0$) and nonreciprocity ($\chi \neq 0$) of the medium, giving the *reciprocal and nonreciprocal material medium*, respectively. In absence of cross-coupling, i.e. $\chi = 0$, the parameters ξ and ζ are reduced to imaginary quantities, and the *bi-isotropic* medium is reduced to a *nonchiral* simple isotropic medium for $\kappa = 0$ and to a *chiral medium* for $\kappa \neq 0$. It is also known as *Pasteur medium*. It supports the left-hand and right-hand circularly polarized waves as the normal modes of propagation. It is a reciprocal medium. For $\kappa = 0$, $\chi \neq 0$ another medium, called *Tellengen medium*, is obtained. It is a nonreciprocal medium. The general bi-isotropic medium has $\chi \neq 0$, $\kappa \neq 0$. It is a nonreciprocal medium.

The gyrotropic medium and bianisotropic medium support left-hand and right-hand circularly polarized EM-waves. However, there is a difference. The gyrotropic medium supports the *Faraday rotation*, i.e. rotation of linearly polarized wave while propagating in the medium, whereas bianisotropic medium does not support it [B.21].

4.2.4 Nondispersive and Dispersive Medium

The medium in which the EM-wave propagates with equal phase velocity at all frequencies is called the *nondispersive* medium. For such a medium, the relative permittivity and relative permeability are real quantities, and they are independent of frequency. However, practically every material medium has losses; and thus their relative permittivity and relative permeability are

Biaxial anisotropic ([ε], [μ]) *(Direction dependence. No magneto – electric coupling)*	Isotropic (ε, μ) *(No-direction dependence. No magneto – electric coupling)*
Bi-isotropic (ε, μ, ξ, ζ) *(No-direction dependence. Magneto – electric coupling)*	Bi-anisotropic ([ε], [μ], [ξ], [ζ]) *(Direction dependence. Magneto – electric coupling)*

(a) Groups of bi-anisotropic materials.

Pasture *(Reciprocal, $\chi = 0$ Chiral, $k \neq 0$)*	Isotropic *(Reciprocal, $\chi = 0$ Non-chiral, $k = 0$)*
Tellengen *(Non-reciprocal, $\chi \neq 0$ Non-chiral, $k = 0$)*	Bi-isotropic *(Non-reciprocal, $\chi \neq 0$ Chiral, $k \neq 0$)*

(b) Groups of bi-isotropic materials.

Figure 4.5 Classification of bianisotropic and bi-isotropic materials.

complex quantities, and these are a function of frequency. Thus, normally a material medium supports frequency-dependent phase velocity. Such a medium is called the *dispersive medium*. This kind of dispersion is known as *material dispersion* [B.2, B.3, B.9, B.10]. The material dispersion in the microwave and the mm-wave ranges is normally negligible for the substrates used in the planar technology. However, a *composite substrate*, made of the layered dielectric media, is dispersive. Likewise, the *artificial metamaterial substrate* is also dispersive. The plasma and conducting media are also dispersive in the microwave range. The material dispersion is an important phenomenon near the optical frequency. The constitutive relation for the dispersive medium is

$$\vec{D}(\omega) = \varepsilon(\omega)\vec{E}(\omega). \tag{4.2.17}$$

Lorentz oscillator model of a dielectric material, discussed in section (6.5) of chapter 6, helps to understand the frequency-dependent origin of the $\varepsilon(\omega)$. When an EM-pulse, like a Gaussian pulse, passes through a dispersive medium, its pulse-width widens due to the separation of different frequency components, as each frequency component travels at a different velocity in the dispersive medium. This is known as the *pulse-spreading* phenomenon. It limits the speed of digital data transmission through the dispersive medium, as the digital bits can overlap each other. However, a dispersive medium can be nonlinear also, where the pulse spreading can be balanced by the nonlinearity. Under such combined dispersion, and nonlinearity, a pulse can propagate without changing its shape. Such robust EM-waves are called the *solitons*. The solitons are useful in optical communication. However, the solitons have also been generated in the microwaves frequency range

4.2.5 Non-lossy and Lossy Medium

All physical dielectric and magnetic material media have some amount of loss. Normally, the substrates used in the microwave planar technology do not have a high loss, except the doped semiconducting substrates. The lossy dielectrics are known as the *imperfect dielectrics*. The losses in the dielectric and magnetic materials change the relative permittivity and permeability into complex quantities:

$$\begin{aligned} \varepsilon_r^* &= \varepsilon_r' - j\varepsilon_r'', && \text{(a)} \\ \mu_r^* &= \mu_r' - j\mu_r'' && \text{(b).} \end{aligned} \tag{4.2.18}$$

The imaginary parts of the relative permittivity and permeability are taken as negative quantities due to our choice of time-harmonic function $e^{j\omega t}$. However, $|\varepsilon_r''|$ and $|\mu_r''|$ are always positive quantities in a passive lossy medium. The lossy capacitor and the lossy inductor can model the complex relative permittivity and permeability of a medium. This is called the *circuit modeling of a material medium*. The modeling of the relative permittivity of a medium by the capacitor shows that both the medium and capacitor can store electric energy. Likewise, the relative permeability of medium and the corresponding inductor model both are the magnetic energy storage components.

4.2.6 Static Conductivity of Materials

Figure (4.6a) shows a cylindrical section of the lossy material of conductivity (σ), i.e. resistivity $\rho = 1/\sigma$. Its cross-sectional area and length are A and h, respectively. The lossy material can be modeled as a resistor R, also shown in Fig. (4.6a). The conduction current density J_c flowing through the conductor, due to the free mobile charges, is $J_c = I_c/A$, where conduction current is $I_c = V/R$. The electric field intensity in the material is $E = V/h$. On combining these expressions, the following relation is obtained:

$$\begin{aligned} &\vec{J}_c = \frac{h}{RA}\vec{E} && \text{(a)} \\ \Rightarrow\ &\vec{J}_c = \sigma\vec{E} && \text{(b)} \\ &\text{where, } \sigma = \frac{h}{RA} && \text{(c).} \end{aligned} \tag{4.2.19}$$

The above expression (4.2.19b) is Ohm's law for a lossy medium. The following expression of the equivalent resistance of a lossy material, in terms of its resistivity, follows from the above equation (4.2.19c):

$$R = \frac{\rho h}{A}. \tag{4.2.20}$$

In general, the Ohm's law for the anisotropic medium is written in the vector form as $\vec{J}_c = [\sigma]\,\vec{E}$. Further, the expression is also obtained below for the conductivity σ of the material in terms of the basic parameters of mobile charges in a lossy material, i.e. in terms of electron charge and its mobility. The expression is also applicable to semiconductors. In the case of a semiconductor, the conduction current is due to both the electrons (negative charges) and holes (positive charges).

Figure (4.6a) considers $h = \Delta x$ length of a cylindrical section of conducting material with a cross-sectional area ΔA. The free charges move in the direction x with a velocity v_x on the application of electric field intensity E_x. The conduction current in the x-direction is the rate

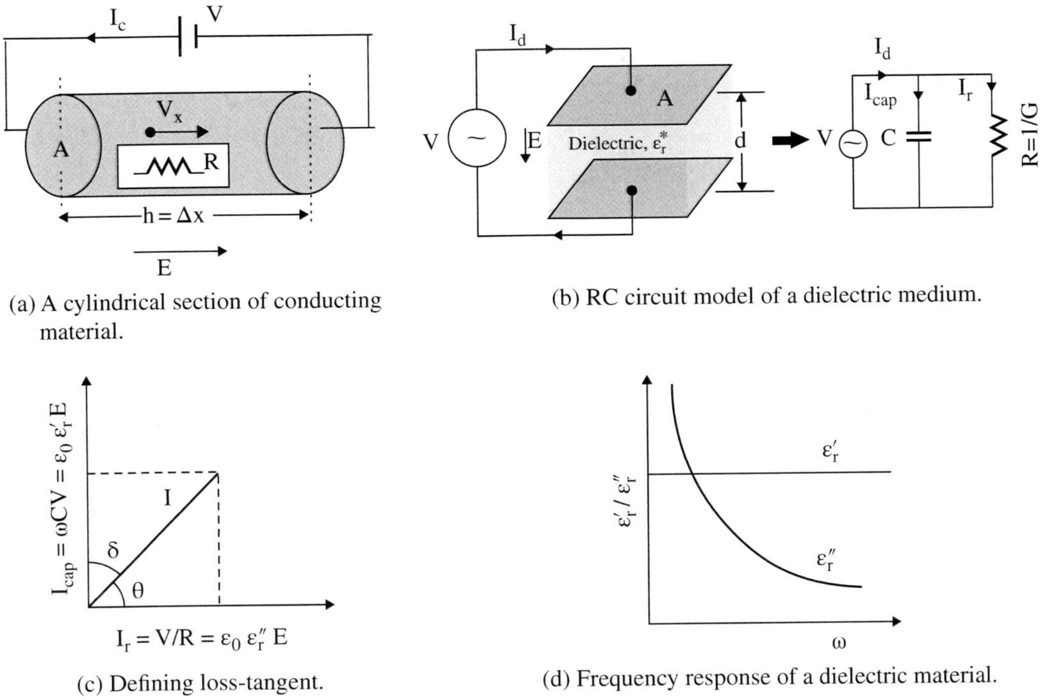

(a) A cylindrical section of conducting material.

(b) RC circuit model of a dielectric medium.

(c) Defining loss-tangent.

(d) Frequency response of a dielectric material.

Figure 4.6 Circuit model, parameters of a dielectric medium.

of flow of total charge ΔQ_e contained in a volume, $\Delta V = \Delta A \times \Delta x$.

$$I_c = \frac{\Delta Q_c}{\Delta t} = \rho_e \frac{\Delta V}{\Delta t} = (\rho_e \, \Delta A) \frac{dx}{dt}. \qquad (4.2.21)$$

In the limiting case, $\vec{v}_x = \underset{\Delta t \to 0}{\mathrm{Lt}} (\Delta x/\Delta t) = dx/dt$.

Thus, $\vec{I}_c = \rho_e \, \Delta A \, \vec{v}_x$, and the conduction current density is

$$\vec{J}_c = \frac{\vec{I}_c}{\Delta A} = \rho_e \, \vec{v}_x. \qquad (4.2.22)$$

In a material, the charge movement is random due to the scattering and so forth. However, an average motion is assumed, giving the drift of charges in the x-direction with a *drift velocity* $\vec{v}_x$. The drift velocity is proportional to the electric field intensity that provides another expression for J_c:

$$\vec{v}_x = \mu_m \vec{E} \qquad (a)$$
$$\vec{J}_c = \rho_e \mu_m \vec{E} \qquad (b). \qquad (4.2.23)$$

where constant μ_m is called the *mobility* of a charge. It is noted that μ is also used as a symbol for permeability. On comparing equations (4.2.19b) and (4.2.23b), the following expression for conductivity is obtained:

$$\sigma = \rho_e \mu_m. \qquad (4.2.24)$$

If N is the number of free charges per unit volume, with charge q on each carrier, the charge density is $\rho_e = Nq$. The equation (4.2.24) of the conductivity is changed to

$$\sigma = N \, q \, \mu_m. \qquad (4.2.25)$$

In the case of a conductor, the charge carrier is electron, i.e. $q = q_e$ (the *electron charge*) and $\mu = \mu_m^e$ (*electron mobility*). However, for a semiconductor, its conductivity σ_s is due to both electrons and holes leading to the following expression:

$$\sigma_s = \sigma_e + \sigma_h = N_e q_e \mu_m^e + N_h q_h \mu_m^h, \qquad (4.2.26)$$

where N_e and N_h are numbers of electrons and holes per unit volume. The charges on electron and holes are equal $q_e = q_h = e = 1.6 \times 10^{-19}$ Coulombs. The electron and hole mobilities, in a semiconductor, are μ_m^e and μ_m^h, respectively. The signs of q_e and μ_m^e are negative, whereas the signs of q_h and μ_m^h are positive. However, the conductivity σ_s of a semiconductor is always positive.

4.3 Circuit Model of Medium

The circuit model helps to understand the electrical property of a medium. It is further useful for simulating the electrical responses of a medium. The electrical property of a dielectric medium is expressed through relative permittivity that shows the electric energy storage ability of the medium. The capacitor also stores electric energy. Therefore, a *dielectric medium is modeled as a capacitor*. The loss in a medium is due to the dissipation of energy that is modeled as a resistor. Similarly, the permeability of a medium, such as an inductor, shows its ability to store magnetic energy. Therefore, the *permeability of a medium is modeled as an inductor* [B.10–B.12].

4.3.1 RC Circuit Model of Lossy Dielectric Medium

Figure (4.6b) shows a parallel-plate capacitor, containing a lossy dielectric medium with complex relative permittivity $\varepsilon_r^* = \varepsilon_r' - j\varepsilon_r''$. It further shows its RC circuit model that is a parallel combination of the capacitor (C) and resistor (R). It is connected to a time-harmonic voltage source $v = v_0 e^{j\omega t}$ that produces a time-harmonic electric field, $E = E_0 e^{j\omega t}$ in the dielectric medium. The displacement current density in the dielectric medium is

$$\begin{aligned} J_d &= \frac{dD}{dt} = \varepsilon_0 \varepsilon_r^* \frac{d}{dt}\left(E_0 e^{j\omega t}\right) \\ \Rightarrow J_d &= j\omega\varepsilon_0\varepsilon_r^* E = j\omega\varepsilon_0\left(\varepsilon_r' - j\varepsilon_r''\right)E \\ \Rightarrow J_d &= \left(j\omega\varepsilon_r' + \omega\varepsilon_r''\right)\varepsilon_0 E \quad \text{(a)} \\ I_d &= J_d \times A = \left(\omega\varepsilon_r'' + j\omega\varepsilon_r'\right)\varepsilon_0 \frac{AV}{d} \quad \text{(b)}. \end{aligned} \tag{4.3.1}$$

The displacement current density has two components with the quadrature phase:

$$\begin{aligned} J_d &= J_r + j\,J_{cap} \Rightarrow J_{cap} = \omega\varepsilon_r'\varepsilon_0 E \quad \text{(a)} \\ J_r &= \omega\varepsilon_r''\varepsilon_0 E \quad \text{(b)}. \end{aligned} \tag{4.3.2}$$

The reactive current density (J_{cap}) flows through ε_r', i.e. through the capacitor. It is shown by the presence of "j." Thus, *the real part of the relative permittivity ε_r' is modeled as a capacitor C*. The resistive current density (J_r), causing a loss in the dielectric medium, flows through ε_r'', i.e. through resistor R. The *imaginary part of relative permittivity ε_r'' is modeled as a resistor R, parallel to the capacitor C*. It is obvious from equation (4.3.1) that the loss caused by ε_r'' is a positive quantity only when the complex relative permittivity ε_r^* is defined as a subtractive combination of its real and imaginary parts mentioned in equation (4.2.18a). The current through the equivalent circuit is

$$\begin{aligned} I &= I_r + j\,I_{cap} = (1/R + j\omega C)V = (G + j\omega C)V \\ \Rightarrow I_r &= V/R \quad \text{(a)} \\ I_{cap} &= \omega C V \quad \text{(b)}. \end{aligned} \tag{4.3.3}$$

Both current components are shown in Fig. (4.6b). The loss-tangent, showing the dissipation factor of the RC circuit, is defined using Fig. (4.6c):

$$\begin{aligned} \text{Dissipation factor, } \tan\delta &= \frac{\text{Loss current}}{\text{Reactive current}} = \frac{I_r}{I_{cap}} \\ \Rightarrow \tan\delta &= \frac{1}{\omega RC}. \end{aligned} \tag{4.3.4}$$

Therefore, the *dissipation factor*, i.e. the *loss tangent* ($\tan\delta$), of lossy dielectric material is defined using equation (4.3.2) as follows:

$$\tan\delta = \frac{I_r}{I_{cap}} = \frac{J_r A}{J_{cap} A} = \frac{\varepsilon_r''}{\varepsilon_r'}, \tag{4.3.5}$$

where A is the area of the parallel-plate capacitor. The loss current is zero, i.e. $I_r = 0$ for a lossless capacitor, and also for a lossless dielectric medium. It leads to $R \to \infty$, and $\varepsilon_r'' = 0$, $\tan\delta = 0$. In Fig. (4.6c), the angle θ is the *power-factor angle*. On comparing equations (4.3.2) and (4.3.3), an equivalence is obtained between the lossy dielectric medium and a lossy capacitor

$$\begin{aligned} \varepsilon_r' &= \frac{C\,d}{\varepsilon_0 A} \quad \text{(a)} \\ \varepsilon_r'' &= \frac{d}{\omega\,\varepsilon_0 R A} \quad \text{(b)}. \end{aligned} \tag{4.3.6}$$

On replacing the dielectric medium of Fig. (4.6b) by the air medium, i.e. $\varepsilon_r = 1$, capacitance C_0 is obtained:

$$C_0 = \frac{\varepsilon_0 A}{d}. \tag{4.3.7}$$

On using the above equations, the real and imaginary parts of a complex relative permittivity and loss tangent are defined in terms of the circuit elements:

$$\begin{aligned} \varepsilon_r' &= \frac{C}{C_0} \quad \text{(a)} \\ \varepsilon_r'' &= \frac{1}{\omega\,RC_0} \quad \text{(b)} \\ \tan\delta &= \frac{\varepsilon_r''}{\varepsilon_r'} = \frac{1}{\omega\,RC} \quad \text{(c)}. \end{aligned} \tag{4.3.8}$$

The above equations provide a practical means to measure the dielectric constant of any dielectric material with the help of a parallel-plate capacitor. Equation (4.3.8a) also gives a practical definition of the relative permittivity of a homogeneous dielectric medium. The relative permittivity is a ratio of capacitances of a parallel-plate capacitor, with a material medium and with the air medium; while keeping the geometry of the parallel-plate capacitor unchanged. We get a homogenized dielectric medium even if the parallel-plate capacitor, as shown in Fig. (4.3a), is made of layered dielectric sheets. The measurement of capacitance C ignores the layered medium and views it as a homogeneous medium. So, *the relative permittivity of a material is the macrolevel homogenization concept that ignores the microlevel discrete composition of a medium.* The concept of homogenization is important to design the *engineered metamaterials* using the discrete metallic and nonmetallic structures embedded in a host medium. It is discussed in section (21.4) of chapter 21.

Figure (4.6d) shows the frequency response of a lossy dielectric medium, as predicted by the RC circuit model. The real part of the permittivity (ε_r') is frequency independent, whereas the imaginary part of the permittivity (ε_r'') decreases hyperbolically with frequency. Some dielectric materials may not exhibit this kind of frequency response. More realistic circuit models may be needed for such a dielectric medium. Chapter 6 discusses a few more circuit models of the dielectric media.

The loss-tangent of a dielectric is also a measurable quantity. Manufacturers provide data on it. However, the loss of a semiconducting substrate is characterized by the conductivity (σ) of a substrate. Even a dielectric material can have some amount of free charge carriers, contributing to its conductivity (σ). The finite conductivity causes a dielectric loss in the material. The imaginary part of the complex relative permittivity (ε_r'') arises due to the damping of oscillation during the polarization process of a dielectric material, under the influence of an externally applied AC electric field discussed in chapter 6. However, it is difficult to distinguish between two sources of the dielectric loss; the contribution of the free charge carriers (*conduction current*) and the contribution of the dielectric polarization (*polarization current*). Therefore, both could be grouped in the total loss-tangent.

The parallel-plate capacitor, shown in Fig. (4.6b), supports two kinds of current densities – the conduction current density, J_c given by equation (4.1.9), and the displacement current density, J_d given by equation (4.3.1a). The total current density is

$$J_t = \sigma E + (j\omega\varepsilon_0 \varepsilon_r' + \varepsilon_0\varepsilon_r'')E$$

$$\Rightarrow J_t = \underbrace{(\sigma + \omega\varepsilon_0\varepsilon_r'')E}_{\text{Loss current}} + \underbrace{j\omega\varepsilon_0\,\varepsilon_r' E}_{\text{Reactive current}} \quad (4.3.9)$$

The total loss-tangent, from equations (4.3.4) and (4.3.9) of dielectric material is

$$\tan\delta_t = \frac{\sigma + \omega\varepsilon_0\varepsilon_r''}{\omega\varepsilon_0\varepsilon_r'}. \quad (4.3.10)$$

Equation (4.3.9), in a changed form, is rewritten as follows:

$$J_t = j\omega\varepsilon_0\left[\varepsilon_r' - j\left(\varepsilon_r'' + \frac{\sigma}{\omega\varepsilon_0}\right)\right]E \quad \text{(a)}$$

$$\Rightarrow J_t = j\omega\varepsilon_0\left[\varepsilon_r' - j\,\varepsilon_{req}''\right]E \quad \text{(b).}$$

(4.3.11)

The equivalent ε_{req}'' of lossy dielectrics, due to the combined effect of polarization and finite conductivity, is

$$\varepsilon_{req}'' = \varepsilon_r'' + \frac{\sigma}{\omega\varepsilon_0}. \quad (4.3.12)$$

The lossy dielectric medium is also described by the concept of the *complex equivalent conductivity* $\sigma_{eq}^* = \sigma_{eq}' + j\,\sigma_{eq}''$. Using the expression $J_t = \sigma_{eq}^* E$ and equation (4.3.9), the complex equivalent conductivity is expressed as follows:

$$\sigma_{eq}^* = \sigma_{eq}' + j\sigma_{eq}'' = (\sigma + \omega\varepsilon_0\varepsilon_r'') + j\omega\varepsilon_0\varepsilon_r'. \quad (4.3.13)$$

The real part of a complex equivalent conductivity causes the dielectric loss in a medium, whereas its imaginary part stores the electric energy of the dielectric medium. Therefore, the imaginary part of a complex equivalent conductivity is related to the relative permittivity of a medium, and its real part is associated with the imaginary part of the complex relative permittivity:

$$\sigma_{eq}'' = \omega\,\varepsilon_0\varepsilon_r' \quad \text{(a)}$$

$$\sigma_{eq}' = \sigma + \omega\,\varepsilon_0\varepsilon_r'' \quad \text{(b).}$$

(4.3.14)

Sometimes, loss due to the polarization causing ε_r'' is ignored, say in a semiconductor, if the loss due to the conduction current is more significant. The loss characteristic of a semiconducting substrate is given by its conductivity. In that case, the loss-tangent is expressed in terms of the conductivity of a medium:

$$\tan\delta = \frac{\varepsilon_r''}{\varepsilon_r'} = \frac{\sigma}{\omega\,\varepsilon_0\varepsilon_r'}. \quad (4.3.15)$$

The above expression helps to convert the conductivity of a substrate to its loss-tangent at each frequency of the required frequency band. It can also be used to convert the loss-tangent to the conductivity at each frequency.

The equivalence between the relative permittivity and capacitance has been obtained by treating both the dielectric and capacitor as electric energy storage devices. Thus, the complex relative permittivity is equivalent to a complex capacitance. Figure (4.6b) provides the admittance of a lossy capacitor:

$$Y = G + j\omega C, \quad \Rightarrow \quad Y = j\omega\left[C - j\frac{G}{\omega}\right] = j\omega C^*, \quad (4.3.16)$$

where the complex capacitance C^* is given by

$$C^* = C' - j\,C''. \tag{4.3.17}$$

The real and imaginary parts of a complex capacitance, and also loss-tangent, are given by the following expressions:

$$C' = C \quad \text{(a)}$$
$$C'' = \frac{G}{\omega} \quad \text{(b)}$$
$$\tan\delta = \frac{C''}{C} = \frac{G}{\omega C} = \frac{1}{\omega RC} \quad \text{(c)}. \tag{4.3.18}$$

The complex relative permittivity is also expressed as follows:

$$\varepsilon_r^* = \varepsilon_r' - j\,\varepsilon_r'' = \frac{C' - j\,C''}{C_0}. \tag{4.3.19}$$

The above expression is helpful in the computation of the real and imaginary parts of the effective relative permittivity of a lossy planar transmission line. The complex line capacitance of a lossy planar transmission line can be numerically evaluated. It also helps to compute the effective loss-tangent of a multilayered planar transmission line by the *variational method*. It is discussed in chapter 14.

4.3.2 Circuit Model of Lossy Magnetic Medium

The magnetic loss in magnetic materials is due to the process of *magnetization* that results in a complex relative permeability, $\mu_r^* = \mu_r' - j\mu_r''$. The relative permeability is defined as a ratio of two inductances; inductance (L^*) of the coil with a magnetic material, and inductance (L_0) of the same coil with the air-core,

$$\mu_r^* = \mu_r' - j\mu_r'' = \frac{L^*}{L_0}. \tag{4.3.20}$$

A magnetic material placed inside a coil, shown in Fig. (4.7a), gets magnetized in the process of *magnetization*. The current causing the magnetization is known as the *magnetization current* (I_m). However, during the magnetization process, some power is lost. It is viewed through the *magnetic loss-current* (I_r). The magnetization of a magnetic material involves the storage of magnetic energy. So, a lossy magnetic material is modeled through the RL equivalent circuit, shown in Fig. (4.7b).

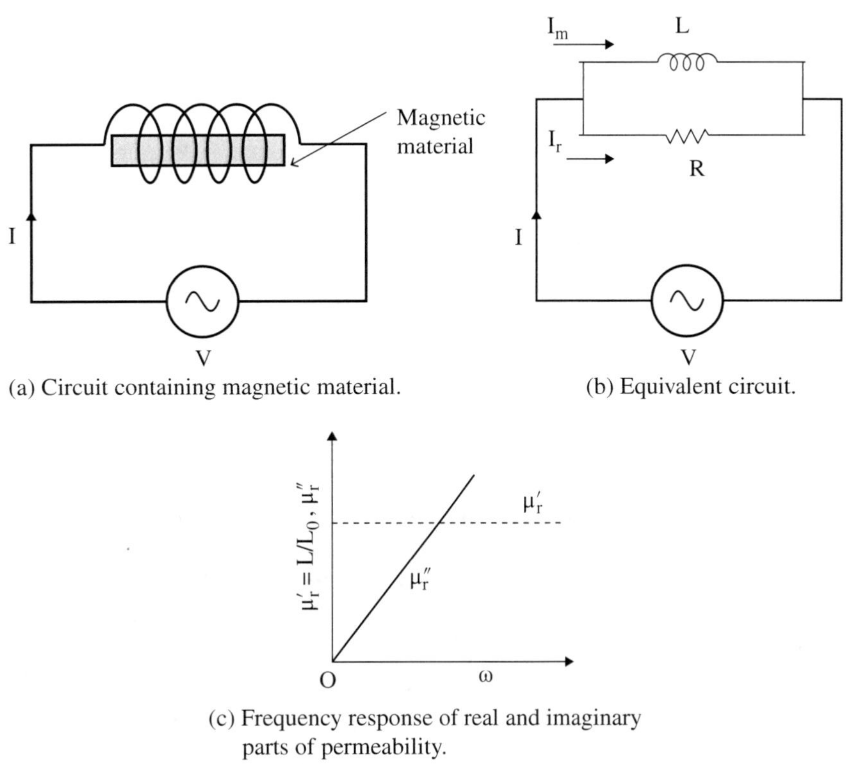

Figure 4.7 Circuit model and frequency response of lossy magnetic material.

The inductor models the stored magnetic energy in a magnetic material, whereas the resistor models its loss.

A time-harmonic current $I = I_0 e^{j\omega t}$ flows through the coil containing the magnetic material. The voltage across the coil and current through it are given below:

$$V = L^* \frac{dI}{dt} = j\omega\mu_r^* L_0 I, \quad \text{(a)}$$

$$I = \frac{V}{j\omega\mu_r^* L_0} = -jV\frac{(\mu_r' + j\mu_r'')}{\omega L_0 (\mu_r'^2 + \mu_r''^2)} \quad \text{(b)}. \tag{4.3.21}$$

Figure (4.7b) shows the current flowing in the parallel RL equivalent circuit. By comparing the above current, with the circuit model current; the expressions for the circuit elements, in terms of the material parameters, are obtained as follows:

$$I = I_m + I_r = -j\frac{V}{\omega L} + \frac{V}{R} \quad \text{(a)}$$

$$L = L_0\frac{(\mu_r'^2 + \mu_r''^2)}{\mu_r'} \quad \text{(b)}$$

$$R = \omega L_0\frac{(\mu_r'^2 + \mu_r''^2)}{\mu_r''} \quad \text{(c)}. \tag{4.3.22}$$

The magnetic loss-tangent of the lossy inductor is defined below that helps to get the magnetic loss-tangent of magnetic material:

$$\tan\delta_m = \frac{I_r}{I_m} = \frac{\omega L}{R} \quad \text{(a)}$$

$$\tan\delta_m = \frac{\mu_r''}{\mu_r'} \quad \text{(b)}. \tag{4.3.23}$$

A lossless magnetic material has $\mu_r'' = 0$, i.e. $L = \mu_r' L_0$, $R \to \infty$. It means an inductor, without the resistor, models the lossless magnetic material. *It shows that the permeability is an ability of a magnetic material to store magnetic energy, just as the permittivity of dielectric material is its ability to store electric energy.* For the case of the low-loss magnetic materials, $\mu_r'' \ll \mu_r'$, the following expressions are obtained from equations (4.3.22b) and (4.3.22c):

$$\mu_r' = \frac{L}{L_0} \quad \text{(a)}$$

$$\mu_r'' = \frac{\omega L}{R}\mu_r' \quad \text{(b)}. \tag{4.3.24}$$

It is observed that for a low-loss magnetic material, μ_r' is independent of frequency, whereas μ_r'' increases linearly with frequency, as shown in Fig. (4.7c). *The above equations show that inductor models the real part of the complex permeability, whereas resistance models its imaginary part.* This kind of frequency response may not be a realistic description of a hysteretic magnetic material.

4.4 Maxwell Equations and Power Relation

The experimental observations on the time-dependent electromagnetic field are described by a set of the vector differential equations, known as the Maxwell equations. These sets of equations form the fundamental laws of nature, governing behaviors of the EM-field in free space, and also in material media. Each part of Maxwell equations shows a definite nature of the EM-field, explored by different investigators; Faraday, Ampere, and Gauss. However, Maxwell has put them in the form of a consistent set of equations using a mathematical format called the *quaternions*. He has also introduced the *concept of displacement current*. On solving the equations, either for the electric field or the magnetic field, the wave equations are obtained for the electric and magnetic fields traveling at a velocity of light in free space. Poynting has determined the power carried by the EM-wave. Finally, Heaviside replaced the quaternions by the vector notations and presented the Maxwell equations in the present form, i.e. in the convenient form of the vector differential equations [B.26–B.28]. Such a formulation of the electromagnetic theory has opened the grand path of modern research and investigation. So, the modern form of Maxwell equations can be truly called *Maxwell–Heaviside equations*. However, in this text, we use normal terminology to follow the current practice.

4.4.1 Maxwell's Equations

The set of Maxwell's equations, given below, consists of four time-dependent vector equations; relating the sources such as *conduction current density* ($\vec{J}_c$), *electric displacement current density* $\left(\vec{J}_d = \partial\vec{D}/\partial t\right)$, and *magnetic displacement current* density $\left(\vec{J}_{d,m} = \partial\vec{B}/\partial t\right)$ to the magnetic field ($\vec{H}$) and the electric field ($\vec{E}$), respectively:

$$\nabla \times \vec{E} = -\frac{\partial \vec{B}}{\partial t} - \vec{M}_{ext} \quad \text{(a)}$$

$$\nabla \times \vec{H} = \vec{J}_c + \frac{\partial \vec{D}}{\partial t} + \vec{J}_{ext} \quad \text{(b)}$$

$$\nabla . \vec{D} = \rho \quad \text{(c)}$$

$$\nabla . \vec{B} = 0 \quad \text{(d).} \qquad (4.4.1)$$

All quantities in the above equations are space-time dependent. The current densities $\vec{J}_c$, $\vec{J}_d$, and $\vec{J}_{d,m}$ are not the power supplying sources to the propagating EM-wave in a medium. These current densities are created by the *externally applied* magnetic and electric current densities $\vec{M}_{ext}$ and $\vec{J}_{ext}$ supplying power to the EM-wave and partly getting absorbed as loss in a conducting medium.

The right-hand sides of the above equations could be treated as the sources (*excitations*) and the left-hand fields as the *responses*. The *force field quantities* $(\vec{E}, \vec{H})$, *flux field quantities* $(\vec{D}, \vec{B})$, and current densities are functions of both the space variables and time variable. Further, in any material medium, the flux field quantities are related to the force field quantities by the *constitutive relations,* given in equation (4.1.7). The conduction current density in a lossy medium is related to the electric field, as given in equation (4.1.9).

In general, ε_r, μ_r, σ are the tensors quantities for an anisotropic medium. However, these are scalar quantities for an isotropic medium. They are also treated as complex quantities to include the losses of a medium. In the case of a complex conductivity σ^*, its real part is responsible for the loss in a medium, whereas its imaginary part accounts for the energy storage. In a dispersive medium, ε_r, μ_r and σ are also frequency-dependent. The characteristics of various kinds of media, such as dielectrics, conductors, plasma, semiconductors, ferrites, and so forth are accounted for in Maxwell's equations through the *constitutive relations* applicable to these physical media. Maxwell's equations, along with the constitutive relations, are the *field equations,* not the *force equations,* i.e. these equations do not express the forces exerted by the fields on stationary or moving charges. This is achieved through *Lorentz's force equation*:

$$\vec{F} = q\left(\vec{E} + \vec{v} \times \vec{B}\right), \qquad (4.4.2)$$

where q is the charge on a mass m that is moving with velocity v. In this case, Lorentz's force acting on a moving charge is equal to Newtonian force:

$$m\frac{d\vec{v}}{dt} = \vec{F} = q\left(\vec{E} + \vec{v} \times \vec{B}\right). \qquad (4.4.3)$$

Lorentz's force has two components: (i) *electric force component* and (ii) *magnetic force component.* The electric force $\vec{F}_e = q\vec{E}$ is exerted either on a moving or on a stationary charge in the static, or in the time-varying electric field. The magnetic force $\vec{F}_m = q\vec{v} \times \vec{B}$ is exerted only on a moving charge in the static, or in the time-varying, magnetic field. In the case of a time-varying electric, or the time-varying magnetic field, both fields are always present and are related through Maxwell's equations (4.4.1a) and (4.4.1b). Thus, both components of Lorentz's force are present on a moving charge in a time-varying EM-field.

It is observed that in the absence of the external sources, in a lossless medium Maxwell's equation (4.4.1a) states that *a time-varying magnetic field creates a time-varying electric field*; and Maxwell's equation (4.4.1b) states that *a time-varying electric field* $(\vec{D} = \varepsilon_0\varepsilon_r\vec{E})$ *creates a time-varying magnetic field.* Thus, Maxwell's equations (4.4.1a) and (4.4.1b) form a set of the *coupled equations,* showing an interdependence of the time-varying electric and magnetic fields. It is like the two-variable simultaneous equations that occur in ordinary algebra. However, in the present case, the field variables $(\vec{E}, \vec{H})$ are vector quantities. The coupled partial differential equations are solved either for the $\vec{E}$ or $\vec{H}$ by following the rules of the vector algebra. The solutions provide the wave equation either for the electric $(\vec{E})$ or for the magnetic $(\vec{H})$ field.

Equation (4.4.1c) of Maxwell's equations is a divergence relation. It shows that the electric field originates from a charge, and ends on another charge. Equation (4.4.1d) shows that the divergence of the magnetic field is zero; meaning thereby that a magnetic charge does not exist in nature. The magnetic field exists as a closed-loop around a current-carrying conductor. However, sometimes a presence of the hypothetical *magnetic charge* is assumed, and the divergence equation (4.4.1d) is modified as $\nabla.\vec{B} = \rho_m$. This assumption maintains the symmetry of Maxwell's equations. Likewise, to maintain the symmetry of Maxwell's equations, a hypothetical *magnetic current* density $(-J_m)$ term is also to be added to equation (4.4.1a). The modified Maxwell equations in the symmetrical form are given below:

$$\nabla \times \vec{E} = -\vec{J}_m - \frac{\partial \vec{B}}{\partial t} - \vec{M}_{ext} \quad \text{(a)}$$

$$\nabla \times \vec{H} = \vec{J}_c + \frac{\partial \vec{D}}{\partial t} + \vec{J}_{ext} \quad \text{(b)}$$

$$\nabla . \vec{D} = \rho \quad \text{(c)}$$

$$\nabla . \vec{B} = \rho_m \quad \text{(d).} \qquad (4.4.4)$$

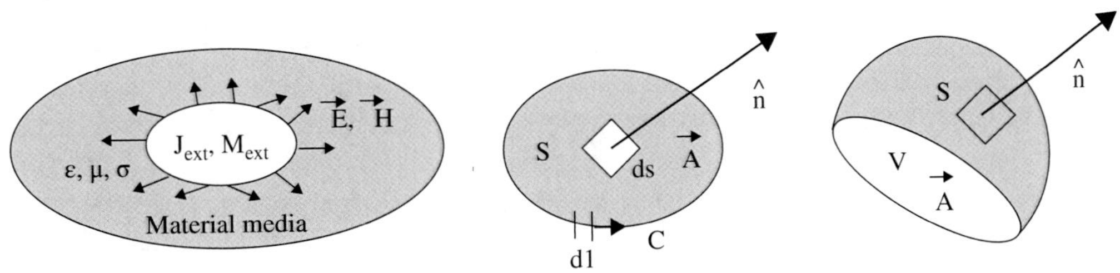

(a) External surface electric and magnetic currents creating a field in outer enclosure. (b) A closed-surface S formed by perimeter C. (c) A volume V enclosed by the surface S.

Figure 4.8 Surfaces and volume used in the integral form of Maxwell equations.

In a real material medium, a current can flow due to the time-dependent electric polarization (P), and also due to the time-dependent magnetic polarization (M). These are known as the *electric and magnetic polarization currents* and are incorporated in Maxwell's equations as the electric and magnetic displacement current densities. Sometimes, the external sources, *magnetic current density* $-\vec{M}_{ext}$ and *electric current density* $\vec{J}_{ext}$ are further added to Maxwell equations (4.4.1a) and (4.4.1b). The external current sources are retained in the modified form of Maxwell equations. Figure (4.8a) shows externally impressed current sources, creating fields in an enclosure. These externally applied current sources supply power to create the electric field and magnetic field in a material medium. They also cause the flow of current in a lossy medium. It is examined in the next section while discussing the energy balance of the electromagnetic field [B.8, B.9]. The lossless free space is treated as a charge-free and source-free medium, i.e. $\vec{M}_{ext} = \vec{J}_{ext} = \vec{J}_c = -\vec{J}_m = \rho = \rho_m = 0$.

A divergence of Maxwell's equations (4.4.1b) in absence of $\vec{J}_{ext}$, and using equation (4.4.1d), leads to the following *continuity equation*:

$$\nabla.\vec{J}_c = -\frac{\partial\rho}{\partial t}. \tag{4.4.5}$$

The above equation shows the conservation of charge. In any enclosed volume, the decrease of charge is associated with a flow of current out of the volume.

Normally, the time-harmonic fields are used in most of the applications. The time-dependent electric field and other field quantities are written in the *phasor form* for the time-harmonic field:

$$\underbrace{\vec{E}(x,y,z,t)}_{\text{Instantaneous time dependent field.}} = \mathrm{Re}\Big[\underbrace{\vec{E}(x,y,z)}_{\text{Space coordinate dependent phasor field.}}\ \underbrace{e^{j\omega t}}_{\text{Time harmonic variation.}}\Big]. \tag{4.4.6}$$

$$\begin{aligned}
\vec{H}(x,y,z,t) &= \mathrm{Re}\left[\vec{H}(x,y,z)e^{j\omega t}\right] && (a)\\
\vec{D}(x,y,z,t) &= \mathrm{Re}\left[\vec{D}(x,y,z)e^{j\omega t}\right] && (b)\\
\vec{B}(x,y,z,t) &= \mathrm{Re}\left[\vec{B}(x,y,z)e^{j\omega t}\right] && (c)\\
\vec{J}(x,y,z,t) &= \mathrm{Re}\left[\vec{J}(x,y,z)e^{j\omega t}\right] && (d)\\
\vec{q}(x,y,z,t) &= \mathrm{Re}\left[\vec{q}(x,y,z)e^{j\omega t}\right] && (e).
\end{aligned} \tag{4.4.7}$$

It is noted that the same notation is used for both the time-space dependent fields $[\vec{E}(x,y,z,t)]$ and only space-dependent fields $[\vec{E}(x,y,z)]$ in the phasor form. The context of the discussion can clarify the situation. In general, the space coordinates based $[\vec{E}(x,y,z)$ etc.$]$ phasor fields are complex quantities, with both the magnitude and phase angle associated with them. The real part $\mathrm{Re}\left[\vec{E}(x,y,z)e^{j\omega t}\right]$ gives actual sinusoidal field quantity with phase relation. $\vec{E}(x,y,z,t)$ is the instantaneous field quantity without any phase term. Further, the following relations are useful:

$$\begin{aligned}
&\frac{d}{dt}\left[E(x,y,z)e^{j\omega t}\right] = j\omega E(x,y,z)e^{j\omega t} && (a)\\
&\int E(x,y,z)e^{j\omega t}dt = -\frac{j}{\omega}E(x,y,z)e^{j\omega t} && (b).
\end{aligned} \tag{4.4.8}$$

Using the above equation for the electric field and similar equation for the magnetic field, Maxwell's equations (4.4.1), in the external source free medium, are rewritten below in the *phasor form* for the time-harmonic field:

$$\begin{aligned}
\nabla\times\vec{E} &= -j\omega\vec{B} && (a)\\
\nabla\times\vec{H} &= \vec{J}_c + j\omega\vec{D} && (b)\\
\nabla.\vec{D} &= \rho && (c)\\
\nabla.\vec{B} &= 0 && (d).
\end{aligned} \tag{4.4.9}$$

where flux density vectors are related to field intensity vectors in an anisotropic medium by the following tensor constitutive relations:

$$\vec{D} = \bar{\bar{\varepsilon}}.\vec{E} \quad (a)$$
$$\vec{B} = \bar{\bar{\mu}}.\vec{H} \quad (b)$$
$$\vec{J}_c = \bar{\bar{\sigma}}.\vec{E} \quad (c). \quad (4.4.10)$$

In the case of an external source-free homogeneous isotropic medium, Maxwell's equations are written in terms of field intensities only:

$$\nabla \times \vec{E} = -j\omega\mu_0\mu_r\vec{H} \quad (a)$$
$$\nabla \times \vec{H} = (\sigma + j\omega\varepsilon_0\varepsilon_r)\vec{E} \quad (b)$$
$$\nabla.\vec{E} = \rho/\varepsilon \quad (c)$$
$$\nabla.\vec{H} = 0 \quad (d), \quad (4.4.11)$$

where the scalar constitutive relations are $\vec{D} = \varepsilon_0\varepsilon_r\vec{E}$, $\vec{B} = \mu_0\mu_r\vec{H}$, $\vec{J}_c = \sigma\vec{E}$. The *differential form* of the above given Maxwell's equations provide relations between the electric field and the magnetic field at any location in the medium. Also, the sources are specified at a point in the medium. The *Maxwell equations (4.4.9) involving the flux densities* $\left(\vec{D}, \vec{B}\right)$ *and field intensities* $\left(\vec{E}, \vec{H}\right)$ *apply to both the isotropic and anisotropic media, whereas Maxwell's equations (4.4.11) involving the field intensities* $\left(\vec{E}, \vec{H}\right)$ *apply to the isotropic medium only.*

The differential form of Maxwell equations does not account for the creation of the fields in the space due to the sources such as the charge or current distributed over a line, surface, or volume. This case is incorporated in Maxwell's equations by converting them to the *integral forms*. It is achieved with the help of two vector identities:

Stokes theorem :
$$\oint_c \vec{A}.d\vec{l} = \iint_S \left(\nabla \times \vec{A}\right).d\vec{s}. \quad (4.4.12)$$

Gauss divergence theorem :
$$\oiint_S \vec{A}.d\vec{s} = \iiint_V \left(\nabla.\vec{A}\right)dv. \quad (4.4.13)$$

Figure (4.8b) shows the existence of a vector $\vec{A}$ over the surface S. Its boundary is enclosed by the perimeter C. Stoke theorem is defined with respect to Fig. (4.8b) and Gauss divergence theorem with respect to Fig. (4.8c). The unit vector $\hat{n}$ shows the direction of a normal to the surface S. Figure (4.8c) shows a vector $\vec{A}$ existing in the whole of the volume V that is enclosed by the surface S.

Maxwell's equations in the integral form, for $\vec{M}_{ext}=0$, are obtained by taking the surface integral of Maxwell's equation (4.4.1a):

$$\iint_S \left(\nabla \times \vec{E}\right).d\vec{s} = -\iint_S \frac{\partial \vec{B}}{\partial t} d\vec{s} = -\frac{\partial}{\partial t}\iint_S \vec{B}.d\vec{s}. \quad (4.4.14)$$

It is assumed that a surface enclosing the magnetic field does not change with time. By using equation (4.4.12), equation (4.4.14) is changed in the following form:

$$\oint_C \vec{E}.d\vec{l} = -\frac{\partial}{\partial t}\iint_S \left(\vec{B}.d\vec{s}\right) = -\frac{\partial \psi_m}{\partial t}, \quad (4.4.15)$$

where ψ_m is the magnetic flux. It is the *Faraday law of induction* that gives the induced voltage V, i.e. the emf, on a conducting loop containing the time-varying magnetic flux, ψ_m:

$$\text{emf} = V = \oint \vec{E}.dl = -\frac{\partial \psi_m}{\partial t}. \quad (4.4.16)$$

Likewise, using Maxwell's equation (4.4.1b) and (4.4.12) for $\vec{J}_{ext}=0$, the second Maxwell's equation is written in the integral form, giving the *modified Ampere's law*:

$$\oint_C \vec{H}.d\vec{l} = \iint_S \vec{J}_c.d\vec{s} + \frac{\partial}{\partial t}\iint_S \vec{D}.d\vec{s} = \iint_S \left(\vec{J}_c + \vec{J}_d\right).d\vec{s}. \quad (4.4.17)$$

In the above equation, J_c and J_d are conduction and displacement current densities creating the magnetic field $\vec{H}$. The above expression is generalized Ampere's law due to Maxwell. The *magnetomotive force, mmf,* is obtained as follows:

$$\text{mmf} = \oint_C \vec{H}.d\vec{l} = \iint_S \left(\vec{J}_c + \vec{J}_d\right).d\vec{s} \quad (a)$$
$$\Rightarrow \oint \vec{H}.d\vec{l} = \iint_S \vec{J}_c.d\vec{s} + \frac{\partial \psi_e}{\partial t} \quad (b). \quad (4.4.18)$$

where ψ_e is the time-dependent electric flux. Equation (4.4.18b) is *Maxwell's induction law* giving the induced mmf due to the time-varying electric field. For the source free medium with $Jc = 0$, it is the complementary induction law of Faraday's law of induction.

4.4.2 Power and Energy Relation from Maxwell Equations

A medium supporting the electromagnetic fields also stores the EM-energy and supports the power flow. The EM-power is supplied to the enclosure by the time-dependent external electric current density J_{ext} and the time-dependent external magnetic current density M_{ext}. They create the time-dependent electric field $(\vec{E})$ and the time-dependent magnetic field $(\vec{H})$ shown in Fig. (4.8a).

The external power, supplied by the source to the medium, is

$$P_{ext} = \iint_S \left(\vec{H}.\vec{M}_{ext} + \vec{E}.\vec{J}_{ext}\right) ds. \quad (4.4.19)$$

The field and source quantities have RMS values, and these are also time-dependent. The power on a transmission line, carrying the voltage and current wave, is $P = VI \cos \phi$, i.e. a scalar product of the voltage and current. The EM-wave is a transverse electromagnetic wave, where the fields $\vec{E}$ and $\vec{H}$ are normal to each other. The power density $\vec{S}$ is defined by a vector product of $\vec{E}$ and $\vec{H}$, known as the *Poynting vector*:

$$\vec{S} = \vec{E} \times \vec{H}. \quad (4.4.20)$$

The divergence of the above equation provides the power entering, or leaving, a location in the space:

$$\nabla.\vec{S} = \nabla.\left(\vec{E} \times \vec{H}\right) = \vec{H}.\nabla \times \vec{E} - \vec{E}.\nabla \times \vec{H}. \quad (4.4.21)$$

The energy contained per unit volume, i.e. the *energy density*, in a dispersive and a nondispersive medium, in the form of the electric and magnetic energy, is given by the following expressions:

$$\text{Nondispersive medium}: W = \frac{1}{2}\varepsilon_0\varepsilon_r\, \vec{E}.\vec{E} + \frac{1}{2}\mu_0\mu_r\, \vec{H}.\vec{H} \quad \text{(a)}$$
$$\text{Dispersive medium}: \quad W = \frac{1}{2}\varepsilon_0 \frac{\partial}{\partial \omega}[\varepsilon_r(\omega)]\, \vec{E}.\vec{E} + \frac{1}{2}\mu_0 \frac{\partial}{\partial \omega}[\mu_r(\omega)]\, \vec{H}.\vec{H} \quad \text{(b).} \quad (4.4.22)$$

A physical medium is dispersive. For a dispersive medium, the modified equation (4.4.22b) is valid [B.2, B.8, B.11–B.15]. The energy density W is a positive quantity, i.e. the following relations must be satisfied:

$$\frac{\partial(\varepsilon_r(\omega))}{\partial \omega} > 0, \quad \frac{\partial(\mu_r(\omega))}{\partial \omega} > 0. \quad (4.4.23)$$

The medium having finite conductivity σ dissipates the EM-energy in the form of heat given by the Joule's law:

$$\vec{P}_d = \vec{E}.\vec{J}_c = \sigma \vec{E}^2. \quad (4.4.24)$$

The total power carried in a medium, in the form of the EM-wave, is

$$P_{wave} = \iint_S \left(\vec{E} \times \vec{H}\right).d\vec{s}. \quad (4.4.25)$$

The integration is carried over the cross-section of the medium carrying the EM-power. The total energy stored in volume V is

$$W_{\substack{energy\\stored}} = \frac{1}{2}\iiint_V \left(\varepsilon_0\varepsilon_r|E|^2 + \mu_0\mu_r|H|^2\right) dv. \quad (4.4.26)$$

The total power dissipated in volume V of the medium is

$$P_{dis} = \iiint_V \left(\sigma|E|^2\right)dv. \quad (4.4.27)$$

Thus, power (P_{ext}) supplied by the external source is balanced by the following equation:

$$P_{ext} = P_{wave} + P_{dis} + \frac{\partial}{\partial t}\left(W_{\substack{energy\\stored}}\right). \quad (4.4.28)$$

Using Maxwell equations, the power balance equation (4.4.28) is evaluated in terms of the field quantities and external sources. The resulting power balance equation identifies the above-mentioned expressions for the power in a wave, energy dissipated in the medium, energy stored in the medium, and also the energy supplied by the external electric and magnetic currents.

The dot product of equation (4.4.1a) is taken with $\vec{H}$, and the dot product of equation (4.4.1b) with $\vec{E}$, and subtract one from another to get the following expression:

$$\left(\vec{H}.\nabla \times \vec{E}\right) - \left(\vec{E}.\nabla \times \vec{H}\right) = -\left(\vec{H}.\vec{M}_{ext} + \vec{E}.\vec{J}_{ext}\right) - \sigma\vec{E}.\vec{E} - \mu_0\left[\vec{H}.\frac{\partial}{\partial t}\left(\mu_r\vec{H}\right)\right] - \varepsilon_0\left[\vec{E}.\frac{\partial}{\partial t}\left(\varepsilon_r\vec{E}\right)\right]. \quad (4.4.29)$$

In the above equation, $\vec{J}_c = \sigma\vec{E}$ is used. In the case of a time-invariant medium, the relative permittivity and relative permeability of a medium are constants. By using an equation (4.4.21), the above equation is rewritten as follows:

$$\nabla.\left(\vec{E}\times\vec{H}\right) = -\left(\vec{H}.\vec{M}_{ext} + \vec{E}.\vec{J}_{ext}\right) - \sigma\,\vec{E}.\vec{E} - \frac{\partial}{\partial t}\frac{1}{2}\left(\varepsilon_0\varepsilon_r|E|^2 + \mu_0\mu_r|H|^2\right). \tag{4.4.30}$$

On taking volume integral of the above equation and further using Gauss divergence theorem (4.4.13), the following expression is obtained:

$$\underbrace{\iiint_V\left(\vec{H}.\vec{M}_{ext} + \vec{E}.\vec{J}_{ext}\right)dv}_{P_{ext}} = -\left[\underbrace{\iint_S\left(\vec{E}\times\vec{H}\right).\vec{ds}}_{P_{wave}} + \underbrace{\iiint_V\sigma|E|^2dv}_{P_{dis}} + \frac{\partial}{\partial t}\frac{1}{2}\underbrace{\iiint_V\left(\varepsilon_0\varepsilon_r E^2 + \mu_0\mu_r|H|^2\right)dv}_{W_{energy\ stored}}\right]. \tag{4.4.31}$$

The power supplied by the external sources is positive. So, the total power in a medium is negative. Out of the total power supplied to a medium, P_{wave} is power carried away by the EM-wave, P_{dis} is power loss in the medium due to the finite conductivity of the medium, and $\frac{\partial}{\partial t}\left(W_{\substack{energy\\stored}}\right)$ is the oscillating electric and magnetic energy in the medium.

4.5 EM-waves in Unbounded Isotropic Medium

Maxwell's coupled vector differential equations (4.4.1a) and (4.4.1b) are solved in an external source-free medium with $\vec{M}_{ext} = \vec{J}_{ext} = 0$ to obtain separate wave equations for the electric and magnetic fields. It is like getting the voltage and current wave equations on a transmission line. Maxwell equations are further presented in the *vector algebraic form.* The present section is concerned with the uniform 1D wave propagation in an unbounded isotropic dielectric medium, also in a conducting medium. The EM-wave propagation in the anisotropic medium is discussed in section (4.7).

4.5.1 EM-wave Equation

Maxwell's coupled equations (4.4.1), in the external source-free medium $\vec{M}_{ext} = \vec{J}_{ext} = 0$, are solved, using the rule of vector algebra, for the electric field intensity $\vec{E}$ by substituting $\vec{H}$ from equation (4.4.1b) to equation (4.4.1a):

$$\begin{aligned}&\nabla\times(\nabla\times\vec{E}) = -\mu_0\mu_r\frac{\partial}{\partial t}\left(\nabla\times\vec{H}\right) \Rightarrow \nabla\times\left(\nabla\times\vec{E}\right) = -\mu_0\mu_r\frac{\partial}{\partial t}\left[\vec{J}_c + \varepsilon_0\varepsilon_r\frac{\partial\vec{E}}{\partial t}\right]\\ &\Rightarrow \nabla\times\left(\nabla\times\vec{E}\right) = -\mu_0\mu_r\frac{\partial}{\partial t}\left[\sigma\vec{E} + \varepsilon_0\varepsilon_r\frac{\partial\vec{E}}{\partial t}\right], \Rightarrow \nabla^2\vec{E} = \mu_0\mu_r\sigma\frac{\partial\vec{E}}{\partial t} + \mu_0\varepsilon_0\varepsilon_r\mu_r\frac{\partial^2\vec{E}}{\partial t^2}.\end{aligned} \tag{4.5.1}$$

In the above equation, the identity $\nabla\times\left(\nabla\times\vec{E}\right) = \nabla\left(\nabla.\vec{E}\right) - \nabla^2\vec{E}$ is used. The above wave equation for the electric field is valid in an isotropic, homogeneous, and lossy medium. In a homogeneous medium, the (μ_r, ε_r) are not a function of position. Further, the medium is taken as charge-free, i.e. $\rho = 0, \nabla.\vec{E} = 0$. Likewise, the following wave equation is obtained for the magnetic field:

$$\nabla^2\vec{H} = \mu_0\mu_r\sigma\frac{\partial\vec{H}}{\partial t} + \mu_0\varepsilon_0\varepsilon_r\mu_r\frac{\partial^2\vec{H}}{\partial t^2}. \tag{4.5.2}$$

To get the time-harmonic field, i.e. $\vec{E}(x,y,z,t) = \vec{E}(x,y,z)\exp(j\omega t)$, the time differential variable is replaced as follows: $\partial/\partial t \to j\omega$ and $\partial^2/\partial t^2 \to -\omega^2$. The above wave equations for the time-harmonic fields are written as,

$$\begin{aligned}&\nabla^2\vec{E} - j\omega\mu_0\mu_r(\sigma + j\varepsilon_0\varepsilon_r\omega)\vec{E} = 0 \quad \text{(a)}\\ &\nabla^2\vec{H} - j\omega\mu_0\mu_r(\sigma + j\varepsilon_0\varepsilon_r\omega)\vec{H} = 0 \quad \text{(b)}.\end{aligned} \tag{4.5.3}$$

The complex propagation constant $\gamma = \alpha + j\beta$ is defined as follows:

$$\gamma^2 = j\omega\mu_0\mu_r(\sigma + j\varepsilon_0\varepsilon_r\omega). \quad (4.5.4)$$

For a lossless medium $\sigma = 0$, and the propagation constant is a real quantity:

$$\gamma = j\omega\sqrt{\mu_0\varepsilon_0}\sqrt{\mu_r\varepsilon_r} \Rightarrow \beta = \omega\sqrt{\mu_0\varepsilon_0}\sqrt{\mu_r\varepsilon_r} = \omega\sqrt{\mu\varepsilon} \quad (a)$$
$$\alpha = 0 \quad (b). \quad (4.5.5)$$

In a homogeneous medium, propagation constant β is also expressed as the *wavenumber* k. In free space, $\mu_r = \varepsilon_r = 1$. The velocity of the EM-wave is equal to the velocity of light (c) in free space:

$$c = \frac{\omega}{\beta_0} = \frac{1}{\sqrt{\mu_0\varepsilon_0}}, \quad (4.5.6)$$

where $\beta_0 = \omega\sqrt{\mu_0\varepsilon_0}$ is the propagation constant, i.e. the wavenumber (k_0) in free space. A lossless material medium is electrically characterized by (ε_r, μ_r). However, it is also characterized by the *refractive index* $n = \sqrt{\mu_r\varepsilon_r}$. In the case of a dielectric medium, it is $n = \sqrt{\varepsilon_r}$. The velocity of the EM-wave propagation in a medium is

$$v_p = \frac{\omega}{\beta} = \frac{c}{\sqrt{\mu_r\varepsilon_r}} \quad (a)$$
$$v_p = \frac{c}{\sqrt{\varepsilon_r}} = \frac{c}{n} \quad (b). \quad (4.5.7)$$

For a lossy medium, the complex propagation constant can be further written as:

$$\gamma^2 = -\omega^2\mu_0\mu_r\varepsilon_0\left(\varepsilon_r - j\frac{\sigma}{\omega\varepsilon_0}\right). \quad (4.5.8)$$

For a lossy dielectric medium, ε_r is defined as a complex quantity:

$$\varepsilon_r^* = \varepsilon_r' - j\varepsilon_r'' \quad (a)$$
$$\text{where, } \varepsilon_r'' = \frac{\sigma}{\omega\varepsilon_0} \quad (b). \quad (4.5.9)$$

It is like the previous discussion on the complex relative permittivity ε_r^* in a lossy dielectric medium, with the following expressions for the loss-tangent and propagation constant:

$$\tan\delta = \frac{\varepsilon_r''}{\varepsilon_r'} = \frac{\sigma}{\omega\varepsilon_0\varepsilon_r} \quad (a)$$
$$\gamma^2 = -\omega^2\mu_0\mu_r\varepsilon_0\varepsilon_r(1 - j\tan\delta) \quad (b)$$
$$\Rightarrow \alpha^2 - \beta^2 + j2\alpha\beta = -\omega^2\mu_0\mu_r\varepsilon_0\varepsilon_r(1 - j\tan\delta) \quad (c). \quad (4.5.10)$$

In the above equation, the real part of the complex relative permittivity is $\varepsilon_r = \varepsilon_r'$.

On separating the real and imaginary parts, the attenuation constant (α) and propagation constant (β) are obtained:

$$\alpha = \omega\left[\frac{\mu_0\mu_r\varepsilon_0\varepsilon_r}{2}\left\{\sqrt{1 + \tan^2\delta} - 1\right\}\right]^{1/2} \quad (a)$$
$$\beta = \omega\left[\frac{\mu_0\mu_r\varepsilon_0\varepsilon_r}{2}\left\{\sqrt{1 + \tan^2\delta} + 1\right\}\right]^{1/2} \quad (b). \quad (4.5.11)$$

The wave equation (4.5.3a) and (4.5.3b) for the $(\vec{E}, \vec{H})$ fields in a lossy and lossless $(\alpha = 0)$ media are rewritten below:

$$\nabla^2\vec{E} - \gamma^2\vec{E} = 0 \quad (a)$$
$$\nabla^2\vec{E} + \beta^2\vec{E} = 0 \quad (b)$$
$$\nabla^2\vec{H} - \gamma^2\vec{H} = 0 \quad (c)$$
$$\nabla^2\vec{H} + \beta^2\vec{H} = 0 \quad (d). \quad (4.5.12)$$

The propagation constant β is also expressed as the wavenumber k of the wavevector $\vec{k}$. Sometimes in place of the complex propagation constant γ, the complex wavevector k is used as a complex propagation constant, i.e. $k^* = \beta - j\alpha$. On using the complex k, the field is written as $E_0\,e^{-jkx} = E_0\,e^{-j(\beta - j\alpha)x} = (E_0e^{-\alpha x})\,e^{-j\beta x}$.

4.5.2 1D Wave Equation

For the wave propagating only in the x-direction, equations (4.5.12a) and (4.5.12b) are reduced to the 1D wave equations:

$$\frac{d^2\vec{E}}{dx^2} - \gamma^2\vec{E} = 0 \quad (a)$$
$$\frac{d^2\vec{H}}{dx^2} - \gamma^2\vec{H} = 0 \quad (b). \quad (4.5.13)$$

Equation (4.5.13a) has the solution, $\vec{E}(x) = \vec{E}_0e^{-\gamma x} = \vec{E}_0e^{-\alpha x}e^{-j\beta x}$. The time-harmonic wave propagating in the x-direction is

$$\vec{E}(x,t) = \text{Re}\left[\vec{E}(x)e^{j\omega t}\right] = \text{Re}\left[\vec{E}_0\,e^{-\alpha x}e^{j(\omega t - \beta x)}\right] \quad (a)$$

Likewise,

$$\vec{H}(x,t) = \text{Re}\left[\vec{H}(x)e^{j\omega t}\right] = \text{Re}\left[\vec{H}_0\,e^{-\alpha x}e^{j(\omega t - \beta x)}\right] \quad (b). \quad (4.5.14)$$

The field equations in the time-domain are also written as follows:

$$\vec{E}(x,t) = \vec{E}_0 e^{-\alpha x}\cos(\omega t - \beta x) \quad (a)$$
$$\vec{H}(x,t) = \vec{H}_0 e^{-\alpha x}\cos(\omega t - \beta x) \quad (b). \qquad (4.5.15)$$

In the case of a lossy medium, equation (4.5.11) shows that both α and β depend on the loss-tangent of a medium. For a lossless medium, $\tan\delta = 0$, leading to $\alpha = 0$, and $\beta = (\omega/c)\sqrt{\mu_r\varepsilon_r}$. In the case of low conductivity, i.e. the low-loss medium, the approximation $\tan\delta \ll 1$, or $(\sigma/\omega\varepsilon_0\varepsilon_r) \ll 1$, can be used. In such a medium $\sigma \ll \omega\varepsilon_0\varepsilon_r$, the contribution of the conduction current is small as compared to that of the displacement current. Such a medium is a dielectric medium with a small loss. On approximating, $[1 + \tan^2\delta]^{1/2} \approx [1 + (1/2)\tan^2\delta]$; the following expression is obtained from equation (4.5.11a):

$$\alpha = \frac{\omega\sqrt{\mu_0\varepsilon_0}\sqrt{\mu_r\varepsilon_r}}{2}\tan\delta = \frac{\omega}{2c}\sqrt{\mu_r\varepsilon_r}\tan\delta. \qquad (4.5.16)$$

The above equation computes the dielectric loss of a low-loss dielectric medium. The approximation $\sqrt{1+\tan^2\delta} + 1 \approx [2 + (\tan^2\delta)/2] \approx 2$ is used to get an approximate value of β for such medium from equation (4.5.11b):

$$\beta = \omega\sqrt{\mu_0\varepsilon_0}\sqrt{\mu_r\varepsilon_r} = \frac{\omega}{c}\sqrt{\mu_r\varepsilon_r} = \beta_0\sqrt{\mu_r\varepsilon_r}. \qquad (4.5.17)$$

For a low-loss dielectric medium *the dielectric loss, due to tan δ, increases linearly with frequency* ω. However, the propagation constant β is dispersionless, giving the frequency-independent phase velocity. The above approximation can also be carried out in a little different way:

$$\sqrt{1+\tan^2\delta} \approx \left(1 + \frac{\sigma^2}{2\,\omega^2\,\varepsilon_0^2\varepsilon_r^2}\right)$$
$$\alpha \approx \omega\left[\frac{\mu_0\mu_r\varepsilon_0\varepsilon_r}{2} \times \frac{\sigma^2}{2\,\omega^2\varepsilon_0^2\varepsilon_r^2}\right]^{1/2}$$
$$\Rightarrow \alpha = \frac{\sigma}{2}\sqrt{\frac{\mu_0\mu_r}{\varepsilon_0\varepsilon_r}} = \frac{\sigma}{2}\eta\sqrt{\frac{\mu_r}{\varepsilon_r}} \quad (a)$$
$$\text{For a dielectric medium:}\ \alpha = \frac{\sigma}{2}\frac{\eta}{\sqrt{\varepsilon_r}} \quad (b). \qquad (4.5.18)$$

In the above equation, $\eta = \sqrt{\mu_0/\varepsilon_0} = 377\Omega$ is the characteristic (*intrinsic*) impedance of free space. A low-loss medium is a mildly dispersive medium, with the frequency-dependent phase velocity:

$$\beta \approx \omega\sqrt{\frac{\mu_0\mu_r\varepsilon_0\varepsilon_r}{2}}\left[\left(1 + \frac{\sigma^2}{2\,\omega^2\varepsilon_0^2\varepsilon_r^2}\right) + 1\right]^{1/2}$$
$$= \omega\sqrt{\mu_0\varepsilon_0}\sqrt{\mu_r\varepsilon_r}\left[1 + \frac{\sigma^2}{8\,\omega^2\varepsilon_0^2\varepsilon_r^2}\right]$$
$$\Rightarrow \beta \approx \beta_0\sqrt{\mu_r\varepsilon_r}\left[1 + \frac{\sigma^2}{8\,\omega^2\varepsilon_0^2\varepsilon_r^2}\right] \quad (a)$$
$$v_p = \frac{\omega}{\beta} \approx v_0\left[1 - \frac{\sigma^2}{8\,\omega^2\varepsilon_0^2\varepsilon_r^2}\right] \quad (b). \qquad (4.5.19)$$

In the above equation, $v_0 = c/\sqrt{\mu_r\varepsilon_r}$ is the velocity of the EM-wave in the lossless dielectric medium. The presence of the loss has decreased the phase velocity, i.e. a lossy medium supports the dispersive *slow-wave* propagation. The use of α and β from equations (4.5.11a) and (4.5.11b) provide more accurate results. The dispersion in a medium is always associated with loss. This fundamental property is further discussed in chapter 6.

4.5.3 Uniform Plane Waves in Linear Lossless Homogeneous Isotropic Medium

Figure (4.9a) shows the propagation of the TEM wave in the x-direction of an unbounded medium. Figure (4.9b) further shows that for the TEM wave, electric and magnetic field components, i.e. the pairs (E_y, H_z), or (E_z, H_y), are transverse to the direction of propagation, i.e. $\pm x$ - direction. For the field pair (E_y, H_z), the wave is y-polarized; and for the field pair(E_z, H_y), it is the z-polarized. The *polarization of an EM-wave* is determined by the direction of the $\vec{E}$ vector. The propagating wave is called the *uniform plane wave*, as amplitudes of electric and magnetic fields are constant over the equiphase surfaces. Figure (4.9c) shows that the phases of the E_y field at any instant of time, over the equiphase surfaces, are either 0° or 180°. The (y-z)-plane is the equiphase surface.

The field components of a uniform plane wave do not change with y and z coordinates, i.e. $\partial/\partial y$(field E or H) $= \partial/\partial z$(field E or H) $= 0$. The field components are a function of x only. So, the field components of the EM-wave propagating in the x-direction can be written as follows:

$$E_i = E_{0i}\, e^{j(\omega t \mp \beta_x x)} \quad (a)$$
$$H_i = H_{0i}\, e^{j(\omega t \mp \beta_x x)} \quad (b) \qquad (4.5.20)$$

where, $i = y, z$.

In the above expressions, the (−) sign shows the wave propagation in the positive x-direction, whereas

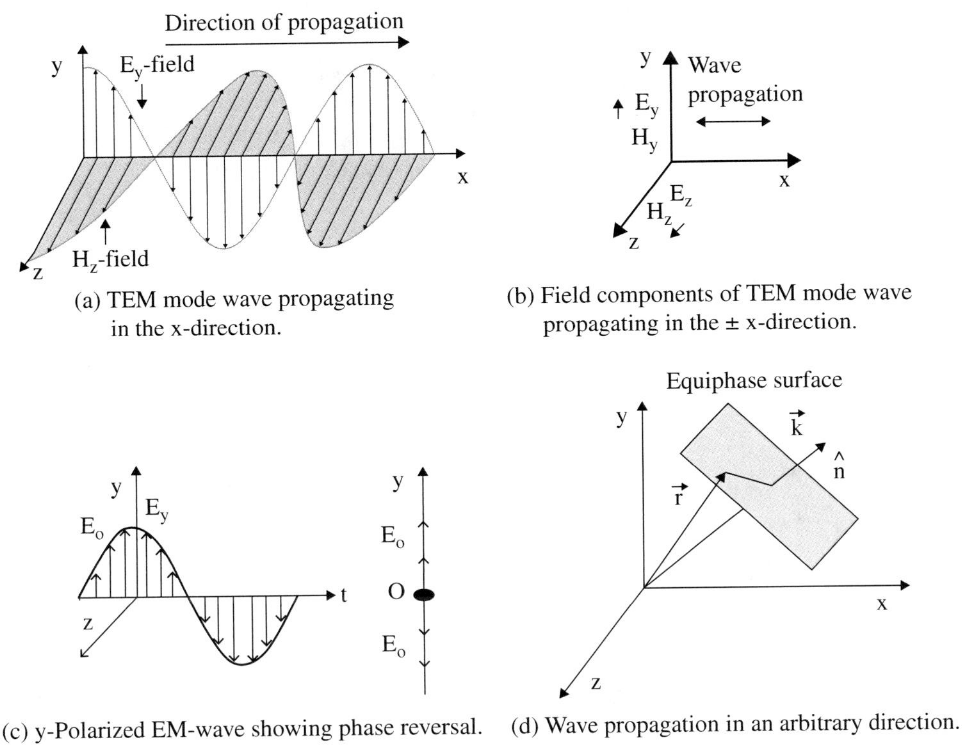

(a) TEM mode wave propagating in the x-direction.

(b) Field components of TEM mode wave propagating in the ± x-direction.

(c) y-Polarized EM-wave showing phase reversal.

(d) Wave propagation in an arbitrary direction.

Figure 4.9 TEM mode wave in an unbounded medium.

the (+) sign is for the wave propagation in the negative x-direction. The wave propagation in the positive x-direction is discussed below.

The above expressions related to a uniform plane wave, in an external source-free $\vec{M}_{ext} = \vec{J}_{ext} = 0$ lossless $\vec{J}_c = 0$ medium, can be applied to the Maxwell equations (4.4.1). In the present case, the del operator is replaced by a derivative with respect to x, i.e. $\nabla \rightarrow \hat{x}(\partial/\partial x)$ as a derivative with respect to y and z are zero. Maxwell first curl equation is reduced to a simpler form:

$$\hat{x}\frac{\partial}{\partial x} \times \left[\hat{x}E_x + \hat{y}E_y + \hat{z}E_z\right] = -\mu_0\mu_r\frac{\partial}{\partial t}\left[\hat{x}H_x + \hat{y}H_y + \hat{z}H_z\right]$$

$$\Rightarrow -\hat{y}\frac{\partial E_z}{\partial x} + \hat{z}\frac{\partial E_y}{\partial x} = -\mu_0\mu_r\left[\hat{x}\frac{\partial H_x}{\partial t} + \hat{y}\frac{\partial H_y}{\partial t} + \hat{z}\frac{\partial H_z}{\partial t}\right]. \tag{4.5.21}$$

On separating each component of the fields, the following expressions are obtained:

$$\begin{aligned} \frac{\partial H_x}{\partial t} &= 0 && \text{(a)}\\ \frac{\partial E_z}{\partial x} &= \mu_0\mu_r\frac{\partial H_y}{\partial t} && \text{(b)}\\ \frac{\partial E_y}{\partial x} &= -\mu_0\mu_r\frac{\partial H_z}{\partial t} && \text{(c).} \end{aligned} \tag{4.5.22}$$

Likewise, the following expressions are obtained from Maxwell's second curl expression (4.4.1b):

$$\begin{aligned} \frac{\partial E_x}{\partial t} &= 0 && \text{(a)}\\ -\frac{\partial H_z}{\partial x} &= \varepsilon_0\varepsilon_r\frac{\partial E_y}{\partial t} && \text{(b)}\\ \frac{\partial H_y}{\partial x} &= \varepsilon_0\varepsilon_r\frac{\partial E_z}{\partial t} && \text{(c).} \end{aligned} \tag{4.5.23}$$

It is seen from the above equations that the E_x and H_x components, in the direction of propagation, are time-independent, i.e. constant. They do not play any role in the wave propagation and can be assumed to be zero, without affecting the wave propagation [B.3]. Only transverse field components play a role in wave propagation. The time-varying H_y component generates the E_z, whereas the time-varying E_y component generates the H_z. It is also true for another time-varying pair (H_z, E_z). Maxwell divergence relations also show $\partial E_x/\partial x = \partial H_x/\partial x = 0$. Again, E_x and H_x components do not show any variation along the direction of propagation that is essential for wave propagation. So, in the TEM mode propagation, the longitudinal field components are zero, i.e. $E_x = H_x = 0$.

By using equation (4.5.20) with the above equations, the time-harmonic fields are rewritten as follows:

$$\frac{\partial H_z}{\partial x} = -j\omega\varepsilon_0\varepsilon_r E_y \quad (a)$$
$$\frac{\partial H_y}{\partial x} = j\omega\varepsilon_0\varepsilon_r E_z \quad (b)$$
$$\frac{\partial E_z}{\partial x} = j\omega\mu_0\mu_r H_y \quad (c)$$
$$\frac{\partial E_y}{\partial x} = -j\omega\mu_0\mu_r H_z \quad (d). \tag{4.5.24}$$

On using equation (4.5.20) with the above equations, the following algebraic expressions are obtained for the EM-wave propagating in the positive x-direction:

$$\beta_x H_z = \omega\varepsilon_0\varepsilon_r E_y \quad (a)$$
$$\beta_x H_y = -\omega\varepsilon_0\varepsilon_r E_z \quad (b)$$
$$\beta_x E_z = -\omega\mu_0\mu_r H_y \quad (c)$$
$$\beta_x E_y = \omega\mu_0\mu_r H_z \quad (d). \tag{4.5.25}$$

The *wave impedance* in free space, or in homogeneous unbounded medium, is defined in a plane normal to the direction of propagation [B.2]. For instance, Fig. (4.9a) shows the direction of propagation is along the x-axis, and wave impedance is defined in the (y-z) plane. It is also called the *intrinsic impedance* η_0 of free space and intrinsic impedance η of material filled homogeneous space. The following expression is obtained for the wave impedance η, for y-polarized waves, from equation (4.5.25a):

$$\vec{\eta} = \frac{E_y}{H_z} = \hat{x}\frac{\beta_x}{\omega\varepsilon_0\varepsilon_r} = \hat{x}\frac{\omega\sqrt{\mu_0\varepsilon_0}\sqrt{\mu_r\varepsilon_r}}{\omega\varepsilon_0\varepsilon_r}$$
$$\Rightarrow \vec{\eta} = \hat{x}\sqrt{\frac{\mu_0\mu_r}{\varepsilon_0\varepsilon_r}} = \hat{x}\,\eta_0\sqrt{\frac{\mu_r}{\varepsilon_r}} \quad (a)$$
$$\vec{\eta} = +\hat{x}\,\eta_0\sqrt{\frac{\mu_r}{\varepsilon_r}} \quad (b). \tag{4.5.26}$$

Equation (4.5.25b) provides the following wave impedance for the z-polarized waves:

$$\vec{\eta} = \frac{-E_z}{H_y} = \hat{x}\frac{\beta_x}{\omega\varepsilon_0\varepsilon_r} \quad (a)$$
$$\vec{\eta} = \hat{x}\frac{-E_z}{H_y} = \hat{x}\eta_0\sqrt{\frac{\mu_r}{\varepsilon_r}} \quad (b). \tag{4.5.27}$$

The positive wave impedance of the (E_y, H_z) or ($-E_z$, H_y) fields corresponds to a wave traveling in the +x direction. However, for the (E_z, H_y) fields, the wave impedance is negative showing the wave propagation in the negative x-direction. Under certain conditions, a medium can have an imaginary value of propagation constant, i.e. $\beta_x = -jp$. In this case, the wave impedance becomes reactive, and there is no wave propagation. Again, the wave equation (4.5.20) is reduced to $E_i = E_{0i}e^{-px}e^{j\omega t}$ and $H_i = H_{0i}\,e^{-px}e^{j\omega t}$ for the wave propagation in the positive x-direction. These are decaying nonpropagating *evanescent mode* waves. These are only decaying oscillations.

The wave equations (4.5.13a) and (4.5.13b), for the E_y and H_z field components, are solved to get the total solution as a superposition of two waves traveling in opposite directions:

$$E_y(x,t) = \underbrace{E_{oy}^{+}\,e^{j(\omega t-\beta x)}}_{\text{Forward moving wave}} + \underbrace{E_{oy}^{-}\,e^{j(\omega t+\beta x)}}_{\text{Backward moving wave}} \quad (a)$$

$$H_z(x,t) = \left[H_{oz}^{+}\,e^{j(\omega t-\beta x)} - H_{oz}^{-}\,e^{j(\omega t+\beta x)}\right]$$
$$\Rightarrow H_z(x,t) = \frac{1}{\eta}\left[E_{oy}^{+}\,e^{j(\omega t-\beta x)} - E_{oy}^{-}\,e^{j(\omega t+\beta x)}\right] \quad (b). \tag{4.5.28}$$

In the above expression, $\sqrt{\mu_0/\varepsilon_0} = \eta_0 = 377\ \Omega$ is the intrinsic or characteristic impedance, of the uniform plane in free space. The power movement is obtained from the Poynting vector $\vec{S} = \vec{E} \times \vec{H} = (\hat{y}E_y) \times (\hat{z}H_z) = (\hat{y} \times \hat{z})E_yH_z = \hat{x}(E_yH_z)$. The power of the forward wave travels in the positive x-direction. *The direction of the power movement is the direction of the group velocity.* In the x-direction, the *direction of the phase velocity is associated with the direction of the propagation vector*, i.e. the wavenumber $\vec{\beta} = \vec{k}$.

Figure (4.9d) shows the propagating EM-wave in an arbitrary direction of the wavevector $\vec{k} = \hat{x}k_x + \hat{y}k_y + \hat{z}k_z$. The wavevector $\vec{k}$ is normal to the equiphase surface. The position vector of a point P at the equiphase surface is $\vec{r} = \hat{x}x + \hat{y}y + \hat{z}z$. The following expressions describe the propagating wave as a solution to the wave equations (4.5.12a) and (4.5.12c):

$$\vec{E}\left(\vec{r},t\right) = \vec{E}_{or}e^{j\left(\omega t - \vec{k}.\vec{r}\right)} \quad (a)$$
$$\vec{H}\left(\vec{r},t\right) = \vec{H}_{or}e^{j\left(\omega t - \vec{k}.\vec{r}\right)} \quad (b)$$
$$\text{where, } \vec{k}.\vec{r} = k_x x + k_y y + k_z z \quad (c)$$
$$\vec{k}.\vec{k} = k^2 = k_x^2 + k_y^2 + k_z^2 \quad (d), \tag{4.5.29}$$

where $|k| = \omega\sqrt{\mu\varepsilon}$. Equation (4.5.29d) is the *dispersion equation*. In the above equations, field quantities show time dependence, i.e. *temporal dependence*, through factor $e^{j\omega t}$ and space dependence, i.e. *spatial dependence*, through factor $e^{-j\vec{k}.\vec{r}}$. It also shows the lagging phase of the propagating wave in the positive direction.

This is the convention adopted by engineers. On several occasions, physicist prefers an alternative sign convention, i.e. $e^{-j\omega t}$ and $e^{j\vec{k}\cdot\vec{r}}$. It leads to a leading phase for propagating waves in a positive direction. A reader must be careful while reading literature from several sources. For y-polarized wave propagating in the x-direction with $k_y = k_z = 0$ and $k_x = \beta_x - j\alpha$, the solution of the wave equation could be written as follows:

$$E_y(x,t) = E_{oy}^{+} e^{-\alpha x} e^{j(\omega t - \beta_x x)} + E_{oy}^{-} e^{\alpha x} e^{j(\omega t + \beta_x x)} \quad (a)$$

$$H_z(x,t) = \frac{1}{\eta}\left[E_{oy}^{+} e^{-\alpha x} e^{j(\omega t - \beta_x x)} - E_{oy}^{-} e^{\alpha x} e^{j(\omega t + \beta_x x)}\right] \quad (b). \tag{4.5.30}$$

The second terms of the above equations show wave propagation in the negative x-direction. The wave equations show the decaying propagating waves. For the case $\alpha = 0$, the above equations are the same as that of equations (4.5.28).

4.5.4 Vector Algebraic Form of Maxwell Equations

Maxwell's equations in the unbounded medium could be also written in the *vector algebraic form*. The del operator can be replaced as follows: $\nabla \rightarrow -j\vec{k}$. Using equation (4.4.9), for the charge-free lossless medium $\rho = \sigma = 0$, two sets of Maxwell equations, for the isotropic and anisotropic media, are written in the following algebraic forms:

Set #I for the isotropic medium:

$$\nabla \times \vec{E} = -j\vec{k} \times \vec{E} = -j\omega\mu\vec{H} \quad \Rightarrow \quad \vec{k} \times \vec{E} = \omega\mu\vec{H} \quad (a)$$
$$\nabla \times \vec{H} = -j\vec{k} \times \vec{H} = j\omega\varepsilon\vec{E} \quad \Rightarrow \quad \vec{k} \times \vec{H} = -\omega\varepsilon\vec{E} \quad (b)$$
$$\nabla.\vec{E} = 0 \quad \Rightarrow \quad \vec{k}.\vec{E} = 0 \quad (c)$$
$$\nabla.\vec{H} = 0 \quad \Rightarrow \quad \vec{k}.\vec{H} = 0 \quad (d).$$

Set #II for the anisotropic medium:

$$\vec{k} \times \vec{E} = \omega\vec{B} \quad (e)$$
$$\vec{k} \times \vec{H} = -\omega\vec{D} \quad (f)$$
$$\vec{k}.\vec{D} = 0 \quad (g)$$
$$\vec{k}.\vec{B} = 0 \quad (h). \tag{4.5.31}$$

Equations (4.5.31a) and (4.5.31b) show that *for the positive values of μ and ε, the triplet $\vec{E}, \vec{H}, \vec{k}$ follows the right-handed orthogonal coordinate system*. Equations (4.5.31c) and (4.5.31d) show that in an isotropic medium, the field vectors $\vec{E}$ and $\vec{H}$ are orthogonal to the wavevector $\vec{k}$. Equations (4.5.31c) and (4.5.31d) directly follow from the first two Maxwell equations by taking their dot product with the wavevector $\vec{k}$. Equation (4.5.31a) further shows that $\vec{H}$ is normal to both $\vec{E}, \vec{k}$ vectors, and equation (4.5.31b) shows that $\vec{E}$ is normal to both $\vec{k}, \vec{H}$ vectors. In brief, the vectors $\vec{E}, \vec{H}, \vec{k}$ are orthogonal to each other, and there is no field component along the wavevector $\vec{k}$, i.e. the wave is a *transverse electromagnetic (TEM) type*. Also, *in an isotropic medium, $\vec{E}$ is parallel to vector $\vec{D}$ and $\vec{H}$ is parallel to vector $\vec{B}$. This statement does not hold for the anisotropic medium.* Maxwell equations (4.5.31e)–(4.5.31h) apply to an anisotropic medium. In an anisotropic medium, equations (4.5.31e) and (4.5.31f) show that $\vec{B}$ is normal to vectors $\vec{k}, \vec{E}$ and $\vec{D}$ is normal to vectors $\vec{k}, \vec{H}$. However, $\vec{B}$ is not parallel to $\vec{H}$. Also, $\vec{D}$ is not parallel to $\vec{E}$. It is discussed in subsection (4.2.3).

Equations (4.5.31a) and (4.5.31b) are solved $\vec{E}$ to get the *vector algebraic form of wave equation* as follows:

$$\vec{k} \times \vec{k} \times \vec{E} = \omega\mu\,\vec{k} \times \vec{H} = -\omega^2\mu\varepsilon\,\vec{E} \quad \Rightarrow \vec{k} \times \vec{k} \times \vec{E} + \omega^2\mu\varepsilon\,\vec{E} = 0 \quad (a)$$
$$\text{On using identity, } \vec{k} \times \vec{k} \times \vec{E} = \vec{k}\left(\vec{k}.\vec{E}\right) - \vec{E}\left(\vec{k}.\vec{k}\right), \quad \left(\vec{k}.\vec{k} - \omega^2\mu\varepsilon\right)\vec{E} = 0 \quad (b). \tag{4.5.32}$$

Likewise, the wave equation for $\vec{D}$ could be written that is useful for the EM-wave propagation in an anisotropic medium. Equation (4.5.32b) is an *eigenvalue equation*, and the nontrivial solution for $E \neq 0$ provides the eigenvalue of k belonging to the propagating waves in the unbounded isotropic medium. The medium supports two numbers of linearly y-polarized waves known as *normal modes* propagating in ±x-directions with same phase velocity (v_p) given below:

$$k = \pm\omega\sqrt{\mu\varepsilon} = \pm k_0\sqrt{\mu_r\varepsilon_r} \quad (a)$$
$$v_p = \pm\frac{\omega}{k} = \pm\frac{c}{\sqrt{\mu_r\varepsilon_r}} \quad (b). \tag{4.5.33}$$

In the above equation, $k_0 = \omega\sqrt{\mu_0\varepsilon_0}$ is the wavenumber in free space. Using the intrinsic impedance with equation (4.5.31a), the magnetic field vector and Poynting vector are obtained below:

$$
\begin{aligned}
&\vec{k} \times \vec{E} = \omega\mu\,\vec{H} \quad \Rightarrow \quad \vec{H} = \frac{k\hat{k} \times \vec{E}}{\omega\mu} = \frac{\sqrt{\varepsilon\mu}}{\mu}\hat{k} \times \vec{E} = \frac{1}{\eta}\hat{k} \times \vec{E} && \text{(a)}\\
&\vec{k} \times \vec{H} = -\omega\varepsilon\,\vec{E} \quad \Rightarrow \quad \vec{E} = -\frac{k\hat{k} \times \vec{H}}{\omega\varepsilon} = -\frac{\sqrt{\varepsilon\mu}}{\varepsilon}\hat{k} \times \vec{H} = -\eta\hat{k} \times \vec{H} && \text{(b)}\\
&\bar{S} = \vec{E} \times \vec{H} = \frac{1}{\eta}\left(\vec{E} \times \hat{k} \times \vec{E}\right) = \frac{1}{\eta}\left[\left(\vec{E}.\vec{E}\right)\hat{k} - \left(\vec{E}.\hat{k}\right)\vec{E}\right] = \frac{E^2}{\eta}\hat{k} && \text{(c)}\\
&\text{Time averaged } S_{av} = \frac{1}{2}\,\text{Re}\left(\vec{E} \times \vec{H}^*\right) = \frac{E^2}{2\eta} && \text{(d).}
\end{aligned}
\tag{4.5.34}
$$

In the case of the propagation of waves *in an isotropic medium, the wavevector* $\vec{k}$ *and Poynting vector* $\vec{S}$ *both are in the same direction*. It provides the phase and group velocities in the same direction.

4.5.5 Uniform Plane Waves in Lossy Conducting Medium

The loss-tangent ($\tan\delta$), given in equation (4.5.10), is much greater than unity, i.e. $\tan\delta \gg 1$ for a highly conducting medium. It means a contribution of the conduction current is much more than that of the displacement current in a conducting medium, i.e. $\left(\sigma/\omega\varepsilon_0\varepsilon_r'\right) \gg 1$. However, the approximation for a low-loss medium is taken differently. The propagation constants of the EM-wave in a highly conducting and also in a low-loss medium are obtained from equation (4.5.4) as follows:

$$
\begin{aligned}
&\text{Highly conducting medium:}\\
&\gamma^2 = -\omega^2\mu_0\mu_r\varepsilon_0\varepsilon_r\left[1 - j\frac{\sigma}{\omega\varepsilon_0\varepsilon_r}\right] \approx j\omega\mu_0\mu_r\sigma && \text{(a)}\\
&\gamma = \alpha + j\beta = \sqrt{j}\sqrt{\omega\mu\sigma} = \left(\frac{1+j}{\sqrt{2}}\right)\sqrt{\omega\mu\sigma}\\
&\Rightarrow \alpha = \beta = \sqrt{\frac{\omega\mu\sigma}{2}} && \text{(b)}\\
&\text{Low-loss medium:}\\
&\gamma = j\frac{\omega}{c}\sqrt{\mu_r\varepsilon_r}\left[1 - j\frac{\sigma}{\omega\varepsilon_0\varepsilon_r}\right]^{1/2} \approx j\frac{\omega}{c}\sqrt{\mu_r\varepsilon_r}\left[1 - j\frac{\sigma}{2\omega\varepsilon_0\varepsilon_r}\right]\\
&\Rightarrow \gamma = \alpha + j\beta = \frac{\sigma\eta}{2} + j\frac{\omega}{c}\sqrt{\mu_r\varepsilon_r} && \text{(c).}
\end{aligned}
\tag{4.5.35}
$$

In a lossy medium, the plane wave propagates in the x-direction with the uniform field components in the (y-z)-plane as shown in Fig. (4.9a). The field components are given by equation (4.5.24), incorporating the conductivity σ of a medium. They are modified as,

$$
\begin{aligned}
&\frac{\partial H_z}{\partial x} = -\sigma E_y - j\omega\varepsilon_0\varepsilon_r E_y && \text{(a)}\\
&\frac{\partial H_y}{\partial x} = \sigma E_z + j\omega\varepsilon_0\varepsilon_r E_z && \text{(b)}\\
&\frac{\partial E_z}{\partial x} = j\omega\mu_0\mu_r H_y && \text{(c)}\\
&\frac{\partial E_y}{\partial x} = -j\omega\mu_0\mu_r H_z && \text{(d).}
\end{aligned}
\tag{4.5.36}
$$

Using the field solutions from equations (4.5.20), the above equations are reduced to the following forms:

$$
\begin{aligned}
&\gamma H_z = (\sigma + j\omega\varepsilon_0\varepsilon_r)E_y && \text{(a)}\\
&-\gamma H_y = (\sigma + j\omega\varepsilon_0\varepsilon_r)E_z && \text{(b)}\\
&\gamma E_z = -j\omega\mu_0\mu_r H_y && \text{(c)}\\
&\gamma E_y = \omega\mu_0\mu_r H_z && \text{(d).}
\end{aligned}
\tag{4.5.37}
$$

In the above equations, the complex propagation constant γ is given by equation (4.5.4).

The conducting medium is highly dispersive, whereas the low-loss medium is nondispersive. Using equations (4.5.35a,b) with equation (4.5.12a), the wave equation and the phase velocity in a conducting medium are given below:

$$
\begin{aligned}
&\frac{d^2\vec{E}}{dx^2} - j\omega\mu\sigma\,\vec{E} = 0 && \text{(a)}\\
&v_p = \frac{\omega}{\beta} = \sqrt{\frac{2\omega}{\mu\sigma}} && \text{(b).}
\end{aligned}
\tag{4.5.38}
$$

It shows that the conducting medium is dispersive, and the phase velocity increases with an increase in frequency.

The characteristic impedance (*intrinsic impedance*) of the low-loss and highly conducting media are obtained from equations (4.5.37) and (4.5.35a) as follows:

$$\text{Low-loss: } \frac{E_y}{H_z} = -\frac{E_z}{H_y} = \eta_c^* = \frac{\gamma}{\sigma + j\omega\varepsilon_0\varepsilon_r}$$

$$\Rightarrow \quad \eta_c^* = \left[\frac{j\omega\mu_0\mu_r}{\sigma + j\omega\varepsilon_0\varepsilon_r}\right]^{\frac{1}{2}} = \frac{\eta}{[1 - j(\sigma/\omega\varepsilon_0\varepsilon_r)]^{1/2}} \approx \eta\left[1 + j\frac{\sigma}{2\omega\,\varepsilon_0\varepsilon_r}\right] \qquad (a) \qquad (4.5.39)$$

$$\text{High-loss: } \eta_c^* = \left[\frac{j\omega\mu_0\mu_r}{\sigma + j\omega\varepsilon_0\varepsilon_r}\right]^{\frac{1}{2}} \approx \left[\frac{j\omega\mu_0\mu_r}{\sigma}\right]^{\frac{1}{2}} = (1+j)\left[\frac{\pi f\,\mu_0\mu_r}{\sigma}\right]^{\frac{1}{2}} \qquad (b).$$

The characteristic impedance, i.e. the *intrinsic impedance* η_c, of a high-loss conducting medium is a complex quantity, with an equal magnitude of real and inductive imaginary parts. The real part of η_c^* is known as the *surface resistance*, R_s incurring an Ohmic loss in the conducting medium; and its imaginary gives the *internal inductance* L_i of a conducting medium:

$$R_s = \sqrt{\frac{\pi\mu f}{\sigma}} \qquad (a)$$
$$\omega L_i = \sqrt{\frac{\pi\mu f}{\sigma}} \qquad (b). \qquad (4.5.40)$$

For the unbounded medium, $|R_s| = |\omega L_i|$, and the internal inductance L_i is due to the penetration of the magnetic field in the medium. It is further discussed in subsection (8.4.2) of chapter 8. The expressions for the magnetic and electric fields and Poynting vector in a conducting medium are given below:

$$E_y = E_0 e^{-\gamma x} e^{j\omega t} = E_0\, e^{-\alpha x}\, e^{j(\omega t - \beta x)} \qquad (a)$$
$$H_z = \frac{E_0}{\eta_c^*} e^{-\alpha x}\, e^{j(\omega t - \beta x)} \qquad (b)$$
$$H_z = H_0 e^{-\alpha x}\, e^{j(\omega t - \beta x)} \qquad (c)$$
$$S_{av} = \text{Re}\,[E_y \times H_z]\,\hat{x} = \frac{|E_0|^2}{2} e^{-2\alpha x}\, \text{Re}\left[\frac{1}{\eta_c^*}\right]\hat{x} \qquad (d). \qquad (4.5.41)$$

The α, β, and η_c for a highly conducting medium are given by equations (4.5.35b) and (4.5.39b). The uniform plane wave in an unbounded conducting medium is still TEM-type. However, field components H_z and E_y are not in-phase. These are in-phase in a dielectric medium shown in Fig. (4.9a). The power transported per unit area, in a conducting medium, in the x-direction is also given by the following expression:

$$S_x = \frac{E_y H_z^*}{2} = \frac{\eta_c^* H_z H_z^*}{2} = \frac{H_0^2}{2}(1+j)\left[\frac{\pi\mu f}{\sigma}\right]^{\frac{1}{2}} e^{-2\alpha x}. \qquad (4.5.42)$$

The power transmitted through the conducting lossy medium is a complex quantity. Its real part gives the power that comes out from the medium of length x, whereas the imaginary part gives stored energy due to the field penetration in the conductor. The input power density available at $x = 0$ is $S_0 = (H_0^2/2)\sqrt{(\pi\mu f/\sigma)}$. The power density after traveling distance x in a highly conducting medium is

$$S_{out} = \frac{H_0^2}{2}\left[\frac{\pi\mu f}{\sigma}\right]^{\frac{1}{2}} e^{-2\alpha x}. \qquad (4.5.43)$$

The field decreases by a factor $e^{-\alpha x}$, whereas the wave travels through a lossy medium. If the wave travels a distance $x = \delta = 1/\alpha$, known as the *skin depth,* the field is decreased by 1/e of its initial strength, i.e. approximately 37% of its initial field strength. However, the power decreases at a faster rate, i.e. by the factor $e^{-2\alpha x}$. If the initial power density is S_0, the power density at distance x is

$$S(x) = S_0 e^{-2\alpha x} \qquad (a)$$
$$\text{Skin-depth, } \delta = \frac{1}{\alpha} = \frac{1}{\sqrt{\pi\mu\sigma f}} \qquad (b). \qquad (4.5.44)$$

The attenuation constant α in the above equation is used from equation (4.5.35b). The power loss of wave traveling a distance x is computed after computing the power loss at unit distance x = 1m:

$$\alpha_{dB} = 10\log_{10}\frac{S(x=1)}{S_0} = [-20\log_{10}(e)]\,\alpha, \Rightarrow \alpha_{dB} = -8.686\,\alpha \;\; (\text{dB/m}) \qquad (a)$$
$$\text{Power after traveling distance x: } \quad \alpha_{dB} = -8.686\,\frac{x}{\delta} \;\; \text{dB} \qquad (b). \qquad (4.5.45)$$

In the above equation, the value of e is 2.71828. The power loss is about 9 dB per skin-depth. The attenuation constant α of a lossy medium is defined by equation (4.5.44a) as follows:

$$\frac{dS(x)}{dx} = -2\,\alpha\, S_0\, e^{-2\alpha x} = -2\alpha\, S(x)$$

$$\Rightarrow |\alpha| = \frac{dS(x)/dx}{2S(x)} = \frac{\text{Power loss per unit length}}{2 \times \text{Power avialable at lenght x}}. \tag{4.5.46}$$

4.6 Polarization of EM-waves

The uniform plane wave in the unbounded medium is the TEM-type wave. The monochromatic EM-wave is characterized by *amplitude*, *phase*, and *polarization states*. The microwave to optical wave devices can appropriately manipulate these characteristics to steer the EM-waves in the desired direction with shaped wavefront. The modern metasurfaces, discussed in sections (22.5) and (22.6) of chapter 22, can achieve such controls on the reflected and transmitted waves.

In general, both $\vec{E}$ and $\vec{H}$ fields have two orthogonal field components in a plane normal to the direction of propagation, the x-direction, as shown in Fig. (4.9a and b). The field components are in the (y-z)-plane. For the TEM mode, it is possible to get either (E_y, H_z) or (E_z, H_y) pair of fields. Both pairs of field components can also exist. The orientation of the electric field component and the movement of the tip of the resultant E-field determine the polarization of the EM-wave. Thus, the (E_y, H_z) pair of the EM-wave is called a y-*polarized wave,* as only the E_y component of wave exists. The (E_z, H_y) pair of the EM-wave is called the *z-polarized wave*. The (y-z)-plane is known as the *plane of polarization*. It is normal to the direction of propagation, i.e. the x-axis. Both these polarizations are *linear polarization*.

Figure (4.9a and c) show that for the EM-wave propagating in the x-direction, the tip of the E_y field component moves along the y-axis from $+E_0$ to 0 to $-E_0$. The movement and rotation of the tip of the $\vec{E}$-vector could be seen using the instantaneous $\vec{E}(x,t)$-vector of the wave propagating in the x-direction. In Fig. (4.9c), the path of the tip of the E_y-field vector, in the plane of polarization, traces a line with respect to time. The linear trace demonstrates the *vertical linear polarization*. However, if both E_y and E_z field components are present, any of the three kinds of wave polarizations can be obtained: (i) *linear polarization,* (ii) *circular polarization,* and (iii) *elliptical polarization.* These polarizations are shown in Fig. (4.10a–c). The type of wave polarization depends upon the magnitude and the relative phase of the orthogonal E_y and E_z field components. The polarization states are briefly discussed below. Finally, the *Jones vector and Jones matrix* descriptions are summarized to describe elegantly the polarization states and their control by the polarizing devices.

4.6.1 Linear Polarization

Figure (4.10a) shows E_y and E_z field components of the EM-wave propagating in the x-direction. The E-electric field vector in the (y-z)-plane could be written in the phasor form as follows:

$$\vec{E}(x) = \left(\hat{y}\, E_{0y} + \hat{z}\, E_{0z}\right) e^{-jk_0 x} \qquad \text{(a)}$$

$$\text{for } x = 0,\ |E| = \sqrt{E_{0y}^2 + E_{0z}^2} \qquad \text{(b)}$$

$$\theta = \tan^{-1}\left(\frac{E_{oz}}{E_{0y}}\right) \qquad \text{(c).} \tag{4.6.1}$$

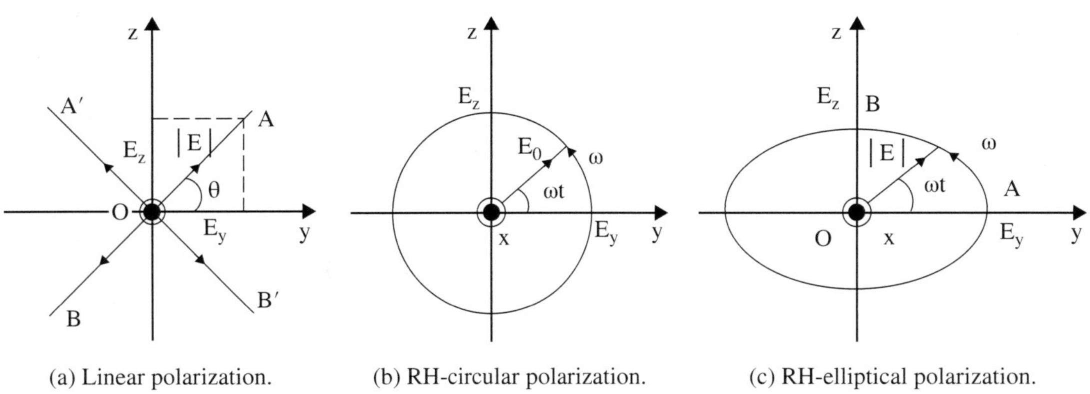

Figure 4.10 Type of polarizations.

The $e^{j\omega t}$ time-harmonic factor is suppressed in the above equation (4.6.1a). Equation (4.6.1b) shows the magnitude of the E-field, and equation (4.6.1c) computes its inclination with respect to the y-axis. For y-polarized wave, $E_{0z} = 0$, and for the z-polarized wave, $E_{0y} = 0$. In general, the field components E_{0y} and E_{0z} are complex quantities. For the in-phase field components, these are expressed as $E_{0y}=|E_{0y}|\,e^{j\phi}$ and $E_{0z}=|E_{0z}|\,e^{j\phi}$. The instantaneous field components are considered to trace the movement of the tip of the $\vec{E}(x,t)$-vector:

$$\vec{E}(x,t) = \mathrm{Re}\left\{ \left(\hat{y}\,|E_{0y}|\,e^{j\phi} + \hat{z}\,|E_{0z}|\,e^{j\phi}\right) e^{j\omega t}\right\} \quad \text{(a)}$$
$$\Rightarrow \vec{E}_y = |E_{0y}|\cos(\omega t + \phi), \qquad \vec{E}_z = |E_{0z}|\cos(\omega t + \phi) \quad \text{(b)}. \qquad (4.6.2)$$

In the above equations, both field components have an identical phase $(\omega t + \phi)$. Figure (4.10a) shows the *slant or inclined linearly polarized wave* with the in-phase E_y and E_z components. The tip of the electric vector $(\vec{E})$ moves along line A-O-B with respect to time. However, the slant angle θ does not change with time. If both the E-field components are either in-phase $(A - O - B)$ or out of phase $(A' - O - B')$ and have the same magnitude, i.e. $E_{0y} = E_{0z} = E_0$, the corresponding inclination angle of the linear polarization trace, with respect to the y-axis, is $\theta = 45°$ and $135°$, respectively. For the linear polarization, the total E-field given by equation (4.6.1a) could also be written as follows:

$$\vec{E}(x) = E_0(\hat{y} + \hat{z})\,e^{-jk_0x} \quad \text{(a)}$$
$$\vec{E}(x) = E_0(\hat{z} - \hat{y})\,e^{-jk_0x} \quad \text{(b)}. \qquad (4.6.3)$$

4.6.2 Circular Polarization

The circular polarization, shown in Fig. (4.10b), is obtained for two orthogonal field components of equal magnitude, and phase in quadrature. So, to get the circular polarization, two electric field components oscillate at the same frequency and meet the following conditions:

- *Equal amplitude: The magnitudes of* E_y *and* E_z *are equal, i.e.* $|E_{0y}| = |E_{0z}| = E_0$.
- *Space quadrature: The* E_y *and* E_z *field components are normal to each other.*
- *Time (phase) quadrature: The phase difference between the* E_y *and* E_z field components are $(\phi = \pm 90°)$, *i.e.* $E_{0y} = E_0$, and $E_{0z} = E_0 e^{\pm\pi/2} = \pm j\,E_0$.

The phasor form of the E-field vector of the circularly polarized waves, meeting the above conditions, could be written from equation (4.6.1a) as follows:

$$\text{LHCP}: \quad \vec{E}(x) = E_0(\hat{y} + j\hat{z})\,e^{-j k_0 x} \quad \text{(a)}$$
$$\text{RHCP}: \quad \vec{E}(x) = E_0(\hat{y} - j\hat{z})\,e^{-j k_0 x} \quad \text{(b)}. \qquad (4.6.4)$$

The $e^{j\omega t}$ time-harmonic factor is suppressed in the above equations. The time-domain forms of the circularly polarized waves, using the instantaneous $\vec{E}(x,t)$-field components, at any location in the positive x-direction and also at the $x = 0$, i.e. in the (y-z)-plane, are expressed as follows:

$$\text{LHCP}: \quad \vec{E}(x,t) = \mathrm{Re}\left[E_0(\hat{y} + j\hat{z})\,e^{j(\omega t - k_0 x)}\right] \quad \text{(a)}$$
$$\text{RHCP}: \quad \vec{E}(x,t) = \mathrm{Re}\left[E_0(\hat{y} - j\hat{z})\,e^{j(\omega t - k_0 x)}\right] \quad \text{(b)}$$
$$\text{LHCP}: \quad \text{At } x = 0,\ \vec{E}(t) = E_0(\hat{y}\cos\omega t - \hat{z}\sin\omega t) \quad \text{(c)}$$
$$\text{RHCP}: \quad \text{At } x = 0,\ \vec{E}(t) = E_0(\hat{y}\cos\omega t + \hat{z}\sin\omega t) \quad \text{(d)}$$
$$\text{LHCP}: \quad \left|\vec{E}(x,t)\right| = E_0,\ \theta = \tan^{-1}[-\tan(\omega t)] = -\omega t \quad \text{(e)}$$
$$\text{RHCP}: \quad \left|\vec{E}(x,t)\right| = E_0,\ \theta = \tan^{-1}[\tan(\omega t)] = \omega t \quad \text{(f)}. \qquad (4.6.5)$$

The *handedness*, i.e. the sense of rotation of the $\vec{E}$-vector of the circularly polarized wave is determined by the direction of rotation of the $\vec{E}(x,t)$-vector in the transverse (y-z)-plane with respect to time, whereas the EM-wave propagates in the x-direction. Therefore, using equation (4.6.5f), Fig. (4.10b) shows the *anti-clock rotation* of the E-field vector tracing a circle of radius E_0 in the (y-z)-plane. The sense of rotation is considered with respect to $\cos(\omega t)$, taking $t = 0$. It shows the wave propagating in the positive x-direction, i.e. the wave coming toward an observer standing on the positive x-axis is the *right-hand circularly polarized wave (RHCP)*. Its field vector is given by equation (4.6.4b). Similarly, equation (4.6.5e) shows the *clockwise rotation*, i.e. the *left-hand circularly polarized wave (LHCP)* propagating in the positive x-direction. The magnetic fields of the RHCP and LHCP wave are obtained using equation (4.6.4) with equation (4.5.34a):

$$\text{RHCP}: \quad \vec{H}(x) = \frac{E_0}{\eta_0}\hat{x} \times (\hat{y} - j\hat{z})\,e^{-jk_0x} = \frac{E_0}{\eta_0}(\hat{z} + j\hat{y})\,e^{-jk_0x} \quad \text{(a)}$$
$$\text{LHCP}: \quad \vec{H}(x) = \frac{E_0}{\eta_0}\hat{x} \times (\hat{y} + j\hat{z})\,e^{-jk_0x} = \frac{E_0}{\eta_0}(\hat{z} - j\hat{y})\,e^{-jk_0x} \quad \text{(b)}. \qquad (4.6.6)$$

In the case of the wave propagation in the negative x-direction, i.e. the wave moves away from the observer standing on the positive x-axis, the role of RHCP and LHCP gets interchanged. Further, the handedness of circular polarization can be reversed by applying 180°

phase-shift to either the y or z component of the $\vec{E}$-field vector. *The circular polarization could be further considered as a linear combination of two linearly polarized waves.* Alternatively, *the linear polarization can also be considered as a linear combination of the left-hand and right-hand circularly polarized waves* [B.9].

4.6.3 Elliptical Polarization

Two orthogonal field components, in the space quadrature, of unequal magnitude and arbitrary phase angle (ϕ) between them, generate the elliptical polarization, i.e. the resultant E-field vector rotates in the plane of polarization such that its tip traces an elliptical path as shown in Fig. (4.10c). The instantaneous orthogonal E-field components in the x = 0 plane are given below:

$$\vec{E}_y(t) = \hat{y}\, E_{0y} \cos\omega t, \qquad \vec{E}_z(t) = \hat{z}\, E_{0z} \cos(\omega t \pm \phi)$$
$$\vec{E}(t) = \hat{y}\, E_{0y} \cos\omega t + \hat{z}\, E_{0z} \cos(\omega t \pm \phi). \tag{4.6.7}$$

Using the above equation and identity $\cos\omega t = E_y(t)/E_{0y}$, $\sin\omega t = \sqrt{1 - \{E_y(t)/E_{0y}\}^2}$, the following equation of the ellipse is obtained:

$$E_z(t) = E_{0z}[\cos\omega t \cos\phi \mp \sin\omega t \sin\phi]$$
$$\Rightarrow \frac{E_z(t)}{E_{0z}} = \frac{E_y(t)}{E_{0y}} \cos\phi \mp \sqrt{1 - \left(\frac{E_y(t)}{E_{0y}}\right)^2}\sqrt{1 - \cos^2\phi}$$
$$\Rightarrow P\,E_y^2 - Q\,E_y E_z + R\,E_z^2 = 1$$
$$\text{where,}\quad P = \frac{1}{E_{0y}^2 \sin^2\phi},\quad Q = \frac{2\cos\phi}{E_{0y}E_{0z}\sin^2\phi},\quad R = \frac{1}{E_{0z}^2 \sin^2\phi}. \tag{4.6.8}$$

The semi-major axis OA, the semi-minor axis OB of the ellipse shown in Fig. (4.10c), and the *axial ratio* (AR) of the polarization ellipse are given below [B.9, B.29]:

$$OA = \left[\frac{1}{2}\left\{E_y^2 + E_z^2 + \left(E_y^4 + E_z^4 + 2E_y^2E_z^2\cos 2\phi\right)^{1/2}\right\}\right]^{1/2} \quad \text{(a)}$$
$$OB = \left[\frac{1}{2}\left\{E_y^2 + E_z^2 - \left(E_y^4 + E_z^4 + 2E_y^2E_z^2\cos 2\phi\right)^{1/2}\right\}\right]^{1/2} \quad \text{(b)} \tag{4.6.9}$$
$$AR = \frac{E_{max}}{E_{min}} = \frac{OA}{OB} = \left[\frac{E_y^2 + E_z^2 + \left(E_y^4 + E_z^4 + 2E_y^2E_z^2\cos 2\phi\right)^{1/2}}{E_y^2 + E_z^2 - \left(E_y^4 + E_z^4 + 2E_y^2E_z^2\cos 2\phi\right)^{1/2}}\right]^{1/2}, 1 \le AR \le \infty \quad \text{(c)}.$$

The *tilt angle* θ of the polarization ellipse, i.e. inclination of the major axis OA with y-axis is [B.9, B.29]:

$$\theta = \frac{1}{2}\tan^{-1}\left[\frac{2E_yE_z}{E_y^2 + E_z^2}\cos\phi\right]. \tag{4.6.10}$$

For $\phi \ne \pi/2$, the polarization ellipse is inclined with respect to the y-axis. The linear and circular polarizations are obtained as special cases from the elliptical polarization. For instance, for $E_z(t) = 0$ the wave is horizontally polarized in the y-direction. For $E_{0y} = E_{0z} = E_0$ and $\phi = \pm\pi/2$, the LHCP/RHCP wave is obtained as equation (4.6.9) is reduced to an equation of a circle with OA = OB. For the linear polarization, AR is infinity. However, for the circular polarization, AR is unity. In the case $E_{0y} = E_{0z} = E_0$ and $\phi \ne \pi/2$, the wave is not circularly polarized and its AR is $\cot\phi/2$. For a practical circularly polarized antenna, the axial ratio is frequency-dependent and its *axial ratio bandwidth* is defined as the frequency band over which AR ≤ 3dB.

4.6.4 Jones Matrix Description of Polarization States

The polarizing devices change the state of polarization. For instance, the polarizing devices could change the rotation of the linear polarization or convert the linear polarization into circular polarization. The Jones matrix method describes and manipulates the polarization states of the EM-wave using a 2×1 column vector, known as the *Jones vector* and transfer matrix of the polarizing device, known as the *Jones matrix* [B.30, B.31]. The Jones matrix is used in chapter-22 with metasurfaces.

Jones Vector

The polarization state of the EM-wave propagation in the x-direction is given by equation (4.6.1). The orthogonal E-field components can be expressed in the form of the following column vector, known as the *Jones vector*:

$$\vec{E}(x) = (\hat{y}\,|E_{0y}|\,e^{j\phi_y} + \hat{z}\,|E_{0z}|\,e^{j\phi_z})\,e^{-jk_0x} = \begin{bmatrix} |E_{0y}| \\ |E_{0z}|\,e^{j\Delta\phi} \end{bmatrix} \quad \text{(a)}$$

$$\text{where, differential - phase}: \Delta\phi = \phi_z - \phi_y \quad \text{(b).} \qquad (4.6.11)$$

In the above expression, the common phase angle has been absorbed in the propagation factor e^{-jk_0x}. The Jones vector describes polarization states of any plane wave field. Some common polarization states are summarized below with respect to Fig. (4.10):

$$\text{Horizontal polarization (HP) along the y-axis}: \begin{bmatrix} 1 \\ 0 \end{bmatrix} \quad \text{(a)}$$

$$\text{Vertical polarization (VP) along the z-axis}: \begin{bmatrix} 0 \\ 1 \end{bmatrix} \quad \text{(b)}$$

$$\text{Linear polarization (LP) at an angle } \theta \text{ with y-axis}: \begin{bmatrix} \cos\theta \\ \sin\theta \end{bmatrix} = \frac{1}{\sqrt{2}}\begin{bmatrix} 1 \\ 1 \end{bmatrix}_{\theta = 45^\circ} \quad \text{(c)} \qquad (4.6.12)$$

$$\text{Right - hand circular polarization (RHCP)}: \frac{1}{\sqrt{2}}\begin{bmatrix} 1 \\ -j \end{bmatrix} \quad \text{(d)}$$

$$\text{Left - hand circular polarization (LHCP)}: \frac{1}{\sqrt{2}}\begin{bmatrix} 1 \\ j \end{bmatrix} \quad \text{(e).}$$

In the above expressions, the normalized magnitude of the E-field components are $|E_{oy}| = |E_{oz}| = 1$.

Jones Matrix

Figure (4.11) shows that the *polarizing device* could be described by a 2×2 transfer matrix, i.e. the Jones matrix [J]. It relates the output of the device to its input. The input could be the incident wave at certain slab/surface, acting as a polarizing device, and the output could be the transmission or reflection of the waves.

$$\vec{E}^{\,out} = \bar{\bar{J}}\,.\,\vec{E}^{\,in}, \ [E]_{out} = [J][E]_{in} \quad \text{(a)}$$

$$\Rightarrow \underset{\text{Outgoing waves}}{\begin{bmatrix} E_y^{out} \\ E_z^{out} \end{bmatrix}} = \underset{\text{Polarizing device}}{\begin{bmatrix} J_{11} & J_{12} \\ J_{21} & J_{22} \end{bmatrix}} \underset{\text{Incoming waves}}{\begin{bmatrix} E_y^{in} \\ E_z^{in} \end{bmatrix}} \quad \text{(b).}$$

where $[E]_{out}$ and $[E]_{in}$ are output and input Jones vectors, and [J] is Jones matrix. (4.6.13)

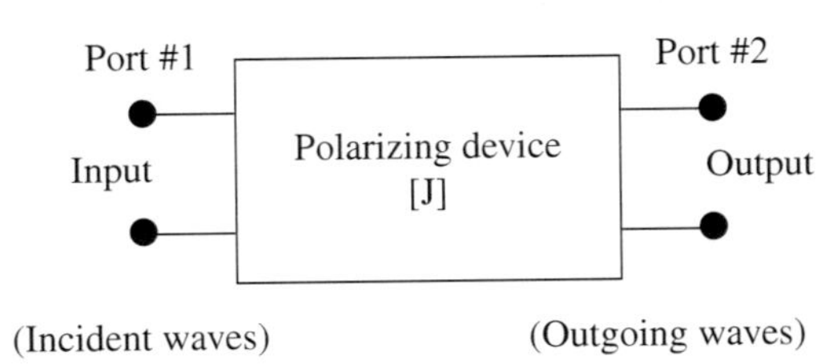

Figure 4.11 Polarizing device described by Jones matrix.

The Jones matrix elements are interpreted in the terms of the co-polarized and cross-polarized outgoing waves after transmission/reflection from a slab/surface:

$$\text{Jones matrix}: [J] = \begin{bmatrix} J_{11} & J_{12} \\ J_{21} & J_{22} \end{bmatrix} = \begin{bmatrix} J_{yy} & J_{yz} \\ J_{zy} & J_{zz} \end{bmatrix}, \qquad (4.6.14)$$

where J_{yy} and J_{zz} are responsible for the *co-polarized* outgoing waves, and J_{yz} and J_{zy} account for the presence of *cross-polarized* waves at the output. The co-polarized output waves have the same polarization as that of the incident input waves. Whereas, the cross-polarized output waves have orthogonal polarization with respect to the polarization of the incident input waves.

Jones Matrix of Linear Polarizer

A *linear polarizer* allows the transmission of the incoming wave only along the transmission axis of the polarizer and blocks the transmission of the orthogonal polarizations. Jones matrices of the linear polarizers are summarized below:

$$\text{Horizontal Polarizer (HP)}: \begin{bmatrix} 1 & 0 \\ 0 & 0 \end{bmatrix} \quad \text{(a)}$$

$$\text{Vertical Polarizer (VP)}: \begin{bmatrix} 0 & 0 \\ 0 & 1 \end{bmatrix} \quad \text{(b)}$$

$$\text{Linear Polarizer}: \begin{bmatrix} p_y & 0 \\ 0 & p_z \end{bmatrix},$$

$$\text{where } 0 \le p_y \le 1;\ 0 \le p_z \le 1 \quad \text{(c).} \qquad (4.6.15)$$

For instance, at an inclination angle $\theta = 45°$, the linearly polarized wave, given by Jones vector of equation (4.6.12c), is incident on the horizontal polarizer. The polarizer will provide a horizontally polarized wave given by equation (4.6.12a). It could be examined with the help of equation (4.6.13).

Jones Matrix of a Linear Polarizer Rotated at Angle θ with the y-Axis

Figure (4.12) shows the (y-z)-coordinate system and also the (e_1-e_2) -coordinate system rotated at an angle θ with respect to the y-axis. The *original polarizer*, located in the (y-z)-coordinate system is described by the Jones matrix [J]; and the *rotated polarizer* at an angle θ is described as the *rotated Jones matrix* $[J_{rot}(\theta)]$. The rotated polarizer is located in the (e_1-e_2) coordinate system. The relation between the Jones matrix [J] of unrotated polarizer and Jones matrix $[J_{rot}(\theta)]$ of the rotated polarizer is obtained by the coordinate transformation through the *rotation Jones matrix* $[R_\theta(\theta)]$ and the *inverse rotation Jones matrix* $[R_\theta(\theta)]^{-1} = [R_\theta(-\theta)]$.

Figure (4.12) shows that the unit vectors of two coordinate systems are related through the following transformations:

$$\left.\begin{aligned} \hat{y} &= \cos\theta\,\hat{e}_1 - \sin\theta\,\hat{e}_2 \\ \hat{z} &= \sin\theta\,\hat{e}_1 + \cos\theta\,\hat{e}_2 \end{aligned}\right\}$$

$$\Rightarrow \begin{bmatrix} \hat{y} \\ \hat{z} \end{bmatrix} = \begin{bmatrix} \cos\theta & -\sin\theta \\ \sin\theta & \cos\theta \end{bmatrix} \begin{bmatrix} \hat{e}_1 \\ \hat{e}_2 \end{bmatrix} \quad \text{(a)}$$

$$\begin{bmatrix} \hat{e}_1 \\ \hat{e}_2 \end{bmatrix} = \begin{bmatrix} \cos\theta & \sin\theta \\ -\sin\theta & \cos\theta \end{bmatrix} \begin{bmatrix} \hat{y} \\ \hat{z} \end{bmatrix} \quad \text{(b).}$$

(4.6.16)

The following *rotation Jones matrix* $[R_\theta(\theta)]$ and its *inverse* $[R_\theta(-\theta)]$ can be defined from the above relations that could be useful to transform the vectors from one co-ordinated system to another:

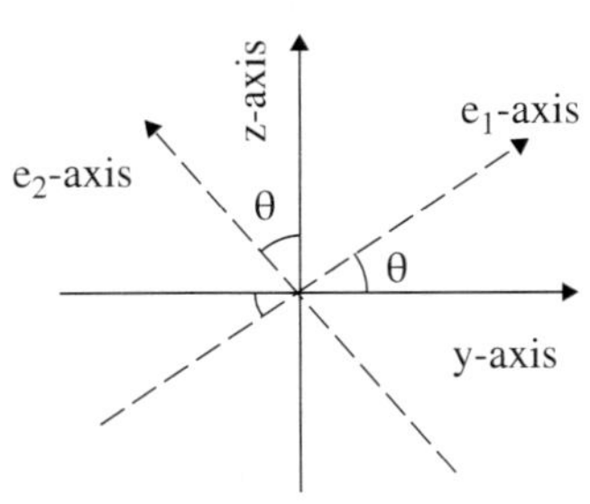

Figure 4.12 The (y−z) and rotated (e_1−e_2) Coordinate systems.

$$[R_\theta(\theta)] = \begin{bmatrix} \cos\theta & \sin\theta \\ -\sin\theta & \cos\theta \end{bmatrix} \quad \text{(a)}$$

$$[R_\theta(\theta)]^{-1} = [R_\theta(-\theta)] = \begin{bmatrix} \cos\theta & -\sin\theta \\ \sin\theta & \cos\theta \end{bmatrix} \quad \text{(b).}$$

(4.6.17)

The above rotation matrices are used to transform both the Jones vector and the Jones matrix from one to another coordinate system. The Jones matrix concept is applied to the polarizing system in two steps:

Transformation of E-vector Components

The *rotation Jones matrix* $[R_\theta(\theta)]$ transforms the vector components from the rotated (e_1-e_2) coordinate system to the (y-z) Cartesian coordinate system. Whereas the *inverse rotation Jones matrix* $[R_\theta(-\theta)]$ transforms the vector components from the (y-z) Cartesian coordinate system back to the rotated (e_1-e_2) coordinate system.

Transformation of Jones Matrix of Polarizer

To perform the transformation of the E-vector components from one to another coordinate system using a polarizer, the Jones matrix describing the polarizer has to be transformed from one to another coordinated system. Thus, the above relations transform the Jones matrix $[J] = [J]_{Car}$ of a linear polarizer, given by equation (4.6.14) in the Cartesian (y-z) coordinate system, to the Jones matrix $[J_{rot}(\theta)] = [J]_e$ of the polarizer in the rotated (e_1-e_2) coordinate system.

The input/output relations of the E-field in both coordinate systems are expressed in terms of their respective Jones matrices:

Cartesian coordinate system:

$$\left[\vec{E}^{\,out}\right]_{Car} = [J]_{Car}\left[\vec{E}^{\,in}\right]_{Car} \quad \text{(a)}$$

Rotated ($e_1 - e_2$) coordinate system:

$$\left[\vec{E}^{\,out}\right]_{e} = [J]_{e}\left[\vec{E}^{\,in}\right]_{e} \quad \text{(b).}$$

(4.6.18)

In the above expression, subscript "Car." with Jones matric stands for the (y-z) Cartesian coordinate system; and the subscript "e" with Jones matric stands for the general (e_1-e_2) coordinate system. In the present case, it is the anticlockwise rotated Cartesian system, as shown in Fig. (4.12).

Using the rotation Jones matrices of equation (4.6.17a), the input and output E-field vectors could be transformed from the rotated (e_1-e_2) coordinate system to the Cartesian (y-z) coordinate system:

$$\begin{aligned} \left[\vec{E}^{in}\right]_{Car} &= [R_\theta(\theta)]\left[\vec{E}^{in}\right]_e && (a) \\ \left[\vec{E}^{out}\right]_{Car} &= [R_\theta(\theta)]\left[\vec{E}^{out}\right]_e && (b). \end{aligned} \quad (4.6.19)$$

On substituting the above equations in equation (4.6.18a), the following expression is obtained:

$$\begin{aligned} &[R_\theta(\theta)]\left[\vec{E}^{out}\right]_e = [J]_{Car}[R_\theta(\theta)]\left[\vec{E}^{in}\right]_e \\ \Rightarrow &\left[\vec{E}^{out}\right]_e = \left\{[R_\theta(\theta)]^{-1}[J]_{Car}[R_\theta(\theta)]\right\}\left[\vec{E}^{in}\right]_e. \end{aligned} \quad (4.6.20)$$

On comparing the above equations against equation (4.6.18b), we get the transformed Jones matrix $[J]_e$ of the rotated polarizer in the (e_1-e_2) coordinate system from the Jones matrix $[J]_{Car}$ of the original polarizer in the Cartesian (y-z) coordinate system:

$$\begin{aligned} &[J]_e = [R_\theta(\theta)]^{-1}[J]_{Car}[R_\theta(\theta)] \\ \Rightarrow &[J]_e = [R_\theta(-\theta)]\,[J]_{Car}[R_\theta(\theta)] && (a) \\ &\text{Alternatively, } [J]_{Car} = [R_\theta(\theta)]\,[J]_e[R_\theta(-\theta)] && (b). \end{aligned} \quad (4.6.21)$$

The transformation (4.6.21b) transforms the Jones matrix $[J]_e$ describing a polarizer in the rotated (e_1-e_2) coordinate system to the Jones matrix $[J]_{Car}$ in the Cartesian coordinate system. The transformation equation (4.6.21b) is obtained by matrix manipulation. However, it could also be obtained independently, as it is done for equation (4.6.21a).

The use of the coordinate transformation for the polarizer is illustrated below by a few illustrative simple examples. Application of Jones matrix to the more complex polarizing system is available in the reference [B.30].

- Examples:

1) The Jones matrix of a linear polarizer given by equation (4.6.15c) is transformed below to the rotated Jones matrix $[J]_e = [J^{LP}(\theta)]$ of a linear polarizer that is rotated at an angle θ:

$$\begin{aligned} [J^{LP}(\theta)]_e &= \begin{bmatrix} \cos\theta & -\sin\theta \\ \sin\theta & \cos\theta \end{bmatrix} \begin{bmatrix} p_y & 0 \\ 0 & p_z \end{bmatrix} \begin{bmatrix} \cos\theta & \sin\theta \\ -\sin\theta & \cos\theta \end{bmatrix} \\ \Rightarrow [J^{LP}(\theta)]_e &= \begin{bmatrix} p_y\cos^2\theta + p_z\sin^2\theta & (p_y - p_z)\sin\theta\cos\theta \\ (p_y - p_z)\sin\theta\cos\theta & p_y\sin^2\theta + p_z\cos^2\theta \end{bmatrix}. \end{aligned} \quad (4.6.22)$$

It is noted that the original linear polarizer in the Cartesian system has no cross-polarization element. However, the rotated linear polarizer has a cross-polarization element. The above transformation can be applied to the *horizontal polarizer* ($p_y = 1$, $p_z = 0$) rotated at an angle θ, and also to the linear polarizer rotated at an angle $\theta = 45°$, to get the following rotated Jones matrices:

Horizontal polarizer:

$$[J^{HP}(\theta)]_e = \begin{bmatrix} \cos^2\theta & \sin\theta\cos\theta \\ \sin\theta\cos\theta & \sin^2\theta \end{bmatrix} \quad (a)$$

$$[J^{HP}(\theta = 45°)]_e = \frac{1}{2}\begin{bmatrix} 1 & 1 \\ 1 & 1 \end{bmatrix} \quad (b)$$

Linear polarizer:

$$[J^{LP}(\theta = 45°)]_e = \frac{1}{2}\begin{bmatrix} (p_y + p_z) & (p_y - p_z) \\ (p_y - p_z) & (p_y + p_z) \end{bmatrix} \quad (c). \qquad (4.6.23)$$

2) The anisotropic polarizer with cross-coupling, in the Cartesian system, is given by equation (4.6.14). The polarizer is rotated by an angle θ. The rotated Jones matrix of the anisotropic polarizer is obtained as follows:

$$[J^{aniso}(\theta)]_e = \begin{bmatrix} \cos\theta & -\sin\theta \\ \sin\theta & \cos\theta \end{bmatrix} \begin{bmatrix} J_{yy} & J_{yz} \\ J_{zy} & J_{zz} \end{bmatrix} \begin{bmatrix} \cos\theta & \sin\theta \\ -\sin\theta & \cos\theta \end{bmatrix}$$

$$\Rightarrow [J^{aniso}(\theta)]_e = \begin{bmatrix} \{J_{yy}\cos^2\theta + J_{zz}\sin^2\theta - (J_{yz} + J_{zy})\sin\theta\cos\theta\} & \{(J_{yy} - J_{zz})\sin\theta\cos\theta + (J_{yz}\cos^2\theta - J_{zy}\sin^2\theta)\} \\ \{(J_{yy} - J_{zz})\sin\theta\cos\theta + (J_{zy}\cos^2\theta - J_{yz}\sin^2\theta)\} & \{J_{yy}\sin^2\theta + J_{zz}\cos^2\theta + (J_{yz} + J_{zy})\sin\theta\cos\theta\} \end{bmatrix}. \quad (4.6.24)$$

Jones Matrix for Retarder (Phase Shifter)

The *wave retarder* also called the *waveplate* alters the relative phase between two orthogonal field components passing through it. In this respect, it is acting as a phase shifter. The waveplates are designed using the birefringent, i.e. anisotropic material with orthogonal *fast-axis* and *slow-axis*. The relative permittivity, also the refractive index, of the anisotropic material, has lower value along the fast-axis and higher value along the slow-axis, causing relatively slower phase velocity of the EM-wave propagation along the slow-axis. The half-wavelength thick slab, called the *half-wave plate*, changes the direction of the linear polarization at its output. Whereas, the *quarter-waveplate*, i.e. a quarter-wavelength thick slab, converts the linearly polarized incident waves into the circularly polarized waves at its output. The waveplates, i.e. the wave retarders, are characterized by the Jones matrices as discussed below.

For the EM-wave propagating in the x-direction, the field components at the output of a *wave retarder* could be written from equation (4.6.11), by normalizing the magnitude of field components to the unity, as follows:

$$\begin{bmatrix} E_y^{out}(x) \\ E_z^{out}(x) \end{bmatrix}_{x=d} = \begin{bmatrix} 1 & 0 \\ 0 & e^{j\Delta\phi} \end{bmatrix} \begin{bmatrix} E_{0y}^{in}(x) \\ E_{0z}^{in}(x) \end{bmatrix}_{x=0} = \begin{bmatrix} \left|E_{0y}^{in}(x)\right| \\ e^{j\Delta\phi}\left|E_{0z}^{in}(x)\right| \end{bmatrix}_{x=d} \quad (a)$$

$$J_R(\Delta\phi) = \begin{bmatrix} 1 & 0 \\ 0 & e^{j\Delta\phi} \end{bmatrix} \quad (b).$$

$$(4.6.25)$$

The $J_R(\Delta\phi)$ is the Jones matrix of a wave retarder (*waveplate*) given by equation (4.6.25b). At the input, the incident wave is linearly polarized and at the output of the retarder slab of thickness d, the differential phase $\Delta\phi = \phi_z - \phi_y$ is the relative phase between the field components $E_y^{out}(d)$ and $E_z^{out}(d)$.

Jones Matrix of Half-waveplate

The Jones matrix of a half-waveplate and the field components at its output are obtained by taking the relative phase $\Delta\phi = \pi$:

$$J_R(\Delta\phi = \pi) = \begin{bmatrix} 1 & 0 \\ 0 & e^{j\pi} \end{bmatrix} = \begin{bmatrix} 1 & 0 \\ 0 & -1 \end{bmatrix} \quad (a)$$

$$\begin{bmatrix} E_y^{out}(x) \\ E_z^{out}(x) \end{bmatrix}_{x=d} = \begin{bmatrix} 1 & 0 \\ 0 & -1 \end{bmatrix} \begin{bmatrix} E_{0y}^{in}(x) \\ E_{0z}^{in}(x) \end{bmatrix}_{x=0} = \begin{bmatrix} E_{0y}^{in}(x) \\ -E_{0z}^{in}(x) \end{bmatrix} \quad (b).$$

$$(4.6.26)$$

It is noted that at the output of the half-wave plate the phase difference between two field components is 180°. Two field components are in-phase at the input of the half-waveplate.

Jones Matrix of Quarter-waveplate

The Jones matrix of a quarter-wave retarder and also the field components at the output are obtained by taking the relative phase $\Delta\phi = -\pi/2$:

$$J_R(\Delta\phi = -\pi/2) = \begin{bmatrix} 1 & 0 \\ 0 & e^{-j\pi/2} \end{bmatrix} = \begin{bmatrix} 1 & 0 \\ 0 & -j \end{bmatrix} \quad (a)$$

$$\begin{bmatrix} E_y^{out}(x) \\ E_z^{out}(x) \end{bmatrix}_{x=d} = \begin{bmatrix} 1 & 0 \\ 0 & -j \end{bmatrix} \begin{bmatrix} E_0 \\ E_0 \end{bmatrix}_{x=0} = \begin{bmatrix} 1 \\ -j \end{bmatrix} E_0 \quad (b).$$

$$(4.6.27)$$

In the above equation, both field components are equal to $E_0 = 1$. It is noted that at the output of the quarter-waveplate, the wave is a right-hand circularly polarized wave. In the case, input wave components are $E_y^{in}(x = 0) = m\cos\phi_y$ and $E_z^{in}(x = 0) = n\cos\phi_y$, the field components at the output of the quarter-waveplate, using equation (4.6.25a) are

$$E_y^{out} = m\cos\phi_y \quad (a)$$

$$E_z^{out} = n\cos(\phi_z) = n\cos\left(\phi_y - \pi/2\right) = n\sin\phi_y \quad (b)$$

$$\Rightarrow \left(\frac{E_y^{out}}{m}\right)^2 + \left(\frac{E_z^{out}}{n}\right)^2 = 1 \quad (c).$$

$$(4.6.28)$$

Equation (4.6.28c) shows that an ellipse is traced by the E-field in the (y-z)-plane. In this case, the quarter-waveplate produces an elliptically polarized wave. However, for the case m = n, it degenerates into the circularly polarized wave. Further, for the rotated retarders the rotated Jones matrices could be obtained, similar to the case of the rotated polarizer, using equation (4.6.21a).

4.7 EM-waves Propagation in Unbounded Anisotropic Medium

Two cases of wave propagations in the uniaxial anisotropic media – without off-diagonal elements and with off-diagonal elements, are considered in this section. The dispersion relation is also discussed leading to the concept of *hypermedia* useful for the realization of *hyperlens* [J.1, J.5–J.7].

4.7.1 Wave Propagation in Uniaxial Medium

The unbounded lossless homogeneous uniaxial medium is considered. The y-axis is the *optical axis*, i.e. the extraordinary axis. In the direction of the optic axis, the permittivity is different as compared to the other two directions. The medium is described by a diagonalized matrix with all off-diagonal elements zero. The permeability of the medium is μ_0 and its permittivity tensor is expressed as follows:

$$\bar{\bar{\varepsilon}} = [\varepsilon] = \varepsilon_0 \begin{bmatrix} \varepsilon_{r\perp} & 0 & 0 \\ 0 & \varepsilon_{r||} & 0 \\ 0 & 0 & \varepsilon_{r\perp} \end{bmatrix}. \qquad (4.7.1)$$

Figure (4.13a) shows the TEM plane wave propagation in the x-direction. The TEM waves have $E_x = 0$, $H_x = 0$; $E_y \neq 0$, $H_y \neq 0$; $E_z \neq 0$, $H_z \neq 0$. Also, the uniform field components in y and z-directions do not vary, i.e. $\partial/\partial y$ (field) $= \partial/\partial z$(field) $= 0$. Under these conditions, the following Maxwell equations provide the transverse field components of electric and magnetic fields:

$$\nabla \times \vec{E} = -j\omega\mu_0 \vec{H} \qquad (a)$$
$$\nabla \times \vec{H} = j\omega\varepsilon_0[\varepsilon_r]\vec{E} \qquad (b). \qquad (4.7.2)$$

On expansion, the above equations provide the following sets of transverse field components:

$$\frac{\partial E_z}{\partial x} = j\omega\mu_0 H_y \qquad (a)$$
$$\frac{\partial E_y}{\partial x} = -j\omega\mu_0 H_z \qquad (b) \qquad (4.7.3)$$

$$-\frac{\partial H_z}{\partial x} = j\omega\varepsilon_0\varepsilon_{r||} E_y \qquad (a)$$
$$\frac{\partial H_y}{\partial x} = j\omega\varepsilon_0\varepsilon_{r\perp} E_z \qquad (b). \qquad (4.7.4)$$

On eliminating H_z and H_y from the above equations, wave equations for the electric field transverse components are obtained. Likewise, the wave equations for magnetic field transverse components are obtained on eliminating E_z and E_y:

$$\frac{d^2E_y}{dx^2} + k_0^2\,\varepsilon_{r||}E_y = 0 \qquad (a)$$
$$\frac{d^2E_z}{dx^2} + k_0^2\,\varepsilon_{r\perp}E_z = 0 \qquad (b)$$
$$\frac{d^2H_y}{dx^2} + k_0^2\,\varepsilon_{r\perp}H_y = 0 \qquad (c)$$
$$\frac{d^2H_z}{dx^2} + k_0^2\,\varepsilon_{r||}H_z = 0 \qquad (d). \qquad (4.7.5)$$

Equation (4.7.5a) is a 1D wave equation of the y-polarized electric field $E_y \neq 0$ and $E_z = 0$. The magnetic field component H_z is along the z-axis. The E_y field component causes polarization in the dielectric medium creating relative permittivity $\varepsilon_{r||}$ along the y-axis. Further, the electric field E_y is transverse to the (x-z)-plane containing the direction of propagation x. Such waves are called the *transverse electric (TE) waves*. The z-polarized electric field with $E_z \neq 0$ and $E_y = 0$ follows the wave equation (4.7.5b). The E_z component generates another polarization in the dielectric medium creating relative permittivity $\varepsilon_{r\perp}$ along the z-axis. In this case, H_y-component is transverse to the (x-z)-plane. These waves are called the *transverse magnetic (TM) waves*. In the present case, also in the case of the oblique incident of the plane waves, the waves are still TEM only. However, the TE and TM terminology is normally used for the *non-TEM* mode waves supported by the waveguides. It is discussed in chapter 7. A reader should be careful in the dual use of terminology TE/TM to show the *mode of propagation*, and also the *polarization of*

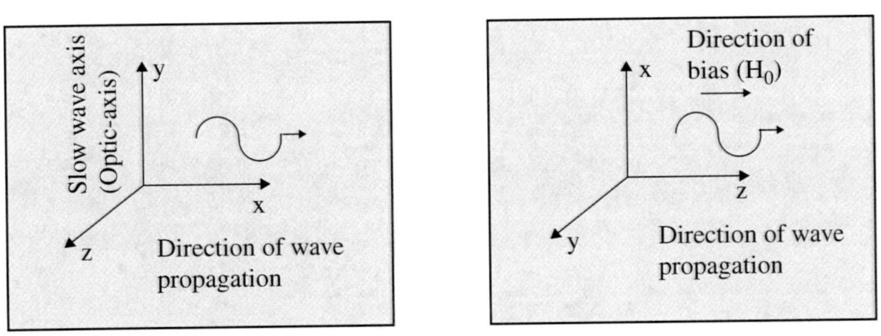

(a) Wave propagation in the uniaxial medium. (b) Wave propagation in the gyroelectric medium.

Figure 4.13 Wave propagation uniaxial media.

propagation. For the guided waves, TE/TM is modes of propagation; and in the open medium, it indicates the polarization of the propagating waves. The second use in the context of obliquely incident EM-waves at the interface of two medias is further discussed in section (5.2) of chapter 5.

On solving the above wave equations, the time-harmonic wave propagating in the positive x-direction is obtained as,

$$\begin{aligned}
\vec{E}(x,t) &= \left[\hat{y}\,E_y(x) + \hat{z}\,E_z(x)\right] e^{j\omega t} \\
&= \left[\hat{y}\,E_{0y} e^{-jk^e x} + \hat{z}\,E_{0z} e^{-jk^o x}\right] e^{j\omega t} && \text{(a)} \\
\text{where, } k^e &= k_0\sqrt{\varepsilon_{r||}} && \text{(b)} \\
k^o &= k_0\sqrt{\varepsilon_{r\perp}} && \text{(c)} \\
v_{pe} &= \frac{c}{\sqrt{\varepsilon_{r||}}} && \text{(d)} \\
v_{po} &= \frac{c}{\sqrt{\varepsilon_{r\perp}}} && \text{(e)} \\
\vec{H}(x,t) &= \left[\hat{y}\,H_y(x) + \hat{z}\,H_z(x)\right] e^{j\omega t} \\
&= \left[\hat{y}\,H_{0y} e^{-jk^e x} + \hat{z}\,H_{0z} e^{-jk^o x}\right] e^{j\omega t} && \text{(f).}
\end{aligned} \tag{4.7.6}$$

In the above equations, k_0 and c are wavenumber and velocity of EM-waves in free space. Also, k^e and k^o are wavenumbers of the *extraordinary waves* and *ordinary waves*, respectively, traveling in the x-direction with phase velocities v_{pe} and v_{po}. The extraordinary waves are y-polarized, i.e. TE-polarized wave viewing the permittivity component $\varepsilon_{r||}$. The ordinary waves are z-polarized, i.e. TM-polarized wave viewing the permittivity component $\varepsilon_{r\perp}$. Thus, an obliquely incident linearly polarized EM-waves, with E_y and E_z components, entering the slab of the anisotropic medium is split into two distinct *normal mode waves* and travel with two different phase velocities. They come out from the slab with a phase difference. This phenomenon is known as *double refraction* or *birefringence*. The dispersion relation for both normal waves is discussed in subsection (4.7.5).

The wave impedances η^e and η^o of the extraordinary waves and ordinary waves propagating in the x-direction are obtained by substituting the field solutions of equation (4.7.6) in equations (4.7.3) and (4.7.4):

$$\begin{aligned}
&\text{TE} - \text{waves } \left(E_y \neq 0, E_z = 0\right): \\
&\eta^e = \frac{E_{0y}}{H_{0z}} = \frac{\omega\mu_0}{k^e} = \frac{\eta_0}{\sqrt{\varepsilon_{r||}}}\hat{x} && \text{(a)} \\
&\text{TM} - \text{waves } \left(E_z \neq 0, E_y = 0\right): \\
&\eta^o = \frac{E_{0z}}{H_{0y}} = -\frac{\omega\mu_0}{k^o} = \frac{\eta_0}{\sqrt{\varepsilon_{r\perp}}}\hat{x} && \text{(b).}
\end{aligned} \tag{4.7.7}$$

The plasma medium could be taken as an example. It is a uniaxial anisotropic medium with $\varepsilon_{r\perp} = 1$ and $\varepsilon_{r||} = \varepsilon_r$. In this case, y-polarized extraordinary waves travel with a slower phase velocity v_{pe} as compared to a phase velocity v_{po} of the z-polarized ordinary waves. So, the extraordinary waves are also known as the *slow-waves* with $\varepsilon_{r||} > \varepsilon_{r\perp}$. The ordinary waves are called *fast-waves*. The optic axis of the uniaxial medium is called the *slow-wave axis* and ordinary axis as the *fast-wave axis*.

Waveplates and Phase Shifters

The incident linearly polarized wave on a slab at 45° has two in-phase E-field components. After traveling a distance d, the field components develop a phase difference $\Delta\phi$. So, for the case $E_{oy} = E_{oz} = E_0$ and a phase difference $\Delta\phi = 90°$ at the output of the slab of thickness x = d, the uniaxial anisotropic dielectric slab converts linearly polarized incident waves into the circularly polarized waves. It is shown below:

$$\begin{aligned}
&\text{Incidence wave: } \quad \vec{E}(x=0) = (\hat{y} + \hat{z})E_0 && \text{(a)} \\
&\text{Phase difference } \Delta\phi \text{ at output } (k^e - k^o)d = \frac{\pi}{2} \\
&\Rightarrow d = \frac{\lambda_0}{4\left(\sqrt{\varepsilon_{r||}} - \sqrt{\varepsilon_{r\perp}}\right)} && \text{(b).}
\end{aligned} \tag{4.7.8}$$

The electric components of the extraordinary and ordinary waves and also the total E-field at the output of the slab are

$$\begin{aligned}
&E_y^e(x=d) = E_0\,e^{-jk^e d} = E_0\,e^{-j\left(k^o d + \pi/2\right)} \\
&\Rightarrow E_y^e(x=d) = -j\,E_0\,e^{-jk^o d} && \text{(a)} \\
&E_z^o(x=d) = E_0\,e^{-jk^o d} && \text{(b)} \\
&\vec{E}(x=d) = (-j\hat{y} + \hat{z})E_0\,e^{-jk^o d} && \text{(c)} \\
&\vec{E}(x=d,t) = \text{Re}\left[(-j\hat{y} + \hat{z})E_0\,e^{-jk^o d} e^{j\omega t}\right] && \text{(d).}
\end{aligned} \tag{4.7.9}$$

The wave at the output of the slab is a left-hand circularly polarized wave. Such a slab converting the incoming linearly polarized wave to the circularly polarized waves is called the *quarter-waveplate*. A waveplate with $(k^e - k^o)d = \pi$, $\Rightarrow d = \lambda_0/\{2\left(\sqrt{\varepsilon_{r||}} - \sqrt{\varepsilon_{r\perp}}\right)$, i.e. the *half-waveplate* rotates the 45° linearly polarized waves by 90° and output electric field component is $\vec{E}(x=d) = (-\hat{y} + \hat{z})E_0$. The phase of the outgoing waves could be controlled by changing the thickness of a slab. The waveplate acts as a *phase shifter*. In the case of uniaxial anisotropic plasma medium, the permittivity components are $\varepsilon_{r||} = \varepsilon_r$ and $\varepsilon_{r\perp} = 1$.

The above characteristics of a slab are also realized by thin *metasurfaces* discussed in subsections (22.5) and (22.6) of chapter 22.

4.7.2 Wave Propagation in Uniaxial Gyroelectric Medium

Figure (4.13b) shows uniform TEM-waves propagation in the z-direction in an unbounded uniaxial gyroelectric medium created by the magnetized plasma on the application of the DC magnetic field H_0 in the z-direction. The permittivity tensor $[\varepsilon_r]$ of the medium is given equation (4.2.11). Due to the presence of off-diagonal matrix elements $\pm j\kappa$ in the permittivity matrix of the gyroelectric medium, the E_x component of the linearly polarized incident wave also generates the E_y component with a time quadrature. It is due to the presence of factor "j." Similarly, the E_y component of an incident wave generates E_x component also with a time quadrature. The presence of two orthogonal E-field components with a time quadrature in a gyroelectric medium creates the left-hand circularly polarized (LHCP) and right-hand circularly polarized (RHCP) waves as the *normal modes* in the uniaxial gyroelectric medium. Both circularly polarized waves travel with two different phase velocities. Thus, the *gyro* medium with the *cross-coupling gyrotropic factor* $\pm j\kappa$ has the ability of polarization conversion.

Maxwell equation (4.7.2a) is expanded in the usual way to get the transverse field components E_x, E_y and H_x, H_y:

$$\frac{\partial E_y}{\partial z} = j\omega\mu_0 H_x, \qquad (a)$$
$$\frac{\partial E_x}{\partial z} = -j\omega\mu_0 H_y \qquad (b). \qquad (4.7.10)$$

However, the Maxwell equation (4.7.2b) in the present case is expanded differently:

$$\begin{bmatrix} \hat{x} & \hat{y} & \hat{z} \\ \frac{\partial}{\partial x} & \frac{\partial}{\partial y} & \frac{\partial}{\partial z} \\ H_x & H_y & 0 \end{bmatrix} = j\omega\varepsilon_0 \begin{bmatrix} \hat{x} & 0 & 0 \\ 0 & \hat{y} & 0 \\ 0 & 0 & \hat{z} \end{bmatrix} \begin{bmatrix} \varepsilon_r & j\kappa & 0 \\ -j\kappa & \varepsilon_r & 0 \\ 0 & 0 & \varepsilon_{r,zz} \end{bmatrix} \begin{bmatrix} E_x \\ E_y \\ 0 \end{bmatrix}$$
$$\Rightarrow -\frac{\partial H_y}{\partial z} = j\omega\varepsilon_0\left[\varepsilon_r E_x + j\kappa E_y\right] \qquad (a)$$
$$\frac{\partial H_x}{\partial z} = j\omega\varepsilon_0\left[-j\kappa E_x + \varepsilon_r E_y\right] \qquad (b). \qquad (4.7.11)$$

For the uniform plane wave propagating in the positive z-direction, $\partial H_y/\partial x = \partial H_x/\partial y = 0$. In the above equations, it is noted that the $\varepsilon_{r,\,zz}$ component of permittivity does not play any role in the TEM mode wave propagation in the z-direction. However, for the wave propagation in the x-direction $E_x = 0$, $E_z \neq 0$ and $\varepsilon_{r,\,zz}$ permittivity component occurs in the wave propagation. Similar is the case for the wave propagation in the y-direction. Further, due to the cross-coupling between E_x and E_y components in the above equations, it is not possible to obtain a single second-order wave equation for either E_x or E_y. However, the solution could be assumed for the field vectors $\vec{E}(z)$ and $\vec{H}(z)$ as follows:

$$\vec{E}(z) = \vec{E}_0 e^{-j\beta_z z} \qquad (a)$$
$$\vec{H}(z) = \vec{H}_0 e^{-j\beta_z z} \qquad (b). \qquad (4.7.12)$$

On substituting the above equations in equations (4.7.10) and (4.7.11), the following sets of equations are obtained:

$$\beta_z E_y = -\omega\mu_0 H_x \qquad (a)$$
$$\beta_z E_x = \omega\mu_0 H_y \qquad (b)$$
$$\beta_z H_y = \omega\varepsilon_0\left[\varepsilon_r E_x + j\kappa E_y\right] \qquad (c)$$
$$\beta_z H_x = -\omega\varepsilon_0\left[-j\kappa E_x + \varepsilon_r E_y\right] \qquad (d). \qquad (4.7.13)$$

On solving the above equations for E_x and E_y, the following characteristics equation is obtained:

$$\begin{bmatrix} \beta_z^2 - \varepsilon_r k_0^2 & -j\kappa k_0^2 \\ j\kappa k_0^2 & \beta_z^2 - \varepsilon_r k_0^2 \end{bmatrix} \begin{bmatrix} E_x \\ E_y \end{bmatrix} = 0, \qquad (4.7.14)$$

where wavenumber in free space is $k_0^2 = \omega^2\mu_0\varepsilon_0$. The det $[\,] = 0$ of the above homogeneous equation provides the nontrivial solutions giving the following two eigenvalues of the propagation constant β_z:

$$\begin{vmatrix} \beta_z^2 - \varepsilon_r k_0^2 & -j\kappa k_0^2 \\ j\kappa k_0^2 & \beta_z^2 - \varepsilon_r k_0^2 \end{vmatrix} = 0, \quad \left(\beta_z^2 - \varepsilon_r k_0^2\right)^2 - \left(\kappa k_0^2\right)^2 = 0$$
$$\Rightarrow \beta_z^2 = k_0^2(\varepsilon_r \pm \kappa) = \left(\beta_z^{\pm}\right)^2$$
$$\Rightarrow \quad \beta_z^+ = k_0\sqrt{\varepsilon_r + \kappa} \qquad (a)$$
$$\beta_z^- = k_0\sqrt{\varepsilon_r - \kappa} \qquad (b). \qquad (4.7.15)$$

It is shown below that the eigenvalue β_z^+ and β_z^- are the propagation constant of two circularly polarized normal mode waves propagating in the z-direction. The wave with propagation constant β_z^+ travels at slower velocity compared to the wave traveling in an isotropic medium with relative permittivity ε_r. The wave with propagation constant β_z^- is a faster traveling wave.

The electric fields, i.e. the eigenvectors $\vec{E}^{+}(z)$ and $\vec{E}^{-}(z)$ for both the normal mode waves are obtained by substituting the eigenvalue $\beta_z = \beta_z^+$ and $\beta_z = \beta_z^-$ in equation (4.7.14), and using equation (4.7.12a):

For β_z^+ normal mode wave: $E_y = -jE_x = -jE_0$. The E-field of wave is,

$$\vec{E}^{+}(z) = (\hat{x}E_{ox} + \hat{y}E_{oy})\, e^{-j\beta_z^+ z}, \quad \Rightarrow \vec{E}^{+}(z) = E_0(\hat{x} - j\hat{y})\, e^{-j\beta_z^+ z} \quad \text{(a)}$$

$$\text{Likewise, for } \beta_z^- \text{ normal mode wave: } \vec{E}^{-}(z) = E_0(\hat{x} + j\hat{y})\, e^{-j\beta_z^- z} \quad \text{(b).} \qquad (4.7.16)$$

Using $\vec{E}^{+}(z,t) = \text{Re}\left[\vec{E}^{+}(z)\, e^{j\omega t}\right]$, equation (4.7.16a) shows the RHCP waves coming toward an observer. Likewise, equation (4.7.16b) is for the LHCP waves coming toward an observer.

Suppose the x-polarized wave with $\vec{E}(z = 0) = \hat{x}E_0$ is incident on the gyroelectric slab of thickness d. At the plane of entry, the linearly polarized electric field can be decomposed into the RHCP and LHCP waves traveling in the positive z-direction. The electric field at any distance inside the slab is a sum of two circularly polarized waves:

$$\text{At input of the slab: } \vec{E}(z = 0) = \hat{x}E_0 = \frac{E_0}{2}(\hat{x} - j\hat{y}) + \frac{E_0}{2}(\hat{x} + j\hat{y}) \quad \text{(a)}$$

$$\text{At distance z inside the slab: } \vec{E}(z) = \frac{E_0}{2}(\hat{x} - j\hat{y})\, e^{-j\beta_z^+ z} + \frac{E_0}{2}(\hat{x} + j\hat{y})\, e^{-j\beta_z^- z} \qquad (4.7.17)$$

$$\Rightarrow \vec{E}(z) = E_0\left[\hat{x}\cos\left(\frac{\beta_z^+ - \beta_z^-}{2}\right) - \hat{y}\sin\left(\frac{\beta_z^+ - \beta_z^-}{2}\right)\right] e^{-j\left(\frac{\beta_z^+ + \beta_z^-}{2}\right)z} \quad \text{(b).}$$

However, the wave is still linearly polarized with a rotation of ϕ with respect to the x-axis. The angle of rotation ϕ at the output of the slab is

$$\phi(z) = \tan^{-1}\left(\frac{E_y}{E_x}\right) = -\left(\frac{\beta_z^+ - \beta_z^-}{2}\right)z \quad \text{(a)}$$

$$\phi(z = d) = -\left(\frac{\beta_z^+ - \beta_z^-}{2}\right)d \quad \text{(b).}$$

$$(4.7.18)$$

The above equation shows that the E-field polarization vector rotates while the wave travels in the medium. For the wave reflected at the end of the slab, the total rotation at the input is 2ϕ. This is known as *Faraday rotation*. It is the characteristic of a gyrotropic medium – gyroelectric, as well as gyromagnetic [B.2–B.4]. The wave propagation in the gyromagnetic medium is obtained similarly [B.3]. Similar to the gyroelectric medium, the gyromagnetic medium also supports the circularly polarized normal modes. The word *gyro* indicates rotation and the *gyro media* supports circularly polarized normal mode wave propagation. They do not support the linearly polarized EM-waves. The analysis of the wave propagation in other complex media- bi-isotropic and bianisotropic is cumbersome. However, it can be followed by consulting more advanced textbooks [B.13, B.17, B.21–B.23].

4.7.3 Dispersion Relations in Biaxial Medium

A biaxial medium could be considered with scalar permeability μ and permittivity tensor $[\varepsilon]$. The off-diagonal elements of the matrix equation (4.2.4a) are zero. Maxwell equations (4.5.31a) and (4.5.31b) are used in the present case with permittivity tensor $[\varepsilon]$ in place of a scalar ε. The wave equation (4.5.32a) is suitably modified to incorporate the tensor $[\varepsilon]$:

$$\vec{k}\left(\vec{k}.\vec{E}\right) - \vec{E}\left(\vec{k}.\vec{k}\right) + \omega^2\mu[\varepsilon]\,\vec{E} = 0, \text{ where } \vec{k}.\vec{k} = k_x^2 + k_y^2 + k_z^2$$

$$\Rightarrow (\hat{x}k_x + \hat{y}k_y + \hat{z}k_z)\left[k_xE_x + k_yE_y + k_zE_z\right] - \left(k_x^2 + k_y^2 + k_z^2\right)(\hat{x}E_x + \hat{y}E_y + \hat{z}E_z)$$

$$+\,\omega^2\mu\begin{bmatrix}\varepsilon_{xx} & 0 & 0\\ 0 & \varepsilon_{yy} & 0\\ 0 & 0 & \varepsilon_{zz}\end{bmatrix}\begin{bmatrix}E_x\\ E_y\\ E_z\end{bmatrix} = 0$$

$$\Rightarrow \left[\begin{bmatrix} -\left(k_y^2+k_z^2\right) & k_x k_y & k_x k_z \\ k_x k_y & -\left(k_x^2+k_z^2\right) & k_y k_z \\ k_x k_z & k_y k_z & -\left(k_x^2+k_y^2\right) \end{bmatrix} + \omega^2\mu \begin{bmatrix} \varepsilon_{xx} & 0 & 0 \\ 0 & \varepsilon_{yy} & 0 \\ 0 & 0 & \varepsilon_{zz} \end{bmatrix}\right] \begin{bmatrix} E_x \\ E_y \\ E_z \end{bmatrix} = 0$$

$$\Rightarrow \begin{bmatrix} \left[\omega^2\mu\varepsilon_{xx} - \left(k_y^2+k_z^2\right)\right] & k_x k_y & k_x k_z \\ k_x k_y & \left[\omega^2\mu\varepsilon_{yy} - \left(k_x^2+k_z^2\right)\right] & k_y k_z \\ k_x k_z & k_y k_z & \left[\omega^2\mu\varepsilon_{zz} - \left(k_x^2+k_y^2\right)\right] \end{bmatrix} \begin{bmatrix} E_x \\ E_y \\ E_z \end{bmatrix} = 0. \quad (4.7.19)$$

$$\text{Let } k^2 - \omega^2\mu\varepsilon_{xx} = m^2; \quad k^2 - \omega^2\mu\varepsilon_{yy} = n^2; k^2 - \omega^2\mu\varepsilon_{zz} = p^2. \quad (4.7.20)$$

Using equation (4.7.20), equation (4.7.19) is rewritten as,

$$\Rightarrow \begin{bmatrix} k_x^2 - m^2 & k_x k_y & k_x k_z \\ k_x k_y & k_y^2 - n^2 & k_y k_z \\ k_x k_z & k_y k_z & k_z^2 - p^2 \end{bmatrix} \begin{bmatrix} E_x \\ E_y \\ E_z \end{bmatrix} = 0. \quad (4.7.21)$$

The nontrivial solution for E_i $(i = x, y, z)$ of the above homogeneous equation is det[ ] = 0, i.e.

$$(npk_x)^2 + \left(mpk_y\right)^2 + (mnk_z)^2 = (mnp)^2$$
$$\Rightarrow \frac{k_x^2}{m^2} + \frac{k_y^2}{n^2} + \frac{k_z^2}{p^2} = 1. \quad (4.7.22)$$

The above dispersion relation is a quadratic equation of any component of k^2. So, there are two solutions for any component of k. Two solutions correspond to two normal modes of propagation in the anisotropic medium. At a fixed frequency, equation (4.7.22) is the equation of an ellipsoid surface in the *k-space (wavevector space)*, i.e. the *normal space*. For an isotropic medium, it is reduced to a sphere. Further, on knowing the wavenumber, the field components E_i $(i = x, y, z)$ can be determined from equation (4.7.21).

4.7.4 Concept of Isofrequency Contours and Isofrequency Surfaces

The discussion of dispersion in the uniaxial medium requires an understanding of the concept of *isofrequency contours* and *isofrequency surfaces* in the 2D and 3D k-space. Figure (4.14a–d) explain the concept of isofrequency contours.

The wave propagation is considered in the isotropic (x-y)-plane. The dispersion relations for the 2D waves propagation in the isotropic medium and also in air medium are expressed as

$$k_x^2 + k_y^2 = \left(\frac{\omega}{v_p}\right)^2 \quad (a)$$
$$k_x^2 + k_y^2 = \left(\frac{\omega}{c}\right)^2 \quad (b). \quad (4.7.23)$$

At a fixed frequency ω, the above equations are equations of circles in the $(k_x\text{-}k_y)$-plane. Figure (4.14a) shows that the radius of the circle increases with an increase in frequency. It forms a *light cone*. Figure (4.14b) further shows the increase in the 2D wavevector at the increasing order of frequencies $\omega_1 < \omega_2 < \omega_3 < \omega_4$. The concentric contours of the wavevector are known as the *isofrequency* contours, displaying the dispersion relation. The light cone of the 2D dispersion diagram is generated by revolving Fig. (2.4) of the 1D dispersion diagram, shown in chapter 2, around the ω-axis. Likewise, 3D *isofrequency surfaces* are obtained. It is discussed in the next section.

The propagating wave in the z-direction is described by equation (4.7.12). The propagation constant $\beta_z = k_z = \sqrt{k^2 - \left(k_x^2 + k_y^2\right)}$ of the waves must be real. So, the propagating waves are obtained only for $\left(k_x^2 + k_y^2\right) < k^2$, i.e. within the light cone. Outside the light cone, i.e. for $\left(k_x^2 + k_y^2\right) > k^2$, the waves are nonpropagating *evanescent waves*. Thus, the light cone divides the k-space into the propagating and evanescent wave regions shown in Fig. (4.14a,c and d).

Further, any point P on the isofrequency contour surface, shown in Fig. (4.14b), connected to the origin O shows the direction of the wavevector. It is also the direction of the phase velocity v_p. The direction of the normal at point P shows the direction of the Poynting vector, i.e. the direction of the group velocity v_g. For an isotropic medium, both the v_p and v_g are in the same direction. It is noted that the wavefront is always normal to the wavevector $\vec{k}$. However, for the anisotropic medium, the phase and group velocities may not in the same direction. It is discussed below.

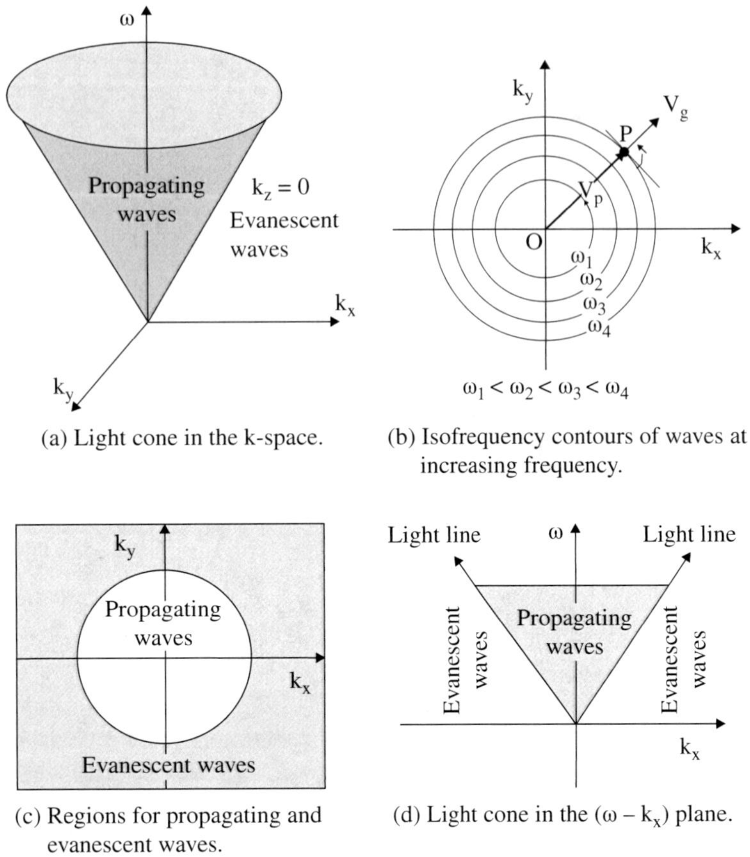

Figure 4.14 Dispersion diagrams of the wave propagating in the z-direction in the isotropic medium.

4.7.5 Dispersion Relations in Uniaxial Medium

This section considers the dispersion relation of a uniaxial anisotropic permittivity medium as a special case of the dispersion relation (4.7.19) of the biaxial medium. The permittivity along the optic axis, i.e. the z-axis is $\varepsilon_{zz} = \varepsilon_{||}$ and permeability of the medium is μ. The (x-y)-plane, with permittivity tensor elements $\varepsilon_{xx} = \varepsilon_{yy} = \varepsilon_{\perp}$, is a transverse plane. So, the medium is isotropic in the (x-y)-plane. The propagation constant along the z-axis is k_z and in the transverse plane, it is k_t, satisfying the relation $k^2 = k_x^2 + k_y^2 + k_z^2 = k_t^2 + k_z^2$. To simplify equation (4.7.19) for the uniaxial medium, the transverse wavevector $\vec{k}_t$ in the (x-y)-plane is aligned such that the wave propagates only along the x-axis, i.e. $k_y = 0$ and $k_x = k_t$. Under such alignment, equation (4.7.19) is reduced to the following dispersion relation:

$$\begin{vmatrix} \omega^2\mu\varepsilon_{\perp} - k_z^2 & 0 & k_t k_z \\ 0 & \omega^2\mu\varepsilon_{\perp} - k^2 & 0 \\ k_t k_z & 0 & \omega^2\mu\varepsilon_{||} - k_t^2 \end{vmatrix} = 0. \tag{4.7.24}$$

The above equation provides the following characteristic equation:

$$\left(k^2 - \omega^2\mu\varepsilon_{\perp}\right)\left[\left(\omega^2\mu\varepsilon_{\perp} - k_z^2\right)\left(\omega^2\mu\varepsilon_{||} - k_t^2\right) - (k_z k_t)^2\right] = 0. \tag{4.7.25}$$

The above expression is also obtained from equation (4.7.19) even for the alignment of the wavevector $\vec{k}_t$ to the y-axis, i.e. $k_x = 0$ and $k_y = k_t$. So, the characteristic equation (4.7.25) gives the following two independent dispersion relations:

$$k^2 = \omega^2\mu\varepsilon_{\perp} \qquad \text{(a)}$$

$$\frac{k_z^2}{\varepsilon_{\perp}} + \frac{k_t^2}{\varepsilon_{||}} = \omega^2\mu \qquad \text{(b)}$$

$$k^2 = k_y^2 + k_z^2 = k_0^2\,\varepsilon_{r\perp} \qquad \text{(c)}$$

$$\frac{k_z^2}{\varepsilon_{r\perp}} + \frac{k_y^2}{\varepsilon_{r||}} = k_0^2 \qquad \text{(d).}$$

(4.7.26)

The above equations (4.7.26a) and (4.7.26b) demonstrate the presence of two normal modes of wave propagation in the 3D (k_x, k_y, k_z) – space. Next, the 3D dispersion relation is reduced to the 2D dispersion relation given by equations (4.7.26c) and (4.7.26d). Equation (4.7.26a) is an equation of sphere in the 3D k-space at a fixed frequency, i.e. at ω = constant. It shows the dispersion relation of the *ordinary waves* with the wavevector $k^o = k_0\sqrt{\varepsilon_{r\perp}}$. At increasing order of frequencies, equation (4.7.26a) provides *isofrequency spherical surfaces*. Likewise, equation (4.7.26b) is an equation of ellipsoid in the 3D k-space at a fixed frequency. It provides the dispersion relation of the *extraordinary waves* in the form of the concentric *isofrequency ellipsoidal surfaces*. The wavevector of *extraordinary wave* k^e is rotation dependent.

The 2D dispersion diagram of both the ordinary and extraordinary waves is considered in the (y-z)-plane with the help of equations (4.7.26c) and (4.7.26d). Equation (4.7.26c) is the equation of an isofrequency dispersion circle, shown in Fig. (4.15a) in the (k_y-k_z) plane. It is the dispersion relation of the *ordinary waves* in the isotropic (y-z)- plane. The wavenumber $\left|\vec{k}^o\right|$ of the ordinary wave is independent of permittivity $\varepsilon_{\|}$ along the z-axis.

Likewise, equation (4.7.26d) is the dispersion relation of the *extraordinary waves*, giving the *isofrequency ellipse* in the (k_y-k_z)-plane, as shown in Fig. (4.15b). In this case, the wavenumber is direction-dependent because the relative permittivity for this case is direction-dependent, $k^e = \sqrt{\varepsilon_r(\theta)}\,k_0$. The wavevector k^e subtends an angle θ with respect to the z-optic axis. In this case, wavevectors components are $k_z = k^e\cos\theta = \sqrt{\varepsilon_r(\theta)}\,k_0\cos\theta$, and $k_y = k^e\sin\theta = \sqrt{\varepsilon_r(\theta)}\,k_0\sin\theta$. Using, these relations with equation (4.7.26d), the following equation of ellipse of permittivity, supporting the extraordinary wave, is obtained:

$$\frac{\cos^2\theta}{\varepsilon_{r\perp}} + \frac{\sin^2\theta}{\varepsilon_{r\|}} = \frac{1}{\varepsilon_r(\theta)} \qquad \text{(a)}$$

$$\Rightarrow \frac{\cos^2\theta}{n_0^2} + \frac{\sin^2\theta}{n_e^2} = \frac{1}{n^2(\theta)} \qquad \text{(b).}$$

$$(4.7.27)$$

The dispersion relation (4.7.27b) is in the term of refractive indices. The refractive index $n(\theta)$ is the direction-dependent, and n_0 and n_e are refractive indices for the ordinary and extraordinary waves. The wave

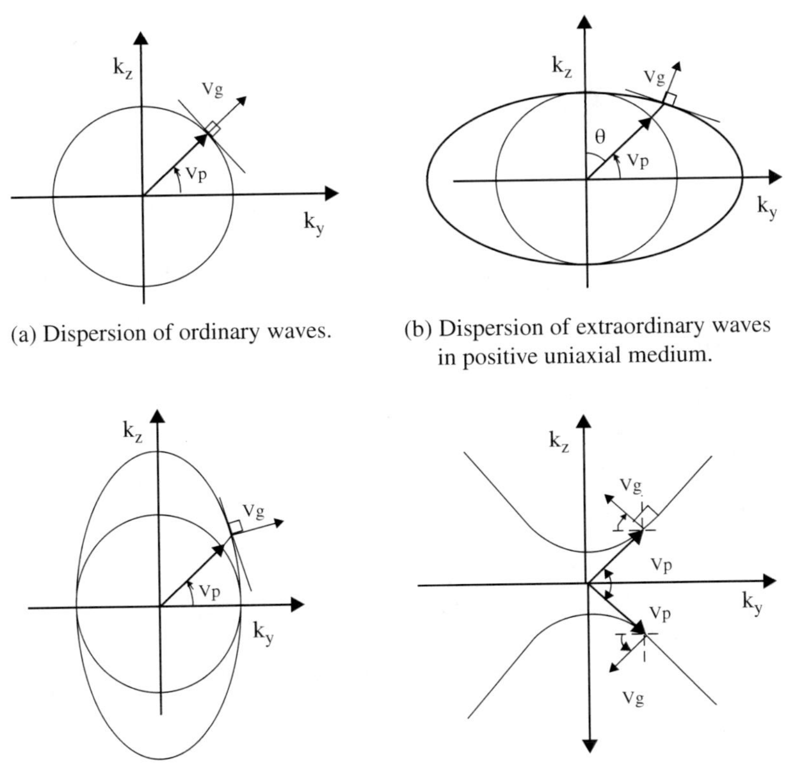

(a) Dispersion of ordinary waves. (b) Dispersion of extraordinary waves in positive uniaxial medium.

(c) Dispersion of extraordinary waves in the negative uniaxial medium. (d) Dispersion of extraordinary waves in the hyperbolic uniaxial medium.

Figure 4.15 Dispersion diagrams in the uniaxial anisotropic medium.

analysis using the refractive index, in place of permittivity and permeability, is commonly used for the optical wave propagation in the uniaxial medium [B.18–B.20].

Figure (4.15a) shows that the ordinary waves have both the phase and group velocities, on the isofrequency contour, in the same direction. Figure (4.15b) shows the elliptical dispersion diagram for the extraordinary waves. In this case, there is a deviation of the direction of the group velocity from the direction of the phase velocity. The inner circle of Fig. (4.15b) shows the dispersion diagram of the ordinary waves with relative permittivity $\sqrt{\varepsilon_{r\perp}} = n_\perp$(refractive index). Along the k_y, the extraordinary waves have relative permittivity $\sqrt{\varepsilon_{r||}} = n_{||}$(refractive index), $\varepsilon_{r||} > \varepsilon_{r\perp}$. Only along the z-optic axis, i.e. along the k_z-axis, the identical relative permittivity for both kinds of waves is obtained. In all other directions, the ordinary wave is a *fast-wave* as compared to the extraordinary wave. The medium is known as the *positive uniaxial medium*. Figure (4.15c) shows the case of the *negative uniaxial medium*. In this case, $\varepsilon_{r||} < \varepsilon_{r\perp}$ and the ordinary wave is a *slow-wave* as compared to the extraordinary wave in all directions except along the y-axis.

The phase velocities v_{p1} and v_{p2} of the ordinary and extraordinary waves using equations (4.7.26a) and (4.7.27a) are expressed through the following relations [B.3]:

$$v_{p1}^2 = (v^o)^2 = \frac{1}{\mu\varepsilon_\perp} \qquad \text{(a)}$$

$$v_{p2}^2 = (v^o)^2 \cos^2\theta + (v^e)^2 \sin^2\theta \qquad \text{(b)}$$

$$\text{where } v^e = \frac{c}{\sqrt{\varepsilon_{r||}}} \qquad \text{(c).}$$

$$(4.7.28)$$

The phase velocity v_{p1} of the ordinary wave is independent of the angle θ. However, the phase velocity v_{p2} of the extraordinary wave is dependent on the angle θ. In Figure (4.15b), both velocities are identical only for the wave propagation along the z-optic axis.

Finally, it is possible to artificially realize a uniaxial anisotropic material with one of the permittivity components as a negative quantity, say $\varepsilon_{r\perp} = -|\varepsilon_{r\perp}|$. In this case, equation (4.7.26d) shows the hyperbolic dispersion relation. This medium is known as the *hypermedium*. It supports the wave propagation with a larger value of wavenumbers and can convert an incident evanescent wave to the propagating waves. Such engineered materials are needed by the hyperlens [J.1, J.7]. Figure (4.15d) shows the dispersion relation of the hypermedium with phase and group velocities. In this case, y-components of v_p and v_g are opposite to each other. So, the hypermedium supports the backward wave propagation, and it is a metamaterial medium. The *hyperlens* is discussed in the subsection (5.5.6) of chapter 5.

References

Books

B.1 Baden-Fuller, A.J.: *Microwaves*, 3rd Edition, Pergamon Press, New York, 1990.

B.2 Jordan, E.C.; Balmain Keith, G.E.: *Electromagnetic Wave and Radiating System*, Prentice-Hall India, New Delhi, 1989.

B.3 Ramo, S.; Whinnery, J.R.; Van Duzer, T.: *Fields, and Waves in Communication Electronics*, 3rd Edition, John Wiley & Sons, Singapore, 1994.

B.4 Collin, R.E.: *Foundations for Microwave Engineering*, 2nd Edition, McGraw-Hill, Inc., New York, 1992.

B.5 Rao, N.N.: *Elements of Engineering Electromagnetics*, 3rd Edition, Prentice-Hall, Englewood Cliff, NJ, 1991.

B.6 Sadiku, M.N.O.: *Elements of Electromagnetics*, 3rd Edition, Oxford University Press, New York, 2001.

B.7 Cheng, D.K.: *Fields and Wave Electromagnetics*, 2nd Edition, Pearson Education, Singapore, 1989.

B.8 Weeks, W.L.: *Electromagnetic Theory for Engineering Applications*, John Wiley & Sons, New York, 1964.

B.9 Balanis, C.A.: *Advanced Engineering Electromagnetics*, John Wiley & Sons, New York, 1989.

B.10 Hipple, A. Von: *Dielectrics, and Waves*, Artech House, Norwood, MA, 1995.

B.11 Plonsey, R.; Collin, R.E.: *Principle and Applications of Electromagnetic Fields*, TMH, New Delhi, 1973.

B.12 Polivanov, K.: *The Theory of Electromagnetic Fields*, Mir Publisher, Moscow, 1975.

B.13 Orfanidis, S.J.: *Electromagnetic Waves and Antenna*, Free Book on the Web.

B.14 Staelin, D.H.; Morgenthaler, A.W.; Kong, J.A.: *Electromagnetic Waves*, Prentice-Hall, Englewood Cliffs, NJ, 1994.

B.15 Landau, L.D.; Lifshitz, E.M.; Pitaevskii, L.P.: *Electrodynamics of Continuous Media*, Pergamon Press, New York, 1984.

B.16 Engheta, N.; Ziolkowski, R.W. (Editors): *Metamaterials: Physics and Engineering Explorations*, Wiley-Interscience, John Wiley & Sons, Inc., Hoboken, NJ, 2006.

B.17 Lakhtakia, A.; Varadan, V.K.; Varadan, V.V.: *Time-Harmonic Electromagnetic Fields in Chiral Media*, Springer-Verlag, Berlin, Heidelberg, 1989.

B.18 Born, M.; Wolf, E.: *Principles of Optics*, 6th Edition, Pergamon Press, New York, 1980.

B.19 Haus, H.A.: *Waves and Fields in Optoelectronics*, Prentice-Hall, Englewood Cliffs, NJ, 1984.

B.20 Yariv, A.; Yeh, P.: *Optical Waves in Crystals*, Wiley, New York, 1984.

B.21 Kong, J.A.: *Electromagnetic Wave Theory*, Wiley, New York, 1986.

B.22 Mackay, T.G.; Lakhtakia, A.: *Electromagnetic Anisotropy and Bianisotropy: A Field Guide*, World Scientific, Singapore, 2010.

B.23 Lindell, I.V.; Sihvola, A.H.; Tretyakov, S.A.; Viitanen, A.J.: *Electromagnetic Waves in Chiral and Bi-anisotropic Media*, Artech House, Boston, MA, 1994.

B.24 Hoffmann, R.: *Microwave Integrated Circuit Handbook*, Artech House, Boston, MA, 1985.

B.25 Capolino, F. (Editor): *Theory and Phenomena of Metamaterials*, CRC Press, Boca Raton, FL, 2009.

B.26 *The Heaviside Centenary Volume*, The Institution of Electrical Engineers, London, 1950.

B.27 Whittaker, E.T.: Oliver Heaviside, In *Electromagnetic Theory*, Vol. 1, Oliver Heaviside, Reprint, Chelsea Pub. Co., New York, 1971.

B.28 Behrend, B.A.: The work of Oliver Heaviside, In *Electromagnetic Theory*, Vol. 1, Oliver Heaviside, Reprint, Chelsea Pub. Co., New York, 1971.

B.29 Kraus, J.D.: *Antenna*, 2nd Edition, McGraw-Hill, 1988.

B.30 Collett, E.: *Polarized Light: Fundamentals and Applications*, Marcel Dekker, Inc., 1993.

B.31 Saleh, B.E.A.; Teich, M.C.: *Fundamental of Photonics*, Wiley, New York, 1991.

Journals

J.1 Jacob, Z.; Alekseyev, L.V.; Narimanov, E.: *Optical hyperlens: far-field imaging beyond the diffraction limit, Opt. Express*, Vol. 14, No. 18, pp. 8247–8256, 2006.

J.2 Lee, H. et al.: *Development of optical hyperlens for imaging below the diffraction limit, Opt. Express*, Vol. 15, pp. 15886–15891, 2007.

J.3 Seddon, N.; Bearpark, T.: *Observation of the inverse Doppler effect, Science*, Vol. 302, pp. 1537–1539, Nov. 2003.

J.4 Mosallaei, H.; Sarabandi, K.: *Magneto-dielectrics in electromagnetics: concept and applications, IEEE Trans. Ant. Propagat.*, Vol. 52, No. 6, pp. 1558–1567, June 2004.

J.5 Alu, A.; Engheta, N.: *Pairing an epsilon–negative slab with a mu-negative slab: anomalous tunneling and transparency, IEEE Trans. Antenna Proag., Special Issue on Metamaterials*, Vol. 51, No. 10, pp. 2558–2570, Oct. 2003.

J.6 Lee, H.; Xiong, Y.; Fang, N.; Srituravanich, N.; Durant, S.; Ambati, M.; Sun, C.; Zhang, X.: *Realization of optical superlens imaging below the diffraction limit, New Journal of Physics*, Vol. 7, No. 255, pp. 1–16, 2005.

J.7 Kim, M.; Rho, J.: *Metamaterials and imaging, Nano Convergence*, Vol. 2, No. 22, pp. 1–16, 2015.

5

Waves in Material Medium-II

(Reflection & Transmission of Waves, Introduction to Metamaterials)

Introduction

The characteristics of material media and EM-waves propagation in unbounded media are discussed in the previous chapter 4. Continuing the topics, this chapter is about the normal and oblique incidence of EM-waves at the interface of two media. The characteristics of both the normal and oblique incident EM-waves are obtained using an analytical method and convenient equivalent transmission line models. Under certain conditions, the interface surface of two media could acquire the property of a perfect electric conductor (PEC), or perfect magnetic conductor (PMC), or even a reactive impedance surface (RIS). These surfaces play a significant role in the modern microwave and antenna technologies. Some electromagnetic characteristics of the engineered composite materials, with negative permittivity and permeability, are also presented in the present chapter. These artificially structured materials are known as *metamaterials*. The realization, circuit modeling, and some applications of the metamaterials and *metasurfaces* are further presented in chapters 21 and 22, respectively. The properties of natural and artificial dielectrics are further examined in chapter 6.

Objectives

- *To discuss the normal and oblique incidence of the EM-waves at the interface of two media.*
- *To present the transmission line model of the normal and oblique incidence of the EM-waves.*
- *To obtain the dispersion diagrams of refracted waves in the isotropic and anisotropic media.*
- *To obtain characteristics of Brewster and the critical angles of incidence.*
- *To discuss the general properties of metamaterials media and their classifications.*
- *To obtain some characteristics of the EM-wave propagation in the metamaterials.*
- *To obtain the circuit models of metamaterials.*
- *To discuss the possibility of obtaining the flat lens, superlens, and hyperlens beyond the diffraction limit.*
- *To discuss the nature of the Doppler effect and Cerenkov radiation in the metamaterials.*
- *To discuss metamaterials as thin microwave absorbers.*

5.1 EM-Waves at Interface of Two Different Media

The EM-waves can strike the interface of two media with different electrical characteristics, either normally or obliquely. If the second medium is a dielectric medium, the waves undergo both reflection and transmission, whereas if the second medium is a perfect conductor, the reflection occurs. The obliquely incident plane waves follow the well-known Snell's law (also called Snell-Descartes law) of reflection and refraction. There are important applications of obliquely incident EM-waves.

5.1.1 Normal Incidence of Plane Waves

Figure (5.1a) shows an interface of two media #1 and #2. The media are electrically characterized by the primary parameters ε_{ri}, μ_{ri}, σ_i; $i = 1, 2$ and also by the secondary parameters, such as the refractive index n_i, and intrinsic or wave impedance η_i. Initially, both media are considered lossless dielectric media. Next, the second medium is treated as a PEC.

Introduction to Modern Planar Transmission Lines: Physical, Analytical, and Circuit Models Approach, First Edition. Anand K. Verma.

The incident wave in the medium #1, propagating in the x-direction with propagation constant k_1, is y-polarized with an electric field component E_y^{inc}. The direction-x is normal to the interface PQ along the *y*-axis; so the incident magnetic field component H_z^{inc} is +z directed. This is a special case of the TM-polarized obliquely incident wave, discussed in section (5.2.2). The incident ray strikes at the location O and gets partly reflected and refracted (i.e. transmitted) at O with the field components $\left(E_y^{ref}, -H_z^{ref}\right)$, and $\left(E_y^{tra}, H_z^{tra}\right)$, respectively. In the medium #2, the propagation constant is k_2. In general, the k is a wavevector with three components. Figure (5.1a) shows that the direction of propagation is decided by the direction of the Poynting vector. The field components are summarized in equation (5.1.1) on suppressing the time-harmonic dependence $e^{j\omega t}$:

Incident plane wave :

$$\vec{E}_y^{\,inc} = \hat{y}E_m^{inc}e^{-jk_1x},\quad \vec{H}_z^{\,inc} = \hat{z}\frac{E_m^{inc}}{\eta_1}e^{-jk_1x} \qquad (a)$$

Reflected plane wave :

$$\vec{E}_y^{\,ref} = \hat{y}E_m^{ref}e^{+jk_1x},\quad \vec{H}_z^{\,ref} = -\hat{z}\frac{E_m^{ref}}{\eta_1}e^{+jk_1x} \qquad (b)$$

Transmitted plane wave :

$$\vec{E}_y^{\,tra} = \hat{y}E_m^{tra}e^{-jk_2x},\quad \vec{H}_z^{\,tra} = \hat{z}\frac{E_m^{tra}}{\eta_2}e^{-jk_2x} \qquad (c),$$

$$(5.1.1)$$

where η_1 and η_2 are the intrinsic impedance of medium #1 and medium #2, respectively. The total tangential components of the electric and magnetic fields in both media are continuous across the interface at $x = 0$:

$$E_y^{inc}(x)\Big|_{x=0^-} + E_y^{ref}(x)\Big|_{x=0^-} = E_y^{tra}(x)\Big|_{x=0^+}, \Rightarrow E_m^{inc} + E_m^{ref} = E_m^{tra} \qquad (a)$$

$$\vec{H}_z^{\,inc}(x)\Big|_{x=0^-} + \vec{H}_z^{\,ref}(x)\Big|_{x=0^-} = \vec{H}_z^{\,tan}(x)\Big|_{x=0^+}, \Rightarrow \frac{E_m^{inc}}{\eta_1} - \frac{E_m^{ref}}{\eta_1} = \frac{E_m^{tra}}{\eta_2} \qquad (b)$$

$$\text{i.e, } 1 + \frac{E_m^{ref}}{E_m^{inc}} = \frac{E_m^{tra}}{E_m^{inc}}, \Rightarrow 1 + \Gamma = \tau \qquad (c)$$

$$\frac{1}{\eta_1}\left[1 - \frac{E_m^{ref}}{E_m^{inc}}\right] = \frac{1}{\eta_2}\frac{E_m^{tra}}{E_m^{inc}}, \Rightarrow 1 - \Gamma = \frac{\eta_1}{\eta_2}\tau \qquad (d).$$

$$(5.1.2)$$

On solving the above last two equations, the *reflection coefficient* Γ, and *the transmission coefficient* τ are computed at the interface:

$$\frac{E_m^{ref}}{E_m^{inc}}\Bigg|_{x=0} = \Gamma = \frac{\eta_2 - \eta_1}{\eta_2 + \eta_1} \qquad (a)$$

$$\frac{E_m^{tra}}{E_m^{inc}}\Bigg|_{x=0} = \frac{\eta_2}{\eta_1}\frac{H_z^{tra}}{H_z^{inc}}\Bigg|_{x=0} = \tau = \frac{2\eta_2}{\eta_2 + \eta_1} \qquad (b).$$

$$(5.1.3)$$

For a lossy medium, the intrinsic impedance is a complex quantity, and reflection and transmission coefficients are also complex quantities. However, the magnitude of the reflection coefficient of a passive medium is always equal to or less than unity, i.e. $|\Gamma| \le 1$. The magnitude and phase of the reflected and transmitted waves are different from that of the incident waves. For a lossless composite medium, the matching is obtained, i.e. no reflection $\Gamma = 0$, for an incident wave at the interface with $\eta_1 = \eta_2$. It could be realized in two ways:

$$\text{for } \mu_1 = \mu_2 = \mu_0 : \ \varepsilon_{r1} = \varepsilon_{r2} \qquad (a)$$

$$\text{for } \mu_1 \neq \mu_2, \varepsilon_{r1} \neq \varepsilon_{r2} : \ \frac{\mu_{r1}}{\varepsilon_{r1}} = \frac{\mu_{r2}}{\varepsilon_{r2}} \qquad (b).$$

$$(5.1.4)$$

In equations (5.1.4a,b), the first case is the usual one, as both media are identical. However, the second case is interesting; as the matching is possible even for dissimilar media. It may not be a practical one with natural materials. However, for the *artificially engineered metamaterials*, both μ_r and ε_r can be tuned independently to achieve the matching condition, It is discussed in subsection (5.5.8). The impedance-matched *Huygens' metasurface* has also been developed by tuning of electric and magnetic inclusions discussed in subsection (22.5.2) of chapter 22.

Sometimes, the concept of the *reflectance* (reflectivity) R and *transmittance* (transmissivity) T are used to show the portion of the reflected and transmitted powers:

$$R = \frac{p^{ref}}{p^{inc}} = \frac{\left(E_m^{ref}\right)^2/2\eta_1}{\left(E_m^{inc}\right)^2/2\eta_1}, \quad \Rightarrow R = |\Gamma|^2 = \left|\frac{\eta_2-\eta_1}{\eta_2+\eta_1}\right|^2 \quad (a)$$

$$T = \frac{p^{tra}}{p^{inc}} = \frac{\left(E_m^{tra}\right)^2/2\eta_2}{\left(E_m^{inc}\right)^2/2\eta_1}, \quad \Rightarrow T = |\tau|^2\frac{\eta_1}{\eta_2} = \frac{4\eta_1\eta_2}{(\eta_1+\eta_2)^2} \quad (b)$$

$$\text{Power balance:} \quad R + T = 1, \Rightarrow |\Gamma|^2 + |\tau|^2\frac{\eta_1}{\eta_2} = 1 \quad (c).$$

(5.1.5)

The total field in medium #1 is a combination of the incident and reflected waves causing interference of waves. The field is uniform in the *x*-direction. However, it creates partly a standing wave, and partly the traveling wave moving in the *x*-direction:

$$\vec{E}^{total}(x) = \hat{y}E_m^{inc}\left[e^{-jk_1x} + \Gamma e^{+jk_1x}\right] \quad (a)$$

$$\vec{H}^{total}(x) = \hat{z}\frac{1}{\eta_1}E_m^{inc}\left[e^{-jk_1x} - \Gamma e^{+jk_1x}\right] \quad (b).$$

(5.1.6)

For $\Gamma > 0$, i.e. for $\eta_2 > \eta_1$, the above expression can be decomposed as follows:

In space:

$$\vec{E}^{total}(x) = \hat{y}E_m^{inc}\left[e^{-jk_1x} - \Gamma e^{-jk_1x} + \Gamma e^{-jk_1x} + \Gamma e^{+jk_1x}\right]$$

$$\Rightarrow \vec{E}^{total}(x) = \hat{y}E_m^{inc}\left[(1+\Gamma)e^{-jk_1x} + j2\Gamma\sin k_1x)\right] \quad (a)$$

In space-time:

$$\vec{E}^{total}(x,t) = \hat{y}E_m^{inc}[(1+\Gamma)\underbrace{\cos(\omega t - k_1x)}_{\text{Traveling wave}} - \underbrace{2\Gamma\sin(k_1x)\sin(\omega t)}_{\text{Standing wave}}] \quad (b).$$

(5.1.7)

In equation (5.1.7), $\vec{E}^{total}(x,t) = \hat{y}\,\mathrm{Re}\left[\vec{E}^{total}(x)e^{j\omega t}\right]$ is taken.

Likewise, for $\Gamma < 0$, i.e. for $\eta_2 < \eta_1$, the wave in the medium can be written as follows:

In space – time:

$$\vec{E}^{total}(x,t) = \hat{y}E_m^{inc}[(1+\Gamma)\underbrace{\cos(\omega t - k_1x)}_{\text{Traveling wave}} - \underbrace{2\Gamma\cos(k_1x)\cos(\omega t)}_{\text{Standing wave}}]$$

(5.1.8)

The combined traveling and standing wave description follow a similar condition on a mismatched line given by equation (2.1.89) of chapter 2. Using the phasor – (5.1.6a), the $E_{max}^{total}(x)$ occurs for $e^{j2k_1x} = +1 = e^{j2n\pi}$, i.e. for $k_1x = n\pi$, and $E_{min}^{total}(x)$ occur for $e^{j2k_1x} = -1 = e^{j(2n+1)\pi}$, i.e. for $k_1x = (2n+1)\pi/2$, where n = 0,1,2, ... The VSWR in the medium is obtained as follows:

$$E_{max} = (1+|\Gamma|)E_m^{inc}, \quad E_{min} = (1-|\Gamma|)E_m^{inc} \quad (a)$$

$$\Rightarrow VSWR = \frac{|E_{max}|}{|E_{min}|} = \frac{(1+|\Gamma|)}{(1-|\Gamma|)} \quad (b).$$

(5.1.9)

The power densities in medium #1 (x<0) and #2 (x>0) are obtained using Poynting vector relation:

$$S^*(x) = \hat{x}\frac{1}{2}E_y^{total}(x)H_z^{*total}(x)$$

$$= \hat{x}\begin{vmatrix} \frac{1}{2\eta_1}\left[1-|\Gamma|^2 + j2|\Gamma|\sin(2k_1x)\right]\left(E_m^{inc}\right)^2, & x<0 \\ \frac{1}{2\eta_2}\tau^2\left(E_m^{inc}\right)^2, & x>0. \end{vmatrix}$$

(5.1.10)

The time-averaged power density is a real part of $S^*(x)$:

$$S_{av}(x) = \hat{x}\begin{vmatrix} \frac{1}{2\eta_1}\left[1-|\Gamma|^2\right]\left(E_m^{inc}\right)^2, & x<0 \\ \frac{1}{2\eta_2}\tau^2\left(E_m^{inc}\right)^2, & x>0. \end{vmatrix}$$

(5.1.11)

In the case of a lossless composite medium, the power balance is maintained. It is seen by separating the incident, reflected, and transmitted power densities:

$$S_{av}^{inc} = \frac{\left(E_m^{inc}\right)^2}{2\eta_1}, \quad S_{av}^{ref} = -|\Gamma|^2\frac{\left(E_m^{inc}\right)^2}{2\eta_1}$$

$$S_{av}^{tra} = \tau^2\frac{\left(E_m^{inc}\right)^2}{2\eta_2} = \tau^2\frac{\eta_1}{\eta_2}S_{av}^{inc} \quad (a)$$

$$\left[1-|\Gamma|^2\right] = \frac{\eta_1}{\eta_2}\tau^2 \quad (b).$$

(5.1.12)

Equation (5.1.12b) has been also obtained in equation (5.1.5c) from the concept of reflectance and transmittance. The above relation shows that the power balance $\Gamma^2 + \tau^2 = 1$ does not hold in the present case. The modified power balance relation is given by equation (5.1.12b). It shows that the transmission coefficient could be even more than unity, as it is multiplied by a factor $\sqrt{\eta_1/\eta_2}$. However, the above-given power balance relation must be maintained [B.1].

5.1.2 The Interface of a Dielectric and Perfect Conductor

Medium #2 could be a conducting medium, with the intrinsic impedance, $\eta_2 = (1+j)/\sigma\delta = (1+j)\sqrt{\omega\mu/2\sigma}$. For a PEC medium, $\sigma \to \infty$, $\eta_2 \to 0$.The PEC has infinite permittivity, i.e. $\varepsilon \to \infty$. Using equation (5.1.3a), the interface provides the total reflection with the reflection coefficient $\Gamma = -1$; i.e. a phase reversal for the reflected E_y^{ref} component. The total tangential electric field at the interface $(x = 0^-)$ is zero. There is no field in the medium #2. The total electric field, using an equation (5.1.6a), in medium #1 is,

$$\vec{E}^{total}(x) = \hat{y}E_m^{inc}\left[e^{-jk_1x} - e^{+jk_1x}\right] = -\hat{y}j2E_m^{inc}\sin(k_1x) \quad \text{(a)}$$

In the space – time :

$$\vec{E}^{total}(x,t) = \hat{y}\,\mathrm{Re}\left[\vec{E}^{total}(x)e^{j\omega t}\right] = \hat{y}2E_m^{inc}\sin(k_1x)\sin(\omega t) \quad \text{(b)}. \tag{5.1.13}$$

The expression (5.1.13) shows the formation of a standing wave in the medium #1, without any traveling wave. The standing wave for the H_z field component can be easily obtained. At the PEC surface, the E_y-field is zero, and H_z-field is at maximum.

The total reflection at the interface also occurs for $\eta_2 \to \infty$, i. e. for $\mu \to \infty$. In this case, medium #2 acts as a PMC, and it offers $\Gamma = +1$. The PMC has infinite permeability, i.e. $\mu \to \infty$. Again, a standing wave is formed in the medium #1, with E_y-field maximum at the interface; while H_z is zero. The PMC is a hypothetical medium. However, it is realized on the periodically loaded surface as an *artificial magnetic conductor* (AMC) over a band of frequencies. The interface can also totally reflect the wave if the interface offers either inductive or capacitive impedance. In this case, the interface is a RIS. The periodic surfaces are discussed in chapter 20. These are widely used in the modern microwave and antenna engineering. The PEC, and PMC surfaces, forming the idealized rectangular waveguides, are discussed in chapter 7.

5.1.3 Transmission Line Model of the Composite Medium

Both the unbounded medium and transmission line supports the 1D wave propagation. So an unbounded medium could be easily modeled, shown in Fig (5.1b), as a transmission line. The propagation constant of wave on the equivalent transmission line could be treated as identical to that of in the medium. This approach is simple and effective in obtaining the impedance transformation and also impedance matching by using the multilayer dielectric medium. On comparing the wave equations for the E_y (x) and H_z (x), given in equation (4.5.13) of chapter 4, against the voltage and current wave equation (2.1.37) of chapter 2, the following equivalences are observed:

$$E_y \equiv V, H_z \equiv I;\ \mu_r \equiv L/\mu_o,\ \varepsilon_r \equiv C/\varepsilon_0;\ \eta \equiv Z_0,$$

where $\eta = \sqrt{\dfrac{\mu_o\mu_r}{\varepsilon_0\varepsilon_r}}$

$$\Rightarrow \eta = \eta_0\sqrt{\frac{\mu_r}{\varepsilon_r}} = \sqrt{\frac{L}{C}} \quad \text{(a)}$$

phase velocity:

$$v_p^{medium} = \frac{\omega}{\beta} = \frac{c}{\sqrt{\varepsilon_r\mu_r}} = \frac{1}{\sqrt{LC}} = v_p^{line} \quad \text{(b)}. \tag{5.1.14}$$

The solution of the voltage and current waves, given in equations (2.1.63) and (2.1.64), can be converted to the

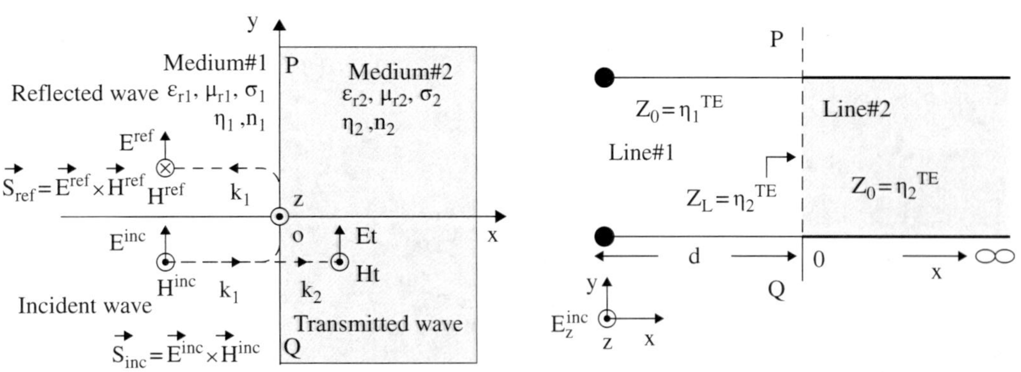

(a) Normal incidence at the interface. (b) Equivalent transmission line model.

Figure 5.1 Normal incidence of TM-polarized plane wave at the interface of two media.

solution of E_y and H_z. Finally, the expression of the input impedance of a lossless line of length d, given in equation (2.2.65) of chapter 2, terminated in a load Z_L, helps to write the following expression of the input impedance of a dielectric slab of thickness d; and also the expressions for the reflection coefficient, and transmission coefficient at the interface PQ:

$$Z_{in}(x) = \eta_{in}(x) = \eta_1 \left[\frac{Z_L + j\eta_1 \tan(\beta d)}{\eta_1 + jZ_L \tan(\beta d)}\right], \quad Z_L = \eta_2 \quad (a)$$
$$\Gamma = \frac{Z_L - \eta_1}{Z_L + \eta_1} \quad (b)$$
$$\tau = \frac{2Z_L}{Z_L + \eta_1} \quad (c). \qquad (5.1.15)$$

In expression (5.1.15), medium #2 has an infinite extent. In medium #1, the incident wave is located at a distance d from the interface. The transmission line model is used for the multilayer dielectric medium also [B.1–B.4].

5.2 Oblique Incidence of Plane Waves

The plane wave of any polarization, obliquely incident at the interface of two dielectric media, can be decomposed to the TE and TM polarizations, so the plane wave of any polarization is a linear combination of the TE and TM polarizations. This section considers the reflection and transmission of both polarizations at the boundary of lossless media. However, losses in the media can also be accounted for [J.1].

5.2.1 TE (Perpendicular) Polarization Case

Figure (5.2a) shows an oblique incidence of the z-polarized plane wave, i.e. TE^z polarized ray at the interface P-O-Q of two media with different electrical characteristics. The interface PQ is along the *y*-axis, and the normal to the interface is along the *x*-axis. The plane x-o-y containing the normal to the interface, and the rays of the obliquely incident, reflected and refracted waves, i.e. the wavevectors ($\vec{k}_1$, $\vec{k}'_1$, $\vec{k}_2$), is called the *plane of incidence*. In the present case, the plane of the paper, i.e. the (x − y)-plane, is the plane of incidence. The incident ray, with the wavevector $\vec{k}_1$, strikes the interface at an angle θ_1. It is partly reflected in the medium #1 at an angle θ'_1 and partly refracted in the medium #2 at an angle θ_2. All angles are measured with respect to the normal o-x. The electric field is perpendicular to the plane of incidence, i.e. in the *z*-direction; so it is called the *TE-polarization*. It is also known as the *perpendicular*, i.e. the *s-polarization* as the E^{inc} field is perpendicular to the plane of incidence, or even the *σ-polarization*. The *letter* "*s*" is from the German word *senkrecht* (perpendicular). It is noted that the polarization of the incident wave is preserved even after the reflection and refraction at the interface of the natural materials. However, the interface created by the engineered metasurfaces controls and alters the polarization states after reflection and refraction. It is discussed in chapter 22. The magnetic field is in the plane of incidence, and its orientation follows the Poynting vector $\vec{S} = \vec{E} \times \vec{H}$ giving the direction of power-flow. The position vector and the wavevectors for the incident, reflected, and transmitted (refracted) waves are summarized below:

Position vector:
$$\vec{r} = \hat{x}x + \hat{y}y + \hat{z}z \quad (a)$$
Incident plane wave:
$$\vec{k}_1 = \hat{x}k_{1x} + \hat{y}k_{1y} = k_1(\hat{x}\cos\theta_1 + \hat{y}\sin\theta_1) \quad (b)$$
Reflected plane wave:
$$\vec{k}'_1 = \hat{x}k'_{1x} + \hat{y}k'_{1y} = k'_1(-\hat{x}\cos\theta'_1 + \hat{y}\sin\theta'_1) \quad (c)$$
Transmitted plane wave:
$$\vec{k}_2 = \hat{x}k_{2x} + \hat{y}k_{2y} = k_2(\hat{x}\cos\theta_2 + \hat{y}\sin\theta_2) \quad (d). \qquad (5.2.1)$$

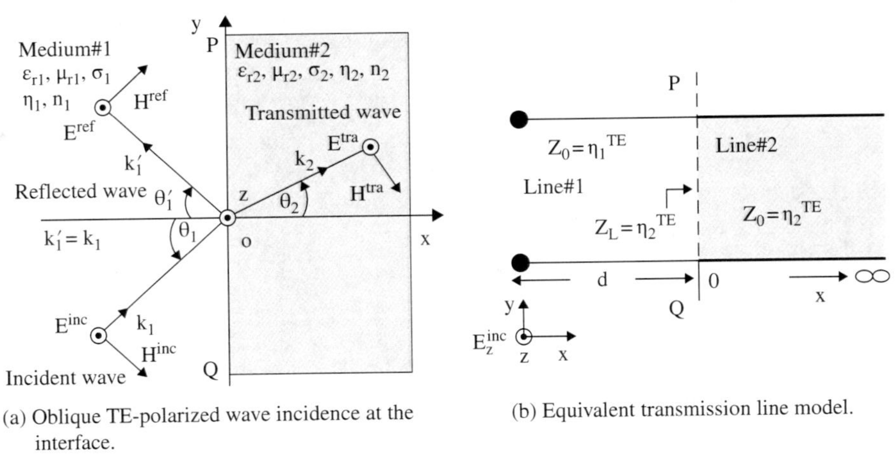

(a) Oblique TE-polarized wave incidence at the interface.

(b) Equivalent transmission line model.

Figure 5.2 Oblique incidence of a plane wave with TE-polarization at the interface of two media.

Using Fig (5.2a), the incident, reflected, and transmitted field components of the TE^z polarized waves are summarized below:

Incident wave:

$$\vec{E}^{inc}(x,y) = \hat{z}E_m^{inc}e^{-j\vec{k}_1\cdot\vec{r}} = \hat{z}E_m^{inc}e^{-jk_1(x\cos\theta_1 + y\sin\theta_1)} \quad (a)$$

$$\vec{H}^{inc}(x,y) = \frac{1}{\eta_1}\hat{k}_1 \times E^{inc}(x,y)$$

$$= \frac{E_m^{inc}}{\eta_1}(\hat{x}\cos\theta_1 + \hat{y}\sin\theta_1) \times \hat{z}e^{-j\vec{k}_1\cdot\vec{r}}$$

$$\Rightarrow \vec{H}^{inc}(x,y) = \frac{E_m^{inc}}{\eta_1}(\hat{x}\sin\theta_1 - \hat{y}\cos\theta_1)e^{-jk_1(x\cos\theta_1 + y\sin\theta_1)} \quad (b). \tag{5.2.2}$$

Reflected wave:

$$\vec{E}^{ref}(x,y) = \hat{z}E_m^{ref}e^{-j\vec{k}_1'\cdot\vec{r}} = \hat{z}E_m^{ref}e^{-jk_1(-x\cos\theta_1' + y\sin\theta_1')} \quad (a)$$

$$\vec{H}^{ref}(x,y) = \frac{E_m^{ref}}{\eta_1}(\hat{x}\sin\theta_1' + \hat{y}\cos\theta_1')e^{-jk_1(-x\cos\theta_1' + y\sin\theta_1')} \quad (b). \tag{5.2.3}$$

Transmitted wave:

$$\vec{E}^{tra}(x,y) = \hat{z}E_m^{tra}e^{-j\vec{k}_2\cdot\vec{r}} = \hat{z}E_m^{tra}e^{-jk_2(x\cos\theta_2 + y\sin\theta_2)} \quad (a)$$

$$\vec{H}^{tra}(x,y) = \frac{E_m^{tra}}{\eta_2}(\hat{x}\sin\theta_2 - \hat{y}\cos\theta_2)e^{-jk_2(x\cos\theta_2 + y\sin\theta_2)} \quad (b). \tag{5.2.4}$$

In equations (5.2.2a,b)–(5.2.4a,b) η_1 and η_2 are the intrinsic impedance of the medium #1 and #2, respectively; and $|k_1| = |k_1'|$ as both the incident and reflected waves are in the same medium #1. The Poynting vectors for the incident, reflected, and transmitted (*refracted*) fields are obtained from the above equations:

$$S_{av}^{inc} = \mathrm{Re}\left[\vec{E}^{inc} \times \vec{H}^{*inc}\right]$$

$$= \mathrm{Re}\left[\left(\hat{z}E_m^{inc}e^{-j\vec{k}_1\cdot\vec{r}}\right) \times \frac{E_m^{inc}}{\eta_1}(\hat{x}\sin\theta_1 - \hat{y}\cos\theta_1)e^{j\vec{k}_1\cdot\vec{r}}\right]$$

$$\Rightarrow S_{av}^{inc} = \frac{\left(E_m^{inc}\right)^2}{2\eta_1}(\hat{x}\cos\theta_1 + \hat{y}\sin\theta_1) \quad (a)$$

$$\text{Likewise,}\quad S_{av}^{ref} = \frac{\left(E_m^{ref}\right)^2}{2\eta_1}(-\hat{x}\cos\theta_1 + \hat{y}\sin\theta_1) \quad (b)$$

$$S_{av}^{tra} = \frac{\left(E_m^{tra}\right)^2}{2\eta_2}(\hat{x}\cos\theta_2 + \hat{y}\sin\theta_2) \quad (c). \tag{5.2.5}$$

Equations (5.2.1) and (5.2.5) show that the wavevector $(\vec{k})$ and Poynting vector $(\vec{S})$ of the incident, reflected, and transmitted waves are in the same direction; so the waves in both media are the *forward waves*. The continuity of the tangential components of the $\vec{E}$ and $\vec{H}$ fields, across the interface at $x = 0$, provides the following expressions:

$$\vec{E}_z^{inc}(x=0,y) + \vec{E}_z^{ref}(x=0,y) = \vec{E}_{t,z}(x=0,y)$$

$$\Rightarrow \vec{E}_m^{inc}e^{-jyk_1\sin\theta_1} + \vec{E}_m^{ref}e^{-jyk_1\sin\theta_1'} = \vec{E}_m^{tra}e^{-jyk_2\sin\theta_2} \quad (a)$$

$$\vec{H}_y^{inc}(x=0,y) + \vec{H}_y^{ref}(x=0,y) = \vec{H}_y^{tra}(x=0,y)$$

$$\Rightarrow \frac{1}{\eta_1}\left[-\vec{E}_m^{inc}\cos\theta_1 e^{-jyk_1\sin\theta_1} + \vec{E}_m^{ref}\cos\theta_1' e^{-jyk_1\sin\theta_1'}\right]$$

$$= -\frac{\vec{E}_m^{tra}}{\eta_2}\cos\theta_2 e^{-jyk_2\sin\theta_2} \quad (b). \tag{5.2.6}$$

The fields are complex quantities on both the left and right-hand sides of the interface. To match the fields at the interface, i.e. along the *y*-axis, both the phase and amplitude matching are needed. The continuity equation, given by equation (5.2.6a), holds at all points along the interface, i.e. along the *y*-axis. To achieve it, the exponential terms, giving phases of the incident, reflected, and refracted waves must be identical. It is known as the *phase matching* at the interface. The phase-matching results in the following well-known *Snell's laws* of reflection and refraction:

Reflection phase matching at the interface :

The phase of incident wave = Phase of the reflected wave

$$\Rightarrow k_1\sin\theta_1 = k_1\sin\theta_1', \quad \Rightarrow \theta_1 = \theta_1' \quad (a)$$

Refraction phase matching at the interface :

The phase of incident wave = Phase of the refracted wave (5.2.7)

$$\Rightarrow k_1\sin\theta_1 = k_2\sin\theta_2, \Rightarrow \frac{\sin\theta_1}{\sin\theta_2} = \frac{k_2}{k_1} = \frac{v_{p1}}{v_{p2}} = \sqrt{\frac{\mu_{r2}\varepsilon_{r2}}{\mu_{r1}\varepsilon_{r1}}} \quad (b)$$

$$\text{or,}\quad \frac{\sin\theta_1}{\sin\theta_2} = \sqrt{\frac{\varepsilon_{r2}}{\varepsilon_{r1}}} = \frac{n_2}{n_1} = \frac{\eta_1}{\eta_2}, \quad (\text{for } \mu_{r2} = \mu_{r1}) \quad (c).$$

Equation (5.2.7b) is used for a *magnetodielectric* medium, whereas equation (5.2.7c) is valid for a dielectric medium. The n_1 and n_2 are the refractive indexes, whereas η_1 and η_2 are the intrinsic impedance of the medium #1, and medium #2, respectively. Moreover, the classical Snell's laws are obtained under the condition of the uniform phase at the interface in the direction of the *y*-axis. However, the phase gradient $d\phi/dy$ can be created on an interface of the *engineered metasurface*. In this case, the classical Snell's laws are modified to obtain the *generalized Snell's* laws. It is discussed in subsection (22.5.4) of chapter 22.

The *amplitude matching* of the tangential components of the E and H-fields at the interface $x = 0$, from equation (5.2.6), provides the following expressions for the *reflection* ($\Gamma_\perp^{TE}$) and *transmission* ($\tau_\perp^{TE}$) coefficients of the perpendicular (TE) polarized obliquely incident plane waves:

$$E_t|_{x=0^+} = E_t|_{x=0^-} \text{ and, } H_t|_{x=0^+} = H_t|_{x=0^-} \text{ leading to}$$

$$\Rightarrow E_m^{inc} + E_m^{ref} = E_m^{tra} \quad (a)$$

$$\frac{1}{\eta_1}\left[E_m^{inc} - E_m^{ref}\right]\cos\theta_1 = \frac{E_m^{tra}}{\eta_2}\cos\theta_2 \quad (b)$$

Reflection coefficient:

$$\Gamma_\perp^{TE} = \frac{E_m^{ref}}{E_m^{inc}} = \frac{\eta_2\cos\theta_1 - \eta_1\cos\theta_2}{\eta_2\cos\theta_1 + \eta_1\cos\theta_2} \quad (c)$$

Transmission coefficient:

$$\tau_\perp^{TE} = \frac{E_m^{tra}}{E_m^{inc}} = \frac{2\eta_2\cos\theta_1}{\eta_2\cos\theta_1 + \eta_1\cos\theta_2} \quad (d). \tag{5.2.8}$$

Equations (5.2c,d) are known as the *Fresnel's Equations* of the TE-polarized waves. They describe the ratio of the reflected and transmitted electric fields to that of the incident electric field. As the reflection and transmission coefficients are complex quantities, they describe both the relative amplitude and phase shifts between the waves. The above equations show that if both media are identical; there is no reflection, $\Gamma_\perp = 0$; and $\eta_1 = \eta_2$, $\theta_1 = \theta_2$, leading to total transmission $\tau_\perp = 1$. It is also noted that $\tau_\perp^{TE} = 1 + \Gamma_\perp^{TE}$.

The total field component in the medium #1 is a summation of the incident and reflected fields, given by equations (5.2.2) and (5.2.3), while in the medium #2, only a refracted field, given by equation (5.2.4) exists. These field equations are summarized below:

Medium #1

$$E_{\perp,z}^{total} = E_m^{inc} e^{-jyk_1\sin\theta_1}\{\underbrace{e^{-jxk_1\cos\theta_1}}_{\text{Traveling wave}} + \underbrace{\Gamma_\perp^{TE} e^{+jxk_1\cos\theta_1}}_{\text{Standing wave}}\} \quad (a)$$

$$H_{\perp,x}^{total} = \frac{E_m^{inc}}{\eta_1}\sin\theta_1 e^{-jyk_1\sin\theta_1}\{e^{-jxk_1\cos\theta_1} + \Gamma_\perp^{TE} e^{+jxk_1\cos\theta_1}\} \quad (b)$$

$$H_{\perp,y}^{total} = \frac{E_m^{inc}}{\eta_1}\cos\theta_1 e^{-jyk_1\sin\theta_1}\{-e^{-jxk_1\cos\theta_1} + \Gamma_\perp^{TE} e^{+jxk_1\cos\theta_1}\} \quad (c). \tag{5.2.9}$$

Medium #2

$$\vec{E}_{\perp,z}^{tra} = \hat{z}\tau_\perp^{TE} E_m^{inc} e^{-jk_2(x\cos\theta_2 + y\sin\theta_2)} \quad (a)$$

$$\vec{H}_\perp^{tra} = \frac{\tau_\perp^{TE} E_m^{inc}}{\eta_2}\cos\theta_1\{\hat{x}\sin\theta_2 - \hat{y}\cos\theta_2\}e^{-jk_2(x\cos\theta_2 + y\sin\theta_2)} \quad (b). \tag{5.2.10}$$

In medium #1, due to the superposition of incident and reflected waves, interference occurs. Following the process used in equation (5.1.7), the computed total wave is partly traveling wave along the *y*-axis, while along the negative *x*-axis, it is partly standing wave. However, the minima of the standing wave do not reach zero levels as $\Gamma_\perp \neq -1$, like a PEC. Although, the wave is traveling along the interface in the *y*-direction, still the wave is not the *surface wave*, as in the x-direction the field is not confined to the surface. Under certain conditions, discussed in subsection (5.3.2), propagation of the surface wave is possible. The surface wave is further discussed in section (7.5) of chapter 7.

The Oblique Incidence on a Perfect Electric Conductor

If the medium #2 is a PEC, i.e. $\eta_2 = 0$, $\varepsilon_{r2} = \infty$, the reflection coefficient is $\Gamma_\perp^{TE} = -1$. The total field components, in the medium #1, are given below from equation (5.2.9); while in the medium #2 there is no field:

$$E_{\perp,z}^{total} = -2jE_m^{inc}\sin(xk_1\cos\theta_1)e^{-jyk_1\sin\theta_1} \quad (a)$$

$$H_{\perp,x}^{total} = \frac{-2jE_m^{inc}\sin\theta_1}{\eta_1}\sin(xk_1\cos\theta_1)e^{-jyk_1\sin\theta_1} \quad (b)$$

$$H_{\perp,y}^{total} = \frac{-2E_m^{inc}\cos\theta_1}{\eta_1}\underbrace{\cos(xk_1\cos\theta_1)}_{\text{Standing wave}}\underbrace{e^{-jyk_1\sin\theta_1}}_{\text{Traveling wave}} \quad (c). \tag{5.2.11}$$

In equation (5.2.11a, b, c) harmonic-time dependence $e^{j\omega t}$ is suppressed. It is seen from equations (5.2.11a, b, c) that the interference produces a total wavefield moving in the y-direction along the interface, while along the direction of the normal, i.e. in the x-direction, it is a standing wave. It shows that the interference field pattern is traveling along the interface.

5.2.2 TM (Parallel) Polarization Case

Figure (5.3a) shows an oblique incidence of TM^z-polarized plane wave ray at the interface of two electrically different media. The electric field of the incident wave is in the (x-y)- plane of incidence, so the incident wave has a parallel polarization as the E^{inc} field is parallel to the plane of incidence. The *parallel polarization*, or *p-polarization*, is also known as the *transverse magnetic (TM) polarization* as the z-oriented magnetic field is normal to the plane of incidence. It is also called *π-polarization*.

The visual inspection of the direction of field vectors in Fig (5.3a) shows that the magnetic field component H_z is normal to the $(x-y)$-plane, i.e. to the plane of polarization, while the electric field components E_x and E_y are in the plane of polarization. The field components of the incident, reflected, and transmitted (*refracted*) TM-polarized obliquely incident wave, as shown in Fig (5.3a), are summarized below:

Incident wave:

$$\vec{H}^{inc}(x,y) = \hat{z}\frac{E_m^{inc}}{\eta_1}e^{-j\vec{k}_1\cdot\vec{r}} = \hat{z}\frac{E_m^{inc}}{\eta_1}e^{-jk_1(x\cos\theta_1 + y\sin\theta_1)} \quad (a)$$

$$\begin{aligned}\vec{E}^{inc}(x,y) &= \eta_1\vec{H}_m^{inc}(x,y)\times\hat{k}_1\\ &= E_m^{inc}[-\hat{x}\sin\theta_1 + \hat{y}\cos\theta_1]e^{-jk_1(x\cos\theta_1 + y\sin\theta_1)} \quad (b).\end{aligned} \tag{5.2.12}$$

Reflected wave:

$$\vec{H}^{ref}(x,y) = -\hat{z}\frac{E_m^{ref}}{\eta_1}e^{-j\vec{k}_1'\cdot\vec{r}} = -\hat{z}\frac{E_m^{ref}}{\eta_1}e^{-jk_1(-x\cos\theta_1' + y\sin\theta_1')} \quad (a)$$

$$\begin{aligned}\vec{E}^{ref}(x,y) &= \eta_1\vec{H}^{ref}(x,y)\times\hat{k}_1'\\ &= E_m^{ref}[\hat{x}\sin\theta_1' + \hat{y}\cos\theta_1']e^{-jk_1(-x\cos\theta_1' + y\sin\theta_1')} \quad (b).\end{aligned} \tag{5.2.13}$$

Transmitted wave:

$$\vec{H}^{tra}(x,y) = \hat{z}\frac{E_m^{tra}}{\eta_2}e^{-j\vec{k}_2\cdot\vec{r}} = \hat{z}\frac{E_m^{tra}}{\eta_2}e^{-jk_2(x\cos\theta_2 + y\sin\theta_2)} \quad (a)$$

$$\begin{aligned}\vec{E}^{tra}(x,y) &= \eta_2\vec{H}^{tra}(x,y)\times\hat{k}_2\\ &= E_m^{tra}[-\hat{x}\sin\theta_2 + \hat{y}\cos\theta_2]e^{-jk_2(x\cos\theta_2 + y\sin\theta_2)} \quad (b).\end{aligned} \tag{5.2.14}$$

The continuity of the tangential electric and magnetic field components, across the interface at $x = 0$, provides the following expression:

$$\begin{aligned}E_m^{inc}\cos\theta_1 e^{-jyk_1\sin\theta_1} + E_m^{ref}\cos\theta_1' e^{-jyk_1\sin\theta_1'} &= E_m^{tra}\cos\theta_2 e^{-jyk_2\sin\theta_2} \quad (a)\\ \frac{E_m^{inc}}{\eta_1}e^{-jyk_1\sin\theta_1} - \frac{E_m^{ref}}{\eta_1}e^{-jyk_1\sin\theta_1'} &= \frac{E_m^{tra}}{\eta_2}e^{-jyk_2\sin\theta_2} \quad (b).\end{aligned} \tag{5.2.15}$$

The Poynting vectors are the same as that of the TE-polarization. The phase matching, from equation (5.2.15), again provides Snell's reflection and refractions laws as given by equation (5.2.7). However, the amplitude matching provides different expressions for the reflection ($\Gamma_{\parallel}^{TM}$), and transmission $\tau_{\parallel}^{TM}$ coefficients of the parallel polarized obliquely incident plane wave:

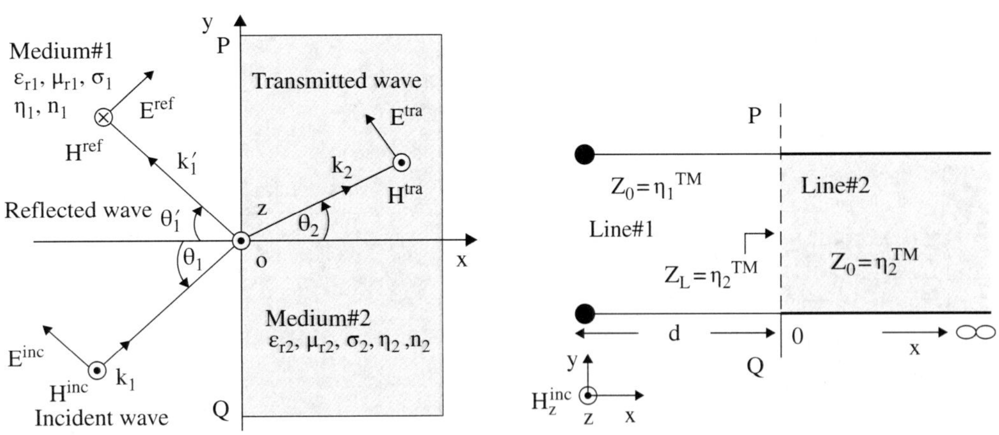

(a) Oblique TM-polarized wave incidence at the interface. (b) Equivalent transmission line model.

Figure 5.3 Oblique incidence of a plane wave with TM-polarization at the interface of two media.

$$
\begin{aligned}
&E_m^{inc}\cos\theta_1 + E_m^{ref}\cos\theta_1 = E_m^{tra}\cos\theta_2 && (a)\\
&\frac{E_m^{inc}}{\eta_1} - \frac{E_m^{ref}}{\eta_1} = \frac{E_m^{tra}}{\eta_2} && (b)\\
&\Rightarrow \Gamma_{ll}^{TM} = \frac{E_m^{ref}}{E_m^{inc}} = \frac{\eta_2\cos\theta_2 - \eta_1\cos\theta_1}{\eta_2\cos\theta_2 + \eta_1\cos\theta_1} && (c)\\
&\tau_{ll}^{TM} = \frac{E_m^{tra}}{E_m^{inc}} = \frac{2\eta_2\cos\theta_1}{\eta_2\cos\theta_2 + \eta_1\cos\theta_1} && (d).
\end{aligned}
\tag{5.2.16}
$$

For normal incidence, equations (5.2.16c,d) reduce to equations (5.1.3a,b). Equations (5.2.16c,d) are known as the *Fresnel's Equations* of the TM-polarized waves. It is noted that $\tau_{ll}^{TM} = (1+\Gamma_{ll}^{TM})\cos\theta_1/\cos\theta_2$. The total fields in both the media are summarized below:

Medium #1

The total field is a summation of the incident and reflected fields:

$$
\begin{aligned}
&E_{ll,x}^{total} = E_m^{inc}\sin\theta_1 e^{-jyk_{1y}}\left[-e^{-jxk_{1x}} + \Gamma_{||}^{TM}e^{+jxk_{1x}}\right] && (a)\\
&E_{ll,y}^{total} = E_m^{inc}\cos\theta_1 e^{-jyk_{1y}}\left[e^{-jxk_{1x}} + \Gamma_{||}^{TM}e^{+jxk_{1x}}\right] && (b)\\
&H_{ll,z}^{total} = \frac{E_m^{inc}}{\eta_1}e^{-jyk_{1y}}\left[e^{-jxk_{1x}} - \Gamma_{||}^{TM}e^{+jxk_{1x}}\right] && (c).
\end{aligned}
$$

↓ Traveling wave ↓ Standing wave

(5.2.17)

Medium #2

$$
\begin{aligned}
&H_{||}^{tra} = \frac{\tau_{||}^{TM}E_m^{inc}}{\eta_2}e^{-j\vec{k}_2\cdot\vec{r}} && (a)\\
&\vec{E}_{||}^{tra} = \tau_{||}^{TM}E_m^{inc}[-\hat{x}\sin\theta_2 + \hat{y}\cos\theta_2]e^{-j\vec{k}_2\cdot\vec{r}} && (b).
\end{aligned}
\tag{5.2.18}
$$

In medium #1, there is a traveling wave along the *y*-axis, whereas, in the direction of normal to the interface, the wave is a standing wave. However, the minima of the standing wave never reach zero levels as $\Gamma_{||}^{TM} \neq 1$, like a PMC. Again, even if the wave travels in the y-direction along the interface of two media, the traveling wave is not a surface wave, as in the transverse x-direction, there is no confinement of field near the interface. However, under certain conditions, discussed in subsection (5.3.2), the surface wave could exist along the interface.

The Oblique Incidence on a Perfect Electric Conductor

If medium #2 is a perfect conductor, $\Gamma_{||}^{TM} = -1,\ \tau_{||} = 0.$ The medium #1 supports the wave with interference, and the fields are written from equation (5.2.17) as follows:

$$
\begin{aligned}
&E_{ll,x}^{total} = -2E_m^{inc}\sin\theta_1\cos(k_{1x}x)e^{-jyk_{1y}} && (a)\\
&E_{ll,y}^{total} = -2jE_m^{inc}\cos\theta_1\sin(k_{1x}x)e^{-jyk_{1y}} && (b)\\
&H_{ll,z}^{total} = \frac{2E_m^{inc}}{\eta_1}\cos(k_{1x}x)e^{-jyk_{1y}} && (c).
\end{aligned}
$$

↓ standing wave ↓ Traveling wave

(5.2.19)

In equation (5.2.19), time-harmonic dependence $e^{j\omega t}$ is suppressed. At the surface of the perfect conductor $x = 0$, the field component $E_{ll,x}^{total}$ forms the standing wave. However, the standing wave pattern travels along the interface in the y-direction.

5.2.3 Dispersion Diagrams of Refracted Waves in Isotropic and Uniaxial Anisotropic Media

The dispersion diagrams of the EM-waves in the isotropic and also uniaxial anisotropic medium displaying Snell's law, given in equation (5.2.7), are represented on the $(k_z - k_y)$-plane. Snell's law was obtained from the phase matching of the incident and refracted waves across the boundary of two media. The dispersion diagrams in the isotropic and uniaxial anisotropic media are obtained by continuing the process discussed in subsections (4.7.4) and (4.7.5) of chapter 4.

Figure (5.4a) shows the geometrical process of constructing the dispersion diagram showing the refraction of the wave at the interface of the air and dielectric media. The incident wavevector $\overrightarrow{PO} = \vec{k}_0$ in the air medium ($\varepsilon_r = 1$) subtends an angle of incidence θ_i. The refracted wavevector $\overrightarrow{OS} = \vec{k}_r = k_0\sqrt{\varepsilon_r}$, in the isotropic dielectric medium with relative permittivity ε_r, subtends an angle of refraction θ_r with respect to the k_z, normal to the interface. Only half of the dispersion circle of each medium is shown. The normal RS intersects the dispersion circle of the isotopic dielectric medium at S. At the interface, the line segments $PQ(k_0\sin\theta_i) = QR(k_r\sin\theta_r) = k_0\sqrt{\varepsilon_r}\sin\theta_r$ show the phase-matching across the interface. It is seen that the directions of the phase velocity v_p and group velocity v_g of the refracted wave are the same.

Likewise, Fig (5.4b) shows the geometrical construction of *double refraction* in the uniaxial anisotropic medium. In this case, the line segments $PQ(k_0\sin\theta_i) = QR(k_{ro}\sin\theta_o = k_{re}\sin\theta_e)$ provide the phase-matching across the interface. The wavenumbers k_{ro} and k_{re} belong to the refracted *ordinary* and *extraordinary waves.* The normal drawn from the location R on the dispersion circle to the interface of air-dielectric medium intersects the dispersion circle of the uniaxial anisotropic medium at S_1 for the ordinary waves. It also intersects at the dispersion

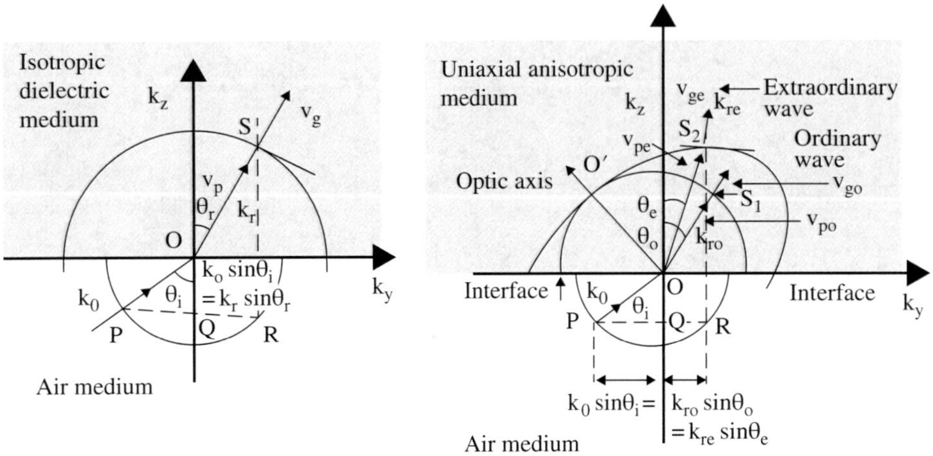

(a) Refraction in the isotropic medium.

(b) Refraction in uniaxial anisotropic medium showing double refraction.

Figure 5.4 Dispersion diagrams of refracted waves.

ellipse at S_2 for the extraordinary waves. The wavevectors OS_1 and OS_2 for the ordinary and extraordinary waves subtend the angle of refraction θ_o and θ_e respectively with k_z-axis. The first refraction is for the ordinary waves with phase velocity v_{po} and group velocity v_{go} in the same direction. The second refraction, for the extraordinary waves, has phase velocity v_{pe} and group velocity v_{ge} in different directions. The angles of refraction are different in both cases. The phase and group velocities for both kinds of waves are identical only along the optic axis OO', where the dispersion circle and dispersion ellipse touch each other.

5.2.4 Wave Impedance and Equivalent Transmission Line Model

The obliquely incident plane wave is partly reflected and partly transmitted at the interface of two electrically dissimilar media. It helps to think the medium #1 and medium #2 as two dissimilar transmission lines with a junction PQ, corresponding to the interface PQ as shown in Figs (5.2a) and (5.3a) for the TE and TM polarization, respectively. Figures (5.2b) and (5.3b) show the *equivalent transmission line networks* of the composite media, supporting the oblique incidence of the TE and TM-polarized plane waves. The source connected to the line #1, with characteristic impedance Z_0, corresponding to the wave impedance of the incident wave in medium #1, is assumed to be located at the left of the junction. The line #2 is of an infinite extent that offers a load Z_L, corresponding to the wave impedance of the transmitted (refracted) wave in medium #2, at the junction. The reflection and transmission coefficients of the incident wave at the physical media interface correspond to the mismatch, causing reflection and transmission, at the junction of two equivalent lines. Our task is to determine the Z_0 and Z_L in terms of the wave impedances of medium #1 and medium #2, respectively, for both the TE and TM polarizations. The correlation between the reflection/transmission coefficient of the TE and TM-polarized waves at the interface of physical media and reflection/transmission coefficient at the junction of equivalent lines is also be considered.

Formulation of Transmission Line Model

At first, the characteristic impedance of equivalent transmission lines, corresponding to wave impedance of both media, is obtained for both the TE and TM-polarized waves. Next, the relations between reflection/transmission coefficient at the interface of the physical media and reflection/transmission coefficient at the junction of equivalent lines are obtained.

Correspondence between Wave Impedance and Characteristic Impedance

The *wave impedance* of the incident, and transmitted *TE waves* in the medium #1 and medium #2, as shown in Fig (5.2a), with respect to the direction of propagations k_1 and k_2 are given below:

$$\text{Medium \#1}: \quad \eta_1^{k_1} = \frac{E_{m,z}^{inc}}{H_m^{inc}} \qquad \text{(a)}$$

$$\text{Medium \#2}: \quad \eta_2^{k_2} = \frac{E_{m,z}^{tran}}{H_m^{tran}} \qquad \text{(b).}$$

$$(5.2.20)$$

However, the interface in the $(y - z)$-plane views both the above-given wave impedances differently due to the oblique incidence. The left $(x = 0^-)$ and right $(x = 0^+)$ side faces view the following x-directed wave impedances, η_1^{TE} and η_2^{TE}, due to the (E_z, H_y) field components:

$$\text{Medium\#1:}\quad \eta_1^x = \eta_1^{TE} = \frac{E_{m,z}^{inc}}{H_{m,y}^{inc}} = \frac{E_{m,z}^{inc}}{-H_m^{inc}\cos\theta_1} = -\frac{\eta_1^{k1}}{\cos\theta_1} = Z_0 \quad (a)$$

$$\text{Medium\#2:}\quad \eta_2^x = \eta_2^{TE} = \frac{E_{m,z}^{tra}}{H_{m,y}^{tra}} = \frac{E_{m,z}^{tra}}{-H_m^{tra}\cos\theta_2} = -\frac{\eta_2^{k2}}{\cos\theta_2} = Z_L \quad (b).$$

(5.2.21)

Figure (5.2b) shows the equivalent transmission line model of the obliquely incident TE-polarized wave. The wave impedances η_1^x, and η_2^x seen by the interface are taken as the characteristic impedance of the equivalent lines#1 and #2, respectively.

Similarly, the wave impedances viewed by the interface with the obliquely incident TM-polarized wave are obtained with reference to Fig (5.3a):

$$\text{Medium\#1}:\quad \eta_1^{k_1} = \frac{E_m^{inc}}{H_{m,z}^{inc}} \quad (a)$$

$$\text{Medium\#2}:\quad \eta_2^{k_2} = \frac{E_m^{tra}}{H_{m,z}^{tra}} \quad (b).$$

(5.2.22)

$$\text{Medium\#1}:\quad \eta_1^x = \eta_1^{TM} = \frac{E_{m,y}^{inc}}{H_{m,z}} = \frac{E_m^{inc}\cos\theta_1}{H_{m,z}} = \eta_1^{k_1}\cos\theta_1 = Z_0 \quad (a)$$

$$\text{Medium\#2}:\quad \eta_2^x = \eta_2^{TM} = \frac{E_{m,y}^{tra}}{H_{m,z}^{tra}} = \frac{E_m^{tra}\cos\theta_2}{H_{m,z}^{tra}} = \eta_2^{k_2}\cos\theta_2 = Z_L \quad (b).$$

(5.2.23)

The superscripts k_1 and k_2, used in equations (5.2.22) and (5.2.23) are dropped in further discussion.

Refection/Transmission Coefficients at Media Interface and Lines Junction

The reflection and transmission coefficients of both the TE and TM-polarized obliquely incident waves at the interface ($x = 0$) of physical media as taken as follow:

$$\textit{Reflection coefficient}:\quad \Gamma_\perp^{TE} \text{ or } \Gamma_{||}^{TM} = \left.\frac{E_m^{ref}}{E_m^{inc}}\right|_{x=0} \quad (a)$$

$$\textit{Transmission coefficient}:\quad \tau_\perp^{TE} \text{ or } \tau_{||}^{TM} = \left.\frac{E_m^{tra}}{E_m^{inc}}\right|_{x=0} \quad (b).$$

(5.2.24)

The reflection and transmission coefficients of TE-polarized at the junction of two equivalent lines, shown in Fig (5.2b), are obtained from E_z-field components of the incident, reflected, and transmitted waves using equations (5.2.2)–(5.2.4):

Line junction Reflection coefficient :

$$\Gamma_{\perp,Line}^{TE} = \frac{E_z^{ref}}{E_z^{inc}} = \frac{E_m^{ref}}{E_m^{inc}} = \Gamma_\perp^{TE} \quad (a)$$

Line junction Transmission coefficient :

$$\tau_{\perp,Line}^{TE} = \frac{E_z^{tra}}{E_z^{inc}} = \frac{E_m^{tra}}{E_m^{inc}} = \tau_\perp^{TE} \quad (b).$$

(5.2.25)

Thus, for the TE-polarized obliquely incident EM-wave, the line reflection, and transmission coefficients correspond to the reflection and transmission coefficients at the interface of two physical media. However, for the TM-polarized case, shown in Fig (5.3b), change occurs for the transmission coefficient. The E_y-field components of the incident, reflected, and transmitted waves using equations (5.2.12)–(5.2.14) are considered to define line junction reflection and transmission coefficients of the TM-polarized case:

Line junction Reflection coefficient :

$$\Gamma_{||,Line}^{TM} = \frac{E_y^{ref}}{E_y^{inc}} = \frac{E_m^{ref}}{E_m^{inc}} = \Gamma_{||}^{TM} \quad (a)$$

Line junction Transmission coefficient :

$$\tau_{||,Line}^{TM} = \frac{E_y^{tra}}{E_y^{inc}} = \frac{E_m^{tra}(\cos\theta_2)}{E_m^{inc}(\cos\theta_1)} \quad (b)$$

$$\Rightarrow \tau_{||,Line}^{TM} = \tau_{||}^{TM}\frac{\cos\theta_2}{\cos\theta_1},$$

$$\Rightarrow \tau_{||}^{TM} = \tau_{||,Line}^{TM}\frac{\cos\theta_1}{\cos\theta_2} \quad (c).$$

(5.2.26)

Computation of Reflection and Transmission Coefficients

At this stage, the reflection and transmission coefficients of the TE/TM-polarized waves at the interface of two physical media could be computed. Using the above discussion, the reflection coefficient $\Gamma_\perp^{TE}$ and the transmission $\tau_\perp^{TE}$ coefficient of an obliquely incident TE (perpendicular) – polarized wave are computed from the transmission line model as follows:

Reflection coefficient :

$$\Gamma_\perp^{TE} = \Gamma_{\perp,Line}^{TE} = \frac{Z_L - Z_0}{Z_L + Z_0} = \frac{\eta_2\cos\theta_1 - \eta_1\cos\theta_2}{\eta_2\cos\theta_1 + \eta_1\cos\theta_2} \quad (a)$$

Transmission coefficient :

$$\tau_\perp^{TE} = \tau_{\perp,Line}^{TE} = \frac{2Z_L}{Z_L + Z_0} = \frac{2\eta_2\cos\theta_1}{\eta_2\cos\theta_1 + \eta_1\cos\theta_2} \quad (b).$$

(5.2.27)

Equation (5.2.27a,b) are identical to equation (5.2.8c,d), The reflection coefficient $\Gamma_{||}^{TM}$ and the transmission coefficient $\tau_{||}^{TM}$ of an obliquely incident TM-polarized wave are further computed using the transmission line model, shown in Fig (5.3b), as follows:

$$\textit{Reflection coefficient}:$$

$$\Gamma_{||}^{TM} = \frac{Z_L - Z_0}{Z_L + Z_0} = \frac{\eta_2 \cos\theta_2 - \eta_1 \cos\theta_1}{\eta_2 \cos\theta_2 + \eta_1 \cos\theta_1} \qquad \text{(a)}$$

$$\textit{Transmission coefficient}:$$

$$\tau_{||}^{TM} = \frac{2Z_L}{Z_L + Z_0}\left(\frac{\cos\theta_1}{\cos\theta_2}\right) = \frac{2\eta_2 \cos\theta_1}{\eta_2 \cos\theta_2 + \eta_1 \cos\theta_1} \qquad \text{(b)}. \qquad (5.2.28)$$

Equations (5.2.23) and (5.2.26) are used to get the above expressions. Equation (5.2.28a, b) are identical to equation (5.2.16c,d). The transmission line models are also used to obtain the reflection and transmission coefficients of both normal and oblique incident plane waves on a multilayered slab medium [B.1, B.3, B.4]

5.3 Special Cases of Angle of Incidence

There are two special cases of the angle of incidence: one for the complete transmission of waves at the interface, and another for the total reflection of the wave at the interface of two media. These are known as *Brewster angle* and *critical angle. Brewster angle is the angle of incidence at which the reflection coefficient is zero and the incident wave is fully transmitted from one medium to another with refraction.* So the Brewster angle corresponds to the impedance matching condition under which reflection at the interface, in the medium #1, is zero, and the incident power is completely transmitted to the medium #2. *At the critical angle of incidence or the incidence angle greater than the critical angle, the incident wave is completely reflected by the interface of two dielectric media.* This kind of total reflection is superior to that of the total reflection from any metallic surface that always introduces loss. We consider these two special cases of the angle of incidence for both the TE and TM polarizations.

5.3.1 Brewster Angle

The Brewster angle of incidence could be obtained for both the TE and TM polarizations.

TE-Polarization

In the case of the *TE-polarization*, at an incident angle $\theta_1 = \theta_B$, i.e. at Brewster angle the reflection coefficient $\Gamma_{\perp}^{TE} = 0$. So from equation (5.2.8c), and also using the Snell's refraction law (5.2.7b), the following expression for Brewster angle is obtained:

$$\eta_2 \cos\theta_1 - \eta_1 \cos\theta_2 = 0,$$

$$\Rightarrow 1 - \sin^2\theta_1 = \left(\frac{\eta_1}{\eta_2}\right)^2 \left[1 - \sin^2\theta_2\right]$$

$$= \left(\frac{\eta_1}{\eta_2}\right)^2 \left[1 - \left(\frac{v_{p2}}{v_{p1}}\right)^2 \sin^2\theta_1\right]$$

$$\Rightarrow \sin\theta_B = \left[\frac{1 - (\mu_{r1}\varepsilon_{r2}/\mu_{r2}\varepsilon_{r1})}{1 - (\mu_{r1}/\mu_{r2})^2}\right]^{1/2} \qquad \text{(a)}$$

$$\text{For } \varepsilon_{r1} = \varepsilon_{r2}\text{: } \sin\theta_B = \left[\frac{\mu_{r2}}{\mu_{r2} + \mu_{r1}}\right]^{1/2} \qquad \text{(b)}. \qquad (5.3.1)$$

The above relation applies to a *magnetodielectric medium*. However, if both media are dielectrics i.e. $\mu_{r1} = \mu_{r2} = 1$ then $\sin\theta_B \rightarrow \infty$. Thus, *the Brewster angle does not exist at the interface of two dielectric media*. The Brewster angle could exit for the case (b). Such a material is difficult to get in practice. However, the situation is different for the TM-polarized plane wave.

TM-Polarization

In the case of the *TM-polarization*, again the reflection is zero, i.e. $\Gamma_{||}^{TM} = 0$, at $\theta_1 = \theta_B$. Using equation (5.2.16c), and the Snell's refraction law; the Brewster angle is obtained:

$$\eta_2 \cos\theta_2 - \eta_1 \cos\theta_1 = 0,$$

$$\Rightarrow \left(1 - (v_{p2}/v_{p1})^2 \sin^2\theta_1\right) = (\eta_1/\eta_2)^2 (1 - \sin^2\theta_1)$$

$$\textit{For magneto} - \text{dielectric}:$$

$$\Rightarrow \sin\theta_B = \left[\frac{1 - (\mu_{r2}\varepsilon_{r1}/\mu_{r1}\varepsilon_{r2})}{1 - (\varepsilon_{r1}/\varepsilon_{r2})^2}\right]^{1/2} \qquad \text{(a)}$$

$$\textit{For dielectric}:$$

$$(\mu_{r1} = \mu_{r2} = 1) \Rightarrow \sin\theta_B = \left[\frac{\varepsilon_{r2}}{\varepsilon_{r1} + \varepsilon_{r2}}\right]^{1/2} \qquad \text{(b)}$$

$$\text{Or, } \cos\theta_B = \left[\frac{\varepsilon_{r1}}{\varepsilon_{r1} + \varepsilon_{r2}}\right]^{1/2} \qquad \text{(c)}$$

$$\tan\theta_B = \sqrt{\frac{\varepsilon_{r2}}{\varepsilon_{r1}}} = \frac{n_2}{n_1} \qquad \text{(d)}. \qquad (5.3.2)$$

Unlike the TE polarization, the Brewster angle of the TM-polarized obliquely incident wave exists even at the interface of the nonmagnetic dielectric medium. Thus, the Brewster angle could be used to separate the TE and TM polarizations from the obliquely incident unpolarized wave at the interface of two dielectric media, as only the TE-polarized wave component will get reflected. A Brewster angle is also called the *polarizing angle*.

5.3.2 Critical Angle

At the critical angle of incidence ($\theta_1 = \theta_c$), complete reflection occurs at the interface. It occurs for both the TE and TM polarizations. The *transmission line model* helps to understand it. In the case of a terminated line, total reflection occurs for the load impedance $Z_L = 0, \infty, \pm jX$. In the first case, the line is short-circuited, i.e. terminated in a PEC with $\varepsilon_r \to \infty$, in the second case, the line is open-circuited, i.e. terminated in a PMC with $\mu_r \to \infty$; in the third case, the line is terminated in a RIS, either inductive or capacitive. The corresponding surface, i.e. the interface, is a PEC, or PMC, or RIS. These surfaces are further discussed in chapter 20 for the artificially engineered periodic surfaces known as the *electromagnetic bandgap (EBG)* surfaces. By varying the angle of incidence of both polarizations with respect to the critical angle of incidence, the reflection and transmission of waves could be controlled.

Figure (5.3) shows that in the case TM - polarized wave, total reflection at the critical angle, is obtained by taking the load impedance at the interface ($x = 0^+$) zero. Likewise, for the TE - polarization shown in Fig (5.2), total reflection at a critical angle is obtained for the infinite load impedance at the interface. Using equations (5.2.23b), (5.2.21b), and (5.2.7b), the following conditions are obtained: for TM-polarization, $Z_L = \eta_2 \cos\theta_2 = 0$; for TE polarization, $Z_L = -\eta_2/\cos\theta_2 = \infty$. Both cases give the following expression for the critical angle $\theta_1 = \theta_c$ and Snell's Law of refraction:

$$\cos\theta_2 = 0 = \left(1 - \sin^2\theta_2\right)^{1/2}$$

$$\Rightarrow \left(1 - (\mu_{r1}\varepsilon_{r1}/\mu_{r2}\varepsilon_{r2})^2 \sin^2\theta_c\right)^{1/2} = 0$$

$$\Rightarrow \textit{for magneto-dielectric media}:$$

$$\sin\theta_c = \sqrt{\frac{\mu_{r2}\varepsilon_{r2}}{\mu_{r1}\varepsilon_{r1}}} \qquad \text{(a)}$$

$$\Rightarrow \textit{for dielectric media}: \sin\theta_c = \sqrt{\frac{\varepsilon_{r2}}{\varepsilon_{r1}}} \qquad \text{(b)}$$

$$\text{Snell's Law}: \ \sin\theta_2 = \frac{\sin\theta_1}{\sin\theta_c} = \sqrt{\frac{\varepsilon_{r1}}{\varepsilon_{r2}}}\sin\theta_1 \qquad \text{(c)}. \qquad (5.3.3)$$

For the real value of the critical angle θ_c, permittivities of media are taken as $\varepsilon_{r1} > \varepsilon_{r2}$. Thus, the critical angle of the interface is a fixed quantity. The angle of refraction θ_2 deciding the direction of wave propagation in medium #2 is a function of angle incidence θ_1. Figure (5.5a–c), applicable to for both the TM and TE polarizations, consider three cases, $\theta_1 < \theta_c$, $\theta_1 = \theta_c$, $\theta_1 > \theta_c$, of propagation for the obliquely incident plane wave.

Case #1: $\theta_1 < \theta_c$
In this case, equation (5.3.3c) provides the real value of the angle θ_2 as $\sin\theta_2 < 1$ corresponding to an angle of incidence θ_1. Figure (5.5a) shows the reflection and transmission of the obliquely incident plane wave at the interface for the case $\varepsilon_{r1} > \varepsilon_{r2}$.

Case#2: $\theta_1 = \theta_c$
Figure (5.5b) shows the case for $\theta_1 = \theta_c$. The angle of refraction θ_2 is obtained on substituting $\sin\theta_1 = \sin\theta_c$ in equation (5.3.3c):

$$\sin\theta_2 = \sqrt{\frac{\varepsilon_{r1}}{\varepsilon_{r2}}}\sin\theta_1\Big|_{\theta_1=\theta_c} = \sqrt{\frac{\varepsilon_{r1}}{\varepsilon_{r2}}} \times \sqrt{\frac{\varepsilon_{r2}}{\varepsilon_{r1}}} = 1$$

$$\Rightarrow \theta_2 = \frac{\pi}{2}. \qquad (5.3.4)$$

Figure (5.5b) shows that the refracted wave travels along the interface $x = 0^+$ in the y-direction and no component of the refracted (transmitted) wave propagates in the medium #2. Equation (5.2.28a) also shows that for the TM-polarized incident wave, the reflection coefficient is $\Gamma_{||}^{TM} = -1$ at the angle of refraction $\theta_2 = \pi/2$. It demonstrates that under the total reflection condition, occurring at the critical angle $\theta_1 = \theta_c$, the interface behaves as a *PEC* for the TM-polarized incident wave. However, equation (5.2.28b) provides also a transmission coefficient $\tau_{||}^{TM} = 2\eta_2/\eta_1$. It shows that the transmitted wave in the medium #2, confined at the interface, travels along the y-axis at the interface $x = 0^+$. The transmitted field components at an angle $\theta_2 = \pi/2$ in the medium #2 are given in equation (5.2.18).

In the case of the TE polarization of the obliquely incident wave, equation (5.2.27a,b) shows the reflection coefficient $\Gamma_{\perp}^{TE} = +1$ and transmission coefficient $\tau_{\perp}^{TE} = \tau_{||}^{TM}$. In this case, the interface acts like a *PMC*. Again, the transmitted wave in the medium #2 propagates in the y-direction at the interface $x = 0^+$. It is seen from the field equation (5.2.10) for $\theta_2 = \pi/2$.

Case#3: $\theta_1 > \theta_c$
Figure (5.5c) shows the case for $\theta_1 > \theta_c$. Equation (5.3.3c) of Snell's refraction law shows that $\sin\theta_2 > 1$. Therefore,

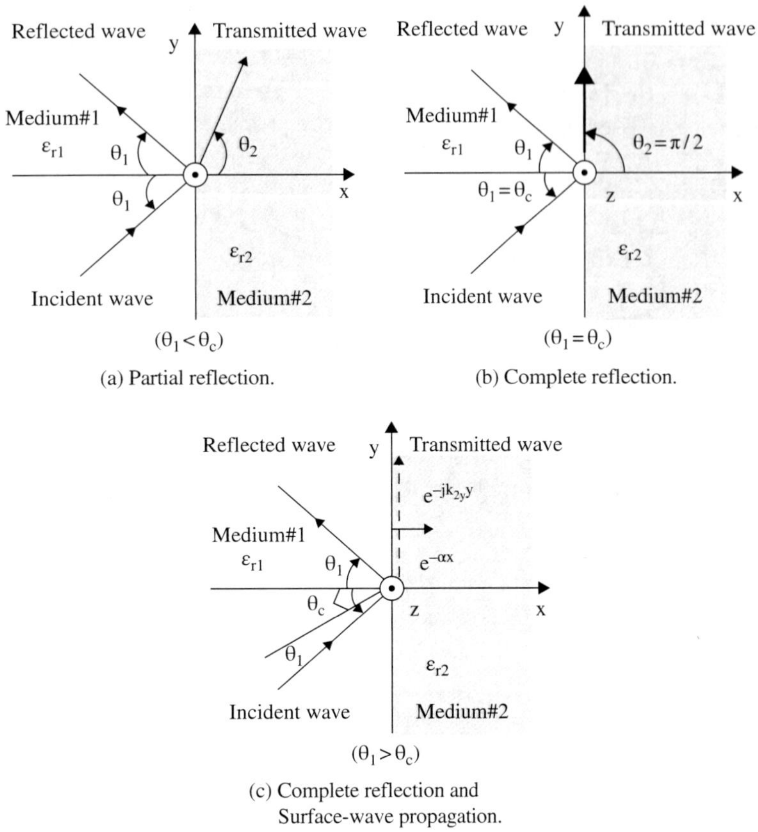

Figure 5.5 Oblique incidence of plane wave at three different angles of incidence.

there is no real solution to the angle θ_2. However, the wave propagation in medium #2 requires $\sin\theta_2$ and $\cos\theta_2$, not the value of the angle of refraction θ_2. The $\sin\theta_2$ and $\cos\theta_2$ could be obtained from Snell's Law given by equation (5.2.7c), also from the wavevector k_2 and its x and y-directed components k_{2x} and k_{2y} using equation (5.2.1):

$$\sin\theta_2 = \sqrt{\frac{\varepsilon_{r1}}{\varepsilon_{r2}}}\sin\theta_1 > 1, \ \text{for } \varepsilon_{r1} > \varepsilon_{r2} \quad \text{(a)}$$

$$\Rightarrow \cos\theta_2 = \pm j\sqrt{\frac{\varepsilon_{r1}}{\varepsilon_{r2}}\sin^2\theta_1 - 1} \quad \text{(b)}$$

$$\sin\theta_2 = \frac{k_{2y}}{k_2}, \quad \Rightarrow k_{2y} = k_2\sqrt{\frac{\varepsilon_{r1}}{\varepsilon_{r2}}}\sin\theta_1 \quad \text{(c)}$$

$$\cos\theta_2 = \frac{k_{2x}}{k_2}, k_{2x} = \pm j\alpha; \quad \text{(d)}$$

$$\text{where, } \alpha = k_2\sqrt{\frac{\varepsilon_{r1}}{\varepsilon_{r2}}\sin^2\theta_1 - 1} \quad \text{(e).}$$

(5.3.5)

The negative (−) value is selected in further computation as it provides exponentially decaying fields with distance from the interface in the medium #2. The choice of positive (+) value results in the exponentially growing field with the distance that is physically not possible in a usual passive medium. The expressions (5.3.5b,d, e) are used for both the TE and TM polarization to compute the reflection and transmission coefficients, and also the refracted fields in medium #2 for the case $\theta_1 > \theta_c$

TE Polarization

Equation (5.2.8c,d) for the reflection and transmission coefficients of the obliquely incident TE-polarized wave, using an equation (5.3.5a) are reduced to the following expression:

$$\Gamma_\perp^{TE} = \frac{(\eta_2/\eta_1)\cos\theta_1 - j\left((\varepsilon_{r1}/\varepsilon_{r2})\sin^2\theta_1 - 1\right)^{1/2}}{(\eta_2/\eta_1)\cos\theta_1 + j\left((\varepsilon_{r1}/\varepsilon_{r2})\sin^2\theta_1 - 1\right)^{1/2}}$$

$$= +1e^{j\phi_\perp} \quad \text{(a)}$$

At $\theta_1 = \theta_c, \ \sin\theta_1 = \sqrt{\varepsilon_{r2}/\varepsilon_{r21}}$

and $\Gamma_\perp^{TE} = +1, \ \theta_2 = \pi/2$ (b)

$$\tau_\perp^{TE} = \frac{2\cos\theta_1}{\cos\theta_1 + (\eta_1/\eta_2)\cos\theta_2}$$

$$= \frac{2\cos\theta_1}{\cos\theta_1 + j(\eta_1/\eta_2)\left((\varepsilon_{r1}/\varepsilon_{r2})\sin^2\theta_1 - 1\right)^{1/2}} \quad \text{(c)}$$

At $\theta_1 = \theta_c, \ \sin\theta_1 = \sqrt{\varepsilon_{r2}/\varepsilon_{r1}}$, and $\tau_\perp^{TE} = 2$ (d).

(5.3.6)

The interface surface acts as a PMC surface for the TE-polarized incident wave under the case $\theta_1 \geq \theta_c$ because $\Gamma_\perp^{TE} = +1$. Over a certain frequency band, it is realized as the *artificial magnetic conductor* (AMC) surface. At the critical angle, $\theta_2 = \pi/2$, the transmitted wave propagates along the y-direction at the interface $x = 0^+$, as shown in Fig (5.5b). For the case $\theta_1 > \theta_c$, shown in Fig (5.5c), the interface supports the *slow-wave type surface wave* propagation. It is demonstrated below.

For the case $\theta_1 > \theta_c$, the electric and magnetic field component and power flow of the TE-polarized wave in the medium #2 is obtained from using equation (5.3.5) with equation (5.2.10a):

$$E_{\perp z}^{tra} = \hat{z}\tau_\perp^{TE}E_m^{inc}e^{-xk_2\left[(\varepsilon_{r1}/\varepsilon_{r2})\sin^2\theta_1 - 1\right]^{1/2}} \times e^{-jk_2 y \sin\theta_1\sqrt{\varepsilon_{r1}/\varepsilon_{r2}}}$$

$$\Rightarrow E_{\perp z}^{tra} = \hat{z}\left[\tau_\perp^{TE}E_m^{inc}e^{-\alpha x}\right]e^{-jk_{2y}y} \qquad \text{(a)}$$

$$H_\perp^{tra} = \left[\frac{\tau_\perp^{TE}E_m^{inc}}{\eta_2}\cos\theta_1(\hat{x}\sin\theta_2 - \hat{y}\cos\theta_2)e^{-\alpha x}\right]e^{-jk_{2y}y} \qquad \text{(b)} \qquad (5.3.7)$$

$$\vec{S}^* = E \times H^* = \left[\frac{\left|\tau_\perp^{TE}\right|^2\left|E_m^{inc}\right|^2}{\eta_2}\cos\theta_1(\hat{y}\sin\theta_2 + \hat{x}\cos\theta_2)\right]e^{-2\alpha x}$$

$$\Rightarrow \vec{S}^* = \left[\frac{\left|\tau_\perp^{TE}\right|^2\left|E_m^{inc}\right|^2}{\eta_2 k_2}\cos\theta_1\left(\hat{y}k_{2y} + \hat{x}j\alpha\cos\theta_2\right)\right]e^{-2\alpha x} \qquad \text{(c).}$$

In equation (5.3.7a,b,c), the y-directed wavevector k_{2y} and attenuation factor α are given by equation (5.3.5). Equation (5.3.7a,b,c) shows that along the direction normal to the interface, i.e. + x-direction, the field is exponentially decaying in medium #2; showing the confinement of field near the interface. However, the wave propagates in the y-direction. It is also evident from the complex Poynting vector giving real power transportation in the y-direction, while imaginary power shows storages of energy in the x-directed *evanescent field*. So the interface supports a *surface wave*, excited at the interface of two media by the obliquely TE-polarized incident plane wave at the angle of incidence $\theta_1 \geq \theta_c$. The surface wave, in more detail, is discussed in chapter 7. Its phase velocity along the y-axis is

$$v_p' = \frac{\omega}{k_{2y}} = \frac{\omega}{k_2\sin\theta_1}\sqrt{\frac{\varepsilon_{r2}}{\varepsilon_{r1}}} = \frac{v_p}{\sin\theta_1}\sqrt{\frac{\varepsilon_{r2}}{\varepsilon_{r1}}}. \qquad (5.3.8)$$

In equation (5.3.8), as $\varepsilon_{r1} > \varepsilon_{r2}$; so $v_p' < v_p$, i.e. the phase velocity of the surface wave is less than that of in the medium #2. Therefore, the surface wave is a slow-wave.

TM Polarization

All three cases of the angle of incidence apply to the TM-polarized obliquely incident plane wave. For $\theta_1 > \theta_c$, the reflection and transmission coefficients of the TM-polarization, given by equation (5.2.28) are reduced to

$$\Gamma_\parallel^{TM} = -\left[\frac{\eta_1/\eta_2\cos\theta_1 - j\left\{(\varepsilon_{r1}/\varepsilon_{r2})\sin^2\theta_1 - 1\right\}^{1/2}}{\eta_1/\eta_2\cos\theta_1 + j\left\{(\varepsilon_{r1}/\varepsilon_{r2})\sin^2\theta_1 - 1\right\}^{1/2}}\right]$$

$$= -1e^{j\phi_\parallel^{TM}} \qquad \text{(a)}$$

$$\tau_\parallel^{TM} = \frac{2\cos\theta_1}{\eta_1/\eta_2\cos\theta_1 + j\left\{(\varepsilon_{r1}/\varepsilon_{r2})\sin^2\theta_1 - 1\right\}^{1/2}} \qquad \text{(b)}$$

At $\theta_1 = \theta_c$, $\theta_2 = \pi/2$, $\sin\theta_1 = \sqrt{\varepsilon_{r2}/\varepsilon_{r1}}$ and $\Gamma_\parallel^{TM} = -1$ $\tau_\parallel^{TM} = 2\eta_2/\eta_1$ (c).

(5.3.9)

The electric and magnetic field components of the TM polarization, also the complex Poynting vector in the medium #2 under $\theta_1 > \theta_c$, could be obtained using equation (5.3.5), from equation (5.2.18). The results are summarized below:

$$\vec{E}_\parallel^{tra} = \left[\frac{\tau_\parallel^{TM}E_m^{inc}}{k_2}\left(-\hat{x}\,k_{2y} + \hat{y}\,j\alpha\right)e^{-\alpha x}\right]e^{-jk_{2y}y} \qquad \text{(a)}$$

$$H_{\parallel,z}^{tra} = \left[\hat{z}\frac{\tau_\parallel^{TM}E_m^{inc}}{\eta_2}e^{-\alpha x}\right]e^{-jk_{2y}y} \qquad \text{(b)} \qquad (5.3.10)$$

$$\vec{S}^* = \vec{E} \times \vec{H}^* = \left[\frac{\left|\tau_\parallel^{TM}\right|^2\left|E_m^{inc}\right|^2}{\eta_2 k_2}\left(\hat{y}\,k_{2y} + \hat{x}\,j\alpha\right)\right]e^{-2\alpha x} \qquad \text{(c).}$$

The expression (5.3.9a) shows that there is a total reflection $\left(\Gamma_{\parallel}^{TM} = -1\right)$ at the interface for $\theta_1 \geq \theta_c$, without any transmission of power from the medium #1 to medium #2. The transmission coefficient shows the presence of the x-directed *evanescent field* in the medium #2. Under such conditions, the interface acts as a perfect electrical conductor (PEC); so the interface surface of two media can act as the *artificial electric conductor (AEC)*. However, there is an exponentially decaying field in the medium #2, and *the interface supports the surface wave propagation* along the interface at $x = 0^+$ in the y-direction. The real power carried in the y-direction along the surface is given by the real part of equation (5.3.10). The imaginary part of the Poynting vector shows the stored energy in the evanescent field. Figure (5.5c) shows the surface wave propagation in the y-direction that also occurs in the case of the obliquely incident TE-polarized waves.

This subsection shows the existence of a surface wave at the interface of natural media. However, *artificially engineered metasurfaces* discussed in subsection (22.5.5) of chapter 2 has additional ability to control the surface wave in the desired manner, and also reradiate it as the leaky wave.

5.4 EM-Waves Incident at Dielectric Slab

The EM-waves can strike the slab embedded in a homogeneous medium both normally and obliquely. Both cases of wave incidence and their transmission line models are discussed below. The analysis can be extended to the multilayer medium.

5.4.1 Oblique Incidence

Figure (5.6a) illustrates the oblique incidence of the TE-polarized waves on a three-layered dielectric medium. It is desired to find overall reflection and transmission coefficients of a dielectric slab of thickness d embedded in a homogeneous medium, while the TE-polarized wave is obliquely incident at the first interface located at the y-axis. The forward and reflected waves are present in both the media #1 and #2 and finally transmitted to the medium #3. The media #1 and #3 have identical electrical properties.

Extending the process given in equations (5.2.1)–(5.2.7), the total E and H-fields in three media are written as follows:

E – field in medium #1:

$$\vec{E}_{m1}^{total}(x, y) = \hat{z}E_m^{inc}e^{-jk_1(x\cos\theta_1 + y\sin\theta_1)} + \hat{z}E_m^{ref}e^{-jk_1(-x\cos\theta_1 + y\sin\theta_1)} \quad (a)$$

E – field in medium #2:

$$\vec{E}_{m2}^{total}(x, y) = \hat{z}E_m^{m2f}e^{-jk_2(x\cos\theta_2 + y\sin\theta_2)} + \hat{z}E_m^{m2b}e^{-jk_2(-x\cos\theta_2 + y\sin\theta_2)} \quad (b) \qquad (5.4.1)$$

E – field in medium #3:

$$\vec{E}_{m3}^{total}(x, y) = \hat{z}E_m^{tra}e^{-jk_1((x-d)\cos\theta_1 + y\sin\theta_1)} \quad (c).$$

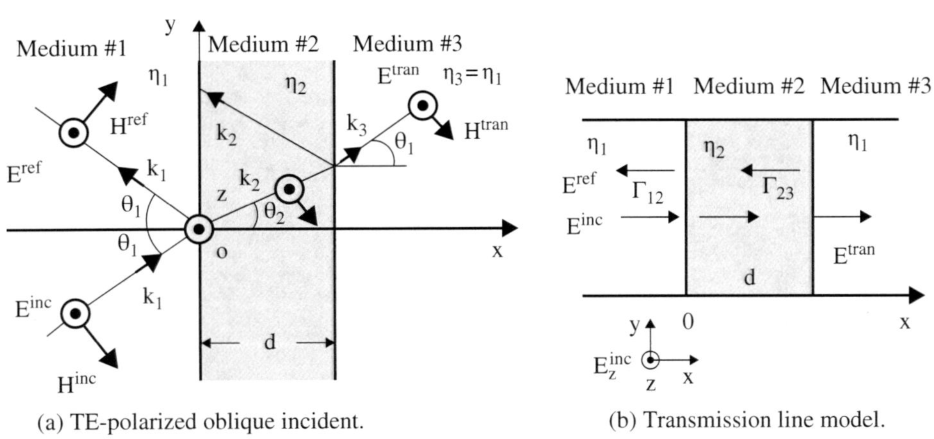

(a) TE-polarized oblique incident.
(b) Transmission line model.

Figure 5.6 Plane-wave incident on a dielectric slab.

H – field in medium #1:

$$\vec{H}^{total}_{m1}(x,y) = \frac{E^{inc}_m}{\eta_1}(\hat{x}\sin\theta_1 - \hat{y}\cos\theta_1)e^{-jk_1(x\cos\theta_1 + y\sin\theta_1)} + \frac{E^{ref}_m}{\eta_1}(\hat{x}\sin\theta_1 + \hat{y}\cos\theta_1)e^{-jk_1(-x\cos\theta_1 + y\sin\theta_1)} \quad (a)$$

H – field in medium #2:

$$\vec{H}^{total}_{m2}(x,y) = \frac{E^{m2f}_m}{\eta_2}(\hat{x}\sin\theta_2 - \hat{y}\cos\theta_2)e^{-jk_2(x\cos\theta_2 + y\sin\theta_2)} + \frac{E^{m2b}_m}{\eta_2}(\hat{x}\sin\theta_2 + \hat{y}\cos\theta_2)e^{-jk_2(-x\cos\theta_2 + y\sin\theta_2)} \quad (b) \qquad (5.4.2)$$

H – field in medium #3:

$$\vec{H}^{total}_{m3}(x,y) = \frac{E^{tra}_m}{\eta_1}(\hat{x}\sin\theta_1 - \hat{y}\cos\theta_1)e^{-jk_1((x-d)\cos\theta_1 + y\sin\theta_1)} \quad (c).$$

In equations (5.4.2a,b,c) superscripts m_1, m_2, and m_3 correspond to medium #1, #2, and #3, respectively. The propagations constant and intrinsic impedance in media are

$$k_i = \sqrt{\varepsilon_i\mu_i} = k_0\sqrt{\varepsilon_{ri}\mu_{ri}} \quad (a), \qquad \eta_i = \sqrt{\frac{\mu_i}{\varepsilon_i}} = \eta_0\sqrt{\frac{\mu_{ri}}{\varepsilon_{ri}}} \quad (b)$$

$$\text{also,}\ \eta_i = \frac{k_i}{\omega\varepsilon_i} \quad (c), \qquad \eta_i = \frac{\omega\mu_i}{k_i} \quad (i = 1,2) \quad (d). \qquad (5.4.3)$$

Further, in equations (5.4.1b) and (5.4.2b), the superscripts m2f and m2b show the forward and backward moving waves in the medium #2. The tangential E_z and H_y field components at the first interface located at $x = 0$ are continuous:

$$E_z - \text{field}: \ \vec{E}^{total}_{m1}(x,y)\Big|_{x=0^-} = \vec{E}^{total}_{m2}(x,y)\Big|_{x=0^+} \quad (a)$$

$$\Rightarrow E^{inc}_m e^{-jk_1\sin\theta_1 y} + E^{ref}_m e^{-jk_1\sin\theta_1 y} = E^{m2f}_m e^{-jk_2\sin\theta_2 y} + E^{m2b}_m e^{-jk_2\sin\theta_2 y} \quad (b). \qquad (5.4.4)$$

$$H_y - \text{field}: \ H^{total}_{m1}(x,y)\big|_{x=0^-} = H^{total}_{m2}(x,y)\big|_{x=0^+} \quad (a)$$

$$\Rightarrow -\frac{E^{inc}_m}{\eta_1}\cos\theta_1 e^{-jk_1\sin\theta_1 y} + \frac{E^{ref}_m}{\eta_1}\cos\theta_1 e^{-jk_1\sin\theta_1 y} = -\frac{E^{m2f}_m}{\eta_2}\cos\theta_2 e^{-jk_2\sin\theta_2 y} + \frac{E^{m2b}_m}{\eta_2}\cos\theta_2 e^{-jk_2\sin\theta_2 y} \quad (b). \qquad (5.4.5)$$

The continuity of the y-components of the field across the interface also provides the phase matching giving the following result:

$$k_{1y} = k_{2y}, \quad \Rightarrow k_1\sin\theta_1 = k_2\sin\theta_2. \qquad (5.4.6)$$

The dispersion relation (4.5.29d) of chapter 4 provides the following expressions for the propagation constant of propagating wave, in the medium #2 and medium #3, in the x-axis direction:

Medium#2:

$$k_{2x}^2 + k_{2y}^2 = \omega^2\mu_2\varepsilon_2, \Rightarrow k_{2x} = \sqrt{\omega^2\mu_2\varepsilon_2 - k_1^2\sin^2\theta_1} \quad (a)$$

Medium#3:

$$k_{3x} = k_{1x} = \sqrt{\omega^2\mu_1\varepsilon_1 - k_{1y}^2} = \sqrt{\omega^2\mu_1\varepsilon_1 - k_1^2\sin^2\theta_1} \quad (b). \qquad (5.4.7)$$

After canceling the phase-matching factors in equations (5.4.4) and (5.4.5), the amplitude matching at the interface #1 provides the following expressions:

$$E^{inc}_m + E^{ref}_m = E^{m2f}_m + E^{m2b}_m \quad (a)$$

$$\frac{\cos\theta_1}{\eta_1}\left(E^{inc}_m - E^{ref}_m\right) = \frac{\cos\theta_2}{\eta_2}\left(E^{m2f}_m - E^{m2b}_m\right) \quad (b). \qquad (5.4.8)$$

The above equations are solved to get the following set of expressions for E^{m2f}_m and E^{m2b}_m:

$$E^{m2f}_m = \frac{1}{2}\{(1+P)E^{inc}_m + (1-P)E^{ref}_m\} \quad (a)$$

$$E^{m2b}_m = \frac{1}{2}\{(1-P)E^{inc}_m + (1+P)E^{ref}_m\} \quad (b)$$

$$\text{where,} \quad P = \frac{\eta_2\cos\theta_1}{\eta_1\cos\theta_2} \quad (c). \qquad (5.4.9)$$

After cancellation of the phase-matching factors at the interface #2 (at $x = d$), the continuity of tangential components E_z and H_y provide the following expressions:

$$E_z-\text{field}: \left.\vec{E}^{\,total}_{m2}(x,y)\right|_{x=d^-} = \vec{E}^{\,total}_{m2} = \left.\vec{E}^{\,total}_{m3}(x,y)\right|_{x=d^+} \quad (a)$$

$$\Rightarrow E_m^{m2f}e^{-j\phi} + E_m^{m2b}e^{+j\phi} = E_m^{tran} \quad (b), \text{ where } \phi = k_2 d\cos\theta_2 \quad (c)$$

$$H_y-\text{field}: \left.\vec{H}^{\,total}_{m1}(x,y)\right|_{x=d^-} = \left.\vec{H}^{\,total}_{m3}(x,y)\right|_{x=d^+} \quad (5.4.10)$$

$$\Rightarrow \frac{\cos\theta_2}{\eta_2}\left(E_m^{m2f}e^{-j\phi} - E_m^{m2b}e^{+j\phi}\right) = \frac{\cos\theta_1}{\eta_1}E_m^{tran},$$

$$\Rightarrow E_m^{m2f}e^{-j\phi} - E_m^{m2b}e^{+j\phi} = PE_m^{tran} \quad (d).$$

Equations (5.4.10a,b,c) are again solved for E_m^{m2f} and E_m^{m2b}:

$$E_m^{m2f} = \frac{1}{2}(1+P)E_m^{tran}e^{+j\phi} \quad (a)$$

$$E_m^{m2b} = \frac{1}{2}(1-P)E_m^{tran}e^{-j\phi} \quad (b). \quad (5.4.11)$$

Equations (5.4.9) and (5.4.11) are equated to get a pair of expressions for the reflection (Γ^{TE}) and transmission (τ^{TE}) coefficients:

$$(1+P) + (1-P)\Gamma^{TE} = (1+P)e^{+j\phi}\tau^{TE} \quad (a)$$

$$(1-P) + (1+P)\Gamma^{TE} = (1-P)e^{-j\phi}\tau^{TE} \quad (b)$$

$$\text{where, } \Gamma^{TE} = \frac{E_m^{ref}}{E_m^{inc}} \quad (c)$$

$$\tau^{TE} = \frac{E_m^{tra}}{E_m^{inc}} \quad (d). \quad (5.4.12)$$

The above equations are solved to get the following expressions for the reflection and transmission coefficients of the obliquely incident TE-polarized wave:

$$\Gamma^{TE} = \frac{(1-P^2)\left(e^{-j\phi} - e^{+j\phi}\right)}{(1+P^2)e^{+j\phi} - (1-P^2)e^{-j\phi}} \quad (a)$$

$$\tau^{TE} = \frac{4P}{(1+P^2)e^{+j\phi} - (1-P^2)e^{-j\phi}} \quad (b)$$

$$\text{where, } P = \frac{\eta_2\cos\theta_1}{\eta_1\cos\theta_2} \quad (c). \quad (5.4.13)$$

For the obliquely incident TM-polarized wave on the three-layered medium, shown in Fig (5.6a), the process can be repeated to get the following expressions, similar to expressions (5.4.14):

$$\Gamma^{TM} = \frac{\left(1-Q^2\right)\left(e^{j\phi} - e^{-j\phi}\right)}{\left(1+Q^2\right)e^{+j\phi} - \left(1-Q^2\right)e^{-j\phi}} \quad (a)$$

$$\tau^{TM} = \frac{4Q}{\left(1+Q^2\right)e^{+j\phi} - \left(1-Q^2\right)e^{-j\phi}} \quad (b)$$

$$\text{where, } Q = \frac{\eta_1\cos\theta_1}{\eta_2\cos\theta_2} \quad (c)$$

$$\phi = k_2 d\cos\theta_2 \quad (d). \quad (5.4.14)$$

The reflection and transmission coefficients of both the TE and TM-polarized incident waves are identical in form, except that parameters P and Q are different.

5.4.2 Normal Incidence

For the case of the normally incident TE-polarized wave, the angles of incidence and refraction are $\theta_1 = \theta_2 = 0°$. Furthermore, $\phi = k_2 d$, $(k_{2x} = k_2)$, and $P = \eta_2/\eta_1$. Under, this condition, equation (5.4.13a,b) for the reflection and transmission coefficients are reduced to the following expressions:

$$\Gamma^{Nor} = \left[\frac{\eta_2-\eta_1}{\eta_2+\eta_1}\right]\left[\frac{1-e^{-j2k_2d}}{1-\{(\eta_2-\eta_1)/(\eta_2+\eta_1)\}^2e^{-j2k_2d}}\right] \quad (a)$$

$$\tau^{Nor} = \left[\frac{4\eta_2\eta_1}{(\eta_2+\eta_1)^2}\right]\left[\frac{e^{-jk_2d}}{1-\{(\eta_2-\eta_1)/(\eta_2+\eta_1)\}^2e^{-j2k_2d}}\right] \quad (b). \quad (5.4.15)$$

The above expressions are valid for the normally incident TM-polarized waves also.

Figure (5.6b) shows the equivalent transmission line model of the three-layered medium. The reflection coefficients at both interfaces can be obtained from the expression, $\Gamma = (Z_L - Z_0)/(Z_L + Z_0)$. At the first interface, corresponding medium impedances are $Z_L \to \eta_2$ and $Z_0 \to \eta_1$, while at the second interface, these are $Z_L \to \eta_1$ and $Z_0 \to \eta_2$. The reflection coefficients Γ_{12} and Γ_{23} are defined at the first and second interfaces. The above equations are rewritten as follows:

$$\Gamma^{Nor} = \frac{\Gamma_{12}\left(1-e^{-j2k_2d}\right)}{1+\Gamma_{12}\Gamma_{23}e^{-j2k_2d}} \quad (a)$$

$$\tau^{Nor} = \left[\frac{4\eta_2\eta_1}{(\eta_2+\eta_1)^2}\right]\frac{e^{-j2k_2d}}{1+\Gamma_{12}\Gamma_{23}e^{-j2k_2d}} \quad (b)$$

$$\Gamma^{Nor} \approx \Gamma_{12}\left(1-e^{-j2k_2d}\right) \quad (c)$$

$$\text{where, } \Gamma_{12} = \left[\frac{\eta_2-\eta_1}{\eta_2+\eta_1}\right], \quad (d)$$

$$\Gamma_{23} = \left[\frac{\eta_1-\eta_2}{\eta_2+\eta_1}\right] = -\Gamma_{12} \quad (e). \quad (5.4.16)$$

The approximate expression (5.4.16c) is used for the small reflection at the interfaces, i.e. for $|\Gamma_{12}\Gamma_{23}| < < 1$.

The above expressions for total reflection and transmission of a slab can also be obtained from the theory of multiple reflections [B.5, J.2]. Using the relation (5.4.3), equation (5.4.16) can be recast in the following format also:

$$\Gamma^{Nor} = \frac{\left[(\mu_2 k_1)^2 - (\mu_1 k_2)^2\right]\left[1 - e^{-j2k_2 d}\right]}{\left[\mu_2 k_1 + \mu_1 k_2\right]^2 - \left[\mu_2 k_1 - \mu_1 k_2\right]^2 e^{-j2k_2 d}} \quad (a)$$

$$\tau^{Nor} = \frac{4\mu_1\mu_2 k_1 k_2 e^{-jk_2 d}}{\left[\mu_2 k_1 + \mu_1 k_2\right]^2 - \left[\mu_2 k_1 - \mu_1 k_2\right]^2 e^{-j2k_2 d}} \quad (b). \qquad (5.4.17)$$

In the case of the obliquely incident TE waves, in the above equations, we replace the wavevectors k_1 and k_2 as $k_1 \to k_{1x} \to k_1 \cos\theta_1$ and $k_2 \to k_{2x} \to k_2 \cos\theta_1$ respectively; where the waves propagate in the x-direction. The waves are confined in the y-direction. As a special limiting case, if the electrical parameters of the slab of thickness d and the host medium are identical, the $k_2 = k_1$, $\mu_2 = \mu_1$. It leads to $\Gamma^{Nor} = 0$, i.e. no reflection at the interface, and $\tau^{Nor} = e^{-jk_2 d}$, i.e. the slab provides a complete transmission with a lagging phase $k_2 d$. This is the case of the matched layer seen in equation (5.4.15a) for $\eta_1 = \eta_2$. The matching applies even to the interface of electrically dissimilar media given by the condition in equation (5.1.4b). The expressions obtained in the present subsection are useful to characterize the metamaterials slab. It is discussed in subsection (21.4.1) of chapter 21.

5.5 EM-Waves in Metamaterials Medium

The electromagnetic properties of metamaterials, propagation of EM – waves in the metamaterials media, and circuit models of metamaterials are presented in this section. Some applications of the metamaterials are also discussed. The realization of engineered metamaterials and their further modeling are discussed in chapter 21.

5.5.1 General Introduction of Metamaterials and Their Classifications

The materials existing in nature, ignoring losses, are characterized electromagnetically by the relative permittivity and relative permeability. Normally, these are positive quantities. Such a medium is called the *double-positive (DPS) medium*. It has a positive real refractive index n. It is placed in the first quadrant of the (μ_r, ε_r)-plane as shown in Fig (5.7). The conductivity or resistivity is added to a medium to account for the losses. Some noble metals, such as silver and gold, have negative permittivity in the visible frequency bands. Drude model, based on the plasma model of metal, is used to model such behavior. It is discussed in chapter 6, section (6.5.2). An identical phenomenon exists in the ionospheres. Below the so-called plasma frequency mentioned briefly in section (3.3.1) of chapter 3, the permittivity of a plasma medium becomes negative. In this case, the medium does not support wave propagation. The waves become oscillatory decaying evanescent mode. The permeability of the plasma medium remains a positive quantity. Under the evanescent mode condition, the plasma medium below the cut-off frequency is called the *epsilon-negative (ENG) medium*. It is shown in the second quadrant in Fig (5.7). Thus, below the plasma frequency, the medium reflects the wave incident on it from the DPS medium; while inside the ENG medium only decaying evanescent mode exists. However, above the plasma frequency, the ENG plasma medium becomes a normal DPS medium. Likewise, a gyrotropic medium, such as a ferrite medium, has negative permeability over a lower microwave frequency range. However, its permittivity is a positive quantity. Such a medium is called the *mu-negative (MNG) medium*. It is shown in the fourth quadrant of Fig (5.7). Both the ENG and MNG media have imaginary refractive index n and support only decaying *evanescent mode,* as shown in Fig (5.7). As a single negative medium, both the ENG and MNG belong to the group of the single negative, i.e. (*SNG*) *medium*. Both the ENG and MNG are described by the following forms of the Drude model:

$$\text{ENG:} \quad \varepsilon_r = 1 - \left(\omega_{ep}/\omega\right)^2 \quad (a)$$

$$\text{MNG:} \quad \mu_r = 1 - \left(\omega_{mp}/\omega\right)^2 \quad (b), \qquad (5.5.1)$$

where ω_{ep} and ω_{mp} are the electric plasma and the magnetic plasma frequency respectively. The magnetic plasma is conceptual as no magnetic charge is available in nature.

The natural dielectric and magneto-dielectric media support the slow-wave propagation as for these media $\varepsilon_r > 1$, $\mu_r > 1$. However, it is possible to engineer a DPS metamaterial for $0 < \varepsilon_r < 1$, $\mu_r \geq 1$ and $0 < \mu_r < 1$, $\varepsilon_r \geq 1$. The first medium is called *epsilon near zero* (ENZ) medium, while the second one is called *mu near-zero* (MNZ) medium. They belong to the group called *near-zero* (NZ) medium. These media support the *fast-wave*. The DNG medium also supports the ENZ and MNZ cases. The NZ medium is shown as a dark circle at the origin in Fig (5.7). These media are obtained just above the plasma frequency. The refractive index of the ENZ and MNZ is numerically less than unity, supporting the fast-wave. Even a zero refractive index could be obtained. It is known as the "*nihility*" [B.6].

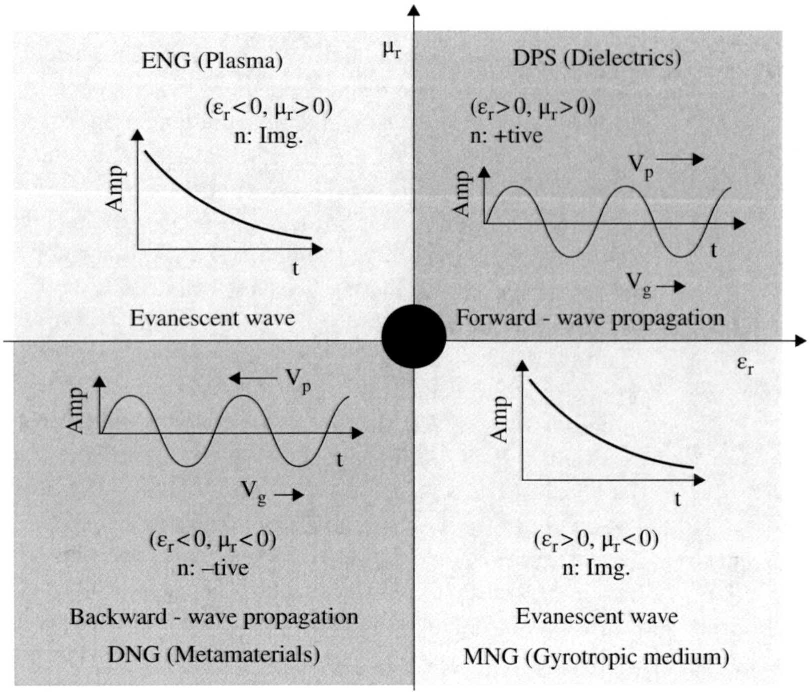

Figure 5.7 Electrical grouping of the materials media in the (μ_r, ε_r)-plane.

Following the above-discussed nomenclature scheme of a material medium, it is natural to envisage an artificial medium with both the permittivity and permeability as negative quantities. Such an engineered medium could be called a *double negative medium (DNG) medium*. It *is* shown in the third quadrant of Fig (5.7). It has a negative refractive index (−n). The DPS medium supports the wave propagation such that both the phase velocity v_p and group velocity v_g are in the same direction, whereas the DNG medium supports wave propagation with phase velocity v_p and group velocity v_g in opposite directions. Such EM-wave is known as the *backward wave*. However, no DNG material is known to exist in nature. The DNG material medium has been synthesized over certain frequency bands, from the microwave to optical frequency ranges.

The DNG materials are commonly known as *metamaterials*. However, artificially engineered ENG, MNG, ENZ, and MNZ media also belong to the metamaterials. Even the DPS could be engineered, and called a metamaterial, to realize the controlled values of permittivity and permeability to meet the specific design requirement. One such requirement is to develop the microwave absorbers discussed in subsection (5.5.8). Several dedicated books are available on the topics of the metamaterials [B.6–B.10].

Depending on their characteristics, several other names have also been proposed for the metamaterials [B.7, B.10]. Veselago proposed the concept and electrodynamics of the metamaterials [J.3]. The metamaterials have been synthesized by embedding either the resonating type inclusions or the C-L type artificial transmission line inclusions in a host medium. The metamaterials could be either *resonant metamaterials* or *nonresonant metamaterials*, so the metamaterials are structured composite materials. Starting with J.C. Bose, such composite materials have a long history [J.4–J.6]. The concept of composite materials and their modeling is discussed in section (6.3) of chapter 6. The C-L line supports the backward wave propagation. It is discussed in section (3.4) of chapter 3. A metamaterial, i.e. the DNG medium, is also called the *backward wave (BW) medium*, or BW material [J.7]. The normal forward wave supporting DPS medium could be called *forward wave (FW) medium*, or FW material.

The physical dimensions of the resonating inclusions, or the C-L line inclusions, are much smaller than the operating wavelengths. The metamaterials could be treated as the *homogenized medium* described by the parameters permittivity and permeability. The inclusions are arranged in the host medium either periodically, or nonperiodically. The periodically arranged inclusions, with a period $\lambda_g/2$, form another group of metamaterials known as the *electronic bandgap (EBG)* materials. It is discussed in chapters 19 and 20. Even the ENG and MNG materials have been artificially developed using such inclusions.

The (μ_r, ε_r)-plane of Fig (5.7) summarizes the above-discussed four-groups of material media with their basic electrical characteristics. The circuit models and further wave characteristics of these media are shown in Fig (5.10). The synthesis, modeling, and some illustrative applications of the metamaterials are discussed in chapter 21. This section only presents some basic electrodynamics properties of the DNG medium.

5.5.2 EM-Waves in DNG Medium

This section considers the EM-wave propagation in the lossless DNG medium, the concept of the negative refractive index, the basic circuit model of a DNG medium, and the lossy DNG medium.

Lossless DNG Medium

Maxwell's equation (4.4.11a,b) of chapter 4 could be written for a lossless ($\sigma = 0$) DNG medium by using ($-|\varepsilon_r|$) and ($-|\mu_r|$) in place of their usual positive values:

$$\nabla \times \vec{E} = -j\omega\mu_0(-|\mu_r|)\vec{H} \quad \text{(a)}$$

$$\nabla \times \vec{H} = j\omega\varepsilon_0(-|\varepsilon_r|)\vec{E} \quad \text{(b)}$$

$$\Rightarrow \nabla \times \vec{E} = -j\omega\mu_0|\mu_r|\left(-\vec{H}\right) \quad \text{(c)}$$

$$\nabla \times \left(-\vec{H}\right) = j\omega\varepsilon_0|\varepsilon_r|\vec{E} \quad \text{(d).}$$

$$(5.5.2)$$

Equations (4.5.31a) and (4.2.3b) of chapter 4 show Maxwell's equations for a DPS medium in terms of the wavevector $\vec{k}$ and the field vectors $\left(\vec{E}, \vec{H}\right)$. They form the *wavevector triplet* ($\vec{E}, \vec{H}, \vec{k}$) relations in the right-hand (RH) coordinate system, shown in Fig (5.8a) for a DPS medium. Normally, the direction of the wavevector $\vec{k}$ determines the direction of the wave propagation, i.e. the direction of the phase velocity v_p. However, the Poynting vector ($\vec{S}$) given by equation (4.4.20) provides another *power-vector triplet* ($\vec{E}, \vec{H}, \vec{S}$) that determines the true direction of wave propagation. It is the direction of the energy flow from the source to a load. So *the corresponding group velocity v_g defines the direction of the wave propagation.*

Figure (5.8c) shows both the sets of the vector triplets of a DPS medium. The field vectors are rotated in cyclic order so that the power flow is maintained from left to right-hand side, i.e. from a source located at the origin O, power flows outwardly in a positive direction. For both the phase and group velocities, the DPS medium follows the right-hand (RH) coordinate system, so a normal DPS medium is called the *right-handed, i.e. the RH-medium.* As the RH-system holds for both triplets, they are combined into one diagram as shown in Fig (5.8c). For a DPS medium, the directions of the vectors $\vec{k}$ and $\vec{S}$ are identical, i.e. they are parallel vectors giving the relation $\vec{k}\cdot\vec{S} > 0$. Therefore, in a DPS medium, both the phase and group velocities are in the same direction giving $\vec{v}_p\cdot\vec{v}_g > 0$. It is shown in the first quadrant of Fig (5.7). *The phase of the propagating EM-wave in the DPS medium lags while traveling in the $\vec{S}$-direction.*

In the case of a DNG medium, both permittivity and permeability are negative. Using Maxwell's equations (5.5.2c,d), and replacing $\nabla \to -jk$ the wavevector triplet- $\vec{E}, \vec{H}, \vec{k}$ relations for the DNG medium are written as follows:

$$\vec{k} \times \vec{E} = \omega|\mu|\left(-\vec{H}\right) \quad \text{(a)}$$

$$\vec{k} \times \vec{H} = \omega|\varepsilon|\vec{E} \quad \text{(b).}$$

$$(5.5.3)$$

However, in the above expression reversal of the direction of the magnetic field involves the reversal of the direction of the power flow toward the source. Physically, it is not possible, so the above equations are rearranged as follows by associating the negative sign with the wavevector $\vec{k}$:

$$\left(-\vec{k}\right) \times \vec{E} = \omega|\mu|\vec{H} \quad \text{(a)}$$

$$\left(-\vec{k}\right) \times \vec{H} = -\omega|\varepsilon|\vec{E} \quad \text{(b).}$$

$$(5.5.4)$$

The above wavevector triplet- $\left(\vec{E}, \vec{H}, \vec{k}\right)$ is shown in Fig (5.8b), i.e. in the left-hand (LH) coordinate system, so the DNG medium is also called the left-handed, *LH-medium.* Both the power-vector and wavevector triplets and their combination are further shown in Fig (5.8d). The field vectors are rotated to maintain the power flow in a positive direction. Figure (5.8d) shows that the phase and group velocities are opposite to each other ($\vec{v}_p\cdot\vec{v}_g < 0$) as for a DNG medium the vectors $\vec{k}$ and $\vec{S}$ are antiparallel, i.e. $\vec{k}\cdot\vec{S} < 0$. The DNG occupies the third quadrant in the (μ_r, ε_r)-plane as shown in Fig (5.7).

In conclusion, a DNG medium supports the *backward wave* propagation, whereas a *forward wave* is supported by the DPS medium. As the phase velocity travels toward the source, while energy is traveling from the source to a load, *a propagating EM-wave in a DNG medium, in the direction of the vector $\vec{S}$, has a leading phase.* This is a unique property of the DNG medium. It significantly influences the EM-wave characteristics of the DNG medium [J.8, B.6, B.10].

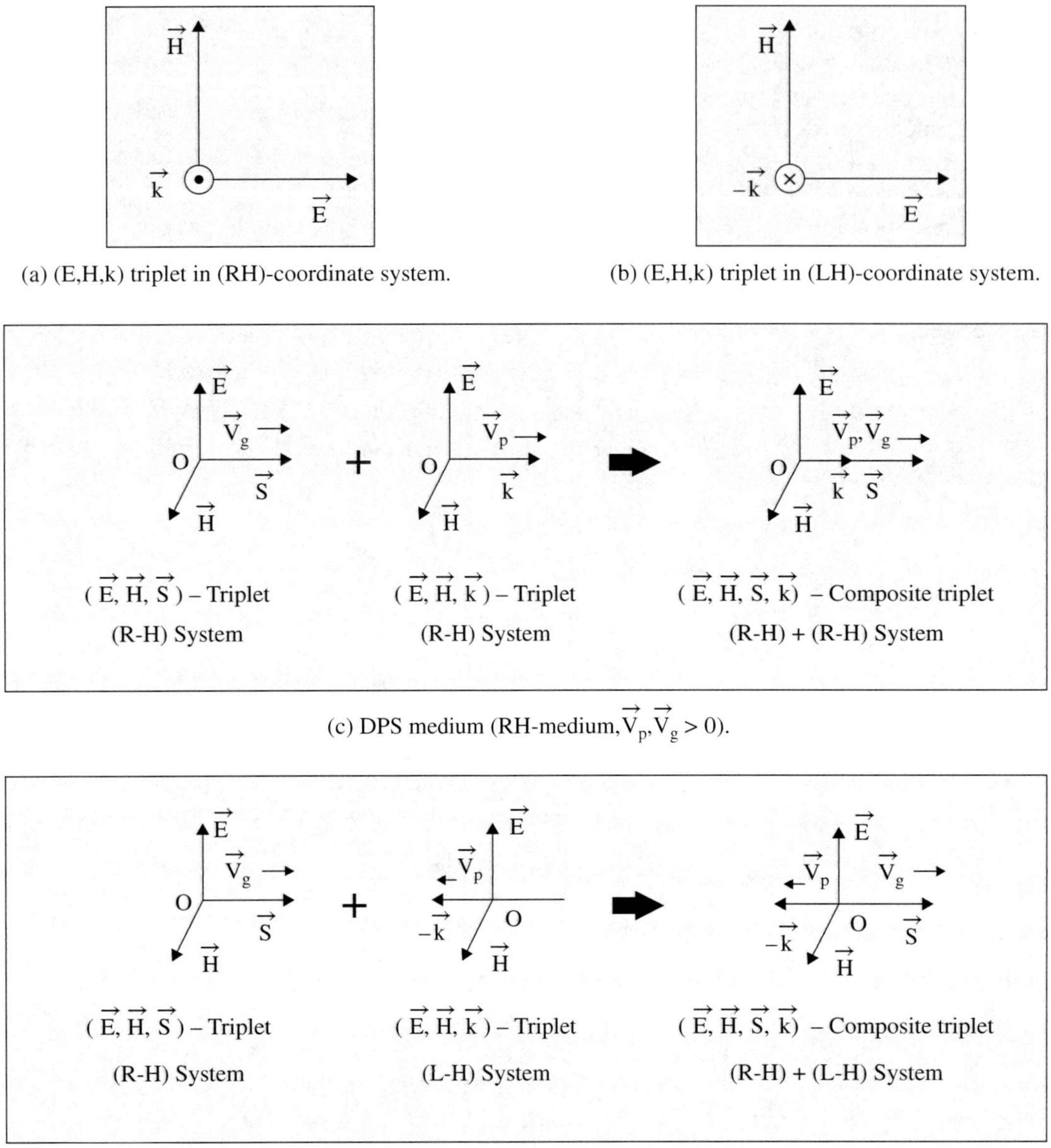

(a) (E,H,k) triplet in (RH)-coordinate system. (b) (E,H,k) triplet in (LH)-coordinate system.

(c) DPS medium (RH-medium, $\vec{V}_p, \vec{V}_g > 0$).

(d) DNS medium (LH-medium, $\vec{V}_p, \vec{V}_g < 0$).

Figure 5.8 RH and LH-coordinate systems for the DPS and DNG media.

Refractive Index of DNG Medium

The above discussion shows that Maxwell's equations in the DNG medium are written in the LH-coordinate system. However, the wave equation (4.5.32) of chapter 4 for the DPS medium remains valid for a lossless ($\sigma = 0$) DNG medium. It provides the following expressions for the propagation constant $\beta = k^{DPS}$ and refraction index of a DPS medium:

$$\vec{k}^{DPS} = k_0\sqrt{\mu_r\varepsilon_r}\,\hat{k} \quad (a), \qquad n^{DPS} = \sqrt{\mu_r\varepsilon_r} \quad (b). \qquad (5.5.5)$$

The evaluation of the square root of negative permeability and negative permittivity is a critical issue in the DNG medium. The negative number (−1) is exp ($\pm j\pi$). However, to meet the physical condition, discussed in subsection (5.5.3), we take $\{-1 = \exp(-j\pi)\}$ [J.8, J.9]. Therefore, the square roots of negative permeability and negative permittivity are obtained as follows:

$$\sqrt{-|\mu_r|} = \sqrt{e^{-j\pi}|\mu_r|} = e^{-j\pi/2}\sqrt{|\mu_r|} = -j\sqrt{|\mu_r|}, \text{ and}$$

$$\sqrt{-|\varepsilon_r|} = e^{-j\pi/2}\sqrt{|\varepsilon_r|} = -j\sqrt{|\varepsilon_r|}.$$

Using the above relations, the refractive index of a DNG medium, and also the propagation constant, are obtained as follows:

$$n^{DNG} = \sqrt{-|\mu_r|} \times \sqrt{-|\varepsilon_r|} = \left(\sqrt{|\mu_r|}e^{-j\pi/2}\right)\left(\sqrt{|\varepsilon_r|}e^{-j\pi/2}\right) = e^{-j\pi}\sqrt{|\mu_r||\varepsilon_r|}$$

$$\Rightarrow n^{DNG} = -\sqrt{|\mu_r||\varepsilon_r|} \quad \text{(a)}$$

$$\vec{k}^{DNG} = \vec{\beta} = -\sqrt{|\mu_r||\varepsilon_r|}\,\vec{k}_0 \quad \text{(b).} \qquad (5.5.6)$$

It is interesting to note that the refractive index for a DPS medium n^{DPS} is a positive quantity, whereas for a DNG medium n^{DNG} is a negative quantity. So *the metamaterials are also known as the negative refractive index materials*, i.e. the *NIM*. Snell's law of refraction for a DNG medium is also modified accordingly. The negative refractive index also shows the reversal of the direction of the phase velocity of the EM-wave. However, first let us discuss the intrinsic impedance, i.e. the wave impedance for the DNG and SNG media.

Wave Impedance of DNG and SNG Media

Following equation (4.5.26b) of chapter 4, the wave impedance η^{DNG} in a DNG medium is written below:

$$\eta^{DNG} = \eta_0\sqrt{\frac{-|\mu_r|}{-|\varepsilon_r|}} = \eta_0\frac{e^{-j\pi/2}}{e^{-j\pi/2}}\sqrt{\frac{|\mu_r|}{|\varepsilon_r|}} = \eta_0\sqrt{\frac{|\mu_r|}{|\varepsilon_r|}} = \eta^{DPS}. \qquad (5.5.7)$$

Like the wave impedance in a DPS medium η^{DPS}, the wave impedance in the DNG medium η^{DNG} is a positive quantity; showing the outward power flow from the source into a DNG medium. However, the wave impedances of the ENG and MNG media are reactive due to nonpropagating evanescent mode:

$$\eta^{ENG} = \eta_0\sqrt{\frac{|\mu_r|}{-|\varepsilon_r|}} = \eta_0 e^{j\pi/2}\sqrt{\frac{|\mu_r|}{|\varepsilon_r|}} = j\eta^{DPS} \quad \text{(a)}$$

$$\eta^{MNG} = \eta_0\sqrt{\frac{-|\mu_r|}{|\varepsilon_r|}} = e^{-j\pi/2}\eta_0\sqrt{\frac{|\mu_r|}{|\varepsilon_r|}} = -j\eta^{DPS} \quad \text{(b).} \qquad (5.5.8)$$

The inductive/capacitive reactive wave impedances of the ENG and MNG media create the reflecting surfaces. The circuit model of the metamaterials, discussed in section (5.5.3), elaborates on the nature of the RIS. Further details of the artificial RIS surface is discussed in section (20.2) of chapter 20. It is noted that the ENG/MNG medium is realized through the nonpropagating evanescent wave. Such an environment is provided by a rectangular waveguide below the cutoff region. It is commented in subsection (7.4.1) of chapter 7.

The propagation constants of EM-waves in the ENG and MNG media are obtained below:

$$\beta^{ENG} = \pm\beta_0\sqrt{|\mu_r|(-|\varepsilon_r|)} = \mp j\beta_0\sqrt{|\mu_r|(|\varepsilon_r|)} \quad \text{(a)}$$

$$\beta^{MNG} = \pm\beta_0\sqrt{|\varepsilon_r|(-|\mu_r|)} = \mp j\beta_0\sqrt{|\mu_r|(|\varepsilon_r|)} \quad \text{(b).} \qquad (5.5.9)$$

Expressions (5.5.9 a,b) show that the ENG and MNG media do not support wave propagation. Figure (5.7) shows that these media are placed in the second and fourth quadrants in the (μ_r, ε_r)-plane. They only support the decaying evanescent mode.

Negative Refraction in DPS-DNG Composite Medium

Figure (5.9a) shows the DPS-DNG composite medium. The incident ray is in the third quadrant of the DPS medium #1. For the DPS medium #2, the refracted ray comes out in the first quadrant; as the angle of refraction (θ_2) is positive. The refraction in the DPS-DPS composite medium follows Snell's law given by equation (5.2.7c). However, if the medium #2 is DNG-type then the angle of refraction $(-\theta_2)$ is negative due to the negative refractive index. It follows from Snell's law:

$$n_1 \sin\theta_1 = -|n_2|\sin\theta_2 = |n_2|\sin(-\theta_2). \qquad (5.5.10)$$

Figure (5.9a) shows that due to the negative angle of refraction $(-\theta_2)$, the refracted ray in the DNG medium #2 emerges from the fourth quadrant. It shows *the reversal of Snell's law* in the DPS-DNG composite medium, as compared to Snell's law in the DPS-DPS composite medium. Figure (5.9b) shows that the wavevector $\vec{k}_2^{DNG}$ in medium #2 must be in the reverse direction to meet the phase-matching condition, $k_{1y}^{DPS} = k_{2y}^{DNG}$, at the interface of the composite medium.

In the case of the TE-polarized incident waves, the wavevectors of the incident, reflected, and refracted rays are given by equations (5.2.1a–d), whereas their Poynting vectors are given in equations (5.2.5a–c). For the DPS-DPS composite medium, both vectors are in the same direction after refraction, showing the presence of the *forward-wave* in medium #2. However, for the DPS-DNG medium, these vectors of the transmitted wave in the DNG medium #2 are given as

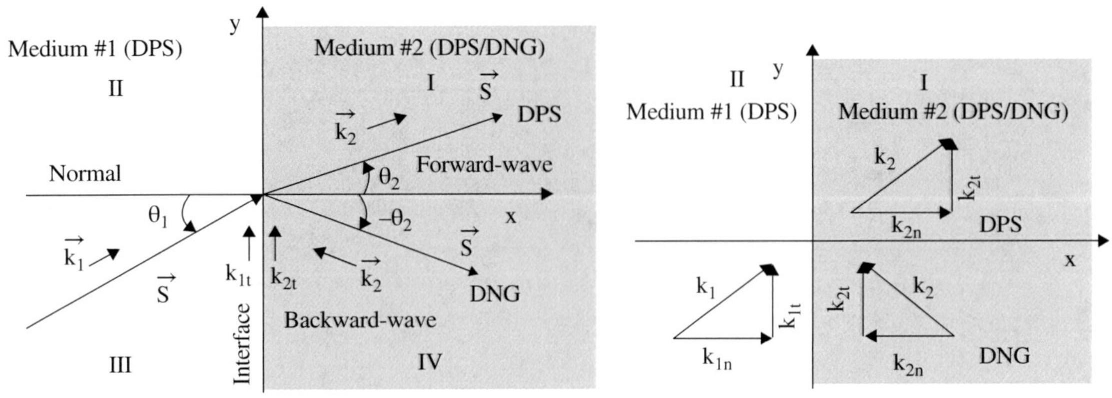

(a) Oblique incidence of EM-wave at DPS-DPS. (b) Wavevector diagrams.

Figure 5.9 Refraction of the obliquely incident EM-wave at the interface of the DPS-DPS and DPS-DNG composite medium.

$$\vec{k}_2^{\,DNG} = \hat{x}k_{2x} + \hat{y}k_{2y} = -\hat{n}k_{2n} + \hat{t}k_{2t} = k_2(-\hat{x}\cos\theta_2 + \hat{y}\sin\theta_2) \quad (a)$$

$$\vec{S}_{av}^{\,tra(DNG)} = \frac{\left(E_m^{tra}\right)^2}{2\eta_2}(+\hat{x}\cos\theta_2 - \hat{y}\sin\theta_2) \quad (b). \qquad (5.5.11)$$

The wavevector $\vec{k}_2 = \vec{k}_2^{\,DNG}$, satisfying the phase-matching at the interface, of the above expressions, is obtained from Fig (5.9b). The Poynting vector is written from the wavevector diagram shown in Fig (5.9a). It is also obtained by using equation (5.2.4) to compute the Poynting vector in the DNG medium for the angle of refraction $(-\theta_2)$. The above expressions further show that the DNG medium #2 supports the *backward-wave* propagation because the vectors $\vec{k}_2^{\,DNG}$ and $\vec{S}_{av}^{\,tra(DNG)}$ are anti-parallel.

5.5.3 Basic Transmission Line Model of the DNG Medium

The unbounded DPS medium and a transmission line both support the forward wave propagation in the TEM mode, so Fig (3.28a) of chapter 3 models a DPS medium by the LC transmission line. The equivalence between the material parameters ε, μ and the circuit parameters C, L is discussed in subsection (3.4.2) of chapter 3. The characteristics impedance and propagation constant of the DPS medium and equivalent transmission lines are summarized below:

$$Z_o^{DPS} = \sqrt{\frac{\mu}{\varepsilon}}, \quad Z_o^{Line} = \sqrt{\frac{L}{C}} \quad (a)$$

$$\beta^{DPS} = \omega\sqrt{\mu\varepsilon}, \quad \beta^{Line} = \omega\sqrt{LC} \quad (b). \qquad (5.5.12)$$

The equivalent LC transmission line is an analog of the DPS medium. Therefore, the medium parameters μ and ε could be treated through the analogous series inductance L, and shunt capacitance C, i.e. $L \rightarrow \mu$, $C \rightarrow \varepsilon$. The expressions $L = Z(\omega)/j\omega$ and $C = Y(\omega)/j\omega$ for series inductance and shunt capacitance, respectively, model the DPS medium parameters as follows:

$$\mu = Z(\omega)/j\omega \quad (a), \qquad \varepsilon = Y(\omega)/j\omega \quad (b). \qquad (5.5.13)$$

In equation (5.5.13), $Z(\omega)$ and $Y(\omega)$ are the series impedance and shunt admittance of the equivalent line.

Following the above discussion, a DNG medium, supporting the backward-wave propagation, could be modeled through the CL-line, shown in Fig (3.28b) of chapter 3. Using the above expressions, the material parameters of a DNG medium could be modeled as follows:

$$|\mu| = \frac{-j/\omega C}{j\omega} = -\frac{1}{\omega^2 C} \quad (a), \qquad |\varepsilon| = \frac{-j/\omega L}{j\omega} = -\frac{1}{\omega^2 L} \quad (b). \qquad (5.5.14)$$

The above expressions show that the CL-line model provides negative values for the permeability and permittivity, as needed for a DNG medium. The characteristic impedance and the propagation constant of the EM-wave in a DNG medium follows from the above equations:

$$Z_o^{DNG} = \sqrt{\frac{|\mu|}{|\varepsilon|}} = \sqrt{\frac{1}{\omega^2 C} \times \omega^2 L} = \sqrt{\frac{L}{C}} \quad (a)$$

$$\beta^{DNG} = \pm\omega\sqrt{|\mu||\varepsilon|} = \mp\sqrt{\frac{1}{\omega^2 C} \times \frac{1}{\omega^2 L}} = \mp\frac{1}{\omega^2\sqrt{LC}} \quad (b). \qquad (5.5.15)$$

Expression (5.5.15b) shows that even *a lossless DNG medium is highly dispersive due to the presence of a term ω^2 in the denominator. It is unlike a DPS medium which is nondispersive for the lossless case.* However, a dispersive medium is always associated with loss to meet the

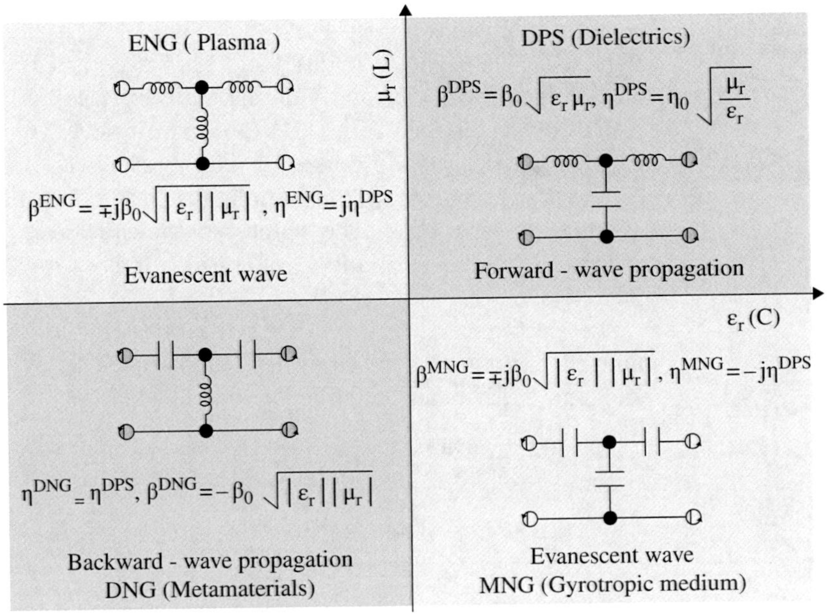

Figure 5.10 Circuit models of four kinds of the medium on the (μ, ε)-plane.

Kramer–Kronig condition of causality. It is discussed in subsection (6.5.4) of chapter 6. The causality fails for a fictitious nondispersive DNG medium. The discussion shows that the DNG medium parameters and wavenumber, i.e. ε_r, μ_r, n, k, are complex quantities [J.8].

The DPS transmission line LC-model can also be extended to the ENG and MNG media. Both these media do not support any EM-wave propagation. Only the decaying evanescent mode is supported by them, as these media are reactive. The ENG medium is obtained for $-\varepsilon_r$ and $+\mu_r$ that correspond to negative shunt capacitance and positive series inductance. Likewise, the MNG medium is obtained for $-\mu_r$ and $+\varepsilon_r$, i.e. for negative series inductance and positive shunt capacitance. However, the negative capacitance and negative inductance in reality correspond to equivalent positive inductance and equivalent positive capacitance, respectively. It is shown below from their reactance:

For negative capacitance :

$$jX_c = \frac{-j}{\omega(-C)} = j\omega L_{eq}, \quad \Rightarrow L_{eq} = \frac{1}{\omega^2 C} \quad \text{(a)}$$

For negative inductance :

$$jX_L = j\omega(-L) = -\frac{j}{\omega C_{eq}}, \quad \Rightarrow C_{eq} = \frac{1}{\omega^2 L} \quad \text{(b).}$$

$$(5.5.16)$$

Figure (5.10) shows the *equivalent T-network* unit cell for all four kinds of media in the (ε_r, μ_r)-plane, i.e. in the corresponding (C, L)-plane. The ε_r(C)-axis and μ_r(L)-axis show the capacitive shunt element and the inductive series element of the equivalent T-network of material media. The DPS medium in the first quadrant is changed to the ENG medium by taking a negative value for shunt capacitance that corresponds to a shunt inductor. The L-L network in the second quadrant corresponds to the ENG medium. The MNG medium is obtained by taking a negative value for the series inductance of the DPS medium. It results in the C-C network in the fourth quadrant. The characteristics impedance of the ENG and MNG equivalent lines are inductive and capacitive respectively. Of course, by taking negative values for both the series inductance and shunt capacitance of the DPS medium, the DNG medium is created. It is shown as a CL unit cell in the third quadrant of Fig (5.10). Individually, the ENG and MNG media do not support EM-wave propagation. However, jointly they form a transparent medium, and EM-wave propagates through the joint medium. It is known as the *tunneling phenomenon* [J.10, J.11].

The above description of circuit modeling is also applicable to a waveguide below the cut-off frequency. The TE-mode waveguide below cut-off frequency provides the inductive load. So it could be viewed as an ENG medium, while the TM-mode waveguide below the cut-off frequency provides the capacitive load and it could be viewed as the MNG medium [B.6, B.8]. The behavior of the modal wave impedance of a rectangular waveguide below the cut-off frequency is discussed in subsection (7.4.1) of chapter 7.

5.5.4 Lossy DPS and DNG Media

A lossy DPS medium is characterized by the complex permittivity and complex permeability $\varepsilon_r^{*DPS} = \varepsilon_r' - j\varepsilon_r''$ and $\mu_r^{*DPS} = \mu_r' - j\mu_r''$, also $\varepsilon_r'' \ll \varepsilon_r'$ and $\mu_r'' \ll \mu_r'$. The complex wavenumber for the EM-wave in a lossy DPS medium is obtained as follows:

$$k^{*DPS} = k_0\sqrt{\varepsilon_r^{*DPS}.\mu_r^{*DPS}} \quad (a)$$

$$\text{where } \sqrt{\varepsilon_r^{*DPS}} \approx \sqrt{\varepsilon_r'}\left(1 - j\frac{\varepsilon_r''}{2\varepsilon_r'}\right)$$

$$= \left(\sqrt{\varepsilon_r'} - j\frac{\varepsilon_r''}{2\sqrt{\varepsilon_r'}}\right) \text{ and, } \sqrt{\mu_r^{*DPS}} \approx \left(\sqrt{\mu_r'} - j\frac{\mu_r''}{2\sqrt{\mu_r'}}\right) \quad (b). \quad (5.5.17)$$

$$\Rightarrow k^{*DPS} \approx k_0\left[\left(\sqrt{\varepsilon_r'\mu_r'} - \frac{\varepsilon_r''\mu_r''}{4\sqrt{\varepsilon_r'\mu_r'}}\right) - j\frac{1}{2}\left(\frac{\varepsilon_r''\mu_r' + \mu_r''\varepsilon_r'}{\sqrt{\varepsilon_r'\mu_r'}}\right)\right]$$
$$\approx k_0\left[\sqrt{\varepsilon_r'\mu_r'} - j\frac{1}{2}\left(\frac{\varepsilon_r''\mu_r' + \mu_r''\varepsilon_r'}{\sqrt{\varepsilon_r'\mu_r'}}\right)\right]. \quad (5.5.18)$$

On separating the real and imaginary parts of a complex wavenumber in the DPS medium, $k^{*DPS} = k' - jk''$, the following expressions are obtained:

$$k' = \text{Re}\left(k^{*DPS}\right) = k_0\sqrt{\varepsilon_r'\mu_r'} \quad (a)$$
$$k'' = \text{Im}\left(k^{*DPS}\right) = \frac{k_0}{2}\left(\frac{\varepsilon_r''\mu_r' + \mu_r''\varepsilon_r'}{\sqrt{\varepsilon_r'\mu_r'}}\right) \quad (b). \quad (5.5.19)$$

The DPS medium has $|\varepsilon_r'| > 0$, $|\varepsilon_r''| > 0$, $|\mu_r'| > 0$, $|\mu_r''| > 0$. So using the above equations, it is obvious that $k' > 0$, $k'' > 0$.

The above expressions are also valid for the DNG medium with some modifications. In the case of a DNG medium, the medium parameters are $|\varepsilon_r'| < 0$, $|\mu_r'| < 0$. However, *even in the DNG medium, the electric and magnetic losses are positive quantities*, i.e. $|\varepsilon_r''| > 0$, $|\mu_r''| > 0$. Using equation (5.5.19), the complex wavenumber in the DNG is given as follows:

$$k^{*DNG} = -k' - jk'' = -(k' + jk''). \quad (5.5.20)$$

The complex DPS medium is also described by the complex refractive index:

$$n^{*DPS} = k^{*DPS}\frac{c}{\omega} = \frac{k^{*DPS}}{k_0} \approx \sqrt{\varepsilon_r'\mu_r'} - \frac{j}{2}\left(\frac{\varepsilon_r''\mu_r' + \mu_r''\varepsilon_r'}{\sqrt{\varepsilon_r'\mu_r'}}\right) \quad (a)$$
$$\Rightarrow n^{*DPS} = n' - jn'', \text{ where}$$
$$n' = \sqrt{\varepsilon_r'\mu_r'},\ n'' = \frac{1}{2}\left(\frac{\varepsilon_r''\mu_r' + \mu_r''\varepsilon_r'}{\sqrt{\varepsilon_r'\mu_r'}}\right) \quad (b). \quad (5.5.21)$$

However, in the case of a complex DNG medium, the complex refractive index is $n^{*DNG} = -(n' + jn'')$; as $\text{Re}(n^{*DNG})$ is a negative quantity and $\text{Im}(n^{*DNG})$ is still a positive quantity. The above discussion is for the propagating waves in the DPS and DNG media. However, in case the waves are nonpropagating (*evanescent*) in both media, the real part of the wavenumber is an imaginary quantity i.e. $k' = -j\alpha'$. The evanescent wave behaves differently in the DPS and DNG media. It is examined below.

The electric fields of the x-directed propagating and nonpropagating EM-waves in the unbounded lossy DPS and DNG media could be expressed as follows:

Propagating wave :

$$E^{DPS} = E_0 e^{+j[\omega t - k^{*DPS}x]} = E_0 e^{+j[\omega t - (k' - jk'')x]},$$
$$\Rightarrow E^{DPS} = E_0 e^{-k''x} e^{j(\omega t - k'x)} \quad (a)$$
$$E^{DNG} = E_0 e^{+j[\omega t - k^{*DNG}x]} = E_0 e^{+j[\omega t + (k' + jk'')x]},$$
$$\Rightarrow E^{DNG} = E_0 e^{-k''x} e^{j(\omega t + k'x)} \quad (b)$$

Evanescent wave($k' = -j\alpha'$) :

$$E^{DPS} = E_0 e^{-(k'' + \alpha')x} e^{j\omega t} \quad (c)$$
$$E^{DNG} = E_0 e^{(\alpha' - k'')x} e^{j\omega t}, \ (\alpha' > k'') \quad (d). \quad (5.5.22)$$

The propagating EM-wave is attenuated while traveling in both the DPS and DNG lossy media $k'' \neq 0$. However, the DPS medium offers a *lagging phase*, whereas the DNG medium offers a *leading phase* to the wave traveling in the positive x-direction. Poynting vector decides the direction of the EM-wave propagation. The $\lambda_g/2$-line resonator could be designed in the DPS-DNG composite, with a length $\lambda_g/4$ in each medium, without any phase-shift at the output. The classical $\lambda_g/2$-line resonator, in a DPS medium, has 180° phase at the output. It is further seen from the above equation that the evanescent wave is decaying with distance x while traveling in the DPS medium. The enhancement of amplitude by the DNG medium could be viewed as the step-up transformer action, whereas it is increasing in amplitude while traveling in the DNG medium. This property is more clearly seen in a lossless medium with $k'' = 0$.

5.5.5 Wave Propagation in DNG Slab

The EM-wave propagating through a normal DPS slab provides the *lagging phase* of the propagating part of the field at the output end of the slab. It is noted that the nonpropagating *decaying evanescent wave* also exists, along with the propagating wavefield components, inside the slab. However, for a propagating wave, the DNG slab provides the *leading phase* at its output. The *increasing evanescent wave* exists inside a DNG slab, along with the propagating wavefield components.

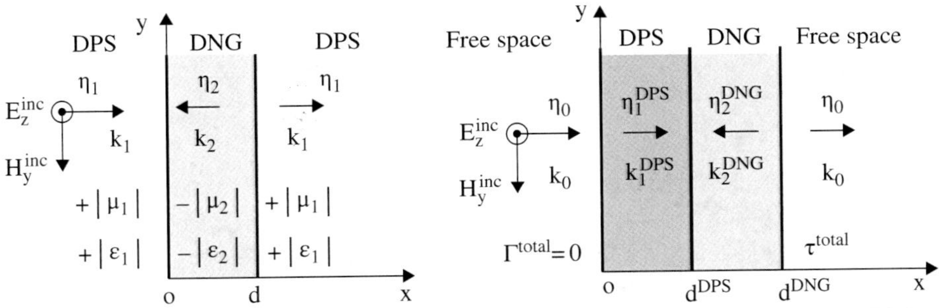

(a) DNG slab in the host DPS medium. (b) Composite DPS - DNG slab in the host DPS medium.

Figure 5.11 EM-wave propagation through the DNG and composite DPS-DNG slabs.

Thus, the DNG slab can act both as *(i) a phase-compensator and (ii) as an amplitude-compensator*, i.e. as a *field amplifier*, without any power amplification. The field amplifier is like a voltage step-up transformer without any power amplification. The first property, in the form of the DGS-DNG composite slab, is helpful in the design of compact *sub-wavelength resonator* [J.12, B.6]. A normal DPS resonator is half-wavelength long. The second property offers an opportunity to design *sub-wavelength resolution superlens* [J.2, J.13–J.21]. This subsection explains both properties of a DNG slab. The *superlens* is discussed briefly in the next subsection.

Phase - Compensation in the DPS-DNG Slab

Figure (5.11a) shows a DNG slab, $\mu_2 = -|\mu_2|$, $\varepsilon_2 = -|\varepsilon_2|$, η_2, of thickness d embedded in the DPS host medium, $\mu_1 = +|\mu_1|$, $\varepsilon_1 = +|\varepsilon_1|$, η_1. The TE-polarized wave is normally incident at the first interface. The DNG slab supports the backward wave with the wavevector k_2 in the opposite direction, as power flows from the left to right. This has important consequences.

Let us assume that in Fig (5.11a), the slab is a DPS type, and it is impedance matched with the host DPS medium, i.e. $\eta_1 = \eta_2$. The reflection and transmission coefficients are obtained from equation (5.4.15):

$$\Gamma_{DPS}^{Nor} = 0 \quad (a), \quad \tau_{DPS}^{Nor} = e^{-jk_2^{DPS}d} = e^{-j|k_2|d} \quad (b). \tag{5.5.23}$$

However, if the DPS slab is replaced by a matched DNG slab, then the following expressions are obtained, on using $k_2^{DNG} = -k_2^{DPS} = -|k_2|$ from equation (5.5.6b):

$$\Gamma_{DNG}^{Nor} = 0 \quad (a), \quad \tau_{DNG}^{Nor} = e^{-jk_2^{DPS}d} = e^{+j|k_2|d} \quad (b). \tag{5.5.24}$$

A complete transmission of waves occurs through the DPS/DNG slab. However, the DPS slab provides the *lagging phase* ($\phi = -|k_2|d$) at the output of the slab, whereas the DNG slab provides the *leading phase* ($\phi = +|k_2|d$). This property of the DNG slab is useful in compensating the lagging phase of a DPS slab in a DPS-DNG composite slab.

Figure (5.11b) of the composite DPS-DNG slab illustrates the application of a DNG slab as a *phase compensator*. Again, the impedance matching of both slabs with the host medium is assumed, i.e. $\eta_1 = \eta_2 = \eta_0$, such that the total reflection coefficient is zero, $\Gamma_{total}^{Nor} = 0$. The total transmission coefficient at the out of the DNG slab is given as follows:

$$\tau_{toal}^{Nor} = \left(e^{-jk_1^{DPS}d_1^{DPS}}\right)\left(e^{+jk_2^{DNG}d_2^{DNG}}\right) = e^{-j\left[k_1^{DPS}d_1^{DPS} - k_2^{DNG}d_2^{DNG}\right]}. \tag{5.5.25}$$

To get the compensated phase, i.e. the zero phase, at the output of the DPS-DNG slab, the following condition, obtained from the above equation, must be met:

$$\begin{aligned} & k_1^{DPS}d_1^{DPS} - k_2^{DNG}d_2^{DNG} = 0 \\ \Rightarrow\ & k_0\sqrt{\mu_{r1}\varepsilon_{r1}}\, d_1^{DPS} = k_0\sqrt{|\mu_{r2}||\varepsilon_{r2}|}\, d_2^{DNG} \\ \Rightarrow\ & \frac{d_1^{DPS}}{d_2^{DNG}} = \frac{n_2^{DNG}}{n_1^{DPS}}, \end{aligned} \tag{5.5.26}$$

where n_1^{DPS} and n_2^{DNG} are refractive indices of the DPS and DNG slabs. The thicknesses of the slabs need not be equal. The dimension of thicknesses could be even in the sub-wavelength. It is useful in designing of very compact *sub-wavelength cavity resonators* and *parallel plate waveguides* [J.12]. Such resonance can be developed even in the compact composite ENG-MNG slab [B.6]. The present concept also finds application in designing properly matched *electrically small dipole antenna* [B.6].

Amplitude-Compensation in the DNG Slab

The resolving power of an optical lens is restricted by the wavelength of a source. This is known as the *diffraction limit* of the lens. Fourier optics, i.e. the wave optics adequately explain it in terms of the propagating waves and decaying evanescent waves generated by the object

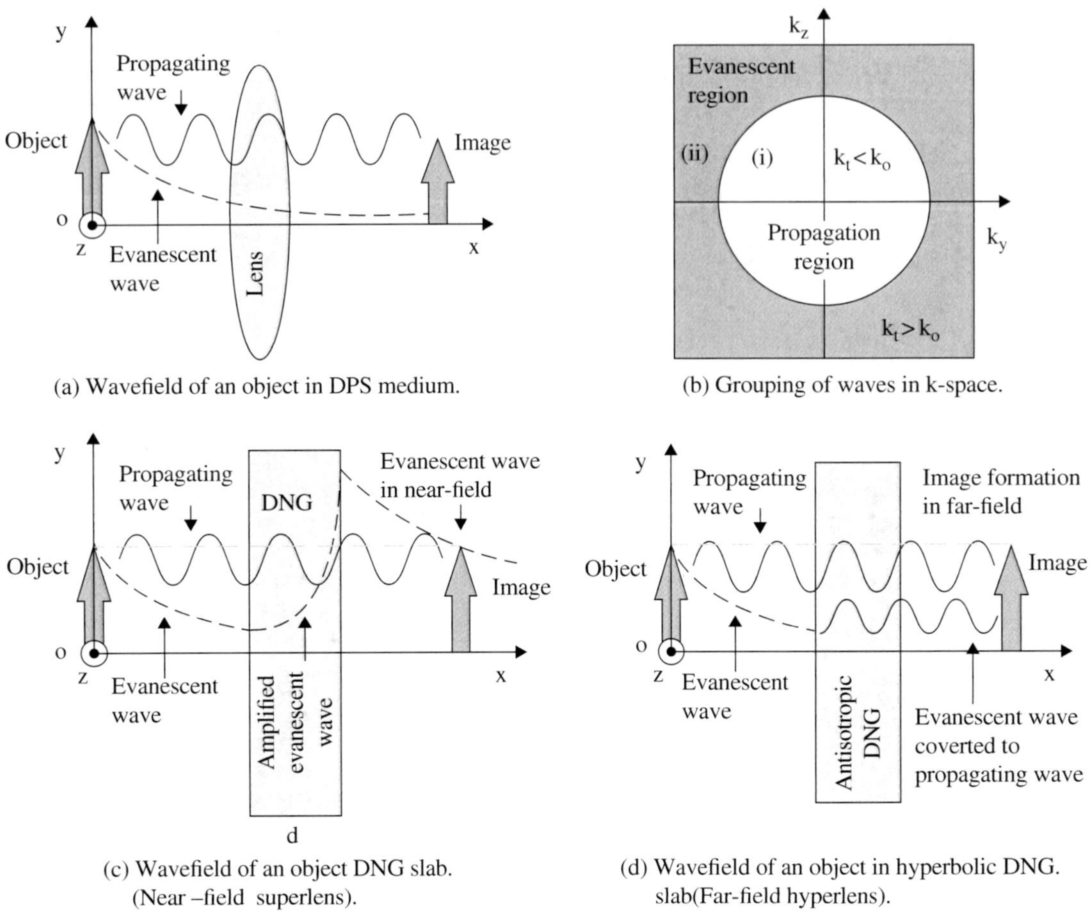

(a) Wavefield of an object in DPS medium.
(b) Grouping of waves in k-space.
(c) Wavefield of an object DNG slab. (Near –field superlens).
(d) Wavefield of an object in hyperbolic DNG. slab(Far-field hyperlens).

Figure 5.12 Creation of image using wave optics.

source shown in Fig (5.12a). The object source located at $x = 0$ generates an arbitrary wave field that can be decomposed in terms of the plane wave spectrum. Their superposition reconstructs the wave field. Thus, the wavefield in the real DPS space, propagating in the x-direction, is composed of 2D Fourier plane wave components in the Fourier domain, i.e. in the k-space (*propagation vector space*). The wave field $\vec{E}^{DPS}(x, y, z, t)$ is expressed through the following 2D Fourier integral:

$$\vec{E}^{DPS}(y, z, x, t) = \frac{1}{4\pi^2}\int_{-\infty}^{\infty}\int_{-\infty}^{\infty}\tilde{E}(k_y, k_z)e^{j\left(\omega t - \vec{k}\cdot\vec{r}\right)}dk_y dk_z$$

$$\Rightarrow \vec{E}^{DPS}(y, z, x, t) = \frac{1}{4\pi^2}\int_{-\infty}^{\infty}\int_{-\infty}^{\infty} \times$$

$$\left[\tilde{E}(k_y, k_z)e^{-j(k_y y + k_z z)}\right]e^{j(\omega t - k_x x)}dk_y dk_z \qquad (a)$$

$$\text{where } k_x^{DPS} = \sqrt{k_0^2 - \left(k_y^2 + k_z^2\right)}, \quad k_o = \omega/c \qquad (b).$$

$$(5.5.27)$$

The wave propagates in the x-direction and the spectral field components are in the $(k_y - k_z)$-plane. The spectral field amplitudes $\tilde{E}(k_y, k_z)$ with phase distribution within the bracket [ ] are Fourier components of the electric field $\vec{E}^{DPS}(y, z, x, t)$. The propagation constant k_x^{DPS} of the plane wave, given by equation (5.5.27b), is obtained from the dispersion relation (4.5.29d) of chapter 4. The propagation constant k_x^{DPS} of propagating plane waves must be a real quantity, whereas imaginary values of k_x^{DPS} provide decaying evanescent waves in the x-direction. Thus, Fourier components of the wave originated by an object could be divided into two groups: *propagating waves* and *nonpropagating evanescent waves*. The grouping of waves is shown in Fig (5.12b). In a DPS medium, the propagation constants for both groups of waves, from equation (5.5.27b), could be written as follows:

Propagating waves :

$$k_x^{DPS} = \sqrt{k_0^2 - \left(k_y^2 + k_z^2\right)} = \beta_x, \quad k_0^2 > \left(k_t^2 = k_y^2 + k_z^2\right) \qquad (a)$$

Evanescent waves :

$$k_x^{DPS} = -j\sqrt{\left(k_y^2 + k_z^2\right) - k_0^2} = -j\alpha_x, \quad k_0^2 < \left(k_t^2 = k_y^2 + k_z^2\right) \qquad (b).$$

$$(5.5.28)$$

The electric fields of both the propagating and evanescent waves in the DPS medium could be written from the above relations as follows:

$$\text{Propagating waves}: \quad \vec{E}^{DPS}(y,z,x,t) = \frac{1}{4\pi^2}\int_{-\infty}^{\infty}\int_{-\infty}^{\infty}\left[\tilde{E}(k_y,k_z)e^{-j(k_y y + k_z z)}\right]e^{j(\omega t - \beta_x x)}dk_y dk_z \quad (a)$$

$$\text{Evanescent waves}: \quad \vec{E}^{DPS}(y,z,x,t) = \frac{1}{4\pi^2}\int_{-\infty}^{\infty}\int_{-\infty}^{\infty}\left[\tilde{E}(k_y,k_z)e^{j[\omega t - (k_y y + k_z z)]}\right]e^{-\alpha_x x}dk_y dk_z \quad (b). \quad (5.5.29)$$

Figure (5.12a) shows both the propagating and exponentially decaying evanescent waves in the DPS medium, generated by an object. The evanescent waves decay fast within a distance under λ. The higher value of transverse components of the wavevector (k_y, k_z) corresponds to the finer spatial details of the object. So at the image-plane, finer spatial details of an image are lost. The DPS based lens cannot recover the lost spatial details in the evanescent waves. The maximum value of the transverse wavevector $\vec{k}_{t,max} = (\hat{y}k_y + \hat{z}k_z)$ at the cut-off wavenumber $k_x = 0$ determines the limit of the finer spatial details, i.e. the diffraction limit:

$$k_x^{DPS} = \sqrt{k_0^2 - \left(k_y^2 + k_z^2\right)} = \sqrt{k_0^2 - k_{t,max}^2} = 0$$
$$\Rightarrow k_{t,max} = \omega/c \quad (a)$$
$$\text{Maximum resolution}(\Delta) = \frac{2\pi}{k_{t,max}} = \lambda \quad (b). \quad (5.5.30)$$

The maximum resolution Δ is the minimum adjacent distance of the finer spatial details of the object. It called the *diffraction limit*. The relation $\Delta \times k_{t,max} = 2\pi$ is obtained in subsection (19.1.1) of chapter 19 from the relation between the *direct space period* (Δ) and the propagation vector in the k-space. The diffraction limit for a very wide lens aperture could be reduced to λ/2 [J.13].

The DNG planar slab acts as a flat lens [J.2, J.13–J.15, J.22]. The DNG slab provides exponentially increasing evanescent waves inside the slab. So the DNG can compensate for the decaying evanescent waves that reach the image plane creating a high-resolution image with finer details. It is shown in Fig (5.12c). It is also examined in the next subsection (5.5.6).

5.5.6 DNG Flat Lens and Superlens

The focal length (f) of a thin lens is related to its radius of curvature (R) by the expression $f = R/(n - 1)$, where n is the refractive index of a lens. For $n = +1$, the lens does not refract, and the focusing of the EM-wave at the image plane does not occur. However, for a DNG lens with $n = -1$, refraction occurs. Veselago [J.3] has shown that due to negative refraction in a DNG slab, even a slab acts as a *flat lens*. Pendry has further shown that such a lens is a *perfect lens* in the *near-field region* as it enables recovery of the decaying evanescent field at the image plane [J.2]. Such recovery of the evanescent waves is not possible with a normal lens, irrespective of the size of its aperture. The perfect lens, also called the *superlens*, has a subwavelength resolution, breaking the diffraction limit barrier. The *superlens* creates an image in the near-field very close to the lens, so it is difficult to use it in practice. However, anisotropic hyperbolic DNG lens creates a perfect image in the *far-field region*, by converting the evanescent waves into the propagating waves. Such a lens is called the *hyperlens*. The proof of concept has been demonstrated experimentally for both the superlens and hyperlens in optical and microwave range. The proper functioning of these lenses is limited by the losses associated with a DNG medium. A brief theory of three lenses is presented in this section.

Veselago Flat Lens

Figure (5.13) shows the ray diagram of a DNG slab acting as a lens due to the negative refraction [J.3]. It has two foci N and Q. The *focal length* f_1 of focus N is located inside the DNG slab of thickness d. The *focal length* f_2 of focus Q is located outside the slab. To avoid mismatch at the interface, the DNG slab has $\mu_r = \varepsilon_r = -1$ and $\eta_1^{DNG} = \eta_0$. Figure (5.13) shows the angles of incidence θ_i and refraction θ_r and other geometrical dimensions. Snell's law for the DNG slab is written below:

$$\sin\theta_r = \frac{1}{n}\sin\theta_i. \quad (5.5.31)$$

The first focal length f_1 is determined as follows:

$$\tan\theta_i = \frac{H_1}{d_1}, \quad \Rightarrow H_1 = d_1\tan\theta_i \quad (a)$$
$$\tan\theta_r = \frac{H_1}{f_1}, \quad \Rightarrow H_1 = f_1\tan\theta_r \quad (b)$$
$$\text{i.e.} \quad f_1 = \frac{d_1\tan\theta_i}{\tan\theta_r} \quad (c). \quad (5.5.32)$$

The incident ray OM and the output ray P_1Q are parallel, i.e. $\theta_i = \theta_o$.

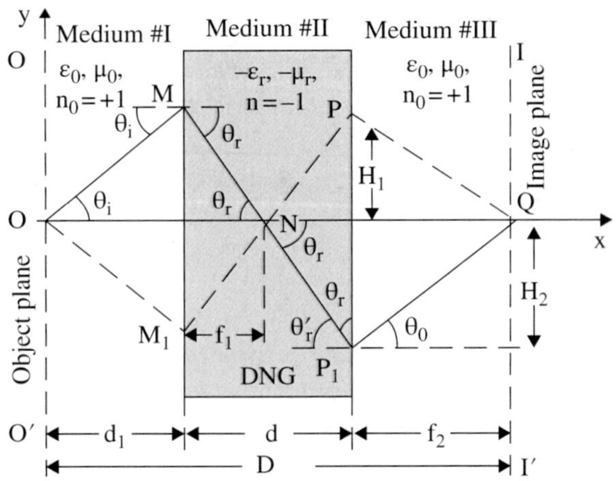

Figure 5.13 Ray diagram of a DNG flat lens.

The second focal length f_2 is computed as follows:

$$\tan\theta_r' = \tan\theta_r = \frac{H_2}{d-f_1}, \quad \Rightarrow H_2 = (d-f_1)\tan\theta_r \quad \text{(a)}$$

$$\tan\theta_0 = \tan\theta_i = \frac{H_2}{f_2}, \quad \Rightarrow H_2 = f_2\tan\theta_i \quad \text{(b)}$$

$$\text{i.e. } f_2 = \frac{(d-f_1)\tan\theta_r}{\tan\theta_i} \quad \text{(c)}. \tag{5.5.33}$$

For $d > f_1$, equations (5.5.32c) and (5.5.33c) provide a relation between two focal lengths:

$$f_2 = \frac{(d-f_1)d_1\tan\theta_i}{f_1\tan\theta_i}, \quad \Rightarrow \frac{d}{f_1} - \frac{f_2}{d_1} = 1. \tag{5.5.34}$$

Equation (5.5.31) shows a matched DNG slab has $\theta_i = \theta_r$. For this case, both focal lengths, and also distance D between the object and image at the second focus are given below:

$$f_1 = d_1 \quad \text{(a)}, \quad f_2 = d - d_1 \quad \text{(b)}, \quad D = 2d \quad \text{(c)}. \tag{5.5.35}$$

Pendry Superlens Lens

In the case of a DNG-based matched flat superlens, the image is reconstructed at the second focus by amplification of the evanescent electric fields. Thus, the DNG slab provides an opportunity to design subdiffraction lens, i.e. practically a lens without any diffraction limit. The refractive index of the DNG, matched to air medium, is $n_2^{DNG} = -1$. The longitudinal wavevector k_x, given by equation (5.5.28), is rewritten for the backward wave supporting DNG slab as follows [J.2, J.13, J.22]:

Propagating waves :

$$k_x^{DNG} = -\sqrt{n_2^2k_0^2 - \left(k_y^2 + k_z^2\right)}$$
$$= -\beta_x, \quad \left(n_2^2k_0^2\right) > \left(k_y^2 + k_z^2\right) \quad \text{(a)}$$

Evanescent waves :

$$k_x^{DNG} = -\left(-j\sqrt{\left(k_y^2 + k_z^2\right) - n_2^2k_0^2}\right)$$
$$= j\alpha_x, \quad k_0^2 < \left(k_y^2 + k_z^2\right) \quad \text{(b)}. \tag{5.5.36}$$

The electric fields of both the propagating and evanescent waves, inside the DNG slab, are obtained using the above equations with equation (5.5.29a,b):

Propagating waves :

$$\vec{E}^{DNG}(y,z,x) = \frac{1}{4\pi^2}\int_{-\infty}^{\infty}\int_{-\infty}^{\infty}$$
$$\left(\tilde{E}(k_y,k_z)e^{-j(k_y y + k_z z)}\right)e^{j(\omega t + \beta_x x)}dk_y dk_z \quad \text{(a)}$$

Evanescent waves :

$$\vec{E}^{DNG}(y,z,x) = \frac{1}{4\pi^2}\int_{-\infty}^{\infty}\int_{-\infty}^{\infty}$$
$$\left(\tilde{E}(k_y,k_z)e^{j[\omega t - (k_y y + k_z z)]}\right)e^{\alpha_x x}dk_y dk_z \quad \text{(b)}. \tag{5.5.37}$$

Equation (5.5.37a) shows that the propagating waves acquire the leading phase ($\beta_x d$) at the end of the DNG slab of thickness $x = d$. The evanescent waves are amplified exponentially by $\alpha_x d$ at the end of the slab. Figure (5.12c) shows such behavior of the DNG based superlens. At the output of the DNG slab, the wave is still evanescent waves in the *near-field region*; although at the image plane its magnitude level is the same as that of at the object plane. So at the second focus, the image has high resolution beyond the diffraction limit. It is noted that the evanescent waves still propagate at the interface along the y-axis in the $(y-z)$-plane as the *surface waves*. The surface wave is bounded in the x-direction. The amplification of the evanescent wave is also treated through the concept of coupled quasi-particles called *plasmon-polariton*, especially in the optical frequency range [J.13, B.11].

HyperLens

The anisotropic DNG medium-based hyperlens converts the evanescent waves to the propagating waves by reducing the values of the transverse wavenumber below the wavenumber in the medium. It is realized by the hyperbolic dispersion relation of the uniaxial anisotropic medium discussed in section (4.7.5) of chapter 4. Thus, a hyperlens projects the high-resolution image in the far-field region. It is shown in Fig (5.12d). The hyperlens has been realized in the cylindrical geometry using very thin alternate curved layers of conducting and dielectric films. The object is placed at the center and the propagating waves are available at the flat end of the cylindrical medium in which the hyperlens is embedded. The normal optical lens, attached with the hyperlens, carries out the optical processing to create the high-resolution image in the far-field region [J.17, J.18].

5.5.7 Doppler and Cerenkov Radiation in DNG Medium

The DNG medium acts inversely on Doppler and Cerenkov radiations. It is examined below.

Doppler Effect

Doppler effect is related to the change in frequency of a source due to the relative motion of the source and receiver. If a source is *moving away, i.e. receding,* from the receiver in a DPS medium, the received frequency is less than the stationary frequency of the source. However, if the source is *moving toward* the receiver, the received frequency is increased.

Let us examine the receding case of a moving source. Figure (5.14a) shows that a source #**S** is moving away from receiver #R with velocity $\mathbf{V_s}$ along the positive x-axis. The source occupies positions P_1 ($x = d_1$) and P_2($x = d_2$) at time t_1 and t_2 ($t_2 > t_1$), respectively. The source S radiates frequency f_s and the radiated spherical wave propagates with the phase velocity V_p in the *DPS medium*. Figure (5.14a) shows the spherical wavefronts at time $t_1,\ldots,t_4$. The phase of the received wave at time t_1 and t_2 could be written as

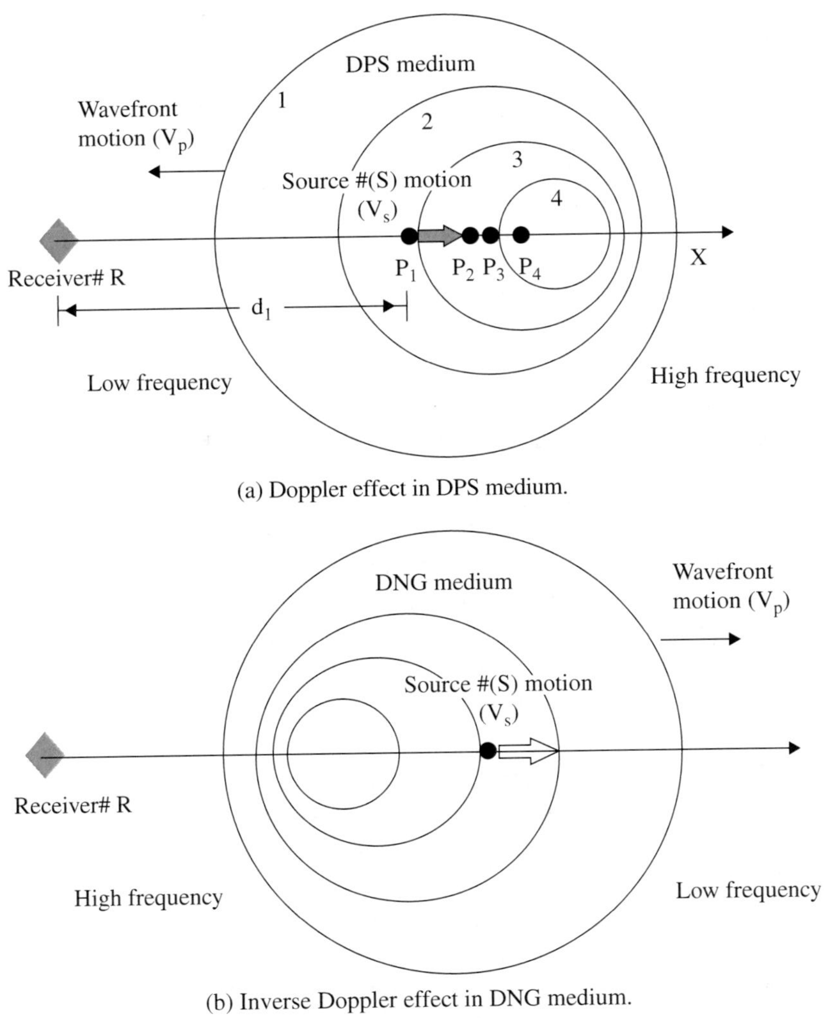

Figure 5.14 Doppler effect in the DPS and DNG media. The source receding the receiver.

$$\begin{aligned}
&\text{Phase at time } t_1: \phi_1 = 2\pi f_s t_1 - \beta d_1 && \text{(a)}\\
&\text{Phase at time } t_2: \phi_2 = 2\pi f_s t_2 - [\beta d_1 + \beta V_s(t_2 - t_1)] && \text{(b)}\\
&\text{Change in phase during the interval } \Delta t = (t_2 - t_1):\\
&\Delta\phi = (2\pi f_s - \beta V_s)\Delta t && \text{(c)}.
\end{aligned} \tag{5.5.38}$$

The received frequency f_R of the wave at the stationary receiver is the rate of change in phase, i.e. $\omega_R = 2\pi f_R = Lt_{\Delta t \to 0}(\Delta\phi/\Delta t)$. The received frequency, from the above expression (5.5.38c) is obtained as

$$\begin{aligned}
&Lt_{\Delta t \to 0}\left\{\frac{\Delta\phi}{\Delta t}\right\} = 2\pi f_R = \left[2\pi f_s - \frac{2\pi f_s}{V_p}V_s\right]\\
&\Rightarrow f_R^{DPS} = f_s\left[1 - \frac{V_s}{V_p}\right], \quad V_s < V_p.
\end{aligned} \tag{5.5.39}$$

Inverse Doppler Effect

The DNG medium, shown in Fig (5.14b), supports the backward wave propagation with opposite phase velocity. In this case, the spherical wavefronts move in the backward direction. Equation (5.5.39) is used to get the frequency of the received waves from a source moving away from the receiver in the DNG medium:

Source moving away from receiver :

$$f_R^{DNG} = f_s\left[1 + \frac{V_s}{V_p}\right], \quad V_s < V_p \qquad \text{(a)}$$

Source moving toward receiver :

$$f_R^{DNG} = f_s\left[1 - \frac{V_s}{V_p}\right], \quad V_s < V_p \qquad \text{(b)}. \tag{5.5.40}$$

The received frequency of the radiated waves from a source, moving towards the receiver, is reduced at the stationary receiver. In the case, the source is moving away from the receiver, the received frequency increases as $f_R^{DPS} = f_s(1 + V_s/V_p)$.

Due to the reversal of the change in the received frequency, the DNG medium supports the *inverse Doppler Effect* [J.6]. It has been experimentally confirmed both in the microwave and optical frequency ranges. The inverse Doppler Effect could be used to design *tunable* and *multifrequency* radiation sources [J.23–J.25].

Cerenkov Radiation

The Cerenkov radiation, also called the Cherenkov (or Cherenkov) radiation, in the DPS medium is generated from the charged particles traveling with a velocity faster than the velocity of the EM-wave in that medium [B.12]. The radiation from a *leaky-wave antenna* closely follows the radiation mechanism of Cerenkov radiation. Likewise, the planar transmission lines radiate the Cerenkov type radiation within a substrate resulting in high substrate loss [B.13]. It is discussed in subsection (9.7.3) of chapter 9. In a DPS medium the directions of the radiated power, i.e. the direction of the Poynting vector, and the wavevector are in the same direction. However, in a DNG medium, these directions are opposite to each other, giving *inverse Cerenkov radiation* [J.3]. Both Cerenkov radiation and inverse Cerenkov radiation are similar to the shock waves of supersonics generating the radiation cone.

Figure (5.15a) shows the generation of the radiation cone of the spherical wavefronts by a moving charge, with velocity V_c, in a DPS medium. The charged particle velocity V_c is greater than the phase velocity V_p of the EM-wave in a DPS medium, i.e. $V_c > V_p(= c/n)$. The positive refractive index is n, and c is the velocity of EM-wave in free space. The locations P_0–P_4, on the positive x-axis, are successive positions of the moving charge at time t_0–t_4. During the time interval $T = (t_4 - t_0)$, the spherical wavefront acquires radius $P_0N = V_pT = (c/n)T$, while the charged particle travels a distance $P_0P_4 = V_cT$. Figure (5.15a) also shows other spherical wavefronts for the time sequence for $t < T$. Following the Huygens principle, the inclined tangential wavefront of the secondary radiation is NP_4. The wavevector (k), i.e. P_0N, is normal to the wavefront NP_4. It forms a *Cerenkov angle* ϕ_c at P_0 or angle θ_c at P_4. It is obtained from the triangle $P_0N\,P_4$:

$$\sin\theta_c = \cos\phi_c = \frac{P_0N}{P_0P_4} = \frac{V_pT}{V_cT} = \frac{c}{nV_c}. \tag{5.5.41}$$

Figure (5.15b) shows the directions of the wavevectors $\left(k_1^{DPS}, k_2^{DPS}\right)$ and Poynting vectors $\left(S_1^{DPS}, S_2^{DPS}\right)$ of the upper and lower radiation from the charged particle moving in the positive x-direction. Cerenkov radiation is a linearly polarized EM-wave with an electric field component parallel to the path of the charged particle, i.e. the x-axis. Cerenkov radiation is directive in the *forward direction* within the cone angle $2\phi_c$. Similarly, a leaky-wave antenna is highly directive. The beam becomes narrower for the higher value of the refractive index n.

Inverse Cerenkov Radiation

Figure (5.15c) shows that the charged particle moving with a velocity V_c ($V_c > V_p$) in a DNG medium with a negative refractive index $-|n|$. It creates *backward radiation* with the negative Cerenkov angle,

$$\sin\theta_c = \frac{c}{-|n|V_c}, \quad \Rightarrow \sin(-\theta_c) = \frac{c}{|n|V_c}. \tag{5.5.42}$$

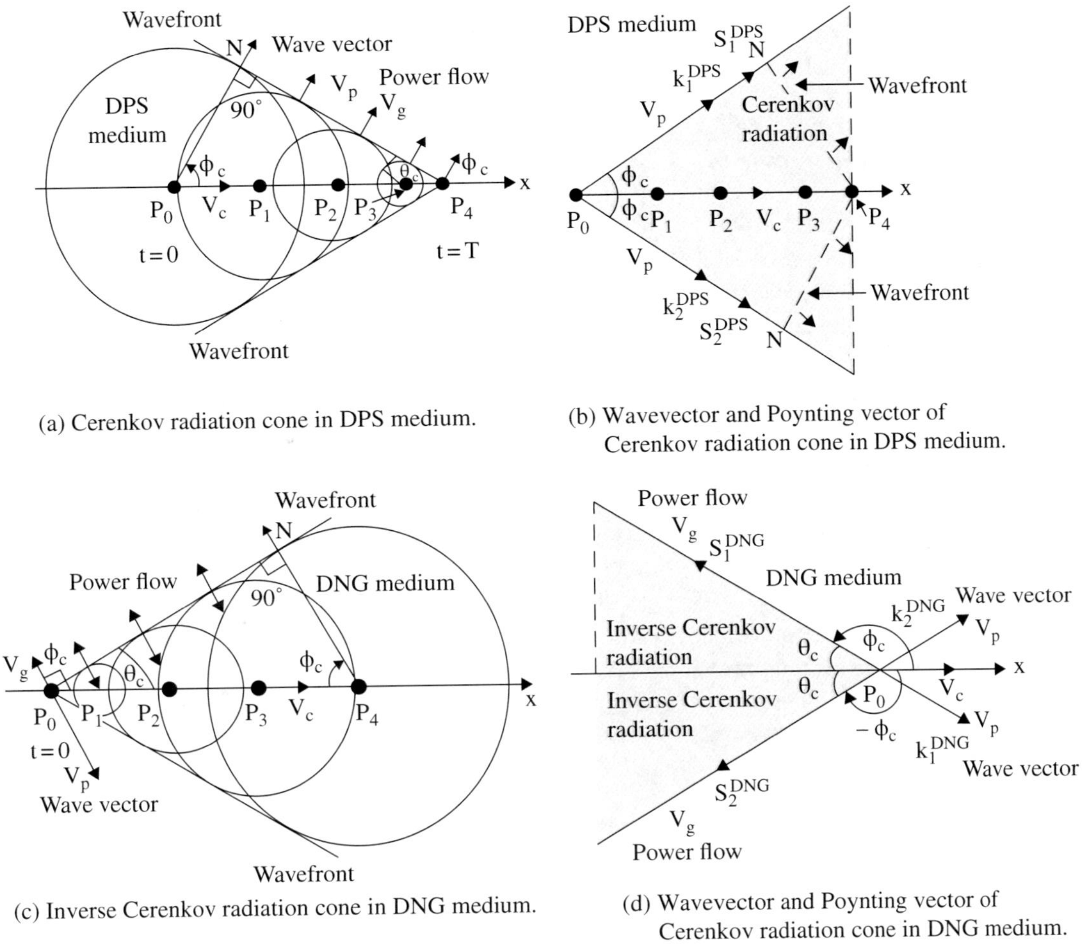

Figure 5.15 Cerenkov radiation in DPS and DNG media.

Figure (5.15d) shows that the wavevectors $\left(k_1^{DNG}, k_2^{DNG}\right)$ and Poynting vectors $\left(S_1^{DNG}, S_2^{DNG}\right)$, for both the upper and lower radiations, are anti-parallel that causes the backward radiation. Thus, a DNG medium supports the reverse, i.e. inverse Cerenkov radiation. The metamaterial medium could be continuously varied from the DPS to DNG forming the forward end-fire to broadside to backward end-fire antenna [J.26, J.27, B.6, B.7]. The inverse Cerenkov radiation has been experimentally confirmed in the microwave frequency range [J.28]. The realization of such a microstrip antenna is illustrated in the subsection (22.4.1) of chapter 22.

5.5.8 Metamaterial Perfect Absorber (MPA)

The absorbing materials are needed to absorb undesired reflected EM-waves and interfering signals. The perfect absorbing materials absorb 100% of incident RF power without any *reflection*, *scattering*, and *transmission*. The frequency-dependent absorbed power $A(\omega)$, i.e. *attenuation*, of an absorber, in absence of scattering and diffraction, is expressed as

$$A(\omega) = 1 - R(\omega) - T(\omega) \qquad (a)$$

$$\text{or } A(\omega) = 1 - |S_{11}(\omega)|^2 - |S_{21}(\omega)|^2 \qquad (b),$$

$$(5.5.43)$$

where $R(\omega)$ is the *reflectivity* (reflectance) related to the reflection coefficient (S_{11}) and $T(\omega)$ is the *transmissivity* (transmittance) of the absorbing sheet [B.14]. The attenuation $A(\omega)$ is also called *absorptivity* (absorptance). For 100% absorption, both $R(\omega)$ and $T(\omega)$ must be zero. To get $R(\omega) = 0$, the absorbing sheet must be matched to the free space impedance $\eta_0 = 377\,\Omega$ over a frequency band and for getting $T(\omega) = 0$ the absorbed power must be dissipated in the absorbing sheet. The transmissivity can be made zero even by placing a conducting sheet behind the absorbing sheet. However, it may result in multiple reflections degrading R(ω) [J.29].

Salisbury Absorber

The classical *Salisbury absorber*, shown in Fig (5.16a), is a narrow band resonant type absorber. The *absorbing resistive screen* has resistive 377Ω/sq impedance so that it is matched to free space. A metallic sheet (*electric-wall*) placed behind the absorbing screen at a distance $\lambda_0/4$ creates the *magnetic-wall*, i.e. the *high impedance surface* (HIS) at the plane of the resistive screen. The equivalent transmission line model of free space shows the absorbing screen as a 377Ω load followed by an open-circuited termination corresponding to the magnetic-wall. The concept of magnetic-wall is discussed in the subsection (7.2.2) of chapter 7. At the magnetic-wall, the total tangential components of the incident and reflected waves provide a high-intensity electric field, i.e. $E_t^{total} = 2E_0^{inc}$. However, the total tangential component of the magnetic field is zero, i.e. $H_t^{total} = 0$. The total electric field induces a current on the resistive screen placed at the magnetic-wall. The absorbed incident RF power is dissipated. Salisbury absorbing sheet is narrowband, thick, and bulky. By placing several absorbing resistive screens above the conducting plane, over a distance λ_0, a broadband absorber, known as *Jaumann absorber*, is obtained [B.15]. Again, it is a thick and bulky absorber. The *metasurface* provides an alternate arrangement for creating the high impedance surface to develop a thin and compact absorber. The realization of the metasurface is discussed in section (22.5) of chapter 22.

Metasurface Absorber

Figure (5.16b) shows the composition of the metasurface-based absorber [J.30, J.31]. A thin dielectric sheet $d << \lambda_0$ is backed by a conducting surface. An inductive surface is created at the dielectric surface. The capacitive grid of lines or patches is constructed on the dielectric surface such that the surface resonance creates the metasurface, i.e. the high impedance surface at the plane of the air-dielectric interface. Like a Salisbury absorber, again 377 Ω/sq resistive screen is placed at the interface to get the impedance matching with free space. The absorbed RF power is dissipated as heat.

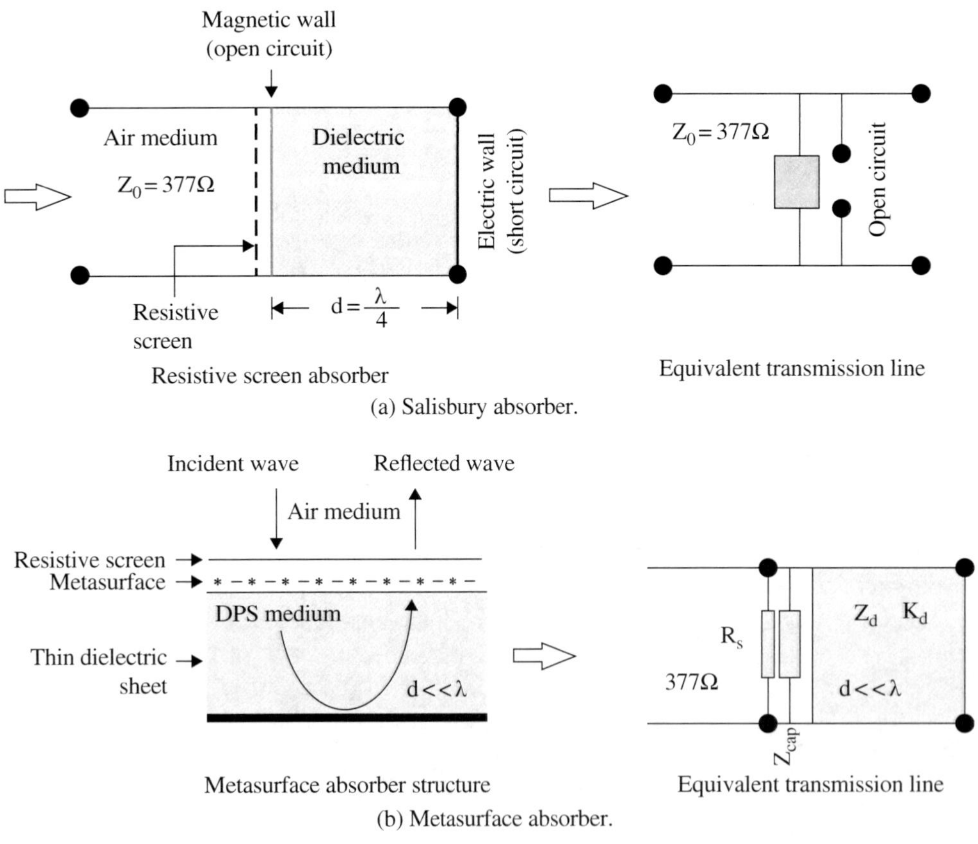

Figure 5.16 Composite surface absorber.

Figure (5.16b) also shows the equivalent transmission line model. The total surface impedance that creates a metasurface is

$$\frac{1}{Z_{surafce}} = \frac{1}{Z_{cap}} + \frac{1}{jZ_d \tan(k_d d)}$$
$$\Rightarrow Z_{surafce} = \frac{jZ_{cap}Z_d \tan(k_d d)}{Z_{cap} + jZ_d \tan(k_d d)}. \quad (5.5.44)$$

where Z_d and k_d are characteristic impedance and wave-vector of the conductor backed dielectric sheet. At a certain resonance frequency, the denominator of the above expression is zero giving the needed value of the surface impedance Z_{cap} of the capacitive grid:

$$Z_{cap} = -jZ_d \tan(k_d d). \quad (5.5.45)$$

The surface parallel resonance creates the high impedance metasurface at which the resistive screen is located. In this case, $Z_{surafce}(\omega_{res}) = R_s(\text{Screen}) = 377\,\Omega$ is obtained to achieve matching condition. The absorbed RF power is dissipated in the resistive screen to get nearly perfect absorber. By using multilayer metasurface, the broadband thin absorber has been developed.

DNG Slab Absorber

The lossy DNG slab can also be used as a perfect absorber, in place of the metasurface and resistive screen combination. The DNG slab absorber is developed around two configurations shown in Fig (5.17). Figure (5.17a) shows the lossy DNG slab with a conductor backing, and Fig (5.17b) shows it without any conductor backing. The working principle of a lossy DNG slab absorber is completely different. However, it is not based on the negative refraction property and backward wave nature of the DNG medium.

The characteristic impedance of any *magneto-dielectric* slab, DPS or DNG, is computed as $Z = \eta_0\sqrt{\mu_r(\omega)/\varepsilon_r(\omega)}$. The impedance matching with free space could be achieved $\mu_r(\omega) = \varepsilon_r(\omega)$. Such natural DPS materials are not available. However, $\mu_r(\omega)$ and $\varepsilon_r(\omega)$ of a DNG slab are engineered using two different structures. So both $\mu_r(\omega)$ and $\varepsilon_r(\omega)$ can be tuned independently to obtain their equalization at the same frequency. Thus, impedance matching could be realized in a DNG slab. If a DNG slab is sufficiently lossy, then absorbed RF power could be dissipated in the DNG slab of appropriate thickness.

Lossy DNG Slab with Conductor Backing

Figure (5.17a) shows the obliquely incident wave on a conductor backed lossy DNG slab. The angle of refraction is negative. The incident wave could be either TE or TM-polarized discussed in section (5.2). Following the results of equations (5.2.8c) and (5.2.16c) the reflection coefficients of both polarizations could be written as follows:

$$\Gamma_\perp^{TE} = \frac{\mu_r \cos\theta - \sqrt{n^2 - \sin^2\theta}}{\mu_r \cos\theta + \sqrt{n^2 - \sin^2\theta}} \quad (a)$$
$$\Gamma_\parallel^{TM} = \frac{\varepsilon_r \cos\theta - \sqrt{n^2 - \sin^2\theta}}{\varepsilon_r \cos\theta + \sqrt{n^2 - \sin^2\theta}} \quad (b), \quad (5.5.46)$$

where $\theta_i = \theta$ is the angle of incidence, $n = \sqrt{\mu_r\varepsilon_r}$ is the refractive index of a DNG slab and (μ_r, ε_r) are material parameters of the DNG medium. The *reflectivity* $R_\perp^{TE/TM}(\omega)$ for the incident TE and TM-polarized waves is defined as $R_\perp^{TE}(\omega) = \left|\Gamma_\perp^{TE}(\omega)\right|^2$ and $R_\parallel^{TM}(\omega) = \left|\Gamma_\parallel^{TM}(\omega)\right|^2$. In the case of normal incidence $\theta = 0°$, the above results provide the following expression for reflectivity:

$$R^{nor} = \left|\frac{\eta^{DNG} - \eta_0}{\eta^{DNG} + \eta_0}\right|^2 = \left|\frac{\mu_r - n}{\mu_r + n}\right|^2 = \left|\frac{\varepsilon_r - n}{\varepsilon_r + n}\right|^2. \quad (5.5.47)$$

To get no reflection, i.e. for $R^{nor} = 0$, the material parameters are $n = \varepsilon_r$, i. e. $\mu_r = \varepsilon_r$. Thus, the DNG slab is

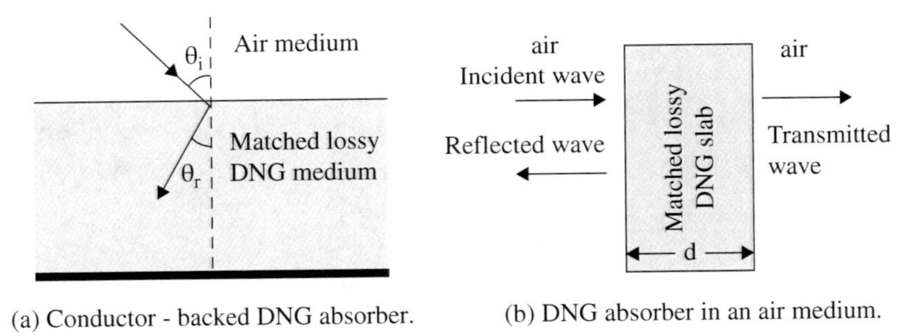

(a) Conductor - backed DNG absorber. (b) DNG absorber in an air medium.

Figure 5.17 Free space matched lossy DNG absorber.

impedance matched i.e. $\eta^{DNG} = \eta_0\sqrt{\mu_r/\varepsilon_r} = \eta_0$. The transmission coefficient τ^{nor} and *transmissivity* T^{nor} of a normally incident wave on the impedance matched DNG slab is obtained from equation (5.4.15):

$$\tau^{nor} = e^{-jk^{DNG}d} = e^{-jn^{DNG}k_0 d} = e^{-j[-(n'+jn'')]k_0 d} = \left[e^{-n''k_0 d}\right]e^{+jnk_0 d} \quad (a)$$

$$T^{nor} = |\tau^{nor}|^2 = e^{-2n''k_0 d} \quad (b). \qquad (5.5.48)$$

The condition $2n''k_0 d_{min} = 1$ provides the minimum propagation depth to attenuate the absorbed RF power to 1/e:

$$d_{min} = \delta = \frac{1}{2n''k_0}. \qquad (5.5.49)$$

The conductor backing reflects the attenuated EM-wave that gets further attenuated by 1/e before appearing as the reflected wave from the air-DNG interface. It degrades the absorption of the absorbing slab. An expression for the reflection coefficient is available, *using the theory of multiple reflections* [J.29]. However, for a thicker slab, multiple reflections can be ignored and simpler expression can be used to get the absorptivity of an absorber:

$$A^{Nor} = 1 - R^{Nor} = 1 - \left|\frac{\eta^{DNG} - \eta_0}{\eta^{DNG} + \eta_0}\right|^2 = 1 - \left|\frac{\mu_r - n}{\mu_r + n}\right|^2 = \frac{4\mu_r n}{(\mu_r + n)^2}. \qquad (5.5.50)$$

Lossy DNG Slab Without Conductor Backing

Figure (5.17b) shows the matched DNG slab without any conductor backing. In this case, the impedance matching of the DNG slab is also obtained for the condition $\mu_r = \varepsilon_r$. The transmissivity of a normally incident wave is given by equation (5.5.48b). A thicker DNG slab is taken ($d >> d_{min}$) to reduce the transmissivity T almost to a negligible value.

Both arrangements can be simulated using the Drude–Lorentz model discussed in chapter 6. We have discussed only the case of a single-layered DNG absorber. It has a limited bandwidth, as $\mu_r = \varepsilon_r$ is obtained at one frequency. However, several thin layers of the lossy DNG could be stacked to get a wideband absorber. The multiple resonance DNG slabs provide multiband absorber also [J.32–J.35].

The metamaterials have several other characteristics and applications. For instance, the DNG medium could be tailored to hide an object from the incident waves. It leads to the concept of *cloaking*. The cloak to hide any object is designed using the concept of the *transformation electromagnetics* [J.36]. The graded anisotropic refractive index between zero and unity is obtained through transformation electromagnetics. The metamaterials are used from microwave to optical frequency ranges, including the THz band [B.16]. Chapter 21 discusses realization and some applications of metamaterial in planar technology.

References

Books

B.1 Balanis, C.A.: *Advanced Engineering Electromagnetics*, John Wiley & Sons, New York, NY, 1989.

B.2 Jordan, E.C.; Balmain Keith, G.E: *Electromagnetic Wave and Radiating System*, Prentice-Hall India, New Delhi, 1989.

B.3 Ramo, S.; Whinnery, J.R.; Van Duzer, T.: *Fields, and Waves in Communication Electronics*, 3rd Edition, John Wiley & Sons, Singapore, 1994.

B.4 Orfanidis, S.J.: *Electromagnetic Waves and Antenna*, Free Book on The Web, ECE Department, Rutgers University, Piscataway, NJ, 2016.

B.5 Collin, R.E.: *Foundations for Microwave Engineering*, IEEE Press, Wiley Student Edition, John Wiley & Sons, Singapore, 2004.

B.6 Engheta, N.; Ziolkowski, R.W. (Editors): *Metamaterials: Physics and Engineering Explorations*, Wiley - Interscience, John Wiley & Sons, Inc., Hoboken, NJ, 2006.

B.7 Christophe Caloz, C.; Itoh, T.: *Electromagnetic Metamaterials: Transmission line Theory and Microwave Applications (The Engineering Approach)*, Wiley- Interscience, John Wiley & Sons, Inc., Hoboken, NJ, 2006.

B.8 Eleftheriades, G.I.; Balmain, K.G. (Editors): *Negative-Refraction Metamaterials: Fundamental Principles and Applications*, Wiley- Interscience, John Wiley & Sons, Inc., Priceton, NJ, 2005.

B.9 Capolino, F. (Editor): *Theory and Phenomena of Metamaterials*, CRC Press, Boca Raton, 2009.

B.10 Marques, R.; Martin, F.; Sorolla, M.: *Metamaterials with Negative Parameters. Theory, Design and Microwave Applications*, Wiley, Hoboken, NJ, 2008.

B.11 Sarychev, A.K.; Shalaev, V.M.: *Electrodynamics of Metamaterials*, World Scientific Publishing, Singapore, 2007.

B.12 Smith, G.S.: *An Introduction to Classical Electromagnetic Radiation*, Cambridge University Press, New York, NY, 1997.

B.13 Rutledge, D.B., Neikirk, D. P.; Kasilingam, D.P.: *Planar Integrated Circuit Antennas*, In *Infrared and Millimeter Waves: Volume 10, Millimeter Components and Techniques Part II*, Chapter-1, K.J. Button, Academic Press, New York, NY, 1983.

B.14 Liao, S.Y.: *Microwave Devices and Circuits*, 2nd Edition, Prentice – Hall of India Pvt. Ltd, New Delhi, 1989.

B.15 Kraus, J.D.: *Antenna*, 2nd Edition, McGraw-Hill International. Editions, New York, NY, 1988.

B.16 Bass, A.F. de (Ed. In-Chief): *Nonostructured Metamaterials*, European Commission, Brussels, 2018.

Journals

J.1 Holmes, J.J.; Balanis, C.A.: *Refraction of the uniform plane wave incident on a plane boundary between two lossy media, IEEE Trans. Antenna Propagat.*, Vol. 26, No. 5, pp. 738–741, Sept. 1978.

J.2 Pendry, J.B.: *Negative refraction makes a perfect lens, Phys. Rev. Lett.*, Vol. 85, No. 18, pp. 3966–3969, Oct. 2000.

J.3 Veselago, V.: *The electrodynamics of substances with simultaneously negative values of ε and μ, Soviet Physics Uspekhi*, Vol. 10, No. 4, pp. 509–514, Jan., Feb. 1968.

J.4 Bose, J.C.: *On the rotation of the plane of polarization of electric waves by a twisted structure, Proc. Poy. Soc.*, Vol. 63, pp. 46–52, 1898.

J.5 Sihvola, A.: *Metamaterials in electromagnetics, Metamaterials*, Vol. 1, pp. 2–11, 2007.

J.6 Shamonina, E.; Solymar, L.: *Metamaterials: How the subject started, Metamaterials*, Vol. 1, pp. 12–18, 2007.

J.7 Lindell, I.V.; Tretyakov, S.A.; Nikoskinen, K.I.; Ilvonen, S.: *BW media –media with negative parameters, capable of supporting backward waves, Microw. Opt. Technol. Lett.*, Vol. 31, No. 2, pp. 129–133, Oct. 2001.

J.8 Ziolkowski, R.W.; Heyman, E.: *Wave propagation in media having negative permittivity and permeability, Phys. Rev. E*, Vol. 64, pp. 1–15, 056625:, 2001.

J.9 Engheta, N.; Ziolkowski, R.W.: *Positive future for double-negative metamaterials, IEEE Trans. Microwave Theory Tech. MTT*, Vol. 53, No. 4, pp. 1535–1556, April 2005.

J.10 Alu, A.; Engheta, N.: *Pairing an epsilon –negative slab with a mu-negative slab: Anomalous tunneling and transparency, IEEE Trans. Antenna Proag.*, Special Issue on Metamaterials, Vol. 51, No.10, pp. 2558–2570, Oct. 2003.

J.11 Fujishige, T.; Caloz, C.; Itoh, T.: *Experimental demonstration of transparency in the ENG-MNG pair in a CRLH transmission-line implementation, Microw. Opt. Technol. Lett.*, Vol. 46, No. 5, pp. 476–48, Sept. 2005.

J.12 Engheta, N.: *An idea for thin subwavelength cavity resonators using metamaterials with negative permittivity and permeability, IEEE Antenna Wireless Propag. Lett.*, Vol. 1, pp. 10–13, 2002.

J.13 Pendry, J.B.; Smith, D.R.: *Reversing light: Negative refraction, Phys. Today*, Vol. 57, No. 6, pp. 37–43, June 2003.

J.14 Zhang, X.; Liu, Z.: *Superlenses to overcome the diffraction limit, Nat. Mater.*, Vol.7, pp. 435–441, June 2008.

J.15 Fang, N.; Liu, Z.; Yen, T.J.; Zhang, X.: *Regenerating evanescent waves from a silver superlens, Opt. Express*, Vol. 11, pp. 682–687, 2003.

J.16 Grbic, A.; Eleftheriades, G.V.: *Overcoming the diffraction limit with a planar left-handed transmission-line lens, Phys. Rev. Lett.*, Vol. 92, 117403, 2004.

J.17 Jacob, Z.; Alekseyev, L.V.; Narimanov, E.: *Optical hyperlens: Far-field imaging beyond the diffraction limit, Opt. Express*, Vol.14, pp. 8247–8256, 2006.

J.18 Lee, H.; Liu, Z.; Xiong, Y.; Sun, C.; Zhang, X.: *Development of optical hyperlens for imaging below the diffraction limit, Opt. Express*, Vol. 15, pp. 15886–15891, 2007.

J.19 Lee, H.; Xiong, Y.; Fang, N.; Srituravanich, N.; Durant, S.; Ambati, M.; Sun, C.; Zhang, X.: *Realization of optical superlens imaging below the diffraction limit, New J. Phys.*, Vol. 7, No. 255, pp. 1–16, 2005.

J.20 Kim, M.; Rho, J.: *Metamaterials and imaging, Nano Convergence*, Vol.2, No. 22, pp. 1–16, 2015.

J.21 Jacob, Z.; Alekseyev, L.V.; Narimanov, E.: *Optical hyperlens: Far-field imaging beyond the diffraction limit, Opt. Express*, Vol. 14, No.18, pp. 8247–8256, 2006.

J.22 Pendry, J.: *New electromagnetic materials emphasize the negative, Phys. World*, pp. 1–5, 2001, Archived 2011-07-17 at the Wayback Machine.

J.23 Leong, K.M.K.H.; Lai, A.; Itoh, T.: *Demonstration of reverse Doppler effect using a left-handed transmission line, Microw. Opt. Technol. Lett.*, Vol. 48, No. 3, pp. 545–547, March 2006.

J.24 Seddon, N.; Bearpark, T.: *Observation of the inverse Doppler effect, Science*, Vol. 302, pp. 1537–1539, Nov. 2003.

J.25 Chen, J.; Wang, Y.; Jia, B.; Geng, T.; Li, X.; Feng, L.; Qian, W.; Liang, B.; Xuanxiong Zhang, X.; Min Gu, M.; Songlin Zhuang, S.: *Observation of the inverse Doppler effect in negative-index materials at optical frequencies, Nat. Photonics*, Vol. 5, pp. 239–242, April. 2011.

J.26 Wu, B.I; Lu, J.; Kong, J.A.; Chen, M.: *Left-handed metamaterial design for Čerenkov radiation*, *J. Appl. Phys.*, Vol. 102, No.114907, pp. 1–5, 2007.

J.27 Duan, Z.Y.; Wu, B.-I.; Xi, S.; Chen, H.S.; Chen, M.: *Research progress in reversed Cherenkov radiation in double-negative metamaterials*, *Prog. Electromagn. Res., PIER*, Vol. 90, pp. 75–87, 2009.

J.28 Xi, S.; Chen, H.; Jiang, T.; Lixin Ran, L.; Huangfu, J.; Wu, B.-I.; Kong, J.A.; Chen, M.: *Experimental verification of reversed Cherenkov radiation in left-handed metamaterial*, *Phys. Rev. Lett., PRL*Vol. 103, No. 194801, pp. 1–4, Nov. 2009.

J.29 Wanghuang, T.; Chen, W.; Huang, Y.; Wen, G.: *Analysis of metamaterial absorber in normal and oblique incidence by using interference theory*, *AIP Adv.*, Vol. 3, No. 102118, pp. 1–9, 2013.

J.30 Engheta, N.: *Thin absorbing screens using metamaterial surfaces*, IEEE Antennas Propagation Int. Symp., vol. 2, pp. 392–395, June 2002.

J.31 Han, Y.; Che, W.Q.; Christopoulos, C.; Chang, Y.M.: *Investigation of thin and broadband capacitive surface-based absorber by the impedance analysis method*, *IEEE Trans. Electromagn. Compat.*, Vol. 57, pp. 22–26, 2015.

J.32 Watts, C.M.; Liu, X.; Padilla, W.J.: *Metamaterial electromagnetic wave absorbers*, *Adv. Opt. Mater.*, Vol. 24, pp. 98–120, 2012.

J.33 Yang, C.; Han Xiong, H.; Li, X.P.: *Investigation of a metamaterial absorber by using reflection theory model*, *Prog. Electromagn. Res. M, PIER*, Vol. 59, pp. 65–73, 2017.

J.34 Zhang, Z.H.; Wang, Z.P.; Wang, L.H.: *Design principle of single or double-layer wave absorbers containing left-handed materials*, *Mater. Des.*, Vol. 30, pp. 3908–3912, 2009.

J.35 Jamilan1, S.; Azarmanesh, M.N.; Zari, D.: *Design and characterization of a dual-band metamaterial absorber based on destructive interferences*, *Prog. Electromagn. Res. C, PIER*, Vol. 47, pp. 95–101, 2014.

J.36 Eleftherriades, G.V.; Selvanayagam, M.: *Transforming electromagnetics using metamaterials*, *IEEE Microw. Mag.*, Vol. 13, No. 2, pp. 26–38, March/April 2012.

6

Electrical Properties of Dielectric Medium

Introduction

The electrical property of a dielectric-medium is characterized by the *macroscopic concept* of relative permittivity, permeability, and conductivity. However, these parameters have their origin in several kinds of polarization or magnetization occurring at the *microscopic level*, i.e. at the level of discrete atoms, molecules, and ions constituting a natural material. The *polarization* normally means "orientation," originated from the Greek word *polos*, meaning the axis of a spinning globe, although all polarizations do not involve the rotation. The macroscopic description ignores the discreteness of atoms, molecules, and ions. It views primarily the discrete materials as the homogenous materials described by the electrical parameters *permittivity, permeability, and conductivity*. It is known as the *homogenization* of a material. The finite time lag for the building up of polarization results in the lossy and dispersive dielectric-medium with a complex relative permittivity $\varepsilon_r^*(f) = \varepsilon_r'(f) - j\varepsilon_r''(f)$ describe by *Kramers–Kronig relations*. Several materials dispersion models, such as *Debye, Lorentz, and Drude models*, as applied to the dielectric and conductors, are developed.

The *artificial dielectrics* are developed using *classical mixture models*. The mixture models are based on the concept of the *host medium* and *inclusion*. The inclusions behave like artificial molecules undergoing polarization, similar to the polarization in natural materials. The concept of homogenization applies to mixture-materials and artificial dielectrics. Thus, the dispersion models mentioned above are applied to artificial dielectrics. These dispersion models, also mixing models, are helpful in understanding and designing the modern engineered *bandgap materials and metamaterials* discussed in chapters 5 and 19–22.

Objectives

- *To understand electrical polarization and magnetization in a material medium.*
- *To discuss in brief properties, models, and realization of some artificial dielectrics.*
- *To discuss the classical mixture models using the concept of the host medium and inclusions.*
- *To explore the frequency-dependent relative permittivity, and also the frequency-dependent conductivity of natural materials, and artificial dielectrics using Debye, Lorentz, Drude, and Maxwell–Wagner models.*
- *To present the circuit models of the dielectric materials.*
- *To explore the causal models of the complex relative permittivity of a substrate.*
- *To present a brief discussion of the metasubstrates.*

6.1 Modeling of Dielectric-Medium

The complex relative permittivity, $\varepsilon_r^* = \varepsilon_r' - j\varepsilon_r''$, of a lossy dielectric-medium is a result of several kinds of material polarizations. Newtonian mechanics, applied to the bound charges of material provides the working models of a dielectric or a conductor. It culminates in the *oscillator model*, due to Lorentz and Drude [B.1–B.10]. The oscillator model explains the general characteristics of dielectrics and conductors.

6.1.1 Dielectric Polarization

The materials at the microscopic level exist in the form of atoms, ions, and molecules. Individually, these basic building blocks of material are electrically neutral. A lossless dielectric material contains mostly the *bound*

Introduction to Modern Planar Transmission Lines: Physical, Analytical, and Circuit Models Approach, First Edition. Anand K. Verma.

charges. The *free charges* are almost nonexistent. However, imperfect dielectrics have free charges also. On the application of an external electric field, the positive and negative bound charges are displaced from each other and the dielectric material gets *polarized*, as shown in Fig (6.1a). The induced *local field* E_{loc} tends to oppose the applied *external field* E_{ext}. The conduction current is not present in a perfect dielectric material. The polarization in most of the material occurs only on the application of the external field. However, some materials are already polarized. Some materials have disoriented polarizations, and they get aligned to exhibit polarization on the application of the external field. These are *ferroelectric* materials (barium titanate, lead titanate, etc.) showing also hysteresis loop-like ferromagnetic materials. Some materials and polymers also retain electric polarization, like molten wax on cooling in the electric field. These are known as the *electrets*.

The electrical polarization involves motion of the charged particles. It could be one of three kinds – (i) *displacement of charges*, (ii) *orientation of charges*, and (iii) *transportation*, or *migration of charges*. The type

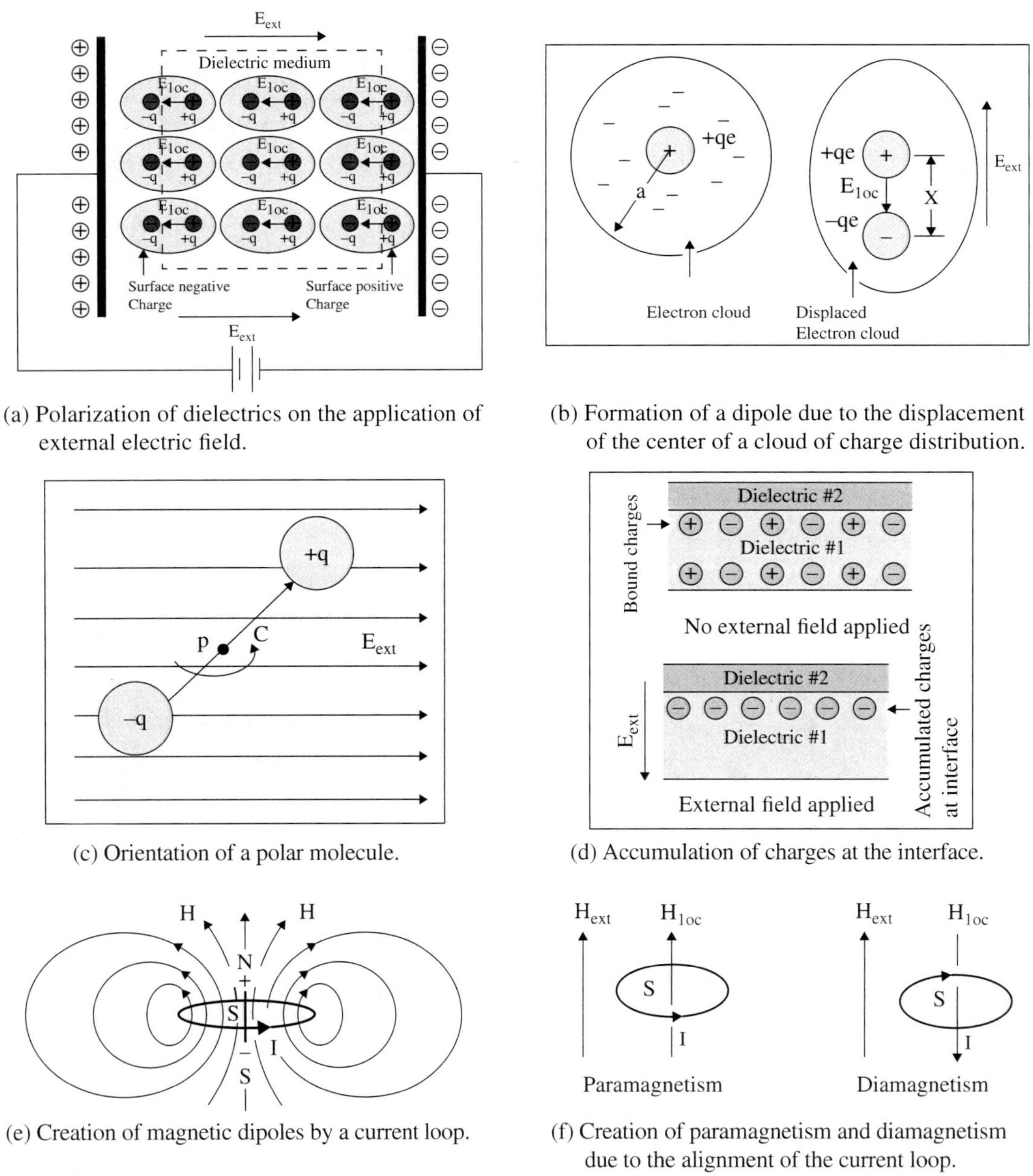

Figure 6.1 Several kinds of electric polarization and magnetization in the material medium.

of motion of the charged particles determines the type of polarization. Materials contain bound electronic charges at the atomic level. At the molecular level, materials can also have bound ionic charges. Some polarized materials at the molecular level have disoriented dipoles, whereas some imperfect dielectrics also contain free charges. Some materials can have all four kinds of polarizations – *electronic polarization, ionic polarization, orientational polarization,* and *space-charge polarization*. The first two polarizations involve displacement of charges. The third polarization involves the orientation of dipoles (charge-pair). The fourth polarization involves transportation or migration of charges. The concept of polarization is the *microlevel* description of the materials, accounting discreteness of a material. The description is also valid for the *artificial dielectric* materials, including engineered *metamaterials*, where embedded metallic inclusions act as mega-molecules.

The building up of any polarization in a material takes a finite amount of time and that is different for different kinds of materials. Thus, polarization is a function of time in the time-domain. In the frequency-domain, it is a frequency-dependent process. The displacement of the charges involves the storage of electrical energy by a material. *At the macroscopic level, the electric energy storing ability of a dielectric material appears as the dielectric constant, i.e. relative permittivity*. As the polarization process is a frequency-dependent process; therefore, the material relative permittivity is also frequency-dependent that leads to the material dispersion. The time lag involved in the building up of polarization results in the complex relative permittivity with the dielectric loss of material.

6.1.1.1 Electronic Polarization

It works at the atomic level. In the classical model, the electron charge is viewed as a spherical cloud of distributed charge around the positive nucleus. Figure (6.1b) shows that in the absence of an external field, the positive nucleus and the center of the spherical negative charge cloud are at the same location. However, under the influence of an external electric field, E_{ext}, both of them get displaced to form a dipole. The dipole is described by its *dipole moment*. As shown in Fig (6.1a), the field of the induced dipole, i.e. *the local field* E_{loc}, *is in the direction opposite to that of the external electric field* [J.1]. The displacement of the electronic charge cloud under the influence of an external electric field is known as the *electronic polarization*.

6.1.1.2 Ionic Polarization

A molecule formed by two different atoms is held either by the *ionic bond* due to the transfer of an electron from one atom to another, or by the *covalent bond* formed due to the sharing of common electrons between them. Depending upon the structure of a molecule, the molecule could either be a *nonpolar* or a *polar*. The nonpolar molecules do not have permanent polarization. However, on the application of an external electric field, there is a displacement of the positive and negative ions at the molecular level, causing the *ionic polarization*. The polarization of a nonpolar material is temperature-independent.

6.1.1.3 Orientational (Dipolar) Polarization

The dipolar polarization is applicable to the polar molecules, possessing the permanent dipole moment in the absence of an external field. In the polar dielectrics, the positive and negative charges of molecules do not coincide, causing the existence of the permanent dipoles. The permanent *dipole moment* is given by

$$\vec{p} = q\vec{x}, \tag{6.1.1}$$

where q is the charge, and x is the separation between the charges. Normally, the dipoles are randomly oriented. Figure (6.1c) shows that under the influence of an external field a couple, given below, is produced:

$$\vec{C} = \vec{p} \times \vec{E}_{ext}. \tag{6.1.2}$$

The torque acting on the dipolar molecules orient, i.e. align them in one direction. Therefore, the orientation of the polar molecules under the influence of an external electric field causes the *orientational (dipolar) polarization*. The polarization of a dipolar material is temperature-dependent, as collisions can destroy the alignment of dipolar molecules.

6.1.1.4 Space Charge (Interfacial) Polarization

In some materials, free charges are present. However, their movement is restricted by some kinds of barriers, like the grain boundaries, or in the case of a heterogeneous mixture by the boundaries of inclusions in the host medium. The layered medium, shown in Fig (6.1d), also comes under this category. When an external field is applied the free charges accumulate at the boundary, resulting in the *space-charge polarization* of the material. This type of polarization occurs in the semiconducting medium, like Metal-Insulator-Semiconductor (MIS) and Schottky's structures causing a large increase in the relative permittivity of the composite structure.

It is also known as the *interfacial* or *Maxwell–Wagner polarization* that is a frequency-dependent phenomenon [B.1, B.8, B.9]. Due to the creation of large relative permittivity, the MIS and Schottky structures support the *slow-wave* propagation at the lower end of microwaves [J.2–J.5].

The above discussion demonstrates that several kinds of polarizations occur in a dielectric-medium, as a response to the applied external electric field. The building up of polarizations always takes a finite amount of time. The electronic polarization builds up very fast, whereas the interfacial polarization takes more time to build up. If the applied field is time-harmonic, the electronic polarization occurs in all materials, at all frequencies up to the visible range of frequency. Only in the ultraviolet range, the time duration is so short that electronic polarization cannot build up properly. The ionic polarization occurs in the ionic materials and its response time is more than the response time of the electronic polarization. Thus, the ionic polarization does not occur above the infrared frequency region. The materials, such as water, $BaTiO_3$, amorphous polymers, have permanent dipoles, and the orientational polarization takes more time to build up. Therefore, the orientational or dipolar polarization does not take place beyond a microwave and mm-wave frequency range. So is the case for the interfacial polarization. It normally occurs only up to a few GHz frequency ranges. The interfacial polarization also takes place in the mixture and composite materials. Such composite materials are the basis for obtaining the engineered artificial dielectrics and metamaterials.

6.1.2 Susceptibility, Relative Permittivity, and Clausius–Mossotti Model

The nonpolar materials do not have a permanent dipole moment. The external electric field $\vec{E}_{ext}$ induces the *polarization* through the mechanism of the induced dipole moment. It is also called the *electric polarization* to distinguish it from the *magnetic polarization*, i.e. the *magnetization*. It is discussed in subsection (6.1.4). Both the applied static E-field and the time-varying external E-field of the incident TEM waves on a material can create polarization in the material medium. *The polarization is the key concept to understand the interaction of fields and waves with materials.* The materials could be a natural one or engineered one, such as artificial dielectrics and metamaterials. The artificial dielectrics are discussed in section (6.2.2). In a linear isotropic dielectric material, the induced *dipole moment* $(\vec{p})$, due to the displacement of charges, is proportional to the *local electric field* $\vec{E}_{loc}$ [B.1, B.8, B.10–B.12]:

$$\vec{p} = \alpha^e \, \vec{E}_{loc}, \tag{6.1.3}$$

where the proportionality constant α^e is a characteristic parameter, known as the *electric polarizability* of a material. It relates the polarization in a material with the local field. The complex local electric field $\vec{E}_{loc}$, near a molecule, is different from the applied external field $\vec{E}_{ext}$. It is described by the *Lorentz model*. In an isotropic medium, the induced dipole moment is in the direction of the local electric field. So polarizability α^e for an isotropic medium is a scalar, whereas for the anisotropic medium it is a tensor $[\alpha]$.

At the atomic level, i.e. for the *electronic polarization*, the dipole moment is $\vec{p} = \vec{p}_e$ and its electric polarizability is $\alpha^e = \alpha_e^e$, i.e. the *electronic polarizability*. In the case of the electronic polarization, the displacement of the electronic charge cloud is limited to the radius of an atom. For the *ionic polarization*, the dipole moment is $\vec{p} = \vec{p}_i$ and the *ionic polarizability* is $\alpha^e = \alpha_i^e$. In the case of the ionic polarization, displacement of the ions is limited to the internuclear distance of atoms forming a molecule. *The dielectric constant, i.e. relative permittivity of a material is due to its polarization.* If the dielectric constant of a nonpolar material is measured at the optical frequency, its value is only due to the electronic polarization, as heavy ions cannot move during a short time interval at the optical frequency. However, the measured dielectric constant at the infrared is due to both the electronic and ionic polarization. It has a higher value of the dielectric constant that is depicted in Fig (6.10a).

The *polarization* $\vec{P}$ of a material is the number of dipole moments $\vec{p}$ per unit volume. A small volume Δv contains $N\vec{p}\,\Delta v$ dipole moments, where N is the number of dipole moments in volume Δv. Thus, the polarization density, i.e. the polarization $\vec{P}$ of the material is expressed as follows:

$$\vec{P} = \frac{\vec{p}}{\Delta v} = N\vec{p} = N\alpha^e \, \vec{E}_{loc}. \tag{6.1.4}$$

The concentration of the dipole moment is $N = 1/\Delta v$. The local field is due to the superposition of applied external electric field and the field produced by the polarized neighboring dipoles. However, as the first approximation, the fields due to induced neighboring dipoles are ignored. In such a case, the polarization is proportional to the external electric field E_{ext}:

$$\vec{P} = \chi \vec{E}_{ext} = \varepsilon_0 \chi_r \vec{E}_{ext}, \quad (6.1.5)$$

where χ and χ_r are the *susceptibility* and *relative susceptibility* of dielectric material, respectively. Lorentz has provided a model relating the local field to an external field. We should be careful as some authors treat χ like relative susceptibility [B.7]. The susceptibility is a measure of the ability of a material to get polarized under the influence of an external electric field. The presence of a polarized medium, i.e. the polarization $\vec{P}$, between two parallel plates of a capacitor, contributes to the local field in the direction opposite to that of the external field. It is shown in Fig (6.1a). So the polarization $\vec{P}$ reduces the flux density $\vec{D}$ inside the parallel-plate capacitor. The phenomenological behavior of the flux density $\vec{D}$ reduction, related to the external field, is expressed as follows:

$$\vec{D} - \vec{P} = \varepsilon_0 \vec{E}_{ext}, \quad \Rightarrow \quad \vec{D} = \varepsilon_0 \vec{E}_{ext} + \vec{P} \quad (a)$$
$$\vec{D} = (\varepsilon_0 + \chi) \vec{E}_{ext}, \quad \Rightarrow \quad \vec{D} = \varepsilon_0 (1 + \chi/\varepsilon_0) \vec{E}_{ext} \quad (b). \quad (6.1.6)$$

where $\chi_r = \chi/\varepsilon_0$ is the *relative susceptibility* of a material. The equation (6.1.5) is used in the above relation. The relative susceptibility is a dimensionless quantity. It helps to compute the *relative permittivity* of a material that relates the flux density $\vec{D}$ to the electric field E_{ext} inside a dielectric-medium:

$$\varepsilon_r = 1 + \frac{\chi}{\varepsilon_0} = 1 + \chi_r \quad (a), \quad \vec{D} = \varepsilon_0 \varepsilon_r \vec{E}_{ext} \quad (b). \quad (6.1.7)$$

The phenomenological equation (6.1.6) is very interesting in the theory of dielectrics. It correlates the measurable *macroscopic* quantity ε_r to the *microscopic* polarization phenomenon $\vec{P}$. Thus, the measurement of the dielectric constant helps to understand the atomic and molecular behavior of material through the *polarizability* α^e at the atomic and molecular level. On assuming $\vec{E}_{loc} = \vec{E}_{ext}$, and using equations (6.1.4) and (6.1.5), the susceptibility $\chi = N\alpha^e$ is obtained. This assumption is valid only for a low-density medium. The expression for the *relative permittivity of a low-density medium* can be written as follows:

$$\varepsilon_r = 1 + \frac{N\alpha^e}{\varepsilon_0}. \quad (6.1.8)$$

Sometimes an expression for the polarizability α^e is premultiplied by ε_o. In this case, the above expression is modified, as the relative permittivity is $\varepsilon_r = 1 + N\alpha^e$. In the case of natural materials, the quantity N is obtained from Avogadro number, i.e. the number of molecule per mole $N = \rho \times 6.023 \times 10^{23}/M$, where ρ is the density in kg/m^3. The molecular weight M is in kg. In the case of an artificial dielectric, discussed in section (6.2.2), one inclusion is located in the volume Δv of the unit cell giving $N = 1/\Delta v$.

In general, the local field is different from the external field. The modeling of the local field is discussed below.

6.1.2.1 Clausius–Mossotti Model

Clausius–Mossotti model of the relative permittivity, in terms of the microscopic parameters, is obtained through the *approximate local electric field model* developed by Lorentz.

6.1.2.2 Lorentz Local Electric Field Model

Lorentz modeled the process of polarization by assuming a *hypothetical spherical cavity* around the polarized molecule. The polarization-electronic or ionic is caused by the action of a *local electric* field in the assumed cavity. The local field E_{loc} in the hypothetical cavity is a combination of the external (i.e. incident) E-field, and the E-field arising out of the induced polarized atoms and molecules around the assumed cavity. So the local field could be expressed through the following empirical relation using a parameter γ [B.2, B.8, B.11]:

$$\vec{E}_{loc} = \vec{E}_{ext} + \left(\frac{\gamma}{\varepsilon_0}\right) \vec{P}. \quad (6.1.9)$$

The parameter γ determines the interaction between the local and external fields in a material medium. The parameter γ is called the *interaction constant* of a material. For a low-density material, like a gas medium, the field interaction could be ignored, i.e. $\gamma = 0$. It gives $\vec{E}_{loc} = \vec{E}_{ext}$. However, to account for such interaction between the fields, Lorentz has taken $\gamma = 1/3$. Onsager has further improved the model for a denser material by accounting the effect of the relative permittivity on the local field. Onsager has taken $\gamma = 1/(2\varepsilon_r + 1)$ [B.2]. Even the Onsager model is not a correct description of the local field for a solid, or a denser, material. However, the value of the parameter γ could be estimated from the experimentally obtained data on the dielectric constants of the materials [B.13, B.14].

On using equations (6.1.5) and (6.1.7a) with the above equation (6.1.9), we get an expression for the local field in terms of the relative permittivity:

$$\vec{E}_{loc} = \vec{E}_{ext} + \left(\frac{\gamma\chi}{\varepsilon_0}\right)\vec{E}_{ext} = (1 + \gamma\chi_r)\vec{E}_{ext}$$

$$\Rightarrow \vec{E}_{loc} = [1 + \gamma(\varepsilon_r - 1)]\vec{E}_{ext} = [\gamma\varepsilon_r + (1 - \gamma)]\vec{E}_{ext}. \tag{6.1.10}$$

On using Lorentz's and Onsager's values for the parameter γ, the following expressions are obtained for the local electric field:

$$\text{Lorentz}: \vec{E}_{loc} = \left[\frac{\varepsilon_r + 2}{3}\right]\vec{E}_{ext} \quad \text{(a)}$$

$$\text{Onsager}: \vec{E}_{loc} = \frac{3\varepsilon_r}{2\varepsilon_r + 1}\vec{E}_{ext} \quad \text{(b)}. \tag{6.1.11}$$

The local electric field in equation (6.1.11) is known as the *Mossotti local electric field* that depends on the relative permittivity of a medium. Normally, Lorentz's local field is used to get the Clausius–Mossotti model. The following expression is obtained for the polarization $\vec{P}$ for such a medium from equations (6.1.4) and (6.1.11a):

$$\vec{P} = N\alpha^e \frac{(\varepsilon_r + 2)}{3}\vec{E}_{ext}. \tag{6.1.12}$$

By using the equations (6.1.5) and (6.1.7a) with the above equation, the following *Clausius–Mossotti* expression is obtained:

$$\chi_r\varepsilon_0 = N\alpha^e \frac{(\varepsilon_r + 2)}{3} \Rightarrow \frac{\varepsilon_r - 1}{\varepsilon_r + 2} = \frac{N\alpha^e}{3\varepsilon_0}. \tag{6.1.13}$$

$$\Rightarrow \varepsilon_r = 1 + \frac{N\alpha^e/\varepsilon_0}{1 - N\alpha^e/3\varepsilon_0} \quad \text{(a)}$$

$$\text{or} \quad \varepsilon_r = 1 + \frac{N\alpha^e/\varepsilon_0}{1 - N\alpha^e\gamma/\varepsilon_0} \quad \text{(b)}. \tag{6.1.14}$$

The above relation computes the macroscopic quantity, i.e. the relative permittivity of a dielectric material, from the microscopic molecular-level parameter, i.e. the polarizability α^e. Equation (6.1.14b) is in terms of the interaction constant γ that accommodates Lorenz, as well as the Onsager local field model. It is obtained by using an equation (6.1.10) with equations (6.1.4) and (6.1.5). The local field is too complicated for a solid dielectric. So the interaction constant γ could be determined experimentally and the empirical curve-fitted expression can be obtained. In the case of artificial dielectric materials, the *interaction constant* γ is computed from the static field theory [B.11].

Both the natural material and artificial dielectric materials, under polarization, constitute an array of induced dipoles in the vacuum. Therefore, the relative permittivity computed in the above equation (6.1.14) could be treated as the effective or equivalent permittivity of the composite medium with the vacuum as a *host medium*, and the dipoles as the *inclusions*. Using such a concept, Clausius–Mossotti expression is used to get the Maxwell Garnett (MG) mixing rule. The concept of host medium and inclusions, and also MG rule, is discussed in section 6.3.

For a low-density medium, usually $\varepsilon_r \approx 1$ is taken. For this case, the above equation (6.1.13) is approximated as follows:

$$(\varepsilon_r - 1) = \frac{N\alpha^e}{\varepsilon_0}. \tag{6.1.15}$$

It is identical to equation (6.1.8), where the local field has been taken as the external field. Clausius–Mossotti model expression is used for both the electronic and ionic polarization. Each kind of polarization has a different expression for the polarizability α^e. It is discussed below. If the material cannot be polarized, $\alpha^e = 0$ and relative permittivity could be treated as unity, $\varepsilon_r = 1$. This "material" acts as the vacuum that cannot be polarized.

6.1.3 Models of Polarizability

The Clausius–Mossotti model could be used to compute the relative permittivity of a material. The models for different kinds of polarizability are needed for this purpose.

6.1.3.1 Electronic Polarizability

The polarizability, in this case, is computed from the spherical uniform charge density model shown in Fig (6.1b) [B.7]. If q_e is the electronic charge distributed uniformly over the sphere of radius a, the charge density is

$$\rho_e = \frac{q_e}{(4/3)\pi a^3}. \tag{6.1.16}$$

The charge on the displaced spherical electron cloud of radius x is

$$q_e' = \frac{4}{3}\pi x^3 \times \frac{3q_e}{4\pi a^3} = \frac{x^3}{a^3}q_e. \tag{6.1.17}$$

Two opposite polarity charges, separated by a distance x, form a dipole. The local electrostatic force acting between them is

$$F_{loc} = \frac{q_e' q_e}{4\pi\varepsilon_0 x^2} = \frac{q_e^2 \times x^3}{4\pi\varepsilon_0 x^2 a^3}, \Rightarrow F = \left[\frac{q_e^2}{4\pi\varepsilon_0 a^3}\right] x. \tag{6.1.18}$$

On getting the displacement x, the dipole moment is computed. It provides the *electronic polarizability* ($\alpha^e = \alpha_e^e$), with the help of the equation (6.1.3):

$$\vec{p} = q_e \vec{x} = q_e \left[\vec{F}_{loc} \frac{4\pi\varepsilon_0 a^3}{q_e^2}\right]$$
$$\Rightarrow \vec{p} = 4\pi\varepsilon_0 a^3 \vec{E}_{loc} \quad (a), \quad \alpha_e^e = 4\pi\varepsilon_0 a^3 \quad (b). \tag{6.1.19}$$

The above expression is used with equations (6.1.8) and (6.1.14a), based on $\vec{E}_{loc} = \vec{E}_{ext}$ and Clausius–Mossotti model, respectively, to compute the relative permittivity of a low-density medium:

$$\varepsilon_r = 1 + 4\pi N a^3 \quad (a), \quad \varepsilon_r = \frac{8\pi N a^3 + 3}{3 - 4\pi N a^3} \quad (b). \tag{6.1.20}$$

The above equation (6.1.20b) is a more accurate expression that uses Lorentz's local field. The ε_r and χ_r linearly depend on the polarized particles density N, and they have a cubic dependence on the radius "a" of the atoms involved in the electronic polarization process. The above expressions for the relative permittivity are approximate ones as the charge distribution of the electron cloud is nonuniform. The polarization, similar to the electronic polarization, also occurs in the artificial dielectrics and metamaterials due to induced dipole moments in these engineered materials.

6.1.3.2 Ionic Polarizability

The ionic polarization, explained with the help of the ionic dipole moment, and associated *ionic polarizability* ($\alpha^e = \alpha_i^e$) of the molecule is given below [B.2]:

$$\vec{p}_i = \alpha_i^e \vec{E}_{loc} \quad (a), \quad \text{where} \quad \alpha_i^e = \frac{64\varepsilon_0 d_a^3}{7(n+1) - 16} \quad (b). \tag{6.1.21}$$

In the above expression, d_a is the interatomic distance between the centers of atoms forming a molecule. The value of n depends on the size of the ion. The local field $\vec{E}_{loc}$ is related to the external field by Lorentz's expression (6.1.11a). The order of magnitude of both the electronic and ionic polarizability is the same. The ionic polarization occurs in natural materials, not in the artificial dielectrics and metamaterials.

6.1.3.3 Orientational (Dipolar) Polarizability

In a polar medium, with permanent dipoles, the polarization under the influence of the local electric field has to be computed differently. The model applies to gases, liquids, and also to some solids. However, it is not needed for artificial dielectrics. The effect of the thermal agitation has to be accounted for, as it disturbs the alignment of the dipoles. The average polarization of a dipolar medium and its *dipolar polarizability*, i.e. $\alpha^e = \alpha_d^e$, is given below [B.2]:

$$P = \frac{N\mu^2 E_{loc}}{3kT} \quad (a), \quad \alpha_d^e = \frac{\mu^2}{3kT} \quad (b), \tag{6.1.22}$$

where μ (not permeability), K, and T are the dipole moment of a permanent dipole, Boltzmann constant, and temperature of a medium in Kelvin, respectively. All three kinds of polarizations – electronic, ionic, and dipolar – are present in a polar material at the lower end of microwave frequency. In the case of the gases and low-density materials, the external and internal fields are almost the same, i.e. $\vec{E}_{ext} \approx \vec{E}_{loc}$. The expressions for the total polarization, total polarizability, and corresponding susceptibility are written as follows:

$$\vec{P} = N\left(\alpha_e^e + \alpha_i^e + \frac{\mu^2}{3kT}\right)\vec{E}_{loc} \quad (a), \quad \alpha_{total}^e = \alpha_e^e + \alpha_i^e + \frac{\mu^2}{3kT} \quad (b)$$
$$\vec{P} = N\alpha_{total}\vec{E}_{loc} \quad (c), \quad \chi_r = \frac{N}{\varepsilon_0}\left(\alpha_e^e + \alpha_i^e + \frac{\mu^2}{3kT}\right) \quad (d). \tag{6.1.23}$$

In the case of the nonpolar low-density materials, polarizations are only due to the electronic polarizability and ionic polarizability, with $\alpha_d^e = 0$.

6.1.4 Magnetization of Materials

Materials acquire magnetic property, through the process of magnetization on the application of external magnetic field H_{ext}. The magnetization of material occurs due to the magnetic dipole, as in the case of a permanent magnet, created by a small current-carrying loop shown in Fig (6.1e). In natural materials, current loops are formed from the orbital motion of electrons in the atoms and also due to their spin. In the case of *artificial magnetic materials*, the external magnetic field induces a circulating current on a metallic object, creating a magnetic dipole. The induced magnetic dipoles have their own local magnetic field H_{loc}. In case the local field is in the direction of the external field, the permeability is enhanced and the material is *paramagnetic*. In case the local field opposes the external field, the material is *diamagnetic*. Figure (6.1f) shows both the cases of paramagnetism and diamagnetism. The ferromagnetic materials, such as ferrite, have permanently magnetized magnetic dipoles. However, on the application of the external magnetic field, they align to increase permeability. The artificial dielectric materials have diamagnetic behavior, creating permeability less than unity [J.6].

The *magnetic dipole moment* $\vec{m}$ of a current loop of area $\vec{S} = S\hat{n}$ and current I, shown in Fig (6.1e) and the *magnetization vector* $\vec{M}$ within a small volume Δv are defined as follows:

$$\vec{m} = I\vec{S} = IS\hat{n} \quad (a), \qquad \text{also } \vec{m} = \alpha^m \vec{H}_{loc} \quad (b)$$

$$\vec{M} = \frac{\vec{m}}{\Delta v} = N\vec{m} = N\alpha^m \vec{H}_{loc} \quad (c). \tag{6.1.24}$$

If each unit cell of the volume Δv contains one magnetic dipole, then the dipole concentration is $N = 1/\Delta v$. The above expression follows the description of the polarization process discussed in subsection (6.1.2). The magnetization, with *magnetic polarizability* α^m, is linearly proportional to the local magnetic field. *A magnetization vector* $\vec{M}$ *is the number of magnetic dipole moments per unit volume.* Further, equations (6.1.24b) and (6.1.24c) are analogous to equations (6.1.3) and (6.1.4) for the electric dipole moment $\vec{p}$ and electric polarization vector $\vec{P}$. Sometimes, the magnetic dipole moment is defined in a little different way, i.e. $\vec{m} = \mu_0 I\vec{S}$ and $\vec{m} = \alpha^m \vec{B}_{loc}$ [B.15, B.16].

The magnetization is caused by the external magnetic field and it is also proportional to $\vec{H}_{ext}$:

$$\vec{M} = \chi_m \vec{H}_{ext} \quad (a), \quad \text{or} \quad \vec{M} = \mu_o \chi_{rm} \vec{H}_{ext} \quad (b), \tag{6.1.25}$$

where χ_m and χ_{rm} are *magnetic susceptibility* and *magnetic relative susceptibility* of the material. The total magnetic flux density $\vec{B}$ due to both the external field and magnetization, following equation (6.1.6a), is obtained from the following expression:

$$\vec{B} = \mu_o \vec{H}_{ext} + \vec{M} \quad (a), \qquad \Rightarrow \vec{B} = \mu_o \vec{H}_{ext} + \chi_m \vec{H}_{ext} \quad (b)$$

$$\Rightarrow \vec{B} = \mu_o \left(1 + \frac{\chi_m}{\mu_o}\right) \vec{H}_{ext} = \mu_o \mu_r \vec{H}_{ext}$$

$$\Rightarrow \mu_r = 1 + \chi_{rm} \quad (c), \quad \chi_m = \mu_o \chi_{rm} \quad (d). \tag{6.1.26}$$

The following expression is obtained under the assumption of low-density magnetic dipoles, i.e. $H_{loc} = H_{ext}$:

$$\vec{B} = \mu_o \vec{H}_{ext} + N\alpha^m \vec{H}_{ext} = \mu_o \left(1 + \frac{N\alpha^m}{\mu_o}\right) \vec{H}_{ext} \quad (a)$$

$$\mu_r = 1 + \frac{N\alpha^m}{\mu_o} \quad (b), \quad \text{or } \mu_r = 1 + N\alpha^m \quad (c). \tag{6.1.27}$$

The magnetic *polarizability* α^m can be either a positive or negative quantity. In the case α^m is positive, $\mu_r > 1$ is obtained creating paramagnetism. For α^m slightly negative, the positive diamagnetism, with $0 < \mu_r < 1$, is obtained. Such materials are known as the *mu near-zero (MNZ)* materials. However, for the strongly negative α^m, the relative permeability μ_r can become negative. Sometimes, the magnetic polarizability α_m is pre-multiplied by μ_0. In this case, the above relation (6.1.27c), i.e. $\mu_r = 1 + N\alpha^m$ is used.

Clausius–Mossotti expression (6.1.14), accounting for the effect of the local field, can also be adopted for the magnetization. It provides the following improved expressions for relative permeability:

$$\mu_r = 1 + \frac{N\alpha^m/\mu_o}{1 - N\alpha^m/3\mu_o} \quad (a), \qquad \Rightarrow \frac{\mu_r - 1}{\mu_r + 2} = \frac{N\alpha^m}{3\mu_o} \quad (b)$$

$$\mu_r = 1 + \frac{N\alpha^m/\mu_o}{1 - N\alpha^m \gamma_m/\mu_o} \quad (c), \text{ or } \mu_r = 1 + \frac{N\alpha^m}{1 - N\alpha^m \gamma_m} \quad (d). \tag{6.1.28}$$

where γ_m is the *interaction constant* due to the presence of neighboring magnetic dipoles. It could be experimentally or numerically determined. For the Lorentz model, the interaction constant is $\gamma_m = 1/3$. Like the case of electric polarization, the above equations could be taken as the equation of the effective relative permeability of the magnetic inclusion, described by the magnetic dipole of polarizability α^m, in the air (vacuum) as the host medium. It leads to the MG model of a mixture of the magnetic material discussed in section (6.3.3). In equation (6.1.28d), the magnetic polarizability α_m is pre-multiplied by μ_0.

Normally, the electric polarization occurs due to the electric field and the magnetization is due to the magnetic field. These processes are described through the electric polarizability and magnetic polarizability, creating the electric and magnetic dipoles, given in the equations (6.1.3) and (6.1.24b). These expressions are rewritten below in more logical and general way [B.16]:

$$\vec{p} = \alpha^{ee}\vec{E}_{loc} \quad (a), \quad \vec{m} = \alpha^{mm}\vec{H}_{loc} \quad (b). \qquad (6.1.29)$$

However, the magnetic field can also cause electric polarization, and the electric field can cause magnetization in any material. They are known as the *cross-polarization* and expressed as follows: $\vec{p} = \alpha^{em}\vec{H}$ and $\vec{m} = \alpha^{me}\vec{E}$. The first letter of the superscript shows electric polarization/magnetization, and the second letter shows the polarizing or magnetizing field. In some *bi-isotropic materials* both polarizations and cross – polarizations occur, giving the following expressions for the joint polarization and magnetization:

$$\vec{p} = \alpha^{ee}\vec{E} + \alpha^{em}\vec{H} \quad (a), \quad \vec{m} = \alpha^{me}\vec{E} + \alpha^{mm}\vec{H} \quad (b)$$
$$\Rightarrow \begin{bmatrix} \vec{p} \\ \vec{m} \end{bmatrix} = \begin{bmatrix} \alpha^{ee} & \alpha^{em} \\ \alpha^{me} & \alpha^{mm} \end{bmatrix} \begin{bmatrix} \vec{E} \\ \vec{H} \end{bmatrix} \quad (c), \quad \alpha^{em} = -\alpha^{me} \quad (d). \qquad (6.1.30)$$

In the above-given expressions, all polarizability are scalar quantities. However, for the *bi-anisotropic materials*, these are the tensor quantities giving the following expression:

$$\begin{bmatrix} \vec{p} \\ \vec{m} \end{bmatrix} = \begin{bmatrix} \overline{\overline{\alpha}}^{ee} & \overline{\overline{\alpha}}^{em} \\ \overline{\overline{\alpha}}^{me} & \overline{\overline{\alpha}}^{mm} \end{bmatrix} \begin{bmatrix} \vec{E} \\ \vec{H} \end{bmatrix} \quad (a)$$
$$[\alpha^{em}] = -[\alpha^{me}]^T \quad (b). \qquad (6.1.31)$$

The bi-isotropic and bi-anisotropic materials are discussed in section (4.2.3) of chapter 4. The above concept of the polarization and cross-polarization, developing bi-anisotropy in the *split ring resonator* (SRR), is discussed in subsection (21.2.4) of chapter 21 on metamaterials.

6.2 Static Dielectric Constants of Materials

The natural dielectric materials are bulky for many applications in microwave and antenna technology. They have relative permittivity $|\varepsilon_r| > 1$. The lightweight artificial dielectric materials are engineered by using metallic inclusions in a host air-medium to get $|\varepsilon_r| > 1$ [B.11, J.7]. They are also engineered to get the fractional relative permittivity $0 < |\varepsilon_r| < 1$ [J.6, J.8]. Such materials are known as the *epsilon-near-zero (ENZ) materials*. Even negative permittivity could be obtained over a certain frequency band that is discussed in chapter 21. In this section, we discuss the dielectric constant of both the natural and artificial dielectric materials.

6.2.1 Natural Dielectric Materials

The models of the relative permittivity of the natural dielectric materials are strongly influenced by the computation of the local electric field that is normally different from the external field. Using the equation (6.1.23), Clausius–Mossotti equation (6.1.13) can be generalized. Further, by including the total polarizability α_{total} to equation (6.1.14b), more general expression for relative permittivity is obtained. The material could be also a homogeneous blend of several kinds of molecules leading to several polarizations. The general expression of relative permittivity can be modified to include a summation of polarization due to all molecules. The different value of the interaction constant γ is used in the Lorentz model and Onsager model. The experimental data-based curve-fitted model of the interaction constant γ can also be used [B.1, B.2, B.14]. The expressions involving single and multiple polarizations are summarized below:

$$\text{General expression}: \varepsilon_r = 1 + \frac{N\alpha_{total}/\varepsilon_o}{1 - N\gamma\alpha_{total}/\varepsilon_o} \quad (a)$$
$$\text{Several polarizations}: \varepsilon_r = 1 + \frac{\sum_n (N_n\alpha_n)/\varepsilon_o}{1 - \gamma\sum_n (N_n\alpha_n)/\varepsilon_o} \quad (b). \qquad (6.2.1)$$

Clausius–Mossotti model could be obtained by taking $\gamma = 1/3$ in the above equations. The Onsager model could be obtained for $\gamma = 1/(2\varepsilon_r + 1)$.

Like a low-density gaseous dielectric-medium, the solid dielectrics also have three kinds of polarization conditions:

- Only electronic polarization.
- Electronic + ionic polarization.
- Electronic + ionic + orientational (dipolar) polarizations.

Normally, the artificial dielectrics could also be treated as a low-density medium with induced dipole kind of polarization [B.11]. However, like many natural polar dielectrics, the induced dipole polarization-based artificial dielectrics are not temperature-dependent. The induced dipole polarization is more akin to the electronic polarization of natural dielectrics. The artificial dielectrics are discussed briefly in subsection (6.2.2). A solid material could be nonpolar or polar material.

6.2.1.1 Nonpolar Solid

The nonpolar solids do not have permanent dipolar polarization. They can acquire either electronic polarization or a combination of electronic and ionic polarization. A suitable value of the interaction parameter γ can be selected to match the computed relative permittivity with the experimentally obtained value. Lorenz and Onsager model can provide an approximate value of permittivity. The measurements of the dielectric constants, both at low frequency and high frequency, are used to know the electronic and ionic polarization ratio that gives information on the relative strength of ionic and electronic polarization. The susceptibility, and also relative permittivity, of this kind of material is also temperature-independent [B.1, B.2].

6.2.1.2 Polar Solid

No theory exists to compute the dipolar polarization P_d of a polar solid. The dipolar polarizability, $\alpha_d = \mu^2/3kT$, is valid for the polar gases and polar liquids. The permanently polarized molecules of gases and liquids can rotate freely. However, the molecules (dipoles) of a solid are restricted by the structural arrangement of a crystal and have the directional character of bonding. It varies from material to material. The rotation of the permanent dipole in a solid is not random under the thermal agitation. Therefore, a factor 3kT associated with α_d is not correct. However, some rotation of polar molecule is available showing the temperature dependence of the relative permittivity, i.e. there is a decrease in ε_r with an increase in the temperature. At lower temperatures, the dipolar polarizability $\alpha_d = 0$ is showing freezing, i.e. decrease of ε_r. Anyway, the total polarization of a natural polar solid is due to all three kinds of polarization.

6.2.2 Artificial Dielectric Materials

Several dielectric materials used in the microwave applications are *heterogeneous*, i.e. they are mixtures of two or more numbers of individual dielectric materials. Such mixtures are also called *two* and *multiphase mixtures*. Normally, materials of some regular or irregular geometrical shapes, called the *inclusions*, are embedded in another material, called the *host*. The dielectric or metallic inclusions, i.e. the metallic particles of several geometrical shapes, can be treated as *mega molecules*. They get polarized under the influence of an external electric field. Traditionally, the mixing formulae have been developed for the dielectric constants of such mixtures for the geophysical, remote sensing, and hydrological research. However, recently a wide range of *artificial dielectric materials* are developed by using a dielectric mixture, dielectric-metal mixture, and dielectric-ferroelectric mixture [B.17, B.18]. Even a porous material can be treated as a mixture of dielectrics and air pockets. For instance, 99.6% alumina (Al_2O_3) ceramic could be treated as a mixture of aluminum oxide as a host material and the air pockets of diameter <1 μm as the inclusion material [B.19]. Even natural dielectrics can be viewed as composite materials with vacuum as a host material and embedded dipoles as inclusions. In this case, the dipoles are held in a vacuum with the help of electrostatic forces. However, in the case of artificial dielectrics, the vacuum as a host holds the metallic inclusion structures either with the support of insulating rods or by embedding the metallic inclusions in a foam medium [J.7]. Depending on the geometry of the metallic inclusions, electric polarization or magnetic polarization (magnetization) can develop in the inclusions. Finally, the polarization (magnetization) leads to the permittivity (permeability) in the engineered medium. The magnetic dipole moments usually create diamagnetism in the artificial dielectrics. Normally, artificial dielectrics develop both electrical and magnetic properties. The relative permittivity ε_r and relative permeability μ_r could be created in the artificial dielectrics. The refractive index of the artificial dielectrics is $n = \sqrt{\varepsilon_r \mu_r}$. The refractive index is more than unity because $\varepsilon_r > 1$. However, due to diamagnetism, permeability is $\mu_r < 1$, so magnetization reduces the value of the refractive index. On suppressing magnetization in inclusions, the refractive index can be enhanced. Sometimes, one of the electrical properties could be predominant, while others can be ignored. In some cases, both properties can be tailored for the realization of engineered metamaterials. Normally such materials are anisotropic.

Over years, several closed-form expressions have been developed to model the heterogeneous materials, both natural and artificial, by treating them as *equivalent homogeneous material* with *equivalent relative permittivity*. This modeling process is called the *homogenization* of the heterogeneous mixture in which the host and inclusions ignore their electric behaviors in a range of operating frequency bands. The inclusions could be randomly distributed in a volume of a host material. Developing such artificial material is cumbersome. However, if inclusions form a regular structure, such as the crystal structure of a solid, or periodically arranged metallic inclusion in the artificial dielectrics, the material could be treated as a homogeneous medium only at the lower frequency. In lower frequency band, the size of the inclusion is much smaller than the half wavelength of the exciting field; and also *lattice constant* (separation between two adjacent inclusion elements) is also smaller than the half wavelength. In this case, the EM-waves interacting with the material undergo refraction. The EM-waves view the material as a homogeneous medium described by the permittivity, permeability, or refractive index.

In the case, the separation between inclusions is half-wavelength, the homogenization does not work, and the Bragg's diffraction patterns are obtained. The EM waves view the material medium as an arrangement of discrete inclusions. Such a material mixture, called *band-gap materials*, could be 1D, 2D, and 3D-medium. They exhibit interesting electromagnetic properties and are known as the *photonic bandgap (PBG)* or *electromagnetic band-gap (EBG)* materials. The 1D and 2D EBG materials are discussed in chapters 19 and 20. Similarly, artificial inclusions can also create so-called *metamaterials*. These can be subjected to the process of homogenization with both their equivalent relative permittivity and relative permeability as the negative quantities. The *metamaterials* are introduced in section (5.5) of chapter 5 and further discussed in the chapter 21. The *equivalent permeability* is obtained only if either the host or the inclusion material is magnetic [J.9]. Further, it is also possible to get the equivalent complex relative permittivity that includes the losses in a mixture.

In brief, the engineered artificial dielectrics have a long history – starting with a composite mixture of natural dielectric inclusions in a host dielectric-medium, to the metallic inclusion of several metallic shapes in a host air-medium or a host low permittivity (foam) medium, to the EBG material, and metamaterials. A few classical illustrative examples of artificial dielectrics and the process of their modeling, using the concept of polarization are presented below. Also, their transmission line modeling is discussed to get the relative permittivity of these media. It helps to understand the realization of 1D and 2D EBG materials and metamaterials.

6.2.2.1 Classical Examples of Artificial Dielectrics

Figure (6.2a–d) shows embedded *metallic spheres* or metallic disks as inclusions in the host low permittivity dielectric-medium, or the host air-medium. A linear periodic arrangement of inclusions creates the 1D artificial dielectric-medium, whereas the 2D lattice structures create the 2D artificial dielectric-medium. Similarly, the 3D lattice structures with such inclusions create artificial dielectrics. If the lattice constant d and diameters of inclusion are much smaller than the wavelength of external field incident on the material, then artificial dielectrics are treated as the homogeneous medium. In the present discussion, the interaction between inclusions is ignored. However, mutual coupling occurs among closely spaced inclusions [B.11]. Otherwise, they form EBG materials. These are discussed in chapter 19. The spherical and disk inclusions are normally embedded, shown in Figure (6.2a and b), in a host foam-medium. However, disks could be either printed or pasted over a dielectric sheet. Figure (6.2c and d) show that metallic spheres are supported by the insulated rods in the air-medium.

Due to the symmetry, the spherical inclusions create the isotropic dielectric-medium, whereas the disk inclusions create an anisotropic dielectric-medium. Figure (6.3a) shows that on the application of the external electric field, the metallic sphere gets polarized, creating electric dipole moment p. However, the external magnetic fields on the surface of the sphere are distorted and induced eddy current creates magnetic dipole with fields opposing the external magnetic field. The diamagnetic effect decreases the permeability of the host medium below unity. The computation of electric and magnetic polarizability is discussed further in this section. Figure (6.3b) shows that the magnetic field on the surface of the disk could be undistorted, and there is no induced eddy current. Therefore, the magnetization is absent in this case.

Figure (6.4a and b) shows the 1D periodic arrangement of the thin metallic strip and metallic rods inclusions in a low permittivity host dielectric-medium creating the *1D artificial dielectrics*. Figure (6.4c and d) shows the 2D periodic arrangement of these inclusions creating the *2D artificial dielectrics*. Figure (6.4a–d) shows the external electric field is normal to the strips and rods. The polarized inclusions create the capacitive effect, and relative permittivity is developed in the artificial dielectrics. However, the created permittivity is small. Although under such polarization, the refractive index is more than unity, and phase velocity is reduced causing a delay in propagation as compared to the propagation in free space. Such dielectrics are known as the *phase delay dielectrics*. However, in case, the electric field is parallel to the strips and rods, it creates the magnetic effect, i.e. inductive loading, in the opposite direction reducing permeability less than unity. The magnetic

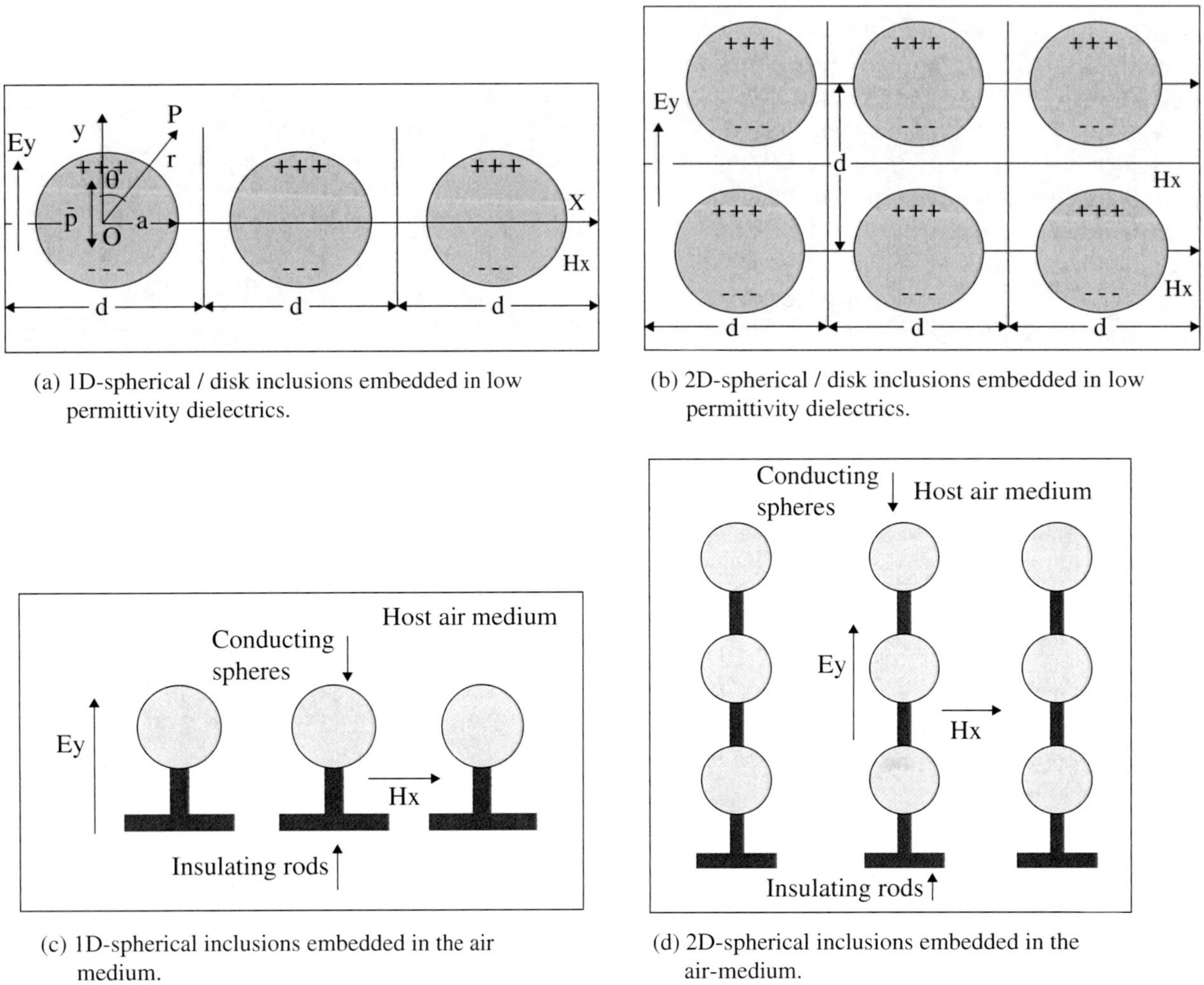

Figure 6.2 1D and 2D artificial media created by spherical inclusions in low permittivity dielectric host medium and host air-medium.

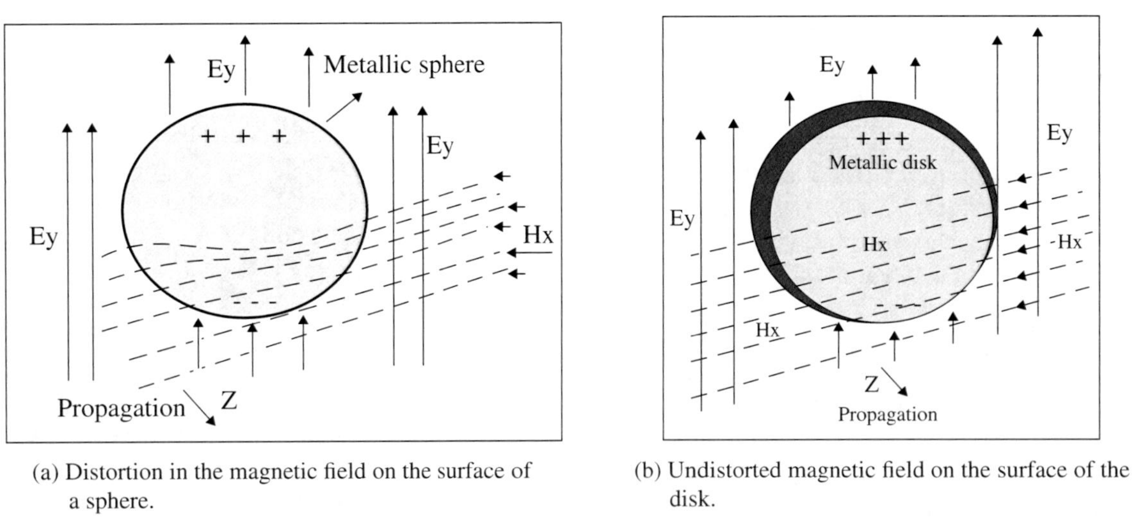

Figure 6.3 Metallic spherical and disk inclusions in low permittivity dielectric host/air host medium.

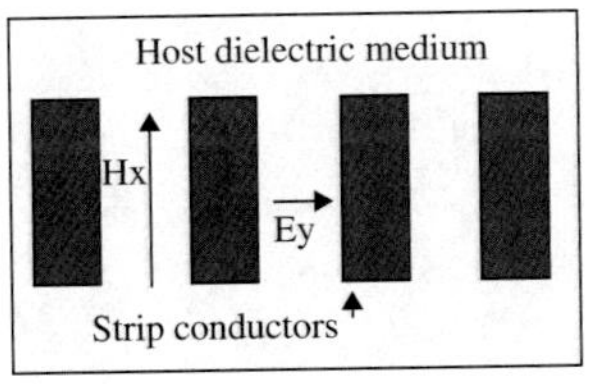

(a) 1D-metallic strips in low permittivity dielectrics.

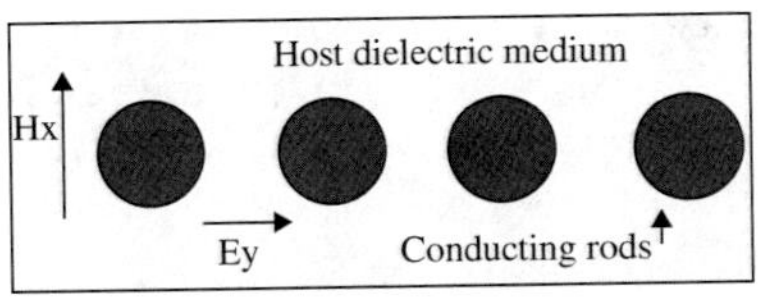

(b) 1D-metallic rods in low permittivity dielectrics.

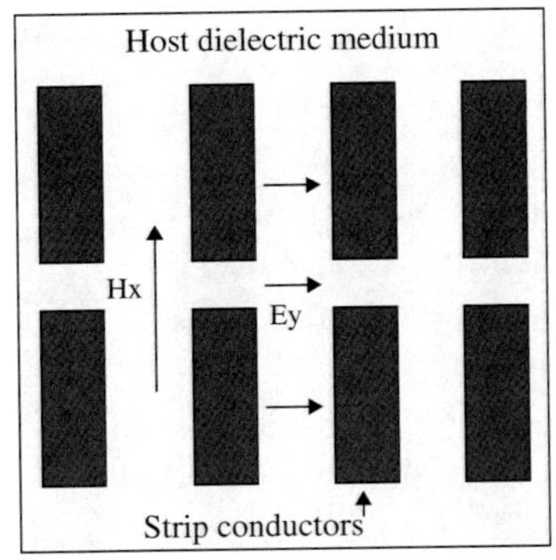

(c) 2D-metallic strips in low permittivity electrics.

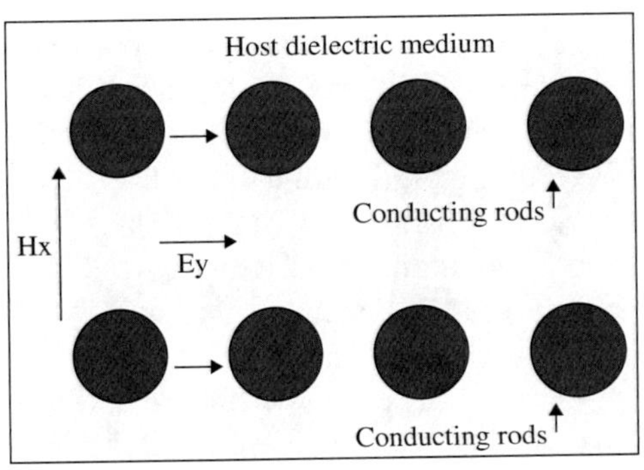

(d) 2D-metallic rods in low permittivity dielectrics.

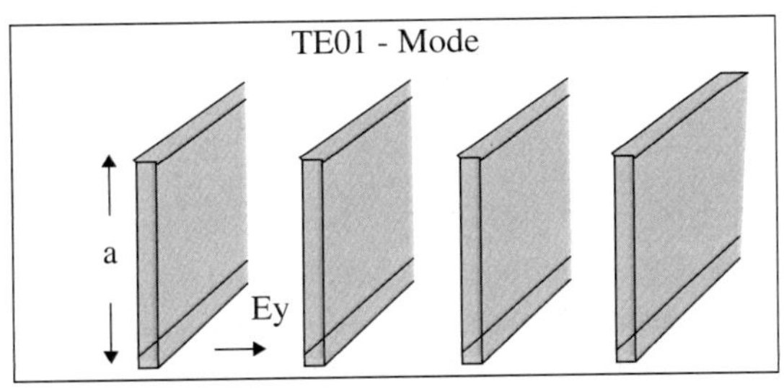

(e) 1D-parallel plate waveguide medium.

Figure 6.4 Metallic strips, conducting rods and parallel-plate waveguide inclusions forming the 1D and 2D media.

effect could depend on the separation of strip conductors. It increases the phase velocity causing a phase advance. Such dielectrics are known as the *phase advance dielectrics* [B.11, J.6, J.7]. In the case of closely spaced rodded or wire medium, with incident electric field parallel to the wire stronger polarization can occur and the wire medium behaves like an *electric plasma medium*. It is discussed in section (21.1) of chapter 21. Again, this medium provides positive permittivity less than unity. The wire medium acts as the phase advance dielectrics. However, below the plasma frequency, the wire medium has negative permittivity.

Figure (6.4e) shows the 1D periodic arrangement of parallel conducting plates supporting the TE_{01} kind of wave propagation. It is like the TE_{10} mode of a rectangular waveguide, discussed in chapter 7. Using expression (7.4.27) of chapter 7, the refractive index and relative permittivity of the parallel-plate artificial dielectrics are computed as follows:

$$n = \frac{c}{v_p} = \frac{\beta^{TE_{10}}}{\beta_0} = \left[1 - \left(\frac{\pi}{a\beta_0}\right)^2\right]^{1/2} = \left[1 - \left(\frac{\lambda_0}{2a}\right)^2\right]^{1/2}$$
$$= \left[1 - \left(\frac{\lambda_0}{\lambda_c}\right)^2\right]^{1/2} \qquad \text{(a)}$$
$$\varepsilon_r = \left[1 - \left(\frac{\lambda_0}{\lambda_c}\right)^2\right] = \left[1 - \left(\frac{f_c}{f}\right)^2\right] \qquad \text{(b).}$$
$$(6.2.2)$$

For $f > f_c$, the refractive index n and relative permittivity are less than unity; so the present medium supports fast-wave propagation. For $f < f_c$, it behaves as an ENG medium, discussed in section (5.5.3) of chapter 5. By comparing the above expression for the relative permittivity against expression (6.5.16d), it is observed that the parallel-plate medium behaves like a plasma medium. Below the cut-off frequency, the medium does not support the wave propagation. The *rodded medium*, called the *wire medium*, also behaves like a plasma medium. It is further discussed in chapter 21 as an essential component for the realization of metamaterials.

6.2.2.2 Computation of Dielectric Constant of Artificial Dielectrics

The polarizability of a metallic sphere, shown in the unit cell of Fig (6.2a), is computed under the assumption that neighboring spherical inclusions do not interact and influence the local field of the dipole created at the center of the sphere. The dipole moment $\vec{p}$ is y-directed due to the static or low-frequency external field E_y. The total electric potential V_{tot} at any location P at the distance r is a summation of the potential V_{ext} due to external E-field and potential V_{dip} due to the dipole moment $\vec{p}$ [J.7]:

$$V_{ext} = -E_{ext}\, r\cos\theta \quad (a), \qquad V_{dip} = \frac{p}{4\pi\varepsilon r^2}\cos\theta \quad (b)$$
$$V_{tot} = V_{ext} + V_{dip}, \Rightarrow V_{tot} = -E_{ext}\, r\cos\theta + \frac{p}{4\pi\varepsilon r^2}\cos\theta \quad (c). \tag{6.2.3}$$

Using equation (6.2.3c) at r = a, total potential on the surface of the metallic sphere of radius "a" is zero, giving an expression for the electric dipole moment p. It is further used with equation (6.1.3) to obtain the expression for the *static or low-frequency electric polarizability* α_e^e (we have used this symbol for the electronic polarization also) of the metallic sphere embedded in low permittivity dielectric ($\varepsilon_r \approx 1$, i. e. $\varepsilon \approx \varepsilon_0$) medium:

$$p = 4\pi\varepsilon_0 a^3 E_{ext} \quad (a), \qquad p = \alpha_e^e E_{ext} \quad (b)$$
$$\text{On comparing equations} - (a) \text{ and } (b): \alpha_e^e = 4\pi\varepsilon_0 a^3 \quad (c). \tag{6.2.4}$$

Equation (6.2.4c) is identical to equation (6.1.19b) of electronic polarization as both equations are due to the induced dipoles under the assumption $E_{loc} = E_{ext}$. The effective relative permittivity of the homogeneous medium, i.e. the *static relative permittivity* of the artificial dielectrics is obtained by substituting equation (6.2.4c) in equation (6.1.8):

$$\varepsilon_r = 1 + 4\pi N a^3. \tag{6.2.5}$$

The above expression is improved by using an equation (6.2.4c) with Lorentz model given in equation (6.1.14a):

$$\varepsilon_r = 1 + \left[\frac{4\pi N a^3}{1 - (4\pi/3)Na^3}\right]. \tag{6.2.6}$$

In equation (6.2.6) the interaction constant (γ) is taken as 1/3. A little different value of the interaction constant γ is obtained by solving the Laplace equation for the 3D lattice of spherical inclusion. For the 2D lattice, modified result for the parameter γ is obtained [B.11]. It is noted that only one inclusion is located within a unit of the 3D lattice. For the cubic lattice of lattice constant d, the density of the inclusion is $N = 1/\Delta v = 1/d^3$.

The time-varying magnetic field of the incident TEM wave causes the magnetic polarization, i.e. the magnetization in the spherical inclusions. On substituting magnetic polarizability α_m of the metallic sphere, given below, in equation (6.1.27b), the relative permeability of the artificial magnetic medium is obtained [J.7]:

$$\alpha_m = -2\pi\mu_0 a^3 \quad (a), \qquad \mu_r = 1 - 2\pi N a^3 \quad (b). \tag{6.2.7}$$

It is noted that the electric polarizability is a positive quantity giving the increased value of relative permittivity, more than unity. However, the magnetic polarizability is a negative quantity, giving permeability less than unity, $0 < \mu_r < 1$. The Lorentz model can also be used to get an improved expression for μ_r, similar to equation (6.2.6). Once the effective values for ε_r and μ_r of the artificial dielectrics with the magnetic property are computed, the refractive index can be obtained from $n = \sqrt{\varepsilon_r \mu_r}$. The reduced value of relative permeability μ_r lowers the refractive index. It is not a desirable feature [J.6, J.7]. The resonating inclusions can also provide a negative value of permittivity and permeability. The Lorentz and Drude dispersion models predict such negative values over a certain range of frequencies. For the metallic sphere periodically arranged in a dielectric host medium, the effective permittivity and permeability of the artificial medium could be computed with the help of MG mixing rule, discussed in section (6.3.3). A modern treatment of frequency-dependent polarizability of some geometrical shapes of inclusion, Lorentz, and Drude dispersion models, and also MG models, from the field theory point of view, is discussed by Bohren and Huffman [B.20].

The polarization response of the inclusions under the high-frequency incident EM-wave is frequency-dependent. It results in the material dispersion, i.e. the frequency dependence of ε_r and refractive index. Lorentz's oscillator model, discussed in section (6.5.1) explains such dispersion. The artificial dielectric-medium could be treated as a loss-less plasma medium. Thus, using equation (6.5.5b) the frequency-dependent relative permittivity of the artificial dielectric-medium is written as follows:

$$\varepsilon_{rc}^*(f) = 1 + \frac{f_p^2}{f_c^2 - f^2} \quad \text{(a)}$$

$$\text{for} \quad f << f_c, \varepsilon_{rc}^*(f) \approx 1 + \frac{f_p^2}{f_c^2} \quad \text{(b)},$$

$$(6.2.8)$$

where f_c is the characteristic resonance frequency of the induced dipoles in the inclusions and f is the operating frequency of the incident EM-waves. The plasma frequency f_p is constant. Equation (6.2.8a) shows dispersion in the artificial dielectrics. However, at a frequency much below the resonance frequency f_c, equation (6.2.8b) demonstrates dispersionless relative permittivity for the artificial dielectrics. The dispersionless artificial dielectrics are important for the development of wideband microwave components and antenna [J.6, J.7].

In the above discussion, the spherical inclusion is treated in detail. The polarizability of inclusions of other geometrical shapes and corresponding relative permittivity of the artificial dielectrics has been obtained [B.11, B.12, J.7]. For instance, these parameters for the metallic circular disks of radius "a" and thin conducting strips of width "w" are summarized below:

$$\text{Circular disks}: \alpha_e = \frac{16\varepsilon_0 a^3}{3} \quad \text{(a)}$$

$$\varepsilon_r = 1 + \frac{N\alpha_e}{\varepsilon_0} = 1 + \frac{16a^3}{3}N \quad \text{(b)}$$

$$\text{Metallic strips}: \alpha_e = \frac{\pi\varepsilon_0 w^2}{4} \quad \text{(c)}$$

$$\varepsilon_r = 1 + \frac{N\alpha_e}{\varepsilon_0} = 1 + \frac{\pi w^2}{4}N \quad \text{(d)}$$

$$\text{Thin metallic rod}: \alpha_e = \frac{4\pi\varepsilon_0 d^3}{3\log(2d/a) - 1} \quad \text{(e)}$$

$$\varepsilon_r = 1 + \frac{4\pi d^3}{3\log(2d/a) - 1} \quad \text{(f)}.$$

$$(6.2.9)$$

The above expressions of relative permittivity are obtained on assumption $E_{loc} = E_{ext}$. In equation (6.2.9e), 2d is the length of the metallic rod and "a" is its radius. The improved results are obtained using an equation (6.1.14a). However, the experimentally obtained value of the parameter γ can further improve the results of computed relative permittivity. The polarizabilities of several kinds of inclusions have also been analytically determined [B.18, B.21].

6.3 Dielectric Mixtures

The permittivity of a mixture is due to the interfacial polarization, discussed in section (6.6) using the circuit models. However, the permittivity of a mixture, or even of the metamaterials, can be obtained in terms of the polarizability of the inclusions, leading again to the equivalent relative permittivity [B.15, J.10, J.11]. This section summarizes a few such expressions for the isotropic host and isotropic inclusion materials of *spherical shapes*. However, some models can also account for the anisotropy in both the host and inclusion materials [J.12].

6.3.1 General Description of Dielectric Mixture Medium

Figure (6.5a) shows the cross-section of a heterogeneous medium. The main continuum material, known as the *host material*, has relative permittivity ε_{rh} in which the *inclusion material*, with relative permittivity ε_{ri}, is randomly scattered. However, in the case of the artificial dielectrics and metamaterials, the inclusions are periodically arranged. To get the homogenized medium, the size of inclusions and lattice constant are much smaller than a wavelength. The dielectric constant of a heterogeneous material is different from the dielectric constant of the host material. It depends on the relative permittivity, concentration, size, and shape of the inclusion material.

The modeling of the heterogeneous mixture is treated in two ways – on the *mass basis* and *volume basis* [B.17, B.22, B.23]. The mass fraction (M) of inclusions and the volume fraction (f) of inclusions in the mixture are expressed as follows:

$$M = \frac{m_i}{m_i + m_h} \quad \text{(a)}, \quad f = \frac{v_i}{v_i + v_h} = \frac{v_i}{v_m} \quad \text{(b)}$$

$$f = \frac{4/3\pi a^3}{d^3} \quad \text{(c)},$$

$$(6.3.1)$$

where m_i and m_h are the mass of inclusions and host respectively. The volume of inclusions and host are v_i and v_h, respectively. The total volume of the mixture is $v_{mi} = v_i + v_h$. In the case of 3D, artificial dielectric material is created by the centrally located spherical inclusion of radius a in the cubic lattice of lattice constant d, the volume fraction of inclusion is given by equation (6.3.1c).

For the density ρ_i of inclusion and density ρ_h of host material, the mass fraction could be rewritten as

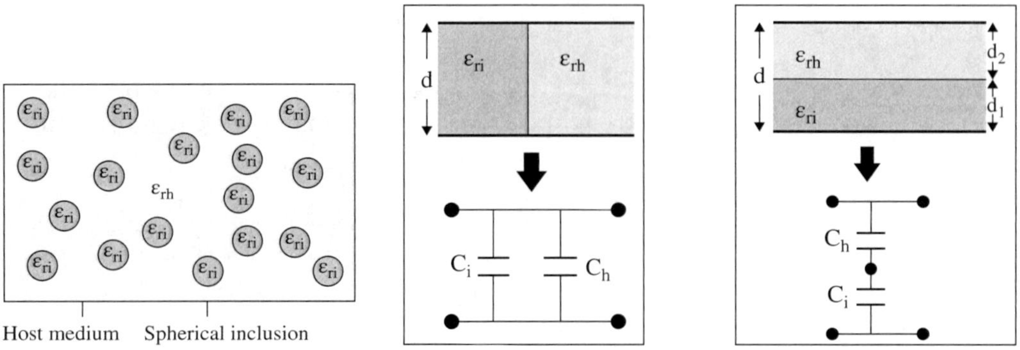

(a) A heterogeneous mixture: Spherical inclusion material embedded in the host material medium.

(b) Parallel arrangement of inclusion and host materials.

(c) Series arrangement of inclusion and host materials.

Figure 6.5 A heterogeneous mixture and equivalent circuit models of the limiting cases.

$$M = \frac{v_i \rho_i}{v_i \rho_i + v_h \rho_h}. \tag{6.3.2}$$

In the case of a porous material, the porosity (P) is defined as [B.19]:

$$P = \frac{\rho_h - \rho_o}{\rho_h}, \tag{6.3.3}$$

where ρ_h is the density of host material without air pockets and ρ_o is the density of the pores with air packets. The porosity P is treated as the volume fraction f in the expression for the equivalent relative permittivity of a mixture.

6.3.2 Limiting Values of Equivalent Relative Permittivity

The equivalent relative permittivity of a mixture is different from that of the host and the inclusion materials. The upper and lower bounds on the equivalent relative permittivity of a mixture could be estimated from the capacitor circuit models, shown in Fig (6.5b and c) [B.1, B.17, J.13]. The capacitor circuit model is known as the *Wiener model* of a mixture [J.14]. The relative permittivities of host and inclusion are treated as the capacitors and their combination represents the equivalent relative permittivity of a mixture medium. The lower and upper bounds on the equivalent relative permittivity are also estimated by the variational method [J.15].

Likewise, the series combination of capacitors corresponds to the minimum value of equivalent relative permittivity of a mixture medium. The average of the minimum and maximum equivalent relative permittivity can be accepted as the equivalent relative permittivity of a mixture medium.

6.3.2.1 Parallel Capacitors Model

The total volume v_m of a heterogeneous mixture is separated into two volumes v_i and v_h, occupied by the inclusion and host materials. Figure (6.5b) shows the parallel capacitor model of a mixture of two materials. It provides the maximum value of the equivalent relative permittivity $\varepsilon_{r,\,eqmax}$ of a mixture medium. The capacitors C_i and C_h corresponding to the inclusion and host materials are given below:

$$C_i = \frac{\varepsilon_0 \varepsilon_{ri} A_i}{d} = \frac{\varepsilon_0 \varepsilon_{ri} A_i d}{d^2} = \frac{\varepsilon_0 \varepsilon_{ri} v_i}{d^2} \quad (a), \quad C_h = \frac{\varepsilon_0 \varepsilon_{rh} v_h}{d^2} \quad (b). \tag{6.3.4}$$

The equivalent capacitance, corresponding to the equivalent relative permittivity of a mixture medium, is

$$C_{eq} = \frac{\varepsilon_0 \varepsilon_{r\,eq\,max} (v_h + v_i)}{d^2} = \frac{\varepsilon_0 \varepsilon_{r\,eq\,max} v_m}{d^2}. \tag{6.3.5}$$

The parallel combination of C_h and C_i is given by $C_{eq} = C_i + C_h$. Therefore, the equivalent relative permittivity is obtained from the above equations in terms of the volume fractions f,

$$\varepsilon_{r\,eq} = \varepsilon_{ri} \frac{v_i}{v_m} + \varepsilon_{rh} \frac{v_h}{v_m} \quad (a)$$

where,

$$v_i + v_h = v_m \Rightarrow \frac{v_i}{v_m} + \frac{v_h}{v_m} = 1 \quad (b)$$

$$\frac{v_i}{v_m} = f \quad (c), \qquad \frac{v_h}{v_m} = 1 - f \quad (d). \tag{6.3.6}$$

The upper limit of the equivalent relative permittivity for a mixture is

$$\varepsilon_{r\,eq\,max} = f\,\varepsilon_{ri} + (1 - f)\,\varepsilon_{rh}. \tag{6.3.7}$$

6.3.2.2 Series Capacitors Model

Figure (6.5c) shows the series capacitor model of a mixture of two materials. It provides the minimum value of the equivalent relative permittivity $\varepsilon_{r\,eq\,min}$ of a mixture medium. The capacitances, C_i, C_h, and C_{eq} corresponding to the inclusion, host, and mixture, respectively, are given by the following expressions:

$$C_i = \frac{\varepsilon_0 \varepsilon_{ri} V_i}{{d_1}^2}, \quad C_h = \frac{\varepsilon_0 \varepsilon_{rh} V_h}{{d_2}^2}, \quad C_{eq} = \frac{\varepsilon_0 \varepsilon_{r\,eq\,min} V_m}{d^2}$$

where, $d = d_1 + d_2$.

$$(6.3.8)$$

The equivalent capacitance of two series-connected inclusion and host media capacitances provide minimum equivalent relative permittivity $\varepsilon_{r\,eq\,min}$ of the mixture:

$$\frac{1}{C_{eq}} = \frac{1}{C_i} + \frac{1}{C_h}, \Rightarrow \frac{d^2}{\varepsilon_{r\,eq\,min} V_m} = \frac{d_1^2}{\varepsilon_{ri} V_i} + \frac{d_2^2}{\varepsilon_{rh} V_h}$$

$$\Rightarrow \frac{1}{\varepsilon_{r\,eq\,min}} = \frac{(d_1/d)}{\varepsilon_{ri}} + \frac{\{1-(d_1/d)\}}{\varepsilon_{rh}}. \tag{6.3.9}$$

The volume fraction is defined as $(v_i/v_m) = f, \Rightarrow Ad_1/Ad = f, \Rightarrow d_1/d = f$. Therefore, the minimum value of the equivalent relative permittivity of a mixture of two materials mixture, giving the lower bound, is

$$\varepsilon_{r\,eq\,min} = \frac{\varepsilon_{ri}\,\varepsilon_{rh}}{f\,\varepsilon_{rh} + (1-f)\varepsilon_{ri}}. \tag{6.3.10}$$

The upper and lower bounds on the equivalent relative permittivity can be extended to the (n−1) layered inclusion materials in a host material. The expressions for the bounds are valid for both $\varepsilon_{ri} > \varepsilon_{rh}$ and $\varepsilon_{rh} > \varepsilon_{ri}$ cases.

6.3.2.3 Variational Model of Bounds on the Equivalent Relative Permittivity of the Mixture

Hashin and Shtrikman have provided the following expressions, using the variational method, for minimum and maximum values of the equivalent relative permittivity $\varepsilon_{r\,eq}$ of the 3D mixture [J.15, J.16]:

$$\varepsilon_{r\,eq\,min} = \varepsilon_{rh} + \frac{f}{\left(\frac{1}{\varepsilon_{ri}-\varepsilon_{rh}}\right) + \left(\frac{1-f}{3\varepsilon_{rh}}\right)} \quad \text{(a)}$$

$$\varepsilon_{r\,eq\,max} = \varepsilon_{ri} + \frac{1-f}{\left(\frac{1}{\varepsilon_{rh}-\varepsilon_{ri}}\right) + \left(\frac{f}{3\varepsilon_{ri}}\right)} \quad \text{(b).}$$

$$(6.3.11)$$

The above expressions are valid for $\varepsilon_{ri} > \varepsilon_{rh}$. For a 2D mixture, the bounds could be written as follows:

$$\varepsilon_{r\,eq\,min} = \varepsilon_{rh} + \frac{f}{\left(\frac{1}{\varepsilon_{ri}-\varepsilon_{rh}}\right) + \left(\frac{1-f}{2\varepsilon_{rh}}\right)} \quad \text{(a)}$$

$$\varepsilon_{r\,eq\,max} = \varepsilon_{ri} + \frac{1-f}{\left(\frac{1}{\varepsilon_{rh}-\varepsilon_{ri}}\right) + \left(\frac{f}{2\varepsilon_{ri}}\right)} \quad \text{(b).}$$

$$(6.3.12)$$

The variational expressions are more accurate as compared to the capacitor models. The variational expressions provide a narrower margin between the minimum and maximum values of $\varepsilon_{r\,eq}$. The average value of the lower and upper bounds provides the more accurate value of the equivalent relative permittivity of a mixture [J.16]. Figure (6.6a) compares the lower and upper bounds, obtained from the capacitor model and also from the variational method, against the finite-difference time-domain (FDTD) results of the equivalent relative permittivity of a mixture of host medium $\varepsilon_{rh} = 1$ and inclusion with $\varepsilon_{ri} = 16$. All FDTD results are within the bounds of both methods. However, bounds of the variational methods are closer to the FDTD results.

6.3.3 Additional Equivalent Permittivity Models of Mixture

Several models of the equivalent relative permittivity of a mixture medium, apart from the capacitor models discussed above, have been reported with a varying degree of accuracy. Three models presented below are valid for the *isotropic host* (ε_{rh}) and *isotropic inclusion* (ε_{ri}) materials. Both materials are individually homogeneous and their mixture is dilute; i.e. the *contrast* $K = \varepsilon_{ri}/\varepsilon_{rh}$ is small, with normally $\varepsilon_{ri} > \varepsilon_{rh}$.

6.3.3.1 Maxwell Garnett Formula

J.C Maxwell Garnett who provided the mixing rule [J.19] was another researcher, not the famous J.C. Maxwell [B.24, J.20]. The effective or equivalent relative permittivity $\varepsilon_{r\,eq}$ of a mixture of the 3D *spherical shape,* and also the *circular cylinder shape,* i.e. the 2D *cylindrical* dielectric inclusion material, in the host dielectric-medium is computed by the following phenomenological MG formulae [B.17, J.16, J.19, J.20]:

3D spherical inclusion :

$$\varepsilon_{r\,eq} = \varepsilon_{rh}\left[1 + \frac{3f(\varepsilon_{ri}-\varepsilon_{rh})}{\varepsilon_{ri} + 2\varepsilon_{rh} - f(\varepsilon_{ri}-\varepsilon_{rh})}\right] \quad \text{(a)}$$

2D cylindrical inclusion :

$$\varepsilon_{req} = \varepsilon_{rh}\left[1 + \frac{2f(\varepsilon_{ri}-\varepsilon_{rh})}{\varepsilon_{ri} + \varepsilon_{rh} - f(\varepsilon_{ri}-\varepsilon_{rh})}\right] \quad \text{(b),}$$

$$(6.3.13)$$

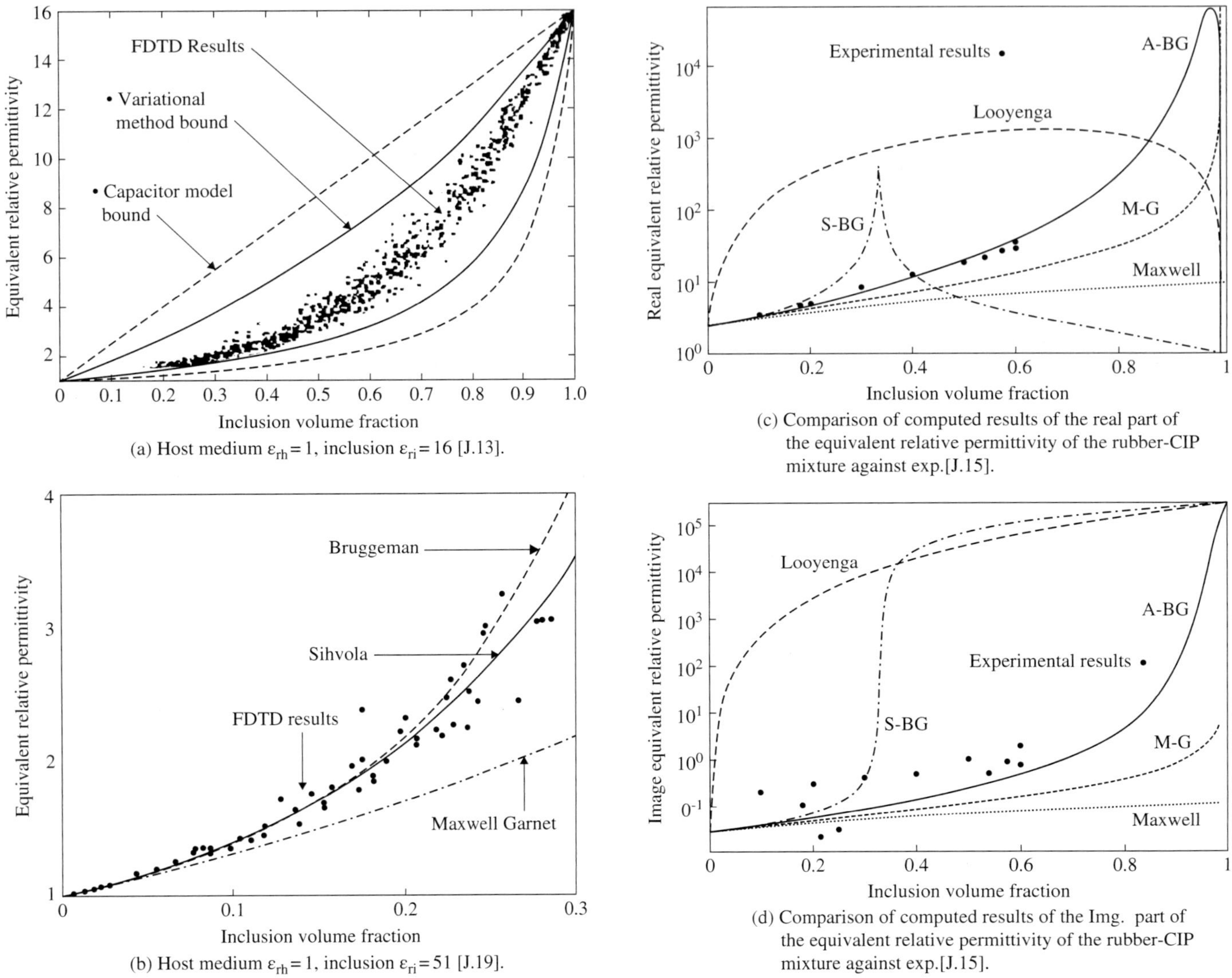

Figure 6.6 Comparisons of several mixture formulae against the FDTD and experimental results. *Source:* Kärkkäinen et al. [J.16], Merrill et al. [J.17], and Kärkkäinen and Sihvola [J.18].

where f is the volume fraction of the inclusion material. In case, the factor $f(\varepsilon_{ri} - \varepsilon_{rh})$ in the denominator of the above equation is dropped, the *Maxwell model* of a mixture is obtained [J.17]. The MG model is more suitable for the well-separated noninteracting inclusion of low-inclusion density. For the higher density of inclusion, the polarized field of inclusions, forming dipoles, interacts with each other. In this case, the MG model underestimates the permittivity of the mixture as shown in Fig (6.6b). The MG model is also suitable to obtain homogenization of the cubic-centered inclusion of artificial dielectrics and metamaterials. This aspect is further discussed in subsection (21.4.2) of chapter 21. The ellipsoidal inclusions can also be incorporated in the mixing rule [J.21].

The phenomenological MG formula can be obtained from the Clausius–Mossotti expression (6.1.14a). To get the MG mixing formula, the air is taken as the host medium and the spherical dielectric inclusion of radius 'a' is replaced by the electric dipole of electric polarizability α_e obtained from the EM Theory [B.20, B.25]:

$$\alpha_e = 4\pi\varepsilon_0\, a^3 \frac{(\varepsilon - \varepsilon_0)}{(\varepsilon + 2\,\varepsilon_0)}, \qquad (6.3.14)$$

where ε is the permittivity of the dielectric sphere. In the case of a metallic sphere $\varepsilon \to \infty$, the above expression is reduced to equation (6.2.4c). Each unit cell of the volume Δv of the artificial material contains one spherical inclusion, i.e. the density (concentration) of inclusion is $N = 1/\Delta v$. Using equation (6.3.14), density N can be obtained in terms of the volume fraction f. The expression for the factor $N\alpha_e/\varepsilon_0$ used in equation (6.1.14a) is obtained as follows:

$$f = \frac{v_i}{\Delta v} = \frac{4\pi}{3}a^3 N, \Rightarrow N = \frac{3f}{4\pi a^3} \quad (a)$$

$$\frac{N\alpha_e}{\varepsilon_0} = \frac{3f(\varepsilon - \varepsilon_0)}{(\varepsilon + 2\varepsilon_0)} \quad (b). \qquad (6.3.15)$$

On substituting equation (6.3.15b) in equation (6.1.14a), the following MG model for the effective relative permittivity ε_r of the artificial medium is obtained:

$$\varepsilon_r = 1 + \frac{3f(\varepsilon - \varepsilon_0)}{(\varepsilon + 2\varepsilon_0) - f(\varepsilon - \varepsilon_0)}. \qquad (6.3.16)$$

Equation (6.3.16) is valid for the spherical dielectric inclusions of permittivity ε and volume fraction f in the air host medium. The above expression is reduced to the MG equation (6.3.13a) for the host medium with relative permittivity ε_{rh} and inclusion with relative permittivity ε_{ri}. To achieve this end, the parameters are replaced as follows: $\varepsilon_0 \to \varepsilon_0\varepsilon_{rh}$, $\varepsilon \to \varepsilon_0\varepsilon_{ri}$, and $\varepsilon_r \to \varepsilon_0\varepsilon_{req}$ in the above equation and multiply the total expression by the factor ε_{rh}.

The MG equation (6.3.13) can also be used for the magnetic inclusion μ_{ri} in the host magnetic medium μ_{rh}:

3D cylindrical inclusion :

$$\mu_{r\,eq} = \mu_{rh}\left[1 + \frac{3f(\mu_{ri} - \mu_{rh})}{\mu_{ri} + 2\mu_{rh} - f(\mu_{ri} - \mu_{rh})}\right] \quad (a)$$

2D cylindrical inclusion :

$$\mu_{r\,eq} = \mu_{rh}\left[1 + \frac{2f(\mu_{ri} - \mu_{rh})}{\mu_{ri} + \mu_{rh} - f(\mu_{ri} - \mu_{rh})}\right] \quad (b). \qquad (6.3.17)$$

The above discussion shows that the Clausius–Mossotti model of homogenization is based on the concept of the polarizability of the inclusion as a dipole, whereas the classical MG model computes the effective (equivalent) permittivity/permeability of the homogenized heterogeneous mixture in terms of the permittivity/permeability of the inclusion and host materials. Both forms of mixing models have been used to develop only dielectric-based metamaterials useful in the THz, infrared, and optical frequency ranges by using the concept of Mie magnetic and electric resonances. It is discussed in section 21.4.2 of chapter 21.

The 3D and 2D MG formula can also be extended to a host medium with two inclusion materials, with relative permittivity ε_{ri1} and ε_{ri2}, and volumetric fractions f_1 and f_2 [J.22]:

$$\varepsilon_{r\,eq} = \varepsilon_{rh}\left[1 + \frac{3(f_1 k_1 + f_2 k_2)}{1 - (f_1 k_1 + f_2 k_2)}\right] \quad (a)$$

$$\varepsilon_{r\,eq} = \varepsilon_{rh}\left[1 + \frac{2(f_1 k_1 + f_2 k_2)}{1 - (f_1 k_1 + f_2 k_2)}\right] \quad (b)$$

$$\text{where,}\ k_1 = \frac{\varepsilon_{ri1} - \varepsilon_{rh}}{\varepsilon_{ri1} + 2\varepsilon_{rh}} \quad \text{and} \quad k_2 = \frac{\varepsilon_{ri2} - \varepsilon_{rh}}{\varepsilon_{ri2} + 2\varepsilon_{rh}} \quad (c). \qquad (6.3.18)$$

If the inclusion material is a *conducting sphere* $\varepsilon_{ri} \to \infty$, both MG formulae of equation (6.3.13), are reduced to

$$\text{3D case}: \varepsilon_{r\,eq} = \varepsilon_{rh}\left[\frac{1 + 2f}{1 - f}\right] \quad (a)$$

$$\text{2D case}: \varepsilon_{r\,eq} = \varepsilon_{rh}\left[\frac{1 + f}{1 - f}\right] \quad (b). \qquad (6.3.19)$$

6.3.3.2 Sihvola Formula

Sihvola has suggested the following more general formulae for the equivalent relative permittivity of a mixture material containing the 3D spherical, and also 2D circular cylindrical, inclusions [B.17, J.16, J.18, J.23–J.25]:

$$\text{3D case}: \frac{\varepsilon_{r\,eq} - \varepsilon_{rh}}{\varepsilon_{r\,eq} + 2\varepsilon_{rh} + a(\varepsilon_{req} - \varepsilon_{rh})} = f\frac{\varepsilon_{ri} - \varepsilon_{rh}}{\varepsilon_{ri} + 2\varepsilon_{rh} + a(\varepsilon_{r\,eq} - \varepsilon_{rh})} \quad (a)$$

$$\text{2D case}: \frac{\varepsilon_{req} - \varepsilon_{rh}}{\varepsilon_{r\,eq} + \varepsilon_{rh} + a(\varepsilon_{r\,eq} - \varepsilon_{rh})} = f\frac{\varepsilon_{ri} - \varepsilon_{rh}}{\varepsilon_{ri} + \varepsilon_{rh} + a(\varepsilon_{r\,eq} - \varepsilon_{rh})} \quad (b). \qquad (6.3.20)$$

On changing the value of the empirical parameter 'a', various classical mixing formulae could be obtained. Thus, for the 3D spherical inclusion case; $a = 0$ gives *Maxwell Garnett formula*, $a = 2$ gives the *3D symmetrical Bruggeman formula* [J.26] and $a = 3$ gives the *coherent potential rule* [J.27]. Likewise, for the 2D cylindrical inclusion case $a = 1$ gives the 2D symmetrical

Bruggeman formula. Further, a = 2 gives the 2D *coherent potential rule*. Both symmetrical Bruggeman formulae (S-BG) are summarized below [B.17]:

$$\begin{aligned}
&\text{3D case}: (1-f)\frac{\varepsilon_{rh}-\varepsilon_{r\,eq}}{\varepsilon_{rh}+2\varepsilon_{r\,eq}}+f\frac{\varepsilon_{ri}-\varepsilon_{r,\,eq}}{\varepsilon_{ri}+2\varepsilon_{r\,eq}}=0 \quad (a)\\
&\text{2D case}: (1-f)\frac{\varepsilon_{rh}-\varepsilon_{r\,eq}}{\varepsilon_{rh}+\varepsilon_{r\,eq}}+f\frac{\varepsilon_{ri}-\varepsilon_{r,\,eq}}{\varepsilon_{ri}+\varepsilon_{r\,eq}}=0 \quad (b).
\end{aligned} \tag{6.3.21}$$

Sihvola formula for a = 1.2 and contrast ratio $K = (\varepsilon_{ri}/\varepsilon_{rh}) > 1$ provides better results for the equivalent relative permittivity against the numerically simulated results. It also gives a more accurate formula for the contrast ratio $K < 1$. The parameter "a" is a function of contrast ratio (K) and volumetric function (f). It can be computed using the following empirical expression [J.18] that is used with Sihvola formula to get the equivalent relative permittivity of a mixture:

$$\begin{aligned}
a(f,K) &= \left(1.27+1.43e^{-0.048K}\right)f^2 \\
&\quad + \left(-2.76-0.09e^{-0.043K}\right)f + 2.35, \text{ for } K>1 \quad (a)\\
a(f,K) &= 1.06f^2 + \left(-1.23+0.44e^{-0.5.95K}\right)f + \\
&\quad 1.7, \text{ for } K<1 \quad (b).
\end{aligned} \tag{6.3.22}$$

The accuracy of MG, Bruggeman, and Sihvola's empirical models are compared in Fig (6.6b) for the host medium $\varepsilon_{rh} = 1$, inclusion, with K = 51, against the FDTD results. Additional comparison against the results of FDTD show that Sihvola's empirical formula has higher accuracy, and the next accurate model is the Bruggeman model.

There are *power-law* [J.16] and *logarithmic–law* [J.28] also, given below, to compute the equivalent relative permittivity $\varepsilon_{r,\,eq}$ of mixtures:

$$\begin{aligned}
&\text{Power law}: \varepsilon^{\beta}_{r,eq} = f\varepsilon^{\beta}_{ri} + (1-f)\varepsilon^{\beta}_{rh} \quad (a)\\
&\text{Logarithmic law}: \ln\left(\varepsilon_{r,eq}\right) = f\ln\varepsilon_{ri} + (1-f)\ln\varepsilon_{rh} \quad (b).
\end{aligned} \tag{6.3.23}$$

The logarithmic law is a special case of the power-law. Likewise, the *Looyenga rule* could be obtained for $\beta = 1/3$ [J.29] and *Birchak rule* for $\beta = 1/2$[J.30]. The popular Bruggeman model is also put in the symmetrical and asymmetrical formats summarized below.

6.3.3.3 Bruggeman Formula

The asymmetric Bruggeman (A-BG) formulae is written as follows [J.17, J.23]:

$$\text{A-BG}: \frac{\varepsilon_{ri}-\varepsilon_{r\,eq}}{\varepsilon_{ri}-\varepsilon_{rh}} = (1-f)\left[\frac{\varepsilon_{r\,eq}}{\varepsilon_{rh}}\right]^{1/3}. \tag{6.3.24}$$

The expression (6.3.21) is a symmetric one, i.e. ε_{ri} and ε_{rh}, also f and (1-f), can be interchanged without affecting equivalent relative permittivity of mixture. This model is more suitable for a random mixture where clustering is allowed [B.17]. The asymmetrical Bruggeman formula (6.3.24) shows better agreement with the available experimental data of Doyle and Jacob [J.17]. If the inclusion material is a *solid conducting sphere*, then $\varepsilon_{ri} \to \infty$. The equations (6.3.21a) and (6.3.24) provide the following expressions:

$$\begin{aligned}
&\text{S-BG}: \varepsilon_{r\,eq} = \frac{\varepsilon_{rh}}{1-3f} \quad (a)\\
&\text{A-BG}: \varepsilon_{r\,eq} = \frac{\varepsilon_{rh}}{(1-f)^3} \quad (b).
\end{aligned} \tag{6.3.25}$$

The above mixture models can be extended to compute the *complex equivalent relative permittivity* of a mixture of materials by replacing $\varepsilon_{rh} = \varepsilon'_{rh} - j\varepsilon''_{rh}$ and $\varepsilon_{ri} = \varepsilon'_{ri} - j\varepsilon''_{ri}$ [J.17]. In case of a conducting inclusion of finite conductivity, we take

$$\varepsilon^{*}_{ri} = 1 - j\frac{\sigma}{\varepsilon_0\,\omega}, \tag{6.3.26}$$

where σ is the conductivity of an inclusion conductor. The result on a $\mathrm{Re}(\varepsilon_{r\,eq})$ and $\mathrm{Img}(\varepsilon_{r\,eq})$ is shown in Fig (6.6c and d) respectively, for a mixture of *carbon iron power* (CIP) inclusion, with $\sigma = 10^4$ at 4GHz, in the host rubber medium with $\varepsilon_{rh} = 2.43 - j0.029$. The asymmetrical Bruggeman model shows better agreement with the measured data.

Hoffman [B.19] has given the following expression to compute the equivalent relative permittivity ε_{req} of a porous material:

$$\varepsilon_{r\,eq} = \varepsilon_{rh}\left[\frac{1-3P+2\varepsilon_{rh}}{1+(2-3P)\,\varepsilon_{rh}}\right]. \tag{6.3.27}$$

where the porosity (P) in a host material is taken as the inclusion. It is determined by equation (6.3.3). For the alumina as a host, the relative permittivity is $\varepsilon_{rh} = 10.22$.

6.4 Frequency Response of Dielectric Materials

The full building up of any kind of polarization, giving the *static permittivity* of a dielectric material, takes a finite amount of time. Thus, the polarization and permittivity are time-dependent, or a frequency-dependent, phenomena causing material dispersion and loss. The dielectric materials show two kinds of responses to the applied external field (i) *relaxation response* (ii) *resonance response*. The dipolar (orientational) polarization is described by *Debye relaxation model*. It also describes the relaxation response of the interfacial polarization. Normally, the relaxation response is operative up to microwave frequency. The atomic and ionic polarizations take place due to the displacement of bound charges under the influence of the external field. These polarizations are described by *Lorentz type harmonic oscillator model*. Lorentz model is also reduced to Drude model, applicable to conductors. Both Debye and Lorentz harmonic oscillator models provide frequency-dependent complex relative permittivity. The real and imaginary parts of the complex permittivity of dielectric material are related to each other by the well-known *Kramers–Krönig relation* [B.5, B.10].

This section first presents the *decay law* of the charges inside a conductor, then the *polarization law* of a linear dielectric-medium, and finally Debye dispersion relation for the complex relative permittivity. Lorentz and Drude harmonic oscillator models are presented in section (6.5). In brief, the present and next sections model the dielectric responses using the microlevel polarization concepts – both the nonresonant and resonant types. Section (6.7) presents the circuit models of Debye and Lorentz polarization processes also.

6.4.1 Relaxation in Material and Decay Law

6.4.1.1 Relaxation in Conducting Medium

Figure (6.7a) shows a conductor of volume V and conductivity σ that contains charge density ρ_v. Under the influence of the electric field $\vec{E}$, current density $\vec{J}$ flows out of volume V so that the charges appear at the surface, satisfying the following condition:

Rate of flow of charges out of surface S =
Rate of decrease of charges in volume V
bounded by surface S.

$$\Rightarrow I = \frac{dQ}{dt} = -\frac{d}{dt}\int_V \rho_v dv$$

$$\Rightarrow \int_S \vec{J}.d\vec{s} = -\int_V \frac{d\rho_v}{dt} dv$$

On using the divergence theorem, we get the following continuity equation:

$$\int_V \nabla.\vec{J}\,dv = -\int_V \frac{d\rho_v}{dt} dv$$

$$\Rightarrow \nabla.\vec{J} = -\frac{d\rho_v}{dt} \Rightarrow \frac{d\rho_v}{dt} + \frac{\sigma}{\varepsilon_0\,\varepsilon_r}\rho_v = 0. \tag{6.4.1}$$

To get the final form of the above equation, we have used Ohm's law $\vec{J} = \sigma\vec{E}$ and relations $\vec{D} = \varepsilon_0\varepsilon_r\vec{E}$ and $\nabla.\vec{D} = \rho_v$.

The above equation is the *decay law*, describing the decay of charges, i.e. the migration of the charges from the inside of volume V to the surface of a conductor. The above equation is solved as follows, to get the exponential decay law under the initial condition $\rho_v(t) = \rho_0$ at time $t = 0$:

$$\int \frac{d\rho_v}{\rho_v} = -\int \frac{\sigma}{\varepsilon_0\varepsilon_r}dt, \Rightarrow \ln(\rho_v) = -\frac{\sigma}{\varepsilon_0\,\varepsilon_r}t + \ln(\rho_0)$$

$$\Rightarrow \rho_v = \rho_0 e^{-\frac{t}{\tau}}, \tag{6.4.2}$$

where $\tau = (\varepsilon_0\varepsilon_r/\sigma)$ is the *relaxation time*. The charges inside the volume exponentially decrease toward the stable condition until all charges migrate to the surface, i.e. $\rho_v \to 0$, $t \to \infty$. On inserting external charges in volume V, the system becomes unstable and it is *relaxed* after reaching the stable, i.e. steady-state condition. During relaxation time duration $t = \tau$, the charges decrease to the value $\rho_v = \rho_0/e$, i.e. 36.8% of its original quantity. The relaxation time is the material-dependent parameter. For instance, the relaxation time of copper is

$$\tau = \frac{\varepsilon_r\varepsilon_0}{\sigma} = \frac{1 \times 8.856 \times 10^{-12}\text{F/m}}{5.8 \times 10^7\text{S/m}}$$
$$= 1.527 \times 10^{-19}\text{Sec.}$$

For the seawater with $\sigma = 4\text{S/m}$ and $\varepsilon_r = 80$, the relaxation time is $\tau = 1.77 \times 10^{-10}$ Sec. It means for a conductor, the relaxation time is very short, and for dielectrics, it is large. Therefore, the dielectrics hold the charges for a longer duration. Say for the glass, with $\sigma = 10^{-12}\text{S/m}$ and $\varepsilon_r = 6$, the relaxation time is $\tau = 53.14$ Sec. ≈ 1 min.

6.4.2 Polarization Law of Linear Dielectric-Medium

The *polarization law* of a dielectric-medium demonstrates its relaxation response, resulting from the applied external electric field. Under the influence of

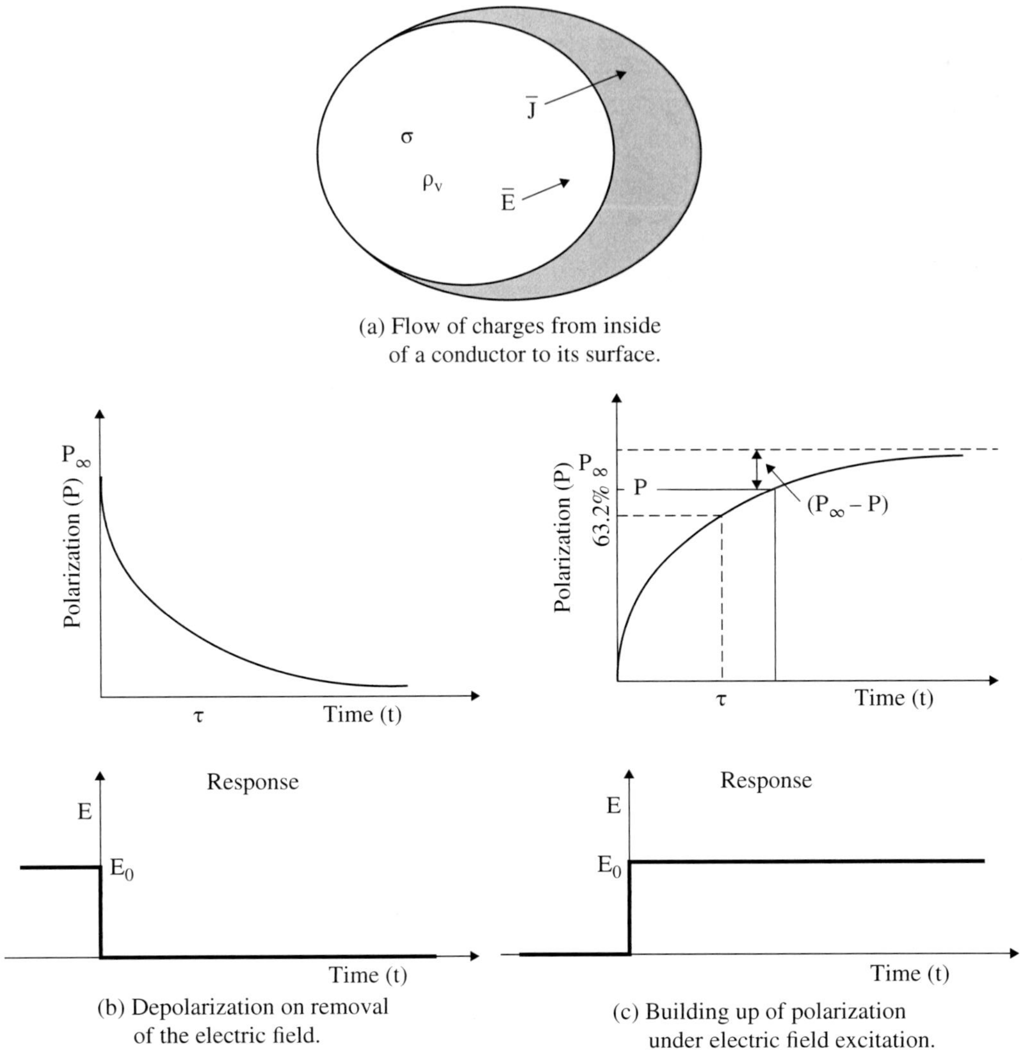

Figure 6.7 Polarization and depolarization process in a dielectric material.

an external field, the atoms, molecules, and dipoles of a linear dielectric-medium start aligning themselves along the direction of the field. Equation (6.1.5) provides a steady-state, i.e. the static polarization, $\vec{P}_\infty = \varepsilon_0\chi_r\vec{E}_{ext}$. The χ_r is the *static susceptibility* of a material. For a nonpolar material, the initial polarization is zero, so the polarization increases from $P = 0$ to $P = P_\infty$. The time-domain polarization law, for a linear dielectric-medium, could be stated as follows:

The rate of increase of polarization (dP/dt) is proportional to the difference between the steady – state value (P_∞) and polarization P at any time t, i.e.

$$\frac{d\vec{P}}{dt} \propto \text{Remaining polarization at time t,}$$

$$\Rightarrow \frac{d\vec{P}}{dt} = C\left[\vec{P}_\infty - \vec{P}\right] \qquad \text{(a)}$$

$$\Rightarrow \tau\frac{d\vec{P}}{dt} + \vec{P} = \varepsilon_0\chi_r\vec{E}_{ext} \qquad \text{(b),} \qquad (6.4.3)$$

where $\tau = 1/C$ is the proportionality constant in the time unit. Equation (6.4.3b) is the *Debye dispersion law in the time-domain*. It is similar to the RC circuit model of material response. The above equation is solved to get the polarization decay law and also its rise law.

6.4.2.1 Polarization Decay Law

Let us assume that the medium is initially fully polarized with polarization P_∞ under the influence of the external field of magnitude E_0. At time t = 0, the external field E_0 is removed, as shown in Fig (6.7b). The depolarization process, i.e. the decay of the polarization, starts. It is described by

$$\tau\frac{d\vec{P}}{dt} + \vec{P} = 0, \Rightarrow \frac{dP}{P} = -\frac{dt}{\tau}, \Rightarrow P = P_\infty e^{-t/\tau}. \quad (6.4.4)$$

The above expression is known as the *polarization decay law*. During the time $t = \tau$, the polarization $P = P_\infty/e$ is decreased by 36.8% of P_∞.

6.4.2.2 Polarization Rise Law

Figure (6.7c) shows that under the application of the step type electric field, the polarization reaches the maximum P_∞. Equation (6.4.3b) is solved under condition $t = 0$, $P = 0$, i.e. no initial polarization, as follows:

$$\int\frac{dP}{(\varepsilon_0\chi_r E_0 - P)} = \int\frac{dt}{\tau}, \Rightarrow P = \varepsilon_0\chi_r E_0 - Ce^{-t/\tau}$$

$$\Rightarrow P = \varepsilon_0\chi_r E_0\left(1 - e^{-t/\tau}\right), \quad (6.4.5)$$

where constant, $C = \varepsilon_0\chi_r E_0 = P_\infty$. The final form of the above equation is called the *polarization rise law*. During the relaxation time, $t = \tau$, 63.2% polarization builds up. Therefore, the *relaxation is a process under which the delayed polarization response, after a finite time interval, appears on the application of excitation*. The charging of the capacitor through a series resistor is also described similarly. Therefore, the RC circuit could be used to model the polarization process.

6.4.3 Debye Dispersion Relation

The first-order phenomenological differential equation (6.4.3) is a statement of the time-domain Debye dispersion relation. To obtain the frequency-domain Debye dispersion relation, instead of the step type field, the dielectric-medium is subjected to the sinusoidal time-harmonic field. The polarization, as a response, also follows the sinusoidal time behavior. Equation (6.4.3) can be rewritten in the frequency-domain using the phasor forms as follows:

$$\text{Phasor form}: \tilde{E}_{ext}(\omega) = E_0 e^{j\omega t}, \tilde{P}(\omega) = P_0 e^{j\omega t} \quad (a)$$

$$P_0 e^{j\omega t} + j\omega\tau P_0 e^{j\omega t} = \varepsilon_0\chi_r E_0 e^{j\omega t}$$

$$\Rightarrow \tilde{P}(\omega) = \varepsilon_0\left[\frac{\chi_r}{1 + j\omega\tau}\right]\tilde{E}_{ext}(\omega) \quad (b). \quad (6.4.6)$$

By using the above equation with equation (6.1.6a) in the frequency-domain, the following expression is obtained:

$$\tilde{D}(\omega) = \varepsilon_0\tilde{E}_{ext}(\omega) + \tilde{P}(\omega),$$
$$\Rightarrow \tilde{D}(\omega) = \varepsilon_0\left[1 + \frac{\chi_r}{1 + j\omega\tau}\right]\tilde{E}_{ext}(\omega). \quad (6.4.7)$$

Therefore, the complex relative permittivity of a dielectric medium is

$$\varepsilon_r^*(\omega) = \varepsilon_r' - j\varepsilon_r'' = \left[1 + \frac{\chi_r}{1 + j\omega\tau}\right] \quad (a)$$
$$\chi_r^*(\omega) = \frac{\chi_r}{1 + j\omega\tau} \quad (b), \quad (6.4.8)$$

where $\chi_r^*(\omega)$ is the complex frequency-dependent susceptibility of a material. The equation (6.4.8a) is *Debye's dispersion relation* for a frequency-dependent dielectric-medium. The real and imaginary parts of the complex relative permittivity $\varepsilon_r^*(\omega)$ are given below:

$$\varepsilon_r' = \frac{1 + \chi_r + (\omega\tau)^2}{1 + (\omega\tau)^2} \quad (a), \quad \varepsilon_r'' = \frac{\chi_r\omega\tau}{1 + (\omega\tau)^2} \quad (b). \quad (6.4.9)$$

For $\omega \to 0$, the static value, or the low-frequency limit of the relative permittivity, is obtained, as $\varepsilon_r' = 1 + \chi_r, \varepsilon_r'' = 0$. For $\omega \to 0$, the frequency is very low and the material medium gets sufficient time to build up to full polarization. At $\omega \to \infty, \varepsilon_r' = 1$, $\varepsilon_r'' = \chi_r/\omega\tau \approx 0$, i.e. at the extremely high frequency, the medium is not able to polarize at all. In this case, the medium behaves as free space. Such behavior is because the expression (6.4.9) does not take into account the presence of other kinds of polarization. However, it accounts for the role of one relaxation in the modeling of the frequency response of the relative permittivity. Therefore, the present model is improved by accounting for various kinds of polarization. At any other frequency ω, the dielectric loss is obtained through ε_r''.

6.4.3.1 Debye's Dispersion Equation for Three Kinds of Polarizations

Three kinds of static polarizations considered in equation (6.1.23) have three relaxation times. However, the electronic and ionic polarization at the optical and infrared frequency are so fast that their relaxation times are taken as zero, i.e. polarizations take place instantaneously without any time lag. *The delayed response, leading to Debye dispersion relation, is obtained only due to the permanent dipolar polarization, with relaxation time* τ_d. Following the dispersion model of equation (6.4.7), the static expression (6.1.23) could be modified to get the following frequency-dependent total polarization:

$$\widetilde{P}(\omega) = N\varepsilon_0\left[\left(\alpha_e^e + \alpha_i^e\right) + \frac{\mu^2}{3kT}\left(\frac{1}{1 + j\,\omega\,\tau_d}\right)\right]\widetilde{E}_{loc}(\omega). \tag{6.4.10}$$

On the assumption $\widetilde{E}_{loc}(\omega) = \widetilde{E}_{ext}(\omega)$, and using equation (6.1.5), equation (6.4.10) provides the frequency-dependent complex relative susceptibility and complex relative permittivity of a low-density medium:

$$\chi_r^*(\omega) = N\left(\alpha_e^e + \alpha_i^e\right) + \frac{N\mu^2}{3\,kT(1 + j\,\omega\,\tau_d)} \quad (a)$$

$$\varepsilon_r^*(\omega) = 1 + N\left(\alpha_e^e + \alpha_i^e\right) + \frac{N\mu^2}{3\,kT\,(1 + j\,\omega\,\tau_d)} \quad (b). \tag{6.4.11}$$

The above Debye's dispersion relation accounts for three kinds of polarization. The limiting values of the relative permittivity at a very high frequency (optical frequency, $\omega \to \infty$), and at low frequency (static, $\omega \to 0$) are obtained as

$$\varepsilon_r^*(\omega \to \infty) = 1 + N\left(\alpha_e^e + \alpha_i^e\right) \quad (a)$$

$$\varepsilon_r^*(\omega \to 0) = 1 + N\left(\alpha_e^e + \alpha_i^e\right) + \frac{N\,\mu^2}{3\,kT} \quad (b). \tag{6.4.12}$$

It is noted that there is no dielectric loss at very high and very low frequency, as the limiting values of the relative permittivity are real quantities, with $\varepsilon_r'' = 0$. The contribution of the dipolar (orientational) polarization to the static relative permittivity is obtained from the above expressions:

$$\frac{N\mu^2}{3kT} = \varepsilon_r'(\omega \to 0) - \varepsilon_r'(\omega \to \infty). \tag{6.4.13}$$

The ε_r' is the same as ε_r. Using equations (6.4.12a) and (6.4.13), the frequency-dependent complex relative permittivity, given by equation (6.4.11b), is rewritten as

$$\varepsilon_r^*(\omega) = \varepsilon_r(\omega \to \infty) + \left[\frac{\varepsilon_r(\omega \to 0) - \varepsilon_r(\omega \to \infty)}{1 + j\omega\tau_d}\right] \quad (a)$$

$$\varepsilon_r^*(\omega) = \varepsilon_r(\omega \to \infty) + \left[\frac{\Delta\varepsilon_r}{1 + j\omega\tau_d}\right] \quad (b). \tag{6.4.14}$$

The above equation is the *Debye dispersion equation*. The difference relative permittivity $\Delta\varepsilon_r$ is a change in the relative permittivity at two steady-state polarization levels. The local field could be replaced with the Lorentz field to improve the above model. The local field can also be replaced by the Onsager field to avoid the Mossotti catastrophe. Even experimentally determined empirical parameter γ can be used to improve the accuracy of the Debye model. However, the nature of the Debye dispersion equation remains unchanged. Only the relaxation time τ_d gets modified. For instance, using the modified relaxation time τ_m for Mossotti field, the real and imaginary parts of the Debye dispersion equation, and also loss-tangent of the medium, are given by the following expressions [B.1]:

$$\varepsilon_r'(\omega) = \varepsilon_r'(\omega \to \infty) + \frac{\left[\varepsilon_r'(\omega \to 0) - \varepsilon_r'(\omega \to \infty)\right]}{1 + \omega^2\tau_m^2} \quad (a)$$

$$\varepsilon_r''(\omega) = \frac{\left[\varepsilon_r'(\omega \to 0) - \varepsilon_r'(\omega \to \infty)\right]\omega\tau_m}{1 + \omega^2\tau_m^2} \quad (b)$$

$$\tan(\delta) = \frac{\varepsilon_r''(\omega)}{\varepsilon_r'(\omega)} = \frac{\left[\varepsilon_r'(\omega \to 0) - \varepsilon_r'(\omega \to \infty)\right]\omega\tau_m}{\varepsilon_r'(\omega \to 0) + \varepsilon_r'(\omega \to \infty)\omega^2\tau_m^2} \quad (c)$$

$$\text{where,} \quad \tau_m = \left\{\frac{\varepsilon_r'(\omega \to 0) + 2}{\varepsilon_r'(\omega \to \infty) + 2}\right\}\tau_d \quad (d). \tag{6.4.15}$$

Figure (6.8) shows Debye dispersion behaviors of a dielectric material. Beyond the *dielectric relaxation frequency*, $f_{re} = 1/\tau_m$, the contribution of the dipolar polarization is reduced and finally, its contribution vanishes. The remaining high-frequency relative permittivity is due to ionic and electronic polarization. In the transition region, the Debye dispersion model provides the *anomalous, i.e. negative, dispersion* in the medium, as in this region, the phase velocity increases with the increase in frequency. The anomalous dispersion is discussed in chapter 3, section (3.3). Furthermore, as there is no relaxation behavior at low and at a very high frequency, the imaginary part of the relative permittivity ε_r'' vanishes at both ends of frequency band giving the maximum value of $\varepsilon_{r\,max}''$, i.e. the absorption peak, near the middle of the transition frequency. The frequency f_{max}, at which the maximum $\varepsilon_{r\,max}''$ occurs, is obtained by differentiating equation (6.4.15b) with respect to ω and equating the expression to zero. $\varepsilon_{r\,max}''$ is computed at frequency f_{max}:

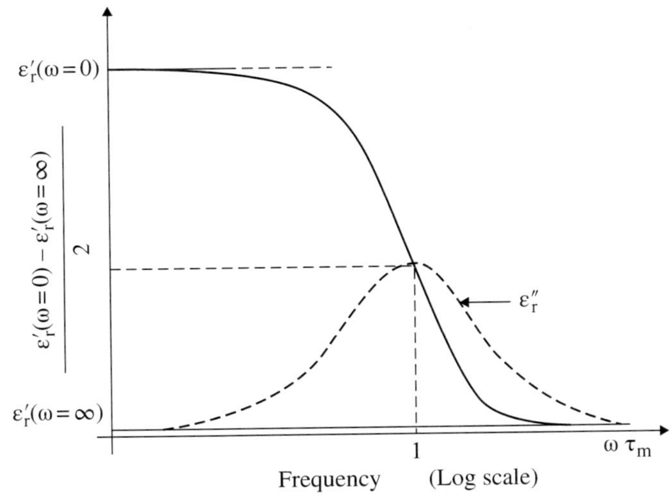

Figure 6.8 Relaxation type Debye dispersion behaviors of a dielectric material.

$$\omega_{max}\tau_m = 1 \Rightarrow f_{max} = \frac{1}{2\pi\tau_m} \quad \text{(a)}$$

$$\varepsilon''_{r\,max} = \frac{\varepsilon'_r(\omega \to 0) - \varepsilon'_r(\omega \to \infty)}{2} \quad \text{(b)}. \tag{6.4.16}$$

The $\varepsilon''_{r\ max}$ is determined in terms of the limiting values of the real parts of the complex relative permittivity. The frequency-dependent relative permittivity $\varepsilon'_r(\omega)$ and tan (δ) are measured experimentally. In practice, the Debye model is curve-fitted to the experimental results of $\varepsilon'_r(\omega)$ and $\varepsilon''_r(\omega)$ by evaluating its parameters $\varepsilon'_r(\omega \to \infty)$, $\Delta\varepsilon'_r$, and τ_m. It is discussed in section (6.8).

6.5 Resonance Response of the Dielectric-Medium

In general, the polarization of material could be due to a combination of several polarizations. The oscillator model discussed in this section, along with the relaxation model, provides a realistic description of dispersion in both the normal and artificial materials. Debye model, applicable to a relaxation type response of the dipolar polarization, is discussed in the previous section. Above the relaxation frequency, the material gets depolarized and its permittivity is decreased. However, most of the practical dielectrics used in the microwave range, say up to 100 GHz, do not show a meaningful change in dielectric constant. It means most of the materials used in the microwave range are nonpolar and the relative permittivity of these materials is due to the ionic and electronic polarization. However, materials show the resonance peaks in relative permittivity and also the absorption peaks in the infrared and visible frequency ranges. At the molecular level, such *polarization resonances* are modeled by the 1D linear harmonic oscillator, known as the *Lorentz model*. This model also is reduced to the conductivity model of Drude. Therefore, it is also called the *Lorentz–Drude model* [B.1, B.10]. The harmonic oscillator model can be reduced to the relaxation type model for the overdamped case. The RLC circuit also provides Lorentz–Drude kind of response of dielectric materials presented in subsection (6.7.5). The Lorentz model is also applicable to the artificial magnetic material. It is expressed by equation (21.2.4) of chapter 21.

The artificial dielectrics developed by introducing the *inclusions*, such as an array of electrically small conducting scatters – spheres, discs, thin wires, metallic strips, etc., in a *host* dielectric-medium, also show resonance peaks. The metallic inclusions of artificial dielectric materials act as macromolecules and exhibit frequency-dependent relative permittivity, even at a lower frequency. They show Lorentz type resonance and anomalous dispersion. The insertion material, i.e. the *inclusion,* could also be a high permittivity material, different from the host medium. Such an artificial dielectrics medium provides interesting dielectric properties for many engineering applications. The dispersion models – both Debye and Lorentz types are important for these materials [B.11, J.31–J.33].

6.5.1 Lorentz Oscillator Model

Each of the polarization in a dielectric material, natural or artificial, discussed above is modeled separately by the harmonic oscillator. Figure (6.9a) shows the displaced bound charges, under the influence of external field E_{ext}, and the spring model of Hook's type restoring force for the *c-kind harmonic oscillator* with the charge q_c and mass m_c. The generic symbol "c" could be the electron, ion, dipole, or inclusion. The external electric field $\vec{E}_{ext}$ displaces the charge by a distance x. The attractive force, i.e. the restoring force, due to the *local electric field* $\vec{E}_{loc}$ between the charges, is modeled by the spring. The motion of a charge, like the motion of a mass attached to the stretched spring, can be written as

$$\frac{d^2x}{dt^2} + \gamma_c\frac{dx}{dt} + \omega_c^2 x = \frac{q_c E_{loc}}{m_c}, \tag{6.5.1}$$

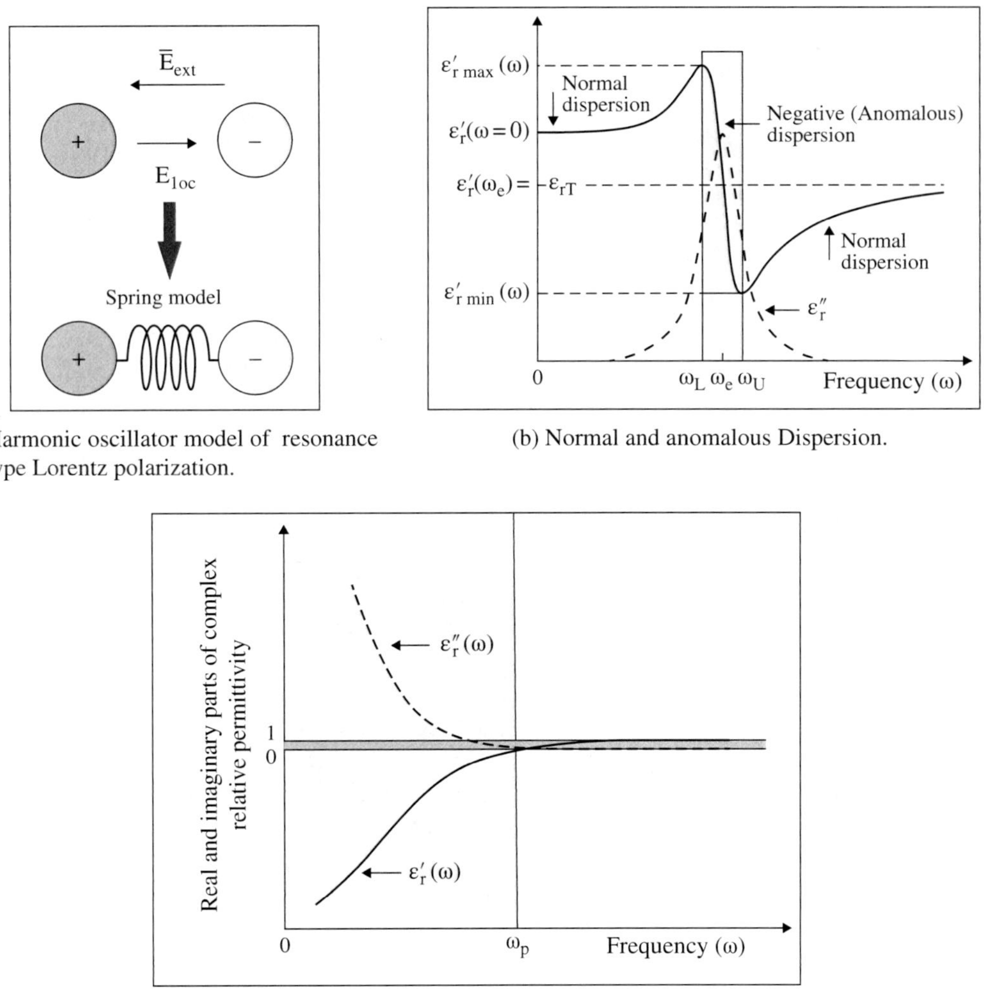

(a) Harmonic oscillator model of resonance type Lorentz polarization.

(b) Normal and anomalous Dispersion.

(c) Response of plasma medium.

Figure 6.9 Harmonic oscillator model of resonance type Lorentz polarization and its frequency response, showing both the positive and negative dispersions. Also the response of the Drude model of the plasma medium.

where γ_c is the damping coefficient, in the frequency unit, of the oscillation indicating losses in materials and ω_c is the characteristic resonance frequency. In absence of losses, the system performs a simple harmonic motion with frequency ω_c. For N_c numbers of the c-type oscillators per unit volume with the dipole moment $p_c = q_c x$, the polarization, using an equation (6.1.4), due to all oscillators is

$$P_c = N_c p_c = N_c\, q_c\, x. \tag{6.5.2}$$

Using Mossotti–Lorentz field, i.e. the Lorentz field, from equation (6.1.9), the following equation of motion for the polarization can be written from the equations (6.5.1) and (6.5.2):

$$\text{Mossotti field : } E_{loc} = E_{ext} + \frac{P_c}{3\varepsilon_0} \qquad \text{(a)}$$

$$\text{Polarization Eq : } \frac{d^2P_c}{dt^2} + \gamma_c \frac{dP_c}{dt} + \left(\omega_c^2 - \frac{N_c q_c^2}{3\,\varepsilon_0\, m_c}\right) P_c = \frac{N_c\, q_c^2}{m_c} E_{ext} \qquad \text{(b)}$$

$$\text{where plasma frequency } \omega_p = \sqrt{\frac{N_c q_c^2}{\varepsilon_0 m_c}}, \qquad \text{(c)}$$

$$\omega'_c = \left[\omega_c^2 - \frac{N_c q_c^2}{3\varepsilon_0 m_c}\right]^{\frac{1}{2}} \qquad \text{(d)}$$

$$\text{Polarization Eq : } \frac{d^2P_c}{dt^2} + \gamma_c \frac{dP_c}{dt} + \omega'^2_c P_c = \varepsilon_0 \omega_p^2 E_{ext} \qquad \text{(e)}. \tag{6.5.3}$$

The characteristics resonance frequency ω_c is reduced to ω_c', due to the Mossotti field. However, for the low-density artificial dielectric material and gases, it could be ignored. Thus, the subsequent discussion assumes $\omega_c' = \omega_c$. For the time-harmonic excitation, polarization (P_c) is also time-harmonic; expressed as $E_{ext} = E_0 e^{j\omega t}$, $P_c = P_0 e^{j\omega t}$. Under the time-harmonic condition, the above-given polarization equation (6.5.3e) is reduced to the following expression:

$$\left(\omega_c^2 - \omega^2 + j\omega\,\gamma_c\right)P_c = \frac{N_c q_c^2}{m_c} E_{ext}. \tag{6.5.4}$$

On comparing the above equation with equation (6.1.5), i.e. $\vec{P}_c = \varepsilon_0 \chi_r \vec{E}_{ext}$, and also using equation (6.1.7a), the following expressions, related to three dispersion models, are obtained for the relative susceptibility and complex relative permittivity of the c-type oscillator:

$$\chi_{rc} = \frac{\left(N_c q_c^2/\varepsilon_0 m_c\right)}{\omega_c^2 - \omega^2 + j\omega\gamma_c} = \frac{\omega_p^2}{\omega_c^2 - \omega^2 + j\omega\gamma_c} \quad \text{(a)}$$

$$\text{Lorentz model}: \varepsilon_{rc}^* = 1 + \frac{\left(N_c q_c^2/\varepsilon_0 m_c\right)}{\omega_c^2 - \omega^2 + j\omega\gamma_c} = 1 + \frac{\omega_p^2}{\omega_c^2 - \omega^2 + j\omega\gamma_c} \quad \text{(b)}$$

$$\text{Drude model}: \varepsilon_{rc}^* = 1 + \frac{\omega_p^2}{-\omega^2 + j\omega\gamma_c}, \quad \text{for } \omega_c = 0 \quad \text{(c)}$$

$$\text{Debye model}: \varepsilon_{rc}^* = 1 + \frac{\omega_p^2/\omega_c^2}{1 + j\omega\left(\gamma_c/\omega_c^2\right)}, \quad \text{for } \left(\omega_c^2 + j\omega\gamma_c\right) >> \omega^2 \quad \text{(d)}. \tag{6.5.5}$$

The c-type oscillators, constituting a dielectric material, are located in the vacuum with $\varepsilon_r = 1$. In the absence of the restoring force, there is no resonance, i.e. $\omega_c = 0$, and the Lorentz model is reduced to the Drude model. Drude model, discussed in the next subsection (6.5.2), applies to the modeling of the conducting medium of metal, and also to the *wire medium* discussed in chapter 21. Thus, the Lorentz and Drude models help to synthesize the artificial resonant and nonresonant types' dispersive metamaterials. Further, the Debye model, in a form little different from equation (6.4.8a), is obtained from the Lorentz model under the condition $\left(\omega_c^2 + j\omega\gamma_c\right) >> \omega^2$.

In equation (6.5.5b), only one kind of resonating element is present. It applies to the artificial dielectric materials and metamaterials with one kind of resonating inclusions. However, in the real material, or a mixture of materials, several kinds of resonating elements are present. The relative permittivity of such a material is due to a linear combination of all noninteracting polarizations:

$$\varepsilon_{rc}^* = 1 + \sum_c \frac{\left(N_c q_c^2/\varepsilon_0 m_c\right)}{\omega_c^2 - \omega^2 + j\omega\gamma_c} = 1 + \sum_c \frac{\omega_{p.c}^2}{\omega_c^2 - \omega^2 + j\omega\gamma_c}. \tag{6.5.6}$$

At the frequency much below the resonance, $\omega << \omega_c$, i.e. $\omega \to 0$, each kind of polarized element adds the nonresonating factor $\Delta\varepsilon_r$ to the total static, i.e. *low-frequency relative permittivity* (ε_{rT}). It is obtained as follows:

$$\Delta\varepsilon_r = \frac{N_c q_c^2}{\varepsilon_0 m_c}\frac{1}{\omega_c^2} \quad \text{(a)}, \qquad \varepsilon_{rT} = 1 + \sum_c \frac{N_c q_c^2}{\varepsilon_0 m_c}\frac{1}{\omega_c^2} \quad \text{(b)}. \tag{6.5.7}$$

To obtain the relative permittivity around the generic resonance frequency ω_e of the e-type resonating inclusions, the total relative permittivity includes the contribution of all nonresonating polarizations below ω_e i.e. for $\omega_c < \omega_e$. The total complex relative permittivity due to nonresonating c-type polarizations and resonating e-type polarization is given below, along with its real and imaginary parts:

$$\varepsilon_r^*(\omega) = 1 + \sum_{\substack{c \neq e \\ c < e}} \frac{N_c q_c^2}{\varepsilon_0 m_c}\frac{1}{\omega_c^2} + \frac{N_e q_e^2}{\varepsilon_0 m_e}\left\{\frac{1}{\omega_e^2 - \omega^2 + j\omega\gamma_e}\right\} = \varepsilon_{rT} + \frac{N_e q_e^2}{\varepsilon_0 m_e}\left\{\frac{1}{\omega_e^2 - \omega^2 + j\omega\gamma_e}\right\} \quad \text{(a)}$$

$$\varepsilon_r'(\omega) = \varepsilon_{rT} + \frac{N_e q_e^2}{\varepsilon_0 m_e}\left\{\frac{\left(\omega_e^2 - \omega^2\right)}{\left(\omega_e^2 - \omega^2\right)^2 + \omega^2\gamma_e^2}\right\} \quad \text{(b)}$$

$$\varepsilon_r''(\omega) = \frac{N_e q_e^2}{\varepsilon_0 m_e}\left\{\frac{\omega\gamma_e}{\left(\omega_e^2 - \omega^2\right)^2 + \omega^2\gamma_e^2}\right\} \quad \text{(c)}. \tag{6.5.8}$$

In case only one kind of polarizing element is present in the host vacuum medium, $\varepsilon_{rT} = 1$. The above relation is examined below at (i) *low frequency*, (ii) *around resonance*, and (iii) *high frequency* above the resonance.

6.5.1.1 Low-Frequency Relative Permittivity

On taking the limiting case, $\omega \to 0$, of equation (6.5.8b), the quasistatic low-frequency value is obtained for relative permittivity:

$$\varepsilon_r'(\omega = 0) = \varepsilon_{rT} + 2\left(\frac{N_e q_e^2}{2\varepsilon_0 m_e \omega_e}\right)\frac{1}{\omega_e}$$

$$\Rightarrow \varepsilon_r'(\omega = 0) = \varepsilon_{rT} + \frac{2\varepsilon_{rE}}{\omega_e} \quad (a)$$

$$\text{where} \quad \varepsilon_{rE} = \frac{N_e q_e^2}{2\varepsilon_0 m_e \omega_e} = \frac{\omega_p^2}{\omega_e} \quad (b)$$

$$\Rightarrow \varepsilon_r'(\omega = 0) = \varepsilon_{rT} + \frac{\omega_p^2}{\omega_e^2} \quad (c).$$

(6.5.9)

In the case of the artificial dielectrics and metamaterials, static relative permittivity $\varepsilon_r'(\omega = 0)$ could be the relative permittivity of a host medium. The host medium could be the vacuum itself with $\varepsilon_r'(\omega = 0) = 1$, i.e. $\varepsilon_{rT} = 1$, $\varepsilon_{rE} = 0$. In this case, no polarizing element is present. In the frequency range below the resonance frequency, $0 < \omega < \omega_e$, the relative permittivity is highly dispersive and $\varepsilon_r'(\omega)$ increases to a peak value.

6.5.1.2 Relative Permittivity at and Near the Resonance Frequency

Near resonance, on using approximations, the approximate complex relative permittivity is obtained from equation (6.5.8a) along with its real and imaginary parts:

$$\text{Approximation}: \Delta\omega = \omega_e - \omega, \omega_e + \omega \approx 2\omega_e, \frac{\omega}{\omega_e} \approx 1 \quad (a)$$

$$\text{Relative permittivity near resonance}: \varepsilon_r^*(\omega) = \varepsilon_{rT} + \frac{\varepsilon_{rE}}{\Delta\omega + j\frac{\gamma_e}{2}} \quad (b)$$

$$\varepsilon_r'(\omega) = \varepsilon_{rT} + \frac{\varepsilon_{rE}\Delta\omega}{(\Delta\omega)^2 + \left(\frac{\gamma_e}{2}\right)^2} = \varepsilon_{rT} + \frac{\varepsilon_{rE}(\omega_e - \omega)}{(\omega_e - \omega)^2 + \left(\frac{\gamma_e}{2}\right)^2} \quad (c)$$

$$\varepsilon_r''(\omega) = \frac{\frac{\gamma_e}{2}\varepsilon_{rE}}{(\Delta\omega)^2 + \left(\frac{\gamma_e}{2}\right)^2} = \frac{\frac{\gamma_e}{2}\varepsilon_{rE}}{(\omega_e - \omega)^2 + \left(\frac{\gamma_e}{2}\right)^2} \quad (d).$$

(6.5.10)

The relative permittivity and absorption peak, i.e. the maximum value of $\varepsilon_r(\omega)$, are obtained at the resonance frequency, $\omega = \omega_e$, $\Delta\omega = 0$:

$$\varepsilon_r'(\omega = \omega_e) = \varepsilon_{rT} \quad (a), \quad \varepsilon''_{r\,max}(\omega = \omega_e) = \frac{2\varepsilon_{rE}}{\gamma_e} \quad (b).$$

(6.5.11)

The maximum and minimum values of the real part of the relative permittivity $\varepsilon_r'(\omega)$, given by equation (6.5.8b), are obtained by taking

$$\frac{d\varepsilon_r'(\omega)}{d(\omega)} = 0, \text{ giving } \Delta\omega = \pm\frac{\gamma_e}{2} \quad (a)$$

$$\text{Upper frequency}, \omega_U = \omega_e + \frac{\gamma_e}{2}$$

$$\text{Lower frequency}, \omega_L = \omega_e - \frac{\gamma_e}{2} \quad (b).$$

(6.5.12)

On substituting the ω_L and ω_U in equation (6.5.10c), the following values of the maximum $\varepsilon'_{r\,max}(\omega)$ and minimum $\varepsilon'_{r\,min}(\omega)$ are obtained:

$$\varepsilon'_{rmax}(\omega = \omega_L) = \varepsilon_{rT} + \frac{\varepsilon_{rE}}{\gamma_e} \quad (a)$$

$$\varepsilon'_{r\,min}(\omega = \omega_U) = \varepsilon_{rT} - \frac{\varepsilon_{rE}}{\gamma_e} \quad (b).$$

(6.5.13)

6.5.1.3 High-Frequency Relative Permittivity Above Resonance

At high frequency above the resonance, $\omega \to \infty$, the steady relative permittivity is obtained from equation (6.5.8b) as,

$$\varepsilon_r'(\infty) = \varepsilon_{rT}. \quad (6.5.14)$$

Figure (6.9b) shows the frequency response of both $\varepsilon_r'(\omega)$ and $\varepsilon_r''(\omega)$ corresponding to material dispersion and loss, respectively. The resonance type Lorentz dispersion model exhibits both the normal and anomalous dispersion in a material. The $\varepsilon_r'(\omega)$ rises to the peak value $\varepsilon'_{r\,max}(\omega)$, from the quasistatic value $\varepsilon_r'(\omega = 0)$, with an increase in frequency. It is given by equation (6.5.13a). The resonance occurs at $\omega = \omega_c$. Further, an increase in frequency causes a decrease in relative permittivity $\varepsilon_r'(\omega)$ from its peak $\varepsilon'_{r\,max}(\omega)$ to the minimum $\varepsilon'_{r\,min}(\omega)$. The decreasing slope of relative permittivity $\varepsilon_r'(\omega)$ is a linear function of frequency. Further increase in frequency, above $\varepsilon'_{r\,min}(\omega)$, increases the relative permittivity $\varepsilon_r'(\omega)$ from $\varepsilon'_{r\,min}(\omega)$ to $\varepsilon'_{rT}(\omega)$.

So for an increase in the frequency from ($\omega = 0$) to ($\omega = \omega_L = \omega_e - \Delta\omega$), and also from ($\omega = \omega_U = \omega_e + \Delta\omega$) to ($\omega \to \infty$), the slope of $\varepsilon_r'(\omega)$ is positive, i.e. $d\varepsilon_r'/d\omega > 0$. Therefore, both in the lower and upper frequency ranges the dispersion is *normal*. However, between the frequency range from ($\omega = \omega_L = \omega_e - \Delta\omega$) to ($\omega = \omega_U = \omega_e + \Delta\omega$), the slope is negative, i.e. $d\varepsilon'/d\omega < 0$. It shows that over a narrow frequency band around the resonance, the dispersion is *anomalous*. Further, at the frequency, $\omega = \omega_e + \Delta\omega$, the value of relative permittivity is less than the steady-state relative permittivity ε_{rT} contributed by the nonresonating elements of the medium. It is interesting to note that if the steady-state relative permittivity of a material is $\varepsilon'_{rT} = 1$, then the highly dispersive relative permittivity can become negative at a frequency $\omega > \omega_e$ over a narrow band. In chapter 3, subsection (3.3.2), it is discussed that the anomalous dispersive medium, under special condition, supports the backward wave propagation. However, losses are very high as shown in Fig (6.9b).

There could be polarization also in heterogeneous dielectrics due to the embedded inclusions in a host medium. This kind of polarization arises due to Maxwell–Wagner interfacial polarization, which is discussed in the next section (6.6). Figure (6.10a and b) shows dispersion and loss behaviors of a *hypothetical material* with several polarizations. In general, the $\varepsilon_r(\omega)$ decreases with an increase in frequency, i.e. the overall material dispersion is anomalous. It is obtained from Debye's relaxation model. Figure (6.10a) shows two extended Debye responses I and II for both Maxwell–Wagner and dipolar polarizations with a noticeable difference in the magnitude of their relative permittivity. However, in practice, such difference may not be visible for two Debye type polarizations. The Maxwell–Wagner polarization provides the extended Debye-I response, and the dipole polarization provides the extended Debye-II response. The magnitude of the relative permittivity of both polarizations tends to the same value $\varepsilon'_r(\omega = \infty)$ at high frequency, when polarizations are lost. The minima and maxima in $\varepsilon''_r(\omega)$ for two Debye type polarizations could be visible in the measurements [B.21]. The Lorentz oscillator model is reduced to the

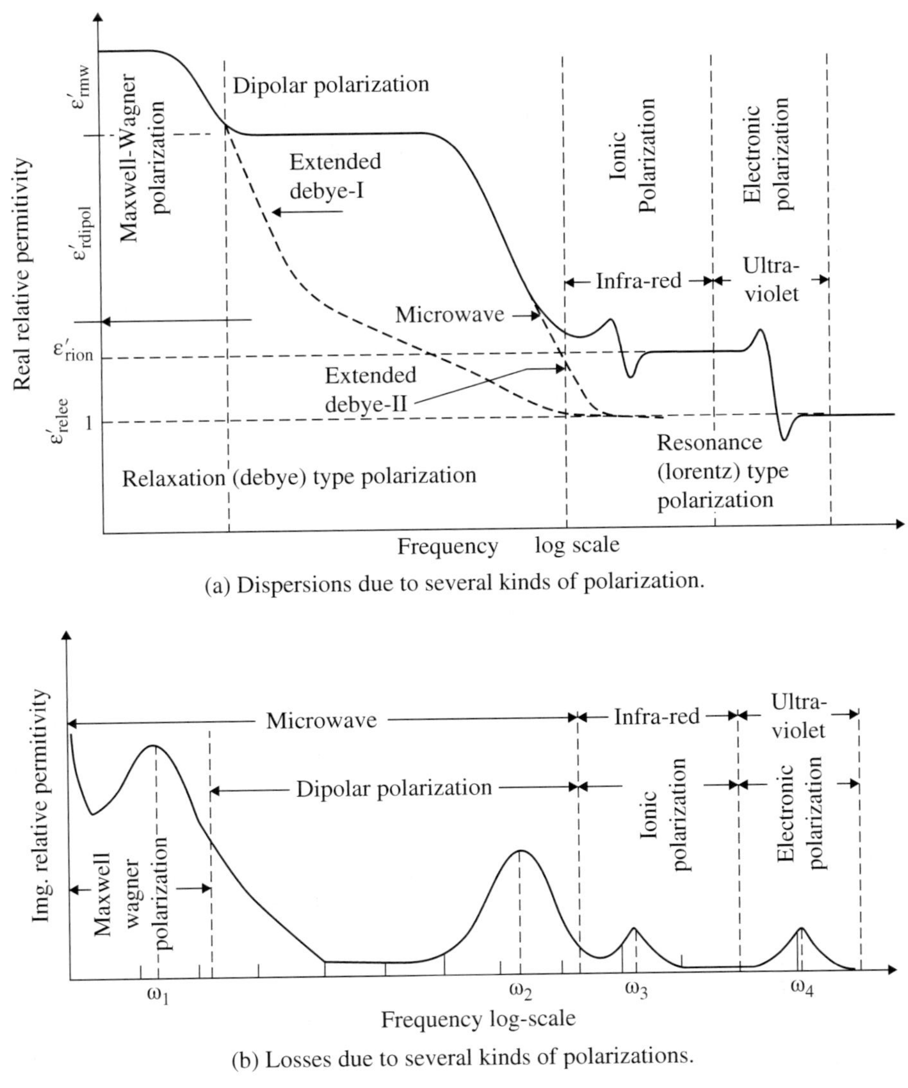

Figure 6.10 Dispersion behavior of a hypothetical material showing two relaxation type dispersion and two resonance type dispersion.

Debye relaxation model by ignoring d^2P_c/dt^2 term in the polarization differential equation (6.5.3e). It can be written as

$$\frac{\gamma_c}{\omega_c^2}\frac{dP_c}{dt} + P_c = \frac{N_c q_c^2}{m_c \omega_c^2} E_{ext}. \qquad (6.5.15)$$

A comparison of equation (6.5.15) with equation (6.4.4) shows that it is a relaxation type dispersion model with relaxation time $\tau = \gamma_c/\omega_c^2$. It is further noted from equation (6.5.8c) that if $\varepsilon_r''(\omega) = 0$, i.e. there is no loss in the material, $\varepsilon_r'(\omega)$ becomes nondispersive. Thus, dispersion in a material is associated with material loss. The material dispersion and loss are related through Kramers–Kronig relations [B.5]. It is summarized in the subsection (6.5.4). If we measure the frequency-dependent real part of relative permittivity $\varepsilon_r'(\omega)$, the measured frequency-dependent loss $\varepsilon_r''(\omega)$ could be recovered using Kramers–Kronig relations.

6.5.2 Drude Model of Conductor and Plasma

Drude model provides the DC and frequency-dependent conductivity of metal under the time-harmonic field excitation. It shows Debye-type relaxation at a frequency below the plasma frequency, as shown in Fig (6.9c). Drude model is a special case of the Lorentz model given by equation (6.5.5c). The metal has free electrons; so in absence of the restoring force, there is no characteristic resonance, i.e. $\omega_c = 0$. The following Drude model expressions can be written from equation (6.5.5c) for the complex susceptibility and complex relative permittivity of the plasma medium of a metal:

$$\chi_r^* = \frac{\omega_p^2}{-\omega^2 + j\omega\gamma} \quad (a), \quad \omega_p^2 = \frac{Ne^2}{\varepsilon_0 m_e} \qquad (b)$$

$$\varepsilon_r^* = 1 + \chi_r^* = 1 - \frac{\omega_p^2}{\omega^2 - j\omega\gamma} \qquad (c)$$

$$\text{for } \gamma_c = 0, \varepsilon_r = 1 - \frac{\omega_p^2}{\omega^2} \qquad (d)$$

$$\Rightarrow \mathrm{Re}\left(\varepsilon_r^*\right) = \varepsilon_r' = \varepsilon_r = 1 - \left[\frac{\omega_p^2}{\omega^2 + \gamma^2}\right] \qquad (e)$$

$$\mathrm{Im}\left(\varepsilon_r^*\right) = \varepsilon_r'' = \frac{\gamma_c \omega_p^2}{\omega\left(\omega^2 + \gamma_c^2\right)} \qquad (f),$$

$$(6.5.16)$$

where ω_p is the plasma frequency of the free electrons of charge e and mass m_e. N is the electron density. The free electrons are confined within a thin layer of conductor form the plasma medium. The propagation constant in the lossy plasma medium is $\beta = \beta_0\sqrt{\varepsilon_r}$. The EM-wave incident, with frequency $\omega^2 < \left(\omega_p^2 - \gamma_c^2\right)$, views the plasma medium of the conductor as an epsilon negative (ENG) medium, discussed in section (5.5) of chapter 5. Even in the case of a lossless plasma medium, the ENG medium is obtained for the frequency $\omega < \omega_p$. Figure (6.9c) shows that the *Drude model provides wideband negative permittivity* below the plasma frequency. Equation (6.5.16f) shows that in the region of negative permittivity, dielectric losses increase with a decrease in frequency. The ENG medium is highly dispersive and highly lossy. However, the Drude model provides negative permittivity over a wider bandwidth, whereas the *resonant Lorentz model provides the negative permittivity over a very narrow bandwidth.*

In the ENG medium, for $\omega < \omega_p$, the incident EM-wave cannot propagate through the plasma medium. The wave propagation occurs only for $\omega > \omega_p$. Thus, the plasma frequency ω_p acts like the cut-off frequency of a waveguide. Therefore, the plasma medium could be modeled as a waveguide discussed in chapter 7. Alternatively, the waveguide can be treated as a plasma medium. In the case of metals, the plasma frequency occurs in the optical and UV frequency ranges. However, the plasma medium has been also created artificially using *the wire-medium*. It brings down the plasma frequency in the GHz frequency range. It is discussed in section (21.1.1) of chapter 21.

The current density is the *polarization current density* due to the time-varying electric field. Using equation (6.1.5), the polarization current is computed. On using equation (6.5.16), the frequency-dependent conductivity is further computed:

$$\vec{J} = \frac{d\vec{P}}{dt} = j\omega\vec{P} = j\omega\varepsilon_0\chi_r\vec{E}_{ext} \qquad (a)$$

$$\text{also} \quad \vec{J} = \sigma(\omega)\vec{E}_{ext} \qquad (b)$$

$$\Rightarrow \sigma^*(\omega) = \frac{j\varepsilon_0\omega_p^2}{(j\gamma - \omega)} \qquad (c)$$

$$\text{Static conductivity}: \sigma(\omega = 0) = \frac{\varepsilon_0\,\omega_p^2}{\gamma} \qquad (d)$$

$$\Rightarrow \sigma^*(\omega) = \frac{\sigma(\omega = 0)}{1 + j\,\omega\,\tau} \qquad (e)$$

$$\text{where relaxation time, } \tau = \frac{1}{\gamma_c} \qquad (f).$$

$$(6.5.17)$$

Equations (6.5.17d) and (6.5.17e) are the static and frequency-dependent Drude models, respectively, for the conductivity of a metal. It shows the Debye-type relaxation, similar to equation (6.4.8b).

6.5.3 Dispersion Models of Dielectric Mixture Medium

Several expressions have been considered in section (6.3) to evaluate the static equivalent relative permittivity of the dielectric mixture medium formed by scattered *inclusion material*, with relative permittivity ε_{ri}, in a *host continuum* with relative permittivity ε_{rh}. We use the terminology *equivalent relative permittivity*, in place of usual *effective relative permittivity* for the corresponding equivalent (effective) homogeneous medium of the mixture, to avoid confusion of similar terminology used in the microstrip description. The inclusion and host materials can have either Debye type or the Lorentz type dispersion behavior. In the case of the metallic inclusion, the plasmonic model for the permittivity could be used to get the Drude type dispersion. The MG mixing formula has been combined with these dispersion models to obtain the dispersion model of a mixture medium [J.34–J.38]. In particular, either the inclusion or host material could be frequency-dependent. In general, both could be frequency-dependent. Maxwell Garnett formula is summarized below incorporating Debye, Lorentz, and Drude models, for the *dispersive inclusion*, $\varepsilon_{ri}(\omega)$ and the *static relative permittivity* ε_{rh} of the host continuum [J.34].

6.5.3.1 Maxwell Garnett–Debye Model

Debye model of equation (6.4.14) has been modified for the mixing medium using the MG formula (6.3.13) for the 3D spherical dispersive inclusions. The inclusion itself follows the Debye model with relaxation time τ_{di}. In the case of a mixture, it is modified to τ_{eq}; as the *mixture is treated as an equivalent homogeneous medium*. Following equation (6.4.14a), Maxwell Garnett–Debye model giving the dispersive complex equivalent relative permittivity of a two-phase mixture is summarized below [J.34].:

$$\varepsilon^*_{req}(\omega) = \varepsilon_{req}(\omega \to \infty) + \left[\frac{\varepsilon_{req}(\omega \to 0) - \varepsilon_{req}(\omega \to \infty)}{1 + j\,\omega\,\tau_{eq}}\right]. \tag{6.5.18}$$

The homogenized mixture could be also treated with the help of the circuit model, discussed in subsection 6.7.3. The low- and high-frequency equivalent relative permittivity of a mixture, containing Debye type frequency-dependent inclusion, are obtained by modifying the MG formula (6.3.13) as follows:

$$\begin{aligned}
\varepsilon'_{req}(\omega \to 0) &= \varepsilon_{rh}\left[1 + \frac{3f\{\varepsilon_{ri}(\omega \to 0) - \varepsilon_{rh}\}}{\varepsilon_{ri}(\omega \to 0) + 2\varepsilon_{rh} - f\{\varepsilon_{ri}(\omega \to 0) - \varepsilon_{rh}\}}\right] \quad &(a)\\
\varepsilon'_{req}(\omega \to \infty) &= \varepsilon_{rh}\left[1 + \frac{3f\{\varepsilon_{ri}(\omega \to \infty) - \varepsilon_{rh}\}}{\varepsilon_{ri}(\omega \to \infty) + 2\varepsilon_{rh} - f\{\varepsilon_{ri}(\omega \to \infty) - \varepsilon_{rh}\}}\right] \quad &(b).
\end{aligned} \tag{6.5.19}$$

Finally, the relaxation time of the mixture is given by

$$\tau_{eq} = \tau_{di}\left[\frac{(1-f)\varepsilon_{ri}(\omega \to \infty) + (2+f)\varepsilon_{rh}}{(1-f)\varepsilon_{ri}(\omega \to 0) + (2+f)\varepsilon_{rh}}\right]. \tag{6.5.20}$$

In equation (6.5.20), f is the volume fraction of a mixture. The real and imaginary parts of the complex equivalent relative permittivity of the mixture could be separated. It helps to obtain its loss-tangent of the mixture.

6.5.3.2 Maxwell Garnett–Lorentz Model

Similarly, Lorentz resonance type model (6.5.5b) could be modified for the two-phase mixture and get the following *Maxwell Garnett–Lorentz model* [J.34]:

$$\varepsilon^*_{req}(\omega) = \varepsilon_{req}(\omega \to \infty) + \frac{\omega^2_{p,eq}}{\omega^2_{0,eq} - \omega^2 + j\omega\,\gamma_{eq}}. \tag{6.5.21}$$

In the present model, inclusion is dispersive and the host-medium is frequency independent. The $\varepsilon_{req}(\omega \to \infty)$ of the mixture is given by equation (6.5.19b). The plasma frequency $\omega_{p,\,eq}$ and resonance frequency $\omega_{0,\,eq}$ of the mixture are expressed in terms of the plasma frequency $\omega_{p,\,i}$ and resonance frequency $\omega_{0,\,i}$ of the inclusion:

$$\begin{aligned}
\omega_{p,eq} &= \sqrt{f}\left[\frac{3\varepsilon_{rh}}{(1-f)\varepsilon_{ri}(\omega \to \infty) + (2+f)\varepsilon_{rh}}\right]\omega_{p,i} \quad &(a)\\
\omega^2_{0,eq} &= \omega^2_{0,i} + \left[\frac{(1-f)}{(1-f)\varepsilon_{ri}(\omega \to \infty) + (2+f)\varepsilon_{rh}}\right]\omega^2_{p,i} \quad &(b).
\end{aligned} \tag{6.5.22}$$

In equation (6.5.22), f is the volume fraction of a mixture. The damping factor of the inclusion remains unchanged in the mixture, i.e. $\gamma_{eq} = \gamma_i$. The resonance frequency of a mixture is higher than that of the resonance frequency of inclusion, and it increases with a low-density mixture. The plasma frequency of a mixture increases for a denser mixture.

6.5.3.3 Maxwell Garnett–Drude Model

Drude model has been discussed above as a special case of the Lorentz model, as applied to the metals with free electrons showing no resonance. However, the metallic inclusions are used in a host medium. The mixture behaves like a dielectric-medium, showing the Lorentz type resonance with resonance frequency $\omega_{0,eq}^{metal}$ given below. For the resonance frequency $\omega_{0,eq}^{metal}$, the MG–Lorentz model (6.5.21) could be treated as the MG–Drude model applicable to the metallic inclusions:

$$\omega_{0,eq}^{metal} = \left[\frac{(1-f)}{(1-f)\varepsilon_{ri}(\omega \to \infty) + (2+f)\varepsilon_{rh}}\right]^{1/2} \omega_{p,i}, \tag{6.5.23}$$

where for a metallic inclusion $\varepsilon_{ri}(\omega \to \infty) \approx 1$. It shows that the mixture behaves like the nonmetallic type Drude model. However, for the usual metallic Drude model $\omega_{0,eq} = 0$ in equation (6.5.21) to get the plasma-medium type response.

If the measurement results for the real and imaginary parts of any mixture medium are known at a few frequencies, the multiterms Debye models could be empirically obtained to get the curve-fit expression of experimental results for the wideband applications. It is discussed in section (6.8).

6.5.4 Kramers–Kronig Relation

The general relation between the frequency-dependent real and imaginary parts of complex relative permittivity $\varepsilon_r'(\omega)$ and $\varepsilon_r''(\omega)$ is expressed by the Kramers–Kronig relation given below [B.5, B.10]:

$$\varepsilon_r'(\omega) = \varepsilon_r(\omega = \infty) + \frac{2}{\pi}\int_0^\infty \frac{\omega'\varepsilon_r''(\omega')}{((\omega')^2 - \omega^2)} d\omega' \quad \text{(a)}$$

$$\varepsilon_r''(\omega) = -\frac{2}{\pi}\int_0^\infty \frac{\omega\left[\varepsilon_r'(\omega') - \varepsilon_r(\infty)\right]}{((\omega')^2 - \omega^2)} d\omega' \quad \text{(b)}. \tag{6.5.24}$$

It can be shown that Debye's equation follows the Kramers–Kronig relation. However, the D.C. conductivity could be the part of the Debye model, although it is not the part of Kramers–Kronig causal relation. If the imaginary part of a dispersive lossy dielectric is measured at one frequency, the first expression of Kramers–Kronig can provide frequency-dependent real relative permittivity at all other frequencies.

6.6 Interfacial Polarization

The interfacial polarization, i.e. *Maxwell–Wagner polarization*, occurs in the layered inhomogeneous dielectric and semiconductor media, such as the metal insulator-semiconductor (MIS) structure, Schottky structure, PN-junction, etc. [B.1, J.2–J.5]. It also occurs in heterogeneous mixing materials, where the contact between the inclusion and host also forms the boundary layer [J.39, J.40]. The interfacial polarization provides *Debye kind of polarization*. Normally, it occurs in the lower range of microwaves. This section presents the polarization equation of *Maxwell–Wagner polarization* for a two-layered structure. The next section discusses circuit modeling.

6.6.1 Interfacial Polarization in Two-Layered Capacitor Medium

Figure (6.11a) shows the two-layered parallel-plate capacitor to examine the interfacial polarization. The lossy dielectric layers have relative permittivity $\varepsilon_{r\,i}$, conductivity σ_i, and substrate thickness h_i (i = 1,2). On the application of a voltage V across the capacitor, the layers support conduction current density, J_i. Once the switch is closed at t = 0, the interfacial polarization process starts with the flow of the *transient polarization current*. The migrated charges, from the interior of the medium, accumulate at the interface, as shown in Fig (6.1d). Finally, the *steady-state condition* is reached. However, initially, there is no accumulated charge at the interface. The time taken for the accumulation of charges is described by the *relaxation time* of the media given by

$$\tau_1 = \frac{\varepsilon_1}{\sigma_1}, \quad \tau_2 = \frac{\varepsilon_2}{\sigma_2}, \text{(where } \varepsilon_i = \varepsilon_0\varepsilon_{r\,i}, i = 1, 2). \tag{6.6.1}$$

The relaxation time is defined in equation (6.4.2).

6.6.1.1 Boundary Conditions

Using Ohm's law, the conduction current densities, normal to the interface of two media, are expressed as

$$J_{n1} = \sigma_1 E_{n1} \quad \text{(a)}, \quad J_{n2} = \sigma_2 E_{n2} \quad \text{(b)}, \tag{6.6.2}$$

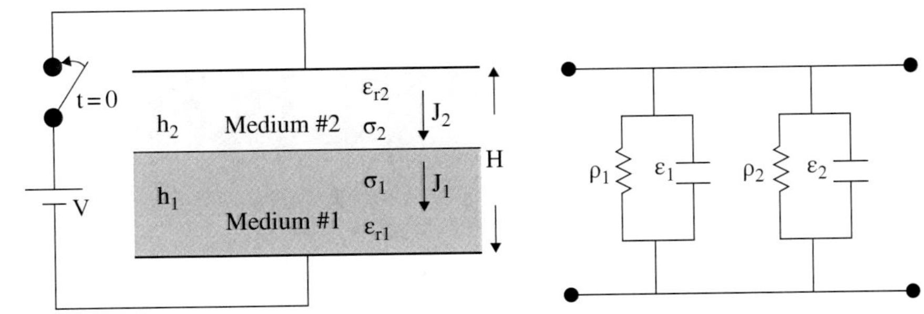

(a) Two-layered lossy capacitor medium exhibiting interfacial polarization at the interface.

(b) Equivalent circuit of two-layered lossy inhomogeneous medium to compute time constant.

Figure 6.11 Interfacial polarization at the interface of two-layered lossy dielectrics.

where E_{n1} and E_{n2} are the normal components of the electric field in the medium #1 and medium #2, respectively. Initially, the current follows across the interface, and there is no accumulation of charges:

Current entering the interface =

Current leaving the interface

$$J_{n1} = J_{n2} \quad (a), \qquad \Rightarrow \frac{E_{n1}}{E_{n2}} = \frac{\sigma_2}{\sigma_1} \qquad (b). \tag{6.6.3}$$

However, during the *transient process*, the charges build up at the interface, giving $J_{n1} \neq J_{n2}$. The acculturated negative surface charge ρ_s provides discontinuity to the normal components of the electric displacement (electric flux density) vectors:

$$D_{n1} - D_{n2} = \rho_s \Rightarrow \varepsilon_{r1}E_{n1} - \varepsilon_{r2}E_{n2} = \frac{\rho_s}{\varepsilon_0}. \tag{6.6.4}$$

On using equations (6.6.1) and (6.6.3b), the accumulated charges at the surface are obtained in terms of the relaxation times:

$$\frac{\rho_s}{\varepsilon_0} = \left[\varepsilon_{r1} - \varepsilon_{r2}\frac{\sigma_1}{\sigma_2}\right]E_{n1}, \Rightarrow \frac{\rho_s}{\varepsilon_0} = \left[\varepsilon_{r1}\frac{\sigma_2}{\sigma_1} - \varepsilon_{r2}\right]E_{n2} \quad (a)$$

$$\Rightarrow \rho_s = (\tau_1 - \tau_2)\sigma_1 E_{n1} \quad (b)$$

For no accumulation of charge at the interface,

$$\rho_s = 0 : \frac{\varepsilon_{r1}}{\varepsilon_{r2}} = \frac{\sigma_1}{\sigma_2} \quad (c). \tag{6.6.5}$$

If $\tau_1 = \tau_2$, $\rho_s = 0$, i.e. there is no interfacial polarization in a homogeneous medium. *The interfacial polarization occurs only when the relaxation times of both media are different*, $\tau_1 \neq \tau_2$. Different relaxation times are obtained for two different materials ($\varepsilon_{r1} \neq \varepsilon_{r2}$), or even for the same material having different conductivity ($\sigma_1 \neq \sigma_2$). The latter case is obtained in a semiconducting material through the doping or by the formation of a depletion layer at the interface. The case $h_1 = h_2 = h$ is considered below. The case $h_1 \neq h_2$ is considered in subsection (6.7.4) using the circuit model.

The double-layered structure of Fig (6.11a) is analyzed under both the steady-state and transient conditions. Under the *initial condition*, at time $t = 0$, there is no surface charge at the interface of two media, $\rho_s(0) = 0$. Under the *transient condition*, the time-dependent surface charge $\rho_s(t)$ accumulates at the interface, and finally, under the *steady-state condition* for $t \to \infty$, the static charges $\rho_s(\infty)$ accumulate at the interface.

6.6.1.2 Steady-State Condition

Under the steady-state condition, the applied voltage V at the capacitor is

$$E_{n1}h + E_{n2}h = V \quad (a)$$

$$\Rightarrow E_{n2}\left(1 + \frac{E_{n1}}{E_{n2}}\right) = \frac{V}{h}, \Rightarrow E_{n2}\left(1 + \frac{\sigma_2}{\sigma_1}\right) = \frac{V}{h}$$

$$\Rightarrow E_{n2} = \frac{\sigma_1}{\sigma_1 + \sigma_2}\frac{V}{h} \quad (b), \quad \text{or} \quad E_{n1} = \frac{\sigma_2}{\sigma_1 + \sigma_2}\frac{V}{h} \quad (c). \tag{6.6.6}$$

The surface charge density at the interface is obtained by substituting E_{n1} from equation (6.6.6c) in equation (6.6.5a):

$$\rho_s = \left[\varepsilon_1 - \frac{\varepsilon_2\sigma_1}{\sigma_2}\right]\frac{\sigma_2}{\sigma_1 + \sigma_2}\frac{V}{h}$$

$$\Rightarrow \rho_s = \left[\frac{\varepsilon_1\sigma_2 - \varepsilon_2\sigma_1}{\sigma_1 + \sigma_2}\right]\frac{V}{h}. \tag{6.6.7}$$

6.6.1.3 Transient Condition

During the transient process $J_{n1} \neq J_{n2}$, and it results in a flow of the interfacial polarization current:

$$\frac{\partial \rho_s}{\partial t} = J_{n2} - J_{n1} \quad (a), \quad \Rightarrow \frac{\partial \rho_s}{\partial t} = E_{n2}\sigma_2 - E_{n1}\sigma_1 \quad (b). \tag{6.6.8}$$

Both field quantities, E_{n1} and E_{n2}, are computed by using equations (6.6.4) and (6.6.6a):

$$E_{n1}(\varepsilon_1/\varepsilon_2) - E_{n2} = \rho_s/\varepsilon_2 \quad (a), \quad E_{n1} + E_{n2} = \frac{V}{h} \quad (b).$$

On adding the above equations (a) and (b), E_{n1} is obtained:

$$\left(\frac{\varepsilon_1}{\varepsilon_2} + 1\right) E_{n1} = \frac{V}{h} + \frac{\rho_s}{\varepsilon_2}$$

$$\Rightarrow E_{n1} = \left(\frac{\varepsilon_2}{\varepsilon_1 + \varepsilon_2}\right)\frac{V}{h} + \frac{\rho_s}{\varepsilon_1 + \varepsilon_2} \quad (a)$$

$$\text{also,} \quad E_{n2} = \frac{V}{h} - E_{n1}$$

$$\Rightarrow E_{n2} = \left(\frac{\varepsilon_1}{\varepsilon_1 + \varepsilon_2}\right)\frac{V}{h} - \frac{\rho_s}{\varepsilon_1 + \varepsilon_2} \quad (b). \tag{6.6.9}$$

The following *interfacial polarization equation* is obtained by substituting E_{n1} and E_{n2} from equation (6.6.9) in equation (6.6.8b):

$$\frac{\partial \rho_s}{\partial t} = -\left[\frac{\sigma_1 + \sigma_2}{\varepsilon_1 + \varepsilon_2}\right]\rho_s + \left[\frac{\sigma_2\varepsilon_1 - \sigma_1\varepsilon_2}{\varepsilon_1 + \varepsilon_2}\right]\frac{V}{h} \quad (a)$$

$$\Rightarrow \frac{\partial \rho_s}{\partial t} = -A\rho_s + B \quad (b)$$

$$\text{where,} \quad A = \frac{\sigma_1 + \sigma_2}{\varepsilon_1 + \varepsilon_2}, B = \frac{\sigma_2\varepsilon_1 - \sigma_1\varepsilon_2}{\varepsilon_1 + \varepsilon_2}\frac{V}{h} \quad (c). \tag{6.6.10}$$

On solving the above equation, the transient surface charge density is obtained at the interface:

$$-A\rho_s(t) + B = De^{-At},$$

$$\Rightarrow \rho_s(t) = Ke^{-At} + \frac{B}{A}, \left(\text{where,} \quad K = -\frac{D}{A}\right)$$

$$\Rightarrow \rho_s(t) = Ke^{-\left[\frac{\sigma_1 + \sigma_2}{\varepsilon_1 + \varepsilon_2}\right]t} + \left[\frac{\sigma_2\varepsilon_1 - \sigma_1\varepsilon_2}{\sigma_1 + \sigma_2}\right]\frac{V}{h}. \tag{6.6.11}$$

The constant K is evaluated from the initial condition, at $t = 0$, $\rho_s(0) = 0$:

$$\rho_s(0) = K + \left[\frac{\sigma_2\varepsilon_1 - \sigma_1\varepsilon_2}{\sigma_1 + \sigma_2}\right]\frac{V}{h} = 0$$

$$\Rightarrow K = -\left[\frac{\sigma_2\varepsilon_1 - \sigma_1\varepsilon_2}{\sigma_1 + \sigma_2}\right]\frac{V}{h}. \tag{6.6.12}$$

The final interfacial polarization, showing the Debye type response, in the time-domain could be written as follows:

$$\rho_s(t) = \frac{V}{h}\left[\frac{\sigma_2\varepsilon_1 - \sigma_1\varepsilon_2}{\sigma_1 + \sigma_2}\right]\left[1 - e^{-t/T_r}\right] \quad (a)$$

$$\text{where} \quad T_r = \left[\frac{\varepsilon_1 + \varepsilon_2}{\sigma_1 + \sigma_2}\right] \quad (b)$$

$$\Rightarrow T_r = \left[\frac{\varepsilon_1 + \varepsilon_2}{(1/\rho_1 + 1/\rho_2)}\right] \quad (c). \tag{6.6.13}$$

The lossy dielectric layers could be viewed as the complex capacitors represented by the RC-circuit model, as shown in Fig (6.11b). The relaxation time constant T_r of the inhomogeneous layered medium could also be obtained from Fig (6.11b), showing the parallel combination of two complex capacitors corresponding to two layers of lossy dielectrics:

$$\frac{1}{R} = \frac{1}{\rho_1} + \frac{1}{\rho_2} \quad (a)$$

$$C = \varepsilon_1 + \varepsilon_2 \quad (b)$$

$$T_r = RC = \left[\frac{\varepsilon_1 + \varepsilon_2}{\sigma_1 + \sigma_2}\right] \quad (c). \tag{6.6.14}$$

The circuit model can be extended to get the relaxation time constant for n numbers of interfaces, corresponding to (n+1) numbers of lossy dielectric layers. It also provides the relaxation frequency of the Debye type polarization response:

$$T_r = \frac{\sum_{i=1}^{n+1} \varepsilon_i}{\sum_{i=1}^{n+1} (1/\rho_i)} \quad (a), \quad f_r = \frac{1}{T_r} \quad (b). \tag{6.6.15}$$

The above analysis can be extended to get the Debye type polarization of the interfacial polarization in the frequency-domain. In subsection (6.7.4), the circuit model is used to get such results. In the case of interfacial polarization, the polarization is not achieved using the concept of polarizability as no dipole formation is involved. The interfacial polarization is achieved by the process of charge accumulation at the interface of heterogeneous media. However, even for the interfacial polarization, the polarizability could be extracted, as done in equation (6.7.36).

6.7 Circuit Models of Dielectric Materials

The dielectric dispersion behavior of a homogeneous material, at the molecular level, is described either by the relaxation type response or by the resonance type response. The homogeneous dielectric materials can be divided into two broad groups: *(i) Debye materials and (ii) Lorentz materials* [J.41]. The heterogeneous dielectrics with the interfacial Maxwell–Wagner polarization can be called *Maxwell–Wagner materials*. However, they also show the relaxation response [B.1]. In previous sections, these responses are considered using the polarization models. The polarization works at the microlevel and is finally responsible for the creation of relative permittivity in a material at the macrolevel, ignoring details of each polarized molecules. It is known as the process of *homogenization*. Likewise, the circuit model elements depict the microlevel polarization response, relaxation, or resonance type, to the applied field. In the end, the circuit models of materials – natural or artificial, help to compute their complex relative permittivity. Thus, the circuit models help to obtain homogenized media. The circuit models are useful in the development of a new kind of artificial dielectric materials with interesting properties. This section presents the modeling of polarization responses, and complex relative permittivity of materials, using the circuit models.

6.7.1 Series RC Circuit Model

The relaxation in a dielectric material is related to its delayed polarization response, under the excitation of an external field. The growth of the polarization is described by the first-order differential equation, given by equation (6.4.3). The delayed polarization, with relaxation time, can also be modeled by the RC series circuit shown in Fig (6.12a). On closing of the switch, the capacitor C_s is charged to Q. The voltage around the circuit is

$$V = V_c + V_R, \Rightarrow IR_s + \frac{Q}{C_s} = V$$
$$\Rightarrow R_s\frac{dQ}{dt} + \frac{Q}{C_s} = V, \Rightarrow \tau\frac{dQ}{dt} + Q = C_sV, \tag{6.7.1}$$

where the time constant of the circuit is $\tau = R_sC_s$. Equation compares well with the polarization equation (6.4.3). Let the solution of equation (6.7.1) is $Q = Q_0e^{j\omega t}$ on the application of a sinusoidal voltage $V = V_0e^{j\omega t}$. On substituting these in the above equation, the expression is obtained for the complex capacitance of the circuit:

$$j\omega\tau Q + Q = C_sV, \Rightarrow C^*(\omega) = \frac{Q}{V} = \frac{C_s}{1 + j\omega\tau}. \tag{6.7.2}$$

The above relation shows that the RC circuit can model the Debye type of polarization of a dielectric material giving the following expressions for complex relative permittivity, along with its limiting values:

$$\varepsilon_r^* = \varepsilon_r' - j\varepsilon_r'' = \frac{C'(\omega) - jC''(\omega)}{C_0} = \frac{C^*(\omega)}{C_0} = \frac{C_s/C_0}{1 + j\omega\tau} \quad \text{(a)}$$
$$\text{For} \quad \omega \to 0, \varepsilon_r^*(0) = \frac{C_s}{C_0} \quad \text{(b)}$$
$$\text{For} \quad \omega \to \infty, \varepsilon_r^*(\infty) = 0 \quad \text{(c)}. \tag{6.7.3}$$

The complex relative permittivity can be separated into the real and imaginary parts:

$$\varepsilon_r'(\omega) = \frac{\varepsilon_r'(0)}{1 + \omega^2\tau^2} \quad \text{(a)}, \quad \varepsilon_r''(\omega) = \frac{\omega\tau\varepsilon_r'(0)}{1 + \omega^2\tau^2} \quad \text{(b)}$$
$$\tan(\delta) = \frac{\varepsilon''}{\varepsilon'} = \omega\tau \quad \text{(c)}. \tag{6.7.4}$$

The above relation can be obtained directly from the admittance of the RC circuit:

$$Y = \frac{j\omega C_s}{1 + j\omega C_s R_s} = \frac{j\omega C_s}{1 + j\omega\tau}. \tag{6.7.5}$$

On using $\tau = R_sC_s$ and $Y = j\omega C^*(\omega)$, the complex capacitance of the circuit is

$$C^*(\omega) = \frac{C_s}{1 + j\omega\tau}. \tag{6.7.6}$$

Equation (6.7.6) is identical to equation (6.7.2). Figure (6.12b) shows the Debye type frequency response of both the $\varepsilon_r'(\omega)$ and $\varepsilon_r''(\omega)$, obtained from the RC circuit model. It is seen that at the optical frequency, i.e. at the higher frequency, $\omega \to \infty$, $\varepsilon_r'(\omega) \to 0$. It is unrealistic, as ε_r' of even the vacuum is unity, not zero. The model could be improved by assuming that the capacitance relation (6.7.2) provides the susceptibility, in place of relative permittivity. Alternatively, a capacitance can be added in parallel to the RC circuit. It is discussed in subsection (6.7.3). The imaginary part of the complex relative permittivity $\varepsilon_r''(\omega)$ is zero, both at $\omega = 0$, and $\omega \to \infty$. Further, an absorption peak is obtained at a frequency, $\omega = \omega_{max} = 1/\tau$ giving $\varepsilon_r''(\omega_{max}) = \varepsilon_r(0)/2$.

On an increase in the frequency of the applied external electric field, polarization is not fully developed, as

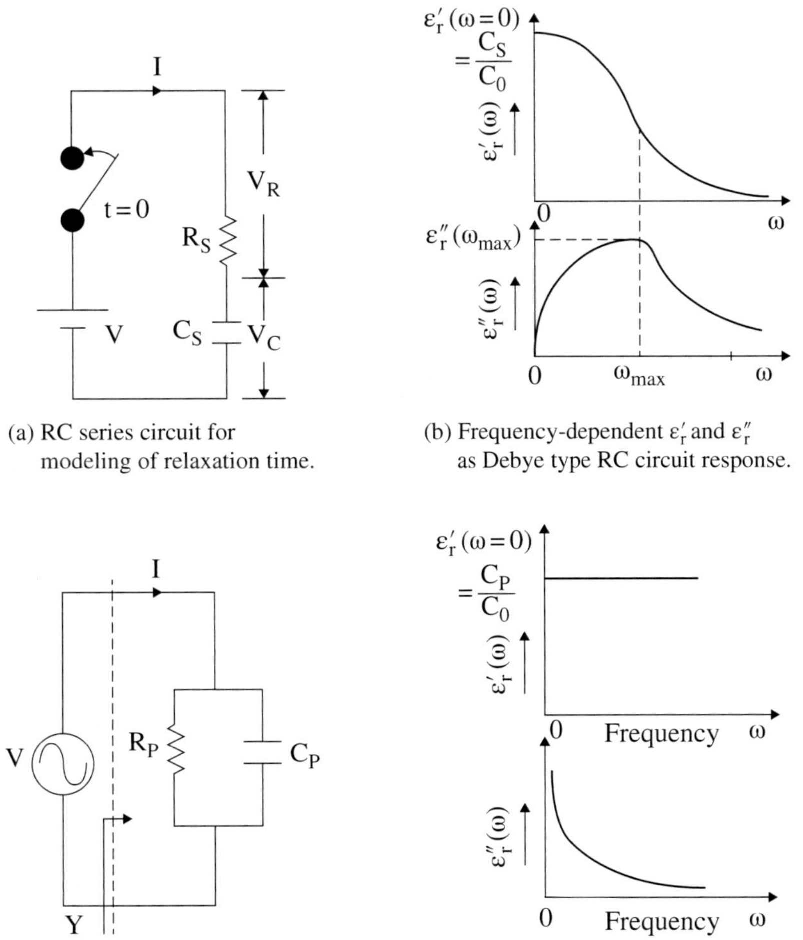

(a) RC series circuit for modeling of relaxation time.

(b) Frequency-dependent ε_r' and ε_r'' as Debye type RC circuit response.

(c) RC parallel circuit to model nonrelaxation type material.

(d) Frequency response of nonrelaxation type of material.

Figure 6.12 Series and parallel equivalent circuits of the relaxation and nonrelaxation type materials.

sufficient time is not available for the full charging of the capacitor. To polarize the material completely, the electric field must be applied for a duration more than the relaxation time τ of the dielectrics, or its circuit model. Once the experimental results on the $\varepsilon_r'(\omega)$ and $\varepsilon_r''(\omega)$ are known, the numerical value of C_s and R_s can be determined to model the material by the R_s C_s series circuit.

6.7.2 Parallel RC Circuit Model

Some dielectrics do not show noticeable variation in the relative permittivity ε_r'. However, the imaginary part of their relative permittivity ε_r'' decreases with frequency. Such material can be modeled by the RC parallel circuit, shown in Fig (6.12c). However, it is a noncausal model. It is analyzed by using the input admittance Y:

$$Y = G_p + j\omega C_p \quad \text{(a)}, \qquad Y = j\omega C^*(\omega) \quad \text{(b)}$$
$$C^*(\omega) = \frac{G_p + j\omega C_p}{j\omega} \quad \text{(c)}. \tag{6.7.7}$$

The complex capacitance provides the complex relative permittivity $\varepsilon_r^*(\omega)$ of a medium:

$$\varepsilon_r^*(\omega) = \varepsilon_r' - j\varepsilon_r'' = \frac{C^*(\omega)}{C_0}$$
$$\Rightarrow \varepsilon_r' = \frac{C_P}{C_0} \quad \text{(a)}, \qquad \varepsilon_r''(\omega) = \frac{1}{\omega\, C_0 R_P} \quad \text{(b)}. \tag{6.7.8}$$

Figure (6.12d) shows the frequency response of real $\varepsilon_r'(\omega)$ and imaginary $\varepsilon_r''(\omega)$ parts of complex relative permittivity. The parallel RC circuit does not show any relaxation and the real $\varepsilon_r'(\omega)$ is frequency independent, while imaginary $\varepsilon_r''(\omega)$ is given by a hyperbolic curve.

The material response modeled around simple series and parallel RC circuits may not correspond to any real material. Still, such simple models have been used to model dispersive materials. For instance, the RC parallel circuit has been used to model the diamond film deposited on the Si substrate [J.42]. The combination of the RC networks can provide more realistic material modeling.

6.7.3 Parallel Series Combined Circuit Model

Figure (6.13a) shows a parallel combination of a capacitor C_p and R_sC_s circuit to model a homogeneous material medium with two kinds of polarizations (i) showing no dispersion, like the electronic and ionic polarizations up to very high frequency; and (ii) showing Debye type dispersion in the lower frequency range. The first kind of polarization of host medium is modeled by the capacitance C_p, while the R_sC_s series circuit, corresponding to lossy inclusion, models the second kind of polarization, like the dipolar polarization. The circuit could also model the artificial dielectric medium. The impedance of the series arm and also input admittance of the circuit are given below:

$$Z_s = R_s + \frac{1}{j\omega\, C_s} \quad \text{(a)}, \qquad Y = j\omega\, C_p + \frac{1}{Z_s} \quad \text{(b)}. \tag{6.7.9}$$

The time constant $\tau_s = R_sC_s$, with relaxation frequency $\omega_s = 1/\tau_s = 1/(R_s\, C_s)$, controls the relaxation response of the circuit. Its input admittance is

$$Y = \frac{\omega^2 C_s\, \tau_s}{1+\omega^2\, \tau_s^2} + j\left[\omega\, C_p + \frac{\omega\, C_s}{1+\omega^2\, \tau_s^2}\right]. \tag{6.7.10}$$

The admittance can be further written using the equivalent complex capacitance,

$$\begin{aligned} &C^*(\omega) = \left(\varepsilon_r' - j\varepsilon_r''\right)C_0 \\ &\Rightarrow Y = j\omega C^*(\omega) = \left(\varepsilon_r'' + j\varepsilon_r'\right)\omega C_0. \end{aligned} \tag{6.7.11}$$

On comparing the above equations, the expressions are obtained for real $\varepsilon_r'(\omega)$, imaginary $\varepsilon_r''(\omega)$, complex $\varepsilon_r^*(\omega)$ and $\tan\delta$ of a lossy material in terms of the circuit elements:

$$\begin{aligned} &\varepsilon_r'(\omega) = \frac{C_p}{C_0} + \frac{C_s}{C_0}\left[\frac{1}{1+\omega^2\tau_s^2}\right] && \text{(a)} \\ &\varepsilon_r''(\omega) = \frac{C_s}{C_0}\left[\frac{\omega\, \tau_s}{1+\omega^2\, {\tau_s}^2}\right] && \text{(b)} \\ &\varepsilon_r^*(\omega) = \varepsilon_r' - j\varepsilon_r'', \\ \Rightarrow\ &\varepsilon_r^*(\omega) = \frac{C_p}{C_0} + \frac{C_s}{C_0}\left[\frac{1}{1+j\omega\, \tau_s}\right] && \text{(c)} \\ &\tan(\delta_s) = \omega\, \tau_s, \quad \text{also, } \tan(\delta) = \frac{\varepsilon_r''}{\varepsilon_r'} \\ \Rightarrow\ &\tan(\delta) = \frac{\tan(\delta_s)}{1+\dfrac{C_p}{C_s}\left(1+\omega^2\tau_s^2\right)} && \text{(d).} \end{aligned} \tag{6.7.12}$$

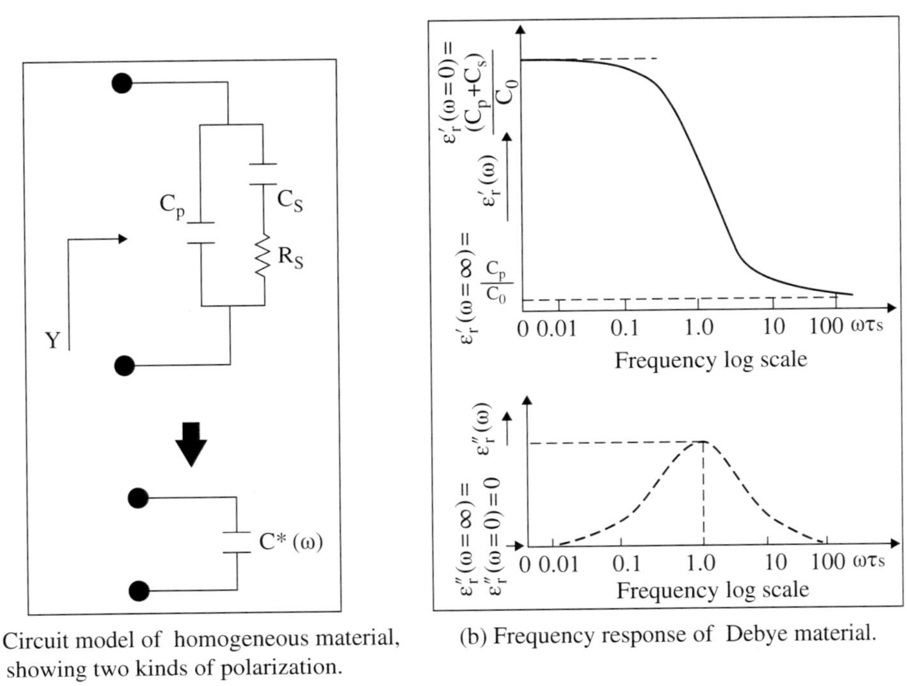

(a) Circuit model of homogeneous material, showing two kinds of polarization.

(b) Frequency response of Debye material.

Figure 6.13 Circuit model and relaxation response of Debye material.

The above equations show that the circuit of Fig (6.13a) models Debye-type relaxation response of a material. For the limiting case $\omega \to 0$, the static relative permittivity is due to both kinds of polarizations, whereas at the optical frequency, $\omega \to \infty$, the dipolar kind of polarization or polarization in inclusions of a mixture, does not exist. The following expressions are obtained for both the limiting cases and also the contribution of the dipolar polarization to the relative permittivity:

$$\text{For } \omega \to 0 : \varepsilon_r'(\omega = 0) = \frac{C_p + C_s}{C_0} \quad \text{(a)}$$

$$\varepsilon_r''(\omega = 0) = 0 \quad \text{(b)}$$

$$\text{For } \omega \to \infty : \varepsilon_r'(\omega \to \infty) = \frac{C_p}{C_0}, \quad \text{(c)}$$

$$\varepsilon_r''(\omega \to \infty) = 0 \quad \text{(d)}$$

Contribution of dipole polarization :

$$\frac{C_s}{C_0} = \varepsilon_r'(\omega = 0) - \varepsilon_{reff}(\omega = \infty) \quad \text{(e).} \qquad (6.7.13)$$

The contribution of the dipolar polarization to the overall polarization of a material decreases at the frequency above the relaxation frequency, i.e. for $\omega > \omega_s$ ($\omega_s \tau_s = 1$). From equations (6.7.12) and (6.7.13), the dispersion expression of the material could be written as

$$\varepsilon_r'(\omega) = \varepsilon_r'(\omega = \infty) + \frac{\varepsilon_r'(\omega = 0) - \varepsilon_r'(\omega = \infty)}{1 + (\omega/\omega_s)^2} \quad \text{(a)}$$

$$\varepsilon_r''(\omega) = \frac{[\varepsilon_r'(\omega = 0) - \varepsilon_r'(\omega = \infty)](\omega/\omega_s)}{1 + (\omega/\omega_s)^2} \quad \text{(b).} \qquad (6.7.14)$$

The above equations could be combined to write the following expression:

$$\varepsilon_r^*(\omega) = \varepsilon_r'(\omega = \infty) + \frac{\varepsilon_r'(\omega = 0) - \varepsilon_r'(\omega = \infty)}{1 + j\omega\tau_s}. \qquad (6.7.15)$$

The above equation is identical to the MG–Debye model, given by equation (6.5.18). Figure (6.13b) shows the usual frequency response for a Debye-type material. The component's values can be computed for the circuit model from the experimental results. The circuit model is useful for the design of components and also to study distortion in a signal propagating through the medium. More general circuit model, useful for the SPICE circuit simulator, are developed for a Debye-type dispersive material [J.43].

6.7.4 Series Combination of RC Parallel Circuit

Figure (6.11a) shows an inhomogeneous lossy medium formed with a two-layered lossy dielectric capacitor. Figure (6.14a) shows its circuit model, using the series

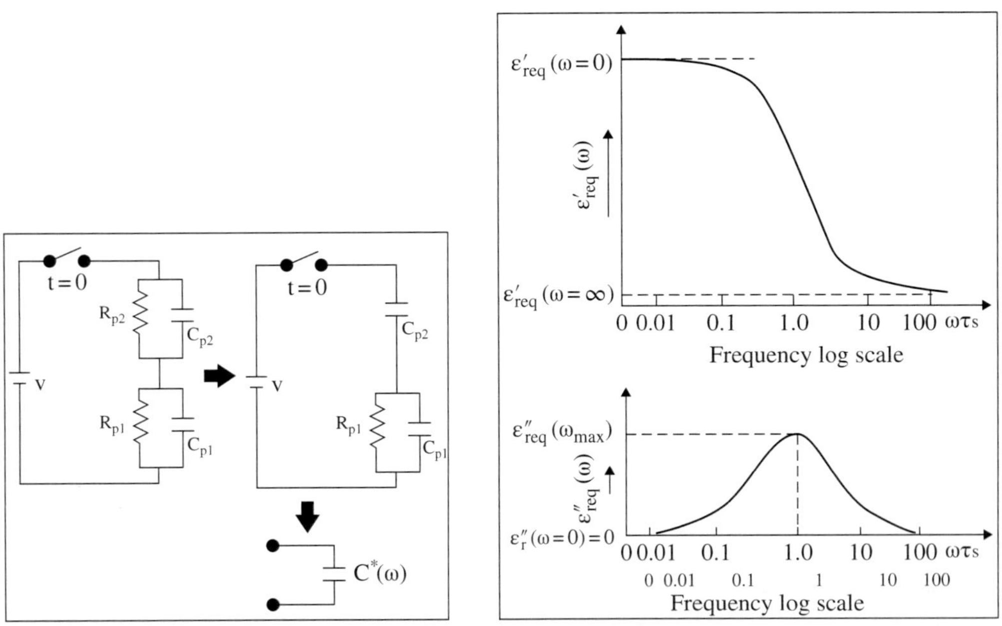

(a) Circuit model of two-layered inhomogeneous medium.

(b) Relaxation type response of interfacial polarization.

Figure 6.14 Circuit model of the two-layered inhomogeneous medium under the interfacial polarization.

combination of two parallel RC circuits. Each parallel circuit is the analog of each layer of the lossy dielectric-medium. Such an inhomogeneous dielectric-medium is encountered in the microstrip line structures based on Si-SiO_2 and Schottky-contact substrates [J.2–J.5]. It involves the Maxwell–Wagner type interfacial polarization. In the case of the Si-SiO_2 structure, the upper SiO_2 layer is a lossless passivation layer with $R_{p2} \to \infty$ and the lower Si- layer is a high conductivity doped silicon layer. Thus, two cases are considered for the two-layered medium shown in Fig (6.11a): (i) *upper layer lossless* and (ii) *upper layer lossy*. The first circuit model of Fig (6.14a) shows the case (ii), and it is reduced to another circuit model for the case (i), that is a series combination of R_{p1} C_{p1} circuit and C_{p2}. The interfacial polarization response of the inhomogeneous medium could be obtained from the response of the circuit models. The Maxwell–Wagner type polarization also occurs in the polycrystalline materials – semiconductor and ceramic [J.40]. The low-resistance grains are molded by one parallel RC circuit and narrow high-resistance grain boundaries are modeled by another parallel RC circuit.

6.7.4.1 Lossless Upper Layer

The input impedance Z and input admittance ($Y = 1/Z$) of the lossless upper layer, shown as the reduced circuit in Fig (6.14a), is determined below

$$Y_1 = \frac{1}{R_{p1}} + j\omega C_{p1}, Z = \frac{1}{j\omega C_{p2}} + \frac{1}{Y_1}$$
$$\Rightarrow Z = \frac{1 + j\omega R_{p1}C_{p1} + j\omega R_{p1}C_{p2}}{j\omega C_{p2}\left(1 + j\omega R_{p1}C_{p1}\right)}. \quad (6.7.16)$$

$$Y = \frac{\omega^2\left[C_{p2}(\tau_1 + C_{p2}R_{p1}) - C_{p2}\tau_1\right] + j\omega C_{p2}\left[1 + \omega^2\tau_1(\tau_1 + R_{p1}C_{p2})\right]}{\left[1 + \omega^2(\tau_1 + R_{p1}C_{p2})^2\right]}, \quad (6.7.17)$$

where the time constant of the circuit is $\tau_1 = R_{p1}C_{p1}$. The whole circuit is treated as an equivalent complex capacitor with a single lossy dielectric-medium:

$$\varepsilon^*_{req} = \varepsilon'_{req} - j\varepsilon''_{req},$$
$$Y = j\left(\varepsilon'_{req} - j\varepsilon''_{req}\right)\omega C_0 = \left(\varepsilon''_{req} + j\varepsilon'_{req}\right)\omega C_0, \quad (6.7.18)$$

where C_0 is the capacitance of the structure with air-medium. On comparing equations (6.7.17) and (6.7.18), the expressions for the real and imaginary parts of ε^*_{req} are obtained in terms of the circuit elements:

$$\varepsilon'_{req}(\omega) = \frac{C_{p2}}{C_0}\left[\frac{1 + \omega^2\tau_1(\tau_1 + R_{p1}C_{p2})}{\left[1 + \omega^2(\tau_1 + R_{p1}C_{p2})^2\right]}\right] \quad (a)$$

$$\varepsilon''_{req}(\omega) = \frac{C_{p2}}{C_0}\frac{\omega R_{p1}C_{p2}}{1 + \omega^2(\tau_1 + R_{p1}C_{p2})^2} \quad (b). \quad (6.7.19)$$

6.7.4.2 Limiting Cases

The frequency $\omega \to 0$ gives the static value of equivalent relative permittivity of the two-layered inhomogeneous medium. Whereas the $\omega \to \infty$ gives its equivalent relative permittivity when interfacial polarization is lost, and relative permittivity is due to the remaining polarizations at a higher frequency. The limiting values of the equivalent relative permittivity are obtained from the equations (6.7.19a,b):

$$\varepsilon'_{req}(\omega = 0) = \frac{C_{p2}}{C_0} \quad (a), \qquad \varepsilon''_{req} = 0 \quad (b)$$

$$\varepsilon'_{req}(\omega \to \infty) = \frac{C_{p2}}{C_0}\left[\frac{1}{1 + \dfrac{R_{p1}C_{p2}}{\tau_1}}\right] = \frac{\varepsilon'_{req}(\omega = 0)}{1 + \dfrac{R_{p1}C_{p2}}{\tau_1}} \quad (c)$$

$$\varepsilon''_{req}(\omega \to \infty) = 0 \quad (d). \quad (6.7.20)$$

The responses of Fig (6.14b) show that the $\varepsilon'_{req}(\omega)$ decreases from $\varepsilon'_{req}(\omega = 0)$ to $\varepsilon'_{req}(\omega \to \infty)$, with an increase in frequency, whereas the $\varepsilon''_{r\,eq}(\omega)$ exhibits a maximum absorption peak in the transition region of $\varepsilon'_{req}(\omega)$. The frequency ω_{max} at which $\varepsilon''_{req,\,max}(\omega)$ occurs is computed on differentiating equation (6.7.19b) with respect to frequency ω and equating it to zero. The $\varepsilon''_{req,\,max}(\omega)$ is obtained at $\omega = \omega_{max}$:

$$\omega_{max} = \frac{1}{\tau_1 + R_{p1}C_{p2}} \quad (a)$$

$$\varepsilon''_{req\,max}(\omega_{max}) = \frac{C_{p2}\,R_{p1}}{2\,C_0}\frac{1}{\tau_1 + R_{p1}C_{p2}},$$

$$\Rightarrow \varepsilon''_{req\,max}(\omega) = \frac{C_{p2}}{2\,C_0}\frac{1}{C_{p1} + C_{p2}} \quad (b).$$

$$(6.7.21)$$

The equivalent circuit model of Fig (6.14a) can be identified with the two-layered medium of Fig (6.11a), that has $\sigma_2 = 0$, as follows:

$$C_{p1} = \frac{A}{h_1}\varepsilon'_1,\ C_{p2} = \frac{A}{h_2}\varepsilon'_2,\ R_{p1} = \frac{h_1}{A\sigma_1},$$

$$R_{p2} = \infty,\ \tau_1 = \frac{\varepsilon'_{r1}}{\sigma_1},\ C_0 = \frac{A}{h}\varepsilon_0,\ h = h_1 + h_2,$$

$$(6.7.22)$$

where A is the area of the parallel-plate capacitor.

6.7.4.3 Lossy Upper Layer

The first circuit model of Fig (6.14a) corresponds to both lossy layers of Fig (6.11a). In this case, the steady-state complex relative permittivity can also be obtained, without ignoring R_{p2}. The admittance function for each parallel circuit is given below:

$$Y_1 = \frac{1}{R_{p1}} + j\omega C_{p1} = \frac{1 + j\omega\tau_1}{R_1} \quad (a)$$

$$Y_2 = \frac{1}{R_{p2}} + j\omega C_{p2} = \frac{1 + j\omega\tau_2}{R_2} \quad (b)$$

with the time constants; $\tau_1 = R_{p1}C_{p1}$ (c)

$$\tau_2 = R_{p2}C_{p2} \quad (d).$$

$$(6.7.23)$$

The inhomogeneous medium of Fig (6.11a) could be replaced by a homogeneous medium with an equivalent complex relative permittivity and an equivalent relaxation time.

The equivalent relaxation time could be computed following the process discussed in the subsection (6.6.1). In this process, two parallel circuits shown in Fig (6.14a) are connected in parallel as shown in Fig (6.11b). Two parallel resistances are combined, and also parallel capacitors are combined to provide the equivalent relaxation time:

$$\tau_{eq} = \left(\frac{R_{p1}R_{p2}}{R_{p1} + R_{p2}}\right)(C_{p1} + C_{p2})$$

$$\Rightarrow \tau_{eq} = \frac{(C_{p1} + C_{p2})}{\frac{1}{R_{p1}} + \frac{1}{R_{p2}}} \quad (a)$$

$$\text{also,}\quad \tau_{eq} = \frac{\varepsilon'_{r1}h_2 + \varepsilon'_{r2}h_1}{\sigma_1 h_2 + \sigma_2 h_1}$$

$$\Rightarrow \tau_{eq} = \frac{\varepsilon'_{r1}/h_1 + \varepsilon'_{r2}/h_2}{\sigma_1/h_1 + \sigma_2/h_2} \quad (b).$$

$$(6.7.24)$$

Equation (6.7.24b) is obtained with the help of equation (6.7.22). For the multilayer lossy dielectric layers medium, the equivalent relaxation time could be computed as follows:

$$\tau_{eq} = \frac{\sum_i C_{pi}}{\sum_i 1/R_{pi}} = \frac{\sum_i(\varepsilon'_{ri}/h_i)}{\sum_i(\sigma_i/h_i)}. \quad (6.7.25)$$

The expressions (6.7.24) and (6.7.25) for the equivalent relaxation time of the two-layered medium and the multilayered medium could be rewritten as,

$$\tau_{eq} = \frac{\tau_1/R_{p1} + \tau_2/R_{p2}}{1/R_{p1} + 1/R_{p2}} = \frac{R_{p1}\tau_2 + R_{p2}\tau_1}{R_{p1} + R_{p2}} \quad (a)$$

$$\tau_{eq} = \frac{\sum_i \tau_i/R_{pi}}{\sum_i 1/R_{pi}} \quad (b).$$

$$(6.7.26)$$

6.7.4.4 Equivalent Relative Permittivity

To obtain the equivalent relative permittivity of the composite medium, the input admittance of the circuit shown in Fig (6.14a) is extended to the multilayer case:

$$\text{Two layers}: Y = \frac{Y_1 Y_2}{Y_1 + Y_2},$$

$$\Rightarrow Y = \frac{1}{(R_{p1} + R_{p2})}\frac{(1 + j\omega\tau_1)(1 + j\omega\tau_2)}{\left(1 + j\omega\tau_{eq}\right)} \quad (a)$$

$$\text{Multilayer}: Y = \frac{\prod_i(1 + j\omega\tau_i)}{\left(1 + j\omega\tau_{eq}\right)\sum_i R_{pi}} \quad (b).$$

$$(6.7.27)$$

The admittance, $Y = j\,\omega\,\varepsilon_0\varepsilon^*_{req}\,C_0$, of the equivalent single-layer complex capacitance is equated to the above two-layered admittance equation to obtain the complex equivalent relative permittivity:

$$
\begin{aligned}
Y &= j\omega\varepsilon_0\varepsilon^*_{req}C_0 = j\omega\varepsilon_0\left(\varepsilon'_{req} - j\varepsilon''_{req}\right)C_0 \\
&= \omega\varepsilon_0\left(\varepsilon''_{req} + j\varepsilon'_{req}\right)C_0 \qquad \text{(a)} \\
j\omega\varepsilon_0\varepsilon^*_{req}C_0 &= \frac{1}{(R_{p1}+R_{p2})}\frac{(1+j\omega\tau_1)(1+j\omega\tau_2)}{\left(1+j\omega\tau_{eq}\right)} \qquad \text{(b),}
\end{aligned} \tag{6.7.28}
$$

where $C_0 = A\varepsilon_0/(h_1 + h_2)$ is the capacitance of the parallel-plate capacitor with air-medium. On separating the real and imaginary parts, the following expressions are obtained:

$$
\begin{aligned}
\varepsilon_0\varepsilon'_{req}(\omega) &= \frac{\tau_1+\tau_2-\tau_{eq}+\omega^2\tau_1\tau_2\tau_{eq}}{C_0(R_{p1}+R_{p2})\left(1+\omega^2\tau^2_{eq}\right)} \qquad \text{(a)} \\
\varepsilon_0\varepsilon''_{req}(\omega) &= \frac{1-\omega^2\tau_1\tau_2+\omega^2\tau_{eq}(\tau_1+\tau_2)}{\omega C_0(R_{p1}+R_{p2})\left(1+\omega^2\tau^2_{eq}\right)} \qquad \text{(b).}
\end{aligned} \tag{6.7.29}
$$

In the case of $\omega \to 0$, the static relative permittivity is computed for both the two-layered and multilayer dielectric media:

$$
\begin{aligned}
\varepsilon_0\varepsilon'_{req}(\omega=0) &= \frac{\tau_1+\tau_2-\tau_{eq}}{C_0(R_{p1}+R_{p2})} \qquad \text{(a)} \\
\varepsilon_0\varepsilon'_{req}(\omega=0) &= \frac{\left(\sum_i \tau_i\right)-\tau_{eq}}{C_0\sum_i R_{pi}} \qquad \text{(b).}
\end{aligned} \tag{6.7.30}
$$

In the case of $\omega \to \infty$, we get the high-frequency relative permittivity of two-layered and multilayered media:

$$
\begin{aligned}
\varepsilon_0\varepsilon'_{req}(\omega\to\infty) &= \frac{\tau_1\tau_2}{C_0(R_{p1}+R_{p2})\tau_{eq}} \\
\Rightarrow \varepsilon_0\varepsilon'_{req}(\omega\to\infty) &= \frac{1}{C_0\left[1/C_{p1}+1/C_{p2}\right]} \qquad \text{(a)} \\
\text{Multilayer}: \varepsilon_0\varepsilon'_{req}(\omega\to\infty) &= \frac{1}{C_0\sum_i(1/C_{pi})} \qquad \text{(b).}
\end{aligned} \tag{6.7.31}
$$

The dispersion relation, given by equations (6.7.29a), can be recast in terms of the limiting values of equivalent relative permittivity:

$$
\begin{aligned}
\varepsilon_0\varepsilon'_{req}(\omega) &= \frac{\tau_1+\tau_2-\tau_{eq}}{C_0\left(R_{p1}+R_{p2}\right)\left(1+\omega^2\tau^2_{eq}\right)} + \frac{\omega^2\,\tau_1\,\tau_2\,\tau_{eq}}{C_0\left(R_{p1}+R_{p2}\right)\left(1+\omega^2\tau^2_{eq}\right)} \times \frac{\tau_{eq}}{\tau_{eq}} \\
\Rightarrow \varepsilon_0\varepsilon'_{req}(\omega) &= \frac{\varepsilon_0\,\varepsilon'_{req}(\omega=0)}{1+\omega^2\,\tau^2_{eq}} + \frac{\omega^2\,\varepsilon_0\,\varepsilon'_{req}(\omega=\infty)\,\tau^2_{eq}}{1+\omega^2\,\tau^2_{eq}} \\
\Rightarrow \varepsilon'_{req}(\omega) &= \frac{\varepsilon'_{req}(\omega=0)+\varepsilon'_{req}(\omega=\infty)\tau^2_{eq}\omega^2+\varepsilon'_{req}(\omega=\infty)-\varepsilon'_{req}(\omega=\infty)}{1+\omega^2\,\tau^2_{eq}} \\
\Rightarrow \varepsilon'_{req}(\omega) &= \varepsilon'_{req}(\omega=\infty) + \frac{\varepsilon'_{req}(\omega=0)-\varepsilon'_{req}(\omega=\infty)}{1+\omega^2\tau^2_{eq}}.
\end{aligned} \tag{6.7.32}
$$

Likewise, the $\varepsilon''_{r\,eq}(\omega)$, given by equation (6.7.29b), can be written as follows:

$$
\begin{aligned}
\varepsilon_0\varepsilon''_{req}(\omega) &= \frac{1}{\omega\,C_0\left(R_{p1}+R_{p2}\right)\left(1+\omega^2\,\tau^2_{eq}\right)} + \frac{\omega\left[\tau_{eq}(\tau_1+\tau_2)-\tau_1\tau_2\right]}{C_0\left(R_{p1}+R_{p2}\right)\left(1+\omega^2\,\tau^2_{eq}\right)} \\
\Rightarrow \varepsilon_0\,\varepsilon''_{req}(\omega) &= \frac{1}{\omega\,C_0\left(R_{p1}+R_{p2}\right)\left(1+\omega^2\tau^2_{eq}\right)} + \frac{\omega\,\varepsilon_0\,\varepsilon'_{req}(\omega=\infty)\,\tau_{eq}\left[\tau_{eq}(\tau_1+\tau_2)-\tau_1\,\tau_2\right]}{\tau_1\,\tau_2\left(1+\omega^2\tau^2_{eq}\right)} \\
\Rightarrow \varepsilon''_{req}(\omega) &= \frac{\varepsilon'_{req}(\omega=\infty)\tau_{eq}}{\omega\tau_1\tau_2} + \frac{\left[\varepsilon'_{req}(\omega=0)-\varepsilon'_{req}(\omega=\infty)\right]\omega\tau_{eq}}{\left(1+\omega^2\tau^2_{eq}\right)}.
\end{aligned} \tag{6.7.33}
$$

Equations (6.7.19a,b) are special case of equations (6.7.32) and (6.7.33) for the lossless upper layer, with $R_{p2} \to 0$. On comparing equations (6.7.32) and (6.7.33) with equation (6.4.15), it is seen that the $\varepsilon'_{req}(\omega)$ follows the Debye relaxation response. However, $\varepsilon''_{req}(\omega)$ has an additional term $\left[\varepsilon'_{req}(\omega = \infty)\,\tau_{eq}\right]/\omega\,\tau_1\,\tau_2$ due to the DC conductivity of the lossy dielectric layers. The equivalent DC conductivity σ_{eq} is given by

$$\sigma_{eq} = \frac{\varepsilon'_{req}(\omega = \infty)\,\tau_{eq}\,\varepsilon_0}{\tau_1\,\tau_2}. \tag{6.7.34}$$

The σ_{eq} is taken as σ in further discussion. The real and imaginary parts of the complex equivalent permittivity $\varepsilon^*_{req}(\omega)$ can be combined to provide the following dispersion equation of Maxwell–Wagner polarization:

$$\varepsilon^*_{req}(\omega) = \varepsilon'_{req}(\omega = \infty) + \frac{\varepsilon'_{req}(\omega = 0) - \varepsilon'_{req}(\omega = \infty)}{1 + j\,\omega\,\tau_{eq}} - j\frac{\sigma}{\omega\,\varepsilon_0}. \tag{6.7.35}$$

The Maxwell–Wagner polarization follows the Debye relaxation response. The above relation with DC conductivity does not satisfy Kramers–Kronig causal relations [J.44]. However, it can be further extended to more numbers of dielectric layers. The polarizability α_{int} of the interfacial polarization can also be extracted as follows:

$$P_{interfacial} = N\alpha_{int}E_{ext}$$
$$= \varepsilon_0\left[\varepsilon^*_{req}(\omega) - \varepsilon'_{req}(\omega = \infty)\right]E_{ext} \qquad (a)$$
$$\Rightarrow \alpha_{int} = \left[\frac{\varepsilon^*_{req}(\omega = 0) - \varepsilon'_{req}(\omega = \infty)}{1 + j\omega\tau_{eq}} - j\frac{\sigma}{\omega\varepsilon_0}\right]\frac{\varepsilon_0}{N} \qquad (b), \tag{6.7.36}$$

where N is the density of mobile charge carrier. Equation (6.7.36a) is obtained by using equations (6.1.4), (6.1.6), and (6.1.7). In the process, the following replacement is also carried out: $\varepsilon_r \to \varepsilon^*_{req}(\omega = 0)$, and $(\varepsilon_r = 1) \to \varepsilon'_{req}(\omega \to \infty)$. Equations (6.1.4) and (6.1.6) are valid for the polarization in a natural material, not for the polarization in the composite heterogeneous considered presently. The interfacial polarizability can be combined with another polarizability to get the total polarizability of an inhomogeneous dielectric-medium.

6.7.5 Series RLC Resonant Circuit Model

Lorentz type dielectric materials can be modeled by the RLC series resonant circuit, shown in Fig (6.15a). The RLC circuit can also be used to model the *capacitor loaded wire-medium* and *split-ring resonator (SRR)* for the realization of the generic model of dispersive effective relative permittivity and permeability. These are discussed in subsections (21.1.1) and (21.2.1) of chapter 21. One type of resonating polarization is modeled by one kind of RLC series circuit. Therefore, the multiple kinds of resonating polarizations in a material are modeled by a series combination of multiple series resonant circuits, shown in Fig (6.15b). At the frequency below their resonance frequency, the resonating elements contribute to the fixed value of relative permittivity. The following *integrodifferential* equation can be written from Kirchhoff' law:

$$L\frac{di}{dt} + Ri + \frac{1}{C}\int idt = v(t) \qquad (a)$$
$$\text{As } i = \frac{dQ}{dt}, \Rightarrow L\frac{d^2Q}{dt^2} + R\frac{dQ}{dt} + \frac{Q}{C} = v(t) \qquad (b)$$
For open switch at $t = 0$, $v(t) = 0$,
$$\text{for} \quad t > 0 : L\frac{d^2Q}{dt^2} + R\frac{dQ}{dt} + \frac{Q}{C} = 0 \qquad (c). \tag{6.7.37}$$

Equation (6.7.37) describes the oscillating behavior of the circuit. Let its solution be $Q(t) = Q_0e^{Kt}$. On substituting it in equation (6.7.37), the following expressions are obtained:

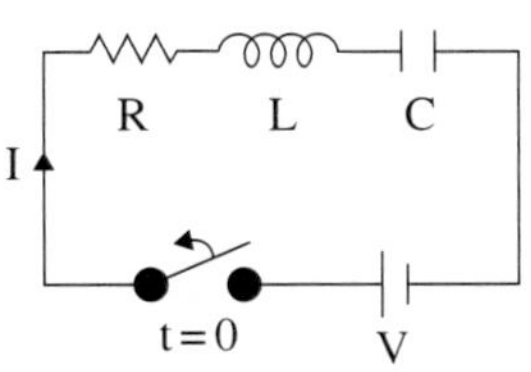

(a) One resonating circuit corresponding to one type of polarization.

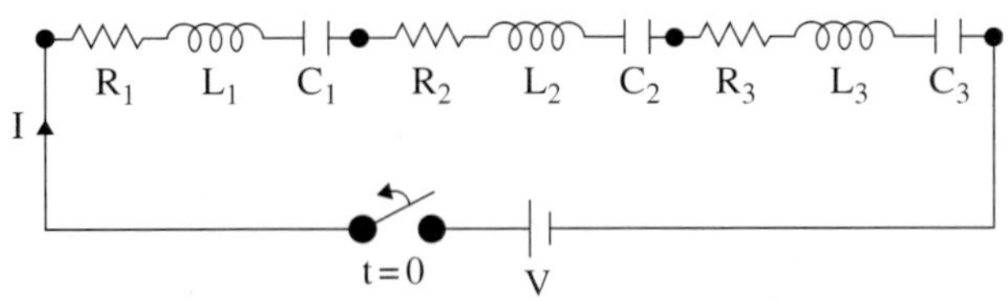

(b) Three resonating circuits corresponding to three types of polarization.

Figure 6.15 Series resonant circuit for modeling of Lorentz type material.

$$LK^2 + RK + \frac{1}{C} = 0 \Rightarrow K^2 + \gamma K + \omega_0^2 = 0 \quad \text{(a)}$$

$$\text{where damping factor}: \gamma = R/L \quad \text{(b)}$$

$$\omega_0 = 1/\sqrt{LC} \quad \text{(c)}$$

$$\text{Two roots of solutions}: K_{1,2} = -\frac{\gamma}{2} \pm j\sqrt{\omega_0^2 - \frac{\gamma^2}{4}} \text{ (d)}, \quad (6.7.38)$$

where ω_0 is the natural frequency of oscillation. Thus, the solution of equation (6.7.37) is

$$Q(t) = Ae^{+K_1 t} + Ae^{+K_2 t} \quad \text{(a)}$$

$$\text{where,} \quad K_{1,2} = -\frac{\gamma}{2} \pm j\omega_d \quad \text{(b)}$$

$$\text{and,} \quad \omega_d = \sqrt{\omega_0^2 - \frac{\gamma^2}{4}} \quad \text{(c).} \quad (6.7.39)$$

In equation (6.7.39), ω_d is the frequency of the damped oscillator. It is less than the natural frequency ω_o of the undamped oscillation. Three cases of oscillation are discussed below:

Case #I: If $\omega_o > \gamma/2$, i.e.$1/\sqrt{LC} > R/2L$, the coefficients K_1 and K_2 are complex quantities, and the oscillatory charge decays as

$$Q(t) = 2A\, e^{-\frac{\gamma}{2}t} \cos(\omega_d t). \quad (6.7.40)$$

Case #II: For $\omega_0 < \gamma/2$, there is no oscillation and the charge decays with two relaxations.

Case #III: For $\omega_0 = \gamma/2$, i.e. $K_1 = K_2$, the charge decay is critically damped for $\omega_0 = \gamma/2$, giving the following expression:

$$Q(t) = Ae^{-\alpha t}, \quad \alpha = \frac{R}{2L} = \frac{2}{RC}. \quad (6.7.41)$$

The above behavior is similar to that of an RC series circuit, giving Debye kind of dispersion response. So the critically damped Lorentz material behaves as Debye material. However, its relaxation time is $\tau = 1/\alpha = RC/2$, instead of RC. On comparing equation (6.7.37) with equation (6.5.3), it is obvious that *the charge Q on the capacitor is an analog of the polarization P*. The damping factor and the natural frequency of the circuit are analog of the damping factor and the natural frequency of the oscillating polarizing element. Thus, the response of the circuit model could provide the relative permittivity of a material. The complex admittance of the circuit shown in Fig (6.15a) is

$$Y^* = \frac{1}{Z^*} = \frac{1}{R + j[\omega L - (1/\omega C)]}$$

$$\Rightarrow Y^* = \frac{j(\omega/L)}{(\omega_0^2 - \omega^2 + j\omega\gamma)}, \quad (6.7.42)$$

where ω and γ are given by equation (6.7.38). An approximation can be used near resonance $\Delta\omega = \omega_0 - \omega$, $\omega \approx \omega_0$, $\omega + \omega_0 \approx 2\omega$ to get equation (6.7.43):

$$Y^* = \frac{j}{2L(\Delta\omega + j\gamma/2)} \quad \text{(a)}$$

$$\Rightarrow Y^* = \frac{\gamma/2}{2L\{(\Delta\omega)^2 + (\gamma/2)^2\}} + \frac{j\,\Delta\omega}{2L\{(\Delta\omega)^2 + (\gamma/2)^2\}} \quad \text{(b).} \quad (6.7.43)$$

The admittance of the parallel-plate capacitor, containing the material with relative permittivity $\varepsilon_r^* = \varepsilon_r' - j\varepsilon_r''$ is $Y^* = j\omega\,\varepsilon_0\,(\varepsilon_r' - j\varepsilon_r'')\,C_0 = \omega\,C_0\,\varepsilon_0(\varepsilon_r'' + j\varepsilon_r')$, where C_0 is the capacitance of the parallel-plate capacitor without material. On comparing admittances of the circuit and material, the following expressions are obtained for the relative permittivity in terms of Lorentz resonator circuit:

$$\omega\varepsilon_0\varepsilon_r' C_0 = \frac{\Delta\omega}{2L[(\Delta\omega)^2 + (\gamma/2)^2]},$$

$$\text{using } K = \frac{1}{2\omega\varepsilon_0 C_0 L} \quad \text{(a)}$$

$$\Rightarrow \varepsilon_r'(\omega) = \frac{K\Delta\omega}{[(\Delta\omega)^2 + (\gamma/2)^2]} \quad \text{(b)}$$

$$\varepsilon_r''(\omega) = \frac{K\gamma}{2[(\Delta\omega)^2 + (\gamma/2)^2]} \quad \text{(c)}$$

$$Q = \frac{\omega_0 L}{R}, \text{ for Lorentz oscillator.} \quad \text{(d).} \quad (6.7.44)$$

The $\varepsilon_r'(\omega)$ and $\varepsilon_r''(\omega)$ obtained from the resonator circuit model is the same as that of obtained from the one-dimensional molecular Lorentz oscillator given by equation (6.5.10). The circuit parameter K corresponds to the material parameter ε_{rE}. The ε_{rT} is added to equation (6.7.44) to get the relative permittivity due to all kinds of polarization processes, as shown in Fig (6.15b). The frequency responses of the real $\varepsilon_r'(\omega)$ and imaginary $\varepsilon_r''(\omega)$ parts of the complex relative permittivity are obtained from the measurements. Using these results, the values of R, L, C, and ω_0 parameters of the resonator circuit model can be extracted. The circuit model also provides the frequency response of $\varepsilon_r'(\omega)$ and $\varepsilon_r''(\omega)$, as shown in Fig (6.9b).

A more general form of Debye and Lorentz dispersion models, using the time derivative of the external field, i.e. the polarizing field, has been developed using the small Hertzian dipoles as the inclusions in the host medium. By further loading, the Hertzian dipoles inclusions with simple resistance, RC and RL circuits, etc., and by loading it with a diode and operational amplifiers new kinds of artificial absorbing materials and smart materials have been reported. These materials are developed using the circuit models for more general polarization of the macromolecules of passive and active inclusions [J.33, J.45, J.46].

6.8 Substrate Materials for Microwave Planar Technology

The static and frequency-dependent relative permittivities are discussed in the previous sections, for both the natural and artificial dielectric materials. The frequency dependence of relative permittivity of natural materials has been perceived using Debye and Lorentz type phenomenological polarization models. Further, these models have also been viewed through the equivalent circuit models. However, the substrate materials used in the microwave do not follow a single relaxation frequency of the Debye model; nor do they follow a single resonance frequency of Lorentz model; a real-life substrate material, say FR-4 of complex composition, can have several relaxations due to the presence of several polarizations. Even in the case of a single relaxation frequency, or a single resonance frequency, it is not easy to evaluate the characteristic parameters of both the models for a specific substrate material, whose frequency-dependent real or complex relative permittivity is known experimentally at few frequencies.

Normally, to design and simulate components and circuits in the microwave planar technology, relative permittivity and loss tangent of a substrate are taken as constant, i.e. frequency-independent. However, it results in the noncausal response in the time-domain. It adversely influences the computations of the pulse propagating on interconnect in such a medium. Further, the noncausal constant permittivity model does not provide the correct group delay response over the broadband. The Debye type causal dispersion models are needed not only for the substrate materials but also for the composites and artificial dielectrics, for their application to the time-domain EM-solvers. The following two kinds of extended Debye models have been developed for their use in the EM-solvers:

- *Multiterm Debye models*
- *Wideband Debye model.*

A summary of both models is presented below. The characteristic parameters of the single-term Debye and Lorentz models are also evaluated using experimental results. The first step to develop a Debye model for any substrate material is to get accurate experimental data on the frequency-dependent relative permittivity. It is normally obtained by using the cavity resonator method or an air-coaxial line. The microstrip line and patch resonator are also used for this purpose with some limitations [J.47–J.51]. The modeling process can be started from the experimental data of manufacturers of substrates.

The engineered artificial dielectric materials, using nonresonant metallic inclusions in a host dielectric-medium, are discussed in subsection (6.2.2). However, resonant metallic inclusions embedded in the dielectric host can create high permittivity and high permeability artificial materials depending upon the type of the resonant inclusions [J.52–J.57]. Such magneto-dielectric materials are useful for the synthesis of engineered substrates, called the *metasubstrates*, useful for the development of compact microstrip based circuits and antenna [J.58]. A brief discussion of metasubstrates is presented at the end of the present section.

6.8.1 Evaluation of Parameters of Single-Term Debye and Lorentz Models

6.8.1.1 Single-Term Debye Model

The Debye model of a lossy dielectric material with single relaxation frequency f_{re} can be written from equation (6.7.35), with some changes, as follows:

$$\varepsilon_r^*(f) = \varepsilon_r'(f = \infty) + \frac{\Delta\varepsilon_r'}{1 + j(f/f_{re})} - j\frac{\sigma}{2\pi f\varepsilon_0} \quad \text{(a)}$$

$$\text{where,} \quad \Delta\varepsilon_r' = \varepsilon_r'(\omega = 0) - \varepsilon_r'(\omega = \infty) \quad \text{(b).} \qquad (6.8.1)$$

In equation (6.8.1), relaxation frequency is $f_{re} = 1/\tau_d$ and σ is DC conductivity of the dielectric-medium. So a lossy and dispersive dielectric material modeled by the Debye model is characterized by four parameters $\Delta\varepsilon_r', f_{re}, \varepsilon_r'(\omega = \infty)$, and σ. For known $\Delta\varepsilon_r'$ and $\varepsilon_r'(\omega = \infty)$, the static relative permittivity $\varepsilon_r'(\omega = 0)$ can be easily obtained. The real and imaginary parts of a complex relative permittivity can be separated. These are evaluated at two frequencies f_1 and f_2 from the known experimental results. This arrangement provides the following four numbers of equations for

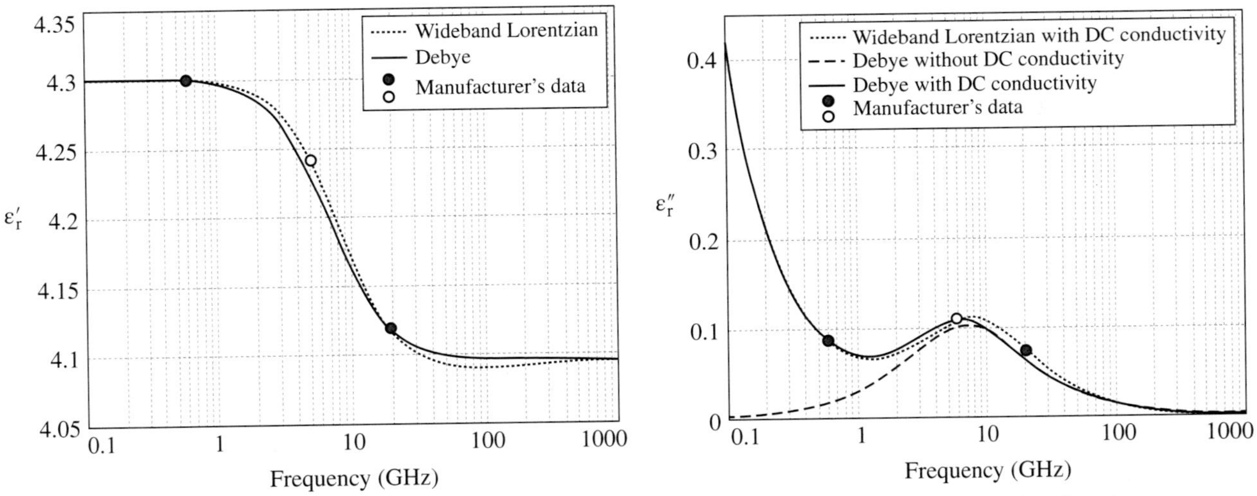

(a) Frequency-dependent real relative permittivity.

(b) Frequency-dependent imaginary relative permittivity.

Figure 6.16 Response of Debye model for FR-4 substrate [J.59]. *Source:* Koledintseva et al. [J.59].

evaluations of four numbers of characteristics parameters [J.48, J.49]:

$$\varepsilon'_{r1}(f_1) = \varepsilon'_r(f = \infty) + \frac{\Delta\varepsilon'_r}{1 + (f_1/f_{re})^2} \quad \text{(a)}$$

$$\varepsilon'_{r2}(f_2) = \varepsilon'_r(f = \infty) + \frac{\Delta\varepsilon'_r}{1 + (f_2/f_{re})^2} \quad \text{(b)}$$

$$\varepsilon''_{r1}(f_1) = \frac{(f_1/f_{re})\Delta\varepsilon'_r}{1 + (f_1/f_{re})^2} - \frac{\sigma}{2\pi f_1 \varepsilon_0} \quad \text{(c)}$$

$$\varepsilon''_{r2}(f_2) = \frac{(f_2/f_{re})\Delta\varepsilon'_r}{1 + (f_2/f_{re})^2} - \frac{\sigma}{2\pi f_2 \varepsilon_0} \quad \text{(d).} \qquad (6.8.2)$$

In equation (6.8.2), if $f_2 > f_1$, then $\varepsilon'_{r1}(f_1) > \varepsilon'_{r2}(f_2)$ as the relative permittivity decreases with increasing frequency. Also, as the dielectric loss increases with frequency, so $\tan\delta(f_1) < \tan\delta(f_2)$, i.e. $\varepsilon''_{r1}(f_1)\varepsilon'_{r2}(f_2) < \varepsilon''_{r2}(f_2)\varepsilon'_{r1}(f_1)$. The experimental data must fulfill these physical conditions. The following expressions are obtained from the above equations, for the characteristics parameters of the Debye model; in terms of the experimentally known real and imaginary parts of the relative permittivity at f_1 and f_2 [J.48, J.49]:

$$f_{re} = \frac{\varepsilon''_{r2}(f_2).f_2 - \varepsilon''_{r1}(f_1).f_1}{\varepsilon'_{r1}(f_1) - \varepsilon'_{r2}(f_2)} \quad \text{(a)}$$

$$\sigma = 2\pi f_1 \varepsilon_0 \left\{ \varepsilon''_{r1}(f_1) + \frac{f_1 f_{re}}{f_1^2 - f_2^2} (\varepsilon'_{r1}(f_1) - \varepsilon'_{r2}(f_2)).\left(1 + \left(\frac{f_2}{f_{re}}\right)^2\right) \right\} \quad \text{(b)} \qquad (6.8.3)$$

$$\varepsilon'_r(f = \infty) = \frac{\varepsilon'_{r1}(f_1) - \varepsilon'_{r2}(f_2) + \varepsilon'_{r1}(f_1).(f_1/f_{re})^2 - \varepsilon'_{r2}(f_2).((f_2/f_{re})^2}{(f_1/f_{re})^2 - ((f_2/f_{re})^2} \quad \text{(c)}$$

$$\varepsilon'_r(f = 0) = \varepsilon'_{r1}(f_1) + \{\varepsilon'_{r1}(f_1) - \varepsilon'_r(f = \infty)\}.(f_1/f_{re})^2 \quad \text{(d).}$$

The above parameters in equation (6.8.3a–d) are evaluated for the FR-4 substrate, using the manufacturer's data. Figure (6.16a) shows Debye's response for both the real and imaginary parts of the FR-4 with one relaxation time. It also shows the manufacture's data. The evaluated Debye model parameters are obtained as follows:

$\varepsilon'_{req}(f=0) = 4.301, \varepsilon'_{req}(f=\infty) = 4.096$, $\sigma = 2.294\times 10^{-3}$S/m and $\tau_d = 2.294\times10^{-11}$s. At the lower end of frequency, a large increase in $\varepsilon''_r(\omega)$ is seen due to the presence of σ. If DC conductivity is ignored $\sigma = 0$, then only three parameters are evaluated for the single-term Debye model, namely $\Delta\varepsilon'_r, f_{re}, \varepsilon'_r(f=\infty)$. For both the measured value of relative permittivity and tanδ at frequency f_1, and the known value of static relative permittivity at $f = 0$, Debye parameters could be evaluated to get the Debye model [J.60]. In the case of slowly varying relative permittivity in the relaxation region, the Cole–Cole and other versions of the Debye model can provide a better fit against the experimental results [B.21]. These variations can be attempted for the substrates and artificial dielectric materials.

6.8.1.2 Single-term Lorentz Model

Normally, the Lorentz model is applicable over a narrow band for dispersive dielectrics, such as a mixture of aluminum inclusions of 10–15 μm diameters in the polymer host with $\varepsilon_r = 2.15$. However, it could apply to the wide-band dielectrics also, such as FR-4 [J.59], if the resonance frequency is very high and the Q-factor is less than unity. The single-term Lorentz model, with the single resonance frequency f_0 ($=\omega_0/2\pi$), can be written, in the modified form, using an equation (6.5.8) and other expressions of section (6.5). The DC conductivity σ of the lossy dielectric material can also be included [J.48, J.49].

$$\varepsilon^*_r(f) = \varepsilon'_r(f=\infty) + \frac{\Delta\varepsilon'_r}{1-(f/f_0)^2 + j(f/f_{re})} - j\frac{\sigma}{2\pi f\varepsilon_0} \quad \text{(a)}$$

$$\text{where,}\quad \Delta\varepsilon'_r = \varepsilon'_r(\omega=0) - \varepsilon'_r(\omega=\infty) \quad \text{(b)}$$

$$f_{re} = f_0^2/\Delta f \quad \text{(c).}$$

(6.8.4)

In equation (6.8.4), the relaxation frequency parameter f_{re} is defined in terms of 3dB bandwidth Δf of Lorentz resonance at f_0. The above expression of complex relative permittivity can be written in terms of the frequency-dependent complex susceptibility; its real and imaginary parts can be separated:

$$\varepsilon^*_r(f) = \varepsilon'_r(f=\infty) + \chi^*_r(f) \quad \text{(a)}$$

$$\chi^*_r(f) = \chi'_r(f) - j\chi''_r(f) \quad \text{(b)}$$

$$\chi'_r(f) = \frac{\Delta\varepsilon'_r\left(1-(f/f_0)^2\right)}{\left\{1-(f/f_0)^2\right\}^2 + (f/f_{re})^2} \quad \text{(c)}$$

$$\chi''_r(f) = \frac{\Delta\varepsilon'_r(f/f_{re})}{\left\{1-(f/f_0)^2\right\}^2 + (f/f_{re})^2} - j\frac{\sigma}{2\pi\varepsilon_0} \quad \text{(d).}$$

(6.8.5)

In general, five numbers of characteristics parameters, namely $\varepsilon'_r(f=\infty), \varepsilon'_r(f=0), f_0, f_{re}$, and σ, are to be determined. The following five numbers of equations are used with experimental results for three numbers of imaginary parts of the susceptibility and two numbers of real parts of the susceptibility at three frequencies: f_1, f_2, and f_3:

$$\chi'_r(f_1) = \frac{\Delta\varepsilon'_r\left(1-(f_1/f_0)^2\right)}{\left\{1-(f_1/f_0)^2\right\}^2 + (f_1/f_{re})^2} \quad \text{(a)}$$

$$\chi'_r(f_2) = \frac{\Delta\varepsilon'_r\left(1-(f_2/f_0)^2\right)}{\left\{1-(f_2/f_0)^2\right\}^2 + (f_2/f_{re})^2} \quad \text{(b)}$$

$$\chi''_r(f_1) = \frac{\Delta\varepsilon'_r(f_1/f_{re})}{\left\{1-(f_1/f_0)^2\right\}^2 + (f_1/f_{re})^2} - j\frac{\sigma}{2\pi f_1\varepsilon_0} \quad \text{(c)}$$

$$\chi''_r(f_2) = \frac{\Delta\varepsilon'_r(f_2/f_{re})}{\left\{1-(f_2/f_0)^2\right\}^2 + (f_2/f_{re})^2} - j\frac{\sigma}{2\pi f_2\varepsilon_0} \quad \text{(d)}$$

$$\chi''_r(f_3) = \frac{\Delta\varepsilon'_r(f_3/f_{re})}{\left\{1-(f_3/f_0)^2\right\}^2 + (f_3/f_{re})^2} - j\frac{\sigma}{2\pi f_3\varepsilon_0} \quad \text{(e).}$$

(6.8.6)

In case the DC conductivity is ignored, i.e. $\sigma = 0$, equation (6.8.6e) could be dropped. The measured susceptibility, both real and imaginary, are needed only at two frequencies. In case, only the susceptibility parameters are needed, three characteristic parameters $\Delta\varepsilon'_r, f_0$, and f_{re} could be determined to get the single-term Lorentz model. A set of three equations for the measured imaginary parts of the susceptibility, at three frequencies f_1, f_2, and f_3, are needed to evaluate these parameters. In this case again $\sigma = 0$, and equations (6.8.6c) and (6.8.6d) could be dropped. The system of nonlinear equations can be solved using the MATLAB program. While evaluating the parameters, these parameters are taken as the positive real quantities and measurement frequencies must be both above and below the resonance frequency, i.e. $f_1 < f_0 < f_2$, or $f_2 < f_0 < f_3$.

If the experimental data are $\chi''_1(5\text{GHz}) = 1.2, \chi''_2(8\text{GHz}) = 2.0$, and $\chi''_3(15\text{GHz}) = 0.8$, then the evaluated parameters are $\Delta\varepsilon'_r = 3.82$, $f_0 = 10.1$ GHz, and $f_{re} = 99.03$ GHz, which corresponds to $\Delta f = 1.03$ GHz. For three numbers of experimental data on FR-4 given in Fig (6.16b), the evaluated parameters are $\varepsilon'_{req}(f=0) = 4.301, \varepsilon'_{req}(f=\infty) = 4.096$, $\sigma = 2.294 * 10^{-3}$S/m, $f_0 = 39.5$GHz, $\Delta f = 200$GHz. These parameters are used to develop the single-term Lorentz model for both the real and imaginary parts of the complex relative permittivity. Figure (6.16) shows the results of the Lorentz model which are almost identical to that of the Debye model. In the present case, due to high resonance frequency and very wide bandwidth, the Q-factor of

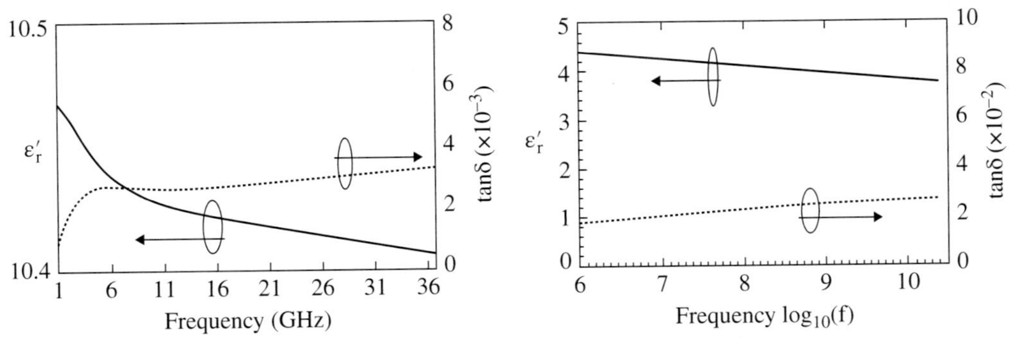

(a) Debye model for Duriod-6010. (b) Debye model for FR-4.

Figure 6.17 8 terms Debye models for relative permittivity and loss-tangent. *Source:* Zhou and Melde [J.61].

resonance is less than unity and the Lorentz model applies to the wideband dispersion [J.59].

6.8.2 Multiterm and Wideband Debye Models

The single-term Debye model is not sufficient to model the frequency-dependent real and imaginary parts, or the loss-tangent, accurately over a wider frequency range, as the substrate materials have several polarizations with different relaxation times. The multiterm or the wideband Debye models are needed to improve the accuracy of the curve-fitting of the measured dispersion results [B.21]. It involves several more characterizing phenomenological parameters and several approaches have been used to evaluate these parameters from the experimental results [J.47–J.49, J.59–J.63]. Such models are also accommodated in the modern EM-Simulators that require only a few experimental data to generate Debye models. A brief introduction to such Debye models is given below.

6.8.2.1 Multiterm Debye Model

Debye model of expression (6.8.1) can be extended to the multirelaxation frequency materials. It provides frequency-dependent relative permittivity; as a superposition of Debye type relaxations due to N number of polarizations:

$$\varepsilon_r^*(\omega) = \varepsilon'_r(\omega = \infty) + \sum_{i=1}^{N} \frac{\Delta\varepsilon'_{ri}}{1 + j(\omega/\omega_{rei})} - j\frac{\sigma}{\omega\varepsilon_0} \quad \text{(a)}$$

$$\text{Real part}: \varepsilon'_r(\omega) = \varepsilon'_r(\omega = \infty) + \sum_{i=1}^{N} \frac{\Delta\varepsilon'_{ri}}{1 + (\omega/\omega_{rei})^2} \quad \text{(b)}$$

$$\text{Imaginary part}: \varepsilon''_r(\omega) = \sum_{i=1}^{N} \frac{(\omega/\omega_{rei})\Delta\varepsilon'_{ri}}{1 + (\omega/\omega_{rei})^2} - \frac{\sigma}{\omega\varepsilon_0} \quad \text{(c)}. \tag{6.8.7}$$

In equation (6.8.7), $\varepsilon'_r(\omega = \infty)$ is the relative permittivity at high frequency. The incremental relative permittivity $\Delta\varepsilon'_{ri}$ and ω_{rei} are the polarization strength and relaxation frequency of the ith Debye type polarization. Further, σ is DC conductivity that influences very low frequency $\varepsilon''_r(\omega)$. It is dropped for many cases or is known from the low-frequency measurement. $N = 2$ to 8 numbers of Debye terms are normally used to accurately curve-fit the experimental results. For the assumed value of N, the system of nonlinear equations is solved in MATLAB to get the parameters $\Delta\varepsilon'_{ri}$ and ω_{rei} [J.61]. However, the empirical approach has also been applied to the FR-4 substrate by taking $N = 8$ [J.37].

The real $\varepsilon'_r(\omega)$ of FR-4 shows a linear dependence of frequency, on the logarithm scale. One-term per decade of frequency could be taken. The total experimentally obtained real part of relative permittivity is divided by N to get the uniform value of $\Delta\varepsilon'_{ri}$ at each of i=1 to 8. Next, each of eight numbers of Debye polarization strength $\Delta\varepsilon'_{ri}$ could be fine-tuned to curve-fit the model to experimental results. However, the genetic algorithm-based optimization process appears to be more efficient and systematic for the evaluation of these parameters. Figure (6.17) shows the results for the eight-term Debye model for both RT–Duriod 6010 and FR-4 substrates, using the genetic algorithm approach. At 10 GHz, these results follow closely the manufacture's results [J.48]. Both dielectric substrates have a substantial difference in their frequency response due to the different nature of polarizations involved.

6.8.2.2 Wideband Debye Model

Both the relative permittivity and loss-tangent, say for FR-4, show slow variations over a very large frequency range. This makes discrete multiterm Debye modeling a laborious process. The relaxation can be treated as a distributed function and replace the summation in the above formulation by an integral in terms of the lower (τ_2) and upper (τ_1) relaxation time [B.21]. In the case of the logarithmic variation, Frohlich distribution is used to get the following Debye expression for the frequency-dependent complex relative permittivity of material with continuously variable relaxation time [J.61]:

$$\begin{aligned}
&\varepsilon_r^*(\omega) = \varepsilon_r'(\omega=\infty) + \frac{\varepsilon_r'(\omega=0)-\varepsilon_r'(\omega=\infty)}{\ln(\tau_1/\tau_2)} \ln\left\{\frac{1+j\omega\tau_1}{1+j\omega\tau_2}\left(\frac{\tau_2}{\tau_1}\right)\right\}, \tau_1 > \tau_2 \quad (a)\\
&\varepsilon_r'(\omega) = \varepsilon_r'(\omega=\infty) + \frac{\varepsilon_r'(\omega=0)-\varepsilon_r'(\omega=\infty)}{\ln(\tau_1/\tau_2)} \ln\left\{\left|\frac{1+j\omega\tau_1}{1+j\omega\tau_2}\left(\frac{\tau_2}{\tau_1}\right)\right|\right\} \quad (b)\\
&\varepsilon_r''(\omega) = \frac{\varepsilon_r'(\omega=0)-\varepsilon_r'(\omega=\infty)}{\ln(\tau_1/\tau_2)} \arg\left\{\frac{1+j\omega\tau_1}{1+j\omega\tau_2}\left(\frac{\tau_2}{\tau_1}\right)\right\} \quad (c).
\end{aligned} \tag{6.8.8}$$

Unlike the multiterms Debye model, the above model evaluates only four characteristic parameters, τ_1, τ_2, $\varepsilon_r'(\omega=0)$, and $\varepsilon_r'(\omega=\infty)$, from the experimental results, using the *genetic algorithm*-based optimization [J.61].

In the place of Frochlich distribution function, hyperbolically decaying relaxation function has been used to get the following Debye model [J.64]:

$$\varepsilon_r^*(\omega) = \varepsilon_r'(\omega=\infty) + a\left[\frac{1}{2}\ln\left(\frac{\tau_2^2(1+\omega^2\tau_1^2)}{\tau_1^2(1+\omega^2\tau_2^2)}\right)\right] - j\left[\tan^{-1}(\omega\tau_2) - \tan^{-1}(\omega\tau_1)\right],\ \tau_2 > \tau_1. \tag{6.8.9}$$

The four characteristic parameters,(τ_1), (τ_2), a, and $\varepsilon_r'(\omega=\infty)$ are evaluated by the trial and error process to curve-fit the measured results, for both the real and imaginary parts of the complex permittivity. In the case of the FR-4 substrate, the curve-fitted parameters are $\varepsilon_r'(\omega=\infty) = 4.2$, $a = 0.06$, $\tau_1 = 1.6$ ps, and $\tau_2 = 1.6$ ms. The model also helps to compute the frequency-dependent loss-tangent with an increasing trend as a function of frequency.

The above expressions are valid for wideband applications. However, the evaluation of four numbers of characteristic parameters from the experimental is not a direct process. Another closed-form wideband Debye model for FR-4 substrate has also been suggested that is simple to use. This model also considers continuously varying relaxation frequency. The polarization strength $\Delta\varepsilon_{ri}'$ is replaced by the average value of the change in the relative permittivity between two assumed frequency limits in the logarithmic scale. The change in relative permittivity at two frequencies is known from the experimental results. The lower and upper frequencies are taken as $\omega_1 = 10^{m_1}$ and $\omega_2 = 10^{m_2}$, $m_2 > m_1$. The average value is $\Delta\varepsilon_{ri}' = \Delta\varepsilon_r'/(m_2 - m_1)$. So the summation over the N relaxations of equation (6.8.7a) can be replaced by an integral, where relaxation frequency is substituted by another frequency variable [J.47]:

On taking $\omega_{rei} = 10^x$, $m_1 \le x \le m_2, \omega_1 << \omega << \omega_2$

$$\sum_{i=1}^{N} \frac{\Delta\varepsilon_{ri}'}{1+j(\omega/\omega_{rei})} \to \frac{\Delta\varepsilon_r'}{(m_2-m_1)}\int_{m_1}^{m_2} \frac{dx}{1+j(\omega/10^x)}$$
$$= \frac{\Delta\varepsilon_r'}{(m_2-m_1)}\left[\frac{\ln\{(\omega_2+j\omega)/(\omega_1+j\omega)\}}{\ln 10}\right]. \tag{6.8.10}$$

Under the above limiting condition, equation (6.8.7) is modified as follows:

$$\begin{aligned}
&\varepsilon_r^*(\omega) = \varepsilon_r'(\omega=\infty) + \frac{\Delta\varepsilon_r'}{(m_2-m_1)}\left[\frac{\ln\{(\omega_2+j\omega)/(\omega_1+j\omega)\}}{\ln 10}\right] - j\frac{\sigma}{\omega\varepsilon_0} \quad (a)\\
&\text{Ignoring } \sigma, \text{i.e.} \sigma = 0\\
&\text{Real part}: \varepsilon_r'(\omega) = \varepsilon_r'(\omega=\infty) + \frac{\Delta\varepsilon_r'}{(m_2-m_1)\ln 10}\left[\frac{1}{2}\ln\left\{\frac{(\omega_2/\omega)^2+1}{(\omega_1/\omega)^2+1}\right\}\right] \quad (b)\\
&\text{Imaginary part}: \varepsilon_r''(\omega) = \frac{\Delta\varepsilon_r'}{(m_2-m_1)}\frac{1}{\ln 10}\left[\tan^{-1}\left(\frac{\omega}{\omega_2}\right) - \tan^{-1}\left(\frac{\omega}{\omega_1}\right)\right] \quad (c).
\end{aligned} \tag{6.8.11}$$

The expressions for both the real and imaginary parts of the complex relative permittivity are approximated by using the limiting values of Debye dispersion at $\omega_1 \to 0$, and $\omega_2 \to \infty$:

$$\varepsilon_r'(\omega) \approx \varepsilon_r'(\omega = \infty) + \frac{\Delta\varepsilon_r'}{(m_2 - m_1)} \frac{\ln(\omega_2/\omega)}{\ln 10} \quad (a)$$

$$\varepsilon_r''(\omega) = \frac{\Delta\varepsilon_r'}{(m_2 - m_1)} \frac{-\pi/2}{\ln 10} \quad (b). \qquad (6.8.12)$$

It noted from equation (6.8.12) that $\varepsilon_r'(\omega)$ decreases linearly with the logarithmic increase in frequency ω, whereas $\varepsilon_r''(\omega)$ remains constant, giving a slow increase in $\tan\delta(\omega)$. FR-4 substrate has been modeled with reasonable accuracy against experimental results for $\varepsilon_r'(\omega = \infty) = 4.27, \sigma = 80\ \text{pS/m}, \omega_1 = 10^4 (m_1 = 4), \omega_2 = 10^{12} (m_2 = 12)$, and $\Delta\varepsilon_r'/(m_2 - m_1) = 0.14$ [J.47].

The noncausal RLCG parameters have been obtained for the microstrip line [J.65]. However, using the Debye model the causal RLCG model has also been developed for its application in the time-domain analysis [J.62].

6.8.3 Metasubstrates

The high permittivity and high permeability substrates, called metasubstrates, can be engineered by embedding resonant metallic structures in the host dielectric-medium. The capacitively loaded wire inclusions create an *electric medium* [J.52, J.53]. It also acts as a superstrate for the patch antenna [J.54]. The resonant loop, metasolonoid, and capacitor loaded metallic loops create the *magnetic medium* [J.55–J.58]. The parallel electric field excited wire inclusions develop electric dipoles creating the electric medium, whereas the magnetic field, normal to the plane of loops, develops magnetic dipoles creating the magnetic medium. However, due to the cross-polarization, the wire medium can create a weaker magnetic response, and the loop medium can also create an electric response. Thus, these media, created in the host dielectric-medium, are *magneto-dielectric-medium*. Such dispersive materials are *Lorentz type materials*. These are also used to engineer the metamaterials and their modeling is discussed in sections (21.1) and (21.2) of chapter 21. The inclusions are periodically arranged. However, the period and size of inclusions are much smaller than the operating wavelength forming the *homogeneous effective medium* characterized by the equivalent permittivity and permeability, again discussed in section (21.4).

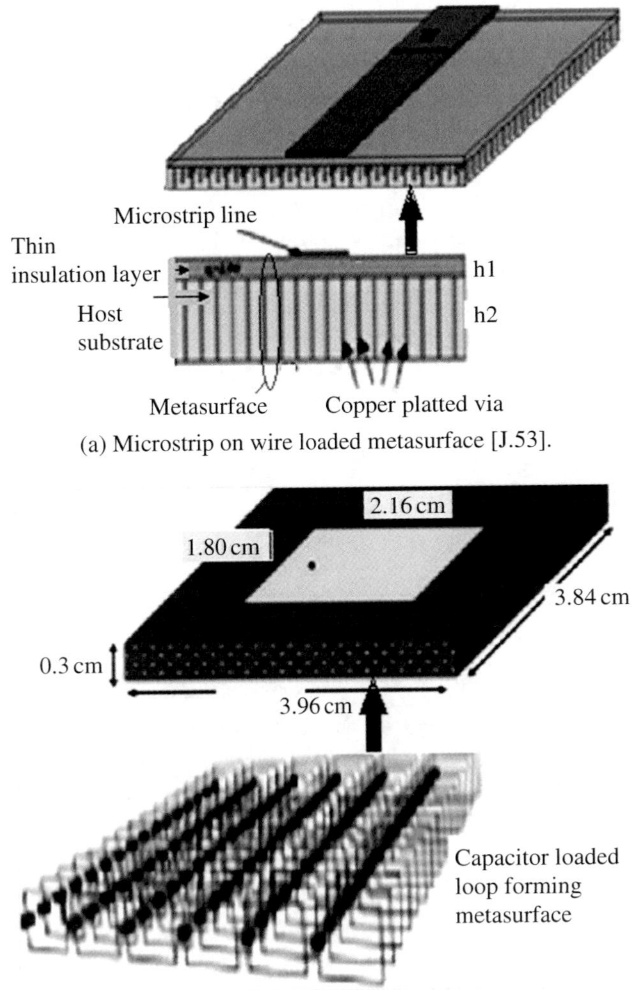

(a) Microstrip on wire loaded metasurface [J.53].

(b) Microstrip patch antenna on loop metasurface [J.58].

Figure 6.18 Wire loaded and loop loaded magneto-dielectric substrates. *Source*: Fernandez et al. [J.53] and Mosallaei and Sarabandi [J.58].

Figure (6.18a and b) illustrates the magneto-dielectric metasurface engineered by the wire loaded and loop loaded host dielectric substrate [J.53, J.58]. For the structure of Fig (6.18a), the host medium is Duroid 6002 with $\varepsilon_r = 2.94$, $h_2 = 0.508\text{mm}$ that is loaded with 200×200 copper-plated via (wires) connected to the ground conductor of the microstrip. The diameter of via (wire) is 0.38 mm, and the period of via is 0.635 mm. For simplicity of fabrication, the same Duriod-6002 is used as the insulating layer between microstrip and wired medium. It creates a capacitance between the via (wire) top and microstrip. Thus, the capacitor loaded wire (via) acts as a resonant circuit creating the Lorentz material with high permittivity below the resonance frequency. However, closely placed wires also create U-shaped loops responsible for the magnetic response. Thus, the metasurface is a magneto-dielectric-medium with $\varepsilon_{r,\,eff} = 4.8$ and $\mu_{r,\,eff} = 1.3$. These parameters are extracted from the S-parameters of the microstrip on the metasurface [J.53].

Figure (6.18b) shows the loop loaded host substrate of relative permittivity and thickness $\varepsilon_r = 4.5$, h = 3 mm. The loop itself is loaded with a capacitor. A microstrip antenna at frequency 1.74 GHz is designed on the metasurface. The metasurface has $\mu_r \approx 2$. This is a compact microstrip antenna with improved bandwidth. The details are available in reference [J.58].

References

Books

B.1 Von Hipple, A.: *Dielectrics, and Waves*, Artech House, Norwood, MA, 1995.

B.2 Elliot, R.S.: *Electromagnetics: History, Theory, and Applications*, IEEE Press, USA, 1993.

B.3 Edward, M.: *Purcell, Electricity, and Magnetism*, McGraw-Hill, New York, 1965.

B.4 Griffiths, D.J.: *Introduction to Electrodynamics*, Prentice-Hall, Englewood Cliffs, NJ, 1999.

B.5 Simon, R., Whinnery John R., Theodore, V.D.: *Fields, and Waves in Communication Electronics*, 3rd Edition, John Wiley & Sons, Singapore, 1994.

B.6 Collin, R.E.: *Foundation for Microwave Engineering*, 2nd Edition, McGraw-Hill, Inc., New York, 1992.

B.7 Jordan, E.C., Balmain Keith G.E.: *Electromagnetic wave and Radiating System*, Prentice-Hall India, New Dehi, 1989.

B.8 Robert, P., Collin Robert, E.: *Principle and Applications of Electromagnetic Fields*, Tata McGraw-Hill, New Delhi, 1973.

B.9 Polivanov, K.: *The Theory of Electromagnetic Fields*, Mir Publisher, Moscow, 1975.

B.10 Balanis, C.A.: *Advanced Engineering Electromagnetics*, John Wiley & Sons, New York, 1989.

B.11 Collin Robert, E.: *Field Theory of Guided Waves*, IEEE Press USA, 1991.

B.12 Smythe: *Static and Dynamic Electricity*, McGraw-Hill, New York, 1939.

B.13 Baden-Fuller, A.J.: *Microwaves*, 3rd Edition, Pergamon Press, New York, 1990.

B.14 Thuery, J.: *Microwaves: Industrial, Scientific, and Medical Applications*, Artech House, USA, 1992.

B.15 Marques, R., Martin, F., and Sorolla, M.: *Metamaterials with Negative Parameters-Theory, Design, and Microwave Applications*, Wiley-Interscience, New Jersey, 2008.

B.16 Solymar, L. and Shamonina, E.: *Waves in Metamaterials*, Oxford University Press, New York, USA, 2009.

B.17 Sihvola, A.: *Electromagnetic Mixing Formulae and Applications*, IEE Electromagnetic Waves Series, 47, Institution of Electrical Engineers, London, UK, 1999.

B.18 Schelkunoff, S.A. and Friis, H.T.: *Antennas: Theory and Practice*, John Wiley, New York, 1952.

B.19 Hoffmann, R.: *Microwave Integrated Circuit Handbook*, Artech House, Boston, 1985.

B.20 Bohren, C.F. and Huffman, D.R.: *Absorption and Scattering of Light by Small Particles*, Wiley-Int. Pub., John Wiley, New York, 1983.

B.21 Tretyakov, S.: *Analytical Modeling on Applied Electromagnetics*, Artech House, Boston, USA, 2000.

B.22 Nyfors, E.G. and Vainikainen, P.: *Industrial Microwave Sensors*, Artech House, Boston, 1989.

B.23 Kraszewski, A.: *Microwave Aquametry: Electromagnetic Wave Interaction with Water-containing Materials*, (Edited), IEEE Press, New York, 1996.

B.24 Maxwell, J. C.: *A Treatise on Electricity and Magnetism-Volume 1*, pp. 450–464, Clarendon Press, Oxford, 3rd Edition, reprint by Dover, 1891.

B.25 Jackson, J.D.: *Classical Electrodynamics*, 3rd Edition, John Wiley, New York, 1999.

Journals

J.1 Herczyński, A.: *Bound charges and currents, Am. J. Phys.*, Vol. 81, No. 3, pp. 202–205, 2013.

J.2 Hasegawa, H. and Furukawa, M., and Yanai, H.: *Properties of the microstrip line on Si-SiO2 system, IEEE Trans. Microw. Theory Tech.*, Vol. 19, pp. 869–881, 1971.

J.3 Jager, D.: *Slow-wave propagation along variable Schottky-contact microstrip line, IEEE Trans. Microw. Theory Tech.*, Vol. 24, pp. 566–573, 1976.

J.4 Verma, A. K.; Nasimuddin: *Propagation Characteristics of Schottky contact suspended slow-wave microstrip line, IEEE Microw. Wirel. Compon. Lett.*, Vol. 11, No. 9, pp. 385–387, 2001.

J.5 Verma, A. K.; Nasimuddin; Sharma, E.K.: *Propagation Characteristics of suspended MIS coupled slow-wave microstrip lines, Microw. Opt. Technol. Lett.*, Vol. 34, No. 6 pp. 403–405, 2002.

J.6 Brown, J.: *Artificial dielectrics having refractive indices less than unity, Proc. IEE, Monograph*, Vol. 100, No. 62R, pp.51–62, May 1953.

J.7 Kock, W.E.: *Metallic delay lenses, Bell Syst. Tech. J.*, Vol. 27, pp. 58–82, 1948.

J.8 Rotman, W.: *Plasma simulation by artificial dielectrics and parallel - plate media, IRE Trans. Antennas Propag.*, Vol. 10, No. 1, pp. 82–95, Jan. 1962.

J.9 Polder, D.; van Santen, J. H.: *The effective permeability of mixtures of solids, Physica*, Vol. 12, No. 5, pp. 257–271, 1946

J.10 Sihvola, A. H.; Lindell, I.V: *Polarizability modeling of heterogeneous media, Progress in Electromagnetic Research, PIERS-6*, pp. 101–151, A. Priou, Ed, Elsevier Science Pub., New York, 1992.

J.11 Sihvola, A.; Kettunen, H.; Wall'en, H.: *Mixtures and composite particles: correspondence of effective description*, Proc. 2013 Int. Symposium on Electromagnetic Theory, IEICE, pp. 908–911, 2013.

J.12 Sihvola, A.: *Homogenization of a dielectric mixture with anisotropic spheres in anisotropic background*, Department of Electromagnetic Theory, Lund Institute of Technology, Sweden, October 1996.

J.13 Qizheng, Y.; Jin, L.: *Effective permittivity of dielectric mixture based on the electric-circuit model, Electromagnetics*, Vol. 21, pp. 341–350, 2001.

J.14 Wiener, O.: *Zur theorie der refraktionskonstanten*, Berichteüber Verhandlungen Königlich-Sächsischen Gesellschaft Wisseschaften, Leipzig, pp. 256–277, 1910.

J.15 Hashin, Z.; Shtrikman, S.: *A variational approach to the theory of effective magnetic permeability of multiphase materials, J. Appl. Phys.*, Vol. 33, pp. 3125–3131, 1962.

J.16 Kärkkäinen, K.K.; Sihvola, A.H.; Nikoskinen, K.I.: *Effective permittivity of mixtures: numerical validation by the FDTD method, IEEE Trans. Geosci. Remote Sens.*, Vol. 38, No. 3, 1303–1308, 2000.

J.17 Merrill, W.M.; Diaz, R.E.; LoRe, M.M.; Squires, M.C.; Alexopoulos, N.G.: *Effective medium theories for artificial materials composed of multiple sizes of spherical inclusions in a host continuum, IEEE Trans. Antennas Propag.*, Vol. 47, No. 1, 142–148, Jan. 1999.

J.18 Kärkkäinen, K.; Sihvola, A.: *Analysis of a three-dimensional dielectric mixture with finite difference method, IEEE Trans. Geosci. Remote Sens.*, Vol. 39, No. 5, pp. 1013–1015, May 2001.

J.19 Maxwell Garnett, J. C.: *Colours in metal glasses and metallic films, Philos. Trans. R. Soc. Lond. A*, Vol. 203, pp. 385–420, 1904.

J.20 Sihvola, A.: *Two main avenues leading to the Maxwell Garnett mixing rule, J. Electromagn. waves Appl.*, Vol. 15, pp. 715–725, 2001.

J.21 Jylha, L.; Sihvola, A.: Equation for the effective permittivity of particle-filled composites for material design applications, *J. Phys. D. Appl. Phys.*, Vol. 40, pp. 4966–4973, 2007.

J.22 Robinson, D.A.; Friedman, S.P.: *The effective permittivity of dense packing of glass beads, quartz sand and their mixtures immersed in different dielectric backgrounds, J. Non-Cryst. Solids*, Vol. 305, No. 1–3. pp. 261–267, 2002.

J.23 Sihvola, A. H.; Kong, J. A.: *Effective permittivity of dielectric mixtures, IEEE Trans. Geosci. Remote Sens.*, Vol. 26, No. 4, 420–429, Jul. 1988.

J.24 Liisi, J.; Sihvola, A. H.: *Decreasing the computational complexity of determination of effective permittivity of a pseudorandom mixture*, Proc. Intern Symposium on Electromagnetic Theory (URSI), (Pisa, Italy), pp. 23–27, 2004.

J.25 Sihvola, A: *Self-consistency aspects of dielectric mixing theories, IEEE Trans. Geosci. Remote Sens.*, Vol. 27, No. 4, pp. 403–415, July 1989.

J.26 Bruggeman, D.A.G.: *Berechnung verschiedener physikalischer konstanten von heterogeneous substanzen, i. dielektrizitätskonstanten und leitfähigkeiten der mischkörper aus isotropen substanzen, Ann. Phys.*, Vol. 416, No. 7, pp. 636–664, 1935.

J.27 Kohler, W. E.; Papanicolaou, G. C: *Some applications of the coherent potential approximation, Multiple Scattering and Waves*, P. L. Kohler and G. C. Papanicolaou, Eds., pp. 199–223, New York: North-Holland, 1981.

J.28 Lichtenecker, K.; Rother, K.: *Die herleitung des logarithmischen mischungsgesetzes aus allgemeinen prinzipien der stationären strömung, Phys. Z.*, Vol. 32, pp. 255–260, 1931.

J.29 Looyenga, H., *Dielectric constants of mixtures, Physica*, Vol. 31, pp. 401–406, 1965.

J.30 Birchak, J. R.; Gardner, L. G.; Hipp, J.W.; Victor, M.: *High dielectric constant microwave probes for sensing soil moisture, Proc. IEEE*, Vol. 62, pp. 93–98, Jan. 1974.

J.31 Kock, W.E.: *Metallic delay lenses, Bell Syst. Tech. J.*, Vol. 27, pp.58–82, Jan. 1948.

J.32 Wickersham, A.F., *Anomalous dispersion in artificial dielectrics, J. Appl. Phys.*, Vol. 29, No. 11, pp. 1537–1542, Nov. 1958.

J.33 Ziolkowski, R.W.: *Time -derivative Lorentz material model-based absorbing boundary condition, IEEE Trans. Antennas Propag.*, Vol. 45, No. 10, pp. 1530–1535, Oct. 1997.

J.34 Kristensson, G.; Rikte, S.: *Mixing formulas in the time-domain, J. Opt. Soc. Am. A*, Vol. 15, No. 5, pp. 1411–1422, May 1998.

J.35 Nisanci, M.H.; Paulis, F.d; Koledintseva, M.Y.; Drewniak, J.L.: Orlandi, A., *From Maxwell Garnett to*

Debye model for electromagnetic simulation of composite dielectrics – Part II: Random cylindrical inclusions, *IEEE Trans. Electromagn. Compat.*, Vol. 54, No. 2, pp. 280–289, April 2012.

J.36 Febo, D.D.; Paulis, F.D.; Orlandi, A.: *Three-dimensional full-wave validation of the Maxwell-Garnett to Debye Model approach*, Proc. 10th Int. Sym. Electromagnetic Compatibility (EMC Europe 2011), (York, UK), Sept. 26–30, 2011.

J.37 Salski, B.; Celuch, M.: *On the equivalence between the Maxwell-Garnett mixing rule and the Debye relaxation formula*, *IEEE Trans. Microw. Theory Tech.*, Vol. 60, No. 8, pp.2352–2358, Aug. 2012.

J.38 Wallén, H.; Kettunen, H.; Qi, J.; Sihvola, A.: *Some Interesting Effects when Homogenizing Plasmonic Composites*, www.ursi.fi/2010/papers/A33_Wallen.

J.39 Wagner, K.W.: *Erkl¨arung der dielekrischen Nachwirkungsvorg¨ange auf Grund Maxwellscher Vorstellungen*, *Archiv für Electrotechnik*, Vol. II, No. 9, pp. 371–387, 1914.

J.40 Billing, E.; Plessner, K.W.: *A note on the dielectric dispersion in polycrystalline materials*, *Proc. Phys. Soc. B*, Vol. 64, pp. 361–363, 1951.

J.41 Ziolkowski, R.W.: *Passive artificial molecule realizations of dielectric materials*, *J.App Phy.*, Vol. 82, No. 7, pp. 3195–3198, Oct. 1997.

J.42 Ye, H.; Sun, C.Q.; Huang, H.; Hing, P.: *Single semicircular response of dielectric properties of diamond films*, *Thin Solid Films*, Vol. 381, pp.52–56, 2001, Elsevier.

J.43 Maradei, F; Ke, H.; Hubing, T.H.: *Fullwave model of frequency-dependent -dispersive media with Debye dispersion relation by circuit-oriented FEM*, *IEEE Trans. Electromagn. Compat.*, Vol. 51, No. 2, pp.312–319, May 2009.

J.44 Baker-Jarvis, J.; Kim, S.: *The interaction of radio-frequency fields with dielectric materials at macroscopic to mesoscopic scale*, *J. Res. Natl. Inst. Stand. Technol.*, Vol. 117, pp. 1–60, 2012.

J.45 Auzanneau, F.; Ziolkowski, R.W: *Microwave signal rectification using artificial composite materials composed of a diode-loaded electrically small dipole antenna*, *IEEE Trans. Microw. Theory Tech.*, Vol. 46, No. 11, pp. 1628–1637, Nov. 1998.

J.46 Auzanneau, F.; Ziolkowski, R.W: *Artificial composite materials* consisting of nonlinearly loaded *electrically small dipole antenna: operational amplifier -based circuits with application to the smart skin*, *IEEE Trans. Microw. Theory Tech.*, Vol. 47, No. 8, pp. 1330–1339, Aug. 1999.

J.47 Djordjevic, A.R.; Biljic, R. M.; Likar-Smiljanic, V.D.; Sarkar, T.K.: *Wideband frequency-domain characterization of FR-4 and time-domain causality*, *IEEE Trans. Microw. Theory Tech.*, Vol. 43, No. 4, pp.662–667, Nov. 2001.

J.48 Koledintseva, M.Y.; Rozanov, K.N.; Orlandi, A.; Drewniak, J.L.: *Extraction of Lorentzian and Debye parameters of dielectric and magnetic dispersive materials for FDTD modeling*, *J. Electr. Eng.*, Vol. 53, No. 9/s, pp.97–100, 2002.

J.49 Koledintseva, M. Y.; Rozanov, K. N.; Fazio, G. D.; Drewniak, J. L.: *Restoration of the Lorentzian and Debye curves of dielectrics and magnetics for FDTDmodeling*, Proc. EMC EUROPE 5th Int. Symp. Electromagnetic Compatibility, (Sorrento, Italy), pp. 687–692. Sep. 2002.

J.50 Verma, A. K.; Nasimuddin: *Determination of dielectric constant and loss tangent of substrate sheet using microstrip patch resonator*, *Microw. Opt. Technol. Lett.*, Vol. 35, No. 3, pp. 175–179, 2002.

J.51 Verma, Y.K.; Verma, A.K.: *Accurate determination of dielectric constant of substrate materials using modified Wolff model*, IEEE Int Microwave Symp Dig MTT-S, 2000.

J.52 Machac, J.: *Microstrip line on an artificial dielectric substrate*, *IEEE Microw. Wirel. Compon. Lett.*, Vol. 16, No. 7, pp. 416–418, July 2006.

J.53 J.M. Fernandez; Sierra-Castañer, M.; Caloz, C.: *Metallic wire substrate (MWS) microstrip structure characterization and application to a patch antenna*, Proc. "EuCAP 2006," Nice, France, 6–10 Nov. 2006 (ESA SP-626, October 2006).

J.54 Ikonen, P.M.T.; Saenz, E.; Gonzalo, R.; Tretyakov, S. A.: *Modeling and Analysis of Composite Antenna: Superstrates Consisting on Grids of Loaded Wires*, pp. 1–9, arXiv: physics/0701035v2 [physics.class-ph] 11 May 2007.

J.55 Kärkkäinen, M.K.; Tretyakov, S. A.; Ikonen, P.: *PIFA with dispersive material fillings*, *Microw. Opt. Technol. Lett.*, Vol. 45, No. 1, pp. 5–8, 2005.

J.56 Buell, K.; Mosallaei, H.; Sarabandi, K.: *A substrate for small patch antennas providing tunable miniaturization factors*, *IEEE Trans. Microw. Theory Tech.*, Vol. 54, No. 1, pp. 135–146, Jan. 2006.

J.57 Ikonen, P. M. T.; Maslowski, S. I.; Simovski, C. R; Tretyakov, S. A.: *On artificial magnetodielectric loading for improving the impedance bandwidth properties of microstrip antennas*, *IEEE Trans. Antennas Propag.*, Vol. 54, No. 6, pp. 1654–1662, Jun. 2006.

J.58 Mosallaei, H.; Sarabandi, K.: *Design and modeling of patch antenna printed on magneto-Dielectric embedded- circuit metasubstrate*, *IEEE Trans. Antennas Propag.*, Vol. 55, No. 1, 45–52, Jan. 2007.

J.59 Koledintseva, M. Y.; Drewniak, J.L.; Pommerenke, D.J.; Antonini, G.; Orlandi, A.; Rozanov, K.N.: *Wide-Band Lorentzian Media in the FDTD Algorithm*, *IEEE Trans. Electromagn. Compat.*, Vol. 47, No. 2, pp. 392–399, May 2005.

J.60 Coperich Branch, K.M.C; Morsey, J.; Cangellaris, A.C.; Ruehli, A.E.: *Physically consistent transmission line models for high-speed interconnects in lossy dielectrics*, *IEEE Trans. Adv. Packag.*, Vol. 5, No. 2, pp. 129–135, May 2002.

J.61 Zhou, Z.; Melde, K. L.: *A comprehensive technique to determine the broadband physically consistent material characteristics of microstrip lines*, *IEEE Trans. Microw. Theory Tech.*, Vol. 58, No. 1, pp.185–194, Jan. 2010.

J.62 Zhang. J.; Drewniak, J.L.; Pommerenke, D.J.; Koledintseva, M.Y.; DuBroff, R.E.; Cheng, W.C.; Yang Z.; Chen, Q.B.; Orlandi, A, *Causal RLGC(f) models for transmission lines from measured S-Parameters*, *IEEE Trans. Electromagn. Compat.*, Vol. 52, No. 1, pp. 189–198, Feb. 2010.

J.63 Mrozowski, M.; Stuchly, M.A.: *Parameterization of media dispersive properties for FDTD*, *IEEE Trans. Antennas Propag.*, Vol. 45, No. 9, 1438–1439, Sept. 1997.

J.64 Svensson, C.; Dermer, G.E.: *Time-domain modeling of lossy interconnects*, *IEEE Trans. Adv. Packag.*, Vol. 24, No. 2, pp. 191–196, May 2001.

J.65 Verma, A.K.; Nasimuddin: *quasistatic RLCG parameters of lossy microstrip line for CAD application*, *Microwave Opt. Tech. Lett.*, Vol. 28, No. 3, pp. 209–212, Feb. 2001.

7

Waves in Waveguide Medium

Introduction

A general class of electromagnetic wave supporting transmission medium is called the *waveguides*. The waveguides are three-dimensional structures, supporting the *non-TEM* modes. This chapter reviews the basic field structures of the non-TEM modes and considers some basic rectangular cross-section waveguides, as these are important structures for the analysis and modeling of the planar transmission lines and resonating patch structures. The analysis of the waveguides is presented in terms of the *electric and magnetic scalar potentials*. The *spectral domain analysis* (SDA) of the planar lines, discussed in chapter 16, is formulated using the scalar electric and magnetic potentials. However, the waveguide structures considered in this chapter are useful in their own right. The waveguides are useful low-loss media at the mm-wave and even at a much higher frequency. In the planar medium format, it has reinvented itself as the *substrate integrated waveguide* (SIW) structures. Several textbooks are available on this topic [B.1–B.12].

Objectives

- *To review the basic classification of the EM-field structures and formulate the EM-field problems in terms of the electric and magnetic scalar potentials.*
- *To present modal solutions of the waveguide structures, including the surface wave waveguides.*
- *To develop circuit models of the waveguide.*
- *To obtain the propagation parameters of several waveguide structures using the transverse resonance method (TRM).*
- *To discuss the basic models and characteristics of the substrate integrated waveguides (SIW).*

7.1 Classification of EM-Fields

The three-dimensional waveguide structures can have a maximum of six field components – three electric and three magnetic field components. Depending on the field components along the longitudinal direction of propagation, the EM field structures, known as *modes*, are classified into four groups

- The *transverse electromagnetic* (TEM) mode: It has no electric and magnetic field component in the direction of wave propagation. All field components are in the transverse plane. The transverse plane is normal to the direction of propagation. Thus if the z-axis in the direction of propagation, the TEM mode has $E_z = H_z = 0$. This mode is supported by the two-conductor transmission line in a homogeneous medium.
- The *transverse magnetic* (TM) mode: It has no magnetic field component in the direction of wave propagation. However, there is an electric field component in the direction of propagation. Thus, the TM-mode has $H_z = 0$, $E_z \neq 0$. Other x and y-field components exist in the transverse plane. This mode is also known as the *E-mode*. This mode is supported by the homogeneous metallic waveguide. The conductor backed dielectric sheet also supports this mode.
- The *transverse electric* (TE) mode: It has no electric field component in the direction of wave propagation. However, there is a magnetic field component in the direction of propagation. Thus, the TE-mode has $E_z = 0$, $H_z \neq 0$. This mode is also known as the *H-mode*. This mode is supported by the homogeneous metallic waveguide and the conductor-backed dielectric sheet.
- The *hybrid* (HE and EH) mode: It has both the electric and magnetic field components in the direction of

Introduction to Modern Planar Transmission Lines: Physical, Analytical, and Circuit Models Approach, First Edition. Anand K. Verma.

wave propagation. Thus, the hybrid mode has $E_z \neq 0$, $H_z \neq 0$. The hybrid modes are expressed in two nomenclature formats. Under the first nomenclature, these are known as the *longitudinal section electric mode* (*LSE*) and the *longitudinal section magnetic mode* (*LSM*) [B.2, B.5]. Under the second nomenclature the hybrid modes are known as the HE and EH modes, depending on the strength of the H_z and E_z longitudinal field components [B.1, B.9]. The inhomogeneous waveguides [B.1, B.2, B.5] and inhomogeneous planar transmission lines support the hybrid mode [B.13, B.15].

It is interesting to note that all the modes can be described by two scalar potentials – the *scalar electric potential* ϕ^e and the *scalar magnetic potential* ϕ^m. These potentials are related to the electric current density and the magnetic current density. Therefore, these potentials are treated as the sources that generate the electromagnetic modal structures.

7.1.1 Maxwell Equations and Vector Potentials

The time-harmonic electric and magnetic fields are not independent quantities. These are dependent on each other, as described by the following coupled curl equations, known as Maxwell's equations:

$$\nabla \times \vec{E} = -j\omega \vec{B} \quad (a), \qquad \nabla \times \vec{H} = j\omega \vec{D} + \vec{J} \quad (b). \tag{7.1.1}$$

Equation (7.1.1b) shows that the magnetic field $\left(\vec{H}\right)$ is a response to two kinds of sources (excitation): (i) Electrical current density $\vec{J}$, that is related to the motion of electrons or other charged particles; (ii) Displacement current density $\vec{J}_d = j\omega\vec{D}$, that is associated with the time changing electric field $\left(\vec{D} = \varepsilon\vec{E}\right)$ at the source point. The time-dependent magnetic field intensity is described by the time-dependent magnetic flux density $\vec{B}$ $\left(\vec{B} = \mu\vec{H}\right)$ and it produces the electric field intensity $(\vec{E})$ through equation (7.1.1a).

A vector quantity is fully described only when both the curl and divergence of the vector are prescribed [B.11]. The curl of the electric and magnetic field vectors are given by equation (7.1.1). Their divergences, in the material medium, are given by Gauss's law.

$$\begin{aligned} &\nabla.\vec{B} = 0 \quad (a), \qquad \nabla.\vec{D} = \rho_e \quad (b) \\ &\text{where } \vec{B} = \mu\vec{H} \quad (c), \qquad \vec{D} = \varepsilon\vec{E} \quad (d). \end{aligned} \tag{7.1.2}$$

$$\nabla.\left(\mu\vec{H}\right) = 0 \quad (a), \qquad \nabla.\left(\varepsilon\vec{E}\right) = \rho_e \quad (b), \tag{7.1.3}$$

where ρ_e is the volumetric electric charge density. If the medium is homogeneous, i.e. its permeability μ and permittivity ε are not functions of the position, the following expressions are obtained:

$$\nabla.\vec{H} = 0 \quad (a), \qquad \nabla.\vec{E} = \frac{\rho_e}{\varepsilon} \quad (b). \tag{7.1.4}$$

7.1.2 Magnetic Vector Potential

Equation (7.1.1b) shows that components of the magnetic field $\vec{H}$ are not directly related to the components of source term $\vec{J}$ or $j\omega\vec{D}$. Another auxiliary vector $\vec{A}$, that is known as the *magnetic vector potential,* is introduced to achieve this end. It has a direct relationship with the source $\vec{J}$, that is the electric current density. The magnetic vector potential is introduced through equation (7.1.4a) by taking its solution as

$$\vec{H} = \nabla \times \vec{A} \quad (a), \qquad \vec{B} = \mu\nabla \times \vec{A} = \nabla \times \left(\mu\vec{A}\right) \quad (b). \tag{7.1.5}$$

The following relation is obtained from equations (7.1.1a) and (7.1.5b):

$$\nabla \times \left(\vec{E} + j\omega\mu\vec{A}\right) = 0. \tag{7.1.6}$$

Vector equation (7.1.6) could be solved through the gradient of the *scalar electric function* ψ^e:

$$\vec{E} + j\omega\mu\vec{A} = -\nabla\psi^e. \tag{7.1.7}$$

A relation between $\vec{A}$ and $\vec{J}$ is obtained by substituting equations (7.1.5a) and (7.1.7) in equation (7.1.1b). The relation is obtained as the *vector wave equation* for the magnetic vector potential:

$$\begin{aligned} &\nabla \times \nabla \times \vec{A} = j\omega\varepsilon\left[-j\omega\mu\vec{A} - \nabla\psi^e\right] + \vec{J} \\ &\Rightarrow \nabla\left(\nabla.\vec{A}\right) - \nabla^2\vec{A} = \omega^2\mu\varepsilon\vec{A} - j\omega\varepsilon\nabla\psi^e + \vec{J} \\ &\Rightarrow \nabla^2\vec{A} + k^2\vec{A} = \nabla\left(\nabla.\vec{A} + j\omega\varepsilon\psi^e\right) - \vec{J}, \end{aligned} \tag{7.1.8}$$

where $k = \omega\sqrt{\mu\varepsilon}$ is called the *wavenumber* of an electromagnetic wave, i.e. EM-wave. This is also the propagation constant β of a wave in the homogeneous medium. The divergence of the magnetic vector potential $\vec{A}$ is obtained by taking

$$\nabla.\vec{A} + j\omega\varepsilon\,\psi^e = 0, \quad \Rightarrow \quad \nabla.\vec{A} = -j\omega\varepsilon\,\psi^e. \tag{7.1.9}$$

This is known as the *Lorentz gauge condition*. The Lorentz gauge condition gives divergence of the vector $\vec{A}$. The curl equation (7.1.5a) and divergence equation (7.1.9) completely define the magnetic vector potential $\vec{A}$. Under the Lorentz gauge condition, the wave equation (7.1.8) is reduced to a simpler form,

$$\nabla^2\vec{A} + k^2\vec{A} = -\vec{J}. \tag{7.1.10}$$

The current density $\vec{J}$ is a source of the EM-wave described by the magnetic vector potential. In the rectangular coordinate system, each component of the magnetic vector potential $\vec{A}$ is directly related to each component of the current density $\vec{J}$. As the vector potential $\vec{A}$ is related to the electric current density, some authors [B.2–B.4] prefer to call it electric vector potential [B.2, B.4], as against the usual classical magnetic vector potential adopted in this chapter [B.5, B.10]. Equation (7.1.10) is solved under the boundary conditions to get $\vec{A}$ for a particular geometry. The EM-field components can be obtained from equations (7.1.5), (7.1.7), and (7.1.9):

$$\begin{aligned} \vec{E} &= -j\omega\mu\vec{A} + \frac{1}{j\omega\varepsilon}\nabla\left(\nabla.\vec{A}\right) && \text{(a)} \\ \vec{H} &= \nabla\times\vec{A} && \text{(b).} \end{aligned} \tag{7.1.11}$$

So the electric current density ($\vec{J}$) generates the magnetic vector potential $\vec{A}$, that, in turn, helps to compute the electric and magnetic fields of the EM-wave. The electrostatic electric potential is a measurable quantity. It also simplifies the computation of the electric fields for the complex arrangement of charges. *The magnetic vector potential $\vec{A}$ and several other potentials introduced in the electromagnetics are not measurable physical quantities*. The potentials help to solve the problems for the computation of the electric and magnetic field quantities. This is the usual approach to solve the EM-field problems. However, modern quantum mechanical views the potentials as the physical reality and considers the E and H-fields as the secondary quantities. This point of view is reviewed in subsection (1.1.2) of chapter 1.

Use of the magnetic vector potential is consolidated below:

$$\begin{aligned} \nabla^2\vec{A} + k^2\vec{A} &= -\vec{J} && \text{(a)} \\ \vec{E} &= -j\omega\mu\vec{A} + \frac{1}{j\omega\varepsilon}\nabla\left(\nabla.\vec{A}\right) && \text{(b)} \\ \vec{H} &= \nabla\times\vec{A} && \text{(c).} \end{aligned} \tag{7.1.12}$$

7.1.3 Electric Vector Potential

The *hypothetical magnetic current*, related to the electric field, also acts as the source for the EM-wave. The concept of the magnetic current density $\vec{M}$ could be introduced in Maxwell's equation (7.1.1a) to maintain its symmetry with respect to equation (7.1.1b):

$$\nabla\times\vec{E} = -j\omega\vec{B} - \vec{M}. \tag{7.1.13}$$

It is discussed in the previous section that the magnetic vector potential $\vec{A}$ corresponds to the electric current density $\vec{J}$. Thus, there could *the electric vector potential* $\vec{F}$ corresponding to the magnetic current density $\vec{M}$. Again, as the electric vector potential $\vec{F}$ is created by the magnetic current, Collin [B.2, B.4] calls it the magnetic vector potential, not the electric vector potential adopted here [B.10]. It is introduced through the divergence relation of equation (7.1.4b). If there is no accumulation of charge in a homogeneous medium, it is written as

$$\nabla.\vec{E} = 0. \tag{7.1.14}$$

The electric field intensity could be solved in term of the electric vector potential $\vec{F}$

$$\vec{E} = -\nabla\times\vec{F}. \tag{7.1.15}$$

The assumed magnetic current $\vec{M}$ generates the electric vector potential $\vec{F}$. A relation is obtained as below between $\vec{F}$ and $\vec{M}$. The medium could be free of source charge, i.e. $\rho_e = 0$, and $\vec{J} = 0$. In the case of a charge-free homogeneous medium, the following expression is obtained from equations (7.1.1b) and (7.1.15):

$$\begin{aligned} &\nabla\times\vec{H} = j\omega\varepsilon\left(-\nabla\times\vec{F}\right) \\ \Rightarrow\quad &\nabla\times\left(\vec{H} + j\omega\varepsilon\vec{F}\right) = 0. \end{aligned} \tag{7.1.16}$$

Equation (7.1.16) can be solved by introducing a gradient of the *scalar magnetic function* ψ^m:

$$\vec{H} + j\omega\varepsilon\vec{F} = -\nabla\psi^m. \tag{7.1.17}$$

The EM-wave equation is obtained for the electric vector potential $\vec{F}$ from equations (7.1.13), (7.1.15), and (7.1.17):

$$\begin{aligned} \nabla\times\nabla\times\vec{F} &= j\omega\mu\vec{H} + \vec{M} \\ &= j\omega\mu\left[-j\omega\varepsilon\vec{F} - \nabla\psi^m\right] + \vec{M} \\ \Rightarrow\quad \nabla\left(\nabla.\vec{F}\right) - \nabla^2\vec{F} &= \omega^2\mu\varepsilon\vec{F} - j\omega\mu\nabla\psi^m + \vec{M} \\ \Rightarrow\quad \nabla^2\vec{F} + k^2\vec{F} &= \nabla\left(\nabla.\vec{F} + j\omega\mu\psi^m\right) - \vec{M}. \end{aligned} \tag{7.1.18}$$

The curl of the vector $\vec{F}$ has already been defined by equation (7.1.15). Its divergence can be defined through another Lorentz gauge condition as follows:

Let

$$\nabla.\vec{F} + j\omega\mu\,\psi^m = 0, \quad \Rightarrow \quad \nabla.\vec{F} = -j\omega\mu\,\psi^m. \tag{7.1.19}$$

Under the Lorentz gauge condition, a simpler form of the wave equation is obtained for the electric vector potential $\vec{F}$:

$$\nabla^2\vec{F} + k^2\vec{F} = -\vec{M}. \tag{7.1.20}$$

The magnetic current density $\vec{M}$ is a source for the EM-wave, described by the wave equation (7.1.20). The wave equation (7.1.20) for the electric vector potential $\vec{F}$ is solved under the boundary conditions to obtain the magnetic field $\vec{H}$ from equations (7.1.17) and (7.1.19):

$$\vec{H} = -j\omega\varepsilon\,\vec{F} + \frac{1}{j\omega\mu}\nabla\left(\nabla.\vec{F}\right). \tag{7.1.21}$$

In brief, the magnetic current $\vec{M}$ also generates the EM-fields described through the electric vector potential $\vec{F}$. A set of equations for the computation of electric and magnetic fields created by the magnetic current is summarized below:

$$\begin{aligned}
&\nabla^2\vec{F} + k^2\vec{F} = -\vec{M} && \text{(a)}\\
&\vec{H} = -j\omega\varepsilon\,\vec{F} + \frac{1}{j\omega\mu}\nabla\left(\nabla.\vec{F}\right) && \text{(b)}\\
&\vec{E} = -\nabla\times\vec{F} && \text{(c).}
\end{aligned} \tag{7.1.22}$$

Equations (7.1.12) and (7.1.22) are key expressions that describe the EM-fields generated by the electric current and magnetic current. If both the electric and magnetic currents are present, the total EM-field, in a linear medium, is the superposition of the fields produced by the individual sources. The magnetic vector potential $(\vec{A})$ is obtained from $\nabla.\vec{H} = 0$ and the electric vector potential $(\vec{F})$ from the expression $\nabla.\vec{E} = 0$. However, some authors [B.1] define $\vec{A}$ and $\vec{F}$ from $\nabla.\vec{B} = 0$ and $\nabla.\vec{D} = 0$ giving

$$\vec{B} = \nabla\times\vec{A} \quad \text{(a)}, \qquad D = -\nabla\times\vec{F} \quad \text{(b)}. \tag{7.1.23}$$

In this case, the magnetic and electric field intensities are

$$\vec{H} = \frac{1}{\mu}\nabla\times\vec{A} \quad \text{(a)}, \qquad \vec{E} = -\frac{1}{\varepsilon}\nabla\times\vec{F} \quad \text{(b)}. \tag{7.1.24}$$

Using the above definitions of $\vec{H}$ and $\vec{E}$, equations (7.1.12) and (7.1.22) can be modified by replacing $\vec{A} \to \frac{\vec{A}}{\mu}$ and $\vec{F} \to \frac{\vec{F}}{\varepsilon}$ to get the following expressions:

$$\begin{aligned}
&\nabla^2\vec{A} + k^2\vec{A} = -\mu\,\vec{J} && \text{(a)}\\
&\vec{E} = -j\omega\,\vec{A} + \frac{1}{j\omega\varepsilon\mu}\nabla\left(\nabla.\vec{A}\right) && \text{(b)}\\
&\vec{H} = \frac{1}{\mu}\nabla\times\vec{A} && \text{(c).}
\end{aligned} \tag{7.1.25}$$

$$\begin{aligned}
&\nabla^2\vec{F} + k^2\vec{F} = -\varepsilon\,\vec{M} && \text{(a)}\\
&\vec{H} = -j\omega\,\vec{F} + \frac{1}{j\omega\mu\varepsilon}\nabla\left(\nabla.\vec{F}\right) && \text{(b)}\\
&\vec{E} = -\frac{1}{\varepsilon}\nabla\times\vec{F} && \text{(c).}
\end{aligned} \tag{7.1.26}$$

However, in this text, equations (7.1.12) and (7.1.22) are preferred.

7.1.4 Generation of EM-Field by Electric and Magnetic Vector Potentials

The previous section has shown that the magnetic and electric vector potentials generate electromagnetic fields. It is shown in this section that *only one component of the vector potentials, i.e. either the scalar electric or scalar magnetic potential, is sufficient to generate the complete EM-field*. However, the nature of the EM-mode depends on the type of the scalar potential. The second term of equations (7.1.12b) and (7.1.22b) can be neglected, as compared to their first term. In that case, $\vec{A} \propto \vec{E} \propto \vec{J}$ and $\vec{F} \propto \vec{H} \propto \vec{M}$. It shows that the longitudinal component of electric current density J_z is parallel to E_z and A_z. Likewise, the longitudinal component of magnetic current density M_z is parallel to H_z and F_z. Figure (7.1a and b) show the field components of the TM^z and TE^z mode waves, in the Cartesian coordinate.

TM^z Mode

Figure (7.1a) shows that the z-axis is in the direction of propagation of the EM-wave, and the TM^z mode is defined with respect to the z-axis. Any of three longitudinal scalar quantities – J_z, E_z, and A_z, could be a source of the TM^z mode with H_x, H_y, $H_z = 0$, and $\vec{F} = 0$ field. The H_x and H_y field components, forming a loop of the magnetic field, are in the transverse (*x–y*)-plane. The magnetic field loop is generated due to the axial conduction current or axial electric displacement current. The transverse magnetic field follows the right-hand rule. In this case, $H_z = 0$, $E_z \neq 0$, and $\vec{F} = 0$. The A_z

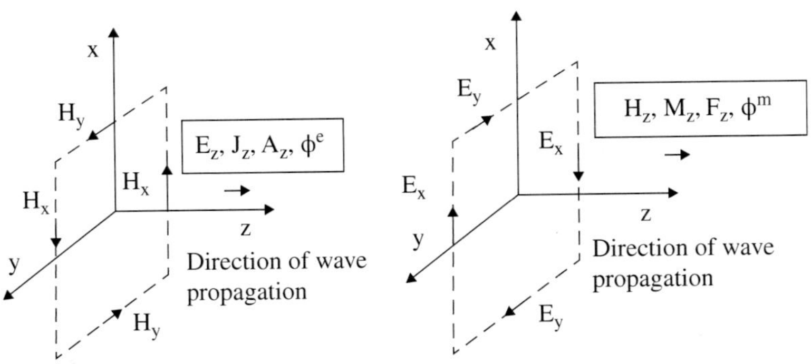

(a) TM^Z - mode generation. (b) TE^Z - mode generation.

Figure 7.1 Generation of TM^z and TE^z modes by the longitudinal field or current, or potential components.

component of the magnetic vector potential $\vec{A}$ could be written as the *z*-directed *scalar electric potential* ϕ^e generating the TM^z mode:

$$\vec{A} = \hat{k}\,A_z = \hat{k}\,\phi^e(x,y,z) = \hat{k}\,\phi^e(x,y)\,e^{-j\beta_z z}. \quad (7.1.27)$$

The TM^z mode wave propagates in the z-direction with propagation constant β_z. It is noted from equations (7.1.9) and (7.1.27) that the scalar electric potential ϕ^e is related to the scalar electric function ψ^e by $\phi^e = (1/\eta)\,\psi^e$, where η is the intrinsic impedance of the medium. Either ϕ^e or ψ^e could be used to obtain the field components of the TM^z mode. However, following Harrington, the function ϕ^e is used in this chapter [B.1].

The field components of the TM^z mode are obtained below in terms of the electric scalar potential $\phi^e(x, y, z)$ from equation (7.1.12b),

$$\hat{i}E_x^e + \hat{j}E_y^e + \hat{k}E_z^e = -j\omega\mu\left(\hat{k}\phi^e\right) + \frac{1}{j\omega\varepsilon}\nabla\left(\frac{\partial\phi^e}{\partial z}\right)$$

$$\Rightarrow \hat{i}E_x^e + \hat{j}E_y^e + \hat{k}E_z^e = -j\omega\mu\left(\hat{k}\phi^e\right) + \frac{1}{j\omega\varepsilon}\left(\hat{i}\frac{\partial^2\phi^e}{\partial x\partial z} + \hat{j}\frac{\partial^2\phi^e}{\partial y\partial z} + \hat{k}\frac{\partial^2\phi^e}{\partial z^2}\right).$$

The superscript "e" indicates that the field components of TM^z mode are generated by the scalar electric potential ϕ^e. On separating each component, the expressions for E_x^e, E_y^e, and E_z^e are obtained. On using equation (7.1.27) with equation (7.1.12c), expressions for H_x^e and H_y^e are also obtained. The field components are summarized below:

$$E_x^e = \frac{1}{j\omega\varepsilon}\frac{\partial^2\phi^e}{\partial x\partial z}, \qquad E_y^e = \frac{1}{j\omega\varepsilon}\frac{\partial^2\phi^e}{\partial y\partial z}$$

$$E_z^e = \frac{1}{j\omega\varepsilon}\left(\frac{\partial^2\phi^e}{\partial z^2} + k^2\phi^e\right)$$

$$H_x^e = \frac{\partial\phi^e}{\partial y}, \qquad H_y^e = -\frac{\partial\phi^e}{\partial x}, \qquad H_z^e = 0, \quad (7.1.28)$$

where $k = \omega\sqrt{\mu\varepsilon}$. The scalar electric potential $\phi^e(x, y, z)$ of equation (7.1.27) satisfies the wave equation (7.1.12a):

$$\nabla^2\phi^e(x,y,z) + k^2\phi^e(x,y,z) = -\vec{J}_z \quad (a)$$

$$\nabla^2\phi^e(x,y,z) + k^2\phi^e(x,y,z) = 0, \quad \vec{J}_z = 0 \quad (b). \quad (7.1.29)$$

Scalar wave equation (7.1.29) for the function $\phi^e(x, y, z)$ could be solved under the boundary conditions.

Equation (7.1.29b) is used for the source-free medium. For the TM^z mode wave propagating in the z-direction, the scalar electric potential $\phi^e(x, y, z)$ is written as

$$\phi^e(x,y,z) = \phi^e(x,y)\,e^{-j\beta_z z}, \quad (7.1.30)$$

where β_z is the propagation constant of the TM^z mode in the *z*-direction. On multiplying expression (7.1.30) by $e^{j\omega t}$, the wave equation is obtained. Using expression (7.1.30), with equation (7.1.28), the field components of the TM^z mode are written as follows:

$$E_x^e = -\frac{\beta_z}{\omega\varepsilon}\frac{\partial\phi^e}{\partial x}, \qquad E_y^e = -\frac{\beta_z}{\omega\varepsilon}\frac{\partial\phi^e}{\partial y}$$

$$E_z^e = \frac{1}{j\omega\varepsilon}\left(k^2 - \beta_z^2\right)\phi^e$$

$$H_x^e = \frac{\partial\phi^e}{\partial y}, \qquad H_y^e = -\frac{\partial\phi^e}{\partial x}, \qquad H_z^e = 0. \quad (7.1.31)$$

For a source-free medium, wave equation (7.1.29b) is reduced to

$$\nabla_t^2\phi^e(x,y) + \left(k^2 - \beta_z^2\right)\phi^e(x,y) = 0. \quad (7.1.32)$$

TE^z Mode

Figure (7.1b) shows that any of the three longitudinal scalar quantities – M_z, H_z, and F_z could be a source of the TE^z mode, with E_x, E_y components, forming a loop of the electric field in the transverse (x–y)-plane. The electric field loop is due to the axial magnetic current or axial magnetic displacement current. The transverse electric field follows the left-hand rule. In this case, the magnetic vector potential is zero, $\vec{A} = 0$, and also $E_z = 0$. The z-directed *scalar magnetic potential* ϕ^m, generating the TE^z mode, is

$$\vec{F} = \hat{k}\, F_z = \hat{k}\, \phi^m(x, y, z) = \hat{k}\, \phi^m(x, y)\, e^{-j\,\beta_z z}. \qquad (7.1.33)$$

The scalar magnetic potential ϕ^m is related to the similar scalar magnetic function ψ^m of equation (7.1.19) by $\phi^m = \eta\, \psi^m$.

The following field components of the TE^z mode are obtained in terms of the magnetic scalar potential by substituting equation (7.1.33) in equation (7.1.22):

$$\begin{aligned} E_x^m &= -\frac{\partial \phi^m}{\partial y}, \quad E_y^m = \frac{\partial \phi^m}{\partial x}, \quad E_z^m = 0 \\ H_x^m &= \frac{1}{j\,\omega\mu}\frac{\partial^2 \phi^m}{\partial x \partial z}, \quad H_y^m = \frac{1}{j\,\omega\mu}\frac{\partial^2 \phi^m}{\partial y \partial z} \\ H_z^m &= \frac{1}{j\,\omega\mu}\left(\frac{\partial^2 \phi^m}{\partial z^2} + k^2 \phi^m\right). \end{aligned} \qquad (7.1.34)$$

Using equation (7.1.33), the above equations could be rewritten as,

$$\begin{aligned} E_x^m &= -\frac{\partial \phi^m}{\partial y}, \quad E_y^m = \frac{\partial \phi^m}{\partial x}, \quad E_z^m = 0 \\ H_x^m &= -\frac{\beta_z}{\omega\mu}\frac{\partial \phi^m}{\partial x}, \quad H_y^m = -\frac{\beta_z}{\omega\mu}\frac{\partial \phi^m}{\partial y} \\ H_z^m &= \frac{1}{j\,\omega\mu}\left(k^2 - \beta_z^2\right)\phi^m, \end{aligned} \qquad (7.1.35)$$

where $\phi^m(x, y, z) = \phi^m(x, y)\, e^{-j\,\beta_z z}$ is a solution of following inhomogeneous or homogeneous wave equation, with and without source, under the boundary conditions:

$$\begin{aligned} &\nabla^2 \phi^m(x, y, z) + k^2 \phi^m(x, y, z) = -\vec{M}_z \qquad &(a) \\ &\nabla^2 \phi^m(x, y, z) + k^2 \phi^m(x, y, z) = 0 \qquad &(b) \\ &\nabla_t^2 \phi^m(x, y) + \left(k^2 - \beta_z^2\right) \phi^m(x, y) = 0 \qquad &(c). \end{aligned} \qquad (7.1.36)$$

Hybrid Mode

Figure (7.2) shows a layered dielectric medium that supports the hybrid mode, propagating in the *z*-direction. In the case of a hybrid mode both E_z and H_z longitudinal components, in the direction of EM-wave propagation, are present. So the hybrid modes are expressed as the HE mode for the stronger H_z component, and as the EH mode for stronger E_z component. Figure (7.2) shows that these field components are generated by *y*-directed F_y and A_y components corresponding to the scalar potentials $\phi^m(x, y, z)$ and $\phi^e(x, y, z)$, respectively. These potentials generate the TE^y and TM^y modes with E_z and H_z longitudinal field components along the direction propagation. The TE^y and TM^y modes are defined with respect to the y-axis that is normal to the interface of multilayered dielectrics. *The hybrid mode propagating in the z-direction is a linear combination of the TE^y and TM^y modes.*

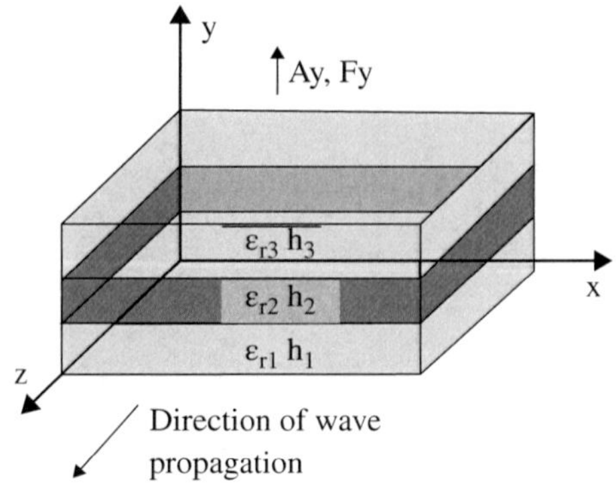

Figure 7.2 Hybrid mode fields in the layered medium.

The TE^y mode is further designated as the *longitudinal section electric (LSE^y) mode* with respect to the y-axis. The longitudinal (z–x)-plane contains the electric field components E_x and E_z. The longitudinal section is normal to the y-axis. Likewise, the TM^y is also called the *longitudinal section magnetic (LSM^y) mode.* In this case, the longitudinal (z–x)-plane contains the magnetic field components H_x and H_z [B.5].

In summary, the hybrid mode, with $\vec{E}_z$ and $\vec{H}_z$ field components, propagating in the *z*-direction is generated by the scalar potentials $\phi^m(x, y, z)$, and $\phi^e(x, y, z)$:

$$\begin{aligned} \vec{F}_y &= \hat{y}\, \phi^m(x, y, z) \qquad &(a) \\ \vec{A}_y &= \hat{y}\, \phi^e(x, y, z) \qquad &(b). \end{aligned} \qquad (7.1.37)$$

The field components of the TE^y and TM^y modes are obtained on substituting equations (7.1.37a and b) in equations (7.1.22) and (7.1.12), respectively. These can be equally obtained from equations (7.1.34) and (7.1.28) by replacing x, y, z in the cyclic order, such as $x \to z,\ y \to x,\ z \to y$.

For TE^Y Mode [LSE^Y Mode]

$$\begin{aligned} E_x^m &= \frac{\partial \phi^m}{\partial z}, \quad E_y^m = 0, \quad E_z^m = -\frac{\partial \phi^m}{\partial x} \\ H_x^m &= \frac{1}{j\,\omega\mu}\frac{\partial^2 \phi^m}{\partial x \partial y}, \quad H_y^m = \frac{1}{j\,\omega\mu}\left(\frac{\partial^2 \phi^m}{\partial y^2} + k^2 \phi^m\right) \\ H_z^m &= \frac{1}{j\,\omega\mu}\frac{\partial^2 \phi^m}{\partial y \partial z}. \end{aligned} \qquad (7.1.38)$$

For TM^y Mode [LSM^y Mode]

$$E_x^e = \frac{1}{j\,\omega\varepsilon}\frac{\partial^2\phi^e}{\partial x\partial y},\quad E_y^e = \frac{1}{j\,\omega\varepsilon}\left(\frac{\partial^2\phi^e}{\partial y^2} + k^2\phi^e\right)$$
$$E_z^e = \frac{1}{j\,\omega\varepsilon}\frac{\partial^2\phi^e}{\partial y\partial z}$$
$$H_x^e = -\frac{\partial\phi^e}{\partial z},\quad H_y^e = 0,\quad H_z^e = \frac{\partial\phi^e}{\partial x}. \tag{7.1.39}$$

The field components of the hybrid mode, with respect to the z-direction, are a linear combination of the fields of the TE^y and TM^y modes:

$$E_x = E_x^e + E_x^m,\quad E_z = E_z^e + E_z^m$$
$$H_x = H_x^e + H_x^m,\quad H_z = H_z^e + H_z^m \tag{7.1.40}$$
$$E_y = E_y^e,\quad H_y = H_y^m.$$

The scalar potentials ϕ^e and ϕ^m are obtained as the solutions of following wave equations under the boundary conditions:

$$\nabla^2\phi^e + k^2\phi^e = 0 \quad \text{(a)}$$
$$\nabla^2\phi^m + k^2\phi^m = 0 \quad \text{(b)}. \tag{7.1.41}$$

7.2 Boundary Surface and Boundary Conditions

The guided electromagnetic fields are confined in two kinds of waveguide structures: the *closed waveguide* and the *open waveguide*. The closed waveguide structures are made of the boundary surfaces – either a *perfect electric conductor* (PEC) type or a *perfect magnetic conductor* (PMC) type. The waveguide structure could also be formed by a combination of the PEC and PMC surfaces. The metallic waveguides structures are treated as the PEC type closed waveguides, whereas the dielectric waveguides are open waveguides and not really the PMC-type waveguides. An open waveguide supports the surface wave propagation and the fields are confined to the surface due to the exponentially decaying fields in the transverse plane of air region. Only in the case of a very high permittivity material, it could be approximated as the PMC type waveguide, ignoring the fringing field. However, the PMC-type waveguide could serve as a modeling tool. Likewise, the combined PEC and PMC structure could be another modeling tool.

In this section, the behavior of the electric and magnetic fields are summarized at these surfaces and also at the interface of two different media. This topic is covered in many well-known books [B.1–B.10]. The behaviors of the EM-field at these surfaces are known as the *surface boundary conditions, or simply as the boundary conditions*. The solutions of Laplace's equation and the wave equation inside the waveguide must meet the boundary conditions at the boundary surfaces. The lateral confinement of the EM-field in the open waveguide takes place due to the *evanescent mode* of the EM-field i.e. due to the exponentially decaying EM-field in the transverse plane of the waveguide. The waveguide medium could also consist of several layers of the dielectrics. In that case, Laplace's equation and wave equation must be solved separately in each layer of the medium. The boundary and continuity conditions at the interface of two layers must be satisfied.

7.2.1 Perfect Electric Conductor (PEC)

The perfect electric conductor surface has infinite conductivity ($\sigma \to \infty$). It is called a sheet of zero impedance or an *electric wall* (EW). The tangential component of an electric field, or the normal component of a magnetic field, is zero at the PEC or EW:

$$E_t = 0,\quad \text{i.e.}\quad \hat{n}\times\vec{E} = 0 \quad \text{(a)}$$
$$H_n = 0,\quad \text{i.e.}\quad \hat{n}.\vec{H} = 0 \quad \text{(b)}, \tag{7.2.1}$$

where $\hat{n}$ is a normal unit vector on the PEC, shown in Fig (7.3a). If there is an accumulation of the surface electric charge, or if the surface current exists on the surface of a conductor, the following boundary conditions are obtained:

$$\rho_{e,s} = \hat{n}.\vec{D} \quad \text{(a)},\qquad \vec{J}_s = \hat{n}\times\vec{H} \quad \text{(b)}. \tag{7.2.2}$$

The subsection (7.1.4) presents the EM-field propagating in the *z*-direction in the form of the TM^z and TE^z modes. These are generated by the z-directed scalar electric potential ϕ^e and scalar magnetic potential ϕ^m, respectively. For both these cases, the boundary conditions given in equation (7.2.1) are rewritten in terms of the potential functions discussed below.

Scalar Electric Potential

Using the boundary conditions in equation (7.2.1) with equation (7.1.31), the following boundary conditions are obtained for the TM^z mode inside a rectangular waveguide with the PEC, i.e. EW, shown in Fig (7.5a):

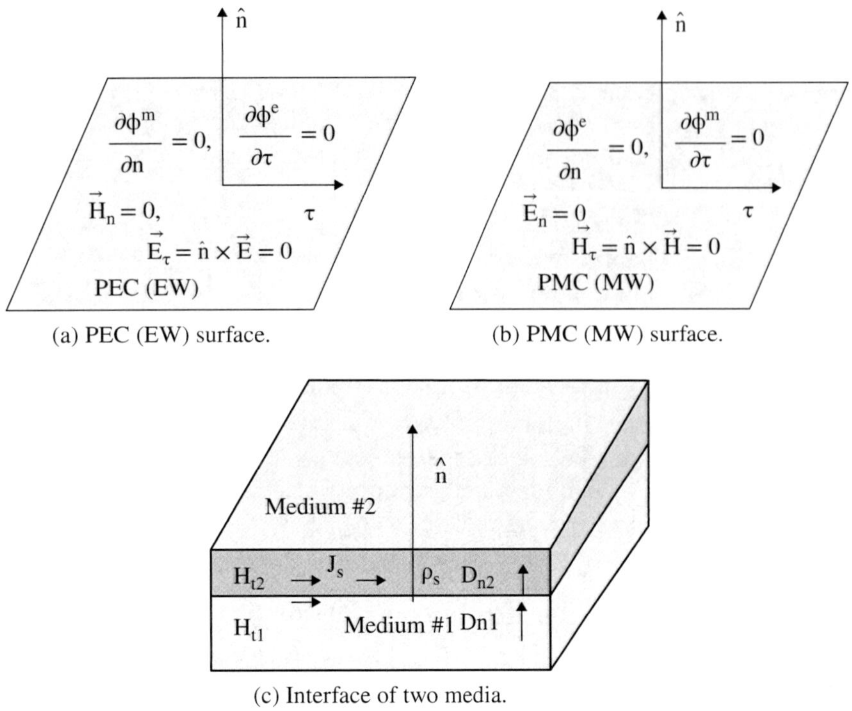

Figure 7.3 The EM-field at (a) PEC surface (b) PMC surface (c) interface of two media.

At the narrow – wall PEC surfaces: $x = 0,\ a$

$$E_z^e = 0, \quad \phi^e = 0; \quad E_y^e = H_x^e = 0, \quad \text{i.e.} \quad \frac{\partial\phi^e}{\partial y} = 0 \quad \text{(a)}$$

At the broad face PEC surfaces : $y = 0,\ b$

$$E_z^e = 0, \quad \phi^e = 0; \quad E_x^e = H_y^e = 0, \quad \text{i.e.} \quad \frac{\partial\phi^e}{\partial x} = 0 \quad \text{(b)}. \tag{7.2.3}$$

Therefore, on the PEC surface, i.e. at the EW, the derivative of the scalar electric potential ϕ^e with respect to the tangent is zero:

$$\frac{\partial\phi^e}{\partial\tau} = 0, \tag{7.2.4}$$

where τ is a tangent to the PEC surface. The tangent τ could be either the *x* or *y* coordinate in the Cartesian coordinate system. It is shown in Fig (7.3a).

Scalar Magnetic Potential

The TE^z mode in the rectangular waveguide, shown in Fig (7.5a), is generated by the scalar magnetic potential ϕ^m. On using the boundary condition (7.2.1) with equation (7.1.35), following boundary conditions are obtained:

At the narrow-wall PEC surface, $x = 0, a$:

$$E_y^m = 0, \quad \text{i.e.} \quad \frac{\partial\phi^m}{\partial x} = 0 \qquad \text{(a)}$$

At the broad-face PEC surface, $y = 0, b$:

$$E_x^m = 0, \quad \text{i.e.} \quad \frac{\partial\phi^m}{\partial y} = 0 \qquad \text{(b)}. \tag{7.2.5}$$

Therefore, on the PEC surface, i.e. at the EW, the derivative of the scalar magnetic potential ϕ^m with respect to the normal n is zero, i.e.

$$\frac{\partial\phi^m}{\partial n} = 0. \tag{7.2.6}$$

Figure (7.3a) shows that the unit vector $\hat{n}$ is normal to the PEC. Figure (7.5a) illustrates the PEC surfaces of a rectangular waveguide with $\hat{x}$ and $\hat{y}$ unit vectors.

7.2.2 Perfect Magnetic Conductor (PMC)

It is a complementary surface that supports the tangential electric field component; however, it does not support any tangential component of the magnetic field. It is like a perfect insulating surface that does not support the electric surface current. The PMC is also known as *the magnetic wall* (MW). The EM-field on the PMC surface follows the following boundary conditions:

$$\vec{H}_t = 0, \quad \text{i.e.} \quad \hat{n} \times \vec{H} = 0 \quad (a)$$
$$E_n = 0, \quad \text{i.e.} \quad \hat{n}.\vec{D} = 0 \quad (b). \qquad (7.2.7)$$

Figure (7.3b) shows the above boundary conditions. However, a PMC can support the *fictitious magnetic surface charge* density $\rho_{m,\,s}$ and the *fictitious magnetic surface current* M_s. These are related to the field quantities as follows:

$$\rho_{m,s} = \hat{n}.\vec{B} \quad (a), \qquad \vec{M}_s = \vec{E} \times \hat{n} \quad (b). \qquad (7.2.8)$$

Again the boundary condition (7.2.7) could be rewritten in terms of the scalar potentials ϕ^e and ϕ^m. Sometimes, we use the symbol $\vec{J}_m$ in place of $\vec{M}_s$.

Scalar Electric Potential

The boundary condition (7.2.7), along with equation (7.1.31), is applied to the rectangular waveguide, shown in Fig (7.5b). The waveguide is constructed with the MWs, i.e. with the PMC surfaces. For a waveguide supporting the TM^z mode, the following boundary conditions are satisfied:

At the narrow-wall of PMC surfaces, $x = 0, a$:

$$H_y^e = 0, \quad \text{i.e.} \quad \frac{\partial \phi^e}{\partial x} = 0 \quad (a)$$

At the broad-face of PMC surfaces, $y = 0, b$:

$$H_x^e = 0, \quad \text{i.e.} \quad \frac{\partial \phi^e}{\partial y} = 0 \quad (b). \qquad (7.2.9)$$

At the PMC surface, the derivative of the scalar electric potential with respect to normal to the surface is zero,

$$\frac{\partial \phi^e}{\partial n} = 0, \qquad (7.2.10)$$

where n is normal to the PMC surface. The symbol n signifies either an x or y component of the rectangular coordinate. Figure (7.3b) shows the boundary condition.

Scalar Magnetic Potential

The PMC-based waveguide, shown in Fig (7.5b), supports TE^z mode. The following boundary condition is obtained for the scalar magnetic potential ϕ^m, on using an equation (7.1.35) along with equation (7.2.7):

At the narrow-wall of PMC surfaces, $x = 0, a$:

$$H_y^m = 0, \quad \text{i.e.} \quad \frac{\partial \phi^m}{\partial y} = 0; \quad H_z = 0, \quad \text{i.e.}\ \phi^m = 0 \quad (a)$$

At the broad-face of PMC surfaces, $y = 0, b$:

$$H_x^m = 0, \quad \text{i.e.} \quad \frac{\partial \phi^m}{\partial x} = 0; \quad H_z = 0, \quad \text{i.e.}\ \phi^m = 0 \quad (b). \qquad (7.2.11)$$

Therefore, at the PMC, surface derivative of the scalar magnetic potential with respect to the tangent to the surface is zero,

$$\frac{\partial \phi^m}{\partial \tau} = 0, \qquad (7.2.12)$$

where, τ is a tangent to the PMC surface. The symbol τ is either an x or y coordinate. Figure (7.3b) shows the above boundary condition.

7.2.3 Interface of Two Media

The boundary conditions of the EM-field at the interface of two media, shown in Fig (7.3c), are expressed as follows:

At the interface containing no charge/magnetic current:

$$\left(\vec{E}_2 - \vec{E}_1\right) \times \hat{n} = 0 \quad (a)$$

At the interface supporting surface current $\vec{J}_s$:

$$\hat{n} \times \left(\vec{H}_2 - \vec{H}_1\right) = \vec{J}_s \quad (b)$$

At the interface supporting surface charge:

$$\hat{n} \bullet \left(\vec{D}_2 - \vec{D}_1\right) = \rho_s \quad (c)$$

At the interface containing no magnetic charge:

$$\hat{n} \bullet \left(\vec{B}_2 - \vec{B}_1\right) = 0 \quad (d), \qquad (7.2.13)$$

where $\hat{n}$ is the unit vector normal at the interface, directed from the medium #1 to medium #2. Figure (7.3c) illustrates the presence of surface current density $\vec{J}_s$ and surface charge density ρ_s causing discontinuities in the tangential component of the magnetic field and also in the normal component of the electric flux density. The interface may contain the magnetic surface current density $\vec{M}_s$ and magnetic charge density ρ_m providing the following additional boundary conditions:

$$\left(\vec{E}_2 - \vec{E}_1\right) \times \hat{n} = \vec{M}_s \quad (e)$$
$$\hat{n} \bullet \left(\vec{B}_2 - \vec{B}_1\right) = \rho_m \quad (f). \qquad (7.2.14)$$

7.3 TEM-Mode Parallel-Plate Waveguide

The finite width parallel-plate waveguide supporting the TEM mode is shown in Fig (7.4). The upper plate of the waveguide is at potential V and the lower plate is at the ground potential. It can be modeled as a two-wire transmission line, supporting the voltage and current waves. The voltage and current for the TEM mode supporting structure could be uniquely defined. However, the next section shows that a parallel-plate waveguide also supports the higher order TE and TM modes.

7.3.1 TEM Field in Parallel-Plate Waveguide

The TEM mode has no longitudinal field components, i.e. $E_z = H_z = 0$. Thus, from equation (7.1.28), the following equation is obtained:

$$\frac{\partial^2 \phi^e}{\partial z^2} + k^2 \phi^e = 0, \quad k = \omega\sqrt{\mu\varepsilon}. \tag{7.3.1}$$

In general, the *scalar electric potential* ϕ^e is a function of x, y, z coordinates. However, the potential function ϕ^e could be separated in the transverse coordinates and longitudinal coordinate-dependent potential functions, i.e. $\phi^e(x, y, z) = \phi^e(x, y)\,\phi^e(z)$. The solution of the above equation could be written as

$$\phi^e(z) = \phi^e(z=0)\, e^{-j\beta_z z} \qquad \text{(a)}$$

$$\text{where propagation constant,} \quad \beta_z = k = \omega\sqrt{\mu\varepsilon} \qquad \text{(b).}$$

$$(7.3.2)$$

Using equation (7.3.1), the wave equation (7.1.29b) for the potential function ϕ^e is reduced to Laplace's equation:

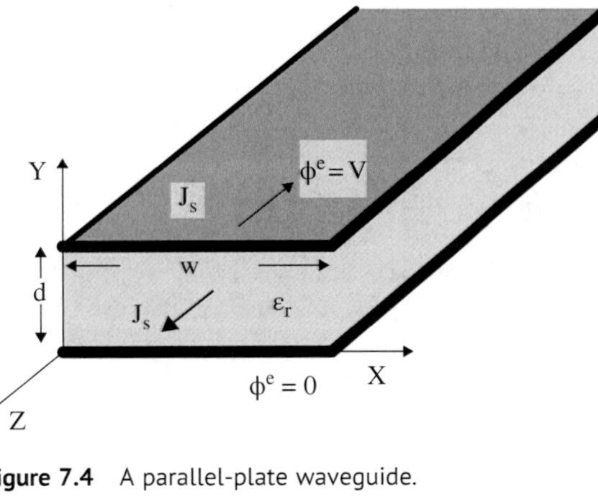

Figure 7.4 A parallel-plate waveguide.

$$\nabla_t^2 \phi^e(x, y) + \frac{\partial^2 \phi^e}{\partial z^2} + k^2 \phi^e = 0$$

$$\Rightarrow \quad \nabla_t^2 \phi^e(x, y) = 0. \tag{7.3.3}$$

Therefore, the TEM mode in the transverse plane is described by Laplace's equation for the potential function $\phi^e(x, y)$. In the case of the uniform field along the width of a waveguide, there is no variation in the potential function with x i.e. $\frac{\partial \phi^e}{\partial x} = 0$. The above Laplace's equation is reduced to $\frac{\partial^2 \phi^e}{\partial y^2} = 0$ and its solution is

$$\frac{\partial \phi^e}{\partial y} = A, \quad \Rightarrow \phi^e = Ay + B.$$

At $y = 0$, potential is zero, i.e. $B = 0$: $\phi^e = Ay$.

At $y = d$, the potential is V, i.e. $A = \frac{V}{d}$: $\phi^e = \frac{V}{d}y$. The potential function in the parallel-plate waveguide is

$$\phi^e(x, y, z) = \frac{V}{d} y\, e^{-j\beta_z z}. \tag{7.3.4}$$

The field components of the waveguide are obtained using equations (7.1.28) and (7.3.4),

$$E_x^e = 0, \quad E_y^e = -\frac{\beta_z V\, e^{-j\beta_z z}}{\omega\,\varepsilon\, d}, \quad E_z^e = 0$$

$$H_x^e = \frac{V\, e^{-j\beta_z z}}{d}, \quad H_y^e = 0, \quad H_z^e = 0. \tag{7.3.5}$$

The *wave impedance* of the TEM mode traveling in the z-direction is given by

$$Z_w^{TEM} = -\frac{E_y^e}{H_x^e} = \frac{\beta_z}{\omega\,\varepsilon} = \frac{\omega\sqrt{\mu\varepsilon}}{\omega\,\varepsilon} = \sqrt{\frac{\mu}{\varepsilon}} = \eta, \tag{7.3.6}$$

where, η is the intrinsic impedance of a homogeneous medium having permittivity ε and permeability μ.

7.3.2 Circuit Relations

The lineal surface current densities on the upper and lower conductor planes of Fig (7.4) are obtained from the magnetic field relation on the conductor, $J_s = \hat{n} \times \vec{H}$ amp/m, where $\hat{n}$ is the direction of normal to the conductor. At the lower conductor plane ($y = 0$) of the waveguide: $J_{s,z} = \hat{y} \times (\hat{x}\, H_x^e) = -\hat{z}\, H_x^e$. Using H_x^e from equation (7.3.5), the z-directed surface current densities on the lower and upper conductor planes are

$$\text{At lower plate:} \quad J_{s,z} = -\frac{V\, e^{-j\beta_z z}}{d} \qquad \text{(a)}$$

$$\text{At upper plate:} \quad J_{s,z} = \frac{V\, e^{-j\beta_z z}}{d} \qquad \text{(b).} \tag{7.3.7}$$

The current flowing over the width dx of the conductor is $dI = J_{s,z}\,dx$. Therefore, the total current flowing over the conductor width w is

$$I(z) = -\int_0^w H_x^e(z)\,dx \Rightarrow I(z) = -H_x^e(z)\,w. \quad (7.3.8)$$

The current density $J_{s,z}$ is uniform over the width of the conductor. The voltage between the lower and upper plates are obtained from the line integral,

$$V(z) = -\int_0^d E_y^e(z)\,dy \Rightarrow V(z) = -E_y^e(z)\,d. \quad (7.3.9)$$

For the TEM mode supporting parallel-plate waveguide, the line integral is independent of the path of integration. Therefore, equations (7.3.8) and (7.3.9) define uniquely the line current and line voltage, respectively, in terms of the field quantities. It also uniquely defines the characteristic impedance of a TEM mode supporting line. Using these relations, the Kelvin–Heaviside transmission line equations could be obtained from Maxwell's equations.

7.3.3 Kelvin–Heaviside Transmission Line Equations from Maxwell Equations

The source-free Maxwell equations are written from equation (7.1.1):

$$\nabla \times \vec{E} = -j\omega\mu\,\vec{H} \quad (a)$$
$$\nabla \times \vec{H} = j\omega\varepsilon\,\vec{E} \quad (b). \quad (7.3.10)$$

The TEM mode parallel-plate waveguide has only E_y and H_x field components. Thus, the above Maxwell's equations are reduced to

$$\frac{\partial E_y(z)}{\partial z} = j\omega\mu H_x, \quad \frac{\partial H_x(z)}{\partial z} = j\omega\varepsilon E_y$$
$$\int_0^d \left[\frac{\partial E_y(z)}{\partial z}\right] dy = j\omega\mu \int_0^d H_x\,dy$$
$$\int_0^w \left[\frac{\partial H_x(z)}{\partial z}\right] dx = j\omega\varepsilon \int_0^w E_y\,dx$$
$$\frac{\partial E_y(z)\,d}{\partial z} = j\omega\mu\,d\,H_x \quad (a)$$
$$\frac{\partial H_x(z)\,w}{\partial z} = j\omega\varepsilon\,w\,E_y \quad (b). \quad (7.3.11)$$

On using equations (7.3.8) and (7.3.9) with above equations, the following Kelvin–Heaviside transmission line equations are obtained for a TEM mode supporting waveguide:

$$\frac{\partial V(z)}{\partial z} = j\omega L I(z) \quad (a)$$
$$\frac{\partial I(z)}{\partial z} = j\omega C V(z) \quad (b). \quad (7.3.12)$$

Therefore, the TEM mode supporting waveguide could be modeled by the transmission line having the line inductance and capacitance *p.u.l.* as

$$L = \frac{\mu d}{w} \quad (a), \qquad C = \frac{\varepsilon w}{d} \quad (b). \quad (7.3.13)$$

Therefore, *permeability and permittivity of a medium could be modeled as the series inductance p.u.l. and shunt capacitance p.u.l.of an equivalent transmission model* supporting the TEM mode propagation. The phase velocity and characteristics impedance of the equivalent transmission line corresponds to that of the TEM mode supporting medium:

$$v_p^{line} = \frac{1}{\sqrt{LC}} = \frac{1}{\sqrt{\mu\varepsilon}} = v_p^{medium} \quad (a)$$
$$Z_0^{line} = \sqrt{\frac{L}{C}} = \frac{d}{w}\sqrt{\frac{\mu}{\varepsilon}} = \frac{d}{w}\eta^{medium} \quad (b). \quad (7.3.14)$$

In the case of $w = d$, $L = \mu$, $C = \varepsilon$, $Z_0^{line} = \eta^{medium}$.

7.4 Rectangular Waveguides

The EM-wave propagation through the homogeneous rectangular waveguide acquires modal field structures described by the TE and TM modes. The rectangular waveguide could be constructed in three forms:

i) All four walls are PEC, i.e. EWs, as shown in Fig (7.5a)
ii) All four walls are PMC, i.e. MWs, as shown in Fig (7.5b)
iii) Two walls are PEC (EW) and two walls are the PMC (MW), as shown in Fig (7.5c)

The modal fields for the TM^z and TE^z modes are determined by equations (7.1.28) and (7.1.34), respectively.

7.4.1 Rectangular Waveguide with Four EWs

Figure (7.5a) shows a rectangular waveguide made of perfect electric conductors (PEC) i.e. the EWs. The EM-wave propagates in the z-direction. The waveguide supports independently both the TE^z and TM^z modes. The waveguide has width "a" along the x-axis and height "b" along the y-axis.

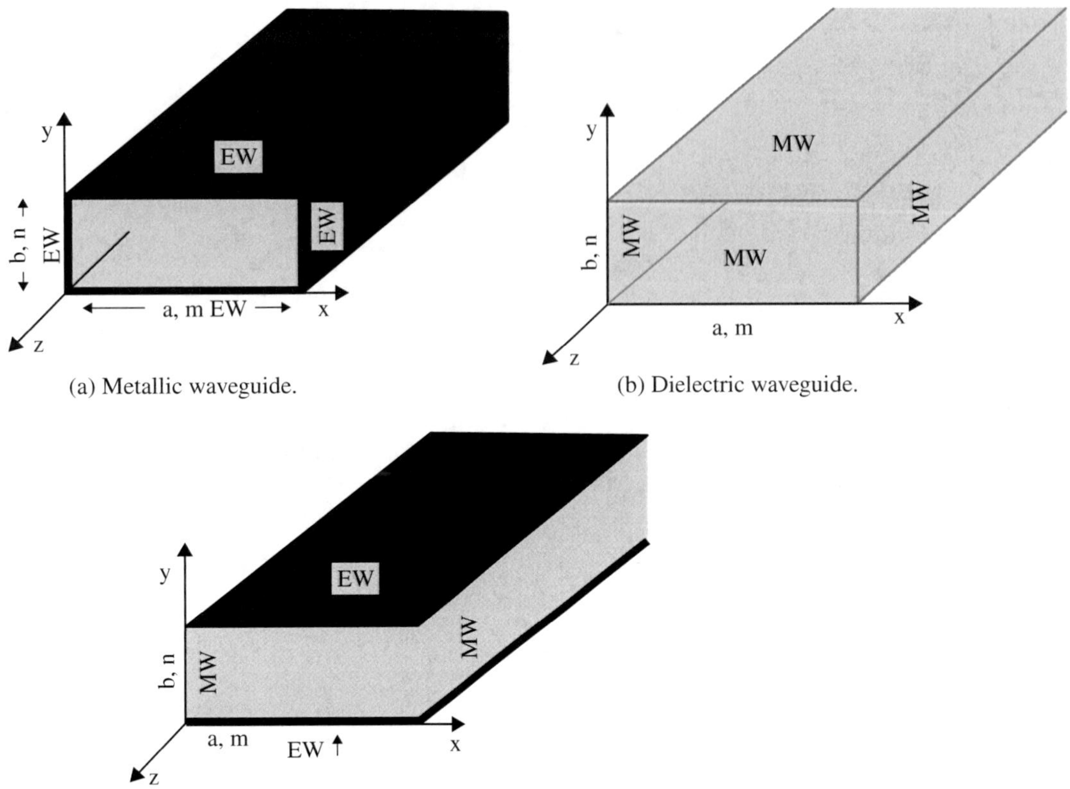

(a) Metallic waveguide.

(b) Dielectric waveguide.

(c) Parallel-plate waveguide.

Figure 7.5 Rectangular waveguide constructed using EW and MWs. "m" is the modal number along the width, and "n" is the modal number along the height of waveguide.

TE^z Modes

The TE^z modes have no electric field component in the z-direction. They have the electric field components only in the transverse (x–y)-plane. In general for the TE^z modes, $E_x \neq 0$, $E_y \neq 0$, $E_z = 0$, $H_x \neq 0$, $H_y \neq 0$, $H_z \neq 0$. Figure (7.1b) shows that the TE^z modes are generated by the scalar magnetic potential $\phi^m(x, y, z)$. For the TE^z modes propagating in the z-direction, it is given by

$$\phi^m(x,y,z) = \phi^m(x,y)e^{-j\beta_z z}. \tag{7.4.1}$$

The wave nature of the potential function $\phi^m(x, y, z)$ could be seen by multiplying the above expression by $\exp(j\omega t)$ and taking the real part of the wave function, i.e. Re $\{\phi^m(x, y)\exp[j(\omega t - \beta_z z)]\}$. For a source free waveguide, the wave equation of the TE^z mode is written on substituting the above expression in equation (7.1.36b):

$$\begin{aligned} &\nabla_t^2\phi^m(x,y) + \left(k^2 - \beta_z^2\right)\phi^m(x,y) = 0 && \text{(a)} \\ &\nabla_t^2\phi^m(x,y) + k_c^2\,\phi^m(x,y) = 0 && \text{(b)}. \end{aligned} \tag{7.4.2}$$

The magnetic scalar potential $\phi^m(x, y)$ in the waveguide is solved under the boundary conditions (7.2.6) by using the method of separation of variables.

Eigenvalue and Eigenfunction

The *cut-off wavenumber*, or the cut-off propagation constant, of the waveguide, is defined as

$$k_c^2 = \left(k^2 - \beta_z^2\right). \tag{7.4.3}$$

The wave equation corresponding to k_c is given in equation (7.4.2b). Once $k_c = 0$, the wave equation (7.4.2) reduces to Laplace's equation $\nabla_t^2\phi^m(x,y) = 0$ and the wave propagation does not exist in the waveguide. The wave propagation occurs in the waveguide for $k_c > 0$. The two-dimensional TE^z wave function $\phi^m(x, y)$ could be separated as two independent one-dimensional partial scalar magnetic potential functions $\phi_1^m(x)$ and $\phi_2^m(y)$:

$$\phi^m(x,y) = \phi_1^m(x)\,\phi_2^m(y). \tag{7.4.4}$$

On substituting equation (7.4.4) in the partial differential equation (7.4.2), two separated ordinary second-order differential equations are obtained:

$$\frac{1}{\phi_1^m(x)}\frac{d^2\phi_1^m(x)}{dx^2} + \frac{1}{\phi_2^m(y)}\frac{d^2\phi_2^m(y)}{dy^2} = -k_c^2. \qquad (7.4.5)$$

Let

$$\frac{1}{\phi_1^m(x)}\frac{d^2\phi_1^m(x)}{dx^2} = -\beta_x^2, \Rightarrow \frac{d^2\phi_1^m(x)}{dx^2} + \beta_x^2\phi_1^m(x) = 0 \quad (a)$$

$$\frac{1}{\phi_2^m(y)}\frac{d^2\phi_2^m(y)}{dy^2} = -\beta_y^2, \Rightarrow \frac{d^2\phi_2^m(y)}{dy^2} + \beta_y^2\phi_2^m(y) = 0 \quad (b). \qquad (7.4.6)$$

The x, y, z-directed propagation constants β_x, β_y, β_z are related to the wave number $k = \omega\sqrt{\mu\varepsilon}$ in a homogeneous medium of the waveguide by the following expression:

$$\beta_x^2 + \beta_y^2 = k_c^2 \quad (a)$$

$$\text{also, } \beta_x^2 + \beta_y^2 + \beta_z^2 = k^2,$$

$$\Rightarrow \beta_z = \pm\sqrt{k^2 - \left(\beta_x^2 + \beta_y^2\right)} \quad (b), \qquad (7.4.7)$$

where β_z is the propagation constant of TE^z propagating in the z-direction. For the *forward traveling* wave in the waveguide, the propagation constant β_z is a positive real quantity. The propagation constant β_z as a negative real quantity gives the *backward traveling* wave. Equation (7.4.1) shows the propagation constant β_z has to be a negative imaginary quantity $\beta_z = -j\alpha_z$ to get the wave decay by a factor $e^{-\alpha_z z}$ in the direction of propagation, i.e. in the z-direction. Such a nonpropagating wave is known as the *evanescent wave*. The general harmonic solutions of equations (7.4.6) are written as

$$\phi_1^m(x) = A_1\sin(\beta_x x) + B_1\cos(\beta_x x) \quad (a)$$

$$\phi_2^m(y) = A_2\sin\left(\beta_y y\right) + B_2\cos\left(\beta_y y\right) \quad (b). \qquad (7.4.8)$$

The constants A_1, A_2, B_1, B_2 are determined by applying the boundary condition (7.2.6) on four EWs. The tangential components of the electric field are zero at the PEC surface.

i) *The boundary condition at the narrow-wall surface*, $x = 0$, $(0 \le y \le b)$: To get $E_y^m = 0$ at the surface of the narrow-wall, the following condition is obtained using equation (7.1.34) or equation (7.2.6):

$$\phi_2^m(y)\frac{d\phi_1^m(x)}{dx} = 0.$$

$$\text{For } \phi_2^m(y) \ne 0, \quad \frac{d\phi_1^m(x)}{dx} = 0, \quad \text{at } x = 0. \qquad (7.4.9)$$

Thus on differentiating equation (7.4.8a) with respect to x and using an equation (7.4.9), the constant A_1 is evaluated as zero, i.e. $A_1 = 0$ and

$$\phi_1^m(x) = B_1\cos(\beta_x x). \qquad (7.4.10)$$

ii) *The boundary condition at narrow-wall surface*, $x = a$, $(0 \le y \le b)$:

$$\frac{d\phi_1^m(x)}{dx} = 0, \quad \text{at } x = a. \qquad (7.4.11)$$

On differentiating equation (7.4.10) with respect to x and using an equation (7.4.11)

$$\sin(\beta_x a) = 0 = \sin(m\pi)$$

$$\Rightarrow \beta_x = \frac{m\pi}{a}, \quad m = 0, 1, 2... \qquad (7.4.12)$$

iii) *The boundary condition at the broad-face surface*, $y = 0$, $(0 \le x \le a)$:

$$E_x^m = -\phi_1^m(x)\frac{d\phi_2^m(y)}{dy} = 0. \text{ For } \phi_1^m(y) \ne 0$$

$$\frac{d\phi_2^m(y)}{dy} = 0, \quad \text{at } y = 0. \qquad (7.4.13)$$

On differentiating equation (7.4.8b) with respect to *y* and using an equation (7.4.13), the constant A_2 is evaluated as zero, i.e. $A_2 = 0$ and

$$\phi_2^m(y) = B_2\cos\left(\beta_y y\right). \qquad (7.4.14)$$

iv) *The boundary condition at the broad-face surface*, $y = b$, $(0 \le x \le a)$:

$$\frac{d\phi_2^m(y)}{dy} = 0, \quad \text{at } y = b. \qquad (7.4.15)$$

On differentiating equation (7.4.14) with respect to *y* and using an equation (7.4.15):

$$\sin\left(\beta_y b\right) = 0 = \sin(n\pi)$$

$$\Rightarrow \beta_y = \frac{n\pi}{b}, \quad n = 0, 1, 2... \qquad (7.4.16)$$

Thus, the partial magnetic scalar potential functions $\phi_1^m(x)$ and $\phi_2^m(y)$ exist in the waveguide only for the discrete values of propagation constants β_x and β_y given by equations (7.4.12) and (7.4.16), respectively. These discrete values are known as *eigenvalues*. The transverse magnetic potential functions $\phi_{mn}^m(x, y)$ corresponding to the *eigenvalues* are obtained from equations (7.4.4), (7.4.10), and (7.4.14). The potential functions $\phi_{mn}^m(x, y)$ are known as the *eigenfunctions* of TE^z modes given below:

$$\phi^m_{mn}(x,y) = B_{mn} \cos\left(\frac{m\pi}{a}x\right) \cos\left(\frac{n\pi}{b}y\right), \quad (7.4.17)$$

where "m" and "n" in the subscript of $\phi^m_{mn}(x,y)$ are the *modal numbers,* indicating the half-wavelength variations of the EM-field along the width and height of the waveguide shown in Fig (7.5a). It is seen from equations (7.4.12) and (7.4.16), $a = m\lambda_g/2$, $b = n\lambda_g/2$. The modal number "m" shown in superscript of the function $\phi^m_{mn}(x,y)$ indicates the magnetic scalar potential generating the TE^z modes. $B_{mn} = B_1B_2$ is the modal constant for the mn^{th} TE^z modes. Both the modal numbers *m* and *n* cannot be zero simultaneously; otherwise, all transverse field components will reduce to zero. It is seen in equations (7.4.19). The presence of both *m* and *n* as the infinite sets of integers shows the presence of infinite numbers of *two-dimensional modes* and their field configurations.

TE^z Modal Fields

The modal scalar magnetic wave functions for the TE^z modes, traveling in the z-direction, and their propagation constants are written from equations (7.4.1) and (7.4.7b):

$$\phi^m_{mn}(x,y,z) = \phi^m_{mn}(x,y)e^{-j\beta_z z} \quad (a)$$
$$\beta_z = \left[k^2 - \left\{\left(\frac{m\pi}{a}\right)^2 + \left(\frac{n\pi}{b}\right)^2\right\}\right]^{1/2} \quad (b). \quad (7.4.18)$$

The five field components of the TE^z modes, i.e. TE^z_{mn} modes are obtained on substituting the modal function $\phi^m_{mn}(x,y,z)$ in equation (7.1.34):

$$E^m_x = \frac{n\pi}{b} B_{mn} \cos\left(\frac{m\pi}{a}x\right) \sin\left(\frac{n\pi}{b}y\right) e^{-j\beta_z z} \quad (a)$$
$$E^m_y = -\frac{m\pi}{a} B_{mn} \sin\left(\frac{m\pi}{a}x\right) \cos\left(\frac{n\pi}{b}y\right) e^{-j\beta_z z} \quad (b)$$
$$E^m_z = 0 \quad (c)$$
$$H^m_x = \frac{m\pi}{a}\frac{B_{mn}\beta_z}{\omega\mu} \sin\left(\frac{m\pi}{a}x\right) \cos\left(\frac{n\pi}{b}y\right) e^{-j\beta_z z} \quad (d)$$
$$H^m_y = \frac{n\pi}{b}\frac{B_{mn}\beta_z}{\omega\mu} \cos\left(\frac{m\pi}{a}x\right) \sin\left(\frac{n\pi}{b}y\right) e^{-j\beta_z z} \quad (e)$$
$$H^m_z = -j\frac{B_{mn}}{\omega\mu}(k^2-\beta_z^2) \cos\left(\frac{m\pi}{a}x\right) \cos\left(\frac{n\pi}{b}y\right) e^{-j\beta_z z} \quad (f). \quad (7.4.19)$$

The superscript "m" in E^m, H^m shows that the field components of the TE^z mode are generated by the scalar magnetic potential ϕ^m as shown in Fig (7.1b).

TM^z Modes

The waveguide of Fig (7.5a) also independently supports the TM^z modes, generated by the scalar electric potential $\phi^e(x,y,z) = \phi^e(x,y)e^{-j\beta_z z}$ that satisfies the wave equation (7.1.32). The wave equation (7.1.32) is rewritten below:

$$\nabla_t^2\phi^e(x,y) + k_c^2\,\phi^e(x,y) = 0. \quad (7.4.20)$$

The cut-off wave number k_c is given by equation (7.4.3). Equation (7.4.20) can be solved to get the eigenvalues and eigenfunctions of the mn^{th} TM^z mode, as in the case of the TE^z mode.

Eigenvalue and Eigenfunction

Using the separation of the variable, the following expression could be written:

$$\phi^e(x,y) = \phi^e_1(x)\,\phi^e_2(y). \quad (7.4.21)$$

Using equation (7.4.21) with equation (7.1.31), we get the following boundary conditions on four EW surfaces.

i) *The boundary condition at the narrow-wall surface*

$$x = 0,\ (0 \le y \le b) \quad \text{and} \quad x = a,\ (0 \le y \le b):$$
$$E^e_y = \phi^e_1(x)\frac{d\phi^e_2(y)}{dy} = 0$$
$$\Rightarrow \text{For } \frac{d\phi^e_2(y)}{dy} \ne 0,\ \phi^e_1(x) = 0 \quad (a). \quad (7.4.22)$$

The condition $\phi^e_1(x) = 0$ also results in $E^e_z = 0$ at the EW of narrow-face.

ii) *The boundary condition at the broad-wall surfaces*

$$y = 0, (0 \le x \le a) \quad \text{and} \quad y = b, (0 \le x \le a):$$
$$E^e_x = \phi^e_2(y)\frac{d\phi^e_1(x)}{dx} = 0$$
$$\Rightarrow \text{For } \frac{d\phi^e_1(x)}{dx} \ne 0,\ \phi^e_2(y) = 0 \quad (b). \quad (7.4.22)$$

The condition $\phi^e_2(y) = 0$ also provides $E^e_z(y) = 0$ at the EW of the broad-face.

The above boundary conditions give the following eigenfunction and propagation constant for the mn^{th} TM^z mode.

$$\phi^e_{mn}(x,y) = A_{mn} \sin\left(\frac{m\pi}{a}x\right) \sin\left(\frac{n\pi}{b}y\right) \quad (m, n = 1, 2, 3...) \quad (a)$$
$$\beta_z = \left[k^2 - \left\{\left(\frac{m\pi}{a}\right)^2 + \left(\frac{n\pi}{b}\right)^2\right\}\right]^{1/2} \quad (b). \quad (7.4.23)$$

It is noted that $m = 0$, $n = 0$ have to be excluded from the solution, as they reduce $\phi^e_{mn}(x,y) = 0$ leading to the nonexistence of the TM^z modal field altogether in the waveguide.

TM^z Modal Fields

On substituting $\phi^e(x,y,z) = \phi^e_{mn}(x,y)e^{-j\beta_z z}$ in equation (7.1.31), the field components of the TM^z modes are obtained:

$$
\begin{aligned}
E^e_x &= -\frac{m\pi}{a} A_{mn}\frac{\beta_z}{\omega\varepsilon}\cos\left(\frac{m\pi}{a}x\right)\sin\left(\frac{n\pi}{b}y\right)e^{-j\beta_z z} && (a)\\
E^e_y &= -\frac{n\pi}{b} A_{mn}\frac{\beta_z}{\omega\varepsilon}\sin\left(\frac{m\pi}{a}x\right)\cos\left(\frac{n\pi}{b}y\right)e^{-j\beta_z z} && (b)\\
E^e_z &= -j\frac{A_{mn}}{\omega\varepsilon}(k^2-\beta_z^2)\sin\left(\frac{m\pi}{a}x\right)\sin\left(\frac{n\pi}{b}y\right)e^{-j\beta_z z} && (c)\\
H^e_x &= \frac{n\pi}{b} A_{mn}\sin\left(\frac{m\pi}{a}x\right)\cos\left(\frac{n\pi}{b}y\right)e^{-j\beta_z z} && (d)\\
H^e_y &= -\frac{m\pi}{a} A_{mn}\cos\left(\frac{m\pi}{a}x\right)\sin\left(\frac{n\pi}{b}y\right)e^{-j\beta_z z} && (e)\\
H^e_z &= 0 && (f).
\end{aligned}
\qquad (7.4.24)
$$

Characteristics of Modes

The propagation constant and the wave impedance of modes are examined below for understanding the behavior of modes.

Modal Propagation Constant

The propagation constants of both the TE^z and TM^z modes are given by

$$\beta_z^{mn} = \pm\left[k^2 - \left\{\left(\frac{m\pi}{a}\right)^2 + \left(\frac{n\pi}{b}\right)^2\right\}\right]^{1/2}. \qquad (7.4.25)$$

For the TE^z modes: $m \neq n = 0$, $m = 0, 1, 2, \ldots$, $n = 0, 1, 2\ldots$, and for the TM^z modes: $m = 1, 2, \ldots$, $n = 1, 2\ldots$ The transverse electric and transverse magnetic modes with respect to the *z*-direction of propagation are written as TE^z_{mn} and TM^z_{mn}, respectively. For any forward-moving modal wave, the propagation constant β_z^{mn} must be a positive real quantity. The negative real β_z^{mn} shows propagation of the wave in the backward direction. For $\beta_z^{mn} = 0$, no wave propagation occurs and the cut-off phenomenon exists at the corresponding *cut-off wave number* $\beta = \beta_c = k_c$. There is a corresponding *cut-off frequency* f_c^{mn} for the mn^{th} mode,

$$
\begin{aligned}
&\text{At } \beta_z = 0,\quad \beta_c = \sqrt{\left(\frac{m\pi}{a}\right)^2 + \left(\frac{n\pi}{b}\right)^2} = \omega_c\sqrt{\mu\varepsilon} && (a)\\
&\Rightarrow f_c^{mn} = \frac{1}{2\pi\sqrt{\mu\varepsilon}}\sqrt{\left(\frac{m\pi}{a}\right)^2 + \left(\frac{n\pi}{b}\right)^2}\\
&\qquad = \frac{c}{2\pi\sqrt{\mu_r\varepsilon_r}}\sqrt{\left(\frac{m\pi}{a}\right)^2 + \left(\frac{n\pi}{b}\right)^2} && (b),
\end{aligned}
\qquad (7.4.26)
$$

where c is the velocity of the EM-wave in free space. Thus, the operating frequency f, exciting the propagating mn^{th} mode signal in a waveguide, must be higher than f_c^{mn}. The modes having identical cut-off frequency are known as the *degenerate modes.* The cut-off behavior of a waveguide could be expressed in terms of the *cut-off guided-wavelength* $\lambda_c = 2\pi/\beta_c$.

In the case of the imaginary value for β_z^{mn} $(-j\,\alpha_z^{mn})$, wave propagation does not occur. The nonpropagating modes, called the *evanescent modes,* are just the exponentially decaying field quantities. The evanescent modes appear for the operating frequency $f < f_c^{mn}$. The cut-off behavior of a waveguide is summarized below:

$$
\beta_z^{mn} = \begin{cases} \pm\beta\sqrt{1-\left(\dfrac{f_c}{f}\right)^2}, & f > f_c\\ 0, & f = f_c\\ \mp j\beta\sqrt{\left(\dfrac{f_c}{f}\right)^2 - 1}, & f < f_c, \end{cases} \qquad (7.4.27)
$$

where $k = \beta$ is the propagation constant of the EM-wave in the homogeneous medium. Under the condition $f > f_c$, the propagation constant $\beta_z^{mn} = +\beta\sqrt{1-(f_c/f)^2}$ is used for the propagating wave in the forward z-direction, whereas under the condition $f < f_c$, the $\beta_z^{mn} = -j\beta\sqrt{(f_c/f)^2 - 1}$ acts an attenuation constant and provides attenuation to the decaying evanescent wave in the forward z-direction. The cut-off occurs at the frequency $f = f_c$. The wavelength of mn^{th} mode λ_g^{mn}, known as the *guided-wavelength*, is given by

$$
\begin{aligned}
\lambda_g^{mn} &= \frac{2\pi}{\beta_z^{mn}} = \frac{2\pi}{\beta\sqrt{1-\left(\dfrac{f_c}{f}\right)^2}} = \frac{\lambda}{\sqrt{1-\left(\dfrac{f_c}{f}\right)^2}} && (a)\\
\lambda &= \frac{\lambda_0}{\sqrt{\varepsilon_r\mu_r}} && (b),
\end{aligned}
\qquad (7.4.28)
$$

where λ is the wavelength of the EM-wave in a homogeneous medium of the waveguide. The guided-wavelength is a real number for the propagating mode and is an imaginary number for the nonpropagating evanescent mode. It is infinite at the cut-off frequency.

Wave Velocities

Following the case of a plasma medium with cut-off phenomenon presented in equations (3.3.8) and (3.3.13) of chapter 3, the phase and group velocities of the wave in a waveguide could be obtained as

$$v_p = v\left[1-\left(\frac{f_c}{f}\right)^2\right]^{-1/2} \quad (a), \quad v_p = \left(\frac{\lambda_g^{mn}}{\lambda}\right)v \quad (b)$$
$$v_g = v\left[1-\left(\frac{f_c}{f}\right)^2\right]^{1/2} \quad (c), \quad v_g = \left(\frac{\lambda}{\lambda_g^{mn}}\right)v \quad (d)$$
$$v_g \times v_p = v^2 \quad (e). \qquad (7.4.29)$$

In the above equations, the velocity of the EM-wave in the homogeneous medium inside to a waveguide is $v = c/\sqrt{\varepsilon_r\mu_r}$, and the corresponding wavelength is $\lambda = \lambda_0/\sqrt{\varepsilon_r\mu_r}$.

Modal Wave Impedance

The concept of the characteristic impedance of a transmission line is discussed in subsection (2.1.3) of chapter 2. It explains the wave phenomenon on a line. It also helps to treat a section of the transmission line as a circuit element. For a similar purpose, the concept of *wave impedance* of the TE^z and TM^z modes, i.e. $Z_w^z(TE_{mn}^z)$ and $Z_w^z(TM_{mn}^z)$, respectively, are introduced. The wave impedance is defined as a ratio of the transverse field components.

For the TE_{mn}^z mode, the wave impedance is

$$Z_w^z(TE_{mn}^z) = \frac{E_x}{H_y} = \frac{E_y}{-H_x} = \frac{\omega\mu}{\beta_z^{mn}}. \qquad (7.4.30)$$

The following expression is written from equations (7.4.27) and (7.4.30):

$$Z_w^z(TE_{mn}^z) = \begin{cases} \dfrac{\eta}{\sqrt{1-\left(\frac{f_c}{f}\right)^2}}, & f > f_c \\ \infty, & f = f_c \\ \dfrac{+j\eta}{\sqrt{\left(\frac{f_c}{f}\right)^2-1}}, & f < f_c, \end{cases} \qquad (7.4.31)$$

where $\eta = \sqrt{\frac{\mu}{\varepsilon}} = \eta_0\sqrt{\frac{\mu_r}{\varepsilon_r}}$ is the intrinsic impedance of the EM-wave propagating in a homogeneous medium of the waveguide and $\eta_0 = 120\pi$ is the intrinsic impedance of the free space. The wave impedance of the propagating TE_{mn}^z, mode under the condition $f > f_c$, is a real quantity providing the propagating wave. The wave impedance is higher than the intrinsic impedance of the medium inside a waveguide. For the evanescent mode under the condition $f < f_c$, the wave impedance $Z_w^z(TE_{mn}^z)$ offers an *inductive reactance* that reflects the power supplied to the waveguide. So the TE mode waveguide, below cut-off frequency, could be viewed as the *ENG metamaterial* medium discussed in the subsection (5.5.3) of chapter 5. At the cut-off $f = f_c$, the wave impedance is infinitely large creating the *artificial magnetic conductor (AMC) surface* with in-phase reflection. The AMC surface is discussed in section (20.2) of chapter 20. Figure (7.6a) shows the impedance behavior of the TE mode waveguide. Figure (7.6b) shows the realization of the *inductive surface* created by an array of TE_{mn}^z mode waveguide for the operation below cut-off frequency $f \le f_c$.

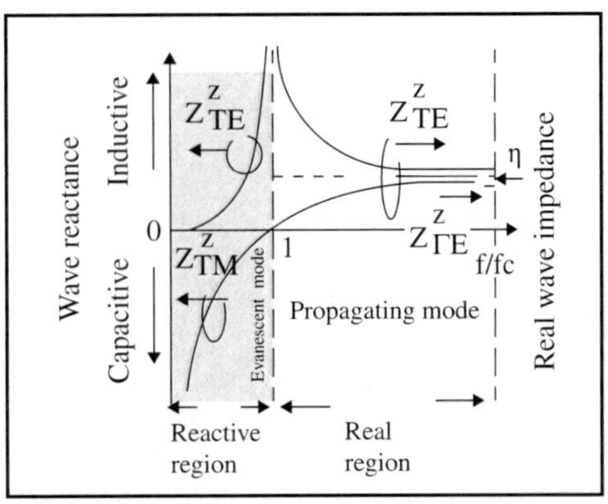

(a) Wave impedance of non-propagating and propagating regions and also at the cut-off frequency.

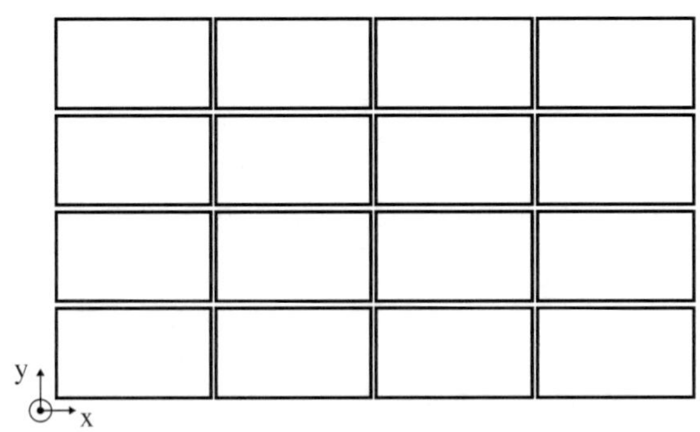

(b) Realization of an inductive surface using an array of TE_{mn}^z mode waveguide below the cut-off frequency.

Figure 7.6 Wave impedance behavior and inductive surface.

Likewise, the following wave impedance of the TM^z_{mn} mode is obtained:

$$Z^z_w\left(TM^z_{mn}\right) = \frac{\beta^{mn}_z}{\omega\,\varepsilon} \qquad \text{(a)}$$

$$Z^z_w\left(TM^z_{mn}\right) = \begin{cases} \eta\sqrt{1-\left(\frac{f_c}{f}\right)^2}, & f > f_c \\ 0, & f = f_c \\ -j\eta\sqrt{\left(\frac{f_c}{f}\right)^2 - 1}, & f < f_c \end{cases} \qquad \text{(b)}. \tag{7.4.32}$$

It is obvious that below the cut-off frequency, i.e. for $f < f_c$ the wave impedance of the TM^z_{mn} mode offers *capacitive reactance*. It reflects the TM^z_{mn} mode. So the TM mode waveguide, below cut-off frequency, could be viewed as the *MNG metamaterial* medium discussed in subsection (5.5.3) of chapter 5. Likewise, at the cut-off frequency $f = f_c$, the wave impedance of TM^z_{mn} mode is zero creating the *artificial electric conductor (AEC) surface* exhibiting out-of-phase reflection. It discussed in section (20.2) of chapter 20. The nature of the TE/TM-mode rectangular waveguide, above and below cut-off frequency, is further validated with the help of the circuit model in subsection (7.6.2).

The TE^z_{mn} and TM^z_{mn} modes show the high pass filtering property. The wave impedance of both modes has opposite behaviors as shown in Fig (7.6a). At the cut-off frequency, the wave impedance for the TE^z_{mn} mode is very large, showing the open circuit behavior, whereas for the TM^z_{mn} mode it is zero, showing the short-circuit behavior. At high frequency, the wave impedances move toward η, i.e. toward the intrinsic impedance of the homogeneous medium inside the waveguide. For the TM^z_{mn} modes, the arrangement of Fig (7.6b) creates a *capacitive surface*.

Modal Fields

Figure (7.7) shows the modal field for a few TE^z_{mn} modes. The integers "m" and "n" denote half guided-wavelength variations of the field along the x and y-axes of a waveguide, i.e. along the broad-face (width) and narrow-face (height) of the waveguide. The modal field pattern repeats itself along the direction of propagation, i.e. in the z-direction, at every $\lambda_g/2$ distance. The upper part of Fig (7.7a) shows the E and H-fields distribution of the fundamental TE^z_{10} mode. The E-field lines are shifted by $\lambda/4$ (90°) with respect to the center of the closed-loop of the H-field lines.

The conduction current and its continuity to the displacement is shown in the lower part of Fig (7.7a). The conduction current flows on the inner surface of the broad-face of a waveguide. Further, electric displacement current, entering toward the lower broad-face of the waveguide, completes the current loop. The electric displacement current, generated by the time-varying electric field, is at the center D of the loop of the magnetic field. It shows $\lambda_g/4$ shifting from the location of maxima of the electric field, because of the presence of "j" with the displacement current. It indicates the $\pi/2$ phase-shift of the displacement current.

Figure (7.7b) shows the E and H-fields distribution of the TE^z_{20} and TE^z_{30} modes. The width "a" accommodates 2 and 3 numbers of TE^z_{10} to form the TE^z_{20} and TE^z_{30} mode, respectively. Each TE^z_{10} mode, constituting TE^z_{20} and TE^z_{30} modes, is separated by the EWs. It is noted that at the plane of the EW, the total electric field is zero and the magnetic field is maximum.

Figure (7.7c–e) show the electric variations of three TE^z_{m0} modes (m = 1,2,3). The location of the maximum of the electric field, along the broad-face, is mode dependent and repeats itself in the z-direction. The maximum of the E-field for the TE_{10} mode is at the center of the waveguide. Also, the cut-off frequency of TE^z_{30} mode is equal to the cut-off frequency of TE^z_{10} mode of the width a/3 of the original waveguide. The TE^z_{10} mode is the *fundamental mode*, with the lowest cut-off frequency. There is no field variation for the TE^z_{m0} mode along the height (y-direction) of a waveguide.

Figure (7.8) shows field lines of the TM^z_{21} mode. The broad-face view shows the presence of a closed-loop of the electric field that is created by the magnetic displacement current at the center of the electric field loop. The E-field lines start at the positive charges on the waveguide wall and after bending end at the negative charges. The pattern is repeated along the z-axis. The magnetic displacement current is generated by the time-varying magnetic field.

Surface Current on the Waveguide

The EM-wave in a waveguide induces a current on its inner surfaces. The surface current density is related to the tangential component of the magnetic field on a surface,

$$\vec{J}_s = \hat{n} \times \vec{H}_t, \tag{7.4.33}$$

where $\hat{n}$ is the unit vector, normal to the conducting surface. The current density is useful for the computation of conductor loss on the walls of a waveguide. It is computed for a perfectly conducting surface. However, an

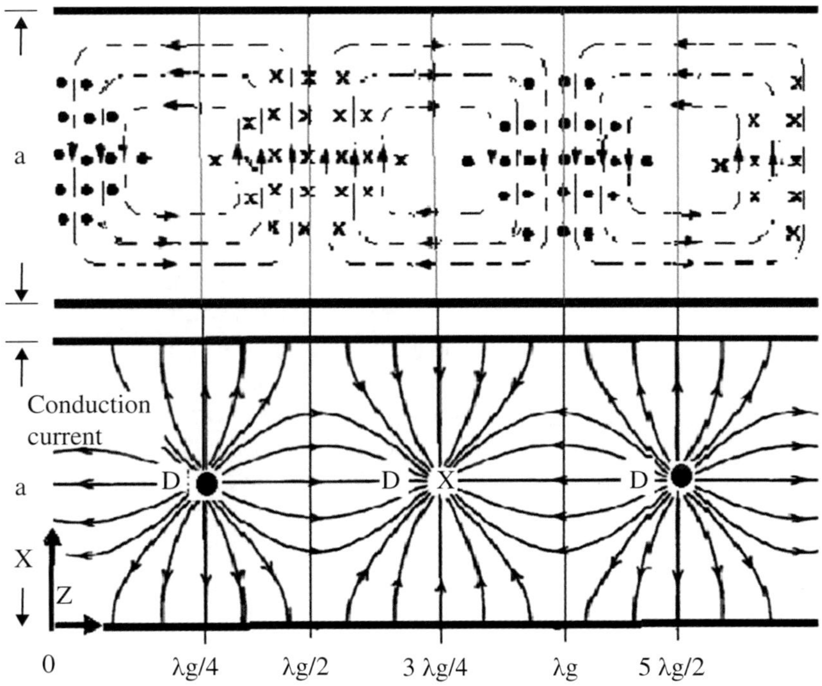

(a) TE10 mode- upper figure --- H-field, (●, ×) E-field arrow. lower figure —— conduction current, (●, ×) displacement current arrow.

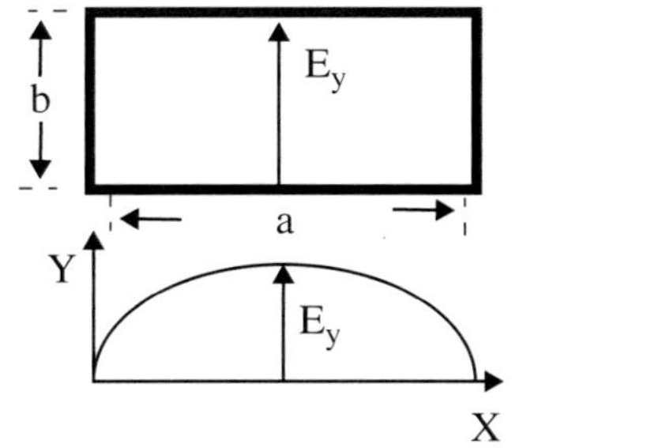

(c) TE_{10} mode
E- field variation in cross-section.

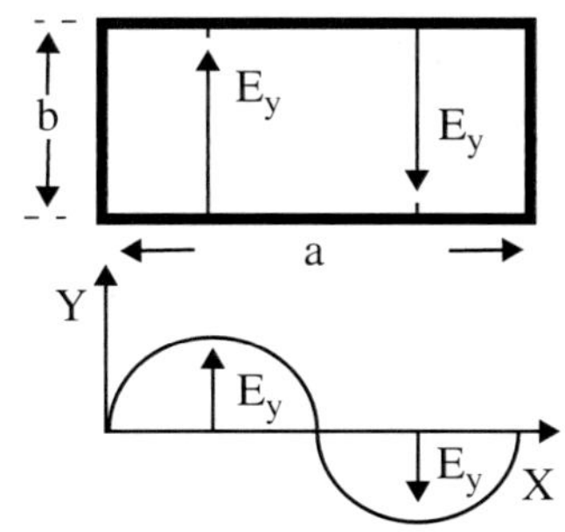

(d) TE20 mode
E-field variation in cross-section.

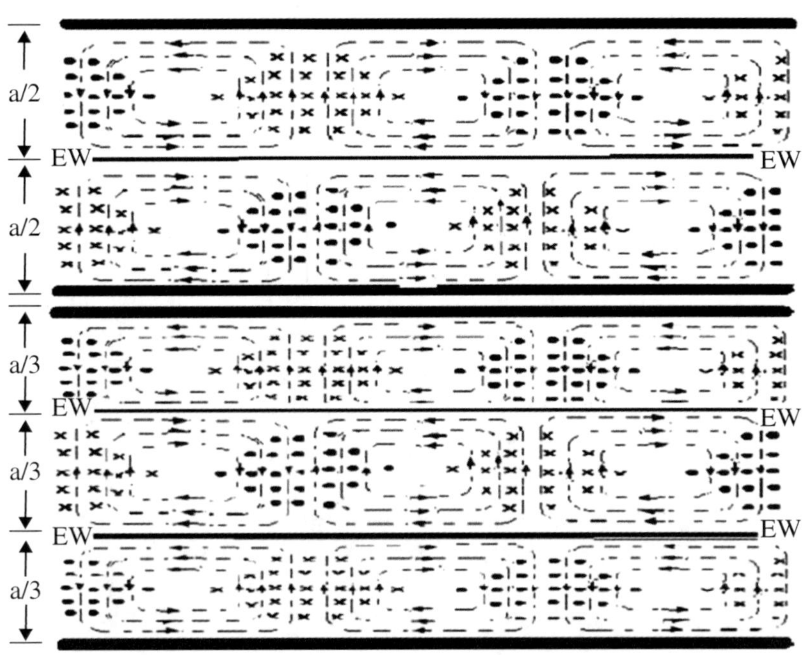

(b) TE20 and TE30 modes.

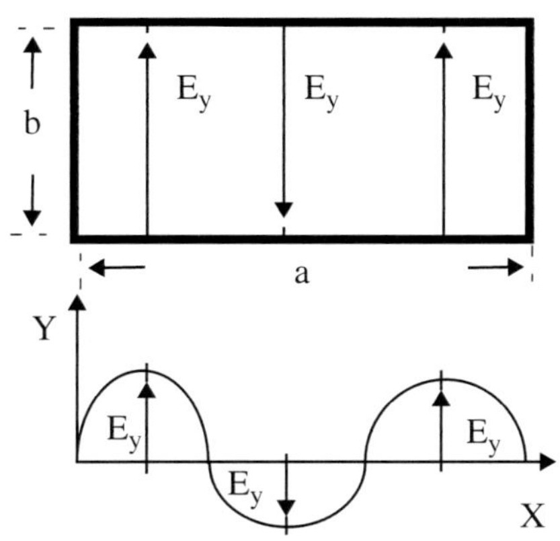

(e) TE30 mode
E- field variation in cross-section.

Figure 7.7 Field and current distribution of TE^z_{m0} modes (m = 1,2,3).

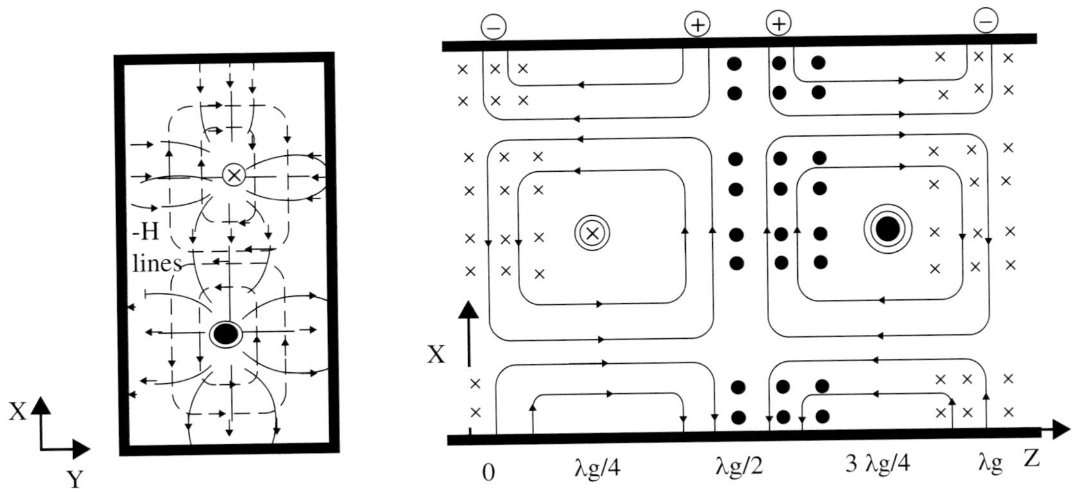

Leftside figure --- H-field lines
⊗ ,●: Electric displacement current

Rightside figure, (•, ×) H-field arrow, — E-field lines
◎,◉ : Magnetic displacement current.

Figure 7.8 Field lines of TM^z_{21} mode.

actual conducting surface has finite conductivity that influences the computation of conductor loss. Let us consider the surface current distribution of the TE^z_{10} fundamental mode, for m = 1, n = 0 on inner surfaces of a perfectly conducting waveguide. The following field components for TE^z_{10} mode are obtained from equation (7.4.19):

$$E^m_x = E^m_z = 0, \qquad E^m_y = \frac{-\pi}{a} B_{10} \sin\left(\frac{\pi}{a}x\right) e^{-j\beta_z z}$$

$$H^m_x = \frac{\pi B_{10} \beta_z}{a\,\omega\mu} \sin\left(\frac{\pi}{a}x\right) e^{-j\beta_z z}$$

$$H^m_y = 0, \quad H^m_z = -j\frac{B_{10}}{\omega\mu}(k^2 - \beta_z^2) \cos\left(\frac{\pi}{a}x\right) e^{-j\beta_z z}$$

$$\text{where } k^2 - \beta_z^2 = k_c^2 = (\pi/a)^2. \tag{7.4.34}$$

The surface current density is easily obtained at all four inner conducting surfaces of the waveguide shown in Fig (7.5a).

Left hand sidewall at x=0:

$$\vec{J}_s(x = 0, 0 \le y \le b) = \hat{x} \times \left[\hat{y}\,H^m_y + \hat{z}\,H^m_z\right]\Big|_{x=0}$$

$$= \left[-\hat{y}\,H^m_z\right]\Big|_{x=0}$$

$$\Rightarrow \quad \vec{J}_s = \hat{y}\,j\frac{B_{10}}{\omega\mu}(k^2 - \beta_z^2)\,e^{-j\beta_z z}$$

$$= \hat{y}\,j\frac{B_{10}}{\omega\mu}\left(\frac{\pi}{a}\right)^2 e^{-j\beta_z z} \qquad \text{(a)}$$

Right-hand sidewall at x=a:

$$\vec{J}_s(x = a, 0 \le y \le b) = -\hat{x} \times \left[\hat{y}\,H^m_y + \hat{z}\,H^m_z\right]\Big|_{x=a}$$

$$= \left[\hat{y}\,H^m_z\right]\Big|_{x=a}$$

$$\Rightarrow \quad \vec{J}_s = -\hat{y}\,j\frac{B_{10}}{\omega\mu}(k^2 - \beta_z^2)\,e^{-j\beta_z z}$$

$$= -\hat{y}\,j\frac{B_{10}}{\omega\mu}\left(\frac{\pi}{a}\right)^2 e^{-j\beta_z z} \qquad \text{(b)}$$

The bottom wall at y=0:

$$\vec{J}_s(0 \le x \le a, y = 0) = \hat{y} \times \left[\hat{x}\,H^m_x + \hat{z}\,H^m_z\right]\Big|_{y=0}$$

$$= \left[-\hat{z}\,H^m_x + \hat{x}\,H^m_z\right]\Big|_{y=0}$$

$$\Rightarrow \quad \vec{J}_s = -\hat{x}\frac{j\,B_{10}}{\omega\mu}\left(\frac{\pi}{a}\right)^2 \cos\left(\frac{\pi}{a}x\right) e^{-j\beta_z z}$$

$$-\hat{z}\frac{B_{10}\,\beta_z}{\omega\mu}\left(\frac{\pi}{a}\right) \sin\left(\frac{\pi}{a}x\right) e^{-j\beta_z z} \qquad \text{(c)}$$

The upper wall at y=b:

$$\vec{J}_s(0 \le x \le a, y = b) = -\vec{J}_s(0 \le x \le a, y = 0) \quad \text{(d)}. \tag{7.4.35}$$

The current distribution on the upper wall of the inner surface is sketched in Fig (7.7a).

EM-wave Power Transfer in Waveguide

The EM-wave power in a waveguide is transported in forms of the power associated with the TE^z_{mn} and

TM^z_{mn} modes. This is computed by using the Poynting vector relation,

$$\vec{S} = \vec{E} \times \vec{H}^*, \tag{7.4.36}$$

where $\vec{H}^*$ is the complex conjugate of the complex vector $\vec{H}$. The powers contained in the TE^z_{mn} and TM^z_{mn} modes are computed separately, as these modes exist independently of each other.

TE^z_{mn} Mode

The power density of TE^z_{mn} mode transported in the z-direction is

$$\begin{aligned} S_z^{TE} &= \frac{1}{2}\,\text{Re}\left(\vec{E} \times \vec{H}^*\right).\hat{z}\big|_{TE^z_{mn}} \\ &= \frac{1}{2}\,\text{Re}\left(E_x^m H_y^{m*} - E_y^m H_x^{m*}\right). \end{aligned} \tag{7.4.37}$$

On using an equation (7.4.19) with equation (7.4.37), the following expression is obtained:

$$S_z^{TE} = \frac{1}{2}\,\text{Re}\begin{bmatrix} \frac{n\pi}{b} B_{mn} \cos\left(\frac{m\pi}{a}x\right) \sin\left(\frac{n\pi}{b}y\right) e^{-j\beta_z z} \\ \times \frac{n\pi}{b} B^*_{mn} \frac{\beta_z}{\omega\mu} \cos\left(\frac{m\pi}{a}x\right) \sin\left(\frac{n\pi}{b}y\right) e^{j\beta_z z} \\ + \frac{m\pi}{a} B_{mn} \sin\left(\frac{m\pi}{a}x\right) \cos\left(\frac{n\pi}{b}y\right) e^{-j\beta_z z} \\ \times \frac{m\pi}{a} B^*_{mn} \frac{\beta_z}{\omega\mu} \sin\left(\frac{m\pi}{a}x\right) \cos\left(\frac{n\pi}{b}y\right) e^{j\beta_z z} \end{bmatrix}$$

$$\Rightarrow \quad S_z^{TE} = \frac{|B_{mn}|^2 \beta_z}{2\,\omega\mu}\left[\left(\frac{n\pi}{b}\right)^2 \cos^2\left(\frac{m\pi}{a}x\right) \sin^2\left(\frac{n\pi}{b}y\right) + \left(\frac{m\pi}{a}\right)^2 \sin^2\left(\frac{m\pi}{a}x\right) \cos^2\left(\frac{n\pi}{b}y\right)\right]. \tag{7.4.38}$$

The TE^z_{mn} mode power crossing the cross-section of the waveguide is

$$\begin{aligned} P_{mn}^{TE} &= \int_0^b \int_0^a S_z^{TE}\, dxdy \\ &= |B_{mn}|^2 \frac{\beta_z}{2\,\omega\mu}\left[\left(\frac{n\pi}{b}\right)^2 \left(\frac{a}{\varepsilon_{0m}}\right)\left(\frac{b}{\varepsilon_{0n}}\right) + \left(\frac{m\pi}{a}\right)^2 \left(\frac{a}{\varepsilon_{0m}}\right)\left(\frac{b}{\varepsilon_{0n}}\right)\right] \\ \Rightarrow \quad P_{mn}^{TE} &= |B_{mn}|^2 \frac{\beta_z}{2\,\omega\mu}\left(\frac{a}{\varepsilon_{0m}}\right)\left(\frac{b}{\varepsilon_{0n}}\right)\left[\left(\frac{m\pi}{a}\right)^2 + \left(\frac{n\pi}{b}\right)^2\right], \end{aligned} \tag{7.4.39}$$

where the following expressions are used in equation (7.4.39):

$$\begin{aligned} &\int_0^a \cos^2\left(\frac{m\pi}{a}x\right)dx = \begin{cases} \frac{a}{2}, & m \neq 0 \\ a, & m = 0 \end{cases} && (a) \\ &\int_0^a \sin^2\left(\frac{m\pi}{a}x\right)dx = \begin{cases} \frac{a}{2}, & m \neq 0 \\ 0, & m = 0 \end{cases} && (b) \\ &\varepsilon_{0p} = \begin{cases} 1, & p = 0 \\ 2, & p \neq 0 \end{cases}, \quad p = m \text{ or } n && (c). \end{aligned} \tag{7.4.40}$$

Using the cut-off wavenumber, $\beta_c^2 = \left(\frac{m\pi}{a}\right)^2 + \left(\frac{n\pi}{b}\right)^2$ and equation (7.4.27), equation (7.4.39) is written as

$$\begin{aligned} P_{mn}^{TE} &= |B_{mn}|^2 \frac{\beta_z}{2\,\omega\mu}\left(\frac{a}{\varepsilon_{0m}}\right)\left(\frac{b}{\varepsilon_{0n}}\right)\beta_c^2 \\ &= |B_{mn}|^2 \frac{\beta}{2\,\omega\mu}\left(\frac{a}{\varepsilon_{0m}}\right)\left(\frac{b}{\varepsilon_{0n}}\right)\beta_c^2\left[1 - \left(\frac{f_c^{mn}}{f}\right)^2\right]^{1/2} \\ \Rightarrow \quad P_{mn}^{TE} &= |B_{mn}|^2 \times \frac{\omega\sqrt{\mu\varepsilon}}{2\,\omega\mu}\left(\frac{a}{\varepsilon_{0m}}\right)\left(\frac{b}{\varepsilon_{0n}}\right)\beta_c^2\left[1 - \left(\frac{f_c^{mn}}{f}\right)^2\right]^{1/2} \\ \Rightarrow \quad P_{mn}^{TE} &= \frac{|B_{mn}|^2}{2\eta}\left(\frac{a}{\varepsilon_{0m}}\right)\left(\frac{b}{\varepsilon_{0n}}\right)\beta_c^2\left[1 - \left(\frac{f_c^{mn}}{f}\right)^2\right]^{1/2}. \end{aligned} \tag{7.4.41}$$

TM^z_{mn} Mode

The power carried by the waveguide in the TM^z_{mn} mode is also computed similarly [B.5],

$$P_{mn}^{TM} = \frac{|A_{mn}|^2 \eta \beta_c^2}{2\mu^2}\left(\frac{a}{2}\right)\left(\frac{b}{2}\right)\left[1 - \left(\frac{f_c^{mn}}{f}\right)^2\right]^{1/2}. \tag{7.4.42}$$

The total power transmitted in the waveguide is a sum of power carried by both the TE^z_{mn} and TM^z_{mn} modes.

Attenuation

The conducting walls of a waveguide are not the perfect ones. They have finite conductivity. The modal EM-fields induce the modal surface currents that cause the conductor loss in a waveguide. Once the conductor loss is computed on the four walls of a waveguide, its attenuation constant α_c, due to the conductor loss, is computed easily. Only the case of the TE^z_{10} fundamental mode is considered below.

Conductor Loss

The total conductor loss per unit length is twice the sum of conductor loss on one broad-face and one narrow-face of the waveguide:

$$P_c^{TE_{10}}(z) = \underbrace{2[P_c(z)|_{x=0}]}_{\text{Loss at narrow-face}} + \underbrace{2[P_c(z)|_{y=0}]}_{\text{Loss at broad-face}}. \quad (7.4.43)$$

The conductor loss of the fundamental mode TE_{10} at the left-hand side narrow-face of the waveguide, shown in Fig (7.5a), is computed using equation (7.4.35a):

$$P_c^{TE_{10}}(z)|_{x=0} = \frac{1}{2}\int_0^b R_s\,|J_s(x=0)|^2 dy = \frac{R_s}{2}\int_0^b \frac{B_{10}^2}{\omega^2\mu^2}\left(\frac{\pi}{a}\right)^4 dy$$

$$\Rightarrow P_c^{TE_{10}}(z)|_{x=0} = \frac{R_s\,B_{10}^2 b}{2\,\omega^2\mu^2}\left(\frac{\pi}{a}\right)^4 \quad (a).$$

The conductor loss at bottom broad-face, using equation (7.4.35c) is

$$P_c^{TE_{10}}(z)|_{y=0} = \frac{1}{2}\int_0^a R_s\,|J_s(y=0)|^2 dx = \frac{R_s}{2}\int_0^a \left[J_s(y=0)\,J_s^*(y=0)\right]dx$$

$$= \frac{R_s}{2}\int_0^a\left[\frac{B_{10}^2}{\omega^2\mu^2}\left(\frac{\pi}{a}\right)^4\cos^2\left(\frac{\pi}{a}x\right) + \frac{B_{10}^2}{\omega^2\mu^2}\left(\frac{\pi}{a}\right)^2\beta_z^2\sin^2\left(\frac{\pi}{a}x\right)\right]dx$$

$$= \frac{R_s\,B_{10}^2}{2\,\omega^2\mu^2}\left(\frac{\pi}{a}\right)^2\int_0^a\left[\left(\frac{\pi}{a}\right)^2\cos^2\left(\frac{\pi}{a}x\right) + \beta_z^2\sin^2\left(\frac{\pi}{a}x\right)\right]dx$$

$$\Rightarrow P_c^{TE_{10}}(z)\,|_{y=0} = \frac{R_s\,B_{10}^2}{2\,\omega^2\mu^2}\left(\frac{\pi}{a}\right)^2\left(\frac{a}{2}\right)\left[\left(\frac{\pi}{a}\right)^2 + \beta_z^2\right] \quad (b). \quad (7.4.44)$$

The total conductor loss *p.u.l.* is

$$P_c^{TE_{10}}(z) = \frac{R_s\,B_{10}^2}{\omega^2\mu^2}\left(\frac{\pi}{a}\right)^2\left[\left(\frac{\pi}{a}\right)^2 b + \frac{a}{2}\left(\frac{\pi}{a}\right)^2 + \frac{a}{2}\beta_z^2\right]$$

$$= \frac{R_s\,B_{10}^2}{\omega^2\mu^2}\left(\frac{\pi}{a}\right)^2\left[\left(\frac{\pi}{a}\right)^2 b + \frac{a}{2}\left(\frac{\pi}{a}\right)^2 + \frac{a}{2}\left(\omega^2\mu\varepsilon - \left(\frac{\pi}{a}\right)^2\right)\right]$$

$$\Rightarrow P_c^{TE_{10}}(z) = \frac{R_s\,B_{10}^2}{\omega^2\mu^2}\left(\frac{\pi}{a}\right)^2\left[\left(\frac{\pi}{a}\right)^2 b + \frac{a}{2}\omega^2\mu\varepsilon\right] = \frac{R_s\,B_{10}^2}{\omega^2\mu^2}\left(\frac{\pi}{a}\right)^4\left[b + \frac{a}{2}\omega^2\mu\varepsilon\left(\frac{a}{\pi}\right)^2\right]. \quad (7.4.45)$$

The following expressions are used to simplify equation (7.4.45) for the conductor loss:

$$\frac{\pi}{a} = \beta_c = \frac{2\pi}{\lambda_c} = \frac{2\pi\,f_c^{10}}{v} = 2\pi\sqrt{\mu\varepsilon}\,f_c^{10} \quad \Rightarrow a\sqrt{\mu\varepsilon} = \frac{1}{2\,f_c^{10}}. \quad (7.4.46)$$

On using equation (7.4.46) with equation (7.4.45), the following expression is obtained:

$$P_c^{TE_{10}}(z) = \frac{R_s\,B_{10}^2}{\omega^2\mu^2}\left(\frac{\pi}{a}\right)^4\left[b + \frac{a}{2}\frac{4\pi^2 f^2}{\pi^2}\frac{1}{4\left(f_c^{10}\right)^2}\right] = \frac{R_s\,B_{10}^2}{\omega^2\mu^2}\left(\frac{\pi}{a}\right)^4\left[b + \left(\frac{f}{f_c^{10}}\right)^2\frac{a}{2}\right]$$

$$\Rightarrow P_c^{TE_{10}}(z) = \frac{R_s\,B_{10}^2}{\omega^2\mu^2}\left(\frac{\pi}{a}\right)^4\frac{a}{2}\left(\frac{f}{f_c^{10}}\right)^2\left[1 + \left(\frac{f_c^{10}}{f}\right)^2\frac{2b}{a}\right]. \quad (7.4.47)$$

$$P_c^{TE_{10}}(z) = \frac{R_s\,B_{10}^2}{\pi^2 f^2\mu}\left(\frac{\pi}{a}\right)^4\frac{a}{2}f^2 a^2\varepsilon\left[1 + \left(\frac{f_c}{f}\right)^2\frac{2b}{a}\right]$$

$$\Rightarrow P_c^{TE_{10}}(z) = R_s\,B_{10}^2\frac{\varepsilon}{\mu}\left(\frac{\pi}{a}\right)^2\frac{a}{2}\left[1 + \left(\frac{f_c}{f}\right)^2\frac{2b}{a}\right]. \quad (7.4.48)$$

The power P_0 at the location $z = 0$ decreases to the P_{mn} (z) at the location z by the following relation:

$$P_{mn}^{TE}(z) = P_0\,e^{-2\alpha_c z}, \quad (7.4.49)$$

where α_c is the attenuation constant due to conductor loss. The loss of transmitted power *p.u.l.* is equal to the conductor loss *p.u.l.* for the TE_{10}^z mode,

$$P_c^{TE_{10}}(z) = -\frac{P_{10}^{TE}(z)}{dz} = 2\alpha_c\,P_{10}^{TE}(z) \quad \Rightarrow \alpha_c = \frac{P_c^{TE_{10}}(z)}{2P_{10}^{TE}(z)}. \quad (7.4.50)$$

The attenuation constant for the TE_{10}^z mode is obtained by substituting equations (7.4.41) and (7.4.48) in equation (7.4.50):

$$\alpha_c^{TE_{10}} = \frac{R_s\,B_{10}^2\frac{\varepsilon}{\mu}\left(\frac{\pi}{a}\right)^2\frac{a}{2}\left[1 + \frac{2b}{a}\left(\frac{f_c^{10}}{f}\right)^2\right]}{\frac{B_{10}^2}{\eta}\frac{a}{2}\,b\left(\frac{\pi}{a}\right)^2\left[1 - \left(\frac{f_c^{10}}{f}\right)^2\right]^{1/2}} = \frac{R_s}{b\,\eta}\frac{\left[1 + \frac{2b}{a}\left(\frac{f_c^{10}}{f}\right)^2\right]}{\left[1 - \left(\frac{f_c^{10}}{f}\right)^2\right]^{1/2}}\ \text{Np/m}, \quad (7.4.51)$$

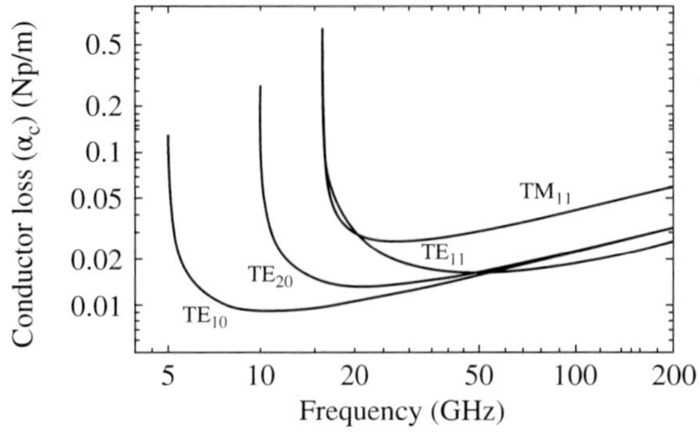

(a) Conductor loss of fundamental and higher-order modes.

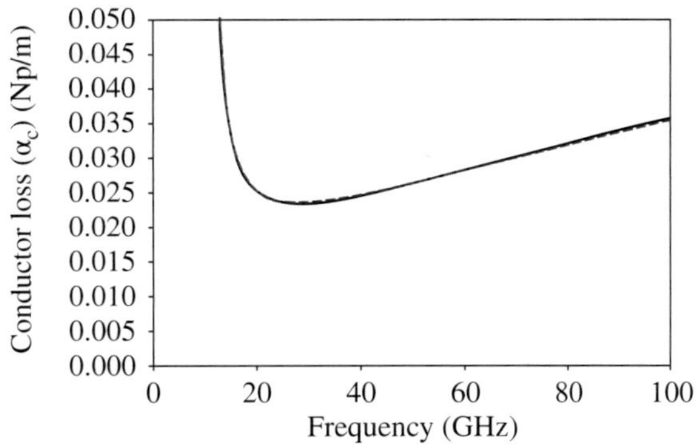

(b) Conductor loss of TE_{10} mode in the propagation region using two methods [J.1].

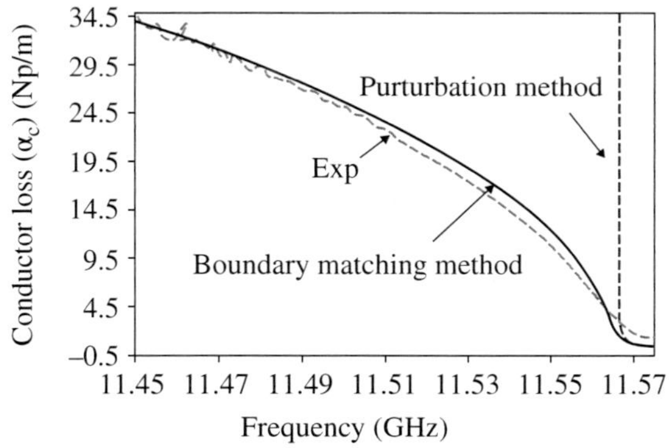

(c) Conductor loss of TE_{10} mode in cut-off region using two methods [J.1].

Figure 7.9 Conductor loss for the TE^z_{10} and higher-order modes. *Source*: From Yeap et al. [J.1]. Public domain.

where the surface impedance of conductor is $R_s = \sqrt{\pi f \mu_{cond}/\sigma_{cond}}$ (for a conductor $\mu_{cond} = 1$). The cut-off frequency of the TE^z_{10} mode is given by equation (7.4.26). The intrinsic impedance of the material medium in a waveguide is $\eta = \eta_0/\sqrt{\varepsilon_r}$.

Likewise, the conductor loss of the TM_{11} mode is obtained:

$$\alpha_c^{TM_{11}} = \frac{2\,Rs}{\eta\, a\, b}\,\frac{\left[\dfrac{b}{a^2}+\dfrac{a}{b^2}\right]}{\left[1-\left(\dfrac{f_c^{11}}{f}\right)^2\right]^{1/2}\left[\dfrac{1}{a^2}+\dfrac{1}{b^2}\right]}\ \text{Np/m.} \tag{7.4.52}$$

The general expression for the conductor loss is obtained for both the TE_{mn} and TM_{mn} modes [B.5]. Figure (7.9a) shows the plot of conductor loss for the TE^z_{10} and TM^z_{11}, and other higher-order modes for a waveguide with a cut-off frequency of the fundamental mode TE^z_{10} about 5 GHz. The fundamental mode TE^z_{10} is least lossy, as compared to other higher-order modes. At the cut-off, all modes show a sharp increase in the loss. Figure (7.9b) confirms the accuracy of the conductor loss computation of the TE^z_{10}-mode of the air-filled waveguide with a = 1.3 cm and b = 0.64 cm, in the propagation region, using the present approximate perturbation method, as against the measurement results and the results of the boundary matching method. However, Fig (7.9c) shows that the perturbation method does not work in the cut-off region. The boundary matching method shows more accurate loss computation in the cut-off region also [J.1–J.3].

Dielectric Loss

If the waveguide is loaded with a lossy dielectric material, with the complex relative permittivity $\varepsilon_r^* = \varepsilon_r' - j\,\varepsilon_r''$, the propagating EM-wave modes suffer the dielectric loss. So the dielectric loss is mode-dependent and shows the cut-off behavior, like the conductor loss. The dielectric loss can be computed from the complex propagation constant using the dispersion relation (7.4.18b). It is reproduced below

$$\beta_z = \sqrt{k^2 - \beta_c^2}, \tag{7.4.53}$$

where the lossy dielectric medium has the wavenumber k:

$$k = k_0\sqrt{\mu_r\varepsilon_r^*} = k_0\sqrt{\mu_r\left(\varepsilon_r' - j\varepsilon_r''\right)}. \qquad (7.4.54)$$

$$\beta_z = \sqrt{k_0^2\mu_r\left(\varepsilon_r' - j\varepsilon_r''\right) - \beta_c^2} = \sqrt{\left(k_0^2\mu_r\varepsilon_r' - \beta_c^2\right) - jk_0^2\mu_r\varepsilon_r''}$$
$$\Rightarrow \beta_z = \sqrt{\left(k'^2 - \beta_c^2\right) - jk_0^2\mu_r\varepsilon_r''};\quad \text{Where,}\quad k' = k_0\sqrt{\mu_r\varepsilon_r'}$$
$$\beta_z = \sqrt{\left(k'^2 - \beta_c^2\right)}\left[1 - j\frac{k_0^2\mu_r\varepsilon_r''}{\left(k'^2 - \beta_c^2\right)}\right]^{1/2}$$
$$\Rightarrow \beta_z \approx \sqrt{\left(k'^2 - \beta_c^2\right)} - j\frac{k_0^2\mu_r\varepsilon_r''}{2\sqrt{\left(k'^2 - \beta_c^2\right)}} = \beta_d - j\alpha_d. \qquad (7.4.55)$$

The propagation constant in the dielectric medium of the waveguide is

$$\beta_d = \text{Re}\ (\beta_z) = \sqrt{\left(k'^2 - \beta_c^2\right)} = k_0\sqrt{\mu_r\varepsilon_r'}\sqrt{1 - \left(\tfrac{f_c}{f}\right)^2}\ \text{rad/m}. \qquad (7.4.56)$$

The attenuation constant, due to the lossy dielectric, in the dielectric medium of the waveguide is

$$\alpha_d = \text{Im}\ (\beta_z) = \frac{k_0^2\mu_r\varepsilon_r''}{2\sqrt{\left(k'^2 - \beta_c^2\right)}} = \frac{k_0^2\mu_r\varepsilon_r''}{2k_0\sqrt{\mu_r\varepsilon_r'}}\frac{1}{\sqrt{1 - \left(\tfrac{f_c}{f}\right)^2}}$$
$$\Rightarrow \alpha_d = \frac{\omega\sqrt{\mu_0\varepsilon_0}\,\varepsilon_r''\sqrt{\mu_r}}{2\sqrt{\varepsilon_r'}}\frac{1}{\sqrt{1 - \left(\tfrac{f_c}{f}\right)^2}}. \qquad (7.4.57)$$

$$\Rightarrow \alpha_d = \frac{\omega\,\varepsilon''}{2}\sqrt{\frac{\mu}{\varepsilon'}}\frac{1}{\sqrt{1 - \left(\tfrac{f_c}{f}\right)^2}} = \frac{\eta'}{2}\frac{\sigma_d}{\sqrt{1 - \left(\tfrac{f_c}{f}\right)^2}}\ \text{Np/m}. \qquad (7.4.58)$$

where σ_d is the conductivity of the imperfect dielectric medium.

7.4.2 Rectangular Waveguide with Four MWs

Figure (7.5b) shows a rectangular waveguide made of high permittivity perfect dielectric slabs. Normally, this waveguide has the field confinement in two transverse directions, as the fields show exponential decay in the air-medium adjacent to the dielectric slab. Marcatili [J.4] has obtained the approximate solution of the field problem related to the dielectric slab structure. The slab structure supports the hybrid mode with respect to the z-direction of propagation. However, our concern is not to discuss the detailed field solutions with the decaying field in the air-medium. We are interested in treating the rectangular dielectric slab waveguide as a complementary structure of the metallic waveguide, shown in Fig (7.5a).

The electromagnetic field is assumed to be confined inside the dielectric waveguide and there is no exponentially decaying field in the air region at the surface of the dielectric waveguide. Such a simplistic field confinement model of the dielectric waveguide is achieved by assuming that the surface of the rectangular waveguides is formed by the MWs, i.e. by the perfect magnetic conductors (PMC), as against the EW (PEC) surfaces of the metallic rectangular waveguide. The TE^z and TM^z modes for the dielectric slab waveguide are obtained below under the condition of four MWs.

TE^z Mode

The TE^z mode has H_z component with $E_z = 0$, and all other field components are generated by scalar magnetic potential $\phi^m(x, y, z) = \phi^m(x, y)e^{-j\beta_z z}$ given by equation (7.4.1).Wave equation (7.4.2) for the potential function $\phi^m(x, y)$ is solved under the following boundary conditions on the MWs, obtained from equation (7.1.35).

i) *Boundary conditions on left and right-hand sidewalls:*

$$H_z = 0,\ \text{i.e.}\ \phi^m = 0\ \text{at}\ x = 0,\ a;\ \ 0 \le y \le b. \qquad (7.4.59)$$

ii) *Boundary condition on the bottom and top walls:*

$$H_z = 0,\ \text{i.e.}\ \phi^m = 0\ \text{at}\ y = 0, b;\ \ 0 \le x \le a. \qquad (7.4.60)$$

Following the method of separation of variables discussed for the metallic rectangular waveguide, the following modal solution is obtained from wave equation (7.4.2) for the scalar magnetic potential $\phi^m(x, y)$:

$$\phi_{mn}^m(x, y) = A_{mn}\sin\left(\frac{m\pi}{a}x\right)\sin\left(\frac{n\pi}{b}y\right);\quad m = 1, 2, 3...;\ \ n = 1, 2, 3..... \qquad (7.4.61)$$

where for the eigenfunction $\phi_{mn}^m(x, y)$, the single superscript "m" indicates the magnetic scalar potential and the paired subscripts ("m, n") show the integral modal numbers, indicating the numbers of half-wavelength variations of the fields along the *x*- and *y*-axes, shown in Fig (7.5b). The modal field components are obtained by using an equation (7.4.61) with equation (7.1.35):

$$
\begin{aligned}
E_x^m &= -A_{mn}\left(\frac{n\pi}{b}\right)\sin\left(\frac{m\pi}{a}x\right)\cos\left(\frac{n\pi}{b}y\right) && \text{(a)}\\
E_y^m &= A_{mn}\left(\frac{m\pi}{a}\right)\cos\left(\frac{m\pi}{a}x\right)\sin\left(\frac{n\pi}{b}y\right) && \text{(b)}\\
H_x^m &= \frac{-\beta_z}{\omega\mu}\frac{\partial\varphi_{mn}^m}{\partial x}\\
&= \frac{-A_{mn}\beta_z}{\omega\mu}\left(\frac{m\pi}{a}\right)\cos\left(\frac{m\pi}{a}x\right)\sin\left(\frac{n\pi}{b}y\right) && \text{(c)}\\
H_y^m &= \frac{-\beta_z}{\omega\mu}\frac{\partial\varphi_{mn}^m}{\partial y}\\
&= \frac{-A_{mn}\beta_z}{\omega\mu}\left(\frac{n\pi}{b}\right)\sin\left(\frac{m\pi}{a}x\right)\cos\left(\frac{n\pi}{b}y\right) && \text{(d)}\\
H_z^m &= \frac{A_{mn}}{j\omega\mu}(k^2-\beta_z^2)\sin\left(\frac{m\pi}{a}x\right)\sin\left(\frac{n\pi}{b}y\right) && \text{(e)},
\end{aligned}
\tag{7.4.62}
$$

where the z-directed propagation constant is given by

$$\beta_z^{mn} = \pm\left[k^2 - \left\{\left(\frac{m\pi}{a}\right)^2 + \left(\frac{n\pi}{b}\right)^2\right\}\right]^{½}. \tag{7.4.63}$$

The cut-off wave number is

$$\beta_c^{mn} = k_c^{mn} = \sqrt{\left(\frac{m\pi}{a}\right)^2 + \left(\frac{n\pi}{b}\right)^2}. \tag{7.4.64}$$

The cut-off frequency of the mnth mode is obtained from $k_c^{mn} = 2\pi f_c^{mn}\sqrt{\mu\varepsilon}$

$$f_c^{mn} = \frac{c}{2\pi\sqrt{\mu_r\varepsilon_r}}\sqrt{\left(\frac{m\pi}{a}\right)^2 + \left(\frac{n\pi}{b}\right)^2}; \quad m = 1, 2, 3.....; , n = 1, 2, 3..., \tag{7.4.65}$$

where c is the velocity of the EM wave in the free space. The dielectric medium has $\mu_r = 1$. The fundamental mode of all MWs waveguide for width $a > b$ is obtained by taking $m = n = 1$. The cut-off frequency for the TE_{11} mode is

$$f_c^{11} = \frac{c}{2\pi\sqrt{\varepsilon_r}}\sqrt{\left(\frac{\pi}{a}\right)^2 + \left(\frac{\pi}{b}\right)^2}. \tag{7.4.66}$$

The field components for the TE_{11} fundamental mode are obtained from equation (7.4.62).

TM^z Mode

The TM^z mode has E_z component and no longitudinal magnetic field component, i.e. $H_z = 0$. All other field components, given by equation (7.1.28), are generated by the electric scalar potential $\phi^e(x, y, z) = \phi^e(x, y)e^{-j\beta_z z}$. The modal function $\phi^e(x, y)$ is a solution of wave equation (7.4.20), under the boundary conditions on the MWs obtained from equation (7.1.28).

i) *Boundary condition on the left and right-hand sidewalls:*

$$H_y^e = 0, \text{ i.e. } \frac{\partial\phi^e(x, y)}{\partial x} = 0 \text{ at } x = 0, a; \quad 0 \le y \le b. \tag{7.4.67}$$

ii) *Boundary condition on the bottom and top walls:*

$$H_x = 0, \text{ i.e. } \frac{\partial\phi^e(x, y)}{\partial y} = 0 \text{ at } y = 0, b; \quad 0 \le x \le a. \tag{7.4.68}$$

The following eigenfunction $\phi_{mn}^e(x, y)$ for the TM^z mode is obtained as a solution of equation (7.4.20) under the above boundary conditions:

$$\left.\begin{aligned}\phi_{mn}^e(x, y) = B_{mn}\cos\left(\frac{m\pi}{a}x\right)\cos\left(\frac{n\pi}{b}y\right);\\ m = 0, 1, 2, ...; n = 0, 1, 2, 3...\\ m = n \neq 0\end{aligned}\right\}. \tag{7.4.69}$$

The field components of the TM^z modes are obtained on substituting equation (7.4.69) in equation (7.1.28)

$$
\begin{aligned}
E_x^e &= \frac{-j\beta_z}{j\omega\varepsilon}\frac{\partial\phi_{mn}^e(x, y)}{\partial x}\\
&= \frac{B_{mn}\beta_z}{\omega\varepsilon}\left(\frac{m\pi}{a}\right)\sin\left(\frac{m\pi}{a}x\right)\cos\left(\frac{n\pi}{b}y\right) && \text{(a)}\\
E_y^e &= \frac{-j\beta_z}{j\omega\varepsilon}\frac{\partial\phi_{mn}^e(x, y)}{\partial y}\\
&= \frac{B_{mn}\beta_z}{\omega\varepsilon}\left(\frac{n\pi}{b}\right)\cos\left(\frac{m\pi}{a}x\right)\sin\left(\frac{n\pi}{b}y\right) && \text{(b)}\\
E_z^e &= \frac{B_{mn}}{j\omega\varepsilon}(k^2-\beta_z^2)\cos\left(\frac{m\pi}{a}x\right)\cos\left(\frac{n\pi}{b}y\right) && \text{(c)}\\
H_x^e &= -B_{mn}\left(\frac{n\pi}{b}\right)\cos\left(\frac{m\pi}{a}x\right)\sin\left(\frac{n\pi}{b}y\right) && \text{(d)}\\
H_y^e &= B_{mn}\left(\frac{m\pi}{a}\right)\sin\left(\frac{m\pi}{a}x\right)\cos\left(\frac{n\pi}{b}y\right) && \text{(e)}.
\end{aligned}
\tag{7.4.70}
$$

In equation (7.4.70), both m and n cannot be simultaneously zero, as it leads to zero fields. All field components should be multiplied by the factor $e^{-j\beta_z z}$ to get the complete space-dependent field components. The time-harmonic field components are obtained by multiplying the above field components by the wave-factor $e^{j(\omega t - \beta_z z)}$. The modal propagation constant is

$$\beta_z^{mn} = \pm\left[k^2 - \left\{\left(\frac{m\pi}{a}\right)^2 + \left(\frac{n\pi}{b}\right)^2\right\}\right]^{½}. \tag{7.4.71}$$

The cut-off wave number is

$$\beta_c^{mn} = k_c^{mn} = \sqrt{\left(\frac{m\pi}{a}\right)^2 + \left(\frac{n\pi}{b}\right)^2}. \quad (7.4.72)$$

On using $k_c^{mn} = 2\pi f_c^{mn}\sqrt{\mu\varepsilon}$, the cut-off frequency of the mnth mode is

$$\left.\begin{aligned} f_c^{mn} &= \frac{c}{2\pi\sqrt{\mu_r\varepsilon_r}}\sqrt{\left(\frac{m\pi}{a}\right)^2 + \left(\frac{n\pi}{b}\right)^2}; \\ m &= 0, 1, 2...; n = 0, 1, 2.. \\ m &= n \neq 0 \end{aligned}\right\}. \quad (7.4.73)$$

The dielectric medium has $\mu_r = 1$. The fundamental mode for width $a > b$ is obtained by taking m=1, n=0. The cut-off frequency for the TM_{10} mode is

$$f_c^{10} = \frac{c}{2a\sqrt{\varepsilon_r}}. \quad (7.4.74)$$

The TM_{10} is the fundamental mode of all MWs based waveguide. The TE_{10} mode does not exist for this waveguide. The cut-off frequency of the TM_{10} mode is lower than the cut-off frequency of the TE_{11} mode. It is opposite to the case of a metallic waveguide. The field components for the fundamental TM_{10} mode are obtained from equation (7.4.70).

7.4.3 Rectangular Waveguide with Composite Electric and MWs

Figure (7.5c) shows a rectangular waveguide with top and bottom perfect conductors, forming the EWs, and the open sides, forming the MWs. This structure is a perfect parallel-plate waveguide structure that has no fringing field at the sidewalls. The electromagnetic energy is confined within the waveguide structure. This section discusses the modal EM-field configuration supported by the structure. The structure supports both the TE^z and TM^z modes. However, its fundamental mode is the *static* TEM mode, static in the (x–y)-plane, obtained as a solution of the Laplace's equation discussed in subsection (7.3.1). A microstrip line could be modeled by such a waveguide [B.13]. This aspect is discussed in chapter 8. The side MWs, over a limited frequency band, can also be realized using the *electronic/photonic bandgap* (EBG/PBG) structures [J.5]. The EBG/PBG is discussed in chapters 19 and 20.

TE^z Mode

The $\phi^m(x, y, z) = \phi^m(x, y)e^{-j\beta_z z}$, generating the TE^z mode with E_z=0 and $H_z \neq 0$, is obtained as a solution of wave equation (7.4.2) under the boundary conditions of section (7.2) and field equation (7.1.35).

i) *Boundary conditions at the bottom and top EWs:*

$$E_x^m = 0, \text{ i.e. } \frac{\partial\phi^m(x, y)}{\partial y} = 0 \text{ at } y = 0, b; \ 0 \le x \le a. \quad (7.4.75)$$

ii) *Boundary conditions at both magnetic sidewalls:*

$$H_z^m = 0, \text{ i.e.} \phi^m(x, y) = 0, \quad \text{at } x = 0, \ a; \ 0 \le y \le b. \quad (7.4.76)$$

The following modal eigenfunction is obtained by solving wave equation (7.4.2) under the above boundary conditions:

$$\phi_{mn}^m = A_{mn}\sin\left(\frac{m\pi}{a}x\right)\cos\left(\frac{n\pi}{b}y\right); \quad m = 1, 2, 3..; n = 0, 1, 2, 3.... \quad (7.4.77)$$

The following field components are obtained on substituting $\phi_{mn}^m e^{-j\beta_z z}$ in equation (7.1.34):

$$\begin{aligned}
E_x^m &= A_{mn}\left(\frac{n\pi}{b}\right)\sin\left(\frac{m\pi}{a}x\right)\sin\left(\frac{n\pi}{b}y\right) && (a)\\
E_y^m &= A_{mn}\left(\frac{m\pi}{a}\right)\cos\left(\frac{m\pi}{a}x\right)\cos\left(\frac{n\pi}{b}y\right) && (b)\\
H_x^m &= \frac{-j\beta_z}{j\omega\mu}\frac{\partial\phi^m}{\partial x} \\
&= \frac{-A_{mn}\beta_z}{\omega\mu}\left(\frac{m\pi}{a}\right)\cos\left(\frac{m\pi}{a}x\right)\cos\left(\frac{n\pi}{b}y\right) && (c)\\
H_y^m &= \frac{-j\beta_z}{j\omega\mu}\frac{\partial\phi^m}{\partial y} \\
&= \frac{A_{mn}\beta_z}{\omega\mu}\left(\frac{n\pi}{b}\right)\sin\left(\frac{m\pi}{a}x\right)\sin\left(\frac{n\pi}{b}y\right) && (d)\\
H_z^m &= \frac{A_{mn}}{j\omega\mu}(k^2 - \beta_z^2)\sin\left(\frac{m\pi}{a}x\right)\cos\left(\frac{n\pi}{b}y\right) && (e).
\end{aligned} \quad (7.4.78)$$

In equations (7.4.78), the modal propagation constant is used $\beta_z \to \beta_z^{mn}$. The modal propagation constant is given by

$$\beta_z^{mn} = \pm\left[k^2 - \left\{\left(\frac{m\pi}{a}\right)^2 + \left(\frac{n\pi}{b}\right)^2\right\}\right]^{1/2}. \quad (7.4.79)$$

The cut-off wave number is

$$\beta_c^{mn} = \sqrt{\left(\frac{m\pi}{a}\right)^2 + \left(\frac{n\pi}{b}\right)^2}; \quad m = 1, 2, 3...; n = 0, 1, 2, 3.... \quad (7.4.80)$$

The cut-off frequency of the mnth mode is

$$f_c^{mn} = \frac{c}{2\pi\sqrt{\mu_r\varepsilon_r}}\sqrt{\left(\frac{m\pi}{a}\right)^2 + \left(\frac{n\pi}{b}\right)^2}. \quad (7.4.81)$$

In the case of dielectric medium, $\mu_r = 1$ and for $a > b$, the first higher-order mode is TE^z_{10} mode with a cut-off frequency

$$f_c^{10} = \frac{c}{2a\sqrt{\varepsilon_r}}, \quad (7.4.82)$$

where c is the velocity of the EM-wave in the free space. The guided-wavelength of TE^z_{mn} mode can be computed. Its wave impedance is $\omega\mu/\beta_z$.

TM^z Mode

The scalar electric potential function $\phi^e(x,y,z) = \phi^e(x,y)e^{-j\beta_z z}$ generating the TM^z modes in the waveguide of Fig (7.5c) is obtained as a solution of the wave equation (7.4.20) under the following boundary conditions, obtained from equation (7.1.28).

i) *Boundary conditions at the bottom and top* EWs

$$E_z^e = 0, \quad \text{i.e. } \phi^e(x,y) = 0 \quad \text{at } y = 0, b; \; 0 \le x \le a. \quad (7.4.83)$$

ii) *Boundary conditions at both magnetic sidewalls:*

$$H_y^e = 0 \text{ i.e. } \frac{\partial \phi^e(x,y)}{\partial x} = 0 \text{ at } x = 0, a; \; 0 \le y \le b. \quad (7.4.84)$$

The modal scalar electric potential function $\phi^e_{mn}(x,y)$ is obtained by solving wave equation (7.4.20) under the above boundary conditions:

$$\phi^e_{mn}(x,y) = B_{mn}\cos\left(\frac{m\pi}{a}x\right)\sin\left(\frac{n\pi}{b}y\right); \quad m = 0,1,2,3\ldots; n = 1,2,3\ldots. \quad (7.4.85)$$

The field components of the TM^z_{mn} mode are obtained by substituting $\phi^e(x,y,z) = \phi^e(x,y)e^{-j\beta_z z}$ in equation (7.1.28):

$$E_x^e = \frac{-j\beta_z}{j\omega\varepsilon}\frac{\partial\phi^e_{mn}(x,y)}{\partial x} = \frac{B_{mn}\beta_z}{\omega\varepsilon}\left(\frac{m\pi}{a}\right)\sin\left(\frac{m\pi}{a}x\right)\sin\left(\frac{n\pi}{b}y\right) \quad (a)$$

$$E_y^e = \frac{-j\beta_z}{j\omega\varepsilon}\frac{\partial\phi^e_{mn}(x,y)}{\partial y} = \frac{-B_{mn}\beta_z}{\omega\varepsilon}\left(\frac{n\pi}{b}\right)\cos\left(\frac{m\pi}{a}x\right)\cos\left(\frac{n\pi}{b}y\right) \quad (b)$$

$$E_z^e = \frac{B_{mn}}{j\omega\varepsilon}(k^2 - \beta_z^2)\cos\left(\frac{m\pi}{a}x\right)\sin\left(\frac{n\pi}{b}y\right) \quad (c)$$

$$H_x^e = B_{mn}\left(\frac{n\pi}{b}\right)\cos\left(\frac{m\pi}{a}x\right)\cos\left(\frac{n\pi}{b}y\right) \quad (d)$$

$$H_y^e = B_{mn}\left(\frac{m\pi}{a}\right)\sin\left(\frac{m\pi}{a}x\right)\sin\left(\frac{n\pi}{b}y\right) \quad (e).$$

(7.4.86)

The propagation constant β_z of the TM^z_{mn} mode is given by

$$\beta_z^{mn} = \pm\left[k^2 - \left\{\left(\frac{m\pi}{a}\right)^2 + \left(\frac{n\pi}{b}\right)^2\right\}\right]^{1/2}. \quad (7.4.87)$$

The cut-off wave number is

$$\beta_c^{mn} = k_c^{mn} = \sqrt{\left(\frac{m\pi}{a}\right)^2 + \left(\frac{n\pi}{b}\right)^2}. \quad (7.4.88)$$

On using $k_c^{mn} = 2\pi f_c^{mn}\sqrt{\mu\varepsilon}$, the cut-off frequency of the TM^z_{mn} mode is

$$f_c^{mn} = \frac{c}{2\pi\sqrt{\mu_r\varepsilon_r}}\sqrt{\left(\frac{m\pi}{a}\right)^2 + \left(\frac{n\pi}{b}\right)^2}; \quad m = 0,1,2,3\ldots; \; n = 1,2,3,\ldots. \quad (7.4.89)$$

For the dielectric medium of the waveguide $\mu_r = 1$, the cut-off frequency of the TM^z_{01} mode is

$$f_c^{01} = \frac{c}{2b\sqrt{\varepsilon_r}}. \quad (7.4.90)$$

As $a > b$, the cutoff frequency f_c^{10} of the TE^z_{10} mode is less than f_c^{01} of the TM^z_{01} mode. However, this structure also supports the TEM mode with no cut-off frequency. Therefore, the TEM mode is the dominant mode of a parallel-plate waveguide and the TE^z_{10} mode is the first higher-order mode.

Current Distribution on the Bottom and Top Conductors

The modal surface current densities, on the upper and lower conducting planes of the composite surface waveguide, are determined from the surface magnetic fields:

$$\vec{J}^{mn} = \hat{n} \times \vec{H}_{mn}, \quad (7.4.91)$$

where $\pm\hat{n}$ is the direction of a unit vector normal to the conducting surface. The surface current flows on the inner surfaces of the upper and lower conductors. The current density is useful for the computation of conductor loss on each plane of the conducting surfaces.

TE^z Mode

At the upper conducting plane, i.e. at $y = b$, the magnetic field components from equation (7.4.78) are

$$H_x^m = \frac{-A_{mn}\beta_z}{\omega\mu}\left(\frac{m\pi}{a}\right)\cos\left(\frac{m\pi}{a}x\right)\cos(n\pi) = \frac{-A_{mn}\beta_z}{\omega\mu}\left(\frac{m\pi}{a}\right)\cos\left(\frac{m\pi}{a}x\right)(-1)^n \quad (a)$$

$$H_y^m = \frac{A_{mn}\beta_z}{\omega\mu}\left(\frac{n\pi}{b}\right)\sin\left(\frac{m\pi}{a}x\right)\sin(n\pi) = 0 \quad (b)$$

$$H_z^m = \frac{A_{mn}}{j\omega\mu}(k^2 - \beta_z^2)\sin\left(\frac{m\pi}{a}x\right)(-1)^n \quad (c).$$

(7.4.92)

Using equation (7.4.91) and (7.4.92), the surface current density at the upper conductor, y = b, is

$$\vec{J}^{mn} = -\hat{y} \times \left[\hat{x}H_x^m + \hat{z}H_z^m\right] = \hat{z}H_x^m - \hat{x}H_z^m.$$

Therefore, the longitudinal current density J_z^{mn} and transverse current density J_x^{mn} are

$$J_z^{mn} = H_x^m = -\frac{A_{mn}\beta_z}{\omega\mu}\left(\frac{m\pi}{a}\right)\cos\left(\frac{m\pi}{a}x\right)(-1)^n \quad (a)$$
$$J_x^{mn} = -H_z^m = -\frac{A_{mn}}{j\omega\mu}(k^2-\beta_z^2)\sin\left(\frac{m\pi}{a}x\right)(-1)^n \quad (b). \quad (7.4.93)$$

The current densities for a thin parallel-plate waveguide $b << \lambda_0$, (n = 0, m = 1, 2, 3...) are

$$J_z^{m0} = H_z^m = -\frac{A_{m0}\beta_z}{\omega\mu}\left(\frac{m\pi}{a}\right)\cos\left(\frac{m\pi x}{a}\right) \quad (a)$$
$$J_x^{m0} = -H_z^m = -\frac{A_{m0}}{j\omega\mu}(k^2-\beta_z^2)\sin\left(\frac{m\pi}{a}x\right) \quad (b). \quad (7.4.94)$$

The current density on the lower conducting plane, y = 0, is identical to equation (7.4.94), expect the change of sign from negative to positive, due to the return current. The path is completed through the displacement current. Figure (7.10a and b) show the magnitude of the longitudinal and transverse current distributions of the TE_{10} mode. To sketch the symmetrical current distribution, the origin is shifted by x = a/2. The longitudinal current shows edge singularities.

The E_y, H_x, H_z field components, in the composite waveguide for the TE^z_{m0} modes, are proportional to $\cos\left(\frac{m\pi x}{a}\right)$, $\cos\left(\frac{m\pi x}{a}\right)$ and $\sin\left(\frac{m\pi x}{a}\right)$, respectively.

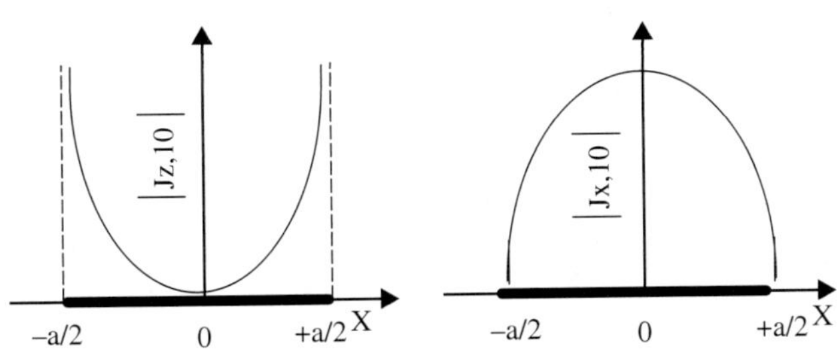

(a) Longitudinal current density distribution $|\cos\{(\pi/a)(x-a/2)\}|$. (b) Transverse current density distribution $|\sin\{(\pi/a)(x-a/2)\}|$.

Figure 7.10 Current distributions for the TE_{10} mode.

TM^z Mode

At the upper conducting plane, i.e. at y=b, the magnetic field components of the TM^z_{mn} modes are obtained from equation (7.4.86),

$$H_x^e = B_{mn}\left(\frac{n\pi}{b}\right)\cos\left(\frac{m\pi}{a}x\right)(-1)^n$$
$$H_y^e = 0,\ H_z^e = 0; \quad m = 0, 1, 2, 3..\ , n = 1, 2, 3...... \quad (7.4.95)$$

For the TM^z_{mn} mode, only the longitudinal surface current density exists,

$$\vec{J}_z^{mn} = -\hat{y} \times \left[\hat{x}H_x^e + \hat{y}H_y^e\right] = \hat{z}H_x^e$$
$$\Rightarrow J_z^{mn} = H_x^e = B_{mn}\left(\frac{n\pi}{b}\right)\cos\left(\frac{m\pi}{a}x\right)(-1)^n. \quad (7.4.96)$$

A thin parallel-plate waveguide has $b << \lambda_0$, (n = 0). However, it is not a valid solution, i.e. for n = 0, TM^z_{mn}-modes do not exist, as $\phi^e_{mn}(x, y) = 0$. The TM^z_{01}-mode is the first mode that comes into existence only at very high frequency. The magnitude of modal longitudinal current distributions for TM^z_{01}, TM^z_{11} and TM^z_{21} modes, with the origin shift by x = a/2, are shown in Fig (7.11)

There is no transverse current for the TM^z_{mn} modes i.e. $J_{x,\,mn} = 0$. The identical current density distribution is obtained on the lower conducting plane, with the reversal of sign. The E_y, E_z, and H_x field distributions for the TM^z_{01} are proportional to $\cos[(\pi/b)y]$, $\sin[(\pi/b)y]$, and $\cos[(\pi/b)y]$, respectively.

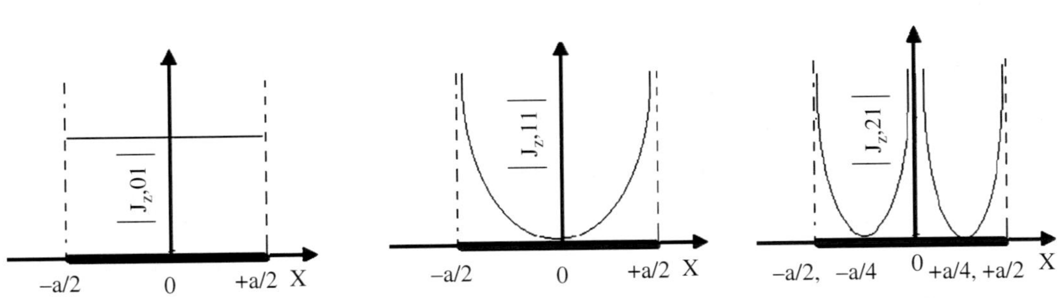

(a) TM^z_{01}-mode. (b) TM^z_{11}-mode. (c) TM^z_{21}-mode.

Figure 7.11 Longitudinal current distributions for the TM^z_{01}, TM^z_{11}, and TM^z_{21} modes.

7.5 Conductor Backed Dielectric Sheet Surface Wave Waveguide

The microstrip line in its limiting condition of zero strip conductor width could be treated as a conductor-backed dielectric waveguide that supports the surface wave propagation. Similarly, two coplanar ground plane portions of the CPW structure could also be regarded as the conductor-backed dielectric sheet. In the case of the usual planar lines, the undesired surface waves are excited on the substrates, adversely influencing the performances of the planar lines and also microstrip antenna. The undesired surface waves on a planar line increase the loss and undesired coupling between lines and components.

Figure (7.12a) shows a conductor-backed dielectric sheet waveguide, supporting the surface waves propagating in the z-direction. The dielectric sheet of thickness h_1 and relative permittivity ε_{r1} has an infinite extent in $\pm x$ direction. Therefore, the field is uniform in the *x*-direction, i.e. the field does not vary with *x*, so $\frac{\partial}{\partial x}(\text{field}) = 0$. The EM-wave propagates in the z-direction in the form of the TE^z and TM^z surface wave modes. Figure (7.12b) shows that physically speaking the field must decay exponentially in the *y*-direction in the air region above the dielectric-air interface. It is needed to achieve the EM-field confinement in the dielectric region of the waveguide structure.

7.5.1 TM^z Surface Wave Mode

The TM^z mode is expressed in terms of the scalar electric potential $\phi^e(x, y, z) = \phi^e(x, y)e^{-j\beta_z z}$ that satisfies wave equation (7.1.29). For the uniform field in the x-direction having $\frac{\partial \phi^e}{\partial x} = 0$, the wave equation for the scalar electric potential $\phi^e(x, y)$ is reduced to the following one-dimensional wave equations in the dielectric and air regions:

Region #1 (Dielectric region):

$$\frac{\partial^2 \phi_1^e(x,y)}{\partial y^2} + \left(\varepsilon_{r1} k_0^2 - \beta_z^2\right) \phi_1^e(x,y) = 0, \quad \text{for } 0 \le y \le h_1 \qquad \text{(a)}$$

Region #2 (Air region):

$$\frac{\partial^2 \phi_2^e(x,y)}{\partial y^2} + \left(k_0^2 - \beta_z^2\right) \phi_2^e(x,y) = 0, \quad \text{for } h_1 \le y \le \infty \qquad \text{(b)}. \tag{7.5.1}$$

To satisfy the phase matching at the interface $y = h_1$, the propagation constant β_z in the z-direction is common in both media. The cut-off wave numbers for two regions are defined as follows:

$$\textit{Region \#1 (Dielectric region)}: k_{c1}^2 = \left(\varepsilon_{r1} k_0^2 - \beta_z^2\right) \qquad \text{(a)}$$

$$\textit{Region \#2 (Air region)}: k_{c2}^2 = \left(\beta_z^2 - k_0^2\right) \qquad \text{(b)}. \tag{7.5.2}$$

The wave equations in both regions are rewritten as follows:

$$\frac{\partial^2 \phi_1^e(x,y)}{\partial y^2} + k_{c1}^2\, \phi_1^e(x,y) = 0, \quad \text{for } 0 \le y \le h_1 \qquad \text{(a)}$$

$$\frac{\partial^2 \phi_2^e(x,y)}{\partial y^2} - k_{c2}^2\, \phi_2^e(x,y) = 0, \quad \text{for } h_1 \le y \le \infty \qquad \text{(b)}. \tag{7.5.3}$$

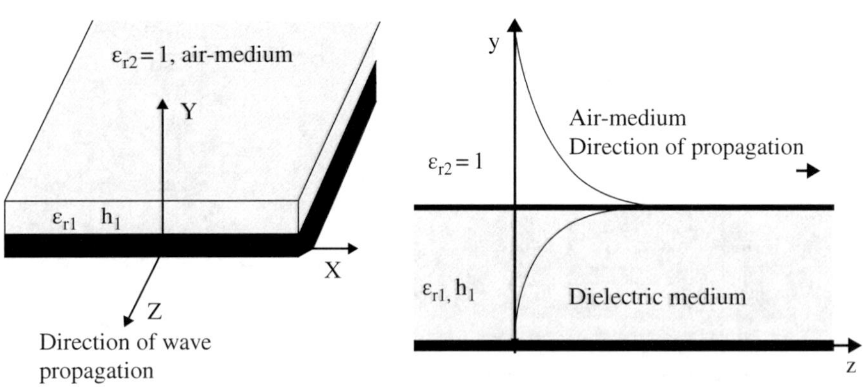

(a) Conductor-backed dielectric waveguide.
(b) Confinement of surface wave

Figure 7.12 Conductor-backed dielectric surface wave waveguide.

The solutions for wave–equation (7.5.3) are written as follows:

Region #1 (Dielectric region):

$$\phi_1^e(x,y) = A\sin(k_{c1}y) + B\cos(k_{c1}y) \quad (a)$$

Region #2 (Air region):

$$\phi_2^e(x,y) = C\,e^{k_{c2}(y-h_1)} + D\,e^{-k_{c2}(y-h_1)} \quad (b). \qquad (7.5.4)$$

The exponentially decaying field in the air-medium is required for the field confinement in the y-direction. It leads to C = 0, as the first term of equation (7.5.4b) does not provide the field confinement in the y-direction. Under the condition for a uniform field in the x-direction, i.e. for $\frac{\partial \phi^e}{\partial x} = 0$, the field components for the TM^z mode are written from equation (7.1.31) as follows:

$$E_x^e = 0, \quad E_y^e = \frac{-\beta_z}{\omega\varepsilon}\frac{\partial\phi^e(x,y)}{\partial y}$$

$$E_z^e = \frac{1}{j\omega\varepsilon}(k^2 - \beta_z^2)\phi^e(x,y) \qquad (7.5.5)$$

$$H_x^e = \frac{\partial\phi^e(x,y)}{\partial y}, \quad H_y^e = 0, \quad H_z^e = 0\cdot$$

Equation (7.5.5) is valid in both regions. It is noted that in the region #1, $\phi(x,y) = \phi_1^e(x,y)$ with the wavenumber $k = k_0\sqrt{\varepsilon_r}$ and in the region #2, $\phi(x,y) = \phi_2^e(x,y)$ with the wavenumber $k = k_0$. In region #1, at y = 0, i.e. at the ground conductor, $E_{z1}^e = 0$ i.e. $\phi_1^e(x,y,z) = 0$ leading B = 0. In region #2 for $y \to \infty$, $\phi_2^e(x, y \to \infty) = 0$ leading to C=0. Thus, equation (7.5.4) is written as

Region #1(Dielectric region):

$$\phi_1^e(x,y) = A\sin(k_{c1}y) \quad (a)$$

Region #2(Air region):

$$\phi_2^e(x,y) = D\,e^{-k_{c2}(y-h_1)} \quad (b). \qquad (7.5.6)$$

The fields in both regions are given below
Region #1(Dielectric region) $(0 \le y \le h_1)$:

$$E_{x1}^e = 0, \quad E_{y1}^e = \frac{-A k_{c1}\beta_z}{\omega\varepsilon_1}\cos(k_{c1}y)$$

$$E_{z1}^e = \frac{A}{j\,\omega\varepsilon_1}(\varepsilon_{r1}k_0^2 - \beta_z^2)\sin(k_{c1}y)$$

$$H_{x1}^e = A k_{c1}\cos(k_{c1}y), \quad H_{y1}^e = 0, \quad H_{z1}^e = 0. \qquad (7.5.7)$$

Region #2 (Air region, $y \ge h_1$):

$$E_{x2}^e = 0, \quad E_{y2}^e = \frac{D k_{c2}\beta_z}{\omega\varepsilon_0}e^{-k_{c2}(y-h_1)}$$

$$E_{z2}^e = -\frac{D}{j\omega\varepsilon_0}(\beta_z^2 - k_0^2)\,e^{-k_{c2}(y-h_1)} \qquad (7.5.8)$$

$$H_{x2}^e = -D k_{c2}e^{-k_{c2}(y-h_1)}$$

$$H_{y2}^e = 0, \quad H_{z2}^e = 0.$$

The continuity condition of the fields, across the dielectric-air interface, provides the following relations between the constants A and D:

At $y = h_1, E_{z1}^e = E_{z2}^e$:

$$\frac{A}{j\omega\varepsilon_{r1}\varepsilon_0}(\varepsilon_{r1}k_0^2 - \beta_z^2)\sin(k_{c1}h_1) = -\frac{D}{j\omega\varepsilon_0}(\beta_z^2 - k_0^2) \quad (a)$$

At $y = h_1, H_{x1}^e = H_{x2}^e$:

$$A k_{c1}\cos(k_{c1}h_1) = -D k_{c2} \quad (b). \qquad (7.5.9)$$

For the nontrivial solution of the constants A and D, the determinant of equation (7.5.9) must be zero,

$$\frac{k_{c1}^2 k_{c2}}{\varepsilon_{r1}}\sin(k_{c1}h) - k_{c2}^2 k_{c1}\cos(k_{c1}h_1) = 0$$

$$\Rightarrow k_{c1}\sin(k_{c1}h_1) = \varepsilon_{r1}k_{c2}\cos(k_{c1}h_1)$$

$$\Rightarrow \tan(k_{c1}h_1) = \frac{\varepsilon_{r1}k_{c2}}{k_{c1}} \qquad (7.5.10)$$

$$\Rightarrow \tan\left[\left\{\varepsilon_{r1} - (\beta_z/k_0)^2\right\}^{1/2}k_0h_1\right] = \varepsilon_{r1}\left[\frac{(\beta_z/k_0)^2 - 1}{\varepsilon_{r1} - (\beta_z/k_0)^2}\right]^{1/2}.$$

For a specified surface wave structure, with known thickness h_1 and relative permittivity ε_{r1}, the above transcendental equation can be solved for (β_z/k_0) at each assumed frequency, i.e. at assumed k_0. If $\theta = \left(\sqrt{\varepsilon_{r1} - (\beta_z/k_0)}\right)k_0h_1$, for the real value of θ, we have

$$1 \le \beta_z/k_0 \le \sqrt{\varepsilon_{r1}}. \qquad (7.5.11)$$

Physically, it means the propagation constant β_z of the surface wave structure is in between the propagation constants of the EM-wave in the air and the homogeneous dielectric media, giving phase velocity of the surface wave less than the velocity of EM-wave in free space. The lateral field confinement in the y-direction is achieved by the exponentially decaying field $e^{-k_{c2}(y-h_1)}$, where k_{c2}

is a real positive wave number in the y-direction. For $k_{c2} = 0$, the field confinement is lost for the TM^z mode. This is the condition for the cut-off phenomenon of the TM mode. Equation (7.5.2) provides the following expressions:

$$k_{c1}^2 + k_{c2}^2 = (\varepsilon_{r1} - 1)\, k_0^2. \tag{7.5.12}$$

Under the cut-off condition in the air-medium $k_{c2} = 0$, i.e. $\beta_z = k_o$ giving the following expression for the cut-off wavenumber of the TM^z surface wave mode:

$$k_{c1} = k_0\sqrt{\varepsilon_{r1} - 1}. \tag{7.5.13}$$

On using the cut-off condition with equation (7.5.10), the expression of the cut-off frequency is obtained for the TM^z mode of the conductor-backed dielectric sheet surface wave waveguide:

$$\begin{aligned} &\tan\left[\left(\sqrt{\varepsilon_{r1}} - 1\right) k_0 h_1\right] = 0 \\ \Rightarrow\ &\frac{2\pi f_c^{TM_n} h_1 \sqrt{\varepsilon_{r1} - 1}}{c} = n\pi \\ \Rightarrow\ &f_c^{TM_n} = \frac{n\,c}{2h_1\sqrt{\varepsilon_{r1} - 1}};\quad (n = 0, 1, 2, 3 \ldots). \end{aligned} \tag{7.5.14}$$

where c is the velocity of the EM-wave in the free space and $f_c^{TM_n}$ is the cut-off frequency of the TM_0, TM_1, TM_2, etc. surface wave modes. The TM_0 is the fundamental mode with no cut-off frequency, as $f_c^{TM_0} = 0$. Thus, the conductor-backed dielectric sheet always supports the TM_0 surface wave mode at any frequency, $f > 0$. At any frequency, equation (7.5.10) is solved to get the frequency-dependent effective relative permittivity,

$$\varepsilon_{r\,eff}^{TM}(f) = \left(\frac{\beta_z}{k_0}\right)^2. \tag{7.5.15}$$

The EM-field components of the TM_n^z surface wave mode obtained from equations (7.5.7) and (7.5.8), are summarized below. The constant D is replaced by using an equation (7.5.9).

Region #1 $(0 \le y \le h_1)$:

$$E_{y1}^e = \frac{-A\,k_{c1}\beta_z}{\omega\varepsilon_1}\cos(k_{c1}y)\,e^{-j\beta_z z} \quad \text{(a1)}$$

$$E_{z1}^e = \frac{A\,k_{c1}^2}{j\omega\varepsilon_1}\sin(k_{c1}y)\,e^{-j\beta_z z} \quad \text{(b1)}$$

$$H_{x1}^e = A\,k_{c1}\cos(k_{c1}y)\,e^{-j\beta_z z} \quad \text{(c1)}$$

Region #2 $(h_1 \le y < \infty)$:

$$E_{y2}^e = \frac{-A\,k_{c1}^2\beta_z}{\omega\varepsilon_1 k_{c2}}\sin(k_{c1}h_1)\,e^{-k_{c2}(y-h_1)}e^{-j\beta_z z} \quad \text{(a2)}$$

$$E_{z2}^e = \frac{A\,k_{c1}^2}{j\omega\varepsilon_1}\sin(k_{c1}h_1)\,e^{-k_{c2}(y-h_1)}e^{-j\beta_z z} \quad \text{(b2)}$$

$$H_{x2}^e = \frac{A\,k_{c1}^2}{\varepsilon_{r1}k_{c2}}\sin(k_{c1}h_1)\,e^{-k_{c2}(y-h_1)}e^{-j\beta_z z} \quad \text{(c2)}\,.$$

(7.5.16)

The field components are maximum at $y = h_1$, which is at the interface of the dielectric and air media. It decays exponentially in the air-medium away from the interface. The EM-wave propagating in the y-direction has energy confinement near the surface of the dielectric sheet, therefore, the wave is known as the *surface wave*. Figure (7.12b) shows the field confinement at the interface of the dielectric and air media.

7.5.2 TE^z Surface Wave Mode

The TE^z surface wave mode is generated by the scalar magnetic potential $\phi^m(x, y, z) = \phi^m(x, y)e^{-j\beta_z z}$. The uniform field in the x-direction has $\frac{\partial\phi^m}{\partial x} = 0$. The function $\phi^m(x, y)$ satisfies the following one-dimensional wave equations, obtained from equation (7.1.36c), in both the regions:

Region #1(Dielectric region):

$$\frac{\partial^2\phi_1^m(x,y)}{\partial y^2} + \left(\varepsilon_{r1}k_0^2 - \beta_z^2\right)\phi_1^m(x,y) = 0 \quad \text{(a)}$$

Region #2(Air region):

$$\frac{\partial^2\phi_2^m(x,y)}{\partial y^2} + \left(k_0^2 - \beta_z^2\right)\phi_2^m(x,y) = 0, \quad \text{(b)}.$$

(7.5.17)

The cut-off wavenumbers for the TE^z modes are defined as

$$\text{Region \#1(Dielectric region)}: k_{c1}'^2 = \left(\varepsilon_{r1}k_0^2 - \beta_z^2\right) \quad \text{(a)}$$

$$\text{Region \#2(Air region)}: k_{c2}'^2 = \left(\beta_z^2 - k_0^2\right) \quad \text{(b)}.$$

(7.5.18)

Wave equation (7.5.17), in both regions, can be rewritten as follows:

Region #1(Dielectric region):

$$\frac{\partial^2\phi_1^m(x,y)}{\partial y^2} + k_{c1}'^2\phi_1^m(x,y) = 0, \quad \text{(a)}$$

Region #2(Air region):

$$\frac{\partial^2\phi_2^m(x,y)}{\partial y^2} - k_{c2}'^2\phi_2^m(x,y) = 0, \quad \text{(b)}.$$

(7.5.19)

The solutions for the scalar magnetic potential in both regions are written as

Region #1(Dielectric region, $0 \le y \le h_1$):

$$\phi_1^m(x,y) = A\sin(k'_{c1}y) + B\cos(k'_{c1}y) \quad (a)$$

Region #2(Air region, $h_1 \le y < \infty$):

$$\phi_2^m(x,y) = C\,e^{k'_{c2}(y-h_1)} + D\,e^{-k'_{c2}(y-h_1)} \quad (b). \tag{7.5.20}$$

On using $\frac{\partial\phi^m}{\partial x} = 0$ as the uniform field is in the x-direction, the field components for the TE^z modes are written from equation (7.1.35):

$$E_x^m = -\frac{\partial\phi^m(x,y)}{\partial y}, \quad E_y^m = 0, \quad E_z^m = 0$$
$$H_x^m = 0, \quad H_y^m = \frac{-\beta_z}{\omega\mu}\frac{\partial\phi^m(x,y)}{\partial y}$$
$$H_z^m = \frac{1}{j\omega\mu}\left(k^2 - \beta_z^2\right)\phi^m(x,y). \tag{7.5.21}$$

In the region# 1, at y=0, $E_x^m = 0$ i.e. $\frac{\partial\phi_1^m(x,y)}{\partial y} = 0$. It results in $A = 0$ from equation (7.5.20a). In region #2 for $y \to \infty$, the constant C of equation (7.5.20b) is $C = 0$. Thus, equation (7.5.20) is reduced to following equation:

Region #1(Dielectric region, $0 \le y \le h_1$):

$$\phi_1^m(x,y) = B\cos(k'_{c1}y) \quad (a)$$

Region #2(Air region, $h_1 \le y < \infty$):

$$\phi_2^m(x,y) = D\,e^{-k'_{c2}(y-h_1)} \quad (b). \tag{7.5.22}$$

The EM-field components in both regions can be written as

Region #1(Dielectric region, $0 \le y \le h_1$):

$$E_{x1}^m = B\,k'_{c1}\sin(k'_{c1}y)$$
$$H_{y1}^m = \frac{B\,k'_{c1}\beta_z}{\omega\mu_1}\sin(k'_{c1}y), \quad H_{z1}^m = \frac{B\,k'^2_{c1}}{j\omega\mu_1}\cos(k'_{c1}y) \quad (a)$$

Region #2(Air region, $h_1 \le y < \infty$):

$$E_{x2}^m = D\,k'_{c2}e^{-k'_{c2}(y-h_1)}$$
$$H_{y2}^m = \frac{D\,k'_{c2}\beta_z}{\omega\mu_0}e^{-k'_{c2}(y-h_1)}$$
$$H_{z2}^m = \frac{-D\,k'^2_{c2}}{j\omega\mu_0}e^{-k'_{c2}(y-h_1)} \quad (b). \tag{7.5.23}$$

The relation between constants B and D is obtained through the continuity condition of the field components at $y = h_1$ giving $E_{x1}^m = E_{x2}^m$ and $H_{z1}^m = H_{z2}^m$. On substituting the field expressions from (7.5.23a,b), the following expressions are obtained

$$B\,k'_{c1}\sin(k'_{c1}h_1) = D\,k'_{c2} \quad (a)$$
$$\frac{B\,k'^2_{c1}}{j\omega\mu_1}\cos(k'_{c1}h_1) = \frac{-D\,k'^2_{c2}}{j\omega\mu_0} \quad (b). \tag{7.5.24}$$

In the case of a dielectric slab $\mu_1 = \mu_0$. For the nontrivial solution of B and D, the determinant of equation (7.5.24) must be zero giving

$$\frac{k'_{c1}k'^2_{c2}\sin(k'_{c1}h_1)}{j\omega\mu_0} + \frac{k'_{c2}k'^2_{c1}\cos(k'_{c1}h_1)}{j\omega\mu_0} = 0$$
$$\Rightarrow \cot(k'_{c1}h_1) = -\frac{k'_{c2}}{k'_{c1}} \quad (a)$$
$$\Rightarrow \cot\left[\left\{\varepsilon_{r1} - (\beta_z/k_0)^2\right\}^{1/2}k_0h_1\right] = -\left[\frac{(\beta_z/k_0)^2 - 1}{\varepsilon_{r1} - (\beta_z/k_0)^2}\right]^{1/2} \quad (b). \tag{7.5.25}$$

For a given substrate, i.e. known substrate thickness and relative permittivity, and assumed frequency, transcendental equation (7.5.25) could be solved for the propagation constant β_z. The propagation constant provides the frequency-dependent effective relative permittivity of the TE^z surface wave modes,

$$\varepsilon_{r\,eff}^{TE}(f) = \left(\frac{\beta_z}{k_0}\right)^2. \tag{7.5.26}$$

Equation (7.5.26) provides the dispersions in the TE_n surface wave modes. At the cut-off frequency, the field confinement in the y-direction is lost and it occurs at $k'_{c2} = 0$. Using this condition in equation (7.5.18), the cut-off wavenumber in the dielectric region #1 could be obtained,

$$k'_{c1} = k_0\sqrt{(\varepsilon_{r1} - 1)}. \tag{7.5.27}$$

The cut-off frequency for the TE^Z mode is obtained from equation (7.5.25)

$$\cot(k'_{c1}h_1) = 0 \quad \Rightarrow k'_{c1}h_1 = (2n-1)\frac{\pi}{2} \quad (a)$$
$$\Rightarrow \frac{2\pi f_c^{TE_n}}{c}\sqrt{(\varepsilon_{r1} - 1)}\,h_1 = (2n-1)\frac{\pi}{2}$$
$$\Rightarrow f_c^{TE_n} = \frac{(2n-1)\,c}{4\,h_1\sqrt{(\varepsilon_{r1} - 1)}}, \quad n = 1,2,3\ldots \quad (b), \tag{7.5.28}$$

where c is the velocity of the EM-wave in the free space. The field components of the TE^z surface wave mode are written in both regions using equation (7.5.23). The constant D could be written in terms of B, using equation (7.5.24). The multilayer substrate is used for several applications. This method can be extended to such cases [B.12]. However, for obtaining the propagation constant a simpler method, discussed in Section (7.7), could be followed.

7.6 Equivalent Circuit Model of Waveguide

A waveguide transports the EM-power in the form of electric and magnetic fields. The transportation of the EM-power is described by the wave impedance or wave admittance that is a ratio of the transverse electric and transverse magnetic fields. This aspect of the guided wave propagation, in the form of the TE and TM modes, is discussed in section (7.4). Each mode of a waveguide transports power independently. Each mode can be described by an *equivalent transmission line* in terms of the *modal voltage* and *modal current*. An equivalent transmission line is described by its characteristic impedance or characteristic admittance. Therefore, the *modal wave impedance* of the original waveguide has to be correlated to the characteristic impedance of an equivalent transmission line. The equivalent transmission line concept is useful in solving complex waveguide problems. Both the TEM mode supporting parallel-plate waveguide, and its equivalent transmission line are discussed in section 7.3. This section discusses the equivalent transmission line of the TE and TM modes supporting waveguide [B.1–B.3, B.10]. The TE/TM equivalent transmission lines are further modeled by the lumped element equivalent circuit models. In this case, the circuit elements are expressed in terms of the original waveguide parameters. Furthermore, the transmission line concept is useful to develop a method called the *transverse resonance method* (TRM) which is a powerful method for the determination of propagation constant of more complex waveguide structures [B.9]. It is discussed in section (7.7).

7.6.1 Relation Between Wave Impedance and Characteristic Impedance

A summary of the wave impedance and the propagation constant for the TE^z and TM^z modes are given below:

$$Z_w^{TE^z} = \frac{\omega\mu}{\beta_z^{mn}} \quad (a), \qquad Z_w^{TM^z} = \frac{\beta_z^{mn}}{\omega\varepsilon} \quad (b)$$
$$\beta_z^{mn} = \sqrt{\omega^2\mu\varepsilon - k_c^2} = \omega\sqrt{\mu\varepsilon}\sqrt{1-(\omega_c/\omega)^2} \quad (c). \tag{7.6.1}$$

The expression for the guided-wavelength is

$$\lambda_g = \frac{2\pi}{\beta_z^{mn}} = \frac{\lambda}{\sqrt{1-(\omega_c/\omega)^2}} = \frac{\lambda}{\sqrt{1-(\lambda/\lambda_c)^2}}$$
$$\Rightarrow \frac{1}{\lambda_g^2} = \frac{1}{\lambda^2} - \frac{1}{\lambda_c^2}. \tag{7.6.2}$$

The waveguide supporting the EM-field propagation is characterized by its propagation constant and the wave impedance. However, it is also modeled as a transmission line. In this case, its characteristics impedance has to be defined. For an air-filled waveguide, $\lambda = \lambda_0$, the characteristic impedance Z_0 of an infinity long transmission line can be defined in three different ways by using voltage, current and power relations,

$$Z_0^{VI} = \frac{V}{I} \quad (a), \quad Z_0^{PI} = \frac{2\,P}{I^2} \quad (b), \quad Z_0^{PV} = \frac{V^2}{2\,P} \quad (c), \tag{7.6.3}$$

where V and I are line voltage and line current. P is the power carried by the line. For a line supporting the TEM mode, all three definitions give the identical value of the characteristic impedance. However, for a non-TEM mode waveguide, all three definitions give different results.

Let us examine the characteristic impedance of the TE_{10} mode rectangular waveguide shown in Fig (7.5a). The voltage between the lower and upper conductors of the waveguide cannot be defined uniquely. It is path-dependent. However, the maximum voltage V_m is defined at the center of the waveguide, i.e. at $x = a/2$. Thus, the maximum voltage for the TE_{10} mode is obtained from equation (7.4.34)

$$V_m = -\int_0^b E_y(\max)dy = \int_0^b \left(\frac{\pi}{a}B_{10}\right)dy = \frac{\pi\,b}{a}B_{10}. \tag{7.6.4}$$

The current density, from equation (7.4.35c), at the lower broad-face of the waveguide is

$$J_z = -H_x = -\frac{\pi}{a}\frac{\beta_z^{10}}{\omega\mu}B_{10}\sin\left(\frac{\pi}{a}x\right).$$

The current flowing at the lower broad-face of the waveguide in the z-direction is

$$I = -\frac{\pi}{a}\frac{\beta_z^{10}}{\omega\mu}B_{10}\int_0^a \sin\left(\frac{\pi}{a}x\right)dx \tag{7.6.5}$$

$$\Rightarrow \quad I = \frac{2\beta_z^{10}}{\omega\mu}B_{10}.$$

The power transported by the TE_{10} mode in the *z*-direction is obtained by using the Poynting vector and equation (7.4.34):

$$P = \frac{1}{2}\int_0^b\int_0^a \left[\frac{\pi}{a}B_{10}\sin\left(\frac{\pi}{a}x\right)\frac{\pi}{a}\frac{\beta_z^{10}}{\omega\mu}B_{10}\sin\left(\frac{\pi}{a}x\right)\right]dxdy$$

$$\Rightarrow \quad P = \frac{1}{2}\left(\frac{\pi}{a}\right)^2 B_{10}^2\frac{\beta_z^{10}}{\omega\mu}\cdot\frac{ab}{2}. \tag{7.6.6}$$

The characteristic impedance of the equivalent transmission line for the TE_{10} mode waveguide could be obtained by using the voltage-current definition. It is also related to the wave impedance of the TE_{10} mode,

$$Z_0^{VI} = \frac{V_m}{I} = \frac{\pi b}{a}B_{10}\times\frac{\omega\mu}{2\beta_z^{10}B_{10}} = \frac{\pi b}{2a}\frac{\omega\mu}{\beta_z^{10}} \qquad (a)$$

$$Z_0^{VI} = \frac{\pi b}{2a}Z_W^{TE^z} = \frac{\pi b\eta}{2a\sqrt{1-(\omega_c/\omega)^2}} \qquad (b)$$

$$\Rightarrow \quad Y_0^{VI} = \frac{2a}{\pi b}Y_W^{TE^z} \qquad (c), \tag{7.6.7}$$

where η is the intrinsic impedance of the homogeneous medium inside the waveguide. Equation (7.6.1a) is used to obtain the above expression that relates the characteristic impedance to wave impedance. Expression (7.6.7) is valid for the TM^z mode also. However, the wave impedance $Z_W^{TE^z}$ of the TE^z mode is replaced by the wave impedance of the TM^z mode, i.e. by taking $Z_W^{TM^z} = \beta_z^{TM}/\omega\varepsilon$. At frequency $\omega < \omega_c$, the wave impedance is an imaginary quantity giving the reactive characteristic impedance Z_0^{VI}. For frequency $\omega > \omega_c$, the characteristic impedance Z_0^{VI} is a real quantity. Therefore, both the waveguide and its equivalent transmission line behave as the high-pass filters.

Let us also compute the characteristic impedance of the equivalent transmission line of the TE_{10} mode waveguide by using the power-current definition. It is achieved with the help of equations (7.6.5) and (7.6.6),

$$Z_0^{PI} = \frac{2P}{I^2} = \frac{2\times\frac{1}{2}\left(\frac{\pi}{a}\right)^2 B_{10}^2\frac{\beta_z^{10}}{\omega\mu}\frac{ab}{2}}{\left(\frac{2\beta_z^{10}}{\omega\mu}\right)^2 B_{10}^2} = \frac{\pi^2 b}{8a}\cdot\frac{\omega\mu}{\beta_z^{10}}$$

$$= \frac{\pi^2 b}{8a}Z_W^{TE^z} = \frac{\pi}{4}Z_0^{VI}. \tag{7.6.8}$$

Likewise, using the power-voltage definition, the characteristic impedance, for the equivalent transmission line of a TE_{10} mode waveguide, is obtained below. Equations (7.6.4) and (7.6.6) are used.

$$Z_0^{PV} = \frac{V_m^2}{2P} = \frac{\left(\frac{\pi b}{a}\right)^2 B_{10}^2}{2\times\frac{1}{2}\left(\frac{\pi}{a}\right)^2 B_{10}^2\frac{\beta_z^{10}}{\omega\mu}\frac{ab}{2}}$$

$$= \frac{2b}{a}\cdot\frac{\omega\mu}{\beta_z^{10}} = \frac{2b}{a}Z_W^{TE^z} = \frac{4}{\pi}Z_0^{VI}. \tag{7.6.9}$$

We note that all three definitions of the characteristic impedance of the equivalent transmission line differ only by a constant. For the non-TEM mode line, the characteristic impedance is suitably defined depending upon the line structure. For instance, in the case of a microstrip line, the power-current definition is more common [B.13, B.15, J.6, J.7], whereas for a coplanar waveguide and slot lines, the power-voltage definition is more acceptable [B.13, B.15, J.8]. The concept discussed in this section is applied to the analysis of the planar slot line [J.9] in chapter 10.

7.6.2 Transmission Line Model of Waveguide

The equivalent transmission line model of a waveguide can be obtained by using the equivalence of power carried by the original waveguide and the equivalent transmission line. Furthermore, characteristic impedance and propagation constant of the equivalent transmission line can be taken the same as the wave impedance and propagation constant of the waveguide modes. The equivalent modal voltage and modal current can be obtained from Maxwell's equation [B.1, B.3, B.10]. The lumped circuit model of the equivalent transmission line could be obtained from the results of the equivalent transmission line with respect to Fig (7.13).

In chapter 2, the following expressions are obtained for the characteristic impedance and propagation constant of the TEM mode transmission line, shown in Fig (7.13a):

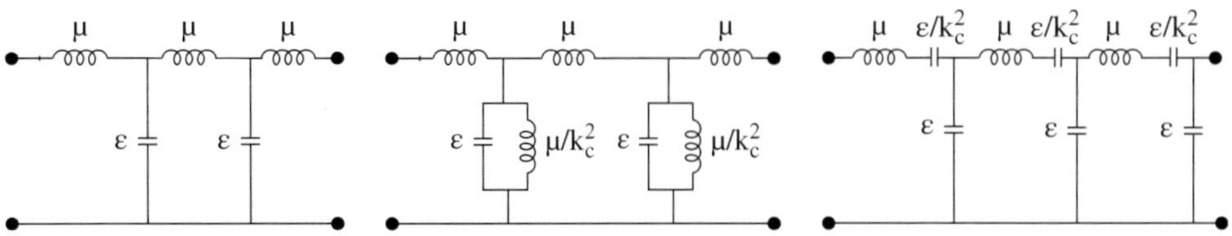

(a) TEM-mode line. (b) TE-mode line. (c) TM-mode line.

Figure 7.13 Lumped element circuit model of the waveguide modes.

$$Z_c = \pm\sqrt{\frac{Z}{Y}} = \pm\sqrt{\frac{X}{B}} \quad \text{(a)}, \quad \gamma = \pm j\sqrt{XB} \quad \text{(b)}. \tag{7.6.10}$$

Equations (7.6.10) (+) sign is used for the forward-moving wave, whereas (−) sign is used for the backward-moving wave. Equations (7.3.13a,b) show correspondence of series L, and shunt C to permittivity and permeability of medium respectively, For the TE-mode waveguide, the characteristic impedance of the equivalent transmission line is equal to the wave impedance $(\omega\mu/\beta^{TE})$ and its propagation constant is β^{TE}. The series reactance X^{TE} *p.u.l.* and shunt susceptance B^{TE} *p.u.l.* of the equivalent TE-mode transmission line are obtained as follows:

$$Z_c^{TE}.\gamma^{TE} = \left[j\sqrt{\frac{X^{TE}}{B^{TE}}}\sqrt{X^{TE}B^{TE}} = jX^{TE}\right]_{\text{line}}$$

$$= \left[j\frac{\omega\mu}{\beta^{TE}}\beta^{TE}\right]_{\text{waveguide}}, \quad \Rightarrow \quad Z = jX^{TE} = j\omega\mu. \tag{7.6.11}$$

$$\text{Also, } B^{TE} = -\frac{(\gamma^{TE})^2}{X^{TE}} = -\frac{(k_c^2 - \omega^2\mu\varepsilon)}{\omega\mu}$$

$$\Rightarrow \quad Y = jB^{TE} = j\left(\omega\varepsilon - \frac{k_c^2}{\omega\mu}\right). \tag{7.6.12}$$

Equations (7.6.11) and (7.6.12) provide the series impedance Z *p.u.l.* and shunt admittance Y *p.u.l.* of the equivalent transmission line model for the TE mode as shown in Fig (7.13b). The series inductance of the equivalent transmission line is $L = \mu$ *p.u.l.* and the shunt connected parallel resonant circuit components *p.u.l.* are $L = \mu/k_c^2$ and $C = \varepsilon$.

For the TM-mode supporting rectangular waveguide, the series reactance X^{TM} *p.u.l.* and shunt susceptance B^{TM} *p.u.l.* of its equivalent TM-mode transmission line are shown in Fig (7.13c). The X^{TM} and B^{TM} are obtained similarly:

$$\left[Z_0^{TM}\gamma^{TM} = \frac{\beta^{TM}}{\omega\varepsilon}j\beta^{TM} = j\frac{\beta^{2TM}}{\omega\varepsilon}\right]_{\text{waveguide}} = \left[jX^{TM}\right]_{\text{line}}$$

$$\Rightarrow Z = jX^{TM} = \frac{j(\omega^2\mu\varepsilon - k_c^2)}{\omega\varepsilon} = j\omega\mu + \frac{k_c^2}{j\omega\varepsilon}. \tag{7.6.13}$$

$$\text{Also, } B^{TM} = -\frac{(\gamma^{TM})^2}{X^{TM}} = \frac{\beta^{2TM}}{\beta^{2TM}/\omega\varepsilon} = \omega\varepsilon$$

$$\Rightarrow \quad Y = jB^{TM} = j\omega\varepsilon. \tag{7.6.14}$$

Equation (7.6.14) provide the equivalent TM-mode transmission line shown in Fig (7.13c). The series arm has series resonant circuit elements $L = \mu$ and $C = \varepsilon/k_c^2$ *p.u.l.* whereas the shunt arm has capacitance $C = \varepsilon$ *p.u.l.*

The reactance and susceptance loaded equivalent transmission lines for both the TE and TM modes behave as the high-pass filters. The cut-off frequency for both cases is $\omega_c = k_c/\sqrt{\mu\varepsilon}$. The characteristic impedance and propagation constant for both the equivalent lines for the propagating waves are obtained as follows:

$$Z_0^{TE} = \sqrt{\frac{X^{TE}}{B^{TE}}} = \sqrt{\frac{j\omega\mu}{j\omega\varepsilon + k_c^2/j\omega\mu}} = \eta\frac{1}{\sqrt{1-(f_c/f)^2}} \quad \text{(a)}$$

$$\gamma^{TE} = \pm j\left[\omega\mu\left(\omega\varepsilon - \frac{k_c^2}{\omega\mu}\right)\right]^{1/2} = \pm j\left[k^2 - k_c^2\right]^{1/2} \quad \text{(b)}$$

$$Z_0^{TM} = \sqrt{\frac{X^{TM}}{B^{TM}}} = \sqrt{\frac{j\omega\mu + k_c^2/j\omega\varepsilon}{j\omega\varepsilon}} = \eta\sqrt{1-(f_c/f)^2} \quad \text{(c)}$$

$$\gamma^{TM} = \pm j\left[\omega\varepsilon\left(\omega\mu - \frac{k_c^2}{\omega\varepsilon}\right)\right]^{1/2} = \pm j\left[k^2 - k_c^2\right]^{1/2} \quad \text{(d)}. \tag{7.6.15}$$

Expressions (7.6.15) are the same as those given in equations (7.4.31) and (7.4.32). Figure (7.6) shows that at the cut-off frequency $Z_0^{TE} \to \infty$ and $Z_0^{TM} \to 0$, i.e. at the cut-off frequency, there is no wave propagation. The equivalent transmission line of TE-mode, shown in Fig (7.13b), at a frequency below cut-off frequency $\omega < \omega_c$ becomes all inductive networks as it is an L–L line. It shows that below the cut-off frequency, the TE-mode waveguide acts as a 1D ENG medium. Above the cut-off frequency $\omega > \omega_c$, the waveguide acts as a 1D DPS medium. Thus, the below cut-off frequency, the TE mode waveguide, axially loaded with negative permeability $(-|\mu_r|)$, acts as the 1D DNG medium.

The negative permittivity ($-|\varepsilon_r|$) is offered by the TE mode waveguide below the cut-off frequency. The negative permeability could be realized with the help of printed split-ring resonators (SRR) [J.10]. The SRR is discussed in the section (21.2) of chapter 21. Likewise, the equivalent transmission line of TM-mode, shown in Fig (7.13c), becomes all capacitive C–C line below the cut-off frequency. Thus, the TM-mode waveguide below cut-off frequency acts as a 1D MNG medium. Such TM – mode waveguide axially loaded with the negative permittivity wire medium can act as a 1D DNG medium. The wire medium is discussed in section (21.1) of chapter 21. The behavior of the L–L line as the ENG plasma type medium and behavior of the C–C line as the MNG type medium are presented in the section (5.5) of chapter 5. The evanescent mode has propagation constant $\beta_z = -j\alpha$ and equations (7.4.31b) and (7.4.32b) provide inductive and capacitive wave impedances for the TE and TM modes, respectively, below the cut-off frequency.

The waveguide in the TE/TM mode acts as 1D medium – DPS, ENG, and MNG medium. So the equivalent permittivity $\left(\varepsilon_{eq}^{TE}, \varepsilon_{eq}^{TM}\right)$ and equivalent permeability $\left(\mu_{eq}^{TE}, \mu_{eq}^{TM}\right)$ of both modes of waveguide medium could be also computed. The propagation constants (β^{TE}, β^{TM}) for both modes are obtained from equations (7.6.15b and d) to compute equivalent permittivity and equivalent permeability:

$$\beta^{TE} = \left[\omega\mu\left(\omega\varepsilon - \frac{k_c^2}{\omega\mu}\right)\right]^{1/2} = \omega\sqrt{\mu\left[\varepsilon\left(1 - \frac{k_c^2}{\omega^2\mu\varepsilon}\right)\right]} = \omega\sqrt{\mu_{eq}^{TE}\varepsilon_{eq}^{TE}} \quad \text{(a)}$$

$$\text{where, } \mu_{eq}^{TE} = \mu,\ \varepsilon_{eq}^{TE} = \varepsilon\left(1 - \frac{k_c^2}{\omega^2\mu\varepsilon}\right) = \varepsilon\left(1 - \frac{\omega_c^2}{\omega^2}\right) \quad \text{(b)}$$

$$\beta^{TM} = \left[\omega\varepsilon\left(\omega\mu - \frac{k_c^2}{\omega\varepsilon}\right)\right]^{1/2} = \omega\sqrt{\varepsilon\left[\mu\left(1 - \frac{k_c^2}{\omega^2\mu\varepsilon}\right)\right]} = \omega\sqrt{\mu_{eq}^{TM}\varepsilon_{eq}^{TM}} \quad \text{(c)}$$

$$\text{where, } \varepsilon_{eq}^{TM} = \varepsilon,\ \mu_{eq}^{TM} = \mu\left(1 - \frac{k_c^2}{\omega^2\mu\varepsilon}\right) = \mu\left(1 - \frac{\omega_c^2}{\omega^2}\right) \quad \text{(d)}.$$

$$(7.6.16)$$

Equations (7.6.16) demonstrate the plasma-like behavior of the waveguide. Below cut-off frequency, the negative permittivity and negative permeability are realized for the TE and TM mode, respectively. They result in the 1D ENG and MNG medium.

7.7 Transverse Resonance Method (TRM)

The transverse resonance method (TRM) is a powerful and simple method to determine the propagation constant of a layered medium waveguide, including the surface wave supporting dielectric slab waveguides [B.2, B.3, B.5, B.9]. The method is also applicable to the traveling wave antenna [B.5, B.14]. The guided wave structures have EM – field confinement in the lateral direction, i.e. in the transverse direction, either due to the presence of metallic walls, as in the case of a standard waveguide, or due to the exponentially decaying field, i.e. the evanescent mode field, as in the case of a planar open surface waveguide. In the transverse plane, transverse to the direction of propagation, these structures could be treated as a *resonating section* of the equivalent transmission line. The resonating line avoids the traveling wave and confines the field in the form of a standing wave. The waveguide in the transverse direction could be treated as the equivalent transmission supporting the resonance.

It is assumed that the characteristic impedance of the equivalent transmission line is equal to the wave impedance of the mode supported by the waveguide. The field is confined, i.e. trapped, in the transverse direction. So a condition of resonance is assumed to exist in the transverse plane of a waveguide. Therefore, the method is called the *transverse resonance method* (TRM). At a suitable location in the transverse plane, the total impedance or total admittance looking toward the left and right-hand sides of the location is either zero or infinity. Alternatively, at any location, under the resonance condition, the total reactance or susceptance is zero. This section uses the concept of TRM to determine the propagation constants of several waveguide structures.

7.7.1 Standard Rectangular Waveguide

Figure (7.14a) shows the cross-section of a standard rectangular waveguide supporting TE_{10} mode. The field is confined along the x-axis due to the conducting walls (PEC) at $x = 0$ and $x = a$. Using equation (7.6.1a), the TE mode wave impedance of the wave propagating in the *z*-direction is

$$Z_w^{TE^z} = \frac{E_y}{-H_x} = \frac{\omega\mu_0}{\beta_z^{10}}. \quad (7.7.1)$$

To use the TRM, a plane wave is assumed with components E_y and H_z, propagating in the x-direction, such that the equivalent transmission is along the x-axis. Its characteristics impedance is

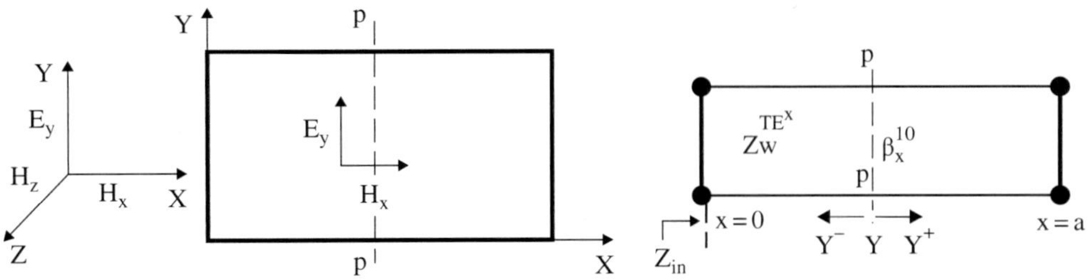

(a) Cross-section of the waveguide. (b) Equivalent transverse transmission line.

Figure 7.14 TRM applied to the rectangular waveguide.

$$Z_0 = Z_w^{TE^x} = \frac{E_y}{H_z} = \frac{\omega\mu_0}{\beta_x^{10}}. \tag{7.7.2}$$

Figure (7.14b) shows both ends shorted equivalent transverse transmission line. It is due to the presence of the conducting walls (PEC) at both ends of a waveguide. At $x = 0$, the input impedance of the short-circuited line of length $x = a$ is zero,

$$\begin{aligned} Z_{in} &= Z_w^{TE^x} \tan\left(\beta_x^{10} a\right) = 0 \\ \Rightarrow \beta_x^{10} &= \frac{m\pi}{a} \quad (m = 1, 2, 3......) \quad (a). \end{aligned} \tag{7.7.3}$$

Therefore, the longitudinal propagation constant β_z of the TE_{m0} mode of the rectangular waveguide in the z-direction is

$$\beta_z^2 = k^2 - \beta_x^2 = k_o^2 - \left(\frac{m\pi}{a}\right)^2 \quad (b), \tag{7.7.3}$$

where k is the wavenumber of EM-wave in the homogeneous medium of a waveguide and β_x is the transverse propagation constant in the x-direction. This result has been obtained in equation (7.4.71) by the field analysis. However, the TRM computes the results with much less effort.

7.7.2 Dielectric Loaded Waveguide

Figure (7.15) shows the cross-sections of horizontally and vertically layered dielectric-loaded rectangular waveguides. The EM-waves propagate in the z-direction. The inhomogeneous medium waveguide supports the hybrid mode that is described by TE and TM modes with respect to a normal to the interface of two dielectric media [J.11–J.18]. Figure (7.15a) supports the TE^y, i.e. the *longitudinal sectional electric (LSE^y) mode*, and also the TM^y, i.e. the *longitudinal sectional magnetic (LSM^y) mode*. Normally the field components and waveguide parameters are determined by solving the wave equation in each medium [B.1, B.5, J.11–J.13]. However, it is desired to determine the propagation constant β_z for both modes by using the TRM. The equivalent transverse transmission line approach applies to both TE^y and TM^y modes. The propagation constant for the hybrid mode satisfies the following dispersion relations in two media:

$$\begin{aligned} &\text{Medium \#1:} \quad \left(\frac{m\pi}{a}\right)^2 + \beta_{y1}^2 + \beta_z^2 = \omega^2\varepsilon_1\mu_1 \quad (a) \\ &\text{Medium \#2:} \quad \left(\frac{m\pi}{a}\right)^2 + \beta_{y2}^2 + \beta_z^2 = \omega^2\varepsilon_2\mu_2 \quad (b), \end{aligned} \tag{7.7.4}$$

where longitudinal propagation constant β_z could be either $\beta_z^{TE^y}\left(\beta_z^{LSE^y}\right)$ or $\beta_z^{TM^y}\left(\beta_z^{LSM^y}\right)$ for the LSE^y or LSM^y mode respectively. The y-directed transverse propagation constants β_{y1} and β_{y2} in both the media are to be computed with the help of the TRM, to finally determine the propagation constant of both modes.

Let us apply the TRM to the horizontally layered waveguide shown in Figure (7.15a) that also shows the equivalent transmission line section for two media in the transverse plane. Both the upper and lower transmission line sections are short-circuited at the ends. Their input impedances at the plane p–p are

$$\begin{aligned} Z_{in}^+ &= j Z_{02} \tan\left[\beta_{y2}(b - b_1)\right] \quad (a) \\ Z_{in}^- &= j Z_{01} \tan\left[\beta_{y1} b_1\right] \quad (b), \end{aligned} \tag{7.7.5}$$

where Z_{01} and Z_{02} are the characteristic impedances and β_{y1}, β_{y2} are the propagation constants of the lower and upper equivalent transmission lines, respectively. Under the resonance condition in the transverse plane of the waveguide, the total input reactance at the plane p–p is zero,

$$\begin{aligned} &Z_{in}^+ + Z_{in}^- = 0 \\ \Rightarrow\ &Z_{02} \tan\left[\beta_{y2}(b - b_1)\right] = -Z_{01} \tan\left[\beta_{y1} b_1\right]. \end{aligned} \tag{7.7.6}$$

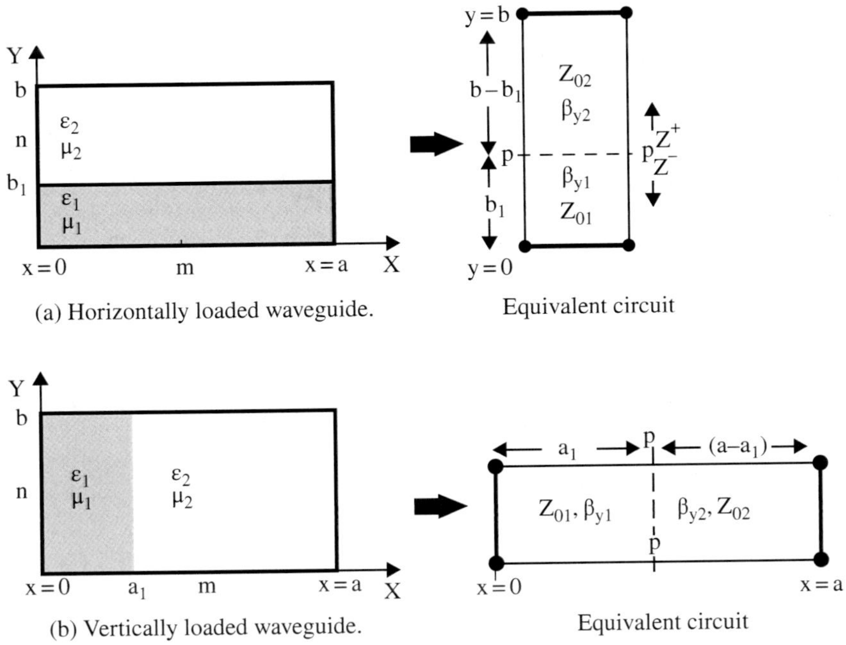

Figure 7.15 Dielectric loaded rectangular waveguide.

TE^y (LSE^y) Hybrid Mode

For the TE^y hybrid mode, characteristic impedances of the upper and lower line sections are obtained from equation (7.6.1a),

$$Z_{02} = Z_{w2}^{TE^y} = \frac{\omega\,\mu_2}{\beta_{y2}^{mn}} \quad (a), \quad Z_{01} = Z_{w1}^{TE^y} = \frac{\omega\,\mu_1}{\beta_{y1}^{mn}} \quad (b). \tag{7.7.7}$$

In superscript "mn", m is an modal integer, and modal number n is the order of roots of equation (7.7.8).

Thus, under the transverse resonance condition, the following dispersion relation is obtained from equations (7.7.6) and (7.7.7):

$$\begin{aligned} &\frac{\omega\mu_2}{\beta_{y2}^{mn}}\tan\left[\beta_{y2}b_2\right] = -\frac{\omega\mu_1}{\beta_{y1}^{mn}}\tan\left[\beta_{y1}b_1\right] && (a)\\ \Rightarrow\ &\frac{\beta_{y1}^{mn}}{\mu_1}\cot\left[\beta_{y1}b_1\right] = -\frac{\beta_{y2}^{mn}}{\mu_2}\cot\left[\beta_{y2}b_2\right] && (b), \end{aligned} \tag{7.7.8}$$

where $b_2 = b - b_1$. On substituting the propagation constants β_{y1} and β_{y2} from equation (7.7.4), equation (7.7.8) could be solved to obtain the propagation constant $\beta_z^{TE^y}$ of the TE^y mode, i.e the LSE^y mode.

TM^y (LSM^y) Hybrid Mode

For the TM^y mode, characteristic impedances of the upper and lower line sections are obtained from equation (7.6.1b),

$$Z_{02} = Z_{w2}^{TM^y} = \frac{\beta_{y2}^{mn}}{\omega\varepsilon_2} \quad (a), \quad Z_{01} = Z_{w1}^{TM^y} = \frac{\beta_{y1}^{mn}}{\omega\varepsilon_1} \quad (b). \tag{7.7.9}$$

In superscript "mn", m is an modal integer, and modal number n is the order of roots of equation (7.7.10).

Again, under the transverse resonance condition, the following dispersion relation of the TM^y mode is obtained:

$$\begin{aligned} &\frac{\beta_{y1}^{mn}}{\omega\varepsilon_1}\tan\left[\beta_{y1}b_1\right] = -\frac{\beta_{y2}^{mn}}{\omega\varepsilon_2}\tan\left[\beta_{y2}b_2\right] && (a)\\ \Rightarrow\ &\frac{\omega\varepsilon_1}{\beta_{y1}^{mn}}\cot\left[\beta_{y1}b_1\right] = -\frac{\omega\varepsilon_2}{\beta_{y2}^{mn}}\cot\left[\beta_{y2}b_2\right] && (b). \end{aligned} \tag{7.7.10}$$

Again, equation (7.7.10), along with equation (7.7.4), can be solved to get the propagation constant $\beta_z^{TM^y}$ for the mode TM^y, i.e. the LSM^y mode. Normally equations (7.7.8) and (7.7.10) are obtained by the field matching approach. However, that is a time-consuming process. Similarly, the propagation constants for LSE^x and LSM^x modes are determined for the vertically loaded waveguide shown in Fig (7.15b), using the TRM method as applied to the transmission line equivalent circuit [B.1, B.5, J.14–J.18].

7.7.3 Slab Waveguide

Figure (7.16a) shows the slab waveguide that is infinite in the y-direction, so that there is no field variation in the y-direction. The z-axis is the direction of propagation of the guided EM-wave. On the symmetry consideration with respect to the plane p–p, the propagating EM-field could be decomposed into the even and odd modes for both the TE and TM modes. Figure (7.16) shows the field components and the direction of propagation for both modes.

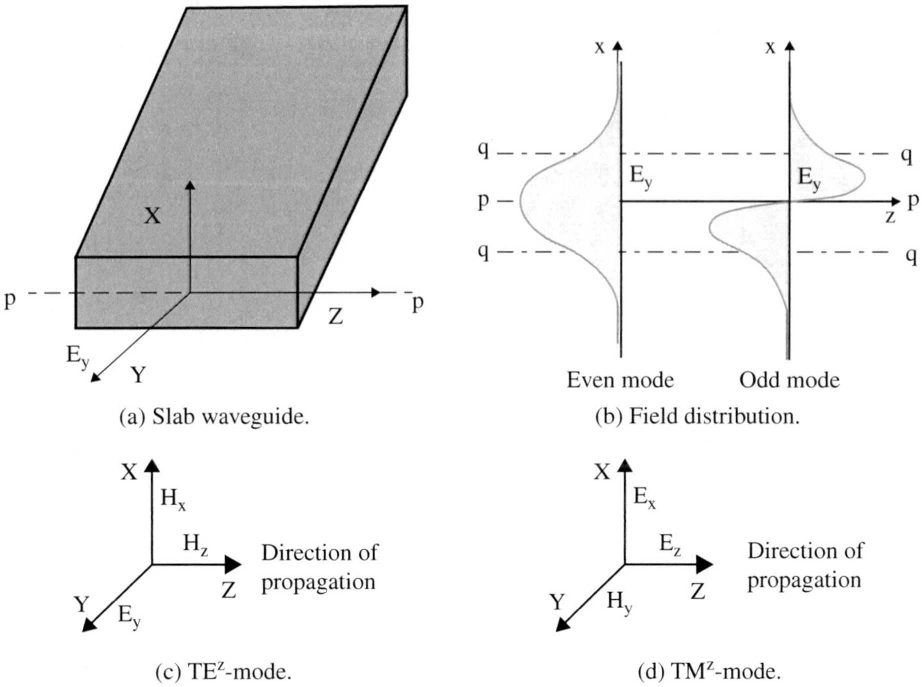

Figure 7.16 Slab waveguide supporting even and odd surface wave modes.

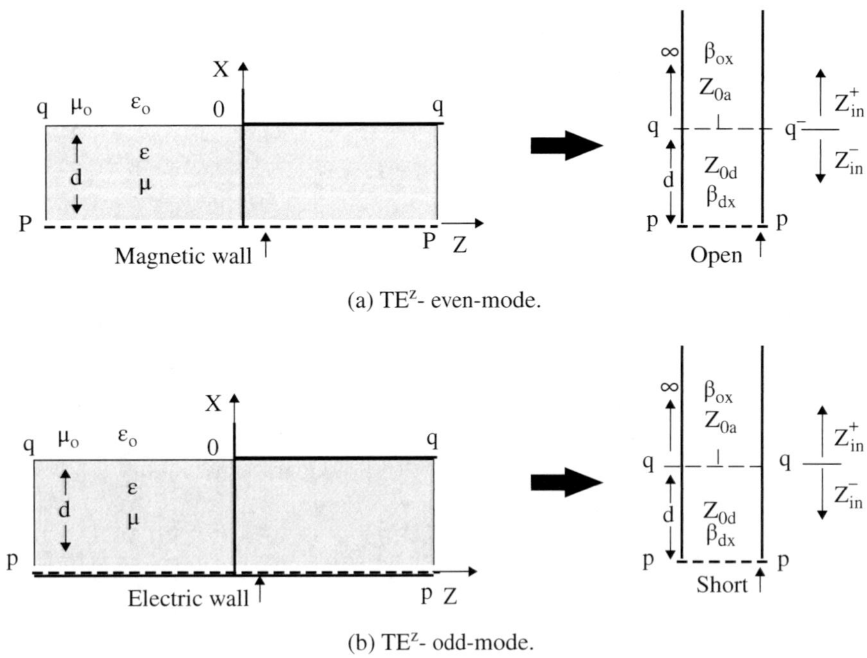

Figure 7.17 TE^z Even and odd modes for slab waveguide.

Figure (7.16b) shows the even and odd mode excitations. The propagation constants for four modes are normally determined by solving the wave equation [B.5, B.9, B.16]. However, the TRM is applied to accomplish this task. Figure (7.17a and b) show that the plane of symmetry p–p can be treated as the MW or EW corresponding to the even and odd mode excitations. The magnetic and EWs arrangement bifurcates the slab waveguide.

TE^z Even Mode

Figure (7.17a) shows one-half of the dielectric slab waveguide ($\mu = \mu_0$), with the MW, supporting the even mode. It also shows the equivalent transmission line in the transverse plane of the waveguide. This structure supports the TE^z surface wave mode. The transverse resonance method is used in the (x–z)-plane as the field is bounded by the MW, along the lower part of the *x*-axis. For analysis, the *x*-axis is shifted to the top surface of the slab waveguide. The even mode E_y field is maximum at the MW plane p–p, showing that the plane p–p is open-circuited. The plane q–q is the interface of the air-dielectric media. The upper side of the plane q–q is air-medium in the cut-off region that provides the field confinement in the x-direction by the evanescent mode. The infinitely long upper section of equivalent TE^x mode transmission under the cut-off condition offers an inductive load, as shown in equation (7.4.31).

Following equation (7.6.1), the characteristic impedance of the lower dielectric and upper airline sections of the TE^x modes are

$$Z_{0d} = Z_{wd}^{TE^x} = \frac{\omega\mu_0}{\beta_{dx}} \quad (a), \qquad Z_{0a} = Z_{wa}^{TE^x} = j\frac{\omega\mu_0}{\beta_{0x}} \quad (b), \tag{7.7.11}$$

where β_{dx} and β_{0x} are the propagation constant of these equivalent lines in the transverse x-direction. The input impedance of the lower and upper sections of the transmission lines at the plane q–q are

$$\begin{aligned} Z_{in}^{-} &= -j\,Z_{0d}\cot\left[\beta_{dx}d\right] \\ &= -j\frac{\omega\mu_0}{\beta_{dx}}\cot\left[\beta_{dx}d\right] \quad (a), \quad Z_{in}^{+} = j\frac{\omega\mu_0}{\beta_{0x}} \quad (b). \end{aligned} \tag{7.7.12}$$

The transverse resonance condition, at the plane-qq, satisfies the following relation:

$$Z_{in}^{+} + Z_{in}^{-} = 0, \quad \Rightarrow \quad \beta_{0x}\cot\left[\beta_{dx}d\right] = \beta_{dx}. \tag{7.7.13}$$

The dispersion relations, using equation (7.5.2), in the slab and air-medium are

$$\begin{aligned} &\text{Lower Slab medium:} \quad \beta_{dx}^2 + \beta_z^2 = \beta_d^2 = \omega^2\varepsilon_r\,\varepsilon_0\mu_0 \quad (a) \\ &\text{Upper Air medium:} \quad -\beta_{0x}^2 + \beta_z^2 = \beta_0^2 = \omega^2\varepsilon_0\mu_0 \quad (b). \end{aligned} \tag{7.7.14}$$

In equation (7.7.14) air-medium, acting as the upper line-section, is in the cut-off mode with x-directed imaginary propagation constant $j\beta_{0x}$. On subtracting the above dispersion equations, the following expression is obtained:

$$\beta_{dx}^2 + \beta_{0x}^2 = (\varepsilon_r - 1)\,k_0^2 \quad \text{where,} \quad k_0^2 = \omega^2\varepsilon_0\mu_0. \tag{7.7.15}$$

The eigenvalue equation (7.7.13) and the dispersion relation (7.7.15) are rewritten as

$$\begin{aligned} &u.\tan(u) = w \quad (a), \qquad u^2 + w^2 = v^2 \quad (b) \\ &\text{where,} \quad u = \beta_{dx}d, \quad w = \beta_{0x}d, \quad v = \sqrt{(1-\varepsilon_r)}\,k_0d \quad (c). \end{aligned} \tag{7.7.16}$$

The parameter v is known for a given slab-waveguide structure at the assumed frequency. Equations (7.7.16a and b) can be solved for u and w simultaneously. Finally, the propagation constant β_z, for TE^z even mode, is obtained from equation (7.7.14). Depending upon the frequency, a given structure can support several TE$_n$ (even) modes, where n numbers of the solutions of equation (7.7.16) give n numbers of modes. The graphical method is also used for the solution of propagation constant β_z [B.5, B.9, B.12].

TE^z Odd Mode

Figure (7.17b) shows that for the TE^z odd mode excitation, the E_y field component is zero at the plane of symmetry p–p, i.e. at the EW. The equivalent transverse line shows the short circuit at the plane p–p. The plane q–q is the interface of two media. The upper air-medium is again in the cut-off region, supporting the evanescent mode in the transverse x-direction for the field confinement. Therefore, the input impedance of the lower and upper sections of the transmission line at the plane q–q are

$$\begin{aligned} Z_{in}^{-} &= j\,Z_{0d}\tan\left[\beta_{dx}d\right] = j\frac{\omega\mu_0}{\beta_{dx}}\tan\left[\beta_{dx}d\right] \quad (a) \\ Z_{in}^{+} &= j\frac{\omega\mu_0}{\beta_{0x}} \quad (b). \end{aligned} \tag{7.7.17}$$

Under the transverse resonance condition, we get the following equation:

$$\begin{aligned} &Z_{in}^{+} + Z_{in}^{-} = 0 \quad (a) \\ \Rightarrow \quad &\beta_{dx}d\,\cot(\beta_{dx}d) = -\beta_{0x}d \quad (b). \end{aligned} \tag{7.7.18}$$

Equation (7.7.18), with the help of equation (7.7.16), is rewritten as

$$\begin{aligned} &u\cot(u) = -w \quad (a), \qquad u^2 + w^2 = v^2 \quad (b) \\ &\text{where,} \quad u = \beta_{dx}d, \quad w = \beta_{0x}d, \quad v = \sqrt{(1-\varepsilon_r)}\,k_0d \quad (c). \end{aligned} \tag{7.7.19}$$

Again, the above equations can be solved for u and w, once the parameter v is known for the given structure at the assumed frequency. Finally, the propagation constant β_z for the TE^z odd mode surface wave is computed at the assumed frequency.

TM^z Even Mode

Figure (7.16d) shows that this mode has even symmetry for the H_y-field, giving maximum H_y-field component at the plane p–p. Therefore, the plane p–p could be replaced by an EW. It shows that the TE^z odd mode equivalent circuit shown in Fig (7.17b) could be used in this case also. However, for this case of TM^z even mode, the characteristic impedance of the equivalent lower and upper transmission line sections correspond to the wave impedance of the TM^x mode. Again, the TM^x mode in the transverse x-direction of air-medium is in the cut-off region. Following expression (7.4.32), it offers capacitive wave impedance. Therefore, on using equations (7.6.1b) and (7.7.17), the characteristic impedances of equivalent TM^x mode transmission lines and their input impedances at the plane q–q are written as

$$Z_{0d} = Z_{wd}^{TM^x} = \frac{\beta_{dx}}{\omega\varepsilon} \quad (a), \quad Z_{0a} = Z_{wa}^{TM^x} = -j\frac{\beta_{0x}}{\omega\varepsilon_0} \quad (b)$$
$$Z_{in}^- = j\frac{\beta_{dx}}{\omega\varepsilon}\tan[\beta_{dx}d] \quad (c), \quad Z_{in}^+ = -j\frac{\beta_{0x}}{\omega\varepsilon_0} \quad (d). \tag{7.7.20}$$

The transverse resonance condition is satisfied as follows:

$$Z_{in}^+ + Z_{in}^- = 0, \quad (a), \quad \Rightarrow \frac{\beta_{dx}d}{\varepsilon_r}\tan[\beta_{dx}d] = \beta_{0x}d \quad (b). \tag{7.7.21}$$

Equation (7.7.21) could be rewritten as follows along with the dispersion relation:

$$u.\tan(u) = \varepsilon_r w \quad (a), \qquad u^2 + w^2 = v^2 \quad (b)$$
$$\text{where,}\ u = \beta_{dx}d, \ w = \beta_{0x}d, \ v = \sqrt{(1-\varepsilon_r)}\,k_0 d \quad (c). \tag{7.7.22}$$

Again, the propagation constant β_z of the TM^z even mode at an assumed frequency is obtained from the above equations and equation (7.7.14).

TM^z Odd Mode

The TM^z odd mode is excited for zero H_y-field component at the plane of symmetry p–p shown in Fig (7.16d). It shows that the plane of symmetry p–p could be replaced by the MW, as is the case for the TE^z even mode. Therefore, the equivalent transmission line shown in Fig (7.17a) is again applicable to this case by using the characteristic impedances of TM mode. The air-medium, in the x-direction, is in the cut-off region offering the capacitive wave impedance. Therefore, the input impedance of the lower and upper transmission lines at the plane q–q is

$$Z_{in}^- = -j\frac{\beta_{dx}}{\omega\varepsilon}\cot[\beta_{dx}d] \quad (a), \quad Z_{in}^+ = -j\frac{\beta_{0x}}{\omega\varepsilon_0} \quad (b). \tag{7.7.23}$$

Under the transverse resonance condition, the following relations are obtained:

$$Z_{in}^+ + Z_{in}^- = 0 \quad (a), \quad \Rightarrow \frac{\beta_{dx}d}{\varepsilon_r}\cot[\beta_{dx}d] = -\beta_{0x}d \quad (b). \tag{7.7.24}$$

Equation (7.7.24) and the dispersion relation could be rewritten as

$$u.\cot(u) = -\varepsilon_r w \quad (a), \qquad u^2 + w^2 = v^2 \quad (b)$$
$$\text{where,}\ u = \beta_{dx}d, \ w = \beta_{0x}d, \ v = \sqrt{(1-\varepsilon_r)}\,k_0 d \quad (c). \tag{7.7.25}$$

For a given structure and frequency, v is a known parameter and equation (7.7.25) can be solved for u and w simultaneously. Finally, the propagation constant β_z is computed for the odd mode of the TM^z surface wave from the dispersion relation (7.7.14).

7.7.4 Conductor Backed Multilayer Dielectric Sheet

Figure (7.18a) shows the double-layer dielectric sheet backed by a conductor. It is a multilayer slow-wave structure. The propagation constants for the TE^z and TM^z modes could be easily obtained by using the TRM. Let us do it for the TM mode. The TRM is applied through the equivalent transmission line shown in Fig (7.18b). The wave propagation in the transverse x-direction in the air-medium is in the cutoff region, offering a capacitive wave impedance and imaginary propagation constant.

The characteristic impedance for three-line sections supporting TM^x are given below:

$$Z_{0d1} = \frac{\beta_{d1x}}{\omega\varepsilon_1} \quad (a), \quad Z_{0d2} = \frac{\beta_{d2x}}{\omega\varepsilon_2} \quad (b), \quad Z_{0a} = -j\frac{\beta_{0x}}{\omega\varepsilon_0} \quad (c). \tag{7.7.26}$$

The propagation constants in three media meet the following dispersion relations:

$$\beta_{d1x}^2 = k_1^2 - \beta_z^2 \quad (a), \quad \beta_{d2x}^2 = k_2^2 - \beta_z^2 \quad (b), \quad \beta_{0x}^2 = \beta_z^2 - k_0^2 \quad (c)$$
$$\text{where,}\ k_i^2 = \varepsilon_{ri}k_0^2 \quad (i = 1, 2) \quad (d). \tag{7.7.27}$$

The input impedance at the plane q–q shown in Fig (7.18b) are given as follows:

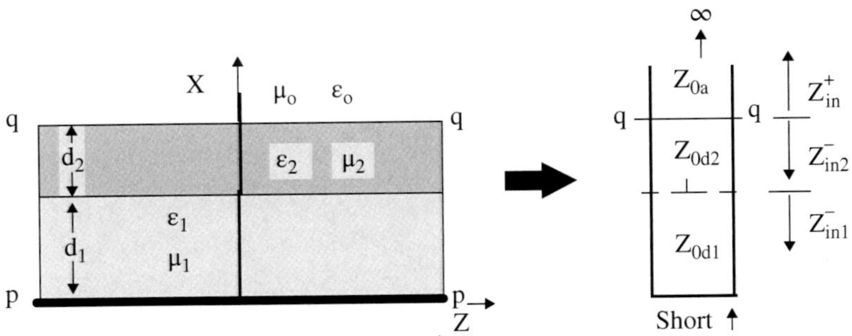

Figure 7.18 Double-layer slow-wave structure.

$$Z^-_{in1} = j\frac{\beta_{d1x}}{\omega\varepsilon_1}\tan(\beta_{d1x}d_1) \qquad \text{(a)}$$

$$Z^-_{in2} = Z_{od2}\left[\frac{Z^-_{in1} + j\,Z_{od2}\tan(\beta_{d2x}d_2)}{Z_{od2} + j\,Z^-_{in1}\tan(\beta_{d2x}d_2)}\right] \qquad \text{(b)}$$

$$Z^+_{in} = Z_{0a} = -j\frac{\beta_{ox}}{\omega\varepsilon_0} \qquad \text{(c)}. \qquad (7.7.28)$$

On applying the transverse resonance condition, $Z^+_{in} + Z^-_{in} = 0$ at the plane q–q, the following dispersion expression is obtained:

$$\left[\frac{b\tan[\beta_{d1x}d_1] + a\tan[\beta_{d2x}d_2]}{a - b\tan[\beta_{d1x}d_1]\tan[\beta_{d2x}d_2]}\right] = b \qquad \text{(a)}$$

where,

$$a = \frac{\varepsilon_{r1}\,\beta_{0x}}{\beta_{d1x}} = \varepsilon_{r1}\sqrt{\left[\frac{(\beta_z/k_0)^2 - 1}{\varepsilon_r - (\beta_z/k_0)^2}\right]} \qquad \text{(b)}$$

$$b = \frac{\varepsilon_{r2}\beta_{0x}}{\beta_{d2x}} = \varepsilon_{r2}\sqrt{\left[\frac{(\beta_z/k_0)^2 - 1}{\varepsilon_{r2} - (\beta_z/k_0)^2}\right]} \qquad \text{(c)}$$

$$\beta_{d1x}d_1 = \frac{2\pi d_1}{\lambda_0}\sqrt{\varepsilon_{r1} - (\beta_z/k_0)^2} \qquad \text{(d)}$$

$$\beta_{d2x}d_2 = \frac{2\pi d_2}{\lambda_0}\sqrt{\varepsilon_{r2} - (\beta_z/k_0)^2} \qquad \text{(e)}\ . \qquad (7.7.29)$$

Elliott [B.15] has obtained equation (7.7.29) by a long process of the field matching at the interface. However, the TRM is a direct method. For $d_2 = 0$, the above equation is reduced to equation (7.7.21). It is the case for the conductor backed single layer dielectric sheet. The TRM could be also applied to compute the propagation constant β_z of the TE^z mode. The method is easily extended to the N layers of the dielectric sheets.

7.8 Substrate Integrated Waveguide (SIW)

The metallic rectangular waveguide, discussed in section (7.4) is popular in the mm-wave frequency range due to its low-loss. In mm-wave and THz ranges the planar lines are lossier. It is difficult to integrate the standard rectangular waveguide with planar technology. However, the synthetic rectangular waveguide has been developed in the PCB and LTCC technology that has been integrated with other planar lines, say microstrip. This structure is known as the *substrate integrated waveguide (SIW)*, also called the *laminated waveguide*. Several components and leaky antenna have been developed using SIW [J.19–J.26]. There are several formats for the SIW. This section briefly presents the basic characteristics and simple modeling of two important structures.

7.8.1 Complete Mode Substrate Integrated Waveguide (SIW)

Figure (7.19a) shows the synthetic rectangular substrate integrated waveguide (SIW) embedded in a substrate of the thickness h and relative permittivity ε_r. The top and bottom walls of the SIW are conducting surfaces, i.e. the EWs, whereas the sidewalls of height h are created by two rows of arrays of the metalized vias of diameter d drilled in the substrate. The vias are separated by the distance S. The center-to-center distance between two rows is W. The structure behaves as a rectangular waveguide filled in with dielectrics of relative permittivity ε_r, although there are some important differences from the standard metallic waveguide. Figure (7.19b) shows that the SIW can be excited by the microstrip and matching is achieved using a linear transition [J.26]. The width W_t and length L_t of the feed microstrip linear taper are optimized on the EM-simulator for wideband matching. Figure (7.19b) shows both the input and output ports.

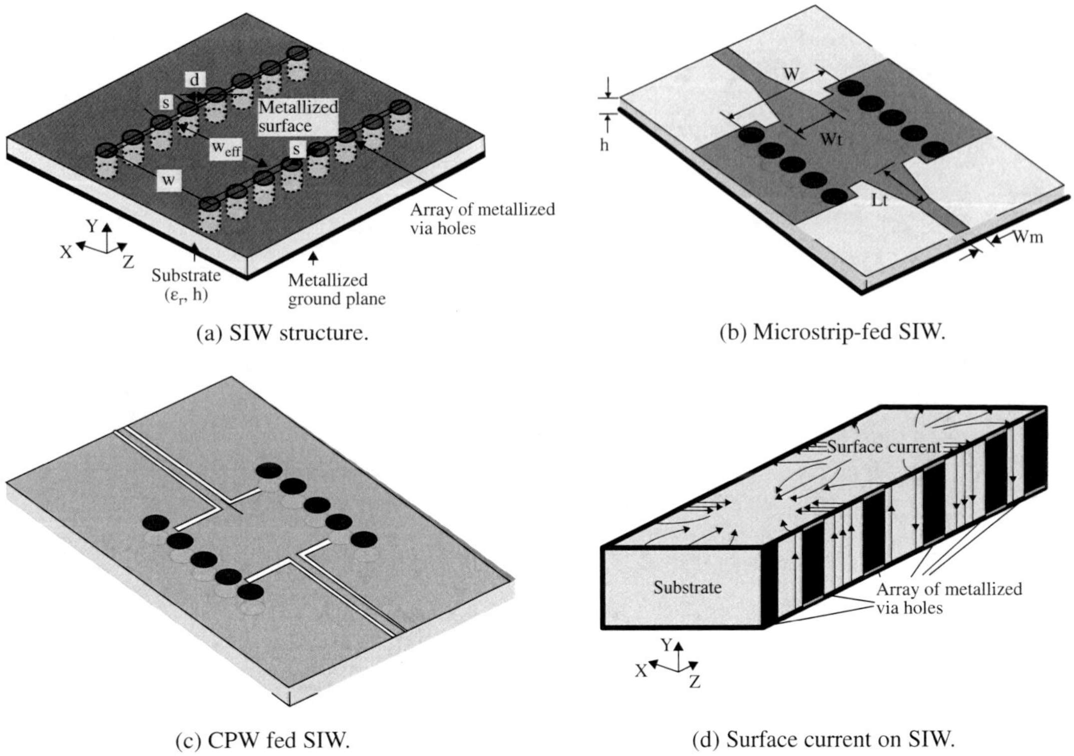

Figure 7.19 Synthetic SIW created through two rows of metalized via holes.

Figure (7.19c) shows that the SIW can also be fed through the CPW [J.26].

The mode supported by a waveguide is related to the surface current of the waveguide. Figure (7.19d) shows the surface currents for the TE_{10} mode in the SIW, it is similar to the standard metallic waveguide. There is no longitudinal surface current on the side-walls. Therefore, any narrow gap, created by the periodic arrangement of the metalized vias holes, will not disturb the TE_{10}, or even TE_{mo} modes. However, TE_{mn} $(n \neq 0)$ and TM_{mn} modes have longitudinal surface currents on the sidewalls and they cannot flow on the SIW sidewalls. Therefore, unlike the standard waveguide, the SIW does not support the TE_{mn} $(n \neq 0)$ and TM_{mn} modes.

The SIW is primarily a periodic structure, due to the periodic array of the metallic post, created through the metalized vias. However, the size of via (d) and its separation (S) are adjusted so that the stopband characteristic is avoided in the useful frequency band. This process is known as the *homogenization* that replaces the periodic discontinuous side-walls by the continuous equivalent metallic walls. The concept of *homogenization* has been used for other kinds of periodic surfaces too [B.6]. The SIW structures have been analyzed using several types of numerical methods, including the commercial EM-Simulators. However, for design, the *equivalent waveguide model* could also be used. The waveguide model uses the closed-form expressions for the effective width w_{eff} of the SIW shown in Fig (7.19a) that is less than W. On replacing the width W of the standard rectangular waveguide by $W_{eff,}$ the expressions presented in section (7.4) is used to compute the cut-off frequency, dispersion, losses and wave impedance of the modes of the SIW. The results of the waveguide model are compared, with satisfactory accuracy, against the results of the numerical methods and also against the experimental results.

The equivalent width of the SIW is dependent on W, S, and d of the SIW. Any of the following expressions can be used [J.6, J.20, J.25]:

$$W_{eff} = W - \frac{d^2}{0.95\,S} \qquad \text{(a)}$$

$$W_{eff} = W - 1.08\frac{d^2}{S} + 0.1\frac{d^2}{W} \qquad \text{(b)}$$

$$W_{eff} = \frac{W}{\sqrt{\begin{array}{l}1 + \{(2W-d)/S\}\{d/(W-d)\}^2 \\ -\left(4W/5S^4\right)\{d^2/(W-d)\}^3\end{array}}} \qquad \text{(c)}$$

$$W_{eff} = W - \frac{d^2}{1.1\,s} - \frac{d^3}{6.6\,S^2} \qquad \text{(d)}, \qquad (7.8.1)$$

where $d < \lambda_g/5$, $S \le 2d$ and $\lambda_g = \lambda_0/\sqrt{\varepsilon_r}$. Equation (7.8.1b) [J.22] and equation (7.8.1c) [J.25] are more

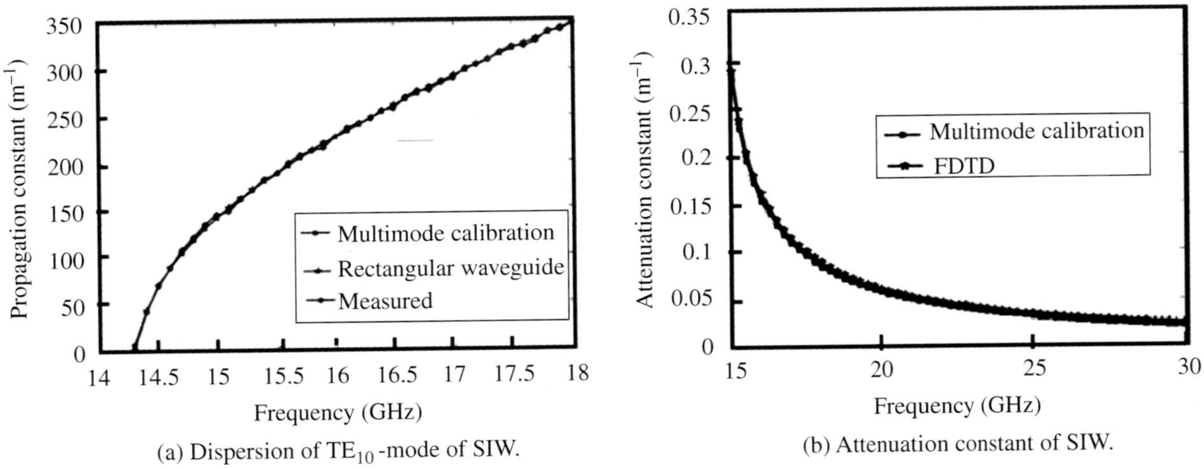

Figure 7.20 Dispersion and loss characteristic of the SIW. *Source*: Xu and Wu [J.22].

accurate expressions. The first one does not show d/w dependence and is acceptable provided (S-d) is small, giving closely spaced via holes. It gives very small leakage loss from the sidewall. The equivalent width is in between W and (W-d). The outer width of the equivalent rectangular waveguide is W and its inner width is (W-d). The above expression can be used to compute the cut-off frequency of the fundamental mode TE_{10}, or even the cut-off frequency of the higher-order mode. Equation (7.8.1d) has been used to determine the cut-off frequency of the TE_{20} mode [J.20]. The cut-off frequency of the TE_{10} mode, using equation (7.8.1a), has accuracy within ±5% and the TE_{20} mode has the accuracy +4 %/ − 9%, using an equation (7.8.1d) for the SIW on a substrate with $\varepsilon_r = 2.2$, h = 20 mil. Figure (7.20a) compares favorably the dispersion characteristic of the TE_{10} mode of the SIW, on the substrate $\varepsilon_r = 2.33$, h = 0.508 mm, using the equivalent waveguide model against the results of both the numerical method and experimental results [J.22].

The accuracy of the above expressions can be tested from the dispersion results of the SIW obtained from the EM-Simulator [J.24]. The EM-simulated S_{21} parameters of two SIW of lengths ℓ_1 and ℓ_2 ($\ell_1 > \ell_2$) provide the phase difference. The propagation constant is evaluated from the $\angle S_{21}^{\ell_2}$ and $\angle S_{21}^{\ell_1}$. The dispersion relation is used to extract the effective width of the SIW:

$$\beta_{SIW} = \frac{\angle S_{21}^{\ell_2} - \angle S_{21}^{\ell_1}}{\ell_1 - \ell_2}\ \text{rad/m} \qquad (a)$$

$$\left(\frac{\pi}{W_{eff}}\right)^2 + (\beta_{SIW})^2 = (k_d)^2$$

$$\Rightarrow W_{eff} = \frac{\pi}{\left[\omega^2\mu_0\varepsilon_0\varepsilon_r - \beta_{SIW}^2\right]} \qquad (b), \qquad (7.8.2)$$

where k_d is the wavenumber in the substrate.

Once the propagation constant β_{SIW} is evaluated using an EM-Simulator, equation (7.8.2b) computes the equivalent width of the SIW. It is used to verify the accuracy of the expressions of equation (7.8.1).

The dielectric loss and conductor losses of the TE_{10} mode SIW can be computed from the standard expressions of the metallic waveguide [J.23, B.2, B.3]:

$$\alpha_d = \frac{k_0^2\,\varepsilon_r\,\tan\delta}{2\,\beta_{SIW}}\ \text{Np/m} \qquad (a)$$

$$\alpha_c = \frac{R_s}{W_{eff}\eta\sqrt{1-(k_c/k_0)^2}}\left[\frac{W_{eff}}{h} + 2\left(\frac{k_c}{k_0}\right)^2\right]\ \text{Np/m} \qquad (b)$$

$$\text{where, } R_s = \sqrt{\omega\mu_0/2\sigma}, \quad \text{and} \quad \eta = \sqrt{\mu_0/\varepsilon_0\varepsilon_r} \qquad (c). \qquad (7.8.3)$$

In equation (7.8.3) conductivity σ of all walls is taken the same, whereas due to gaps between vias holes in the side-walls the conductivity of the equivalent homogeneous surface gets reduced.

There is a leakage loss from the side-walls, as the RF power is coupled to the adjacent TEM mode supporting the parallel-plates guide. The total loss of the SIW can be evaluated from the S_{21} of SIW obtained from the EM-Simulator. The total attenuation coefficient, α_{total}, due to conductor, dielectric and leakage losses, can also be evaluated from the S_{21} of two unequal line lengths of the SIW:

Total loss in SIW

$$= 10\log_{10}\left(1 - \left\{|S_{11}|^2 + |S_{21}|^2\right\}\right)\ \text{dB} \qquad (a)$$

$$\alpha_{total} = \frac{\left[\left(|S_{11}^{\ell_1}|^2 + |S_{21}^{\ell_1}|^2\right) - \left(|S_{11}^{\ell_2}|^2 + |S_{21}^{\ell_2}|^2\right)\right]}{\ell_1 - \ell_2}\ \text{Np/m}, \quad (\ell_1 > \ell_2) \qquad (b). \qquad (7.8.4)$$

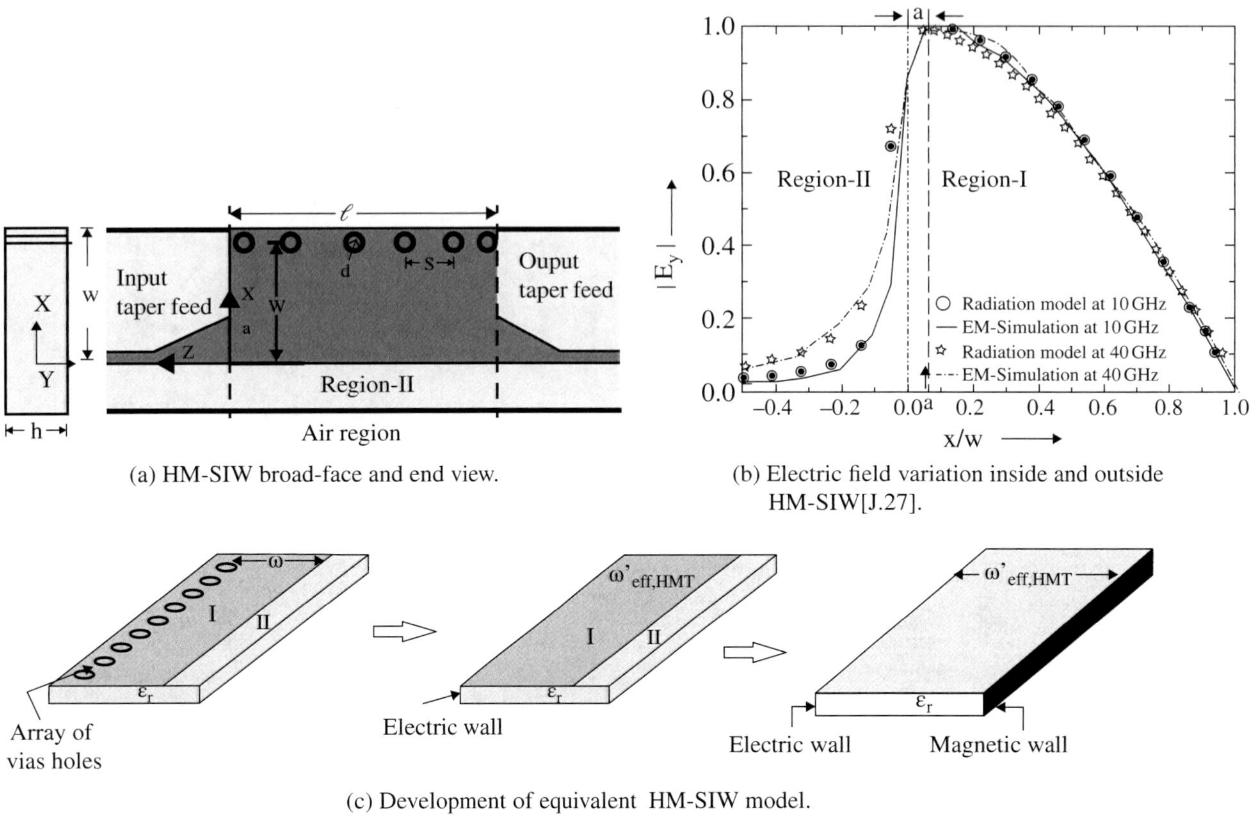

(a) HM-SIW broad-face and end view.

(b) Electric field variation inside and outside HM-SIW[J.27].

(c) Development of equivalent HM-SIW model.

Figure 7.21 Modeling of HM-SIW. *Source*: Modified from Lai et al. [J.27].

The above EM-Simulation process can also be used to evaluate the conductor, dielectric and leakage losses and their corresponding attenuation coefficients separately. The attenuation in the SIW can be also computed using the FDTD and multimode calibration. Figure (7.20b) illustrates the attenuation of the SIW on the substrate $\varepsilon_r = 2.33$, $h = 0.508$ mm. It is noted that the leakage loss is high at the lower end of the frequency band [J.22].

The leakage loss of the SIW can be effectively controlled by controlling the sizes of the vias holes and their separation. Further, it is also desirable to avoid any stop-band appearance in the useful bandwidth of SIW. These are achieved through the following conditions [J.26]:

$$d < 0.25\lambda_c \quad (a), \qquad d < S \leq 2d \quad (b). \tag{7.8.5}$$

7.8.2 Half-Mode Substrate Integrated Waveguide (SIW)

The SIW discussed is a low-profile and low-cost structure, easily integrated with the planar technology. However, the SIW is a wide width structure, occupying more areas of the substrate. The half-mode SIW (HM-SIW), shown in Fig (7.21a), with half of the conducting plane width of the SIW, offers a more compact waveguide structure that has been used for the development of filters, 3 dB coupler, directional couplers, etc. [J.28–J.31]. The HM-SIW can be simulated and analyzed using the EM-Simulators. However, in this section design equations are presented, as these are useful for the development of components [J.27]. Again, the HM-SIW is fed through a microstrip line and a microstrip linear taper, optimized by the EM-Simulator.

Figure (7.21a) shows that one sidewall of the HM-SIW has an EW, realized with help of an array of metalized vias holes, while the other sidewall is open to the dielectric substrate with ε_r, h. Again, beyond the substrate edge, there is an infinite air-medium. The HM-SIW structure has two distinct regions – *region-I*, inside the waveguide supporting TE_{mo} type mode, called quasi-$TE_{m'o}$ half mode (HM), and *region-II* supporting the evanescent radiation mode. In the case of the *quasi*-HM, the open sidewall is approximately located at the center of the standard rectangular waveguide. It results in the cosine type E_y-field distribution. So $m' = 0.5, 1.5, 2.5, \ldots$. The HM-SIW supports a bound mode, propagating in the Z-direction with propagation constant β_z, with one side field confinement by the EW and another side field confinement by the evanescent E_y-field. Figure (7.21b) shows EM-Simulated cosine type E_y-field distribution across the cross-section of the HM-SIW. At $x = w$, E_y-field is zero as usual. However, the maximum of E_y-field is shifted inside the HM-SIW by the distance "a." Figure (7.21b) shows the location of the maximum and location of distance "a." The HM-SIW has been modeled using two models – *model-I, the radiation model*, and *model-II, the equivalent waveguide model* [J.27].

Model-I, Radiation Model

The model-I helps to compute the evanescent E_y-field outside the HM-SIW, i.e. the radiation field. It is computed by obtaining the magnetic current on the open sidewall. The magnetic current is computed through the E_y-field component of the *quasi*-$TE_{m'o}$ half mode by following the standard TE_{mn} mode discussed in section 7.4. The magnetic scalar potential and the field components of the *quasi*-$TE_{m'o}$ following equations (7.4.17) and (7.4.34), can be written as follows for fundamental mode ($m' = 0.5$):

$$\begin{aligned}
&\phi^{m}_{m'0}(x,y,z) = B_{m'0}\cos\{\beta^{x}_{m'0}(w-x)\}\,e^{-j\beta_z z}, 0 \le x \le w \quad (a)\\
&E^{m}_{y} = -B_{m'0}\beta^{x}_{m'0}\sin\{\beta^{x}_{m'0}(w-x)\}\,e^{-j\beta_z z} \quad (b)\\
&H^{m}_{x} = B_{m'0}\beta^{x}_{m'0}(\beta_z/\omega\mu)\sin\{\beta^{x}_{m'0}(w-x)\}\,e^{-j\beta_z z} \quad (c)\\
&H^{m}_{z} = -B_{m'0}\beta^{x}_{m'0}(\beta^{x}_{m'0}/\omega\mu)(k^2-\beta_z^2)\\
&\qquad \times\cos\{\beta^{x}_{m'0}(w-x)\}\,e^{-j\beta_z z} \quad (d),
\end{aligned} \tag{7.8.6}$$

where the directional propagation constant, wavenumber, dispersion relation, and wave impedance are given by the following expression for the fundamental *quasi*-$TE_{m'o}$ half mode, ($m' = 0.5$):

$$\beta^{x}_{m'0} = \frac{\pi}{2(W-a)} \quad (a), \quad k = k_0\sqrt{\varepsilon_r} \quad (b)$$
$$\beta_z = \left[k^2 - \beta^{x2}_{m'0}\right]^{1/2} \quad (c), \quad Z^{TE}_{w} = \frac{\omega\mu}{\beta_z} \quad (d). \tag{7.8.7}$$

In equation (7.8.7), the origin has been shifted to the location of E_y maximum by the distance a, as shown in Fig. (7.21b). The *z*-directed magnetic current is obtained from the E_y component,

$$\begin{aligned}
M_z &= -2(-\hat{x}\times\hat{y})\,E_y\\
&= \hat{z}\,\frac{B_{m'o}\pi}{(W-a)}\sin\left[\frac{\pi}{2(W-a)}(W-x)\right]e^{-j\beta_z z}.
\end{aligned} \tag{7.8.8}$$

In expression (7.8.8) image current has been accounted for. Using the standard radiation integral, the E_y-field component can be evaluated in the homogeneous open region-II that has effective relative permittivity between 1 and ε_r. However, for a thin substrate, one can take the effective relative permittivity ≈1 [J.27]. Fig (7.21b) shows that the evanescent mode E_y-field, computed by the radiation model, agrees satisfactory with results obtained from the EM-simulation.

Model-II, Equivalent Waveguide Model

The equivalent waveguide model of the HM-SIW is based on the similar model of the SIW discussed previously. However, it has one sidewall as the EW, whereas its other sidewall is the MW. Figure (7.21c) shows the stepwise development of the composite walls equivalent waveguide model of the HM-SIW. If the width of a SIW is 2W and width of the HW-SIW w, the effective width ($W'_{eff,HM}$) of the HM-SIW, using an equation (7.8.1b), can be computed. However, there is a fringing field at the open side of the HM-SIW that can be accounted by further extending the width by ΔW and placing a MW, giving the final *equivalent waveguide mode* of width $W_{eff,\,HM}$ [J.27]:

$$\begin{aligned}
&W'_{eff,HM} = \frac{1}{2}\left(2W - 1.08\frac{d^2}{S} + 0.1\frac{d^2}{2W}\right) \quad (a)\\
&W_{eff,HM} = W'_{eff,HM} + \Delta W \quad (b)\\
&\text{where,}\quad \frac{\Delta W}{h} = \left(0.05 + \frac{0.3}{\varepsilon_r}\right)\\
&\ell n\left(0.79\frac{W'^2_{eff,HM}}{h^3} + \frac{104W'_{eff,HM} - 261}{h^2} + \frac{38}{h} + 2.77\right) (c).
\end{aligned} \tag{7.8.9}$$

Equation (7.8.9) is valid for the range $2.2 \le \varepsilon_r \le 15$, 0.254 mm ≤ h ≤ 2.54 mm, 2.5 mm ≤ w ≤10 mm. It covers the frequency range from 2 GHz to 60 GHz. On considering the width of HM-SIW as an $\lambda_g/4$ open-end resonator, the cut-off frequency and the dispersion relations are obtained:

$$\begin{aligned}
&f_{c,TE_{0.5,0}} = \frac{3\,c}{4\sqrt{\varepsilon_r}W_{eff,HM}} \quad (a)\\
&\beta_{Z,TE_{0.5,0}} = \left[\varepsilon_r k_0^2 - \left(\frac{\pi}{2W_{eff,HM}}\right)^2\right]^{\frac{1}{2}} \quad (b).
\end{aligned} \tag{7.8.10}$$

Figure (7.22a) compares satisfactorily dispersion of the fundamental $TE_{0.5,0}$ mode of the HM-SIW, using the equivalent waveguide model-II, against the EM-Simulation and experimental results [J.27]. Figure (7.22b) compares the attenuation of the quasi-$TE_{0.5,0}$ mode HM-SIW against the attenuations of the SIW and 50 Ω microstrip line. The width of 40 mm, 80 mm, and 100 mm long SIW is 2W = 5.25 on a substrate ε_r = 2.2, tan δ= 0.001, h = 0.254 mm and strip conductivity σ = 5×10^7 S/m. The diameter of via d = 0.05 mm and the separation S between two via is

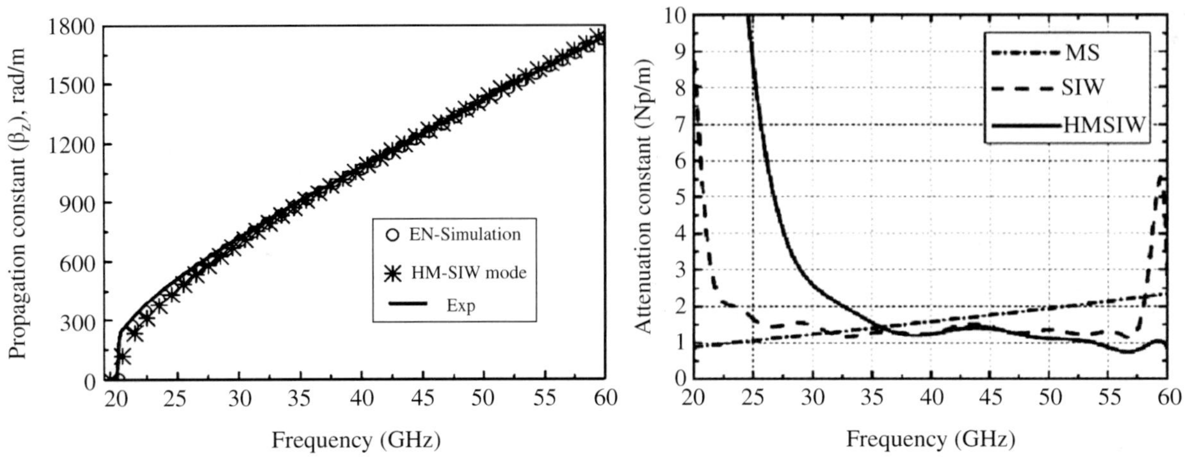

(a) Dispersion of quasi-$TE_{0.5,0}$-mode.

(b) Attenuation of microstrip, SIW, HM-SIW.

Figure 7.22 Dispersion and attenuation characteristics of -$TE_{0.5,\,0}$-mode HM-SIW. *Source*: Lai et al. [J.27].

0.6 mm. The attenuation of the microstrip increases almost linearly with frequency from 20 GHz to 60 GHz. The attenuation in microstrip is less than the attenuation of both SIW and HM-SIW at frequencies below 35 GHz. The SIW shows a sudden increase in attenuation at 58 GHz, due to the bandstop of the periodic nature of the SIW. The HM-SIW shows a higher cut-off frequency. Its loss is decreased below both these structures above 35 GHz and no bandstop occurs up to 60 GHz. However, as compared to a microstrip line, both the SIW and HM-SIW occupy more area on a PCB. The HM-SIW has wider single-mode operation bandwidth, as compared to that of the SIW. In the case of the HM-SIW, the cut-off frequency of next higher quasi-$TE_{1.5,0}$ mode occurs at a frequency that is three times of the cut-off frequency of the fundamental quasi-$TE_{0.5,0}$ mode. The single-mode operation bandwidth of the HM-SIW is two times the bandwidth of the SIW.

References

Books

B.1 Harrington, R.F.: *Time-Harmonic Electromagnetic Fields*, McGraw-Hill Book Company, New York, 1961

B.2 Collin, Robert E.: *Field Theory of Guided Waves*, IEEE Press, New York, 1991.

B.3 Collin, Robert E.: *Foundations for Microwave Engineering*, 2nd Edition, McGraw-Hill, Inc., New York, 1992

B.4 Collin, Robert E.: *Antenna and Radio Wave Propagation*, McGraw-Hill, Inc., New York, 1985.

B.5 Balanis, C.A.: *Advanced Engineering Electromagnetics*, John Wiley & Sons, New York, 1989

B.6 Ramo Simon, Whinnery John R., Van Duzer Theodore: *Fields and Waves in Communication Electronics*, 3rd Edition, John Wiley & Sons, Singapore, 1994.

B.7 Sadiku, N.O.: *Elements of Electromagnetics*, 3rd Edition, Oxford University Press, UK,2004.

B.8 Chang, D.K.: *Field and Waves Electromagnetics*, Pearson Education Asia, Delhi, 2001.

B.9 Rozzi, T.; Mongiardo, M.: *Open Electromagnetic Waveguides*, IEE Press, U.K., 1997.

B.10 Jordan, Edward C., Balmain Keith G. E.: *Electromagnetic Wave and Radiating System*, Prentice-Hall, New Delhi India, 1989.

B.11 Plonsey Robert, Collin, Robert E.: *Principle and Applications of Electromagnetic Fields*, Tata McGraw-Hill, New Delhi, 1973.

B.12 Elliott, R.S.: *An Introduction to Guided Waves and Microwave Circuits*, Englewood Cliffs, NJ: Prentice-Hall, 1993

B.13 Itoh Tatsuo (Editor): *Numerical Techniques for Microwave and Millimetre-Wave Passive Structures*, Wiley-Interscience, New York, 1989.

B.14 Walter, C. H.: *Traveling Wave Antennas*, McGraw-Hill, 1965, Dover, 1970.

B.15 Mirshekar-Syahkal, D.: *Spectral Domain Method for Microwave Integrated Circuits*, ResearchStudies Press Ltd., John Wiley & Sons, N.Y., 1990.

B.16 Koul, S. K.: *Millimeter Wave and Optical Dielectric Integrated Guides and Circuits*, Wiley-Interscience, New York, 1997.

Journals

J.1 Yeap, K.K.; Tham, C.Y.; Yassin, G.; Yeong, K.C.: *Attenuation in rectangular waveguides with finite conductivity walls, Radioengineering*, Vol. 20, No. 2, pp. 472–478, June 2011.

J.2 Yeap, K.K.; Tham, C.Y.; Yassin, G.; Yeong, K.C.; Propagation in lossy in rectangular waveguides, *Electromagnetic Waves Propagation in Complex Matter*, Ed. InTech Pub., 2011

J.3 Lee, C.S.; Lee, S.W.; Chuang, S. L.: *Plot of modal field distribution in rectangular and circular waveguides, IEEE Trans. Microwave Theory Tech.*, Vol. 33, No. 3, pp. 271–274, 1985.

J.4 Marcatili, E.A.J.: *Dielectric rectangular waveguide and directional coupler for integrated optics, Bell Syst. Tech. J., BSTJ*, Vol. 48, No. 7, pp. 2071–2102, Sept. 1969.

J.5 Chang, C.C.; Qian, Y.; Itoh, T.: *Analysis, and applications of uniplanar compact photonic bandgap structures, Prog. Electromagn.Res., PIER*, Vol. 41, pp. 211–235, 2003.

J.6 Krischnining, M.; Jansen, R.H.; Koster, N.H.L.: *New aspects concerning the definition of microstrip characteristic impedance as a function of frequency*, IEEE., MTT-S, Int MicrowaveSymp. Dig., pp. 305–307, 1982.

J.7 Bianco, B.; Panini, L.; Parodi, M.; Ridella, S.: *Some considerations about the frequency dependence of the characteristic impedance of uniform microstrip, IEEE Trans. Microwave Theory Tech.*, Vol. MTT- 26, No. 3, pp. 182–185, Mar. 1978.

J.8 Knorr, Jeffrey B.; Kuchler, Klaus-Dieter: *Analysis of coupled slots and coplanar strips on dielectric substrate, IEEE Trans. Microwave Theory Tech.*, Vol. MTT- 23, pp. 541–548, July 1975.

J.9 Cohn, S.B.: *Slot line on a dielectric substrate, IEEE Trans. Microwave Theory Tech.*, Vol. MTT- 17, No. 10, pp. 768–778, Oct. 1969.

J.10 Marque, R.; Martel, J.; Mesa, F.; Medina, F.: *Left-handed-media simulation and transmission of EM waves in subwavelength split-ring-resonator-loaded metallic waveguides, Phys. Rev. Lett.*, Vol. 89, No. 18, pp. 183901-1–183901-4, Oct. 2002.

J.11 Pincherle, L.: *Electromagnetic waves in metal tubes filled longitudinally with two dielectrics, Phys. Rev.* Vol. 66, No. 5&6, pp. 118–130, Sept. 1944.

J.12 Weeks, W.L.: *Propagation constants in rectangular waveguide partially filled with dielectric, IRE Trans. Microwave Theory Tech.*, Vol. 7, No. 2, p. 294, April 1959.

J.13 Gardial, F.E.; Vander Vorst, A.S.: *Wave propagation in a rectangular waveguide loaded with an H-plane dielectric slab, IEEE Trans. Microwave Theory Tech.*, Vol. MTT, No.1, p. 56, Jan. 1969.

J.14 Vartanian, P. H.; Ayres, W. P.; Efelgesson, A. L.: *Propagation in dielectric slab loaded rectangular waveguide, IRS Trans. Microwave Theory Tech.*, Vol. MTT- 15, No.5, pp. 215–222, April 1958.

J.15 Eberhardt, N.: *Propagation in the off center E-plane dielectrically loaded waveguide, IEEE Trans. Microwave Theory Tech.*, Vol. MTT, No.15,pp. 282–289, May 1967.

J.16 Seckelmann, R.: *Propagation of TE modes in dielectric loaded waveguides, IEEE Trans. Microwave Theory Tech.*, Vol. MTT, 4, No.11, pp. 518–527, Nov. 1966.

J.17 Witt, H.R.; Biss, R.E.; Price, E.L.: *Propagation constants of a waveguide containing parallel sheets of finite conductivity, IEEE Trans. Microwave Theory Tech.*, Vol. MTT-15, No.4, pp. 232–239, April 1967.

J.18 Gardiol, F.E.: *Higher-order modes in dielectrically loaded rectangular waveguides, IEEE Trans. Microwave Theory Tech.*, Vol. MTT- 16, No.11, pp. 919–924, Nov. 1968.

J.19 Deslandes, D.; Wu, K.: *Integrated microstrip and rectangular waveguide in planar form, IEEE Microwave Wireless Compon. Lett.*, Vol.11, No.2, pp. 68–70, Feb. 2001.

J.20 Cassivi, Y.; Perregrini, L.; Arcioni, P.; Bressan, M.; Conciauro, G.: *Dispersion characteristics of substrate integrated rectangular waveguide, IEEE Microwave Wireless Compon. Lett.*, Vol.12, No.8, pp. 333–335, Sept. 2002.

J.21 Deslandes, D.; Wu, K.: *Single-substrate integration technique of planar circuits and waveguide filters, IEEE Trans. Microwave Theory Tech.*, Vol. 51, No. 2, pp. 593–596, Feb. 2003.

J.22 Xu, F.; Wu, K.: *Guided wave and leakage characteristics of substrate integrated waveguide, IEEE Trans. Microwave Theory Tech.*, Vol. 53, No. 1, pp. 66–73, Jan. 2005.

J.23 Deslandes, D.; Wu, K.: *Accurate modeling, wave mechanisms, and design considerations of a substrate integrated waveguide, IEEE Trans. Microwave Theory Tech.*, Vol. 54, No. 6, pp. 2516–2526, June 2006.

J.24 Deslandes, D.; Wu, K.: *Design consideration and performance analysis of substrate integrated waveguide components*, Proc. 32th Eur. Microw. Conf., Milan, Italy, vol. 2, pp. 881–884, Sept. 2002.

J.25 Salehi, M.; Mehrshahi, E.: *A closed-form formula for dispersion characteristics of fundamental SIW mode*, *IEEE Microwave Wireless Compon. Lett.*, Vol. 21, No. 1, pp. 4–6, Jan. 2011.

J.26 Wu, K.; Deslandes, D.; Cassivi, Y.: *The substrate integrated circuits – a new concept for high-frequency electronics and optoelectronics*, TELSIKS 2003 6th Int. Conf., vol. 1, pp. P-III–P-X, October 2003.

J.27 Lai, Q.; Fumeaux, C.; Hong, W.; Vahldieck, R.: *Characterization of the propagation properties of the half-mode substrate integrated waveguide*, *IEEE Trans. Microwave Theory Tech.*, Vol. 57, No. 8, pp. 1996–2004, Aug. 2009.

J.28 Hong, W.; Liu, B.; Wang, Y. Q.; Lai, Q. H.; Wu, K.: *Half mode substrate integrated waveguide: A new guided wave structure for microwave and millimeter-wave application*, Proc. Joint 31st Int. Infrared Millimeter Wave Conf./14th Int. Terahertz Electron. Conf., pp. 18–22, Shanghai, China, Sep. 2006.

J.29 Liu, B.; Hong, W.; Wang, Y. Q.; Lai, Q. H.; Wu, K.: *Half mode substrate integrated waveguide (HMSIW) 3 dB coupler*, *IEEE Microw. Wireless Compon. Lett.*, Vol. 17, No. 1, pp. 22–24, Jan. 2007.

J.30 Wang, Y.; Hong, W.; Dong, Y. D.; Liu, B. et al.: *Half mode substrate integrated waveguide (HMSIW) bandpass filter*, *IEEE Microw. Wireless Compon. Lett.*, Vol. 17, No. 4, pp. 256–267, Apr. 2007.

J.31 Liu, B.; Hong, W.; Zhang, Y.; Tang, H. J.; Yin, X. X.; Wu, K.: *Half mode substrate integrated waveguide 180 3-dB directional couplers*, *IEEE Trans. Microw. Theory Tech.*, Vol. 55, No. 12, pp. 2586–2592, Dec. 2007.

8

Microstrip Line

Basic Characteristics

Introduction

The microstrip line contributes about ninety percent toward the MIC and MMIC technology, as against other planar transmission lines, such as coplanar waveguide (CPW), slot lines, etc. It has found its applications in the frequency range 1–100 GHz and beyond up to a few THz. Unlike a coaxial cable and a two-wire transmission line, embedded in a *homogeneous medium*, the microstrip line is constructed in an *inhomogeneous medium*. The basic line parameters; such as propagation constant, attenuation factor, and characteristic impedance, of a coaxial cable and a two-wire transmission line are frequency independent, i.e. these lines are *nondispersive*. However, the basic line parameters of microstrip and several other planar transmission lines are frequency-dependent, i.e. the planar transmission lines are *dispersive*. The main emphasis in this chapter is to understand the basic parameters and characteristics of the microstrip line. It can help us to compute its propagation characteristics, characteristic impedance, and losses in terms of the physical parameters of a microstrip line.

Objectives

- *To explain the general nature of the mode of propagation of the EM-wave on a microstrip line, including the quasi-static nature of wave propagation.*
- *To understand Wheeler's transformation that transforms the inhomogeneous medium microstrip to the homogenous medium microstrip through the concept of the effective relative permittivity.*
- *To present closed-form static models, to compute effective relative permittivity and the characteristic impedance of a microstrip under several kinds of physical conditions.*
- *To understand the nature of microstrip dispersion and to develop dispersion models.*
- *To model the dielectric and conductor losses of a microstrip line.*
- *To present a circuit model of a lossy microstrip line.*

8.1 General Description

Figure (8.1) shows the cross-sectional views of a few basic microstrip structures. Figure (8.1a) shows the standard form of an *open microstrip line*. It is also called just the *microstrip* or the microstrip line. The strip conductor width of the microstrip is w and the thickness of the strip conductor is t. The microstrip is printed on a dielectric sheet, called the *substrate*, with relative permittivity ε_{r1} and thickness h_1. The open space above the microstrip is unbounded medium. Therefore, the microstrip forms an *inhomogeneous* transmission medium for the EM-wave propagation, as the strip conductor is located at the interface of two dielectrics media – the substrate with relative permittivity ε_{r1} and the air-medium with $\varepsilon_{r2} = 1$. For simplicity, relative permittivity and thickness of a substrate are taken as ε_r and h, respectively.

The real-life microstrip-based components are always enclosed in a metallic enclosure, shown in Fig (8.1b), to protect the circuits against the environment. The metallic enclosure also provides the EM-shielding to the microstrip circuits and also the mechanical strength to the substrate. It further acts as a heat-sink to the high-power microwave devices. The presence of a metallic enclosure, i.e. the top shield and two sidewalls, has a significant influence on the propagation characteristics and the characteristic impedance of an open microstrip line. Therefore, either their presence has to be accounted for at the design stage itself, or the top shield, and sidewalls should be kept far away from the strip conductor, so that they do not alter the characteristics of an open microstrip line. The *top-shielded microstrip line*, shown in Fig (8.1c), has the sidewalls far from the strip conductor. Therefore, the sidewalls are not shown in this case. The upper dielectric layer of relative permittivity ε_{r2} and thickness h_2, i.e. the dielectric cover to the microstrip, is called the *superstrate*. Figure (8.1a–c) do not show any superstrate. The upper dielectric is just an air-medium.

In the case of the shielded and the enclosed microstrip lines, both the lower (substrate) and the upper (superstrate)

Introduction to Modern Planar Transmission Lines: Physical, Analytical, and Circuit Models Approach, First Edition. Anand K. Verma.

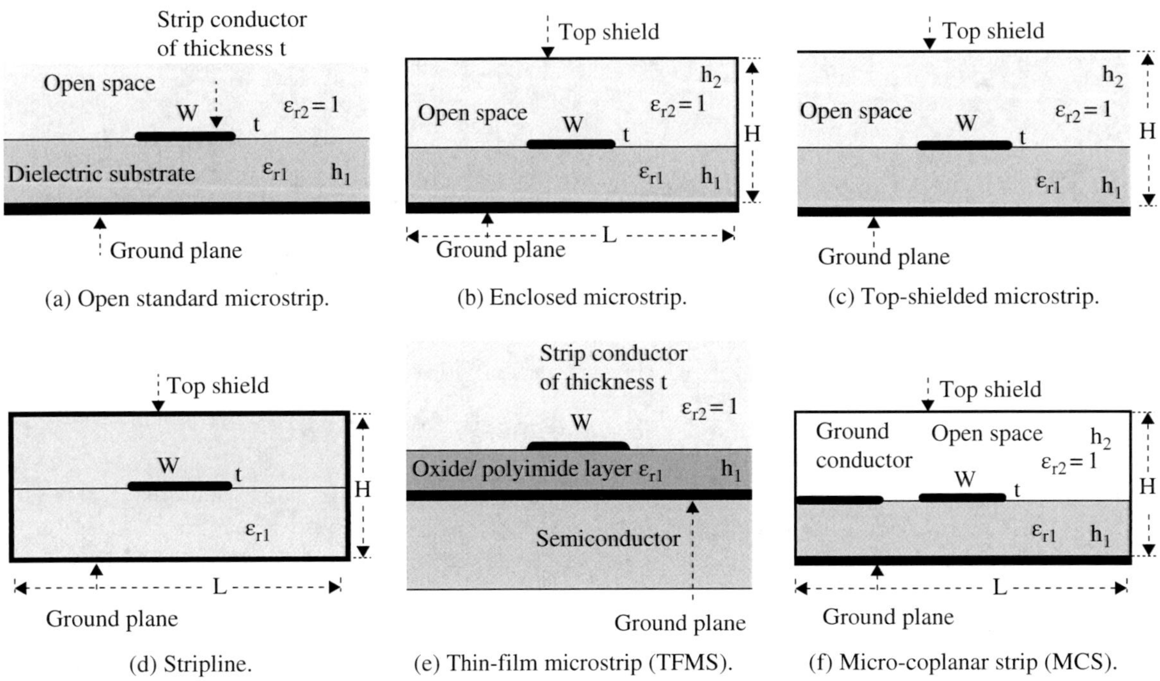

Figure 8.1 Cross-sections of some microstrip structures.

dielectric layers are present. However, layers can have identical relative permittivity and thickness, i.e. $\varepsilon_{r2} = \varepsilon_{r1}$ with $h_2 = h_1$. This special structure, shown in Fig (8.1d), is the *strip* transmission line. The stripline is developed in a *homogeneous* dielectric-medium. The stripline can also be asymmetrical. This line is also used in MIC technology. However, it is not much popular like a microstrip; as the strip conductor of a strip transmission line is not approachable for mounting the components. There are many more interesting and useful variations in the basic microstrip line [B.1–B.7, J.1, J.2]. One such variation, shown in Fig (8.1e), is known as the *thin film microstrip* (TFMS) line. The TFMS is constructed on a semiconductor substrate. The ground plane conductor is deposited on a supporting semiconductor that need not be expensive semi-insulating type. The substrate is a thin layer (a few micron thicknesses) of the oxide or polyimide on which the strip conductor of microstrip is deposited. The thin-film microstrip line is a useful structure for MMIC technology. Normally, it is difficult to ground any device in the microstrip, as it may be brittle substrates such as alumina and semiconducting substrates. Drilling a hole in these substrates is a difficult and critical task. Figure (8.1f) shows the micro-coplanar strip (MCS) structure. It is a microstrip line with an additional coplanar ground conductor strip that helps to ground a device in the microstrip technology [J.3].

Figure (8.2a) shows the 3D view of $\lambda_g/2$-section of an infinite-length microstrip line. The microwave signal is applied between the strip conductor and the ground plane conductor. It also shows two sets of transverse electric and magnetic field lines, separated by half wavelength, in the direction of propagation. In general, the substrate of a microstrip could be made of dielectric material, magnetic material, semiconductor, ferroelectric material, etc. However, a dielectric sheet is commonly used as the substrate. Normally, the substrates are lossy materials and they cause the *dielectric loss* in a microstrip line. The strip and ground conductors can be made of copper, aluminum, gold, and even superconductors, depending upon the particular technological requirement. The conducting strip and ground plane are usually several skin-depth thick conductors. The ground plane is taken as an infinitely thick conductor; however, the thickness of the strip conductor is either neglected or taken as finite. We have to account for the influence of the strip conductor thickness on the propagation characteristics and the characteristic impedance of a microstrip line. Figure (8.2b) demonstrates the cross-section of a thick conductor (thickness t) microstrip with more field lines in the air-region.

The strip conductor is printed on a substrate either through the process of *etching* on a premetallized substrate or by the process of *deposition* of a conducting layer on a substrate. In the process of etching, the unwanted parts of the pre-coated conducting sheet are removed either by the *chemical etching* or by the *mechanical milling*. The fabrication processes of the microstrip, and other planar lines

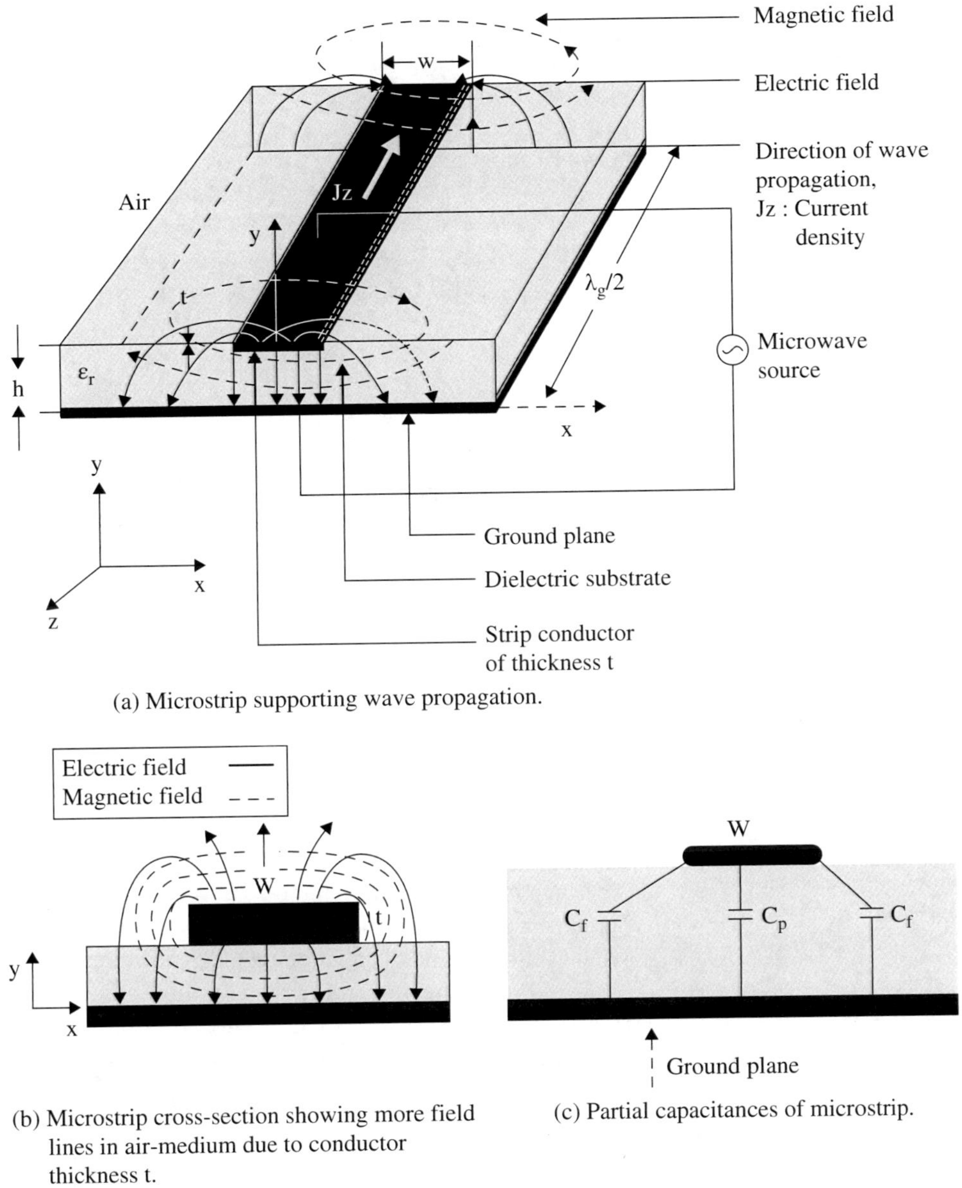

Figure 8.2 The fields and wave of a microstrip.

also, are discussed in chapter 13. The strip and ground conductors are not perfect conductors. They have a finite conductivity that causes loss to the EM-wave propagating on a microstrip line. This loss is called the *conductor loss.*

The microstrip and its several variants are primarily two-conductor transmission lines. A two-conductor transmission line, in a *homogeneous* dielectric-medium, supports the *TEM-mode* wave propagation. Figure (8.2a) shows that in the case of a microstrip line, the electric and magnetic fields are in two different dielectric media. Thus, a microstrip line is located in an *inhomogeneous* dielectric-medium. Figure (8.2b) further shows the field distribution of a microstrip line in the transverse plane. The electric field is not totally inside the dielectric-medium, between the strip and ground conductors. A part of the total electric field, forming the *fringing field*, is away from the central region of the strip conductor. Figure (8.2c) shows that the central capacitance C_p *p.u.l.* of microstrip is due to the main electric field, whereas both side fringe capacitance C_f *p.u.l.* is due to the fringing field. The total line capacitance of microstrip is due to both C_p and C_f. The next section demonstrates, with the help of Maxwell's equations, that a microstrip line in an inhomogeneous medium cannot support the pure TEM mode wave propagation, as there are always electric and magnetic field components along the direction of propagation. In general, a microstrip supports the so-called *hybrid mode*, discussed in chapter 7. However, at the lower frequency, the hybrid mode could be approximated by the TEM kind of mode that is known as the *quasi-TEM mode*. For the quasi-TEM mode, the magnitude of the longitudinal field components can be neglected, as compared to the magnitude of the transverse field components. Figure (8.3) exhibits the conceptual evolution of a microstrip from the coaxial line. It helps to appreciate the TEM nature of a microstrip line.

8.1.1 Conceptual Evolution of Microstrip Lines

Figure (8.3) demonstrates the evolution of a microstrip line from a coaxial line. The circular coaxial line supports the pure TEM mode propagation as the fundamental mode. At higher frequency, the higher-order modes can also appear. However, these can be ignored, as normally, the single fundamental mode propagation is more important. The circular coaxial line is gradually deformed into the elongated elliptical coaxial line that too supports the pure TEM mode propagation. The elongated sides can be replaced by the vertical sidewalls. Thus, a stripline structure is obtained that again supports the pure TEM mode propagation. Both sidewalls and the materials above the strip conductor can be removed. It results in a microstrip structure. The microstrip appears to support the TEM mode, as it has evolved from the TEM mode supporting the coaxial line. However, due to the presence of fields in two different dielectric media, the microstrip does not support the *pure* TEM mode. It supports the *quasi*-TEM mode.

8.1.2 Non-TEM Nature of Microstrip Line

Maxwell's equations are used below to demonstrate that technically a microstrip line cannot support the TEM mode propagation, i.e. a mode with no electric and magnetic field components along the direction of propagation. Figure (8.2a) shows that the electric field E undergoes the refraction, i.e. bending of its path, when it enters from the air-medium to the dielectric substrate medium. At the interface of two media, the tangential components of electric field E_x in both media must be continuous:

$$E_x|_{dielectric} = E_x|_{air}. \tag{8.1.1}$$

As there is no free charge at the interface of two media, Maxwell's equation could be written as

$$\nabla \times \vec{H} = j\varepsilon_0\varepsilon_r\omega\,\vec{E}. \tag{8.1.2}$$

On using the above expressions in both media, the following expression is obtained at the interface:

$$\frac{1}{j\varepsilon_0\varepsilon_r\omega}\left(\nabla \times \vec{H}\right)_x\Big|_{dielectric} = \frac{1}{j\varepsilon_0\omega}\left(\nabla \times \vec{H}\right)_x\Big|_{air}$$

$$\Rightarrow \left(\frac{\partial H_z}{\partial y} - \frac{\partial H_y}{\partial z}\right)\Big|_{dielectric} = \varepsilon_r\left(\frac{\partial H_z}{\partial y} - \frac{\partial H_y}{\partial z}\right)\Big|_{air}. \tag{8.1.3}$$

The normal component of the magnetic field H_y is also continuous at the interface of two media:

$$\frac{\partial H_y}{\partial z}\Big|_{dielectric} = \frac{\partial H_y}{\partial z}\Big|_{air}. \tag{8.1.4}$$

The following expression is obtained from the above two equations:

$$\varepsilon_r\frac{\partial H_z}{\partial y}\Big|_{air} - \frac{\partial H_z}{\partial y}\Big|_{dielectric} = (\varepsilon_r - 1)\frac{\partial H_y}{\partial z}. \tag{8.1.5}$$

For an inhomogeneous microstrip medium, $\varepsilon_r \neq 1$. Therefore, the H_z component of the magnetic field exists, as the right-hand side of equation (8.1.5) is nonzero. However, for a homogeneous medium, say for the air dielectric-medium, $\varepsilon_r = 1$ and the right-hand side of the above equation is zero, leading to $H_z = 0$. Likewise, the second Maxwell's equation can be used to demonstrate that for an inhomogeneous microstrip medium, the E_z component of the electric field also exists. Thus, a microstrip line supports the hybrid mode with $E_z \neq 0$ and $H_z \neq 0$. However, the longitudinal field components, E_z and H_z, are significant only at a higher frequency. At the lower end of the microwave,

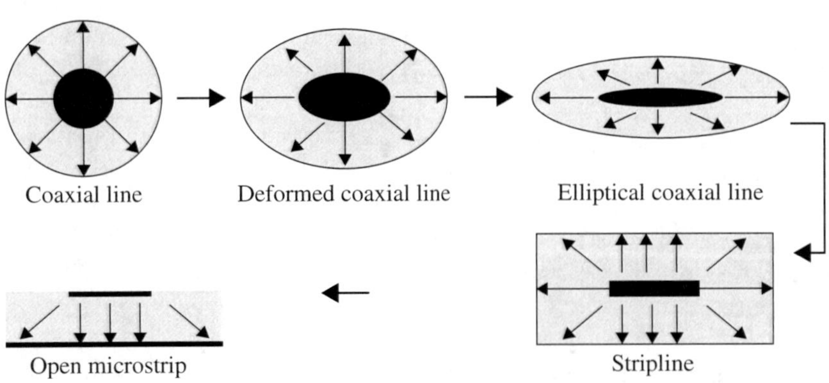

Figure 8.3 Evolution of stripline and microstrip line from the TEM-mode coaxial line.

say at a frequency below 4 GHz, these field components can be neglected as compared to the dominant transverse field components. It leads to approximately TEM mode. The *approximate TEM* mode is called the *quasi-TEM* mode. Even at the low frequency, the metallic ground plane backed dielectric substrate supports the fundamental *TM_0 surface wave mode.* The surface wave mode is discussed in chapter 7. The TEM mode couples with the TM_0 surface wave mode producing the dispersive nature of a microstrip line. *The quasi-TEM mode can be regarded as a combination of the TEM mode and the TM_0 surface wave mode* [J.4]. Thus, the pure TEM mode supporting line is nondispersive one, whereas the quasi-TEM mode supporting line is dispersive. The material dispersion in a substrate is negligible in the microwave and mm-wave ranges.

8.1.3 Quasi-TEM Mode of Microstrip Line

The separation of variable method, applicable to the metallic waveguides, is not applicable to solve the wave equation for a microstrip line in an inhomogeneous medium. For the case of a microstrip line, and other planar transmission lines, several numerical methods, called the *full-wave methods* have been developed to obtain the frequency-dependent hybrid mode field solutions. We present below the limiting case at low frequency. In this case, the wave equation could be reduced to Laplace's equation, showing the static nature of microstrip fields in the transverse plane [B.8]. Thus, the quasi-TEM mode of a microstrip could be dealt with the Laplace's equation. Its solution is much simpler as compared to the solution of Helmholtz's wave equation. The wave equation for the electric and magnetic field is given below:

$$\nabla^2 \begin{Bmatrix} \vec{E} \\ \vec{H} \end{Bmatrix} + \omega^2 \varepsilon\mu \begin{Bmatrix} \vec{E} \\ \vec{H} \end{Bmatrix} = 0. \tag{8.1.6}$$

Figure (8.2a) shows that the strip conductor of a microstrip line supports the time-varying current density $\vec{J}_z(h, z)$ located at the distance h from the ground plane. The retarded magnetic vector potential at any location below the strip conductor in the transverse (x-y)-plane is [B.9]:

$$\vec{A}(r) = \frac{\mu}{4\pi}\int \frac{\vec{J}_z(h, z)e^{-j\omega\sqrt{\mu\varepsilon}|r-r'|}}{|r-r'|}ds'. \tag{8.1.7}$$

In the above equation, ds' is the length of the current element and r, r' are radial vectors. For the field just below the conducting strip, i.e. at $|r - r'| = h$, the magnetic vector potential is

$$\vec{A}(r) = \frac{\mu}{4\pi}\int \frac{\vec{J}_x(h, z)e^{-j2\pi h/\lambda}}{|r-r'|}ds'. \tag{8.1.8}$$

If the substrate thickness h is much smaller than the wavelength λ, $(2\pi h/\lambda \ll 1)$, the exponential factor showing the phase variation of magnetic vector potential can be neglected and it is given by

$$\vec{A}(r) = \frac{\mu}{4\pi}\int \frac{\vec{J}_z(h, z)}{|r-r'|}ds'. \tag{8.1.9}$$

The above magnetic vector potential of a *magnetostatic field* is valid for the substrate thickness, $h \ll \lambda/2\pi$. In this case, the time retardation in the magnetic vector potential is ignored. It shows that the transverse field of a microstrip at a longer wavelength, with limiting case of $\omega \to 0$, could be treated as the static field in the transverse plane. Although, the time-dependent wave still propagates along the microstrip line as the quasi-TEM mode. Thus, the magnetostatic field expressed as below, models the quasi-TEM mode of a microstrip line:

$$\nabla \times \vec{H} = 0. \tag{8.1.10}$$

The electric field of the TEM mode is related to the magnetic field through Maxwell's equation:

$$\nabla \times \vec{E} = -j\mu_0\mu_r\omega\vec{H}. \tag{8.1.11}$$

The following Laplace equation is obtained from the above two equations:

$$\nabla^2 \vec{E} = 0. \tag{8.1.12}$$

In the transverse (x-y)-plane, Laplace equation could be further written in terms of the electric potential V:

$$\nabla^2 V = 0, \tag{8.1.13}$$

where $\vec{E} = -\hat{y}V/h$. Therefore, the quasi-TEM mode assumption requires the solution of Laplace's equation in the transverse (x-y)-plane of a microstrip line. The static field solution is useful only in the lower microwave frequency range, say around 4 GHz. The static microstrip line parameters are improved empirically to incorporate the frequency-dependent behavior of a microstrip line. In summary, a microstrip and other planar transmission lines are analyzed by following three approaches:

- *Full-wave methods:* These methods, such as spectral domain analysis (SDA), integral equation method, method of lines, finite element method (FEM), finite difference time domain (FDTD), mode-matching method, etc., numerically solve the wave equation. They provide frequency-dependent line parameters. The SDA is discussed in chapter 16. The SDA is a semi-analytical method involving more analytical processing and less computational time.

- *Static methods:* There are several static methods, such as the variational method in the Fourier domain and the space domain, conformal mapping methods, method of moment (MoM), finite difference method, etc. The conformal mapping method and the variational method are discussed in chapters 9 and 14, respectively.
- *Closed-form models:* Several *static closed-form models*, for the frequency-independent microstrip line parameters, and several *dynamic closed-form models*, for the frequency-dependent microstrip line parameters, have been reported. A few of these models are discussed in this chapter.

The above methods and models have been summarized in several books [B.1–B.7, B.10]. The full-wave methods involve the rigorous solution of the wave equation (8.1.6) that is analytically demanding and numerically inefficient, i.e. it takes a long time on a computer to get the results. Therefore, these methods are not suitable for the computer-aided design (CAD) that is the core of the modern circuit simulators, such as ADS, Microwave Office, etc. However, the full-wave methods can provide accurate results on the propagation behavior and also for the field and current distributions of a microstrip line and other planar transmission lines. The full-wave methods are suitable for the detailed investigations of the planar transmission lines and passive components. These methods are the foundation block of the modern 2D and 3D *EM-field simulators* – Momentum, Microwave Studio, HFSS, LINMIC, Sonnet, Empire, etc. The EM-Simulators are useful for the *numerical experimentations* without conducting the real physical experimentations on the planar transmission lines and components. The real-life experimental investigations are expensive and time-consuming. These methods are also used to generate numerical data for the multidimensional curve-fitting to obtain numerically efficient CAD-oriented closed-form models. The closed-form models are the foundation blocks of the modern microwave circuit simulators mentioned above. Most of the practical design works are carried out on these *circuit simulators*. The static methods are also numerical methods, but they are analytically much simpler and computationally much faster. The static methods are combined with the closed-form dynamic models, discussed in chapter 15, to provide the design-oriented dispersion and loss models for the multilayer microstrip and other planar lines.

8.1.4 Basic Parameters of Microstrip Line

The above discussion shows that a microstrip line supports the hybrid mode of propagation that can be approximated by the TEM kind of wave propagation. Thus, a microstrip line could be modeled as *a parallel conductor transmission line* that we have discussed in chapter 2. The microstrip and other planar transmission lines, such as the TEM mode supporting two-conductor transmission lines, are described by the *secondary line parameter* – the complex propagation constant γ and the characteristic impedance Z_0. These secondary line parameters are described by the *primary line constants* – R, L, C, G, i.e. the resistance, inductance, capacitance, and conductance per unit length (*p.u.l.*) of a microstrip line. The secondary parameters are directly measurable quantities. These are useful in the design of microwave components. Therefore, a microstrip and other planar transmission lines are characterized in terms of the secondary line parameters.

Careful consideration is needed for the modeling of a microstrip line, supporting the hybrid mode wave propagation, by the TEM mode. The static solution of Laplace's equation provides the frequency-independent secondary line parameters: characteristic impedance $Z(f = 0)$, propagation constant $\beta(f = 0)$, and attenuation constant $\alpha(f = 0)$. Finally, the frequency-independent phase velocity $v_p(f = 0)$ is obtained. The microstrip line is described as a line without any dispersion. However, this is not the correct description of microstrip and other planar transmission lines. The microstrip and other planar transmission lines are primarily dispersive transmission lines. Their characteristic impedance $Z(f)$, propagation constant $\beta(f)$, and attenuation constant $\alpha(f)$ are frequency-dependent parameters. Moreover, the fundamental TEM mode is supplemented by the appearance of *higher-order guided modes*. Further, *the radiation modes*, viewed as the *leaky modes*, also exist. Figure (8.4) shows the results of the full-wave analysis, using the *eigenmode* solution, on frequency-dependent effective relative permittivity, with respect to the normalized wavelength h/λ, of a top-shielded microstrip line on an alumina substrate ($w/h_1 = 4.72$, $\varepsilon_r = 9.8$, $h_2 = 5h_1$) for the quasi-TEM mode [B.1, J.5, J.6]. In the hybrid mode format, the quasi-TEM mode is the HE_0 hybrid mode. It also shows the appearance of higher-order hybrid modes, i.e. the HE_1 and HE_2 hybrid modes, with an increase in frequency; as the normalized substrate thickness (h/λ) increases with frequency. The higher-order modes appear above their cutoff frequencies. These are the bound guided modes. Other full-wave methods are also used to get higher-order modes [J.7]. Figure (8.4) further shows the presence of the unbound radiation mode at all frequencies. It is significant at a higher frequency. Thus, in the case of an open microstrip line operating in the higher-order mode, the *leaky mode* could be obtained. It is useful to design the leaky-wave antenna. The useful frequency range of the microstrip line, as a wave-propagating medium, is limited by the appearance of the first higher-order mode HE_1.

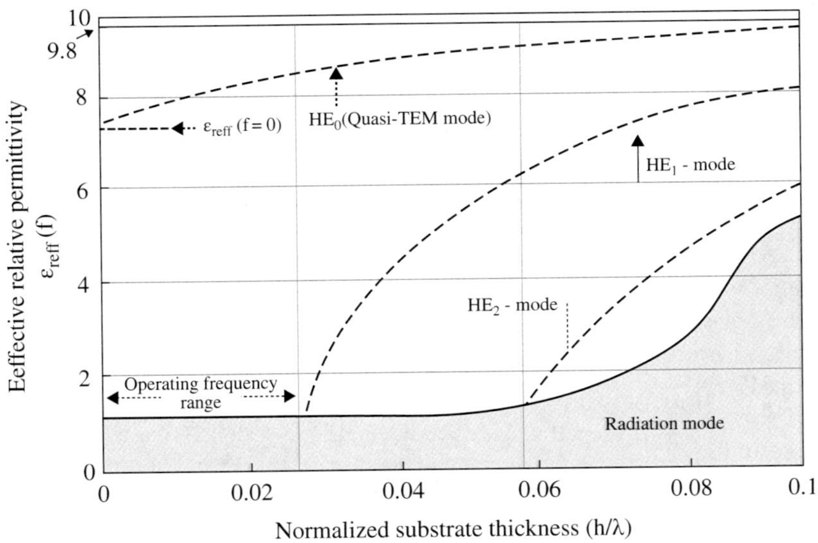

Figure 8.4 The quasi-TEM mode and higher-order modes of a microstrip line on the alumina substrate [J.5]. *Source:* Modified from Ermert [J.5].

The cutoff frequency of the first higher-order mode should be known, to estimate the frequency range of the quasi-TEM operation. The microstrip line can tolerate, without any damage, only a finite amount of microwave power. Therefore, the power handling capability of a microstrip line should also be estimated. It is important for the high-power application of a microstrip line [B.2]. Of course, for the usual low-power application, this information is normally not needed.

In brief, our primary concern is to model a microstrip line by the static closed-form models. Next, the models are improved by incorporating the frequency-dependent behavior in it. Finally, the closed-form frequency-dependent models should be able to accommodate several variations in the microstrip line, shown in Fig (8.1). It is considered in chapter 15. At present, the microstrip is treated as a lossless transmission line with $R = G = 0$ that is embedded in an inhomogeneous dielectric-medium. The process of obtaining the equivalent homogeneous dielectric-medium for a primarily inhomogeneous medium microstrip line is discussed in the next section. From chapter 2, the characteristic impedance and phase constant of a lossless line are given below:

$$Z_0 = \sqrt{\frac{L}{C}} \quad (a), \quad \beta = \omega\sqrt{LC} \quad (b). \tag{8.1.14}$$

The phase velocity of the TEM-wave propagating on a transmission line is

$$v_p = \frac{\omega}{\beta} = \frac{1}{\sqrt{LC}}. \tag{8.1.15}$$

The phase velocity of a microstrip line in the air-medium is the same as that of the velocity of light in the free space. The phase velocity of the EM-wave in the nonmagnetic dielectric-medium with $\mu_r = 1$ and $\varepsilon_r \neq 1$ is

$$v_p = \frac{\omega}{\beta} = \frac{1}{\sqrt{\mu_0 \varepsilon_0}} \frac{1}{\sqrt{\varepsilon_r}} = \frac{c}{\sqrt{\varepsilon_r}}. \tag{8.1.16}$$

The wavelength λ_0 of the signal in the free space is decreased in a dielectric-medium. The guided wavelength (λ_g) of the TEM-wave in a dielectric-medium is

$$\lambda_g = \frac{\lambda_0}{\sqrt{\varepsilon_r}}. \tag{8.1.17}$$

The guided-wave propagation constant (β_g) of the EM-wave propagating in a dielectric-medium is related to the propagation constant (β_0) in the free space by

$$\beta_g = \beta_0 \sqrt{\varepsilon_r}. \tag{8.1.18}$$

The characteristic impedance can be written in terms of the line capacitance C *p.u.l.*:

$$Z_0(\varepsilon_r) = \frac{1}{v_p C} = \frac{Z_0(\varepsilon_r = 1)}{\sqrt{\varepsilon_r}}. \tag{8.1.19}$$

If the relative permittivity of a medium is frequency-dependent, i.e. if the medium is dispersive, then the phase velocity, propagation constant, guided wavelength, and characteristic impedance associated with the transmission line are also frequency-dependent.

A parallel-plate transmission line with the homogeneous dielectric-medium inside it also represents a microstrip line. The line capacitances *p.u.l.* for a parallel-plate transmission line in a homogeneous dielectric-medium and the air-medium are given by

$$C(\varepsilon_r) = \frac{\varepsilon_0 \varepsilon_r W}{h} \quad (a), \quad C_o = C(\varepsilon_r = 1) = \frac{\varepsilon_0 W}{h} \quad (b), \tag{8.1.20}$$

where w is the width of the parallel-plate line, and h is the separation between two parallel-plates. The relative permittivity of a homogeneous medium is taken as a ratio of the above given two capacitances:

$$\varepsilon_r = \frac{C(\varepsilon_r)}{C(\varepsilon_r = 1)}. \tag{8.1.21}$$

The above expression is very useful both from the experimental and theoretical points of view. The relative permittivity of a dielectric sheet is measured at low frequency with help of a capacitance meter, by measuring the capacitances with and without a dielectric sheet. During the process, the same separation between the plates, for both the measurements, is maintained. Likewise, once an expression is available to determine the line capacitance of a microstrip, the relative permittivity of a microstrip medium is computed by calculating the line capacitances with and without dielectric substrate, while keeping the w/h-ratio unchanged. The relative permittivity obtained is called the *effective relative permittivity* of an equivalent homogeneous medium of microstrip, as all field lines are not confined in the dielectric-medium, some fringing field is always present in the air-medium. The value of the effective relative permittivity is always more than unity and less than the relative permittivity of a substrate. The line capacitance can be obtained by solving Laplace's equation discussed above. The characteristic impedance of a TEM-mode transmission line can also be expressed in terms of these two line capacitances

$$Z_0(\varepsilon_r) = \frac{1}{c\sqrt{C(\varepsilon_r)\,C(\varepsilon_r = 1)}} = \frac{1}{c\sqrt{C\,C_0}}, \tag{8.1.22}$$

where c is the velocity of EM-wave in the air-medium. Sometimes, the parametric description of quantities is dropped for the sake of simplicity, as it has been done above for the line capacitance.

8.2 Static Closed-Form Models of Microstrip Line

This section introduces the concept of *Wheeler's transformation, filling-factor,* and *effective relative permittivity.* Wheeler's transformation is used to convert the inhomogeneous medium microstrip line to an equivalent homogeneous medium microstrip. The closed-form models are also presented to compute the static effective relative permittivity and the static characteristic impedance of a microstrip line with zero strip conductor thickness. Subsequently, the effects of finite conductor thickness, top shield, and the sidewalls are accounted for on the effective relative permittivity and the static characteristic impedance. These closed-form models are important for the MIC design.

8.2.1 Homogeneous Medium Model of Microstrip Line (*Wheeler's Transformation*)

Figure (8.5) shows the steps involved to obtain *Wheeler's transformation* that provides the equivalent homogeneous medium microstrip line from the original microstrip line structure in an inhomogeneous medium. The *concept of partial capacitance* is used in the process.

- **Step (a):** The microstrip line is on the substrate of thickness h with relative permittivity $\varepsilon_{r1} = \varepsilon_r$. The upper-medium is an infinite block of air with relative permittivity $\varepsilon_{r2} = 1$. The strip conductor has width w, with zero

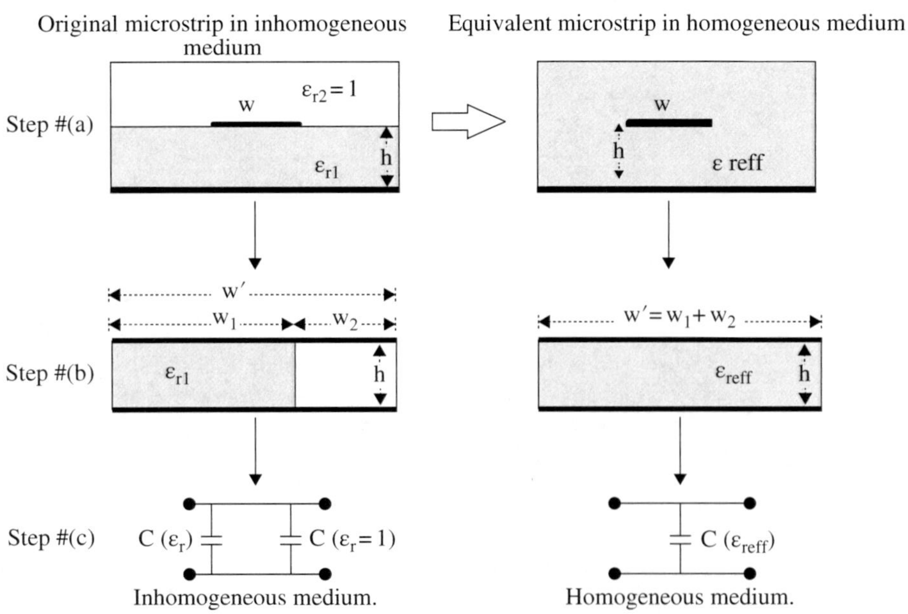

Figure 8.5 Equivalent homogeneous dielectric-medium for a microstrip line.

thickness. The *equivalent homogeneous medium* microstrip line is located in the dielectric-medium of the infinite extent with a relative permittivity called the *effective relative permittivity* ($\varepsilon_r = \varepsilon_{reff}$). The effective relative permittivity depends on the relative permittivity (ε_r) and w/h ratio of the original microstrip line. It also depends, to some extent, on the thickness of the strip conductor that is neglected at present.

Following the definition of the relative permittivity of a homogeneous dielectric-medium given by equation (8.1.21), an expression can be written for the effective relative permittivity of a microstrip line in terms of its line capacitances:

$$\varepsilon_{reff}\left(\varepsilon_r, \frac{w}{h}\right) = \frac{C\left(\varepsilon_r, \frac{w}{h}\right)}{C\left(\varepsilon_r = 1, \frac{w}{h}\right)}. \tag{8.2.1}$$

The line capacitance of a microstrip line is obtained by solving Laplace's equation. Once the effective relative permittivity of microstrip is known, the guided-wavelength and the characteristic impedance are obtained by replacing ε_r for ε_{reff} equations (8.1.17) and (8.1.19):

$$\lambda_g = \frac{\lambda_0}{\sqrt{\varepsilon_{reff}}} \quad \text{(a)}, \quad Z_0\left(\varepsilon_r, \frac{w}{h}\right) = \frac{Z_0(\varepsilon_r = 1)}{\sqrt{\varepsilon_{reff}}} \quad \text{(b)}. \tag{8.2.2}$$

- **Step (b):** The microstrip is treated as an equivalent parallel-plate capacitor with two parallel dielectric media. The material dielectric with relative permittivity $\varepsilon_{r1} = \varepsilon_r$ occupies the width w_1 and the air-medium occupies the width w_2. The total equivalent strip width $w' = w_1 + w_2$ of a microstrip remains unchanged. For the equivalent homogeneous medium, the strip width w' is the same and relative permittivity is ε_{reff}.
- **Step (c):** The microstrip line is described by two partial parallel-plate capacitors $C_1(\varepsilon_r)$ and $C_2(\varepsilon_r = 1)$ corresponding to the dielectric and air media, respectively. The equivalent homogeneous medium microstrip line is described by the effective line capacitance $C_{eff}(\varepsilon_{reff})$. *Wheeler's transformation is a static process in which the electrostatic energy remains unchanged.* It is expressed as

$$\frac{1}{2}C_{eff}V^2 = \frac{1}{2}C_1V^2 + \frac{1}{2}C_2V^2, \tag{8.2.3}$$

where V is the potential between the strip conductor and the ground plane. The above equation provides

$$C_{eff} = C_1 + C_2. \tag{8.2.4}$$

The *filling-factor* (q) is defined as a ratio of the electrostatic energy *p.u.l.* in the dielectric-medium to the total electrostatic energy *p.u.l.* in the microstrip line:

$$q = \frac{\frac{1}{2}C_1V^2}{\frac{1}{2}C_1V^2 + \frac{1}{2}C_2V^2}, \quad \Rightarrow \quad q = \frac{C_1}{C_1 + C_2}. \tag{8.2.5}$$

The partial capacitances *p.u.l.* could be written as

$$C_1 = \frac{\varepsilon_0\varepsilon_r w_1}{h}, C_2 = \frac{\varepsilon_0 w_2}{h}, \text{and } C_{eff} = \frac{\varepsilon_0\varepsilon_{reff}(w_1 + w_2)}{h}. \tag{8.2.6}$$

On substituting expressions of equation (8.2.6) in equation (8.2.4), the following expression is obtained:

$$\varepsilon_{reff} = \varepsilon_r\frac{w_1}{w_1 + w_2} + \frac{w_2}{w_1 + w_2}. \tag{8.2.7}$$

Likewise, from equations (8.2.5) and (8.2.6), the filling-factor is expressed as

$$q = \frac{\varepsilon_r w_1}{\varepsilon_r w_1 + w_2} = \frac{\varepsilon_r}{\varepsilon_r + \frac{w_2}{w_1}}. \tag{8.2.8}$$

The above expression shows that the filling-factor q (<1) is a function of both the w_2/w_1-ratio and the relative permittivity of a substrate. The w_2/w_1-ratio is an indicator of the division of electrostatic energy between the air and dielectric media. It is assumed that the division of the electrostatic energy is independent of the relative permittivity of a substrate. On this assumption, $\varepsilon_r = 1$ could be taken and the filling-factor becomes approximately independent of relative permittivity.

$$q = \frac{w_1}{w_1 + w_2} \quad \text{(a)}, \quad 1 - q = \frac{w_2}{w_1 + w_2} \quad \text{(b)}. \tag{8.2.9}$$

For w/h>>1, i.e. in the case of a *very wide width* microstrip, total electric field lines are confined in the dielectric substrate region w_1, giving $w_2 = 0$ and $q = 1$ as the limiting case. For the case of equal division of electrostatic energy between the air and dielectric-media, $C_1 = C_2$ and the ratio of widths is $w_2/w_1 = \varepsilon_r$. Under this limiting condition, the filling-factor is $q = 1/2$. It applies to a microstrip with a *very narrow strip width*. On substituting equation (8.2.9) in equation (8.2.7), the following expression is obtained [J.8, J.9]:

$$\varepsilon_{reff}(f = 0) \Leftrightarrow 1 + q(\varepsilon_r - 1). \tag{8.2.10}$$

The above expression is called *Wheeler's transformation*. The transformation is bi-directional. If the right-hand side of equation (8.2.10) is known, then it transforms the known inhomogeneous medium microstrip line to a homogeneous medium microstrip. It is assumed that the expression for the filling-factor in terms of w/h-ratio is known. Wheeler has obtained such expression for the filling-factor from the conformal mapping method. The result is given below. If microstrip is embedded in a homogeneous dielectric-medium of relative permittivity $\varepsilon_{reff}(f=0)$, i.e. the left-hand side of equation (8.2.10) is known, the equation (8.2.10) transforms it to the standard open microstrip line in an inhomogeneous medium. The reverse medium transformation could be called *inverse Wheeler's transformation*. This concept is useful for the modeling of multilayer microstrip and multilayer CPW structures. It is discussed in chapter 15.

The filling-factor has a range $0.5 \le q \le 1$. Under limiting cases, the effective relative permittivity of the microstrip line could be separated into two groups:

$$\begin{aligned} &\text{For } q = 0.5, w \ll h, \text{very narrow strip conductor}: \\ &\varepsilon_{reff}(f=0) = \frac{\varepsilon_r + 1}{2} \qquad (a) \\ &\text{For } q = 1.0, w \gg h \text{ very wide strip conductor}: \\ &\varepsilon_{reff}(f=0) = \varepsilon_r \qquad (b). \end{aligned} \tag{8.2.11}$$

On knowing the expression for the filling-factor, equation (8.2.10) computes the effective relative permittivity of a microstrip line for any w/h-ratio. In general, the filling-factor is a function of both w/h-ratio and ε_r of the substrate. It is expressed in the functional form using the structure-dependent function $F(w/h, \varepsilon_r)$:

$$q(w/h, \varepsilon_r) = \frac{F(w/h, \varepsilon_r) + 1}{2}. \tag{8.2.12}$$

The dependence of the filling-factor on the relative permittivity ε_r is estimated, from the expression (8.2.8), as follows [J.10, J.11]:

$$\begin{aligned} q(W/h, \varepsilon_r) &= \frac{1}{1 + \dfrac{1}{\varepsilon_r}\dfrac{W_2}{W_1}} \approx \left(1 + \frac{1}{\varepsilon_r}\right)^{-1} q(W/h) \\ &\approx \left(1 - \frac{1}{\varepsilon_r}\right) q(W/h). \end{aligned} \tag{8.2.13}$$

In the above expression, we first assumed a special case for $w_1 = w_2$ and then multiplied it by the w/h dependent filling-factor, q(w/h). For the air-substrate, $q(w/h, \varepsilon_r) = 0$ leading to $\varepsilon_{reff}(f=0) = 1$. For a very high permittivity substrate, the dependence of the filling-factor on permittivity ε_r can be ignored. Wheeler [B.1] has given the following approximate expression for the structure-dependent function $F(w/h, \varepsilon_r)$ that is independent of ε_r:

$$\begin{aligned} F(w/h, \varepsilon_r) &= \left[1 + \frac{12}{w/h}\right]^{-\frac{1}{2}}, \text{for } \frac{w}{h} > 1 \qquad (a) \\ F\left(\frac{w}{h}, \varepsilon_r\right) &= \left[1 + \frac{12}{w/h}\right]^{-\frac{1}{2}} + 0.04\left[1 - \frac{w}{h}\right]^2, \text{for } \frac{w}{h} \le 1 \quad (b). \end{aligned} \tag{8.2.14}$$

The following more accurate expression for the function $F(w/h, \varepsilon_r)$ is obtained from the expressions given by Hammerstad–Jensen [J.12]:

$$\begin{aligned} F(w/h, \varepsilon_r) &= \left[1 + \frac{10}{w/h}\right]^{-a.b} \qquad (a) \\ a &= 1 + \frac{1}{49}\ln\left[\frac{\left(\frac{w}{h}\right)^4 + \left(\frac{1}{52}\frac{w}{h}\right)^2}{\left(\frac{w}{h}\right)^4 + 0.432}\right] \\ &\quad + \frac{1}{18.7}\ln\left[1 + \left(\frac{1}{18.1}\frac{w}{h}\right)^3\right] \qquad (b) \\ b &= 0.564\left[\frac{\varepsilon_r - 0.9}{\varepsilon_r + 3}\right]^{0.053} \qquad (c). \end{aligned} \tag{8.2.15}$$

In the above expressions, the parameters "a" and "b" are functions of w/h ratio and ε_r, respectively.

8.2.2 Static Characteristic Impedance of Microstrip Line

The static characteristic impedance of a lossless microstrip line, with zero conductor thickness, is obtained below, in terms of the characteristic impedance of a microstrip line with zero conductor thickness on the air-substrate. The static characteristic impedance is the dc characteristic impedance, i.e. for the case frequency $f \to 0$. The equation (8.1.14) helps to write the following expression:

$$Z_0(w/h, \varepsilon_{r1} = \varepsilon_r, \varepsilon_{r2} = 1) = Z_0(w/h, \varepsilon_{reff}) = \sqrt{\frac{L(w/h, \varepsilon_{reff})}{C(w/h, \varepsilon_{reff})}}. \tag{8.2.16}$$

However, for a nonmagnetic dielectric-medium, the line inductance does not depend on the relative permittivity. For a microstrip in an equivalent homogeneous dielectric-medium, the expressions for its line inductance and line capacitance are

$$\begin{aligned} L(w/h, \varepsilon_{reff}) &= L(w/h, \varepsilon_{reff} = 1) \qquad (a) \\ C(w/h, \varepsilon_{reff}) &= \frac{\varepsilon_0 \varepsilon_{reff} w}{h} = \varepsilon_{reff} C(w/h, \varepsilon_{reff} = 1) \quad (b). \end{aligned} \tag{8.2.17}$$

The expression for the static characteristic impedance of a lossless microstrip line is obtained from the above two equations:

$$Z_0(w/h, \varepsilon_{r1} = \varepsilon_r, \varepsilon_{r2} = 1, f = 0) = \frac{Z_0(w/h, \varepsilon_{r1} = \varepsilon_{r2} = 1)}{\sqrt{\varepsilon_{reff}(f=0)}}. \tag{8.2.18}$$

The static nature of the line characteristics is shown by writing frequency $f = 0$ as the parameter. It is dropped subsequently, unless it is needed to emphasize the static nature. Several closed-form expressions, with different accuracy, have also been proposed [B.1]. The closed-form expressions of Hammerstad–Jensen are summarized below for the characteristic impedance $Z_0(w/h, \varepsilon_{r1} = \varepsilon_{r2} = 1)$ of a microstrip line on the air-substrate [J.12]:

$$Z_0(w/h, \varepsilon_{r1} = \varepsilon_{r2} = 1) = 60 \ln \left[\frac{F}{w/h} + \sqrt{1 + \left(\frac{2}{w/h} \right)^2} \right] \quad \text{(a)}$$

$$\text{where, } F = 6.0 + 0.2832 \exp \left[- \left(\frac{30.666}{w/h} \right)^{0.7528} \right] \quad \text{(b)}. \qquad (8.2.19)$$

The above expression has error about 0.01% for $w/h < 1$ and error 0.03% for $w/h < 1000$ with respect to the exact conformal mapping method. The static effective relative permittivity of a microstrip line with conductor thickness $t = 0$ is computed from expressions (8.2.10), (8.2.12), and (8.2.15).

8.2.3 Results on Static Parameters of Microstrip Line

The filling-factor (q) is primarily a function of the w/h ratio. However, it also depends to some extent on ε_r [J.10, J.11]. It is incorporated in equation (8.2.15). Figure (8.6a) shows its variation from 0.5 to 1 corresponding to the narrow- and wide-strip conductors for permittivity $\varepsilon_r = 2.5$.

The effective relative permittivity varies from $(\varepsilon_r + 1)/2$, for a very narrow-strip conductor, to permittivity ε_r, for a very wide-strip conductor. Figure (8.6b) shows an increase in $\sqrt{\varepsilon_{reff}(f = 0)}$ with an increase in the w/h-ratio. It is due to the confinement of more electric field in the dielectric region below the strip conductor with an increase in w/h-ratio. Table (8.1a) shows some computed results for $\varepsilon_{reff}(f = 0)$ on the plastic substrate ($\varepsilon_r = 2.3$) and alumina substrate ($\varepsilon_r = 9.8$) using Hammerstad–Jensen (H-J) and Wheeler's (Wh.) closed-form expressions. Results are also

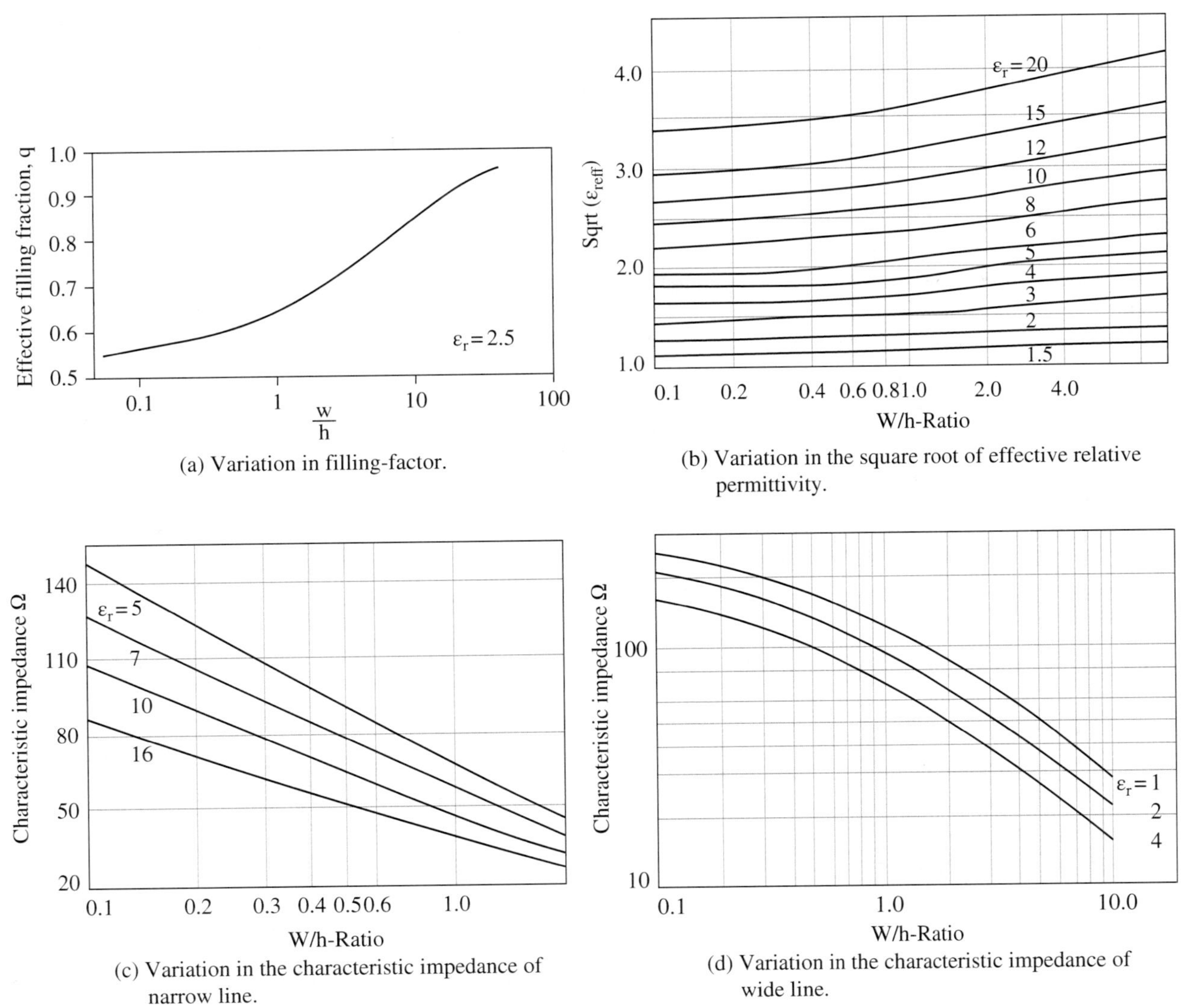

(a) Variation in filling-factor.

(b) Variation in the square root of effective relative permittivity.

(c) Variation in the characteristic impedance of narrow line.

(d) Variation in the characteristic impedance of wide line.

Figure 8.6 Characteristics of microstrip lines. *Source:* Maloratsky [J.1].

Table 8.1a Comparison of computed $\varepsilon_{reff}(f = 0)$ by several numerical methods and closed-form models.

ε_r	w/h	MOL	SDA	Var	H-J	Wh.
2.3	0.100	1.738	1.758	1.738	1.737	1.700
	1.000	1.838	1.836	1.835	1.836	1.822
	10.00	2.098	2.096	2.034	2.100	2.092
9.8	0.100	5.926	5.986	6.150	5.929	5.739
	1.000	6.574	6.563	6.555	6.579	6.563
	10.00	8.382	8.406	8.345	8.389	8.395

Comparison of $\varepsilon_{reff}(f = 0)$ against experimental results

	w/h	Exp.	SDA	Var	H-J	Wh.
$\varepsilon_{r\perp} = 9.4$	0.100	6.465	6.459	6.431	6.448	6.405
εrII=11.6	1.259	7.614	7.601	7.606	7.658	7.607
h = 0.5 ± 0.05 mm	9.140	9.787	9.775	9.758	9.725	9.702

Source: Edwards and Owens [J.13] and Hammerstad`and Jensen [J.12].

Table 8.1b The $\varepsilon_{reff}(f = 0)$ and characteristic impedance of microstrip lines.

ε_r	w/h	ε_{reff}	Z_0 Ω
2.2	0.50	1.93	120
	1.00	1.99	89.4
	1.50	2.03	73.2
	2.50	2.10	54.1
	4.00	2.60	9.30
4.8	0.50	3.28	92.1
	1.00	3.43	68.1
	1.50	3.53	55.5
	2.50	3.69	40.8
	4.00	3.85	29.5

Source: Edwards and Owens [J.13] and Hammerstad`and Jensen [J.12].

presented using the numerical methods; such as the method of the line (MOL), variational (var.) method, and SDA at 1 GHz. As compared to the MOL, the results of the SDA have a maximum deviation of 1.5%, the variational method and Wheeler's expression have above 3% deviation for the narrow microstrip; whereas the H-J model has deviation within 0.5%. Table (8.1a) also shows experimental results on the anisotropic substrate (Sapphire) to compare the accuracy of models against them [J.13, J.14].

Table (8.1b) shows a few computed results for both static effective permittivity $\varepsilon_{reff}(f = 0)$ and characteristic impedance of a microstrip line. Figure (8.6c and d) further shows a decrease in the characteristic impedance of a microstrip line with the increase in relative permittivity of the substrate and also with an increase in w/h-ratio. It is because the characteristic impedance of a microstrip line is inversely proportional to the line capacitance that increases with the w/h-ratio and also with the relative permittivity of a substrate.

8.2.4 Effect of Conductor Thickness on Static Parameters of Microstrip Line

Depending upon the substrate and the fabrication process, a microstrip line has strip conductor thickness in the range, t = 1 μm–35 μm, which is usually t/h < 0.3. Several numerical methods, including the conformal mapping methods, have been used to account for the effect of conductor thickness on microstrip line parameters, including losses. Figure (8.2b) shows that the conductor thickness t increases the content of the electric field lines in the air-medium. It results in a decrease in the effective relative permittivity of a microstrip line. It further decreases the characteristic impedance of a microstrip due to the increase in the effective width of the strip conductor. Wheeler presented a method to compute the *equivalent strip conductor width*, $w_{eq}(t/h, w/h, \varepsilon_r)$, of microstrip with finite conductor thickness [J.8, J.9].

The equivalent conductor width, with zero conductor thickness, is more than the physical width of the strip conductor. It helps to obtain expressions for the effective relative permittivity and the characteristic impedance of a microstrip line. In general, the equivalent strip conductor width is a function of t/h, w/h, and ε_r. It could be written in the form of the differential increase in the width of a strip conductor:

$$w_{eq}(t/h, w/h, \varepsilon_r) = w + \Delta w(t/h, w/h, \varepsilon_r). \tag{8.2.20}$$

Wheeler [J.8], using the conformal mapping method, has obtained the following equation for the differential increase in the width for a microstrip line on the air-substrate:

$$\Delta w(t/h, w/h, \varepsilon_r = 1) = \begin{cases} \dfrac{t}{h}\left[1 + \ln\left(\dfrac{2}{t/h}\right)\right], & \text{for } w/h > 0.16 \\ \dfrac{t}{\pi}\left[1 + \ln\left(\dfrac{4\pi}{t/h}\right)\right], & \text{for } w/h \le 0.16. \end{cases} \tag{8.2.21}$$

The above expression applies to a conductor of small thickness, i.e. for t/h < 0.1. Wheeler [J.9] has also obtained another expression that applies to all widths of thick conductor microstrip line on the air-substrate, in the range 0 < w/h < ∞, t/w < 1, and t/h < 1:

$$\Delta w(t/h, w/h, \varepsilon_r = 1) = \frac{t}{\pi} \ln\left[\frac{10.872}{\sqrt{(t/h)^2 + \left\{\dfrac{1/\pi}{(w/h) + 1.10}\right\}^2}}\right]. \tag{8.2.22}$$

The equation (8.2.22) is to be modified for a microstrip line on a dielectric substrate. While replacing the original microstrip of a finite conductor thickness t by an equivalent microstrip with $t = 0$, the characteristic impedance of the original microstrip, with known w/h and t/h, and the characteristic impedance of the equivalent microstrip, with equivalent strip conductor width $w_{eq}(t/h, w/h, \varepsilon_r)$ and $t = 0$, should remain unchanged. The line inductance and line capacitance of a microstrip view the differential conductor width in different ways. The line inductance does not depend on the relative permittivity of a substrate. Therefore, the differential conductor width for the line inductance remains unchanged for a microstrip on the air and the dielectric substrate:

$$w_{eq}(t/h, w/h, \varepsilon_r)_{Induc} = w + \Delta w(t/h, w/h, \varepsilon_r = 1). \quad (8.2.23)$$

However, with the increasing ε_r of a substrate, more electric field lines concentrate in the dielectric region just below the width of the strip. It means the line capacitance experiences linearly decrease in the differential conductor width with increasing ε_r. It is assumed that the differential width due to the fringe field of a strip conductor decreases with an increase in relative permittivity ε_r:

$$\Delta w(t/h, w/h, \varepsilon_r)_{Cap} = \frac{\Delta w(t/h, w/h, \varepsilon_r = 1)}{\varepsilon_r}. \quad (8.2.24)$$

The strip conductor thickness has extended the width and increased the line capacitance that decreases the characteristic impedance. To maintain the constant characteristic impedance, the line inductance must be increased with differential width. Therefore, the relative changes in line capacitance, impedance, and inductance can be written as follows:

$$\frac{\Delta C}{C} = \frac{\Delta w(t/h, w/h, \varepsilon_r)_{Cap}}{w} = \frac{\Delta w(t/h, w/h, \varepsilon_r = 1)}{w\varepsilon_r} \quad (a)$$

$$\frac{\Delta Z_0}{Z_0} = \frac{\Delta w(t/h, w/h, \varepsilon_r)_{Z_0}}{w} \quad (b)$$

$$\frac{\Delta L}{L} = \frac{\Delta w(t/h, w/h, \varepsilon_r = 1)_{Indc}}{w} = \frac{\Delta w(t/h, w/h, \varepsilon_r = 1)}{w} \quad (c).$$

(8.2.25)

The characteristic impedance of microstrip is a function of L and C, i.e. $Z_0 = Z_0(L, C)$. Keeping in view decrease in Z_0 with line capacitance, the differential change in Z_0 with respect to differentials of L and C is

$$dZ_0 = \frac{\partial Z_0}{\partial L} dL - \frac{\partial Z_0}{\partial C} dC \quad (a)$$

Additionally, using $Z_0 = \sqrt{L/C}$, we have following:

$$\frac{\partial Z_0}{\partial L} = \frac{1}{\sqrt{C}} \frac{d}{dL}\left(L^{\frac{1}{2}}\right) = \frac{1}{2}\frac{1}{\sqrt{LC}} \quad (b)$$

$$\text{and } \frac{\partial Z_0}{\partial C} = \sqrt{L}\frac{d}{dC}\left(C^{-\frac{1}{2}}\right) = -\frac{1}{2}\sqrt{\frac{L}{C}}\frac{1}{C} \quad (c)$$

Therefore, differential change in Z_0 is

$$\Delta Z_0 = \frac{1}{2}\sqrt{\frac{L}{C}}\left[\frac{\Delta L}{L} + \frac{\Delta C}{C}\right]$$

$$\Rightarrow \frac{dZ_0}{Z_0} = \frac{1}{2}\left[\frac{\Delta L}{L} + \frac{\Delta C}{C}\right] \quad (d).$$

(8.2.26)

On substituting equations (8.2.25a)–(8.2.25c) in the above equation, the following expression is obtained for the differential width $\Delta w(t/h, w/h, \varepsilon_r)_{Z_0}$, as viewed by the characteristic impedance of a microstrip line on the dielectric substrate:

$$\Delta w(t/h, w/h, \varepsilon_r)_{Z_0} = \frac{1}{2}\left[1 + \frac{1}{\varepsilon_r}\right]\Delta w(t/h, w/h, \varepsilon_r = 1). \quad (8.2.27)$$

Hammerstad–Jensen [J.12] have given the following set of closed-form expressions for the differential width of microstrip over air-substrate:

$$\Delta w(t/h, w/h, \varepsilon_r = 1) = \frac{t}{\pi}\ln\left[\frac{10.872}{t/h\,\coth^2\left(\sqrt{6.517(w/h)}\right)}\right] \quad (a)$$

$$\Delta w(t/h, w/h, \varepsilon_r) = \frac{1}{2}\left[\frac{1}{\coth\left(\sqrt{\varepsilon_r - 1}\right)}\right]\Delta w(t/h, w/h, \varepsilon_r = 1) \quad (b).$$

(8.2.28)

The next task is to use the expression (8.2.20) to compute the characteristic impedance $Z_0(t/h, w/h, \varepsilon_r)$ and effective relative permittivity $\varepsilon_{reff}(t/h, w/h, \varepsilon_r)$ of a microstrip line with strip conductor thickness. Two equivalent schemes are used for the computation of the characteristic impedance of a microstrip. In the first scheme, the extension in width $w_{eq}(t/h, w/h, \varepsilon_r = 1)$ is obtained on the air-substrate, and then the characteristic impedance with conductor thickness is computed from the following equation:

$$Z_0(t/h, w/h, \varepsilon_r) = \frac{Z_0\left(t = 0, \varepsilon_r = 1, w_{eq}(t/h, w/h, \varepsilon_r = 1)\right)}{\sqrt{\varepsilon_{reff}(t/h, w/h, \varepsilon_r)}}. \quad (8.2.29)$$

The effective relative permittivity $\varepsilon_{reff}(t/h, w/h, \varepsilon_r)$ in the above equation is unknown. It could be computed from the following expression due to Bahl and Garg [J.17]:

$$\varepsilon_{reff}(t/h, w/h, \varepsilon_r) = \varepsilon_{reff}(t = 0, w/h, \varepsilon_r) - \frac{\varepsilon_r - 1}{4.6}\frac{t/h}{\sqrt{w/h}}. \quad (8.2.30)$$

In the second scheme, we first obtain the extension in width on the dielectric substrate, i.e. $w_{eq}(t/h, w/h, \varepsilon_r)$, and use previously discussed expressions to compute characteristic impedance on air-substrate with $t = 0$ and effective relative permittivity with t = 0. Finally, the characteristic impedance of microstrip with finite conductor thickness, i.e. $t \neq 0$, is computed from the expression given below:

$$Z_0(t/h, w/h, \varepsilon_r) = \frac{Z_0(t=0, \varepsilon_r=1, w_{eq}(t/h, w/h, \varepsilon_r))}{\sqrt{\varepsilon_{reff}(t=0, w_{eq}(t/h, w/h, \varepsilon_r), \varepsilon_r)}}. \tag{8.2.31}$$

The above expression does not require equation (8.2.30) for the computation of characteristic impedance. On equating equations (8.2.29) and (8.2.31), the following expression is obtained to compute effective relative permittivity of microstrip with finite conductor thickness:

$$\varepsilon_{reff}(t/h, w/h, \varepsilon_r) = \varepsilon_{reff}(t=0, w_{eq}(t/h, w/h, \varepsilon_r), \varepsilon_r) \times \left[\frac{Z_0(t=0, \varepsilon_r=1, w_{eq}(t/h, w/h, \varepsilon_r=1)}{Z_0(t=0, \varepsilon_r=1, w_{eq}(t/h, w/h, \varepsilon_r)}\right]^2. \tag{8.2.32}$$

The above expression is more accurate as compared to equation (8.2.30). Figure (8.7a and b) illustrates variations in the effective relative permittivity [J.15] and characteristic impedance [B.11, J.16] with the conductor thickness. Both the effective relative permittivity and characteristic impedance decrease with an increase in the conductor thickness. The characteristic impedance also decreases with an increase in the w/h-ratio.

8.2.5 Effect of Shield on Static Parameters of Microstrip Line

Figure (8.8a) shows a shielded microstrip line in a metallic box. The substrate thickness is h_1 (original h) and its relative permittivity is ε_{r1} (original ε_r). The top shield height is h_2 and the sidewall is at the distance "a" from the edge of the strip. The shield is used to avoid radiation from the circuit and also to protect the circuit from outside interference and the environment. Sometimes it is also used for the post-fabrication adjustment of the designed circuit. For the far placed top shield and side shielding walls, i.e. $h_2/h_1 > 10$ and $a/w > 20$, the influence of shield on the characteristic impedance and effective relative permittivity could be disregarded. If the sidewalls are far away, the enclosed microstrip is reduced to the top-shielded microstrip, shown in Fig (8.8b). The line capacitance of an open microstrip C_{mic} increases due to additional shunt capacitance C_{sh} introduced by the shield. These capacitances are shown in the circuit model of Fig (8.8c). Following equations (8.1.21) and (8.1.22), the effective relative permittivity and characteristic impedance of the unshielded and shielded microstrip lines could be obtained:

$$\text{Unshielded line}: \varepsilon_{reff} = \frac{C_{mic}(\varepsilon_r)}{C_{mic}(\varepsilon_r=1)} \tag{a}$$

$$Z_0 = \frac{1}{c\sqrt{C_{mic}(\varepsilon_r)C_{mic}(\varepsilon_r=1)}} \tag{b}$$

$$\text{Shielded line}: \varepsilon_{reff}^{Shield} = \frac{C_{mic}(\varepsilon_r) + C_{sh}(\varepsilon_r=1)}{C_{mic}(\varepsilon_r=1) + C_{sh}(\varepsilon_r=1)} \tag{c}$$

$$Z_0^{Shield} = \frac{1}{c\sqrt{[C_{mic}(\varepsilon_r) + C_{sh}(\varepsilon_r=1)][C_{mic}(\varepsilon_r=1) + C_{sh}(\varepsilon_r=1)]}} \text{ (d).} \tag{8.2.33}$$

The above equations show that the characteristic impedance of an open microstrip is reduced in the presence of the shield, $Z_0^{Shield} < Z_0$. The effective relative permittivity of microstrip is also reduced due to the presence of a shield,

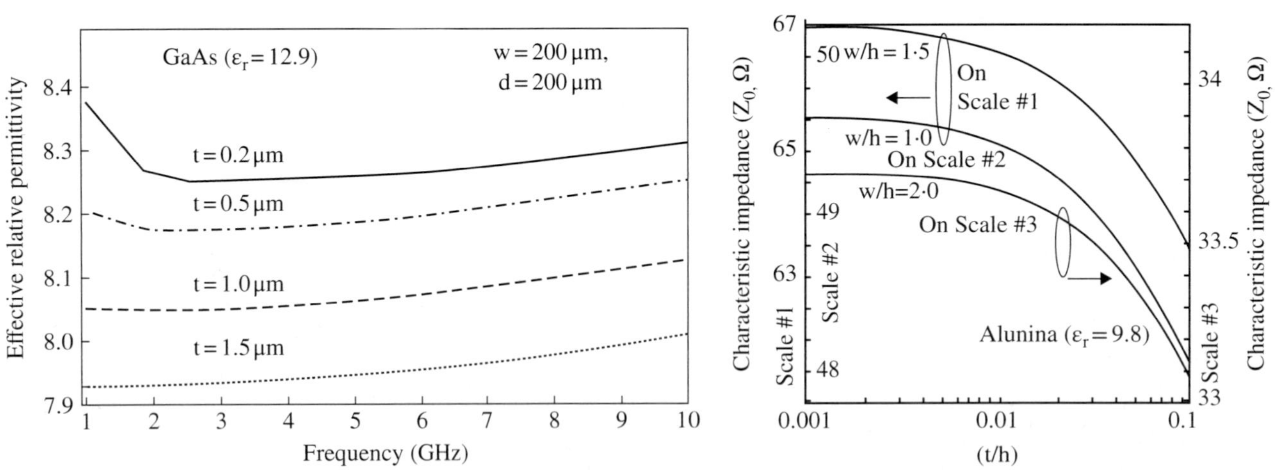

(a) Effect of conductor thickness on ε_{reff} (f) [J.15].

(b) Effect of conductor thickness on Z_0 [J.16].

Figure 8.7 Variation in microstrip parameters with conductor thickness. *Source:* Farina and Rozzi [J.15] and Gunsto and Weale [J.16].

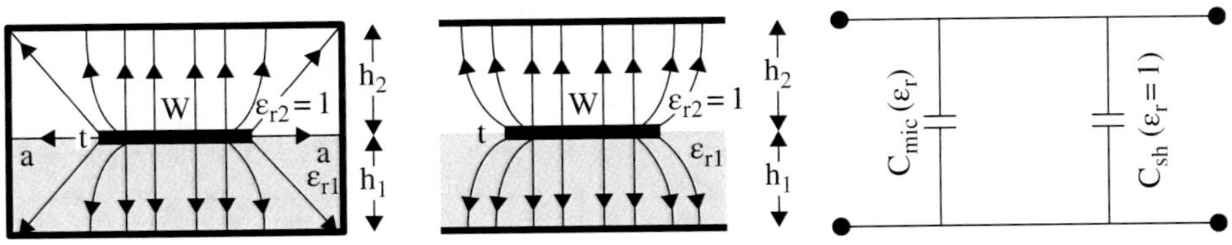

(a) Enclosed microstrip. (b) Top shielded microstrip. (c) Circuit model.

Figure 8.8 Shielded microstrip line and its circuit model. *Source:* Hoffmann [B.1] and March [J.18].

$\varepsilon_{reff}^{Shield} < \varepsilon_{reff}$. The electric field of the microstrip also moves from the dielectric of the substrate to the air-region in the presence of a shield that reduces the effective relative permittivity of an open microstrip.

The characteristic impedance and effective relative permittivity of a shielded microstrip can be determined by the *static and dynamic* SDA methods presented in chapters 14 and 16, respectively. Bahl [J.17] and March [J.18] have reported the closed-form models for this purpose with higher error as compared to the results of the SDA. The *modified March's model* [J.19] is summarized below. It has a deviation of <0.8% for the relative permittivity $\varepsilon_{r1} \leq 140$. The characteristic impedance of a shielded microstrip line on the air-substrate is computed by using the following expressions [B.1, J.18]:

$$\begin{aligned}
&Z_0^{shield}(\varepsilon_{r1}=1) = Z_0^{open}(\varepsilon_{r1}=1) - \Delta Z_0 && \text{(a)}\\
&\Delta Z_0 = 270\left[1.0 - \tanh\left(1.192 + 0.706\sqrt{1 + \frac{h_2}{h_1} - \frac{1.389}{1 + h_2/h_1}}\right)\right]E && \text{(b)}\\
&E = \left[1.0109 - \tanh^{-1}\left\{\frac{0.012(w/h_1) + 0.177(w/h_1)^2 - 0.027(w/h_1)^3}{(1 + h_2/h_1)^2}\right\}\right] && \text{(c).}
\end{aligned} \quad (8.2.34)$$

The $\tanh^{-1}(x)$ used in the above equation is computed by the following expression:

$$\tanh^{-1}(X) = \frac{1}{2}\ln\left[\frac{1+X}{1-X}\right]. \quad (8.2.35)$$

The static effective relative permittivity of a top-shielded microstrip line is computed using the filling-factor with a shield:

$$\varepsilon_{reff}(f=0, w/h_1, h_2/h_1) = 1 + q(\text{shield})\ (\varepsilon_{r1} - 1). \quad (8.2.36)$$

The shielded filling-factor, in terms of the filling-factor q (open) of an open microstrip line and another parameter p, that modifies the q (open), is given by

$$q(\text{shield}) = 0.5 + p[q(\text{open}) - 0.5]. \quad (8.2.37)$$

The filling-factor of the open microstrip line, q (open), is obtained from

$$q(\text{open}) = \frac{[\varepsilon_{reff}(f=0, w/h_1) - 1]}{[\varepsilon_{r1} - 1]}. \quad (8.2.38)$$

The effective relative permittivity of the open microstrip line, $\varepsilon_{reff}(f=0, w/h_1)$, is obtained by either Wheeler's or Hammerstad–Jensen's expression given in subsection (8.2.1). The parameter p is given by

$$\begin{aligned}
&p = \tanh\left[0.922 + 0.121\left(1 + \frac{h_2}{h_1}\right) - 1.164\left(\frac{h_1}{h_2}\right)\right]K && \text{(a)}\\
&K = 1 - \alpha\left(\frac{h_1/w}{(h_2/h_1)^{1.8}}\right) && \text{(b)}\\
&\alpha = \begin{cases} 0.0663 + 0.0576\,w/h_1 & ,0.1 \leq w/h_1 \leq 0.6 \\ -0.2504 + 0.5684\,w/h_1 & ,0.6 \leq w/h_1 \leq 3.0 \\ 10^y & 3.0 \leq w/h_1 \leq 10.3 \end{cases} && \text{(c)}\\
&y = -49 + 1.485\log_{10}(w/h_1) && \text{(d).}
\end{aligned} \quad (8.2.39)$$

In the case of a shielded microstrip with finite conductor thickness, the following expression is used for the shielded filling-factor:

$$\begin{aligned}
&q(\text{shield}) = p[q(\text{open}) - q(t/h_1)] && \text{(a)}\\
&q(t/h_1) = \frac{2\ln(2)}{\pi}\frac{(t/h_1)}{\sqrt{w/h_1}} && \text{(b).}
\end{aligned} \quad (8.2.40)$$

Table 8.2 Comparison of computed $\varepsilon_{reff}(f = 0)$ of the shielded microstrip by numerical methods and closed-form model.

ε_r	w/h	$h_2/h_1 = 2$			$h_2/h_1 = 6$		
		SDA	Var	Mod. March	SDA	Var	Mod. March
2.2	0.6	1.677	1.677	1.667	1.718	1.729	1.727
	5.0	1.759	1.760	1.762	1.899	1.900	1.901
	10.0	1.778	1.771	1.779	1.954	1.933	1.967
9.8	0.6	5.944	5.934	5.944	6.281	6.267	6.259
	5.0	6.562	6.565	6.554	7.545	7.551	7.535
	10.0	6.711	6.641	6.692	7.968	8.143	8.034
40	0.6	22.83	22.78	22.86	24.25	25.18	24.22
	5.0	25.63	25.62	25.57	29.94	29.92	30.15
	10.0	26.26	26.14	26.19	31.83	31.57	32.11

Finally, using equations (8.2.34) and (8.2.36), the characteristics impedance of a shielded microstrip with t = 0 is computed by

$$Z_o^{\text{shielded}}(\varepsilon_{r1}) = \frac{Z_o^{\text{shielded}}(\varepsilon_{r1} = 1)}{\sqrt{\varepsilon_{\text{reff}}(f = 0, h_2/h_1)}}. \tag{8.2.41}$$

The above equation is taken from equation (8.2.2b). However, the effect of the shield on the characteristics impedance and effective relative permittivity of a microstrip line is accounted for.

Table (8.2) shows the computed effective relative permittivity of a shielded microstrip on several substrates. Table (8.2) also shows results computed using the SDA and variational method. Figure (8.9a and b) shows a decrease in the effective relative permittivity and the characteristic impedance of a microstrip line on an alumina substrate with the nearness of the top shield to a microstrip line [J.1].

8.2.6 Microstrip Line on Anisotropic Substrate

The anisotropy in a substrate is discussed in chapter 4. Some materials, such as sapphire [$\varepsilon_{r\|} = 11.6$, $\varepsilon_{r\perp} = 9.4$], α–quartz [$\varepsilon_{r\|} = 4.637$, $\varepsilon_{r\perp} = 4.52$], pyrolytic boron nitride [$\varepsilon_{r\|} = 3.40$, $\varepsilon_{r\perp} = 5.12$], etc., used as the MIC substrates, exhibit dielectric anisotropy. Anisotropy also exists in a small quantity in alumina Al_2O_3 and glass fiber-reinforced PTFE substrates. Figure (8.10) shows the orientation of the crystallographic axis (ξ, η, ζ) of a uniaxial substrate with respect to the physical axes (x,y,z) of a microstrip. The angle θ is the *misalignment angle* between the substrate's crystal axis and the physical axis.

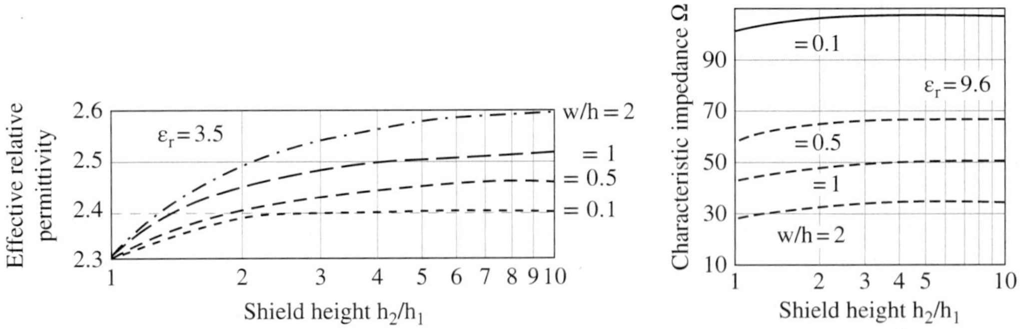

(a) Variation in $\varepsilon_{reff}(f=0)$ with shield height.

(b) Variation in Z_0 with shield height.

Figure 8.9 Effect of shielded on the microstrip line. *Source:* Modified from Maloratsky [J.1].

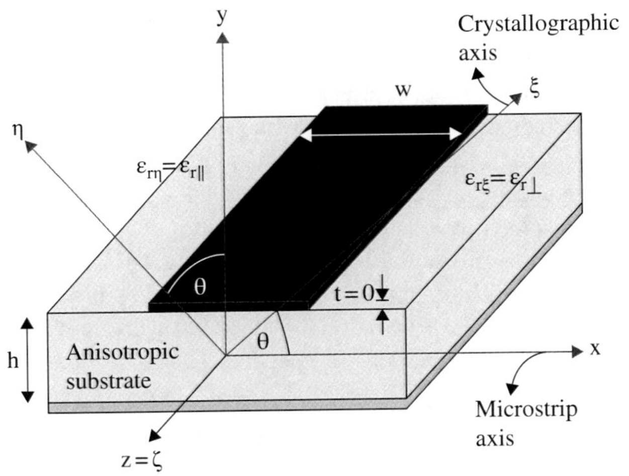

Figure 8.10 Orientation of crystallographic axes (ξ, η, ζ) of uniaxial substrate with respect to physical axes (x,y,z) of a microstrip.

The anisotropy is expressed through the *anisotropic-ratio* n. For the isotropic case, n = 1. For the accurate design of the MIC-based components, the anisotropy in a substrate should be accounted for. Normally, the anisotropic substrate is transformed into an electrically equivalent isotropic substrate [J.13, J.20–J.24]. A set of transformation equations is summarized below. However, this simple transformation is not accurate, as the transformed equivalent isotropic substrate does not account for the w/h-dependence. The correct transformation has been obtained by using the finite difference method [J.25], SDA [J.26], and also by the partial capacitance method [J.27]. The *partial capacitance method* is presented below.

Figure (8.11) shows the two-step process of getting the w/h-dependent equivalent isotropic relative permittivity $\varepsilon^{eq}_{r,iso}(w/h)$ of an anisotropic substrate. The *first step* converts an anisotropic substrate to a substrate with equivalent isotropic relative permittivity $\varepsilon_{r,eq}$ and equivalent substrate thickness h_{eq}, without consideration of the w/h-ratio. It is achieved through the following expressions [B.1, J.28]:

$$\varepsilon_{req} = \sqrt{\varepsilon_{rxx}\varepsilon_{ryy} - \left(\varepsilon_{rxy}\right)^2} \quad (a), \qquad h_{eq} = h\sqrt{\left[\frac{\varepsilon_{rxx}}{\varepsilon_{ryy}} - \left(\frac{\varepsilon_{rxy}}{\varepsilon_{ryy}}\right)^2\right]} \quad (b)$$

$$\text{where, } \varepsilon_{rxx} = \varepsilon_{r\varsigma}\cos^2\theta + \varepsilon_{r\eta}\sin^2\theta \quad (c), \qquad \varepsilon_{ryy} = \varepsilon_{r\varsigma}\sin^2\theta + \varepsilon_{r\eta}\cos^2\theta \quad (d)$$

$$\varepsilon_{rxy} = \varepsilon_{ryx} = \left(\varepsilon_{r\varsigma} - \varepsilon_{r\eta}\right)\sin\theta\cos\theta \quad (e), \qquad \varepsilon_{r\varsigma} = \varepsilon_{r\perp}, \ \varepsilon_{r\eta} = \varepsilon_{r}, \quad (f)$$

$$\text{Anisotropic ratio, } n = \sqrt{\frac{\varepsilon_{rxx}}{\varepsilon_{ryy}}} \quad (g). \tag{8.2.42}$$

The *second step* provides, with the help of the *method of partial capacitance*, the w/h-dependent equivalent isotropic relative permittivity $\varepsilon^{eq}_{r,iso}(w/h)$ with original substrate thickness h. The width of the microstrip is extended to width $(w + \Delta w)$ due to the fringe capacitance. The main central parallel-plate capacitance is due to the component ε_{ryy}. It is assumed that the fringe capacitance of the original structure is approximately the same as that of the microstrip in step #1, on a substrate with the relative permittivity $\varepsilon_{r,\,eq}$ and substrate thickness $h_{eq} \approx h$. Thus, the total line capacitance C_t *p.u.l.* on the equivalent isotropic substrate could be written as a combination of the fringe capacitance C_f and the central capacitance C_p:

$$C_t = C_f + C_p$$

$$\Rightarrow \frac{\varepsilon_0\,\varepsilon^{eq}_{r,iso}(w/h)(w + \Delta w)}{h} = \frac{\varepsilon_0\,\varepsilon_{r,eq}\Delta w}{h} + \frac{\varepsilon_0\varepsilon_{r,yy}w}{h}$$

$$\Rightarrow \varepsilon^{eq}_{r,iso}(w/h) = \frac{\varepsilon_{r,eq}\Delta w + \varepsilon_{r,yy}w}{w + \Delta w}. \tag{8.2.43}$$

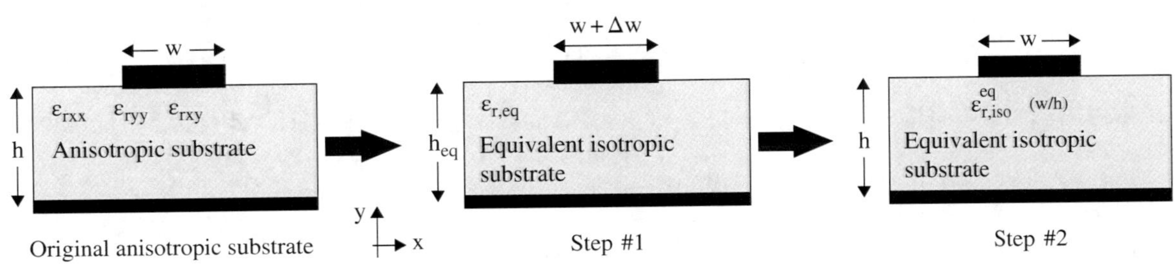

Figure 8.11 Process of obtaining w/h-dependent equivalent isotropic substrate. *Source:* Modified from Verma et al. [J.27].

However, as the line capacitance *p.u.l.* is proportional to the width of a strip conductor, i.e. fringe capacitance $C_f \propto \Delta w$, central parallel-plate capacitance $C_p \propto w$, and the total line capacitance $C_t \propto (w + \Delta w)$, the above equation gives the following expression for the w/h-dependent equivalent isotropic relative permittivity $\varepsilon_{r,iso}^{eq}(w/h)$:

$$\varepsilon_{r,iso}^{eq}(w/h) = \frac{\varepsilon_{r,eq}C_f + \varepsilon_{r,yy}C_p}{C_t} = \frac{\varepsilon_{r,eq}C_t + (\varepsilon_{r,yy} - \varepsilon_{r,eq})C_p}{C_t}, \tag{8.2.44}$$

where $C_t = C_p + C_f$. Using the *equivalent substrate* of step #1 and the ε_{ryy} component of the anisotropic substrate, the C_t and C_p *p.u.l.* are obtained as follows:

$$C_t = \frac{\sqrt{\varepsilon_{reff}(w, h_{eq}, \varepsilon_{req})}}{c\ Z_0(w, h_{eq}, \varepsilon_{req} = 1)} \quad (a), \qquad C_p = \varepsilon_{ryy}\varepsilon_0\frac{w}{h} \quad (b), \tag{8.2.45}$$

where c is the velocity of EM-wave in the free space. The characteristic impedance $Z_0(w, h_{eq}, \varepsilon_{req} = 1)$ on the air-substrate and the effective relative permittivity on the equivalent isotropic substrate, with relative permittivity ε_{req} and substrate thickness h_{eq}, are obtained from the closed-form expressions of Hammerstad–Jensen.

The w/h-dependent equivalent isotropic relative permittivity $\varepsilon_{r,iso}^{eq}(w/h)$ of an anisotropic substrate can also be computed using the following set of closed-form expressions [J.27]:

$$\varepsilon_{r,iso}^{eq}(w/h) = K.Y.\varepsilon_{ryy} \quad (a)$$

$$\text{where, } Y = -7.0 \times 10^{-5}(w/h)^4 + 0.0017(w/h)^3 - 0.0149(w/h)^2 + 0.0568(w/h) + 0.9099 \quad (b)$$

$$K = \begin{cases} 1.1256\sqrt{\dfrac{\varepsilon_{rxx}}{\varepsilon_{ryy}}} - 0.0151; \varepsilon_{rxx} \neq \varepsilon_{ryy}, & 0 < w/h \le 1 \\ 1 & w/h > 1 \end{cases} \quad (c). \tag{8.2.46}$$

The w/h-dependent equivalent isotropic relative permittivity $\varepsilon_{r,iso}^{eq}(w/h)$ with substrate thickness h and the strip-width w can be used to compute the correct static effective relative permittivity, and the characteristic impedance of the original microstrip on the anisotropic substrate by using the closed-form expression of Hammerstad–Jensen. The process of obtaining the equivalent isotropic substrate has a 0.5% deviation against the SDA results.

8.3 Dispersion in Microstrip Line

The microstrip line is a dispersive transmission line with frequency-dependent phase velocity. Thus, its effective relative permittivity $\varepsilon_{r\,eff}(f)$ is also frequency-dependent which is computed by the full-wave methods [B.6, B.8, B.9, J.5, J.6, J.29–J.36]. Normally, the frequency dependence is significant above 4 GHz. The frequency-dependent effective relative permittivity $\varepsilon_{r\,eff}(f)$ of a microstrip increases with frequency and its phase velocity decreases with the increasing frequency. It has significant implications for both the digital and RF analog signals on a microstrip. On a dispersive interconnect, frequency components of digital signal travel with different velocities. It results in a distorted pulse due to pulse spreading, ringing, etc. The distorted pulse limits the rate of high-speed data transmission. Likewise, a complex RF analog signal traveling on a microstrip also gets distorted due to dispersion. The guided wavelength $\lambda_g = \lambda_0/\sqrt{\varepsilon_{r\,eff}(f)}$ is also frequency-dependent. Thus, microstrip-based filters, couplers, and matching networks become frequency sensitive and these devices are frequency band limited. The characteristic impedance of microstrip is also frequency-dependent that creates a problem for broadband matching. A microwave circuit designer cannot work only with the static models of the microstrip transmission line parameters. Several closed-form models are reported in the literature [J.10, J.11, J.37–J.46].

Some closed-form dispersion models are based on the curve-fitting of dispersion data, obtained from the rigorous field-theoretic formulations. A few models are based on the coupling between the TEM mode and the lowest order surface wave TM_0 mode. Some authors have considered the dispersion due to the coupling between TE_1 surface wave mode and TEM mode. The experimental and numerical investigations have shown dispersion in a microstrip at a frequency below the cutoff frequency of the TE_1 surface wave mode. Therefore, models based on the coupling between the TM_0 surface wave mode and TEM mode appear to be more appropriate, as the TM_0 surface wave mode has no cutoff frequency. This section presents a discussion on the nature of dispersion and closed-form models to compute the frequency-dependent effective relative permittivity and the characteristic impedance of a microstrip line.

8.3.1 Nature of Dispersion in Microstrip

The increase in the effective relative permittivity $\varepsilon_{reff}(f)$ with frequency is due to the coupling between the TEM parallel-plate mode and the TM_0 surface wave mode [B.2]. During the process of field coupling, the electric field lines move from the air-region to the dielectric-region increasing the effective relative permittivity. The surface wave mode is supported by a metal-coated dielectric sheet. It is discussed in the section (7.5) of chapter 7. Thus, the *quasi-TEM* mode is viewed as a combination of these two modes:

$$\textit{Quasi-TEM} \rightarrow \textit{TEM parallel-plate mode} + TM_0\ \textit{surface wave mode}. \tag{8.3.1}$$

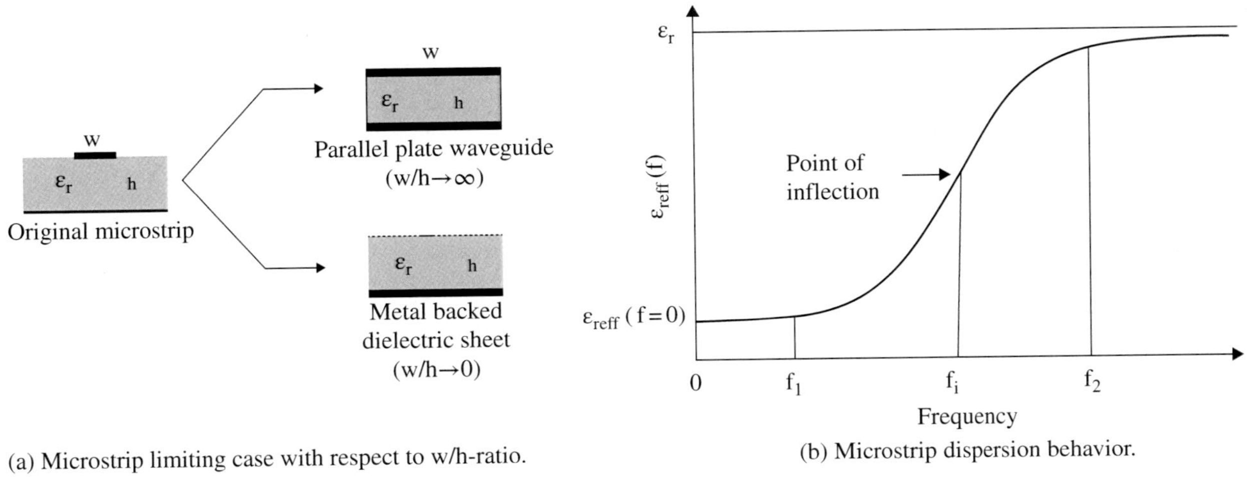

Figure 8.12 Limiting cases of microstrip with respect to w/h-ratio and frequency.

The above description of the quasi-TEM mode of a microstrip line leads to two limiting cases shown in Fig (8.12a). For $w/h \to \infty$, microstrip tends to a parallel-plate waveguide supporting the TEM fundamental mode. This case gives $\varepsilon_{r\,eff}(f = 0) \to \varepsilon_r$, whereas for $w/h \to 0$, the microstrip tends to metal-supported dielectric waveguide that supports the TM_0 surface wave mode. This case leads to $\varepsilon_{reff}(f = 0) \to (\varepsilon_r + 1)/2$. Through the process of mode coupling, the electric field lines of a dispersive microstrip move from the air-medium to the dielectric-medium, so that the $\varepsilon_{reff}(f)$ increases from its lower limiting value $\varepsilon_{r\,eff}(f = 0)$ to the upper limiting value ε_r for $f \to \infty$. The frequency-dependent nature of $\varepsilon_{reff}(f)$ of a microstrip line in the frequency range $0 \le f < \infty$ is shown in Fig (8.12b).

Figure (8.12b) illustrates that $\varepsilon_{reff}(f)$ is a monotonically increasing function of frequency, i.e. $\partial\varepsilon_{r\,eff}(f)/\partial f > 0$ in the range $\varepsilon_{r\,eff}(f = 0) \le \varepsilon_{r\,eff}(f) < \varepsilon_r$. However, the frequency-dependent effective relative permittivity $\varepsilon_{reff}(f)$ has an inflection at the frequency f_i, known as the *inflection frequency*. Schneider [J.39] states the following physical conditions that should be satisfied by the frequency-dependent function of the effective relative permittivity:

$$\text{i. Initial conditions}: \varepsilon_{reff}(f) \to \varepsilon_{reff}(f = 0), \text{as } f \to 0 \qquad \text{(i a)}$$

$$\frac{\partial\varepsilon_{reff}(f)}{\partial f} \to 0, \text{as } f \to 0 \qquad \text{(i b)}$$

$$\text{ii. Final condition}: \varepsilon_{reff}(f) \to \varepsilon_r, \text{as } f \to \infty \qquad \text{(ii a)}$$

$$\frac{\partial\varepsilon_{reff}(f)}{\partial f} \to 0, \text{as } f \to \infty \qquad \text{(ii b)}$$

$$\text{iii. Inflection condition}: \frac{\partial^2\varepsilon_{reff}(f)}{\partial f^2} = 0, \text{at } f = f_i \qquad \text{(iii)}.$$

$$(8.3.2)$$

The first initial condition states that with decreasing frequency, $\varepsilon_{reff}(f)$ tends to a static value of the effective relative permittivity of microstrip and the second initial condition states that there is no initial jump (discontinuity) in $\varepsilon_{reff}(f)$ at $f = 0$. The first final condition states that $\varepsilon_{reff}(f)$ saturates to the value ε_r with $f \to \infty$ and the second final condition states about the smooth saturation of $\varepsilon_{reff}(f)$ with increasing frequency. The presence of the inflection frequency f_i shows that below it the change in $\varepsilon_{reff}(f)$ is accelerated one, and above it the change in $\varepsilon_{reff}(f)$ is de-accelerated, leading to its saturation to relative permittivity ε_r.

All dispersion models must follow these physical conditions. However, the dispersion models following Schneider's condition may not have identical accuracy. Figure (8.12b) shows that a microstrip line is less dispersive, both in the lower frequency range from 0 to f_1 and in the upper-frequency range from f_2 to $(f \to \infty)$. It shows a rapid change in $\varepsilon_{reff}(f)$ during the frequency range f_1–f_2, around the inflection frequency, leading to higher dispersion [J.10, J.11]. Therefore, if a complex signal has all its frequency components, fundamental and several higher harmonics, below f_1, the signal may not be much distorted due to dispersion. Similar is the case for the distortion in a signal if its fundamental and higher harmonics occupy the frequency spectrum $f_2 < f < \infty$. However, if frequency components of a complex signal occupy the frequency spectrum $f_1 < f < f_2$, the signal undergoes a severe distortion due to the significant dispersion in the microstrip.

The confinement of the electric field between the strip conductor and the ground plane suggests that a microstrip line can be treated with the help of the *waveguide model*. The waveguide model of a microstrip has been developed through the efforts of many investigators [J.46–J.48]. The model has limited accuracy. However; it is a simple one and it provides several characteristics of a microstrip line. Moreover, it has been used to study the behavior of several kinds of discontinuities in a microstrip [B.12, J.49]. A consolidated formulation of the waveguide model is presented below.

8.3.2 Waveguide Model of Microstrip

Figure (8.13a) illustrates the equivalent parallel-plate waveguide model with the equivalent width $w_{eq} = w + \Delta w$ of a microstrip so that it contains all the electric field lines between two conducting strips [B.12, J.47, J.49]. The top and bottom are the *electric walls* (EW), whereas the sidewalls are the *magnetic walls* (MW). This waveguide is discussed in chapter 7. It supports the fundamental TEM mode with zero cutoff frequency, which is also supported by the original microstrip line. The next higher-order mode is TE_{10}. The waveguide model is used to obtain expressions for the frequency-dependent effective relative permittivity and the characteristic impedance of a microstrip line. It is also used to obtain the cutoff frequency of the higher-order mode of a microstrip line. The cutoff frequency of the waveguide model is

$$f_c = \frac{1}{2\sqrt{\mu\varepsilon}}\left[\left(\frac{m}{w_{eq}}\right)^2 + \left(\frac{n}{h}\right)^2\right]^{1/2}. \quad (8.3.3)$$

For the TE_{10} mode, the modal numbers are m = 1, n = 0. The cutoff frequency f_T of the TE_{10} mode is

$$f_c|_{10} = f_T = \frac{c}{2w_{eq}\sqrt{\mu_r\varepsilon_r}}, \quad (8.3.4)$$

where c is the velocity of the EM-wave in the free space. A dielectric substrate has $\mu_r = 1$. For equivalence, the line capacitances of the original microstrip line and the parallel-plate waveguide should be identical:

$$\underbrace{C(w/h, \varepsilon_r) = \frac{\varepsilon_0\varepsilon_r w_{eq}(f=0)}{h}}_{\text{Waveguide medium}} \equiv \underbrace{\frac{1}{v_p Z(w/h, \varepsilon_r)} = \frac{\sqrt{\varepsilon_{reff}(0)}}{cZ(w/h, \varepsilon_r)}}_{\text{Microstrip line}}. \quad (8.3.5)$$

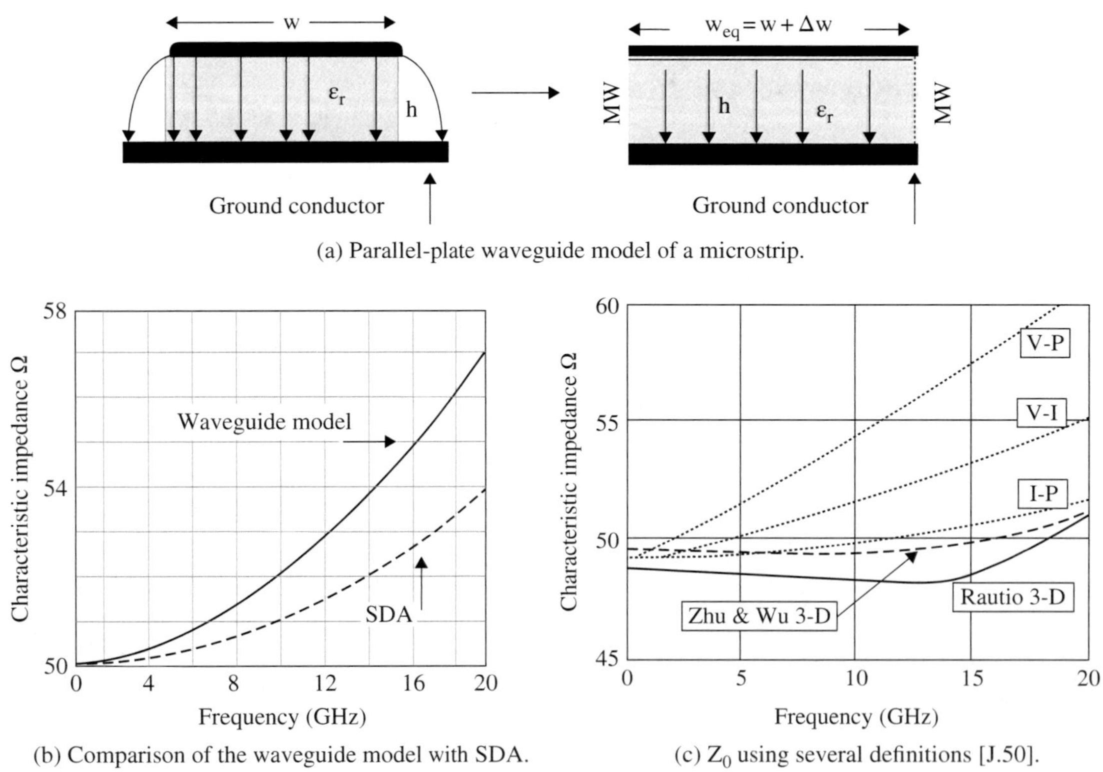

Figure 8.13 Parallel-plate waveguide model of a microstrip. *Source*: Zhu and Wu [J.50].

The above equation gives an expression of the static equivalent width of the parallel-plate waveguide:

$$w_{eq}(f=0) = \frac{h\sqrt{\varepsilon_{reff}(0)}}{\varepsilon_0\varepsilon_r c Z_0(w/h,\varepsilon_r)} = \frac{h\varepsilon_{reff}(0)}{\varepsilon_0\varepsilon_r c Z_0(w/h,\varepsilon_r=1)}. \quad (8.3.6)$$

The static effective relative permittivity and characteristic impedance, used in the above equation, are obtained from the closed-form expressions of Hammerstad–Jensen, or Wheeler, presented in section (8.2). The following empirical relation can be written for the frequency-dependent effective relative permittivity that is known as *Getsinger's model* [J.45]:

$$\varepsilon_{reff}(f) = \varepsilon_r - \frac{\varepsilon_r - \varepsilon_{reff}(0)}{1+(f/f_T)^2}. \quad (8.3.7)$$

The above expression meets *Schneider's physical conditions* mentioned in the previous subsection. It is noted that the operating frequency f is normalized by the cutoff frequency of the TE_{10} waveguide mode. The cutoff frequency f_T is modified by taking into account the fringe field through a parameter G, called the *Getsinger's parameter*. The modified expression of cutoff frequency f_T of a microstrip on a dielectric substrate ($\mu_r = 1$) is obtained in terms of the Getsinger's parameter G. For this purpose, the equivalent width w_{eq}, from equation (8.3.6), is substituted in equation (8.3.4), to get the following expression:

$$f_T = f_t\sqrt{\frac{\varepsilon_r}{\varepsilon_{reff}(0)}} = f_t\frac{1}{\sqrt{G}}. \quad (8.3.8)$$

The normalized frequency f_t and parameter G are expressed as follows:

$$f_t = \frac{Z_0(w/h,\varepsilon_r)}{2\mu_0 h} \quad (a), \quad G = \frac{\varepsilon_{reff}(0)}{\varepsilon_r} \quad (b). \quad (8.3.9)$$

On substituting equation (8.3.8) in equation (8.3.7), the Getsinger's dispersion model is obtained:

$$\varepsilon_{reff}(f) = \varepsilon_r - \frac{\varepsilon_r - \varepsilon_{reff}(0)}{1+G(f/f_t)^2}. \quad (8.3.10)$$

The accuracy of the above model compares well against several available models [J.4]. The waveguide model is also used to get an expression for the frequency-dependent characteristic impedance of a microstrip line [J.47, J.48, J.51]. To do so, we have to obtain an expression for the frequency-dependent equivalent width of the waveguide model.

Frequency-Dependent Equivalent Width

The physical width of a microstrip strip conductor is enlarged due to the fringing electric field. The electric field lines move below the strip conductor in the dielectric region with an increase in frequency. Such enhanced confinement of the electric field in the dielectric region is due to an increase in the coupling, with increasing frequency, between the TEM mode and the TM_0 surface wave mode. It results in the monotonic decrease in width, i.e. $\partial w_{eqf}(f)/\partial f < 0$, of the waveguide model toward the physical width w of a microstrip. Schneider's physical conditions are adapted to the frequency-dependent equivalent width as follows [J.46]:

i.Initial condition :

$$w_{eq}(f) \to w_{eq}(f=0) = w + \Delta w, \text{ as } f \to 0 \quad (i\,a)$$

$$\frac{\partial w_{eq}(f)}{\partial f} \to 0, \text{ as } f \to 0 \quad (i\,b)$$

ii. Final condition : $w_{eq}(f) \to w$, as $f \to \infty$ (ii a)

$$\frac{\partial w_{eq}(f)}{\partial f} \to 0, \text{ as } f \to \infty \quad (ii\,b)$$

iii.Inflection condition : $\dfrac{\partial^2 w_{eq}(f)}{\partial f^2} = 0$, at $f = f_i$ (iii). (8.3.11)

The dispersion model of Getsinger's can be empirically modified to model the frequency-dependent behavior of the equivalent width:

$$w_{eq}(f) = w + \frac{w_{eq}(f=0) - w}{1+G(f/f_t)^2}. \quad (8.3.12)$$

Frequency-Dependent Characteristic Impedance

The following replacements are done in equation (8.3.6) to get an expression for the frequency-dependent characteristic impedance of a microstrip line:

$$\varepsilon_r \to \varepsilon_{reff}(f=0) \to \varepsilon_{reff}(f), \; w_{eq}(f=0) \to w_{eq}(f),$$

$$c = \frac{1}{\sqrt{\mu_0\varepsilon_0}}, \; \eta_0 = \sqrt{\frac{\mu_0}{\varepsilon_0}} = 377\,\Omega$$

$$Z_0(w/h,\varepsilon_r,f) = \frac{377h}{w_{eq}(f)\sqrt{\varepsilon_{reff}(f)}} \quad (a)$$

alternatively, $Z_0(w/h,\varepsilon_r,f)$

$$= Z_0(f=0)\left(\frac{\varepsilon_{reff}(f)-1}{\varepsilon_{reff}(0)-1}\right)\sqrt{\frac{\varepsilon_{reff}(0)}{\varepsilon_{reff}(f)}} \quad (b), \quad (8.3.13)$$

where $\varepsilon_{reff}(f)$ and $w_{eq}(f)$ are obtained from equations (8.3.10) and (8.3.12), respectively. The above expression shows an increase in the characteristic impedance of a microstrip line with frequency. Figure (8.13b) compares the frequency-dependent characteristic impedance of a 50Ω microstrip line on a substrate $\varepsilon_r = 10.0$, h = 0.635 mm, w/h = 1.0, as computed by the waveguide model, against the results of the SDA [B.1, J.52]. The computed results from the waveguide model are not accurate. However, the deviation in the computed characteristic impedance is within 2% for the frequency up to 10 GHz. It is interesting to note that this simple model shows that the dispersion behavior, i.e. the increase in characteristic impedance with frequency, follows the power-current definition of characteristic impedance. Figure (8.13c) shows the full-wave results, using the P-V, V-I, P-I, and 3D port definitions, of the characteristic impedance of microstrip on an alumina substrate, $\varepsilon_r = 9.7$, $h = 0.635$ mm, $w/h = 1.0$ [J.36, J.50]. There is no unique definition of hybrid-mode characteristic impedance of a microstrip line [J.51]. However, power-current (P-I) is more acceptable and even the 3D port definitions from both sources are close to the P-I definition; although these definitions are *noncasual*. A *causal* definition of the characteristic impedance has also been proposed [J.53]. Krischning–Jansen model provides a set of closed-form expressions to compute the frequency-dependent characteristic impedance of a microstrip line that follows the power-current definition [J.54]. Subsection (8.3.5) summarizes the expressions. At this stage, it is also noted that the $Z_o(\varepsilon_r, f) = Z_o(\varepsilon_r, f = 0)/\sqrt{\varepsilon_{reff}(f)}$ definition is popular among microwave designers, although it shows a decrease in the characteristic impedance of a microstrip line with frequency. However, equation (8.3.13b) suggests alternate empirically obtained expression for the characteristic impedance that exhibits increasing characteristic impedance, similar to the P-I-based model, with an increase in the frequency [J.14]. The effect of the conductor thickness can be incorporated in this expression by using $w_{eq}(f)$ in place of the strip width w, and by modifying the expression for the $\varepsilon_{reff}(0)$ and $\varepsilon_{reff}(f)$ suitably.

Higher-Order Mode

Using expression (8.3.4) and the replacement discussed above, the cutoff frequency of higher-order modes of a microstrip can be estimated by the following expression:

$$f_c(m, 0) = \frac{mc}{2w_{eq}(f)\sqrt{\varepsilon_{reff}(f)}}. \quad (8.3.14)$$

It has been demonstrated that m = 0, 1, 2... approximately corresponds to the hybrid modes HE_0 ($HE_{0,0}$), HE_1 ($HE_{1,0}$), HE_2 ($HE_{2,0}$), etc. [B.12, J.49]. The fundamental mode HE_0 is the quasi-static mode. The effective relative permittivities of the higher-order modes HE_1, HE_2, etc., are dispersive and can be computed using the waveguide model. The guided wave propagation constant of the HE_m mode obtained from the waveguide model is

$$\beta_g^m = k\sqrt{1 - \left(\frac{f_c(m, 0)}{f}\right)^2} \quad (a)$$

where

$$\beta_g^m = \beta_0\sqrt{\varepsilon_{r\,eff}^m(f)} \quad (b)$$

$$k = \beta_0\sqrt{\varepsilon_{r\,eff}(f = 0)} \quad (c). \quad (8.3.15)$$

In the above equation, k is the wavenumber in the equivalent homogeneous medium. The following dispersion equation for the higher-order modes are obtained from the above equations:

$$\varepsilon_{reff}^m(f) = \varepsilon_{reff}(f = 0)\left[1 - \left(\frac{f_c(m, 0)}{f}\right)^2\right]. \quad (8.3.16)$$

8.3.3 Logistic Dispersion Model of Microstrip (*Dispersion Law of Microstrip*)

Figure (8.12b) shows the phenomenological dispersion behavior of a microstrip. It is used to develop the logistic dispersion model (LDM) [J.4]. The LDM is based on the coupling of the TEM mode and TM_0 surface wave mode to determine the inflection frequency f_i. The dispersion in a microstrip line is due to the growth of the effective relative permittivity $\varepsilon_{r\,eff}(f)$ from its static value $\varepsilon_{r\,eff}(f = 0)$ to the relative permittivity ε_r of a substrate with the increase in frequency, in the range $0 \le f \le \infty$. The following statement can be made on the *dispersion law* for the growth of the effective relative permittivity of a microstrip line with increasing frequency:

> *"The rate of increase of the effective relative permittivity with frequency α [Effective relative permittivity at given frequency] × [Remaining fractional relative permittivity of the substrate]"*

The above statement of the dispersion law can be written in the form of the first-order differential equation:

$$\frac{d\varepsilon_{reff}(f)}{df} = K\varepsilon_{reff}(f)\left[\frac{\varepsilon_r - \varepsilon_{reff}(f)}{\varepsilon_r}\right], \quad (8.3.17)$$

where K is the proportionality constant. The constant K is structure-dependent, i.e. it depends on the w/h-ratio and relative permittivity ε_r of a substrate. The K could be estimated and expressed in terms of the static effective relative permittivity $\varepsilon_{reff}(f = 0)$ and ε_r. For the sake of convenience, the K is taken on the natural logarithms as follows:

$$K = \ln\left[\frac{3\,\varepsilon_{reff}(0) - \varepsilon_r}{\varepsilon_{reff}(0)}\right]. \quad (8.3.18)$$

The solution of the differential equation (8.3.17) leads to the following logistic dispersion equation:

$$\varepsilon_{reff}(f) = \frac{\varepsilon_r}{1 + e^{-(Kf + C)}}, \quad (8.3.19)$$

where C is the constant of integration. Using Schneider's initial condition (8.3.2), equation (8.3.18), and $e^{-c} = M$, constant M is obtained from equation (8.3.19):

$$M = \frac{\varepsilon_r - \varepsilon_{reff}(0)}{\varepsilon_{reff}(0)}. \quad (8.3.20)$$

The operating frequency in equation (8.3.19) is normalized by the inflection frequency f_i. The final form of the logistic dispersion equation is

$$\varepsilon_{r\,eff}(f) = \frac{\varepsilon_r}{1 + M\,e^{-K\,(f/f_i)}}. \tag{8.3.21}$$

The above equation meets Schneider's conditions (8.3.2). However, it does not meet the condition (8.3.2–i b). The *inflection frequency* is determined from the coupling frequency $f_{k,\,TM}$ [B.2, J.10]:

$$f_{k,TM} = c\frac{\tan^{-1}\left[\varepsilon_r\sqrt{\dfrac{\varepsilon_{r\,eff}(0) - 1}{\varepsilon_r - \varepsilon_{r\,eff}(0)}}\right]}{2\pi h\sqrt{\varepsilon_r - \varepsilon_{r\,eff}(0)}}. \tag{8.3.22}$$

The static effective relative permittivity $\varepsilon_{r\,eff}(f = 0)$ is obtained from the model of Hammerstad–Jensen [J.14]. The c is the velocity of the EM-wave in the free space and h is the substrate thickness. The effect of w/h-ratio on the inflection frequency could be accounted for through the following modified empirical expression [J.4, J.10]:

$$f_i = \frac{f_{k,TM}}{\sqrt{3}\left(1 + B\dfrac{W}{h}\right)^A}. \tag{8.3.23}$$

The parameters A and B are determined empirically after a comparison of the LDM against the dispersion data obtained from the SDA. Both the parameters are w/h dependent. However, A is also dependent on ε_r of the substrate. The curve-fitted expressions for parameter A are given below:

$$\text{For } 1.05 \le \varepsilon_r \le 10: \quad A = ax + b, \quad x = \log_{10}\left(\frac{W}{h}\right). \tag{8.3.24}$$

where a and b are determined from the following set of expressions in a different range of x and relative permittivity ε_r:

$$\begin{aligned}
&\text{For } -1 \le x \le 0, \ \text{i.e.}(0.1 \le w/h \le 1):\\
&\quad a = -0.1122\varepsilon_r + 1.428, \ b = 0.0649\varepsilon_r + 0.3136\\
&\text{For } 0 < x \le 0.7, \ \text{i.e.}(1 < w/h \le 5):\\
&\quad a = -0.0927\varepsilon_r + 0.9081, \ b = 0.0648\varepsilon_r + 0.3142\\
&\text{For } 0.7 < x \le 1, \ \text{i.e.}(w/h \ge 5): A = 0.95\\
&\text{For } 10 \le \varepsilon_r \le 20, \ x \ge -1, \ \text{i.e.}(w/h \ge 0.1):\\
&\quad A = 0.95.
\end{aligned} \tag{8.3.25}$$

$$\text{For } 20 \le \varepsilon_r \le 30, \ w/h \ge 0.6: \quad A = \left[0.1164\left(\frac{W}{h}\right)^{-0.1435} - 0.095\right][\varepsilon_r - 20] + 0.95. \tag{8.3.26}$$

$$\text{For } 30 \le \varepsilon_r \le 200, \ w/h \ge 0.6: \quad A = 1.1639\left(\frac{W}{h}\right)^{-0.1435}. \tag{8.3.27}$$

The curve-fitted expressions for parameter B are given below:

$$\text{For } -1 \le x < 0, \ \text{i.e. } (0.1 \le w/h \le 1): \quad B = e^{-1.7047x}. \tag{8.3.28}$$

$$\text{For } 0 \le x \le 0.3, \ \text{i.e. } (1 \le w/h \le 2): \quad B = 0.5\left(0.8 + e^{-1.7047x}\right). \tag{8.3.29}$$

$$\text{For } x > 0.3, \ \text{i.e. } (w/h > 2): B = 0.8. \tag{8.3.30}$$

Correction Factor in LDM

The LDM expression (8.3.21) does not meet Schneider's condition (8.3.2-i b) that results in higher value for the effective relative permittivity, as compared to results of the SDA. A correction factor is incorporated into the expression of the LDM. The form of correction term is obtained from the first derivative of equation (8.3.21) in which ε_r is replaced by $(\varepsilon_r - \varepsilon_{reff}(0))$ to avoid a significant change in the nature of the LDM. The correction factor $\Delta\varepsilon_{r\,eff}(f)$ is,

$$\Delta\varepsilon_{r\,eff}(f) = \begin{cases} \dfrac{M\,K(\varepsilon_r - \varepsilon_{r\,eff}(0))^{-Q}}{P}\left(\dfrac{f}{f_i}\right), & f \le f_i\\[2ex] \dfrac{M\,K(\varepsilon_r - \varepsilon_{r\,eff}(0))^{-Q}}{P}\left(\dfrac{f_i}{f}\right), & f > f_i \end{cases} \tag{8.3.31}$$

where $Q = k\dfrac{f_i}{f}, R = K\left(\dfrac{f}{f_i}\right), P = \left(1 + M\,e^{-R}\right)^2.$

The corrected LDM is

$$\begin{aligned}
&\varepsilon_{r\,eff}(f) = \frac{\varepsilon_r}{1 + M\,e^{-K\,(f/f_i)}} - E\Delta\varepsilon_r(f) && \text{(a)}\\
&\text{where}: E = \begin{cases} 0, & 0.1 \le w/h < 5\\ 1, & w/h \ge 5 \end{cases}, \quad \text{for } 1 \le \varepsilon_r \le 20 && \text{(b)}\\
&\text{and} \quad E = \begin{cases} 1, & 0.5 \le w/h < 1\\ 0, & w/h \ge 1 \end{cases}, \quad \text{for } 20 < \varepsilon_r \le 200 && \text{(c)}.
\end{aligned} \tag{8.3.32}$$

The finite strip conductor thickness can also be incorporated in the LDM. It is achieved by calculating the strip thickness (t)-dependent effective relative permittivity $\varepsilon_{reff}(t,\,0)$ that replaces $\varepsilon_{r\,eff}(0)$ in the above expressions. Likewise, w/h-ratio is replaced by the $w_{eq}(t,f)$/h-ratio. Such replacement should be carried out even in equations (8.3.18)–(8.3.22). The expressions for $\varepsilon_{r\,eff}(t,\,0)$ and $w_{eq}(t,f)$/h are given below:

$$\varepsilon_{r\,eff}(t, 0) = \varepsilon_{r\,eff}(t = 0, f = 0) - \frac{(\varepsilon_r - 1)(t/h)}{4.6\sqrt{w/h}}. \tag{8.3.33}$$

$$\begin{aligned}
&w_{eq}(t, f) = w + \frac{w_{eq}(t, 0) - w}{1 + (f/f_0)^2} && \text{(a)}\\
&f_0 = \frac{c}{2w_{eq}(t, 0)\sqrt{\varepsilon_{r\,eff}(t, 0)}} && \text{(b)},
\end{aligned} \tag{8.3.34}$$

where c is the velocity of the EM-wave in free space. The static equivalent width w_{eq} (t,0) is determined from equation (8.2.21). Unlike other dispersion models, the LDM is an open model, in a sense that a designer can himself curve-fit the parameters A and B to suit his experimental data or simulation data for a specific substrate, such as GaAs, Si, etc.

8.3.4 Kirschning–Jansen Dispersion Model

A popular dispersion model due to Kirschning–Jansen is summarized below [J.44]:

$$\varepsilon_{r\,eff}(f) = \varepsilon_r - \frac{\varepsilon_r - \varepsilon_{r\,eff}(0)}{1 + P(\varepsilon_r, f\,h, w/h)}$$

$$P(\varepsilon_r, f\,h, w/h) = P_1 P_2 ((0.1844 + P_3 P_4) f\,h)^{1.5763}$$

$$P_1 = 0.27488 + \frac{w}{h}\left(0.6315 + \left(\frac{0.525}{(1 + 0.157 f\,h)^{20}}\right) - 0.065683 \exp\left(\frac{-8.7513 w}{h}\right)\right)$$

$$P_2 = 0.33622(1 - \exp(-0.03442\varepsilon_r))$$

$$P_3 = 0.0363 \exp(-4.6 w/h)\left(1 - \exp\left(\frac{-f\,h}{38.7}\right)^{4.97}\right)$$

$$p_4 = 1 + 2.751\left[(1 - \exp(-\varepsilon_r/15.916)^8\right]$$

$$\text{where } f\,h \text{ in GHz} - \text{mm}. \tag{8.3.35}$$

The static effective relative permittivity $\varepsilon_{r\,eff}(0)$ is computed from the model of Hammerstad–Jensen [J.14]. The above expression of the Kirschning–Jansen [K-J] model has an average accuracy of 0.6% over the range, $0.1 \le w/h \le 100$, $1 \le \varepsilon_r \le 20$, and $0 \le h/\lambda_0 \le 0.13$. The original K-J model does not account for the effect of strip conductor thickness t. However, the effect of the conductor thickness on the dispersion could be accounted for by carrying out the following replacement: $\varepsilon_{reff}(0) \to \varepsilon_{reff}(w/h, t)$ and $W \to W_{eq}(t, f)$. It is called the modified K-J, i.e. MKJ model. The MKJ model has accuracy within 1% up to 1.5 THz frequency for the same range of parameters of the original K-J model with $0 \le t/h \le 0.2$ [J.55, J.56].

For the narrow line, w/h = 0.1 the inaccuracy is above 2%. The LDM has average accuracy < 1% and the model applies to any substrate, any w/h-ratio, and frequency for the fundamental mode. These models are valid for $f < f_{TE}; f_{TE} = c/4h\sqrt{\varepsilon_r - 1}$. The accuracy of several available dispersion models is compared in the literature [J.4, J.57]. The results of both the LDM and MKJ, incorporating the conductor thickness, have been compared up to 10 THz against the results obtained using the EM-Simulator, HFSS. The average deviation of the MKJ is within 0.5%, whereas for the LDM, it is within 1.5%. Moreover, the LDM works for the conductor thickness up to $t/h \le 0.05$, whereas the MKJ model works for a thicker strip conductor $t/h \le 0.2$ [J.55].

8.3.5 Improved Model of Frequency-Dependent Characteristic Impedance

The waveguide model to compute the frequency-dependent characteristic impedance of a microstrip line has limited accuracy. Kirschning–Jansen [B.1, J.54] have reported a more accurate model using the power-current definition of the characteristic impedance. However, the original model does not take into account the finite thickness of the strip conductor. However, to account for the effect of conductor thickness, the conductor width w is replaced by $w_{eq}(t, f)$ in the set of expressions of Kirschning–Jansen model summarized below:

$$Z_0(f) = Z_0(f = 0)\left(\frac{R_{13}}{R_{14}}\right)^{R_{17}}, \tag{8.3.36}$$

where $Z_0(0)$ is the static characteristic impedance of microstrip. Other parameters R_1 to R_{17}, using $u = w_{eq}(t, f)/h$ and $f_n = f\,h$, are defined as follows:

$$\begin{aligned}
&R_1 = 0.03891\varepsilon_r^{1.4}, R_2 = 0.2671u^7, R_3 = 4.766 \exp(-3.228u^{0.641})\\
&R_4 = 0.016 + (0.0514\varepsilon_r)^{4.524}, R_5 = \left(\frac{f_n}{28.843}\right)^{12}\\
&R_6 = 22.2u^{1.92}, R_7 = 1.206 - 0.3144 \exp(-R_1)[1 - \exp(-R_2)]\\
&R_8 = 1 + 1.275\left\{1 - \exp\left[-0.004625 R_3\, \varepsilon_r^{1.674}\left(\frac{f_n}{18.365}\right)^{2.745}\right]\right\}\\
&R_9 = 5.086 R_4 \frac{R_5}{0.3838 + 0.386 R_4} \frac{\exp(-R_6)}{1 + 1.2992 R_5} \frac{(\varepsilon_r - 1)^6}{1 + 10(\varepsilon_r - 1)^6}\\
&R_{10} = 0.00044\varepsilon_r^{2.136} + 0.0184, R_{11} = \frac{\left(\frac{f_n}{19.47}\right)^6}{1 + 0.0962\left(\frac{f_n}{19.47}\right)^6}, R_{12} = \frac{1}{1 + 0.00245u^2}\\
&R_{13} = 0.9408\varepsilon_{r\,eff}(f_n)^{R_8} - 0.9603, R_{14} = (0.9408 - R_9)\left(\varepsilon_{r\,eff}(0)\right)^{R_8} - 0.9603\\
&R_{15} = 0.707 R_{10}\left(\frac{f_n}{12.3}\right)^{1.097}, R_{16} = 1 + 0.0503\varepsilon_r^2 R_{11}\left\{1 - \exp\left[-\left(\frac{u}{15}\right)^6\right]\right\}\\
&R_{17} = R_7\left[1 - 1.1241\frac{R_{12}}{R_{16}} \exp\left(-0.026 f_n^{1.15656} - R_{15}\right)\right].
\end{aligned} \tag{8.3.37}$$

The above model has accuracy within 1% against the results of the SDA for $0.1 \le w/h \le 10$, $1 \le \varepsilon_r \le 18$ and $0 \le f\,h \le 30$, where f-h is in GHz-mm. The values of ε_r and w/h should be restricted in such a way that R_1, R_2, and R_3 should be less than 20 to avoid the problem of overflow on the computer.

8.3.6 Synthesis of Microstrip Line

The design of microstrip components requires the computation of the w/h-ratio for the known characteristic impedance on a given substrate. The synthesis of a microstrip line for $0 < w/h < \infty$ and $1 \le \varepsilon_r < \infty$ is done with the help of the following frequency-independent closed-form expression [B.1]:

$$\frac{w}{h} = \frac{8\left[\{G-1\}\left\{\frac{7+(4/\varepsilon_r)}{11}\right\} + \left\{\frac{1+(1/\varepsilon_r)}{0.81}\right\}\right]^{0.5}}{G-1} \quad \text{(a)}$$

$$\text{where, } G = \exp\left[\frac{Z_0(w/h, \varepsilon_r)\sqrt{(\varepsilon_r+1)}}{42.4}\right] \quad \text{(b).}$$

(8.3.38)

8.4 Losses in Microstrip Line

There are four mechanisms of losses, causing attenuation of the signal on a microstrip, i.e. the dielectric loss, conductor loss, radiation loss, and surface wave loss. The *dielectric loss*, i.e. the *substrate loss*, is caused by imperfect lossy substrate material. The *conductor loss* is caused by the finite conductivity of the strip and ground plane conductors. For a low-loss substrate, like PTFE and alumina, the conductor loss is usually a predominant factor. For instance, the dielectric loss of a 50-ohm microstrip on an alumina substrate, at 10 GHz, is 0.04 dB/m, while its conductor loss is 1.9 dB/m. The conductor loss depends on the skin-depth and the surface resistivity of a conductor.

A microstrip circuit has several kinds of discontinuities, such as open, step, gap, bend, etc. [B.1–B.6]. They radiate power to the outer space causing loss to a microstrip section. This is known as *radiation loss*. The radiation from a microstrip occurs due to the generation of higher-order modes at the discontinuity and under certain conditions, the higher-order modes are radiating modes. The radiation modes constitute a continuous spectrum [B.13]. It is shown in Fig (8.4). The radiation loss is high for a microstrip on a low permittivity substrate, like PTFE ($\varepsilon_r = 2.5$), and is low for a microstrip on high permittivity substrate, like alumina ($\varepsilon_r = 9.8$). A discontinuity also generates the surface wave on a conductor-backed substrate of a microstrip. The power loss in a surface wave is known as the *surface wave loss* of a microstrip. A part of the signal traveling on the substrate, as surface waves, interacts and couples with components located on the substrate, causing degradation in the performance of a circuit. The surface wave loss is significant in a microstrip on a thick substrate and also for the substrate of high relative permittivity. However, if the operating frequency is below the coupling frequency of the TEM mode and TM_0 surface wave mode given in equation (8.3.22), the surface wave loss is small. The dielectric and conductor losses of a microstrip line are computed below.

8.4.1 Dielectric Loss in Microstrip

A lossy substrate is characterized through its complex relative permittivity, finite resistivity, loss-tangent, and complex capacitance. These concepts are summarized in this section.

EM-Wave in a Lossy Dielectric-Medium

It is assumed that the x-polarized uniform TEM wave propagates in the z-direction in a lossy homogeneous dielectric-medium. Maxwell's equations of the x-polarized plane wave, using equation (4.4.11) of chapter 4, are given below:

$$\frac{dE_x}{dz} = -j\,\omega\mu_0 H_y \quad \text{(a)}, \qquad \frac{dH_y}{dz} = -(\sigma + j\,\omega\varepsilon_0\varepsilon_r)E_x \quad \text{(b)}, \tag{8.4.1}$$

where σ and ε_r are conductivity and relative permittivity of a lossy dielectric-medium. The 1D wave equation is obtained from the above equations:

$$\frac{d^2E_x}{dz^2} = -\omega^2\mu_0\varepsilon_0\varepsilon_r\left(1 - j\frac{\sigma}{\omega\varepsilon_0\varepsilon_r}\right)E_x$$
$$= -\omega^2\mu_0\varepsilon_0\varepsilon_r(1 - j\,\tan\delta)E_x \tag{8.4.2}$$
$$\Rightarrow \frac{d^2E_x}{dz^2} + \omega^2\mu_0\varepsilon_0\varepsilon_r^*\,E_x = 0.$$

Thus, a lossy dielectric-medium is expressed through the loss-tangent, $\tan\delta$, and also using the complex relative permittivity ε_r^*:

$$\varepsilon_r^* = \varepsilon_r(1 - j\,\tan\delta) = \varepsilon_r - j\,\varepsilon_r\tan\delta = \varepsilon_r' - j\varepsilon_r''. \tag{8.4.3}$$

From the above consideration, the following expressions are obtained:

$$\text{Re}\left(\varepsilon_r^*\right) = \varepsilon_r' = \varepsilon_r \quad \text{(a)}, \qquad \text{Im}\left(\varepsilon_r^*\right) = \varepsilon_r'' = \varepsilon_r\tan\delta \quad \text{(b)}. \tag{8.4.4}$$

$$\tan\delta = \frac{\varepsilon_r''}{\varepsilon_r'} = \frac{\sigma}{\omega\varepsilon_0\varepsilon_r} \quad \text{(a)}, \qquad \varepsilon_r'' = \frac{\sigma}{\omega\varepsilon_0} \quad \text{(b)}. \tag{8.4.5}$$

The imaginary part of a complex relative permittivity indicates the loss in a dielectric-medium. The conductivity of a medium σ can be converted to the loss-tangent at each frequency of interest by equation (8.4.5a). Normally, $\tan\delta$ and σ of a substrate are taken frequency independent, i.e. the substrate has usually no noticeable material dispersion in the microwave range. However, over a wide range of frequency, both the ε_r' and ε_r'' are frequency-dependent and follow the causal Krammer–Kronig relation or the Debye model discussed in chapter 6.

Representation of Lossy Dielectric-Medium by a Lossy Capacitor

Figure (8.14a) shows the cross-section of a lossy homogeneous medium parallel-plate transmission line. Figure (8.14b) represents the *equivalent circuit model* expressed through the admittance *p.u.l.*:

$$Y = G + j\,\omega C. \tag{8.4.6}$$

The conductance (G) represents the loss in a dielectric-medium that is related to the loss-tangent through the phasor diagram, shown in Fig (8.14c):

$$\tan\delta = \frac{(1/\omega C)}{R} = \frac{G}{\omega C}. \tag{8.4.7}$$

Therefore, the admittance of a lossy transmission line *p.u.l.* is

$$Y = \omega C(\tan\delta + j). \tag{8.4.8}$$

The conductance G is related to the conductivity σ of a lossy medium. Under the influence of an external voltage V, the leakage transverse current I flow from the upper conductor to lower conductor through the dielectric-medium. If A is an area of the parallel-plate capacitor, the current density through a lossy dielectric-medium is

$$\vec{J} = \frac{\vec{I}}{A} = \frac{\vec{V}}{AR} = \frac{\vec{E}h}{RA}. \tag{8.4.9}$$

The electric field $\vec{E}$ and current density $\vec{J}$ in a lossy dielectric-medium are related through Ohm's law:

$$\vec{J} = \sigma\,\vec{E}. \tag{8.4.10}$$

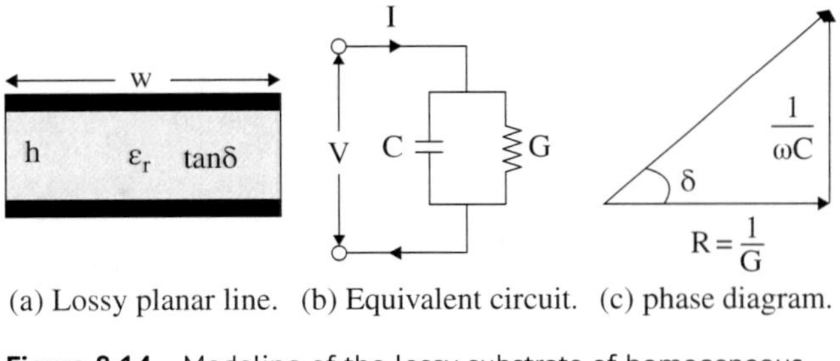

(a) Lossy planar line. (b) Equivalent circuit. (c) phase diagram.

Figure 8.14 Modeling of the lossy substrate of homogeneous medium transmission line.

An expression for conductance, $G = 1/R$ *p.u.l.* is obtained from the above equations,

$$G = \frac{\sigma A}{h} = \frac{\sigma w}{h}. \tag{8.4.11}$$

As the capacitance C *p.u.l.* of a parallel-plate line is $\varepsilon_0\varepsilon_r w/h$, the loss tangent of the substrate is expressed as

$$\tan\delta = \frac{G}{\omega C} = \frac{\sigma}{\omega\,\varepsilon_0\varepsilon_r}. \tag{8.4.12}$$

It is identical to equation (8.4.5a) from the field point of view. It shows the equivalence of field and circuit treatment of a homogeneous lossy dielectric-medium. The circuit model shown in Fig (8.14b) can also be represented by the *complex capacitance* C^*. The admittance of the equivalent circuit, given by expression (8.4.6), can be rewritten as

$$Y = j\,\omega\left[C - j\frac{G}{\omega}\right] = j\omega\,(C' - j\,C''). \tag{8.4.13}$$

The complex capacitance C^* of a lossy transmission line is

$$\begin{aligned} &C^* = j\omega(C' - jC'') \quad (a), \quad C' = C \quad (b), \quad C'' = \frac{G}{\omega} \quad (c) \\ &\tan\delta = \frac{C''}{C'} = \frac{G}{\omega C} \quad (d). \end{aligned} \tag{8.4.14}$$

The complex line capacitance of a lossy homogeneous medium is further related to its complex relative permittivity ε_r^*:

$$\varepsilon_r^* = \varepsilon_r' - j\varepsilon_r'' = \frac{C^*(\varepsilon_r^*)}{C_0(\varepsilon_r^* = 1)} = \frac{C'}{C_0} - j\frac{C''}{C_0}. \tag{8.4.15}$$

Likewise, the complex effective relative permittivity of an inhomogeneous microstrip line on a lossy substrate is related to its complex line capacitance:

$$\begin{aligned} &\varepsilon_{r\,\mathrm{eff}}^*(f = 0) = \frac{C^*(\varepsilon_r^*)}{C_0(\varepsilon_r^* = 1)} \\ &\Rightarrow \varepsilon_{r\,\mathrm{eff}}'(f = 0) - j\varepsilon_{r\,\mathrm{eff}}''(f = 0) = \frac{C'}{C_0} - j\frac{C''}{C_0}. \end{aligned} \tag{8.4.16}$$

The above expression is useful for the determination of the effective loss-tangent needed in the computation of the dielectric loss of a microstrip.

The Dielectric Loss of EM-Wave in a Homogeneous Lossy Dielectric-Medium

The following expression for the attenuation of a low-loss homogeneous medium transmission line is obtained in equation (2.5.0a) of chapter 2:

$$\alpha = \frac{R}{2Z_0(\varepsilon_r)} + \frac{G\,Z_0(\varepsilon_r)}{2}. \tag{8.4.17}$$

The first term of equation (8.4.17) is the conductor loss of the line and the second term shows its dielectric loss:

$$\text{Conductor loss}: \alpha_C = \frac{R}{2Z_0(\varepsilon_r)}\ \text{Np/length}. \tag{8.4.18}$$

$$\text{Dielectric loss}: \alpha_d = \frac{G\,Z_0(\varepsilon_r)}{2}\ \text{Np/length}. \tag{8.4.19}$$

Using equation (8.4.12), with the above equation, the dielectric loss of a parallel-plate line could be written as

$$\begin{aligned}
\alpha_d &= \frac{\omega C Z_0(\varepsilon_r)\tan\delta}{2} = \left(\frac{2\pi f\, w\varepsilon_0\varepsilon_r}{2\ \ h}\right)\left(\frac{Z_0(\varepsilon_r=1)}{\sqrt{\varepsilon_r}}\right)\tan\delta \\
&= \left(\frac{\pi f w\varepsilon_0\varepsilon_r}{h}\right)\left(\frac{1}{cC_0\sqrt{\varepsilon_r}}\right)\tan\delta \\
&= \left(\frac{\pi f w\varepsilon_0\varepsilon_r}{h}\right)\left(\frac{1}{c\varepsilon_0 w/h\sqrt{\varepsilon_r}}\right)\tan\delta \\
\Rightarrow \alpha_d &= \frac{\pi f}{c}\sqrt{\varepsilon_r}\tan\delta\ \ \text{Np/length} \quad (a) \\
\Rightarrow \alpha_d &= \frac{\pi}{\lambda_0}\sqrt{\varepsilon_r}\tan\delta\ \ \text{Np/length} \quad (b) \\
\Rightarrow \alpha_d &= \frac{27.3}{\lambda_0}\sqrt{\varepsilon_r}\tan\delta\ \ \text{dB/length} \quad (c).
\end{aligned} \tag{8.4.20}$$

The dielectric loss can also be expressed in terms of the conductivity σ of the medium, instead of the loss-tangent:

$$\begin{aligned}
\alpha_d &= \frac{\eta_0}{2}\frac{\sigma}{\sqrt{\varepsilon_r}}\ \ \text{Np/length} \quad (a) \\
\Rightarrow \alpha_d &= 4.34\frac{\sigma}{\sqrt{\varepsilon_r}}\ \ \text{dB/length} \quad (b),
\end{aligned} \tag{8.4.21}$$

where $\eta_0 = 120\pi\ \Omega$ is the intrinsic impedance of free space.

The Dielectric Loss of a Microstrip Line

The waveguide model of a microstrip forms a parallel-plate line with equivalent width w_{eq} and a medium of the effective relative permittivity $\varepsilon_{r\,eff}(0)$. Likewise, for a lossy microstrip, the loss-tangent of the medium of the waveguide model is the *effective loss-tangent* $\tan\delta_{eff}$. The expression for $\tan\delta_{eff}$ is obtained using Wheeler's transformation, given by equation (8.2.10), in the complex form:

$$\begin{aligned}
&\varepsilon^*_{reff}(0) = 1 + q(\varepsilon^*_r - 1) \\
&\Rightarrow \varepsilon'_{r\,eff}(0) - j\varepsilon''_{r\,eff}(0) = 1 + q(\varepsilon'_r - 1) - jq\varepsilon''_r.
\end{aligned} \tag{8.4.22}$$

On separating the real and imaginary parts of the above expression, the following expressions are obtained:

$$\begin{aligned}
\varepsilon_{r\,eff}(0) &= \varepsilon'_{r\,eff}(0) = 1 + q(\varepsilon'_r - 1) \quad (a) \\
\varepsilon''_{r\,eff}(0) &= q\varepsilon''_r \quad (b).
\end{aligned} \tag{8.4.23}$$

The expression for the effective loss-tangent of a microstrip line used in the equivalent waveguide model is

$$\tan\delta_{eff} = \frac{\varepsilon''_{r\,eff}(0)}{\varepsilon'_{r\,eff}(0)} = \frac{q\varepsilon''_r}{1 + q(\varepsilon'_r - 1)} = \frac{q\,\varepsilon_r\tan\delta}{1 + q(\varepsilon_r - 1)}, \tag{8.4.24}$$

where $\varepsilon'_r = \varepsilon_r$ is the relative permittivity of a substrate and $\tan\delta$ is its loss-tangent. Using equation (8.2.10), the filling-factor q is expressed as

$$q = \frac{\varepsilon_{r\,eff}(0) - 1}{\varepsilon_r - 1}. \tag{8.4.25}$$

On using $\varepsilon_{r\,eff}(0)$, or more accurately $\varepsilon_{r\,eff}(f)$ and $\tan\delta_{eff}$ in equation (8.4.20) in place of ε_r and $\tan\delta$, the following expression is obtained for the dielectric loss of a microstrip line:

$$\begin{aligned}
\alpha_d &= \frac{\pi}{\lambda_0}\sqrt{\varepsilon_{r\,eff}(0)}\tan\delta_{eff} = \frac{\pi\sqrt{\varepsilon_{r\,eff}(0)}}{\lambda_0}\frac{q\,\varepsilon_r}{1 + q(\varepsilon_r - 1)}\tan\delta \\
\Rightarrow \alpha_d &= \frac{\pi[\varepsilon_{r\,eff}(0) - 1]}{\lambda_0[\varepsilon_r - 1]}\frac{\varepsilon_r\tan\delta}{\sqrt{\varepsilon_{r\,eff}(0)}}\ \ \text{Np/Length} \quad (a) \\
\alpha_d &= \frac{27.3[\varepsilon_{r\,eff}(f) - 1]}{\lambda_0[\varepsilon_r - 1]}\frac{\varepsilon_r\tan\delta}{\sqrt{\varepsilon_{r\,eff}(f)}}\ \ \text{dB/Length} \quad (b).
\end{aligned} \tag{8.4.26}$$

The above equation shows that the dielectric loss increases linearly with frequency. It is the same as the final results of Schneider [J.59]. It computes the dielectric loss that is about 1% higher than the value computed by the numerical method of Simpson and Tseng [J.58]. Table (8.3) compares the normalized dielectric loss as computed by both the methods for a few substrates. Normalization is done by the guided wavelength $\lambda_g = \lambda_0/\sqrt{\varepsilon_{reff}(0)}$. Usually, the dielectric loss computation is satisfactory up to 100 GHz. However, by using the empirically obtained frequency-dependent loss-tangent, the model can compute the dielectric loss up to 1 THz, and beyond [J.55].

8.4.2 Conductor Loss in Microstrip

The RF signal also gets attenuated due to the finite conductivity of the strip and ground conductors of a microstrip line. The conducting surface presents a *surface impedance* to the RF field penetrating a conductor. Its real part is used to compute the conductor loss, while its imaginary inductive part shows the penetration of the surface magnetic field and corresponding surface current density inside a conductor. An expression is obtained below for the surface impedance of the EM-wave propagating over a block of infinite extent conductor.

Table 8.3 Normalized dielectric loss of microstrip line in dB/λ_g.

	ε_r = 2.22		ε_r = 3.82		ε_r = 9.00		ε_r = 11.70	
W/h	**Schneider**	**Simpson**	**Schneider**	**Simpson**	**Schneider**	**Simpson**	**Schneider**	**Simpson**
0.4	743.95	743.80	596.94	594.56	406.81	405.03	359.69	357.60
0.6	764.34	760.20	611.33	606.02	415.07	411.89	366.76	363.51
0.8	780.85	774.70	622.59	616.18	421.73	417.97	372.46	368.7
1.0	794.87	787.91	632.28	625.48	427.36	423.44	377.27	373.67
5.0	921.10	916.37	718.58	714.34	477.24	475.12	419.97	418.14

Source: Modified from Simpson and Tseng [J.58].

Using Maxwell's equations in a conducting medium, i.e. for $\sigma >> j\,\omega\varepsilon$, the wave equation (4.5.3) of chapter 4 is written as the *diffusion equation* in three forms:

$$\nabla^2 \vec{E} = j\omega\mu\sigma \vec{E} \quad (a), \quad \nabla^2 \vec{J} = j\omega\mu\sigma \vec{J} \quad (b)$$
$$\nabla^2 \vec{H} = j\omega\mu\sigma \vec{H} \quad (c). \tag{8.4.27}$$

Figure (8.15a) shows that the longitudinal current density (Amp/m^2) J_z, associated with the EM-wave, flowing in the z-direction on the surface of a conductor. The current I is supplied by the voltage source V, connected across the line of length L. The conductor width in the x-direction is infinite, leading to the *uniform* current flow on a strip. The strip conductor section offers a *series of impedance* Z_{series} to the source. It is evaluated below. Figure (8.15b) shows that while the surface current density J_z, at the top of the conductor, flows in the z-direction, it also penetrates inside the conductor in the y-direction. Such penetration of the surface current is responsible for the conductor loss in a microstrip line.

The penetration of longitudinal J_z in the y-direction is described by the 1D diffusion equation:

$$\frac{d^2J_z}{dy^2} = j\omega\mu\sigma J_z = \Gamma^2 J_z. \tag{8.4.28}$$

The propagation constant Γ is

$$\Gamma = \sqrt{j}\sqrt{2\pi f\mu\sigma} = (1 + j)\sqrt{\pi f\mu\sigma} \quad (a)$$
$$\Rightarrow \Gamma = (1 + j)/\delta \quad (b). \tag{8.4.29}$$

The parameter δ is known as the *skin-depth* inside a planar thick conductor. It is given by

$$\delta = \frac{1}{\sqrt{\pi f\mu\sigma}}, \tag{8.4.30}$$

where inside a conductor $\mu = \mu_0$. If J_0 is the surface current density at the top surface of a conductor, then the solution of equation (8.4.28) is written as

$$J_z(y) = J_0 e^{-\Gamma y} = \underbrace{J_0\, e^{-\frac{y}{\delta}}}_{\text{Amplitude}} \times \underbrace{e^{-j\frac{y}{\delta}}}_{\text{Phase of current density}}. \tag{8.4.31}$$

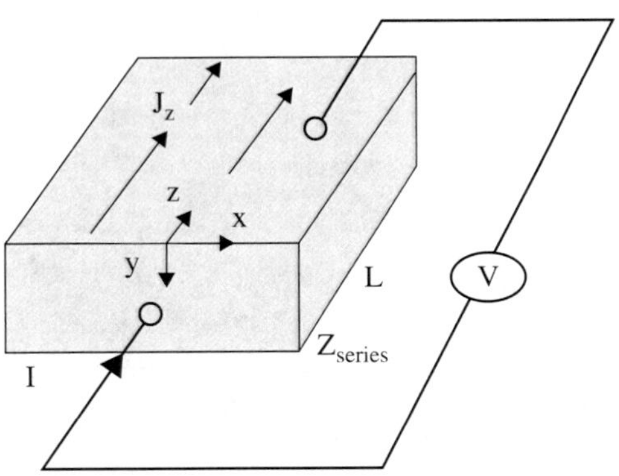

(a) Longitudinal current flowing on the surface of the conductor.

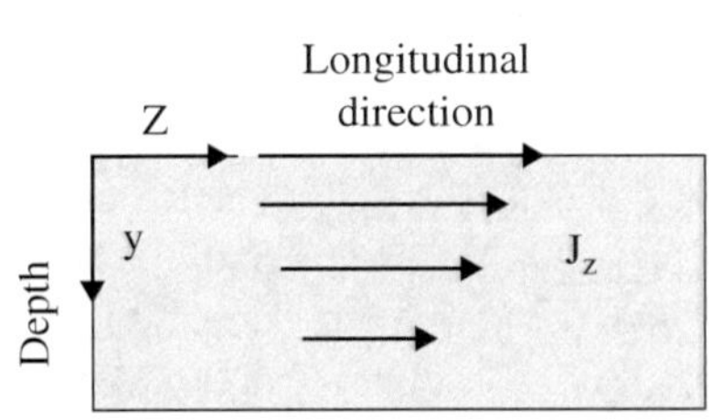

(b) Penetration of surface current inside the metal strip.

Figure 8.15 Surface current penetration in the thick metal strip.

The *lineal current density per unit width (Amp/m)*, flowing in the z-direction, is

$$J_{zw} = \int_0^\infty J_z(y)dy = \int_0^\infty J_0 e^{-\frac{(1+j)}{\delta}y}dy = \frac{J_0\delta}{1+j} = \frac{J_0\delta}{\sqrt{2}}\angle -45^\circ. \quad (8.4.32)$$

Therefore, the z-directed lineal current density is 45° out of phase, with respect to the surface current density at the surface of the conductor. The current decays exponentially inside a conductor. At the surface of the finite conductivity conductor, the electric field intensity E_0 is related to the voltage drop, $J_0 = \sigma E_0$. The lineal current density per unit width is

$$J_{zw} = \frac{\delta\sigma}{1+j}E_0. \quad (8.4.33)$$

The above equation gives an expression for the surface impedance:

$$Z_s = R_s + j\omega L_i = \frac{E_0}{J_{zw}} = \frac{1+j}{\sigma\delta} \quad (a)$$
$$\Rightarrow R_s + j\omega L_i = \frac{1}{\sigma\delta} + j\omega\mu\frac{\delta}{2} \quad (b), \quad (8.4.34)$$

where equation (8.4.34b) is obtained using equation (8.4.30). The real part R_s of the above equation is known as the *surface resistance*. It is responsible for the conductor loss. Its imaginary part forms the *internal inductance* of a conductor, due to the penetration of the magnetic field. At the field penetration $y = \delta/2$, the surface resistance and surface reactance are numerically equal, i.e. $\omega L_s = R_s$. Figure (8.15a) shows the line current I *(Amp)* passing through the conductor of width w. It is computed as follows:

$$I = \int_0^w J_{zw}(x)\,dx = J_{zw}w. \quad (8.4.35)$$

In the above equation, $J_{zw}(x)$ is assumed to be *uniform* with respect to x. It is an unrealistic assumption, as the current distribution across the width of a conducting strip has an edge singularity. However, at present the voltage drop across a length of the conductor L is

$$V = E_0 L. \quad (8.4.36)$$

Therefore, the *series impedance* Z_{series} offered by the block of a strip conductor connected to a source is obtained as follows:

$$Z_{series} = \frac{V}{I} = \frac{E_0 L}{J_{zw}w} = Z_s\left(\frac{L}{w}\right) = \left(\frac{1+j}{\sigma\delta}\right)\left(\frac{L}{w}\right). \quad (8.4.37)$$

For $L = w$, $Z_{seies} = Z_s$, i.e. *any size of a square conductor offers the same series impedance to the source.* The real part of the series impedance is resistance R_{series} or simply R, i.e. the line resistance. The line resistance and the internal reactance *p.u.l.*, i.e. for L = 1, are

$$R = \frac{1}{\sigma\delta w} = \frac{R_s}{w} \quad (a), \quad X_i = \omega L_i = \frac{R_s}{w} \quad (b). \quad (8.4.38)$$

The power loss in a conductor of length L and width w, carrying the *uniform* surface current J_{zw} per unit width, is

$$P = \frac{1}{2}\mathrm{Re}\,[V I^*] = \frac{1}{2}\mathrm{Re}\,[(E_0 L)(J_{zw}w)^*]$$
$$= \frac{1}{2}\mathrm{Re}\,[(R_s J_{zw} L)(J_{zw}w)^*] = \frac{1}{2}R_s|J_{zw}|^2 Lw.$$

The power loss per unit surface area is $p = (1/2)R_s|J_{zw}|^2 = (1/2)R_s|J_s|^2$, where $J_{zw} = J_s$ is the linear surface density per unit width. It is related to the surface magnetic field by the expression $|J_s| = |H_t|$. The power loss over the conducting surface is

$$P = \frac{1}{2}\iint_S R_s|J_s|^2 ds. \quad (8.4.39)$$

To compute the conductor loss of a microstrip line, the above equation is applied to both the strip and ground conductors. This method is known as the *perturbation method* [B.9]. It requires determination of the surface current density J_s over both the strip and ground conductors. It has been done by numerical methods [J.60]. The closed-form Maxwell's distribution function takes care of the edge singularity [J.61, J.62]. The closed-form expressions for the current distribution over the ground conductor have also been obtained [J.63]. However, for simplicity, the *uniform current density* is *assumed* over the strip conductor and uses the parallel-plate waveguide model to compute the conductor loss of a microstrip line. It is used to compute the conductor loss of a waveguide discussed in chapter 7. Next, the results are summarized for the perturbation method that takes care of Maxwell's distribution function of surface current distribution showing the edge singularity.

Waveguide Model for Computation of Conductor Loss of Microstrip Line

The waveguide model of a microstrip line is a parallel-plate transmission line of conductor width w and conductor thickness t. Two conducting plates are separated by a distance h, i.e. the thickness of the substrate. The parallel-plate line contains a dielectric of relative permittivity $\varepsilon_{r\,eff}(f)$. Using equation (8.4.38a), the RF line resistance *p.u.l.* is

$$R = \frac{2}{\sigma\delta_s w}. \quad (8.4.40)$$

The factor 2 in the above equation is due to two strip conductors, lower and upper ones, forming the parallel-plate line. The δ_s is the skin-depth of a strip conductor. However, if the strip and ground conductors are of different materials with different conductivities, the line resistance *p.u.l* is

$$R = R_{ss} + R_{sg} = \frac{1}{\sigma\delta_s w} + \frac{1}{\sigma\delta_g w}, \qquad (8.4.41)$$

where δ_g is the skin-depth of the ground conductor. In the above equation, R_{ss} and R_{sg} are the surface resistance of strip and ground conductors, respectively. The conductor loss *p.u.l.* is given by equation (8.4.18):

$$\alpha_C = \frac{R}{2Z_0(\varepsilon_r)} = \frac{R_{ss} + R_{sg}}{2Z_0(\varepsilon_r)} \text{ Np/length.} \qquad (8.4.42)$$

As the characteristic impedance of a parallel-plate waveguide is $Z_o(\varepsilon_r) = (120\pi\, h/w)\sqrt{1/\varepsilon_r} = (h/w)\sqrt{\mu_0/\varepsilon_0\varepsilon_r}$, the conductor loss of a microstrip line using equations (8.4.18), (8.4.30), and (8.4.40) is

$$\alpha_C = \frac{\pi\delta_s\sqrt{\varepsilon_{r\,eff}(f)}}{\lambda_0\, h} \text{ Np/length} \qquad (a)$$

$$\alpha_C = \frac{27.3\delta_s\sqrt{\varepsilon_{r\,eff}(f)}}{\lambda_0\, h} \text{ dB/length} \qquad (b)$$

$$\alpha_C = 27.3\sqrt{\frac{\varepsilon_0}{\pi\sigma}}\frac{\sqrt{\varepsilon_{r\,eff}(f)}}{h}\sqrt{f} \text{ dB/length} \qquad (c).$$

$$(8.4.43)$$

The waveguide model of microstrip helps to replace $\varepsilon_r \rightarrow \varepsilon_{r\,eff}(f)$ in the above expressions. For the strip and ground plane conductors, with different conductivities, the conductor loss is

$$\alpha_C = \frac{\sqrt{\pi\mu_r\mu_0}}{2Z_0(\varepsilon_r)w}\left[\frac{1}{\sqrt{\sigma_s}} + \frac{1}{\sqrt{\sigma_g}}\right]\sqrt{f} \text{ Np/length.} \qquad (8.4.44)$$

The nonmagnetic conductors have $\mu_r = 1$. The various forms of the above expressions to compute the conductor loss are valid only for a wide microstrip with $w/h \rightarrow \infty$. For $w/h \leq 2$, these expressions show about 80% higher conductor loss, as compared to the experimental results [J.61, J.62]. *The conductor loss increases with the square root of frequency*, provided $\varepsilon_{r\,eff}(f)$ does not change much with frequency. Although the expression for the surface resistance is obtained for an infinitely thick conductor, however it is valid for thickness $t \geq 3\delta_s$. The above expressions show that at nearly zero frequency, $f \rightarrow 0$, the conductor loss is zero. In practice, this not the case. The skin-depth becomes larger than the conductor thickness with decreasing frequency. For the case of the conductor thickness less than the skin-depth, i.e. for $t < \delta_s$ corresponding to lower frequency limit f_{c1}, the RF resistance is replaced by the dc resistance given below:

$$R = \frac{2}{\sigma t w}. \qquad (8.4.45)$$

For the above case, the current is uniformly distributed throughout the strip thickness and conductor loss is frequency independent. At high frequency, above f_{c2}, corresponding to $t \geq 2\delta_s$, the conductor loss follows the square root law given by (8.4.43c). These frequency limits are given by the following expressions [J.64, J.65]:

$$\text{Lower frequency limit}: f_{c1} = \frac{R}{2\pi L} \qquad (a)$$

$$\text{Upper frequency limit}: f_{c2} = \frac{4}{\pi\mu\sigma t^2} \qquad (b),$$

$$(8.4.46)$$

where $R = 1/\sqrt{\sigma w t}$ and L are resistance and inductance *p.u.l.* of a microstrip. It is also obvious from equation (8.4.43c) that the conductor loss increases with increasing permittivity of the substrate. Equation (8.4.44) shows that the conductor loss decreases with increasing strip width w. In the above equations, conductivity of a conductor is treated as constant, i.e. independent of the frequency. However, the computed conductor loss deviates from the experimental results with increasing frequency. The following empirical expression for the frequency-dependent conductivity improves the computed frequency-dependent conductor loss [J.66]:

$$\sigma(f) = \sigma(0)\sqrt{1 + C_0 f}, \qquad (8.4.47)$$

where $\sigma(0)$ is the given or measured dc conductivity of a conductor, f is the frequency in GHz and for a microstrip line the constant $C_0 = 0.045$. Drude model, discussed in chapter 6, section (6.5), shows a decrease in the conductivity with frequency. However, the experimentally obtained conductor loss follows the above relation. Possibly the conductor–dielectric interface morphology is responsible for such behavior.

Figure (8.16a) compares the computed conductor loss by the present method against the experimental results of a microstrip [J.66]. It shows good agreement of the computed and measured conductor loss. For the conductor thickness much more than the skin-depth, $t >> \delta_s$, the conductor loss shows a linear behavior on the logarithmic frequency scale. In the lower frequency range, $t < \delta_s$, the conductor loss deviates from it. However, if the expression for the surface impedance, given by equation (8.4.34), is replaced by the thickness-dependent surface impedance, more accurate computed results are obtained. It is shown in Fig (8.16a). Such expression is presented later on. The minimum conductor loss, i.e. the loss smaller than the loss of the infinitely thick conductor, is obtained for a conductor thickness in the range $1.142\,\delta_s < t < 2.851\,\delta_s$.

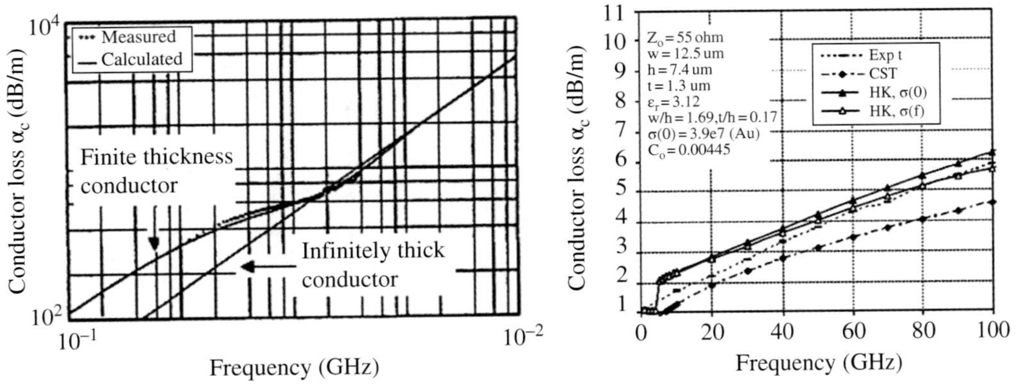

(a) $w = 20\,\mu m$, $t = 0.65\,\mu m$, $h = 2.8\,\mu m$, $\varepsilon_r = 3.0$, $\tan\delta = 0.01$, $\sigma = 3.6 \times 10^7$ s/m.[J.66].

(b) Conductor loss on TFML(55 Ω) [J.67].

Figure 8.16 Conductor of the microstrip line. *Source:* Konno [J.66] and Bansal et al. [J.67].

Perturbation Method Using Surface Current with Edge Singularity

A narrow microstrip line has a concentration of current at the edges, resulting in higher conductor loss. The waveguide model does not treat the edge singularity of the current distribution on a strip conductor and the nonuniform current distribution on a ground plane. A summary of two methods that take care of the edge singularity of the surface current is given below.

Conformal Mapping Method

Collin [B.8] has corrected the expression for a surface resistance of both the strip and ground conductors given in equation (8.4.34) by using the conformal mapping method. The results are summarized below:

$$\text{For strip conductor}: R_{ss} = R_s\left[\frac{1}{\pi} + \frac{1}{\pi^2}\ln\left(\frac{4\pi w}{t}\right)\right]LR \quad (a)$$

where for $w/h \le 0.5 : LR = 1$;

$0.5 < w/h \le 10 : LR = 0.94 + 0.132w/h - 0.0062(w/h)^2$

For ground conductor :

$$R_{sg} = R_s\left[\frac{w/h}{w/h + 5.8 + 0.03h/w}\right], 0.1 \le w/h \le 10 \quad (b). \qquad (8.4.48)$$

where R_{ss} and R_{sg} are surface resistances of the strip and ground conductors, respectively. In the above expression, the effective width of the ground plane is $w + 5.8h$ of the uniform current density. Collins has taken surface resistance $R_s = \sqrt{(\omega\mu/\sigma)}$. Liou [J.68] has suggested another multiplying factor for correct expression of the surface impedance that accounts for w and t of a conductor:

$$Z_s\,(\text{Improved}) = Z_s\left[1 + \frac{2t}{w}\left(1 - e^{-\frac{w}{\delta_s}}\right)\left(1 - e^{-\frac{t}{\delta_s}}\right)\right]. \qquad (8.4.49)$$

It is seen that for $w \to \infty$, $t/w \to 0$, $Z_s(\mathit{Improved}) \to Z_s$.

Perturbation Method

The perturbation method is used to compute the conductor loss of both the thin and thick strip conductors. The conductor thickness of a thin conductor is less than the skin-depth. However, the method is applied separately to the strip conductor and the ground conductor. In the case, the strip current distribution is expressed through Maxwell's function, $J_z = I/\sqrt{(w/2)^2 - x^2}$, equation (8.4.39) shows the logarithmic divergence. To avoid it, the integration omits the endpoints of the width and stops at $\pm(w/2 - \Delta)$, where Δ is known as the *stopping distance* [J.69]. For an isolated strip, Holloway and Kuester have given the values of the stopping distance in a tabular form [J.70]. However, the following closed-form expressions for the reciprocal of normalized stopping distance Δ, as a function of normalized strip conductor thickness, are also available for the CAD application [J.71]:

$$y = \begin{cases} -60.89\cdot x^4 + 282.17\cdot x^3 - 27.764\cdot x^2 + 0.5103\cdot x + 9.1907 & \text{for } 0.03 \le x < 0.64 \\ 711.01\cdot x^6 - 2120.2\cdot x^5 - 1038.3\cdot x^4 + 9881.7\cdot x^3 - \\ 12798\cdot x^2 + 6935.8\cdot x - 1372.7 & \text{for } 0.64 \le x < 1.5 \\ -183.05\cdot x^4 + 1583\cdot x^3 - 4981.9\cdot x^2 + 6618.6\cdot x - 2807.6 & \text{for } 1.5 \le x < 2.76 \\ 0.2953\cdot x^4 - 7.7733\cdot x^3 + 73.991\cdot x^2 - 293.88\cdot x + 581.98 & \text{for } 2.76 \le x < 8.0 \\ -0.0198\cdot x^3 + 0.9848\cdot x^2 - 12.104\cdot x + 240.01 & \text{for } 8.0 \le x < 16. \end{cases} \qquad (8.4.50)$$

The normalized parameters used in the above expressions are given below:

$$\text{Reciprocal normalized stopping distance}: y = t/\Delta \quad (a)$$
$$\text{Normalized strip thickness}: x = \frac{t}{2\cdot 0\delta_s} \quad (b). \qquad (8.4.51)$$

Using the concept of stopping distance, Holloway and Kuester (HK) have computed the conductor loss of a microstrip line as a summation of the conductor loss on the strip (α_{cs}) and conductor loss on the ground plane (α_{cg}), i.e. $\alpha_c = \alpha_{cs} + \alpha_{cg}$ [J.69, J.72]. The expressions are summarized below:

Loss on strip :

$$\alpha_{cs} = \frac{R_{sm}}{2\pi^2 w Z_0(w_{eff}, h, t, \varepsilon_r, f)} \ln\left(\frac{w-\Delta}{\Delta}\right) \text{Np/m} \quad (a)$$

$$\text{where,} \quad R_{sm} = \mu_0 \omega t \; \text{Im}\left[\frac{\cot(k_c t) + \csc(k_c t)}{(k_c t)}\right] \quad (b)$$

$$k_c = \omega\sqrt{\mu_0\varepsilon_0}\left[1 - j\frac{\sigma}{\omega\varepsilon_0}\right]^{1/2} \quad (c)$$

Loss on ground plane :

$$\alpha_{cg} = \frac{R_{sm}}{\pi\cdot w\cdot Z_0(w_{eff}, h, t, \varepsilon_r, f)} \cdot \left\{\tan^{-1}\left(\frac{w}{2h}\right) - \frac{h}{w}\cdot\ln\left[1 + \left(\frac{w}{2h}\right)^2\right]\right\} \quad (d). \qquad (8.4.52)$$

In the above equation, "Im" stands for imaginary part, R_{sm} is the surface impedance and k_c is the wavenumber in a conductor. In the above expression, Holloway and Kuester have taken the DC bulk conductivity of the conductor. However, equation (8.4.47) suggests frequency-dependent conductivity. Furthermore, for a thin film microstrip line (TFML), shown in Fig (8.1e), the conductivity is thickness dependent and is less than the conductivity of a bulk metal. Figure (8.16b) shows the conductor loss, up to 100 GHz, of a 55Ω TFML constructed with the gold film, where thin-film conductivity of gold is taken as 3.9e7 S/m, instead of the bulk conductivity of gold that is 4.1e7 S/m. The Holloway and Kuester model with frequency-dependent, conductivity $\sigma(f)$ follows the experimental results more closely, as compared to their model with $\sigma(0)$. Figure (8.16b) also shows the results of the commercial EM-Simulator CST, with more deviation against the experimental results [J.67].

In the above discussion, a single layer of metal for a strip conductor is considered. However, a practical microstrip line has composite layers of metals to improve its adhesion to the dielectric substrate. The fabrication process is discussed in chapter 13. A more general expression for the surface impedance, accounting for the composite layer of metals, is obtained below.

Surface Impedance of Composite Layers Conductor

Figure (8.17) shows a two-layered conductor of an infinite extent. It has uniform current distribution at the upper conducting layer of thickness t, conductivity σ_1, and permeability μ_1. The lower conductor is infinitely thick. Its conductivity and permeability are σ_2 and μ_2, respectively. The 1D diffusion equation (8.4.28) is solved for the current density in both the conducting media. The more appropriate solution involves the solution of the 2D diffusion equation that also accounts for the width of the conducting strip [J.73].

$$\text{Medium – I}: \quad J_{z1} = A\sinh(\Gamma_1 y) + B\cosh(\Gamma_1 y), 0 < y \le t \quad (a)$$

$$\text{Medium – II}: J_{z2} = C\,e^{-\Gamma_2(y-t)}, y \ge t \quad (b)$$

$$\text{where} \quad \Gamma_1 = \frac{1+j}{\delta_{s1}} = (1+j)\sqrt{\pi f \mu_1 \sigma_1} \text{ and}$$
$$\Gamma_2 = \frac{1+j}{\delta_{s2}} = (1+j)\sqrt{\pi f \mu_2 \sigma_2} \quad (c). \qquad (8.4.53)$$

In both media, the longitudinal current density obeys Ohm's law:

$$E_{zi} = \frac{J_{zi}}{\sigma_i} \quad (i = 1, 2 \text{ showing medium}). \qquad (8.4.54)$$

The tangential magnetic field components are obtained from Maxwell's equation:

$$H_{xi} = \frac{1}{j\omega\mu_i}\frac{dE_{zi}}{dy} = \frac{\sigma_i}{\Gamma_i^2}\frac{dE_{zi}}{dy} \quad (i = 1, 2). \qquad (8.4.55)$$

Equation (8.4.29) is used to get the final form of the above equations. The electric and magnetic fields in both media are obtained from the above equations:

$$\text{Medium I}: \quad E_{z1} = \frac{1}{\sigma_1}[A\sinh(\Gamma_1 y) + B\cosh(\Gamma_1 y)] \quad (a)$$
$$H_{x1} = \frac{1}{\Gamma_1}[A\cosh(\Gamma_1 y) + B\sinh(\Gamma_1 y)] \quad (b). \qquad (8.4.56)$$

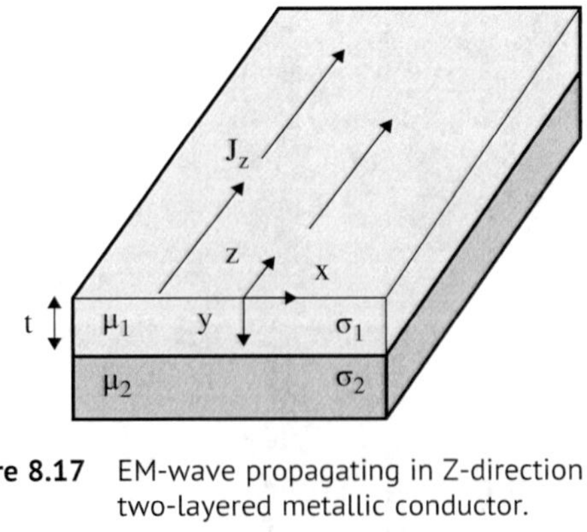

Figure 8.17 EM-wave propagating in Z-direction of the two-layered metallic conductor.

$$\text{Medium II}: \quad \begin{aligned} E_{z2} &= \frac{C}{\sigma_2} e^{-\Gamma_2 (y-t)} \quad (a) \\ H_{x2} &= \frac{C}{\Gamma_2} e^{-\Gamma_2 (y-t)} \quad (b). \end{aligned} \tag{8.4.57}$$

At the interface of both conductors, the following continuity conditions are used:

$$E_{z1} = E_{z2}, \quad H_{x1} = H_{x2}. \tag{8.4.58}$$

The above expressions provide a ratio of two constants:

$$\frac{B}{A} = -\frac{[\Gamma_1\sigma_2/\Gamma_2\sigma_1]}{[\Gamma_1\sigma_2/\Gamma_2\sigma_1]} \frac{\left[\sinh(\Gamma_1 t) + \left\{\frac{\Gamma_2\sigma_1}{\Gamma_1\sigma_2}\right\}\cosh(\Gamma_1 t)\right]}{\left[\cosh(\Gamma_1 t) + \left\{\frac{\Gamma_2\sigma_1}{\Gamma_1\sigma_2}\right\}\sinh(\Gamma_1 t)\right]} \tag{8.4.59}$$

$$\text{where,} \quad \begin{aligned} \frac{\Gamma_1}{\sigma_1} &= \frac{1+j}{\delta_{s1}\sigma_1} = (1+j)R_{s_1} \quad (a) \\ \frac{\Gamma_2}{\sigma_2} &= \frac{1+j}{\delta_{s2}\sigma_2} = (1+j)R_{s_2} \quad (b). \end{aligned} \tag{8.4.60}$$

$$\frac{B}{A} = -\frac{\left[\sinh(\Gamma_1 t) + \left\{\frac{Rs_2}{Rs_1}\right\}\cosh(\Gamma_1 t)\right]}{\left[\cosh(\Gamma_1 t) + \left\{\frac{Rs_2}{Rs_1}\right\}\sinh(\Gamma_1 t)\right]}. \tag{8.4.61}$$

At the top surface of the conductor I, i.e. at $y = 0$, the surface current density is

$$J_{sz} = -H_x|_{y=0}. \tag{8.4.62}$$

The surface impedance of the conductor I is

$$\begin{aligned} Z_s(y=0) &= \frac{E_z|_{y=0}}{J_{sz}} = -\frac{E_{z1}}{H_{x1}}\bigg|_{y=0} \\ &= -\frac{\Gamma_1}{\sigma_1}\frac{B}{A} = -(1+j)R_{s1}\frac{B}{A}. \end{aligned} \tag{8.4.63}$$

Using the above two equations, an expression is obtained for the surface impedance of conductor I having finite thickness t. The conductor I is supported by another conductor of an infinite extent.

$$\begin{aligned} &Z_s(y=0) \\ &= (1+j)R_{s1}\left[\frac{\sinh(\Gamma_1 t) + \{R_{s_2}/R_{s_1}\}\cosh(\Gamma_1 t)}{\cosh(\Gamma_1 t) + \{R_{s_2}/R_{s_1}\}\sinh(\Gamma_1 t)}\right] \quad (a) \\ &\text{where,} \quad Z_s = R_s + j\omega L_{is} \quad (b). \end{aligned} \tag{8.4.64}$$

In the above equation, R_s and L_{is} are the *surface resistance* and *internal surface inductance* of the composite conductor's system. The following special cases are considered:

- *Medium II* is a perfect conductor, i.e. $\sigma_2 \to \infty$, leading to $R_{s2} \to 0$. Equation (8.4.64) is reduced to

$$\begin{aligned} Z_s(y=0) &= (1+j)R_{s1}\tanh(\Gamma_1 t) \\ &= (1+j)R_{s1}\tanh\left[(1+j)\frac{t}{\delta_{s1}}\right] \quad (a) \\ R_s(t) &= \operatorname{Re}\{Z_s(y=0)\} = R_{s1}\tanh[1+j(t/\delta_{s1})] \quad (b). \end{aligned} \tag{8.4.65}$$

- *Medium II* is a perfect dielectric insulator, i.e. $\sigma_2 \to 0$, leading to $R_{s2} \to \infty$.

Equation (8.4.64) is reduced to

$$\begin{aligned} Z_s(y=0) &= (1+j)R_{s1}\coth\left[(1+j)\frac{t}{\delta_{s1}}\right] \quad (a) \\ R_s(t) &= \operatorname{Re}\{Z_s(y=0)\} = R_{s1}\left\{\coth\left[(1+j)\frac{t}{\delta_{s1}}\right]\right\} \quad (b). \end{aligned} \tag{8.4.66}$$

The above equation provides *Leontovich surface impedance* of a conductor of finite thickness [J.63]. This is an important case involving either a thin ground conductor over an air-medium or a thin ground conductor over the dielectric substrate. These conditions arise in a microstrip. The above expression is also used for different conductors of the strip and the ground plane. The surface resistance obtained from the above equation is

$$\begin{aligned} R_s(t) &= R_{s1}\frac{\left[\sinh\left(\frac{2t}{\delta_{s1}}\right) + \sin\left(\frac{2t}{\delta_{s1}}\right)\right]}{\left[\cosh\left(\frac{2t}{\delta_{s1}}\right) - \cos\left(\frac{2t}{\delta_{s1}}\right)\right]} \quad (a) \\ RF &= \frac{[\sinh(y) + \sin(y)]}{[\cosh(y) - \cos(y)]}, \text{where} \quad y = 2t/\delta_{s1} \quad (b). \end{aligned} \tag{8.4.67}$$

The above expression is used in equation (8.4.42) to compute the conductor loss of microstrip with conductor thickness less than, and also more than, the skin-depth. The finite conductor thickness, through the *resistance factor* (RF), modifies the surface impedance $R_{s1} = R_s$ of the infinitely thick conductor. For a minimum loss, the RF of equation (8.4.67b) must be minimized with respect to the thickness t. It gives the optimum conductor thickness for the minimum conductor loss [J.66].

Wheeler's Incremental Inductance Rule

Wheeler's *incremental inductance rule* determines the line resistance from the internal *inductance* of a conductor to compute the conductor loss of a microstrip line for the strip conductor thickness more than the skin-depth, $t \geq 1.1\,\delta_s$ [J.61, J.62, J.74, J.75]. Equation (8.4.38b) provides the

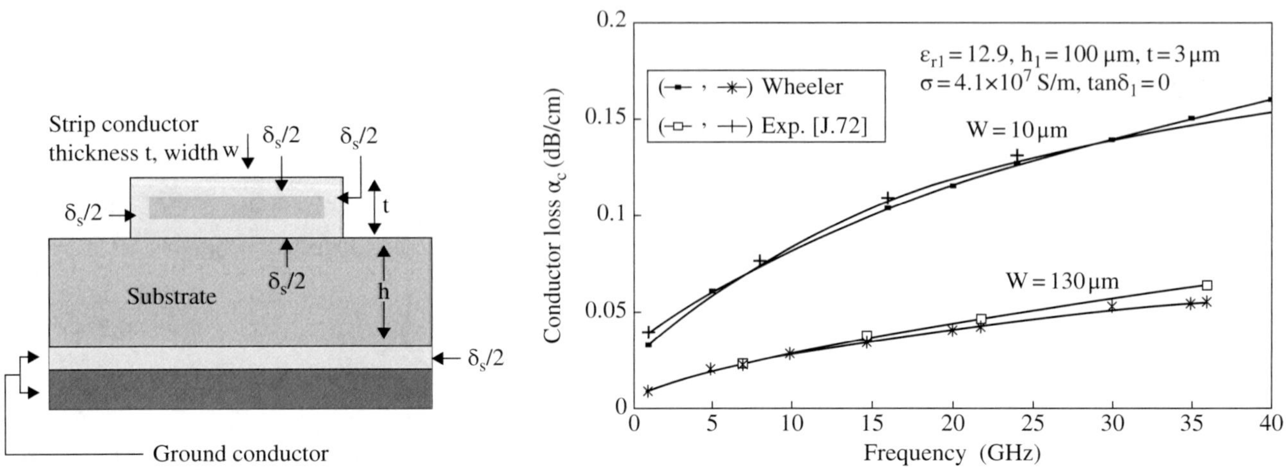

(a) $\delta_s/2$ - δ penetration of magnetic field/current around strip conductor and in the ground.

(b) Comparison of Wheeler's method against experimental results [J.75].

Figure 8.18 Application Wheeler's incremental rule to compute the conductor loss of a microstrip line. *Source:* From Verma and Bhupal [J.75].

following expression for the line resistance in terms of the internal inductance:

$$R = X_i = \frac{R_s}{w} = \omega\, L_i. \tag{8.4.68}$$

Equation (8.4.34b) shows that the internal inductance $L_i = \mu\delta_s/2$ is due to the half skin-depth ($\delta_s/2$) penetration of the magnetic field in a conductor. Figure (8.18a) shows the magnetic field penetration in a conducting strip from all directions, whereas in the ground conductor, it is in one direction. After the field penetration, the effective width and thickness of the strip conductor are reduced to $w - \delta_s$ and $t - \delta_s$, respectively. The substrate thickness is increased to the thickness $h + \delta_s$.

The dielectric substrate does not influence the line inductance. Therefore, *the internal inductance due to the field penetration is equal to the incremental change in the line inductance*, $L_i = \Delta L$. The incremental line inductance can be easily computed from the total and external line inductances. The external line inductance L_{ext} of microstrip is computed without any field penetration, i.e. from w, t, and h of a microstrip line. Whereas the total line inductance L_{tot} is computed from $w - \delta_s$, $t - \delta_s$ and $h + \delta_s$ that takes care of the field penetration in the strip and ground plane conductors. The line inductance of a microstrip line is obtained from $L = Z_0(\text{air})/c$, where c is the velocity of EM-wave in the free space. Thus, the internal inductance is

$$\begin{aligned} L_i = \Delta L &= \frac{\Delta Z_0}{c} \\ &= \frac{1}{c}[Z_0(w - \delta_s, h + \delta_s, t - \delta_s, \varepsilon_r = 1) - Z_0(w, h, t, \varepsilon_r = 1)]. \end{aligned} \tag{8.4.69}$$

The computation of characteristic impedance of a microstrip line with conductor thickness using the expressions of Hammerstad–Jensen has already been discussed in subsection (8.2.2). The line resistance is obtained from equations (8.4.68) and (8.4.69):

$$\begin{aligned} R &= \omega L_i \\ &= \frac{\omega}{c}[Z_0(w - \delta_s), h + \delta_s, t - \delta_s, \varepsilon_r = 1) - Z_0(w, h, t, \varepsilon_r = 1)] \\ &= \beta \Delta Z_0. \end{aligned} \tag{8.4.70}$$

The conductor loss for a microstrip line is computed by the following expression, obtained by substituting the above equation in equation (8.4.42):

$$\begin{aligned} \alpha_C &= \frac{\beta \Delta Z_0}{2Z_0(w, h, t, \varepsilon_r)} = \frac{\pi}{\lambda_0} \frac{\Delta Z_0}{Z_0(w, h, t, \varepsilon_r)} \quad \text{Np/length} \quad (a) \\ \alpha_C &= \frac{27.3 \Delta Z_0}{\lambda_0 Z_0(w, h, t, \varepsilon_r)} \quad \text{dB/length} \quad (b), \end{aligned} \tag{8.4.71}$$

where ΔZ_0 is the differential change in the characteristic impedance of a microstrip line, due to the penetration of the magnetic field. The above method is accurate for the conductor thickness $t \geq 1.1\,\delta_s$ against the experimental results [J.75, J.76]. Figure (8.18b) shows the accuracy of the present formulation of Wheeler's inductance rule against the experimental results. The classical formulation of Wheeler's inductance rule is not as accurate [J.61, J.62]. Moreover, the present formulation applies to a microstrip in a layered medium, discussed in chapter 15. It is also applicable to a coupled microstrip in the layered dielectric-medium [J.77]. Normally, the conductor surface is not smooth. The *uneven* conductor surface causes an increase in the conductor loss; as the surface current has to travel a longer path. The surface roughness can increase the conductor loss by 50%–60% [J.1]. The conductor loss models discussed in this section normally work up to 100 GHz. However, the working range of the models could be extended up to 1 THz, and beyond, considering the empirically obtained frequency-dependent model of the strip and ground conductors' conductivity [J.55].

8.5 Circuit Model of Lossy Microstrip Line

The characteristic impedance of the lossy microstrip is a complex quantity. Moreover, the microstrip line also shows low-frequency dispersion, say below 1 GHz, due to the field penetration in the conducting strip and ground planes with finite conductivity [J.15]. The above-discussed models ignore such important aspects. The quasi-static RLCG circuit model of the lossy microstrip, shown in Fig (8.19a), computes the complex characteristic impedance. However, it does not provide low-frequency dispersion [J.78]. The alternate computation of the dynamic RLCG parameters is discussed below.

The frequency-dependent primary line parameters R(f), L(f), C(f), G(f) can be approximately determined from the dispersion relation, frequency-dependent characteristic impedance, dielectric loss, and conductor loss models discussed in this chapter. Several models working in the different frequency ranges have been combined to get the *integrated model* of each of these parameters of a lossy microstrip working up to THz frequency ranges [J.55]. These models can be used to obtain the R(f), L(f), C(f), G(f) parameters from the following set of equations:

$$\begin{aligned}
R(f) &= 2\,Z_0\,(w_{eff}, h, t, \varepsilon_r, f)\,\alpha_C && \text{(a)}\\
L(f) &= Z_0\,(w_{eff}, h, t, \varepsilon_r, f)\sqrt{\varepsilon_{reff}(w_{eff}, h, t, \varepsilon_r, f)}/c && \text{(b)}\\
C(f) &= \sqrt{\varepsilon_{reff}(w_{eff}, h, t, \varepsilon_r, f)}/c_0 Z_0(w_{eff}, h, t, \varepsilon_r, f) && \text{(c)}\\
G(f) &= 2\alpha_d/Z_0\,(w_{eff}, h, t, \varepsilon_r, f) && \text{(d)},
\end{aligned} \tag{8.5.1}$$

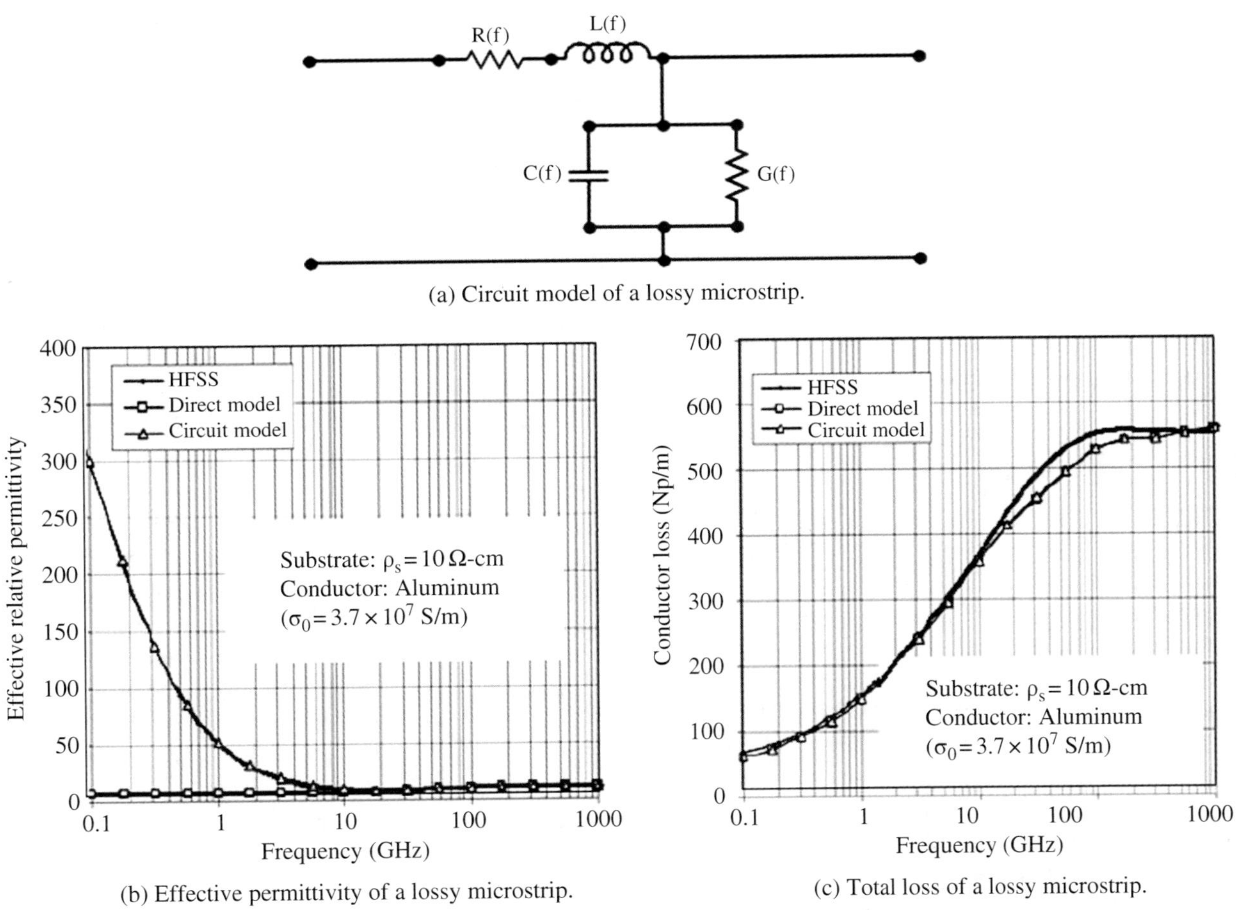

Figure 8.19 Dispersion and total loss of lossy microstrip using the circuit model. *Source:* Awasthi et al. [J.55].

where c is the velocity of the EM-wave. Using the above expression, the complex characteristic impedance and complex propagation constant of a lossy quasi-TEM microstrip are computed from the following well-known equations:

$$Z_0^*(f) = \sqrt{\frac{R(f) + j\omega L(f)}{G(f) + j\omega C(f)}} \quad (a), \quad \gamma(f) = \alpha_T(f) + j\beta(f) \quad (b)$$

$$\gamma(f) = \sqrt{\{R(f) + j\omega L(f)\}\{G(f) + j\omega C(f)\}} \quad (c), \qquad (8.5.2)$$

where $\alpha_T(f) = \alpha_d(f) + \alpha_c(f)$ is the total loss as a combination of the dielectric and conductor losses. The circuit model provides interaction between individual line parameters and improves the computation of characteristic impedance, dispersion, and losses. The dispersion and loss results are presented in Fig (8.19b and c), respectively, for a microstrip on the Si-substrate: $\varepsilon_r = 11.9$, $h = 450\,\mu m$, $t = 0.5\,\mu m$, $w/h = 0.67$. The lossy Si-substrate has resistivity $\rho_s = 10\,\Omega - cm$ and aluminum conductor has conductivity $\sigma_0 = 3.7\,10^7 S/m$ [J.55].

Figures (8.19b,c) compare the results of the circuit model against the results of the EM-Simulator, HFSS. Both results are almost identical. The strong low-frequency dispersion is noted.

References

Books

B.1 Hoffmann, R.: *Microwave Integrated Circuit Handbook*, Artech House, Boston, 1985.

B.2 Gupta, K.C.; Ramesh, G.; Bahl I.; Bhartia P.: *Microstrip Lines and Slot Lines*, 2nd Edition, Artech House, Boston, 1996.

B.3 Edward, T.C.: *Foundations for Microstrip Circuit Design*, John Wiley & Sons, New York, 1987.

B.4 Edward, T.C., Steer M.B.: *Foundations* of Interconnects and Microstrip Circuit Design, 3rd Edition, John Wiley & Sons, New York, 2000.

B.5 Fooks, E.H.: *Zakarevicius:* Microwave Engineering Using Microstrip Circuits, Prentice-Hall PTR, New Jersey, 1989.

B.6 Paolo, D.: *Franco:* Network, and Devices Using Planar Transmission Lines, CRC Press, New York, 2000.

B.7 Wadell Brian, C.: *Transmission Line Design Handbook*, Artech House, Boston, 1991.

B.8 Collin, R.E., *Foundations for Microwave Engineering*, 2nd Edition, McGraw-Hill, Inc., New York 1992.

B.9 Jordan, E.C., Balmain Keith, G. E.: *Electromagnetic Wave and Radiating System*, Prentice Hall India, New Dehi, 1989.

B.10 Gunter, K.: *Practical Microstrip Design and Applications*, Artech House, Boston, 2005.

B.11 Frey, J.; Bhasin, K.B.: *Microwave Integrated Circuits*, Artech House, Dedham, MA, 1985.

B.12 Wolff, I.: *The waveguide model for the analysis of microstrip discontinuities*, In *Numerical Techniques for Microwave and Millimeter-Wave Passive Structures*, Itoh Tatsuo, John Wiley & Sons, New York., 1989.

B.13 Rozzi, T.; Mongiardo, M.: *Open Electromagnetic Waveguides*, IEE, London, 1997.

Journals

J.1 Maloratsky, L.G.: *Reviewing the Basics of Microstrip Lines*, Microwave & RF, pp. 79–88, Mar 2000.

J.2 Schnieder, F.; Heinrich, W.: *Model of thin-film microstrip line for circuit design., IEEE Trans. Microw. Theory Tech.*, Vol. 49, No. 1, pp. 104–110, 2001.

J.3 Yamashita, E.; Ke Ren, L.; Suzuki, Y.: *Characterization method and simple design formulas of MCS lines proposed for MMIC's, IEEE Trans. Microw. Theory Tech.*, Vol. 35, No. 12, pp. 1355–1362, Dec. 1987.

J.4 Verma, A.K.; Kumar, R.: A *new Dispersion model for microstrip line, IEEE Trans. Microw. Theory Tech.*, Vol. MTT-46, No. 8, pp. 1183–1186, Aug. 1998.

J.5 Ermert, H.: *Guiding and radiation characteristics of planar waveguides, IEE J. Microwaves, Opt. & Acoust*, Vol. 3, pp. 59–62, March 1979.

J.6 Ermert, H.: *Guided modes and radiation characteristics of covered microstrip lines, Arch. Elektron. Ubertragungstechn*, Vol. 30, No. 2, pp. 65–70, 1976.

J.7 Mittra, R.; Itoh, T.: *A new technique for the analysis of the dispersion characteristics of microstrip lines, IEEE Trans. Microw. Theory Tech.*, Vol. MTT-19, pp. 47–56, Jan. 1971.

J.8 Wheeler, H.A.: *Transmission-line properties of parallel strips separated by a dielectric sheet, IEEE Trans. Microw. Theory Tech.*, Vol. MTT-13, No. 2, pp. 172–185, March 1965.

J.9 Wheeler, H.A.: *Transmission-line properties of a strip on a dielectric sheet on a plane, IEEE Trans. Microw. Theory Tech.*, Vol. MTT-25, No. 8, pp. 631–647, 1977.

J.10 Kobayashi, M.: *Important role of inflection frequency in the dispersive properties of microstrip lines, IEEE Trans. Microw. Theory Tech.*, Vol. MTT-30, No. 11, pp. 2057–2059, 1982.

J.11 Kobayashi, M.: *A dispersion formula satisfying recent requirements in microstrip CAD, IEEE Trans. Microw. Theory Tech.*, Vol. MTT-36, No. 8, pp. 1246–1250, Aug. 1986.

J.12 Hammerstad, E.; Jensen, O.: *Accurate models for microstrip computer aided design*, IEEE., MTT-S, Int. Microwave Symposium Digest, pp. 407–409, 1980.

J.13 Edwards, T.C.; Owens, R.P.: *2-18 GHz dispersion Measurements on 10-100 Ω Microstrip Lines on Sapphire, IEEE Trans. Microw. Theory Tech.*, Vol. MTT-25, pp. 172–185, March 1965.

J.14 Hammerstad, E.O.; Jensen Ø.: *Accurate models for microstrip computer-aided design*, IEEE MTT-S Int. Microwave. Symp. Dig., pp. 407–409, 1980.

J.15 Farina, M.; Rozzi, T.: *Spectral domain approach to 2D-modelling of open planar structures with thick lossy conductors, IEE Proc. -Microw. Antenna Propagat.*, Vol. 147, No. 5, pp. 321–324, Oct. 2000.

J.16 Gunsto, M.A.R.; Weale, J.R.: *Variation of microstrip impedance with strip thickness, Electron. Lett.*, Vol. 5, No. 26, pp. 697–698, Dec. 1969.

J.17 Bahl, I.J.; Garg, R.: *Simple and accurate formulas for a microstrip with finite conductor thickness, Proc. IEEE*, Vol. 65, No. 11, pp. 1611–1612, 1977.

J.18 March, S.: *Microstrip packaging. Watch the last step, Microwaves*, Vol. 20, No. 12, pp. 83–94, 1981.

J.19 Verma, A.K.; Kumar, R.: *New empirical unified dispersion model for shielded, suspended, and composite-substrate microstrip line for microwave and mm-wave applications, IEEE Trans. Microw. Theory Tech.*, Vol. MTT-46, No. 8, pp. 1187–1192, Aug. 1998.

J.20 Yamashita, E.; Atsuki, K.; Mori, T.: *Application of MIC formula to a class of integrated – optics modulator analysis, IEEE Trans. Microw. Theory Tech.*, Vol. MTT-25, pp. 146–150, 1977.

J.21 Szentkuti, B.T.: *Simple analysis of anisotropic microstrip lines by a transform method, Electron. Lett.*, Vol. 12, No. 25, pp. 672–673, 1976.

J.22 Szentkuti, B.T.: *Corrections to simple analysis of anisotropic microstrip lines by a transform method, Electron. Lett.*, Vol. 13, No. 3, pp. 92, 1977.

J.23 Kusase, S.; Terakado, R.: *Mapping of two-dimensional anisotropic regions, Proc. IEEE*, Vol. 67, No. 1, pp. 171–172, 1979.

J.24 Kobayashi, M.; Terakado, R: *New view on an anisotropic medium and its application to the transformation from anisotropic to isotropic problems, IEEE Trans. Microw. Theory Tech.*, Vol. MTT-27, No. 9. pp. 769–775, 1978.

J.25 Owens, R.P.; Aitken, J.E.; Edwards T.C.: *Quasi-static characteristics of microstrip on an anisotropic sapphire substrate, IEEE Trans. Microw. Theory Tech.*, Vol. MTT-24, pp. 499–505, Aug. 1976.

J.26 Fritsch U. and Wolff, I.: *Characterization of anisotropic substrate materials for microwave applications*, IEEE. MTT-S Digest, pp. 1131–1134, 1992.

J.27 Verma, A.K.; Awasthi, Y.K.; Singh, H.: *Equivalent isotropic relative permittivity of microstrip on multilayer anisotropic substrate, Int. J. Electron.*, Vol. 96, No. 8, pp. 865–875, Aug. 2009.

J.28 Alexopoulos, N.G.: *Integrated-circuit structures on anisotropic substrates, IEEE Trans. Microw. Theory Tech.*, Vol. MTT-33, pp. 847–881, Oct. 1985.

J.29 Zysman, G.I.; Voron, B.: *Wave propagation in microstrip transmission line*, IEEE., MTT-S, Int. Microwave Symposium Digest, pp. 3–9, May 1969.

J.30 Yamashita, E.; Atsuki, K.: *Analysis of microstrip-like transmission lines by non-uniform discretization of integral equation, IEEE Trans. Microw. Theory Tech.*, Vol. MTT-24, pp. 195–200, April 1976.

J.31 Kowalski, G.; Pregla, R.: *Dispersion characteristics of shielded microstrips with finite thickness, Arch. Elektron. Ubertragungstechn*, Vol. 25, pp. 193–196, April 1971.

J.32 Hornsby, J.S.; Gopinath, A.: *Numerical analysis of the dielectric loaded waveguide with microstrip line – finite difference methods, IEEE Trans. Microw. Theory Tech.*, Vol. MTT-17, pp. 684–690, Sept. 1969.

J.33 Zang, X.; Fang, J.; Mei, K.K.; Tiu, Y.: *Calculation of dispersion characteristics of microstrips by time-domain finite difference method, IEEE Trans. Microw. Theory Tech.*, Vol. MTT-36, pp. 263–276, Feb. 1988.

J.34 Schultz, U; Pregla, R.: *A new technique for analysis of the dispersion characteristics of planar waveguides, Arch. Elektron. Ubertragungstechn*, Vol. 34, pp. 169–173, April 1980.

J.35 Jansen, R.H.: *A novel CAD tool and concept compatible with the requirement of multilayer GaAs MMIC technology*, IEEE., MTT-S, Int. Microwave Symposium Digest, pp. 711–714, 1985.

J.36 Rautio, J.C.: *Comments on Revisiting characteristic impedance and its definition of microstrip line with a self – calibrated 3-D MOM scheme, IEEE Trans. Microw. Theory Tech.*, Vol. MTT-47, No. 1. pp. 115–119, Jan. 1999.

J.37 Jain, O.P.; Makious, V.; Chudobiak, W.J.: *Coupled mode model for dispersion in microstrip, Electron. Lett.*, Vol. 7, No. 25, pp. 405–407, 1971.

J.38 Chudobiak, W.J.; Jain, O.P.; Makious, V.: *Dispersion in microstrip, IEEE Trans. Microw. Theory Tech.*, Vol. MTT-19, No. 9, pp. 783–784, 1971.

J.39 Schneider, M.V.: *Microstrip dispersion, Proc. IEEE*, Vol. 20, pp. 144–146, Jan. 1972.

J.40 Carlin, H.J.; *A simplified circuit model for microstrip, IEEE Trans. Microw. Theory Tech.*, Vol. MTT-21, No. 8, pp. 589–591, Sept. 1973.

J.41 Nobel, O.F.; Carlin, H.J.: *Circuit properties of coupled dispersive transmission lines, IEEE Trans. Circuit Theory*, Vol. CT-20, pp. 56–65, 1973.

J.42 Bhartia P.; Pramanick, P.: *An accurate description of dispersion in microstrip, Microw. J.*,Vol. 26, No. 12, pp. 89–92, Dec. 1983.

J.43 Yamashita, E.; Atsuki, K.; Veda, T.: *An accurate dispersion formula in microstrip line for computer-aided design of* *microwave integrated circuits, IEEE Trans. Microw. Theory Tech.*, Vol. MTT-27, pp. 1036–1038, Dec. 1979.

J.44 Krischnining, M.; Jansen, R.H.: *Accurate model for effective dielectric constant with validity up to millimeter-wave frequency, Electron. Lett.*, Vol. 18, pp. 272–273, Jan. 1982.

J.45 Getsinger, W.J: *Microstrip dispersion model, IEEE Trans. Microw. Theory Tech.*, Vol. MTT-21, pp. 34–39, Jan. 1973.

J.46 Bhartia P.; Pramanick, P.: *A new microstrip dispersion model, IEEE Trans. Microw. Theory Tech.*, Vol. MTT-32, No. 10, pp. 1379–1384, Oct. 1984.

J.47 Kompa G.; Mehran, R: *Planar waveguide model for calculating microstrip components, Electron. Lett.*, Vol. 11, pp. 459–460, 1975.

J.48 Owens, R.P, *Predicted frequency dependence of microstrip characteristics impedance using planar-waveguide model, Electron. Lett.*, Vol. 12, pp. 269–270, May 1976.

J.49 Menzel, W.; Wolff, I.: *A method for calculating the frequency-dependent properties of microstrip discontinuities, IEEE Trans. Microw. Theory Tech.*, Vol. MTT-25, No.2, pp. 107–112, Feb. 1977.

J.50 Zhu, L.; Wu, K.: *Revisiting characteristic impedance and its definition of a microstrip line with a self – calibrated 3-D MOM scheme, IEEE Microw. Guid. Wave Lett.*, Vol. 4, No. 2, pp. 87–89, Feb. 1998.

J.51 Bianco, B; Panini, L.; Parodi, M.; Ridella, S.: *Some considerations about the frequency dependence of the characteristic impedance of uniform microstrip, IEEE Trans. Microw. Theory Tech.*, Vol. MTT-26, No. 3, pp. 182–185, Mar. 1978.

J.52 Knorr, J.B.; Tufekcioglu, A.: *Spectral domain calculation of microstrip characteristic impedance, IEEE Trans. Microw. Theory Tech.*, Vol. MTT-23, pp. 725–728, Sept. 1975.

J.53 Williams, D. F.; Alpert, B.K.; Arz, U.; Walker, D.K.; Grabinski, H.: *Causal characteristic impedance of planar transmission lines, IEEE Trans. Adv. Packag.*, Vol. 26, No. 2, pp. 165–171, May 2003.

J.54 Krischnining, M.; Jansen, R.H: *Argument and an accurate model for the power-current formulation of microstrip characteristic impedance, Aeue. Arciv Fuer Elecktronik und Uebertrangungstechik*, Vo. 37, No. 3–4, pp.108–112, 1983.

J.55 Awasthi, Y.K; Singh, H.; Sharma, M.; Kumari, S.; Verma, A.K.: *Computer-aided design-based circuit model of microstrip line for terahertz interconnects technology, J. Eng, (IET Pub.)*, doi:10.1049/joe.2017.0078, Vol. 2017, No. 9, pp. 512–526, 2017.

J.56 Bansal, R.; Awasthi, Y.K.; Singh, P.; Verma, A.K.: *Accurate dispersion model for microstrip line up to terahertz frequency range*, Proc. SPIE 8760, International Conference on Communication and Electronics System Design, 87601A (28 January 2013); doi:10.1117/12.2012151.

J.57 Sadiku Matthew N.O.; Musa, M.S.: *Comparison of dispersion formula for microstrip lines*, IEEE, 2004.

J.58 Simpson, T.L; Tseng, B.: *Dielectric loss in microstrip line, IEEE Trans. Microw. Theory Tech.*, Vol. MTT-24, No. 2, pp. 106–108, Feb. 1976.

J.59 Schneider, M.V.: *Dielectric loss in integrated microwave circuits, Bell Syst. Tech. J.*, Vol. 48, pp. 35–332, Sept. 1969.

J.60 Vakanas, L.P. and Cangellaris, A.C.: *A parametric study of the attenuation constant of microstrip lines, IEEE Trans. Microw. Theory Tech.*, Vol. MTT-38, No. 8, pp. 1136–1139, Aug. 1990.

J.61 Pucel, R.A.; Masse, D.J; Hartwig, C.P.: *Losses in microstrip, IEEE Trans. Microw. Theory Tech.*, Vol. MTT-16, pp. 342–350, 1968.

J.62 Pucel, R.A.; Masse, D.J; Hartwig, C.P.: *Correction to - Losses in microstrip, IEEE Trans. Microw. Theory Tech.*, Vol. MTT-16, pp. 1064, 1968.

J.63 Holloway, C.L.; Hufford, G.A.: *Internal inductance and conductor loss associated with the ground plane of a microstrip line, IEEE Trans. Electromagn. Compat.*, Vol. 39, No. 2, pp. 73–78. May 1997.

J.64 Rautio, J.C.: *An investigation of microstrip conductor loss, IEEE Microw. Mag.*, Vol. 1, No. 4, pp. 60–67, Dec. 2000.

J.65 Rautio, J.C. and Demir, V.: *Microstrip conductor loss models for electromagnetic analysis, IEEE Trans. Microw. Theory Tech.*, Vol. MTT-51, No. 3, pp. 915–921, Mar. 2003.

J.66 Konno, M.: *Conductor loss in thin-film transmission lines, Electron. Commun. Jpn. Pt. 2*, Vol. 82, No. 10, pp. 83–99, 1999.

J.67 Bansal, R.; Singh, P.; Verma, A.K., *Experimental verification of conductor loss model for thin film microstrip line*, 3rd Int. Conf. Advanced Computing & Comm. Tech., 2013.

J.68 Liou, J.C.; Lau, K.M.: *Analysis of slow-wave transmission lines on a multi-layered semiconductor structures including conductor loss, IEEE Trans. Microw. Theory Tech.*, Vol. MTT-41, No. 5, pp. 824–829, May 1993.

J.69 Holloway, C.L.; Kuester, E.F.: *Edge shape effects and quasi-closed form expressions for the conductor loss of microstrip lines, Radio Sci.*, Vol. 29, No. 3, pp. 539–559, May-June 1994.

J.70 Holloway, C.L.; Kuester, E.F.: *A quasi-closed form expression for the conductor loss of cpw lines, with an investigation of edge shape effects, IEEE Trans. Microw. Theory Tech.*, Vol. MTT-43, No. 12, pp. 2695–2701, Dec. 1995.

J.71 Majumdar, P.; Verma, A.K.: *Accurate CAD model of stopping distance to compute conductor loss of CPW, Int. J. Electron. Commun. (AEU)*, Vol. 64, pp. 1157–1166, 2010.

J.72 Booth, J.C.; Holloway, C.L.: *Conductor loss in superconducting planar structures: calculations and measurements, IEEE Trans. Microw. Theory Tech.*, Vol. MTT-47, No. 6, pp. 769–774, June 1999.

J.73 Faraji-Dana, R.; Chow, Y.L.: *The current distribution and AC resistance of a microstrip structure, IEEE Trans. Microw. Theory Tech.*, Vol. MTT-38, No. 9, pp. 1268–1277, Sept. 1990.

J.74 Wheeler H.A.: *Formula for the skin effect, Proc. IRE*, Vol. 30, No. 9, pp. 412–424, Sept. 1942.

J.75 Verma, A.K.; Bhupal, A.: *Conductor loss of multilayer microstrip line using the SLR formulation, Microw. Opt. Technol. Lett.*, Vol. 19, No. 1, pp. 20–24, Sept. 1998.

J.76 Goldfarb, M.E.; PLatzker, A.: Losses in GaAs microstrip, *IEEE Trans. Microw. Theory Tech.*, Vol. MTT-38, pp. 1957–1963, 1990.

J.77 Verma, A.K.; Nasimuddin: *Determination of conductor loss of multilayer-coupled microstrip lines for CAD application, Microw. Opt. Technol. Lett.*, Vol. 38, No. 5, pp. 409–415, Sept. 2003.

J.78 Verma, A.K.; Nasimuddin: *Quasi-static RLCG parameters of lossy microstrip lines for CAD application, Microw. Opt. Technol. Lett.*, Vol. 28, No. 3, pp. 209–212, Feb. 2001.

9

Coplanar Waveguide and Coplanar Stripline

Basic Characteristics

Introduction

The coplanar waveguide (CPW), coplanar stripline (CPS), and the slot line are *uniplanar* transmission lines with all conductors, the signal and ground conductors, on one side of the substrate [J.1]. A microstrip requires via-holes for the shunt connections of the components. It is difficult to drill a semiconductor or a ceramic substrate. The vias also add parasitic inductance to the FET or HEMT devices. A thin semiconducting substrate, down to 100 μm or less, is used to get a suitable microstrip line width, even for a 50 Ω line. The backside of a wafer requires thick conductor plating to provide mechanical strength to it. Furthermore, the line characteristics of a microstrip are influenced by the variation in thickness of a substrate. However, the microstrip-based circuit has a common ground to all components and devices.

The CPW provides both the shunt and series connections of the components, without making any via-hole in the substrate. However, the bond-wires are used in a CPW to connect two ground planes. These are inconvenient in an MMIC fabrication. The CPW supports the elliptically or circularly polarized magnetic field in the slot region that is needed to develop the nonreciprocal magnetic devices.

The CPS does not need the bonding wire. Also, the CPS takes much less space, as compared to the CPW. The relatively large range of characteristic impedance can be obtained with the CPS. Both the CPS and CPW are less dispersive as compared to a microstrip line. These lines are also less sensitive to the substrate thickness and no backside processing of the substrate is required. However, the main drawback of a CPS is the lack of shielding that causes stray coupling to other lines. This drawback could be avoided by adding the coplanar ground planes on both sides of the CPS line [J.2].

Objectives

- *To present a general description of the uniplanar transmission lines.*
- *To review the conformal mapping method as applicable to the CPW and CPS.*
- *To get analytical expressions of several CPW and CPS structures using the conformal mapping method.*
- *To understand the nature of modes and present dispersion relations for the CPW and CPS.*
- *To model dielectric and conductor losses of the CPW and CPS.*
- *To present synthesis expressions for the CPW and CPS.*

9.1 General Description

The CPW and CPS have been proposed by C.P. Wen [J.1]. S. B. Cohn has proposed the slot line that is discussed in chapter 10. Figure (9.1a–c) shows these *uniplanar* transmission lines, both in the 3-D and in the cross-sectional view. The ground conductors (G) of a CPW, shown in Fig (9.1a), are infinite in lateral dimension. Normally, the central strip (S) is symmetrically placed with respect to the pair of ground conductors. The slot width (W) separates the central strip conductor (S) from the ground conductor (G). The signal conductor can also be placed asymmetrically, giving the asymmetrical CPW (ACPW). Figure (9.1b) shows the CPS. It is a *complementary line* to the CPW in the sense that conductor strips and dielectric gaps of the CPW are interchanged by the dielectric gaps and conductor strips in a CPS. In the case of a slot line, shown in Fig (9.1c), both the ground conductors have infinite lateral dimensions. Once their widths are finite, the slot line is reduced to a CPS. Similarly, the slot line could be viewed as a limiting case of a CPW with zero central strip width.

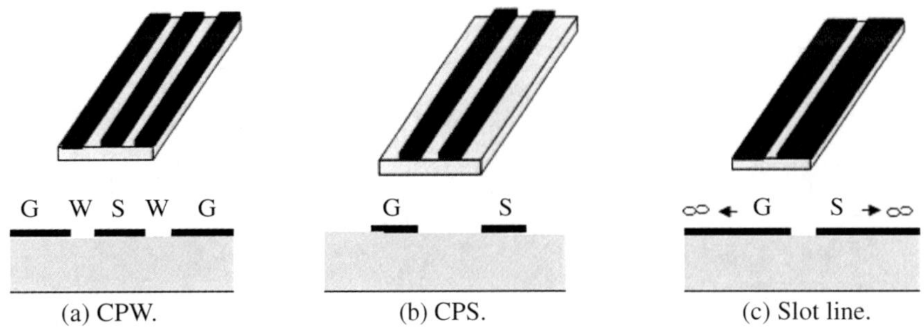

Figure 9.1 Some uniplanar transmission lines with cross-sectional views.

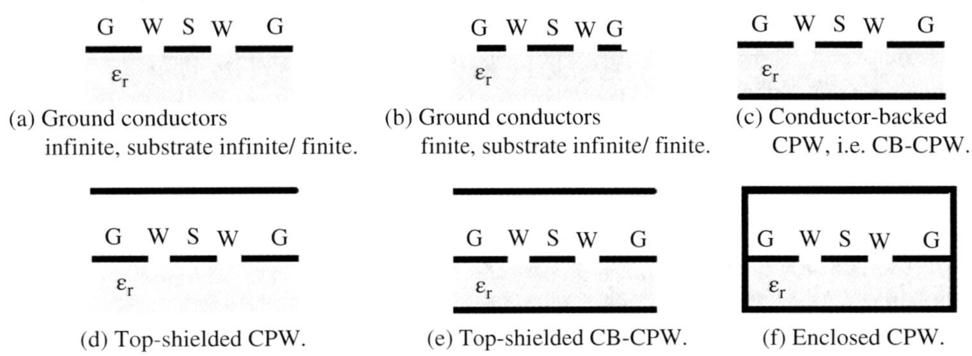

Figure 9.2 Some CPW structures.

There are several forms of the standard CPW structure, such as infinite extent CPW, finite substrate CPW, finite ground plane CPW, top-shielded CPW, conductor-backed CPW (CB-CPW), enclosed CPW, etc. Figure (9.2a–f) shows these structures. A CPW structure supports the quasi-TEM mode that has no cutoff frequency for the fundamental quasi-TEM mode. The CPW structures are analyzed by the static methods, like the conformal mapping method [B.1, B.2, B.3, J.1, J.3–J.8], the variational method [J.9], etc. They are also analyzed by the full-wave methods [B.1, B.4, B.5]. However, the static conformal mapping method is very popular, as it provides simple CAD-oriented closed-form expressions for the characteristic impedance, propagation constant, and losses. Under the usual environment of the MIC, the static expressions can be used up to 20 GHz. This chapter presents the conformal mapping method for the analysis of the CPW and CPS structures. The *variational method* and the *spectral domain method* are discussed in chapters 14 and 16, respectively.

9.2 Fundamentals of Conformal Mapping Method

This section briefly reviews the basics of the complex variable that leads to the conformal mapping method, especially with the help of *Schwarz–Christoffel (SC)-transformation*. Once applied to the CPW and CPS structures, the SC-transformation requires the *elliptic sine function* that is less familiar. An elementary discussion on the elliptic sine function is presented. Several mathematical books cover this topic [B.2, B.3, B.6–B.9].

9.2.1 Complex Variable

A *dependent* complex variable $w = u + jv$ is a function of another *independent* complex variable $z = x + jy$. Figure (9.3a,b) jointly defines these functions using two Argand diagrams.

$$w = F(z) = u(x, y) + jv(x, y). \tag{9.2.1}$$

The complex variable z is defined in the complex z-plane, also called (x–y plane), shown in Fig (9.3a) and the w is defined in the complex w-plane, also called (u–v plane), shown in Fig (9.3b). In the case of a *single-valued* complex variable function, for each value of a complex number, say P(x, y) in the z-plane, there is only one corresponding complex number, say Q(u, v) in the w-plane. The variables u and v, i.e. the real and imaginary parts of the complex variable w, are functions of independent variables x and y. If a curve C is traced by the point P(x, y) in the z-plane, then due to the functional dependence, the point Q(u,v) also traces another curve C' in the w-plane. The curve C' is called the

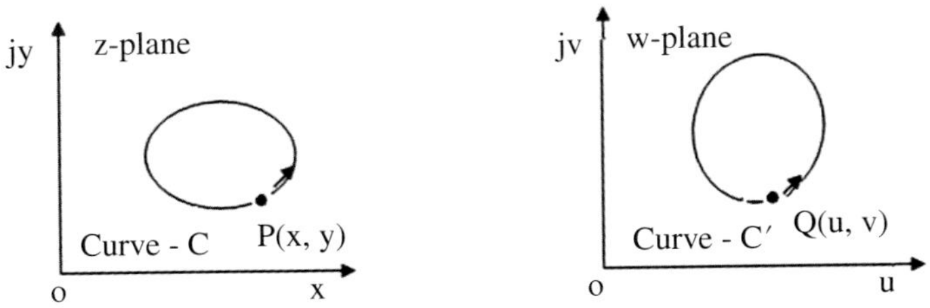

(a) Z, i.e. (x, y)-plane. (b) W, i.e. (u,v)-plane.

Figure 9.3 Mapping of point P from Z-plane to W-plane as point Q.

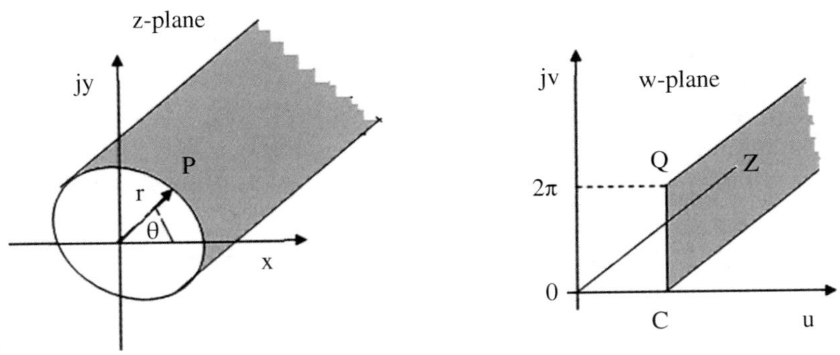

(a) Circular cylinder in z-plane. (b) Parallel plane in w-plane.

Figure 9.4 Transformation of the circular cylinder to a parallel plane.

mapping or the *image* of the curve C. The function F(z) is called the *mapping function*. In the process of mapping, the geometrical shape can get changed. Let us see it through an example.

Figure (9.4) shows that the mapping function $w(z) = \ln(z)$ transforms a circular cylinder located in the z-plane to a parallel plate in the w-plane. The independent variable z can be written in the polar coordinate as $z = re^{j\theta}$. It describes a circular cross-section of radius r, while *argument* θ changes from 0 to 2π. The mapping function in the polar coordinate is

$$w(z) = \ln(z) = \ln(r) + j(\theta + 2\pi k), k = 0, 1, 2, 3...$$

Therefore,

$$\text{Re}(w) = u(x, y) = \ln(r) = \frac{1}{2}\ln\left(x^2 + y^2\right) = C \quad \text{(a)}$$
$$\text{Im}(w) = v(x, y) = \theta + 2\pi k = \tan^{-1}\frac{y}{x} + 2\pi k \quad \text{(b)}, \tag{9.2.2}$$

where k = 0, 1, 2... For the principal value, k = 0.

Let us interpret the above transformation (*mapping*) graphically, as shown in Fig (9.4). The point P(x, y) in the z-plane traces a circle of radius r, while *argument* θ changes from 0 to 2π radians. The point Q(u, v), in the w-plane, is located at distance $u = \ln(r) = C$ moves along a line CQ parallel to the v-axis from θ = 0 to 2π. In this process, a circle in the z-plane is transformed into a straight line in the w-plane, i.e. *the circular cylinder is transformed into a parallel plate* as there is no change in the z-direction after the conformal mapping.

9.2.2 Analytic Function

The above discussion is related to the mapping of points, lines, and the enclosed area. However, attention is to be paid to the *transformation of an angle* between two lines, or between two curves. It involves the derivative of a complex function, i.e. dw/dz. The uniqueness of the derivative is satisfied only by a special class of the complex functions, known as the *analytic function*. An analytic function satisfies the following Cauchy–Riemann conditions [B.8, B.10]:

$$\frac{\partial u}{\partial x} = \frac{\partial v}{\partial y} \quad \text{(a)}$$
$$\frac{\partial u}{\partial y} = -\frac{\partial v}{\partial x} \quad \text{(b)}. \tag{9.2.3}$$

The following Laplace's equations could be obtained from the above equations:

$$\frac{\partial^2 u}{\partial x^2} + \frac{\partial^2 u}{\partial y^2} = 0 \quad \text{(a)}$$
$$\frac{\partial^2 v}{\partial x^2} + \frac{\partial^2 v}{\partial y^2} = 0 \quad \text{(b)}. \tag{9.2.4}$$

It shows that an electrostatic potential function is adequately described by the analytic function. The functions u(x, y) and v(x, y) are known as the *conjugate functions*. If u (x, y) represents the *electrostatic potential*, then v(x, y) would correspond to the *electrostatic flux*. The reverse representation is also applicable.

In summary, the complex analytic function, w = F(z), involves two coordinate systems – the z-plane, i.e. [*(x-y)-system*] and the w-plane, i.e. *(u-v)-system*. The analytic function can be used for the mapping of points, lines, and enclosed regions from the (x-y)-system to the (u-v)-coordinate system with a unique derivative of the function, i.e. dw/dz is independent of direction. The change in the angle between two curves, located in the (x-y)-system, is further examined for their transformation in the (u-v)-system, with the help of an analytic function.

Figure (9.5a) shows that a point P in the z-plane moving along the curve C from z_0 to $z_0 + \Delta z$. Figure (9.5b) shows the corresponding movement of the point Q, from the location w_0 to $w_0 + \Delta w$, in the w-plane along the path C'. Let the curves be described by the parameter t. The path C of the curve in the z-plane is

$$z = z(t) \ \text{i.e.} \ \ x = x(t), \ y = y(t). \tag{9.2.5}$$

The path C' of the curve in the w-plane is

$$w = w(t) \ \text{i.e.} \ \ u = u(t), \ v = v(t). \tag{9.2.6}$$

The derivative dz/dt is tangent to the curve C at z_0 in the z-plane. The derivative dw/dt is tangent to the curve C' at w_o in the w-plane. Two derivatives are related as follows:

$$\frac{dw}{dt} = \frac{dw}{dz}\frac{dz}{dt} = F'(z)\frac{dz}{dt}. \tag{9.2.7}$$

The function F(z) is analytic; therefore, its derivative at z_0 is unique:

$$\frac{dw}{dt}\Big|_{w = w_0} = F'(z_0)\frac{dz}{dt}\Big|_{z = z_0}. \tag{9.2.8}$$

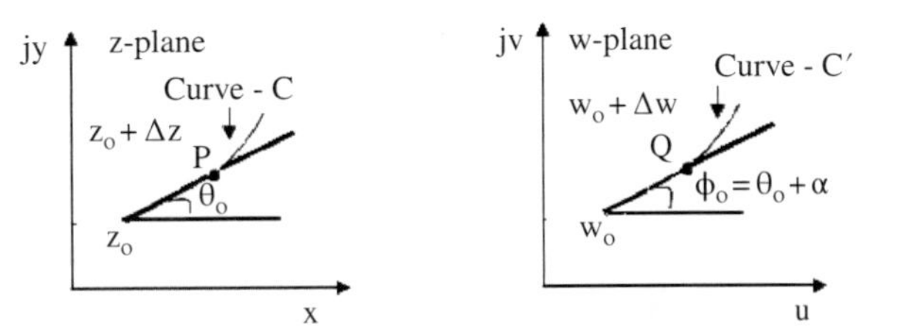

(a) Tangent at P in Z-plane. (b) Rotation of tangent at Q in W-plane.

Figure 9.5 Rotation of curve during conformal transformation.

The complex derivatives could be expressed in the polar forms,

$$\frac{dw}{dt}\Big|_{w = w_0} = \rho_0 e^{j\phi_0}, \quad \frac{dz}{dt}\Big|_{z = z_0} = r_0 e^{j\theta_0} \\ \text{and } F'(z_0) = Re^{j\alpha}, \tag{9.2.9}$$

where r_0, ρ_0; θ_0, and ϕ_0 correspond to z_0P and w_0Q. The R and α are related to dw/dz. The substitution of the above equation (9.2.9) in equation (9.2.8) provides $\rho_0 e^{j\phi_0} = r_0 \, R \, e^{j(\theta_0 + \alpha)}$. On separating the real and imaginary parts, the following expressions are obtained:

$$\phi_0 = (\theta_0 + \alpha) = \theta_0 + \arg[F'(z_0)] \quad (a) \\ \rho_0 = r_0 R \quad (b). \tag{9.2.10}$$

The above equations show that on *transformation (mapping), the curve in the z-plane is rotated by an angle α in the w-plane*. The angle of rotation α is equal to the argument of the derivative dw/dz at z_0. *The magnitude of the derivative dz/dt, i.e. the length z_0P in the z-plane, is modified by the factor R on mapping*. Figure (9.5b) demonstrates the rotation of the curve in the w-plane. If two curves in the z-plane intersect at z_0, both curves rotate by the angle α, as both curves have identical $\frac{dw}{dt}\Big|_{z = z_0}$. Therefore, the angle between them remains unchanged on their mapping from the z-plane to the w-plane. As the angle between two curves in the w-plane confirms the angle in the z-plane on the transformation (mapping), such geometrical transformation using the analytic function is called the *conformal mapping*.

9.2.3 Properties of Conformal Transformation

Some more properties of the conformal mapping transformations are summarized below. A reader can consult references for their proofs [B.3, B.7, B.11, B.12].

Under the conformal mapping transformation, from the z-plane to the w-plane, the angle between two intersecting curves remains unchanged. The z-plane, i.e. the (x,y) coordinate system, is an *orthogonal* one. The (u,v) coordinate system forming the w-plane for an analytical function is also an orthogonal system. This property can be demonstrated using Cauchy–Riemann conditions.

Figure (9.6a) shows two orthogonal surfaces, u(x, y) = constant and v(x, y) = constant, in the z-plane. As the plane, u(x,y) = constant and plane v(x,y) = constant are described by their normal, is obtained from grad (u) = ∇u and grad (v) = ∇v, respectively, the following expression is obtained:

$$\nabla u \cdot \nabla v = (\hat{x}u_x + \hat{y}u_y)\cdot(\hat{x}v_x + \hat{y}v_y) = u_x v_x + u_y v_y. \tag{9.2.11}$$

The Cauchy–Riemann condition (9.2.3) in the above equation provides

$$\nabla u \cdot \nabla v = 0. \tag{9.2.12}$$

The above expression shows that the normal of surfaces u = constant and v = constant are orthogonal to each other. Therefore, the (u–v) coordinate system forms an orthogonal coordinate system. Figure (9.6b) shows that u = constant and v = constant curves, in the z-plane, are transformed to the parallel planes in the (u − v) coordinate system.

In the z-plane, the constant u and constant v curves could have any geometrical form such as circle, parabola, hyperbola, ellipse, straight line, etc. All of them can be conformally transformed to the parallel lines in the (u − v) coordinate system, on using a suitable analytic mapping function. Thus, *if the system of curved surfaces, in the z-plane, are equipotential surfaces, then they are transformed to a parallel-plate capacitor in the w-plane*, where capacitance can be easily computed.

Normally, the conformal mapping process is used to solve the electrostatic problems. The *electrostatic potential* (ϕ) is expressed through the variable u; and the *electric flux* (ψ), corresponding to the charges on a conductor, is expressed by the variable v. The variables v and u can be interchanged for the description of the potential and electric flux. Laplace's equation remains invariant on the transformation from the z-plane to the w-plane. Further, the electrostatic energy in both the z-plane and w-plane remains unchanged. Therefore, under the conformal transformation of the arrangement of conductor and enclosed space, the capacitance remains invariant in both the z-plane and the w-plane. *In conclusion, we state that in the process of the conformal mapping, only the geometry of the arrangement of conductors is changed to an analytically convenient geometry, while the capacitance remains invariant in both the z- and w-planes.* It helps to evaluate the line capacitances of several complicated transmission lines.

The conformal mapping is just a *geometrical transformation process*. It is not a *physical transformation process*. The physical properties and composition of the enclosed space, such as the electrical (permittivity, conductivity), magnetic (permeability), mechanical properties, etc., remain invariant under the conformal mapping transformation. The physical nature of the boundaries, such as the *electric wall* (EW) and *magnetic wall* (MW), also remain invariant under the conformal mapping transformation from the z-plane to the w-plane.

An application of the conformal mapping is demonstrated below, by computing the line capacitance of a coaxial line of inner conductor radius r = a and radius of outer conductor r = b. Figure (9.7a) shows the coaxial line in the z-plane.

Let us take the following mapping function that transforms the coaxial line to a parallel plate line, shown in Fig (9.7b):

$$w(z) = \ln(z) = u + jv. \tag{9.2.13}$$

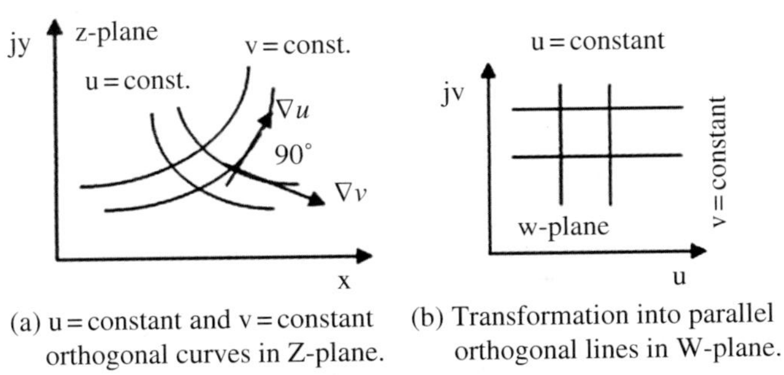

(a) u = constant and v = constant orthogonal curves in Z-plane. (b) Transformation into parallel orthogonal lines in W-plane.

Figure 9.6 Transformation of curves into parallel lines.

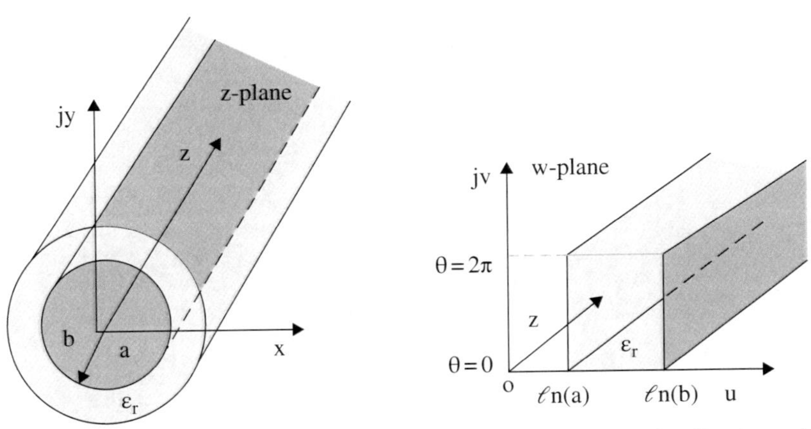

(a) Coaxial line in z-plane. (b) Transformed parallel plate line in w-plane.

Figure 9.7 Conformal mapping of the coaxial line. Dark surfaces are conductors.

The complex variable z of the above equation, in the polar coordinate, describes a circle in the (x–y)-plane:

$$z = x + jy = re^{j\theta}, \quad r^2 = x^2 + y^2$$
$$\theta = \tan^{-1}\frac{y}{x}. \tag{9.2.14}$$

The transformation function ln(z) in the polar coordinate is

$$w(z) = \ln\left(re^{j\theta}\right) = \ln(r) + j\theta. \tag{9.2.15}$$

Thus, the transformed variables in the (u–v)-plane are

$$u = \ln(r) \quad (a)$$
$$v = \theta \quad (b). \tag{9.2.16}$$

The u = constant and v = constant planes (lines) in the $(u - v)$-system form a parallel plate transmission line. The r = constant, i.e. r = a, b, are the circular conducting surfaces. Therefore, $u_1 = \ln(a)$ and $u_2 = \ln(b)$ are also the conducting surfaces in the (u–v)-plane. In the z-plane, $\theta = 0,\ 2\pi$ forms two circles with the dielectric medium between them. Therefore, in the (u–v)-plane, $v_1 = \theta = 0$ and $v_2 = \theta = 2\pi$; further $u = \ell n(a)$ and $\ell n(b)$ enclose dielectric medium between two conducting parallel planes. Figure (9.7b) shows that the width of the parallel plate conductor in the w-plane is $w = 2\pi$ and the separation of the parallel plates is $h = \ln(b) - \ln(a)$. Thus, the line capacitance *p.u.l.* of the original coaxial line obtained from the transformed parallel plate transmission line is

$$C = \frac{2\pi\varepsilon_0\varepsilon_r}{\ln(b) - \ln(a)} = \frac{2\pi\varepsilon_0\varepsilon_r}{\ln(b/a)}. \tag{9.2.17}$$

9.2.4 Schwarz–Christoffel (SC) Transformation

Figure (9.8a) shows an arrangement of the *line segments*, $\widehat{x_1x_2}$, $\widehat{x_2 0}$, $\widehat{0, x_3 \ldots, x_N}$, etc., along the real axis in the z-plane. The SC-transformation transforms them in the form of a *polygon* in the w-plane, shown in Fig (9.8b) [B.3, B.6, B.12]. In the process, the upper half of the open space in the z-plane is also transformed into an enclosed space within the polygon in the w-plane, of course without changing the relative permittivity of the space. The opposite can be carried out with the help of the SC-transformation, i.e. a polygon with enclosed space can be transformed into the line segments with open half-space. Figure (9.8) illustrates the SC-transformation.

The SC-transformation is written in the form of the derivative of the complex function w, as follows:

$$F'(w) = \frac{dw}{dz} = A(z - x_1)^{-k_1}(z - x_2)^{-k_2}\ldots(z - x_n)^{-k_N}$$
$$= A\prod_{i=1}^{N}(z - x_i)^{-k_i}. \tag{9.2.18}$$

The points $x_1 \langle x_2 \langle x_3 \ldots \langle x_N$, forming line segments in the z-plane, are located on the x-axis. We move on the x-axis from the *left to right* in such a manner that points $w_1, w_2, w_3, \ldots, w_N$ create an *anticlockwise* polygon in the w-plane. In the above equation, z is a real variable x, as $y = 0$. The constant A is a complex quantity. Its magnitude determines the size and its angle (argument), i.e. $\angle A$, determines the orientation of a polygon in the w-plane. The parameters $k_1, k_2, \ldots, k_N$ are determined by the exterior angles $\theta_1, \theta_2, \ldots, \theta_N$ of the polygon. The angle of the SC-transformation is

$$\angle\Gamma_0 = \angle\frac{dw}{dz} = \angle F'(Z) = \angle A - \sum_{i=1}^{N}[k_i \angle z - x_i], \tag{9.2.19}$$

where the angle $\sum_{i=1}^{N}[k_i \angle z - x_i]$ is the total external angle of the polygon. The angle $\angle z - x_i$ along the line acquires the value π or 0, depending on the position of z(=x) with respect to points x_i. Figure (9.9a,b) illustrates the process. Figure (9.9a) shows that the variable x crossing a fixed point x_i from the left to right undergoes the change of angle by π radian. Figure (9.9b) shows the movement of the variable x after crossing the point x_i does not involve any change in angle.

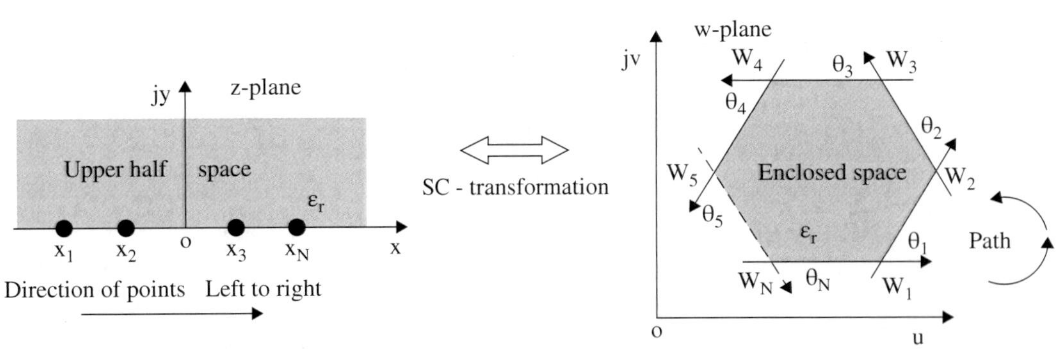

Figure 9.8 Formation of a polygon in w-plane by transforming collinear points in z-plane.

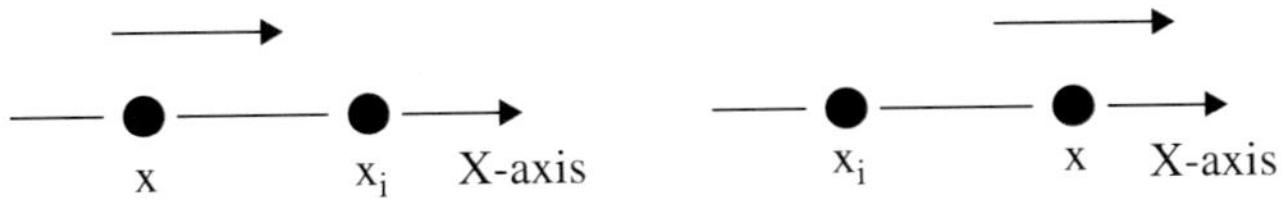

(a) $x < x_i$ (b) $x > x_i$

Figure 9.9 Angle change while crossing a fixed point.

$$\angle(x - x_i) = \pi, \text{ for } x \langle x_i \quad ; \quad (i = 1, 2, \ldots N) \quad (a)$$
$$\angle(x - x_i) = 0, \text{ for } x \rangle x_i \quad ; \quad (i = 1, 2, \ldots N) \quad (b). \tag{9.2.20}$$

The crossing points $x_1, x_2, x_3, \ldots, x_N$ in the z-plane appear as the vertices of a polygon in the w-plane. The external angle at each vertex of the polygon is given by

$$\theta_i = k_i\pi, \quad \Rightarrow \quad k_i = \frac{\theta_i}{\pi}; \quad (i = 1, 2, \ldots N). \tag{9.2.21}$$

To determine the parameter k_i, *we assume the shape of the polygon* in which line segments of the z-plane is to be transformed. The following relation holds for the interior angles $\alpha_1, \alpha_2, \ldots, \alpha_N$ of a polygon:

$$\alpha_i + \theta_i = \pi, \quad \alpha_i + k_i\pi = \pi$$
$$\Rightarrow k_i = \frac{\pi - \alpha_i}{\pi}, \quad (i = 1, 2, \ldots N). \tag{9.2.22}$$

For the interior angle of the SC-transformation, equation (9.2.18) is modified into

$$F'(w) = \frac{dw}{dz} = A(z - x_1)^{\beta_1}(z - x_2)^{\beta_2} \ldots (z - x_n)^{\beta_N}$$
$$\text{where}, \beta_1 = \left(\frac{\alpha_1}{\pi} - 1\right), \beta_2 = \left(\frac{\alpha_2}{\pi} - 1\right), \beta_N = \left(\frac{\alpha_N}{\pi} - 1\right). \tag{9.2.23}$$

Thus, the SC-transformation is available in two forms – using the external angle or using the internal angle of a polygon. The sum of external angles of any polygon is

$$\sum_{i=1}^{N} k_i\pi = 2\pi \Rightarrow \sum_{i=1}^{N} k_i = 2. \tag{9.2.24}$$

The above condition imposed on k_i can be used as a check of its correct determination for a polygon.

9.2.5 Elliptic Sine Function

The analytic functions involved in the SC-transformation are generally expressed with the help of the definite integrals. Sometimes these integrals are not solved in terms of the trigonometric and algebraic functions. The special kind of function, known as the *elliptic sine function,* is used for such cases [B.3, B.10, B.13].

1st Form of the 1st Kind of the Elliptic Integral

The following definite integral of the variable t is known as the 1st kind of the elliptic integral:

$$z = \int_0^w \frac{dt}{\sqrt{(1 - t^2)(1 - k^2t^2)}}, \quad \text{modulus} \quad \begin{matrix} |k| \le 1 \\ |w| \le 1, \end{matrix} \tag{9.2.25}$$

where the upper limit w is a real quantity with $0 \le t \le w$ and k is a *parameter (modulus)* such that $|k| \le 1$. Further, the quantity w is located on the real axis Re(t) of the t-plane such that $|w| \le 1$. The value z of the above elliptic integral, for the known w and k, could be obtained using the numerical methods. The above elliptic integral also defines the inverse sine function as a special case:

$$\text{For } k = 0, \quad z = \sin^{-1}(w) = \int_0^w \frac{dt}{\sqrt{(1 - t^2)}}. \tag{9.2.26}$$

The above function z is the *sine inverse* function, $\sin^{-1}(w)$ with $|w| \le 1$. At present, the sine inverse function is defined with the help of a definite integral. Normally, $\sin^{-1}(w)$ is defined through geometrical means. For $k \neq 0$, the integral given by equation (9.2.25) defines the *elliptical sine inverse function* that is a generalized form of the sine inverse function. It also helps to get the *elliptic sine function.*

$$\text{Elliptic sine inverse function}: \quad z = sn^{-1}(w, k) = \int_0^w \frac{dt}{\sqrt{(1 - t^2)(1 - k^2t^2)}}, \quad \begin{matrix} |k| \le 1 \\ |w| \le 1 \end{matrix} \quad (a)$$
$$\text{Elliptic sine function}: \quad sn(z, k) = (w, k) \quad (b). \tag{9.2.27}$$

$$\text{For } k = 0, \quad \text{inverse sine function},$$
$$sn^{-1}(w, 0) = \sin^{-1}(w) = z \quad (a)$$
$$\text{For } k \neq 0 \text{ elliptic sine function}, \quad sn(z, k) = (w, k) \quad (b)$$
$$\text{For } k = 0, \quad \text{sine function}, \quad sn(z, 0) = \sin(z) = w \quad (c). \tag{9.2.28}$$

The *elliptic sine inverse function* and thus the *elliptic sine function* sn(z, k) are obtained with the help of equation (9.2.27) using the following process:

$$\underset{\substack{\downarrow \\ \text{The numerically calculated value} \\ \text{of z in z – plane using elliptic integrals}}}{sn(z, k)} = \overset{\substack{\text{Assumed parameter} \\ \uparrow}}{\underset{\substack{\downarrow \\ \text{Known value of w located in t – plane}}}{(w, k)}} \tag{9.2.29}$$

The elliptic cosine function cn(z) and another function dn(z) are defined as follows:

$$\begin{aligned} cn(z) &= \sqrt{1-sn^2(z)} \quad (a) \\ dn(z) &= \sqrt{1-k^2 sn^2(z)} \quad (b). \end{aligned} \tag{9.2.30}$$

The parameter "k" is dropped in the above functions cn(z) and dn(z).

2nd Form of the 1st Kind of the Elliptic Integral

The 1st kind of the elliptic integral could be arranged in a different format. Let $t = \sin\theta$ and $w = \sin\phi$. We can change the limits $t = 0$ and $t = w = \phi$ of the integral (9.2.25), and the following integral expression could be obtained:

$$\begin{aligned} z &= \int_0^{\phi} \frac{d\theta}{\sqrt{1-k^2\sin^2\theta}} \quad (a) \\ \text{For} \quad \phi &= \frac{\pi}{2}, \quad z = \int_0^{\pi/2} \frac{d\theta}{\sqrt{1-k^2\sin^2\theta}} \quad (b). \end{aligned} \tag{9.2.31}$$

As $w = \sin\frac{\pi}{2} = 1$, the integral shown in equation (9.2.27) is

$$z = sn^{-1}(1,k) = K(k) = \int_0^1 \frac{dt}{\sqrt{(1-t^2)(1-k^2t^2)}}, \quad \text{for } 0 \le t \le 1. \tag{9.2.32}$$

In the above equation, K(k) is the value of the *elliptic inverse sine function* for the parameter k (modulus). The K(k) can be computed only for $0 \le t \le 1$. However, the *complimentary elliptic sine inverse function* can be obtained for $1 \le t \le 1/k$ as follows:

$$K'(k) = \int_1^{1/k} \frac{dt}{\sqrt{(t^2-1)(1-k^2t^2)}}, \quad 1 \le t \le \frac{1}{k}. \tag{9.2.33}$$

For the change in a variable $t = (1 - k'^2 t'^2)^{-1/2}$ with the *complementary modulus* (k') as $k'^2 = 1 - k^2$, the above complementary integral is written as

$$K(k') = \int_0^1 \frac{dt'}{\sqrt{(1-t'^2)(1-k'^2t'^2)}}, \quad 0 \le t' \le 1. \tag{9.2.34}$$

It is noted that $K(k') = K'(k)$.

Complete Elliptic Integral

The *complete elliptic integral,* in the whole range of variable t, is defined by combining equations (9.2.32) and (9.2.33):

$$z = \int_0^{1/k} \frac{dt}{\sqrt{(1-t^2)(1-k^2t^2)}}, \quad 0 \le t \le 1/k. \tag{9.2.35}$$

For the case $0 \le t \le 1$, z is a real quantity K(k). For the case, $t > 1$, i.e. for $1 \le t \le 1/k$, z is an imaginary quantity jK(k). However, for $0 \le t \le 1/k$, z is a complex quantity given by the following expression:

$$\begin{aligned} z = \pm K(k) + jK(k') &= \pm \int_0^1 \frac{dt}{\sqrt{(1-t^2)(1-k^2t^2)}} + j\int_1^{1/k} \frac{dt}{\sqrt{(t^2-1)(1-k^2t^2)}} \\ \Rightarrow \pm K(k) + jK(k') &= \pm \int_0^1 \frac{dt}{\sqrt{(1-t^2)(1-k^2t^2)}} + j\int_0^1 \frac{dt'}{\sqrt{(1-t'^2)(1-k'^2t'^2)}}. \end{aligned} \tag{9.2.36}$$

To define the elliptic function geometrically, two complex planes are used: the t-plane (or the z-plane) with t as a real number on the x-axis, i.e. $x = t$, $y = 0$, and the z-plane (or the w-plane). Figure (9.10) shows the correlation, i.e. the transformation between these planes.

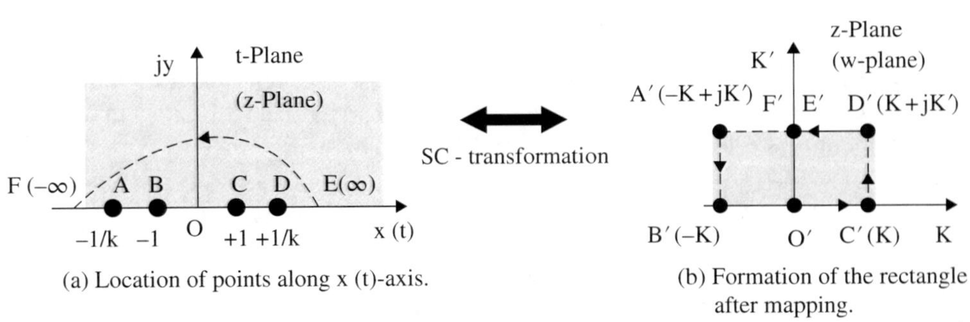

(a) Location of points along x (t)-axis.
(b) Formation of the rectangle after mapping.

Figure 9.10 Formation of the rectangle using SC-transformation.

The transformation of points located on the real x-axis of the z-plane to the points located in the complex w-plane is achieved with the help of the 1st kind of elliptic integral. The transformation process is explained below

$$\begin{array}{ccc} \text{The parameter of a fixed value} & & \\ \uparrow & & \\ \mathrm{sn}(z,k) & = & t \\ \downarrow & & \downarrow \\ \text{Value of integral shown as} & & \text{Point on the x – axis of z – plane(t – plane)} \\ \text{a point on the z i.e.w – plane} & & \text{appears as an upper limit of the integral} \end{array} \quad (9.2.37)$$

Figure (9.10) shows the SC-transformation using the elliptic function sn(z, k) = t that transforms the points located on the real axis in the t-plane (i.e. the z-plane) to the vertices of a rectangle in the complex z-plane (i.e. the complex w-plane). There are two important aspects of the SC-transformation:

- The line segments along the x-axis are transformed into the sides of a rectangle in the complex w-plane.
- In the t-plane, movement is from the left to right, whereas in the complex w-plane, the rectangle formation follows the anticlockwise path.

The transformation of points from the t-plane to z-plane, through the *mapping function* t = sn(z,k), are summarized in Table (9.1).

Approximation for the Elliptic Integral

The following approximate expressions are used to evaluate the elliptic integral [B.3, B.14, J.6]:

For the parameter $k << 1$ and $k' << 1$

$$K(k) = K(k') = \begin{cases} \frac{\pi}{2}\left[1 + \frac{k^2}{4} + \frac{9}{64}k^4\right], & \text{for} \quad k \le 0.4 \\ \frac{2}{1+k}K\left[\frac{1-k}{1+k}\right], & \text{for} \quad 0.4 \le k \le 0.65 \\ \frac{\pi}{2}\left[1 + \frac{k'^2}{4} + \frac{9}{64}k'^4\right]\ln\left(\frac{4}{k'}\right) - \left[\frac{k'^2}{4} + \frac{21}{168}k'^4\right], & \text{for} \quad k \ge 0.65. \end{cases} \quad (9.2.38)$$

where k and k′ can be interchanged.

$$\frac{K(k)}{K(k')} = \begin{cases} \left\{\frac{1}{\pi}\ln\left[2\frac{1+\sqrt{k'}}{1-\sqrt{k'}}\right]\right\}^{-1}, & \text{for} \quad 0 \le k \le 0.7 \\ \frac{1}{\pi}\ln 2\left[\frac{1+\sqrt{k}}{1-\sqrt{k}}\right], & \text{for} \quad 0.7 \le k \le 1 \end{cases}$$

where, $k' = \sqrt{k^2 - 1}$. (9.2.39)

3rd Form of the 1st Kind of Elliptic Integral

Sometimes the SC-transformation can have the following *scaled form* that is useful in the analysis of a CPW structure:

$$\frac{dw}{dz} = \frac{A}{\sqrt{(z^2 - a_1^2)(z^2 - b_1^2)}}, \quad (9.2.40)$$

where point $\pm a_1$ and point $\pm b_1$ are on the real axis of the z-plane, shown in Fig (9.11a). Equation (9.2.40) can be integrated as follows:

Table 9.1 Conformal transformation using t = Sn(z,k) mapping function.

Points in the t(z)-plane	Mapping function Sn(z,k) = t	Points in the z(w)-plane
$O = 0$	$Sn(0, k) = 0$	$O' = Sn(0, k) = 0$
$C = +1$	$Sn(K, k) = 1$	$C' = K(k)$
$D = \frac{1}{k}$	$Sn(K + jK', k) = \frac{1}{k}$	$D' = K(k) + jK'(k)$
$E \to \infty, F \to -\infty$	$Sn(0 + jK', k) = \pm\infty$	$E', F' \to jK'(k)$
$A = \frac{-1}{k}$	$Sn(-K + jK', k) = -\frac{1}{k}$	$A' = -K(k) + jK'(k)$
$B = -1$	$Sn(-K, k) = -1$	$B' = -K(k)$

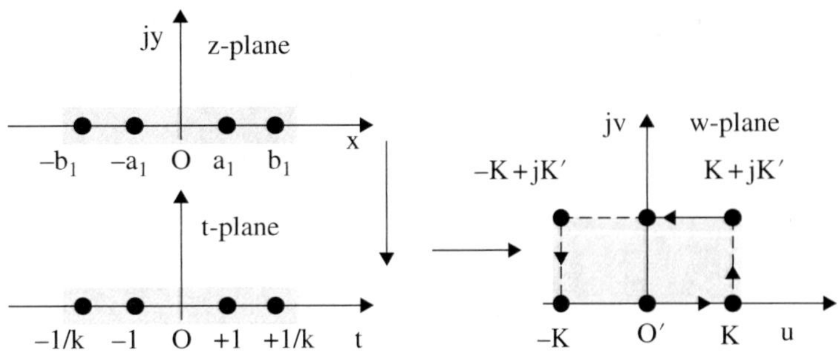

(a) Change in a variable. (b) SC-transformation.

Figure 9.11 SC-transformation of scaled locations of points.

$$w(z) = u + jv = \int_0^{b1} \frac{A\ dz}{\sqrt{(z^2 - a_1^2)(z^2 - b_1^2)}}. \tag{9.2.41}$$

Equation (9.2.41) is transformed to the standard elliptic function of the 1st kind by using the following change of variable and limits of integration: $t = z/a_1$, $z = 0 \rightarrow t = 0$, $z = b_1 \rightarrow t = b_1/a_1 = 1/k$. Figure (9.11a) shows the points located in the z-plane and their corresponding points in the t-plane.

$$\begin{aligned} u + jv &= \frac{A}{b_1}\int_0^{1/k} \frac{dt}{\sqrt{(1-t^2)(1-k^2t^2)}} \\ &= A'\int_0^{1/k} \frac{dt}{\sqrt{(1-t^2)(1-k^2t^2)}} = K(k) + jK(k') \quad (a) \\ &\text{where aspect} - \text{ratio},\ k = \frac{a_1}{b_1} \text{ and } k' = \sqrt{1-k^2} \quad (b). \end{aligned} \tag{9.2.42}$$

Figure (9.11b) shows that equation (9.2.42) achieves the SC-transformation. It transforms points located along the x- or t-axis into the vertices of a rectangle in the complex w-plane, as discussed previously and shown in Fig (9.10).

9.3 Conformal Mapping Analysis of Coplanar Waveguide

Figure (9.2) shows several forms of the CPW. The closed-form expressions for static effective relative permittivity and the characteristic impedance of most of these structures are obtained with the help of the conformal mapping method. The method is applied below to a few important CPW structures.

9.3.1 Infinite Extent CPW

Figure (9.12a) shows a CPW, with infinitely wide two ground conductors, on an infinitely thick substrate [J.1]. The width of the central signal conductor and two symmetrical gaps are s and w, respectively. The conductor thickness t is zero.

The CPW structure is located in the z-plane. The line segments AB, BC, CD, DE, EF forming the ground conductors (*electric walls, EW*), dielectric gaps (*magnetic walls, MW*), and signal conductor (*EW*) are located on the x-axis. The SC-transformation maps line segments on the x-axis to form a rectangle in the w-plane, shown in Fig (9.12b). The $A(-\infty)$ and $F(\infty)$ are taken at the origin of the w-plane, and the mapped points are vertices of a rectangle. The mapping sequence of points $A \rightarrow B \rightarrow C \rightarrow D \rightarrow E \rightarrow F$ on the Re(z)-axis in the z-plane is from the left to right, and the corresponding vertices $A' \rightarrow B' \rightarrow C' \rightarrow D' \rightarrow E' \rightarrow F'$ of the rectangle in the complex w-plane are created. In the w-plane, two infinite ground conductors, (AB and EF), of a CPW appear as the ground conductor $E'B'(A'B' + E'F')$ of the parallel-plate capacitor. The signal conductor CD appears as the conductor $C'D'$. Two gaps BC and DE appear as the MW $B'C'$ and $E'D'$ of the parallel-plate capacitor in the w-plane. *The following SC-transformation could be written to achieve the geometrical transformation of a CPW structure to the parallel-plate capacitor:*

$$\frac{dw}{dz} = A(z - x_1)^{-k_1}(z - x_2)^{-k_2}(z - x_3)^{-k_3}(z - x_4)^{-k_4}, \tag{9.3.1}$$

where, shown in Fig (9.12a), $x_1 = -b$, $x_2 = -a$, $x_3 = a$, and $x_4 = b$ are points B, C, D, E, respectively, defining the conducting strips of a CPW in the z-plane. The width of signal strip = CD = 2a = s and width of the symmetrical gaps = BC = DE = b − a = w.

Therefore,

$$a = \frac{s}{2} \text{ and } b = w + \frac{s}{2}. \tag{9.3.2}$$

All external angles, shown in Fig (9.12b), are $\pi/2$ for the transformed rectangle. Using equation (9.2.21), we get $k_1\pi = \pi/2 \Rightarrow k_1 = 1/2$. Likewise, $k_1 = k_2 = k_3 = k_4 = 1/2$. Hence, the needed SC-transformation is

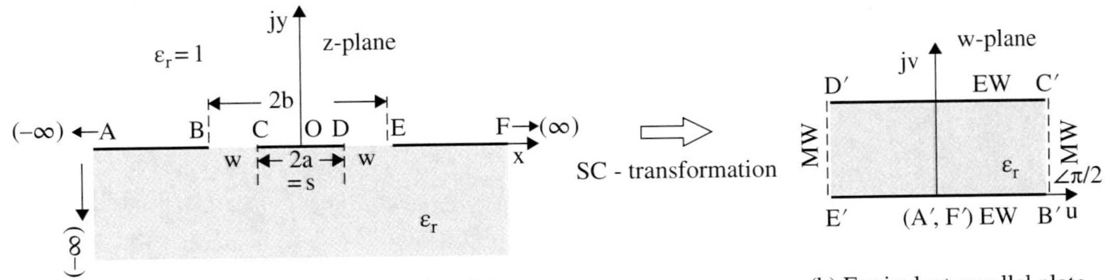

(a) Infinite width ground plane and infinite substrate thickness CPW. Movement along the x-axis, with the anticlockwise turn, covers the upper half-space with $\varepsilon_r = 1$.

(b) Equivalent parallel plate line. $\varepsilon_r = 1$ for the upper half space.

Figure 9.12 Infinite extent CPW.

$$w = \int_0^b \frac{A\,dz}{\sqrt{(z^2 - a^2)(z^2 - b^2)}}. \quad (9.3.3)$$

Figure (9.12a) shows that with anticlockwise turn from F to A, the SC-transformation applies to the upper half of the original structure, i.e. for space with an air dielectric $\varepsilon_r = 1$ in the z-plane. Due to the symmetry, the SC-transformation is applied separately to the upper and lower space of the CPW in the z-plane. For the lower space, the inside space of the rectangle in the w-plane is the dielectric medium with relative permittivity ε_r. Using the change in a variable $(z/a = t)$, the SC-transformation is written as

$$u + jv = A' \int_0^{1/k} \frac{dt}{\sqrt{(1 - t^2)(1 - k^2 t^2)}} = K(k) + jK(k'). \quad (9.3.4)$$

The parameter (modulus) $k = k_1$ is aspect − ratio :

$$k_1 = \frac{a}{b} = \frac{s}{s + 2w} \quad (a)$$

and the complimentary modulus is $k' = k_1'$. It is given as,

$$k_1' = \sqrt{1 - \left(\frac{s}{s + 2w}\right)^2} \quad (b). \quad (9.3.5)$$

In the w-plane, the width of the parallel-plate capacitor and separation between the parallel plates are $D'C' = E'B' = 2K(k_1)$ and $B'C' = D'E' = K(k_1')$, respectively, giving the line capacitances *p.u.l.* for the upper and lower half-spaces:

$$\text{Upper air − filled half − space :} \quad C_1 = 2\varepsilon_0 \frac{K(k_1)}{K(k_1')} \quad (a)$$

$$\text{Lower dielectric − filled half − space :} \quad C_2 = \frac{2\varepsilon_0 \varepsilon_r K(k_1)}{K(k_1')} \quad (b). \quad (9.3.6)$$

The total line capacitances *p.u.l.* of the CPW on dielectric and air-substrates are

$$C(\varepsilon_r) = C_1 + C_2 = 2\varepsilon_0(\varepsilon_r + 1)\frac{K(k_1)}{K(k_1')} \quad (a)$$

$$C_0(\varepsilon_r = 1) = 4\varepsilon_0 \frac{K(k_1)}{K(k_1')} \quad (b). \quad (9.3.7)$$

The effective dielectric constant of the CPW on an infinite extent dielectric substrate, shown in Fig (9.12a), and the filling-factor are given below:

$$\varepsilon_{r\,eff} = \frac{C(\varepsilon_r)}{C_0(\varepsilon_r = 1)} = \frac{(\varepsilon_r + 1)}{2} \quad (a)$$

$$\varepsilon_{r\,eff} = 1 + q(\varepsilon_r - 1) \quad (b), \quad q = \frac{1}{2} \quad (c). \quad (9.3.8)$$

The $q = 1/2$ shows the equal division of the electric fields in the upper and lower half-spaces for the present structure. The characteristic impedance of a CPW on the air and dielectric substrates are

$$Z_0(\varepsilon_r = 1) = \frac{1}{cC} = \sqrt{\mu_0 \varepsilon_0} \times \frac{K(k_1')}{4\varepsilon_0 K(k_1)}$$

$$\Rightarrow Z_0(\varepsilon_r = 1) = 30\pi \frac{K(k_1')}{K(k_1)} \quad (a)$$

$$Z(\varepsilon_r) = \frac{30\pi}{\sqrt{\varepsilon_{r\,eff}}} \times \frac{K(k_1')}{K(k_1)} \quad (b). \quad (9.3.9)$$

The ratio $K(k_1')/K(k_1)$ is calculated by using equation (9.2.39). In practical circumstances, the width of the ground planes is finite, as well as the substrate has a finite thickness. These are considered below.

9.3.2 CPW on Finite Thickness Substrate and Infinite Ground Plane

Figure (9.13a) shows a CPW on a substrate with finite thickness h and relative permittivity ε_r. However, the ground planes are infinite in width. Exact conformal

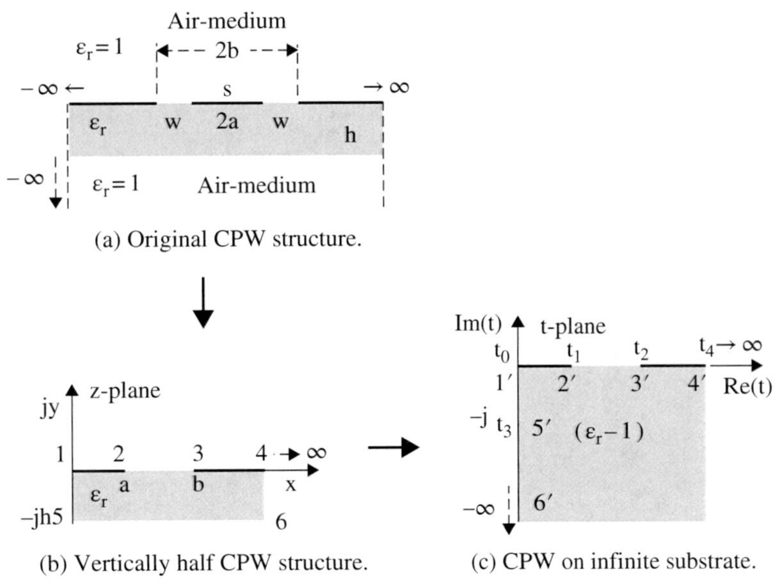

(a) Original CPW structure.

(b) Vertically half CPW structure.

(c) CPW on infinite substrate.

Figure 9.13 Infinite ground plane CPW on finite thickness substrate.

mapping of this structure involves mapping of the dielectric–air interface in the lower half-space. Such transformation is difficult [J.11]. However, based on the following assumption, an approximate conformal mapping could be achieved [J.3]:

The line capacitance of a CPW structure on the substrate of thickness h = Line capacitance of the CPW line on air substrate (both upper and lower half-spaces) + The line capacitance of the lower half-space of the CPW on an infinitely thick substrate of the relative permittivity (ε_r-1).

The results, based on the above assumption, provide exact results only for the limiting cases, $h \to \infty$ and $h \to 0$. The results are approximate for the cases in between the limits. In the present case, the conformal mapping is a two-step process:

First Step: At this stage, half of the original structure is considered and shown in Fig (9.13b). The half structure is transformed to Fig (9.13c) that is a CPW on the infinitely thick substrate, with relative permittivity $(\varepsilon_r - 1)$. The transformation from the z-plane to t-plane is carried out with the help of the following mapping function and the mapping points are shown in Table (9.2):

$$t = \sinh\left(\frac{\pi z}{2h}\right). \tag{9.3.10}$$

Second Step: Using the results of subsection (9.3.1), the SC-transformation is carried out to transform Fig (9.13c) to a rectangle in the w-plane with the following aspect-ratio:

Table 9.2 Transformation using $t = \sinh(\pi z/2h)$.

Points in z-plane	Points in the t-plane
1) $z = 0$	$1'$. $t_0 = 0$
2) $z = a$	$2'$. $t_1 = \sinh\left(\frac{\pi a}{2h}\right)$
3) $z = b$	$3'$. $t_2 = \sinh\left(\frac{\pi b}{2h}\right)$
4) $z \to \infty$	$4'$. $t_4 \to \infty$
5) $z = -jh$	$5'$. $t_5' = \sinh\left(\frac{\pi}{2h}(-jh)\right) = -j$
6) $z \to \infty - jh$	$6'$. $t_6 \to -j\infty$

$$k_2 = \frac{t_1}{t_2} = \frac{\sinh(\pi a/2h)}{\sinh(\pi b/2h)} \quad \text{(a)}$$
$$k_2' = \sqrt{1 - k_2^2} \quad \text{(b).} \tag{9.3.11}$$

The line capacitance of the upper half of the structure is

$$C_1 = 2\varepsilon_0 \frac{K(k_1)}{K(k_1')}, \quad \text{where } k_1 = \frac{a}{b}, \; k_1' = \sqrt{1 - k_1^2}. \tag{9.3.12}$$

If C_2 is the line capacitance of one-half of the structure on the substrate with relative permittivity $(\varepsilon_r - 1)$, then using equation (9.3.6b), the following expression is obtained:

$$C_2 = 2\varepsilon_0(\varepsilon_r - 1)\frac{K(k_2)}{K(k_2')}. \tag{9.3.13}$$

In the view of our assumption, the total line capacitance *p.u.l.* of a CPW on the finite thickness substrate is written as follows:

$$C(\varepsilon_r) = C_{1U} + C_{1L} + C_2, \tag{9.3.14}$$

where C_{1U} and C_{1L} are the line capacitances of the upper and lower half of the CPW on the air-substrate. In the present case, the line capacitances are equal, i.e. $C_{1U} = C_{1L} = C_1$.

The limiting cases are examined: For $h \to 0$, Fig (9.13a) indicates that the lower half-space is filled with an air medium ($\varepsilon_r = 1$). Thus, from equation (9.3.13), $C_2 = 0$ and using equation (9.3.14), $C(\varepsilon_r = 1) = 2C_1$, i.e. the CPW is totally in the air medium. Another limiting case is examined for $h \to \infty$:

$$\sinh\left(\frac{\pi a}{2h}\right) \approx \left(\frac{\pi a}{2h}\right), \quad \text{and} \quad \sinh\left(\frac{\pi b}{2h}\right) \approx \left(\frac{\pi b}{2h}\right)$$

$$\text{also for } h \to \infty, \quad k_2 = \frac{\pi a/2h}{\pi b/2h} = \frac{a}{b} = k.$$

This is the case for a CPW on the infinitely thick substrate. Therefore, the assumption is valid for limiting cases. The effective relative permittivity and the filling-factor of the CPW structure, shown in Fig (9.13a), are given as follows:

$$\varepsilon_{r\,eff} = \frac{C(\varepsilon_r)}{C(\varepsilon_r = 1)} = \frac{2C_1 + C_2}{2C_1} = 1 + \frac{1}{2}\left(\frac{C_2}{C_1}\right)$$

$$= 1 + \frac{(\varepsilon_r - 1)}{2}\frac{K(k_2)}{K(k_2')}\cdot\frac{K(k_1')}{K(k_1)} \qquad \text{(a)}$$

$$\varepsilon_{r\,eff} = 1 + q(\varepsilon_r - 1) \qquad \text{(b)}$$

$$q = \frac{1}{2}\frac{K(k_2)}{K(k_2')}\cdot\frac{K(k_1')}{K(k_1)} \qquad \text{(c)}.$$

$$\tag{9.3.15}$$

In the case of the CPW on a finite thickness substrate, the following expression of characteristic impedance is obtained:

$$Z_o(\varepsilon_r) = \frac{30\pi}{\sqrt{\varepsilon_{r\,eff}}} \times \frac{K(k_1')}{K(k_1)}. \tag{9.3.16}$$

9.3.3 CPW with Finite Ground Planes

Figure (9.14a) shows a CPW of finite width ground conductor on an infinite thickness substrate [J.10]. Figure (9.14b) shows another CPW structure with a finite width ground conductor on a finitely thick substrate. The aspect-ratio, in both cases, is modified by the finite size of ground conductors as follows:

$$k_1^G = k_1 F_1 = \frac{a}{b} F_1 \qquad \text{(a)}$$

$$\text{where } F_1 = \left[\frac{1 - (b/c)^2}{1 - (a/c)^2}\right]^{\frac{1}{2}} \qquad \text{(b)}.$$

$$\tag{9.3.17}$$

$$k_2^G = k_2 F_2 = \frac{\sinh(\pi a/2h)}{\sinh(\pi b/2h)} F_2 \qquad \text{(a)}$$

$$\text{where } F_2 = \left[\frac{1 - \left(\frac{\sinh(\pi b/2h)}{\sinh(\pi c/2h)}\right)^2}{1 - \left(\frac{\sinh(\pi a/2h)}{\sinh(\pi c/2h)}\right)^2}\right]^{\frac{1}{2}} \qquad \text{(b)}.$$

$$\tag{9.3.18}$$

In the above equations, k_1^G and k_2^G are the aspect-ratios of the infinitely and finitely thick substrates with finite width ground planes [J.4]. The corrections are obtained in the process of the SC-transformation discussed below. In both cases for $c \to \infty$ (infinitely wide ground planes), $F_1 \to 1$ and $F_2 \to 1$.

The CPW structure of Fig (9.14b) is analyzed in two steps.

Step #I

The step I, shown in Fig (9.15a), considers the numbered vertically half of the original CPW. The upper part, on the infinite air substrate, is mapped into the upper half of the t-plane, shown in Fig (9.15b), using the following mapping function:

$$t = z^2. \tag{9.3.19}$$

The mapping points are summarized in Table (9.3). Figure (9.15c) shows the mapping of conducting strip (*EW*) and the gap (*MW*) to form a parallel plate

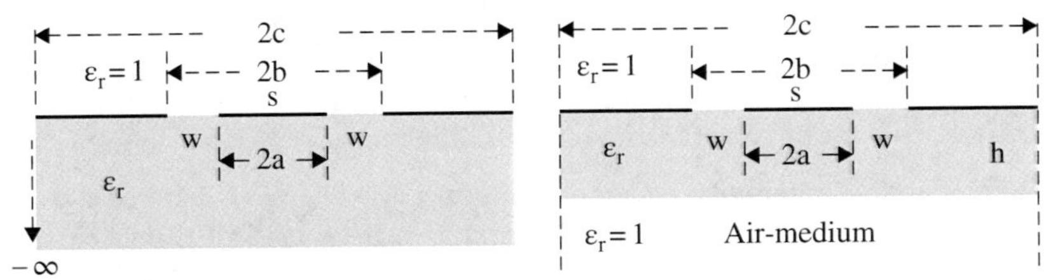

(a) Finite width CPW on infinite thickness substrtae.

(b) Finite width CPW on finite thickness substrtae.

Figure 9.14 Finite ground plane width CPW.

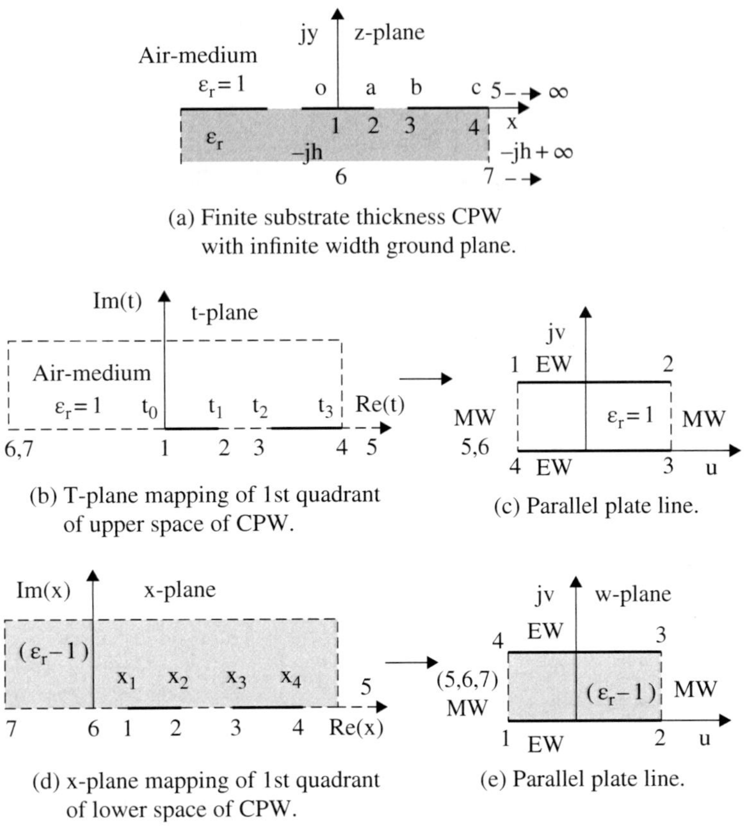

Figure 9.15 SC-transformation of finite substrate thickness CPW with the infinite ground plane.

Table 9.3 Transformation using $t = z^2$.

Points in the z-plane	Points in the t-plane
1) $z = o$	1) $t = 0$
2) $z = a$	2) $t_1 = a^2$
3) $z = b$	3) $t_2 = b^2$
4) $z = c$	4) $t_3 = c^2$
5) $z \to x(x \to \infty)$	5) $t \to \infty$
6) $z = -j\,y(y \to \infty)$	6) $t \to -\infty$
7) $z \to (-jh + \infty)$	7) $t \to -\infty$

rectangular structure in the w-plane by using the following SC-transformation:

$$u + jv = w(t) = \int_{t_0}^{t} \frac{dt}{\sqrt{t(t-t_1)(t-t_2)(t-t_3)}}. \qquad (9.3.20)$$

The above integral is transformed to a standard elliptic integral of the 1st kind by the following change of variable [B.15]:

$$p = \frac{(\beta-\delta)(t-\alpha)}{(\alpha-\delta)(t-\beta)} \qquad (a)$$

$$\int_{\delta}^{t} \frac{dt}{\sqrt{(t-\alpha)(t-\beta)(t-\gamma)(t-\delta)}} = N\int_{0}^{\frac{1}{k}} \frac{dp}{\sqrt{(1-p^2)(1-k^2p^2)}} \qquad (b)$$

$$k^2 = \frac{(\alpha-\beta)(\gamma-\delta)}{(\alpha-\gamma)(\beta-\delta)} \qquad (c)$$

$$k'^2 = \frac{(\alpha-\delta)(\beta-\gamma)}{(\alpha-\gamma)(\beta-\delta)} \qquad (d),$$

$$(9.3.21)$$

where $\alpha > \beta > \gamma > \delta$. Sometimes the role of the modulus k and complimentary modulus k' could be interchanged [J.12]. Thus, $\alpha = t_3 = c^2, \beta = t_2 = b^2, \gamma = t_1 = a^2, \delta = t_0 = 0$. Thus, the aspect-ratio of the CPW on an infinite substrate with thick finite extent coplanar ground planes is

$$k = k_3 = k_1^G = \frac{a}{b}\sqrt{\frac{(c^2-b^2)}{(c^2-a^2)}}. \qquad (9.3.22)$$

The aspect-ratio, $k_1 = a/b$, applies to the infinite width ground planes, and the correction factor, $F_1 = \sqrt{(c^2-b^2)/(c^2-a^2)}$, mentioned in equation (9.3.17b), is due to the finite width of ground planes. Following the discussion of subsection (9.3.1), the line capacitance of a CPW on the air-substrate is

$$C_1(\varepsilon_r = 1) = 2\varepsilon_0\frac{K(k_3)}{K(k_3')}. \qquad (9.3.23)$$

Step #II

Using Figure (9.15a), step II considers the numbered vertically half of the original CPW in the lower half-plane that is a finite thickness substrate backed by the air-substrate. To compute the line capacitance for this case, the dielectric–air interface is replaced by a MW, and relative permittivity $(\varepsilon_r - 1)$ is used in the infinite extent lower half of the x-plane, shown in Fig (9.15d). The transformation from the z-plane of Fig (9.15a) to the x-plane of Fig (9.15d) is carried out using the following mapping function:

$$x = \cosh^2\left(\frac{\pi z}{2h}\right). \quad (9.3.24)$$

Using the above equation, the transformed locations are summarized in Table (9.4). The following SC-transformation maps structure of Fig (9.15d) to a parallel-plate capacitor in the w-plane, shown Fig (9.15e):

$$u + jv = w = \int_{x_1}^{x} \frac{dx}{\sqrt{(x-x_1)(x-x_2)(x-x_3)(x-x_4)}}. \quad (9.3.25)$$

The above integral is transformed into a standard elliptic integral of the 1st kind with the help of equation (9.3.21):

$$x_1 = \delta = 1,\ x_2 = \gamma = \cosh^2\left(\frac{\pi a}{2h}\right),$$
$$x_3 = \beta = \cosh^2\left(\frac{\pi b}{2h}\right),\ x_4 = \alpha = \cosh^2\left(\frac{\pi c}{2h}\right). \quad (9.3.26)$$

Using equation (9.3.21c), the aspect-ratio is

$$k^2 = k_4^2 = \left[\frac{\cosh^2(\pi c/2h) - \cosh^2(\pi b/2h)}{\cosh^2(\pi c/2h) - \cosh^2(\pi a/2h)}\right] \times \left[\frac{\cosh^2(\pi a/2h) - 1}{\cosh^2(\pi b/2h) - 1}\right]$$
$$\Rightarrow k_4 = k_2^G = \left[\frac{\sinh^2(\pi c/2h) - \sinh^2(\pi b/2h)}{\sinh^2(\pi c/2h) - \sinh^2(\pi a/2h)}\right]^{\frac{1}{2}} \times \left[\frac{\sinh(\pi a/2h)}{\sinh(\pi b/2h)}\right]. \quad (9.3.27)$$

Table 9.4 Transformation using $x = \cosh^2(\pi z/2h)$.

Points in the z-plane	Points in the x-plane
1) $z = 0$	$x = x_1 = 1$
2) $z = a$	$x = x_2 = \cosh^2\left(\frac{\pi a}{2h}\right)$
3) $z = b$	$x = x_3 = \cosh^2\left(\frac{\pi b}{2h}\right)$
4) $z = c$	$x = x_4 = \cosh^2\left(\frac{\pi c}{2h}\right)$
5) $z = x(x \to \infty)$	$x \to \infty$
6) $z = -jh$	$x = 0$
7) $z = (x - jh)(x \to \infty)$	$x \to -\infty$

The above equation provides the correction factor F_2 of equation (9.3.18b). The line capacitance of the lower half of the CPW with the dielectric layer is

$$C_2 = 2\varepsilon_0(\varepsilon_r - 1)\frac{K(k_4)}{K(k_4')}. \quad (9.3.28)$$

Using equation (9.3.14), the total line capacitance *p.u.l.* of CPW with the finite substrate and finite ground conductors is

$$C_2(\varepsilon_r) = 2C_1 + C_2 = 4\varepsilon_0\frac{K(k_3)}{K(k_3')} + 2\varepsilon_0(\varepsilon_r - 1)\frac{K(k_4)}{K(k_4')}. \quad (9.3.29)$$

The effective relative permittivity, filling-factor, and characteristic impedance of the considered CPW are given below:

$$\varepsilon_{r\,eff} = \frac{C(\varepsilon_r)}{C(\varepsilon_r = 1)} = \frac{4\varepsilon_0\frac{K(k_3)}{K(k_3')} + 2\varepsilon_0(\varepsilon_r - 1)\frac{K(k_4)}{K(k_4')}}{4\varepsilon_0\frac{K(k_3)}{K(k_3')}}$$

$$\Rightarrow \varepsilon_{r\,eff} = 1 + \frac{1}{2}(\varepsilon_r - 1)\frac{K(k_4)\,K(k_3')}{K(k_4')\,K(k_3)} \quad (a)$$

$$q = \frac{1}{2}\frac{K(k_4)\,K(k_3')}{K(k_4')\,K(k_3)} \quad (b). \quad (9.3.30)$$

$$Z_o(\varepsilon_r) = \frac{30\pi}{\sqrt{\varepsilon_{r\,eff}}}\frac{K(k_3')}{K(k_3)}. \quad (9.3.31)$$

9.3.4 Static Characteristics of CPW

Infinite Dielectric Thickness

The effective relative permittivity of a CPW on an infinitely thick substrate, shown in Fig (9.12), does not depend on the a/b aspect-ratio. However, its characteristic impedance depends on a/b-ratio. For a fixed value of b, the separation between two grounds is fixed and an increase in the width of the central strip conductor ($s = 2a$) narrows the gap w. It results in an increase in the line capacitance that decreases the characteristic impedance of a CPW. Figure (9.16) shows such a decrease in the characteristic impedance with an increase in a/b-ratio [B.16]. Several values of the strip width (2a) and separation of the ground planes (2b) can provide identical value for the characteristic impedance. *Thus, the dimensions (2a) and (2b) are selected for the compactness of the CPW, and also for the minimum conductor loss.*

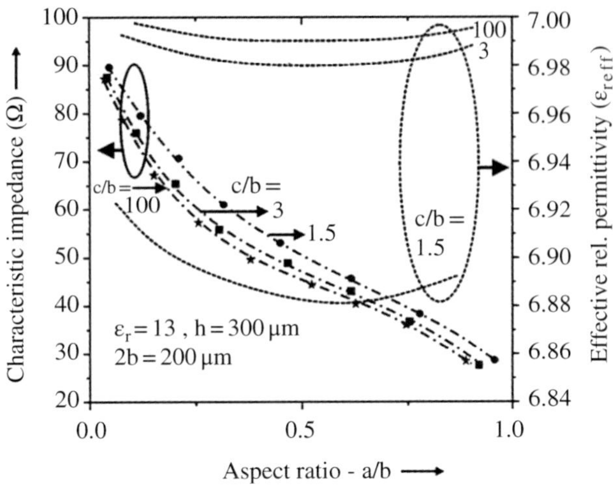

Figure 9.16 Changes in CPW characteristics with aspect ratio-(a/b) and c/b-ratio. *Source:* Gupta et al. [B.16]. © 1996, Horizon House Publications, Inc.

Finite Dielectric Thickness

The characteristic impedance and effective relative permittivity of a CPW are influenced to some extent by the finite size of the substrate thickness and finite width of the ground planes. At a fixed aspect-ratio a/b, the decrease in h/b results in more electric fields in the air medium of the lower half-plane and less electric fields in the dielectric medium. This causes a decrease in line capacitance for the lower part of CPW. It increases its characteristic impedance and also decreases its effective relative permittivity $\varepsilon_{r\,eff}$. The finite ground conductor width c decreases the line capacitance, as some of the electric field lines move to the air regions. The characteristic impedance of CPW increases and its effective relative permittivity decreases with a decrease in c/b ratio. These changes are shown in Fig (9.16) for c/b-ratio 1.5,3,100 [B.16, J.3]. For a thick substrate, $h/b \geq 3$, and for the wide ground planes $c/b \geq 3$ the effect of finite substrate thickness and effect of the finite width of ground planes on the characteristic impedance and effective relative permittivity could be ignored.

9.3.5 Top-Shielded CPW

Figure (9.17a) shows a top-shielded CPW structure on a finite thickness (h) substrate with relative permittivity ε_r. However, the ground conductors have an infinite width and the top shield is located at the height h_1. The medium between the CPW strip conductors and the top shield is air, i.e. $\varepsilon_r = 1$. The top shield is used to protect a CPW against the environment. It is also useful for the post-fabrication adjustment of the line parameters. The conformal mapping method is applied separately to the upper and lower spaces of the structure. Both slots are modeled by the MW and the conducting strips are treated as the EW.

The line capacitance *p.u.l.* of the present CPW is $C_5(\varepsilon_r)$ = line capacitance $C_3(\varepsilon_r = 1)$ of the upper half of shielded space + line capacitance $C_2(\varepsilon_r)$ of the lower half of finite thickness substrate, i.e.

$$C_5(\varepsilon_r) = C_3(\varepsilon_r = 1) + C_2(\varepsilon_r) \quad \text{(a)}$$
$$C_2(\varepsilon_r) = C_{1L}(\varepsilon_r = 1) + C_2 \quad \text{(b),} \qquad (9.3.32)$$

where C_{1L} is given by equation (9.3.12) and C_2 is given by equation (9.3.13). We have to obtain the expression only for $C_3(\varepsilon_r = 1)$.

The first transformation, using the function given below, maps all conductors of the upper half-space, shown in Fig (9.17a), on the real t-axis, shown in Fig (9.17b):

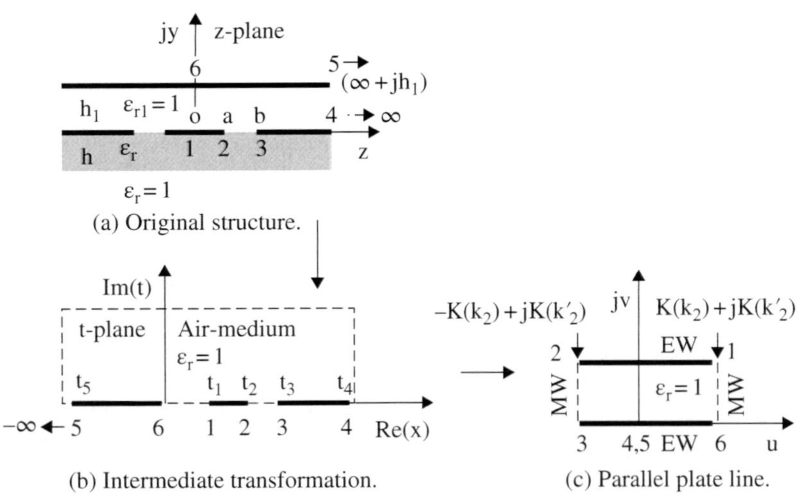

Figure 9.17 Conformal mapping of the upper half of the top-shielded CPW on the finite thickness.

Table 9.5 Transformation using $x = \cosh^2(\pi z/2h_1)$.

Points in the z-plane	Points in the t-plane
1) $x = 0$	$t_1 = 1$
2) $x = a$	$t_2 = \cosh^2\left(\frac{\pi a}{2h_1}\right)$
3) $x = b$	$t_3 = \cosh^2\left(\frac{\pi b}{2h_1}\right)$
4) $x = x(x \to \infty)$	$t_4 \to \infty$
5) $x = (x + jh_1)(x \to \infty)$	$t_5 \to -\infty$
6) $z = jh_1$	$t_6 = 0$

$$t = \cosh^2\left(\frac{\pi x}{2h_1}\right). \qquad (9.3.33)$$

Table (9.5) shows the transformed locations of the strip edges in the t-plane.

Next, the SC-transformation converts the strips and gaps located on Re(t)-axis into a rectangle in the w-plane, shown in Fig (9.17c). The mapping function for this transformation is written as follows:

$$W = \int_{t_0}^{t} \frac{dt}{\sqrt{t(t-t_1)(t-t_2)(t-t_3)}}. \qquad (9.3.34)$$

The above integral can be put in the standard form of the elliptical integral of the 1st kind by the change of the variable expressed through equation (9.3.21). The aspect ratio K_5 is given by

$$K_5^2 = \frac{(\alpha-\delta)(\beta-\gamma)}{(\alpha-\gamma)(\beta-\delta)}, \Rightarrow \quad K_5 = \frac{\tanh(\pi a/2h_1)}{\tanh(\pi b/2h_1)} \qquad \text{(a)}$$

$$K_5'^2 = 1 - K_5^2 \qquad \text{(b)}$$

$$\text{where } \alpha = \cosh^2\frac{\pi b}{2h}, \quad \beta = \cosh^2\frac{\pi a}{2h}, \quad \gamma = 1, \quad \delta = 0 \qquad \text{(c)}.$$

(9.3.35)

Figure (9.17c) shows the transformation of the locations from the t-plane to w-plane forming the parallel-plate capacitor. The line capacitance $C_3(\varepsilon_r = 1)$ of the upper half of the shielded CPW is

$$C_3(\varepsilon_r = 1) = 2\varepsilon_0\frac{K(k_5)}{K(k_5')}. \qquad (9.3.36)$$

The line capacitance *p.u.l.* of the shielded CPW on a finitely thick substrate with the top shield is

$$C_5(\varepsilon_r) = \underbrace{2\varepsilon_0\frac{K(k_5)}{K(k_5')}}_{\text{Upper half - space}} + \underbrace{\left[2\varepsilon_0\frac{K(k_3)}{K(k_3')} + 2\varepsilon_0(\varepsilon_r - 1)\frac{K(k_4)}{K(k_4')}\right]}_{\text{Lower half - space}}. \qquad (9.3.37)$$

The line capacitance *p.u.l.* of the shielded CPW in the air medium is

$$C_5(\varepsilon_r = 1) = 2\varepsilon_0\frac{K(k_5)}{K(k_5')} + 2\varepsilon_0\frac{K(k_3)}{K(k_3')}. \qquad (9.3.38)$$

The effective relative permittivity and the filling-factor q of the shielded CPW are

$$\varepsilon_{r\,eff} = \frac{C_5(\varepsilon_r)}{C_5(\varepsilon_r = 1)} = 1 + (\varepsilon_r - 1)\left[\frac{\frac{K(k_4)}{K(k_4')}}{\frac{K(k_5)}{K(k_5')} + \frac{K(k_3)}{K(k_3')}}\right] \qquad \text{(a)}$$

$$q = \frac{\frac{K(k_4)}{K(k_4')}}{\frac{K(k_5)}{K(k_5')} + \frac{K(k_3)}{K(k_3')}} \qquad \text{(b)}.$$

(9.3.39)

The following aspect-ratios, from equations (9.3.5a), (9.3.11a), and (9.3.35a), are used in the above equations for the infinitely wide ground planes:

$$K_3 = \frac{a}{b}, \quad K_4 = \frac{\sinh(\pi a/2h)}{\sinh(\pi b/2h)}$$

$$K_5 = \frac{\tanh(\pi a/2h_1)}{\tanh(\pi b/2h_1)}, \quad K_j'^2 = 1 - K_j^2. \qquad (9.3.40)$$

The characteristic impedance of the top-shielded CPW on the air and dielectric substrates are

$$Z_0(\varepsilon_r = 1) = \frac{60\pi}{\frac{K(k_5)}{K(k_5')} + \frac{K(k_3)}{K(k_3')}} \qquad \text{(a)}$$

$$Z_0(\varepsilon_r) = \frac{Z_0(\varepsilon_r = 1)}{\sqrt{\varepsilon_{r\,eff}}} \qquad \text{(b)}. \qquad (9.3.41)$$

Only the infinitely wide ground planes are considered above. However, for the finite width of the ground planes separated by distance 2c, shown in Fig (9.14b), the aspect-ratios K_3, K_4, K_5 should be modified according to expressions (9.3.17) and (9.3.18) as follows:

$$K_3(\text{mod}) = K_3\left[\frac{c^2 - b^2}{c^2 - a^2}\right]^2 \qquad \text{(a)}$$

$$K_4(\text{mod}) = K_4\left[\frac{\sinh^2(\pi c/2h) - \sinh^2(\pi b/2h)}{\sinh^2(\pi c/2h) - \sinh^2(\pi a/2h)}\right] \qquad \text{(b)}$$

$$K_5(\text{mod}) = K_5\left[\frac{\sinh^2(\pi c/2h) - \sinh^2(\pi b/2h)}{\sinh^2(\pi c/2h) - \sinh^2(\pi a/2h)}\right] \qquad \text{(c)}$$

$$K_i'^2(\text{mod}) = 1 - K_i^2(\text{mod}), \quad (i = 3, 4, 5,) \qquad \text{(d)}.$$

(9.3.42)

9.3.6 Conductor-Backed CPW

Figure (9.2c) shows the CB-CPW structure. The conductor backing provides mechanical support to the thin and fragile semiconductor and quartz substrates. It also acts as a heat sink to the active microwave devices. The CB-CPW is a mixed structure of the CPW and microstrip structures. This structure can support the unwanted microstrip mode that is the dominant mode of a thin substrate and wide slot width, $w/h > 2$, CB-CPW. For a moderate aspect-ratio, such as $s/2h = 1/3$ and $w/h = 2/3$, the structure is less dispersive, as compared to a microstrip line of $w/h = 2/3$. In this section, the conformal mapping method is applied to obtain the static effective relative permittivity and the characteristic impedance of a top-shielded CB-CPW, shown in Fig (9.2e).

The line capacitance of the top-shielded CB-CPW is a sum of the line capacitances of the upper half $C_3(\varepsilon_r = 1)$ and the lower half $C_2(\varepsilon_r)$ of the structure:

$$C(\varepsilon_r) = C_3(\varepsilon_r = 1) + C_2(\varepsilon_r), \tag{9.3.43}$$

where $C_3(\varepsilon_r = 1)$ is given by equation (9.3.36) for height h_1 of the top shield. Following the same process, $C_2(\varepsilon_r)$ is given by

$$C_2(\varepsilon_r) = 2\varepsilon_0\varepsilon_r\frac{K(k_6)}{K(k_6')} \quad \text{(a)}$$

$$\text{where aspect ratio}: \quad K_6 = \frac{\tanh(\pi a/2h)}{\tanh(\pi b/2h)} \quad \text{(b)}$$

$$K_6'^2 = 1 - K_6^2 \quad \text{(c)}. \tag{9.3.44}$$

In the above expression, h is the thickness of a conductor-backed substrate. The line capacitance of the top shield CB-CPW on the dielectric and the air substrates are

$$C(\varepsilon_r) = 2\varepsilon_0\frac{K(k_5)}{K(k_5')} + 2\varepsilon_0\varepsilon_r\frac{K(k_6)}{K(k_6')} \quad \text{(a)}$$

$$C(\varepsilon_r = 1) = 2\varepsilon_0\frac{K(k_5)}{K(k_5')} + 2\varepsilon_0\frac{K(k_6)}{K(k_6')} \quad \text{(b)}. \tag{9.3.45}$$

The effective relative permittivity and filling-factor q of the shielded CB-CPW structure are obtained from the above equations:

$$\varepsilon_{r\,\text{eff}} = \frac{C(\varepsilon_r)}{C(\varepsilon_r = 1)} = \frac{\dfrac{K(k_5)}{K(k_5')} + \varepsilon_r\dfrac{K(k_6)}{K(k_6')}}{\dfrac{K(k_5)}{K(k_5')} + \dfrac{K(k_6)}{K(k_6')}} \quad \text{(a)}$$

$$q = \frac{\dfrac{K(k_6)}{K(k_6')}}{\dfrac{K(k_5)}{K(k_5')} + \dfrac{K(k_6)}{K(k_6')}} \quad \text{(b)}. \tag{9.3.46}$$

The characteristic impedance of the structure is

$$Z(\varepsilon_r) = \frac{60\pi}{\sqrt{\varepsilon_{r\,\text{eff}}}}\frac{1}{\dfrac{K(k_5)}{K(k_5')} + \dfrac{K(k_6)}{K(k_6')}}. \tag{9.3.47}$$

For $h_1 = h$, $\varepsilon_{r\,\text{eff}} = (\varepsilon_r + 1)/2$. For the shield height $h_1 \rightarrow \infty$, $K_5 \rightarrow a/b$, the expressions for simple CB-CPW are obtained. The presence of a top or a bottom shield reduces the characteristic impedance of a CPW. The characteristic impedance also decreases with an increase in a/b-ratio, shown in Fig (9.18a). The nearness of the conductor backing to the strip conductors increases $\varepsilon_{r\,\text{eff}}$ [J.4]. It is shown in Fig (9.18b).

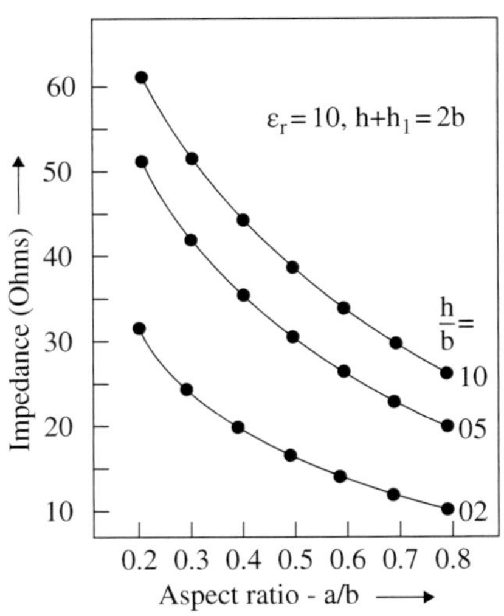

(a) Variation in characteristic impedance.

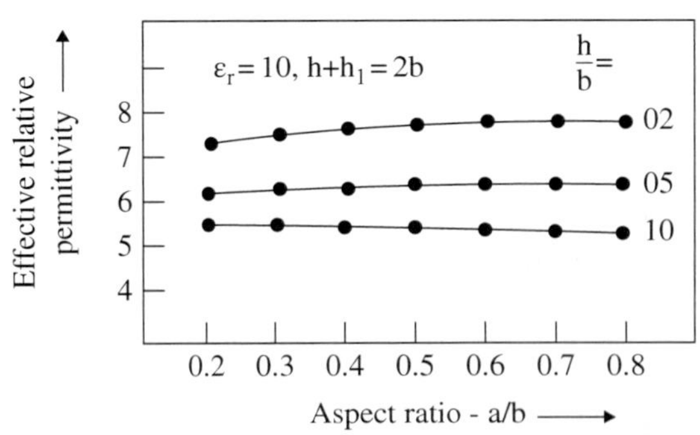

(b) Variation in the effective relative permittivity.

Figure 9.18 Variation in the line parameters of top-shielded conductor-backed CPW with aspect-ratio. Substrate thickness is a parameter. *Source:* Ghione and Naldi [J.4].

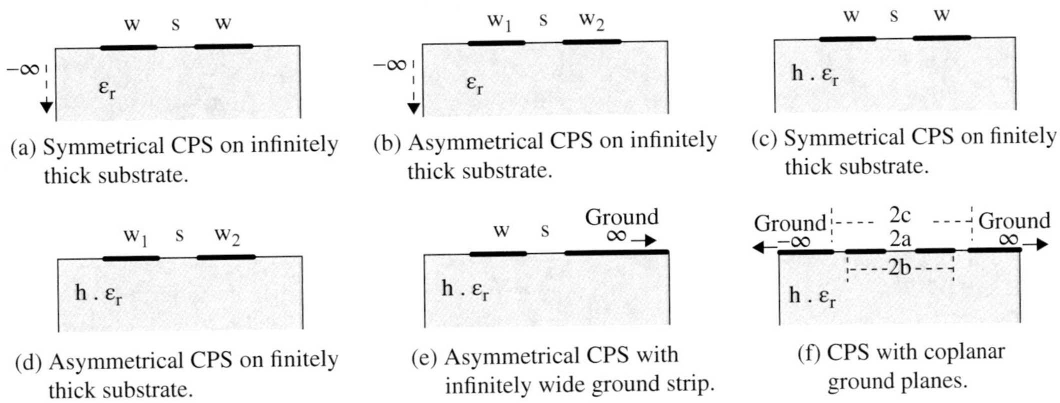

Figure 9.19 Some CPS structures.

9.4 Coplanar Stripline

Figure (9.19) shows several CPS structures. It consists of two strip conductors of width w separated by a slot-gap of width s. The CPS line also supports the quasi-TEM mode and is less dispersive, as compared to a microstrip line. To reduce the losses on the CPS line, normally a wider strip width, i.e. $w \geq 4s$, is maintained [J.13]. However, the strip width (w) has to be kept small to exclude the parasitic non-TEM modes and also to reduce the MMIC chip area. The CPS line further supports the TM_0 and TE_0 dielectric slab parasitic modes. The TM_0 surface wave mode has no cutoff frequency [J.14, J.15]. This section discusses the conformal mapping method to get the static line parameters of the CPS structures. The dispersion and losses are also considered in sections (9.6) and (9.7).

9.4.1 Symmetrical CPS on Infinitely Thick Substrate

Figure (9.20a) shows the zero conductor thickness symmetrical CPS on a finitely thick substrate. It also shows the TEM field lines. Figure (9.20b) further shows the current distributions across both the strips. The charges of opposite polarity on two conductors attract each other to produce the increased current density, i.e. the edge singularity, at the inner edges of a CPS line.

The CPS line is a *complementary structure* of the CPW structure shown in Fig (9.12a), i.e. the slots and conductor strips of the CPW are interchanged by the conductor strips (w) and slot (s) to get the CPS structure. The conformal mapping, shown in Fig (9.12b), is also applicable to a CPS by interchanging the MW to the EW, and also EW to the MW. The CPS, after mapping in the w-plane, is shown in Fig (9.20c). It forms a parallel-plate capacitor. The line capacitance C_1 for the upper half of the CPS structure with air medium ($\varepsilon_r = 1$) and the line capacitance C_2 for the lower half of the CPS with dielectric medium (ε_r) are computed as follows:

$$C_1 = \frac{\varepsilon_0 K(k_1')}{2K(k_1)} \quad \text{(a)}$$
$$C_2 = \frac{\varepsilon_r \varepsilon_0 K(k_1')}{2K(k_1)} \quad \text{(b)}. \qquad (9.4.1)$$

The modulus k_1, giving the aspect-ratio and complementary modulus k_1', is given by equation (9.3.5). The total line capacitance *p.u.l.* of symmetrical CPS on an infinitely thick dielectric substrate and air substrate is

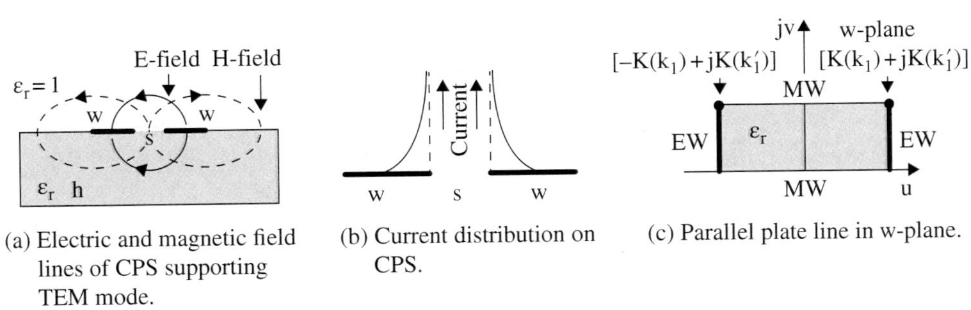

Figure 9.20 Field and current distribution on the CPS line.

$$C(\varepsilon_r) = C_1 + C_2 = \frac{(\varepsilon_r + 1)\varepsilon_0}{2}\frac{K(k_1')}{K(k_1)} \quad \text{(a)}$$
$$C(\varepsilon_r = 1) = \frac{\varepsilon_0 K(k_1')}{K(k_1)} \quad \text{(b).} \qquad (9.4.2)$$

The effective dielectric constant and the filling-factor of symmetrical CPS line on an infinitely thick substrate are obtained from the above equations:

$$\varepsilon_{r\,eff} = \frac{C(\varepsilon_r)}{C(\varepsilon_r = 1)} = \frac{\varepsilon_r + 1}{2} \quad \text{(a)}$$
$$q = \frac{1}{2} \quad \text{(b).} \qquad (9.4.3)$$

Like a CPW on an infinitely thick substrate, the above equation shows the equal division of field in the air and dielectric regions for a CPS structure also. The characteristic impedances of the CPS on the air and dielectric substrates are

$$Z(\varepsilon_r = 1) = \frac{1}{v_p C} = \sqrt{\mu_0\varepsilon_0} \times \frac{K(k_1)}{\varepsilon_0 K(k_1')} = 120\pi\frac{K(k_1)}{K(k_1')} \quad \text{(a)}$$
$$Z(\varepsilon_r) = \frac{120\pi}{\sqrt{\varepsilon_{r\,eff}}} \times \frac{K(k_1)}{K(k_1')} \quad \text{(b).} \qquad (9.4.4)$$

9.4.2 Asymmetrical CPS (ACPS) on Infinitely Thick Substrate

Figure (9.21a) shows an asymmetrical CPS (ACPS) structure in the z-plane [J.14]. It has two strip conductors of width w_1 and w_2 separated by a slot gap s. The SC-transformation, given by equation (9.3.21b), is used to get the parallel plate line in the w-plane shown in Fig (9.21b). Locations of conducting strips and slot edges in the z-plane are marked by points $\alpha, \beta, \gamma, \delta$, such that $\alpha > \beta > \gamma > \delta$.

The modulus, given by equation (9.3.21c), could be either k or complementary modulus k' depending upon the way the SC-transformation (9.3.21a) is applied to get the rectangular figure in the w-plane. Figure (9.21a) shows the ACPS with dimensions $\delta = -w_1$, $\gamma = 0$, $\beta = s$, $\alpha = s + w_2$. The SC-transformation from equation (9.3.21) is

$$w = A\int_{\delta}^{z} \frac{dt}{\sqrt{(z + w_1)z(z - s)(z - s - w_2)}}. \qquad (9.4.5)$$

The above function maps the strip conductor line segments (1)–(2) and (3)–(4) in the z-plane as the line segments (1′)–(2′) and (3′)–(4′) (EW) along the jv-axis in the w-plane. The slot segments (2)–(3) and (4–5–6–1) are mapped as (2′)–(3′) and (4′, 5′, 6′, 1′) MW. The modulus k given by equation (9.3.21) is the complementary modulus for the present case:

$$k_7' = \left[\frac{w_1 w_2}{(s + w_1)(s + w_2)}\right]^{\frac{1}{2}} \quad \text{(a)}$$
$$k_7 = \left[1 - k_7'^2\right]^{\frac{1}{2}} = \left[\frac{s(w_1 + w_2 + s)}{(s + w_1)(s + w_2)}\right]^{\frac{1}{2}} \quad \text{(b).} \qquad (9.4.6)$$

The line capacitances C_1 for ACPW with the air-filled upper half-space and the line capacitance C_2 for the dielectric-filled lower half-space is given as follows:

$$C_1 = \varepsilon_0\frac{K(k_7')}{K(k_7)} \quad \text{(a)}$$
$$C_2 = \varepsilon_0\varepsilon_r\frac{K(k_7')}{K(k_7)} \quad \text{(b).} \qquad (9.4.7)$$

The total line capacitances *p.u.l.* of the ACPS line with dielectric and air-substrates are

$$C(\varepsilon_r) = C_1(\varepsilon_r = 1) + C_2(\varepsilon_r) = (\varepsilon_r + 1)\varepsilon_0\frac{K(k_7')}{K(k_7)} \quad \text{(a)}$$
$$C(\varepsilon_r = 1) = 2\varepsilon_0\frac{K(k_7')}{K(k_7)} \quad \text{(b).} \qquad (9.4.8)$$

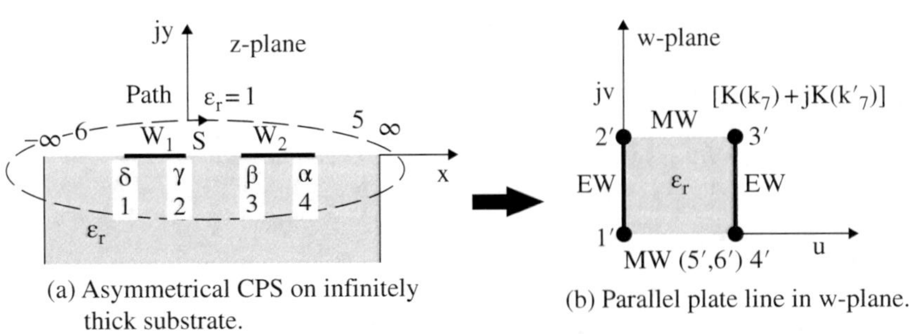

Figure 9.21 The conformal mapping of the asymmetrical CPS line on the infinitely thick substrate.

The effective relative permittivity, filling-factor, and characteristic impedances of the ACPS on the air and infinitely thick substrates are

$$\varepsilon_{r\,eff} = \frac{C(\varepsilon_r)}{C(\varepsilon_r = 1)} = \frac{\varepsilon_r + 1}{2} \quad \text{(a)}$$

$$q = \frac{1}{2} \quad \text{(b)}$$

$$Z(\varepsilon_r = 1) = \frac{1}{v_p C(\varepsilon_r = 1)} = \sqrt{\mu_0\varepsilon_0}\frac{K(k_7)}{2\varepsilon_0 K(k_7')} = 60\pi\frac{K(k_7)}{K(k_7')} \quad \text{(c)}$$

$$Z(\varepsilon_r) = \frac{60\pi}{\sqrt{\varepsilon_{r\,eff}}}\frac{K(k_7)}{K(k_7')} \quad \text{(d).}$$

(9.4.9)

The SC-transformation process presently discussed is equally applicable to the ACPW structure with two different slot gaps.

9.4.3 Symmetrical CPS on Finite Thickness Substrate

The symmetrical CPS on a finite thickness substrate, shown in Fig (9.22a), can be treated as a complementary structure of the CPW on a finite thickness substrate [J.16]. However, the assumption of a complementary structure is valid only for a thick substrate, or more correctly only for a CPS line on the air-substrate [J.17]. In any case, it is easier to implement the concept.

The line capacitance C_1 of a CPS for the upper half-space, with air medium, is given by equation (9.4.1a). Following the discussion in section (9.3.2), the line capacitance of a CPS for the lower half-space, with an inhomogeneous medium, is

$$C_2(\varepsilon_r) = C_1(\varepsilon_r = 1) + C_2'(\varepsilon_r - 1), \quad (9.4.10)$$

where C_1 is the line capacitance of a CPS for the upper half-space and C_2' is the line capacitance of a CPS with the lower half-space filled in with infinitely thick dielectric medium having relative permittivity $(\varepsilon_r - 1)$.

The capacitance C_2' is obtained by following the process given in section (9.3.2) and shown in Fig (9.22). It is noted that Fig (9.22a) is a complementary structure of Fig (9.13a). In this process, the strip conductors of Fig (9.13a) are replaced by the MW, and the slots are replaced by the strip conductors, i.e. by the EW. Thus, as shown in Fig (9.22a), the slot gap is $s = 2a$ and the strip width (w) is related to $2b = s + 2w$. The mapped Fig (9.22b) provides us the line capacitance C_2':

$$C_2'(\varepsilon_r - 1) = \frac{\varepsilon_0(\varepsilon_r - 1)K(k_2')}{2K(k_2)} \quad \text{(a)}$$

Modulus k_2, from equation – (9.3.11) :

$$k_2 = \frac{\sinh\left(\frac{\pi s}{4h}\right)}{\sinh\left[\frac{\pi}{4h}(s + 2w)\right]} \quad \text{(b)}$$

$$k_2' = \sqrt{1 - k_2^2} \quad \text{(c).}$$

(9.4.11)

Using Fig (9.22b), the line capacitance C_1 of the upper region in the air medium could be written as:

$$C_1 = \frac{\varepsilon_0 K(k_1')}{2K(k_1)} \quad \text{(a)}$$

$$\text{where,}\ k_1 = \frac{2a}{2b} = \frac{s}{s + 2w} \quad \text{(b)}$$

$$k_1' = \sqrt{1 - k_1^2} \quad \text{(c).}$$

(9.4.12)

The total line capacitance *p.u.l.* of a CPS on the finite thickness substrate and the air substrates are

$$C(\varepsilon_r) = 2C_1(\varepsilon_r = 1) + C_2'(\varepsilon_r - 1),$$

$$\Rightarrow C(\varepsilon_r) = \varepsilon_0\frac{K(k_1')}{K(k_1)} + \frac{\varepsilon_0(\varepsilon_r - 1)K(k_2')}{2K(k_2)} \quad \text{(a)}$$

$$C(\varepsilon_r = 1) = \frac{\varepsilon_0 K(k_1')}{K(k_1)} \quad \text{(b).}$$

(9.4.13)

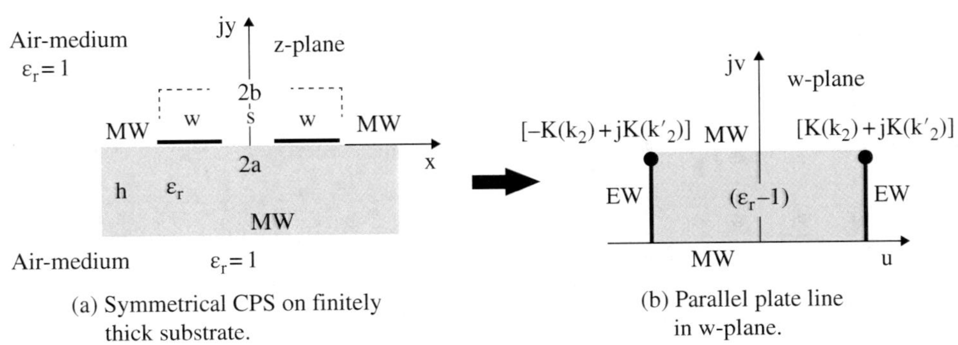

Figure 9.22 The symmetrical CPS line on the finite thickness substrate.

The effective relative permittivity and the filling-factor of a CPS line on the finite thickness substrate are

$$\varepsilon_{r\,eff} = \frac{C(\varepsilon_r)}{C(\varepsilon_r = 1)} = 1 + \frac{1}{2}(\varepsilon_r - 1)\frac{K(k_2')}{K(k_2)}\cdot\frac{K(k_1)}{K(k_1')} \quad (a)$$

$$q = \frac{1}{2}\frac{K(k_2')}{K(k_2)}\cdot\frac{K(k_1)}{K(k_1')} \quad (b). \qquad (9.4.14)$$

The characteristic impedance of the CPS on the finite thickness substrate is obtained from equation (9.4.4) with $\varepsilon_{r\,eff}$ taken from the above equation (9.4.14).

Let us examine the limiting case, $h \to 0$. The physical condition suggests that the substrate is gradually replaced by the air-substrate. Therefore, for $h \to 0$, ε_{reff} should decrease and the characteristic impedance should increase. However, as $h \to 0$, $(\pi s/4h) \to \infty$ and also $[\pi(s + 2w)/4h] \to \infty$ that gives $k_2 \to 0$. It means for $h \to 0$, $K(k_2')/K(k_2)$ increases gradually [B.14] and equation (9.4.14) provides the contradictory results with the increasing value of $\varepsilon_{r\,eff}$ and the decreasing value of the characteristic impedance. Such nonphysical results are due to the assumption of the CPS as a complementary structure of the CPW [J.16]. To remove the inconsistency, Ghione and Naldi [J.18] have assumed that the phase velocities of the CPW on the infinitely thick substrate and its complementary CPS structure are identical. Under this assumption, the effective relative permittivity $\varepsilon_{r\,eff}$ of the CPS can be written from equation (9.3.15a) as

$$\varepsilon_{r\,eff} = 1 + \frac{1}{2}(\varepsilon_r - 1)\frac{K(k_1')}{K(k_1)}\cdot\frac{K(k_2)}{K(k_2')}. \qquad (9.4.15)$$

For a CPS line under the condition $h \to 0$, the factor $K(k_2)/K(k_2')$ decreases giving a decrease in the value of $\varepsilon_{r\,eff}$ and an increase in the value of the characteristic impedance. However, this solution to the problem is artificial [B.14, J.19, J.20]. The failure of the complementary model of the CPS, with respect to a CPW structure on the finitely thick substrate, suggests the need for a new conformal mapping transformation. It is discussed below.

Alternate Conformal Mapping of the CPS Line on Finitely Thick Substrate

The total line capacitance $C(\varepsilon_r)$ of a CPS on a finitely thick substrate is still given by equation (9.4.13a). However, C_2', i.e. the line capacitance on the $(\varepsilon_r - 1)$ substrate, has to be recomputed. For this purpose, the CPS is redrawn in Fig (9.23a) and an EW of length (1)–(6) is placed along the y-axis at the center of the slot. It is assumed that the capacitance C_2' is due to the series combination of two identical sections of capacitance C_2'' in the third and fourth quadrants, i.e. $C_2' = C_2''/2$. The conformal mapping to compute the line capacitance C_2'' is achieved in the following two steps:

Step #1

The EW (1)–(6) of the z-plane is transformed as a coplanar strip $(t_6 - t_1)$ in the t-plane, shown in Fig (9.23b), on an infinitely thick substrate with relative permittivity $(\varepsilon_r - 1)$ by using the following mapping function:

$$t = \cosh^2\left[\frac{\pi z}{2h}\right]. \qquad (9.4.16)$$

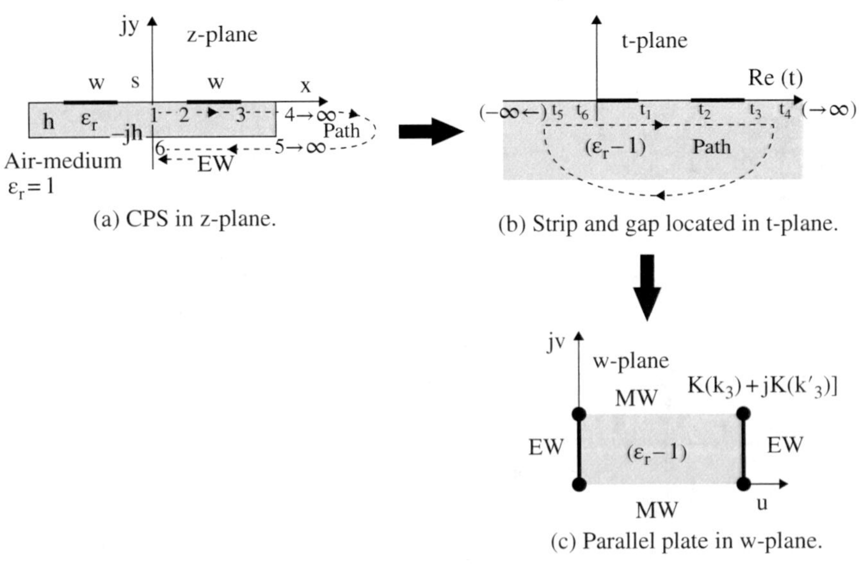

Figure 9.23 Lower half-space mapping of symmetrical CPS on finite substrate thickness.

Table 9.6 Transformation using $t = \cosh^2[\pi z/2h]$.

Points in z-plane	Points in the t-plane
1) $z = 0$	$t_1 = 1$
2) $z = \frac{s}{2}$	$t_2 = \cosh^2\left[\frac{\pi s}{4h}\right]$
3) $z = w + \frac{s}{2} = \frac{2w+s}{2}$	$t_3 = \cosh^2\left[\frac{\pi}{4h}(2w+s)\right]$
4) $z \to \infty$	$t_4 = \infty$
5) $z = (-jh + x),\ x \to \infty$	$t_5 = -\infty$
6) $z = -jh$	$t_6 = 0$

The path of transformation is shown on both the z- and t-planes. The locations (1)–(6) on the z-plane are transferred as the corresponding locations t_1, t_2, t_3, t_4, t_5, t_6 on the t-plane, as shown in Table (9.6).

Step #2

The coplanar strip structure from the t-plane is transformed as a rectangle in the w-plane. It is shown in Fig (9.23c). The mapping is achieved through the following SC-transformation:

$$w = u + jv = \int_{t_6}^{t} \frac{dt}{\sqrt{(t-t_6)(t-t_1)(t-t_2)(t-t_3)}} = K(k_3) + jK(k_3'). \tag{9.4.17}$$

The modulus k_3 is obtained from equation (9.3.21d) [J.12, J.20]:

$$\begin{aligned}
k_3^2 &= \frac{(\alpha-\delta)(\beta-\gamma)}{(\alpha-\gamma)(\beta-\delta)}; \text{ where } \delta = t_6 = 0,\\
\gamma &= t_1,\ \beta = t_2 = \cosh^2\left[\frac{\pi s}{4h}\right]\\
\alpha &= t_3 = \cosh^2\left[\frac{\pi}{4h}(s+2w)\right]\\
\Rightarrow k_3^2 &= \frac{\left[\cosh^2\left(\frac{\pi s}{4h}\right) - 1\right]\cosh^2\left\{\frac{\pi}{4h}(s+2w)\right\}}{\left[\cosh^2\left\{\frac{\pi}{4h}(s+2w)\right\} - 1\right]\cosh^2\left(\frac{\pi s}{4h}\right)}\\
&= \frac{\sinh^2\left(\frac{\pi s}{4h}\right)\cosh^2\left\{\frac{\pi}{4h}(s+2w)\right\}}{\sinh^2\left\{\frac{\pi}{4h}(s+2w)\right\}\cosh^2\left(\frac{\pi s}{4h}\right)}\\
\Rightarrow k_3 &= \frac{\tanh\left(\frac{\pi s}{4h}\right)}{\tanh\left\{\frac{\pi}{4h}(s+2w)\right\}}.
\end{aligned} \tag{9.4.18}$$

The partial line capacitance C_2'' for the one-half of the CPS, and also C_2' for the complete CPS, on the finitely thick substrate is

$$\begin{aligned}
C_2'' &= \varepsilon_0(\varepsilon_r - 1)\frac{K(k_3')}{K(k_3)} && \text{(a)}\\
C_2' &= \frac{C_2''}{2} = \frac{\varepsilon_0(\varepsilon_r-1)}{2}\frac{K(k_3')}{K(k_3)} && \text{(b).}
\end{aligned} \tag{9.4.19}$$

The total line capacitance of a CPS on a finitely thick dielectric and air substrates are

$$\begin{aligned}
&C(\varepsilon_r) = 2C_1 + C_2',\\
&\Rightarrow C(\varepsilon_r) = \varepsilon_0\frac{K(k_1')}{K(k_1)} + \frac{\varepsilon_0(\varepsilon_r-1)}{2}\frac{K(k_3')}{K(k_3)} && \text{(a)}\\
&C(\varepsilon_r = 1) = \varepsilon_0\frac{K(k_1')}{K(k_1)} && \text{(b).}
\end{aligned} \tag{9.4.20}$$

Finally, the effective relative permittivity and the filling-factor of the CPS are

$$\begin{aligned}
\varepsilon_{r\,eff} &= \frac{C(\varepsilon_r)}{C(\varepsilon_r = 1)} = 1 + \frac{1}{2}(\varepsilon_r - 1)\frac{K(k_3')}{K(k_3)}\cdot\frac{K(k_1)}{K(k_1')} && \text{(a)}\\
q &= \frac{1}{2}\frac{K(k_3')}{K(k_3)}\cdot\frac{K(k_1)}{K(k_1')} && \text{(b).}
\end{aligned} \tag{9.4.21}$$

The characteristic impedance of the CPS is obtained from equation (9.4.4) on using equation (9.4.21a) for $\varepsilon_{r\,eff}$. The presence of tanh in place of sinh in the modulus alters the limiting conditions. For the case $h \to \infty$, $k_3 \to k_1 = s/(2w+s)$ and for $h \to 0$, $k_3 \to 1$. The decreasing value of $K(k_3')/K(k_3)$ is obtained for the substrate thickness $h \to 0$. Therefore, for $h \to 0$, $\varepsilon_{r\,eff}$ decreases and the characteristic impedance increases for a CPS line. It is the physically correct description. It is noted that the above expression is obtained without using the complementary condition for the CPS line. Figure (9.24) shows the effective relative permittivity computed using the CPW-based expression (9.4.15) and also using the expression (9.4.21). The complementary condition holds only for the thick substrates, i.e. for $2h/(2w+s) > 2$ and low permittivity substrate.

The CPS structure could be provided with the additional floating ground plane at the backside of the substrate [J.21]. The backside metallization of a CPS provides mechanical strength and also heat sink facility for the active devices. The presence of the backside conductor increases the effective relative permittivity of a CPS. The CPS with the bottom conductor supports two fundamental modes – *the CPS mode and the parasitic microstrip mode*. The parasitic microstrip mode is significant for a thin substrate with a wide strip conductor and wide slot width. Both the dielectric and conductor losses of the line structure are high, as compared to the normal CPS line without conductor backing.

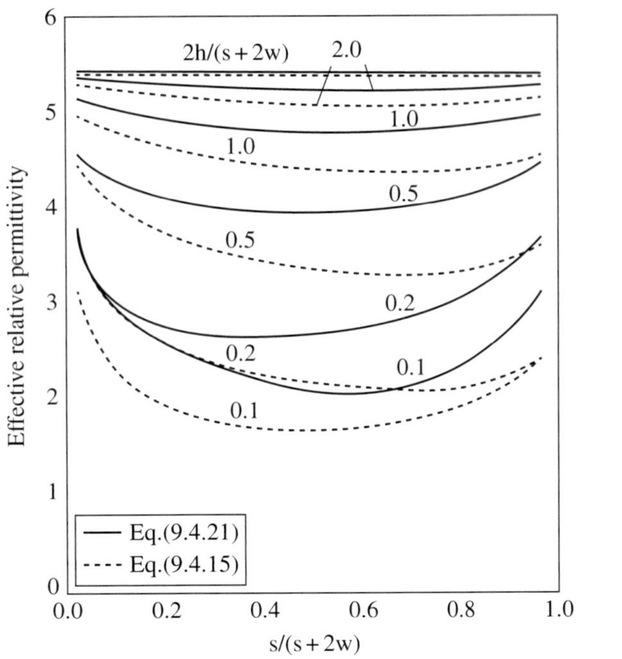

Figure 9.24 Validation of concept of complementary structure (ε_r = 9.9). *Source:* Zhu and Wang [J.20]

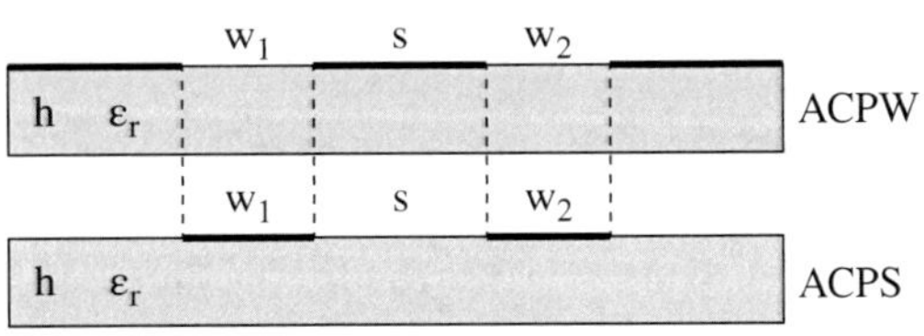

Figure 9.25 ACPW and ACPS as complementary pairs.

9.4.4 Asymmetrical CPW (ACPW) and Asymmetrical CPS (ACPS) on Finite Thickness Substrate

Figure (9.25) shows the ACPW and ACPS as the complementary pairs [J.12, J.22]. It is seen in the previous section that the complementary pair's concept does not hold for a thin and high permittivity substrate. It gives an error in the computation of the line parameters.

ACPW Line

Figure (9.26) shows that the conformal mapping of the ACPW structure is carried out in two steps. *Step* #1 transforms the ACPW with lower half-space of the inhomogeneous medium to the ACPW with lower half-space filled in with the infinitely thick homogeneous medium having assumed relative permittivity $(\varepsilon_r - 1)$. *Step* #2 transforms the CPW structure, on the infinitely thick substrate, of the step #1 to a parallel-plate capacitor in the w-plane.

The line capacitance of the ACPW structure is

$$C(\varepsilon_r) = 2C_1(\varepsilon_r = 1) + C_2(\varepsilon_r - 1), \tag{9.4.22}$$

where C_1 is the line capacitance of the upper/lower half-space of the ACPW on the air-substrate. C_2 is the line capacitance of the ACPW with lower half-space filled with infinite extent dielectric having relative permittivity $(\varepsilon_r - 1)$.

Step #1

The following mapping function transforms the lower half-space of the ACPW located in the z-plane to an ACPW on the infinite extent substrate in the t-plane:

$$t = \sinh\left(\frac{\pi}{2h} z\right). \tag{9.4.23}$$

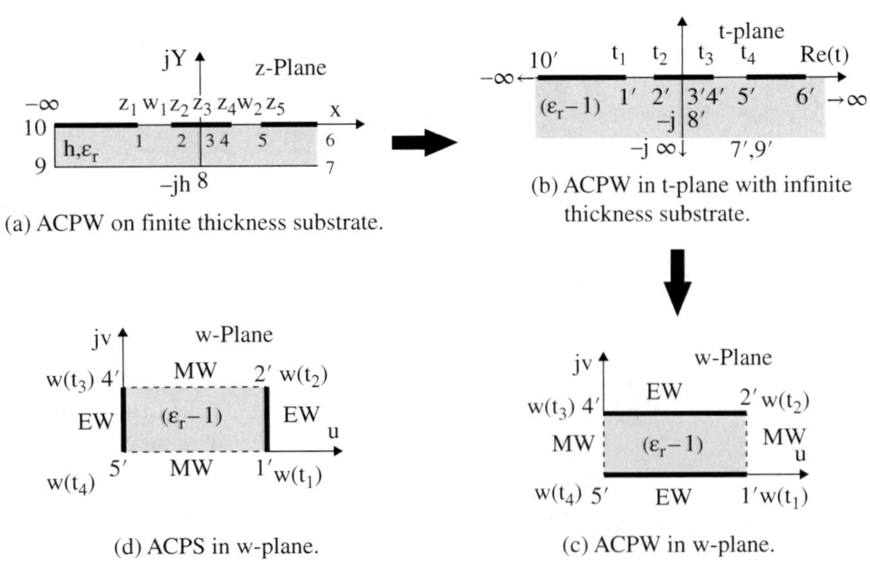

Figure 9.26 Mapping of asymmetrical CPW/CPS of finite thickness substrate.

The transformation is shown in Fig (9.26a,b). The transformation rotates lines (8)–(7) and (8)–(9) about jy-axis such that the closed finite space of the substrate thickness h in the z-plane, opens to the infinite lower half-space in the t-plane. The t-plane transformation is not needed for an ACPW in the air medium. The above mapping function maps the locations z_1, z_2, z_3, z_4, etc. in the z-plane to the corresponding locations in the t-plane. These are summarized in Table (9.7).

Step #2

Figure (9.26c) shows that the following SC-transformation converts the line segments of the lower half-space in the t-plane to a rectangle in the w-plane:

$$\begin{aligned} w &= u + jv = K(k) + jK(k') \\ &= A\int_t \frac{dt}{\sqrt{(t-t_1)(t-t_2)(t-t_4)(t-t_5)}} && \text{(a)} \\ \text{where,}\ k &= \left[\frac{(t_4-t_2)(t_5-t_1)}{(t_5-t_2)(t_4-t_1)}\right]^{\frac{1}{2}} && \text{(b)} \\ k' &= \left[\frac{(t_5-t_4)(t_2-t_1)}{(t_5-t_2)(t_4-t_1)}\right]^{\frac{1}{2}} && \text{(c).} \end{aligned} \tag{9.4.24}$$

Table 9.7 Transformation using $t = \sinh(\pi z/2h)$.

Points in the z-plane	Points in the t-plane
$z_1 = -\left(w_1 + \frac{s}{2}\right)$	$t_1 = -\sinh\left[\frac{\pi}{4h}(s + 2w_1)\right]$
$z_2 = -\frac{s}{2}$	$t_2 = -\sinh\left[\frac{\pi}{4h}s\right]$
$z_3 = 0$	$t_3 = 0$
$z_4 = \frac{s}{2}$	$t_4 = \sinh\left[\frac{\pi}{4h}s\right]$
$z_5 = w_2 + \frac{s}{2}$	$t_5 = \sinh\left[\frac{\pi}{4h}(s + 2w_2)\right]$
$z_6 \to \infty$	$t_6 \to \infty$
$z_7 \to -jh + \infty$	$t_7 = -j\infty$
$z_8 \to -jh$	$t_8 = -j$
$z_9 \to -jh - \infty$	$t_9 = -j\infty$
$z_{10} \to -\infty$	$t_{10} \to -\infty$

The role of the modulus k and complimentary modulus k′ gets interchanged. The modulus k and k′ are obtained from equation (9.3.21). On substituting the variable t from Table (9.7) in the above equation, the modulus $k = k_8, k' = k'_8$ is obtained.

$$\begin{aligned} k_8^2 &= \frac{2\sinh\left(\frac{\pi s}{4h}\right)\left[\sinh\left\{\frac{\pi}{4h}(s+2w_1)\right\} + \sinh\left\{\frac{\pi}{4h}(s+2w_2)\right\}\right]}{\left[\sinh\left\{\frac{\pi}{4h}(s+2w_1)\right\} + \sinh\left(\frac{\pi s}{4h}\right)\right]\left[\sinh\left(\frac{\pi s}{4h}\right) + \sinh\left\{\frac{\pi}{4h}(s+2w_2)\right\}\right]} && \text{(a)} \\ k_8'^2 &= \frac{\left[\sinh\left\{\frac{\pi}{4h}(s+2w_1)\right\} - \sinh\left(\frac{\pi s}{4h}\right)\right]\left[\sinh\left(\frac{\pi s}{4h} + 2w_2\right) - \sinh\left\{\frac{\pi s}{4h}\right\}\right]}{\left[\sinh\left\{\frac{\pi}{4h}(s+2w_1)\right\} + \sinh\left(\frac{\pi s}{4h}\right)\right]\left[\sinh\left(\frac{\pi s}{4h}\right) + \sinh\left\{\frac{\pi}{4h}(s+2w_2)\right\}\right]} && \text{(b).} \end{aligned} \tag{9.4.25}$$

In the limiting case of infinite extent substrate, i.e. for $h \to \infty$, the above equations for k_8 and k'_8 reduce to equation (9.4.6a,b), applicable to k_7 and k'_7. The line capacitance of the lower/upper space of the ACPW on the air-substrate, total line capacitance of the ACPW on the air-substrate, and total line capacitance on the dielectric substrate are

$$\begin{aligned} C_1 &= \varepsilon_0 \frac{K(k_7)}{K(k'_7)} && \text{(a)} \\ C(\varepsilon_r = 1) &= 2C_1 = 2\varepsilon_0 \frac{K(k_7)}{K(k'_7)} && \text{(b)} \\ C(\varepsilon_r) &= 2\varepsilon_0 \frac{K(k_7)}{K(k'_7)} + \varepsilon_0(\varepsilon_r - 1)\frac{K(k_8)}{K(k'_8)} && \text{(c).} \end{aligned} \tag{9.4.26}$$

The effective relative permittivity and filling-factor of the ACPW are

$$\begin{aligned} \varepsilon_{r\,eff} &= \frac{C(\varepsilon_r)}{C(\varepsilon_r = 1)} = 1 + \frac{1}{2}(\varepsilon_r - 1)\frac{K(k'_7)}{K(k_7)} \cdot \frac{K(k_8)}{K(k'_8)} && \text{(a)} \\ q &= \frac{1}{2}\frac{K(k'_7)}{K(k_7)} \cdot \frac{K(k_8)}{K(k'_8)} && \text{(b).} \end{aligned} \tag{9.4.27}$$

The characteristic impedance of the ACPW on the air and finite extent dielectric substrates are

$$\begin{aligned} Z(\varepsilon_r = 1) &= \frac{1}{v_p C(\varepsilon_r = 1)} = \sqrt{\mu_0\varepsilon_0}\frac{K(k'_7)}{2\varepsilon_0 K(k_7)} \\ \Rightarrow Z(\varepsilon_r = 1) &= 60\pi \frac{K(k'_7)}{K(k_7)} && \text{(a)} \\ Z(\varepsilon_r) &= \frac{60\pi}{\sqrt{\varepsilon_{r\,eff}}}\frac{K(k'_7)}{K(k_7)} && \text{(b).} \end{aligned} \tag{9.4.28}$$

The effective relative permittivity $\varepsilon_{r\,eff}$ could be examined under the limiting cases, $h \to \infty$ and $h \to 0$. For

$h \to \infty$, the ACPW line is on an infinite extent substrate leading to $k_8 \to k_7, k'_8 \to k'_7$, $q = 1/2$. For this case, $\varepsilon_{r\,eff} = (\varepsilon_r + 1)/2$ that shows an equal distribution of the field in the upper and lower media. For $h \to 0$, $k_8 \to 0$, i.e. $K(k_8)/K(k'_8) \to 0$, i.e. $q \to 0$, so $\varepsilon_{r\,eff} \to 1$. It is physically correct, as $h \to 0$ removes the substrate and the lower half-space is filled with air-medium.

ACPS Line

The ACPS could be treated through complementary Fig (9.26d). However, it is better to assume identical phase velocity for both the ACPW structure and its complementary ACPS structure [J.18]. Thus, equation (9.4.27) for $\varepsilon_{r\,eff}$ and q are also applicable to an ACPS. On using the effective relative permittivity $\varepsilon_{r\,eff}$ with equation (9.4.9), the characteristic impedance of the ACPS line on the finite extent substrate is computed. The alternate expressions are also available in the literature for the modulus k_8 and k'_8 [B.14, B.16, J.12].

9.4.5 Asymmetric CPS Line with Infinitely Wide Ground Plane

The asymmetric CPS line with an infinitely wide ground plane, shown in Fig (9.19e), is commonly used in the electro-optics modulators [J.23, J.24]. It is also known as the *micro coplanar line* (MCL). The line to line coupling of the CPS structure is reduced in the MCL. The TM_0 dielectric slab waveguide mode that exists with the CPS is eliminated by the ground plane of the MCL. The structure could be treated as a limiting case of the ACPS structure with $w_2 \to \infty$. Thus, for an infinitely thick substrate ($h \to \infty$), the modulus k_9 is obtained from equation (9.4.6) for $w_2 \to \infty$, $w_1 = w$:

$$k_9 = \left[\frac{s}{s+w}\right]^{\frac{1}{2}} \quad \text{(a)}$$

$$k'_9 = \left[\frac{w}{s+w}\right]^{\frac{1}{2}} \quad \text{(b).}$$

(9.4.29)

The characteristic impedance of the MCL obtained from equation (9.4.9d) is

$$Z(\varepsilon_r) = \frac{60\pi}{\sqrt{\varepsilon_{r\,eff}}} \frac{K(k_9)}{K(k'_9)} \quad \text{(a)}$$

$$\text{where,} \quad \varepsilon_{r\,eff} = \frac{\varepsilon_r + 1}{2} \quad \text{(b).}$$

(9.4.30)

However, for a finite substrate thickness, there is another modulus also. For the lower half-space, the strip widths are $w_1 = w$, $w_2 \to \infty$. The modulus k_{10} is obtained from equation (9.4.25):

$$k_{10}^2 \approx \frac{2\sinh\left(\frac{\pi s}{4h}\right)\left[\sinh\left\{\frac{\pi}{4h}(s+2w)\right\} + \exp\left\{\frac{\pi}{4h}(s+2w_2)\right\}\right]}{\left[\sinh\left\{\frac{\pi}{4h}(s+2w)\right\} + \sinh\left(\frac{\pi s}{4h}\right)\right]\left[\sinh\left(\frac{\pi s}{4h}\right) + \exp\left\{\frac{\pi}{4h}(s+2w_2)\right\}\right]}$$

$$\approx \frac{2\sinh\left(\frac{\pi s}{4h}\right)\exp\left\{\frac{\pi}{4h}(s+2w_2)\right\}}{\left[\sinh\left\{\frac{\pi}{4h}(s+2w)\right\} + \sinh\left(\frac{\pi s}{4h}\right)\right]\exp\left\{\frac{\pi}{4h}(s+2w_2)\right\}}$$

$$\Rightarrow k_{10}^2 = \frac{2\sinh\left(\frac{\pi s}{4h}\right)}{\left[\sinh\left\{\frac{\pi}{4h}(s+2w)\right\} + \sinh\left(\frac{\pi s}{4h}\right)\right]} \quad \text{(a)}$$

$$k'^2_{10} = \sqrt{1 - k_{10}^2} \quad \text{(b).}$$

(9.4.31)

Another expression for modulus k_{10} is also available [J.12, B.14, B.16].

The effective relative permittivity and filling-factor from equation (9.4.27) are

$$\varepsilon_{r\,eff} = 1 + \frac{1}{2}(\varepsilon_r - 1)\frac{K(k'_9)}{K(k_9)} \cdot \frac{K(k_{10})}{K(k'_{10})} \quad \text{(a)}$$

$$q = \frac{K(k'_9)}{K(k_9)} \cdot \frac{K(k_{10})}{K(k'_{10})} \quad \text{(b).}$$

(9.4.32)

The MCL could be provided with another ground plane at the back of the substrate. Such MCL could be viewed as a microstrip line with an additional coplanar ground plane that helps the mounting of shunt components in the microstrip technology. This structure, also called *micro-coplanar strip* (MCS), was proposed and analyzed by Yamashita et al. [J.25].

9.4.6 CPS with Coplanar Ground Plane [CPS–CGP]

The CPS with two infinitely wide *coplanar ground planes* forms the (CPS–CGP) line. It is shown in Fig (9.27) [J.2, J.20, J.26]. The presence of ground planes reduces the

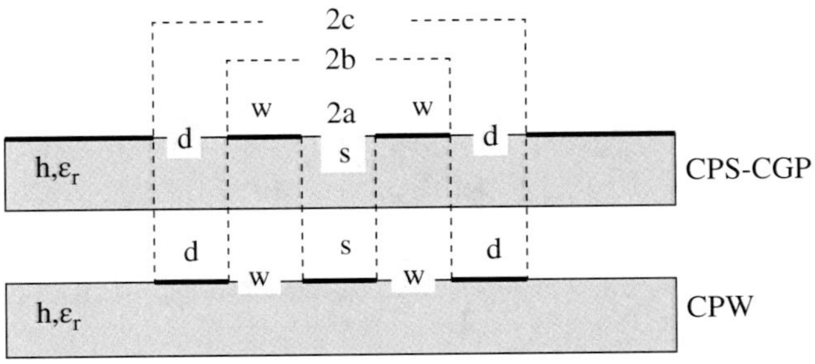

Figure 9.27 The CPS with the coplanar ground plane and its complimentary CPW.

line to line coupling that is a drawback of the conventional CPS line. The TM_0 dielectric slab mode, supported by the CPS line, is eliminated. The spacing between the signal lines and ground planes is changed to modify the characteristic impedance and effective relative permittivity of the CPS–CGP line. Mclean and Itoh have carried out both the full-wave and quasi-static analysis of the structure [J.2]. The CPS–CGP structure can also be treated as a complementary structure of the finite ground plane CPW on a finitely thick substrate. It is shown in Fig (9.27).

The characteristic impedance of the CPS–CGP structure on the air-substrate is given by equation (9.4.4) for the standard CPS. However, the modulus has to be modified to account for the lateral ground planes. The CPS–CGS line is treated as a complementary structure of the CPW with the finite ground plane. Following the discussion of previous sections, the phase velocities of both structures are taken as identical [J.18]. The modulus k_{11} and k'_{11} are obtained from equation (9.3.17):

$$k_{11} = \frac{a}{b}\left[\frac{c^2 - b^2}{c^2 - a^2}\right]^{\frac{1}{2}} = \frac{s}{2w + s}\left[\frac{(d + 2w + s)d}{(d + w + s)(d + w)}\right] \quad \text{(a)}$$

$$k'_{11} = \sqrt{1 - k_{11}^2} \quad \text{(b).}$$

where, $2a = s,\ 2b = 2w + s,\ 2c = 2d + 2w + s$

$$(9.4.33)$$

Likewise, the modulus k_{12} and k'_{12} in the presence of the finitely thick substrate are obtained from equation (9.3.18):

$$k_{12} = \frac{\sinh\left(\frac{\pi a}{2h}\right)}{\sinh\left(\frac{\pi b}{2h}\right)}\left[\frac{\sinh^2\left(\frac{\pi c}{2h}\right) - \sinh^2\left(\frac{2\pi b}{2h}\right)}{\sinh^2\left(\frac{\pi c}{2h}\right) - \sinh^2\left(\frac{2\pi a}{2h}\right)}\right]^{\frac{1}{2}} \quad \text{(a)}$$

$$k'_{12} = \sqrt{1 - k_{12}^2} \quad \text{(b).}$$

$$(9.4.34)$$

The effective relative permittivity and the characteristic impedance of the CPS–CGP structure are

$$\varepsilon_{r\,eff} = 1 + \frac{1}{2}(\varepsilon_r - 1)\frac{K(k'_{11})}{K(k_{11})}\cdot\frac{K(k_{12})}{K(k'_{12})}. \quad (9.4.35)$$

$$Z(\varepsilon_r = 1) = 120\pi\frac{K(k_{11})}{K(k'_{11})} \quad \text{(a)}$$

$$Z(\varepsilon_r) = \frac{120\pi}{\sqrt{\varepsilon_{r\,eff}}}\frac{K(k_{11})}{K(k'_{11})} \quad \text{(b).} \quad (9.4.36)$$

Zhu and Wang [J.20] have shown that the concept of the complementary pair is satisfactory for $h/b \geq 0.5$. For the thinner substrates, on using the direct conformal mapping transformation, they have obtained the following improved expression for the modulus k_{13}:

$$k_{13} = \left[\frac{\tanh^2\left(\frac{\pi c}{4h}\right) - \tanh^2\left[\frac{\pi(b + a)}{4h}\right]\tanh^2\left[\frac{\pi(b - a)}{4h}\right]}{\tanh^2\left(\frac{\pi c}{4h}\right)\left\{1 - \tanh^2\left[\frac{\pi(b + a)}{4h}\right]\tanh^2\left[\frac{\pi(b - a)}{4h}\right]\right\}}\right]^{\frac{1}{2}} \quad \text{(a)}$$

$$k'_{13} = \sqrt{1 - k_{13}^2} \quad \text{(b).}$$

$$(9.4.37)$$

The k_{12} is replaced by k_{13} in equation (9.4.35) to compute the improved value of the effective relative permittivity for the CPS–CGP structure.

9.4.7 Discussion on Results for CPS

Figure (9.28) shows the variation in the effective relative permittivity and the characteristic impedance of the symmetrical CPW and symmetrical CPS on alumina ($\varepsilon_r = 10$) substrate [J.18]. The CPS is treated as the complementary structure of the CPW and their phase velocities are taken as identical, i.e. their effective relative permittivities are identical. The normalized substrate thickness (h/b) is taken as the parameter.

Figure (9.28a) shows a decrease in the effective relative permittivity of CPW/CPS for a thinner substrate. It causes an increase in characteristic impedance for both the CPS and CPW for a thin substrate. These results on the characteristics impedance for the CPS and CPW are shown in Fig (9.28b,c), respectively. However, the CPS line provides a much higher value of the characteristic impedance. The characteristic impedance of the CPS increases, while that of the CPW decreases with an increasing aspect-ratio (a/b). Therefore, the characteristic impedance can be controlled both by varying the substrate thickness and the slot width for a fixed width of the strip. The strip width could be decided by the space consideration on a substrate and by considering the conductor loss [J.12]. The conformal mapping results are compared against the results of the variational method for both the upper and lower bounds. In conclusion, it is seen that the characteristic impedance of

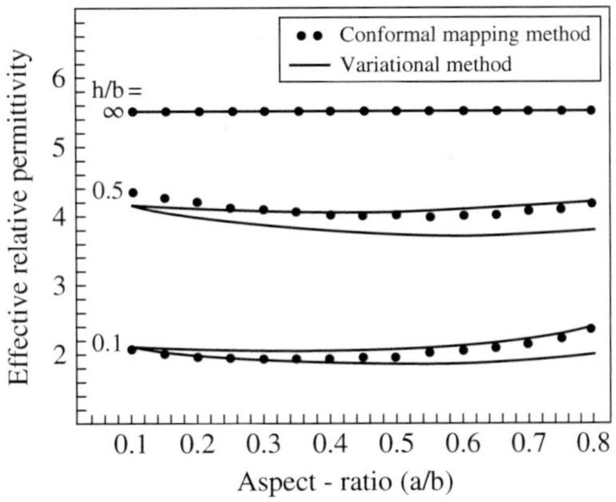

(a) Effective relative permittivity of CPS and CPW.

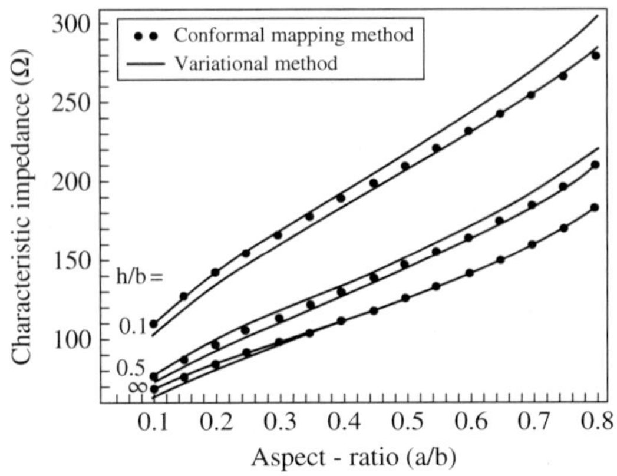

(b) Characteristic impedance of CPS.

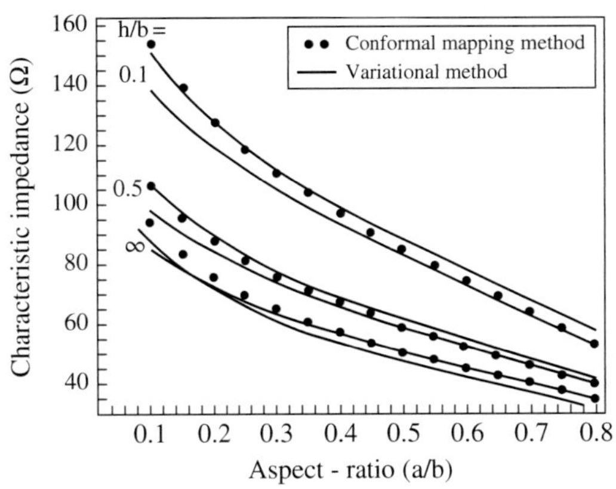

(c) Characteristic impedance of CPW.

Figure 9.28 Characteristics of CPS and CPW on the finitely thick substrate. *Source:* Ghione and Naldi [J.18]. © IET.

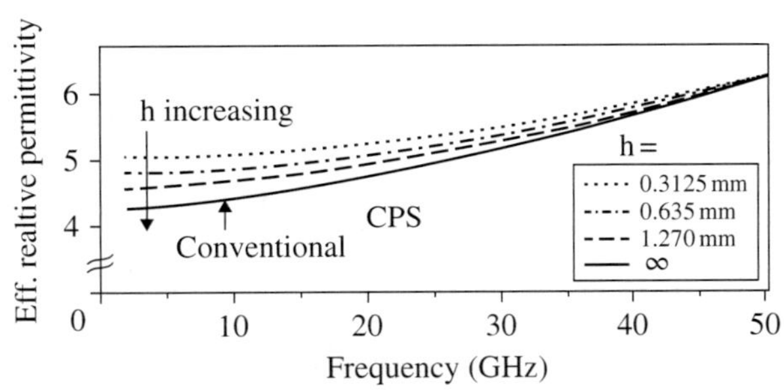

(a) Effective relative permittivity of CPS – CGP line [J.26].

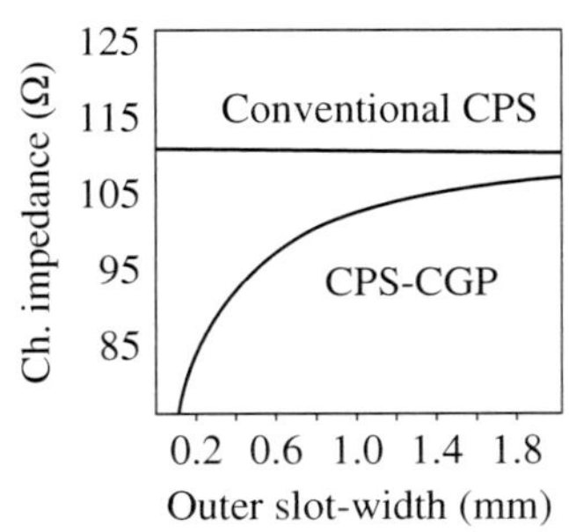

(b) Characteristic impedance of CPS – CGP line [J.2].

Figure 9.29 Characteristic of CPS–CPG line on an alumina substrate (ε_r = 9.9, h = s = w = 0.635 mm). *Source:* McLean and Itoh [J.2]. © IEEE. Svacina [J.26]. © IEEE.

the asymmetric CPS and CPW are significantly controlled by varying the strip width ratio and the ratio of the slot gaps, respectively [J.22]. However, impedance control in the case of ACPS is more as compared to the control in ACPW.

In the case of a CPS–CPG line on a finitely thick substrate, Fig 9.29a shows that the addition of coplanar ground increases the effective relative permittivity of the CPS, as more field lines are confined in the dielectric region. Furthermore, the CPS–CPG line is less dispersive, as compared to the standard CPS line. Figure (9.29b) shows that the coplanar ground plane strips reduce the characteristic impedance of CPS significantly. The coplanar ground plane strips can be used as an additional means to control the characteristic impedance of a CPS structure.

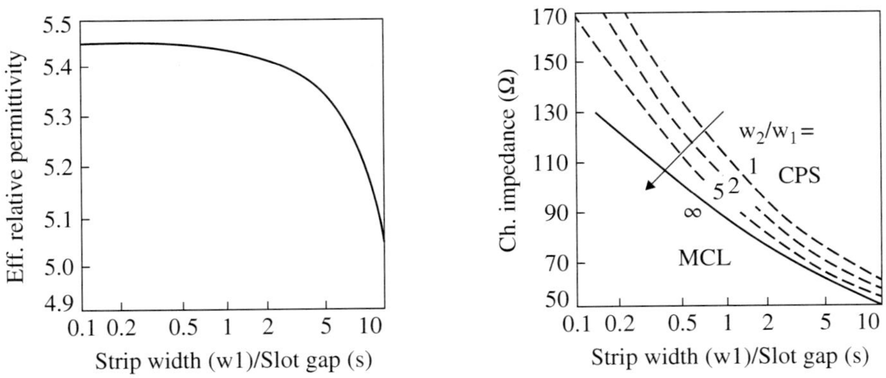

(a) `Effective relative permittivity of MCL. (b) Characteristic impedance of MCL.

Figure 9.30 Characteristic of MCL structure on a finitely thick substrate (ε_r = 9.9, s/h = 0.1). *Source:* Svacina [J.26]. © IEEE.

Figure (9.30a,b) shows the variation in the effective relative permittivity and characteristic impedance of the MCL structure on a finitely thick alumina substrate ($\varepsilon_r = 9.7$) [J.26]. The nearness of the large ground plane, i.e. a decrease in slot width, reduces the effective relative permittivity. It also reduces the characteristic impedance. Figure (9.30b) also shows the gradual approach of the CPS line to the MCL, as one of the strip-conductors increases to infinity. It shows that the characteristic impedance of the MCL is less than that of the CPS for the same slot width.

9.5 Effect of Conductor Thickness on Characteristics of CPW and CPS Structures

The conductor thickness of the CPW and CPS structures, like a microstrip line, has a noticeable influence on their effective relative permittivity, characteristic impedance, and conductor loss. Such influences are also frequency-dependent and are obtained using the full-wave methods and also incorporated empirically in the results of the conformal mapping method [B.16, B.17, J.27–J.29].

9.5.1 CPW Structure

The conductor thickness (t) causes more confinement of the electric fields in the air-medium of two slot regions causing a decrease in the effective relative permittivity. Figure (9.31a) shows the full-wave method results of a CPW for the frequency-dependent slowing-factor $\left(1/\sqrt{\varepsilon_{r\,eff}(f,t)}\right)$. It also demonstrates an increase in the slowing-factor with the increase in normalized conductor thickness (t/w) strip conductors. Figure (9.31b) further shows the decrease in the frequency-dependent characteristic impedance with an increase in the normalized conductor thickness [J.28]. The increase in conductor thickness increases the equivalent widths of the strip conductors that reduce the slot gap width causing an increase in the CPW line capacitance. The enhanced value of the line capacitance reduces the characteristics impedance $Z_0 = 1/\left[c\sqrt{C(\varepsilon_r)C(\varepsilon_r = 1)}\right]$ of the CPW. For t/w = 0.02, an increase in the guided wavelength is about 1% and a decrease in the characteristic impedance is about 1.5% with respect to the results of zero conductor thickness.

To account for the effect of conductor thickness on the CPW line parameters, the physical width(s) of the central signal strip is replaced by the equivalent width [B.16]. Two empirical relations are used for this purpose. Effectively, conductor thickness decreases the slot width w:

$$s_{eq} = s + \Delta s \quad \text{(a)}$$

$$w_{eq} = w - \Delta s \quad \text{(b)}$$

$$\text{where,}\quad \Delta s = \frac{t}{2\pi\varepsilon_r}\left[1 + \ln\frac{8\pi w}{t}\right] \quad \text{(c)}$$

Alternatively,

$$\Delta s = \frac{t}{\pi}\left[1 + \ln 4 - 0.5\ \ln\left[\left(\frac{t}{h}\right)^2 + \left(\frac{t}{\pi s}\right)^2 \frac{8\pi w}{t}\right]\right] \quad \text{(d).}$$

(9.5.1)

The equation (9.5.1d) is taken from the reference [B.17]. The aspect-ratio parameter $k_{1,t}$ of the CPW, on infinite thickness substrate, with conductor thickness is obtained from equation (9.3.5):

$$k_{1,t} = \frac{s_{eq}}{s_{eq} + 2w_{eq}} \quad \text{(a)}$$

$$k'_{1,t} = \sqrt{1 - k_{1,t}^2} \quad \text{(b).}$$

(9.5.2)

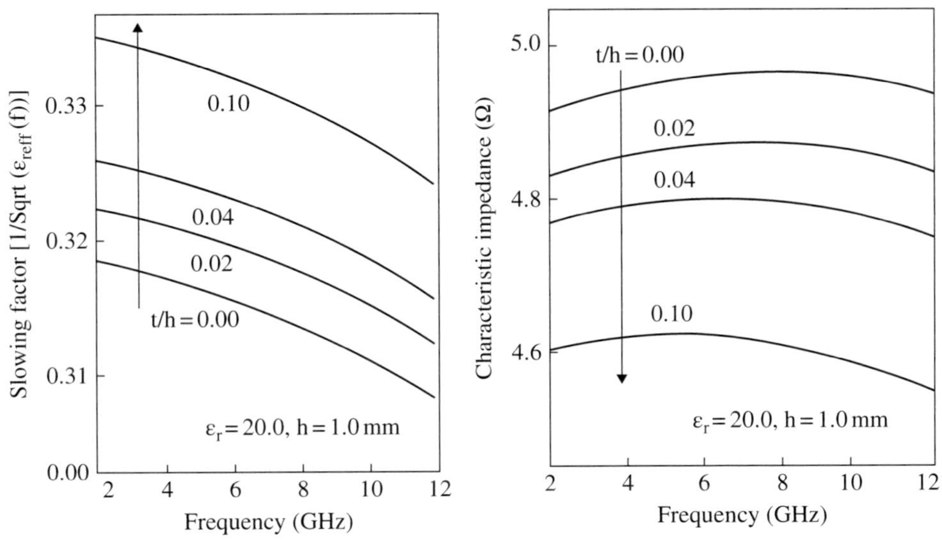

(a) Slowing factor. (b) Characteristic impedance.

Figure 9.31 Effect of conductor thickness on the line parameters of CPW. *Source:* Kitazawa et al. [J.28]. © IEEE.

Likewise, the aspect-ratio parameter $k_{2,t}$ of the CPW, on *finite* thickness substrate, with conductor thickness could be obtained from equation (9.3.11) on using s_{eq} and w_{eq}. The effective relative permittivity and characteristic impedance are computed by the following empirical relation [B.16]:

$$\varepsilon_{r\,eff}(t) = \varepsilon_{r\,eff}(t=0) - \frac{0.7(\varepsilon_{r\,eff}(t=0)-1)t/w}{[K(k)/K(k')] + 0.7t/w} \quad (a)$$

$$Z(t) = \frac{30\pi}{\sqrt{\varepsilon_{r\,eff}(t)}}\frac{K(k_t')}{K(k_t)} \quad (b).$$

$$(9.5.3)$$

The above expressions are approximate and do not provide good results for a narrow slot width CPW.

9.5.2 CPS Structure

The full-wave results, shown in Fig (9.32), demonstrate that the effective relative permittivity and the characteristic impedance of a CPS decrease with an increase in the conductor thickness [J.29]. Results are shown for a CPS on GaAs ($\varepsilon_r = 12.9$) substrate with slot width $(b-a)/h = w/h = 0.5$. The normalized conductor thickness $t/(b-a) = t/w$ is a parameter.

The characteristic impedance, using the power–voltage definition, decreases, and the effective relative permittivity increases with increasing frequency. The changes, due to the conductor thickness, in the characteristic impedance, especially in the higher frequency range, are more as compared to the changes in the effective relative permittivity. For a change in the normalized strip thickness from 0 to 0.1, the $\varepsilon_{r\,eff}(f)$ is changed by 6%, whereas the $Z_0(f)$ is changed by 9 and 22% at $h/\lambda = 0.01$ and $h/\lambda = 0.1$, respectively.

9.6 Modal Field and Dispersion of CPW and CPS Structures

The quasi-static descriptions of the CPW and CPS structures are not sufficient in the mm-wave range. The realistic descriptions of the fields, modes, and dispersion are only possible with the help of the full-wave analysis [B.1, J.30]. This section discusses the results obtained from the full-wave analysis to appreciate the modal fields and dispersion. The closed-form expressions for dispersion are also presented.

9.6.1 Modal Field Structure of CPW

Figure (9.2) presents several structural forms of the CPW line. Each structure has a different field configuration and supports a separate modal field. Thus, the dispersion behavior of the CPW line is different for each structure. The finite thickness substrate CPW, and also the CB-CPW, is discussed below.

CPW on a Finite Substrate

Figure (9.33a) shows the electric and magnetic field lines of a CPW for the quasi-static case, supporting the TEM-type mode. Both the ground planes are at zero potential. The identical potential on both ground conductors is

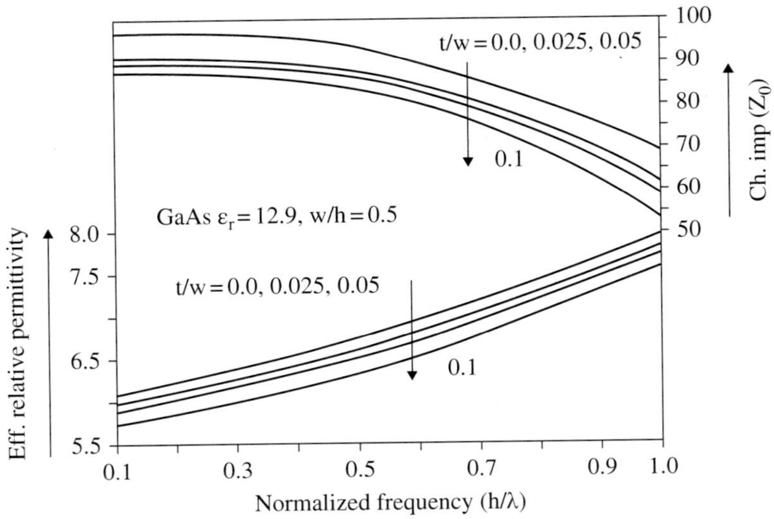

Figure 9.32 Effect of conductor thickness on effective relative permittivity and the characteristic impedance of a CPS. *Source:* Rahman and Nguyen [J.29]. © IEEE.

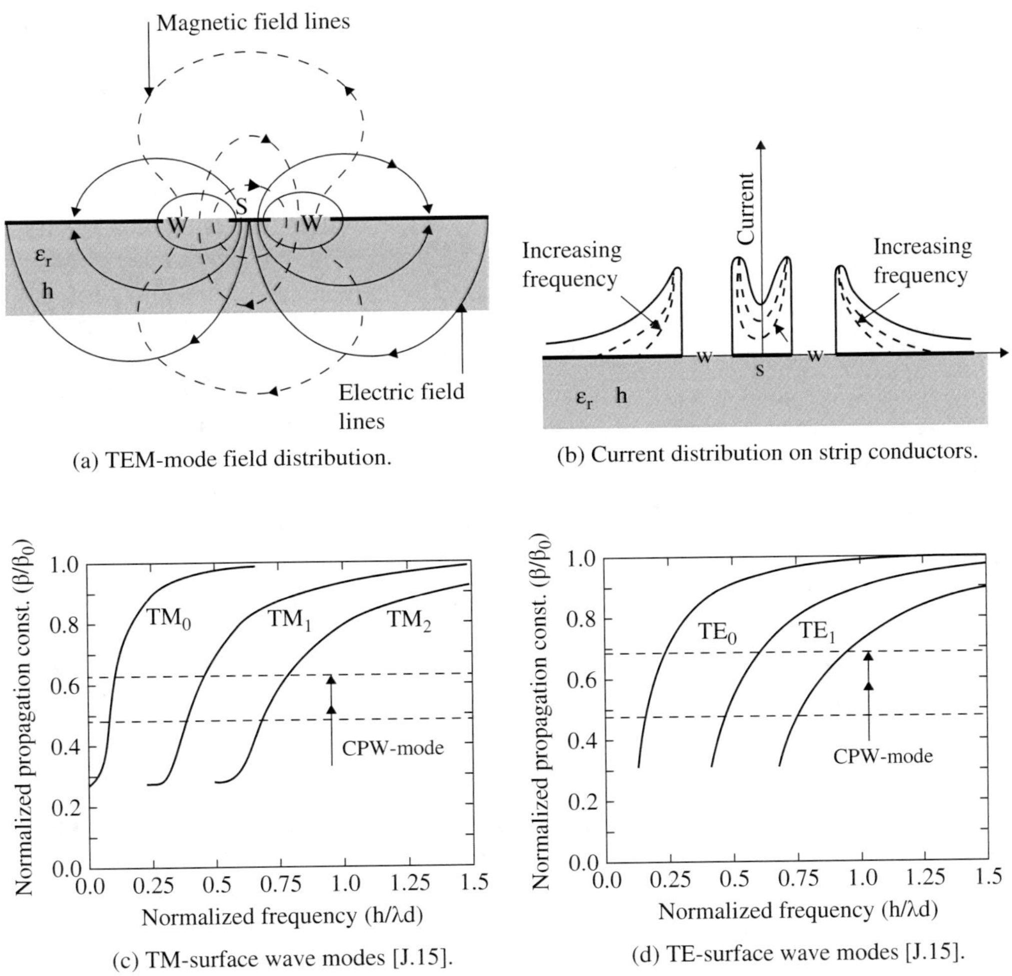

Figure 9.33 Fields, current, and modes on finite thickness substrate CPW. *Source:* Riaziat et al. [J.15]. © IEEE.

maintained with the help of the air-bridge, (not shown in the figure), connecting them. Such CPW supports the *even mode* [J.4]. If both ground conductors are at different potentials, the CPW supports the *odd mode* and a slot line mode could be excited. In the case of the even-mode operation, electric field lines originate at the central conductor and terminate at the ground planes. The magnetic fields surround the central conductor. One can place any other material in the slot region, or just below the central strip, for their interaction with CPW fields. It helps to characterize the material using CPW. Figure (9.33b) demonstrates the longitudinal current distribution across the width of strips, showing maximum value, i.e. the singularity, at the edges. The current decreases away from the slot edges. The decrease is faster at a higher frequency. The *edge singularity* of the longitudinal current is associated with the excess charge concentration at the edges of strip conductors.

A CPW without the conductor backing supports the quasi-static mode that is a combination of the TEM-mode (static CPW mode) and the surface wave mode. The surface wave modes are supported by a conductor-backed dielectric sheet. It is discussed in section (7.5) of chapter 7. In the case of a CPW with a finite thickness substrate, the structure supports both the TM- and TE-type surface wave modes shown in Fig (9.33c,d) [J.15]. These modes are excited by the slot fields of a CPW.

Figure (9.33c) shows the range of normalized propagation constant for the CPW mode (*dashed line*), corresponding to ($0.48 \le \varepsilon_{r\,eff}/\varepsilon_r \le 0.64$). The TM_0 surface wave mode has no cutoff frequency. It is similar to the CPW mode (TEM-mode). At lower frequency, phase velocities of the TM_0 surface mode and CPW mode are different. So coupling between these two modes is weak, leading to a small dispersion. However, at high frequency, close to the point of intersection, both modes have the same phase velocity leading to a strong coupling that results in a large dispersion. Thus, in the case of a dispersive CPW, without conductors backing, the dispersion is due to the coupling between the TEM mode (CPW mode) and the surface wave mode. The CPW should be used below the first point of intersection of modes, shown in Fig (9.33c). The operating frequency should meet the condition $h \le 0.15\,\lambda_d$, where λ_d is the wavelength in a dielectric medium. Above the point of mode intersection, the loss in a CPW increases with frequency due to the surface wave loss. The TE_0 surface wave mode is excited at the substrate thickness $2h \approx 0.25\,\lambda_d$. Figure (9.33d) shows that near $2h \approx 0.35\,\lambda_d$, the phase velocities of TE_0 surface wave mode and CPW mode are almost identical. For $h \approx 0.12\,\lambda_d$, the TE_0 mode can be avoided, and also dispersion due to coupling between the TM_0 mode and CPW mode can be reduced.

Conductor-Backed CPW (CB-CPW)

Figure (9.34) shows the electric field configurations of different modes of the CB-CPW structure with the finite width coplanar ground planes. It is noted that the CB-CPW structure can support normal CPW even mode, microstrip mode, first higher-order microstrip mode, and image guide mode, depending on the structural parameters and the operating frequency. However, in practice, the structural parameters are selected such that only the CPW and microstrip modes are supported by the structure. The back conductor is at the floating potential; still, it supports the undesired microstrip mode.

The CB-CPW also behaves like a parallel plate waveguide. Figure (9.34e) demonstrates that its fundamental mode is the TEM parallel-plate mode and it also supports other TE/TM higher-order modes. The phase velocity of the TEM parallel-plate mode is less than the phase velocity of the TEM-type CPW mode. It causes a loss of power in a CPW in the absence of lateral confinement of fields. The TE_1/TM_1 modes appear at a higher frequency, above $h = 0.12\ \lambda_d$. The coupling between the CPW mode and TE_1/TM_1 parallel plate modes causes power loss and dispersion. The loss and dispersion can be controlled, if $h \le 0.12\ \lambda_d$. The lateral confinement of fields in a CB-CPW is achieved by connecting the ground planes of the upper layer, using an array of vias, to the conductor backing. In this case, the structure also supports the rectangular waveguide modes. The coupling between the waveguide mode and the CPW mode also causes dispersion. This is again avoided by operating the CPW below the cutoff frequency of the waveguide modes.

9.6.2 Modal Field Structure of CPS

A CPS line also supports the quasi-TEM mode. Figure (9.35a) shows its field lines. At higher frequency, it also supports the higher-order modes. Figure (9.35b) presents the propagation constants of the fundamental TEM mode and the first higher-order mode [J.31]. The first higher-order modes appear around 9 GHz and the next higher-order mode appears at 40 GHz not shown in Fig (9.35b). The CPS line on an infinitely thick substrate does not support any parasitic surface wave mode or the dielectric slab mode. However, it supports the leaky-wave

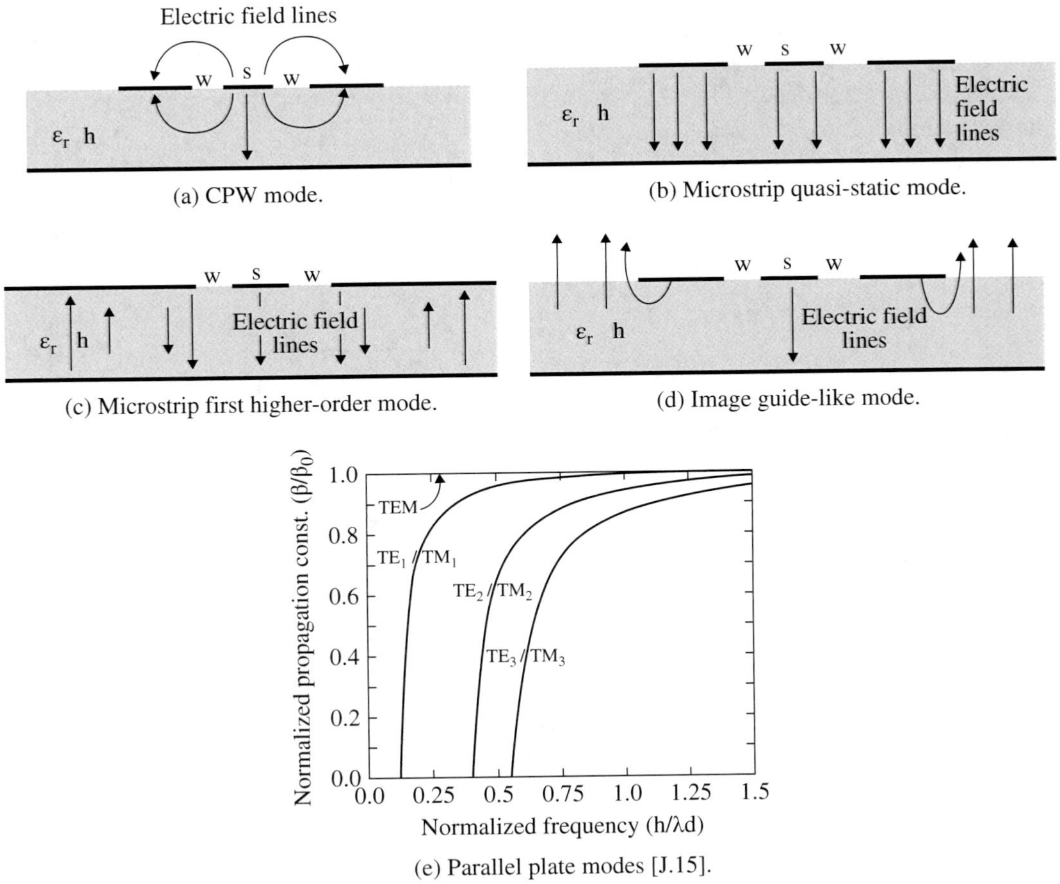

Figure 9.34 The modes on finite width conductor-backed CPW. *Source:* Riaziat et al. [J.15]. © IEEE.

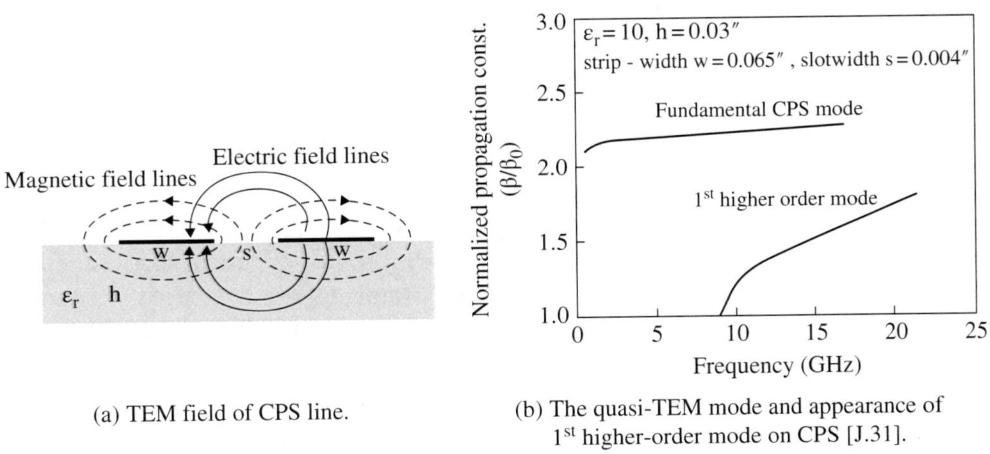

Figure 9.35 Field and modes of CPS line. *Source:* Goverdhan et al. [J.31].

radiation mode that causes the power loss to the CPS. This aspect of power loss, both in the CPW and CPS structures, is further discussed in the next section [J.32]. The CPS on a finite thickness substrate can excite the TM- and TE-type dielectric waveguide modes both in the even and odd mode formats. The TM_0 and TE_0 modes have no cutoff frequency. Again the power of a CPS is lost to these parasitic guided modes, as these modes become the leaky modes. A narrow strip width CPS excites the TM-type substrate mode, whereas the wide strip excites both the TM- and TE-type substrate modes.

The conductor-backed CPS, i.e. CB-CPS, supports two fundamental modes – *the CPS mode and the parasitic microstrip mode* [J.33]. It also supports parallel-plate modes. The parasitic microstrip mode is significant for a thin substrate and wide strip dimension. The CB-CPS further supports the surface wave mode due to the conductor-backed dielectric. The higher-order modes appear above the critical frequency given by

$$f_c = \frac{1}{h\sqrt{2\mu_0\varepsilon_0(\varepsilon_r - 1)}}. \quad (9.6.1)$$

By using a thin substrate, the critical frequency could be pushed to a very high value, so that the CPS line could be used without the appearance of higher-order modes.

9.6.3 Closed-Form Dispersion Model of CPW

The closed-form dispersion model of the CPW, on a finitely thick substrate, is presented below. However, coplanar ground conductors are infinitely wide. The conductors are treated as the perfect conductors. Next, the dispersion model is improved to take into account the finite conductivity and finite thickness of the strip conductor.

CPW on a Finitely Thick Substrate

Hasnain *et al.* [J.34] have adopted the dispersion expression of a microstrip line proposed by Yamashita *et al.* [J.35] to model the dispersion behavior of a CPW on a finitely thick substrate. The dispersion model is given below:

$$\varepsilon_{r\,eff}(f) = \left[\sqrt{\varepsilon_{r\,eff}(0)} + \frac{\sqrt{\varepsilon_r} - \sqrt{\varepsilon_{r\,eff}(0)}}{(1 + aF^{-b})}\right]^2. \quad (9.6.2)$$

In the above equation, F is the normalized frequency with respect to the cutoff frequency of the TE_1 surface wave mode:

$$F = f/f_c^{TE} \quad (a)$$
$$f_c^{TE} = \frac{c}{4h\sqrt{\varepsilon_r - 1}} \quad (b).$$
$$(9.6.3)$$

The expression for f_c^{TE} is taken from equation (7.5.28) of chapter 7. The static effective relative permittivity $\varepsilon_{r\,eff}(0)$ of a CPW is obtained from equation (9.3.15). Empirically obtained constants a and b are dependent on the CPW structure:

$$b = 1.8 \quad (a)$$
$$a \approx \exp\left[u \log\left(\frac{s}{w}\right) + v\right] \quad (b)$$
$$u = 0.54 - 0.64q + 0.015q^2 \quad (c)$$
$$v = 0.43 - 0.86q + 0.54q^2 \quad (d)$$
$$q = \log\left(\frac{s}{h}\right) \quad (e).$$
$$(9.6.4)$$

The model has 5% accuracy for the ranges $0.1 < s/w < 5$, $0.1 < s/h < 5$, $1.5 < \varepsilon_r < 50$, $0 < f/f_{TE} < 10$. Figure (9.36a) compares the dispersion in a 50Ω CPW, on the GaAs substrate ($\varepsilon_r = 13$, $h = 100$ μm) with central conductor width $s = 85$ μm, slot width $w = 50$ μm, against a 50 Ω microstrip on the same substrate with strip width $w = 73$ μm. The static $\varepsilon_{r\,eff}(0)$ of a CPW is lower compared to $\varepsilon_{r\,eff}(0)$ of a microstrip line. At very high frequency, for both lines, $\varepsilon_{r\,eff}(f)$ approaches to ε_r. Therefore, for a very wide bandwidth signal, a CPW is more dispersive than a microstrip. However, below f_c^{TE}, a CPW is less dispersive as compared to a microstrip line. Jackson *et al.* [J.37] have shown that the static $\varepsilon_{r\,eff}(0)$ can be used for the design of the GaAs-based MMIC up to 40 GHz. The small dispersion in a CPW is due to the confinement of fields in the slot region that is to a large extent independent of frequency. The nature of dispersion in the CPW, shown in Fig (9.36a), suggests that the CPW dispersion can also be modeled using the LDM of chapter 8. However, a set of parameters applicable to the CPW has to determined from the data obtained using an EM-Simulator.

Improved Dispersion Model

The above model ignores the thickness and conductivity of the strip conductors. The finite conductivity of a strip conductor permits penetration of the magnetic field inside the strip that creates the internal inductance [J.38]. The line capacitance increases with an increase in the finite thickness of the strip conductor. Both parameters are combined to modify the static effective relative permittivity, $\varepsilon_{r\,eff}(f = 0)$. So, the $\varepsilon_{r\,eff}(f = 0)$ used in equation (9.6.2) could be replaced by the conductor thickness and skin-depth-dependent effective relative permittivity $\varepsilon_{r\,eff}(t, \delta, f = 0)$. The $\varepsilon_{r\,eff}(t, \delta, f = 0)$ is computed below.

The field penetrates through all sides of the strip and ground conductors by the skin-depth, $\delta = 1/\sqrt{\mu_0 \pi f \sigma}$, where σ is the conductivity of conductors. It is shown in Fig (9.38). In the process, the strip width s is reduced and gap width w is increased. Finally, the aspect-ratio $k_{1\delta}$ due to the skin-depth is changed. Using the equality

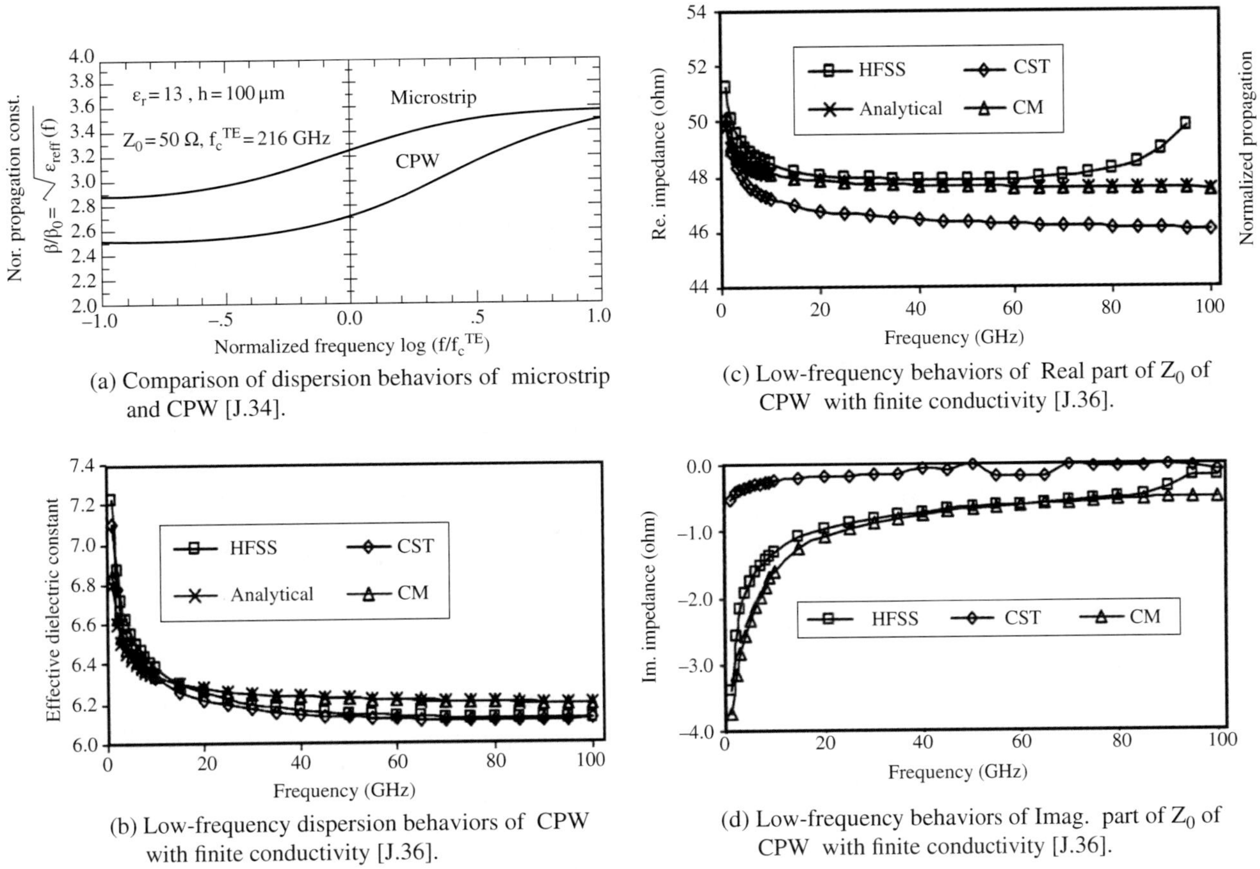

(a) Comparison of dispersion behaviors of microstrip and CPW [J.34].

(c) Low-frequency behaviors of Real part of Z_0 of CPW with finite conductivity [J.36].

(b) Low-frequency dispersion behaviors of CPW with finite conductivity [J.36].

(d) Low-frequency behaviors of Imag. part of Z_0 of CPW with finite conductivity [J.36].

Figure 9.36 Dispersion behaviors of CPW. CM: Circuit model. *Source:* Hasnai et al. [J.34]. © IEEE. Verma et al. [J.36]. © John Wiley & Sons.

of the phase velocity of the EM-wave on the CPW considered on the air-substrate and the EM-wave in the air-medium and also using equation (9.3.7), the total line inductance *p.u.l.* is obtained as follows:

$$L(\delta)C(\varepsilon_r = 1, \delta) = \varepsilon_0\mu_0, \quad \Rightarrow L(\delta) = \frac{\mu_0}{4}\frac{K(k'_{1,\delta})}{K(k_{1,\delta})} \qquad (a)$$

$$\text{where,} \quad k_{1,\delta} = \frac{s-\delta}{s+2w+2\delta} \qquad (b)$$

$$k'_{1,\delta} = \sqrt{1-K_{1\delta}^2} \qquad (c).$$

(9.6.5)

Using equation (9.3.7b), line capacitance *p.u.l.* of a CPW, on the air-substrate, with conductor thickness t is given below:

$$C(\varepsilon_r = 1, t) = 4\varepsilon_0\frac{K(k_{1,t})}{K(k'_{1,t})}. \qquad (9.6.6)$$

The modified aspect-ratio $k_{1,t}$ is given in equation (9.5.2). The conductor thickness-dependent total line capacitance *p.u.l.* can be written as follows:

$$C(\varepsilon_r, t) = C(\varepsilon_r = 1, t)\ \varepsilon_{r\,eff}(t, f = 0). \qquad (9.6.7)$$

The phase velocity of the EM-wave on the CPW is

$$v_p = \frac{1}{\sqrt{L(\delta)C(\varepsilon_r, t)}} = \frac{c}{\sqrt{\varepsilon_{r\,eff}(\delta, t, f=0)}}$$

$$\Rightarrow \varepsilon_{r\,eff}(\delta, t, f=0) = c^2 L(\delta)C(t, \varepsilon_r)$$

$$\Rightarrow \varepsilon_{r\,eff}(\delta, t, f=0) = c^2\left[\frac{\mu_0}{4}\frac{K(k'_{1,\delta})}{K(k_{1\delta})}\right]\left[4\varepsilon_0\frac{K(k_{1,t})}{K(k'_{1,t})}\right]\varepsilon_{r\,eff}(t, f=0)$$

$$\Rightarrow \varepsilon_{r\,eff}(\delta, t, f=0) = \left[\frac{K(k'_{1,\delta})}{K(k_{1\delta})}\right]\left[\frac{K(k_{1,t})}{K(k'_{1,t})}\right]\varepsilon_{r\,eff}(t, f=0)$$

$$\Rightarrow \varepsilon_{r\,eff}(\delta, t, f=0) = P\,\varepsilon_{r\,eff}(t, f=0) \qquad (a)$$

$$\text{where}\ P = \left[\frac{K(k'_{1,\delta})}{K(k_{1\delta})}\right]\left[\frac{K(k_{1,t})}{K(k'_{1,t})}\right] \qquad (b).$$

(9.6.8)

The above equation gives the conductor thickness and skin-depth-dependent relative permittivity $\varepsilon_{r\,eff}(t, \delta, f = 0)$ in terms of factor P due to the skin-depth and conductor thickness. On replacing the $\varepsilon_{r\,eff}(f = 0)$ used in equation (9.6.2) by $\varepsilon_{r\,eff}(\delta, t, f = 0)$, the final dispersion expression is obtained for the CPW with finite strip conductivity and finite strip thickness:

$$\varepsilon_{r\,eff}(f, t, \delta) = \left[\sqrt{P\varepsilon_{r\,eff}(t,0)} + \frac{\sqrt{\varepsilon_r} - \sqrt{P\varepsilon_{r\,eff}(t,0)}}{(1 + aF^{-b})}\right]^2. \qquad (9.6.9)$$

The above model is valid for frequency, $f > 1/(\mu_0 \sigma\, t\, s)$. Below this frequency, the current distribution is almost uniform across the strip and the line inductance has to be computed as the DC inductance [J.39].

Frequency-Dependent Characteristic Impedance

The characteristic impedance for a CPW can be computed from

$$Z_0(f, \varepsilon_r, t, \delta) = \frac{Z_0(f = 0, \varepsilon_r = 1, t, \delta)}{\sqrt{\varepsilon_{r\,\mathrm{eff}}(f, t)}}. \quad (9.6.10)$$

The characteristic impedance, $Z_0(f = 0, \varepsilon_r = 1, t, \delta)$, of a finite conductivity CPW on the air-substrate is obtained as follows:

$$Z_0(f = 0, \varepsilon_r = 1, t, \delta) = \sqrt{\frac{L(\delta)}{C(\varepsilon_r = 1, t)}}$$

$$= \left[\frac{\mu_0}{4}\frac{K(k'_{1,\delta})}{K(k_{1,\delta})}\cdot\frac{1}{4\varepsilon_0}\frac{K(k'_{1,t})}{K(k_{1,t})} \times \frac{K(k_{1,t})}{K(k'_{1,t})}\cdot\frac{K(k'_{1,t})}{K(k_{1,t})}\right]^{1/2}$$

$$\Rightarrow Z_0(f = 0, \varepsilon_r = 1, t, \delta) = 30\pi\frac{K(k'_{1,t})}{K(k_{1,t})}\sqrt{P}$$

$$= Z_0(f = 0, \varepsilon_r = 1, t)\sqrt{P}. \quad (9.6.11)$$

Using the above equations, we can compute the characteristic impedance of a CPW of strip conductors with finite conductivity and finite thickness from the following expression:

$$Z_0(f, \varepsilon_r, t, \delta) = \frac{Z_0(f = 0, \varepsilon_r = 1, t)}{\sqrt{\varepsilon_{r\,\mathrm{eff}}(f, t, P = 1)}}\sqrt{P}. \quad (9.6.12)$$

The characteristic impedance $Z_0(f = 0, \varepsilon_r = 1, t)$, frequency-dependent effective relative permittivity, and factor P are obtained for various CPW structures from our previous discussions. *The effect of penetration of the magnetic field inside the strip conductor of finite conductivity can also be viewed as the creation of effective permeability in a CPW and the factor P is related to it.* This aspect is accommodated in the frequency-dependent modeling of the multilayer CPW structure using the SLR process [J.36]. It is discussed in the section (15.3) of chapter 15.

Figure (9.36b–d) shows the effect of finite conductivity on a CPW designed on GaAs substrate, $\varepsilon_r = 12.9$, $\tan\delta_2$, $h = 380\,\mu m$, $s = 16\,\mu m$, $w = 12\,\mu m$, $t = 3\,\mu m$, $\sigma = 4.1 \times 10^7$ S/m over the frequency range 1–100 GHz. There is a significant increase in the relative permittivity and real part of characteristic impedance at the lower end of frequency due to the dispersive behavior. It is due to frequency-dependent magnetic field penetration. The imaginary part of the characteristic impedance also shows such strong dispersion. The present analytical models show similar behavior. The complex impedance is obtained using the circuit model (CM), discussed in the section (9.8.1).

9.6.4 Dispersion in CPS Line

The full-wave methods, such as the SDA [J.40], variational [J.13], and mode matching [J.29], have been used to obtain the frequency-dependent characteristic impedance and effective relative permittivity of a CPS line. In a lower frequency range, the CPS line behaves as a quasi-TEM mode line. However, in a higher frequency range, its hybrid nature appears. The voltage–power definition [J.29, J.40] shows an increase in the characteristic impedance of a CPS line with increasing frequency. The power–current definition shows an initial decrease and then an increase in characteristic impedance with increasing frequency [J.13]. The effective relative permittivity $\varepsilon_{r,\,\mathrm{eff}}(f)$ shows an increase with an increase in frequency.

Closed-Form Dispersion Model of CPS Line

The closed-form expression, discussed in this section, to compute the frequency-dependent effective relative permittivity $\varepsilon_{r\,\mathrm{eff}}(f)$ of the CPW structure, has also been used for the CPS line up to mm-wave range [J.32, J.34, J.41, J.42]. The effect of conductor thickness on the static effective relative permittivity and characteristic impedance should be accounted for. The following expressions, for the CPS on an infinitely thickness substrate, compute the thickness-dependent line parameters [B.16, J.42]:

$$\varepsilon_{r\,\mathrm{eff}}(t) = \varepsilon_{r\,\mathrm{eff}}(t = 0) - \frac{1.4(\varepsilon_{r\,\mathrm{eff}}(t = 0) - 1)t/s}{[K(k'_1)/K(k_1)] + 1.4t/s} \quad (a)$$

$$Z(t) = \frac{120\,\pi}{\sqrt{\varepsilon_{r\,\mathrm{eff}}(t)}}\frac{K(k'_{1,t})}{K(k_{1,t})} \quad (b).$$

$$(9.6.13)$$

For a finite thickness substrate, $k_{1,\,t}$ is replaced by $k_{2,\,t}$. While calculating the aspect-ratios $k_{1,\,t}$ and $k_{2,\,t}$, using equations (9.3.5) and (9.3.11), the modified dimensions $s_{eq} = s - \Delta w$, $w_{eq} = w + \Delta w$ should be used in place of the physical dimensions s and w. Equation (9.5.1c) is used to compute Δw. The finite conductivity can also be accommodated in the computation. However, its effect, showing an increase in both the effective relative permittivity and characteristic impedance at the lower end of frequency, is also accounted for using the CM discussed in section (9.8.1).

Figure (9.37a) presents the dispersive behavior of a CPS on GaAs substrate, $\varepsilon_r = 12.9$, $\tan\delta_2$, $h = 670\,\mu m$, $s = 7\,\mu m$, $w = 7\,\mu m$, $\sigma = 4.1 \times 10^7$ S/m, over the

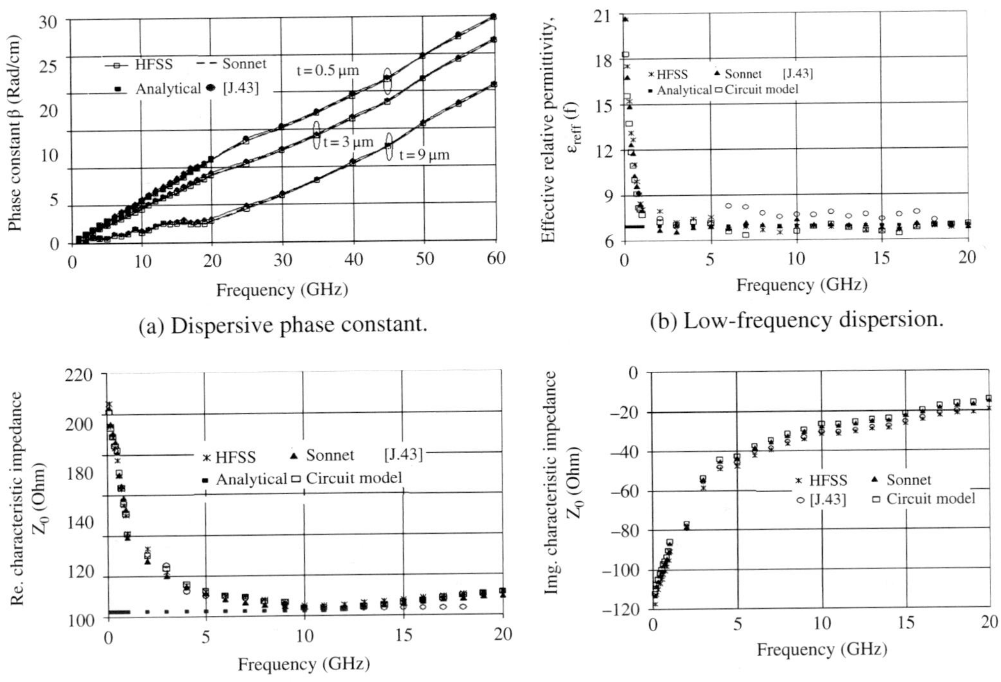

(a) Dispersive phase constant.

(b) Low-frequency dispersion.

(c) Low-frequency dispersion in Real Z_0.

(d) Low-frequency dispersion in Img. Z_0.

Figure 9.37 Dispersion behavior of CPS. *Source:* Majumdar and Verma [J.42]. © Taylor & Francis Ltd.

frequency range 1–60 GHz. Figure (9.37b–d) shows the dispersion of CPS for s = 8 μm, w = 4 μm, $\sigma = 4.1 \times 10^7$ S/m over 1–20 GHz. The present analytical model does not account for the finite conductivity. So, it does not show low-frequency dispersion. The circuit model (CM) accounts for the effect of both the finite conductivity and loses on dispersion. It shows the low-frequency dispersion for effective relative permittivity and also for the real and imaginary parts of characteristic impedance. The results follow closely the results of two EM-Simulators, HFSS and Sonnet [J.42], and also the experimental results [J.43].

9.7 Losses in CPW and CPS Structures

There are three prime mechanisms for the power loss in both the CPW and CPS structures – *conductor loss, dielectric loss,* and *radiation loss*. Under normal circumstances, conductor loss, due to the finite conductivity of conducting strip conductors, is the main source of power loss. The dielectric loss is small due to low loss-tangent of the substrate materials. The radiation loss is also small at the microwave and the lower end of mm-wave. However, under certain circumstances, it becomes a major source of loss. This section discusses the mechanism of losses and presents closed-form expressions to compute them.

9.7.1 Conductor Loss

The conductor loss in the CPW and CPS is decided by the finite conductivity of the strip conductors and geometry of the structures. It also strongly depends on the thickness of strip conductors. It increases with a decrease in the normalized conductor thickness t/δ_s, where δ_s is the skin-depth. Several quasi-static and full-wave methods have been reported for the computation of conductor loss [J.44–J.46]. The quasi-static methods, such as the conformal mapping methods [B.3, J.12, J.17, J.47], and numerical methods, like MOM, [B.18, J.48] are based on the perturbation approach. This method assumes that the magnetic fields tangential to the conductor surfaces do not change significantly when the conductivity is changed from infinite to a finite practical value. However, the assumption is not correct at the low frequency; where magnetic field penetration in a practical conductor is significant, leading to an increase in the internal inductance of a line [J.39]. The penetration of the magnetic field influences not only the conductor loss but also the dispersion characteristics and characteristic impedance [J.49, J.50]. It is discussed in subsection (9.6.3). The current density on the strip conductor shows the *edge-singularity* that causes the integral, used in the perturbation method, to diverge. However, the problem of divergence has been overcome by the concept of *stopping distance* while integrating the current density on the strip conductor [J.51, J.52].

The quasi-static *Wheeler's incremental inductance method* discussed below, can be used to compute the conductor loss in both the CPW and CPS structures [B.16, J.25, J.53]. However, this method is applicable only for the conductor thickness greater than the skin-depth, i.e. for $t > 1.1\,\delta_s$ [J.54]. The *perturbation method with stopping distance* concept is also presented. It applies to a conductor thickness both greater and less than the skin-depth.

Wheeler's Incremental Inductance Method

Wheeler's incremental inductance method normally uses the partial derivative of the characteristic impedance of a CPW [B.16]. However, this process is not very accurate [J.12]. A more accurate and general formulation of Wheeler's inductance rule, in terms of the fractional characteristic impedance, is discussed in subsection (8.4.2) of chapter 8 [J.53, J.55]. It is also applicable to the CB-CPW structures and the CPW with top shield, shown in Fig (9.38). The expression presented here is also applicable to the multilayer case.

The expression (8.4.71) of chapter 8, applicable to a microstrip line, is given below for the CPW case:

$$\alpha_c = \frac{27.3}{\lambda_0}\sqrt{\varepsilon_{r\,eff}^{CPW}(\varepsilon_r, w, s, h, t)} \times \frac{\Delta Z_0(\varepsilon_r = 1, w, s, h, t, \delta_s)}{Z_0(s + \Delta s, w - \Delta s, h, \varepsilon_r = 1)}\ \text{dB/m}. \quad (9.7.1)$$

The above expression is written for a CPW on the finite thickness substrate. Figure (9.38b) shows that the conductor thickness increases the strip width (s) of the central conductor to the enlarged width $s + \Delta s$ and decreases the slot width (w) to $(w - \Delta s)$. The characteristic impedance $Z_0(s + \Delta s,\ w - \Delta s,\ h,\ \varepsilon_r = 1)$ with conductor thickness t is for the CPW on the air-substrate. It is, along with $\varepsilon_{r\,eff}^{CPW}(\varepsilon_r, w, s, h, t)$, obtained from equation (9.5.3). Figure (9.38c) shows that the strip width and conductor thickness decrease due to skin-depth while the slot width increases. The incremental change in width (Δs) due to the conductor thickness and skin-depth is computed using equation (9.5.1d). The difference characteristic impedance, due to the skin-depth, ΔZ_0, is defined as follows:

$$\Delta Z_0(\varepsilon_r, w, s, h, t, \delta_s) = Z_0\left(s + \Delta' s - \delta_s, w - \Delta' s + \delta_s, h + \frac{\delta_s}{2}, \varepsilon_r = 1\right) - Z_0(s + \Delta s, w - \Delta s, h, \varepsilon_r = 1). \quad (9.7.2)$$

On the right-hand side, the first characteristic impedance takes care of both the conductor thickness and the skin-depth δ_s in the strip conductor. Figure (9.38c) shows the field penetration $\delta_s/2$ all around the strip conductor. The effective thickness of the strip conductor is reduced to $t' = t - \delta_s$, where $\delta_s = 1/\sqrt{\mu_0 \pi f \sigma}$. The substrate thickness is increased to the thickness $h + \delta_s/2$. The incremental increase in the strip width $(\Delta' s)$ due to the effective conductor thickness t' is given by

$$\Delta' s = (t'/\pi)\left[1 + \ln(4) - 0.5\ \ln\left\{(t'/h)^2 + (t'/\pi s)^2\right\}\right]. \quad (9.7.3)$$

The effective strip conductor width due to both effective conductor thickness t' and skin-depth δ_s is $(s + \Delta' s - \delta_s)$ and the effective slot width is $(w - \Delta' s + \delta_s)$. The characteristic impedance corresponding to the first term of equation (9.7.2) is computed for the effective strip width $(s + \Delta' s - \delta_s)$, effective slot width $(w - \Delta' s + \delta_s)$, and the effective substrate thickness $(h + \delta_s/2)$ with relative permittivity ε_r, using expression (9.3.16). The characteristic impedance, corresponding to the second term, is also computed without skin-depth using the same expression.

The above-discussed process is also applicable to a CB-CPW. In this case, $\varepsilon_{r\,eff}^{CPW}(t = 0)$ has to be computed from equation (9.3.46). However, in this case, the substrate thickness with skin-depth is $(h + \delta_s)$. The difference characteristic impedance in for CB-CPW is

$$\Delta Z = Z_0(s + \Delta' s - \delta_s, w - \Delta' s + \delta_s, h + \delta_s, \varepsilon_r = 1) - Z_0(s + \Delta s, w - \Delta s, h, \varepsilon_r = 1). \quad (9.7.4)$$

The characteristic impedance of the CB-CPW is computed from equation (9.3.47).

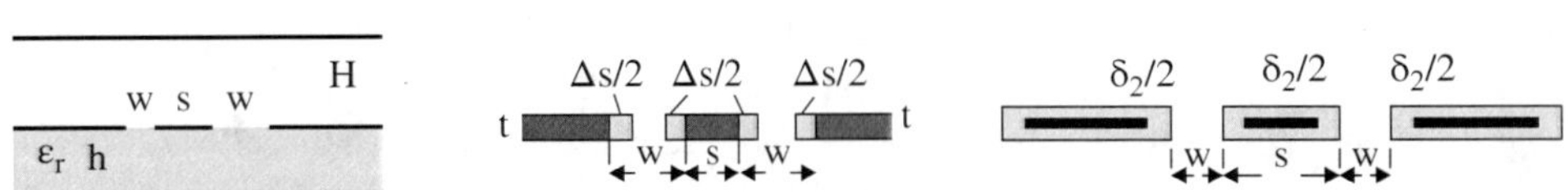

(a) Top-shielded CPW. (b) Extension of strip width due to conductor thickness. (c) δ/2 - field penetration around strip-conductor.

Figure 9.38 Application of Wheeler's inductance rule to compute conductor loss.

The conductor loss of a CB-CPW, with the top shield, could be also computed. The suitable expression for $\varepsilon_{reff}^{CPW}(t = 0)$ is used for the CB-CPW with a top shield. The shield height with skin-depth is $(H + \delta_s)$. The difference characteristic impedance for this case is

$$\Delta Z = Z_0(s + \Delta' s - \delta_s, w - \Delta' s + \delta_s, h + \delta_s, H + \delta_s, \varepsilon_r = 1) - Z_0(s + \Delta s, w - \Delta s, h, H, t, \varepsilon_r = 1). \quad (9.7.5)$$

Modified Perturbation Method

Wheeler's incremental inductance method presented above does not apply to the case of a strip conductor thickness less than the skin-depth, i.e. for $t < 1.1\,\delta_s$ [J.53]. Even for the conductor thickness slightly more than the skin-depth, it is not accurate. The modified perturbation method, using the concept of the *stopping distance* (Δ), computes conductor loss of the CPW and CPS structures without any restriction on the conductor thickness [J.56, J.57]. The closed-form expression of the conductor loss is obtained by modifying the classical perturbation method in which the integration limit excludes the strip edge singularity by excluding a small distance, called the stopping distance (Δ), from both edges of the strip conductor. The conductor losses for both the CPW and CPS are computed by the following modified perturbation method [J.56]:

$$\alpha_c = \frac{R_{sm}}{4Z_0^{CPW/CPS}}\int_{\Delta}\left(\frac{J}{I}\right)^2 dl, \quad (a)$$

$$\text{For CPW}: \quad \alpha_c = \frac{R_{sm}}{4Z_0^{CPW}}\int_{-a+\Delta}^{a-\Delta}\left(\frac{J}{I}\right)^2 dx + 2\int_{b+\Delta}^{\infty}\left(\frac{J}{I}\right)^2 dx \quad (b). \quad (9.7.6)$$

The modified surface impedance R_{sm}, that takes care of both the upper and lower surfaces of the strip conductor, is given by the following expression [J.58, J.59]:

$$R_{sm} = \omega\mu_c t\,\mathrm{Im}\left[\frac{\cot(k_c t) + \mathrm{cosec}(k_c t)}{k_c t}\right]. \quad (9.7.7)$$

In the above equations, Im[ ] stands for the imaginary part of the expression under the bracket [ ]. For a conductor, $\mu_c = \mu_0$, the wavenumber in a conductor is given by

$$k_c = \omega\sqrt{\mu_c\varepsilon_c} = \omega\sqrt{\mu_0\left(\varepsilon_0 - j\frac{\sigma}{\omega}\right)} = \omega\sqrt{\mu_0\varepsilon_0}\left[1 - j\frac{\sigma}{\omega\varepsilon_0}\right]^{\frac{1}{2}}. \quad (9.7.8)$$

CPW Structure

Figure (9.39a) shows the CPW structure on a finite thickness substrate and with infinite extent ground conductors. However, to avoid the edge singularity, the stopping distance Δ is considered at all conductor edges. The strip can acquire a trapezoidal shape in the fabrication process. The current distribution (J) on the central strip conductor is taken as follows, showing singularity at $x = \pm a, b$ [J.17]:

$$J = \frac{A}{\sqrt{(a^2 - x^2)(b^2 - x^2)}}, \quad |x| < a \quad (a)$$

$$\text{where, constant } A = \frac{bI}{2K(k)} \quad (b)$$

$$\text{aspect-ratio}, \quad k = \frac{a}{b} \quad (c). \quad (9.7.9)$$

In the above equation, I is the current on the strip conductor. The current density on the ground plane of CPW is

$$J = \frac{A}{\sqrt{(x^2 - a^2)(x^2 - b^2)}}; \quad |x| > b. \quad (9.7.10)$$

Equation (9.7.6b) for a CPW is written below by taking into account the stopping distance Δ and symmetry of the CPW while evaluating the integral for the conductor loss computation:

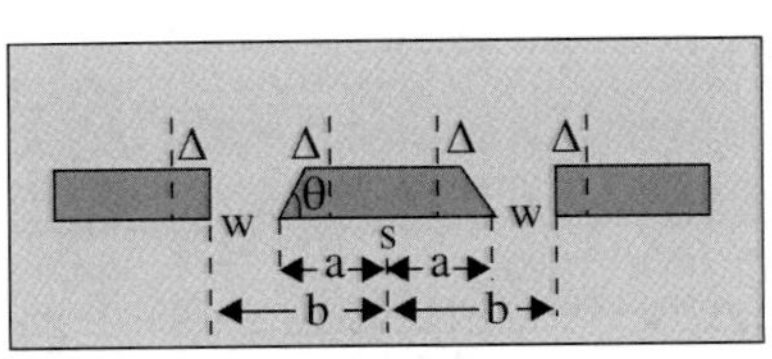

(a) CPW structure.

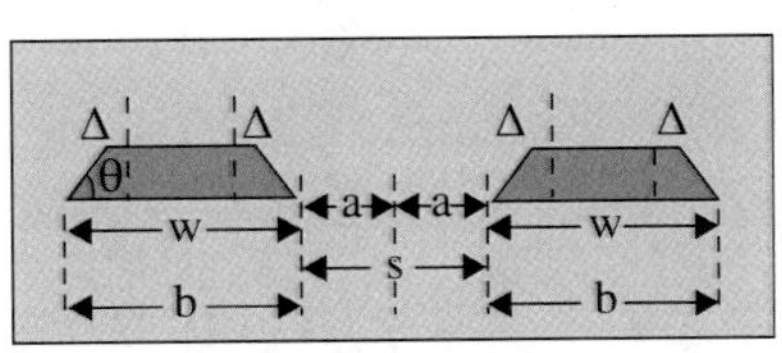

(b) CPS structure.

Figure 9.39 CPW and CPS strip conductors with stopping distance (Δ).

$$\alpha_c = \frac{R_{sm} b^2}{8Z_0^{CPW} K^2(k)} \left[\int_0^{a-\Delta} \frac{dx}{(a^2-x^2)(b^2-x^2)} + \int_{b+\Delta}^{\infty} \frac{dx}{(x^2-a^2)(x^2-b^2)} \right]$$

$$\Rightarrow \alpha_c = \frac{R_{sm} b^2}{16Z_0^{CPW}(f,t) K^2(k)(b^2-a^2)} \left[\frac{1}{a} \ln\left\{ \left(\frac{2a}{\Delta}-1\right)\left(\frac{b-a+\Delta}{b+a+\Delta}\right) \right\} + \frac{1}{b} \ln\left\{ \left(\frac{2b}{\Delta}+1\right)\left(\frac{b-a+\Delta}{b+a-\Delta}\right) \right\} \right]. \quad (9.7.11)$$

The stopping distance Δ is small as compared to a, b, and (a − b). So the above expression can be written as

$$\alpha_c = \frac{R_{sm} b^2}{16Z_0^{CPW}(f,t) K^2(k)(b^2-a^2)} \left[\frac{1}{a} \ln\left\{ \left(\frac{2a}{\Delta}\right)\left(\frac{b-a}{b+a}\right) \right\} + \frac{1}{b} \ln\left\{ \left(\frac{2b}{\Delta}\right)\left(\frac{b-a}{b+a}\right) \right\} \right]. \quad (9.7.12)$$

In the above equation, $Z_0^{CPW}(f,t)$ is the dispersive characteristic impedance of a finite substrate, infinite ground plane CPW with conductor thickness t. The effect of the skin-depth penetration on $Z_0^{CPW}(f,t)$ could also be accounted for. The process can be applied to the finite width ground conductor also [J.58]. It is also applicable to an ACPW structure.

The stopping distance (Δ) is computed from following closed-form expressions applicable to the isolated single strip with rectangular and trapezoidal-shaped conductor cross-section [J.58, J.59]:

- Expressions for rectangular 90° cross-sections :

$$y = \begin{cases} -60.89x^4 + 282.17x^3 - 27.764x^2 + 0.5103x + 9.1907, & \text{for } 0.03 \le x < 0.64 \\ 711.01x^6 - 2120.2x^5 - 1038.3x^4 + 9881.7x^3 - 12798x^2 + 6935.8x - 1372.7, & \text{for } 0.64 \le x < 1.5 \\ -183.05x^4 + 1583.9x^3 - 4981.9x^2 + 6618.6x - 2807.6, & \text{for } 1.5 \le x < 2.76 \\ 0.2953x^4 - 7.7733x^3 + 73.991x^2 - 293.88x + 581.98, & \text{for } 2.76 \le x < 8.0 \\ -0.0198x^3 + 0.9848x^2 - 12.104x + 240.01 & \text{for } 8.0 \le x < 16.0. \end{cases} \quad (9.7.13)$$

- Expressions for trapezoidal 45° cross-section

$$y = \begin{cases} -234.52x^5 + 231.86x^4 + 140.04x^3 - 20.231x^2 + 1.1217x + 6.5491, & \text{for } 0.03 \le x < 0.64 \\ 589.31x^6 + 2175.7x^5 + 1049x^4 + 4836.5x^3 - 7660.5x^2 + 4439.8x - 905.16, & \text{for } 0.64 \le x < 1.5 \\ -206.38x^4 + 1848.3x^3 - 6025.2x^2 + 8396.8x - 3888.4, & \text{for } 1.5 \le x < 2.76 \\ -1.0366x^4 + 24.614x^3 - 210.34x^2 + 776.85x - 726.6, & \text{for } 2.76 \le x < 8.0 \\ 0.4653x^3 - 10.904x^2 + 113.16x - 63.032 & \text{for } 8.0 \le x < 16.0. \end{cases} \quad (9.7.14)$$

In the above expressions, the stopping distance is $y = (t/\Delta)$ and normalized conductor thickness is $x = (t/2\delta_s)$. The following empirical expression, obtained from the experimental results of the conductor loss of the CPW, is more useful over the range $0.045 < x < 2.0$.

$$\begin{aligned} &\text{For } 0.10 \le s/2b \le 0.4 \\ &y = 1668x^4 - 1504.1x^3 + 595.85x^2 - 18.823x + 0.6874, \quad \text{for } 0.045 \le x < 0.7 \\ &\text{For } 0.4 < s/2b \le 0.73: \\ &y = \begin{cases} -2308.6x^4 + 2314.6x^3 - 595.39x^2 + 65.242x - 1.6152, & \text{for } 0.045 \le x < 0.35 \\ -8702.8x^4 + 18054x^3 - 13118x^2 + 4109.8x - 462.55, & \text{for } 0.35 \le x < 0.7 \\ -43937x^4 + 170137x^3 - 233814x^2 + 138995x - 30454, & \text{for } 0.7 \le x < 1.4. \end{cases} \end{aligned} \quad (9.7.15)$$

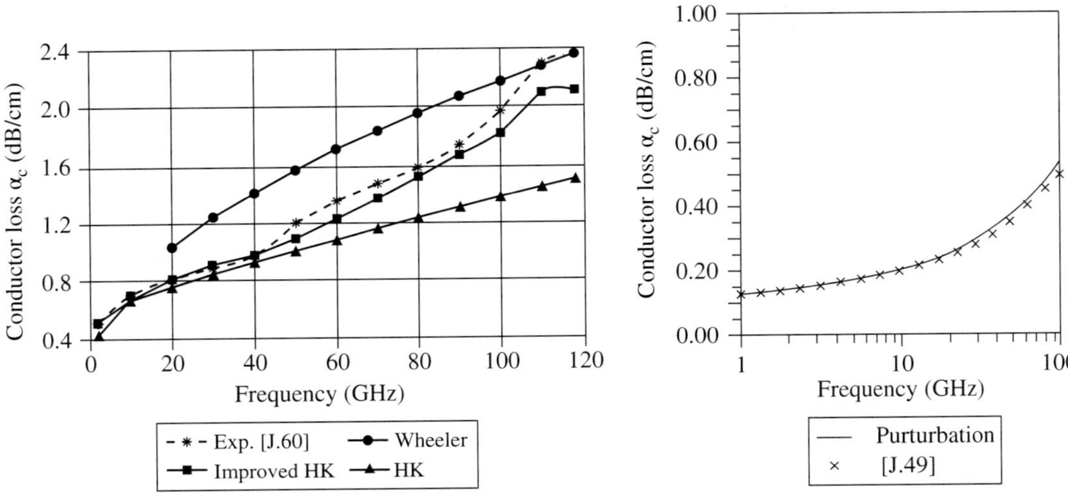

(a) Conductor loss of CPW.

(b) Conductor loss of CPS.

Figure 9.40 Comparison of conductor loss of CPW and CPS.

Figure (9.40a) compares the accuracy of the Wheeler method and perturbation method, using expressions (9.7.13) and (9.7.15), against the experimental results of the conductor loss of the CPW on quartz ($\varepsilon_r = 3.8$) substrate up to 120 GHz. The first one is called Holloway and Kuester (HK) model and the next one is called improved HK (IHK) model. The IHK shows better agreement, within 5%, as compared to other models [J.60–J.62].

CPS Structure

Figure (9.39b) shows the CPS structure with stopping distance. The current distribution on a CPS line is given by [J.42, J.57]:

$$J = \frac{A}{\sqrt{(x^2 - a^2)(b^2 - x^2)}}; \quad a < x < b, \tag{9.7.16}$$

where constant A and aspect-ratio k are given by equation (9.7.9). The characteristic impedance of the CPS line is

$$Z_0^{CPS} = \frac{120\pi}{\sqrt{\varepsilon_{r\,eff}}} \frac{K(k)}{K(k')}. \tag{9.7.17}$$

For the thick substrate h >> (b – a), $\varepsilon_{reff}^{CPS} = (\varepsilon_r + 1)/2$. The conductor loss α_c for the CPS line is obtained as follows, on substituting equation (9.7.16) in equation (9.7.6) and carrying integration with respect to Fig (9.39b):

$$\begin{aligned}
\alpha_c &= \frac{2R_{sm}b^2}{4Z_0^{CPS}K^2(k')} \int_{a+\Delta}^{b-\Delta} \frac{dx}{(x^2 - a^2)(b^2 - x^2)} \\
&= \frac{R_{sm}b^2}{4Z_0^{CPS}K^2(k')(b^2 - a^2)} \left[\frac{1}{a} \ln\left\{ \left(\frac{2a}{\Delta} + 1\right)\left(\frac{b - a - \Delta}{b + a - \Delta}\right) \right\} \right. \\
&\quad \left. + \frac{1}{b} \ln\left\{ \left(\frac{2b}{\Delta} - 1\right)\left(\frac{b - a - \Delta}{b + a + \Delta}\right) \right\} \right] \\
\Rightarrow Z_0^{CPS} &\approx \frac{R_{sm}b^2}{4Z_0^{CPS}K^2(k')(b^2 - a^2)} \\
&\quad \left[\frac{1}{a} \ln\left\{ \left(\frac{2a}{\Delta}\right)\left(\frac{b - a}{b + a}\right) \right\} + \frac{1}{b} \ln\left\{ \left(\frac{2b}{\Delta}\right)\left(\frac{b - a}{b + a}\right) \right\} \right].
\end{aligned} \tag{9.7.18}$$

The expressions (9.7.13) and (9.7.14) for the stopping distance Δ are valid for the CPS also. However, the presence of a ground plane disturbs its value. Figure (9.40b) shows that the conductor loss of the CPS computed by the present method is almost identical to the results obtained using the mode-matching method [J.45, J.62, J.63]. The CPS is on the GaAs substrate $\varepsilon_r = 12.9$, a = 10 μm, b = 25 μm, t = 1.5 μm, and $\sigma = 3 \times 10^7$ S/m.

9.7.2 Dielectric Loss

The CPW and CPS lines are fabricated on the low-loss substrates. Therefore, the dielectric loss is much smaller than the conductor loss. The dielectric loss for a CPW on a finite thickness substrate, with and without conductor backing, is computed by the *single-layer reduction (SLR) formulation*. The method also includes the top shield. Figure (9.41) shows the shielded finite thickness substrate CPW with the steps involved in the SLR process.

This method is also applicable to the CPW with a multilayer dielectric medium [J.64]. It is discussed in section (15.3) of chapter 15. The dielectric loss of the CPW structure on the finite thickness substrate of Fig (9.40a) is computed by the following expression [J.64, J.65]:

$$\alpha_d^{CPW} = \frac{\pi}{\lambda_0}\frac{\varepsilon_{r,eq}}{\varepsilon_{r\,eff}}\cdot\frac{(\varepsilon_{r\,eff}-1)}{(\varepsilon_{r,eq}-1)}\tan\delta_{eq}\ \text{Np/m} \qquad \text{(a)}$$

$$\Rightarrow \alpha_d^{CPW} = 27.29\frac{\varepsilon_{r,eq}}{\varepsilon_{r\,eff}}\cdot\frac{(\varepsilon_{r\,eff}-1)}{(\varepsilon_{r,eq}-1)}\frac{\tan\delta_{eq}}{\lambda_0}\ \text{dB/m} \qquad \text{(b)}, \tag{9.7.19}$$

where ε_{reff} of the CPW structure, without conductor backing, with conductor backing, and with/without a top shield, is computed from expressions presented previously.

The effect of the conductor thickness and frequency, in the computation of $\varepsilon_{r\,eff}$ could also be included in the computation. The real permittivity ε_r used in these expressions is replaced by the complex relative permittivity $\varepsilon_r^* = \varepsilon_r' - j\varepsilon_r''$, $\tan\delta = \varepsilon_r''/\varepsilon_r'$ of a lossy substrate. The complex effective relative permittivity $\varepsilon_{r\,eff}^*$ is computed to place the CPW in a homogeneous lossy dielectric medium, as shown in Fig (9.41b). Figure (9.41c) shows the next step of the SLR process that reduces the CPW structure, located in the homogeneous medium, to a CPW on the infinitely thick substrate with equivalent relative permittivity $\varepsilon_{r,\,eq}$ and equivalent loss-tangent $\tan\delta_{eq}$. The $\varepsilon_{r,\,eq}$ and $\tan\delta_{eq}$, i.e. the complex equivalent relative permittivity $\varepsilon_{r,eq}^*$, take care of finite thickness of a substrate, conductor backing, and also top shield. The complex equivalent relative permittivity $\varepsilon_{r,eq}^*$ is computed as follows:

$$\varepsilon_{r,eff}^* = 1 + q\left(\varepsilon_{r,eq}^* - 1\right)$$
$$\text{for,}\quad q = 1/2,\quad \varepsilon_{r,eq}^* = 2\varepsilon_{r,eff}^* - 1. \tag{9.7.20}$$

The real part of the above equation gives $\varepsilon_{r,\,eq}$ of the equivalent infinitely thick substrate. The $\tan\delta_{eq}$ of the equivalent substrate is

$$\tan\delta_{eq} = \frac{\text{Im}\left(\varepsilon_{r,eq}^*\right)}{\text{Re}\left(\varepsilon_{r,eq}^*\right)}. \tag{9.7.21}$$

Figure (9.42a) presents the dielectric loss of a CPW, using the SLR and MOM-based software LINPAR [B.18], on a GaAs substrate with and without conductor backing for the slot width w = 0.02 mm, 0.2 mm. The presence of a conductor backing increases the dielectric loss, as the field lines are confined in the lossy dielectric region. For the same reason, the dielectric loss increases with a strip width of a CB-CPW, whereas it decreases with strip width for the CPW without conductor backing. Both methods show good agreement. Figure (9.42b) further shows the effect of the top shield on the CB-CPW. The proximity of the top shield increases the dielectric loss. However, for the shield height more than the substrate thickness, its influence can be neglected. Figure (9.42b) also shows that for the same loss-tangent and the same CPW structure, the dielectric loss increases with an increase in the relative permittivity of a substrate. Again, it is due to more field confinement in the dielectric region of a high permittivity substrate.

9.7.3 Substrate Radiation Loss

The fundamental modes of the CPW and CPS, under the certain condition, can become the *leaky mode* and radiate power into the substrate. This kind of radiation loss is very strong, as compared to both the dielectric and conductor losses. The mechanism of such substrate radiation is related to *Cherenkov-type radiation* [B.19]. Cherenkov has shown experimentally that if a charged particle is moving in a dielectric medium with a uniform velocity v_q greater than the EM-wave velocity v_m in that medium, it radiates the EM shock wave forming a radiation cone, shown in Fig (9.43a). For the Cherenkov radiation, also called the *Cerenkov radiation*, power increases with f^3 and is sensed by a photographic film. The mechanism of the Cherenkov radiation is discussed in subsection (5.5.7) of chapter 5.

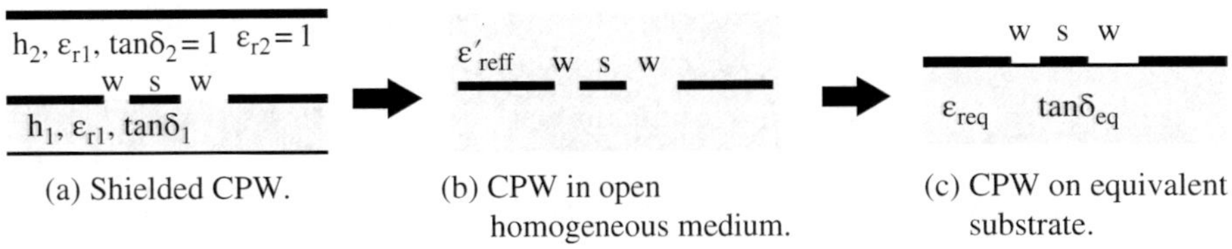

Figure 9.41 The SLR process to compute the dielectric loss of conductor-backed CPW with a top shield. *Source:* Verma et al. [J.64]. Verma and Bhupal [J.65].

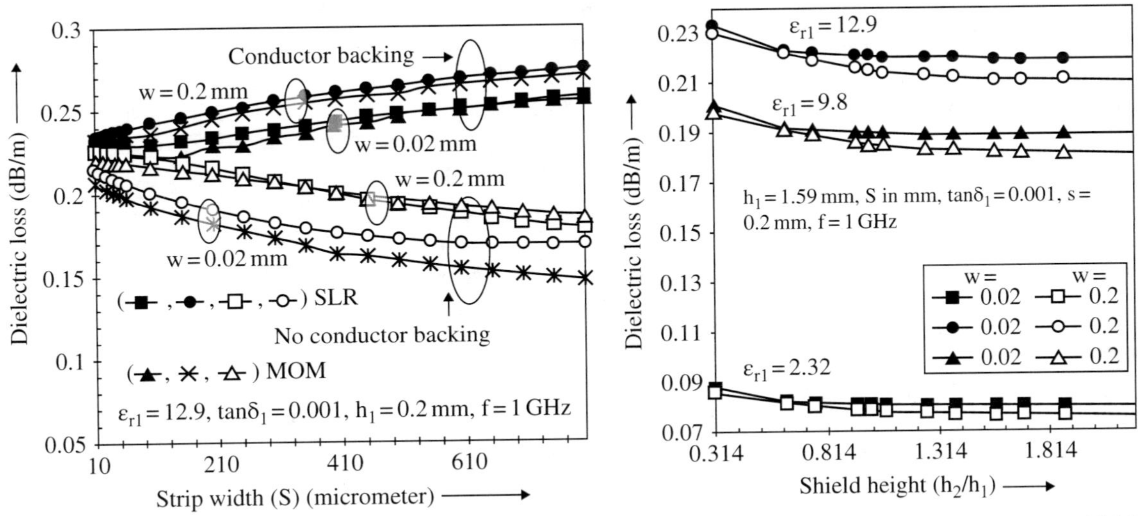

(a) Dielectric loss with/without conductor backing. (b) Dielectric loss of CPW with a top shield.

Figure 9.42 Comparison of computation of dielectric loss CPW structures. *Source:* Verma et al. [J.64].

Figure (9.43b) shows the wave propagating on a CPS structure. The propagating wave induces the dipolar charges on the CPS. The dipolar charges move on the conductor faster than the phase velocity of the EM-wave in the surrounding dielectric substrate medium. Thus, a signal traveling on the CPS structure radiates power in the substrate in the form of Cherenkov-type radiation, causing a serious loss in the signal. The radiation in

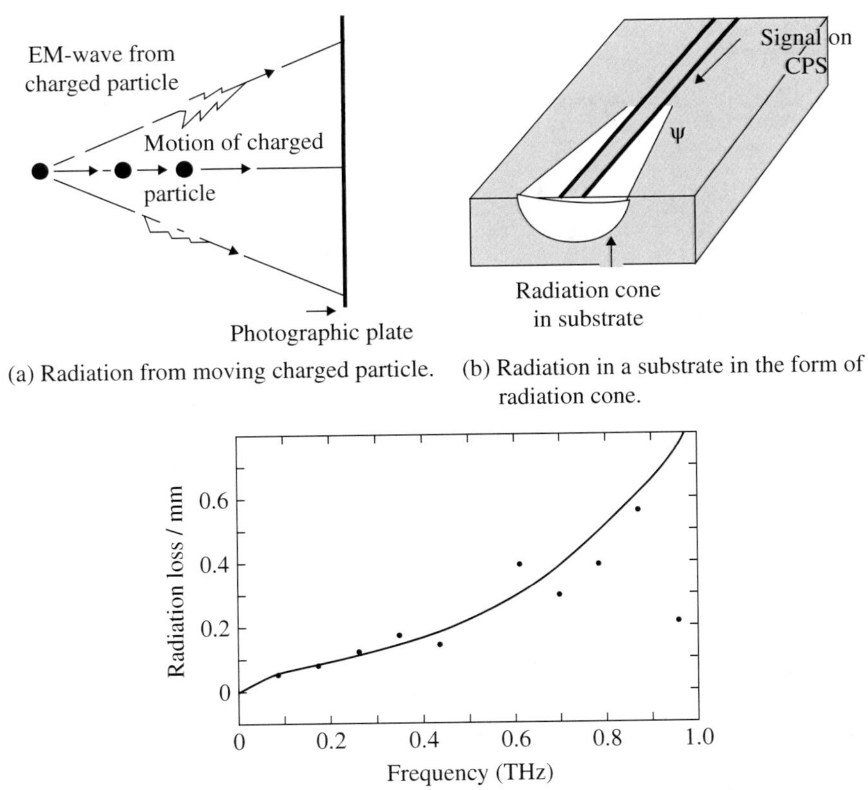

(a) Radiation from moving charged particle. (b) Radiation in a substrate in the form of radiation cone.

(c) Radiation loss for CPS [J.66].

Figure 9.43 Cherenkov-type radiation loss. *Source:* Grischkowsk et al. [J.66]. © APS.

the substrate is at an angle ψ forming the radiation cone. Figure (9.43c) experimentally shows the strong f^3-dependent radiation loss from the 150 Ω CPS taken on silicon on the sapphire (SOS) substrate. The CPS has strip width (s) 5 μm, strip gap (w) 15 μm, and aluminum conductor thickness (t) 0.5 μm [B.20, J.66–J.68]. The dots show the experimental loss of the CPS line from DC to 1 THz that is approximately described by the following curve-fitted expression:

$$\alpha_{rad} = 0.2\sqrt{f} + 0.65f^3 \ \text{Np/mm}. \tag{9.7.22}$$

The first term of the above expression is the conductor loss of the aluminum conductor-based CPS that is small. It increases slowly as the square root of frequency in the THz range. The second term is the strong radiation loss due to Cherenkov-type leaky mode radiation. Rutledge *et al.* have obtained analytic expressions to model this unique radiation loss on both the CPW and CPS structures [B.20]. Their expressions have been modified by other investigators [J.68]. A summary of their expressions is presented below.

Figure (9.43a) shows the cone angle of Cherenkov-type radiation of a moving charge by the following expression:

$$\cos\psi = \frac{v_m}{v_q}; \quad v_q > v_m, \tag{9.7.23}$$

where v_m is the phase velocity of the EM-wave in the surrounding medium and v_q is a uniform velocity of the moving charge. In the case of a CPS, shown in Fig (9.43b), the charge velocity v_q is identified to the movement of the dipoles on a strip conductor that is related to the velocity of the fundamental mode with propagation constant k_z. The EM-wave, radiated in the substrate, has the substrate mode velocity v_m, with the propagation constant k_d. Thus, the velocities are $v_m = c/k_d$ and $v_q = c/k_z$. The radiation cone angle for the coplanar line is

$$\begin{aligned} &\text{In dielectric medium}: \quad \cos\psi = k_z/k_d = \left(\frac{\varepsilon_{r\,eff}}{\varepsilon_r}\right)^{1/2} && \text{(a)} \\ &\text{In air medium}: \quad \cos\psi = k_z/k_0 && \text{(b)}. \end{aligned} \tag{9.7.24}$$

The above relation also shows the phase matching, at the interface, of the radiated wave in a dielectric medium. The substrate leaky-mode radiation loss occurs only for $k_z < k_d$. Similarly, the radiation in the air-medium occurs, if $k_z < k_0$. This condition is needed for working of a leaky-wave antenna [J.69, J.70]. The bound-mode wave propagation on a transmission line, without any radiation, occurs if $k_z > k_d > k_0$. The radiation in the substrate occurs for the condition $k_d > k_z > k_0$. However, radiation in air-medium occurs for the condition $k_0 > k_z$.

For the CPS and CPW on a thick substrate, wavenumber is $k_d = \sqrt{\varepsilon_r}k_0$ for the radiated wave inside the infinitely thick substrate. The wavenumber of the fundamental quasi-static mode, on both the line structures, is $k_z = \sqrt{(\varepsilon_r + 1)/2}k_0$. For $k_d > k_z$, the fundamental quasi-static mode becomes leaky, and radiates in the substrate. The CPS and CPW structures, on the infinitely thick substrate, are unconditionally leaky with radiation in the substrate. The expression for the radiation loss follows the cubic frequency dependence under the quasi-static approximation [J.15, J.41, J.66, J.67]:

$$\begin{aligned} \alpha_{sw}^{CPS} &= \left(\frac{\pi}{2}\right)^5 \frac{\sqrt{2}(3-\sqrt{8})(1-1/\varepsilon_r)^2 w^2\sqrt{1+1/\varepsilon_r}\,\varepsilon_r^{3/2}}{c^3K(k')K(k)} f^3 && \text{(a)} \\ \alpha_{sw}^{CPW} &= \left(\frac{\pi}{2}\right)^5 \frac{1}{\sqrt{2}}\frac{(1-1/\varepsilon_r)^2}{\sqrt{1+1/\varepsilon_r}}\frac{(s+2w)\varepsilon_r^{3/2}}{c^3K(k')K(k)} f^3 && \text{(b)}. \end{aligned} \tag{9.7.25}$$

where aspect-ratio is $k = s/(s+2w)$, $k' = \sqrt{1-k^2}$ and c is the velocity of the EM-wave in the free space. In the case of a CPW, s is the width of central strip and w is the slot width, whereas for a CPS, w is the strip width and s is the slot gap. The CB-CPW line is also unconditionally leaky [B.16].

For the CPS and CPW lines on a finitely thick substrate, Frankel *et al.* [J.41] have observed that substrate radiation loss does not follow the cubic frequency dependence. It follows frequency dependence less than the exponent factor 3. This can be accommodated by taking $\cos\psi = \sqrt{\varepsilon_{r\,eff}(f)/\varepsilon_r}$ and by modifying the above loss expression as follows:

$$\begin{aligned} \alpha_{sw}^{CPS} &= \pi^5 \frac{(3-\sqrt{8})}{2}\sqrt{\frac{\varepsilon_{r\,eff}(f)}{\varepsilon_r}}\left(1-\frac{\varepsilon_{r\,eff}(f)}{\varepsilon_r}\right)^2 \frac{(s+2w)^2\varepsilon_r^{3/2}}{c^3K(k')K(k)} f^3 && \text{(a)} \\ \alpha_{sw}^{CPW} &= \left(\frac{\pi}{2}\right)^5 2\left[\frac{\left(1-\frac{\varepsilon(f)}{\varepsilon_r}\right)^2}{\sqrt{\frac{\varepsilon_{r\,eff}(f)}{\varepsilon_r}}}\right]\frac{(s+2w)^2\varepsilon_r^{3/2}}{c^3K(k')K(k)} f^3 && \text{(b)}. \end{aligned} \tag{9.7.26}$$

The finitely thick substrate supports the TE/TM surface waves. Therefore, the surface wave loss for the TE/TM polarization could be computed separately

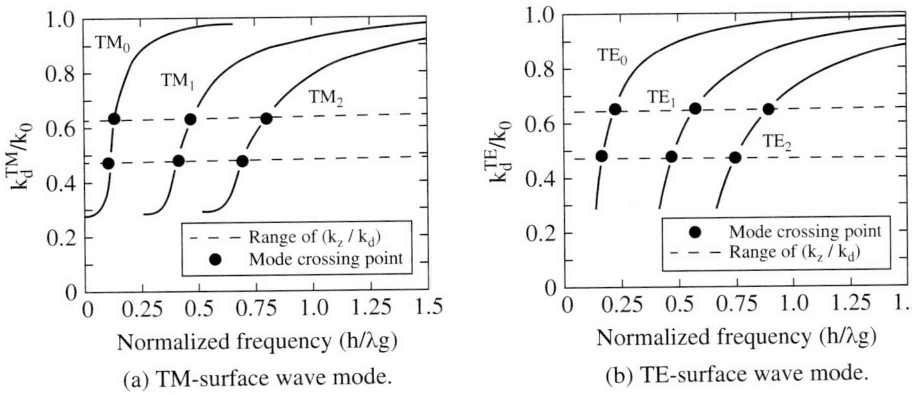

Figure 9.44 Mode crossing of surface wave modes in a conductor-backed dielectric slab. *Source:* Riaziat et al. [J.15]. © IEEE.

[J.68]. The above expressions ignore this aspect. The computation of frequency-dependent effective relative permittivity of both the CPW and CPS lines, on the finitely thick substrate, has already been presented. The expressions are valid for $0.1 < s/w < 10$ and $h > 3w$ and for wavelength $\lambda_0 > s + 2w$.

The CB-CPW supports the TM and TE surface wave modes on the conductor-backed dielectric substrate of thickness h. If the wavenumbers of the TM/TE surface wave modes exceed the wavenumber of the CPW line, the quasi-TEM mode of CPW becomes leaky and radiates power in that mode in a substrate. Figure (9.44a, b) shows the mode crossings of both the TM and TE surface wave modes [J.15]. The broken lines show the $\varepsilon_{r\,eff}(f)/\varepsilon_r$ of the CPW line in the range 0.48–0.64. It is determined by the CPW structure. The surface wave propagation constant is normalized by $k_d = \sqrt{\varepsilon_r}k_0$. For the wideband operation of the CPW, without excessive substrate radiation loss, the substrate thickness should be such that the fundamental surface wave mode does not cross the normalized wave number of the quasi-TEM mode of a CPW. This is achieved by taking substrate thickness $h < 0.15\,\lambda_d$, where $\lambda_d = \lambda_0/\sqrt{\varepsilon_r}$. This condition is equally applicable to a microstrip line that also supports the spurious TM_0-mode surface wave.

9.8 Circuit Models and Synthesis of CPW and CPS

The computation of the characteristic impedance, effective relative permittivity, conductor loss, and dielectric loss of the CPW and CPS structures is discussed in the previous sections. In these computations, the structural parameters on a given substrate are known quantities. This is known as the *analysis* of the CPW or CPS. However, in designing RF circuits, it is required to determine the structural dimensions of the CPW/CPS on a specified substrate for the assigned characteristic impedance. This process is known as the *synthesis* of CPW or CPS. This section presents the final expressions of the synthesis of the CPW and CPS structures. Even in the case of the analysis, the losses also influence Z_0 and $\varepsilon_{r\,eff}(f, t)$. Such influence can be accommodated through the RLCG-circuit model discussed in this section.

9.8.1 Circuit Model

The lossy CPW, CPS, and other quasi-TEM planar lines are described either by the secondary line parameters, such as the complex propagation constant γ and the complex characteristic impedance Z_0^*, or by the primary line constants R, L, C, G *p.u.l.* In the case of the quasi-TEM planar lines, the primary line constants are frequency dependent. Figure (9.45) shows the equivalent TEM-type transmission line model of the lossy quasi-TEM planar lines.

The frequency-dependent and also conductor thickness-dependent primary line constants, R(f,t), L(f,t), C(f,t), G(f,t), shown in Fig (9.45), can be evaluated from the known frequency and conductor thickness-dependent secondary line parameters, $Z_0(f, t)$, $\varepsilon_{r\,eff}(f, t)$, $\alpha_c(f, t)$, and $\alpha_d(f, t)$ as follows [J.50]:

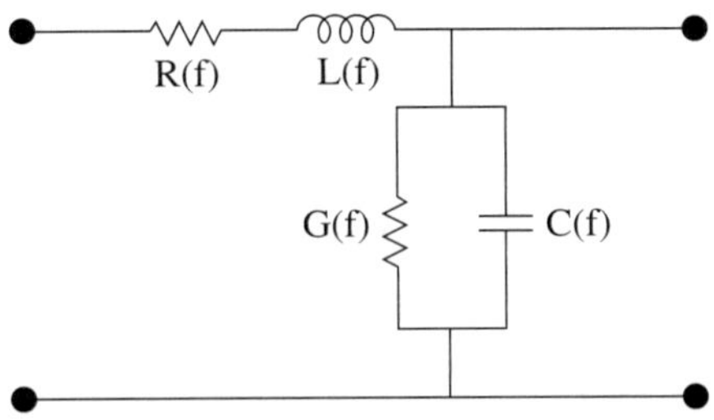

Figure 9.45 Equivalent transmission line model of lossy quasi-TEM planar lines.

$$R(f,t) = 2Z_0(f,t)\alpha_c(f) \quad (a)$$
$$L(f) = \frac{Z_0(f,t)\sqrt{\varepsilon_{r\,eff}(f,t)}}{c} \quad (b)$$
$$C(f,t) = \frac{\sqrt{\varepsilon_{r\,eff}(f,t)}}{cZ_0(f,t)} \quad (c)$$
$$G(f,t) = \frac{2\alpha_d(f,t)}{Z_0(f,t)} \quad (d), \qquad (9.8.1)$$

where c is the velocity of the EM-wave in the free space. The secondary line parameters are computed, for the CPW and CPS structures, using the analytical expressions presented in this chapter and for the microstrip line, using expressions presented in chapter 8. The complex propagation constant γ and the complex characteristic impedance Z_0^* of the planar lines are computed from the following expressions:

$$Z_0^*(f,t) = Z_0(f,t) + jZ_0'(f,t) = \left[\frac{R(f,t) + j\omega L(f,t)}{G(f,t) + j\omega C(f,t)}\right]^{1/2} \quad (a)$$
$$\gamma(f,t) = \alpha_T(f,t) + j\beta(f,t) = [(R(f,t) + j\omega L(f,t))(G(f,t) + j\omega C(f,t))]^{1/2} \quad (b). \qquad (9.8.2)$$

The characteristic impedance $Z_0(f, t)$ and $\varepsilon_{r\,eff}(f, t)$ are evaluated from the above expressions. At this stage, effect of the losses and the magnetic field penetration on $Z_0(f, t)$ and $\varepsilon_{r\,eff}(f, t)$ are included. Besides, the imaginary part of $Z_0^*(f,t)$ is also computed. These are not possible with analytical expressions presented in this chapter. Figures (9.36) and (9.37) show the results of circuit model (CM) for the CPW and CPS. These results show the low-frequency dispersion in the line parameters. The above-computed R(f,t), L(f,t), C(f,t), G(f,t), can be further used to develop closed-form models of the primary line constants. Such expressions are available for a standard microstrip line [J.71].

9.8.2 Synthesis of CPW

The characteristic impedance of a microstrip line depends on a single structural parameter, i.e. the w/h-ratio. However, the characteristic impedance of a CPW depends on two structural parameters, i.e. on the w/h-ratio and s/h-ratio. Thus, the CPW of a given value of characteristic impedance can be synthesized by several combinations of w/h-ratio and s/h-ratio [J.72–J.75]. There is no unique choice of the w/h-ratio and s/h-ratio to achieve the given characteristic impedance of a CPW. Therefore, two sets of synthesis expressions are given below – one for the assumed s/h (s: strip width) ratio and another for the assumed w/h (w: slot width) ratio. Usually, the slot width ratio (w/h) is less than 1 for a compact CPW. This choice also gives small dispersion, as the electric field is confined to the slot region and the so-called microstrip mode is suppressed. The CPW has negligible dispersion up to 60 GHz. Thus, the slot width in the range $0.1 \le w/h \le 1$ and corresponding to the strip width in the range $0.1 \le s/h \le 8$ are used to achieve characteristic impedance of a CPW between 25 and 100 Ω on GaAs substrate, $\varepsilon_r = 12.9$, h = 0.2 mm [J.72].

Synthesis Expression Set - I (Known w/h-Ratio)

The synthesis expressions, given by Deng [J.72], are summarized below. These are valid for a substrate with any relative permittivity. This set of expressions assumes w/h (slot width/substrate thickness)-ratio. For the given characteristic impedance $Z_0(\varepsilon_r)$, on a known substrate (ε_r, h), it computes the s/h (strip width/substrate thickness)-ratio:

$$\frac{s}{h} = \frac{w}{h} G_1(\varepsilon_r, Z_0, w/h). \qquad (9.8.3)$$

For $Z_0 < \dfrac{\eta_0}{\sqrt{2(\varepsilon_r + 1)}}$: $G_1(\varepsilon_r, Z_0, w/h)$

$$= \left[\frac{1}{4}\exp\left(\frac{\pi}{4\sqrt{\varepsilon_{r\,eff}(0)}}\frac{\eta_0}{Z_0}\right) + \exp\left(-\frac{\pi}{4\sqrt{\varepsilon_{r\,eff}(0)}}\frac{\eta_0}{Z_0}\right) - 1\right] \quad (a)$$

For $Z_0 \ge \dfrac{\eta_0}{\sqrt{2(\varepsilon_r + 1)}}$:

$$G_1(\varepsilon_r, Z_0, w/h) = \left[\frac{1}{8}\exp\left(2\pi\sqrt{\varepsilon_{r\,eff}(0)}\frac{Z_0}{\eta_0}\right) - \frac{1}{2}\right]^{-1} \quad (b). \qquad (9.8.4)$$

The $\varepsilon_{r\,eff}(0)$ is the static effective dielectric constant that is a function of ε_r, Z_0 and the assumed w/h-ratio. It can be computed by the following expression:

$$\varepsilon_{r\,eff}(0) = \varepsilon_{r\,eff}(\varepsilon_r, Z_0, w/h) = A(Z_0, \varepsilon_r) \quad B(\varepsilon_r, Z_0, w/h) \qquad \text{(a)}$$

$$A(Z_0, \varepsilon_r) = 1 + \sqrt{2}(\varepsilon_r - 1)\sqrt{\varepsilon_r + 1}\frac{Z_0}{\eta_0}\frac{K(k)}{K(k')} \qquad \text{(b)}$$

$$B(\varepsilon_r, Z_0, w/h) = \text{sech}\left\{\frac{\varepsilon_r^5}{4\pi(\varepsilon_r + 1)^6}\left(\frac{\eta_0}{Z_0}\right)^2 \exp\left[\left(1 + 0.0016\varepsilon_r Z_0 \frac{w}{h}\right)\ln\left(0.6 + \frac{w}{h}\right)\right]\right\} \qquad \text{(c)}$$

$$\text{where aspect} - \text{ratio } k: \ k = \frac{\exp\left[\frac{[\pi(1+g)]}{2}\frac{w}{h}\right] - \exp\left[\frac{\pi}{2}\frac{w}{h}\right]}{\max\left[\frac{[\pi(2+g)]}{2}\frac{w}{h}\right] - 1} \quad \text{(d)}, \quad k' = \sqrt{1-k^2} \qquad \text{(e)} \qquad (9.8.5)$$

$$g = \frac{\varepsilon_r + 1}{2} \qquad \text{(f)}$$

$$\eta_0 = 120\pi \qquad \text{(g)}.$$

The approximate expression for $K(k)/K(k')$ is given by equation (9.2.39). The above expressions are valid for $w/h \le 10/[3(1 + \ln \varepsilon_r)]$ and $s/h \le 80/[3(1 + \ln \varepsilon_r)]$. The above expressions also compute the static effective relative permittivity in the process of synthesis. Usually, its value is less than the value computed by the static method discussed previously or by the full-wave methods. For a practical range of parameters, its accuracy is 1.5% that gets degraded for the extreme case of the w/h- and s/h-ratios. The above expressions are complicated. However, once w/h- and s/h-ratios of a CPW are known, $\varepsilon_{r\,eff}(0)$ can be accurately computed using equation (9.3.15). Another set of expressions to compute G_1, that does not involve computation of $\varepsilon_{r\,eff}(0)$, is also available [J.75]. It is valid for the substrates with relative permittivity range $6 \le \varepsilon_r \le 100$.

Synthesis Expression Set - II (Known s/h-Ratio)

The slot width (w/h) ratio of a CPW, for known, ε_r, Z_0, h can be computed if s/h-ratio is assumed **[J.72]**:

$$\frac{w}{h} = \frac{s/h}{G_2(\varepsilon_r, Z_0, s/h)}. \qquad (9.8.6)$$

On replacing w/h-ratio by s/h-ratio in the expression (9.8.4), the expression of G_1 is used for G_2. The w/h-ratio is replaced by the s/h-ratio. However, $\varepsilon_{r\,eff}(0)$ has to be computed as follows:

$$\varepsilon_{r\,eff}(0) = \varepsilon_{r\,eff}(\varepsilon_r, Z_0, s/h) = A{\cdot}B \qquad \text{(a)}$$

$$B = 1 + \tanh\left(\frac{\varepsilon_r^7}{4\pi^2(\varepsilon_r + 1)^8}\left(\frac{\eta_0}{Z_0}\right)^2 Q_1\right) \qquad \text{(b)}$$

$$Q_1 = \exp\left[\left(1 + 0.0004\,\varepsilon_r \frac{Z_0}{g}\frac{s}{h}\right)\ln\left(\frac{s}{hg}\right)\right] \qquad \text{(c)}. \qquad (9.8.7)$$

The parameter A of the above equation is computed from the expression (9.8.5b). The aspect-ratio k is obtained from the following expression:

$$k = \frac{\exp\left[\pi\frac{(1+1/g)}{2}\frac{s}{h} - \exp\left(\frac{\pi}{2g}\frac{s}{h}\right)\right]}{\exp\left[\frac{\pi(1+2/g)}{2}\frac{s}{h} - 1\right]}. \qquad (9.8.8)$$

The above set of equations also computes $\varepsilon_{r\,eff}(0)$. However, its value is more than the results obtained from the analysis formula. It can be computed accurately using the expression (9.3.15). An alternate set of expression, for $6 \le \varepsilon_r \le 100$, is also available to compute G_2 without computing $\varepsilon_{r\,eff}(0)$ [J.75].

9.8.3 Synthesis of CPS

The expressions for the analysis of the CPS can be taken in the iterative process to synthesize the CPS. The synthesis expression for the CPS on the infinitely thick substrate is available [B.14]. Three different sets of the

Table 9.8 The values for the coefficient $\alpha_1 - \alpha_9$.

Set	α_1	α_2	α_3	α_4	α_5	α_6	α_7	α_8	α_9
Set #1: Low impedance	−1.0083	0.4261	2.3613	1.1966	9.7854	4.0479	3.7066	2.0667	4.2467
High impedance	−0.4394	0.1753	−0.0848	−0.0524	1.8679	1.8822	1.2015	2.1294	4.2530
Set#2: Low impedance	−1.0050	0.6229	4.7557	2.3716	8.4331	4.2339	3.5531	2.1547	4.4647
High impedance	−0.9255	0.4156	3.1556	1.3762	1.6588	2.0995	2.2560	1.8277	3.8805

closed-form expressions have also been proposed to synthesize the symmetrical CPS on a finitely thick substrate [J.76–J.78]. Two sets of accurate expressions are given below [J.76].

Set #1 (Known s/h-ratio): In this case, the substrate details (ε_r,h) and characteristic impedance $Z_0(\varepsilon_r)$ of CPS are specified. Further, the slot width (s) is assumed appropriately. The strip width (w) is computed as follows:

$$w = \frac{s}{G[\varepsilon_r, Z_0(\varepsilon_r), X]} \quad \text{(a)}$$
$$X = s/h \quad \text{(b)}. \qquad (9.8.9)$$

Set #2 (Known w/h-ratio): In this case also, the substrate details (ε_r,h) and characteristic impedance $Z_0(\varepsilon_r)$ of CPS are specified. However, the strip width (w) is assumed appropriately and we calculate the slot width (s) as follows:

$$s = w \cdot G[\varepsilon_r, Z_0(\varepsilon_r), X] \quad \text{(a)}$$
$$X = w/h \quad \text{(b)}. \qquad (9.8.10)$$

The curve-fitted expression for the factor $G[\varepsilon_r, Z_0(\varepsilon_r), X]$ is given below:

$$G[\varepsilon_r, Z_0(\varepsilon_r), X] = G_1 + G_2 + G_3 \quad \text{(a)}$$
$$\text{where } G_1 = \alpha_1 \cdot \exp\left[X^{\alpha_2} \cdot \varepsilon_r^{\alpha_3}\left(\frac{Z_0(\varepsilon_r)}{120\pi}\right)^{\alpha_4}\right] \quad \text{(b)}$$
$$G_2 = \exp\left[\alpha_5\left(\frac{Z_0(\varepsilon_r)}{120\pi}\right)^{\alpha_6}\right] \quad \text{(c)}$$
$$G_3 = \alpha_7 \cdot \varepsilon_r^{\alpha_8}\left(\frac{Z_0(\varepsilon_r)}{120\pi}\right)^{\alpha_9} \quad \text{(d)}. \qquad (9.8.11)$$

The expression for the factor $G(\varepsilon_r, Z_0(\varepsilon_r), X)$ is common for both the sets and the value of the X is suitably assumed. The values for the coefficient $\alpha_1 - \alpha_9$, given in Table (9.8) for both sets of expressions, apply to the low impedance and high impedance CPS. For the low impedance lines, $Z_0(\varepsilon_r) < \left(120\pi/\sqrt{2(\varepsilon_r + 1)}\right)$, and for the high impedance line, the relation is $Z_0(\varepsilon_r) > \left(120\pi/\sqrt{2(\varepsilon_r + 1)}\right)$. The formulas are valid for the range $s/h \le \{10/[3(1 + \ln(\varepsilon_r))]\}$, $w/h \le \{10/(1 + \ln(\varepsilon_r))\}$ and $2.2 \le \varepsilon_r \le 50$. It has accuracy 1.56% against the results of the analysis formula.

References

Books

B.1 Simon, R.N.: *Coplanar Waveguide Circuits Components and Systems*, John Wiley & Sons, New York, NY, 2001.

B.2 Collin, R.E.: *Field Theory of Guided Waves*, 2nd Edition, Wiley-IEEE Press, Piscataway, New Jersey, 1990.

B.3 Collin, R.E.: *Foundations for Microwave Engineering*, 2nd Edition, McGraw-Hill, Inc., New York, NY, 1992.

B.4 Wolff, I.: *Coplanar Microwave Integrated Circuits*, John Wiley & Sons, New York, NY, 2006.

B.5 Kinayman, N.; Aksun, M.I.: *Modern Microwave Circuits*, Artech House, Boston, MA, 2003.

B.6 Plonsey, R.; Collin, R.E.: *Principles and Applications of Electromagnetic Fields*, 2nd Edition, McGraw-Hill, Inc., New York, NY, 1982.

B.7 Dworsky, L.N.: *Modern Transmission Line Theory and Applications*, John Wiley & Sons, New York, NY, 1997.

B.8 Pipes, L.A.; Harvill, L.R.: *Applied Mathematics for Engineers and Physicist*, International Edition, McGraw-Hill, Inc., New York, NY, 1971.

B.9 Ramo, S.; Whinnery, J.R.; Van Duzer, T.: *Fields and Waves in Communication Electronics*, 3rd Edition, John Wiley & Sons, Singapore, 1994.

B.10 Kreyszig, E.: *Advanced Engineering Mathematics*, 8th Edition, John Wiley & Sons, New York, NY, 1999.

B.11 Weeks, W.L.: *Electromagnetic Theory for Engineering Applications*, John Wiley & Sons, New York, NY, 1964.

B.12 Nehari, Z.: *Conformal Mapping*, Dover Publications, New York, NY, 1975.

B.13 Spiegel, M.R.: *Theory and Problems of Complex Variables with an Introduction to Conformal Mapping and its Applications*, Schaum's Publishing Company, New York, NY, 1964.

B.14 Hoffmann, R.: *Microwave Integrated Circuit Handbook*, Artech House, Boston, MA, 1985.

B.15 Abramowitz, M.; Stegun, I.A.: *Handbook of Mathematical Functions with Formulas, Graphs, and Mathematical Tables*, Martino Fine Books, Eastford, CT, 2014.

B.16 Gupta, K.C.; Garg, R.; Bahl, I.; Bhartia, P.: *Microstrip Lines and Slot Lines*, 2nd Edition, Artech House, Boston, MA, 1996.

B.17 Kneppo, I.; Fabian, J.; Bezousek P.; Hrnicko, P.; Pavel, M.: *Microwave Integrated Circuits*, Chapman & Hall, London, 1994.

B.18 Sarkar, T.K.; Harrington, R.F.: *LINPAR – Software*, Artech House, Boston, 1999.

B.19 Smith, G.S.: *An Introduction to Classical Electromagnetic Radiation*, Cambridge University Press, NewYork, 1997.

B.20 Rutledge, D.B., et al.: *Millimeter Components and Techniques, Infrared and Millimeter Waves Part II, Chapter-1*, In *Planar Integrated Circuit Antennas* Vol. 10, K.J. Button, Academic Press, New York, NY, 1983

Journals

J.1 Wen, C.P.: *Coplanar waveguide: A surface strip transmission line suitable for non-reciprocal gyromagnetic device application, IEEE Trans. Microw. Theory Tech.*, Vol. MTT- 17, No. 12, pp. 1087–1090, Dec. 1969.

J.2 McLean, J.S.; Itoh, T.: *Analysis of a new configuration of coplanar stripline, IEEE Trans. Microw. Theory Tech.*, Vol. MTT- 40, pp. 772–774, April 1992.

J.3 Veyres, C.; Hanna, V.F.; *Extension of the application of conformal mapping techniques to coplanar lines with finite dimensions, Int. J. Electron.*, Vol. 48, No. 1, pp. 47–56, Jan. 1980.

J.4 Ghione, G.; Naldi, C.U.: *Coplanar waveguides for MMIC applications: Effect of upper shielding, conductor backing, finite extent ground planes, and line-to-line coupling, IEEE Trans. Microw. Theory Tech.*, Vol. MTT- 35, No. 3, pp. 260–267, March 1987.

J.5 Bedair, S.S.; Wolff, I.: *Fast, accurate and simple approximate analysis formulas for calculating the parameters of supported coplanar waveguides, IEEE Trans. Microw. Theory Tech.*, Vol. MTT- 40, No. 1, pp. 41–48, Jan. 1992.

J.6 Hilberg, W.: *From approximations to exact relation for characteristics impedances, IEEE Trans. Microw. Theory Tech.*, Vol. MTT- 17, pp. 259–265, 1969.

J.7 Ghione, G.; Naldi, C.: *Parameters of coplanar waveguides with the lower ground, Electron. Lett.*, Vol. 19, pp. 734–735, 1983.

J.8 Crampagane, R.; Khoo, G.: *Synthesis of certain transmission lines employed in microwave integrated circuits, IEEE Trans. Microw. Theory Tech.*, Vol. MTT- 25, pp. 440–442, 1977.

J.9 Yamashita, E.; Atsuki, K.: *Analysis of microstrip- like transmission lines by non-uniform discretization of integral equations, IEEE Trans. Microw. Theory Tech.*, Vol. MTT- 24, pp. 195–200, 1976.

J.10 Ponchak, G.E.; Katehi, L.P.B.; Tentzeris, E.M.: Finite Ground Coplanar (FGC) Waveguide: It's Characteristics and Advantages For Use In RF and Wireless Communication Circuits," 3rd Int. Wireless Commun. Conf. (WCC '98) Dig., pp. 75–83, San Diego, CA, Nov. 1–3, 1998.

J.11 Davis, M.E.; Williams, E.W.; Celestini, A.C.: *Finite-boundary corrections to the coplanar waveguide analysis, IEEE Trans. Microw. Theory Tech.*, Vol. MTT- 21, No. 9, pp. 594–596, Sept. 1973.

J.12 Ghione, G.: *A CAD-oriented analytical model for the losses of general asymmetric coplanar lines in hybrid and monolithic MICs, IEEE Trans. Microw. Theory Tech.*, Vol. MTT- 41, pp. 1499–1510, Sept. 1993.

J.13 Pintzos, S.G.: *Full-wave spectral-domain analysis of coplanar strips, IEEE Trans. Microw. Theory Tech.*, Vol. MTT- 39, pp. 239–246, Feb. 1991.

J.14 Kneppo, I.; Gotzrnan, J.; Cesta, I.: *Basic parameters of nonsymmetrical coplanar line, IEEE Trans. Microw. Theory Tech.*, Vol. MTT- 25, pp. 718, Aug. 1977.

J.15 Riaziat, M.; Majidi-Ahy, R.; Feng, I.-J.: *Propagation modes and dispersion characteristics of coplanar waveguides, IEEE Trans. Microw. Theory Tech.*, Vol. MTT- 38, pp. 245–251, March 1990.

J.16 Fouad Hanna, V.: *Finite boundary corrections to coplanar stripline analysis, Electron. Lett.*, Vol. 16, No. 15, pp. 604–606, July 1980.

J.17 Owyang, G.H.; Wu, T.T.: *The approximate parameters of slot lines and their complement, IRE Trans. Antennas Propag.*, Vol. 4, pp. 49–55, June 1958.

J.18 Ghione, G.; Naldi, C.: *Analytical formulas for coplanar lines in hybrid and monolithic MICS, Electron. Lett.*, Vol. 20, No. 4, pp. 179–181, Feb. 1984.

J.19 Gevorgian, S.; Berg, H.: *Line capacitance and impedance of coplanar-strip waveguides on substrates with multiple dielectric layers*, Proc. Eu. Micro. Conf., London, 2001

J.20 Zhu, N.H.; Wang, Z.Q.: *On the accuracy of analytical expressions for calculating the parameters of coplanar strips on a finitely thick substrate, Microw. Opt. Technol. Lett.*, Vol. 8, No. 3, pp. 160–164, Feb. 1995.

J.21 Song, Y.K.; Lee, C.C.: *RF modeling and design of flip-chip configurations of microwave devices on PCBs*, Proc. 54th Electronic Components and Technology Conference, Vol. 2, pp. 1837–1842, 2004.

J.22 Flint, J.A.; Vardaxoglou, J.C.: *Exploitation of non-radiating modes in asymmetric coplanar strip folded dipoles, IEEE Proc. Microw. Antennas Propag.*, Vol. 151, pp. 307–310, 2004.

J.23 Chung, H.; Chang, W.S.C.; Adler, E.J.: *Modelling and optimization of traveling-wave LiNbO3 interferometric modulators, IEEE J. Quantum Electron.*, Vol. 27, No. 3, pp. 608–617, Mar. 1991.

J.24 Donnelly, J.P.; Gopinath, A.: *A comparison of power requirements of traveling-wave LiNbO3 optical couplers end interferometric modulators, IEEE J. Quantum Electron.*, Vol. QE-23, No. 1, pp. 30–41, Jan. 1987.

J.25 Yamashita, E.; Li, K.R.; Suzuki, Y.: *Characterization method and simple design formulas of MCS lines proposed for MMIC'S, IEEE Trans. Microw. Theory Tech.*, Vol. MTT- 35, pp. 1355–1362, Dec. 1987.

J.26 Svacina, J.: *Special types of coplanar transmission lines suitable up to mm-wave bands*, Conf. IEEE Electrical Performance of Electronic Packaging, 1997, Austin, TX, pp. 99–102.

J.27 Kitazawa, T; Hayashi, Y.: *Quasi-static characteristics of a coplanar waveguide with thick metal coating, Proc. IEEE, Part-H*, Vol. 133, No. 1, pp. 18–20, Feb. 1986.

J.28 Kitazawa, T.; Hayashi, Y.; Suzuki, M.: *A coplanar waveguide with a thick metal coating, IEEE Trans. Microw. Theory Tech.*, Vol. MTT- 24, No. 9, pp. 604–608, Sept. 1976.

J.29 Rahman, K.M.; Nguyen, C.: *Full-wave analysis of coplanar strips considering the finite strip metallization thickness, IEEE Trans. Microw. Theory Tech.*, Vol. MTT- 42, pp. 2177–2179, Nov. 1999.

J.30 Kitazawa, T.; Itoh, T.: *Propagation characteristics of coplanar type transmission lines with lossy media, IEEE Trans. Microw. Theory Tech.*, Vol. MTT- 39, pp. 1994–1700, 1991.

J.31 Goverdhan, K.; Simon, R.N.; Katchi, L.P.B.: *Coplanar stripline propagation characteristics and a bandpass filter, IEEE Trans. Microw. Theory Tech.*, Vol. MTT- 7, pp. 214–216, Aug. 1997.

J.32 Phatak, D.S.; Defonzo, A.P.: *Dispersion characteristics of optically excited coplanar striplines: Pulse propagation, IEEE Trans. Microw. Theory Tech.*, Vol. MTT- 38, pp. 654–661, May 1990.

J.33 Song, Y.K.; Lee, C.C.: *Characterization of coplanar strip lines on dielectric boards for RF and microwave applications, Int. J. RF Microw. Comput. Aided Eng.*, Vol. 18, pp. 1–10, Wiley Interscience, 2007.

J.34 Hasnain, G.; Dienes, A.; Whinnery, J.R.: *Dispersion of picosecond pulses in coplanar transmission lines, IEEE Trans. Microw. Theory Tech.*, Vol. MTT- 34, pp. 738–741, June 1986.

J.35 Yamashita, E.; Atsuki, K.; Veda, T.: *An accurate dispersion formula in microstrip line for computer-aided design of microwave integrated circuits, IEEE Trans. Microw. Theory Tech.*, Vol. MTT- 27, pp. 1036–1038, Dec. 1979.

J.36 Verma, A.K.; Singh, P.; Bansal, R.: *Computation of static and frequency dependent line parameters of multilayer CPW using static SDA and single layer reduction method, Int. J. RF Microw. Comput. Aided Eng.*, Vol. 24, No. 1, pp. 18–29, Jan. 2014.

J.37 Jackson, R.W.: *Considerations in the use of coplanar waveguide for millimeter-wave integrated circuits, IEEE Trans. Microw. Theory Tech.*, Vol. MTT- 34, pp. 1450–1456, 1986.

J.38 Gevorgian, S.S.; Martinsson, T.; Deleniv, A.; Kollberg, E.; Vendik, I.: *Simple and accurate dispersion the expression for the effective dielectric constant of coplanar waveguide, Proc. IEEE, Part-H*, Vol. 144, No. 2, pp. 145–148, April 1997.

J.39 Heinrich, W.: *Quasi-TEM description of MMIC coplanar lines including conductor loss effects, IEEE Trans. Microw. Theory Tech.*, Vol. MTT- 41, No. 1, pp. 45–52, Jan. 1993.

J.40 Knorr, J.B.; Kuchler, K.-D.: *Analysis of coupled slots and coplanar strips on dielectric substrate, IEEE Trans. Microw. Theory Tech.*, Vol. MTT- 23, pp. 541–548, July 1975.

J.41 Frankel, M.Y.; Gupta, S.; Valdmanis, J.A.; Mourou, G. A.: *Terahertz attenuation and dispersion characteristics of coplanar transmission lines, IEEE Trans. Microw. Theory Tech.*, Vol. MTT- 39, pp. 910–916, June 1991.

J.42 Majumdar, P.; Verma, A.K.: *CAD models of losses for symmetric elliptical and circular cylindrical coplanar strip lines, Int. J. Electron.*, Vol. 98, No. 7, pp. 863–882, July 2011.

J.43 Kiziloglu, K.; Dagli, N.; Matthaei, G.L.; Long, S.I.: *Experimental analysis of transmission line parameters in high –speed GaAs digital circuit interconnects, IEEE Trans. Microw. Theory Tech*, Vol. MTT- 39, pp. 1361–1367, Aug. 1991.

J.44 Heinrich, W.: *Full-wave analysis of conductor losses on MMIC transmission lines, IEEE Trans. Microw. Theory Tech.*, Vol. MTT- 38, pp. 1468–1472, Oct. 1990.

J.45 Heinrich, W.: *Mode-matching approach for superconducting planar transmission lines, IEEE Microw. Guid. Wave Lett.*, Vol. 1, pp. 294–296,1991.

J.46 Costache, G.I.: *Finite-element method applied to skin-effect problems in strip transmission lines, IEEE Trans. Microw. Theory Tech.*, Vol. MTT- 35, pp. 1009–1013, Nov. 1987.

J.47 Tuncer, E.; Lee, B.-T.; Islam, M.S.; Neikirk, D.P.: *Quasi-static conductor loss calculation in transmission lines using a new conformal mapping technique, IEEE Trans. Microw. Theory Tech.*, Vol. MTT- 42, No. 9, pp. 1807–1815, Sept. 1994.

J.48 Gopinath, A.: *Losses in the coplanar waveguide, IEEE Trans. Microw. Theory Tech.*, Vol. MTT- 30, pp. 1101–1104, 1982.

J.49 Jager, D.: *Slow-wave propagation along variable Schottky- contact microstrip line, IEEE Trans. Microw. Theory Tech.*, Vol. MTT- 24, No. 9, pp. 566–573, Sept. 1976.

J.50 Verma, A.K.; Nasimuddin; Sharma, E.K.: *Analysis and circuit model of multilayer semiconductor slow-wave microstrip line, IEEE Proc. Microw. Antenna Propag.*, Vol. 151, No. 5, pp. 441–449, Oct. 2004.

J.51 Barsotti, E.L.; Kuester, E.F.; Dunn, J.M.: *A simple method to account for edge shape in the conductor the loss in microstrip, IEEE Trans. Microw. Theory Tech.*, Vol. MTT- 39, pp. 98–106, Jan. 1991.

J.52 Lewin, L.: *A method of avoiding the edge current divergence in perturbation loss calculations, IEEE Trans. Microw. Theory Tech.*, Vol. MTT- 32, pp. 717–719, July 1984.

J.53 Verma, A.K.; Bhupal, A.: *Conductor loss of multilayer microstrip line using the SLR formulation, Microw. Opt. Technol. Lett.*, Vol. 19, No. 1, pp. 20–24, Sep. 1998.

J.54 Verma, A.K.; Bhupal, A.: *Conductor loss of multilayer microstrip line using the SLR formulation, Microw. Opt. Technol. Lett.*, Vol. 19, No. 1, pp. 20–24, Sept. 1998.

J.55 Verma, A.K.; Nasimuddin; Singh, H.: *Conductor loss of the coplanar waveguide with conductor backing and top shield*, Asia-Pacific Microwave Conference, Dec. 2004, New Delhi, India.

J.56 Holloway, C.L.; Kuester, E.F.: *A quasi- closed-form expression for the conductor loss of CPW lines with the investigation of edge shape effects, IEEE Trans. Microw. Theory Tech.*, Vol. MTT- 43, No. 12, pp. 2695–2701, Dec. 1995.

J.57 Holloway, C.L.: *Expressions for the conductor loss of stripline, and coplanar strip (CPS) structures, Microw. Opt. Technol. Lett.*, Vol. 25, No. 3, pp. 162–168, May 2000.

J.58 Majumdar, P.; Verma, A.K.: *Accurate CAD model of stopping distance to compute conductor loss of CPW, AEU*, Vol. 64, pp. 1157–1166, 2010.

J.59 Holloway, C.L.; Kuester, E.F.: *Edge shape effects and quasi- closed-form expressions for the conductor loss of microstrip lines, Radio Sci.*, Vol. 29, pp. 539–559, 1994.

J.60 Papapolymerou, J.; East, J.; Katehi Linda, P.B.: *GaAs versus quartz FGC lines for MMIC applications, IEEE Trans. Microw. Theory Tech.*, Vol. 46, No. 11, pp. 1790–1793, 1998.

J.61 Majumdar, P.; Verma, A.K.: *Accurate closed-form model for computation of conductor loss of coplanar waveguide, Int. J. RF Microw. Comput.-Aided Eng.*, Vol. 20, No.1, pp. 6–14, Jan. 2010.

J.62 Ponchak, G.E.; Matloubian, G.; Katehi, L.P.B.: *A measurement-based design equation for the attenuation of MMIC-compatible coplanar waveguides, IEEE Trans. Microw. Theory Tech.*, Vol. 47, pp. 241–243, 1999.

J.63 Ponchak, G.E.; Katehi, L.P.B.: *Measured attenuation of coplanar waveguide on CMOS grade silicon Substrate lines with polyimide interface layer, Electron. Lett.*, Vol. 13, pp. 1327–1329, June 1998.

J.64 Verma, A.K.; Nasimuddin; Singh, H.: *Dielectric loss of multilayer coplanar waveguide using the single layer reduction (SLR) formulation*, Asia-Pacific Microwave Conference, 2005, Japan.

J.65 Verma, A.K.; Bhupal, A.: *Dielectric loss of multilayer microstrip lines, Microw. Opt. Technol. Lett.*, Vol. 17, pp. 368–370, 1998.

J.66 Grischkowsky, D.; Duling, I.N.; Chen, J.C.; Chi, C.C.: *Electromagnetic shock waves from transmission lines, Phys. Rev. Lett.*, Vol. 59, No. 15, pp. 1663–1666, Oct. 1987.

J.67 Fattinger, C.; Grischkowsky, D.: *Observation of electromagnetic shock waves from propagating surface-dipole distribution, Phys. Rev. Lett.*, Vol. 62, No. 25, pp. 2961–2964, June 1989.

J.68 Kasilingam, D.P.; Rutledge, D.B.: *Surface-wave loss of coplanar lines*, IEEE MTT(s) Digest, pp. 113–116, 1083.

J.69 Lin, Y.-D; Sheen, U.W.: *Mode distinction and radiation –efficiency analysis of planar leaky-wave line source, IEEE Trans. Microw. Theory Tech.*, Vol. MTT- 45, No. 2, pp. 1672–1680, Oct. 1997.

J.70 Lin, Y.D.; Itoh, T.: *Leakage phenomenon in multilayered conductor-backed coplanar waveguide, IEEE Microw. Guid. Wave Lett.*, Vol. 3, pp. 426–427, 1993.

J.71 Verma, A.K.; Nasimuddin: *Quasistatic RLCG parameters of lossy microstrip line for CAD application, Microw. Opt. Tech. Lett.*, Vol. 28, No. 3, pp. 209–212, Feb. 2001.

J.72 Deng, T.: *CAD model* for *coplanar waveguide synthesis, IEEE Trans. Microw. Theory Tech.*, Vol. MTT- 44, No. 10, pp. 1733–1738, Oct. 1996.

J.73 Deng, T.Q.; Leong, M.S.; Kooi, P.S.; Yeo, T.S.: *Synthesis formula simplify coplanar waveguide design, Microw. RF*, Vol. 36, No. 3, pp. 84–98, March 1997.

J.74 Deng, T.Q.; Leong, M.S.; Kooi, P.S.: *Accurate and simple closed-form formulas for coplanar waveguide synthesis, Electron. Lett.*, Vol. 31, No. 23, pp. 2017–2019, Nov. 1995.

J.75 Yildiz, C.; Turkmen, M.: *New and very simple models for coplanar waveguide synthesis*, *Microw. Opt. Technol. Lett.*, Vol. 41, pp. 49–53, April 2004.

J.76 Yildiz, C.; Akdagli, A.; Turkmen, M.: *Simple and accurate synthesis formulas obtained by using a differential evolution algorithm for coplanar striplines*, *Microw. Opt. Technol. Lett.*, Vol. 48, No. 6, pp. 1133–1137, June 2006.

J.77 Yildiz, C.: *New and very simple synthesis formulas for coplanar strip line*, *Microw. Opt. Technol. Lett.*, Vol. 44, No. 2, pp. 199–202, Jan. 2005.

J.78 Deng, T.Q.; Leong, M.S.; Kooi, P.S.; Yeo, T.S.: *Synthesis formulas for coplanar lines in hybrid and monolithic MIC*, *Electron. Lett.*, Vol. 32, pp. 2253–2254, Nov. 1996.

10

Slot Line

Basic Characteristics

Introduction

S.B. Cohn proposed the basic slot line structure and, along with others, also demonstrated its application [B.1, B.2, J.1–J.6]. It is truly a *uniplanar* transmission line, as there is no backside metallization. Normally, it is used along with the microstrip and CPW structures. At the mm-wave frequency, the microstrip conductor losses are high and its dimensions are very small. So it requires specialized fabrication technology. The slot line and coplanar line structures are useful in the mm-wave and sub-mm-wave ranges [B.3], [J.7, J.8]. To achieve the EM-field confinement in the slot region, normally a substrate of high permittivity, i.e. $\varepsilon_r \geq 9$, is used. A significant part of the electric field is in the air region, above the conducting plane. The effective relative permittivity of a slot line is about 15% lower than that of a microstrip. Its characteristic impedance is between 40 Ω and 200 Ω. A slot line mounted in a rectangular waveguide forms the *fin line* that is another useful structure for the applications in the mm-wave ranges [B.4].

The slot line is primarily a non-TEM mode structure that can be accurately analyzed only by using the full-wave methods, such as the SDA discussed in chapter 14 [B.5–B.7]. However, Cohn has modeled it as a rectangular waveguide supporting the *quasi-TE_{10} mode* [J.1]. This model shows a cutoff frequency that is an unrealistic description of a practical slot line. Still, it is interesting to view the 2D geometry of a slot line, as a 3D rectangular waveguide. Furthermore, the static conformal mapping method has been also used with some modifications to compute the line parameters of a slot line [J.9–J.11].

Objectives

- *To present several geometrical forms of the slot line.*
- *To discuss the magnetic current model of a slot line.*
- *To present the equivalent waveguide model of standard slot line.*
- *To extend the waveguide model to a multilayer slot line.*
- *To present the closed-form integrated model of slot line to compute its line parameters.*

10.1 Slot Line Structures

The slot line could be used on single layer and multilayer substrates. It could be a conductor-backed structure that could be also covered with a shield. A brief description of some of these structures is presented below.

10.1.1 Structures of the Open Slot Line

A few configurations of the open slot line structure are shown in Fig (10.1). The lateral conducting planes are infinitely large. Figure (10.1a) is a basic standard slot line. The effective relative permittivity, or the slowing-factor given by λ_g/λ_0, and characteristic impedance are determined by the relative permittivity of the substrate, slot-width, and operating frequency. The slot line could be considered as a *complementary structure* of a microstrip line. The central strip conductor of microstrip is replaced by a slot and two lateral very large bare substrate portions are replaced by the infinitely large two lateral conducting planes of a slot line. Further, the ground conductor of the microstrip is replaced by the bare face of its dielectric substrate. However, even though a slot line is a two-conductor line, like a CPS, it does not support the TEM-mode. For infinitely wide conductors of a slot line, its line capacitance is $C \rightarrow \infty$. Therefore, the TEM-mode characteristic impedance of a slot line vanishes, $Z_0 = (1/v_pC) \rightarrow 0$. However, a practical slot line is fabricated with large, but finitely wide strip conductors. In that case, the slot line mode could be approximated by the quasi-TEM mode with zero cutoff frequency. However, Cohn has modeled it as a rectangular waveguide supporting the *quasi-TE_{10} mode* with a cutoff frequency.

Introduction to Modern Planar Transmission Lines: Physical, Analytical, and Circuit Models Approach, First Edition. Anand K. Verma.

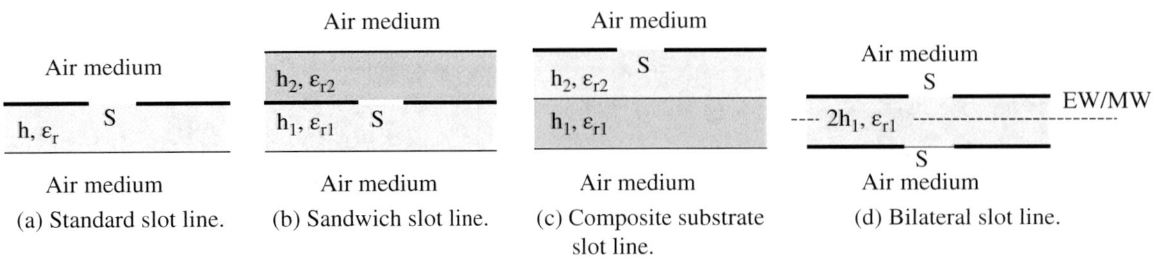

Figure 10.1 Some open slot-line structures.

Figure (10.1b) shows that the *sandwich slot line* is located at the interface of two dielectric media [J.2, J.9, J.12–J.14]. For the identical lower and upper dielectric layers, it reduces to an *embedded (buried)* slot line [J.13]. The sandwich slot line has a higher effective relative permittivity and lower characteristic impedance compared to a standard slot line on the same substrate with an identical s/h ratio. Therefore, an overlay, i.e. the superstrate, serves the purpose of controlling the line parameters. It further acts as a protective cover that also reduces radiation loss.

The *composite substrate slot line*, shown in Fig (10.1c), is another useful multilayer structure [J.7, J.9] that also controls the line characteristics and distortion in a signal. For the case $\varepsilon_{r1} > \varepsilon_{r2}$, it diverts the EM-wave energy from the conducting planes to the dielectric region that reduces the conductor loss of a slot line [J.7]. The substrate #1 also provides mechanical support to substrate #2. The composite substrate slot line has higher characteristic impedance, as compared to a standard slot line. It is again due to the diversion of the EM-field away from the surface conductors [J.15].

Figure (10.1d) shows the double-sided slot line, also called the *bilateral slot line* [J.15–J.17]. The structure is a broadside coupled slot lines. Due to the symmetry, it supports both the *even* and *odd* modes. In the case of the odd-mode excitation, the tangential components of electric fields in both the slot regions are equal and opposite that form the electric wall (*EW*) at the plane of symmetry. The structure could be reduced to a *conductor-backed slot line* (CBSL), shown in Fig (10.2a). However, in the case of the even-mode excitation, the tangential components of the electric field in the slot regions are equal and in the same direction that forms the magnetic wall (*MW*) at the plane of symmetry. The even-mode excitation provides nearly a standard slot line. The characteristic impedance of the odd-mode excited bilateral slot line is always much less than that of a standard slot line on a substrate of identical thickness and identical slot-width. On several occasions, the characteristic impedance of a bilateral slot line is one-half to that of the characteristic impedance of a slot line. It means to get fixed characteristic impedance for a given substrate; wider slot-width could be used in case of a bilateral slot line. Thus, a bilateral slot line overcomes the fabrication difficulties associated with the narrow slot of a standard slot line. Due to the symmetrical field distribution, a bilateral slot line is less dispersive than a standard slot line. The bilateral slot line is more suitable as a feed line to a tapered slot-radiator for the wideband antenna application [J.5, J.15, J.18].

10.1.2 Shielded Slot Line Structures

The slot line structures are also used with a conductor-backing and top shield [J.15]. The slot line could be fully enclosed in a metallic shield too. Fig (10.2) shows a few of such structures. The CBSL appears to be a useful structure, as it permits mounting of the circuits on a metallic ground plane. The conductor backing adds mechanical strength to a slot line. It also provides a good heat-sink and dc biasing to the active devices. However, the structure has a serious problem of power-leakage, due to the leaky-mode, in the substrate region. It gets coupled to the neighboring circuits on the same substrate, degrading their performances. The CBSL structure is normally not suitable as a transmission medium. The wide-slot CBSL is commonly used as the leaky-wave antenna [J.19–J.24]. Figure (10.2b) shows that the CBSL could be modified by adding a superstrate to the structure. For the case $\varepsilon_{r1} > \varepsilon_{r2}$, it has identical behavior to that of CBSL, shown in Fig (10.2a). However, for the case $\varepsilon_{r2} > \varepsilon_{r1}$, known as the *inverted conductor-backed slot line* (ICBSL), the structure has more useful features – both with respect to the signal transmission and upward radiation in the outer space [J.21]. In the lower frequency range, it is possible to obtain a frequency band that is suitable for the guided wave propagation without the leaky-mode and radiation. However, radiation can be observed in some frequency ranges.

The standard slot line is an open structure that causes excessive radiation at the discontinuity and also cross-coupling to the neighboring circuits. Figure (10.2c) shows added top shield to form the *shielded suspended slot line*. The top shield is also added to other slot lines. At higher

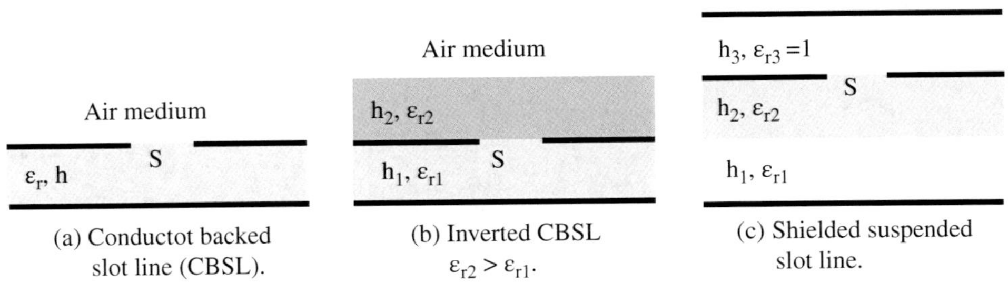

Figure 10.2 Some conductor-backed shielded slot lines.

frequency, a shielded slot line could be as efficient as a finline and much easier to integrate with circuits [J.8]. The shielded suspended slot line structure permits propagation of a band of frequency without radiation. The lower end of frequency is limited by the cutoff phenomenon and the higher frequency is limited by the radiation. The lower ground plane has a stronger influence on its performance. The practical planar circuits are always mounted in the shielded enclosures, to protect them from the environment and RF interferences. For conducting walls very near to the slot region, the slot mode moves to the cutoff, where field confinement in the slot is less [J.13].

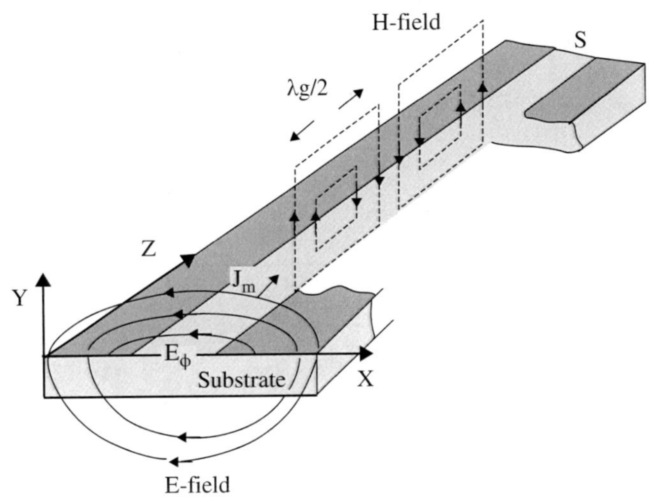

Figure 10.3 Approximate 3D view of the dominant mode of a slot line with a cut to show field lines.

10.2 Analysis and Modeling of Slot Line

Cohn suggested two approximate models for a slot line – *the magnetic current model* and the *equivalent waveguide model* [J.1]. The magnetic current model, discussed in this section, demonstrates the EM-field confinement in the slot region, due to the exponential decay of the field away from the slot. It also demonstrates the existence of the elliptically polarized wave in the slot region. However, the model does not compute the frequency-dependent slot line parameters. It is achieved with the help of the equivalent waveguide model, discussed in the next section. The equivalent waveguide model has been further extended to the multilayer shielded slot lines and the coupled slot lines [J.13, J.25, J.26]. The spectral-domain analysis (SDA) of the slot line is discussed in chapter 16. A summary of the results of the conformal mapping method as well as the closed-form models of the slot line to compute the losses and dispersion are also presented subsequently.

10.2.1 Magnetic Current Model

Figure (10.3) illustrates an approximate 3D-view of the field lines of a slot line in the cylindrical coordinate, with E_ϕ, H_r, and H_z field components. The electric and magnetic fields are shown separately for better clarity, although they are orthogonal to each other. The magnetic field pattern along the slot line repeats itself at every $\lambda_g/2$ distance, where λ_g is the guided wavelength of a slot line. The current flowing on the surface of the strip conductors causes the conductor loss. The *fictitious magnetic current* J_m in the slot region is assumed. It is related to the slot electric field through expression $\vec{J}_m = \vec{E} \times \hat{n}$. The magnetic current model of the slot field is valid for a narrow slot, i.e. $s/\lambda_0 << 1$. The slot-width is s and λ_0 is operating wavelength in free space. A slot line with an infinitely wide pair of conductors supports the quasi-TE mode in the cylindrical coordinate excited by the magnetic current J_m. Maxwell's equations in the cylindrical coordinate give the transverse field components, summarized below, in terms of the H_z-field component [B.8]:

$$E_r = -\frac{j}{k_c^2}\frac{\omega\mu}{r}\frac{\partial H_z}{\partial \phi}, \quad E_\phi = +\frac{j\omega\mu}{k_c^2}\frac{\partial H_z}{\partial r}, \quad E_z = 0$$
$$H_r = -\frac{j\beta_z}{k_c^2}\frac{\partial H_z}{\partial r}, \quad H_\phi = -\frac{j}{k_c^2}\frac{\beta_z}{r}\frac{\partial H_z}{\partial \phi}, \tag{10.2.1}$$

where the cutoff wavenumber k_c, given below, is related to the wavenumber k_0 in the free space and the propagation constant β_z of the EM-wave on the slot line in the z-direction. The wave propagation is described by $e^{j(\omega t - \beta z)}$.

$$k_0 = \frac{2\pi}{\lambda_0}, \quad \beta_z = \frac{2\pi}{\lambda_g}, \quad k_c^2 = k_0^2 - \beta_z^2$$
$$\Rightarrow k_c = k_0\sqrt{1-\left(\frac{\beta_z}{k_0}\right)^2} = j\frac{2\pi}{\lambda_0}\sqrt{\left(\frac{\lambda_0}{\lambda_g}\right)^2 - 1}. \qquad (10.2.2)$$

For the zeroth-order solution, i.e. the *quasi-static solution* of the wave propagation, it is assumed that the EM-wave propagates in the *equivalent isotropic medium* of a slot line on an infinitely thick substrate with relative permittivity $(\varepsilon_r + 1)/2$. The guided wavelength for such wave propagation and the cutoff wavenumber are given as follows:

$$\lambda_g = \frac{\lambda_o}{\sqrt{\varepsilon_{reff}(f=0)}} = \frac{\lambda_o}{\sqrt{(\varepsilon_r+1)/2}} \qquad (a)$$
$$k_c = \frac{2\pi}{\lambda_0}j\left[\left(\frac{\lambda_0}{\lambda_g}\right)^2 - 1\right]^{\frac{1}{2}} = \frac{2\pi}{\lambda_0}\left[\frac{\varepsilon_r+1}{2} - 1\right]^{\frac{1}{2}}$$
$$= j\frac{2\pi}{\lambda_0}\sqrt{\frac{\varepsilon_r - 1}{2}} \qquad (b).$$
$$(10.2.3)$$

The relative permittivity of a substrate is always more than unity, i.e. $\varepsilon_r > 1$. Therefore, k_c is an imaginary quantity.

For a narrow slot, the field component H_z does not change across the slot-width, i.e. $\partial H_z/\partial\phi = 0$. Thus, the transverse field components of the TE-mode, from equation (10.2.1), are

$$E_r = H_\phi = 0, \quad E_\phi = j\frac{\omega\mu}{k_c^2}\frac{\partial H_z}{\partial r}, \quad H_r = -j\frac{\beta_z}{k_c^2}\frac{\partial H_z}{\partial r}. \qquad (10.2.4)$$

The H_z-field component satisfies the wave equation in the cylindrical coordinate system:

$$\frac{\partial^2 H_z}{\partial r^2} + \frac{1}{r}\frac{\partial H_z}{\partial r} + \frac{1}{r^2}\frac{\partial^2 H_z}{\partial\phi^2} + k_c^2 H_z = 0. \qquad (10.2.5)$$

On using the method of separation of variable, the solution of the above equation is written as follows:

$$H_z = R(r)\cdot F(\phi) \qquad (a)$$
$$F\frac{d^2R}{dr^2} + \frac{F}{r}\frac{dR}{dr} + \frac{1}{r^2}R\frac{d^2F}{d\phi^2} + k_c^2 RF = 0$$
$$\Rightarrow \frac{r^2}{R}\frac{d^2R}{dr^2} + \frac{r}{R}\frac{dR}{dr} + k_c^2 r^2 = -\frac{1}{F}\frac{d^2F}{d\phi^2} = n^2 \qquad (b).$$
$$(10.2.6)$$

On separating equations for R(r) and F(ϕ), the following two equations are obtained:

$$\frac{d^2F}{d\phi^2} + n^2F = 0 \ (a), \qquad \frac{r^2}{R}\frac{d^2R}{dr^2} + \frac{r}{R}\frac{dR}{dr} + r^2k_c^2 - n^2 = 0$$
$$\Rightarrow \frac{d^2R}{dr^2} + \frac{1}{r}\frac{dR}{dr} + \left(k_c^2 - \frac{n^2}{r^2}\right)R = 0 \qquad (b).$$
$$(10.2.7)$$

The solution of equation (10.2.7a) is the linear combination of harmonic functions. The solution of the Bessel equation of nth order (10.2.7b) is written using a linear combination of the first and second kinds of Hankel function [B.9], as it provides radial outgoing wave from the magnetic current source.

$$F(\phi) = A\cos(n\phi) + B\sin(n\phi) \quad (a)$$
$$R(r) = C\,H_n^{(1)}(k_c r) + D\,H_n^{(2)}(k_c r) \quad (b). \qquad (10.2.8)$$

The approximate uniform field distribution with respect to ϕ shows the *zeroth*-order solution by taking $n = 0$. The first and the second kind of Hankel functions for the *zeroth*-order are approximated by the following expressions [B.9, B.10]:

$$\underset{r\to\infty}{H_0^{(1)}}(k_c r) = \sqrt{\frac{2}{\pi k_c r}}\exp\left\{j\left(k_c r - \frac{\pi}{4}\right)\right\} \qquad (a)$$
$$\underset{r\to\infty}{H_0^{(2)}}(k_c r) = \sqrt{\frac{2}{\pi k_c r}}\exp\left\{-j\left(k_c r - \frac{\pi}{4}\right)\right\} \qquad (b).$$
$$(10.2.9)$$

On multiplying the above equations by $e^{j\omega t}$, equation (10.2.9a) shows the *inward moving wave* and equation (10.2.9b) shows the *outward moving wave* from the slot region. Thus, equation (10.2.9b) shows the radiation from the slot that is always present in an open structure. The radiation could be both in the free space, forming the *space-wave* and also in the dielectric substrate forming the *surface wave*. However, the present discussion is not about radiation. The present discussion is concerned with the field confinement in the slot region, forming the guided wave on a slot line. The radiation term of the solution (10.2.8) is discarded by taking $D = 0$, giving the zeroth-order $(n = 0)$ H_z-field for the guided wave on the slot line:

$$H_z = A\,H_0^{(1)}(k_c r). \qquad (10.2.10)$$

For getting the above expression, the constant $A \times C$ is again denoted by A. It does not influence the solution. Equation (10.2.3b) shows that the argument $(k_c r)$ of the above Hankel function is an imaginary quantity. On using the identity given below [B.9], the azimuthal electric field E_ϕ and the radial magnetic field H_r components are written from equations (10.2.4) and (10.2.10):

$$\frac{d}{dx}\left[H_0^{(1)}(x)\right] = -H_1^{(1)}(x) \quad (a)$$

$$E_\varphi = -\eta A\left(\frac{\lambda_g}{\lambda_0}\right)\frac{H_1^{(1)}(k_c r)}{\sqrt{1-(\lambda_g/\lambda_0)^2}} \quad (b) \qquad (10.2.11)$$

$$H_r = \frac{A\,H_1^{(1)}(k_c r)}{\sqrt{1-(\lambda_g/\lambda_0)^2}} \quad (c).$$

As the argument of the Hankel function is an imaginary quantity, Hankel function $H_n^{(1)}(jx)$ is given by the modified Bessel function K_n [B.9, B.10]:

$$K_n(x) = \frac{\pi}{2}(j)^{n+1}H_n^{(1)}(jx), \qquad (10.2.12)$$

where n is the order of functions. There is a 90° phase difference between H_z and H_r magnetic field components. Therefore, these orthogonal magnetic field components may produce the *circular polarization* at a certain distance $r = r_{cp}$, where their magnitudes must be the same:

$$\left|\frac{H_r}{H_z}\right| = 1 \;\Rightarrow\; \left|\frac{H_1^{(1)}(k_c r_{cp})}{H_0^{(1)}(k_c r_{cp})}\right| = \sqrt{1-\left(\frac{\lambda_g}{\lambda_0}\right)^2}. \qquad (10.2.13)$$

The right-hand side of the above expression is <1, i.e. $\left|H_1^{(1)}(k_c r_{cp})\right| < \left|H_0^{(1)}(k_c r_{cp})\right|$. However, a table of the Hankel function [B.11] shows that it is not possible. Thus, our assumption that $|H_z| = |H_r|$ at distance $r = r_{cp}$ is not possible. Therefore, two orthogonal magnetic field components produce the *elliptical polarization* in (r–z)-plane. It is used for the development of the nonreciprocal devices on the ferrite substrates [J.6].

It is shown below that away from the slot region, the field decays very fast enabling the field confinement in a slot region. The ratio of voltage V(r), along with a semicircular path at the constant radius, and voltage V across the slot region is

$$\frac{V(r)}{V} = \frac{|k_c r|\,H_1^{(1)}(j\,|k_c r|)}{\underset{|x|\to 0}{Lt}\left[|x|\,H_1^{(1)}(j\,|x|)\right]}$$

$$\Rightarrow \frac{V(x)}{V} = \frac{\pi}{2}|k_c r|\left|H_1^{(1)}(j\,|k_c r|)\right| = (|k_c r|)\,K_1(|k_c r|), \qquad (10.2.14)$$

where $K_1(k_c r)$ is the modified Bessel function given by equation (10.2.12). The field decay is illustrated by the numerical computation of the above given voltage ratio. The slot line is considered at f = 3 GHz, i.e. λ_0 = 10 cm on a substrate with ε_r = 16. Using equation (10.2.3), $|k_c r|$ = 1.721r, where r is in cm. At r = 0 cm, 1.778 cm, 2.54 cm, and 3.302 cm, the voltage ratio V(r)/V is computed by getting $K_1(k_c r)$ from the table given in reference [B.12]. For $k_c r$ = 0, 3.1, 4.4, 5.7, $K_1(k_c r)$ is 0.03563, 0.00792, 0.0019 and V(r)/V = 1, 0.1110, 0.03486, and 0.011. The computations correspond to 0 dB, −19.14 dB, −29.15 dB, and −39.3 dB field at 0 cm, 1.778 cm, 2.54 cm, and 3.302 cm away from the slot region in any direction in the transverse plane. It shows faster decay of the field away from the slot region.

At a larger distance, $k_c r > 4$, or $r/\lambda_0 > 1/\sqrt{\varepsilon_r - 1}$, Hankel function used in equation (10.2.14) is approximated as

$$H_1^{(1)}(j\,|k_c r|) = -\frac{2}{\pi}\sqrt{\frac{\pi}{2|k_c r|}}\,e^{-|k_c r|}. \qquad (10.2.15)$$

The voltage ratio, in this case, is approximated by the following expression on using equation (10.2.3):

$$\frac{V(x)}{V} = \pi\sqrt{\frac{r}{\lambda_0}}\left[\frac{\varepsilon_r - 1}{2}\right]^{\frac{1}{4}}\left[e^{-\left\{\frac{\pi r}{\lambda_0}\sqrt{2(\varepsilon_r - 1)}\right\}}\right]. \qquad (10.2.16)$$

The above expression has error <8%, as compared to computation with the table of the modified Bessel function. However, it demonstrates confinement of the exponentially decaying field in the slot region and around it. The field, with distance from the slot region, decreases fast both with an increase in the relative permittivity of the substrate and also with an increase in frequency.

The magnetic current model of the slot line demonstrates the field confinement in the slot region to achieve the guided wave propagation. However, the guided mode is not bound. The EM-field is also present both in the air region and in the substrate-region, producing the undesired space wave and surface wave, respectively. The propagation constant obtained from the *zero*th-order solution shows the dispersionless behavior of a slot line. However, it is not a correct description. Thus, a more accurate model of the slot line is needed to provide a dispersion model for the propagation constant and the characteristic impedance. Cohn has achieved it through the equivalent waveguide model.

10.3 Waveguide Model

Cohn has modeled the standard slot line as a combination of the *iris-loaded rectangular waveguide* and the *air-filled waveguide sections* [J.1–J.3]. Next, the *transverse resonance method* (TRM) is applied at the junction of two equivalent waveguides to determine the guided wavelength of the slot line [B.10, B.11]. The waveguide model is also extended to the multilayer slot line.

10.3.1 Standard Slot Line

Figure (10.4) shows two-step development of the waveguide model of a standard slot line.

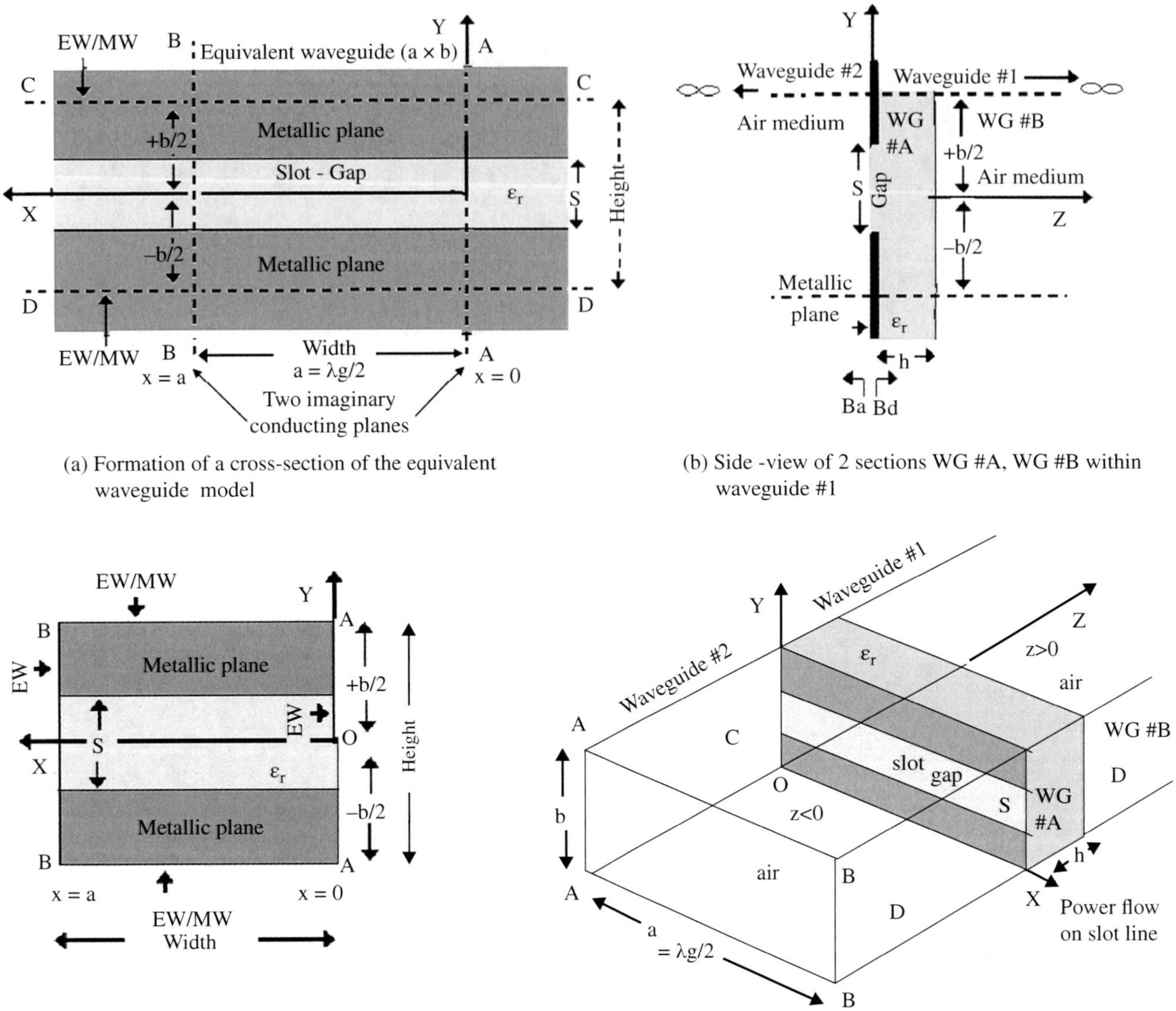

(a) Formation of a cross-section of the equivalent waveguide model

(b) Side -view of 2 sections WG #A, WG #B within waveguide #1

(c) Waveguide cross-section with iris

(d) 3D- view of waveguide model. A section of the slot line appears as an iris at z = 0.

Figure 10.4 Development of equivalent waveguide model of the slot line.

Step #1: Formation of Width (a) of the Equivalent Waveguide

Figure (10.4a) shows the top view of a slot line in the (x–y)-plane. The EM-wave propagates along the x-axis. Two infinitely wide strip conductors of the slot line are separated by a gap *s* in the y-direction. The substrate has a relative permittivity ε_r and the substrate thickness h is in the z-direction. It is shown in the side view of Fig (10.4b). The guided wavelength λ_g of the slot line is in the x-direction. *Two positions, x = 0 and x = a = $\lambda_g/2$, are selected on the slot line to form the broadside width of the equivalent waveguide.* The λ_g is the guided wavelength of the slot line and $\lambda_g = \lambda_0/\sqrt{\varepsilon_{reff}(f)}$, where $\varepsilon_{reff}(f)$ is the frequency-dependent effective relative permittivity of a slot line. At these locations, the transverse component of the E-field and the normal component of the H-field are zero. Thus, the *artificial EW*, i.e. the perfect conductors, can be placed at the planes AA (x = 0) and BB (x = a) extending in the x-direction. The slot line sections, right to the plane AA and also left to the plane BB, are removed. The planes AA and BB are the narrow-face of the equivalent waveguide.

Step # 2: Formation of Height (b) of the Equivalent Waveguide

The EM-field of a slot line also spreads in the ±y-direction on the metallic surface. The previous section has shown that the field decays very fast away from the slot region.

It helps us to place the planes CC and DD, located at +b/2 and –b/2, as the boundary walls. The conductors of the slot line are removed beyond these walls. The *distance CD = b, forming the height of the equivalent waveguide*, is several times of the slot gap s. The planes CC and DD are the broad-face of the equivalent waveguide. *These planes could be either the electric walls (EW) or the magnetic walls (MW).* These walls also extend in the ±z-direction. In the case the planes CC and DD are the EW, a rectangular waveguide model is achieved that has cross-section $a \times b$ in the x–y plane and it extends in the ±z-direction. Figure (10.4b) shows the side view of the equivalent waveguide model and Fig (10.4c) shows its cross-section at $z = 0$, containing *the iris* formed from the slot line.

Figure (10.4d) shows the 3D view of the equivalent waveguide, loaded with the metallic iris of the gap s supported on a substrate of thickness h. The iris-loaded substrate of thickness h forms the waveguide section WG #A. The plane of the iris is located at $z = 0$, i.e. at the junction of two waveguides – #1 and #2. The air-filled space exists in the ±z-directions, both above and below the physical slot line. The fields exponentially decay in the air-filled space so the fields are confined in the slot region of a slot line.

The air-filled spaces of the slot line appear as two air-filled waveguide sections WG #1 and WG #B in Fig (10.4b and d). The air-filled waveguide sections are in the cutoff region to achieve the field confinement in the slot region. Thus, the propagating fields are confined in the strips-loaded dielectric region of WG #A. Such field confinement is required for the guided wave propagation on a slot line. The propagation characteristic and characteristic impedance of a slot line are obtained by applying the *TRM* at the junction of $z = 0$ of the waveguides #1 and #2. The TRM is discussed in section (7.7) of chapter 7.

Modes in Waveguide

Let us consider the case of all-EW waveguide. Figure (10.4b) divides the waveguide model into two equivalent waveguides #1 and #2. The *equivalent waveguide #1* contains the dielectric-loaded **WG Section-A (WG #A)** of thickness h followed by an infinite extent air medium **WG Section-B (WG #B)**. The *equivalent waveguide #2* contains an infinite extent of air medium. The waveguides #1 and #2 form a junction at $z = 0$. The *capacitive iris* formed out of the slot line section is also located at the junction $z = 0$. In the slot region of the slot line, the EM-wave has a $= \lambda_g/2$ variation in the x-direction. Figure (10.4d) shows the 3D waveguide model of the slot line. Further, *it is assumed that in the equivalent waveguide, all modes have $\lambda_g/2$ variation in the x-direction.* Due to the symmetry of slot-structure with respect to the x-axis, located at mid of the slot gap s, the E-fields for all modes must have a maximum at the center of the slot. Therefore, only TE_{10}, TE_{12}, TE_{14}... and TM_{12},TM_{14},..., i.e. $TE_{1,2n}$ ($n = 0,1,2,...$) and $TM_{1,2n}$ ($n = 1,2,...$), modes exist in the structure.

In the slot region of the waveguide #1, the guided wavelength of the slot line wave is $\lambda_g < \lambda_0$. Therefore, the width of the equivalent waveguide, carved out of the slot line, is $a = \lambda_g/2$. Under such formulation, the width of the air-filled waveguide #2 and WG #B is less than half wavelength, i.e. $\{(a = \lambda_g/2) < \lambda_0/2\}$. It helps to achieve the cutoff behavior in the air-filled waveguides for the TE_{10} and all other higher-order modes. These modes are non-propagating in the air-regions of the waveguide #2 and WG #B section of waveguide #1. *Thus, both air regions provide reactive loading to the waveguide sections.* In the dielectric region of WG #A, the TE_{10} and a few more modes are propagating modes depending on the dimension "b" along the y-axis. The iris on substrate-loaded waveguide #1 offers *capacitive susceptance* B_d at the junction. The air-filled iris-loaded waveguide #2 also offers *capacitive susceptance* B_a at the junction. The field confinement in the slot region of WG #A, sandwiched between the reactive sections of the waveguide #2 and WG #B, could be treated through the transverse resonance in the (y–z)-plane. Hence, at $z = 0$, the *total susceptance* $B_t = B_d + B_a$ is zero satisfying the transverse resonance condition. It is shown in Fig (10.4b). The primary task is to compute the susceptance B_d and B_a at the plane of iris, i.e. at $z = 0$ for the following two cases:

Case A: All electric wall (EW–EW) waveguide.

Case B: Combined electric and magnetic walls (EW–MW) waveguide.

Both Fig (10.4b) and (10.5d) show that the iris is located at $z = 0$, i.e. at the junction of the waveguides #1 and #2. The waveguide #1, in the direction of $z > 0$, has two sections – the **WG #A** and **WG #B**. The WG #A is the iris-supported dielectric section of thickness h and the WG #B is an air-filled waveguide of infinite length in the cutoff region. The waveguide #2, in the direction of $z < 0$, is the air-filled waveguide of infinite length, also in the cutoff region. The waveguide #2 and WG #B, with the width a $< \lambda_0/2$, load the junction reactively. *The nature of the reactive loading, i.e. the inductive or capacitive, depends on the TE/TM modes in the cutoff regions.*

Summary of Expressions used in Waveguide

The analysis of the rectangular waveguide with all-EW is discussed in chapter 7. A summary of expressions is presented below for immediate reference. The expressions for the TE and TM modes are given with some changes. The constant $B = -j\,(B_{mn}/\omega\mu)\,\beta_c^2$ is used in the expression (7.4.19), and the constant $A = -j\,(A_{mn}/\omega\varepsilon)\,\beta_c^2$ in the expression (7.4.24), to get the following expressions of the TE/TM modes [B.9]:

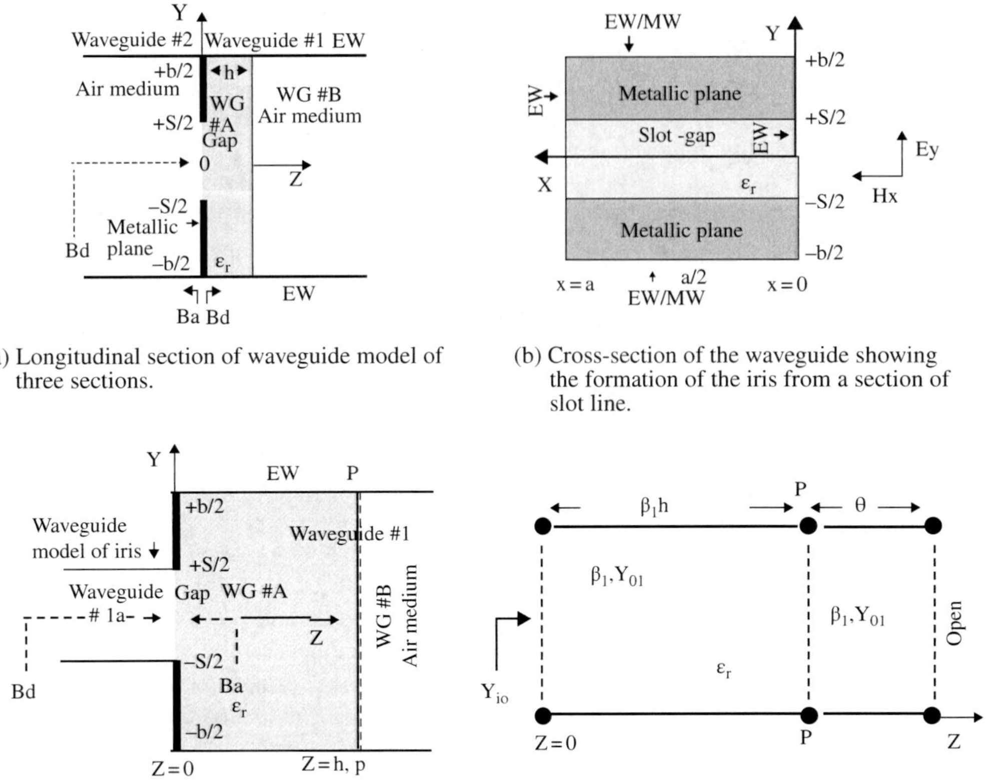

(a) Longitudinal section of waveguide model of three sections.

(b) Cross-section of the waveguide showing the formation of the iris from a section of slot line.

(c) Waveguide model of slot region. Plane PP is the reactive plane, either EW/MW.

(d) Transmission line model of a waveguide section.

Figure 10.5 Analysis of the slot line as a waveguide.

TE-modes:

$$H_z = B\cos(\beta_x x)\cos(\beta_y y)$$
$$E_x = j\frac{\eta\,\beta_y f}{\beta_c f_c} B\cos(\beta_x x)\sin(\beta_y y)$$
$$E_y = -j\frac{\eta\,\beta_x f}{\beta_c\,f_c} B\sin(\beta_x x)\cos(\beta_y y) \qquad (10.3.1)$$
$$H_x = -\frac{E_y}{Z_{TE}},\quad H_y = \frac{E_x}{Z_{TE}}.$$

TM-modes:

$$E_z = A\sin(\beta_x x)\sin(\beta_y y)$$
$$H_x = j\frac{\beta_y f}{\beta_c \eta f_c} A\sin(\beta_x x)\cos(\beta_y y)$$
$$H_y = -j\frac{\beta_x f}{\beta_c \eta f_c} A\cos(\beta_x x)\sin(\beta_y y) \qquad (10.3.2)$$
$$E_x = Z_{TM}\,H_y,\qquad E_y = -Z_{TM}\,H_x.$$

The waveguide model considers the EM-wave propagation in the z-direction with the propagation factor $\exp\{j(\omega t - \gamma z)\}$. The wave impedances for the TE and TM modes, i.e. Z_{TE} and Z_{TM} from equations (7.4.31) and (7.4.32) of chapter 7, are given below:

$$Z_{TE} = \eta\left[1-\left(\frac{f_c}{f}\right)^2\right]^{-1/2} \quad \text{(a)}$$
$$Z_{TM} = \eta\left[1-\left(\frac{f_c}{f}\right)^2\right]^{1/2} \quad \text{(b)}. \qquad (10.3.3)$$

The intrinsic impedance of the waveguide medium is $\eta = \sqrt{\mu/\varepsilon}$ and for the free space it is $\eta_0 = 377\,\Omega$, such that $\eta = \eta_0/\sqrt{\varepsilon_r}$ for a dielectric medium. The transverse wavenumbers β_x, β_y, and the cutoff wavenumber are

$$\beta_x = \frac{m\pi}{a},\quad \beta_y = \frac{n\pi}{b},\quad \beta_c = \sqrt{\beta_x^2+\beta_y^2} = \sqrt{\left(\frac{m\pi}{a}\right)^2+\left(\frac{n\pi}{b}\right)^2},\quad \beta_z = \left[k^2-\beta_c^2\right]^{1/2} \qquad (10.3.4)$$

also, $\gamma = j\beta_z = j\left[k^2-\beta_c^2\right]^{1/2}$, $k = k_0\sqrt{\varepsilon_r}$, $\lambda = \lambda_0/\sqrt{\varepsilon_r}$, $\lambda_g = 2\pi/\beta_z$, $\lambda_0 = 2\pi/k_0$.

In the above equation, a is the dimension of broad-face, and b is the dimension of the narrow-face of the equivalent waveguide. For the TE-mode, either of two *modal numbers* *m* and *n* could be zero. However, for the TM-mode neither of two modal numbers m and n is zero, i.e. for TM-mode $m, n \geq 1$. If either m or n is zero for the TM-modes, all field components are zero. The modal numbers m, n are integers. The cutoff wavelength and cutoff frequency for both the TE and TM modes are given as follows:

$$\lambda_{c,mn} = \frac{2\pi}{\beta_c} = \frac{2}{\sqrt{\left(\frac{m}{a}\right)^2 + \left(\frac{n}{b}\right)^2}} \quad \text{(a)}$$

$$f_{c,mn} = \frac{\beta_c}{2\pi\sqrt{\mu\varepsilon}} = \frac{1}{2\sqrt{\mu\varepsilon}}\sqrt{\left(\frac{m}{a}\right)^2 + \left(\frac{n}{b}\right)^2} \quad \text{(b)}. \tag{10.3.5}$$

Determination of Susceptances B_d and B_a for Case-A

Figure (10.5a and b) shows the longitudinal and cross-sectional views of the (EW–EW) waveguide model. It is assumed that both the waveguides #1 and #2 are excited by the aperture field of an iris, located at the junction $z = 0$ of the waveguides. At the plane of the iris at the right-hand side of $z = 0$, the susceptance B_d is present due to WG #A and WG #B in the cutoff region for all modes. At the same iris plane, from the left-hand side of $z = 0$, the susceptance B_a is present due to the WG #2 again in the cut-off region.

At first, the susceptance B_d at $z = 0$, offered by the waveguide #1, is computed. The iris slot-gap s is modeled by the air-filled *waveguide #1a* with a cross-section ($a \times s$) that excites waveguide #1. It is shown in Fig (10.5c). The waveguide #1 has width "a" and height "b" and is divided into two waveguides – **WG#A** filled with strip-slot-loaded dielectric substrate of thickness h and **WG#B** filled with air medium and is under the cutoff region. The WG#A could be treated as terminated in either the inductive reactance or capacitive reactance created by the air-filled waveguide WG#B in the cutoff region. *The TE mode WG#B in the cutoff region provides the inductive reactance at x = h and, whereas the TM mode in cutoff provides the capacitive reactance*. Further, the capacitive reactance could be replaced by the extended length of WG#A terminated in the open-circuit, and the inductive reactance is replaced by the extended length of WG#A terminated in the short-circuit. The *input admittance* Y_i at plane $z = 0$, looking toward waveguide #1 containing dielectric of relative permittivity ε_r, is a *susceptance* B_d, as the dielectric is to be taken as lossless.

To get an expression for the B_d ($Y_i = j\,B_d$), let us consider the waveguide #1 excited by the aperture field of the iris. It has been noted earlier that, due to symmetry, only $TE_{1,2n}$ ($n = 0,1,2,\ldots$) and $TM_{1,2n}$ ($n = 1,2,\ldots$) modes are present in waveguide #1. The field expressions (10.3.1) of the TE-modes in the transverse x–y plane show that the E_y and H_x field components have the following proportionality relations:

$$E_y \propto \sin\left(\frac{\pi x}{a}\right)\cos\left(\frac{2\pi n y}{b}\right) \quad \text{(a)}$$

$$H_x \propto \sin\left(\frac{\pi x}{a}\right)\cos\left(\frac{2\pi n y}{b}\right) \quad \text{(b)}. \tag{10.3.6}$$

At $z = 0$ and $x = a/2$, the total E_y and H_x are given as the summation of fields of all modes:

$$E_y = R_0 + \sum_{n=1,2,\ldots} R_n \cos\left(\frac{2\pi n y}{b}\right) \quad \text{(a)}$$

$$H_x = -y_{i,0}R_0 - \sum_{n=1,2,\ldots} y_{i,n}R_n \cos\left(\frac{2\pi n y}{b}\right) \quad \text{(b)}, \tag{10.3.7}$$

where R_0 and R_n are constants. The input *wave admittance* for the TE_{10} mode is $y_{i,0}$ and for the hybrid mode, it is $y_{i,n}$. The linear combination of other remaining modes $TE_{1,2n}$ ($n \geq 1$) and $TM_{1,2n}$ ($n \geq 1$) with input wave admittance $y_{i,n}$ constitute the hybrid modes. The wave admittances of the TE_{10} mode and hybrid modes are expressed as follows:

$$y_{i,0} = -\left[\frac{H_x}{E_y}\right]_{TE_{10}} \quad \text{(a)}$$

$$y_{i,n} = -\left[\frac{(H_x)_{TE_{1,2n}} + (H_x)_{TM_{1,2n}}}{(E_y)_{TE_{1,2n}} + (E_y)_{TM_{1,2n}}}\right] \quad \text{(b)}. \tag{10.3.8}$$

A constant field E_y in the capacitive iris slot is assumed to excite the TE_{10} and other higher-order modes, mentioned above, in the waveguide #1. *The iris slot is modeled as the waveguide #1a*, shown in Fig (10.5c), and its assumed field components at $z = 0$ are written as follows:

$$E_y = \begin{cases} C_0, & \text{for } |y| \leq \frac{s}{2} \\ 0, & \text{for } \frac{s}{2} < |y| \leq \frac{b}{2} \end{cases} \quad \text{(a)}$$

$$H_x = -C_0 y_i', \text{ for } |y| \leq \frac{s}{2} \quad \text{(b)}. \tag{10.3.9}$$

In the above equation, y_i' is the *input wave admittance* of waveguide #1a of height s supporting only TE_{10} mode. As the length of the exciting waveguide #1a, i.e. the thickness of iris is zero, therefore y_i' is also the *input wave admittance* of the waveguide #1 that is related to its input admittance Y_i. Let us define the *fractional slot gap* δ as

$$\delta = \frac{s}{b}, \qquad s \ll b. \tag{10.3.10}$$

The assumed slot-field of equation (10.3.9) ignores higher-order modes and is valid for a narrow slot of fractional width $\delta \le 0.15$. The constant slot field E_y could be expressed as follows, in the form of Fourier series, giving its modal composition:

$$E_y = C_0 = R_0 + \sum_{n=1,2,\ldots} R_n \cos\left(\frac{2\pi n}{b} y\right). \tag{10.3.11}$$

The Fourier coefficients R_0 and R_n are determined by using the orthogonality relation. Thus, the coefficient R_0 is obtained by integrating both sides of the equation over $-s/2 \le y \le s/2$ and coefficient R_n is obtained by the same integration process after multiplying both sides of the above equation by $\cos\{(2\pi m/b)y\}$:

$$R_0 = \frac{1}{b}\int_{-s/2}^{s/2} C_0\, dy = C_0 \frac{s}{b} = C_0\delta \tag{10.3.12}$$

$$\int_{-b/2}^{b/2} C_0 \cos\left(\frac{2\pi m}{b} y\right) dy = R_0 \int_{-b/2}^{b/2} C_0 \cos\left(\frac{2\pi m}{b} y\right) dy$$

$$+ \sum_n R_n \int_{-b/2}^{b/2} C_0 \cos\left(\frac{2\pi n}{b} y\right) \cos\left(\frac{2\pi m}{b} y\right) dy$$

$$\text{For } m = n \ : \ R_n = 2C_0\delta\left[\frac{\sin(n\pi\delta)}{n\pi\delta}\right]. \tag{10.3.13}$$

Next, it is desired to obtain an expression for the input wave admittance y_i' of waveguide #1a at $z = 0$ in terms of the modal input wave admittances $y_{i,0}$ and $y_{i,n}$ of waveguide #1. To achieve it, equation (10.3.7b) is substituted in equation (10.3.9b) and integrated over slot height $\pm s/2$:

$$\int_{-s/2}^{s/2} C_0 y_i' dy = \int_{-s/2}^{s/2} \left[y_{i,0} R_0\, dy + \sum_{n=1,2,\ldots} y_{i,n} R_n \cos\left(\frac{2\pi n}{b} y\right) dy \right]$$

$$\Rightarrow C_0 y_i' = y_{i,0} R_0 + \sum_{n=1,2,\ldots} y_{i,n} R_n \frac{\sin(n\pi\delta)}{n\pi\delta}$$

On substitution of R_0 and R_n from equations (10.3.12) and (10.3.13) in the above equation, the following expression is obtained:

$$\left(\frac{b}{s}\right) y_i' = y_{i,0} + 2\sum_{n=1,2,\ldots} y_{i,n} \left[\frac{\sin(n\pi\delta)}{n\pi\delta}\right]^2. \tag{10.3.14}$$

So far the *wave admittance* has been used in the discussion. It is defined as a ratio of the transverse magnetic field component and transverse electric field component given by equation (10.3.1). It can be replaced by the waveguide *characteristic admittance* defined for the TE_{10}-mode on the voltage-power basis [B.8]. It is discussed in section (7.6) of chapter 7. The change of admittance is needed, as the slot line is described through its characteristic impedance and not through the wave impedance.

The input power at $z = 0$ to the waveguide #1 is obtained using the assumed field expressions in the slot by equation (10.3.9):

$$P_i = \frac{1}{2}\int_{-s/2}^{s/2}\int_0^a E_y H_x\, dy\, dx$$
$$\Rightarrow P_i = \frac{1}{2}\int_{-s/2}^{s/2}\int_0^a C_0^2 y_i'\, dy\, dx = \frac{1}{2} C_0^2 y_i' s\, a. \tag{10.3.15}$$

The voltage over the slot is $V = \int_{-s/2}^{s/2} E_y\, dy = s\, E_y$.

$$\tag{10.3.16}$$

Using the above given two equations, the characteristic admittance Y_i of the waveguide #1a, in terms of its wave admittance, is obtained as follows:

$$P_i = Y_i V^2 = Y_i (E_y s)^2 = Y_i (C_0 s)^2 = \frac{1}{2} C_0^2 y_i' s\, a$$
$$\Rightarrow Y_i = \frac{a}{2s} y_i' \qquad (a). \tag{10.3.17}$$

Likewise, the input waveguide admittances Y_{i0} and Y_{in} corresponding to the TE_{10} and higher-order modes of the waveguide #1 for the height b are

$$Y_{i0} = \frac{a}{2b} y_{i,0} \quad (b), \qquad Y_{in} = \frac{a}{2b} y_{i,n} \quad (c). \tag{10.3.18}$$

On substituting the y_i' from equation (10.3.17), and y_{i0} and y_{in} from equation (10.3.18) in equation (10.3.14), the following expression for the input admittance of the waveguide #1 at $z = 0$ is obtained:

$$Y_i = j B_d = Y_{i0} + 2\sum_{n=1,2,\ldots} \left[Y_{in} \left(\frac{\sin(\pi n \delta)}{\pi n \delta}\right)^2 \right]. \tag{10.3.19}$$

The *input admittance is capacitive*. It has two components. The first component $Y_{i,0}$ is due to the TE_{10} mode in the waveguide #1 containing dielectric substrate, followed by the air-filled infinite waveguide. The second component of the input admittance, under the $\sum$ sign, is contributed by the higher-order $TE_{1,2n}$ and $TM_{1,2n}$ $n \ge 1$ modes in the waveguide #1. The next task is the computation of the $Y_{i,0}$ and $Y_{i,n}$.

Computation of Input Admittance $Y_{i,0}$ due to the TE_{10} Mode

The input admittance $Y_{i,0}$ due to the TE_{10} mode is obtained by using the transmission line model of the waveguide #1 with two waveguide sections WG #A and WG #B.

Figure (10.5d) shows the process to compute the input admittance Y_{i0} due to the TE_{10} mode in waveguide #1. At $z = h$, shown in Fig (10.5c), the TE_{10} mode in the air-filled waveguide section WG #B is evanescent. Further, the waveguide section WG #B could be treated as a finite length line terminated in an open-circuit, offering capacitive load admittance Y_0 to the waveguide section WG #A. In the case a slot line is enclosed in a metallic box, the air-filled equivalent waveguide section WG #B could be treated as a line terminated in the short-circuit. This case is discussed in subsection (10.3.3).

The capacitive load Y_0 terminated at the plane PP can be further replaced by the extended open line section of the length θ. *It is noted that the infinite length WG #B of characteristics admittance Y_0 is replaced by the extended length θ of the WG #A* having characteristic admittance Y_{01} and propagation constant β_1. Hence, the total electrical length of the waveguide section WG #A is $(\beta_1 h + \theta)$ that is terminated in the open-circuited load $Z_L = \infty$ or $Y_L = 0$. The TE_{10} mode in the WG #A is a lossless propagating mode. Therefore, the input admittance of the extended line length θ at the plane PP is

$$Y_{in}|_{PP} = Y_0 = jY_{01}\tan\theta,$$
$$\Rightarrow \theta = \tan^{-1}\left(\frac{Y_0}{jY_{01}}\right), \quad \beta_1 = \frac{\gamma_1}{j} = \frac{2\pi}{\lambda_g}. \tag{10.3.20}$$

The guided wavelength λ_{g1} is for the TE_{10} mode in the dielectric-filled WG #A. At $z = 0$, using equation (2.1.78) of chapter 2, the input admittance of the waveguide #1 is computed by replacing it with the extended WG #A of finite length $(\beta_1 h + \theta)$ terminated in the open. The input admittance of the waveguide #1 seen by the TE_{10} mode is

$$Y_{i,0} = jY_{01}\tan(\beta_1 h + \theta),$$
$$\Rightarrow Y_{i,0} = jY_{01}\tan\left[\beta_1 h + \tan^{-1}\left(\frac{Y_0}{jY_{01}}\right)\right]. \tag{10.3.21}$$

The characteristic admittances Y_0 and Y_{01}, defined on the power-voltage basis, are related to the wave admittance of the TE mode. The wave admittance is given in terms of the physical parameters of a waveguide. It is discussed in chapter 7. Using equation (7.4.30) of chapter 7, the wave admittance of the TE mode supporting air-filled waveguide is given below [B.8]:

$$Y_w^{TE} = \frac{\beta}{\omega\mu} = \frac{\beta}{\eta_0 k_0}. \tag{10.3.22}$$

The intrinsic impedance of free space is $\eta_0 = \sqrt{\mu_0/\varepsilon_0} = 377\ \Omega$. Using equation (7.6.9) of chapter 7, the characteristic admittance on the power-voltage basis is related to the TE mode wave admittance Y_w^{TE} by the following expression:

$$Y_0^{PV} = \frac{a}{2b}Y_w^{TE}. \tag{10.3.23}$$

Therefore, using the above two equations, the characteristic admittance Y_0 of the air-filled waveguide section WG #B for the TE_{10} mode is

$$Y_0 = Y_0^{PV} = \frac{a\gamma}{j\,2b\eta_0 k_0}. \tag{10.3.24}$$

The parameter $\gamma = j\beta$ is the z-directed propagation constant of the TE_{10} mode air-filled waveguide section WG #B. Likewise, the characteristic admittance Y_{01} of the dielectric-filled waveguide section WG #A for the TE_{10} mode is

$$Y_{01} = Y_0^{PV} = -j\frac{a\gamma_1}{2b\eta_1 k_1} = \frac{a}{2b\eta_0}\frac{\lambda_0}{\lambda_{g1}}$$
$$\text{where, } \eta_1 = \eta_0/\sqrt{\varepsilon_r},\ k_0 = 2\pi/\lambda_0,\ k_1 = \frac{2\pi\sqrt{\varepsilon_r}}{\lambda_0} \tag{10.3.25}$$
$$\gamma_1 = j\beta_1,\ \beta_1 = 2\pi/\lambda_{g1}.$$

Computation of Input Admittance $Y_{i,n}$ Due to Higher-order Modes

At $z = 0$, the input admittance $Y_{i,n}$, for $n > 0$, of the combined $TE_{1,2n}$ and $TM_{1,2n}$ modes, i.e. the *hybrid mode* is obtained from equation (10.3.8b) as follows:

$$Y_{i,n} = -\left[\frac{\dfrac{(H_x)_{TE_{1,2n}}}{(E_y)_{TM_{1,2n}}} + \dfrac{(H_x)_{TM_{1,2n}}}{(E_y)_{TM_{1,2n}}}}{\dfrac{(E_y)_{TE_{1,2n}}}{(E_y)_{TM_{1,2n}}} + 1}\right]$$
$$\Rightarrow Y_{i,n} = \left[\frac{Y_i^{TM^n} + Y_i^{TE^n}D_n}{1 + D_n}\right] \quad \text{(a)} \tag{10.3.26}$$
$$\text{where,}\quad D_n = \frac{(E_y)_{TE_{1,2n}}}{(E_y)_{TM_{1,2n}}} \quad \text{(b)}.$$

The expressions for the field components of both the $TE_{1,2n}$ and $TM_{1,2n}$ modes, in waveguide #1, are needed to evaluate D_n. At $z = 0$, i.e. at the plane of the iris, shown in Fig (10.4a), only TE_{10} mode with constant E_y field in the slot region is considered. The slot field is given by equation (10.3.9a). Therefore, the total E_x field component of both the TE and TM higher-order modes must be zero at $z = 0$. So, only the TE_{10} mode field is retained at $z = 0$:

$$E_x\big|_{TE_{1,2n}} + E_x\big|_{TM_{1,2n}} = 0. \quad (10.3.27)$$

On substituting E_x field components from equations (10.3.1) and (10.3.2) in the above equation, the following expression is obtained:

$$\eta_0\, \beta_y B = A\, Z_{TM} \frac{\beta_x}{\eta_0}. \quad (10.3.28)$$

Again, on substituting E_y, for both the TE and TM modes, from equations (10.3.1) and (10.3.2) in equation (10.3.26b); and using the resulted expression with equation (10.3.28), the following relation is obtained:

$$D_n = \frac{\beta_x^2}{\beta_y^2} = \left(\frac{m\pi/a}{n\pi/b}\right)^2. \quad (10.3.29)$$

As discussed previously, due to the symmetry of the structure, the waveguide #1 has m = 1 and n → 2n. Hence, the above equation is written as follows:

$$D_n = \left(\frac{b}{2na}\right)^2. \quad (10.3.30)$$

Following the method adopted above to obtain expression (10.3.21) for the input admittance of the TE_{10} mode, the expressions for the input admittance $Y_i^{TE_n}$ and $Y_i^{TM_n}$ of $TE_{1,2n}$ and $TM_{1,2n}$ ($n \geq 1$) modes could be obtained. The $TE_{1,2n}$ mode of the WG #B in the cutoff region offers an inductive load to the WG #A at x = h, whereas the $TM_{1,2n}$ mode offers a capacitive load. These considerations help to compute the input admittance for both the higher-order modes. The derivation of the results could be followed from the steps given for the *shielded slot line on the suspended composite substrate* in Fig (10.7). The results summarized below, and also the expression (10.3.21), are limiting cases of the relevant equations, for $h_3 = 0$, $h_1 \to \infty$, in subsection (10.3.3). The equations to compute the input admittance $Y_i^{TE_n}$ and $Y_i^{TM_n}$ of the present structure are given below [J.1]:

$$Y_i^{TE_n} = Y_{01}^{TE_n} \coth\left[\gamma_{n1} h + \coth^{-1}\left(\frac{Y_0^{TE_n}}{Y_{01}^{TEn}}\right)\right] \quad (a)$$

$$Y_i^{TM_n} = Y_{01}^{TM_n} \tanh\left[\gamma_{n1} h + \tanh^{-1}\left(\frac{Y_0^{TM_n}}{Y_{01}^{TMn}}\right)\right] \quad (b),$$

(10.3.31)

where $Y_{01}^{TE_n}$ and $Y_{01}^{TM_n}$ are the characteristic admittance of the dielectric-filled waveguide section WG #A for the $TE_{1,2n}$ and $TM_{1,2n}$ modes, respectively. $Y_0^{TE_n}$ and $Y_0^{TM_n}$ are the characteristic admittances for the air-filled waveguide section WG #B. The characteristic admittances for the TE and TM modes defined with respect to power–voltage ratio are related to their respective wave admittances [B.8]. It is discussed in chapter 7.

TE Mode

For dielectric – filled waveguide :

$$Y_{01}^{TE_n} = \frac{a}{2b} \times Y_w^{TE} = \frac{a}{2b} \times \frac{\beta_{n1}}{\omega\mu} = \left(\frac{a}{2b}\right)\left(\frac{\beta_{n1}}{\eta_1 k_1}\right) \quad (a)$$

$$\text{For air – filled waveguide : } Y_0^{TE_n} = \left(\frac{\beta_n}{\eta_0 k_0}\right)\left(\frac{a}{2b}\right) \quad (b).$$

(10.3.32)

TM Mode

For dielectric – filled waveguide :

$$Y_{01}^{TM_n} = \frac{a}{2b} \times Y_w^{TM} = \frac{a}{2b} \times \frac{\omega\varepsilon}{\beta} = \left(\frac{a}{2b}\right)\left(\frac{k_1}{\eta_1 \beta_{n1}}\right) \quad (a)$$

$$\text{For air – filled waveguide : } Y_0^{TM_n} = \left(\frac{a}{2b}\right)\left(\frac{k_0}{\eta_0 \beta_n}\right) \quad (b).$$

(10.3.33)

In the above equations, the propagation constants $\gamma_{n1} = j\,\beta_{n1}$ and $\gamma_n = j\,\beta_n$ are for both modes in the dielectric-filled WG #A and air-filled WG #B, respectively. The propagation constants for the dielectric-filled waveguide section A and the air-filled waveguide section B are obtained from equation (10.3.4) on using m = 1, n → 2n:

$$\gamma_{n1} = \left[\left(\frac{\pi}{a}\right)^2 + \left(\frac{2n\pi}{b}\right)^2 - \left(\frac{2\pi}{\lambda_0}\right)^2 \varepsilon_r\right]^{\frac{1}{2}} = \frac{2\pi\, n}{b}\left[1 - \left(\frac{b}{2\pi\, n}\right)^2 \left\{\left(\frac{2\pi}{\lambda_0}\right)^2 \varepsilon_r - \left(\frac{\pi}{a}\right)^2\right\}\right]^{\frac{1}{2}}$$

$$= \frac{2\pi\, n}{b}\left[1 - \left(\frac{b}{n}\right)^2 \left\{\frac{\varepsilon_r}{\lambda_0^2} - \left(\frac{1}{2a}\right)^2\right\}\right]^{\frac{1}{2}} \Rightarrow \gamma_{n1} = \frac{2\pi\, n}{b}\left[1 - \left(\frac{b}{n}\right)^2 \left\{\frac{1}{\lambda^2} - \frac{1}{\lambda_c^2}\right\}\right]^{\frac{1}{2}}. \quad (10.3.34)$$

where $\lambda = \lambda_0/\sqrt{\varepsilon_r}$. In the above equation, the cutoff wavelength of the $TE_{1,2n}$ mode is $\lambda_c = 2a$. It is related to the following expression [B.8]:

$$\frac{1}{\lambda_g^2} = \frac{1}{\lambda^2} - \frac{1}{\lambda_c^2}. \tag{10.3.35}$$

The above equation follows from equation (10.3.4). Using equations (10.3.34) and (10.3.35), the propagation constant γ_{n1} for $n \geq 1$ mode, in the dielectric-filled waveguide section WG #A, is obtained. Likewise, the propagation constant γ_n for the air-filled waveguide section WG #B is obtained using $\gamma = j\beta$, $\lambda_g = 2\pi/\beta$. Both expressions are given below:

$$\gamma_{n1} = \frac{2\pi n}{b}\left[1 - \left(\frac{b}{n\lambda_{g1}}\right)^2\right]^{\frac{1}{2}} \quad \text{(a)}$$
$$\gamma_n = \frac{2\pi n}{b}\left[1 + \left(\frac{b\gamma}{2\pi n}\right)^2\right]^{\frac{1}{2}} \quad \text{(b).} \tag{10.3.36}$$

Two more parameters q_n and r_n are defined for equation (10.3.31) as

$$q_n = \gamma_{n1}h + \coth^{-1}\left[\frac{Y_0^{TE_n}}{Y_{01}^{TE_n}}\right] \quad \text{(a)}$$
$$r_n = \gamma_{n1}h + \tanh^{-1}\left[\frac{Y_0^{TM_n}}{Y_{01}^{TM_n}}\right] \quad \text{(b).} \tag{10.3.37}$$

On using equations (10.3.32) and (10.3.33) for the $TE_{1,2n}$ and $TM_{1,2n}$ modes, the ratio of characteristic admittances can be evaluated for both modes, in terms of their propagation constants:

$$\frac{Y_0^{TE_n}}{Y_{01}^{TE_n}} = \frac{\gamma_n \eta_1 k_1}{\gamma_{n1}\eta_0 k_0} = \frac{\gamma_n}{\gamma_{n1}} \quad \text{(a)}$$
$$\frac{Y_0^{TM_n}}{Y_{01}^{TM_n}} = \frac{\gamma_{n1}\eta_1 k_0}{\gamma_n \eta_0 k_1} = \frac{\gamma_{n1}}{\varepsilon_r \gamma_n} \quad \text{(b).} \tag{10.3.38}$$

Thus,

$$q_n = \gamma_{n1}h + \coth^{-1}\left[\frac{\gamma_n}{\gamma_{n1}}\right] \quad \text{(a)}$$
$$r_n = \gamma_{n1}h + \tanh^{-1}\left[\frac{\gamma_{n1}}{\varepsilon_r \gamma_n}\right] \quad \text{(b).} \tag{10.3.39}$$

On using the above equations for q_n and r_n in equation (10.3.31), the input admittances of the $TE_{1,2n}$ and $TM_{1,2n}$ modes can be expressed as

$$Y_i^{TE_n} = Y_{01}^{TE_n}\coth(q_n) \quad \text{(a)}, \qquad Y_i^{TM_n} = Y_{01}^{TM_n}\tanh(r_n) \quad \text{(b).} \tag{10.3.40}$$

On substituting the characteristic admittances from equations (10.3.32a) and (10.3.33a) for the $TE_{1,2n}$ and $TM_{1,2n}$ $(n \geq 1)$ modes, respectively, the above equations are written as

$$Y_i^{TE_n} = \frac{a\gamma_{n1}}{j2b\eta_1 k_1}\coth(q_n) \quad \text{(a)}$$
$$Y_i^{TM_n} = \frac{ja k_1}{2b\eta_1\gamma_{n1}}\tanh(r_n) \quad \text{(b).} \tag{10.3.41}$$

On substituting equations (10.3.41) and (10.3.30) in equation (10.3.26a), the input admittance of combined $TE_{1,2n}$ and $TM_{1,2n}$ modes, i.e. the hybrid mode, at $z = 0$ is

$$Y_{i,n} = \frac{\dfrac{ja k_1}{2b\eta_1\gamma_{n1}}\tanh(r_n) + \dfrac{a\gamma_{n1}}{j2b\eta_1 k_1}\coth(q_n)\left(\dfrac{b}{2an}\right)^2}{1 + \left(\dfrac{b}{2an}\right)^2}$$
$$= \frac{ja k_0}{2b\eta_0\gamma_{n1}}\left[\frac{\varepsilon_r\tanh(r_n) - \left(\dfrac{b\gamma_{n1}}{2a\,nk_0}\right)^2\coth(q_n)}{1 + \left(\dfrac{b}{2a\,n}\right)^2}\right]. \tag{10.3.42}$$

A few more parameters are defined below to recast the above equation in the more convenient form:

$$p = \frac{\lambda_0}{2a} = \frac{\lambda_0}{\lambda_c}, \text{ also } \lambda_g = \lambda_c = 2a,\ \lambda = \frac{\lambda_0}{\sqrt{\varepsilon_r}} \quad \text{(a)}, \quad u = \frac{\lambda_0}{\lambda_{g1}} = \frac{2\pi}{k_0} \times \frac{\beta_1}{2\pi} = \frac{\gamma_1}{jk_0} \quad \text{(b)}$$
$$\left(\frac{1}{\lambda_{g1}}\right)^2 = \left(\frac{1}{\lambda}\right)^2 - \left(\frac{1}{\lambda_c}\right)^2 = \varepsilon_r\left(\frac{1}{\lambda_0}\right)^2 - \left(\frac{1}{\lambda_c}\right)^2, \ \Rightarrow \frac{1}{\lambda_{g1}^2} = \frac{1}{\lambda_0^2}(\varepsilon_r - p^2) \quad \text{(c)}$$
$$\text{So, } u = \frac{\lambda_0}{\lambda_{g1}} = \sqrt{\varepsilon_r - p^2} = \frac{\gamma_1}{jk_0} \quad \text{(d)}, \quad v = \frac{\gamma}{k_0} = \sqrt{p^2 - 1}, \quad u = \frac{v}{j} \quad \text{(e),} \tag{10.3.43}$$

where $\gamma = \gamma_0$ is the propagation constant of the air-filled waveguide. The propagation constants, given by equations (10.3.36a) and (10.3.36b), can be modified as follows, by using parameters p, u, and v:

$*$ For dielectric – filled waveguide section WG#A :

$$F_{n1} = \frac{b}{2\pi n}\gamma_{n1} = \left[1 - \left(\frac{b}{n\lambda_{g1}}\right)^2\right]^{\frac{1}{2}} = \left[1 - \left(\frac{bu}{n\lambda_0}\right)^2\right]^{\frac{1}{2}} = \left[1 - \left(\frac{bu}{2a\,np}\right)^2\right]^{\frac{1}{2}} \quad (a)$$

$*$ For air – filled waveguide section#B :

$$F_n = \frac{b}{2\pi n}\gamma_n = \left[1 + \left(\frac{b\gamma}{2\pi n}\right)^2\right]^{\frac{1}{2}} = \left[1 + \left(\frac{b\,v\,k_0}{2\pi n}\right)^2\right]^{\frac{1}{2}} = \left[1 + \left(\frac{b\,v}{2\,anp}\right)^2\right]^{\frac{1}{2}} \quad (b). \tag{10.3.44}$$

Let us change two factors involved in equation (10.3.42) in terms of parameters F_{n1} and p:

$$\frac{b\gamma_{n1}}{2ank_0} = \frac{b\gamma_{n1}}{2\pi n}\cdot\frac{\pi}{ak_0} = F_{n1}\cdot\frac{2\pi}{2ak_0} = F_{n1}\cdot\frac{\lambda_0}{2a} = F_{n1}\cdot p \quad (a)$$

$$\frac{ak_0}{2b\gamma_{n1}} = \frac{a\pi}{\lambda_0 b\gamma_{n1}} = \frac{a\pi}{2apb\gamma_{n1}} = \frac{1}{2p\left(\frac{b\gamma_{n1}}{\pi}\right)} = \frac{1}{2p\left[\frac{b\gamma_{n1}}{2\pi n}\right]2n} = \frac{1}{4p\,nF_{n1}} \quad (b). \tag{10.3.45}$$

Equation (10.3.42) could be rewritten as

$$Y_{i,n} = j\frac{1}{\eta_0(4pnF_{n1})}\left[\frac{\varepsilon_r\tanh(r_n) - (pF_{n1})^2\coth(q_n)}{1 + \left(\frac{b}{2an}\right)^2}\right]. \tag{10.3.46}$$

The input admittances at $z = 0$, due to all higher modes, $TE_{1,2n}$ and $TM_{1,2n}$ for $n \geq 1$, are obtained from equations (10.3.19) and (10.3.46) as follows:

$$2\sum_{n=1,2....}\left[Y_{i,n}\left(\frac{\sin\pi n\delta}{\pi n\delta}\right)^2\right] = j\frac{1}{2p\eta_0}\sum\left[\frac{\varepsilon_r\tanh(r_n) - (pF_{n1})^2\coth(q_n)}{\left\{1 + \left(\frac{b}{2an}\right)^2\right\}F_{n1}}\cdot\left(\frac{\sin\pi n\delta}{\pi n\delta}\right)^2\frac{1}{n}\right]. \tag{10.3.47}$$

The above capacitive input susceptance is due to the non-propagation higher-order modes. At $z = 0$, the input admittance Y_{i0} given by equation (10.3.21), could be written in terms of the variables p, u, and v. To do so, the characteristic admittance Y_{01}, Y_0/jY_{01} and electrical length ($\beta_1 h$) are expressed in terms of p, u, and v. Equations (10.3.24), (10.3.25), and (10.3.43) are used to get the following expressions:

$$Y_{01} = \frac{a}{2b\eta_0}\times\frac{\lambda_0}{\lambda_{g1}} = \frac{au}{2b\eta_0} \quad (a)$$

$$\beta_1 h = \frac{2\pi h}{\lambda_{g1}} = u\,k_0 h = \frac{2\pi}{\lambda_0}uh = \frac{\pi\,uh}{ap} \quad (b)$$

$$\frac{Y_0}{jY_{01}} = -\frac{\gamma}{k_0}\frac{\lambda_{g1}}{\lambda_0} = -\frac{v}{u} \quad (c). \tag{10.3.48}$$

Using variables of equation (10.3.48) in equation (10.3.21), the input admittance of TE_{10} mode is obtained as

$$Y_{i,0} = j\frac{au}{2b\eta_0}\tan\left[\frac{\pi\,uh}{ap} - \tan^{-1}\left(\frac{v}{u}\right)\right]. \tag{10.3.49}$$

Finally, the input capacitive susceptance B_d, given by equation (10.3.19), at $z = 0$ is rewritten with the help of equations (10.3.47) and (10.3.49) as

$$\eta_0 B_d = \frac{au}{2b}\tan\left[\frac{\pi\,uh}{ap} - \tan^{-1}\left(\frac{v}{u}\right)\right] + \frac{1}{2p}\sum_{n=1,2,...}\left\{\left[\frac{\varepsilon_r\tanh(r_n) - p^2F_{n1}^2\coth(q_n)}{\left[1 + \left(\frac{b}{2an}\right)^2\right]F_{n1}}\right]\frac{\sin^2(\pi\,n\delta)}{n(\pi\,n\delta)^2}\right\}. \tag{10.3.50}$$

The above capacitive susceptance is due to the discontinuity offered by the waveguide #1 that has an iris at $z = 0$ supported on the dielectric-loaded WG #A section followed by the WG Section B in the cutoff mode. The *first term* of the above equation is the susceptance due to TE_{10} mode and the *second term* is the susceptance due to combined higher-order modes $TE_{1,2n}$ and $TM_{1,2n}$ ($n = 1,2,...$) in a section of WG #B. The rate of convergence of the series shown in the above equation is slow. It requires improvement in convergence. To do so, the limiting value of the term under the sign $\sum$ is obtained.

$$\text{Let } S_n = \left[\frac{\varepsilon_r\tanh(r_n) - p^2F_{n1}^2\coth(q_n)}{\left[1 + \left(\frac{b}{2an}\right)^2\right]F_{n1}}\right]\frac{\sin^2(\pi\,n\delta)}{n(\pi\,n\delta)^2}. \tag{10.3.51}$$

Under the limiting condition, $h \to \infty$ and $f \to 0$, i.e. $\lambda_0 = \frac{c}{f} \to \infty$, $\lambda_g \to \infty$, $a \to \infty$, also $\underset{\lambda_g\lambda_{g1}\to\infty}{Lt}(F_{n1}) \to 1$, $\underset{\lambda_g\lambda_{g1}\to\infty}{Lt}(\tanh r_n) \to 1$, $\underset{\lambda_g\lambda_{g1}\to\infty}{Lt}(\coth q_n) \to 1$ the following expression is obtained:

$$S_n \to S_n' = (\varepsilon_r - p^2)\frac{\sin^2(\pi\,n\delta)}{n(\pi\,n\delta)^2} = u^2\frac{\sin^2(\pi\,n\delta)}{n(\pi\,n\delta)^2}. \tag{10.3.52}$$

The series summation of equation (10.3.50) is rewritten as follows:

$$\sum_{n=1,2,\ldots} S_n = \sum_{n=1,2,\ldots} (S_n - S'_n) + \sum_{n=1,2,\ldots} S'_n. \quad (10.3.53)$$

$$\sum_{n=1,2,\ldots} S'_n = u^2 \sum_{n=1,2,\ldots} \left[\frac{\sin^2(\pi n\delta)}{n(\pi n\delta)^2}\right] \approx u^2\left[\ln\left(\frac{1}{2\pi\delta}\right) + \frac{3}{2}\right]$$
$$= u^2\left[\ln\left(\frac{1}{\delta}\right) - 0.3379\right]. \quad (10.3.54)$$

The above summation has been evaluated by expressing the sine function in the form of exponentials and also using the identity given by Collin in Appendix-A.2 [B.11]. Thus, under the condition $\delta \to 0$, the field is confined in the slot region providing the capacitive susceptance B_c that exists on both sides of the iris. Hence, from equations (10.3.54) and (10.3.50), the following expression is obtained:

$$\eta_0 B_c \approx \frac{u^2}{p}\left[\ln\left(\frac{1}{\delta}\right) - 0.3379\right]. \quad (10.3.55)$$

However, a more accurate formula in the limiting case $\lambda_{g1} \to \infty$ is [J.1, B.13],

$$\eta_0 B_c \approx \frac{u^2}{p}\ln\left(\frac{2}{\pi\delta}\right) = \frac{u^2}{p}\left[\ln\left(\frac{1}{\delta}\right) - 0.452\right]. \quad (10.3.56)$$

The difference in B_c is due to our assumption of the constant E_y and H_x fields accross the iris slot. The slot field is a function of y, not a constant as assumed in equation (10.3.9). Using the above expression, equation (10.3.53) is improved as

$$\sum_{n=1,2,\ldots} S_n = \sum_{n=1,2,\ldots} (S_n - S'_n) + u^2 \ln\left(\frac{2}{\pi\delta}\right). \quad (10.3.57)$$

At $z = 0$, the susceptance B_d looking into the waveguide #1 given by equation (10.3.51) is rewritten by using equations (10.3.51), (10.3.52), and (10.3.57):

$$\eta_0 B_d = \frac{a\,u}{2b}\tan\left[\frac{\pi\,u\,h}{a\,p} - \tan^{-1}\left(\frac{v}{u}\right)\right] + \frac{u^2}{2p}\ln\left(\frac{2}{\pi\delta}\right)$$
$$+ \frac{1}{2p}\sum_{n=1,2,\ldots}\left\{\left[\frac{\varepsilon_r \tanh(r_n) - p^2F_{n1}^2\coth(q_n)}{\left[1+\left(\frac{b}{2a\,n}\right)^2\right]F_{n1}} - u^2\right]\frac{\sin^2(\pi\,n\delta)}{n\,(\pi\,n\delta)^2}\right\}. \quad (10.3.58)$$

The *first term* of the above susceptance is due to the TE_{10} mode in waveguide #1 with WG sections – WG #A and WG #B. The *second term* is the capacitive reactance of the slot iris. The *third term* is the susceptance contributed by the higher-order modes in waveguide #1.

At $z = 0$, the susceptance B_a offered by the air-filled waveguide #2 in the region $z < 0$ is obtained from the above equation by using $\varepsilon_r = 1$ or $h = 0$:

$$\eta_0 B_a = -\frac{av}{2b} - \frac{v^2}{2p}\ln\left(\frac{2}{\pi\delta}\right)$$
$$+ \frac{1}{2p}\sum_{n>1}\left\{\left[\frac{1 - p^2F_{n1}^2}{\left[1+\left(\frac{b}{2a\,n}\right)^2\right]F_{n1}} - u^2\right]\frac{\sin^2(\pi n\delta)}{n\,(\pi\,n\delta)^2}\right\}. \quad (10.3.59)$$

It is seen that for $\varepsilon_r = 1$, $\gamma_{n1} = \gamma_n$, so $\tanh(r_n) = \coth(q_n) = 1$, also $F_{n1} = \sqrt{1 - \left(\frac{bu}{2anp}\right)^2}$. Thus, the expression under the $\sum$ sign is

$$\frac{1 - p^2F_{n1}^2}{\left[1+\left(\frac{b}{2an}\right)^2\right]F_n} - u^2 = \frac{1 - p^2\left[1 - \left(\frac{b}{2an}\right)^2\frac{u^2}{p^2}\right]}{\left[1+\left(\frac{b}{2an}\right)^2\right]F_n} - u^2 = \frac{1 - p^2 + \left(\frac{b}{2an}\right)^2u^2}{\left[1+\left(\frac{b}{2an}\right)^2\right]F_n} - u^2$$
$$= \frac{u^2\left[1+\left(\frac{b}{2an}\right)^2\right]}{\left[1+\left(\frac{b}{2an}\right)^2\right]F_n} - u^2 = u^2\left(\frac{1}{F_n} - 1\right) = v^2\left(1 - \frac{1}{F_n}\right). \quad (10.3.60)$$

At $z = 0$, the susceptance B_a, looking toward the air-filled waveguide #2 is

$$\eta_0 B_a = -\frac{av}{2b} - \frac{v^2}{2p}\ln\left(\frac{2}{\pi\delta}\right) + \frac{1}{2p}\sum_{n=1,2,\ldots}\left[v^2\left(1-\frac{1}{F_n}\right)\frac{\sin^2(\pi n\delta)}{n(\pi n\delta)^2}\right]. \quad (10.3.61)$$

At $z = 0$, the total susceptance $B_t = B_d + B_a$ is obtained, for the **case-A** of the (EW–EW) waveguide, from equations (10.3.58) and (10.3.61):

$$\eta_0 B_t = \frac{a}{2b}\left[-v + u\tan\left\{\frac{\pi u h}{ap} - \tan^{-1}\left(\frac{v}{u}\right)\right\}\right] + \frac{1}{p}\left\{\left(\frac{\varepsilon_r+1}{2}\right) - p^2\right\}\ln\left(\frac{2}{\pi\delta}\right) + \frac{1}{2}\sum_{n=1,2,\ldots}\left\{\left[v^2\left(1-\frac{1}{F_n}\right) + M_n\right]\frac{\sin^2(\pi n\delta)}{n(\pi n\delta)^2}\right\}. \quad (10.3.62)$$

where M_n for real F_{n1} is;

$$M_n = \frac{\varepsilon_r\tanh(r_n) - p^2F_{n1}^2\coth(q_n)}{\left[1+\left(\frac{b}{2an}\right)^2\right]F_{n1}} - u^2 \quad (a)$$

$$r_n = \frac{2\pi nhF_{n1}}{b} + \tanh^{-1}\left(\frac{F_{n1}}{\varepsilon_r F_n}\right) \quad (b)$$

$$q_n = \frac{2\pi nhF_{n1}}{b} + \coth^{-1}\left(\frac{F_n}{F_{n1}}\right) \quad (c). \quad (10.3.63)$$

M_n for F_{n1} imaginary is;

$$M_n = \frac{\varepsilon_r\tan(r_n') - p^2|F_{n1}|^2\cot(q_n')}{\left[1+\left(\frac{b}{2an}\right)^2\right]|F_{n1}|} - u^2 \quad (a)$$

$$r_n' = \frac{2\pi nh|F_{n1}|}{b} + \tan^{-1}\left(\frac{|F_{n1}|}{\varepsilon_r F_n}\right) \quad (b)$$

$$q_n' = \frac{2\pi nh|F_{n1}|}{b} + \cot^{-1}\left(\frac{F_n}{|F_{n1}|}\right) \quad (c). \quad (10.3.64)$$

The above expression for the total susceptance B_t is valid for the narrow slot, $\delta = s/b \le 0.15, s \le \lambda_0/(4\sqrt{\varepsilon_r})$.

Case B: Magnetic Wall at y = ±b/2

In this case, two broad-faces, separated by height b in Fig (10.4) and (10.5), are treated as the MW. The equivalent waveguide model, with the EW–MW surfaces, is formed. The rest of the analysis follows case A. However, this waveguide does not support the propagating TE_{10} mode. Therefore, its contribution to the susceptance B_d, through the first term of equation (10.3.50), is dropped. Also keeping in view the MW and symmetry, TE_{11}, TE_{13}, TE_{15},... and TM_{11},TM_{13}, TM_{15},..., i.e. $TE_{1,2n}$ and $TM_{1,2n}$ modes are excited, where n = 1/2, 3/2, 5/2, etc. Under this value for n, the following limiting value is obtained:

$$\underset{\delta\to 0}{Lt}\sum_{n=\frac{1}{2},\frac{3}{2},\ldots}\frac{\sin^2(\pi n\delta)}{n(\pi n\delta)^2} = \ln\left(\frac{4}{\delta}\right) - 0.3379 \approx \frac{u^2}{p}\ln\left(\frac{8}{\pi\delta}\right). \quad (10.3.65)$$

The above expression provides the capacitive susceptance offered by the iris in the (EW–MW) waveguide. Therefore, the susceptance B_d and B_a for the case of MW at $y = \pm b/2$ is written, from equations (10.3.58), (10.3.61), and (10.3.65), as follows:

$$\eta_0 B_d = \frac{u^2}{2p}\ln\left(\frac{8}{\pi\delta}\right) + \frac{1}{2p}\sum_{n=\frac{1}{2},\frac{3}{2},\ldots}\left\{\left[\frac{\varepsilon_r\tanh(r_n) - p^2F_{n1}^2\coth(q_n)}{\left[1+\left(\frac{b}{2an}\right)^2\right]F_{n1}} - u^2\right]\frac{\sin^2(\pi n\delta)}{n(\pi n\delta)^2}\right\}. \quad (10.3.66)$$

$$\eta_0 B_a = -\frac{v^2}{2p}\ln\left(\frac{8}{\pi\delta}\right) + \frac{1}{2p}\sum_{n=\frac{1}{2},\frac{3}{2},\ldots}\left\{v^2\left(1-\frac{1}{F_n}\right)\frac{\sin^2(\pi n\delta)}{n(\pi n\delta)^2}\right\}. \quad (10.3.67)$$

The first term of the above equations shows the capacitive susceptance offered by the iris in the (EW–MW) waveguide. The *second term* is the capacitive susceptance due to the higher-order modes. At $z = 0$, the total susceptance $B_t = B_d + B_a$, for **case B** of the (EW–MW) waveguide, is obtained by adding the above two equations:

$$\eta_0 B_t = \frac{1}{p}\left\{\left[\left(\frac{\varepsilon_r+1}{2}\right) - p^2\right]\ln\left(\frac{8}{\pi\delta}\right) + \frac{1}{2}\sum_{n=\frac{1}{2},\frac{3}{2},\ldots}\left[v^2\left(1-\frac{1}{F_n}\right) + M_n\right]\frac{\sin^2(\pi n\delta)}{n(\pi n\delta)^2}\right\}. \quad (10.3.68)$$

The above expression is a modification of equation (10.3.62). The first term of equation (10.3.62), due to the TE_{10} mode, is dropped. The parameters M_n for the real and imaginary F_{n1} are given by equations (10.3.63) and (10.3.64), respectively.

Slowing-Factor

The slowing-factor $1/\sqrt{\varepsilon_{reff}(f)}$ is determined by using the *transverse resonance condition*, i.e. $B_t = 0$ at $z = 0$. For a given slot line, i.e. known substrate thickness h and relative permittivity ε_r and assumed b, equations (10.3.62) and (10.3.68) are solved for the parameter p at a given frequency. At the transverse resonance, $a = \lambda_g/2$, i.e. $p = \lambda_0/\lambda_g$, where λ_g is the guided wavelength of the slot line. The value of the parameter p is in the range $1 < p < \sqrt{\varepsilon_r}$ corresponding to $\lambda_{g1} \to \lambda_0$, $\lambda_{g1} \to \infty$. The solution provides $1/p = \lambda_g/\lambda_0 = 1/\sqrt{\varepsilon_{reff}(f)}$, i.e. the dispersive slowing-factor. For the correct choice of dimension b, both equations (8.3.62) and (8.3.68) provide almost the same values for the slowing-factor. Thus, at each assumed frequency, the effective relative permittivity $\varepsilon_{reff}(f)$ of a slot line is determined, once the parameter p is evaluated. It should be independent of parameter b. Usually, b is several times of the slot-width, $b > 10\,s$.

Characteristic Impedance

The characteristic impedance of a slot line is defined based on voltage–power ratio, as follows:

$$Z_0 = \frac{V_+^2}{2P_+}, \tag{10.3.69}$$

where P_+ is the average power flow of the wave traveling in + x-direction on a slot line, shown in Fig (10.4a, d) and (10.5b) and V_+ is the peak amplitude of the voltage across the slot. The energy U *p.u.l.* travels with the group velocity v_g, i.e.

$$U = \frac{2P_+}{v_g}. \tag{10.3.70}$$

In the case of the resonant line length $x = \lambda_g/2$, two oppositely traveling waves form the standing wave, such that the total power on the line is $2P_+$. The stored energy in the $\lambda_g/2$ line length is $W_t = (2P_+/v_g) \times (\lambda_g/2) = P_+\lambda_g/v_g$. Using phase velocity $v_p = \lambda_g c/\lambda_0$ from equation (7.4.29b) of chapter 7, the stored energy is

$$W_t = \frac{P_+}{v_g} \times v_p \frac{\lambda_0}{c} = \frac{2\pi P_+}{\omega} \frac{v_p}{v_g}. \tag{10.3.71}$$

At the center of the slot ($x = \lambda_g/2$), the voltage is maximum:

$$V_0 = 2V_+. \tag{10.3.72}$$

The characteristic impedance of the slot line, from equations (10.3.69) to (10.3.71), is

$$Z_0 = \frac{\pi V_0^2}{4\,\omega W_t} \times \frac{v_p}{v_g}. \tag{10.3.73}$$

The stored energy in a cavity, at resonance, is given by [J.1]:

$$W_t = \frac{V_0^2}{4}\frac{dB}{d\omega}, \tag{10.3.74}$$

where B is the susceptance, at the port of the cavity. In the present case, the susceptance is B_t at the port $z = 0$ giving the following expression:

$$Z_0 = \frac{\pi}{\omega} \cdot \frac{1}{(d(\eta_0 B_t)/d\omega)} \cdot \frac{v_p}{v_g}. \tag{10.3.75}$$

However, $\omega = 2\pi c/\lambda_0$, at the transverse resonance $p = \lambda_0/2a = \lambda_0/\lambda_g$, $\omega = 2\pi\,c/2a\,p = \pi\,c/a\,p$, $d\omega = -(\pi\,c\,dp)/ap^2 = -\omega(dp/p)$. Therefore, the above characteristic impedance is written as

$$Z_0 = 377\,\frac{\pi}{p} \cdot \left[\frac{\Delta p}{-\Delta(\eta_0 B_t)}\right] \cdot \frac{v_p}{v_g}. \tag{10.3.76}$$

An expression for v_p/v_g, i.e. ratio of the phase and group velocities for a slot line, is needed to compute its characteristic impedance. The group velocity is related to the propagation constant β_x of the slot line by

$$v_g = \frac{d\omega}{d\beta_x} = \frac{1}{\frac{d\beta_x}{d\omega}}, \quad \text{where,} \quad \beta_x = \frac{2\pi}{\lambda_g}. \tag{10.3.77}$$

$$v_g = \frac{1}{\frac{d}{d\omega}\left(\frac{2\pi}{\lambda_g}\right)} = \frac{1}{\frac{2\pi}{2\pi}\frac{d}{df}\left(\frac{1}{\lambda_g}\right)} = -\frac{1}{\frac{1}{\lambda_g^2}\frac{d\lambda_g}{df}} = -\frac{\lambda_g^2}{\frac{d\lambda_g}{df}}$$

$$\Rightarrow \frac{\lambda_g}{v_g} = -\frac{1}{\lambda_g}\frac{d\lambda_g}{df}. \tag{10.3.78}$$

Using equation (7.4.29b) of chapter 7, the phase velocity is also expressed as follows:

$$v_p = \frac{\lambda_g}{\lambda_0}c = f{\cdot}\lambda_g. \tag{10.3.79}$$

$$\text{Further, } p = \frac{\lambda_0}{\lambda_g}, \quad \frac{dp}{df} = \frac{d}{df}\left(\frac{\lambda_0}{\lambda_g}\right) = \frac{1}{\lambda_g}\frac{d\lambda_0}{df} - \frac{\lambda_0}{\lambda_g^2}\frac{d\lambda_g}{df}.$$

$$\text{However, } \lambda = \frac{c}{f} \text{ i.e. } \frac{d\lambda_0}{df} = -\frac{c}{f^2} = -\frac{\lambda_0}{f}.$$

$$\text{Therefore, } \frac{d(\lambda_0/\lambda_g)}{df} = -\frac{\lambda_0}{\lambda_g}\left(\frac{1}{f} + \frac{1}{\lambda_g}\frac{d\lambda_g}{df}\right)$$

$$\Rightarrow \frac{d\lambda_g}{df} = -\lambda_g\left(\frac{1}{f} + \frac{\lambda_g}{\lambda_0}\frac{d(\lambda_0/\lambda_g)}{df}\right). \tag{10.3.80}$$

From equations (10.3.78) and (10.3.79),

$$\frac{\lambda_g}{v_g} = \frac{1}{f} + (\lambda_g/\lambda_0)\frac{\Delta(\lambda_0/\lambda_g)}{\Delta f}. \tag{10.3.81}$$

On substituting λ_g from equation (10.3.79) in the above equation, the following relation is obtained:

$$\frac{v_p}{v_g} = 1 + \frac{f}{(\lambda_0/\lambda_g)}\frac{\Delta(\lambda_0/\lambda_g)}{\Delta f}. \tag{10.3.82}$$

The $\Delta(\lambda_0/\lambda_g)$, i.e. Δp and Δf, is computed from two separate solutions of $\eta_0 B_t = 0$ for the fixed values of ε_r, s, h, and b and for two slightly different values of $a = \lambda_g/2 = \lambda_0/2p$, incremented plus and minus from the defined parameter "a" at the frequency f. The frequency f is taken midway in the Δf interval. Once v_p/v_g is obtained, it can be substituted in equation (10.3.76) to compute the characteristic impedance of a slot line.

10.3.2 Sandwich Slot Line

The waveguide model discussed above is extended to a sandwich slot line, shown in Fig (10.1b) [J.12–J.14]. The method is also applicable to the coupled slot lines [J.25, J.26]. Following the discussion of the previous section, the waveguide model is developed for a sandwich slot line, shown in Fig (10.6a). Figure (10.6b) shows its equivalent waveguide of width $a = \lambda_g/2$, in the direction of the x-axis and height $\pm b/2$, in the direction of the y-axis. The width is not shown in Fig (10.6b). Similar to Fig (10.4c), the broadface of the present equivalent waveguide could be either an EW or a MW. A pair of strip conductors of a slot line appears as the iris located at the interface of two dielectric sheets. The $z = 0$ plane of Fig (10.6b) is a junction of two modeled waveguides – the waveguide #1 for $z \geq 0$ and the waveguide #2 for $z \leq 0$.

For the *case of EW* at $y = \pm b/2$ and on using equation (10.3.58), the susceptance B_{d1} and B_{d2}, i.e. B_{di} $(i = 1,2)$ for the $z > 0$ and $z < 0$ sides are written below:

$$\eta_0 B_{di} = \frac{a\,u_i}{2b}\tan\left[\frac{\pi\,u_i h_i}{ap} - \tan^{-1}\left(\frac{v_i}{u_i}\right)\right] + \frac{u_i^2}{2p}\ln\left(\frac{2}{\pi\delta}\right) + \frac{1}{2p}\sum_{n=1,2,3..}\left\{\left[\frac{\varepsilon_{ri}\tanh(r_{ni}) - p^2F_{ni}^2\coth(q_{ni})}{\left[1+\left(\frac{b}{2an}\right)^2\right]F_{ni}} - u_i^2\right]\frac{\sin^2(\pi\,n\delta)}{n\,(\pi\,n\delta)^2}\right\}. \quad (10.3.83)$$

Let,
$$M_{ni} = \frac{\varepsilon_{ri}\tanh(r_{ni}) - p^2F_{ni}^2\coth(q_{ni})}{\left[1+\left(\frac{b}{2an}\right)^2\right]F_{ni}} - u_i^2 \quad (i = 1, 2). \quad (10.3.84)$$

For waveguide #1:

$$\eta_0 B_{d1} = \frac{au_1}{2b}\tan\left[\frac{\pi\,u_1h_1}{ap} - \tan^{-1}\left(\frac{v_1}{u_1}\right)\right] + \frac{u_1^2}{2p}\ln\left(\frac{2}{\pi\delta}\right) + \frac{1}{2p}\sum_{n=1,2,\ldots}\left\{M_{n1}\frac{\sin^2(\pi n\delta)}{n(\pi n\delta)^2}\right\}. \quad (10.3.85)$$

Likewise, for waveguide #2:

$$\eta_0 B_{d2} = \frac{au_2}{2b}\tan\left[\frac{\pi\,u_2h_2}{ap} - \tan^{-1}\left(\frac{v_2}{u_2}\right)\right] + \frac{u_2^2}{2p}\ln\left(\frac{2}{\pi\delta}\right) + \frac{1}{2p}\sum_{n=1,2,\ldots}\left\{M_{n2}\frac{\sin^2(\pi\,n\delta)}{n\,(\pi\,n\delta)^2}\right\}. \quad (10.3.86)$$

The total susceptance at plane $z = 0$:
$$B_t = B_{d1} + B_{d2}. \quad (10.3.87)$$

$$\eta_0 B_t = \frac{a}{2b}\left[u_1\tan\left[\frac{\pi\,u_1h_1}{ap} - \tan^{-1}\left(\frac{v_1}{u_1}\right)\right] + u_2\tan\left[\frac{\pi u_2h_2}{ap} - \tan^{-1}\left(\frac{v_2}{u_2}\right)\right]\right] + \frac{1}{2p}\left[u_1^2 + u_2^2\right]\ln\left(\frac{2}{\pi\delta}\right) + \frac{1}{2p}\sum_{n=1,2,3..}\left[(M_{n1} + M_{n2})\frac{\sin^2(\pi\,n\delta)}{n\,(\pi\,n\delta)^2}\right], \quad (10.3.88)$$

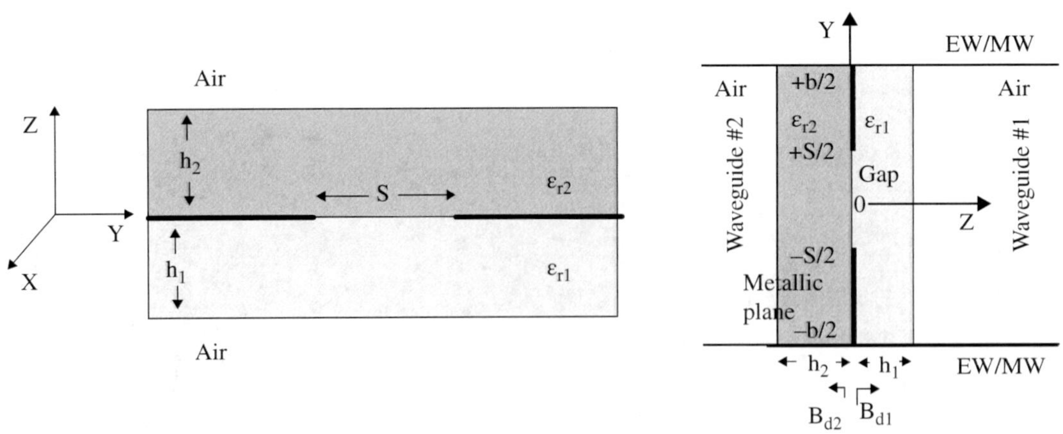

(a) Sandwich slot line. (b) Equivalent waveguide model.

Figure 10.6 Sandwich slot line and its equivalent waveguide model.

$$\text{where} \quad u_i = \sqrt{\varepsilon_{ri} - p^2} \quad (a), \quad F_{ni} = \sqrt{1 - \left(\frac{bu_i}{2anp}\right)^2}, \; i = 1, 2 \quad (b)$$
$$v_1 = v_2 = v = \sqrt{p^2 - 1} \quad (c), \quad F_n = \sqrt{1 + \left(\frac{bv}{2anp}\right)^2} \quad (d). \tag{10.3.89}$$

The parameter M_{ni} for the real F_{ni} and imaginary F_{ni} ($i = 1, 2$) are obtained from equations (10.3.63) and (10.3.64), respectively. Using equation (10.3.58), the total susceptance B_t, at $z = 0$, is

$$\eta_0 B_t = \frac{a}{2b}\left[u_1 \tan\left[\frac{\pi u_1 h_1}{ap} - \tan^{-1}\left(\frac{v}{u_1}\right)\right] + u_2 \tan\left[\frac{\pi u_2 h_2}{ap} - \tan^{-1}\left(\frac{v}{u_2}\right)\right]\right] + \frac{1}{2p}\left[\varepsilon_{r1} + \varepsilon_{r2} - 2p^2\right] \ln\left(\frac{2}{\pi\delta}\right) + \frac{1}{2p}\sum_{n=1,2,3..}\left[(M_{n1} + M_{n2})\frac{\sin^2(\pi n\delta)}{n(\pi n\delta)^2}\right]. \tag{10.3.90}$$

Likewise, for the *case of the MW* at $y = \pm b/2$, using equation (10.3.66), the total susceptance B_t at $z = 0$ is written as follows:

$$\eta_0 B_t = \frac{1}{2p}\left[\varepsilon_{r1} + \varepsilon_{r2} - 2p^2\right] \ln\left(\frac{8}{\pi\delta}\right) + \frac{1}{2p}\sum_{n=\frac{1}{2},\frac{3}{2},\ldots}\left[(M_{n1} + M_{n2})\frac{\sin^2(\pi n\delta)}{n(\pi n\delta)^2}\right]. \tag{10.3.91}$$

In the case of the MW at $y = \pm b/2$, the first two terms of equation (10.3.90), due to the TE_{10} mode, are dropped in the above equation.

In the present case, the equivalent waveguides #1 and #2 are partly occupied with dielectric slabs corresponding to the substrate and superstrate of the sandwiched slot line. If the two substrates are widely different, either u_1 or u_2 can be imaginary. In that case, $u_i^2 = -|u_i|^2$ and $u_i = j|u_i|$. For the u_i imaginary, the first term of equation (10.3.90) is replaced by another factor as follows:

$$u_i \tan\left[\frac{\pi h_i u_i}{ap} - \tan^{-1}\left(\frac{v}{u_i}\right)\right] \rightarrow |u_i| \tanh\left[\frac{\pi h_i |u_i|}{ap} - \tanh^{-1}\left(\frac{v}{|u_i|}\right)\right] \quad (i = 1, 2). \tag{10.3.92}$$

At the assumed frequency, the parameter $p = (\lambda_0/\lambda_g) = \sqrt{\varepsilon_{reff}(f)}$ is obtained from the solution of either equation (10.3.90) or equation (10.3.91), for a large value of b, $b > 10\,s$. The characteristic impedance is obtained from equation (10.3.76). The zeroth-order solution for the guided wavelength is

$$\frac{\lambda_g}{\lambda_0} = \sqrt{\frac{2}{(\varepsilon_{r1} + \varepsilon_{r2})}}. \tag{10.3.93}$$

The above analysis is used for the dielectric substrate–superstrate. However, the permeability of the substrate–superstrate can be easily accounted for in the analysis [J.12].

10.3.3 Shielded Slot Line

Figure (10.7a) shows a *suspended composite substrate slot line*, totally enclosed inside a metallic box [J.13]. Figure (10.7b) shows its equivalent waveguide model. Figure (10.7c) shows the cross-section of the equivalent waveguide. The EM-wave on the slot line propagates in the x-direction. In the present case, the EW, due to the side shield conductors, are placed at $y = \pm b/2$. The lower region, below the strip conductors, comprising of three dielectric media, is converted to the right-hand side *waveguide #1* shown in Fig (10.7b). The WG #1 is terminated in the conducting wall at $z = (h_1 + h_2 + h_3)$. The upper region of the structure, with air medium, appears to the left-hand side *waveguide #2* terminated in the conducting wall at $z = -h_4$. Hence, at the plane $z = 0$, a junction is created by two waveguides – waveguide #1 and waveguide #2. The metallic iris is located at the junction of two waveguides. Fig (10.7c) shows the cross-sectional view of the iris corresponding to the $a = \lambda_g/2$ section of a slot line.

The structural details are taken into account, while computing the capacitive susceptance B_d at $z = 0$, using the expressions of subsection (10.3.1). The total E_y and H_x fields at $z = 0$ plane and $x = a/2$ are still given by equation (10.3.7). The TE_{10} mode and all higher-order modes are at the cutoff or non-propagating in the air region of thickness h_1 and h_4. However, in the dielectric regions, the TE_{10} mode propagates. The first few higher-order modes may be in the cutoff, depending upon height "b" of the equivalent

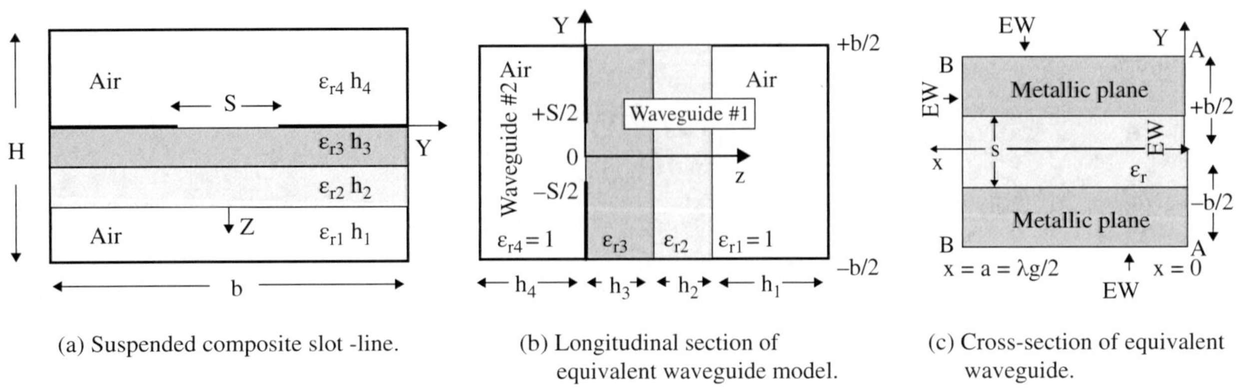

Figure 10.7 Multilayer shielded slot line and its equivalent waveguide model.

waveguide. The input wave admittance of the fundamental TE_{10} mode $y_{i,0,}$ and the input wave admittance, $y_{i,n}$ of the combined n^{th} $TE_{1,2n,}$ and $TM_{1,2n}$ modes, i.e. the hybrid modes, are given by equation (10.3.8). The slot field E_y, given by equation (10.3.9), is assumed to be constant in the slot region. The analysis is valid for the narrow slot $\delta = s/b \leq 0.15$ [J.1].

Computation of B_d at z = 0 for Waveguide #1

The input admittance Y_i, related to the capacitive susceptance B_d, at the plane z = 0 for the waveguide section #1, is given by equation (10.3.19). The Y_i has two components – (i) *input admittance* $Y_{i,0}$ *of the* TE_{10} *mode,* and (ii) *input admittance* $Y_{i,n}$ *contributed by the higher-order modes* $TE_{1,2n}$ *and* $TM_{1,2n}$. The following expressions are obtained for them:

Input Admittance $Y_{i,0}$ of the TE_{10} Mode

The input admittance of waveguide #1 is computed in three steps with reference to Fig (10.8). The input admittance of the equivalent transmission line of the waveguide section of length h terminated in load Y_L is

$$Y_{in} = Y_0 \frac{Y_L + Y_0 \tanh(\gamma h)}{Y_0 + Y_L \tanh(\gamma h)}. \quad (10.3.94)$$

Figure (10.8a) shows the *1st step*. The last section of the waveguide #1 is terminated in the EW forming the short-circuited load $Y_L \to \infty$.

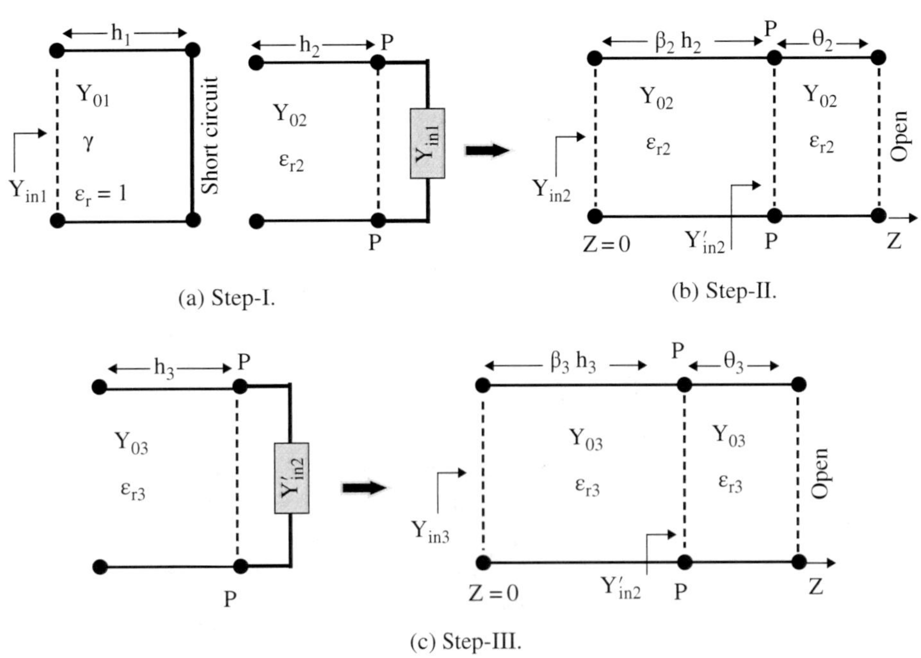

Figure 10.8 Computation of the input admittance.

$$Y_{in1} = Y_{01} \coth(\gamma h_1), \tag{10.3.95}$$

where Y_{01} is the characteristic admittance of the waveguide section h_1. For the TE_{10} mode, Y_{01} is given by equation (10.3.24) and γ is the propagation constant of the waveguide.

The *2nd step* of Fig (10.8b) shows that Y_{in1} acts as a load to dielectric (ε_{r2})-filled waveguide section h_2 with characteristic admittance Y_{02}. This load is replaced by the extended open-end line of length θ_2. Its input admittance is

$$Y'_{in1} = jY_{02} \tan\theta_2 = Y_{in1} = Y_{01}\coth(\gamma h_1)$$
$$\Rightarrow \tan\theta_2 = \frac{Y_{01}\coth(\gamma h_1)}{j Y_{02}}. \tag{10.3.96}$$

The following expression is obtained from equation (10.3.48c):

$$\frac{Y_0}{jY_{02}} = -\frac{v}{u_2}. \tag{10.3.97}$$

The input admittance $Y_{in,2}$ of the waveguide section h_2 with extended length θ_2 is

$$Y_{in,2} = jY_{02} \tan(\beta_2 h_2 + \theta_2). \tag{10.3.98}$$

The *3rd step* of Fig (10.8c) shows that $Y_{in,2}$ acts as a load to the dielectric (ε_{r3})-loaded waveguide section h_3 with the characteristic admittance Y_{03}. The load $Y_{in,2}$ is again replaced by an open circuit line section of length θ_3. Its input admittance is

$$Y'_{in,2} = jY_{03} \tan\theta_3 = jY_{02} \tan(\beta_2 h_2 + \theta_2)$$
$$\Rightarrow \quad \tan\theta_3 = \frac{Y_{02}}{Y_{03}} \tan(\beta_2 h_2 + \theta_2), \tag{10.3.99}$$

where using an equation (10.3.48a) : $\dfrac{Y_{02}}{Y_{03}} = \dfrac{u_2}{u_3}$. (10.3.100)

$$\tan\theta_3 = \frac{u_2}{u_3} \tan\left[\beta_2 h_2 - \tan^{-1}\left\{\frac{v}{u_2}\coth(\gamma h_1)\right\}\right] \tag{10.3.101}$$

where $v = (p^2-1)^{1/2}, \gamma = v\,k_0,\ k = 2\pi/\lambda_0,$
$u_2 = (\varepsilon_{r2} - p^2)^{1/2}, u_3 = (\varepsilon_{r3} - p^2)^{1/2}.$ (10.3.102)

The input admittance of the section h_3 with line extension θ_3 is

$$Y_{in,3} = jY_{03} \tan(\beta_3 h_3 + \theta_3). \tag{10.3.103}$$

On using equations (10.3.48a) and (10.3.48b), the above equation is written as the input admittance $Y_{i,0}$ of waveguide #1 for the TE_{10} mode:

$$Y_{i,0} = Y_{in,3} = j\frac{a\,u_3}{2b\eta_0}\tan\left[\frac{\pi u_3 h_3}{ap} + \tan^{-1}\left\{\frac{u_2}{u_3}\tan\left(\frac{\pi u_2 h_2}{ap} - \tan^{-1}\left(\frac{v}{u_2}\coth(\gamma h_1)\right)\right)\right\}\right]. \tag{10.3.104}$$

Input Admittance $Y_{i,n}$ of Higher-order Modes

To compute Y_{in} for the waveguide #1, the input admittances $Y_i^{TE_n}$ and $Y_i^{TM_n}$, of the $TE_{1,2n}$ and $TM_{1,2n}$ ($n \geq 1$) modes, are obtained. The process is identical to the previous case for the TE_{10} mode.

$Y_i^{TE_n}$ for the $TE_{1,2n}$ Mode

Figure (10.9) shows the three-step process of obtaining $Y_i^{TE_n}$ at $z = 0$ for waveguide #1. In the air-filled waveguide section of length h_1, the $TE_{1,2n}$ higher-order modes are in the cutoff region, giving the *imaginary inductive characteristic impedance*. Hence, the waveguide #1 is terminated in the *inductive load* at $z = (h_2 + h_3)$. The inductive load at $z = (h_2 + h_3)$ is also achieved if the $TE_{1,2n}$ mode supporting the waveguide section of length h_1 is terminated in a short-circuit [J.1]. Figure (10.9a) shows the *1st step* to compute the input admittance of section h_1 for load $Y_L \to \infty$:

$$Y_{in1}^{TEn} = Y_{01}^{TEn}\coth(\gamma_{n1} h_1). \tag{10.3.105}$$

Figure (10.9b) of the *2nd step* shows that the Y_{in1}^{TEn} is connected as a load to the output of section h_2. It can be expressed as a short-circuited line of extension θ_2. The input admittance Y'^{TEn}_{in1} due to line length θ_2 is

$$Y'^{TEn}_{in1} = Y_{02}^{TEn}\coth\theta_2 = Y_{01}^{TEn}\coth(\gamma_{n1} h_1)$$
$$\Rightarrow \coth\theta_2 = \frac{Y_{01}^{TEn}}{Y_{02}^{TEn}}\coth(\gamma_{n1} h_1). \tag{10.3.106}$$

However, using equations (10.3.38) and (10.3.44), the ratio of the characteristic admittances are

$$\frac{Y_{01}^{TEn}}{Y_{02}^{TEn}} = \frac{Y_n}{Y_{n2}} = \frac{F_n}{F_{n2}} \quad \text{(a)}$$
$$\gamma_n h_1 = \gamma_{n1} h_1 = \frac{2\pi n}{b} F_n h_1 \quad \text{(b)}. \tag{10.3.107}$$

The input admittance Y_{in2}^{TEn} for the line length $(\gamma_{n2} h_2 + \theta_2)$ is

$$Y_{in\,2}^{TEn} = Y_{02}^{TEn}\coth(\gamma_{n2} h_2 + \theta_2) \quad \text{(a)}$$
$$\text{where,} \quad \gamma_{n2} h_2 = \frac{2\pi\,n}{b} F_{n2} h_2 \quad \text{(b)}. \tag{10.3.108}$$

The 3rd step of Fig (10.9c) shows that Y_{in2}^{TEn} is a load connected to line section h_3. The load Y_{in2}^{TEn} is replaced by a

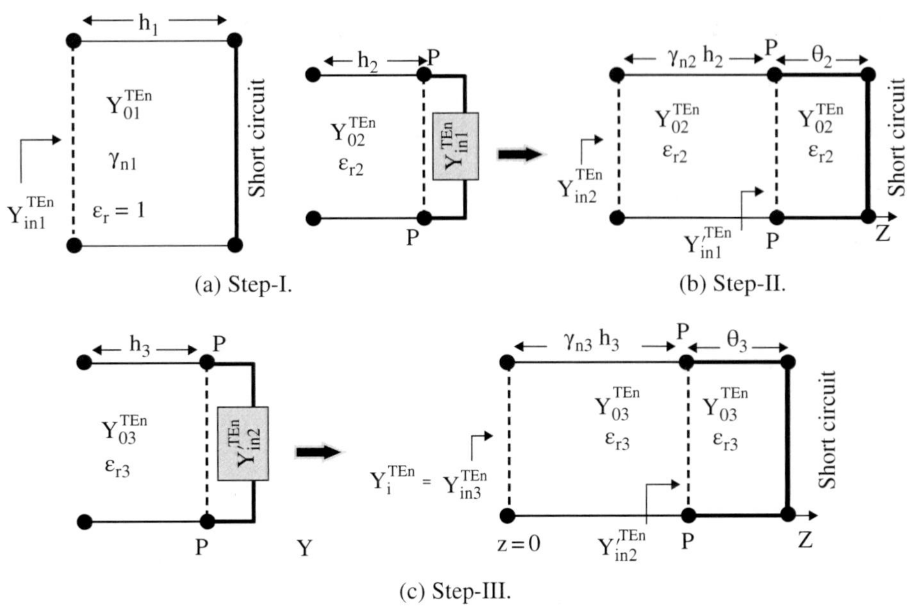

Figure 10.9 Computation of the input admittance $Y_i^{TE_n}$ of the equivalent transmission line of $TE_{1,2n}$ mode.

short-circuited line of length θ_3. The input admittance of line length θ_3 is

$$Y'^{TEn}_{in\,2} = Y^{TEn}_{03}\coth(\theta_3) = Y^{TEn}_{02}\coth(\gamma_{n2}h_2 + \theta_2)$$

$$\Rightarrow \coth\theta_3 = \frac{Y^{TEn}_{02}}{Y^{TEn}_{03}}\coth(\gamma_{n2}h_2 + \theta_2) \qquad (a)$$

$$\text{where,} \quad \frac{Y^{TEn}_{02}}{Y^{TEn}_{03}} = \frac{F_{n2}}{F_{n3}} \qquad (b). \qquad (10.3.109)$$

Figure (10.9c) shows that finally the input admittance Y^{TEn}_{in3} is

$$Y^{TEn}_{in\,3} = Y^{TEn}_{03}\coth(\gamma_{n3}h_3 + \theta_3) \qquad (a)$$

$$\text{where} \quad Y^{TEn}_{03} = \frac{a\gamma_{n3}}{j\,2b\eta_3 k_3} \qquad (b). \qquad (10.3.110)$$

The input admittance Y^{TEn}_{03} is obtained from equation (10.3.32a). The θ_2 and θ_3 are replaced to get the expression for the input admittance Y^{TEn}_{i} of the waveguide #1 for the $TE_{1,2n}$ mode at $z = 0$.

$$Y^{TEn}_{i} = Y^{TEn}_{in\,3} = Y^{TEn}_{03}\coth\left[\frac{2\pi\, nF_{n3}h_3}{b} + \coth^{-1}\left\{\frac{F_{n2}}{F_{n3}}\coth\left(\frac{2\pi nF_{n2}h_2}{b} + \coth^{-1}\left\{\frac{F_n}{F_{n2}}\coth\left(\frac{2\pi\, nF_n}{b}h_1\right)\right\}\right)\right\}\right] \qquad (10.3.111)$$

$$\text{where,} \quad F_{n1} = F_n = \sqrt{1 + \left(\frac{bv}{2anp}\right)^2}, \quad F_{n2} = \sqrt{1 - \left(\frac{bu_2}{2anp}\right)^2}, \quad F_{n3} = \sqrt{1 - \left(\frac{bu_3}{2anp}\right)^{-2}}. \qquad (10.3.112)$$

The variables v, u_2, and u_3 are given in equation (10.3.102).

Input Admittance Y_i^{TMn}

The similar three-step process, shown in Fig (10.10), is followed to obtain the input admittance of higher-order $TM_{1,2n}$ modes, i.e. Y^{TMn}_{in1} of the waveguide #1. Again, $TM_{1,2n}$ higher-order modes are in cutoff for the waveguide #1 section h_1 that is short-circuited at the end. It is the *1st step* and shown in Fig (10.10a). However, the *characteristic admittance* of the line section h_1, supporting the evanescent $TM_{1,2n}$ modes, is *capacitive*. Hence, the line section h_1 offers a *capacitive load*, Y^{TMn}_{in1}, to the line section h_2, instead of the inductive load, as in the case of the evanescent $TE_{1,2n}$ modes discussed previously. The capacitive load Y^{TMn}_{in1} is given by

$$Y^{TMn}_{in1} = Y^{TMn}_{01}\coth(\gamma_{n1}h_1). \qquad (10.3.113)$$

Figure (10.10b) of the *2nd step* shows that the input admittance Y^{TMn}_{in1} is connected as a load to the line section h_2. The capacitive load Y^{TMn}_{in1} is further treated as an open-ended line of length θ_2, i.e. $Y_L \to 0$ using equation (10.3.94).

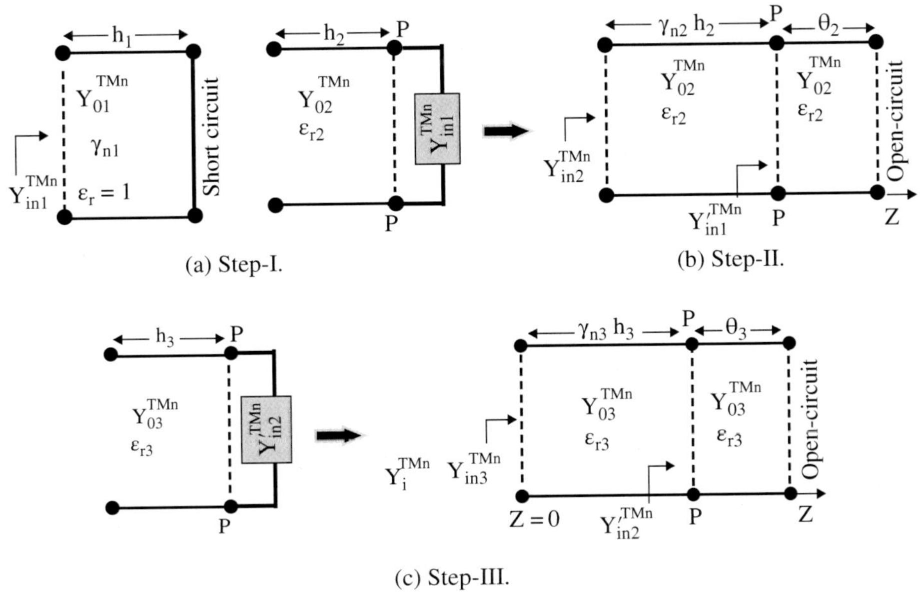

Figure 10.10 Computation of the input admittance of the equivalent transmission line of TM_n mode.

$$Y'^{TMn}_{in1} = Y^{TMn}_{02}\tanh(\theta_2) = Y^{TMn}_{in1} = Y^{TMn}_{01}\coth(\gamma_{n1}h_1)$$
$$\Rightarrow \tanh\theta_2 = \frac{Y^{TMn}_{01}}{Y^{TMn}_{02}}\coth(\gamma_{n1}h_1). \tag{10.3.114}$$

Following equations (10.3.33) and (10.3.44), the following expression is obtained:

$$\frac{Y^{TMn}_{01}}{Y^{TMn}_{02}} = \frac{Y^{TMn}_{0}(\text{For air}-\text{filled waveguide})}{Y^{TMn}_{02}} = \frac{1}{\varepsilon_{r2}}\frac{F_{n2}}{F_n}. \tag{10.3.115}$$

$$\tanh\theta_2 = \frac{1}{\varepsilon_{r2}}\frac{F_{n2}}{F_n}\coth(\gamma_{n1}h_1), \tag{10.3.116}$$

where $\gamma_{n1}h_1$ is given by equation (10.3.107b). Figure (10.10b) shows the input admittance Y^{TMn}_{in2} of the open-circuited line of length $(\gamma_{n2}h_2 + \theta_2)$:

$$Y^{TMn}_{in2} = Y^{TMn}_{02}\tanh(\gamma_{n2}h_2 + \theta_2). \tag{10.3.117}$$

Again, Fig (10.10c) of the *3rd step* shows that admittance Y^{TMn}_{in2} is connected as a load to the line section h_3. The Y^{TMn}_{in2} is replaced by the open-circuited line section θ_3. Its input admittance Y'^{TMn}_{in2} is

$$Y'^{TMn}_{in2} = Y^{TMn}_{03}\tanh(\theta_3) = Y^{TMn}_{02}\tanh(\gamma_{n2}h_2 + \theta_2)$$
$$\Rightarrow \tanh(\theta_3) = \frac{Y^{TMn}_{02}}{Y^{TMn}_{03}}\tanh(\gamma_{n2}h_2 + \theta_2), \tag{10.3.118}$$

where following equation (10.3.115), the below expression is written:

$$\frac{Y^{TMn}_{02}}{Y^{TMn}_{03}} = \frac{\varepsilon_{r2}}{\varepsilon_{r3}}\frac{F_{n3}}{F_{n2}}. \tag{10.3.119}$$

$$\tanh(\theta_3) = \frac{\varepsilon_{r2}}{\varepsilon_{r3}}\frac{F_{n3}}{F_{n2}}\tanh(\gamma_{n2}h_2 + \theta_2). \tag{10.3.120}$$

The electrical length of line section h_2, $\gamma_{n2}h_2$, is given by equation (10.3.108b). Finally, the input admittance of line section h_3 at $z = 0$ is given below. The characteristic admittance Y^{TMn}_{03} is obtained from equations (10.3.33) and (10.3.45b).

$$Y^{TMn}_{in} = Y^{TMn}_{in3} = Y^{TMn}_{03}\tanh(\gamma_{n3}h_3 + \theta_3) \qquad \text{(a)}$$
$$Y^{TMn}_{03} = j\frac{\varepsilon_{r3}}{4pnF_{n3}\eta_0} \qquad \text{(b)}. \tag{10.3.121}$$

$$Y^{TMn}_{in} = \frac{j\varepsilon_{r3}}{4pn\eta_0 F_{n3}}\tanh\left[\frac{2\pi nF_{n3}h_3}{b} + \tanh^{-1}\left\{\frac{\varepsilon_{r2}}{\varepsilon_{r3}}\frac{F_{n3}}{F_{n2}}\tanh\left(\frac{2\pi\, nF_{n2}h_2}{b}\tanh^{-1}\left(\frac{1}{\varepsilon_{r2}}\frac{F_{n2}}{F_n}\coth\left(\frac{2\pi nF_n h_1}{b}\right)\right)\right)\right\}\right]. \tag{10.3.122}$$

The expressions (10.3.111) and (10.3.122) are rewritten below in terms of arguments q_n and r_n for n^{th} $TE_{1,2n}$ and $TM_{1,2n}$ modes.

$$Y_i^{TEn} = \frac{a\gamma_{n3}}{j2b\eta_0 k_0}\coth(q_n) \quad (a)$$
$$Y_i^{TMn} = \frac{j\varepsilon_{r3}}{4pnF_{n3}\eta_0}\tanh(r_n) \quad (b) \qquad (10.3.123)$$

where,

$$q_n = \frac{2\pi n F_{n3} h_3}{b} + \coth^{-1}\left\{\frac{F_{n2}}{F_{n1}}\coth\left(\frac{2\pi n}{b}F_{n2}h_2\right) + \coth^{-1}\left\{\frac{F_n}{F_{n2}}\coth\left(\frac{2\pi n F_n}{b}h\right)\right\}\right\} \quad (a)$$
$$r_n = \frac{2\pi n F_{n3} h_3}{b} + \tanh^{-1}\left\{\frac{\varepsilon_{r2}}{\varepsilon_{r3}}\frac{F_{n3}}{F_{n2}}\tanh\left(\frac{2\pi n F_{n2} h_2}{b}\right) + \tanh^{-1}\left(\frac{1}{\varepsilon_{r2}}\frac{F_{n2}}{F_n}\coth\left(\frac{2\pi n}{b}F_n h_1\right)\right)\right\} \quad (b). \qquad (10.3.124)$$

The capacitive susceptance B_d of the iris, supported on a dielectric substrate, due to the TE_{10}, $TE_{1,2n}$, and $TM_{1,2n}$ modes looking into the waveguide #1, is obtained by modifying equation (10.3.58) with help of equations (10.3.114) and (10.3.123):

$$\eta_0 B_d = Y_{i,0} + \frac{u_3^2}{2p}\ln\left(\frac{2}{\pi\delta}\right) + \frac{1}{2p}\sum_{n=1,2,\dots}\left\{\left[\frac{\varepsilon_{r3}\tanh r_n - p^2F_{n3}^2\coth q_n}{\left[1+\left(\frac{b}{2\pi n}\right)^2\right]F_{n3}} - u_3^2\right]\times\frac{\sin^2(\pi n\delta)}{n(\pi n\delta)}\right\}. \qquad (10.3.125)$$

↓ Due to TE_{10} mode, ↓ Due to iris, ↓ Due to $TE_{1,2n} + TM_{1,2n}$ modes

The input admittance $Y_{i,0}$ of the waveguide #1 for the TE_{10} mode is given by equation (10.3.104).

Computation B_a at z = 0 for Waveguide #2

Further, the capacitive susceptance B_a of iris, in the air-filled waveguide #2, is obtained by taking $h_2 = h_3 = 0$, and $h_1 = h_4$, and by modifying equation (10.3.58):

$$\eta_0 B_a = -\frac{av}{2b}\coth(vk_0h_4) - \frac{v^2}{2p}\ln\left(\frac{2}{\pi\delta}\right) + \frac{1}{2}\sum_{n=1,2,\dots}\left[\left[v^2\left\{1-\frac{\coth(2\pi nF_nh_4/b)}{F_n}\right\}\right]\frac{\sin^2(\pi n\delta)}{n(\pi n\delta)^2}\right]. \qquad (10.3.126)$$

At $z = p$, the total susceptance $B_t = B_d + B_a$ is

$$\eta B_t = \frac{a}{2b}\left\{-v\coth(vk_0h_4) + u_3\tan\left[\frac{\pi h_3 u_3}{ap} + \tan^{-1}\left\{\frac{u_2}{u_3}\tan\left(\frac{\pi h_2 u_2}{ap} - \tan^{-1}\left(\frac{v}{u_2}\coth(vk_0h_1)\right)\right)\right\}\right]\right\}$$
$$+\frac{1}{p}\left\{\left(\frac{\varepsilon_r+1}{2} - p^2\right)\ln\frac{2}{\pi\delta} + \frac{1}{2},\sum_{n=1,2,\dots}^{v}\left[v^2\left(1-\frac{\coth(2\pi nF_nh_4/b)}{F_n}\right) + M_n\right]\frac{\sin^2(\pi n\delta)}{n(\pi n\delta)^2}\right\} \qquad (10.3.127)$$

where, $$M_n = \frac{\varepsilon_{r3}\tanh(r_n) - p^2F_{n3}^2\coth(q_n)}{\{1+(b/2an)^2\}F_{n3}} - u_3^2. \qquad (10.3.128)$$

Under the transverse resonance condition, $B_t = 0$. At assumed frequency, the variable p is determined for a slot line. The value of the parameter "p" leads to the determination of frequency-dependent effective relative permittivity. The frequency-dependent characteristic impedance of the present slot line structure is determined by using equations (10.3.76) and (10.3.82).

10.3.4 Characteristics of Slot Line

The slowing-factor $\lambda_g/\lambda_0 = 1/\sqrt{\varepsilon_{reff}(f)}$, or the effective relative permittivity $\varepsilon_{reff}(f)$, and the characteristic impedance of a slot line are dependent on their physical parameters: relative permittivity (ε_r), a substrate thickness (h), slot-width (s), and frequency, even if the strip conductor

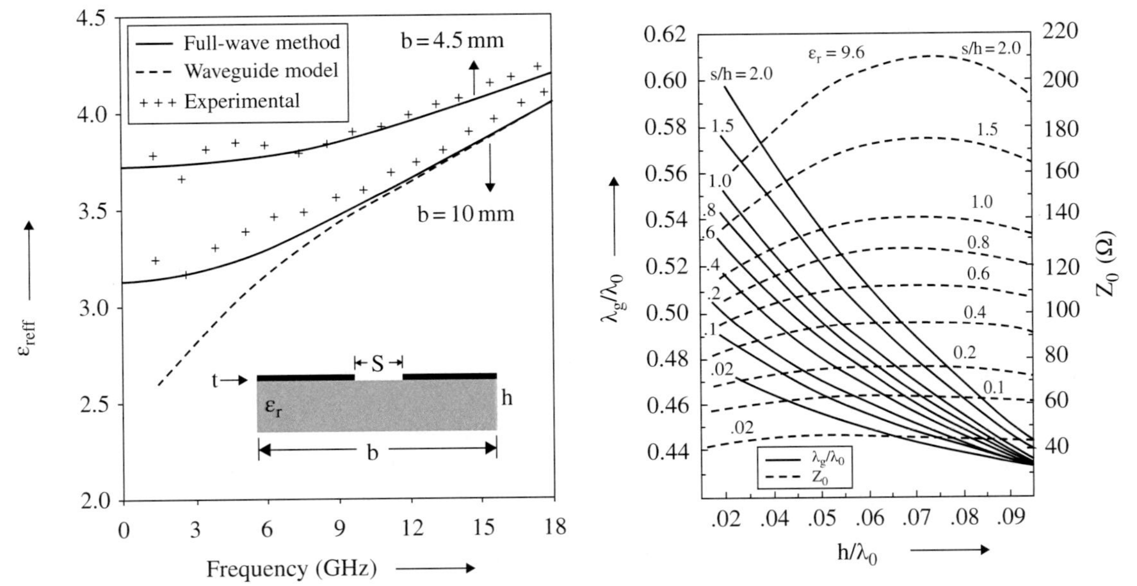

(a) Comparison of results of the full-wave method and waveguide model against experimental results [J.27].

(b) Computation of slowing-factor and Z_0 using waveguide model [J.2].

Figure 10.11 Nature of dispersion in the slot line. *Source*: Boremann [J.27]. © IEEE. Mariani et al. [J.2]. © IEEE.

thickness and losses are ignored. The waveguide model provides information on the line parameters. The modal behavior of a slot line using the waveguide model could be compared against the results obtained from the full-wave analysis of Boremann, using the scattering-type TRM [J.27]. Figure (10.11a) compares such results of $\varepsilon_{reff}(f)$ for a slot line, on the substrate $\varepsilon_r = 9.7$, $h = 0.635$, $t = 7.03$ μm, slot-width $s = 0.5$ mm, and line width b, against the experimental results. The *equivalent waveguide model* does not correctly describe the dispersion behavior of a slot line at the lower frequency, due to the presence of the cutoff phenomenon. The full-wave results show no cutoff phenomenon and follow the experimental results. It demonstrates the *quasi-TEM*-type mode of a finite width slot line, as the $\varepsilon_{reff}(f)$ tends toward the static $\varepsilon_{reff}(f \to 0)$. Therefore, the *quasi-TE$_{10}$* mode, shown by the waveguide model, is not a correct physical description of the finite width practical slot line. However, much above the cutoff frequency, the results of the waveguide model follow both the full-wave and experimental results.

Figure (10.11b) shows the results for the slowing-factor (λ_g/λ_0) and the characteristic impedance of a slot line on the substrate with $\varepsilon_r = 9.6$ using Cohn's equivalent waveguide model [J.1, J.2]. The λ_g/λ_0 curves for various slot-widths, i.e. s/h-ratio as a parameter, converge toward $h/\lambda_0 \approx 1/\left(4\sqrt{(\varepsilon_r - 1)}\right)$. The normalized substrate thickness, $h/\lambda_0 \approx 1/\left(4\sqrt{(\varepsilon_r - 1)}\right)$, corresponds to the cutoff frequency of the TE_1 surface wave mode that occurs nearly at $h = \lambda_g/4$. Thus, the slot line model is useful for the substrate thickness $h/\lambda_0 < 1/\left(4\sqrt{(\varepsilon_r - 1)}\right)$. It shows that effective relative permittivity is $\varepsilon_{reff} < (\varepsilon_r - 1)/2$. The behavior of characteristic impedance with respect to h/λ_0 is more complicated. It firstly increases, next remains flat, and finally decreases with frequency. However, its increase is obvious with an increase in the slot-width.

Figure (10.12a and b), on the substrate with $\varepsilon_r = 16$, demonstrates that *a slot line forms the complementary pair with a microstrip line*. It is obvious that the microstrip characteristic impedance decreases with an increase in strip width ($s = w$), i.e. w/h-ratio, whereas for a slot line, it increases with an increase in the slot-width ratio, i.e. s/h. Thus, the slot line is a reciprocal structure of a microstrip. An increase in the slot-width results in the slot field confinement in the air medium that increases the characteristic impedance. Figure (10.12b) shows that the effective relative permittivity of a slot line decreases with s/h-ratio. However, as a complementary pair, the effective relative permittivity of a microstrip increases with the increase of s(=w)/h-ratio. The effective relative permittivity of a slot line is always less than that of a microstrip line for the same s/h-ratio and it increases with frequency.

Kitazawa and Itoh have further compared, using the SDA, the $\varepsilon_{reff}(f)$ and the characteristic impedance of the slot line against the symmetrical CPW, on a substrate with $\varepsilon_r = 12.8$, $h = 0.1$ mm, $\tan\delta = 0.0006$, at 60 GHz [J.28].

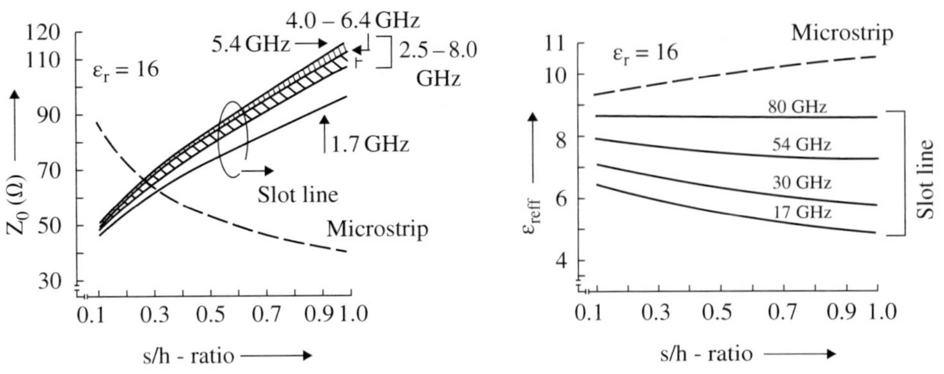

(a) Z_o of a slot line and microstrip. (b) ε_{reff} of a slot line and microstrip.

Figure 10.12 Slot line and microstrip as the complementary pairs. *Source*: Mariani et al. [J.2]. © IEEE.

Figure (10.13a) shows frequency-dependent characteristic impedance of CPW using three definitions – *power–voltage (pv), power–current (pi), and voltage–current (vi)*. The power–voltage definition, applicable to a slot line also, shows a small increase in the $Z_0(f)$ of the CPW with frequency. In the case of a slot line, the increase in the $Z_0(f)$ is very significant, indicating the highly dispersive nature of a slot line. Figure (10.13b) further compares the effective relative permittivity and Z_0 of a CPW and slot line for increasing slot-width. The characteristic impedance of CPW is always less than that of a slot line. As a matter of the fact, for a wide-slot (s = 0.4 mm) CPW, the characteristic impedance is one-half of the characteristic impedance of a slot line. Likewise, ε_{reff} of a slot line is always less than that of the CPW structure.

Simon [J.13] has shown that in the case of the suspended slot line, the nearness of the top shield to the slot line decreases its effective relative permittivity and increases its characteristic impedance. The presence of the conductor backing increases the field concentration of a slot line in the dielectric region that increases its effective relative permittivity and decreases its characteristic impedance. Boremann has obtained the frequency-dependent effective relative permittivity and characteristic impedance of the CBSL, along with the appearance of high-order mode [J.27].

10.4 Closed-form Models

A slot line of finite width conductors could be viewed as a CPS structure, supporting the *quasi-static* TEM mode. The quasi-static nature of a slot line is mentioned by Cohn and others [J.1, J.2, J.9, J.29]. So the method of moment (MOM) has been used to obtain the characteristic impedance of a slot line of finite width [J.29]. Likewise, the variational method is used to get the line parameters of the CPS [J.30]. The conformal mapping method, along with empirical expressions for the *virtual heights*, is also reported to compute the frequency-dependent effective relative permittivity and

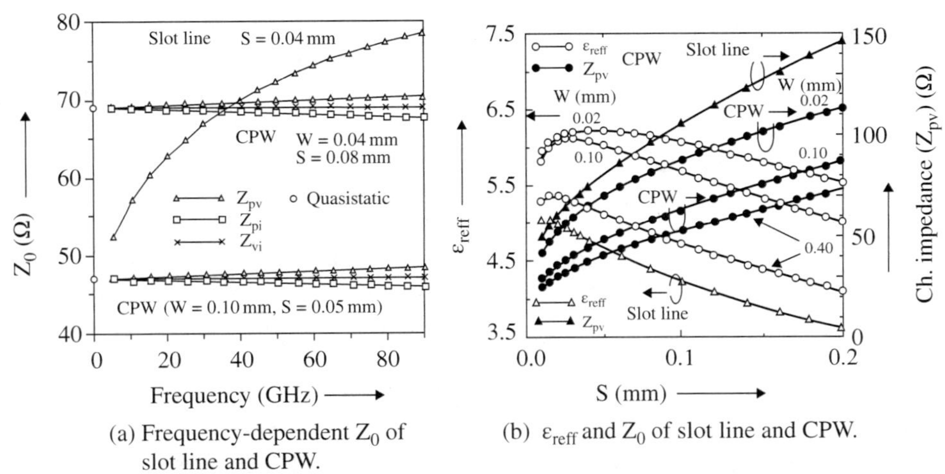

(a) Frequency-dependent Z_0 of slot line and CPW. (b) ε_{reff} and Z_0 of slot line and CPW.

Figure 10.13 Comparison of slot line against the CPW. *Source*: Kitazawa and Itoh [J.28]. © IEEE.

the characteristic impedance of three kinds of slot lines: the *standard, sandwich,* and *composite substrate* slot lines, shown in Fig (10.1) [J.9]. Krowne has given a set of closed-form expressions to compute both the frequency-dependent slowing-factor (λ_g/λ_0) and the characteristic impedance of a slot line over the range $2.2 \le \varepsilon_r \le 20$ [J.29]. Garg and Gupta have reported the curve-fitted expression to compute the slowing-factor (λ_g/λ_0) and frequency-dependent characteristic impedance of a narrow gap slot line on high permittivity substrate, over the range $9.7 \le \varepsilon_r \le 20.0$, $0.02 \le s/h \le 2.0$, as $0.01 \le h/\lambda_0 \le 1/(4\sqrt{\varepsilon_r - 1})$ [J.31]. This model has about 2% accuracy against the results of Cohn's waveguide model. Janaswamy and Schaubert [J.32, J.33] have provided another set of the closed-form expressions of a wide slot-width slot line on the low permittivity substrate, over the range $2.2 \le \varepsilon_r \le 9.8$, $0.006 \le h/\lambda_0 \le 0.06$ and $0.005 \le s/\lambda_0 \le 2.0$. This model also has about 2% accuracy against the results of SDA.

The above-mentioned closed-form models do not account for the strip conductor thickness. Moreover, the low-frequency dispersion occurs in the slot line due to the finite conductivity of the strip conductor. Such low-frequency dispersion is also not accounted for by these models. An *integrated model* of the slot line has also been reported, over the range $2.2 \le \varepsilon_r \le 20$ [J.10, J.11, J.33, J.34]. This model is a combination of the models of Garg and Gupta, and Janaswamy and Schaubert. Further, the integrated model incorporates the effect of the conductor thickness on the slot line parameters. It computes the dielectric loss, as well as the conductor loss. Next, the integrated model helps to get the *circuit model* with R(f), L(f), C(f), G(f) parameters. These line parameters compute the low-frequency dispersion of a slot line, due to the finite conductivity of strip conductors. A summary of the closed-form models is given below.

10.4.1 Conformal Mapping Method

The characteristic impedance of the slot line structures, shown in Fig (10.1a–c), is computed by using the following result of the conformal mapping [J.9]:

$$Z_0 = \frac{60\pi}{\sqrt{\varepsilon_{reff}}} \frac{K(k_0)}{K(k_0')}. \tag{10.4.1}$$

In the above equation, $K(k_0)$ and $K(k_0')$ are the *complete elliptic integrals* of the first kind and their modulus k_0 and complimentary modulus k_0' are given by

$$k_0^2 = 2 \cdot \frac{\alpha_0}{1 + \alpha_0} \quad (a), \qquad k_0'^2 = 1 - k_0^2 \quad (b)$$
$$\alpha_0 = \tanh\left[\frac{\pi s}{2h_0}\right] \quad (c). \tag{10.4.2}$$

The h_0 is the hypothetical *"virtual height"* below and above the slot, forming a region in which most of the EM-energy is confined. The virtual height h_0 is obtained empirically for three cases of the slot line.

Standard Slot Line

Figure (10.1a) shows a finite substrate thickness slot line. The virtual height is proportional to the substrate thickness h and inversely proportional to the relative permittivity ε_r of the substrate. An empirical expression for the frequency-independent virtual height is

$$h_0 = h \cdot \left[1 + \frac{133}{\varepsilon_r + 2}\right]. \tag{10.4.3}$$

The effective relative permittivity of a standard slot line is obtained from the following expression:

$$\varepsilon_{reff} = 1 + \frac{\varepsilon_r - 1}{2} \frac{K(k_1')}{K(k_1)} \frac{K(k_0)}{K(k_0')}. \tag{10.4.4}$$

The complete elliptical integral of the 1st kind is computed by the approximate expression discussed in chapter 9, section (9.2). The modulus is given by

$$k_1^2 = 2 \cdot \frac{\alpha_1}{1 + \alpha_1} \quad (a), \qquad k_1'^2 = 1 - k_1^2 \quad (b)$$
$$\alpha_1 = \tanh\left[\frac{\pi s}{2h}\right] \quad (c). \tag{10.4.5}$$

This method has accuracy within 2% for the computation of effective relative permittivity over the following range: substrate thickness $h/\lambda_0 \le 0.01$, slot-width $0.02 \le s/h \le 1$, and relative permittivity $2.22 \le \varepsilon_r \le 20$.

The *frequency-dependent effective relative permittivity* is obtained, using the above equations, by taking the virtual height h_0 as a frequency-dependent parameter through the following empirical relations:

$$h_0 = h \cdot \left[1 + \frac{0.0133}{\varepsilon_r + 1}\left(\frac{\lambda_0}{h}\right)^2\right], \tag{10.4.6}$$

where λ_0 is the free space wavelength and the slowing-factor is $\lambda_g/\lambda_0 = 1/\left(\sqrt{\varepsilon_{eff}(f)}\right)$. Once ε_{reff} is obtained, the static or frequency-dependent characteristic impedance is computed using equation (10.4.1). The model is valid for $9.7 \le \varepsilon_r \le 20$, $0.02 \le s/h \le 1.0$, and $0.01 \le h/\lambda_0 \le 0.25/\sqrt{\varepsilon_r - 1}$.

Sandwich Slot Line

Figure (10.1b) shows the sandwich slot line structure. In this case, there are two frequency-dependent virtual heights: h_{01} below the slot and h_{02} above the slot. These are given by the following empirical relations:

$$h_{01} = h_1\cdot\left[1 + \frac{0.0133}{\varepsilon_{r1} + 2}\left(\frac{\lambda_0}{h_1}\right)^2\right] \quad \text{(a)}$$

$$h_{02} = h_2\cdot\left[1 + \frac{0.0133}{\{\varepsilon_{r2} + (\varepsilon_{r1} - \varepsilon_{r2})\, h_1/h_2 + 2\}}\left(\frac{\lambda_0}{h_2}\right)^2\right] \quad \text{(b).} \tag{10.4.7}$$

The h_1 and h_2 are the thicknesses of the substrate and the superstrate. The frequency-dependent effective relative permittivity of the sandwich slot line is obtained as follows:

$$\varepsilon_{reff} = 1 + \left[(\varepsilon_{r1} - 1)\frac{K(k_1')}{K(k_1)} + (\varepsilon_{r2} - 1)\frac{K(k_2')}{K(k_2)}\right]\left[\frac{K(k_{01}')}{K(k_{01})} + \frac{K(k_{02}')}{K(k_{02})}\right]^{-1}. \tag{10.4.8}$$

Where, for $[i = 1, 2]$:

$$k_i^2 = 2\cdot\frac{\alpha_i}{1 + \alpha_i} \text{ (a)}, \quad k_i'^2 = 1 - k_i^2 \text{ (b)}, \quad \alpha_i = \tanh\left[\frac{\pi s}{2h_i}\right] \text{ (c).} \tag{10.4.9}$$

also for $[i = 1, 2]$:

$$k_{0i}^2 = 2\cdot\frac{\alpha_{0i}}{1 + \alpha_{0i}} \text{ (a)}, \quad k_{0i}'^2 = 1 - k_{0i}^2 \text{ (b)}, \quad \alpha_{0i} = \tanh\left[\frac{\pi s}{2h_{0i}}\right] \text{ (c).} \tag{10.4.10}$$

The Composite Substrate Slot Line

Figure (10.1c) shows the structure. The frequency-dependent virtual height in this case is

$$h_0 = h_2\cdot\left[1 + \frac{0.0133}{\varepsilon_{r1} + (\varepsilon_{r2} - \varepsilon_{r1})\,(h_1/H) + 2}\left(\frac{\lambda_0}{h_1}\right)^2\right]$$

$$H = h_1 + h_2. \tag{10.4.11}$$

The effective relative permittivity for this case is computed as follows:

$$\varepsilon_{reff} = 1 + \frac{\varepsilon_{r2} - \varepsilon_{r1}}{2}\cdot\left\{\frac{K(k_1')}{K(k_1)}\frac{K(k_0)}{K(k_0')}\right\} + \frac{\varepsilon_{r2} - 1}{2}\cdot\left\{\frac{K(k_2')}{K(k_2)}\frac{K(k_0)}{K(k_0')}\right\}, \tag{10.4.12}$$

where k_i and k_i' $(i = 1,2)$ are given by equation (10.4.9). The k_0 and k_0' are computed from equation (10.4.2). However, for computing α_0, equation (10.4.11) is used in equation (10.4.2c). The above expression has accuracy of 2%–3% for the sandwich slot line against the results of the waveguide model [J.1].

10.4.2 Krowne Model

Krowne [J.35] has given the closed-form expressions to compute the slowing-factor $\left(\lambda_g/\lambda_0 = 1/\sqrt{\varepsilon_{reff}(f)}\right)$ and the characteristic impedance of a slot line shown in Fig (10.1a). The model has been tested for $9.6 \le \varepsilon_r \le 20$. However, it is also suitable for the substrate of lower permittivity, such as $\varepsilon_r = 2.22$. For the wider slot-width, $s/h \ge 0.2$, the error in the computation of characteristic impedance is within 4% and for the narrow slot-width, the error goes up to 14.5%. For the computation of λ_g/λ_0, the error is within ±3.7% against the results given by Mariani et al. [J.2] for $2.22 \le \varepsilon_r \le 20$, $0.02 \le s/h \le 2.0$, and $0.015 \le h/\lambda_0 \le 0.08$.

Slowing-Factor λ_g/λ_0

The slowing-factor is computed by the following expression:

$$\frac{\lambda_g}{\lambda_0} = f_1(\varepsilon_r)\left[f_2\left(\frac{s}{h}\right) + f_3\left(\frac{s}{h}\right)\left(\frac{h}{\lambda_0}\right)^{f_4\left(\frac{s}{h}\right)} + f_5\left(\frac{s}{h}\right)\right]. \tag{10.4.13}$$

The parameters f_1 to f_5 of the above equation are given as follows:

$$\begin{aligned}
&f_1(\varepsilon_r) = 3.549\varepsilon_r^{-0.56} \quad \text{(a)}, \qquad f_2(s/h) = 0.5632\,(s/h)^{0.104(s/h)^{0.266}} && \text{(b)}\\
&f_3(s/h) = -0.8777\,(s/h)^{0.81} + 0.4233\,(s/h) - 0.2492 && \text{(c)}\\
&f_4(s/h) = -1.269\times10^{-2}[\ln(50\,(s/h))]^{1.7} + 0.0674\ln[(50\,(s/h))] + 0.2 && \text{(d)}\\
&f_5(s/h) = 1.906\times10^{-3}[\ln(50\,(s/h))]^{2.9} - 7.203\times10^{-3}\ln(50\,(s/h)) + 0.1223 && \text{(e).}
\end{aligned} \tag{10.4.14}$$

Characteristic Impedance

The characteristic impedance of a slot line is computed by the following expression:

$$Z_0 = g_1(\varepsilon_r, s/h, h/\lambda_0)\, g_2(s/h, h/\lambda_0). \tag{10.4.15}$$

The parameters of the above equation are given as follows:

$$\begin{aligned}
&g_1(\varepsilon_r, s/h, h/\lambda_0) = [11/\varepsilon_r]^T, \text{where } T = p_1\left(\frac{s}{h}, \frac{h}{\lambda_0}\right) && (a)\\
&p_1(s/h, h/\lambda_0) = [a_1(s/h)]\,(h/\lambda_0)^2 + [a_2(s/h)]\,(h/\lambda_0) + a_3(s/h) && (b)\\
&a_1(s/h) = -30.21\ \ln(s/h) - 46.03 && (c)\\
&a_2(s/h) = 0.5073\ \ln(s/h) + 3.358\,(s/h) + 6.442 && (d)\\
&a_3(s/h) = -2.013 \times 10^{-2}\ \ln(s/h) - 0.1374\,(s/h) + 0.2365 && (e).
\end{aligned} \tag{10.4.16}$$

$$\begin{aligned}
&g_2(s/h, h/\lambda_0) = [p_2(s/h)]\,(h/\lambda_0)^2 + [p_3(s/h)]\,[h/\lambda_0] + p_4(s/h) && (a)\\
&p_2(s/h) = -1.176 \times 10^4 (s/h)^{0.502} - 6.311 \times 10^3 (s/h) - 162.7 && (b)\\
&p_3(s/h) = 900.5\,(s/h)^{0.28} + 1262\,(s/h) - 123.8 && (c)\\
&p_4(s/h) = 1.637\ \ln[50\,s/h] + 40.99\,(s/h)^{0.46} + 30.96 && (d).
\end{aligned} \tag{10.4.17}$$

10.4.3 Integrated Model

The integrated model of a slot line is a closed-form model with four stages of development [J.10, J.11, J.33, J.34]. At the *first stage,* it has combined the closed-form models of Garg-Gupta [J.31], and Janaswamy and Schaubert [J.32, J.36] to compute the frequency-dependent effective relative permittivity and characteristic impedance. At this stage, conductor thickness of strips and losses are not accounted for. The combined model is written in the following parametric form:

$$\begin{aligned}
&\varepsilon_{reff}(f, t = 0) = \left(\frac{\lambda_0}{\lambda_g}\right)^2\\
&\lambda_g = \lambda_0(A_1 + B_1{\cdot}f_1 + C_1{\cdot}f_2 + D_1{\cdot}f_3 + E_1 + F_1 + G_1) && (a)\\
&Z_0(f, t = 0) = A_2 + B_2{\cdot}g_1 + C_2{\cdot}g_2 + D_2{\cdot}g_3 + E_2 + F_2 && (b).
\end{aligned} \tag{10.4.18}$$

The set of parameters involved in the above expressions, in terms of the physical parameters of the slot line, have been obtained from the models of Garg-Gupta and Janaswamy and Schaubert. These are available in the Appendices I and II at the end of chapter 10 [J.10, J.11]. The combined model is valid over a wider range of parameters, as compared to individual models, i.e. over the range $2.22 \le \varepsilon_r \le 20$, $0.02 \le w/h \le 1.0$ (i.e. $2 \times 10^{-4} \le w/\lambda_0 \le 0.25/\sqrt{(\varepsilon_r - 1)}$), $0.01 \le h/\lambda_0 \le 0.25/\sqrt{(\varepsilon_r - 1)}$. The combined model has an average accuracy of 2% against the full-wave results. Figure (10.14a and b) illustrates one such comparison, for three closed-form models, against the full-wave results [J.11, J.37]. The integrated model follows more closely the full-wave results. The model of Svacina for the characteristic impedance does not follow the full-wave results.

At the *second stage* of development, the integrated model accounts for the effect of the conductor thickness on frequency-dependent effective relative permittivity and characteristic impedance. The strip conductor thickness reduces both the effective relative permittivity and the characteristic impedance of a slot line. However, its effect, in the integrated model, is accounted for in different ways. The frequency- and thickness-dependent effective relative permittivity, i.e. $\varepsilon_{reff}(f, t)$, is computed as follows:

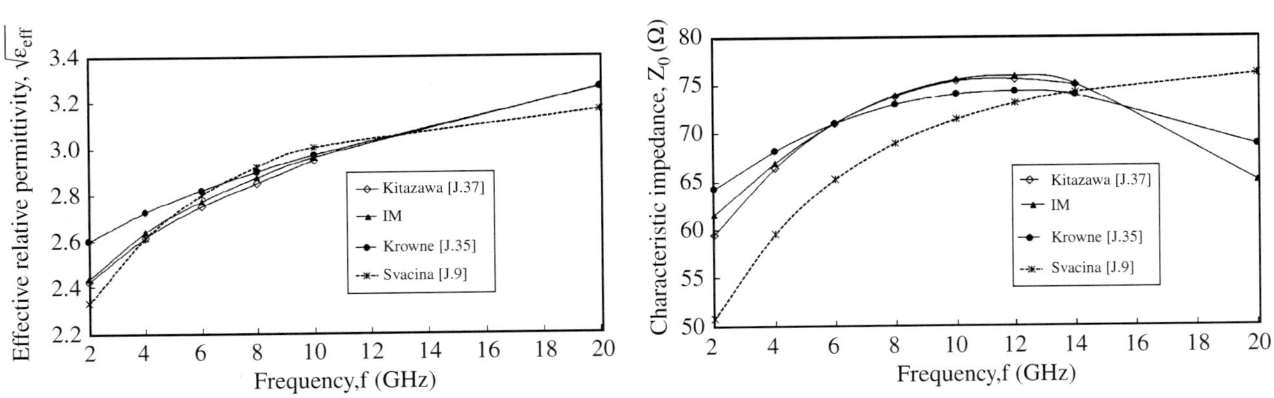

(a) Frequency-dependent ε_{eff} (f).

(b) Frequency-dependent Z_0 (f).

Figure 10.14 Comparison of three models of slot line, ε_r=20, s/h=0.5, t=0, against the full-wave results. *Source*: Majumdar and Verma [J.11]. © Springer.

$$\varepsilon_{reff}(f,t) = \begin{cases} \varepsilon_{reff}(f,t=0) - \left(\dfrac{\varepsilon_r - 1}{4.6}\dfrac{t/h}{\sqrt{s/h}}\right), & \text{for } 4\times10^{-5} \le \dfrac{t}{\lambda_0} < 6.67\times10^{-4} \quad \text{(a)} \\ \varepsilon_{reff}(f,t=0) - \left\{\left(\dfrac{\varepsilon_r - 1}{4.6}\dfrac{t/h}{\sqrt{s/h}}\right) - \left(\dfrac{t}{\lambda_0}\right)^{p_0}\right\}, & \text{for } 6.67\times10^{-4} \le \dfrac{t}{\lambda_0} < 3.3\times10^{-3} \end{cases} \quad (10.4.19)$$

where, $p_0 = 0.0006f^2 - 0.0369f + 0.7714$ (b).

In the above expression, the frequency-dependent effective relative permittivity $\varepsilon_{reff}(f, t = 0)$ is computed from equation (10.4.18).

To compute the frequency- and thickness-dependent characteristic impedance, i.e. $Z_0(f, t)$, reduced size of the *equivalent slot gap* $s_{eq} = s - \Delta s$ is used, in place of the physical gap s in equation (10.4.18a). The same process can be used with the models of Krowne and also Svacina discussed it previously. The incremental reduction in the slot-gap can be computed from the following empirical relations:

$$\Delta s = \begin{cases} \dfrac{t}{2\pi}\left[1 + \dfrac{1}{\varepsilon_r}\right] \times \ln\left\{\dfrac{10.872}{(t/\lambda_0)^{q_0} + \left(\dfrac{1/\pi}{(s/h)+1.1}\right)^2}\right\}, & \text{for } 4\times10^{-5} \le \dfrac{t}{\lambda_0} < 3.6\times10^{-4} \\ \dfrac{t}{2\pi}\left[1 + \dfrac{1}{\cosh\left(\sqrt{\varepsilon_r - 1}\right)}\right] \times \ln\left\{\dfrac{10.872}{(t/\lambda_0)\coth^2\left(\sqrt{6.517(s/h)}\right)}\right\}, & \text{for } 3.6\times10^{-4} \le \dfrac{t}{\lambda_0} < 6\times10^{-4} \quad \text{(a)} \\ \dfrac{t}{\pi}\left[1 + \ln(4) - 0.5\ln\left(\left[\dfrac{t}{\lambda_0}\right]^{q_0} + \left[\dfrac{t}{\pi s}\right]^2\right)\right], & \text{for } 6\times10^{-4} \le \dfrac{t}{\lambda_0} \le 3.33\times10^{-3} \end{cases}$$

$$\text{where, } q_0 = \begin{cases} 2, \text{ for } f < 18 \text{ GHz} \\ 0.092\left(f^{0.4184}\right), f \ge 18 \text{ GHz} \end{cases} \quad \text{(b)}. \qquad (10.4.20)$$

Figure (10.15a–d) compares results of the integrated model, for the frequency- and thickness-dependent effective relative permittivity and the characteristic impedance of a slot line, against the full-wave results from the EM-simulators (HFSS, Sonnet). The HFSS excites a slot line through the *wave-port* (WP). However, Fig (10.15a and c) demonstrates that the results of HFSS (WP) do not follow the full-wave results of the SDA [J.37]. So, the slot line is considered as a limiting case of the CPW with a central strip conductor $w \to 0$; i. e. $w = 0.1$ μm for minimum slot-width $s = 10$ μm. The HFSS (CPW) provides a correct result for the slot line [J.10]. Normally, the results of the integrated model are in between the results of two EM-Simulators.

At the *third stage* of the development of the integrated model, computations of the dielectric and conductor losses are incorporated in the integrated model. These are due to the imperfect dielectric substrate and imperfect conductors. At the mm-wave range, the radiation loss and leakage loss through surface wave mode also become important [J.38]. However, it is not included in the integrated model. The dielectric loss of a slot line is computed using the following expression, obtained in chapter 8:

$$\alpha_d = \frac{\pi}{\lambda_0}\frac{\varepsilon_r[\varepsilon_{reff}(f,t) - 1]}{\sqrt{\varepsilon_{reff}(f,t)}\cdot(\varepsilon_r - 1)}\tan\delta \text{ Np/m.} \qquad (10.4.21)$$

The $\varepsilon_{reff}(f, t)$ is computed from equation (10.4.19). Figure (10.16a) compares the dielectric loss of a slot line between 2 GHz and 30 GHz, computed by the integrated model against two full-wave results [J.38, J.39]. The average and maximum deviations of the integrated model are 0.011 Np/m and 0.041 Np/m, respectively.

The integrated model computes the conductor loss of a slot line using the *perturbation method* with the concept of stopping distance Δ. The method applies to the microstrip and the CPW also. It is discussed in chapter 9. A slot line could be treated as a limiting case of the CPW structure by reducing the width of its central conductor to zero. Under this limiting case, the conductor loss of a slot line is computed from the following expression [J.10]:

$$\alpha_c \approx \frac{R_{sm}}{16\, Z_0(f)\, K^2(k_0)\, s}\ln\left(\left(\frac{2s}{\Delta} + 1\right)\left(\frac{s+\Delta}{s-\Delta}\right)\right) \text{ Np/m,} \qquad (10.4.22)$$

where modulus k_0 is given by equation (10.4.2).

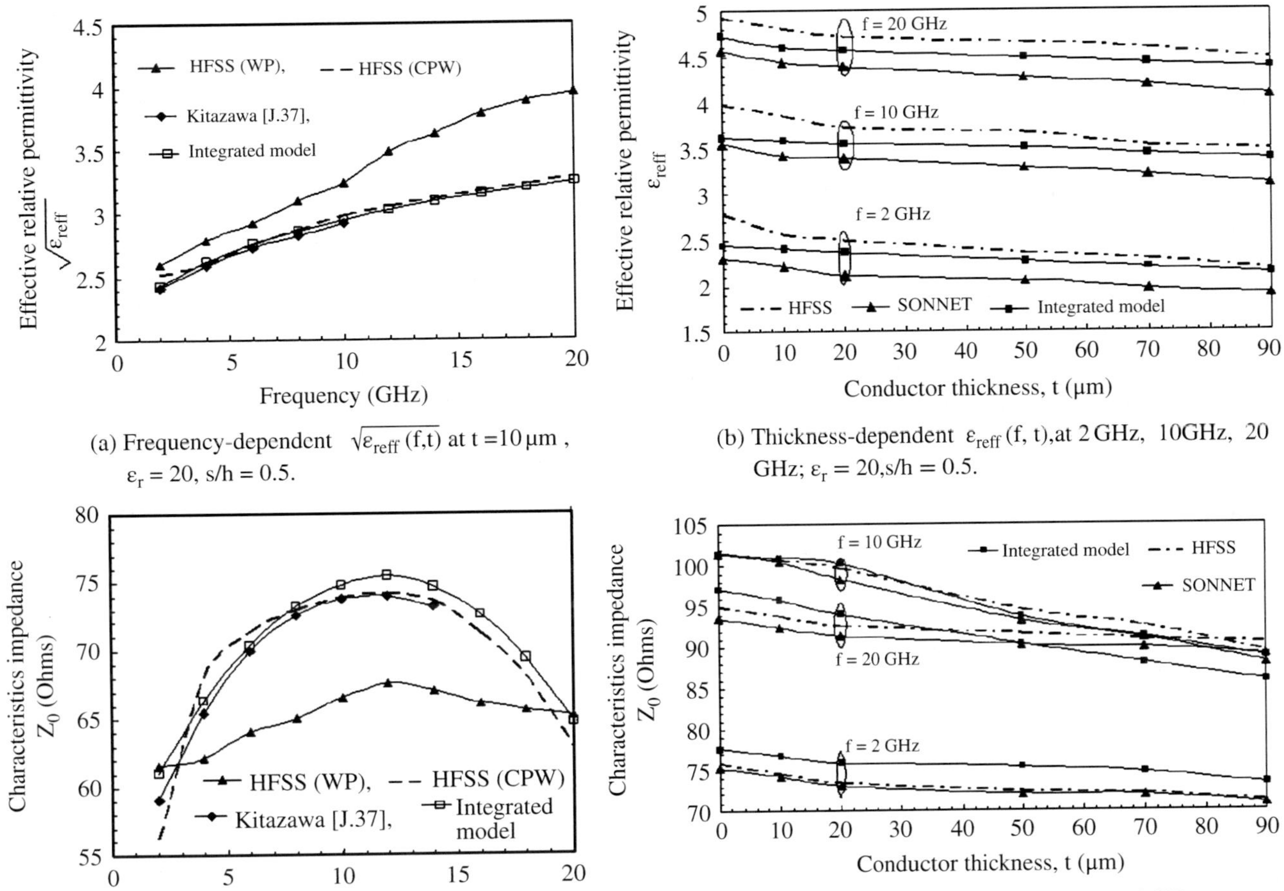

(a) Frequency-dependent $\sqrt{\varepsilon_{reff}}$ (f,t) at t = 10 μm, $\varepsilon_r = 20$, s/h = 0.5.

(b) Thickness-dependent ε_{reff} (f, t), at 2 GHz, 10GHz, 20 GHz; $\varepsilon_r = 20$, s/h = 0.5.

(c) Frequency-dependent Z_0 (f, t) at t = 20 μm, $\varepsilon_r = 20$, s/h = 0.5.

(d) Thickness-dependent Z_0 (f, t) at 2 GHz, 10 GHz, 20 GHz; $\varepsilon_r = 9.8$, s/h = 0.5.

Figure 10.15 Comparison of the integrated model against HFSS and Sonnet. *Source*: Majumdar and Verma [J.10]. © IET.

The characteristic impedance $Z_0(f)$ is computed using equation (10.4.18b). The surface impedance R_{sm} of the strip conductor of thickness t and conductivity σ_c in S/m is computed from the following expression:

$$R_{sm} = \omega\mu_c t \,\mathrm{Im}\left(\frac{\cot(k_c t) + \csc(k_c t)}{k_c t}\right) \quad (a)$$

$$k_c = \omega\sqrt{\mu_0\varepsilon_0}\left[1 - j\frac{\sigma_c}{\omega\varepsilon_0}\right]^{1/2} \quad (b). \qquad (10.4.23)$$

Finally, the stopping distance for a slot line is computed from the following empirical relation:

$$y = T_1 e^{T_2 x} \ (a), \qquad \text{where, } T_1 = 12.093e^{(-25.101w)} \quad (b)$$

$$T_2 = -50w^4 + 30w^3 + 30w^2 - 0.691w + 1.0109 \quad (c)$$

$$\text{where, } y \equiv (t/\Delta) \text{ and } x \equiv (t/2\delta_s);\ \ 0.816 < x < 7.3 \quad (d). \qquad (10.4.24)$$

In the above equation, the slot-width s is in mm.

Wheeler's incremental inductance rule, discussed in chapter 8, can also be used to compute the conductor loss of a slot line [J.11]. However, it does not apply to the conductor thickness near and less than the skin-depth. Figure (10.16b) compares the conductor loss computations, using the integrated model and Wheeler's model, against the full-wave results [J.39]. The results of two EM-simulators are also used in comparison. Results of the integrated model, full-wave method, and Sonnet have average and maximum deviations (4.91%, 7.37%), (15.1%, 27.1%), and (2.07%, 4.49%), respectively, against the results of HFSS.

At the *final stage* of the integrated model, the secondary line parameters are used to obtain the R(f), L(f), C(f), G(f) parameters of a circuit model of the slot line. The process is discussed in section (9.8) of chapter 9 for a CPW. The circuit model helps us to get low-frequency dispersion for both the effective relative permittivity and characteristic impedance. It also provides the imaginary part of the complex characteristic impedance. The accuracy of the circuit model is

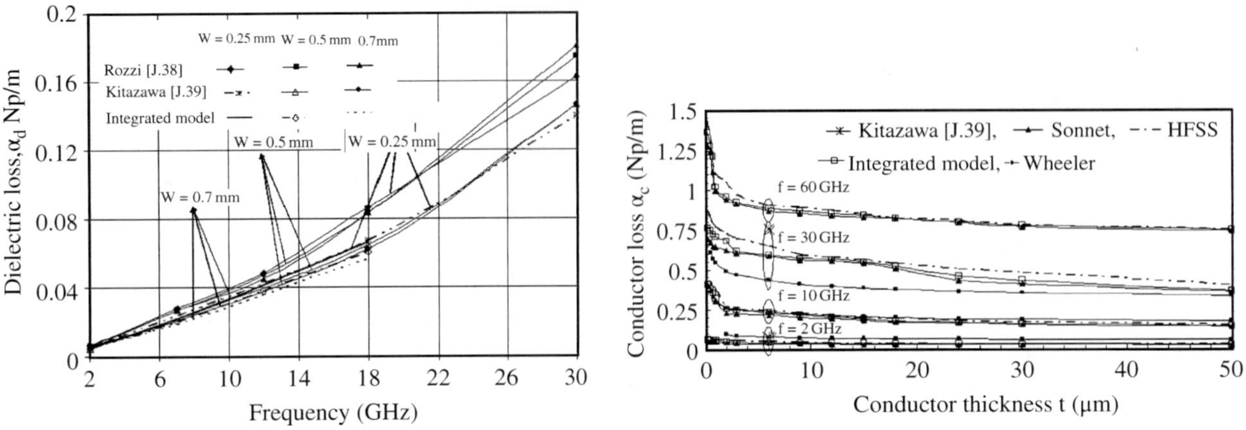

(a) Comparison of computed dielectric loss for $\varepsilon_r = 9.8$, $t = 6\ \mu m$, $h = 0.635\ mm$[J.11].

(b) Comparison of computed conductor loss for $\varepsilon_r = 9.8$, $t = 6\ \mu m$, $h = 0.635\ mm$[J.10].

Figure 10.16 Comparison of the integrated model, for loss computation, against full-wave methods. *Source*: Majumdar and Verma [J.11] © Springer. Majumdar and Verma [J.10]. © IET.

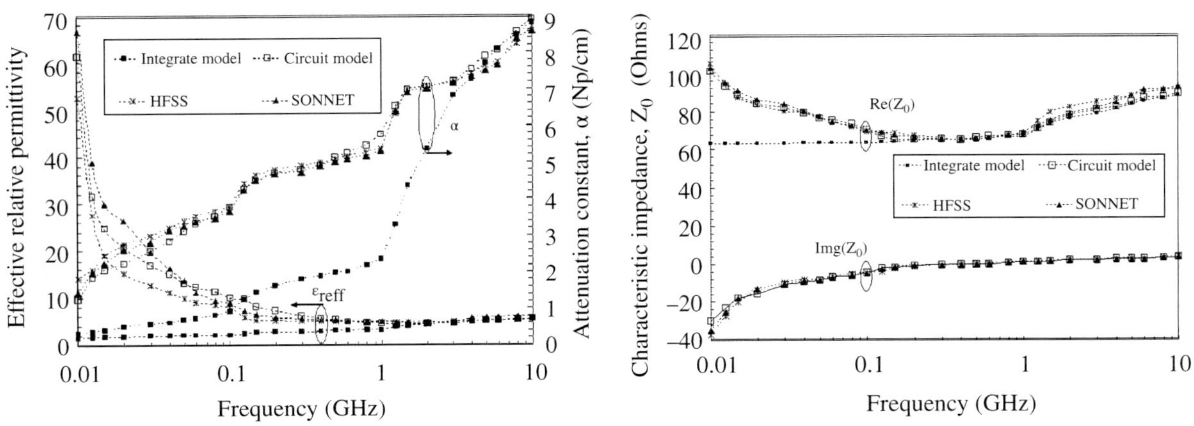

(a) $\varepsilon_{reff}(f, t)$ and total attenuation.

(b) $ReZ_0(f, t)$ and total Img $Z_0(f, t)$.

Figure 10.17 Low-frequency dispersion in a slot line ($\varepsilon_r = 12.9$, $t = 6\ \mu m$, $s/h = 0.5$).

higher for computation of line parameters, including losses, compared to the accuracy of individual components of the integrated model. Figure 10.17a and b compares the results of the slot line parameters obtained from both the integrated model and circuit model against two EM-Simulators.

References

Books

B.1 Hoffmann, R.: *Microwave Integrated Circuit Handbook*, Artech House, Boston, 1985.

B.2 Gupta, K.C.; Garg, R.; Bahl, I.; Bhartia, P.: *Microstrip Lines and Slot Lines*, 2nd Edition, Artech House, Boston, 1996.

B.3 Di Paolo, F.: *Network, and Devices Using Planar Transmission Lines*, CRC Press, New York, 2000.

B.4 Bhat, B.; Koul, S.K.: *Analysis Design and Applications of Fin-Lines*, Artech House, Boston, 1987.

B.5 Rozzi, T.; Mongiardo, M.: *Open Electromagnetic Waveguides*, IEE Press, UK, 1997.

B.6 Mirshekar-Syahkal, D.: *Spectral Domain Method for Microwave Integrated Circuits*, Research Studies Press Ltd., John Wiley & Sons, New York., 1990.

B.7 Itoh, T. (Editor): *Numerical Techniques for Microwave and Millimetre-Wave Passive Structures*, John Wiley & Sons, New York, 1989.

B.8 Jordan, E. C.; Balmain, K.G.: *Electromagnetic Wave and Radiating System*, Prentice-Hall India, New Delhi, 1989.

B.9 Ramo, S., Whinnery, J. R., Van Duzer, T.: *Fields, and Waves in Communication Electronics*, 3rd Edition, John Wiley & Sons, Singapore, 1994.

B.10 Balanis, C.A.: *Advanced Engineering Electromagnetics*, John Wiley & Sons, New York, 1989.

B.11 Collin, R.E.: *Field Theory of Guided Waves*, IEEE Press, New York, 1991.

B.12 Abramowitz, M.; Stegun, I.A.: *Handbook of Mathematical Functions with Formula, Graphs, and Mathematical Tabless*, Martino Publishing, Eastford, CT, 2014.

B.13 Marcuvitz, N.: *Waveguide Handbook*, M.I.T. Radiation Laboratory Series, Vol. 10, pp. 218–219, McGraw-Hill, New York, 1951.

Journals

J.1 Cohn, S.B.: *Slot line on a dielectric substrate, IEEE Trans. Microwave Theory Tech.*, Vol. MTT- 17, No. 10, pp. 768–778, Oct. 1969.

J.2 Mariani, E.O.; Heinzman, C.P.; Agrios, J.P.; Cohn, S.B.: *Slot line characteristics, IEEE Trans. Microwave Theory Tech.*, Vol. MTT- 17, No. 12, pp. 1091–1096, Dec. 1969.

J.3 Hunton, J.K.: *Novel contributions to microwave circuit design*, IEEE, MTT-S, Int. Microwave Symp. Dig., pp. 753–755, 1989.

J.4 Donn, C.: *A miniature, circularly polarized cross-slot line antenna*, IEEE AP-S Int. Symp. Dig., (San Jose, CA), pp. 1328–1331, June 1989.

J.5 Povinelli, M.J.; Johnson, J.A.: *Design and performance of wideband dual-polarized stripline notch arrays*, IEEE AP-S Int. Symp. Dig., (Syracuse, NY), pp. 200–203, June 1988.

J.6 Robinson, G.H.; Allen, J.I.: *Slot line application to miniature ferrite devices, IEEE Trans. Microwave Theory Tech.*, Vol. Vol. MTT- 17, No. 12, pp. 1097–1101, Dec. 1969.

J.7 Samardzija, N.; Itoh, T.: *Double-layered slot line for millimeter-wave integrated circuits, IEEE Trans. Microwave Theory Tech.*, Vol. MTT- 24, No. 11, pp. 827–831, Nov. 1976.

J.8 Abdel-Moniem, A.; El-Sherbiny: *Millimeter-wave performance of shielded slot-lines, IEEE Trans. Microwave Theory Tech.*, Vol. MTT- 30, No. 5, pp. 750–756, May 1982.

J.9 Svacina, J.: *Dispersion characteristics of multilayered slot lines—a simple approach, IEEE Trans. Microwave Theory Tech.*, Vol. MTT- 20, No. 2, pp. 1826–1829, Sept. 1999.

J.10 Majumdar, P.; Verma, A.K.: *Integrated closed-form model and circuit model of lossy slot line, IET Microw. Antennas Propag* Vol. 5, No. 14, pp. 1763–1772, 2011.

J.11 Majumdar, P.; Verma, A.K.: *Modified closed-form dispersion and loss models of slot-line with conductor thickness, J Infrared Milli Terahz Waves*, Vol. 31, pp. 271–287, 2010.

J.12 Cohn, S.B. *Sandwich slot line, IEEE Trans. Microwave Theory Tech.*, Vol. MTT- 19, No. 9, pp. 773–774, Sept. 1971.

J.13 Simons, R.: *Suspended slot line using double-layer dielectric, IEEE Trans. Microwave Theory Tech.*, Vol. MTT- 29, No. 10, pp. 1102–1107, Oct. 1981.

J.14 Han, K.J.; Kim, J.P.: *Guide wavelength estimation of sandwich-type double-layer slot line, Microwave Opt. Technol. Lett.*, Vol. 43, No. 3, pp. 261–264, Nov. 2004.

J.15 Coetzee, J.C.; Malherbe, J.A.G.: *Characteristic impedance for double-sided slot lines, Microwave Opt Technol Lett.*, Vol. 3, No. 3, pp. 85–88, Mar. 1990.

J.16 Coetzee, J.C.; Malherbe, J.A.G.: *Dispersion characteristics for double-sided slot lines, Electron. Lett.*, Vol. 25, No. 20, pp. 1383–1385, Sept. 1989.

J.17 Janaswamy, R.: *Even-mode characteristics of the bilateral slot line, IEEE Trans. Microwave Theory Tech.*, Vol. MTT- 38, No. 6, pp. 760–765, June 1990.

J.18 Simons, R.N.; Taub, S.; Lee, R.Q.; Young, P.G.: *Microwave characterization of slot line and coplanar silicon for a slot antenna feed network, Microwave Opt Technol. Lett.*, Vol. 7, No. 11, pp. 489–494, Aug. 1994.

J.19 Shigesawa, H.; Tsuji, M.; Oliner, A.A.: *Conductor backed slot line and coplanar waveguide: dangers and full-wave analysis*, IEEE MTT-S Int. Microwave Symp. Dig., vol. 1, pp. 199–202, June 1988.

J.20 Zehentner, J.; Machac, J.; Mrkvica, J: *Novel selected modes on the conductor-backed slot line*, IEEE MTT-S Int. Microwave Symp. Dig., pp. 961–964, Seattle, June 2002.

J.21 Zehentner, J.; Machac, J.; Mrkvica, J.; Tuzi, C.: *The inverted conductor-backed slot line – a challenge to the antenna and circuit design*, Proc. 33rd European Microwave Conf., (Munich), pp. 73–76, 2003.

J.22 Zehentner, J.; Machac, J.; Mrkvica, J.; Tuzi, C.: *Modes on the standard and inverted conductor-backed slot line*, IEEE MTT-S Int. Microwave Symp. Dig., vol. 2, pp. 677–680, June 2003.

J.23 Zehentner, J.; Machac, J.; Mrkvica, J.: *Even and odd modes on a conductor backed slot line*, Proc. of 32nd European Microwave Conf., vol. 2, pp. 609–612, Sept. 2002.

J.24 Das, N.K.; Pozar, D.M.: *Full-wave spectral-domain computation of material, radiation, and guided wave losses in infinite multilayered printed transmission lines, IEEE Trans. Microwave Theory Tech.*, Vol. MTT- 39, No.1, pp. 54–63, Jan. 1991.

J.25 Simons, R.: *Suspended coupled slot line using double-layer dielectric, IEEE Trans. Microwave Theory Tech.*, Vol. MTT-29, No. 2, pp. 162–165, Feb. 1981.

J.26 Simons, R.: *Suspended broadside- coupled slot line with overlay, IEEE Trans. Microwave Theory Tech.*, Vol. MTT- 30, No. 1, pp. 76–81, Jan. 1982.

J.27 Boremann, J.: *A scattering-type transverse resonance formulation and its application to open, conductor-backed and shielded slot line (M) MIC structures*, IEEE MTT-S Int. Microwave Symp. Digest, pp. 695–698, June 1991.

J.28 Kitazawa, T; Itoh, T.: *Propagation characteristics of co-planar type transmission lines with lossy media, IEEE Trans. Microwave Theory Tech.*, Vol. MTT- 39, No. 10, pp. 1694–1700, Oct. 1991.

J.29 Lee, J.J.: *Slot line impedance, IEEE Trans. Microwave Theory Tech.*, Vol. MTT- 39, No. 4, pp. 666–672, April 1991.

J.30 Yamashita, E.; Yamazaki, S.: *Parallel-strip line embedded in or printed on a dielectric sheet, IEEE Trans. Microwave Theory Tech.*, Vol. MTT- 16, No. 11, pp. 792–793, Nov. 1968.

J.31 Garg, R.; Gupta, K.C.: *Expression for wavelength and impedance of slot line, IEEE Trans. Microwave Theory Tech.*, Vol. MTT- 24, No. 8, pp. 532, Aug. 1976.

J.32 Janaswamy, R.; Schaubert, D.H.: *Characteristic impedance of a wide slot line on low-permittivity substrates, IEEE Trans. Microwave Theory Tech.*, Vol. MTT- 34, No. 8, pp. 900–902, Aug. 1986.

J.33 Majumdar, P; Verma, A K.: *Propagation characteristics of slot – line in lossy media with conductor thickness*, Proc. Int. Symp. Antenna & Propagat. ISAP, Taiwan, Oct. 2008.

J.34 Majumdar, P.; Verma, A.K.: *Conductor thickness* based *propagation characteristics and losses of slot-line, Proc. Asia-Pacific Microwave Conf., APMC*, Hong-Kong, China, Dec. 2008.

J.35 Krowne, C.M.: *Approximations to hybrid mode slot line behavior, Electron. Lett.*, Vol. 14, No. 8, pp. 258–259, April 1978.

J.36 Janaswamy, R.; Schaubert, D.H.: *Dispersion characteristics for wide slot lines on low-permittivity substrates, IEEE Trans. Microwave Theory Tech.*, Vol. MTT- 33, No. 8, pp. 723–726, Aug. 1985.

J.37 Kitazawa, T.; Fujiki, Y.; Hayashi, Y.; Suzuki, M.: *Slot line with a thick metal coating, IEEE Trans. Microwave Theory Tech.*, Vol. MTT- 21, No. 9, pp. 580–582, Sept. 1973.

J.38 Rozzi, T.; Moglie, F.; Morini, A.; Marchionna, E.; Politi, M.: *Hybrid modes, substrate leakage, and losses of slotline at millimeter-wave frequencies, IEEE Trans. Microwave Theory Tech.*, Vol. MTT- 38, No. 8, pp. 1069–1078, Aug. 1990.

J.39 Kitazawa, T.; Kuo, C.W.; Kong, K.S.; Itoh, T.: *Planar transmission lines with finitely thick conductors and lossy substrates*, IEEE MTT-S Int. Microwave Symp. Digest, pp. 769–772, June, 1991.

Appendix – I

The coefficients A_1 – G_1 and parameters f_1 – f_3 for equation (8.4.18a).

1) $2.22 \le \varepsilon_r \le 3.8$

A) $0.0015 \le w/\lambda_0 \le 0.075$

$$f_1 = \ln \varepsilon_r \quad \text{(i)}, \quad f_2 = \frac{w}{h} \quad \text{(ii)}, \quad f_3 = \ln\left(\frac{h}{\lambda_0}\right) \quad \text{(iii)}, \; A_1 = 1.045 \quad \text{(iv)}, \; B_1 = -0.365 \quad \text{(v)}$$

$$C_1 = \left[\frac{6.3\varepsilon_r^{0.945}}{238.64 + 100w/h}\right] \quad \text{(vi)}, \; D_1 = \left[0.0599 - \frac{8.3695}{100\varepsilon_r}\right] \quad \text{(vii)}, \quad E_1 = 0 \text{ (viii)}, \quad F_1 = 0 \quad \text{(ix)}, \; G_1 = 0 \quad \text{(x)}.$$

B) $0.075 \le w/\lambda_0 \le 1.0$

$$f_1 = \ln \varepsilon_r \quad \text{(i)}, \; f_2 = \frac{w}{h} \quad \text{(ii)}, \; f_3 = \ln\left(\frac{h}{\lambda_0}\right) \quad \text{(iii)}, \; A_1 = 1.194 \quad \text{(iv)}, \; B_1 = -0.24 \quad \text{(v)}, \; C_1 = 0 \quad \text{(vi)},$$

$$D_1 = \left[0.0561 - \frac{0.1234}{\varepsilon_r}\right] \quad \text{(vii)}, \; E_1 = \left[\frac{0.621\varepsilon_r^{0.835}(w/\lambda_0)^{0.48}}{(1.344 + w/h)}\right] \text{(viii)}, \; F_1 = 0 \text{ (ix)}, \; G_1 = 0 \quad \text{(x)}.$$

2) $3.8 \le \varepsilon_r \le 9.8$

A) $0.0015 \le w/\lambda_0 \le 0.075$

$$f_1 = \ln \varepsilon_r \quad \text{(i)}, \quad f_2 = \frac{w}{h} \text{ (ii)}, \quad f_3 = \ln\left(\frac{h}{\lambda_0}\right) \quad \text{(iii)}, \; A_1 = 0.9217 \quad \text{(iv)}, \; B_1 = -0.277 \quad \text{(v)},$$

$$C_1 = \left[\frac{1.03684 \times 10^{-3}\varepsilon_r}{w/h + 0.435}\right]^{\frac{1}{2}} \quad \text{(vi)}, \; D_1 = \left[0.046 - \frac{0.0365}{\varepsilon_r^2\sqrt{w/\lambda_0}(9.06 - 100w/\lambda_0)}\right] \quad \text{(vii)}, \; E_1 = 0 \text{ (viii)},$$

$$F_1 = 0 \quad \text{(ix)}, \; G_1 = 0 \quad \text{(x)}.$$

B) $0.075 \le w/\lambda_0 \le 1.0$

$$f_1 = \ln \varepsilon_r \quad (i), \quad f_2 = \frac{w}{h} \quad (ii), \quad f_3 = \ln\left(\frac{h}{\lambda_0}\right) \quad (iii), \ A_1 = 1.05 \quad (iv), \ B_1 = 0 \quad (v), \ C_1 = 0 \ (vi),$$

$$D_1 = 0.139(1 + 0.52\varepsilon_r \ln(14.7 - \varepsilon_r))\left(\frac{h}{\lambda_0}\right) \quad (vii),$$

$$E_1 = \left[1.411 \times 10^{-2}(\varepsilon_r - 1.421)\cdot \ln\{w/h - 2.012(1 - 0.146\varepsilon_r)\}\right] \ (viii),$$

$$F_1 = -\left(0.04 + 0.04063\sqrt{w/\lambda_0}\right)\varepsilon_r \quad (ix), \ G_1 = 0.111\sqrt{w/\lambda_0} \quad (x).$$

3) $9.7 \le \varepsilon_r \le 20$

A) $0.02 \le w/h \le 0.2$

$$f_1 = \log_{10}\varepsilon_r \quad (i), \quad f_2 = \frac{w}{h} \quad (ii), \quad f_3 = \log_{10}\left(\frac{h}{\lambda_0}\right) \ (iii), \quad A_1 = 0.829 \quad (iv),$$

$$B_1 = -0.448 \ (v), \ C_1 = -0.38 \quad (vi), \ D_1 = -\left(0.29\left(\frac{w}{h}\right) + 0.047\right) \quad (vii), \ E_1 = 0 \ (viii),$$

$$F_1 = 0 \ (ix), G_1 = 0 \ (x).$$

B) $0.2 \le w/h \le 1$

$$f_1 = \log_{10}\varepsilon_r \quad (i), \ f_2 = \frac{w}{h} \quad (ii), \quad f_3 = \log_{10}\left(\frac{h}{\lambda_0}\right) \quad (iii), \ A_1 = 0.745 \quad (iv), \ B_1 = -0.483 \quad (v),$$

$$C_1 = -0.077 \quad (vi), \ D_1 = -\left(0.094\left(\frac{w}{h}\right) + 0.121\right) \quad (vii),$$

$$E_1 = \left(-0.0022\left(\frac{w}{h}\right) + 0.0032\log_{10}\left(\frac{h}{\lambda_0}\right) + 0.0064\right)\varepsilon_r \quad (viii), \ F_1 = 0 \ (ix), \qquad G_1 = 0 \ (x).$$

Appendix - II

The coefficients A_2 – F_2 and parameters g_1 – g_3 for equation (8.4.18b).

1) $2.22 \le \varepsilon_r \le 3.8$

A) $0.0015 \le w/\lambda_0 \le 0.075$

$$g_1 = \ln \varepsilon_r \quad (i), \quad g_2 = \frac{w}{h} \quad (ii), \quad g_3 = \ln\left(\frac{h}{\lambda_0}\right) \quad (iii), \ A_2 = 60 \quad (iv),$$

$$B_2 = 133.5\sqrt{w/\lambda_0} - 0.1423 \ \varepsilon_r - 131.1\sqrt{h/\lambda_0} + \frac{2.246w/h}{\sqrt{\varepsilon_r - 2.06 + 0.85(w/h)^2}} \qquad (v),$$

$$C_2 = 12.94 - 0.6378\varepsilon_r + \frac{12.48}{\sqrt{\varepsilon_r - 2.06 + 0.85(w/h)^2}} \quad (vi),$$

$$D_2 = 2.89w/h - \varepsilon_r(0.1385w/h + 0.0309\ln\varepsilon_r) \quad (vii),$$

$$E_2 = 134.77\sqrt{h/\lambda_0} + 307.39\sqrt{w/\lambda_0} \quad (viii), \ F_2 = 3.69\sin\left[\frac{(\varepsilon_r - 2.22)\pi}{2.36}\right].(ix).$$

B) $0.075 \le w/\lambda_0 \le 1.0$

$$g_1 = \varepsilon_r \quad (i), g_2 = \frac{w}{h} \quad (ii), \ g_3 = \frac{h}{\lambda_0} \quad (iii), \ A_2 = 166.5 \quad (iv),$$

$$B_2 = 10.34\varepsilon_r + \left(\frac{w}{\lambda_0}\right)^2(13.34\varepsilon_r - 45.56) - \left(\frac{h}{\lambda_0}\right)(684.48\varepsilon_r + 1848) - 37.224 \qquad (v),$$

$$C_2 = \left[\frac{1}{w/h}\left(1 + \frac{2.32\varepsilon_r - 0.56}{w/h}\right)\cdot\{(32.5 - 6.67\varepsilon_r)(100h/\lambda_0)^2 - 1\}\right]^{1/2}\cdot\left[2.87\varepsilon_r^2 - 9.08\varepsilon_r + 15.675\right] \ (vi),$$

$$D_2 = -1247.4 + 39.23\left(\frac{w}{h}\right)^2\left(\frac{h}{\lambda_0}\right) \quad (vii), \ E_2 = 0 \ (viii), F_2 = 0 \quad (ix).$$

2) $3.8 \le \varepsilon_r \le 9.8$

A) $0.0015 \le w/\lambda_0 \le 0.075$

$$g_1 = \varepsilon_r \quad \text{(i)},\ g_2 = \frac{w}{h} \quad \text{(ii)},\ g_3 = \ln\left(\frac{h}{\lambda_0}\right) \quad \text{(iii)},\ A_2 = 73.6 \quad \text{(iv)},$$

$$B_2 = -\left[2.15 + 31.37(w/\lambda_0)^{0.6} + \left(0.753(h/\lambda_0)/\sqrt{w/\lambda_0}\right)\right] \quad \text{(v)},$$

$$C_2 = \left[4.9789 + 2.3485\varepsilon_r + \frac{\left(36.23\sqrt{\varepsilon_r^2 + 41}\right) - 225}{(w/h + 0.876\varepsilon_r - 2)}\right] \quad \text{(vi)},\ D_2 = \left(0.51\varepsilon_r\frac{w}{h}\right) + \left(1.0812\frac{w}{h}\right) \quad \text{(vii)},$$

$$E_2 = 638.9(w/\lambda_0)^{0.6} \quad \text{(viii)}, \qquad F_2 = 0 \quad \text{(ix)}.$$

B) $0.075 \le w/\lambda_0 \le 1.0$

$$g_1 = \ln\varepsilon_r \quad \text{(i)},\ g_2 = \frac{w}{h} \quad \text{(ii)},\ g_3 = \frac{h}{\lambda_0} \quad \text{(iii)},\ A_2 = 100.479 \quad \text{(iv)},$$

$$B_2 = 7.892 - 3.6144\varepsilon_r + \left[\left(\frac{w}{\lambda_0}\right)\left(\frac{w}{\lambda_0} + 0.66\right)(72.471 - 33.1908\varepsilon_r)\right] \quad \text{(v)}$$

$$C_2 = (w/h)^{(p-1)}.50\left[\tan^{-1}(2\varepsilon_r) - 0.8\right].\ln\left[100h/\lambda_0 + \sqrt{(100h/\lambda_0)^2 + 1}\right] \text{(vi)}$$

where, $p = [1.11 + (0.132(\varepsilon_r - 27.7)/(100h/\lambda_0 + 5))]$,

$$D_2 = 154.74 - 70.872\varepsilon_r + \left[\left(\frac{w}{\lambda_0}\right)\left(\frac{w}{\lambda_0} + 0.66\right)(1421 - 650.8\varepsilon_r)\right] \quad \text{(vii)},$$

$$E_2 = \left[\left(\frac{w}{\lambda_0}\right)\left(\frac{w}{\lambda_0} + 0.66\right)(-186.151 + 85.25\varepsilon_r)\right] + 5.543\varepsilon_r \quad \text{(viii)},\ F_2 = 0 \quad \text{(ix)}.$$

3) $9.7 \le \varepsilon_r \le 20$

A) $0.02 \le w/h \le 0.2$

$$g_1 = \log_{10}\varepsilon_r \quad \text{(i)},\ g_2 = \frac{w}{h} \quad \text{(ii)},\ g_3 = \frac{h}{\lambda_0} \quad \text{(iii)},\ A_2 = 169.475 \quad \text{(iv)},$$

$$B_2 = -131.16 - 19.58\log_{10}\left(\frac{w}{h}\right) - (11.78\log_{10}\varepsilon_r - 48.33)\log_{10}\varepsilon_r + \frac{h}{\lambda_0}\left(-3200\frac{h}{\lambda_0} - 388.48\log_{10}\varepsilon_r + 863.14\right) \quad \text{(v)},$$

$$C_2 = -137.14 + \frac{0.1}{(w/h)^2} + \log_{10}\varepsilon_r\left(60.23 + 691.4\frac{h}{\lambda_0}\right) - (100h/\lambda_0)^2(1.07\log_{10}\varepsilon_r + 1.44) + (\log_{10}\varepsilon_r)^2\left(-39.41\log_{10}\varepsilon_r - 1298.9\frac{h}{\lambda_0} + 95.02\right) \quad \text{(vi)},$$

$$D_2 = 1100\left(\frac{h}{\lambda_0}\right) - 250.8 + 3283.2\left(\frac{w}{h}\right) \quad \text{(vii)},\ E_2 = 44.28\log_{10}(w/h) \quad \text{(viii)},\ F_2 = 0 \quad \text{(ix)}.$$

B) $0.2 \le w/h \le 1$

$$g_1 = \log_{10}\varepsilon_r \quad \text{(i)},\ g_2 = \frac{w}{h} \quad \text{(ii)},\ g_3 = \frac{h}{\lambda_0} \quad \text{(iii)},\ A_2 = 93.431 \quad \text{(iv)},$$

$$B_2 = -62.335 - 365.69\left(\frac{w}{h}\right) + 19.83\log_{10}\varepsilon_r + 116.01\left(\frac{w}{h}\right)^2\log_{10}\varepsilon_r + 321.5\left(\frac{h}{\lambda_0}\right) - 5.75(\log_{10}\varepsilon_r)^2 - 128.81\left(\frac{w}{h}\right)^2 - 13.841\left(\frac{w}{h}\right)^3 - 2300\left(\frac{h}{\lambda_0}\right)^2 \quad \text{(v)},$$

$$C_2 = 243.762 + 13.35\left(\frac{w}{h}\right) - 1556.5\left(\frac{h}{\lambda_0}\right) + 241.31(\log_{10}\varepsilon_r)^2 + 3.484\left(\frac{w}{h}\right)^2 + 7900\left(\frac{h}{\lambda_0}\right)^2 - 55.01(\log_{10}\varepsilon_r)^3\ 20700\log_{10}\varepsilon_r\left(\frac{h}{\lambda_0}\right)^2 \quad \text{(vi)},$$

$$D_2 = 307.5 - 1500\left(\frac{h}{\lambda_0}\right) + 5087.5\log_{10}\varepsilon_r\left(\frac{w}{h}\right) - 230(\log_{10}\varepsilon_r)^2 - 331.8\left(\frac{w}{h}\right)^2 - 2135.32\left(\frac{w}{h}\right)(\log_{10}\varepsilon_r)^2$$

$$- 587.88\left(\frac{w}{h}\right)^2(\log_{10}\varepsilon_r)^2 + 1093.76\left(\frac{w}{h}\right)^2\log_{10}\varepsilon_r \quad \text{(vii)},$$

$$E_2 = -4.174\left(\frac{w}{h}\right)^3(\log_{10}\varepsilon_r)^3 + 13.932\left(\frac{w}{h}\right)^3(\log_{10}\varepsilon_r)^2 - 29.858\left(\frac{w}{h}\right)^2(\log_{10}\varepsilon_r) \quad \text{(viii)}, \quad F_2 = 0 \quad \text{(ix)}.$$

11

Coupled Transmission Lines

Basic Characteristics

Introduction

The coupled transmission line structures are important for the design of several microwave components such as directional couplers, filters, phase shifters, delay lines, 3dB-hybrids, matching networks, etc. The coupling between lines is a desirable feature of the coupled lines. However, the multiple pairs of telephone lines, multi-conductor strips on the PCB boards and IC chips have an undesired coupling, called the *cross-talk*. It leads to the distortion and degradation in the signals and malfunctioning of the digital circuits and networks. Such undesirable couplings should be either avoided or at least controlled to the minimum acceptable limits. This chapter covers the basic theoretical backgrounds needed for the design of the coupled lines [B.1–B.25]. The theoretical formulations are used to analyze the planar coupled transmission line structures in chapter 12.

Objectives

- *To review some coupled line structures.*
- *To introduce basic concepts of coupled lines.*
- *To develop the circuit models of coupling.*
- *To present even–odd mode analysis of symmetrical coupled lines.*
- *To discuss capacitance and inductance matrices of multi-conductor lines.*
- *To obtain the wave equations for asymmetrical and symmetrical coupled lines.*
- *To discuss the C and π-modes of asymmetrical coupled lines.*

11.1 Some Coupled Line Structures

The proximity of two transmission lines with a common ground conductor, shown in Fig (11.1), provides an electromagnetic coupling between them. It results in the transfer of RF power from one line to another line. The coupling of the RF power is viewed as the coupling of the electric and magnetic fields of the TEM or non-TEM mode, supported by the line structures. The coupling is also viewed through the mutual capacitance and mutual inductance between the coupled lines. Thus, the coupling is described by the *capacitive (electric)* coupling and the *inductive (magnetic)* coupling. The coupling can also exist on the multiconductor lines, and is described through Maxwell's capacitance matrix and inductance matrix [B.1–B.3, J.1–J.5].

The planar transmission lines – the strip lines, microstrip lines, CPW structures, etc. – are used to form the coupled transmission lines. These lines are either the *edge coupled* lines as shown in Fig (11.2a–c), or the *broadside coupled lines,* shown in Fig (11.3a–c). In the case of the coupled strip lines, both upper and lower dielectric layers are identical. Normally, the edge-coupled lines have a *weak coupling* (15–25 dB), or a *medium coupling* (6–15 dB), whereas the broadside coupled lines provide a *strong coupling* (3–6 dB) [B.4–B.15].

The coupling between two coupled transmission lines, mentioned above, is a *continuous coupling* throughout the proximity range of two lines. There is another group of coupling known as the *discrete coupling* between two lines, or between two waveguides. It is achieved through the coupling slots, known as the *aperture coupling* between two transmission lines or between two waveguide structures. Fig (11.4a–c) shows three cases of the aperture coupling: the broadside coupling of two waveguides [B.10, B.11], the aperture coupled two microstrip lines located on the opposite sides of a common ground plane [J.6], and the microstrip line-fed image dielectric waveguide [J.7]. The discrete coupling is also achieved by interconnecting two transmission lines at discrete locations, using the *stubs* [B.5–B.10, B.13, B.14]. Figure (11.5a and b) show the stub coupled transmission lines (*hybrids*) for the waveguide and microstrip structures. Presently, we are only interested in the

Introduction to Modern Planar Transmission Lines: Physical, Analytical, and Circuit Models Approach, First Edition. Anand K. Verma.

continuously coupled planar lines: strip lines, microstrips, and CPW. However, the reader can consult available textbooks and published works for the aperture coupling, and the stub coupling forming the hybrids.

The planar coupled line structures, shown in Figs (11.2 and 11.3), could also be fabricated in the multilayered media: dielectric, ferrite, semiconducting, and other media. The coupling between the lines is strongly influenced by the medium and dimensions of the strip conductors. In the case of the coupled strip lines, the medium is *homogeneous*. However, two strip widths could be different. Such coupled line structure is called the *asymmetrical* coupled lines. The coupled lines with identical widths form the *symmetrical* coupled lines. This chapter is concerned with only the uniformly coupled transmission lines, giving the *uniform coupling* between the lines. The nonuniform transmission lines, say exponential coupled microstrips, could be used to get the *nonuniform coupling* [J.8, J.9]. Such a structure provides stronger coupling and wider bandwidth. The microstrip and CPW coupled lines are the *inhomogeneous* media coupled lines, as the medium is *inhomogeneous*. However, they are either symmetrical or asymmetrical coupled lines. The symmetrical edge coupled microstrip structure is more popular for the design of several microwave components [B.4–B.10, B.15]. The asymmetrical coupled microstrip structure is the less used one. However, they provide both the coupling and impedance transformation simultaneously [B.16, B.17, J.10–J.13].

The planar coupled transmission lines are analyzed by three methods:

- *Even–odd mode analysis.*
- *Coupled transmission line equations.*
- *Coupled-mode analysis.*

The first method is simple and direct one; however, applicable to only symmetrical lines [B.4, B.6–B.10, B.15]. The method decouples the symmetrical coupled transmission lines into two independent transmission lines, supporting two independent modes called the *even-mode* and the *odd-mode*. These individual modes are the *normal modes* of the coupled lines. The normal modes of the coupled lines are further discussed in

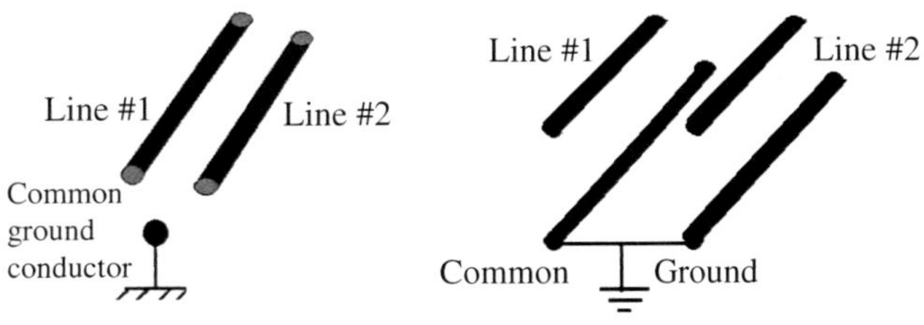

Figure 11.1 Two-wire transmission lines.

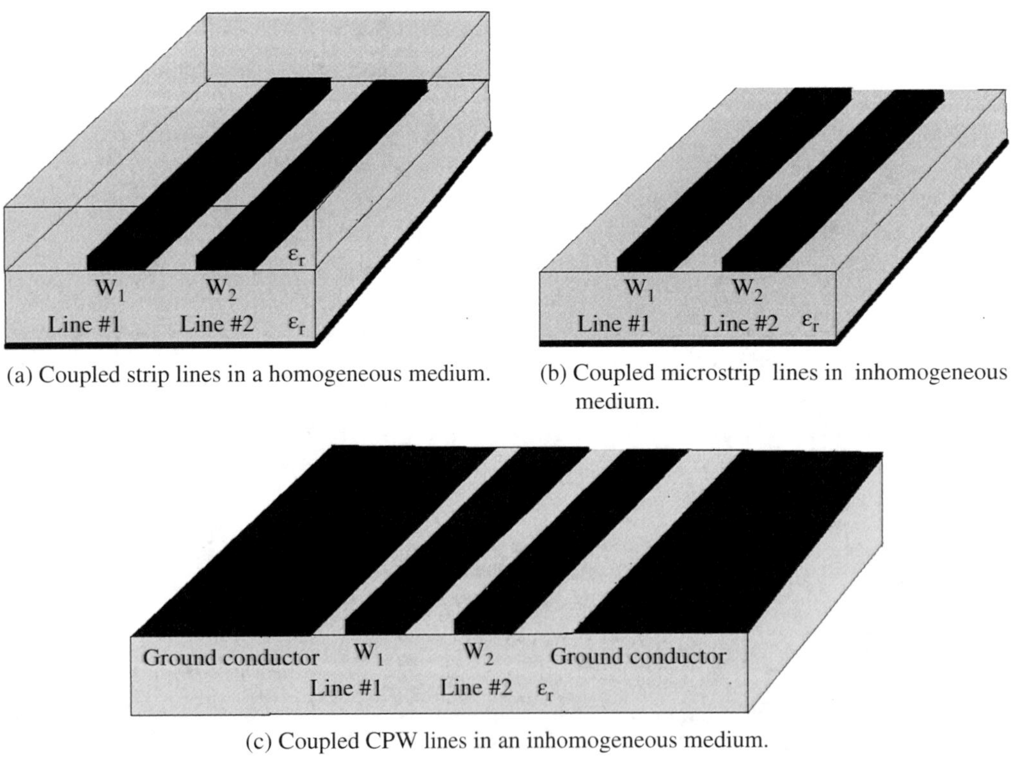

Figure 11.2 Edge coupled planar transmission lines.

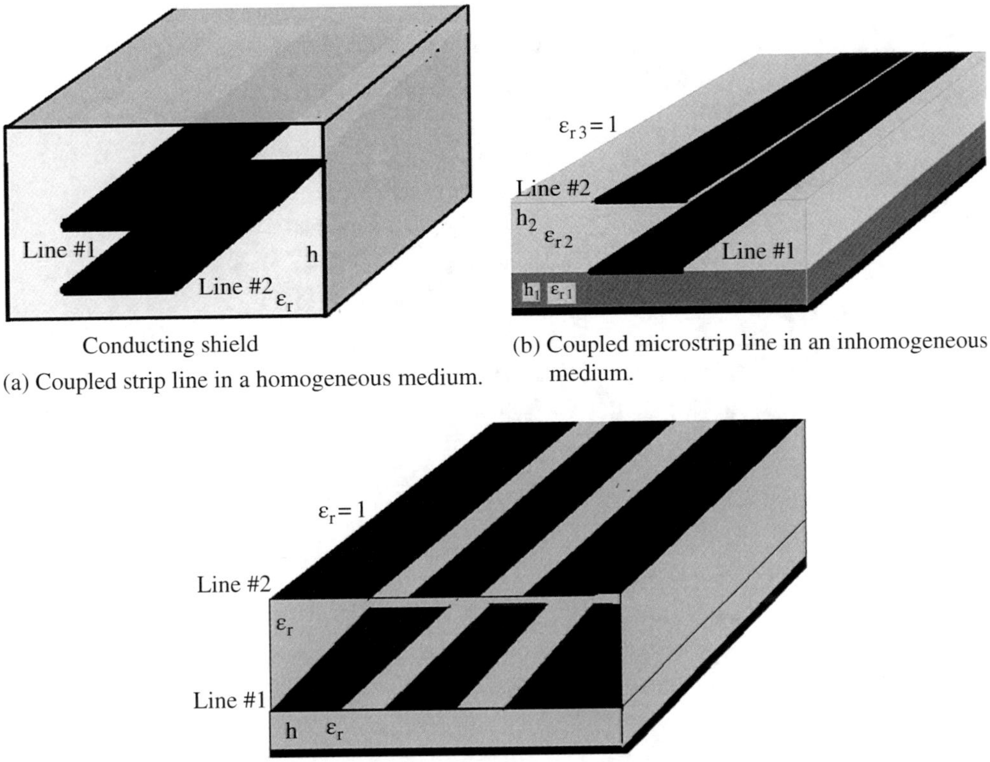

(a) Coupled strip line in a homogeneous medium.

(b) Coupled microstrip line in an inhomogeneous medium.

(c) Coupled CPW line in an inhomogeneous medium.

Figure 11.3 Broadside coupled planar transmission lines.

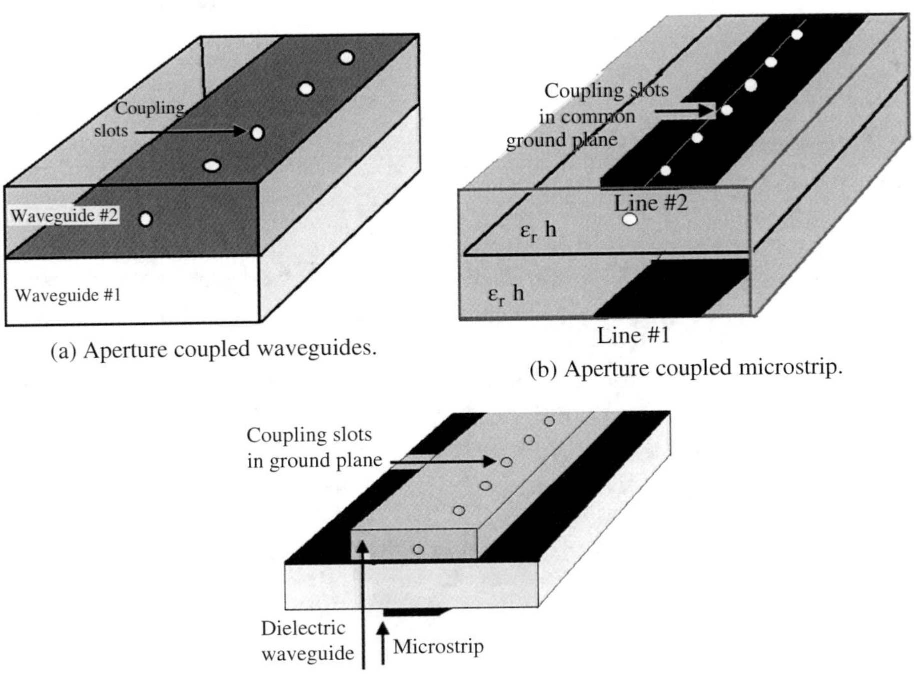

(a) Aperture coupled waveguides.

(b) Aperture coupled microstrip.

(c) Aperture coupled dielectric waveguide and microstrip.

Figure 11.4 Apertures (slot) coupled transmission line structures. *Source:* Collin [B.10, B.11].

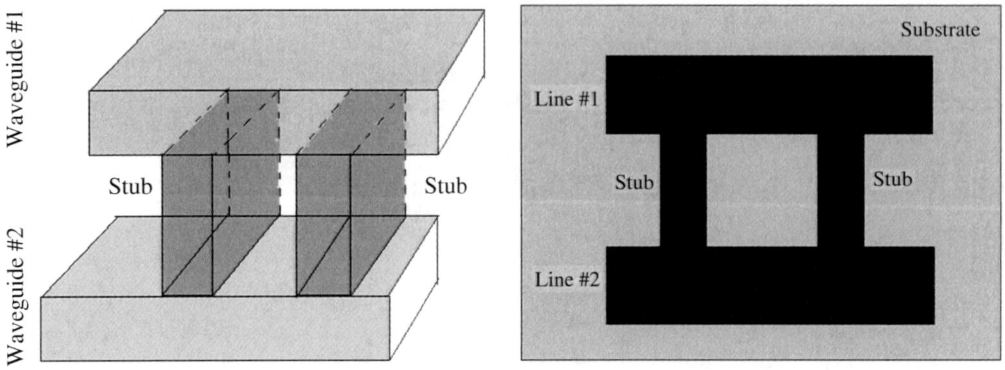

(a) Stubs coupled waveguide. (b) Stubs coupled microstrip/strip lines.

Figure 11.5 Stubs coupled transmission line structures.

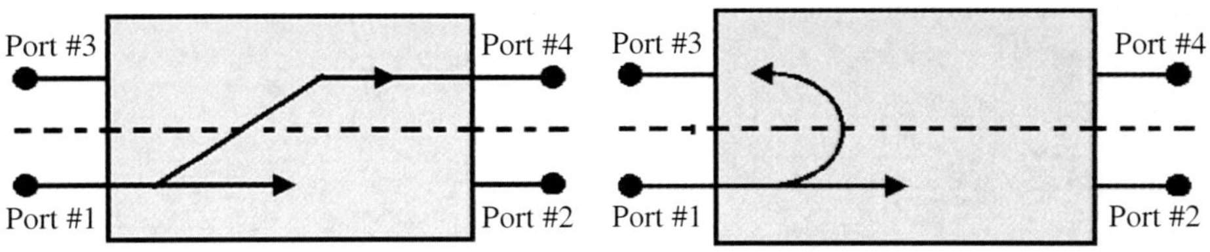

(a) Co directional coupler. (b) Counter-directional coupler.

Figure 11.6 Schematic diagrams of couplers.

subsection (11.5.2). The second method applies to both the symmetrical and asymmetrical coupled lines, supporting the TEM and *quasi*-TEM modes [B.1–B.4, B.15 J.10, J.14]. The *C and π-modes* are the normal modes of the asymmetrical coupled lines. The coupled transmission line equations are obtained through the equivalent coupled lumped circuit model. The third method is more general and helps to understand the nature of coupling [B.18, J.1, J.9, J.15]. However, the method is more involved. The last two methods are also applicable to the multiconductor coupled line structures. The present chapter discusses only the first two methods applicable to a pair of coupled transmission lines.

11.2 Basic Concepts of Coupled Transmission Lines

The basic nature of coupling and the basic definition of parameters to characterize the coupled transmission lines are discussed in this section. The next section presents the circuit model of the coupled lines.

11.2.1 Forward and Reverse Directional Coupling

A pair of coupled transmission lines forms the 4-port network. Ideally, the coupling between two lines could be either the *codirectional,* i.e. the *forward directional coupling* or the *counter directional,* i.e. the *reverse directional coupling*. Figure (11.6a and b) shows the schematic diagrams of both types of couplings. The upper pairs of ports, ports-(3) and (4), are associated with the *coupled line,* while the lower pairs of ports, ports-(1) and (2), are associated with the *mainline*. In the case of a codirectional (forward) coupling, the direction of the coupled power at port #4 is the same as that of the direction of the input power flowing in the mainline, from port #1 to port #2. Port #3 is the *isolated port,* and ideally, no coupled power is available there. In the case of a counter-directional (reverse) coupling, the direction of the coupled power, at port # 3, is in the direction opposite to that of the direction of input power in the mainline. Table (11.1) shows the designation of the ports of a 4-port coupler.

Table 11.1 Designation of the ports of a 4-port coupler.

Name of ports	Codirectional coupler	Counter-directional coupler
Input port	Port #1	Port #1
Coupled port	Port #4	Port #3
Through port	Port #2	Port #2
Isolated port	Port #3	Port #4

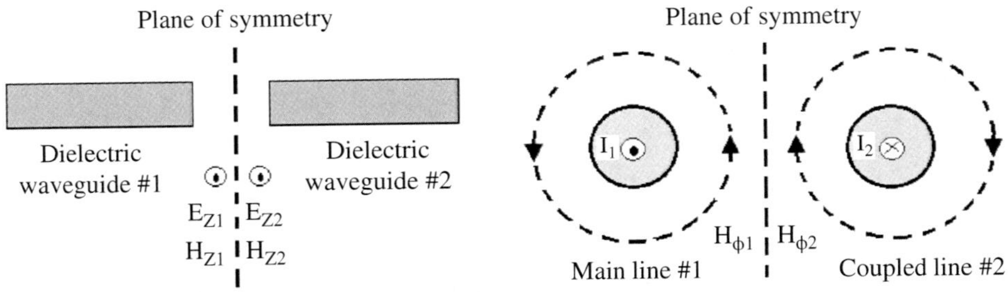

(a) Co directional coupled dielectric waveguide. (b) Counter-directional coupled two-wire lines.

Figure 11.7 Coupled lines structure.

The coupled transmission lines supporting the non-TEM modes, i.e. the TE, TM, and hybrid modes, provide the *codirectional coupling*. Such coupled transmission line structures are formed by the metallic waveguide, dielectric waveguide, etc. The codirectional coupling is physically explained by the mode coupling, shown in Fig (11.7a), of the hybrid mode coupled dielectric waveguides. The longitudinal field components E_Z and H_z of the main guide #1 are in the direction of the wave prorogation. The direction of field components, E_z and H_z, determines the direction of the power flow in a waveguide. At the plane of symmetry, the tangential field components of both the waveguides #1 and #2 are continuous, i.e. $\vec{E}_{z1} = \vec{E}_{z2}$ and $\vec{H}_{z1} = \vec{H}_{z2}$. Therefore, the directions of the field components in both the coupled waveguide #2 and the main waveguide #1 are the same, i.e. the powers in both the waveguides flow in the same direction. So, a *non-TEM mode coupler is a codirectional (forward) coupler.*

A pair of transmission lines in a homogenous medium supporting a *pure TEM mode*, i.e. a mode with no longitudinal field component, forms a *counter-directional (reverse)* coupler, shown in Fig (11.7b). The current (I_1), in the mainline #1, comes out of the plane of the paper, showing the direction of power flow in the mainline #1. The magnetic field $H_{\Phi 1}$ around mainline #1 is in the anticlockwise direction. At the plane of symmetry, the magnetic fields of both the lines are continuous, i.e. $\vec{H}_{\phi 1} = \vec{H}_{\phi 2}$. Consequently, the magnetic field $\vec{H}_{\phi 2}$ around the coupled line #2 is in the clockwise direction. The induced current I_2 in line #2 enters the plane of the paper. It is in the direction opposite to the current I_1 in line #1. So, the direction of the coupled power in the coupled line #2 is opposite to that of the power flow in the mainline #1. *In conclusion, the TEM-type coupled transmission lines create a counter-directional (reverse) coupler.*

Strictly speaking, a microstrip line supports the hybrid mode. The hybrid mode, i.e. *quasi*-TEM mode, of a microstrip line is a linear combination of the TEM mode and non-TEM (TM_0 surface wave) mode [J.16]. The main mode is the TEM one. Therefore, the coupled microstrip lines, or even the coupled CPW, have the counter-directional coupling property. However, the presence of the TM_0 surface wave modes also gives some codirectional coupling at the isolated port that degrades the isolation and directivity of a microstrip coupler. Further, the strength of the TM_0 mode increases with an increase in frequency. It causes more degradation in the directivity of a microstrip coupler at a higher frequency. However, a wideband high directivity codirectional coupler has been developed in microstrip [J.17, J.18].

11.2.2 Basic Definitions

The basic characteristics of the *symmetrical* directional coupler are defined using the *scattering parameters*, discussed in subsection (3.1.4) of chapter 3 [B.3, B.9, B.10, B.15, B.19]. Figure (11.8) shows the schematic diagram of a 4-port codirectional coupler that is used to define its scattering parameters. It is a coupler showing the coupling from port #1 to port #4 or from port #2 to port #3. Ideally, there is no coupling between port #1 and port #3, and vice versa. Also, there is no coupling between port #2 and port #4 and vice-versa. The general scattering matrix of the 4-port network is given below:

$$[S] = \begin{bmatrix} S_{11} & S_{12} & S_{13} & S_{14} \\ S_{21} & S_{22} & S_{23} & S_{24} \\ S_{31} & S_{32} & S_{33} & S_{34} \\ S_{41} & S_{42} & S_{43} & S_{44} \end{bmatrix}. \tag{11.2.1}$$

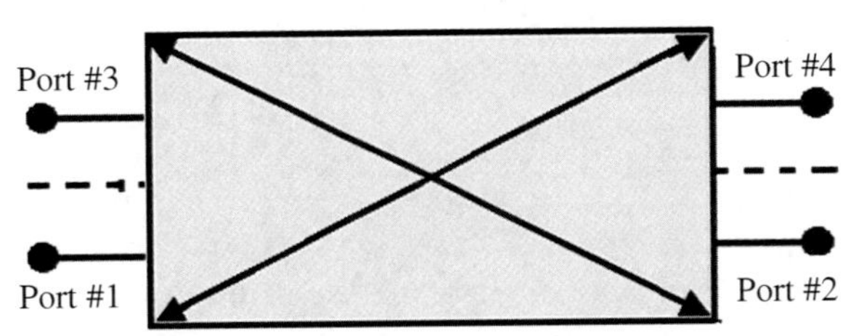

Figure 11.8 Symmetrical codirectional coupler.

If all ports are matched terminated, normally to the 50 Ω, then $S_{11} = S_{22} = S_{33} = S_{44} = 0$, as there is no reflection at these ports. Also, due to the ideal isolation of ports, discussed above, $S_{13} = S_{31} = S_{24} = S_{42} = 0$. The voltage *coupling coefficient* k_c is defined as

$$|S_{14}| = |S_{32}| = k_c. \tag{11.2.2}$$

The *through transmission* coefficient is given as

$$|S_{12}| = |S_{34}| = k_t = \sqrt{1-k_c^2}. \tag{11.2.3}$$

For no dissipation in a coupler, the power balance for the unit input power is written as follows:

$$k_c^2 + k_t^2 = 1, \tag{11.2.4}$$

where k_c^2 and k_t^2 are the coupled and through powers, respectively. For a loose, i.e. weak, coupling, $k_c \approx 0.1$ and $k_t \approx 1$. For a *quadrature coupler*, two output signals are 90^0 phase apart, i.e. $S_{14} = S_{32} = k_c$ and $S_{12} = S_{34} = j\,k_t$. The scattering matrix for the matched reciprocal quadrature coupler could be written as follows:

$$[S] = \begin{bmatrix} 0 & jk_t & 0 & k_c \\ jk_t & 0 & k_c & 0 \\ 0 & k_c & 0 & jk_t \\ k_c & 0 & jk_t & 0 \end{bmatrix}. \tag{11.2.5}$$

The *coupling factor* ($K_c = P_c/P_i$) in dB is defined as

$$\text{Coupling (dB)} = 10\log_{10}\left(\frac{P_c}{P_i}\right) = 10\log_{10}K_c = -20\log_{10}|S_{41}| = -20\log_{10}k_c, \tag{11.2.6}$$

where P_i is the input power at port #1 and P_c is the coupled power at port #4. *The coupling is a measure of the power transferred from the mainline (waveguide) to the coupled line (waveguide) under the matched condition.* This is also known as the *insertion loss* between port #1 and port #4.

The *transmission factor* ($K_t = P_t/P_i$), i.e. the *insertion loss (IL)*, in dB is a measure of the power appearing at the through (*direct*) port #2 of the mainline:

$$\text{IL (dB)} = 10\log_{10}\left(\frac{P_t}{P_i}\right) = 10\log_{10}K_t = -20\log_{10}|S_{21}| = -20\log_{10}\left|\sqrt{1-k_c^2}\right|, \tag{11.2.7}$$

where P_t is the power available at the through port #2. The *reflection factor* in dB, defined below, also known as the *return-loss*, is measured at the *input port* #1, when all other ports are terminated in the matched loads:

$$\text{Return loss} = -20\log_{10}|S_{11}|. \tag{11.2.8}$$

The return loss is a measure of the reflected power at the input port.

Ideally, no power should be transferred to the *isolated (decoupled)* port #3. However, in a practical coupler, some undesired power is always transferred to the isolated port #3. Thus, out of the coupled power, from the mainline to the coupled line, a major portion of the power is transferred to the coupled port #4. However, some undesired power is also available at the isolated (decoupled) port #3. The *isolation factor* ($K_d = P_d/P_i$) of port #3 in dB is defined as

$$\text{Isolation (dB)} = 10\log_{10}\left(\frac{P_d}{P_i}\right) = -20\log_{10}|S_{31}|, \tag{11.2.9}$$

where P_d is the power at the isolated port #3.

Out of the coupled power, the maximum power is desired at the coupled port #4 and minimum power at the isolated port #3. This characteristic is defined by the *directivity of a coupler*:

$$\text{Directivity (dB)} = 10\log_{10}\left(\frac{P_d}{P_c}\right), \tag{11.2.10}$$

where $K_{dir} = P_d/P_c$ is the *directivity factor*. Thus, *the directivity of a coupler is a measure of the isolation of a coupled port from the isolated port.* It is also seen that the isolation (dB) is a summation of directivity (dB) and coupling (dB), i.e. $K_d(\text{dB}) = K_{dir}(\text{dB}) + K_c(\text{dB})$. In the above definitions, the S-parameters in dB are negative quantities, as the power available at all ports is less than the incident (input) power.

11.3 Circuit Models of Coupling

The electromagnetic coupling between two transmission lines takes place through the process of electric and magnetic inductions [B.4, B.10]. *Electrostatic induction* is a familiar concept. Once uncharged conductor #A is brought near a charged conductor #B, the charges of opposite polarity are induced on the conductor #A and capacitance is formed between two conductors. Thus, two originally isolated conductors are *coupled* through the mutual capacitance between them. Likewise, the *law of dynamic induction* (Faraday law of induction) is another familiar concept. A time-varying electric current in the conductor #A creates a time-varying magnetic field around it that couples with the conductor #B to induce a potential on it. The magnetic field coupling is represented by the mutual inductance between two conductors. In brief, the coupling between the two transmission lines is due to two mechanisms:

- *Capacitive coupling, i.e. the electric coupling.*
- *Inductive coupling, i.e. the magnetic coupling.*

In this section, the circuit models of both the couplings are obtained. The models are based on *direct capacitances* and *direct inductances*. These lumped elements are evaluated in terms of the elements of Maxwell's *capacitance matrix*, and *inductance matrix*. In the case of the symmetrical coupled planar lines, the concept of even–odd modes capacitance and inductance is useful.

11.3.1 Capacitive Coupling – Even and Odd Mode Basics

Figure (11.9a) shows the cross-section of a 3-conductor coupled transmission lines. A positive potential is applied on the conductor #1 that terminates its electric field lines partly on the ground conductor #3, and partly on the conductor #2. In the process, it induces the negative charges on the conductor #2 and establishes the capacitive coupling through the *direct mutual coupling capacitance* C^d_{12} *p.u.l.* between two conductors. The potentials at the conductors #1 and #2 further create *direct self-capacitances* C^d_{11} and C^d_{22} *p.u.l.* with respect to the ground conductor #3. Figure (11.9b) shows that the capacitive coupling is modeled by the π-network of the direct capacitances.

For the asymmetrical coupled lines, shown in Fig (11.9a), both conductors have different sizes, giving unequal self-capacitance, i.e. $C^d_{11} \neq C^d_{22}$ between the conductors and ground. However, for symmetrical coupled lines, both conductors are physically identical, giving identical self-capacitance, i.e. $C^d_{11} = C^d_{22}$. The characteristic impedance of a single conductor TEM line could be obtained from its line capacitance. In the case of the coupled TEM transmission lines, such direct computation is applicable. The symmetrically coupled TEM modes are treated as a superposition of two *normal* TEM modes. These normal modes are called the *even mode* and *odd mode*. However, for the asymmetrical coupled lines, the normal modes are the *C-mode* and *π-mode*. These normal modes could be excited on the coupled transmission lines. The *even–odd mode analysis method*, applicable to the symmetrical coupled lines only, separates the coupled lines into two uncoupled lines. Likewise, the coupled asymmetrical lines are separated into the C-mode supporting line and the π-mode supporting line.

Even-Mode Excitation

Figure (11.10a) shows the even-mode excitation on the symmetrical coupled lines by applying identical voltages of the same polarity, i.e. $V_1 = V_2 = +1V$, on both the conductors. The electric field lines of both the conductors are terminated in the same direction on the ground conductor. At the plane of symmetry-AA′, the magnetic fields cancel each other forming a *magnetic wall* with $H_t = 0$, i.e. there is no surface-current on the magnetic wall. Therefore, from the circuit point of view, *the magnetic wall is an open-circuited structure.* Figure (11.10b) shows the rearranged original π-equivalent circuit of Fig (11.9b) with a plane of symmetry. Figure (11.10c) shows the separated even-mode equivalent circuit with the open-circuited plane of symmetry. The even-mode line capacitance is obtained as follows:

$$C_e = C^d_{11}. \tag{11.3.1}$$

Odd-Mode Excitation

Figure (11.11a) shows the odd-mode excitation of the symmetrical coupled lines by applying equal voltages of opposite polarity on two coupled conductors. At the plane of symmetry-AA′, the electric field lines cancel each other; forming an *electric wall* with zero tangential

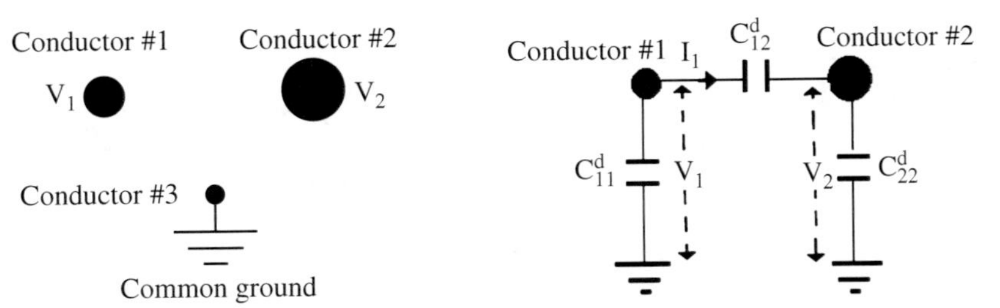

(a) Cross-section of two-conductor lines. (b) Equivalent direct capacitances π-network.

Figure 11.9 The capacitive coupling between two conductors.

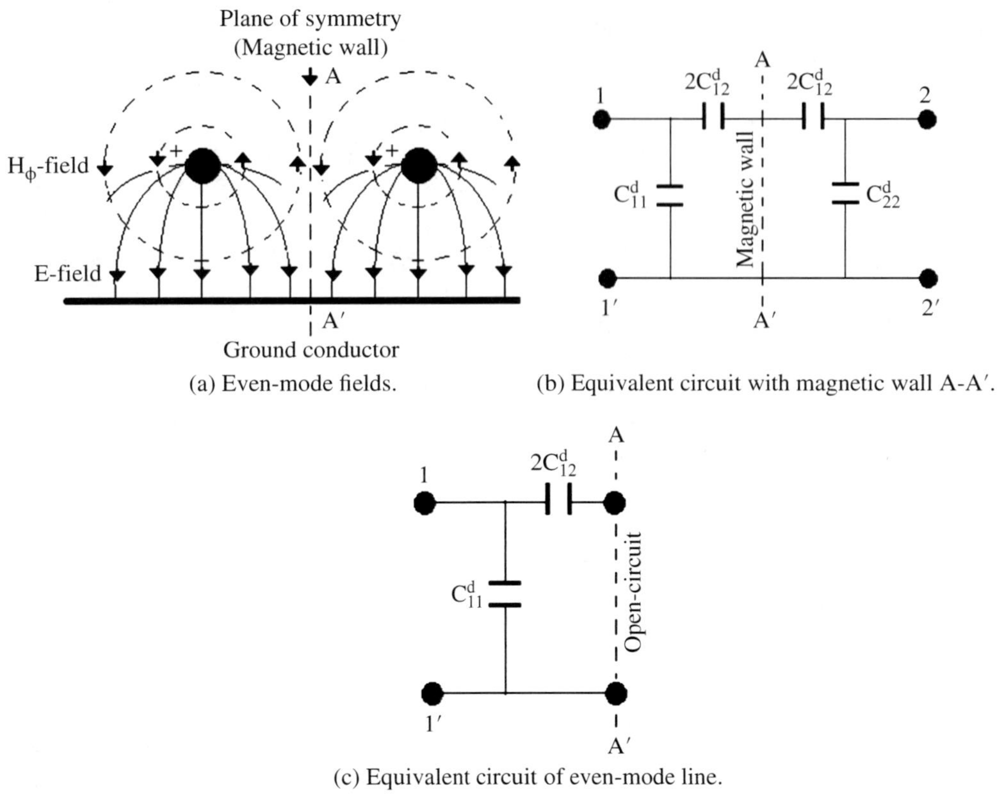

Figure 11.10 The even-mode excitation of the symmetrical coupled lines.

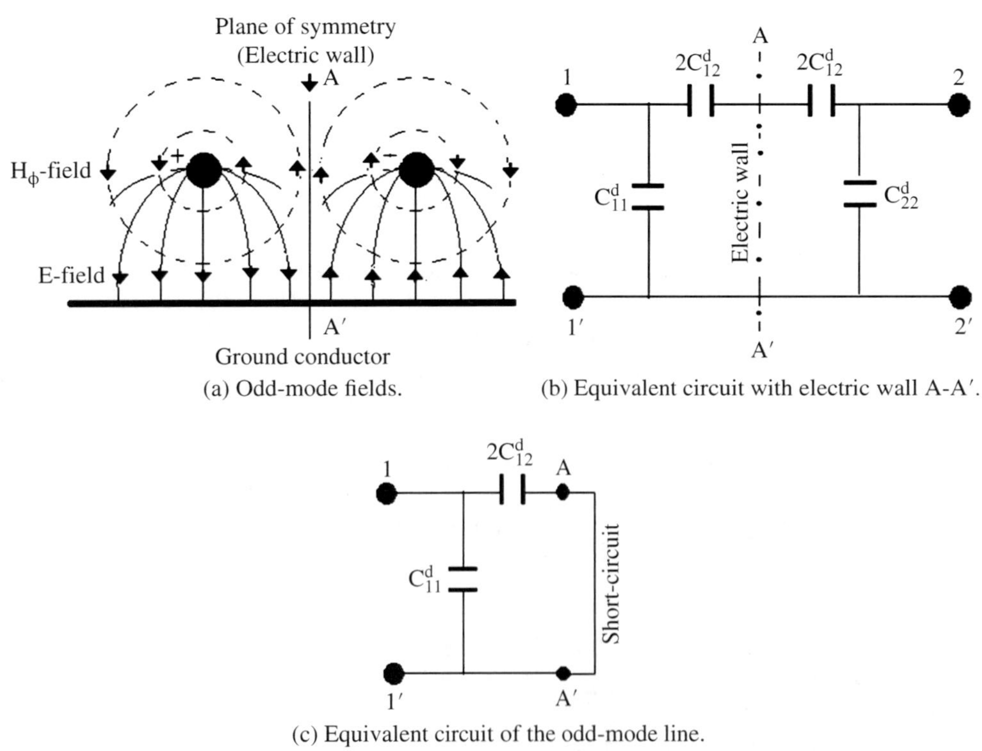

Figure 11.11 The odd-mode excitation of the symmetrical coupled lines.

electric field component, i.e. $E_t = 0$. However, the magnetic fields of both conductors add together. Thus, from the circuit point of view, *the electric wall is a short-circuited structure*. Figure (11.11b) shows the rearranged original π-equivalent circuit with the plane of symmetry-AA' treated as an electric wall. Finally, Fig (11.11c) shows the separated odd-mode supporting line. The odd-mode line capacitance is obtained by short-circuiting the output terminals:

$$C_0 = C_{11}^d + 2C_{12}^d. \tag{11.3.2}$$

It is possible to compute the even- and odd-mode line capacitances of the symmetrical coupled strip lines, microstrip lines, and CPW, using either the closed-form models or numerical methods [B.4–B.9, B.12, B.13, B.16, B.17, B.20, J.10, J.11, J.19–J.21]. The equations (11.3.1) and (11.3.2) are used to obtain the direct capacitances C_{11}^d and C_{12}^d of the symmetrical circuit model from the even- and odd mode-line capacitances.

To excite the C-mode on the asymmetrical coupled lines, two unequal voltages with zero (*cipher*) phase difference are applied to both the conductors. It appears that the *nomenclature C-mode* is derived from the word-**C**ipher. To excite the π-mode, both the conductors are excited with out of phase unequal voltages, i.e. with 180° (π) phase difference that gives the *nomenclature π-mode*.

The above-given description is about the capacitive coupling in the terms of the direct capacitances, forming the π-model. However, the direct capacitances appear due to the electrostatic induction. Therefore, the electric coupling is also described by *Maxwell's capacitance matrix* of the multiconductor transmission lines [B.21, J.2, J.3, J.22]. The elements of the capacitance matrix are used to determine the direct capacitances of the π-model.

11.3.2 Forms of Capacitive Coupling

The capacitive coupling of the symmetrical coupled transmission lines is expressed in either of three ways: (i) *the coefficients of capacitance matrix*, (ii) the *direct capacitances of the π-circuit model*, and (iii) *the even–odd mode capacitances*. The choice depends on the ease of their computations. These three forms of capacitive coupling are examined below.

Capacitance Matrix

Figure (11.9a) of the (2 + 1) conductors can be replaced by a system of the (N + 1) conductor transmission lines. Where N is the number of transmission lines and the additional line is the common ground conductor. If charges $q_1, q_2, \ldots q_N$ are placed on the line conductors #1, 2, 3, ..., N, apart from retaining self-charges, these charges on respective conductors induce opposite polarity charges on all other conductors.

Let us consider the induced charge on the conductor #1. However, if the original charge (self-charge) on the conductor #1 is q_{11}, the total charge (q_1) on it is a summation of self-charge and all other induced charges from the conductors 2 to N:

$$q_1 = q_{11} + q_{12} + q_{13} + \ldots q_{1N}. \tag{11.3.3}$$

The self-charge on conductor #1 is proportional to the potential on it, i.e. $q_{11} = C_{11}V_1$. The induced charge on conductor #1 is proportional to the potential V_j on the jth conductor, i.e. the induced charge on conductor #1 is $q_{1j} = C_{1j}V_j$. Therefore, if the conductors #1 to N are charged with the positive potentials $V_1 \ldots V_j \ldots V_N$, the following expression is obtained for the total charges $q_1 \ldots q_N$ on the N numbers of conductors:

$$\begin{aligned} q_1 &= C_{11}V_1 + C_{12}V_2 + C_{13}V_3. --- C_{1N}V_N \\ q_2 &= C_{21}V_1 + C_{22}V_2 + C_{23}V_3 --- C_{2N}V_N \\ -- &\quad ---- \quad ----- \quad ---- \quad -- \quad --- \\ q_N &= C_{N1}V_1 + C_{N2}V_2 + C_{N3}V_3. - - C_{NN}V_N \end{aligned} \qquad \text{(a)}$$

$$\Rightarrow [q] = [C][V] \qquad \text{(b)}. \tag{11.3.4}$$

The [C] is Maxwell's capacitance matrix of the coefficients of capacitance *p.u.l*:

$$[C] = \begin{bmatrix} C_{11} & C_{12} & . & C_{1N} \\ C_{21} & C_{22} & . & C_{2N} \\ . & . & . & . \\ C_{N1} & C_{N2} & . & C_{NN} \end{bmatrix}. \tag{11.3.5}$$

The coefficients of the capacitance matrix are the short-circuit parameters. The C_{ii} is the *self-coefficient* of the capacitance of the ith conductor due to a potential V_i on it, while all other conductors are grounded. C_{ij} is the coefficient of capacitance matrix showing the *mutual capacitance* between the ith and j^{th} conductors. The potential V_j is applied on the jth conductor, while all other conductors are grounded. *The mutual capacitances* (C_{ij}), *i.e. the off-diagonal elements of the capacitance matrix, are always negative as the induced charges have opposite polarity.* Thus, C_{ii} and C_{ij} are obtained as follows:

$$C_{ii} = \left.\frac{q_i}{V_i}\right|_{V_j=0} \quad \text{(a)}, \qquad C_{ij} = -\left.\frac{q_i}{V_j}\right|_{V_i=0} \quad \text{(b)}; \quad j \neq i. \tag{11.3.6}$$

The coupled lines of Fig (11.9a) illustrate an application of the capacitance matrix. The potentials V_1 and

V_2 are applied to two conductors. The charges on the conductors, in terms of the coefficients of capacitance matrix, are:

$$\begin{bmatrix} q_1 \\ q_2 \end{bmatrix} = \begin{bmatrix} C_{11} & -C_{12} \\ -C_{21} & C_{22} \end{bmatrix} \begin{bmatrix} V_1 \\ V_2 \end{bmatrix} \quad \text{(a)}$$

$$\Rightarrow C_{11}|_{V_2=0} = \frac{q_1}{V_1},\ C_{21}|_{V_2=0} = -\frac{q_2}{V_1}$$
$$C_{22}|_{V_1=0} = \frac{q_2}{V_2},\ C_{12}|_{V_1=0} = -\frac{q_1}{V_2} \quad \text{(b)}. \tag{11.3.7}$$

In the above equation, the *mutual capacitance* C_{12} *is negative* and the reciprocity relation $C_{21} = C_{12}$ holds for the coupled lines within an isotropic medium.

Direct Capacitances

Figure (11.9b) shows the direct capacitance, $C^d_{12} = C^d_{21}$, between conductors, and the direct capacitances, C^d_{11} and C^d_{22}, between conductors and the ground. Following expressions could be written for the total charges on each conductor of the coupled transmission lines:

$$q_1 = C^d_{11}V_1 + C^d_{12}(V_1 - V_2) \Rightarrow q_1 = (C^d_{11} + C^d_{12})V_1 - C^d_{12}V_2$$
$$q_2 = C^d_{12}(V_2 - V_1) + C^d_{22}V_2 \qquad q_2 = -C^d_{12}V_1 + (C^d_{22} + C^d_{12})V_2$$
$$\Rightarrow \begin{bmatrix} q_1 \\ q_2 \end{bmatrix} = \begin{bmatrix} C^d_{11} + C^d_{12} & -C^d_{12} \\ -C^d_{12} & C^d_{22} + C^d_{12} \end{bmatrix} \begin{bmatrix} V_1 \\ V_2 \end{bmatrix}. \tag{11.3.8}$$

On comparing the above equation (11.3.8) against equation (11.3.5) for N = 2 conductors, the following relations are obtained between the coefficients of capacitance matrix and the direct capacitances:

$$\begin{aligned} &\text{Self capacitance of line\#1:} && C_{11} = C^d_{11} + C^d_{12} \quad \text{(a)} \\ &\text{Self capacitance of line\#2:} && C_{22} = C^d_{22} + C^d_{12} \quad \text{(b)} \\ &\text{Mutual capacitance between lines:} && C_{12} = -C^d_{12} \quad \text{(c)}. \end{aligned} \tag{11.3.9}$$

However, the off-diagonal coefficients of the capacitance matrix given by equation (11.3.7a) are negative giving $C_{12} = C^d_{12}$. In the case of symmetrical structure ($C_{11} = C_{22}$), the even mode is excited for $V_1 = V_2$ and the odd mode is excited for $V_1 = -V_2$. Using these conditions in equation (11.3.8), the even- and odd-mode line capacitances are obtained in terms of the direct capacitances. Using equation (11.3.9), they can be further expressed in terms of the coefficients of the capacitance matrix. On several occasions, such as the use of the variational method discussed in chapter 14, the even- and odd-mode line capacitances could be easily computed for the symmetrical coupled microstrip lines. The coefficients of capacitance matrix are computed from the even- and odd-mode line capacitances:

$$\begin{aligned} &\text{Even–mode line capacitance}: C_e = q_1/V_1 = C^d_{11} && \text{(a)}, \quad C_e = C_{11} - C_{12} && \text{(b)} \\ &\text{Odd–mode line capacitance}: C_o = q_1/V_1 = C^d_{11} + 2C^d_{12} && \text{(c)}, \quad C_o = C_{11} + C_{12} && \text{(d)} \\ &\text{Coefficient of capacitance}: C_{11} = (C_e + C_o)/2 && \text{(e)}, \quad C_{12} = (C_o - C_e)/2 && \text{(f)}. \end{aligned} \tag{11.3.10}$$

Capacitive Coupling Coefficient

The voltage coupling is due to electric field coupling. It is defined by the *capacitive coupling coefficient* k_c, $[0 \le k_c \le 1]$. For $k_c = 0$, there is no electric field coupling between lines #1 and #2. So, $k_c = 1$ provides full coupling between two lines. However, it is not possible.

In Fig (11.9b), the time-varying voltage V_1 is applied to the input port #1. The current flowing in the circuit is

$$I_1 = \frac{j\omega C^d_{22} C^d_{12}}{C^d_{22} + C^d_{12}} V_1.$$

The coupled voltage V_2 across the output port of the conductor #2 is

$$V_2 = \frac{I_1}{j\omega C^d_{22}} = \frac{C^d_{12}}{C^d_{22} + C^d_{12}} V_1.$$

On using equation (11.3.9), the voltage coupling k_{c2} from the port #1 to port #2 and also the coupling k_{c1} from the conductor #2 to conductor #1 are obtained as follows:

$$k_{c2} = \frac{V_2}{V_1} = \frac{C^d_{12}}{C^d_{22} + C^d_{12}} = \frac{C_{12}}{C_{22}} \quad \text{(a)}$$
$$k_{c1} = \frac{V_1}{V_2} = \frac{C^d_{12}}{C^d_{11} + C^d_{12}} = \frac{C_{12}}{C_{11}} \quad \text{(b)}. \tag{11.3.11}$$

The voltage coupling factor k_c between the *asymmetric coupled lines* is defined as a geometric mean of k_{c1} and k_{c2}:

$$k_c = \sqrt{K_{c1}K_{c2}} = \frac{C_{12}}{\sqrt{C_{11}C_{22}}}. \tag{11.3.12}$$

As for the *symmetrical coupled lines* $C_{11} = C_{22}$, so the voltage coupling k_c, using equation (11.3.10), is

$$k_c = \frac{C_{12}}{C_{11}} = \frac{C_o - C_e}{C_o + C_e}. \quad (11.3.13)$$

11.3.3 Forms of Inductive Coupling

The inductive coupling of the symmetrical coupled transmission lines is also expressed in either of three ways: (i) *the coefficients of the inductance matrix*, (ii) the *direct inductances of the T-circuit model*, and (iii) *the even–odd mode inductances*. Again the choice depends on the ease of their computations. These three forms of inductive coupling are examined below.

Inductance Matrix

The inductive coupling, among N numbers of multiconductor coupled lines, is properly described in terms of the coefficients of *inductance matrix* L_{ij} *p.u.l* [B.10, B.22, J.2, J.3, J.22]. The flux linkage ψ_j in the jth line is proportional to the current in the ith line, i.e.$\psi_j \propto I_i$ or $\psi_j = L_{ji}I_i$. So, the current I_1 in the conductor #1 creates its *self-inductance* L_{11}, while currents in all other conductors are zero. The currents $I_2, I_3, \ldots I_N$ in the neighboring conductors also provide the flux linkage from these conductors to the conductor #1 through *mutual inductances*. The total flux linkage to conductor #1 is $\psi_1 = L_{11}I_1 + L_{12}I_2 + \ldots + L_{1N}I_N$. The flux linkages in each one of N conductors, due to the currents in all N conductors, are given by the following set of linear equations:

$$\begin{aligned}
\psi_1 &= L_{11}I_1 + L_{12}I_2 + \ldots + L_{1N}I_N \\
\psi_2 &= L_{21}I_1 + L_{22}I_2 + \ldots + L_{2N}I_N \\
&- \quad -- \quad -- \quad -- \qquad \text{(a)}\\
&- \quad -- \quad -- \quad -- \\
\psi_N &= L_{N1}I_1 + L_{2N}I_2 + \ldots + L_{NN}I_N
\end{aligned}$$

$$\Rightarrow [\psi] = [L][I] \qquad \text{(b)}. \quad (11.3.14)$$

where the inductance matrix *p.u.l.* is given as

$$[L] = \begin{bmatrix} L_{11} & L_{12} & \cdots & L_{1N} \\ L_{21} & L_{22} & \cdots & L_{2N} \\ \vdots & \vdots & \vdots & \vdots \\ L_{N1} & L_{N2} & \cdots & L_{NN} \end{bmatrix}. \quad (11.3.15)$$

The coefficients of the inductance matrix are defined as the open-circuited parameters. Thus, the self-inductance L_{ii} at the port #i, due to the current I_i, is obtained, while all ports are open-circuited, i.e. I_j=0 ($i \neq j$). Likewise, the mutual inductance L_{ij} between the ith and j^{th} ports, when current flows only in the jth line, is obtained, while all other ports are open-circuited:

$$L_{ii} = \frac{\psi_i}{I_i}\bigg|_{I_j=0} \text{ (a)}, \quad L_{ij} = \frac{\psi_i}{I_j}\bigg|_{I_i=0} \text{ (b)}; \quad i \neq j. \quad (11.3.16)$$

11.3.3.1 Direct Inductances

Figure (11.12a) shows the T-type equivalent circuit, in terms of the *direct inductances* of the inductively coupled asymmetrical coupled lines. A relation exists between the direct inductances and coefficients of the inductance matrix. The total flux linkages in line #1 and #2, due to

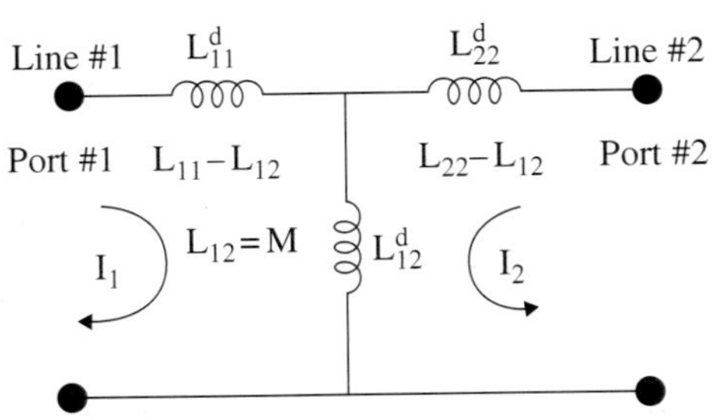

(a) Equivalent T - circuit model of asymmetrical coupled line.

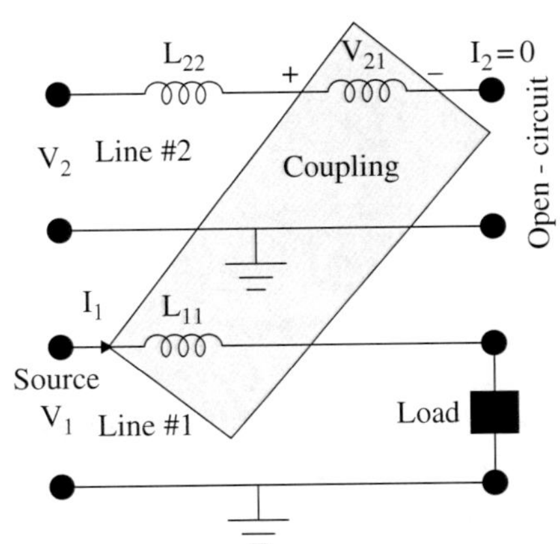

(b) Inductively coupled lines with induced voltage.

Figure 11.12 Inductive coupling of the coupled lines.

currents I_1 and I_2 in the conductors #1 and #2, are given as follows:

$$\psi_1 = L_{11}^d I_1 + L_{12}^d (I_1 + I_2), \quad \psi_1 = \left(L_{11}^d + L_{12}^d\right) I_1 + L_{12}^d I_2$$

$$\psi_2 = L_{22}^d I_2 + L_{21}^d (I_1 + I_2), \quad \psi_2 = L_{21}^d I_1 + \left(L_{22}^d + L_{21}^d\right) I_2$$

$$\begin{bmatrix} \psi_1 \\ \psi_2 \end{bmatrix} = \begin{bmatrix} L_{11}^d + L_{12}^d & L_{12}^d \\ L_{21}^d & L_{22}^d + L_{21}^d \end{bmatrix} \begin{bmatrix} I_1 \\ I_2 \end{bmatrix}. \tag{11.3.17}$$

Due to the reciprocity relation, in the above equation, we have $L_{21}^d = L_{12}^d$. The flux linkages in terms of the coefficients of inductance matrix are written from equation (11.3.14) as follows:

$$\begin{bmatrix} \psi_1 \\ \psi_2 \end{bmatrix} = \begin{bmatrix} L_{11} & L_{12} \\ L_{12} & L_{22} \end{bmatrix} \begin{bmatrix} I_1 \\ I_2 \end{bmatrix}. \tag{11.3.18}$$

On comparing the above two matrix equations, the following relations are obtained between the direct inductances of the equivalent T-circuit model and coefficients of inductance matrix:

$$\begin{aligned} L_{11} &= \left(L_{11}^d + L_{12}^d\right) && \text{(a)} \\ L_{22} &= \left(L_{22}^d + L_{12}^d\right) && \text{(b)} \\ L_{12} &= L_{21}^d = M && \text{(c).} \end{aligned} \tag{11.3.19}$$

The symmetrical coupled lines have $L_{11}^d = L_{22}^d$. The matrix equation (11.3.17) provides the following even- and odd-mode inductances, in terms of the direct inductances, under the even-mode ($I_1 = I_2$) and odd-mode excitations ($I_1 = -I_2$):

$$\begin{aligned} \psi_1 &= \left(L_{11}^d + 2L_{12}^d\right) I_1, && \Rightarrow L_e = (\psi_1/I_1) = \left(L_{11}^d + 2L_{12}^d\right) && \text{(a)} \\ \psi_1 &= L_{11}^d I_1, && \Rightarrow L_o = (\psi_1/I_1) = L_{11}^d && \text{(b).} \end{aligned} \tag{11.3.20}$$

Likewise, the matrix equation (11.3.18) provides the coefficients of inductance matrix for the even-mode and odd-mode line inductances:

$$\begin{aligned} L_e &= L_{11} + L_{12} && \text{(a)} \\ L_0 &= L_{11} - L_{12} && \text{(b).} \end{aligned} \tag{11.3.21}$$

The above equations can also be obtained from Fig (11.12a) on using the open- and short-circuit conditions at the plane of symmetry under the even- and odd-mode excitations. To create the plane of symmetry, the shunt-connected mutual inductance L_{12}^d is replaced with a parallel combination of two inductances of value $2L_{12}^d$. The plane of symmetry (magnetic/electric wall) is placed in the middle of two inductances.

The coefficients of inductance matrix for the symmetrical coupled lines are computed in terms of the even- and odd-mode inductances:

$$\begin{aligned} L_{11} &= \frac{L_e + L_O}{2} && \text{(a)} \\ L_{12} &= \frac{L_e - L_O}{2} && \text{(b).} \end{aligned} \tag{11.3.22}$$

Inductive Coupling Coefficient

Figure (11.12b) shows a pair of inductively coupled transmission lines with the induced voltage V_{21} in the line #2, due to the current I_1 in the line #1, while the line #2 is open-circuited. The open-circuited voltage at terminals of line #2 is $V_2 = V_{21}$. The inductive voltage coefficients, k_{L2} from line #1 to line #2 and k_{L1} from line #2 to line #1, are obtained as follows:

For line #1 with short- circuited load, $V_1 = j\,\omega L_{11} I_1$, and the induced voltage in line #2 is $V_2 = V_{21} = j\,\omega L_{12} I_1$, i.e.

$$\begin{aligned} & V_2 = V_{21} = j\omega L_{12} \times \frac{V_1}{j\omega L_{11}} = \frac{L_{12}}{L_{11}} V_1, \Rightarrow k_{L2} = \frac{V_2}{V_1} = \frac{L_{12}}{L_{11}} && \text{(a)} \\ & \text{similarly,} \quad k_{L1} = \frac{L_{21}}{L_{22}} && \text{(b).} \end{aligned} \tag{11.3.23}$$

The reciprocity relation provides $L_{12} = L_{21}$. The inductive voltage coupling coefficient k_L for the asymmetrical coupled lines is defined as a geometric mean of k_{L1} and k_{L2}. It also provides an expression of k_L for symmetrical coupled lines ($L_{22} = L_{11}$):

$$\begin{aligned} k_L &= \sqrt{k_{L1} k_{L2}} = \frac{L_{12}}{\sqrt{L_{11} L_{22}}} && \text{(a)} \\ k_L &= \frac{L_{12}}{L_{11}} = \frac{L_e - L_o}{L_e + L_o} = \frac{L_{12}^d}{L_{11}^d + L_{12}^d} && \text{(b).} \end{aligned} \tag{11.3.24}$$

If the transmission lines are coupled both electrically and magnetically, the coupling coefficient C is defined as [J.15, J.23],

$$C = \sqrt{k_C\, k_L}. \tag{11.3.25}$$

11.4 Even–Odd Mode Analysis of Symmetrical Coupled Lines

Levy [B.15, J.24] developed the *even–odd mode analysis method* to obtain the reflection coefficient, transmission coefficient, coupling, and directivity of the TEM mode supporting symmetrical coupled transmission lines (strip lines) structure in a homogeneous medium. However, the method is also applied to the analysis of coupled microstrip and CPW lines, supporting the *quasi*-TEM mode [B.4–B.15]. The method has also

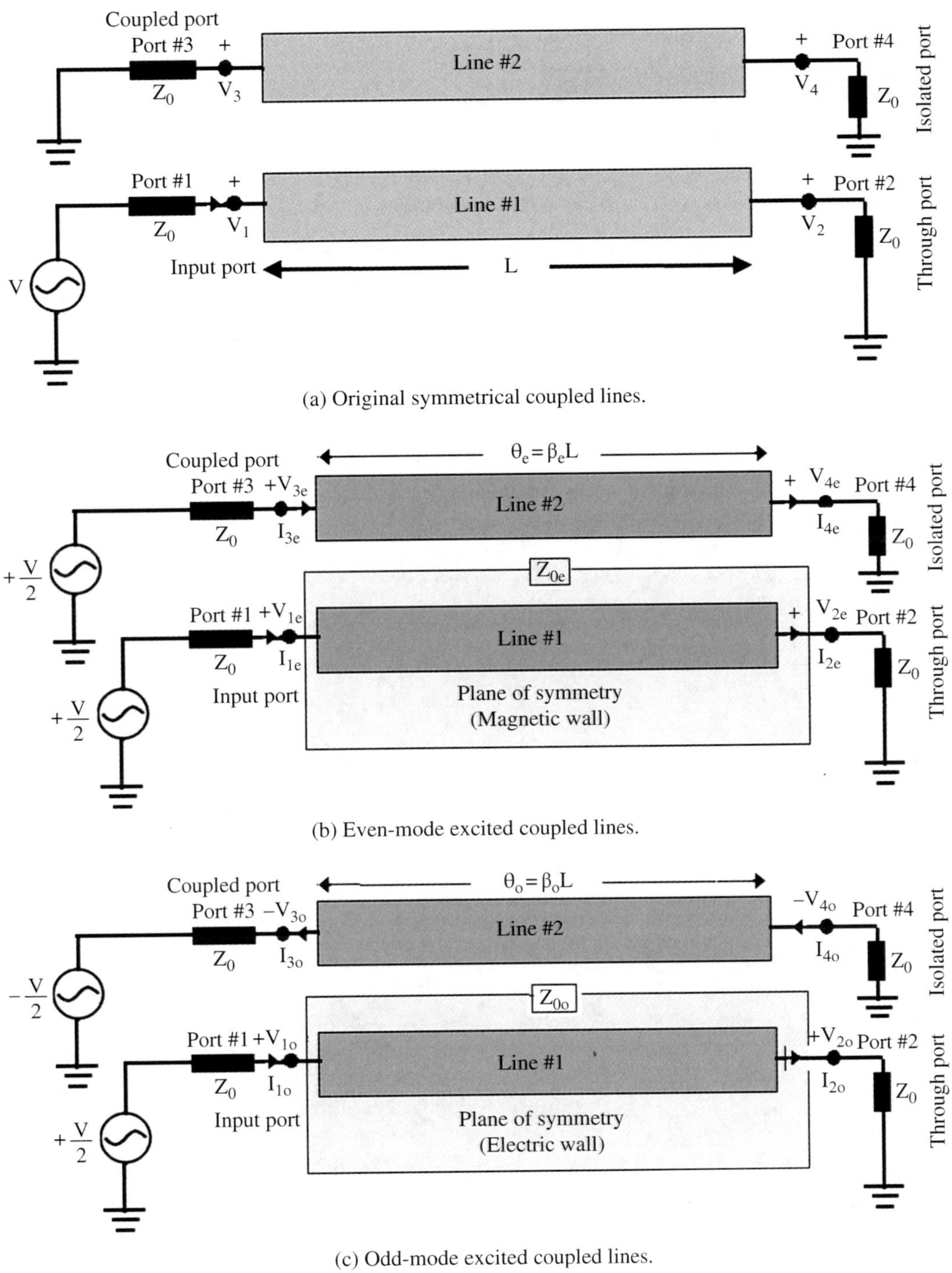

Figure 11.13 Coupled lines resolved in two separate lines supporting the even and odd modes.

been used to the coupled waveguide structure [B.15]. Figure (11.13) shows symbolically the 2-conductor coupled lines. The RF input voltage V is applied to the *input port* #1, whereas port #2, port #3, and port #4 are the *through*, *coupled*, and *isolated ports*, respectively.

The coupled symmetrical lines, shown in Fig (11.13a), support two independent normal modes, called the even mode and odd mode discussed in the previous section. The phase velocities of the even and odd modes of the coupled microstrip and CPW structures are different. Their characteristic impedances are also different. In the case of coupled strip lines, the even and odd modes have identical phase velocities, due to homogeneous medium. However, the even and odd mode

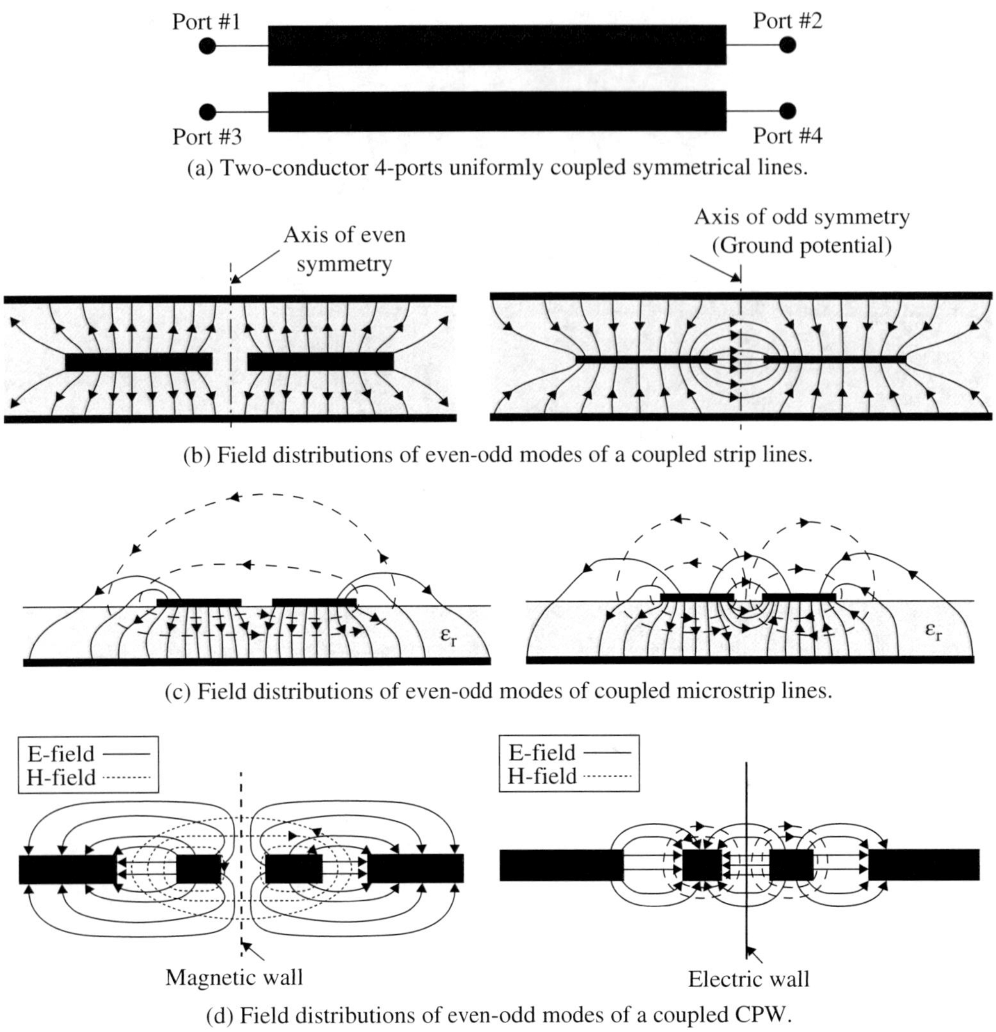

Figure 11.14 Cross-sections of coupled strip lines, microstrip, and CPW showing even–odd mode field distributions.

characteristic impedances of a coupled strip line are still different, due to different field distributions for both the modes. Figure (11.14a–d) shows the coupled lines section, the field distributions of both modes for the edge coupled strip lines, microstrip lines, and CPW. The even and odd modes analysis is equally valid to the symmetrical broadside coupled transmission lines.

11.4.1 Analysis Method

The even–odd mode analysis is applied to the coupled microstrip lines. The coupled strip line structure is treated as its special case. Figure (11.13a) shows the original coupled strip lines excited by the voltage source V at its input port #1. The source has an internal impedance Z_0. The port #2, port #3, and port #4 are terminated in the load Z_0. Under proper excitations, the twin conductors support two normal modes – the even and odd modes [J.25]. Figure (11.13b) shows that the even mode is excited by the application of identical polarity voltages +V/2 at port #1 and port #3. The plane of symmetry, between two conductors, is a *magnetic wall*. Likewise, the opposite polarity voltages ±V/2 applied at these ports excite the odd mode. In this case, the plane of symmetry is an *electric wall*. The currents and voltages of the even and odd modes, at all the 4-ports, are given below:

$$\text{Even mode}: \quad \begin{array}{ll} I_{1e} = I_{3e}, I_{2e} = I_{4e} & \text{(a)} \\ V_{1e} = V_{3e}, V_{2e} = V_{4e} & \text{(b)}. \end{array} \tag{11.4.1}$$

$$\text{Odd mode}: \quad \begin{array}{ll} I_{1o} = -I_{3o}, \ I_{2o} = -I_{4o} & \text{(a)} \\ V_{1o} = -V_{3o}, V_{2o} = -V_{4o} & \text{(b)}. \end{array} \tag{11.4.2}$$

In summary, the concept of the even and odd modes decouples the original pair of coupled lines into two

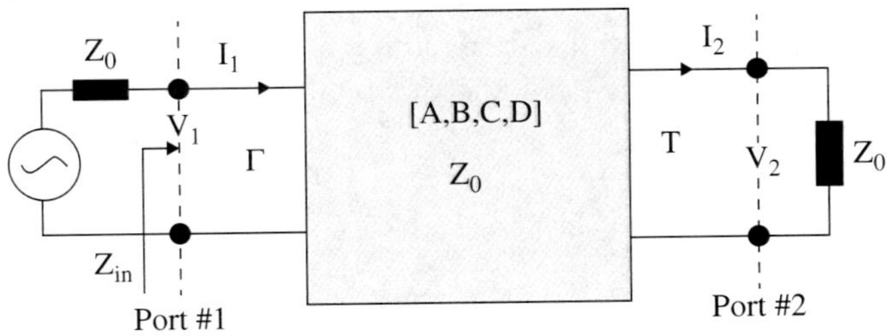

Figure 11.15 2-Port network for even- or odd-mode line section.

isolated single lines, supporting the even and odd modes. So, the even-mode supporting line is just a single conductor TEM line in presence of the magnetic wall, whereas the odd-mode supporting line is another single conductor TEM line in the presence of the electric wall. The Z_{oe} and Z_{oo} are the even- and odd-mode characteristic impedances of these single lines. Thus, a section of the 4-port coupled lines structure is partitioned into two 2-port line sections – the even-mode and odd-mode line sections, shown in Fig (11.13). Figure (11.15) shows that the 2-port line section could be treated as the 2-port network. It is characterized by the ABCD matrix discussed in subsection (3.1.3) of chapter 3.

The input voltage–current (V_1,I_1) are related to the output voltage–current (V_2,I_2) as follows [B.10, B.23]:

$$\begin{bmatrix} V_1 \\ I_1 \end{bmatrix} = \begin{bmatrix} A & B \\ C & D \end{bmatrix} \begin{bmatrix} V_2 \\ I_2 \end{bmatrix} = \begin{bmatrix} \cos\theta & jZ_o \sin\theta \\ j\dfrac{\sin\theta}{Z_o} & \cos\theta \end{bmatrix} \begin{bmatrix} V_2 \\ I_2 \end{bmatrix}, \tag{11.4.3}$$

where the characteristic impedance Z_o could be either Z_{oe} or Z_{oo} for the even- or odd-mode line section, respectively. The above expression is given in equation (3.1.25) of chapter 3. Likewise, the electrical length $\theta_e = \beta_e L$, $\theta_o = \beta_o L$ corresponds to the even and odd modes. However, for the coupled strip line propagation constants $\beta_e = \beta_o = \beta$, giving $\theta_e = \theta_o = \theta$. However, the coupled microstrip lines have $\theta_e \neq \theta_o$. The physical length of the coupled lines section is L. The input impedance of the line section, terminated in a matched load, is computed. It helps to obtain the reflection coefficient (Γ), showing a mismatch at the input port, and the transmission coefficient (T), showing the available signal at the output port [B.23]:

$$Z_{in} = \frac{AZ_o + B}{CZ_o + D} \quad \text{(a)}$$
$$\Gamma = \frac{A + B/Z_o - CZ_o - D}{A + B/Z_o + CZ_o + D} \quad \text{(b)}$$
$$T = \frac{2}{A + B/Z_o + CZ_o + D} \quad \text{(c)}. \tag{11.4.4}$$

The above expressions are used for both the even and odd modes giving (Γ_e) and (T_e), and also (Γ_o) and (T_o).

Following equation (11.4.3), the ABCD matrix descriptions for both the even- and odd-mode lines could be written as follows:

$$\begin{bmatrix} V_{1e} \\ I_{1e} \end{bmatrix} = \begin{bmatrix} \cos\theta_e & j Z_{oe}\sin\theta_e \\ j\dfrac{\sin\theta_e}{Z_{oe}} & \cos\theta_e \end{bmatrix} \begin{bmatrix} V_{2e} \\ I_{2e} \end{bmatrix} \quad \text{(a)}$$
$$\begin{bmatrix} V_{1o} \\ I_{1o} \end{bmatrix} = \begin{bmatrix} \cos\theta_o & j Z_{oo}\sin\theta_o \\ j\dfrac{\sin\theta_o}{Z_{oo}} & \cos\theta_o \end{bmatrix} \begin{bmatrix} V_{2o} \\ I_{2o} \end{bmatrix} \quad \text{(b)}. \tag{11.4.5}$$

S-Parameters of Coupled Microstrip Lines

To get the S-parameters, the port voltages are computed. For the input voltage V = 1 at port #1, shown in Fig (11.13a), the even- and odd-mode voltages at port #1 are ½. The even- and odd-mode reflected signals, A_{1e} and A_{1o}, at port #1 are obtained as follows:

$$\text{Reflected signal} = \text{Reflection coefficient} \times \text{Input signal}$$
$$\Rightarrow \text{For the even mode}: A_{1e} = \Gamma_e \times \frac{1}{2} \quad \text{(a)}$$
$$\text{For the odd mode}: A_{1o} = \Gamma_o \times \frac{1}{2} \quad \text{(b)}. \tag{11.4.6}$$

The total reflected signal A_1 at port #1 of the original coupled lines is a combination of the even and odd modes reflected signals. It is written in terms of the S-parameter giving the *return-loss*:

$$S_{11} = A_1 = A_{1e} + A_{1o} = \frac{1}{2}(\Gamma_e + \Gamma_o) \quad \text{(a)}$$
$$\text{Return–loss}: RL(dB) = -20\log_{10}|S_{11}| \quad \text{(b)}. \tag{11.4.7}$$

Likewise, the even- and odd-mode transmitted signals A'_{1e} and A'_{1o} at the output port#2 are obtained as

$$\text{Transmitted signal} = \text{Transmission coefficient} \times \text{Input signal}$$
$$\Rightarrow \text{For the even mode}: A'_{1e} = T_e \times \frac{1}{2} \quad \text{(a)}$$
$$\text{For the odd mode}: A'_{1o} = T_o \times \frac{1}{2} \quad \text{(b)}. \tag{11.4.8}$$

The total transmitted signal A_2, at the through port #2 of the original coupled lines, helps to obtain the S_{21} and *insertion loss*:

$$S_{21} = A_2 = \frac{1}{2}(T_e + T_o) \quad \text{(a)}$$
$$\text{Insertion loss, I L(dB)} = -20\log_{10}|S_{21}| \quad \text{(b)}. \tag{11.4.9}$$

Likewise, the coupled signal A_3 at the coupled port #3 computes the *coupling*:

$$S_{31} = A_3 = \frac{1}{2}(\Gamma_e - \Gamma_o) \quad \text{(a)}$$
$$\text{Coupling, } C(\text{dB}) = -20\log_{10}|S_{31}| \quad \text{(b)}. \tag{11.4.10}$$

Finally, the signal at the isolated port #4 provides the isolation:

$$S_{41} = A_4 = \frac{1}{2}(T_e - T_o) \quad \text{(a)}$$
$$\text{Isolation, } I(\text{dB}) = -20\log_{10}|S_{41}| \quad \text{(b)}. \tag{11.4.11}$$

The *directivity* of a coupler is

$$D\,(\text{dB}) = -20\log_{10}\left|\frac{A_4}{A_3}\right|. \tag{11.4.12}$$

The scattering parameters are complex quantities. Therefore, their magnitudes are taken for the determination of the return-loss, insertion-loss, coupling, isolation, and directivity. The even and odd modes' electrical lengths of the coupled microstrip lines, in an inhomogeneous medium, are different:

$$\theta_e = \beta_e L = \beta_0\sqrt{\varepsilon_{r\,\text{effe}}(f)}L \quad \text{(a)}$$
$$\theta_o = \beta_o L = \beta_0\sqrt{\varepsilon_{r\,\text{effo}}(f)}L \quad \text{(b)}, \tag{11.4.13}$$

where β_0 is the propagation constant in the free space. The effective even- and odd-mode relative permittivity, $\varepsilon_{r\,\text{effe}}(f)$ and $\varepsilon_{r\,\text{effo}}(f)$, are frequency-dependent.

The above S-parameters are obtained in terms of the reflection and transmission coefficients for the even and odd modes. A section of the coupled line in an isotropic medium forms the reciprocal network with A = D. Using equations (11.4.4) and (11.4.5), the even- and odd-mode reflection and transmission coefficients are computed as

$$\Gamma_e = \frac{j\left(\frac{Z_{oe}}{Z_o} - \frac{Z_o}{Z_{oe}}\right)\sin\theta_e}{\sum_e} \quad \text{(a)}$$
$$\Gamma_o = \frac{j\left(\frac{Z_{oo}}{Z_o} - \frac{Z_o}{Z_{oo}}\right)\sin\theta_o}{\sum_o} \quad \text{(b)}, \tag{11.4.14}$$

where

$$\sum_e = 2\cos\theta_e + j\left(\frac{Z_{oe}}{Z_o} + \frac{Z_o}{Z_{oe}}\right)\sin\theta_e \quad \text{(a)}$$
$$\sum_o = 2\cos\theta_o + j\left(\frac{Z_{oo}}{Z_o} + \frac{Z_o}{Z_{oo}}\right)\sin\theta_o \quad \text{(b)}. \tag{11.4.15}$$

Likewise, using equations (11.4.4c) and (11.4.5), the following expressions are obtained for the transmission coefficients of the even and odd modes:

$$T_e = \frac{2}{\sum_e} \quad \text{(a)}, \qquad T_o = \frac{2}{\sum_o} \quad \text{(b)}. \tag{11.4.16}$$

The input port #1 must be matched to the source impedance Z_0. The condition for input impedance matching is obtained below.

Input Impedance Matching

The port voltages and port currents (V_1, I_1) and (V_2, I_2) at the port #1 and port #2, respectively, are computed as a linear combination of the even and odd voltages:

$$\text{At port\#1}: \begin{bmatrix} V_1 \\ I_1 \end{bmatrix} = \begin{bmatrix} V_{1e} \\ I_{1e} \end{bmatrix} + \begin{bmatrix} V_{1o} \\ I_{1o} \end{bmatrix} \quad \text{(a)}$$
$$\text{At port\#2}: \begin{bmatrix} V_2 \\ I_2 \end{bmatrix} = \begin{bmatrix} V_{2e} \\ I_{2e} \end{bmatrix} + \begin{bmatrix} V_{2o} \\ I_{2o} \end{bmatrix} \quad \text{(b)}. \tag{11.4.17}$$

Equations (11.4.5) and (11.4.17) provide

$$\begin{bmatrix} V_1 \\ I_1 \end{bmatrix} = \begin{bmatrix} (\cos\theta_e + \cos\theta_o) & j(Z_{oe}\sin\theta_e + Z_{oo}\sin\theta_o) \\ j\left(\frac{\sin\theta_e}{Z_{oe}} + \frac{\sin\theta_o}{Z_{oo}}\right) & (\cos\theta_e + \cos\theta_o) \end{bmatrix} \begin{bmatrix} V_2 \\ I_2 \end{bmatrix}. \tag{11.4.18}$$

The input impedance of coupled lines under matching at port #1 is

$$\Gamma = \frac{Z_{in} - Z_0}{Z_{in} + Z_0} = 0, \Rightarrow Z_{in} = Z_0 = \frac{AZ_0 + B}{CZ_0 + D}. \tag{11.4.19}$$

For a reciprocal network with A = D, the above matching condition is reduced to

$$(CZ_0^2 + Z_0A) = AZ_0 + B \Rightarrow \frac{B}{Z_0} = Z_oC. \tag{11.4.20}$$

On substituting the matrix elements B and C of equation (11.4.18) in the above equation, the following matching condition is obtained:

$$\frac{Z_{oe}}{Z_o}\sin\theta_e + \frac{Z_{oo}}{Z_o}\sin\theta_o = \frac{Z_o}{Z_{oe}}\sin\theta_e + \frac{Z_o}{Z_{oo}}\sin\theta_o$$
$$\Rightarrow \frac{Z_{oe}Z_{oo}}{Z_0^2} = \frac{Z_{oo}\sin\theta_e + Z_{oe}\sin\theta_o}{Z_{oe}\sin\theta_e + Z_{oo}\sin\theta_o}. \tag{11.4.21}$$

The perfect matching, at the input port, is satisfied at only one center frequency with a narrow frequency band, as θ_e,θ_o and Z_{oe}, Z_{oo} are frequency-dependent, due to the frequency-dependent nature of $\varepsilon_{r\,\text{effe}}(f)$ and $\varepsilon_{r\,\text{effo}}(f)$.

Isolation Between Ports

The parameter S_{41} determines the isolation between ports #1 and #4. However, the ports #2 and #4 should also be isolated from each other. It is needed to improve directivity. Using equations (11.4.9a) and (11.4.11a), the *isolation* between the ports #2 and #4 is obtained as follows [J.24]:

$$\text{Isolation} = \frac{A_4}{A_2} = \frac{T_e - T_o}{T_e + T_o} = \frac{2\left(\cos\theta_o - \cos\theta_e\right) + j\left(r + \dfrac{1}{r}\right)\left(\sin\theta_o - \sin\theta_e\right)}{2\left(\cos\theta_o + \cos\theta_e\right) + j\left(r + \dfrac{1}{r}\right)\left(\sin\theta_o + \sin\theta_e\right)}$$

$$\Rightarrow \frac{A_4}{A_2} = \tan\left(\frac{\theta_e - \theta_o}{2}\right)\left[\frac{2\sin\left(\dfrac{\theta_e + \theta_o}{2}\right) - j\left(r + \dfrac{1}{r}\right)\cos\left(\dfrac{\theta_e + \theta_o}{2}\right)}{2\cos\left(\dfrac{\theta_e + \theta_o}{2}\right) + j\left(r + \dfrac{1}{r}\right)\sin\left(\dfrac{\theta_e + \theta_o}{2}\right)}\right] \quad (a) \qquad (11.4.22)$$

$$\text{where, } r = \sqrt{\frac{Z_{oe}}{Z_{oo}}}, \text{ and } Z_0 \approx \sqrt{Z_{0e}Z_{0o}} \quad (b).$$

It is noted that for the coupled strip lines, in a homogeneous medium, $\theta_e = \theta_o$, giving the perfect isolation as $A_4 = 0$. However, in the case of coupled microstrip lines $\theta_e \neq \theta_o$; therefore, the perfect isolation is not possible for a microstrip-based coupler. For a loose coupling case, $Z_{oe} \approx Z_{oo}$ i.e. $r + \frac{1}{r} \approx 2$, giving

$$\left|\frac{A_4}{A_2}\right| = \left|\tan\left[\frac{\theta_e - \theta_o}{2}\right]\right| \Rightarrow A_4 \approx \left|\frac{\theta_e - \theta_o}{2}\right|. \qquad (11.4.23)$$

For a loose coupling $|A_2| \approx 1$, and $|(\theta_e - \theta_o)/2|$ is small. Both the directivity and isolation degrade for loose coupling ($\leq$10 dB). The isolation also degrades with an increase in frequency.

Coupled Strip Lines

The symmetric coupled strip lines are in a *homogeneous medium* with equal $\varepsilon_{r\,effe}(f) = \varepsilon_{r\,effo}(f) = \varepsilon_r$, giving $\theta_e = \theta_o = \theta$. Therefore, for coupled strip lines the input matching condition, given by equation (11.4.21), is reduced to a simpler form:

$$Z_o = \sqrt{Z_{oe}Z_{oo}}. \qquad (11.4.24)$$

The above expression is also used for the design of the microstrip-based coupler. To obtain the frequency response of a coupler, the variable θ is expressed in terms of the normalized frequency f/f_0, where f_0 is the central design frequency of a coupler for maximum coupling.

Let the length of the coupled strip lines section is $L = (\lambda_g/4)$. It gives the electrical length as $\theta_e = \theta_o = \theta = \beta_g L = (2\pi/\lambda_g)\times(\lambda_g/4) = (\pi/2)$. The length of the coupled section at the center frequency ω_0 is $L = (\pi/2)\times(v_p/\omega_0)$. The electrical length of the coupled section at any other frequency is

$$\theta = \beta L = \frac{\omega}{v_p}\left(\frac{\pi v_p}{2\omega_o}\right), \Rightarrow \theta = \frac{\pi}{2}\times\frac{f}{f_o}. \qquad (11.4.25)$$

Using the input matching condition (11.4.24), the even–odd modes reflection and transmission coefficients of the coupled strip lines are obtained from equations (11.4.14)–(11.4.16):

$$\Gamma_e = -\Gamma_o = \frac{j\left(\sqrt{\dfrac{Z_{oe}}{Z_{oo}}} - \sqrt{\dfrac{Z_{oo}}{Z_{oe}}}\right)\sin\theta}{\Sigma}. \qquad (11.4.26)$$

$$T_e = T_o = \frac{2}{\Sigma} \qquad (a)$$

$$\text{where, } \Sigma = 2\cos\theta + j\left(\sqrt{\frac{Z_{oe}}{Z_{oo}}} + \sqrt{\frac{Z_{oo}}{Z_{oe}}}\right)\sin\theta \qquad (b). \qquad (11.4.27)$$

Thus, using the above relation and equations (11.4.7)–(11.4.11), the output signals, i.e. the S-parameters, at all the four ports are given below:

$$\text{Input matched}: A_1 = 0, \qquad \text{Coupling}: A_3 = \Gamma_e$$
$$\text{Perfect isolation}: A_4 = 0, \qquad \text{Through port}: A_2 = T_e. \qquad (11.4.28)$$

The coupled signal at port #3 is

$$A_3 = j\left(\frac{Z_{oe} - Z_{oo}}{Z_{oe} + Z_{oo}}\right)\left[\frac{\sin\theta}{\dfrac{2Z_o}{Z_{oe} + Z_{oo}}\cos\theta + j\sin\theta}\right]. \qquad (11.4.29)$$

The maximum A_3, showing the maximum coupling (C), is obtained for the electrical length $\theta = \frac{\pi}{2}$:

$$C = A_3 = \left(\frac{Z_{oe} - Z_{oo}}{Z_{oe} + Z_{oo}}\right) \quad (a), \qquad \sqrt{1 - C^2} = \frac{2Z_o}{Z_{oe} + Z_{oo}} \quad (b). \tag{11.4.30}$$

Finally, under the matched condition, the following expressions are obtained for the coupled and through signals at the ports #3 and 2 of a stripline coupler:

$$A_3 = \Gamma_e = j\frac{C \sin\theta}{\sqrt{1 - C^2}\cos\theta + j\sin\theta} \quad (a)$$

$$A_2 = T_e = \frac{\sqrt{1 - C^2}}{\sqrt{1 - C^2}\cos\theta + j\sin\theta} \quad (b). \tag{11.4.31}$$

The two output signals, at port #2 and port #3, are in the quadrature-phase. The expression for θ, given by equation (11.4.25), is used to obtain the frequency response at all ports, i.e. the frequency-dependent return-loss, insertion loss, coupling, isolation, and directivity of a microstrip and strip lines couplers. The expressions for the S-parameters are also applicable to the coupled CPW lines. To compute the frequency response at all ports, expressions are needed for $\varepsilon_{r\,effe}(f)$, $\varepsilon_{r\,effo}(f)$, Z_{oe}, and Z_{oo} of the coupled microstrip lines, coupled strip lines, and coupled CPW. The expressions are presented in chapter 12. Chapter 14 also presents the variational method to determine the static $\varepsilon_{r\,effe}(f = 0)$ and $\varepsilon_{r\,effo}(f = 0)$. The present method is valid only for the symmetrical coupled lines. The next section (11.5) discusses the *coupled wave equations approach* that is also used for the *asymmetrical coupled lines*.

11.4.2 Coupling Coefficients

Equation (11.4.30) for the coupling is strictly valid only for the symmetrical coupled strip lines in a homogeneous medium. More accurate expressions for the symmetrical coupled microstrip lines, involving using both the capacitive and inductive couplings, are obtained below.

These expressions for coupling, in terms of the even- and odd-mode line parameters, are also applicable to the symmetrical coupled CPW.

Using equations (11.3.1) and (11.3.20a), the even-mode characteristic impedance Z_{oe}, phase velocity v_{pe}, and even-mode effective relative permittivity $\varepsilon_{r\,effe}$ are obtained as follows:

$$Z_{oe} = \sqrt{\frac{L_e}{C_e}} = \left[\frac{L_{11}^d + 2L_{12}^d}{C_{11}^d}\right]^{1/2} = \left[\frac{L_{11} + L_{12}}{C_{11} - C_{12}}\right]^{1/2} \quad (a)$$

Alternatively,

$$Z_{oe} = \frac{1}{v_{p1}C_e} = \frac{\sqrt{\varepsilon_{r\,effe}}}{cC_e} = \frac{1}{c\sqrt{C_e(\varepsilon_r)C_e(\varepsilon_r = 1)}}, \Rightarrow Z_{oe} = \frac{\eta_o\varepsilon_o}{\sqrt{C_e(\varepsilon_r)C_e(\varepsilon_r = 1)}} \quad (b) \tag{11.4.32}$$

$$v_{pe} = \frac{c}{\sqrt{\varepsilon_{r\,effe}}} = \frac{1}{\sqrt{L_eC_e}} = \frac{1}{\sqrt{(L_{11}^d + 2L_{12}^d)C_{11}^d}} = \frac{1}{\sqrt{(L_{11} + L_{12})(C_{11} - C_{12})}} \quad (c)$$

$$\Rightarrow \varepsilon_{r\,effe} = c^2[(L_{11} + L_{12})(C_{11} - C_{12})] \quad (d),$$

where c is the velocity of EM-wave in free space and $\eta_o = 120\pi$ is the intrinsic impedance of the free space. Likewise, the odd-mode characteristic impedance Z_{oo}, phase velocity v_{po}, and effective relative permittivity $\varepsilon_{r\,effo}$ are also given below:

$$Z_{oo} = \sqrt{\frac{L_o}{C_o}} = \left[\frac{L_{11}^d}{C_{11}^d + 2C_{12}^d}\right]^{1/2} = \left[\frac{L_{11} - L_{12}}{C_{11} + C_{12}}\right]^{1/2} \quad (a)$$

Alternatively,

$$Z_{oo} = \frac{\sqrt{\varepsilon_{r\,effo}}}{cC_o} = \frac{1}{c\sqrt{C_o(\varepsilon_r)C_o(\varepsilon_r = 1)}}, \Rightarrow Z_{oo} = \frac{\eta_o\varepsilon_o}{\sqrt{C_o(\varepsilon_r)C_o(\varepsilon_r = 1)}} \quad (b) \tag{11.4.33}$$

$$v_{po} = \frac{c}{\sqrt{\varepsilon_{r\,effo}}} = \frac{1}{\sqrt{L_oC_o}} = \frac{1}{\sqrt{L_{11}^d(C_{11}^d + 2C_{12}^d)}} = \frac{1}{\sqrt{(L_{11} - L_{12})(C_{11} + C_{12})}} \quad (c)$$

$$\Rightarrow \varepsilon_{r\,effo} = c^2[(L_{11} - L_{12})(C_{11} + C_{12})] \quad (d).$$

The closed-form expressions for the characteristic impedance and effective relative permittivity are presented in section (12.1) of chapter 12. Therefore, the direct circuit constants L_{11}^d, L_{12}^d, C_{11}^d, and C_{12}^d are expressed in terms of the even- and odd-mode secondary parameters as follows:

$$Z_{oe}v_{pe} = \frac{1}{C_{11}^d}, \qquad \Rightarrow C_{11}^d = \frac{1}{v_{pe}Z_{oe}} = \frac{\sqrt{\varepsilon_{r\,effe}}}{cZ_{oe}} \quad (a)$$

$$Z_{oo}v_{po} = \left[\frac{1}{C_{11}^d + 2C_{12}^d}\right], \Rightarrow C_{12}^d = \frac{1}{2c}\left(\frac{\sqrt{\varepsilon_{r\,effo}}}{Z_{oo}} - \frac{\sqrt{\varepsilon_{r\,effe}}}{Z_{oe}}\right) \quad (b). \tag{11.4.34}$$

$$\frac{Z_{oe}}{v_{pe}} = L_{11}^d + 2L_{12}^d \quad (a), \qquad \frac{Z_{oo}}{v_{po}} = L_{11}^d \quad (b). \tag{11.4.35}$$

$$L_{11}^d + L_{12}^d = \frac{1}{2}\left[\frac{Z_{oe}}{v_{pe}} + \frac{Z_{oo}}{v_{po}}\right] = \frac{1}{2c}\left[Z_{oe}\sqrt{\varepsilon_{r\,effe}} + Z_{oo}\sqrt{\varepsilon_{r\,effo}}\right] \quad (a)$$

$$L_{12}^d = \frac{1}{2}\left[\frac{Z_{oe}}{v_{pe}} - \frac{Z_{oo}}{v_{po}}\right] = \frac{1}{2c}\left[Z_{oe}\sqrt{\varepsilon_{r\,effe}} - Z_{oo}\sqrt{\varepsilon_{r\,effo}}\right] \quad (b). \tag{11.4.36}$$

The capacitive (k_C) and inductive (k_L) coupling coefficients are obtained as follows:

$$k_C = \frac{C_o - C_e}{C_o + C_e} = \frac{C_{12}^d}{C_{11}^d + C_{12}^d}, \quad \Rightarrow k_C = \frac{Z_{oe}\sqrt{\varepsilon_{r\,effo}} - Z_{oo}\sqrt{\varepsilon_{r\,effe}}}{Z_{oe}\sqrt{\varepsilon_{r\,effo}} + Z_{oo}\sqrt{\varepsilon_{r\,effe}}} \quad (a)$$

$$k_L = \frac{L_o - L_e}{L_o + L_e} = \frac{L_{12}^d}{L_{11}^d + L_{12}^d}, \quad \Rightarrow k_L = \frac{Z_{oe}\sqrt{\varepsilon_{r\,effe}} - Z_{oo}\sqrt{\varepsilon_{r\,effo}}}{Z_{oe}\sqrt{\varepsilon_{r\,effe}} + Z_{oo}\sqrt{\varepsilon_{r\,effo}}} \quad (b). \tag{11.4.37}$$

In the case of a homogeneous medium coupled strip lines $\varepsilon_{r\,effe} = \varepsilon_{r\,effo} = \varepsilon_r$, and the inductive and capacitive coupling coefficients $k_C = k_L = C$ are given by equation (11.4.30a). The coupling coefficient of the coupled microstrip lines, or the coupled CPW, in the inhomogeneous media is computed from the following expression [J.15, J.23]:

$$C = \sqrt{k_C k_L}. \tag{11.4.38}$$

For the symmetrical coupled strip lines, $k_C = k_L$ and the above equations are reduced to equation (11.4.30a).

11.5 Wave Equation for Coupled Transmission Lines

A pair of coupled *asymmetrical* transmission lines of length dx, located in an isotropic homogeneous/inhomogeneous medium, is shown in Fig (11.16a). Figure (11.16b) shows its equivalent circuit using the coefficients of capacitance and inductance matrices. The coupling of power from one line to another line takes places through thc following two mechanisms:

- *Coupling due to the time-varying magnetic field, i.e. through the inductive coupling.*
- *Coupling due to the time-varying electric field, i.e. through the capacitive coupling.*

The equivalent circuit, shown in Fig (11.16b), takes into account both coupling mechanisms. The coupling capacitance, i.e. the mutual capacitance, is a negative quantity, $C_{12} = -C_m$, whereas the inductive coupling, i.e. the mutual inductance, is a positive quantity, $L_{12} = M$.

The self-inductance, L_{11} and L_{22}, and the self-capacitance, C_{11}, and C_{22}, of line #1 and line #2 are the elements of the inductance and capacitance matrices discussed in section (11.3). On the application of the basic *induction law of Faraday* for the magnetic coupling and *induced displacement current law of Maxwell* for the electric coupling, a wave equation is obtained for the asymmetrical coupled transmission lines. It can be easily extended to N numbers of the coupled transmission lines. The matrix form of the formulation is a compact one, and easier to manipulate. The time-harmonic variation is assumed for the phasor voltage and phasor current [B.4–B.6, B.21, B.22, J.5, J.10, J.17, J.23, J.26].

11.5.1 Kelvin–Heaviside Coupled Transmission Line Equations

The Kelvin–Heaviside transmission line equations of subsection (2.1.3) of chapter 2 can be extended to the coupled transmission line also. In view of two coupling mechanisms, the equivalent circuit of the coupled lines is obtained in two steps. Figure (11.17a) shows the inductive coupling through the *induced emf* in line #1, $L_{12}\,dx\,(\partial i_2/\partial t)$, caused by the time-varying current i_2 in line #2. The polarity of the induced voltage follows Lenz's law. Figure (11.17b) shows the capacitive coupling, through the *induced displacement current*, $C_{12}dx(\partial v_2/\partial t)$, due to

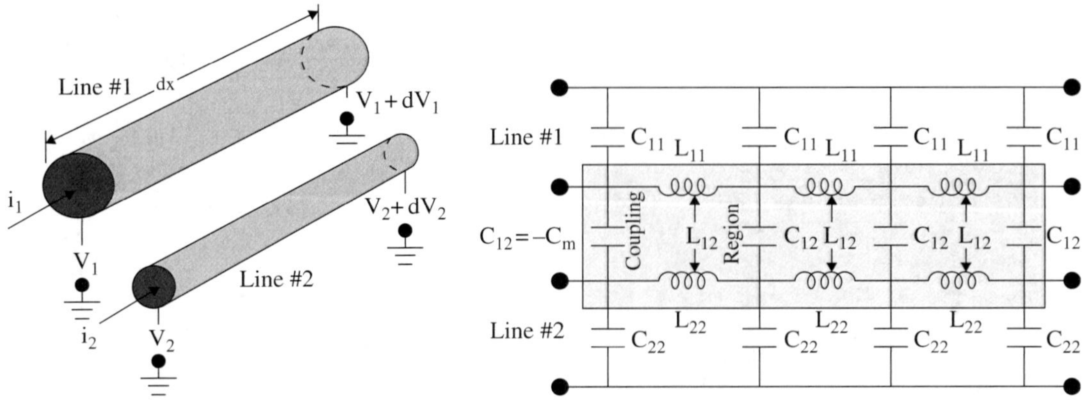

(a) Coupled asymmetrical lines.

(b) Equivalent circuit of coupled asymmetrical lines.

Figure 11.16 Coupled asymmetrical lines in an inhomogeneous medium.

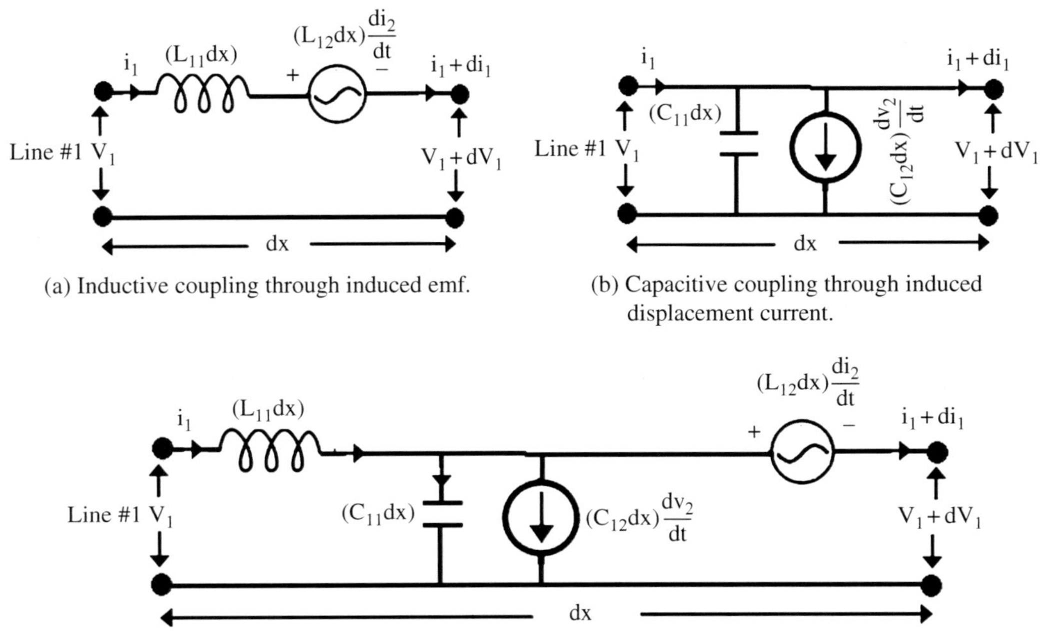

(a) Inductive coupling through induced emf.

(b) Capacitive coupling through induced displacement current.

(c) Combined capacitive and inductive couplings.

Figure 11.17 The coupling mechanisms in the asymmetrical coupled transmission lines.

the time-varying electric field between lines #1 and #2. Figure (11.17c) shows both couplings simultaneously.

The following loop and node equations are written, from Fig (11.17c), for line #1 due to the couplings from the line #2:

$$\text{Loop equation}: v_1 - (L_{11}dx)\frac{\partial i_1}{\partial t} - (L_{12}dx)\frac{\partial i_2}{\partial t} - (v_1 + dv_1) = 0$$

$$\Rightarrow -\frac{\partial v_1}{\partial x} = L_{11}\frac{\partial i_1}{\partial t} + L_{12}\frac{\partial i_2}{\partial t} \quad \text{(a)}$$

$$\text{Node equation}: i_1 - (C_{11}dx)\frac{\partial v_1}{\partial t} - (C_{12}dx)\frac{\partial v_2}{\partial t} - (i_1 + di_1) = 0$$

$$\Rightarrow -\frac{\partial i_1}{\partial x} = C_{11}\frac{\partial v_1}{\partial t} + C_{12}\frac{\partial v_2}{\partial t} \quad \text{(b).}$$

(11.5.1)

Likewise, the voltage and current equations for the line #2, due to the coupling from line #1 to line #2, are written:

$$-\frac{\partial v_2}{\partial x} = L_{22}\frac{\partial i_2}{\partial t} + L_{21}\frac{\partial i_1}{\partial t} \quad \text{(a)}$$

$$-\frac{\partial i_2}{\partial t} = C_{22}dx\frac{\partial v_2}{\partial t} + C_{21}\frac{\partial v_1}{\partial t} \quad \text{(b).}$$

(11.5.2)

Using the phasor notation, the above Kelvin–Heaviside coupled equations are rewritten below in terms of the impedances and admittances:

$$-\frac{dv_1}{dx} = Z_{11}i_1 + Z_{12}i_2 \quad \text{(a)}, \qquad -\frac{dv_2}{dx} = Z_{21}i_1 + Z_{22}i_2 \quad \text{(b).}$$

(11.5.3)

$$-\frac{di_1}{dx} = Y_{11}v_1 + Y_{12}v_2 \quad (a), \qquad -\frac{di_2}{dx} = Y_{21}v_1 + Y_{22}v_2 \quad (b). \tag{11.5.4}$$

The above equations, in the *phasor form*, are generalized to include the dielectric and conductor losses through the conductance G and resistance R, respectively. The phasor voltages and currents are written without the tilde sign $\tilde{v}_1, \tilde{i}_1$, etc. The self-impedance, mutual impedance, self-admittance, and mutual admittance of the coupled lines *p.u.l.* are given by the following expressions:

$$\begin{aligned} Z_{11} &= R_{11} + j\omega L_{11} \quad (a), & Y_{11} &= G_{11} + j\omega C_{11} \quad (b)\\ Z_{12} &= R_{12} + j\omega L_{12} \quad (c), & Y_{12} &= G_{12} + j\omega C_{12} \quad (d)\\ Z_{22} &= R_{22} + j\omega L_{22} \quad (e), & Y_{22} &= G_{22} + j\omega C_{22} \quad (f). \end{aligned} \tag{11.5.5}$$

The L_{11}, C_{11}, and L_{22}, C_{22} pairs are the self-inductance and self-capacitance of the conductors #1 and #2. The pairs R_{11}, G_{11}, and R_{22}, G_{22} account for the conductor loss and the leakage loss of the conductors #1 and #2. The G_{12} is the leakage loss between the conductors #1 and #2 through the leakage current in the lossy dielectric medium. The R_{12} is caused by the proximity effect of two conductors [J.10, J.26]. For an *isotropic* and *linear* medium, the following reciprocity relations hold [B.22, B.23]:

$$Z_{12} = Z_{21} \quad (a), \qquad Y_{12} = Y_{21} \quad (b). \tag{11.5.6}$$

The coupled line equations for the isotropic and linear medium are written below, in more compact matrix notation:

$$\begin{aligned} -\frac{d}{dx}\begin{bmatrix} v_1 \\ v_2 \end{bmatrix} &= \begin{bmatrix} Z_{11} & Z_{12} \\ Z_{12} & Z_{22} \end{bmatrix}\begin{bmatrix} i_1 \\ i_2 \end{bmatrix} = j\omega\begin{bmatrix} L_{11} & L_{12} \\ L_{12} & L_{22} \end{bmatrix}\begin{bmatrix} i_1 \\ i_2 \end{bmatrix} + \begin{bmatrix} R_{11} & R_{12} \\ R_{12} & R_{22} \end{bmatrix}\begin{bmatrix} i_1 \\ i_2 \end{bmatrix} \quad (a)\\ -\frac{d}{dx}\begin{bmatrix} i_1 \\ i_2 \end{bmatrix} &= \begin{bmatrix} Y_{11} & Y_{12} \\ Y_{12} & Y_{22} \end{bmatrix}\begin{bmatrix} v_1 \\ v_2 \end{bmatrix} = j\omega\begin{bmatrix} C_{11} & C_{12} \\ C_{12} & C_{22} \end{bmatrix}\begin{bmatrix} v_1 \\ v_2 \end{bmatrix} + \begin{bmatrix} G_{11} & G_{12} \\ G_{12} & G_{22} \end{bmatrix}\begin{bmatrix} v_1 \\ v_2 \end{bmatrix} \quad (b). \end{aligned} \tag{11.5.7}$$

$$\Rightarrow -\frac{d}{dx}[V] = [Z][I] \quad (a), \qquad -\frac{d}{dx}[I] = [Y][V] \quad (b), \tag{11.5.8}$$

where [V] and [I] are the column matrices for the voltage and current on the coupled transmission lines. The matrix form easily accommodates N numbers of conductors [J.2, J.25, J.26]. The [Z] and [Y] are the $n \times n$ order impedance and admittance matrices for N numbers of coupled lines. The above matrix equations are solved to obtain the following voltage wave equation for coupled transmission lines:

$$\begin{aligned} -\frac{d^2}{dx^2}[V] &= [Z]\frac{d}{dx}[I] = -[Z][Y][V]\\ \Rightarrow \frac{d^2}{dx^2}[V] &= [Z][Y][V]. \end{aligned} \tag{11.5.9}$$

$$[Z]\,[Y] = \begin{bmatrix} Z_{11}Y_{11} + Z_{12}Y_{12} & Z_{11}Y_{12} + Z_{12}Y_{22} \\ Z_{12}Y_{11} + Z_{22}Y_{12} & Z_{12}Y_{12} + Z_{22}Y_{22} \end{bmatrix}. \tag{11.5.10}$$

$$\Rightarrow [Z]\,[Y] = \begin{bmatrix} a_1 & b_1 \\ b_2 & a_2 \end{bmatrix} \tag{11.5.11}$$

where,

$$\begin{aligned} a_1 &= Z_{11}Y_{11} + Z_{12}Y_{12}, & a_2 &= Z_{12}Y_{12} + Z_{22}Y_{22}\\ b_1 &= Z_{11}Y_{12} + Z_{12}Y_{22}, & b_2 &= Z_{12}Y_{11} + Z_{22}Y_{12}. \end{aligned} \tag{11.5.12}$$

Similarly, the current wave equation is also obtained for the coupled transmission lines.

11.5.2 Solution of Coupled Wave Equation

The following form of the solution is assumed for the voltage wave propagation in the x-direction on the coupled lines:

$$V = V_0\, e^{-\gamma x}, \tag{11.5.13}$$

where γ is the complex propagation constant. On substituting the above solution in the wave equation (11.5.9), the following expression is obtained:

$$\gamma^2[V] = [Z][Y][V], \Rightarrow [V]\{[Z][Y] - \gamma^2[U]\} = 0, \tag{11.5.14}$$

where [U] is the unit matrix. The eigenvalues of the above equation are obtained from the following characteristics equation:

$$\text{Det}\left|\begin{bmatrix} a_1 & b_1 \\ b_2 & a_2 \end{bmatrix} - \begin{bmatrix} \gamma^2 & 0 \\ 0 & \gamma^2 \end{bmatrix}\right| = 0.$$
$$\Rightarrow \text{Det}\begin{vmatrix} a_1 - \gamma^2 & b_1 \\ b_2 & a_2 - \gamma^2 \end{vmatrix} = 0 \tag{11.5.15}$$

$$\Rightarrow \gamma^4 - \gamma^2(a_1 + a_2) + (a_1a_2 - b_1b_2) = 0$$
$$\Rightarrow \gamma^2 = \frac{(a_1 + a_2)}{2} \pm \frac{1}{2}\left[(a_1 - a_2)^2 + 4b_1b_2\right]^{\frac{1}{2}}. \tag{11.5.16}$$

Equation (11.5.16) provides two solutions for the propagation constants, indicating the existence of two linearly independent modes known as the *normal modes*. Two normal modes are two independent solutions of wave equations. In the present case, two normal modes are supported by a pair of coupled lines. Each normal mode has ± propagation constants, indicating the presence of the *forward* and *backward* traveling modal voltage waves. Likewise, the *structure of the N-coupled line supports N numbers of normal modes.* Each normal voltage mode has a distinct propagation constant. The phase velocity of the N normal modes of the N-coupled lines is given by $v_{p,i} = 1/\sqrt{[L][C]} = c/\sqrt{\varepsilon_{reff,i}}(i = 1, 2, ..., N)$.

The coupled lines could be either an *asymmetrical* or *symmetrical* structure. The medium could also be either an *inhomogeneous* medium or a *homogeneous* medium. Further, a medium could be isotropic or anisotropic. However, this subsection only considers the coupled transmission lines in the isotropic medium with four cases:

- *Asymmetrical coupled lines in an inhomogeneous medium.*
- *Asymmetrical coupled lines in a homogeneous medium.*
- *Symmetrical coupled lines in an inhomogeneous medium.*
- *Symmetrical coupled lines in a homogeneous medium.*

Case # 1: Asymmetrical Coupled Lines in the Inhomogeneous Medium

In the case of the asymmetrical coupled lines in an inhomogeneous medium, such as the *asymmetrically coupled microstrips and CPW*, the self-impedance and the self-admittance of two conductors are unequal, i.e. $Z_{11} \neq Z_{22}$ and $Y_{11} \neq Y_{22}$. The C-mode and π-mode are the *distinct normal modes* of asymmetrical coupled lines. The normal modes have distinct propagations constants and distinct characteristic impedances.

The propagation constants for the C and π-modes are obtained from equation (11.5.16):

$$\gamma_{1,2} = \pm\gamma_c \text{ (a)}, \qquad \gamma_{3,4} = \pm\gamma_\pi \text{ (b)}, \tag{11.5.17}$$

where (+) indicates the forward-moving C and π-modes and (−) indicates the backward moving C and π-modes. The propagation constants of these modes are given below:

$$\gamma_c^{\ 2} = \frac{(a_1 + a_2)}{2} + \frac{1}{2}\left[(a_1 - a_2)^2 + 4b_1b_2\right]^{\frac{1}{2}} \quad \text{(a)}$$
$$\gamma_\pi^{\ 2} = \frac{(a_1 + a_2)}{2} - \frac{1}{2}\left[(a_1 - a_2)^2 + 4b_1b_2\right]^{\frac{1}{2}} \quad \text{(b).} \tag{11.5.18}$$

The propagation constants of a lossless asymmetrical coupled transmission lines, in the isotropic inhomogeneous medium, can be obtained in terms of the primary line constants. For this purpose, the following expressions are used:

$$Z_{11} = j\omega L_{11}, \ Z_{12} = j\omega L_{12}, \quad Z_{22} = j\omega L_{22} \quad \text{(a)}$$
$$Y_{11} = j\omega C_{11}, \ Y_{12} = -j\omega C_{12}, \ Y_{22} = j\omega C_{22}$$
$$\frac{(a_1 + a_2)}{2} = \frac{j^2\omega^2}{2}[(L_{11}C_{11} + L_{22}C_{22}) - 2L_{12}C_{12}], (a_1 - a_2)^2$$
$$= j^4\omega^4\left[(L_{11}C_{11} - L_{22}C_{22})^2\right] \quad \text{(b)}$$
$$4b_1b_2 = 4j^4\omega^4[(L_{12}C_{11} - L_{22}C_{12})(L_{12}C_{22} - L_{11}C_{12})] \quad \text{(c).} \tag{11.5.19}$$

In the above equation, the negative value is used for the mutual capacitance (C_{12}). Normally, the line constants of microstrip and CPW kind of structures are frequency-dependent [J.4]. The following expressions for the propagation constants β_c and β_π of the lossless C and π-modes, $\gamma_{c,\pi} = j\beta_{c,\pi}$, are obtained on using the above equations with equation (11.5.18):

$$\beta_c = \frac{\omega}{\sqrt{2}}[(L_{11}C_{11} + L_{22}C_{22} - 2L_{12}C_{12}) + F]^{1/2} \quad \text{(a)}$$
$$\text{where, } F = \sqrt{(L_{11}C_{11} - L_{22}C_{22})^2 + 4(L_{12}C_{11} - L_{22}C_{12})(L_{12}C_{22} - L_{11}C_{12})} \quad \text{(b).} \tag{11.5.20}$$

$$\beta_\pi = \frac{\omega}{\sqrt{2}}\left[(L_{11}C_{11} + L_{22}C_{22} - 2\,L_{12}C_{12}) - F\right]^{1/2}. \quad (11.5.21)$$

The phase velocities and effective relative permittivities for C and π-modes are given below:

$$v_c = \frac{\omega}{\beta_c} = \frac{c}{\sqrt{\varepsilon_{\text{reff c}}}} \quad (a), \quad v_\pi = \frac{\omega}{\beta_\pi} = \frac{c}{\sqrt{\varepsilon_{\text{reff}\,\pi}}} \quad (b)$$
$$\Rightarrow \sqrt{\varepsilon_{\text{reff c}}} = \frac{c\beta_c}{\omega} \quad (c), \quad \sqrt{\varepsilon_{\text{reff}\,\pi}} = \frac{c\beta_\pi}{\omega} \quad (d). \quad (11.5.22)$$

For $L_{11} = L_{22}$ and $C_{11} = C_{22}$, the asymmetrical coupled lines structure is reduced to the symmetrical coupled lines. Also, the C and π-mode effective relative permittivities $\varepsilon_{\text{reff c}}$ and $\varepsilon_{\text{reff}\,\pi}$ are reduced to the even- and odd-mode effective relative permittivities $\varepsilon_{\text{reff e}}$ and $\varepsilon_{\text{reff o}}$ given by equations (11.4.32d) and (11.4.33d).

Case #2: Asymmetrical Coupled Lines in a Homogeneous Medium

In a homogeneous medium, the phase velocities of the EM-wave for both the transmission lines are identical:

$$\frac{1}{\sqrt{L_{11}C_{11}}} = \frac{1}{\sqrt{L_{22}C_{22}}}. \quad (11.5.23)$$

$$\omega^2 L_{11}C_{11} = \omega^2 L_{22}C_{22} \Rightarrow Z_{11}Y_{11} = Z_{22}Y_{22}. \quad (11.5.24)$$

Using the above expressions, the following expressions are obtained from equations (11.5.12):

$$a = a_1 = a_2 = Z_{11}Y_{11} + Z_{12}Y_{12} \quad (a)$$
$$b_1 = Z_{11}Y_{12} + Z_{12}Y_{22} \quad (b)$$
$$b_2 = Z_{12}Y_{11} + Z_{22}Y_{12} \quad (c). \quad (11.5.25)$$

The condition $b_1 \neq b_2$, ($Z_{11} \neq Z_{22}$, $Y_{11} \neq Y_{22}$), is valid for the asymmetrical coupled lines in a homogeneous medium. The following expressions for the propagation constants of the C and π-modes in a homogeneous medium are obtained, on using the above equations with equation (11.5.18):

$$\gamma_c^2 = (Z_{11}Y_{11} + Z_{12}Y_{12}) + \left[(Z_{11}Y_{12} + Z_{12}Y_{22})(Z_{12}Y_{11} + Z_{22}Y_{12})\right]^{\frac{1}{2}}. \quad (11.5.26)$$

$$\gamma_\pi^2 = (Z_{11}Y_{11} + Z_{12}Y_{12}) - \left[(Z_{11}Y_{12} + Z_{12}Y_{22})(Z_{12}Y_{11} + Z_{22}Y_{12})\right]^{\frac{1}{2}}. \quad (11.5.27)$$

For lossless asymmetrical coupled lines in a homogeneous medium, the propagation constants β_c and β_π are computed from the above equations.

Case # 3: Symmetrical Coupled Lines in an Inhomogeneous Medium

Both lines are identical for the symmetrical coupled lines in an inhomogeneous medium, i.e. $Z_{11} = Z_{22}$ and $Y_{11} = Y_{22}$. Equation (11.5.12) provides $a_1 = a_2 = a$ and $b_1 = b_2 = b$. Equation (11.5.18) gives the following expressions of the propagation constants for the C and π-modes:

$$\gamma_c^2 = a + b \Rightarrow \gamma_c = \pm\sqrt{a+b} \quad (a)$$
$$\gamma_\pi^2 = a - b \Rightarrow \gamma_\pi = \pm\sqrt{a-b} \quad (b). \quad (11.5.28)$$

In the case of the symmetrical coupled lines, the γ_c corresponds to a *sum mode*, also known as the *even mode, i.e.* $\gamma_c = \gamma_e$, while the γ_π corresponds to a *difference mode*, i.e. the *odd mode* leading to $\gamma_\pi = \gamma_o$. Thus, the even and odd modes are the *normal modes* of the symmetrical coupled lines. The propagation constants of these modes are further written in terms of the line parameters of the symmetrical coupled lines:

- *The propagation constant of the even mode (sum mode):*

$$\gamma_e = \pm\left[Z_{11}(Y_{11} + Y_{12}) + Z_{12}(Y_{11} + Y_{12})\right]^{\frac{1}{2}}$$
$$\Rightarrow \gamma_e = \pm\left[(Z_{11} + Z_{12})(Y_{11} + Y_{12})\right]^{\frac{1}{2}}. \quad (11.5.29)$$

- *The propagation constant of the odd mode (difference mode):*

$$\gamma_o = \pm\left[Z_{11}(Y_{11} - Y_{12}) - Z_{12}(Y_{11} - Y_{12})\right]^{\frac{1}{2}}$$
$$\Rightarrow \gamma_o = \pm\left[(Z_{11} - Z_{12})(Y_{11} - Y_{12})\right]^{\frac{1}{2}}. \quad (11.5.30)$$

For lossless symmetrical coupled lines, the even- and odd-mode propagation constants are reduced to the following expressions:

$$\beta_e = \pm\,\omega\sqrt{\left[(L_{11} + L_{12})(C_{11} - C_{12})\right]}$$
$$\Rightarrow \beta_e = \pm\,\omega\sqrt{\left[(L_{11}^d + 2L_{12}^d)C_{11}^d\right]} \quad (a)$$
$$\beta_o = \pm\,\omega\sqrt{\left[(L_{11} - L_{12})(C_{11} + C_{12})\right]}$$
$$\Rightarrow \beta_o = \pm\,\omega\sqrt{\left[(C_{11}^d + 2C_{12}^d)L_{11}^d\right]} \quad (b). \quad (11.5.31)$$

The above equations have already been obtained in the form of equations (11.4.32c) and (11.4.33c) by using the even–odd mode analysis method. It is noted that

$\beta_e = \beta_o\sqrt{\varepsilon_{r\,effe}} = (\omega/c)\sqrt{\varepsilon_{r\,effe}}$ and $\beta_o = \beta_o\sqrt{\varepsilon_{r\,effo}} = (\omega/c)\sqrt{\varepsilon_{r\,effo}}$. For an isolated line, the above equations are reduced to $\beta_e = \beta_o = \beta = \omega\sqrt{L_{11}C_{11}}$.

Case #4: Symmetrical Coupled Lines in the Homogeneous Medium

The even-mode and odd-mode phase velocities are equal in a homogeneous medium, their propagation constants are also equal, i.e. $\gamma_e = \gamma_o$. Using equations (11.5.29) and (11.5.30), the following condition is obtained:

$$Z_{12}Y_{11} = -Z_{11}Y_{12}, \Rightarrow L_{12}C_{11} = -L_{11}C_{12} \Rightarrow L_{12}C_{11} = L_{11}C_m, \quad (11.5.32)$$

where C_m is the capacitive mutual coupling, $C_m = -C_{12}$.

11.5.3 Modal Characteristic Impedance and Admittance

The characteristic impedances of both the C and π-modes, computed in this section, are normally expressed in terms of the *voltage-ratios* (R_c, R_π) of the C and the π-modes:

$$R_c = \left.\frac{v_2}{v_1}\right|_c \quad (a), \quad R_\pi = \left.\frac{v_2}{v_1}\right|_\pi \quad (b). \quad (11.5.33)$$

The C or π-mode phasor voltage traveling waves in the x-direction, on each of two conductors, is written as:

$$v_1 = v_{01}e^{-\gamma x} \quad (a), \quad v_2 = v_{02}e^{-\gamma x} \quad (b). \quad (11.5.34)$$

The coupled voltage waves, on the coupled transmission lines, are described by equation (11.5.9). It is rewritten below:

$$\frac{d^2}{dx^2}\begin{bmatrix} v_1 \\ v_2 \end{bmatrix} = \begin{bmatrix} a_1 & b_1 \\ b_2 & a_2 \end{bmatrix}\begin{bmatrix} v_1 \\ v_2 \end{bmatrix}. \quad (11.5.35)$$

$$\Rightarrow \frac{d^2v_1}{dx^2} - a_1v_1 - b_1v_2 = 0 \quad (a)$$
$$\frac{d^2v_2}{dx^2} - a_2v_2 - b_2v_1 = 0 \quad (b). \quad (11.5.36)$$

On substituting equations (11.5.34) in the above equations, the following characteristic equations are obtained:

$$\gamma^2v_1 - a_1v_1 - b_1v_2 = 0 \quad (a)$$
$$\gamma^2v_2 - a_2v_2 - b_2v_1 = 0 \quad (b). \quad (11.5.37)$$

Two values of the propagation constants, corresponding to the C or π-mode, are obtained. The voltage-ratio of either of the modes is

$$\frac{v_2}{v_1} = \frac{\gamma^2 - a_1}{b_1} = \frac{b_2}{\gamma^2 - a_2}. \quad (11.5.38)$$

The following expressions, for the voltage ratios of the C or π–mode, i.e. for R_c and R_π, are written by using equation (11.5.18) with the above equation:

$$R_c = \frac{1}{2b_1}\left[(a_2 - a_1) + \left\{(a_2 - a_1)^2 + 4b_1b_2\right\}^{\frac{1}{2}}\right] \quad (a)$$
$$R_\pi = \frac{1}{2b_1}\left[(a_2 - a_1) - \left\{(a_2 - a_1)^2 + 4b_1b_2\right\}^{\frac{1}{2}}\right] \quad (b). \quad (11.5.39)$$

Using equations (11.5.12) and (11.5.19), the above equations could be rewritten in terms of the equivalent circuit elements as follows:

$$R_c = \frac{(L_{22}C_{22} - L_{11}C_{11}) + F}{2(L_{12}C_{22} - L_{11}C_{12})} \quad (a), \quad R_\pi = \frac{(L_{22}C_{22} - L_{11}C_{11}) - F}{2(L_{12}C_{22} - L_{11}C_{12})} \quad (b)$$
$$\text{where, } F = \sqrt{(L_{11}C_{11} - L_{22}C_{22})^2 + 4(L_{12}C_{11} - L_{22}C_{12})(L_{12}C_{22} - L_{11}C_{12})} \quad (c). \quad (11.5.40)$$

The circuit elements of the asymmetrical coupled lines could be determined from the static analysis. The closed-form expressions for these circuit elements of the asymmetrically coupled microstrips are presented in subsection (12.2.1) of chapter 12.

The symmetrical coupled lines have $a_1 = a_2$ and $b_1 = b_2$. In this case, the above voltage-ratios are reduced to

$$R_c = +1, \text{corresponding to the even mode} \quad (a)$$
$$R_\pi = -1, \text{corresponding to the odd mode} \quad (b). \quad (11.5.41)$$

Therefore, to excite the C-mode with $R_c \neq +1$, the voltages v_1 and v_2 must have the same polarity; and to excite the π–mode with $R_\pi \neq -1$, the voltages v_1 and v_2 must have opposite polarity. The coupled lines phasor equations, in terms of the Z-matrix, are given by equation (11.5.7a). Substitution of the assumed solutions (11.5.34) in equation (11.5.7a), using a similar solution for the current waves, and taking $\gamma = \gamma_c$ for the C mode, provides the following expressions:

$$\gamma_c v_1 = Z_{11} i_1 + Z_{12} i_2 \quad (a)$$
$$\gamma_c v_2 = Z_{22} i_2 + Z_{12} i_1 \quad (b), \qquad \Rightarrow i_2 = \frac{\gamma_c v_2 - Z_{12} i_1}{Z_{22}}. \tag{11.5.42}$$

On substituting i_2 in equation (11.5.42a),

$$\gamma_c Z_{22} v_1 = Z_{11} Z_{22} i_1 + Z_{12}(\gamma_c v_2 - Z_{12} i_1)$$
$$\Rightarrow \gamma_c \left(Z_{22} - Z_{12} \frac{v_2}{v_1} \right) = \left(Z_{11} Z_{22} - Z_{12}^2 \right) \frac{i_1}{v_1}. \tag{11.5.43}$$

The characteristics admittance of the line #1 for the C-mode is $Y_{c1} = i_1/v_1$ and its voltage-ratio is $R_c = v_2/v_1$. The Y_{c1} for line #1 follows from the above equation:

$$Y_{c1} = \gamma_c \frac{(Z_{22} - Z_{12} R_c)}{\left(Z_{11} Z_{22} - Z_{12}^2 \right)} = \frac{1}{Z_{c1}}, \tag{11.5.44}$$

where Z_{c1} is the characteristics impedance of the C-mode supporting line #1. The symmetrical coupled lines, supporting the even mode, has $Z_{11} = Z_{22}$, $R_c = +1$, and $\gamma_c = \gamma_e$. The even-mode characteristics admittance follows from the above equation:

$$Y_e = \frac{\gamma_e}{(Z_{11} + Z_{12})} = \frac{1}{Z_e}. \tag{11.5.45}$$

Using equation (11.5.29), the characteristics impedance of the even mode is further written below in terms of the primary line constants of the symmetrical coupled lines:

$$Z_e = \frac{(Z_{11} + Z_{12})}{\gamma_e} = \frac{(Z_{11} + Z_{12})}{\sqrt{(Z_{11} + Z_{12})(Y_{11} + Y_{12})}}$$
$$= \sqrt{\frac{(Z_{11} + Z_{12})}{(Y_{11} + Y_{12})}} \quad (a)$$
$$\Rightarrow Z_e = \sqrt{\frac{(L_{11} + L_{12})}{(C_{11} - C_{12})}} \quad (b). \tag{11.5.46}$$

The above relation is already incorporated in equation (11.4.32a) by using the even–odd mode analysis method. In the above equation, a negative sign is taken for C_{12}. Further, the characteristics admittance for the π-mode is

$$Y_{\pi 1} = \gamma_\pi \frac{(Z_{22} - Z_{12} R_\pi)}{\left(Z_{11} Z_{22} - Z_{12}^2 \right)} = \frac{1}{Z_{\pi 1}}. \tag{11.5.47}$$

The characteristics admittance of the odd mode follows from the above equations for $R_\pi = -1$, $Z_{11} = Z_{22}$ and $\gamma_\pi = \gamma_o$:

$$Y_o = \frac{\gamma_o}{(Z_{11} - Z_{12})} = \frac{1}{Z_o}. \tag{11.5.48}$$

Using equation (11.5.30), the characteristics impedance for the odd mode can be further written in terms of the primary line constants of the symmetrical coupled lines:

$$Z_o = \frac{(Z_{11} - Z_{12})}{\gamma_o} = \frac{(Z_{11} - Z_{12})}{\sqrt{(Z_{11} - Z_{12})(Y_{11} - Y_{12})}}$$
$$= \sqrt{\frac{(Z_{11} - Z_{12})}{(Y_{11} - Y_{12})}} \quad (a)$$
$$\Rightarrow Z_o = \sqrt{\frac{(L_{11} - L_{12})}{(C_{11} + C_{12})}} \quad (b). \tag{11.5.49}$$

Again, the above relation is incorporated in equation (11.4.33a) by using the even–odd mode analysis method.

The C and π-modes on line #2 of the asymmetrical coupled lines could be considered to obtain its characteristic admittance for both modes. The current i_1 is solved from equation (11.5.42a) to get the relation $i_1 = (\gamma_c v_1 - Z_{12} i_2)/Z_{11}$. On substituting i_1 in equation (11.5.42b), the following expression is obtained:

$$\gamma_c Z_{11} v_2 = Z_{11} Z_{22} i_2 + Z_{12}(\gamma_c v_1 - Z_{12} i_2)$$
$$\Rightarrow \gamma_c (Z_{11} v_2 - Z_{12} v_1) = \left(Z_{11} Z_{22} - Z_{12}^2 \right) i_2$$
$$\Rightarrow \gamma_c \left(Z_{11} - Z_{12} \frac{v_1}{v_2} \right) = \left(Z_{11} Z_{22} - Z_{12}^2 \right) \frac{i_2}{v_2}. \tag{11.5.50}$$

The characteristics admittance of line #2 of a coupled transmission lines for the C-mode is $Y_{c2} = i_2/v_2$ and its voltage-ratio is $R_c = v_2/v_1$. The Y_{c2} for the line #2, from the above equation, is obtained as:

$$Y_{c2} = \frac{\gamma_c}{R_c} \frac{(Z_{11} R_c - Z_{12})}{\left(Z_{11} Z_{22} - Z_{12}^2 \right)} = \frac{1}{Z_{c2}}. \tag{11.5.51}$$

Likewise, the characteristics admittance of line #2 of the π-mode is

$$Y_{\pi 2} = \frac{\gamma_\pi}{R_\pi} \frac{(Z_{11} R_\pi - Z_{12})}{\left(Z_{11} Z_{22} - Z_{12}^2 \right)} = \frac{1}{Z_{\pi 2}}. \tag{11.5.52}$$

The characteristic impedance/admittance of both modes is related through the voltage-ratio of both modes given by equation (11.5.33). The C and π-modes are excited by the identical pair of unequal voltages. To excite the C-mode, both voltages have the same polarity. However, in the case, the π-mode one voltage on the strip conductor has the opposite polarity. It helps to define the voltage-ratio of one mode in terms of current-ratio of another mode as follows [B.17, J.10]:

$$R_c = \frac{V_{2c}}{V_{1c}} = -\frac{I_{1\pi}}{I_{2\pi}} \quad (a), \quad R_\pi = \frac{V_{2\pi}}{V_{1\pi}} = -\frac{I_{1c}}{I_{2c}} \quad (b). \tag{11.5.53}$$

Using the relations $V_{\pi i} = Z_{\pi i} I_{\pi i} (i = 1, 2)$ and $V_{c i} = Z_{c i} I_{c i} (i = 1, 2)$, the following expressions are obtained:

$$R_c = -\frac{I_{1\pi}}{I_{2\pi}} = -\frac{V_{1\pi}/Z_{1\pi}}{V_{2\pi}/Z_{2\pi}} = -\frac{Z_{2\pi}}{Z_{1\pi}}\frac{1}{R_\pi} \quad (a)$$

$$R_\pi = -\frac{I_{1c}}{I_{2c}} = -\frac{V_{1c}/Z_{1c}}{V_{2c}/Z_{2c}} = -\frac{Z_{2c}}{Z_{1c}}\frac{1}{R_c} \quad (b)$$

$$\text{Therefore,} \quad \frac{Y_{c1}}{Y_{c2}} = \frac{Y_{\pi 1}}{Y_{\pi 2}} = -R_c R_\pi \quad (c). \tag{11.5.54}$$

For the symmetrical coupled lines, even-mode characteristics admittances for both the lines are equal. So is the case of the odd-mode characteristics admittances. However, for asymmetrical coupled lines, the C and π-mode characteristics admittances, for line #1 and line #2, are unequal:

$$\text{For symmetrical coupled lines}: Y_{e1} = Y_{e2}, Y_{o1} = Y_{o2} \quad (a)$$
$$\text{For asymmetrical coupled lines}: Y_{c1} \neq Y_{c2}, Y_{\pi 1} \neq Y_{\pi 2} \quad (b). \tag{11.5.55}$$

The capacitive and inductive coupling coefficients, k_C and k_L, and the final coupling coefficient C of the asymmetrical coupled lines are obtained from equations (11.3.12), (11.3.24), and (11.3.25).

The concepts and formulations of this chapter are used further in Chapter 12 to analyze the coupled microstrips and coupled CPW structures.

References

Books

B.1 Fache, N.; Frank, O.; Daniel, D.Z.: *Electromagnetic and Circuit Modelling of Multiconductor Transmission Lines*, Clarendon Press, Oxford, NY, 1993.

B.2 Bewley, L.V.: *Travelling Waves on Transmission Systems*, Wiley, New York, 1951.

B.3 Noyan, K.; Aksun M.I.: *Modern Microwave Circuits*, Artech House, Boston, 2003.

B.4 Hoffmann, R.: *Microwave Integrated Circuit Handbook*, Artech House, Boston, 1985.

B.5 Gupta, K.C.; Garg R.; Bahl I.; Bhartia P.: *Microstrip Lines and Slot Lines*, 2nd Edition, Artech House, 1996.

B.6 Edward, T.C.: *Foundations for Microstrip Circuit Design*, John Wiley & Sons, New York, 1987.

B.7 Fooks, E.H.; Zakarevicius, R.A.: *Microwave Engineering Using Microstrip Circuits*, Prentice-Hall PTR, New Jersey, 1989.

B.8 Di Paolo, F.: *Network, and Devices Using Planar Transmission Lines*, CRC Press, New York, 2000.

B.9 Edward, T.C.; Steer, M.B.: *Foundations of Interconnects and Microstrip Circuit Design*, 3rd Edition, John Wiley & Sons, New York, 2000.

B.10 Collin, R.E.: *Foundations for Microwave Engineering*, 2nd Edition, McGraw-Hill, Inc., New York, 1992.

B.11 Collin, R.E.: *Field Theory of Guided Waves*, IEEE Press, New York, 1991.

B.12 Simon, R.N.: *Coplanar Waveguide Circuits Components and Systems*, John Wiley, New York, 2001.

B.13 Wolff, I.: *Coplanar Microwave Integrated Circuits*, John Wiley & Sons, New York, 2006.

B.14 Young, L. (Editor): *Advances in Microwaves*, Vol. 8, Academic Press, 1974.

B.15 Young, L. (Editor): *Advances in Microwaves*, Vol. 1, Academic Press, 1966.

B.16 Kneppo, I.; Fabian J.; Bezousek P.; Hrnicko P.; Pavel M.: *Microwave Integrated Circuits*, Chapman & Hall, London, 1994.

B.17 Mongia, R.; Bhartia, P.; Bahl, I.J.: *R. F. and Microwave Coupled-Line Circuits*, 2nd Edition, Artech-House, Boston, 1999.

B.18 Hung-Chia, H.: *Coupled Mode Theory- As Applied to Microwaves and Optical Transmission*, VNU Science Press, the Netherland, 1984.

B.19 Dworsky, L.N.: *Modern Transmission Line Theory and Applications*, John Wiley, New York, 1979.

B.20 Costa, L.; Valtonen, M.: *Implementation of Single and Coupled Microstrip Lines in APLAC*, CT-33, ISBN 951-222-3881-0, Helsinki University, Finland, Dec. 1997.

B.21 Djordjevic, A.R; Harrington, R.F.; Sarkar, T.K.; Bazdar, M.B.: Matrix Parameters for Multiconductor Transmission Lines – *Software and User's Manual*, Norwood Artech House, 1989.

B.22 Mattick, R.E.: *Transmission Lines For Digital and Communication Networks*. IEEE Press, New York, 1995

B.23 Rizzi, P.A.: *Microwave Engineering- Passive Circuits*, Prentice-Hall Int. Ed., New Jersey, 1988.

B.24 Orfanidis, S.J.: *Electromagnetic Waves and Antenna*, Free Book on Web, ECE Department, RutgersUniversity, Piscataway, NJ, 2016.

B.25 Kuznetsov, P.I.; Stratonovich, R.L.: *The Propagation of Electromagnetic Waves in Multiconductor Transmission Lines*, Pergamon Press, London, 1964.

Journals

J.1 Yasumoto, K.: *Coupled – mode formulation of multilayered and multiconductor transmission lines, IEEE Trans. Microw. Theory Tech.*, Vol. MTT-44, No. 4, pp. 585–590, April. 1996.

J.2 Pipes, L.A.: *Matrix theory of multiconductor transmission lines, Philos. Mag.*, Vol. 24, No. 159, pp. 97–11, 1937.

J.3 Bazdar, M.B.; Djordjevic, A.R.; Harrington, R.F.; Mautz, J.R.; Sarkar, T.K.: *Evaluation of quasi- static matrix parameters for multiconductor transmission lines using Galerkin's method, IEEE Trans. Microw. Theory Tech.*, Vol. MTT- 42, No. 7, pp. 1223–1228, July 1984.

J.4 Tripathi, V.K.: *A dispersion model for coupled microstrips, IEEE Trans. Microw. Theory Tech.*, Vol. MTT-34, No. 1, pp. 66–67, 1986.

J.5 Marx, K.D.: *Propagation modes, equivalent circuits, and characteristic terminations for multiconductor transmission lines with inhomogeneous dielectrics, IEEE Trans. Microw. Theory Tech.*, Vol. MTT-21, No. 7, pp. 450–457, July 1973.

J.6 Rao, J.S.; Joshi, K.K.; Das, B.N.: *Analysis of small aperture coupling between the rectangular waveguide and microstrip line, IEEE Trans. Microw. Theory Tech.*, Vol. MTT-29, No. 2, pp. 150–154, Feb 1981.

J.7 Miao, J.-F.; Itoh, T.: *Coupling between microstrip and image guide through a small aperture in the common ground plane, IEEE Trans. Microw. Theory Tech.*, Vol. MTT-31, No. 4, pp. 361–363, April 1983.

J.8 Sobhy, M.I.; Hosny, E.A.: *The design of directional couplers using exponential lines in inhomogeneous media, IEEE Trans. Microw. Theory Tech.*, Vol. MTT-30, No. 1, pp. 71–76, Jan 1982.

J.9 Adair, J.E.; Haddad, G.I.: *Coupled-mode analysis of non-uniform coupled transmission lines, IEEE Trans. Microw. Theory Tech.*, Vol. MTT-17, No. 10, pp. 746–752, Oct. 1969.

J.10 Tripathi, V.K.: *Asymmetric coupled transmission lines in an inhomogeneous medium, IEEE Trans. Microw. Theory Tech.*, Vol. MTT-23, No. 9, pp. 734–739, 1975.

J.11 Allan, J.L.: *Non-symmetrical coupled lines in an inhomogeneous dielectric medium, Int. J. Electron.*, Vol. 38, pp. 337–347, 1975.

J.12 Tripathi, V.K.; Chang, C.L.: *Quasi-TEM parameters of non-symmetrical coupled microstrip lines, Int. J. Electron.*, Vol. 45, No. 2, pp. 215–223, 1978.

J.13 Tripathi, V.K.: *Equivalent circuits and characteristics of inhomogeneous non-symmetrical coupled line two-port circuits, IEEE Trans. Microw. Theory Tech.*, Vol. MTT-25, No. 2, pp. 140–142, Feb. 1977.

J.14 Jones, E.M.T.; Bolljahn, T.: *Coupled strip transmission line filters and directional couplers, IEEE Trans. Microw. Theory Tech.*, Vol. MTT-4, No. 2, pp. 75–81, April., 1956.

J.15 Krage, M.K; Haddad, G.I.: *Characteristics* of *coupled microstrip transmission lines-I: Coupled mode formulation of inhomogeneous lines, IEEE Trans. Microw. Theory Tech.*, Vol. MTT-18, No. 4, pp. 217–222, 1970.

J.16 Verma, A.K.; Kumar, R.: *A new dispersion model for microstrip line, IEEE Trans. Microw. Theory Tech.*, Vol. MTT-46, No. 8, pp. 1183–1187, 1998.

J.17 Ikalainen, P.K.; Matthaei, G.L.: *Wide-band, forward – coupling microstrip hybrid with high directivity, IEEE Trans. Microw. Theory Tech.*, Vol. MTT-35, No. 8, pp. 719–725, Aug. 1987.

J.18 Verma, A. K.; Nasimuddin, N.; Ojha, R.K.: *Synthesis and analysis of optimum directivity suspended microstrip coupler*, Asia-Pacific Microwave Conference, APMC -Dec. 2000, Sydney, Australia.

J.19 Bryant, T.G.; Weiss, J.A.: *Parameters of microstrip transmission lines and coupled pairs of microstrip lines, IEEE Trans. Microw. Theory Tech.*, Vol. MTT-16, No. 12, pp. 1021–1027, 1968.

J.20 Hammerstad, E.; Jensen, O.: *Accurate models for Microstrip Computer-Aided Design*, IEEE., MTT- S, Int. Microwave Symposium Digest, pp. 407–409, 1980.

J.21 Garg, R.; Bahl, I.J.: *Characteristics* of *coupled microstrip lines, IEEE Trans. Microw. Theory Tech.*, Vol. MTT-27, No. 7, pp. 700–705, 1979. Also corrections in, Vol. MTT- 28, No., pp. 272, 1980.

J.22 Wei, C.; Harrington, R.F.; Mautz, J.R.; Sarkar, T.K.: *Multiconductor transmission lines multilayered dielectric media, IEEE Trans. Microw. Theory Tech.*, Vol. MTT-32, No. 4, pp. 439–450, April 1984.

J.23 Krage, M.K; Haddad, G.I.: *Characteristics* of *coupled microstrip transmission lines-II: Evaluation of coupled lines parameters, IEEE Trans. Microw. Theory Tech.*, Vol. MTT-18, No. 4, pp. 222–228, 1970.

J.24 Levy, R.: *Transmission line directional couplers for very broad-band operation, Proc. IEE*, Vol. 112, No. 3, March 1965.

J.25 Carson, J.R.; Hoyt, R.S.: *Propagation of periodic currents over a system of parallel wires, Bell Syst. Tech. J.*, Vol. 6, pp.495–545, 1927.

J.26 Rice, S.O.: *Steady-state solutions of transmission line equations, Bell Syst. Tech. J.*, Vol. 20, No. 2, pp. 131–178, April 1941.

12

Planar Coupled Transmission Lines

Introduction

The basic theoretical concepts and formulations of the coupled lines are presented in chapter 11. This chapter is concerned with their applications to the planar coupled transmission lines, such as the coupled microstrips and CPW. These planar coupled lines are needed for the design of important microwave components such as directional couplers, filters, phase shifters, delay lines, 3 dB-hybrids, matching networks, etc. The circuits and components based on the coupled lines are discussed in several books [B.1–B.11].

Objectives

- *To compute static and dynamic parameters of symmetrical coupled microstrip lines.*
- *To compute static and dynamic parameters of asymmetrical coupled microstrip lines.*
- *To compute static parameters of coupled CPW.*
- *To obtain the 4-ports and 2-ports [Z], [Y], and [ABCD] parameters of coupled strip lines.*
- *To obtain the 4-ports and 2-ports [Z], [Y], and [ABCD] parameters of coupled microstrip lines.*
- *To obtain the [ABCD] parameters of the C-mode and π-mode of asymmetrical coupled microstrip lines.*

12.1 Line Parameters of Symmetric Edge Coupled Microstrips

Figure (12.1) shows coupled microstrips in the inhomogeneous media. The medium could be homogeneous also. This section presents both the static and frequency-dependent closed-form models to compute the line parameters of symmetrical *edge coupled* microstrips in terms of physical dimensions. It also accounts for the effects of the top shield, and the finite conductor thickness in the models [B.1]. Bryant and Weiss [J.1] and also Krage and Haddad [J.21] have provided the static numerical results on the parameters of the symmetrically coupled microstrips. The *method of moment*-based software is available for such computation [B.2, J.2, J.3]. A summary of the frequency-dependent models, developed by Kirschning and Jansen is also presented [J.4], for the symmetrical coupled microstrip lines. The methods are also available for computation of line parameters for the *broadside coupled* microstrips [J.5, J.6]. In chapter 14, the variational method, applicable to the microstrip lines in the layered dielectric medium, is presented to compute the static coupled line parameters more systematically.

12.1.1 Static Models for Even- and Odd-Mode Relative Permittivity and Characteristic Impedances of Edge Coupled Microstrips

Figure (12.1) shows shielded asymmetrical coupled microstrips of strip widths (w_1,w_2), separation s, and strip conductor thickness t. The substrate is having relative permittivity ε_r, thickness h_1, and is covered with a shield height h_2. However, in the case of symmetrically coupled microstrips, strip conductors have width $w_1 = w_2 = w$. The presence of a top shield creates additional capacitance in the air medium, between the top shield and strip conductors, thereby reducing the effective relative permittivity and characteristics impedance of coupled microstrips. The physical dimensions of the coupled microstrip are normalized with respect to the substrate thickness:

$$u = w/h_1, g = s/h_1, t_h = t/h_1, h_s = h_2/h_1. \tag{12.1.1}$$

The expressions summarized below have accuracy within 1% in the range $0.1 \le u \le 10$, $0.1 \le g \le 10$, and $1 \le \varepsilon_r \le 18$. These expressions are used in the commercial circuit simulator for the microwave design applications [B.1, B.3, J.4, J.7].

Even-Mode Effective Relative Permittivity

Following the process used in subsection (8.2.1) of chapter 8 applicable to the single conductor microstrip, an

Introduction to Modern Planar Transmission Lines: Physical, Analytical, and Circuit Models Approach, First Edition. Anand K. Verma.

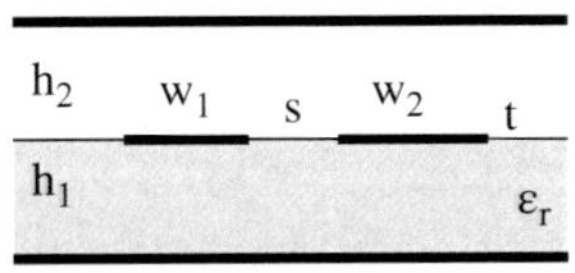

Figure 12.1 Asymmetrical coupled microstrip lines with a top shield. (For the symmetrical coupled lines, $w_1 = w_2 = w$.)

expression is written below for the even-mode effective relative permittivity of the symmetrically coupled microstrips, in terms of the *even-mode filling-factor* q_e [J.7]:

$$\varepsilon_{r\,effe}(f=0) = \left(\frac{\varepsilon_r+1}{2}\right) + \left(\frac{\varepsilon_r-1}{2}\right)F_e$$

$$\Rightarrow \varepsilon_{r\,effe}(f=0) = 1 + q_e(\varepsilon_r - 1) \qquad \text{(a)}$$

Where even mode filling – factor :

$$q_e = \frac{F_e+1}{2} \qquad \text{(b).} \qquad (12.1.2)$$

The even-mode function F_e is dependent on the structural parameters u, g, and ε_r. The even-mode function F_e, taking care of the conductor thickness and the top shield height, is

$$F_e = (F_{\infty,e} - F_{t,e})F_{s,e} \qquad \text{(a)}$$

$$F_{\infty,e} = \left[1 + \frac{10}{p}\right]^{-a_e(p)b_e(\varepsilon_r)} \qquad \text{(b)}$$

$$F_{t,e} = \frac{2\ln(2)}{\pi}\frac{t_h}{\sqrt{u}} \qquad \text{(c)}$$

$$F_{s,e} = \begin{cases} \tanh\left(1.626 + 0.107h_s - \dfrac{1.733}{\sqrt{h_s}}\right), & \text{for } h_s \le 39 \\ 1 & \text{for, } h_s > 39 \end{cases} \qquad \text{(d).} \qquad (12.1.3)$$

where $F_{\infty,\,e}$ is the even-mode-related function of open coupled microstrips, i.e. for the case $h_2 \to \infty$. The conductor thickness is accounted for by using a correction factor $F_{t,e,}$ and the shield height is accounted for by another correction factor $F_{s,e}$ of March [J.8]. The more accurate expression is also available [J.9]. The structure-dependent variable p is given below. It determines the parameter $a_e(p)$ of equation (12.1.3b). The parameter $b_e(\varepsilon_r)$ is permittivity dependent. The closed-form expressions for parameters $a_e(p)$ and $b_e(\varepsilon_r)$ are summarized below:

$$a_e(p) = 1 + \frac{1}{49}\ln\left[\frac{p^4 + \left(\frac{p}{52}\right)^2}{p^4 + 0.432}\right] + \frac{1}{18.7}\ln\left[1 + \left(\frac{p}{18.1}\right)^3\right] \qquad \text{(a)}$$

$$b_e(\varepsilon_r) = 0.564\left(\frac{\varepsilon_r - 0.9}{\varepsilon_r + 3.0}\right)^{0.053} \qquad \text{(b)}$$

$$p = \frac{u\,(20 + g^2)}{10 + g^2} + g.\exp(-g) \qquad \text{(c).} \qquad (12.1.4)$$

The above equations are a modification of equation (8.2.15) given in chapter 8.

Odd-Mode Effective Relative Permittivity

The improved Hammerstad and Jensen model [B.1, J.4, J.7], to compute the static odd-mode effective relative permittivity, in terms of the odd-mode filling-factor, is given below:

$$\varepsilon_{r\,effo}(f=0) = \left[\frac{\varepsilon_r+1}{2} + a_o(u,\varepsilon_r) - \varepsilon_{r\,eff}(f=0)\right]q_o + \varepsilon_{r\,eff}(f=0), \qquad (12.1.5)$$

where $\varepsilon_{r\,eff}(f=0)$ is the static effective relative permittivity of a *single isolated microstrip line*, given in subsection (8.2.1) of chapter 8. The parameter a_o (u,ε_r) is given below:

$$a_o(u,\varepsilon_r) = 0.7287\left[\varepsilon_{r\,eff}(f=0) - \frac{\varepsilon_r+1}{2}\right] \times [1 - \exp(-0.179\,u)]. \qquad (12.1.6)$$

The following modified *odd-mode filling-factor* q_o takes into account the correction factors $q_{t,\,o}$ and $q_{s,o}$, due to the conductor thickness and shield height, respectively:

$$q_o = (q_{\infty,o} - q_{t,o})\,q_{s,o} \quad \text{(a)}, \quad \text{where, } q_{t,o} = \frac{2\ell n\,2}{\pi}\frac{t_h}{\sqrt{u}} \qquad \text{(b)}$$

$$q_{\infty,o} = \exp\left[-c_o\,g^{d_o}\right], \quad c_o = b_o(\varepsilon_r) - [b_o(\varepsilon_r) - 0.207]\exp[-0.414\,u]$$

$$b_o(\varepsilon_r) = \frac{0.747\,\varepsilon_r}{0.15 + \varepsilon_r}, \quad d_o = 0.593 + 0.694.\exp(-0.562\,u) \qquad \text{(c)} \qquad (12.1.7)$$

$$q_{s,o} = \begin{cases} \tanh\left(\dfrac{9.575}{7 - h_s} - 2.965 + 1.68h_s - 0.311h_s^2\right), & \text{for } h_s < 7 \\ 1 & \text{for } h_s \ge 7 \end{cases} \qquad \text{(d),}$$

where $q_{\infty,o}$ is the filling-factor of the open structure, i.e. for the case $h_2 \to \infty$. In the case of the odd mode, the electric field confinement in the line structure is stronger. So, the odd mode case has $h_s>7$, whereas for the even mode, due to weak field confinement, $h_s> 39$ for taking $q_{s,e} = q_{s,o}$.

Even-Mode Characteristic Impedance

The closed-form expressions to compute the static even-mode characteristic impedance of open coupled microstrips of zero conductor thickness, due to Kirschning and Jansen [J.4], are summarized below:

$$Z_{oe,\infty}(\varepsilon_r, f=0) = \frac{Z_o(f=0)\sqrt{\varepsilon_{r,eff}(f=0)/\varepsilon_{reff,e}(f=0)}}{1-[Z_o(f=0)/\eta_0]\sqrt{\varepsilon_{r,eff}(f=0)}\,Q_4}, \tag{12.1.8}$$

where the static characteristic impedance and effective relative permittivity, $Z_o(f=0)$ and $\varepsilon_{r,\,eff}(f=0)$, of a single isolated microstrip, are computed in subsection (8.2.2) of chapter 8. The even mode $\varepsilon_{reff,e}(f=0)$ is determined from equation (12.1.2). The effect of the top shield and conductor thickness on the even-mode characteristic impedance $Z_{oe,\,\infty}(f=0,\ \varepsilon_r)$ is accounted for separately. The parameter Q_4 of equation (12.1.8) is evaluated from the following set of equations:

$$\begin{aligned}
Q_4 &= \frac{2Q_1/Q_2}{\exp(-g)\,u^{Q_3} + [2-\exp(-g)]\,u^{-Q_3}} && \text{(a)}\\
Q_3 &= 0.1975 + \left[16.6 + (8.4/g)^6\right]^{-0.387} + \frac{1}{241}\ln\left[\frac{g^{10}}{1+(g/3.4)^{10}}\right] && \text{(b)}\\
Q_2 &= 1 + 0.7519\,g + 0.189\,g^{2.31} && \text{(c)}\\
Q_1 &= 0.8695\,u^{0.194} && \text{(d).}
\end{aligned} \tag{12.1.9}$$

Odd-Mode Characteristic Impedance

The expressions to compute the static odd-mode characteristic impedance are summarized below [J.4]:

$$Z_{oo,\infty}(\varepsilon_r, f=0) = \frac{Z_o(f=0)\sqrt{\varepsilon_{reff}(f=0)/\varepsilon_{reff,o}(f=0)}}{1-[Z_o(f=0)/\eta_0]\sqrt{\varepsilon_{reff}(f=0)}\,Q_{10}}. \tag{12.1.10}$$

The odd mode $\varepsilon_{r,effo}(f=0)$ is determined from equation (12.1.5), and the parameter Q_{10} is computed as follows:

$$\begin{aligned}
Q_{10} &= \left[Q_2\,Q_4 - Q_5\,\exp\left(\ln(u)\,Q_6\,u^{-Q_9}\right)\right]/Q_2 && \text{(a)}\\
Q_9 &= \ln(Q_7)[Q_8 + (1/16.5)] && \text{(b)}\\
Q_8 &= \exp\left[-6.5 - 0.95\ln(g) - (g/0.15)^5\right] && \text{(c)}\\
Q_7 &= \left[10 + 190g^2\right]/\left[1 + 82.3g^3\right] && \text{(d)}\\
Q_6 &= 0.2305 + (1/281.3)\ln\left[g^{10}/\left(1+(g/5.8)^{10}\right)\right] \\
&\quad + (1/5.1)\ln\left[1 + 0.598\,g^{1.154}\right] && \text{(e)}\\
Q_5 &= 1.794 + 1.14\ln\left[1 + 0.638/\left(g + 0.517g^{2.43}\right)\right] && \text{(f).}
\end{aligned} \tag{12.1.11}$$

The effect of the top shield on the even/odd mode characteristics impedance is accounted below [B.4, J.8].

Effect of the Top Shield on Even-Mode Characteristic Impedance

The proximity of the top shield to strip conductors reduces the characteristic impedance of open coupled microstrips. It is accounted for by using the following even-mode impedance correction factor ΔZ_{oe}:

$$\begin{aligned}
Z_{oe}(f=0,\varepsilon_r,u,h_s) &= Z_{oe,\infty}(f=0,\varepsilon_r,u,h_s\to\infty) \\
&\quad - \Delta Z_{oe}(f=0,\varepsilon_r,u,h_s) && \text{(a)}\\
\text{where, } \Delta Z_{oe}(f=0,\varepsilon_r,u,h_s) &= \frac{\Delta Z_{oe}(f=0,\varepsilon_r=1,u,h_s)}{\sqrt{\varepsilon_{r,effe}(f=0,\varepsilon_r,u)}} && \text{(b).}
\end{aligned} \tag{12.1.12}$$

For the static even mode, $\varepsilon_{r,effe}(f = 0,\ \varepsilon_r,\ u)$ and $Z_{oe,\infty}(f = 0,\ \varepsilon_r,\ u,\ h_s \to \infty)$ are computed using equations (12.1.2) and (12.1.8). The even-mode impedance correction factor on the air-substrate is obtained from the following set of equations [B.4]:

$$\begin{aligned}
&\Delta Z_{oe}(f=0,\varepsilon_r=1,u,h_s) = f_e(u,h_s).g_e(g,h_s)\\
&f_e(u,h_s) = 1-\tanh^{-1}[A+(B+C.u)u]\\
&A = -4.351/(1+h_s)^{1.842},\ B = 6.639/(1+h_s)^{1.861}\\
&C = -2.291/(1+h_s)^{1.90}\\
&g_e(g,h_s) = 270\left[1-\tanh\left(D+E\sqrt{1+h_s}-F/(1+h_s)\right)\right]\\
&D = 0.747/\sin(0.5.\pi.x),\ E = 0.725.\sin(0.5.\pi.y)\\
&\log_{10}F = 0.11-0.0947.g\\
&\log_{10}x = 0.103.g-0.159,\ \log_{10}y = 0.0492.g-0.073.
\end{aligned} \tag{12.1.13}$$

Effect of the Top Shield on Odd-Mode Characteristic Impedance

The expression for the odd-mode characteristic impedance of the top-shielded coupled microstrips, using the correction factor ΔZ_{oo}, is written below:

$$Z_{oo}(f=0,\varepsilon_r,u,h_s)=Z_{oo,\infty}(f=0,\varepsilon_r,u,h_s\to\infty)-\Delta Z_{oo}(f=0,\varepsilon_r,u,h_s) \quad (a)$$

$$\text{where, } \Delta Z_{oo}(f=0,\varepsilon_r,u,h_s)=\frac{\Delta Z_{oo}(f=0,\varepsilon_r=1,u,h_s)}{\sqrt{\varepsilon_{r,effo}(f=0,\varepsilon_r,u)}} \quad (b). \tag{12.1.14}$$

For the static odd mode, $\varepsilon_{r,\ effo}(f = 0,\ \varepsilon_r,\ u)$ and $Z_{oo}(f = 0,\ \varepsilon_r,\ u,\ h_s \to \infty)$ are computed using equations (12.1.5) and (12.1.10). The odd-mode impedance correction factor, on the air-substrate, is obtained from the following set of equations [B.4]:

$$\Delta Z_{oo}(f=0,\varepsilon_r=1,u,h_s)=f_o(u,h_s).g_o(g,h_s)$$

$$f_o(u,h_s)=u^J, J=\tanh\left[(1+h_s)^{1.585}/6\right]$$

$$g_o(g,h_s)=270\left[1-\tanh\left(G+K\sqrt{1+h_s}-L/(1+h_s)\right)\right]$$

$$G=2.178-0.796.g,$$

$$K=\begin{cases}\log_{10}(20.429.g^{0.174}), & g>0.858\\ 1.30, & g\le 0.858\end{cases}$$

$$L=\begin{cases}2.51.g^{-0.462}, & g>0.873\\ 2.674, & g\le 0.873.\end{cases} \tag{12.1.15}$$

Effect of Conductor Thickness on Microstrip

The conductor thickness increases the normalized effective even- and odd-mode strip conductor widths as follows:

$$u_{t,e}=u+\Delta u[1-0.5.\exp(-0.69\Delta u/\Delta t)] \quad (a)$$

$$u_{t,o}=u_{t,e}+\Delta t \quad (b)$$

$$\text{where, } \Delta t=\frac{t_h}{g.\varepsilon_r} \quad (c)$$

$$\Delta u=\frac{1.25}{\pi}t_h\left[1+\ln\left(\frac{2}{t_h}+\frac{(4\pi u-2)/t_h}{1+\exp\{-100(u-1/2\pi)\}}\right)\right] \quad (d). \tag{12.1.16}$$

The Δu is a change in the strip width of a single conductor [J.10].

To account for the effect of the strip conductor thickness, the normalized strip-width u is replaced by the thickness-dependent strip-width $u_{t,e}$, and $u_{t,o}$, for the even mode and odd mode, in the expressions of even and odd modes effective relative permittivity and characteristic impedance.

Another Simple Closed-Form Expression

The even- and odd-mode line capacitances can be computed from the following expressions obtained using the conformal mapping method [J.11]:

$$C_e=2\varepsilon_0\left[\varepsilon_r\frac{K(k_1)}{K'(k_1)}+\frac{K(k_2)}{K'(k_2)}\right] \quad (a)$$

$$C_o=2\varepsilon_0\left[\varepsilon_r\frac{K(k_3)}{K'(k_3)}+\frac{K(k_4)}{K'(k_4)}\right] \quad (b)$$

$$\text{where, } k_1=\tanh(\pi w/4h)\tanh(\pi\{w+s\}/4h)$$

$$k_2=\tanh(\pi w/4\{h+\pi w\})\tanh(\pi\{w+s\}/4\{h+\pi w\})$$

$$k_3=\tanh(\pi w/4h)\coth(\pi\{w+s\}/4h), k_4=w/(w+s). \tag{12.1.17}$$

The K(k) is the first kind elliptical integral and $K'(k)=K\left(\sqrt{1-k^2}\right)$. The even- and odd-mode characteristic impedance are computed using equations (11.4.32) and (11.4.33) of chapter 11. Likewise, the even- and odd-mode effective relative permittivities are computed as $\varepsilon_{r,eff,\ e/o}=C_{e,\ o}(\varepsilon_r)/C_{e,\ o}(\varepsilon_r=1)$. Figure (12.2a and b) compares the results of the even- and odd-mode characteristic impedance computed by the present method against other closed-form models [J.4], Galerkin's method, and results of the EM-simulator, HFSS. The method has accuracy within 5% even for a very tight coupling

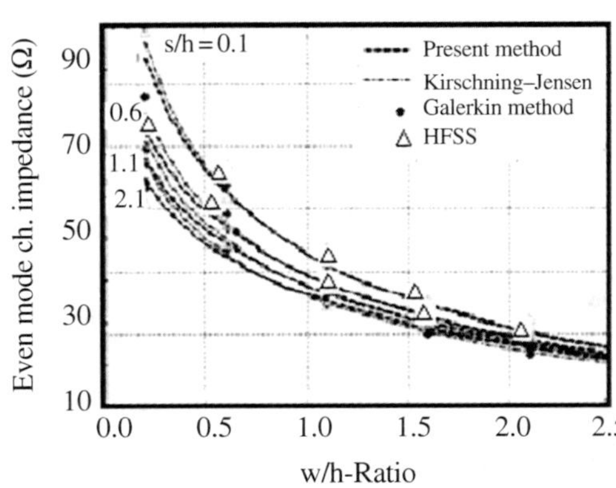

(a) Even-mode characteristic impedance.

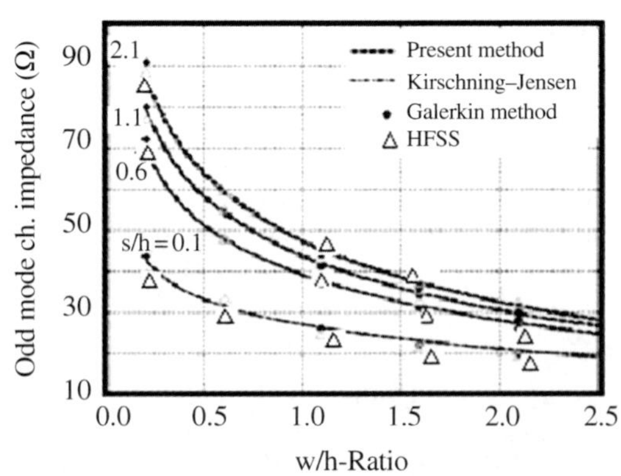

(b) Odd-mode characteristic impedance.

Figure 12.2 Even-and odd-mode characteristic impedance of symmetrical couple microstrips on an alumina substrate with ε_r = 9.8. *Source*: Abbosh [J.11]. © 2009 The Institution of Engineering and Technology.

(s/h = 0.0001), whereas the Kirschning–Jansen model has error 20% for s/h = 0.1.

The effect of a conductor thickness on the line parameters can be accounted for by computing the normalized effective line widths for both the modes [B.1]:

$$\frac{w_{eff,e}}{h_1} = \frac{w}{h_1} + \frac{\Delta w}{h_1}\left[1 - 0.5\exp\left(-0.69\frac{\Delta(w/h_1)}{\Delta(t/h_1)}\right)\right] \quad (a)$$

$$\frac{w_{eff,o}}{h_1} = \frac{w_{eff,e}}{h_1} + \frac{\Delta t}{h_1} \quad (b)$$

$$\text{where} \quad \frac{t}{h_1} = \frac{1}{\varepsilon_r}\frac{t/h_1}{s/h_1} \quad (c). \qquad (12.1.18)$$

Bedair has also provided the closed-form expressions to compute the odd-mode capacitance [J.42].

12.1.2 Frequency-Dependent Models of Edge Coupled Microstrip Lines

At high frequency, normally above 4 GHz, the even- and odd-mode effective relative permittivity and characteristic impedance are frequency-dependent. Both modes have different field distributions, resulting in different frequency dependence. Getsinger has extended his model to the symmetrically coupled microstrip [J.12]. However, it has lower accuracy. Garg and Bahl [J.13] have also given simpler but less accurate expressions for the frequency-dependent characteristic impedance of both modes. The Getsinger's model has the advantage of its applicability to the asymmetrically coupled microstrips also. It is discussed in the next section. Kirschning and Jansen [J.4] have provided a set of more accurate closed-form expressions to compute the frequency-dependent line parameters of edge-coupled symmetrical microstrip, shown in Fig (12.1). These expressions are summarized below. The dispersion model has an accuracy better than 2.5% over the normalized frequency f_n up to 20; and better than 1.5% for $f_n \le 15$, and $\varepsilon_r \le 12.9$, where $f_n = f.\,h$ GHz – mm.

Even-Mode Effective Relative Permittivity

The even-mode dispersion is given by the following expression, in the format of Getsinger's model [J.4]:

$$\varepsilon_{reff,e}(f_n) = \varepsilon_r - \frac{\varepsilon_r - \varepsilon_{reff,e}(f_n = 0)}{1 + F_e(f_n)}. \qquad (12.1.19)$$

The static even-mode effective relative permittivity is computed using equation (12.1.2), and the parameter $F_e(f)$ is evaluated using the following set of equations:

$$F_e(f_n) = P_1P_2[(P_3P_4 + 0.1844P_7)f_n]^{1.5763}$$
$$P_1 = 0.27488 + u\left[0.6315 + \left\{0.525/(1 + 0.0157.f_n)^{20}\right\}\right] - 0.065683\exp(-8.7513u)$$
$$P_2 = 0.33622[1 - \exp(-0.03442\varepsilon_r)]$$
$$P_3 = 0.0363\exp(-4.6u)\left[1 - \exp\left\{-(f_n/38.7)^{4.97}\right\}\right]$$
$$P_4 = 1 + 2.751\left[1 - \exp\left\{-(\varepsilon_r/15.916)^8\right\}\right]$$
$$P_5 = 0.334\left[\exp\left\{-3.3(f_n/15)^3\right\}\right] + 0.746$$
$$P_6 = P_5\left[\exp\left\{-(f_n/18)^{0.368}\right\}\right]$$
$$P_7 = 1 + 4.069P_6.g^{0.479}\exp\left[-1.347g^{0.595} - 0.17g^{2.5}\right]. \qquad (12.1.20)$$

Even-Mode Characteristics Impedance

It is computed from the following expression [J.4]:

$$Z_{o,e}(\varepsilon_r, f_n) = Z_{o,e}(\varepsilon_r, f_n = 0)\left[\frac{\left(0.9408\{\varepsilon_{r,eff}(f=0)\}^{C_e} - 0.9603\right)^{Q_0}}{\left((0.9408 - d_e)\{\varepsilon_{r,eff}(f=0)\}^{C_e} - 0.9603\right)^{Q_0}}\right], \qquad (12.1.21)$$

where the static even-mode characteristic impedance $Z_{0,e}(\varepsilon_r, f_n = 0)$ is given by equation (12.1.8). The effective relative permittivity $\varepsilon_{r,\,eff}(f_n = 0)$ for the single isolated microstrip has been computed in subsection (8.2.1) of chapter 8. The parameters C_e and d_e are computed from the following set of expressions:

$$C_e = 1 + 1.275\left[1 - \exp\left\{-0.004625p_e(\varepsilon_r)^{1.674}(f_n/18.365)^{2.745}\right\}\right] - Q_{12} + Q_{16} - Q_{17} + Q_{18} + Q_{20}$$
$$d_e = 5.086.q_e[r_e/(0.3838 + 0.386q_e)].\left[\exp(-22.2u^{1.92})/(1 + 1.299r_e)\right] \times \left[(1-\varepsilon_r)^6/\left\{1 + 10(1-\varepsilon_r)^6\right\}\right]$$
$$p_e = 4.766.\exp(-3.228u^{0.641}), \quad q_e = 0.016 + (0.0514\varepsilon_rQ_{21})^{4.524}, \quad r_e = (f_n/28.843)^{12}. \qquad (12.1.22)$$

The expressions for the variables Q_{11}–Q_{21} are further given below:

$$Q_{11} = 0.893[1 - 0.3/\{1 + 0.7(\varepsilon_r - 1)\}]$$
$$Q_{12} = 2.121\left[(f_n/20)^{4.91}/\left\{1 + Q_{11}(f_n/20)^{4.91}\right\}\right]\exp(-2.87g)g^{0.902}$$
$$Q_{13} = 1 + 0.038.(\varepsilon_r/8)^{5.1}$$
$$Q_{14} = 1 + 1.203.\left[(\varepsilon_r/15)^4/\left\{1 + (\varepsilon_r/15)^4\right\}\right]$$
$$Q_{15} = \left[1.887.\exp(-1.5.g^{0.84}).g^{Q_{14}}/\left[1 + 0.41\left((f_n/15)^3\left\{u^{2/Q_{13}}/(0.125 + u^{1.626/Q_{13}})\right\}\right]\right]$$

$$
\begin{aligned}
Q_{16} &= \left[1 + 9/\left\{1 + 0.403.(\varepsilon_r - 1)^2\right\}\right].Q_{15} \\
Q_{17} &= 0.394\left[1 - \exp\left\{-1.47(u/7)^{0.672}\right\}\right] \times \\
&\quad \left[1 - \exp\left\{-4.25(f_n/20)^{1.87}\right\}\right] \\
Q_{18} &= 0.61.\left[\left\{1 - \exp\left\{-2.13(u/8)^{1.593}\right\}\right\}/\right. \\
&\quad \left.\left\{1 + 6.544g^{4.17}\right\}\right] \\
Q_{19} &= 0.21g^4/\left[\left(1 + 0.18g^{4.9}\right).\left(1 + 0.1u^2\right)\left\{1 + (f_n/24)^3\right\}\right] \\
Q_{20} &= \left[0.09 + 1/\left\{1 + 0.1.(\varepsilon_r - 1)^{2.7}\right\}\right].Q_{19} \\
Q_{21} &= \left[1 - \left[42.54.g^{0.133}\exp\{-0.812g\}.u^{2.5}/\right.\right. \\
&\quad \left.\left.\left\{1 + 0.033u^{2.5}\right\}\right]\right].
\end{aligned} \tag{12.1.23}
$$

Odd-Mode Effective Relative Permittivity

The odd-mode dispersion expression, in the format of Getsinger's model, is given below [J.4]:

$$
\varepsilon_{r,effo}(f_n) = \varepsilon_r - \frac{\varepsilon_r - \varepsilon_{r,effo}(f=0)}{1 + F_o(f)}. \tag{12.1.24}
$$

The static odd-mode effective relative permittivity is computed using equation (12.1.5), and the parameter $F_o(f)$ is evaluated using the following set of equations:

$$
\begin{aligned}
F_o(f_n) &= P_1P_2\left[(P_3P_4 + 0.1844).f_nP_{15}\right]^{1.5763} \\
P_8 &= 0.7168.\left[1 + \left[1.076/\left\{1 + 0.0576(\varepsilon_r - 1)\right\}\right]\right] \\
P_9 &= P_8 - 0.7913\left[1 - \exp\left\{-(f_n/20)^{1.424}\right\}\right] \times \\
&\quad \tan^{-1}\left[2.481(\varepsilon_r/8)^{0.946}\right] \\
P_{10} &= 0.242(\varepsilon_r - 1)^{0.55} \\
P_{11} &= 0.6366.\left[\exp(-0.3401f_n) - 1\right]\tan^{-1}\left[1.263(u/3)^{1.629}\right] \\
P_{12} &= P_9 + \left[\{1 - P_9\}/\left\{1 + 1.183u^{1.376}\right\}\right] \\
P_{13} &= (1.695P_{10})/(0.414 + 1.605P_{10}) \\
P_{14} &= 0.8928 + 0.1072\left[1 - \exp\left\{-0.42(f_n/20)^{3.215}\right\}\right] \\
P_{15} &= \left|1 - [0.8928/P_{14}].(1 + P_{11}).P_{12}.\exp\left\{-P_{13}(g)^{1.092}\right\}\right|.
\end{aligned} \tag{12.1.25}
$$

The coefficients P_1–P_4 are taken from equation (12.1.20). The absolute value is taken in the last expression of equation (12.1.25).

Odd-Mode Characteristics Impedance

The frequency-dependent odd-mode characteristics impedance is computed from the following expression:

$$
Z_{o,o}(\varepsilon_r, f_n) = Z_o(\varepsilon_r, f_n) + \left[\frac{Z_{o,o}(\varepsilon_r, f=0)\left(\varepsilon_{r,eff,o}(f)/\varepsilon_{r,eff,o}(f=0)\right)^{Q_{22}} - Z_o(\varepsilon_r, f)Q_{23}}{1 + Q_{24} + (0.46g)^{2.2}Q_{25}}\right], \tag{12.1.26}
$$

where $Z_o(\varepsilon_r, f_n)$, the frequency-dependent characteristic impedance of a single isolated microstrip, is computed by equation (8.3.36) of chapter 8. Equation (12.1.10) computes the odd-mode characteristic impedance $Z_{o,o}(\varepsilon_r, f_n = 0)$. The Q-variables are obtained from the following set of expressions:

$$
\begin{aligned}
Q_{22} &= \left[0.925(f_n/Q_{26})^{1.536}\right]/\left[1 + 0.3(f_n/30)^{1.536}\right] \\
Q_{23} &= 1 + \left[0.005f_nQ_{27}/\left[\left\{1 + 0.812(f_n/15)^{1.9}\right\}\left(1 + 0.025u^2\right)\right]\right] \\
Q_{24} &= 2.506Q_{28}u^{0.894}\left[(1 + 1.3u)(f_n/99.25)\right]^{4.29}/\left[3.575 + u^{0.894}\right] \\
Q_{25} &= \left[0.3f_n^2/\left(10 + f_n^2\right)\right].\left[1 + \left\{2.333(\varepsilon_r - 1)^2/\left(5 + (\varepsilon_r - 1)^2\right)\right\}\right] \\
Q_{26} &= 30 - 22.2\left[\left((\varepsilon_r - 1)/13\right)^{12}/\left\{1 + 3\left((\varepsilon_r - 1)/13\right)^{12}\right\}\right] - Q_{29} \\
Q_{27} &= 0.4g^{0.84}\left[1 + \left\{\left(2.5(\varepsilon_r - 1)^{1.5}\right)/\left(5 + (\varepsilon_r - 1)^{1.5}\right)\right\}\right] \\
Q_{28} &= \left[0.149(\varepsilon_r - 1)^3\right]/\left[94.5 + 0.038(\varepsilon_r - 1)^3\right] \\
Q_{29} &= 15.16/\left[1 + 0.196(\varepsilon_r - 1)^2\right].
\end{aligned} \tag{12.1.27}
$$

12.2 Line Parameters of Asymmetric Coupled Microstrips

This section presents the closed-form models to compute the static and frequency-dependent parameters of the asymmetrically coupled microstrips. The static parameters are computed by two closed-form models – one computes the direct capacitances of the π-circuit model, shown in Fig (11.9b) of chapter 11, using the results of the symmetrically coupled microstrips [J.14], while another computes these capacitances by the *partial capacitance method* [J.15]. The more accurate former model is used in a CAD-package also [B.1]. The *network analog method* is also used to compute the capacitance matrix of the asymmetrically coupled microstrips [J.16]. The dispersion in the asymmetrically coupled microstrips is accounted for by extending the Getsinger's model [J.16, J.17].

12.2.1 Static Parameters of Asymmetrically Coupled Microstrips

The expressions for the propagation constants and characteristic impedances of two *normal modes*, i.e. the C and π-modes, of an asymmetrically coupled microstrips have been obtained in section (11.5), chapter 11. These are also obtained in terms of the elements of the capacitance and inductance matrices. The inductance of the microstrips on a dielectric substrate is independent of the relative permittivity. Therefore, the inductance matrix is determined from the capacitance matrix on the air-substrate:

$$C(\varepsilon_r = 1) = \begin{bmatrix} C_{11}(\varepsilon_r = 1) & -C_{12}(\varepsilon_r = 1) \\ -C_{12}(\varepsilon_r = 1) & C_{22}(\varepsilon_r = 1) \end{bmatrix} \quad (a)$$

$$L = \mu_0\varepsilon_o \begin{bmatrix} C_{11}(\varepsilon_r = 1) & -C_{12}(\varepsilon_r = 1) \\ -C_{12}(\varepsilon_r = 1) & C_{22}(\varepsilon_r = 1) \end{bmatrix}^{-1} \quad (b). \qquad (12.2.1)$$

The following expressions are obtained, from equations (11.5.20)–(11.5.22) of chapter 11, on using $(\varepsilon_{r,\ eff,\ i} = (\beta_i/\beta_o)^2, i = c, \pi)$, for the *static* effective relative permittivity of the C and π-modes:

$$\varepsilon_{r,eff,c}(f = 0) = \frac{c^2}{2}[L_{11}C_{11} + L_{22}C_{22} - 2L_{12}C_{12} + F] \quad (a)$$

$$\varepsilon_{r,eff,\pi}(f = 0) = \frac{c^2}{2}[L_{11}C_{11} + L_{22}C_{22} - 2L_{12}C_{12} - F] \quad (b)$$

$$\text{where, } F = [4(L_{12}C_{11} - L_{22}C_{12}).(L_{12}C_{22} - L_{11}C_{12}) + (L_{11}C_{11} - L_{22}C_{22})^2]^{1/2} \quad (c). \qquad (12.2.2)$$

where c is the velocity of EM-wave in free space. The static modal characteristics impedance Z_{c1} and $Z_{\pi1}$ of the line #1 for the C and π-modes are given by equations (11.5.44) and (11.5.47) of chapter 11. These equations could be rewritten in terms of the circuit parameters. The expression for the Z_{c1} of the C-mode is obtained below. The elements of the [L]-matrix are related to the elements of the [C]-matrix through the phase velocity of the C-mode:

$$[L][C] = \frac{\varepsilon_{reff\ c}}{c^2}, \Rightarrow \begin{bmatrix} L_{11} & L_{12} \\ L_{12} & L_{22} \end{bmatrix} = \frac{\varepsilon_{reff\ c}}{c^2}\begin{bmatrix} C_{11} & -C_{12} \\ -C_{12} & C_{22} \end{bmatrix}^{-1}$$

$$\Rightarrow \begin{bmatrix} L_{11} & L_{12} \\ L_{12} & L_{22} \end{bmatrix} = \frac{\varepsilon_{reff\ C}}{c^2\left[C_{11}C_{22} - C_{12}^2\right]}\begin{bmatrix} C_{22} & C_{12} \\ C_{12} & C_{11} \end{bmatrix}$$

$$\Rightarrow L_{11} = \left[\frac{\varepsilon_{reff\ c}}{c^2\left[C_{11}C_{22} - C_{12}^2\right]}\right]C_{22} \quad (a)$$

$$L_{12} = \left[\frac{\varepsilon_{reff\ c}}{c^2\left[C_{11}C_{22} - C_{12}^2\right]}\right]C_{12} \quad (b)$$

$$L_{22} = \left[\frac{\varepsilon_{reff\ c}}{c^2\left[C_{11}C_{22} - C_{12}^2\right]}\right]C_{11} \quad (c). \qquad (12.2.3)$$

On using $Z_{11} = j\omega L_{11}$, $Z_{22} = j\omega L_{22}$, $Z_{12} = j\omega L_{12}$, equation (11.5.44) for the characteristics impedance of C-mode on line #1 is written as

$$Z_{c1} = \frac{c}{\sqrt{\varepsilon_{reff,c}}}\left[\frac{L_{11}L_{22} - L_{12}^2}{L_{22} - L_{12}\,R_c}\right]. \qquad (12.2.4)$$

Using equation (12.2.3) with the above equations, the following expression for the Z_{c1} of the C-mode is obtained:

$$Z_{c1} = \frac{\sqrt{\varepsilon_{reff\ c}}}{c}\left[\frac{1}{C_{11} - C_{12}\,R_c}\right] = \frac{1}{Y_{C1}}. \qquad (12.2.5)$$

The voltage ratio R_c of C-mode is given by equation (11.5.40a) of chapter 11. The above equation of Z_{c1} can also be expressed in terms of the elements of [L]-matrix of asymmetrical coupled lines and the voltage ratio R_π of the π-mode. To do so, the relation between R_c and R_π in terms of the circuit elements is needed. Using equations (11.5.44), (11.5.51), and (11.5.53) of chapter 11, the following relations are obtained:

$$R_c = -\frac{1}{R_\pi}\left(\frac{Z_{c2}}{Z_{c1}}\right) \quad (a)$$

$$\left(\frac{Z_{c2}}{Z_{c1}}\right) = R_c\frac{(Z_{22} - Z_{12}R_c)}{(Z_{11}R_c - Z_{12})} \quad (b). \qquad (12.2.6)$$

The above equations provide the following identities:

$$R_c = \frac{Z_{12}R_\pi - Z_{22}}{Z_{11}R_\pi - Z_{12}} = \frac{L_{12}R_\pi - L_{22}}{L_{11}R_\pi - L_{12}} \quad (a)$$

$$R_\pi = \frac{Z_{12}R_c - Z_{22}}{Z_{11}R_c - Z_{12}} = \frac{L_{12}R_c - L_{22}}{L_{11}R_c - L_{12}} \quad (b). \qquad (12.2.7)$$

Using equation (12.2.7a), equation (12.2.4) for Z_{c1} could be rewritten, in terms of the voltage ratio R_π, as follows:

$$Z_{c1} = \frac{c}{\sqrt{\varepsilon_{reff,c}}}\left[L_{11} - \frac{L_{12}}{R_\pi}\right]. \qquad (12.2.8)$$

Following the above process, expressions for the static characteristic impedance $Z_{\pi1}$ of the π-mode of line #1 can also be obtained. The static characteristic impedance of C and π-modes for both line #1 and line #2 are summarized below [B.5, J.18–J.20]:

$$\text{Conductor \#1: } Z_{c1} = \frac{\sqrt{\varepsilon_{r,eff,c}(f=0)}}{c}\left[\frac{1}{(C_{11} - C_{12}R_c)}\right] = \frac{c}{\sqrt{\varepsilon_{r,eff,c}(f=0)}}(L_{11} - L_{12}/R_\pi) = \frac{1}{Y_{c1}} \quad (a)$$

$$Z_{\pi1} = \frac{\sqrt{\varepsilon_{r,eff,\pi}(f=0)}}{c}\left[\frac{1}{(C_{11} - C_{12}R_\pi)}\right] = \frac{c}{\sqrt{\varepsilon_{r,eff,\pi}(f=0)}}(L_{11} - L_{12}/R_c) = \frac{1}{Y_{\pi1}} \quad (b)$$

$$\text{Conductor \#2: } Z_{c2} = -R_c.R_\pi Z_{c1} \quad (c)$$

$$Z_{\pi2} = -R_c.R_\pi Z_{\pi1} \quad (d). \qquad (12.2.9)$$

To use the above expressions, the elements of the capacitance matrix are computed. At first, the direct capacitances are computed, using the closed-form expressions; and next, the capacitance matrix is obtained. The *method of moments* computes all elements of a capacitance matrix more effectively [B.2, J.2, J.3].

The Method Based on Symmetrically Coupled Microstrips

Figure (11.9b) shows the direct capacitances of the π-circuit model of the asymmetrically coupled microstrips. The direct capacitances of the asymmetrically coupled microstrips are obtained below on modifying the direct capacitances of the symmetrically coupled microstrips [J.14]. The following normalized dimensions are used:

$$u_1 = w_1/h, \quad u_2 = w_2/h, \quad g = s/h. \qquad (12.2.10)$$

Equation (11.4.34) of chapter 11 compute the direct capacitances $C^d_{11,Sym}(u, g, \varepsilon_r)$ and $C^d_{12,Sym}(u, g, \varepsilon_r)$ for the symmetrically coupled microstrips of strip widths, $w_1 = w_2 = w$. The mutual coupling capacitance, $C^d_{12,asym}(u_1, u_2, g, \varepsilon_r)$, of the asymmetrically coupled microstrips is obtained from the following average value of two symmetric coupled microstrips, each having strip widths $u = u_1$ and $u = u_2$, respectively:

$$C^d_{12,asym}(u_1, u_2, g, \varepsilon_r) = \left[\frac{2}{\left\{C^d_{12,sym}(u_1, g, \varepsilon_r)\right\}^{-K_m} + \left\{C^d_{12,sym}(u_2, g, \varepsilon_r)\right\}^{-K_m}}\right]^{\frac{1}{K_m}}. \qquad (12.2.11)$$

The exponent K_m is determined from the empirical expression:

$$K_m = 0.95 + 0.33\ln(u_1.u_2) - 0.4\ln(g). \qquad (12.2.12)$$

Two direct capacitances, $C^d_{11,asym}(u_1, u_2, g, \varepsilon_r), C^d_{22,asym}(u_1, u_2, g, \varepsilon_r)$ of the asymmetrically coupled microstrips, between the strip conductors and ground, are obtained by modifying the direct capacitances of two symmetric coupled microstrips; $C^d_{11,Sym}(u_1, g, \varepsilon_r)$, $C^d_{11,Sym}(u_2, g, \varepsilon_r)$ corresponding to the strip widths $u=u_1$ and $u=u_2$, respectively. The direct capacitances of the asymmetrically coupled microstrips are obtained as follows:

$$\begin{aligned} C^d_{11,asym}(u_1, u_2, g, \varepsilon_r) &= C^d_{11,sym}(u_1, g, \varepsilon_r) + \Delta C_1 \quad (a)\\ C^d_{22,asym}(u_1, u_2, g, \varepsilon_r) &= C^d_{22,sym}(u_2, g, \varepsilon_r) + \Delta C_2 \quad (b). \end{aligned} \qquad (12.2.13)$$

The capacitance correction factors ΔC_1 and ΔC_2 are computed from the following expressions:

$$\begin{aligned} \Delta C_1 &= [1 - K_s(\varepsilon_r)].\left[C^d_{12,sym}(u_1, g, \varepsilon_r) - C^d_{12,asym}(u_1, u_2, g, \varepsilon_r)\right] \quad (a)\\ \Delta C_2 &= [1 - K_s(\varepsilon_r)].\left[C^d_{12,sym}(u_2, g, \varepsilon_r) - C^d_{12,asym}(u_1, u_2, g, \varepsilon_r)\right] \quad (b). \end{aligned} \qquad (12.2.14)$$

The relative permittivity-dependent parameter $K_s(\varepsilon_r)$ is given by

$$\begin{aligned} K_s(\varepsilon_r) &= \frac{K_s(\varepsilon_r = 1)}{1 + 0.58\left(1 - 1/\sqrt{\varepsilon_r}\right)} \quad (a)\\ K_s(\varepsilon_r = 1) &= [0.21 - 0.023\ln(u_1.u_2)]\\ &\quad \exp[-g\{1.56 + 0.22\ln(u_1.u_2)\}] \quad (b). \end{aligned} \qquad (12.2.15)$$

The elements of the following capacitance matrix of the asymmetrically coupled microstrips are obtained, using equation (11.3.9) of chapter 11; from the direct capacitances $C^d_{11,asym}(u_1, u_2, g, \varepsilon_r)$, $C^d_{22,asym}(u_1, u_2, g, \varepsilon_r)$, and $C^d_{12,asym}(u_1, u_2, g, \varepsilon_r)$:

$$C(\varepsilon_r) = \begin{bmatrix} C_{11}(\varepsilon_r) & -C_{12}(\varepsilon_r)\\ -C_{12}(\varepsilon_r) & C_{22}(\varepsilon_r) \end{bmatrix}. \qquad (12.2.16)$$

To compute the inductance matrix, using equation (12.2.1b), the capacitance matrix $[C(\varepsilon_r = 1)]$, on air-substrate is evaluated by taking $\varepsilon_r = 1$ in the above equation. Next, the matrix $[L(\varepsilon_r = 1)]$ is evaluated from the expression $[L(\varepsilon_r = 1)] = \sqrt{\mu_0\varepsilon_0}[C(\varepsilon_r = 1)]^{-1}$. Finally, elements of the capacitance and inductance matrices, $[C(\varepsilon_r)]$ and $[L(\varepsilon_r = 1)]$, are substituted in equations (12.2.2) and (12.2.9) to compute the static *modal* effective relative permittivity and modal characteristic impedances of the C and π-modes.

12.2.2 Frequency-Dependent Line Parameters of Asymmetrically Coupled Microstrips

Tripathi [J.17] developed the *waveguide dispersion model* to compute the frequency-dependent modal effective relative permittivity and the modal characteristic impedance of the asymmetrically coupled microstrips, by obtaining the frequency-dependent capacitance and inductance matrices. The model has been adapted in a CAD package [B.1]. The waveguide model of the coupled microstrip lines could be called the "*Tripathi model.*" Simpler but less accurate dispersion model, by extending Getsinger's model of a single microstrip, is also reported [J.12, J.16].

Tripathi Waveguides Dispersion Model

Figure (12.3a and b) shows the asymmetrically coupled microstrips and its equivalent waveguide model. The

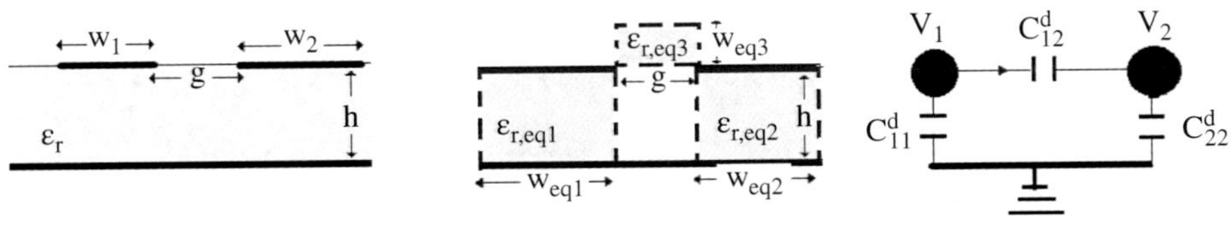

(a) Asymmetric coupled microstrips. (b) Waveguide model. (c) π-circuit model.

Figure 12.3 Asymmetric coupled microstrips and its equivalent circuit model.

waveguide model treats two strip conductors of the widths w_1 and w_2 as the parallel plate waveguides of height h. It also treats the coupling gap region g as another waveguide of height g. Figure (12.3c) further shows the π-capacitance model of the coupled lines structure. The waveguides have *static* equivalent relative permittivity, $\varepsilon_{r,\,eq1}(\varepsilon_r, f = 0)$, $\varepsilon_{r,\,eq2}(\varepsilon_r, f = 0)$, and $\varepsilon_{r,\,eq3}(\varepsilon_r, f = 0)$. Their *static* equivalent widths are $w_{eq1}(f = 0)$, $w_{eq2}(f = 0)$, and $w_{eq3}(f = 0)$.

Equation (11.3.9) of chapter 11 relates the *static* direct capacitances, $C^d_{11}, C^d_{22}, C^d_{12}$ of the π-model to the capacitance matrix elements, C_{11}, C_{12}, C_{22}. The *static* equivalent relative permittivities of three equivalent waveguides are computed as follows:

$$\varepsilon_{r,eq1}(\varepsilon_r, f=0) = \frac{C^d_{11}(\varepsilon_r)}{C^d_{11}(\varepsilon_r=1)} = \frac{C_{11}(\varepsilon_r)-C_{12}(\varepsilon_r)}{C_{11}(\varepsilon_r=1)-C_{12}(\varepsilon_r=1)} \quad \text{(a)}$$

$$\varepsilon_{r,eq2}(\varepsilon_r, f=0) = \frac{C^d_{22}(\varepsilon_r)}{C^d_{22}(\varepsilon_r=1)} = \frac{C_{22}(\varepsilon_r)-C_{12}(\varepsilon_r)}{C_{22}(\varepsilon_r=1)-C_{12}(\varepsilon_r=1)} \quad \text{(b)}$$

$$\varepsilon_{r,eq3}(\varepsilon_r, f=0) = \frac{C^d_{12}(\varepsilon_r)}{C^d_{12}(\varepsilon_r=1)} = \frac{C_{12}(\varepsilon_r)}{C_{12}(\varepsilon_r=1)} \quad \text{(c).}$$

$$(12.2.17)$$

The above equations take into account the negative value of the coupling capacitance. The *static* equivalent widths of the waveguides, for the air-substrate, are also computed as follows from the direct capacitances:

$$w_{eq1}(f=0) = \frac{h\, C^d_{11}(\varepsilon_r = 1)}{\varepsilon_0} \quad \text{(a)}$$

$$w_{eq2}(f=0) = \frac{h\, C^d_{22}(\varepsilon_r = 1)}{\varepsilon_0} \quad \text{(b)}$$

$$w_{eq3}(f=0) = \frac{g\, C^d_{12}(\varepsilon_r = 1)}{\varepsilon_0} \quad \text{(c).} \qquad (12.2.18)$$

Subsection (12.2.1) presents the closed-form expressions for the static elements of the capacitance matrix and direct capacitances of π-model of asymmetrical coupled microstrip lines. The method of moments (MOM) is also used to compute the static capacitance matrix [B.2, J.2, J.3]. In the case of the symmetrically coupled microstrips, $C^d_{11} = C^d_{22}$ and these capacitances can also be computed by the variational method presented in chapter 14. In practice, the equivalent widths and equivalent relative permittivity are frequency-dependent parameters. Getsinger's model is used to estimate its frequency dispersion:

$$\varepsilon_{r,eq,j}(\varepsilon_r, f) = \varepsilon_r - \frac{\varepsilon_r - \varepsilon_{r,eq,j}(\varepsilon_r, f=0)}{1 + G\left(f/f_p\right)^2}, j = 1,2,3 \quad \text{(a)}$$

$$w_{eq,j}(f) = w_j - \frac{w_j - w_{eq,j}(f=0)}{1 + \left(f/f_p\right)^2}, \quad j = 1,2,3 \quad \text{(b)}$$

$$\text{where,} \quad G = 0.6 + 0.009\left(2\mu_0 h f_p\right) \quad \text{(c).}$$

$$(12.2.19)$$

The above expression for G is optimized for the alumina substrate. It could be corrected for other substrates to get improved results. Another expression for the parameter $G = \varepsilon_{req,\,j}/\varepsilon_r$, ($j = 1, 2, 3$) given in equation (8.3.9) of chapter 8 can also be used.

At this stage, the task is to compute the inflection frequency f_p. It is obtained by using the *transverse resonance method*, shown in Fig (12.4a and b). Three lossless interconnected equivalent waveguides are treated as the open-end line sections of lengths w_j, $j = 1, 2, 3$. Their characteristic impedances and propagation constants are obtained from the following expressions:

$$Z_j = \sqrt{\mu_0/\varepsilon_0}.\left[h_j/\sqrt{\varepsilon_{r,eq,j}(f=0)}\right], \begin{cases} j=1,2,3 \\ h_1 = h_2 = h, h_3 = g \end{cases} \quad \text{(a)}$$

$$\beta_j = \omega\sqrt{\mu_0 \varepsilon_{r,eq,j}(f=0)}, \qquad j = 1,2,3 \quad \text{(b).}$$

$$(12.2.20)$$

Under the *transverse resonance* condition, the total admittance at the plane X–X is zero:

$$Y'_1 + Y'_2 = 0; \Rightarrow (1/Z_1)\tan\left[\beta_1 w_{eq,1}(f=0)\right] + Y'_2 = 0 \quad \text{(a)}$$

$$Z^{in}_2 = Z_2\cot\left\{\beta_2 w_{eq,2}(f=0)\right\}, Z^{in}_3 = Z_3\cot\left\{\beta_3 w_{eq,3}(f=0)\right\} \quad \text{(b)}$$

$$Y'_2 = \frac{1}{Z'_2} = \frac{1}{Z^{in}_2 + Z^{in}_3} \quad \text{(c).}$$

$$(12.2.21)$$

Usually, $w_{eq,3}(f = 0)$ is much smaller than the strip conductor widths. So, the coupling capacitance C_{12} can be treated as frequency independent. The open stub

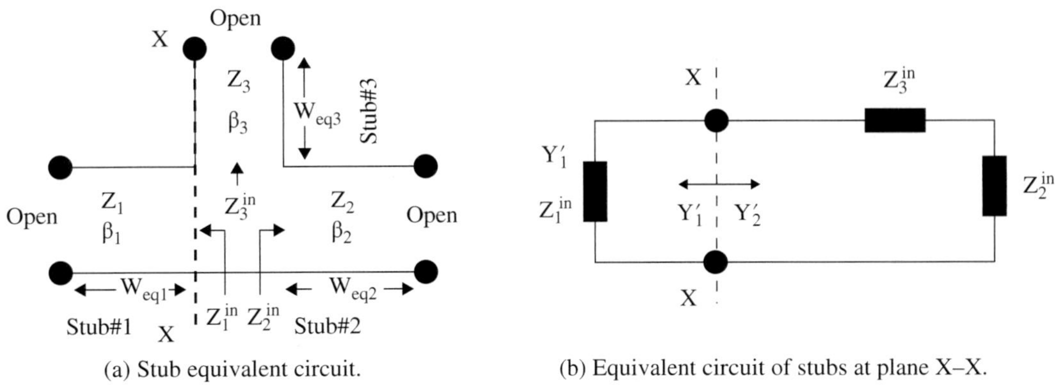

Figure 12.4 Application of the transverse resonance method.

#3 is replaced by the impedance $Z_3^{in} = 1/\omega C_{12}$. Thus, transverse resonance condition at the plane X–X is modified as follows:

$$Z_1^{in} + \left[Z_2^{in} + Z_3^{in}\right] = 0$$
$$\Rightarrow Z_1 \cot\left\{\beta_1 w_{eq,1}(f = 0)\right\} + \left[Z_2 \cot\left\{\beta_2 w_{eq,2}(f = 0)\right\} + 1/\omega C_{12}\right] = 0. \qquad (12.2.22)$$

The above equations (12.2.21) or (12.2.22) is solved numerically for the first transverse resonance frequency, i.e. for the first root ω_1. It is the inflection frequency f_p of the coupled microstrips. It helps to compute the frequency-dependent equivalent widths $w_{eq,j}$ (f), and equivalent relative permittivity, $\varepsilon_{r,eq,j}$ (ε_r, f), j = 1, 2, 3. The frequency-dependent capacitance matrix elements are computed as follows:

$$C_{11}(\varepsilon_r, f) = C_{11}^d(\varepsilon_r, f) + C_{12}^d(\varepsilon_r, f) \qquad (a)$$
$$C_{22}(\varepsilon_r, f) = C_{22}^d(\varepsilon_r, f) + C_{12}^d(\varepsilon_r, f) \qquad (b)$$
$$\text{also}[C(\varepsilon_r, f)] = \begin{bmatrix} C_{11}(\varepsilon_r, f) & -C_{12}(\varepsilon_r, f) \\ -C_{12}(\varepsilon_r, f) & C_{22}(\varepsilon_r, f) \end{bmatrix} \qquad (c).$$
$$(12.2.23)$$

where

$$C_{11}(\varepsilon_r, f) = \left[\varepsilon_{r,eq,1}(f).w_{eq,1}(f)/h\right] + \left[\varepsilon_{r,eq,3}(f).w_{eq,3}(f)/g\right] \quad (a)$$
$$\Rightarrow C_{22}(\varepsilon_r, f) = \left[\varepsilon_{r,eq,2}(f).w_{eq,2}(f)/h\right] + \left[\varepsilon_{r,eq,3}(f).w_{eq,3}(f)/g\right] \quad (b)$$
$$C_{12}(\varepsilon_r, f) = \left[\varepsilon_{r,eq,3}(f).w_{eq,3}(f)/g\right] \quad (c).$$
$$(12.2.24)$$

The frequency-dependent inductance matrix elements are determined by evaluating the above capacitances on the air-substrate:

$$[L(f)].[C(\varepsilon_r = 1, f)] = \frac{1}{c^2}, \Rightarrow [L(\varepsilon_r = 1, f)] = \frac{1}{c^2}\begin{bmatrix} C_{11}(f) & -C_{12}(f) \\ -C_{12}(f) & C_{22}(f) \end{bmatrix}^{-1}. \qquad (12.2.25)$$

Equation (12.2.2) normally computes the *static* C and π- effective relative permittivity, as the capacitance and inductance matrix elements are frequency independent. On using the frequency-dependent elements of both [C(ε_r, f)] and [L(ε_r = 1, f)] matrices, equation (12.2.2) computes the dispersion, $\varepsilon_{r,\,eff,\,c}(f)$, $\varepsilon_{r,\,eff,\,\pi}(f)$, of both the C and π-modes, whereas equation (11.5.31), or equations (11.4.32d) and (11.4.33d), computes $\varepsilon_{r,\,eff,\,e}(f)$, $\varepsilon_{r,\,eff,\,o}(f)$ of the symmetrically coupled microstrips. Figure (12.5a and b) shows the illustrative results using the dispersion model of Tripathi against the results of Getsinger's model for both the symmetrical and asymmetrical coupled microstrips [J.17].

On using the frequency-dependent elements of both [C(ε_r, f)] and [L(ε_r = 1, f)] matrices, equation (12.2.9) computes the dispersive characteristic impedances, $Z_{c1}(f)$, $Z_{\pi1}(f)$, $Z_{c2}(f)$, $Z_{\pi2}(f)$ of both the C and π-modes for two asymmetric strip conductors. Likewise, on using the frequency-dependent elements of both [C(ε_r, f)] and [L(ε_r = 1, f)] matrices, equations (11.5.46) and (11.5.49) of chapter 11 compute the $Z_{oe}(f)$, $Z_{oo}(f)$ for the even and odd modes of the symmetrically coupled microstrips. Figure (12.6) shows the illustrative results using Tripathi's model for the even- and odd-mode characteristics impedance, $Z_{oe}(f)$, $Z_{oo}(f)$, of the symmetrically coupled microstrips [J.17]. It also compares the results against the results of Jansen [J.4].

Extended Getsinger's Dispersion Models for C and π-Modes

Getsinger's dispersion model applicable to a single microstrip, discussed in subsection (8.3.2) of chapter 8, is further extended to get the dispersion relations to compute the $\varepsilon_{r,\,eff,\,c}$ (ε_r, f) and $\varepsilon_{r,\,eff,\,\pi}$ (ε_r, f) of both the C and π-modes of the asymmetrically coupled microstrips, shown in Fig (12.7a and b) [J.16]:

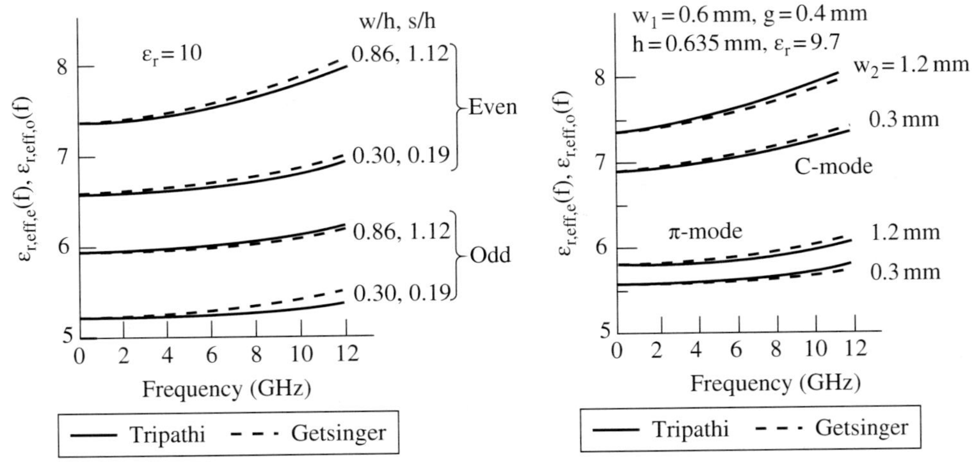

(a) Dispersion of even and odd modes.

(b) Dispersion of C and π-modes.

Figure 12.5 Dispersion of symmetrical and asymmetrical coupled microstrip. *Source:* From Tripathi [J.17]. © 1986 IEEE.

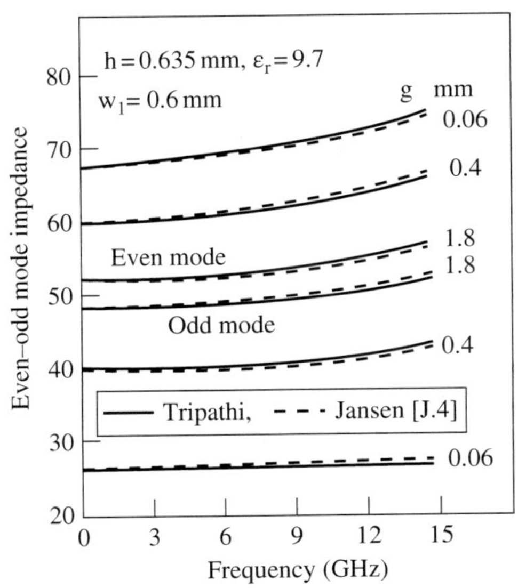

Figure 12.6 Dispersive characteristic impedances of even and odd modes. *Source*: From Tripathi [J.17]. © 1986 IEEE.

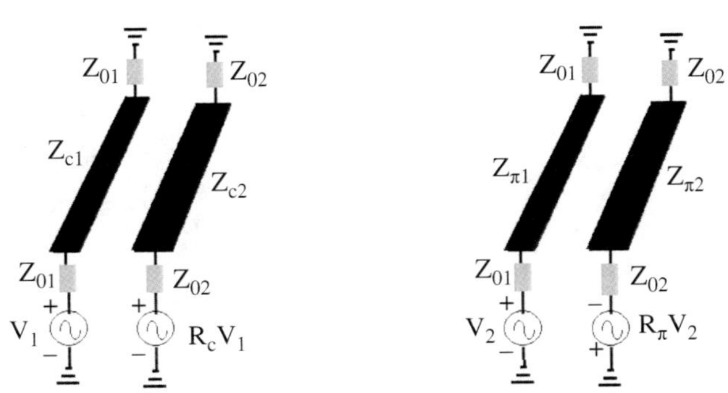

(a) C-mode excitations.

(b) π-mode excitations.

Figure 12.7 C- and π-mode excitations of the asymmetrically coupled microstrips.

$$\text{For C}-\text{mode}: \varepsilon_{r,eff,c}(\varepsilon_r, f) = \varepsilon_r - \frac{\varepsilon_r - \varepsilon_{r,eff,c}(\varepsilon_r, f=0)}{1 + G_c\left(f/f_p^C\right)^2} \quad \text{(a)}$$

$$\text{where}, G_c = 0.6 + 0.009Z_{0,eq}^c(\varepsilon_r, f=0)$$

$$f_p^c = Z_{0,eq}^C(\varepsilon_r, f=0)/(2\mu_0 h) \quad \text{(b)}$$

$$\text{For } \pi-\text{mode}: \varepsilon_{r,eff,\pi}(\varepsilon_r, f) = \varepsilon_r - \frac{\varepsilon_r - \varepsilon_{r,eff,\pi}(\varepsilon_r, f=0)}{1 + G_\pi\left(f/f_p^\pi\right)^2} \quad \text{(c)}$$

$$\text{where}, G_\pi = 0.6 + 0.009\,Z_{0,eq}^\pi(\varepsilon_r, f=0)$$

$$f_p^\pi = Z_{0,eq}^\pi(\varepsilon_r, f=0)/(2\mu_0 h) \quad \text{(d)}. \quad (12.2.26)$$

The static modal effective relative permittivities, $\varepsilon_{r,\,eff,\,c}(\varepsilon_r, f = 0)$ and $\varepsilon_{r,\,eff,\,\pi}(\varepsilon_r, f = 0)$, are computed using equation (12.2.2). The above *waveguide dispersion model* requires numerical evaluation of the inflection frequency f_p of the coupled microstrips. However, Getsinger's model computes the inflection frequencies, f_p^c and f_p^π, separately for both C and π-modes in terms of the equivalent static mode impedances, $Z_{0,eq}^c(\varepsilon_r, f = 0)$ and $Z_{0,eq}^\pi(\varepsilon_r, f = 0)$. The equivalent mode impedances of the C and π-modes excited in the coupled microstrips, shown in Fig (12.7a and b), are to be evaluated.

Figure (12.7a and b) shows that the voltage pair ($+V_1$, $+R_cV_1$) excites the C-mode, and the voltage pair ($+V_2$, $-R_\pi V_2$) excites the π-mode. Figure (12.8a and b) further shows that the mode voltage ratios R_c and R_π modify the corresponding source, characteristic impedance, and load impedance of each of the C and π-modes. Figure (12.8a) demonstrates the parallel combination of two conductors of the C-mode, as both conductors are at the same potential. It provides the equivalent C-mode characteristic impedance. Likewise, Fig (12.8b) demonstrates the series connection of two conductors of the π-mode, due to the

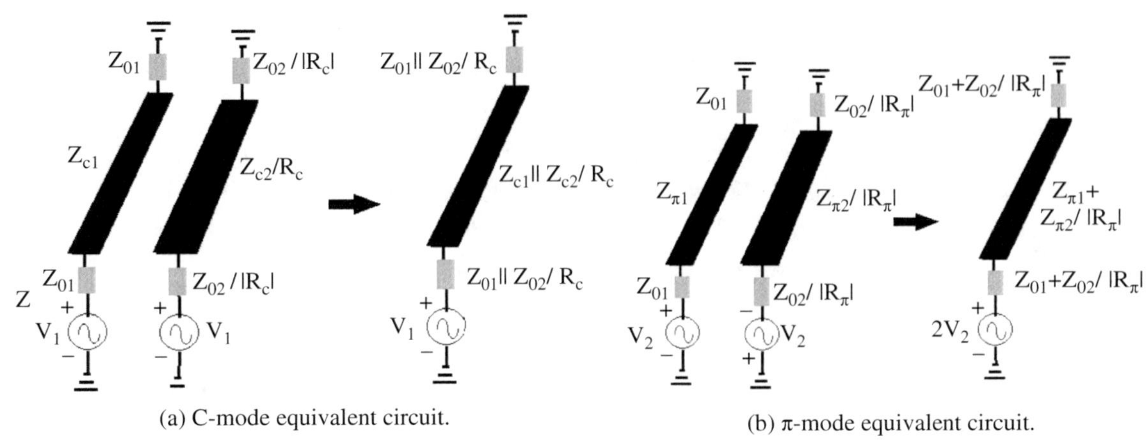

Figure 12.8 Equivalent C- and π-mode impedances.

opposite potential on both conductors. It gives the equivalent π-mode characteristic impedance. Both equivalent modal impedances are given below:

$$Z^{c}_{0,eq}(\varepsilon_r, f = 0) = \frac{Z_{o,c1} \cdot Z_{o,c2}/R_c}{Z_{o,c1} + Z_{o,c2}/R_c} \quad (a)$$

$$Z^{\pi}_{0,eq}(\varepsilon_r, f = 0) = Z_{o,\pi 1} + Z_{o,\pi 2}/R_\pi \quad (b). \quad (12.2.27)$$

The characteristic impedance of the C and π-modes are computed from equation (12.2.9a and b). The voltage ratios R_c and R_π are computed from equation (11.5.40) of chapter 11. The above model is also applicable to the symmetrically coupled microstrip giving dispersion model for the even and odd modes [J.12, J.16].

12.3 Line Parameters of Coupled CPW

This section summarizes results on the line parameters of the symmetric edge coupled CPW and the line parameters of a shielded broadside coupled CPW. Normally, the closed-form results are obtained using the conformal mapping methods [J.22–J.24].

12.3.1 Symmetric Edge Coupled CPW

Wen [J.25] presented an analysis of the symmetrical edge coupled CPW on an infinite thickness substrate. However, using the conformal mapping method, Keung and Cheng [J.22] have obtained the closed-form expressions for the even- and odd-mode capacitances of the symmetrical edge coupled CPW. It is shown in Fig (12.9), on a finite thickness substrate, with thickness h and relative permittivity ε_r. Its coupling gap is 2a. The 2b is a summation of the coupling gap and two strip widths, whereas the separation between two coplanar ground planes is 2c.

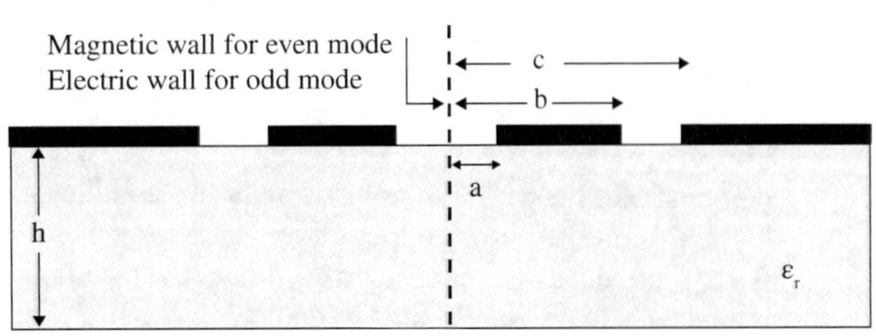

Figure 12.9 Symmetrical edge-coupled CPW on finite thickness substrate. *Source*: From Keung and Cheng [J.22]. © 1996 IEEE.

Even-Mode CPW Capacitance

The even-mode excitation has a *magnetic wall* (MW) at the plane of symmetry of the edge-coupled CPW. It separates the coupled lines into two quasi-TEM even-mode supporting lines. The even-mode line capacitance, $C_e(\varepsilon_r)$ *p.u.l.* is given by the following expression:

$$C_e(\varepsilon_r) = 2\varepsilon_0 \frac{K(k_{e1})}{K'(k_{e1})} + (\varepsilon_r - 1)\varepsilon_0 \frac{K(k_{e2})}{K'(k_{e2})} \quad (a)$$

$$\text{where } k^2_{e1} = \frac{b^2 - a^2}{c^2 - a^2} \quad (b)$$

$$k^2_{e2} = \frac{\sinh^2\left(\frac{\pi b}{2h}\right) - \sinh^2\left(\frac{\pi a}{2h}\right)}{\sinh^2\left(\frac{\pi c}{2h}\right) - \sinh^2\left(\frac{\pi a}{2h}\right)} \quad (c). \quad (12.3.1)$$

In the above expressions, the K(k) and K′(k) are the *complete elliptic integrals* of the first kind and its complement. The ratio K(k)/K′(k) is evaluated by the closed-form expression [B.7, J.26]. It is also summarized in subsection (9.2.4) of chapter 9.

Odd-Mode CPW Capacitance

The odd-mode excitation has an *electric wall* (EW) at the plane of symmetry. It separates the symmetrical coupled CPW in two quasi-TEM odd-mode supporting lines. The odd-mode line capacitance, $C_o(\varepsilon_r)$ *p.u.l.*, is given by the following expression:

$$C_o(\varepsilon_r) = 2\varepsilon_0 \frac{K(k_{o1})}{K'(k_{o1})} + (\varepsilon_r - 1)\varepsilon_0 \left[\frac{K(k_{o2})}{K'(k_{o2})} + \frac{K(k_{o3})}{K'(k_{o3})}\right] \quad \text{(a)}$$

$$\text{where,} \quad k_{o1}^2 = \frac{c^2(b^2-a^2)}{b^2(c^2-a^2)} \quad \text{(b)}. \qquad (12.3.2)$$

The parameters k_{o2} and k_{o3} are computed in terms of the *incomplete elliptic integral* of the first kind in Jacobian notation F(ϕ, k) using the following expressions:

$$\frac{F\left[\sin^{-1}\left(\frac{k_{o2}}{k_{c2}}\right), k_{c2}\right]}{K(k_{c2})} = \frac{\beta_1}{\beta_3} \quad \text{(a)}$$

$$\frac{F\left[\sin^{-1}\left(\frac{k_{o3}}{k_{c3}}\right), k_{c3}\right]}{K(k_{c3})} = \frac{1-\beta_2}{1-\beta_3} \quad \text{(b)}$$

$$\beta_3 = \frac{\beta_1 + \beta_2}{2} \quad \text{(c)}. \qquad (12.3.3)$$

The parameters β_1,β_2, and another parameter α, in terms of structural dimensions, are given by the following expressions:

$$\alpha = \frac{K(k_{c1})}{K'(k_{c1})} \quad \text{(a)}$$

$$K_{c1}^2 = \left[\frac{\sinh^2\left(\frac{\pi c}{2h}\right)}{\sinh^2\left(\frac{\pi b}{2h}\right)}\right]\left[\frac{\sinh^2\left(\frac{\pi b}{2h}\right) - \sinh^2\left(\frac{\pi a}{2h}\right)}{\sinh^2\left(\frac{\pi c}{2h}\right) - \sinh^2\left(\frac{\pi a}{2h}\right)}\right] \quad \text{(b)}$$

$$\beta_1 = \frac{F\left[\sin^{-1}\left\{\sqrt{\frac{\sinh^2\left(\frac{\pi c}{2h}\right) - \sinh^2\left(\frac{\pi a}{2h}\right)}{\sinh^2\left(\frac{\pi c}{2h}\right)\cosh^2\left(\frac{\pi a}{2h}\right)}}\right\}, k_{c1}\right]}{K(k_{c1})} \quad \text{(c)}$$

$$\beta_2 = F\frac{F\left[\sin^{-1}\left\{\sqrt{\frac{\sinh^2\left(\frac{\pi c}{2h}\right) - \sinh^2\left(\frac{\pi a}{2h}\right)}{\sinh^2\left(\frac{\pi c}{2h}\right)}}\right\}, k_{c1}\right]}{K(k_{c1})} \quad \text{(d)}. \qquad (12.3.4)$$

The parameters k_{c2} and k_{c3} are further computed using the following expressions:

$$\frac{K(k_{c2})}{K'(k_{c2})} = \alpha\beta_3 \quad \text{(a)}$$

$$\frac{K(k_{c3})}{K'(k_{c3})} = \alpha(1-\beta_3) \quad \text{(b)}$$

$$\text{For} \frac{K(k)}{K'(k)} = p;$$

$$k = \begin{cases} \left[(e^{\pi p}-2)/(e^{\pi p}+2)\right]^2, & 1 < p < \infty \\ \sqrt{1-\left[(e^{\pi/p}-2)/(e^{\pi/p}+2)\right]^4}, & 0 < p < 1 \end{cases} \quad \text{(c)}. \qquad (12.3.5)$$

The right-hand sides of equation (12.3.5) are known and the k_{c2} and k_{c3} are computed by using the approximate expression for K(k)/K′(k) given above [B.7, J.26]. Also, the ratio k_{o2}/k_{c2} and k_{o3}/k_{c3} are obtained from equation (12.3.3), as the right-hand sides are known. The values for k_{o2} and k_{o3} are obtained. These are used in expression (12.3.2a) to compute the odd-mode line capacitance.

Finally, the even- and odd-mode effective relative permittivity and characteristic impedances are computed from the even- and odd-mode line capacitances as follows:

$$\varepsilon_{r\,eff\,e/o} = \frac{C_{e/o}(\varepsilon_r)}{C_{e/o}(\varepsilon_r = 1)}. \qquad (12.3.6)$$

$$Z_{o\,e/o} = \frac{1}{c\sqrt{C_{e/o}(\varepsilon_r)C_{e/o}(\varepsilon_r = 1)}}. \qquad (12.3.7)$$

The above expressions are used either for the even mode or the odd mode. The c is the velocity of EM-wave in free space. The effect of the top shielding can be accounted for by using the expressions of Ghione and Naldi [J.24]. Figure (12.10a and b) shows the comparisons of the computed even- and odd-mode characteristic impedances using the present method, again the results of the SDA, for $\varepsilon_r = 12.9$. Figure (12.10c) further presents the results for the even- and odd-mode effective permittivity.

The even- and odd-mode characteristic impedance can also be computed by the following empirical expressions [J.22]:

$$\text{Even mode: } Z_{oe} = \frac{120\pi\frac{K(k_{e1})}{K'(k_{e1})}}{\sqrt{2[2+f_e(\varepsilon_r-1)]}} \quad \text{(a)}$$

$$f_e = 1 - \frac{\varepsilon_r}{\varepsilon_r + 0.641}\left\{-0.0175\left(\frac{c}{h}\right) + 0.18\left(\frac{c}{h}\right)^2 - 0.05\left(\frac{c}{h}\right)^3\right\} \quad \text{(b)}. \qquad (12.3.8)$$

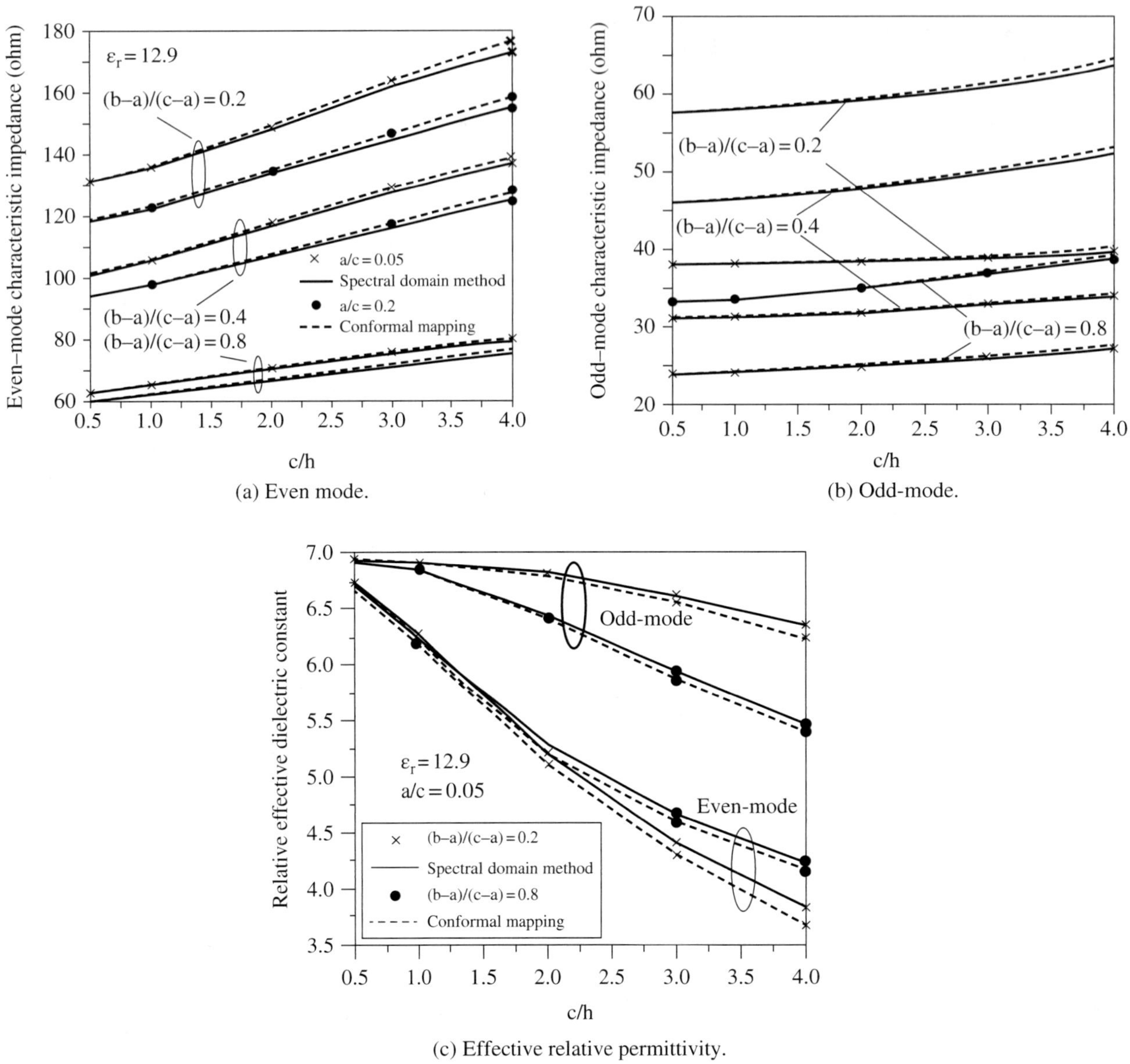

Figure 12.10 Characteristic impedance and effective relative permittivity of coupled CPW for even and odd modes. *Source*: From Keung and Cheng [J.22]. © 1996 IEEE.

$$\text{Odd mode: } Z_{oo} = \frac{120\pi \dfrac{K(k_{o1})}{K'(k_{o1})}}{\sqrt{2\left[2+f_o(\varepsilon_r-1)\right]}} \quad \text{(a)}$$

$$f_o = 1 - \frac{\varepsilon_r}{\varepsilon_r + 1.345}\left\{-0.0177\left(\frac{c}{h}\right) + 0.0264\left(\frac{c}{h}\right)^2\right\} \quad \text{(b)}. \qquad (12.3.9)$$

where k_{e1} and k_{o1} are given by equations (12.3.1) and (12.3.2). The above expressions have 2% deviations against the results of the SDA for (c/b) <2 [J.22, J.27]. The even- and odd-mode effective relative permittivity, $\varepsilon_{r\,eff\,e}$ and $\varepsilon_{r\,eff\,o}$, are computed from the Z_{oe} and Z_{oo}.

12.3.2 Shielded Broadside Coupled CPW

Bedair and Wolff [J.23] have given the following expressions for the odd- and even-mode line capacitances of the shielded broadside coupled CPW, shown in Fig (12.11a).

Odd Mode

For the odd-mode excitation, the plane of symmetry shown in Fig (12.11a) is an EW. The odd-mode line capacitance C_{o1} is due to the dielectric medium between the strip and upper shield and odd-mode line capacitance C_{o2} is due to the dielectric medium between the strip and EW. The total odd-mode line capacitance is

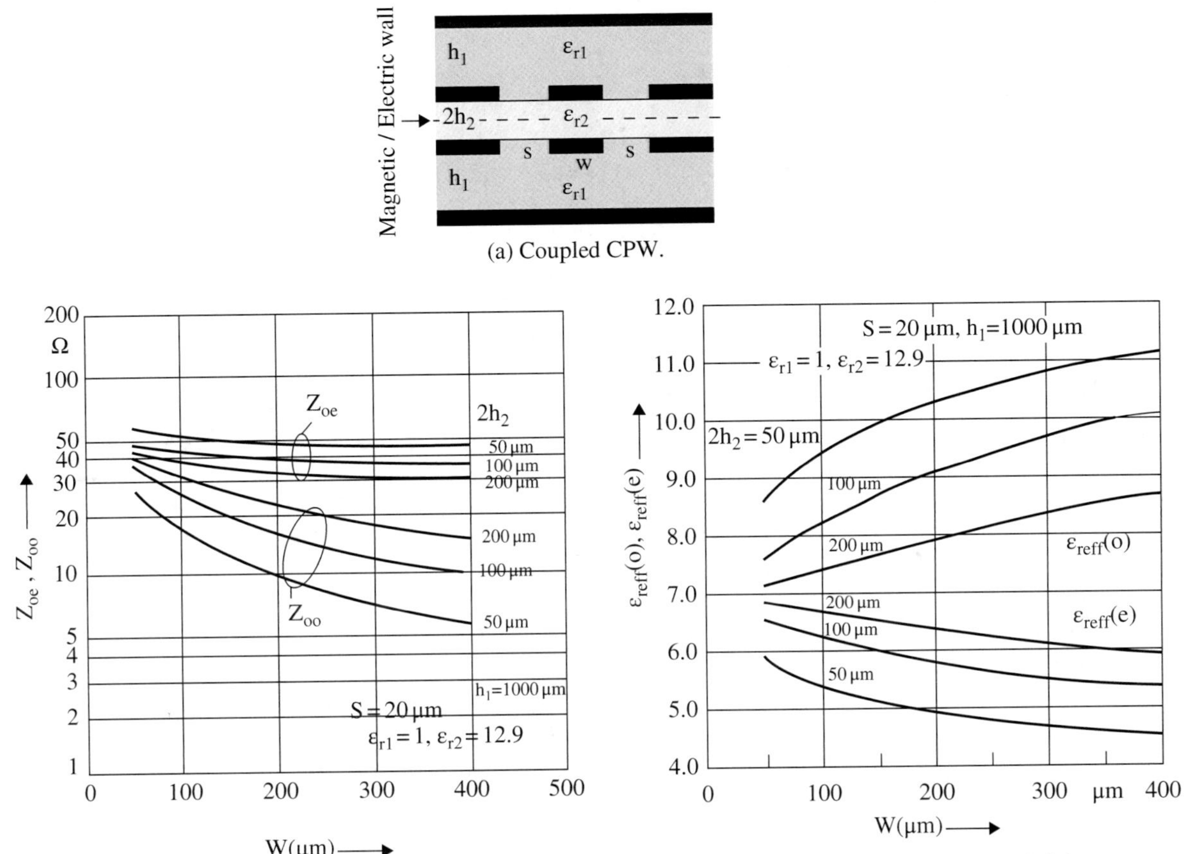

(a) Coupled CPW.

(b) Characteristic impedance

(c) Effective relative permittivity.

Figure 12.11 Shielded broadside coupled CPW and its even–odd mode characteristic impedance and effective relative permittivity. *Source*: From Bedair and Wolff [J.23]. © 1996 IEEE.

$$C_o = C_{o1} + C_{o2} \qquad \text{(a)}$$

$$\text{where} \quad C_{oi} = 2\varepsilon_o\varepsilon_{r\,i}\frac{K(k_{oi})}{K'(k_{oi})}, i = 1, 2 \qquad \text{(b)}$$

$$K_{oi} = \frac{\tanh\left(\dfrac{\pi w}{4h_i}\right)}{\tanh\left(\dfrac{\pi(w + 2s)}{4h_i}\right)}, i = 1, 2 \qquad \text{(c).}$$

$$(12.3.10)$$

The ratio $K(k)/K'(k)$ is computed from the following approximate relations [B.7, J.26]:

$$\frac{K(k)}{K'(k)} = \begin{cases} (1/\pi)\ln\left[2\left(1+\sqrt{k}\right)/\left(1-\sqrt{k}\right)\right], & 0.5 \le k^2 \le 1 \\ \pi/\ln\left[2\left(1+\sqrt{k'}\right)/\left(1-\sqrt{k'}\right)\right], & 0.0 \le k^2 \le 0.5 \end{cases}. \qquad (12.3.11)$$

Even Mode

For the even-mode excitation, the plane of symmetry shown in Fig (12.11a) is a MW. The even-mode line capacitances C_{e1} and C_{e2} are due to the upper and lower dielectric media. The total even-mode line capacitance is

$$C_e = C_{e1} + C_{e2} \qquad \text{(a)}$$

$$\text{where} \quad C_{e1} = C_{o1}\,\text{and}\, C_{e2} = 2\varepsilon_o\varepsilon_{r2}\frac{K(k_{e2})}{K'(k_{e2})} \qquad \text{(b)}$$

$$k_{e2} = \frac{\sinh\left(\dfrac{\pi w}{4h_2}\right)}{\sinh\left(\dfrac{\pi(w + 2s)}{4h_2}\right)} \qquad \text{(c).}$$

$$(12.3.12)$$

The even- and odd-mode characteristic impedance and effective relative permittivity could be computed from the above relations. Figure (12.11b and c) shows the variation in the computed even–odd mode characteristic impedance, and relative permittivity, with respect to strip width w.

12.4 Network Parameters of Coupled Line Section

The coupled transmission line section is a building block of the microwave circuits, such as filters, couplers, matching networks, etc. It is a 4-port network. However, for the circuit applications, it is reduced to a 2-port network. In that case, a pair of the port is taken as the input/output port, while the remaining pair of the port could be either short-circuited or open-circuited. Jones and Bolljahn have considered ten circuit configurations of the *4-port symmetrical coupled strip lines in a homogeneous medium* [J.28]. Zysman and Johnson have extended these configurations to the *symmetrical coupled microstrip lines in an inhomogeneous medium* [J.29]. Levy has improved the circuit model of symmetrically coupled microstrip [J.30]. Tripathi [J.18, J.33], Speciale [J.19], and Allan [J.31, J.32] have further extended the analysis to the *asymmetrical coupled microstrip lines in an inhomogeneous medium*.

This section obtains the [Z], [Y], and [ABCD] parameter representations of the symmetrical coupled lines section. However, their applications to the design of the filter are not considered here. The readers can refer to other available sources [B.6–B.10]. The asymmetrical coupled microstrip lines network is discussed in the next section.

12.4.1 Symmetrical Coupled Line in Homogeneous Medium

The analysis is carried out in two steps – first, the [Z] parameters is determined for the 4-port coupled transmission lines section; and in the next step, the 2-port network, between the input and output port, is obtained for a symmetrical coupled transmission lines section, while other two ports are properly terminated, or connected to the next stage.

[Z] Parameters Description of 4-Port Coupled Lines

Figure (12.12) shows a section of the 4-port symmetrical coupled transmission lines in a homogeneous medium, along with the *port currents* $I_{p1}...I_{p4}$, and the *port voltages* $V_{p1}...V_{p4}$. These port currents and voltages are related to the [Z] parameters:

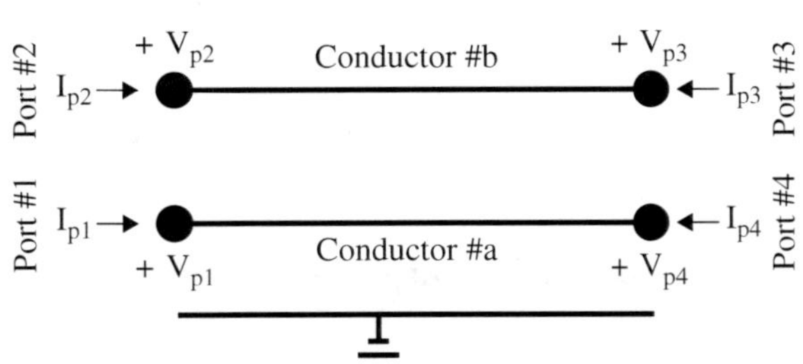

Figure 12.12 4-Port symmetrical coupled transmission line section.

$$\begin{bmatrix} V_{p1} \\ V_{p2} \\ V_{p3} \\ V_{p4} \end{bmatrix} = \begin{bmatrix} Z_{11} & Z_{12} & Z_{13} & Z_{14} \\ Z_{21} & Z_{22} & Z_{23} & Z_{24} \\ Z_{31} & Z_{32} & Z_{33} & Z_{34} \\ Z_{41} & Z_{42} & Z_{43} & Z_{44} \end{bmatrix} \begin{bmatrix} I_{p1} \\ I_{p2} \\ I_{p3} \\ I_{p4} \end{bmatrix}. \qquad (12.4.1)$$

The even- and odd-mode analysis is used to determine elements of the [Z] parameters. The coupled lines structure could be excited in the *even mode* by applying the same polarity equal potentials $V_{p1} = V_{p2}$ at port #1 and #2, and also the same polarity equal potentials $V_{p3} = V_{p4}$ at port #3 and #4. Figure (12.13a) shows that it results in positive *even-mode excitation currents*, $i_{ex1} = i_{ex2}$, and $i_{ex3} = i_{ex4}$, entering the ports. The plane of symmetry between the conductor #a and conductor #b is a MW. Likewise, Fig (12.13b) shows the odd-mode excitation, providing an EW at the plane of symmetry. In this case, the equal and opposite polarity potentials are applied at the pair of ports #1–2 and ports #3–4. It results in the *odd-mode excitation currents* i_{ex2} and i_{ex4}. At the ports #1 and #4 of *conductor #a*, excitation currents i_{ex2} and i_{ex4} enter into the ports, while at the ports #2 and #3 of *conductor #b*, excitation currents i_{ex2} and i_{ex4} leave the ports. The MW and EW reduce the symmetrical coupled lines into the single conductor even- and odd-mode supporting lines, with the even-/odd-mode characteristic impedances Z_{oe} and Z_{oo}. The even or odd mode supporting a single conductor line has one side MW/EW, whereas no wall is present at another side. In the case of a single conductor standard microstrip, there is no EW/MW at all on either side of the strip conductor.

The port currents are the superposition of the even- and odd-mode excitation currents:

$$\begin{aligned} I_{p1} &= i_{ex1} + i_{ex2} && \text{(a)} \\ I_{p2} &= i_{ex1} - i_{ex2} && \text{(b)} \\ I_{p3} &= i_{ex3} - i_{ex4} && \text{(c)} \\ I_{p4} &= i_{ex3} + i_{ex4} && \text{(d).} \end{aligned} \qquad (12.4.2)$$

In the above equation, the current entering a port is taken as positive, while the current leaving a port is negative. Figure (12.14a) shows the isolated even mode supporting conductor# b, with port #2 and #3. The length of the line section is ℓ. The other half of the coupled lines, i.e. the conductor #a, along with port #1 and port #4, could also be taken. Port #3 and #4 are considered open-circuited to evaluate the [Z] parameters, i.e. the open-circuited parameters [B.7, B.9]. The Z-parameter is discussed in subsection (3.1.1) of chapter 3.

The time-harmonic voltage V^+ is applied to port #1. The incoming wave at the point P, located at distance z, is

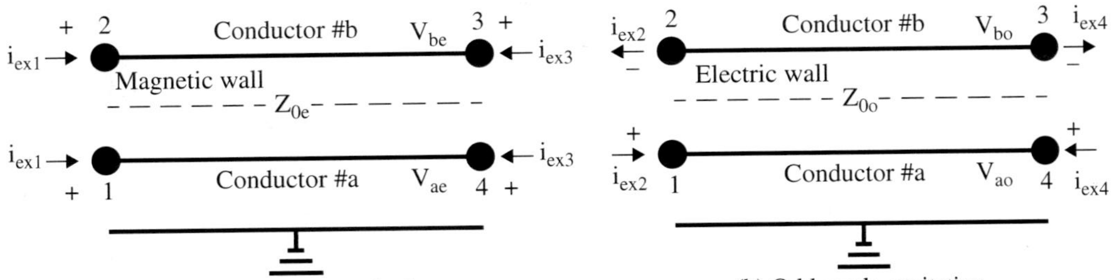

(a) Even-mode excitation. (b) Odd-mode excitation.

Figure 12.13 Even and odd modes excitation of the symmetrical coupled lines.

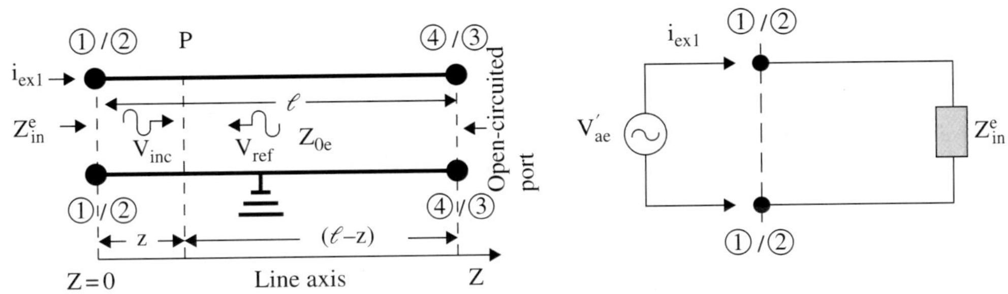

(a) Even-mode supporting conductor either #b/a. (b) Equivalent circuit at the input port.

Figure 12.14 Even-mode analysis of the symmetrical coupled lines.

$$v(z) = V^+ e^{-j\beta z}. \quad (12.4.3)$$

The time-harmonic dependent factor $e^{j\omega t}$ is suppressed in the above equation. In a homogeneous medium, the even- and odd-mode propagation constants are equal, i.e. $\beta = \beta_e = \beta_o$. The incoming voltage at port #4 on conductor #a is

$$v_{inc} = V^+ e^{-j\beta \ell}. \quad (12.4.4)$$

The voltage wave is reflected from the open-circuited port #4, with the voltage reflection coefficient +1. Therefore, the reflected voltage at port #4 is

$$v_{ref} = V^+ e^{-j\beta \ell}. \quad (12.4.5)$$

Even-Mode Excitation

At the location P, on the conductor #a shown in Fig (12.14a), the total even-mode line voltage $v_{ae}(z)$ is

$$\begin{aligned} v_{ae}(z) &= v_{inc}|_{at\ P} + v_{ref}|_{at\ P} \\ \Rightarrow v_{ae}(z) &= V^+ e^{-j\beta z} + \left(V^+ e^{-j\beta \ell}\right) e^{-j\beta(\ell - z)} \\ \Rightarrow v_{ae}(z) &= V^+ e^{-j\beta \ell}\left[e^{j\beta(\ell - z)} + e^{-j\beta(\ell - z)}\right]. \end{aligned} \quad (12.4.6)$$

Assuming $|V^+ e^{-j\beta\ell}| = V_e^+$ as the amplitude of the even-mode line voltage, the above equation is rewritten as follows:

$$\begin{aligned} v_{ae}(z) &= v_{be}(z) = V_e^+\left[e^{-j\beta(\ell - z)} + e^{j\beta(\ell - z)}\right] \\ \Rightarrow v_{ae}(z) &= v_{be}(z) = 2V_e^+ \cos \beta(\ell - z), \end{aligned} \quad (12.4.7)$$

where v_{be} is the voltage at conductor #b and v_{ae} is the voltage at conductor #a. The $2V_e^+$ is determined in terms of the even-mode excitation current i_{ex1}. The even-mode input impedance at the port #1, or the port #2 located at the conductor #b, is

$$Z_{in}^e = -j Z_{oe} \cot \beta\ell. \quad (12.4.8)$$

Figure (12.14b) shows the equivalent circuit of the even-mode excited line. The above input impedance is terminated at the port # 1/port # 2, where input port voltage # $v_{ae}(z = 0)/ v_{be}(z = 0)$ is available, giving the even-mode excitation current i_{ex1}. Using the above expressions, the even-mode voltage at the port #1/port #2, i.e. at z = 0, is

$$\begin{aligned} v'_{ae} &= v'_{be} = 2V_e{}^+ \cos\beta\ell = i_{ex1} Z_{in}^e = -jZ_{oe}\cot\beta\ell\, i_{ex1} \\ \Rightarrow 2V_e{}^+ &= -j\frac{Z_{oe}}{\sin\beta\ell} i_{ex1}. \end{aligned} \quad (12.4.9)$$

The voltage on the isolated even mode supporting conductor #a, due to the excitation current i_{ex1}, is v_{ae}^1. On the conductor #b, it is v_{be}^1. The superscript "1" with voltage indicates the voltage on conductors due to the excitation current i_{ex1}. Using equations (12.4.7) and (12.4.9), the voltage on the conductor #a/conductor #b, due to the even-mode excitation current i_{ex1}, at the port #1/port #2, is given below:

$$v_{ae}^1(z) = v_{be}^1(z) = -j\,Z_{oe}\frac{\cos\beta(\ell - z)}{\sin\beta\ell}i_{ex1}. \qquad (12.4.10)$$

Figure (12.13a) shows that the excitation current i_{ex3} also drives the coupled lines in the even mode. However, i_{ex3} is applied at the port #4/port #3, while the port #1/port #2 is open-circuited. Thus, the voltage on the conductor #a/conductor #b, due to the excitation current i_{ex3}, is

$$v_{ae}^3(z) = v_{be}^3(z) = -j\,Z_{oe}\frac{\cos\beta z}{\sin\beta\ell}i_{ex3}. \qquad (12.4.11)$$

The total even-mode voltage on the conductor # a / conductor # b is due to both excitation currents i_{ex1} and i_{ex3}. Therefore, from the above two equations, the following total even-mode line voltages are determined:

$$\begin{aligned}&\text{On the conductor \#a: } v_{a,even}(z) = v_{ae}^1(z) + v_{ae}^3(z) \quad (a)\\&\text{On the conductor \#b: } v_{b,even}(z) = v_{be}^1(z) + v_{be}^3(z) \quad (b).\end{aligned} \qquad (12.4.12)$$

Odd-Mode Excitation

Figure (12.13b) shows the odd-mode excitation is due to the excitation current $\pm\, i_{ex2}$ at port #1 and #2. For the open-circuited port #4/port #3, the odd-mode input impedance, at the port #1/port #2, is

$$Z_{in}^o = -j\,Z_{oo}\cot\beta\ell. \qquad (12.4.13)$$

The voltage on the isolated odd mode supporting conductor #a, due to the excitation current i_{ex2}, is v_{ao}^2. On line conductor# b, it is v_{bo}^2. However, for the odd-mode excitation, the potential at the conductor# b is negative. These potentials could be written as

$$\begin{aligned}&v_{ao}^2(z) = -v_{bo}^2(z) = V_o^+\left[e^{-j\beta(\ell - z)} + e^{j\beta(\ell - z)}\right]\\&\Rightarrow v_{ao}^2(z) = 2V_o^+\cos\beta(\ell - z).\end{aligned} \qquad (12.4.14)$$

The equivalent circuit for the odd mode is obtained, like the even mode case shown in Fig (12.14b). The odd-mode voltage at port #1/ port #2, i.e. at z=0, is

$$\begin{aligned}&v_{ao}^2(z=0) = -v_{bo}^2(z=0)\\&\Rightarrow v_{ao}^2(z=0) = 2V_o^+\cos\beta\ell = i_{ex2}\,Z_{in}^o = -jZ_{oo}\cot\beta\ell\, i_{ex2}\\&\Rightarrow 2V_o^+ = -j\frac{Z_{oo}}{\sin\beta\ell}i_{ex2}.\end{aligned} \qquad (12.4.15)$$

The odd-mode voltage on the conductor #a and conductor # b, due to the odd-mode excitation current i_{ex2}, is obtained from above two equations:

$$v_{ao}^2(z) = -v_{bo}^2(z) = -j\,Z_{oo}\frac{\cos\beta(\ell - z)}{\sin\beta\ell}i_{ex2}. \qquad (12.4.16)$$

Similarly, the odd-mode voltage on the conductor #a and conductor #b, due to the odd-mode excitation current i_{ex4}, is

$$v_{ao}^4(z) = -v_{bo}^4(z) = -j\,Z_{oo}\frac{\cos\beta z}{\sin\beta\ell}i_{ex4}. \qquad (12.4.17)$$

The total odd-mode voltages on the conductor #a and conductor #b, due to the odd-mode excitation currents i_{ex2} and i_{ex4}, are given below:

$$\begin{aligned}&\text{On conductor \#a: } v_{a,odd}(z) = v_{ao}^2(z) + v_{ao}^4(z) \quad (a)\\&\text{On conductor \#b: } v_{b,odd}(z) = -\left[v_{bo}^2(z) + v_{bo}^4(z)\right] \quad (b).\end{aligned} \qquad (12.4.18)$$

Finally, the port voltage V_{p1} at the port #1, located on the conductor #a, is a superposition of total even-mode voltage and total odd-mode voltage, given by equations (12.4.12) and (12.4.18):

$$\begin{aligned}&V_{p1} = v_{a,even}(z=0) + v_{a\,odd}(z=0)\\&\Rightarrow V_{p1} = v_{ae}^1(0) + v_{ao}^2(0) + v_{ae}^3(0) + v_{ao}^4(0)\\&\qquad\qquad\ \downarrow\qquad\ \downarrow\qquad\ \downarrow\qquad\ \downarrow\\&\text{Modes} \rightarrow \text{even}\quad \text{odd}\quad \text{even}\quad \text{odd}\end{aligned} \qquad (12.4.19)$$

$$\begin{aligned}\Rightarrow V_{p1} = &-j\left(Z_{oe}\,i_{ex1} + Z_{oo}i_{ex2}\right)\cot\theta\\&-j(Z_{oe}\,i_{ex3} + Z_{oo}\,i_{ex4})\,\mathrm{cosec}\,\theta,\end{aligned}$$

where $\beta\,\ell = \theta$ is the electrical length of the symmetrical coupled lines in a homogeneous medium. The even- and odd-mode excitation currents are related to the port currents and are obtained from equation (12.4.2):

$$\begin{aligned}i_{ex1} &= \frac{1}{2}\left(I_{p1} + I_{p2}\right) \quad (a)\\i_{ex2} &= \frac{1}{2}\left(I_{p1} - I_{p2}\right) \quad (b)\\i_{ex3} &= \frac{1}{2}\left(I_{p3} + I_{p4}\right) \quad (c)\\i_{ex4} &= \frac{1}{2}\left(I_{p4} - I_{p3}\right) \quad (d).\end{aligned} \qquad (12.4.20)$$

On substituting the above mode excitation currents in equation (12.4.19), the port voltage at the port #1 is obtained in terms of the port currents:

$$\begin{aligned}V_{p1} = &\frac{-j}{2}\left[Z_{oe}I_{p1} + Z_{oe}I_{p2} + Z_{oo}I_{p1} - Z_{oo}I_{p2}\right]\cot\theta\\&-\frac{j}{2}\left[Z_{oe}I_{p3} + Z_{oe}I_{p4} + Z_{oo}I_{p4} - Z_{oo}I_{p3}\right]\mathrm{cosec}\theta\\\Rightarrow V_{p1} = &\frac{-j}{2}[(Z_{oe} + Z_{oo})\cot\theta]I_{p1} - \frac{j}{2}[(Z_{oe} - Z_{oo})\cot\theta]I_{p2}\\&-\frac{j}{2}[(Z_{oe} - Z_{oo})\,\mathrm{cosec}\theta]I_{p3} - \frac{j}{2}[(Z_{oe} + Z_{oo})\,\mathrm{cosec}\theta]I_{p4}.\end{aligned} \qquad (12.4.21)$$

On comparing equation (12.4.21) against the first row of equation (12.4.1), and also taking into account the symmetry of the coupled line structure, elements of the [Z]-matrix are written as follows:

$$Z_{11} = Z_{22} = Z_{33} = Z_{44} = \frac{-j}{2}[Z_{oe} + Z_{oo}]\cot\theta \quad \text{(a)}$$
$$Z_{12} = Z_{21} = Z_{34} = Z_{43} = -\frac{j}{2}[Z_{oe} - Z_{oo}]\cot\theta \quad \text{(b)}$$
$$Z_{13} = Z_{24} = Z_{31} = Z_{42} = -\frac{j}{2}[Z_{oe} - Z_{oo}]\,\text{cosec}\,\theta \quad \text{(c)}$$
$$Z_{14} = Z_{23} = Z_{32} = Z_{41} = -\frac{j}{2}[Z_{oe} + Z_{oo}]\,\text{cosec}\,\theta \quad \text{(d)}. \qquad (12.4.22)$$

The 2-Port Network of Coupled Transmission Lines

Out of the 4 ports of a homogeneous medium symmetrical coupled lines section, one can take any two ports as the input port and the output port. The remaining two ports could be open-circuited (O.C.), or short-circuited (S.C.), or ports could be connected together. Jones and Bolljahn [J.28] have discussed ten different kinds of circuit configurations, shown in Fig (12.15a–j), to develop several kinds of filters. These are used in the design of several filters. The open-circuited port is easy to implement in the strip and microstrip lines. The configuration-3, shown in Fig (12.15c), is popular for the bandpass filter design [B.7, B.9–B.11]. Let us examine the [Z]-matrix for the 2-port structure.

For the 2-port network of Fig (12.15c), port #1 is the *input port* (I.P.), and port #3 is the *output port* (O.P.). Ports #2 and # 4 are *open-circuited* (O.C.). The strip/microstrip lines coupler is a reverse (backward) directional coupler. So, the coupled power is available at port #2. However, reflected power at the open-circuited port #2 appears at the output port #3. Under the terminal conditions, shown in Fig (12.15c), the port currents $I_{p2} = I_{p4} = 0$, the matrix equation (12.4.1) is rewritten as follows:

$$\begin{bmatrix} V_{p1} \\ V_{p2} \\ V_{p3} \\ V_{p4} \end{bmatrix} = \begin{bmatrix} Z_{11} & Z_{12} & Z_{13} & Z_{14} \\ Z_{21} & Z_{22} & Z_{23} & Z_{24} \\ Z_{31} & Z_{32} & Z_{33} & Z_{34} \\ Z_{41} & Z_{42} & Z_{43} & Z_{44} \end{bmatrix} \begin{bmatrix} I_{p1} \\ 0 \\ I_{p3} \\ 0 \end{bmatrix}$$
$$\Rightarrow \begin{bmatrix} V_{p1} \\ V_{p3} \end{bmatrix} = \begin{bmatrix} Z_{11} & Z_{13} \\ Z_{31} & Z_{33} \end{bmatrix} \begin{bmatrix} I_{p1} \\ I_{p3} \end{bmatrix}. \qquad (12.4.23)$$

It is noted that the original 4-port network of a coupled lines is reduced to the 2-port network described by the following [Z] matrix:

$$[Z] = \begin{bmatrix} Z_{11} & Z_{13} \\ Z_{31} & Z_{33} \end{bmatrix}. \qquad (12.4.24)$$

Figure (12.16a and b) shows the 4-port and equivalent 2-port network. Equation (12.4.22) provides elements of the above [Z]-matrix, in terms of the even–odd mode characteristic impedances. The 2-port network acts as a filter. The filter characteristic is obtained by computing the *image impedance* that is identical at port #1 and #3 due to the symmetry. In Fig (12.16b), port #3 is terminated with the image impedance Z_I. The input impedance at port #1 is also equal to the image impedance Z_I. For the symmetrical structure, Z_{11}=Z_{33}, the input impedance is computed at port #1 as follows [B.9, B.10, J.28]:

$$Z_{in} = Z_I = \sqrt{Z_{11}^2 - Z_{13}^2} \quad \text{(a)}$$
$$\Rightarrow Z_{in} = Z_I = \frac{1}{2}\left[(Z_{oe} - Z_{oo})^2 \,\text{cosec}^2\theta - (Z_{oe} + Z_{oo})^2 \cot^2\theta\right]^{\frac{1}{2}} \quad \text{(b)}. \qquad (12.4.25)$$

At $\theta = \frac{\pi}{2}$, i.e. for the coupled section length $\ell = \lambda_g/4$, the input impedance is

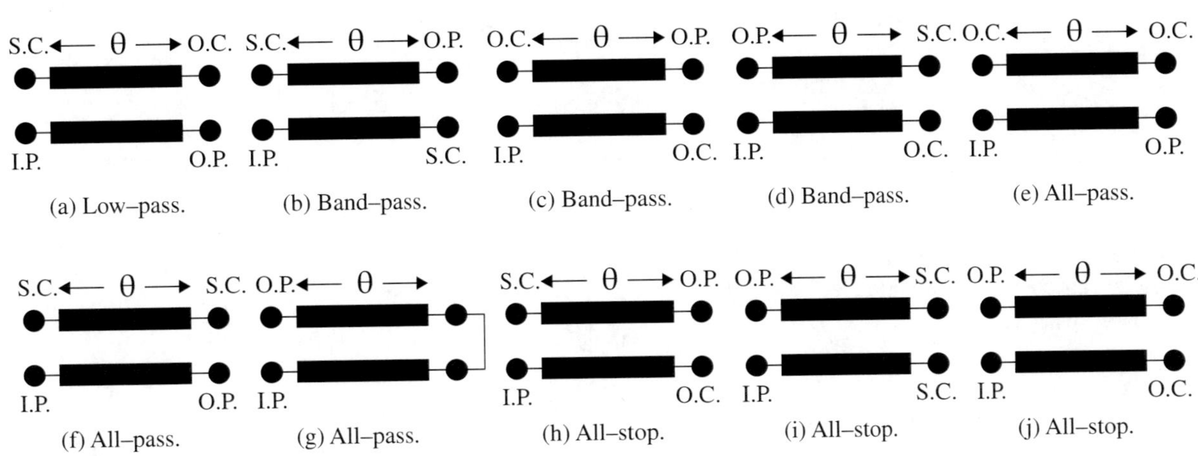

Figure 12.15 Ten configurations of the symmetrical coupled lines. *Source*: Modified from [J.28].

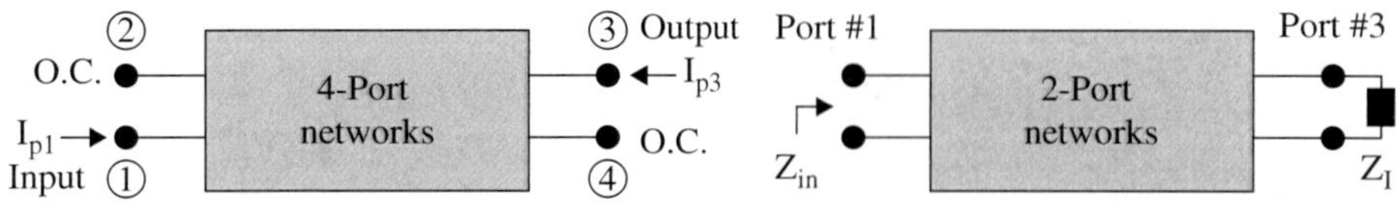

(a) 4-Port network with 2 port open-circuited.

(b) Equivalent 2-port network.

Figure 12.16 The 4-port network and equivalent 2-port with terminated load at port #3.

$$Z_{in} = Z_I = \frac{1}{2}(Z_{oe} - Z_{oo}). \quad (12.4.26)$$

The Z_{in} is a positive real quantity for the case $Z_{oe} > Z_{oo}$. The frequency band, over which it is true, provides the bandpass characteristics. At the cut-off frequency Z_{in}=0, giving two ends, θ_1 and θ_2, of the pass-band as follows:

$$\cos\theta = \pm \left[\frac{Z_{oe} - Z_{oo}}{Z_{oe} + Z_{oo}}\right] \quad (a)$$

$$\Rightarrow \cos\theta_1 = \left[\frac{Z_{oe} - Z_{oo}}{Z_{oe} + Z_{oo}}\right] \text{ and } \cos\theta_2 = -\left[\frac{Z_{oe} - Z_{oo}}{Z_{oe} + Z_{oo}}\right] \quad (b)$$

$$\cos\theta_2 = -\cos\theta_1 = \cos(\pi - \theta_1) \Rightarrow \theta_2 = (\pi - \theta_1) \quad (c). \quad (12.4.27)$$

It shows that the Z_{in} is positive real for $\theta_1 \langle\, \theta\, \langle \theta_2(\pi - \theta_1)$. The propagation constant is obtained from the following expression [B.9]:

$$\cos\beta\ell = \frac{Z_{11}}{Z_{12}} = \frac{Z_{oe} + Z_{oo}}{Z_{oe} - Z_{oo}} \cos\theta. \quad (12.4.28)$$

Jones and Bolljahn [J.28] have also discussed the image impedance and propagation constants of symmetrical coupled lines.

12.4.2 Symmetrical Coupled Microstrip Line in An Inhomogeneous Medium

Unlike the symmetrical coupled strip lines in a *homogeneous* medium, the symmetrical coupled microstrips structure is always in an *inhomogeneous* dielectric medium [J.29]. However, the even–odd mode analysis is still valid. In the case of the symmetrically coupled microstrips, both the characteristic impedance and phase velocities for the even and odd modes are different, i.e. $Z_{oe} \neq Z_{oo}, v_{pe} \neq v_{po}$, $\beta_e \neq \beta_o$. The [Z] and [Y] parameters, also [ABCD] parameters, are obtained below for the symmetrically coupled microstrips.

[Z] Parameters

To get the [Z] parameters of coupled microstrips, the relation between the port voltages and port currents is obtained. The even–odd mode excitations are considered below for this purpose.

Even–odd Mode Excitation

Figure (12.12) shows the port voltages, $V_{p1}...V_{p4}$, and port currents, $I_{p1}...I_{p4}$, of the coupled microstrips. These are related through the [Z]-matrix of equation (12.4.1). Figure (12.13a) shows the even-mode excitation by applying the excitation currents, i_{ex1} at port #1/#2, and i_{ex3} at port #4/#3. Figure (12.13b) shows the odd-mode excitation by the excitation currents i_{ex2} between port #1 and port #2 and also by i_{ex4} between port #3 and port #4. Port #1 and #4 are located on the *conductor* #a, whereas port #2 and #3 are on the *conductor #b*. The port current $I_{p1}...I_{p4}$ is a superposition of the excitation currents, $i_{ex1}...i_{ex4}$, given by equation (12.4.2). Due to the symmetry, the coupled microstrips are separated in the even mode and odd mode supporting single conductor lines. The even-mode line has the even-mode characteristics impedance Z_{oe}, and the odd-mode line has odd-mode characteristics impedance Z_{oo}.

As discussed in the previous subsection, each conductor #a and #b support a linear combination of the even-mode voltage, and odd-mode voltage wave traveling in the z-direction. In the present symmetrical coupled lines case, the conductors #a and #b are identical. Therefore, the relation between the port voltage and port current on the conductor #a is sufficient to provide all elements of the [Z]-matrix. It is given in the previous subsection. However, presently both conductors are considered separately. This process is useful to consider the asymmetrical coupled lines, discussed in the next section (12.5).

On each conductor #a and conductor #b there are four *modal* voltage waves – two even and two odd modes, excited by four excitation currents i_{ex1}, i_{ex2}, i_{ex3}, i_{ex4}. The superposition of four modal voltages provides the following total voltage $V_a(z)$ and $V_b(z)$ on the conductors #a and #b:

$$V_a(z) = V^1_{a,e}(z) + V^3_{a,e}(z) + V^2_{a,o}(z) + V^4_{a,o}(z) \quad (a)$$

Modes → even even odd odd

$$V_b(z) = V^1_{b,e}(z) + V^3_{b,e}(z) + V^2_{o}(z) + V^4_{b,o}(z) \quad (b).$$

Modes → even even odd odd

(12.4.29)

In the above expressions, the paired subscripts (a,e) and (a,o) correspond to the even- and odd-mode voltage on the conductor #a, whereas (b,e) and (b,o) correspond to the mode voltage on the conductor #b. The superscripts 1...4 correspond to the mode excitation currents $i_{ex1}...i_{ex4}$. Thus, the $V^1_{a,e}(z)$ is the even-mode voltage on the conductor #a, due to the even-mode excitation current i_{ex1}. The mode voltage $V^i_{p,k}(z)$ (p = a, b; k = e,o, and i = 1...4) are obtained in terms of the line length ℓ, propagation constants β_e and β_o, and characteristic impedances Z_{oe} and Z_{oo} from the circuit analysis. It has been done in the previous subsection. However, at present, it is done in a little different way. First, the expression for the voltage wave on a transmission section is obtained. It is excited by the current sources I_1 and I_2 at port #1 and #2. Next, it is applied to the analysis of coupled microstrips.

Voltage Wave on a Line Section Excited by Current Sources at Both Ends

Equation (2.1.79) of chapter 2 presents voltage and current waves on a 2-port transmission line section with characteristic impedance Z_0 and propagation constant β. Suppressing the time-harmonic term, the governing equations are reproduced below:

$$V(z) = V^+ e^{-j\beta z} + V^- e^{j\beta z} \quad (a)$$
$$I(z) = \frac{1}{Z_o}\left(V^+ e^{-j\beta z} - V^- e^{j\beta z}\right) \quad (b). \tag{12.4.30}$$

Let us consider the line section of length ℓ that is excited by the ideal current sources I_1 at the port #1, located at $z = 0$, and I_2 at the port #2, located at $z = \ell$. Our task is to determine the line voltage due to current sources I_1 and I_2. It is computed by using the *principle of superposition*. First, let us take $I_2 = 0$, i.e. port #2 open-circuited $I_2 = I(z = \ell) = 0$. It leads to

$$V^+ e^{-j\beta\ell} = V^- e^{j\beta\ell}, \Rightarrow V^- = V^+ e^{-2j\beta\ell} \tag{12.4.31}$$

and the voltage on the line is

$$V(z) = V^+ \left(e^{-j\beta z} + e^{j\beta z} e^{-2j\beta\ell}\right)$$
$$\Rightarrow V(z) = V^+ e^{-j\beta\ell}\left(e^{j\beta(\ell - z)} + e^{-j\beta(\ell - z)}\right). \tag{12.4.32}$$

The constant V^+ is evaluated, using equation (12.4.30b). The port condition, at the input port z=0, is $I(z = 0) = I_1$. The line voltage from the above equation is obtained as follows:

$$I_1 Z_o = V^+ - V^+ e^{-j2\beta\ell}, \quad \Rightarrow V^+ = \frac{I_1 Z_o}{1 - e^{-j2\beta\ell}}$$
$$V(z) = I_1 Z_o e^{-j\beta\ell} \frac{\left[e^{j\beta(\ell - z)} + e^{-j\beta(\ell - z)}\right]}{e^{-j\beta\ell}\left[e^{j\beta\ell} - e^{-j\beta\ell}\right]}$$
$$\Rightarrow V(z) = -jI_1 Z_o \frac{\cos[\beta(\ell - z)]}{\sin\beta\ell}. \tag{12.4.33}$$

Next, consider the current source is $I(z = \ell) = I_2$ at port #2. The port #2 is located at $z = \ell$. Port #1 is open-circuited, i.e. $I(z = 0) = I_1 = 0$. In this case, the line voltage due to current I_2 is

$$V(z) = -j\, I_2 Z_o \frac{\cos\beta z}{\sin\beta\ell}. \tag{12.4.34}$$

Finally, if the line is excited by both I_1 and I_2 at both the port #1 and #2 simultaneously, the total voltage on the line is a superposition of the voltages given by equations (12.4.33) and (12.4.34):

$$V(z) = -j\left[\frac{Z_o \cos[\beta(\ell - z)]}{\sin\beta\ell} I_1 + \frac{Z_o \cos\beta z}{\sin\beta\ell} I_2\right]. \tag{12.4.35}$$

The above expression is now applied to the even and odd mode supporting lines on the conductor #a and conductor #b separately.

The Voltage on Conductor #a

Figure (12.13a) shows that the even-mode voltage $V_{ae}(z)$ on the *conductor #a* is excited by the excitation currents i_{ex1} at the port #1 and by i_{ex3} at the port #4. On using the above equation, the expression for the total even-mode voltage on conductor #a is written:

$$V_{ae}(z) = -j\left[\frac{Z_{oe}\cos[\beta_e(\ell - z)]}{\sin\beta_e\ell} i_{ex1} + \frac{Z_{oe}\cos\beta_e z}{\sin\beta_e\ell} i_{ex3}\right]. \tag{12.4.36}$$

Likewise, Fig (12.13b) shows that the odd-mode voltage $V_{ao}(z)$ on the *conductor #a*, excited by the excitation currents i_{ex2} at the port #1 and i_{ex4} at port #4, is

$$V_{ao}(z) = -j\left[\frac{Z_{oo}\cos[\beta_o(\ell - z)]}{\sin\beta_o\ell} i_{ex2} + \frac{Z_{oo}\cos\beta_o z}{\sin\beta_o\ell} i_{ex4}\right]. \tag{12.4.37}$$

The total voltage on the *conductor #a* is a superposition of the even- and odd-mode voltages due to the excitation currents $i_{ex1}, \ldots, i_{ex4}$:

$$V_a(z) = V_{a,e}(z) + V_{a,o}(z). \tag{12.4.38}$$

The port voltage V_{p1} at the port #1, i.e. at $z = 0$, on the conductor #a is

$$V_{p1} = -j\left[Z_{oe}\cot(\beta_e\ell)\, i_{ex1} + Z_{oe}\operatorname{cosec}(\beta_e\ell)\, i_{ex3}\right]$$
$$-j\left[Z_{oo}\cot(\beta_o\ell)\, i_{ex2} + Z_{oo}\operatorname{cosec}(\beta_o\ell)\, i_{ex4}\right]. \tag{12.4.39}$$

On substituting the excitation currents i_{ex1} to i_{ex4} from equation (12.4.20) in the above equation, the following expression for V_{p1}, due to 4-port currents I_{p1} to I_{p4}, is written as:

$$V_{p1} = -\frac{j}{2}[Z_{oe}\cot(\beta_e\ell)(I_{p1}+I_{p2}) + Z_{oe}\operatorname{cosec}(\beta_e\ell)(I_{p3}+I_{p4})] - \frac{j}{2}[Z_{oo}\cot(\beta_o\ell)(I_{p1}-I_{p2}) + Z_{oo}\operatorname{cosec}(\beta_o\ell)(I_{p4}-I_{p3})]$$

$$\Rightarrow V_{p1} = -\frac{j}{2}[Z_{oe}\cot(\beta_e\ell) + Z_{oo}\cot(\beta_o\ell)]I_{p1} - \frac{j}{2}[Z_{oe}\cot(\beta_e\ell) - Z_{oo}\cot(\beta_o\ell)]I_{p2} - \frac{j}{2}[Z_{oe}\operatorname{cosec}(\beta_e\ell) - Z_{oo}\operatorname{cosec}(\beta_o\ell)]I_{p3} - \frac{j}{2}[Z_{oe}\operatorname{cosec}(\beta_e\ell) + Z_{oo}\operatorname{cosec}(\beta_o\ell)]I_{p4}. \quad (12.4.40)$$

At the port #4, located at $z = \ell$, the port voltage V_{p4}, on the conductor # a, is obtained from equations (12.4.36)–(12.4.38):

$$V_{p4} = -j[Z_{oe}\operatorname{cosec}(\beta_e\ell)\, i_{ex1} + Z_{oe}\cot(\beta_e\ell)\, i_{ex3}] - j[Z_{oo}\operatorname{cosec}(\beta_o\ell)\, i_{ex2} + Z_{oo}\cot(\beta_o\ell)\, i_{ex4}]$$

$$\Rightarrow V_{p4} = -\frac{j}{2}[Z_{oe}\operatorname{cosec}(\beta_e\ell) + Z_{oo}\operatorname{cosec}(\beta_o\ell)]I_{p1} - \frac{j}{2}[Z_{oe}\operatorname{cosec}(\beta_e\ell) - Z_{oo}\operatorname{cosec}(\beta_o\ell)]I_{p2} - \frac{j}{2}[Z_{oe}\cot(\beta_e\ell) - Z_{oo}\cot(\beta_o\ell)]I_{p3} - \frac{j}{2}[Z_{oe}\cot(\beta_e\ell) + Z_{oo}\cot(\beta_o\ell)]I_{p4}. \quad (12.4.41)$$

The Voltage on Conductor #b

Equation (12.4.35) is also used to compute the voltage on the conductor #b as a superposition of the even- and odd-mode voltages:

$$V_b(z) = V_{b,e}(z) + V_{b,o}(z). \quad (12.4.42)$$

Figure (12.13a) shows that the even-mode voltage $V_{be}(z)$ is excited by the excitation currents i_{ex1} and i_{ex3} at port #2 and #3. Figure (12.13b) shows that the odd-mode voltage is excited by the excitation currents $-i_{ex2}$ and $-i_{ex4}$ at port #2 and #3. At the port #2 and #3, the excitation currents i_{ex2} and i_{ex4} are leaving the port, so these are taken as negative. Following equation (12.4.39), the port voltage V_{p2}, at z=0, on the conductor #b is

$$V_{p2}(z=0) = -j[Z_{oe}\cot(\beta_e\ell)\, i_{ex1} + Z_{oe}\operatorname{cosec}(\beta_e\ell)\, i_{ex3}] - j[-Z_{oo}\cot(\beta_o\ell)\, i_{ex2} - Z_{oo}\operatorname{cosec}(\beta_o\ell)\, i_{ex4}]. \quad (12.4.43)$$

On substituting the excitation currents i_{ex1} to i_{ex4} from equation (12.4.20) in the above equation, the port voltage V_{p2}, due to the 4-port currents, is obtained:

$$V_{p2} = -\frac{j}{2}[Z_{oe}\cot\beta_e\ell\,(I_{p1}+I_{p2}) + Z_{oe}\operatorname{cosec}\beta_e\ell\,(I_{p3}+I_{p4})] + \frac{j}{2}[Z_{oo}\cot\beta_o\ell\,(I_{p1}-I_{p2}) + Z_{oo}\operatorname{cosec}\beta_o\ell\,(I_{p4}-I_{p3})]$$

$$\Rightarrow V_{p2} = -\frac{j}{2}[Z_{oe}\cot(\beta_e\ell) - Z_{oo}\cot(\beta_o\ell)]I_{p1} - \frac{j}{2}[Z_{oe}\cot(\beta_e\ell) + Z_{oo}\cot(\beta_o\ell)]I_{p2} - \frac{j}{2}[Z_{oe}\operatorname{cosec}(\beta_e\ell) + Z_{oo}\operatorname{cosec}(\beta_o\ell)]I_{p3} - \frac{j}{2}[Z_{oe}\operatorname{cosec}(\beta_e\ell) - Z_{oo}\operatorname{cosec}(\beta_o\ell)]I_{p4}. \quad (12.4.44)$$

Again, for the even mode, the exciting currents i_{ex1} and i_{ex3} are positive, while for the odd mode, the exciting currents i_{ex2} and i_{ex4} are negative. At the port #3, located at $z = \ell$, on the conductor #b, the port voltage V_{p3} is obtained from equations (12.4.35) and (12.4.42):

$$V_{p3}(z=\ell) = -j[Z_{oe}\operatorname{cosec}(\beta_e\ell)\, i_{ex1} + Z_{oe}\cot(\beta_e\ell)\, i_{ex3}] - j[-Z_{oo}\operatorname{cosec}(\beta_o\ell)\, i_{ex2} - Z_{oo}\cot(\beta_o\ell)\, i_{ex4}]$$

$$\Rightarrow V_{p3}(z=\ell) = \frac{-j}{2}[Z_{oe}\operatorname{cosec}(\beta_e\ell) - Z_{oo}\operatorname{cosec}(\beta_o\ell)]I_{p1} - \frac{j}{2}[Z_{oe}\operatorname{cosec}(\beta_e\ell) + Z_{oo}\operatorname{cosec}(\beta_o\ell)]I_{p2} - \frac{j}{2}[Z_{oe}\cot(\beta_e\ell) + Z_{oo}\cot(\beta_o\ell)]I_{p3} - \frac{j}{2}[Z_{oe}\cot(\beta_e\ell) - Z_{oo}\cot(\beta_o\ell)]I_{p4}. \quad (12.4.45)$$

The port voltages V_{p1}, V_{p2}, V_{p3}, and V_{p4} are arranged in the matrix form to get the following elements of the Z-matrix:

$$Z_{11} = Z_{22} = Z_{33} = Z_{44} = -\frac{j}{2}[Z_{oe}\cot(\beta_e\ell) + Z_{oo}\cot(\beta_o\ell)]$$

$$Z_{12} = Z_{21} = Z_{34} = Z_{43} = -\frac{j}{2}[Z_{oe}\cot(\beta_e\ell) - Z_{oo}\cot(\beta_o\ell)]$$

$$Z_{13} = Z_{31} = Z_{24} = Z_{42} = -\frac{j}{2}[Z_{oe}\operatorname{cosec}(\beta_e\ell) - Z_{oo}\operatorname{cosec}(\beta_o\ell)]$$

$$Z_{14} = Z_{41} = Z_{23} = Z_{32} = -\frac{j}{2}[Z_{oe}\operatorname{cosec}(\beta_e\ell) + Z_{oo}\operatorname{cosec}(\beta_o\ell)]. \quad (12.4.46)$$

The ports are symmetrical due to the structural symmetry. Equation (12.4.46) could have been written from the expression of the port voltage V_{p1} itself without writing the expressions for the remaining port voltages, i.e. V_{p2}, V_{p3}, V_{p4}. For the symmetrical coupled lines in a homogeneous medium, the even- and odd-mode propagation constants are equal, i.e. $\beta_e = \beta_o = \beta$. In that case, equation (12.4.46) is reduced to the expression (12.4.22). Again by taking any two-port as the input and output port, and the remaining two-port either the open-circuited or the short-circuited, or connecting two output ports together, ten numbers of the configurations of the 2-port networks, similar to Fig (12.15a–j), are obtained.

Y Parameters

The [Z] parameters of the symmetrical coupled lines, discussed above, are suitable for the open-circuited port, normally used for the stripline and the microstrip line configurations. However, for the short-circuited port, the [Y] parameter description of a coupled line section is more appropriate. The [Y] parameters can be obtained by inverting the [Z] parameters. However, inverting a 4×4 matrix is a difficult process. Alternatively, following the above method, expressions are derived for the [Y] parameters of a 4-port network. It is defined below:

$$\begin{bmatrix} I_{p1} \\ I_{p2} \\ I_{p3} \\ I_{p4} \end{bmatrix} = \begin{bmatrix} Y_{11} & Y_{12} & Y_{13} & Y_{14} \\ Y_{21} & Y_{22} & Y_{23} & Y_{24} \\ Y_{31} & Y_{32} & Y_{33} & Y_{34} \\ Y_{41} & Y_{42} & Y_{43} & Y_{44} \end{bmatrix} \begin{bmatrix} V_{p1} \\ V_{p2} \\ V_{p3} \\ V_{p4} \end{bmatrix}. \qquad (12.4.47)$$

The [Y] parameters are the short-circuit parameters, defined by taking the port voltages as the independent parameters. Therefore, the even–odd mode excitation is achieved through the *even–odd mode excitation voltages*, v_{ex1}, v_{ex2}, v_{ex3}, v_{ex4}, not through the mode excitation currents, as it has been done for the [Z] parameters. The port voltages are a linear combination of the excitation voltages. The line currents on both conductors are generated by the excitation voltages, giving the port currents. The [Y] parameters are evaluated, once the matrix relation between the port currents and port voltages for a section of coupled lines is obtained. At first, an expression is obtained for the current wave on a transmission section, shown in Figure (12.17). The current wave is excited by the voltage sources V_1 and V_2 at port #1 and #2. The result is applied below to a coupled microstrip section to obtain the [Y] matrix.

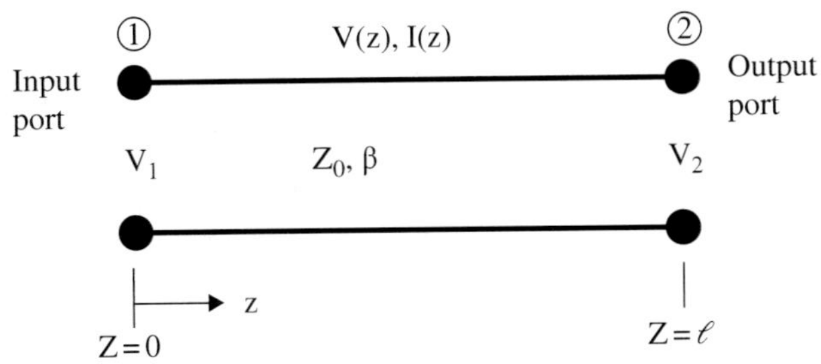

Figure 12.17 Two-port single conductor transmission line section.

Current Wave on a Line Section Excited by Voltage Sources at Both Ends

Figure (12.17) shows a single line section of length ℓ, excited at both ends by the ideal voltage sources V_1 and V_2 at port #1 and #2, located at $z = 0$ and $z = \ell$, respectively. The total voltage and current on the line section are computed by the *superposition method*. Using equation (12.4.30), the following expressions for the line current $I^1(z)$ and line voltage $V^1(z)$, due to the voltage source V_1, are computed, while the port #2, located at $z = \ell$, is short-circuited, i.e. $V_2=0$:

$$V^1(z=\ell) = V_2 = 0 = V^+ e^{-j\beta\ell} + V^- e^{j\beta\ell}$$
$$\Rightarrow V^- = -V^+ e^{-2j\beta\ell} \text{ and}$$
$$V^1(z) = V^+ e^{-j\beta\ell}\left[e^{j\beta(\ell-z)} - e^{-j\beta(\ell-z)}\right]$$
$$\Rightarrow V^1(z) = j\,2V^+ e^{-j\beta\ell}[\sin\beta(\ell-z)] \qquad (a)$$
$$I^1(z) = \frac{V^+}{Z_0} e^{-j\beta\ell}\left[e^{j\beta(\ell-z)} + e^{-j\beta(\ell-z)}\right]$$
$$\Rightarrow I^1(z) = \frac{2V^+}{Z_0} e^{-j\beta\ell}[\cos\beta(\ell-z)] \qquad (b)$$
$$\text{At } z=0,\ V^1(z=0) = V_1 = j\,2V^+ e^{-j\beta\ell}\sin\beta\ell,$$
$$\Rightarrow 2V^+ = -je^{j\beta l}[\operatorname{cosec}\beta\ell]\,V_1 \qquad (c). \qquad (12.4.48)$$

The above equations provide the following expressions for the line voltage and current due to the voltage source V_1, while port #2 is short-circuited (V_2=0):

$$V^1(z) = \frac{\sin\beta(\ell-z)}{\sin\beta\ell}V_1 \qquad (a)$$
$$I^1(z) = -jY_0\frac{\cos\beta(\ell-z)}{\sin\beta\ell}V_1 \qquad (b). \qquad (12.4.49)$$

Next, for the voltage source, V_2 is connected to the port #2 and V_1 =0 at z=0, i.e. the port #1 is short-circuited. Using equation (12.4.30), the line voltage $V^2(z)$ and line current $I^2(z)$, due to V_2, are given as follows:

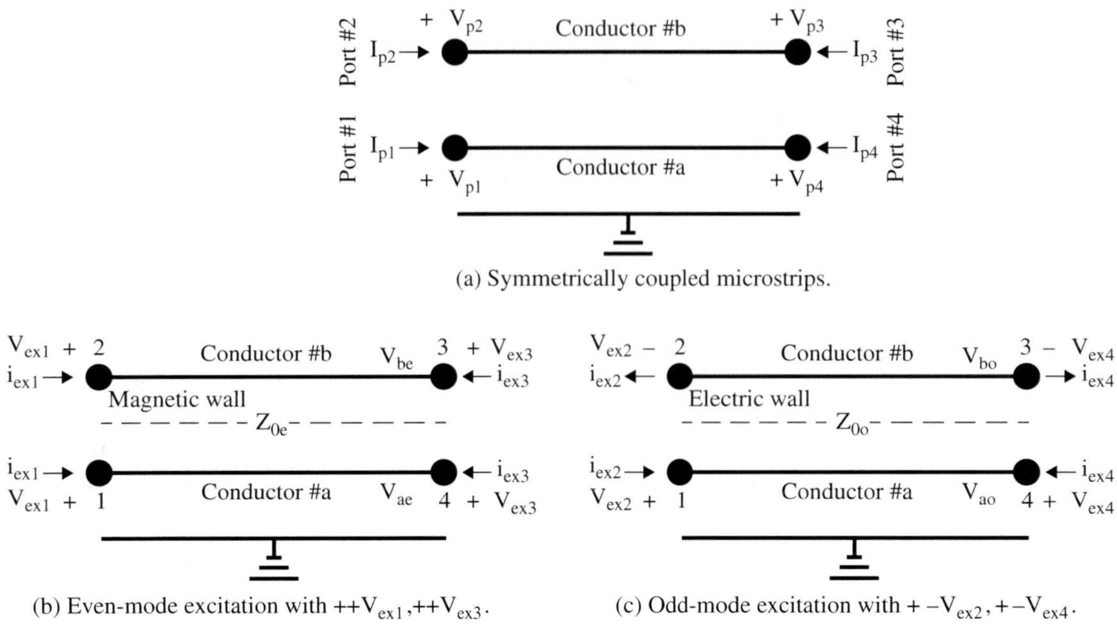

Figure 12.18 Even–odd analysis of the symmetrically coupled microstrips with port voltages and current.

$$V^2(z) = -j2V^+ \sin\beta z \quad (a)$$
$$I^2(z) = \frac{2V^+}{Z_0}\cos\beta z \quad (b)$$
$$\text{At } z = \ell, V^2(z) = V_2 = -j2V^+ \sin\beta\ell \Rightarrow 2V^+ = jV_2 \operatorname{cosec}\beta\ell \quad (c). \tag{12.4.50}$$

The above equations provide the following expressions for the line voltage and current, due to the voltage source V_2, while port #1 is short-circuited ($V_1 = 0$):

$$V^2(z) = \frac{\sin\beta z}{\sin\beta\ell}V_2 \quad (a)$$
$$I^2(z) = jY_0\frac{\cos\beta z}{\sin\beta\ell}V_2 \quad (b). \tag{12.4.51}$$

Finally, the total line voltage and current, due to the voltages V_1 and V_2 at both the ports, are obtained as a superposition of the line voltages and currents given in equations (12.4.49) and (12.4.51):

$$V(z) = V^1(z) + V^2(z)$$
$$\Rightarrow V(z) = \left[\frac{\sin\beta(\ell - z)}{\sin\beta\ell}V_1 + \frac{\sin\beta z}{\sin\beta\ell}V_2\right] \quad (a)$$
$$I(z) = I^1(z) + I^2(z)$$
$$\Rightarrow I(z) = -jY_0\left[\frac{\cos\beta(\ell - z)}{\sin\beta\ell}V_1 - \frac{\cos\beta z}{\sin\beta\ell}V_2\right] \quad (b). \tag{12.4.52}$$

The above equation is applied below for the even–odd analysis of the symmetrically coupled microstrips to obtain the [Y] matrix.

Even–Odd Mode Excitations

Figure (12.18a) shows the port voltages V_{p1}, V_{p2}, V_{p3}, V_{p4} of symmetrically coupled microstrips, whereas Fig (12.18b and c) shows the even- and odd-mode excitation of the coupled lines by the excitation voltage pairs $(++v_{ex1}, ++v_{ex3})$ and $(+-v_{ex2}, +-v_{ex4})$. The polarity of these excitation voltages to excite the even and odd modes is shown in Fig (12.18b and c), respectively. The port voltages of Fig (12.18a) are the superposition of the even- and odd-mode excitation voltages:

$$V_{p1} = v_{ex1} + v_{ex2}, \quad V_{p2} = v_{ex1} - v_{ex2}$$
$$V_{p3} = v_{ex3} - v_{ex4}, \quad V_{p4} = v_{ex3} + v_{ex4} \quad (a)$$
$$\Rightarrow v_{ex1} = \frac{V_{p1} + V_{p2}}{2}, \quad v_{ex2} = \frac{V_{p1} - V_{p2}}{2}$$
$$v_{ex3} = \frac{V_{p4} + V_{p3}}{2}, \quad v_{ex4} = \frac{V_{p4} - V_{p3}}{2} \quad (b). \tag{12.4.53}$$

Current on Conductor #a

The even- and odd-mode excitations decouple both conductor #a and conductor #b. The excitation voltage pairs $(++v_{ex1}, ++v_{ex3})$ generate the even-mode line currents $I^1_{ae}(z)$ and $I^3_{ae}(z)$, giving currents $I_{ae}(z) = I^1_{ae}(z) + I^3_{ae}(z)$

on the conductor #a. Similarly, the excitation voltage pairs $(+-v_{ex2}, +-v_{ex4})$ generate odd-mode line currents $I_{ao}^2(z)$ and $I_{ao}^4(z)$, giving $I_{ao}(z) = I_{ao}^2(z) + I_{ao}^4(z)$ on the conductor #a. The total line current $I_a(z)$ on the conductor #a is due to the superposition of the even- and odd-mode line currents, $I_{ae}(z), I_{ao}(z)$. Both these currents are obtained from equation (12.4.52b) on replacing the voltages V_1 and V_2 as follows:

$$\begin{aligned}
&\text{Even-mode}: V_1 \rightarrow v_{ex1},\ V_2 \rightarrow v_{ex3},\ Y_o \rightarrow Y_{oe},\ \beta \rightarrow \beta_e \quad &(a)\\
&\text{Odd-mode}: V_1 \rightarrow v_{ex2},\ V_2 \rightarrow v_{ex4},\ Y_o \rightarrow Y_{oo},\ \beta \rightarrow \beta_o \quad &(b).
\end{aligned} \tag{12.4.54}$$

The total line current on the conductor #a:

$$\begin{aligned}
I_a(z) &= I_{a,e}^1(z) + I_{a,e}^3(z) + I_{a,o}^2(z) + I_{a,o}^4(z) \qquad (a)\\
\text{Modes} &\rightarrow \text{even} \quad \text{even} \quad \text{odd} \quad \text{odd}\\
I_a(z) &= I_{a,e}(z) + I_{a,o}(z) \qquad (b).
\end{aligned} \tag{12.4.55}$$

$$\begin{aligned}
I_a(z) = &-j\left[Y_{oe}\frac{\cos\beta_e(\ell-z)}{\sin\beta_e\ell}v_{ex1} + Y_{oo}\frac{\cos\beta_o(\ell-z)}{\sin\beta_o\ell}v_{ex2}\right] +\\
&j\left[Y_{oe}\frac{\cos\beta_e z}{\sin\beta_e\ell}v_{ex3} + Y_{oo}\frac{\cos\beta_o z}{\sin\beta_o\ell}v_{ex4}\right].
\end{aligned} \tag{12.4.56}$$

Therefore, by using equation (12.4.53), the port current I_{p1}, at $z = 0$, i.e. $I_{p1} = I_a(z = 0)$, in terms of the port voltages, $V_{p1}, V_{p2}, V_{p3}, V_{p4}$ is obtained as follows:

$$\begin{aligned}
I_{p1} &= -j[Y_{oe}\cot(\beta_e\ell)\,v_{ex1} + Y_{oo}\cot(\beta_o\ell)\,v_{ex2}] + j[Y_{oe}\operatorname{cosec}(\beta_e\ell)\,v_{ex3} + Y_{oo}\operatorname{cosec}(\beta_o\ell)\,v_{ex4}]\\
\Rightarrow I_{p1} &= -j\left[\frac{Y_{oe}}{2}\cot(\beta_e\ell)(v_{p1}+v_{p2}) + \frac{Y_{oo}}{2}\cot(\beta_o\ell)(v_{p1}-v_{p2})\right]\\
&\quad + j\left[\frac{Y_{oe}}{2}\operatorname{cosec}(\beta_e\ell)(v_{p4}+v_{p3}) + \frac{Y_{oo}}{2}\operatorname{cosec}(\beta_o\ell)(v_{p4}-v_{p3})\right]\\
\Rightarrow I_{p1} &= -j\left[\frac{Y_{oe}\cot(\beta_e\ell) + Y_{oo}\cot(\beta_o\ell)}{2}\right]v_{p1} - j\left[\frac{Y_{oe}\cot(\beta_e\ell) - Y_{oo}\cot(\beta_o\ell)}{2}\right]v_{p2}\\
&\quad + j\left[\frac{Y_{oe}\operatorname{cosec}(\beta_e\ell) - Y_{oo}\operatorname{cosec}(\beta_e\ell)}{2}\right]v_{p3} + j\left[\frac{Y_{oe}\operatorname{cosec}(\beta_e\ell) + Y_{oo}\operatorname{cosec}(\beta_o\ell)}{2}\right]v_{p4}.
\end{aligned} \tag{12.4.57}$$

Likewise, by using equation (12.4.56), the following expression is obtained for the port current I_{p4}, leaving the port #4, i.e. $I_a(z = \ell) = -I_{p4}$:

$$\begin{aligned}
-I_{p4} &= -j[Y_{oe}\operatorname{cosec}(\beta_e\ell)v_{ex1} + Y_{oo}\operatorname{cosec}(\beta_o\ell)v_{ex2}] + j[Y_{oe}\cot(\beta_e\ell)v_{ex3} + Y_{oo}\cot(\beta_o\ell)\,v_{ex4}]\\
\Rightarrow I_{p4} &= j\frac{Y_{oe}}{2}\operatorname{cosec}(\beta_e\ell)(v_{p1}+v_{p2}) + j\frac{Y_{oo}}{2}\operatorname{cosec}(\beta_o\ell)(v_{p1}-v_{p2})\\
&\quad - j\frac{Y_{oe}}{2}\cot(\beta_e\ell)(v_{p4}+v_{p3}) - j\frac{Y_{oo}}{2}\cot(\beta_o\ell)(v_{p4}-v_{p3})\\
\Rightarrow I_{p4} &= j\left[\frac{Y_{oe}\operatorname{cosec}(\beta_e\ell) + Y_{oo}\operatorname{cosec}(\beta_o\ell)}{2}\right]v_{p1} + j\left[\frac{Y_{oe}\operatorname{cosec}(\beta_e\ell) - Y_{oo}\operatorname{cosec}(\beta_o\ell)}{2}\right]v_{p2}\\
&\quad - j\left[\frac{Y_{oe}\cot(\beta_e\ell) - Y_{oo}\cot(\beta_o\ell)}{2}\right]v_{p3} - j\left[\frac{Y_{oe}\cot(\beta_e\ell) + Y_{oo}\cot(\beta_o\ell)}{2}\right]v_{p4}.
\end{aligned} \tag{12.4.58}$$

Current on Conductor #b

Figure (12.18a) shows the line current $I_b(z)$ on conductor #b. It is a linear combination of four modal line currents. These currents are due to the even-mode excitation voltages $(+v_{ex1}, +v_{ex3})$ and the odd mode excitation voltages $(-v_{ex2}, -v_{ex4})$. The mode currents at the conductor #b are obtained from equation (12.4.52b) by replacing the voltages V_1 and V_2 with excitation voltages: for even mode $V_1 \rightarrow v_{ex1}, V_2 \rightarrow v_{ex3}, Y_O \rightarrow Y_{oe}, \beta \rightarrow \beta_e$; for odd mode $V_1 \rightarrow -v_{ex2}, V_2 \rightarrow -v_{ex4}, Y_O \rightarrow Y_{oo}, \beta \rightarrow \beta_o$. The total current on conductor #b is

$$\begin{aligned}
I_b(z) &= I_{be}^1(z) + I_{be}^3(z) + I_{bo}^2(z) + I_{bo}^4(z) \qquad (a)\\
\text{Modes} &\rightarrow \text{even} \quad \text{even} \quad \text{odd} \quad \text{odd}\\
\Rightarrow I_b(z) &= -j\left[Y_{oe}\frac{\cos\beta_e(\ell-z)}{\sin\beta_e l}v_{ex1} - Y_{oo}\frac{\cos\beta_o(\ell-z)}{\sin\beta_o l}v_{ex2}\right]\\
&\quad + j\left[Y_{oe}\frac{\cos\beta_e(z)}{\sin\beta_e\ell}v_{ex3} - Y_{oo}\frac{\cos\beta_o(z)}{\sin\beta_o\ell}v_{ex4}\right] \qquad (b).
\end{aligned} \tag{12.4.59}$$

The port current at port #2, located at $z = 0$, is $I_b(z = 0) = I_{p2}$. It is evaluated, first in terms of the excitation voltages, $v_{ex1}...v_{ex4}$, and then in terms of the port voltages $v_{p1}...v_{p4}$ from the above equation:

$$\begin{aligned}
I_{p2} &= -j[Y_{oe}\cot(\beta_e\ell)v_{ex1} - Y_{oo}\cot(\beta_o\ell)v_{ex2}] + j[Y_{oe}\operatorname{cosec}(\beta_e\ell)v_{ex3} - Y_{oo}\operatorname{cosec}(\beta_o\ell)v_{ex4}]\\
\Rightarrow I_{p2} &= -jY_{oe}\cot(\beta_e\ell)\left(\frac{v_{p1}+v_{p2}}{2}\right) + j\frac{Y_{oo}}{2}\cot(\beta_o\ell)\frac{(v_{p1}-v_{p2})}{2}\\
&\quad + j\frac{Y_{oe}}{2}\operatorname{cosec}(\beta_e\ell)\left(\frac{v_{p4}+v_{p3}}{2}\right) - j\frac{Y_{oo}}{2}\operatorname{cosec}(\beta_o\ell)\left(\frac{v_{p4}-v_{p3}}{2}\right)\\
\Rightarrow I_{p2} &= -j\left[\frac{Y_{oe}\cot(\beta_e\ell) - Y_{oo}\cot(\beta_o\ell)}{2}\right]v_{p1} - j\left[\frac{Y_{oe}\cot(\beta_e\ell) + Y_{oo}\cot(\beta_o\ell)}{2}\right]v_{p2}\\
&\quad + j\left[\frac{Y_{oe}\operatorname{cosec}(\beta_e\ell) + Y_{oo}\operatorname{cosec}(\beta_o\ell)}{2}\right]v_{p3} + j\left[\frac{Y_{oe}\operatorname{cosec}(\beta_e\ell) - Y_{oo}\operatorname{cosec}(\beta_o\ell)}{2}\right]v_{p4}.
\end{aligned} \tag{12.4.60}$$

Likewise, the port current $I_b(z = \ell) = -I_{p3}$ at port #3 is also computed using equation (12.4.59):

$$\begin{aligned}
I_b(z=\ell) = -I_{p3} &= -j[Y_{oe}\operatorname{cosec}(\beta_e\ell)v_{ex1} - Y_{oo}\operatorname{cosec}(\beta_o\ell)v_{ex2}]\\
&\quad + j[Y_{oe}\cot(\beta_e\ell)v_{ex3} - Y_{oo}\cot(\beta_o\ell)v_{ex4}]\\
\Rightarrow I_{p3} &= jY_{oe}\operatorname{cosec}(\beta_o\ell)\left(\frac{v_{p1}+v_{p2}}{2}\right) - j\frac{Y_{oo}}{2}\operatorname{cosec}(\beta_e\ell)\frac{(v_{p1}-v_{p2})}{2}\\
&\quad - j\frac{Y_{oe}}{2}\cot(\beta_e\ell)\left(\frac{v_{p4}+v_{p3}}{2}\right) + j\frac{Y_{oo}}{2}\cot(\beta_o\ell)\left(\frac{v_{p4}-v_{p3}}{2}\right)\\
\Rightarrow I_{p3} &= j\left[\frac{Y_{oe}\operatorname{cosec}(\beta_e\ell) - Y_{oo}\operatorname{cosec}(\beta_e\ell)}{2}\right]v_{p1} + j\left[\frac{Y_{oe}\operatorname{cosec}(\beta_e\ell) + Y_{oo}\operatorname{cosec}(\beta_o\ell)}{2}\right]v_{p2}\\
&\quad - j\left[\frac{Y_{oe}\cot(\beta_e\ell) + Y_{oo}\cot(\beta_o\ell)}{2}\right]v_{p3} - j\left[\frac{Y_{oe}\cot(\beta_e\ell) - Y_{oo}\cot(\beta_o\ell)}{2}\right]v_{p4}.
\end{aligned} \tag{12.4.61}$$

The port currents I_{p1}, I_{p2}, I_{p3}, I_{p4}, given by equations (12.4.57)–(12.4.61), are rearranged, and compared to the matrix (12.4.47). It provides the following elements of a [Y] matrix:

$$\begin{aligned}
Y_{11} = Y_{22} = Y_{33} = Y_{44} &= -j\left[\frac{Y_{oe}\cot\theta_e + Y_{oo}\cot\theta_o}{2}\right]\\
Y_{12} = Y_{21} = Y_{34} = Y_{43} &= -j\left[\frac{Y_{oe}\cot\theta_e - Y_{oo}\cot\theta_o}{2}\right]\\
Y_{13} = Y_{31} = Y_{24} = Y_{42} &= j\left[\frac{Y_{oe}\operatorname{cosec}\theta_e - Y_{oo}\operatorname{cosec}\theta_o}{2}\right]\\
Y_{14} = Y_{41} = Y_{23} = Y_{32} &= j\left[\frac{Y_{oe}\operatorname{cosec}\theta_e + Y_{oo}\operatorname{cosec}\theta_o}{2}\right].
\end{aligned} \tag{12.4.62}$$

The [Y] matrix elements could be written from the expression of I_{p1} itself by consideration of the symmetry of the conductors and ports. Zysman and Johnson [J.29] have provided the above results; that is reduced to a [Y] matrix of the symmetrical coupled lines in a homogeneous medium. However, admittance parameters Y_{12} and Y_{14} do not agree with the results of Jones and Bolljahn [J.28].

12.4.3 ABCD Matrix of Symmetrical Coupled Transmission Lines

The coupled transmission lines are used to develop microwave circuits by cascading several sections. The even–odd mode analysis of symmetrical coupled lines in a homogeneous medium has been discussed in section (11.4) of chapter 11. In the present section, a more versatile analysis is presented, using the [ABCD] matrix of a 4-port network. The 4-port network can be reduced to 10 numbers of two-port configurations shown in Fig (12.19). It is based on data provided by Zysman and Johnson [J.29]. Figure (12.18a and c) shows the 4-port network and also even–odd mode exciting voltages, v_{ex1}, v_{ex2}, v_{ex3}, v_{ex4}, and exciting currents, i_{ex1}, i_{ex2}, i_{ex3}, i_{ex4}. Equation (12.4.53) relates the port voltages to the exciting voltages. Similarly, the port currents are related to the exciting currents as follows:

Structure	Equivalent circuits	[ABCD] Parameters
 Short-circuited section.	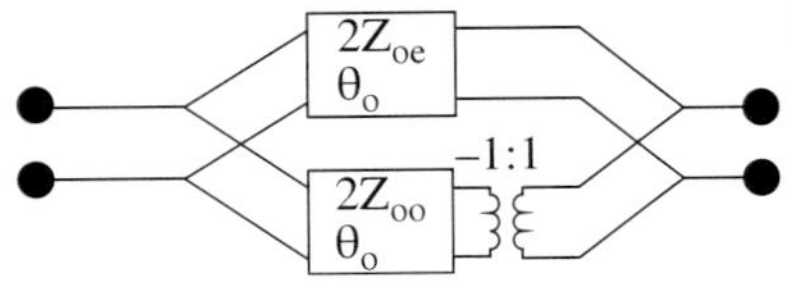	$A=\dfrac{Y_{oe}\cot\theta_e+Y_{oo}\cos\theta_o}{Y_{oe}\csc\theta_e-Y_{oo}\csc\theta_o}=D$ $B=\dfrac{2j}{Y_{oe}\csc\theta_e-Y_{oo}\csc\theta_o}$ $C=\dfrac{j}{2}\,\dfrac{Y_{oe}^2+Y_{oo}^2-2Y_{oe}Y_{oo}(\cot\theta_e\cot\theta_o+\csc\theta_e\csc\theta_o)}{Y_{oe}\csc\theta_e-Y_{oo}\csc\theta_o}.$
 Meander line section.	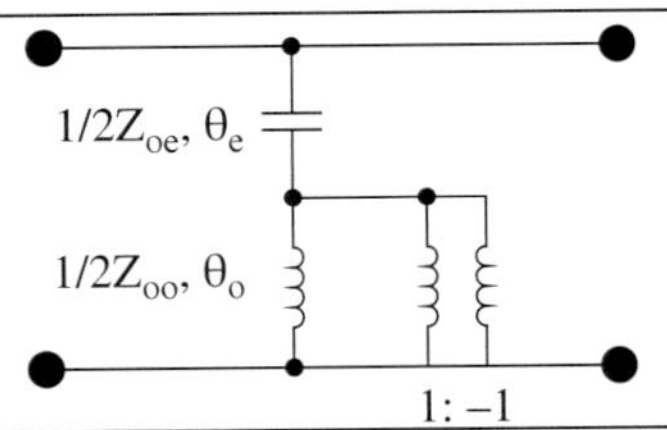	$A=\dfrac{Z_{oe}\cot\theta_e-Z_{oo}\tan\theta_o}{Z_{oe}\cot\theta_e+Z_{oo}\tan\theta_o}=D$ $B=\dfrac{2j\,Z_{oe}Z_{oo}\cot\theta_e\tan\theta_o}{Z_{oe}\cot\theta_e+Z_{oo}\tan\theta_o}$ $C=\dfrac{2j}{Z_{oe}\cot\theta_e+Z_{oo}\tan\theta_o}.$
 Comb line section.	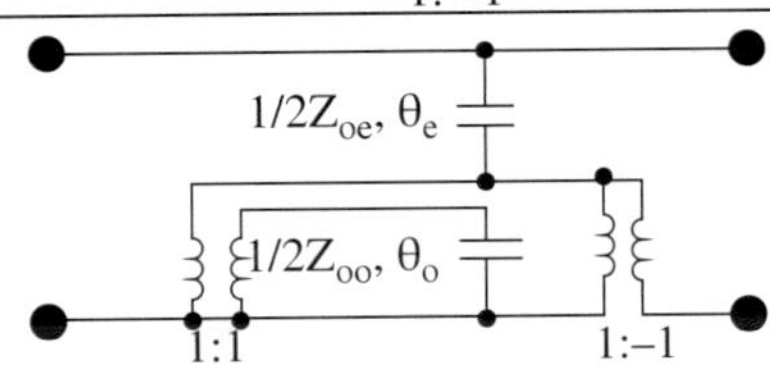	$A=\dfrac{Z_{oe}\cot\theta_e+Z_{oo}\cot\theta_o}{Z_{oe}\cot\theta_e-Z_{oo}\cot\theta_o}=D$ $B=-j\,\dfrac{2Z_{oe}Z_{oo}\cot\theta_e\cot\theta_o}{Z_{oe}\cot\theta_e-Z_{oo}\tan\theta_o}$ $C=j\,\dfrac{2}{Z_{oe}\cot\theta_e-Z_{oo}\cot\theta_o}.$
 Dual comb line section.	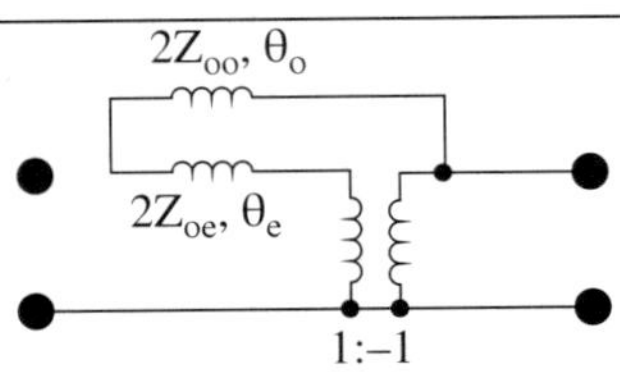	$A=\dfrac{Y_{oo}\cot\theta_o+Y_{oe}\cot\theta_e}{Y_{oo}\cot\theta_o-Y_{oe}\cot\theta_e}=D$ $B=j\,\dfrac{2}{Y_{oo}\cot\theta_o-Y_{oe}\cot\theta_e}$ $C=-j\,\dfrac{2Y_{oo}Y_{oe}\cos\theta_o\cot\theta_e}{Y_{oo}\cot\theta_e-Y_{oe}\cot\theta_e}.$
 Open-circuited symmetric section.	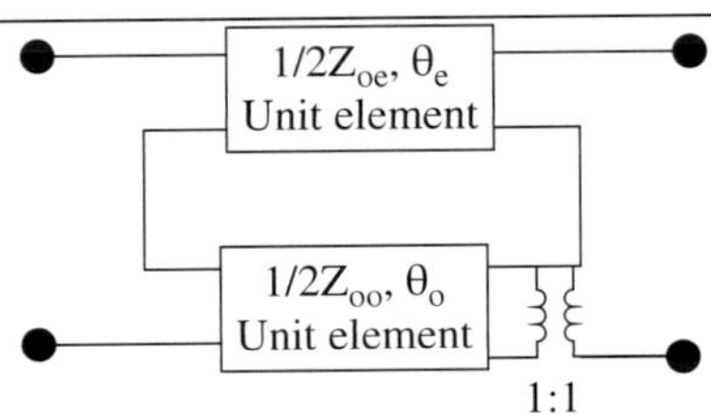	$A=\dfrac{Z_{oe}\cot\theta_e+Z_{oo}\cot\theta_o}{Z_{oe}\csc\theta_e-Z_{oo}\csc\theta_o}=D$ $B=j\,\dfrac{Z_{oe}^2+Z_{oo}^2-2Z_{oe}Z_{oo}(\csc\theta_e\csc\theta_o-\cot\theta_e\cot\theta_o)}{2(Z_{oe}\csc\theta_e-Z_{oo}\csc\theta_o)}$ $C=j\,\dfrac{2}{Z_{oe}\csc\theta_e-Z_{oo}\csc\theta_o}.$
 Asymmetrical section		$A=\dfrac{2\cos\theta_o\cos\theta_e}{\cos\theta_o-\cos\theta_e}$, $B=j\,\dfrac{Z_{oo}\sin\theta_o\cos\theta_e+Z_{oe}\cos\theta_o\sin\theta_e}{\cos\theta_o-\cos\theta_e}$ $C=j\,\dfrac{Y_{oo}\sin\theta_o\cos\theta_e+Y_{oe}\cos\theta_o\sin\theta_e}{\cos\theta_o-\cos\theta_e}$, $D=\dfrac{2(\cos\theta_o\cos\theta_e-1)-\sin\theta_o\sin\theta_e(Y_{oo}Z_{oe}+Z_{oo}Y_{oe})}{2(\cos\theta_o-\cos\theta_e)}.$
 Asymmetrical section		$A=\dfrac{2\cos\theta_o\cos\theta_e}{\cos\theta_o+\cos\theta_e}$, $B=j\,\dfrac{Z_{oo}\sin\theta_o\cos\theta_e+Z_{oe}\cos\theta_o\sin\theta_e}{\cos\theta_o+\cos\theta_e}$ $C=j\,\dfrac{Y_{oo}\sin\theta_o\cos\theta_e+Y_{oe}\cos\theta_o\sin\theta_e}{\cos\theta_o+\cos\theta_e}$, $D=\dfrac{2(\cos\theta_o\cos\theta_e+1)-\sin\theta_o\sin\theta_e(Y_{oo}Z_{oe}+Z_{oo}Y_{oe})}{2(\cos\theta_o+\cos\theta_e)}.$
 Asymmetrical section		$A=\dfrac{Z_{oe}^2+Z_{oo}^2-2Z_{oe}Z_{oo}\csc\theta_e\csc\theta_o(\cos\theta_e\cos\theta_o+1)}{Z_{oe}^2-Z_{oo}^2}$, $B=-j\,\dfrac{2Z_{oe}Z_{oo}(Z_{oo}\cot\theta_e+Z_{oe}\cot\theta_o)}{Z_{oe}^2-Z_{oo}^2}$ $C=-j\,\dfrac{2(Z_{oe}\cot\theta_e+Z_{oo}\cot\theta_o)}{Z_{oe}^2-Z_{oo}^2}$, $D=\dfrac{Z_{oe}^2+Z_{oo}^2-2Z_{oe}Z_{oo}\csc\theta_e\cos\theta_o(\cos\theta_e\cos\theta_o-1)}{Z_{oe}^2-Z_{oo}^2}.$

Figure 12.19 Several two-port networks, with ABCD parameters, of coupled microstrips. *Source*: Modified from Zysman and Johnson [J.29].

$I_{p1} = i_{ex1} + i_{ex2}, I_{p2} = i_{ex1} - i_{ex2}, I_{p3} = i_{ex3} - i_{ex4}, I_{p4} = i_{ex3} + i_{ex4}.$

$$\text{For even mode}: i_{ex1} = \frac{I_{p1}+I_{p2}}{2}, i_{ex3} = \frac{I_{p3}+I_{p4}}{2} \quad (a)$$
$$\text{For odd mode}: i_{ex2} = \frac{I_{p1}-I_{p2}}{2}, i_{ex4} = \frac{I_{p4}-I_{p3}}{2} \quad (b). \qquad (12.4.63)$$

The coupled transmission lines are separated into two independent even- and odd-mode supporting lines, with the characteristic impedance Z_{oe} and Z_{oo}, respectively. The input even-mode exciting voltage and current (v_{ex1}, i_{ex1}) are related to its output (v_{ex3}, i_{ex3}) through the even mode $[ABCD]_{even}$ matrix. Likewise, the input odd-mode exciting voltage and current (v_{ex2}, i_{ex2}) are related to its output (v_{ex4}, i_{ex4}) through the odd mode $[ABCD]_{odd}$ matrix. These relations are given below:

$$\begin{bmatrix} v_{ex1} \\ i_{ex1} \end{bmatrix} = \begin{bmatrix} A & B \\ C & D \end{bmatrix}_{even} \begin{bmatrix} v_{ex3} \\ -i_{ex3} \end{bmatrix} \quad (a)$$
$$\begin{bmatrix} v_{ex2} \\ i_{ex2} \end{bmatrix} = \begin{bmatrix} A & B \\ C & D \end{bmatrix}_{odd} \begin{bmatrix} v_{ex4} \\ -i_{ex4} \end{bmatrix} \quad (b). \qquad (12.4.64)$$

The output port currents for both the modes are taken as negative, following the definition of the [ABCD] matrix given in subsection (3.1.3) of chapter 3. The sets of equations given below are obtained from the above relations. Figure (12.18b) shows that the positive even-mode exciting currents (i_{ex1}, i_{ex3}) enter the port, while the negative odd-mode exciting currents (i_{ex2}, i_{ex4}) leave the port. The matrix relations of both modes are given in equation (11.4.5) of chapter 11.

Using equations (12.4.53) and (12.4.63), the above relations can be expressed in terms of the port voltages and port currents:

$$\begin{bmatrix} \left(\frac{V_{p1}+V_{p2}}{2}\right) \\ \left(\frac{I_{p1}+I_{p2}}{2}\right) \end{bmatrix} = \begin{bmatrix} \cos\theta_e & jZ_{oe}\sin\theta_e \\ jY_{oe}\sin\theta_e & \cos\theta_e \end{bmatrix} \begin{bmatrix} \left(\frac{V_{p4}+V_{p3}}{2}\right) \\ -\left(\frac{I_{p4}+I_{p3}}{2}\right) \end{bmatrix} \quad (a)$$
$$\begin{bmatrix} \left(\frac{V_{p1}-V_{p2}}{2}\right) \\ \left(\frac{I_{p1}-I_{p2}}{2}\right) \end{bmatrix} = \begin{bmatrix} \cos\theta_o & jZ_{oo}\sin\theta_o \\ jY_{oo}\sin\theta_o & \cos\theta_o \end{bmatrix} \begin{bmatrix} \left(\frac{V_{p4}-V_{p3}}{2}\right) \\ -\left(\frac{I_{p4}-I_{p3}}{2}\right) \end{bmatrix} \quad (b). \qquad (12.4.65)$$

$$\Rightarrow \begin{aligned} \frac{V_{p1}+V_{p2}}{2} &= \frac{\cos\theta_e}{2}(V_{p4}+V_{p3}) - j\frac{Z_{oe}\sin\theta_e}{2}(I_{p4}+I_{p3}) \quad (a) \\ \frac{I_{p1}+I_{p2}}{2} &= j\frac{Y_{oe}\sin\theta_e}{2}(V_{p4}+V_{p3}) - \frac{\cos\theta_e}{2}(I_{p4}+I_{p3}) \quad (b) \\ \frac{V_{p1}-V_{p2}}{2} &= \frac{\cos\theta_o}{2}(V_{p4}-V_{p3}) - j\frac{Z_{oo}\sin\theta_o}{2}(I_{p4}-I_{p3}) \quad (c) \\ \frac{I_{p1}-I_{p2}}{2} &= j\frac{Y_{oo}\sin\theta_o}{2}(V_{p4}-V_{p3}) - \frac{\cos\theta_o}{2}(I_{p4}-I_{p3}) \quad (d). \end{aligned} \qquad (12.4.66)$$

The input-port voltages (V_{p1}, V_{p2}) and the input currents (I_{p1}, I_{p2}) are obtained from the above:

$$\begin{aligned} V_{p1} &= \frac{1}{2}(\cos\theta_e + \cos\theta_o)V_{p4} + \frac{1}{2}(\cos\theta_e - \cos\theta_o)V_{p3} \\ &\quad + \frac{j}{2}(Z_{oe}\sin\theta_e + Z_{oo}\sin\theta_o)(-I_{p4}) \\ &\quad + \frac{j}{2}(Z_{oe}\sin\theta_e - Z_{oo}\sin\theta_o)(-I_{p3}) \quad (a) \\ V_{p2} &= \frac{1}{2}(\cos\theta_e - \cos\theta_o)V_{p4} + \frac{1}{2}(\cos\theta_e + \cos\theta_o)V_{p3} \\ &\quad + \frac{j}{2}(Z_{oe}\sin\theta_e - Z_{oo}\sin\theta_o)(-I_{p4}) \\ &\quad + \frac{j}{2}(Z_{oe}\sin\theta_e + Z_{oo}\sin\theta_o)(-I_{p3}) \quad (b) \\ I_{p1} &= \frac{j}{2}(Y_{oe}\sin\theta_e + Y_{oo}\sin\theta_o)(V_{p4}) \\ &\quad + \frac{j}{2}(Y_{oe}\sin\theta_e - Y_{oo}\sin\theta_o)(V_{p3}) \\ &\quad + \frac{1}{2}(\cos\theta_e + \cos\theta_o)(-I_{p4}) \\ &\quad + \frac{1}{2}(\cos\theta_e - \cos\theta_o)(-I_{p3}) \quad (c) \\ I_{p2} &= \frac{j}{2}(Y_{oe}\sin\theta_e - Y_{oo}\sin\theta_o)(V_{p4}) \\ &\quad + \frac{j}{2}(Y_{oe}\sin\theta_e + Y_{oo}\sin\theta_o)(V_{p3}) \\ &\quad + \frac{j}{2}(\cos\theta_e - \cos\theta_o)(-I_{p4}) \\ &\quad + \frac{j}{2}(\cos\theta_e + \cos\theta_o)(-I_{p3}) \quad (d). \end{aligned} \qquad (12.4.67)$$

The above equation is rewritten in the 4-port [ABCD] matrix form:

$$\begin{bmatrix} V_{p1} \\ V_{p2} \\ \cdots \\ I_{p1} \\ I_{p2} \end{bmatrix} = \begin{bmatrix} a_{11} & a_{12}: & b_{11} & b_{12} \\ a_{21} & a_{22}: & b_{21} & b_{22} \\ \cdots & \cdots & \cdots & \cdots \\ c_{11} & c_{12}: & d_{11} & d_{12} \\ c_{21} & c_{22}: & d_{21} & d_{22} \end{bmatrix} \begin{bmatrix} V_{p4} \\ V_{p3} \\ \cdots \\ -I_{p4} \\ -I_{p3} \end{bmatrix}, \qquad (12.4.68)$$

where

$$\begin{aligned} a_{11} = a_{22} = d_{11} = d_{22} &= \frac{1}{2}(\cos\theta_e + \cos\theta_o) \\ a_{12} = a_{21} = d_{12} = d_{21} &= \frac{1}{2}(\cos\theta_e - \cos\theta_o) \\ b_{11} = b_{22} &= \frac{j}{2}(Z_{oe}\sin\theta_e + Z_{oo}\sin\theta_o) \\ b_{12} = b_{21} &= \frac{j}{2}(Z_{oe}\sin\theta_e - Z_{oo}\sin\theta_o) \\ c_{11} = c_{22} &= \frac{j}{2}(Y_{oe}\sin\theta_e + Y_{oo}\sin\theta_o) \\ c_{12} = c_{21} &= \frac{j}{2}(Y_{oe}\sin\theta_e - Y_{oo}\sin\theta_o), \end{aligned} \qquad (12.4.69)$$

$$\text{and}, \theta_e = \beta_e \ell_e = \frac{\pi}{2}\frac{f}{f_{oe}} \quad \text{(a)}$$
$$\theta_o = \beta_o \ell_o = \frac{\pi}{2}\frac{f}{f_{oo}} \quad \text{(b)}. \qquad (12.4.70)$$

The f_{oe} and f_{oo} are frequencies at which the coupled lines, for each mode, are $\ell_e = \lambda_{oe}/4$ and $\ell_o = \lambda_{oo}/4$, respectively. The physical coupled line length is an average of both lengths.

The above equations show the symmetry of the port. The matrix elements a_{ij}, b_{ij}, c_{ij}, and d_{ij} correspond to A, B, C, and D of the 2-port [ABCD] matrix, and these are the dimensionless *voltage ratio*, *trans-impedance*, *transconductance*, and a *current ratio* of the two-port network. Like normal 2-port ABCD parameters, discussed in chapter 3; the elements [a] and [c] are determined by the output open-circuit condition $[I_{p4} = I_{p3} = 0]$. The elements [b] and [d] are determined by the output short-circuit condition $[V_{p4} = V_{p3} = 0]$. The combinations of the open-circuit, short-circuit cases and connecting output ports together provide 10 numbers of 2-port networks [J.29]. Their ABCD parameters, along with their equivalent circuits, are listed in Fig (12.19). These ABCD parameters for the 2-port networks can be obtained from equations (12.4.68) and (12.4.69). These networks are useful for the design of coupled line-based circuits. They are equally applicable to the symmetrical coupled lines in the homogeneous medium [J.28, J.29]. Alternative equivalent circuits have also been suggested [J.30].

12.5 Asymmetrical Coupled Lines Network Parameters

Figure (12.20) shows a section of the asymmetrical coupled transmission lines in an inhomogeneous

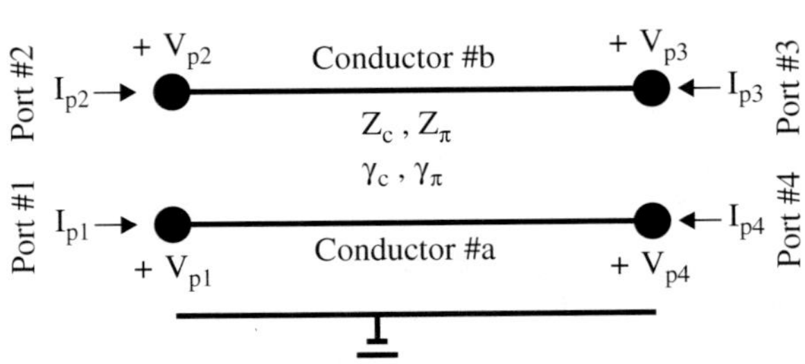

Figure 12.20 Asymmetrical coupled lines in an inhomogeneous medium.

medium. The wave theory of the coupled transmission lines is discussed in subsection (11.5.2) of chapter 11. The asymmetrical coupled lines structure supports two *normal* modes; namely, the C-mode and the π-mode. The voltage ratios on two conductors for both modes, given below, are defined by equation (11.5.33):

$$\text{For C - mode: } R_C = \frac{V_{2c}}{V_{1c}} \quad \text{(a)}$$
$$\text{For } \pi \text{ - mode: } R_\pi = \frac{V_{2\pi}}{V_{1\pi}} \quad \text{(b)}. \qquad (12.5.1)$$

In the case of the symmetrical coupled lines, the C-mode corresponds to the even mode, $R_C = +1$, and the π–mode corresponds to the odd mode, $R_\pi = -1$. Under certain congruent conditions, even the asymmetrical coupled lines can support the even voltage and odd current mode, or even current and odd voltage mode. These modes are the special cases of the C-mode and π-mode [J.18, J.19, J.34, J.35]. The propagation constants γ_c and γ_π of the C-mode and π-mode, for a lossy coupled lines, are complex quantities. Likewise, Y_{c1} and $Y_{\pi 1}$ are the characteristic admittances of these modes on the line #a, and their corresponding modal characteristic impedances are Z_{c1} and $Z_{\pi 1}$. Similarly, the modal characteristic admittances and impedances of line #b are Y_{c2} and $Y_{\pi 2}$, Z_{c2} and $Z_{\pi 2}$ [J.18]. Subsection (11.5.2) of chapter 11 discusses the coupled lines secondary parameters expressed in terms of the primary line constants. Figure (12.20) shows the port voltages, V_{p1}, V_{p2}, V_{p3}, V_{p4} and port currents I_{p1}, I_{p2}, I_{p3}, I_{p4}. Under the normal mode excitations, both lines are decoupled and support two independent C and π–modes. So, the line voltage or the current on line #a is a linear combination of the modal voltages v_{c1} and $v_{\pi 1}$, or modal currents i_{c1} and $i_{\pi 1}$. Our task in this section is to obtain the ABCD parameters of 4-port asymmetrical coupled lines. The ABCD parameters of the 2-port network are discussed in subsection (3.1.3) of chapter 3. For the 2-port network, the ABCD elements are complex scalar quantities. However, for the 4-port network, each element of the [ABCD] matrix is described by the 2×2 matrix.

12.5.1 [ABCD] Parameters of the 4-Port Network

The 4-port [ABCD] matrix parameters of the symmetrical coupled lines, shown in Figure (12.18a), are described by the following equation:

$$\begin{bmatrix} V_{p1} \\ V_{p2} \\ \cdots \\ I_{p1} \\ I_{p2} \end{bmatrix} = \begin{bmatrix} a_{11} & a_{12} : & b_{11} & b_{12} \\ a_{21} & a_{22} : & b_{21} & b_{22} \\ \cdots & \cdots & \cdots & \cdots \\ c_{11} & c_{12} : & d_{11} & d_{12} \\ c_{21} & c_{22} : & d_{21} & d_{22} \end{bmatrix} \begin{bmatrix} V_{p4} \\ V_{p3} \\ \cdots \\ -I_{p4} \\ -I_{p3} \end{bmatrix}$$

$$\Rightarrow \begin{bmatrix} V_{in} \\ I_{in} \end{bmatrix} = \begin{bmatrix} [A] & [B] \\ [C] & [D] \end{bmatrix} \begin{bmatrix} V_{out} \\ -I_{out} \end{bmatrix} \qquad (a)$$

$$\text{where,} \quad V_{in} = \begin{bmatrix} V_{p1} \\ V_{p2} \end{bmatrix}, I_{in} = \begin{bmatrix} I_{p1} \\ I_{p2} \end{bmatrix}$$

$$\text{and,} \quad V_{out} = \begin{bmatrix} V_{p4} \\ V_{p3} \end{bmatrix}, I_{out} = \begin{bmatrix} I_{p4} \\ I_{p3} \end{bmatrix} \qquad (b). \qquad (12.5.2)$$

Like a standard 2-port [A, B, C, D], the [A], [B], [C], [D] parameters of a 4-port network are determined under the *open-circuit* and *short-circuit* conditions at the output port #3 and #4.

Open Circuit Condition At Output Port [$I_{p3} = I_{p4} = 0$]

The input port voltages and input currents, at the port #1and #2, due to the output open-circuited port voltages, V_{p4} and V_{p3}, at the port #4 and #3 are given below:

$$[V_{in}] = [A][V_{out}] \text{ i.e.}$$

$$V_{p1} = a_{11}V_{p4} + a_{12}V_{p3} \qquad (a)$$

$$V_{p2} = a_{21}V_{p4} + a_{22}V_{p3} \qquad (b) \qquad (12.5.3)$$

$$\text{where,} \quad [A] = \begin{bmatrix} a_{11} & a_{12} \\ a_{21} & a_{22} \end{bmatrix} \qquad (c).$$

$$[I_{in}] = [C][I_{out}] \text{ i.e.}$$

$$I_{1p} = c_{11}V_{p4} + c_{12}V_{p3} \qquad (a)$$

$$I_{p2} = c_{21}V_{p4} + c_{22}V_{p3} \qquad (b) \qquad (12.5.4)$$

$$\text{where,} \quad [C] = \begin{bmatrix} c_{11} & c_{12} \\ c_{21} & c_{22} \end{bmatrix} \qquad (c).$$

In the above equations, the A-parameter is the voltage ratio and the C-parameter has a dimension of admittance. The elements of both the [A] and [C] matrices are determined below.

Determination of [A] Matrix

Each of the conductors #a and #b support the C and π-modes. The port voltages for the conductors #a and #b are expressed as a superposition of the C and π-modal voltages at these ports:

$$\text{Conductor \#a:} \begin{matrix} \text{At input port}: & V_{p1} = v_{c1} + v_{\pi 1} \\ \text{At output port}: & V_{p4} = v_{c4} + v_{\pi 4} \end{matrix} \qquad (a)$$

$$\text{Conductor \#b:} \begin{matrix} \text{At input port}: & V_{p2} = v_{c2} + v_{\pi 2} \\ \text{At output port}: & V_{p3} = v_{c3} + v_{\pi 3} \end{matrix} \qquad (b). \qquad (12.5.5)$$

The input port #1 and the output port #4 are located on the conductor #a. The open-circuited C and π-mode voltages at the output port #4 are V_{c4} and $V_{\pi 4}$. Therefore, at the input port #1, the modal voltages are expressed as $v_{c1} = v_{c4}\cosh\gamma_c\ell$ and $v_{\pi 1} = v_{\pi 4}\cosh\gamma_\pi\ell$. Likewise, on the conductor #b, the modal voltages at the input port #2 are obtained as $v_{c2} = v_{c3}\cosh\gamma_c\ell$ and $v_{\pi 2} = v_{\pi 3}\cosh\gamma_\pi\ell$. These voltages are due to the open-circuited modal voltages v_{c3} and $v_{\pi 3}$ at the output port #3. Using the above equations, the total input port voltages on the conductors #a and #b are given as follows:

$$V_{p1} = v_{c4}\cosh\gamma_c\ell + v_{\pi 4}\cosh\gamma_\pi\ell \qquad (a)$$

$$V_{p2} = v_{c3}\cosh\gamma_c\ell + v_{\pi 3}\cosh\gamma_\pi\ell \qquad (b). \qquad (12.5.6)$$

The above equations, for both the C and π modes, could be written by using equation (2.1.77) of chapter 2. Also, in the homogeneous medium, the propagation constants γ_c and γ_π are identical. In subsection (11.5.2) of chapter 11, both γ_c and γ_π are computed in terms of the primary line constants. These are different for the lines located in an inhomogeneous medium. However, in the case of the asymmetrical coupled lines, even in a homogeneous medium, the characteristic impedances Z_c, Z_π and characteristic admittances Y_c, Y_π are different. To compute the elements of the matrix [A], the modal voltages, ($V_{c4}, V_{\pi 4}$) and (V_{c3}, $V_{\pi 3}$) at the output port #4 and #3, are further expressed in terms of the port voltages V_{p4}, and V_{p3}. The voltage ratio, R_c and R_π, for both the modes, are defined as follows:

$$\text{Mode \#C:} \ \frac{V_{c3}}{V_{c4}} = R_c \qquad (a)$$

$$\text{Mode \#}\pi\text{:} \ \frac{V_{\pi 3}}{V_{\pi 4}} = R_\pi \qquad (b). \qquad (12.5.7)$$

In subsection (11.5.3) of chapter 11, R_c and R_π are computed in terms of the primary line constants. In the above equation, the voltage ratio R_c and R_π are taken as positive. However, in actual computation R_π becomes negative. On using equations (12.5.5) and (12.5.7), the following expressions are obtained for the modal voltages in terms of the port voltages:

π-Mode Voltage

$$\begin{aligned} v_{c4} &= V_{p4} - v_{\pi 4} = V_{p4} - \frac{v_{\pi 3}}{R_\pi} && \text{(a)} \\ v_{c3} &= V_{p3} - v_{\pi 3} = V_{p3} - R_\pi v_{\pi 4} && \text{(b).} \end{aligned} \qquad (12.5.8)$$

$$\begin{aligned} R_c &= \frac{V_{p3} - v_{\pi 3}}{V_{p4} - v_{\pi 4}} = \frac{V_{p3} - v_{\pi 3}}{V_{p4} - v_{\pi 3}/R_\pi} \\ \Rightarrow R_c\left[V_{p4} - \frac{v_{\pi 3}}{R_\pi}\right] &= V_{p3} - v_{\pi 3} \\ \Rightarrow v_{\pi 3} &= \frac{R_\pi\left[V_{p3} - R_c V_{p4}\right]}{R_\pi - R_c}. \end{aligned} \qquad (12.5.9)$$

$$\begin{aligned} \text{Again, } R_c &= \frac{V_{p3} - v_{\pi 3}}{V_{p4} - v_{\pi 4}} = \frac{V_{p3} - R_\pi v_{\pi 4}}{V_{p4} - v_{\pi 4}} \\ \Rightarrow V_{\pi 4} &= \frac{V_{p3} - R_c V_{p4}}{R_\pi - R_c}. \end{aligned} \qquad (12.5.10)$$

C-mode Voltage

On using equations (12.5.8) and (12.5.10), the following expressions are obtained:

$$\begin{aligned} v_{c4} &= V_{p4} - v_{\pi 4} = V_{p4} - \left[\frac{V_{p3} - R_c V_{p4}}{R_\pi - R_c}\right] \\ \Rightarrow v_{c4} &= \frac{R_\pi V_{p4} - V_{p3}}{R_\pi - R_c} \end{aligned} \qquad (12.5.11)$$

$$\begin{aligned} v_{c3} &= V_{p3} - v_{\pi 3} = V_3 - \left[\frac{R_\pi V_{p3} - R_\pi R_c V_{p4}}{R_\pi - R_c}\right] \\ \Rightarrow v_{c3} &= \frac{R_\pi R_c V_{p4} - R_c V_{p3}}{R_\pi - R_c}. \end{aligned} \qquad (12.5.12)$$

On substituting the modal voltages (v_{c4}, $v_{\pi 4}$) and (v_{c3}, $v_{\pi 3}$) in equation (12.5.6), the following expressions are obtained for the input port voltages:

$$\begin{aligned} V_{p1} &= \cosh\gamma_c\ell\left[\frac{R_\pi V_{p4} - V_{p3}}{R_\pi - R_c}\right] + \cosh\gamma_\pi\ell\left[\frac{V_{p3} - R_c V_{p4}}{R_\pi - R_c}\right] \\ V_{p2} &= \cosh\gamma_c\ell\left[\frac{R_\pi R_c V_{p4} - R_c V_{p3}}{R_\pi - R_c}\right] + \cosh\gamma_\pi\ell\left[\frac{R_\pi V_{p3} - R_\pi R_c V_{p4}}{R_\pi - R_c}\right] \\ \Rightarrow V_{p1} &= \left[\frac{R_\pi\cosh\gamma_c\ell - R_c\cosh\gamma_\pi\ell}{R_\pi - R_c}\right]V_{p4} + \left[\frac{\cosh\gamma_\pi\ell - \cosh\gamma_c\ell}{R_\pi - R_c}\right]V_{p3} && \text{(a)} \\ V_{p2} &= \left[\frac{R_\pi R_c\cosh\gamma_c\ell - R_\pi R_c\cosh\gamma_\pi}{R_\pi - R_c}\right]V_{p4} + \left[\frac{R_\pi\cosh\gamma_\pi\ell - R_c\cosh\gamma_c\ell}{R_\pi - R_c}\right]V_{p3} && \text{(b).} \end{aligned} \qquad (12.5.13)$$

On comparing the above equations against equation (12.5.3), the elements of the [A]-matrix are obtained as follows:

$$\begin{aligned} a_{11} &= \frac{R_\pi\cosh\gamma_c\ell - R_c\cosh\gamma_\pi\ell}{R_\pi - R_c} && \text{(a)} \\ a_{12} &= \frac{\cosh\gamma_\pi\ell - \cosh\gamma_c\ell}{R_\pi - R_c} && \text{(b)} \\ a_{21} &= \frac{R_\pi R_c\cosh\gamma_c\ell - R_\pi R_c\cosh\gamma_\pi\ell}{R_\pi - R_c} && \text{(c)} \\ a_{22} &= \frac{R_\pi\cosh\gamma_\pi\ell - R_c\cosh\gamma_c\ell}{R_\pi - R_c} && \text{(d).} \end{aligned} \qquad (12.5.14)$$

The [A] matrix parameters of the *symmetrical coupled lines*, in the *inhomogeneous* medium, are evaluated from the above equations. The C and π-modes are reduced to the even and odd modes with $\gamma_c \to \gamma_e$ and $\gamma_\pi \to \gamma_o$, and also $R_c = +1$ (even mode) and $R\pi = (-1)$ (odd mode).

$$\begin{aligned} a_{11} &= \frac{\cosh\gamma_e\ell + \cosh\gamma_o\ell}{2} && \text{(a)} \\ a_{12} &= \frac{\cosh\gamma_e\ell - \cosh\gamma_o\ell}{2} && \text{(b)} \\ a_{21} &= \frac{\cosh\gamma_e\ell - \cosh\gamma_o\ell}{2} && \text{(c)} \\ a_{22} &= \frac{\cosh\gamma_e\ell + \cosh\gamma_o\ell}{2} && \text{(d).} \end{aligned} \qquad (12.5.15)$$

For a lossless line, $\gamma = \alpha + j\beta = 0 + j\beta$, the above equation is reduced to

$$\begin{aligned} a_{11} &= \frac{\cos\beta_e\ell + \cos\beta_o\ell}{2} && \text{(a)} \\ a_{12} &= \frac{\cos\beta_e\ell - \cos\beta_o\ell}{2} && \text{(b)} \\ a_{21} &= \frac{\cos\beta_e\ell - \cos\beta_o\ell}{2} && \text{(c)} \\ a_{22} &= \frac{\cos\beta_e\ell + \cos\beta_o\ell}{2} && \text{(d).} \end{aligned} \qquad (12.5.16)$$

For the symmetrical coupled lines, the [A] matrix is symmetrical and reciprocal, i.e. $a_{11} = a_{22}$, $a_{12} = a_{21}$. However, for the asymmetrical coupled lines, the [A] matrix is neither symmetrical nor reciprocal, i.e. the $a_{11} \neq a_{22}$, $a_{12} \neq a_{21}$. For the symmetrical coupled line in the homogeneous medium, $\beta_e = \beta_o = \beta$, leading to $a_{12} = a_{21} = 0$, $a_{11} = a_{22} = \cos\beta\,\ell$. Even for an asymmetrical coupled lines in a homogeneous medium, $\gamma_\pi = \gamma_c = \gamma$; resulting in $a_{12} = a_{21} = 0$ and $a_{11} = a_{22} = \cosh\gamma\ell$. It shows that a_{12} and a_{21} are due to inhomogeneity of medium, and these are unequal for asymmetrical conductors.

Determination of [C] Matrix

The elements of the [C] matrix are also obtained from the open-circuited output port #3 and #4, with $I_{p3} = I_{p4} = 0$. Like the port voltages, the port currents I_{p1}

and I_{p2} on the conductor #a and conductor #b are a combination of the C and π-mode currents given as follows:

$$\text{Conductor\#1: } I_{p1} = \frac{v_{c4}}{Z_{c1}} \sinh\gamma_c\ell + \frac{v_{\pi4}}{Z_{\pi1}} \sinh\gamma_\pi\ell \quad \text{(a)}$$

$$\text{Conductor\#2: } I_{p2} = \frac{v_{c3}}{Z_{c2}} \sinh\gamma_c\ell + \frac{v_{\pi3}}{Z_{\pi2}} \sinh\gamma_\pi\ell \quad \text{(b)}.$$

$$\downarrow \qquad \downarrow \qquad \downarrow$$

$$\text{Modal current} \quad \text{C - mode} \quad \pi\text{ - mode} \qquad (12.5.17)$$

The modal voltages ($v_{\pi4}$, v_{c4}) and ($v_{\pi3}$, v_{c3}), from equations (12.5.9)–(12.5.12), are substituted to get the following port currents I_{p1} and I_{p2}, in the terms of the output port voltages, (V_{p4}, V_{p3}):

$$I_{p1} = \frac{\sinh\gamma_c\ell}{Z_{c1}}\left[\frac{R_\pi V_{p4} - V_{p3}}{R_\pi - R_c}\right] + \frac{\sinh\gamma_\pi\ell}{Z_{\pi1}}\left[\frac{V_{p3} - R_c V_{p4}}{R_\pi - R_c}\right]$$

$$\text{and } I_{p2} = \frac{\sinh\gamma_c\ell}{Z_{c2}}\left[\frac{R_\pi R_c V_{p4} - R_c V_{p3}}{R_\pi - R_c}\right] + \frac{\sinh\gamma_\pi\ell}{Z_{\pi2}}\left[\frac{R_\pi V_{p3} - R_c R_\pi V_{p4}}{R_\pi - R_c}\right].$$

$$\Rightarrow I_{p1} = \left[\frac{R_\pi \sinh\gamma_c\ell}{Z_{c1}(R_\pi - R_c)} - \frac{R_c \sinh\gamma_\pi\ell}{Z_{\pi1}(R_\pi - R_c)}\right]V_{p4} + \left[\frac{\sinh\gamma_\pi\ell}{Z_{\pi1}(R_\pi - R_c)} - \frac{\sinh\gamma_c\ell}{Z_{c1}(R_\pi - R_c)}\right]V_{p3} \quad \text{(a)}$$

$$I_{p2} = \left[\frac{R_\pi R_c \sinh\gamma_c\ell}{Z_{c2}(R_\pi - R_c)} - \frac{R_c R_\pi \sinh\gamma_\pi\ell}{Z_{\pi2}(R_\pi - R_c)}\right]V_{p4} + \left[\frac{R_\pi \sinh\gamma_\pi\ell}{Z_{\pi2}(R_\pi - R_c)} - \frac{R_c \sinh\gamma_c\ell}{Z_{c2}(R_\pi - R_c)}\right]V_{p3} \quad \text{(b)}. \qquad (12.5.18)$$

On comparing the above equations against equations (12.5.4a) and (12.5.4b), the elements of the [C] matrix are obtained:

$$c_{11} = \frac{R_\pi Z_{\pi1} \sinh\gamma_c\ell - Z_{c1} R_c \sinh\gamma_\pi\ell}{Z_{\pi1} Z_{c1}(R_\pi - R_c)} \quad \text{(a)}$$

$$c_{12} = \frac{Z_{c1} \sinh\gamma_\pi\ell - Z_{\pi1} \sinh\gamma_c\ell}{Z_{\pi1} Z_{c1}(R_\pi - R_c)} \quad \text{(b)}$$

$$c_{21} = \frac{R_\pi R_c Z_{\pi2} \sinh\gamma_c\ell - R_c R_\pi Z_{c2} \sinh\gamma_\pi\ell}{Z_{\pi2} Z_{c2}(R_\pi - R_c)} \quad \text{(c)} \qquad (12.5.19)$$

$$c_{22} = \frac{R_\pi Z_{c2} \sinh\gamma_\pi\ell - R_c Z_{\pi2} \sinh\gamma_c\ell}{Z_{\pi2} Z_{c2}(R_\pi - R_c)} \quad \text{(d)}.$$

The expressions for the elements of [C] matrix are obtained for the even and odd modes of the symmetrical coupled lines in an *inhomogeneous* medium: $R_c = +1$, $\gamma_c \to \gamma_e$, $Z_{c1} = Z_{c2} = Z_{oe}$ and $R_\pi = -1$, $\gamma_\pi \to \gamma_o$, $Z_{\pi1} = Z_{\pi2} = Z_{oo}$. For even and odd mode, the above equations are reduced to the following expressions:

$$c_{11} = \frac{Y_{oe} \sinh\gamma_e\ell + Y_{oo} \sinh\gamma_o\ell}{2} \quad \text{(a)}$$

$$c_{12} = \frac{Y_{oe} \sinh\gamma_e\ell - Y_{oo} \sinh\gamma_o\ell}{2} \quad \text{(b)}$$

$$c_{21} = \frac{Y_{oe} \sinh\gamma_e\ell - Y_{oo} \sinh\gamma_o\ell}{2} \quad \text{(c)} \qquad (12.5.20)$$

$$c_{22} = \frac{Y_{oe} \sinh\gamma_e\ell + Y_{oo} \sinh\gamma_o\ell}{2} \quad \text{(d)}.$$

Using $(Y_{c1}/Y_{c2}) = (Y_{\pi1}/Y_{\pi2}) = -R_c R_\pi$ from equation (11.5.54) of chapter 11, equation (12.5.19) is further simplified:

$$c_{11} = \frac{R_\pi Y_{c1} \sinh\gamma_c\ell - R_c Y_{\pi1} \sinh\gamma_\pi\ell}{(R_\pi - R_c)} \quad \text{(a)}$$

$$c_{12} = c_{21} = \frac{Y_{\pi1} \sinh\gamma_\pi\ell - Y_{c1} \sinh\gamma_c\ell}{(R_\pi - R_c)} \quad \text{(b)} \qquad (12.5.21)$$

$$c_{22} = \frac{R_\pi Y_{\pi2} \sinh\gamma_\pi\ell - R_c Y_{c2} \sinh\gamma_c\ell}{(R_\pi - R_c)} \quad \text{(c)}.$$

For symmetrical coupled lines in an inhomogeneous medium, $c_{12} = c_{21}$, $c_{11} = c_{22}$. In the case of a lossless medium, equation (12.5.20) is reduced to

$$c_{11} = c_{22} = \frac{j}{2}[Y_{oe} \sin\beta_e\ell + Y_{oo} \sin\beta_o\ell] \quad \text{(a)}$$

$$c_{12} = c_{21} = \frac{j}{2}[Y_{oe} \sin\beta_e\ell - Y_{oo} \sin\beta_o\ell] \quad \text{(b)}. \qquad (12.5.22)$$

The above results agree with the previously obtained results for the [C] matrix elements of equation (12.4.69).

Short-Circuit Condition At Outputs ($V_{p4} = V_{p3} = 0$)

The short-circuited output port #4 and #3 help us to determine elements of the [B] and [D] matrices of an asymmetrical coupled lines as follows.

Determination of Elements of [B] Matrix

The input port voltages and port currents are obtained from equation (12.5.2), for $V_{p4} = V_{p3} = 0$:

$$[V_{in}] = [B][-I_{out}] \text{ i.e.}$$

$$V_{p1} = b_{11}(-I_{p4}) + b_{12}(-I_{p3}) \quad \text{(a)} \quad (12.5.23)$$

$$V_{p2} = b_{21}(-I_{p4}) + b_{22}(-I_{p3}) \quad \text{(b)}.$$

$$[I_{in}] = [D][-I_{out}] \text{ i.e.}$$

$$I_{p1} = d_{11}(-I_{p4}) + d_{12}(-I_{p3}) \quad \text{(a)} \quad (12.5.24)$$

$$I_{p2} = d_{21}(-I_{p4}) + d_{22}(-I_{p4}) \quad \text{(b)}.$$

The input port voltages V_{p1} and V_{p2}, on conductor #a and #b, using a superposition of the modal voltages, are created by the output short-circuited modal port currents (i_{c4}, $i_{\pi 4}$) and (i_{c3}, $i_{\pi 3}$):

$$V_{p1} = Z_{c1}\sinh\gamma_c\ell(-i_{c4}) + Z_{\pi 1}\sinh\gamma_\pi\ell(-i_{\pi 4}) \quad (a)$$
$$V_{p2} = Z_{c2}\sinh\gamma_c\ell(-i_{c3}) + Z_{\pi 2}\sinh\gamma_\pi\ell(-i_{\pi 3}) \quad (b). \qquad (12.5.25)$$

The above equations, for both the C and π modes, are written by using equation (2.1.75) of chapter 2. Further, the short-circuited output port currents are also a superposition of the modal currents:

$$I_{p4} = i_{c4} + i_{\pi 4} \quad (a), \qquad I_{p3} = i_{c3} + i_{\pi 3} \quad (b). \qquad (12.5.26)$$

Using equations (12.5.7) and (11.5.53) of chapter 11, the *modal current ratios* are calculated:

$$\text{C - mode}: \frac{i_{c3}}{i_{c4}} = R_c\frac{Y_{c2}}{Y_{c1}} = -\frac{1}{R_\pi} \quad (a)$$
$$\pi\text{ - mode}: \frac{i_{\pi 3}}{i_{\pi 4}} = R_c\frac{Y_{\pi 2}}{Y_{\pi 1}} = -\frac{1}{R_c} \quad (b). \qquad (12.5.27)$$

The C- mode currents at output port are computed from the above equations:

$$i_{c4} = I_{p4} - i_{\pi 4} = I_{p4} + R_c i_{\pi 3} \quad (a)$$
$$i_{c3} = I_{p3} - i_{\pi 3} = I_{p3} + \frac{i_{\pi 4}}{R_c} \quad (b). \qquad (12.5.28)$$

The above C and π-mode current ratios are used to obtain the π and C-mode output currents at port # 4 and #3 as follows:

$$\frac{i_{c4}}{i_{c3}} = \frac{I_{p4} - I_{\pi 4}}{I_{p3} + \frac{I_{\pi 4}}{R_c}}, \Rightarrow -R_\pi = \frac{R_c I_{p4} - R_c i_{\pi 4}}{R_c I_{p3} + i_{\pi 4}}$$

$$\Rightarrow i_{\pi 4} = \frac{R_c I_{p4} + R_c R_\pi I_{p3}}{(R_c - R_\pi)} \quad (a)$$

$$\text{also}, -R_\pi = \frac{R_c I_{p4} + R_c^2 i_{\pi 3}}{R_c I_{p3} - R_c i_{\pi 3}}, \Rightarrow i_{\pi 3} = -\frac{(I_{p4} + R_\pi I_{p3})}{(R_c - R_\pi)} \quad (b)$$

$$\text{Likewise}, : i_{c3} = \frac{(I_{p4} + R_c I_{p3})}{(R_c - R_\pi)} \quad (c)$$

$$i_{c4} = \frac{-(R_\pi I_{p4} + R_c R_\pi I_{p3})}{(R_c - R_\pi)} \quad (d). \qquad (12.5.29)$$

Using the above equations with equation (12.5.25), the port voltage V_{p1} and V_{p2} are obtained at port #1 and #2:

$$V_{p1} = Z_{c1}\sinh\gamma_c\ell\left[\frac{R_\pi I_{p4} + R_c R_\pi I_{p3}}{R_c - R_\pi}\right] - Z_{\pi 1}\sinh\gamma_\pi\ell\left[\frac{R_c I_{p4} + R_c R_\pi I_{p3}}{R_c - R_\pi}\right]$$

$$\Rightarrow V_{p1} = \left[\frac{R_c Z_{\pi 1}\sinh\gamma_\pi\ell - Z_{c1}R_\pi\sinh\gamma_c\ell}{R_c - R_\pi}\right](-I_{p4}) + \left[\frac{Z_{\pi 1}R_c R_\pi\sinh\gamma_\pi\ell - Z_{c1}R_c R_\pi\sinh\gamma_c\ell}{R_c - R_\pi}\right](-I_{p3})$$

$$\text{Likewise}, V_{p2} = \left[\frac{Z_{c2}\sinh\gamma_c\ell - Z_{\pi 2}\sinh\gamma_\pi\ell}{R_c - R_\pi}\right](-I_{p4}) + \left[\frac{Z_{c2}R_c\sinh\gamma_c\ell - Z_{\pi 2}R_\pi\sinh\gamma_\pi\ell}{R_c - R_\pi}\right](-I_{p3}). \qquad (12.5.30)$$

On comparing the above equation against equation (12.5.23), the following expressions for the elements of [B] matrix are obtained:

$$b_{11} = \left[\frac{R_c Z_{\pi 1}\sinh\gamma_\pi\ell - Z_{c1}R_\pi\sinh\gamma_c\ell}{R_c - R_\pi}\right] \quad (a)$$
$$b_{12} = \left[\frac{R_c R_\pi Z_{\pi 1}\sinh\gamma_\pi\ell - Z_{c1}R_c R_\pi\sinh\gamma_c\ell}{R_c - R_\pi}\right] \quad (b)$$
$$b_{21} = \left[\frac{Z_{c2}\sinh\gamma_c\ell - Z_{\pi 2}\sinh\gamma_\pi\ell}{R_c - R_\pi}\right] \quad (c)$$
$$b_{22} = \left[\frac{Z_{c2}R_c\sinh\gamma_c\ell - Z_{\pi 2}R_\pi\sinh\gamma_\pi\ell}{R_c - R_\pi}\right] \quad (d). \qquad (12.5.31)$$

The use of equation (11.5.54) of chapter 11 with equation (12.5.31b) provides $b_{12} = [Z_{c2}\sinh\gamma_c\ell - Z_{\pi 2}\sinh\gamma_\pi\ell]/R_c - R_\pi]$, and also $b_{12} = b_{21.}$ The symmetrical coupled lines in an inhomogeneous medium support the even and odd modes: $R_c = +1, \gamma_c \to \gamma_e, Z_{c1} = Z_{c2} = Z_{oe}$ and $R_\pi = -1, \gamma_\pi \to \gamma_o, Z_{\pi 1} = Z_{\pi 2} = Z_{oo}$; and the elements of the [B] matrix, in this case, are given below:

$$b_{11} = b_{22} = \frac{Z_{oo}\sinh\gamma_o\ell + Z_{oe}\sinh\gamma_e\ell}{2} \quad (a)$$
$$b_{12} = b_{21} = \frac{Z_{oe}\sinh\gamma_e\ell - Z_{oo}\sinh\gamma_o\ell}{2} \quad (b). \qquad (12.5.32)$$

It is noted that for a symmetrical coupled lines in an inhomogeneous medium, $b_{11} = b_{22}$, $b_{12} = b_{21}$. For the lossless case, the following expressions could be written:

$$b_{11} = b_{22} = \frac{j}{2}[Z_{oe}\sin\beta_e\ell + Z_{oo}\sin\beta_o\ell] \quad (a)$$
$$b_{12} = b_{21} = \frac{j}{2}[Z_{oe}\sin\beta_e\ell - Z_{oo}\sin\beta_o\ell] \quad (b). \qquad (12.5.33)$$

The above equations agree with equation (12.4.69).

Determination of Elements of [D] Matrix

The short-circuited output port conditions ($V_{p4} = V_{p3} = 0$) help us to determine elements of [D] matrix. The input port currents at the port #1 and #2 are written as the linear combination of the C and π current modes:

$$\begin{aligned} I_{p1} &= -i_{c4}\cosh\gamma_c\ell - i_{\pi4}\cosh\gamma_\pi\ell \quad (a)\\ I_{p2} &= -i_{c3}\cosh\gamma_c\ell - i_{\pi3}\cosh\gamma_\pi\ell \quad (b).\\ &\downarrow \qquad\qquad \downarrow \qquad\qquad \downarrow\\ &\text{Current mode,}\ C-\text{mode} \qquad \pi-\text{mode} \end{aligned} \tag{12.5.34}$$

The above equations are obtained by using equation (2.1.75) of chapter 2. The C and π-mode short-circuited currents, from equation (12.5.29), are substituted to get the following expression for the port currents I_{p1} and I_{p2}:

$$\begin{aligned} I_{p1} &= \cosh\gamma_c\ell\left[\frac{R_\pi I_{p4} + R_c R_\pi I_{p3}}{R_c - R_\pi}\right] - \cosh\gamma_\pi\ell\left[\frac{R_c I_{p4} + R_c R_\pi I_{p3}}{R_c - R_\pi}\right] \quad (a)\\ I_{p2} &= -\cosh\gamma_c\ell\left[\frac{I_{p4} + R_c I_{p3}}{R_c - R_\pi}\right] + \cosh\gamma_\pi\ell\left[\frac{I_{p4} + R_\pi I_{p3}}{R_c - R_\pi}\right] \quad (b)\\ I_{p1} &= \left[\frac{R_c\cosh\gamma_\pi\ell - R_\pi\cosh\gamma_c\ell}{R_c - R_\pi}\right](-I_{p4}) + \left[\frac{R_\pi R_c\cosh\gamma_\pi\ell - R_c R_\pi\cosh\gamma_c\ell}{R_c - R_\pi}\right](-I_{p3}) \quad (c)\\ \Rightarrow\quad I_{p2} &= \left[\frac{\cosh\gamma_c\ell - \cosh\gamma_\pi\ell}{R_c - R_\pi}\right](-I_{p4}) + \left[\frac{R_c\cosh\gamma_c\ell - R_\pi\cosh\gamma_\pi\ell}{R_c - R_\pi}\right](-I_{p3}) \quad (d). \end{aligned} \tag{12.5.35}$$

On comparing the above equations against equation (12.5.24), the following elements of the [D] matrix are obtained:

$$\begin{aligned} d_{11} &= \left[\frac{R_c\cosh\gamma_\pi\ell - R_\pi\cosh\gamma_c\ell}{R_c - R_\pi}\right] \quad (a)\\ d_{12} &= \left[\frac{R_c R_\pi\cosh\gamma_\pi\ell - R_c R_\pi\cosh\gamma_c\ell}{R_c - R_\pi}\right] \quad (b)\\ d_{21} &= \left[\frac{\cosh\gamma_c\ell - \cosh\gamma_\pi\ell}{R_c - R_\pi}\right] \quad (c)\\ d_{22} &= \left[\frac{R_c\cosh\gamma_c\ell - R_\pi\cosh\gamma_\pi\ell}{R_c - R_\pi}\right] \quad (d). \end{aligned} \tag{12.5.36}$$

For symmetrical coupled lines in an inhomogeneous medium, the modes reduce to the even and odd modes: $R_c = +1, \gamma_c \to \gamma_e; R_\pi = -1, \gamma_\pi \to \gamma_o$. The elements of [D] matrix are written as follows:

$$\begin{aligned} d_{11} = d_{22} &= \left[\frac{\cosh\gamma_e\ell + \cosh\gamma_o\ell}{2}\right] \quad (a)\\ d_{12} = d_{21} &= \left[\frac{\cosh\gamma_e\ell - \cosh\gamma_o\ell}{2}\right] \quad (b). \end{aligned} \tag{12.5.37}$$

For a lossless symmetrical coupled lines structure, the above equation is written as

$$\begin{aligned} d_{11} = d_{22} &= \left[\frac{\cos\beta_e\ell + \cos\beta_o\ell}{2}\right] \quad (a)\\ d_{12} = d_{12} &= \left[\frac{\cos\beta_e\ell - \cos\beta_o\ell}{2}\right] \quad (b). \end{aligned} \tag{12.5.38}$$

The above equations agree with the results shown in equation (12.4.69). The symmetrical coupled lines have [A] = [D]. The [Z] and [Y] parameters can also be obtained for asymmetrical coupled lines [J.18]. The [ABCD], [Z], and [Y] parameters are used to obtain the equivalent 2-port networks of the asymmetrical coupled lines [J.36]. The scattering parameters of the asymmetrical coupled lines have also been obtained [J.37, J.38]. The even–odd mode concept has been further extended to the differential and common mode transmission lines [J.39] The method discussed above has been also extended to 3-conductors coupled lines to design the six-port microstrip directional coupler [J.40, J.41].

References

Books

B.1 Costa, L.; Valtonen, M.: *Implementation of Single and Coupled Microstrip Lines in APLAC*, CT-33, ISBN 951-222-3881-0, Helsinki University, Finland, Dec. 1997.

B.2 Djordjevic, A.R; Harrington, R.F.; Sarkar, T.K.; Bazdar, M.B.: *Matrix Parameters for Multiconductor Transmission Lines – Software and User's Manual*, Norwood Artech House, Boston, 1989.

B.3 Hoffmann, R.: *Microwave Integrated Circuit Handbook*, Artech House, Boston, 1985.

B.4 Kneppo, I.; Fabian, J.; Bezousek, P.; Hrnicko, P.; Pavel, M.: *Microwave Integrated Circuits*, Chapman & Hall, London, 1994.

B.5 Gupta, K.C.; Garg, R.; Bahl, I.; Bhartia, P.: *Microstrip Lines and Slot Lines*, 2nd Edition, Artech House, Boston, 1989, 1996.

B.6 Young, L. (Editor): *Advances in Microwaves*, Vol. 8, Academic Press, New York, 1974.

B.7 Collin, R.E.: *Foundations for Microwave Engineering*, 2nd Edition, McGraw-Hill, Inc., New York. 1992.

B.8 Mongia, R.; Bhartia, P.; Bahl, I.J.: *R. F. and Microwave Coupled-Line Circuits*, Artech-House, Boston, 1999.

B.9 Pozar, D.M.: *Microwave Engineering*, 2nd Edition, John Wiley & Sons, Singapore, 1999.

B.10 Matthaei, G; Young, L.; Jones, E.M.T.: *Microwave Filters, Impedance-Matching Networks, and Coupling Structures*, Norwood Artech House, Boston, 1980.

B.11 Edward, T.C.: *Foundations for Microstrip Circuit Design*, John Wiley & Sons, New York, 1987.

Journals

J.1 Bryant, T.G.; Weiss, J.A.: *Parameters of microstrip transmission lines and coupled pairs of microstrip lines, IEEE Trans. Microw. Theory Tech.*, Vol. MTT-16, No. 12, pp. 1021–1027, 1968.

J.2 Krage, M.K; Haddad, G.I.: *Characteristics* of *coupled microstrip transmission lines-I: Coupled mode formulation of inhomogeneous lines, IEEE Trans. Microw. Theory Tech.*, Vol. MTT-18, No. 14, pp. 217–222, 1970.

J.3 Bazdar, M.B.; Djordjevic, A.R.; Harrington, R.F.; Mautz, J.R.; Sarkar, T.K.: *Evaluation of quasi- static matrix parameters for multiconductor transmission lines using Galerkin's method, IEEE Trans. Microw. Theory Tech.*, Vol. MTT-42, No. 7, pp. 1223–1228, July 1984.

J.4 Kirschning, M.; Jansen, R.H.: *Accurate wide range design equations* for *parallel-coupled microstrip lines, IEEE Trans. Microw. Theory Tech.*, Vol. MTT-32, No. 1, pp. 83–90, 1984. Also corrections in Vol. MTT- 33, No.1 pp. 288, Jan. 1985.

J.5 Allen, J.J.; Estes, M.F.: *Broadside coupled strip in a layered dielectric medium, IEEE Trans. Microw. Theory Tech.*, Vol. MTT-20, No. 10, pp. 662–669, Oct. 1972.

J.6 Bedair, S.S; Wolff, I.: *Fast and accurate analytic formulas for calculating the parameters of a general broadside coupled coplanar waveguide for (M) MIC applications, IEEE Trans. Microw. Theory Tech.*, Vol. MTT-37, No. 5, pp. 843–850, May. 1989.

J.7 Hammerstad, E.; Jensen, O.: *Accurate models for microstrip computer-aided design, IEEE., MTT-S, Int. Microwave Symposium Digest*, pp. 407–409, 1980.

J.8 March, S.: *Microstrip packaging: watch the last step, Microwaves*, Vol. 220, No. 13, pp.83–94,1981.

J.9 Verma, A.K.; Kumar, R.: *New empirical unified dispersion model for shielded -, suspended-, and composite – substrate microstrip line for microwave and mm-wave applications, IEEE Trans. Microw. Theory Tech.*, Vol. MTT-46, No. 8, pp. 1187–1192, Aug. 1998.

J.10 Atwater, H.A.: *Simplified design equations for microstrip line parameters, Microw. J.*, Vol. 32, No. 11 pp.109–115, Nov. 1989.

J.11 Abbosh, A.M.: *Analytical closed-form solutions for different configurations of parallel-coupled microstrip lines, IET Microw. Antenna Propag.*, Vol. 3, No. 1, pp.137–147, 2009.

J.12 Getsinger, W.J.: *Dispersion of parallel-coupled microstrip, IEEE Trans. Microw. Theory Tech.*, Vol. MTT-21, No. 3, pp. 144–1145, 1973.

J.13 Garg, R.; Bahl, I.J.: *Characteristics of coupled microstrip lines, IEEE Trans. Microw. Theory Tech.*, Vol. MTT-27, No. 7, pp. 700–705, 1979. Also corrections in, Vol. MTT- 28, No. 3, pp. 272, 1980.

J.14 Sellberg, F.: *Simple and accurate algorithms to include Lange and asymmetric microstrip couplers in CAD-packages*, Proc. 18th Eu. Microwave Conf., pp. 1157–1162, Sept. 1988.

J.15 Bedair, S.S.: *Characteristics of some asymmetrical coupled transmission lines IEEE Trans. Microw. Theory Tech.*, Vol. MTT-32, No. 1, pp. 108–110, Jan. 1984.

J.16 El-Deeb, N.A.; Abdallah, E.A.F.; Saleh, M.B.: *Design parameters of inhomogeneous asymmetrical coupled transmission lines, IEEE Trans. Microw. Theory Tech.*, Vol. MTT-33, No. 7, pp. 592–596, 1983.

J.17 Tripathi, V.K.: *A dispersion model for coupled microstrips, IEEE Trans. Microw. Theory Tech.*, Vol. MTT-34, No. 1, pp. 66–71, Jan. 1986.

J.18 Tripathi, V.K.: *Asymmetric coupled transmission lines in an inhomogeneous medium, IEEE Trans. Microw. Theory Tech.*, Vol. MTT-23, No. 9, pp. 734–739, 1975.

J.19 Speciale, R.A.: *Even and odd- mode waves for non-symmetrical coupled lines in on-homogeneous media, IEEE Trans. Microw. Theory Tech.*, Vol. MTT, No. 11, pp. 897–908, 1975.

J.20 Chang, F.Y.: *Transient analysis of lossless coupled transmission lines in a non-homogeneous dielectric medium, IEEE Trans. Microw. Theory Tech.*, Vol. MTT-18, No. 9, pp. 616–626, 1970.

J.21 Krage, M.K.; Haddad, G.I.: *Characteristics* of *coupled microstrip transmission lines-II: Evaluation of coupled lines parameters, IEEE Trans. Microw. Theory Tech.*, Vol. MTT-18, No. 4, pp. 222–228, 1970.

J.22 Keung, K.; Cheng, M.:*Analysis and synthesis of coplanar-coupled lines on a substrate of finite thickness, IEEE Trans. Microw. Theory Tech.*, Vol. MTT-44, No. 4, pp. 636–643, April. 1996.

J.23 Bedair, S.S.; Wolff, I.: *Fast and accurate analytic formulas for calculating the parameters of a general* *broadside coupled coplanar waveguide for (M) MIC applications*, IEEE Trans. Microw. Theory Tech., Vol. MTT-37, No. 5, pp. 843–850, May 1989.

J.24 Ghione, G.; Naldi, C.: *Coplanar waveguides for MMIC applications: effect of upper shielding, conductor backing, finite-extent ground plane, and line-to-line coupling*, IEEE Trans. Microw. Theory Tech., Vol. MTT-35, No. 3, pp. 260–267, March 1987.

J.25 Wen, C.P.:*Coplanar waveguide directional coupler*, IEEE Trans. Microw. Theory Tech., Vol. MTT-3, No. 6, pp. 318–322, June. 1970.

J.26 Hilberg, W.: *From approximation to exact relation for characteristic impedances*, IEEE Trans. Microw. Theory Tech., Vol. MTT-17, No. 5, pp. 259–265, 1969.

J.27 Kitazawa, T.; Mittra, R.: *Quasi-static chrematistic of asymmetrical and coupled coplanar - type transmission lines*, IEEE Trans. Microw. Theory Tech., Vol. MTT-33, pp. 771–778, Sept. 1985.

J.28 Jones, E.M.T.; Bolljahn, T.: *Coupled strip transmission line filters and directional couplers*, IEEE Trans. Microw. Theory Tech., Vol. MTT-4, No. 2, pp. 75–81, April. 1956.

J.29 Zysman, G.I.; Johnson, A.K.: *Coupled transmission line networks in an inhomogeneous dielectric medium*, IEEE Trans. Microw. Theory Tech., Vol. MTT-17, No. 10, pp. 753–759, Oct. 1969.

J.30 Levy, R.: *New equivalent circuits for inhomogeneous coupled lines with synthesis applications*, IEEE Trans. Microw. Theory Tech., Vol. MTT-36, No. 6, pp. 1087–1094, June. 1988.

J.31 Allan, J.L.: *Non-symmetrical coupled lines in an inhomogeneous dielectric medium*, Int. J. Electron., Vol. 38, pp. 337–347, 1975.

J.32 Allen, J.J.; Estes, M.F.: *Broadside coupled strip in a layered dielectric medium*, IEEE Trans. Microw. Theory Tech., Vol. MTT-20, No. 10, pp. 662–669, Oct. 1972.

J.33 Tripathi, V.K.; Chang, C.L.: *Quasi-TEM parameters of non-symmetrical coupled microstrip lines*, Int. J. Electron., Vol. 45, No. 2, pp. 215–223, 1978.

J.34 Allen, J.L.: *Non – symmetrical coupled lines in an inhomogeneous dielectric medium*, Int. J. Electron., Vol. 38, No. 3, pp. 337–347, 1975.

J.35 Speciale, R.A.; Tripathi, V.K.: *Wave modes and parameters matrices of non – symmetrical coupled lines in a non- homogeneous medium*, Int. J. Electron., Vol. 40, No. 4, pp. 371–375, 1976.

J.36 Tripathi, V.K.: *Equivalent circuits and characteristics of inhomogeneous non-symmetrical coupled line two-port circuits*, IEEE Trans. Microw. Theory Tech., Vol. MTT-25, No. 2, pp. 140–142, Feb. 1977.

J.37 Sellberg, F.: *Formulas useful for the synthesis and optimization of general, uniform contra-directional couplers*, IEEE Trans. Microw. Theory Tech., Vol. MTT-38, No. 8, pp. 1000–1010, Aug. 1990.

J.38 Ikalainen, P.K.; Matthaei, G.L.: *Wide-band, forward – coupling microstrip hybrid with high directivity*, IEEE Trans. Microw. Theory Tech., Vol. MTT-35, No. 8, pp. 719–725, Aug. 1987.

J.39 Bockelman, D.E.; Eisenstadt, W.R.: *Combined differential and common mode scattering parameters: theory and simulation*, IEEE Trans. Microw. Theory Tech., Vol. MTT-43, No. 7, pp. 1530–1539, July 1995.

J.40 Yu, L.; Rawat, B.: *Coupled-mode analysis of an ideal microstrip six-port directional coupler on an anisotropic substrate*, Int. J. Electron., Vol. 83, No. 5, pp. 599–614, 1997.

J.41 Wang, A.; Jin, T.; Wu, Y.: *The ABCD parameters for symmetric coupled three-line in an inhomogeneous medium*, Int. J. Electron., Vol. 91, No. 5, pp. 289–299, May 2004.

J.42 Bedair, S.S.: *On the odd mode capacitance of coupled microstrip lines*, IEEE Trans. Microw. Theory Tech., Vol. MTT-28, No. 11,pp. 1225–1227, 1980.

13

Fabrication of Planar Transmission Lines

Introduction

At present there are following four important technologies involved in the fabrication of planar transmission lines, such as microstrips, coplanar waveguide (CPW), slot lines and related components, and circuits:

- *Microwave integrated circuit (MIC),*
- *Monolithic microwave integrated circuit (MMIC),*
- *Microelectro-mechanical systems (MEMS),*
- *Low temperature cofired ceramic (LTCC).*

The MIC is normally a single dielectric layer, called the *laminate, printed circuit board* (PCB) technology. However, the multilayer microstrip lines structures, such as the suspended microstrip, inverted microstrip, are also realized in the MIC technology. The combination of the passive components, made out of microstrips, etc., and discrete active components on the same substrate is known as the *hybrid MIC (HMIC)*. The *MMIC* and *MEMS* are the semiconducting substrates-based technology, whereas the *LTCC* is a ceramic tape-based technology. The planar lines and passive components are also developed around multidielectric medium, and one of the layers could be of ferrite, ferroelectric, etc.

The analytical modeler and designer of components and circuits in the planar technology should have some basic information on the planar line fabrication processes, and visualization of the layout of the planar lines based passive components in the above mentioned four technologies.

Objectives

- *To review the fabrication process of the MIC, MMIC, MEMS, and LTCC technologies.*
- *To discuss the physical parameters and materials involved in the fabrication of line structures.*
- *To familiarize some planar line structures and passive components in these technologies.*

13.1 Elements of Hybrid MIC (HMIC) Technology

The single-layer microstrip, CPW, and other planar transmission lines are already discussed in earlier chapters. Figure (13.1a–i) shows a few examples of common types of multilayer microstrip structures. Figure (13.1a) is a *suspended microstrip* line, with an air-gap. It is a composite substrate microstrip line. Figure (13.1b) shows the *inverted microstrip*, again with an air-gap. However, in this case, the microstrip is on the air-substrate and the dielectric layer is a *superstrate* or an *overlay*. These lines are more suitable for mm-wave applications. Similarly, the other designated lines, with more number of dielectric layers or overlays, are also used. In the place of microstrips, the layered medium also supports the CPW, slot lines, etc. The dielectric substrate, or the superstrate, is also replaced by the nondielectric substrates/superstrate, such as the ferrites, semiconductors, ferroelectric materials.

The planar transmission lines such as microstrip, CPW, and slot lines are the printed conducting strips or slots either on a *plastic substrate* or a *ceramic substrate*. In the case of a plastic substrate, the conducting strips and slots are fabricated by the *etching process*. In the MIC fabrication process, the unwanted copper coating is removed from the copper-cladded substrate by a chemical process. The etching is also achieved by the computer-controlled mechanical milling machine. In the case of the plastic substrate-based hybrid MIC technology, the lumped elements such as a capacitor, resistor, and active devices are added to the MIC circuit, whereas in case of the ceramic substrate-based hybrid MIC technology, the *thin film* and the *thick film deposition technologies* are used to integrate the lumped elements to a MIC circuit. In this case, the planar transmission lines and the passive components based on them, along with the film resistors are deposited on the ceramic substrates. Sometimes even a capacitor is

Introduction to Modern Planar Transmission Lines: Physical, Analytical, and Circuit Models Approach, First Edition. Anand K. Verma.

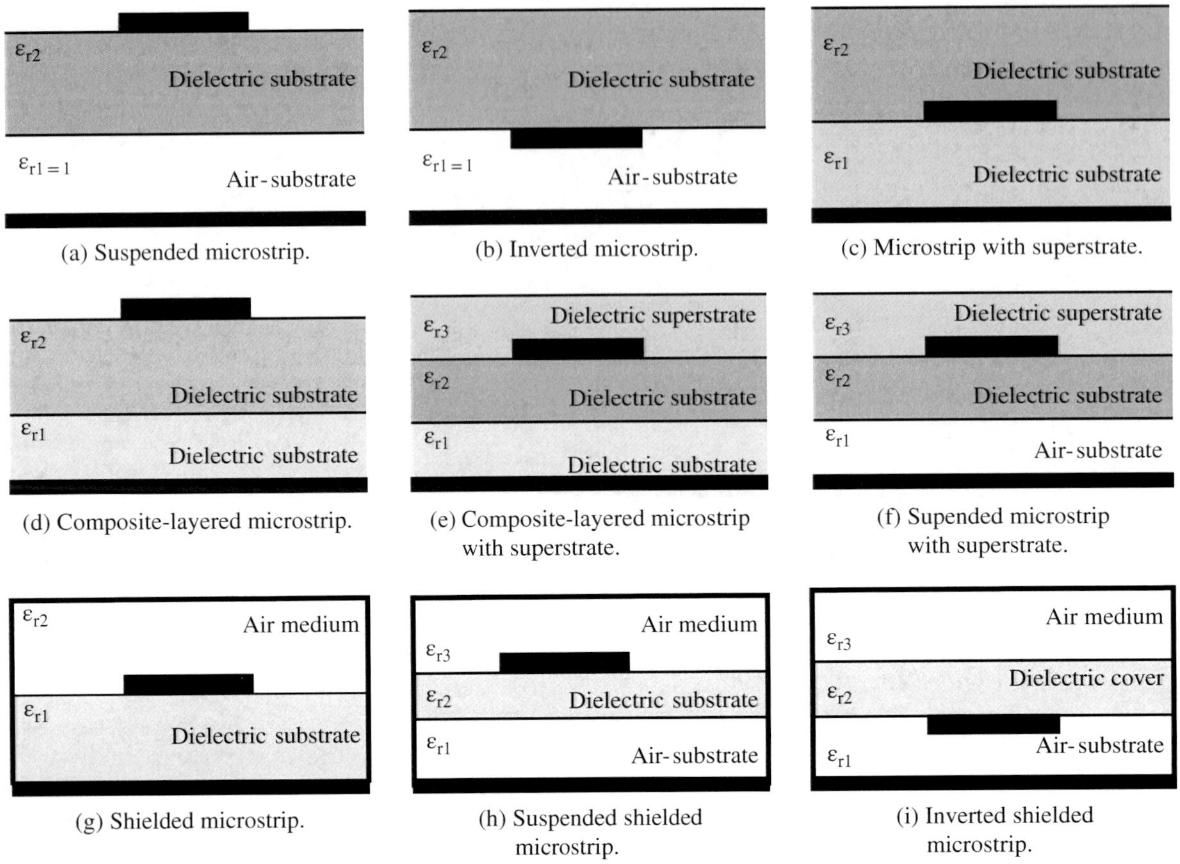

Figure 13.1 Some common multilayer microstrip lines.

deposited in this process. However, normally capacitor and active elements are added separately to a substrate that forms the hybrid MIC, i.e. the HMIC. This chapter is concerned only with the planar transmission lines, and not with the full hybrid MIC technology, used for the fabrication of components and devices. Figure (13.2) gives a summary of the substrate and fabrication process of the planar transmission lines in the hybrid MIC technology. The basic fabrication is described in this section briefly. Familiarity with it could be useful in designing and modeling of planar lines and circuits.

13.1.1 Substrates

The hybrid MIC uses two kinds of substrates – the *plastic substrate* and the *ceramic substrate*. However, there is another kind of composite substrate – the semiconductor (silicon) on sapphire. It is known as *SOS*, i.e. *the silicon on sapphire* technology. It is an intermediate technology between HMIC and MMIC. The planar lines are constructed on the lossless sapphire substrate, and the thin film small islands of Si or GaAs are deposited on the sapphire to develop active devices such as Schottky diodes, bipolar transistors, FET, and MOSFET [J.1–J.3].

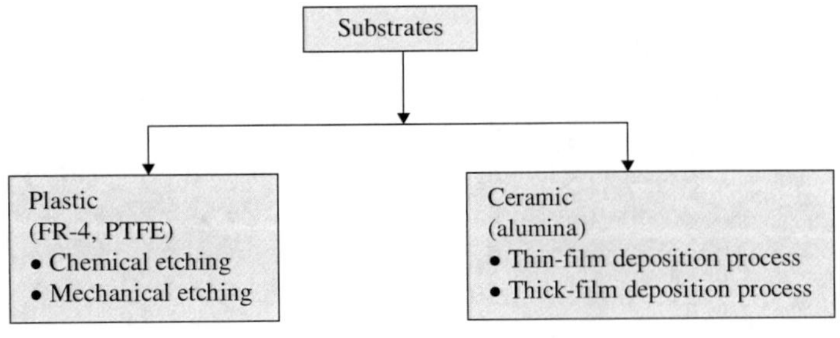

Figure 13.2 Substrates and fabrication process of HMIC.

The substrate for the HMIC is decided on a consideration of its mechanical, thermal, chemical, and electrical properties [B.1–B.3]. Normally, the plastic substrates, i.e. the organic material based substrates are soft, and a hole, i.e. *via*, is easily drilled in it to short-circuit the microstrip line. The hole is drilled at a temperature of about 250°C and also by using the chemical etching process. The dielectric constant of a plastic substrate is usually low, i.e. in the range $\varepsilon_r = 2–3$. It has a low loss-tangent, $\tan\delta \sim 0.001$. However, less expensive FR-4, $\varepsilon_r \sim 4–5$ and $\tan\delta \sim 0.01$, is a lossy plastic substrate [J.4]. Its metal foil, usually about 35 μm thick, is bonded to the plastic substrate by adhesive bonding. It is normally used for the analog and digital circuits at low frequency. It is an inexpensive substrate, so it is used to develop the prototype microwave circuit up to 6 GHz.

The organic materials available in the pure plastic form (*polyolefin, polystyrene*) have $\varepsilon_r \sim 2.3–2.6$, and these are the isotropic materials. These are mechanically and thermally not stable as substrates. So the glass–fiber-reinforced PTFE or polystyrene, i.e. the glass–fiber reinforced plastic, is used. The PTFE has $\varepsilon_r \sim 2.4–2.6$ and it is slightly anisotropic. The woven glass-reinforced plastic, i.e. PTFE with glass weave is also used as a substrate. Some anisotropy is associated with it. Finally, sometimes PTFE with ceramic powder is used as a substrate that has $\varepsilon_r \sim 2–20$ with some anisotropy [B.4, J.5, J.6]. The copper coating is normally 17, 35, or 70 μm thick. The data-sheet should be consulted for details.

The aluminum oxide ceramic, known as alumina (Al_2O_3), is a popular substrate for the thin and thick film MIC technologies. It is a polycrystalline material and has $\varepsilon_r \sim 10$. In the monocrystalline form, it is known as sapphire $\varepsilon_r = 9.4–10.6$. It is an anisotropic substrate, and has high-surface quality, low-loss $\tan\delta$, as well as low variation in ε_r from location to location on a substrate. The variation in substrate thickness is 0.051 mm. Other ceramic substrates are beryllium oxide (BeO), ($\varepsilon_r = 6.3$) and titanium oxide TiO_2, i.e. rutile, ($\varepsilon_r > 80$). As BeO has high thermal conductivity, it could be useful to high-power devices. However, its dust is toxic. So careful handling of BeO is needed. TiO_2 is a highly temperature-dependent material. The glass as fused quartz ($\varepsilon_r = 3.75$) acts as a good substrate. Quartz is a very low-loss substrate up to mm-wave range. Table 13.1 gives a summary of some useful parameters for commonly used plastic and ceramic substrates. Detailed information can be obtained from the handbooks and data-sheets [B.1–B.3, J.7–J.9]. Most of the PTFE substrates have some amount of anisotropy that has to be accounted for in the design of components.

13.1.2 Hybrid MIC Fabrication Process

The PTFE plastic substrates are copper-cladded on both sides. Similarly, alumina substrates are also available with a chrome-gold metallization, or a chrome–copper–gold metallization on both sides. The metallization of the alumina substrate is done by the *thin film process*. In both cases, the planar transmission lines are printed on a substrate by removing the unwanted metallization on the substrate that leaves the metal lines on a substrate. The process of removal of the unwanted metal from the surface of the substrate is called the *etching*. There are two kinds of etching – (i) the *mechanical etching* and (ii) the *chemical etching*. The mechanical etching is carried out by a numerically controlled milling machine to fabricate the line structures on a PTFE substrate. It consumes a lot of the PTFE substrate. However, for the microwave circuit application, it generates the prototype with more ease.

The chemical etching applies to both the PTFE and ceramic substrates. It is an accurate process for commercial applications. However, unlike mechanical etching, it requires more complicated processing. Its basic steps are shown in Fig (13.3) [B.1, B.2, B.5, B.6, J.10].

The first step is the preparation of a planar circuit to make an artwork. We draw the initially designed circuit with all its dimensions and next draw the layout on a paper with good black and white contrast. It is normally done using the software. The computer-generated artwork is used to generate a *positive film (mask)* with the help of the photographic process. The circuit/transmission line is black that does not permit the passage of the UV-light. The remaining area of the film is transparent that permits the passage of the UV-light.

The metalized substrate is cleaned properly by removing dust and grease from its surface. The copper-cladded lamination should not be touched with fingers. Nonmetallic tweezers should be used. High purity alcohol is used for cleaning. After cleaning, the substrate is baked. The *positive photoresist* is applied on the substrate, either by spraying or by the process of spin-coating. The coating is dried at about 70°C for nearly 15 minutes. Next, the *positive mask* is placed on the surface of a photoresist-coated substrate. The UV-light is illuminated on the photoresist-coated substrate. The complete process is known as the substrate development process.

The *positive photolithography process* is shown in Fig (13.4a). In the positive masking method, the strip conductor part is not illuminated, and the photoresist of the remaining area of the substrate is illuminated. The exposed unwanted area of the photoresist-coated substrate gets hardened. The hardened area of photoresist should be removed. It is achieved by first developing

Table 13.1 Some commonly used plastic, ceramic, and semiconducting substrates.

Substrate	Relative permittivity	Tan δ/ resistivity (Ω-cm)	Substrate thickness (mm)	Some other characteristics
Plastic substrates				
RT-Duriod 5870	2.33 ± 0.02	0.0012	0.127, 0.254, ..., 3.175	PTFE glass fiber. Other thicknesses are available.
Ultralam 2000	2.20 ± 0.02	0.0019	0.101, 0.256, ..., 1.524	PTFE Woven Glass. Other thicknesses available.
RT-Duriod 6002	2.94 ± 0.04	0.0012	0.127, 0.254, ..., 3.048	PTFE ceramic. Other thicknesses available.
RT-Duriod 6006	6.15 ± 0.15	0.0019	0.254, 0.635, ..., 2.54	PTFE ceramic. Other thicknesses available.
RO4003C	3.38 ± 0.05	0.0027	0.381, 0.508, ..., 1.524	Hydrocarbon ceramic. Other thicknesses available.
RO3010	10.2 ± 0.3	0.0023	0.254, 0.635, 1.27	Hydrocarbon ceramic.
FR-4	4.1, 4.35, 4.5–4.8 ± 10%	0.018 0.022	0.8, 1.2, 1.59, 2.0	Epoxy glass, anisotropic.
LCP	3.0	0.003	0.025	Suitable for implantable device, thin film process.
Polyimide	2.5–3.2	0.001–0.05		Used as an insulating layer.
BCB	2.65, 3	0.001–0.025	0.011, 0.045	THz application, Used as insulating layer.
Ceramic substrates				
Alumina, Al_2O_3 (99.5%) (96%)	9.8–9.9	0.0001 0.001	0.251, 0.381, ..., 1.207	Al_2O_3 ceramic, Some anisotropy.
Sapphire	⊥9.39; ‖11.58	0.0001	0.002, 0.004, 0.025–0.500	2″, 4″, 6″ diameter wafer Ceramic, anisotropic.
Quartz glass	3.78	0.0001	0.2, 0.6, 1.0	Glass.
Coring glass	5.75	0.0036	0.4–1.1	Glass.
Beryllium oxide (98%) BeO	6.3; ⊥6.94, ‖7.65	0.006	0.254, 0.635	Ceramic, high thermal conductivity.
Aluminum nitride AlN	8.0, 8.5, 9.14	0.0002	0.254, 0.635	Ceramic, 2″ × 2″, 6. 5″ × 4. 5″.
Titanium oxide TiO_2	85	0.004		Ceramic.
Lanthanum aluminate, $LaAlO_3$	24.5–25	0.0004	00.5, 1.0, 2.8	Useful for HTC superconductors 10 × 3, 10 × 10, 20 × 20 (mm × mm) etc.
Magnesium oxide, MgO	9.5–9.8	0.009	0.5, 1.0	
Strontium titanate, $SrTiO_3$	300	0.0026	0.5, 1.0	
SiON	5	0.001–0.025		Used as an insulating layer.

the UV-light exposed substrate. The UV-light exposed substrate is immersed in the *positive developer*. It removes the hardened photoresist from the unwanted metalized area of the substrate, while maintaining the photoresist on the stripline. Finally, to *etch out* the unwanted metal from the substrate, it is immersed in the ferric chloride that is an etching medium. The ferric chloride dissolves the copper from the unwanted surface

Chemical etching process

⇓

- ❖ Preparation of circuit
- ❖ Mask making.
- ❖ Cleaning of substrate:
  - ▪ Solvent bath
  - ▪ Rinsing
  - ▪ Baking.
- ❖ Application of positive photo resist on the substrate. (By immersion/spraying/spin coating).
- ❖ Artwork, mask placement on the substrate.
- ❖ Illumination of the substrate in ultraviolet light.
- ❖ Development of the substrate.
- ❖ Etching of the unwanted metallization from substrate. (By immersion/spraying of the echant).

Figure 13.3 Steps of the chemical etching process.

area of the substrate while leaving the photoresist-coated conducting strip line on it. However, some ferric chloride may enter from the lateral sides of the strip conductor giving it a trapezoidal shape as shown in Fig (13.4b).On removal of photoresist, finally, a microstrip line is realized.

13.1.3 Thin-Film Process

The thin-film process is a deposition of the metal film on a ceramic substrate like alumina, either by the *thermal evaporation process* or by the *sputtering process,* shown in Fig (13.5). Both these processes are known as the Physical Vapor Deposition (PVD) process. [B.1, B.3, B.7–B.9]. The thin-film metal deposition on a ceramic substrate, like alumina (Al_2O_3), is carried out to get the high-quality hybrid microwave circuits. The alumina ceramic substrate has 99.5% purity, and its common thickness is either 0.6325 mm (25 mils) or 1.27 mm (50 mils). It has some amount of anisotropy that is much less than anisotropy in the sapphire substrate.

The thickness of the deposited metal film is on a ceramic substrate is less than 1 μm. It is difficult to deposit high-conductivity conductors such as aluminum, silver, copper, and gold on the alumina substrate. Thus, a thin adhesion layer of Nickel–chromium (NiCr), i.e. nichrome ~0.05 μm is first deposited on the alumina substrate. The NiCr is a poor conductor, but it has good adhesion to the alumina substrate. Over the NiCr layer, 0.1 μm thick gold layer is deposited. Finally, a 5–10 μm thick conducting layer of gold is deposited by the electroplating process. The gold as a conductor for the strip lines is a better choice as compared to the copper because it does not oxidize and corrode. Sometimes another thin layer (~0.05 μm) of Pt or Pd (platinum or palladium) is deposited as the barrier layer between the nichrome layer and the gold layer, to avoid the diffusion between nichrome and gold. Thus, the metallization on the alumina substrate is a *composite metallic structure* of thickness more than three times the skin depth. This process is also used to deposit the thin-film resistors and the thin-film capacitors, apart from the conducting strip lines. The premetalized, single-sided, or double-sided alumina substrates in sizes $1'' \times 1''$–$3'' \times 3''$ are available from the commercial market. The microstrip, CPW, slot line and the circuits constructed out of these planar lines are etched over the alumina substrate by the photolithography process, as it is done for the PTFE substrate.

Thermal Evaporation Process

Figure (13.5a) shows the basic thermal evaporation process [B.1, B.3, B.7]. A vacuum enclosure is made of glass or stainless steel. A provision is made for a quartz window for observation. Normally, the vacuum level obtained using the turbo/diffusion pump is in the range

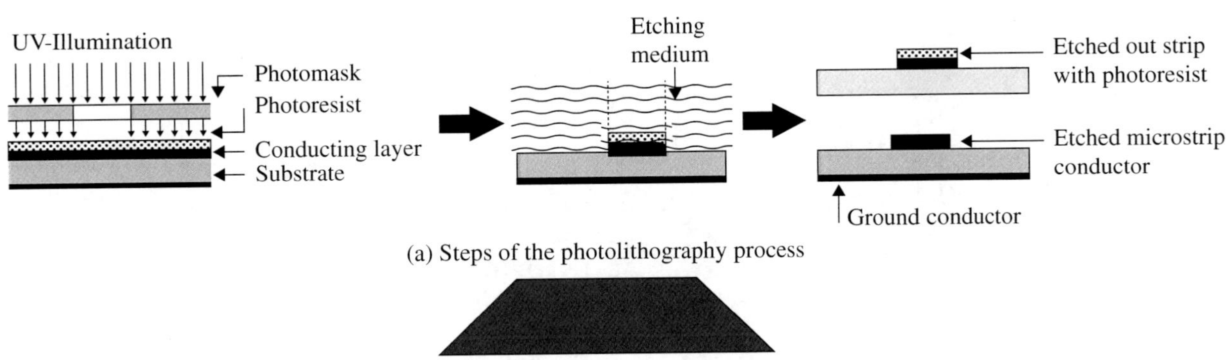

(a) Steps of the photolithography process

(b) Trapezoidal shaped cross-section of the copper conductor due to the etching process.

Figure 13.4 Positive photolithography process.

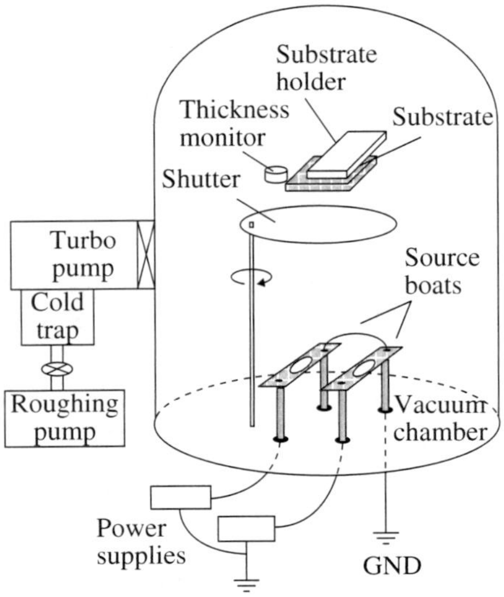

(a) Thermal vapor evaporation process.

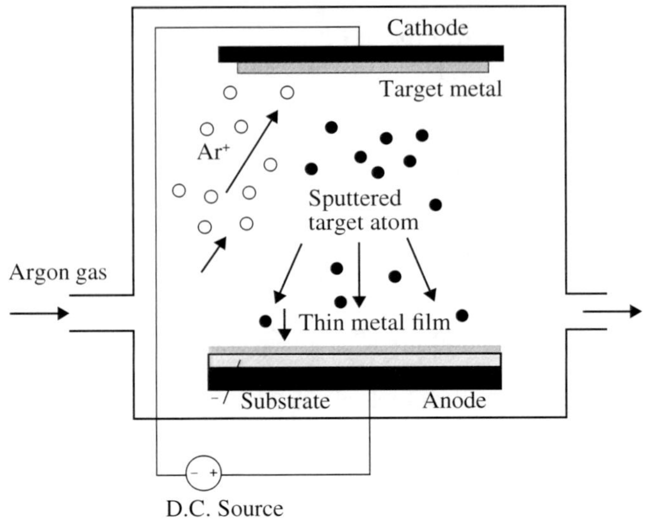

(b) Sputtering process.

Figure 13.5 Physical vapor deposition (PVD) process.

10^{-3} to 10^{-10} Torr. The basic thermal evaporation is a three steps process, mentioned below:

- *Step#1:* The metal is vaporized at a high temperature in the vacuum environment. The deposited materials, such as nichrome, gold, are kept in a pot called the *boat*. The boat is made of the refractory metal like molybdenum or tungsten that is heated to over 1000°C by applying the electrical current from power supplies.
- *Step#2:* Next, the metal vapor in the vacuum is transported from the boat to the substrate surface through a shutter that controls amount of metal deposition.
- *Step#3:* Finally, the metal vapor condenses as a metal film, nichrome, or gold and gets deposited as a film on the substrate.

Sputtering Process

Figure (13.5b) shows the basic D.C. sputtering process [B.3]. It is a plasma-assisted process. The vacuum chamber is filled with Argon gas at above 10^{-3} Torr. On the application of high voltage, 1–3 KV, the gas discharge is produced and the plasma medium is formed that releases high energy (>10 KeV) Ar^+ ions. The Ar^+ ions hit the target metal that is to be deposited on the substrate. The target metal is mounted on the cathode. The sputtered gold atoms released from the target are deposited on the substrate mounted on the anode. The sputtering process is suitably controlled one, as compared to the evaporation process. Several materials could be used in this process. Further sophistication in the process leads to the RF and D.C. magnetron sputtering process. There are many more commercial thin-film deposition processes [B.8, B.10].

13.1.4 Thick-Film Process

The *thick-film process* of creating the planar transmission lines and the hybrid MIC on a *ceramic substrate* is a cheaper alternative to the expensive thin-film process. It has acquired importance due to the rapid growth of the *MCM (Multichip Module)* technology. This technology is used to produce only passive elements, such as strip conductors, overlays, resistors, and capacitors. There is no sharp distinction between the thin film and the thick film based on the thicknesses of deposited conductor films on a ceramic substrate. The thickness of a thick-film conductor is at least ten times greater than the *mean free path* of an electron in a filmmaking material. Normally, it is between 0.2 mils (5.08 μm) and 2 mils (50.8 μm), whereas the thickness of the *thin film* is less than 1.0 μm. The mean free path is the average distance an electron travels before it is scattered in the deposited film. It governs the resistivity of a metallic film. The most important difference between the thick and thin film arises due to the film deposition process. The thick-film deposition is done both on the ceramic substrates and LTCC substrates [J.11]. The thick film is deposited on a substrate by a process called *screen printing*. This has two versions [B.1, B.3, B.11, J.10, J.12–J.21]:

- Conventional thick-film technology.
- Photoimageable thick-film technology.

The thick-film technology requires a suitable substrate and metallic paste also called *ink* in both kinds of technology.

- **Substrate:** Two kinds of substrates, i.e. ceramics, and glasses, are used in thick-film technology. The substrate materials should not react with the deposited film. However, it should provide mechanical strength and adhesion to the deposited film. The 94–96% pure alumina (Al_2O_3) is commonly used as a substrate. However, 99.6% alumina is preferable for microwave applications, as it has 20% less dielectric loss at microwave frequency [J.15, J.16]. Apart from alumina, beryllia (BeO), aluminum nitride (AlN), magnesia, etc. substrates are also used. The thermal conductivity of beryllia is about 12 times higher than that of alumina and is also about 12 times more expensive. The thermal conductivity of aluminum nitride substrate is nearly the same as beryllia, but their production costs are lower than of beryllia. Other types of ceramic substrates are also used in the LTCC process. It is discussed in section (13.4).
- **Metallic paste:** In the thick-film technology, the conducting materials are used in the form of a paste. The *metallic paste* or *slurry, called ink,* is deposited on the ceramic substrate through a screen made of either silk (nylon) or stainless steel. The thick-film metallic paste has the following three ingredients:

  i) *Host material:* It is also known as the *functional phase* that is a fine particle powder of metal or metal alloys, such as silver, gold, palladium–gold, having the particle size about 5 Å. For printing a resistor, the powder of metal oxides (RuO_2, $Bi_2 Ru_2O_7$, Pd, Ag) is used. Likewise, for the dielectric layer, glass-ceramic, glass, or barium titanate ($BaTiO_3$) powder are used.
  ii) *Binder material:* It is also known as the *binder phase*. It is a glass frit, usually lead-bismuth borosilicate ($PbBiBSiO_2$), or oxides, e.g. CuO or Bi_2O_3, that holds the paste to the ceramic substrate. During the high-temperature *firing*, it merges with the ceramic.
  iii) *Vehicle material:* The vehicle is an organic system and that acts as a carrier for the powders and also a *plasticizer*. It controls, to some extent, the desired printing characteristics. It is composed of both volatile (solvents) and nonvolatile (polymers) organics. These materials evaporate and burn off during the early stages of drying and firing, respectively.

In short, the paste (ink) is a blend of metal and glass powders mixed in the organic vehicle like paint thinner. The usual composition of the metal particles in the paste is 50–70%, the vehicle occupies 15–35% and the rest is the binder.

The thick-film conductor should have low sheet resistance, <5 mΩ/sq. The metallic paste of gold (Au) and silver (Ag), or their alloys with palladium (Pd) and/or platinum (Pt) have following sheet resistances: Au (3 mΩ/sq.), Ag (3 − 5 mΩ/sq.), Ag/Pt (2 mΩ/sq.), Ag/Pd (30 mΩ/sq.), Au/Pd (50 mΩ/sq.). The silver paste, with high conductivity, creates an excellent conducting line pattern of a good definition. However, it is susceptible to tarnishing and has undesirable high migration. The gold-based paste is insensitive to an atmosphere and has higher stability. However, it is expensive compared to the silver-based paste. The 125 μm lines, with 250/125 μm separation could be printed. It is difficult to produce line width and separation less than 100 μm due to unreliable lines. However, manufacturers have claimed even 35 μm lines for the *photoimageable technology* [J.12, J.14]. The thickness of the printed conductor is in the range of $8 - 12 \pm 2.5$ μm. The copper paste is normally less preferred, as it is not compatible with the process using other conductors. It requires nitrogen firing. The alloy-based paste has a higher resistivity, as compared to a pure metal paste, so these are unsuitable for microwave applications.

Conventional Thick-Film Technology

Figure (13.6a) shows that a *sieve* is used in the screen printing process. It is made of either nylon or stainless steel wire mesh. It has normally 50–150 wires/cm stretched over a frame. A *stencil* showing the conductor pattern is put on the screen of the sieve. The stencil is cut for the metallic strip conductor that we desire to print on the substrate.

Alternatively, a pattern of the conductor shape to be printed is made on the screen by the *photosensitive emulsion*. The emulsion covered screen is exposed through a film mask shown in Fig (13.6b). After exposure, the emulsion is washed out by *developing* the screen. The emulsion is removed from the portion where thick-film metal conductor is to be formed on the substrate shown Fig (13.6c). Now the screen is mounted on the substrate and the paste is applied through the opening on the screen or in the stencil as shown in Fig (13.6d). The metallic paste passes through the screen/ stencil and the metallic strip line is printed on the substrate. Finally, the substrate containing the thick-film line is dried at temperature 160–200°C and is *fired* for about 10 minutes at the temperature between 600 and 1000°C depending on the metal paste. The thick-film conductor has a

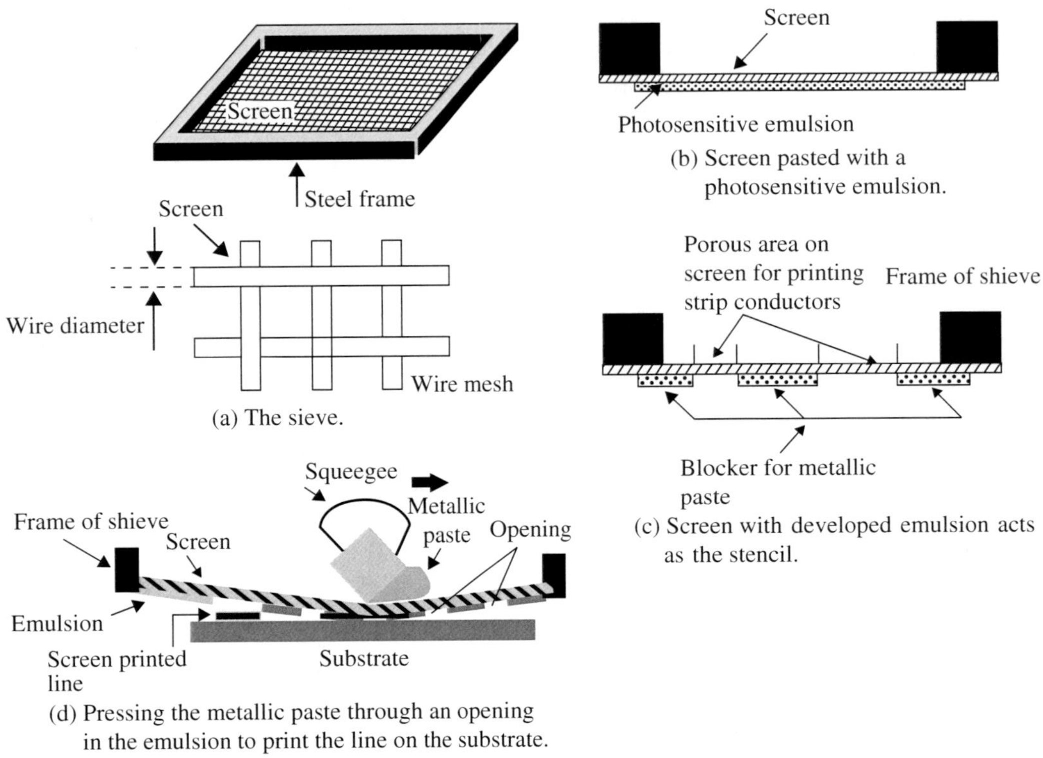

Figure 13.6 Schematic diagram of the thick film process using screen printing.

thickness about 25 μm and its minimum width could be 100–200 μm with edge accuracy 20–30 μm.

Photo-Imageable Thick-Film Technology

The conventional thick-film screen printing technology does not provide narrow width and narrow spacing printing of the conductors on the ceramic substrate. The microwave and mm-wave applications require high impedance and low-loss lines [J.18]. Photo-defined technology is the *photo-imageable thick film technology*. It is an extension of standard thick-film technology that offers fine lines and small spacing down to 25 μm with ±5 μm tolerance. The fine pattern geometry lines have a stable width that is comparable to those obtained by the thin-film technology. The microwave lines, coupled-lines, and circuits have been tested from 4 to 18 GHz [J.16]. This technology is a combination of conventional thick-film technology and some processes typical for thin-films. Figure (13.7) shows the process of the photo-imageable thick-film technology [J.16]. At first, a layer of photosensitive paste is deposited on the substrate, leveled, and dried at the temperature of 80°C. After drying, this layer is exposed to UV light through the negative photomask and then developed to create final circuits.

13.2 Elements of Monolithic MIC (MMIC) Technology

Microwave community conceived the idea of a monolithic microwave integrated circuit-*monolithic MIC*, i.e. MMIC on the semi-insulating GaAs semiconductor substrate in 1962 [B.12–B.14]. Since then the MMIC (chip size about $3.5 \times 1.5\,\text{mm}^2$) technology has moved in many directions, such as the incorporation of new devices, an extension of frequency range from VHF to UHF to microwave to mm-wave and beyond, use of silicon as a substrate and also novel use of silicon–germanium (SiGe), indium phosphate (InP), etc., as new substrates. The design of planar transmission lines in the MMIC environment has contributed significantly toward the growth of the MMIC technology. The development of MMIC on the silicon substrate has acquired a new name – the *radio frequency IC (RFIC)* [B.4, B.15, B.16]. Likewise, MMIC at 30 GHz and beyond, i.e. in the millimeter range, called *the monolithic millimeter-wave integrated circuits (MMWIC or M^3IC)*, has also been developed [B.17]. However, the term "MMIC" could be used to represent all these cases. The MMIC is a specific fabrication technology for the microwave circuits – such as an amplifier, oscillator, mixer, switches, phase-shifter,

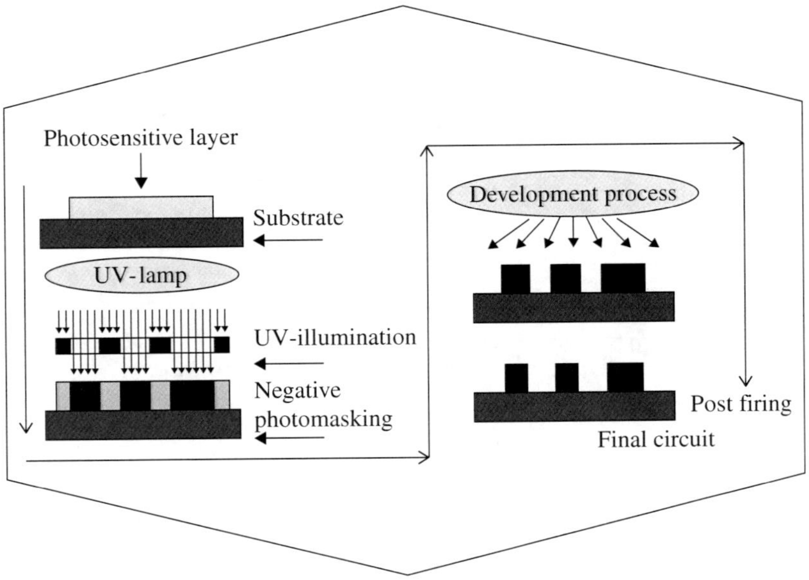

Figure 13.7 The photo-imageable thick film technology process.

on a single semiconductor substrate. Its normal operating frequency range is from 1 to 100 GHz. However, the frequency range is moving upward in the THz ranges with the development of technology [B.17, B.18]. The term "*monolithic*" indicates that all passive and active components of a circuit are deposited and grown on the same substrate, as against the discrete component-based hybrid MIC. Despite such integration, the MMIC is only used in the RF, microwave, mm-Wave, and THz applications. It does not apply to the digital circuits.

The present-day technology is moving toward the *system-on-chip (SOC)* that combines both the digital and the analog circuits with the MEMS components, optical components, and transceivers on a single substrate. However, it is difficult to develop all kinds of electronic devices on one kind of substrate. Therefore, the development of a true SOC is a difficult proposition. However, it is necessary due to the technological requirement of the high-demand and high-volume communication system technology, especially for the mobile telephony, Local Multipoint Distribution Service (LMDS), satellite communication, etc. The *multichip module (MCM)* technology has been developed as a solution to the problems associated with the SOC [B.19]. It is a newer approach to SOC. The SOC divides a complete electronic system into sections. Each section is designed and developed with the best performing components for that section, such as CMOS for the digital circuit, HEMT for the mm-wave application, MEMS for switches and passive components and planar structure. Finally, all these sections (chips) are combined through *interconnects* to form a single module for the MCM.

The MCM depends on *three main technologies* – (i) the thin-film deposition technique resulting in the MCM-D, (ii) the ceramic-based cofired technology resulting in the MCM-C and (iii) the lamination-based technology resulting in the MCM-L. The MCM-D is the Si and GaAs-based technology, the MCM-C is LTCC-based technology and the MCM-L is an advanced form of the multilayer PCB technology.

MMIC technology consists of three items – *substrate, active devices, passive components, and interconnects.* GaAs and Si are the two most commonly used semiconductor substrates. However, InP, SiGe, GaN, and SiC are also finding applications. Table 13.2 shows the basic characteristics of some semiconductor substrates. SiGe semiconductor is grown as an epitaxial layer on the Si-substrate to provide high-electron mobility in transistors. InP is a more brittle substrate, as compared against GaAs-substrate. On several occasions, InP is grown as an epitaxial layer on a GaAs-substrate to develop InP transistors. Silicon carbide (SiC) and gallium nitride (GaN) are wide band-gap semiconductors, and they work at higher junction temperature and also have higher breakdown voltages [B.20]. These are useful substrates to develop high-power amplifiers.

The passive components and interconnects occupy a very large chip area. These are used as the impedance matching network, biasing network, filters, attenuators, phase-shifters, etc. These are also the

Table 13.2 Some commonly used semiconducting substrates.

Substrate	Working freq. (GHz)	Electron mobility (cm/Vs)	Relative permittivity	Tan δ/resistivity (Ω-cm)	Substrate thickness (mm)	Comments
GaAs	>75 GHz	5500–7000	12.9–13.1	0.0006 (10^7–10^9)	0.0005, 0.001, 0.35, 0.5, 0.625	4″ or 6″ diameter wafer
GaN	20–30	400–1600	∥10.4, ⊥9.5 8.9	0.0027	0.001, 0.003–0.420	Anisotropic, deposited on Sapphire and other substrates
InP	>115	10^4–12 000	12.4	0.001 (10^7)	0.35, 0.60, 0.625	Used at THz
Si (high resistivity)	<20	900–1100	11.7–12.9	0.005–015 (10^3–10^5)	0.279, 0.281, 0.525, 0.625	1″ to 12″ diameter wafer
Si on Sapphire			11.6	10^{14}		
SiGe	10–40	2000–3 × 10^5				6″ wafer
SiC	15–20	500–1000				

lossy components. The loss depends on the semiconductors, the layered structure of a substrate, and the type of transmission line adopted in the circuit development. The microstrip, CPW, and slot lines are used with some modification. The "*thin film microstrip*" *(TFMS)* is a popular choice [B.17, J.19]. The contemporary focus is on the planar transmission line structures for the MMIC, etc. The interested readers can follow the MMIC technology and design from the available sources given in references at the end of the chapter. The electrodes on the MESFET, HEMT, etc., that support the traveling wave for the mm-wave applications can also be treated as a transmission line [B.21]. Thus, a planar transmission line in the MMIC structure encounters the complex layered medium for its application in the design of interconnects passive components and even modeling of the active devices. Some familiarization of the MMIC fabrication process is useful for the modeling of the transmission line structures in the MMIC technology. This section briefly discusses the fabrication process and also the fabrication of planar transmission lines in the MMIC technology.

13.2.1 Fabrication Process

Four fabrication methods for the *islands of active devices* on the semi-insulating GaAs-substrate are presented in brief. The process is also used for the Si-substrate. These methods are – (i) selective epitaxial growth, (ii) mesa etching method, (iii) ion implantation method, and (iv) insulation implantation method. Figure (13.8a–d) illustrates these fabrication methods [B.1–B.3, B.13].

Epitaxial Growth Method

The word 'epitaxy', from the Greek roots 'epi', means 'above', and 'taxis' means in 'orderly manner'. Thus, the *epitaxy* is a process of growing the single-crystal film on top of a crystalline substrate. It is an important process of the thin-film semiconductor device technology. The *homoepitaxy* process has the same material for both the grown thin film and substrate. For instance, Si is grown on the Si-substrate. However, the grown film is purer than the substrate material, and it can be suitably doped. In the case of the *heteroepitaxy* process, the grown film and substrate are of a different material. For instance, AlAs is grown on the GaAs-substrate. In this case, there can be unmatched lattice parameters, leading to changes in the electrical, optical, thermal, and mechanical properties of the films. The metal-semiconductor structures are also used for the electrode contact applications. For high-frequency devices (1–1000 GHz), higher-electron mobility compound semiconductors are used. The heterojunction nature also increases the gain and efficiency of the amplifiers developed from these devices.

Figure (13.8a) shows the steps involved in the epitaxial growth process. The process initiated with *a semi-insulating GaAs-substrate* of typical thickness is 150 μm. Before the selective epitaxial growth, a 0.4 μm thick *masking* layer of SiO_2/Si_3N_4 is deposited on the substrate. The gaps are etched out in the masking insulating layer

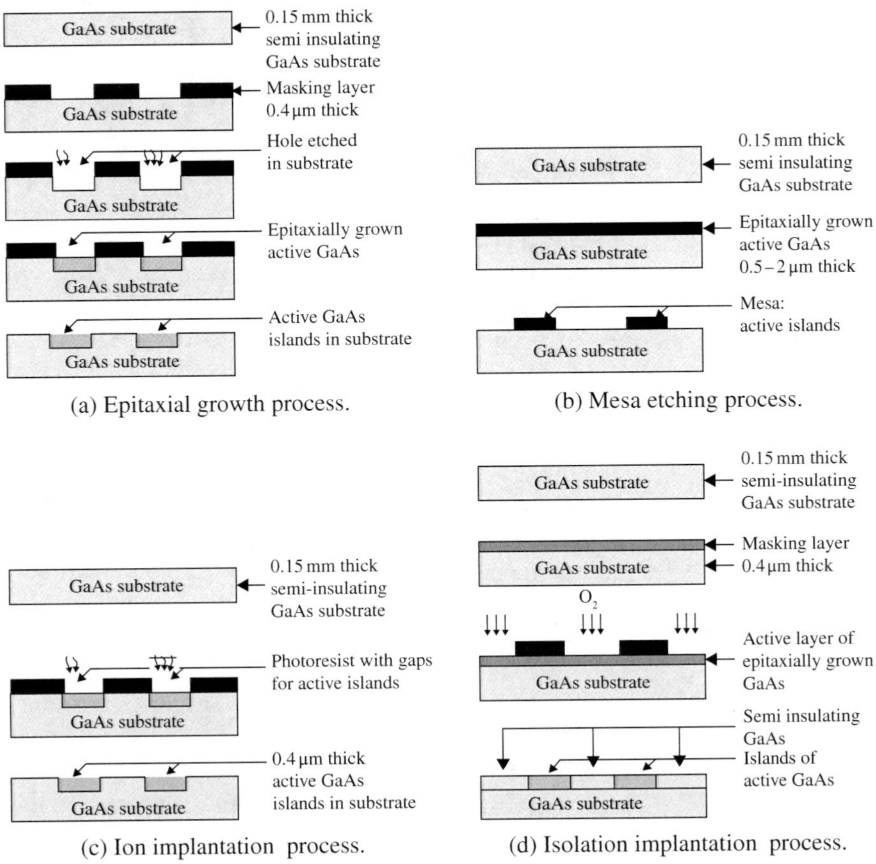

Figure 13.8 MMIC fabrication process.

to create the islands for the active devices. In the next step, a hole of few microns is further etched on the GaAs-substrate. The doped monocrystalline layer of an *active GaAs* is epitaxially grown in the hole. Finally, the masking material is removed, and the *active GaAs island* remains on the semi-insulating GaAs.

Mesa Etching Method

Figure (13.8b) shows the mesa etching process. An active (doped) layer of GaAs 0.5–2 μm thickness is epitaxially grown on the semi-insulating GaAs-substrate. The *mesa* island of active GaAs is left on the substrate by removing the unwanted part of epitaxially grown GaAs, using the photolithographic method. The mesa is projected above the substrate.

Ion Implantation Method

This is the most commonly used method and is shown in Fig (13.8c). The semi-insulating GaAs-substrate is covered with photoresist, and the gaps in the photoresist are created by etching out the photoresist from the gap area. The GaAs-substrate in the gap area is doped by implanting the ion of a doping element. It has a typical depth of 0.4–1 μm. Finally, the photoresist is removed, and the islands of the active GaAs, i.e. the doped GaAs, are created.

Isolation Implantation Method

Figure (13.8d) shows the process. On the semi-insulating GaAs-substrate, a thick layer, about 1 μm, of active GaAs (doped GaAs) is epitaxially grown. Next, the photoresist layer is deposited over the epitaxially grown active GaAs. The photo-resist mask is retained over the area, where islands of the active GaAs are grown epitaxially. The remaining surface without photoresist, is implanted with the oxygen atoms that create the semi-insulating GaAs between the active islands. Finally, the photoresist is removed to get the semi-insulating GaAs-substrate with active islands.

Several processes, including the thin-film method, are available for fabricating the planar passive components and interconnect. The planar strip lines and lumped elements are also developed by the vapor deposition method, sputtering, and electrolysis processes. The diodes and transistors are developed by controlling the doping of active islands. The Ohmic and Schottky contacts are also deposited as needed. Usually, a high-doping concentration $1 \times 10^{18}\ cm^{-3}$ or more, known as the n^+ layer, together with an AuGe or other suitable metal layer is used for the Ohmic contact. The high doping concentration results in the tunneling effect that reduces the resistivity of the Ohmic contact. The metal-semiconductor contacts are usually annealed or alloyed after the initial deposition of the metal to further improve the contact resistivity. The Schottky contact is obtained by the deposition of TiPtAu metal layer on a semiconductor with a lower doping concentration level of 1×10^{17} cm^{-3}. The proper choice of metal on the semiconductor (n or p), taking into account the electron affinity and bandgap energy, is required to develop an appropriate contact. Moreover, for the MMIC, the overbridge and the underpasses are also created for connecting the active devices with the line structure.

13.2.2 Planar Transmission Lines in MMIC

The MMIC has three distinct design strategies – *(i) all-transistor circuits, (ii) lumped element circuits, and (iii) distributed element circuits*. The choice of an MMIC circuit is determined by the operating frequency. All transistor circuits operate approximately up to 5 GHz, and the lumped element circuits operate up to 20 GHz. The high-end microwave and mm-wave MMIC are developed around the distributed elements, i.e. around the microstrip, CPW and slot lines, and their judicious combinations. Even the lumped inductor and lumped capacitor are developed using the microstrip and CPW technology. The planar transmission lines-based circuits in the MMIC environment operate up to 100 GHz on GaAs-substrate, even beyond up to 1.5 THz on InP substrate, depending upon the choice of substrate, device, and fabrication process [B.16, B.17].

In the MMIC environment, the planar transmission lines have two broad kinds of applications – a medium for the development of passive components, such as filters, couplers, balun, matching networks, interconnects [B.4, B.14–B.17], and as a structure to develop the electrodes of active devices [B.21]. Both applications are separately discussed below for the appreciation of the planar transmission lines in the MMIC technology.

Planar Transmission Lines for Passive Components and Interconnect

The microstrip, CPW, coplanar strips, and slot lines are the basic transmission line structures, and their combinations adapted to the MMIC. Microstrip is the most popular one, followed by the CPW. For microstrip lines, the GaAs-substrate is thinned to 100 or 200 μm so that strip conductor width is in the range of 6 to 120 μm, providing 40–120 Ω characteristic impedance. The CPW does not require thinning of a substrate. We use 600 μm thick substrate, without any processing or metallization of its back surface. The signal line and two ground conductors of a CPW are on the same side of the substrate. No vias are needed for the shunt connection of components. However, the shunt connection of a component requires a via in the microstrip. Of course, an overbridge, or underbridge, is added to connect both ground conductors of a CPW. Because of less dispersion, also smaller losses, the CPW is a preferred transmission medium above 40 GHz. The slot lines are used normally to incorporate balun for needed applications. A judicious combination of microstrip and CPW is also used in the MMIC.

However, the planar lines are not always fabricated directly on the GaAs or Si and other semiconducting substrates, primarily due to high-dielectric losses. The dielectric losses increase with an increase in the operating frequency. The line structures face a new multilayer environment to improve their performance with respect to loss and dispersion. The standard closed-form expressions are not available to compute the multilayered transmission line parameters. Normally, a circuit designer has to depend on the *EM-Simulators*. Chapter 15 presents another approach, the *single-layer reduction* (SLR) approach, to compute multilayered line parameters.

The total loss of a planar transmission line has three components – dielectric loss in the substrate, conductor loss in the strip conductor, and the ground plane, and radiation loss in the substrate through the surface wave and leaky wave. The substrate radiation also causes crosstalk to a highly dense MMIC by the increased level of undesired coupling between the adjacent lines on the same substrate. All the three mechanisms of losses need to be addressed with respect to the semi-insulating GaAs and the high-resistivity Si-substrate. The dielectric loss of the planar lines on these substrates is small as compared to the conductor loss in the lines. However, the resistivity of the Si-substrate is mostly determined by the device technology. For instance, in the CMOS process, resistivity of the substrate is small, in the range 1 – 20 Ω - cm, leading to a high dielectric loss. It creates a problem with the design of the high-Q passive components and matching networks. Further, the low-resistivity Si-substrate gives higher parasitic capacitances that degrade the performance of an active device.

Gold is the preferred metal for conductors in GaAs- and InP-based MMIC, whereas aluminum is used in the Si and SiGe technologies. To reduce the conductor loss, normally, the conductor thickness is about four to five times the skin depth.

The following types of line structures are usually involved in the MMIC.

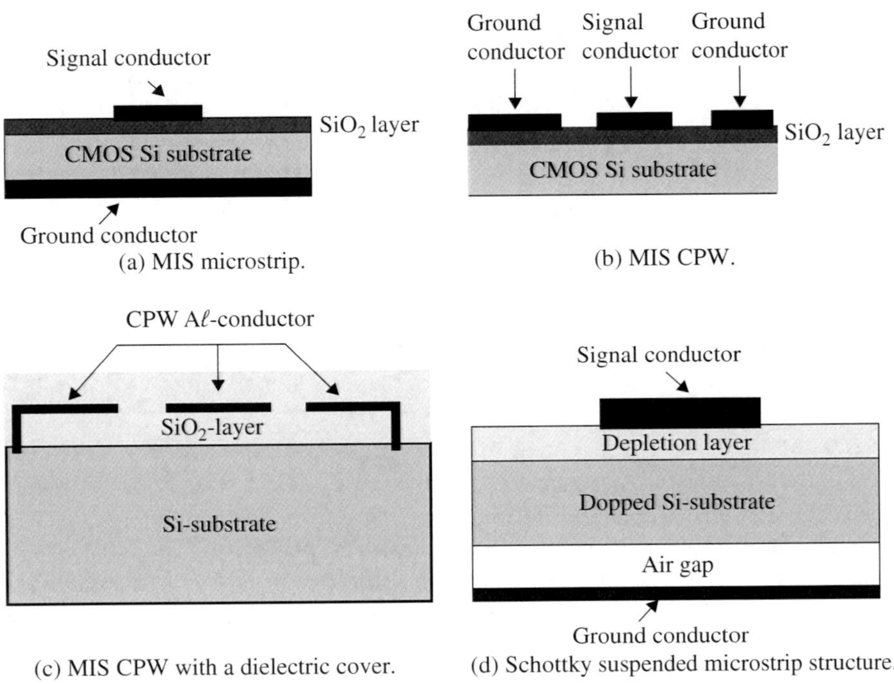

Figure 13.9 MIS and Schottky structures.

Si/GaAs-Substrates

In this case, the microstrip and CPW are fabricated using the photolithographic process directly on the high-resistivity Si and semi-insulating substrates. The losses are mainly due to conductor losses, as the dielectric loss is small.

MIS/MOS Structure

This is a two-layer, i.e. composite substrate structure used in the CMOS technology. Approximately, the 25–1500 nm thick insulating layer of polyimide, SiO_2, etc., is deposited on the Si/GaAs-substrate, as shown in Fig (13.9a–c), for the microstrip and CPW. There could also be an insulating overlay layer. There is an increasing demand for the RF and microwave components in the CMOS technology that uses the low-resistivity (1.0 Ω-cm) Si-substrate. The SiO_2 passivation layer reduces the loss. The MIS kind of microstrip/CPW supports complex frequency-dependent wave propagation. It also involves the slow-wave propagation at lower microwave frequency [J.10, J.20–J.22].

Schottky Structure

The metal-semiconductor contact could be Schottky type that creates a depletion layer, as shown in Fig (13.9d), for a suspended microstrip. The depletion layer, shown in Fig (13.9d), is not a separate material layer. It is shown only for analytical modeling. In reality, the depletion layer is formed within the layer of the doped Si-substrate itself. For a normal Schottky microstrip, the air-gap thickness is zero. However, the suspended Schottky structure provides further control over the line parameters. The lines characteristics of a Schottky line are voltage-controlled, as the depletion layer is voltage-controlled. This structure is suitable for the development of voltage-controlled components. The structure is suitable for the nonlinear transmission also [J.23, J.24].

Thin-Film Structure

Figure (13.10) shows that a thin-film microstrip (TFMS) line is a simple microstrip line created over the GaAs/Si/alumina substrate [J.19]. In this case, the ground plane is

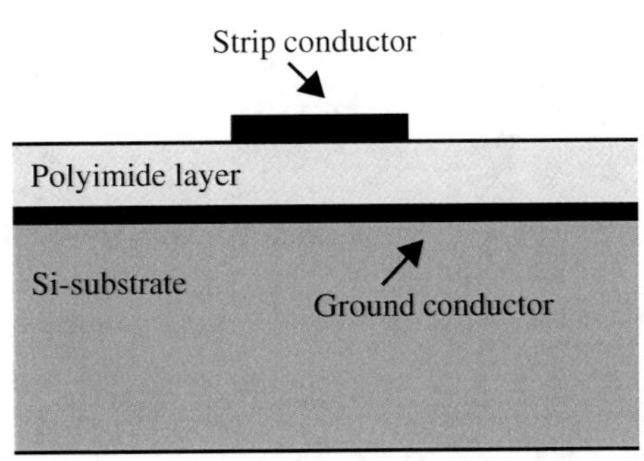

Figure 13.10 Thin-film microstrip. Source: Based on Ponchak and Downey [J.19].

not on the backside of the substrate, rather it is printed on the front face of the semiconductor substrate over which a dielectric thin-film is deposited.

The substrate thickness, for a standard Si 4–12 Ω-cm substrate, is in the range 100–500 μm, and a thin dielectric film is in the range 1–25 μm. The typical dielectrics are polyimide, BCB (*benzo-cyclo-butene*), or SiON with $\varepsilon_r = 2.65–3.0$, 5.0 respectively. The conductor thickness is $t \leq 1.5$ μm. Gold is a preferred conductor. The first 200 Å thick Ti adhesion layer is deposited on the Si-substrate. Next, 2.5 μm thick Au is deposited on the Ti adhesion layer, and the composite metal forms the ground plane. The real advantage of the TFMS is its compact line-width, between 5 and 35 μm, to obtain standard values of the characteristics impedances. Unlike normal microstrip, the TFMS has very small via made of the thin dielectric film that has negligible inductance. The low ε_r of a dielectric substrate is needed for high-speed digital circuits. The TFMS has insignificant dispersion up to 25 GHz, and it is useful up to 110 GHz. However, the conductor loss is a dominant factor for the TFMS lines [J.25].

Flexible Polyethersulfone (PES) Substrate

The MMIC on a flexible substrate is important for flexible display, smart tag, and wearable electronics. The integration of RF components on the transparent flexible substrate is useful for mobile communication devices. The polyethersulfone (PES) substrate is used for a flexible monolithic microwave integrated circuit (MMIC) [J.26]. The fishbone-type transmission line (FTTL) [J.26] has been fabricated on the PES substrate to operate from the direct current, i.e. DC to 50 GHz. The PES substrate thickness is 200 μm and the thickness of the Au/Ti conductor is 2 μm. The FTTL, shown in Fig (13.11), is primarily a CPW line, where a central conductor is periodically loaded with stubs. The FTTL gives ε_{reff} 7.2–9.5 at frequency 5–50 GHz, whereas CPW on PES gives ε_{reff} in the range of 2.45–2.97 in the same frequency range. The higher value of the ε_{reff} for the FTTL type CPW is due to the periodic loading of the original CPW. The FTTL also provides a higher value of ε_{reff}, as against the value of ε_{reff} of a CPW, on the Si-substrate at a frequency above 10 GHz.

Multilayer MMIC Lines

The above mentioned lines have more than one layer of dielectrics/semiconductors. Thus, technically speaking, these are multilayer planar transmission lines. However, Fig (13.12a) shows some other kinds of multilayer microstrip lines for the MMIC applications. These are suggested by Ogawa et al. [J.27]. The four multilayer configurations are – *microstrip line with overlay, inverted microstrip line, trapezoid microstrip line, and valley microstrip line.* The microstrips are fabricated on the GaAs-substrate and then the dielectric layers of 10 μm thick polyimide films deposited on it. Ogawa et al. studied the loss behaviors of 50 Ω, 1.4 mm long lines at 20 GHz. The trapezoid microstrip line has the least loss, and maximum width as the microstrip line is on the low-loss, low permittivity polyimide substrate. Each layer of polyimide is 2.5 μm thick. Figure (13.12b) shows that these structures could also be used as the coupled microstrip lines.

Buried Microstrip Line

The *buried microstrip line* (BMSL) is also known as the *embedded microstrip line.* Figure (13.13a) shows it in a PCB laminate format. It is a standard microstrip line with overlay and a metallic shield can also be added [J.28]. The signal carrying strip line is buried (embedded) at the interface of two dielectric media of relative permittivity ε_{r1} and ε_{r2}. However, the microstrip

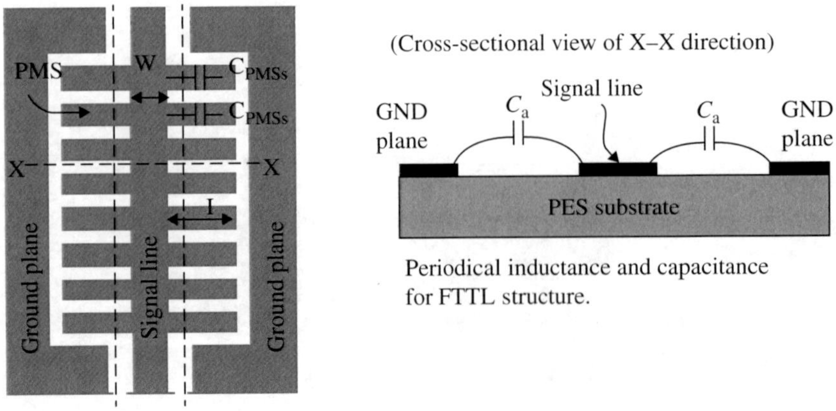

Figure 13.11 Fishbone-type stub loaded line on PES substrate.

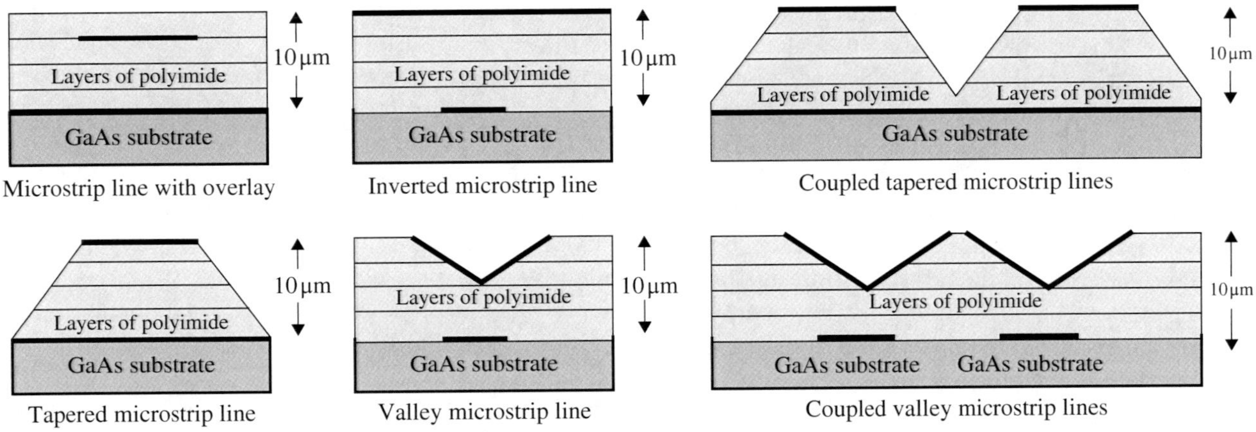

(a) Multilayered microstrip line.

(b) Coupled multilayered microstrip line.

Figure 13.12 Multilayered microstrip and coupled microstrip lines for MMIC applications.

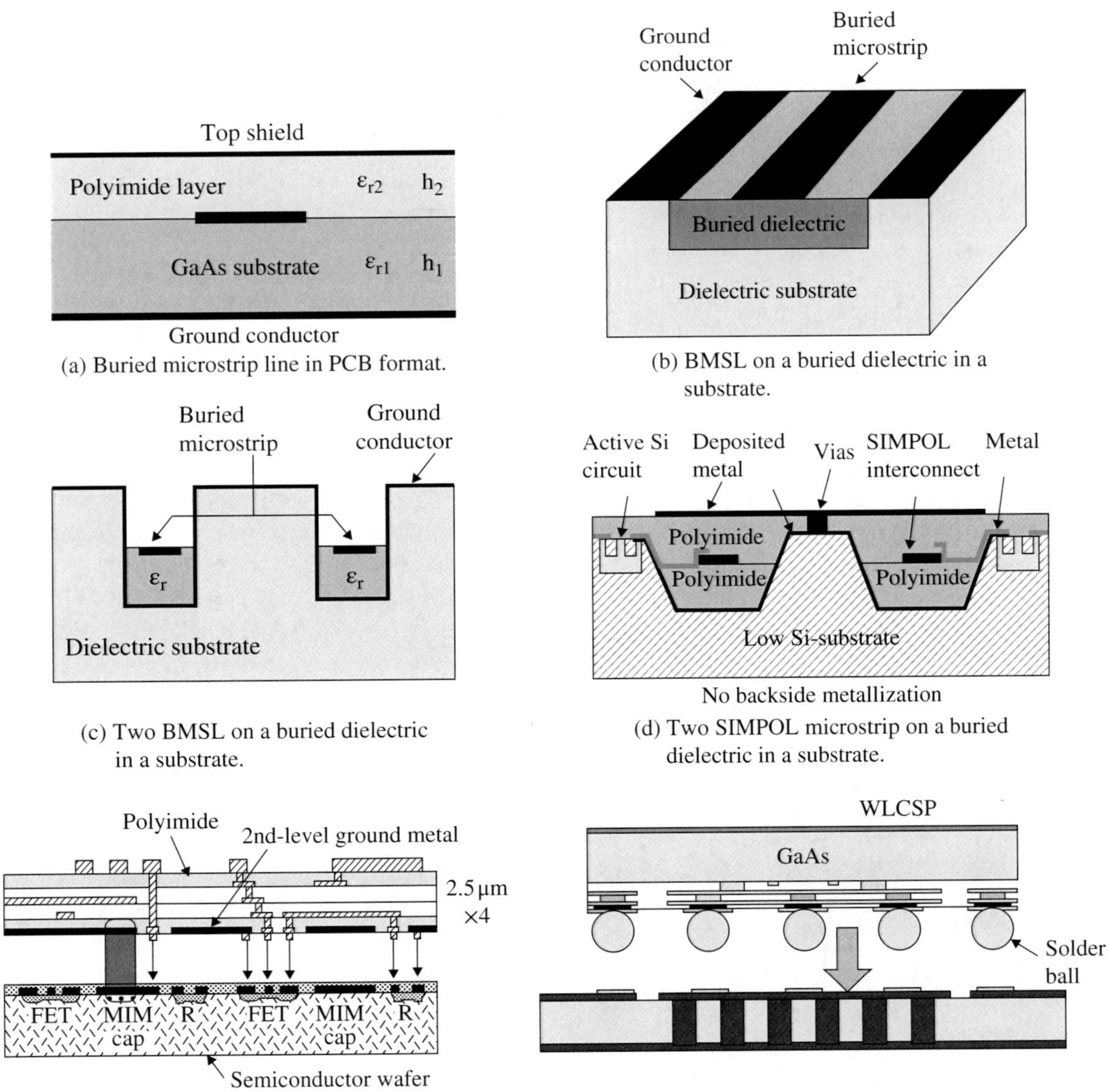

(a) Buried microstrip line in PCB format.

(b) BMSL on a buried dielectric in a substrate.

(c) Two BMSL on a buried dielectric in a substrate.

(d) Two SIMPOL microstrip on a buried dielectric in a substrate.

(e) Basic 3D-MMIC WLCPS structure.

(f) WLCSP assembly process.

Figure 13.13 Some buried microstrip lines.

line could be buried in a uniform medium also. The BMSL on the Si-substrate is shown in Fig (13.13b) [J.29]. The substrate could be any semiconductor or a ceramic substrate on which a groove is cut. The groove is filled in with dielectrics and strip conductor is printed over it. This microstrip line is isolated from the substrate. Thus, unlike a stripline on the CMOS grade Si-substrate, the BMSL is not having a high-loss. The number of BMSL can be fabricated on the same substrate as shown in Fig (13.13c). It has a low coupling, i.e. low crosstalk. The low crosstalk is due to the electromagnetic shielding provided by the side metallic walls.

There are other suggestions also for the 3D-MMIC using the embedded microstrip line, say in 600 μm GaAs-substrate [J.11]. Figure (13.13d) shows another Silicon-metal-polyimide (SIMPOL) 3D MMIC technology on a 375-μm-thick CMOS-grade silicon wafer, by forming a cavity in the Si-substrate [J.30]. The cavity is metalized mostly on all sides and is filled in with the low-loss polyimide. The liquid polyimide (DuPont 2808) is used to deposit an 8.7 μm-thick polyimide layer over which microstrip lines, 8 μm wide and 0.85 μm thick, are created with Ti/Al metallization. The lines are covered with another polyimide layer. The structure is a buried microstrip with additional self-packaging features. This structure is extremely effective in reducing the noise crosstalk, above 100 dB at frequencies up to 50 GHz and beyond, as well as the overall size of MMIC chips.

Thus in 3D MMIC structures, the conductors are also deposited on several levels of the dielectric layers. The microstrip lines are buried both in a homogenous and in an inhomogeneous medium. The high-performance air-gap transmission lines have also been developed on the insulating substrate at 40 GHz. These lines have a small loss and insignificant dispersion [J.11, J.30]. Finally, Fig (13.13e) shows the 3D MMIC Wafer Level Chip Size Package (WLCSP) [J.31]. It is low-cost and surface-mount technology compatible with MMIC. The transistors, resistors, and lower electrodes of MIM capacitors are formed on the substrate and arranged in a matrix to construct a footprint of the master slice MMICs. It has narrow TFMS with small spacing between lines. The structure also has the coupling-free crossing at several polyimide layers and adjustable locations of common ground. The multilayer arrangement, with tiny vias, provides a large area for the passive microwave circuitry fabricated with TFMS lines. The 3D integration structure is suitable for both the semi-insulating GaAs and CMOS grade Si-substrate. Figure (13.13f) further demonstrates the complete WLCSP concept. The 3D MMIC, with common ground, is at the top surface and the solder balls are on to the I/O pads connecting to ground on wafer-like conventional SMT packages. The die assembly is finally mounted on a PCB. Thus, in this format of 3D MMIC, several circuit functions are effectively integrated into a small die.

Electrodes of Active Devices as Planar Transmission Lines

The MESFET is the most common microwave transistor. However, the HEMT (MODFET) and the HBT are important transistors for the mm-wave applications. These transistors involve multilayer structures with three electrodes, shown in Fig (13.14a) for a GaAs-based MESFET. Normally, these transistors are modeled by the equivalent circuit without consideration of the wave propagation on a CPW kind of the electrodes. However, it is important to consider wave propagation on the electrodes of a FET for its high-frequency applications [B.12, J.32]. In the active region, the traveling wave gains energy from its coupling with the Gunn domain on a GaAs sample that is induced by a proper biasing. The wave propagation is the slow-wave in nature. The multilayer structure also supports passive CPW. The full-wave analysis for the HEMT (MODFET) in the 1–10

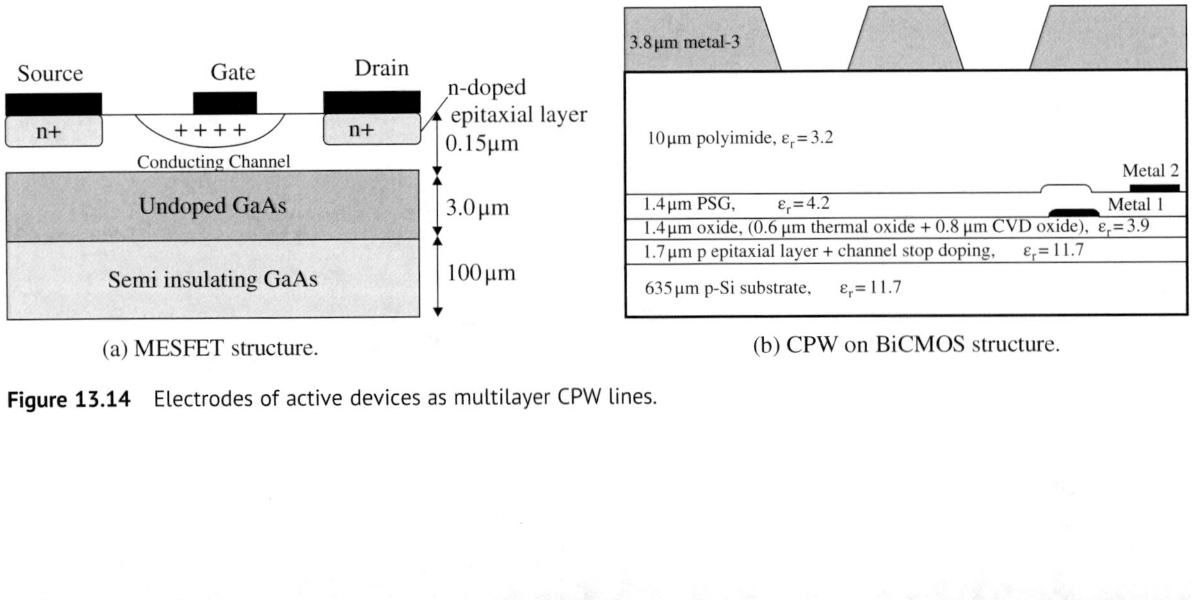

(a) MESFET structure.

(b) CPW on BiCMOS structure.

Figure 13.14 Electrodes of active devices as multilayer CPW lines.

GHz range has been used to control the attenuation and the slowing-factor [B.21]. Likewise, the BiCMOS structure can also have a CPW structure as shown in Fig (13.14b) [B.12]. This structure has been modeled in the 1–10 GHz range by the simple circuit model on the quasi-TEM basis. These structures also support the slow-wave.

13.3 Micromachined Transmission Line Technology

The micromachining is a chemical process and not a mechanical machining process. The microstrip and CPW structures are fabricated on the thin dielectric membrane, created by removing the silicon wafer material from the back of the wafer. The micromachined transmission lines are the silicon-based MEMS transmission lines. The micromachining fabrication method applies to any semiconductor substrate. The choice of silicon is due to its cheap and proven technology. It also facilitates the integration of the MEMS components with SiGe and CMOS circuits useful to modern wireless communications [J.33]. The high-resistivity silicon substrate has comparable electrical, thermal, and mechanical properties to that of a ceramic substrate. This is also suitable for 3D circuit integration that combines several circuit functions on a single substrate. However, system-level packaging became a limiting factor that led to the growth of the *multichip module (MCM)*. The MCM has high loss interconnects and passive components. The coupling between components and the package resonances became another degrading factor for the MCM [B.19]. The silicon MEMS-based transmission lines and components are compact and have high performance in the frequency range from 1 GHz to 1 THz (1000 GHz) region. These lines are the building blocks for the development of low-loss, low-dispersion filters, couplers, switches, resonators, voltage-controlled oscillators, and phase-shifters, etc. [B.22–B.24, J.34].

In this section, attention is focused on the planar transmission lines, primarily on the microstrip and the CPW structures, in the MEMS environment. The microstrip and CPW in the MEMS environment are practically fabricated on the air-substrate and the lines support almost pure TEM mode up to the 1000 GHz, almost without any dielectric loss and any noticeable dispersion. So the passive components developed around these lines have a high Q-factor. However, as there is no wavelength reduction factor of $\sqrt{\varepsilon_r}$, so the lines and components have a larger size. The MEMS technology is more suitable for the mm-wave range and beyond. A brief review of the MEMS fabrication process is presented. Next, a description of some MEMS-based transmission lines is also provided.

13.3.1 MEMS Fabrication Process

The MEMS transmission lines are fabricated on Si-substrate by two processes: *bulk micromachining* and *surface micromachining*. There are important variations in both fabrication processes [B.22–B.24].

Bulk Micromachining

The name "*bulk*" signifies that the micromachining is carried out on the bulk semiconductor substrate, i.e. on the bulk of a single crystal silicon wafer for the creation of a micromechanical structure. The name "*machining*" has nothing to do with the mechanical machining process, such as milling, drilling, turning, grinding, cutting, forging, molding. It is primarily a *chemical process*. The MEMS devices are fabricated using the IC fabrication process – photolithography and thin-film deposition techniques, such as the chemical vapor deposition (CVD) and the physical vapor deposition (PVD), the epitaxial growth and oxidation, doping and diffusion. The bulk micromachining creates the mechanical devices, such as a *cavity*, *membrane*, by etching deeply in the Si-substrate. In the etching process, the Si wafer material is selectively removed. Primarily, there are two kinds of etching process – the *wet etching* and the *dry etching*. The *wet* and *dry* terms are used for the etchants [J.35].

Wet Etching

Figure (13.15) shows both the wet and dry etching process of Si-substrate. In the wet etching process, the liquid etchants, like a solution of potassium hydroxide (KOH), ethylenediamine pyrocatechol (EDP), tetramethylammonium hydroxide (TMAH) and hydrazine water are used. The etching rate is influenced by both the temperature and concentrations of etchant. For the diffusion-determined etching, the agitation produces uniform well-shaped etching. The etching could be *anisotropic* or *isotropic*. The anisotropic etchant etches the silicon substrate at different rates in different directions of the crystallographic planes. In certain crystallographic planes, the etchant changes slowly. These are called the *stop-planes*. Normally, the wet etching is carried out from the backside of a wafer. However, the front-side is also used for the fabrication of CPW lines. In the etching process, selectivity and directionality of the etched area are important. The etching rate is controlled and

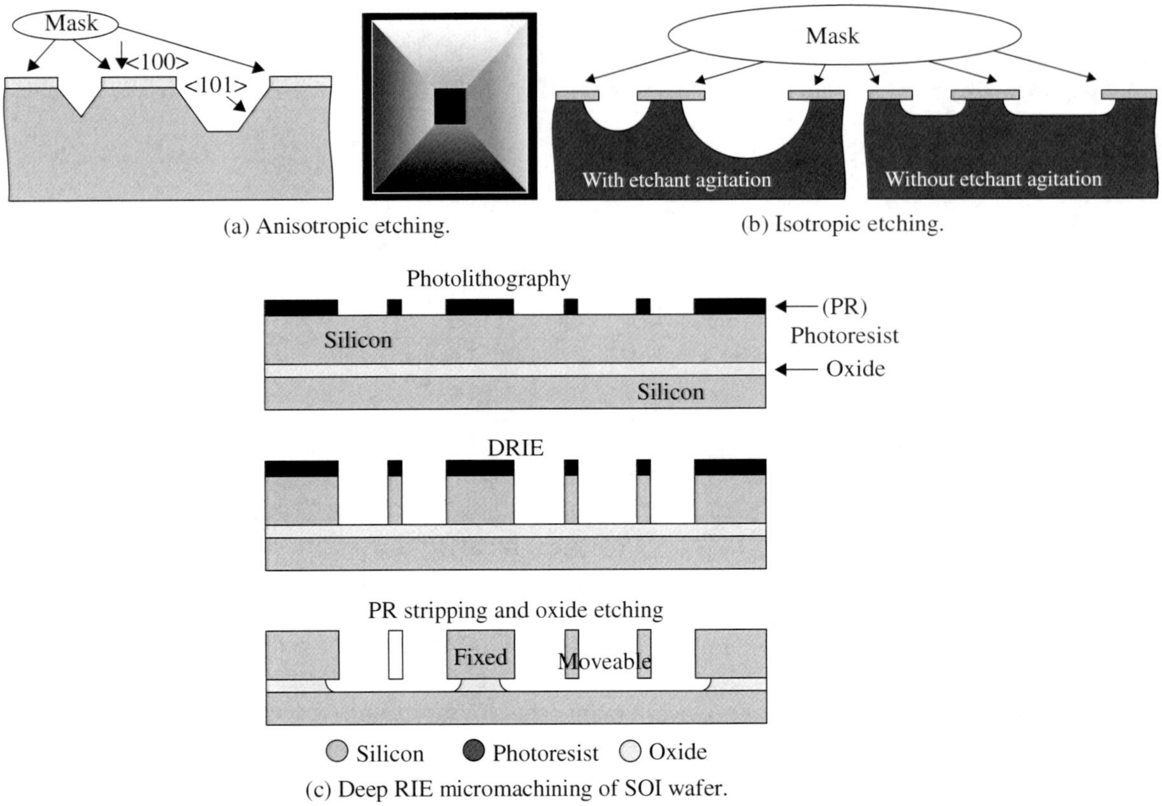

Figure 13.15 Wet and dry etching of Si-substrate.

can be stopped by doping the silicon and by creating the PN junction at which etching, the wet or dry, slows down or stops. It is called the *etch-stop*.

The *silicon membranes* are created by the etch-stop that is developed by a thin heavily boron doped silicon layer. The layer is either epitaxially grown or formed by the diffusion/implantation of boron in the silicon substrate. The *anisotropic etching* creates the V-grooves, pyramids, and channels in the silicon substrate, as shown in Fig (13.15a). The alkali etchants dissolve the unmasked Si-substrate. Sometimes the mask itself gets dissolved creating a problem of getting the correct pattern. A proper choice of etchants is important. The *isotropic etching* removes the silicon from all directions equally and creates a rounded depression-like hemisphere and cylinder shown in Fig (13.15b).

Dry Etching

The *plasma etching* is a dry etching technique that is normally applied at the front-side of the silicon wafer. The *reactive ion etching* (RIE) is another popular dry etching process. The very deep silicon microstructures are fabricated using the *deep RIE* (DRIE) process. The typical RIE gases for Si are CF_4, SF_6, and $BC_{12} + C_{12}$. The DRIE process provides the straight-walled structures as shown in Fig (13.15c). The photoresist (PR) is coated on the SOI wafer, and the pattern is created on the photoresist by the photolithographic process. The DRIE process removes the silicon by etching it and creates straight-walled structures. Further, the photoresist and oxide layers are removed leaving the MEMS structure.

Surface-Micromachining

The surface-micromachining for the creation of the MEMS structure is more suitable for the IC process. It builds the structures on the substrates by creating a layer of the structures. The surface micromachining involves three layers – *substrate, sacrificial layer,* and *structural layer* [B.22, B.25].

The MEMS structures are created with the help of the *structural layer* on a silicon substrate. The *sacrificial layer* is an intermediate layer between the substrate and the structural layer. The sacrificial layer is removed by using any suitable wet etchant to release the structural layer. The substrate finally supports the MEMS structure without the presence of the sacrificial layer that has been "*sacrificed*" during the MEMS fabrication process. Several kinds of structural materials, sacrificial

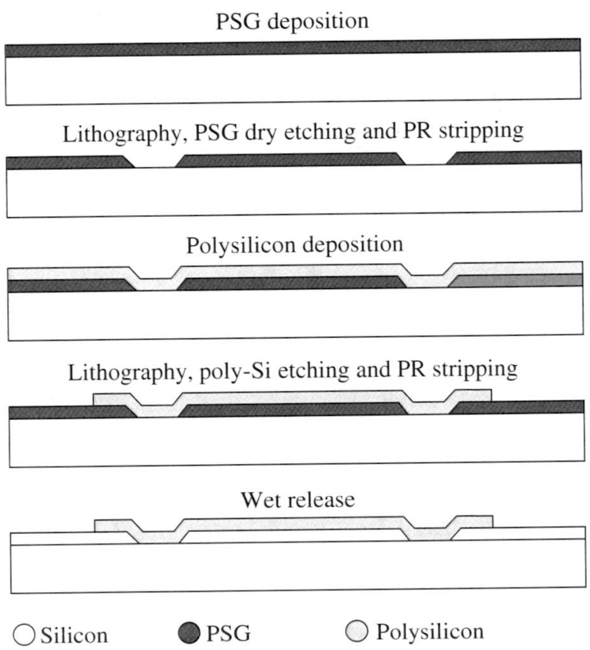

Figure 13.16 Basic steps for creation of microbridge by the surface micromachining.

materials, and the etchants are in use. The aluminum MEMS structure is formed on the silicon substrate. To achieve it, the sacrificial layer of photoresist (PR) is deposited on the silicon substrate, and over it, 1–5 μm thick structural layer of aluminum is deposited. The sacrificial layer is removed by acetone and the silicon substrate finally supports the aluminum structure.

Figure (13.16) shows an example of creating the polysilicon microbridge on a silicon substrate. On the silicon substrate, a sacrificial layer of phosphor silicate glass (PSG) is deposited by the LPCVD process. Next, the photolithography, dry etching of PSG, and removal of photoresist creates a pattern on the substrate. In the third step, the polysilicon layer is deposited on the PSG sacrificial layer. In the fourth step, the photolithography is used to remove partly the polysilicon to create the required shape of microbridge. Finally, the PSG sacrificial layer is removed by using the HF solution and the microbridge on the silicon substrate is created.

The silicon micromachining creates 2D microstructures. However, the wafer bonding with bulk micromachining is used to create the 3D MEMS structures like the cavity and the waveguide. LIGA is another important MEMS process. It is a German acronym for X-ray lithography (Lithographic), electroforming (Galvanoforming), and molding (Abforming). The LIGA creates very high aspect ratio (height to width ratio) 3D microstructures by creating the molds. However, LIGA is a complex and expensive process.

13.3.2 MEMS Transmission Line Structures

The MEMS transmission lines, suitable for the microwave, mm-wave and terahertz technologies, have structural shapes of microstrip, CPW, coplanar strip, slot line, strip line (rectangular coaxial line) and metallic waveguide. These conventional structures are modified in shapes during the MEMS fabrication process. The MEMS planar transmission lines are based on the following technology:

- *Membrane technology*
- *Microshield technology*
- *V-groove CMOS technology*
- *LIGA technology*
- *MEMS waveguide technology*

A brief review is presented below for the fabrication process of the transmission line structures in these technologies. All these structures are created on a composite layer of substrates, with or without a presence of the bottom and top shielding.

Membrane Technology

Figure (13.17) shows several types of microstrip lines fabricated on a low dielectric constant thin membrane. To form the membrane over the silicon substrate, the $SiO/Si_3N_4/SiO_2$ (5000 Å/3000 Å/4000 Å) film layers of dielectric are deposited by thermal growth, i.e. LPCVD and LCCVD processes [B.21, J.36–J.40]. Sometimes membrane thickness is 1.4 μm. So, a membrane is fabricated out of the thin composite dielectric layers. On the top side, the microstrip line is constructed first by depositing a seed layer of gold via evaporation, and then it is made thicker by depositing additional gold through electroplating. The silicon is etched out from the backside of the silicon substrate by using the bulk micromachining technique with the help of the etchant either KOH or EDP. A 500 μm wide microstrip on 350 μm thick substrate supports the TEM mode from low frequency to 320 GHz, whereas a microstrip line of identical dimension on the silicon substrate has a higher-order mode at 62 GHz. The membrane-based microstrip supports the operating frequency up to 1 THz [J.36, J.37].

Figure (13.17a) shows the MEMS-created-suspended microstrip line on the Si-membrane cut out of a Si-substrate that is bonded to an additional metallic ground plane. The membrane-supported, suspended microstrip line is a high impedance line. Its dielectric loss is very small even at several hundreds GHz. Figure (13.17b) further shows that the ground plane could be deposited on another Si-substrate that is bonded to the microstrip supporting Si-based MEMS. Sometimes this scheme to

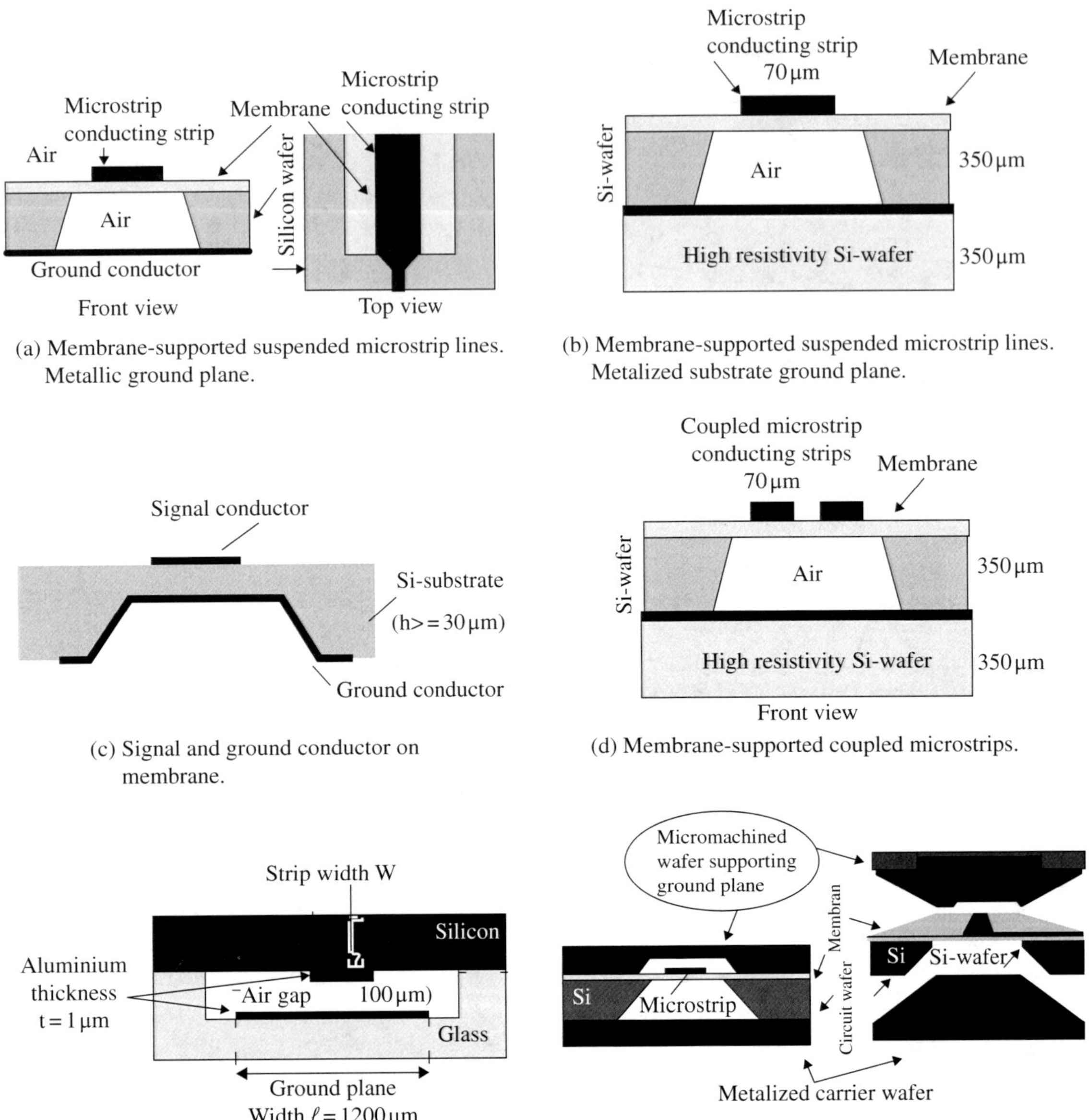

(a) Membrane-supported suspended microstrip lines. Metallic ground plane.

(b) Membrane-supported suspended microstrip lines. Metalized substrate ground plane.

(c) Signal and ground conductor on membrane.

(d) Membrane-supported coupled microstrips.

(e) Inverted microstrip line.

(f) Suspended membrane-supported microstrip with a top shield.

Figure 13.17 Some membrane supported microstrip type of MEMS lines.

provide the ground plane is more suitable for the on-chip integration of the MEMS structure.

Figure (13.17c) shows another membrane-supported microstrip. The ground plane is fabricated on another side of the membrane. The microstrip on the membrane is a low impedance line. Figure (13.17d) shows the *coplanar strip lines* fabricated on the membrane. By adding the ground plane, it has been converted to a *suspended coupled MEMS microstrip lines* on the membrane. Figure (13.17e) shows the fabrication of an *inverted microstrip* on the glass substrate. Of course, the signal strip conductor is fabricated on the silicon wafer itself. The suspended and inverted microstrip lines in the MEMS format can also be provided with the top shield as shown in Fig (13.17f) [J.39].

Figure (13.18) shows some more variants of MEMS strip lines. The line structure in Fig (13.18a) is on low-resistivity Si-substrate. However, a passivation layer of SiO_2 improves, the loss performance of the line. The Spin-On-Glass (SOG) layer is used for another strip. The thickness of the SOG layer is in a range of 20–100 μm. Figure (13.18b) shows the same line

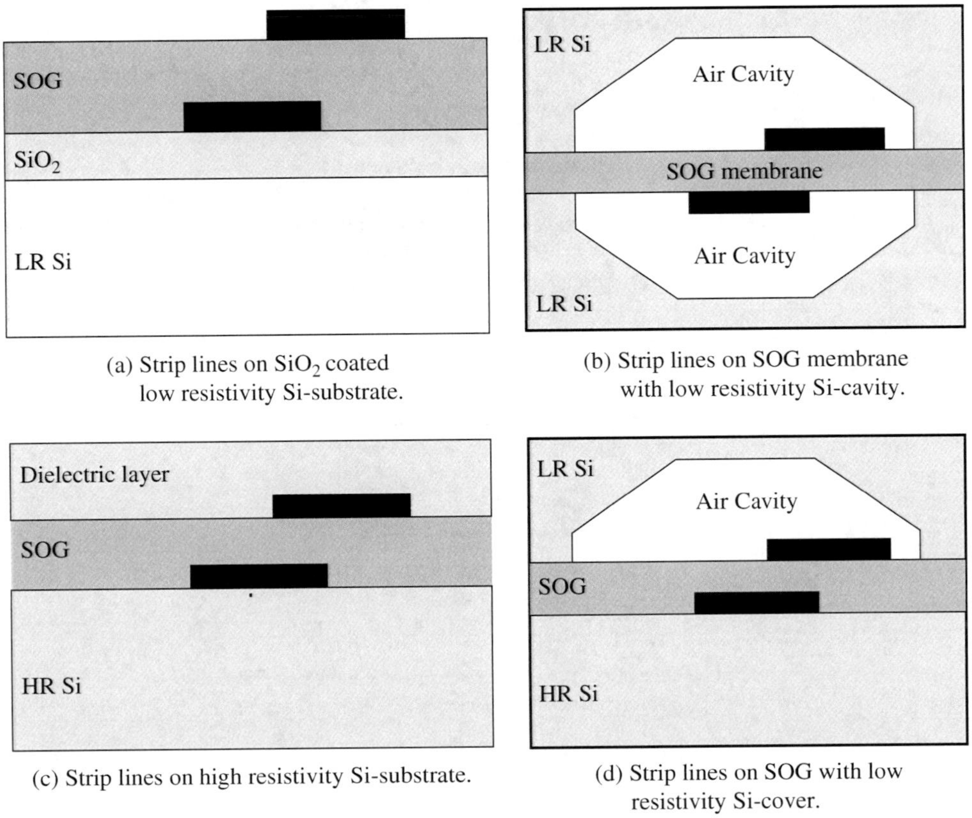

Figure 13.18 Some variants of MEMS strip lines on low and high resistivity substrates.

structure on the SOG membrane that is supported inside a cavity of the low-resistivity Si-substrate.

Figure (13.18c) shows the same line structure, where the stripline is directly printed on the high-resistivity Si-substrate. The second line is printed on the SOG layer that is covered with a dielectric layer. In Fig (13.18d), the dielectric cover is replaced by the low-resistivity Si-substrate cavity. The metallic ground planes can be added in both the silicon cavities of Fig (13.18b, d) such that the structure work as an off-set broadside coupled MEMS microstrip lines.

The microstrip line is also constructed just on the air-substrate by removing even the thin dielectric membrane. Figure (13.19a) shows the metallic membrane-based microstrip line on the air-substrate. Of course, dielectric pillars are created on the semi-insulating GaAs-substrate to support the strip conductor. The steps involved in the fabrication process are shown in Fig (13.19b) [J.41].

The CPW has also been constructed in the MEMS configuration. Figure (13.20a) shows a finite ground CPW (FGC) on a high-resistivity Si-substrate with a SiO_2 passivation layer. Figure (13.20b) further shows that the MEMS technology itself can provide the Si-packaging with a cavity to the CPW. Figure (13.20c) shows that after etching out the Si-substrate, the SiO_2 membrane supported FGC, i.e. CPW with a cavity backing, is obtained [J.42]. Figure (13.20d) further shows that the structure can be backed by the metalized cavity on another substrate [J.40].

Microshield Technology

In the membrane-based MEMS transmission lines, discussed above, the *ground plane* is normally created on another substrate that is bonded to the dielectric membrane and *signal conductor* supporting substrate. However, the ground plane can also be fabricated inside the cavity of a substrate. Such a cavity-backed line structure is called the *micro-shield* structure that is available for both the microstrip and a CPW. The micro-shielded microstrip structures are shown in Fig (13.21). Figure (13.21a) shows the *membrane-supported, suspended microstrip* line. The metalized cavity in the substrate, Si/GaAs, provides the ground plane [J.43]. Figure (13.21b) shows the *substrate-supported, inverted microstrip* line. Again, the metalized cavity in the

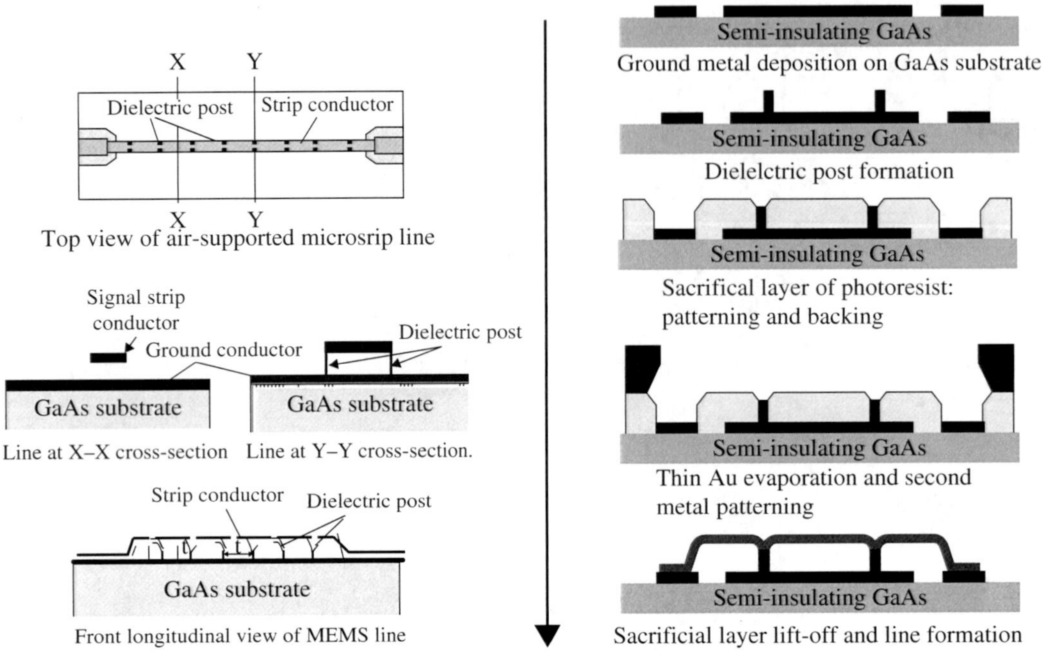

(a) MEMS microstrip at air substrate fabricated on GaAs-substrate.

(b) Fabrication process of air substrate microstrip on semi-insulating GaAs-substrate.

Figure 13.19 Microstrip on air-substrate. *Source:* From Lee et al. [J.41]. © 2004 John Wiley & Sons.

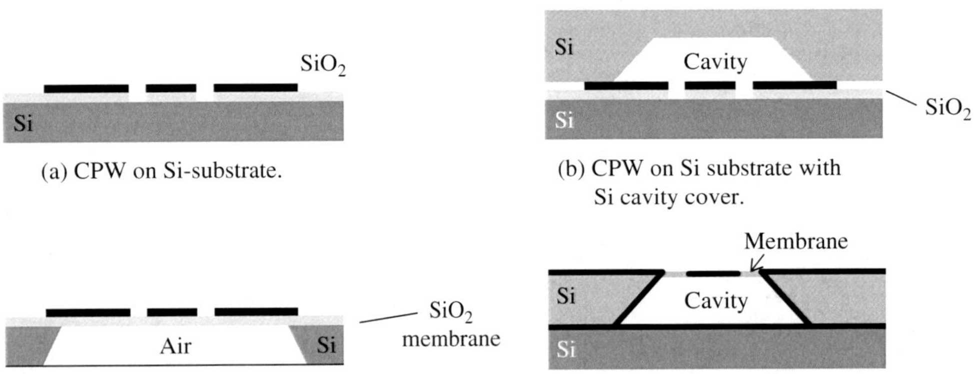

(a) CPW on Si-substrate.

(b) CPW on Si substrate with Si cavity cover.

(c) CPW on SiO_2 membrane.

(d) Cavity-backed CPW.

Figure 13.20 MEMS CPW on high resistivity Si-substrate.

substrate provides the ground plane. The metalized cavity in the substrate also minimizes the signal crosstalk between the adjacent lines, as it isolates the substrate from the line by controlling the radiation to the substrate. The ground plane is available at the top, so there is no need for via hole in the micro-shield microstrip lines. Finally, Fig (13.21c) shows the micro-shielded microstrip line with both the upper and lower ground cavities [J.42]. It is a three-wafer structure. The microstrip is supported on a membrane.

The CPW structure has also been fabricated in the *micro-shield format*. Figure (13.22) shows the micro-shielded cavity-backed CPW structures [J.38, J.45–J.47]. Figure (13.22a) shows a metalized cavity in the upper wafer. The ground plane is constructed on another wafer that is bonded to the upper wafer. The coplanar ground conductors on the membrane are connected to the bottom ground plane through the walls of the metalized cavity. Thus, the present CPW structure avoids the air-bridge. The structure looks like a

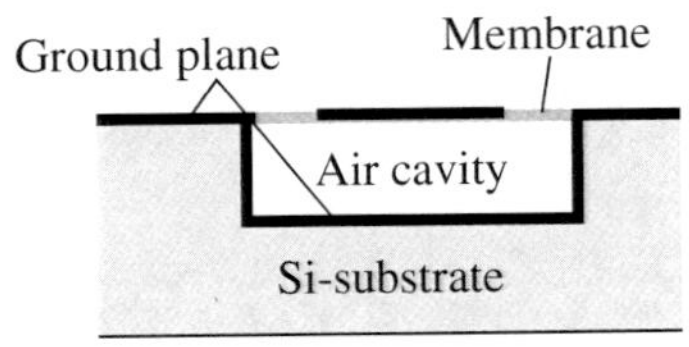

(a) Cavity-backed membrane-based suspended microstrip line.

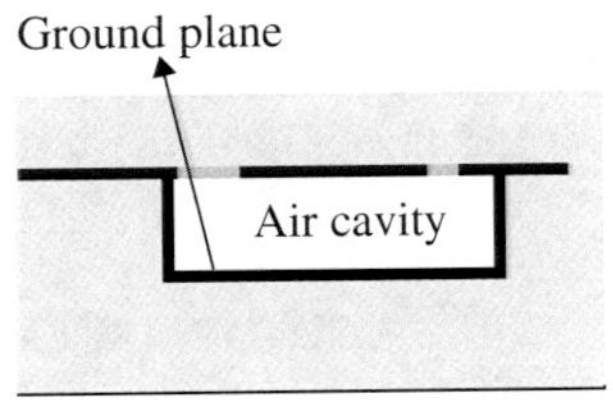

(b) Cavity-backed inverted microstrip line.

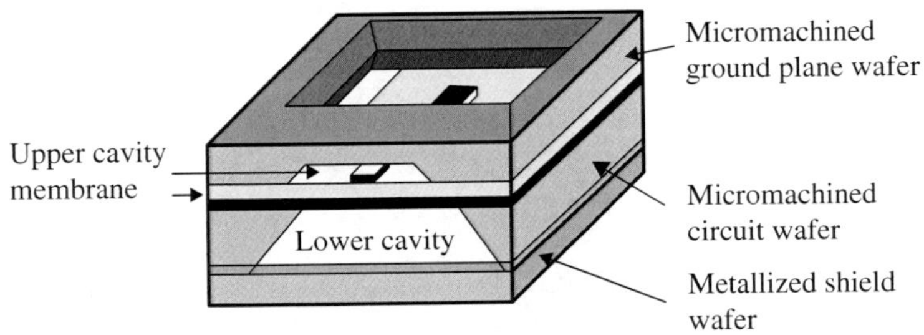

(c) Membrane supported suspended microstrip line with both the upper and lower metalized Cavities[J.42]. (Reproduce from [J.42], ©1998 IEEE).

Figure 13.21 MEMS microshielded microstrip line. Source: From Herrick et al. [J.42]. © 1998 IEEE.

conductor-backed suspended CPW. The self-packaging is added to the upper cavity, as shown in Fig (13.22b). In the case of Fig (13.22b), the lower cavity is filled in with the substrate material. The white strips are three coplanar conductors of a CPW.

V-Groove CMOS Technology

The above-discussed fabrication of the MEMS transmission lines is mostly for the semi-insulating GaAs-substrate and the high-insulating Si-substrate. It is achieved by several photolithographic steps, wafer bonding, and backside etching. This process is not compatible with the CMOS process. Another bulk micromachining process from the front-side of the CPW line structure is suggested by Milanovic et al. [J.44]. It is suitable for the CMOS process. Two cases are considered below.

On the CMOS-grade Si-substrate, the SiO_2 layer is deposited that acts as the *supporting membrane* for the CPW line structure. For front etching, two openings in SiO_2 are provided. In the *first case*, shown in Fig (13.23a, b) for the front-view and top view, the openings are at the end of the finite ground CPW (FGC). In the *second case*, shown in Fig (13.23c, d), two openings are in between the ground conductor and central strip conductor. The CPW structure is patterned over the SiO_2 membrane and the Si-substrate etching is carried out from the

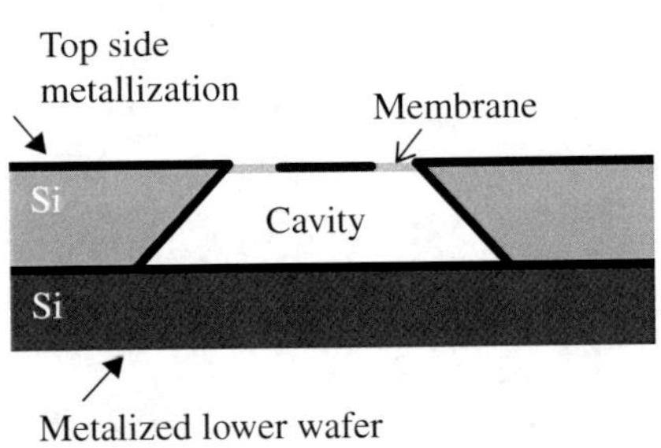

(a) Cavity-backed MEMS CPW.

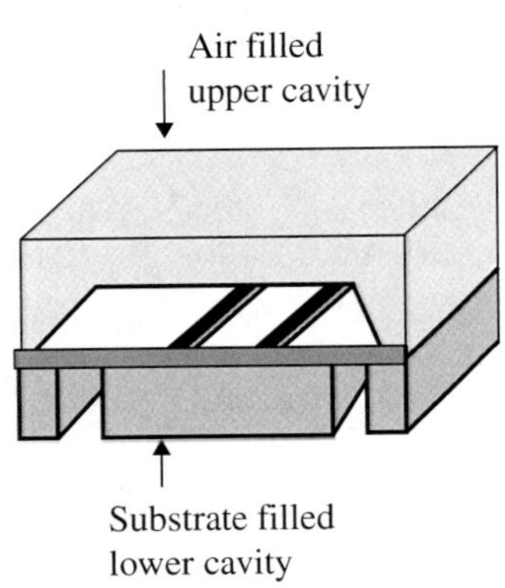

(b) Self-packaged MEMS CPW.

Figure 13.22 Micro-shielded MEMS CPW. *Source:* Based on Milanovic et al. [J.44].

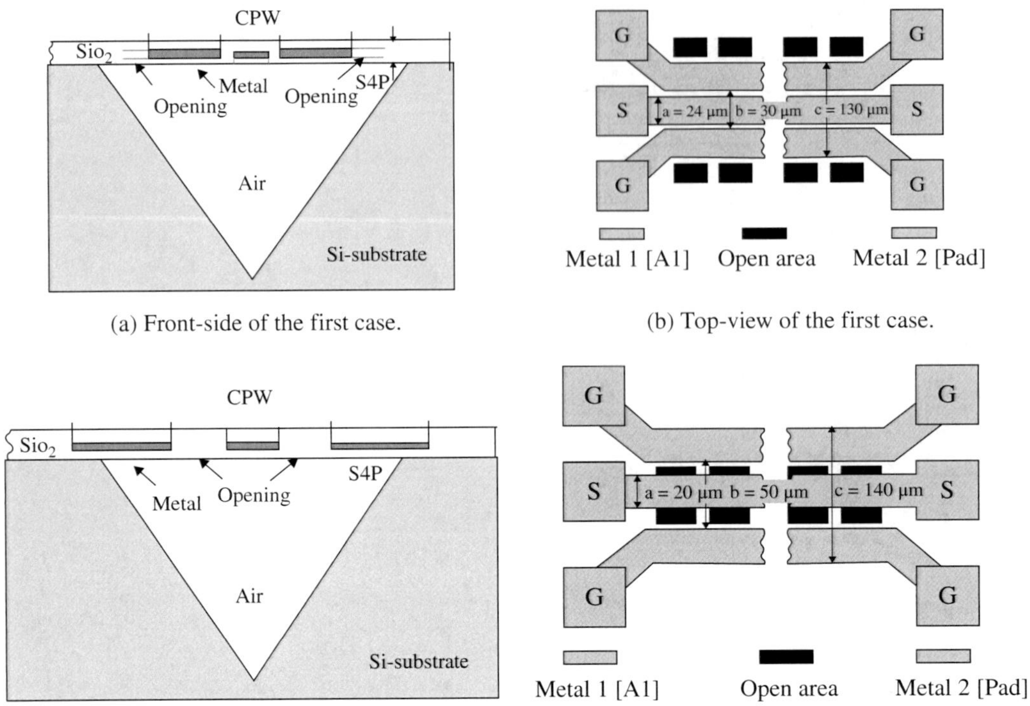

(a) Front-side of the first case.

(b) Top-view of the first case.

(c): Front-side of the second case.

(d) Top-view of the first case.

Figure 13.23 MEMS CPW in CMOS technology. *Source:* Based on Milanovic et al. [J.44]. © 1997 IEEE.

front openings. The bulk micromachining creates the V-groove in the substrate due to the anisotropic etching. This structure is used up to 40 GHz. The V-groove microstrip line can be extended to the coupled microstrip structures also [J.48].

LIGA Technology

This is a 3D molding process to create the micromechanical microwave and mm-wave transmission lines, filters, etc. This process provides the large aspect-ratio metal structure of 1 μm–1 mm thickness that is needed for the high-power MMIC applications.

The microstrip and CPW structures fabricated by the LIGA process are shown in Fig (13.24a, b), respectively [J.49]. Figure (13.24c) further shows the *coupled microstrip lines* fabricated on a fused quartz substrate by the LIGA process. The LIGA is a German acronym for the *Lithographie, Galvanoformung, Abformung* (Lithography, Electroplating, and Molding) fabrication technology. The LIGA process controls the thickness of a conductor that is an additional design parameter. Figure (13.25) shows the LIGA fabrication process for the coupled microstrip. The process is started with the metalized quartz substrate of 120 μm thickness on which 300 μm sheet resist is deposited. The sheet resist is exposed to the X-ray through the Si_3N_4/SiO_2 mask to create the coupled microstrip that is electroplated with 200 μm Ni or Cu.

Waveguide Technology

The MEMS technology has been also used to develop THz range metallic waveguide structures. Its process is shown in Fig (13.26). A Si-substrate is a double-side coated with the silicon nitride layer of 1000 Å. At the step-I, an opening is created by removing the silicon nitride through which the etchant KOH is applied to create half of the waveguide. At step-II, a channel is etched out in the Si-substrate. At the step-III, the waveguide-containing wafer is bonded to the unetched wafer. At the step-IV, the channel is coated with the chrome (250 Å) and gold (5000 Å) metal composite to create 1/2 waveguide. Finally, at the step-V, two such channel sections are combined to form a rectangular waveguide [J.50] in MEMS technology.

13.4 Elements of LTCC

The substrate technology could be categorized into three groups – *laminates based PCB technology, ceramic-based technology,* and *semiconductor-based technology.* The PCB and semiconductor-based technologies are discussed in the previous sections. The ceramic-based interconnect technology has the following forms – a thick

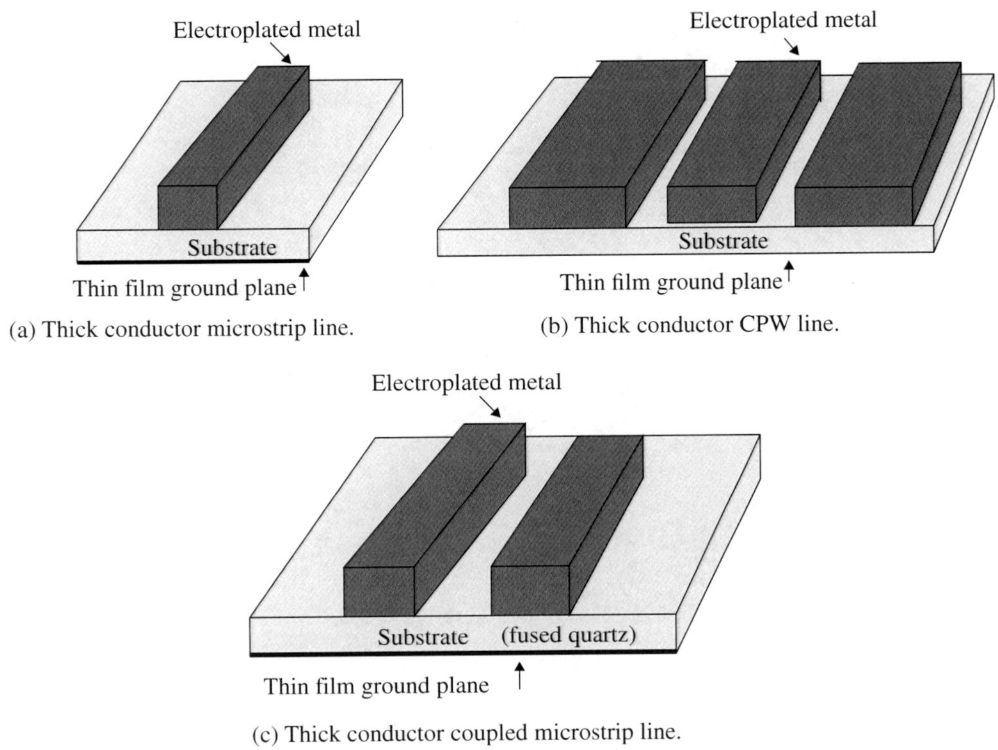

Figure 13.24 LIGA planar transmission lines.

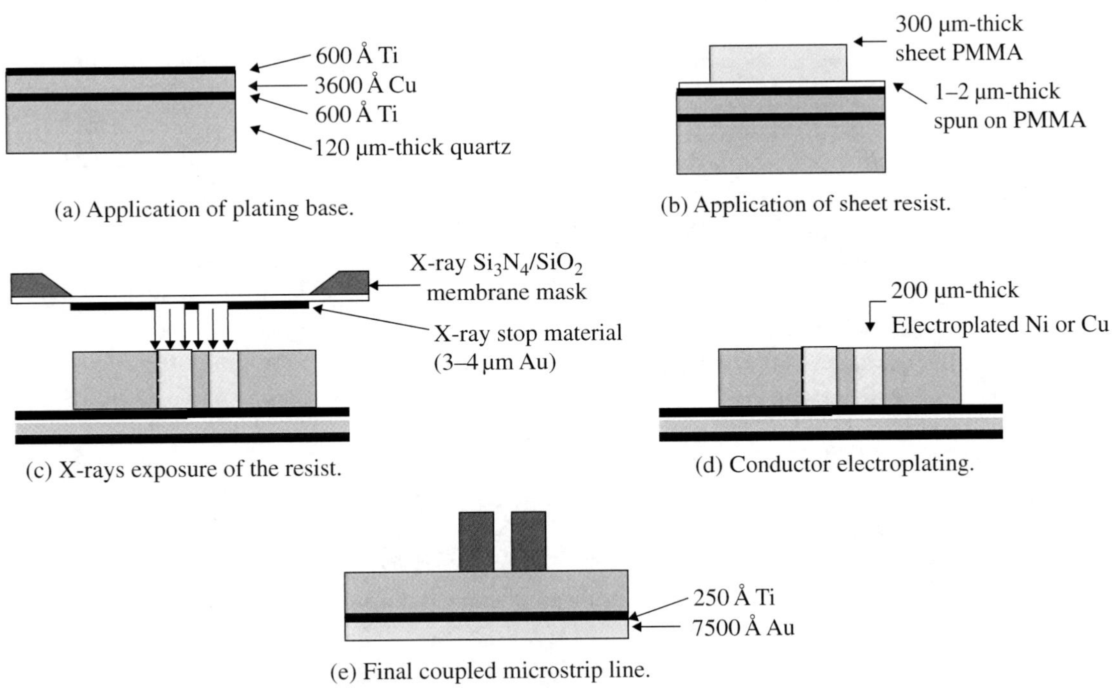

Figure 13.25 LIGA fabrication process.

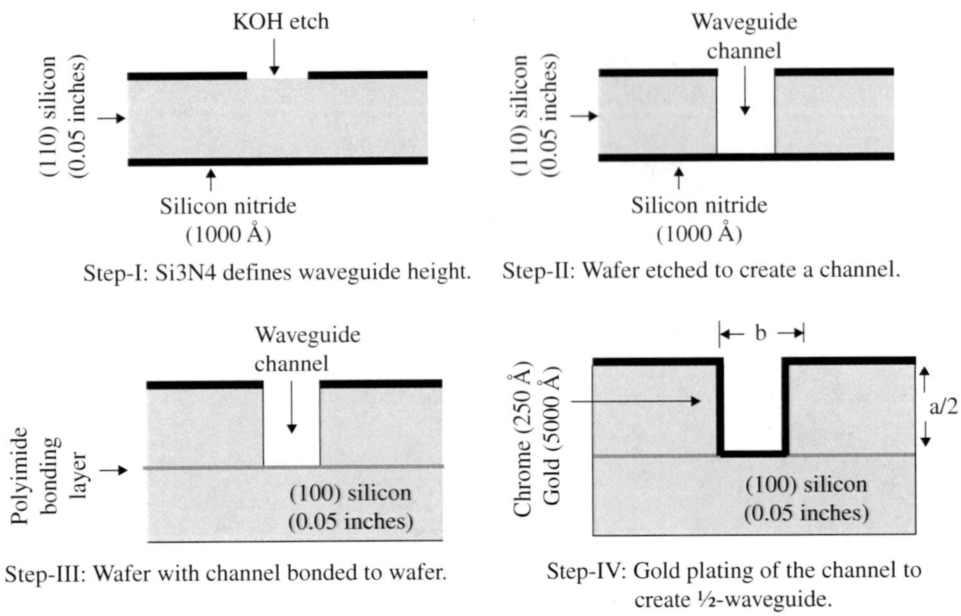

Figure 13.26 Fabrication process of MEMS waveguide.

film on the ceramic substrate, copper plating on the ceramic substrate, thin film on the ceramic substrate, *high-temperature cofired ceramic*, i.e. *HTCC technology* and the *low-temperature cofired ceramic*, i.e. *LTCC technology*. The thick and thin-film technologies have already been considered for the fabrication of transmission lines and the related structures. Primarily, these are single-layer technologies and are adapted to the multilayer structures. However, each ceramic layer has to be processed separately, and they are bonded together. This circuit configuration is not very compact and also not cost-effective. Thus, there is a need for the *vertical integration* of passive components and interconnects, as well as an integration of the active components in a modular form. The multilayer ceramic technology offers a solution to high-density circuit integration. In the case of the HTCC technology, the embedded passive components and interconnects are fabricated on several layers of the ceramic substrates. Finally, the assembled layered structure, i.e. a bundle of all ceramic layers is burnt (*fired*) at a temperature over 1500°C in an oven/furnace. However, the HTCC process requires high-temperature melting point conductors such as tungsten and other refractory metals. These metals are poor conductors and are also expensive. Thus, the HTCC packaging technology is not a popular one.

The *ceramic tape*-based LTCC has become a more popular technology, as the firing temperature of the LTCC is in the range 800–900°C [J.51]. Gold and silver are used as conductors in the LTCC process. These conductors have good electrical conductivity. The LTCC offers significant advantages over other technologies for the MCM used in the RF and microwave applications. It also accommodates both the digital and analog circuits in the module. The LTCC has become an important RF technology that accommodates antenna, filters, couplers, and other passive components along with active devices mounted on the surface and in the cavities. A typical LTCC multilayer structure, with embedded passive components, such as capacitor and inductor, and strip lines, and also surface-mounted SMD component, resistor, and RFIC is shown in Fig (13.27a) [J.52]. Figure (13.27b) shows a stack of multilayer dielectric tapes, called the *green-tape*, on which these components are fabricated/mounted.

This section discusses the LTCC fabrication process and its layered medium environment for the development of the planar transmission lines, including waveguide. The detailed description of the LTCC technology could be read from the books and reports [B.26, B.27, J.52].

13.4.1 LTCC Materials and Process

The LTCC technology is developed around the thin ceramic dielectric tapes ($0.005'' - 0.008''$). These are cut into equal pieces and stacked together to form a multilayer structure, as shown in Fig (13.27b). It shows seven dielectric layers with eight metallic surfaces for the construction of components and line structures. The

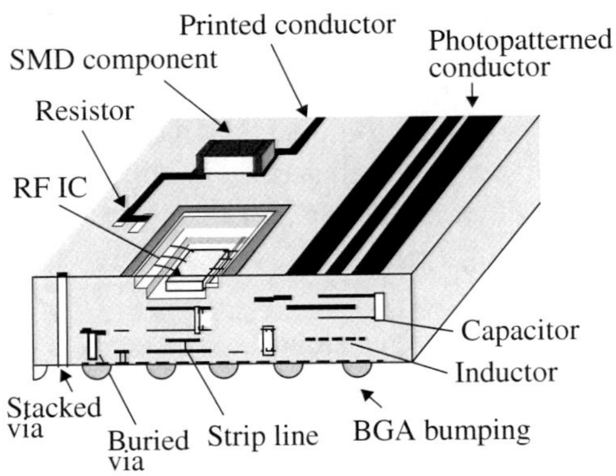

(a) A typical LTCC structure.

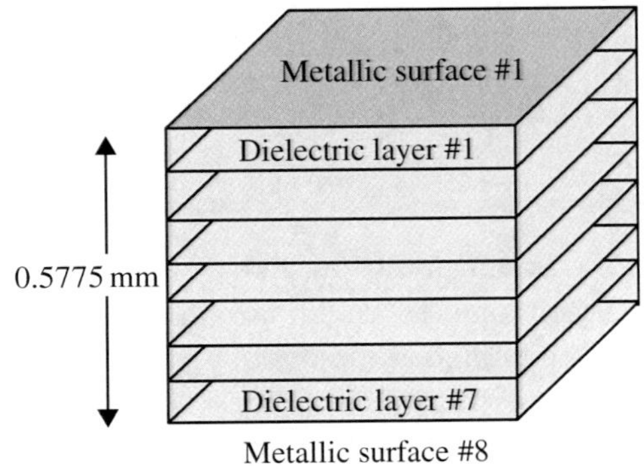

(b) A stacked layer of green dielectric tape.

Figure 13.27 A typical LTCC structure with embedded and surface mounted components.

number of layers could be more, say up to 70–80. The last layer is the metal layer that is fabricated by thick film technology. The unfired thickness of the tape is in the range of 50 –250 μm. There is about 10% shrinkage after firing. The typical dielectric constant of tape is about 7.4 and loss-tangent 0.002. More information can be obtained from the Dupont tape selection guide.

The photolithography is adapted to print the conductors for constructing the line structures, electrodes, etc. The dielectric tapes, i.e. the LTCC tapes, are made of glass-ceramic composite materials – alumina ceramic filler 40%, glass (silica) 45%, and organic binder, i.e. plasticizer and dispersant (deflocculant) 15%. The tape casting process, i.e. the manufacturing of the LTCC tape is adapted from the plastic and paper industries. The base LTCC tape is called "*the green-tape*," as it is soft before baking. It is formed by spreading slurry consisting of ceramic powders mixed in an organic (or water) vehicle, together with appropriate binder and dispersant. The flowchart of the tape casting process is shown in Fig (13.28a). The binder, dispersant (deflocculant), plasticizer, and organic vehicle determine the property of a slurry. The dispersant breaks apart the agglomerates, into individual particles. The binder creates the polymer matrix that holds the chemicals for further processing. The binder determines the mechanical property of dielectric tape. Finally, the plasticizer makes the green-tape bendable (soft) without breaking during the process of punching, cutting, rolling, and laminating. The solvents should have good solubility for the binder and plasticizer. These should also have a moderate evaporation rate after the tape casting [J.53–J.55].

The homogenous slurry is poured on a Mylar (ceramic) tape carrier that is passed under the doctor blade shown in Fig (13.28b) to produce uniform tapes of desired thickness. After drying, the tapes become the ceramic-filled "green-tapes." It is kept in rolls. The "green-tape" is the trade name of Dupont for the dielectric tapes. It is also manufactured by other manufacturers. The tape characteristics from different suppliers

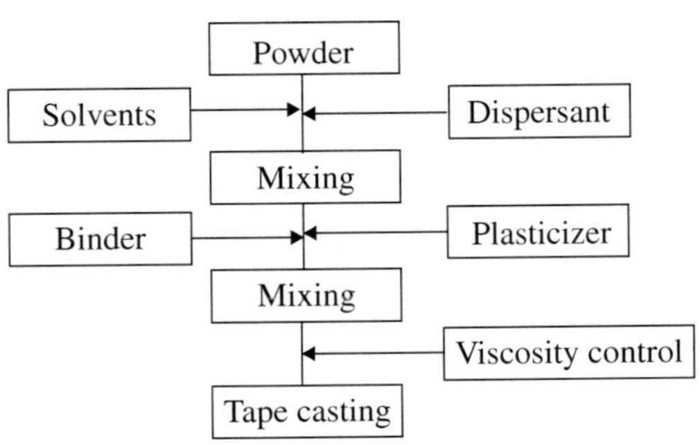

(a) Flow-diagram of LTCC green-tape casting.

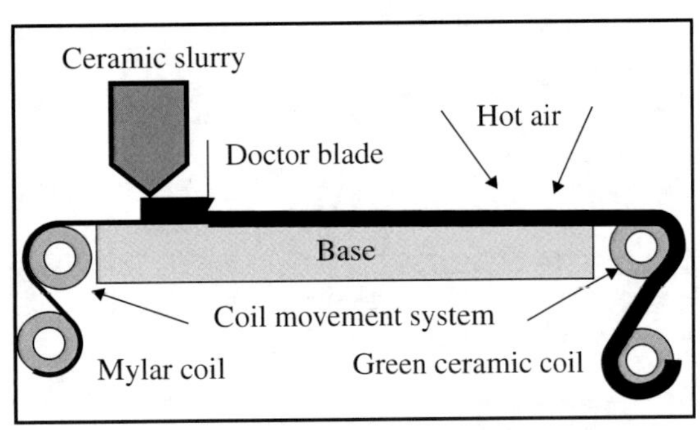

(b) Green-tape casing on the conveyor belt.

Figure 13.28 Casting process of green-tape.

are different. The conductor strips are made of the paste of gold, silver, silver–copper, Pt–Au or Pd–Au, etc. The metallic paste is also used to fill in via holes.

13.4.2 LTCC Circuit Fabrication

The passive components such as resistor, inductor, and capacitor in the LTCC fabrication process could be embedded components, i.e. they are located at the inner layers of the LTCC dielectric green-tape. However, the discrete and the surface-mount components are located at the uppermost layers and attached to the LTCC at the end of the cofiring process known as the *postcofiring* process.

The microwave passive components designed around the planar transmission lines could be located partially on more than one dielectric layer. The sectionalized vertical arrangement of circuits makes the arrangement very compact. The interconnections between parts of the transmission lines and embedded components and ground planes are achieved by vias. The vias are holes in the dielectric layers. The vias are also used to bring terminals of a circuit to the surface of an LTCC. The arrays of vias are used to create the imperfect walls of a waveguide and also a cavity embedded in the dielectric layers. It is similar to the *substrate integrated waveguide (SIW)* discussed in section (7.8) of chapter 7. The ground plane could be located at the end of the dielectric stack, called LTCC metal (LTCC-M). The ground planes, or grounded conductors, can exist at several layers, and they are interconnected by using vias.

Figure (13.29) shows a typical stack of six dielectric layers with several arrangements of vias, buried resistors, capacitors, and conductors. It also shows the postfired (attached after the cofired process) resistor. The vias are terminated on a cover pad and are filled in with metallic paste before the cofiring process. The important steps of the LTCC manufacturing process are shown in Fig (13.30) [J.54–J.56]. The process assumes that the circuit design has already been completed using the design tools like a Circuit Simulator that contains libraries of LTCC embedded components, via and many transmission lines (microstrip, stripline, and CPW) circuits. The complete LTCC circuit can also be simulated on the 3D EM-simulators.

The green dielectric tape roll (i.e. unfired dielectric tape) is cut into pieces and preconditioned in the furnace. The *registration holes,* used for the alignment of all layers in the vertical stacking, are punched or burnt. The Mylar-tape is removed. The *blanking die* is used to create the orientation marks and the final working dimension of the green-tape. Next, the vias are made by the laser tools, CNC or punching tool. The vias are filled with metal paste, using the thick film screen printer with a stencil metal mask.

The *screen printing* process of via filling is shown in Fig (13.31a). A CNC machine is used to create a cavity in the dielectric layer, or to cut a partial depth in the dielectric layer. Figure (13.31b) shows the process. It is filled with a high permittivity dielectric paste for fabricating small size-embedded capacitance. The next stage is the *printing of the conductors* for fabricating the transmission line-based circuits and interconnects. It is shown in Fig (13.30). The thick film screen printing process is used at the surface of the green-tape. The circuit and interconnect printing is partially done on several dielectric layers. The conductors of different layers forming the circuit are connected by the metal-filled vias.

At this stage, the circuit, with embedded components in several dielectric layers of the green-tapes, is ready for cofiring. A stack is made of the green-tape containing

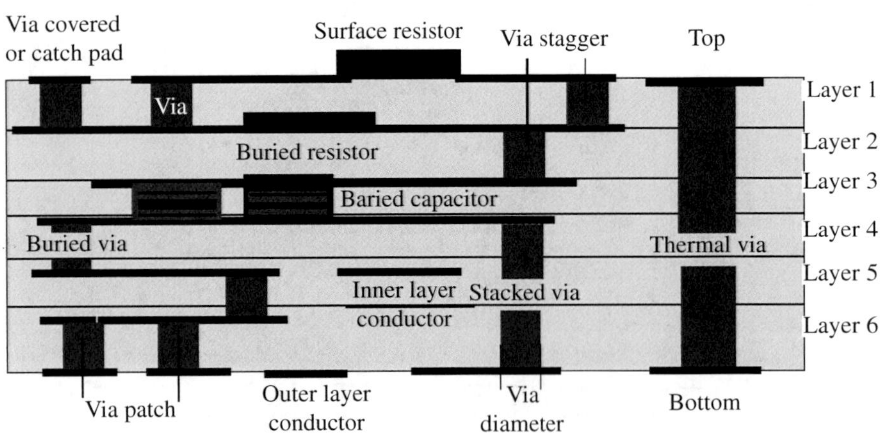

Figure 13.29 A typical arrangement of vias, passive components, and conductors in the dielectric stack.

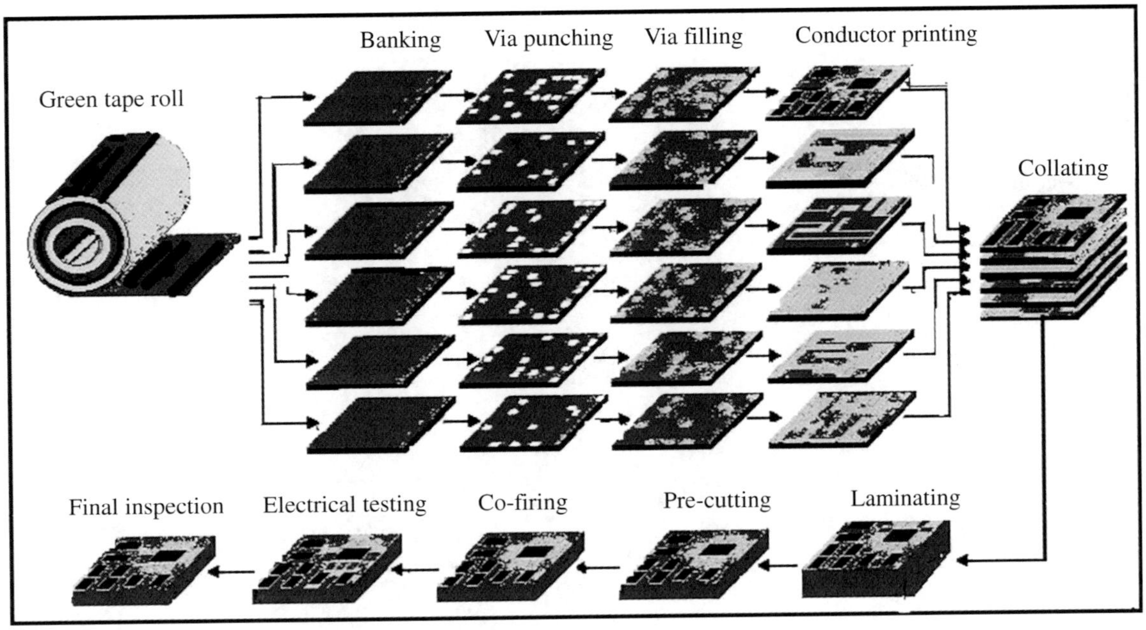

Figure 13.30 Steps of LTCC circuit fabrication process.

partial circuits before the cofiring. It is known as the *collating shown* in Figs (13.30) and (13.31c). The green-tape layers collated and stacked with special tools are fixed with collating pins through the registration holes on the dielectric layers. The whole stack of the green-tapes is laminated by a thermal process at 100°C and 3000 PSI pressure. It is known as the *thermo-compression* process that sticks all the dielectric tape layers to form a composite, i.e. the monolithic single layer. Figure (13.31d) shows the process. The laminated layers are cut to the proper size. Finally, the laminated layers are burnt in the sintering furnace at the peak temperature of 850°C. The temperature control for the sintering process is shown in Fig (13.31e). The typical cycle time is three hours. As all the layers are sintered at the same time, the process is called *cofiring*. During the sintering process, the glass filler of the green-tape melts and spreads to form the host medium for the alumina particles. The sintering process model is shown in Fig (13.31f). The cofired LTCC circuit is electrically tested for the open and short-circuit tests. The SMT components, like chip capacitor, resistor or IC chips are mounted on the outer surface of the LTCC as shown in Figs (13.27a) and (13.29). The microstrip-matching network is also screen printed on the outer surface. These are postcofired processes.

Familiarity with the basic LTCC process helps to appreciate the advantage offered by it, as against the traditional thick film process and HTCC. Some characteristics of the LTCC process are summarized below:

- The green-tape dielectrics are produced in parallel that results in the cost-effective process and increased yield.
- It is one single-firing operation to develop the complete circuit. It reduces production time and cost. The firing operation takes about three hours.
- The partial circuit on each dielectric layer can be inspected separately before stacking. Thus, the defects can be repaired that increases the reliability and yield.
- Multiple dielectric layers permit innovative compact circuits of filter, balun, resonator, patch antenna, etc. The *ferroelectric layer* can be also incorporated for the voltage-controlled phase shifter, filter, delay line, etc.
- The LTCC circuit can effectively operate under the harsh environment. It can withstand high temperatures even at 500°C.
- The LTCC is having a better coefficient of thermal expansion (CTE); so it is better matched to the silicon substrate, as compared to organic or thick-film ceramic (alumina) substrate. Thus, it improves the reliability of the flip-chip components in the thermal cycle testing.
- The LTCC has thermal conductivity 3 W/m°K that is of course less than thermal conductivity (24.7 W/m°K) for

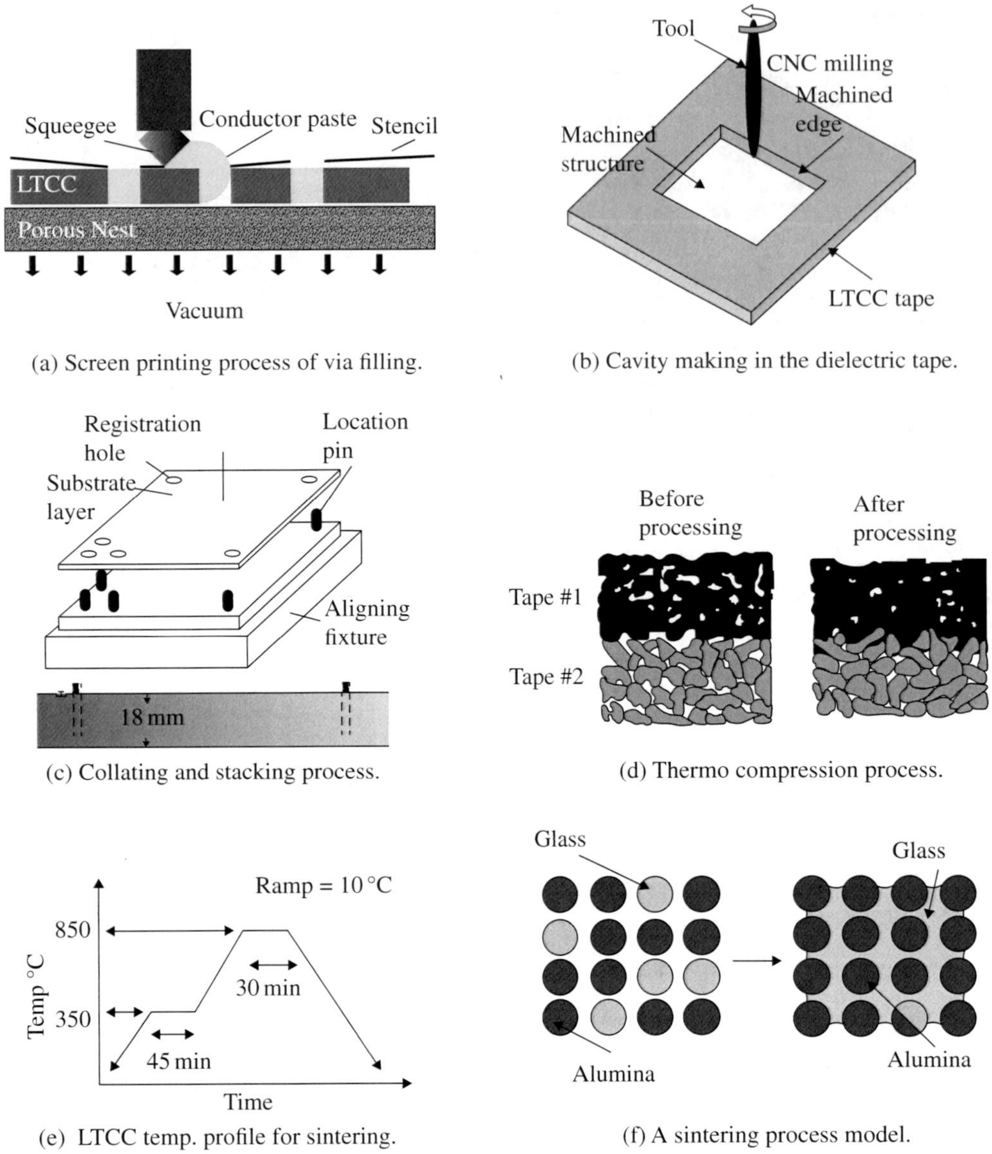

Figure 13.31 Some LTCC circuit fabrication processes.

96% alumina. However, it is better than the thermal conductivity of FR-4 (1.7 W/m°K). The thermal management of the LTCC is improved by their thermal via shown in Fig (13.29). Normally, several thermal vias are provided from the bottom of the heat-generating active circuits to the metallic ground, acting as the heat spreader. Sometimes, the vias and channels below the active components carry fluid for heat removal.

The real difficulty with the LTCC process is associated with the *tape shrinkage* (12–16%) in the planar X–Y direction. The shrinkage of the tape to be accounted for at the design stage itself. However, zero shrinkage tape has been developed as the LTCC-M with the metal ground that also removes the tape shrinking in the planar X–Y direction, but not in the Z-direction.

13.4.3 LTCC Planar Transmission Line and Some Components

The planar transmission lines in the LTCC structure are used both as interconnects and as a medium for the development of distributed circuit elements. The common planar transmission line structures – microstrip, CPW, stripline, slot line, are easily accommodated in the LTCC. These can also be printed on the outer layer of the LTCC after the postfiring process. In this case, both thick and thin-film printing technologies are used.

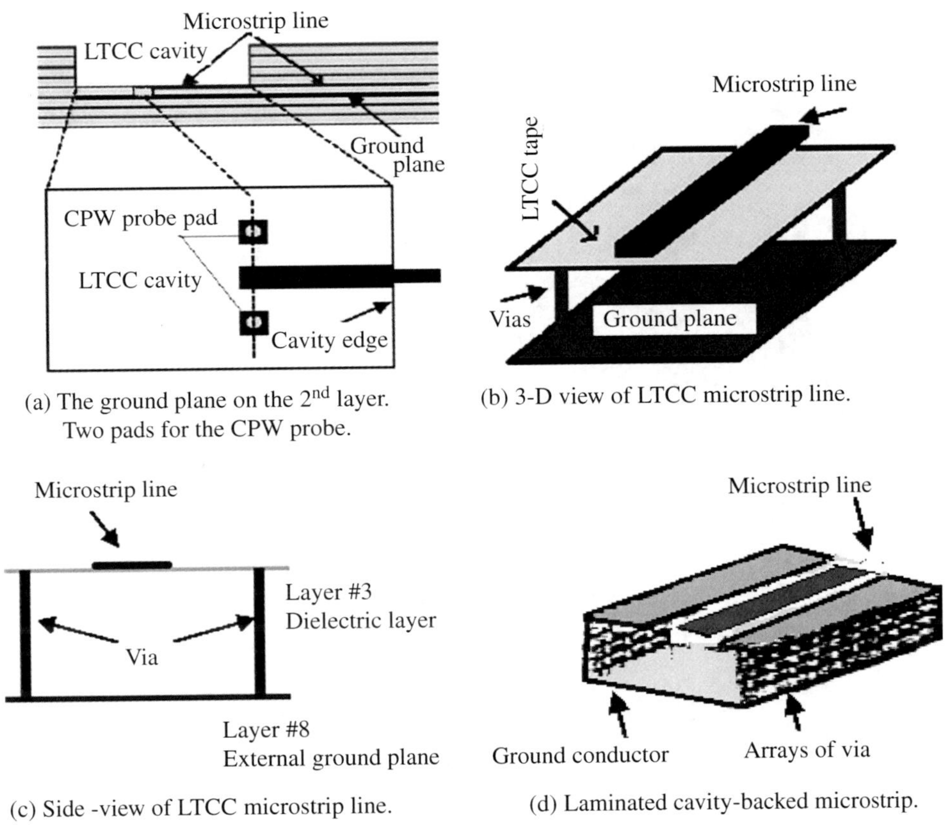

Figure 13.32 Microstrip lines in LTCC.

However, thin-film printing requires a very smooth surface that is available only after polishing the outer surface. It is an expensive exercise, although very narrow lines with narrow spacing could be printed on the LTCC substrate. Normally, thick-film printing is used at the outer and inner dielectric layers, with a minimum line width of about 3–4 mils. The line structures operate at the frequency 1–40 GHz and above 100 GHz [J.55–J.57]. Several distributed components such as filter, coupler, and balun have been fabricated with the planar transmission lines on one dielectric layer or the multiple dielectric layers.

Microstrip

The microstrip in the LTCC format has a more flexible characteristic impedance. The characteristic impedance of a surface-mounted microstrip can be controlled by changing the location of a ground plane at any dielectric layer, as shown in Fig (13.32a, b). In Fig (13.32a), the ground conductor is located on the second layer and next could be located at the third layer. The ground plane of the microstrip is connected, using vias, to the external ground of the LTCC located at the eighth layer. It is shown in Fig (13.32c). The embedded microstrip line can be approached from the top for measurement purposes. Figure (13.32a) also shows two pads for the CPW probe. The *buried (embedded) microstrip* line is embedded in a homogeneous medium. Therefore, it is less dispersive than the surface-mounted microstrip line. Figure (13.32d) further shows that the microstrip line can also be located in a cavity formed within the LTCC [J.58]. Such microstrip lines have better isolation. The sidewalls of the cavity are fabricated by the double column of vias. A top ground plane can also be added to form the TEM mode supporting *the stripline*. The microstrip on the green-tapes can have low-loss, 0.08 dB/m even at 35 GHz [J.59].

CPW to Microstrip Transition Structure

The shielded microstrip in a homogenous medium is a stripline. This line is a preferred structure in the LTCC technology, because it is a broadband nondispersion line. However, the wafer probe uses the CPW launcher. In the LTCC technology, a transition is needed. A basic form of the transition is shown in Fig (13.33) [J.60–J.62]. The microstrip is located at the upper dielectric and the ground plane is located at the next dielectric layer. The CPW is printed at the lower dielectric layer and its two

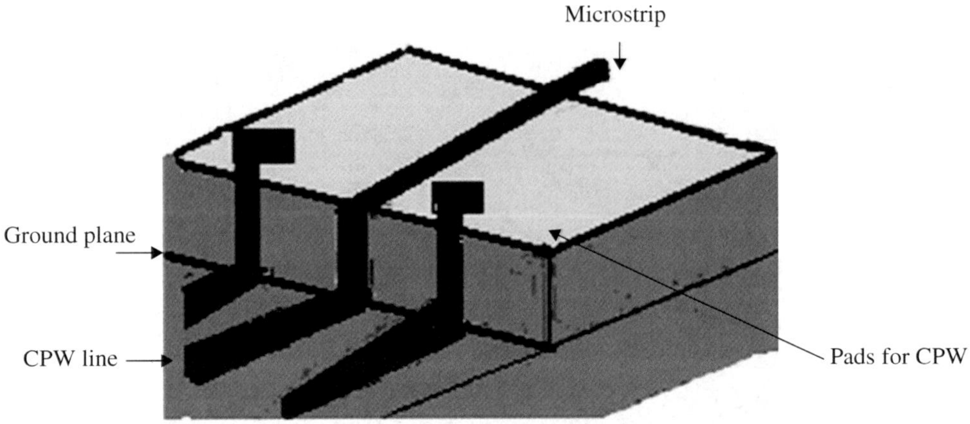

Figure 13.33 Microstrip CPW transition in the LTCC.

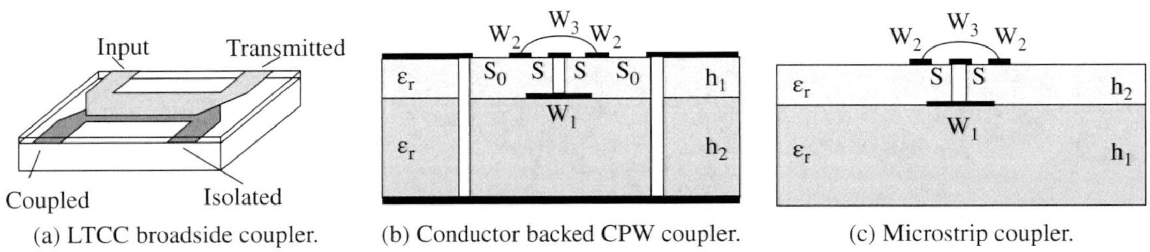

Figure 13.34 Coupler in LTCC technology.

coplanar ground conductors are connected through via to the ground plane of microstrip. A transition, on 5 mils "*green-tape*" works from 10 MHz to 100 GHz.

Coupled Microstrip Line

Figure (13.34a) shows that the *broadside microstrip coupler* with very tight coupling, i.e. 3 dB coupling. The broadside coupler can be easily accommodated in the LTCC. Two microstrip conductors are located at two different dielectric layers. The thickness of the dielectric layers between two strip conductors could be adjusted to meet the design requirement. Moreover, as the structure is buried inside a homogeneous medium, the even and odd mode phase velocities are equalized to achieve a high directivity and less dispersion. The structure has been used as the power combiner and splitter in the frequency range 8–14 GHz [J.63]. Figure (13.34b, c) show other coupler structures in LTCC using the CPW and microstrip line. A wide range of coupling from −10 to −2.6 dB in the frequency range of 0.4–4.0 GHz has been reported [J.64].

Branch-line Coupler

A compact *branch-line coupler,* shown in Fig (13.35a), has been realized in the LTCC environment. The transmission line section is replaced by the equivalent T-network shown in Fig (13.35b). Its LTCC realization is shown in Fig (13.35c) [J.65].

Bandpass Filter

Figure (13.36a) shows a stripline resonator in the LTCC environment. It is useful in developing the LTCC bandpass filter. Figure (13.36b) shows another type of bandpass filter using line resonators in LTCC. The passive components such as inductors, capacitors, and resonator are realized in the LTCC to develop compact filters [J.66–J.69].

Figure (13.37a) further shows a hybrid π-bandpass filter. Figure (13.37b) shows its corresponding LTCC structure in the multilayer format [J.70]. The filter is designed at 2.5 GHz. Four sections are cascaded to achieve four transmissions zero, at 1.90, 1.70, 4.25, 5.25 GHz for getting high rejection at both the lower and upper bands. The passive lumped components, L and C, and the step-impedance resonator are all constructed using the planar line structures at different levels of the LTCC. The partial structures are connected through vias to fabricate a compact filter.

The LTCC structures are complex 3D structures. This kind of structure can be analyzed using the 3D/2.5D EM

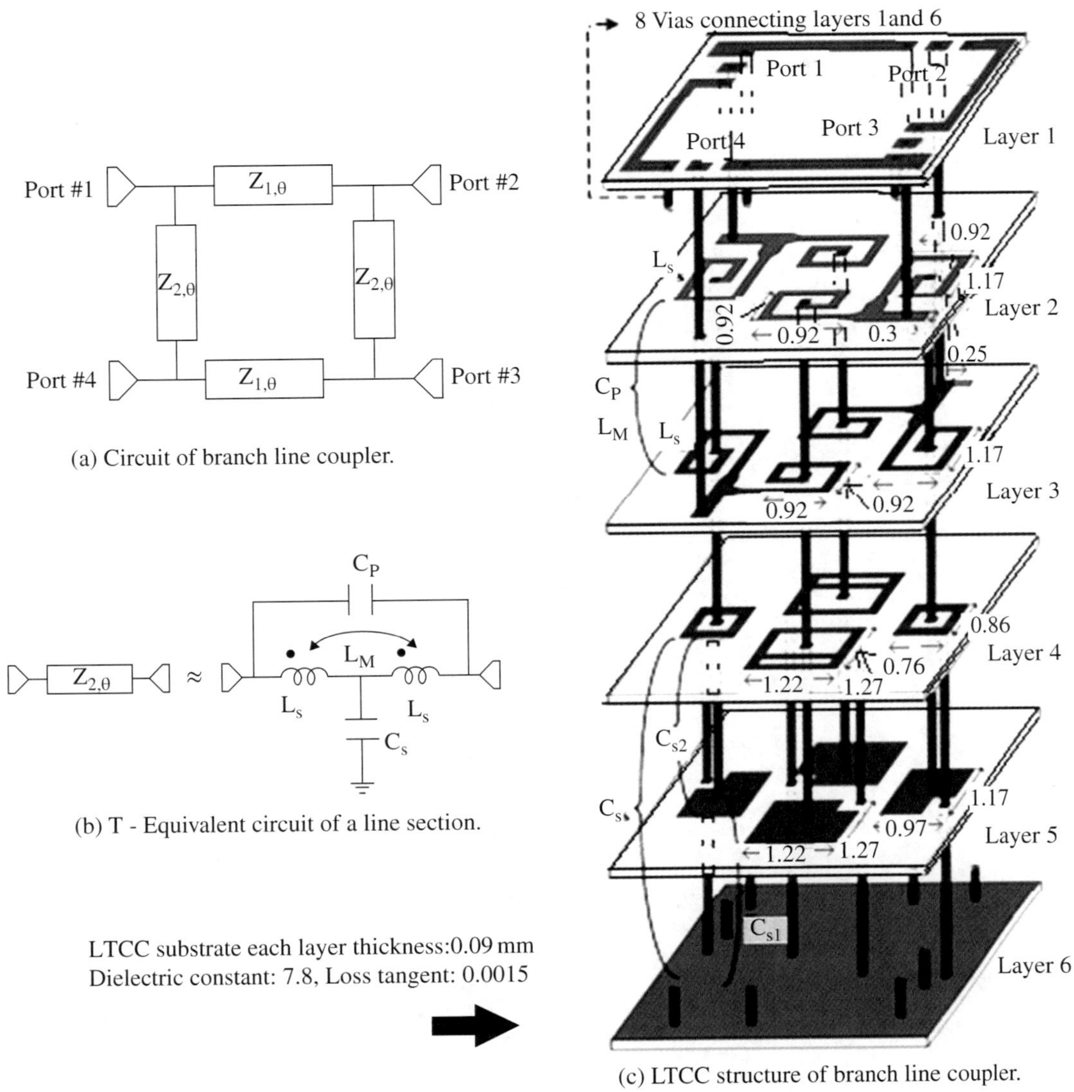

Figure 13.35 Branch line coupler in the LTCC. *Source:* Kuo et al. [J.65]. © 2006 IEEE.

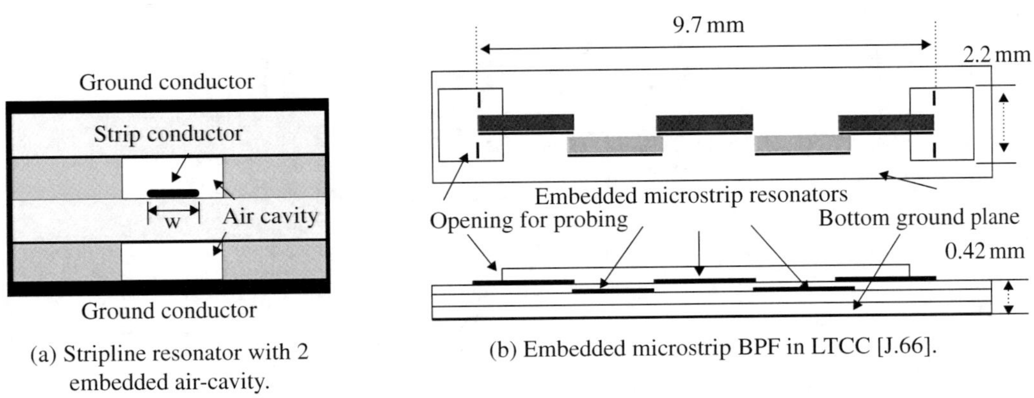

Figure 13.36 The embedded microstrip bandpass filter in LTCC. *Source:* Modified from Choi et al. [J.66]. © 2003 IEEE.

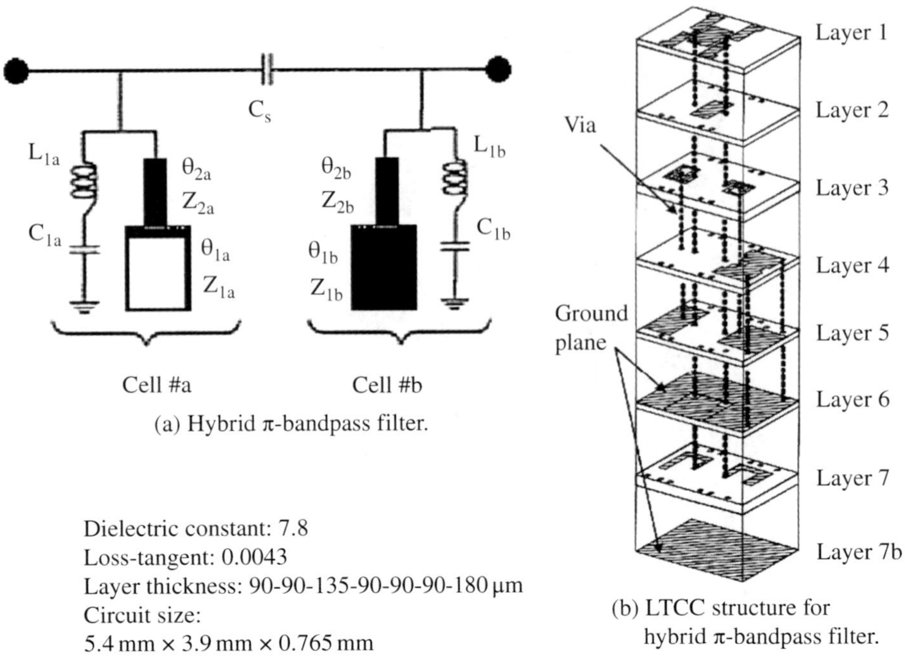

Figure 13.37 Bandpass filter in LTCC. *Source*: From Jeng et al. [J.70]. © 2006 IEEE.

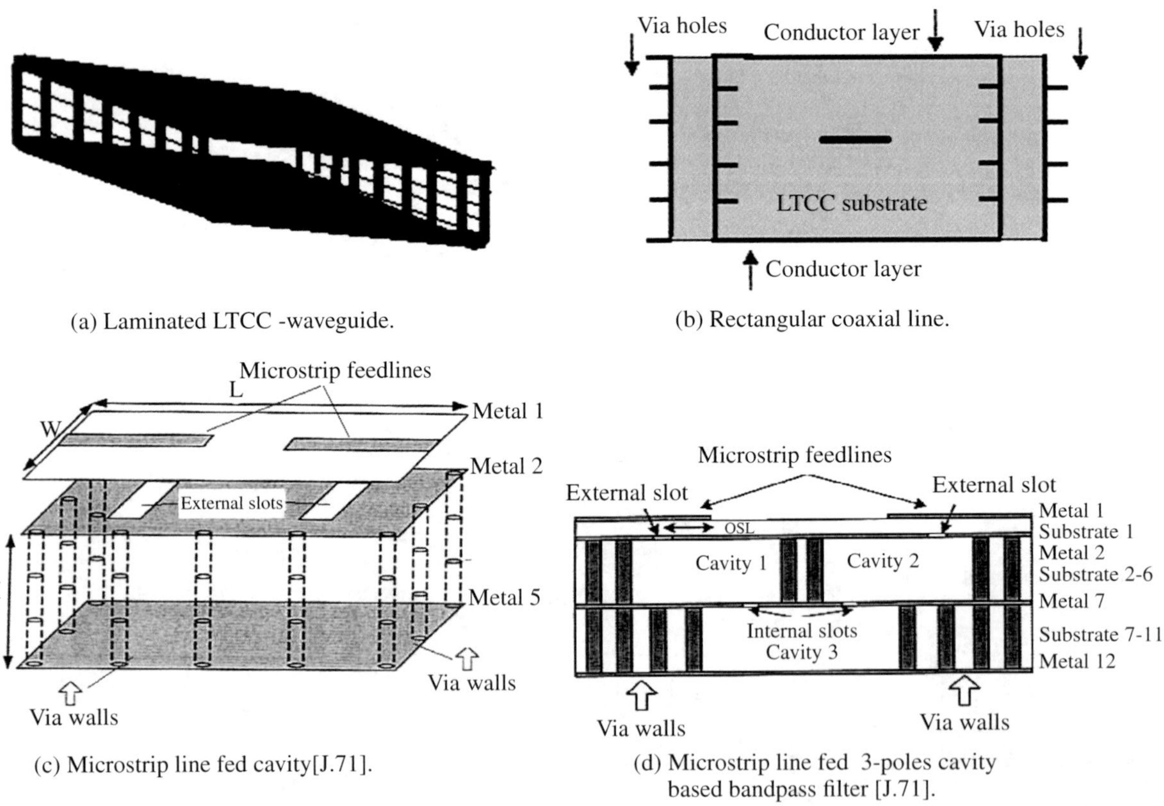

Figure 13.38 Waveguide and cavity structures in the LTCC. *Source:* From Lee et al. [J.71]. © 2005 IEEE.

simulators. However, most of the structures could be broken in parts and could be modeled separately as the lumped elements or as the transmission line sections. The final results could also be obtained by a combination of these blocks. Such modeling is approximate, and the performance of a model should be compared and improved by comparing its results against the results of the EM-Simulator. An analytical model is helpful for a fast design process.

13.4.4 LTCC Waveguide and Cavity Resonators

The multilayer LTCC is a very versatile medium. The low-loss *waveguide structure*, air-filled or the dielectric-filled could be developed in the LTCC. The sidewalls are fabricated by the double column of the closely spaced vias. The top and bottom conductors of the rectangular waveguide could be deposited on two dielectric layers. Figure (13.38a) shows the rectangular laminated waveguide. The *rectangular coaxial line* (*stripline*) is also created in the LTCC. It is shown in Fig (13.38b) [J.58].

Similarly, a *cavity resonator* is also created by using vias for the sidewalls. Figure (13.38c) shows a cavity fed by the aperture-coupled microstrip line. The cavity has a Q factor 372 at 60 GHz. It is further used to develop three poles LTCC bandpass filters as shown in Fig (13.8d) [J.71].

References

Books

B.1 Hoffmann, R.: *Microwave Integrated Circuit Handbook*, Artech House, Boston, 1985.

B.2 Laverghetta, T.S.: *Microwave Materials and Fabrication Techniques*, Artech House, Boston, 1991.

B.3 Gupta, T., *Handbook of Thick- and Thin-film Hybrid Microelectronics*, John Wiley & Sons, NY, 2003.

B.4 Rogers, J.; Plett, C.: *Radio Frequency Integrated Circuit Design*, Artech House, Inc., 2003, USA.

B.5 James, J.R.; Hall, P.S. (Editors): *Handbook of Microstrip Antennas*, Vol. **II**, IEE Pub., 1989.

B.6 Edwards, T.C.; *Foundations for Microstrip Circuit Design*, 2nd Edition, John Willy & Sons, 1992.

B.7 Powell, C.F.; Oxley, J.H.; Blocher Jr, J.M.: *Vapor Deposition*, Wiley, New-York, 1967.

B.8 Westwood, W.D.; *Sputter Deposition*, AVS Education Committee Book Series, Vol. **2**, New York Education Committee, 2003.

B.9 Maissel, L.I.; Glang, R.: *Handbook of Thin Film Technology*, McGraw-Hill, 1970.

B.10 Mattox, D.M.: *Handbook of Physical Vapor Deposition (PVD) Processing: Film Formation, Adhesion, Surface Preparation, and Contamination Control*, Noyes Publications, New Jersey, 1998.

B.11 Haskard, M.; Pitt, K.: *Thick-film Technology and Applications*, Electrochemical Publications Ltd, UK, 1997.

B.12 Robertson, I.D.; Lucyszyn, S. (Editors): *RFIC and MMIC Design and Technology*, IEE, London, 2001.

B.13 Goyal, R.: *Monolithic Microwave Integrated Circuits: Technology & Design*, Artech House, Norwood, MA, 1989.

B.14 Marsh, S.: *Practical MMIC Design*, Artech House, Inc., USA, 2006.

B.15 Nguyen, C.: *Radio-Frequency Integrated-Circuit Engineering*, John Wiley & Sons Inc., 2015.

B.16 Steer, M.: *Microwave and RF Design: A Systems Approach*, SciTech Publishing, 2010, USA.

B.17 Minin, I. (Editor): *Microwave and millimeter-wave Technologies from Photonic Bandgap Devices to Antenna and Applications*, InTech Pub., Mar. 2010.

B.18 Lee, Yun-Shik: *Principles of Terahertz Science and Technology, Springer*, 2009.

B.19 Kuh, Ernest S.: *Multichip Modules*, World Scientific Press, Singapore, 1992

B.20 Medjdoub, F. (Editor): *Gallium Nitride (GaN): Physics, Devices, and Technology*, CRC Press, 2016.

B.21 Rossi, T. and M. Farina: *Advanced Electromagnetic Analysis of Passive and Active Planar Structures*, IEE Pub., UK, 1999.

B.22 Santos Hector J. De Los: *Introduction to Micromechanical Microwave System*, 2, Artech House, 2004.

B.23 Varadan V.K.; Jose, K.A.: *RF MEMS, and Their Applications*, John Willey Ltd., 2003.

B.24 Rebeiz, G.M.: *RF MEMS: Theory, Design, and Technology*, Wiley-Interscience, 2003.

B.25 Johnstone R.W.; Parameswaran, A.: *An Introduction to Surface-Micromachining*, Springer, New York, 2012. ISBN:9781402080203

B.26 Ivanka, Y.: *Multilayered Low-Temperature Cofired Ceramics (LTCC) Technology*, Springer, USA, 2005.

B.27 Kulke, R.: *LTCC – An Introduction and Overview*, IMST GmbH, Germany, 2001.

Journals

J.1 Imthurn, G.: The history of silicon-on-sapphire, *Peregrine Semiconductor*, San Diego, http://www.admiral-microwaves.co.uk.

J.2 Nakamura, T.; Matsuhashi, H.; Nagatomo, Y.: *Silicon on sapphire (SOS) device technology, Oki Tech. Rev.*, Vol. **71**, No. 4, Issue 200, pp. 66–69, Oct. 2004.

J.3 Johnson, R.A.; De la Houssaye, P.R.; Chang, C.E.; Chen, P.-F.; Wood, M.E.; Garcia, G.A.; Lagnado, I.; Asbeck, P.M.; *Advanced thin-film silicon-on-sapphire technology: microwave circuit applications, IEEE Trans. Electron Dev.*, Vol. **ED-45**, No. 5, pp. 1047–1053, May 1998.

J.4 Gardial, F.E.: *Design and layout of microstrip structures, IEE Proc.*, Vol. **135**, No. 3, June 1988.

J.5 Coonrod, J.: *Understanding when to use FR-4 or high-frequency laminates, On Board Technol.*, pp. 26–30, Sept. 2011.

J.6 *High frequency circuit materials product selection guide*, Rogers cor., USA.

J.7 *Thin film ceramic design guide*, CoorsTek, USA.

J.8 Product Guide: *Applied Thin-Film Products (ATP), USA.*

J.9 *RT/duroid® 5870/5880 High Frequency Laminates*, Rogers Corporation, USA.

J.10 Hasegawa, H.; Furukawa, M.; Yanai, H.: *Properties of the microstrip line on SiSiO2 system, IEE Trans. Microwave Theory Tech.*, Vol. **MTT-19**, No. 11, pp. 869–881, Nov. 1971.

J.11 Tseng, H.Q.; Saunier, P.; Ketterson, A.; Witkowski, L. C.; Jones, T.: *Embedded transmission line (ETL) MMIC for low-cost high-density wireless communication applications, IEEE Trans. Microwave Theory Tech.*, Vol. **MTT-45**, No. 12, pp. 2540–2548, Dec. 1997.

J.12 *Fine line thick film solution*, API Technologies Corp., USA.

J.13 Guillaume, B.; Mabire, J.; Maurin, P.; Reffet, D.; Chirol, F.; Polzer, E.K.: *Microwave applications using photo-imageable technology; Microwave J.*, Vol. **40**, No. 3, pp. 102–106, Mar. 1997.

J.14 *The Basics of Screen Printing Thick Film Inks*, Du Pont, USA.

J.15 *Process of Thick Film Technology*, http://www.ett.bme.hu/vlab/lab.

J.16 Crnojević-Bengin, V.; Budimir, D.: *Integrated waveguide bandpass filters using thick-film technology, Microwave Rev.*, Vol. **10**, No. 1, pp. 9–13, June 2004.

J.17 Pan, J.; Tonkay, G.L.; Quintero, A.: *Screen printing process design of experiments for fine line printing of thick film ceramic substrates, Proc. Inter. Sym. Microelectronics*, pp. 264–269, Nov. 1998, San Diego, USA.

J.18 Dziurdzia, D.; Novak, S.; Gregorczyk, W.; CieŜ, M.: *Photo patterned thick film resonators, 14th Int. Conf. Microwave Radar Wirel. Commun.*, MIKON-2002, Vol. **2**, pp. 467–471, May 2002.

J.19 Ponchak, G.E.; Downey, A.N.: *Characterization of thin film microstrip lines on polyimide, IEEE Trans. Comput. Packg. Manfact. Technol. Part-B*, Vol. **21**, No. 2, pp. 171–176, May 1998.

J.20 Malnovic, V.; Ozgur, M. Degroot, D.C.; Jargon, J.A.; Gaitan, M; Zaghloul, M.E.: *Characterization of broadband transmission for coplanar waveguides on CMOS silicon substrates, IEEE Trans. Microwave Theory Tech.*, Vol. **MTT-46**, No. 5, pp. 632–640, May 1998.

J.21 Williams, D.F.: Metal–insulator–semiconductor transmission lines, *IEEE Trans. Microwave Theory Tech.* Vol. **47**, pp. 176–181, Feb. 1999.

J.22 Verma, A.K.; Nasimuddin; Sharma, E.K.: *Analysis and circuit model of a multilayer semiconductor slow-wave microstrip line, IEE Proc.-Microwave Antennas Propag.*, Vol. **151**, No. 5, pp. 441–449, Oct. 2004.

J.23 Jäger, D.: *Slow-wave propagation along variable Schottky-contact microstrip line, IEEE Trans. Microwave Theory Tech.*, Vol. **MTT-24**, No. 9, pp. 566–573, Sept. 1976.

J.24 Verma, A.K.; Nasimuddin, Sharma, E.K.: *Propagation characteristics of Schottky contact suspended slow-wave microstrip line, IEEE Microwave Wirel. Comp. Lett.*, Vol. **11**, No. 9, pp. 385–387, 2001.

J.25 Konno, M.: *Conductor loss in thin-film transmission lines, Electron. Commun. Japan Part-2*, Vol. **82**, No. 10, pp. 83–91, 1999.

J.26 Yun, Y.; Jeong, J.H.; Kim, H.S.; Jang, N.: *Basic RF characteristics of fishbone-type transmission line employing comb-type ground plane (FTLCGP) on PES substrate for use in flexible passive circuits, Electron. Telecommun. Res. Inst. ETRI J.*, Vol. **37**, No. 1, pp. 128–137, Feb. 2015.

J.27 Ogawa, H.; Hasegawa, T.; Banba, S.; Nakamoto, H.: *MMIC transmission lines for multi-layered MMIC's, IEEE MTT-S Int. Microwave Symp. Dig.*, pp. 1067–1070, June 1991.

J.28 Darwish, A.; Ezzeddine, A.; Huang, H.C.; Mah, M; *Analysis of three-dimensional embedded transmission lines (ETL's), IEEE Microwave Guided Wave Lett.*, Vol. **9**, No. 11, pp. 447–449, Nov. 1999.

J.29 Ishigawa, T.; Yamashita, E.: *Characterization of buried microstrip lines for constructing high- density microwave integrated circuits, IEEE Trans. Microwave Theory Tech.*, Vol. **44**, pp. 840–847, June 1996.

J.30 Kim, J.; Qian, Y.; Feng, G.; Ma, P.; Judy, J.; Chang, M. F.; Itoh, T.: *A novel low-loss low crosstalk interconnect for broad-band mixed-signal silicon MMIC's, IEEE Trans. Microwave Theory Tech.*, Vol. **47**, No. 9, pp. 1830–1835, Sept. 1999.

J.31 Tokumitsu, T.: *Three-dimensional MMIC and its evolution to WLCSP technology*, *SEI Tech. Rev.*, Vol. No. **72**, pp. 34–41, Apr. 2011

J.32 Heinrich, W.; Hartangel, H.L.: *Field-theoric analysis of wave propagation on FET electrodes including losses and small signal amplification*, *Int. J. Electron.*, Vol. **58**, No. 4, pp. 413–627, 1985.

J.33 Nguyen, C.; Katehi, L.P.B.; Rebeiz, G.M.: *Micromachined devices for wireless communications*, *Proc. IEEE*, Vol. **86**, No. 8, pp. 1756–1768, Aug. 1998.

J.34 Katehi, L.P.B.; Harvey, J.F.; Herrick, K.J.: *3-D integration of RF circuits using Si micromachining*, *IEEE Microw. Mag.*, Vol. **2**, No.1, pp. 30–39, Mar. 2001.

J.35 Dennis, J.O.; Ahmad, F.; Khir, M.H.: CMOS compatible bulk micromachining, chapter-5, *Advances in Micro/Nano Electromechanical Systems and Fabrication Technologies*, InTech Pub., May 1013.

J.36 Weller, T.M.; Katehi, L.P.B.; Herman, M.I.; Wamhof, P.D.; Lee, K.; Kolawa, E.A.; Tai, B.H.: *New results using membrane-supported circuits: a Ka-band power amplifier and survivability testing*, *IEEE Trans. Microwave Theory Tech.*, Vol. **44**, No. 9, pp. 1603–1606, Sept. 1996.

J.37 Wellert, T.M.; Katehi, L.P.B.; Herman, M.I.; Wamhof, P.D.: *Membrane technology (MIST-T) applied to microstrip: a 33 GHz Wilkinson power divider*, IEEE Int. Microw. Symp. Dig., pp. 911–914, San Diego, CA, 1994.

J.38 Robertson, S.V.; Katehi, L.P.B.; Rebeiz, G.M.: *W-band microshield low-pass filters*, IEEE Int. Microw. Sym. Dig., pp. 625–628, San Diego, CA, 1994.

J.39 Stephen V.; Robertson, S.V.; Katehi, L.P.B.; Rebeiz, G. M.: *Micromachined W-band filters*, *IEEE Trans. Microwave Theory Tech.*, Vol. **MTT-44**, No. 4, pp. 598–606, Apr. 1996.

J.40 Katehi, L.P.B.; Rebeiz, G.M.; Weller, T.M.; Drayton, R. F.; Cheng, H.J, Whitaker, J.F.: *Micromachined circuits for millimeter- and sub-millimeter-wave applications*, *IEEE Antennas Propag. Mag.*, Vol. **35**, No. 5, pp. 9–17, Oct. 1993.

J.41 Lee, H.S.; Kim, S.C.; Lim, B.O; Kim, K.M.; Uhm, W.Y.; Chun, Y.H.; Shin, D.H.; Kim, S.K.; Park, H.C.; Rhee, J.K.: *New micromachined microstrip transmission lines for application in millimeter-wave circuits*, *Microwave Opt. Tech. Lett.*, Vol. **40**, No. 1, pp. 6–9, Jan. 2004.

J.42 Herrick, K.J.; York, J.G.; Katehi, L.P.B.: *Microtechnology in the development of three-dimensional circuits*, *IEEE Trans. Microwave Theory Tech.*, Vol. **46**, No. 11, pp. 1832–1844, Nov. 1998.

J.43 Dib, N.I; Harokopus, W.P.; Katehi, P.B.; Ling, C.C.; Rebeiz, G.M.: *Study of a novel planar transmission line*, IEEE Int. Microw. Sym. Dig., pp. 623–626, Boston, MA, 1991.

J.44 Milanovic, V.; Gaitan, M.; Bowen, E.D.; Zaghlou, M.E.: *Micromachined microwave transmission lines in CMOS technology*, *IEEE Trans. Microwave Theory Tech.*, Vol. **MTT-45**, No. 5, pp. 630–635, May 1997.

J.45 Drayton, R.F.; Katehi, L.P.B.: *Development of miniature microwave circuit components using micromachining techniques*, IEEE Int. Microw. Sym. Dig., pp. 225–228, San Diego, CA, 1994.

J.46 Weller, T.M.; Katehi, L.P.B.; Rebeiz, G.M.: *High-performance microshield line components*, *IEEE Trans. Microwave Theory Tech.*, Vol. **MTT-43**, No. 3, pp. 534–543, Mar. 1995.

J.47 Drayton, R.F.; Katehi, L.P.B.: *Development of self-packaged high-frequency circuits using micromachining techniques*; *IEEE Trans. Microwave Theory Tech.*, Vol. **MTT-43**, No. 9, pp. 2073–2080, Sept. 1995.

J.48 Ashesh, C.B.; Bhattacharya, D.; Garg, R.: *Characterization of V-groove coupled microshield line*, *IEEE Microwave Wirel. Comput. Lett.*, Vol. **15**, No. 2, pp. 110–112, Feb. 2005.

J.49 Fu, J.; Wu, Q.; Gu, X.; Shi, H.; Lee, J.: *MEMS-based transmission lines for microwave applications*, *Proc. SPIE Int. Soc. Opt. Engg.*, Vol. **5116 II**, pp. 744–750, 2003.

J.50 McGrath, W.R.; Walker, C.; Yap, Y.; Tai, Y.C.: *Silicon micromachined waveguides for millimeter-wave and submillimeter-wave frequencies*, *IEEE Microwave Guided Wave Lett.*, Vol. **3**, No. 3, pp. 61–63, Mar. 1993.

J.51 Scranton, C.Q.; Lawson, J.C.: *LTCC technology: where we are and we are going?* **IEEE** Symp.Technol. Wireless Appl., pp. 193–200, New York, 1999.

J.52 Clénet, M.: *Design and analysis of a Yagi-like antenna element buried in LTCC material for AEHF communication systems*, Technical Memorandum, Defence Research & Dev., Canada, Apr. 2005.

J.53 Jantunen, H.: *A novel low temperature co-firing ceramic (LTCC) material for Telecommunication Devices*, Dept. EE and Infotech, University of Oulu, Oulu, 2001.

J.54 Hu, T.: *BST-based Low-Temperature Co-Fired Ceramic (LTCC) Modules for Microwave Tunable Components*, ISBN:951-42-7292-7, Dept. EE and Infotech, University of Oulu, Oulu, 2004.

J.55 Gongora-Rubio, M.R.: Non-packaging applications of LTCC technology, *Micro-Electro-Mechanical System*, Panamerican Advanced Studies Institute, Patagonia, Argentina, Mar 31, 2003.

J.56 *Design guidelines for LTCC: Preliminary Guideline*, Release 1.0, Circuit Materials Division MIT0703.3, Heralockò HL2000 Materials System.

J.57 Chai, L.; Shaikh, A.; Stygar, V.: *LTCC for wireless and photonic packaging applications, IEEE Int. Sym. Electron. Mater. Packg.*, pp. 381–385, Taiwan, 2002.

J.58 Wu, K.L.; Huang, Y: *LTCC technology and its applications in high-frequency front end modules*, 6th Int. Symp. Antennas, Propagat. EM Theory Proc., IEEE, pp. 730–734, Oct. 2003.

J.59 Brokmeier, A.: *LTCC -technology for miniaturized Ka-band frontends*, 33rd European Microwave Conf., pp. 631–634, Munich, 2003.

J.60 Muller, J; Perrone, R.; Rentsch, S.: *Technology & material challenges for future LTCC microwave*, BMBF Centre for Innovation Competence, Technical Univ. Ilmenaur, Ilmenaur, Germany.

J.61 Lautzenhiser, F.; Barnwell, P.; Wood, J.: *Microwave Module Design with HeraLockTM HL2000 LTCC*, Proc. Int. Conf. Advanced Packaging and Systems (ICAPS), pp. 143–149, Reno, Nevada, 2002.

J.62 Kumar, D.; Nair, M; McKinzie, W.E.: *A 10 MHz to 100 GHz LTCC CPW-to-Stripline Vertical Transition*, IEEE IMS, 2013.

J.63 Gipprich, J.W.; Dickens, L.E.; Hayes, R.E.; Restivo, M. A.: *A compact LTCC multilayer multiport stripline coupler network for wideband low-loss power combining/splitting*, IEEE 3rd Topical Meeting Electrical Performance of Electronic Packaging, pp. 167–169, 1994.

J.64 Sawicki, A.; Sachse, K.: *Novel coupled line conductor-backed coplanar and microstrip directional couplers for PCB PCB and LTCC applications, IEEE Trans. Microwave Theory Tech.*, Vol. **MTT-51**, No. 61, pp. 1743–1751, June 2003.

J.65 Kuo, T.N.; Lin, Y.S.; Wang, C.H.; Chen, C.H.: *A compact LTCC branch line coupler using modified -T equivalent circuit model for transmission line, IEEE Microwave Comput. Lett.*, Vol. **16**, No. 2, pp. 90–93, Feb. 2006.

J.66 Choi, B.G.; Stubbs, M.G.; Park, C.S.: *A Ka-band narrow bandpass filter using LTCC technology, IEEE Microwave Comput. Lett.*, Vol. **13**, No. 3, pp. 388–89, Sept. 2003.

J.67 Liang, H.; Sutono, A.; Laskar, J.; Smth, W.R.: *Material parameter characterization of multilayer LTCC and implementation of high Q-resonators, IEEE MTT(S) Dig.*, Vol. **4**, pp. 1901–1904, 1999.

J.68 Eurskens, W.; Wersing, W.; Gohlke, S.; Wannemnacher, V.; Hild, P.; Weigel, R.: *Design and performance of UHF band inductors, capacitors and resonators using LTCC technology for mobile communication systems, IEEE Int. Microw. Symp. Dig.*, pp. 1285–1288, Baltimore, MD, 1998.

J.69 Lee, Y.C.; Park, C.H.: *A novel high Q LTCC strip line resonator for millimeter wave applications, IEEE Microwave Comput. Lett.*, Vol. **13**, No. 12, pp. 499–501, Dec. 2003.

J.70 Jeng, Y.H.; Chang, S.F.R; Lin, H.K.: *A high stopband-rejection LTCC filter with multiple transmission zero, IEEE Trans. Microwave Theory Tech.*, Vol. **MTT-54**, No. 2, pp. 633–638, Feb. 2006.

J.71 Lee, J.H.; Pinel, S.; Papapolymerou, J.; Laskar, J.; Tentzeris, M.M.: *Low-loss LTCC cavity filters using the system-on-package technology at 60 GHz, IEEE Trans. Microwave Theory Tech.*, Vol. **MTT-53**, No. 12, pp. 3817–3823, Dec. 2005.

14

Static Variational Methods for Planar Transmission Lines

Introduction

The variational method is a powerful tool to compute the line capacitance and other line parameters of microstrip and coupled microstrips on a single-layer substrate, as well as, on the multilayer substrate. The *Galerkin's method*, a sophisticated version of the variational method, is also discussed. Galerkin's method computes the line parameters of both the microstrip and CPW structures. The method is used both in the *space domain* and the Fourier transformed domain, i.e. in the *Fourier domain*. The latter method is also known as the *static spectral domain analysis* (SDA). The static results of this chapter are used in chapter 15 to develop the closed-form models for the multilayer microstrip and coupled microstrip lines to compute the frequency-dependent propagation constant, characteristic impedance, conductor and dielectric losses.

Objectives

- *To present variational formulations of line capacitance, both in space and Fourier domains.*
- *To compute line parameters of the boxed single-layer and multilayer microstrip in the space domain.*
- *To compute line parameters of the open single-layer and multilayer microstrip in the Fourier domain.*
- *To compute line parameters of the coupled multilayer microstrip in space and Fourier domains.*
- *To compute line parameters of the multilayer boxed microstrip using the discrete Fourier transform.*
- *To compute line parameters of the multilayer boxed CPW using the discrete Fourier transform.*

14.1 Variational Formulation of Transmission Line

The variational method computes the *upper* and the *lower* bounds on the line capacitance. It means the line capacitance can be computed with a relatively small error for an assumed charge distribution function on a conductor, or an assumed potential distribution around the conductor. The variational method is based on the *calculus of variation* that can be studied from the standard textbooks [B.1–B.4]. It is like the maximum and minimum problem of ordinary calculus. However, some familiarity with the nomenclature of the calculus of variation is useful to follow the variational method.

14.1.1 Basic Concepts of Variation

The differential equation, with known boundary conditions, is normally used to formulate a physical problem. The physical problem could also be reformulated in the form of a definite integral. Its lower and upper limits are associated with the boundary or range. The integrand of the definite integral is known as the *Lagrangian* of the variational problem. The variational method is concerned with obtaining the unknown function of the integrand under the *stationary conditions*, i.e. under the maximum/minimum conditions for the variational integral, known as the *functional*. The stationary conditions give the *lower* and *upper bounds* of the functional.

Let us consider the formulation of a physical problem that is constructed through following definite integral:

$$I(y) = \int_a^b F\left(x, y, \frac{dy}{dx}\right) dx, \tag{14.1.1}$$

where integral I(y) is known as the *functional*. The integrand F(x,y,dy/dx), i.e Lagrangian function, is a function of independent variable x, dependent variable y(x), and derivative dy/dx. We have to determine the function y(x) that provides the maximum/minimum, i.e. the *stationary value* for the integral i.e. functional I(y), within the range $a < x < b$. It is noted that in the case of a simple function of differential calculus, the local maximum and minimum of the function are obtained at the *discrete stationary points*, whereas in the case of the functional of the variational calculus, the extrema of the functional are obtained for the *stationary functions*. Thus, the

Introduction to Modern Planar Transmission Lines: Physical, Analytical, and Circuit Models Approach, First Edition. Anand K. Verma.

domain of a simple function is a region of *coordinate space*, whereas the domain of a functional is a set of the admissible functions in the *functional space*.

The problem of the variational calculus is dealt with by an operator called *variation* δ. It works like a differential operator. Thus, the variation δy of a function $y(x)$ is a small change in y for a fixed value of independent variable x, i.e. for no variation in x, i.e. $\delta x = 0$. The values of the function $y(x)$ are given at $x = a$, $x = b$, i.e. $y(x = a) = A$, $y(x = b) = B$. These are the boundary conditions, such that there is no variation in y, $\delta y = 0$ at $x = a$, $x = b$.

To obtain the unknown function y, i.e. $y(x)$ under the stationary condition of the functional $I(y)$, the variation in $I(y)$ must be zero, i.e. $\delta I(y) = 0$. If the second variation $\delta^2 I(y)$ is positive, the function $y(x)$ provides the minimum value of the integral functional $I(y)$. However, for $\delta^2 I(y)$ negative, it provides the maximum value of the integral functional $I(y)$. Figure (14.1a and b) show the minimum and maximum values of the *variational functional* $I(y)$. Figure (14.1c) shows the *nonvariational functional* $I(y)$. It is noted that the vertical axis of Fig (14.1) shows the values of the functional for the admissible functions y_0, y_1, etc., along the horizontal axis. In the case of a variational expression, the minimum functional $I_{min}(y_0)$, shown in Fig (14.1a), is obtained at $y = y_0$. The value $I_{min}(y_0)$ is known as the *true value* of the functional for the correct solution y_0. However, even for the approximate solution $y = y_1$, the value of the functional does not deviate significantly from its true value $I_{min}(y_0)$, although $I(y_1) > I_{min}(y_0)$. Thus, any admissible approximate or assumed solution $y(x)$ will have the value of functional more than its true value. The assumed solution should be optimized so that the value of functional approaches to the true value $I_{min}(y_0)$. This is the minimization of the functional $I(y)$ with respect to the assumed solution.

Similar is the case for the maximum of the functional $I_{max}(y_0)$, shown in Fig (14.1b). In this case, the functional value has $I(y_1) < I_{max}(y_0)$ for $y = y_1$. Thus, any admissible approximate or assumed solution $y(x)$ will have the value of functional less than its true value. The assumed solution should be maximized so that the value of the functional approaches to the true value $I_{max}(y_0)$. This is the maximization of the functional $I(y)$ with respect to the assumed solution. However, if the functional $I(y)$ is a nonvariational, the value of the functional $I(y_1)$ is greatly different from its true value of $I(y)$ at $y = y_0$. It is shown in Fig (14.1c).

The *stationary conditions* for the minimum and maximum of the *functional* $I(y)$ are summarized below:

$$\begin{aligned}
&\text{First variation}: \quad \delta I(y) = 0 && \text{(a)}\\
&\text{For the minimum condition, the } 2^{nd} \text{ variation positive}: \quad \delta^2 I(y) = +\text{tive} && \text{(b)}\\
&\text{For the maximum condition, the } 2^{nd} \text{ variation negative}: \quad \delta^2 I(y) = -\text{tive} && \text{(c).}
\end{aligned} \tag{14.1.2}$$

The minimum condition provides the *upper bound* of the functional $I(y)$, i.e. the variational value of $I(y)$, for $y \neq y_0$ is always higher than the true value of $I(y)$ at y_0. The relation $I(y_1) > I_{min}(y_0)$ is shown in Fig (14.1a) for the assumed function of $y = y_1$. This variational formulation is applied in section (14.1.2) to compute the line capacitance C using the *energy method based variational expression*. It provides the *upper bound* on the line capacitance. Further, in this variational formulation, the line capacitance C is a *functional,* and potential function ϕ is an *unknown function*. The approximate form of the unknown trial function could be assumed based on physical consideration.

Likewise, the maximum condition, shown in Fig (14.1b), provides the *lower bound* of the functional $I(y)$, i.e. the variational value of $I(y)$, for $y \neq y_0$ is always lower than the

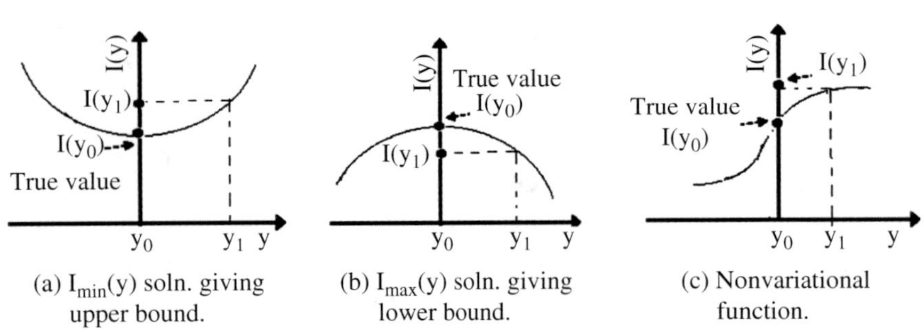

Figure 14.1 Nature of variational (a and b) and nonvariational (c) functionals.

true value of I(y) at $y = y_0$ i.e. $I(y_1) < I_{max}(y_0)$. The assumed function $y_1(x)$, the solution of a given problem, is known as the *trial function*. The variational method of Fig (14.1b) is applied in section (14.1.3), using the *Green's function method-based variational expression*, to compute the *lower bound* of the line capacitance. In this case, the charge density distribution is an unknown function.

14.1.2 Energy Method-Based Variational Expression

Collin [B.1] has provided the variational expressions for both the upper and lower bounds of the line capacitance of a transmission line. The expression for the upper bound on the line capacitance, i.e. the lower bound on the characteristics impedance Z_0, is based on the electrostatic energy. The variational expression for the lower bound of the line capacitance, i.e. the upper bound on the characteristic impedance, is based on Green's function approach. The Green's function method is discussed in subsection (14.1.3). A summary of both expressions is presented below. For a full discussion, the reader should consult Collin.

The electrostatic energy *p.u.l.* stored in the field surrounding a transmission line, shown in Fig (14.2a), is

$$W_c = \frac{\varepsilon_0\varepsilon_r}{2}\iint \vec{E}_t\cdot\vec{E}_t^{\,*}\,dxdy = \frac{\varepsilon_0\varepsilon_r}{2}\iint -(\nabla_t\phi)\cdot(-\nabla_t\phi)dxdy$$

$$\Rightarrow W_c = \frac{\varepsilon_0\varepsilon_r}{2}\iint\left[\left(\frac{\partial\phi}{\partial x}\right)^2 + \left(\frac{\partial\phi}{\partial y}\right)^2\right]dxdy, \tag{14.1.3}$$

where $\vec{E}_t$ is the transverse electric field in the transverse (x–y)-plane, and $\phi(x, y)$ is the potential function giving potentials ϕ_1 and ϕ_2 at conductor #1 and conductor #2 respectively. The potential difference between both the conductors is V_0. For a grounded conductor #1, $\phi_1 = 0$. If C is the line capacitance *p.u.l.*, the electrostatic energy *p.u.l.* is

$$W_c = \frac{1}{2}C(\phi_2 - \phi_1)^2 = \frac{1}{2}CV_0^2. \tag{14.1.4}$$

Using the equations (14.1.3) and (14.1.4), the expression for the line capacitance is obtained as

$$C = \frac{\varepsilon_0\varepsilon_r}{V_0^2}\iint(\nabla_t\phi\cdot\nabla_t\phi)dx.dy = \frac{\varepsilon_0\varepsilon_r}{V_0^2}\iint_S(\nabla_t\phi)^2 ds \qquad \text{(a)}$$

$$\text{where,}\quad V_0 = \int_{Cond\#1}^{Cond\#2}\nabla_t\phi\cdot d\ell \qquad \text{(b).}$$

$$\tag{14.1.5}$$

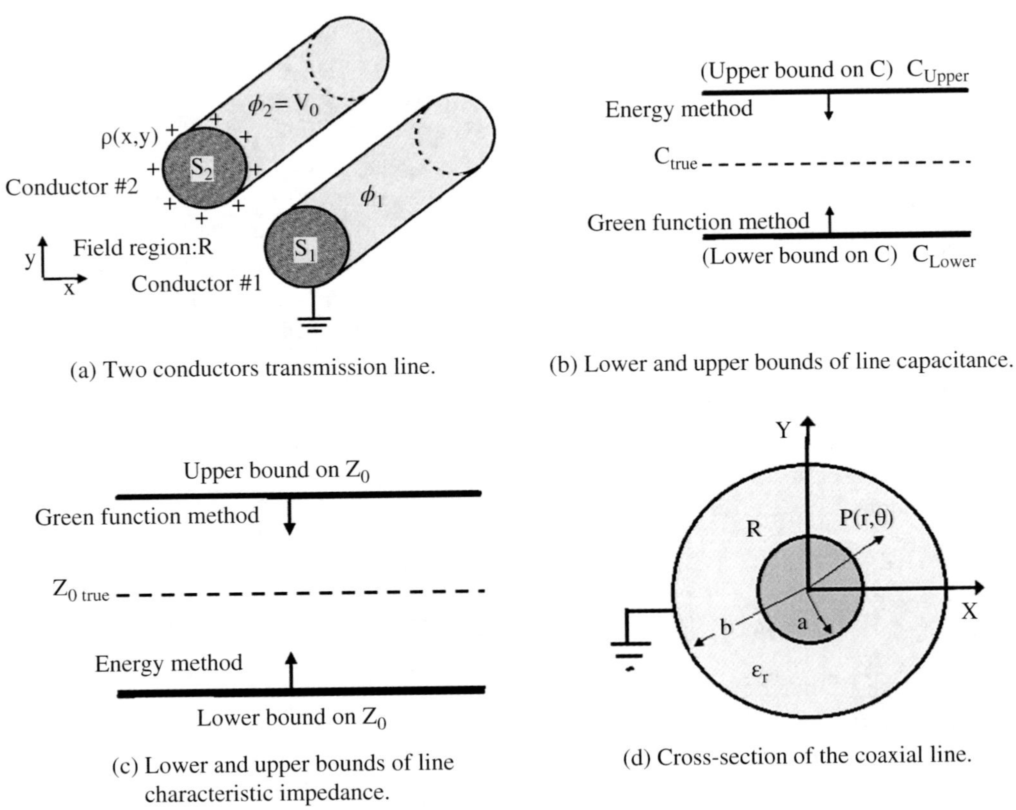

(a) Two conductors transmission line.

(b) Lower and upper bounds of line capacitance.

(c) Lower and upper bounds of line characteristic impedance.

(d) Cross-section of the coaxial line.

Figure 14.2 Transmission lines and variational bounds.

In equation (14.1.5), the differential length dl is along any path between conductor #1 and conductor #2.

The integration is carried out over the cross-sectional region (R) occupied by the electrostatic field. *The potential function* $\phi(x, y)$ *is unknown*. We assume the form of solution for the potential function $\phi(x, y)$ and compute the line capacitance. However, the accuracy of the computed result is uncertain. If expression (14.1.5), i.e. the energy method is variational, in that case, the first-order error in our assumption of the potential function $\phi(x, y)$ results in less error, i.e. the second-order error in the computation of the line capacitance. Therefore, it is to be demonstrated that expression (14.1.5) is a variational expression. For this purpose, the variation in the electrostatic energy is zero:

$$\delta W_c = \delta \int\int_S \left[\left(\frac{\partial\phi}{\partial x}\right)^2 + \left(\frac{\partial\phi}{\partial y}\right)^2\right] dxdy \qquad (14.1.6)$$

$$= -\int\int_R \left[\frac{\partial^2\phi}{\partial x^2} + \frac{\partial^2\phi}{\partial y^2}\right]\delta\phi dxdy = 0.$$

The integral is zero, due as $\nabla^2\phi = 0$. The reader should follow the details in Collin [B.1]. Moreover, Collin shows that $\delta^2 W_c > 0$. *Hence, expression (14.1.5) i.e. the energy method provides the minimum value, i.e. the upper bound for the line capacitance*, shown in Fig (14.2b). The characteristic impedance of a line is given by

$$\frac{1}{Z_0} = \frac{\int\int(\nabla_t\phi\cdot\nabla_t\phi)dx.dy}{\sqrt{\mu/\varepsilon}V_0^2}. \qquad (14.1.7)$$

The potentials on the conductors are $\phi_1 = 0$, $\phi_2 = V_0$. Equation (14.1.7) is also a variational expression that provides the lower bound on the characteristic impedance, shown in Fig (14.2c). An application of the variational formulation is illustrated below by computing the line capacitance of a coaxial cable.

Ex. 14.1

Calculate the line capacitance of a coaxial cable of radii a and b shown in Fig (14.2d). Use the energy method-based variational expression for this purpose.

Solution: Let us assume that the potential at P(r, θ) between the inner and outer conductors is expressed by the following polynomial function:

$$\phi = \phi_0 + \sum_{i=1}^{N} \alpha_i r^i, \quad (i = 1, 2, 3, \ldots) \qquad \text{(i)},$$

where r is a radial distance and α_i is the ith coefficient of the assumed polynomial. For the uniform field, potential is independent of θ. Let us approximate the potential by one term i.e. the order of the polynomial is N = 1:

$$\phi = \phi_0 + \alpha_1 r. \qquad \text{(ii)}$$

The potential at the inner conductor r = a is $\phi = V_0$ and so

$$\nabla_t\phi = \hat{r}\frac{\partial\phi}{\partial r} + \hat{\theta}\frac{\partial\phi}{r\partial\theta} = \alpha_1. \qquad \text{(iii)}$$

The variational line capacitance using equation (14.1.5) is

$$C_v = \frac{\varepsilon_0\varepsilon_r\int\int|\nabla_t\phi|^2 dxdy}{\int_{s1}^{s2}(\nabla_t\phi\cdot dl)^2} = \frac{\varepsilon_0\varepsilon_r\int_0^{2\pi}\int_a^b \alpha_1^2 r\, dr d\theta}{\left[\int_a^b \alpha_1 dr\right]^2}$$

$$= \frac{\varepsilon_0\varepsilon_r 2\pi\frac{1}{2}(b^2 - a^2)\alpha_1^2}{\alpha_1^2(b - a)^2} = \varepsilon_0\varepsilon_r\pi\left[\frac{b + a}{b - a}\right]. \qquad \text{(iv)}$$

However, the *true value* of the line capacitance of a coaxial line is

$$C_0 = \frac{2\pi\varepsilon_0\varepsilon_r}{\ln(b/a)} \approx \pi\varepsilon_0\varepsilon_r\left[\frac{b + a}{b - a}\right]$$

$$\text{where, for } b > a;\ \ln\left(\frac{b}{a}\right) = 2\left[\left(\frac{b - a}{b + a}\right) + \frac{1}{3}\left(\frac{b - a}{b + a}\right)^3 + \cdots\right]$$

$$\approx 2\left[\frac{b - a}{b + a}\right]. \qquad \text{(v)}$$

The above expression is identical to that of the expression (iv). It shows that the variational expressions are approximate. Thus, error in the variational expression occurs due to neglecting the higher-order terms of the logarithmic expansion. If b = 2a, then $C_v \approx 3\pi\varepsilon_0\varepsilon_r$; however, its true value is $C_0 \approx 2.885\pi\varepsilon_0\varepsilon_r$. The variational method has a 3.97% error in C_v. The value of the variational capacitance is higher than the true value of the line capacitance. The error in the computed variational line capacitance is reduced by taking another term of the expression for the potential function, i.e. N = 2.

$$\phi = \alpha_0 + \alpha_1 r + \alpha_2 r^2 \text{ giving } \nabla_t\phi = \alpha_1 + 2\alpha_2 r. \qquad \text{(vi)}$$

$$C_v = \frac{\varepsilon_0\varepsilon_r\int_0^{2\pi}\int_a^b(\alpha_1 + 2\alpha_2 r)^2 r dr\, d\theta}{\left[\int_a^b(\alpha_1 + 2\alpha_2 r)dr\right]^2}$$

$$= \frac{2\pi\varepsilon_0\varepsilon_r\left[\frac{b^2 - a^2}{2}\alpha_1^2 + \frac{b^3 - a^3}{3}4\alpha_1\alpha_2 + (b^4 - a^4)\alpha_2^2\right]}{(b - a)^2\alpha_1^2 + (b^2 - a^2)(b - a)2\alpha_1\alpha_2 + (b^2 - a^2)^2\alpha_2^2}$$

$$\Rightarrow \frac{C_v}{\pi\varepsilon_0\varepsilon_r}\left[(b - a)^2\alpha_1^2 + (b^2 - a^2)(b - a)2\alpha_1\alpha_2 + (b^2 - a^2)^2\alpha_2^2\right]$$

$$- \left[(b^2 - a^2)\alpha_1^2 + \frac{8}{3}(b^3 - a^3)\alpha_1\alpha_2 + 2(b^4 - a^4)\alpha_2^2\right] = 0. \qquad \text{(vii)}$$

The variational line capacitance C_v is a function of two parameters, i.e. coefficients α_1 and α_2. We know that C_v is always higher than the true line capacitance. Thus, C_v has to be minimized with respect to α_1 and α_2, so that it approaches the true value of the line capacitance.

$$\frac{dC_V}{d\alpha_1} = 0, \quad \frac{dC_V}{d\alpha_2} = 0. \tag{viii}$$

Using equations (vii) and (viii), the following matrix equation is obtained:

$$\begin{bmatrix} \frac{2C_V}{\pi\varepsilon_0\varepsilon_r}(b-a)^2 - 2(b^2-a^2) & \frac{2C_V}{\pi\varepsilon_0\varepsilon_r}(b-a)(b^2-a^2) - \frac{8}{3}(b^3-a^3) \\ \frac{2C_V}{\pi\varepsilon_0\varepsilon_r}(b-a)(b^2-a^2) - \frac{8}{3}(b^3-a^3) & \frac{2C_V}{\pi\varepsilon_0\varepsilon_r}(b^2-a^2)^2 - 4(b^4-a^4) \end{bmatrix} \begin{bmatrix} \alpha_1 \\ \alpha_2 \end{bmatrix} = 0. \tag{ix}$$

The variational value of the line capacitance C_v is obtained from the nontrivial solution of the above equation. For the nontrivial solution, the determinant of the above equation is zero. For b=2a, it computes the $C_V = 2.889\pi\varepsilon_0\varepsilon_r$. The value of C_v for N = 2 is less than that of N = 1. The error in the computed value of C_v is only 0.14% higher than the true value of the line capacitance.

In the above example, we have adopted the *Rayleigh–Ritz method* to minimize the line capacitance to determine its value. Thus, if the variational line capacitance is C = N/D, its minimization with respect to the coefficients $\alpha_i(i = 1, 2, ...)$ requires the following condition:

$$\begin{aligned} \frac{\partial C}{\partial \alpha_i} &= \frac{1}{D}\left[\frac{\partial N}{\partial \alpha_i} - \frac{N}{D}\frac{\partial D}{\partial \alpha_i}\right] = 0 \\ \Rightarrow \frac{\partial N}{\partial \alpha_i} &- C\frac{\partial D}{\partial \alpha_i} = 0, \quad i = 1, 2, ..., n. \end{aligned} \tag{14.1.8}$$

The ratio of numerator function $N(\alpha_i)$ and denominator function $D(\alpha_i)$ defines the line capacitance $C(\alpha_i)$. Condition (14.1.8) provides n number of linear homogeneous equations. The solution of the system of linear equations determines the line capacitance. The coefficients $\alpha_i(i = 1, 2, ..., n)$ of the assumed polynomial function can also be determined by solving the set of homogeneous linear equations.

14.1.3 Green's Function Method-Based Variational Expression

Green's function formulation of the variational expression for the line capacitance provides the lower bound of the line capacitance, i.e. the upper bound on its characteristic impedance. It is illustrated in Fig (14.2). *The charge distribution on the line conductor is an unknown function.* The assumed charge distribution function, treated as the *trial function*, is based on physical consideration. We summarize below the Green's function method. The reader can follow Collin for more details [B.1].

Figure (14.2a) shows that conductor #1 is grounded and conductor #2 is positively charged to the potential $\phi = \phi_2 = V_0$, such that there is a charge distribution $\rho(x', y')$ on its surface. Alternatively, a charge distribution $\rho(x', y')$ placed on the surface of a conductor can create the potential function $\phi(x, y)$ around the conductor. The potential function $\phi(x, y)$ describes the electric field around the conductor. It satisfies Poisson's equation:

$$\nabla_t{}^2\phi(x,y) = -\frac{1}{\varepsilon_0\varepsilon_r}\rho(x',y'), \tag{14.1.9}$$

where the locations (x', y') and (x, y) are the *source point* and *field point*, respectively. The potential $\phi(x, y)$ could be taken as a *response* of the charge density $\rho(x', y')$ located on a conductor at the source point (x', y'). The Dirac's delta function $\delta(x-x')\delta(y-y')$ describes the unit charge density *excitation* on the surface of a conductor. The *response* of the unit charge density at the field point (x, y) is the potential *Green's function* $G(x, y|\, x', y')$. It also satisfies Poisson's equation and meets the boundary conditions given below:

$$\nabla_t^2 G(x,y|x',y') = \begin{cases} -\dfrac{1}{\varepsilon_r\varepsilon_0}\delta(x-x')\delta(y-y') \\ 0, \quad x \neq x', \; y \neq y' \end{cases} \tag{a}$$

where $G(x,y|x',y') = 0$ at the conductor#1, and also at ∞ (b).

(14.1.10)

Therefore, for a given problem, Green's function is obtained by solving the above homogeneous partial differential equation under the boundary conditions. The potential at any field location is a superposition of the potential responses due to the charges. For a distributed charge, it is given by the following integral:

$$\phi(x,y) = \oint_{C2} G(x,y|x',y')\rho(x',y')d\ell'. \tag{14.1.11}$$

The potential $\phi(x, y)$ at the surface of conductor #2, that is charged to potential V_0, is

$$\phi(x,y) = V_0 = \oint_{C2} G(x,y|x',y')\rho(x',y')d\ell'. \qquad (14.1.12)$$

In equation (14.1.12), the charge on the conductor #2 is described by the distribution function $\rho(x', y')$. Equation (14.1.12) is the *integral equation*, where V_0 is known and the charge distribution function $\rho(x', y')$ is unknown. By solving it, we can obtain the charge distribution on the conductor. In the space domain, this process is called *the method of moments*, where the integral equation is converted to a matrix equation [B.5, B.6]. The line capacitance *p.u.l.* is

$$C = \frac{Q}{V_0} \ \text{(a), where charge p.u.l.,} \ Q = \oint_{C2} \rho(x',y')dl' \ \text{(b).} \qquad (14.1.13)$$

The variables x', y', and dl' are dummy variables. Expression (14.1.13a) for the line capacitance is not variational in nature. To make the expression variational, we rewrite it as follows, in terms of Green's function:

$$\frac{1}{C_V} = \frac{V_0 Q}{Q^2} \Rightarrow \frac{1}{C_V} = \frac{\oint\oint_{C_2} G(x,y|x',y')\rho(x,y)\rho(x',y')dldl'}{\left[\oint_{C2}\rho(x',y')dl'\right]^2}. \qquad (14.1.14)$$

It could be shown that $\delta(1/C_v) = 0$, for a $\delta\{\rho(x', y')\}$ variation in the charge density $\rho(x, y)$, and $\delta^2(1/C_v) > 0$ [B.1]. Hence, the inverse variational line capacitance $1/C_v$ provides the minimum value for the line capacitance. The variational expression for the line capacitance, in term of the potential function, is

$$\frac{1}{C_V} = \frac{1}{Q^2}\oint_{C_2} \phi(x,y)\rho(x',y')dl'. \qquad (14.1.15)$$

For any charge distribution, the variational $1/C_v$ is higher than the true value $1/C_{true}$, $1/C_V > 1/C_{true}$, i.e. $C_V < C_{true}$. Hence, the value of variational line capacitance is always less than the true value of line capacitance. Green's function method gives the lower bound on a line capacitance, shown in Fig (14.2b). It means *Green's function method, using the unknown charge distribution, gives the upper bound on the characteristic impedance*, shown in Fig (14.2c). It is noted that the free space permittivity ε_0 is not incorporated in expression (14.1.15) for the line capacitance. It has been subsequently incorporated in the expression for the Green's function.

Two variational expressions for the line capacitance have been summarized above. The *energy method,* with an unknown potential function, gives the *upper bound* on the line capacitance. In this case, the parameters (coefficients) of the trial potential function are optimized to get the minimum value of variational line capacitance to bring it closer to the true value of the line capacitance. Likewise, *Green's function* method, with unknown charge distribution function, gives the *lower bound* on the line capacitance. The parameters (coefficients) of the trial charge distribution function are optimized to get the maximum value of the line capacitance. The maximized line capacitance approaches to its true value. Figure (14.2b and c) show the upper and lower bounds on the line capacitance and the characteristic impedance respectively. The average value of line capacitances, obtained from the lower and upper bounds, is more accurate. However, it is a time-consuming process. Thus, we have to decide one variational expression. The decision depends upon the numerical accuracy and efficiency of the numerical process. For a shielded microstrip line, Green's function method is more commonly used.

14.2 Variational Expression of Line Capacitance in Fourier Domain

The potential Green's function, in the *space domain*, for a boxed, i.e. shielded, microstrip is obtained in terms of the slowly convergent infinite series. Thus, the computation of the line capacitance of a shielded microstrip takes a longer time, if the sidewalls and top shield are kept at large distances to simulate an open microstrip structure. The open microstrip and the microstrip with top shield, however, without sidewalls, are analyzed more efficiently in the *Fourier domain.* In the Fourier domain, the second-order PDE is reduced to the second-order ODE and the closed-form Green's function is obtained. The usual Fourier domain method does not apply to a shielded microstrip line, as the Fourier transform is not defined for the boxed structure. However, the Fourier method, using the *discrete Fourier transform* (DFT), applies to a boxed microstrip line. The DFT formulation is known as the *static spectral domain analysis* (SDA). Moreover, the DFT method is also applicable to the multiple strip planar structures, such as the CPW lines, etc. All three methods are discussed in this chapter.

14.2.1 Transformation of Poisson Equation in Fourier Domain

Fourier transform of a function $\phi(x, y)$, with respect to a continuous variable x in the range $(-\infty, \infty)$, and the inverse Fourier transform $\tilde{\phi}(\beta, y)$ to recover the function $\phi(x, y)$ from it, are defined as follows:

$$\tilde{\phi}(\beta, y) = \int_{-\infty}^{\infty} \phi(x, y) e^{j\beta x} dx \quad \text{(a)}$$
$$\phi(x, y) = \frac{1}{2\pi}\int_{-\infty}^{\infty} \tilde{\phi}(\beta, y) e^{-j\beta x} d\beta \quad \text{(b)}, \qquad (14.2.1)$$

where β is a Fourier variable. Fourier transform is symmetrical with respect to β if the function $\phi(x, y)$ is symmetrical with respect to variable x,

$$\phi(x, y) = \phi(-x, y) \quad \text{(a)}, \quad \tilde{\phi}(\beta, y) = \tilde{\phi}(-\beta, y) \quad \text{(b)}. \qquad (14.2.2)$$

Fourier transform of a derivative of a function $\phi(x, y)$ is obtained as follows:

$$\frac{\partial \phi}{\partial x} = \frac{1}{2\pi}\int_{-\infty}^{\infty} (-j\beta)\tilde{\phi}(\beta, y) e^{-j\beta x} d\beta \quad \text{(a)}$$
$$\frac{\partial^2 \phi}{\partial x^2} = \frac{1}{2\pi}\int_{-\infty}^{\infty} \left[-\beta^2 \tilde{\phi}(\beta, y) e^{-j\beta x} d\beta\right] \quad \text{(b)}. \qquad (14.2.3)$$

Therefore, Fourier transform F of $\frac{\partial^2 \phi}{\partial x^2}$ is obtained from equations (14.2.1a) and (14.2.3b) as,

$$F\left[\frac{\partial^2 \phi}{\partial x^2}\right] = \int_{-\infty}^{\infty} \frac{\partial^2 \phi}{\partial x^2} e^{j\beta x} dx = -\beta^2 \tilde{\phi}(\beta, y). \qquad (14.2.4)$$

The derivative of the function $\phi(x,y)$ with respect to y is

$$\frac{\partial \phi}{\partial y} = \frac{1}{2\pi}\int_{-\infty}^{\infty} \frac{\partial \tilde{\phi}(\beta, y)}{\partial y} e^{-j\beta x} dx \quad \text{(a)}$$
$$\frac{\partial^2 \phi}{\partial y^2} = \frac{1}{2\pi}\int_{-\infty}^{\infty} \frac{\partial^2 \tilde{\phi}(\beta, y)}{\partial y^2} e^{-j\beta x} dx \quad \text{(b)}. \qquad (14.2.5)$$

Thus, $F\left[\frac{\partial^2 \phi}{\partial y^2}\right] = \int_{-\infty}^{\infty} \frac{\partial^2 \phi}{\partial y^2} e^{j\beta x} dx = \frac{d^2 \tilde{\phi}(\beta, y)}{\partial y^2}.$ (14.2.6)

Using equations (14.1.1a), Fourier transform of Poisson's equation (14.1.9) is

$$\int_{-\infty}^{\infty} \left(\frac{\partial^2 \phi}{\partial x^2} + \frac{\partial^2 \phi}{\partial y^2}\right) e^{j\beta x} dx = -\frac{1}{\varepsilon_0 \varepsilon_r}\int_{-\infty}^{\infty} f(x)\delta(y - y') e^{j\beta x} dx. \qquad (14.2.7)$$

On a thin strip conductor, located at distance y', the charge distribution f(x) is a function of x only. Hence, the charge distribution function $\rho(x', y')$ has been written, in the above equation, as

$$\rho(x', y') = f(x)\,\delta(y - y'). \qquad (14.2.8)$$

Poisson's equation in the Fourier domain is obtained from equation (14.2.7), on using equations (14.2.4) and (14.2.6):

$$\frac{d^2 \tilde{\phi}(\beta, y)}{dy^2} - \beta^2 \tilde{\phi}(\beta, y) = -\frac{1}{\varepsilon_0 \varepsilon_r}\tilde{f}(\beta)\delta(y - y'), \qquad (14.2.9)$$

where $\tilde{\phi}(\beta, y)$ and $\tilde{f}(\beta)$ are the potential and charge distribution function, respectively, in the Fourier domain. Equation (14.2.9) is an ODE not a PDE, like the original Poisson's equation.

14.2.2 Transformation of Variational Expression of Line Capacitance in Fourier Domain

To avoid taking the inverse Fourier transform of Green's function or potential function, the variational expression of the line capacitance is expressed in terms of Fourier transformed variables. Parseval theorem is helpful in this process. It expresses the definite integral of a product of two functions in the range $(-\infty, \infty)$ in terms of their Fourier transforms. It is established for integral I as follows:

$$I = \int_{-\infty}^{\infty} f(x)\phi(x, y') dx = \int_{-\infty}^{\infty} f(x) dx \left[\frac{1}{2\pi}\int_{-\infty}^{\infty} \tilde{\phi}(\beta, y') e^{-j\beta x} d\beta\right]$$
$$= \frac{1}{2\pi}\int_{-\infty}^{\infty} \tilde{\phi}(\beta, y') d\beta \left[\int_{-\infty}^{\infty} f(x) e^{j(-\beta)x} dx\right]$$
$$\Rightarrow \int_{-\infty}^{\infty} f(x)\phi(x, y') dx = \frac{1}{2\pi}\int_{-\infty}^{\infty} \tilde{\phi}(\beta, y')\tilde{f}(\beta) d\beta. \qquad (14.2.10)$$

Using equation (14.2.10), the variational expression of the line capacitance, given by equation (14.1.15), in the Fourier domain is

$$\frac{1}{C} = \frac{1}{C_v} = \frac{1}{2\pi Q^2}\int_{-\infty}^{\infty} \tilde{\phi}(\beta, y')\tilde{f}(\beta) d\beta. \qquad (14.2.11)$$

In the above equation, y' is the location of strip conductor containing charge distribution. Using Parseval theorem, the potential function, given by equation (14.1.11), in the Fourier domain, is

$$\tilde{\phi}(\beta, y') = \tilde{G}(\beta, y')\tilde{f}(\beta), \qquad (14.2.12)$$

where $\tilde{G}(\beta, b)$ is Green's function in the Fourier domain. Using the symmetrical charge distribution, i.e. $\tilde{f}(-\beta) = \tilde{f}(\beta)$ on a conductor, expression (14.2.12) is rewritten as

$$\frac{1}{C_V} = \frac{1}{\pi}\int_0^{\infty} \left[\frac{\tilde{f}(\beta)}{Q}\right]^2 \tilde{G}(\beta, y') d\beta. \qquad (14.2.13)$$

Again, it is noted that ε_0 is not incorporated in equation (14.2.13). However, the permittivity of medium is normally used with the expression for the Green's function.

Expression (14.2.13) requires assumed trial charge distribution function, $\left[\tilde{f}(\beta)/Q\right]$, in the Fourier domain. Therefore, before preceding further, some popular charge distribution functions on the strip conductor, and their Fourier transforms, are present below.

14.2.3 Fourier Transform of Some Charge Distribution Functions

The charge on a strip conductor distributes itself symmetrically in such a way that it has a singularity (sudden large increase) at the edges of a strip conductor. *Maxwell's distribution function*, obtained using the conformal mapping method, describes such a charge distribution on the strip conductor. Mitra and Itoh have also computed the charge distribution, showing singularity, on a strip conductor using the *function theoretic method* [J.1]. However, simpler functions are also used in practice, as the accuracy of the variational method is not very sensitive toward the assumed charge distribution function. Some symmetrical charge distribution functions on a conducting strip, shown in Fig (14.3), and their Fourier transforms are presented below.

Constant Charge Distribution

Figure (14.3a) shows a symmetrically placed strip conductor of width w. Figure (14.3b) shows the constant charge distribution f(x) on a strip conductor of width w [J.2]. It does not account for the edge singularity.

$$f(x) = \begin{cases} 1, & -w/2 \le x \le w/2 \\ 0, & |x| > w/2 \end{cases}. \tag{14.2.14}$$

Fourier transform of the above function is

$$\tilde{f}(\beta) = \int_{-\infty}^{\infty} f(x)e^{j\beta x}dx = \int_{-w/2}^{w/2} e^{j\beta x}dx$$
$$= \frac{2}{\beta}\frac{\left[e^{j\beta w/2} - e^{-j\beta w/2}\right]}{j2} = w\left[\frac{\sin(\beta w/2)}{\beta w/2}\right]. \tag{14.2.15}$$

The total charge *p.u.l* on the strip conductor is $Q = \int_{-w/2}^{w/2} dx = w$. Therefore, the normalized charge distribution function in the Fourier domain is

$$\frac{\tilde{f}(\beta)}{Q} = \left[\frac{\sin(\beta w/2)}{\beta w/2}\right]. \tag{14.2.16}$$

Linear Charge Distribution

Figure (14.3c) shows the linear charge distribution on a strip conductor [J.2]. It is a better description of the charge distribution on a strip conductor, although it does not show the edge singularity. However, it shows an increase in charge accumulation at the edges of the strip conductor:

$$f(x) = \begin{cases} |x|, & -w/2 \le x \le w/2 \\ 0, & |x| > w/2 \end{cases}. \tag{14.2.17}$$

Fourier transform of function (14.2.17) is obtained as follows:

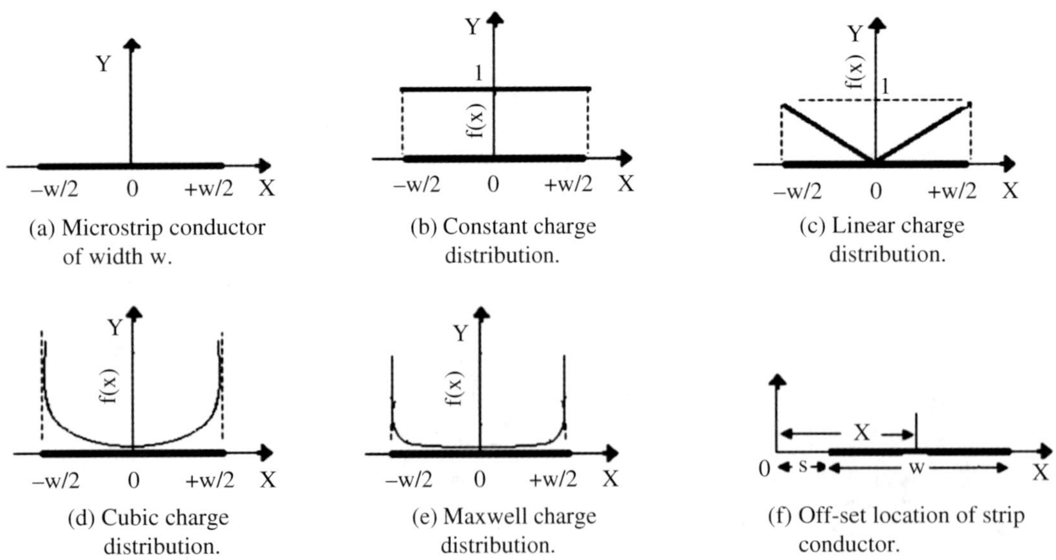

(a) Microstrip conductor of width w. (b) Constant charge distribution. (c) Linear charge distribution. (d) Cubic charge distribution. (e) Maxwell charge distribution. (f) Off-set location of strip conductor.

Figure 14.3 A few common charge distribution functions and potential distribution.

$$\tilde{f}(\beta) = \int_{-\infty}^{\infty} |x| e^{j\beta x} dx = \int_{-\frac{w}{2}}^{\frac{w}{2}} |x|[\cos \beta x + j \sin \beta x] dx$$
$$= 2\int_0^{\frac{w}{2}} x \cos \beta x dx$$
$$\Rightarrow \tilde{f}(\beta) = \frac{w^2}{2}\left[\frac{\sin(\beta w/2)}{\beta w/2}\right] - \frac{w^2}{4}\left[\frac{\sin(\beta w/4)}{\beta w/4}\right]^2. \qquad (14.2.18)$$

The function |x| sin βx an odd function; so its integral is zero. The function |x| cos βx is an even function. The total charge *p.u.l* on the strip conductor and also the normalized charge distribution function in the Fourier domain are given below:

$$Q = \int_{\frac{-w}{2}}^{\frac{w}{2}} |x| dx = 2\int_0^{\frac{w}{2}} x dx = \frac{w^2}{4} \qquad (a)$$
$$\frac{\tilde{f}(\beta)}{Q} = 2\left[\frac{\sin(\beta w/2)}{\beta w/2}\right] - \left[\frac{\sin(\beta w/4)}{\beta w/4}\right]^2 \qquad (b).$$
$$(14.2.19)$$

Cubic Charge Distribution

Figure (14.3d) shows a more realistic cubic charge distribution on a strip conductor [J.3]:

$$f(x) = \begin{cases} 1 + A\left|\frac{x}{w/2}\right|^3, & -w/2 \le x \le w/2 \\ 0, & |x| > w/2. \end{cases} \qquad (14.2.20)$$

Rayleigh–Ritz optimization process is used to evaluate constant A. However, normally, we use A = 1. Fourier transform of the function (14.2.20) is obtained as follows:

$$\tilde{f}(\beta) = \int_{-\infty}^{\infty} \left[1 + A\left|\frac{2x}{W}\right|^3\right] e^{j\beta x} dx$$
$$= \int_{-\frac{w}{2}}^{\frac{w}{2}} e^{j\beta x} dx + A\int_{-w/2}^{w/2} \left|\frac{2x}{w}\right|^3 e^{j\beta x} dx$$
$$\Rightarrow \tilde{f}(\beta) = \frac{2}{\beta}(1 + A)\sin\left(\frac{\beta w}{2}\right) + \frac{12A}{w\beta^2}$$
$$\left[\cos\frac{\beta w}{2} - \frac{4}{\beta w}\sin\frac{\beta w}{2} + \frac{16}{(\beta w)^2}\sin^2\left(\frac{\beta w}{4}\right)\right]. \qquad (14.2.21)$$

Both the total charge *p.u.l* on the strip conductor and the normalized charge distribution function in the Fourier domain are given below:

$$Q = \int_{-\frac{w}{2}}^{\frac{w}{2}} \left[1 + A\left|\frac{2x}{w}\right|^3\right] dx = w + \frac{16A}{w^3}\int_0^{w/2} x^3 dx$$
$$= \frac{w}{4}[4 + A] \qquad (a)$$
$$\frac{\tilde{f}(\beta)}{Q} = \frac{4(A+1)}{A+4}\left\{\frac{\sin(\beta w/2)}{\beta w/2}\right\} + \frac{12A}{(A+4)(\beta w/2)^2}$$
$$\left[\cos\left(\frac{\beta w}{2}\right) - 2\left\{\frac{\sin(\beta w/2)}{\beta w/2}\right\} + \left\{\frac{\sin(\beta w/4)}{\beta w/4}\right\}^2\right] \qquad (b).$$
$$(14.2.22)$$

Maxwell's Charge Distribution Function

Figure (14.3e) shows Maxwell's charge distribution function [J.4]. It shows the edge singularity at $x = \pm w/2$.

$$f(x) = \begin{cases} \dfrac{1}{\sqrt{1 - \left(\frac{x}{w/2}\right)^2}}, & -w/2 \le x \le w/2 \\ 0 & |x| > 0. \end{cases} \qquad (14.2.23)$$

Fourier transform of function (14.2.23), in terms of the zeroth-order Bessel's function, is obtained below from the integral form of Bessel's function of nth order [B.7]:

$$J_n(\beta) = \frac{1}{\pi}\int_{-\pi/2}^{\pi/2} e^{j(\beta\sin\theta - n\theta)} d\theta \qquad (a)$$
$$\text{For } n = 0;\ J_0(\beta) = \frac{1}{\pi}\int_{-\pi/2}^{\pi/2} e^{j(\beta\sin\theta)} d\theta \qquad (b)$$

On substitution of $\sin\theta = t$,

$$J_0(\beta) = \frac{1}{\pi}\int_{-1}^{1} \frac{e^{j\beta t}}{\sqrt{1-t^2}} dt \qquad (c).$$
$$(14.2.24)$$

Fourier transform of equation (14.2.23) is

$$\tilde{f}(\beta) = \int_{-\infty}^{\infty} \frac{1}{\sqrt{1-(2x/w)^2}} e^{j\beta x} dx$$
$$= \int_{-w/2}^{w/2} \frac{1}{\sqrt{1-(2x/w)^2}} e^{j\beta x} dx. \qquad (14.2.25)$$

On using the substitution $2x/w = t$ and equation (14.2.24c), we get

$$\tilde{f}(\beta) = \frac{w}{2}\int_{-1}^{1} \frac{e^{j\left(\frac{\beta w}{2}\right)t}}{\sqrt{1-t^2}} dt = \frac{\pi w}{2} J_0\left(\frac{\beta w}{2}\right). \qquad (14.2.26)$$

Both the total charge *p.u.l* on the strip conductor and the normalized charge distribution function are

$$Q = \int_{-w/2}^{w/2} \frac{1}{\sqrt{1-(2x/w)^2}} dx = \frac{\pi w}{2} \qquad \text{(a)}$$

$$\frac{\tilde{f}(\beta)}{Q} = J_0(\beta w/2) \qquad \text{(b).}$$

$$(14.2.27)$$

Modified Maxwell's Distribution Function

Expression (14.2.23) is modified to obtain the following modified Maxwell's distribution function that is suitable to generate a sequence of functions, showing the edge singularity [B.8]:

$$\rho_i(x) = \begin{cases} \dfrac{\cos\left[(i-1)\pi\left\{\frac{x-s}{w}\right\}\right]}{\sqrt{1-\left[\frac{2(x-s)-w}{w}\right]^2}}, & (s \le x \le s+w) \quad i = 1,2,3,\ldots \\ 0, & \text{otherwise.} \end{cases} \tag{14.2.28}$$

The above charge distribution function $\rho_i(x)$ is used with the DFT method discussed in the section (14.6). The location of the strip conductor is shown in Fig (14.3f). The strip conductor of width w is located at $x = s$ and the center of the strip is at $x = s + w/2$. Thus, the charge distribution density function $\rho_i(x)$ is singular at $x = s$ and $s + w$. The factor $\cos[(i-1)\pi\{(x-s)/w\}]$ helps to generate the *trial basis functions* for $i = 1,2,3, \ldots$ useful in Galerkin's method. The total charge *p.u.l* distributed over the width of the strip conductor is

$$Q_i = \int_s^{s+w} \rho_i(x)dx, \Rightarrow Q_i = \int_s^{s+w} \frac{\cos\left[(i-1)\pi\left\{\frac{x-s}{w}\right\}\right]}{\left[1-\left\{\frac{2(x-s)}{w}-1\right\}^2\right]^{1/2}} dx. \tag{14.2.29}$$

Let $(x-s)/w = \sin(t)$; $dx = w\cos t\ dt$. For $x = s, s+w$; $t = 0, \pi/2$.

$$Q_i = \int_0^{\pi/2} \frac{\cos[(i-1)\pi \sin t]}{\left[1-\{2\sin t - 1\}^2\right]^{1/2}} w\cos t\, dt. \tag{14.2.30}$$

Let $2\sin(t) - 1 = p$, $\cos(t)\,dt = 0.5\,dp$. For $t = 0, \pi/2$; $p = -1, +1$.

$$Q_i = \frac{w}{2}\int_{-1}^{1} \frac{\cos\left[(i-1)\frac{\pi}{2}(1+p)\right]}{[1-p^2]^{1/2}} dp. \tag{14.2.31}$$

The cos [ ] could be expressed in the form of exponentials and then equation (14.2.24c) could be used to evaluate the above integral in term of the zeroth-order Bessel's function to get the following expression:

$$Q_i = \frac{w\pi}{2} J_0\left[(i-1)\frac{\pi}{2}\right] \cos\left[(i-1)\frac{\pi}{2}\right], \quad i = 1,2,3,\ldots \tag{14.2.32}$$

Discrete Fourier Transform of $\rho_i(x)$

The discrete Fourier transform of function $\rho_i(x)$, over the range $0 \le x \le L$, is defined as follows:

$$\tilde{\rho}_i(\beta_n) = \int_0^L \rho_i(x)\sin(\beta_n x)dx, \tag{14.2.33}$$

where $\beta_n = n\pi/L$, $n = 1, 2, \ldots$ is the discrete Fourier variable. Therefore, the discrete Fourier transform of equation (14.2.28) is

$$\tilde{\rho}_i(\beta_n) = \int_s^{s+w} \frac{\cos\left[(i-1)\pi\frac{x-s}{w}\right]}{\left[1-\left\{\frac{2(x-s)}{w}-1\right\}^2\right]^{1/2}} \sin(\beta_n x)dx. \tag{14.2.34}$$

Let $2((x-s)/w) - 1 = p$. Thus, $dx = (w/2)dp$ and for $x = s, (s+w)$; $p = -1, +1$. The above integral is written as

$$\tilde{\rho}_i(\beta_n) = \frac{w}{2}\int_{-1}^{1} \frac{\cos\left[(i-1)\frac{\pi}{2}(1+p)\right]\sin\left[\beta_n\left\{s+\frac{w}{2}(1+p)\right\}\right]}{[1-p^2]^{\frac{1}{2}}} dp. \tag{14.2.35}$$

Again, cos [ ] and sin [ ] could be expressed in the form of exponentials and then equation (14.2.24c) could be used to evaluate the above integral giving the following expression:

$$\begin{aligned} \tilde{\rho}_i(\beta_n) = {} & \frac{w\pi}{j8}\left[e^{j\left\{(i-1)\frac{\pi}{2}+\beta_n\left(s+\frac{w}{2}\right)\right\}} J_0\left\{\frac{\beta_n w}{2}+(i-1)\frac{\pi}{2}\right\} + e^{-j\left\{(i-1)\frac{\pi}{2}-\beta_n\left(s+\frac{w}{2}\right)\right\}} J_0\left\{\frac{\beta_n w}{2}-(i-1)\frac{\pi}{2}\right\}\right] \\ & - \frac{w\pi}{j8}\left[e^{j\left\{(i-1)\frac{\pi}{2}-\beta_n\left(s+\frac{w}{2}\right)\right\}} J_0\left\{\frac{\beta_n w}{2}-(i-1)\frac{\pi}{2}\right\} + e^{-j\left\{(i-1)\frac{\pi}{2}+\beta_n\left(s+\frac{w}{2}\right)\right\}} J_0\left\{\frac{\beta_n w}{2}+(i-1)\frac{\pi}{2}\right\}\right]. \end{aligned} \tag{14.2.36}$$

For $M_i = (i-1)\frac{\pi}{2} + \beta_n\left(s + \frac{w}{2}\right)$ and $N_i = -(i-1)\frac{\pi}{2} + \beta_n\left(s + \frac{w}{2}\right)$, expression (14.2.36) is written as

$$\widetilde{\rho}_i(\beta_n) = \frac{w\pi}{4}\left[\sin(M_i)J_0(M_i - \beta_n s) + \sin(N_i)J_0(N_i - \beta_n s)\right],\quad i = 1, 2, 3, \ldots. \tag{14.2.37}$$

Finally, using equations (14.2.32) and (14.2.37) the normalized charge distribution function is

$$\frac{\widetilde{\rho}_i(\beta_n)}{Q_i} = \frac{\sin(M_i)J_0(M_i - \beta_n s) + \sin(N_i)J_0(N_i)}{2J_0\left[(i-1)\frac{\pi}{2}\right]\cos\left[(i-1)\frac{\pi}{2}\right]},\quad i = 1, 2, \ldots. \tag{14.2.38}$$

14.3 Analysis of Microstrip Line by Variational Method

This section presents the space domain-based variational method to compute the line capacitance of a shielded microstrip line called the *boxed microstrip line*, shown in Fig (14.4a) [B.1, B.8, B.10–B.12]. It also presents the Fourier domain variational method to compute the line capacitance of an *open microstrip line*, shown in Fig (14.5a). Both the variational methods are further applicable to a microstrip line under the multilayer environment [B.13, B.14]. The *transverse transmission line* (TTL) technique conveniently provides Green's function for the multilayer microstrip and CPW structures [J.5–J.8]. The TTL method is discussed in section (14.4).

14.3.1 Boxed Microstrip Line (Green's Function Method in Space Domain)

Figure (14.4a) shows a boxed microstrip line of width w, printed on the isotropic substrate of relative permittivity ε_{r1} and substrate thickness h_1 [B.9, J.3]. The height of the top shield, i.e. the electric wall (EW), from the ground conductor is H. The separation between the sidewalls (EW) is L. The *region*-I, $-L/2 \le x \le L/2,\ 0 \le y \le h_1$, contains the substrate of relative permittivity ε_{r1} and thickness h_1. The region-II, $-L/2 \le x \le L/2,\ h_1 \le y \le h_2(H - h_1)$, contains the air medium with $\varepsilon_{r2} = 1$ and thickness h_2.

The charge density $\rho(x', y')$ on the strip conductor creates the potential field, $\phi_i(x, y)$, in both the regions. It is obtained from the solution of 2D-Poisson's equation:

$$\frac{\partial^2\phi_i}{\partial x^2} + \frac{\partial^2\phi_i}{\partial y^2} = -\frac{\rho(x', y')}{\varepsilon_i},\quad (i = 1, 2). \tag{14.3.1}$$

The functions $\phi_1(x, y)$ and $\phi_2(x, y)$ are potential functions in the regions-I and -II, respectively. Four walls are the electric walls, such that $\phi_i(x,y) = 0$ on them. Equation (14.3.1), in terms of the Green's function $G_i(x, y|x', y')$, is rewritten as

$$\frac{\partial^2 G_i}{\partial x^2} + \frac{\partial^2 G_i}{\partial y^2} = \begin{cases} -\dfrac{1}{\varepsilon_i}\delta(x - x')\delta(y - y'), & x = x', y = y' \\ 0, & x \ne x', y \ne y', \end{cases} \tag{14.3.2}$$

where (x', y') is the location of the unit charge on the strip conductor of width w and (x, y) is the location of a potential field point. The Green's function $G(x, y|x', y')$ is a potential response of the system, i.e. the microstrip structure under excitation of the unit charge source. For an infinitely thin conductor located at $y = h_1$, it is described by Dirac's delta function:

$$\rho(x', y') = \rho(x')\delta(y - h_1). \tag{14.3.3}$$

The total potential at any point (x, y), due to the $\rho(x', y')$ charge distribution, is

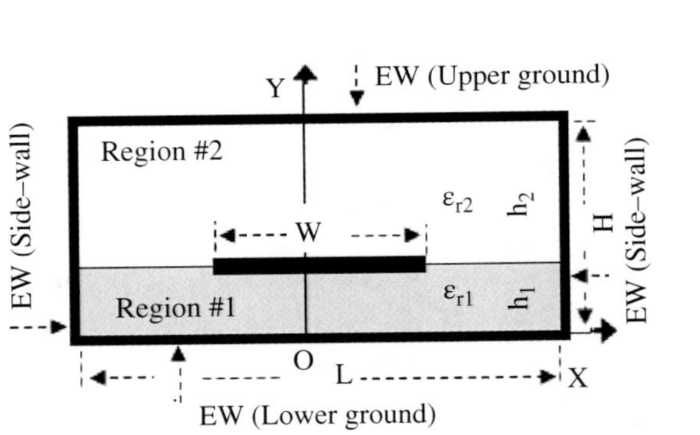

(a) Boxed Microstrip Line.

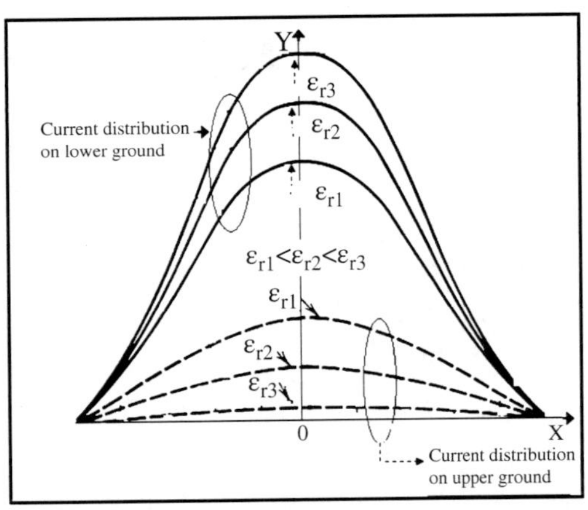

(b) Current distribution on lower and upper ground conductors. [B.9].

Figure 14.4 Boxed microstrip and current distributions on its ground planes. *Source:* Dworsky and Lawrence [B.9].

$$\phi_i(x,y) = \int_{-\frac{w}{2}}^{\frac{w}{2}}\int_{y'=h_1-0}^{y'=h_1+0} G_i(x,y|x',h_1)\rho(x')\delta(y-h_1)dy'dx'$$
$$\Rightarrow \phi_i(x,y) = \int_{-\frac{w}{2}}^{\frac{w}{2}} G_i(x,y|x',h_1)\rho(x')dx'. \tag{14.3.4}$$

For $x \neq x'$, $y \neq y'$, Laplace's equation of Green's function is obtained:

$$\frac{\partial^2 G_i}{\partial x^2} + \frac{\partial^2 G_i}{\partial y^2} = 0, \quad (i = 1, 2). \tag{14.3.5}$$

On using the method of separation of variables, Green's function $G_i(x, y|x', y')$ is written as a product of the *partial Green's functions*, $G_i^x(x|x')$ and $G_i^y(y|y')$, and equation (14.3.6) is separated as two ODE:

$$G_i(x,y|x',y') = G_i^x(x|x')G_i^y(y|y'), \tag{14.3.6}$$

$$\frac{d^2G_i^x(x)}{dx^2} + k^2G_i^x(x) = 0 \quad (a), \qquad \frac{d^2G_i^y(y)}{dy^2} - k^2G_i^y(y) = 0 \quad (b). \tag{14.3.7}$$

The sign of constant k^2 is selected to get the harmonic solution (sine and cosine) with respect to the variable x in both the regions. The solution of equation (14.3.7a) is

$$G_i^x = A\cos kx + B\sin kx, \quad (i = 1, 2). \tag{14.3.8}$$

At $x = \pm L/2$, on the electric wall, $G^x(x) = 0$. The plane of symmetry OY is a magnetic wall so that at $x = 0$, $dG_i^x/dx = 0$. The following expression is obtained under the boundary conditions:

$$B = 0,\ k = \frac{n\pi}{L}, \quad n = 1, 3, 5.... \quad (a)$$
$$G_{i\,n}^x(x) = A_n\cos\left(\frac{n\pi}{L}x\right) \quad (b). \tag{14.3.9}$$

Therefore, the partial and complete Green's functions are

$$G_{i\,n}^x(x) = A_n\cos\left(\frac{n\pi}{L}x\right), \quad (i = 1, 2) \quad (a)$$
$$G_{i\,n}(x,y) = \sum_{n=1,3,5...}^{\infty} A_nG_i^y(y)\cos\left(\frac{n\pi}{L}x\right) \quad (b). \tag{14.3.10}$$

Green's function is a symmetrical function with respect to x. Let us neglect constant A_n that can be absorbed in the solution of partial Green's function $G_i^y(y)$. The following expression is obtained on substituting equation (14.3.10) in equation (14.3.2):

$$\sum_{n,\,odd}\left[-\left(\frac{n\pi}{L}\right)^2G_{i\,n}^y(y) + \frac{d^2G_{i\,n}^y(y)}{dy^2}\right]\cos\left(\frac{n\pi}{L}x\right)$$
$$= -\frac{1}{\varepsilon_i}\delta(x-x')\delta(y-y'), \quad (i = 1, 2). \tag{14.3.11}$$

Equation (14.3.11) is the cosine Fourier series expansion of the Dirac's delta function $\delta(x-x')$ of strength $[-\{\delta(y-y')\}/\varepsilon_i]$. The terms inside the bracket [ ] of equation (14.3.11) is the coefficient of expansion. The coefficients of expansion are evaluated in terms of the orthogonal functions $\cos(n\pi/L)$. To achieve it, expression (14.3.11) is multiplied by $\cos(m\pi/L)$ and integrated over the range $-L/2 \leq x \leq L/2$:

$$\sum_{n,\,odd}\left[-\left(\frac{n\pi}{L}\right)^2G_{i\,n}^y(y) + \frac{d^2G_{in}^y(y)}{dy^2}\right]\int_{-\frac{L}{2}}^{\frac{L}{2}}\cos\left(\frac{n\pi}{L}x\right)\cos\left(\frac{m\pi}{L}x\right)dx = -\frac{\delta(y-y')}{\varepsilon_i}\int_{-\frac{L}{2}}^{\frac{L}{2}}\cos\left(\frac{m\pi}{L}x\right)\delta(x-x')dx \quad (a)$$
$$\text{For } m \neq n, \int_{-\frac{L}{2}}^{\frac{L}{2}}\cos\left(\frac{n\pi}{L}x\right)\cos\left(\frac{m\pi}{L}x\right)dx = 0 \quad (b), \quad \text{for } m = n, \int_{-\frac{L}{2}}^{\frac{L}{2}}\cos^2\left(\frac{n\pi}{L}x\right)dx = \frac{L}{2} \quad (c). \tag{14.3.12}$$

Equation (14.3.12a) is reduced to

$$\left[-\left(\frac{n\pi}{L}\right)^2G_{i\,n}^y(y) + \frac{d^2G_{i\,n}^y(y)}{dy^2}\right]\frac{L}{2}$$
$$= -\frac{1}{\varepsilon_i}\cos\left(\frac{n\pi}{L}x'\right)\delta(y-y'). \tag{14.3.13}$$

For $y \neq y'$, equation (14.3.13) is reduced to the following homogeneous equation:

$$\frac{d^2G_{i\,n}^y(y)}{dy^2} - \left(\frac{n\pi}{L}\right)^2G_{i\,n}^y(y) = 0. \tag{14.3.14}$$

- **For Region-I:** ($i = 1$, $0 \leq y \leq h_1$). The solution of equation (14.3.14) in the region-I is

$$G_{1n}^y(y) = C_1\sinh\left(\frac{n\pi}{L}y\right) + D_1\cosh\left(\frac{n\pi}{L}y\right). \tag{14.3.15}$$

Boundary condition: At $y = 0$, $G^y_{1n}(y) = 0$, giving $D_1 = 0$. The y-dependent partial Green's function in the region-I is

$$G^y_{1n}(y) = C_1 \sinh\left(\frac{n\pi}{L}y\right), \quad 0 \le y \le h_1. \quad (14.3.16)$$

- **For Region-II:** ($i = 2$, $h_1 \le y \le H$). The solution of equation (14.3.14) in the region-II is

$$G^y_{2n}(y) = C_2 \sinh\left[\frac{n\pi}{L}(H-y)\right] + D_2 \cosh\left[\frac{n\pi}{L}(H-y)\right]. \quad (14.3.17)$$

Boundary condition: At $y = H$, $G^y_{in}(y) = 0$, giving $D_2 = 0$. The y-dependent partial Green's function in the region-II is

$$G^y_{2n}(y) = C_2 \sinh\left[\frac{n\pi}{L}(H-y)\right], \quad h_1 \le y \le H. \quad (14.3.18)$$

The constants C_1 and C_2 are determined from the source conditions:

i) At $y = h_1$, the potential function $G^y_{in}(y)$ is continuous, i.e.

$$G^y_{1n}(x, h_1 - 0) = G^y_{2n}(x, h_1 + 0). \quad (14.3.19)$$

On substituting equations (14.3.16) and (14.3.18) in equation (14.3.19),

$$C_1 \sinh\left(\frac{n\pi}{L}h_1\right) = C_2 \sinh\left[\frac{n\pi}{L}(H-h_1)\right]. \quad (14.3.20)$$

ii) At the source (charge) plane, i.e. at $y = h_1$, a derivative of the potential function, i.e. Green's function $G^y_{in}(y)$, is discontinuous. To see it, let us integrate equation (14.3.13) from the limits $h_1 - 0$ to $h_1 + 0$:

$$\frac{\varepsilon_i L}{2}\int_{h_1-0}^{h_1+0}\left[-\left(\frac{n\pi}{L}\right)^2 G^y_{in}(y) + \frac{d^2G^y_{in}(y)}{dy^2}\right]dy$$
$$= -\cos\left(\frac{n\pi x'}{L}\right)\int_{h_1-0}^{h_1+0}\delta(y-y')\,dy, \quad (i = 1, 2).$$

On taking into account, $G^y_{in}(h_1 + 0) = G^y_{in}(h_1 - 0)$ at the source point,

$$\frac{L}{2}\left\{\left[\varepsilon_2\frac{dG^y_{2n}(y)}{dy}\right]_{y=h_1+0} - \left[\varepsilon_1\frac{dG^y_{1n}(y)}{dy}\right]_{y=h_1-0}\right\}$$
$$= -\cos\left(\frac{n\pi x'}{L}\right). \quad (14.3.21)$$

On substituting equations (14.3.16) and (14.3.18) in equation (14.3.21),

$$\frac{n\pi}{2}\left[\varepsilon_2 C_2 \cosh\left(\frac{n\pi}{L}(H-h_1)\right) + \varepsilon_1 C_1 \cosh\left(\frac{n\pi}{L}h_1\right)\right]$$
$$= \cos\left(\frac{n\pi}{L}x'\right), \quad (14.3.22)$$

where $\varepsilon_1 = \varepsilon_0\varepsilon_{r1}$ and $\varepsilon_2 = \varepsilon_0\varepsilon_{r2}$. The constants C_1 and C_2 are evaluated from equations (14.3.20) and (14.3.22):

$$C_1 = \frac{\frac{2}{n\pi}\cos\left(\frac{n\pi}{L}x'\right)\sinh\left[\frac{n\pi}{L}(H-h_1)\right]}{\varepsilon_1\cosh\left(\frac{n\pi}{L}h_1\right)\sinh\left[\frac{n\pi}{L}(H-h_1)\right] + \varepsilon_2\sinh\left(\frac{n\pi}{L}h_1\right)\cosh\left[\frac{n\pi}{L}(H-h_1)\right]}. \quad (14.3.23)$$

$$\text{Let } H - h_1 = h_2, \quad C_1 = \frac{\frac{2}{n\pi}\cos\left(\frac{n\pi}{L}x'\right)\sinh\left(\frac{n\pi}{L}h_2\right)}{\varepsilon_1\cosh\left(\frac{n\pi}{L}h_1\right)\sinh\left(\frac{n\pi}{L}h_2\right) + \varepsilon_2\sinh\left(\frac{n\pi}{L}h_1\right)\cosh\left(\frac{n\pi}{L}h_2\right)} \quad \text{(a)}$$

$$\text{Likewise, } C_2 = \frac{\frac{2}{n\pi}\cos\left(\frac{n\pi}{L}x'\right)\sinh\left(\frac{n\pi}{L}h_1\right)}{\varepsilon_1\cosh\left(\frac{n\pi}{L}h_1\right)\sinh\left(\frac{n\pi}{L}h_2\right) + \varepsilon_2\sinh\left(\frac{n\pi}{L}h_1\right)\cosh\left(\frac{n\pi}{L}h_2\right)} \quad \text{(b)}. \quad (14.3.24)$$

Finally, Green's functions in two regions are obtained from equations (14.3.10b), (14.3.16), and (14.3.18):

$$G_{1n}(x,y) = \sum_{n=1,3,\ldots}^{\infty} C_1 \cos\left(\frac{n\pi}{L}x\right) \sinh\left(\frac{n\pi}{L}y\right), \quad y \le h_1 \qquad \text{(a)}$$

$$G_{2n}(x,y) = \sum_{n=1,3,\ldots}^{\infty} C_2 \cos\left(\frac{n\pi}{L}x\right) \sinh\left\{\frac{n\pi}{L}(H-y)\right\}, \quad y \ge h_1 \qquad \text{(b)}. \tag{14.3.25}$$

The constants C_1 and C_2 are given by equation (14.3.24). Green's function on the central conductor located at $y = h_1$ is

$$G_{1n}(x, h_1|x', h_1) = \sum_{n=1,3,\ldots} \frac{\frac{2}{n\pi}\cos\left(\frac{n\pi}{L}x'\right)\sinh\left(\frac{n\pi}{L}h_2\right)\cos\left(\frac{n\pi}{L}x\right)\sinh\left(\frac{n\pi}{L}h_1\right)}{\varepsilon_1\cosh\left(\frac{n\pi}{L}h_1\right)\sinh\left(\frac{n\pi}{L}h_2\right) + \varepsilon_2\sinh\left(\frac{n\pi}{L}h_1\right)\cosh\left(\frac{n\pi}{L}h_2\right)}$$

$$G_{1n}(x, h_1|x', h_1) = \sum_{n=1,3,\ldots} \frac{2}{n\pi Y}\cos\left(\frac{n\pi}{L}x'\right)\cos\left(\frac{n\pi}{L}x\right) \qquad \text{(a)}$$

$$\text{where, parameter } Y = \varepsilon_1\coth\left(\frac{n\pi}{L}h_1\right) + \varepsilon_2\coth\left(\frac{n\pi}{L}h_2\right) \qquad \text{(b)}. \tag{14.3.26}$$

The parameter Y is the *admittance function* of the *equivalent transverse transmission line (TTL)* of a given microstrip structure. The equivalent transmission line concept is useful in obtaining Green's function of a multilayer microstrip line. It is discussed in section (14.4).

The Green's function in the present case is the potential response due to the unit charge density excitation. The elemental charge density $\rho(x')dx'$ on the strip conductor creates an elemental potential $dV_0 = G_{1n}(x, h_1|x', h_1)\rho(x')dx'$. Thus, the potential at the central conductor due to the charge density distributed over the strip width $-w/2 \le x' \le w/2$ is obtained as

$$V_0 = \int_{-\frac{w}{2}}^{\frac{w}{2}} G_{1n}(x, h_1|x', h_1)\rho(x')dx'. \tag{14.3.27}$$

$$V_0 Q = \int_{-\frac{w}{2}}^{\frac{w}{2}} V_0\rho(x)dx = \int_{-\frac{w}{2}}^{\frac{w}{2}}\int_{-\frac{w}{2}}^{\frac{w}{2}} G_{1n}(x, h_1|x', h_1)\rho(x')\rho(x)dxdx'$$

$$\Rightarrow V_0 Q = \sum_{n=1,3,\ldots} \frac{2\ 4}{n\pi Y}\left[\int_0^{\frac{w}{2}} \cos\left(\frac{n\pi}{L}x\right)\rho(x)dx\right]^2. \tag{14.3.28}$$

The variables x' and x, on the strip conductor, are interchangeable, as x' is a dummy variable. The cubic charge distribution could be considered as the *trial function* [J.9]. However, for simplicity, let us take the linear charge distribution function as a trial function:

$$\rho(x) = \frac{1}{w}\left[1 + \frac{A|x|}{w}\right], \quad |x| \le \frac{w}{2}. \tag{14.3.29}$$

where constant A could be 1. The total charge *p.u.l* on the central conductor is

$$Q = \int_{-\frac{w}{2}}^{\frac{w}{2}} \rho(x)dx = \frac{2}{w}\int_0^{\frac{w}{2}}\left(1 + \frac{A|x|}{w}\right)dx. \tag{14.3.30}$$

The variational expression for the line capacitance is $C = Q^2/V_0Q$, i.e.

$$C = \frac{\left[\frac{2}{w}\int_0^{\frac{w}{2}}\left\{1 + \frac{A|x|}{w}\right\}dx\right]^2}{\sum_{n=1,3,\ldots}\frac{2\times 4}{n\pi Y}\left[\int_0^{\frac{w}{2}}\cos\left(\frac{n\pi x}{L}\right)\frac{1}{w}\left\{1 + \frac{A|x|}{w}\right\}dx\right]^2}$$

$$= \frac{\left[\frac{1}{w}\int_0^{\frac{w}{2}}\left\{1 + \frac{A|x|}{w}\right\}dx\right]^2}{\sum_{n=1,3,\ldots}\frac{2}{n\pi Y}\left[\frac{1}{w}\int_0^{\frac{w}{2}}\cos\left(\frac{n\pi x}{L}\right)\left\{1 + \frac{A|x|}{w}\right\}dx\right]^2}. \tag{14.3.31}$$

On evaluating the integrals of equation (14.3.31),

$$\Rightarrow C = \frac{\frac{1}{4}\left(1 + \frac{A}{4}\right)^2}{\sum_{n=1,3\ldots} g_n(h_1)\left(d_n^1 + Ad_n^2\right)^2}. \tag{14.3.32}$$

$$\text{where, } g_n(h_1) = \frac{2}{n\pi Y} \quad \text{(a)}, \quad d_n^1 = \frac{L}{wn\pi}\sin\left(\frac{n\pi w}{2L}\right) \quad \text{(b)}$$

$$d_n^2 = \frac{L}{2wn\pi}\sin\left(\frac{n\pi w}{2L}\right) + \left(\frac{L}{wn\pi}\right)^2\left[\cos\left(\frac{n\pi w}{2L}\right) - 1\right] \quad \text{(c)}. \tag{14.3.33}$$

The constant A is obtained by maximizing the line capacitance C [B.13, B.9, J.9], as the present variational expression always calculates line capacitance lower than that of the true value:

$$\frac{dC}{dA} = 0. \text{ It gives us, } A = \frac{-\sum_{n=1,3,\ldots} g_n(h_1)d_n^1(d_n^1 - 4d_n^2)}{\sum_{n=1,3,\ldots} g_n(h_1)d_n^2(d_n^1 - 4d_n^2)}. \quad (14.3.34)$$

The following expressions compute the characteristic impedance, effective relative permittivity, and guided wavelength of a boxed microstrip:

$$Z_0(\varepsilon_r) = \frac{1}{v_0\sqrt{C(\varepsilon_r)C(\varepsilon_r = 1)}} \quad (a)$$

$$\varepsilon_{reff} = \frac{C(\varepsilon_r)}{C(\varepsilon_r = 1)} \quad (b), \quad \lambda_g = \frac{\lambda_0}{\sqrt{\varepsilon_{reff}}} \quad (c). \quad (14.3.35)$$

Yamashita and Atsuki [J.9] have extended the present analysis to three dielectric layers microstrip line with the cubic trial function for the charge distribution.

The Potential at the Central Conductor

Equation (14.3.27), with equation (14.3.26), provides the potential at the interface of two media, i.e. at $y = h_1$:

$$V(x, h_1) = \int_{-\frac{w}{2}}^{\frac{w}{2}} \rho(x')\, G_{1n}(x, h_1|x', h_1)dx'$$

$$V(x, h_1) = \int_{-\frac{w}{2}}^{\frac{w}{2}} \rho(x') \sum_{n=1,3,\ldots} g_n(h_1)\cos\left(\frac{n\pi x}{L}\right)\cos\left(\frac{n\pi x'}{L}\right)dx'$$

$$\Rightarrow V(x, h_1) = \frac{2}{w}\sum\left[g_n(h_1)\cos\left(\frac{n\pi x}{L}\right)\int_0^{\frac{w}{2}}\left\{1 + \frac{Ax'}{w}\right\} \times \cos\left(\frac{n\pi x'}{L}\right)dx'\right]. \quad (14.3.36)$$

The equation (14.3.36) provides the potential distribution along the plane $y = h_1$, including potential at the strip conductor. However, it is not variational expression. Therefore, the choice of the trial function for the charge distribution influences the accuracy of the computation of the potential at the interface. Itoh and Mitra have computed more accurately potential distribution by using the *function theoretic method* [J.10]. The potential V(x,y) at any point in the region is also determined similarly. It provides the electric field components, E_x and E_y:

$$E_x(x, y) = -\frac{\partial V}{\partial x} \quad (a), \quad E_y(x, y) = -\frac{\partial V}{\partial y} \quad (b)$$

$$\text{Magnitude of field: } E(x, y) = \sqrt{E_x^2 + E_y^2} \quad (c)$$

$$\text{Direction of field: } \phi = \tan^{-1}\left(\frac{E_x}{E_y}\right) \quad (d), \quad (14.3.37)$$

where the direction ϕ of the electric field is taken with respect to the *y*-axis.

Current Distribution on the Bottom and Top Ground Planes

An approximate distribution of current can be determined on the top shield, as well as on the bottom ground plane, for the assumed charge distribution on the strip conductor. The computed current distribution is approximate, as the expression for the electric field is not variational in nature. However, it helps to understand the nature of current distributions on the top and bottom ground planes. The current distribution on the ground plane is proportional to the charge distribution on it. The charge distribution itself is proportional to the normal component of the displacement vector D:

$$E_y = -\frac{\partial V}{\partial y}, \quad D_y = -\varepsilon\frac{\partial V}{\partial y}$$

$$= -\varepsilon_i\frac{\partial}{\partial y}\left[\int_{-\frac{w}{2}}^{\frac{w}{2}} \rho(x')G(x, y)dx'\right], \quad (i = 1, 2). \quad (14.3.38)$$

In the region-I, the permittivity is $\varepsilon = \varepsilon_1$ and in the region-II, $\varepsilon = \varepsilon_2$.

Using equation (14.3.25) with equation (14.3.38), the following expressions are obtained for the normal component of the displacement vector D at the bottom and top conductors:

$$\text{At the bottom conductor } y = 0: \ D_y = -\varepsilon_1\int_{-\frac{w}{2}}^{\frac{w}{2}} \rho(x') \sum_{n=1,3,\ldots} C_1\left(\frac{n\pi}{L}\right)\cos\left(\frac{n\pi}{L}x\right)dx' \quad (a)$$

$$\text{At the top conductor } y = H: \ D_y = \varepsilon_2\int_{-\frac{w}{2}}^{\frac{w}{2}} \rho(x') \sum_{n=1,3,\ldots} C_2\left(\frac{n\pi}{L}\right)\cos\left(\frac{n\pi}{L}x\right)dx' \quad (b). \quad (14.3.39)$$

The constants C_1 and C_2 are taken from equation (14.3.24). Figure (14.4b) shows the current distributions on the bottom and top ground conductors for three relative permittivities of a substrate. The current at the ground plane is maximum just below the center of the strip conductor and it is larger for a higher value of relative permittivity ε_r of the substrate. This is due to more concentration of the electric field in the substrate region. Therefore, for a higher ε_r substrate, the current at the top conductor is minimum. *To reduce loss in a microstrip circuit, the grounding is done at the point just below the central conductor.* We have seen that the strip conductor current is high at the edges. It is the reserve of the case for the current distribution on the ground conductor.

14.3.2 Open Microstrip Line (Green's Function Method in Fourier Domain)

Yamashita and Mitra [J.2] applied the variational method in the Fourier domain to compute the line capacitance and other static line parameters of an open microstrip line, shown in Fig (14.5a). Yamashita has further extended the analysis to the top shielded microstrip line with three dielectric layers [J.3]. An analysis of a simple open microstrip line is presented. The next section treats the multilayer microstrip line.

The strip conductor is assumed to be infinitely thin, i.e. strip thickness $t \approx 0$ and substrate thickness is much less than the operating wavelength, i.e. $h_1 << \lambda_0$. In this case, the microstrip supports the *quasi-TEM* mode. The unknown surface charge density on the strip conductor, $\rho(x', y') = f(x')\delta(y - h_1)$ creates the potential field $\phi(x, y)$, around the strip conductor, in the (x–y)-plane. Poisson's equation for the potential function $\phi(x, y)$ is

$$\nabla_t^2\phi(x,y) = -\frac{1}{\varepsilon_0\varepsilon_r}\rho(x')\delta(y-h_1). \qquad (14.3.40)$$

In section (14.2), equation (14.3.40) has been converted into the Fourier domain. Equation (14.2.9) is reproduced below to obtain Green's function in the Fourier domain:

$$\frac{d^2\tilde{\phi}(\beta,y)}{dy^2} - \beta^2\tilde{\phi}(\beta,y) = -\frac{1}{\varepsilon_0\varepsilon_r}\tilde{f}(\beta)\delta(y-h_1), \qquad (14.3.41)$$

where β is Fourier variable, defined over the range $[-\infty, \infty]$, corresponding to the space variable x, defined over the range $[-\infty, \infty]$. The substrate has a thickness

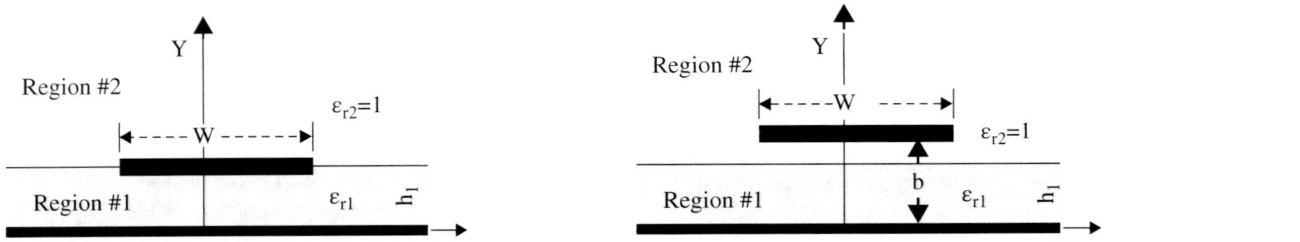

(a) Original microstrip line.

(b) Strip conductor elevated to apply continuity condition at the interface.

(c) Variation in line capacitance[J.2].

(d) Variation in characteristic impedance[J.2].

Figure 14.5 Open microstrip line in the inhomogeneous medium. *Source:* Modified from Yamashita and Mitra [J.2]. © 1968 IEEE.

$h = h_1$ and relative permittivity $\varepsilon_r = \varepsilon_{r1}$. The functions $\tilde{\phi}(\beta, y)$ and $\tilde{f}(\beta)$ are the potential and charge distribution functions in the Fourier domain. Equation (14.2.13), reproduced below, gives the variational line capacitance in the Fourier domain:

$$\frac{1}{C} = \frac{1}{\pi}\int_0^{\infty}\left[\frac{\tilde{f}(\beta)}{Q}\right]^2 \tilde{G}(\beta, h_1)d\beta, \qquad (14.3.42)$$

where $\tilde{G}(\beta, h_1)$ is Green's function in the Fourier domain at strip conductor. The strip conductor is located at $y = h_1$. Section (14.2.4) summarizes several normalized charge distribution functions $\tilde{f}(\beta)/Q$ located at $y = h_1$. The Green's function of an open microstrip line in the Fourier domain is obtained by solving equation (14.3.41). At the location $y \neq h_1$, equation (14.3.41) is reduced to the following Laplace's equation:

$$\frac{d^2\tilde{\phi}(\beta, y)}{dy^2} - \beta^2\tilde{\phi}(\beta, y) = 0. \qquad (14.3.43)$$

To apply the continuity condition at the interface of two media, the strip conductor is elevated to the height $y = b$, as shown in Fig (14.5b). For the original case, shown in Fig (14.5a), $b \to h$. The boundary and continuity conditions in the Fourier domain are given below:

$$\begin{aligned}
&\text{At ground plane } (y = 0), \text{ potential is zero}: \ \tilde{\phi}(\beta, 0) = 0 && \text{(a)}\\
&\text{At infinity } (y \to \infty), \text{ potential is zero}: \ \tilde{\phi}(\beta, \infty) = 0 && \text{(b)}\\
&\text{At interface of two dielectric media } (y = h_1),\\
&\text{Potential function is continuous}: \ \tilde{\phi}(\beta, h_1 + 0) = \tilde{\phi}(\beta, h_1 - 0) && \text{(c)}\\
&\text{At interface } (y = h) \text{ containing no free charge, normal component of } \vec{D} \text{ is continuous}:\\
&D_n|_{h+0} = D_n|_{h-0} \Rightarrow \varepsilon_0 E_n|_{y=h+0} = \varepsilon_{r1}\varepsilon_0 E_n|_{y=h-0}\\
&\Rightarrow \frac{d\tilde{\phi}(\beta, y)}{dy}\Big|_{y=h+0} = \varepsilon_{r1}\frac{d\tilde{\phi}(\beta, y)}{dy}\Big|_{y=h-0} && \text{(d)}\\
&\text{At the interface } (y = b) \text{ containing charge, normal component of } \vec{D} \text{ is discontinuous}:\\
&D_n|_{y=b+0} - D_n|_{y=b-0} = -\tilde{f}(\beta) \Rightarrow E_n|_{y=b+0} - \varepsilon_{r1}E_n|_{y=b-0} = -\frac{\tilde{f}(\beta)}{\varepsilon_0}\\
&\Rightarrow \frac{d\tilde{\phi}(\beta, y)}{dy}\Big|_{y=h+0} - \varepsilon_{r1}\frac{d\tilde{\phi}(\beta, y)}{dy}\Big|_{y=h-0} = -\frac{\tilde{f}(\beta)}{\varepsilon_0}, \ b \to h && \text{(e).}
\end{aligned} \qquad (14.3.44)$$

The solution of equation (14.3.43) is written in both regions as follows:

Region-I:

$$\tilde{\phi}_I(\beta, y) = Ae^{-\beta y} + Be^{\beta y}, \quad 0 \le y \le h_1. \qquad (14.3.45)$$

On using boundary condition (14.3.44a), B=–A, and equation (14.3.45) is written as,

$$\tilde{\phi}_I(\beta, y) = A\left[e^{-\beta y} - e^{\beta y}\right]. \qquad (14.3.46)$$

Region-II: The solution of equation (14.3.43), satisfying the boundary condition (14.3.44b), is given below:

$$\tilde{\phi}_{II}(\beta, y) = Ce^{-\beta y}, \quad y \ge h_1. \qquad (14.3.47)$$

On using the continuity condition (14.3.44c) at $y = h_1$,

$$Ce^{-\beta h} = A\left[e^{-\beta h_1} - e^{\beta h_1}\right]. \qquad (14.3.48)$$

The derivatives of potential functions at $y = h_1$ from the sides of both regions are obtained as follows:

$$\begin{aligned}
&\text{Region I}: \ \frac{d\tilde{\phi}(\beta, y)}{dy}\Big|_{y=h-0} = -\beta A\left[e^{-\beta h_1} + e^{\beta h_1}\right] && \text{(a)}\\
&\text{Region II}: \ \frac{d\tilde{\phi}(\beta, y)}{dy}\Big|_{y=h+0} = -\beta Ce^{-\beta h_1} && \text{(b).}
\end{aligned} \qquad (14.3.49)$$

On substituting equations in the discontinuity condition (14.3.44e):

$$-\beta Ce^{-\beta h_1} + \beta A\varepsilon_{r1}\left[e^{-\beta h_1} + e^{\beta h_1}\right] = -\frac{\tilde{f}(\beta)}{\varepsilon_0}. \qquad (14.3.50)$$

The constants A and C are evaluated from equations (14.3.48) and (14.3.50):

$$\begin{aligned}
A &= \frac{-\tilde{f}(\beta)}{\beta\varepsilon_0(e^{\beta h_1} - e^{-\beta h_1})[1 + \varepsilon_{r1}\coth(\beta h_1)]} && \text{(a)}\\
C &= \frac{\tilde{f}(\beta)e^{\beta h_1}}{\beta\varepsilon_0[1 + \varepsilon_{r1}\coth(\beta h_1)]} && \text{(b).}
\end{aligned} \qquad (14.3.51)$$

Both the potential distribution function and its corresponding Green's function, in the region-I, are obtained from equations (14.3.46) and (14.3.51a):

$$\tilde{\phi}_I(\beta, y) = \frac{\tilde{f}(\beta)\left(e^{\beta y} - e^{-\beta y}\right)}{\beta\varepsilon_0(e^{\beta h_1} - e^{-\beta h_1})[1 + \varepsilon_{r1}\coth(\beta h_1)]} \quad (a)$$

$$\tilde{G}_I(\beta, y) = \frac{\left(e^{\beta y} - e^{-\beta y}\right)}{\beta\varepsilon_0(e^{\beta h_1} - e^{-\beta h_1})[1 + \varepsilon_{r1}\coth(\beta h_1)]} \quad (b). \tag{14.3.52}$$

Similarly, the potential and Green's functions in region-II are obtained from equations (14.3.47) and (14.3.51b):

$$\tilde{\phi}_{II}(\beta, y) = \frac{\tilde{f}(\beta)e^{-|\beta|(y-h_1)}}{\varepsilon_0\beta[1 + \varepsilon_{r1}\coth\beta h_1]} \quad (a)$$

$$\tilde{G}_{II}(\beta, y) = \frac{e^{-|\beta|(y-h_1)}}{\varepsilon_0\beta[1 + \varepsilon_{r1}\coth\beta h_1]} \quad (b). \tag{14.3.53}$$

The potential and Green's function at the strip conductor $y = h_1$ are obtained from equations,

$$\tilde{\phi}(\beta, h) = \frac{\tilde{f}(\beta)}{\beta\varepsilon_0[1 + \varepsilon_{r1}\coth(\beta h_1)]} \quad (a)$$

$$\tilde{G}(\beta, h) = \frac{1}{\beta\varepsilon_0[1 + \varepsilon_{r1}\coth(\beta h_1)]} \quad (b). \tag{14.3.54}$$

Equation (14.3.54) can be used to determine the variation of potential, at the interface $y = h_1$, with respect to variable x. Finally, the line capacitance for the open microstrip is obtained, using equation (14.3.54) with equation (14.3.42), as follows:

$$\frac{1}{C} = \frac{1}{\pi\varepsilon_0}\int_0^\infty \frac{\left[\tilde{f}(\beta)/Q\right]^2}{\beta[1 + \varepsilon_{r1}\coth\beta h_1]}d\beta. \tag{14.3.55}$$

The linear trial charge distribution function, given by equation (14.2.17), has been used in equation (14.3.55) [J.2]. However, the cubic charge distribution function for $A = 1$, given by equation (14.2.20), provides more accurate results for the line capacitance and other line parameters [J.3]. The optimized constant A further improves the result. The computed line capacitance and the characteristic impedance of the microstrip line with respect to w/h ratio are shown in Fig (14.5c and d), respectively. Figure (14.5d) also compares the variational method, against the conformal mapping results of Wheeler [J.11]. The results of both methods follow each other closely.

Conductor Thickness

The strip conductor thickness t is ignored in the above discussion. However, it can be accounted for by taking exponential decay of Green's function on the upper face of strip conductor, i.e. at $y = h_1 + t$,

$$\tilde{G}(\beta, h_1 + t) = e^{-|\beta|t}\,\tilde{G}(\beta, h_1). \tag{14.3.56}$$

The expression of the line capacitance is modified as follows by taking the average Green's function at both faces of the strip conductor:

$$\frac{1}{C} = \frac{1}{\pi\varepsilon_0}\int_0^\infty \left[\frac{1 + e^{-|\beta|t}}{2}\right]\frac{\left[\tilde{f}(\beta)/Q\right]^2}{[1 + \varepsilon_r\coth\beta h_1](\beta h_1)}d(\beta h_1). \tag{14.3.57}$$

The effective relative permittivity and characteristic impedance are computed from the following expressions:

$$\varepsilon_{reff} = \frac{C}{C_0} \quad (a), \quad Z_0(\varepsilon_{r1}) = \frac{1}{c\sqrt{C_0 C}} \quad (b)$$

$$Z_0(\varepsilon_{r1}) = \frac{Z_0(\varepsilon_{r1} = 1)}{\sqrt{\varepsilon_{reff}}} \quad (c), \tag{14.3.58}$$

where c is the velocity of EM-wave in free space. The accuracy of computation of the line capacitance C depends on the accuracy of the evaluation of the improper integral of equation (14.3.55). However, ε_{reff} is computed more accurately. More error occurs in the computation of the characteristic impedance $Z_0(\varepsilon_{r1})$ by using equation (14.3.58b). The accuracy of computation of $Z_0(\varepsilon_{r1})$ improves significantly on using equation (14.3.58c), where $Z_0(\varepsilon_r = 1)$ is obtained from the closed-form expression of Hammerstad–Jansen given in section (8.2.2) of chapter 8. The effective relative permittivity and the characteristic impedance of a microstrip line decrease with an increase in t/h ratio [J.9]. The effect of t/h ratio is more significant on the narrow line for $w/h < 1$. The strip conductor of width w and thickness t can also be replaced by the equivalent width $w_{eq}(w, t)$ of zero conductor thickness $t = 0$, in expression used in equations (14.3.31) and (14.3.55). The expression of the equivalent width $w_{eq}(w, t)$ is given in section (8.2.4) of chapter 8. Table (14.1) shows a comparison of the computed results of the characteristic impedance with conductor thickness by the variational method, also by the numerical method of Silvester, against the experimental results of Kaupp [J.9]. Both methods show good agreement between them and also with the experimental results.

14.3.3 Open Microstrip Line (Energy Method in Fourier Domain)

This method provides the upper bound on the line capacitance [J.12]. In this case, the potential function

Table 14.1 Comparison of the computed value of Z_0 of microstrip against experimental results.

Dimensions					Characteristic impedance Ω		
t (mil)	w (mil)	h (mil)	w/h	ε_r	Exp	Num. method	Var. method
2.8	10	67	0.15	2.9	150	153.0	153.0
2.8	22	19	1.16	4.3	65	64.4	62.8
2.8	15	14	1.07	4.7	61	62.0	61.8
2.8	20	31	0.65	4.7	81	81.3	79.2
2.8	15	31	0.48	4.7	90.5	88.1	87.8
2.8	5	31	0.16	4.7	116	118.5	118.4

Source: Based on Yamashita and Atsuki [J.9].

V(x) at the interface of two media is an unknown trial function for an open microstrip line, shown in Fig (14.5a). The potential function $\phi_i(x, y)$ is obtained in terms of V(x). The dielectric region-I is indicated by $i = 1$ and $i = 2$ shows the air region –II. The potential functions in both regions and the *trial potential function* V(x) at $y = h_1$ in the Fourier domain are

$$\begin{aligned}\tilde{\phi}_i(\beta, y) &= \int_{-\infty}^{\infty}\phi_i(x, y)e^{j\beta x}dx, \quad [i = 1, 2] &\text{(a)}\\ \tilde{V}(\beta) &= \int_{-\infty}^{\infty}V(x)e^{j\beta x}dx &\text{(b)}.\end{aligned} \tag{14.3.59}$$

The potential on the strip conductor is constant, while it decreases away from it in the dielectric region. Let us assume the following trial function for the potential at the interface of two dielectric media:

$$V(x) = \begin{cases} 1, & |x| \le w/2, \ \text{potential on the strip} \\ \dfrac{1}{|x|}, & |x| > w/2, \ \text{potential in the dielectric medium.}\end{cases} \tag{14.3.60}$$

The single-term basis function does not give an accurate result for the line capacitance of a narrow microstrip. Therefore, another trial function for the potential is suggested to obtain more accurate results [J.12].

Laplace's equation in the Fourier domain is

$$\frac{d^2\tilde{\phi}_i(\beta, y)}{dy^2} - \beta^2\tilde{\phi}_i(\beta, y) = 0, \quad [i = 1, 2]. \tag{14.3.61}$$

The solution of the potential function in the region-I is

$$\begin{aligned}&\tilde{\phi}_1(\beta, y) = A\cosh\beta y + B\sinh\beta y &\text{(a)}\\ &\text{At } y = 0, \ \tilde{\phi}_1(\beta, y = 0) = 0: \ A = 0; \ \tilde{\phi}_1(\beta, y) = B\sinh\beta y &\text{(b)}\\ &\text{At the interface, } y = h: \ \tilde{\phi}_1(\beta, h_1^+) = \tilde{\phi}_1(\beta, h_1^-)\\ &\Rightarrow \tilde{V}(\beta) = B\sinh\beta h_1; \ \Rightarrow B = \frac{\tilde{V}(\beta)}{\sinh\beta h_1} &\text{(c)}\\ &\text{The potential in the region - I}: \ \tilde{\phi}_1(\beta, y) = \tilde{V}(\beta)\frac{\sinh|\beta|y}{\sinh|\beta|h_1} &\text{(d)}.\end{aligned} \tag{14.3.62}$$

Likewise, the solution of the potential function in the region-II is

$$\tilde{\phi}_2(\beta, y) = \tilde{V}(\beta)e^{-|\beta|(y-h_1)}; \quad y \ge h_1. \tag{14.3.63}$$

The following identities are used to obtain the electrostatic energy W_{e1} and W_{e2} in both regions:

$$\begin{aligned}&\text{Green's indentity}: \iint(\nabla_t\phi)^2ds\\ &= \oint\phi\frac{\partial\phi}{\partial n}d\ell - \iint\phi\nabla_t^2\phi ds &\text{(a)}\\ &\text{Parseval's indentity}: \int_{-\infty}^{\infty}\phi(x)\psi(x)dx\\ &= \frac{1}{2\pi}\int_{-\infty}^{\infty}\tilde{\phi}(\beta)\tilde{\psi}(\beta)d\beta &\text{(b)}.\end{aligned} \tag{14.3.64}$$

Region-I: For $\left[\nabla_t^2\phi_1 = 0\right]$,

$$\begin{aligned}W_{e1} &= \frac{\varepsilon}{2}\int_0^h\int_{-\infty}^{\infty}(\nabla_t\phi_1)^2dxdy = \frac{\varepsilon}{2}\int_{-\infty}^{\infty}\left[\phi_1\frac{\partial\phi_1}{\partial y}\right]\Big|_{y\le h_1}dx,\\ \Rightarrow W_{e1} &= \frac{\varepsilon_{r1}}{2}\frac{1}{2\pi}\int_{-\infty}^{\infty}\left[\tilde{\phi}_1(\beta, y)\frac{d\tilde{\phi}_1(\beta, y)}{dy}\right]\Big|_{y\le h_1}d\beta\\ \Rightarrow W_{e1} &= \frac{\varepsilon_0\varepsilon_{r1}}{4\pi}\int_{-\infty}^{\infty}\left|\tilde{V}(\beta)\right|^2|\beta|\coth(|\beta|h_1)d\beta.\end{aligned} \tag{14.3.65}$$

Equation (14.3.65) also provides the electrostatic energy in the region-II, by taking $\varepsilon_r = 1$ and $h \to \infty$:

Region-II:

$$W_{e2} = \frac{\varepsilon_0}{4\pi}\int_{-\infty}^{\infty}\left|\tilde{V}(\beta)\right|^2|\beta|d\beta. \tag{14.3.66}$$

The total electric energy, $W_e = W_{e1} + W_{e2}$, in both regions around the strip conductor gives the following variational line capacitance:

$$C = \frac{2W_e}{V^2} \Rightarrow C = \frac{\varepsilon_0}{2\pi V^2}\int_{-\infty}^{\infty} \left|\tilde{V}(\beta)\right|^2 |\beta|[1 + \varepsilon_{r1}\coth(|\beta|h_1)]d\beta, \tag{14.3.67}$$

where V = 1 Volt is potential at the strip conductor. The effective relative permittivity and the characteristic impedance of a microstrip line are obtained from equation (14.3.58). The computed results are not accurate due to the one-term potential basis function. However, the following trial potential function improves the accuracy of the computed results:

$$V(x) = \frac{a_1}{x^1} + \frac{a_2}{x^2} + \cdots + \frac{a_{N+1}}{x^{N+1}}. \tag{14.3.68}$$

The Rayleigh–Ritz procedure determines the coefficients a_i. The $V(x = 1) = 1$ is satisfied for

$$a_1 + a_2 + \cdots + a_{N+1} = 1. \tag{14.3.69}$$

From the above two equations (14.3.68) and (14.3.69), the trial potential function is obtained as

$$V(x) = a_1\left(\frac{1}{x} - \frac{1}{x^{N+1}}\right) + \cdots + a_N\left(\frac{1}{x^N} - \frac{1}{x^{N+1}}\right) + \frac{1}{x^{N+1}}. \tag{14.3.70}$$

Naito and Araki have provided the Fourier transform of expression (14.3.70) [J.12]. They have shown that for N = 4 terms of the trial function, error in the computation of the line capacitance is less than 2% for $w/h = 0.1$, $\varepsilon_r = 2.65$, 8.9. Horno has also used this method for a microstrip on an anisotropic substrate [J.13].

14.4 Analysis of Multilayer Microstrip Line

Yamashita analyzed a microstrip line in three dielectric layers environments, both in the space domain and in the spectral domains [J.3, J.9, J.14–J.16]. The lower-bound variational expression for line capacitance remains unchanged under the multilayer environment. However, Green's function is to be obtained for the multilayer microstrip structures. It involves the solution of Laplace's equation in each dielectric layer with two unknown constants. The problem of analytically evaluating 2N constants is cumbersome. Chang and Chang [J.8] proposed a simple and elegant method for the determination of Green's function of the multilayer microstrip using the transmission line analogy. This method is popularly known as the *TTL* technique [J.5, J.6, J.8]. The TTL method, discussed in this section, is applicable in both the space domain and in the Fourier domain for the lossless dielectric layers. However, it can easily be adapted to the lossy substrates [J.17]. The lossy layer case is discussed section (15.1.4) of chapter 15.

14.4.1 Space Domain Analysis of Multilayer Microstrip Structure

Figure (14.6a) shows the boxed microstrip line with four dielectric layers – two layers above the conducting strip and two layers below it. Several dielectric layers are easily accommodated in the TTL formulation. The microstrip has width w and conductor thickness t. Each dielectric layer has a relative permittivity ε_{r1}, ε_{r2}, ε_{r3}, ε_{r4} and thickness h_1, h_2, h_3, h_4. The top and bottom walls, AB and OC, are normally electric wall (EW), i.e. the perfect electric conductor (PEC); however, either of walls could be the magnetic wall (MW), i.e. the perfect magnetic conductor (PMC) or the upper wall could be even open to free space with $\varepsilon_{r4} = 1$ and $h_4 \rightarrow \infty$. The TTL formulation accounts for the boundary conditions on these walls. The sidewalls, AO and BC, can also be either the electric wall or the magnetic wall. There could be three different cases for the sidewalls:

- *Case-I, Fig (14.6b): Both end walls are the electric wall.*
- *Case-II, Fig (14.6c): One end wall is an electric wall, another is a magnetic wall.*
- *Case-III, Fig (14.6d): Both end walls are the magnetic wall.*

The Green function of the multilayer microstrip, shown in Fig (14.6a), satisfies the following Poisson's equation with boundary:

$$\nabla^2 G(x, y|x', y') = \begin{cases} -\dfrac{1}{\varepsilon_0\varepsilon_r}\delta(x-x')\delta(y-y') \\ 0; \quad x \neq x', \ y \neq y' \end{cases} \qquad (a)$$

Boundary conditions

∗Dirichlet boundary condition :

$$\text{On electric wall, } G(x, y) = 0 \qquad (b)$$

∗Neumann boundary condition :

$$\text{On magnetic wall, } \frac{\partial G(x, y)}{\partial x} = 0 \qquad (c). \tag{14.4.1}$$

Equation (14.4.1a) is solved, under the boundary conditions (14.4.1b) and (14.4.1c), by using the method of separation of variables:

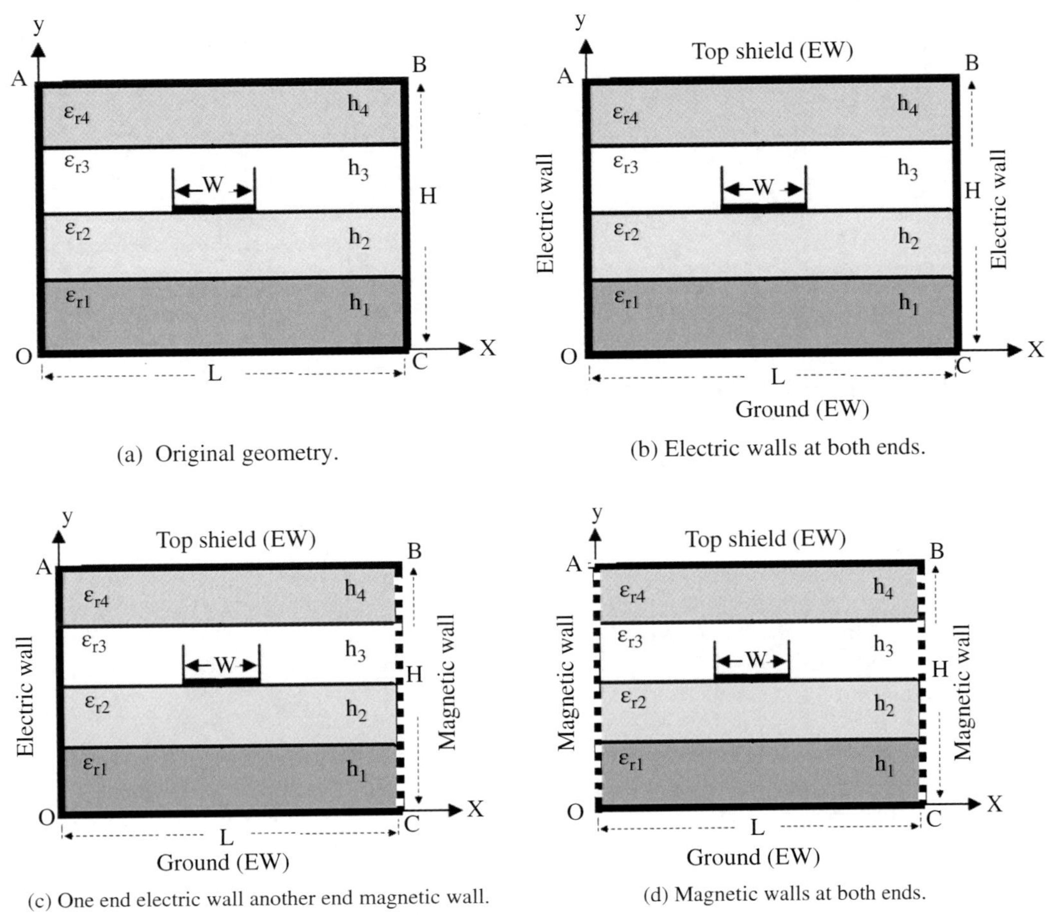

Figure 14.6 Different cases of the multilayer microstrip line with electric/magnetic walls.

$$
\begin{aligned}
&G(x,y) = \sum_{n=1}^{\infty} G_n^x(x) G_n^y(y) && \text{(a)}\\
&\frac{d^2 G_n^x(x)}{dx^2} + k^2 G_n^x(x) = 0 && \text{(b)}\\
&\frac{d^2 G_n^y(y)}{dy^2} - k^2 G_n^y(y) = 0 && \text{(c)},
\end{aligned}
\tag{14.4.2}
$$

where k is a positive constant. The functions $G_n^x(x)$ and $G_n^y(y)$ are the partial Green's functions of x and y, respectively.

CASE-I: Figure (14.6b) shows that both the end walls are electric walls at $x = 0$, $x = L$. The solution of equation (14.4.2b) under boundary condition (14.4.1b) is

$$
\begin{aligned}
&G_n^x(x) = A\cos(kx) + B\sin(kx) && \text{(a)}\\
&G_n^x(x) = 0\big|_{x=0} = 0,\ \ A = 0:\ \ G_n^x(x) = B\sin kx && \text{(b)}\\
&\text{Ignoring constant B},\ \ G_n^x(x) = 0\big|_{x=L} = 0:\ \ \sin(kL) = 0 \Rightarrow k = \frac{n\pi}{L} && \text{(c)}\\
&G_n^x(x) = \sin\left(\frac{n\pi}{L}x\right),\quad n = 0, 1, 2\ldots && \text{(d)}.
\end{aligned}
\tag{14.4.3}
$$

CASE-II: Figure (14.6c) shows that the electric wall is at $x = 0$ and the magnetic wall is at $x = L$. The solution of equation (14.4.2b) under boundary conditions (14.4.1b,c) is

$$G_n^x(x) = A\cos(kx) + B\sin(kx) \quad (a)$$
$$G_n^x(x) = 0|_{x=0} = 0,\ A = 0:\ G_n^x(x) = B\sin(kx) \quad (b)$$
$$\frac{dG_n^x(x)}{dx} = 0|_{x=L} = 0:\ \cos(kL) = 0 \Rightarrow k = (2n+1)\frac{\pi}{2L} \quad (c)$$
$$\text{Ignoring constant, B}:\ G_n^x(x) = \sin\left(\frac{(2n+1)\pi}{2L}x\right),\ n = 0, 1, 2\ldots \quad (d).$$
$$(14.4.4)$$

CASE-III: Figure (14.6d) shows that both end walls are the magnetic walls at $x = 0$, $x = L$. The solution of equation (14.4.2b) under boundary conditions (14.4.1c) is obtained from equation (14.4.3a):

$$\frac{dG_n^x(x)}{dx} = -A\,k\sin(kx) + B\,k\cos(kx) \quad (a)$$
$$\left.\frac{dG_n^x(x)}{dx}\right|_{x=0} = 0,\ B = 0;\ G_n^x(x) = A\cos(kx) \quad (b)$$
$$\frac{dG_n^x(x)}{dx} = 0|_{x=L} = 0:\ \sin(kL) = 0 = \sin n\pi \Rightarrow k = \frac{n\pi}{L} \quad (c)$$
$$\text{Ignoring constant (Ak)}:\ G_n^x(x) = \cos\frac{n\pi}{L}x,\ n = 0, 1, 2, 3\ldots. \quad (d).$$
$$(14.4.5)$$

Using the partial Green's functions $G_n^x(x)$, the Green's functions for three cases are summarized as follows:

$$\text{Case 1}:\ G(x,y) = \sum_n \sin\left\{\frac{n\pi}{L}x\right\}G_n^y(y) \quad (a)$$
$$\text{Case 2}:\ G(x,y) = \sum_n \sin\left\{\frac{(2n+1)\pi}{2L}x\right\}G_n^y(y) \quad (b)$$
$$\text{Case 3}:\ G(x,y) = \sum_n \cos\left\{\frac{n\pi}{L}x\right\}G_n^y(y) \quad (c).$$
$$(14.4.6)$$

The partial Green functions $G_n^y(y)$ for three cases are determined below:

Case 1: On substituting equation (14.4.6a) in Poisson's equation (14.4.1), we get

$$\sum_{n=0}^{\infty}\left\{\left[\frac{d^2G_n^y}{dy^2} - \left(\frac{n\pi}{L}\right)^2 G_n^y(y)\right]\sin\left(\frac{n\pi}{L}\right)x\right\} = -\frac{1}{\varepsilon}\delta(x-x')\delta(y-y'). \quad (14.4.7)$$

The left side of equation (14.4.7) is the Fourier series expansion of the Dirac's delta function $\delta(x-x')$. The expression in the bracket [ ] of equation (14.4.7) is the coefficient of $\sin\{(n\pi/L)x\}$. By using the orthogonality property, the following differential equation is obtained:

$$\frac{d^2G_n^y(y)}{dy^2} - \beta_n^2 G_n^y(y) = -\frac{2}{\varepsilon_0\varepsilon_r L}\sin(\beta_n x')\delta(y-y') \quad (a)$$
$$\beta_n = \frac{n\pi}{L} \quad (b).$$
$$(14.4.8)$$

Similarly, for the case-II and case-III, the following expressions are obtained:

$$\textbf{Case – II}:\ \frac{d^2G_n^y(y)}{dy^2} - \beta_n^2 G_n{}^y(y) = -\frac{2}{\varepsilon_0\varepsilon_r L}\sin(\beta_n x')\delta(y-y') \quad (a)$$
$$\beta_n = \frac{(2n+1)\pi}{2L} \quad (b).$$
$$(14.4.9)$$

$$\textbf{Case – III}:\ \frac{d^2G_n^y(y)}{dy^2} - \beta_n^2 G_n{}^y(y) = -\frac{2}{\varepsilon_0\varepsilon_r L}\cos(\beta_n x')\delta(y-y') \quad (a)$$
$$\beta_n = \frac{n\pi}{L} \quad (b).$$
$$(14.4.10)$$

Equations (14.4.8)–(14.4.10) are solved under the following continuity and boundary conditions, to get the partial Green's functions $G_n^y(y)$ for three cases:

$$G_n^y(x, h_i - 0) = G_n^y(x, h_i + 0) \quad (a)$$
$$\varepsilon_{ri}\frac{\partial G_n^y(x, h_i - 0)}{\partial y} = \varepsilon_{ri+1}\frac{\partial G_n^y(x, h_i + 0)}{\partial y} \quad (b)$$
$$\text{At top/bottom electric wall}:\ G_n^y(x, y = 0) = 0,\ G_n^y(x, y = H) = 0 \quad (c)$$
$$\text{At the infinity}:\ G_n^y(x, y \to \infty) \to 0 \quad (d)$$
$$\text{At top/bottom magnetic wall}:\ \frac{\partial G_n^y(x,y)}{\partial y}\Big|_{y=0,H} = 0 \quad (e).$$
$$(14.4.11)$$

Away from the strip conductor, i.e. for $y \neq y'$, equations (14.4.8)–(14.4.10) are written in the following form:

$$\frac{d^2G_n^y(y)}{dy^2} - \beta_n^2 G_n^y(y) = 0, \quad (14.4.12)$$

where β_n for cases-I, II, and III are given in their respective equations (14.4.8)–(14.4.10). Equation (14.4.12), for $G_n^y(y)$, is solved in dielectric regions (1–4) under the boundary conditions (14.4.11). At $y = 0$ and H the electric wall is present.

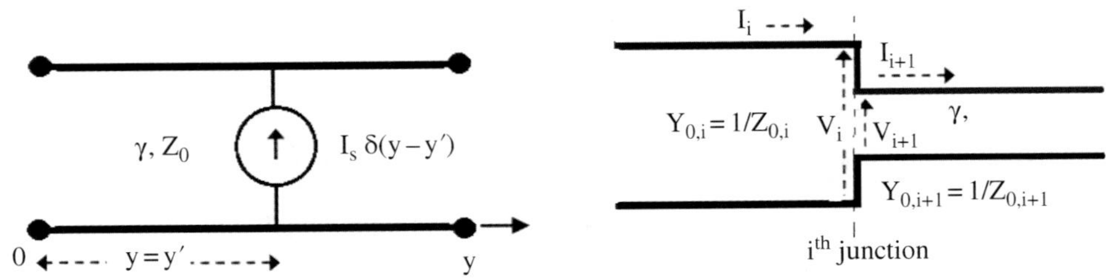

(a) Location of shunt current source I_s at $y=y'$.

(b) Continuity of voltage and current at the ith junction.

Figure 14.7 Equivalent transmission line model used in TTL technique.

$$G_n^y(y) = A_n^1 \sinh\left(\frac{n\pi}{L}y\right), \qquad 0 \le y \le h_1 \quad (a)$$
$$G_n^y(y) = A_n^2 \sinh\left(\frac{n\pi}{L}y\right) + B_n^2 \cosh\left(\frac{n\pi}{L}y\right), \qquad h_1 \le y \le (h_1 + h_2) \quad (b)$$
$$G_n^y(y) = A_n^3 \sinh\left(\frac{n\pi}{L}y\right) + B_n^3 \cosh\left(\frac{n\pi}{L}y\right), \qquad (h_1 + h_2) \le y \le (h_1 + h_2 + h_3) \quad (c)$$
$$G_n^y(y) = A_n^4 \sinh\frac{n\pi}{L}[H - y], \qquad (h_1 + h_2 + h_3) \le y \le H \quad (d),$$
$$(14.4.13)$$

where $H = h_1 + h_2 + h_3 + h_4$. The constants could be evaluated by the boundary conditions. However, such algebraic manipulation is cumbersome [J.9, J.16]. The TTL technique, presented below, easily obtains Green's function of a multilayer microstrip line.

The TTL Method

It treats the multilayer microstrip through an equivalent multisection transmission line to obtain the partial Green's function $G_n^y(y)$ of the original multilayer microstrip. In the equivalent transmission line, each dielectric layer appears as a transmission line section, and the charge on the microstrip conductor appears as a shunt connected current source. Figure (14.7a) shows a two-wire transmission line with propagation constant γ, characteristic impedance Z_0, and current source I_s located at $y = y'$. Kelvin–Heaviside transmission line equation with a current source is discussed in section (2.1.6) of chapter 2. These equations and the voltage wave equation are reproduced below:

$$\frac{dV}{dy} = -\gamma Z_0 I \quad (a), \qquad \frac{dI}{dy} = -\frac{\gamma}{Z_0}V + I_s\delta(y - y') \quad (b)$$
$$\frac{d^2V}{dy^2} - \gamma^2 V = -\gamma Z_0 I_s \delta(y - y') \quad (c).$$
$$(14.4.14)$$

Figure (14.7b) shows a junction of two transmission line sections i^{th} and $(i+1)^{th}$ sections. The following continuity conditions are satisfied at the junction of two transmission lines with two different characteristic impedances and identical propagation constant:

$$V_i = V_{i+1} \quad (a), \quad I_i = I_{i+1} \Rightarrow Y_{0i}\frac{\partial V_i}{\partial y} = Y_{0\,i+1}\frac{\partial V_{i+1}}{\partial y} \quad (b).$$
$$(14.4.15)$$

Case - I: On comparing equations (14.4.8a) and (14.4.14), the following equivalence is obtained:

$$V \equiv G_n^y(y) \quad (a), \quad \gamma \equiv \beta_n = \frac{n\pi}{L} \quad (b), \quad Y_{o\,i} \equiv \varepsilon_i \quad (c)$$
$$-\gamma Z_0 I_s \delta(y - y') = -\frac{2}{\varepsilon L}\sin(\beta_n x')\delta(y - y')$$
$$\text{as} \quad Z_0 = \frac{1}{Y} = \frac{1}{\varepsilon}, \quad \Rightarrow I_s = \frac{2}{n\pi}\sin\left\{\left(\frac{n\pi}{L}\right)x'\right\} \quad (d).$$
$$(14.4.16)$$

Equation (14.4.8a) for Green's function $G_n^y(y)$ is similar to the voltage wave equation (14.4.14c). Under the above equivalence, the multilayer microstrip, shown in Fig (14.6a), is reduced to a multisection equivalent transmission line, shown in Fig (14.8a). The voltage V on the multisection equivalent transmission line is the Green function $G_n^y(y)$ of the original multilayer microstrip. *For case-I*, the propagation constant γ of the equivalent line is $(n\pi/L)$. The thickness of a dielectric layer appears as the length of the line section. The characteristic admittance (Y_{0i}, $i = 1, 2, 3, 4$) of a line section represents the permittivity ($\varepsilon_i = \varepsilon_0\varepsilon_{ri}$, $i = 1, 2, 3, 4$) of the dielectric layer of the original structure. The current source I_s, located at $y \to y'$, corresponds to charge density on the strip conductor.

The bottom and top walls of the originally boxed multilayer microstrip could be either the electric wall (*short circuit*) or the magnetic wall (*open circuit*). At the left-hand side of I_s, located at $y = y'$, Figure (14.8b) shows the effective load admittance Y^- due to the line sections 2 and 1.To the right-hand side of I_s, the effective load is

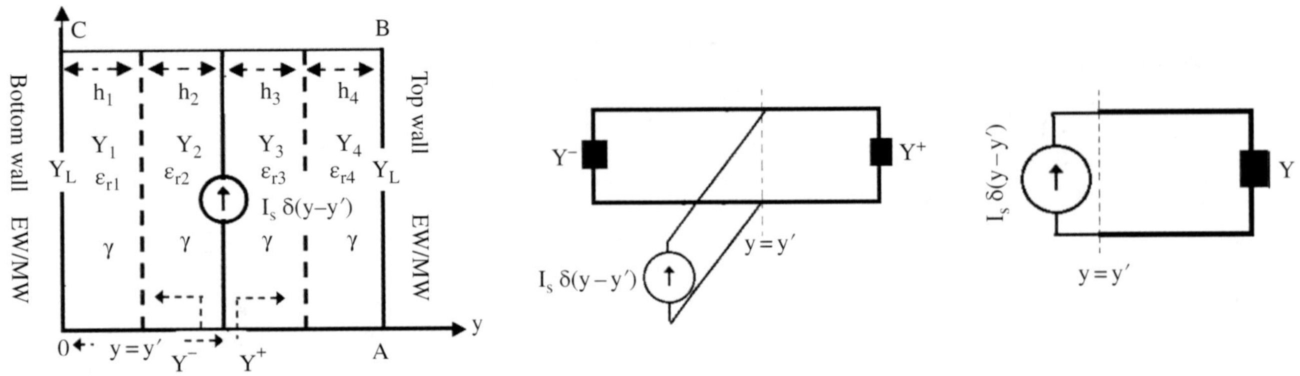

(a) Equivalent transmission with boundary condition and a current source.

(b) Circuit with a current source, and left and right loads.

(c) Current source with equivalent load.

Figure 14.8 Equivalent transmission line model of the four-layered microstrip line. ($Y_L \to \infty$ for the electric wall (short circuit) and $Y_L \to 0$ for the magnetic wall (open circuit).

Y^+ due to the line sections 3 and 4. The total load admittance Y appearing across the current source I_s is

$$Y = Y^- + Y^+. \tag{14.4.17}$$

Figure (14.8c) helps to compute the potential V, corresponding to Green's function, at the plane of the current source as follows:

$$VY = I_s \Rightarrow V = \frac{I_s}{Y} \Rightarrow G_n^y(y) = \frac{I_s}{Y}. \tag{14.4.18}$$

The partial Green's function $G_n^y(y)$ of the multilayer microstrip line and its complete Green's function for case-I, at the plane of the strip conductor, is obtained on using equations (14.4.16d), (14.4.6a) and (14.4.2a):

$$G_n^y(y)\Big|_{y=h_1+h_2} = \frac{2}{n\pi Y}\sin\left(\frac{n\pi}{L}x'\right) \qquad \text{(a)}$$

$$G(x,y|x',y') = \sum_n \frac{2}{n\pi Y}\sin\left(\frac{n\pi}{L}x\right)\sin\left(\frac{n\pi}{L}x'\right) \qquad \text{(b)}. \tag{14.4.19}$$

Case-II: Similarly, the equivalent current source I_s, $G_n^y(y)$ and the complete Green's function for the case-II are obtained:

$$I_s = \frac{4}{(2n+1)\pi}\sin(\beta_n x') \qquad \text{(a)}$$

$$G_n^y(y) = \frac{4}{(2n+1)\pi Y}\sin(\beta_n x') \qquad \text{(b)}$$

$$G(x,y|x',y') = \sum_n \frac{4}{(2n+1)\pi Y}\sin\left\{\frac{(2n+1)\pi}{2L}x\right\}\times \sin\left\{\frac{(2n+1)\pi}{2L}x'\right\} \qquad \text{(c)}. \tag{14.4.20}$$

Case-III: Again, from the equivalent current source I_s, we obtain $G_n^y(y)$ and the complete Green's function for case-III:

$$I_s = \frac{2}{n\pi}\cos\left\{\left(\frac{n\pi}{L}\right)x'\right\} \qquad \text{(a)}$$

$$G_n^y(y) = V = \frac{I_s}{Y} = \frac{2}{n\pi Y}\cos\left\{\left(\frac{n\pi}{L}\right)x'\right\} \qquad \text{(b)}$$

$$G(x,y|x',y') = \sum_n \frac{2}{n\pi Y}\cos\left\{\left(\frac{n\pi}{L}\right)x\right\}\times \cos\left\{\left(\frac{n\pi}{L}\right)x'\right\} \qquad \text{(c)}. \tag{14.4.21}$$

Determination of the Variational Line Capacitance

On an infinitesimally thin central strip conductor, the charge distribution is

$$\rho(x,y) = f(x)\delta(y-y'). \tag{14.4.22}$$

The expression for the variational line capacitance, given by equation (14.1.14), is modified for the thin strip conductor as

$$\frac{1}{C_V} = \frac{\int\int_s G(x,y|x',y')f(x)f(x')dxdx'}{\left[\int_s f(x)dx\right]^2}. \tag{14.4.23}$$

The integration is taken along strip width w. The variable x′ is the dummy variable.

Case-I: The variational line capacitance, for case-I, is obtained from equations (14.4.19) and (14.4.23):

$$\frac{1}{C_V} = \frac{\int\int_s \sum_{n=1}^{\infty} \frac{2}{n\pi Y} \sin\left\{\left(\frac{n\pi}{L}\right)x\right\} \sin\left\{\left(\frac{n\pi}{L}\right)x'\right\} f(x)f(x')dxdx'}{\left[\int_s f(x)dx\right]^2}$$

$$\Rightarrow \frac{1}{C_V} = \frac{\sum_{n=1}^{\infty} \frac{2}{n\pi Y}\left[\int_s \sin\left\{\left(\frac{n\pi}{L}\right)x\right\} f(x)dx \int_s \sin\left\{\left(\frac{n\pi}{L}\right)x'\right\} f(x')dx'\right]}{\left[\int_s f(x)dx\right]^2} \tag{14.4.24}$$

$$\Rightarrow \frac{1}{C_V} = \frac{\sum_{n=1}^{\infty} \frac{2}{n\pi Y}\left[\int_s \sin\left\{\left(\frac{n\pi}{L}\right)x\right\} f(x)dx\right]^2}{\left[\int_s f(x)dx\right]^2}.$$

The variational line capacitances for case-II and III, using equations (14.4.20), (14.4.21), and (14.4.23) are summarized below:

$$\textbf{Case – II}: \frac{1}{C_V} = \frac{\sum_{n=1}^{\infty} \frac{4}{(2n+1)\pi Y}\left[\int_s \sin\left\{\frac{(2n+1)\pi}{2L}x\right\} f(x)dx\right]^2}{\left[\int_s f(x)dx\right]^2}. \tag{14.4.25}$$

$$\textbf{Case – III}: \frac{1}{C_v} = \frac{\sum_{n=1}^{\infty} \frac{2}{n\pi Y}\left[\int_s \cos\left\{\frac{n\pi}{L}x\right\} f(x)dx\right]^2}{\left[\int_s f(x)dx\right]^2}. \tag{14.4.26}$$

In the above expressions (14.4.25 and 14.4.26), admittance parameter Y is determined at the plane of charge, $y = y' = h_1 + h_2$. The Y is determined from the input admittance of the equivalent transmission line. The bottom and top surfaces could be either electric wall or magnetic wall, Y_L is the wall admittance. The electric wall of the original structure appears as a short circuit, $Y_L \rightarrow \infty$, in the equivalent transmission line. The magnetic wall appears as an open circuit, $Y_L = 0$. The input admittance of a line section of length ℓ, with characteristic admittance Y_0 and propagation constant γ, connected to the load Y_L, is

$$Y_{in} = Y_0 \frac{Y_L + Y_0 \tanh\gamma\ell}{Y_0 + Y_L \tanh\gamma\ell}. \tag{14.4.27}$$

Let us take the *case-I* with both top and bottom conductors as the electric walls. For this case, shown in Fig (14.8a), the left-hand side admittance Y^- is obtained as follows:

$$Y_L \rightarrow \infty,\ Y_{L1};\ Y_{in}|_{\ell=h_1} = Y_{L1} = Y_1 \coth \gamma h_1 \quad \text{(a)}$$

$$\text{where}, Y_0 = Y_1 = \varepsilon_0\varepsilon_{r1} \text{ and } \gamma = \frac{n\pi}{L} \quad \text{(b)}$$

Y^- at the plane of current source :

$$Y^- = Y_2 \frac{Y_{L1} + Y_2 \tanh\gamma h_2}{Y_2 + Y_{L1}\tanh\gamma h_2}, \text{ where } Y_2 = \varepsilon_0\varepsilon_{r2} \quad \text{(c)}$$

$$\Rightarrow Y^- = \varepsilon_0\varepsilon_{r2}\left[\frac{\varepsilon_{r1}\coth\left(\frac{n\pi}{L}h_1\right) + \varepsilon_{r2}\tanh\left(\frac{n\pi}{L}h_2\right)}{\varepsilon_{r2} + \varepsilon_{r1}\coth\left(\frac{n\pi}{L}h_1\right)\tanh\left(\frac{n\pi}{L}h_2\right)}\right] \quad \text{(d).} \tag{14.4.28}$$

Likewise, the following expression for the Y^+, at the right-hand side of the source, is obtained under the short-circuited load ($Y_L \rightarrow \infty$):

$$Y_{L4} = Y_L \coth\gamma h_4 = \varepsilon_0\varepsilon_{r4}\coth\left(\frac{n\pi}{L}h_4\right) \quad \text{(a)}$$

$$Y^+ = \varepsilon_0\varepsilon_{r3}\left[\frac{\varepsilon_{r4}\coth\left(\frac{n\pi}{L}h_4\right) + \varepsilon_{r3}\tanh\left(\frac{n\pi}{L}h_3\right)}{\varepsilon_{r3} + \varepsilon_{r4}\coth\left(\frac{n\pi}{L}h_4\right)\tanh\left(\frac{n\pi}{L}h_3\right)}\right] \quad \text{(b).} \tag{14.4.29}$$

The admittance at the plane containing the charge is $Y = Y^- + Y^+$. Nature of the sidewalls changes γ only i.e. for the *case - I and case - III*, $\gamma = n\pi/L$, whereas for the *case-II*, $\gamma = [(2n+1)\pi]/2L$. If the bottom and top surfaces are the magnetic walls, then they form open-circuit loads at the ends of the equivalent transmission line. In this case, $Y_L = 0$ and from equation (14.4.27)

$$Y_{in}|_{\ell=h_1} = Y_{L1} = Y_0 \tanh\gamma\ell = \varepsilon_0\varepsilon_{r1}\tanh((n\pi/L)h_1) \quad \text{(a)}$$
$$Y_{L4} = \varepsilon_0\varepsilon_{r4}\tanh((n\pi/L)h_4) \quad \text{(b).} \tag{14.4.30}$$

The expressions for Y^- and Y^+ are modified accordingly. Any number of dielectric layers can be incorporated above and below the central strip conductor and the TTL process can be used to obtain Green's function for such a multilayer structure. The cubic charge distribution trial function f(x) given in equation (14.2.20) is used in the computation of the line capacitance. However, it is modified as follows to account for the presence of the symmetrical sidewalls:

$$f(x) = \begin{cases} \frac{1}{w}\left[1 + A\left|\frac{(x - L/2)}{w/2}\right|^3\right], & (L-w)/2 \le x \le (L+w)/2 \\ 0, & x > (L \pm w)/2. \end{cases} \tag{14.4.31}$$

Figure (14.6a) shows that the strip conductor of width w is centrally located at $x = L/2$. For simplicity, one can take constant $A = 1$ [J.3]. However, by using the Rayleigh–Ritz process, the capacitance given by equation (14.4.23) is maximized by taking $dC/dA = 0$. It provides the following expression of A for three cases [B.13]:

$$A = -\frac{\sum_{n,odd} \{[L_n - 4M_n]P_n L_n/Y\}}{\sum_{n,odd} \{[L_n - 4M_n]P_n M_n/Y\}} \quad (a); \text{ where, } L_n = \sin\frac{\gamma w}{2}; \text{ for case – I, III}: \ \gamma = \frac{n\pi}{L}$$

$$\text{for case – II, } \gamma = \frac{(2n+1)}{L}\frac{\pi}{2} \quad (b), \quad P_n = \left(\frac{2}{n\pi}\right)\left(\frac{2}{\gamma w}\right)^2 \quad (c) \qquad (14.4.32)$$

$$M_n = \left(\frac{2}{\gamma w}\right)^3 \left[3\left\{\left(\frac{\gamma w}{2}\right)^2 - 2\right\}\cos\left(\frac{\gamma w}{2}\right) + \frac{\gamma w}{2}\left\{\left(\frac{\gamma w}{2}\right)^2 - 6\right\}\sin\left(\frac{\gamma w}{2}\right) + 6\right] \quad (d).$$

The variational expression (14.4.23) for the cubic charge distribution function is written as

$$C_V = \frac{(1 + 0.25A)^2}{\sum_{n,odd} [L_n + AM_n]^2 P_n/Y}. \qquad (14.4.33)$$

Anisotropic Dielectric Layers

The expressions for the variational capacitances are equally applicable to the anisotropic dielectric layers [J.7, J.18, J.19]. Each anisotropic dielectric layer can be replaced by an *equivalent isotropic dielectric layer* discussed in subsection (8.2.6) of chapter 8. The equivalent isotropic relative permittivity and equivalent thickness of the ith layer are

$$\varepsilon_{eqri} = \left[\varepsilon_{rxxi}\varepsilon_{ryyi} - \left(\varepsilon_{rxyi}\right)^2\right]^{1/2} \qquad (a)$$

$$h_{eq\,i} = h_i\left[\frac{\varepsilon_{rxxi}}{\varepsilon_{ryyi}} - \left(\frac{\varepsilon_{rxyi}}{\varepsilon_{ryyi}}\right)^2\right]^{1/2} \qquad (b).$$

$$(14.4.34)$$

The conductor thickness can be accounted for either by the factor $(1 + e^{-\gamma t})/2$, as done in equation (14.3.57) [B.13], or by adopting Wheeler's expression for the equivalent width of zero conductor thickness discussed in subsection (8.2.4) of chapter 8. The expression (14.4.34) for the equivalent isotropic relative permittivity does not account for the effect of w/h-ratio on it. Using the concept of the partial capacitance, a more accurate w/h-dependent expression of the equivalent isotropic relative permittivity could be obtained [J.20].

14.4.2 Static Spectral Domain Analysis of Multilayer Microstrip

This method applies to the multilayer microstrip line without sidewalls, shown in Fig (14.9a) [J.10, J.15]. Figure (14.9b) shows its equivalent transmission line sections, as it was obtained for the space domain case in the previous section. The bottom/top surface could be the electric/magnetic wall, corresponding to the short circuit ($Y_L \to \infty$)/open circuit ($Y_L \to 0$) at the ends of the equivalent line sections. For a microstrip, without the top shield, we take $\varepsilon_{r4} = 1$, $h_4 \to \infty$. The solution of Poisson's equation (14.3.41) in the Fourier domain provides

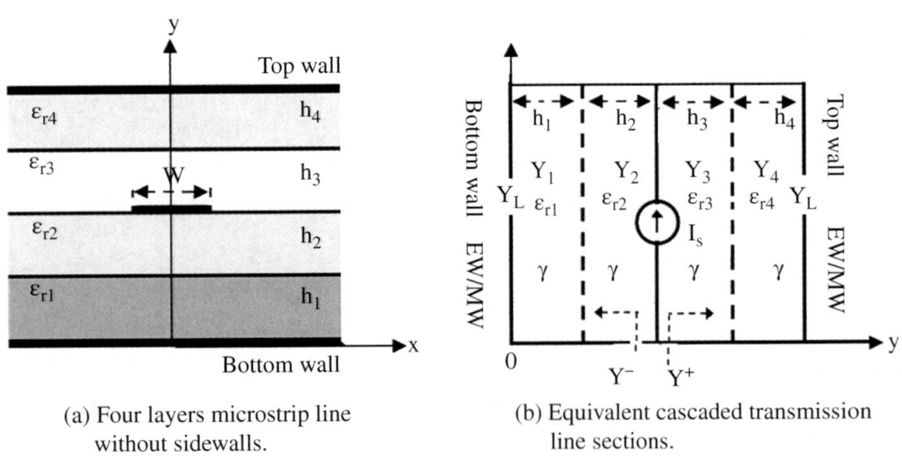

(a) Four layers microstrip line without sidewalls.

(b) Equivalent cascaded transmission line sections.

Figure 14.9 TTL method for the open-sided multilayer microstrip line.

Table 14.2 Equivalence of variables.

Original variables in Fourier domain			Corresponding variables Equivalent transmission line
Potential:	$\tilde{\phi}(\beta, y)$	$\equiv$	Voltage: V
Fourier variable:	β	$\equiv$	Propagation constant: γ
Relative permittivity: $\varepsilon_0\varepsilon_{ri}$		$\equiv$	Characteristic admittance: Y_{0i}
Charge:	$\tilde{f}(\beta)/\beta$	$\equiv$	Current source: I_s

the potential distribution $\tilde{\phi}(\beta, y)$. The following continuity conditions are satisfied at each ith interface of the dielectric layers:

$$\tilde{\phi}(\beta, y)\Big|_{y = h_i - 0} = \tilde{\phi}(\beta, y)\Big|_{y = h_i + 0} \quad (a)$$

$$\varepsilon_{r\,i}\frac{\partial\tilde{\phi}(\beta, y)}{\partial y}\Big|_{y = h_i - 0} = \varepsilon_{r\,i+1}\frac{\partial\tilde{\phi}(\beta, y)}{\partial y}\Big|_{y = h_i + 0} \quad (b). \qquad (14.4.35)$$

Poisson's equation (14.3.41) looks like the transmission line equation (14.4.14). On comparing these two equations, we obtain the equivalence of the variables summarized in Table (14.2).

Each dielectric layer of the multilayer microstrip line, shown in Fig (14.9b), appears as a section of line length h_i ($i = 1, 2, \ldots$) with characteristic admittance Y_{0i}, corresponding to the relative permittivity $\varepsilon_{r\,i}$ of the original multilayered medium. Fourier variable β appears as the propagation constant γ of equivalent transmission line. The charge density on the strip conductor appears as a current source I_s at $y' = h_1 + h_2$. At $y = y'$ the voltage and its corresponding potential, giving Green's function, are

$$V\Big|_{y = y'} = \frac{I_s}{Y} \quad (a), \Rightarrow \tilde{\phi}(\beta, y)\Big|_{y = y'} = \frac{\tilde{f}(\beta)}{\beta Y} \quad (b)$$

$$\tilde{G}(\beta, y) = \frac{1}{\beta Y} \quad (c). \qquad (14.4.36)$$

The above discussion follows the previous discussion of the space domain analysis. Using equation (14.4.36) with equation (14.2.13), the variational expression of line capacitance of the multilayer microstrip is

$$\frac{1}{C_V} = \frac{1}{\pi\varepsilon_0}\int_0^{\infty}\frac{\left[\tilde{f}(\beta)/Q\right]^2 d\beta}{\beta Y}. \qquad (14.4.37)$$

The Y-admittance parameter for the four-layer case, shown in Fig (14.9b), with the perfect bottom and top conductor surfaces, i.e. the electric walls, is obtained as follows:

$$Y_L = \varepsilon_{r1}\coth(\beta h_1) \quad (a)$$

$$Y^- = \varepsilon_{r2}\left[\frac{Y_L + Y_2\tanh(\beta h_2)}{Y_2 + Y_L\tanh(\beta h_2)}\right], \Rightarrow Y^- = \varepsilon_{r2}\left[\frac{\varepsilon_{r1}\coth(\beta h_1)\coth(\beta h_2) + \varepsilon_{r2}}{\varepsilon_{r1}\coth(\beta h_1) + \varepsilon_{r2}\coth(\beta h_2)}\right] \quad (b) \qquad (14.4.38)$$

$$\text{Likewise, } Y^+ = \varepsilon_{r3}\left[\frac{\varepsilon_{r4}\coth(\beta h_4)\coth(\beta h_3) + \varepsilon_{r3}}{\varepsilon_{r3}\coth(\beta h_3) + \varepsilon_{r4}\coth(\beta h_4)}\right] \quad (c), \quad Y = Y^- + Y^+ \quad (d).$$

Following equation (14.3.56), the conductor thickness is incorporated in Green's function. The anisotropy of each layer can be accounted for by obtaining the equivalent isotropic relative permittivity and equivalent substrate thickness on using expression (14.4.34). The four layered structure could be reduced to several technologically important cases, such as the suspended, inverted, microstrip with superstrate. Using equation (14.3.35), the characteristic impedance, and effective relative permittivity of a multilayer microstrip without sidewalls could be computed. Due to the approximate evaluation of the improper integral of the line capacitance, the characteristic impedance computed by equation (14.3.35) is not accurate. However, the following expression can be used for better numerical accuracy:

$$Z_0(\varepsilon_r) = \frac{Z_0(\varepsilon_r = 1)}{\sqrt{\varepsilon_{r\,eff}}}, \qquad (14.4.39)$$

where the closed-form expressions of $Z_0(\varepsilon_r = 1)$ for the open microstrip is given in subsection (8.2.2) of chapter 8. The expression for characteristic impedance on-air substrate $Z_0(\varepsilon_r = 1)$ of a shielded microstrip line is given in section (8.2.5) of chapter 8. The computation of $\varepsilon_{r\,eff}$ of a multilayer microstrip using the variational method is more accurate, as it is a ratio of two-line capacitances computed by the same method [J.21].

14.5 Analysis of Coupled Microstrip Line in Multilayer Dielectric Medium

The coupled microstrip line is an important structure for the development of several kinds of microwave circuit elements – such as couplers, filters [J.22, J.23]. The multilayer coupled microstrip line, like the suspended and inverted ones, is important in the mm-wave range, as

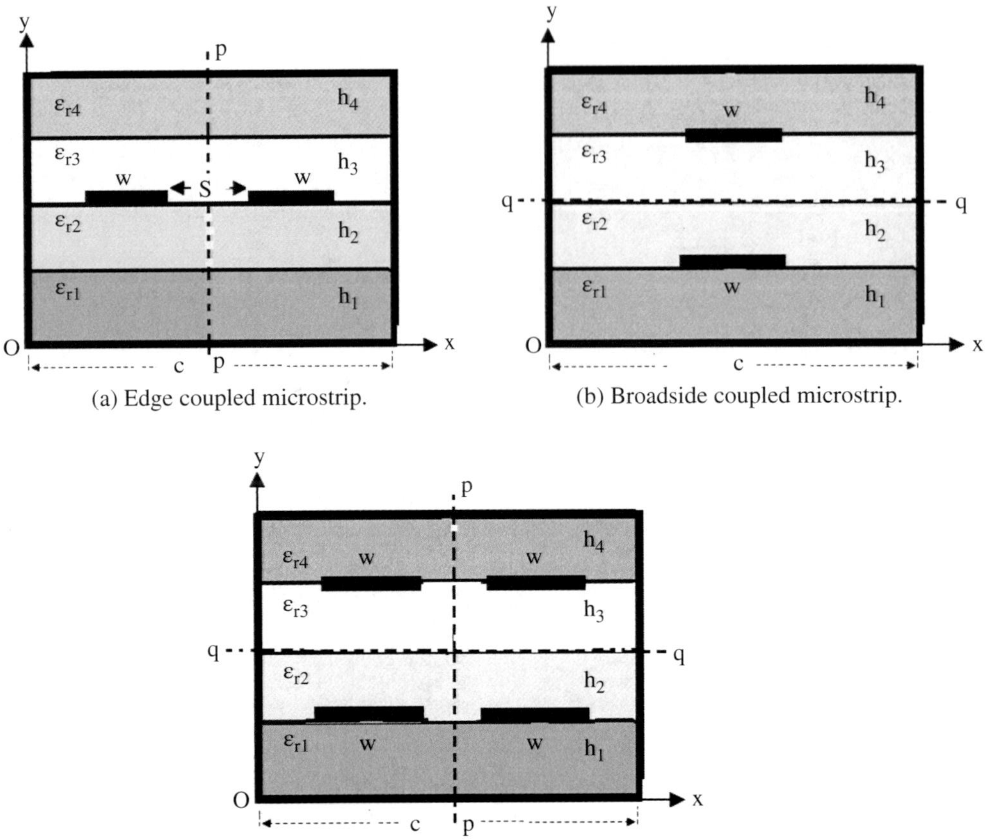

(a) Edge coupled microstrip.

(b) Broadside coupled microstrip.

(c) Edge-broadside coupled microstrip.

Figure 14.10 Multilayer coupled microstrip line structures.

they offer smaller circuit losses and less stringent dimensional tolerances, as compared to the standard coupled microstrip line. These structures are also adopted in MEMS technology. Both the composite substrate coupled microstrip line and the coupled microstrip with superstrate (overlay) are important in the MMIC technology. The effect of the top shield and sidewalls on the coupled, single-layer, or multilayer microstrip line could be accounted for. The anisotropy in the dielectric layers is also to be accounted for, as it influences the design of a coupler. Several methods are used to get the transmission line parameters of a coupled microstrip line [J.22, J.24–J.27]. In this section, the variational method, both in space and Fourier domains, is extended to obtain the *even* and *odd mode* line parameters of the symmetrical coupled microstrip structures, shown in Fig (14.10a–c). The TTL method for the determination of Green's function is also applicable to a multilayer-coupled microstrip line.

14.5.1 Space Domain Analysis

Three groups of the symmetrical coupled microstrip line under the multilayer environment are shown in Fig (14.10): (a) *the edge-coupled microstrip*, (b) *the broadside coupled microstrip*, and (c) *the edge broadside coupled microstrip*. The dashed lines show the planes of symmetry. Figure (14.10a and b) show a single plane of symmetry pp and qq, respectively, whereas Fig (14.10c) shows the dual-plane of symmetry pp, and qq. For the case – (b) and case – (c), the symmetry principle is only applicable for ($\varepsilon_{r1} = \varepsilon_{r4}, \varepsilon_{r2} = \varepsilon_{r3};\ h_1 = h_4, h_2 = h_3$). The concept of the even *mode* and *odd mode* is applied to these structures. Under the even and odd mode excitations, the coupled multilayer microstrip lines are decoupled to the multilayer single microstrip line with the magnetic/electric walls. It is emphasized that (i) the x-component of Green's function of a microstrip structure depends on the nature of sidewalls, and (ii) the nature of bottom and top surfaces are incorporated in the y-component of the Green's function through the TTL technique. The first two cases are considered below. The last one is left as an exercise.

Case A: Multilayer Edge Coupled Microstrip Line

Figure (14.11a), with respect to Fig (14.10a), shows the single isolated microstrip with the plane of

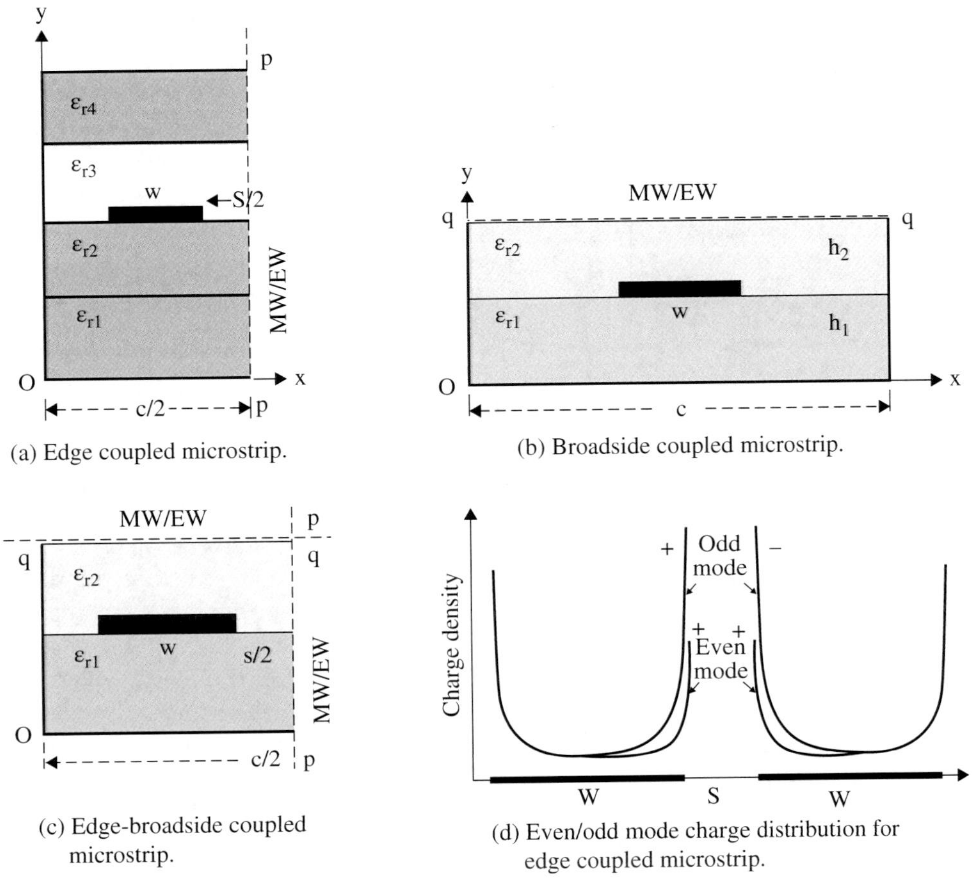

Figure 14.11 Decoupled single microstrip under even/odd mode excitation and charge distribution for the edge coupled microstrip.

symmetry-pp as the magnetic/electric wall corresponding to the even/odd mode excitation. Three sides of the decoupled single microstrip are the electric walls (EW).

On using the MW L = c/2, the even mode line capacitance is obtained from equation (14.4.25):

$$\frac{1}{C_e} = \frac{\sum_{n=0}^{\infty} \frac{4}{(2n+1)\pi Y}\left[\int_s f_e(x)\sin\left\{\left(\frac{2n+1}{c}\right)\pi x\right\}dx\right]^2}{\left[\int_s f_e(x)dx\right]^2}$$

$$\Rightarrow \frac{1}{C_e} = \frac{\sum_{n(odd)}^{\infty} \frac{4}{n\pi Y}\left[\int_s f_e(x)\sin\left\{\frac{n\pi}{c}x\right\}dx\right]^2}{\left[\int_s f_e(x)dx\right]^2}. \quad (14.5.1)$$

where, $f_e(x)$ is the *even mode charge distribution* on the strip conductor. The summation index is taken as $m = (2n+1)$, and the index m is replaced by n with $n = 0, 1, 3, 5, \ldots$ in the final expression.

For the *odd mode charge distribution* on the strip conductor, the electric wall is placed at L = c/2. The odd mode line capacitance is obtained from equation (14.4.24):

$$\frac{1}{C_o} = \frac{\sum_{n=1}^{\infty} \frac{2}{n\pi Y}\left[\int_s f_o(x)\sin\left\{\frac{2n\pi}{c}x\right\}dx\right]^2}{\left[\int_s f_o(x)dx\right]^2}$$

$$\Rightarrow \frac{1}{C_o} = \frac{\sum_{n(even)}^{\infty} \frac{4}{n\pi Y}\left[\int_S f_o(x)\sin\left\{\frac{n\pi}{c}x\right\}dx\right]^2}{\left[\int_S f_o(x)dx\right]^2}. \quad (14.5.2)$$

In the final expression (14.5.2), the summation index is taken as $m = 2n$ and the index m is replaced by n with $n = 2, 4, 6, \ldots$.

The charge distributions for both modes are sketched in Fig (14.11d). Bryant and Weiss [J.22], and also Itoh and Mitra [J.1], have computed the charge distribution on the microstrip line. The admittance parameter Y, signifying the y-component of Green's function, is the same for both the even and odd mode excitations. The integration $\int_S$ is carried over the width of the strip conductor. The conductor thickness is accounted for by replacing the term $(4/n\pi Y)$ in the equations (14.5.1) and (14.5.2) by $(4/n\pi Y)t_n$ [J.24],

$$t_n = \frac{1}{2}\left[1 + \frac{\sinh \frac{n\pi}{c}(p-t)}{\sinh \frac{n\pi}{c} p}\right]. \tag{14.5.3}$$

For the inverted microstrip structure, the parameter p is taken as $p = h_1$ with $h_2 = 0$, whereas for the suspended and standard microstrip structures, $p = h_3 + h_4$ and $p = h_2 + h_1$, respectively. For the open structure, $h_4 \to \infty$. The anisotropy in each layer is accounted for on replacing the anisotropic dielectric layer by an equivalent isotropic dielectric layer, using equation (14.4.34). The following combined closed-form expressions for the even and odd mode charge distributions are used with the above variational expressions of the line capacitances [B.13]:

$$f_{\substack{e\\o}}(x) = \begin{cases} \frac{1}{w}\left[1 + A_{\substack{e\\o}}\left|\frac{2}{w}\left(x - \frac{c-s-w}{2} - g_{\substack{e\\o}}\right)\right|^3\right], & \text{for } \frac{c-s}{2} - w \le x \le \frac{c-s}{2} \quad \text{(a)} \\ 0, & \text{Otherwise} \end{cases} \tag{14.5.4}$$

$$\text{where, for even mode}: \ 0 \le g_e \le \frac{w}{2} \ \text{(b)}, \quad \text{for odd mode}: \ -\frac{w}{2} \le g_o \le 0 \quad \text{(c)}.$$

For a loosely coupled microstrip line, the charge distributions for the even and odd modes are approximately symmetrical and $g_o = g_e = 0$ could be taken. For a strong coupling case, g_o and g_e accounts for the degree of asymmetry and some assumed values could be attempted to improve the computation. The constant A_e and A_o can be optimized numerically for the maximum value of the even and odd mode line capacitances. Alternatively, one can use the optimized expressions of Koul and Bhat [B.13] to compute C_e and C_o, and corresponding values of effective even/odd mode relative permittivity and the even/odd mode characteristic impedances:

$$\varepsilon_{\substack{e\\o}} = \frac{C_{\substack{e\\o}}(\varepsilon_r)}{C_{\substack{e\\o}}(\varepsilon_r = 1)} \quad \text{(a)}$$

$$Z_{\substack{e\\o}} = \frac{1}{v\sqrt{C_{\substack{e\\o}}(\varepsilon_r) C_{\substack{e\\o}}(\varepsilon_r = 1)}} \quad \text{(b)}$$

$$\text{or } Z_{\substack{e\\o}} = \frac{Z_{\substack{e\\o}}(\varepsilon_r = 1)}{\sqrt{\varepsilon_{r\substack{e\\o}}}} \quad \text{(c)}. \tag{14.5.5}$$

The closed-form even and odd mode characteristic impedance expressions on the air-substrate are given by Kneppo et al. [B.15]. The effect of the sidewalls is ignored for the separation $c/(2w+s) \ge 6$, and the effect of the top shield is ignored for the height $h_4/(h_1+h_2) \ge 14$. The parameters $A_e = 2.8$ and $A_o = 5.8$ in the charge distribution function could be taken for all cases, except for $c/(2w+s) \le 3$. For simplicity, one can also take $A_e = A_o = 1$. The variational method results show good agreement with the results of the method of moments [J.22, J.24].

Case B: Multilayer Broadside Coupled Microstrip Line

The broadside coupled microstrip structure, shown in Fig (14.10b), provides stronger coupling, as compared to an edge coupled microstrip line. It also shows the plane of symmetry qq for the electric/magnetic wall, corresponding to odd/even mode excitation. These are accommodated in the admittance parameter Y as the short-circuited/open-circuited termination. Following equation (14.4.24), the line capacitances, both for the even and odd mode excitations, are given below:

$$\frac{1}{C_{\substack{e\\o}}} = \frac{\sum_{n=1}^{\infty} \frac{2}{n\pi Y_{\substack{e\\o}}}\left[\int_S f_{\substack{e\\o}}(x)\sin\left(\frac{n\pi}{c}x\right)dx\right]^2}{\left[\int_S f_{\substack{e\\o}}(x)\right]^2}. \tag{14.5.6}$$

For the even mode, the admittance parameter Y_e is computed with the top wall as a magnetic wall, i.e. an open-circuit termination in Fig (14.11b). For the odd mode, the admittance parameter Y_o is computed with the short-circuit termination. Using the function applied to a single strip case, the charge distributions for both modes are given as follows:

$$f_{\substack{e\\o}}(x) = \begin{cases} \frac{1}{w}\left[1 + A_{\substack{e\\o}}\left|\frac{2}{w}\left(x - \frac{c}{2}\right)\right|^3\right], & \text{for } \frac{c-w}{2} \le x \le \frac{c+w}{2} \\ 0, & \text{otherwise.} \end{cases} \tag{14.5.7}$$

For simplicity, one can take $A_e = A_o = 1$, or it can be determined by maximizing C_e or C_o, respectively. The even and odd mode characteristic impedances of two special cases of the broadside-coupled microstrip line are also compared against other methods, showing good agreement [J.28, J.29].

14.5.2 Spectral Domain Analysis

The TTL technique presented in subsection (14.4.2) is adapted to a coupled multilayer microstrip structure, shown in Fig (14.12). Due to the symmetry of the structure, the even and odd mode concept is applicable. The planes of symmetry pp and qq could be either a magnetic wall (even mode) or an electric wall (odd mode). For the edge coupled microstrip line, equation (14.4.38) provides the admittance parameter that is identical for both the even and odd modes. The even/odd mode line capacitance of the multilayer edge coupled microstrip line, from equation (14.4.37) is

$$\frac{1}{C_{\substack{e\\o}}} = \frac{1}{\pi\varepsilon_0}\int_0^\infty \left|\frac{\tilde{f}_{\substack{e\\o}}(\beta)}{Q_{\substack{e\\o}}}\right|^2 \frac{1}{\beta Y}\,d\beta. \tag{14.5.8}$$

where for weak coupling, i.e. for a large strip conductor spacing $\{s/(h_1 + h_2)\} > 0.2$, the symmetrical charge distribution is identical for both the even and odd modes. It is given by the following expression [B.13]:

$$f_{\substack{e\\o}}(x) = \begin{cases} \left[1 + B_{\substack{e\\o}}\left|\frac{2}{w}\left(x - \frac{s}{2} - \frac{w}{2}\right)^3\right|\right], & \frac{s}{2} \le x \le \frac{s}{2} + w \\ 0, & \text{otherwise.} \end{cases} \tag{14.5.9}$$

The narrow spacing of strip conductor,$\{s/(h_1 + h_2)\} \le 0.2$, gives stronger coupling. It results in the asymmetrical charge distributions for both the even and odd modes expressed as follows:

$$f_e(x) = \begin{cases} \left[1 + B_e\left|\frac{1}{w}\left(x - \frac{s}{2}\right)^3\right|\right], & \frac{s}{2} \le x \le \frac{s}{2} + w \\ 0, & \text{Otherwise} \end{cases} \quad (a)$$

$$f_o(x) = \begin{cases} \left[1 + B_o\left|\frac{1}{w}\left(w - x + \frac{s}{2}\right)^3\right|\right], & \frac{s}{2} \le x \le \frac{s}{2} + w \\ 0, & \text{Otherwise} \end{cases} \quad (b). \tag{14.5.10}$$

Fourier transform of even and odd mode charge distributions are obtained from the following expressions:

$$\tilde{f}_e(\beta) = 2\int_{s/2}^{s/2+w} f_e(x)\cos(\beta x)dx \quad (a)$$

$$\tilde{f}_o(\beta) = -2j\int_{s/2}^{s/2+w} f_o(x)\sin(\beta x)dx \quad (b). \tag{14.5.11}$$

Fourier transforms of equation (14.5.9), for the even and odd modes, are as follows:

$$\tilde{f}_e(\beta) = 2w\cos\left\{\beta\left(\frac{s+w}{2}\right)\right\}\left[(1 + B_e)\left\{\frac{\sin(\beta w/2)}{(\beta w/2)}\right\} + \frac{3B_e}{(\beta w/2)^2}\left\{\cos(\beta w/2) - 2\left\{\frac{\sin(\beta w/2)}{(\beta w/2)}\right\} + \left\{\frac{\sin(\beta w/4)}{(\beta w/4)}\right\}^2\right\}\right] \quad (a)$$

$$\tilde{f}_o(\beta) = -2jw\sin\left\{\beta\left(\frac{s+w}{2}\right)\right\}\left[(1 + B_o)\left\{\frac{\sin(\beta w/2)}{(\beta w/2)}\right\} + \frac{3B_o}{(\beta w/2)^2}\left\{\cos(\beta w/2) - 2\left\{\frac{\sin(\beta w/2)}{(\beta w/2)}\right\} + \left\{\frac{\sin(\beta w/4)}{(\beta w/4)}\right\}^2\right\}\right] \quad (b). \tag{14.5.12}$$

Even and odd mode Fourier transforms of the even/odd mode charge densities given by equation (14.5.10a and b) are summarized below:

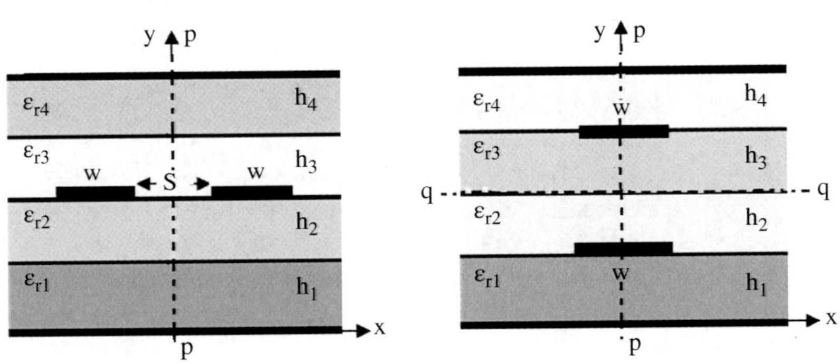

(a) Edge–coupled. (b) Broadside coupled.

Figure 14.12 Multilayer coupled microstrip line without the sidewalls.

$$\tilde{f}_e(\beta) = \frac{2w}{(\beta w)^3}\left[\begin{array}{l}\left\{((2+B_e)((\beta w)^2-6))\sin\left(\frac{\beta w}{2}\right)+3B_e(\beta W)\cos\left(\frac{\beta w}{2}\right)\right\}\cos\left\{\beta\left(\frac{s+w}{2}\right)\right\}\\ +B_e\left\{((\beta w)^2-6)\cos\left(\frac{\beta w}{2}\right)-\frac{3}{\beta w}((\beta w)^2-4)\sin\left(\frac{\beta w}{2}\right)\right\}\sin\left\{\alpha\left(\frac{s+w}{2}\right)\right\}\end{array}\right] \quad \text{(a)}$$

$$\tilde{f}_o(\beta) = \frac{-2jw}{(\beta w)^3}\left[\begin{array}{l}\left\{((2+B_o)(\beta w)^2-6)\sin\left(\frac{\beta w}{2}\right)+3B_o(\beta w)\cos\left(\frac{\beta w}{2}\right)\right\}\sin\left\{\beta\left(\frac{s+w}{2}\right)\right\}\\ +B_o\left\{((\beta w)^2-6)\cos\left(\frac{\beta w}{2}\right)-\frac{3}{\beta W}((\beta w)^2-4)\sin\left(\frac{\beta w}{2}\right)\right\}\cos\left\{\beta\left(\frac{s+w}{2}\right)\right\}\end{array}\right] \quad \text{(b).}$$

(14.5.13)

The total charges *p.u.l* on the strip conductor, for the even and odd modes, are

$$Q_{\substack{e\\o}} = \int_{s/2}^{s/2+w} f_{\substack{e\\o}}(x)dx = [1+0.25]B_{\substack{e\\o}}w. \quad (14.5.14)$$

The conductor thickness t is accounted for on replacing Y parameter by $Y/\tilde{h}(\beta,t)$, where $\tilde{h}(\beta,t)$ is given below [J.30]:

Inverted coupled microstrip :

$$\tilde{h}(\beta,t) = \frac{1}{2}\left[1+\frac{\sinh(\beta h-\beta t)}{\sinh(\beta h)}\right] \quad \text{(a)}$$

Composite/suspended coupled microstrip:

$$\tilde{h}(\beta,t) = \frac{1}{2}\left[1+e^{-\beta t}\right] \quad \text{(b).}$$

(14.5.15)

The anisotropy in the substrate is accounted for by replacing each anisotropic dielectric layer by an equivalent isotropic layer through equation (14.4.34). The following expressions compute the even and odd modes effective relative permittivity and characteristic impedances:

$$\varepsilon_{reff_{\substack{e\\o}}} = \frac{C_{\substack{e\\o}}(\varepsilon_{r1}\ldots\varepsilon_{rn})}{C_{\substack{e\\o}}(\varepsilon_{r1}\ldots\varepsilon_{rn}=1)} \quad \text{(a)}$$

$$Z_{0_{\substack{e\\o}}} = \frac{1}{v\sqrt{C_{\substack{e\\o}}(\varepsilon_{r1}\ldots\varepsilon_{rn})C_{\substack{e\\o}}(\varepsilon_{r1}\ldots\varepsilon_{rn}=1)}} \quad \text{(b)}$$

$$\text{Alternatively,}\quad Z_{0_{\substack{e\\o}}} = \frac{Z_{0_{\substack{e\\o}}}(\varepsilon_{r1}\ldots\varepsilon_{rn}=1)}{\sqrt{\varepsilon_{r\,eff_{\substack{e\\o}}}(\varepsilon_{r1}\ldots\varepsilon_{rn})}} \quad \text{(c).}$$

(14.5.16)

where $Z_{0_{\substack{e\\o}}}(\varepsilon_{r1}\ldots\varepsilon_{rn}=1)$ are the closed-form expressions of the even/odd mode characteristic impedance on the air-substrate. It is given by Kneppo et al. [B.15]. At this stage, we should be careful about the case with a top shield and without the top shield. Hammerstad and Jansen provide the closed-form expression for $Z_{0_{\substack{e\\o}}}$ on the air-substrate without top shield (open substrate) [J.31]. Harneko and Pavel provide the expression for $Z_{0_{\substack{e\\o}}}$ on the air-substrate with a top shield [B.15]. Ideally, both expressions (14.5.16b) and (14.5.16c), for the characteristic impedance, should give identical results. However, the difficulty arises due to the evaluation of the improper integral involved in line capacitances. If we use the Gauss–Laguerre method, the numerical integration is fast, but not very accurate. However, expression (14.5.16c) provides numerically more correct values. This scheme is suitable for CAD applications. The constant $B_e = B_o = 1$ could be assumed for evaluating the line capacitances C_e, C_o. However, line capacitance or the characteristic impedance could be numerically optimized to obtain the optimum values for B_e and B_o, corresponding to the minimum values of Z_{0e}, Z_{0o}. The improved results are obtained by taking $B_e = B_o = 3$ for all cases in the range $0.1 \le w/h \le 5.0$ and $0.2 \le s/h \le 4.0$.

Table (14.3) compares the even and odd modes static effective relative permittivity computed by the SDA [B.14], closed-form model of Krischning–Jansen (KJ Method) [J.25], and the present variational method against the dynamical variational method of Kowalski and Pregla [J.27]. The comparison is done for the coupled microstrip line on an alumina substrate with $\varepsilon_r = 9.8$, for $0.05 \le s/h \le 4.0$, $0.4 \le w/h \le 4.0$. The RMS

Table 14.3 Comparison of results of $\varepsilon_{reffe}(0)$ and $\varepsilon_{reffo}(0)$ for the coupled microstrips.

w/h_1	s/h_1	Ref. data [J.27]	SDA $\varepsilon_{reffe/o}(0)$	KJ $\varepsilon_{reffe/o}$	Var. $\varepsilon_{reffe/o}$
Even mode (ε_r = 9.8)					
0.4	0.05	6.491	6.480	6.492	6.478
	4.00	6.361	6.356	6.359	6.337
1.0	0.05	7.054	7.019	7.015	7.036
3.5	0.05	8.142	8.119	8.094	8.059
	4.00	7.741	7.711	7.780	7.810
Odd mode (ε_r = 9.8)					
0.4	0.05	5.428	5.482	5.447	5.440
	4.00	6.022	6.045	6.019	6.037
1.0	0.05	5.542	5.734	5.561	5.604
3.5	0.05	6.260	6.643	6.209	6.562
	4.00	7.269	7.260	7.204	7.202

deviations of SDA, KJ model, and present variational method are 0.082, 0.062, 0.08%, for the even mode, and 0.16, 0.079, 0.09% for the odd mode, respectively. For a narrow separation and wide strip width, the deviation is high, i.e. 6% for the SDA and 4.2% for the variational method.

Figure (14.13a) shows the phase velocities and characteristic impedances of both the even and odd modes. Figure (14.13a) shows that the even and odd mode phase velocities could be equalized by controlling the anisotropy. Figure (14.13b) shows that they could also be equalized by controlling the air-gap of a suspended coupled microstrip line. Equalization of even and odd mode phase velocities is important for the design of a high-directivity microstrip coupler [J.23]. The results on the characteristic impedances are also compared against other sources [J.32].

14.6 Discrete Fourier Transform Method

Two methods are discussed in previous sections for determination of variational line capacitance of a microstrip line – the *space domain method* applicable to a boxed microstrip line and the *Fourier domain method* applicable to a microstrip without sidewalls. The first method involves two-dimensional Green's function expressed through slowly convergent series and the second method does not apply to a boxed microstrip line. This section presents a numerically efficient discrete Fourier transform method, applicable to a boxed microstrip line, as well as to a boxed CPW. In this case, the Green's function is obtained in the closed-form.

14.6.1 Discrete Fourier Transform

A summary of the discrete Fourier transform is presented. The discrete Fourier transform of a periodic function f(x) comes from the Fourier series expansion of the periodic function f(x), defined over the period $-L \le x \le L$. Fourier series gives the *inverse discrete Fourier transform,* whereas its coefficient of expansion provides the *discrete Fourier transform of* the function f(x).

The following complex Fourier series [B.16] is used to define the discrete Fourier transform pairs:

$$f(x) = \sum_{-\infty}^{\infty} c_n e^{j(n\pi/L)x}, \text{ where } c_n = \frac{1}{2L}\int_{-L}^{L} f(x)e^{-j(n\pi/L)x}.$$

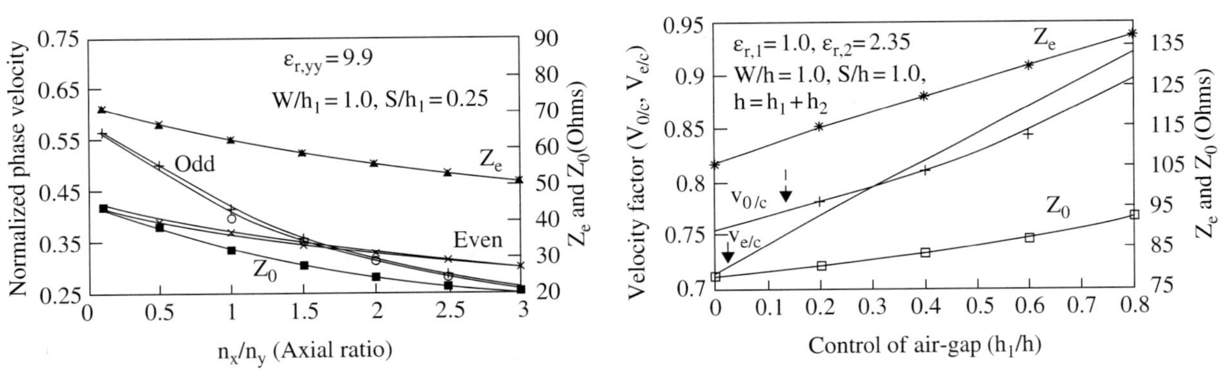

(a) Control of phase velocities and characteristics impedance by axial ratio.

(b) Control of phase velocities and characteristics impedance by air-gap.

Figure 14.13 Control of phase velocities and characteristic impedances of coupled microstrip on the anisotropic substrate and the suspended substrate.

The coefficient of expansion c_n is taken as the discrete Fourier transform $\tilde{f}(\beta_n)$ with a change in the sign of exponential. The discrete Fourier transform pairs are defined as follows [B.3, B.14]:

$$\text{Discrete Fourier transform}: \ \tilde{f}(\beta_n) = \frac{1}{2L}\int_{-L}^{L} f(x)e^{j\beta_n x}dx \qquad (a)$$

$$\text{Inverse Discrete Fourier transform}: \ f(x) = \sum_{-\infty}^{\infty} \tilde{f}(\beta_n)e^{-j\beta_n x} \qquad (b) \qquad (14.6.1)$$

$$\text{where discrete Fourier variable}: \ \beta_n = \frac{n\pi}{L} \qquad (c).$$

Sometimes, the discrete Fourier transform pair is also defined as follows [B.17]:

$$\tilde{f}(\beta_n) = \int_{-L}^{L} f(x)e^{j\beta_n(x)}dx \qquad (a)$$

$$f(x) = \frac{1}{2L}\sum_{-\infty}^{\infty} \tilde{f}(\beta_n)e^{-j\beta_n x} \qquad (b).$$

$$(14.6.2)$$

Parseval's theorem could be defined with respect to both sets of Fourier transform pairs. Parseval's theorem helps to express integral of a product of two functions f(x) and g(x), defined over [−L, L], in terms of their Fourier transforms,

On using equation (14.6.1) :

$$\int_{-L}^{L} f(x)g(x)dx = 2L\sum_{n=-\infty}^{\infty} \tilde{f}(\beta_n)\tilde{g}^*(\beta_n). \qquad (14.6.3)$$

On using equation (14.6.2) :

$$\int_{-L}^{L} f(x)g(x)dx = \frac{1}{2L}\sum_{n=-\infty}^{\infty} \tilde{f}(\beta_n)\tilde{g}^*(\beta_n). \qquad (14.6.4)$$

Sine and Cosine Discrete Fourier Transforms

The *discrete sine Fourier transform* $\tilde{f}_s(\beta_n)$ defines the discrete Fourier transform of odd function f(x). In this case, the function f(x) is defined over the half-range [0, L] with its odd extension over [−L, 0]. However, if the even extension is taken over [−L, 0], the discrete Fourier transform is defined through the *discrete cosine Fourier transform*, $\tilde{f}_c(\beta_n)$. Both the sine and cosine discrete Fourier transform are obtained from Fourier series as follows [B.16]:

$$f(x) = a_0 + \sum_{n=1}^{\infty}\left[a_n\cos\frac{n\pi}{L}x + b_n\sin\frac{n\pi}{L}x\right] \qquad (a)$$

$$n = 1, 2...$$

$$\text{where, } a_n = \frac{1}{L}\int_{-L}^{L} f(x)\cos\frac{n\pi}{L}xdx \qquad (b)$$

$$b_n = \frac{1}{L}\int_{-L}^{L} f(x)\sin\frac{n\pi}{L}xdx \qquad (c)$$

$$a_0 = \frac{1}{2L}\int_{-L}^{L} f(x)dx \qquad (d).$$

$$(14.6.5)$$

The coefficients a_n and b_n define cosine and sine discrete Fourier transform for the f(x) even function and f(x) odd function, respectively.

f(x) Odd Function

Using equation (14.6.5), the inverse discrete sine Fourier transform of the odd function f(x) is obtained as follows [B.8, B.16]:

$$\text{For odd function}: \ a_0 = a_n = 0, \ f(x) = \sum_{n=1}^{\infty}[b_n\sin(\beta_n x)] \Rightarrow f(x) = \sum_{n=1}^{\infty}\left[\tilde{f}_s(\beta_n)\sin(\beta_n x)\right] \qquad (a)$$

$$\text{Discrete sine Fourier transform for } f(x) \text{ odd function}: \ b_n = \tilde{f}_s(\beta_n) = \frac{2}{L}\int_0^L [f(x)\sin(\beta_n x)]dx \qquad (b)$$

$$\text{Parseval's theorem}: \ \int_0^L f(x)g(x)dx = \frac{L}{2}\sum_{n=1}^{\infty}\tilde{f}_s(\beta_n) = \tilde{g}_s(\beta_n) \quad (c), \quad \beta_n = \frac{n\pi}{L} \qquad (d).$$

$$(14.6.6)$$

Some authors [J.17] define the discrete sine Fourier transforms pairs for f(x) odd function as

$$\text{Inverse discrete sine Fourier transform}: \; f(x) = \frac{2}{L}\sum_{n=1}^{\infty}\left[\tilde{f}_s(\beta_n)\sin(\beta_n x)\right] \quad \text{(a)}$$

$$\text{Discrete sine Fourier transform}: \; \tilde{f}_S(\beta_n) = \int_0^L [f(x)\sin(\beta_n x)]dx \quad \text{(b)} \qquad (14.6.7)$$

$$\text{Corresponding Parseval's theorem}: \; \int_0^L f(x)g(x)dx = \frac{2}{L}\sum_{n=1}^{\infty}\left[\tilde{f}_s(\beta_n)\tilde{g}_s(\beta_n)\right] \quad \text{(c)}.$$

f(x) Even Function

Similarly, the inverse discrete cosine Fourier transform is defined for the even function f(x) from equation (14.6.1) [B.16]:

$$\text{For even function}: \; b_n = 0, \; f(x) = \sum_{n=0}^{\infty}[a_n\cos(\beta_n x)] = a_0 + \sum_{n=1}^{\infty}[a_n\cos(\beta_n x)] \quad \text{(a)}$$

$$\text{where}, \; \beta_n = \frac{(2n-1)}{L}\frac{\pi}{2}, \; n = 1, 2... \quad \text{(b)}, \quad a_n = \tilde{f}_c(x) = \frac{2}{L}\int_0^L [f(x)\cos(\beta_n x)]dx \quad \text{(c)} \qquad (14.6.8)$$

$$\text{Constant is obtained for } n = 0, \; \beta_n = 0 : a_0 = \frac{1}{L}\int_0^L [f(x)]dx \quad \text{(d)}$$

$$\text{Corresponding Parsevals theorem}: \; \int_0^L f(x)g(x)dx = \frac{L}{2}\sum_{n=1}^{\infty}\left[\tilde{f}_c(\beta_n)\tilde{g}_c(\beta_n)\right] \quad \text{(e)}.$$

To define the inverse discrete cosine Fourier transform by f(x), the constant a_0 is ignored and Parseval's theorem is obtained.

Again, some authors [J.14] define discrete cosine Fourier transform pairs for f(x) even function as

$$\text{Inverse discrete cosine Fourier transform}: \; f(x) = \frac{2}{L}\sum_{n=1}^{\infty}\left[\tilde{f}_c(\beta_n)\cos(\beta_n x)\right] \quad \text{(a)}$$

$$\text{Discrete cosine Fourier transform}: \; \tilde{f}_c(\beta_n) = \int_0^L [f(x)\cos(\beta_n x)]dx \quad \text{(b)} \qquad (14.6.9)$$

$$\text{Corresponding Parseval's theorem}: \; \int_0^L f(x)g(x)dx = \frac{2}{L}\sum_{n=1}^{\infty}\left[\tilde{f}_c(\beta_n)\tilde{g}_c(\beta_n)\right] \quad \text{(c)}.$$

The sine and cosine discrete Fourier transforms are also defined as

$$b_n = \tilde{f}_s(\beta_n) = \frac{2}{L}\int_C^{C+L} [f(x)\sin(\beta_n x)]dx. \qquad (14.6.10)$$

$$a_n = \tilde{f}_c(\beta_n) = \frac{2}{L}\int_C^{C+L} [f(x)\cos(\beta_n x)]dx. \qquad (14.6.11)$$

In a special case c = 0. The function f(x) has a period 2L

Fourier Transform of the Derivatives

If $f(x), \frac{df(x)}{dx}, \ldots, \frac{d^{m-1}f(x)}{dx^{m-1}} = 0$ at $x = \pm L$, then the discrete Fourier transform (FT) of $\frac{d^m f(x)}{dx^m}$ is

$$\mathrm{FT}\left[\frac{d^m f(x)}{dx^m}\right] = [-j\beta_n]^m \tilde{f}(\beta_n). \tag{14.6.12}$$

14.6.2 Boxed Microstrip Line

Figure (14.14a) shows the geometry of a boxed microstrip line. In this case, widths of the box and the strip conductors are 2L and 2w, respectively. All sides are the electric walls. Our task is twofold, first to determine Green's function of the structure in the discrete Fourier domain, and second to determine the line capacitance in terms of Green's function by using *Galerkin's process* [B.8, B.14, B.18, J.14, J.15, J.13, J.33, J.34].

Determination of Green's Function in Discrete Fourier Domain

The static potential function, in both dielectric regions, satisfies Poisson's equation. On an infinitely thin strip conductor, the charge distribution density is $\rho(x, y) = f(x)\delta(y - h_1)$ and Poisson's equation is written as follows:

$$\frac{\partial^2 \phi_i}{\partial x^2} + \frac{\partial^2 \phi_i}{\partial y^2} = -\frac{f(x)\delta(y-y')}{\varepsilon_0 \varepsilon_r} \quad \text{(a)}$$

For $y \neq y'$;

$$\frac{\partial^2 \phi_i}{\partial x^2} + \frac{\partial^2 \phi_i}{\partial y^2} = 0, \quad i = 1, 2 \quad \text{(b).}$$

$$\tag{14.6.13}$$

Using the discrete Fourier transform, equation (14.6.13) is transformed as

$$\frac{\partial^2 \tilde{\phi}(\beta_n)}{\partial y^2} - \beta_n^2 \tilde{\phi}(\beta_n) = 0. \tag{14.6.14}$$

Using the method of separation of variable and the boundary conditions on the electric walls, the discrete Fourier variable β_n is obtained from equation (14.6.13):

$$\beta_n = \frac{(2n+1)\pi}{2L}, n = 0, 1, 2, 3\ldots \quad \text{(a)}$$

$$\text{Aternatively,} \quad \beta_n = \frac{(2n-1)\pi}{2L}, \quad n = 1, 2, 3\ldots \quad \text{(b).}$$

$$\tag{14.6.15}$$

The solution of equation (14.6.14) in each region is given below:

Region-I: $0 \le y \le h_1$

$$\tilde{\phi}_1(\beta_n, y) = A_1 \sinh(\beta_n y) + B_1 \cosh(\beta_n y)$$

Boundary condition: At $y = 0$

$$\phi_1(\beta_n, 0) = 0, \text{ giving } B_1 = 0.$$

Region-II: $h_1 \le y \le H(h_1 + h_2)$

$$\tilde{\phi}_2(\beta_n, y) = A_2 \sinh[\beta_n(H - y)] + B_2 \cosh[\beta_n(H - y)]$$

Boundary condition: At $y = H$

$$\phi_2(\beta_n, H) = 0, \text{ giving } B_2 = 0.$$

The potentials, in the discrete Fourier domain, in both regions are

$$\tilde{\phi}_1(\beta_n, y) = A_1 \sinh(\beta_n y) \quad 0 \le y \le h_1 \quad \text{(a)}$$

$$\tilde{\phi}_2(\beta_n, y) = A_2 \sinh[\beta_n(h_1 + h_2 - y)], \quad h_1 \le y \le H \quad \text{(b).}$$

$$\tag{14.6.16}$$

The constants A_1 and A_2 are evaluated from the following continuity conditions:

At dielectric interface $(y = h_1)$:

$$\tilde{\phi}_2(\beta_n, y) = \tilde{\phi}_1(\beta_n, y) \quad \text{(a)}$$

$$\text{At strip conductor:} \quad \varepsilon_0\varepsilon_{r2}\frac{d\tilde{\phi}_2(\beta_n, y)}{dy} - \varepsilon_0\varepsilon_{r1}\frac{d\tilde{\phi}_1(\beta_n, y)}{dy} = -\tilde{\rho}(\beta_n) \quad \text{(b),}$$

$$\tag{14.6.17}$$

where $\tilde{\rho}(\beta_n)$ is the discrete Fourier transform of the charge density $\rho(x)$ at the strip conductor. The potential at the interface is divided into two regions. The potential

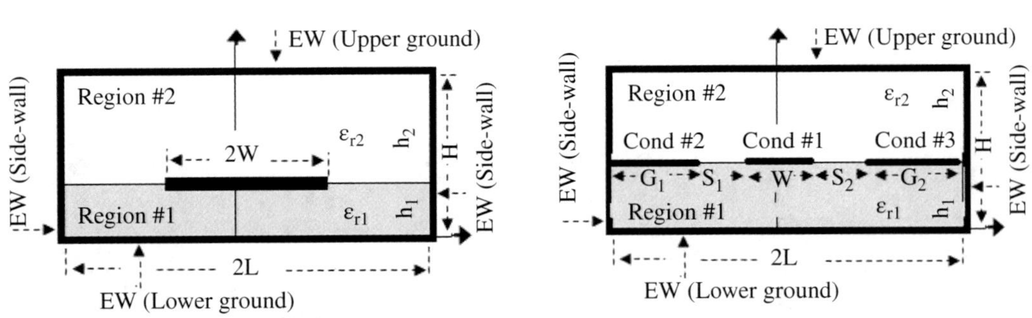

(a) Boxed microstrip line.

(b) Boxed CPW structure.

Figure 14.14 Boxed microstrip and CPW structures.

$V_C(x)$ is at the strip conductor, over $(-w \le x \le w)$ and potential $V_d(x)$ is at the dielectric interface, over $(-L \le x \le -w)$ and $(w \le x \le L)$. The potentials $V_C(x)$ and $V_d(x)$ in Fourier domains are given below:

$$\begin{aligned}
&\text{Potential at the interface: } \tilde{\phi}_1 = \tilde{\phi}_2 = \tilde{V}_C + \tilde{V}_d && \text{(a)}\\
&\text{Potential at strip conductor: } \tilde{V}_C = \int_{-w}^{w} V_C(x) e^{j\beta_n x} dx && \text{(b)}\\
&\text{Potential at dielectric interface: } \tilde{V}_d = \left[\int_{-L}^{-w} V_d(x) e^{j\beta_n x} dx + \int_{w}^{L} V_d(x) e^{j\beta_n x} dx\right] && \text{(c).}
\end{aligned} \quad (14.6.18)$$

On the strip conductor, $V_d = 0$, whereas on the dielectric interface, $V_c = 0$. The strip conductor and dielectric interface form a pair of *complementary surfaces* [J.14]. Using the continuity condition (14.6.17a) at the interface of two dielectrics, and equation (14.6.16), the constant A_1 is evaluated [J.15]:

$$A_1 \sinh(\beta_n h_1) = A_2 \sinh(\beta_n h_2) \Rightarrow A_1 = \frac{A_2 \sinh(\beta_n h_2)}{\sinh(\beta_n h_1)}.$$

Using equations (14.6.16) and (14.6.17b), the following expressions are obtained:

$$A_1 \beta_n \varepsilon_{r1} \cosh(\beta_n h_1) + A_2 \beta_n \varepsilon_{r2} \cosh(\beta_n h_2) = \frac{\tilde{\rho}(\beta_n)}{\varepsilon_0}$$

On solving the above two equations, expressions for A_1 and A_2 are obtained:

$$\begin{aligned}
A_1 &= \frac{\tilde{\rho}(\beta_n)}{\varepsilon_0 \beta_n \sinh(\beta_n h_1)[\varepsilon_{r1}\coth(\beta_n h_1) + \varepsilon_{r2}\coth(\beta_n h_2)]} && \text{(a)}\\
A_2 &= \frac{\tilde{\rho}(\beta_n)}{\varepsilon_0 \beta_n \sinh(\beta_n h_2)[\varepsilon_{r1}\coth(\beta_n h_1) + \varepsilon_{r2}\coth(\beta_n h_2)]} && \text{(b).}
\end{aligned} \quad (14.6.19)$$

The Fourier transformed potentials in two regions are obtained by substituting equations (14.6.19) in equation (14.6.16):

$$\begin{aligned}
\tilde{\phi}_1(\beta_n, y) &= \frac{\sinh(\beta_n y)\tilde{\rho}(\beta_n)}{\varepsilon_0 \beta_n \sinh(\beta_n h_1)[\varepsilon_{r1}\coth(\beta_n h_1) + \varepsilon_{r2}\coth(\beta_n h_2)]}, \quad 0 \le y \le h_1 && \text{(a)}\\
\tilde{\phi}_2(\beta_n, y) &= \frac{\sinh[\beta_n(h_1 + h_2 - y)]\tilde{\rho}(\beta_n)}{\varepsilon_0 \beta_n \sinh(\beta_n h_2)[\varepsilon_{r1}\coth(\beta_n h_1) + \varepsilon_{r2}\coth(\beta_n h_2)]}, \quad h_1 \le y \le H && \text{(b).}
\end{aligned} \quad (14.6.20)$$

Using equation (14.2.12), the potential at any location is related to the Green's function at that location as follows:

$$\tilde{\phi}_1(\beta_n, y) = \tilde{G}(\beta_n, y)\tilde{\rho}(\beta_n). \quad (14.6.21)$$

Therefore, Green's functions in two regions are

$$\begin{aligned}
\tilde{G}_1(\beta_n, y) &= \frac{\sinh(\beta_n y)}{\varepsilon_0 \beta_n \sinh(\beta_n h_1)[\varepsilon_{r1}\coth(\beta_n h_1) + \varepsilon_{r2}\coth(\beta_n h_2)]}, \quad 0 \le y \le h_1 && \text{(a)}\\
\tilde{G}_2(\beta_n, y) &= \frac{\sinh[\beta_n(h_1 + h_2 - y)]}{\varepsilon_0 \beta_n \sinh(\beta_n h_2)[\varepsilon_{r1}\coth(\beta_n h_1) + \varepsilon_{r2}\coth(\beta_n h_2)]}, \quad h_1 \le y \le H && \text{(b)}\\
&\text{Green function at the interface, :}\\
\tilde{G}(\beta_n, h_1) &= \tilde{G}_1(\beta_n, h_1) = \tilde{G}_2(\beta_n, h_2) = \frac{1}{\varepsilon_0 \beta_n[\varepsilon_{r1}\coth(\beta_n h_1) + \varepsilon_{r2}\coth(\beta_n h_2)]} && \text{(c).}
\end{aligned} \quad (14.6.22)$$

Line Capacitance Using Galerkin's Method

To compute the line capacitance, the potential and charge distribution, on the strip conductor, are required. The charge distribution on the strip conductor is an unknown function. However, the unknown charge distribution $\rho(x)$ can be expanded as a linear combination of *assumed basis functions*, $\rho_{bi}(i = 1, 2, ..., N)$. The basis functions are transformed into the discrete Fourier domain. The constants of expansion, $a_i(i = 1, 2, ..., N)$, are determined by *Galerkin's method*. In the space domain, this process is known as the *method of moment* (MOM) [B.5].

$$\rho(x) = \sum_{i=1}^{N} a_i \rho_{bi}(x) \quad (a)$$

$$\tilde{\rho}(\beta_n) = \begin{cases} \sum_{i=1}^{N} a_i \tilde{\rho}_{bi}(\beta_n), & \text{for } |x| \le W \\ 0, & \text{for } |x| > W \end{cases} \quad (b). \qquad (14.6.23)$$

To implement Galerkin's method, the potential at the interface, given by equation (14.6.21), is written as the sum of potentials on the strip conductor and the dielectric interface:

$$\tilde{G}(\beta_n, h_1)\tilde{\rho}(\beta_n) = \tilde{V}_c(\beta_n) + \tilde{V}_d(\beta_n). \qquad (14.6.24)$$

The charge distribution and potential at the interface are divided into two groups,

$$\tilde{\rho}(\beta_n) = \begin{cases} \text{nonzero on strip conductor,} & -w \le x \le w \\ 0, \text{ on dielectric layer,} & |x| > w \end{cases} \quad (a)$$

$$\tilde{V}_c(\beta_n) = \begin{cases} \text{nonzero on strip conductor,} & -w \le x \le w \\ 0, \text{ on dielectric layer} & |x| > w \end{cases} \quad (b)$$

$$\tilde{V}_d(\beta_n) = \begin{cases} 0, \text{ on strip conductor,} & -w \le x \le w \\ \text{nonzero on dielectric layer,} & |x| > w \end{cases} \quad (c). \qquad (14.6.25)$$

The potential in the Fourier domain, due to N number of basis functions of the charge distribution, is obtained from equations (14.6.23) and (14.6.24):

$$\tilde{G}(\beta_n, h_1)\sum_{i=1}^{N} a_i \tilde{\rho}_{bi}(\beta_n) = \tilde{V}_c(\beta_n) + \tilde{V}_d(\beta_n). \qquad (14.6.26)$$

In equation (14.6.26), the coefficient of expansion $a_i (i = 1, 2, ..., N)$ of the charge distribution function is unknown. It is determined by Galerkin's process. The Galerkin's process requires the concept of the *inner product of two functions* $\tilde{\psi}_1(\beta_n)$ and $\tilde{\psi}_2(\beta_n)$. Using Parseval's theorem (14.6.4), it is defined as follows:

$$\langle \tilde{\psi}_1(\beta_n), \ \tilde{\psi}_2(\beta_n) \rangle = \sum_{n=1}^{\infty} \tilde{\psi}_1(\beta_n)\tilde{\psi}_2^*(\beta_n) = L\int_{-L}^{L} \psi_1(x)\tilde{\psi}_2^*(x)dx. \qquad (14.6.27)$$

At this stage, Galerkin's method is used in the Fourier domain. It gives us a matrix equation to determine the coefficients $a_1, a_2, ...a_N$ of the basis functions describing the charge density distribution on the strip conductor. The above-stated Parseval's theorem is also used in the present process. On taking the inner product of equation (14.6.26) with the *testing charge distribution functions,* $\tilde{\rho}_{tj}(\beta_n)$, the following expression is obtained:

$$\sum_{i=1}^{N}\left[a_i\left\langle \tilde{G}(\beta_n, h_1)\tilde{\rho}_{bi}(\beta_n), \tilde{\rho}_{tj}(\beta_n)\right\rangle\right] = \left\langle \tilde{V}_c(\beta_n), \tilde{\rho}_{tj}(\beta_n)\right\rangle + \left\langle \tilde{V}_d(\beta_n), \tilde{\rho}_{tj}(\beta_n)\right\rangle, \quad j = 1, 2, 3, ..., M, \qquad (14.6.28)$$

where $\left\langle \tilde{V}_d(\beta_n), \tilde{\rho}_{tj}(\beta_n)\right\rangle = L\int_{-L}^{L} V_d(x), \rho_{tj}(x)dx = 0$, as on the dielectric interface $\rho_{tj}(x) = 0$ and $V_d(x) = 0$ on the strip conductor. Hence,

$$\sum_{i=1}^{N}\left[a_i\left\langle \tilde{G}(\beta_n, h_1)\tilde{\rho}_{bi}(\beta_n), \tilde{\rho}_{tj}(\beta_n)\right\rangle\right] = \left\langle \tilde{V}_c(\beta_n), \tilde{\rho}_{tj}(\beta_n)\right\rangle, j = 1, 2, 3, ..., M. \qquad (14.6.29)$$

Let us define, $H_j = \left\langle \tilde{V}_c(\beta_n), \tilde{\rho}_{tj}(\beta_n)\right\rangle = \sum_{n=1}^{\infty}\left[\tilde{V}_c(\beta_n)\tilde{\rho}^*_{tj}(\beta_n)\right]$. On using Parseval's theorem (14.6.27),

$$H_j = L\int_{-L}^{L} V_c(x)\cdot\rho_{tj}^*(x)dx = L\int_{-W}^{W} V_c(x)\cdot\rho_{tj}^*(x)dx, \quad j = 1, 2, 3, ..., M. \qquad (14.6.30)$$

Using Parseval's theorem, the following relation is defined:

$$C_{j,i} = \left\langle \tilde{G}(\beta_n, h_1)\tilde{\rho}_{bi}(\beta_n), \tilde{\rho}_{tj}(\beta_n)\right\rangle = \sum_{n=1}^{\infty}\left[\tilde{G}(\beta_n, h_1)\tilde{\rho}_{bi}(\beta_n)\tilde{\rho}^*_{tj}(\beta_n)\right], \quad j = 1, 2, ..., M. \qquad (14.6.31)$$

Thus, equation (14.6.29) is written as

$$\sum_{i=1}^{N} C_{j,i}a_i = H_j, \begin{cases} \text{No of terms of basis function}: & i = 1, 2, ..., N \\ \text{No of terms of testing function}: & j = 1, 2, ..., M \end{cases}. \qquad (14.6.32)$$

Each testing function (j = 1, 2,...,M) tests the potential due to i (=1, 2,...,N) charge distribution basis functions on the strip conductor. Equation (14.6.32) is expanded as follows:

$$C_{11}a_1 + C_{12}a_2 + \cdots + C_{1N}a_N = H_1$$
$$C_{21}a_1 + C_{22}a_2 + \cdots + C_{2N}a_N = H_2$$
$$------------------------$$
$$C_{M1}a_1 + C_{M2}a_2 + \cdots + C_{MN}a_N = H_M$$
$$\Rightarrow \begin{bmatrix} C_{11} & C_{12} & C_{13} \\ C_{21} & C_{22} & C_{23} \\ --- & --- & --- \\ C_{M1} & C_{M2} & C_{MN} \end{bmatrix} \begin{bmatrix} a_1 \\ a_2 \\ - \\ a_3 \end{bmatrix} = \begin{bmatrix} H_1 \\ H_2 \\ - \\ H_M \end{bmatrix}. \tag{14.6.33}$$

Normally, $N = M$, i.e. the number of terms of the basis and the testing functions are taken identical. Using equation (14.6.23a), the line capacitance C *p.u.l.* is determined from the following expression for $V_c(x) = 1$ Volt on the strip conductor,

$$C = \frac{Q}{V_C(x)} = \sum_{i=1}^{N} a_i \int_{-w}^{w} \rho_{bi}(x)dx. \tag{14.6.34}$$

The coefficients a_i are determined from equation (14.6.33). For instance, in the case of *one term basis function* i.e. $N = 1$, the above line capacitance is reduced to

$$C = a_1 \int_{-w}^{w} \rho_b(x)dx. \tag{14.6.35}$$

From equation (14.6.33), $C_{11}a_1 = H_1, \Rightarrow a_1 = H_1/C_{11}$, as $a_2 = \dots = a_N = 0$. For $V_c(x) = 1$ Volt, equation (14.6.30) is written as $H_1 = L\int_{-w}^{w}\rho_{t1}(x)dx$. Usually, identical charge distribution functions are taken for both the basis and testing functions, i.e. $\rho_{b1}(x) = \rho_{t1}(x) = \rho_s(x)$. The matrix coefficient C_{11} from equation (14.6.31) is $C_{11} = \sum_{n=1}^{\infty} \tilde{G}(\beta_n, h_1)[\tilde{\rho}_s(\beta_n)]^2$. Thus, the line capacitance, using equation (14.6.34), for one term basis function is

$$C = \frac{H_1}{C_{11}} \int_{-W}^{W} \rho_S(x)dx,$$
$$\Rightarrow \frac{1}{C} = \frac{\sum_{n=1}^{\infty} \left\{ \tilde{G}(\beta_n, h_1)[\tilde{\rho}_S(\beta_n)]^2 \right\}}{L\left[\int_{-W}^{W}\rho_S(x)dx\right]^2},$$

$$\text{For} \quad Q = \int_{-W}^{W} \rho_S(x)dx \tag{a}$$
$$\Rightarrow \frac{1}{C} = \frac{1}{L}\sum_{n=1}^{\infty}\left\{ \tilde{G}(\beta_n, h_1)\left[\frac{\tilde{\rho}_S(\beta_n)}{Q}\right]^2 \right\} \tag{b}$$
$$\text{where } \beta_n = \left(\frac{2n-1}{L}\right)\frac{\pi}{2} \tag{c}$$
$$(14.6.36)$$

Using the definition of discrete cosine Fourier transform pair and corresponding Parseval's theorem given by equation (14.6.8e), equation (14.6.36) for the shielded microstrip line capacitance is modified as

$$\frac{1}{C} = \frac{L}{2}\sum_{n=1}^{\infty}\left\{ \tilde{G}(\beta_n, h_1)\left[\frac{\tilde{\rho}_S(\beta_n)}{Q}\right]^2 \right\}. \tag{14.6.37}$$

The Green's function $\tilde{G}(\beta_n, h_1)$ at the plane of the strip conductor is taken from equation (14.6.22c) to evaluate the line capacitance of the boxed microstrip. The discrete cosine Fourier transform pair is further used in section (14.6.3) to obtain the line capacitance of a boxed CPW structure.

Any charge distribution function, discussed in subsection (14.2.3), could be used. However, the *modified Maxwell distribution function* is more appropriate as it accounts for the edge singularity and any number of basis functions can be easily generated using this function. The same set of discrete Fourier transforms should be used to get Parseval's theorem and discrete Fourier transform of the charge distribution function. Once the line capacitance of a boxed microstrip is determined, the effective relative permittivity and characteristic impedance, using equation (14.3.35), are easily computed. The above expression is identical to the variational line capacitance given in equation (14.2.13). The Fourier integral is used in equation (14.2.13) as it applies to the open structures. Thus, equations (14.6.36) and (14.6.37) are the variational expressions with a one-term basis function approximation of Galerkin's method. By increasing the number of terms of the basis function, say $N = 3$ in equation (14.6.33); the accuracy of the computed line capacitance can be improved. The effect of the conductor thickness on the microstrip line parameters could be included by modifying the Green's function as follows:

$$\tilde{G}_t(h_1, \beta_n) = h(\beta_n)\tilde{G}(h_1, \beta_n) \tag{a}$$
$$\text{where,} \quad h(\beta_n) = \frac{1}{2}\left[1 + \frac{\sinh\{\beta_n(h_1 - t)\}}{\sinh\{\beta_n h_1\}}\right] \tag{b}$$
$$(14.6.38)$$

14.6.3 Boxed Coplanar Waveguide

Figure (14.14b) shows a shielded conductor backed CPW structure. Its line capacitance is determined by using Galerkin's method in the discrete Fourier domain. The solution of equation (14.6.14) gives Green's function of a CPW structure that is the same as obtained for a microstrip line. Analytically, the shielded microstrip line and CPW are not much different. In the case of a microstrip, two side conductors of the CPW are replaced by the dielectric interfaces with potential distribution; however, without any surface charge.

The central strip conductor of a CPW is maintained at potential $V_0 = 1$ Volt. Two outer conductors are ground conductors. Three *unknown* charge distributions ρ_{s1}, ρ_{s2}, and ρ_{s3} located on the central strip conductor #1, left-side ground conductor #2 and the right-side ground conductor #3, respectively, create the potential at any location in the bounded region. The unknown charge distributions, on each strip conductor, are expanded in terms of the assumed basis functions. However, for simplicity at present, only *one term of the basis functions* is taken, and subsequently, N-terms basis functions are considered.

One Term Basis Function

Figure (14.15) illustrates the Galerkin's process using the basis and testing functions for the charge density on the strip conductors. The one-term *basis functions* of the charge density, in the discrete Fourier domain, $\tilde{\rho}_{b,s1}(\beta_n), \tilde{\rho}_{b,s2}(\beta_n)$ and $\tilde{\rho}_{b,s3}(\beta_n)$, located on the conductors #1, #2, and #3, are confined to their respective regions: $(G_1 + S_1 \le x' \le (G_1 + S_1 + W))$, $0 \le x' \le G_1$ and $(G_1 + S_1 + W + S_2) < x' < G_1 + S_1 + W + S_2 + G_2$, respectively. It is the case of the ACPW. However, for the symmetrical CPW, $G_1 = G_2$ and $S_1 = S_2$. The conductors are placed at the dielectric interface located at $y = h_1$. At this plane Green's function is $\tilde{G}(\beta_n, h_1)$. For simplicity, it is written as $\tilde{G}(\beta_n)$. Green's function in the Fourier domain $\tilde{G}(\beta_n, h_1)$ remains the same for any charge location along the *x*-axis. However, in the space domain, it is *x*-dependent. *The Green's function* $\tilde{G}(\beta_n, h_1)$ *is independent of the number of strip conductors on the same plane.* In the range $[-L, L]$, the potential at $y = h_1$ is a linear combination of the individual potential created by the basis functions $\tilde{\rho}_{b,s1}(\beta_n), \tilde{\rho}_{b,s2}(\beta_n)$ and $\tilde{\rho}_{b,s3}(\beta_n)$. This potential, in the range $[-L, L]$, is also equal to a sum of the potentials of three-strip conductors, and the potentials on two dielectric slot-regions. The following potential expression is written at the interface $y = h_1$:

$$a_1\tilde{G}(\beta_n)\tilde{\rho}_{s1}(\beta_n) + b_1\tilde{G}(\beta_n)\tilde{\rho}_{s2}(\beta_n) + c_1\tilde{G}(\beta_n)\tilde{\rho}_{s3}(\beta_n) = \tilde{V}_c(\beta_n) + \tilde{V}_d(\beta_n). \tag{14.6.39}$$

The coefficients a_1, b_1, and c_1 associated with the charge densities, on the conductors #1, #2, and #3, respectively, are determined using Galerkin's method. Normally, the basis and testing functions are taken identical. The inner product of equation (14.6.39), with the *testing functions* $\tilde{\rho}_{t,1s}(\beta_n)$, $\tilde{\rho}_{t,2s}(\beta_n)$, and $\tilde{\rho}_{t,3s}(\beta_n)$, is taken one by one. In Fig (14.15), the testing functions are shown as $\tilde{\rho}_{t,1k}(\beta_n)$, $\tilde{\rho}_{t,2k}(\beta_n)$, and $\tilde{\rho}_{t,3k}(\beta_n)$, where $k = s$.

The inner product is a process of testing each potential at plane $y = h_1$ by the testing charge, i.e. the testing function. Normally, the identical basis and testing functions are used, i.e. $\tilde{\rho}_{b,1s}(\beta_n) = \tilde{\rho}_{t,1s}(\beta_n) = \tilde{\rho}_{s1}(\beta_n)$, $\tilde{\rho}_{b,2s}(\beta_n) = \tilde{\rho}_{t,2s}(\beta_n) = \tilde{\rho}_{s2}(\beta_n)$, and $\tilde{\rho}_{t,3s}(\beta_n) = \tilde{\rho}_{b,3s}(\beta_n) = \tilde{\rho}_{s3}(\beta_n)$. Thus, the inner product of the expression (14.6.39), taken by the testing functions $\tilde{\rho}_{s1}(\beta_n)$, $\tilde{\rho}_{s2}(\beta_n)$ and $\tilde{\rho}_{s3}(\beta_n)$, are

$$a_1\left\langle \tilde{G}(\beta_n)\tilde{\rho}_{s1}(\beta_n), \tilde{\rho}_{s1}(\beta_n)\right\rangle + b_1\left\langle \tilde{G}(\beta_n)\tilde{\rho}_{s2}(\beta_n), \tilde{\rho}_{s1}(\beta_n)\right\rangle + c_1\left\langle \tilde{G}(\beta_n)\tilde{\rho}_{s3}(\beta_n), \tilde{\rho}_{s1}(\beta_n)\right\rangle = \left\langle \tilde{V}_C(\beta_n), \tilde{\rho}_{s1}(\beta_n)\right\rangle + \left\langle \tilde{V}_d(\beta_n), \tilde{\rho}_{s1}(\beta_n)\right\rangle \quad \text{(a)}$$

$$a_1\left\langle \tilde{G}(\beta_n)\tilde{\rho}_{s1}(\beta_n), \tilde{\rho}_{s2}(\beta_n)\right\rangle + b_1\left\langle \tilde{G}(\beta_n)\tilde{\rho}_{s2}(\beta_n), \tilde{\rho}_{s2}(\beta_n)\right\rangle + c_1\left\langle \tilde{G}(\beta_n)\tilde{\rho}_{s3}(\beta_n), \tilde{\rho}_{s2}(\beta_n)\right\rangle = \left\langle \tilde{V}_C(\beta_n), \tilde{\rho}_{s2}(\beta_n)\right\rangle + \left\langle \tilde{V}_d(\beta_n), \tilde{\rho}_{s2}(\beta_n)\right\rangle \quad \text{(b)}$$

$$a_1\left\langle \tilde{G}(\beta_n)\tilde{\rho}_{s1}(\beta_n), \tilde{\rho}_{s3}(\beta_n)\right\rangle + b_1\left\langle \tilde{G}(\beta_n)\tilde{\rho}_{s2}(\beta_n), \tilde{\rho}_{s3}(\beta_n)\right\rangle + c_1\left\langle \tilde{G}(\beta_n)\tilde{\rho}_{s3}(\beta_n), \tilde{\rho}_{s3}(\beta_n)\right\rangle = \left\langle \tilde{V}_C(\beta_n), \tilde{\rho}_{s3}(\beta_n)\right\rangle + \left\langle \tilde{V}_d(\beta_n), \tilde{\rho}_{s3}(\beta_n)\right\rangle \quad \text{(c).}$$

(14.6.40)

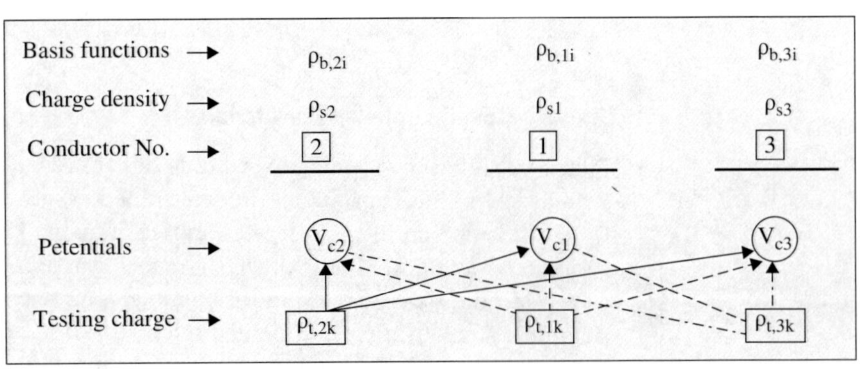

Figure 14.15 Galerkin's process of taking the inner product of three testing charge functions.

Using equation (14.6.27), the inner product elements of the above matrix equation are evaluated as follows:

$$\widetilde{P}_{ji}(\beta_n) = \left\langle \widetilde{G}(\beta_n)\widetilde{\rho}_{b,si}(\beta_n), \widetilde{\rho}_{t,sj}(\beta_n) \right\rangle = \sum_{n=1}^{\infty} \widetilde{G}(\beta_n)\widetilde{\rho}_{b,si}(\beta_n)\widetilde{\rho}_{t,sj}(\beta_n) \begin{cases} i = 1,2,3 \\ j = 1,2,3, \end{cases} \quad (14.6.41)$$

where index "i" is the number of conductors supporting the main charge using one term *basis function* and index "j" is the number of conductors supporting one term *testing function*. In equation (14.6.41), an identical basis and testing functions are assumed i.e. $\widetilde{\rho}_{b,si}(\beta_n) = \widetilde{\rho}_{t,sj}(\beta_n)$. On the dielectric interface, i.e. in two slot-regions S_1 and S_2, there is no charge, i.e. $\rho_{s1} = \rho_{s2} = \rho_{s3} = 0$. Also on the strip conductors $V_d = 0$. Thus, using equation (14.6.8e), the inner product is taken as

$$\left\langle \widetilde{V}_d, \widetilde{\rho}_{si} \right\rangle = \sum_{n=1}^{\infty} \left[\widetilde{V}_d(\beta_n)\widetilde{\rho}_{si}(\beta_n) \right] = \frac{2}{L}\int_0^L V_d(x)\rho_{si}(x)dx = 0, \quad i = 1,2,3. \quad (14.6.42)$$

The above inner product is based on the discrete cosine Fourier transform pair given in equation (14.6.8).

The potential, in the Fourier domain, on three conducting strips in the range $[-L, L]$ is $\widetilde{V}_c(\beta_n) = \widetilde{V}_{c1} + \widetilde{V}_{c2} + \widetilde{V}_{c3}$, where V_{c1}, V_{c2} and V_{c3} are the potentials on the conductors #1, #2, and #3, respectively. The conductors #2 and #3 are grounded, i.e. $V_{c2} = V_{c3} = 0$. It is also noted that at conductor #1, $\rho_{s2} = \rho_{s3} = 0$, as these charge densities exist only on the conductors #2 and #3, respectively. On using equation (14.6.8e) for the Parveral's theorem, the following inner products of equation (14.6.40) are evaluated:

$$\left\langle \widetilde{V}_c(\beta_n), \widetilde{\rho}_{s2}(\beta_n) \right\rangle = \sum_{n=1}^{\infty} \left[\widetilde{V}_c(\beta_n)\widetilde{\rho}_{s2}(\beta_n) \right] = \frac{2}{L}\int_0^L V_c(x)\,\rho_{s2}(x)dx = 0 \quad (a)$$

$$\left\langle \widetilde{V}_c(\beta_n), \widetilde{\rho}_{s3}(\beta_n) \right\rangle = \sum_{n=1}^{\infty} \left[\widetilde{V}_c(\beta_n)\widetilde{\rho}_{s3}(\beta_n) \right] = \frac{2}{L}\int_0^L V_c(x)\rho_{s3}(x)dx = 0 \quad (b)$$

$$\left\langle \widetilde{V}_c(\beta_n), \widetilde{\rho}_{s1}(\beta_n) \right\rangle = \sum_{n=1}^{M} \left[\widetilde{V}_c(\beta_n)\widetilde{\rho}_{s1}(\beta_n) \right] = \frac{2V_0}{L}\int_{G_1+S_1}^{G_1+S_1+w} \rho_{S1}(x)dx = R_1 \quad (c). \quad (14.6.43)$$

Equation (14.6.40) is reduced to

$$a_1\left\langle \widetilde{G}(\beta_n)\widetilde{\rho}_{s1}(\beta_n), \widetilde{\rho}_{s1}(\beta_n) \right\rangle + b_1\left\langle \widetilde{G}(\beta_n)\widetilde{\rho}_{s2}(\beta_n), \widetilde{\rho}_{s1}(\beta_n) \right\rangle + c_1\left\langle \widetilde{G}(\beta_n)\widetilde{\rho}_{s3}(\beta_n), \widetilde{\rho}_{s1}(\beta_n) \right\rangle = R_1$$

$$a_1\left\langle \widetilde{G}(\beta_n)\widetilde{\rho}_{s1}(\beta_n), \widetilde{\rho}_{s2}(\beta_n) \right\rangle + b_1\left\langle \widetilde{G}(\beta_n)\widetilde{\rho}_{s2}(\beta_n), \widetilde{\rho}_{s2}(\beta_n) \right\rangle + c_1\left\langle \widetilde{G}(\beta_n)\widetilde{\rho}_{s3}(\beta_n), \widetilde{\rho}_{s2}(\beta_n) \right\rangle = 0$$

$$a_1\left\langle \widetilde{G}(\beta_n)\widetilde{\rho}_{s1}(\beta_n), \widetilde{\rho}_{s3}(\beta_n) \right\rangle + b_1\left\langle \widetilde{G}(\beta_n)\widetilde{\rho}_{s2}(\beta_n), \widetilde{\rho}_{s3}(\beta_n) \right\rangle + c_1\left\langle \widetilde{G}(\beta_n)\widetilde{\rho}_{s3}(\beta_n), \widetilde{\rho}_{s3}(\beta_n) \right\rangle = 0$$

$$\Rightarrow \begin{bmatrix} \left\langle \widetilde{G}(\beta_n)\widetilde{\rho}_{s1}(\beta_n), \widetilde{\rho}_{s1}(\beta_n) \right\rangle & \left\langle \widetilde{G}(\beta_n)\widetilde{\rho}_{s2}(\beta_n), \widetilde{\rho}_{s1}(\beta_n) \right\rangle & \left\langle \widetilde{G}(\beta_n)\widetilde{\rho}_{s3}(\beta_n), \widetilde{\rho}_{s1}(\beta_n) \right\rangle \\ \left\langle \widetilde{G}(\beta_n)\widetilde{\rho}_{s1}(\beta_n), \widetilde{\rho}_{s2}(\beta_n) \right\rangle & \left\langle \widetilde{G}(\beta_n)\widetilde{\rho}_{s2}(\beta_n), \widetilde{\rho}_{s2}(\beta_n) \right\rangle & \left\langle \widetilde{G}(\beta_n)\widetilde{\rho}_{s3}(\beta_n), \widetilde{\rho}_{s2}(\beta_n) \right\rangle \\ \left\langle \widetilde{G}(\beta_n)\widetilde{\rho}_{s1}(\beta_n), \widetilde{\rho}_{s3}(\beta_n) \right\rangle & \left\langle \widetilde{G}(\beta_n)\widetilde{\rho}_{s2}(\beta_n), \widetilde{\rho}_{s3}(\beta_n) \right\rangle & \left\langle \widetilde{G}(\beta_n)\widetilde{\rho}_{s3}(\beta_n), \widetilde{\rho}_{s3}(\beta_n) \right\rangle \end{bmatrix} \begin{bmatrix} a_1 \\ b_1 \\ c_1 \end{bmatrix} = \begin{bmatrix} R_1 \\ 0 \\ 0 \end{bmatrix}. \quad (14.6.44)$$

Matrix equation (14.6.44) can be arranged in the following compact form:

$$\begin{bmatrix} \widetilde{P}_{11}(\beta_n) & \widetilde{P}_{12}(\beta_n) & \widetilde{P}_{13}(\beta_n) \\ \widetilde{P}_{21}(\beta_n) & \widetilde{P}_{22}(\beta_n) & \widetilde{P}_{23}(\beta_n) \\ \widetilde{P}_{31}(\beta_n) & \widetilde{P}_{32}(\beta_n) & \widetilde{P}_{33}(\beta_n) \end{bmatrix} \begin{bmatrix} a_1 \\ b_1 \\ c_1 \end{bmatrix} = \begin{bmatrix} R_1 \\ 0 \\ 0 \end{bmatrix}. \quad (14.6.45)$$

Matrix elements in equation (14.6.45) correspond to the matrix elements of equation (14.6.44). The *matrix elements* $\widetilde{P}_{ji}(\beta_n)$ are inner products. For an identical basis and testing functions, the off-diagonal matrix elements are symmetrical. The cosine Fourier transform of the charge distribution function is discussed in section (14.6.1). The coefficients a_1, a_2, a_3 are evaluated by inverting the matrix equation 14.6.45.

The line capacitance of the CPW is

$$\frac{Q_1[\text{charge on the central conductor \# 1}]}{V_0} = a_1\frac{\int_{G_1+S_1}^{G_1+S_1+W}\rho_{s1}(x)dx}{V_0}. \quad (14.6.46)$$

From equation (14.6.45) : $a_1\tilde{P}_{11}(\beta_n) + b_1\tilde{P}_{12}(\beta_n) + c_1\tilde{P}_{13}(\beta_n) = R_1$

$$\Rightarrow a_1 = \frac{R_1 - \left(b_1\tilde{P}_{12} + c_1\tilde{P}_{13}\right)}{\tilde{P}_{11}} \quad (a)$$

From equation (14.6.43c) : $R_1 = \frac{2V_0}{L}\int_{G_1+S_1}^{G_1+S_1+w}\rho_{s1}(x)dx$

$$\Rightarrow Q_1 = \int_{G_1+S_1}^{G_1+S_1+W}\rho_{s1}(x)dx = \frac{R_1 L}{2V_0} \quad (b). \quad (14.6.47)$$

The line capacitance of the CPW from the above two equations (14.4.46) and (14.4.47) is

$$C = \frac{R_1 - \left(b_1\tilde{P}_{12}(\beta_n) + c_1\tilde{P}_{13}(\beta_n)\right)}{V_0\tilde{P}_{11}(\beta_n)} \times \frac{R_1 L}{2V_0}$$
$$= \frac{R_1^2 L}{2{V_0}^2\tilde{P}_{11}(\beta_n)}\left[1 - \frac{b_1\tilde{P}_{12}(\beta_n) + c_1\tilde{P}_{13}(\beta_n)}{R_1}\right]$$
$$\Rightarrow C = \frac{L}{2{V_0}^2\tilde{P}_{11}(\beta_n)} \times \frac{4{V^2}_0}{L^2}\left[\int_{G_1+S_1}^{G_1+S_1+w}\rho_{S1}(x)dx\right]^2 \left[1 - \frac{b_1\tilde{P}_{12}(\beta_n) + c_1\tilde{P}_{13}(\beta_n)}{R_1}\right]. \quad (14.6.48)$$

The line capacitance *p.u.l.* of a CPW structure, for one-term basis function on each conductor and $V_0 = 1$, is

$$C = \frac{2}{L}\frac{Q_1^2}{\tilde{P}_{11}(\beta_n)}\left[1 - \left\{\frac{b_1\tilde{P}_{12}(\beta_n) + c_1\tilde{P}_{13}(\beta_n)}{R_1}\right\}\right]. \quad (14.6.49)$$

where using equation (14.6.41) the inner products are

$$\tilde{P}_{11}(\beta_n) = \left\langle\tilde{G}(\beta_n, h_1)\tilde{\rho}_{s1}(\beta_n), \tilde{\rho}_{s1}(\beta_n)\right\rangle = \sum_{n=1}^{\infty}\left[\tilde{G}(\beta_n, h_1)\tilde{\rho}_{s1}(\beta_n), \tilde{\rho}_{s1}(\beta_n)\right]$$
$$\tilde{P}_{12}(\beta_n) = \left\langle\tilde{G}(\beta_n, h_1)\tilde{\rho}_{s2}(\beta_n), \tilde{\rho}_{s1}(\beta_n)\right\rangle = \sum_{n=1}^{\infty}\left[\tilde{G}(\beta_n, h_1)\tilde{\rho}_{s2}(\beta_n), \tilde{\rho}_{s1}(\beta_n)\right] \quad (14.6.50)$$
$$\tilde{P}_{13}(\beta_n) = \left\langle\tilde{G}(\beta_n, h_1)\tilde{\rho}_{s3}(\beta_n), \tilde{\rho}_{s1}(\beta_n)\right\rangle = \sum_{n=1}^{\infty}\left[\tilde{G}(\beta_n, h_1)\tilde{\rho}_{s3}(\beta_n), \tilde{\rho}_{s1}(\beta_n)\right].$$

In the case of a *microstrip*, the conductors #2 and #3 are dielectrics that do not support any charge density. Therefore, $\tilde{\rho}_{s2} = 0, \tilde{\rho}_{s3} = 0$ that leads to $\tilde{P}_{12}(\beta_n) = \tilde{P}_{13}(\beta_n) = 0$. Hence, for a microstrip, the line capacitance equation (14.6.49) is reduced to the following expression:

$$\frac{1}{C} = \frac{L}{2}\frac{\tilde{P}_{11}(\beta_n)}{Q_1^2} = \frac{L\left\langle\tilde{G}(\beta_n)\tilde{\rho}_{s1}(\beta_n), \tilde{\rho}_{s1}(\beta_n)\right\rangle}{2\varepsilon_0 Q_1^2}$$
$$\Rightarrow \frac{1}{C} = \frac{L}{2}\sum_{n=1}^{\infty}\tilde{G}(\beta_n, h_1)\{\tilde{\rho}_{s1}(\beta_n)/Q_1\}^2. \quad (14.6.51)$$

It is the same as the expression (14.6.37) obtained earlier for the microstrip line. In the case of the conductor-backed CPW structure, equation (14.6.49) shows that the presence of the coplanar-grounded conductors reduces the line capacitance of a corresponding microstrip. Once the line capacitance of the boxed CPW is known, the effective relative permittivity and characteristic impedance can be computed using equation (14.3.58).

N-Terms Basis Function

The above Galerkin's process can be extended to the N-terms basis functions [B.8]. The potential at the plane $y = h_1$, in terms of the Green's function, is still given by equation (14.6.39). The total charge *p.u.l* density on three-strip conductors is

$$\rho_s^T(x) = \sum_{i=1}^{3}\rho_{si}(x); \text{ Index i indicates 3-strip conductors, } i = 1, 2, 3. \quad (14.6.52)$$

The charge density $\rho_{si}(x)$ on each strip conductor can be expanded in terms of N number of the basis functions. The charge density on the ith conductor is

$$\rho_{si}(x) = \sum_{\ell=1}^{N} d_{i\ell}\rho_{b,i\ell}(x); \quad i = 1, 2, 3; \quad \ell = 1, 2, ..., N, \quad (14.6.53)$$

where "i" is the *index number for the strip conductors*, and "ℓ" is the *index number for the terms of the basis functions* of charge density. Equation (14.6.53) $d_{i\ell}$ is the unknown coefficient of ℓth of expansion term of the charge density $\rho_{b,\,i\ell}$ on the ith strip conductor. The total charge density on three-strip conductors located at the plane $y = h_1$ is

$$\rho_s^T(x) = \sum_{i=1}^{3}\left[\sum_{\ell=1}^{N} d_{i\ell}\rho_{b,i\ell}(x)\right]. \quad (14.6.54)$$

In the Fourier domain, the total charge density is

$$\tilde{\rho}_s^T(\beta_n) = \sum_{i=1}^{3}\left[\sum_{\ell=1}^{N} d_{i\ell}\tilde{\rho}_{b,i\ell}(\beta_n)\right]. \qquad (14.6.55)$$

The *potential at the plane* $y = h_1$ from equations (14.6.39) and (14.6.55) is

$$\tilde{G}_S(\beta_n, h_1)\sum_{i=1}^{3}\tilde{\rho}_{si}(\beta_n) = \tilde{V}_c(\beta_n) + \tilde{V}_d(\beta_n)$$

$$\Rightarrow \tilde{G}_S(\beta_n, h_1)\sum_{i=1}^{3}\left[\sum_{\ell=1}^{N} d_{i\ell}\tilde{\rho}_{b,i\ell}(\beta_n)\right] = \tilde{V}_c(\beta_n) + \tilde{V}_d(\beta_n). \qquad (14.6.56)$$

Using Galerkin's method $\tilde{V}_d$ can be eliminated from equation (14.6.56) and three equations are obtained to determine three coefficients $d_{i\ell}(i = 1, 2, 3;\ \ell = 1, 2, \ldots, N)$ associated with each term of the basis function $\rho_{b,\,i\ell}$. For one-term ($\ell = 1$) basis function, three equations are obtained with coefficients, $d_{11} = a_1$, $d_{21} = b_1$ and $d_{31} = c_1$ as noted earlier in equation (14.6.39). The coefficients could be determined by taking the inner product of equation (14.6.56) with the *test charge density function,* $\tilde{\rho}_{t,jk}$ on the three conductors [j = 1, 2, 3]. The index k, [k = 1, 2, 3, ..., N], shows the terms of testing function. Equation (14.6.56) is tested, i.e. the inner product is taken with test charge one by one. As there are three test charge densities associated with three conductors [j = 1, 2, 3], each one will test the potential on all three conductors [i = 1, 2, 3]. Figure (14.15) illustrates the inner product process. It results in three expressions that are written in compact form as follows:

$$\left\langle \sum_{i=1}^{3}\sum_{\ell=1}^{N}\tilde{G}(\beta_n)d_{i\ell}\tilde{\rho}_{b,i\ell}, \tilde{\rho}_{t,jk}\right\rangle = \left\langle \tilde{V}_c(\beta_n), \tilde{\rho}_{t,jk}\right\rangle + \left\langle \tilde{V}_d(\beta_n), \tilde{\rho}_{t,jk}\right\rangle, [j = 1, 2, 3]; [k = 1, 2, 3, \ldots, N]. \qquad (14.6.57)$$

The above compact inner product equation (14.6.57) is examined in some details. It is shown in Fig (14.15). Let us take the inner product, i.e. test the potential, on all three conductors (i = 1, 2, 3) with the testing charge $\tilde{\rho}_{t,jk}$ located on the first conductor for j = 1 (One).

Testing of Potential on conductor #1 by test charge $\tilde{\rho}_{t,1k}$	Testing of Potential on conductor #2 by test charge $\tilde{\rho}_{t,1k}$	Testing of Potential on conductor #3 by test charge $\tilde{\rho}_{t,1k}$
↓	↓	↓

$$\sum_{\ell=1}^{N} d_{1\ell}\left\langle \tilde{G}(\beta_n)\tilde{\rho}_{b,1\ell}, \tilde{\rho}_{t,1k}\right\rangle + \sum_{\ell=1}^{N} d_{2\ell}\left\langle \tilde{G}(\beta_n)\tilde{\rho}_{b,2\ell}, \tilde{\rho}_{t,1k}\right\rangle + \sum_{\ell=1}^{N} d_{3\ell}\left\langle \tilde{G}(\beta_n)\tilde{\rho}_{b,3\ell}, \tilde{\rho}_{t,1k}\right\rangle = \qquad (a)$$

$$\left\langle \tilde{V}_C(\beta_n), \tilde{\rho}_{t.1k}\right\rangle + \left\langle \tilde{V}_d(\beta_n), \tilde{\rho}_{t.1k}\right\rangle, \quad [k = 1, 2, \ldots, N]. \qquad (14.6.58)$$

Likewise, testing of the potential on three conductors with the testing charges $\tilde{\rho}_{t,2k}$ and $\tilde{\rho}_{t,3k}$, located at the second conductor for j = 2 and the third conductor for j = 3, are expressed as follows:

$$\sum_{\ell=1}^{N} d_{1\ell}\left\langle \tilde{G}(\beta_n)\tilde{\rho}_{b,1\ell}, \tilde{\rho}_{t,2k}\right\rangle + \sum_{\ell=1}^{N} d_{2\ell}\left\langle \tilde{G}(\beta_n)\tilde{\rho}_{b,2\ell}, \tilde{\rho}_{t,2k}\right\rangle + \sum_{\ell=1}^{N} d_{3\ell}\left\langle \tilde{G}(\beta_n)\tilde{\rho}_{b,3\ell}, \tilde{\rho}_{t,2k}\right\rangle =$$

$$\left\langle \tilde{V}_c(\beta_n), \tilde{\rho}_{t,.2k}\right\rangle + \left\langle \tilde{V}_d(\beta_n), \tilde{\rho}_{t,.2k}\right\rangle, \quad k = 1, 2, \ldots, N] \qquad (b)$$

$$\sum_{\ell=1}^{N} d_{1\ell}\left\langle \tilde{G}(\beta_n)\tilde{\rho}_{b,1\ell}, \tilde{\rho}_{t,3k}\right\rangle + \sum_{\ell=1}^{N} d_{2\ell}\left\langle \tilde{G}(\beta_n)\tilde{\rho}_{b,2\ell}, \tilde{\rho}_{t,3k}\right\rangle + \sum_{\ell=1}^{N} d_{3\ell}\left\langle \tilde{G}(\beta_n)\tilde{\rho}_{b,3\ell}, \tilde{\rho}_{t,3k}\right\rangle =$$

$$\left\langle \tilde{V}_C(\beta_n), \tilde{\rho}_{t,3k}\right\rangle + \left\langle \tilde{V}_d(\beta_n), \tilde{\rho}_{t,3k}\right\rangle, \quad [k = 1, 2, \ldots, N] \qquad (c).$$

(14.6.58)

Equation (14.6.58a,b,c) are rewritten in more compact form as

$$\text{Testing charge, } j = 1: \sum_{\ell=1}^{N} d_{1\ell}\tilde{P}_{11}^{k\ell} + \sum_{\ell=1}^{N} d_{2\ell}\tilde{P}_{12}^{k\ell} + \sum_{\ell=1}^{N} d_{3\ell}\tilde{P}_{13}^{k\ell} = \tilde{Q}_1, \text{ on cond.\#1} \qquad (a)$$

$$\text{Testing charge, } j = 2: \sum_{\ell=1}^{N} d_{1\ell}\tilde{P}_{21}^{k\ell} + \sum_{\ell=1}^{N} d_{2\ell}\tilde{P}_{22}^{k\ell} + \sum_{\ell=1}^{N} d_3\tilde{P}_{23}^{k\ell} = \tilde{Q}_2 \text{ on cond.\#2} \qquad (b)$$

$$\text{Testing charge, } j = 3: \sum_{\ell=1}^{N} d_{1\ell}\tilde{P}_{31}^{k\ell} + \sum_{\ell=1}^{N} d_{2\ell}\tilde{P}_{32}^{kl} + \sum_{\ell=1}^{N} d_{3\ell}\tilde{P}_{33}^{k\ell} = \tilde{Q}_3 \text{ on cond.\#3} \qquad (c).$$

(14.6.59)

where inner products are defined as

$$P_{ji}^{k\ell} = \left\langle \widetilde{G}(\beta_n)\widetilde{\rho}_{b,i\ell}, \widetilde{\rho}_{t,jk} \right\rangle; \quad \left| \begin{array}{ll} i = 1,2,3; & j = 1,2,3 \\ k = 1,2...,N; & \ell = 1,2,...,N \end{array} \right. . \tag{14.6.60}$$

The indices j, i, k,ℓ are grouped as a subscript (ji) with j = 1, 2, 3, indicating the conductor's number of the *test charge density*, and i = 1,2,3 indicates the conductor's number, supporting the main charge using the *basis functions*. The superscript ($k\ell$) indicates the number of expansion terms (k = 1, 2, ...,N) of the test charge density at the jth conductor, and the number of expansion terms (ℓ = 1, 2, ...,N) of the basis function at the ith conductor. Using equation (14.6.57), the $\widetilde{Q}_j$ of equation (14.6.59) is defined as

$$\widetilde{Q}_i = \left\langle \widetilde{V}_c(\beta_n), \widetilde{\rho}_{t,jk} \right\rangle + \left\langle \widetilde{V}_d(\beta_n), \widetilde{\rho}_{t,jk} \right\rangle, \quad j = 1,2,3. \tag{14.6.61}$$

The set of equation (14.6.59) is written below in the compact form:

$$\sum_{i=1}^{3} \left[\sum_{\ell=1}^{N} d_{i\ell} \widetilde{P}_{ji}^{k\ell} \right] = \widetilde{Q}_j,$$

$$\text{For testing function}: j = 1,2,3; \quad k = 1,2,...,N. \tag{14.6.62}$$

The inner products (14.6.60) and (14.6.61) are evaluated as

$$\widetilde{P}_{ji}^{k\ell}(\beta_n) = \left\langle \widetilde{G}(\beta_n)\widetilde{\rho}_{b,i\ell}, \widetilde{\rho}_{t,jk} \right\rangle = \sum_{n=1}^{\infty} \left[\widetilde{G}(\beta_n)\widetilde{\rho}_{b,i\ell}, \widetilde{\rho}_{t,jk} \right];$$

$$\left\{ \begin{array}{l} \text{For basis function}: \ i = 1,2,3; \quad \ell = 1,2,3,...,N \\ \text{For testing function}: \ j = 1,2,3; \ k = 1,2,3,...,N. \end{array} \right. \tag{14.6.63}$$

$$\begin{aligned} \widetilde{Q}_j &= \left\langle \widetilde{V}_c(\beta_n), \widetilde{\rho}_{t,jk} \right\rangle + \left\langle \widetilde{V}_d(\beta_n), \widetilde{\rho}_{t,jk} \right\rangle \\ &= \underbrace{\sum_{n=1}^{\infty} \left[\widetilde{V}_c(\beta_n), \widetilde{\rho}_{t,jk} \right]}_{\widetilde{Q}_{cj}} + \underbrace{\sum_{n=1}^{\infty} \left[\widetilde{V}_d(\beta_n), \widetilde{\rho}_{t,jk} \right]}_{\widetilde{Q}_{dj}}. \end{aligned} \tag{14.6.64}$$

On using Parseval's theorem (14.6.8e), equation (14.6.64) is transformed into the space domain. Let us take up the potential on the conductors:

$$\begin{aligned} \widetilde{Q}_{c,j} &= \sum_{n=1}^{\infty} \left[\widetilde{V}_c(\beta_n), \widetilde{\rho}_{t,jk} \right] \\ &= \frac{2}{L} \int_0^L V_c(x), \rho_{t,jk}(x) dx; \quad \left| \begin{array}{l} j = 1,2,3 \\ k = 1,2,...,N \end{array} \right. . \end{aligned} \tag{14.6.65}$$

The central conductor #1, shown in Fig (14.14b) has potential $V_c(x) = V_0$, i.e.

$$\widetilde{Q}_{c1} = \frac{2V_0}{L} \int_{G_1 + S_1}^{G_1 + S_1 + W} \rho_{t,jk}(x) dx, \quad k = 1,2,3,...,N \tag{a}$$

$$\text{On ground cond.\#2}: \ V_c(x) = 0; \ 0 \le x \le G_1; \ \widetilde{Q}_{c2} = 0 \tag{b}$$

$$\text{On ground cond.\#3}: \ V_c(x) = 0; \quad G_1 + s_1 + w \le x \le G_1 + s_1 + w + G_2; \quad \widetilde{Q}_{c3} = 0 \tag{c}$$

(14.6.66)

Let us take up the potential on the dielectric interface s_1,s_2.

$$\widetilde{Q}_{d,j} = \sum_{n=1}^{\infty} \left[\widetilde{V}_d(\beta_n), \widetilde{\rho}_{t,jk} \right] = \frac{2}{L} \int_0^L V_d(x), \rho_{t,jk}(x) dx = 0. \tag{14.6.67}$$

The testing charge, $\rho_{t,jk}(x) = 0$ on the dielectric interface s_1, s_2. Equation (14.6.59) is written as

$$\text{For } j = 1: \ \sum_{\ell=1}^{N} d_{1\ell} \widetilde{P}_{11}^{k\ell} + \sum_{\ell=1}^{N} d_{2\ell} \widetilde{P}_{12}^{k\ell} + \sum_{\ell=1}^{N} d_{3\ell} \widetilde{P}_{3\ell}^{k\ell} = \frac{2V_0}{L} \int_{G_1 + S_1}^{G_1 + S_1 + w} \rho_{t,1k}(x) dx \tag{a}$$

$$\text{For } j = 2: \ \sum_{\ell=1}^{N} d_{1\ell} \widetilde{P}_{21}^{k\ell} + \sum_{\ell=1}^{N} d_{2\ell} \widetilde{P}_{22}^{k\ell} + \sum_{\ell=1}^{N} d_{3\ell} \widetilde{P}_{23}^{k\ell} = 0 \tag{b}$$

$$\text{For } j = 3: \ \sum_{\ell=1}^{N} d_{1\ell} \widetilde{P}_{31}^{k\ell} + \sum_{\ell=1}^{N} d_{2\ell} \widetilde{P}_{32}^{k\ell} + \sum_{\ell=1}^{N} d_{3\ell} \widetilde{P}_{33}^{k\ell} = 0 \tag{c}$$

(14.6.68)

Normally, the basis function and the testing function are identical in the Galerkin's process, i.e.

$$\rho_{t,jk} = \rho_{b,i\ell} \quad [i = j = 1,2,3 \ \text{and} \ k = \ell = 1,2,3,...,N]. \tag{14.6.69}$$

For one term basis and one term testing function, equation (14.6.68) is reduced to equation (14.6.45). The line capacitance *p.u.l.* of the CPW is

$$\begin{aligned} C &= \frac{Q_0[\text{charge on central conductor \#1}]}{V_0[\text{potential on central conductor \#1}]} \\ &= \frac{1}{V_0} \sum_{\ell=1}^{N} \left[d_{1\ell} \int_{G_1 + S_1}^{G_1 + S_1 + W} \rho_{b,1\ell}(x) dx \right]. \end{aligned} \tag{14.6.70}$$

The right-hand side of equation (14.6.68a) is defined as

$$\begin{aligned} R_l &= \frac{2V_0}{L} \int_{G_1 + S_1}^{G_1 + S_1 + w} \rho_{b,1\ell}(x) \, dx, \ (\ell = 1,2,...,N) \\ &\Rightarrow \int_{G_1 + S_1}^{G_1 + S_1 + w} \rho_{b,1\ell}(x) dx = \frac{L}{2V_0} R_l. \end{aligned} \tag{14.6.71}$$

The line capacitance *p.u.l.* of the CPW is rewritten as

$$C = \frac{L}{2V_0^2}\sum_{\ell=1}^{N} d_{1\ell}R_1 \quad (\ell = 1, 2, \ldots, N). \tag{14.6.72}$$

The coefficients $d_{1\ell}$ and R_1 are to be determined. The coefficients $d_{1\ell}$ are associated with the charge distribution basis function on the conductor #1, i.e. on the central strip conductor. It is achieved by solving the set of equation (14.6.68a–c). For one term $\ell = 1$, the coefficients of the basis and testing functions are $d_{11} = a_1$, $d_{21} = b_1$, $d_{31} = c_1$ and equation (14.6.68) is reduced to the matrix equation (14.6.45). A similar process is repeated for other coefficients for the next expansion term of the basis and testing functions, say for $\ell = 2$.

Equation (14.6.72) is the variational expression, giving the lower bounded on the CPW line capacitance. The following charge distribution functions on the central and ground conductors could be used [B.8]:

$$\text{Cenral strip}: \ \rho_{b,s11}(x) = \frac{\cos\left[(\ell-1)\pi\dfrac{(x - S_1 - G_1)}{W}\right]}{\sqrt{1 - \left[\frac{2(x - S_1 - G_1) - W}{W}\right]^2}} \quad \text{(a)}$$

$$\text{Left strip}: \ \rho_{b,s21}(x) = \frac{\cos\left[\left(\ell - \dfrac{1}{2}\right)\pi\dfrac{(x - G_1)}{G_1}\right]}{\sqrt{1 - \left[\frac{x}{G_1}\right]^2}}$$

$$\text{where,} \quad \ell = 1, 2, \ldots \quad \text{(b)}$$

$$\text{Right strip}: \ \rho_{b,s31}(x) = \frac{\cos\left[\left(\ell - \dfrac{1}{2}\right)\pi\dfrac{(x - L + G_2)}{G_2}\right]}{\sqrt{1 - \left[\frac{L-x}{G_2}\right]^2}} \quad \text{(c)}. \tag{14.6.73}$$

The Fourier transform of the above basis functions are given below:

$$\begin{aligned}
\tilde{\rho}_{b,11}(\beta_n) = \frac{2W}{2L}&\left[J_0\left\{\left|\frac{(\ell-1)\pi + \beta_n W}{2}\right|\right\}\right]\sin\left[\frac{(\ell-1)\pi}{2} + \beta_n\left\{S_1 + G_1 + \frac{W}{2}\right\}\right] \\
&- J_0\left\{\left|\frac{(\ell-1)\pi - \beta_n W}{2}\right|\right\}\sin\left[\frac{(\ell-1)\pi}{2} - \beta_n\left\{S_1 + G_1 + \frac{W}{2}\right\}\right] \quad \text{(a)} \\
\tilde{\rho}_{b,21}(\beta_n) = (-1)^l\frac{\pi G_1}{2L}&\left[J_0\left\{\left|\left(\ell - \frac{1}{2}\right)\pi + \beta_n G_1\right|\right\} - J_0\left\{\left|\left(\ell - \frac{1}{2}\right)\pi - \beta_n G_1\right|\right\}\right] \quad \text{(b)} \\
\tilde{\rho}_{b,31}(\beta_n) = (-1)^{n+1+l}\frac{\pi G_2}{2L}&\left[J_0\left\{\left|\left(\ell - \frac{1}{2}\right)\pi + \beta_n G_2\right|\right\} - J_0\left\{\left|\left(\ell - \frac{1}{2}\right)\pi - \beta_n G_2\right|\right\}\right] \quad \text{(c)}.
\end{aligned} \tag{14.6.74}$$

The above expressions can generate ℓ numbers of the basis functions on three conductors. However, even one term of basis function provides a sufficiently accurate result for the computation of effective relative permittivity and the characteristic impedance of a CPW. The variational method has also been extended to the multiconductor microstrip line [J.35].

References

Books

B.1 Collin, R.E.: *Field Theory of Guided Waves*, IEEE Press, New York, NY, 1991.

B.2 Pipes, L.A.; Harvill, L.R.: *Applied Mathematics for Engineers and Physicists*, 3rd Edition, McGraw-Hill Book Company, New York, NY, 1971.

B.3 Spiegel, M.R.: *Theory and Problem of Advanced Mathematics for Engineers & Scientist*, Schaum's Outline Series, McGraw-Hill Book Company, New York, NY, 1983.

B.4 Elsgolts, L.: *Differential Equations and the Calculus of Variational Method*, MIR Publishers, Moscow, 1977.

B.5 Harrington, R.F.: *Field Computation by Moment Methods*, Krieger Publishing Co, Malabar, FL, 1968.

B.6 Booton, R.C. Jr.: *Computational Methods for Electromagnetics and Microwaves*, John Wiley & Sons, New York, NY, 1992.

B.7 Reichel, A.: *Special Function*, Science Press, Sydney, 1966.

B.8 Nguyen, C.: *Analysis Methods for RF, Microwave, and Millimetre – Wave Planar Transmission Lines*, John Wiley & Sons, New York, NY, 2000.

B.9 Dworsky, L.N.: *Modern Transmission Line Theory and Applications*, John Wiley & Sons, New York, NY, 1979.

B.10 Goel, A.K.: *High-Speed VLSI Interconnects Modelling, Analysis, and Simulation*, John Wiley & Sons, New York, NY, 1994.

B.11 Kinayman, N.; Aksun, M.I.: *Modern Microwave Circuits*, Artech House, Boston, 2003.

B.12 Sadiku Mathew, N.O.: *Numerical Techniques in Electromagnetics*, 2 nd Edition, CRC Press, New York, NY, 2001.

B.13 Bhat, B.; Koul, S.K.: *Stripline-Like Transmission Lines for Microwave Integrated Circuits*, Wiley Eastern Ltd., New Delhi, 1989.

B.14 Mirshekar-Syahkal, D.: *Spectral Domain Method for Microwave Integrated Circuits*, Research Studies Press Ltd., John Wiley & Sons, New York, NY, 1990.

B.15 Kneppo, I.; Fabian, J.; Bezousek, P.; Hrnicko, P.; Pavel, M.: *Microwave Integrated Circuits*, Chapman & Hall, London, 1994.

B.16 Kreyszig, E.: *Advanced Engineering Mathematics*, John Wiley & Sons, New York, NY, 1999.

B.17 Sneddon, I.N.: *Fourier Transforms*, McGraw-Hill Book Co., New York, NY.

B.18 Young, L.; Sobol, H.: *Advances in Microwaves*, Vol. 8, Academic Press, New York, NY, 1974.

Journals

J.1 Itoh, T; Mitra, R.: *An accurate method for calculating charge and potential distributions in coupled microstrip lines, Proc. IEEE*, Vol. 59, pp. 332–334, Feb. 1971.

J.2 Yamashita, E.; Mitra, R.: *Variational method for analysis of microstrip lines, IEEE Trans. Microw. Theory Tech.*, Vol. MTT-16, No. 4, pp. 251–256, April 1968.

J.3 Yamashita, E.: *Variational method for analysis of microstrip – like transmission lines, IEEE Trans. Microw. Theory Tech.*, Vol. MTT-16, No. 8, pp. 529–535, Aug. 1968.

J.4 Denlinger, E.J.L.: *A frequency-dependent solution for microstrip transmission lines., IEEE Trans. Microw. Theory Tech.*, Vol. MTT-19, No. 1, pp. 30–39, Jan. 1971.

J.5 Crampgane, R.; Ahmadpanah, M.; Guiraud, J.L.: *A simple method for determining the Green's function for a large class of MIC lines having multilayered dielectric structures, IEEE Trans. Microw. Theory Tech.*, Vol. MTT-26, No. 2, pp. 82–87, Feb. 1978.

J.6 Bhat, B.; Koul, S.K.: *Unified approach to solve a class of strip and microstrip – like transmission lines, IEEE Trans. Microw. Theory Tech.*, Vol. MTT-30, pp. 679–686, May 1982. Also correction in IEEE Trans. Microwave Theory Tech., Vol. MTT- 30, No., pp. 2067, Nov. 1982.

J.7 Koul, S.K.; Bhat, B.: *Generalized analysis of microstrip – like transmission lines and coplanar strips with anisotropic substrates for MIC, electro-optic modulator and SAW application, IEEE Trans. Microw. Theory Tech.*, Vol. MTT-31, pp. 1051–1059, Dec. 1983.

J.8 Chang, Y.; Chang, I.E.: *Simple method for the variational analysis of a generalized N- dielectric layer transmission line, Electron. Lett.*, Vol. 6, No.3, pp. 49–50, Feb. 1970.

J.9 Yamashita, E.; Atsuki, K.: *Stripline with rectangular outer conductor and three dielectric layers, IEEE Trans. Microw. Theory Tech.*, Vol. MTT-18, No. 5, pp. 238–244, May 1970.

J.10 Mitra, R.; Itoh, T.: *Charge and potential distributions in the shielded strip lines, IEEE Trans. Microw. Theory Tech.*, Vol. MTT-18, pp. 149–156, March 1970.

J.11 Wheeler, H.A.: *Transmission-line properties of parallel strips separated by a dielectric sheet, IEEE Trans. Microw. Theory Tech.*, Vol. MTT-13, No. 3, pp. 172–185, March 1965.

J.12 Araki, K.; Naito, Y.: *Upper bound calculations on capacitance of microstrip line using variational method and Spectral domain approach, IEEE Trans. Microw. Theory Tech.*, Vol. MTT-26, pp.506–509, July 1978.

J.13 Horno, M.: *Quasi-static characteristics of microstrip on arbitrary anisotropic substrates, Proc. IEEE*, Vol. 68, Pt. H, No. 8, pp. 1033–1034, Aug. 1980.

J.14 Itoh, T.; Herbert, A.S.: *A generalized spectral domain analysis for coupled suspended microstrip lines with tuning septum, IEEE Trans. Microw. Theory Tech.*, Vol. MTT-26, pp. 820–826, Oct. 1978.

J.15 Itoh, T.: *A generalized spectral domain analysis for multiconductor printed lines and its application to tunable suspended microstrips, IEEE Trans. Microw. Theory Tech.*, Vol. MTT-26, pp. 983–987, Dec. 1978.

J.16 Kobayashi, M.; Terakodo, R.: *General form of Green's function microstrip structures with any kind of geometry for multilayer microstrip line with rectangular side walls, IEEE Trans. Microw. Theory Tech.*, Vol. MTT-4, pp. 626–628, 1976.

J.17 Verma, A.K.; Nasimuddin; Bhupal, A.: *Dielectric loss of multilayer coupled microstrip line using SLR formulation, Microw. Opt. Technol. Lett.*, Vol. 31, No.1, pp. 65–67, Oct. 2001.

J.18 Verma, A.K.; Singh, P.; Matekovits, L.: *Strip-width and slot-gap dependent equivalent isotropic substrate and dispersion characteristics of asymmetric coplanar waveguide, symmetric coplanar waveguide and micro-coplanar strip line on anisotropic substrates, Microw. Theory Tech.*, Vol. MTT-62, No. 10, pp. 2232–2241, Oct. 2014.

J.19 Verma, A.K.; Bhupal, A.: *Resonance frequency of anisotropic multilayer rectangular microstrip patch using MWM, Electron. Lett.*, Vol. 34, No.4, pp. 323–325, Feb 1998.

J.20 Verma, A.K.; Awasthi, Y.K.; Singh, H.: *Equivalent isotropic relative permittivity of microstrip on multilayer anisotropic substrate, Int. J. Electron.*, Vol. 96, No. 8, pp. 865–875, Aug. 2009.

J.21 Verma, A.K.; Rostamy, Z.: *Resonant frequency of shielded rectangular and circular patch resonators, Int. J. Microw. Millim.- Wave. Comput.-Aided Eng.*, Vol. 5, pp. 13–19, 1995.

J.22 Bryant, T.G.; Weiss, J.A.: *Parameters of microstrip transmission lines and of coupled pairs of microstrip lines, IEEE Trans. Microw. Theory Tech.*, Vol. MTT-16, pp. 1021–1027, 1968.

J.23 Verma, A.K.; Nasimuddin; Ojha, R.K.: *Synthesis, and analysis of optimum directivity suspended microstrip coupler*, Asia-Pacific Microwave Conf., APMC, pp. 1178–1182, 2000, Australia.

J.24 Koul, S.K.; Bhat, B.: *Propagation parameters of coupled microstrip-like transmission lines for millimeter-wave applications, IEEE Trans. Microw. Theory Tech.*, Vol. MTT-29, No. 12, pp. 1364–1370, Dec. 1981.

J.25 Krischnining, M.; Jansen, R.H.: *Accurate model for effective dielectric constant with validity up to millimeter-wave frequency, Electron. Lett.*, Vol. 18, pp. 272–273, Jan. 1982.

J.26 Garg, R.; Bah, I.J.: *Characteristics of coupled microstrip lines, IEEE Trans. Microw. Theory Tech.*, Vol. MTT-27, pp. 700–705, 1979. Also corrections in, Vol. MTT- 28, No., pp. 272, 1980.

J.27 Kawalski, G.; Pregala, R.: *Dispersion characteristics of single and coupled microstrips, A.E.U*, Vol. 26, pp. 276–280, 1972.

J.28 Bahl, I.J.; Bhartia, P.: *Characteristics of inhomogeneous –coupled stripline, IEEE Trans. Microw. Theory Tech.*, Vol. MTT-28, No. 6, pp. 529–535, June 1980.

J.29 Allen, J.L.; Estes, M.F.: *Broadside-coupled strips in a layered dielectric medium, IEEE Trans. Microw. Theory Tech.*, Vol. MTT-20, No. 10, pp. 662–669, Oct. 1972. Correction MTT- 23, No. 9, pp. 779, Sept. 1975.

J.30 Tounsi, M.L.; Halheit, H.; Yagoub, M.C.E.; Khodja, A.: *Analysis of shielded planar circuits by a mixed variational-spectral method*, IEEE CAS Int. Circuit Syst. Symp., pp. 65–68, 2003.

J.31 Hammerstad, E.; Jensen, O.: *Accurate models for microstrip computer aided design, IEEE. MTT-S, Int. Microw. Symp. Dig.*, pp. 407–409, Washington, DC, 1980.

J.32 Horno, M.: *Upper and lower bounds on capacitances of coupled microstrip lines with anisotropic substrates, IEE Proc.*, Vol. 129, Pt. H, No. 3, pp. 89–93, June 1982.

J.33 Darwish, A.; Ezzeddine, A.; Huang, H.C.; Misoon, M.: *Effective dielectric constant of the embedded transmission line (ETL)-a multilayer stripline-like structure, IEEE MTT-S Dig.*, Vol. 3, pp. 1249–1252, 1999

J.34 Darwish, A.; Ezzeddine, A.; Huang, H.C.; Misoon, M.: *Analysis of three-dimensional embedded transmission lines (ETL's), IEEE Microw. Guid. Wave Lett.*, Vol. 9, No. 11, pp.447–449, Nov. 1999.

J.35 Sawicki, A.; Sachse, K.: *Lower and upper bound calculations on the capacitance of multiconductor printed transmission line using the spectral-domain approach and variational method, IEEE Trans. Microw. Theory Tech.*, Vol. MTT-34, No. 2, pp. 236–244, Feb. 1986.

15

Multilayer Planar Transmission Lines

SLR Formulation

Introduction

The closed-form dispersion and loss models of the standard single-layer microstrip, and CPW lines without any shielding are discussed in chapters 8 and 9 respectively. However, the shielded and multilayered line structures, such as suspended, composite, inverted, line structures with superstrate, etc., are important in modern microwave technology. It is difficult to obtain the curve-fitted closed-form dispersion and loss models for the multilayer planar transmission lines because of the involvement of more numbers of independent variables [J.1]. Therefore, several full-wave electromagnetic methods are used for the analysis and development of EM-Simulators [B.1–B.4]. One such method, the dynamic *spectral domain analysis (SDA)* is discussed in chapter 16. However, the full-wave methods are not suitable for the circuit simulators and computer-aided design (CAD) applications. The synthesis of the line structures requires iterative use of the closed-form analytical expressions.

The *single-layer reduction* (SLR) formulation is a practical engineering process to compute the dispersion, frequency-dependent characteristic impedance, dielectric, and conductor losses of the multilayer microstrips, CPW, and slot line structures. The SLR process is developed around the static variational method and Galerkin method, and also around the conformal mapping method. The available frequency-dependent closed-form models of the standard single-layer line structures are also used in the formulation. The present chapter discusses the SLR formulation applicable to the multilayer planar transmission lines.

Objectives

- *To present the basic formulation of the SLR applied to the multilayer microstrip lines to compute dispersion, dispersive characteristic impedance, dielectric loss, and conductor loss.*
- *To present the SLR process for the coupled multilayer microstrip lines to compute dispersion, dispersive characteristic impedance, dielectric loss, and conductor loss.*
- *To present the basic formulation of the SLR applied to the multilayer ACPW/CPW structures to compute dispersion, dispersive characteristic impedance, dielectric loss, and conductor loss.*

15.1 SLR Process for Multilayer Microstrip Lines

Figure (15.1a) shows the four-layered lossy multilayer microstrip line in the inhomogeneous medium. Figure (15.1b-f) also shows some special cases. The concept of the effective relative permittivity converts the inhomogeneous medium microstrip to an equivalent isotropic homogenous medium microstrip line. Such a medium transformation, discussed in subsection (8.2.1) of chapter 8, is known as the *Wheeler's transformation*. In the next step, the equivalent isotropic homogenous medium microstrip is further transformed, using the *inverse Wheeler's transformation*, to an equivalent single-layer microstrip with equivalent relative permittivity ε_{req}, equivalent dielectric loss-tangent δ_{eq}, and equivalent substrate thickness h_{eq}. The parameters of the *equivalent single layer microstrip* depend on the relative permittivity, loss-tangent, and thickness of each layer of the original multilayer microstrip. The process can also take care of the presence of the shield, conductor thickness, and anisotropy in the dielectric layers [J.2–J.5]. Such medium transformation is static in nature. The method can be also applied to a magnetic substrate [J.13]. This section presents the static SLR-formulation. The existing closed-form models of a microstrip line, discussed in chapter 8, are used over the equivalent single-layer microstrip to compute the dispersion,

Introduction to Modern Planar Transmission Lines: Physical, Analytical, and Circuit Models Approach, First Edition. Anand K. Verma.

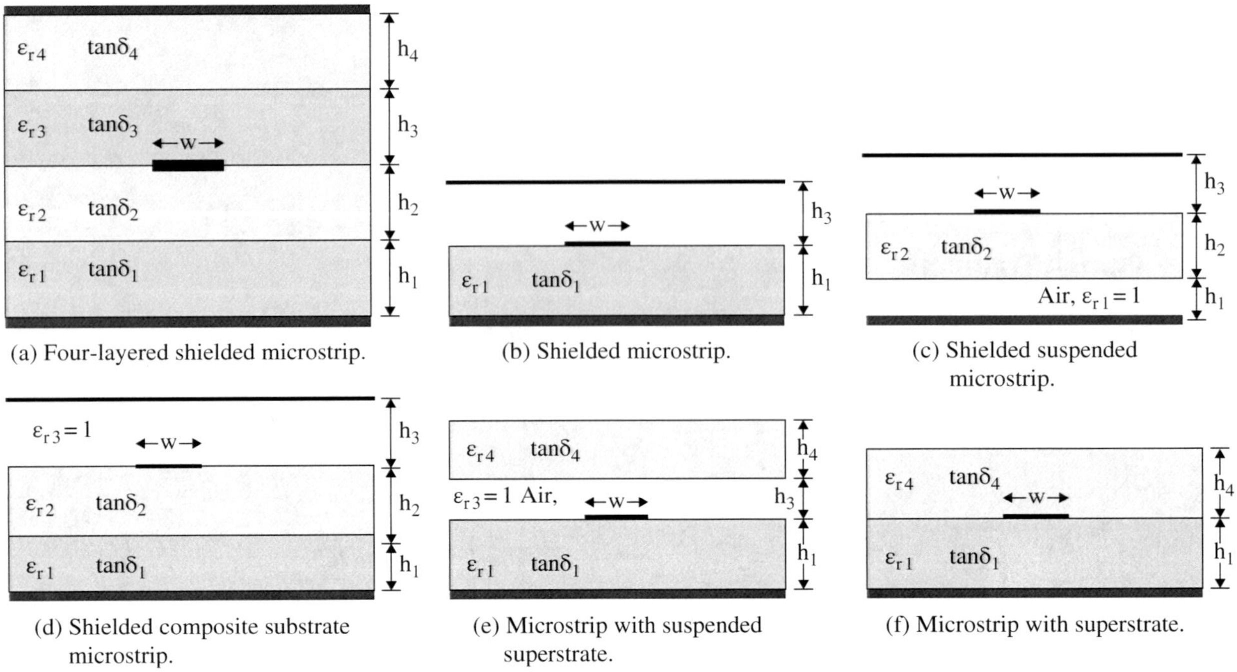

Figure 15.1 Four-layered microstrip and some special cases.

frequency-dependent characteristic impedance, and losses in the multilayer microstrip.

15.1.1 SLR-Process for Lossy Multilayer Microstrip Lines

Figure (15.2) shows the two-step SLR, i.e. the single-layer reduction, process applied to a shielded four-layered lossy microstrip line. The structure can also accommodate more numbers of dielectric layers [J.2, J.3, J.6].

- **Step-1:** It transforms the lossy multilayer inhomogeneous dielectric medium of microstrip to a lossy homogeneous dielectric medium. The process is called *Wheeler's transformation*. The relative permittivity of the complex equivalent homogeneous medium is equal to the complex relative effective permittivity of the four-layered microstrip medium. The complex relative effective permittivity of the multilayer microstrip is computed by the variational method, combined with the transverse transmission line (TTL) technique in the Fourier domain. It is discussed in subsection (14.4.2) of chapter 14. The real relative permittivities in equation (14.4.38) are replaced by the complex relative permittivities to account for the dielectric losses in each layer. The space domain method, discussed in subsection (14.4.1) of chapter 14, can also be used for an enclosed multilayer microstrip line. More general discrete Fourier transformed-based Galerkin method, discussed in subsection (14.6.2), can also be applied to the multilayer shielded microstrip line [B.1, B.4].

At this step, the relative permittivities, loss-tangents, and substrate's thickness of each layer, and also the strip width W are known. For the static case, i.e. at f = 0, the complex effective dielectric constant of a multilayer microstrip is obtained as

$$\begin{aligned}\varepsilon^*_{reff}(f=0) &= \varepsilon'_{reff}(f=0) - j\varepsilon''_{reff}(f=0)\\ &= \frac{C^*_{v,d}\left(\varepsilon^*_{r1},\ldots,\varepsilon^*_{4},h_1,\ldots,h_4,w\right)}{C_{v,o}(\varepsilon^*_{r1},\ldots,=\varepsilon^*_4=1,h_1,\ldots,h_4,w)} \qquad (a)\\ \tan\delta_{eff} &= \frac{\varepsilon''_{reff}(f=0)}{\varepsilon'_{reff}(f=0)} \qquad (b).\end{aligned} \tag{15.1.1}$$

The variational expression (14.4.37) of chapter 14 is used to compute the complex line capacitance $C^*_{v,d}\left(\varepsilon^*_{r1},\ldots,\varepsilon^*_{r4},h_1,\ldots,h_4,w\right)$ of a lossy structure with lossy dielectric layers, and also lossless line capacitance $C_{v,o}\left(\varepsilon^*_{r1},\ldots\varepsilon^*_{r4}=1,h_1,\ldots,h_4,w\right)$ using the air substrates. The complex relative permittivity of each layer is $\varepsilon^*_{ri}=\varepsilon'_{ri}-j\varepsilon''_{ri}=\varepsilon'_{ri}-j\varepsilon'_{ri}\tan\delta_i(i=1,\ldots,4)$. The real relative permittivity of each dielectric layer, given in equation (14.4.38) of chapter 14, is replaced by the complex relative permittivity of corresponding layer.

- **Step-2:** This step is based on *inverse Wheeler's transformation*. It transforms the microstrip line in the lossy homogeneous medium to a microstrip line in the equivalent lossy single-layer microstrip in an inhomogeneous medium. The equivalent single layer has equivalent substrate thickness h_{eq}, equivalent relative

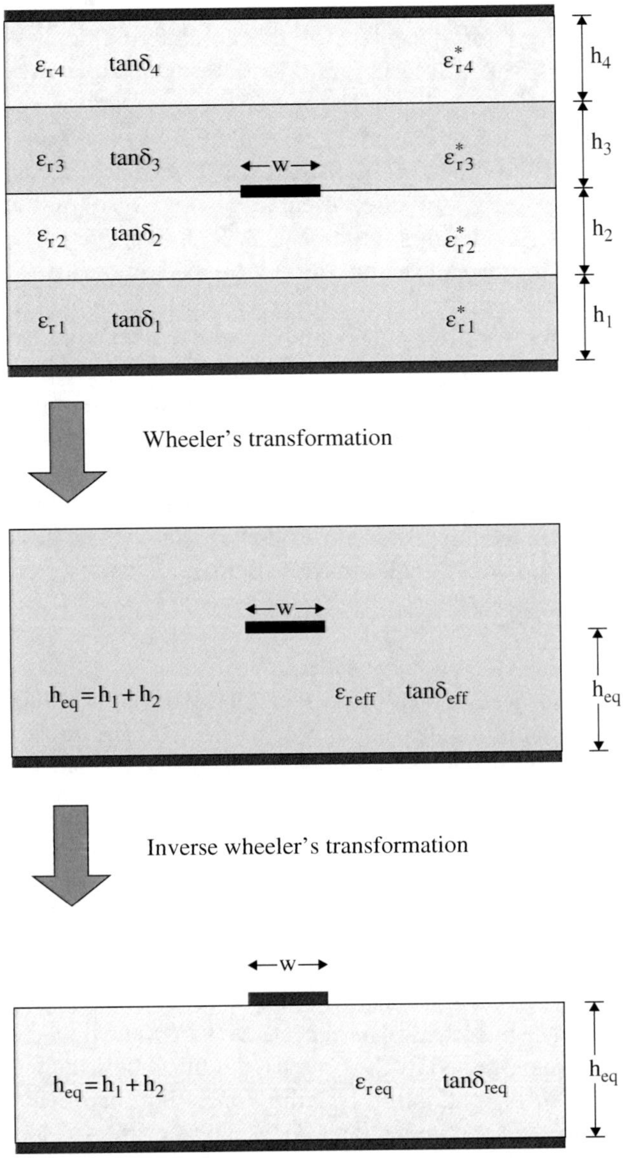

Figure 15.2 The two-step SLR-formulation process. *Source:* Verma and Sadr [J.2], Verma et al. [J.3], and Verma and Bhupal [J.6].

permittivity $\varepsilon'_{req} = \varepsilon_{req}$, and equivalent $\tan\delta_{eq}$. These parameters are obtained by following the process of equations (8.4.22) to (8.4.24) of chapter 8:

$$\varepsilon'_{req} = \varepsilon_{req} = \frac{[\varepsilon'_{reff}(f=0)-1]}{q} + 1 \qquad (a)$$

$$\tan\delta_{eq} = \frac{\varepsilon''_{req}}{\varepsilon'_{req}} = \frac{\varepsilon''_{reff}(f=0)}{\varepsilon'_{reff}(f=0)+(q-1)} \qquad (b)$$

Equivalent substrate thickness, $h_{eq} = h_1 + h_2$ (c).

(15.1.2)

The $\varepsilon'_{reff}(f=0)$ and $\varepsilon''_{reff}(f=0)$ of a multilayer microstrip are computed from variational expression (15.1.1a). During the SLR process, the strip width w remains unchanged. The equivalent substrate thickness h_{eq} is a sum of dielectric layers between the conducting strip and bottom ground conductor. In the above equations, the filling-factor q is an unknown quantity; as for a single-layer microstrip line, it is a function of the w/h_{eq} and ε_{req} of the equivalent substrate. The quantity ε_{req} is unknown at this stage. However, the filling-factor – q is a strong function of w/h_{eq} ratio. It is weakly dependent on the dielectric constant ε_r, i.e. ε_{req} of the substrate. So, an *iterative approach* is used to compute filling-factor q accurately.

To implement the iterative approach; initially, the filling-factor is assumed to be independent of the equivalent relative permittivity ε_{req} of the equivalent substrate. Thus, the *approximate filling-factor* q_{app} is obtained from the following equation [J.2]:

$$q_{app} = \frac{1}{2}(1+p) \qquad (a)$$

where,

$$p = \begin{cases} \left[1 + \dfrac{12h_{eq}}{w}\right]^{-1/2}, & \text{for } \dfrac{w}{h_{eq}} > 1 \\ \left[1 + \dfrac{12h_{eq}}{w}\right]^{-1/2} + 0.04\left[1 - \dfrac{w}{h_{eq}}\right]^2, & \text{for } \dfrac{w}{h_{eq}} \le 1 \end{cases} \qquad (b).$$

(15.1.3)

Using the above given approximate filling-factor q_{app}, an approximate value of ε_{req}, i.e. $\varepsilon_{req} \approx \varepsilon_{req}^{app}$ is computed from equation (15.1.2a). It helps to obtain a more accurate value of the filling- factor from the following set of expressions [J.7]:

$$q = \frac{(\varepsilon_{req}^{app}+1) + (\varepsilon_{req}^{app}-1)T - 2}{2(\varepsilon_{req}^{app}-1)} \qquad (a)$$

$$\text{where, } T = \left(1 + \frac{10}{w/h_{eq}}\right)^{-a.b} \qquad (b)$$

$$b = 0.564\left[\frac{(\varepsilon_{req}^{app}-0.9)}{(\varepsilon_{req}^{app}+3.0)}\right]^{0.053} \qquad (c)$$

$$a = 1 + \frac{1}{49}\ln\left[\frac{(w/h_{eq})^4 + (w/52h_{eq})^2}{(w/h_{eq})^4 + 0.432}\right] + \frac{1}{18.7}\ln\left[1 + \left(\frac{w}{18.1h_{eq}}\right)^3\right] \qquad (d).$$

(15.1.4)

The filling-factor, q, of the above equation (15.1.4a) is further used to compute more accurate value of ε_{req} with help of equation (15.1.2a). It can be further substituted in equation (15.1.4a) to get a more accurate iterated value of q that helps to compute ε_{req} with improved accuracy. In place of equation (15.1.4), equations (8.2.12) and (8.2.15) can also be used in the iteration process to compute the filling-factor q more accurately. The accuracy of the method is tested by considering the single-layer microstrip with the known relative permittivity and the SLR process is used to recover the known relative permittivity. The % deviation in the recovered relative permittivity is used to correct the computed value of ε_{req}. Using the accurate value of q, the equivalent loss-tangent $\tan\delta_{eq}$ is computed from equation (15.1.2b).

Once the equivalent single-layer microstrip with its relative permittivity, loss-tangent, and substrate thickness (ε_{req}, $\tan\delta_{eq}$, h_{eq}) are known, the dispersion, dispersive characteristic impedance, dielectric loss, and conductor loss of a lossy multilayer microstrip are computed by using the standard expressions presented in chapter 8.

15.1.2 Dispersion Model of Multilayer Microstrip Lines

The variational method computes the static complex relative effective permittivity $\varepsilon^*_{reff}(0)$, and the static SLR formulation provides an equivalent single-layer substrate with thickness h_{eq}, relative permittivity ε_{req}, and $\tan\delta_{eq}$. However, a few available approximate closed-form models with a lesser accuracy, to avoid the variational method, can also be used in the SLR process [J.2, J.8, J.9]. Any of the available dispersion models, for instance, the Krischning–Jansen (K–J) model or Logistic Dispersion Model (LDM), could be used to get the dispersion in a multilayer. These models are discussed in chapter 8. The dispersion model of the multilayer microstrip could be called the *SLR-dispersion model*. It is available in two forms.

SLR-Dispersion Model

The SLR-dispersion model, also called the *unified dispersion model*, is developed around the (K–J) dispersion model of the standard single-layer microstrip lines [J.2]. It has been used to examine the dispersion in the shielded microstrip, composite substrate microstrip, suspended substrate microstrip, microstrip with superstrate, etc.

Using equation (8.3.35) of chapter 8, the frequency-dependent effective relative parameter of a multilayer microstrip line is computed from the following modified (K–J) expression:

$$\varepsilon_{reff}(f) = \varepsilon_{req} - \frac{\varepsilon_{req} - \varepsilon_{reff}(0)}{1 + P(\varepsilon_{req}, fh_{eq}, w/h_{eq})}. \tag{15.1.5}$$

For the structure shown in Fig (15.1a), the equivalent substrate thickness is $h_{eq} = h_1 + h_2$. The parameter P is obtained by using ε_{req} in place of ε_r in the set of expressions of equation (8.3.35) of chapter 8. Moreover, to account for the effect of finite conductor thickness, the physical strip width w is also replaced by a frequency-dependent equivalent strip width $w_{eq}(t, f)$ given by equation (8.3.34). Even the static ε_{req} can be modified to account for the effect of the conductor thickness [J.10–J.12]. The SLR-dispersion model, using ε_{req}, works well only over a limited range of the line parameters. However, due to its simplicity and ability to accommodate any numbers of the dielectric layers, the model has been used by the researchers, and also in the commercial circuit simulator [B.5, J.13–J.16, J.17]. The model is not satisfactory for a shielded microstrip if the shield height is in the range of substrate thickness. Again for a multilayer microstrip with a thicker superstrate of high permittivity, the results of the model significantly deviate from the results of the full-wave methods. In this case, the confinement of power shifts from the substrate to the superstrate with an increase in frequency. The SLR-dispersion model does not account for the inversion of the power flow direction. The basic SLR-dispersion model requires improvements.

Improved SLR-Dispersion Model

The improved SLR (ISLR)-dispersion model is also called the new unified dispersion model [J.18]. The *numerical experimentations* on the shielded, composite, and suspended microstrip lines, using the full-wave SDA [B.3], show that with an increase in frequency, the effective relative permittivities of these structures do not approach toward any real permittivity of the multilayer microstrip. The frequency-dependent effective relative permittivity $\varepsilon_{reff}(f)$ of a multilayer microstrip also does not approach the ε_{req}. Instead, the $\varepsilon_{reff}(f)$ of these structures approach toward a hypothetical *virtual relative permittivity* $\varepsilon_{rv}(f)$ that is determined by the physical parameters of a microstrip line and also by the operating frequency. Figure (15.3a) demonstrates such nature of the shielded $\varepsilon_{r1} = 9.8$, $\varepsilon_{r2} = 1.0$, composite $\varepsilon_{r1} = 3.5$, $\varepsilon_{r2} = 12.95$, and suspended $\varepsilon_{r1} = 1.0$, $\varepsilon_{r2} = 9.8$ microstrip lines. Thus, the (K–J) dispersion model should be used for the microstrip on an equivalent substrate of virtual relative permittivity, i.e. in equation (15.1.5), ε_{req} is replaced with $\varepsilon_{rv}(f)$. The dispersion in virtual relative permittivity could be treated as the material dispersion. The material dispersion could be modeled by the Debye model. Further, the Cole–Cole dispersion diagram has been developed for the dispersive three-layered microstrip line [J.19].

Figure (15.3a) suggests that the frequency-dependent *virtual relative permittivity* $\varepsilon_{rv}(f)$ of the equivalent substrate has both its static ε_{req} and additional frequency-dependent $\varepsilon_{radd}(f)$ components:

$$\varepsilon_{rv}(f) = \varepsilon_{req} + \varepsilon_{radd}(f), \tag{15.1.6}$$

where the equivalent relative permittivity for the shielded microstrip is $\varepsilon_{req} = \varepsilon_{r1}$. For the composite/

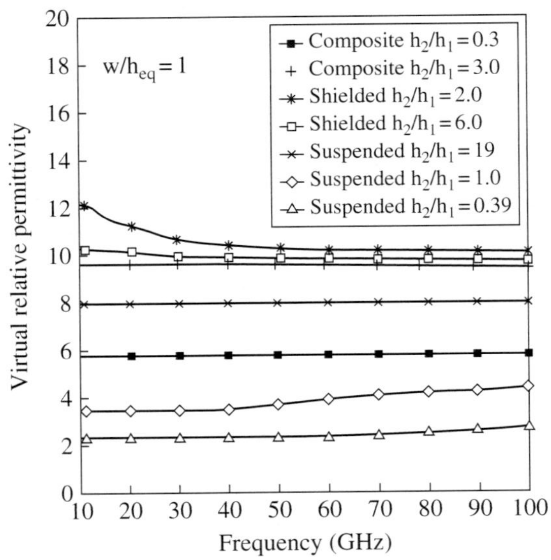

(a) Dispersive virtual relative permittivity[J.18].

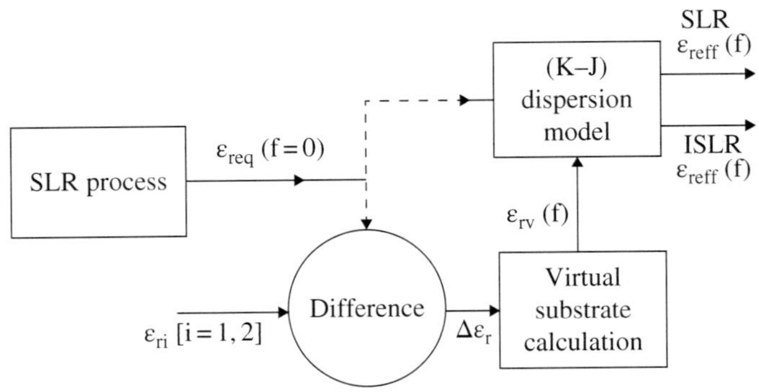

(b) Schematic diagram of SLR and improved SLR dispersion models.

Figure 15.3 The virtual relative permittivity-based SLR-dispersion models. *Source:* From Verma and Kumar [J.18]. © 1998 IEEE.

suspended substrate, the ε_{req} is computed using the static SLR formulation discussed in subsection (15.1.1). The increasing frequency moves the field lines toward a substrate of higher permittivity, bridging the differential gap between the ε_{req} and real permittivity of the substrate. Thus, the additional permittivity $\varepsilon_{radd}(f)$ is modeled around the *differential relative permittivity* $\Delta\varepsilon_r$.

For a *shielded microstrip*, the differential relative permittivity is $\Delta\varepsilon_r = \varepsilon_{r1} - \varepsilon_{req}$. For the *composite* and *suspended* substrates with $\varepsilon_{r2} > \varepsilon_{r1}$, it is $\Delta\varepsilon_r = \varepsilon_{r2} - \varepsilon_{req}$. However, for the *composite* microstrip with $\varepsilon_{r1} > \varepsilon_{r2}$, the differential relative permittivity is $\Delta\varepsilon_r = \varepsilon_{r1} - \varepsilon_{req}$. Satisfying the parametric dependence, the empirical relation for $\varepsilon_{radd}(f)$ could be rewritten as follows [J.18]:

$$\varepsilon_{radd}(f) = \sqrt{\gamma\Delta\varepsilon_r}\,\frac{(1+\alpha w/h_{eq})}{H}(f_n)^{\beta} \quad \text{(a)}$$

$$\text{for shielded microstrip } f_n = \frac{f_p}{\tau f} \quad \text{(b)}$$

$$\text{for composite/suspended microstrip } f_n = \frac{f}{\tau f_p} \quad \text{(c)}$$

$$\text{and } f_p = \frac{Z_0}{2\mu_0 h_{eq}\sqrt{\varepsilon_{reff}(0)}} \quad \text{(d).} \qquad (15.1.7)$$

In the above equation, H is a parameter dependent on the ratio of thicknesses of dielectric layers, and shield height. For each structure, the empirical expression for H is obtained separately. The equivalent substrate thickness is $h_{eq} = h_1 + h_2$ and f_n is the normalized frequency. The inflection frequency f_p has its origin in the waveguide model of the microstrip discussed in subsection (8.3.2) of chapter 8. For a *shielded microstrip*, Z_0 is determined by using the modified March model, given in subsection (8.2.5) of chapter 8. For the composite/suspended substrate, it is obtained from equation (14.4.39) of the chapter 14 using the variational method. For each microstrip structure, the parameters α, β, and γ are determined empirically.

Figure (15.3b) shows a schematic diagram of both the *SLR dispersion model and improved SLR, i.e. ISLR dispersion model*. The static ε_{req}-based SLR dispersion model, using equation (15.1.5), is shown in the dotted lines. It computes the frequency-dependent relative permittivity $\varepsilon_{reff}^{SLR}(f)$. The ISLR dispersion computes the $\varepsilon_{reff}^{ISLR}(f)$ using the following (K–J)–model-based dispersion model:

$$\varepsilon_{reff}^{ISLR}(f) = \varepsilon_{rv}(f) - \frac{\varepsilon_{rv}(f) - \varepsilon_{reff}(0)}{1 + P(\varepsilon_{rv}(f), fh_{eq}, w/h_{eq})}. \qquad (15.1.8)$$

Again, the above equation is obtained by replacing ε_r in equation (8.3.35) of chapter 8 with $\varepsilon_{rv}(f)$.

For a shielded microstrip $h_{eq} = h_1$, and for the composite/suspended substrate $h_{eq} = h_1 + h_2$. The curve-fitted closed-form expressions for computation of the parameters H, α, β, and γ of some important microstrip structures are summarized below.

Shielded microstrip line

The four-layered microstrip, shown in Fig (15.1a), is reduced to a shielded microstrip of Fig (15.1b) for the case $\varepsilon_{r2} = \varepsilon_{r3} = \varepsilon_{r4} = 1$ and $h_2 \to 0$, $h_4 \to 0$. The virtual relative permittivity is obtained by taking the following parameters [J.18]:

$$H = h_3/h_1, \tau = 3, f_n = \begin{cases} f_p/3f, & f > f_p \\ 1, & f \le f_p/3 \end{cases}. \quad (15.1.9)$$

In the lower frequency range, the frequency dependence of the virtual relative permittivity $\varepsilon_{rv}(f)$ is disregarded. The value of the normalized frequency f_n in between the ranges could be linearly interpolated. In this frequency range, the effect of w/h_1 and h_3/h_1 is dominant. The additional permittivity $\varepsilon_{radd}(f)$ of a shielded microstrip also meets the following asymptotic requirements:

$$\begin{aligned} &\varepsilon_{radd}(f) \to 0, \varepsilon_{rv}(f) \to \varepsilon_{r1}, \text{as } h_3/h_1 \to \infty, && \text{(a)} \\ &\varepsilon_{radd}(f) \to 0, \varepsilon_{rv}(f) \to \varepsilon_{r1}, \text{as } f \to \infty && \text{(b)}. \end{aligned} \quad (15.1.10)$$

Figure (15.3a) shows that at the lower end of the operating frequency, the nearness of the top shield increases the value of $\varepsilon_{rv}(f)$. However, $\varepsilon_{rv}(f)$ comes down to ε_{r1} with an increase in frequency. For a shielded microstrip, the empirical relations of the parameters α, β, and γ are summarized below:

$$\begin{aligned} &\alpha = 1.4738(w/h_1)^{0.7758}, \quad 0.1 \le w/h_1 \le 10 && \text{(a)} \\ &\beta = \begin{cases} 1.5731(w/h_1)^{-0.2308}, & 0.1 \le w/h_1 \le 0.6 \\ 1.2604(w/h_1)^{-0.2535}, & 0.6 \le w/h_1 \le 5 \\ 0.5, & 5.0 \le w/h_1 \le 10 \end{cases} && \text{(b)} \\ &\gamma = \begin{cases} -0.9869 + 1.1304\varepsilon_{r1} & 1.05 \le \varepsilon_{r1} \le 2.2 \\ 0.8228 + 0.3286\varepsilon_{r1}, & 2.2 \le \varepsilon_{r1} \le 12.95 \\ -13.5 + 1.4285\varepsilon_{r1}, & 12.95 \le \varepsilon_{r1} \le 14 \\ 0.3342 + 0.3542\varepsilon_{r1} & 14 \le \varepsilon_{r1} \le 20 \end{cases} && \text{(c)}. \end{aligned} \quad (15.1.11)$$

In the range $h_3/h_1 \ge 2$, $0.1 \le w/h_1 \le 10$, and $1.05 \le \varepsilon_{r1} \le 20$, the above model has maximum deviation ≤1.6% against the results of the SDA. Other forms of empirical relations are also available [J.20]. Figure (15.4a) shows the dispersion behavior of a shielded microstrip, as computed by the SLR-model and ISLR-model. The results are compared with some full-wave methods [J.1, J.21–J.23]. The ISLR-model follows the results of the SDA [B.3, J.21] more closely. The results of various full-wave methods also show much deviation, i.e. 6.7% among themselves. It could be due to the choice and number of *basis functions* used in the full-wave methods.

Suspended Microstrip Line

Figure (15.1c) shows the shielded suspended microstrip line. It is reduced to an open suspended microstrip line for $h_3 \to \infty$. The following parameters are taken for the open suspended microstrip [J.18]:

$$\begin{aligned} &H = \exp\left\{-5.1128\left(\frac{h_2}{h_1}\right)^{0.0544}\right\} \\ &\tau = \frac{1}{4}, f_n = \begin{cases} 4f/3f_p, & f > f_p/4 \\ 1, & f \le f_p/4. \end{cases} \end{aligned} \quad (15.1.12)$$

The additional permittivity $\varepsilon_{radd}(f)$ of a suspended microstrip should meet the following asymptotic requirements:

$$\begin{aligned} &\Delta\varepsilon_r \to 0, \varepsilon_{radd}(f) \to 0, \varepsilon_{rv}(f) \to 1, \text{as } h_2 \to 0, && \text{(a)} \\ &\Delta\varepsilon_r \to 0, \varepsilon_{radd}(f) \to 0, \varepsilon_{rv}(f) \to \varepsilon_{r2}, \text{ as } h_1 \to 0 && \text{(b)}. \end{aligned} \quad (15.1.13)$$

The empirically determined curve-fitted parameters α, β, and γ are summarized below:

$$\begin{aligned} &\alpha = 1.5\exp\{-(w/0.7213h_{eq})\}, h_{eq} = h_1 + h_2, \quad 0.1 \le w/h_{eq} \le 10 && \text{(a)} \\ &\beta = 3.928(0.8118)^{A_1}\left\{0.848(1.0248)^{A_1}\right\}^{w/h_{eq}}, \quad A_1 = h_2/h_1 && \text{(b)} \\ &\gamma = 0.345A_2A_3(A_4)^{A_5}(\varepsilon_{r2}/6.0)^{1.35}, \quad 0.1 \le w/h_{eq} \le 10 && \text{(c)} \\ &\text{where,} \\ &A_2 = \begin{cases} \exp\left\{1.3911(0.4236)^{w/h_{eq}} + 0.6861(w/h_{eq})^{-1.888} - 1\right\}, & 0.6 \le w/h_{eq} \le 4 \\ 0.116(1.3474)^{w/h_{eq}}, & 4 < w/h_{eq} \le 10 \end{cases} && \text{(d)} \\ &A_3 = h_2/2h_1, \text{for } h_2/h_1 < 1, \quad \text{otherwise}, \quad A_3 = 1 && \text{(e)} \\ &A_4 = w/h_{eq}, \text{for } 0.6 \le w/h_{eq} \le 5, \quad A_4 = h_{eq}/w, \text{for } 5 < w/h_{eq} \le 10 && \text{(f)} \\ &A_5 = \begin{cases} 3.4266 - 4.2256w/h_{eq}, & 0.1 < w/h_{eq} \le 0.6 \\ 1.0153 - 0.2123w/h_{eq}, & 0.6 < w/h_{eq} \le 5 \\ -0.2339 + 0.0555w/h_{eq}, & 5 < w/h_{eq} \le 10 \end{cases} && \text{(g)}. \end{aligned} \quad (15.1.14)$$

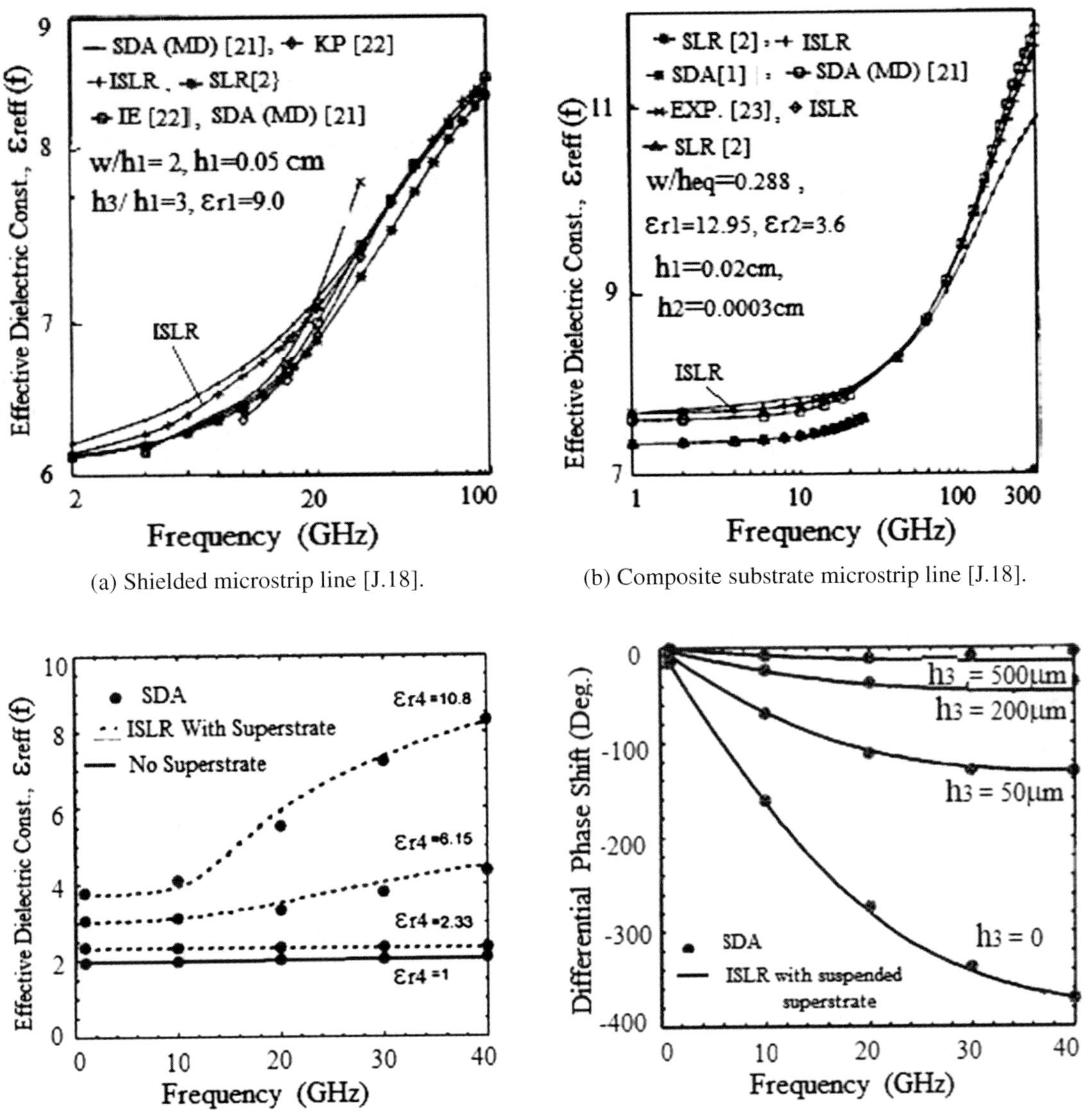

(a) Shielded microstrip line [J.18].

(b) Composite substrate microstrip line [J.18].

(c) Microstrip line with superstrate [J.14].

(d) Microstrip line with suspended superstrate [J.14].

Figure 15.4 Validation of SLR and ISLR-dispersion models against results of SDA. *Source:* From Verma and Kumar [J.18]. © 1998 IEEE. From Yun and Chang [J.14] © 2002.

To improve the accuracy of computation of dispersion for the suspended microstrip, the $\varepsilon_{radd}(f)$ given by equation (15.1.7a) is multiplied by the following correction factor:

$$K_{cor} = 1 - A_6 \exp\left[-\{(f_1 - 60)/20\}^2\right], \quad \text{where } f_1 \text{ in GHz}$$

$$A_6 = \begin{cases} A_7, & h_2/h_1 \le 1, \quad 0.1 \le w/h_{eq} \le 5 \\ A_7/A_1, & 1 < h_2/h_1 \le 3, \quad 0.1 \le w/h_{eq} \le 5 \\ A_7, & 0.05 < h_2/h_1 \le 3, \quad 5 < w/h_{eq} \le 10 \end{cases} \quad \text{(a)}$$

where,

$$A_7 = \begin{cases} -1.6938 + 0.3503 w/h_{eq}, & 0.1 \le w/h_{eq} \le 5 \\ -0.25 + 0.05 w/h_{eq}, & 5 \le w/h_{eq} \le 8 \\ -0.825 + 0.125 w/h_{eq}, & 8 \le w/h_{eq} \le 10 \end{cases} \quad \text{(b)}. \qquad (15.1.15)$$

Figure (15.3a) shows a small variation in $\varepsilon_{rv}(f)$ with frequency. The model has an RMS deviation within 0.8% and a maximum deviation of 2% against the results of the SDA for the following range of parameters, $1 < \varepsilon_{r2} \le 13$, $0 < h_1/h_2 \le 3$, $0.1 \le w/h_{eq} \le 10$, f. $h_{eq} = 2$ GHz-cm

The Composite Substrate Microstrip Line

Figure (15.1d) shows the shielded composite substrate microstrip line [J.18]. It is reduced to an open structure for $h_3 \to \infty$. It has two different cases – $\varepsilon_{r2} > \varepsilon_{r1}$ and $\varepsilon_{r1} > \varepsilon_{r2}$.

Case: $\varepsilon_{r2} > \varepsilon_{r1}$

This case could be used for a suspended microstrip also for a limited range of parameters, i.e. for a thicker and also thinner $h_1/19 \le h_2 \le h_1/3$ substrate with permittivity in the range $2 \le \varepsilon_{r2} \le 6$. Figure (15.3a) shows that the frequency dependence of the virtual permittivity $\varepsilon_{rv}(f)$ is insignificant. In other cases, modeling of $\varepsilon_{rv}(f)$ is needed. The following parameters are taken for an open composite substrate microstrip line:

$$H = \exp\left\{-4.4299\left(\frac{h_2}{h_1}\right)^{0.1115}\right\}, \tau = 1, f_n = \begin{vmatrix} f/f_p, & f > f_p \\ 1, & f \le f_p \end{vmatrix}. \tag{15.1.16}$$

$$\alpha = 0.6 \quad (a), \quad \beta = C_1\left(w/h_{eq}\right)^{-C_2}, \quad h_{eq} = h_1 + h_2 \quad (b), \tag{15.1.17}$$

where C_1 and C_2 are given below:

$$C_1 = \begin{cases} 2.6888 + 0.3566(h_2/h_1), & 1/3 \le h_2/h_1 \le 1 \\ 3.4093 - 0.3639(h_2/h_1), & 1 \le h_2/h_1 < 3 \\ 2.8959 - 0.1261(h_2/h_1), & 3 \le h_2/h_1 < 19 \end{cases} \quad (a)$$

$$C_2 = \begin{cases} 0.3573 - 0.0575(h_2/h_1), & 1/3 \le h_2/h_1 \le 1 \\ 0.3375 - 0.0377(h_2/h_1), & 1 \le h_2/h_1 < 3 \\ 0.266 - 0.014(h_2/h_1), & 3 \le h_2/h_1 < 19 \end{cases} \quad (b). \tag{15.1.18}$$

$$\gamma = \begin{cases} 0.3285(1.2012)^{\varepsilon_{r2}}, & 2 \le \varepsilon_{r2} \le 9.8 \\ 0.0655(1.4173)^{\varepsilon_{r2}}, & 9.8 < \varepsilon_{r2} < 13 \end{cases}. \tag{15.1.19}$$

The above model has RMS deviation within 1.8% against the results of the SDA for the following range of parameters: $2 < \varepsilon_{r2} \le 13$, $\varepsilon_{r2}/\varepsilon_{r1} \le 5$, $0 < h_1/h_2 \le 3$, $0.1 \le w/h_{eq} \le 10$, and f. h_{eq} = 4 GHz-cm. It has the maximum deviation within 2% at $w/h_{eq} = 10$, $h_2/h_1 = 1$, $\varepsilon_{r1} = 3.5$, $\varepsilon_{r2} = 12.95$.

Case: $\varepsilon_{r1} > \varepsilon_{r2}$

For this case, the parameters are taken as follows:

$$H = \frac{2h_2}{h_1}, \tau = \frac{1}{2}, f_n = \begin{cases} 2f/f_p, & f > f_p/2 \\ 1, & f \le f_p/2 \end{cases}. \tag{15.1.20}$$

$$\alpha = 1.0, \quad 0.1 \le w/h_{eq} \le 10, \quad 0.01 \le h_2/h_1 \le 0.05 \quad (a)$$
$$\beta = 0.8028\left(w/h_{eq}\right)^{-0.4288} \quad (b), \quad \gamma = \text{Eq.}-(15.1.19) \quad (c). \tag{15.1.21}$$

The above model has the maximum deviation within 2% against the results of the SDA for $f.h_{eq} \le 4$ GHz-cm. Figure (15.4b) compares the $\varepsilon_{reff}(f)$ computed by both the SLR and ISLR models up to 300 GHz, against the results of the SDA from two sources [J.1, J.21], and also against the experimental results [J.23]. The SLR model has deviation <4%, whereas the ISLR-model has a reduced deviation <1%. The accuracy of the model within 0.5% has been further confirmed up to 100 GHz for the composite substrate microstrip lines on several types of substrates, used in the thin-film MCM-D and MCM-L technologies [J.16].

Shielded Composite Substrate Microstrip Line

Figure (15.1d) shows the shielded microstrip line on a composite substrate. The virtual permittivity meets asymptotic requirements with respect to the top shield height, i.e. $\varepsilon_{rv}(f) \to \varepsilon_{req}$ for $h_3/h_{eq} \to \infty$. Again, it has two different cases: – $\varepsilon_{r1} > \varepsilon_{r2}$ and $\varepsilon_{r2} > \varepsilon_{r1}$.

Case: $\varepsilon_{r1} > \varepsilon_{r2}$

The following parameters are taken for the shielded composite substrate microstrip lines:

$$H = \sqrt{\frac{h_3}{h_{eq}}} \quad (a)$$
$$f_n = \begin{cases} f_p/\{2.5(f - \exp(0.035f))\}, & f.h_{eq} \le 0.8\,\text{GHz-cm} \\ f_p/\{2.0(f - \exp(0.035f))\}, & 0.8 < f.h_{eq} \le 2\,\text{GHz-cm} \\ f_p/133.78, & 2 < f.h_{eq} \le 6\,\text{GHz-cm} \end{cases} \quad (b). \tag{15.1.22}$$

In the above equation, frequency f and f_p are in GHz. Further, in the present case, the model has $\beta = \gamma = 0.5$. The parameter α is obtained from the following set of equations:

$$\alpha = C_3\left(w/h_{eq}\right)^{-C4}, \qquad 0.6 \le w/h_{eq} \le 10 \quad (a)$$

where

$$C_3 = \begin{cases} 2.7392 - 14.587(h_2/h_1), & h_2/h_1 \le 0.05 \\ 1.9715, & 0.05 < h_2/h_1 \le 0.33 \\ 2.1986 - 0.6821(h_2/h_1), & 0.33 < h_2/h_1 \le 1 \end{cases} \quad (b)$$

$$C_4 = \begin{cases} 0.4758 - 0.9732(h_2/h_1), & h_2/h_1 \le 0.05 \\ 0.4246, & 0.05 < h_2/h_1 \le 0.33 \\ 0.4378 - 0.0396(h_2/h_1), & 0.33 < h_2/h_1 \le 1 \end{cases} \quad (c). \tag{15.1.23}$$

For $h_3/h_{eq} > 8$, the present model could be used to compute the dispersion $\varepsilon_{reff}^{ISLR}(f)$ of the open composite substrate microstrip lines.

Case: $\varepsilon_{r2} > \varepsilon_{r1}$

In this case, the following parameters are used to model the additional relative permittivity $\varepsilon_{radd}(f)$ of a shielded composite substrate microstrip line:

$$H = \sqrt{\frac{h_3}{h_{eq}}} \quad (a)$$

$$f_n = \begin{cases} f_p/\{2.2(f - \exp(0.035f))\}, & f.h_{eq} \le 0.8 \ \text{GHz.cm} \\ f_p/\{2.0(f - \exp(0.035f))\}, & 0.8 < f.h_{eq} \le 1.2\,\text{GHz.cm} \\ f_p/125, & 1.2 < f.h_{eq} \le 2.8 \ \text{GHz.cm} \end{cases} \quad (b). \tag{15.1.24}$$

The empirical relations for the parameters α, β, and γ are summarized below:

$$\alpha = C_5(w/h_{eq})^{-C6}, \quad 0.1 \le w/h_{eq} \le 10, \quad h_2/h_1 \le 19 \quad (a)$$

where

$$C_5 = 0.6431(h_2/h_1)^{-0.2506}, \quad C_6 = 0.2977(h_2/h_1)^{-0.0265} \quad (b)$$
$$\beta = 0.5, \quad (c)$$
$$\gamma = 0.5(1 + C_7(h_2/h_1))^{-1}, \quad C_7 = 1.4134(h_2/h_1)^{-1.6936} \quad (d). \tag{15.1.25}$$

In the case of $\varepsilon_{r1} > \varepsilon_{r2}$, the model has a maximum deviation of 2% for $\varepsilon_{r1} = 12.95$, $\varepsilon_{r2} = 3.5$ within the range $f.h_{eq} \le 4$ GHz. cm, $0.1 \le w/h_{eq} \le 10$, and $0 \le h_2/h_1 \le 3$, $h_3/h_1 \ge 2$. In the case of $\varepsilon_{r2} > \varepsilon_{r1}$, the deviation of the model against the results of the SDA is within 3% for the range $0 \le h_2/h_1 \le 3$, $h_3/h_{eq} \ge 2$, $f.h_{eq} \le 2.6$ GHz. cm, $0.1 \le w/h_{eq} \le 10$.

Microstrip Line with Superstrate

Figure (15.1e) shows the microstrip line with a suspended superstrate that is reduced to a microstrip with a superstrate shown in Fig (15.1f). To get the structure of Fig (15.1e), the 5th dielectric layer is added to the structure shown in Fig (15.1a) and also taken $\varepsilon_{r5} = 1$, $h_5 \to \infty$. The variational method is used to obtain $\varepsilon_{reff}(0)$ and static ε_{req} using the SLR formulation. The additional relative permittivity $\varepsilon_{radd}(f)$ of the virtual relative permittivity $\varepsilon_{rv}(f)$ is again obtained from the following empirical expression [J.14]:

$$\varepsilon_{radd}(f) = -\left\{\frac{(\varepsilon_{req} - \varepsilon_r^{hi})}{0.95 + (f_p/1.2f)^D}\right\}$$

$$\text{where } D = \frac{\varepsilon_r^{hi}}{\varepsilon_r^{lo}}\sqrt[4]{\frac{h_4}{2h_1}} \quad (a)$$

$$\text{for Fig.(15.1e) } h_3 \ne 0, \quad f_p = \frac{Z_0(0)}{2\mu_0 h_1\sqrt{\varepsilon_{reff}(0)}} - \frac{h_3}{h_1}f^{1.2} \quad (b)$$

$$\text{for Fig.(15.1f) } h_3 = 0, \quad f_p = \frac{Z_0(0)}{2\mu_0 h_1\sqrt{\varepsilon_{reff}(0)}} \quad (c). \tag{15.1.26}$$

In the above equation, frequency f and f_p are in GHz and the ε_r^{hi} and ε_r^{lo} are higher valued and lower valued relative permittivity, respectively, selected from ε_{r1} and ε_{r4}. In the case of the suspended superstrate, the inflection frequency f_p gets modified to account for the effect of the air-gap h_3. However, for a simple superstrate, there is no modification.

Unlike the SLR-dispersion model of equation (15.1.5), the present model accurately computes the dispersion in microstrip with either simple superstrate or suspended superstrate. The model has an RMS deviation within 1.65% against the results of SDA. Its validity range is $2.33 \le \varepsilon_{r1}$, $\varepsilon_{r2} \le 10.8$, $0.5 \le w/h_1 \le 1.55$, $h_4/h_1 \ge 0.6$, $f < 40$ GHz.

For the case of the microstrip line with a superstrate, the SLR-model, given by equation (15.1.5), shows more deviation in results of $\varepsilon_{reff}(f)$ as compared against the results of the method of moments [J.24]. However, the model still works satisfactorily for thinner dielectric layers. It has been used for the design of wideband sensors up to 1.5 THz. The sensors are used to characterize the polar liquids [J.15]. The accuracy of the SLR-model for the composite substrate microstrip with superstrate, obtained from Fig (15.1a) by taking $\varepsilon_{r4} = 1$, $h_4 \to \infty$, has been confirmed up to 100 GHz. However, the thicknesses of the dielectric layers are within 50 μm. Such line structures are used for MCM-D and MCM-L technologies [B.5, J.16]. The ISLR model for the dielectric covered microstrip is more accurate, as shown in Fig (15.4c). The microstrip width and length are w = 22 mil and $\ell = 1.03$ in. on a substrate with $\varepsilon_{r1} = 2.33$, $h_1 = 20$ mil. In this case, there is no air-gap $h_3 = 0$ and superstrate of relative permittivity $\varepsilon_{r4} = 2.33/6.15/10.8$ with corresponding thicknesses $h_4 = 125/50/50$ mil, respectively, are used in the structure. Figure (15.4d) further compares the differential phase shift for a microstrip line with a suspended superstrate. It is computed by the ISLR and SDA. The experimental results are also shown for four values of air-gaps, i.e. $h_3 = 0$, 50, 200, 500 μm. The strip width is w = 22 mil and its length is L = 1.0 in. on a substrate with $\varepsilon_{r1} = 10.8$, $h_1 = 25$ mil. The superstrate has $\varepsilon_{r4} = 10.8$, $h_4 = 50$ mil. The differential phase is defined as the phase difference of the microstrip line of a fixed length with superstrate and without superstrate [J.14].

15.1.3 Characteristic Impedance and Synthesis of Multilayer Microstrip Lines

Once the frequency-dependent effective relative permittivity of a multilayer microstrip line is computed, its frequency-dependent characteristic impedance is obtained using the standard equation (8.2.31), or more accurate equation (8.3.13b) of chapter 8. The $\varepsilon_{reff}(f)$ computed by the SLR and ISLR models could be used with these expressions to compute the characteristic impedance. The power–current (P–I) based

Kirschning–Jansen–Kostar model of the standard microstrip lines, discussed in subsection (8.3.5) of chapter 8, could also be used to compute the frequency-dependent characteristic impedance of a multilayer microstrip line [B.6, J.25]. However, the relative permittivity ε_r should be replaced by the static equivalent relative permittivity ε_{req} for the SLR model or by the virtual relative permittivity $\varepsilon_{rv}(f)$ for the ISLR model. The model also accounts for the effect of conductor thickness t by using $w_{eqv}(t, f)$ in place of the strip width w. Another expression to compute the frequency-dependent characteristic impedance of a suspended microstrip is also available [J.26]. However, its utility is limited.

Finally, a multilayer microstrip line can be synthesized to realize the needed characteristic impedance at a known frequency by using the SLR model and the model for the characteristic impedance [J.27]. Figure (15.5) shows the flow-chart of the synthesis process for the four-layered structure shown in Fig (15.1a). The variational improper integral could be evaluated either by the Gauss–Legendre scheme or the Gauss–Laguerre scheme. The former one is more time taking. Its numerical evaluation requires numerical convergence using several subdomain integrals. The Gauss–Laguerre scheme can provide the result for the entire domain. The synthesis process is based on a combination of the *linear and golden-section optimization* to get the w/h_{eq} for a given structure and the assigned value of characteristic impedance at a certain frequency [B.7]. The SLR process could be operated in the analysis mode to get the frequency-dependent response, both for the effective relative permittivity and characteristic impedance, of a multilayer microstrip line. The synthesis algorithm for several kinds of multilayer microstrip lines such as the suspended, composite, shielded, dielectric covered microstrip lines has been tested with a deviation within 0.5% [J.27].

15.1.4 Models of Losses in Multilayer Microstrip Lines

The multilayer microstrip lines, like a standard single-layer microstrip line, have losses from three sources – dielectric loss, conductor loss, and surface wave's loss. The losses in single-layer microstrip are discussed in section (8.4) of chapter 8. The formulations developed for the computation of the dielectric and conductor losses in a microstrip line are also applicable to the multilayer microstrip lines by using the SLR and ISLR

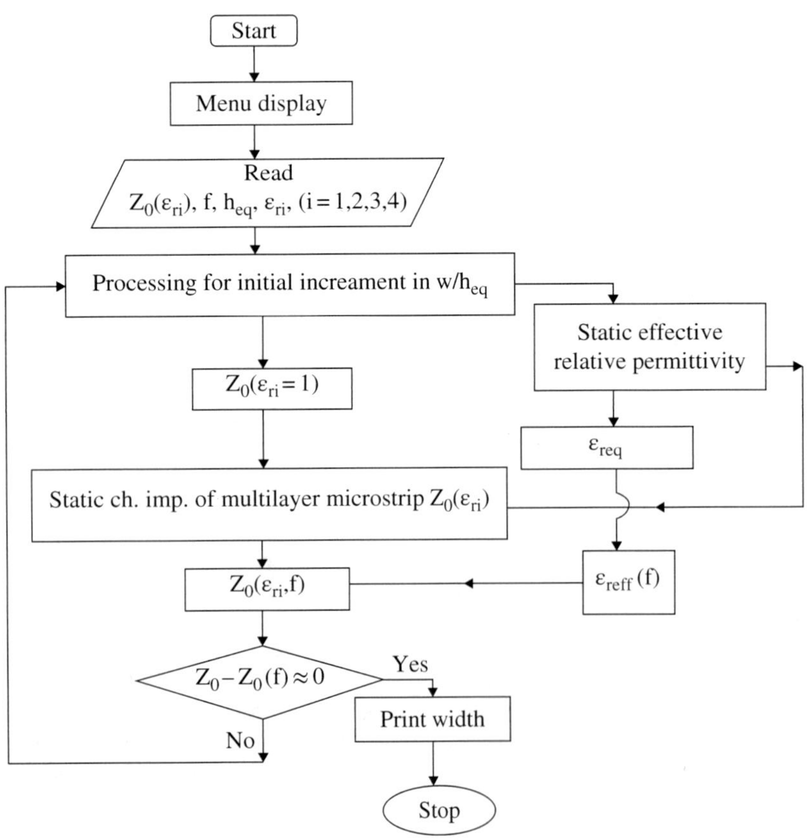

Figure 15.5 Flow-chart of the synthesis process for a multilayered microstrip line.

process. The full-wave methods and the commercial EM-Simulator compute these losses in the multilayer structures. However, SLR and ISLR-based methods are more efficient for CAD applications.

Dielectric Loss

The dielectric loss of a multilayer microstrip, shown in Fig (15.1a), can be computed by modifying equation (8.4.26b) of chapter 8 as [J.6],

$$\alpha_d = \frac{27.3\left[\varepsilon_{r\,eff}(w/h_{eq}, f, t) - 1\right]}{\lambda_0\left[\varepsilon_{req} - 1\right]}\frac{\varepsilon_{req}\tan\delta_{eq}}{\sqrt{\varepsilon_{r\,eff}(w/h_{eq}, f, t)}}\ \text{dB/length}, \tag{15.1.27}$$

where λ_0 is the free space wavelength corresponding to the operating frequency. In the above equation, the static equivalent relative permittivity ε_{req} of the multilayer microstrip is computed using the SLR formulation given by equation (15.1.2a) and $\varepsilon_{r\,eff}(w/h_{eq}, f, t)$ is computed with the help of equation (15.1.5). The effect of the conductor thickness t can also be accommodated in equation (15.1.5). For some of the structures, more accurate ISLR formulation can be used. The computation of the dielectric loss also requires the equivalent loss-tangent $\tan\delta_{eq}$. It is computed with the help of the variational method given by equation (15.1.2b).

Schneider has suggested an alternate method to compute $\tan\delta_{eq}$ using the following expression [J.28]:

$$\tan\delta_{eq} = \frac{1}{\varepsilon_{r\,eff}}\sum_{n=1}^{N}\frac{\partial\varepsilon_{reff}}{\partial\varepsilon_{rn}}\tan\delta_n, \tag{15.1.28}$$

where n = 1,...,N is the number of dielectric layers, $\varepsilon_{r\,eff}$ is the effective relative permittivity of a multilayer microstrip, and $\partial\varepsilon_{r\,eff}/\partial\varepsilon_{rn}$ is the layer-wise derivative of $\varepsilon_{r\,eff}$. Thus, implementation of the above simple-looking expression is a difficult task. It has been implemented using a combination of the SDA and FDTD methods [J.29, J.30]. Thus, this method is not suitable for CAD purposes.

Figure (15.6a) compares the dielectric loss of a shielded microstrip, as computed by the SLR method, for $0.6 \le w/h_1 \le 4$, $\varepsilon_{r1} = 10.0$, $\tan\delta = 0.0002$, $h_1 = 1.0$mm, $h_4 = 9.0$ mm at 1.0 GHz, against the results of the SDA, FEM, and integral equation (IE) methods [J.6, J.21, J.31, J.32]. The agreement of results is within 0.05 dB/cm. Figure (15.6b) compares, up to 100 GHz, the SLR-computed dielectric loss against the results of the SDA for a shielded Si-GaAs composite substrate microstrip line: Si $\varepsilon_{r1} = 11.8$, $\sigma_{Si} = 2000\ \Omega$-cm, $h_1 = 0.254$ mm and GaAs $\varepsilon_{r2} = 12.9$, $\sigma_{GaAs} = 1000\ \Omega$-cm, $h_2 = 4.0\ \mu$m. The strip width is 0.205 mm and the shield height is 5.254 mm [J.33]. The deviation in results is within

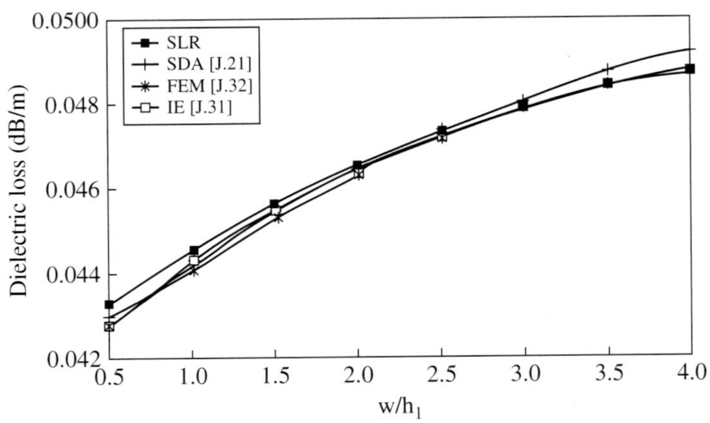

(a) Dielectric loss in a shielded microstrip.

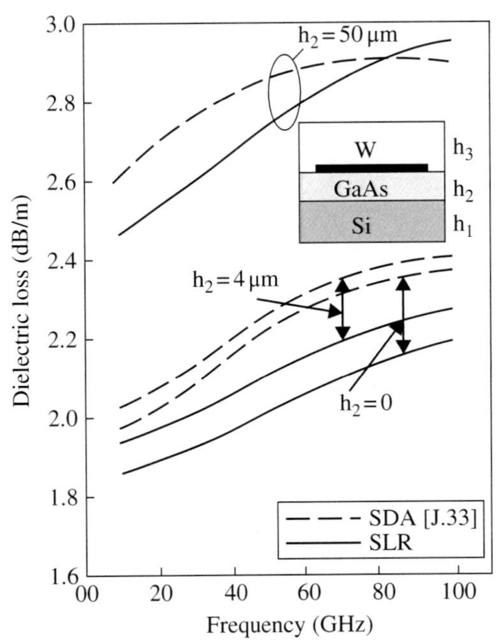

(b) Dielectric loss in a shielded composite substrate microstrip.

Figure 15.6 Dielectric loss in the shielded and composite substrate microstrip lines. *Source:* From Verma and Bhupal [J.6]. © 1998 John Wiley & Sons.

0.03 dB/cm, i.e. within 2%. However, above 100 GHz frequency dependence of $\tan\delta_{eq}$ should be considered to improve the computed dielectric loss using the SLR [J.34]. The ISLR model may provide improved results.

Conductor Loss

The conductor loss of a multilayer microstrip line, shown in Fig (15.1a), can be determined by three methods – *waveguide model, perturbation method*, and *Wheeler's incremental rule*, discussed in subsection (8.4.2) of chapter 8. For using the waveguide model and perturbation method, the characteristic impedance of a single-layer microstrip should be replaced by that of the multilayer microstrip. The perturbation method is useful for both the thin and thick conductors, i.e. both for $t \le \delta_s$ and $t > \delta_s$. The Wheeler's incremental rule is only applicable to a thick strip conductor $t \ge 1.1\,\delta_s$ [J.35]. Using the SLR formulation, equation (8.4.71) of chapter 8 is written below for the multilayer microstrip lines:

$$\alpha_C = \frac{27.3}{\lambda_0}\sqrt{\varepsilon_{reff}(\varepsilon_{r1},\ldots\varepsilon_{r4},w,h_1,\ldots h_4,t)} \times \left[\frac{\Delta Z_0(\varepsilon_{r1}\ldots\varepsilon_{r4}=1,w,h_1,\ldots h_4,t,\delta_s)}{Z_0(\varepsilon_{r1}\ldots\varepsilon_{r4}=1,w,h_1,\ldots h_4,t)}\right], \qquad (15.1.29)$$

where ΔZ_0 is the differential change in the characteristic impedance of a multilayer microstrip line, due to the penetration of the magnetic field. The static effective relative permittivity in the above equation is computed by the following expression:

$$\varepsilon_{reff}(\varepsilon_{r1},\ldots\varepsilon_{r4},h_1,\ldots h_4,w,t) = \varepsilon_{reff}(\varepsilon_{r1},\ldots\varepsilon_{r4},h_1,\ldots h_4,w,t=0) - \frac{(\varepsilon_{req}-1)}{4.6}\frac{t/h_{eq}}{\sqrt{w/h_{eq}}}, \quad \text{where, } h_{eq}=h_1+h_2. \qquad (15.1.30)$$

The equivalent relative permittivity is obtained from equation (15.1.2a) of the SLR process. The static effective relative permittivity used in equation (15.1.29) can also be replaced by the frequency-dependent SLR model of equation (15.1.5), or by the ISLR model of equation (15.1.8).

The differential characteristic impedance of a multilayer microstrip line is computed using the following equation:

$$\begin{aligned}\Delta Z_0(\varepsilon_{r1},\ldots\varepsilon_{r4}=1,w,h_1,\ldots h_4,t,\delta_s) &= Z_0(\varepsilon_{r1},\ldots\varepsilon_{r4}=1,w-\delta_s,h_{eq}+\delta_s,t-\delta_s) \\ &\quad - Z_0(\varepsilon_{r1},\ldots\varepsilon_{r4}=1,w,h_{eq},\ldots t,) \qquad (a)\\ \text{where, skin-depth, } \delta_s &= 1/\sqrt{\pi f\mu_0\sigma} \qquad (b).\end{aligned} \qquad (15.1.31)$$

At higher frequency, in place of static conductivity of a strip conductor, the frequency-dependent conductivity $\sigma(f)$ can be used to compute the conductor loss more accurately [J.34]. The frequency-dependent conductivity $\sigma(f)$ is computed using the empirical equation (8.4.47) of chapter 8.

Figure (15.7a) compares, over the range, $0.5 \le w/h_1 \le 5.0$, the SLR-computed conductor loss of a shielded microstrip line against the results of a few full-wave methods [J.21, J.31, J.32]. The shielded microstrip is considered on a substrate with $\varepsilon_{r1} = 10$, $h_1 = 1.0$ mm, $h_4 = 9.0$ mm, $t/h_1 = 0.0078$ operating at 1.0 GHz. Figure (15.7b) further compares, over the range, $0.1 \le w/h_1 \le 5.1$, the SLR-computed conductor loss of a dielectric-covered microstrip against the results of another method [J.37]. The conductor loss increases with an increase in the permittivity of the superstrate. The dielectric-covered microstrip is considered on a substrate with $\varepsilon_{r1} = 2.32$, $h_1 = 0.159$ mm, $t = 0.035$ mm and dielectric superstrate $\varepsilon_{r3} = 50$, $\sigma = 5.8 \times 10^7$ S/m. The operating frequency is 2.45 GHz. The average deviation in the results is within 5%. It is within the variation in results among full-wave methods.

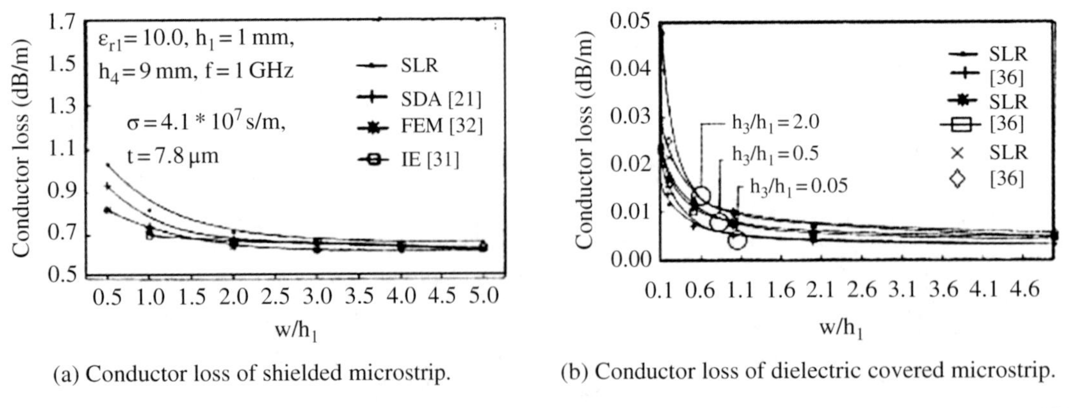

Figure 15.7 Conductor loss in the shielded and dielectric covered microstrip lines. *Source:* From Verma and Bhupal [J.35]. © 1998 John Wiley & Sons.

15.1.5 Circuit Model of Multilayer Microstrip Lines

We have discussed the computation of the frequency-dependent relative permittivity, characteristic impedance, dielectric loss, and conductor loss of a multilayer microstrip line with and without the top shield. These parameters are individually computed using the SLR or ISLR methods. However, the multilayer microstrip is a quasi-TEM line structure that can be modeled by the frequency-dependent RLCG parameters of a circuit model (CM). The RLCG parameters can be evaluated from the computed values of $\varepsilon_{reff}(f)$, $Z_0(f)$, $\alpha_d(f)$, $\alpha_c(f)$ and the CM of a multilayer microstrip line could be developed. It is discussed in section (8.5) of chapter 8. More accurate $\varepsilon_{reff}(f)$, $Z_0(f)$, $\alpha_d(f)$, $\alpha_c(f)$ parameters of a lossy multilayer microstrip are recomputed from the complex propagation constant $\gamma(f)$ and complex characteristic impedance $Z_0^*(f)$ of the CM. The CM-based results can also produce the low-frequency dispersion due to the field penetration in a finite conductivity conducting strip and ground plane.

15.2 SLR Process for Multilayer Coupled Microstrip Lines

The SLR process is also applicable to reduce the multilayer lossy symmetrically coupled microstrip lines to an equivalent single-layer substrate. The equivalent substrate could be a common substrate for both the even and odd modes. Alternatively, two separate equivalent substrates, the even-mode equivalent substrate and the odd-mode substrate, could be obtained to compute the dispersion, frequency-dependent characteristic impedance, dielectric and conductor losses of both the even and odd modes. The present section examines the computation of these parameters.

15.2.1 Equivalent Single-Layer Substrate

Figure (15.8a) shows the top shielded four-layered symmetrically coupled lossy microstrip lines. Figure (15.8b-f) shows some special cases. The complex static even- and odd-mode effective relative permittivities $\varepsilon^*_{reff,e}(0)$ and $\varepsilon^*_{reff,0}(0)$, and also the complex static even- and odd-mode characteristic impedances $Z^*_{0,e}(0)$ and $Z^*_{0,o}(0)$ are computed using the variational method in the Fourier domain. It is discussed in subsection (14.5.2) of chapter 14. The enclosed multilayer coupled microstrip lines could be treated in the space domain as discussed in subsection (14.5.1). The SLR process for the multilayer coupled microstrips is similar to the process applicable to the multilayered microstrip line discussed in section (15.1).

The SLR process is applied in two ways to the multilayer coupled microstrip lines. *Method #1* computes

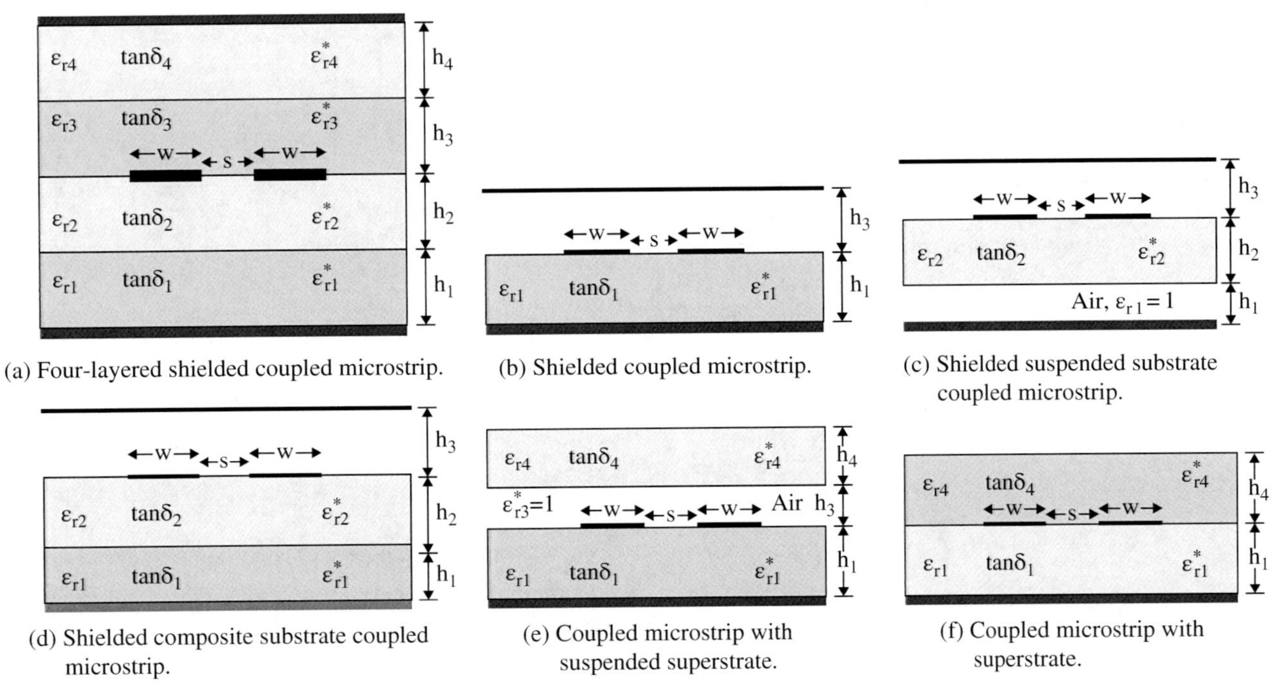

Figure 15.8 Four-layered coupled microstrip and some special cases.

the common equivalent single-layer substrate for both the even and odd modes. *Method #II* computes *separate single-layer even and odd modes equivalent substrates.* In the first step, the SLR process requires the computation of the even- and odd-mode complex effective relative permittivities $\varepsilon^*_{reff,e}(0)$ and $\varepsilon^*_{reff,o}(0)$. It is achieved by computing the complex line capacitances, with and without dielectric layers, for both modes using the variational method in the Fourier domain:

Even mode:

$$\varepsilon^*_{reff,e}(f=0) = \frac{C^*_e(\varepsilon^*_{r1},\ldots,\varepsilon^*_{r4},W)}{C_{o,e}(\varepsilon^*_{r1},\ldots,\varepsilon^*_{r4}=1,W)} \quad (a)$$

$$\tan\delta_{eff,e} = \frac{\varepsilon''_{reff,e}(f=0)}{\varepsilon'_{reff,e}(f=0)} \quad (b). \qquad (15.2.1)$$

Odd mode:

$$\varepsilon^*_{reff,o}(f=0) = \frac{C^*_o(\varepsilon^*_{r1},\ldots,\varepsilon^*_{r4},W)}{C_{o,o}(\varepsilon^*_{r1},\ldots,\varepsilon^*_{r4}=1,W)} \quad (a)$$

$$\tan\delta_{eff,o} = \frac{\varepsilon''_{reff,o}(f=0)}{\varepsilon'_{reff,o}(f=0)} \quad (b). \qquad (15.2.2)$$

The effect of the conductor thickness on the line parameters can be accounted for by replacing the width w with the effective width $w_{eff,e}$ and $w_{eff,o}$ for the even and odd modes, respectively, given by equations (12.1.18a) and (12.1.18b) of chapter 12.

Method – I: Common Equivalent Single-layer Substrate

The first method of the SLR-process determines *the common equivalent single-layer substrate* with *equivalent relative permittivity* ε_{req} and *equivalent loss-tangent* $\tan\delta_{eq}$ used for both the even and odd modes. The coupled line is replaced by a single-strip microstrip. Its equivalent effective relative permittivity and effective loss-tangent are approximately obtained from the mean value of the square root of above-computed parameters:

$$\varepsilon_{reff}(f=0) \approx \left[\frac{\sqrt{\varepsilon_{reff,e}(f=0)} + \sqrt{\varepsilon_{reff,o}(f=0)}}{2}\right]^2 \quad (a)$$

$$\tan\delta_{reff} \approx \left[\frac{\sqrt{\tan\delta_{reff,e}} + \sqrt{\tan\delta_{reff,o}}}{2}\right]^2 \quad (b). \qquad (15.2.3)$$

The approximate ε_{req} and $\tan\delta_{eq}$ of the equivalent common single-layer substrate, corresponding to a multilayer coupled microstrip lines, are recovered from the above equations as follows:

$$\varepsilon_{req} \approx \frac{\varepsilon_{reff}(f=0)-1}{q} \quad (a)$$

$$\tan\delta_{eq} \approx \frac{[1+q(\varepsilon_{req}-1)]\tan\delta_{reff}}{q\varepsilon_{req}} \quad (b). \qquad (15.2.4)$$

The above expressions are obtained from equation (8.4.24) of chapter 8 by replacing the pair (ε_r, $\tan\delta$) with (ε_{req}, $\tan\delta_{eq}$). The filling-factor q of a single conductor microstrip is computed from equations (8.2.12), (8.2.14), and (8.2.15). The process is given in subsection (15.1.1).

Method – II: Separate Equivalent Even-/Odd-Mode Single-layer Substrate

Using the even- and odd-mode filling-factors q_e and q_o, the *inverse Wheeler's transformation* converts the multilayer even- and odd-mode structures to the single-layer even- and odd-mode equivalent microstrip line. Following equation (15.1.2), the even- and odd-mode equivalent relative permittivity and loss-tangents are as given below [J.45]:

Even mode:

$$\varepsilon_{req,e} = \frac{[\varepsilon'_{reff,e}(f=0)-1]}{q_e} + 1 \quad (a)$$

$$\tan\delta_{eq,e} = \frac{\varepsilon''_{req,e}}{\varepsilon'_{req,e}} = \frac{\varepsilon''_{reff,e}}{\varepsilon'_{reff,e} + (q_e-1)} \quad (b)$$

Equivalent substrate thickness, $h_{eq} = h_1 + h_2$. (15.2.5)

Odd mode:

$$\varepsilon_{req,o} = \frac{[\varepsilon_{reff,o}(f=0)-1]}{q_o} + 1 \quad (a)$$

$$\tan\delta_{eq,o} = \frac{\varepsilon''_{req,o}}{\varepsilon'_{req,o}} = \frac{\varepsilon''_{reff,o}}{\varepsilon'_{reff,o} + (q_o-1)} \quad (b). \qquad (15.2.6)$$

Following equation (8.2.12) of chapter 8, the even- and odd-mode filling-factors are expressed as follows:

$$Even-mode: q_e = \frac{[F_e(u,g,\varepsilon_{req})+1]}{2} \quad (a)$$

$$Odd-mode: q_o = \frac{[F_0(u, g, \varepsilon_{req}) + 1]}{2} \quad (b). \tag{15.2.7}$$

where $u = w/(h_1 + h_2)$, $g = s/(h_1 + h_2)$. The structure-dependent even-mode variable F_e is obtained from equations (12.1.3b) and (12.1.4) of chapter 12 for the multilayer microstrip without the top-shield. In the process, ε_r of the substrate is replaced by the approximate ε_{req} of equation (15.2.4a). Alternatively, we use the iterative method with an initial value $\varepsilon_{req,\,e} = 1$ that computes the approximate value of the q_e. It helps to get an approximate value of $\varepsilon_{req,e}$. It is further used with equations (12.1.3b), (12.1.4), and (15.2.7a) to get the improved value of the filing-factor q_e. Using it, a more accurate value of $\varepsilon_{req,e}$ is obtained. Two or three iterations can provide an accurate value $\varepsilon_{req,e}$. The strip conductor thickness and the effect of the top-shield on q_e could be accounted for by using equation (12.1.3).

The odd mode $F_0(u, g, \varepsilon_{req})$ factor is obtained by modifying the available expressions in the open literature [J.38, J.39]:

$$\begin{aligned}
&F_o(u, g, \varepsilon_{req}) = K_0(1 + 10/u)^{-a_o(u,\varepsilon_{req})b_0(\varepsilon_{req})} && (a)\\
&\text{where, } K_0 = 1 + [-1 + 0.7287\exp(-0.179u)]\exp[-C_0 g^{d_o}] && (b)\\
&a_o(u, \varepsilon_{req}) = 0.7287[\varepsilon_{reff}(0) - 0.5(\varepsilon_{req} + 1)][1 - \exp(-0.179u)] && (c)\\
&b_0(\varepsilon_{req}) = 0.747\left[\frac{\varepsilon_{req}}{0.15 + \varepsilon_{req}}\right] && (d)\\
&C_0 = b_0(\varepsilon_{req}) - [b_0(\varepsilon_{req}) - 0.207]\exp(-0.414u) && (e)\\
&d_0 = 0.593 + 0.694\exp(-0.562u), \quad (f), \quad u = w/h_1, g = s/h && (g).
\end{aligned} \tag{15.2.8}$$

In the above equations, the effective relative permittivity of the multilayer single-strip conductor microstrip is obtained from the variational method, and the approximate value of ε_{req} is taken from equation (15.2.4a). However, for the odd mode, the single-layer equivalent substrate with relative permittivity $\varepsilon_{req,o}$ is considered in the above equation. It is determined by the iterative process starting with the initial value $\varepsilon_{req,o} = 1$.

In the process of getting the equivalent permittivities ε_{req}, $\varepsilon_{req,e}$, $\varepsilon_{req,o}$, an error occurs in the SLR due to the approximate nature of the formulation. The correction factors can be generated by taking the coupled microstrip on a single-layer substrate with relative permittivity $\varepsilon_r = \varepsilon_{req}$, $\varepsilon_{req,e}$, or $\varepsilon_{req,o}$. The complete SLR process using the variational method is repeated. Ideally, we get back the known value of the relative permittivity ε_r of the substrate. However, any deviation in computation can be treated as an error in the SLR-process and the originally computed ε_{req}, $\varepsilon_{req,e}$, $\varepsilon_{req,o}$ should be corrected accordingly to get more accurate relative permittivity of the single-layer substrate, even-mode substrate, and odd-mode substrate.

15.2.2 Dispersion Model of Multilayer Coupled Microstrip Lines

Following the process of modeling the dispersion in a multilayer microstrip [J.2], the even- and odd-mode dispersion of the multilayer coupled microstrip are computed by using Kirschning and Jansen (K–J) model of single-substrate coupled microstrip lines [J.38]: The K–J model is discussed in subsection (12.1.2) of chapter 12. The dispersion models for both the even and odd modes, as applicable on the common equivalent single-layer substrate, are given below:

$$\begin{aligned}
&Even-mode: \varepsilon_{reff,e}(f) = \varepsilon_{req} - \frac{\varepsilon_{req} - \varepsilon_{reff,e}(f=0)}{1 + F_e(f)} && (a)\\
&Odd-mode: \varepsilon_{reff,o}(f) = \varepsilon_{req} - \frac{\varepsilon_{req} - \varepsilon_{reff,o}(f=0)}{1 + F_0(f)} && (b).
\end{aligned} \tag{15.2.9}$$

In the above equations, the common equivalent substrate ε_{req} is taken from equation (15.2.4a). Further, in place of a common ε_{req} for both the modes, the equivalent even and odd modes relative permittivities $\varepsilon_{req,e}$ and $\varepsilon_{req,o}$ can be used in the above equations to compute the even- and odd-mode dispersions. The even- and odd-mode static relative permittivities $\varepsilon_{reff,e}$ and $\varepsilon_{reff,o}$ are computed from the variational method. The dispersion parameters $F_e(f)$ and $F_o(f)$ are taken from equations (12.1.20) and (12.1.25) of chapter 12. Again, in these equations ε_r is replaced by ε_{req}; alternatively, by $\varepsilon_{req,e}$ and $\varepsilon_{req,o}$ for the even mode and odd mode, respectively.

The SLR-based dispersion results for a shielded coupled microstrip on the substrate with $\varepsilon_r = 10$ have deviation within 3% as against the full-wave results of

Krage and Haddad [J.40, J.41]. In the SLR-based computation, ε_{req} is used for both modes. Figure (15.9) compares the SLR-computed frequency-dependent $\varepsilon_{reff,e}(f)$ and $\varepsilon_{reff,o}(f)$ of the coupled microstrip lines on a composite substrate against the results obtained from two sources of the SDA [B.8, J.42]. The first layer of the composite substrate is RT Duriod with $\varepsilon_{r1} = 2.2$, $h_1 = 0.53$ mm and the 2nd layer is the alumina substrate with $\varepsilon_{r2} = 9.7$, $h_2 = 0.97$ mm. The strip width and slot gap between the strips are $w = h_{eq} = 1.52$mm, $s = 0.53$ mm. The SLR results have about a 3% deviation from the results of the SDA. Finally, Fig (15.10) compares the SLR results against the full-wave results of Kitazawa and Hayashi [J.43] for the coupled microstrips with a superstrate. In this case, the substrate is anisotropic sapphire with $\varepsilon_{rxx1} = 9.4$, $\varepsilon_{ryy1} = 11.6$, $w/h_1 = 1.0$, $s/h_1 = 0.5$. The superstrate is alumina layer with $\varepsilon_{r3} = 9.6$, $h_3/h_1 = 0.1$. Using equation (8.2.42) of chapter 8, the anisotropic substrate is converted to an equivalent isotropic substrate before using the SLR process. In the present case, $\varepsilon_{req,e}$ and $\varepsilon_{req,o}$ are used in the (K–J) model to compute the even- and odd-mode dispersion. Again, the deviation in the SLR-based model is about 3%. The model can be improved by developing the models for the even- and odd-mode virtual equivalent permittivities.

15.2.3 Characteristic Impedance and Synthesis of Multilayer Coupled Microstrips

The general forms of the static even- and odd-mode characteristic impedances of the multilayer coupled microstrip are given below:

$$Z_{0,e} = \frac{Z_{0,e}(\varepsilon_{r1}, \ldots \varepsilon_{r4} = 1)}{\sqrt{\varepsilon_{reff,e}(\varepsilon_{r1}, \ldots \varepsilon_{r4})}} \quad \text{(a)}$$

$$Z_{0,o} = \frac{Z_{0,o}(\varepsilon_{r1}, \ldots \varepsilon_{r4} = 1)}{\sqrt{\varepsilon_{reff,o}(\varepsilon_{r1}, \ldots \varepsilon_{r4})}} \quad \text{(b).}$$

$$(15.2.10)$$

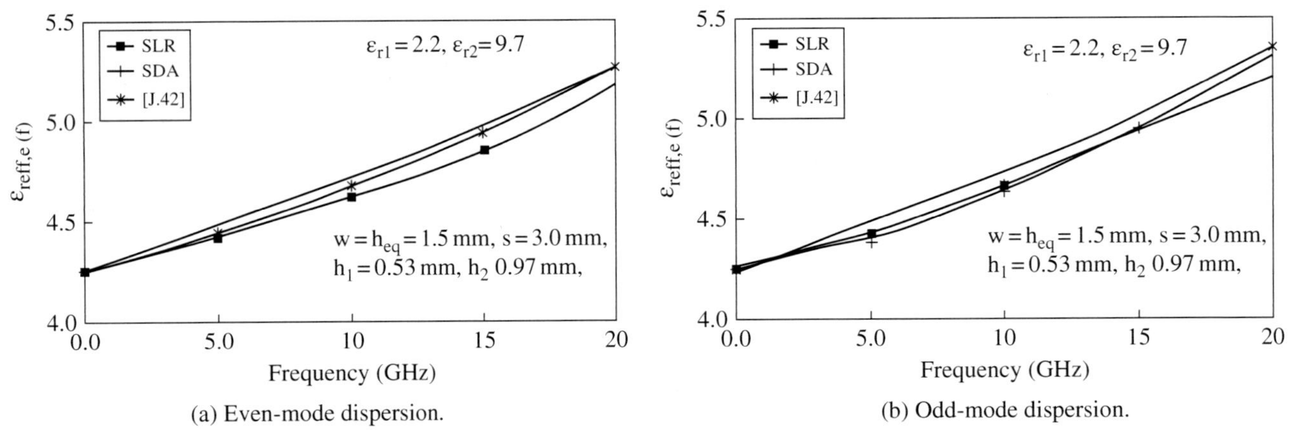

Figure 15.9 Even- and odd-mode dispersion of the coupled microstrip lines on a composite substrate.

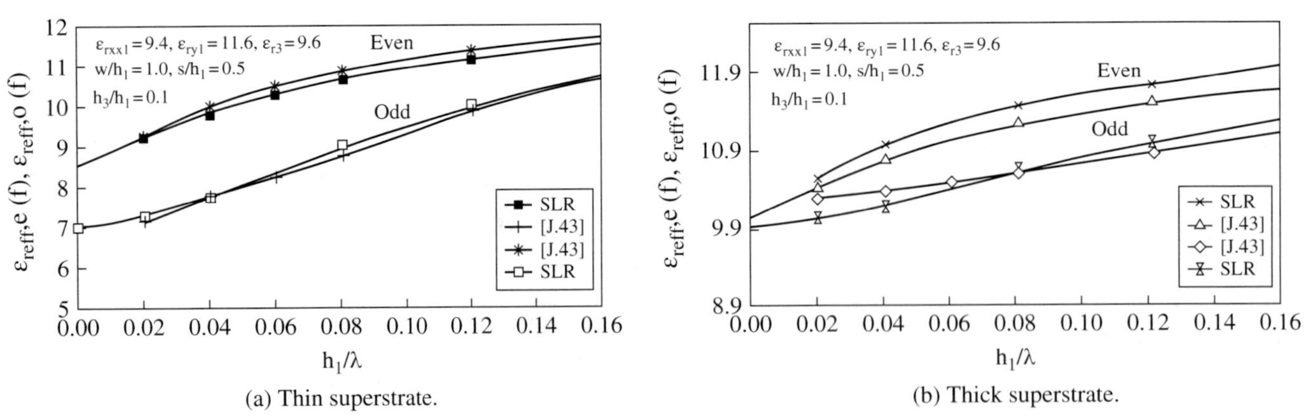

Figure 15.10 Even and odd dispersion characteristics of a coupled microstrip lines with superstrate.

In the above equation $Z_{0,e}(\varepsilon_{r1}, ...\varepsilon_{r4} = 1)$ and $Z_{0,\ o}(\varepsilon_{r1}, ...\varepsilon_{r4} = 1)$ are the even- and odd-mode characteristic impedances on the air-substrate. For the open multilayer microstrips, equations (12.1.8) and (12.1.10) of chapter 12 are used to compute both impedances. The effect of the top shields could be further incorporated using equations (12.1.12)–(12.1.15). Alternatively, for the shielded coupled microstrips on the air-substrate, the expressions of Harneko and Pavel could also be used [B.9]. The static effective relative permittivities $\varepsilon_{reff,\ e}(\varepsilon_{r1}, ...\varepsilon_{r4})$ and $\varepsilon_{reff,\ o}(\varepsilon_{r1}, ...\varepsilon_{r4})$ of the multilayer coupled microstrips are computed using the variational method discussed in section (14.5) of chapter 14. The effect of the conductor thickness on both the even and odd modes is incorporated in the variational formulation using the effective even- and odd-mode widths given by equation (12.1.16) or (12.1.18). In the case of an anisotropic multilayer substrate, each layer is converted to the equivalent isotropic substrate using equation (8.2.42) of chapter 8.

Equations (12.1.21)–(12.1.23), and (12.1.26)–(12.1.27) for the frequency-dependent even- and odd-mode characteristic impedances of a single-layer substrate are accommodated in the SLR formulation to compute the dispersive impedance of the multilayer coupled microstrip lines. However, the relative permittivity ε_r of the substrate requires the replacement as $\varepsilon_r \rightarrow \varepsilon_{req}$ applicable to both the even and odd modes. Alternatively, the replacement is done separately for the even- and odd-mode cases, i.e. $\varepsilon_r \rightarrow \varepsilon_{req,\ e}$ and $\varepsilon_r \rightarrow \varepsilon_{req,\ o}$.

Following the case of the multilayer microstrip, the synthesis of the coupled microstrip can also be achieved by using the two-variable golden section optimization scheme [B.7]. Such a synthesis method has been used to design a high directivity-suspended microstrip coupler [J.44].

15.2.4 Loss Models of Multilayer Coupled Microstrip Lines

The even- and odd-mode dielectric and conductor losses of the multilayer coupled microstrip lines of Fig (15.1) can be computed using the SLR formulation.

Dielectric Losses

The dielectric loss of a multilayer coupled microstrip lines could be estimated from the dielectric loss in the equivalent single-layer substrate with equivalent permittivity ε_{req} and $\tan\delta_{eq}$ given by equation (15.2.4). However, it is more accurate to compute the dielectric loss of the even and odd modes of the coupled multilayer microstrip lines separately. The multilayer coupled microstrips structure is separately reduced to the equivalent single-layer even- and odd-mode lossy substrate with equivalent even- and odd-mode permittivities $\varepsilon_{req,e}$, $\varepsilon_{req,o}$ and $\tan\delta_{eq,e}$, $\tan\delta_{eq,o}$. Thus, using equations (15.2.5) and (15.2.6) with, the following expressions are obtained to compute the even- and odd-mode dielectric losses [J.45].

$$\textit{Even}-\textit{mode}:\alpha_{d,e} = \frac{27.3\left[\varepsilon_{r\ eff,e}(w/h_{eq}, f, t) - 1\right]}{\lambda_0\left[\varepsilon_{req,e} - 1\right]}\frac{\varepsilon_{req,e}\tan\delta_{eq,e}}{\sqrt{\varepsilon_{reff,e}(w/h_{eq}, f, t)}}\ \mathrm{dB/length} \quad \text{(a)}$$

$$\textit{Odd}-\textit{mode}:\alpha_{d,o} = \frac{27.3\left[\varepsilon_{reff,o}(w/h_{eq}, f, t) - 1\right]}{\lambda_0\left[\varepsilon_{req,o} - 1\right]}\frac{\varepsilon_{req,o}\tan\delta_{eq,0}}{\sqrt{\varepsilon_{reff,o}(w/h_{eq}, f, t)}}\ \mathrm{dB/length} \quad \text{(b)}$$

$$\text{Total dielectric loss}:\alpha_d = \alpha_{d,e} + \alpha_{d,o}\ \mathrm{dB/length} \quad \text{(c)}. \qquad (15.2.11)$$

Figure (15.11) illustrates the computation of dielectric loss using the SLR formulation for two cases – coupled microstrip on a composite substrate and a composite substrate with a superstrate. Figure (15.11a) shows the even and odd equivalent loss-tangents and the dielectric loss of the coupled microstrips on a composite substrate with RT Duriod $\varepsilon_{r1} = 2.2, \tan\delta_1 = 0.0012$ and alumina $\varepsilon_{r2} = 9.8, \tan\delta_2 = 0.0023$, $h_1 = h_2 = 0.3175$ mm, w = 0.6 mm, s = 1.2 mm. The dielectric loss is computed at 1.0 GHz for different thicknesses of the 1st substrate while keeping the $h_{eq} = h_1 + h_2$ constant. For the case $h_1/h_{eq} < 0.15$, the even-mode dielectric loss is higher as compared to that of the odd-mode loss. However, it gets interchanged for $h_1/h_{eq} > 0.15$. Figure (15.11b) shows the losses at 1.0 GHz on the composite substrate with a superstrate with $\varepsilon_{r1} = 2.2, \tan\delta_1 = 0.0009$, $\varepsilon_{r2} = 9.8, \tan\delta_2 = 0.0023$, $\varepsilon_{r3} = 9.8, \tan\delta_3 = 0.0023$, and $\varepsilon_{r4} = 1$, $h_4 \rightarrow \infty$; $h_1 = h_2 = 0.3175$ mm, w = 0.6 mm, s = 1.2 mm. The odd-mode losses are always higher than the even-mode losses. It increases with an increase in the thickness of the superstrate. The odd-mode field lines are confined mostly in the lossy superstrate.

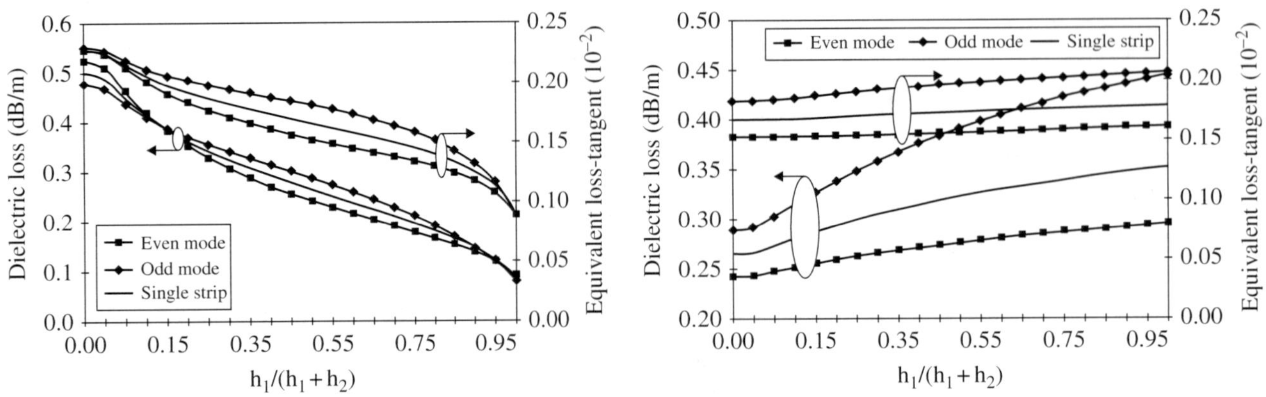

(a) Dielectric losses and eq. loss-tangents of even and odd mode on the composite substrate.

(b) Dielectric loss and eq. loss-tangent of even and odd mode on composite substrate with superstrate.

Figure 15.11 Dielectric loss of coupled microstrip on a composite substrate without and with superstrate. *Source:* Verma and Nasimuddin [J.45]. © 2001 John Wiley & Sons.

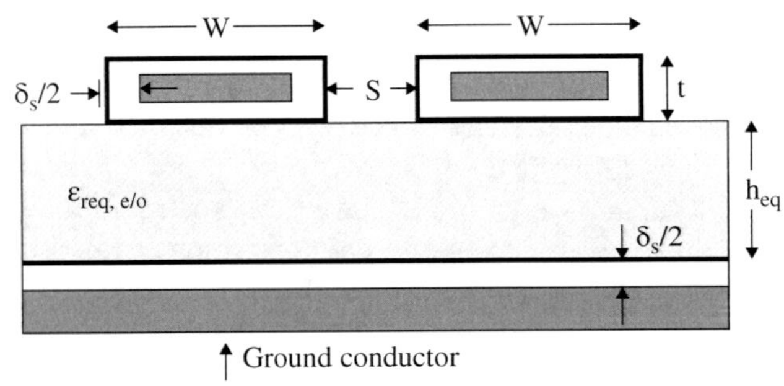

(a) Coupled microstrip with field penetration on the equivalent even/odd mode substrate.

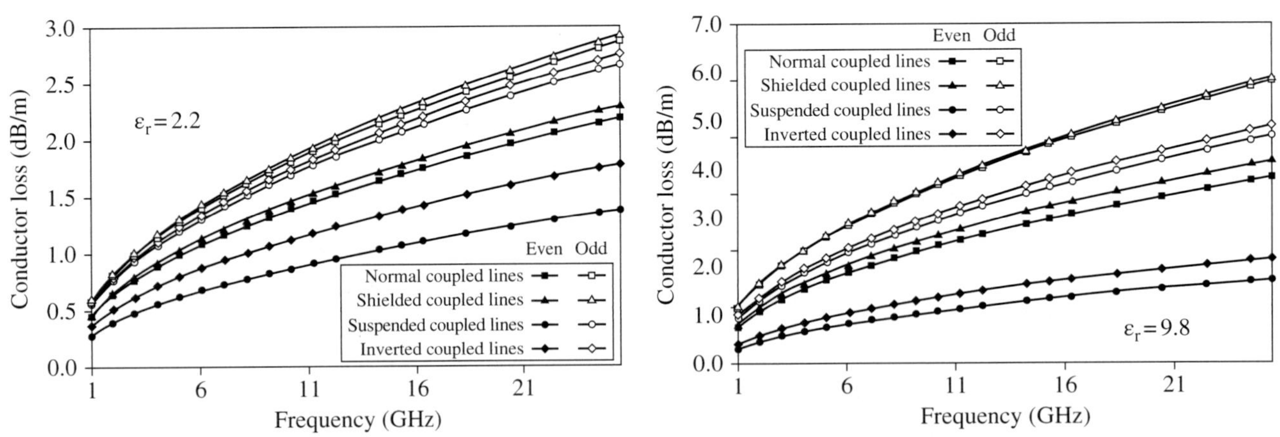

(b) Even/odd mode conductor losses on a low permittivity dielectric sheet [J.36].

(c) Even/odd mode conductor losses on a high permittivity dielectric sheet [J.36].

Figure 15.12 Conductor loss of the coupled microstrip on several types of substrates. *Source:* Modified from Verma and Nasimuddin [J.36]. © 2003 John Wiley & Sons.

Conductor Losses

The conductor loss of the multilayer coupled microstrips, shown in Fig (15.1), can be determined by using *Wheeler's inductance rule* for the even and odd mode separately. Figure (15.12a) shows $\delta_s/2$ field penetration in the conducting strips and also in the ground conductor. It changes the physical dimensions of the coupled line as $w \rightarrow (w - \delta_s)$, $s \rightarrow (s + \delta_s)$, $h_{eq} = h_1 + h_2 \rightarrow (h_{eq} + \delta_s)$, and also $H = h_3 + h_4 \rightarrow (H + \delta_s)$. However, change in the strip gap could be taken empirically as $s \rightarrow (s + 2\delta_s)$

to account for the complex nature of the field penetration in the strip edges of the gap region [J.36]. The Wheeler's inductance rule formulation of multilayer microstrip, given by equations (15.1.29) and (15.1.31), could be used to compute the conductor losses of both the even and odd modes:

Even mode :

$$\alpha_{C,e} = \frac{27.3\sqrt{\varepsilon_{reff,e}(w, t, \delta_s, s, h_1 ... h_4, \varepsilon_{r1}, ... \varepsilon_{r4} = 1)}}{\lambda_0} \times \left[\frac{\Delta Z_{0e}(w, t, \delta_s, s, h_1 ... h_4, \varepsilon_{r1}, ... \varepsilon_{r4} = 1)}{Z_{0e}(w, t, \delta_s, s, h_1 ... h_4, \varepsilon_{r1}, ... \varepsilon_{r4} = 1)}\right] \text{dB/length} \tag{15.2.12}$$

$$\text{where,}\quad \Delta Z_{0e} = \left[Z_{0e}\left[(w-\delta_s), (s+2\delta_s), \left(h_{eq}+\delta_s\right), (t-\delta_s), \varepsilon_r = 1\right] - Z_{0e}(w, s, h, t, \varepsilon_r = 1)\right].$$

Odd mode:

$$\alpha_{C,o} = \frac{27.3\sqrt{\varepsilon_{reff,o}(w, t, \delta_s, s, h_1 ... h_4, \varepsilon_{r1}, ... \varepsilon_{r4} = 1)}}{\lambda_0} \times \left[\frac{\Delta Z_{0o}(w, t, \delta_s, s, h_1 ... h_4, \varepsilon_{r1}, ... \varepsilon_{r4} = 1)}{Z_{0o}(w, t, \delta_s, s, h_1 ... h_4, \varepsilon_{r1}, ... \varepsilon_{r4} = 1)}\right] \text{dB/length} \tag{15.2.13}$$

$$\text{where,}\quad \Delta Z_{0o} = \left[Z_{0o}\left[(w-\delta_s), (s+2\delta_s), \left(h_{eq}+\delta_s\right), (t-\delta_s), \varepsilon_r = 1\right] - Z_{0o}(w, s, h, t, \varepsilon_r = 1)\right].$$

The even- and odd-mode characteristic impedances and the differential characteristic impedance $\Delta Z_{0,\ e/o}$ are computed using the SLR process. It is discussed in subsection (15.2.3). The total conductor loss is $\alpha_c = \alpha_{c,e} + \alpha_{c,,o}$.

The SLR-computed conductor losses show satisfactory agreement with other full-wave methods [J.36]. Figure (15.12b, c) shows the conductor losses of both the even and odd modes for the standard, shielded, suspended, and inverted coupled microstrips on both the low and high permittivity substrates. The dimensions used in the computations are taken as follows: for the shielded structure, $h_1 = 1$ mm, shield height $h_4 = 3$ mm; for the suspended structure, $h_1 = h_2 = 1$ mm; and for the inverted structure, $h_1 = h_3 = 1$ mm; for all structures, $w = 2.88$ mm, $s = 0.5$ mm, $t = 0.5$ μm; and conductivity of the strips and ground conductors are $\sigma = 5.8 \times 10^7$ S/m. The presence of the top shield increases the conductor loss and the air-gap decreases it. The minimum conductor loss occurs to the inverted coupled microstrip lines. Wheeler's inductance rule is applicable for the thick conductors only, i.e. for the case $t \geq 1.1\,\delta_s$. The perturbations method, using the *concept of the stopping distance* discussed in subsection (8.4.2) of chapter 8, can be applied to the thin conductor case. However, the accuracy of the SLR process, using the perturbations method, for the multilayer coupled microstrip has to be determined by comparing the results against the results from the EM-Simulators.

15.3 SLR Process for Multilayer ACPW/CPW

Figure (15.13a, c) shows the shielded lossy multilayer asymmetric CPW (ACPW) line structure with a finite conductor thickness t, and the SLR-process to reduce the structure into two cases. For *case #A*, the lossy multilayer ACPW is reduced to a lossy ACPW on an equivalent single-layer substrate of finite substrate thickness $h_{eq} = h_1 + h_2$. Whereas, *case #B* reduces a lossy multilayer ACPW line to a lossy ACPW line on equivalent single-layer substrate with infinite substrate thickness. Case #A is used to compute the dispersion in both the CPW and ACPW structures, as the dispersion relation of the single-layer CPW is available only for the finite substrate symmetrical CPW [J.46]. Case #B is used to compute the dielectric loss of the multilayer symmetrical CPW [J.47].

The present section discusses the results of the isotropic lossy multilayer CPW only, although the SLR process is also applicable to the anisotropic and lossy ACPW [J.7]. The SLR process requires the computation of the static effective relative permittivity and the static characteristic impedance of the multilayer ACPW/CPW with the finite conductor thickness. Any static method is applicable for this purpose, including the conformal mapping method discussed in chapter 9 [J.48]. However, the present section is based on Galerkin's method in the discrete Fourier domain discussed in subsection (14.6.3) of chapter 14 [B.4].

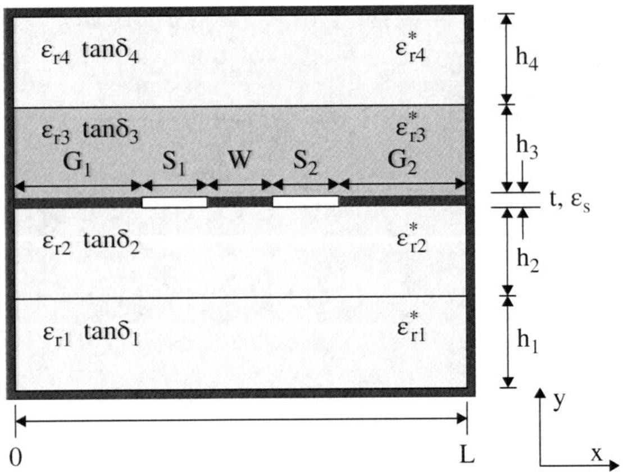

(a) Multilayer ACPW with finite conductor thickness.

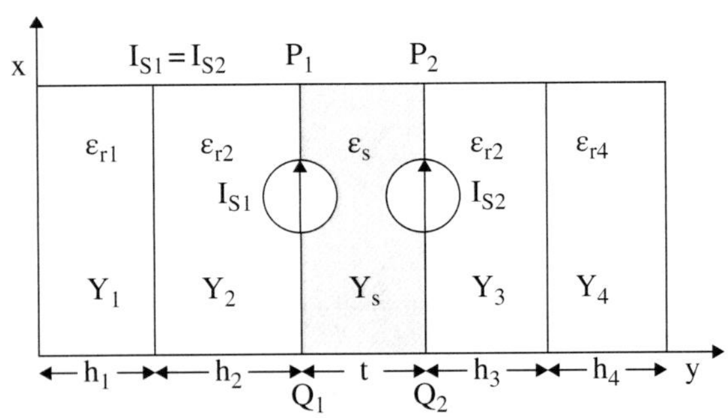

(b) Equivalent transmission line with two-current sources corresponding to the charge distributions over the lower and upper surface of the finite thickness strip conductor.

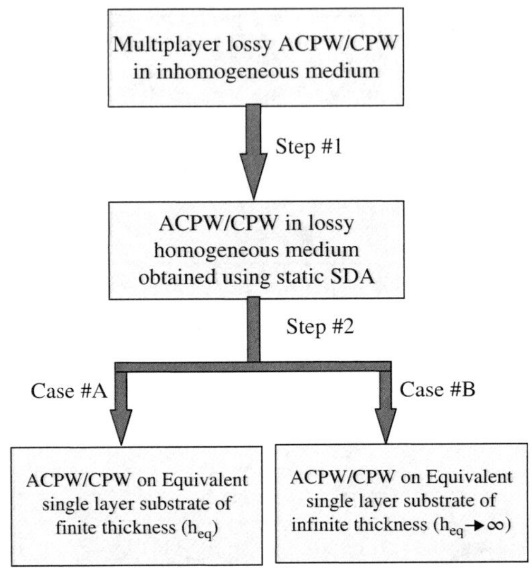

(c) Equivalent single-layer ACPW with finite conductor thickness.

Figure 15.13 Multilayer ACPW and equivalent single-layer substrate using the SLR process.

15.3.1 Single-Layer Reduction (SLR) Process for Multilayer ACPW/CPW

The SLR process applied to a lossy multilayer ACPW/CPW line structure is the same as the process discussed in subsection (15.1.1) for a multilayer microstrip structure. Figure 15.13c shows the SLR process for the lossy multilayer ACPW/CPW structure. The multilayer ACPW/CPW is an inhomogeneous isotropic lossy dielectric medium.

Step #1 converts the inhomogeneous medium to an equivalent homogeneous medium by computing the complex effective relative permittivity. The static spectral domain method (SDA), using Galerkin's method, computes the complex effective relative permittivity. It provides the real effective relative permittivity $\varepsilon^*_{r,eff}(f=0)$ and the effective loss-tangent:

$$\varepsilon^*_{r,eff}(f=0) = \varepsilon'_{reff} - j\varepsilon''_{reff} = \frac{C^*_{v,d}\left(\varepsilon^*_{r1}, \ldots, \varepsilon^*_4, W, S_1, S_2, G_1, G_2, h_1 \ldots, h_4, t\right)}{C_{v,o}(\varepsilon^*_{r1}, = \ldots, = \varepsilon^*_4 = 1, W, S_1, S_2, G_1, G_2, h_1 \ldots, h_4, t)} \quad \text{(a)}$$

$$\tan\delta_{eff} = \frac{\varepsilon''_{r,eff}(f=0)}{\varepsilon'_{r,eff}(f=0)} \quad \text{(b).} \qquad (15.3.1)$$

The static SDA computes the variational line complex capacitance $C^*_{v,d}$ of an ACPW/CPW line with dielectric layers and line capacitance $C_{v,o}$ on the air-substrate.

Step #2, using the *inverse Wheeler's transformation*, further converts the ACPW/CPW in a homogeneous medium to an equivalent single-layer substrate of thickness h_{eq} with equivalent relative permittivity ε_{req} and $\tan\delta_{eq}$ given by equation (15.1.2). However, step #2 has two cases. Case #A computes the equivalent single-layer substrate of finite thickness, i.e. $h_{eq} = h_1 + h_2$. Whereas, case #B computes the equivalent single-layer substrate of infinite thickness, i.e. $h_{eq} \to \infty$. The filling-factor q of the symmetrical CPW for case #A and case #B are obtained from equations (9.3.15c) and (9.3.8c) of chapter 9, respectively [B.10]. The filling-factor of ACPW for the case#A is given by equation (9.4.27) of chapter 9, and for the case#B it is 0.5. These results are summarized below with an interchange in the notation of the strip width (W) and slot width (S) applicable to the CPW structure:

For CPW structure:

$$\text{Case\#A}: q = \frac{1}{2}\frac{K(k_1')}{K(k_1)}\frac{K(k_2)}{K(k_2')} \quad \text{(a)}, \quad \text{Case\#B}: q = \frac{1}{2} \quad \text{(b)}$$

$$\text{where,}\ k_1 = \frac{W}{W+2S} = \frac{a}{b}, \quad k_1' = \sqrt{1-k_1^2}, \quad \text{(c)}$$

$$k_2 = \frac{\sinh\left(\dfrac{\pi W}{4h_1}\right)}{\sinh\left(\dfrac{\pi(W+2S)}{4h_1}\right)}, \quad k_2' = \sqrt{1-k_2^2} \quad \text{(d)}. \qquad (15.3.2)$$

For *ACPW structure*:

$$\text{Case\#A}: q = \frac{1}{2}\frac{K(k_1')}{K(k_1)}\frac{K(k_2)}{K(k_2')} \quad \text{(a)}, \quad \text{Case\#B}: q = \frac{1}{2} \quad \text{(b)}$$

$$\text{where} \quad k_1 = \sqrt{\frac{2a(b_1+b_2)}{(b_1+a)(b_2+a)}}, \quad k_1' = \sqrt{1-k_1^2} \quad \text{(c)}$$

$$k_2 = \sqrt{\frac{2\sinh\left(\dfrac{\pi a}{2h_{eq}}\right)\left[\sinh\left(\dfrac{\pi b_2}{2h_{eq}}\right)+\sinh\left(\dfrac{\pi b_1}{2h_{eq}}\right)\right]}{\left[\sinh\left(\dfrac{\pi b_1}{2h_{eq}}\right)+\sinh\left(\dfrac{\pi a}{2h_{eq}}\right)\right]\left[\sinh\left(\dfrac{\pi b_2}{2h_{eq}}\right)+\sinh\left(\dfrac{\pi a}{2h_{eq}}\right)\right]}}, \quad k_2' = \sqrt{1-k_2^2} \quad \text{(d)}. \qquad (15.3.3)$$

In the above equation, the variables a, b_1, and b_2 are related to the strip width (W) and two slot gap widths S_1 and S_2 as $W = 2a$, $b_1 = S_1 + a$, $b_2 = S_2 + a$.

Our main task is to compute the complex capacitance of a multilayer ACPW/CPW with a finite conductor thickness t using the static SDA. Like the case of a microstrip line, the effect of finite conductor thickness t on the computation of the line parameters of the multilayer ACPW/CPW could be accounted for by replacing the strip width, slot-gaps, and width of the ground conductor as follows: $W \to W_{eq} = (W + \Delta W)$, $S_1 \to S_{1eq} = (S_1 - \Delta S_1)$, $S_2 \to S_{2eq} = (S_2 - \Delta S_2)$, $G_1 \to G_{1eq} = (G_1 + \Delta G_1)$, $G_2 \to G_{2eq} = (G_2 + \Delta G_2)$, $\Delta S_1 = (\Delta W/2 + \Delta G_1)$, $\Delta S_2 = (\Delta W/2 + \Delta G_2)$. The changes in the widths of strip conductor and ground conductors can be approximately computed from the empirical relations given by equation (9.5.1) of chapter 9. The static SDA with Galerkin's method, discussed in subsection (14.6.3), can be applied to the multilayer ACPW/CPW. The Green's function of the multilayer structure can be obtained from the TTL method discussed in section (14.4). It is noted that the Green's function is independent of the shape of strip conductors. However, in the present section, we discuss another more accurate approach, i.e. the *two-layer conductor model*, to account for the effect of the conductor thickness t on the line parameters of the multilayer ACPW/CPW [J.49–J.52].

15.3.2 Static SDA of Multilayer ACPW/CPW Using Two-Conductor Model

Figure (15.13a) shows the four-layered ACPW/CPW structure of conductor thickness t. The strip conductor thickness forms another dielectric layer of relative permittivity ε_{rs}. The potential at any layer of a CPW is created by the Fourier transformed *charge density basis functions* $\tilde{\rho}_{s1i}(\beta_n)$, $\tilde{\rho}_{s2i}(\beta_n)$ and $\tilde{\rho}_{s3i}(\beta_n)$ at the width w of the central strip and the ground conductors G_1 and G_2. The charge densities exist only at the conducting strips. In the slot regions, the charge densities are zero. Equations (14.6.73a–c) of chapter 14 provide the expressions for the charge densities and their Fourier transforms are given by equations (14.6.74a–c). In the case of zero conductor thickness, equation (14.6.18) of chapter 14 describes the Fourier domain potential at the plane containing strip conductors. In the range [0, L], the potential at the plane containing the strip conductors is a combination of potential on the conducting strips $\tilde{V}_c(\beta_n)$ and potential in the slot regions $\tilde{V}_d(\beta_n)$.

The *two-layer strip conductor model* accounts for the finite thickness t of the strip conductor. In this case, the charge densities reside at both the *lower surface #1* and *upper surface #2* of the strip conductors. In the present case, the potentials at both the lower and upper interfaces, containing the strip conductors, are treated as the Green's function $\widetilde{G}(\beta_n)$ of the lossy multilayer structure. Due to two layers of charge densities, Green's function is dyadic, and the total charge density $\widetilde{\rho}_s(\beta_n)$ of both layers is described by a column vector. The potentials, dyadic Green's function, and total charge density are given below:

$$\widetilde{V}_c(\beta_n)+\widetilde{V}_d(\beta_n)=\begin{bmatrix}\widetilde{V}_c^1(\beta_n)+\widetilde{V}_d^1(\beta_n)\\ \widetilde{V}_c^2(\beta_n)+\widetilde{V}_d^2(\beta_n)\end{bmatrix} \quad \text{(a)}$$
$$\widetilde{G}(\beta_n)=\begin{bmatrix}\widetilde{G}_{11}(\beta_n) & \widetilde{G}_{12}(\beta_n)\\ \widetilde{G}_{21}(\beta_n) & \widetilde{G}_{22}(\beta_n)\end{bmatrix} \quad \text{(b)} \qquad (15.3.4)$$
$$\widetilde{\rho}_s(\beta_n)=\begin{bmatrix}\sum_{i=1}^{N} a_i^1\widetilde{\rho}_{s1i}(\beta_n)+\sum_{i=1}^{N} b_i^1\widetilde{\rho}_{s2i}(\beta_n)+\sum_{i=1}^{N} c_i^1\widetilde{\rho}_{s3i}(\beta_n)\\ \sum_{i=1}^{N} a_i^2\widetilde{\rho}_{s1i}(\beta_n)+\sum_{i=1}^{N} b_i^2\widetilde{\rho}_{s2i}(\beta_n)+\sum_{i=1}^{N} c_i^2\widetilde{\rho}_{s3i}(\beta_n)\end{bmatrix} \quad \text{(c).}$$

The index number in the suffix of the above equations shows the lower/upper surface of the strip conductors. Using the above equations with equation (14.6.24) of chapter 14, the following relations are obtained for the potentials at both the lower and upper interface:

$$\begin{bmatrix}\widetilde{G}_{11}(\beta_n) & \widetilde{G}_{12}(\beta_n)\\ \widetilde{G}_{21}(\beta_n) & \widetilde{G}_{22}(\beta_n)\end{bmatrix}\begin{bmatrix}\sum_{i=1}^{N} a_i^1\widetilde{\rho}_{s1i}(\beta_n)+\sum_{i=1}^{N} b_i^1\widetilde{\rho}_{s2i}(\beta_n)+\sum_{i=1}^{N} c_i^1\widetilde{\rho}_{s3i}(\beta_n)\\ \sum_{i=1}^{N} a_i^2\widetilde{\rho}_{s1i}(\beta_n)+\sum_{i=1}^{N} b_i^2\widetilde{\rho}_{s2i}(\beta_n)+\sum_{i=1}^{N} c_i^2\widetilde{\rho}_{s3i}(\beta_n)\end{bmatrix}=\begin{bmatrix}\widetilde{V}_c^1(\beta_n)+\widetilde{V}_d^1(\beta_n)\\ \widetilde{V}_c^2(\beta_n)+\widetilde{V}_d^2(\beta_n)\end{bmatrix}. \qquad (15.3.5)$$

Each row of the above equation, corresponding to the lower and upper surfaces of the strip conductors, can be written separately:

$$\widetilde{G}_{11}(\beta_n)\left[\sum_{i=1}^{N} a_i^1\widetilde{\rho}_{s1i}(\beta_n)+\sum_{i=1}^{N} b_i^1\widetilde{\rho}_{s2i}(\beta_n)+\sum_{i=1}^{N} c_i^1\widetilde{\rho}_{s3i}(\beta_n)\right]$$
$$+\widetilde{G}_{12}(\beta_n)\left[\sum_{i=1}^{N} a_i^2\widetilde{\rho}_{s1i}(\beta_n)+\sum_{i=1}^{N} b_i^2\widetilde{\rho}_{s2i}(\beta_n)+\sum_{i=1}^{N} c_i^2\widetilde{\rho}_{s3i}(\beta_n)\right]=\widetilde{V}_c^1(\beta_n)+\widetilde{V}_d^1(\beta_n) \quad \text{(a)}$$
$$\widetilde{G}_{21}(\beta_n)\left[\sum_{i=1}^{N} a_i^1\widetilde{\rho}_{s1i}(\beta_n)+\sum_{i=1}^{N} b_i^1\widetilde{\rho}_{s2i}(\beta_n)+\sum_{i=1}^{N} c_i^1\widetilde{\rho}_{s3i}(\beta_n)\right]$$
$$+\widetilde{G}_{22}(\beta_n)\left[\sum_{i=1}^{N} a_i^2\widetilde{\rho}_{s1i}(\beta_n)+\sum_{i=1}^{N} b_i^2\widetilde{\rho}_{s2i}(\beta_n)+\sum_{i=1}^{N} c_i^2\widetilde{\rho}_{s3i}(\beta_n)\right]=\widetilde{V}_c^2(\beta_n)+\widetilde{V}_d^2(\beta_n) \quad \text{(b).}$$
$$(15.3.6)$$

The above equations are rearranged as follows:

$$\widetilde{P}_{11}(\beta_n)+\widetilde{P}_{12}(\beta_n)+\widetilde{P}_{21}(\beta_n)+\widetilde{P}_{22}(\beta_n)+\widetilde{P}_{31}(\beta_n)+\widetilde{P}_{32}(\beta_n)=\widetilde{V}_c^1(\beta_n)+\widetilde{V}_d^1(\beta_n) \quad \text{(a)}$$
$$\widetilde{P}_{41}(\beta_n)+\widetilde{P}_{42}(\beta_n)+\widetilde{P}_{51}(\beta_n)+\widetilde{P}_{52}(\beta_n)+\widetilde{P}_{61}(\beta_n)+\widetilde{P}_{62}(\beta_n)=\widetilde{V}_c^2(\beta_n)+\widetilde{V}_d^2(\beta_n) \quad \text{(b),} \qquad (15.3.7)$$

where the $\widetilde{P}(\beta_n)$ parameters are defined as

$$
\left.
\begin{aligned}
&\widetilde{P}_{11}(\beta_n) = \widetilde{G}_{11}(\beta_n)\sum_{i=1}^{N} a_i^1\widetilde{\rho}_{s1i}(\beta_n),\ \widetilde{P}_{12}(\beta_n) = \widetilde{G}_{12}(\beta_n)\sum_{i=1}^{N} a_i^2\widetilde{\rho}_{s1i}(\beta_n)\\
&\widetilde{P}_{21}(\beta_n) = \widetilde{G}_{11}(\beta_n)\sum_{i=1}^{N} b_i^1\widetilde{\rho}_{s2i}(\beta_n),\ \widetilde{P}_{22}(\beta_n) = \widetilde{G}_{12}(\beta_n)\sum_{i=1}^{N} b_i^2\widetilde{\rho}_{s2i}(\beta_n) \qquad \text{(a)}\\
&\widetilde{P}_{31}(\beta_n) = \widetilde{G}_{11}(\beta_n)\sum_{i=1}^{N} c_i^1\widetilde{\rho}_{s3i}(\beta_n),\ \widetilde{P}_{32}(\beta_n) = \widetilde{G}_{12}(\beta_n)\sum_{i=1}^{N} c_i^2\widetilde{\rho}_{s3i}(\beta_n)\\
&\widetilde{P}_{41}(\beta_n) = \widetilde{G}_{21}(\beta_n)\sum_{i=1}^{N} a_i^1\widetilde{\rho}_{s1i}(\beta_n),\ \widetilde{P}_{42}(\beta_n) = \widetilde{G}_{22}(\beta_n)\sum_{i=1}^{N} a_i^2\widetilde{\rho}_{s1i}(\beta_n)\\
&\widetilde{P}_{51}(\beta_n) = \widetilde{G}_{21}(\beta_n)\sum_{i=1}^{N} b_i^1\widetilde{\rho}_{s2i}(\beta_n),\ \widetilde{P}_{52}(\beta_n) = \widetilde{G}_{22}(\beta_n)\sum_{i=1}^{N} b_i^2\widetilde{\rho}_{s2i}(\beta_n) \qquad \text{(b).}\\
&\widetilde{P}_{61}(\beta_n) = \widetilde{G}_{21}(\beta_n)\sum_{i=1}^{N} c_i^1\widetilde{\rho}_{s3i}(\beta_n),\ \widetilde{P}_{62}(\beta_n) = \widetilde{G}_{22}(\beta_n)\sum_{i=1}^{N} c_i^2\widetilde{\rho}_{s3i}(\beta_n)
\end{aligned}
\right\} \qquad (15.3.8)
$$

The charge density basis functions at both the surfaces of the strip conductors are confined only on the strip conductors and are zero outside the strips. The unknown coefficients $a_i^p, b_i^p, c_i^p; p = 1, 2$ associated with the expansions of the charge distribution functions at lower surface $p = 1$ and upper surface $p = 2$ of the strip conductors are determined using the Galerkin's process discussed in subsection (14.6.3) of chapter 14. For the simplicity in computation, the *testing functions* $\widetilde{\rho}_{t1j}(\beta_n)$, etc., and *basis functions* $\widetilde{\rho}_{s1i}(\beta_n)$, etc., are taken identical, i.e. $\widetilde{\rho}_{t1j}(\beta_n) = \widetilde{\rho}_{s1i}(\beta_n)$, $\widetilde{\rho}_{t2j}(\beta_n) = \widetilde{\rho}_{s2i}(\beta_n)$, $\widetilde{\rho}_{t3j}(\beta_n) = \widetilde{\rho}_{s3i}(\beta_n)$. The six numbers of basis functions, three for the lower surface and three for the upper surface of the strip conductors, are taken as the testing functions. Resulting six equations are solved with the help of Parseval's identity to get the charge distribution expansion coefficients $a_i^p, b_i^p, c_i^p; p = 1, 2$. The elements of the dyadic Green's function used in the above equation are evaluated using the TTL techniques.

Elements of Dyadic Green's Function

The evaluation of Green's function of Fig (15.13a), using the TTL technique for the case of zero conductor thickness, is discussed in subsection (14.4.2) of chapter 14. Figure (15.13b) shows the equivalent line sections, corresponding to each dielectric layer of the multilayer ACPW/CPW, with two identical current sources located at the lower and upper surfaces of the strip conductors. The current sources located at the planes $(P_1 - Q_1)$ and $(P_2 - Q_2)$ are separated by a distance t. These are due to identical charge densities over the surfaces of the strip conductors. The dyadic Green's function could be treated with the help of the 2-port network of the impedance matrix. The potentials at the ports #1 and port #2 correspond to the elements $\widetilde{G}_{11}(\beta_n)$ and $\widetilde{G}_{22}(\beta_n)$ of the dyadic Green's function. These are due to the current sources I_{s1} and I_{s2} ($I_{s1} = I_{s2} = I_s$). The elements of the dyadic Green's function are obtained as follows:

$$
\left.
\begin{aligned}
&\widetilde{G}_{11}(\beta_n) = \frac{1}{\beta_n\left(Y_{11}^{+} + Y_{11}^{-}\right)} \quad \text{(a)},\quad Y_{11}^{+}(\beta_n) = \varepsilon_{r2}\left[\frac{\varepsilon_{r2} + \varepsilon_{r1}\coth(\beta_n h_1)\coth(\beta h_2)}{\varepsilon_{r1}\coth(\beta_n h_1) + \varepsilon_{r2}\coth(\beta_n h_2)}\right] \qquad \text{(b)}\\
&Y_{11}^{-}(\beta_n) = \varepsilon_s\left[\frac{\varepsilon_s + \varepsilon_{r3}\left[\dfrac{\varepsilon_{r3} + \varepsilon_{r4}\coth(\beta_n h_3)\coth(\beta_n h_4)}{\varepsilon_{r3}\coth(\beta_n h_3) + \varepsilon_{r4}\coth(\beta_n h_4)}\right]\coth(\beta_n t)}{\varepsilon_s\coth(\beta_n t) + \varepsilon_{r3}\left[\dfrac{\varepsilon_{r3} + \varepsilon_{r4}\coth(\beta_n h_3)\coth(\beta_n h_4)}{\varepsilon_{r3}\coth(\beta_n h_3) + \varepsilon_{r4}\coth(\beta_n h_4)}\right]}\right] \qquad \text{(c).}
\end{aligned}
\right\} \qquad (15.3.9)
$$

$$\tilde{G}_{22}(\beta_n) = \frac{1}{\beta_n\left(Y_{22}^{+} + Y_{22}^{-}\right)} \quad \text{(a)}, \quad Y_{22}^{-}(\beta_n) = \varepsilon_{r3}\left[\frac{\varepsilon_{r3} + \varepsilon_{r4}\coth(\beta_n h_3)\coth(\beta_n h_4)}{\varepsilon_{r3}\coth(\beta_n h_3) + \varepsilon_{r4}\coth(\beta_n h_4)}\right] \quad \text{(b)}$$

$$Y_{22}^{+}(\beta_n) = \varepsilon_s\left[\frac{\varepsilon_s + \varepsilon_{r2}\left[\dfrac{\varepsilon_{r2} + \varepsilon_{r1}\coth(\beta_n h_1)\coth(\beta_n h_2)}{\varepsilon_{r1}\coth(\beta_n h_1) + \varepsilon_{r2}\coth(\beta_n h_2)}\right]\coth(\beta_n t)}{\varepsilon_s\coth(\beta_n t) + \varepsilon_{r2}\left[\dfrac{\varepsilon_{r2} + \varepsilon_{r1}\coth(\beta_n h_1)\coth(\beta_n h_2)}{\varepsilon_{r1}\coth(\beta_n h_1) + \varepsilon_{r2}\coth(\beta_n h_2)}\right]}\right] \quad \text{(c)}. \quad (15.3.10)$$

The cross elements of the Green's function $\tilde{G}_{12}(\beta_n)$ and $\tilde{G}_{21}(\beta_n)$ at the port #1 and port #2, due to the current sources I_{s2} and I_{s1} at the ports # 2 and 1 are obtained as

$$\tilde{G}_{12}(\beta_n) = \tilde{G}_{22}(\beta_n)\left[\frac{\varepsilon_{rs}/\sinh(\beta_n t)}{\varepsilon_{rs}\coth(\beta_n t) + Y_{11}^{+}(\beta_n)}\right] \quad \text{(a)}$$

$$\tilde{G}_{21}(\beta_n) = \tilde{G}_{11}(\beta_n)\left[\frac{\varepsilon_{rs}/\sinh(\beta_n t)}{\varepsilon_{rs}\coth(\beta_n t) + Y_{22}^{-}(\beta_n)}\right] \quad \text{(b)}. \quad (15.3.11)$$

Static Line Parameters of Multilayer ACPW/CPW

The line capacitance *per unit length (p.u.l.)* of a multilayer ACPW/CPW given by equation (14.6.70) of chapter 14 is modified to accommodate charges on both the lower and upper surfaces of the strip conductors [B.4, J.49]:

$$\frac{C}{\varepsilon_0} = \sum_{i=1}^{N}\left(a_i^1 Q_i + a_i^2 Q_i\right). \quad (15.3.12)$$

The total charges Q_i at each surface of the conducting strips are computed from equation (14.2.32) of chapter 14. Further, the above-given line capacitance is used to compute the complex relative effective permittivity and the characteristic impedance of the ACPW/CPW structure:

$$\varepsilon_{reff}^{*} = \frac{C_d^{*}(\varepsilon_r^{*})}{C_a(\varepsilon_r = 1)} \quad \text{(a)}$$

$$Z_0^{*} = \frac{1}{c\sqrt{C_d^{*}(\varepsilon_r^{*})C_a(\varepsilon_r = 1)}} \quad \text{(b)}, \quad (15.3.13)$$

where c is the velocity of EM-wave in the free space. The above equations incorporate the effect of finite strip conductor thickness on ε_{reff}^{*} and Z_0^{*}.

In the above formulation, the strip conductors are treated as perfect conductors. However, the strip conductors have finite conductivity. At the lower end of the microwaves, the magnetic field penetrates the finite conductivity strip conductors. The penetration of the magnetic field creates the *artificial relative effective permeability* $\mu_{reff}(\delta_s)$ in a dielectric medium [J.50]. The effective permeability due to the skin-depth is computed on the air-substrate as follows:

$$\mu_{reff}(\delta_s) = \frac{L_0(\delta_s)}{L_0(\delta_s = 0)} = \frac{C_0(\delta_s = 0, \varepsilon_r = 1)}{C_0(\delta_s, \varepsilon_r = 1)}, \quad (15.3.14)$$

where $L_0(\delta_s)$ and $L_0(\delta_s = 0)$ are the line inductance on the air-substrate with and without skin-depth, respectively. The corresponding line capacitances are $C_0(\delta_s, \varepsilon_r = 1)$ and $C_0(\delta_s = 0, \varepsilon_r = 1)$. The skin-depth penetration reduces the effective width of central strip conductor and ground conductors to $(W - \delta_s)$,$(G_1 - \delta_s/2)$, and $(G_2 - \delta_s/2)$. It increases the slot gaps to $S_1 + \delta_s$ and $S_2 + \delta_s$. Thus, penetration of the magnetic field decreases the line capacitance. At lower frequency, reduction in the line capacitance is more, resulting in $\mu_{reff}(\delta_s) >> 1$. Finally, penetration of the magnetic field increases both the ε_{reff}^{*} and Z_0^{*} of the line structure with a decrease in the operating frequency. Equations (15.3.13a, b) compute the quasi-static ε_{reff}^{*} and Z_0^{*} at the lower end of microwaves:

$$\varepsilon_{reff}^{*}(\delta_s) = \frac{C_d^{*}(\varepsilon_r^{*})}{C_a(\varepsilon_r = 1)}\mu_{reff}(\delta_s) \quad \text{(a)}$$

$$Z^{*}(\delta_s) = \frac{1}{c}\sqrt{\frac{\mu_{reff}(\delta_s)}{C_d^{*}(\varepsilon_r^{*})C_a(\varepsilon_r = 1)}} \quad \text{(b)}. \quad (15.3.15)$$

The skin-effect phenomenon is responsible for the low-frequency dispersion. At higher frequency, its effect gradually disappears and inhomogeneity-dependent dispersion appears.

15.3.3 Dispersion Models of Multilayer ACPW/CPW

The dispersion model of the standard CPW structure, given by equations (9.6.2)–(9.6.4) of chapter 9, can be used to get the dispersion in the multilayer CPW/ACPW structures. The relative permittivity ε_r and substrate thickness h are replaced by the equivalent relative permittivity ε_{req} and equivalent substrate thickness $h_{eq} = h_1 + h_2$ using the concept of the SLR formulation. To get dispersion in the ACPW, it should be converted to an equivalent CPW by replacing the unequal slot-gaps S_1 and S_2 with an equivalent slot-gap S_{eq}. The dispersion models for both the multilayer CPW and ACPW are summarized below [J.50, J.51]:

$$\begin{aligned}
&\varepsilon_{reff}(f, t, \mu_{reff}(\delta_s)) \\
&= \left[\sqrt{\mu_{reff}(\delta_s)\varepsilon_{reff}(f=0,t)} + \frac{\sqrt{\varepsilon_{req}} - \sqrt{\mu_{reff}(\delta_s)\varepsilon_{reff}(f=0,t)}}{(1 + aF^{-b})}\right]^2 && \text{(a)} \\
&\text{where,} \quad F = \frac{f}{f_{TE}}, \quad f_{TE} = \frac{c_0}{4h_{eq}\sqrt{\varepsilon_{req} - 1}}, \quad b = 1.8 && \text{(b)} \\
&\log_e(a) = u \log_e\left(\frac{W}{S}\right) + v && \text{(c)} \\
&u = 0.54 - 0.64q + 0.015q^2, v = 0.43 - 0.86q + 0.540q^2 && \text{(d)} \\
&q = \log_e\left(\frac{W}{h_{eq}}\right) && \text{(e).}
\end{aligned} \tag{15.3.16}$$

The accuracy of the model is within 5% for a range of parameters discussed in subsection (9.6.3) of chapter 9. The dispersive characteristic impedance is computed by using the following expression:

$$\begin{aligned}
&Z_0(f, \varepsilon_{req}, h_{eq}) \\
&= Z_0(f = 0, \varepsilon_{req} = 1, h_{eq})\sqrt{\frac{\mu_{reff}(\delta_s)}{\varepsilon_{reff}(f, \mu_{reff}(\delta_s) = 1)}}.
\end{aligned} \tag{15.3.17}$$

The quasi-static characteristic impedance of a CPW on the air-substrate $Z_0(f = 0, \varepsilon_{req} = 1, h_{eq})$ is computed by using equation (15.3.15b).

To use the above equations to a multilayer ACPW, the slot-gap S is replaced by the equivalent slot-gap S_{eq} [J.7, J.50].

$$\begin{aligned}
&S_{eq} = WF_1 \times S_1 + WF_2 \times S_2 && \text{(a)} \\
&\text{where,} \quad WF_1 = \frac{C_{01}(S_1, \varepsilon_r = 1)}{C_{01}(S_1, \varepsilon_r = 1) + C_{02}(S_2, \varepsilon_r = 1)} && \text{(b)} \\
&\qquad\quad WF_2 = \frac{C_{02}(S_2, \varepsilon_r = 1)}{C_{01}(S_1, \varepsilon_r = 1) + C_{02}(S_2, \varepsilon_r = 1)} && \text{(c).}
\end{aligned} \tag{15.3.18}$$

In the above equations, WF_1 and WF_2 are the *weighting-factors* corresponding to the unequal slot-gaps S_1 and S_2. The line capacitances $C_{01}(S_1, \varepsilon_r = 1)$ and $C_{02}(S_2, \varepsilon_r = 1)$ are computed by replacing all layers with the air-substrate and ACPW is treated as the CPW for both the slot-gap width S_1 and S_2, respectively.

15.3.4 Loss Models of Multilayer ACPW/CPW

The case #B of SLR process, with filling-factor $q = 1/2$, is used to get the equivalent single-layer ACPW/CPW and the dielectric loss is computed using the following expression [J.53, J.54]:

$$\alpha_d = \frac{8.686\pi}{\lambda_0}\frac{\varepsilon'_{req}}{\sqrt{\varepsilon'_{reff}}}\frac{(\varepsilon'_{reff} - 1)}{(\varepsilon'_{req} - 1)}\tan\delta_{eq}. \tag{15.3.19}$$

The conductor losses of a multilayer ACPW/CPW can be computed using the Wheeler's inductance rule:

$$\begin{aligned}
&\alpha_c = \frac{\pi}{\lambda_0}\sqrt{\varepsilon'_{reff}}\frac{\Delta Z_0}{Z_0(\varepsilon_{r1} = \varepsilon_{r1} = \cdots = 1, h_1, \ldots h_4, W, \delta_s, t)} && \text{(a)} \\
&\Delta Z_0 = Z_0(\varepsilon_{r1} = \varepsilon_{r2} = \cdots = 1, W - \delta_s, S_1 + \delta_s, S_2 + \delta_s, t - \delta_s) \\
&\qquad - Z_0(\varepsilon_{r1} = \varepsilon_{r2} = \cdots = 1, W, S_1, S_2, t) && \text{(b),}
\end{aligned} \tag{15.3.20}$$

where δ_s is the skin-depth of the strip conductor. The characteristic impedance and also differential characteristic impedance ΔZ_0 of the ACPW are computed on the air-substrate. For the symmetrical CPW, the perturbation method is also used to compute the conductor loss [J.51].

The SLR does not compute the imaginary part of the characteristic impedance $Z_0^*(f)$. The circuit model (CM) is used to compute more accurately $\varepsilon_{reff}(f), Z_0^*(f)$, $\alpha_d(f), \alpha_c(f)$ for a multilayer lossy ACPW/CPW. The process of developing the CM by using the RLCG parameters is discussed in section (8.5) of chapter 8. The results of the SLR process and the CM are compared against the EM-Simulators. Table (15.1) shows the % average deviation in the results obtained from the EM-Simulator CST,

Table 15.1 % Average deviation in CPW dynamic line parameters computed by the CST, SLR, and circuit model (CM) against the reference results of HFSS ($W = 16\ \mu m$, $S = 12\ \mu m$, $\sigma = 4.1 \times 10^7$ S/m, 1–100 GHz).

	ε_{eff}	Z_0	α_d	α_c	α_T
Standard CPW					
CST	0.56	3.63	0.92	11.61	11.58
SLR	1.49	1.50	4.94	9.97	–
CM	1.41	1.45	–	–	9.86
Composite substrate CPW					
CST	0.56	2.15	0.13	14.57	14.57
SLR	1.84	0.71	4.22	12.63	–
CM	1.78	0.66	–	–	12.46
Suspended substrate CPW					
CST	0.56	1.96	0.29	15.47	15.32
SLR	1.82	0.88	4.24	12.61	–
CM	1.71	0.81	–	–	12.44

SLR, and CM against the reference results of the HFSS. The results of the characteristic impedance and conductor loss, using the SLR and CM, show less deviation as compared to the results of CST, another EM-Simulator. Figure (15.14a–d) further compares the results of $\varepsilon_{reff}(f)$, $\mathrm{Re}\,Z_0^*(f), \mathrm{Im}Z_0^*(f)$, and total loss $\alpha_T(f)$ against the results of EM-Simulators, HFSS, and CST for a shielded CPW on the composite substrate: $\varepsilon_{r1} = 9.8$, $\varepsilon_{r2} = 12.9$, $\varepsilon_{r4} = 1$, $\tan\delta_1 = 0.0001$, $\tan\delta_2 = 0.0003$, $h_1 = 127\,\mu m$, $h_2 = 254\,\mu m$, $t = 3\,\mu m$, $h_3 = 0$, $h_4 = 10h_{eq}$, $h_{eq} = h_1 + h_2$ $W = 16\,\mu m$, $S = 12\,\mu m$, $\sigma = 4.1 \times 10^7$ S/m, $t = 3\,\mu m$. The computation and simulation, using both HFSS and CST, are performed over a frequency range up to 100 GHz. The variation in results of the CST and CM is within 2.0%.

15.4 Further Consideration of SLR Formulation

The above-discussed SLR formulation of the multilayer microstrip and CPW structures considers only lossy isotropic substrates. The anisotropy in the multilayer structures has been also accounted for both the microstrip and CPW by computing the equivalent isotropic substrates for each layer. Next, the usual SLR process is applied over the equivalent isotropic multilayer line structures [J.50]. However, more accurate conversion of the anisotropic layer to an equivalent isotropic layer requires an accounting of W/h dependence for the microstrip case, and W/h, S/h dependence for the CPW case [J.7].

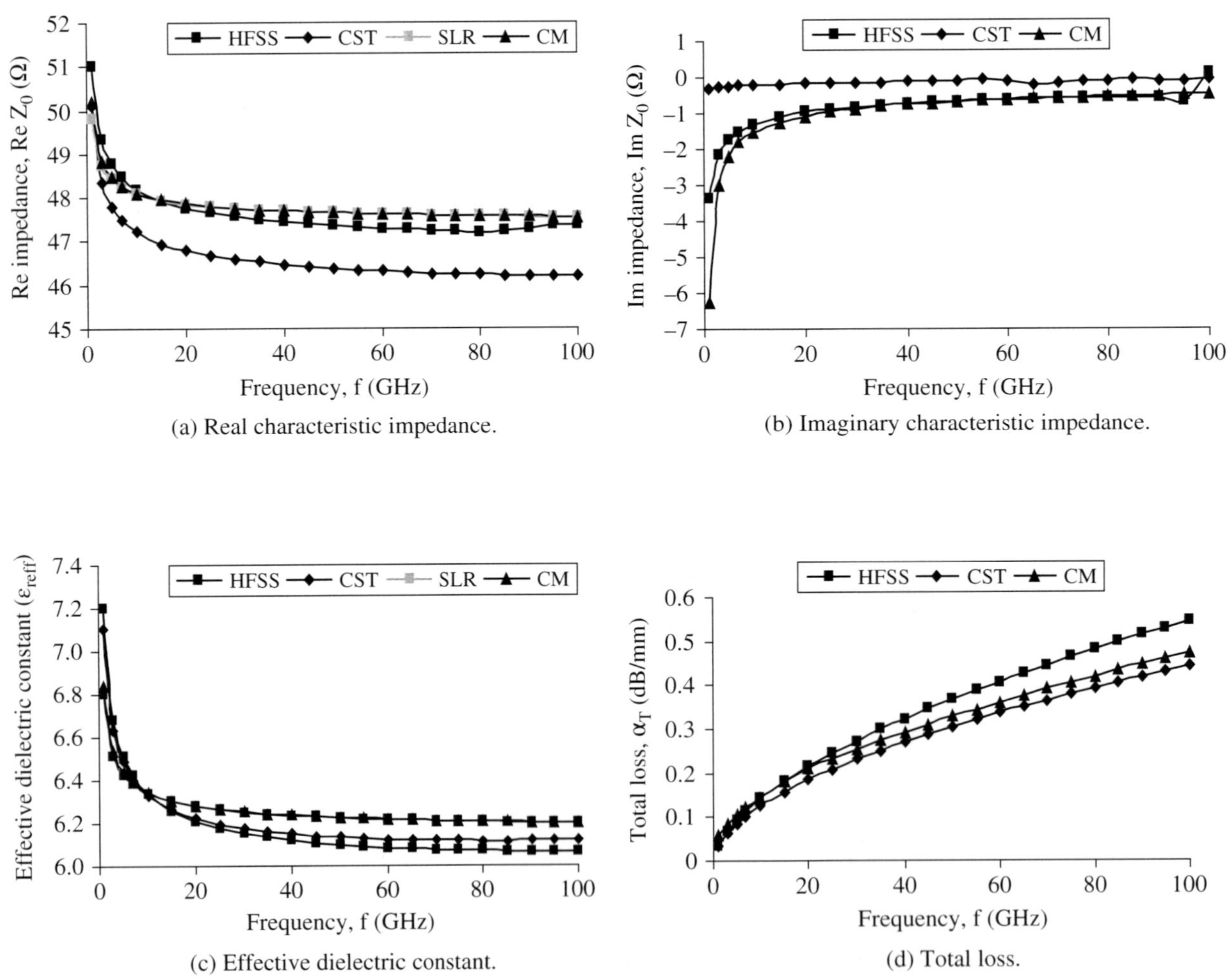

Figure 15.14 Computed and simulated line parameters of shielded CPW on the composite substrate.

The multilayer structure can also contain the layers of magnetic material. For instance, a leaky-wave antenna, using a coupled CRLH transmission line on YIG substrate with a three-layered RT Duriod superstrate, has been developed. The SLR process has been applied to such structure by calculating both the equivalent permittivity and equivalent permeability. The equivalent permeability is computed by replacing ε_r with $1/\mu_r$ in the SLR formulation. The permeability of the ferrite layer is computed by Polder's formula [J.13]:

$$\mu = \mu_0\left[1 + \frac{\omega_m\omega_0}{\omega_0^2 - \omega^2}\right],$$
$$\text{where } \omega_m = \gamma M_s, \omega_0 = \gamma H_0, \text{ and } \frac{\gamma}{2\pi} = 2.8\,\text{MHz/Oe}. \tag{15.4.1}$$

The calculated values of ε_{reff} and μ_{reff} help to determine the characteristic impedance and phase constant of the microstrip line on the magneto-electric substrate:

$$Z_0 = Z_0(\varepsilon_r = 1, \mu_r = 1, h_{eq}, W)\sqrt{\frac{\mu_{reff}}{\varepsilon_{reff}}} \quad \text{(a)}$$
$$\beta = 2\pi f\sqrt{\varepsilon_{reff}\mu_{reff}} \quad \text{(b).} \tag{15.4.2}$$

The dispersion and losses in the line can be computed using the SLR process.

The SLR process has also been used to model the MIS, Schottky, and other semiconductor-based line structures [J.55–J.57]. Furthermore, the SLR process has been used to compute microstrip patch capacitance, and to model the resonating patches under the layered medium condition. Such multilayer resonating patches are modeled by the improved cavity model, called the *modified Wolff model* (MWM) [J.58–J.60]. It is briefly discussed in subsection (18.6.2) of chapter 18. The SLR process has also been used to model the step discontinuity and open-end discontinuity in the multilayer microstrip lines [J.61, J.62]. It shows that the SLR is a simple and effective process to model several planar lines structures, discontinuities, etc., under the multilayer conditions. Such an approach could be useful to develop CAD process of the multilayer substrates.

References

Books

B.1 Tatsuo, T. (Editor): The waveguide model for the analysis of microstrip discontinuities, *Numerical Techniques for Microwave and Millimeter-Wave Passive Structures*, John Wiley & Sons, New York, 1989.

B.2 Gupta, K.C.; Ramesh, G.; Bahl, I.; Bhartia, P.: *Microstrip Lines and Slot Lines*, 2nd Edition, Artech House, Boston, 1996.

B.3 Mirshekar-Syahkal, D.: *Spectral Domain Method for Microwave Integrated Circuits*, Research Studies Press Ltd., John Wiley & Sons, New York, 1990.

B.4 Cam, N.: *Analysis Methods for RF, Microwave, and Millimetre –Wave Planar Transmission Lines*, John Wiley & Sons, New York, 2000.

B.5 Grzyb, J.: Millimeter-wave interconnects, *Advanced Millimeter-wave Techniques*, Editors: Liu, D; Gaucher, B.; Pfeiffer, U.; Grzyb, J., John Wiley & Sons, Inc., 2009.

B.6 Hoffmann, R.: *Microwave Integrated Circuit Handbook*, Artech House, Boston, 1985.

B.7 Roslinec, S.: *Algorithms for Computer-Aided Design of Linear Microwave Circuits*, Artech House, London, 1990.

B.8 Syahkal, M.: *Spectral-domain Methods*, Research Studies Press Ltd., John Wiley & Sons INc., New York, 1990.

B.9 Kneppo, I.; Fbian, J.; Bezousek, P.; Hrnicko, P.; Pavel, M.: *Microwave Integrated Circuits*, Chapman & Hall, London, 1994.

B.10 Simon, R.N.: *Coplanar Waveguide Circuits Components and Systems*, John Wiley & Sons, New York, 2001.

Journals

J.1 Jansen, R.H.: *A novel CAD tool and concept compatible with the requirement of multilayer GaAs MMIC technology*, IEEE MTT-S Microwave Symp. Dig., pp. 711–714, 4–6, June 1985, St. Louis, MO.

J.2 Verma, A.K.; Sadr, G.H.: *Unified dispersion model for multilayer microstrip line, IEEE Trans. Microwave Theory Tech.*, Vol. 40, No. 7, pp. 1587–1591, July 1992.

J.3 Verma, A.K.; Bhupal, A.; Rostamy, Z.; Srivastava, G.P.: *Analysis of rectangular patch antenna with dielectric cover, IEICE Trans.*, Vol. E74, pp. 1270–1276, May 1991.

J.4 Verma, A.K.; Bhupal, A.: *Resonance frequency of anisotropic multilayer rectangular microstrip patch using MWM, Electron. Lett.*, Vol.34, No. 4, pp. 323–325, 1998.

J.5 Verma, A.K.; Awasthi, Y.K.; Singh, H.: *Equivalent isotropic relative permittivity of microstrip on the multilayer anisotropic substrate. Int. J. Electron.*, Vol. 96, No.8, pp. 865–875, Aug. 2009.

J.6 Verma, A.K.; Bhupal, A.: *Dielectric loss of multilayer microstrip line, Microw. Opt. Technol. Lett.*, vol. 17, pp. 368–370, Apr. 1998.

J.7 Verma, A.K.; Singh, P.; Matekovits, L.: *Strip-width and slot-gap dependent equivalent isotropic substrate and dispersion characteristics of asymmetric coplanar waveguide, symmetric coplanar waveguide, and micro-coplanar strip line on anisotropic substrates, IEEE Trans. Microwave Theory Tech.*, Vol. 62, No. 10, pp. 2232–2240, Oct. 2014.

J.8 Svacina, J.: *Analysis of multilayer microstrip lines by a conformal mapping method, EEE Trans. Microwave Theory Tech.*, Vol. MTT-40, No.4, pp. 169–172, Apr. 1992.

J.9 Schellenberg, J.M.: *CAD models for suspended and inverted microstrip, EEE Trans. Microwave Theory Tech.*, Vol. MTT-43, No.6, pp. 1241–1252, June 1995.

J.10 Verma, A.K.; Kumar, R.: *A new dispersion model for microstrip line, IEEE Trans. Microwave Theory Tech.*, Vol 46, No.8, pp. 1183–1187, Aug. 1998.

J.11 Awasthi, Y.K; Singh, H.; Sharma, M.; Kumari, S.; Verma, A.K.: *Computer-aided design-based circuit model of microstrip line for terahertz interconnects technology, J. Eng. (IET Pub.)*, doi:10.1049/joe.2017.0078, Vol. 2017, No. 9, pp. 512–526, 2017.

J.12 Bansal, R.; Awasthi, Y.K.; Singh, P.; Verma, A.K.: *Accurate dispersion model for microstrip line up to terahertz frequency range*, Proc. SPIE 8760, International Conference on Communication and Electronics System Design, 87601A (28 January 2013); doi:10.1117/12.2012151.

J.13 Apaydin, N.; Sertel, K.; Volakis, J.L.: *Nonreciprocal and magnetically scanned leaky-wave antenna using coupled CRLH lines, IEEE Trans. Antenna Propagat.*, Vol. 62, No. 6, pp. 2954–2961, June 2014.

J.14 Yun, T.Y.; Chang, K.: *Analysis and optimization of a phase shifter controlled by a piezoelectric transducer, IEEE Trans. Microwave Theory Tech.*, Vol. 50, No. 1, pp. 101–111, Jan. 2002.

J.15 Ohkubo, T.; Onuma, M.; Kitagawa, J.; Kadoya, Y.: *Micro-strip-line-based sensing chips for characterization of polar liquids in the terahertz regime, Appl. Phys. Lett.*, Vol. 88, pp. 212511-1–212511-3, 2006. doi:10.1063/1.2207989.

J.16 Grzyb, J.; Ruiz, I.; Cottet, D.; Troster, G.: *An Investigation of the material and process parameters for thin-film MCM-D and MCM-L technologies up to lOO GHz*, IEEE Electronic Components and Technology Conf., pp. 478–486, 27–30 May 2003, New Orleans, Louisiana, USA.

J.17 *Suspended Substrate Module of ADS*, Advanced Design System ADS Version 15.01, 2015, Keysight EEsof EDA,Southern California.

J.18 Verma, A.K.; Kumar, R.: *New empirical unified dispersion model for shielded-, suspended-, and composite-substrate microstrip line for microwave and mm-wave applications, IEEE Trans.Microwave Theory Tech.*, Vol. 46, No.8, pp. 1187–1191, Aug. 1998.

J.19 Settapong Malisuwan, S.; Sivaraks, J.: *Design of three-layer shielded microstrip line for microwave circuit applications by applying Cole-Cole diagrams, Int. J. Future Comput. Commun.*, Vol. 2, No. 5, pp. 433–436, Oct. 2013. doi:10.7763/IJFCC.2013.V2.200.

J.20 Verma, A.K.; Kumar, R.: *Closed-form dispersion expressions for a shielded microstrip line for millimeter-wave applications, Electron. Lett.*, Vol. 32, No. 15, pp. 1378–1380, Jul. 1996.

J.21 Syahkal, D.M.; Davies, J.B.: *Accurate solution of microstrip and coplanar structures for dispersion and dielectric and conductor losses, IEEE Trans. Microwave Theory Tech.*, Vol. MTT-27, pp. 694–699, July 1979.

J.22 Yamashita, E.; Atsuki, A.: *Analysis of microstrip like transmission lines by non-uniform discretization of integral equations, IEEE Trans. Microwave Theory Tech.*, Vol. MTT-24, pp. 195–200, Apr. 1976.

J.23 Finlay, H.J.; Jansen, R.H.; Jenkins, J.A.; Eddison, I.G.: *Accurate characterization and modeling of transmission lines for GaAs MMIC, IEEE Trans. Microwave Theory Tech.*, Vol. 36, pp. 961–967, June 1988.

J.24 Jackson, D.R.; Alexopoulos, N.G.: *Analysis of planar strip geometries in a substrate-superstrate configuration, IEEE Trans. Antennas. Propagat.*, Vol. AP-34, pp. 1430–1438, Dec. 1986.

J.25 Krischnining, M.; Jansen, R.H.; Koster, N.H.L.: *New aspects concerning the definition of microstrip characteristic impedance as a function of frequency*, IEEE., MTT-S, Int. Microwave Symposium Digest, pp. 305–307, 15–17 June 1982, Dallas, TX, USA.

J.26 Tomar, R.; Antar, Y.M.M; Bhartia, P.: *Matlab-based accurate dispersion models for unshielded suspended microstrip*,10th Inter. Symp. on Antenna Techn. and Applied Electromagnetics and URSI Conference, 2004, pp. 1–4, 2004.

J.27 Verma, A.K.; Sadr, G.H.: *CAD process for synthesis of the multilayer microstrip line*, 23rd European Microwave Conf, pp. 717–719, 4–6 Oct. 1993, Madrid, Spain.

J.28 Schneider, M.V.: *Dielectric loss in integrated microwave circuits, Bell Syst. Tech. J.*, Vol. 48, pp. 2325–2332, Sept. 1969.

J.29 Gilb, J.P.K.; Balanis, C.A.: *Accurate and efficient computation of dielectric losses in multi-level, multi-conductor microstrip for CAD applications, EEE Trans. Microwave Theory Tech.*, Vol. MTT-41, No. 3, pp. 521–530, Mar. 1993.

J.30 Gilb, J.P.K.; Balanis, C.A.: *Fast and accurate computation of dielectric microstrip structures losses in multilayer, multi-conductor*, IEEE., MTT-S, Int. Microwave Symposium Digest, pp. 385–387, 1–5 June 1992, Albuquerque, NM, USA.

J.31 Deventer, V.T.E.; Katehi, P.B.; Cangellaris, A.C.: *High-frequency conductor and dielectric losses in the shielded microstrip lines*, MTT-S, Int. Microwave Symposium Digest, pp. 919–922, 13–15 June 1989, Long Beach, CA, USA.

J.32 Pantice, Z.; Mitra, R.: *Finite Element Matrices for loss calculation in quasi-TEM analysis of microwave transmission lines, Microw. Opt. Technol. Lett.*, Vol. 1, pp. 142–146, June 1988.

J.33 Aksum, M.S.; Morkok, H.: *GaAs on Si substrate for microwave and mm-wave monolithic integration, IEEE Trans. Microwave Theory Tech.*, Vol. 36, 36, pp. 160–162, Jan. 1988.

J.34 Awasthi, Y.K; Singh, H.; Sharma, M.; Kumari, S.; Verma, A.K.: *Computer-aided design-based circuit model of microstrip line for terahertz interconnect technology, IET J. Eng.*, Vol. 2017, No. 9, pp. 512–526, 2017.

J.35 Vermal, A.K.; Bhupal, A.: *Conductor loss of multilayer microstrip line using the single-layer reduction formulation, Microw. Opt. Technol. Lett.*, Vol. 19, No. 1, pp. 20–24, Sept. 1998.

J.36 Verma, A.K.; Nasimuddin Determination of *conductor loss of multilayer coupled microstrip lines for CAD application, Microw. Opt. Technol. Lett.*, Vol. 38, No. 5, pp. 409–418, Sept. 2003.

J.37 Bahl, I.J.; Stuchly, S.S.: *Analysis of microstrip covered with a lossy dielectric, IEEE Trans. Microwave Theory Tech.*, Vol. MTT-28, pp. 104–109, Feb. 1980.

J.38 Kirschning, M.; Jansen, R.H.: *Accurate wide range design equations for parallel-coupled microstrip lines, IEEE Trans. Microwave Theory Tech.*, Vol. MTT-32, No. 1, pp. 83–90, Jan. 1984. Also corrections in Vol. MTT-33, No., pp. 288, 1985.

J.39 Hammerstad, E.; Jensen, O.: *Accurate models for Microstrip Computer-Aided Design*, IEEE., MTT-S, Int. Microwave Symposium Digest, pp. 407–409, 28–30 May 1980, Washington, DC, USA.

J.40 Verma, A.K.; Bhupal, A.: *Unified dispersion model for a composite substrate shielded parallel coupled microstrip lines using SLR technology*, IEEE MTT-S Int. Conf., pp. 711–714, 24–27 July 1995, Rio de Jena Rio, Brazil.

J.41 Krage, M.K.; Haddad, G.I.: *Frequency-dependent characteristic of the microstrip transmission line, IEEE Trans. Microwave Theory Tech.*, Vol. MTT-20, No. 10, pp. 678–688, 1972.

J.42 Gilb, J.P.; Balanis, C.A.: Pulse *distortion on multilayer coupled microstrip lines, IEEE Trans. Microwave Theory Tech.*, Vol. MTT-37, No.10, Oct. 1989.

J.43 Kitazawa, T.; Hayashi, Y.: *Propagation characteristics of strip lines with multilayered anisotropic media, IEEE Trans. Microwave Theory Tech.*, Vol. MTT-31, pp. 429–433, June 1983.

J.44 Verma, A.K.; Nasimuddin, Ojha, R. K.: *Synthesis and analysis of optimum directivity suspended microstrip coupler*, Asia Pacific Microwave Conf, Sydney, pp. 1–4, Dec. 2000.

J.45 Verma, A.K.; Nasimuddin, Bhupal, A.: *Dielectric loss of multilayer coupled microstrip line using SLR formulation, Microw. Opt. Technol. Lett.*, Vol. 31, No. 1, pp. 65–67, Oct. 2001.

J.46 Hasnain, G.; Dienes, A.; Whinnery, J.R.: *Dispersion of picosecond pulses in a coplanar transmission lines, IEEE Trans. Microwave Theory Tech.*, Vol. MTT-34, pp. 738–741, 1986.

J.47 Verma, A.K.; Nasimuddin, Singh, H.: *Dielectric loss of multilayer coplanar waveguide using the single-layer reduction (SLR) formulation, 2005 Asia-Pacific Microwave Conference Proceedings*, 4–7 December 2005, Suzhou, China.

J.48 Majumdar, P.; Verma, A.K.: *CAD models of dispersion and losses for multilayer coplanar the waveguide*, Asia-Pacific Microwave Conference, APMC, N. Delhi, 2016.

J.49 Tripathi, V.K.; Kollipara, R.T.: *Quasi-TEM spectral domain analysis of thick microstrip for microwave and digital integrated circuits, Electron. Lett.*, Vol. 25, No. 18, pp. 1253–1254, Aug. 1989.

J.50 Verma, A.K.; Singh, P.; Ritu Bansal, R.: *Modeling for dispersion and losses of multilayer asymmetric CPW on iso/anisotropic substrate, Progr. Electromagn. Res. B (PIER-B)*, Vol. 48, pp. 395–419, Feb. 2013.

J.51 Verma, A.K.; Singh, P.; Ritu Bansal, R.: *Computation of static and frequency-dependent line parameters of multilayer CPW using static SDA and single-layer reduction method, Int. J. RF Microwave Computer-Aided Eng.*, Vol. 24, No. 1, pp. 18–29, Jan. 2014, doi:10.1002/mmce.20708.

J.52 Singh, P.; Verma, A.K.; Ritu Bansal, R.: *Analysis of multilayer coplanar waveguide with finite conductor thickness, Int. J. Microwave Optical Technol. (IJMOT)*, Vol. 7, No. 6, pp. 380–387, Nov. 2012.

J.53 Singh, P.; Verma, A.K.: *Dielectric loss computation of multilayer coplanar waveguide*, International

Conference on Communications and Signal Processing (ICCSP), 10–12 February 2011, National Institute of Technology Calicut, Kerala, India.

J.54 Verma, A.K.; Singh, P.; Ritu Bansal, R.: *Loss computation of multilayer coplanar waveguide using single-layer reduction method, Int. J. Electromagnetics Appl. (IJEA)*, Vol. 2, No. 6, pp. 174–181, Nov. 2012.

J.55 Verma, A.K.; Nasimuddin; Sharma, E.K.: *Propagation characteristics of suspended MIS coupled slow wave microstrip lines, Microw. Opt. Technol. Lett.*, Vol. 34, No. 6, pp. 403–405, Sept. 2002.

J.56 Verma, A.K.; Nasimuddin: *Propagation characteristics of Schottky contact suspended slow-wave microstrip line, IEEE Microwave Wireless Comp. Lett.*, Vol. 11, No. 9, pp. 385–387, Sept. 2001.

J.57 Verma, A.K.; Nasimuddin; Sharma, E.K.: *Analysis and circuit model of a multilayer semiconductor slow-wave microstrip line, IEE Proc.-Microwave Antennas Propag.*, Vol. 151, No. 5, pp. 441–449, Oct. 2004.

J.58 Verma, A.K.; Rostamy, Z.: *Static capacitance of some multilayered microstrip capacitors IEEE Trans. Microwave Theory Tech.*, Vol. MTT-43, No. 5, pp. 1144–1152, May 1995.

J.59 Verma, A.K.; Rostamy, Z.: *Resonance frequency of the uncovered and covered rectangular microstrip patch using modified Wolff model*, *IEEE Trans. Microwave Theory Tech.*, Vol. 41, No.1, pp. 109–116, 1993.

J.60 Verma, A.K.; Nasimuddin: *Input impedance of rectangular microstrip patch antenna with iso/anisotropic substrate-superstrate, IEEE Microwave Wireless Comp. Lett.*, Vol. 11, No. 11, pp. 456–458, 2001.

J.61 Verma, A.K.; Sadr, H.: *Determination of the open-end discontinuity and resonance frequency of a suspended microstrip using the SLR technique, Microw. Opt. Technol. Lett.*, Vol. 7, No. 11, pp. 532–535, Aug. 1994.

J.62 Verma, A.K.; Singh, H.; Awasthi, Y.K. *Circuit model of multilayer microstrip steps discontinuity using single – layer reduction formulation, Electromagnetics*, Vol. 29, No. 6, pp. 483–498, Aug. 2000.

16

Dynamic Spectral Domain Analysis

Introduction

The microstrip, slot line, and CPW structures support the hybrid mode wave propagation. The hybrid mode is treated as a linear combination of the LSE and LSM modes. It is equally treated as a linear combination of the TE and TM modes. The TE and TM modes are defined with respect to an axis that is normal to the interface of an inhomogeneous medium. It is discussed in chapter 7. Several approaches, known as the *full-wave analysis*, both in the space domain and in the Fourier domain have been used to analyze the hybrid modes supporting microstrip and CPW structures. Itoh and Mittra [B.1, J.1–J.6] developed the full-wave method, called the *spectral domain analysis* (SDA*)*. It transforms all field components and wave equations in the Fourier domain. At the final stage, it uses the *Galerkin's method* to obtain the propagation constants of the hybrid modes. The *dynamic spectral domain analysis* is a more general method, as against the *static spectral domain analysis* discussed in chapter 14. It provides the frequency-dependent line parameters. However, it does not account for the conductor thickness and conductivity of the strip conductors. It is used to solve the diverse field problems of the planar transmission line structures, planar antennas, and scattering, etc. [B.1–B.9]. The SDA is an analytically simpler and computationally efficient technique. In the process of SDA, the space domain convolution type coupled integral equation is replaced by the simple algebraic equation. However, considerable analytical processing is needed to get Green's function in the Fourier domain.

Objectives

- *To present a general discussion on the SDA.*
- *To obtain Green's function of the boxed planar lines.*
- *To obtain propagation constant using Galerkin's method*
- *To obtain propagation constants of the symmetrically coupled microstrip.*
- *To apply the immittance approach to the multilayer planar lines.*

16.1 General Discussion of SDA

Figure (16.1a–f) shows the shielded single and coupled CPW and microstrip structures. Figure (16.1a) is a conductor-backed CPW (CB-CPW) structure. The strip conductors are located at the interface of two isotropic media, and the conductor thickness is zero. The Y-axis is the plane of symmetry. For the fundamental mode of propagation, it is the *magnetic wall*; as the electric field lines that originated from the central conductor are symmetrically terminated on the ground conductors. Figure (16.1b) is a boxed microstrip that is a special case of the CPW with coplanar ground conductors width G = 0.

Figure (16.1c,d) shows the *edge coupled* CB-CPW and the edge coupled microstrip lines. Again, all strip conductors are located at the interface of the two isotropic media. The *y*-axis is the plane of symmetry. It could be either a magnetic wall (MW) for the even hybrid mode or an electric wall (EW) for the odd hybrid mode. Figure (16.1e, f) shows the *broadside coupled* CB-CPW and the broadside-coupled microstrip lines. In this case, the strip conductors are located at two different levels. There are two planes of symmetries – one with respect to the *y*-axis and another with respect to the *x*-axis. The (x–x′)-plane could be either a MW for the even hybrid mode or an EW for the odd hybrid mode. Likewise, (y–y′)-plane could be either a MW or the EW. On the application of the concept of symmetry, all the abovementioned structures could be reduced to a slot line kind of the generic structure shown in Fig (16.2a).

The generic *shielded slot line* of Fig (16.2a) could be treated as the half-structure of the structure shown in Fig (16.1a). To get the Fig 16.2a, the *y*-axis of Fig (16.1a) is taken as the *MW*. Thus, the segment OC of Fig (16.2a) is a *MW* and remaining three walls, i.e. the segments OA, AB, BC are considered as the EWs. The conductor DE (=W) of is the half-width strip conductor and FG (=G) is the ground conductor of Fig (16.1a). The EF

Introduction to Modern Planar Transmission Lines: Physical, Analytical, and Circuit Models Approach, First Edition. Anand K. Verma.

(=S), i.e. the slot-width, is treated as a MW. The strip and ground conductors are perfect conductors.

Figure (16.2a) also, treats the *shielded microstrip* line of Fig (16.1b). In this case, all four external walls are the EWs. The dielectric interface DE = FG forms the MW and EF = (S) is the strip conductor of width 2W. The coupled structures can be put in the format of Fig (16.2a). However, the edge-coupled CB-CPW requires an additional strip conductor in the generic structure. In the case of the broadside-coupled structure of Fig (16.1e, f), the plane OA along *x*-axis, shown in Fig (16.2a), is the MW for the even hybrid mode. However, it is treated as the EW for the odd hybrid mode. In the next section, the dynamic SDA analysis is carried out for the generic structure shown in Fig (16.2a).

The *quasi-TEM* description of the microstrip line, presented in chapter 14, ignores the transverse current density J_x on the strip conductor. Likewise, the quasi-TEM description of the CPW equally ignores the longitudinal electric field component E_z in the slot region. However, for a realistic description of the microstrip and CPW, the transverse current density J_x and longitudinal electric field component E_z cannot be ignored. The transverse current and longitudinal electric field components generate the hybrid mode on the microstrip and CPW structures. The hybrid mode nature becomes significant when the transverse dimension of the line structure becomes comparable to the guided wavelength.

The SDA considers the current and field components in the Fourier domain, shown by a tilde sign (~). In our discussion, the longitudinal current density $\tilde{J}_z$ and the transverse current density $\tilde{J}_x$, also the longitudinal electric field component $\tilde{E}_z$ and the transverse electric field component $\tilde{E}_x$, are located in the same plane, i.e. at the interface of two media. The hybrid modes are generated in two ways. In the *first case*, a pair of the current densities $(\tilde{J}_z, \tilde{J}_x)$ on the strip conductor generates a pair of

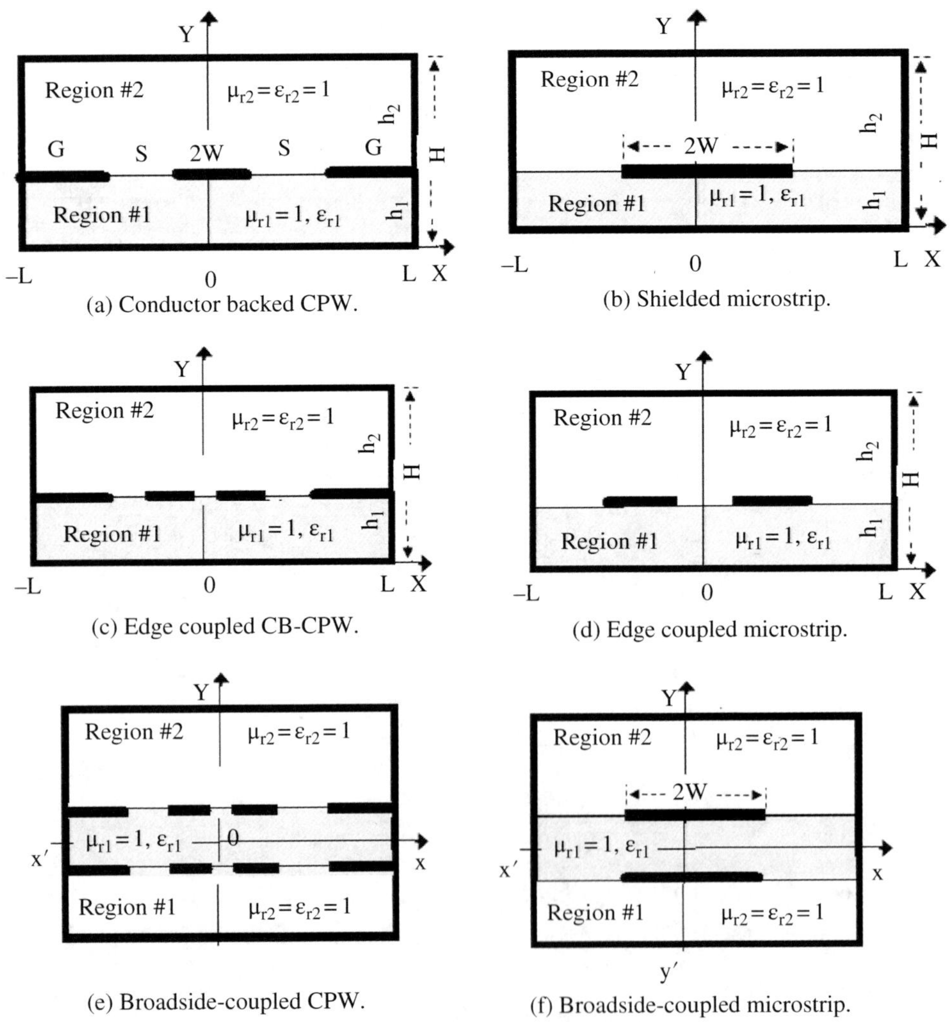

Figure 16.1 Shielded CPW and microstrip structures.

the tangential electric field components $\left(\tilde{E}_z, \tilde{E}_x\right)$ at the dielectric interface. Therefore, in terms of the network description, shown in Fig (16.2b), the current density pair $\left(\tilde{J}_z, \tilde{J}_x\right)$ is *excitation* and the field pair $\left(\tilde{E}_z, \tilde{E}_x\right)$ is the *response* connected through the [Z]-matrix parameters. The voltage pair at a port corresponds to the electric field pair and the current pair at another port corresponds to the current density pair. The [Z]-matrix description could be applied to a microstrip.

In the *second case*, a pair of the electric field components $\left(\tilde{E}_z, \tilde{E}_x\right)$, located in the slot region, is the source for the hybrid mode that produces current densities $\tilde{J}_z$ and $\tilde{J}_x$ response on the strip conductor. In this case, the field pair $(\tilde{E}_z, \tilde{E}_x)$ is *excitation* and the current density pair $\left(\tilde{J}_z, \tilde{J}_x\right)$ is *response* connected through [Y]-matrix parameters. This case applies to the CPW and slot line structures. In the Fourier domain, the surface current density pair $(\tilde{J}_z, \tilde{J}_x)$ on the strip conductors is related to the tangential components of the electric field pair $(\tilde{E}_z, \tilde{E}_x)$ at the dielectric interface, through the two-dimensional *dyadic Green's function*. Hence, *the Green's function is a description of a transmission line medium that can be modeled as the two-port network either by the Z-matrix or by the Y-matrix*, shown in Fig (16.2b).

For the microstrip structure shown in Fig (16.1b), the z-matrix of the two-port network, and the corresponding dyadic Green's function, are written below in the matrix form:

$$\begin{bmatrix} \tilde{Z}_{11} & \tilde{Z}_{12} \\ \tilde{Z}_{21} & \tilde{Z}_{22} \end{bmatrix} \begin{bmatrix} \tilde{I}_1 \\ \tilde{I}_2 \end{bmatrix} = \begin{bmatrix} \tilde{V}_1 \\ \tilde{V}_2 \end{bmatrix}. \qquad (16.1.1)$$

$$\begin{bmatrix} \tilde{G}_{11} & \tilde{G}_{12} \\ \tilde{G}_{21} & \tilde{G}_{22} \end{bmatrix} \begin{bmatrix} \tilde{J}_x \\ \tilde{J}_z \end{bmatrix} = \begin{bmatrix} \tilde{E}_x \\ \tilde{E}_z \end{bmatrix}. \qquad (16.1.2)$$

Thus, the impedance matrix for the microstrip corresponds to the dyadic [G]-type Green's function:

$$\begin{bmatrix} \tilde{G}_{11} & \tilde{G}_{12} \\ \tilde{G}_{21} & \tilde{G}_{22} \end{bmatrix} = \begin{bmatrix} \tilde{Z}_{11} & \tilde{Z}_{12} \\ \tilde{Z}_{21} & \tilde{Z}_{22} \end{bmatrix}. \qquad (16.1.3)$$

The Y-parameter of the two-port network and the corresponding Green's function are

$$\begin{bmatrix} \tilde{Y}_{11} & \tilde{Y}_{12} \\ \tilde{Y}_{21} & \tilde{Y}_{22} \end{bmatrix} \begin{bmatrix} \tilde{V}_1 \\ \tilde{V}_2 \end{bmatrix} = \begin{bmatrix} \tilde{I}_1 \\ \tilde{I}_2 \end{bmatrix}. \qquad (16.1.4)$$

$$\begin{bmatrix} \tilde{H}_{11} & \tilde{H}_{12} \\ \tilde{H}_{21} & \tilde{H}_{22} \end{bmatrix} \begin{bmatrix} \tilde{E}_x \\ \tilde{E}_z \end{bmatrix} = \begin{bmatrix} \tilde{J}_x \\ \tilde{J}_z \end{bmatrix}. \qquad (16.1.5)$$

In the above case, a pair of the slot fields $(\tilde{E}_z, \tilde{E}_x)$ is the source and a pair of the current densities on strip conductor $\left(\tilde{J}_x, \tilde{J}_z\right)$ is the response. The *dyadic [H]-type Green's function applicable to the CPW and the slot-line*, shown in Fig (16.1a), is given by the admittance matrix parameters as follows:

$$\begin{bmatrix} \tilde{H}_{11} & \tilde{H}_{12} \\ \tilde{H}_{21} & \tilde{H}_{22} \end{bmatrix} = \begin{bmatrix} \tilde{Y}_{11} & \tilde{Y}_{12} \\ Y_{21} & \tilde{Y}_{22} \end{bmatrix} \quad (a)$$
$$\begin{bmatrix} G_{11} & G_{12} \\ G_{21} & G_{22} \end{bmatrix} = \begin{bmatrix} H_{11} & H_{12} \\ H_{21} & H_{22} \end{bmatrix}^{-1} \quad (b). \qquad (16.1.6)$$

Two kinds of Green's functions are reciprocal of each other. It also shows that the Green's function does not depend on the current supporting strip conductors, or the field supporting slots. It depends only on the physical parameters of the inhomogeneous dielectric medium. The dynamic SDA involves the following three steps:

- Determination of the dyadic Green's function in the Fourier domain.

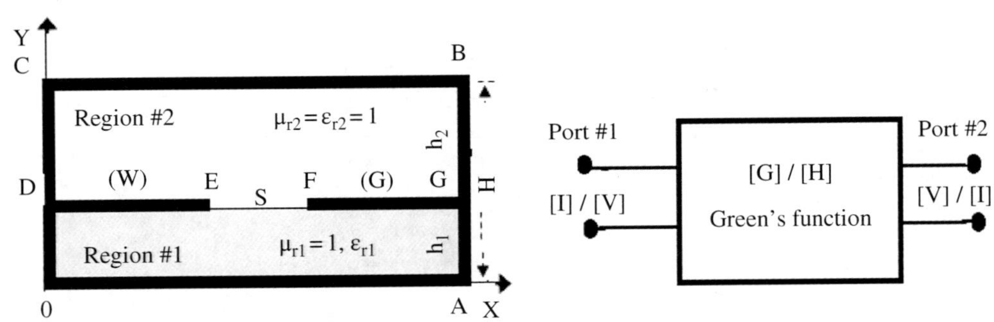

(a) Generic slot line structure. (b) 2-port description of Green's function.

Figure 16.2 Generic structure supporting hybrid mode and its two-port circuit description.

- Application of the Galerkin's method in the Fourier domain to solve either equation (16.1.2) or equation (16.1.5) to get *the characteristic equation.*
- The numerical solution of the characteristic equation determines the propagation constants of the fundamental and higher-order hybrid modes.

The dyadic Green's function in the Fourier domain is determined either as the boundary value problem or using the *immittance approach.*

16.2 Green's Function of Single-Layer Planar Line

Green's function, suitable for the CPW, slot line, and microstrip structures is determined in this section through the hybrid mode analysis of the planar line in the Fourier domain. Attention is paid to the physical symmetry of the modes.

16.2.1 Formulation of Field Problem

Figure (16.1a) shows a boxed single-layer CPW structure. The enclosure is EWs. The OY axis is the plane of symmetry, and it is a MW for the dominant mode. Figure (16.3a) shows one-half of the CPW structure for the present analysis. It supports the hybrid mode with respect to the *z*-axis that is the direction of propagation of the EM-wave. The hybrid mode is decomposed into (*i*) TE^y mode and (*ii*) TM^y mode as discussed in subsection (7.1.4) of chapter 7. The *y*-axis is in the direction of the inhomogeneity of the dielectric medium.

The hybrid modes are described by the *scalar electric potential* $\phi^e(x, y, z)$ and *scalar magnetic potential* $\phi^m(x, y, z)$. These potentials are responsible for the generation of TM^y mode and TE^y mode, respectively. The time-harmonic scalar potential functions, for the EM-wave propagating in the *z*-direction with the propagation constant β_z, are written as follows:

$$\begin{aligned} \phi^e(x,y,z,t) &= \phi^e(x,y)e^{j(\omega t-\beta_z z)} \quad &(a)\\ \phi^m(x,y,z,t) &= \phi^m(x,y)e^{j(\omega t-\beta_z z)} \quad &(b). \end{aligned} \tag{16.2.1}$$

The wave equations (7.1.32) and (7.1.36c) of chapter 7, are reproduced below:

$$\nabla_t^2\phi_i^e(x,y) + (k_i^2-\beta_z^2)\phi_i^e(x,y) = 0 \quad (a)$$

$$\nabla_t^2\phi_i^m(x,y) + (k_i^2-\beta_z^2)\phi_i^m(x,y) = 0 \quad (b)$$

$$\text{where, } \nabla_t^2\phi_i^q = \frac{\partial^2\phi_i^q}{\partial x^2} + \frac{\partial^2\phi_i^q}{\partial y^2}; \ (q = e, m),$$

$$k_i = \omega\sqrt{\mu_i\varepsilon_i}, \ (i = 1,2: \text{ indicating medium } \#1, \#2) \quad (c). \tag{16.2.2}$$

The EM-fields in the medium #1 and medium #2 are described by the scalar electric potentials $\phi_1^e(x,y)$ and $\phi_2^e(x,y)$, and the scalar magnetic potentials $\phi_1^m(x,y)$ and $\phi_2^m(x,y)$. The wavenumbers in two media are k_1 and k_2 given by equation (16.2.2c). Two media have permeability μ_1, μ_2, and permittivity ε_1, ε_2. The space domain field components of the TE^y and TM^y modes, given by equations (7.1.38) and (7.1.39), are summarized below:

$$\textbf{TE}^y\ \textbf{mode}:\ E_x^m = -j\beta_z\phi^m(x,y),\ E_y^m = 0,\ E_z^m = -\frac{\partial\phi^m(x,y)}{\partial x}$$

$$H_x^m = \frac{1}{j\omega\mu}\frac{\partial^2\phi^m(x,y)}{\partial x\partial y},\ H_y^m = \frac{1}{j\omega\mu}\left[\frac{\partial^2\phi^m(x,y)}{\partial y^2} + k^2\phi^m(x,y)\right]$$

$$H_z^m = \frac{-\beta_z}{\omega\mu}\frac{\partial\phi^m(x,y)}{\partial y}. \tag{16.2.3}$$

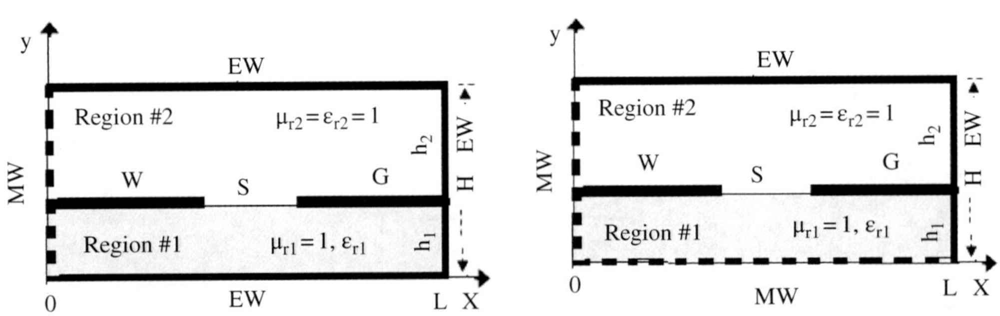

(a) Three electric walls (EW) and 1magnetic wall (MW).

(b) Two electric walls (EW) and two magnetic walls (MW).

Figure 16.3 One-half of the CPW structure.

TM^y mode: $E_x^e = \dfrac{1}{j\omega\varepsilon}\dfrac{\partial^2\phi^e(x,y)}{\partial x\partial y}$

$$\begin{aligned} E_y^e &= \frac{1}{j\omega\varepsilon}\left[\frac{\partial^2\phi^e(x,y)}{\partial y^2} + k^2\phi^e(x,y)\right] \\ E_z^e &= \frac{-\beta_z}{\omega\varepsilon}\frac{\partial\phi^e(x,y)}{\partial y}, \quad H_x^e = j\beta_z\phi^e(x,y) \\ H_y^e &= 0, \quad H_z^e = \frac{\partial\phi^e(x,y)}{\partial x}. \end{aligned} \tag{16.2.4}$$

The *wave-factor* $e^{j(\omega t-\beta_z z)}$ is dropped from equations (16.2.3) and (16.2.4). The scalar potentials $\phi^e(x, y)$ and $\phi^m(x, y)$ are obtained as the solution of the wave equations (16.2.2). To apply the dynamic spectral domain analysis (SDA), the wave equations (16.2.2) and field equations are transformed in the Fourier domain with the help of the following discrete Fourier transform:

$$\tilde{\phi}^q(\alpha_n,y) = \int_{-L}^{L}\phi^q(x,y)e^{j\alpha_n x}dx, \tag{16.2.5}$$

where q = e, m stands for the electric and magnetic fields. The discrete Fourier transform pair is taken from equation (14.6.2) of chapter 14. It is noted that in equation (16.2.5), α_n is taken as the discrete Fourier variable in place of β_n used in chapter 14. The variable β is taken as the propagation constant. The space variable x is transformed into the discrete Fourier variable α_n. Consequently, the space domain electric/magnetic potential function $\phi^q(x, y)$ is transformed as the spectral domain electric/magnetic potential $\tilde{\phi}^q(\alpha_n, y)$. However, once the spectral domain potential is known, the space domain potential is recovered using the inverse discrete Fourier transform from equation (14.6.2) of chapter 14:

$$\phi^q(x,y) = \frac{1}{2L}\sum_{n=-\infty}^{\infty}\tilde{\phi}^q(\alpha_n,y)e^{-j\alpha_n x}. \tag{16.2.6}$$

Normally, the potential function $\phi^q(x, y)$ is either an odd function or an even function of x. Under such cases, the Fourier transforms are obtained using the sine and cosine functions, respectively:

$$\begin{aligned} &\text{For } \phi^q(x,y) \text{ odd function of } x: \begin{cases} \tilde{\phi}^q(\alpha_n,y) = \dfrac{2}{L}\displaystyle\int_0^L \phi^q(x,y)\sin(\alpha_n x)dx & \text{(a)} \\ \alpha_n = \dfrac{n\pi}{L} \quad n = 0,1,2\ldots & \text{(b)} \end{cases} \\ &\text{For } \phi^q(x,y) \text{ even function of } x: \begin{cases} \tilde{\phi}^q(\alpha_n,y) = \dfrac{2}{L}\displaystyle\int_0^L \phi^q(x,y)\cos(\alpha_n x)dx & \text{(c)} \\ \alpha_n = \dfrac{(2n-1)\pi}{2L} \quad n = 1,2\ldots & \text{(d).} \end{cases} \end{aligned} \tag{16.2.7}$$

The inverse sine and inverse cosine transform are given below

$$\begin{aligned} \phi^q(x,y) &= \sum_{n=1}^{\infty}\left[\tilde{\phi}^q(\alpha_n,y)\sin(\alpha_n,x)\right] \quad \text{(a)} \\ \phi^q(x,y) &= \sum_{n=1}^{\infty}\left[\tilde{\phi}^q(\alpha_n,y)\cos(\alpha_n,x)\right] \quad \text{(b).} \end{aligned} \tag{16.2.8}$$

The above sine and cosine discrete Fourier transforms are obtained from equations (14.6.6) and (14.6.8) of chapter 14. Using equation (16.2.5), Fourier transforms of equations (16.2.3) and (16.2.4) are obtained. The hybrid mode field components, in the Fourier domain, are written as the summation of individual field components of the TE^y and TM^y modes:

$$\begin{aligned} &\tilde{E}_{x,i} = -j\frac{\alpha_n}{\hat{y}_i}\frac{\partial\tilde{\phi}_i^e}{\partial y} - j\beta_z\tilde{\phi}_i^m, \quad \tilde{E}_{y,i} = \frac{1}{\hat{y}_i}\left(\frac{\partial^2}{\partial y^2} + k_i^2\right)\tilde{\phi}_i^e \\ &\tilde{E}_{z,i} = -j\frac{\beta_z}{\hat{y}_i}\frac{\partial\tilde{\phi}_i^e}{\partial y} + j\alpha_n\tilde{\phi}^m \qquad \text{(a)} \\ &\tilde{H}_{x,i} = j\beta_z\tilde{\phi}_i^e - j\frac{\alpha_n}{\hat{z}_i}\frac{\partial\tilde{\phi}_i^m}{\partial y}, \quad \tilde{H}_{y,i} = \frac{1}{\hat{z}_i}\left(\frac{\partial^2}{\partial y^2} + k_i^2\right)\tilde{\phi}_i^m \\ &\tilde{H}_{z,i} = -j\alpha_n\tilde{\phi}_i^e - j\frac{\beta_z}{\hat{z}}\frac{\partial\tilde{\phi}_i^m}{\partial y} \qquad \text{(b),} \\ &\text{where, } \hat{y}_i = j\omega\varepsilon_i, \ \hat{z}_i = j\omega\mu_i, \ k_i^2 = \omega^2\mu_i\varepsilon_i, \ i = 1,2. \end{aligned} \tag{16.2.9}$$

The wave equation (16.2.2) is also transformed in the Fourier domain. In the process, the PDE is reduced to a second-order ODE:

$$\frac{d^2\tilde{\phi}_i^q}{dy^2} - \gamma_i^2\tilde{\phi}_i^q(\alpha_n, y) = 0, \quad \text{(a)}$$

$$\text{where, } i = 1,2; \quad q = e, m \text{ and } \gamma_i^2 = \beta_z^2 + {\alpha^2}_n - k_i^2 \quad \text{(b)}, \qquad (16.2.10)$$

where γ_i is the propagation constant in the medium $i = 1,2$. The *y*-component of wavenumber k_{yi} is the propagation constant in the *y*-direction and β_z is the propagation constant of the hybrid mode, as the hybrid mode wave propagates in the *z*-direction on the planar line. The SDA determines the propagation constant β_z of the fundamental hybrid mode of a planar line. It helps to compute its frequency-dependent effective relative permittivity, $\varepsilon_{reff}(f) = [\beta_z(f)/\beta_0]^2$; where β_0 is the propagation constant in the free space.

16.2.2 Case #1: CPW and Microstrip Structures

Figure (16.3a) shows the structure for case #1 that is one-half of the conductor backed CPW, shown in Fig (16.1a). Figure (16.3a) is a slot line structure. On replacing the strip conductors by the dielectric interface and the slot by a strip conductor, the slot line could be treated as a microstrip line, shown in Fig (16.1b). It has the MW–EW side walls and the EW - EW bottom-top walls. The solutions of wave equation (16.2.10) in two dielectric regions are written from inspection:

Region #1 (lower medium, $0 \le y \le h_1$):

$$\begin{aligned} &\tilde{\phi}_1^e(\alpha_n, y) = A^e \sinh[\gamma_1 y] + C^e \cosh[\gamma_1 y] \quad \text{(a)} \\ &\tilde{\phi}_1^m(\alpha_n, y) = A^m \sinh[\gamma_1 y] + C^m \cosh[\gamma_1 y] \quad \text{(b).} \end{aligned} \qquad (16.2.11)$$

Region #2 (upper medium, $h_1 \le y \le H$):

$$\begin{aligned} &\tilde{\phi}_2^e(\alpha_n, y) = B^e \sinh[\gamma_2(H-y)] + D^e \cosh[\gamma_2(H-y)] \quad \text{(a)} \\ &\tilde{\phi}_2^m(\alpha_n, y) = B^m \sinh[\gamma_2(H-y)] + D^m \cosh[\gamma_2(H-y)] \quad \text{(b),} \end{aligned} \qquad (16.2.12)$$

where $H = h_1 + h_2$. The constants of equations (16.2.12) are evaluated below in three steps: *(i)* application of the boundary condition in both the regions, *(ii)* application of the continuity condition at the interface of two media, and *(iii)* assumption of the electric field in the slot-region.

Step #1: Application of Boundary Condition

The boundary condition is applied with the help of equation (16.2.9) as follows:

Region - I (Lower medium):

$$\text{At } y = 0: \ \tilde{E}_{x1} = 0; \ \text{ i.e. } \tilde{\phi}_1^m = 0, \ \frac{\partial\tilde{\phi}_1^e}{\partial y} = 0 \qquad (16.2.13)$$

From equation (16.2.11), $C^m = 0$, $A^e = 0$.

Region - II (Upper medium):

$$\text{At } y = H: \ \tilde{E}_{x2} = 0; \ \text{ i.e. } \tilde{\phi}_2^m = 0, \ \frac{\partial\tilde{\phi}_2^e}{\partial y} = 0 \qquad (16.2.14)$$

From equation (16.2.12), $D^m = 0$, $B^e = 0$.

Equations (16.2.11) and (16.2.12) are now written as

$$\begin{aligned} \text{Lower Medium: } &\tilde{\phi}_1^e(\alpha_n, y) = A^e \cosh[\gamma_1 y] \\ &\tilde{\phi}_1^m(\alpha_n, y) = A^m \sinh[\gamma_1 y] \quad \text{(a)} \\ \text{Upper Medium: } &\tilde{\phi}_2^e(\alpha_n, y) = B^e \cosh[\gamma_2(H-y)] \\ &\tilde{\phi}_2^m(\alpha_n, y) = B^m \sinh[\gamma_2(H-y)] \quad \text{(b).} \end{aligned} \qquad (16.2.15)$$

In equation (16.2.15), a new set of constants is introduced to maintain the symmetry of expressions. The constants A^e and A^m are used for the field solutions in the lower medium #1, and the constants B^e and B^m are used with the field solutions in the upper-medium #2. To determine these constants, the tangential field components of the hybrid mode are written below, in both regions on substituting equation (16.2.15) in equation (16.2.9):

$$\begin{aligned} \text{Lower Medium: } \tilde{E}_{x1} &= -j\alpha_n \frac{\gamma_1}{\hat{y}_1} A^e \sinh[\gamma_1 y] - j\beta_z A^m \sinh[\gamma_1 y] \\ \tilde{H}_{x1} &= j\beta_z A^e \cosh[\gamma_1 y] - j\frac{\alpha_n}{\hat{z}_1}\gamma_1 A^m \cosh[\gamma_1 y] \quad \text{(a)} \\ \tilde{E}_{z1} &= -j\beta_z \frac{\gamma_1}{\hat{y}_1} A^e \sinh[\gamma_1 y] + j\alpha_n A^m \sinh[\gamma_1 y] \\ \tilde{H}_{z1} &= -j\alpha_n A^e \cosh[\gamma_1 y] - j\beta_z \frac{\gamma_1}{\hat{z}_1} A^m \cosh[\gamma_1 y] \\ \text{Upper Medium: } \tilde{E}_{x2} &= j\alpha_n \frac{\gamma_2}{\hat{y}_2} B^e \sinh[\gamma_2(H-y)] - j\beta_z B^m \sinh[\gamma_2(H-y)] \\ \tilde{H}_{x2} &= j\beta_z B^e \cosh[\gamma_2(H-y)] + j\frac{\alpha_n}{\hat{z}_2}\gamma_2 B^m \cosh[\gamma_2(H-y)] \quad \text{(b).} \\ \tilde{E}_{z2} &= j\beta_z \frac{\gamma_2}{\hat{y}_2} B^e \sinh[\gamma_2(H-y)] + j\alpha_n B^m \sinh[\gamma_2(H-y)] \\ \tilde{H}_{z2} &= -j\alpha_n B^e \cosh[\gamma_2(H-y)] + j\beta_z \frac{\gamma_2}{\hat{z}_2} B^m \cosh[\gamma_2(H-y)] \end{aligned} \qquad (16.2.16)$$

To simplify the above expression (16.2.16), the following normalized y-directed propagation constants are assumed:

$$\gamma_{yi} = \frac{\gamma_i}{\hat{y}_i} \text{ and } \gamma_{zi} = \frac{\gamma_i}{\hat{z}_i}; i = 1, 2, \quad (16.2.17)$$

where $\gamma_i = k_{y_i}$ is the propagation constant in the y-direction. The above normalization is done with respect to the permittivity and permeability parameters of the medium. The parameters $\hat{y}_i$ and $\hat{z}_i$ are given in equation (16.2.9).

Step #2: Application of Continuity Condition

The continuity conditions for the *x* and *z*-components of the electric field at the interface of two media, i.e. at $y = h_1$, are applied as follows:

$$\left.\tilde{E}_{x1}\right|_{y=h_1} = \left.\tilde{E}_{x2}\right|_{y=h_1} \Rightarrow -j\alpha_n\gamma_{y1}A^e \sinh[\gamma_1 h_1] - j\beta_z A^m \sinh[\gamma_1 h_1] = j\alpha_n\gamma_{y2}B^e \sinh[\gamma_2 h_2] - j\beta_z B^m \sinh[\gamma_2 h_2]. \quad (16.2.18)$$

$$\left.\tilde{E}_{z1}\right|_{y=h_1} = \left.\tilde{E}_{z2}\right|_{y=h_1} \Rightarrow -j\beta_z\gamma_{y1}A^e \sinh[\gamma_1 h_1] + j\alpha_n A^m \sinh[\gamma_1 h_1] = j\beta_z\gamma_{y2}B^e \sinh[\gamma_2 h_2] + j\alpha_n B^m \sinh[\gamma_2 h_2]. \quad (16.2.19)$$

On multiplying equation (16.2.18) by α_n, and equation (16.2.19) by β_z, constants A^e and A^m, in terms of constants B^e and B^m, are obtained below:

$$-j\alpha_n^2\gamma_{y1}A^e \sinh[\gamma_1 h_1] - j\alpha_n\beta_z A^m \sinh[\gamma_1 h_1] = j\alpha_n^2\gamma_{y2}B^e \sinh[\gamma_2 h_2] - j\alpha_n\beta_z B^m \sinh[\gamma_2 h_2] \quad (a)$$

$$-j\beta_z^2\gamma_{y1}A^e \sinh[\gamma_1 h_1] + j\beta_z\alpha_n A^m \sinh[\gamma_1 h_1] = j\beta_z^2\gamma_{y2}B^e \sinh[\gamma_2 h_2] + j\beta_z\alpha_n B^m \sinh[\gamma_2 h_2] \quad (b)$$

On adding above two equations:

$$-j(\alpha_n^2 + \beta_z^2)\gamma_{y1}A^e \sinh[\gamma_1 h_1] = j(\alpha_n^2 + \beta_z^2)\gamma_{y2}B^e \sinh[\gamma_2 h_2]$$

$$\Rightarrow A^e(\text{Lower medium}) = -\frac{\gamma_{y2}\sinh(\gamma_2 h_2)}{\gamma_{y1}\sinh(\gamma_1 h_1)} B^e(\text{Upper medium}) \quad (c). \quad (16.2.20)$$

On multiplying equation (16.2.18) by β_z and equation (16.2.19) by α_n:

$$-j\beta_z\alpha_n\gamma_{y1}A^e \sinh[\gamma_1 h_1] - j\beta_z^2 A^m \sinh[\gamma_1 h_1] = j\beta_z\alpha_n\gamma_{y2}B^e \sinh[\gamma_2 h_2] - j\beta_z^2 B^m \sinh[\gamma_2 h_2] \quad (a)$$

$$-j\beta_z\alpha_n\gamma_{y1}A^e \sinh[\gamma_1 h_1] + j\alpha_n^2 A^m \sinh[\gamma_1 h_1] = j\beta_z\alpha_n\gamma_{y2}B^e \sinh[\gamma_2 h_2] + j\alpha_n^2 B^m \sinh[\gamma_2 h_2] \quad (b)$$

On subtracting the above two equations:

$$-j(\alpha_n^2 + \beta_z^2)A^m \sinh[\gamma_1 h_1] = -j(\alpha_n^2 + \beta_z^2) \times B^m \sinh[\gamma_2 h_2]$$

$$\Rightarrow A^m(\text{Lower medium}) = \frac{\sinh[\gamma_2 h_2]}{\sinh[\gamma_1 h_1]} \times B^m(\text{Upper medium}) \quad (c). \quad (16.2.21)$$

Step #3: Assumption of the Slot-Field

The lower-medium and upper-medium fields are correlated through equations (16.2.20c) and (16.2.21c). The constants, B^e and B^m, are evaluated in the terms of *assumed electric field components,* $\tilde{E}_x(\alpha_n)$ and $\tilde{E}_z(\alpha_n)$, in a slot region, located at the interface $y = h_1$. Fourier transformed electric field components, $\tilde{E}_x(\alpha_n)$ and $\tilde{E}_z(\alpha_n)$, in the slot-region are expressed through the *assumed basis functions*. The choice of the basis function to express the electric field in the slot region is discussed separately in section (16.4). Finally, the tangential field components in the upper region #2, at $y = h_1$, are written from equation (16.2.16b) as

Known i.e. assumed electric field in slot

↑

$$\tilde{E}_{x2} = \tilde{E}_x(\alpha_n) = j\alpha_n\gamma_{y2}B^e \sinh(\gamma_2 h_2) - j\beta_z B^m \sinh(\gamma_2 h_2) \quad (a)$$

$$\tilde{E}_{z2} = \tilde{E}_z(\alpha_n) = j\beta_z\gamma_{y2}B^e \sinh(\gamma_2 h_2) + j\alpha_n B^m \sinh(\gamma_2 h_2) \quad (b).$$

↓

Known i.e. assumed electric field in slot.

(16.2.22)

Equations (16.2.22a,b) are solved to get the constants B^e and B^m in terms of the assumed slot - fields, $\tilde{E}_x(\alpha_n)$ and $\tilde{E}_z(\alpha_n)$.

$$B^m = -j\left[\frac{\alpha_n\tilde{E}_z(\alpha_n, h_1) - \beta_z\tilde{E}_x(\alpha_n, h_1)}{(\beta_z^2 + \alpha_n^2)\sinh(\gamma_2 h_2)}\right] \quad (a)$$

$$B^e = -j\left[\frac{\alpha_n\tilde{E}_x(\alpha_n, h_1) - \beta_z\tilde{E}_z(\alpha_n, h_1)}{\gamma_{y2}(\beta_z^2 + \alpha_n^2)\sinh(\gamma_2 h_2)}\right] \quad (b). \quad (16.2.23)$$

Determination of Strip Current Densities

The assumed slot-electric fields induce the currents on the strip conductors. The slot-electric field and the current density on the strip conductors are related through Green's function. The computation of the strip current density provides the *admittance type Green's function/ [H]-type* given by equations (16.1.4) and (16.1.5).

Longitudinal Current Density Giving Green's Function Components $\widetilde{H}_{21}$ and $\widetilde{H}_{22}$

The longitudinal current density on strip conductor shows a discontinuity in the tangential components of the transverse magnetic field at the interface $y = h_1$

$$\widetilde{H}_{x2} - \widetilde{H}_{x1} = \widetilde{J}_z \text{ at } y = h_1. \tag{16.2.24}$$

$$j\beta_z B^e \cosh(\gamma_2 h_2) + j\alpha_n \gamma_{z2} B^m \cosh(\gamma_2 h_2) - j\beta_z A^e \cosh(\gamma_1 h_1) + j\alpha_n \gamma_{z1} A^m \cosh(\gamma_1 h_1) = \widetilde{J}_z(\alpha_n). \tag{16.2.25}$$

On substituting A^e and A^m from equations (16.2.20c) and (16.2.21c),

$$j\beta_z\left[\cosh(\gamma_2 h_2) + \frac{\gamma_{y2}}{\gamma_{y1}}\frac{\sinh(\gamma_2 h_2)}{\tanh(\gamma_1 h_1)}\right]B^e + j\alpha_n\left[\gamma_{z2}\cosh(\gamma_2 h_2) + \gamma_{z1}\frac{\sinh(\gamma_2 h_2)}{\tanh(\gamma_1 h_1)}\right]B^m = \widetilde{J}_z(\alpha_n). \tag{16.2.26}$$

$$\text{Let, } M_1 = \beta_z\left[\cosh(\gamma_2 h_2) + \frac{\gamma_{y2}}{\gamma_{y1}}\frac{\sinh(\gamma_2 h_2)}{\tanh(\gamma_1 h_1)}\right] \quad (a)$$

$$N_1 = \alpha_n\left[\gamma_{z2}\cosh(\gamma_2 h_2) + \gamma_{z1}\frac{\sinh(\gamma_2 h_2)}{\tanh(\gamma_1 h_1)}\right] \quad (b). \tag{16.2.27}$$

On substituting equations (16.2.27), and also B^e and B^m from equation (16.2.23), in equation (16.2.26),

$$\left[\frac{M_1\alpha_n}{\gamma_{y2}(\beta_z^2+\alpha_n^2)\sinh(\gamma_2 h_2)} - \frac{\beta_z N_1}{(\beta_z^2+\alpha_n^2)\sinh(\gamma_2 h_2)}\right] \times \widetilde{E}_x(\alpha_n, h_1) + \left[\frac{M_1\beta_z}{\gamma_{y2}(\beta_z^2+\alpha_n^2)\sinh(\gamma_2 h_2)} + \frac{\alpha_n N_1}{(\beta_z^2+\alpha_n^2)\sinh(\gamma_2 h_2)}\right] \times \widetilde{E}_z(\alpha_n, h_1) = \widetilde{J}_z(\alpha_n). \tag{16.2.28}$$

$$\text{Let } P_1 = \left[\frac{M_1\alpha_n}{\gamma_{y2}(\beta_z^2+\alpha_n^2)\sinh(\gamma_2 h_2)} - \frac{\beta_z N_1}{(\beta_z^2+\alpha_n^2)\sinh(\gamma_2 h_2)}\right] \quad (a)$$

$$Q_1 = \left[\frac{M_1\beta_z}{\gamma_{y2}(\beta_z^2+\alpha_n^2)\sinh(\gamma_2 h_2)} + \frac{\alpha_n N_1}{(\beta_z^2+\alpha_n^2)\sinh(\gamma_2 h_2)}\right] \quad (b). \tag{16.2.29}$$

The simplified discontinuity equation, at the interface $y = h_1$, is written as

$$P_1\widetilde{E}_x(\alpha_n, h_1) + Q_1\widetilde{E}_z(\alpha_n, h_1) = \widetilde{J}_z(\alpha_n). \tag{16.2.30}$$

The P_1 and Q_1 could be further rewritten on the substitution of M_1 and N_1

$$P_1 = \frac{\alpha_n\beta_z}{\gamma_{y2}({\alpha^2}_n+\beta_z^2)\sinh(\gamma_2 h_2)}\left[\cosh(\gamma_2 h_2) + \frac{\gamma_{y2}}{\gamma_{y1}}\frac{\sinh(\gamma_2 h_2)}{\tanh(\gamma_1 h_1)}\right] - \frac{\alpha_n\beta_z}{({\alpha^2}_n+\beta_z^2)\sinh(\gamma_2 h_2)}\left[\gamma_{z2}\cosh(\gamma_2 h_2) + \frac{\gamma_{z1}\sinh(\gamma_2 h_2)}{\tanh(\gamma_1 h_1)}\right]$$

$$\Rightarrow P_1 = \frac{\alpha_n\beta_z}{({\alpha^2}_n+\beta_z^2)}\left[\left(\frac{1}{\gamma_{y1}\tanh(\gamma_1 h_1)} + \frac{1}{\gamma_{y2}\tanh(\gamma_2 h_2)}\right) - \left(\frac{\gamma_{z1}}{\tanh(\gamma_1 h_1)} + \frac{\gamma_{z2}}{\tanh(\gamma_2 h_2)}\right)\right]. \tag{16.2.31}$$

$$Q_1 = \frac{\beta_z}{\gamma_{y2}(\alpha_n^2+\beta_z^2)\sinh(\gamma_2 h_2)} \times \beta_z\left[\cosh(\gamma_2 h_2) + \frac{\gamma_{y2}}{\gamma_{y1}}\frac{\sinh(\gamma_2 h_2)}{\tanh(\gamma_1 h_1)}\right] + \frac{\alpha_n}{(\alpha_n^2+\beta_z^2)\sinh(\gamma_2 h_2)} \times \alpha_n\left[\gamma_{z2}\cosh(\gamma_2 h_2) + \frac{\gamma_{z1}\sinh(\gamma_2 h_2)}{\tanh(\gamma_1 h_1)}\right]$$

$$\Rightarrow Q_1 = \frac{1}{(\alpha_n^2+\beta_z^2)}\left[\left(\frac{\beta_z^2}{\gamma_{y1}\tanh(\gamma_1 h_1)} + \frac{\beta_z^2}{\gamma_{y2}\tanh(\gamma_2 h_2)}\right) + \left(\frac{\alpha_n^2\gamma_{z1}}{\tanh(\gamma_1 h_1)} + \frac{\alpha_n^2\gamma_{z2}}{\tanh(\gamma_2 h_2)}\right)\right]. \tag{16.2.32}$$

Let

$$\widetilde{H}_e = \frac{1}{\gamma_{y1}\tanh(\gamma_1 h_1)} + \frac{1}{\gamma_{y2}\tanh(\gamma_2 h_2)} \quad (a)$$

$$\widetilde{H}_m = \frac{\gamma_{z1}}{\tanh(\gamma_1 h_1)} + \frac{\gamma_{z2}}{\tanh(\gamma_2 h_2)} \quad (b). \tag{16.2.33}$$

Equations for P_1 and Q_1 are rewritten as

$$P_1 = \widetilde{H}_{21} = \frac{\alpha_n\beta_z}{({\alpha^2}_n+\beta_z^2)}\left[\widetilde{H}_e - \widetilde{H}_m\right] \quad (a)$$

$$Q_1 = \widetilde{H}_{22} = \frac{1}{(\alpha_n^2+\beta_z^2)}\left[\beta_z^2\widetilde{H}_e + \alpha_n^2\widetilde{H}_m\right] \quad (b). \tag{16.2.34}$$

Finally, equation (16.2.30) is rewritten in the following form, describing one of equations of the admittance type/[H]-type Green's function given by equation (16.1.5):

$$\tilde{H}_{21}\tilde{E}_x(\alpha_n, h_1) + \tilde{H}_{22}\tilde{E}_z(\alpha_n, h_1) = \tilde{J}_z(\alpha_n). \tag{16.2.35}$$

Transverse Current Density Giving Green's Function Components $\tilde{H}_{11}$ and $\tilde{H}_{12}$

The discontinuity in the longitudinal magnetic field components in two media is related to the transverse current density $\tilde{J}_x$ on the strip conductor at the interface $y = h_1$. The magnetic field components are obtained from equation (16.2.16).

$$\tilde{H}_{z1} - \tilde{H}_{z2} = \tilde{J}_x(\alpha_n)\big|_{\text{at } y=h_1} \tag{16.2.36}$$

$$-j\alpha_n A^e \cosh(\gamma_1 h_1) - j\beta_z \gamma_{z1} A^m \cosh(\gamma_1 h_1) + j\alpha_n B^e \cosh(\gamma_2 h_2) - j\beta_z \gamma_{z2} B^m \cosh(\gamma_2 h_2) = \tilde{J}_x(\alpha_n). \tag{16.2.37}$$

On substituting A^e and A^m from equations (16.2.20) and (16.2.21) in the above equation,

$$j\alpha_n \left\{ \frac{\gamma_{y2}}{\gamma_{y1}} \frac{\sinh(\gamma_2 h_2)}{\tanh(\gamma_1 h_1)} + \cosh(\gamma_2 h_2) \right\} B^e - j\beta_z \left\{ \frac{\gamma_{z1} \sinh(\gamma_2 h_2)}{\tanh(\gamma_1 h_1)} + \gamma_{z2} \cosh(\gamma_2 h_2) \right\} B^m = \tilde{J}_x(\alpha_n). \tag{16.2.38}$$

Let

$$M_2 = \left\{ \frac{\gamma_{y2}}{\gamma_{y1}} \frac{\sinh(\gamma_2 h_2)}{\tanh(\gamma_1 h_1)} + \cosh(\gamma_2 h_2) \right\} \quad \text{(a)}$$

$$N_2 = \left\{ \frac{\gamma_{z1} \sinh(\gamma_2 h_2)}{\tanh(\gamma_1 h_1)} + \gamma_{z2} \cosh(\gamma_2 h_2) \right\} \quad \text{(b).} \tag{16.2.39}$$

Equation (16.2.38) is written as follows:

$$j\alpha_n M_2 B^e - j\beta_z N_2 B^m = \tilde{J}_x. \tag{16.2.40}$$

On substituting B^e and B^m from equation (16.2.23) in the above equation,

$$\left[\frac{\alpha_n^2}{\gamma_{y2}(\alpha^2{}_n + \beta_z^2)\sinh(\gamma_2 h_2)} M_2 + \frac{\beta_z^2}{(\alpha_n^2 + \beta_z^2)\sinh(\gamma_2 h_2)} N_2 \right] \times \tilde{E}_x(\alpha_n, h_1) + \left[\frac{\alpha_n \beta_z}{\gamma_{y2}(\alpha_n^2 + \beta_z^2)\sinh(\gamma_2 h_2)} M_2 - \frac{\alpha_n \beta_z}{(\alpha_n^2 + \beta_z^2)\sinh(\gamma_2 h_2)} N_2 \right] \times \tilde{E}_z(\alpha_n, h_1) = \tilde{J}_x(\alpha_n). \tag{16.2.41}$$

$$P_2 \tilde{E}_x(\alpha_n, d) + Q_2 \tilde{E}_z(\alpha_n, d) = \tilde{J}_x(\alpha_n) \tag{16.2.42}$$

where,

$$P_2 = \frac{\alpha_n^2}{\gamma_{y2}(\alpha^2{}_n + \beta_z^2)\sinh(\gamma_2 h_2)} M_2 + \frac{\beta_z^2}{(\alpha_n^2 + \beta_z^2)\sinh(\gamma_2 h_2)} N_2$$

$$\Rightarrow P_2 = \frac{1}{(\alpha_n^2 + \beta_z^2)} \times \left[\alpha_n^2 \left\{ \frac{1}{\gamma_{y1}\tanh(\gamma_1 h_1)} + \frac{1}{\gamma_{y2}\tanh(\gamma_2 h_2)} \right\} + \beta_z^2 \left\{ \frac{\gamma_{z1}}{\tanh(\gamma_1 h_1)} + \frac{\gamma_{z2}}{\tanh(\gamma_2 h_2)} \right\} \right]. \tag{16.2.43}$$

On using equation (16.2.33) with equation (16.2.43) and assuming $P_2 = \tilde{H}_{11}$

$$P_2 = \tilde{H}_{11} = \frac{1}{(\alpha_n^2 + \beta_z^2)} \left[\alpha_n^2 \tilde{H}_e + \beta_z^2 \tilde{H}_m \right]. \tag{16.2.44}$$

Likewise,

$$Q_2 = \left[\frac{\alpha_n \beta_z}{\gamma_{y2}(\alpha_n^2 + \beta_z^2)\sinh(\gamma_2 h_2)} M_2 - \frac{\alpha_n \beta_z}{(\alpha_n^2 + \beta_z^2)\sinh(\gamma_2 h_2)} N_2 \right]$$

$$\Rightarrow Q_2 = \frac{\alpha_n \beta_z}{(\alpha_z^2 + \beta_z^2)} \left[\left\{ \frac{1}{\gamma_{y1}\tanh(\gamma_1 h_1)} + \frac{1}{\gamma_{y2}\tanh(\gamma_2 h_2)} \right\} - \left\{ \frac{\gamma_{y1}}{\tanh(\gamma_1 h_1)} + \frac{\gamma_{y2}}{\tanh(\gamma_2 h_2)} \right\} \right]. \tag{16.2.45}$$

On using equation (16.2.33) with equation (16.2.45) and assuming $Q_2 = \tilde{H}_{12}$

$$Q_2 = \tilde{H}_{12} = \frac{\alpha_n \beta_z}{(\alpha_n^2 + \beta_z^2)} \left[\tilde{H}_e - \tilde{H}_m \right]. \tag{16.2.46}$$

On using expression (16.2.46), equation (16.2.42) can be rewritten as follows:

$$\tilde{H}_{11}\tilde{E}_x(\alpha_n, d) + \tilde{H}_{12}\tilde{E}_z(\alpha_n, d) = \tilde{J}_x. \tag{16.2.47}$$

Admittance Type Green's Function of CPW and Slotline

Equations (16.2.35) and (16.2.47) are summarized below to write the admittance type Green's function, applicable to the slot type planar transmission line structures, such as CPW and slot line. Green's function is obtained at the plane $y = h_1$, containing the strip and slot. In this case, *the slot electric field components* $\tilde{E}_x(\alpha_n, h_1)$ *and* $\tilde{E}_z(\alpha_n, h_1)$ *in the Fourier domain are treated as known functions through the assumed basis functions.*

$$\tilde{H}_{11}\tilde{E}_x(\alpha_n, h_1) + \tilde{H}_{12}\tilde{E}_z(\alpha_n, h_1) = \tilde{J}_x(\alpha_n)$$
$$\tilde{H}_{21}\tilde{E}_x(\alpha_n, h_1) + \tilde{H}_{22}\tilde{E}_z(\alpha_n, h_1) = \tilde{J}_z(\alpha_n). \tag{16.2.48}$$

Dimensionally the $[\tilde{H}]$- dyadic Green's function is an admittance matrix -$[\tilde{Y}_{11}]$, i.e.

$$\begin{bmatrix} \tilde{H}_{11} & \tilde{H}_{12} \\ \tilde{H}_{21} & \tilde{H}_{22} \end{bmatrix} = \begin{bmatrix} \tilde{Y}_{11} & \tilde{Y}_{12} \\ \tilde{Y}_{21} & \tilde{Y}_{22} \end{bmatrix} \tag{a}$$

where,

$$\tilde{Y}_{11} = P_2, \quad \tilde{Y}_{12} = \tilde{Y}_{21} = P_1 = Q_2, \quad \tilde{Y}_{22} = Q_1$$

$$\tilde{Y}_{11} = \frac{1}{(\alpha_n^2 + \beta_z^2)}\left[\alpha_n^2\left\{\frac{1}{\gamma_{y1}\tanh(\gamma_1 h_1)} + \frac{1}{\gamma_{y2}\tanh(\gamma_2 h_2)}\right\} + \beta_z^2\left\{\frac{\gamma_{z1}}{\tanh(\gamma_1 h_1)} + \frac{\gamma_{z2}}{\tanh(\gamma_2 h_2)}\right\}\right] \tag{16.2.49}$$

$$\tilde{Y}_{12} = \tilde{Y}_{21} = \frac{\alpha_n\beta_z}{(\alpha_z^2 + \beta_z^2)}\left[\left\{\frac{1}{\gamma_{y1}\tanh(\gamma_1 h_1)} + \frac{1}{\gamma_{y2}\tanh(\gamma_2 h_2)}\right\} - \left\{\frac{\gamma_{y1}}{\tanh(\gamma_1 h_1)} + \frac{\gamma_{y2}}{\tanh(\gamma_2 h_2)}\right\}\right]$$

$$\tilde{Y}_{22} = \frac{1}{(\alpha_n^2 + \beta_z^2)}\left[\left\{\frac{\beta_z^2}{\gamma_{y1}\tanh(\gamma_1 h_1)} + \frac{\beta_z^2}{\gamma_{y2}\tanh(\gamma_2 h_2)}\right\} + \left\{\frac{\alpha_n^2\gamma_{z1}}{\tanh(\gamma_1 h_1)} + \frac{\alpha_n^2\gamma_{z2}}{\tanh(\gamma_2 h_2)}\right\}\right] \tag{b}$$

Impedance Type Green's Function of Microstrip

The SDA is applied to a microstrip line, shown in Fig (16.1b). *In the case of a microstrip, the unknown current distribution components are the sources for the electric field components of the hybrid mode.* Therefore, the field-current relation in the Fourier domain, applicable to the microstrip line, could be written on inverting equation (16.2.48):

$$\begin{bmatrix} \tilde{H}_{11} & \tilde{H}_{12} \\ \tilde{H}_{21} & \tilde{H}_{22} \end{bmatrix}^{-1}\begin{bmatrix} \tilde{J}_x \\ \tilde{J}_z \end{bmatrix} = \begin{bmatrix} \tilde{E}_x \\ \tilde{E}_z \end{bmatrix}$$
$$\Rightarrow \begin{bmatrix} \tilde{Z}_{11} & \tilde{Z}_{12} \\ \tilde{Z}_{21} & \tilde{Z}_{22} \end{bmatrix}\begin{bmatrix} \tilde{J}_x \\ \tilde{J}_z \end{bmatrix} = \begin{bmatrix} \tilde{E}_x \\ \tilde{E}_z \end{bmatrix}. \tag{16.2.50}$$

where the impedance matrix-$[\tilde{Z}]$ is the dyadic Green's function $[\tilde{G}]$ applicable to the microstrip,

$$[\tilde{G}] = \begin{bmatrix} \tilde{G}_{11} & \tilde{G}_{12} \\ \tilde{G}_{21} & \tilde{G}_{22} \end{bmatrix} = \begin{bmatrix} \tilde{Z}_{11} & \tilde{Z}_{12} \\ \tilde{Z}_{21} & \tilde{Z}_{22} \end{bmatrix} = \begin{bmatrix} \tilde{H}_{11} & \tilde{H}_{12} \\ \tilde{H}_{21} & \tilde{H}_{22} \end{bmatrix}^{-1}$$
$$= \frac{1}{\Delta\tilde{H}}\begin{bmatrix} \tilde{H}_{11} & -\tilde{H}_{12} \\ -\tilde{H}_{21} & \tilde{H}_{22} \end{bmatrix}. \tag{16.2.51}$$

The $\Delta\tilde{H}$ is evaluated as follows:

$$\Delta\tilde{H} = \tilde{H}_{11}\tilde{H}_{22} - \tilde{H}_{12}\tilde{H}_{21}$$

$$\Rightarrow \Delta\tilde{H} = \left[\frac{\alpha_n\beta_z}{(\alpha_n^2 + \beta_z^2)}\right]^2\left[\tilde{H}_e - \tilde{H}_m\right]^2 - \frac{1}{(\alpha_n^2 + \beta_z^2)}\left[\beta_z^2\tilde{H}_e + \alpha_n^2\tilde{H}_m\right] \times \frac{1}{(\alpha_n^2 + \beta_z^2)}\left[\alpha_n^2\tilde{H}_e + \beta_z^2\tilde{H}_m\right]$$

$$= \left(\frac{1}{(\alpha_n^2 + \beta_z^2)}\right)^2\left[\alpha_n^2\beta_z^2\left(\tilde{H}_e^2 + \tilde{H}_m^2 - 2\tilde{H}_e\tilde{H}_m\right) - \left\{\alpha_n^2\beta_z^2\tilde{H}_e^2 + \alpha_n^4\tilde{H}_e\tilde{H}_m + \beta_z^4\tilde{H}_e\tilde{H}_m + \alpha_n^2\beta_z^2\tilde{H}_m^2\right\}\right] \tag{16.2.52}$$

$$= \left(\frac{1}{(\alpha_n^2 + \beta_z^2)}\right)^2\left[\alpha_n^2\beta_z^2\tilde{H}_m^2 - 2\alpha_n^2\beta_z^2\tilde{H}_e\tilde{H}_m - \alpha_n^4\tilde{H}_e\tilde{H}_m - \beta_z^4\tilde{H}_e\tilde{H}_m - \alpha_n^2\beta_z^2\tilde{H}_m^2\right].$$

$$= \left(\frac{1}{(\alpha_n^2 + \beta_z^2)}\right)^2\left[-2\alpha_n^2\beta_z^2\tilde{H}_e\tilde{H}_m - \alpha_n^4\tilde{H}_e\tilde{H}_m - \beta_z^4\tilde{H}_e\tilde{H}_m\right].$$

$$\Rightarrow \Delta\tilde{H} = -\left(\frac{1}{(\alpha_n^2 + \beta_z^2)}\right)^2\left[2\alpha_n^2\beta_z^2 + \alpha_n^4 + \beta_z^4\right]\tilde{H}_e\tilde{H}_m = -\left(\frac{1}{(\alpha_n^2 + \beta_z^2)}\right)^2\left[\alpha_z^2 + \beta_z^2\right]^2\tilde{H}_e\tilde{H}_m. \tag{16.2.53}$$

$$\Delta\widetilde{H} = -\widetilde{H}_e\widetilde{H}_m = -\left[\frac{1}{\gamma_{y1}\tanh(\gamma_1 h_1)} + \frac{1}{\gamma_{y2}\tanh(\gamma_2 h_2)}\right]\left[\frac{\gamma_{z1}}{\tanh(\gamma_1 h_1)} + \frac{\gamma_{z2}}{\tanh(\gamma_2 h_2)}\right]$$
$$= -\left[\frac{j\omega\varepsilon_1}{\gamma_1\tanh(\gamma_1 h_1)} + \frac{j\omega\varepsilon_2}{\gamma_2\tanh(\gamma_2 h_2)}\right]\times\left[\frac{\gamma_1}{j\omega\mu_1\tanh(\gamma_1 h_1)} + \frac{\gamma_2}{j\omega\mu_2\tanh(\gamma_2 h_2)}\right]$$
$$\Rightarrow \Delta\widetilde{H} = -[\varepsilon_2\gamma_1\tanh(\gamma_1 h_1) + \varepsilon_1\gamma_2\tanh(\gamma_2 h_2)]$$
$$\times\frac{1}{\gamma_1\gamma_2\tanh(\gamma_1 h_1)\tanh(\gamma_2 h_2)}\times[\gamma_2\mu_1\coth(\gamma_2 h_2) + \gamma_1\mu_2\coth(\gamma_1 h_1)]\times\frac{1}{\mu_1\mu_2}. \tag{16.2.54}$$

Equations (16.2.33), (16.2.46), and (16.2.51) provide the matrix elements of the impedance type Green's function:

$$\widetilde{G}_{12} = \widetilde{Z}_{12} = -\frac{\widetilde{H}_{12}}{\Delta\widetilde{H}} = \frac{-\alpha_n\beta_z}{\Delta\widetilde{H}(\alpha_n^2 + \beta_n^2)}\left[\widetilde{H}_e - \widetilde{H}_m\right]$$
$$= \frac{-\alpha_n\beta_z}{\Delta\widetilde{H}(\alpha_n^2 + \beta_n^2)}\left[\frac{1}{\gamma_{y1}\tanh(\gamma_1 h_1)} + \frac{1}{\gamma_{y2}\tanh(\gamma_2 h_2)} - \frac{\gamma_{z1}}{\tanh(\gamma_1 h_1)} - \frac{\gamma_{z2}}{\gamma_{y2}\tanh(\gamma_2 h_2)}\right]$$
$$= \frac{-\alpha_n\beta_z}{\Delta\widetilde{H}(\alpha_n^2 + \beta_n^2)}\left[\left(\frac{1}{\gamma_{y1}} - \gamma_{z1}\right)\frac{1}{\tanh(\gamma_1 h_1)} + \left(\frac{1}{\gamma_{y2}} - \gamma_{z2}\right)\frac{1}{\tanh(\gamma_2 h_2)}\right]$$
$$= \frac{-\alpha_n\beta_z}{\Delta\widetilde{H}(\alpha_n^2 + \beta_n^2)}\left[\left(\frac{j\omega\varepsilon_1}{\gamma_1} - \frac{\gamma_1}{\omega\mu_1}\right)\frac{1}{\tanh(\gamma_1 h_1)} + \left(\frac{j\omega\varepsilon_2}{\gamma_2} - \frac{\gamma_2}{\omega\mu_2}\right)\frac{1}{\tanh(\gamma_2 h_2)}\right]$$
$$= \frac{-\alpha_n\beta_z}{\Delta\widetilde{H}(\alpha_n^2 + \beta_n^2)}\left[\left(\frac{-\omega^2\mu_1\varepsilon_1 - \gamma_1^2}{j\gamma_1\omega\mu_1}\right)\coth(\gamma_1 h_1) + \left(\frac{-\omega^2\mu_2\varepsilon_2 - \gamma_2^2}{j\gamma_2\omega\mu_2}\right)\coth(\gamma_2 h_2)\right]$$
$$= \frac{\alpha_n\beta_z}{\Delta\widetilde{H}(\alpha_n^2 + \beta_n^2)}\frac{1}{j\omega}\left[\left(\frac{\gamma_1^2 + k_1^2}{\gamma_1\mu_1}\right)\coth(\gamma_1 h_1) + \left(\frac{\gamma_2^2 + k_2^2}{\gamma_2\mu_{21}}\right)\coth(\gamma_2 h_2)\right]$$
$$= \frac{\alpha_n\beta_z(\alpha_n^2 + \beta_z^2)}{j\omega\Delta\widetilde{H}(\alpha_n^2 + \beta_z^2)}\left[\frac{\coth(\gamma_1 h_1)}{\gamma_1\mu_1} + \frac{\coth(\gamma_2 h_2)}{\gamma_2\mu_2}\right]$$
$$= \frac{\alpha_n\beta_z}{j\omega\Delta\widetilde{H}}\left[\frac{\coth(\gamma_1 h_1)}{\gamma_1\mu_1} + \frac{\coth(\gamma_2 h_2)}{\gamma_2\mu_2}\right]$$
$$= \frac{\alpha_n\beta_z}{j\omega\Delta\widetilde{H}}\left[\frac{\gamma_1\mu_1\tanh(\gamma_1 h_1) + \gamma_2\mu_2\tanh(\gamma_2 h_2)}{\gamma_1\gamma_2\mu_1\mu_2\tanh(\gamma_1 h_1)\tanh(\gamma_2 h_2)}\right]$$
$$= -\frac{\alpha_n\beta_z}{j\omega\varepsilon_1\mu_1}\frac{[\gamma_1\mu_1\tanh(\gamma_1 h_1) + \gamma_2\mu_2\tanh(\gamma_2 h_2)]}{\left[\gamma_2\tanh(\gamma_2 h_2) + \frac{\varepsilon_2}{\varepsilon_1}\gamma_1\tanh(\gamma_1 h_1)\right]\left[\gamma_2\coth(\gamma_2 h_2) + \frac{\mu_2}{\mu_1}\gamma_1\coth(\gamma_1 h_1)\right]}. \tag{16.2.55}$$

$$\widetilde{Z}_{12} = \frac{j\alpha_n\beta_z\omega}{k_1^2\Delta\widetilde{Z}}[\gamma_1\mu_1\tanh(\gamma_1h_1) + \gamma_2\mu_2\tanh(\gamma_2h_2)], \quad (16.2.56)$$

where

$$\Delta\widetilde{Z} = \left[\gamma_2\tanh(\gamma_2h_2) + \frac{\varepsilon_2}{\varepsilon_1}\gamma_1\tanh(\gamma_1h_1)\right]\times \left[\gamma_2\coth(\gamma_2h_2) + \frac{\mu_2}{\mu_1}\gamma_1\coth(\gamma_1h_1)\right]. \quad (16.2.57)$$

$$\widetilde{G}_{11} = \widetilde{Z}_{11} = \frac{\widetilde{H}_{11}}{\Delta\widetilde{H}}, \quad (16.2.58)$$

where $\widetilde{H}_{11}$ is taken from equation (16.2.44) as

$$\widetilde{H}_{11} = \frac{1}{(\alpha_n^2+\beta_z^2)}\left[\alpha_n^2\widetilde{H}_e + \beta_z^2\widetilde{H}_m\right]. \quad (16.2.59)$$

Using equation (16.2.33),

$$\begin{aligned}
\Rightarrow \widetilde{H}_{11} &= \frac{1}{(\alpha_n^2+\beta_z^2)}\left[\alpha_n^2\left(\frac{1}{\gamma_{y2}\tanh(\gamma_2h_2)} + \frac{1}{\gamma_{y1}\tanh(\gamma_1h_1)}\right) + \beta_z^2\left(\frac{\gamma_{z2}}{\tanh(\gamma_2h_2)} + \frac{\gamma_{z1}}{\tanh(\gamma_1h_1)}\right)\right]\\
&= \frac{1}{(\alpha_n^2+\beta_z^2)}\left[\left(\frac{\alpha_n^2}{\gamma_{y2}} + \beta_z^2\gamma_{z2}\right)\frac{1}{\tanh(\gamma_2h_2)} + \left(\frac{\alpha_n^2}{\gamma_{y1}} + \beta_z^2\gamma_{z1}\right)\frac{1}{\tanh(\gamma_1h_1)}\right]\\
&= \frac{1}{(\alpha_n^2+\beta_z^2)}\left[\left(\frac{\alpha_n^2}{\gamma_2}\times j\omega\varepsilon_2 + \beta_z^2\frac{\gamma_2}{j\omega\mu_2}\right)\frac{1}{\tanh(\gamma_2h_2)} + \left(\frac{\alpha_n^2}{\gamma_1}\times j\omega\varepsilon_1 + \beta_z^2\frac{\gamma_1}{j\omega\mu_1}\right)\frac{1}{\tanh(\gamma_1h_1)}\right]\\
&= \frac{1}{(\alpha_n^2+\beta_z^2)}\left[\left(\frac{-\omega^2\mu_2\varepsilon_2\alpha_n^2 + \beta_z^2\gamma_2^2}{j\omega\mu_2\gamma_2\tanh(\gamma_2h_2)}\right) + \left(\frac{-\omega^2\mu_1\varepsilon_1\alpha_n^2 + \beta_z^2\gamma_1^2}{j\omega\mu_1\gamma_1\tanh(\gamma_1h_1)}\right)\right]\\
\Rightarrow \widetilde{H}_{11} &= \frac{1}{j\omega(\alpha_n^2+\beta_z^2)}\left[\left(\frac{-\alpha_n^2k_2^2 + \beta_z^2(\beta_z^2+\alpha_n^2-k_2^2)}{\mu_2\gamma_2\tanh(\gamma_2h_2)}\right) + \left(\frac{-\alpha_n^2k_1^2 + \beta_z^2(\beta_z^2+\alpha_n^2-k_1^2)}{\mu_1\gamma_1\tanh(\gamma_1h_1)}\right)\right].
\end{aligned} \quad (16.2.60)$$

Let us simplify the following expressions, $I = -\alpha_n^2k_2^2 + \beta_z^2(\beta_z^2+\alpha_n^2-k_2^2) = -k_2^2(\alpha_n^2+\beta_z^2) + \beta_z^2(\alpha_n^2+\beta_z^2) = (\alpha_n^2+\beta_z^2)(\beta_z^2-k_2^2)$

Similarly, $II = -\alpha_n^2k_1^2 + \beta_z^2(\beta_z^2+\alpha_n^2-k_1^2) = (\alpha_n^2+\beta_z^2)(\beta_z^2-k_1^2)$. (16.2.61)

$$\begin{aligned}
\widetilde{H}_{11} &= \frac{1}{j\omega(\alpha_n^2+\beta_z^2)}(\alpha_n^2+\beta_z^2)\left[\frac{\beta_z^2-k_2^2}{\mu_2\gamma_2\tanh(\gamma_2h_2)} + \frac{\beta_z^2-k_1^2}{\mu_1\gamma_1\tanh(\gamma_1h_1)}\right]\\
\Rightarrow \widetilde{H}_{11} &= \frac{1}{j\omega}\left[\frac{(\beta_z^2-k_2^2)\mu_1\gamma_1\tanh(\gamma_1h_1) + (\beta_z^2-k_1^2)\mu_2\gamma_2\tanh(\gamma_2h_2)}{\mu_1\mu_2\gamma_1\gamma_2\tanh(\gamma_1h_1)\tanh(\gamma_2h_2)}\right].
\end{aligned} \quad (16.2.62)$$

Using equations (16.2.54) and (16.2.62), the following expression is obtained from equation (16.2.58):

$$\begin{aligned}
\widetilde{G}_{11} = \widetilde{Z}_{11} = \frac{\widetilde{H}_{11}}{\Delta\widetilde{H}} &= \left\{\frac{-j[\mu_1\mu_2\gamma_1\gamma_2\tanh(\gamma_1h_1)\tanh(\gamma_2h_2)]}{-\omega[\varepsilon_2\gamma_1\tanh(\gamma_1h_1) + \varepsilon_1\gamma_2\tanh(\gamma_2h_2)][\gamma_2\mu_1\coth(\gamma_2h_2) + \gamma_1\mu_2\coth(\gamma_1h_1)]}\right\}\times\\
&\quad \left\{\frac{(\beta_z^2-k_2^2)\mu_1\gamma_1\tanh(\gamma_1h_1) + (\beta_z^2-k_1^2)\mu_2\gamma_2\tanh(\gamma_2h_2)}{\mu_1\mu_2\gamma_1\gamma_2\tanh(\gamma_1h_1)\tanh(\gamma_2h_2)}\right\}\\
&= \frac{j\left[(\beta_z^2-k_2^2)\mu_1\gamma_1\tanh(\gamma_1h_1) + (\beta_z^2-k_1^2)\mu_2\gamma_2\tanh(\gamma_2h_2)\right]}{\omega\varepsilon_1\mu_1\left[\gamma_2\tanh(\gamma_2h_2) + \frac{\varepsilon_2}{\varepsilon_1}\gamma_1\tanh(\gamma_1h_1)\right]\left[\gamma_2\coth(\gamma_2h_2) + \frac{\mu_2}{\mu_1}\gamma_1\coth(\gamma_1h_1)\right]}\\
\Rightarrow \widetilde{G}_{11} = \widetilde{Z}_{11} &= \frac{-j\omega\left[(k_2^2-\beta_z^2)\mu_1\gamma_1\tanh(\gamma_1h_1) + (k_1^2-\beta_z^2)\mu_2\gamma_2\tanh(\gamma_2h_2)\right]}{k_1^2\Delta\widetilde{Z}},
\end{aligned} \quad (16.2.63)$$

where $\Delta\widetilde{Z}$ is given by equation (16.2.57).

Determination of Z_{22}

The following expression is obtained by using equation (16.2.51):

$$\widetilde{G}_{22} = \widetilde{Z}_{22} = \frac{\widetilde{H}_{22}}{\Delta\widetilde{H}}. \tag{16.2.64}$$

The following expression is obtained from equation (16.2.34):

$$\widetilde{Z}_{22} = \frac{1}{(\alpha_n^2+\beta_z^2)\Delta\widetilde{H}}\left[\beta_z^2\widetilde{H}_e + \alpha_n^2\widetilde{H}_m\right]. \tag{16.2.65}$$

Using equation (16.2.34), the expression of matrix component $\widetilde{H}_{22}$ is simplified as follow:

$$\begin{aligned}
\widetilde{H}_{22} &= \frac{1}{(\alpha_n^2+\beta_z^2)}\left[\beta_z^2\widetilde{H}_e + \alpha_n^2\widetilde{H}_m\right]\\
&= \frac{1}{(\alpha_n^2+\beta_z^2)}\left[\beta_z^2\left(\frac{1}{\gamma_{y1}\tanh(\gamma_1 h_1)} + \frac{1}{\gamma_{y2}\tanh(\gamma_2 h_2)}\right) + \alpha_n^2\left(\frac{\gamma_{z1}}{\tanh(\gamma_1 h_1)} + \frac{\gamma_{z2}}{\tanh(\gamma_2 h_2)}\right)\right]\\
&= \frac{1}{(\alpha_n^2+\beta_z^2)}\left[\left(\frac{\beta_z^2}{\gamma_{y1}} + \alpha_n^2\gamma_{z1}\right)\frac{1}{\tanh(\gamma_1 h_1)} + \left(\frac{\beta_z^2}{\gamma_{y2}} + \alpha_n^2\gamma_{z2}\right)\frac{1}{\tanh(\gamma_2 h_2)}\right]\\
&= \frac{1}{(\alpha_n^2+\beta_z^2)}\left[\left(\frac{\beta_z^2}{\gamma_1}\times j\omega\varepsilon_1 + \alpha_n^2\frac{\gamma_1}{j\omega\mu_1}\right)\frac{1}{\tanh(\gamma_1 h_1)} + \left(\frac{\beta_z^2}{\gamma_2}\times j\omega\varepsilon_2 + \alpha_n^2\frac{\gamma_2}{j\omega\mu_2}\right)\frac{1}{\tanh(\gamma_2 h_2)}\right]\\
&= \frac{1}{(\alpha_n^2+\beta_z^2)}\left[\left(\frac{-\omega^2\mu_1\varepsilon_1\beta_z^2 + \alpha_n^2\gamma_1^2}{j\,\omega\mu_1\gamma_1\tanh(\gamma_1 h_1)}\right) + \left(\frac{-\omega^2\mu_2\varepsilon_2\beta_z^2 + \alpha_n^2\gamma_2^2}{j\omega\mu_2\gamma_2\tanh(\gamma_2 h_2)}\right)\right].
\end{aligned} \tag{16.2.66}$$

$$\begin{aligned}
\text{Let } I &= -\omega^2\mu_1\varepsilon_1\beta_z^2 + \alpha_n^2\gamma_1^2 = -\beta_z^2 k_1^2 + \alpha_n^2(\beta_z^2+\alpha_n^2-k_1^2)\\
&= -k_1^2(\beta_z^2+\alpha_n^2) + \alpha_n^2(\beta_z^2+\alpha_n^2) = (\beta_z^2+\alpha_n^2)(\alpha_n^2-k_1^2)\\
II &= -\omega^2\mu_2\varepsilon_2\beta_z^2 + \alpha_n^2\gamma_2^2 = (\beta_z^2+\alpha_n^2)(\alpha_n^2-k_2^2).
\end{aligned} \tag{16.2.67}$$

$$\begin{aligned}
\widetilde{H}_{22} &= \frac{(\alpha_n^2+\beta_z^2)}{j\omega(\alpha_n^2+\beta_z^2)}\left[\left(\frac{(\alpha_n^2-k_1^2)}{\mu_1\gamma_1\tanh(\gamma_1 h_1)}\right) + \left(\frac{(\alpha_n^2-k_2^2)}{\mu_2\gamma_2\tanh(\gamma_2 h_2)}\right)\right]\\
\Rightarrow \widetilde{H}_{22} &= \frac{1}{j\omega}\left[\frac{\mu_2\gamma_2(\alpha_n^2-k_1^2)\tanh(\gamma_2 h_2) + \mu_1\gamma_1(\alpha_n^2-k_2^2)\tanh(\gamma_1 h_1)}{\mu_1\mu_2\gamma_1\gamma_2\tanh(\gamma_1 h_1)\tanh(\gamma_2 h_2)}\right].
\end{aligned} \tag{16.2.68}$$

$$\widetilde{G}_{22} = \widetilde{Z}_{22} = \frac{1}{j\omega\Delta\widetilde{H}}\left[\frac{\mu_2\gamma_2(\alpha_n^2-k_1^2)\tanh(\gamma_2 h_2) + \mu_1\gamma_1(\alpha_n^2-k_2^2)\tanh(\gamma_1 h_1)}{\mu_1\mu_2\gamma_1\gamma_2\tanh(\gamma_1 h_1)\tanh(\gamma_2 h_2)}\right]. \tag{16.2.69}$$

On substituting $\Delta\widetilde{H}$ from equation (16.2.54) in the above equation (16.2.69), the following expression is obtained:

$$\begin{aligned}
\widetilde{Z}_{22} &= \frac{1}{j\omega}\left[\frac{\mu_2\gamma_2(\alpha_n^2-k_1^2)\tanh(\gamma_2 h_2) + \mu_1\gamma_1(\alpha_n^2-k_2^2)\tanh(\gamma_1 h_1)}{\mu_1\mu_2\gamma_1\gamma_2\tanh(\gamma_1 h_1)\tanh(\gamma_2 h_2)}\right]\\
&\quad\times\frac{\mu_1\mu_2\gamma_1\gamma_2\tanh(\gamma_1 h_1)\tanh(\gamma_2 h_2)}{-[\varepsilon_2\gamma_1\tanh(\gamma_1 h_1) + \varepsilon_1\gamma_2\tanh(\gamma_2 h_2)][\mu_2\gamma_1\coth(\gamma_1 h_1) + \mu_1\gamma_2\coth(\gamma_2 h_2)]}\\
\widetilde{Z}_{22} &= \frac{j}{\omega\mu_1\varepsilon_1}\left[\frac{\mu_2\gamma_2(\alpha_n^2-k_1^2)\tanh(\gamma_2 h_2) + \mu_1\gamma_1(\alpha_n^2-k_2^2)\tanh(\gamma_1 h_1)}{\left\{\gamma_2\tanh(\gamma_2 h_2) + \frac{\varepsilon_2}{\varepsilon_1}\gamma_1\tanh(\gamma_1 h_1)\right\}\left\{\gamma_2\coth(\gamma_2 h_2) + \frac{\mu_2}{\mu_1}\gamma_1\coth(\gamma_1 h_1)\right\}}\right]\\
\Rightarrow \widetilde{Z}_{22} &= \frac{-j\omega}{k_1^2}\left[\frac{\mu_2\gamma_2(k_1^2-\alpha_n^2)\tanh(\gamma_2 h_2) + \mu_1\gamma_1(k_2^2-\alpha_n^2)\tanh(\gamma_1 h_1)}{\Delta\widetilde{Z}}\right].
\end{aligned} \tag{16.2.70}$$

The elements of Green's function in the impedance matrix form, useful for the microstrip analysis, are summarized below:

$$\tilde{G}_{11} = \tilde{Z}_{11} = \frac{-j\omega}{k_1^2 \Delta\tilde{Z}}\left[(k_1^2 - \beta_z^2)\gamma_2\mu_2 \tanh(\gamma_2 h_2) + (k_2^2 - \beta_z^2)\gamma_1\mu_1 \tanh(\gamma_1 h_1)\right] \quad (a)$$

$$\tilde{G}_{12} = \tilde{G}_{21} = \tilde{Z}_{12} = \tilde{Z}_{21} = \frac{j\alpha_n\beta_z\omega}{k_1^2 \Delta\tilde{Z}}[\gamma_2\mu_2 \tanh(\gamma_2 h_2) + \gamma_1\mu_1 \tanh(\gamma_1 h_1)] \quad (b)$$

$$\tilde{G}_{22} = \tilde{Z}_{22} = \frac{-j\omega}{k_1^2 \Delta\tilde{Z}}\left[(k_1^2 - \alpha_n^2)\gamma_2\mu_2 \tanh(\gamma_2 h_2) + (k_2^2 - \alpha_n^2)\gamma_1\mu_1 \tanh(\gamma_1 h_1)\right] \quad (c) \qquad (16.2.71)$$

where,

$$\Delta\tilde{Z} = \left[\gamma_2 \tanh(\gamma_2 h_2) + \frac{\varepsilon_2}{\varepsilon_1}\gamma_1 \tanh(\gamma_1 h_1)\right]\left[\gamma_2 \coth(\gamma_2 h_2) + \frac{\mu_2}{\mu_1}\gamma_1 \coth(\gamma_1 h_1)\right] \quad (d).$$

16.2.3 Case #II – Sides: MW – EW, Bottom: MW, Top: EW

Figure (16.3b) shows the slot line structure for the case-II. The left-hand side of the slot line is the MW and the right-hand side is the EW. Its bottom is the MW, and the top is the EW. The structure is one fourth of the broadside coupled CPW, shown in Figure (16.1e), and could also serve to obtain Green's function of a broadside coupled microstrip, shown in Figure (16.1f). The $\left[\tilde{H}\right]$-type Green's function in the Fourier domain, i.e. the admittance $\left[\tilde{Y}\right]$, is obtained below. The case #II could be applied to a *slot excited symmetrical broadside coupled CPW*, shown in Fig (16.1e). On inverting it, the impedance $\left[\tilde{Z}\right]$, i.e. the $\left[\tilde{G}\right]$-type Green's function is obtained. It could be applied to a *current excited symmetrical broadside coupled microstrip*, shown in Fig (16.1f). The potential functions $\tilde{\phi}^e(\alpha_n, y)$ and $\tilde{\phi}^m(\alpha_n, y)$ in Fourier domain follow equation (16.2.10). Their solutions in the lower region #1 and the upper region #2 are given by equations (16.2.11) and (16.2.12), respectively. The boundary condition is applied in both regions.

Region #1 (Lower Region)

The bottom plane $y = 0$ is the MW with $H_z = H_x = 0$. We get following boundary conditions, from equation (16.2.9), for both the potentials on the bottom MW:

$$\tilde{\phi}^e(\alpha_n, y = 0) = 0, \quad \left.\frac{d\tilde{\phi}_1^m}{dy}\right|_{y=0} = 0. \qquad (16.2.72)$$

On using the above boundary conditions with equation (16.2.11), $C^e = A^m = 0$. The solutions for the potential functions in the lower region are

$$\tilde{\phi}_1^e(\alpha_n, y) = A^e \sinh[\gamma_1 y], \quad \tilde{\phi}_1^m(\alpha_n, y) = A^m \cosh[\gamma_1 y]. \qquad (16.2.73)$$

In the above equations (16.2.73), the constants are changed without any change in the potential function. It is done to maintain the symmetry in the above equations.

Region #2 (Upper Region)

The plane $y = H$, i.e. the top shield is the EW with $E_z = E_x = 0$. The following boundary conditions are written from equation (16.2.9):

$$\tilde{\phi}^m(\alpha_n, y = H) = 0, \quad \left.\frac{d\tilde{\phi}^e}{dy}\right|_{y=H} = 0. \qquad (16.2.74)$$

The above boundary conditions (16.2.74), used with equation (16.2.12), provide $B^e = D^m = 0$. The solutions for the potential functions in the upper region are

$$\begin{aligned} \phi_2^e(\alpha_n, y) &= B^e \cosh[\gamma_2(H - y)] \\ \phi_2^m(\alpha_n, y) &= B^m \sinh[\gamma_2(H - y)]. \end{aligned} \qquad (16.2.75)$$

Again, the constants in the above equations (16.2.75) are changed without changing the potential functions.

The Tangential Field Components in Region #1

The following field components are obtained on substituting the potential functions given by equation (16.2.73) in an equation (16.2.9):

$$\begin{aligned} \tilde{E}_{x1} &= -j\alpha_n\gamma_{y1}A^e \cosh[\gamma_1 y] - j\beta_z A^m \cosh[\gamma_1 y] \\ \tilde{H}_{x1} &= j\beta_z A^e \sinh[\gamma_1 y] - j\alpha_n\gamma_{z1}A^m \sinh[\gamma_1 y] \\ \tilde{E}_{z1} &= -j\beta_z\gamma_{y1}\, A^e \cosh[\gamma_1 y] + j\alpha_n A^m \cosh[\gamma_1 y] \\ \tilde{H}_{z1} &= -j\alpha_n A^e \sinh[\gamma_1 y] - j\beta_z\gamma_{z1}A^m \sinh[\gamma_1 y] \end{aligned} \qquad (16.2.76)$$

where, $\gamma_{yi} = \dfrac{\gamma_i}{\tilde{y}_i}$, $\gamma_{zi} = \dfrac{\gamma_i}{\tilde{z}_i}$ $(i = 1, 2)$.

The Tangential Field Components in Region #2

Likewise, the components in the upper region #2 are written as follows:

$$\tilde{E}_{x2} = j\alpha_n\gamma_{y2}B^e \sinh[\gamma_2(H-y)] - j\beta_z B^m \sinh[\gamma_2(H-y)]$$
$$\tilde{H}_{x2} = j\beta_z B^e \cosh[\gamma_2(H-y)] + j\alpha_n\gamma_{z2}B^m \cosh[\gamma_2(H-y)]$$
$$\tilde{E}_{z2} = j\beta_z\gamma_{y2}B^e \sinh[\gamma_2(H-y)] + j\alpha_n B^m \sinh[\gamma_2(H-y)]$$
$$\tilde{H}_{z2} = -j\alpha_n B^e \cosh[\gamma_2(H-y)] + j\beta_z\gamma_{z2}B^m \cosh[\gamma_2(H-y)]. \quad (16.2.77)$$

The *continuity condition* at the boundary, $y = h_1$ of two media is applied:

$$\tilde{E}_{x1}\Big|_{y=h_1} = \tilde{E}_{x2}\Big|_{y=h_1}, \quad \tilde{E}_{z1}\Big|_{y=h_1} = \tilde{E}_{z2}\Big|_{y=h_1}. \quad (16.2.78)$$

$$\left\{-j\alpha_n\gamma_{y1}\cosh[\gamma_1 h_1]\right\}A^e + \{-j\beta_z\cosh(\gamma_1 h_1)\}A^m =$$
$$\left\{j\alpha_n\gamma_{y2}\sinh[\gamma_2 h_2]\right\}B^e + \{-j\beta_z\sinh(\gamma_2 h_2)\}B^m \quad (a)$$
$$\left\{-j\beta_z\gamma_{y1}\cosh[\gamma_1 h_1]\right\}A^e + \{j\alpha_n\cosh(\gamma_1 h_1)\}A^m =$$
$$\left\{j\beta_z\gamma_{y2}\sinh[\gamma_2 h_2]\right\}B^e + \{j\alpha_n\sinh(\gamma_2 h_2)\}B^m \quad (b). \quad (16.2.79)$$

The following relations are obtained, by solving the above two equations:

$$A^e = -\frac{\gamma_{y2}}{\gamma_{y1}}\frac{\sinh(\gamma_2 h_2)}{\cosh(\gamma_1 h_1)}B^e \quad (a)$$
$$A^m = \frac{\sinh(\gamma_2 h_2)}{\cosh(\gamma_1 h_1)}B^m \quad (b). \quad (16.2.80)$$

By assuming the electric field components, $\tilde{E}_{x2}(\alpha_n, y)\big|_{y=h_1}$ and $\tilde{E}_{z2}(\alpha_n, y)\big|_{y=h_1}$, in the slot, a region at $y = h_1$, the constants B^e and B^m are evaluated in terms of these field components using equation (16.2.77):

$$\tilde{E}_{x2}(\alpha_n, h_1) = \left\{j\alpha_n\gamma_{y2}\sinh[\gamma_2 h_2]\right\}B^e + \{-j\beta_z\sinh(\gamma_2 h_2)\}B^m \quad (a)$$
$$\tilde{E}_{z2}(\alpha_n, h_1) = \left\{j\beta_z\gamma_{y2}\sinh[\gamma_2 h_2]\right\}B^e + \{j\alpha_n\sinh(\gamma_2 h_2)\}B^m \quad (b). \quad (16.2.81)$$

$$B^e = -j\frac{\alpha_n\tilde{E}_{x2}(\alpha_n, h_1) + \beta_z\tilde{E}_{z2}(\alpha_n, h_1)}{\gamma_{y2}(\alpha_n^2 + \beta_z^2)\sinh(\gamma_2 h_2)} \quad (a)$$
$$B^m = -j\frac{\alpha_n\tilde{E}_{z2}(\alpha_n, h_1) - \beta_z\tilde{E}_{x2}(\alpha_n, h_1)}{(\alpha_n^2 + \beta_z^2)\sinh(\gamma_2 h_2)} \quad (b). \quad (16.2.82)$$

The assumed slot-field components $\tilde{E}_{x2}(\alpha_n, h_1)$ and $\tilde{E}_{z2}(\alpha_n, h_1)$ induce electric currents $\tilde{J}_x(\alpha_n, h_1)$ and $\tilde{J}_z(\alpha_n, h_1)$ on the strip conductors. The field components and current components are related through dyadic Green's function. It is obtained by applying the *discontinuity condition* for H_x and H_z field components on the strip conductors. The discontinuity in the field component H_x is considered below:

$$\left(\tilde{H}_{x2} - \tilde{H}_{x1}\right)\Big|_{y=h_1} = \tilde{J}_z\Big|_{y=h_1}$$
$$\{j\beta_z\cosh[\gamma_2 h_2]\}B^e + \{j\alpha_n\gamma_{z2}\cosh(\gamma_2 h_2)\}B^m -$$
$$\{j\beta_z\sinh[\gamma_1 h_1]\}A^e + \{j\alpha_n\gamma_{z1}\sinh(\gamma_1 h_1)\}A^m = \tilde{J}_z(\alpha_n, h_1). \quad (16.2.83)$$

On substituting A^e and A^m from equation (16.2.80) in the above equation (16.2.83):

$$\{j\beta_z\cosh(\gamma_2 h_2)\}B^e + \{j\alpha_n\gamma_{z2}\cosh(\gamma_2 h_2)\}B^m$$
$$+ \{j\beta_z\sinh(\gamma_1 h_1)\}\left\{\frac{\gamma_{y2}}{\gamma_{y1}}\frac{\sinh(\gamma_2 h_2)}{\cosh(\gamma_1 h_1)}\right\}B^e$$
$$+ \{j\alpha_n\gamma_{z1}\sinh(\gamma_1 h_1)\}\left\{\frac{\sinh(\gamma_2 h_2)}{\cosh(\gamma_1 h_1)}\right\}B^m = \tilde{J}_z$$
$$\Rightarrow j\beta_z\left[\cosh(\gamma_2 h_2) + \frac{\gamma_{y2}\sinh(\gamma_2 h_2)}{\gamma_{y1}\coth(\gamma_1 h_1)}\right]B^e$$
$$+ j\alpha_n\left[\gamma_{z2}\cosh(\gamma_2 h_2) + \frac{\gamma_{z1}\sinh(\gamma_2 h_2)}{\coth(\gamma_1 h_1)}\right]B^m = \tilde{J}_z. \quad (16.2.84)$$

$$\Rightarrow (jM_1)B^e + (jN_1)B^m = \tilde{J}_z(\alpha_n, h_1) \quad (16.2.85)$$

where,

$$M_1 = \beta_z\left[\cosh(\gamma_2 h_2) + \frac{\gamma_{y2}\sinh(\gamma_2 h_2)}{\gamma_{y1}\coth(\gamma_1 h_1)}\right] \quad (a)$$
$$N_1 = \alpha_n\left[\gamma_{z2}\cosh(\gamma_2 h_2) + \frac{\gamma_{z1}\sinh(\gamma_2 h_2)}{\coth(\gamma_1 h_1)}\right] \quad (b). \quad (16.2.86)$$

On substituting B^e and B^m from equation (16.2.82) in equation (16.2.85),

$$(jM_1)\left[-j\frac{\alpha_n\tilde{E}_{x2}(\alpha_n, h_1) + \beta_z\tilde{E}_{z2}(\alpha_n, h_1)}{\gamma_{y2}(\alpha_n^2 + \beta_z^2)\sinh(\gamma_2 h_2)}\right]$$
$$+ (jN_1)\left[-j\frac{\alpha_n\tilde{E}_{z2}(\alpha_n, h_1) - \beta_z\tilde{E}_{x2}(\alpha_n, h_1)}{(\alpha_n^2 + \beta_z^2)\sinh(\gamma_2 h_2)}\right] = \tilde{J}_z(\alpha_n, h_1). \quad (16.2.87)$$

$$\left[\frac{M_1\alpha_n}{\gamma_{y2}(\alpha_n^2+\beta_z^2)\sinh(\gamma_2 h_2)} - \frac{\beta_z N_1}{(\alpha_n^2+\beta_z^2)\sinh(\gamma_2 h_2)}\right] \times \tilde{E}_{x2}(\alpha_n, h_1)$$
$$+\left[\frac{M_1\beta_z}{\gamma_{y2}(\alpha_n^2+\beta_z^2)\sinh(\gamma_2 h_2)} + \frac{N_1\alpha_n}{(\alpha_n^2+\beta_z^2)\sinh(\gamma_2 h_2)}\right] \times \tilde{E}_{z2}(\alpha_n, h_1) = \tilde{J}_z(\alpha_n, h_1)$$
$$\Rightarrow P_1\tilde{E}_{x2}(\alpha_n, d) + Q_1\,\tilde{E}_{z2}(\alpha_n, d) = \tilde{J}_z(\alpha_n, h_1). \tag{16.2.88}$$

where,

$$P_1 = \left[\frac{M_1\alpha_n}{\gamma_{y2}} - N_1\beta_z\right] \times \frac{1}{(\alpha_n^2+\beta_z^2)\sinh(\gamma_2 h_2)} \quad (a)$$
$$Q_1 = \left[\frac{M_1\beta_z}{\gamma_{y2}} + N_1\alpha_n\right] \times \frac{1}{(\alpha_n^2+\beta_z^2)\sinh(\gamma_2 h_2)} \quad (b). \tag{16.2.89}$$

The parameters P_1 and Q_1 are simplified below:

$$P_1 = \left[\frac{\alpha_n}{\gamma_{y2}}\beta_z\left[\cosh(\gamma_2 h_2) + \frac{\gamma_{y2}\sinh(\gamma_2 h_2)}{\gamma_{y1}\coth(\gamma_1 h_1)}\right] - \beta_z\alpha_n\left[\gamma_{z2}\cosh(\gamma_2 h_2) + \frac{\gamma_{z1}\sinh(\gamma_2 h_2)}{\coth(\gamma_1 h_1)}\right]\right] \times \frac{1}{(\alpha_n^2+\beta_z^2)\sinh(\gamma_2 h_2)}$$
$$\Rightarrow P_1 = \frac{\beta_z\alpha_n}{(\alpha_n^2+\beta_z^2)}\left\{\left[\frac{1}{\gamma_{y2}\tanh(\gamma_2 h_2)} + \frac{1}{\gamma_{y1}\coth(\gamma_1 h_1)}\right] - \left[\frac{\gamma_{z2}}{\tanh(\gamma_2 h_2)} + \frac{\gamma_{z1}}{\coth(\gamma_1 h_1)}\right]\right\}. \tag{16.2.90}$$

$$Q_1 = \left[\frac{\beta_z^2}{\gamma_{y2}}\left[\cosh(\gamma_2 h_2) + \frac{\gamma_{y2}\sinh(\gamma_2 h_2)}{\gamma_{y1}\coth(\gamma_1 h_1)}\right] + \alpha_n^2\left[\gamma_{z2}\cosh(\gamma_2 h_2) + \frac{\gamma_{z1}\sinh(\gamma_2 h_2)}{\coth(\gamma_1 h_1)}\right]\right] \times \frac{1}{(\alpha_n^2+\beta_z^2)\sinh(\gamma_2 h_2)}$$
$$\Rightarrow Q_1 = \frac{1}{(\alpha_n^2+\beta_z^2)}\left\{\left[\frac{1}{\gamma_{y2}\tanh(\gamma_2 h_2)} + \frac{1}{\gamma_{y1}\coth(\gamma_1 h_1)}\right]\beta_z^2 + \left[\frac{\gamma_{z2}}{\tanh(\gamma_2 h_2)} + \frac{\gamma_{z1}}{\coth(\gamma_1 h_1)}\right]\alpha_n^2\right\}. \tag{16.2.91}$$

Let,

$$\tilde{H}_e = \frac{1}{\gamma_{y2}\tanh(\gamma_2 h_2)} + \frac{1}{\gamma_{y1}\coth(\gamma_1 h_1)} \quad (a)$$
$$\tilde{H}_m = \frac{\gamma_{z2}}{\tanh(\gamma_2 h_2)} + \frac{\gamma_{z1}}{\coth(\gamma_1 h_1)} \quad (b). \tag{16.2.92}$$

Using equations (16.2.90)–(16.2.92), and (16.2.34), two elements of the admittance type of dyadic Green's function are written as follows:

$$P_1 = \tilde{H}_{21} = \tilde{Y}_{21} = \frac{\beta_z\alpha_n}{(\alpha_n^2+\beta_z^2)}\left\{\tilde{H}_e - \tilde{H}_m\right\} \quad (a)$$
$$Q_1 = \tilde{H}_{22} = \tilde{Y}_{22} = \frac{1}{(\alpha_n^2+\beta_z^2)}\left\{\tilde{H}_e\beta_z^2 + \tilde{H}_m\alpha_n^2\right\} \quad (b). \tag{16.2.93}$$

In terms of the above components (16.2.93) of the admittance type dyadic Green's function, equation (16.2.88) is rewritten as follows:

$$\tilde{H}_{21}(\alpha_n)\,\tilde{E}_{x2}(\alpha_n, h_1) + \tilde{H}_{22}(\alpha_n)\tilde{E}_{z2}(\alpha_n, h_1) = \tilde{J}_z(\alpha_n, h_1). \tag{16.2.94}$$

Two more matrix elements of the admittance type dyadic Green's function are obtained from another boundary condition for the field H_z on the strip conductor located at $y = h_1$. The relation is written from equations (16.2.36) and (16.2.48):

$$\left(\tilde{H}_{z1} - \tilde{H}_{z2}\right)\Big|_{y=h_1} = \tilde{J}_x\Big|_{y=h_1}$$
$$\Rightarrow \tilde{H}_{11}(\alpha_n)\tilde{E}_{x2}(\alpha_n) + \tilde{H}_{12}(\alpha_n)\tilde{E}_{z2}(\alpha_n) = \tilde{J}_x(\alpha_n). \tag{16.2.95}$$

The following expressions are written by using the symmetry in Green's function for case #1, i.e. from equation (16.2.44):

$$\tilde{H}_{12} = \tilde{H}_{21}, \text{ and } \tilde{H}_{11}(\alpha_n) = \frac{1}{(\alpha_n^2+\beta_z^2)}\left\{\tilde{H}_e\alpha_n^2 + \tilde{H}_m\beta_z^2\right\}. \tag{16.2.96}$$

The slot field components $\tilde{E}_{x1}(\alpha_n)$ and $\tilde{E}_{z1}(\alpha_n)$ at $y = h_1$ can be suitably assumed,

$$\tilde{E}_{x1}(\alpha_n) = \tilde{E}_{x2}(\alpha_n) = \tilde{E}_x(\alpha_n),\ \tilde{E}_{z1}(\alpha_n) = \tilde{E}_{z2}(\alpha_n) = \tilde{E}_z(\alpha_n). \tag{16.2.97}$$

The induced current on the strip conductors are related to the assumed slot electric fields $\tilde{E}_x(\alpha_n)$, $\tilde{E}_{z1}(\alpha_n)$ by Green's function:

$$\widetilde{H}_{11}(\alpha_n)\widetilde{E}_x(\alpha_n) + \widetilde{H}_{12}(\alpha_n)\widetilde{E}_z(\alpha_n) = \widetilde{J}_x(\alpha_n)$$
$$\widetilde{H}_{21}(\alpha_n)\widetilde{E}_x(\alpha_n) + \widetilde{H}_{22}(\alpha_n)\widetilde{E}_z(\alpha_n) = \widetilde{J}_z(\alpha_n)$$

$$\Rightarrow \begin{bmatrix} \widetilde{H}_{11}(\alpha_n) & \widetilde{H}_{12}(\alpha_n) \\ \widetilde{H}_{21}(\alpha_n) & \widetilde{H}_{22}(\alpha_n) \end{bmatrix} \begin{bmatrix} \widetilde{E}_x(\alpha_n) \\ \widetilde{E}_z(\alpha_n) \end{bmatrix} = \begin{bmatrix} \widetilde{J}_x(\alpha_n) \\ \widetilde{J}_z(\alpha_n) \end{bmatrix}. \quad (16.2.98)$$

where $\widetilde{H}_{21} = \widetilde{H}_{12}$, $\widetilde{H}_{22}$ and $\widetilde{H}_{11}$ are given by equations (16.2.93) and (16.2.96), respectively. The admittance type Green's function is obtained from the $\widetilde{H}$ Green's function, i.e. $\left[\widetilde{H}\right] = \left[\widetilde{Y}\right]$. Equation (16.2.98) is useful for the slot kind of the structure, like CPW. However, for a microstrip, the admittance type Green's function is inverted to get the impedance kind of Green's function $\left[\widetilde{G}\right] = \left[\widetilde{Z}\right] = \left[\widetilde{H}\right]^{-1}$. The Green's function $\left[\widetilde{G}\right]$ is useful for the analysis of the broadside-coupled microstrip line shown in Fig (16.1f).

16.3 Solution of Hybrid Mode Field Equations (Galerkin's Method in Fourier Domain)

Galerkin's method in the Fourier domain, to solve the static problem of microstrip and CPW, is discussed in section (14.6) of chapter 14. In the present case, we are interested in obtaining the z-directed propagation constant β_z of the hybrid mode. It is computed for both the microstrip and CPW.

16.3.1 Microstrip

At the plane containing the microstrip conductor, shown in Fig (16.1b), the transverse and longitudinal current components on the strip conductor are related to the transverse and longitudinal components of the electric field components at the dielectric interface through the impedance kind of dyadic Green's function:

$$\begin{bmatrix} \widetilde{Z}_{11}(\alpha_n, \beta_z) & \widetilde{Z}_{12}(\alpha_n, \beta_z) \\ \widetilde{Z}_{21}(\alpha_n, \beta_z) & \widetilde{Z}_{22}(\alpha_n, \beta_z) \end{bmatrix} \begin{bmatrix} \widetilde{J}_x(\alpha_n) \\ \widetilde{J}_z(\alpha_n) \end{bmatrix} = \begin{bmatrix} \widetilde{E}_x(\alpha_n) \\ \widetilde{E}_z(\alpha_n) \end{bmatrix}. \quad (16.3.1)$$

At this stage, we have a system of two equations with four unknowns, i.e. $\widetilde{J}_x(\alpha_n)$, $\widetilde{J}_z(\alpha_n)$, $\widetilde{E}_x(\alpha_n)$ and $\widetilde{E}_z(\alpha_n)$. In the previous section, expressions for the elements of impedance kind dyadic Green' function are obtained. In equation (16.3.1), two unknowns $\widetilde{E}_x(\alpha_n)$ and $\widetilde{E}_z(\alpha_n)$ can be easily eliminated through Galerkin's method. The current density components, $\widetilde{J}_x(x)$ and $\widetilde{J}_z(x)$, are nonzero only on the strip conductor and are zero at the interface of two dielectric media, i.e. in the slot regions. Likewise, the tangential electric field components $\widetilde{E}_x(x)$ and $\widetilde{E}_z(x)$ are zero on the perfect strip conductor, while these are nonzero at the dielectric interface, i.e. in the slot region. Thus, the strip-slot regions provide *complementary boundary conditions.*

For applying Galerkin's method, the first step is to *expand the unknown surface current density* $\widetilde{J}_x(\alpha_n)$ *and* $\widetilde{J}_z(\alpha_n)$ *in terms of the known, i.e. the assumed basis functions* $\widetilde{J}_{xm}(\alpha_n)$ and $\widetilde{J}_{zm}(\alpha_n)$:

$$\begin{aligned} \widetilde{J}_x(\alpha_n) &= \sum_{m=1}^{M} c_m \widetilde{J}_{xm}(\alpha_n) \quad &(a) \\ \widetilde{J}_z(\alpha_n) &= \sum_{m=1}^{M} d_m \widetilde{J}_{zm}(\alpha_n) \quad &(b), \end{aligned} \quad (16.3.2)$$

where constants c_m and d_m are the coefficients of expansion for m = 1, 2,...M terms of the *basis functions.* On substituting equation (16.3.2) in equation (16.3.1), the following expressions are obtained:

$$\widetilde{Z}_{11}(\alpha_n, \beta_z) \sum_{m=1}^{M} c_m \widetilde{J}_{xm}(\alpha_n) + \widetilde{Z}_{12}(\alpha_n, \beta_z) \sum_{m=1}^{M} d_m \widetilde{J}_{zm}(\alpha_n) = \widetilde{E}_x(\alpha_n) \quad (a)$$

$$\widetilde{Z}_{21}(\alpha_n, \beta_z) \sum_{m=1}^{M} c_m \widetilde{J}_{xm}(\alpha_n) + \widetilde{Z}_{22}(\alpha_n, \beta_z) \sum_{m=1}^{M} d_m \widetilde{J}_{zm}(\alpha_n) = \widetilde{E}_z(\alpha_n) \quad (b). \quad (16.3.3)$$

An inner product of equation (16.3.3a) can be taken with the *testing functions,* $\widetilde{J}_{x1}(\alpha_n)$, $\widetilde{J}_{x2}(\alpha_n)$, ..., $\widetilde{J}_{xk}(\alpha_n)$, [k = 1, 2, ...M]. We also get the inner products of equation (16.3.3b) with the testing functions, $\widetilde{J}_{z1}(\alpha_n), \widetilde{J}_{z2}(\alpha_n), ..., \widetilde{J}_{z\ell}(\alpha_n)$, [$\ell$ = 1, 2, ...N]:

$$\sum_{m=1}^{M} c_m \left\langle \widetilde{Z}_{11}(\alpha_n, \beta_z)\widetilde{J}_{xm}(\alpha_n), \widetilde{J}_{xk}(\alpha_n) \right\rangle + \sum_{m=1}^{M} d_m \left\langle \widetilde{Z}_{12}(\alpha_n, \beta_z)\widetilde{J}_{zm}(\alpha_n), \widetilde{J}_{xk}(\alpha_n) \right\rangle = \left\langle \widetilde{E}_x(\alpha_n), \widetilde{J}_{xk}(\alpha_n) \right\rangle \quad (a)$$

$$\sum_{m=1}^{M} c_m \left\langle \widetilde{Z}_{21}(\alpha_n, \beta_z)\widetilde{J}_{xm}(\alpha_n), \widetilde{J}_{z\ell}(\alpha_n) \right\rangle + \sum_{m=1}^{M} d_m \left\langle \widetilde{Z}_{22}(\alpha_n, \beta_z)\widetilde{J}_{zm}(\alpha_n), \widetilde{J}_{z\ell}(\alpha_n) \right\rangle = \left\langle \widetilde{E}_z(\alpha_n), \widetilde{J}_{z\ell}(\alpha_n) \right\rangle \quad (b). \quad (16.3.4)$$

where m indicates the number of terms of the basis function for both current densities $\tilde{J}_x(\alpha_n)$ and $\tilde{J}_z(\alpha_n)$. The k and ℓ are the number of terms of the testing functions for $\tilde{J}_x(\alpha_n)$ and $\tilde{J}_z(\alpha_n)$. Normally, $m = k = \ell = M$. For convenience, identical functions are used for both the basis and testing functions. The right-hand side of the above equation is zero on the application of Parseval's theorem:

$$\left\langle \tilde{E}_x(\alpha_n), \tilde{J}_{xk}(\alpha_n) \right\rangle = \sum_{n=1}^{\infty} \left[\tilde{E}_x(\alpha_n), \tilde{J}_{xk}(\alpha_n) \right] = \frac{1}{2L} \int_{-L}^{L} E_x(x) J_{xk}^*(x) dx$$

$$= \underbrace{\int_{-L}^{-W} E_x(x) J_{xk}^*(x) dx}_{\Downarrow\ 0} + \underbrace{\int_{-W}^{W} E_x(x) J_{xk}^*(x) dx}_{\Downarrow\ 0} + \underbrace{\int_{W}^{L} E_x(x) J_{xk}^*(x) dx}_{\Downarrow\ 0} \quad \text{(a)}$$

Dielectric Interface, Strip conductor, Dielectric Interface

$$\text{Likewise, } \left\langle \tilde{E}_z(\alpha_n), \tilde{J}_{z\ell}(\alpha_n) \right\rangle = 0 \quad \text{(b).} \quad (16.3.5)$$

Equation (16.3.4) is rewritten as

$$\sum_{m=1}^{M} K_{km}^{1,1} c_m + \sum_{m=1}^{M} K_{km}^{1,2} d_m = 0, \quad k = 1, 2, \dots M \quad \text{(a)}$$

$$\sum_{m=1}^{M} K_{\ell m}^{2,1} c_m + \sum_{m=1}^{M} K_{\ell m}^{2,2} d_m = 0 \quad \ell = 1, 2, \dots N \quad \text{(b)}$$

$$\Rightarrow \begin{bmatrix} \sum_{m=1}^{M} K_{km}^{1,1} & \sum_{m=1}^{M} K_{km}^{1,2} \\ \sum_{m=1}^{M} K_{\ell m}^{2,1} & \sum_{m=1}^{M} K_{\ell,m}^{2,2} \end{bmatrix} \begin{bmatrix} c_m \\ d_m \end{bmatrix} = \begin{bmatrix} 0 \\ 0 \end{bmatrix} \quad \text{(c),} \quad (16.3.6)$$

where,

$$K_{km}^{1,1} = \left\langle \tilde{Z}_{11}(\alpha_n, \beta_z) \tilde{J}_{xm}(\alpha_n), \tilde{J}_{xk}(\alpha_n) \right\rangle = \sum_{n=1}^{\infty} \left[\tilde{Z}_{11}(\alpha_n, \beta_z) \tilde{J}_{xm}^*(\alpha_n), \tilde{J}_{xk}(\alpha_n) \right] \quad \text{(a)}$$

$$K_{km}^{1,2} = \left\langle \tilde{Z}_{12}(\alpha_n, \beta_z) \tilde{J}_{zm}(\alpha_n), \tilde{J}_{xk}(\alpha_n) \right\rangle = \sum_{n=1}^{\infty} \left[\tilde{Z}_{12}(\alpha_n, \beta_z) \tilde{J}_{zm}^*(\alpha_n), \tilde{J}_{xk}(\alpha_n) \right] \quad \text{(b)}$$

$$K_{\ell m}^{2,1} = \left\langle \tilde{Z}_{21}(\alpha_n, \beta_z) \tilde{J}_{xm}(\alpha_n), \tilde{J}_{z\ell}(\alpha_n) \right\rangle = \sum_{n=1}^{\infty} \left[\tilde{Z}_{21}(\alpha_n, \beta_z) \tilde{J}_{xm}^*(\alpha_n), \tilde{J}_{z\ell}(\alpha_n) \right] \quad \text{(c)} \quad (16.3.7)$$

$$K_{\ell m}^{2,2} = \left\langle \tilde{Z}_{22}(\alpha_n, \beta_z) \tilde{J}_{zm}(\alpha_n), \tilde{J}_{z\ell}(\alpha_n) \right\rangle = \sum_{n=1}^{\infty} \left[\tilde{Z}_{22}(\alpha_n, \beta_z) \tilde{J}_{zm}^*(\alpha_n), \tilde{J}_{z\ell}(\alpha_n) \right] \quad \text{(d).}$$

and, $k = 1, 2, \dots M, \ \ell = 1, 2 \dots N$

For the nontrivial solution of equation (16.3.6c), its determinant (ΔK) is zero, i.e. $\Delta K = 0$. This is a transcendental equation, i.e. the *characteristic equation*. It can be solved for the propagation constant β_z at any frequency. If the root search of the propagation constant β_z is done along the real axis, we get only the real value of β_z showing normal dispersion. However, for higher permittivity lossless substrate at higher frequency complex root, giving the complex propagation constant can exist showing the presence of the *complex modes* [J.7, J.8]. The complex root can also provide the *leaky mode* for a microstrip line. In that case, power can leak in the form of the surface wave mode to the substrate [J.9]. Such leaky mode should be controlled as it results in the cross-talk between the lines. One has to be careful to avoid the spurious solution, indicating the presence of the spurious modes [J.10, J.11]. The number of discrete Fourier terms n is selected for converging value of the propagation constant β_z. It could be about 100–500. The number of the basis functions and type of the basis function influences the accuracy of the computed propagation constant β_z [J.12, J.13]. The reader can use the available FORTRAN codes [B.1, B.2]. Let us consider simple cases.

One-Term Basis Function for Longitudinal Current $J_z(x)$

The transverse current on the strip conductor is associated with higher-order modes. To compute the propagation constant β_z of the *fundamental mode,* we take the transverse current, $J_x = 0$. Only the longitudinal current $J_z(x)$, expressed by a one-term basis function, i.e. $M = 1$, is used. The following characteristic equation is obtained from equations (16.3.6c) and (16.3.7):

$$K_{km}^{1,1} = K_{km}^{1,2} = K_{lm}^{2,1} = 0, \ \sum_{m=1}^{M=1} K_{1,1}^{2,2} d_m = 0$$

$$\Rightarrow K_{1,1}^{2,2} d_1 = 0; \ \text{as } d_1 \neq 0; \ K_{1,1}^{2,2} = 0 \quad (16.3.8)$$

$$\Rightarrow \sum_{n=1}^{\infty} \left[\tilde{Z}_{22}(\alpha_n, \beta_z) \left| \tilde{J}_{z1}(\alpha_n) \right|^2 \right] = 0.$$

The dyadic Green's function component $\tilde{Z}_{22}(\alpha_n, \beta_z)$ for the given geometrical parameter of microstrip is obtained from equation (16.2.71). The assumed basis function $\tilde{J}_{z1}(\alpha_n)$ is discussed in the next section. The above transcendental equation can be solved numerically for the propagation constant β_z at any frequency.

One-Term Basis Function for Both Transverse and Longitudinal Currents

The transverse current is needed to compute the propagation constant of the higher-order modes. For $M = N = 1$, one-term basis expansion for both $\tilde{J}_x(x)$ and $\tilde{J}_z(x)$, the following expression is written from equation (16.3.6):

$$K_{1,1}^{1,1}c_1 + K_{1,1}^{1,2}d_1 = 0,\quad K_{1,1}^{2,1}c_1 + K_{1,1}^{2,2}d_1 = 0$$

$$\Rightarrow \begin{bmatrix} K_{1,1}^{1,1} & K_{1,1}^{1,2} \\ K_{1,1}^{2,1} & K_{1,1}^{2,2} \end{bmatrix} \begin{bmatrix} c_1 \\ d_1 \end{bmatrix} = 0. \qquad (16.3.9)$$

For the nontrivial solution of equation (16.3.9), its determinant is zero, $\Delta K = 0$, i.e

$$K_{1,1}^{1,1} \times K_{1,1}^{2,2} - K_{1,1}^{1,2} \times K_{1,1}^{2,1} = 0 \qquad (16.3.10)$$

where,

$$\begin{aligned} K_{1,1}^{1,1} &= \sum_{n=1}^{\infty}\left[\tilde{Z}_{11}(\alpha_n,\beta_z)\left|\tilde{J}_{x1}(\alpha_n)\right|^2\right] \\ K_{1,1}^{1,2} &= \sum_{n=1}^{\infty}\left[\tilde{Z}_{12}(\alpha_n,\beta_z)\tilde{J}_{z1}(\alpha_n)\tilde{J}_{x1}^{*}(\alpha_n)\right] \\ K_{1,1}^{2,1} &= \sum_{n=1}^{\infty}\left[\tilde{Z}_{21}(\alpha_n,\beta_z)\tilde{J}_{x1}(\alpha_n)\tilde{J}_{z1}^{*}(\alpha_n)\right] \\ K_{1,1}^{2,2} &= \sum_{n=1}^{\infty}\left[\tilde{Z}_{22}(\alpha_n,\beta_z)\left|\tilde{J}_{z1}(\alpha_n)\right|^2\right]. \end{aligned} \qquad (16.3.11)$$

Equation (16.3.10) is solved numerically to get the propagation constant β_z, by assuming the basis functions for $\tilde{J}_z(\alpha_n)$ and $\tilde{J}_x(\alpha_n)$.

16.3.2 CPW Structure

Figure (16.1a) shows the conductor backed CPW (CB-CPW) structure with the top shield. It is described by the admittance kind of Green's function:

$$\begin{bmatrix} \tilde{Y}_{11}(\alpha_n) & \tilde{Y}_{12}(\alpha_n) \\ \tilde{Y}_{21}(\alpha_n) & \tilde{Y}_{22}(\alpha_n) \end{bmatrix} \begin{bmatrix} \tilde{E}_x(\alpha_n) \\ \tilde{E}_z(\alpha_n) \end{bmatrix} = \begin{bmatrix} \tilde{J}_x(\alpha_n) \\ \tilde{J}_z(\alpha_n) \end{bmatrix}. \qquad (16.3.12)$$

The dyadic Green's function $\left[\tilde{Y}(\alpha_n)\right] = \left[\tilde{H}(\alpha_n)\right]$ components are given by equation (16.2.49). Again, we have two equations with four unknowns. In the present case, the slot electric field components, $\tilde{E}_x(\alpha_n)$ and $\tilde{E}_z(\alpha_n)$, are expanded in terms of the known, i.e. the assumed, basis functions. The components of the surface current density, $\tilde{J}_x(\alpha_n)$ and $\tilde{J}_z(\alpha_n)$, are eliminated in the Galerkin process using Parseval's theorem. The resulting matrix equation is used to determine the propagation constant β_z of the hybrid modes of the CB-CPW.

$$\begin{aligned} \tilde{E}_x(\alpha_n) &= \sum_{m=1}^{M} c_m\tilde{E}_{xm}(\alpha_n) \qquad (a) \\ \tilde{E}_z(\alpha_n) &= \sum_{m=1}^{M} d_m\tilde{E}_{zm}(\alpha_n) \qquad (b), \end{aligned} \qquad (16.3.13)$$

where m is the number of terms of the basis functions, $\tilde{E}_{xm}(\alpha_n)$ and $\tilde{E}_{zm}(\alpha_n)$. The following equations are obtained, by substituting equation (16.3.13) in an equation (16.3.12):

$$\begin{aligned} &\tilde{Y}_{11}(\alpha_n,\beta_z)\sum_{m=1}^{M} c_m\tilde{E}_{xm}(\alpha_n) + \\ &Y_{12}(\alpha_n,\beta_z)\sum_{m=1}^{M} d_m\tilde{E}_{zm}(\alpha_n) = \tilde{J}_x(\alpha_n) \qquad (a) \\ &\tilde{Y}_{21}(\alpha_n,\beta_z)\sum_{m=1}^{\infty} c_m\tilde{E}_{xm}(\alpha_n) + \\ &Y_{22}(\alpha_n,\beta_z)\sum_{m=1}^{M} d_m\tilde{E}_{zm}(\alpha_n) = \tilde{J}_z(\alpha_n) \qquad (b). \end{aligned} \qquad (16.3.14)$$

On taking the inner product of equation (16.3.14a) with each term of the testing function $\tilde{E}_{xk}(\alpha_n)$, [k = 1, 2, 3, ...M], and the inner product of equation (16.3.14b) with the testing function $\tilde{E}_{z\ell}(\alpha_n)$ [$\ell = 1, 2, 3, ...N$], the following set of equations are obtained:

$$\begin{aligned} &\sum_{m=1}^{M} K_{\ell m}^{1,1}c_m + \sum_{m=1}^{M} K_{km}^{1,2}d_m = 0, \quad [k=1,2,3,...M] \qquad (a) \\ &\sum_{m=1}^{M} K_{\ell m}^{2,1}c_m + \sum_{m=1}^{M} K_{\ell m}^{2,2}d_m = 0, \quad [\ell=1,2,3,...N] \qquad (b) \\ &\Rightarrow \begin{bmatrix} \sum_{m=1}^{M} K_{\ell m}^{1,1} & \sum_{m=1}^{M} K_{km}^{1,2} \\ \sum_{m=1}^{M} K_{\ell m}^{2,1} & \sum_{m=1}^{M} K_{\ell m}^{2,2} \end{bmatrix} \begin{bmatrix} c_m \\ d_m \end{bmatrix} = \begin{bmatrix} 0 \\ 0 \end{bmatrix} \qquad (c). \end{aligned} \qquad (16.3.15)$$

Expression (16.3.15), m = 1, 2, ...M is the number of terms of the basis function, whereas k and ℓ indicate the number of terms of the testing functions. The identical and equal number of functions are used for both the basis and testing functions. The right-hand side of equation (16.3.14) is zero, as $\{\tilde{J}_x(\alpha_n), \tilde{J}_z(\alpha_n)\}$ is zero on the dielectric interface and $\{\tilde{E}_x(\alpha_n), \tilde{E}_z(\alpha_n)\}$ is zero on the strip conductor. The matrix elements of the above equation are given below:

$$
\begin{aligned}
K_{km}^{1,1} &= \left\langle \tilde{Y}_{11}(\alpha_n, \beta_z)\tilde{E}_{xm}(\alpha_n), \tilde{E}_{xk}(\alpha_n)\right\rangle \\
&= \sum_{n=1}^{\infty} \tilde{Y}_{11}(\alpha_n, \beta_z)\tilde{E}_{xm}^{*}(\alpha_n), \tilde{E}_{xk}(\alpha_n) \\
K_{km}^{1,2} &= \left\langle \tilde{Y}_{12}(\alpha_n, \beta_z)\tilde{E}_{zm}(\alpha_n), \tilde{E}_{xk}(\alpha_n)\right\rangle \\
&= \sum_{n=1}^{\infty} \tilde{Y}_{12}(\alpha_n, \beta_z)\tilde{E}_{zm}^{*}(\alpha_n), \tilde{E}_{xk}(\alpha_n) \\
K_{\ell,m}^{2,1} &= \left\langle \tilde{Y}_{21}(\alpha_n, \beta_z)\tilde{E}_{xm}(\alpha_n), \tilde{E}_{z\ell}(\alpha_n)\right\rangle \\
&= \sum_{n=1}^{\infty} \tilde{Y}_{21}(\alpha_n, \beta_z)\tilde{E}_{xm}^{*}(\alpha_n), \tilde{E}_{z\ell}(\alpha_n) \\
K_{\ell,m}^{2,2} &= \left\langle \tilde{Y}_{22}(\alpha_n, \beta_z)\tilde{E}_{zm}(\alpha_n), \tilde{E}_{z\ell}(\alpha_n)\right\rangle \\
&= \sum_{n=1}^{\infty} \tilde{Y}_{22}(\alpha_n, \beta_z)\tilde{E}_{zm}^{*}(\alpha_n), \tilde{E}_{z\ell}(\alpha_n).
\end{aligned} \tag{16.3.16}
$$

For the nontrivial solution of equation (16.3.15c), its determinant ΔK is zero that gives us the characteristic equation. The propagation constant β_z of the CPW is computed by solving the characteristic equation numerically. In this process, a suitable basis function is assumed for the slot electric field components. It is discussed in the next section.

16.4 Basis Functions for Surface Current Density and Slot Field

The nature of the longitudinal and transverse current density is considered on the strip conductor of a microstrip line. Next, the discussion is on the nature of the longitudinal and the transverse electric field components in the slot regions of a slot line and a CPW structure.

16.4.1 Nature of the Field and Current Densities

The surface current density $J_x(x)$ and $J_z(x)$ on the conducting strip and the electric fields $E_x(x)$ and $E_z(x)$ in the slot region encounter a sharp conductor edge. The behavior of the field at the conducting edge, known as the *edge condition*, is formulated by the *Meixner condition* [B.2, J.14]. At the conducting edge of the wedge $(r \to 0)$, the transverse $\left(\vec{E}_t, \vec{H}_t\right)$ and longitudinal $\left(\vec{E}_z, \vec{H}_z\right)$ electric and magnetic fields show the following *singularity* and *regular behavior*:

$$\vec{E}_t, \vec{H}_t \to \frac{1}{\sqrt{r}} \quad \text{(a)}, \qquad E_z, H_z \to \sqrt{r} \quad \text{(b)}. \tag{16.4.1}$$

For a planar structure, $r \to x$ and the edge conditions are expressed as follows:

$$E_x, H_x \text{ or } (J_z) \to \frac{1}{\sqrt{x}} \quad \text{(a)}, \quad E_z, H_z \text{ or } (J_x) \to \sqrt{x} \quad \text{(b)}. \tag{16.4.2}$$

In the case of a microstrip, on the surface of a perfect strip conductor, the electric field components are $E_x = E_z = 0$. For $x \to 0$, i.e. at the edge of the conductor, the longitudinal current J_z shows the *singularity*, i.e. $J_z \to \infty$ (a sharp increase of J_z), whereas for $x \to 0$, the transverse current shows a *regular behavior*, i.e. $J_x \to 0$. Likewise, in the case of the slot field of the slot line and CPW, the transverse electric field component in the slot region shows the edge singularity, i.e. for $x \to 0$, $E_x \to \infty$, whereas the longitudinal field component at $x \to 0$, $E_z \to 0$ is regular. Denlinger has also come to the same edge condition for a microstrip line [J.12].

If the selected basis functions satisfy the physical behavior (singularity) of J_z or E_x, even one-term expansion of the current or field can provide a satisfactory result for the propagation constant. The proper choice of the basis functions, meeting the physical behavior, improves both the computational efficiency and accuracy of Galerkin's method. It also simplifies the analytical complexities. If the assumed basis function belongs to the *complete set*, the undesirable effect called the relative convergence, i.e. converging to the nonphysical solution, can be avoided. To improve the computational efficiency, the closed-form Fourier transform of the basis functions should be obtained. The basis functions should have twice differentiability of function to eliminate the spurious solution [B.6].

16.4.2 Basis Functions and Nature of Hybrid Modes

Microstrip

The rectangular metallic waveguide, discussed in section (7.4) of chapter 7, shows that each mode has a distinct current pattern on the inner surfaces of the conducting walls. If the current distributions on the conducting walls are known, they can serve as the source for generating the EM-field of different modes of a waveguide. Similarly, the assumed current distribution of the basis function for the current density on a strip conductor influences the nature of the hybrid modes supported by the microstrip line. Capelle and Luypaert [J.15] have obtained approximate five longitudinal current distributions by solving the integral equation and have demonstrated that these current distributions correspond to five hybrid modes of a microstrip. Figure (16.4a) shows the even and odd longitudinal current distributions, and

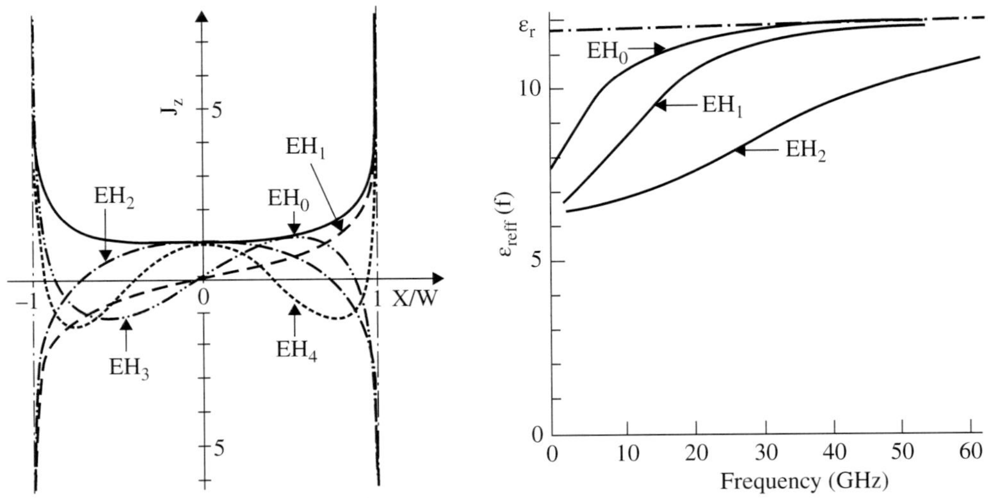

(a) Current distributions and corresponding hybrid modes [J.15].

(b) Dispersion of the first three hybrid modes.

Figure 16.4 Hybrid modes on a microstrip line [J.15]. *Source*: Van de Capelle and Luypaert [J.15]. © 1973, Institution of Engineering and Technology. Reproduced by permission of IET.

their corresponding modes, EH_0, EH_1, EH_2, EH_3, EH_4 for a microstrip on an alumina substrate with $\varepsilon_r = 11.7$, $h = 3.17$ mm, $2w/h = 0.96$. The fundamental quasi-TEM mode is the EH_0 mode. The EH_2 and EH_4 modes are even modes, whereas the EH_1 and EH_3 modes are the odd modes. Mirshakar and Davies [B.2, J.16] have further shown that these current distributions approximately collaborate with the current distributions expressed in terms of Legendre polynomials $P_m(x)$ of the mth order. Figure (16.4b) further shows the dispersion behaviors of the first three microstrip hybrid modes. The higher-order modes travel at higher phase velocity as compared to lower-order modes. So, higher order modes could become the radiating modes, as shown in Figure (8.4) of chapter 8.

Several kinds of basis functions for the current distributions on a microstrip have been used. Three commonly used basis functions, along with their Fourier transforms, are summarized below.

Legendre Polynomials

Legendre polynomials have been used to expand the current density $J_z(x)$ and the slot field $E_x(x)$. However, the trigonometric function is used for the regular current distribution $J_x(x)$ and regular slot field $E_z(x)$ [B.2, J.11, J.16–J.19]. For the *even mode* of a microstrip line of width w, the following basis functions are used:

$$J_{x,p}(x) = \sin\left(\frac{p\pi x}{w}\right), \quad |x| < w/2 \qquad (a)$$
$$J_{z,q}(X) = P_{z(q-1)}(X), \quad X = x/w \qquad (b), \qquad (16.4.3)$$

where $P_{z(q-1)}(x)$ is Legendre polynomial of even degree obtained from the following recursion formula:

$$P_{q+1}(x) = \frac{2q+1}{q+1} x P_q(x) - \frac{q}{q+1} P_{q-1}(x) \qquad (16.4.4)$$
$$P_0(x) = 1, \quad P_1(x) = x.$$

Fourier transform of Legendre polynomial is given below [J.2, J.20]:

$$\widetilde{P}_n(\alpha) = (j)^n \sqrt{\frac{2\pi}{\alpha}} J_{n+\frac{1}{2}}(\alpha) \qquad (a)$$

$$\text{where Bessel function: } J_m(\alpha) = \sum_{k=0}^{\infty} \frac{(-1)^k (\alpha/2)^{2k+m}}{k!\,\Gamma(k+m+1)} \qquad (b). \qquad (16.4.5)$$

However, Legendre polynomial is not a very good choice for the current basis function $J_z(x)$, as it does not show the edge singularity [J.21].

Singular Sinusoidal Basis Functions

Subsection (14.2.3) of chapter 14 shows that Maxwell's distribution function provides edge singularity for the charge distribution on a strip conductor. So $J_z(x) = 1/\sqrt{w^2 - x^2}$ and $J_x(x) = x\sqrt{w^2 - x^2}$ have been used as one-term basis functions for the longitudinal and transverse current density on microstrip [J.4]. However, Maxwell's singularity can be also combined with the sinusoidal basis function to account for the edge condition of the current density as follows [B.1, J.9]:

$$J_{zp}(x) = \frac{\cos\left[2(p-1)\dfrac{\pi x}{w}\right]}{\sqrt{1-(2x/w)^2}} \quad \text{(a)}$$
$$J_{xp}(x) = \frac{\sin\left[2p\dfrac{\pi x}{w}\right]}{\sqrt{1-(2x/w)^2}} \quad p = 1, 2, \ldots \quad \text{(b)}. \tag{16.4.6}$$

Figure (16.5a) shows a few basis functions for the current density distribution of $J_{zp}(x)$ and $J_{xp}(x)$ on a microstrip of width w. The basis function $J_{zp}(x)$ shows the edge singularity, whereas the $J_{xp}(x)$ shows zeros at the conductor edge. The Fourier transforms of the above functions are given below:

$$\tilde{J}_{zp}(\alpha_n) = \frac{\pi w}{4}\left[J_0\left\{\left|\frac{w\alpha_n}{2} + (p-1)\pi\right|\right\} + J_0\left\{\left|\frac{w\alpha_n}{2} - (p-1)\pi\right|\right\}\right] \quad \text{(a)}$$
$$\tilde{J}_{xp}(\alpha_n) = \frac{\pi w}{4j}\left[J_0\left\{\left|\frac{w\alpha_n}{2} + p\pi\right|\right\} - J_0\left\{\left|\frac{w\alpha_n}{2} - p\pi\right|\right\}\right] \quad \text{(b)}. \tag{16.4.7}$$

It is interesting to note that for any order of the basis functions, Fourier transform is always in terms of the zero^{th}-order Bessel's function. The continuous Fourier variable, i.e. $\alpha_n = \alpha$, is used for a shielded box with open sides. The basis function may also be chosen so that the integrals of $J_z(x)$ and $J_x(x)$ over the strip width w are equal to unity. In this case, for p = 1, the basis functions could be written as follows:

$$J_{z1}(x) = \frac{2}{w\pi}\frac{1}{\sqrt{1-\left(\frac{2x}{w}\right)^2}} \quad \text{(a)}$$
$$J_{x1}(x) = \frac{4j}{\pi w}\left[\frac{\sin\left[\dfrac{2\pi x}{w}\right]}{\sqrt{1-(2x/w)^2}}\right] \quad \text{(b)}. \tag{16.4.8}$$

The Fourier transforms of the above functions are given below:

$$\tilde{J}_{z1}(\alpha_n) = J_0\left(\left|\frac{w\alpha_n}{2}\right|\right) \quad \text{(a)}$$
$$\tilde{J}_{x1}(\alpha_n) = \left[J_0\left\{\left|\frac{w\alpha_n}{2} + \pi\right|\right\} - J_0\left\{\left|\frac{w\alpha_n}{2} - \pi\right|\right\}\right] \quad \text{(b)}$$
$$\text{where,} \int_{-w/2}^{w/2} J_{z1}(x)dx = 1, \quad \int_{-w/2}^{w/2} J_{x1}(x)dx = 1 \quad \text{(c)}. \tag{16.4.9}$$

The higher-order, even and odd, modes are properly described by the above basis function [B.1, J.9]. However, some investors have preferred Chebyshev polynomial with the edge singularity for faster convergence [J.22, J.23].

Singular Chebyshev Polynomial Basis Functions

The longitudinal and transverse current densities are defined using the first and second kind Chebyshev polynomials, $T_p(x)$, and $U_p(x)$, respectively. The $T_p(x)/U_p(x)$ is the even/odd function for p an even number. It is the odd/even function for p an odd number. The $T_p(x)$ is multiplied by Maxwell's distribution function to get the edge singularity for the longitudinal current $J_z(x)$. It is even symmetric, with respect to the *y*-axis of Fig (16.1b), for the fundamental mode. The following expressions provide the longitudinal and transverse current basis functions:

$$J_{x,p}(x) = U_{2p}\left(\frac{2x}{w}\right) \quad \text{(a)}, \quad J_{z,p}(x) = \frac{T_{2(p-1)}\left(\frac{2x}{w}\right)}{\sqrt{1-(2x/w)^2}} \quad \text{(b)}. \tag{16.4.10}$$

The above function (16.4.10) is defined over $|x| \le w/2$, and it is zero for $|x| > w/2$. Figure (16.5b) shows the behavior of some of the terms of both functions. Figure (16.5b) shows a more rapid edge singularity for the $J_{z,p}(x)$, compared to the sine type function. Likewise, $J_{x,p}(x)$ component also shows rapid decreases at the edge.

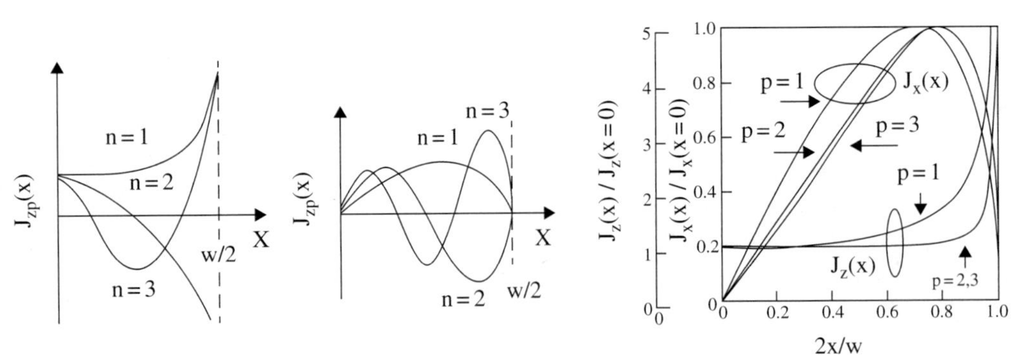

(a) Singular sinusoidal basis functions. (b) Chebyshev basis functions.

Figure 16.5 Basis functions for current density on microstrip.

The current density distribution saturates for $p > 2$. So Chebyshev type basis functions provide faster convergence, even for one or two-term basis functions [J.20, J.22–J.24]. Their Fourier transforms are summarized below:

$$\tilde{J}_{x,p} = j(-1)^{(p-1)/2}\frac{p\pi}{\alpha_n}J_p\left\{\frac{|\alpha_n|w}{2}\right\} \quad (a)$$
$$\tilde{J}_{z,p} = (-1)^{(p-1)/2}\frac{p\pi}{2}J_{p-1}\left\{\frac{|\alpha_n|w}{2}\right\} \quad (b), \qquad (16.4.11)$$

where α_n is the discrete Fourier variable for the boxed microstrip. The continuous Fourier variable α is used for the side open microstrip. The basis functions are defined by number p. It also determines the order of the Bessel function $J_p(\alpha_n)$. In this case, the computer code becomes more involved, unlike the sinusoidal based basis function mentioned above, that has the zeroth-order Bessel function for all order of basis functions. However, using Chebyshev polynomial, frequency-dependent current distribution has been obtained that provides a more accurate computation of frequency-dependent effective relative permittivity of the microstrip [J.22].

Slotline

The microstrip line, shown in Fig (16.1b), can be treated as a boxed slot line also, where strip conductor is replaced by the slot-width w, and both dielectric interfaces are replaced by the conducting strips. The considered slot line structure is symmetrical with respect to the *y*-axis. For the fundamental mode, an EW is in the direction of the *y*-axis by treating one conductor at the positive potential and another conductor at the negative potential. The edge conditions also require that the E_z field is a regular function with $E_z = 0$ at the inner edges of the conductors, and E_x has a singularity (maxima) at the edges. The singular sinusoidal functions are popular as the basis functions for the slot-field components [B.1, J.9]. The electric field components $E_x(x)$ and $E_z(x)$ are the even and odd function, respectively, with respect to x. Such functional behavior could be described by the following basis functions to generate the *EH-fundamental* and *even higher-order modes.*

$$E_{xn}(x) = \frac{\cos\left[2(n-1)\frac{\pi x}{w}\right]}{\sqrt{1-(2x/w)^2}} \quad (a)$$
$$E_{zn}(x) = \frac{\sin\left[2n\frac{\pi x}{w}\right]}{\sqrt{1-(2x/w)^2}};\ n = 1, 2, \ldots \quad (b). \qquad (16.4.12)$$

Fourier transform of the above function (16.4.12) is the same as that of equation (16.4.6). The odd higher-order hybrid mode could be obtained from the following basis functions [J.9]:

$$E_{x\,n}(x) = \frac{\sin\left[(2n-1)\frac{\pi x}{w}\right]}{\sqrt{1-(2x/w)^2}} \quad (a)$$
$$E_{zn}(x) = \frac{\cos\left[(2n-1)\frac{\pi x}{w}\right]}{\sqrt{1-(2x/w)^2}};\ n = 1, 2, \ldots \quad (b), \qquad (16.4.13)$$

where w is the slot-width and the field components. These are zero elsewhere. The solution of the characteristic equation of a slot line can provide both the bound fundamental and higher-order modes. However, due to the complex propagation constant, space leaky mode could be generated. Figure (16.6a) shows the even transverse electric field $E_x(x)$, in relative unit i.e. normalized field, in the slot region of a slot line, on a substrate with $\varepsilon_r = 2.6$, $h = 1.2$ mm, $w = 60$ mm, for the fundamental even mode. It provides the bound fundamental mode, shown in Fig (16.6b). The first odd $E_x(x)$ field generates the first higher-order bound mode and also the first space leaky odd mode at about 1.4 GHz. These are also shown in Fig (16.6b). Space leaky mode decreases at a higher frequency [J.9].

CPW Structure

Figure (16.1a) shows the CPW structure that supports the fundamental, even, and odd hybrid modes. For the *even hybrid mode*, the $E_z(x)$ field component, with respect to x is even and $H_z(x)$ is odd. The *odd hybrid mode* has $E_z(x)$ odd and $H_z(x)$ even, as respect to the variable x. The plane of symmetry, located at $x = 0$, is the MW of the fundamental modes and higher-order even mode with $H_z(x) = 0$ on it. For the odd mode, the plane of symmetry is an EW, with $E_z = 0$ at $x = 0$. Several kinds of basis functions have been used as the basis functions to expand the slot fields [B.2, J.1–J.4, J.7–J.10, J.25, J.26]. However, the singular sinusoidal basis functions appear to be more appropriate. Figure (16.7a) shows the location of slots of width s at $x = \pm c$, where $c = w + s/2$. The CPW could also be treated as the coupled slot line. The total slot fields, $E_{x,m}(x)$ and $E_{z,m}(x)$, in both the slot regions, for the fundamental and even modes, and also for the odd modes are expressed by the following singular sinusoidal basis functions:

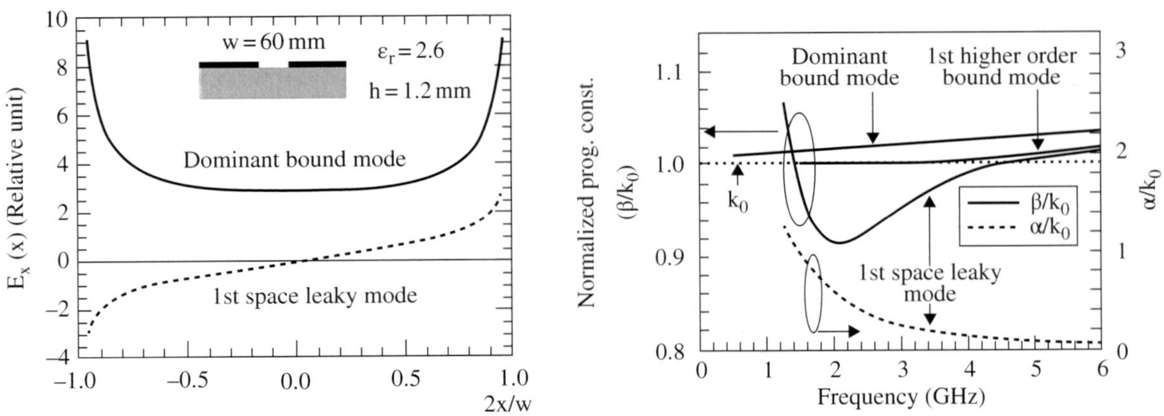

(a) Even/odd $E_x(x)$ field in slot region.

(b) Existence of bound and leaky modes.

Figure 16.6 Bound and leaky mode on a slot-line. *Source*: Zehentner et al. [J.9]. © 2004, Microwave Review. Reproduced by permission of Microwave Review.

$$\text{Even mode: } E_{xm}(x) = \frac{\cos[m\pi(x+c)/s]}{\sqrt{1-[2(x+c)/s]^2}} - \frac{\cos[m\pi(x-c)/s]}{\sqrt{1-[2(x-c)/s]^2}}, \quad m = 0, 2, 4\ldots \quad \text{(a)}$$

$$E_{zm}(x) = \frac{\cos[m\pi(x+c)/s]}{\sqrt{1-[2(x+c)/s]^2}} + \frac{\cos[m\pi(x-c)/s]}{\sqrt{1-[2(x-c)/s]^2}}, \quad m = 1, 3, \ldots \quad \text{(b)}$$

$$\text{Odd mode: } E_{xm}(x) = \frac{\sin[m\pi(x+c)/s]}{\sqrt{1-[2(x+c)/s]^2}} + \frac{\sin[m\pi(x-c)/s]}{\sqrt{1-[2(x-c)/s]^2}}, \quad m = 1, 3, \ldots \quad \text{(c)}$$

$$E_{zm}(x) = \frac{\sin[m\pi(x+c)/s]}{\sqrt{1-[2(x+c)/s]^2}} - \frac{\sin[m\pi(x-c)/s]}{\sqrt{1-[2(x-c)/s]^2}}, \quad m = 2, 4\ldots \quad \text{(d)}. \tag{16.4.14}$$

Both the $E_{x,m}(x)$, and $E_{z,m}(x)$ are zero on the perfectly conducting strips. In the above equations (16.4.14a–d), the first term shows the center of the slot at $x = -c$ in the region $x < 0$, and the second term shows the center of the slot at $x = c$ in the region $x > 0$. We obtain below the Fourier transforms of the above field distributions for the SDA application. The characteristic equation provides both the real and complex propagation constants for an open CPW. Figure (16.7b) shows the bound fundamental mode, and also generation of the first surface leaky mode at $h/\lambda_o = 0.12$ for the CPW on a substrate with $\varepsilon_r = 10$, $h = 1$ mm, $s = 0.6$ mm, $2w = 0.25$ mm. The propagation constants are normalized by the free space wavenumber k_0. Figure (16.7b) also shows the first two surface wave modes, k_{TM0} and k_{TE1}.

To obtain Fourier transform of the slot-field, the following definition is used:

$$\tilde{f}(\alpha_n, y) = \int_{-L}^{L} f(x, y) e^{j\alpha_n x} dx, \quad \text{where,} \quad \alpha_n = \left(\frac{2n-1}{L}\right)\frac{\pi}{2}. \tag{16.4.15}$$

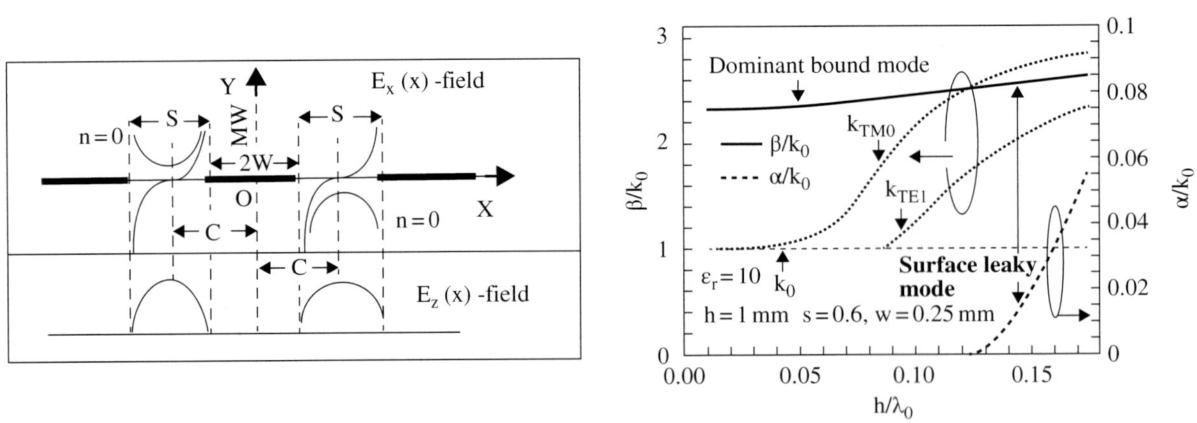

(a) Slot E_x and E_z field distribution.

(b) Fundamental and surface leaky mode.

Figure 16.7 Bound and leaky mode on a CPW. *Source*: Zehentner et al. [J.9]. © 2004, Microwave Review. Reproduced by permission of Microwave Review.

The function f(x, y) is an even function, for the x variable, with respect to the y-axis.

Fourier Transform of $E_{zm}(x)$ Even Mode

Let us obtain the Fourier transform of equation (16.4.14b) as follows:

$$\widetilde{E}_{zm}(\alpha_n) = \int_{-a}^{a} E_{zm}(x)dx = \underbrace{\int_{-\left(c+\frac{s}{2}\right)}^{-\left(c-\frac{s}{2}\right)} \frac{\cos\left[m\pi\left(\frac{x+c}{s}\right)\right]}{\sqrt{1-\left[\frac{2(x+c)}{s}\right]^2}} e^{j\alpha_n x}dx}_{I} + \underbrace{\int_{\left(c-\frac{s}{2}\right)}^{\left(c+\frac{s}{2}\right)} \frac{\cos\left[m\pi\left(\frac{x-c}{s}\right)\right]}{\sqrt{1-\left[\frac{2(x-c)}{s}\right]^2}} e^{j\alpha_n x}dx.}_{II} \quad (16.4.16)$$

In equation (16.4.16), the first integral-I is evaluated over the left-hand side slot shown in Fig (16.7a) and the second integral-II is evaluated over the right-hand side slot.

Let $\frac{2(x+c)}{s} = u, \quad \Rightarrow x = \frac{s}{2}u - c, \; dx = \frac{s}{2}du$

For $x = -\left(\frac{s}{2}+c\right), \quad u = \frac{2}{s}\left(-c - \frac{s}{2} + c,\right) = -1;$

for $x = -\left(c - \frac{s}{2}\right), \quad u = \frac{2}{s}\left(-c + \frac{s}{2} + c,\right) = +1$

$$I = \frac{s}{2}\int_{-1}^{+1} \frac{\cos\left(\frac{m\pi u}{2}\right)}{\sqrt{1-u^2}} \times e^{j\alpha_n\left(\frac{s}{2}u - c\right)}du$$

$$= \frac{s}{2}e^{-jc\alpha_n}\frac{1}{2}\int_{-1}^{+1} \frac{e^{j\left(\frac{\alpha_n s}{2}+\frac{m\pi}{2}\right)u} + e^{j\left(\frac{\alpha_n s}{2}-\frac{m\pi}{2}\right)u}}{\sqrt{1-u^2}}du.$$

On using the Bessel function integral [B.10], $J_0(\alpha) = \frac{1}{\pi}\int_{-1}^{1} \frac{e^{j\alpha t}}{\sqrt{1-t^2}}dt$, the integral-I is evaluated:

$$I = \frac{\pi s}{4}e^{-jc\alpha_n}\left[J_0\left\{\frac{\alpha_n s}{2}+\frac{m\pi}{2}\right\} + J_0\left\{\frac{\alpha_n s}{2}-\frac{m\pi}{2}\right\}\right], \; m = 1,3,5\ldots \quad (16.4.17)$$

The second integral, over the right-hand side of Fig (16.7a), is evaluated as follows:

$$II = \int_{c-\frac{s}{2}}^{c+\frac{s}{2}} \frac{\cos\left[\frac{m\pi(x-c)}{s}\right]e^{j\alpha_n x}}{\sqrt{1-\left[\frac{2(x-c)}{s}\right]^2}}dx.$$

Let $\frac{2(x-c)}{s} = u \Rightarrow x = \frac{s}{2}u + c$ and $dx = \frac{s}{2}du.$

For $x = c - \frac{s}{2}, \quad u = \frac{2}{s}\left[c - \frac{s}{2} - c\right] = -1,$ and

for $x = \left[\frac{s}{2}+c\right], \quad u = \frac{2}{s}\left[c + \frac{s}{2} - c\right] = +1.$

$$II = \frac{s}{2}\int_{-1}^{+1} \frac{\cos\left[\frac{m\pi u}{2}\right]e^{j\alpha_n\left(\frac{s}{2}u + c\right)}}{\sqrt{1-u^2}}du$$

$$= \frac{s}{2}e^{jc\alpha_n}\frac{1}{2}\left[\int_{-1}^{+1} \frac{e^{j\left\{\frac{\alpha_n s}{2}+\frac{m\pi}{2}\right\}u} + e^{j\left\{\frac{\alpha_n s}{2}-\frac{m\pi}{2}\right\}u}}{\sqrt{1-u^2}}du\right]$$

$$\Rightarrow II = \frac{\pi s}{4}e^{jc\alpha_n}\left[J_0\left\{\frac{\alpha_n s}{2}+\frac{m\pi}{2}\right\} + J_0\left\{\frac{\alpha_n s}{2}-\frac{m\pi}{2}\right\}\right], \; m = 1,3,5\ldots. \quad (16.4.18)$$

Fourier transform of $E_{zm}(x)$ of equation (16.4.14b) is

$$\widetilde{E}_{zm}(\alpha_n) = I + II = = \frac{s\pi}{2}\left[\frac{e^{(-j\alpha_n c)} + e^{(j\alpha_n c)}}{2}\right]$$
$$\left[J_0\left\{\left|\frac{\alpha_n s}{2}+\frac{m\pi}{2}\right|\right\} + J_0\left\{\left|\frac{\alpha_n s}{2}-\frac{m\pi}{2}\right|\right\}\right]$$

$$\Rightarrow \widetilde{E}_{zm}(\alpha_n) = \frac{s\pi}{2}\cos(\alpha_n c)$$
$$\left[J_0\left\{\left|\frac{\alpha_n s}{2}+\frac{m\pi}{2}\right|\right\} + J_0\left\{\left|\frac{\alpha_n s}{2}-\frac{m\pi}{2}\right|\right\}\right], \quad m = 1,3,5\ldots \quad (16.4.19)$$

Fourier Transform of $E_{zm}(x)$ Odd Mode

Let us obtain the Fourier transform of equation (16.4.14d).

$$\widetilde{E}_{zm}(\alpha_n) = \underbrace{\int_{-\left(c+\frac{s}{2}\right)}^{-\left(c-\frac{s}{2}\right)} \frac{\sin\left[\frac{m\pi(x+c)}{s}\right]e^{j\alpha_n x}}{\sqrt{1-\left[\frac{2(x+c)}{s}\right]^2}}dx}_{I'} - \underbrace{\int_{c-\frac{s}{2}}^{c+\frac{s}{2}} \frac{\sin\left[\frac{m\pi(x-c)}{s}\right]e^{j\alpha_n x}}{\sqrt{1-\left[\frac{2(x-c)}{s}\right]^2}}dx}_{II'}. \quad (16.4.20)$$

Let us evaluate the above integral in two parts

$$I' = \int_{-(c+\frac{s}{2})}^{-(c-\frac{s}{2})} \frac{\sin\left[\frac{m\pi(x+c)}{s}\right] e^{jx\alpha_n}}{\sqrt{1-\left[\frac{2(x+c)}{s}\right]^2}} dx.$$

Let $\frac{2(x+c)}{s} = u.$

$$\Rightarrow I' = \frac{s}{2}\int_{-1}^{+1} \frac{\sin\left[\frac{m\pi u}{2}\right] e^{j\alpha_n\left(\frac{s}{2}u - c\right)}}{\sqrt{1-u^2}} du$$

$$= \frac{s}{4j} e^{-jc\alpha_n}\left[\int_{-1}^{+1} \frac{e^{j\left(\frac{\alpha_n s}{2}u + \frac{m\pi}{2}\right)} - e^{j\left(\frac{\alpha_n s}{2}u - \frac{m\pi}{2}\right)}}{\sqrt{1-u^2}} du\right]$$

$$= \frac{\pi s}{4j} e^{-jc\alpha_n}\left[J_0\left\{\left|\frac{\alpha_n s}{2} + \frac{m\pi}{2}\right|\right\} - J_0\left\{\left|\frac{\alpha_n s}{2} - \frac{m\pi}{2}\right|\right\}\right]. \tag{16.4.21}$$

Likewise, $II' = \frac{\pi s}{4j} e^{j\alpha_n c}\left[J_0\left\{\left|\frac{\alpha_n s}{2} + \frac{m\pi}{2}\right|\right\} - J_0\left\{\left|\frac{\alpha_n s}{2} - \frac{m\pi}{2}\right|\right\}\right].$ (16.4.22)

Therefore, Fourier transform of equation (16.4.14d) is

$$\widetilde{E}_{zm}(\alpha_n) = I' - II' = \frac{s\pi}{2}\left[\frac{e^{-j\alpha_n c} - e^{j\alpha_n c}}{2j}\right] \times \left[J_0\left\{\left|\frac{\alpha_n s}{2} + \frac{m\pi}{2}\right|\right\} - J_0\left\{\left|\frac{\alpha_n s}{2} - \frac{m\pi}{2}\right|\right\}\right]$$

$$\widetilde{E}_{zm}(\alpha_n) = -\frac{s\pi}{2}\sin(\alpha_n c) \times \left[J_0\left\{\left|\frac{\alpha_n s}{2} + \frac{m\pi}{2}\right|\right\} - J_0\left\{\left|\frac{\alpha_n s}{2} - \frac{m\pi}{2}\right|\right\}\right], \quad m = 2, 4..... \tag{16.4.23}$$

Fourier Transform of $E_{xm}(x)$ Even Mode

Using equations (16.4.17) and (16.4.18), Fourier transform of equation (16.4.14a) is determined:

$$\widetilde{E}_{xm}(\alpha_n) = I - II = \frac{s\pi}{2}\left[\frac{e^{(-j\alpha_n c)} - e^{(j\alpha_n c)}}{2}\right] \times \left[J_0\left\{\left|\frac{\alpha_n s}{2} + \frac{m\pi}{2}\right|\right\} + J_0\left\{\left|\frac{\alpha_n s}{2} - \frac{m\pi}{2}\right|\right\}\right]$$

$$\Rightarrow \widetilde{E}_{xm}(\alpha_n) = -\frac{js\pi}{2}\sin(\alpha_n c) \times \left[J_0\left\{\left|\frac{\alpha_n s}{2} + \frac{m\pi}{2}\right|\right\} + J_0\left\{\left|\frac{\alpha_n s}{2} - \frac{m\pi}{2}\right|\right\}\right], \quad m = 0, 1, 2, ... \tag{16.4.24}$$

Fourier Transform of $E_{xm}(x)$ Odd Mode

Using equations (16.4.21) and (16.4.22), Fourier transform of equation (16.4.14c) obtained:

$$\widetilde{E}_{xm}(\alpha_n) = I' + II' = \frac{s\pi}{2j}\left[\frac{e^{-j\alpha_n c} + e^{j\alpha_n c}}{2}\right] \times \left[J_0\left\{\left|\frac{\alpha_n s}{2} + \frac{m\pi}{2}\right|\right\} - J_0\left\{\left|\frac{\alpha_n s}{2} - \frac{m\pi}{2}\right|\right\}\right]$$

$$\Rightarrow \widetilde{E}_{xm}(\alpha_n) = -\frac{js\pi}{2}\cos(\alpha_n c) \times \left[J_0\left\{\left|\frac{\alpha_n s}{2} + \frac{m\pi}{2}\right|\right\} - J_0\left\{\left|\frac{\alpha_n s}{2} - \frac{m\pi}{2}\right|\right\}\right], \quad m = 1, 3, ... \tag{16.4.25}$$

Fourier transforms of equations (16.4.14a–d) are summarized below:

$$\begin{aligned}
&\text{Even mode:}\\
&\widetilde{E}_{xm}(\alpha_n) = -j\frac{s\pi}{2}\sin(\alpha_n c)\left[J_0\left\{\left|\frac{\alpha_n s}{2} + \frac{m\pi}{2}\right|\right\} + J_0\left\{\left|\frac{\alpha_n s}{2} - \frac{m\pi}{2}\right|\right\}\right], \quad m = 0, 2, 4... \quad (a)\\
&\widetilde{E}_{zm}(\alpha_n) = \frac{s\pi}{2}\cos(\alpha_n c)\left[J_0\left\{\left|\frac{\alpha_n s}{2} + \frac{m\pi}{2}\right|\right\} + J_0\left\{\left|\frac{\alpha_n s}{2} - \frac{m\pi}{2}\right|\right\}\right], \quad m = 1, 3, 5... \quad (b)\\
&\text{Odd mode:}\\
&\widetilde{E}_{xm}(\alpha_n) = -j\frac{s\pi}{2}\cos(\alpha_n c)\left[J_0\left\{\left|\frac{\alpha_n s}{2} + \frac{m\pi}{2}\right|\right\} - J_0\left\{\left|\frac{\alpha_n s}{2} - \frac{m\pi}{2}\right|\right\}\right] \quad m = 1, 3, 5... \quad (c)\\
&\widetilde{E}_{zm}(\alpha_n) = -\frac{s\pi}{2}\sin(\alpha_n c)\left[J_0\left\{\left|\frac{\alpha_n s}{2} + \frac{m\pi}{2}\right|\right\} - J_0\left\{\left|\frac{\alpha_n s}{2} - \frac{m\pi}{2}\right|\right\}\right] \quad m = 0, 2, 4... \quad (d).
\end{aligned} \tag{16.4.26}$$

16.5 Coplanar Multistrip Structure

The previous sections have discussed the single strip microstrip, slot line, and CPW structures. The CPW structure has two slots in the same plane containing the strip conductors. There are occasions when two, three, four, etc., microstrip conductors are located on the same plane, i.e. at the interface of the substrate and air dielectric media. These structures are useful for the design of coupler, filter, etc. The multiple strip

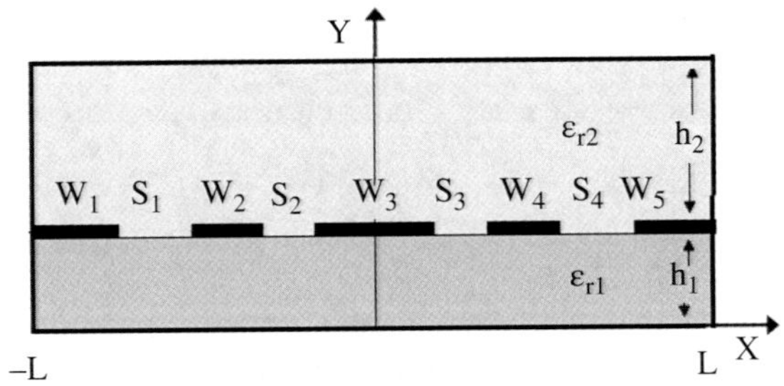

(a) Multistrip and multislot planar line.

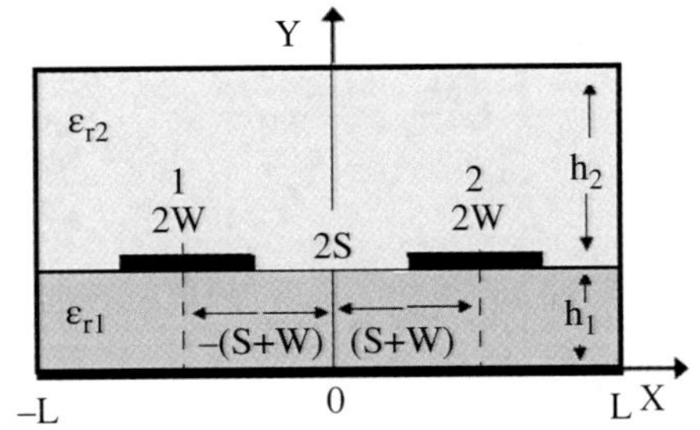

(b) Symmetrical coupled microstrip line.

Figure 16.8 Same level multiconductors and multislots.

conductors are also used as interconnects in the VLSI technology that has pushed the clock-rate in the GHz range. A generic form of such structure is shown in Fig (16.8a). The SDA has been used to obtain the propagation constant of all modes supported by this structure [J.27, J.28]. The symmetrically coupled microstrip line, with identical conductors shown in Fig (16.8b), is a more popular structure. Its physical behavior and equivalent circuit line are discussed in section (11.3) of chapter 11. In this section, the SDA is used to obtain the coupling between the lines. The discussion is limited to the symmetrically coupled microstrip. The SDA has also been used to the asymmetrically coupled microstrip with unequal strip widths [B.2]. The mathematical process used below follows the discussion of section (14.6) of chapter 14.

16.5.1 Symmetrical Coupled Microstrip Line

The impedance kind of Green's function given by equation (16.3.1) in the Fourier domain is applied to the microstrip structure containing any number of coplanar strips, shown in Fig (16.8a). The surface current density components in the Fourier domain $\tilde{J}_x(\alpha_n)$ and $\tilde{J}_z(\alpha_n)$, i.e. $J_x(x)$ and $J_z(x)$ in the space domain, are the total current density on all strip conductors in the range, $-L$ to L. Likewise, the electric field components $\tilde{E}_x(\alpha_n)$ and $\tilde{E}_z(\alpha_n)$, i.e. $E_x(x)$ and $E_z(x)$, are the total electric fields in all slots in the range, $-L$ to L. For the N number of strip conductors, equation (16.3.2), for the current density, can be written as follows:

$$\begin{aligned} \tilde{J}_x(\alpha_n) &= \sum_{t=1}^{N}\left[\sum_{m=1}^{M} c_m^t \tilde{J}_{xm}^t(\alpha_n)\right] && \text{(a)} \\ \tilde{J}_z(\alpha_n) &= \sum_{t=1}^{N}\left[\sum_{m=1}^{M} d_m^t \tilde{J}_{zm}^t(\alpha_n)\right] && \text{(b),} \end{aligned} \qquad (16.5.1)$$

where t is the number of strip conductors, i.e. $t = 1, 2, \ldots$ N; m = 1, 2, ...M is the number of terms of basis function for both the transverse and longitudinal current density $\tilde{J}_x(\alpha_n)$ and $\tilde{J}_z(\alpha_n)$. For the two-conductor symmetrical coupled microstrip line, shown in Fig (16.8b), equation (16.5.1a and b) are reduced to

$$\begin{aligned} \tilde{J}_x(\alpha_n) &= \sum_{m=1}^{M} c_m^1 \tilde{J}_{xm}^1(\alpha_n) + \sum_{m=1}^{M} c_m^2 \tilde{J}_{xm}^2(\alpha_n) && \text{(a)} \\ \tilde{J}_z(\alpha_n) &= \sum_{m=1}^{M} d_m^1 \tilde{J}_{zm}^1(\alpha_n) + \sum_{m=1}^{M} d_m^2 \tilde{J}_{zm}^2(\alpha_n) && \text{(b),} \end{aligned} \qquad (16.5.2)$$

where $\tilde{J}_{xm}^1(\alpha_n)$ and $\tilde{J}_{zm}^1(\alpha_n)$ are Fourier transformed basis functions of current density on the *conductor #1*. The Fourier transformed basis functions $\tilde{J}_{xm}^2(\alpha_n)$ and $\tilde{J}_{zm}^2(\alpha_n)$ are current densities on *conductor #2*. Equation (16.5.1) is substituted in equation (16.3.1) to get the following system of equations:

$$\begin{aligned} &\tilde{Z}_{11}(\alpha_n)\left[\sum_{t=1}^{N}\sum_{m=1}^{M} c_m^t \tilde{J}_{xm}^t(\alpha_n)\right] + \tilde{Z}_{12}(\alpha_n)\left[\sum_{t=1}^{N}\sum_{m=1}^{M} d_m^t \tilde{J}_{zm}^t(\alpha_n)\right] = \tilde{E}_x(\alpha_n) && \text{(a)} \\ &\tilde{Z}_{21}(\alpha_n)\left[\sum_{t=1}^{N}\sum_{m=1}^{M} c_m^t \tilde{J}_{xm}^t(\alpha_n)\right] + \tilde{Z}_{22}(\alpha_n)\left[\sum_{t=1}^{N}\sum_{m=1}^{M} d_m^t \tilde{J}_{zm}^t(\alpha_n)\right] = \tilde{E}_z(\alpha_n) && \text{(b).} \end{aligned} \qquad (16.5.3)$$

The inner product of equation (16.5.3a) is taken with the testing function $\tilde{J}_{xk}^t(\alpha_n)(k = 1, 2, \ldots)$ and of equation (16.5.3b) with the testing function $\tilde{J}_{z\ell}^t(\alpha_n)$, ($\ell = 1$, 2, ...) such that the right-hand side of both equations is zero over the range $-L$ to L, due to the complementary nature of perfect strip conductors and slots. Under the inner product process, the above equation (16.5.3) is reduced to the following system of equations:

$$\sum_{t=1}^{N}\sum_{m=1}^{M} c_m^t P_{im}^t(\alpha_n) + \sum_{t=1}^{N}\sum_{m=1}^{M} d_m^t Q_{im}^t = 0 \qquad \text{(a)}$$

testing function $k = i = 1, 2, ...M$

$$\sum_{t=1}^{N}\sum_{m=1}^{M} c_m^t R_{im}^t(\alpha_n) + \sum_{t=1}^{N}\sum_{m=1}^{M} d_m^t S_{im}^t = 0 \qquad \text{(b)}$$

testing function $\ell = i = 1, 2, ...M$

$$\Rightarrow \begin{bmatrix} \sum_{t=1}^{N}\sum_{m=1}^{M} P_{im}^t(\alpha_n) & \sum_{t=1}^{N}\sum_{m=1}^{M} Q_{im}^t \\ \sum_{t=1}^{N}\sum_{m=1}^{M} R_{im}^t(\alpha_n) & \sum_{t=1}^{N}\sum_{m=1}^{M} S_{im}^t \end{bmatrix} \begin{bmatrix} c_m^t \\ d_m^t \end{bmatrix} = \begin{bmatrix} 0 \\ 0 \end{bmatrix} \qquad \text{(c)},$$

$$(16.5.4)$$

where the number of strip conductors is, $t = 1, 2...N$ and

$$P_{i\,m}^t(\alpha_n) = \sum_{n=1}^{\infty} \tilde{J}_{x\,m}^{t*}(\alpha_n)\tilde{Z}_{11}(\alpha_n)\tilde{J}_{x\,i}^t(\alpha_n)$$

$$Q_{i\,m}^t(\alpha_n) = \sum_{n=1}^{\infty} \tilde{J}_{z\,m}^{t*}(\alpha_n)\tilde{Z}_{12}(\alpha_n)\tilde{J}_{x\,i}^t(\alpha_n)$$

$$R_{i\,n}^t(\alpha_n) = \sum_{n=1}^{\infty} \tilde{J}_{x\,m}^{t*}(\alpha_n)\tilde{Z}_{21}(\alpha_n)\tilde{J}_{z\,i}^t(\alpha_n) \qquad (16.5.5)$$

$$S_{i\,n}^t(\alpha_n) = \sum_{n=1}^{\infty} \tilde{J}_{z\,m}^{t*}(\alpha_n)\tilde{Z}_{22}(\alpha_n)\tilde{J}_{z\,i}^t(\alpha_n).$$

In equation (16.5.5), the suffix $(*)$ is used for the complex conjugate of a function. The above expressions are applied to the symmetrical coupled microstrip line shown in Fig (16.8b). Let us consider the case when transverse current on the strip could be neglected, i.e. $J_x = 0$, and $\tilde{J}_z(\alpha_n)$ has a one-term basis function. This assumption is valid for the fundamental coupled mode. If two strips of width 2w are symmetrically located, its Fourier transform at the center is $\tilde{J}_{z1}(\alpha_n)$. The Fourier transforms of conductor # 1 and conductor # 2 are obtained from the shifting theorem as follows:

$$\text{Conductor \#1}: \quad \tilde{J}_z^1(\alpha_n) = e^{-j\alpha_n(s+w)}\tilde{J}_{z1}(\alpha_n) \qquad \text{(a)}$$

$$\text{Conductor \#2}: \quad \tilde{J}_z^2(\alpha_n) = e^{+j\alpha_n(s+w)}\tilde{J}_{z2}(\alpha_n) \qquad \text{(b)}.$$

$$(16.5.6)$$

For the *even-mode excitation*, currents on both strip conductors are equal in magnitude with the same phase. For this case, the total current density $\tilde{J}_z^t(\alpha_n)$ is

$$\tilde{J}_z^e(\alpha_n) = \left[e^{-j\alpha_n(s+w)} + e^{+j\alpha_n(s+w)}\right]\tilde{J}_{z1}(\alpha_n) = 2\cos\alpha_n(s+w)\tilde{J}_{z1}(\alpha_n). \qquad (16.5.7)$$

For the *odd mode excitation*, currents on both strip conductors are equal in magnitude and opposite in phase. For this case, the total current density $\tilde{J}_z^t(\alpha_n)$ is

$$\tilde{J}_z^o(\alpha_n) = \frac{2j\left[e^{+j\alpha_n(s+w)} - e^{-j\alpha_n(s+w)}\right]\tilde{J}_{z1}(\alpha_n)}{2j} = 2j\sin\alpha_n(s+w)\tilde{J}_{z1}(\alpha_n). \qquad (16.5.8)$$

The expression for Fourier transformed current density $\tilde{J}_{z1}(\alpha_n)$ has already been obtained in section (16.4). For the case considered, the characteristic equation, from equation (16.5.5), is

$$\sum_{n=1}^{\infty} \tilde{Z}_{22}(\alpha_n)\left[\tilde{J}_{z\,i}^t(\alpha_n)\right]^2 = 0. \qquad (16.5.9)$$

In equation (16.5.9), the condition $\tilde{J}_{z\,i}^t(\alpha_n) = \tilde{J}_z^e(\alpha_n)$ is used for the even mode, and the condition $\tilde{J}_{z\,i}^t(\alpha_n) = \tilde{J}_z^o(\alpha_n)$ for the odd mode. Also, the same function is used as the basis and testing function. The numerical solution of the characteristic equation, giving the fundamental even and odd mode propagation constants, for the known physical parameters of the coupled microstrip line at the assumed frequency is obtained from the characteristic equation (16.5.9). In the case of more number of basis function for both the longitudinal and transverse currents, the even and odd mode propagation constants are obtained from equation (16.5.4c) by taking its det = 0. The complex propagation constants could also be obtained, showing the presence of the leaky modes.

16.6 Multilayer Planar Transmission Lines

Figure (16.9a–f) show some multilayer planar transmission line structures used in the MIC, MMIC, MEMS, LTCC technology. The dielectric layers could be isotropic or anisotropic. Generally, these are lossy dielectric layers. Sometimes they could be treated as lossless. Three cases are considered in this section.

i) *Single strip microstrip and single slot slot-line*
Figure (16.9a, b) shows the microstrip line and the slot-line under four-layered dielectric medium. The structures could be easily reduced to several useful cases, such as shielded, dielectric covered, suspended, inverted structures, etc.
ii) *Coplanar multistrip microstrip lines*
Figure (16.9c,d) shows the edge coupled microstrip lines and CPW structures respectively. Multiple

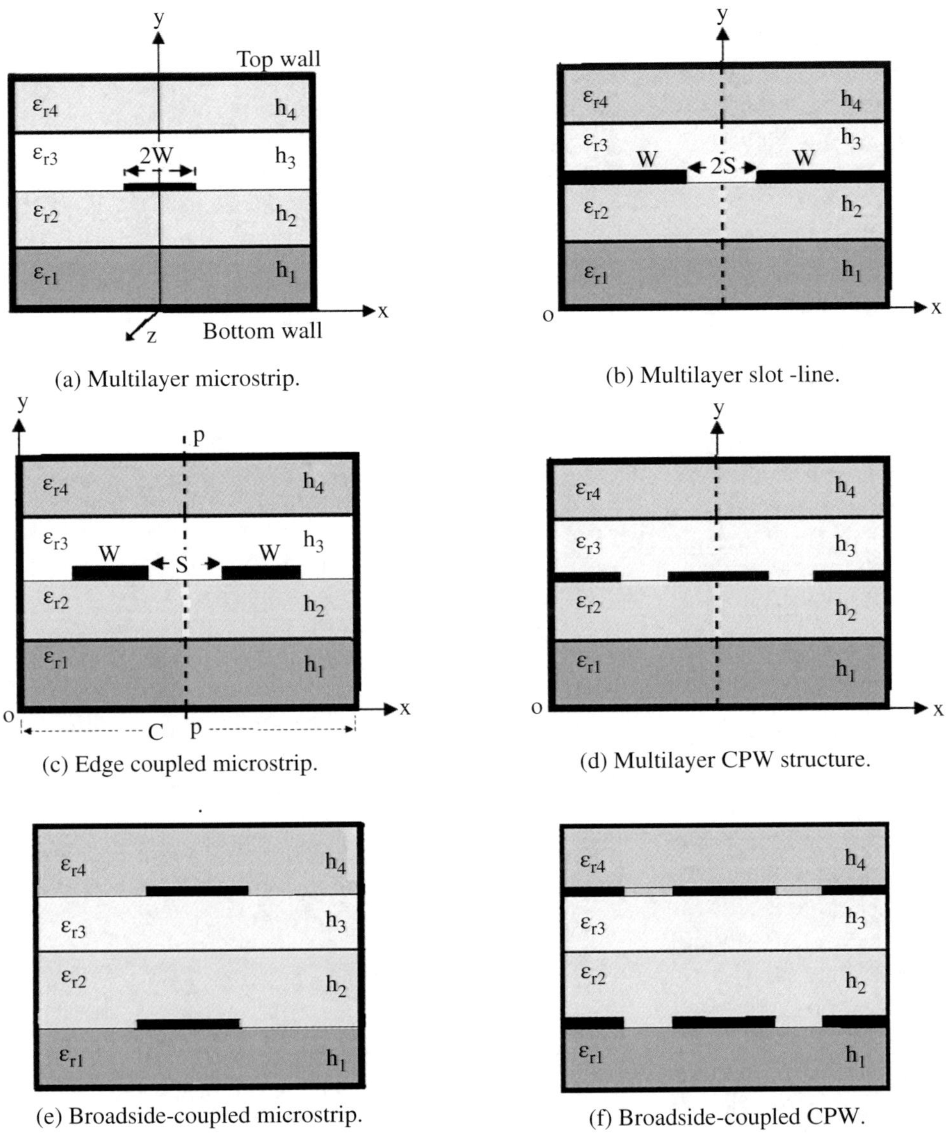

Figure 16.9 Multilayer planar transmission line structures.

conducting strips are located at the same level. Again these structures are under four-layered dielectric medium.

iii) *Multilevel multiconductor microstrip*

Figure (16.9e,f) shows the broadside-coupled microstrip and the broadside coupled CPW structure under four-layered dielectric medium. In this case, the conducting strips are located at two levels. Normally, the second and third layers are identical. In that case, it is a three-layered structure. These structures could also be reduced to several technologically useful cases.

The above mentioned structures are typical. Several dielectric layers and a number of the strip conductors can be added at different levels to meet practical requirements. The single or multilayer strips at any one level are considered by the total surface current density at that level. Likewise, the multiple slots at any one level are considered by the total slot fields at that level. The real task before us is to find the *impedance kind Green's function* for the microstrip type structures used in the multilayer condition and the *admittance kind Green's function* for the slot kind of the multilayer structure. Both kinds of Green's functions are the inverse of each other. In general, two approaches are used to obtain the Green's function of a multilayer planar transmission line: (i) *immittance approach* developed by Itoh [B.1, J.4, J.29–J.32] and *transfer matrix approach* developed by Davies and Syahkal [B.2, B.4, J.11, J.16–J.19]. However, this section considers only the immittance approach due to its simplicity. There are other approaches also to obtain the Green's function of the multilayer planar transmission lines. These could be followed in the open literature after having acquaintance with one approach.

16.6.1 Immittance Approach for Single-Level Strip Conductors

The immittance approach avoids the solution of the wave equation in each layer of a multilayer dielectric medium for the hybrid mode wave propagation. Each layer involves four unknown constants for the electric and magnetic scalar potentials. It is a very cumbersome process to determine the constants of the multilayer case. This difficulty is faced even while solving Laplace's equation for the *quasi-static* case of the planar lines in the layered medium. The transverse transmission line (TTL) method is used in section (14.4) of chapter 14 to obtain Green's function of the quasi-static multilayer microstrip and CPW structures. *The immittance approach is an extension of the TTL technique applied to the dynamic SDA.* The quasi-static method involves one Laplace's equation for one electric potential. Thus, it involves only one equivalent transmission line network to represent the multilayer microstrip line. However, the dynamic analysis of a microstrip line, or a CPW structure, involves the *hybrid mode* that is a linear combination of the LSE (TE^y) and LSM (TM^y) modes described by the wave equation (16.2.2) for the electric and magnetic potentials. Therefore, two equivalent transmission lines are needed – one for the LSE (TE^y) mode and another for the LSM (TM^y) mode. Figure (16.9a) shows that the *y*-axis is the direction of inhomogeneity of the dielectric layers and the hybrid mode propagates in the *z*-direction. This section at first obtains the *transverse transmission line (TTL) equivalence circuit* for the layered medium, shown in Fig (16.9a–f), for the LSE (TE^y) mode and LSM (TM^y) mode, respectively. Next, it is applied to the TTL models to obtain Green's functions of several multilayered planar transmission line structures.

TTL Equivalent Circuits for LSE and LSM Modes

The LSE (TE^y) and LSM (TM^y) modes of the structure shown in Fig (16.9a) are described by magnetic potential $\phi^m(x, y)e^{j(\omega t - \beta z)}$ and electric potential $\phi^e(x, y)e^{j(\omega t - \beta z)}$, respectively. These potentials satisfy the wave equation (16.2.2) both for the TE^y mode ($E^m_y = 0$) and TM^y mode ($H^e_y = 0$). The five field components for the TE^y mode and TM^y mode are given by equations (16.2.3) and (16.2.4), respectively. These field components in the Fourier domain are summarized below:

TE^y Mode

$$\begin{aligned}
&\tilde{E}_x(\alpha) = -j\beta_z\tilde{\phi}^m(\alpha), && \tilde{H}_x(\alpha) = -\frac{j\alpha}{j\omega\mu}\frac{\partial\tilde{\phi}^m(\alpha)}{\partial y} = -\frac{\alpha}{\omega\mu}\frac{\partial\tilde{\phi}^m(\alpha)}{\partial y}\\
&\tilde{E}_y(\alpha) = 0, && \tilde{H}_y(\alpha) = \frac{1}{j\omega\mu}\left(\frac{\partial^2}{\partial y^2} + k^2\right)\tilde{\phi}^m(\alpha)\\
&\tilde{E}_z(\alpha) = j\alpha\tilde{\phi}^m(\alpha), && \tilde{H}_z(\alpha) = -\frac{j\beta_z}{j\omega\mu}\frac{\partial\tilde{\phi}^m(\alpha)}{\partial y} = -\frac{\beta_z}{\omega\mu}\frac{\partial\tilde{\phi}^m(\alpha)}{\partial y}.
\end{aligned} \tag{16.6.1}$$

TM^y Mode

$$\begin{aligned}
&\tilde{E}_x(\alpha) = -\frac{\alpha}{\omega\varepsilon}\frac{\partial\tilde{\phi}^e(\alpha)}{\partial y}, && \tilde{H}_x(\alpha) = j\beta_z\tilde{\phi}^e(\alpha)\\
&\tilde{E}_y(\alpha) = \frac{1}{j\omega\varepsilon}\left(\frac{\partial^2}{\partial y^2} + k^2\right)\tilde{\phi}^e(\alpha), && \tilde{H}_y(\alpha) = 0\\
&\tilde{E}_z(\alpha) = -\frac{j\beta_z}{j\omega\varepsilon}\frac{\partial\tilde{\phi}^e(\alpha)}{\partial y} = -\frac{\beta_z}{\omega\varepsilon}\frac{\partial\tilde{\phi}^e(\alpha)}{\partial y}, && \tilde{H}_z(\alpha) = -j\alpha\tilde{\phi}^e(\alpha).
\end{aligned} \tag{16.6.2}$$

Fourier transforms of the potential and field components are obtained for the laterally open structure by

$$\tilde{\phi}^{e,m}(\alpha, y) = \int_{-\infty}^{\infty} \phi^{e,m}(x, y)e^{j\alpha x}dx. \tag{16.6.3}$$

For the shielded structure, the continuous Fourier variable α is replaced by the discrete Fourier variable α_n. The inverse Fourier transform is

$$\phi^{e,m}(x, y) = \frac{1}{2\pi}\int_{-\infty}^{\infty} \tilde{\phi}^{e,m}(\alpha, y)e^{-j(\alpha x + \beta_z z)}d\alpha. \tag{16.6.4}$$

Equation (16.6.4) shows that the TE^y and TM^y waves could be treated as a superposition of the *spectral plane waves* described by the $\tilde{\phi}^m(\alpha, y)e^{-j(\alpha x + \beta_z z)}$ and $\tilde{\phi}^e(\alpha, y)e^{-j(\alpha x + \beta_z z)}$. The spectral plane waves travel in the (x–z)-plane, with components of propagation constants β_z and α along the *z*-axis and *x*-axis, respectively. Figure (16.10) shows the spectral waves travel in the v direction. The direction of propagation of the spectral waves, i.e. v, is θ degree inclined with the z-axis, such that

$$\cos\theta = \frac{\beta_z}{\sqrt{\alpha^2 + \beta_z^2}}, \qquad \sin\theta = \frac{\alpha}{\sqrt{\alpha^2 + \beta_z^2}}. \tag{16.6.5}$$

The modes TE^y and TM^y have five field components in the (x, y, z) coordinate system. A new coordinate system (u, v, y) is considered for the propagation of the *spectral waves*. The variables of both coordinate systems in the transverse plane, with respect to the *y*-axis, are related by the following coordinate transformation:

$$\begin{bmatrix} u \\ v \end{bmatrix} = \begin{bmatrix} \sin\theta & -\cos\theta \\ \cos\theta & \sin\theta \end{bmatrix}\begin{bmatrix} z \\ x \end{bmatrix}. \tag{16.6.6}$$

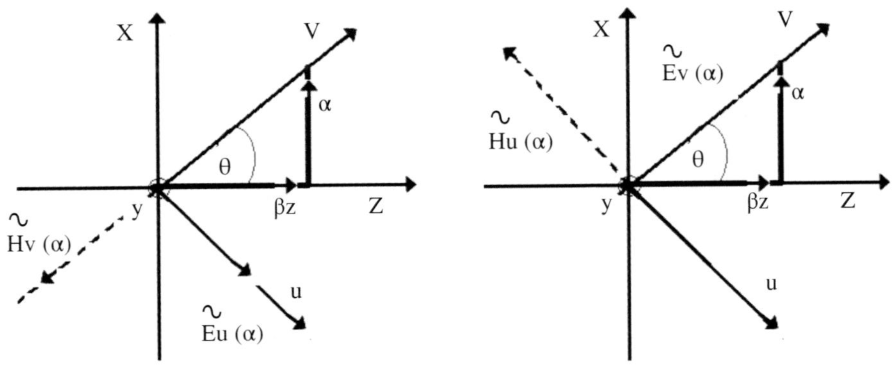

(a) $TE^y \rightarrow TE^v$ mode change. (b) $TM^y \rightarrow TM^v$ mode change.

Figure 16.10 Spectral wave propagation.

Equivalent Transverse Transmission Line for LSE (TE^y) Mode

It is interesting to note that five field components of the TE^y mode in the (x, y, z) coordinate system are reduced to three field components of the TE^v in the (u, v, y) coordinate system. Using equation (16.6.6), the electric field components $\left(\tilde{E}_z, \tilde{E}_x\right)$ and magnetic field components $\left(\tilde{H}_z, \tilde{H}_x\right)$, from the (x, y, z) coordinate system are transformed into the $\left(\tilde{E}_u, \tilde{E}_v\right)$ and $\left(\tilde{H}_u, \tilde{H}_v\right)$ field components in the (u, v, y) coordinate system as follows:

$\tilde{E}_u = \sin\theta\tilde{E}_z - \cos\theta\tilde{E}_x$. Using equations (16.6.1), (16.6.4) and (16.6.5), the field components in the (u, v, y) coordinate system are

$$\tilde{E}_u(\alpha) = \frac{\alpha}{\sqrt{\alpha^2+\beta_z^2}} \times \left(j\alpha\tilde{\phi}^m\right) - \frac{\beta_z}{\sqrt{\alpha^2+\beta_z^2}} \times \left(-j\beta\tilde{\phi}^m\right)$$

$$\Rightarrow \tilde{E}_u(\alpha) = j\sqrt{\alpha^2+\beta_z^2}\,\tilde{\phi}^m(\alpha) \qquad \text{(a)}$$

$$\tilde{E}_v(\alpha) = \cos\theta\tilde{E}_z + \sin\theta\tilde{E}_x = \frac{\beta_z}{\sqrt{\alpha^2+\beta_z^2}} \times \left(j\alpha\tilde{\phi}^m\right) + \frac{\alpha}{\sqrt{\alpha^2+\beta_z^2}} \times \left(-j\beta_z\tilde{\phi}^m\right) = 0 \qquad \text{(b)}$$

$$\tilde{H}_u(\alpha) = \sin\theta\tilde{H}_z - \cos\theta\tilde{H}_x = \frac{\alpha}{\sqrt{\alpha^2+\beta_z^2}} \times \left(-\frac{\beta_z}{\omega\mu}\frac{d\tilde{\phi}^m}{dy}\right) - \frac{\beta_z}{\sqrt{\alpha^2+\beta_z^2}} \times \left(-\frac{\alpha}{\omega\mu}\frac{d\tilde{\phi}^m}{dy}\right) = 0 \qquad \text{(c)}$$

$$\tilde{H}_v(\alpha) = \cos\theta\tilde{H}_z + \sin\theta\tilde{H}_x = \frac{\beta_z}{\sqrt{\alpha^2+\beta_z^2}}\left(-\frac{\beta_z}{\omega\mu}\frac{d\tilde{\phi}^m}{dy}\right) + \frac{\alpha}{\sqrt{\alpha^2+\beta_z^2}}\left(-\frac{\alpha}{\omega\mu}\frac{d\tilde{\phi}^m}{dy}\right) \Rightarrow \tilde{H}_v(\alpha) = -\frac{\sqrt{\alpha^2+\beta_z^2}}{\omega\mu}\frac{d\tilde{\phi}^m}{dy} \qquad \text{(d).}$$

(16.6.7)

Thus, in the (y, u, v) coordinate system, the transformed TE^y mode has three field components $\left(\tilde{H}_y, \tilde{E}_u, \tilde{H}_v\right)$ and three field components are zero $\left(\tilde{E}_y = \tilde{E}_v = \tilde{H}_u = 0\right)$. In the (y, u, v) coordinate system, the usual TE^y mode is transformed into the TE^v mode. The TE^v mode is defined with respect to the *v*-axis, $\tilde{E}_v = 0$. Thus, the original TE^y mode with five field components is now TE^v mode with only three field components. In the original structure, it is assumed that the modal field propagates in the *y*-direction to define the TE^y mode. However, the actual hybrid mode propagates in the *z*-direction with the propagation constant β_z. The TE^v mode could also be treated as the mode propagating in the *y*-direction with constant propagation γ, i.e. $\tilde{\phi}^m(\alpha, y)e^{\gamma y}$. Therefore, TE^v mode is supported by a transmission line of the propagation constant γ. Its *characteristic admittance* Y_{TE} is the *wave admittance* of the TE^v mode given by the following expression:

$$Y_{TE} = -\frac{\tilde{H}_v(\alpha)}{\tilde{E}_u(\alpha)} = -\frac{\left(-\sqrt{\alpha^2+\beta_z^2}\right)\frac{d\tilde{\phi}^m}{dy}}{\omega\mu\left(j\sqrt{\alpha^2+\beta_z^2}\right)\tilde{\phi}^m} = \frac{1}{j\omega\mu}\left[\frac{\frac{d\left\{\tilde{\phi}^m(\alpha)e^{\gamma y}\right\}}{dy}}{\tilde{\phi}^m(\alpha, y)e^{\gamma y}}\right] = \frac{\gamma}{j\omega\mu} \qquad \text{(a)}$$

$$\text{For the } i^{th} \text{ layer of dielectric}: \quad Y_{TEi} = \frac{\gamma_i}{j\omega\mu_i} \qquad \text{(b).}$$

(16.6.8)

The (−) sign of the Y_{TE}, with respect to Fig (16.10), is to maintain the Poynting vector in the positive $\hat{y}$ direction. It is noted that the $\tilde{H}_v(\alpha)$ is in the negative direction along the v-axis. The $\tilde{E}_u(\alpha)$-field component is in the direction of the $\tilde{J}_u(\alpha)$ component of surface current density on the strip conductor. Thus, the $\tilde{J}_u(\alpha)$ current component is taken as the source of the TE^v mode. Now, the *equivalent transverse transmission line (TTL)* could be drawn for the TE^y mode in the (y, u, v) coordinate system as shown in Fig (16.11a). Each dielectric layer of thickness h_i is represented by a transmission line section of length h_i. The characteristic admittance of the line section is Y_{TEi}, and the propagation constant γ_i is given by

$$\gamma_i = \sqrt{\alpha^2 + \beta_z^2 - \varepsilon_i k^2}, \quad i = 1, 2, 3, 4. \qquad (16.6.9)$$

The surface current $\tilde{J}_u(\alpha)$ is located at the plane containing the strip conductor. The lower and uppermost transmission lines are short-circuited, as $\tilde{E}_u(\alpha) = 0$ at both the bottom and top perfect conductors shown in Fig (16.11a).

Equivalent Transverse Transmission Line for LSM (TM^y) Mode

Using equation (16.6.6), the five field components of the LSM, i.e. TM^y mode, in the (z, x, y) coordinate system can also be reduced to three field components in the (u, v, y) coordinate system as shown in Fig (16.10b). The expression (16.6.5) is applied to the field components of the LSM (TM^y) – mode as given by equation (16.6.2):

$\tilde{E}_u = \sin\theta\tilde{E}_z - \cos\theta\tilde{E}_x$. Using $\tilde{E}_x$ and $\tilde{E}_z$ from equation (16.6.2), the following transformations are obtained:

$$\tilde{E}_u = \frac{\alpha}{\sqrt{\alpha^2+\beta^2}} \times \left(-\frac{\beta_z}{\omega\varepsilon}\frac{d\tilde{\phi}^e}{dy}\right) - \frac{\beta_z}{\sqrt{\alpha^2+\beta_z^2}} \times \left(-\frac{\alpha}{\omega\varepsilon}\frac{d\tilde{\phi}^e}{dy}\right) = 0 \quad (a)$$

$$\tilde{E}_v(\alpha) = \cos\theta\tilde{E}_z + \sin\theta\tilde{E}_x = \frac{\beta_z}{\sqrt{\alpha^2+\beta_z^2}}\left(-\frac{\beta_z}{\omega\varepsilon}\frac{d\tilde{\phi}^e}{dy}\right) + \frac{\alpha}{\sqrt{\alpha^2+\beta_z^2}}\left(-\frac{\alpha}{\omega\varepsilon}\frac{d\tilde{\phi}^e}{dy}\right)$$

$$\Rightarrow \tilde{E}_v(\alpha) = \left(-\frac{\sqrt{\alpha^2+\beta_z^2}}{\omega\varepsilon}\frac{d\tilde{\phi}^e(\alpha)}{dy}\right) \quad (b).$$

$$(16.6.10)$$

$$\tilde{H}_u(\alpha) = \sin\theta\tilde{H}_z - \cos\theta\tilde{H}_x = \frac{\alpha}{\sqrt{\alpha^2+\beta_z^2}} \times \left(-j\alpha\tilde{\phi}^e\right) - \frac{\beta_z}{\sqrt{\alpha^2+\beta_z^2}} \times \left(j\beta_z\tilde{\phi}^e\right) = -j\sqrt{\alpha^2+\beta_z^2}\;\tilde{\phi}^e \quad (a)$$

$$\tilde{H}_v = \cos\theta\tilde{H}_z + \sin\theta\tilde{H}_x = \frac{\beta_z}{\sqrt{\alpha^2+\beta_z^2}} \times \left(-j\alpha\tilde{\phi}^e\right) + \frac{\alpha}{\sqrt{\alpha^2+\beta_z^2}} \times \left(j\beta_z\tilde{\phi}^e\right) = 0 \quad (b).$$

$$(16.6.11)$$

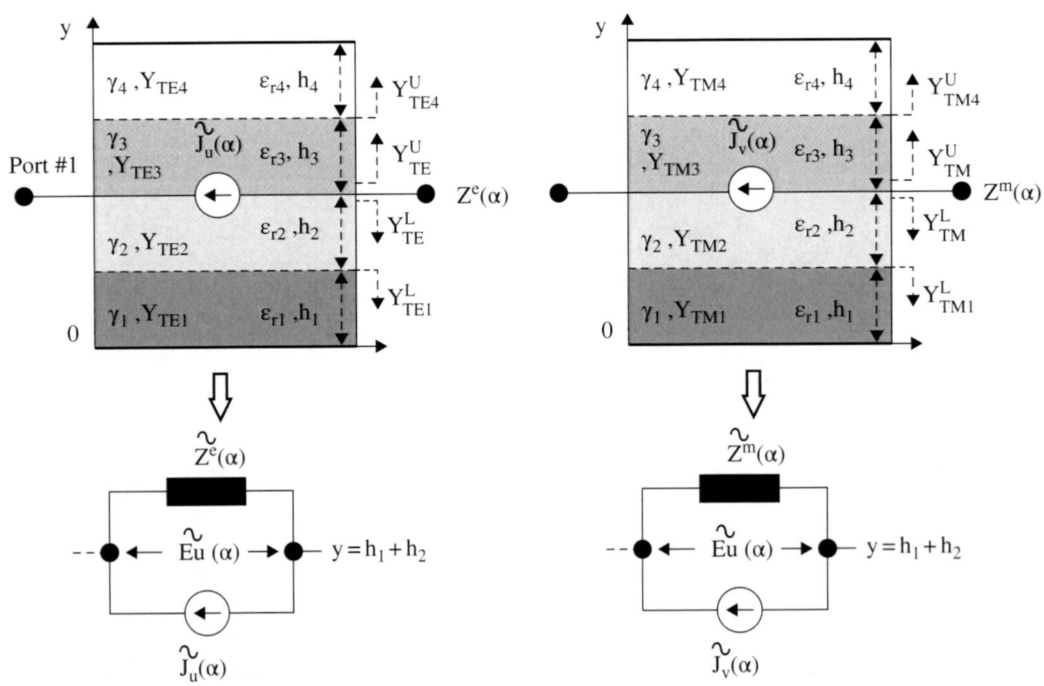

(a) Equivalent circuit of TE^y -mode. (b) Equivalent circuit of TM^y -mode.

Figure 16.11 Equivalent transverse transmission line (TTL) for TE^y and TM^y – modes in (u, v, y) coordinate system and equivalent circuit at $y = h_1 + h_2$ plane.

Thus, the (TM^y)-mode has three field components in the (u, v, y) coordinate system – $\tilde{E}_v(\alpha)$, $\tilde{H}_u(\alpha)$, $\tilde{E}_y(\alpha)$, as shown in Fig (16.10b). It shows that these are the field components of the TM^v mode. Hence, the TM^y mode, $\tilde{\phi}^e(\alpha, y)e^{\gamma y}$ propagating in the y-direction, with five field components in the (z, x, y) coordinate system is transformed in the TM^v mode in the (u, v, y) coordinate system with three field components. Figure (16.11b) shows a representation of the TM^v mode by the transverse transmission line (TTL) sections, with characteristic admittance Y_{TMi}:

$$Y_{TMi} = \frac{\tilde{H}_u(\alpha)}{\tilde{E}_v(\alpha)} = -\frac{\left(-j\sqrt{\alpha^2+\beta_z^2}\right)\tilde{\phi}^e(\alpha)}{-\left(\sqrt{\alpha^2+\beta_z^2}\right)\dfrac{d\tilde{\phi}^e(\alpha)}{dy}} \times \omega\varepsilon_i$$

$$\Rightarrow Y_{TM_i} = \frac{j\omega\varepsilon_i}{\gamma_i} \quad i = 1, 2, 3, 4. \tag{16.6.12}$$

Equation (16.6.9) gives us the propagation constant γ_i in the y-direction. The TM^v mode has the $\tilde{E}_v(\alpha)$ electric field component that is generated by the $\tilde{J}_v(\alpha)$ surface current density, as the strip is located in the (u, v, y) coordinate system. Figure (16.11b) shows that the current source $\tilde{J}_v(\alpha)$ is located at $y = h_1 + h_2$. The propagation constant γ_i is common to both TE^v and TM^v modes, as they are not the independent modes. They are the components of the hybrid mode having propagation constant γ_i in the y-direction. Figure (16.11b) shows that the bottom and top transmission lines are short-circuited. So at both the bottom and top conducting surfaces, the electric field's v-component is $\tilde{E}_v(\alpha) = 0$.

The above-discussed equivalent TTL can be applied to the circuits of the hybrid mode to determine the Green's function in the Fourier domain at the plane y = h1 + h2 containing the strip conductor.

Determination of the [Z] – Type Green's Function

Figure (16.11) shows that at the plane $y = h_1 + h_2$, the current sources, $\tilde{J}_u(\alpha)$ and $\tilde{J}_v(\alpha)$, are present. They generate TE^v and TM^v mode, respectively. The transmission line sections, above and below current sources, present impedances $\tilde{Z}^e(\alpha, h_1 + h_2)$ and $\tilde{Z}^m(\alpha, h_1 + h_2)$ at the port #1, i.e. at the plane $y = h_1 + h_2$ for TE^v and TM^v modes respectively. The equivalent circuits of Fig (16.11) provide the corresponding electric fields, $\tilde{E}_u(\alpha, h_1 + h_2)$ and $\tilde{E}_v(\alpha, h_1 + h_2)$, at the plane $y = h_1 + h_2$ as follows:

$$\begin{aligned} &TE^u \text{ mode: } \tilde{E}_u(\alpha) = \tilde{Z}^e(\alpha)\tilde{J}_u(\alpha) && \text{(a)} \\ &TM^v \text{ mode: } \tilde{E}_v(\alpha) = \tilde{Z}^m(\alpha)\tilde{J}_v(\alpha) && \text{(b).} \end{aligned} \tag{16.6.13}$$

The port impedances $\tilde{Z}^e(\alpha)$ and $\tilde{Z}^m(\alpha)$ are obtained from the port admittances Y_{TE}^L, Y_{TE}^U and Y_{TM}^L, Y_{TM}^U of the transmission line sections looking downward (L) and upward (U):

$$\begin{aligned} \tilde{Z}^e(\alpha) &= \frac{1}{Y_{TE}^L(\alpha) + Y_{TE}^U(\alpha)} && \text{(a)} \\ \tilde{Z}^m(\alpha) &= \frac{1}{Y_{TM}^L(\alpha) + Y_{TM}^U(\alpha)} && \text{(b).} \end{aligned} \tag{16.6.14}$$

Equation (16.6.13) indicates that the hybrid mode of the actual structure in the (z, x, y) coordinate system is decoupled in two independent modes, TE^u and TM^v, in the (u, v, y) coordinate system. The $\tilde{Z}^e(\alpha)$ and $\tilde{Z}^m(\alpha)$ are the *impedance kind of Green's functions* in the (u, v, y) system for the TE^u and TM^v modes, respectively. However, the impedance kind of Green's function is to be obtained in the (z, x, y) coordinate system, belonging to the hybrid mode propagation. The *port admittances* $\tilde{Y}_{TE}^L(\alpha)$ and $\tilde{Y}_{TE}^U(\alpha)$, also $\tilde{Y}_{TM}^L(\alpha)$ and $\tilde{Y}_{TM}^U(\alpha)$, are obtained on using the following input admittance equation:

$$Y_{in} = Y_0\frac{Y_L + Y_0 \tanh \gamma d}{Y_0 + Y_L \tanh \gamma d}, \tag{16.6.15}$$

where Y_{in} is the input admittance of a line section of length d terminated in the load Y_L. The short-circuited line #1 of Fig (16.11a) has $Y_L = \infty$. Using Fig (16.11a), the lower port admittance $\tilde{Y}_{TE}^L(\alpha)$ is obtained in terms of the *layer admittance* $\tilde{Y}_{TE1}^L$ of the first dielectric layer #1:

$$\begin{aligned} \tilde{Y}_{TE1}^L(\alpha) &= \tilde{Y}_{TE1}(\alpha)\coth(\gamma_1 h_1) && \text{(a)} \\ \tilde{Y}_{TE}^L(\alpha) &= \tilde{Y}_{TE2}(\alpha)\frac{\tilde{Y}_{TE1}^L(\alpha) + \tilde{Y}_{TE2}(\alpha)\tanh(\gamma_2 h_2)}{\tilde{Y}_{TE2}(\alpha) + \tilde{Y}_{TE1}^L(\alpha)\tanh(\gamma_2 h_2)} && \text{(b),} \end{aligned} \tag{16.6.16}$$

where $\tilde{Y}_{TEi}(\alpha)$, $i = 1, 2, 3, 4$ is the characteristic admittance of the ith section of the equivalent transmission lines. In the above equation (16.6.16), $\tilde{Y}_{TE1}^L(\alpha)$ is a load to the line section #2. Likewise, the admittance of the upper port admittance $\tilde{Y}_{TE}^U(\alpha)$ is obtained in terms of the *layer admittance* $\tilde{Y}_{TE4}^U(\alpha)$ of the short-circuited fourth dielectric layer #4:

$$\begin{aligned} \tilde{Y}_{TE4}^U(\alpha) &= \tilde{Y}_{TE4}(\alpha)\coth(\gamma_4 h_4) && \text{(a)} \\ \tilde{Y}_{TE}^U(\alpha) &= \tilde{Y}_{TE3}(\alpha)\frac{\tilde{Y}_{TE4}^U(\alpha) + \tilde{Y}_{TE3}(\alpha)\tanh(\gamma_3 h_3)}{\tilde{Y}_{TE3}(\alpha) + \tilde{Y}_{TE4}^U(\alpha)\tanh(\gamma_3 h_3)} && \text{(b).} \end{aligned} \tag{16.6.17}$$

Similarly, using Fig (16.11b), the following expressions are obtained for the $\tilde{Y}^L_{TM}(\alpha)$ and $\tilde{Y}^U_{TM}(\alpha)$:

$$\tilde{Y}^L_{TM1}(\alpha) = \tilde{Y}_{TM1}(\alpha)\coth(\gamma_1 h_1) \quad (a)$$
$$\tilde{Y}^L_{TM}(\alpha) = \tilde{Y}_{TM2}(\alpha)\frac{\tilde{Y}^L_{TM1}(\alpha) + \tilde{Y}_{TM2}(\alpha)\tanh(\gamma_2 h_2)}{\tilde{Y}_{TM2}(\alpha) + \tilde{Y}^L_{TM1}(\alpha)\tanh(\gamma_2 h_2)} \quad (b). \tag{16.6.18}$$

$$\tilde{Y}^L_{TM4}(\alpha) = \tilde{Y}_{TM4}(\alpha)\coth(\gamma_4 h_4) \quad (a)$$
$$\tilde{Y}^U_{TM}(\alpha) = \tilde{Y}_{TM3}(\alpha)\frac{\tilde{Y}^L_{TM4}(\alpha) + \tilde{Y}_{TM4}(\alpha)\tanh(\gamma_3 h_3)}{\tilde{Y}_{TM3}(\alpha) + \tilde{Y}^L_{TM4}(\alpha)\tanh(\gamma_3 h_3)} \quad (b). \tag{16.6.19}$$

At this stage, equation 16.6.13 is to be transformed from the (u, v, y) coordinate system to the (z, x, y) coordinate system by the change of variables from $\tilde{E}_u(\alpha)$, $\tilde{E}_v(\alpha)$ and $\tilde{J}_u(\alpha)$, $\tilde{J}_v(\alpha)$ to $\tilde{E}_z(x)$, $\tilde{E}_x(x)$, $\tilde{J}_z(x)$, and $\tilde{J}_x(x)$. To get the coordinate transformation, equation (16.6.13) could be written in the matrix form:

$$\begin{bmatrix} \tilde{E}_u(\alpha) \\ \tilde{E}_v(\alpha) \end{bmatrix} = \begin{bmatrix} \tilde{Z}^e(\alpha) & 0 \\ 0 & \tilde{Z}^m(\alpha) \end{bmatrix} \begin{bmatrix} \tilde{J}_u(\alpha) \\ \tilde{J}_v(\alpha) \end{bmatrix}. \tag{16.6.20}$$

On using the coordinate transformation (16.6.6), the above equation (16.6.20) is written as follows:

$$\begin{bmatrix} \sin\theta & -\cos\theta \\ \cos\theta & \sin\theta \end{bmatrix} \begin{bmatrix} \tilde{E}_z(\alpha) \\ \tilde{E}_x(\alpha) \end{bmatrix} = \begin{bmatrix} \tilde{Z}^e(\alpha) & 0 \\ 0 & \tilde{Z}^m(\alpha) \end{bmatrix} \begin{bmatrix} \sin\theta & -\cos\theta \\ \cos\theta & \sin\theta \end{bmatrix} \begin{bmatrix} \tilde{J}_z(\alpha) \\ \tilde{J}_x(\alpha) \end{bmatrix}$$
$$= \begin{bmatrix} \tilde{Z}^e(\alpha)\sin\theta & -\tilde{Z}^e(\alpha)\cos\theta \\ \tilde{Z}^m(\alpha)\cos\theta & \tilde{Z}^m(\alpha)\sin\theta \end{bmatrix} \begin{bmatrix} \tilde{J}_z(\alpha) \\ \tilde{J}_x(\alpha) \end{bmatrix} \tag{16.6.21}$$
$$\Rightarrow \begin{bmatrix} \tilde{E}_u(\alpha) \\ \tilde{E}_v(\alpha) \end{bmatrix} = \begin{bmatrix} \sin\theta & -\cos\theta \\ \cos\theta & \sin\theta \end{bmatrix}^{-1} \begin{bmatrix} \tilde{Z}^e(\alpha)\sin\theta & -\tilde{Z}^e(\alpha)\cos\theta \\ \tilde{Z}^m(\alpha)\cos\theta & \tilde{Z}^m(\alpha)\sin\theta \end{bmatrix} \begin{bmatrix} \tilde{J}_z(\alpha) \\ \tilde{J}_x(\alpha) \end{bmatrix}.$$

Since $\begin{bmatrix} \sin\theta & -\cos\theta \\ \cos\theta & \sin\theta \end{bmatrix}^{-1} = \begin{bmatrix} \sin\theta & \cos\theta \\ -\cos\theta & \sin\theta \end{bmatrix}$ the above equation (16.6.21) is written as follows:

$$\begin{bmatrix} \tilde{E}_z(\alpha) \\ \tilde{E}_x(\alpha) \end{bmatrix} = \begin{bmatrix} \sin\theta & \cos\theta \\ -\cos\theta & \sin\theta \end{bmatrix} \begin{bmatrix} \tilde{Z}^e(\alpha)\sin\theta & -\tilde{Z}^e(\alpha)\cos\theta \\ \tilde{Z}^m(\alpha)\cos\theta & \tilde{Z}^m(\alpha)\sin\theta \end{bmatrix} \begin{bmatrix} \tilde{J}_z(\alpha) \\ \tilde{J}_x(\alpha) \end{bmatrix}$$
$$= \begin{bmatrix} \tilde{Z}^e(\alpha)\sin^2\theta + \tilde{Z}^m(\alpha)\cos^2\theta & -\tilde{Z}^e(\alpha)\sin\theta\cos\theta + \tilde{Z}^m(\alpha)\sin\theta\cos\theta \\ -\tilde{Z}^e(\alpha)\sin\theta\cos\theta + \tilde{Z}^m(\alpha)\sin\theta\cos\theta & \tilde{Z}^e(\alpha)\cos^2\theta + \tilde{Z}^m(\alpha)\sin^2\theta \end{bmatrix} \begin{bmatrix} \tilde{J}_z(\alpha) \\ \tilde{J}_x(\alpha) \end{bmatrix}$$
$$\begin{bmatrix} \tilde{E}_z(\alpha) \\ \tilde{E}_x(\alpha) \end{bmatrix} = \begin{bmatrix} \tilde{Z}^e(\alpha)\sin^2\theta + \tilde{Z}^m(\alpha)\cos^2\theta & \left(\tilde{Z}^m - \tilde{Z}^e\right)\sin\theta\cos\theta \\ \left(\tilde{Z}^m - \tilde{Z}^e\right)\sin\theta\cos\theta & \tilde{Z}^e(\alpha)\cos^2\theta + \tilde{Z}^m(\alpha)\sin^2\theta \end{bmatrix} \begin{bmatrix} \tilde{J}_z(\alpha) \\ \tilde{J}_x(\alpha) \end{bmatrix}. \tag{16.6.22}$$

The above equations (16.6.22) give the *impedance type Green's function* of the structure, shown in Fig (16.9a), in the Fourier domain with the following components:

$$\tilde{Z}_{11}(\alpha_n) = \tilde{Z}^e(\alpha_n)\sin^2\theta + \tilde{Z}^m(\alpha_n)\cos^2\theta \quad (a)$$
$$\tilde{Z}_{12}(\alpha_n) = \tilde{Z}_{21}(\alpha_n) = \left(\tilde{Z}^m - \tilde{Z}^e\right)\sin\theta\cos\theta \quad (b) \tag{16.6.23}$$
$$\tilde{Z}_{22}(\alpha_n) = \tilde{Z}^e(\alpha_n)\cos^2\theta + \tilde{Z}^m(\alpha_n)\sin^2\theta \quad (c).$$

The matrix elements of equation (16.6.23) are defined at the plane $y = h_1 + h_2$. For an enclosed structure, a discrete Fourier variable $\alpha = \alpha_n$ is used in the above equations. The Green's function is also applicable to the symmetrical coupled microstrip, separated in the even and odd mode supporting structures.

Determination of [Y] Type Green's Function for Slot Line and CPW

Figure (16.9b,d) shows the multilayer slot line and CPW. The admittance kind of Green's function is more suitable for these structures. The equivalent TTL network representation of TE^y and TM^y modes could be applied to the slot case also. At the plane $y = h_1 + h_2$, containing the slots and conducting strips, the following expressions are written from equation (16.6.13):

$$\begin{aligned} &TE^u \text{ mode}: \ \tilde{J}_u(\alpha) = \tilde{Y}^e(\alpha)\tilde{E}_u(\alpha) && (a) \\ &TM^v \text{ mode}: \ \tilde{J}_v(\alpha) = \tilde{Y}^m(\alpha)\tilde{E}_v(\alpha) && (b), \end{aligned} \qquad (16.6.24)$$

where the admittance parameters $\tilde{Y}^e(\alpha)$ and $\tilde{Y}^m(\alpha)$ are the decoupled Y-kind Green's functions at the plane $y = h_1 + h_2$. These are obtained from the equivalent TTL network shown in Fig (16.11) as follows:

$$\begin{aligned} &\tilde{Y}^e(\alpha) = \tilde{Y}^L_{TE}(\alpha) + \tilde{Y}^U_{TE}(\alpha) && (a) \\ &\tilde{Y}^m(\alpha) = \tilde{Y}^L_{TM}(\alpha) + \tilde{Y}^U_{TM}(\alpha) && (b). \end{aligned} \qquad (16.6.25)$$

The determination of the above admittance parameters has already been discussed. Equation (16.6.24) is written in the matrix form:

$$\begin{bmatrix} \tilde{J}_u(\alpha) \\ \tilde{J}_v(\alpha) \end{bmatrix} = \begin{bmatrix} \tilde{Y}^e(\alpha) & 0 \\ 0 & \tilde{Y}^m(\alpha) \end{bmatrix} \begin{bmatrix} \tilde{E}_u(\alpha) \\ \tilde{E}_v(\alpha \end{bmatrix}. \qquad (16.6.26)$$

The above expression (16.6.26) is transformed from the (u, v, y) coordinate system to the (z, x, y) coordinate system:

$$\begin{bmatrix} \tilde{J}_z(\alpha) \\ \tilde{J}_x(\alpha) \end{bmatrix} = \begin{bmatrix} \tilde{Y}^e(\alpha)\sin^2\theta + \tilde{Y}^m(\alpha)\cos^2\theta & \left(\tilde{Y}^m - \tilde{Y}^e\right)\sin\theta\cos\theta \\ \left(\tilde{Y}^m - \tilde{Y}^e\right)\sin\theta\cos\theta & \tilde{Y}^e(\alpha)\cos^2\theta + \tilde{Y}^m(\alpha)\sin^2\theta \end{bmatrix} \begin{bmatrix} \tilde{E}_z(\alpha) \\ \tilde{E}_x(\alpha) \end{bmatrix}. \qquad (16.6.27)$$

The above equation (16.6.27) could be written directly from equation (16.2.22). The above expression provides us the *admittance type Green's function* in the Fourier domain with the following components:

$$\begin{aligned} &\tilde{Y}_{11}(\alpha_n) = \tilde{Y}^e(\alpha_n)\sin^2\theta + \tilde{Y}^m(\alpha_n)\cos^2\theta && (a) \\ &\tilde{Y}_{12}(\alpha_n) = \tilde{Y}_{21}(\alpha_n) = \left(\tilde{Y}^m - \tilde{Y}^e\right)\sin\theta\cos\theta && (b) \\ &\tilde{Y}_{22}(\alpha_n) = \tilde{Y}^e(\alpha_n)\cos^2\theta + \tilde{Y}^m(\alpha_n)\sin^2\theta && (c). \end{aligned} \qquad (16.6.28)$$

Equations (16.6.23) and (16.6.28) are independent of the number of strip conductors and slots at one level, so both Green's functions could be applied to any number of strips and slots at one level.

16.6.2 Immittance Approach for Multilevel Strip Conductors

The conducting strips could be located at any level of dielectric layers. Figure (16.9e,f) shows the broadside coupled microstrip and the broadside coupled CPW structure respectively. Normally, $\varepsilon_{r2} = \varepsilon_{r3}$ and $h_2 = h_3$, i.e. the second and third dielectric layers could be combined into a single layer. Following the discussion of the previous section, the equivalent TTL networks can be drawn. These are shown in Fig (16.12a,b), for the LSE (TE^y) and LSM (TM^y) modes with two current sources for each of the conductor planes.

Figure (16.12a) shows that the planes $y = h_1$ and $y = h_1 + h_2 + h_3$, containing the strip conductors, could be treated as port #2 and port #1, respectively, where current sources $\tilde{J}_{u2}(\alpha)$ and $\tilde{J}_{u1}(\alpha)$ are located. They generate the TE^y mode. Likewise, Fig (16.12b) shows that the current sources $\tilde{J}_{v2}(\alpha)$ and $\tilde{J}_{v1}(\alpha)$ generate the TM^y mode. The port-voltages appear as the responses at port #1 and port #2 due to the excitation currents, so the port voltages and currents are related through the equivalent [Z]-matrix. In the case of the TE^y mode, the port voltages correspond to the field components, $\tilde{E}_{u2}(\alpha)$ and $\tilde{E}_{u1}(\alpha)$. The excitation current sources $\tilde{J}_{u2}(\alpha)$ and $\tilde{J}_{u1}(\alpha)$ are applied at port #2 and port #1, respectively. In the case of the TM^y mode, the response $\tilde{E}_{v2}(\alpha)$ and $\tilde{E}_{v1}(\alpha)$ are considered as the port voltages at port #2 and port #1, and the excitation current sources $\tilde{J}_{v2}(\alpha)$ and $\tilde{J}_{v1}(\alpha)$ are applied at the ports. Using the concept of the superposition, the following Z-matrix formulations are obtained for both the modes:

$$TE^y \text{ mode}: \left| \begin{aligned} &\tilde{E}_u(\alpha, h_1 + h_2 + h_3) = \tilde{E}_{u1}(\alpha) = Z^e_{11}\tilde{J}_{u1}(\alpha) + Z^e_{12}\tilde{J}_{u2}(\alpha) && (a) \\ &\tilde{E}_u(\alpha, h_1) = \tilde{E}_{u2}(\alpha) = Z^e_{21}\tilde{J}_{u1}(\alpha) + Z^e_{22}\tilde{J}_{u2}(\alpha) && (b). \end{aligned} \right. \qquad (16.6.29)$$

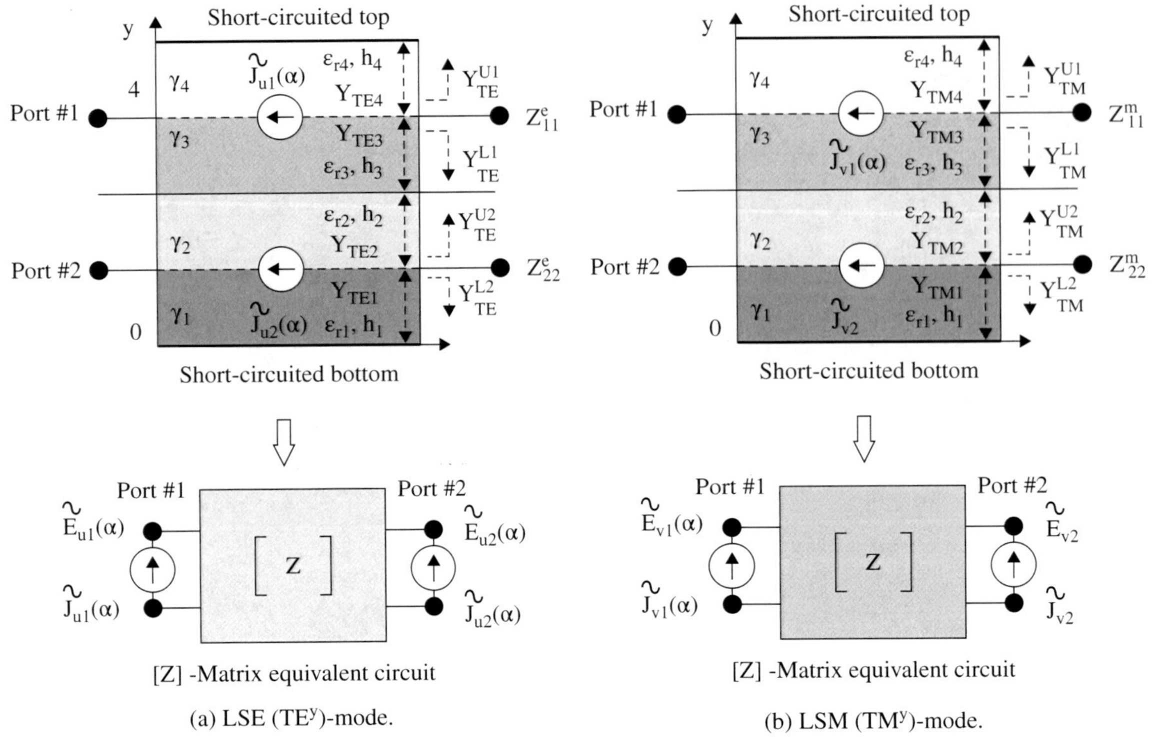

Figure 16.12 Equivalent TTL network for two-level strip conductors in (u, v, y) – coordinate system.

$$\text{TM}^y \text{ mode}: \left| \begin{array}{ll} \tilde{E}_v(\alpha, h_1+h_2+h_3) = \tilde{E}_{v1}(\alpha) = Z^m_{11}\tilde{J}_{v1}(\alpha) + Z^m_{12}\tilde{J}_{v2}(\alpha) & \text{(a)} \\ \tilde{E}_v(\alpha, h_1) = \tilde{E}_{v2}(\alpha) = Z^m_{21}\tilde{J}_{v1}(\alpha) + Z^m_{22}\tilde{J}_{v2}(\alpha) & \text{(b)}, \end{array} \right. \tag{16.6.30}$$

where Z^e_{11} and Z^e_{22} are the driving point, i.e. the *input impedance*, at the port #1 and port #2, respectively, for the TE^y mode. The $\tilde{Z}^e_{12}$ and $\tilde{Z}^e_{21}$ are the *transfer impedances*, i.e. the impedance at port #1 and port #2 due to the current sources at port #2 and port #1 respectively. Likewise, Z^m_{11}, Z^m_{22}, Z^m_{12} and Z^m_{21} are defined for the TM^y mode. The Z^e_{11}, Z^e_{22}, $\tilde{Z}^e_{12}$ and $\tilde{Z}^e_{21}$ for the TE^y mode are determine below, using the TTL network shown in Fig (16.12a). The upper end of the line section #4 and the lower end of the line section #1 are short-circuited.

Determination of Z^e_{11}

It is determined from equation (16.13b). The input port impedance Z^e_{11} is determined, from equation (16.6.29a), in terms of the upper and lower port admittance Y^{U1}_{TE} and Y^{L1}_{TE} at the port #1, while the port #2 is open-circuited, i.e. $\tilde{J}_{u2}(\alpha) = 0$:

$$Z^e_{11} = \frac{1}{Y^{U1}_{TE} + Y^{L1}_{TE}} \quad \text{(a)}$$
$$Z^e_{22} = \frac{1}{Y^{U2}_{TE} + Y^{L2}_{TE}} \quad \text{(b)}. \tag{16.6.31}$$

Therefore, the following expressions are obtained, using the transmission line theory, for the port admittance Y^{U1}_{TE} and Y^{L1}_{TE}:

$$\text{At } y = (h_1 + h_2 + h_3)^+ \;:\; Y^{U1}_{TE} = Y_{TE4}\coth(\gamma_4 h_4). \tag{16.6.32}$$

$$\begin{aligned} &\text{At } y = h_1^-,\ \text{load to line section \#2}: && Y^{L2}_{TE} = Y^{TE}_1 = Y_{TE1}\coth(\gamma_1 h_1) && \text{(a)} \\ &\text{At } y = (h_1 + h_2)^-\ \text{line section \#2}: && Y^{TE}_2 = Y_{TE2}\frac{Y^{TE}_1 + Y_{TE2}\tanh(\gamma_2 h_2)}{Y_{TE2} + Y^{TE}_1\tanh(\gamma_2 h_2)} && \text{(b)} \\ &\text{At } y = (h_1 + h_2 + h_3)^-\ \text{line section \#3}: && Y^{L1}_{TE} = Y_{TE3}\frac{Y^{TE}_2 + Y_{TE3}\tanh(\gamma_3 h_3)}{Y_{TE3} + Y^{TE}_2\tanh(\gamma_3 h_3)} && \text{(c)}. \end{aligned} \tag{16.6.33}$$

Once at the plane $y = h_1 + h_2 + h_3$, Y^{U1}_{TE} and Y^{L1}_{TE} are known, we can get the function Z^e_{11} from equation (16.6.31a).

Determination of Z^e_{12}

Figure (16.13a) shows that the Z^e_{12} is related to the field response $\tilde{E}_{u1}(\alpha)$ at the open-circuited port #1, due to the current source $\tilde{J}_{u2}(\alpha)$ applied at port # 2. It follows from equation (16.6.29a) for $\tilde{J}_{u1}(\alpha) = 0$. The potential $\tilde{E}_{u1}(\alpha)$ is obtained in terms of the current source $\tilde{J}_{u2}(\alpha)$ by using the process discussed in section (2.2) of chapter 2. The process starts with replacing the current source $\tilde{J}_{u2}(\alpha)$ by the voltage source at the port #2 as follows:

$$V_s = \frac{\tilde{J}_{u2}(\alpha)}{Y^{L2}_{TE} + Y^{U2}_{TE}}. \tag{16.6.34}$$

The potential, corresponding to the electric field at the interface of h_2 and h_3, i.e. at $y = h_2$, is

$$V_{h_2} = \frac{V_s(1 + \Gamma_{h_2})}{e^{\gamma_2 h_2} + \Gamma_{h_2} e^{-\gamma_2 h_2}}, \tag{16.6.35}$$

where Γ_{h_2} is the reflection coefficient at the plane $y = h_1 + h_2$. The above relation is obtained from equation (2.2.5) of chapter 2. The reflection coefficient is given by

$$\Gamma_{h_2} = \frac{Y_{TE2} - Y^U_{TE2}}{Y_{TE2} + Y^U_{TE2}} \qquad \text{(a)}$$

$$\text{where,} \quad Y^U_{TE2} = Y_{TE3}\frac{Y^{U1}_{TE} + Y_{TE3}\tanh(\gamma_3 h_3)}{Y_{TE3} + Y^{U1}_{TE}\tanh(\gamma_3 h_3)} \qquad \text{(b)}. \tag{16.6.36}$$

In equation (16.6.36), Y_{TE2} and Y_{TE3} are characteristic admittances of line #2 and line #3, respectively. The admittance Y^U_{TE2} is a load to line #2 at $y = (h_1 + h_2)^-$ due to the line sections #3 and #4. Y^{U1}_{TE} is given by equation (16.6.32). Again, following equation (2.2.5) of chapter 2 the potential at $y = h_1 + h_2 + h_3$, i.e. at port #1, is

$$V_1 = \tilde{E}_{u1}(\alpha) = \frac{V_{h_2}(1 + \Gamma_{h_3})}{e^{\gamma_3 h_3} + \Gamma_{h_3} e^{-\gamma_3 h_3}}. \tag{16.6.37}$$

where Γ_{h_3} is the reflection coefficient at $y = h_1 + h_2 + h_3$ obtained from

$$\Gamma_{h_3} = \frac{Y_{TE3} - Y^{U1}_{TE}}{Y_{TE3} + Y^{U1}_{TE}}. \tag{16.6.38}$$

Thus, the potential at the port #1 is

$$\begin{aligned}
\tilde{E}_{u1}(\alpha) &= \frac{V_{h_2}\left(1 + \dfrac{Y_{TE3} - Y^{U1}_{TE}}{Y_{TE3} + Y^{U1}_{TE}}\right)}{e^{\gamma_3 h_3} + \left(\dfrac{Y_{TE3} - Y^{U1}_{TE}}{Y_{TE3} + Y^{U1}_{TE}}\right)e^{-\gamma_3 h_3}} \\
&= \frac{V_{h_2}\left(Y_{TE3} + Y^{U1}_{TE} + Y_{TE3} - Y^{U1}_{TE}\right)}{e^{\gamma_3 h_3}\left(Y_{TE3} + Y^{U1}_{TE}\right) + \left(Y_{TE3} - Y^{U1}_{TE}\right)e^{-\gamma_3 h_3}} \\
&= \frac{2V_{h_2}Y_{TE3}}{Y_{TE3}(e^{\gamma_3 h_3} + e^{-\gamma_3 h_3}) + Y^{U1}_{TE}(e^{\gamma_3 h_3} - e^{-\gamma_3 h_3})} \\
&= \frac{V_{h_2}Y_{TE3}}{Y_{TE3}\cosh\gamma_3 h_3 + Y^{U1}_{TE}\sinh\gamma_3 h_3}
\end{aligned}$$

$$\Rightarrow \tilde{E}_{u1}(\alpha) = \frac{V_{h_2}(Y_{TE3})}{\sinh(\gamma_3 h_3)\left[Y^{U1}_{TE} + Y_{TE3}\coth(\gamma_3 h_3)\right]}. \tag{16.6.39}$$

The V_{h_2} is obtained from equation (16.6.35) as

$$V_{h_2} = \frac{V_s(Y_{TE2})}{\sinh(\gamma_2 h_2)\left[Y^U_{TE2} + Y_{TE2}\coth(\gamma_2 h_2)\right]}. \tag{16.6.40}$$

The following expressions are also obtained from equations (16.6.34), (16.6.39), and (16.6.40):

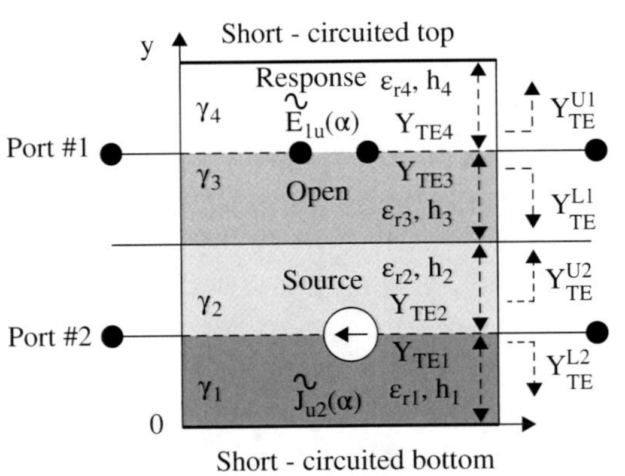

(a) Source at port #2 and response at port #1.

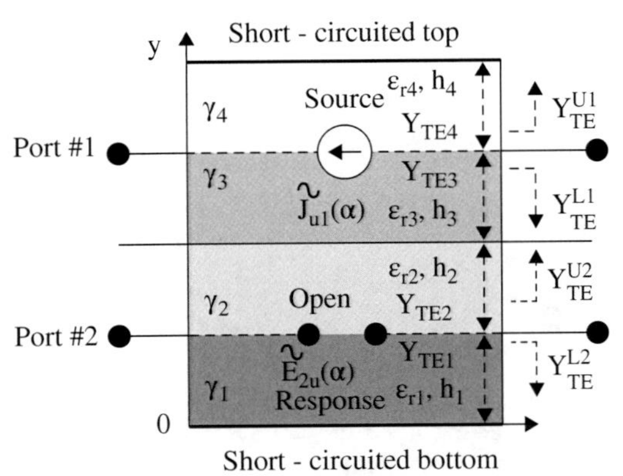

(b) Source at port #1 and response at port #2.

Figure 16.13 Determination of the transfer (mutual) impedance of the LSE (TE^y) – mode.

$$\tilde{E}_{u1}(\alpha) = \frac{\tilde{J}_{u2}(\alpha) Y_{TE2}\, Y_{TE3}}{\left[Y_{TE}^{L2} + Y_{TE}^{U2}\right] \sinh(\gamma_2 h_2) \sinh(\gamma_3 h_3)} \times \frac{1}{\left[Y_{TE2}^{U} + Y_{TE2}\coth(\gamma_2 h_2)\right]\left[Y_{TE}^{U1} + Y_{TE3}\coth(\gamma_3 h_3)\right]}. \quad (16.6.41)$$

Therefore, the transfer impedance from equation (16.6.41) and equation (16.6.29a) showing the component of Green's function is

$$\tilde{Z}^{e}_{12}(\alpha) = \frac{Y_{TE2} Y_{TE3}}{\left[Y_{TE}^{L2} + Y_{TE}^{U2}\right] \sinh(\gamma_2 h_2) \sinh(\gamma_3 h_3)} \times \frac{1}{\left[Y_{TE2}^{U} + Y_{TE2}\coth(\gamma_2 h_2)\right]\left[Y_{TE}^{U1} + Y_{TE3}^{U}\coth(\gamma_3 h_3)\right]}. \quad (16.6.42)$$

In equation (16.6.42), Y_{TE2}^{U} is given by equation (16.6.36b). For three-layered structure, $h_2 = 0$, equation (16.6.42) is reduced to

$$\tilde{Z}^{e}_{12}(\alpha) = \frac{Y_{TE3}}{\left[Y_{TE}^{L2} + Y_{TE}^{U2}\right] \sinh(\gamma_3 h_3)} \times \frac{1}{\left[Y_{TE}^{U} + Y_{TE3}^{U}\coth(\gamma_3 h_3)\right]}. \quad (16.6.43)$$

The above expression (16.6.43) is the same as the Green's function of the three-layered structure obtained by Itoh [J.4].

Determination of $\tilde{Z}^{e}_{22}(\alpha)$

The input impedance $\tilde{Z}^{e}_{22}(\alpha)$ is obtained at port #2, while the port #1 is open-circuited. It is given by equation (16.6.31b). The upper and lower parts of the port admittances Y_{TE}^{L2} and Y_{TE}^{U2}, shown in Fig (16.13b), are determined as follows:

$$Y_{TE}^{L2} = Y_{TE1}\coth(\gamma_1 h_1). \quad (16.6.44)$$

$$Y_{TE}^{U1} = Y_{TE4}\coth(\gamma_4 h_4) \quad (a)$$

$$Y_{3}^{TE} = Y_{TE3}\frac{Y_{TE}^{U1} + Y_{TE3}\tanh(\gamma_3 h_3)}{Y_{TE3} + Y_{TE}^{U1}\tanh(\gamma_3 h_3)} \quad (b). \quad (16.6.45)$$

Finally at $y = h_1^+$:

$$Y_{TE}^{U2} = Y_{TE2}\frac{Y_{3}^{TE} + Y_{TE2}\tanh(\gamma_2 h_2)}{Y_{TE2} + Y_{3}^{TE}\tanh(\gamma_3 h_3)}. \quad (16.6.46)$$

Likewise, $\tilde{Z}^{e}_{21}(\alpha)$ is obtained from Fig (16.13b) by applying $\tilde{J}_{u1}(\alpha)$ current at port #1, while port #2 is open-circuited. It can be written from equation (16.6.42) by using the symmetry:

$$\tilde{Z}^{e}_{21}(\alpha) = \frac{Y_{TE2}\, Y_{TE3}}{\left[Y_{TE}^{L1} + Y_{TE}^{U1}\right] \sinh(\gamma_2 h_2) \sinh(\gamma_3 h_3)} \times \frac{1}{\left[Y_{TE3}^{U} + Y_{TE3}\coth(\gamma_3 h_3)\right]\left[Y_{TE}^{L2} + Y_{TE2}\coth(\gamma_2 h_2)\right]}. \quad (16.6.47)$$

The expressions for $\tilde{Z}^{e}_{11}$, $\tilde{Z}^{e}_{22}$, $\tilde{Z}^{e}_{12}$ and $\tilde{Z}^{e}_{21}$ are obtained from equations (16.2.31), (16.6.42), and (16.6.47). In these equations, the characteristic admittances for the TE^y mode are taken from equation (16.6.8) and propagation constants from equation (16.6.9). These equations are used for getting the $\tilde{Z}^{m}_{11}$, $\tilde{Z}^{m}_{22}$, $\tilde{Z}^{m}_{12}$ and $\tilde{Z}^{m}_{21}$ on replacing TE^y mode by TM^y mode everywhere, so for the TM^y mode the characteristic admittances are taken from equation (16.6.12) and propagation constants are the same as that of the TE^y mode.

The real structure is in the (z, x, y) coordinate system, supporting the hybrid mode. However, equations (16.6.29) and (16.6.30) are in the (u, v, y) coordinate system. To convert them in the (z, x, y) coordinate, these equations are arranged in the matrix form:

$$\begin{bmatrix} \tilde{E}_{u1}(\alpha) \\ \tilde{E}_{v1}(\alpha) \\ \tilde{E}_{u2}(\alpha) \\ \tilde{E}_{v2}(\alpha) \end{bmatrix} = \begin{bmatrix} \tilde{Z}^{e}_{11} & 0 & \tilde{Z}^{e}_{12} & 0 \\ 0 & \tilde{Z}^{m}_{11} & 0 & \tilde{Z}^{m}_{12} \\ \tilde{Z}^{e}_{21} & 0 & \tilde{Z}^{e}_{22} & 0 \\ 0 & \tilde{Z}^{m}_{21} & 0 & \tilde{Z}^{m}_{22} \end{bmatrix} \begin{bmatrix} \tilde{J}_{u1}(\alpha) \\ \tilde{J}_{v1}(\alpha) \\ \tilde{J}_{u2}(\alpha) \\ \tilde{J}_{v2}(\alpha) \end{bmatrix}. \quad (16.6.48)$$

The coordinate transformation equation (16.6.6) can be written as follows for two sets of variables $(u_1, v_1) \to (z_1, x_1)$ and $(u_2, v_2) \to (z_2, x_2)$:

$$\begin{bmatrix} u_1 \\ v_1 \\ u_2 \\ v_2 \end{bmatrix} = \underbrace{\begin{bmatrix} \sin\theta & -\cos\theta & 0 & 0 \\ \cos\theta & \sin\theta & 0 & 0 \\ 0 & 0 & \sin\theta & -\cos\theta \\ 0 & 0 & \cos\theta & \sin\theta \end{bmatrix}}_{[u]} \begin{bmatrix} z_1 \\ x_1 \\ z_2 \\ x_2 \end{bmatrix}. \quad (16.6.49)$$

Using the above transformation (16.6.49), the field variables $\tilde{E}_{u1}(\alpha)$, etc., and current variables $\tilde{J}_{u1}(\alpha)$, etc., in the (u, v, y) coordinate system are transformed to the field variables $\tilde{E}_{z1}(\alpha)$ ... and current variables $\tilde{J}_{z1}(\alpha)$... in the (z, x, y) coordinate system. The coordinate transformation is applied to equation (16.6.48) to get the following transformed relation:

$$[u]\begin{bmatrix}\tilde{E}_{z1}(\alpha)\\ \tilde{E}_{x1}(\alpha)\\ \tilde{E}_{z2}(\alpha)\\ \tilde{E}_{x2}(\alpha)\end{bmatrix}=\begin{bmatrix}\tilde{Z}^{e}_{11} & 0 & \tilde{Z}^{e}_{12} & 0\\ 0 & \tilde{Z}^{m}_{11} & 0 & \tilde{Z}^{m}_{12}\\ \tilde{Z}^{e}_{21} & 0 & \tilde{Z}^{e}_{22} & 0\\ 0 & \tilde{Z}^{m}_{21} & 0 & \tilde{Z}^{m}_{22}\end{bmatrix}\begin{bmatrix}\sin\theta & -\cos\theta & 0 & 0\\ \cos\theta & \sin\theta & 0 & 0\\ 0 & 0 & \sin\theta & -\cos\theta\\ 0 & 0 & \cos\theta & \sin\theta\end{bmatrix}\begin{bmatrix}\tilde{J}_{z1}(\alpha)\\ \tilde{J}_{x1}(\alpha)\\ \tilde{J}_{z2}(\alpha)\\ \tilde{J}_{x2}(\alpha)\end{bmatrix}.$$

$$\Rightarrow [u]\begin{bmatrix}\tilde{E}_{z1}(\alpha)\\ \tilde{E}_{x1}(\alpha)\\ \tilde{E}_{z2}(\alpha)\\ \tilde{E}_{x2}(\alpha)\end{bmatrix}=\begin{bmatrix}\tilde{Z}^{e}_{11}\sin\theta & -\tilde{Z}^{e}_{11}\cos\theta & \tilde{Z}^{e}_{12}\sin\theta & -\tilde{Z}^{e}_{12}\cos\theta\\ \tilde{Z}^{m}_{11}\cos\theta & \tilde{Z}^{m}_{11}\sin\theta & \tilde{Z}^{m}_{12}\cos\theta & \tilde{Z}^{m}_{12}\sin\theta\\ \tilde{Z}^{e}_{21}\sin\theta & -\tilde{Z}^{e}_{21}\cos\theta & \tilde{Z}^{e}_{22}\sin\theta & -\tilde{Z}^{e}_{22}\cos\theta\\ \tilde{Z}^{m}_{21}\cos\theta & \tilde{Z}^{m}_{21}\sin\theta & \tilde{Z}^{m}_{22}\cos\theta & \tilde{Z}^{m}_{22}\sin\theta\end{bmatrix}\begin{bmatrix}\tilde{J}_{z1}(\alpha)\\ \tilde{J}_{x1}(\alpha)\\ \tilde{J}_{z2}(\alpha)\\ \tilde{J}_{x2}(\alpha)\end{bmatrix} \qquad (16.6.50)$$

$$\Rightarrow \begin{bmatrix}\tilde{E}_{z1}(\alpha)\\ \tilde{E}_{x1}(\alpha)\\ \tilde{E}_{z2}(\alpha)\\ \tilde{E}_{x2}(\alpha)\end{bmatrix}=\begin{bmatrix}\sin\theta & \cos\theta & 0 & 0\\ -\cos\theta & \sin\theta & 0 & 0\\ 0 & 0 & \sin\theta & \cos\theta\\ 0 & 0 & -\cos\theta & \sin\theta\end{bmatrix}\begin{bmatrix}\tilde{Z}^{e}_{11}\sin\theta & -\tilde{Z}^{e}_{11}\cos\theta & \tilde{Z}^{e}_{12}\sin\theta & -\tilde{Z}^{e}_{12}\cos\theta\\ \tilde{Z}^{m}_{11}\cos\theta & \tilde{Z}^{m}_{11}\sin\theta & \tilde{Z}^{m}_{12}\cos\theta & \tilde{Z}^{m}_{12}\sin\theta\\ \tilde{Z}^{e}_{21}\sin\theta & -\tilde{Z}^{e}_{21}\cos\theta & \tilde{Z}^{e}_{22}\sin\theta & -\tilde{Z}^{e}_{22}\cos\theta\\ \tilde{Z}^{m}_{21}\cos\theta & \tilde{Z}^{m}_{21}\sin\theta & \tilde{Z}^{m}_{22}\cos\theta & \tilde{Z}^{m}_{22}\sin\theta\end{bmatrix}\begin{bmatrix}\tilde{J}_{z1}(\alpha)\\ \tilde{J}_{x1}(\alpha)\\ \tilde{J}_{z2}(\alpha)\\ \tilde{J}_{x2}(\alpha)\end{bmatrix}$$

$$\begin{bmatrix}\tilde{E}_{z1}(\alpha)\\ \tilde{E}_{x1}(\alpha)\\ \tilde{E}_{z2}(\alpha)\\ \tilde{E}_{x2}(\alpha)\end{bmatrix}=\begin{bmatrix}\tilde{Z}^{e}_{11}\sin^2\theta+\tilde{Z}^{m}_{11}\cos^2\theta & -\tilde{Z}^{e}_{11}\sin\theta\cos\theta+\tilde{Z}^{m}_{11}\sin\theta\cos\theta & \tilde{Z}^{e}_{12}\sin^2\theta+\tilde{Z}^{m}_{12}\cos^2\theta & -\tilde{Z}^{e}_{12}\cos\theta\sin\theta+\tilde{Z}^{m}_{12}\sin\theta\cos\theta\\ -\tilde{Z}^{e}_{11}\sin\theta\cos\theta+\tilde{Z}^{m}_{11}\sin\theta\sin\theta & \tilde{Z}^{m}_{11}\sin^2\theta+\tilde{Z}^{e}_{11}\cos^2\theta & -\tilde{Z}^{e}_{12}\cos\theta\sin\theta+\tilde{Z}^{m}_{12}\sin\theta\cos\theta & \tilde{Z}^{e}_{12}\cos^2\theta+\tilde{Z}^{m}_{12}\sin^2\theta\\ \tilde{Z}^{e}_{21}\sin^2\theta+\tilde{Z}^{m}_{21}\cos^2\theta & -\tilde{Z}^{e}_{21}\cos\theta\sin\theta+\tilde{Z}^{m}_{21}\sin\theta\cos\theta & \tilde{Z}^{e}_{22}\sin^2\theta+\tilde{Z}^{m}_{22}\cos^2\theta & -\tilde{Z}^{e}_{22}\cos\theta\sin\theta+\tilde{Z}^{m}_{22}\sin\theta\cos\theta\\ -\tilde{Z}^{e}_{21}\cos\theta\sin\theta+\tilde{Z}^{m}_{21}\sin\theta\cos\theta & \tilde{Z}^{m}_{21}\sin^2\theta+\tilde{Z}^{e}_{21}\cos^2\theta & -\tilde{Z}^{e}_{22}\cos\theta\sin\theta+\tilde{Z}^{m}_{22}\sin\theta\cos\theta & \tilde{Z}^{e}_{22}\cos^2\theta+\tilde{Z}^{m}_{22}\sin^2\theta\end{bmatrix}\begin{bmatrix}\tilde{J}_{z1}(\alpha)\\ \tilde{J}_{x1}(\alpha)\\ \tilde{J}_{z2}(\alpha)\\ \tilde{J}_{x2}(\alpha)\end{bmatrix}. \qquad (16.6.51)$$

The Z-kind of the Green's function relates the surface current density $\tilde{J}_z(\alpha)$ and $\tilde{J}_x(\alpha)$ to the electric field components $\tilde{E}_z(\alpha)$ and $\tilde{E}_x(\alpha)$ at two layers designated as two ports shown in Fig (16.13).

$$\begin{bmatrix}\tilde{E}_{z1}(\alpha)\\ \tilde{E}_{x1}(\alpha)\\ ----\\ \tilde{E}_{z2}(\alpha)\\ \tilde{E}_{x2}(\alpha)\end{bmatrix}=\begin{bmatrix}\tilde{Z}^{11}_{zz} & \tilde{Z}^{11}_{zx} & \tilde{Z}^{12}_{zz} & \tilde{Z}^{12}_{zx}\\ \tilde{Z}^{11}_{xz} & \tilde{Z}^{11}_{xx} & \tilde{Z}^{12}_{xz} & \tilde{Z}^{12}_{xx}\\ -- & -- & -- & --\\ \tilde{Z}^{21}_{zz} & \tilde{Z}^{21}_{zx} & \tilde{Z}^{22}_{zz} & \tilde{Z}^{22}_{zx}\\ \tilde{Z}^{21}_{xz} & \tilde{Z}^{21}_{xx} & \tilde{Z}^{22}_{xz} & \tilde{Z}^{22}_{xx}\end{bmatrix}\begin{bmatrix}\tilde{J}_{z1}(\alpha)\\ \tilde{J}_{x1}(\alpha)\\ ---\\ \tilde{J}_{z2}(\alpha)\\ \tilde{J}_{x2}(\alpha)\end{bmatrix},$$

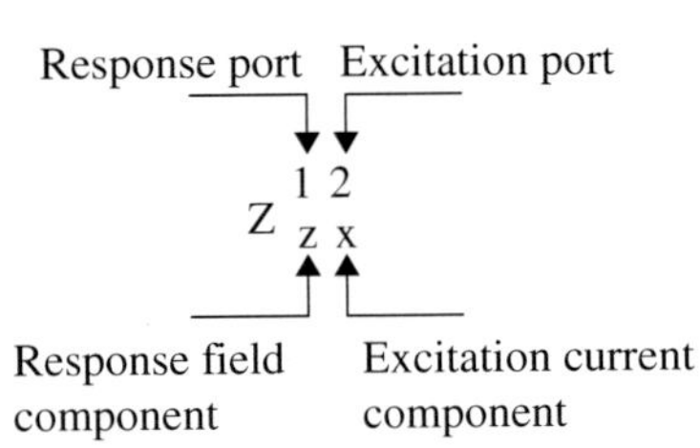

(16.6.52)

where the matrix element Z^{12}_{zx} is explained above. The x-current component at the port #2, i.e. J_{x2}, creates the z-electric field component at port #1. From the above two equations, the elements of the [Z]-kind of dyadic Green's function are given below:

$$\begin{aligned}
&\tilde{Z}^{11}_{zz}=\tilde{Z}^{e}_{11}\sin^2\theta+\tilde{Z}^{m}_{11}\cos^2\theta, && \tilde{Z}^{11}_{zx}=\left(\tilde{Z}^{m}_{11}-\tilde{Z}^{e}_{11}\right)\sin\theta\cos\theta\\
&\tilde{Z}^{12}_{zz}=\tilde{Z}^{e}_{12}\sin^2\theta+\tilde{Z}^{m}_{12}\cos^2\theta, && \tilde{Z}^{12}_{zx}=\left(\tilde{Z}^{m}_{12}-\tilde{Z}^{e}_{12}\right)\sin\theta\cos\theta\\
&\tilde{Z}^{11}_{zx}=\tilde{Z}^{11}_{xz}=\left(\tilde{Z}^{m}_{21}-\tilde{Z}^{e}_{21}\right)\sin\theta\cos\theta, && \tilde{Z}^{11}_{xx}=\tilde{Z}^{e}_{11}\cos^2\theta+\tilde{Z}^{m}_{11}\sin^2\theta\\
&\tilde{Z}^{12}_{xz}=\tilde{Z}^{12}_{zx}=\left(\tilde{Z}^{m}_{12}-\tilde{Z}^{e}_{12}\right)\sin\theta\cos\theta, && \tilde{Z}^{12}_{xx}=\tilde{Z}^{e}_{12}\cos^2\theta+\tilde{Z}^{m}_{12}\sin^2\theta\\
&\tilde{Z}^{21}_{zz}=\tilde{Z}^{e}_{21}\sin^2\theta+\tilde{Z}^{m}_{21}\cos^2\theta, && \tilde{Z}^{21}_{zx}=\left(\tilde{Z}^{m}_{21}-\tilde{Z}^{e}_{21}\right)\sin\theta\cos\theta\\
&\tilde{Z}^{22}_{zz}=\tilde{Z}^{e}_{22}\sin^2\theta+\tilde{Z}^{m}_{22}\cos^2\theta, && \tilde{Z}^{22}_{zx}=\left(\tilde{Z}^{m}_{22}-\tilde{Z}^{e}_{22}\right)\sin\theta\cos\theta\\
&\tilde{Z}^{21}_{xz}=\tilde{Z}^{21}_{zx}=\left(\tilde{Z}^{m}_{21}-\tilde{Z}^{e}_{21}\right)\sin\theta\cos\theta, && \tilde{Z}^{21}_{xx}=\tilde{Z}^{e}_{21}\cos^2\theta+\tilde{Z}^{m}_{21}\sin^2\theta\\
&\tilde{Z}^{22}_{xz}=\tilde{Z}^{22}_{zx}=\left(\tilde{Z}^{m}_{22}-\tilde{Z}^{e}_{22}\right)\sin\theta\cos\theta, && \tilde{Z}^{22}_{xx}=\tilde{Z}^{e}_{22}\cos^2\theta+\tilde{Z}^{m}_{22}\sin^2\theta.
\end{aligned} \qquad (16.6.53)$$

In each block of equation (16.6.52), the cross diagonal elements are equal, i.e. $\tilde{Z}^{11}_{zx}=\tilde{Z}^{11}_{xz}$, $\tilde{Z}^{12}_{zx}=\tilde{Z}^{12}_{xz}$, $\tilde{Z}^{21}_{zx}=\tilde{Z}^{21}_{xz}$ and $\tilde{Z}^{22}_{zx}=\tilde{Z}^{22}_{xz}$.

In the case of the coupling between the slots, the [Y]-kind of Green's function is used:

$$\begin{bmatrix}\tilde{J}_{z1}(\alpha)\\ \tilde{J}_{x1}(\alpha)\\ \tilde{J}_{z2}(\alpha)\\ \tilde{J}_{x2}(\alpha)\end{bmatrix}=[Y]\begin{bmatrix}\tilde{E}_{z1}(\alpha)\\ \tilde{E}_{x1}(\alpha)\\ \tilde{E}_{z2}(\alpha)\\ \tilde{E}_{x2}(\alpha)\end{bmatrix} \qquad (a) \qquad (16.6.54)$$

where, $[Y]=[Z]^{-1}$ (b).

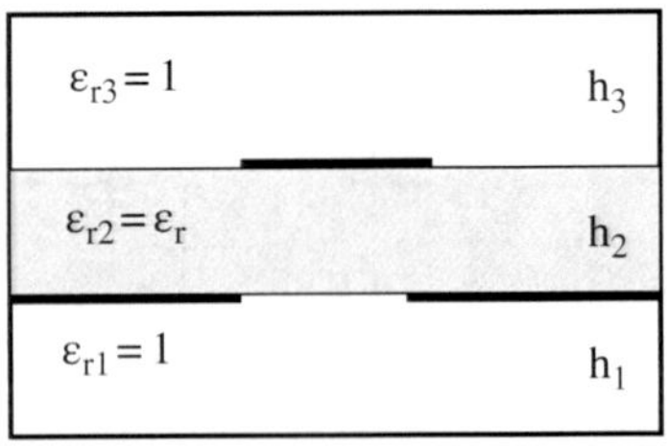

(a) Microstrip–slot line coupling.

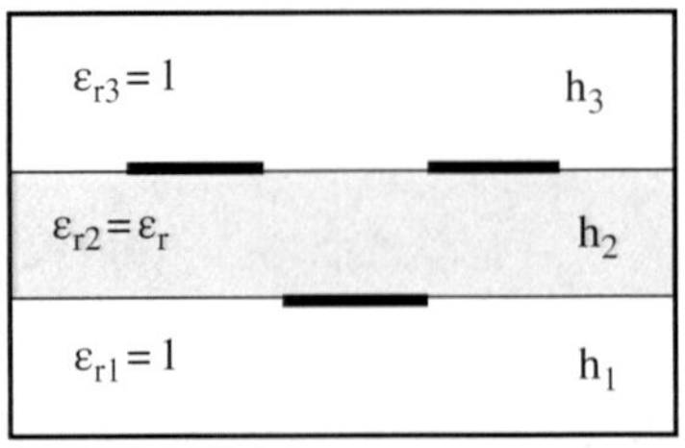

(b) Coupled microstrip with septum.

Figure 16.14 Two levels coupled structures.

The 4×4 matrix is to be inverted. Alternatively, starting from equations (16.6.29) and (16.2.30), the admittance form is expressed as follows:

$$\begin{aligned} \text{TE}^y \text{ mode}: &\left| \begin{array}{l} \tilde{J}_{u1}(\alpha) = \tilde{Y}^e_{11}\tilde{E}_{u1}(\alpha) + \tilde{Y}^e_{12}\tilde{E}_{u2}(\alpha) \\ \tilde{J}_{u2}(\alpha) = \tilde{Y}^e_{21}\tilde{E}_{u1}(\alpha) + \tilde{Y}^e_{22}\tilde{E}_{u2}(\alpha) \end{array} \right. \\ \text{TM}^y \text{ mode}: &\left| \begin{array}{l} \tilde{J}_{v1}(\alpha) = \tilde{Y}^m_{11}\tilde{E}_{v1}(\alpha) + \tilde{Y}^m_{12}\tilde{E}_{v2}(\alpha) \\ \tilde{J}_{v2}(\alpha) = \tilde{Y}^m_{21}\tilde{E}_{v1}(\alpha) + \tilde{Y}^m_{22}\tilde{E}_{v2}(\alpha). \end{array} \right. \end{aligned} \tag{16.6.55}$$

where, using equation (16.2.31), we have

$$\begin{aligned} \tilde{Y}^e_{11} &= \tilde{Y}^{U1}_{TE} + \tilde{Y}^{L1}_{TE} \quad & \tilde{Y}^e_{22} &= \tilde{Y}^{U2}_{TE} + \tilde{Y}^{L2}_{TE} \\ \tilde{Y}^m_{11} &= \tilde{Y}^{U1}_{TM} + \tilde{Y}^{L1}_{TM} \quad & \tilde{Y}^m_{22} &= \tilde{Y}^{U2}_{TM} + \tilde{Y}^{L2}_{TM}. \end{aligned} \tag{16.6.56}$$

The rest of the process for obtaining the admittance type Green's function of the two-level slot structures could be repeated like the computation of the elements of the [Z]- matrix.

The method can be extended to more number of dielectric layers and more levels containing strip conductors and slots [B.2]. However, it is analytically more cumbersome. If the conductors or slots on both the levels are symmetrical, as shown in Fig (16.9e,f), the present analysis is not needed. The MW/EW, parallel to the *x*-axis, can be placed in between two levels. It reduces the structure to the single-level case for the even/odd mode excitation. The present method is more general and is useful for the asymmetrical structure, shown in Fig (16.14a, b) [B.9, J.4, J.5]. The dynamic SDA has been used for more involved cases, even with the anisotropic substrates [J.30]. Only computation of the propagation constants has been discussed. However, the method has also been used to compute the frequency-dependent characteristic impedance [B.1, B.2, J.33, J.28]. The method has been further used for the analysis of the patch resonators, and microstrip antenna [B.3, J.32].

References

Books

B.1 Itoh, T. (Editor): *Numerical Techniques for Microwave and Millimeter-Wave Passive Structures*, John Wiley & Sons, New York, NY, 1989.

B.2 Mirshekar-Syahkal, D.: *Spectral Domain Method for Microwave Integrated Circuits*, Research Studies Press Ltd., John Wiley & Sons, New York, NY, 1990.

B.3 Garg, R.; Bhartia, P.; Behl, I.; Ittipiboon, A.: *Microstrip Antenna Design Handbook*, Artech House, Boston, 2000.

B.4 Bhat, B.; Koul, S.K.: *Stripline-Like Transmission Lines for Microwave Integrated Circuits*, Wiley Eastern Ltd., New Delhi, 1989.

B.5 Goel, A.K.: *High-Speed VLSI Interconnects Modelling, Analysis, and Simulation*, John Wiley & Sons, New York, NY, 1994.

B.6 Nguyen, C.: *Analysis Methods for RF, Microwave, and Millimetre –Wave Planar Transmission Lines*, John Wiley & Sons, New York, NY, 2000.

B.7 Bhattacharyya, A.K.: *Electromagnetic Fields in Multilayered Structures*, Artech House, Boston, 1994.

B.8 Scott, C.R.: *The Spectral Domain Method in Electromagnetics*, Artech House, Boston, 1989.

B.9 Kinayman, N.; Aksun, M.I.: *Modern Microwave Circuits*, Artech House, Boston, 2003.

B.10 Ryshik, I.M.; Gradstein, I.S.: *Tables of Series, Products and Integrals*, 4th Edtion, Academic Press, New York, NY, 1965.

Journals

J.1 Itoh, T.; Mittra, R.: *Spectral-domain approach for calculating the dispersion characteristics of microstrip lines, IEEE Trans. Microw. Theory Tech.*, Vol. MTT-21, pp. 496–499, July 1973.

J.2 Itoh, T.; Mittra, R.: *A technique for computing dispersion characteristics of shielded microstrip lines, IEEE Trans. Microw. Theory Tech.*, Vol. MTT-22, pp. 896–898, Oct. 1974.

J.3 Itoh, T.; Mittra, R.: *Dispersion characteristics of slotlines, Electron. Lett.*, Vol. 7, pp. 364–365, 1971.

J.4 Itoh, T.: *Spectral domain immittance approach for dispersion characteristics of generalized printed transmission lines, IEEE Trans. Microw. Theory Tech*, Vol. MTT-28, pp. 733–736, 1980.

J.5 Itoh, T.; Hebert, A.S.: *A generalized spectral domain analysis for coupled suspended microstrip lines with tuning septum, IEEE Trans. Microw. Theory Tech.*, Vol. MTT-26, No. 10, pp. 820–826, Oct. 1978.

J.6 Itoh, T.: *Generalized spectral domain method for multiconductor printed lines and its application to tunable suspended microstrips, IEEE Trans. Microw. Theory Tech.*, Vol. MTT- 26, pp. 983–987, 1978.

J.7 Huang, W.-X.; Itoh, T.: *Complex modes in lossless shielded microstrip lines, IEEE Trans. Microw. Theory Tech.*, Vol. MTT-36, No. 1, pp. 163–165, 1988.

J.8 Railton, C.J.; Rozzi, T.: *Complex modes in boxed microstrip, IEEE Trans. Microw. Theory Tech.*, Vol. MTT-36, No. 5, pp. 865–873, 1988.

J.9 Zehentner, J.; Mrkvica, J.; Macháč, J.: *Spectral domain analysis of open planar transmission lines, Mikrotalasna Revija (Microw. Rev.)*, Vol. 10, pp. 36–46, Nov. 2004.

J.10 Mittra, R.; Itoh, T.: *A new technique for the analysis of the dispersion characteristics of microstrip lines, IEEE Trans. Microw. Theory Tech.*, Vol. MTT-19, pp. 47–56, Jan. 1971.

J.11 Davies, J.B.; Mirshekar-Syahkal, D.: *Spectral domain solution of arbitrary coplanar transmission line with a multilayer substrate, IEEE Trans. Microw. Theory Tech.*, Vol. MTT-25, pp. 143–146, 1977.

J.12 Denlinger, E.J.: *A frequency dependent solution for microstrip transmission lines, IEEE Trans. Microw. Theory Tech.*, Vol. MTT-19, pp. 30–39, Jan. 1971.

J.13 Jansen, R.H.: *High-speed computation of single and coupled microstrip parameters including dispersion, high-order modes, loss and finite strip thickness, IEEE Trans. Microw. Theory Tech.*, Vol. MTT-26, pp. 75–82, 1978.

J.14 Meixner, J.: *The behavior of electromagnetic fields at edges, IEEE Trans. Antenna Propagat.*, Vol. AP-20, No. 4, pp. 442–446, July 1972.

J.15 Van de Capelle, A.R.; Luypaert, P.J.: *Fundamental- and higher order modes in open microstrip lines, Electron. Lett.*, Vol. 9, pp. 345–346, July 1973.

J.16 Mirshekar-Syahkal, D.; Davies, J.B.: *Accurate solution of microstrip and coplanar structures for dispersion and dielectric and conductor loss, IEEE Trans. Microw. Theory Tech.*, Vol. MTT-23, pp. 694–699, 1979.

J.17 Mirshekar-Syahkal, D.; Davies, J.B.: *Accurate analysis of coupled strip-fin line structure for phase constant, characteristic impedance, dielectric and conductor losses, IEEE Trans. Microw. Theory Tech.*, Vol. MTT-30, pp. 906–910, 1982.

J.18 Mirshekar-Syahkal, D.; Davies, J.B.: *An accurate unified solution to various fin-line structures, of phase constant, characteristic impedance, and attenuation, IEEE Trans. Microw. Theory Tech.*, Vol. MTT-30, pp. 1854–1861, 1982.

J.19 Mirshekar-Syahkal, D.: *An accurate determination of dielectric loss effect in MMIC'S including microstrip and coupled microstrip lines, IEEE Trans. Microw. Theory Tech.*, Vol. MTT-31, pp. 950–954, 1983.

J.20 Fokas, A.S.; Smitheman, S.A.: *The Fourier Transforms of the Chebyshev and Legendre Polynomials*, arXiv:1211.4943v1 [math.NA], Nov. 2012.

J.21 Krage, M.K.; Haddad, G.I.: *Frequency-dependent characteristics of microstrip transmission lines, IEEE Trans. Microw. Theory Tech.*, Vol. MTT-20, pp. 678–688, Oct. 1972.

J.22 Kobayashi, M.; Iijima, T.: *Frequency dependent characteristics of current distributions on microstrip lines, IEEE Trans. Microw. Theory Tech.*, Vol. MTT-37, No. 4, pp. 799–801, April 1989.

J.23 Kitazawa, T.; Hayashi, Y.: *Propagation characteristics of striplines with multilayered anisotropic media, IEEE Trans. Microw. Theory Tech.*, Vol. MTT-31, pp. 429–433, June 1983.

J.24 Dixit, A.; Jiu, L.; Moll, V.H.; Vignat, C.: *The finite Fourier transform of classical polynomials, J. Aust. Math. Soc.*, Vol.98, pp. 145–160, 2015.

J.25 Schmidt, L.P.; Itoh, T.: *Spectral domain analysis of dominant and higher order modes in fin lines, IEEE Trans. Microw. Theory Tech*, Vol. MTT-28, pp. 981–985, 1980.

J.26 Jansen, R.H.: *Fast accurate hybrid mode computation of nonsymmetrical coupled microstrip characteristics*, 7th European Microwave Conference, Copenhagen, Denmark, 5–8 Sept. 1977.

J.27 Kitazawa, T.; Kao, C.W.; Itoh, T.: *Planar transmission lines with finitely thick conductors and lossy substrates*, IEEE MTT-S Int. Microwave Sym. Dig., pp. 769–772, Boston, MA, 10–14 July 1991.

J.28 Knorr, J.B.; Kuchler, K.D.: *Analysis of coupled slots and coplanar strips on a dielectric substrate, IEEE Trans. Microw. Theory Tech.*, Vol. MTT-23, pp. 541–548, 1975.

J.29 Jansen, R.H.: *The spectral-domain approach for microwave integrated circuits, IEEE Trans. Microw. Theory Tech*, Vol. MTT-33, No. 10, pp. 1043–1056, Oct. 1985.

J.30 Tounsi, M.L.; Touhami, R.; Khodja, A.; Yagoub, M.C. E.: *Analysis of the mixed coupling in bilateral microwave circuits including anisotropy for MICs and MMICs applications, Prog. Electromagn. Res., PIER,* Vol. 62, pp. 281–315, 2006.

J.31 Menzel, W.: *A new interpretation of the spectral domain immittance matrix approach, IEEE Microw. Guid. Wave Lett.*, Vol. 3, No. 9, pp. 305–306, Sept. 1993.

J.32 Itoh, T.; Menzel, W.: *A full-wave analysis method for open microstrip structures, IEEE Trans. Antenna Propagat.*, Vol. AP-29, No. 1, pp. 63–68, Jan. 1981.

J.33 Knorr, J.B.; Tufekcioglu, A.: *Spectral-domain calculation of microstrip characteristic impedance, IEEE Trans. Microw. Theory Tech.*, Vol. MTT-23, pp. 725–728, Sept. 1975.

17

Lumped and Line Resonators

Basic Characteristics

Introduction

The resonator is one of the most basic circuit elements in the microwave and other communication systems. They have a primary role either to accept or to reject a band of frequencies. It has three basic characteristics parameters – *resonance frequency*, *unloaded Q-factor*, and *coupling coefficient*. These basic characteristics apply to all kinds of resonators. The resonators are used in the design of filters, frequency discriminators, duplexers, wave meters, oscillators, tuned amplifiers, etc. The resonators are also used as the sensors to measure material properties and other physical parameters. They further act as antennas to radiate power in the free space and also in the material bodies. This chapter presents the basic characteristics of lumped and line resonators. They act as the circuit models for the planar resonators discussed in chapter 18.

- *To review the basic characteristics of lumped resonators.*
- *To discuss the one-port, two-port, and reaction-type resonators.*
- *To understand the coupling of lumped resonators to the external circuits.*
- *To discuss the circuit models of transmission line resonators.*
- *To present the modal analysis of transmission line resonators.*

17.1 Basic Resonating Structures

A section of the open- or short-circuited transmission line supports the standing wave, not the usual traveling wave. Such a standing wave supporting line section forms the basic distributed line resonator. There is no transportation of electrical energy from one location on a resonating line to another location. Only there is an exchange of energy from the electric field to the magnetic field. It is like the exchange of energy from the potential energy to the kinetic energy of a vibrating string or an oscillating pendulum [B.1].

The physical dimensions of a resonator are expressed in terms of the wavelength, corresponding to the resonance frequency. It is known as the *normalized dimension*. The resonating structures, from the dimension point of view, are grouped into the following four categories:

- *Zero-dimensional resonator,*
- *One-dimensional resonator,*
- *Two-dimensional resonator,*
- *Three-dimensional resonator.*

All microwave resonator circuits follow the above classification [B.2]. These four resonating structures are shown in Fig (17.1).

A *zero-dimensional resonator* has its all dimension very small ($<\lambda/50$) as compared to the resonant wavelength. Even the $\lambda/10$ is treated as the zero-dimension resonator. Figure (17.1a) shows a lossless zero-dimensional resonator that is constructed using the lumped capacitor and lumped inductor. This resonator has a fixed resonance frequency that is determined by the values of the capacitance (C) and inductance (L). It behaves like an oscillating pendulum that has a fixed fundamental resonance frequency for a fixed string length. The zero-dimensional resonators are useful to develop our understanding of the basic characteristics, *resonance frequency*, *unloaded Q-factor,* and *coupling coefficient,* of the resonators. These are also helpful to develop the models for the line resonators around the resonance frequency. The lumped resonators are normally used below the VHF frequency band, i.e. below 300 MHz. At the microwave frequency, the LC circuit-based resonators have the following disadvantages:

Introduction to Modern Planar Transmission Lines: Physical, Analytical, and Circuit Models Approach, First Edition. Anand K. Verma.

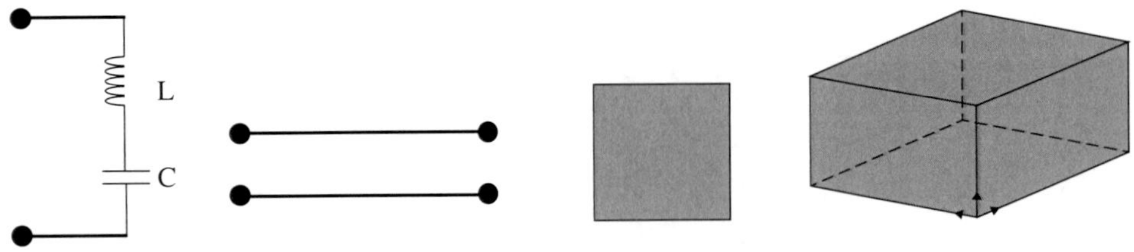

(a) Zero-dimension lumped resonator. (b) One-dimension line resonator. (c) Two-dimentional patch resonator. (d) Three-dimension cavity resonator.

Figure 17.1 Some resonating structures.

- The physical dimensions of an inductor and a capacitor become a significant part of the wavelength. Further, it becomes difficult to define and measure the voltage and current.
- It is difficult to obtain high Q-factor with a lumped inductor and lumped capacitor at high frequency.
- The radiation loss from an LC resonator is high.

Despite the above limitations, the lumped element-based resonators are used in the planar technology in the lower microwave frequency range, as they are compact, and thus suitable for the MMIC design. Presently, the planar lumped resonators and microwave components are available up to 5 GHz, and even up to mm-wave [B.3–B.5].

Figure (17.1b) shows a transmission line section that forms the *one-dimensional resonator*. It could be short-circuited or open-circuited at one end or both ends, thus forming either $\lambda/4$ or $\lambda/2$ resonator. The ends could also be loaded with a reactance, either inductance or capacitance, to reduce the physical size of a one-dimensional line resonator. It is a distributed circuit resonator that works like a stretched vibrating string [B.1]. It could resonate at several harmonically related frequencies; unlike a lumped LC resonator that resonates at only one frequency. Each resonating frequency has its unique field/voltage/current distribution along the line, known as the *resonating mode*. Thus, each resonator mode has a unique resonance frequency. At the lowest resonating frequency, the mode is known as the *fundamental* or *dominant resonant mode*. The line resonator could be realized through any physical line structure, such as the two-wire transmission line, coaxial line, microstrip, CPW, slot line, fin line, etc. However, any physical line could be modeled as a two-wire line structure with specific characteristic impedance and propagation constant. The two-wire line structure itself is modeled using the lumped circuit elements.

Figure (17.1c) shows the *two-dimensional resonator* that is a microstrip planar patch. It is available in several geometrical shapes such as rectangle, square, circle, triangle, ellipse, hexagon, etc. These resonating planar patch resonators and their ring-forms are shown in Fig (17.2) [B.6]. In the case of the microstrip patch

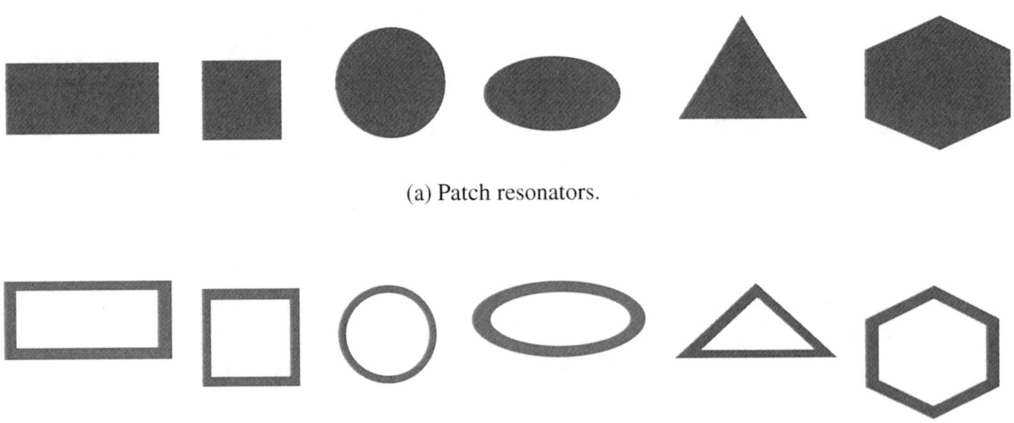

(a) Patch resonators.

(b) Ring resonators.

Figure 17.2 Some patch and ring resonators.

and ring resonators, these structures are printed on a substrate backed by a conducting surface. These shapes can also be realized in the CPW and slot line geometry.

The patch and ring could be loaded with the reactive elements, realized through the slots, and conducting strips. The loading could be also done using the lumped capacitor. Even a varactor diode could be used to get a voltage-controlled resonator. The slot/strip-loaded planar resonator could be called the *complex planar resonator*. These are very important structures for the development of compact planar circuits in the microwave frequency range. At present, the microwave community has shown significant interest in these structures. The embedded complex resonating structures are the backbone of the *metamaterials* that have some unusual properties [B.7, B.8]. The metamaterials are discussed in chapters 5 and 21.

The two-dimensional patch resonators behave like a stretched membrane of the rectangular or circular shape. They have discrete two-dimensional resonating modes with the distinct field distributions and distinct resonance frequencies. Of course, two modes, with different field distributions, can have the same resonance frequency. Such modes are known as *degenerate modes*. The modal field has two-dimensional variation in a plane.

Figure (17.1d) shows a *three-dimensional resonator* that is realized by a closed metallic waveguide section, normally having a rectangular or circular shape cross-section [B.9–B.12]. Such resonators are known as the *cavity resonators*, in which field variation takes place in a volume, i.e. in the three-dimensional space. They have very large Q-factors. However, these are large and bulky structures. Normally, these are not suitable for planar technology. However, at high frequencies, these are realized using the *substrate integrated waveguide* (SIW) and LTCC technology [J.1, J.2]. If there is no field variation in the third dimension, we treat the cavity resonator as a two-dimensional resonator. This concept is useful to model the planar patch resonator as a two-dimensional cavity resonator.

17.2 Zero-Dimensional Lumped Resonator

The series and parallel resonant circuits, using lumped elements, are two basic forms of the zero-dimensional lumped resonator. The series and parallel resonant circuits are complementary to each other and they are treated identically. The combinations of series and parallel resonators are also used to design filters and other circuits. Our objective in this section is to understand the basic working of the series and parallel resonators and their connection to the external circuits – source and load. The resonators used in circuits are characterized by the three kinds of the *Q-factors – unloaded, external, and loaded Q-factor*. The coupling of a resonator to the external circuit is described by the external Q-factor, also by the concept of the *coupling coefficient*. These parameters, and also the resonance frequency, are determined experimentally, or by using an EM-Simulator. These parameters are also used to obtain the circuit model, by evaluating the values of RLC circuit elements, of a resonating planar structure [B.9–B.13].

17.2.1 Lumped Series Resonant Circuit

Figure (17.3a) shows the LC series resonant circuit with an RF voltage source V and its internal impedance R_g. The internal loss of the resonator is expressed through resistance R. On taking $R_g = 0$, the input impedance and current through the resonator are obtained as follows:

$$Z_{in} = R + jX = R + j\left(L\omega - \frac{1}{\omega C}\right) \quad (a)$$

$$I = \frac{V}{R + j\left(L\omega - \frac{1}{\omega C}\right)} \quad (b).$$

$$(17.2.1)$$

Figure (17.3b) shows the current response. At the resonance frequency, $\omega = \omega_r$, the reactance $X = 0$ and current is maximum:

$$\omega_r = \frac{1}{\sqrt{LC}} \quad (a), \quad f_r = \frac{1}{2\pi\sqrt{LC}} \quad (b), \quad I_{max} = \frac{V}{R}. \qquad (17.2.2)$$

The power supplied to the circuit at resonance is P_{res}. Figure (17.3b) shows that at the off-resonance frequencies, f_1 and f_2, symmetrical with respect to f_r, the power through circuit drops to the half of its maximum value,

$$\frac{P_{res}}{2} = \frac{I_{res}^2 R}{2} = \left(\frac{I_{res}}{\sqrt{2}}\right)^2 R. \qquad (17.2.3)$$

At the half-power frequencies, f_1 and f_2, the current drops to $1/\sqrt{2}$ of its maximum value I_{max} at the resonance frequency. The current response is controlled by the variation in the input impedance, whose magnitude and phase are shown in Fig (17.3c). At the half-power frequencies, the following relations are obtained:

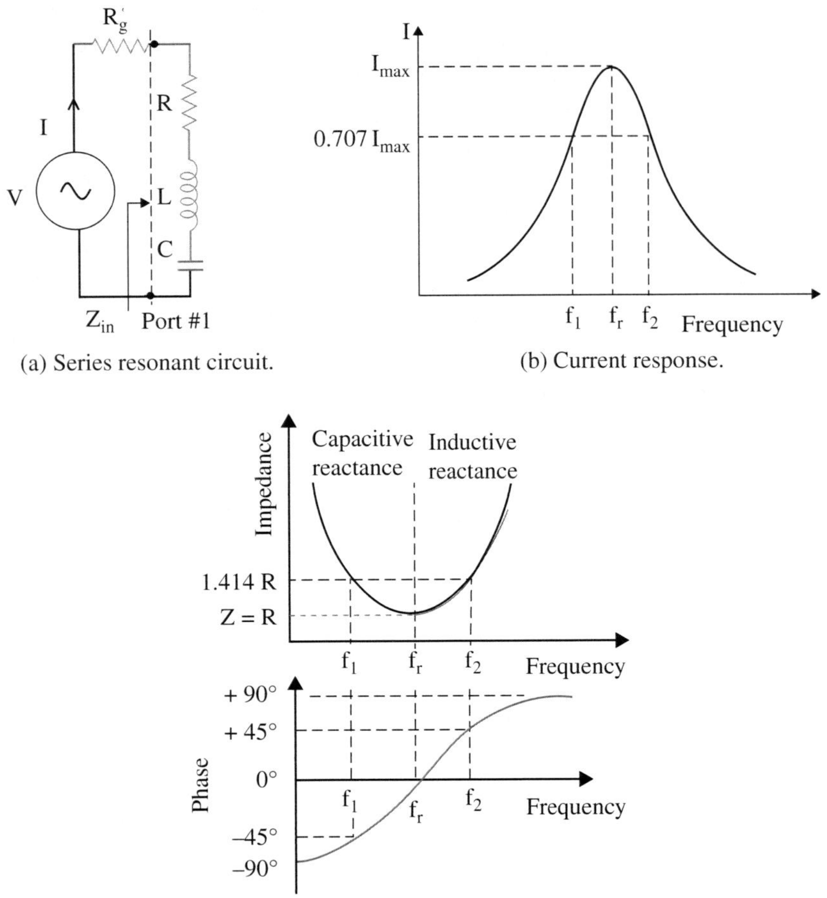

(a) Series resonant circuit.

(b) Current response.

(c) Input impedance magnitude and phase variation with frequency.

Figure 17.3 Behavior of series resonant circuit.

$$I = \frac{I_{max}}{\sqrt{2}} = \frac{1}{\sqrt{2}}\frac{V}{R} = \frac{V}{\sqrt{R^2 + X^2}}$$

$$\Rightarrow R\sqrt{2} = \sqrt{R^2 + X^2}$$

$$\Rightarrow R = \mp X, \quad \Rightarrow R = |X|. \qquad (17.2.4)$$

In the above equation, reactance X is an imaginary quantity. At the *half-power frequencies, the numerical value of reactance is equal to the resistance* and the input impedance is $Z_{in} = \sqrt{2}\,R$. Figure (17.3c) shows that at the frequency $f < f_r$ the LC series resonant circuit behaves as a capacitor, and at the frequency, $f > f_r$, it behaves as an inductor. The phase of the circuit current at frequencies f_1 and f_2 are $\tan\phi = X/R = \mp 1$, i.e. $\phi = \mp 45°$. The frequency difference $(f_2 – f_1)$ is known as the half-power or *3-dB bandwidth* of the circuit. It depends on the internal resistance R of a circuit and is related to another characteristic parameter called the *quality-factor*, i.e. Q-factor, of the resonant circuit.

The Q-factor of a resonant circuit due to its internal loss, without any consideration of external load, is called the *unloaded* Q_U*-factor*. It is defined as follows:

$$Q_U = \omega_r \frac{\text{Average energy stored}}{\text{Average power loss}}\bigg|_{\omega = \omega_r}. \qquad (17.2.5)$$

The inductor and capacitor are devices to store the magnetic and electric energy, respectively. In a resonant circuit, the energy is transferred from an inductor to a capacitor and back. At the resonance, the magnetic and electric energies are equal. The average stored magnetic energy, electric energy and average power loss in a circuit, over the time period T, are

$$W_m = \frac{1}{4}LI_{max}^2 \quad (a), \qquad W_c = \frac{1}{4}CV_{max}^2 \quad (b)$$

$$P_l = \frac{1}{2}RI_{max}^2 \quad (c). \qquad (17.2.6)$$

At the resonance, the total stored energy in a circuit is $W_T = W_m + W_c$ and in the inductor, $W_T = 2W_m$ or in the capacitor, $W_T = 2W_c$, giving the following expressions for the unloaded Q-factor of a series resonant circuit:

$$Q_U = \frac{\omega_r W_T}{P_l} = \frac{2\omega_r W_m}{P_l} = \frac{\omega_r (1/2) L I_{max}^2}{(1/2) R I_{max}^2} = \frac{\omega_r L}{R} \quad \text{(a)}$$

$$\text{or,}\ Q_U = \frac{1}{\omega_r RC} \quad \text{(b).}$$

$$(17.2.7)$$

The input impedance of a series resonant circuit, around its resonance frequency, is further expressed in terms of the Q-factor. The frequency around ω_r is defined by the *frequency detuning parameter* ε_f, and the input impedance is rearranged as follows:

$$Z_{in} = R + j[\omega L - (1/\omega C)]$$

$$= R + j\sqrt{L/C}\left[\omega\sqrt{LC} - \left(1/\omega\sqrt{LC}\right)\right]$$

$$\Rightarrow Z_{in} = R + j\sqrt{\frac{L}{C}}\varepsilon_f \quad \text{(a)}$$

$$\text{where,}\ \varepsilon_f = \frac{\omega}{\omega_r} - \frac{\omega_r}{\omega} \quad \text{(b).}$$

$$(17.2.8)$$

Another variable, i.e. the *normalized deviation* of frequency from the resonance frequency, is defined as follows:

$$\delta = \frac{\Delta\omega}{\omega_r} = \frac{f - f_r}{f_r} = \frac{\omega - \omega_r}{\omega_r} = \frac{\omega}{\omega_r} - 1$$

$$\Rightarrow \frac{\omega}{\omega_r} = 1 + \delta. \quad (17.2.9)$$

The input impedance around the resonance frequency, using equation (17.2.8), could be expressed in terms of the unloaded Q-factor:

$$Z_{in} = R\left[1 + jQ_U\left\{(1+\delta) - (1+\delta)^{-1}\right\}\right]$$

$$\approx R[1 + jQ_U\{(1+\delta) - (1-\delta)\}]$$

$$\Rightarrow Z_{in} = R[1 + j2\delta Q_U] \quad \text{(a)}$$

$$Z_{in} = R\left[1 + j2Q_U\frac{\Delta\omega}{\omega_r}\right] \quad \text{(b)}$$

$$\text{alternatively,}\ Z_{in} = j\frac{2RQ_U}{\omega_r}\left[\omega - \omega_r\left(1 + \frac{j}{2Q_U}\right)\right] \quad \text{(c),}$$

$$(17.2.10)$$

where $\Delta\omega = \omega - \omega_r$ and $\varepsilon_f \approx 2\delta$. The above approximate expression for the input impedance is valid around the resonance frequency. It is useful in the modeling of a line resonator as a *lumped* series resonant circuit. The expression (17.2.10b) shows that the Z_{in} of a lossy resonant circuit can be computed from the Z_{in} of a lossless circuit by replacing the resonance frequency ω_r of the lossless circuit with a complex resonance frequency, $\omega_r(1 + j/2\,Q_U)$.

At the half-power frequencies f_1 and f_2, the real and imaginary parts of Z_{in} are equal, i.e. $R = |X|$. Under this condition, the 3 dB bandwidth of a resonant circuit is obtained from equation (17.2.10b):

$$\text{At half-power frequencies}\quad f_1 \text{ and } f_2;$$

$$R = |X|,\ \Rightarrow R = 2R(\Delta\omega/\omega_r)Q_U$$

$$\Rightarrow \text{Bandwidth}: \omega_2 - \omega_1 = 2\Delta\omega = \frac{\omega_r}{Q_U} \quad \text{(a)}$$

$$BW = f_2 - f_1 = \frac{f_r}{Q_U} \quad \text{(b).}$$

$$(17.2.11)$$

To get a very narrow bandwidth, the Q-factor must be large, i.e. the internal resistance R of a series resonant circuit, signifying the circuit loss, must be very small.

Another useful general expression for the Q-factor is also obtained in terms of the reactance slope of a series resonant circuit:

$$X = \left[\omega L - \frac{1}{\omega C}\right],\ \Rightarrow \frac{dX}{d\omega} = L + \frac{1}{\omega^2 C}$$

$$\Rightarrow \left.\frac{dX}{d\omega}\right|_{\omega=\omega_r} = 2L \quad \text{(a);}\quad \text{as}\ L = \frac{1}{\omega_r^2 C}$$

$$\text{and,}\quad Q_U = \frac{\omega_r L}{R},\quad \Rightarrow Q_U = \frac{1}{2}\frac{\omega_r}{R}\left.\frac{dX}{d\omega}\right|_{\omega=\omega_r} \quad \text{(b).}$$

$$(17.2.12)$$

17.2.2 Lumped Parallel Resonant Circuit

Figure (17.4a) shows a parallel resonant circuit for the computation of the resonance frequency ω_r. The source internal impedance is ignored, i.e. $R_g = 0$. The input admittance of the parallel resonant circuit is

$$Y_{in} = \frac{1}{R} + j\left(\omega C - \frac{1}{\omega L}\right). \quad (17.2.13)$$

Figure (17.4b) shows the voltage response across R with a voltage maximum at the resonance frequency. The voltage maximum at the resonance, $V_{max} = \{R/(R + R_g)\}V$, is the terminal voltage for a source with internal impedance R_g. Figure (17.4b) also shows the drop in the terminal voltage to $V_{max}/\sqrt{2}$ at the half-power frequencies f_1 and f_2. The average power loss P_ℓ in the resistance

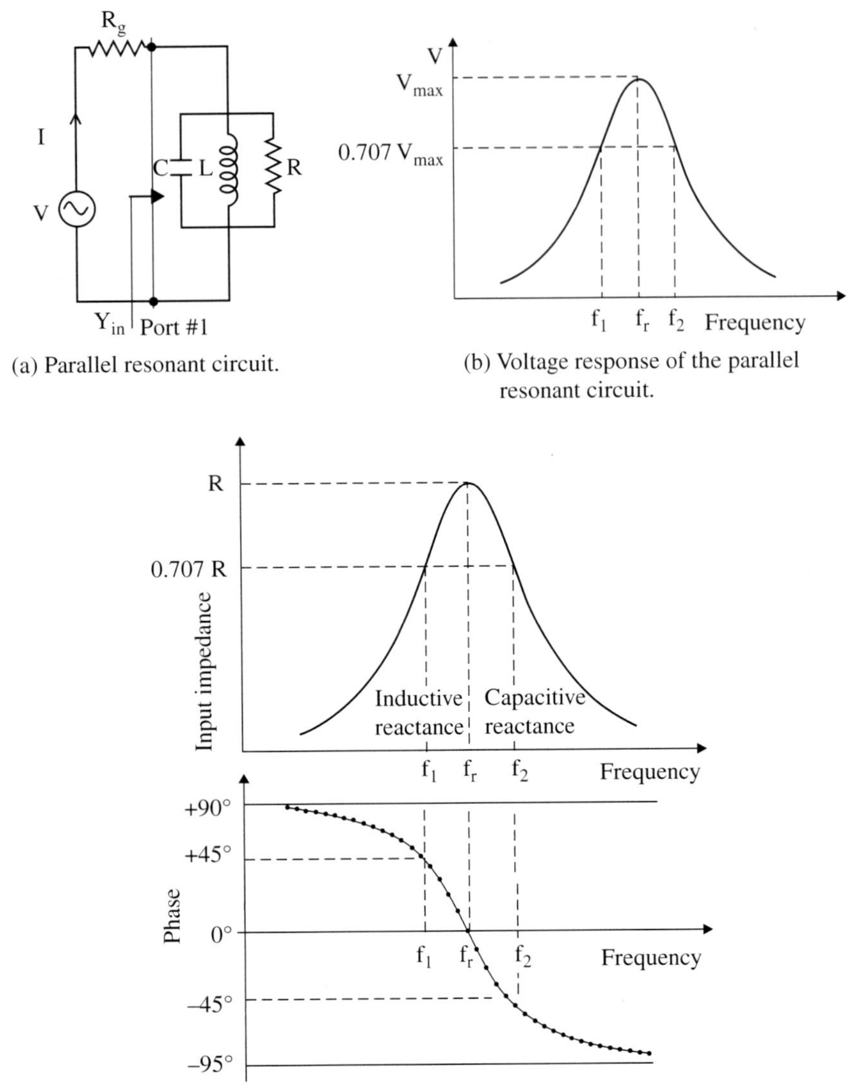

(a) Parallel resonant circuit.

(b) Voltage response of the parallel resonant circuit.

(c) Input impedance magnitude and phase variation with frequency.

Figure 17.4 Behavior of the parallel resonant circuit.

R, electric energy w_c stored in the capacitor, and magnetic energy w_m stored in the inductor are given below:

$$P_l = \frac{1}{2}\frac{|V_{max}|^2}{R} \quad \text{(a)}, \quad w_c = \frac{1}{4}|V_{max}|^2 C \quad \text{(b)}$$

$$w_m = \frac{1}{4}|I_{L,max}|^2 L = \frac{1}{4}\frac{|V_{max}|^2}{\omega_r^2 L} \quad \text{(c)}, \tag{17.2.14}$$

where $I_{L,max}$ is the maximum of current I_L through the inductor. At the resonance frequency, the susceptance part of Y_{in} is zero giving,

$$\omega_r = \frac{1}{\sqrt{LC}} \quad \text{(a)}, \quad f_r = \frac{1}{2\pi\sqrt{LC}} \quad \text{(b)}. \tag{17.2.15}$$

At the resonance $w_c = w_m$, i.e. the total stored energy is $w_T = 2w_m$. Following definition of equation (17.2.5), the unloaded Q-factor is obtained as follows:

$$Q_U = 2 \times \omega_r \frac{1/4\left(|V_{max}|^2/\omega_r^2 L\right)}{1/2\left(|V_{max}|^2/R\right)} = \frac{R}{\omega_r L}$$

$$\Rightarrow Q_U = \omega_r RC = \frac{\omega_r C}{G}. \tag{17.2.16}$$

Figure (17.4c) shows a variation of the input impedance, both magnitude and phase, with frequency. *Below resonance, the input impedance is inductive, while above the resonance, it is capacitive*. The input susceptance and its derivative are

$$B = \omega C - \frac{1}{\omega L} \quad (a), \quad \left.\frac{dB}{d\omega}\right|_{\omega = \omega_r} = 2C \quad (b). \tag{17.2.17}$$

An alternate expression for the unloaded Q-factor of a parallel resonant circuit is written from the above expressions:

$$Q_U = \frac{1}{2}\frac{\omega_r}{G}\left.\frac{dB}{d\omega}\right|_{\omega = \omega_r}, \tag{17.2.18}$$

where $G = 1/R$ is the conductance. Following the previous discussion on series resonance, the input admittance of a parallel resonant circuit, around the resonance frequency, and also its 3 dB bandwidth, are obtained:

$$Y_{in} = G\left[1 + j2Q_U\frac{\Delta\omega}{\omega_r}\right] \quad (a)$$

$$Z_{in} = \frac{1}{Y_{in}} = \frac{R}{1 + j2Q_U\frac{\Delta\omega}{\omega_r}} \quad (b)$$

$$3\text{ dB Bandwidth} = f_2 - f_1 = \frac{f_r}{Q_U} \quad (c). \tag{17.2.19}$$

17.2.3 Resonator with External Circuit

A resonant circuit, series or parallel, never works in isolation. It is always connected to an external circuit. Figure (17.5) shows the following three basic configurations of the resonators with external circuits:

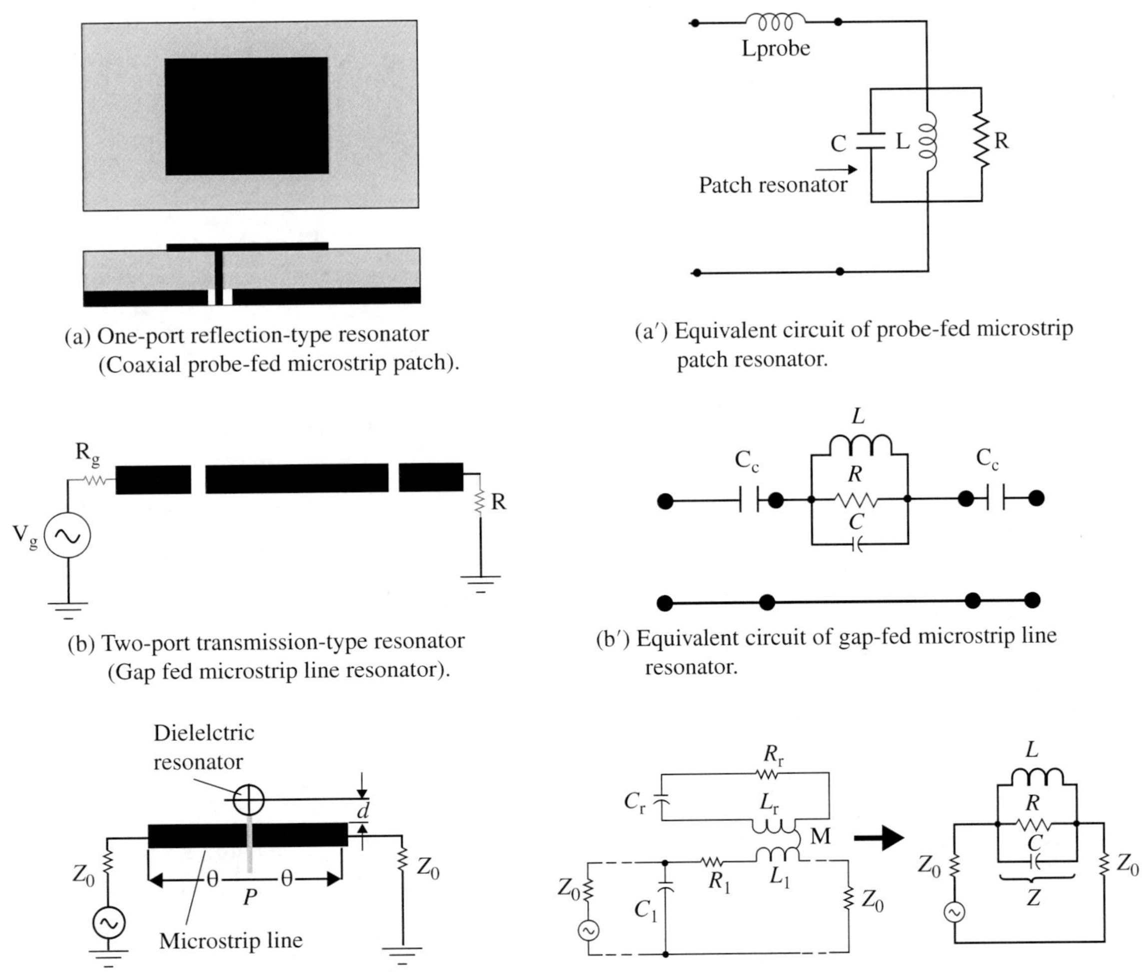

(a) One-port reflection-type resonator (Coaxial probe-fed microstrip patch).

(a′) Equivalent circuit of probe-fed microstrip patch resonator.

(b) Two-port transmission-type resonator (Gap fed microstrip line resonator).

(b′) Equivalent circuit of gap-fed microstrip line resonator.

(c) Two-port reaction-type resonator (Microstrip coupled dielectric resonator).

(c′) Equivalent circuit of microstrip line coupled dielectric resonator.

Figure 17.5 Basic types of planar resonators with equivalent circuits.

- *One-port reflection type*
- *Two-port transmission type*
- *Two-port reaction type.*

Figure (17.5) also show the equivalent circuits of these configurations. The one-port resonator reflects back power to the source. In the case of a two-port transmission-type resonator, at resonance, the power is partially coupled (transmitted) from the source to a resonator; partially reflected to the source. Next, it is coupled from the resonator to an external load. In the case of a microstrip-based reaction-type resonator, shown in Figure (17.5c), the input and output ports are its external circuits. The dielectric resonator reacts to the coupling and reflects the power. Figure (17.5c) also shows that the resonator-loaded line section acts like a band-reject parallel resonant circuit [J.3]. The coupling between a resonator and an *external circuit* is a fundamental characteristic of a practical resonator. The coupling is expressed through the concept of the *external Q-factor* (Q_E), and also through the *coupling coefficient* (β) between the resonator and external circuit. In the discussion to follow, the emphasis is placed on three fundamental characteristics of a practical resonator – *resonance frequency, unloaded Q-factor,* and *coupling coefficient.* Once these parameters are known, either experimentally or by using an EM-Simulator, the R, L, C elements of the circuit model of a resonating structure can be evaluated.

17.2.4 One-Port Reflection-Type Resonator

Figures (17.3a and 17.4a) show the one-port series and parallel resonant circuits act as the reflection-type resonators. This section presents the basic concept of coupling through the external Q-factor, and the coupling coefficient for both types of reflection resonators.

Reflection-Type Series Resonator

Figure (17.3a) shows a lossy series resonant circuit connected to a source with internal resistance R_g. This circuit is also obtained from Fig (17.7a) by taking $R_L = 0$. The coupling is treated as the power loss in the external resistance, $R_{ex} = R_g$. The *external Q-factor* is defined as follows:

$$Q_E = \omega_r \frac{\text{Average energy stored in resonant circuit}}{\text{Power loss in external circuit}}\bigg|_{\omega = \omega_r}. \tag{17.2.20}$$

Following the same method to derive equations (17.2.7) and (17.2.12), the above definition is used to obtain the following expressions for the external Q-factor of a *series resonant circuit* at the input port #1:

$$Q_{E,in} = \frac{\omega_r L}{R_g} \quad (a), \quad Q_{E,in} = \frac{X(\text{slope})|_{\omega=\omega_r}}{R_g} \quad (b)$$

$$\text{where, } X(\text{slope})|_{\omega=\omega_r} = \frac{\omega_r}{2}\frac{dX}{d\omega}\bigg|_{\omega=\omega_r} \quad (c). \tag{17.2.21}$$

The above equation (17.2.21b) expresses the external Q-factor in terms of the reactance slope parameter, X(slope), at the resonance. It is applicable to both the lumped and distributed, i.e. transmission line-type resonator. Equation (17.2.21a) could be obtained from equation (17.2.21b). The total Q-factor of the *reflection-type series resonator,* with external source resistance $R_{ex} = R_g$, is known as the *loaded Q-factor,* Q_L. It is computed by using the total resistance in series with LC circuit,

$$Q_L = \frac{\omega_r L}{R_g + R}. \tag{17.2.22}$$

It clearly shows that the *loaded* Q-factor is less than the *unloaded* Q-factor given by equation (17.2.12). From the above equation, the following relation, between the loaded, unloaded, and external Q-factors, is obtained:

$$\frac{1}{Q_L} = \frac{R}{\omega_r L} + \frac{R_g}{\omega_r L} \Rightarrow \frac{1}{Q_L} = \frac{1}{Q_U} + \frac{1}{Q_E}. \tag{17.2.23}$$

Figure (17.3a) is a *direct coupled* series resonant circuit. It has a low Q-factor. The coupling of a resonator to an external circuit is defined by the *coupling coefficient* β:

$$\beta = \frac{\text{Power loss in external circuit } (P_{ex})}{\text{Power loss in resonator circuit } (P_L)}\bigg|_{\omega=\omega_r}. \tag{17.2.24}$$

In the case of a *reflection-type* series resonant circuit, the *coupling coefficient* β is computed as follows:

$$\beta = \frac{(1/2)R_g I_{max}^2}{(1/2)R I_{max}^2} = \frac{R_g}{R} \quad (a)$$

$$\Rightarrow \beta = \frac{Q_U}{Q_E} = \frac{\text{Unloaded Q - factor}}{\text{External Q - factor}} \quad (b). \tag{17.2.25}$$

The relation between the Q-factors and the coupling coefficient is obtained from equation (17.2.23):

$$\frac{1}{Q_L} = \frac{1}{Q_U} + \frac{\beta}{Q_U} = \frac{1+\beta}{Q_U} \quad (a)$$

$$Q_U = Q_L(1+\beta) = Q_E\beta \quad (b). \tag{17.2.26}$$

The input impedance of a series resonator, given by equation (17.2.10b), can be rewritten as follows, in terms of the coupling coefficient and the loaded Q-factor:

$$
\begin{aligned}
Z_{in} &= R + j2\frac{RQ_U\Delta\omega}{\omega_r} && \text{(a)}\\
\Rightarrow Z_{in} &= R\left[1 + j2\frac{Q_L(1+\beta)\Delta\omega}{\omega_r}\right] && \text{(b)}\\
\Rightarrow Z_{in} &= R\left[1 + j2\,Q_L\,(1+\beta)\delta\right] && \text{(c)}\\
Z_{in} &= R[1 + j2Q_E\beta\delta] && \text{(d)}
\end{aligned}
$$

$$
\text{Normalized input impedance}: \overline{Z}_{in} = \frac{Z_{in}}{R_g} = \frac{R}{R_g}[1 + j2Q_E\beta\delta] = \frac{[1 + j2Q_E\beta\delta]}{\beta} \quad \text{(e)}, \tag{17.2.27}
$$

where $\delta = \Delta\omega/\omega_r$ is the *normalized frequency deviation*. In the above equation, the system impedance is $Z_0 = R_g$. At resonance $\delta = 0$ and $Z_{in} = R$ leading to $\beta = R_g/Z_{in}$, i.e. $Z_{in} = R_g/\beta$. The condition $\beta = 1$, i.e. the *critical coupling* between a resonator and the external circuit, is needed for the matching of a resonator at the resonance to the source. Under the condition of critical coupling, the S-parameter at the input port-1, $S_{11} = 0$. However, equal power is dissipated in the resonator and the external circuit. The critical coupling is achieved for the case $Q_U = 2Q_L$. The condition $\beta < 1$ ($Q_U < 2Q_L$) results in the *under-coupling*, whereas $\beta > 1(Q_U > 2Q_L)$ provides the *over-coupling*. It is obvious that in the case of under-coupling, the loaded Q-factor is higher as compared to its value at the critical coupling. It provides reduced 3dB bandwidth for an under-coupled resonator. The coupling is weak with a higher value of external Q-factor, i.e. for $Q_E > Q_U$. Therefore, more power is dissipated in the resonator. In the case of over-coupling, the loaded Q-factor is low, as compared to its value at the critical coupling, thus giving a larger 3 dB bandwidth. Also, the coupling is stronger with a lower value of the Q_E, i.e. for $Q_E < Q_U$. Therefore, more power is dissipated in the external load.

The coupling coefficient of a series resonant circuit to the source ($R_g = Z_0$) is computed from the known value of $|S_{11}|$ for a series resonator:

$$
\begin{aligned}
S_{11} &= \frac{Z_{in} - Z_0}{Z_{in} + Z_0}\\
&= \frac{R(1 + j2Q_U\delta) - Z_0}{R(1 + j2Q_U\delta) + Z_0} = \frac{R/R_g(1 + j2Q_U\delta) - 1}{R/R_g(1 + j2Q_U\delta) + 1}\\
\Rightarrow S_{11} &= \frac{\{(1 + j2Q_U\delta)/\beta\} - 1}{\{(1 + j2Q_U\delta)/\beta\} + 1} = \frac{(1-\beta) + j2\,Q_U\delta}{(1+\beta) + j2Q_U\delta} \quad \text{(a)}
\end{aligned}
$$

At the resonace, $\delta = 0$, and

$$
S_{11}\big|_{\omega=\omega_r} = S_{11}^{Res} = \frac{1-\beta}{1+\beta} \quad \text{(b)}. \tag{17.2.28}
$$

Equation (17.2.28a) defines the circles in the Γ-plane of the Smith chart. It provides information on the input impedance given by equation (17.2.27a). The *unloaded 3dB bandwidth, at the half-power frequencies*, is obtained from the expression (17.2.10b) under the condition $R = |X|$:

$$
R = 2\frac{RQ_U\Delta\omega}{\omega_r}, \Rightarrow BW = \omega_2 - \omega_1 = 2\Delta\omega = \frac{\omega_r}{Q_U}. \tag{17.2.29}
$$

The magnitude of the reflection coefficient $|S_{11}|$ gives the return loss (RL) at the resonance. The RL and phase $\angle S_{11}$ are obtained from equation (17.2.28):

$$
\begin{aligned}
RL &= -20\log|S_{11}| = -20\log\left[\frac{(1-\beta)^2 + (2Q_U\delta)^2}{(1+\beta)^2 + (2Q_U\delta)^2}\right] && \text{(a)}\\
|S_{11}^{Res}| &= 10^{-RL(dB)/20}, |S_{11}| \le 1 && \text{(b)}\\
\angle S_{11} &= \tan^{-1}\left(\frac{2Q_U\delta}{1-\beta}\right) - \tan^{-1}\left(\frac{2Q_U\delta}{1+\beta}\right) && \text{(c)}.
\end{aligned} \tag{17.2.30}
$$

The RL at resonance is measured or the EM-simulated quantity giving $|S_{11}^{Res}|$. The coupling conditions are summarized below:

$$
\begin{aligned}
&\text{For critical coupling}: \beta = 1, \text{i.e. } |S_{11}^{Res}| = 0, R = R_g && \text{(a)}\\
&\text{For under-coupling}: \beta < 1,\ \beta = \frac{1 - |S_{11}^{Res}|}{1 + |S_{11}^{Res}|}, R > R_g && \text{(b)}\\
&\text{For over-coupling}: \beta > 1,\ \beta = \frac{1 + |S_{11}^{Res}|}{1 - |S_{11}^{Res}|}, R < R_g && \text{(c)}.
\end{aligned} \tag{17.2.31}
$$

The expression for the over-coupling is an inversion of the expression for the under-coupling. The requirement $\beta > 1$ is met, while retaining $|S_{11}^{Res}| < 1$ for both the under- and over-coupling cases. For an under-coupled series resonant circuit, $\beta < 1$, i.e. $R > R_g$, whereas for a critically coupled case $\beta = 1$, i.e. $R = R_g$. The over-coupled resonator has $\beta > 1$, i.e. $R < R_g$.

Figure (17.6a) shows the input impedance variation of a reflection-type series resonant circuit with frequency, for three coupling conditions. At lower frequency f_1, i.e. below resonance f_r, the input impedance is capacitive. At the resonance $Z_{in} = R$. It is marked on the horizontal axis of the Smith chart. At frequency f_2 above the resonance, the input impedance is inductive. Normally, the system impedance is $Z_0 = R_g = 50\Omega$. The nature of coupling, i.e. the critical-, under-, or over-coupling, is determined from the position of the locus of the input impedance on the Smith chart. *The points A, B, and C show the under-coupling, critical-coupling, and over-coupling, respectively*. The coupling coefficient (β) can

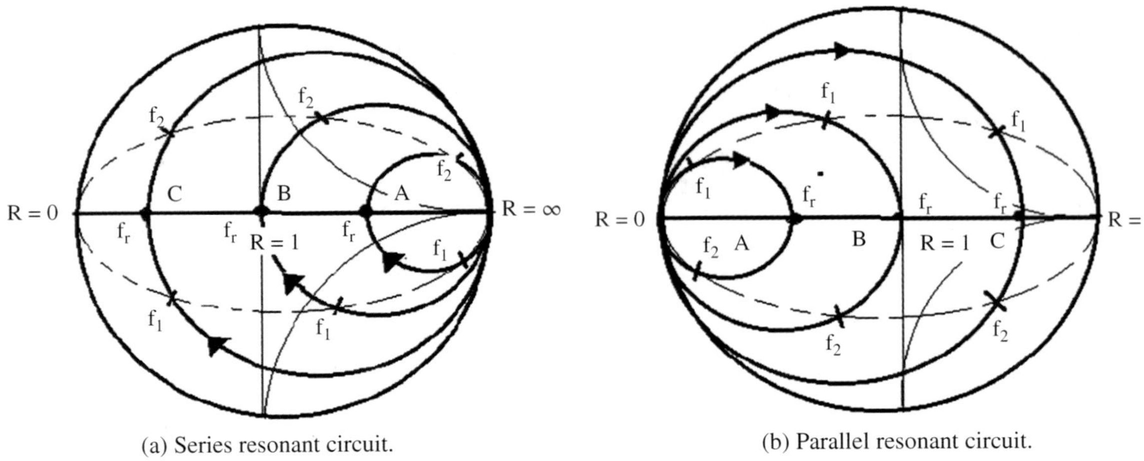

Figure 17.6 Input impedance variation of the one-port reflection-type resonators with frequency. (A) Under-coupling, (B) Critical-coupling, (C) Over-coupling.

be located on the horizontal axis of the Smith chart. Furthermore, Fig (17.6a) also shows the dotted *Q_U-circles*, i.e. the unloaded Q-factor circles, at which $R = |X|$. The intersections of the impedance circles with the Q_U-circles, for three cases of coupling, provide us the lower and upper 3 dB frequency, f_1 and f_2. It helps us to determine the unloaded Q-factor:

$$Q_U = \frac{\omega_r}{\omega_2 - \omega_1} = \frac{f_r}{f_2 - f_1} \quad \text{(a)}$$

$$\text{For } Q_U = (1+\beta)Q_L; \quad Q_L = \frac{f_r}{(1+\beta)(f_2 - f_1)} \quad \text{(b).} \tag{17.2.32}$$

The loaded Q-factor could be also determined from the 3 dB bandwidth of the return-loss response. In this case, the Q_U is further obtained from the *loaded* Q-factor Q_L. Once ω_r, β, and Q_U are known for any resonating structure, either from the experiment or from the EM-Simulation, the R, L, C components of a series resonator are determined from the following expressions:

$$R = \frac{R_g}{\beta} \quad \text{(a)}, \quad L = \frac{RQ_U}{\omega_r} \quad \text{(b)}$$

$$C = \frac{1}{\omega_r R Q_U} \quad \text{(c).} \tag{17.2.33}$$

Reflection-Type Parallel Resonator

Figure (17.4a) shows a lossy parallel reflection-type resonator circuit, connected to a source with internal resistance R_g. Figure (17.7b) is also reduced to Fig (17.4a) for

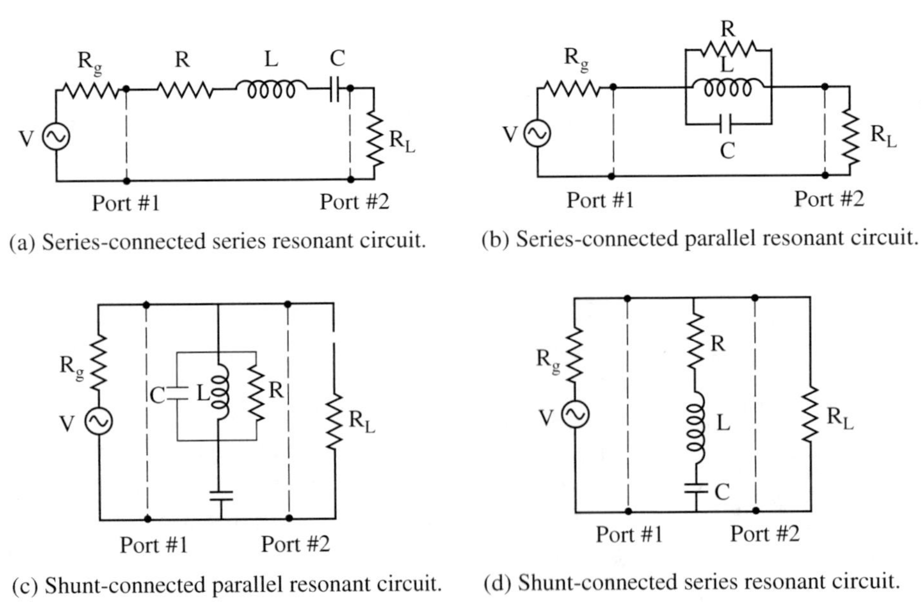

Figure 17.7 Circuit configurations of series and parallel resonators.

the output short-circuited, i.e. $R_L = 0$. The Q-factors, coupling coefficient, and input impedance are determined. The external Q-factor is obtained using the definition given in equation (17.2.20):

$$Q_{E,in} = \frac{\omega_r C}{G_g} \quad (a), \qquad Q_{E,in} = \frac{B(slope)|_{\omega=\omega_r}}{G_g} \quad (b)$$

$$\text{where,} \qquad B(slope)|_{\omega=\omega_r} = \frac{\omega_r}{2}\frac{dB}{d\omega}\bigg|_{\omega=\omega_r} \quad (c). \qquad (17.2.34)$$

In the above equation, $G_{ex} = G_g = 1/R_g$ is the external conductance. The susceptance slope parameter at resonance, $B(slope)|_{\omega=\omega_r}$, provides a more general definition of the external Q-factor. It is applicable to both the lumped and distributed line-type parallel resonant circuit. The loaded Q-factor of the parallel resonant circuit is obtained by taking total conductance across the LC circuit:

$$Q_L = \frac{\omega_r C}{G_g + G}. \qquad (17.2.35)$$

Equation (17.2.31), showing the relation between S_{11}^{Res} and the coupling coefficient, is valid for a parallel resonator also. Of course, equation (17.2.25a) for the coupling coefficient β is modified as follows:

$$\beta = \frac{Q_U}{Q_E} = \frac{R}{\omega_r L}\frac{\omega_r L}{R_g} = \frac{R}{R_g} \quad (a)$$

$$\text{For the under-coupling}: \beta < 1, \text{i.e. } R < R_g \quad (b)$$

$$\text{For the critical coupling}: \beta = 1, \text{i.e. } R = R_g \quad (c)$$

$$\text{For the over-coupling}: \beta > 1, \text{i.e. } R > R_g \quad (d). \qquad (17.2.36)$$

Equation (17.2.31) for β is valid for a reaction-type parallel resonator also. However, the above-given conditions are taken into account.

The input impedance of a parallel resonator, given by equation (17.2.19b), is rewritten as

$$Z_{in} = \frac{R}{1 + j2Q_U\delta} = \frac{R}{1 + j2Q_L(1+\beta)\delta} \quad (a)$$

$$Z_{in} = \frac{R}{1 + j2Q_E\beta\delta} \quad (b). \qquad (17.2.37)$$

The normalized input impedance of a parallel reflection-type resonator is

$$\frac{Z_{in}}{R_g} = \frac{R/R_g}{1 + j2Q_U\delta} = \frac{R/R_g}{1 + j2Q_L(1+\beta)\delta}$$

$$\Rightarrow \overline{Z}_{in} = \frac{\beta}{1 + j2Q_U\delta} = \frac{\beta}{1 + j2Q_L(1+\beta)\delta}. \qquad (17.2.38)$$

Once input impedance is known, an expression is obtained to compute the S_{11} that is related to the coupling coefficient.

$$S_{11} = \frac{\overline{Z}_{in} - 1}{\overline{Z}_{in} + 1} = \frac{(\beta - 1) - j\,2\,Q_U\delta}{(\beta + 1) + j\,2\,Q_U\delta} \quad (a)$$

$$RL = -20\log|S_{11}| = -20\log\left[\frac{(\beta-1)^2 + (2\,Q_U\delta)^2}{(\beta+1)^2 + (2\,Q_U\delta)^2}\right] \quad (b)$$

$$\angle S_{11} = -\tan^{-1}\left(\frac{2Q_U\delta}{1-\beta}\right) - \tan^{-1}\left(\frac{2Q_U\delta}{1+\beta}\right) \quad (c)$$

$$\text{At the resonace, } \delta = 0, \text{ and } S_{11}|_{\omega=\omega_r} = S_{11}^{Res} = \frac{1-\beta}{1+\beta} \quad (d)$$

$$\text{Also, } |S_{11}^{Res}| = 10^{-RL(dB)/20}, \quad |S_{11}| \le 1 \quad (e). \qquad (17.2.39)$$

Figure (17.6b) shows the normalized input impedance variation of a parallel resonant circuit, with respect to the frequency deviation δ, on a Smith chart. The impedances of the series and parallel resonant circuits show the inverse relationship. The *unloaded* Q-factor is computed after obtaining the f_1, f_2, and f_r from the Smith chart. The under-, critical-, and over-couplings are shown on the Smith chart. The coupling coefficient β is also obtained at the horizontal axis of the Smith chart, and Q_L from the 3dB points of the $|S_{11}|$ response. Once Q_L and β are known, the *unloaded* Q-factor of the resonator is determined from equation (17.2.26). On using these parameters, along with known ω_r, the circuit elements of a parallel resonator are extracted from the following expressions:

$$R = \beta\ R_g \quad (a)$$

$$C = \frac{(G + G_g)Q_L}{\omega_r} = \left(\frac{G + G_g}{\omega_r}\right)\frac{Q_U}{(1+\beta)} \quad (b)$$

$$L = \frac{R_g}{\omega_r\,Q_E} = \frac{R_g\,\beta}{\omega_r\,Q_U} \quad (c). \qquad (17.2.40)$$

17.2.5 Two-Port Transmission-Type Resonator

The series and parallel resonant circuits can be connected as a two-port network, forming the transmission-type resonators, both in the *series configuration* and in the *shunt configuration*. Figure (17.7) shows four circuit configurations. The circuit arrangements exhibit either the bandpass filter (BPF) or the bandstop filter (BSF) response. The resonators are *directly coupled* to the source and an external load. We first discuss the external Q-factor, and the coupling coefficient. Next, the S-parameter response of the circuit is presented. The measured S-parameter response of a resonator circuit is obtained from the Network Analyzer. It can also be obtained using an EM-Simulator.

Case-1: Series-Connected Series Resonant Circuit

Figure (17.7a) shows the *transmission-type series resonator* connected in series with the source and load. Figure (17.8a) shows that the circuit acts as a BPF. The external Q-factors are computed at both the input and the output ports. Normally in a transmission system, $R_g = R_L = Z_0$, where $Z_0 = 50\Omega$ is the system impedance. The total external load, in the series with the LC circuit, is $2Z_0$. The total load, including the internal resistance of the resonator, is $R + 2Z_0$. The external and loaded Q-factors are written as follows:

$$Q_E = \frac{\omega_r L}{2Z_0} \quad (a), \qquad Q_L = \frac{\omega_r L}{R + 2Z_0} \quad (b). \tag{17.2.41}$$

The series-connected series resonant circuit, shown in Fig (17.7a), is analyzed below using the following matrix expression for the S-parameters [B.9, B.10, B.13]:

$$[S] = \begin{bmatrix} Z/(Z + 2Z_0) & 2Z_0/(Z + 2Z_0) \\ 2Z_0/(Z + 2Z_0) & Z/(Z + 2Z_0) \end{bmatrix}, \tag{17.2.42}$$

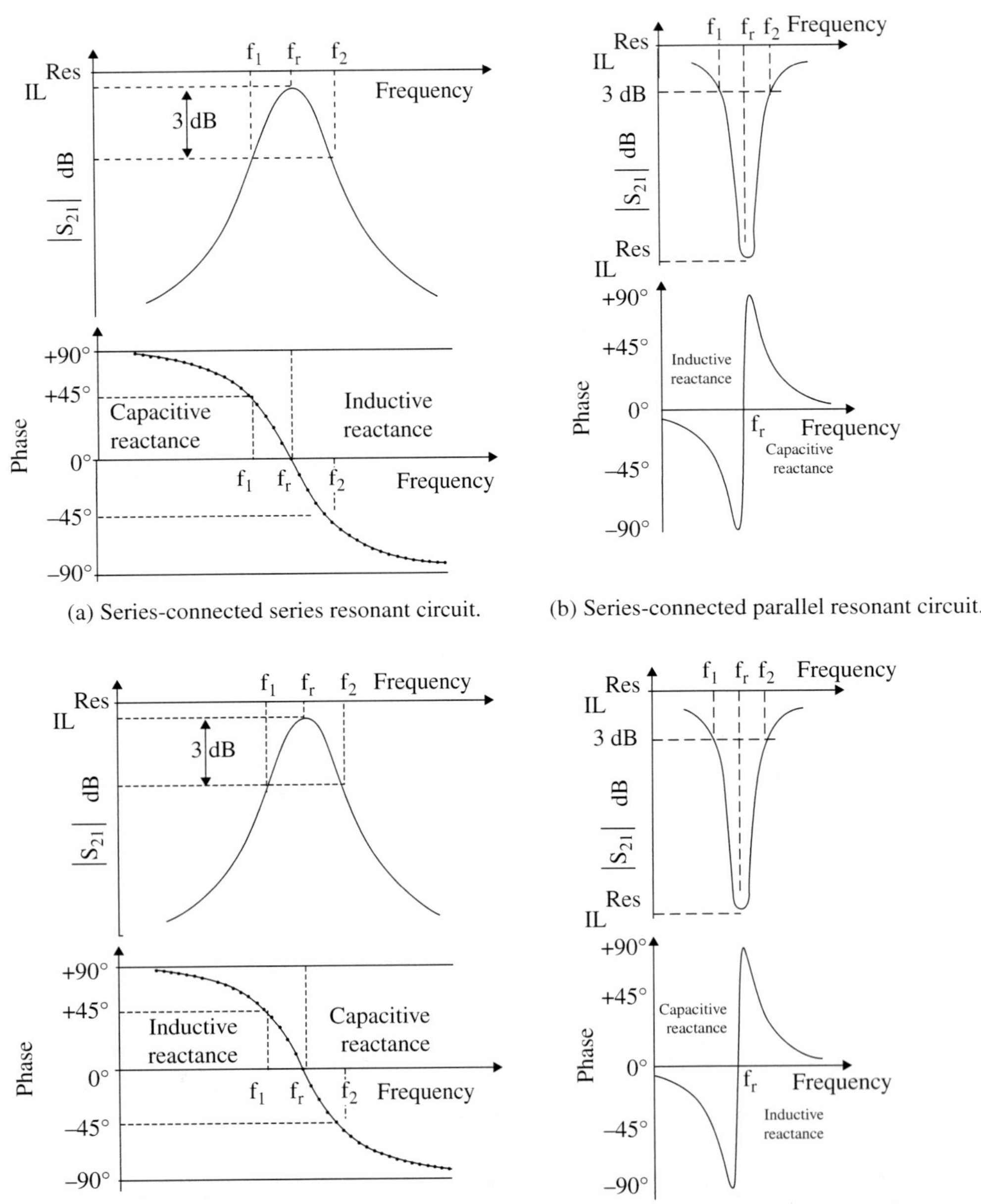

Figure 17.8 Sketched insertion loss and phase response $\angle S_{21}(f)$ of series/shunt-connected series and parallel resonant circuits.

where Z_0 is the system impedance and Z is the impedance of the series resonant RLC circuit near resonance frequency, $Z = R + j(\omega L - 1/\omega C) = R + j2RQ_U\delta$. The above [S]-matrix is obtained in Ex. 3.9 of chapter 3. The insertion loss (IL) and phase shift from the S_{21} are obtained below. The source impedance is a real quantity, and the ports are symmetrical, i.e. $Z_0 = R_G = R_L$.

$$S_{21}(f) = \frac{2Z_0}{Z + 2Z_0} = \frac{2Z_0}{(R + 2Z_0) + j2RQ_U\delta} \quad \text{(a)}$$

$$|S_{21}(f)|^2 = \frac{1}{\left(\frac{2Z_0 + R}{2Z_0}\right)^2 \left[1 + \left(\frac{2RQ_U\delta}{2Z_0 + R}\right)^2\right]} \quad \text{(b)},$$

(17.2.43)

where $\delta = (\Delta\omega/\omega_r)$. The denominator is simplified as follows:

$$\text{Deno} = \frac{\left[1 + \left(\frac{2RQ_U\delta}{(2Z_0 + R)}\right)^2\right]}{\left(\frac{2Z_0}{R + 2Z_0}\right)^2}$$

$$= \frac{\left[1 + \left(\frac{2RQ_U\delta Q_L}{\omega_r L}\right)^2\right]}{\left(1 - \frac{R/\omega_r L}{(R + 2Z_0)/\omega_r L}\right)^2} = \frac{\left[1 + (2\delta Q_L)^2\right]}{\left(1 - \frac{Q_L}{Q_U}\right)^2}.$$

(17.2.44)

The condition for *the half-power provides the 3dB BW*, from which the Q_L can be determined. Next, Q_U is also determined using the value of $|S_{21}^{Res}(f_r)|$. At resonance, the normalized frequency deviation is zero $\delta = 0$. The $|S_{21}(f)|$-response and 3 dB bandwidth from it are obtained as follows:

$$S_{21}^{Res} = 1 - Q_L/Q_U, \Rightarrow Q_U = \frac{Q_L}{1 - S_{21}^{Res}} \quad \text{(a)}$$

$$|S_{21}(f)|^2 = \frac{|S_{21}^{Res}|^2}{\left[1 + (2\delta Q_L)^2\right]^2} \quad \text{(b)}$$

$$\text{For} \quad 2\delta Q_L = 1, \Rightarrow Q_L = \frac{f_r}{f_2 - f_1} \quad \text{(c)},$$

(17.2.45)

where $S_{21}^{Res} = 10^{-IL(dB)/20}$ is IL (dB). It is a measured or EM-Simulated positive quantity. The IL and phase shifts are obtained as follows:

$$\begin{aligned} IL(dB) &= -10 \log_{10}\left[|S_{21}(f)|^2\right] \\ &= -20 \log_{10}[1 - (Q_L/Q_U)] \\ &\quad + 10 \log_{10}\left[1 + (2Q_L\delta)^2\right] \end{aligned} \quad \text{(a)}$$

At resonance, $\delta = 0$, and $IL|_{f = f_r}$

$$= -20 \log_{10}[1 - (Q_L/Q_U)] \quad \text{(b)}$$

$$\phi = -\tan^{-1}\left[\frac{\omega L - 1/\omega C}{R + 2Z_0}\right] = -\tan^{-1}\left[\frac{2RQ_U\delta}{R + 2Z_0}\right] \quad \text{(c)}$$

In limit, for $\omega \to 0$, $\phi \to +\pi/2$;

$$\omega = \omega_r, \delta = 0, \phi = 0; \ \omega \to \infty, \ \phi \to -\pi/2 \quad \text{(d)}.$$

(17.2.46)

Figure (17.8a) shows sketches of the magnitude and phase responses of the series-connected series resonant circuit. The series resonant circuit behaves as a series capacitor at a frequency below the resonance frequency, giving positive phase response. At a frequency above the resonance frequency, it acts as a series inductor, giving negative phase response. The 3 dB frequency points, f_1 and f_2, are marked on both responses. These responses are obtained either experimentally on a *Vector Network Analyzer* or an *EM-simulator*. The Q_L and Q_U are determined from equations (17.2.45c) and (17.2.46b), respectively.

The coupling coefficient (β) is also computed from S_{21}^{Res}. In the present case of the equal coupling at both the input and output, the following relations hold between the Q-factors:

$$\frac{1}{Q_L} = \frac{1}{Q_U} + \frac{2}{Q_E} \Rightarrow Q_L = \frac{Q_U Q_E}{Q_E + 2Q_U} \quad \text{(a)}$$

$$\beta = \frac{2Q_U}{Q_E} \quad \text{(b)}.$$

(17.2.47)

Because of equal coupling at the input and output, the β mentioned above is twice of the coupling coefficient only at the input. The following relation is obtained between S_{21}^{Res} and β at the resonance:

$$\begin{aligned} S_{21}^{Res} &= 1 - \frac{Q_L}{Q_U} = 1 - \frac{1}{1 + (2Q_U/Q_E)} \\ &= 1 - \frac{1}{1 + \beta} = \frac{\beta}{1 + \beta} \end{aligned} \quad \text{(a)}$$

alternatively, the coupling coefficient, $\beta = \dfrac{|S_{21}^{Res}|}{1 - |S_{21}^{Res}|}$ (b).

(17.2.48)

The RL, and also unloaded Q-factor of the circuit, are computed as follows:

$$S_{11} = \frac{Z}{Z+2Z_0} = \frac{R + j(\omega L - 1/\omega C)}{(2Z_0 + R) + j(\omega L - 1/\omega C)}$$
$$= \frac{R + j2RQ_U\delta}{(2Z_0 + R) + j2RQ_U\delta}$$
$$\Rightarrow S_{11} = \frac{Q_L}{Q_U}\left[\frac{1 + j2Q_U\delta}{1 + j2Q_L\delta}\right] \quad (a)$$
$$\text{At resonace,}\ S_{11}^{Res} = \frac{Q_L}{Q_U} \quad (b)$$
$$\text{Phase},\ \angle S_{11} = \tan^{-1}(2Q_U\delta) - \tan^{-1}(2Q_L\delta) \quad (c)$$
$$RL(dB)|_{\omega=\omega_r} = -20\log_{10}\left(\frac{Q_L}{Q_U}\right) \quad (d)$$
$$Q_U = Q_L 10^{RL(dB)/20} \quad (e). \quad (17.2.49)$$

Further, using equation (3.1.64) of chapter 3 the coupling coefficient at the resonance is also obtained:

$$S_{11}^{Res} + S_{21}^{Res} = 1$$
$$\Rightarrow S_{11}^{Res} = 1 - \frac{\beta}{1+\beta} = \frac{1}{1+\beta} \Rightarrow \beta = \frac{1 - |S_{11}^{Res}|}{S_{11}^{Res}}. \quad (17.2.50)$$

The S-parameters S_{21}^{Res} and S_{11}^{Res} are obtained either from the measurement or from the EM-simulation. Once the coupling coefficient, Q-factors, and the resonance frequency of a two-port series-connected series resonant circuit are known; the component values of a series resonator are obtained as follows:

$$R = 2Z_0/\beta \quad (a), \quad L = RQ_U/\omega_r \quad (b), \quad C = 1/\omega_r^2 L \quad (c). \quad (17.2.51)$$

Case- 2: Series-Connected Parallel Resonant Circuit

Figure (17.7b) shows the two-port series-connected parallel resonant circuit. Primarily it is a BSF. Figure (17.8b) shows the stopband response. The analysis is carried out by computing the S-parameters. The admittance of the parallel resonator, around resonance frequency, is

$$Y = G + j\left(\omega C - \frac{1}{\omega L}\right)$$
$$= G\left(1 + j2Q_U\frac{\Delta\omega}{\omega_r}\right) = G(1 + j2\delta Q_U). \quad (17.2.52)$$

The resonator, with admittance Y, is loaded at both ends by the external loads, $R_g = R_L = Z_0$, i.e. by the admittance $1/(2Z_0) = 0.5\ Y_0$. Accounting for the loss of a resonator, loss in the external circuit, and the total loss both in the resonator and external circuit, the expressions for the unloaded, external and loaded Q-factors are written below:

$$Q_U = \frac{\omega_r C}{G} \quad (a), \quad Q_E = \frac{\omega_r C}{0.5Y_0} \quad (b)$$
$$Q_L = \frac{\omega_r C}{G + 0.5Y_0} \quad (c). \quad (17.2.53)$$

The $S_{21}(f)$ is obtained using equation (17.2.42):

$$S_{21}(f) = \frac{2Z_0}{Z + 2Z_0} = \frac{2Y}{Y_0 + 2Y}. \quad (17.2.54)$$

$$\frac{1}{S_{21}(f)} = 1 + \frac{Y_0}{2Y} = 1 + \frac{0.5Y_0}{G(1 + j2\delta Q_U)}$$
$$= \frac{(G + 0.5Y_0) + j2\delta G Q_U}{G(1 + j2\delta Q_U)}$$
$$\Rightarrow \frac{1}{S_{21}(f)} = \left(\frac{G + 0.5Y_0}{G} \times \frac{\omega_r C}{\omega_r C}\right)\left(\frac{1 + j2\delta Q_U[G/(G + 0.5Y_0)] \times (\omega_r C/\omega_r C)}{1 + j2\delta Q_U}\right) \quad (17.2.55)$$
$$\Rightarrow \frac{1}{S_{21}(f)} = \frac{Q_U}{Q_L}\left[\frac{1 + j2\delta Q_U \times (Q_L/Q_U)}{1 + j2\delta Q_U}\right] = \frac{Q_U}{Q_L}\left[\frac{1 + j2\delta Q_L}{1 + j2\delta Q_U}\right].$$

The IL and phase are obtained as follows:

$$S_{21}(f) = \frac{Q_L}{Q_U}\left[\frac{1 + j2\delta Q_U}{1 + j2\delta Q_L}\right] \quad (a)$$
$$\Rightarrow |S_{21}(f)|^2 = \left(\frac{Q_L}{Q_U}\right)^2\left[\frac{1 + (2\delta Q_U)^2}{1 + (2\delta Q_L)^2}\right] \quad (b)$$
$$\phi = \tan^{-1}(2\delta Q_U) - \tan^{-1}(2\delta Q_L) \quad (c). \quad (17.2.56)$$

$$\text{where,}\quad 2\delta \approx \varepsilon_f = \frac{f}{f_r} - \frac{f_r}{f}.$$
$$\text{For}\ f \to 0, \varepsilon_f \to -\infty, \phi \to 0^- \quad (17.2.57)$$
$$f = f_r, \varepsilon_f = 0, \quad \phi = 0$$
$$f \to \infty, \varepsilon_f \to \infty, \phi \to 0^+.$$

Figure (17.8b) shows the phase plot of $S_{21}(f)$. For $f < f_r$ and $Q_U > Q_L$, the phase ϕ is negative, i.e. a parallel resonant circuit behaves as an inductor at a frequency below the resonance frequency. Likewise, for $f > f_r$ and $Q_U > Q_L$, a parallel resonant circuit behaves as a capacitor at a frequency above the resonance frequency.

$$IL = -10\log_{10}|S_{21}(f)|^2$$
$$= -20\log_{10}\left(\frac{Q_L}{Q_U}\right) - 10\log_{10}\left[\frac{1+(2\delta Q_U)^2}{1+(2\delta Q_L)^2}\right] \quad (a)$$
$$\text{At resonance, } \delta = 0, IL^{Res} = -20\log_{10}\left(\frac{Q_L}{Q_U}\right)$$
$$\Rightarrow Q_U = Q_L 10^{\left(IL^{Res}/20\right)} \quad (b). \quad (17.2.58)$$

The second term of the equation (17.2.58a) shows the frequency-dependent bandstop behavior, as shown in Fig (17.8b). As $Q_L < Q_U$, so the IL response $S_{21}(f)$ on the logarithmic scale is a positive quantity. equation (17.2.58b) extracts the unloaded Q-factor. Following Fig (17.6b), the Q_U can also be determined from the corresponding bandwidth obtained from the Smith chart. The Smith chart can also be used to determine the Q_L [J.3]. Normally, one attempts to determine the Q_L from the 3 dB bandwidth shown in Fig (17.8b). However, the above IL equation defines the bandwidth relating to Q_L that is dependent on the coupling ($Q_U = \beta Q_E$). The 3dB location satisfactorily works only if $IL^{Res}(f_r)$ is more than 18 dB [J.3]. However, the Q_L could be also determined from the return-loss response using $S_{11}(f)$:

$$S_{11}(f) = 1 - S_{21}(f) = 1 - \frac{Q_L}{Q_U}\left[\frac{1+j\,2\delta Q_U}{1+j\,2\delta Q_L}\right]$$
$$= \frac{Q_U - Q_L}{Q_U(1+j\,2\delta Q_L)} \quad (a)$$
$$\left|S_{11}^{Res}(f_r)\right|^2 = \left(\frac{Q_U - Q_L}{Q_U}\right)^2 \quad (b)$$
$$|S_{11}(f)|^2 = \frac{\left|S_{11}^{Res}(f_r)\right|^2}{1+(2\delta Q_L)^2} \quad (c)$$
$$\text{Half-power corresponds to } 2\delta Q_L = 1,$$
$$Q_L = \frac{f_r}{f_2 - f_1} \quad (d). \quad (17.2.59)$$

The resonance frequency f_r and 3 dB lower f_1, and upper f_2 frequencies are obtained from the experimental or EM-Simulated $|S_{21}(f)|$ response. Finally, the circuit components G, L, C for a parallel resonant circuit connected in series are obtained as follows:

$$\frac{Q_L}{Q_U} = \frac{G}{\omega_r C} \times \frac{\omega_r C}{G + 0.5Y_0} = \frac{G}{G + 0.5Y_0}$$
$$G = \frac{0.5Y_0 Q_L}{Q_U - Q_L} \quad (a), \quad C = \frac{GQ_U}{\omega_r} \quad (b), \quad L = \frac{1}{C\omega_r^2} \quad (c). \quad (17.2.60)$$

Case-3: Shunt-Connected Parallel Resonator

Figure (17.7c) shows the shunt-connected parallel resonator that has admittance $Y = [G + j(\omega C - 1/\omega L)] = (G + j2\delta\, Q_U G)$, where $G = 1/R$ shows the internal loss of a parallel resonator. The resonator is loaded at both ports with $R_g = R_L = Z_0$. Figure (17.8c) shows that it behaves as a BPF. Taking care of total losses in the internal conductance G, and both end external conductance Y_0, the Q-factors are written as follows:

$$Q_U = \frac{\omega_r C}{G} \quad (a), \quad Q_E = \frac{\omega_r C}{2Y_0} \quad (b)$$
$$Q_L = \frac{\omega_r C}{G + 2Y_0} \quad (c). \quad (17.2.61)$$

An expression for the IL is obtained from the S-parameter of the circuit, given by equation (3.1.71) of chapter 3:

$$[S] = \begin{bmatrix} -\dfrac{YZ_0}{2+YZ_0} & \dfrac{2}{2+YZ_0} \\ \dfrac{2}{2+YZ_0} & -\dfrac{YZ_0}{2+YZ_0} \end{bmatrix}. \quad (17.2.62)$$

$$S_{21}(f) = \frac{2}{2+YZ_0}$$
$$\Rightarrow \frac{1}{S_{12}(f)} = 1 + \frac{YZ_0}{2} = 1 + \frac{G}{2Y_0}(1 + j2\delta Q_U). \quad (17.2.63)$$

The coupling coefficients at both ports are identical. Equation (17.2.36) gives $\beta_1 = \beta_2 = \beta = Y_0/G$, and the above expression is rewritten as

$$\frac{1}{S_{21}(f)} = 1 + \frac{1}{2\beta}(1 + j\,2\delta Q_U). \quad (17.2.64)$$

However, 2β is the total coupling coefficient at both ports that relates Q_L and Q_U through equation (17.2.26), i.e. $2\beta = (Q_U - Q_L)/Q_L$. On substituting 2β in the above equation, the following expressions are obtained:

$$S_{21}(f) = \frac{Q_U - Q_L}{Q_U}\left[\frac{1}{(1 + j2\delta\, Q_L)}\right] \quad (a)$$
$$\text{At resonace } \delta = 0; \left|S_{21}^{Res}(f_r)\right|^2 = \left(\frac{Q_U - Q_L}{Q_U}\right)^2 \quad (b)$$
$$|S_{21}(f)|^2 = \frac{\left|S_{21}^{Res}(f_r)\right|^2}{1 + (2\delta\, Q_L)^2} \quad (c)$$
$$\text{At Half-power, } 2\delta\, Q_L = 1, \Rightarrow Q_L = \frac{f_r}{f_2 - f_1} \quad (d)$$
$$\text{Phase, } \angle S21 = \phi = -\tan^{-1}(2\delta Q_L) \quad (e). \quad (17.2.65)$$

The IL of the circuit is

$$IL(dB) = -10 \log |S_{21}(f)|^2$$
$$= -20 \log \left(\frac{Q_U - Q_L}{Q_U}\right) + 10 \log\left[1 + (2\delta Q_L)^2\right] \quad (a)$$

At resonance $\delta = 0$,

$$IL^{Res}(dB) = -20 \log \left(\frac{Q_U - Q_L}{Q_U}\right) \quad (b). \tag{17.2.66}$$

Figure (17.8c) shows the sketched amplitude and phase responses of $S_{21}(f)$. The Q_L is determined from the 3 dB BW. For $f < f_r$, δ is negative that leads to positive φ and the resonator acts a shunt-connected inductor. At $f = f_r$, $\delta = 0$ leading to $\phi = 0$. For $f > f_r$, δ is positive that leads to negative ϕ, and the resonator acts as a shunt-connected capacitor. The $IL^{res}(dB)$ at the resonance frequency is a measurable positive quantity. However, it appears as a negative quantity, in Figure (17.8c), with respect to no loss case. It is also available from the EM-simulator. The unloaded Q-factor is determined as follows:

$$\frac{Q_U - Q_L}{Q_U} = 10^{-(IL^{res}dB)/20} = |S_{21}(f_r)| \tag{17.2.67}$$
$$\Rightarrow Q_U = \frac{Q_L}{1 - |S_{21}(f_r)|}.$$

In the above equation, the insertion loss $IL^{res}dB$ is taken at the resonance frequency. After evaluation of the resonance frequency, Q_U and Q_L, finally R, L, C are computed from equation (17.2.61):

$$G = \frac{2Y_0 Q_L}{Q_U - Q_L} \quad (a), \quad C = \frac{GQ_U}{\omega_r} \quad (b)$$
$$L = \frac{1}{C\omega_r^2} \quad (c). \tag{17.2.68}$$

Case-4: Shunt-Connected Series Resonance

Figure (17.7d) shows the shunt-connected series resonator. The circuit acts as a BSF. Three Q-factors, and also the impedance of the series resonant circuit, are written as follows:

$$Q_U = \frac{\omega_r L}{R} \quad (a), \quad Q_E = \frac{\omega_r L}{Z_0/2} \quad (b)$$
$$Q_L = \frac{\omega_r L}{R + Z_0/2} \quad (c), \quad Z = R(1 + j2Q_U\delta) \quad (d). \tag{17.2.69}$$

Again, its S-parameters responses, IL, and phase response are obtained from the $S_{21}(f)$ of the shunt-connected admittance:

$$S_{21}(f) = \frac{2}{2 + YZ_0} = \frac{2R(1 + j2Q_U\delta)}{2R(1 + j2Q_U\delta) + Z_0}$$
$$= \frac{(1 + j2Q_U\delta)}{[1 + (Z_0/2)/R) \times (\omega_r L/\omega_r L) + j2Q_U\delta]}$$
$$\Rightarrow S_{21}(f) = \frac{(1 + j2Q_U\delta)}{(1 + Q_U/Q_E) + j2Q_U\delta}$$
$$= \frac{Q_L}{Q_U}\left[\frac{(1 + j2Q_U\delta)}{(1 + j2Q_L\delta)}\right] \quad (a)$$
$$\text{Phase, } \angle S_{21} = \phi = \tan^{-1}(2Q_U\delta) - \tan^{-1}(2Q_L\delta) \quad (b)$$
$$IL(dB) = -20 \log_{10}(Q_L/Q_U) - 10 \log_{10}\left[1 + (2Q_U\delta)^2\right]$$
$$+ 10 \log_{10}\left[1 + (2Q_L\delta)^2\right] \quad (c). \tag{17.2.70}$$

Figure (17.8d) shows the sketched amplitude and phase responses of $S_{21}(f)$. The 3 dB frequencies, f_1 and f_2, are indicated on the response. However, the half-power frequency does not provide correctly the needed bandwidth to get the loaded Q-factor Q_L, as it depends on the coupling coefficient also. However, the internal resistance R of a resonator is evaluated from the insertion-loss at the resonance:

$$S_{21}^{Res}(f_r) = \frac{Q_E}{Q_U + Q_E} = \frac{2\omega_r L/Z_0}{2\omega_r L/Z_0 + \omega_r L/R}$$
$$\Rightarrow S_{21}^{Res}(f_r) = \frac{2R}{2R + Z_0} \quad (a)$$
$$\Rightarrow R = \frac{(Z_0/2)\, S_{21}^{Res}(f_r)}{1 - S_{21}^{Res}(f_r)} \quad (b). \tag{17.2.71}$$

The unloaded Q-factor is related to Q_L, $Q_U = Q_L/S_{21}^{Res}(f_r)$. To compute the unloaded Q-factor Q_U, the Q_L can be evaluated from the power response of $S_{11}(f)$ as follows:

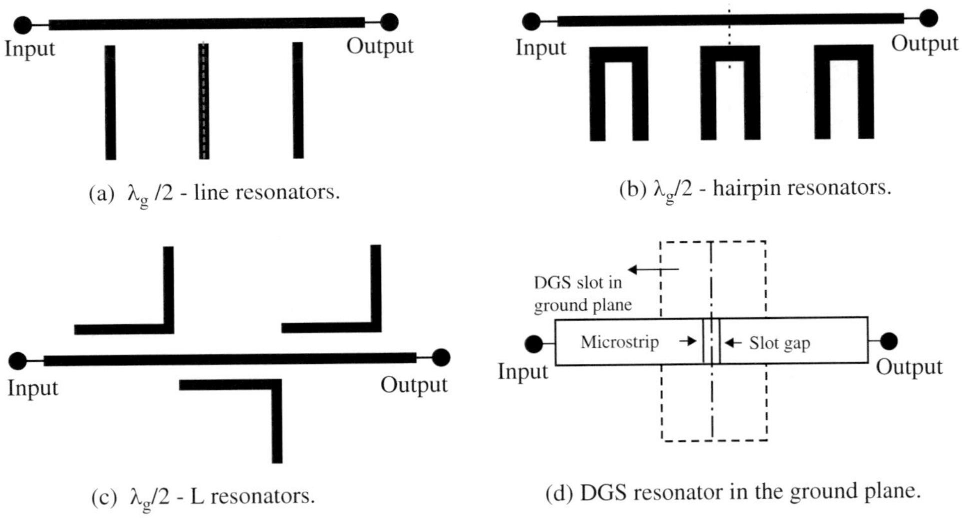

Figure 17.9 Some reaction-type resonators.

$$\begin{aligned}
S_{11}(f) &= -\frac{YZ_0}{2+YZ_0} = \frac{-Z_0}{Z_0+2R(1+j2Q_U\delta)} \\
&= \frac{-Z_0}{(Z_0+2R)+j4RQ_U\delta]} \\
\Rightarrow S_{11}(f) &= -\frac{\omega_r L}{\omega_r L}\times\frac{Z_0/2}{R+Z_0/2}\times \\
&\frac{1}{[1+(2Q_U\delta R)/\{R+Z_0/2\})\times(\omega_r L/\omega_r L)]} \\
\Rightarrow S_{11}(f) &= -\frac{Q_L}{Q_E}\frac{1}{1+j2Q_L\delta} \\
&= -\left[\frac{Q_U-Q_L}{Q_U}\right]\frac{1}{1+j2Q_L\delta} && \text{(a)} \\
\text{At resonace } \delta = 0, &\left|S_{11}^{Res}(f_r)\right|^2 = \left[\frac{Q_U-Q_L}{Q_U}\right]^2 && \text{(b)} \\
|S_{11}(f)|^2 &= \frac{\left|S_{11}^{Res}(f_r)\right|^2}{1+(2Q_L\delta)^2} && \text{(c)} \\
\text{For half} - \text{power}, &\ 2Q_L\delta = 1, Q_L = \frac{f_r}{f_2-f_1} && \text{(d)} \\
\text{Phase}, &\ \angle S_{11} = \phi = \pi - \tan^{-1}(2Q_L\delta) && \text{(e).}
\end{aligned}$$
(17.2.72)

Once R, ω_r, and Q_U are evaluated, the L and C of a resonator can be computed.

17.2.6 Two-Port Reaction-Type Resonator

Figure (17.5c) shows the dielectric resonator-based reaction-type resonator, magnetically coupled to a microstrip line. The coupling with the dielectric resonator loads the microstrip line with the series-connected parallel resonant circuit. Thus, a reaction-type resonator acts as a band-reject circuit. It can also be described by the shunt-connected series resonant circuit. Fig (17.9) shows some more electrically or magnetically coupled reaction-type resonators. These are commonly used to develop BSFs. The resonators in Fig (17.9a and d) are electrically coupled to the microstrip. The defected ground structure (DGS) is a λ_g/2-slot-type resonator that is modeled by a shunt-connected series resonant circuit. However, due to the coupling capacitor, it appears as a series-connected parallel resonant circuit to the microstrip line, like any other reaction-type resonator [J.4]. Normally it is used to develop high rejection low-pass filters [J.5]. The resonators in Fig (17.9b and c) are magnetically coupled to the microstrip line. However, *the nature of the coupling is decided by the predominance of electric or magnetic fields in the coupling region.* At the ends of the line resonators, the electric field is dominant, and around the middle region of the line resonators, the magnetic field is dominant. There can be many variations in the physical structures of these resonators. Some of these could provide more compact planar filters. The coupled resonators are further discussed in section (18.5) of chapter 18.

17.3 Transmission Line Resonator

The transmission line resonators are basic building blocks of the RF, microwave, and mm-wave circuits, as the lumped element-based resonators are normally not suitable for the frequency above 3 GHz. However, in some cases, the lumped element-based circuits are reported even up to mm-wave ranges. A section of the transmission line, such as a two-wire line, coaxial, microstrip, CPW, slot line, fin line, etc., acts as a resonator. It is also

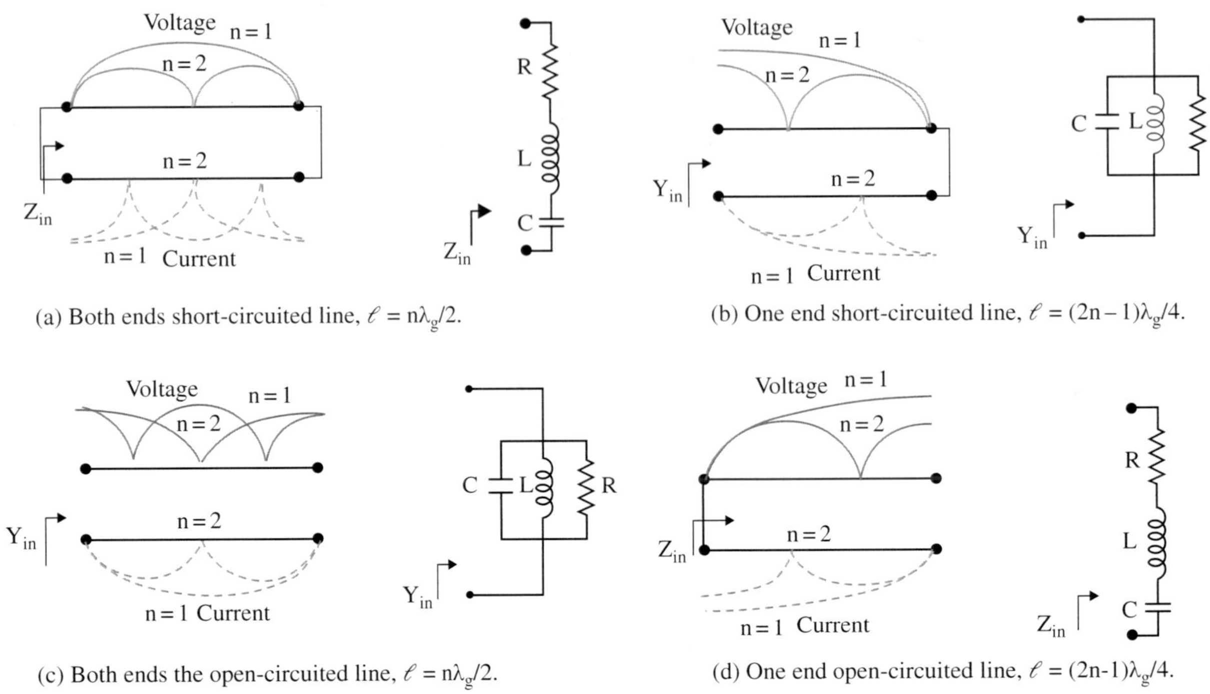

Figure 17.10 Basic transmission line resonators, showing voltage and current distribution for the fundamental and first harmonic and also their lumped equivalent circuits.

used for the realization of capacitors and inductors at microwave frequency. This section presents the modeling of a section of the transmission line as a lumped resonator. The section further presents a complete *modal description* of the $\lambda_g/2$ line resonators.

17.3.1 Lumped Resonator Modeling of Transmission Line Resonator

The output end of a line section is either short-circuited or open-circuited. A fixed line length can support a large number of discrete resonating modes. However, at a fixed frequency, the line lengths could be either even multiple of $\lambda_g/2$ or an odd multiple of $\lambda_g/4$ showing series or parallel type of resonances. Figure (17.10) shows four basic structures of the line resonators with voltage and current distribution. It also shows equivalent lumped resonator models around the resonance frequency.

Case-I: Both Ends Short-Circuited $\lambda_g/2$ Line

The $\lambda_g/2$-line section, short-circuited at both ends, is shown in Fig (17.10a). It also shows the magnitude of the voltage and current standing-waves for the first two modes n = 1, 2. As both ends are short-circuited, the voltage is zero at both ends, whereas the current is maximum at both ends. At the fundamental resonance frequency (n = 1), the line length is $\ell = \lambda_g/2$. In the case of the TEM mode line in the air medium, the guided wavelength $\lambda_g = \lambda_0$; otherwise, λ_g is influenced by the material medium, and also by the geometry of the planar line. The left-hand side port is excited by a current loop, creating the magnetic field that induces the exciting current at the input.

The input impedance of a short-circuited lossy line section is obtained from subsection (2.1.7) of chapter 2:

$$\begin{aligned} Z_{in} &= Z_0[\tanh(\alpha + j\beta)\ell] \\ &= Z_0\left[\frac{\sinh(\alpha\ell)\cosh(j\beta\ell) + \cosh(\alpha\ell)\sinh(j\beta\ell)}{\cosh(\alpha\ell)\cosh(j\beta\ell) + \sinh(\alpha\ell)\sinh(j\beta\ell)}\right]. \end{aligned} \tag{17.3.1}$$

On using the identity, $\cosh(jx) = \cos x$, $\sinh(jx) = j\sin x$, the above expression is reduced to

$$Z_{in} = Z_0\left[\frac{\tanh(\alpha\ell) + j\tan(\beta\ell)}{1 + j\tanh(\alpha\ell)\tan(\beta\ell)}\right]. \tag{17.3.2}$$

A section of the *lossless* line, with $\alpha = 0$, gives the reactive input impedance, or reactive input susceptance. Figure (17.11) shows their behaviors.

$$\begin{aligned} Z_{in} &= jZ_0\tan(\beta\ell) = jX & \text{(a)} \\ Y_{in} &= -jY_0\cot(\beta\ell) & \text{(b)}, \end{aligned} \tag{17.3.3}$$

where β is the propagation constant of the line and ℓ is its length.

For a short-circuited line resonator, the first resonance occurs at $\beta\ell = \pi/2$, i. e. $\ell = \lambda_g/4$. Figure (17.11a) shows

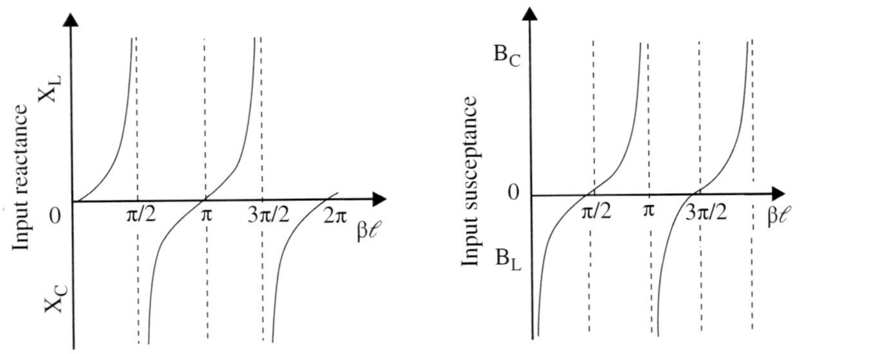

(a) Input reactance w.r.t. electrical length. (b) Input susceptance w.r.t. electrical length.

Figure 17.11 Reactance variation of short-circuited line with frequency $\beta\ell = (2\pi\ell/c)\,f$.

that at the first resonance, the input reactance of the lossless short-circuited line is infinite, $Z_{in} = \infty$. Figure (17.11b) shows that the input susceptance of the $\ell = \lambda_g/4$ line is zero, $Y_{in} = 0$. The behaviors repeat at the odd multiple of $\lambda_g/4$ line length. It is also obvious from the voltage and current distributions on the $\lambda_g/4$-line, shown in Fig (17.10b). To maintain the $Z_{in} = \infty$ at the input, the input must be open-circuited, forming the $\lambda_g/4$ line resonator. The $\lambda_g/4$ line, short-circuited at the output, behaves like a parallel resonant circuit, shown in Figure (17.10b). For the line length $\beta\ell<\pi/2$, i.e. $\ell < (\lambda_g/4)$, $\tan(\beta\ell)$ is positive. Therefore, the short-circuited $\lambda_g/4$-line appears as an *inductor* at the input port.

Figure (17.11a and b) shows that for $\beta\ell = \pi$, i. e. $\ell = \lambda_g/2$, the line resonator has zero input reactance, or infinite input susceptance, i.e. $Z_{in} = 0, Y_{in} = \infty$. The reactance and susceptance repeat themselves at the even multiple of $\lambda_g/2$ line length. Thus, Fig (17.10a) shows that the $\lambda_g/2$ line, shorted at both the input and output, behaves as a *series resonant circuit.* Also for the case $\lambda_g/4 < \ell < \lambda_g/2$, the line section acts as a capacitor.

Let us investigate the behavior of Z_{in} of both ends short-circuited $\lambda_g/2$ line at a frequency slightly away from the resonance frequency ω_r, i.e. at $\omega = \omega_r + \Delta\omega$. The electric length of the line at ω is

$$\beta\ell = n\frac{\omega\ell}{v_p} = n\frac{\omega_r\ell}{v_p} + n\frac{\Delta\omega\ell}{v_p} \quad (n = 1, 2, 3.......). \tag{17.3.4}$$

However, for $\ell = \dfrac{\lambda_g}{2} = \dfrac{\pi}{\beta} = \dfrac{\pi\, v_p}{\omega_r}$

therefore, $\beta\ell = \dfrac{n\,\omega_r}{v_p}\dfrac{\pi v_p}{\omega_r} + \dfrac{n\Delta\omega}{v_p}\dfrac{\pi v_p}{\omega_r}$

$$\Rightarrow \quad \beta\ell = n\pi + n\pi\frac{\Delta\omega}{\omega_r} \qquad \text{(a)}$$

$$\tan\beta\ell = \tan\left(n\pi + n\pi\frac{\Delta\omega}{\omega_r}\right) \tag{17.3.5}$$

$$= \tan\left(n\pi\frac{\Delta\omega}{\omega_r}\right) \approx n\pi\frac{\Delta\omega}{\omega_r} \qquad \text{(b).}$$

Using the above relation, the input impedance of a short-circuited low-loss $[\tanh(\alpha\ell) \approx \alpha\ell]$ line at $\omega = \omega_r + \Delta\omega$ is obtained from equation (17.3.2):

$$Z_{in} = Z_0\left[\frac{\alpha\ell + j(n\pi(\Delta\omega/\omega_r))}{1 + j(n\pi(\Delta\omega/\omega_r))\alpha\ell}\right]. \tag{17.3.6}$$

However, as $n\pi(\Delta\omega/\omega_r)(\alpha\ell) << 1$, so the input impedance of a short-circuited $\lambda_g/2$-line around the resonance frequency is

$$Z_{in} \approx Z_0\,\alpha\ell + jZ_0 n\pi\frac{\Delta\omega}{\omega_r}. \tag{17.3.7}$$

The above expression has the form of the input impedance of a series resonant circuit given by equation (17.2.10a). *Thus, for the line length $\ell = (n\lambda_g/2)$ with both ends short-circuited, the line could be modeled as a series resonant circuit around the resonance frequency.* It is shown in Fig (17.10a). On comparing equations (17.3.7) and (17.2.10a), the equivalent resistance, unloaded Q-factor, equivalent inductance, and capacitance, at the resonance of the $(n\lambda_g/2)$-short-circuited line resonator, are obtained as follows:

$$R = Z_0\alpha\ell \Rightarrow, \quad R = nZ_0\alpha\frac{\lambda_g}{2} \qquad \text{(a)}$$

$$2\frac{\Delta\omega}{\omega_r}RQ_U = nZ_0\pi\frac{\Delta\omega}{\omega_r}$$

$$\Rightarrow Q_U = \frac{nZ_0\pi}{2R} = \frac{nZ_0\pi}{2nZ_0\alpha(\lambda_g/2)} = \frac{\pi}{2\alpha\ell}$$

$$\text{At resonance,} \quad \beta\ell = \pi, \quad Q_U = \frac{\beta}{2\alpha} \qquad \text{(b)}$$

$$L = \frac{Q_U R}{\omega_r} = \frac{\beta}{2\alpha\omega_r}(nZ_0\alpha\ell) = \frac{n\beta\ell}{2\omega_r}Z_0$$

$$\Rightarrow L = n\pi\frac{Z_0}{2\omega_r} \qquad \text{(c)}$$

$$C = \frac{2}{n\pi\omega_r Z_0},\ n = 1, 2, 3... \qquad \text{(d).}$$

(17.3.8)

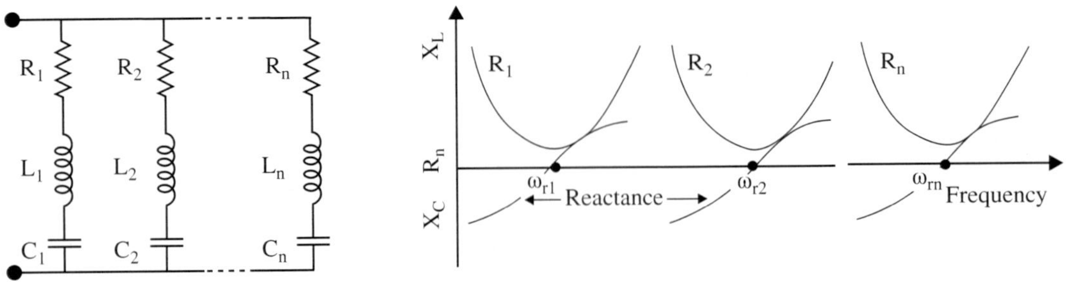

(a) Equivalent circuit. (b) Resonance nature of discrete higher-order modes.

Figure 17.12 Equivalent circuit and higher-order modes of both end short-circuited line resonator.

In the above expression, the Q_U is independent of the modal number (n); although resistance R increases with the modal number. Furthermore, for a fixed-line length ℓ, both ends short-circuited, the line section resonates at n numbers of distinct modes with modal frequency $\omega_{r,n} = (n\,\pi\,v_p/\ell)$. The frequency difference between two modes for a *dispersionless* line is $(\pi\,v_p/\ell)$. Figure (17.12a) shows the equivalent circuit of a line resonator considering n modes. The frequency-dependent input impedances, for discrete modes, are sketched in Fig (17.12b). For each mode loss is accounted for by R_n. The non-resonating modes collectively offer a capacitive reactance. For a lossless line, this is further discussed in the next section (17.3.2). However, for a lossless line section a better understanding is obtained from a complete set of the resonating higher-order modes. It is achieved using the following *Mittag–Leffler* expansion for $\cot\pi x$ [B.14]:

$$\cot\pi x = \frac{1}{\pi x} + \frac{2x}{\pi}\sum_{n=1}^{\infty}\frac{1}{x^2 - n^2}. \qquad (17.3.9)$$

Equation (17.3.3) could be rewritten on replacing $\beta\ell = \omega/v_p \times \lambda_g/2 = \pi\omega/\omega_{r1}$:

$$Z_{in} = \frac{1}{Y_{in}} \qquad (a)$$

$$\text{where} \quad Y_{in} = \frac{\cot[\pi(\omega/\omega_{r1})]}{jZ_0}$$

$$= \frac{1}{jZ_0}\left[\frac{\omega_{r1}}{\pi\omega} + \frac{2\omega}{\pi\omega_{r1}}\sum_{n=1}^{\infty}\frac{1}{(\omega/\omega_{r1})^2 - n^2}\right]$$

$$\Rightarrow \quad Y_{in} = \frac{1}{j\left(\dfrac{\pi Z_0}{\omega_{r1}}\right)\omega} + \frac{1}{j\sum_{n=1}^{\infty}\left[\left(\dfrac{\pi Z_0}{2\omega_{r1}}\right)\omega - \left(\dfrac{n^2\pi Z_0\omega_{r1}}{2}\right)\dfrac{1}{\omega}\right]} \qquad (b).$$

(17.3.10)

The expression (17.3.10a) can be identified with the shunt-connected series resonant circuit, and an additional shunt-connected inductor:

$$Y_{in} = \frac{1}{Z_{ext}} + \frac{1}{Z_n} = \frac{1}{j\omega L_0} + \frac{1}{j(\omega L_n - 1/\omega C_n)}. \qquad (17.3.11)$$

On comparing the above two equations, the following circuit elements of the lumped equivalent circuit are obtained:

$$L_0 = \frac{\pi Z_0}{\omega_{r1}} \quad (a), \quad L_n = \frac{\pi Z_0}{2\omega_{r1}} \quad (b)$$

$$C_n = \frac{2}{n^2\pi Z_0\omega_{r1}} \quad (c) \quad (17.3.12)$$

$$\omega_{rn} = \frac{1}{\sqrt{L_n C_n}} = n\omega_{r1} \quad (d).$$

The above results are compared against equation (17.3.8) that is applicable to the single-mode operation and has been extended to the nth mode by introducing modal number n. However, equation (17.3.8) does not provide harmonically related resonance frequency of the higher-order modes. Otherwise, expressions for the L and C for the fundamental mode (n = 1) are identical to those given in the above equation. The above equation introduces an additional shunt inductance L_0 that is external to the resonant circuit. Normally it is associated with the Foster-type circuit [B.11]. It could be absorbed in the feed-network of the resonator. The loss resistance, in series with the series resonant circuit, could be added and computed using equation (17.3.8a). The *Mittag–Leffler* expansion can be further applied to the other three cases also.

Case-II: Short-Circuited ($\lambda_g/4$) Transmission Line

Figure (17.10b) shows the equivalent circuit of the output-end short-circuited $\lambda_g/4$-line resonator. It resonates while the input is open-circuited. The resonator is excited by a voltage source, i.e. through the electric field coupling at the input. The fundamental mode oscillation occurs at the line length $\ell = \lambda_g/4$. The line resonator of higher-order resonant modes has electrical length $\ell = (2n-1)\lambda_g/4$. The input impedance of a lossy line is written from equation (17.3.2):

$$Z_{in} = Z_0\left[\frac{1 - j\tanh(\alpha\ell)\cot(\beta\ell)}{\tanh(\alpha\ell) - j\cot(\beta\ell)}\right] \quad (a)$$

$$\text{For a lossless line } (\alpha = 0): Z_{in} = jZ_0\tan(\beta\ell) \quad (b). \qquad (17.3.13)$$

At the resonance frequency ω_r,

$$\ell = (2n-1)\lambda_g/4, \text{ and } Z_{in} = \frac{Z_0}{\tanh(\alpha\ell)} \approx \frac{Z_0}{\alpha\ell}. \qquad (17.3.14)$$

For a low-loss line, $\tanh(\alpha l) \approx \alpha l$ and the lossless line has $\alpha = 0$; $Z_{in} \to \infty$. Figure (17.10b) shows that the short-circuited $\lambda_g/4$-line behaves like a *parallel resonant circuit.* Additional $\lambda_g/4$-line length converts the short-circuited $\lambda_g/4$-line to a series resonant circuit of the case-1. The input impedance around ω_r, i.e. at $\omega = \omega_r + \Delta\omega$, is

$$\beta\ell = \frac{\omega_r\ell}{v_p} + \frac{\Delta\omega\,\ell}{v_p} = (2n-1)\left[\frac{\pi}{2} + \frac{\pi}{2}\frac{\Delta\omega}{\omega_r}\right] \quad (a)$$

$$\cot(\beta\ell) = \cot\left[(2n-1)\frac{\pi}{2} + (2n-1)\frac{\pi}{2}\frac{\Delta\omega}{\omega_r}\right]$$

$$\Rightarrow \cot(\beta\ell) \approx -\tan\left[(2n-1)\frac{\pi}{2}\frac{\Delta\omega}{\omega_r}\right]$$

$$\approx -(2n-1)\frac{\pi}{2}\frac{\Delta\omega}{\omega_r} \quad (b). \qquad (17.3.15)$$

$$Z_{in} = Z_0\left[\frac{(1 + j(2n-1)\alpha\ell(\pi\Delta\omega/2\omega_r)}{\alpha\ell + j(2n-1)(\pi\Delta\omega/2\omega_r)}\right]$$

$$\text{as } \frac{(\alpha\ell)\pi\Delta\omega}{2\omega_r} << 1 \qquad (17.3.16)$$

$$\Rightarrow Z_{in} = \frac{Z_0/\alpha\ell}{1 + j(2n-1)(\pi\Delta\omega/2\alpha\ell\omega_r)}.$$

On comparing equations (17.2.19b) and (17.3.16), and on using $\beta\ell = (2n-1)(\pi/2)$ at resonance, the components R, C, L of the *equivalent parallel resonant circuit* are obtained as follows:

$$R = \frac{Z_0}{\alpha\ell} = \frac{4Z_0}{(2n-1)\alpha\ell\lambda_g} \quad (a)$$

$$2Q_U\frac{\Delta\omega}{\omega_r} = (2n-1)\frac{\pi}{2}\frac{1}{\alpha\ell}\frac{\Delta\omega}{\omega_r}$$

$$\Rightarrow Q_U = (2n-1)\frac{\pi}{4}\frac{1}{\alpha\ell} = \frac{\beta}{2\alpha} \quad (b)$$

$$C = \frac{Q_U}{\omega_r R} = \frac{(2n-1)\pi}{4\omega_r Z_0} \quad (c)$$

$$L = \frac{1}{\omega_r^2 C} = \frac{4Z_0}{(2n-1)\pi\omega_r} \quad (d). \qquad (17.3.17)$$

Case-III: Both Ends Open-Circuited $\lambda_g/2$ Line

Figure (17.10c) shows both ends open-circuited line resonator. It also shows the voltage and current distributions for the first two modes. The voltage is maximum at both ends, while the current is zero at both ends for all resonating modes. The resonator is voltage excited, i.e. the electric field excited. For the open-circuited output port, the input impedance is

$$Z_{in} = Z_0\coth(\alpha\ell + j\beta\ell) \quad (a)$$

$$\text{For a lossless line}: Z_{in} = -jZ_0\cot(\beta\ell) \quad (b). \qquad (17.3.18)$$

At the ω_r, line length is $\ell = (n\lambda_g/2)$ and for a low-loss line, $\tanh(\alpha\ell) \approx \alpha\ell$. At frequency $\omega = \omega_r + \Delta\omega$, the Z_{in} is evaluated as follows:

$$\text{On using,} \quad \beta\ell = n\left(\pi + \pi\frac{\Delta\omega}{\omega_r}\right)$$

$$\Rightarrow \tan(\beta\ell) \approx n\pi\frac{\Delta\omega}{\omega_r}, \quad Z_{in} = \frac{Z_0/\alpha\ell}{1 + j\left[\frac{n\pi}{\alpha\ell}\frac{\Delta\omega}{\omega_r}\right]}. \qquad (17.3.19)$$

On comparing equations (17.3.19) and (17.2.19b), the following expressions of the circuit elements of an equivalent *parallel resonant* circuit, shown in Fig (17.10c), are obtained:

$$R = \frac{Z_0}{\alpha\ell} = \frac{2Z_0}{n\alpha\lambda_g} \quad (a), \quad 2Q_U\frac{\Delta\omega}{\omega_r} = \frac{n\pi}{\alpha\ell}\frac{\Delta\omega}{\omega_r}$$

At resonance $\beta\ell = n\pi$,

$$Q_U = \frac{n\pi}{2\alpha\ell} = \frac{\beta\ell}{2\alpha\ell} = \frac{\beta}{2\alpha} \quad (b)$$

$$C = \frac{Q_U}{\omega_r R} = \frac{n\pi}{2\omega_r Z_0} \quad (c), \quad L = \frac{1}{\omega_r^2 C} = \frac{2Z_0}{n\pi\omega_r} \quad (d). \qquad (17.3.20)$$

In the range $\ell < (\lambda_g/4)$, the open-circuited line behaves as a capacitor. In the range $(\lambda_g/4 < \ell < \lambda_g/2)$, it behaves as an inductor.

Case-IV: Open-Circuited $\lambda_g/4$ Line

Figure (17.10d) shows the $\lambda_g/4$-line resonator with its output end open-circuited and the input end short-circuited. This line resonator could be excited by the external magnetic field. It also shows the voltage and current distribution. Its input impedance is written from equations (17.3.18a) and (17.3.15b),

$$Z_{in} = Z_0\left[\frac{\tanh\alpha\ell - j\cot\beta\ell}{1-j\tanh\alpha\ell\ \cot\beta\ell}\right] \approx Z_0\left[\frac{\alpha\ell - j\cot\beta\ell}{1-j\alpha\ell\cot\beta\ell}\right]$$

$$\approx Z_0\left[\frac{\alpha\ell + j[(2n-1)\pi/2]\Delta\omega/\omega_r}{1-j\alpha\ell[(2n-1)\pi/2]\Delta\omega/\omega_r}\right]$$

$$\text{As } (2n-1)\frac{\pi\alpha\ell}{2}\frac{\Delta\omega}{\omega_r} << 1$$

$$Z_{in} = \alpha\ell\, Z_0\left[1 + j(2n-1)\frac{\pi}{2\alpha\ell}\frac{\Delta\omega}{\omega_r}\right]. \tag{17.3.21}$$

The above expression shows that the open-circuited line resonator, around the resonance frequency, $\omega = \omega_r + \Delta\omega$, could be modeled by the *lumped series resonant circuit*. On using $\beta\ell = (2n-1)\pi/2$ at the resonance, the circuit components of the equivalent series resonant circuit, shown in Fig (17.10d), are obtained below on comparing the above equation against equations (17.2.10a):

$$R = \alpha\ell Z_0 = \left(\frac{2n-1}{4}\right)Z_0\alpha\lambda_g \qquad (a)$$

$$2Q_U R\frac{\Delta\omega}{\omega_r} = \left(\frac{2n-1}{2}\right)\frac{\pi}{2}\frac{\Delta\omega}{\omega_r}Z_0$$

$$Q_U = \pi\left(\frac{2n-1}{4}\right)\frac{1}{\alpha\ell Z_0}Z_0 = \frac{\beta}{2\alpha} \qquad (b)$$

$$L = \frac{Q_U R}{\omega_r}, \quad \Rightarrow \quad L = \frac{(2n-1)\pi}{4}\frac{Z_0}{\omega_r} \qquad (c)$$

$$C = \frac{1}{\omega_r^2 L} = \frac{4}{(2n-1)\pi}\frac{1}{\omega_r Z_0} \qquad (d).$$

$$(17.3.22)$$

For lossless line $\alpha = 0$, and the reactive input impedance is

$$Z_{in} = Z_0\coth(j\beta\ell) = -j Z_0\cot(\beta\ell). \tag{17.3.23}$$

For $\ell < \lambda_g/4$, Z_{in} is capacitive. Below resonance, the $\ell < \lambda_g/4$ long open-circuited line resonator behaves as a *capacitor*.

17.3.2 Modal Description of $\lambda_g/2$ Short-Circuited Line Resonator

This subsection presents a systematic *modal analysis* of the $\ell = n\lambda_g/2$ short-circuited line resonator, shown in Fig (17.13) [B.15]. The resonator is excited by an external impressed voltage V^i, applied at the input port located at $z = 0$. The sum of the impressed voltage V^i and line voltage $V(z)$ is zero at $z = 0$ leading to the short-circuited condition at the input. The present resonator is the case-I of the previous subsection. The nth mode resonates at $\omega_{r,n} = n\omega_r$, where ω_r is the fundamental mode resonance frequency. However, the modal frequencies are not harmonically related, i.e. $\omega_{r,n} \neq n\omega_r$, for a dispersive line resonator. The previous discussion does not account for the capacitive effect of the non-resonating modes toward the computation of the input impedance Z_{in}. The modal analysis accounts for it.

The resonator is governed by the Kelvin–Heaviside transmission line equations. The related voltage and current wave equations are discussed in subsection (2.1.4) of chapter 2. For a lossless line, equations are summarized below:

$$\frac{dV}{dz} = -j\omega LI \quad (a), \qquad \frac{dI}{dz} = -j\omega CV \quad (b)$$

$$\frac{d^2V}{dz^2} + \beta^2 V = 0 \quad (c), \qquad \frac{d^2I}{dz^2} + \beta^2 I = 0 \quad (d), \tag{17.3.24}$$

where C and L are line capacitance and line inductance *p.u.l.* and β is the propagation constant, $\beta = \omega\sqrt{LC} = \omega/v_p$. The voltage wave equation, also the current wave equation, is solved under the boundary conditions, $V(z = 0) = V(z = \ell) = 0$ as both ends of the line are short-circuited. The solution is expressed as a summation of the discrete voltage and discrete current modes using the Fourier expansion:

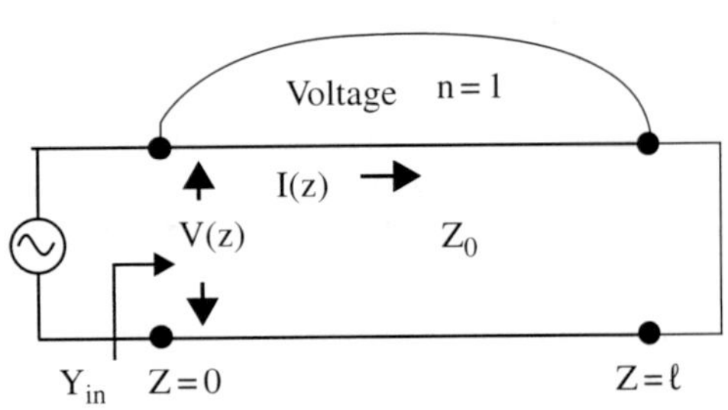

(a) Output short-circuited. In absence of V^i input is short-circuited.

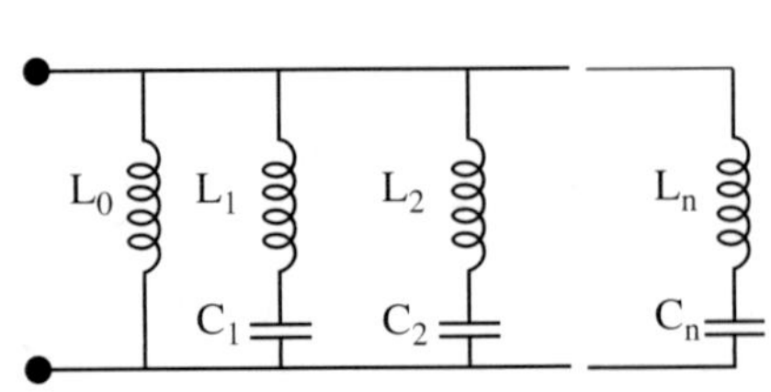

(b) Modal equivalent circuit.

Figure 17.13 Both ends short-circuited line with an equivalent circuit.

$$V(z) = \sum_{n=1}^{\infty} V_n \sin\left(\frac{n\pi}{\ell} z\right) \quad \text{(a)}$$

$$I(z) = I_0 + \sum_{n=1}^{\infty} I_n \cos\left(\frac{n\pi}{\ell} z\right), \quad 0 \le z \le \ell \quad \text{(b)}$$

$$\text{where } \beta = \frac{2\pi}{\lambda_g} = \frac{\pi}{\lambda_g/2} = \frac{\pi}{\ell} \quad \text{(c).}$$

(17.3.25)

The amplitudes of current and voltage modes are,

$$I_0 = \frac{1}{\ell}\int_0^{\ell} I(z)dz \quad \text{(a)}, \quad I_n = \frac{2}{\ell}\int_0^{\ell} I(z)\cos\left(\frac{n\pi}{\ell} z\right)dz \quad \text{(b)}$$

$$V_n = \frac{2}{\ell}\int_0^{\ell} V(z)\sin\left(\frac{n\pi}{\ell} z\right)dz \quad \text{(c).}$$

(17.3.26)

As impressed voltage V_i is applied at $z = 0$, the voltage at the input port is not zero, i.e. $V(0) = V_i$. The above coefficients of Fourier series are evaluated as follows, on substituting $I(z)$ and $V(z)$ from equation (17.3.24):

$$I_0 = -\frac{1}{j\omega L\ell}\int_0^{\ell} \frac{dv}{dz} dz = -\frac{1}{j\omega L\ell}[V(\ell) - V(0)]$$

$$\Rightarrow I_0 = \frac{V_i}{j\omega L\ell} \quad \text{(a)}$$

$$I_n = -\frac{2}{j\omega L\ell}\int_0^{\ell} \frac{dv}{dz}\cos\left(\frac{n\pi}{\ell} z\right)dz \quad \text{(b).}$$

(17.3.27)

On using integration by parts,

$$\Rightarrow I_n = -\frac{2}{j\omega L\ell}\left[\{V(\ell)\cos n\pi - V(0)\} + \frac{n\pi}{\ell}\int_0^{\ell} V(z)\sin\left(\frac{n\pi}{\ell} z\right)dz\right] = \frac{2}{j\omega L\ell}\left[V_i - \frac{n\pi}{2}V_n\right]. \quad (17.3.28)$$

$$\text{Likewise, } V_n = -\frac{2}{j\omega C\ell}\int_0^{\ell} \frac{dI}{dz}\sin\left(\frac{n\pi}{\ell} z\right)dz$$

$$\Rightarrow V_n = -\frac{2}{j\omega C\ell}\left[\{I(\ell)\sin(n\pi) - I(0)\sin(0)\} - \frac{n\pi}{\ell}\int_0^{\ell} I(z)\cos\left(\frac{n\pi}{\ell} z\right)dz\right] = \frac{n\pi I_n}{j\omega C\,\ell}. \quad (17.3.29)$$

The amplitude of the modal current, excited by the *impressed* voltage source V_i, is obtained from the above equations:

$$I_n = \frac{V_i}{(j\omega L\ell/2) + (n^2\pi^2/2j\omega C\ell)}, \quad n = 1, 2, 3, \ldots$$

(17.3.30)

The input current is evaluated at $z = 0$ to get the input admittance due to all excited modes:

$$I(z=0) = \sum_{n=0}^{\infty} I_n = I_0 + \sum_{n}^{=} I_n$$

$$= \frac{V_i}{j\omega L\,\ell} + \sum_{n=1}^{\infty}\left[\frac{V_i}{(j\omega L\ell/2) + (n^2\pi^2/2j\omega C\ell)}\right].$$

(17.3.31)

The input admittance is,

$$Y_{in} = \frac{I(z=0)}{V_i} = \frac{1}{j\omega L\,\ell} + \sum_{n=1}^{\infty}\left[\frac{1}{j\left\{\omega\left(\frac{L\ell}{2}\right) - \left(\frac{n^2\pi^2}{2C\ell}\right)\frac{1}{\omega}\right\}}\right].$$

(17.3.32)

The first term is the admittance of an inductance $L\ell$ that is external to the resonator. The next term shows the modal inductance and modal capacitance for the equivalent series resonant circuit connected in the shunt. Hence, Y_{in} is the admittance of the lossless resonating network, as shown in Fig (17.13b). The above equation provides the following circuit elements of the modal equivalent network:

$$L_0 = L\ell \quad (a), \quad L_n = \frac{1}{2} L\ell \quad (b)$$

$$C_n = \frac{2}{n^2\pi^2} C\ell \quad (c). \tag{17.3.33}$$

It could be shown that the above equation is identical to equation (17.3.12). At the modal resonance frequency, the input impedance is zero, i.e. the input admittance is infinite. The following expressions for the modal frequencies are obtained:

$$\omega_n = \frac{1}{\sqrt{L_n C_n}} = \frac{n\pi}{\ell\sqrt{LC}} = n\omega_1 \quad (a)$$

$$\ell = n\frac{\lambda_g}{2} \quad (b). \tag{17.3.34}$$

The expression for the Y_{in} could be rewritten in terms of the modal frequencies:

$$Y_{in} = \frac{1}{j\omega L\ell} + \sum_{n=1}^{\infty} \frac{2j\omega}{L\ell(\omega_{r,n}^2 - \omega^2)}. \tag{17.3.35}$$

The above series is slowly convergent. Its convergence can be improved by recasting the above equation. For this purpose, the input admittance is determined at a frequency below the fundamental resonance frequency by taking the limiting case:

$$\text{For } \omega \to 0, Y_{in} \to \frac{1}{j\omega L\ell} + \frac{2j\omega}{L\ell}\sum_{n=1}^{\infty}\frac{1}{\omega_{r,n}^2}$$

$$= \frac{1}{j\omega L\ell} + \frac{2j\omega C\ell}{\pi^2}\sum_{n=1}^{\infty}\frac{1}{n^2} \quad (a)$$

$$\text{Since, } \sum_{n=1}^{\infty}(1/n^2) = \pi^2/6,$$

$$Y_{in} = \frac{1}{j\omega L\ell} + \frac{j\omega C\ell}{3}, \quad \text{for } \omega \to 0. \quad (b)$$

$$\text{i.e. } \frac{j\omega C\ell}{3} = \frac{2j\omega}{L\ell}\sum_{n=1}^{\infty}\frac{1}{\omega_{r,n}^2} \quad (c). \tag{17.3.36}$$

Because of the above, equation (17.3.35) is rewritten as follows:

$$Y_{in} = \frac{1}{j\omega L\ell} + \frac{j\omega C\ell}{3} + \frac{2j\omega}{L\ell}\sum_{n=1}^{\infty}\left[\frac{1}{\omega_{r,n}^2 - \omega^2} - \frac{1}{\omega_{r,n}^2}\right]$$

$$\Rightarrow Y_{in} = \frac{1}{j\omega L\ell} + \frac{j\omega C\ell}{3} + \sum_{n=1}^{\infty}\left[\frac{2j\omega^3}{L\ell\,\omega_{r,n}^2(\omega_{r,n}^2 - \omega^2)}\right]. \tag{17.3.37}$$

The above expression for Y_{in} is a fast convergent series. A loss resistance R is introduced in the modal equivalent circuit, as discussed previously. The method can be extended to the other three cases of line resonators.

References

Books

B.1 Waldron: *Waves and Oscillations*, D.Van Nostrand Co., Princeton, 1964.

B.2 Okoshi, T.: *Planar Circuits for Microwaves and Lightwaves*, Springer-Verlag Berlin Heidelberg, 1985.

B.3 Sharma, A.; Khanna, A.P.S.: Resonators, *Microwave Solid State Circuit Design*, Bahl Inder, Bhatia P., John Wiley & Sons, New York, 1989.

B.4 Bahl, I.J.: *Lumped Elements for RF and Microwave Circuits*, Artech House, Boston, 2003.

B.5 Nguyen, C.: *Radio-Frequency Integrated-Circuit Engineering*, John Wiley & Sons, New York, 2015.

B.6 Chang, K.; Hsieh, L.H.: *Microwave Ring Circuits and Related Structures*, 2nd Edition, John Wiley & Sons, New York, 2004.

B.7 Caloz, C.; Itoh, T.: *Electromagnetic Metamaterials: Transmission Line Theory and Microwave Applications*, John Wiley & Sons, New York, 2006.

B.8 Marqués, R.; Martín, F.; Sorolla, M.: *Metamaterials with Negative Parameters: Theory, Design and Microwave Applications*, John Wiley & Sons, New York, 2007.

B.9 Rizzi, P.A.: *Microwave Engineering-Passive Circuits*, Prentice-Hall International Edition, Englewood Cliff, NJ, 1988.

B.10 Pozar, D.M.: *Microwave Engineering*, 2nd Edition, John Wiley & Sons, Singapore, 1999.

B.11 Simon, R.; Whinnery John R.; Theodore, V.D.: *Fields, and Waves in Communication Electronics*, 3rd Edition, John Wiley & Sons, Singapore, 1994.

B.12 Collin, R.E.: *Foundations for Microwave Engineering*, 2nd Edition, McGraw-Hill, Inc., New York, 1992.

B.13 Ludwig, R., Bretchko, P.: *RF Circuit Design*, Pearson Education Asia, New Delhi, 2000.

B.14 Weber, H.J.; Arfken, G.B.: *Essential Mathematical Methods for Physicist*, Academic Press, NewYork, 2003.

B.15 Schelkunoff, S.A.: *Electromagnetic Field*, Blaisdell Publishing Co., New York, 1963.

Journals

J.1 Smith, N.A.: *Microstrip-fed circular substrate integrated waveguide (SIW) cavity resonator and antenna*, IEEE Antenna, and Propagation. Society Inter. Sym.; APSURSI, 2010.

J.2 El-Tager, A.; Bray, J.; Roy, L.: *High-Q LTCC resonators for millimeter wave applications*, IEEE MTT-S International Microwave Sym. Digest, Vol. 3, 2003.

J.3 Khanna, A.P.S.; Garault, Y.: *Determination of loaded, unloaded and external Quality factors of a dielectric resonator coupled to a microstrip line, IEEE Trans. Microw. Theory Tech.*, Vol. **MTT-31**, No. 3, pp. 261–263, 1983.

J.4 Chang, C.C.; Caloz, C.; Itoh, T.: *Analysis of a compact slot resonator in the ground plane for microstrip structures*, Proc. APMC, pp. 1100–1103, Taipei, Taiwan, 2001.

J.5 Verma, A.K.; Kumar, A.: *Synthesis of microstrip lowpass filter using defected ground structures, IET Microw. Antenna Propagat.*, Vol. **5**, No. 12, pp. 1431–1439, Sept. 2011.

18

Planar Resonating Structures

Introduction

The basic theory and characteristics of the lumped element resonators and the line resonators are discussed in chapter 17. The resonating structures are characterized using four important parameters – *compactness, resonance frequency, Q-factor, and appearance of next resonating mode*. These are emphasized in this chapter while discussing various forms of the planar resonating structures. The planar resonators, in the formats of the microstrip, CPW, and slot line resonators, are very important circuit elements of the planar microwave technology. The microwave planar resonators have been developed in all planar technologies – PCB-based MIC, MMIC, MEMS, and LTCC. These structures are rigorously analyzed using the full-wave methods, and also using the EM-Simulators. However, for the design purpose, these are also modeled using the lumped resonators and line resonators, discussed in chapter 17. Apart from their usual applications in the design of filters, frequency discriminators, duplexers, oscillators, etc., these are also used to develop the DGS, PBG, and metamaterials structures. These topics are discussed in chapters 19–22.

This chapter presents some forms of the planar line resonators and their modeling. It also discusses in brief the 2D-patch resonators, fractal resonators, and dual-mode resonators. The circuit modeling of these planar resonators and their characteristics are presented.

Objectives

- *To discuss the characteristics and modeling of the microstrip, CPW, and slot line resonators.*
- *To present modeling of the step impedance, hairpin, and ring resonators.*
- *To discuss the coupling of the resonators with the external circuits.*
- *To discuss in brief the characteristics of the 2D-patch resonators of several geometrical shapes.*
- *To discuss in brief the basic properties of the fractal geometry.*
- *To discuss modeling of the line and patch fractal resonators.*
- *To present modeling of the dual-mode resonators.*

18.1 Microstrip Line Resonator

The microstrip line resonators are commonly used as the filter elements. These are also used to measure the relative permittivity of a substrate [J.1, J.2]. Figure (18.1a–i) shows some forms of the microstrip line resonators. Figure (18.1a) shows a very common *open-ended* $\lambda_g/2$-*microstrip uniform impedance resonator (UIR)*. Both open ends radiate power, thereby reducing its Q-factor. The Q-factor is improved by using the *shorting pins* at both ends, as shown in Fig (18.1b) [J.3]. The shorting pins connect the microstrip to the ground plane. However, the shorting pins, or *vias*, add inductances at both ends. Figure (18.1c) shows the $\lambda_g/4$-microstrip line resonator. It is developed by shorting *one end* of the microstrip with the help of a pin. The $\lambda_g/4$-resonator is used in the interdigital and comb-line filter designs. Several modifications in a microstrip line resonator are available. Figure (18.1d) shows a *spur microstrip line resonator*. The spur, i.e. a microstrip with a slot, acts as a stub. It behaves as a notch filter to suppress higher-order harmonic component with the wavelength approximately equal to four times of the stub length.

Figure (18.1e) shows another modification of the microstrip line *uniform impedance resonator* (UIR). It is called the *step impedance resonator* (SIR). If both ends are open-circuited, resonance occurs at the line length $\lambda_g/2$. One end of the SIR could be short-circuited to get the $\lambda_g/4$-resonator. The SIR pushes the higher-order spurious resonance to the higher frequency end, thereby improving the out-of-band frequency response of a

Introduction to Modern Planar Transmission Lines: Physical, Analytical, and Circuit Models Approach, First Edition. Anand K. Verma.

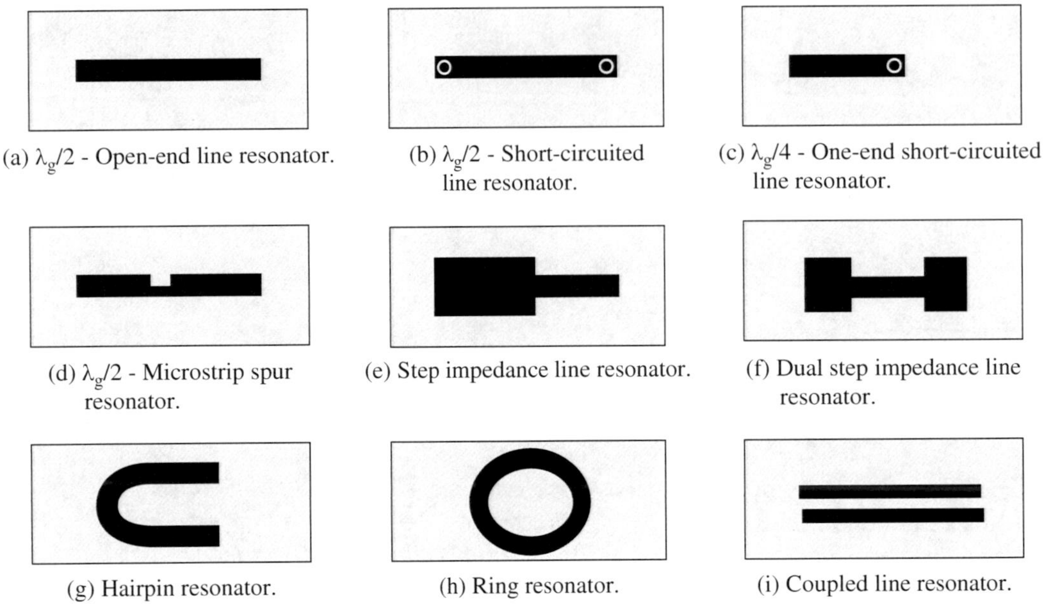

Figure 18.1 Some popular forms of microstrip resonators.

bandpass filter. However, it increases the in-band insertion loss, mainly due to the increased conductor loss of the narrow line section. Figure (18.1f) shows that the step impedances can be added to both ends of a microstrip line to form the *dual step impedance resonator*. The structure can be treated as a *capacitor-loaded* microstrip resonator. It reduces the line length of a $\lambda_g/2$-uniform microstrip resonator. Figure (18.1g) shows that to reduce the size of a straight $\lambda_g/2$-microstrip resonator, it could be bent into the U-shape, forming the microstrip *hairpin resonator*. The hairpin resonator has a higher Q-factor. It is popular in the filter design. The slot and step discontinuities can also be added to the hairpin resonator to develop more compact and versatile filters.

The microstrip line is also bent to form a *ring resonator* [B.1], as shown in Fig (18.1h). Figure (17.2) of chapter 17 shows that a microstrip ring can acquire several shapes. There is no end effect in this case and the radiation loss is less as compared to the open-ended straight microstrip resonator. So, a microstrip ring resonator has a Q-factor higher than a straight microstrip resonator. It is used for the measurement of the relative permittivity of a substrate. Figure (18.1i) shows the *microstrip edge-coupled resonator*. It is useful in the design of a filter. The microwave literature is full of unending modifications in the microstrip resonators. The responses of the complex microstrip resonators and other planar resonators are normally obtained on the EM-simulator to develop new designs. Only a few basic resonating structures are considered below.

18.1.1 $\lambda_g/2$ Open-end Microstrip Resonator

Figure (18.2a) shows a $\lambda_g/2$-open-ended microstrip resonator on a substrate with relative permittivity ε_r, and substrate thickness h [J.1]. Its width is small compared to the guided wavelength, i.e. $w << \lambda_g$, so there is no field variation along the width of the resonator. The fringing electric fields at both ends extend beyond the microstrip resonator, thus increasing its effective length from ℓ to $(\ell + 2\Delta\ell)$.

The approximate resonance frequency $f_{r\,n}$ of the nth mode is computed, ignoring the line extension $2\Delta\ell$, due to the fringing electric fields, by the following expression:

$$\ell = n\frac{\lambda_g}{2} = n\frac{\lambda_r}{2\sqrt{\varepsilon_{r\,eff}(f_r)}} \qquad \text{(a)}$$

$$f_{r\,n} = \frac{n\,c}{2\ell\sqrt{\varepsilon_{r\,eff}(f_r)}}, \quad (n = 1, 2, \ldots) \qquad \text{(b)},$$

(18.1.1)

where n is the modal number, c is the velocity of the EM-wave in the free space, and λ_r is the wavelength at resonance frequency f_r. The effective relative permittivity $\varepsilon_{r\,eff}(f_r)$ is computed at resonance. At first, the approximate resonance frequency is determined for the static effective relative permittivity $\varepsilon_{r\,eff}(f = 0)$. Next, the relative permittivity $\varepsilon_{r\,eff}(f_r)$ is computed at the approximate resonance frequency using the dispersion relation presented in section (8.3) of chapter 8. Once $\varepsilon_{r\,eff}(f_r)$ is obtained, the more accurate resonance frequency is computed for the microstrip line resonator of a given length. To compute resonance frequency more accurately, the extended length, due to the fringing electric fields, is used in replacing the ℓ by $(\ell + 2\Delta\ell)$:

$$f_{rn} = \frac{n\,c}{2\,(\ell + 2\Delta\ell)\sqrt{\varepsilon_{r\,eff}(f_r)}} \quad (a)$$

where,

$$\frac{\Delta\ell}{h} = 0.412\left[\frac{[\varepsilon_{r\,eff}(f=0) + 0.3][(w/h) + 0.262]}{[\varepsilon_{r\,eff}(f=0) - 0.258][(w/h) + 0.813]}\right] \quad (b). \qquad (18.1.2)$$

In the above equation, the line extension $\Delta\ell$ is computed using the expression due to Hammerstad [J.4]. More accurate, but a little complex, expression due to Kirschning, Jansen, and Koster [J.5] can be used for better accuracy of the computed resonance frequency. For the dispersionless case, the frequencies of the higher-order modes are harmonically related, i.e. $f_{rn} = nf_{r1}$, where f_{r1} is the resonance frequency of the fundamental resonating mode at $\ell \approx \lambda_g/2$. However, for a dispersive microstrip resonator, such harmonic relation does not hold.

Figure (18.2b) shows a better *transmission line model* of a microstrip line resonator. It accounts for the radiating open-end through a lossy capacitor, i.e. through a parallel combination of C_f and G_{rad}. The radiation conductance, radiation resistance, fringe capacitance, and fringe susceptance at the open ends are given by the following expression [J.2, J.6, B.2]:

$$G_{rad} = \frac{1}{R_{rad}} \quad (a)$$

$$R_{rad} = \frac{120\,\lambda_r}{1 - (\beta_{ro}h)^2/24} \quad (b)$$

$$C_f = \Delta\ell\left[\frac{\sqrt{\varepsilon_{r\,eff}(f=0)}}{c\,Z_0}\right] \quad (c)$$

$$B_f = \frac{\beta_{ro}\Delta\ell\sqrt{\varepsilon_{r\,eff}(f=0)}}{Z_0} \quad (d), \qquad (18.1.3)$$

where β_{ro} is the wavenumber of a microstrip line at the resonance frequency ω_r, and Z_0 is the characteristic impedance of the microstrip, computed in subsection (8.2.2) of chapter 8. The input admittance at the input port is

$$Y_{in} = G_{rad} + jB_f + Y_0\left[\frac{(G_{rad} + j\,B_f) + j\,Y_0\,\tan\,\beta_g\ell}{Y_0 + j\,(G_{rad} + j\,B_f)\,\tan\,\beta_g\ell}\right] \quad (a)$$

$$\text{where, } \beta_g = \beta_{ro}\sqrt{\varepsilon_{r\,eff}(f=0)}, \quad Y_0 = 1/Z_0 \quad (b). \qquad (18.1.4)$$

At the resonance, the imaginary part of the input admittance $\mathrm{Im}(Y_{in}) = 0$. It leads to the following equation:

$$\tan\beta_g\ell = \frac{2Y_0B_f}{B_f^2 + G_{rad}^2 - Y_0^2}. \qquad (18.1.5)$$

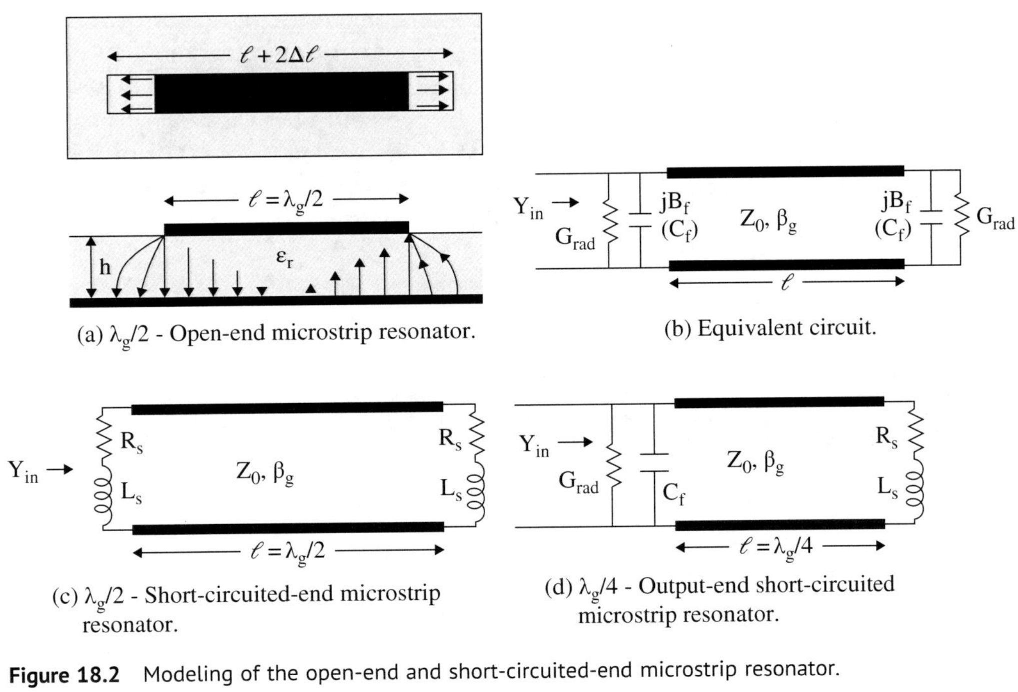

(a) $\lambda_g/2$ - Open-end microstrip resonator.

(b) Equivalent circuit.

(c) $\lambda_g/2$ - Short-circuited-end microstrip resonator.

(d) $\lambda_g/4$ - Output-end short-circuited microstrip resonator.

Figure 18.2 Modeling of the open-end and short-circuited-end microstrip resonator.

For a given line length, the above transcendental equation is solved numerically to get β_g, giving the resonance frequency [B.2, J.6]. In equation (18.1.4), the complex propagation constant γ_g can be used in place of real β_g. The propagation constant γ_g is expressed through the tan-hyperbolic function to include the effect of the attenuation-factor α, due to the conductor and dielectric losses of a microstrip line. The imaginary part of input admittance, Img (Yin) = 0 compute the resonance frequency; whereas the real part of Y_{in} gives the resonant resistance R_{res} at the resonance frequency. The method is also used to find the resonance frequency, and resonant resistance of a $\lambda_g/4$-microstrip resonator.

The total Q-factor of an open-ended $\lambda_g/2$-microstrip resonator is computed by obtaining the individual Q-factors due to the conductor loss (Q_c), dielectric loss (Q_d), surface-wave loss (Q_{sw}), and radiation loss (Q_{rad}). The combined Q-factor, $Q_{d,c}$, due to the combined dielectric and conductor losses, $\alpha_{d,c} = \alpha_d + \alpha_c$, is computed using equation (17.3.22b) of chapter 17. The attenuation factors α_d, α_c and propagation constant β_g are computed from the expressions given in sections (8.4) and (8.2) of chapter 8. The radiation Q-factor Q_{rad} is obtained using the following expression [J.2]:

$$Q_{rad} \approx \frac{3Z_0\varepsilon_{reff}(f_r)}{6032(h/\lambda_r)^2}. \qquad (18.1.6)$$

The quality factor Q_{sw} due to the TM_0 surface-wave mode can be estimated from the following expression [J.7]:

$$Q_{sw} = Q_{rad}\frac{P_{sp}}{P_{sw}}, \qquad (18.1.7)$$

where P_{sp} and P_{sw} are the space-wave radiated power and the power in the surface-wave, respectively. These are computed by the following expressions [J.8]:

$$P_{sp} = \frac{Z_0\beta_{ro}^2(\beta_{ro}h)^2}{3\pi}\left(1-\frac{1}{\varepsilon_r}+\frac{2}{5\varepsilon_r^2}\right) \qquad \text{(a)}$$

$$P_{sw} = \frac{Z_0\beta_{ro}^2}{4}$$

$$\frac{\varepsilon_r(x_0^2-1)}{\varepsilon_r\left[\frac{1}{\sqrt{(x_0^2-1)}}+\frac{\sqrt{(x_0^2-1)}}{\varepsilon_r-x_0^2}\right]+\left[1+\frac{\varepsilon_r^2(x_0^2-1)}{\varepsilon_r-x_0^2}\right](\beta_{ro}h)} \qquad \text{(b)}$$

$$\text{where, } x_0 = 1+\frac{-\varepsilon_r^2+\alpha_0\alpha_1+\varepsilon_r\sqrt{\varepsilon_r^2-2\alpha_0\alpha_1+\alpha_0^2}}{(\varepsilon_r^2-\alpha_1^2)} \qquad \text{(c)}$$

$$\alpha_0 = s\tan(\beta_{ro}hs),\ \alpha_1 = \frac{-1}{s}\left[\tan(\beta_{ro}hs)+\frac{\beta_{ro}hs}{\cos^2(\beta_{ro}hs)}\right] \qquad \text{(d)}$$

$$s = \sqrt{\varepsilon_r-1},\ Z_0 = \sqrt{\mu_0/\varepsilon_0},\ \beta_{ro} = 2\pi/\lambda_r \qquad \text{(e)}$$

$$h \le c/4f_r\sqrt{\varepsilon_r-1} \qquad \text{(f)}.$$

$$(18.1.8)$$

The wavelength λ_r is at the resonance frequency. The total Q-factor is determined by combining all Q-factors, i.e. $1/Q_{total} = 1/Q_d + 1/Q_c + 1/Q_{sur} + 1/Q_{rad}$. The higher-order surface-waves are avoided by the selection of appropriate substrate thickness to satisfy equation (18.1.8f). For a given substrate and characteristic impedance, the Q-factor of a $\lambda_g/2$-resonator can be optimized for its thickness. The optimized Q-factor is low for a high impedance line, due to an increase in the conductor loss. For a given substrate and line impedance, the Q-factor of a line resonator is high at the lower resonance frequency. For instance, it is reduced from about 200 to 50 for an increase in the resonance frequency from 8 GHz to 96 GHz [J.2].

Figure (18.3a and b) shows two interesting variations in the $\lambda_g/2$-open-end microstrip resonators. The first one shows a loop (ring) microstrip resonator, along with its capacitively loaded version. As compared to a standard $\lambda_g/2$-resonator of the same physical size, the loaded loop microstrip resonator has a much lower resonance frequency, depending on the amount of loading. For instance, the resonance frequency of a microstrip

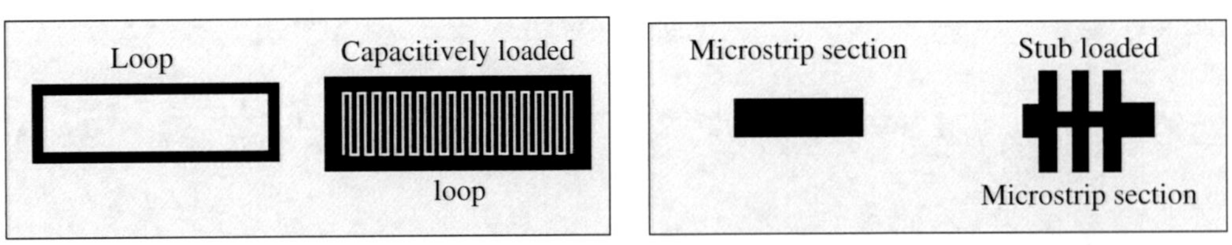

(a) Loop resonator and the capacitively loaded version.

(b) Standard microstrip resonator and its periodically loaded stub version.

Figure 18.3 Periodically loaded microstrip $\lambda_g/2$ resonator.

resonator, on an RT-Duroid substrate with $\varepsilon_r = 2.2$, h = 1.57 mm, is reduced from 4.9 GHz to 3.7 GHz. Thus, the capacitively loaded loop resonator is a compact resonator with a slowing factor 1.33. The capacitive loading of a loop resonator also increases the Q-factor approximately by a factor of 2 [J.9]. Figure (18.3b) shows another variation of a periodically stub-loaded microstrip resonator. On an alumina substrate, the 50 Ω microstrip resonator resonates at 2.5 GHz with an unloaded Q-factor 208. After the periodic loading with stub, the resonance frequency is reduced to 1.94 GHz, with unloaded Q-factor 1372 [J.10].

18.1.2 $\lambda_g/2$ and $\lambda_g/4$ Short-circuited Ends Microstrip Resonator

Figure (18.2c and d) shows the equivalent circuits of the short-circuited $\lambda_g/2$ and $\lambda_g/4$-microstrip resonators. The physical microstrip short-circuited $\lambda_g/2$ and $\lambda_g/4$-microstrip resonators are shown in Fig (18.1b and c), respectively. The shorting-pin is approximately modeled by a lossy shunt inductor. The following expressions are used to evaluate the L_s and R_s in series [B.3]:

$$\frac{L_s}{h} = 0.2\left[\ln\left(\frac{4h}{d}\right) + \left(\frac{d}{4h}\right) - 1\right] \quad \text{nH/mm} \quad (a)$$
$$R_s = 2.63 \times 10^{-3}\frac{h}{d}\sqrt{f_r\frac{\rho}{\rho_{Cu}}} \quad \Omega \quad (b), \tag{18.1.9}$$

where d is the diameter of a shorting-pin, h is the substrate thickness, ρ is the specific resistance of the shorting-pin conductor in Ω-cm. The frequency f_r, in GHz, is approximately computed by using equation (18.1.2a) with $\Delta\ell = 0$. The specific resistance of copper is $\rho_{cu} = 1.72\times10^{-6}$Ω-cm. The above expression is valid for a very small diameter shorting-pin $d/w << 1$, where w is the width of the microstrip conductor. The open-ends can also be short-circuited by adding a conducting strip between the microstrip and the ground plane [B.1].

At this stage, the equivalent circuit models, shown in Fig (18.2c and d), are used to compute the Y_{in} of the $\lambda_g/2$ and $\lambda_g/4$-microstrip resonators. The transcendental equation, obtained from Im $(Y_{in}) = B_{in} = 0$, is solved to get the resonance frequency. The expression Re $(Z_{in} = 1/Y_{in}) = R_{in}$ gives the input resistance of the series/parallel resonant circuits for $\lambda_g/2$ or $\lambda_g/4$-microstrip resonators.

The Q-factors, due to the conductor and dielectric losses, and also due to the surface-wave loss, are computed as discussed above. The radiation Q-factor is computed by the following expression [B.3]:

$$Q_{rad} \approx \frac{15Z_0\varepsilon_{reff}^2(f_r)}{32\eta_0(h/\lambda_r)^2\left[1 + \frac{2}{7\varepsilon_{reff}(f_r)}\right]}, \tag{18.1.10}$$

where η_0 is the intrinsic impedance of free space. The Q-factor of both ends short-circuited microstrip resonator is much higher than that of the Q-factor of the corresponding resonator with the open-circuited ends. For instance, the Q-factor of a 50 Ω open-ends microstrip resonator, at 4.8 GHz on an alumina (Al_2O3) substrate ($\varepsilon_r = 9.9$, $\tan\delta = 0.00025$, $h = 0.05''$), is 185. It increases to 480 for the short-circuited ends [J.3]. Due to its high Q-factor, the short-circuited $\lambda_g/2$-resonator is suitable to measure the dispersion in a microstrip line [J.11].

18.1.3 Microstrip Ring Resonator

The microstrip ring resonators, with several modifications, are widely used as the circuit elements for the development of microwave planar circuits [B.4]. The ring resonators are used to measure the dispersion in a microstrip line [J.12]. Figure (18.1h) shows a microstrip ring resonator. As there is no open-end for a ring resonator, so there is no fringe field or radiation from the open-end [J.13]. It results in a higher unloaded Q-factor for a ring resonator, as compared to that of an open-end linear microstrip resonator. However, the radiation loss is high for a smaller ring. as the radiated power is proportional to $h^2/\lambda_r^2 Z_0$. The high Q-factor can be maintained only for a larger loop [J.14]. To avoid the effect of curvature, the circumference of a ring resonator is at least 5 wavelengths long [J.1].

The nth mode resonance occurs for the resonator length $n\lambda_g$. The resonance frequency of the nth mode is computed below for the mean circumference r_m of a ring resonator:

$$n\lambda_g = 2\pi\, r_m, \Rightarrow \frac{n\,c}{f_{r\,n}\sqrt{\varepsilon_{reff}(f_{r\,m})}} = 2\pi\, r_m$$
$$\Rightarrow f_{r\,n} = \frac{n}{2\pi r_m}\frac{c}{\sqrt{\varepsilon_{r\,eff}(f_{r\,m})}} \quad (a)$$
$$\text{where, } r_m = \frac{r_a + r_b}{2},$$
$$r_a \text{ and } r_b \text{ are inner and outer radius of ring} \quad (b). \tag{18.1.11}$$

In the above equation, c is the velocity of the EM-wave in free space. The fundamental resonance occurs for $n = 1$ at the mean circumference of length $\ell_m = 2\pi r_m = \lambda_g$. At the experimentally known higher-order resonance frequency $f_{r\,n}$, the frequency-dependent effective relative permittivity $\varepsilon_{r\,eff}(f_{r\,n})$ is determined. The above relation provides acceptable results for resonance frequency only for a large radius and narrow width ring, ignoring the effect of curvature. The cavity model, or the full-wave analysis, is used to get more accurate results. The inner and the outer circumferences of a ring resonator provide a longer length of the fringing field. It helps in getting an enhanced interaction of the ring resonator with the externally loaded material. Thus, the ring resonator acts as a better *sensor device* [J.15]. The ring resonator also supports dual-mode operation that is useful for the design of a compact bandpass filter [B.1]. The *dual-mode resonator* is discussed in section (18.8) at the end of this chapter.

Equivalent Circuit Model of Ring Resonator

The transmission line model is used to develop the equivalent circuit model of the *two-port ring resonator*, shown in Fig (18.4a) [J.16–J.18]. It also accommodates the single-port ring resonator. The ring resonator is excited at port #1 using the capacitively coupled microstrip feed. The model, shown in Fig (18.4b), helps to compute the input impedance at the capacitively coupled microstrip feed at the port #1. *The ring is excited at the maximum of the electric field on the ring, and another maximum occurs at the diametrically opposite output port #2.* The port #2 is conceptually treated, using the magnetic-wall (MW), as an open-circuited port. Thus, the ring could be decomposed into two semi-circular parallel-connected rings with output open-circuited. Both rings have the identical characteristic impedance Z_0 and propagation constant β_g. Furthermore, the parallel-connected semi-circular rings at the input port #1, ignoring the effect of curvature, could be treated as the parallel-connected two linear lines of length $\ell_m/2 = \pi r_m$ with characteristic impedance Z_0. The parallel-connected lines form a single line with characteristic impedance $Z_0/2$. The input impedance at the port #1, of the output open-circuited equivalent lossy line, is given below:

$$Z_{in} = \frac{Z_0}{2}\coth\left(\alpha\ell_m/2 + j\,\beta_g\ell_m/2\right)$$
$$= \frac{Z_0}{2}\left[\frac{1 + j\tanh(\alpha\ell_m/2)\tan\left(\beta_g\ell_m/2\right)}{\tanh(\alpha\ell_m/2) + j\tan\left(\beta_g\ell_m/2\right)}\right]. \quad (18.1.12)$$

In the above description, the ring resonator has been finally modeled as parallel-connected two open-circuited transmission lines. The resonance condition of the open-circuit line determines the resonance in the ring resonator. For the line length $\ell_m/2 = (2n-1)\lambda_g/4$, the *series resonance* occurs on the line, and for the line length $\ell_m/2 = n\lambda_g/2$, *parallel resonance* occurs on the line. Therefore, depending on the operating frequency, the mean circumference ℓ_m of the microstrip ring resonator in the above equation could be either odd or even multiple of $\lambda_g/2$, corresponding to the series resonance and parallel resonance, giving the following relations:

$$\text{Series resonance:} \quad \ell_{m\,S} = (2n-1)\lambda_g/2 \quad (a)$$
$$\text{Parallel resonance:} \quad \ell_{m\,P} = n\lambda_g \quad (b). \quad (18.1.13)$$

For the *first case*, the ring resonator behaves like a series resonant circuit, and for the *second case*, it behaves like a parallel resonant circuit. However, at the microstrip feed line located at the port #1, the impedance inversion occurs due to the series-connected gap-capacitance that leads to the inversion of resonance, i.e. the parallel/series resonance in the ring appears as the series/parallel resonance at the source end of coupling gap.

For *the case-I* of the series resonance, the electrical line length is $\beta_g\ell_m/2 = \beta_g\ell_{S\,m}/2 = (2\pi/\lambda_g)\times(2n-1)\lambda_g/4 = (2n-1)\pi/2$, and the Z_{in} is computed from equation (18.1.12):

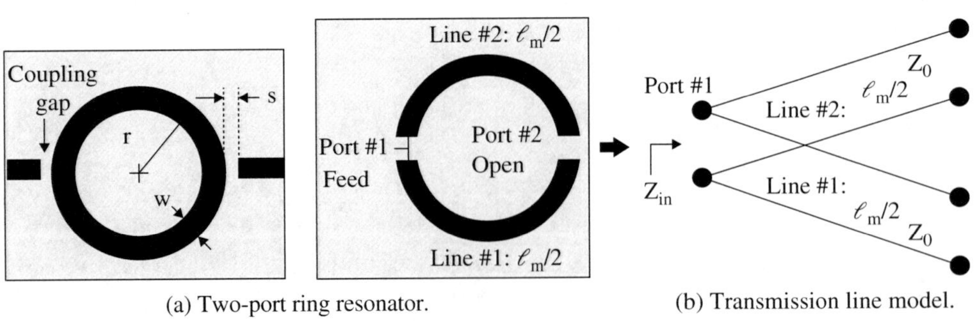

Figure 18.4 Transmission line model of the ring resonator.

$$Z_{in} = \frac{Z_0}{2}\left[\frac{\tanh(\alpha\ell_m/2) - j\cot\left(\beta_g\ell_m/2\right)}{1 - j\tanh(\alpha\ell_m/2)\cot\left(\beta_g\ell_m/2\right)}\right]$$

$$\approx \frac{Z_0}{2}\left[\frac{\alpha\ell_m/2 + j[(2n-1)\pi/2]\,\Delta\omega/\omega_r}{1 - j(\alpha\ell_m/2)\,[\{(2n-1)\pi/2\}\,\Delta\omega/\omega_r]}\right]$$

$$\Rightarrow Z_{in} \approx \frac{Z_0\alpha\ell_m}{4} + j\frac{(2n-1)\pi Z_0}{4}\frac{\Delta\omega}{\omega_r}. \quad (18.1.14)$$

The approximation for cotangent function shown in equation (17.3.15b) of chapter 17 is used in the above equation.

In the case of series resonance, the input impedance is zero. *So, the series resonance for the case-I is not excited by the gap-fed ring resonator.* Following equation (17.3.22) of chapter 17 and carrying out the replacement $\ell \to \ell_m/2$ and $Z_0 \to Z_0/2$, the following expressions are obtained for the R, L, C elements of the equivalent circuit model and also unloaded Q-factor Q_U of the ring resonator supporting the series resonance:

$$R = \frac{Z_0\alpha_c\ell_m}{4} \;(a), \qquad L = \frac{(2n-1)\pi Z_0}{8\omega_r} \;(b)$$

$$C = \frac{1}{\omega_r^2 L} \;(c), \qquad Q_U = \frac{\beta_g}{2\alpha_T} \;(d). \quad (18.1.15)$$

The total attenuation in a ring is due to the conductor, dielectric, radiation, and surface-wave losses, i.e. $\alpha_T = \alpha_c + \alpha_d + \alpha_{rad} + \alpha_{sur}$. The attenuation is taken as only α_c, due to the conductor loss, to compute the resistance R. The losses can be approximately computed by treating the mean circumference of the ring as the linear length of a microstrip. The width of the microstrip line and the ring width are identical. The α_{rad} and α_{sur} can also be computed from the radiation Q-factor of the short-circuited microstrip resonator. More accurate computation is carried using the cavity model [B.5]. However, an EM-Simulator can also be used to determine these parameters from the S-parameter response:

$$\alpha_T = \frac{n\,\pi}{Q_U\ell} \;(a), \quad Q_U = \frac{Q_L}{1 - S_{21}^{res}} \;(b), \quad Q_L = \frac{f_r}{BW|_{3dB}} \;(c). \quad (18.1.16)$$

The *case-II* corresponds to the parallel resonance. In this case, the transmission line model has both ends open-circuited. Under the low-loss condition, following equation (17.3.19) of chapter 17, the approximate expression for the Z_{in}, around the resonance could be written as

$$Z_{in} = \frac{Z_0/2}{\alpha\ell_m/2 + jn\pi\Delta\omega/\omega_r} = \frac{Z_0/\alpha\ell_m}{(1 + j\,(2n\pi/\alpha\ell_m)\,\Delta\omega/\omega_r}. \quad (18.1.17)$$

Again, following equation (17.3.20) of chapter 17 and carrying out the replacement $\ell \to \ell_m/2$ and $Z_0 \to Z_0/2$, the following expressions are obtained for the R, L, C elements of the equivalent circuit model, and also unloaded Q-factor Q_U of the ring resonator supporting the parallel resonance:

$$R = \frac{Z_0}{n\alpha_c\ell_m} \;(a), \quad C = \frac{n\pi}{Z_0\omega_r} \;(b), \quad L = \frac{1}{\omega_r^2 C} \;(c)$$

$$Q_U = \frac{\beta_g}{2\alpha_T} \;(d). \quad (18.1.18)$$

The case-II, i.e. the parallel resonance, is useful in the filter design. However, the input is coupled through the capacitive gap that inverts the parallel resonator to a series resonator. Thus, from the port point of view, the ring resonator of the case-II acts as a series-connected series resonant circuit showing the bandpass behavior. It can also be seen on the Smith chart that provides impedance circle such as shown in Fig (17.6a) of chapter 17. The coupling can be controlled by adjusting the feed gap s, shown in Fig (18.4a).

18.1.4 Microstrip Step Impedance Resonator

Previously discussed *uniform impedance resonator* (UIR) is not a compact line resonator. The harmonic frequencies of a nondispersive $\lambda_g/2$-UIR are $f_{r\,n} = n\,f_r$, $(n = 2, 3, 4, \ldots)$, i.e. the harmonics are separated by f_r. In the case of a $\lambda_g/4$-non-dispersive UIR, the harmonics are $f_{r\,n} = (2n + 1)f_r$, $(n = 1, 2, 3, 4, \ldots)$, i.e. they are separated by $2f_r$. Therefore, the bandpass filter, designed around the $\lambda_g/4$-UIR, is more compact, and its spurious passband is shifted to $2f_r$, as compared to only f_r shifting of the spurious passband of a $\lambda_g/2$-UIR. The design of a compact filter requires a smaller wavelength λ_g. A high relative permittivity substrate is useful for this purpose. It is also achieved by the slow-wave structures, discussed in chapter 19. The SIR has two attractive features as compared against the UIR – (i) compact structure and (ii) pushing upward the spurious harmonic resonance frequencies.

The single-end and both end *step impedance resonators* (SIRs), shown in Fig (18.1e and f) respectively, have lower fundamental resonance frequencies. They also push the spurious resonance frequency up to $5f_r$ [J.19]. Figure (18.5) shows three cases of the SIR, and also two more variations [J.20, J.21]. More variations in the SIR ring resonators are available. The SIR is also used

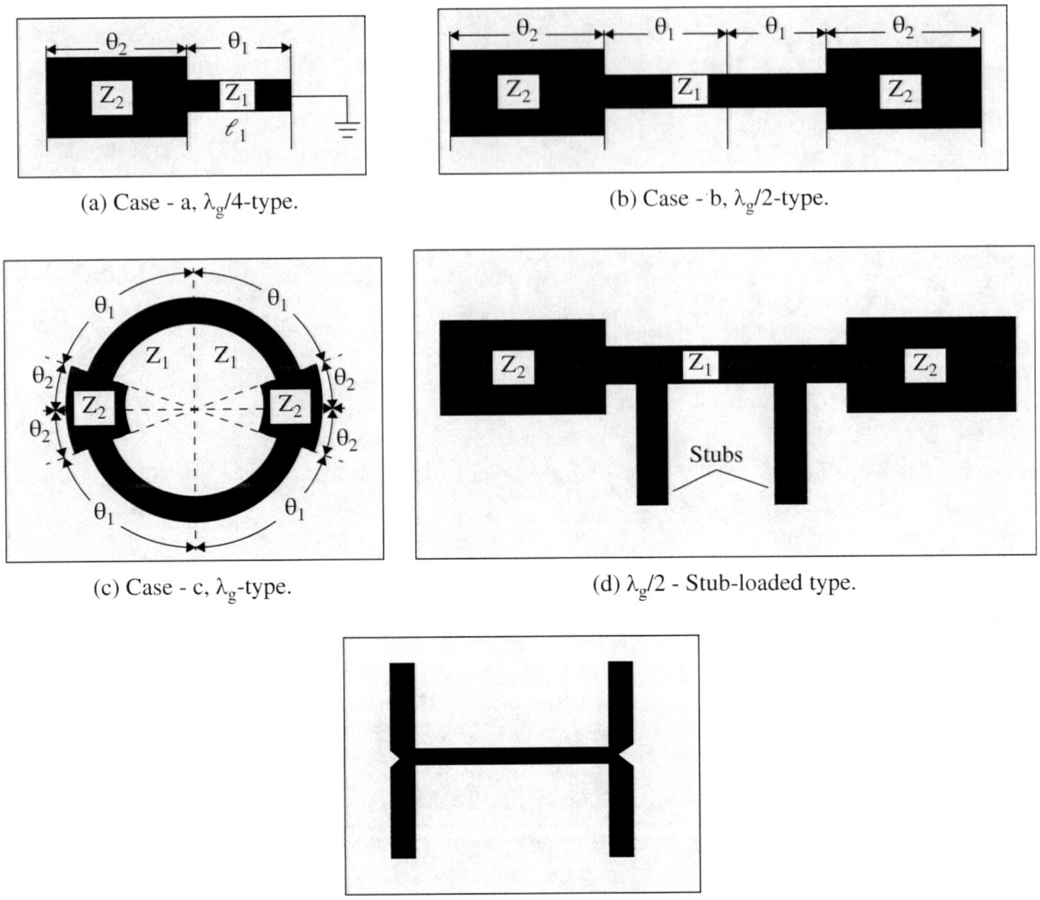

(a) Case - a, $\lambda_g/4$-type. (b) Case - b, $\lambda_g/2$-type. (c) Case - c, λ_g-type. (d) $\lambda_g/2$ - Stub-loaded type. (e) $\lambda_g/2$ - Both ends split type.

Figure 18.5 Step impedance resonators.

with the hairpin resonators. The use of the SIR results in the compact and high-performance microstrip bandpass filters [J.19, J.22–J.26]. The multielement microstrip SIR has also been used to develop a high magnetic field volume RF coil for the MRI applications [J.27].

The $\lambda_g/2$-SIR has both ends open-circuited; however, both ends could also be short-circuited. Figure (18.5b) shows that the $\lambda_g/2$ straight SIR is composed of 2 units of the $\lambda_g/4$-type SIR. Whereas, the λ_g ring-type SIR, shown in Figure (18.5c), is composed of 4 units of the $\lambda_g/4$-type SIR. Figure (18.5d) shows a variation in the $\lambda_g/2$-SIR with two stubs loading. It further improves the compactness of a SIR structure to nearly $1/3^{rd}$ of the UIR resonator. It could be folded to get a compact SIR-type hairpin resonator. A high performance very compact BPF has been developed at 900 MHz using this structure [J.21]. Figure (18.5e) shows yet another variation in the $\lambda_g/2$-SIR, where both end sections are split into half and bent upward. Again this modification provides further compactness, and also adds transmission zeros that attenuate two lowest spurious resonance frequencies [J.28]. The approximate resonance condition can be obtained for all three cases by ignoring the step discontinuities and treating the SIR as cascading of line sections of different characteristic impedances. The simple transmission line model helps to understand the impact of *impedance ratio* (K) of the SIR.

Figure (18.5a) shows the one end short-circuited $\lambda_g/4$-SIR. It is primarily a parallel resonant-type resonator. Its resonance condition is obtained by neglecting the step and open-end discontinuities. The first SIR section, length θ_1, is terminated in a short-circuit and its input admittance is $Y_{in\,1}$. The input admittance of a complete lossless SIR, at the end of the length θ_2, is Y_{in}. These are given below:

$$Y_{in1} = -jY_{01}\cot\theta_1 \qquad \text{(a)}$$

$$Y_{in} = j\frac{[Y_{02}\tan\theta_1\tan\theta_2 - Y_{01}]}{[Y_{01}\tan\theta_2 + Y_{02}\tan\theta_2]} \qquad \text{(b)},$$

(18.1.19)

where electrical lengths of two sections are $\theta_i = \beta_{gi}\ell_i = \beta_0\ell_i\sqrt{\varepsilon_{r\,\text{effi},i}}\ (i = 1, 2)$. Their characteristic admittances and impedances are $Y_{0i} = 1/Z_{0i}$, $Z_{0i} = Z_{0i}(\varepsilon_r = 1)/\sqrt{\varepsilon_{r\,\text{effi},i}}$. At this stage, a lossless SIR structure is considered. However, for a lossy SIR structure, the complex propagation constant $\gamma_g = \alpha + j\beta_g$ is used in place of β_g, and the admittance relation is in the hyperbolic form. At the resonance, Im $(Y_{in}) = 0$, giving the following *odd-mode resonance* condition, corresponding to the line length of odd multiples of the short-circuited $\lambda_g/4$ line:

$$\tan\theta_1 \tan\theta_2 = \frac{Y_{01}}{Y_{02}} = \frac{Z_{02}}{Z_{01}} = K, \qquad (18.1.20)$$

where K is the impedance ratio defined above. For a normal uniform impedance line resonator (UIR), $K = 1$. For $K < 1$, i.e. $Z_{02} < Z_{01}$, the first line is narrow, and the second line is wide. Whereas for $K > 1$, i.e. $Z_{02} > Z_{01}$, the first line is wide, and the second line is narrow. The total electrical line length of the SIR at the fundamental resonance frequency, for the case #a, shown in Fig (18.5a), is

$$\theta_{TA} = \theta_1 + \theta_2 \Rightarrow \quad \theta_{TA} = \theta_1 + \tan^{-1}\left(\frac{K}{\tan\theta_1}\right). \qquad (18.1.21)$$

At the fundamental resonance frequency, the total electrical line lengths for the case #b and case #c, shown in Fig (18.5b and c) are given below:

$$\theta_{TA} = \theta_{TB}/2 \quad (a), \qquad \theta_{TA} = \theta_{TC}/4 \quad (b), \qquad (18.1.22)$$

where θ_{TB} and θ_{TC} are the total electrical lengths of the SIR for case #b and case #c, respectively. The electrical line length of the case #a at the resonance is controlled by the physical length ℓ_1 of line #1, and also by the impedance ratio K. To determine the resonance frequency, another variable u is defined below:

$$u = \frac{\theta_2}{\theta_1 + \theta_2}, \quad \Rightarrow \quad \theta_2 = \frac{u}{1-u}\theta_1. \qquad (18.1.23)$$

The *variable u* is a fractional electrical length of line #2. Equation (18.1.20) is modified as follows:

$$\tan\theta_1 \tan\left[\frac{u}{1-u}\theta_1\right] = K \qquad (a)$$

$$\text{where, } \theta_1 = \frac{2\pi f_r}{c}\sqrt{\varepsilon_{r\,\text{eff}1}(f=0)}\,\ell_1 \qquad (b). \qquad (18.1.24)$$

For the case $u = 0$, $\theta_2 = 0$, the SIR is reduced to UIR with the characteristic impedance Z_1 and length θ_1. For $u = 1$, $\theta_1 = 0$, the SIR is reduced to UIR with the characteristic impedance Z_2 and length θ_2. Figure (18.6) shows the normalized *undesired spurious harmonic resonance frequencies* (f_{spi}, i = 1,2,3) for K = 0.3, 0.59, 1, 1.4 [J.21]. The normalization is with respect to the fundamental resonance frequency f_r. The resonances shown in Fig (18.6) are discussed below.

For the UIR with $K = 1$, the normalized 1st, 2nd, and 3rd harmonic resonance frequencies are given as $(f_{sp\,i}/f_r) = 3, 5, 7$, $(i = 1, 2, 3)$. For $K > 1$, say for $K = 1.4$, the harmonic resonance frequencies of the SIR move below the $f_{sp\,i}/f_r$ of the UIR. However, the deviations in $f_{sp\,i}/f_r$ of SIR are still closer to $f_{sp\,i}/f_r$ of the UIR. For $K < 1$, they move away from the $f_{sp\,i}/f_r$ of the UIR. For K=0.3, the 1st and 3rd spurious harmonics have maximum separation from the spurious frequency (harmonics) $f_{sp\,i}/f_r$ of the UIR. In the middle of the response, the 2nd spurious harmonics of SIR exhibits a dip with the least separation from the $f_{sp\,i}/f_r$ of the UIR. However, it has two other side

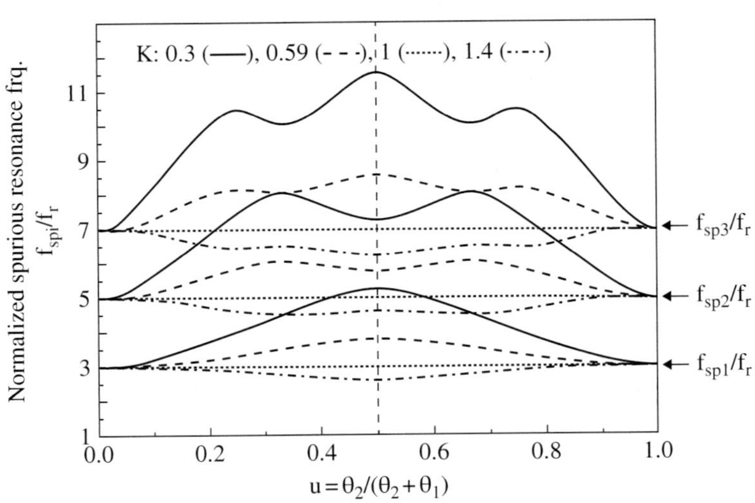

Figure 18.6 Harmonic frequency control of the $\lambda_g/4$– SIR. *Source:* From Lin et al. [J.21]. © 2006, IEEE.

maxima away from the $f_{sp\,i}/f_r$ of the UIR. By further lowering, the value of K, larger separation of the harmonic frequency could be realized. However, a narrower line#1 causes more conductor loss. It also results in greater step discontinuity, again causing more radiation and surface-wave losses. The net result is a degradation of the unloaded Q-factor. *In conclusion, K < 1 is desirable to enlarge the stopband characteristic of a BPF.* Moreover, the SIR-based BPF suppresses the harmonic contents of the amplifier and oscillator.

Control of Spurious Harmonic Frequency

The analytical evaluation of the spurious harmonic frequencies is considered below under the condition of electrically equal length of both sections, i.e. $\theta_1 = \theta_2 = \theta_0$ for the $\lambda_g/4$-SIR. It is *case #a* shown in Fig (18.5a). The resonance condition given by equation (18.1.20) is reduced to

$$\tan\theta_0 = \pm\sqrt{K} \quad (a)$$

for the fundamental resonance frequency:

$$\theta_0 = \tan^{-1}\sqrt{K} \quad (b), \quad \beta_{g1}\ell_1 = \beta_{g2}\ell_2 \qquad (c).(18.1.25)$$

$$\text{The SIR total length: } \ell = \ell_1 + \ell_2 = \ell_1\left[1 + \frac{\beta_{g1}}{\beta_{g2}}\right]$$

$$= \ell_1\left[1 + \sqrt{\frac{\varepsilon_{r\,eff1}}{\varepsilon_{r\,eff2}}}\right] \qquad (a)$$

$$\theta_{TA} = 2\theta_0, \text{ where } \theta_0 = \tan^{-1}\sqrt{K} = \beta_{g1}\ell_1$$

$$\Rightarrow \ell_1 = \frac{1}{\beta_0\sqrt{\varepsilon_{r\,eff1}}}\tan^{-1}\sqrt{K} \qquad (b).$$

$$(18.1.26)$$

Equation (18.1.24b) provides the radian length of a $\lambda_g/4$-SIR at the fundamental resonance frequency. The spurious harmonic frequencies, ignoring the dispersion in the SIR, are obtained from the periodic nature of $\tan\theta$ function as follows:

$$\tan\theta_{sp1} = \tan(\theta_0 - \pi) = -\sqrt{K}, \Rightarrow \frac{\theta_{sp1}}{\theta_0} = \frac{\pi}{\theta_0} - 1$$

$$\Rightarrow f_{sp1} = \left(\frac{\pi}{\tan^{-1}\sqrt{K}} - 1\right)f_r \qquad (a)$$

$$\text{also, } \tan\theta_{sp2} = \tan(\theta_0 + \pi) = \sqrt{K}$$

$$\frac{\theta_{sp2}}{\theta_0} = \frac{(\pi - \theta_0)}{\theta_0} + 2, \Rightarrow f_{sp2} = f_{sp1} + 2f_r \qquad (b)$$

$$\text{and, } \tan\theta_{sp3} = \tan(\theta_0 - 2\pi) = -\sqrt{K}$$

$$\frac{\theta_{sp3}}{\theta_0} = \frac{2(\pi - \theta_0)}{\theta_0} + 1, \Rightarrow f_{sp3} = 2f_{sp1} + f_r \qquad (c).$$

$$(18.1.27)$$

In the above equations, the electrical line length $\theta_{spn}(n = 1, 2, 3)$ corresponds to the spurious resonance frequency f_{spn} (n = 1,2,3). Also, in the above equations $\theta_{sp1}/\theta_0 = f_{sp1}/f_r$, etc., are obtained for the nondispersive microstrip SIR resonator from the relation $\theta_{spn} = \beta_g\ell = \omega_{spn}\ell/v_p$, i.e. $\theta_0 = \omega_r\ell/v_p$, $\theta_{sp1} = \omega_{sp1}\ell/v_p$, etc.

Figure (18.7a) shows three normalized spurious resonant frequencies (f_{spn}/f_r, n = 1,2,3) of the $\lambda_g/4$-SIR. For K = 1, i.e. for the UIR, the spurious resonance occurs at a separation of 2.0. Whereas, for K < 1, i.e. for the SIR, the separation of the spurious resonance increases with the lowering of the value of K. The structure is also used to design the BPF in the coaxial line technology [J.26].

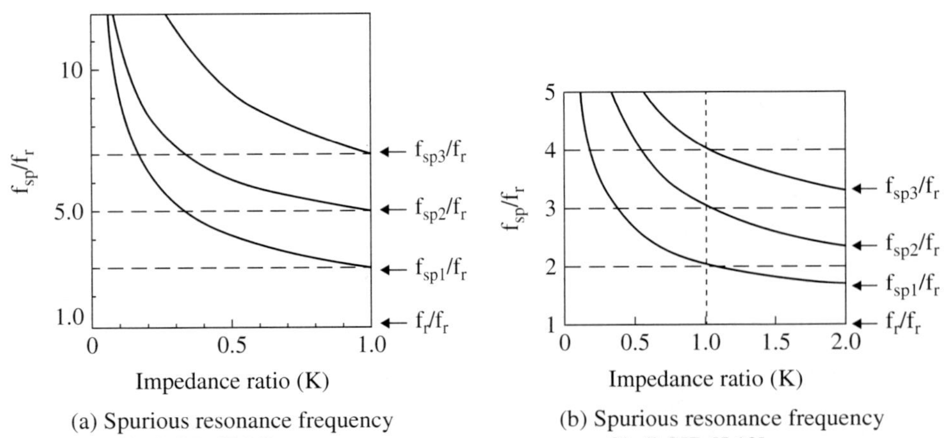

(a) Spurious resonance frequency of $\lambda_g/4$ SIR [J.26].

(b) Spurious resonance frequency of $\lambda_g/2$ SIR [J.19].

Figure 18.7 Control of spurious resonance by the impedance ratio K. *Source:* From Makimoto et al. [J.19]. © 1980, IEEE. From Sagawa et al. [J.26]. © 1985, IEEE.

Likewise, the following expressions are also obtained for the spurious harmonic frequencies for *case #b* shown in Fig (18.5b) [J.19]:

$$f_{sp1} = \frac{\pi}{2\tan^{-1}\sqrt{K}} f_r \text{ (a)}, \quad f_{sp2} = 2f_{sp1} - f_r \text{ (b)}$$

$$f_{sp3} = 2f_{sp1} \qquad \text{(c)}. \qquad (18.1.28)$$

For the *case #b*, both ends are open-circuited and the fundamental frequency is again given by $\theta_0 = \tan^{-1}(\sqrt{K})$. Figure (18.7b) shows f_{spn}/f_r, (n = 1,2,3) for the $\lambda_g/2$-SIR. In this case, for K = 1, the spurious resonance occurs at the separation of distance 1.0. The lower value of K (K < 1) increases the separation of spurious resonances of a SIR.

In conclusion, the spurious harmonic resonance frequencies can be effectively controlled by impedance-ratio K.

Optimum Resonant Length

The effect of K is considered below on the *normalized resonant length* L_n of the SIR at the fundamental resonance frequency. It is examined for the optimum case, $\theta_1 = \theta_2 = \theta_0$, giving the maximum shifting of the spurious harmonic frequency of the SIR from the harmonic frequency of UIR. The optimum case provides both the *minimum and maximum normalized resonant lengths* L_{nm}, depending upon the value of K. The normalized lengths of the SIR for three cases, mentioned above, are given below [J.25]:

$$\text{For } \theta_1 \neq \theta_2: \quad L_n = \frac{\theta_{TA}}{\pi/2} = \frac{\theta_{TB}}{\pi} = \frac{\theta_{TC}}{2\pi} \qquad \text{(a)}$$

$$\text{For } \theta_1 = \theta_2 = \theta_0: L_{nm} = \frac{4\theta_0}{\pi} = \frac{4\tan^{-1}(\sqrt{K})}{\pi} \qquad \text{(b)},$$

(18.1.29)

where normalization is carried out by the radian resonant lengths $\pi/2$, π, 2π for the UIR corresponding to case #a, b, and c, respectively. Figure (18.8a) shows the dependence of the normalized length L_n on the length $\theta_1 = \beta_{g1}\ell_1$ of the line section #1, and for the impedance-ratio K. The L_n is reduced, for K < 1, as compared to the resonant length $L_n = 1$ for the UIR (K = 1). *Thus, K < 1 makes the SIR physically a compact resonator, as against the UIR.* Also for a given value of K < 1, there is the minimum normalized length L_n. Of course, there is a maximum L_n also for a value of K > 1. The min/max of L_n, called L_{nm}, occurs for $\theta_1 = \theta_2 = \theta_0$. Figure (18.8b) shows the effective control of min/max normalized length by the impedance-ratio K.

The above investigation carried on the SIR is approximate. It provides only general characteristics of the SIR. For the proper design of a SIR, an EM-Simulator is used.

18.1.5 Microstrip Hairpin Resonator

Some common type microstrip hairpin resonators are shown in Fig (18.9a–d). The basic structure of the *microstrip hairpin resonator*, shown in Fig (18.9a), is primarily a $\lambda_g/2$ resonator. Its each arm length is $\lambda_g/4$ with respect to the plane of symmetry. An approximate resonance frequency is obtained from its mean length, i.e.

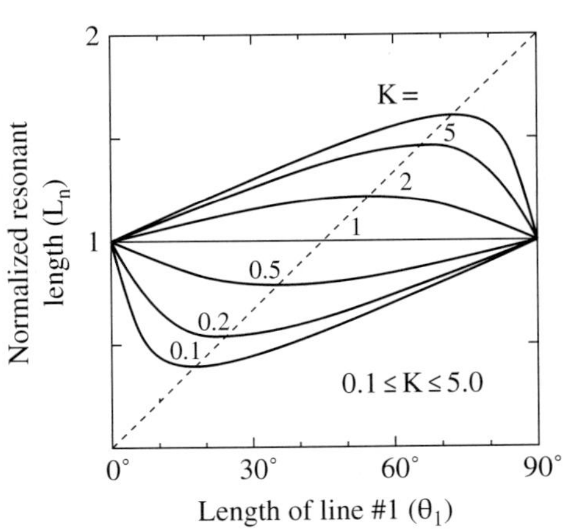

(a) Variation in normalized resonant length with the length of line #1. Impedance ratio K is a parameter.

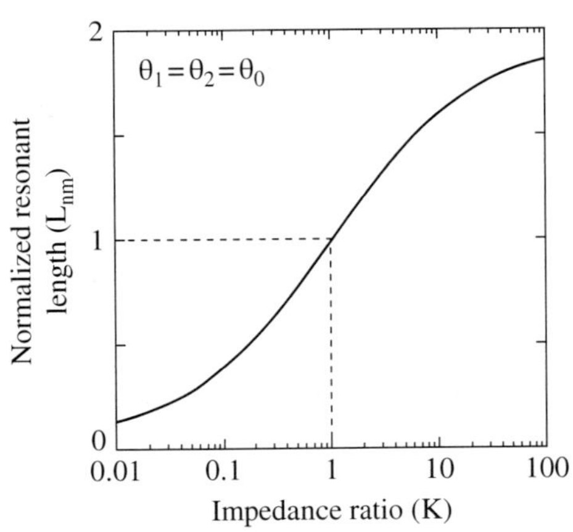

(b) Variation in min/max normalized resonant length with impedance ratio K.

Figure 18.8 *Control of length of the SIR by the length of line #1 and impedance ratio K. Source:* Sagawa et al. [J.25]. © 2000, IEEE.

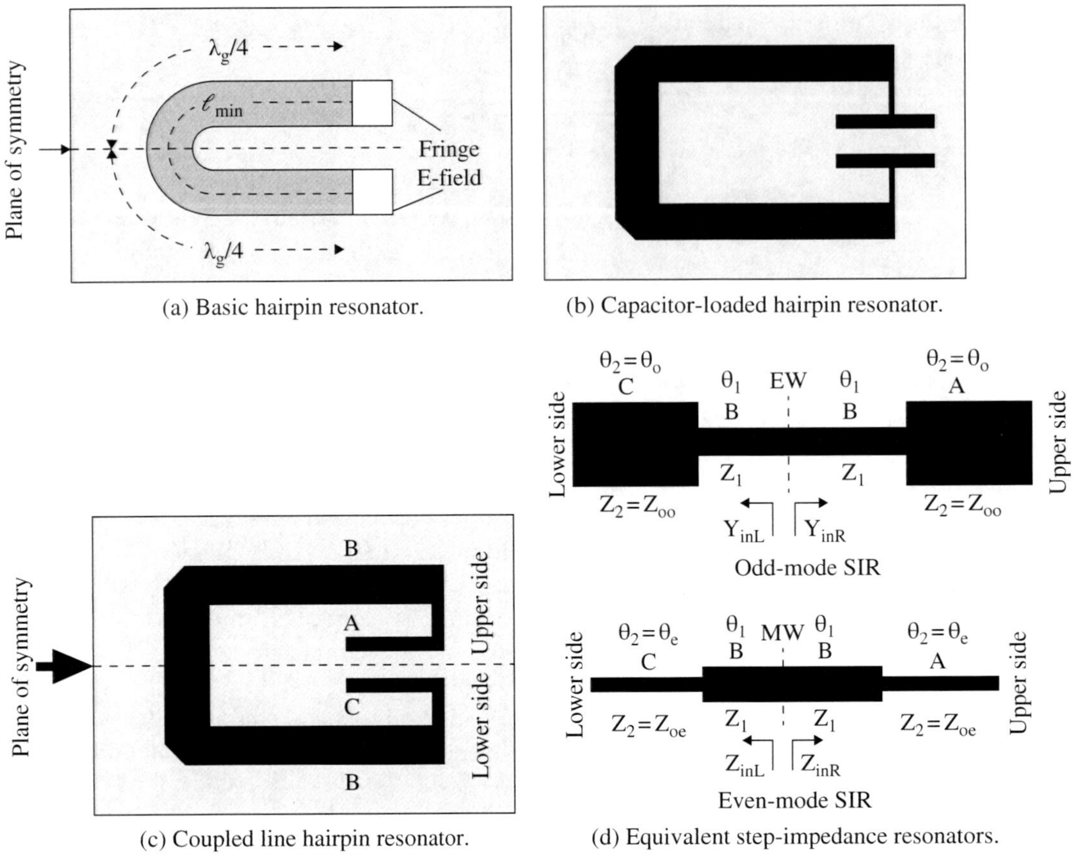

Figure 18.9 Some microstrip hairpin resonator.

$\ell_m = (\lambda_g/2)$. The electric fields at the open ends, located on the same side, have opposite phases due to the 180° phase difference of the electric fields at two ends. Therefore, the radiation is less, as compared to the radiation from a straight open $\lambda_g/2$-microstrip resonator. So, its Q-factor is high, about 220 for the 50 Ω hairpin resonator on a low loss 0.5 mm sapphire substrate. For a smaller separation between two arms, the Q-factor is reduced to 183. However, it is still high compared to the Q-factor (150) of a straight open $\lambda_g/2$-microstrip resonator [J.13]. Due to its small linear length, and also due to the higher Q-factor, the structure is a popular choice for designing the filters, especially BPF [J.20, J.22, J.23, B.6].

Although the hairpin microstrip resonator is a compact element, still it is physically large in the L-band. Figure (18.9b) shows the *capacitor-loaded hairpin resonator*. Its size is reduced to almost half, without degrading the Q-factor. However, getting the chip capacitor at a high frequency is difficult. Figure (18.9c) shows the replacement of the capacitor by a coupled line section, without enlarging the size of the structure. This structure is called the *folded hairpin resonator*. The folded and mainline sections of the folded hairpin resonator are shown as A, B, C, D in Fig (18.9c). Figure (18.9c) also shows the lower and upper sections with respect to the plane of symmetry. For a narrow spacing between two arms, the structure could also be viewed as a coupled $\lambda_g/4$ resonator. However, for a large spacing between the folded arms, the coupling is small.

For a stronger coupling, the folded hairpin resonator forms a miniaturized hairpin resonator. The coupled line section of the microstrip hairpin resonator supports the odd and even resonances. The lower resonance frequency, i.e. the fundamental resonance frequency, is due to the odd-mode excitation [J.22, J.23]. The even-mode frequency is treated as the first spurious higher-order mode.

The structure of Fig (18.9c) is analyzed by getting an equivalent SIR resonator for both the odd and even modes, as shown in Fig (18.9d) [J.22]. The approximate odd- and even-mode resonance frequencies are computed from the symmetry consideration. The plane of symmetry is the electric-wall (EW) for the odd-mode excitation, and it is the MW for the even-mode excitation. The resonance conditions, under symmetry, i.e. ($Z_{inR} = Z_{inL}$) and ($Y_{inR} = Y_{inL}$), for the even and odd mode, respectively, are obtained below:

$$Z_{in} = Z_{inR} + Z_{inL} = 0$$

$$\Rightarrow Z_{in} = -j2Z_{01}\left[\frac{K - \tan\theta_1 \tan\theta_2}{K\tan\theta_1 + \tan\theta_2}\right] = 0$$

$$\Rightarrow \cot\theta_1 = -K\cot\theta_2$$

$$\Rightarrow \tan\theta_1 = K\cot\theta_2, \text{(for odd mode)} \qquad \text{(a)}$$

$$Y_{in} = Y_{inR} + Y_{inL} = 0$$

$$\Rightarrow Y_{in} = j2Y_{01}\left[\frac{K\tan\theta_1 + \tan\theta_2}{K - \tan\theta_1 \tan\theta_2}\right] = 0$$

$$\Rightarrow \cot\theta_1 = -K\cot\theta_2, \text{(for even mode)} \qquad \text{(b)}$$

$$\text{where impedance ratio } K = Z_{02}/Z_{01} = Y_{01}/Y_{02} \qquad \text{(c).}$$

$$(18.1.30)$$

Equation (2.1.65c) of chapter 2 is used to get the Z_{inR}/Y_{inR} of the cascaded line-sections θ_2 and θ_1. The end of the line section θ_2 is open-circuited.

In the above equation, the odd mode has $Z_2 = Z_{oo}$; $\theta_2 = \theta_o$; and for the even mode the characteristic impedance is $Z_2 = Z_{oe}$; $\theta_2 = \theta_e$. These parameters are determined for the given coupled line section. It is discussed in section (12.1) of chapter 12. Figure (18.10a and b) shows normalized frequency of the odd- and even-mode resonances as the function of normalized odd- and even-mode line lengths. The frequency normalization is with respect to the odd- and even-mode frequency of a UIR, and lengths are normalized by θ_1.

The odd-mode frequency should be lowered for a compact resonator. Figure (18.10a) shows that it is achieved by lowering the value of K below 1, $K < 1$. It is easily obtained as the characteristic impedance of the odd mode is always less than that of a single line. It is further brought down by the length ratio. Figure (18.10b) shows that the even-mode frequency, i.e. the spurious undesired resonance, is pushed up by increasing the value of K. The value of K is naturally higher, as the even-mode characteristic impedance is always more than that of a single line. The length ratio further controls the even-mode resonance frequency. By controlling the gap between coupled line sections, both the odd- and even-mode characteristic impedances are controlled. A proper choice of the characteristic impedance of the central single line is also important. The resonator has been used to develop compact bandpass filters [J.20, J.22, J.23].

18.2 CPW Resonator

Figure (18.11a–i) shows a few basic forms of the CPW resonators, without any connection to the external source and load. These structures are used for the development of filters and diplexers. Again, the CPW resonators are classified as $\lambda_g/2$ and $\lambda_g/4$-uniform impedance resonator (UIR), SIR, and the ring resonator. The CPW resonators are also used in the hairpin form to reduce its linear length. As usual, the SIRs are more compact structures. There could be many more modifications to these basic structures. The ring shape could be a rectangle, triangle, etc. The CPW resonator could be

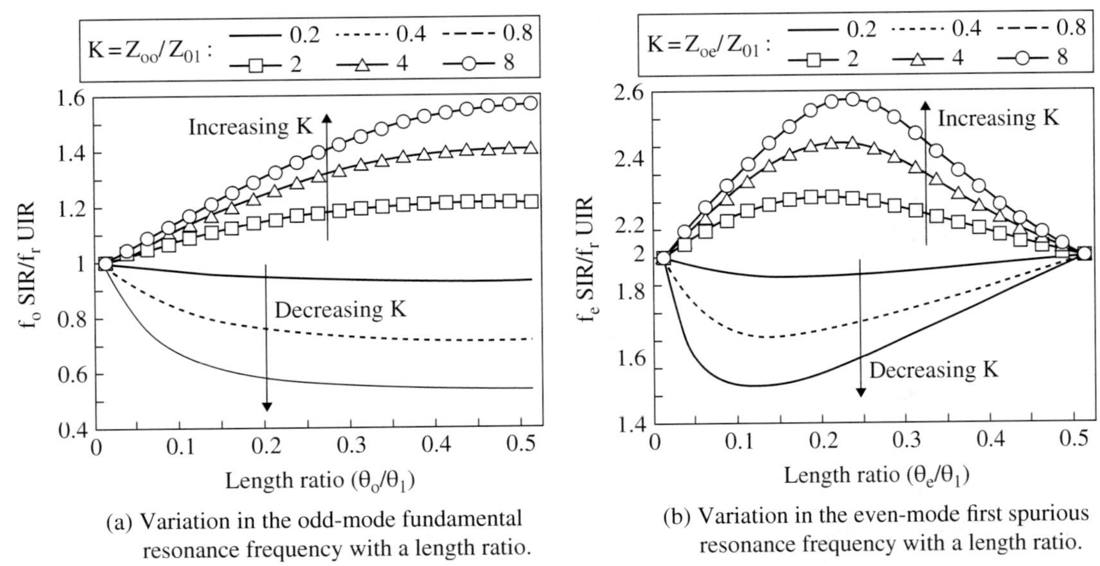

(a) Variation in the odd-mode fundamental resonance frequency with a length ratio.

(b) Variation in the even-mode first spurious resonance frequency with a length ratio.

Figure 18.10 Control of fundamental and first spurious resonance frequencies by the impedance ratio and length ratio. *Source:* From Lee et al. [J.23]. © 2000, IEEE.

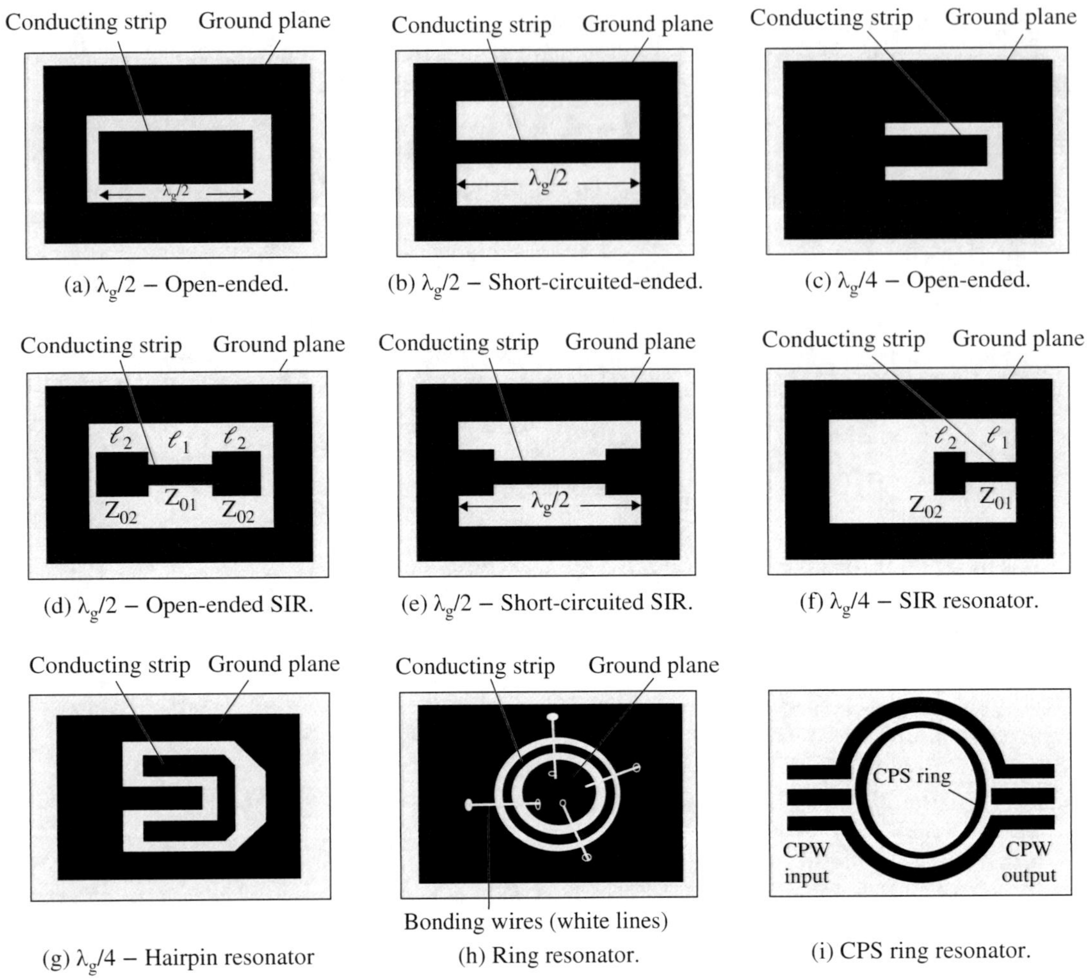

Figure 18.11 CPW resonators.

loaded by the slot and stub to reduce the size of a CPW resonator.

The CPW resonators could be treated with the help of the equivalent transmission line models. The open-end fringing electric fields are modeled either by the extension of the strip conductor or by a lumped capacitor. Likewise, the short-circuited end is modeled by a lumped inductor. Their values can be approximately obtained from the closed-form expression [B.7]. Thus, Fig (17.10a and c) of chapter 17 models the $\lambda_g/2$-CPW resonators, shown in Fig (18.11a and b), respectively. Figure (17.10b) of chapter 17 models $\lambda_g/4$-one end open-circuited CPW, shown in Fig (18.11c). Both the $\lambda_g/2$ and $\lambda_g/4$-CPW SIRs, shown in Fig (18.11d–f), can also be analyzed by using the multistep transmission line models. The Z_0 and $\varepsilon_{r\,eff}(f = 0)$ of the CPW structures are computed using the closed-form expressions given in chapter 9. The CPW could also be the conductor-backed structure. A hairpin CPW, shown in Fig (18.11g), could be just a folded $\lambda_g/2$ resonator for the weak coupling. Alternatively, it could be treated as coupled CPW, as discussed for the microstrip counterpart. Figure (18.11h) shows a CPW ring resonator with bonding wires to equalize the ground potentials, and to suppress the unwanted modes. The impedance steps can also be added to the CPW ring resonator. Finally, Fig (18.11i) shows a CPS ring resonator connected to the CPW line that has been fabricated in the Si-MEMS technology at 10.62 GHz, with Q-factor 53.5 and the first spurious resonance at 20.93 GHz [J.29]. The transmission line models are approximate. So, an EM-Simulator has to be used for the accurate design of the CPW resonators, and circuits based on them.

Numerical studies of the CPW resonators on a substrate (ε_r = 10.5, h = 1.6 mm) at 2 GHz, shown in Fig (18.11a–e), have been carried out using the FDTD method [J.30]. The radiation Q-factor Q_{rad} for a $\lambda_g/2$-CPW resonator is half, as compared to that of the $\lambda_g/4$-CPW resonator. It is due to the radiation from two ends of the $\lambda_g/2$-CPW resonator. All losses are less for a $\lambda_g/4$-resonator. Thus, the unloaded Q-factor of a $\lambda_g/2$-open CPW resonator is about 100, whereas for the unloaded Q-factor of a $\lambda_g/4$-CPW resonator is about 150. The unloaded Q-factor has been computed for the line width from 1 mm to 3.5 mm. The Q-factors of CPW $\lambda_g/2$-SIR and $\lambda_g/4$-SIR resonators are almost the same, about 150. It is not much influenced by the impedance-ratio K and line width.

Figure (18.12a) illustrates the realization of a CPW ultra wideband BPF [J.31]. The end sections are adjusted to $\lambda_g/4$, while the middle section is adjusted to $\lambda_g/2$. Figure (18.12b) shows the cascaded transmission line equivalent circuit.

The resonance frequencies are obtained from Im $(Z_{in}) = 0$, where the input impedance is

$$Z_{in} = jZ_2 \frac{2\,K \tan\theta_1 \tan\theta_2 [K \tan\theta_1 + \tan\theta_2]}{K\,(1 - \tan^2\theta_1)\,(1 - \tan^2\theta_2) - 2(1 + K^2) \tan\theta_1 \tan\theta_2}. \tag{18.2.1}$$

Normally, $\theta_1 = \theta_2 = \theta$ is taken. It simplifies the design. The condition Im $(Z_{in}) = 0$ gives three resonance frequencies of the series resonant circuits, resulting in the UWB response for the BPF. The first and third frequencies are dependent on the impedance-ratio K.

$$\theta(f_1) = \tan^{-1}\sqrt{K} \quad \text{(a)}, \quad \theta(f_2) = \pi/2 \quad \text{(b)}$$
$$\theta(f_3) = \pi - \tan^{-1}\sqrt{K} \quad \text{(c)}, \quad \text{where, } K = (Z_{01}/Z_{02}) \quad \text{(d)}$$
$$\theta_1 = \beta_1 \ell_1, \quad \theta_2 = \beta_2 \ell_2 \quad \text{(e)}. \tag{18.2.2}$$

Figure (18.12c) shows the response of the filter is controlled by the gap d. So, the multiple mode resonances of the CPW SIR provide a 3.1 GHz–10.6 GHz passband filter.

18.3 Slot Line Resonator

The slot line resonators are made in the ground plane of a PCB. Some common types of slot resonators, etched in the ground plane of the microstrip, are shown in Fig (18.13a–i). These resonators are fed by the microstrip line. The rectangular slot, for instance, and also the ring-shaped slot have been used as a radiating element [B.2, B.5]. These are also popular elements of the frequency selective surfaces (FSS) [B.1, B.8], and useful in the design of the filters. They are the basis of the defected ground structures (DGS). There is a large variety of the DGS shapes [J.32]. These structures are discussed in chapter 19.

The slots are also made in the ground plane of the CPW structures [J.33]. The DGS slots inductively load the microstrip, or a CPW to form the slow-wave line structures and compact resonators. The slot in the ground plane further acts as a coupling element between the line and a resonator, or a microstrip antenna, located at different dielectric layer levels of the multilayer structure [B.1, B.2]. The slot embedded in a patch resonator reduces its size and controls the polarization of a patch resonator. The slot resonators are analyzed by the full-wave methods. These are also simulated on the EM-simulators. For the simpler geometrical shapes, the circuit model can also be used to analyze approximately most of the structures shown in Fig (18.13a–i).

The slot resonators are also made in the *fin line structures*. The single and coupled slots, shown in Fig (18.13a–i), have been analyzed both in the microstrip and fin line structures by using the spectral domain method [J.29, J.32, J.34–J.36]. Primarily, these are the $\lambda_g/2$-resonators. So these can be modeled by the series

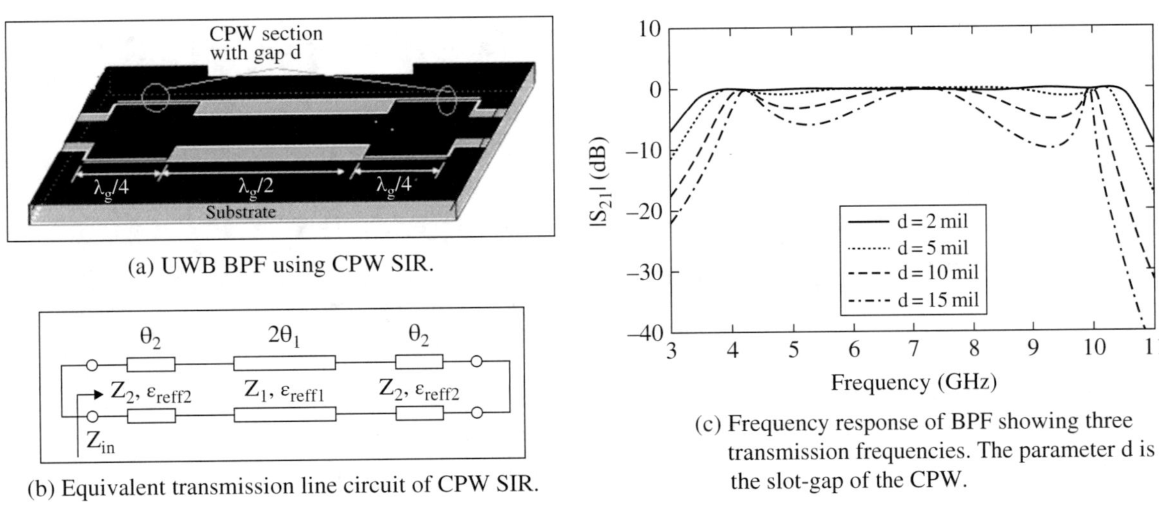

(a) UWB BPF using CPW SIR.

(b) Equivalent transmission line circuit of CPW SIR.

(c) Frequency response of BPF showing three transmission frequencies. The parameter d is the slot-gap of the CPW.

Figure 18.12 UBPF using CPW SIR. *Source:* From Gao et al. [J.31]. © 2006, IEEE.

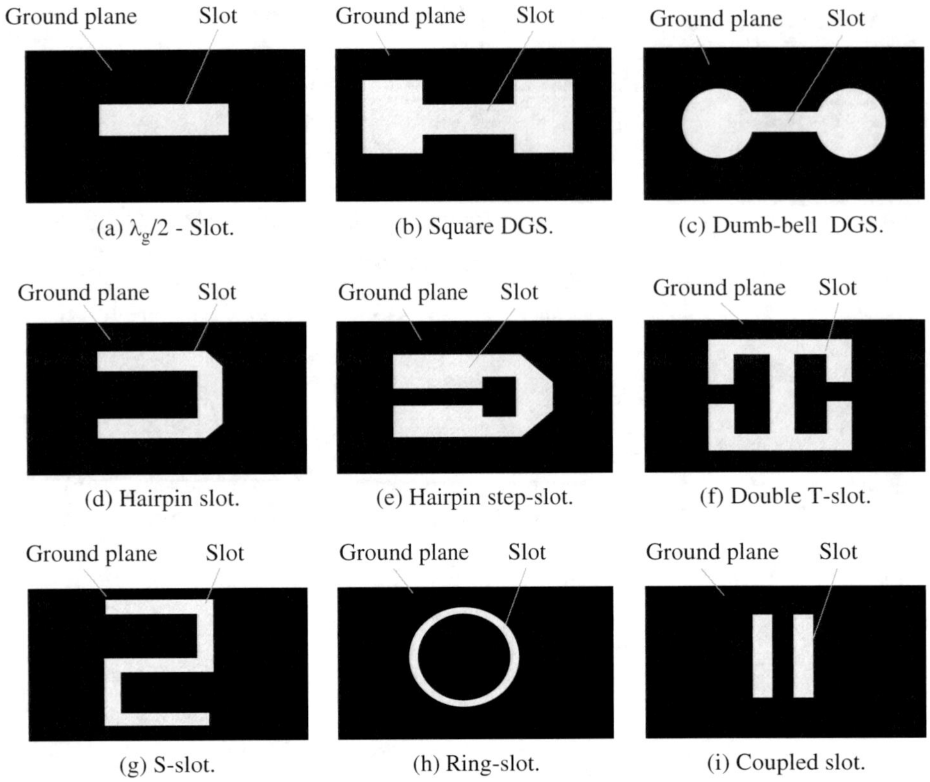

Figure 18.13 Some simple slot line resonators.

RLC circuit. For instance, a DGS structure is primarily a series resonant circuit, as the slot resonator is short-circuited at both ends. However, the coupling capacitance between the microstrip line and a slot inverts it to the series-connected parallel resonant circuit. Of course, the DGS can be also modeled as the shunt-connected series resonant circuit. This approach has been taken to model the double-T-resonator, or the folded slot resonator, shown in Fig (18.13f) [J.35]. It is a compact resonator, useful at the lower end of the microwave frequency. The double-T-resonator exhibits two resonance frequencies. The first resonance corresponds to the physical length of the center slot, combined with the folded slots. The second resonance is due to the resonance in the folded slots. The resonance frequency can be computed by the transmission line model. The compact slot resonator can also be obtained by folding the $\lambda_g/2$-slot resonator in the S-form, shown in Fig (18.13g).

Both ends of the $\lambda_g/2$-slot resonator are treated as perfectly short-circuits. However, in the transmission line model, the discontinuities due to the short-circuiting are modeled as inductors at both ends. The inductor-loaded model can provide better results for the resonance frequency. The DGS, shown in Fig (18.13b and c), can also be treated as the slot-based SIR. So these structures can be modeled as the SIR. The SIR concept is also useful to model the hairpin slot resonator and the hairpin step slot resonators, shown in Fig (18.13d and e).

18.4 Coupling of Line Resonator to Source and Load

The physical line resonators, such as the microstrip, CPW, and slot line resonators, are always coupled to the source and load. Such a connection of a resonator with an external circuit is described by the *external Q-factor*. It is also described by the *coupling coefficient*, as it has been done for a lumped resonator. It is discussed in subsection (17.2.3) of chapter 17. The external circuit has a strong influence on the bandwidth. It weakly influences the resonance frequency of a resonator. The source is connected to a line resonator in following manner [B.9–B.17]:

- *Direct-coupled resonator.*
- *Capacitively/inductivity coupled resonator.*
- *Tapped line resonator.*

18.4.1 Direct-coupled Resonator

Fig (18.14a and b) shows the short-circuited and open-circuited line resonators directly connected a source with an internal resistance R_g. In the case of a short-circuited $\lambda_g/2$-line resonator, the series resonances are obtained for $\beta_g \ell = n\pi$, (n = 1, 2, 3,...), and for the one end short-circuited $\lambda_g/4$-resonator, the parallel resonances occur for $\beta_g \ell = (2n-1)\pi/2$, (n = 1, 2, 3,...). For both cases the output port #2 could be open-circuited also. These cases of line terminations are discussed in section (17.3) of chapter 17.

Short-circuited $\lambda_g/2$-line Resonator

The input impedance of the short-circuited line, at port #1, is given by equation (17.3.2) of chapter 17. Using $\tanh(\alpha l) \approx \alpha l$, and also ignoring the terms with $(\alpha\ell)^2$ and $\tan^2\beta_g\ell << 1$, for a small loss line, the input impedance is evaluated as

$$Z_{in} = Z_0 \left[\frac{(\alpha\ell + j\tan\beta_g\ell)}{(1 + j\alpha\ell\tan\beta_g\ell)} \right]$$

$$\Rightarrow Z_{in} = Z_0 \left[\frac{(\alpha\ell + j\tan\beta_g\ell) - j(\alpha\ell)^2\tan\beta_g\ell + \alpha\ell\tan^2\beta_g\ell}{\{1 + (\alpha\ell)^2\tan^2\beta_g\ell\}} \right]$$

$$\Rightarrow Z_{in} = Z_0\alpha\ell + jZ_0\tan\beta_g\ell. \tag{18.4.1}$$

The short-circuited line offers the following reactance and resistance at the input:

$$X = Z_0\tan\beta_g\ell, \quad \beta_g = \omega/v_p \quad (a), \qquad R = Z_0\alpha\ell \quad (b). \tag{18.4.2}$$

Incorporating the source impedance R_g in equation (17.2.12b) of chapter 17, the loaded Q-factor of the series resonant circuit is

$$Q_L = \frac{\omega_r}{2(R_g + R)} \left. \frac{dX}{d\omega} \right|_{\omega = \omega_r}. \tag{18.4.3}$$

In the case of $R_g = 0$, $Q_L \to Q_U$, i.e. without any loading by an external circuit, the loaded Q-factor is reduced to the unloaded Q-factor.

The derivative of the reactance is evaluated at the resonance by taking $\beta\ell = n\pi$. It helps to compute the unloaded Q-factor Q_U of a *direct-coupled line resonator*:

$$\frac{dX}{d\omega} = Z_0\sec^2\beta_g\ell \times (\ell/v_p), \quad \left.\frac{dX}{d\omega}\right|_{\omega=\omega_r} = Z_0\frac{\ell}{v_p} \quad (a)$$

$$Q_U = \frac{\omega_r}{2Z_0\alpha\ell}\frac{Z_0\ell}{v_p} = \frac{\beta_g}{2\alpha} \quad (b). \tag{18.4.4}$$

Short-circuited $\lambda_g/4$-line Resonator

The short-circuited line resonator of length $\ell = (2n-1)\lambda_g/4$ exhibits the parallel circuit resonance. Using equation (17.2.18) of chapter 17, the loaded Q-factor of a parallel resonant circuit is

$$Q_L = \frac{\omega_r}{2(G_g + G)}\left.\frac{dB}{d\omega}\right|_{\omega=\omega_r}. \tag{18.4.5}$$

The following approximate expression of the input admittance Y_{in}, using equation (18.4.1), is determined from the low-loss case near resonance, i.e. for $\alpha\ell << 1$, and $\tan\beta_g\ell \to \infty (\tan\beta_g\ell >> 1)$:

$$Y_{in} = Y_0\left[\frac{(1 + j\alpha\ell\tan\beta_g\ell)}{(\alpha\ell + j\tan\beta_g\ell)}\right]$$

$$= Y_0\left[\frac{\alpha\ell(1 + \tan^2\beta_g\ell) + j\{(\alpha\ell)^2 - 1\}\tan\beta_g\ell}{(\alpha\ell)^2 + \tan^2\beta_g\ell}\right] \quad (a)$$

$$\Rightarrow Y_{in} \approx Y_0\alpha\ell - jY_0\cot\beta_g\ell \quad (b). \tag{18.4.6}$$

Thus, the susceptance and the conductance offered by the $(2n-1)\lambda_g/4$ line resonator are

$$B = -Y_0\cot\beta_g\ell \quad (a), \qquad G = G_0 = Y_0\alpha\ell \quad (b). \tag{18.4.7}$$

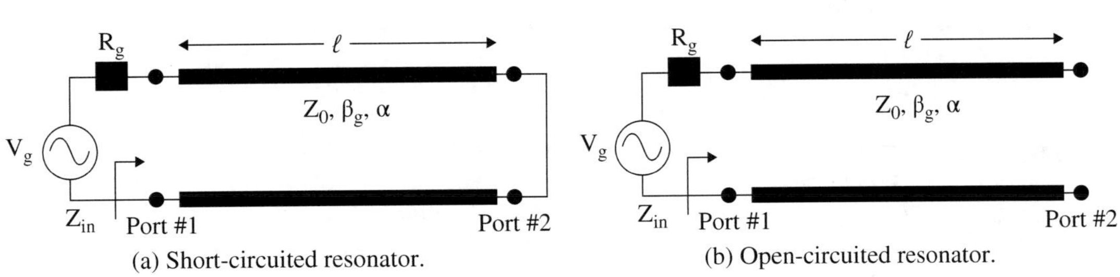

Figure 18.14 Direct-coupled line resonator.

In the absence of the source conductance, the external circuit is not present. The loaded Q-factor is reduced to the unloaded Q-factor, i.e. for $G_g = 0$, $Q_L \to Q_U$:

$$\left.\frac{dB}{d\omega}\right|_{\omega=\omega_r} = \frac{Y_0\ell}{v_p} \quad \text{(a)}, \quad Q_U = \frac{\omega_r}{2\,G}\frac{Y_0\ell}{v_p} = \frac{\beta_g Y_0 \ell}{2Y_0\alpha\ell} = \frac{\beta_g}{2\alpha} \quad \text{(b)}. \tag{18.4.8}$$

The unloaded Q-factors for $\lambda_g/2$ and $\lambda_g/4$-line resonators are identical. A similar expression is obtained for the open-circuited line resonators also. The external Q-factor for $\lambda_g/2$ and $\lambda_g/4$-line resonators are obtained from expressions (18.4.3) and (18.4.5), respectively, on consideration of the lossless line resonators with $R = G = 0$:

$\frac{\lambda_g}{2}$ resonator $\left(\beta_g\ell = n\pi\right)$:

$$Q_E = \frac{\omega_r}{2R_g}\frac{Z_0\ell}{v_p} = \frac{\beta_g Z_0 \ell}{2R_g} = \frac{n\pi}{2}\frac{Z_0}{R_g}. \tag{18.4.9}$$

$\frac{\lambda_g}{4}$ resonator $\left(\beta_g\ell = (2n-1)\frac{\pi}{2}\right)$:

$$Q_E = \frac{\omega_r}{2G_g}\frac{Y_0\ell}{v_p} = \frac{\beta_g\ell\, R_g}{2Z_0} = \frac{(2n-1)\pi}{4}\frac{R_g}{Z_0}. \tag{18.4.10}$$

Usually, the source impedance is $R_g = 50$, and for the matching purpose, $Z_0 = 50$. It provides lower value of loaded Q-factor for matched line resonator $Z_0 \approx R_g$: $Q_E \approx Q_L = \pi/2$ and $\pi/4$ for the direct coupled $\lambda_g/2$ and $\lambda_g/4$-line resonators, respectively. *Therefore, a narrow bandwidth filter cannot be obtained using the direct-coupled resonators.* Some improvement is possible by increasing or decreasing the characteristic impedance of line resonator in the range 120 Ω–20 Ω. However, the external Q-factor Q_E can be improved either by the inductive or capacitive coupling between the source and the resonator.

18.4.2 Reactively Coupled Line Resonator

The source could be either inductively or capacitively coupled to a short-circuited or an open-circuited line resonator. Figure (18.15a) shows the coupling of a source to the short-circuited line resonator through the shunt-connected coupling inductance L_c. The coupling inductance could be an equivalent circuit representation of the coupling mechanism between the source and a resonator through the magnetic field. Likewise, the series-connected coupling capacitance C_c, shown in Fig (18.15b), could also be an equivalent circuit representation of the coupling mechanism between a source and the resonator through the electric field.

Inductive Coupling

A lossless short-circuited line, for the line length $\ell < \lambda_g/4$, offers an inductive susceptance $-jY_0 \cot\beta_g\ell$ at the plan a′–a shown in Fig (18.15a). The total susceptance at the input port #1 is

$$B_{in} = -\frac{1}{\omega L_c} - Y_0 \cot\beta_g\ell. \tag{18.4.11}$$

For the parallel resonance, the resonance condition is $B_{in} = 0$. It provides the following expression for the line length ℓ of a resonator:

$$\frac{1}{\omega_r L_c} = -Y_0\cot\frac{\omega_r\ell}{v_p}, \quad \Rightarrow \quad \tan\frac{\omega_r\ell}{v_p} = -\frac{\omega_r L_c}{Z_0} \quad \text{(a)}$$

$$\Rightarrow \frac{2\pi\ell}{\lambda_g} = \tan^{-1}\left[-\frac{\omega_r L_c}{Z_0}\right] = \pi - \tan^{-1}\left[\frac{\omega_r L_c}{Z_0}\right] \quad \text{(b)}. \tag{18.4.12}$$

The above expression shows that $\ell \approx \lambda_g/2$, i.e. the resonator length is a little less than $\lambda_g/2$. The unloaded line resonator is a series resonator at the plane a′–a of Fig (18.15a). However, the shunt coupling inductance L_c inverts it to a parallel resonant circuit at the input port #1. Equation (18.4.12a) can be solved graphically, as shown in Fig (18.16a), to get the fundamental and harmonic frequencies.

The external Q-factor of a parallel resonant circuit at the input port #1 is computed by using equation (18.4.5) for $G = 0$:

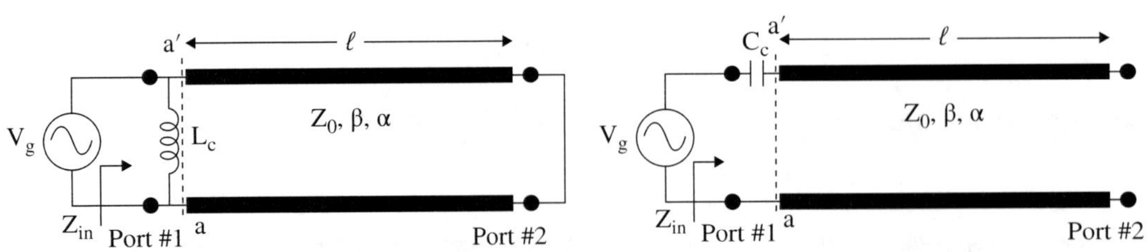

(a) Short-circuited resonator $\ell = \lambda_g/2$ inductively coupled.

(b) Open-circuited resonator $\ell = \lambda_g/2$ capacitively coupled.

Figure 18.15 Reactively coupled line resonators.

$$Q_E = \frac{\omega_r}{2G_g}\frac{dB_{in}}{d\omega}|_{\omega=\omega_r}, \quad \text{where } G_g = \frac{1}{R_g}. \tag{18.4.13}$$

The derivative of the susceptance is obtained from equation (18.4.11) and also using equation (18.4.12a):

$$\frac{dB_{in}}{d\omega} = -\frac{d}{d\omega}\left[\frac{1}{\omega L_c} + Y_0 \cot\frac{\omega\ell}{v_p}\right]$$

$$\Rightarrow \left.\frac{dB_{in}}{d\omega}\right|_{\omega=\omega_r} = \left[\frac{1}{\omega^2 L_c} + Y_0\left(1+\cot^2\beta_r\ell\right)\left(\frac{\ell}{v_p}\right)\right]_{\omega=\omega_r}$$

$$\Rightarrow \left.\frac{dB_{in}}{d\omega}\right|_{\omega=\omega_r} = \frac{Y_0}{\omega_r}\left[\frac{1}{\omega_r L_c Y_0} + \left\{1+\left(\frac{1}{\omega_r L_c Y_0}\right)^2\right\}\beta_r\ell\right]. \tag{18.4.14}$$

Thus, the external Q-factor is

$$Q_E = \frac{Y_0 R_g}{2}\left[\frac{1}{\omega_r L_c Y_0} + \left\{1+\left(\frac{1}{\omega_r L_c Y_0}\right)^2\right\}\beta_r\ell\right], \tag{18.4.15}$$

where ω_r is resonance frequency, Y_0 is characteristic admittance, and β_r is propagation constant of the line resonator at resonance. At the resonance, $\beta_r\ell \approx \pi$ and ω_r are known. The coupling inductance L_c could be adjusted to get the required external Q-factor Q_E. The solution of transcendental equation (18.4.12a) is obtained from Figure (18.16a). It shows that Q_E is large near $\phi \approx \pi$, as $dB/d\omega$ is high near ϕ_r. The first spurious resonance is nearly an octave away from the fundamental resonance frequency. The coupling also reduces the resonance frequency a little.

Capacitive Coupling

Figure (18.15b) shows the capacitive coupling of the open-circuited line resonator to a source. At the plane a′–a, capacitive reactance of the open-circuited line is $(-Z_0 \cot\beta_g\ell)$, $(\beta = \omega/v_p)$. The input reactance X_{in} is computed at the port #1, and series resonance occurs for $X_{in} = 0$, giving the resonant length of the line resonator:

$$X_{in} = -\frac{1}{\omega C_c} - Z_0\cot\beta_g\ell = 0, \Rightarrow \tan\beta_g\ell = -\frac{\omega C_c}{Y_0} \quad (a)$$

$$\Rightarrow \text{where, } \phi_r = \frac{2\pi\ell}{\lambda_g} = \pi - \tan^{-1}\frac{\omega_r C_c}{Y_0} \quad (b). \tag{18.4.16}$$

At resonance $\beta_r\ell \approx \pi$, the line length is $\ell = \lambda_g/2$. The $\ell = \lambda_g/2$ open-circuited line resonator is a parallel resonant circuit. However, the coupling capacitor inverts it to a series resonant circuit at the input port #1. The above transcendental equation is solved graphically, as shown in Fig (18.16b). The external Q-factor of the series resonant circuit is obtained from equation (18.4.3) for a lossless line resonator, i.e. R = 0:

$$Q_E = \frac{\omega_r}{2R_g}\frac{dX_{in}}{d\omega}|_{\omega=\omega_r}. \tag{18.4.17}$$

The derivative of the input reactance, with respect to frequency, is evaluated at the resonance frequency and rearranged using equation (18.4.16a):

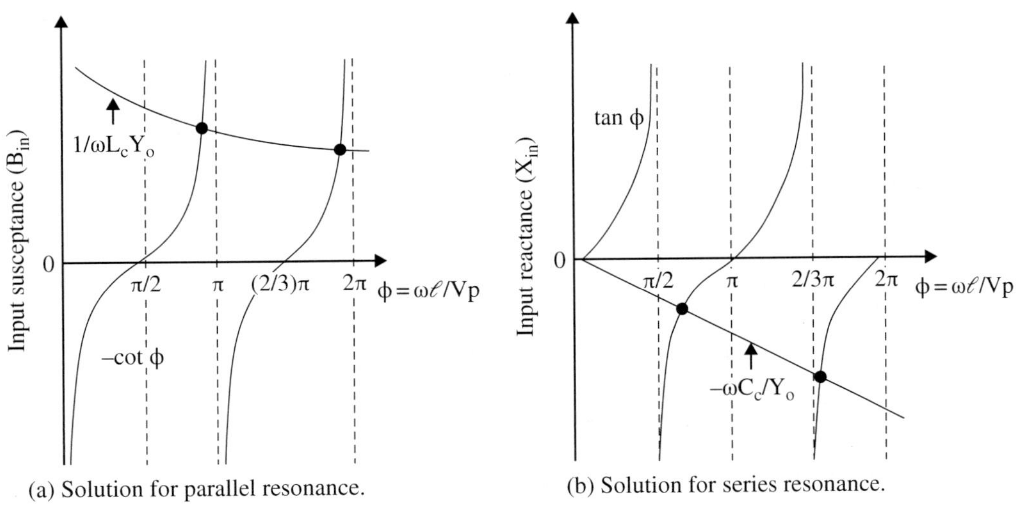

Figure 18.16 Graphical solution of the transcendental equations.

$$\left.\frac{dX_{in}}{d\omega}\right|_{\omega=\omega_r} = \left[\frac{1}{\omega^2 C_c} + Z_0 \operatorname{cosec}^2\beta_g \ell \, \frac{1}{v_p}\right]_{\omega=\omega_r}$$
$$= \left[\frac{1}{\omega_r^2 C_c} + Z_0\left(1 + \cot^2\beta_r\ell\right)\frac{\beta_r\ell}{\omega_r}\right]$$
$$\Rightarrow \left.\frac{dX_{in}}{d\omega}\right|_{\omega=\omega_r} = \frac{1}{\omega_r}\left(\frac{1}{\omega_r C_c} + Z_0\left\{1 + \left(\frac{Y_0}{\omega_r C_c}\right)^2\right\}\beta_r\ell\right). \tag{18.4.18}$$

The external Q-factor is obtained as

$$Q_E = \frac{Z_0}{2R_g}\left[\left(\frac{Y_0}{\omega_r C_c} + \left\{1 + \left(\frac{Y_0}{\omega_r C_c}\right)^2\right\}\beta_r\ell\right)\right]. \tag{18.4.19}$$

Again, a high value of the Q_E is obtained near the resonance for $\beta_r\ell \approx \pi$. Once the resonance frequency ω_r is known, the coupling capacitance C_c can be adjusted to get a desired value of the external Q-factor.

18.4.3 Tapped Line Resonator

Figure (18.17a and b) shows two cases of the line resonator of length ℓ, the *case-I*: one end short-circuited, and the *case-II* for both ends open-circuited. The line has characteristic impedance Z_0 or characteristic admittance Y_0, and propagation constant β. They are fed by the transversely connected voltage source V_g. The source is connected at the distance x from the open end, near the current maxima, to achieve a high valued loaded Q-factor. The shunt-connected source of frequency ω is also located at a distance x' from the maximum of the current on the line.

In the case of the $\lambda_g/2$-resonator of Fig (18.17b), the source is connected at distance x from the left-hand side of the resonator. The source has an internal impedance R_g. It is connected to the resonator using a transmission line of characteristic admittance $Y_s = 1/R_g$. The connecting line is matched to the source. Usually, R_g is 50 Ω. At the junction a–a′, the total admittance, due to two-open-circuited line sections, is computed as follows:

$$Y_{in} = jY_0 \tan\frac{\omega x}{v_p} + jY_0 \tan\frac{\omega(\ell - x)}{v_p} + Y_s. \tag{18.4.20}$$

Both ends open-circuited line supports the parallel resonance. Therefore, at the plane a–a′, the line section acts as a parallel resonator. Under the resonance condition, the total susceptance at the plane a–a′ is zero $B_{in}|_{\omega=\omega_r} = 0$. It helps to compute the resonating length:

$$\tan\frac{\omega_r x}{v_p} = -\tan\frac{\omega_r(\ell - x)}{v_p} = \tan\left[n\pi - \frac{\omega_r(\ell - x)}{v_p}\right]$$
$$\Rightarrow \frac{\omega_r x}{v_p} = \left[n\pi - \frac{\omega_r(\ell - x)}{v_p}\right] \Rightarrow \frac{\omega_r\ell}{v_p} = n\pi. \tag{18.4.21}$$

The length of the resonator is

$$\ell = n\pi\frac{v_p}{\omega_r} = n\frac{2\pi}{2\beta} = n\frac{\lambda_g}{2}. \tag{18.4.22}$$

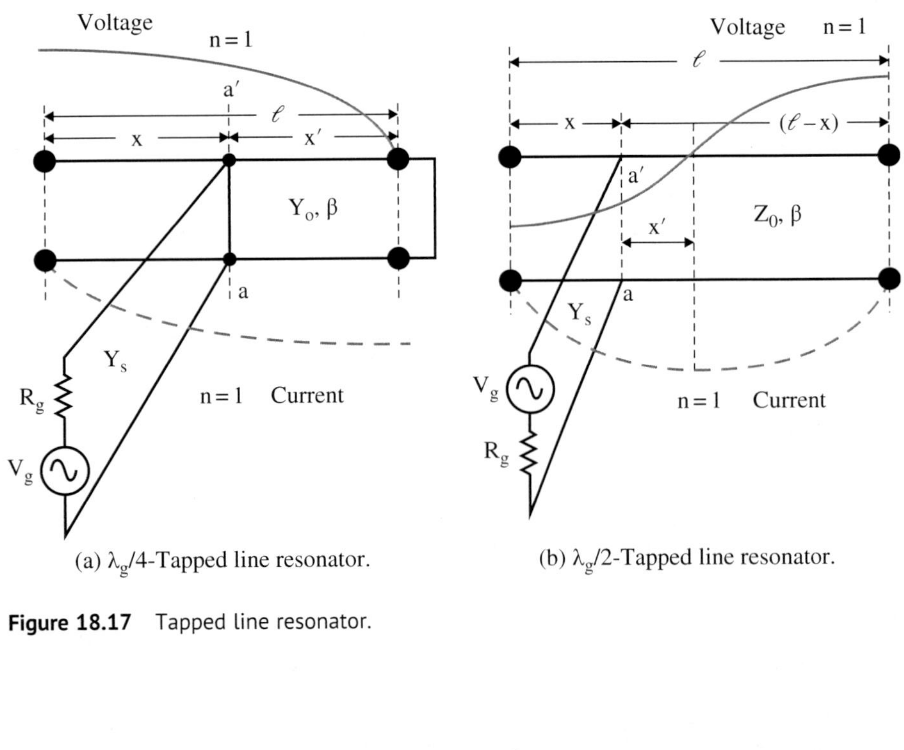

Figure 18.17 Tapped line resonator.

Thus, the tapped line is still a $\lambda_g/2$-type of resonator. The open-end discontinuity has been neglected in this discussion. The fundamental resonance frequency occurs for n = 1. The resonance frequency is given by the total length of the resonator. It does not depend on the feed location. The external Q-factor of a tapped line resonator is computed below, on neglecting the resonator loss, i.e. by taking G = 0 in equation (18.4.5):

$$Q_E = \frac{\omega_r}{2G_g}\frac{dB_{in}}{d\omega}\Big|_{\omega=\omega_r}. \tag{18.4.23}$$

The derivative of the input susceptance given by equation (18.4.20) is obtained as follows:

$$\frac{dB_{in}}{d\omega} = Y_0 \sec^2\left[\frac{\omega x}{v_p}\right]\frac{x}{v_p} - Y_0 \sec^2\left[\omega\left(\frac{\ell - x}{v_p}\right)\right]\frac{(\ell - x)}{v_p}. \tag{18.4.24}$$

At the resonance, $\frac{\omega_r \ell}{v_p} = \beta_g \ell = \pi$ and $\cos\left[\pi - \frac{\omega_r x}{v_p}\right] = -\cos\frac{\omega_r x}{v_p}$.

Therefore, the derivative of the input susceptance at the resonance is

$$\frac{dB_{in}}{d\omega}\Big|_{\omega=\omega_r} = \frac{Y_0 \ell}{v_p}\sec^2\left[\frac{\omega_r x}{v_p}\right] = \frac{\ell}{v_p Z_0 \cos^2(\omega_r x/v_p)}. \tag{18.4.25}$$

Finally, the external Q-factor is computed as follows:

$$Q_E = \frac{\omega_r R_g \ell}{2v_p Z_0 \cos^2(\omega_r x/v_p)} = \frac{\omega_r \lambda_g/2}{2v_p}\frac{R_g}{Z_0 \cos^2(\omega_r x/v_p)}$$

$$\Rightarrow Q_E = \frac{\pi}{2}\frac{R_g}{Z_0 \cos^2(\omega_r x/v_p)}. \tag{18.4.26}$$

However, for the distance of feed x', measured from the current maxima, and using $\beta = \omega/v_p$, $\beta(x + x') = \pi/2$, i. e. $\beta x = (\pi/2 - \beta x')$, the above equation could be rewritten as

$$Q_E = \frac{\pi}{2}\frac{R_g}{Z_0 \sin^2(\omega_r x'/v_p)}. \tag{18.4.27}$$

The above expression is also applicable to the $\lambda_g/4$-resonator, shown in Fig (18.17a), with a little change

$$Q_E = \frac{\pi}{4}\frac{R_g}{Z_0 \sin^2(\omega_r x'/v_p)}. \tag{18.4.28}$$

For the $\ell = \lambda_g/4$-resonator, the external Q-factor is one-half of the external Q-factor of the $\lambda_g/2$-resonator. On choosing x' near current maxima, a large value of Q_E could be obtained. However, in practice, the loaded Q-factor may not be more than 100, as the unloaded Q-factor is low near the current maxima [B.9].

18.4.4 Feed to Planar Transmission Line Resonator

The external circuits at the input and output of a line resonator determine the loaded Q-factor that determines the bandwidth of a resonator. We have considered the loading of the external circuits both for the lumped resonator and for the two-wire line resonator. The analysis of the previous section is also applicable to the planar line resonators.

The feed to several reflection types, and transmission type microstrip and CPW resonators, are shown in Fig (18.18 and 18.19), respectively. The feeds could be modeled by the transmission line and lumped circuit models. The step, gap, and transitions can be modeled by the available closed-form expressions. Alternatively, these can be modeled using the T and π-networks, and values of network elements could be determined with the help of an EM-Simulator [B.1, B.6, B.9, B.13, B.16, B.18]. Finally, the filters, using these resonators, are developed more effectively on an EM-Simulator.

18.5 Coupled Resonators

The planar coupled resonators are key circuit elements for the development of high-performance bandpass filters. The coupling can exist between two identical resonators with the same isolated frequencies, or between two resonators with different isolated resonance frequencies. The first one is known as the *synchronously* tuned coupled resonator and the second one is known as the *asynchronously* tuned coupled resonator [J.37, J.38, B.6]. The coupling between the resonators can take place through any of the three mechanisms – (i) *electric field, i.e. the capacitive coupling,* (ii) *magnetic field, i.e. the inductive coupling, and* (iii) the *mixed couplings* using both the electric and magnetic fields, i.e. both the capacitive and inductive coupling simultaneously. The basic theory of coupling is discussed in chapter 11 for coupled transmission lines. It applies to the coupled resonators also. The coupled planar resonators can be in the microstrip, CPW, or slot line environment, and can be in any of the several shapes. This section considers three cases of the coupled microstrip resonators, and one case of the CPW-fed slot line coupled resonator.

18.5.1 Coupled Microstrip Line Resonator

Figure (18.20a) shows a gap-fed transmission-type $\lambda_g/2$-coupled microstrip resonator. Whereas, Fig (18.20b) shows one end short-circuited reflection-type $\lambda_g/4$-coupled microstrip resonator. The symmetrically coupled resonator supports the *even-mode and odd-mode*

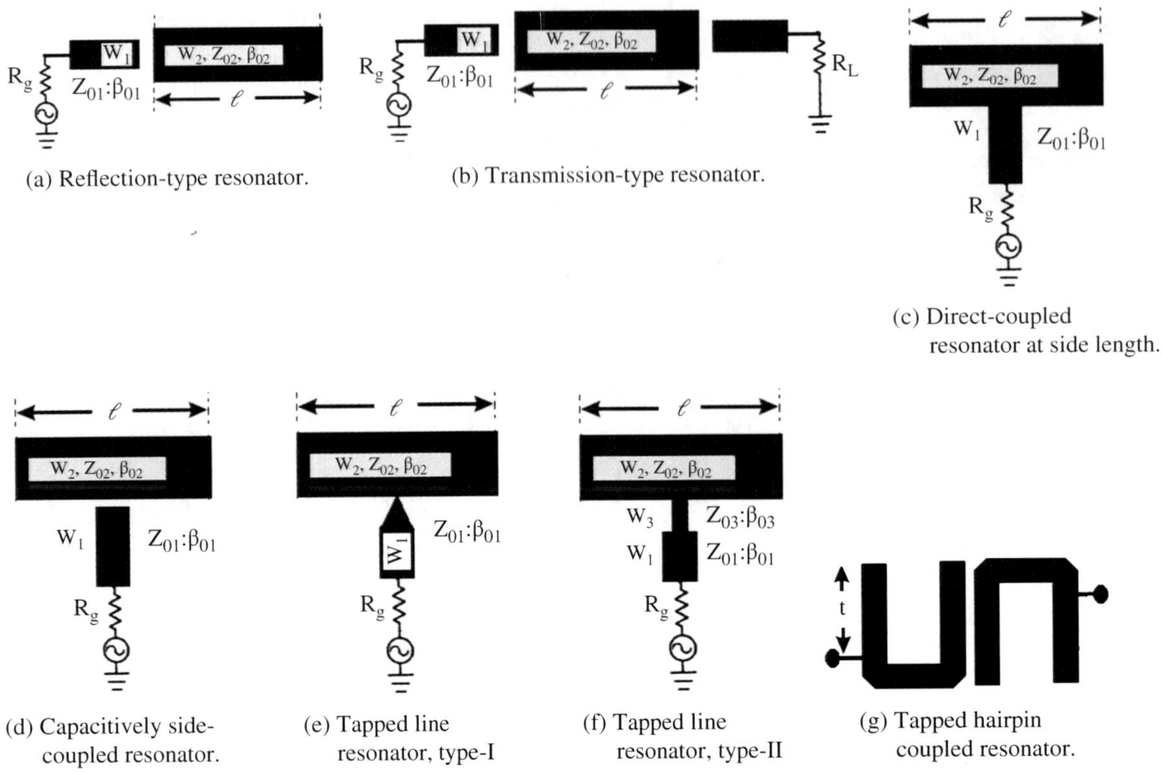

(a) Reflection-type resonator.
(b) Transmission-type resonator.
(c) Direct-coupled resonator at side length.
(d) Capacitively side-coupled resonator.
(e) Tapped line resonator, type-I
(f) Tapped line resonator, type-II
(g) Tapped hairpin coupled resonator.

Figure 18.18 Feeding arrangement of $\lambda_g/2$ microstrip line resonators.

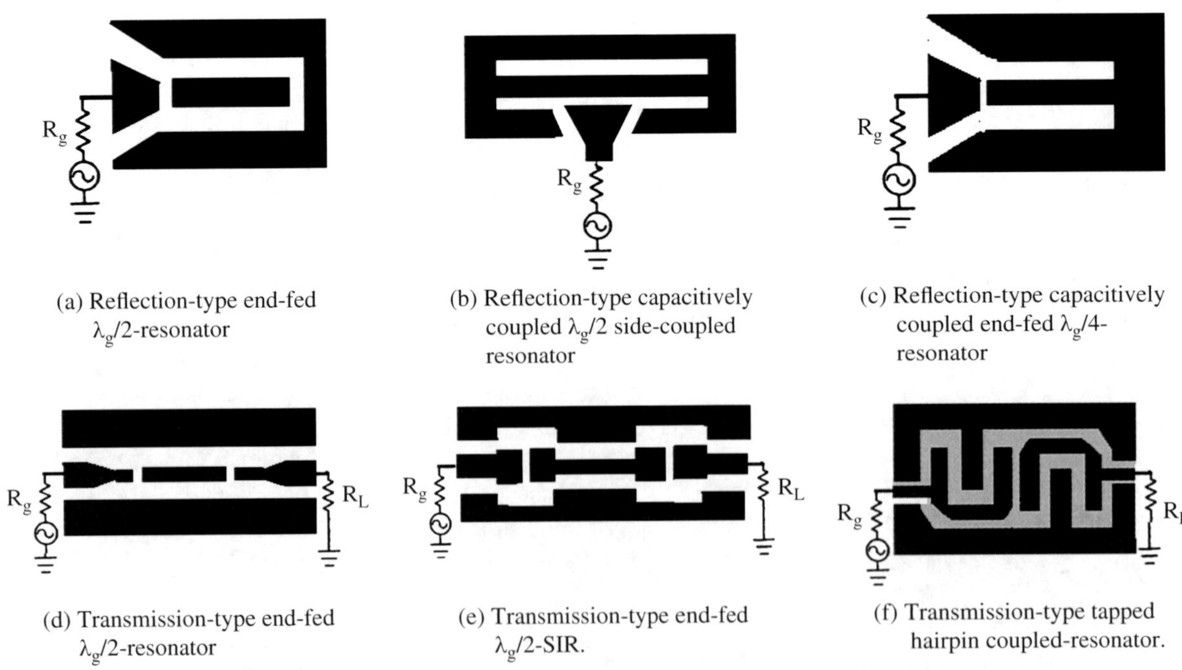

(a) Reflection-type end-fed $\lambda_g/2$-resonator
(b) Reflection-type capacitively coupled $\lambda_g/2$ side-coupled resonator
(c) Reflection-type capacitively coupled end-fed $\lambda_g/4$-resonator
(d) Transmission-type end-fed $\lambda_g/2$-resonator
(e) Transmission-type end-fed $\lambda_g/2$-SIR.
(f) Transmission-type tapped hairpin coupled-resonator.

Figure 18.19 Feeding arrangement of CPW line resonators.

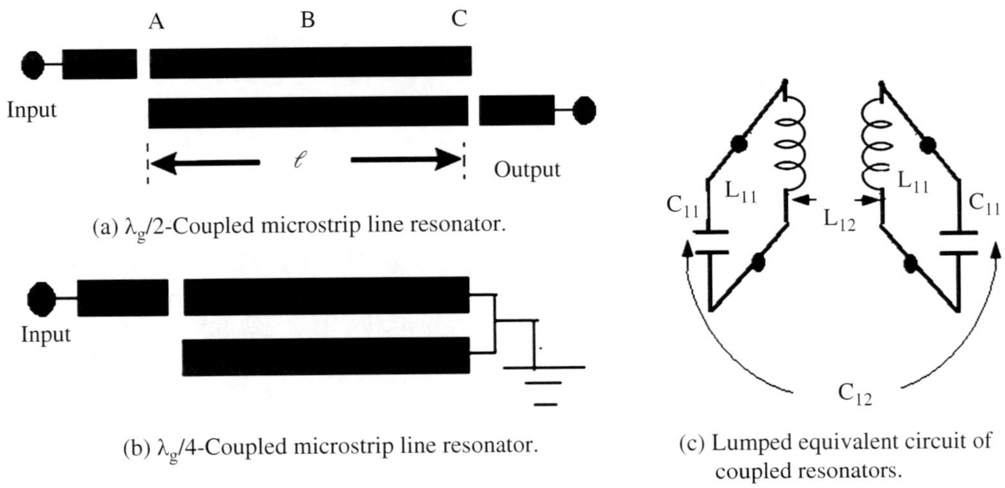

(a) $\lambda_g/2$-Coupled microstrip line resonator.

(b) $\lambda_g/4$-Coupled microstrip line resonator.

(c) Lumped equivalent circuit of coupled resonators.

Figure 18.20 Coupled microstrip line resonators.

resonances, in place of one resonance of an isolated line resonator. Using the concepts of the even–odd mode analysis, discussed in section (11.4) of chapter 11, the coupled line resonators can be decomposed into two isolated line resonators, supporting the even- and odd-mode resonance separately. The following conditions can be written for the even- and odd-mode resonance for the decomposed resonators:

$$\lambda_g/2-\text{resonator: Even mode: } \beta_e(\ell + 2\Delta\ell_e) = n\pi \quad \text{(a)}$$
$$\text{: Odd mode: } \beta_o(\ell + 2\Delta\ell_o) = n\pi \quad \text{(b)}$$
$$\lambda_g/4-\text{resonator: Even mode: } \beta_e(\ell + \Delta\ell_e) = (2n-1)\pi/2 \quad \text{(c)}$$
$$\text{: Odd mode: } \beta_o(\ell + \Delta\ell_o) = (2n-1)\pi/2 \quad \text{(d)}, \qquad (18.5.1)$$

where n = 1,2,.... The odd- and even-mode propagation constants of the coupled microstrip lines are β_o and β_e. The coupled line length is ℓ; $\Delta\ell_e$ and $\Delta\ell_o$ are the line extensions of the even- and odd-mode open-end discontinuity. The above expressions provide the even- and odd-mode resonance frequencies in terms of the even- and odd-mode effective relative permittivity:

$$\lambda_g/2-\text{resonator: Even/Odd mode:}$$
$$f_{re/o} = \frac{nc}{2\sqrt{\varepsilon_{reff,e/o}}(\ell + 2\Delta\ell_{e/o})} \quad \text{(a)}$$
$$\lambda_g/4-\text{resonator: Even/Odd mode:}$$
$$f_{re/o} = \frac{(2n-1)c}{4\sqrt{\varepsilon_{reff,e/o}}(\ell + 2\Delta\ell_{e/o})} \quad \text{(b)}, \qquad (18.5.2)$$

where c is the velocity of the EM-wave in free space. The even-/odd-mode effective relative permittivity, $\varepsilon_{reff,\ e/o}$, can be computed either from the closed-form expressions given in section (12.1) of chapter 12, or using the variational method discussed in section (14.5) of chapter 14. The open-end extension $\Delta\ell_{e/o}$ in the even-/odd-mode microstrip lines can again be computed from the closed-form expressions given in equation [J.39]. The above expressions can also be expressed in terms of the coupled lines primary constants, i.e. the self and mutual capacitances, and the self and mutual inductance presented in equations (11.3.10) and (11.3.21) of chapter 11. The even and odd frequencies of the coupled $\lambda_g/2-$resonator are computed as follows:

$$\text{Even mode: } f_{re} = \frac{nv_{pe}}{2(\ell + 2\Delta\ell_e)} = \frac{n}{2\sqrt{C_e L_e}(\ell + 2\Delta\ell_e)}$$
$$\Rightarrow \qquad f_{re} = \frac{n}{2\sqrt{(C_{11}-C_{12})(L_{11}+L_{12})}\ (\ell + 2\Delta\ell_e)} \quad \text{(a)}$$
$$\text{Odd mode: } f_{ro} = \frac{n}{2\sqrt{(C_{11}+C_{12})(L_{11}-L_{12})}\ (\ell + 2\Delta\ell_o)} \quad \text{(b)}. \qquad (18.5.3)$$

The above expressions can be easily modified for the coupled $\lambda_g/4-$ resonator. The above expressions involve both the capacitive and inductive couplings.

18.5.2 Circuit Model of Coupled Microstrip Line Resonator

Following the lumped circuit modeling of a resonating line section, the coupled microstrip line resonator can also be modeled by the equivalent coupled parallel resonant circuit, shown in Fig (18.20c). It shows two identical parallel resonant circuits corresponding to each of the coupled lines. In general, both the electric and magnetic couplings are involved in the coupling process of the resonators. The electric coupling is through the coupling capacitor C_{12}, while the magnetic coupling is through the coupling inductor L_{12}. The C_{11} and L_{11} are the self-capacitance and self-inductance of both resonators. The electric and magnetic couplings are considered below individually [J.37, J.38, B.6].

Electric (Capacitive) Coupling

Figure (18.20a) shows the open-circuited coupled resonator with the voltage maxima, i.e. the electric field maxima, at the locations A and C. The maximum current, i.e. the maximum magnetic field, exists at the location B. So, the input and output lines are assumed to couple with the resonator only electrically (capacitively), without any magnetic coupling. However, at location B, a loop (not shown) can couple with resonator magnetically or inductively. Figure (18.20c) describes the electric coupling only through the coupling capacitor C_{12}, with $L_{12} = 0$. The coupled line resonator supports both the even- and odd-mode resonance frequencies. These frequencies are computed below; using the even- and odd-mode capacitances and self-inductance given by equations (11.3.10b) and (11.3.10d) of chapter 11:

$$\text{Even mode:}\quad f^{e}_{r,e} = \frac{1}{2\pi\sqrt{L_{11}C_e}} = \frac{1}{2\pi\sqrt{L_{11}(C_{11} - C_{12})}} = \frac{f_r}{\sqrt{(1 - k_C)}} \qquad \text{(a)}$$

$$\text{Odd mode:}\quad f^{e}_{r,o} = \frac{1}{2\pi\sqrt{L_{11}C_o}} = \frac{1}{2\pi\sqrt{L_{11}(C_{11} + C_{12})}} = \frac{f_r}{\sqrt{(1 + k_C)}} \qquad \text{(b),}$$

(18.5.4)

where the self-resonance frequency of the uncoupled resonator f_r, and the capacitive coupling coefficients k_C are defined as follows:

$$f_r = \frac{1}{2\pi\sqrt{L_{11}C_{11}}} \quad \text{(a)}, \qquad k_C = \frac{C_{12}}{C_{11}}, 0 \le k_C < 1 \quad \text{(b)}.$$

(18.5.5)

As $(1 - k_C) < 1$ and $(1 + k_C) > 1$, the even-mode frequency is higher than the odd-mode frequency. The electric, i.e. the capacitive, coupling coefficient k_C can be expressed in terms of the even- and odd-mode resonance frequencies as follows:

$$\left(\frac{f^{e}_{r,e}}{f^{e}_{r,o}}\right)^2 = \frac{1 + k_C}{1 - k_C}, \qquad \Rightarrow k_C = \frac{\left(f^{e}_{r,e}\right)^2 - \left(f^{e}_{r,o}\right)^2}{\left(f^{e}_{r,e}\right)^2 + \left(f^{e}_{r,o}\right)^2}.$$

(18.5.6)

Equation (18.5.6) is a practically useful expression to determine the k_C. The even- and odd-mode resonance frequencies of an electrically coupled microstrip can be obtained on an EM-simulator from the S_{21} frequency response. Figure (18.21a) shows an S_{21} response of the electric coupling of a coupled resonator for three separations. Two resonance peaks, corresponding to the odd-mode and even-mode excitations, move away from each other for a decrease in the separation of the resonators, i.e. with an increase in the electric coupling k_C. Thus, *a larger separation of two resonance peaks, with a larger dip in transmission (>10 dB) at the center frequency, shows a stronger coupling with a higher value of k_C*. The value of k_C for a weakly edge-coupled microstrip resonator is $k_C < 0.1$.

Magnetic (Inductive) Coupling

The even- and odd-mode resonance frequencies of only magnetically coupled resonators, with $L_{12} \neq 0$ and $C_{12} = 0$, are obtained as

$$\text{Even mode:}\quad f^{m}_{r,e} = \frac{1}{2\pi\sqrt{L_eC_{11}}} = \frac{1}{2\pi\sqrt{(L_{11} + L_{12})C_{11}}} = \frac{f_r}{\sqrt{(1 + k_L)}} \qquad \text{(a)}$$

$$\text{Odd mode:}\quad f^{m}_{r,o} = \frac{1}{2\pi\sqrt{L_oC_{11}}} = \frac{1}{2\pi\sqrt{(L_{11} - L_{12})C_{11}}} = \frac{f_r}{\sqrt{(1 - k_L)}} \qquad \text{(b),}$$

(18.5.7)

where the magnetic coupling is defined as $k_L = L_{12}/L_{11}$, $0 \le k_L < 1$. The above equations are obtained using equation (11.3.21) of chapter 11. Once the even- and odd-mode resonance frequencies of magnetically coupled resonators are known from the EM-Simulation of the S_{21}-frequency response, the magnetic coupling is determined from the following equations:

$$\left(\frac{f^{m}_{r,e}}{f^{m}_{r,o}}\right)^2 = \frac{1 - k_L}{1 + k_L}, \qquad \Rightarrow k_L = \frac{\left(f^{m}_{r,o}\right)^2 - \left(f^{m}_{r,e}\right)^2}{\left(f^{m}_{r,o}\right)^2 + \left(f^{m}_{r,e}\right)^2}.$$

(18.5.8)

In this case, the above results show that the odd-mode frequency is higher than the even-mode frequency, as $(1 - k_L) <1$ and $(1 + k_L)>1$. Figure (18.21b) shows the S_{21} response of the magnetic coupling of coupled resonators

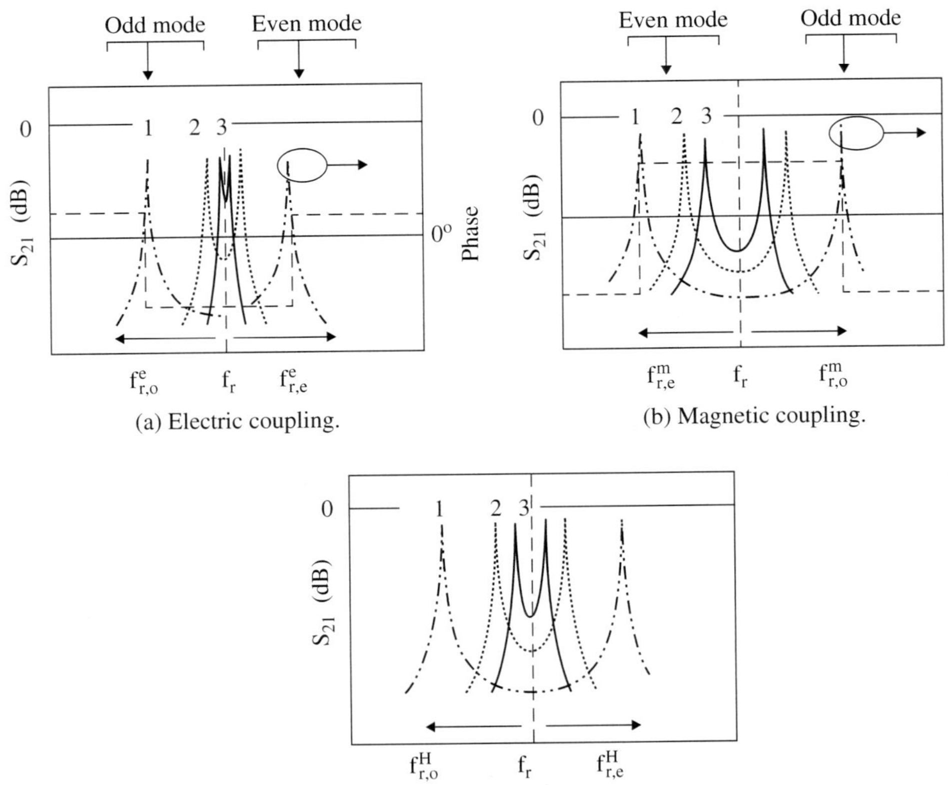

(c) Hybrid coupling.

Figure 18.21 S_{21} response of coupled microstrip resonator.

for three separation gaps of the resonators. Again, a larger separation of two resonance peaks indicates stronger coupling. Also, the separation of peaks, for the same coupling distance, is more for the magnetic coupling, as compared against the electric coupling of Fig (18.21a). It demonstrates that the *magnetic coupling is stronger than the electric coupling*, i.e. $k_L > k_C$. It indicates weak coupling for the edge-coupled microstrip resonators, whereas the broadside-coupled microstrip resonators have stronger coupling. Further, the phases of the electric coupling and magnetic coupling are the reverses of each other. So, if the electric coupling is taken as a *positive coupling*, then the magnetic coupling is a *negative coupling*.

Hybrid Coupling – Both Capacitive and Inductive Coupling

In general, the coupled microstrip resonators have both the electric and magnetic couplings. Such mixed-coupling is called a *hybrid coupling*. The pure electric and pure magnetic couplings are an idealization of the coupling mechanism. The electric coupling and magnetic coupling, discussed above, are the *dominant* electric coupling and *dominant* magnetic coupling. The mixed-coupling is due to the presence of both C_{12} and L_{12}. The even- and odd-mode resonance frequencies of the hybrid coupling are computed as follows:

$$\text{Even mode:}\ f^H_{r,e} = \frac{1}{2\pi\sqrt{L_e C_e}} = \frac{1}{2\pi\sqrt{(L_{11}+L_{12})(C_{11}-C_{12})}} = \frac{f_r}{\sqrt{(1+k_L)(1-k_C)}} \quad \text{(a)}$$

$$\text{Odd mode:}\ f^H_{r,o} = \frac{1}{2\pi\sqrt{L_o C_o}} = \frac{1}{2\pi\sqrt{(L_{11}-L_{12})(C_{11}+C_{12})}} = \frac{f_r}{\sqrt{(1-k_L)(1+k_C)}} \quad \text{(b).} \tag{18.5.9}$$

Following the previous discussion, the hybrid coupling coefficient k_H is defined as follows:

$$\left(\frac{f^H_{r,o}}{f^H_{r,e}}\right)^2 = \frac{(1+k_L)(1-k_C)}{(1-k_L)(1+k_C)}$$

$$\Rightarrow \frac{\left(f^H_{r,o}\right)^2 - \left(f^H_{r,e}\right)^2}{\left(f^H_{r,o}\right)^2 + \left(f^H_{r,e}\right)^2} \approx k_L - k_C \quad \text{(a)}$$

$$\text{Define,}\ k_H = \frac{\left(f^H_{r,o}\right)^2 - \left(f^H_{r,e}\right)^2}{\left(f^H_{r,o}\right)^2 + \left(f^H_{r,e}\right)^2} \quad \text{(b)}$$

$$k_H = k_L - k_C \quad \text{(c).} \tag{18.5.10}$$

The above equation (18.5.10a) is obtained ignoring $(k_L.k_C) < 1$. Equation (18.5.10c) shows that magnetic and electric couplings are out-of-phase. In this case, the electric and magnetic couplings cancel each other. However, if the mutual inductance or the mutual capacitance change sign, both couplings enhance the hybrid coupling [B.6].

The above equations also show the relation among three coupling coefficients. The case of only electric and magnetic couplings follows from the case of the combined coupling. Figure (18.21c) shows the S_{21} frequency response of the edge-coupled microstrip resonators for the combined coupling. For the same separation of resonators, two peaks of the combined coupling are less apart, as compared to that of the magnetic coupling, whereas it is more than that of the electric coupling, i.e. $k_L > k_H > k_C$. It is possible that at a specific frequency, called the *null frequency* f_{null}, the transmission of a coupled resonator could be practically zero. It is related to the uncoupled resonator frequency as follows [J.40]:

$$f_{null} = f_r\sqrt{\frac{k_L}{k_C}}. \tag{18.5.11}$$

It shows that for $k_L > k_C$, the null occurs in at a frequency above f_r in the upper stopband, and in the case of $k_C > k_L$, the null occurs at the frequency below f_r lower stopband. This is very useful for the design of a BPF with the transmission zeros near the lower and upper cutoff frequencies of the filter. However, it requires independent control of the electric and magnetic couplings [J.40]. Such applications are illustrated below.

18.5.3 Some Structures of Coupled Microstrip Line Resonator

The $\lambda_g/2$-end coupled microstrip, or CPW resonators, used in the design of a BPF, are electrically coupled resonators. Figure (18.22a–d) shows four cases of the coupling arrangements of the $\lambda_g/2$-microstrip coupled open square resonators [J.37, J.41, B.6]. The first case shows the electric field coupling between the open ends, where the electric field is dominant. The second case shows the magnetic coupling between two arms, where the magnetic field is dominant, due to the maximum of current at the center of the resonators. The electric field decays faster than the magnetic field, providing stronger magnetic coupling. Figure (18.21a and b) shows responses for the first and second cases of Figure (18.22). The third and fourth cases are for the mixed coupling. Figure (18.21c) shows the response of the mixed coupling. The coupling is controlled by changing the separation between the resonators. The zero coupling is obtained at some distance that produces a transmission zero at the null frequency. These

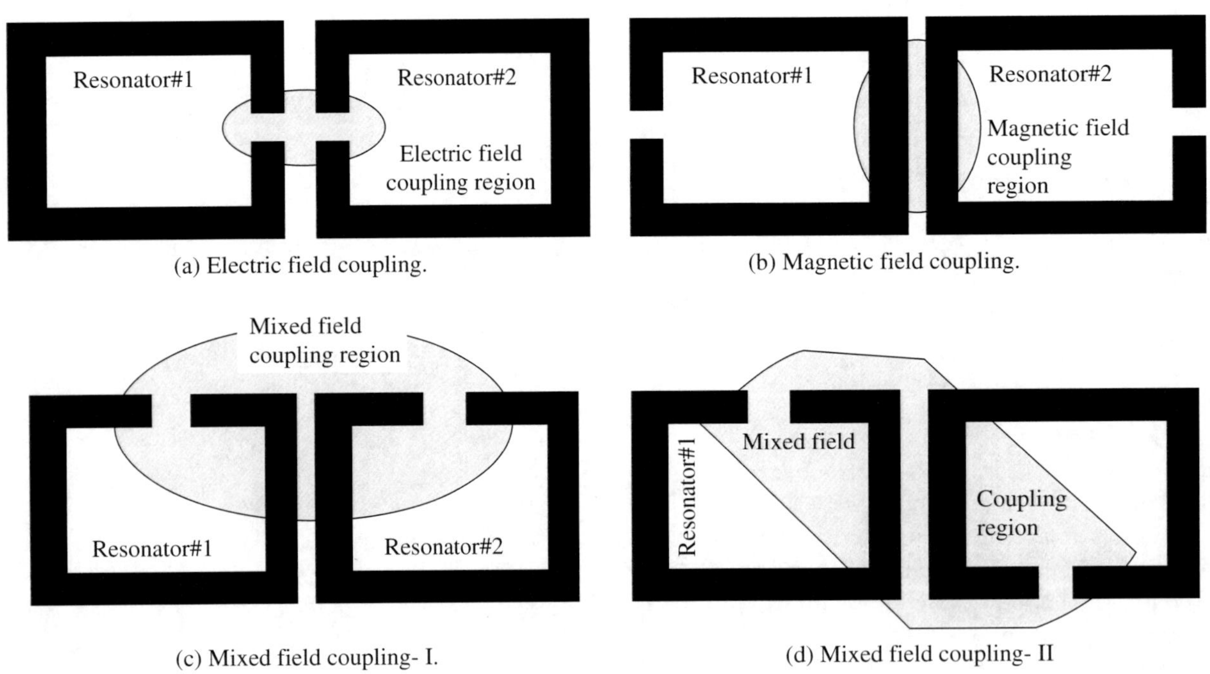

Figure 18.22 Coupling arrangements between open square resonators.

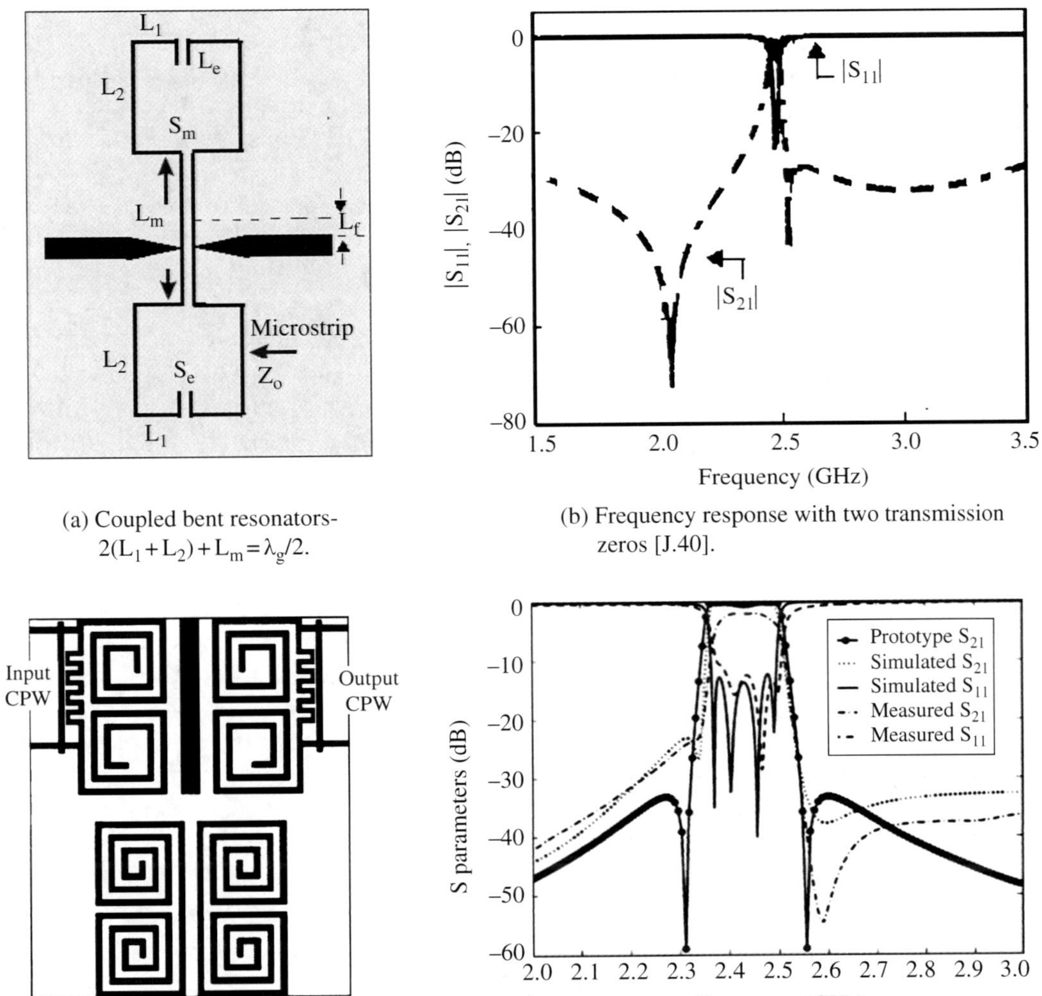

(a) Coupled bent resonators- $2(L_1+L_2)+L_m=\lambda_g/2$.

(b) Frequency response with two transmission zeros [J.40].

(c) Slot line-based coupled spiral resonators forming 4-pole BPF.

(d) Response of BPF showing both sides transmission zeros [42].

Figure 18.23 Coupling arrangements of bent microstrip resonators and slot line resonator. *Source:* From Tang et al. [J.40]. © 2009, IEEE. From Azadegan and Sarabandi [J.42]. © 2004, IEEE.

structures have been used to develop the multipole BPF using the *coupling matrix* for the coupled multiple resonators.

In the case of the microstrip coupled open-square resonators, it is difficult to control the electric and magnetic couplings, almost independent of each other. However, Fig (18.23a) shows a bent form of $\lambda_g/2$-coupled resonator [J.40] that provides better independent control of the electric and magnetic couplings. The longer coupled lengths L_m and their separation S_m control the magnetic coupling k_L, whereas the bent lengths, near open ends L_e, and their separation S_e control the electric coupling k_C. The coupling-ratio k_L/k_C can be adjusted to place the transmission zero at the left-hand side of the passband of 2-pole BPF. The input and output of the 2-pole BPF of Fig (18.23a) are tapped fed. Their location L_f controls another transmission zero of BPF. The resonators are designed around high-impedance ($Z_0 = 110\ \Omega$) microstrip, as it provides a larger tuning range for the coupling coefficients. The frequency response of Fig (18.23b) demonstrates both transmission zeros.

Figure (18.23c) shows the slot-line-based, CPW-fed, coupled resonator 4-pole BPF. To reduce the size of the BPF, the resonators are spiral loaded at both ends [J.42]. It is a series resonant-type structure that involves both the electric and magnetic couplings. By controlling these couplings, the transmission zero can be placed on either left or right of the passband. Figure (18.23d) shows a response to the 4-pole BPF with the transmission zeros on both sides.

The inductive and capacitive couplings to the microstrip resonators, for instance, the SIR type, can also be achieved with the help of the short-circuited stubs [J.43, J.44]. These coupling arrangements provide high-performance filters. There are several other coupling arrangements for the microstrip resonators.

18.6 Microstrip Patch Resonators

The two-dimensional microstrip resonators can be arranged in three groups – (i) *patch and ring resonators*, (ii) *2D space-filling fractal resonators*, and (iii) *dual-mode resonators*. These structures are important for both the microwave circuits and antenna applications. This section summarizes the basic properties of patch and ring resonators. The fractal resonators and dual-mode resonators are discussed in the next sections (18.7) and (18.8), respectively.

Figure (17.2) of chapter 17 shows several geometrical shapes of microstrip resonators – the patch and rings. These have been extensively studied and used for both the circuit and antenna applications [B.2–B.6, B.9, B.12, B.19]. Four structures, shown in Fig (18.24), are discussed in this section. Using the full-wave analysis methods, these resonating structures are analyzed to obtain the resonant frequencies, Q-factors, and field structures. However, the *cavity model* of these structures has a special attraction, as it helps visualization of the resonating modes [J.45–J.49]. It is a simple and effective method to get the resonant frequencies and Q-factors. The wave equation in the enclosure, made of the EW and MW, is solved like all metallic cavity resonators [B.10–B.12, B.15, B.16]. The cavity resonators are just the enclosed waveguide sections. The metallic waveguide of a rectangular or circular cross-section supports TE^Z_{mn} and TM^Z_{mn} modes, defined with respect to the Z-direction of propagation. The reflecting metallic surfaces are placed at two ends of a waveguide section to form a cavity resonator with TE^Z_{mnp} and TM^Z_{mnp} modes, where p is an integer, showing the number of half-wavelength variations along the Z-axis. The *method of separation of variables*, along with boundary conditions, discussed in chapter 7, is used to get the modal solutions. This method has been used to get the cavity model of the patch resonator. Below the results are presented, without going into details of their derivations. The expressions are not presented for the fields inside a cavity. This information is available in the standard books [B.2, B.10–B.12, B.16].

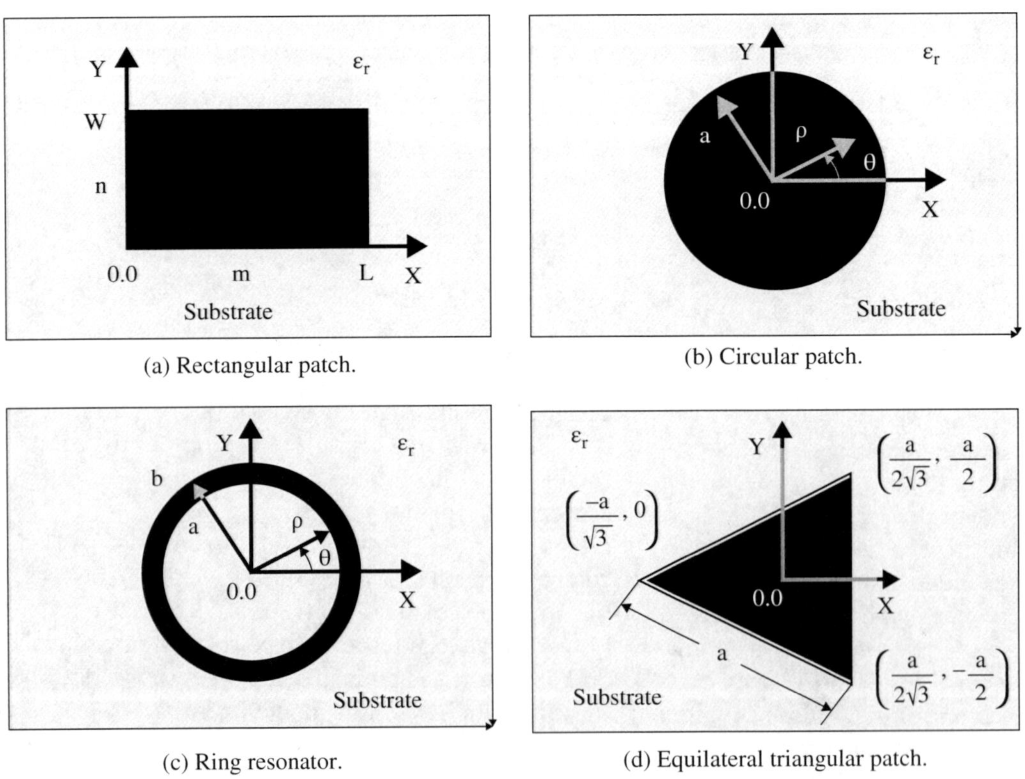

Figure 18.24 Some microstrip patch resonators. Thickness h of the substrate is along Z-axis.

To model the patch resonators as the cavity resonators, the bottom and top surfaces are taken as the *EWs*, whereas the sidewalls are treated as the *MWs*. The substrate thickness h is the separation between two EWs. The substrate thickness is much less than the half-wavelength, i.e. $p = 0$. Normally, the integer p is dropped from the resonating-mode description of a patch resonator. As there is an E-field component in the Z-direction, the patch resonators support the TM^z_{mn} modes. Following the discussion in chapter 7, these modes could be described by the electric scalar potential eigenfunction $\phi^e(x, y, z) = \phi^e_{mn}(x, y)e^{-j\beta_z Z}$ for a rectangular cross-section; and by $\phi^e(x, y, z) = \phi^e_{mn}(\theta, \rho)e^{-j\beta_z Z}$ for the circular disc and circular ring cross-sections. The wave equation is solved in an appropriate coordinate system to get the eigenfunction of a patch resonator [B.2, B.12]. The case of the equilateral triangular patch resonator is more involved. However, Schelkunoff has provided the eigenfunction of equilateral triangular cross-section waveguide [B.20]. It is used for the cavity model of a triangular patch resonator [J.48, J.49]. The wave equations, in both the rectangular and cylindrical coordinate systems, for the electric scalar potential function, are given below:

$$\frac{\partial^2\phi^e(x,y)}{\partial x^2} + \frac{\partial^2\phi^e(x,y)}{\partial y^2} + k_c^2\phi^e(x,y) = 0 \qquad \text{(a)}$$

$$\frac{1}{\rho}\frac{\partial}{\partial\rho}\left(\rho\frac{\partial\phi^e}{\partial\rho}\right) + \frac{1}{\rho^2}\frac{\partial^2\phi^e}{\partial\theta^2} + k_c^2\phi^e(\rho,\theta) = 0 \qquad \text{(b)}$$

$$\text{where,}\ k_c^2 = k^2 - \beta_z^2 \quad \text{(c)}, \quad k = \omega\sqrt{\mu\varepsilon} \qquad \text{(d)}. \qquad (18.6.1)$$

The MW surfaces are located just at the edges of the conducting patches, so that relative permittivity ε_r is maintained inside the cavity. Under the MW boundary condition $(\partial\phi^e/\partial\hat{n} = 0)$, the electric scalar potential eigenfunction and also the modal cutoff wavenumber, $k_c = k_{mn}$, are computed. Under the resonance constraint, the modal resonance frequencies are obtained as

$$f_{mn} = \frac{k_{mn}}{2\pi\sqrt{\mu\varepsilon}} = \frac{ck_{mn}}{2\pi\sqrt{\varepsilon_r}}. \qquad (18.6.2)$$

18.6.1 Rectangular Patch

Figure (18.24a) shows a rectangular patch resonator. Its eigenfunction and cutoff wavenumber are obtained by solving equation (18.6.1a):

$$\phi^e_{mn} = \cos\left(\frac{m\pi}{L}\right)\cos\left(\frac{n\pi}{W}\right) \qquad \text{(a)}$$

$$k_{mn} = \left[\left(\frac{m\pi}{L}\right)^2 + \left(\frac{n\pi}{W}\right)^2\right]^{1/2} \qquad \text{(b)}, \qquad (18.6.3)$$

where, $m, n = 0, 1, 2, 3, \ldots$.

Using the above eigenfunction, the electric and magnetic field components are obtained from equation (7.1.28) of chapter 7 with a detailed solution in subsection (7.4.2). Figure (18.25a–c) shows the E_z-field distribution of a few TM_{mn}-modes. The magnetic currents (M), over the MWs, are also shown. At the location of $E_z = 0$, the input impedance is zero; and at the edge, where E_z is maximum, it is maximum. The feed can be suitably located to achieve the matching. The magnetic currents are determined at the apertures of substrate thickness. They help to compute the radiated power. It is important for antenna applications [B.2, B.5, B.21]. It is also important for evaluating the radiation Q-factor of a resonator. The radiation is less for a patch resonator on a thin and high permittivity substrate, such as on an alumina substrate. Following the process applicable to a rectangular waveguide, the conductor loss and dielectric losses of various modes are computed. Finally, the total Q-factor and the bandwidth $\approx$1% are obtained.

The cavity model, in its simplest form, ignores the fringe fields at the edges. So, the computed resonance frequency of a patch deviates from the measured one by several percent. The fringe fields increase both the length and width of a patch. In that case, the MWs should be placed at the end of the effective length L_{eff} and effective width W_{eff}. It helps to include the total field within the cavity model. Likewise, the fringe fields also decrease the relative permittivity of a patch resonator. The L_{eff} and W_{eff} are computed using the expression of open-end discontinuity [B.3], and the effective relative permittivity is obtained from closed-form expression discussed in subsection (8.2.1) of chapter 8. This arrangement improves the computation of resonance frequency. However, improvement is not adequate. Wolff and Knoppik [J.47] have noted that the fringe fields are mode dependent, and the *dynamic relative permittivity* of a patch resonator could be evaluated for more accurate computation of its resonance frequency. Their model has been generalized as the *modified Wolff model (MWM)* to improve the results further. The MWM applies to the patch resonators of several shapes under the multilayer conditions. The multilayer substrates could be lossless, lossy, isotropic, and anisotropic [J.50–J.60]. The MWM process, for a rectangular patch resonator, is summarized below.

18.6.2 Modified Wolff Model (MWM)

Figure (18.26a) shows a shielded multilayer microstrip rectangular patch resonator. The structure can be reduced to several important cases, including the single-substrate open-patch resonator. The MWM computes the resonance frequency of the fundamental mode with accuracy 0.5%, and the resonance frequencies

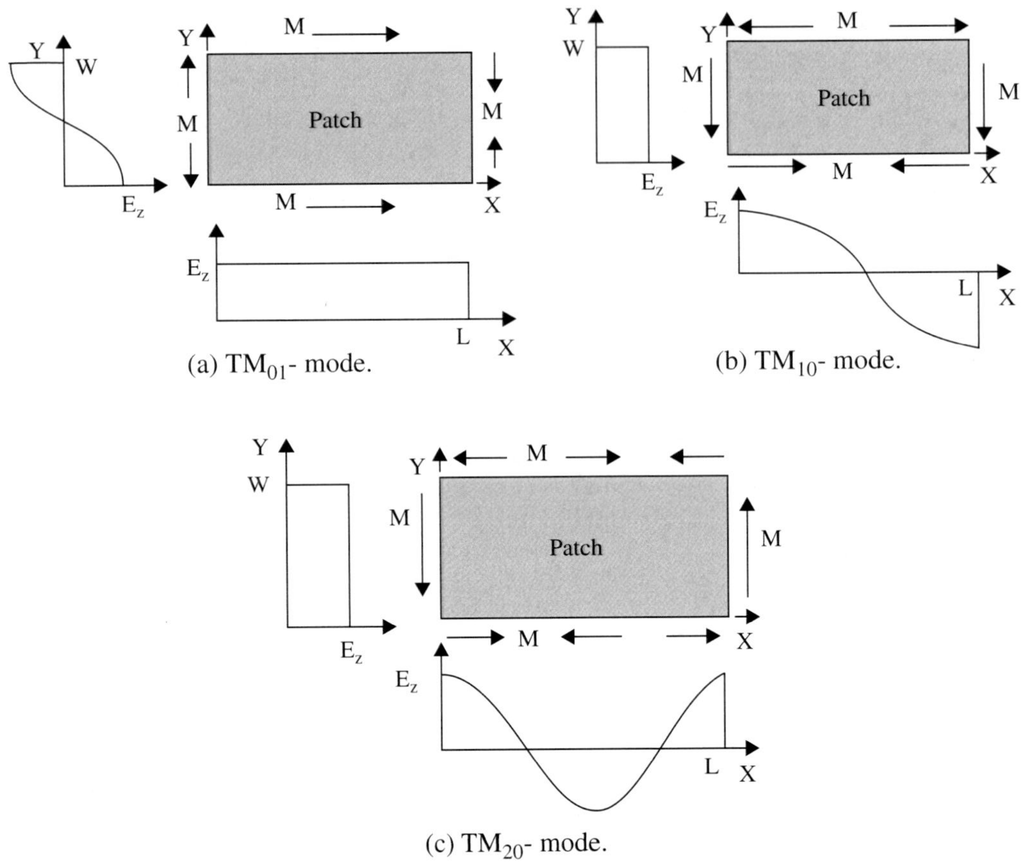

Figure 18.25 Ez-field distribution of a few TM$_{mn}$-modes. (M: Magnetic current at magnetic walls.)

of the higher-order modes with an accuracy within 1.7% [J.50, J.52]. Using the concept of the *mode-dependent dynamic relative permittivity* $\varepsilon_{r,dyn}$, the effective length (L_{eff}), and effective width (W_{eff}), equation (18.6.3) provides the following expression to compute the modal resonance frequency:

$$f_{mn} = \frac{k_{mn}}{2\pi\sqrt{\mu\varepsilon}} = \frac{c}{2\sqrt{\varepsilon_{r,dyn}}}\left[\left(\frac{m}{L_{eff}}\right)^2 + \left(\frac{n}{W_{eff}}\right)^2\right]^{1/2}. \quad (18.6.4)$$

The mode-dependent dynamic relative permittivity of a complete structure is computed by computing the dynamic capacitance of a structure with dielectric layers, and the dielectric layers replaced by an air medium:

$$\varepsilon_{r,dyn} = \frac{C_{dyn}(\varepsilon_{r1}, \varepsilon_{r2}, \varepsilon_{r3}, W)}{C_{dyn}(\varepsilon_{r1} = \varepsilon_{r2} = \varepsilon_{r3} = 1, W)} \quad (a)$$

$$\text{where, } C_{dyn}(\varepsilon_{r1}, \varepsilon_{r2}, \varepsilon_{r3}, W) = C_{0,dyn}(\varepsilon_{r1}) + 2C_{e1,dyn}(\varepsilon_{r1}, \varepsilon_{r2}, \varepsilon_{r3}, W) + 2C_{e2,dyn}(\varepsilon_{r1}, \varepsilon_{r2}, \varepsilon_{r3}, L) \quad (b). \quad (18.6.5)$$

In the above equation, $C_{0,dyn}(\varepsilon_{r1})$ is the central dynamic patch capacitance that is not influenced by the dielectric cover and shield. The capacitances $C_{ej,dyn}(\varepsilon_{r1}, \varepsilon_{r2}, \varepsilon_{r3}, W \text{ or } L)$, $(j = 1, 2)$ are the fringe capacitance, along the length and width of the patch. These capacitances are evaluated as follows [J.47, J.50]:

$$C_{o,dyn}(\varepsilon_{r1}) = \frac{C_{o,stat}(\varepsilon_{r1})}{\gamma_n\gamma_m} \quad (a)$$

$$C_{o,stat}(\varepsilon_{r1}) = \frac{\varepsilon_0\varepsilon_{r1}LW}{h_1} \quad (b), \quad \gamma_i = \begin{cases} 1, \text{for } i = 0 \\ 2 \text{ for } i \neq 0. \end{cases} \quad (18.6.6)$$

$$C_{e1,dyn}(\varepsilon_{r1}, \varepsilon_{r2}, \varepsilon_{r3}) = \frac{C_{e1,stat}(\varepsilon_{r1}, \varepsilon_{r2}, \varepsilon_{r3}, W)}{\gamma_n} \quad (a)$$

$$C_{e2,dyn}(\varepsilon_{r1}, \varepsilon_{r2}, \varepsilon_{r3}) = \frac{C_{e2,stat}(\varepsilon_{r1}, \varepsilon_{r2}, \varepsilon_{r3}, L)}{\gamma_m} \quad (b). \quad (18.6.7)$$

In the above equation (18.6.6), i = n, m is the modal number. The static fringe capacitance of the length-side of the patch is computed after subtracting the central capacitance from the total capacitance of a patch in the layered medium:

$$C_{e1,stat}(\varepsilon_{r1}, \varepsilon_{r2}, \varepsilon_{r3}, W) = \frac{1}{2}[C_tL - C_{o,stat}(\varepsilon_{r1})], \quad (18.6.8)$$

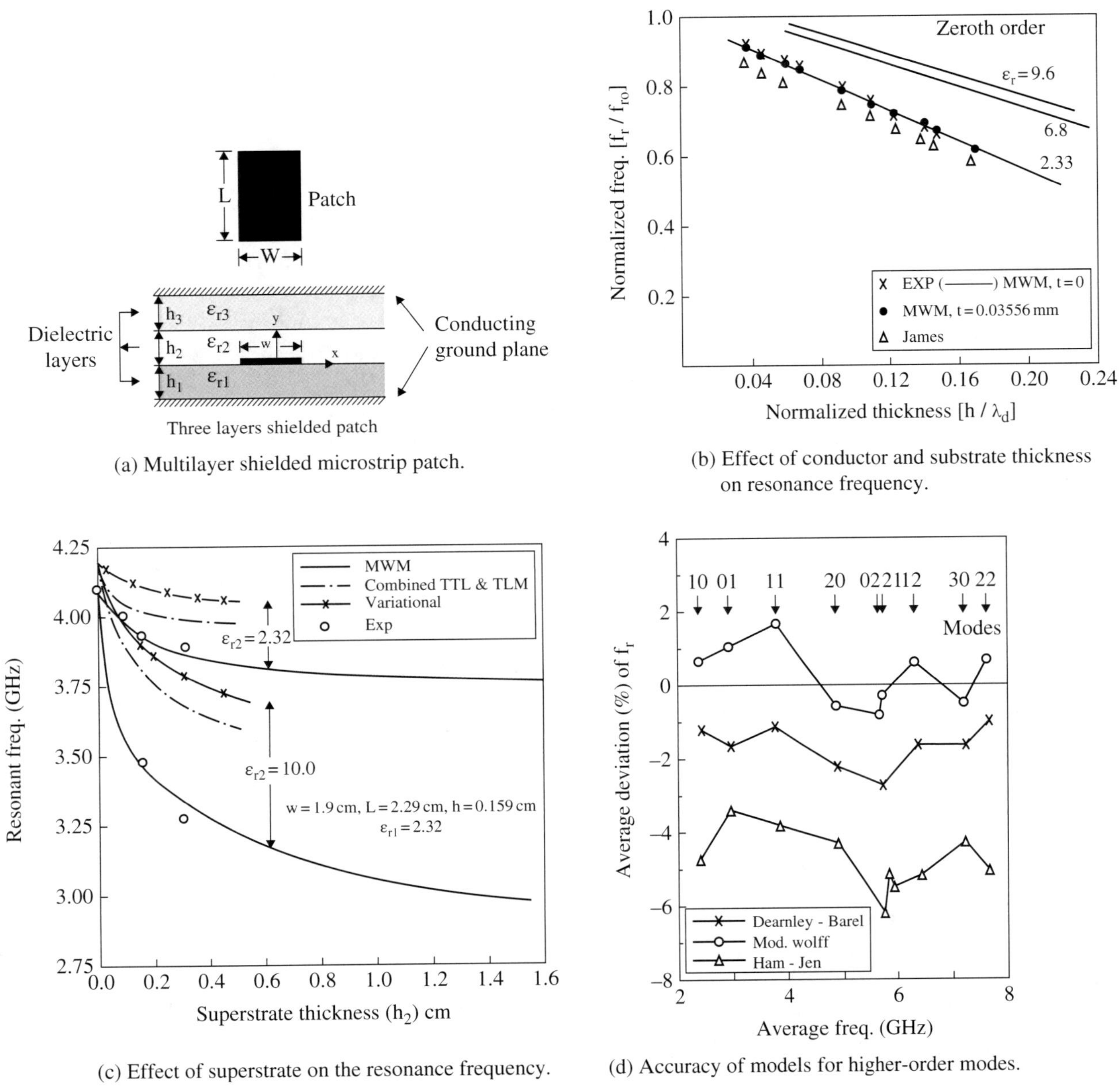

Figure 18.26 Shielded multilayer microstrip rectangular patch resonator. *Source:* From Verma and Rostamy [J.52]. © 1993, IEEE.

where the total line capacitance (C_t) *p.u.l.* of the layered medium patch is computed using the variational method in the Fourier domain, discussed in subsection (14.4.2) of chapter 14. The result, applicable to the present structure, is given below:

$$\frac{1}{C_t} = \frac{1}{\pi\varepsilon_0}\int_0^{\infty}\frac{\left[\tilde{f}(\beta)/Q\right]^2}{\beta h_1}\frac{1}{Y}d(\beta h_1) \qquad \text{(a)}$$

$$\text{where, } Y = \varepsilon_{r1}\coth(\beta h_1) + \varepsilon_{r2}\left[\frac{\varepsilon_{r3}\coth(\beta h_3) + \varepsilon_{r2}\tanh(\beta h_2)}{\varepsilon_{r2} + \varepsilon_{r3}\tanh(\beta h_2)\coth(\beta h_3)}\right] \qquad \text{(b).}$$

(18.6.9)

A layer can be eliminated by taking the thickness of a substrate to zero. Further, $h_3 \to \infty$ and $\varepsilon_{r3} = 1$ eliminate the top shield. So, the structure can be reduced to several interesting cases. Several layers can be accommodated by getting additional terms in the *admittance parameter* Y. In the above equation, the function $\tilde{f}(\beta)$ is assumed to be Fourier domain cubic charge density function along the width W of the patch, and Q is the total charge *p.u.l.* on the patch. The details are presented in chapter 14. The process is repeated to determine the static fringe capacitance $C_{e2,stat}(\varepsilon_{r1}, \varepsilon_{r2}, \varepsilon_{r3})$ of the width side of a patch.

Next, the L_{eff} is computed, using the equivalent width (W_{eq}), from the following expressions:

$$L_{eff} = L + \left[\frac{W_{eq} - W}{2}\right]\left[\frac{\varepsilon_{reff}(W) + 0.3}{\varepsilon_{reff}(W) - 0.258}\right] \quad (a)$$

$$W_{eq} = \frac{120\pi h_1}{Z(\varepsilon_{r1}, \varepsilon_{r2}, \varepsilon_{r3}, W)\,\varepsilon_{reff}(W)} \quad (b). \qquad (18.6.10)$$

The effective relative permittivity $\varepsilon_{reff}(W)$ and the characteristic impedance $Z(\varepsilon_{r1}, \varepsilon_{r2}, \varepsilon_{r3})$ are determined, from the width-side of the patch, using the variational expression of the line capacitance:

$$\varepsilon_{reff}(W) = \frac{C(\varepsilon_{r1}, \varepsilon_{r2}, \varepsilon_{r3}, W)}{C(\varepsilon_{r1} = \varepsilon_{r2} = \varepsilon_{r3} = 1)} \quad (a)$$

$$Z(\varepsilon_{r1}, \varepsilon_{r2}, \varepsilon_{r3}, W) = \frac{1}{c\sqrt{C(\varepsilon_{r1}, \varepsilon_{r2}, \varepsilon_{r3}, W)C(\varepsilon_{r1} = \varepsilon_{r2} = \varepsilon_{r3} = 1, W)}} \quad (b). \qquad (18.6.11)$$

The accuracy of the evaluation of the integral, used in equation (18.6.9a), determines the accuracy of characteristic impedance $Z(\varepsilon_{r1}, \varepsilon_{r2}, \varepsilon_{r3}, W)$. Similarly, the W_{eff} is determined by exchanging W by L and L by W in the above expressions.

The effect of the conductor thickness on the resonance frequency could also be accounted for [J.52]. However, it is not significant. Figure (18.26b) compares the results, using MWM, and other methods, for the normalized resonance frequency of a standard patch resonator against the experimental results. It shows a monotonic decrease in the resonance frequency with increasing normalized substrate thickness. The resonance frequency f_r is normalized with respect to the zeroth-order resonance frequency, $f_{ro} = 15/L\sqrt{\varepsilon_{r1}}$ GHz. The patch length L is in cm. Figure (18.26c) compares the results of a dielectric covered patch. Figure (18.26d) shows the accuracy of the MWM for the computations of higher-order modes. In general, the MWM has an accuracy of 0.7% and 1.7% for the fundamental and higher-order modes, respectively. The MWM has also been used to compute the Q-factors due to the dielectric, conductor, radiation, and surface-wave losses [J.54, J.55]. It helps to determine the bandwidth and input impedance of a multilayer patch resonator. The patches could be the circular disc, annular ring, and triangle also. The anisotropy in the substrate is also accounted for in the MWM [J.53, J.54]. Due to its high accuracy, the MWM is used to develop a sensor to measure the complex relative permittivity of material in the sheet, paste, or liquid form [J.56].

18.6.3 Circular Patch

Figure (18.24b) shows the geometry of a circular patch resonator of the radius a, on a substrate with relative permittivity ε_r, and thickness h. The circular patch could be under the multilayer environment also [J.51, J.55, J.57]. The cavity is formed with the *MW* just at the circumference of the patch. Equation (18.6.1b) is solved for the potential function ϕ^e_{mn}, under the boundary condition $\partial\phi^e/\partial\rho = 0$, at $\rho = a$, where "a" is the radius of a circular patch. The solution for the potential function ϕ^e_{mn} is expressed as a linear combination of two solutions in terms of Bessel functions of the first and second kinds. However, the solution using the second kind of Bessel function is infinite at $\rho = 0$. So, it is dropped as a nonphysical solution, giving the following expression of the eigenfunction and wavenumber of the TM^z_{nm} mode [B.11, J.45]:

$$\phi^e_{mn} = J_m(k_{mn}\rho)e^{jm\varphi} \quad (a)$$

$$J'_m(k_{mn}a) = 0, \quad \Rightarrow \quad k_{mn} = \chi'_{mn}/a \quad (b)$$

where, the first 4 roots of Bessel function :

$$\chi'_{11} = 1.8412, \quad \chi'_{21} = 3.0542, \quad \chi'_{01} = 3.8318, \quad \chi'_{31} = 4.2012 \quad (c). \qquad (18.6.12)$$

The subscript n is the *radial modal number.* It shows the radial variation of the modal field of the TM^z_{mn} mode. Another *modal number m* shows the angular (ϕ) variation, over the range $0 \le \phi \le 2\pi$, of the modal field. The modal number n is also the order of Bessel function, giving the decaying cosine radial field, along ρ, $0 \le \rho \le a$, with n roots. The subscript n shows the number of roots of the derivative of Bessel function with respect to ρ.

For the modal number n = 1, the first root of m = 0–3 order Bessel functions provides the $TM^z_{01}, TM^z_{11}, TM^z_{21}, TM^z_{31}$ resonating modes. Once roots χ_{mn} are known, the resonance frequency is computed by using $k_{mn} = \chi'_{mn}/a$ in equation (18.6.2). However, for accurate computation of the resonance frequency, the MWM is used; in that case, the radius "a" is replaced by the effective radius a_{eff} and ε_r is replaced by $\varepsilon_{r,dyn}$:

$$f_{mn} = \frac{c\chi'_{mn}}{2\pi a_{eff}\sqrt{\varepsilon_{r,dyn}}} \quad (a)$$

$$a_{eff} = a\left[1 + \frac{2h}{\pi a\varepsilon_r}\left\{\ln\left(\frac{a}{2h}\right) + (1.41\varepsilon_r + 1.77) + \frac{h}{a}(0.268\varepsilon_r + 1.65)\right\}\right]^{1/2} \quad (b). \qquad (18.6.13)$$

In the above equation, the dynamic relative permittivity is computed by converting the circular patch to an equivalent rectangular patch, on taking the width of the rectangle 2a. Its length is computed to maintain the area of the rectangular patch the same as that of the circular patch [J.51, J.55]. In place of equation (18.6.13b), another more accurate expression is also available to compute the a_{eff} for a wide range of parameters [J.61]. The MWM is also applicable for computing resonance frequency, various Q-factors, input impedance, and bandwidth of a multilayer lossy circular patch [J.55].

The values of roots χ'_{mn} show that the TM^z_{11}-mode is the fundamental mode. Once ϕ^e_{mn} is known, the field equations are obtained in the cylindrical coordinate [B.11]. Figure (18.27) shows the field and current patterns of four modes [J.46]. The entering and leaving displacement current provide continuity to the conduction current on the patch. The symmetric mode TM^z_{01} has no angular (azimuthal) field variation. Whereas for the other three modes, the electric field, i.e. the displacement currents, in pairs are obtained, corresponding to m = 1, 2, 3.

18.6.4 Ring Patch

Figure (18.24c) shows the ring-type patch resonator, with inner radius a and outer radius b. The concentric cylindrical MWs are located at ρ = a and ρ = b. The potential function, in terms of the first and second kind Bessel functions, within the MWs, is again a solution of

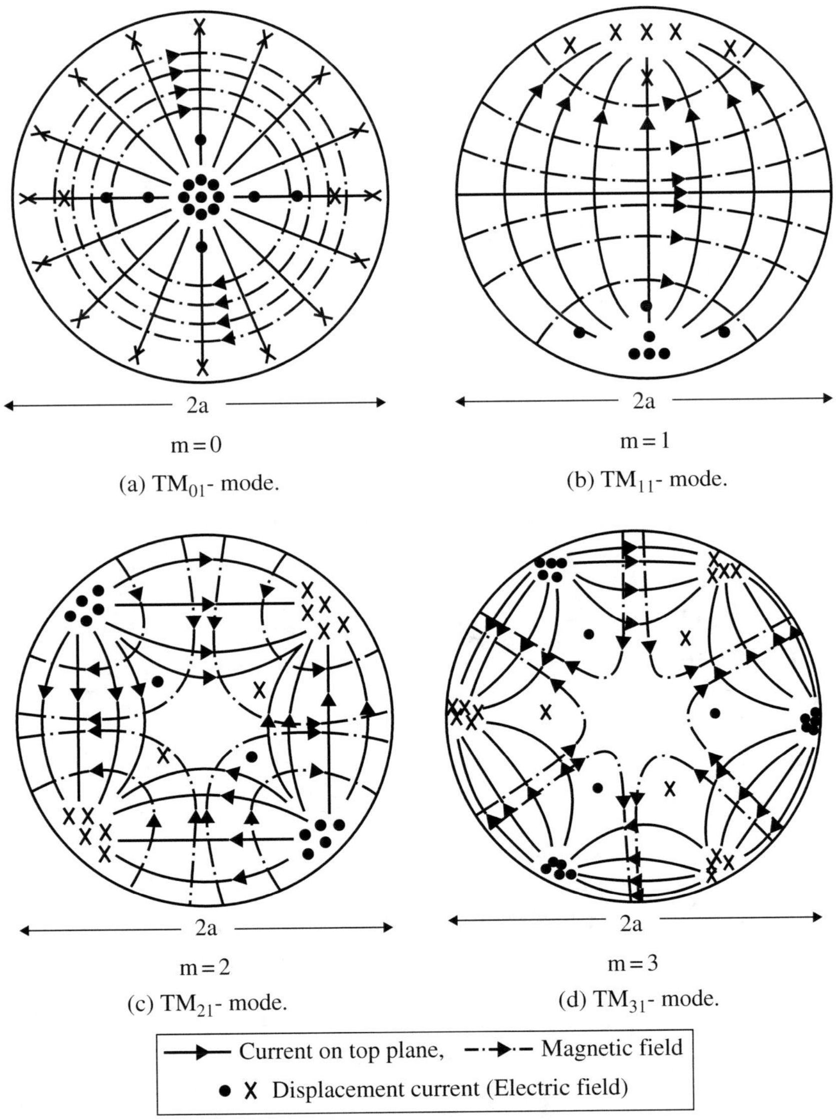

Figure 18.27 Field and current patterns of a few TM_{mn} modes of microstrip circular patch resonator. *Source:* Watkins [J.46]. © 1969, Institution of Engineering and Technology.

equation (18.6.1b). It also provides a constraint condition. Its solutions give the n number of roots χ_{mn} for the mth-order Bessel functions. These results are shown below [B.2, B.11, B.19, J.46]:

$$\phi^e_{mn} = \left[J_m(k_{mn}\rho)Y'_m(k_{mn}\rho) - J'_m(k_{mn}\rho)Y_m(k_{mn}\rho)\right]e^{jm\phi} \quad (a)$$

$$J'_m(k_{mn}a)Y'_m(k_{mn}b) - J'_m(k_{mn}b)Y'_m(k_{mn}a) = 0 \quad (b)$$

$$\text{For } C = b/a \text{ ratio, and } \chi_{mn} = k_{mn}a, \quad k_{mn} = \chi_{mn}/a \quad (c)$$

$$J'_m(C\chi_{mn})Y'_m(\chi_{mn}) - J'_m(\chi_{mn})Y'_m(C\chi_{mn}) = 0 \quad (d)$$

For $C = 2$, the first 3 roots are,

$$\chi_{11} = 0.6773, \chi_{21} = 1.3406, \chi_{31} = 1.9789$$

$$\text{additional root, } \chi_{02} = 3.1966, \ \chi_{12} = 3.2825 \quad (e). \quad (18.6.14)$$

The calculation of resonance frequency requires replacement of parameters: $a \to a_{eff}$, $b \to b_{eff}$ and $\varepsilon_r \to \varepsilon_{reff}$. The required expressions are summarized below:

$$\varepsilon_{reff} = \frac{1}{2}(\varepsilon_r + 1) + \frac{1}{2}(\varepsilon_r - 1)\left(1 + \frac{10}{W/h}\right)^{-1/2}$$

$$\text{where } W = (b - a) \quad (a)$$

$$b_{eff} = b + [W_e(f) - W]/2 \quad (b)$$

$$a_{eff} = a - [W_e(f) - W]/2 \quad (c)$$

$$\text{where, } W_e(f) = W + \frac{W_e(f=0) - W}{1 + (f/f_p)^2} \quad (d)$$

$$W_e(f=0) = \frac{120\pi h}{Z_0\sqrt{\varepsilon_{reff}}}, \quad f_p = \frac{Z_0}{2\mu h} \quad (e). \quad (18.6.15)$$

The factor $[W_e(f) - W]/2$ in the above equation is an increase in one side width of the strip conductor. It reduces the inner radius a, and increases the outer radius b. The following simpler equations for $a_{eff,}$ and b_{eff} are also used:

$$a_{eff} = a - (3/4)h \ (a), \quad b_{eff} = b + (3/4)h \ (b). \quad (18.6.16)$$

Using $\chi_{mn} = k_{mn}/a$, the resonance frequency is calculated from equation (18.6.2):

$$f_{mn} = \frac{c\chi_{mn}}{2\pi\, a_{eff}\sqrt{\varepsilon_{reff}}}. \quad (18.6.17)$$

In the place of the above equation, again the MWM can be used. It applies to the multilayer ring resonator also [J.59].

Once the potential function is known, the field components can be computed. Figure (18.28a–d) shows the field and current distributions on the patch surface of a ring resonator for a few TM^z_{mn} modes [B.19].

18.6.5 Equilateral Triangular Patch

Figure (18.24d) shows the geometry of an equilateral triangular patch resonator with side length a, on a substrate with thickness h and relative permittivity ε_r. Other triangular shapes have also been considered [B.19]. The $TM^z_{m,n,\ell}$ field components have been obtained from the following eigenfunction [J.48, J.49]:

$$\begin{aligned}\phi^e_{mn} = &\cos\left(\frac{2\pi\ell}{3b}\right)\left[\frac{u}{2} + b\right]\cos\left(\frac{\pi(m+n)(v-w)}{9b}\right)\\ &+ \cos\left(\frac{2\pi m}{3b}\right)\left[\frac{u}{2} + b\right]\cos\left(\frac{\pi(n-\ell)(v-w)}{9b}\right)\\ &+ \cos\left(\frac{2\pi n}{3b}\right)\left[\frac{u}{2} + b\right]\cos\left(\frac{\pi(\ell-m)(v-w)}{9b}\right) \quad (a)\end{aligned}$$

$$\text{where, } \ell + (m+n) = 0, \ u = \frac{\sqrt{3}}{2}x + \frac{1}{2}y,$$

$$v - w = -\frac{\sqrt{3}}{2}x + \frac{3}{2}y, \ b = \frac{a}{2\sqrt{3}} \quad (b)$$

$$k^2_{mn} = \left(\frac{4\pi}{3a}\right)^2(m^2 + n^2 + mn) \quad (c). \quad (18.6.18)$$

The integers m, n, ℓ do not represent the half-wavelength variation along the arms of a triangle. However, the boundary value problem is solved under the constraint $m + n + \ell = 0$. The resonance frequency is obtained on substituting k_{mn} in equation (18.6.2) [J.49, B.20]:

$$f_{mn} = \frac{2c}{3a_{eff}\sqrt{\varepsilon_r}}(m^2 + n^2 + mn)^{1/2} \quad (a)$$

$$a_{eff} = a + \frac{h}{\sqrt{\varepsilon_r}} \quad (b). \quad (18.6.19)$$

The mode $TM^z_{1,0,-1}$ is the fundamental mode of an equilateral triangular patch resonator. The modes $TM^z_{1,0,-1}$, $TM^z_{-1,0,1}$, and $TM^z_{0,-1,1}$ are the degenerate modes. Figure (18.29a and b) shows the field patterns of the fundamental and next higher-order modes, i.e. the $TM^z_{1,0,-1}$ and $TM^z_{1,1,2}$ [J.49]. The conduction currents on the patch are at a right angle to the magnetic field lines. These are terminated in the displacement currents, shown as the dot and cross. The fringe field is accounted through the empirically obtained effective side length a_{eff}, while keeping ε_r unchanged. It is

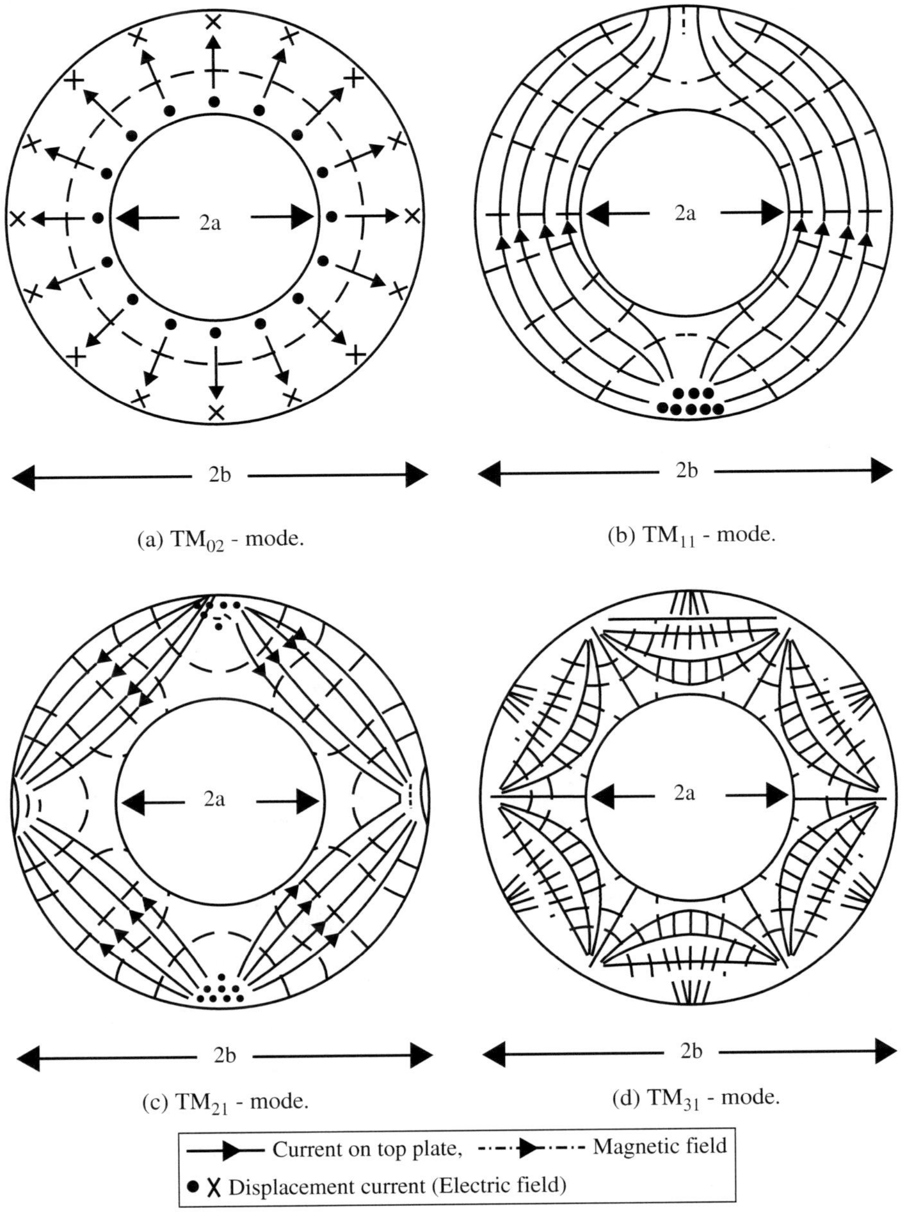

Figure 18.28 Field and current patterns of a few TM_{nm}-modes of microstrip ring patch resonator. *Source:* Sharma et al. [B.19]. © 1989, John Wiley & Sons.

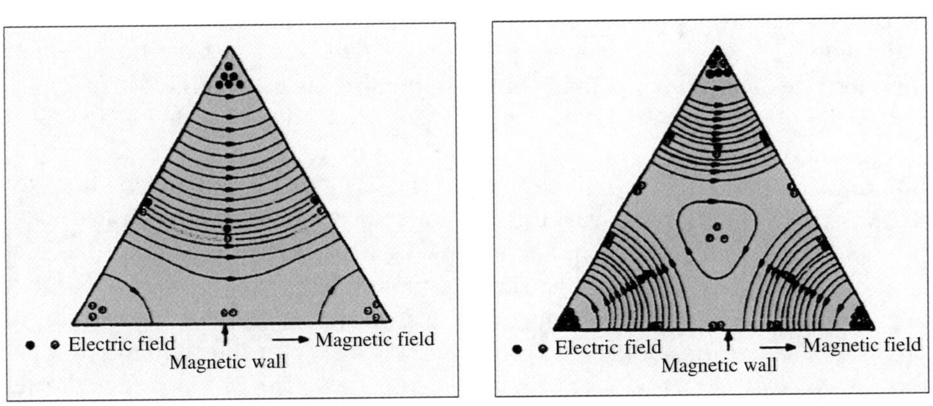

(a) $TM_{1,0,-1}$ - mode.

(b) $TM_{1,1,-2}$ - mode.

Figure 18.29 Field patterns of two TM_{nml}-modes of microstrip equilateral triangle patch resonator. *Source:* Helszajn and James [J.49]. © 1978, IEEE.

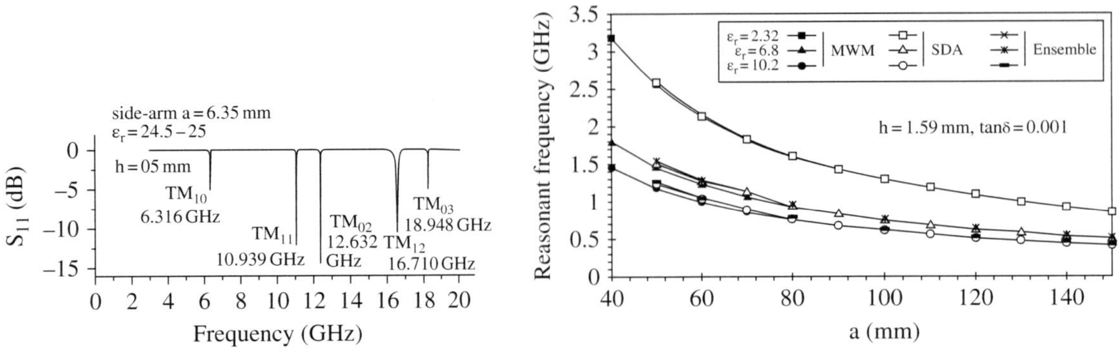

(a) First five modes of an equilateral triangular patch.

(b) Fundamental mode resonance frequency using MWM.

Figure 18.30 Resonance frequency of the equilateral triangular patch.

unrealistic. Again, the MWM computes the resonance frequency more accurately through the effective length and $\varepsilon_{r,dyn}$ [J.58]. Moreover, it applies to the lossy multi-layer cases also, giving the Q-factors and bandwidth [J.60]. Figure (18.30a) shows the EM-simulated resonance frequencies of the first five modes of an equilateral triangular patch resonator on the $LaAl_2O_3$ substrate (ε_r = 24.5 − 25) of thickness 0.5 mm, and side arm a = 6.35 mm. Figure (18.30b) shows a variation of the resonance frequency of the fundamental mode, with respect to the side-arm length [J.62]. It further compares the accuracy of the MWM against the results of the SDA [J.63], and also against the full-wave EM-simulator Ensemble.

18.7 2D Fractal Resonators

The miniaturization of the planar antenna, filters, and other microwave components are very important for the front-ends of the modern wireless and satellite communication systems. The antenna itself is a narrow band resonating element. The physical size of the microstrip resonator can be reduced by folding the line or increasing the relative permittivity of a substrate. The size reduction of a line is also achieved by the inductive or capacitive loading. The loading is realized through the stubs or by cutting slots in the line conductor, such as the defected microstrip structure (DMS), or by the slots in the ground plane, forming the DGS. These loadings also reduce the size of a resonating patch.

This section introduces a different approach, i.e. the *planar fractal technology,* to the problem of miniaturization of the planar resonators. It has also been extensively used with the antenna and filter technologies. The fractal technology is based on *fractal geometry.* In this geometrical arrangement, the planar metallic surface of a microstrip resonator is created not by the 2D metallic patch but by filling the required 2D-space with an infinitely long line, arranged in a specific manner. Several kinds of curves are available to achieve the objective of the space-filling that creates a plane sheet with a very long line. In the place of a metallic line, one can equally take the slot line and can create a slot area in the metallic ground plane of a microstrip line. The slots are also used repeatedly to defect a metallic patch or a line resonator. Before discussing the fractal resonator, it is useful to appreciate some important curves of the *fractal geometry.* These curves do not follow the integral dimensional arrangements, like 0, 1, 2, and 3 dimensions, of the Euclidean space. The *fractal curves* have the fractional dimension between 1 and 2.

18.7.1 Fractal Geometry

The fractal object is defined through the concept of the *non-Euclidean* dimension, called the *similarity dimension.* There are also other kinds of *non-Euclidean* fractal dimensions [B.22]. The fractal dimension could be viewed as the *space-filling* ability of the fractal shapes. Mandelbrot has initiated the present interest in the fractal geometry [B.23]. He emphasized that shapes of the objects in nature are more effectively described by the fractal geometry, i.e. the geometry of broken lines. However, the natural shapes of objects are created by the complex forces of nature. By just adopting the fractal geometrical description of an object, we cannot modify or control the natural forces. So, just by adopting the fractal geometry, we cannot get the optimum or unique electrodynamic properties of an object, although useful changes can be derived. Yet, the effort has been made to achieve something like the *fractal electrodynamics*

[B.24]. Despite its widespread innovative use in the microwave and antenna technologies, some limitations will be obvious by the end of this section.

The fractal curves, used in the planar microwave technology, are normally described with the help of the similarity dimension. In the case of the Euclidean 2D plane, all shapes, such as a square, circle, etc., have a fixed dimensional number, i.e. 2. These Euclidean shapes completely fill in the 2D Euclidean space. Likewise, the line and cube fill in the 1D and 3D Euclidean spaces, respectively. Even though several 2D Euclidean objects have the ability to fill in the 2D-space, their electrical characteristics as resonators, filters, and antennas are different. Different 2D conventional geometrical shapes, even for the same area, or the same length of the perimeter, can provide different performances of the resonators, filters, and antennas. A similar situation holds for various fractal curves. The fractal curve with a higher similarity dimension, or with identical curve length, or identical enclosed area, may not provide identical electrical performances, when used as the resonators, filters, and antenna. The space-filling curves, creating the non-Euclidean planes, have their own similarity dimensions. Before proceeding further, it is important to define the *similarity dimension,* so that it could be applied to the fractal curves.

Similarity Dimension and Basic Characteristics of Fractal Structures

A larger object can be broken into smaller similar objects, called the *scaled copies* of the original large object. On assembly of several numbers of the scaled copies, in a particular sequence or format, a larger object can be created. The copies and the large object are *similar in shape*; they are different only in the *scale*. Let us apply this concept to 1, 2, and 3-dimensional objects of the Euclidean geometry, i.e. to a geometrical line, square, and cube, shown in Fig (18.31a–c).

In Fig (18.31a), a unit length line is broken into 3 equal similar segments. The size of each smaller segment, called the *copy*, is 1/3. For n number of copies, the *scale-factor* is r = 1/n. If the *self-similar* copy is repeated along a line, without changing the direction, the 1D space is *filled in* and a long line is created.

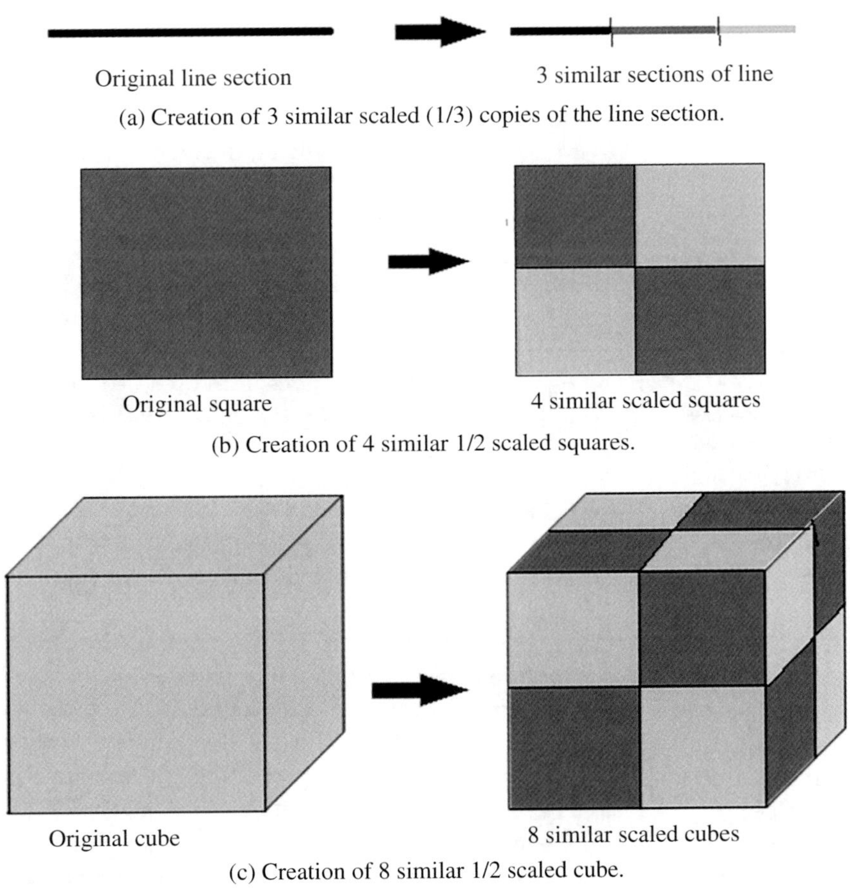

Figure 18.31 Creation of scaled copies of objects in Euclidean space.

Figure (18.31b) shows four similar segments (copies) of the original unit square. If it is broken into n number of copies, the *scale-factor* of a copy is $r = 1/\sqrt{n}$. Thus, for $n = 4$, $r = 1/2$, i.e. sides of the original unit square are reduced by a factor 1/2. Finally, Figure (18.31c) shows eight similar cubic segments (copies) of a unit cube. The *scale-factor* of a cube copy is $r = 1/\sqrt[3]{n}$. So, for $n = 8$, $r = 1/2$, i.e. each side of the original unit cube is reduced by a factor 1/2. Using the *induction method*, the following expression is written for the *scale-factor* (**r**), relating it to the number of *self-similar segments (copies,* **n**), and space *dimension* **d**:

$$\text{Scale} - \text{factor}, r = \frac{1}{\sqrt[d]{n}} \text{ (a)}, \quad \text{No.of copies}, \ n = \frac{1}{r^d} \text{ (b)}$$
$$\text{Dimension}, \ d = \frac{\ln(n)}{-\ln(r)} \qquad \text{(c)}. \tag{18.7.1}$$

In the above equation, the scale-factor r is a fraction giving a positive number for the dimension. The above equation provides the similarity dimension of an object, if its scale-factor and the number of copies involved in its creation are known. In generating a copy, the rule of *self-similarity* has to be followed. So, the dimension obtained is known as the *similarity dimension*. The self-similar, i.e. the similarity, dimension **d** can be defined using the following more general expression, involving more number of variables that occur in some fractal curves [J.64]:

$$n_1\left(\frac{1}{r_1}\right)^d + n_2\left(\frac{1}{r_2}\right)^d + \cdots + n_m\left(\frac{1}{r_m}\right)^d = 1, \tag{18.7.2}$$

where m is the number of scales, used in the *initiator* of the fractal curve. The above expression shows that *a fractal curve can have more than one scale-factor.* The scale-factor r_m is an integer, and a dimension is a positive number. For only one kind of scale-factor $m = 1$, the above equation is reduced to equation (18.7.1). It is noted that the scale-factor r in equation (18.7.1) is a fractional number, whereas the scale-factor r_m of equation (18.7.2) is an integer. Both the scale-factors are reciprocal to each other. Both forms are used in further discussion.

The above concept is applied to a regular closed geometrical area to get a finite area, with infinite broken irregular boundary, i.e. the *fractal boundary*. The irregular *fractal curve* can also be generated using the same process. Likewise, the 2D-space is filled with the broken curve of an infinite length to get 2D-like surface. Alternatively, the filled 2D geometrical space can be depleted by the self-similar copies. These are all irregular broken fractal shapes used for the development of the compact planar resonators, filters, antennas, etc.

The following four popular fractal constructions are considered in further discussion:

- *Koch curves and Koch islands.*
- *Minkowski curves and Minkowski islands.*
- *Hilbert curves.*
- *Sierpinski triangles and Sierpinski carpet.*

The above fractal geometrical shapes can be generated using MATLAB codes. The coordinate details of the fractal curve generated in MATLAB are transferred to the EM-Simulator for the development of the fractal components [J.65].

The *self-similarity* and *space-filling*, or *space consumption*, and *scaling* are the important basic characteristics for the creation of the fractal curves and fractal areas. *The fractal curve is generated by an infinite repetition of the self-similar copy, with changing scale-factor.* In practice, very fine structures are ignored, and the recursive process of fractal curve generation is terminated with 2–6 *iterations*. This method of fractal curve generation is known as the *iterated function system* (IFS). It is based on a series of affine transformation, where the fractal elements undergo the rotation, shearing, translation, and scaling [J.64, J.66, J.67]. The basic characteristics of the fractal shapes have the following implication in the microwave fractal technology:

- The self-similarity helps to design the multiband antenna. In the case of filters, it provides better stop-band rejection and generates the transmission zeros to improve the selectivity.
- The space-filling ability helps to develop an electrically small antenna and resonators. The area fractals and boundary fractals can provide high directivity antenna and antenna-array with low-side levels.
- The space-filling curves have been used to develop the high-impedance EBG surfaces and metasurfaces.

Koch Curves and Koch Islands

The equilateral triangular shape is a *generator* that acts on an initial line segment of unit length to create the Koch curve. In this case, the initial line is the *initiator* of the Koch curve. Figure (18.32) explains the process. A unit line length is divided into three segments, creating *self-similar* three copies of a line, with the reduced *scale-factor* ($r = 1/3$). During the first iteration, the generator replaces the middle part of the line segment with an equilateral triangle without a base. So, the original line segment containing 3 self-similar copies is replaced by the four self-similar copies. The segment numbers are marked on the figure. *During the second and subsequent iterations, each line section is replaced by four self-similar*

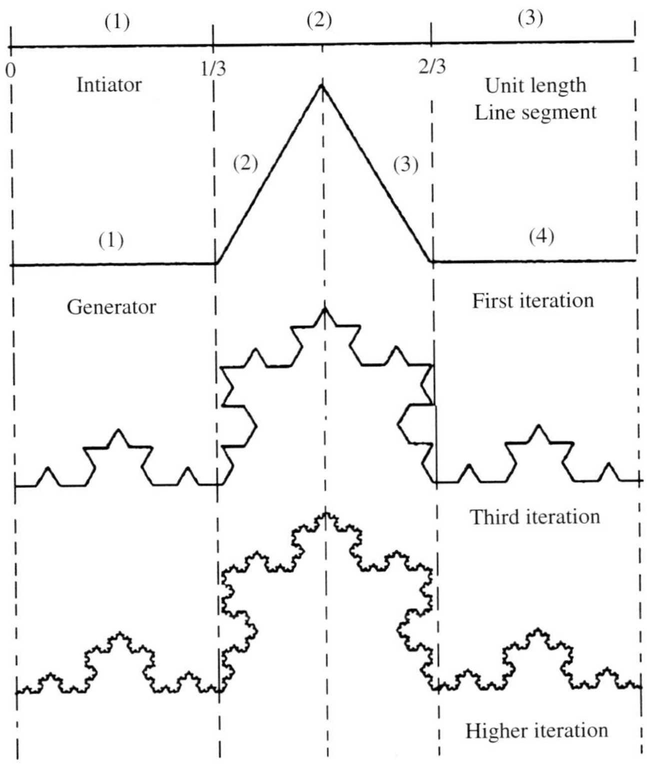

Figure 18.32 Creation of a Koch curve from a line segment.

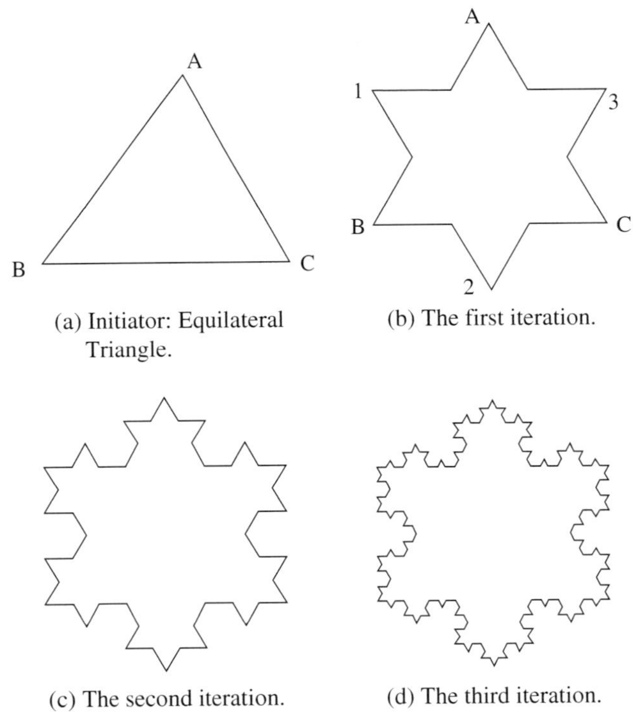

Figure 18.33 Creation of *Koch island* or *Koch Snowflake* from an equilateral triangle.

copies by applying the generator shown in Figure (18.32). It results in the Koch curve, created using the broken lines. Figure (18.32) shows the third and higher-order Koch curve. It is a non-differentiable curve.

It is interesting to note that within the 1D dimension of the original line, the Koch curve has an infinitely large number of line sections, although its 1D Euclidian dimension is fixed. The Koch curve has a peculiar property – *with an increasing number of iterations, the curve length increases to infinity, while its enclosed area is finite.* After n^{th} iteration, the perimeter ℓ and enclosed area (A) of the Koch curve are given by the following expressions:

$$\ell = \left(\frac{4}{3}\right)^{n} \quad (a), \quad A = 1 + \left(\frac{4}{9}\right)^{n-1} \quad (b). \tag{18.7.3}$$

The limiting value of the enclosed area is 1.8. However, after a few iterations, say 3, the Koch structure becomes so fine that the iteration can be terminated for any practical application. The terminated fractal is known as the *pre-fractal.* The pre-fractal ignores the fine structure of the arbitrarily small scale. The Koch curve has the number of *copies* n = 4 and the *scale-factor* r = 1/3. So, its *dimension* d, using equation (18.7.1), is 1.2619.

In the above discussion, the Koch curve generation process is applied to a straight line. It can also be applied to an *initiator* equilateral triangle and other polygons. It generates the *Koch island* or Koch *Snowflake.* Figure (18.33a–d) demonstrates it through three iterations for an equilateral triangle, where the middle part of each arm is replaced by a triangle without a base. The equilateral triangle without a base acts as a *generator.*

The standard Koch curve is obtained using a triangular generator. However, the triangle can be replaced by a square of size w. The parameter w is called the *indentation length.* In Fig (18.34), the modified version of Koch generator is applied to all four sides of a square. The resulted loop, or island, is also known as the *Minkowski island.* Its perimeter, after n^{th} iterations, is given by the following expression:

$$\ell_n = \left(1 + \frac{2}{3}w\right)^{n} \ell_{n-1}. \tag{18.7.4}$$

Minkowski Curves

Figure (18.35) shows the recursive process for the generation of the 1st, 2nd, and 3rd iteration of the Minkowski curve, from a line *initiator.* The eight-sided double square acts as a *generator.* The generator itself is the first iteration, that is obtained by getting eight copies of a line, with the reduced *scale-factor* 4.

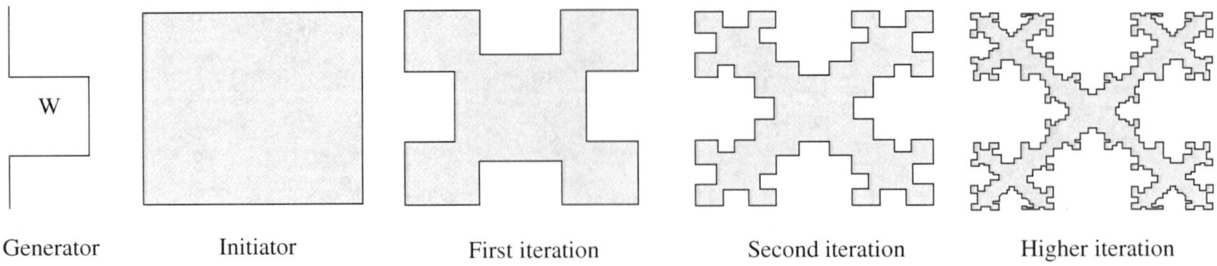

Figure 18.34 Creation of square *Koch island* or square *Koch Snowflake* from a square.

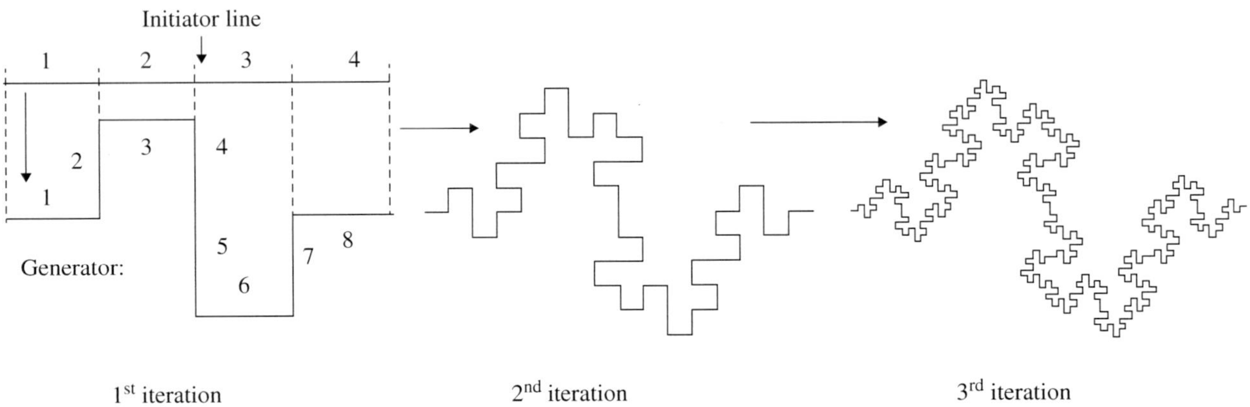

Figure 18.35 Generation of Minkowski curve from the 8-sided pulse generator.

The initiator line is divided into four sections, and sections 2 and 3 are replaced by the squares without a base to get the first iteration of the Minkowski curve. Subsequently, each arm of the Minkowski curve is replaced by the eight-sided Minkowski generator of the previous iteration step. At each iteration step, the line segment length is reduced by a factor 4, and to get total perimeter, it is multiplied by 8. In each step, the total curve length is increased by a factor of 2. So, the length of the Minkowski curve increases to infinite with increasing order of iterations. Its dimension d, with $n = 8$ and $r = 4$, is 1.5. It has better space-filling ability, as compared to the Koch curve that has dimension 1.2619. Likewise, one can use the square Sierpinski fractal curves to fill a given area. It looks like Minkowski fractal with a 2 : 3 ratio between the inner and outer squares [J.68].

In place of a line section, a planar area such as a triangle, or square, can be taken as the initiator in the Minkowski process. It results in the *Minkowski islands* [J.69]. The single-sided square [J.65], or a rectangular, generator is also used to generate the Minkowski island. Its square format, shown in Figure (18.34), is also known as the Koch island. In Figure (18.36), the five-sided Minkowski square or rectangular self-similar generator is applied to each of the four sides of a square patch resonator [B.25]. The rectangular pulse has height ℓ, known as the *indentation width*. The pulse width is g. Figure (18.36) shows the results of the first two iterations. The initiator rectangle can also be replaced by a square ring initiator [J.64]. In the case of a rectangular generator, two scale-factors are involved due to three horizontal segments ($n_1 = 3$) with size g ($r_1 = 3$), and two vertical segments ($n_2 = 2$) with size ℓ ($r_2 = 3/\ell$). The ℓ-dependent dimension of Minkowski island is computed by taking two terms of equation (18.7.2). In the case of a square generating pulse, the dimension is $d = \ln 5/\ln 3 = 1.465$. For ℓ in between 0 and 0.9, the dimension is in between 1.0 and 1.413. The perimeter of the Minkowski island increases with each iteration, and for the mth iteration, it is given by the following expression:

$$p_m = \left(1 + \frac{2}{3}\ell\right)^m p_{m-1}, \quad p_0 = 4, \tag{18.7.5}$$

where m is the number of iteration. Normally, the *pre-fractal* Minkowski islands, up to 3–4 iterations, are used in the microwave circuits and antenna applications.

Hilbert Curves

Hilbert curve is a space-filling curve. It is a continuous 1D line, formed by joining the broken U-shaped line structure. It fills in the prescribed 2D area with several iterations. In the process, the perimeter of the Hilbert curve increases infinitely. Figure (18.37a–d) shows a few iterations of the Hilbert curve, within a $S \times S$ square.

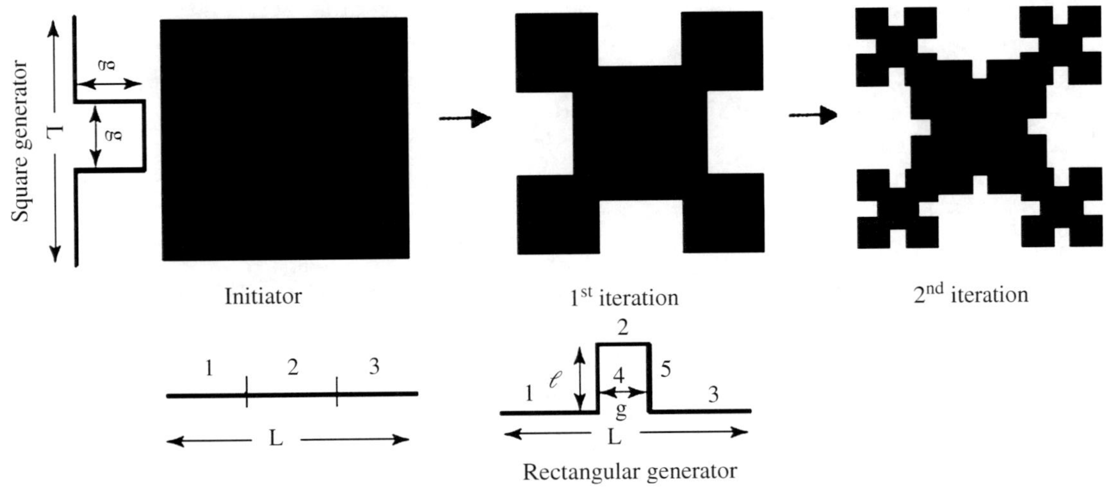

Figure 18.36 Generation of Minkowski island from the 5-sided square/rectangular generator.

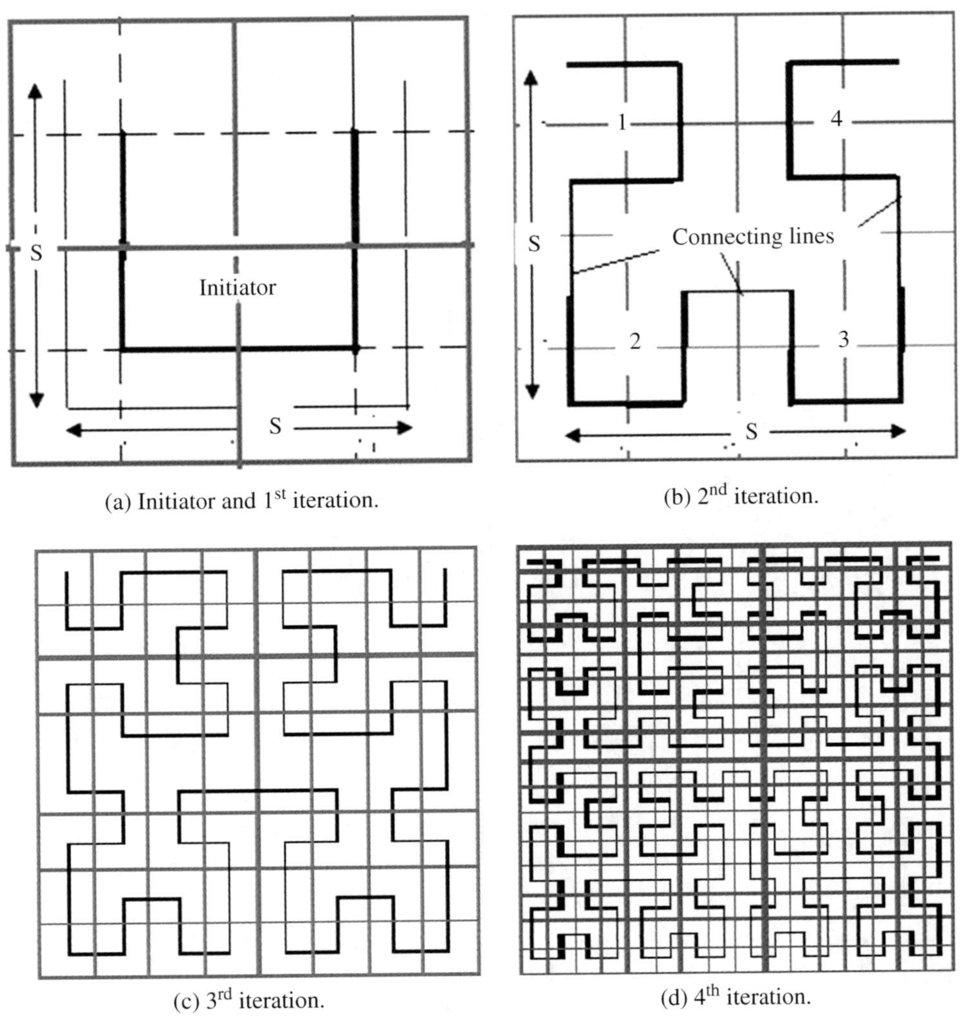

Figure 18.37 Generation of Hilbert curves using U shape generator.

The bent U- shape line is an *initiator* of the Hilbert curve, and the *generator* is also a reduced scaled U-shape with 1/2 of vertical and horizontal lengths. The total length of a scaled U-shape generator is 1/2 of the length of the initiator. So, the *scale-factor* (r) of the Hilbert curve is 2.

The first iteration, sometimes taken as the zeroth-order iteration, is just the enlarged U-shape of the size S × S. Figure (18.37b) shows that during the second iteration at each corner of the S × S size U-shape, four *copies* of the scaled U-shapes (thick dark lines) are placed, with proper orientation. Four U-shaped copies are connected using additional three lines. For n = 4, the *dimension* of the Hilbert curve is d = log 4/log 2 = 2. It shows that the Hilbert curve has the ability to fill in completely the 2D Euclidean space with a higher order of iterations. Hilbert curve is created by joining centers of the background grids at each stage of iteration. The grid size G(m) reduces with m number of iterations. During the iteration process, the length L(m) of the curve also increases. These are obtained using the following expressions:

$$G(m) = \frac{S}{2^m - 1} \quad (a), \qquad L(m) = (2^m + 1)\,S \quad (b). \qquad (18.7.6)$$

The similarity ratio can also be changed, like a Minkowski rectangular generator, by changing the width of the U-shaped generator. Sometimes *Peano curves* are also used as the space-filling curve to realize compact microstrip filters [J.70]. The Peano-curve algorithm has a relatively higher compression rate than the Hilbert-curve algorithm to fill-in a 2-D region. So, for the same area of a fractal resonator, it has a lower resonance frequency.

Sierpinski Triangles and Sierpinski Carpet

Sierpinski Triangles

Sierpinski triangles are also self-similar geometrical structures. These structures have been used for the development of compact multiband planar monopoles. However, it is not a space-filling structure. Normally, it is used to create sieve-type structures with an iterative process, shown in Fig (18.38a–e). There is some modification of the standard Sierpinski triangles.

In the present case, the initiator is an equilateral triangle of side length **a**. During the first iteration, three half-sized black triangles are created, and the central, white triangular area is removed. During the second iteration, every three black triangles create another half-sized three numbers of black triangles, while dropping central white triangles. At this stage, nine black triangles are created. Likewise, during the 3rd and 4th iterations, 27 and 81 triangles are created. During each step of

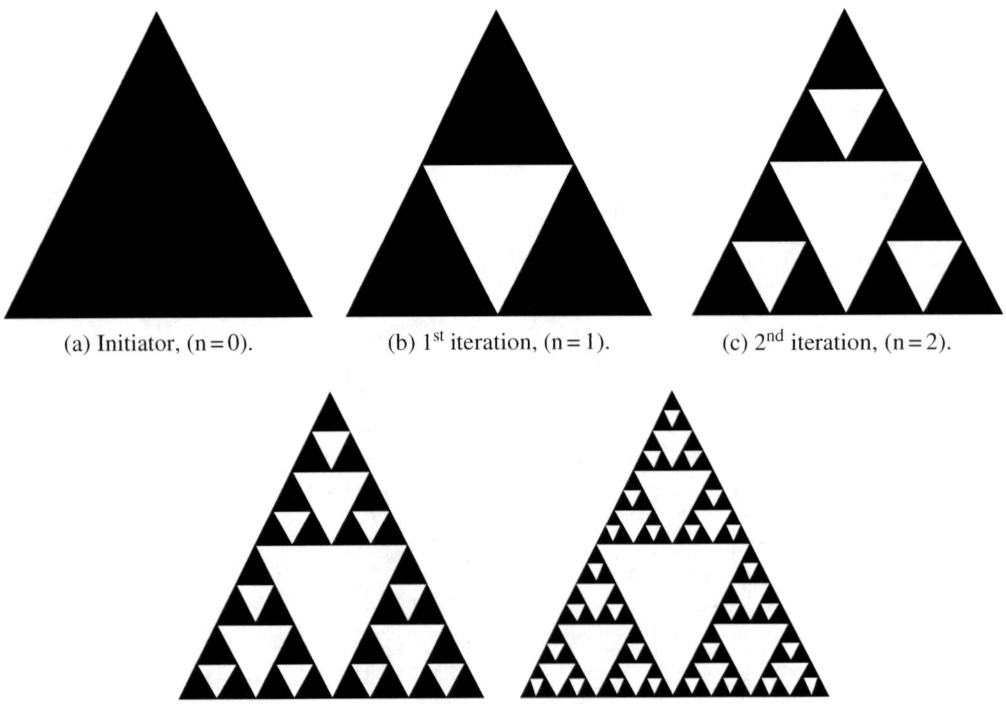

(a) Initiator, (n = 0). (b) 1st iteration, (n = 1). (c) 2nd iteration, (n = 2). (d) 3rd iteration, (n = 3). (e) 4th iteration, (n = 4).

Figure 18.38 Generation of Sierpinski triangles.

the iteration, 1, 3, $9 \ldots 3^{(n-1)}$ central triangles are dropped. It forms a geometric series, and the total number of triangles removed at the nth iteration is $(3^n - 1)/2$. This formation is known as the *Sierpinski triangle or Sierpinski gasket.* All removed triangles are inverse (top-down) with respect to the initial triangle. At the nth iteration, the sum total of removed triangular area, and the total perimeter of retained triangles are given by the following expressions:

$$\text{Area removed} = \left(a^2 \frac{\sqrt{3}}{4}\right)\left[1 - \left(\frac{3}{4}\right)^n\right] \quad \text{(a)}$$

$$\text{Total perimeter} = 9\left[\left(\frac{3}{2}\right)^n - 1\right] \quad \text{(b)}. \qquad (18.7.7)$$

It is seen from the above that for $n \to \infty$ the removed area is $a^2\sqrt{3}/4$ that is the same as the area of the initial triangle. Thus, after the infinite number of iterations, the area of the Sierpinski triangle is reduced to zero, whereas its perimeter increases to infinite. However, in practice, 2–3 iterations are used to develop a compact multiband monopole antenna. The multiband operation is due to the self-similarity of the triangle at the half-scale, i.e. *scale-factor* $(r = 2)$ with $n = 3$ *copies* at each stage of iteration. It gives us the *dimension* of the Sierpinski triangle as, $d = \ln(3)/\ln(2) = 1.585$.

Sierpinski Carpet

Sierpinski carpet is another popular fractal structure. It is used to design the microstrip array with a low side-lobe level [J.67].

Figure (18.39a–d) shows the generation of a Sierpinski carpet from a blank square. During the first iteration, the blank square is divided into nine squares through three horizontal and three vertical divisions. The eight numbers of the white square of 1/3 size are retained, giving the *dimension* of Sierpinski square as $\ln 8/\ln 3 = 1.8928$. The 1/9 black square is removed. During the second iteration, each of 8 blank squares is further divided into 8 white squares giving 64 white squares. The process continues and generates the square Sierpinski carpet. Likewise, the pentagonal and hexagonal Sierpinski carpets are also obtained. Similar to the previous case, with infinite iterations, the area of Sierpinski square is reduced to zero, while its perimeter becomes infinitely large.

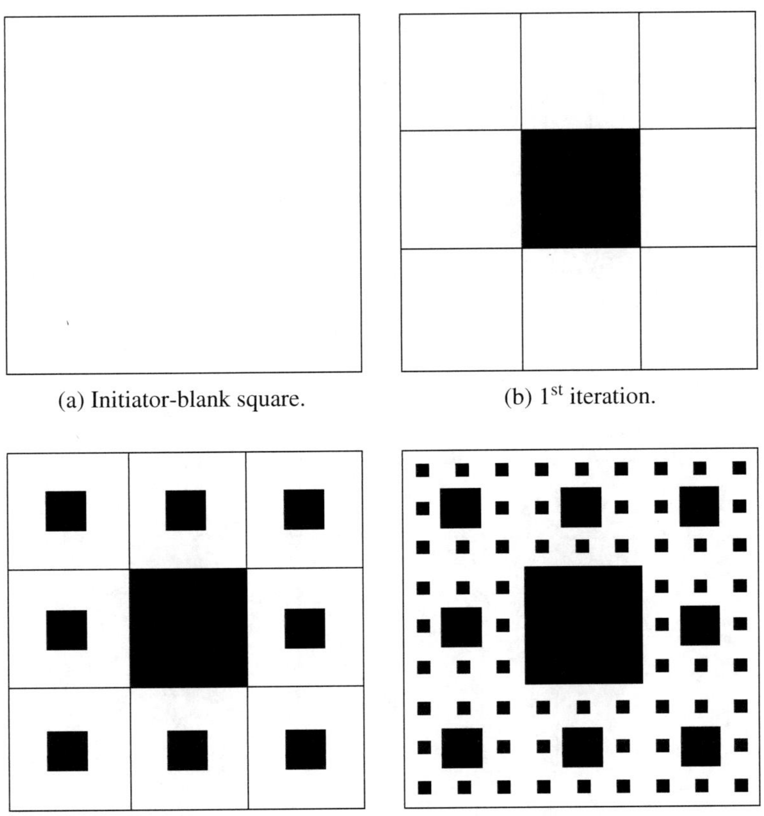

(a) Initiator-blank square. (b) 1st iteration. (c) 2nd iteration. (d) 3rd iteration.

Figure 18.39 Generation of Sierpinski carpet.

18.7.2 Fractal Resonator Antenna

The above-discussed fractal geometrical structures have been used to develop compact multiband antenna, antenna array, resonators, filters, and multiband PBG structures. *Space-filling ability* provides compactness to the structure, while increasing the resonating length. The *self-similarity* provides multiband behavior. The monopole, dipole, and loop antennas are primarily resonating structures to which fractal geometrical arrangements have been applied. Their use provides a reduction in the resonant frequency. Thus, we can get more compact structures, as compared to their Euclidian counterparts. However, such a reduction is not unique to fractal geometry. Other kinds of folded structures also provide identical compactness, or sometimes even better compactness. We summarize below the results of some of the important fractal geometries in respect of a reduction in the resonance frequency only. Details could be followed through the published literature [J.64, J.66, J.67, J.71–J.73].

Koch fractal geometry has been used for the development of compact resonant monopoles. For the same overall height, Koch monopole shows lower resonance frequency, as compared to the identical height of a simple wire monopole, i.e. the Euclidean monopole. It means to maintain a fixed resonance frequency, obtained from the Euclidean monopole, the height of Koch monopole should be reduced. Both the resonance frequency and 2 : 1 VSWR bandwidth decrease with an increase in the order of iterations of Koch monopole. However, this characteristic is not unique to Koch geometry. Better compactness, i.e. more decrease in the resonant frequency, could be obtained on using the *meander line* and *normal mode helix* monopoles. Of course, their 2 : 1 VSWR BW is less than that of Koch monopole. In all the cases of folding, the wire length increases with an increase in the number of iterations. It results in a decrease in resonance frequency. Figure (18.40) shows five iterations of three kinds of monopoles. K_0 is the Euclidian wire monopole. The resonance frequency for them decreases with an increase in the number of iterations [J.74]. It is noted that up to the second iteration much difference is not seen for a reduction in the resonance frequency for all three kinds of monopoles.

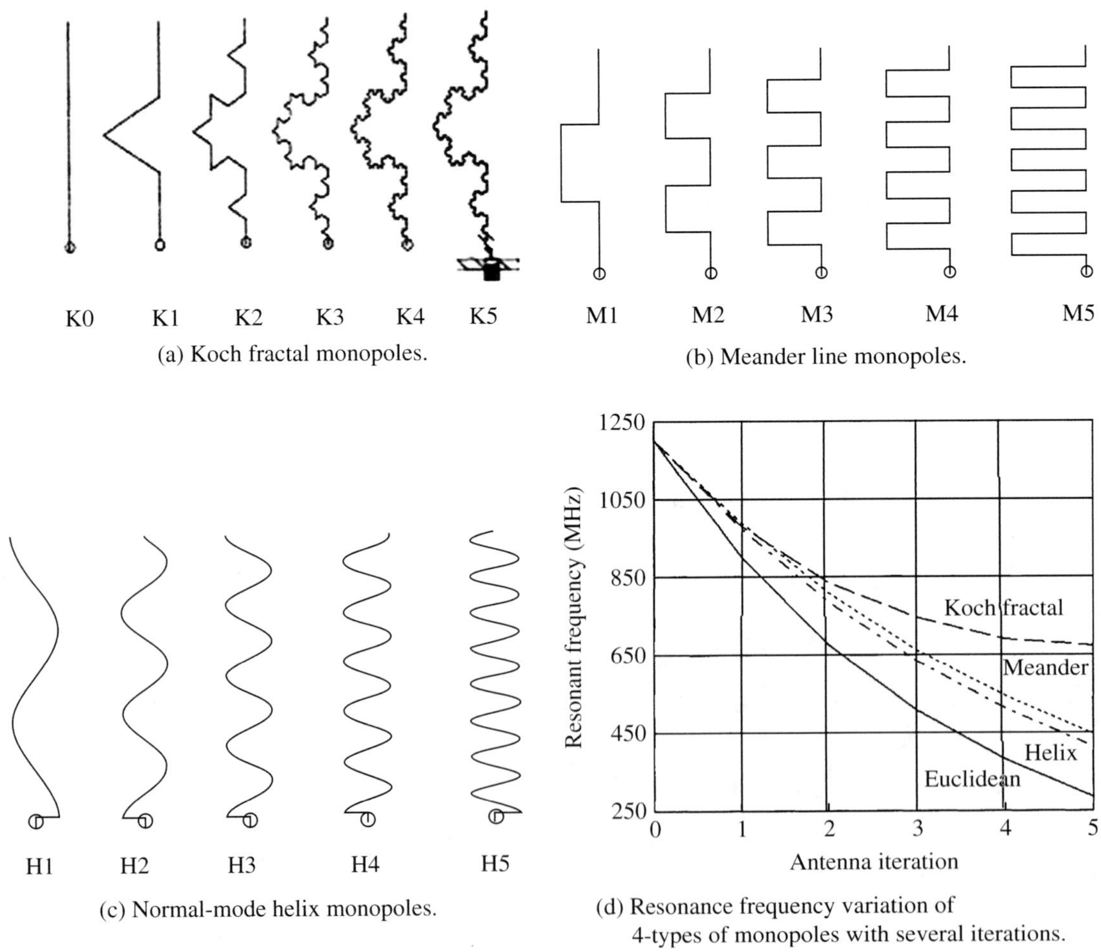

Figure 18.40 Resonance frequency behavior of Koch and other folded monopoles. *Source:* From Best [J.71]. © 2003, IEEE.

However, above the 4th iteration, frequency reduction gets saturated for Koch fractals, whereas in the case of the meander, and normal helix monopole further frequency reduction can be achieved. The Euclidian monopole provides more decreases.

Koch geometry has also been used to get the compact *dipole printed antenna*. The Q-factor decreases with an increase in the number of iterations [J.67]. Even Hilbert fractals have been used to get the compact dipole antenna [J.73]. However, a meander dipole provides more reduction in the frequency as compared to that of Hilbert fractals. In the case of Hilbert fractals geometry, the currents in adjacent wires are in opposite directions, reducing the effective inductance of the Hilbert fractals loops.

Minkowski fractal, shown in Fig (18.36), has been used to get a compact loop antenna, as compared to that of a circular or rectangular loop antenna [J.64, J.66, J.72]. The decrease in the resonance frequency is more for a higher number of iterations. However, it saturates after 2–3 iterations. The indentation width also controls the lowering of the resonance frequency. However, such behavior is not unique to Minkowski fractal or any other fractal. Other kinds of non-fractal folded geometry can provide even more or identical resonance frequency reduction.

The self-similar Sierpinski gasket has also been used to get the multiband antenna [J.71]. The multiband response increases with an increase in the self-similar iterations. However, it has been noted that a significant portion of the standard self-similar Sierpinski gasket can be either removed or altered, without changing the multiband nature. It shows that the standard Sierpinski gasket is not an optimum or unique geometrical arrangement to get the multiband antenna.

18.7.3 Fractal Resonators

The planar fractal resonators have been used to develop high-performance compact filters. The case of a microstrip fractal resonator is more complicated than that of the fractal monopole in the air medium. The fractal iterations increase the microstrip line length in a fixed area. It results in a lowering of the resonance frequency, with increasing iterations. So, the length and area of a resonator can be reduced significantly to maintain the same prescribed frequency. Due to the longer microstrip line, packed in a smaller area, the energy storage ability of a fractal microstrip resonator is more. It increases the Q-factor of a fractal $\lambda_g/2$-resonator 10 times, as compared to a conventional resonator. However, with increasing iterations, to pack longer microstrip lengths, the conductor width is reduced. It increases the conductor loss. On reducing the spacing between the conductor tracks, the conductor loss is further increased due to the proximity effect. Normally, after 3 iterations, the Q-factor of a microstrip fractal resonator is reduced. Also, due to the increase in the discontinuities with iterations, both the radiation and surface-wave losses increase, leading to a reduction in the Q-factor [J.65, J.75].

The resonating structures are characterized using four parameters – *compactness, resonance frequency, Q-factor, and appearance of next resonating mode*. The fractal geometry has been used both for the patch and line resonators. We summarize below results on a few planar fractal resonators with the main focus on the Minkowski and Hilbert resonators. The details can be followed from the available literature [B.25, J.65, J.68–J.70, J.75–J.80].

Figure (18.36) shows the 5-sided Minkowski prefractal *generator* as applied to a square patch resonator. Figure (18.41a) shows the change in the resonance

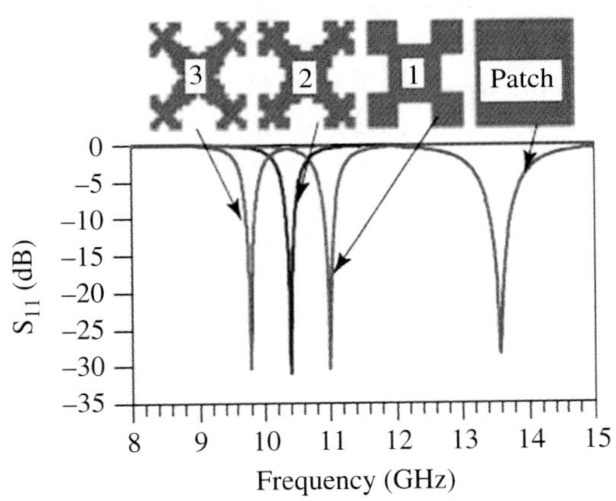

(a) Effect of increasing iterations on the resonance frequency of a patch resonator[B.25].

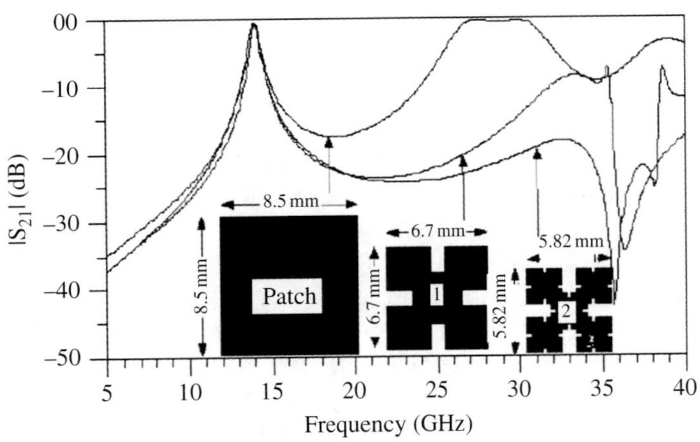

(b) Effect of increasing iterations on 2nd passband [J.76].

Figure 18.41 Behavior of microstrip Minkowski fractal patch resonator. *Source:* From Jarry et al. [B.25]. © 2009, John Wiley & Sons. From Hanna et al. [J.76]. © 2006, IEEE.

Table 18.1 Resonance frequency compression of fractal planar resonators.

Fractal/freq.	f_0 (GHz)	f_1 (GHz)	f_2 (GHz)	f_3 (GHz)	$\Delta f_1/f_0$ (%)	$\Delta f_2/f_0$ (%)	$\Delta f_3/f_0$ (%)	Dimension
Sierpinski	13.8	10.18	9.60	9.46	26.2	4.2	1.02	1.893
Minkowski	13.6	11.00	10.40	9.80	19.1	4.4	4.4	1.465
Hilbert	4.3	2.90	1.85	–	32.6	24	–	2.000
Koch	3.7	2.85	2.58	–	23.0	7.3	–	1.262

frequencies of Minkowski pre-fractal square microstrip patch resonator for 3 iterations. During each of the iteration processes, a portion of the patch is removed, while maintaining the overall patch area. The initial patch resonance frequency 13.6 GHz is reduced at each stage of iteration. The results are shown in Table (18.1). The first iteration reduces the patch frequency to 11 GHz (19.1%), whereas the 2nd and 3rd iterations only add another 4.4% reduction in the resonance frequency. The saturation in the frequency reduction is reached at the 2nd iteration. If it is desired to maintain a constant resonance frequency, with Minkowski fractal geometry, the patch area is reduced, and a compact resonator is obtained.

Figure (18.41b) demonstrates this aspect of Minkowski fractal patch resonator, using the shielded suspended microstrip medium. The resonator is designed on an RT-Duroid 5880 substrate with $\varepsilon_r = 2.2$, $\tan\delta = 0.0009$, $h = 0.254$ mm [J.76]. The unperturbed patch resonance frequency 14 GHz is maintained during both steps of iterations, by reducing the patch area from 8.54×8.54 mm^2 to 6.76×6.76 mm^2, and to 5.82×5.82 mm^2. However, the perimeter increases from 34.16 mm to 40.48 mm, and next to 48.56 mm. In the process, the unloaded Q-factor increases from 353.84 to 424.3, and next to 425.8. At the first iteration, the patch area is reduced by 37.34%, while the perimeter is increased by 18.5%. At the second iteration, the patch area is reduced by an additional 16.21%, and the perimeter is further increased by 23.65%. However, the Q-factor increases by 70.46 at the first iteration, whereas it further increases only by 1.5 at the second iteration.

Thus, it is noted that the rate of area saving gets reduced at the second iterations. Despite a long perimeter, the Q-factor comes to almost saturation. Any further iteration reduces the Q-factor. For instance, the 3rd and 4th iterations reduce the area to 5.7×5.7 mm^2 and 4.74×4.74 mm^2, and the corresponding Q-factors reduce to 403.4 and 344.1 [B.25]. So, Minkowski fractal patch resonator is useful only up to the second iteration.

Figure (18.41b) also shows suppression of the 2nd harmonics for both iterations, and further appearance of a transmission zero at the second iteration. These features are useful to the design of the bandpass filters. In the WLAN band, the patch area reduction up to 64% is also reported for the Minkowski fractal patch resonator [J.81].

Sierpinski square fractal has also been used to reduce the patch area. Table (18.1) shows the results. Sierpinski square fractal has better frequency compression, as compared to that of Minkowski fractal patch resonator. Koch fractal has been applied to a circular patch resonator to achieve a 60% reduction in the area, while its unloaded Q-factor has increased by 30% [J.75]. It is due to the localization of energy in Koch fractal.

Hilbert and Koch's fractals have also been used to miniaturize the microstrip line resonators. Figure (18.42a) shows the decreasing resonance frequency of *the U-type line resonator* on applications of Hilbert fractal iterations. The frequency lowering does not saturate for two iterations. Table (18.1) summarizes the results for both Hilbert and Koch fractals. It also shows the dimensions of four kinds of fractal geometries. Let us recall that the *dimension is a space-filling ability*. The frequency reduction, during the 1st iteration of three cases, almost follows the increasing order of the dimension. The 2nd iteration has more exceptions. However, the increasing dimensional order works for the patch and line resonators differently. Figure (18.42b and c) shows a decrease in the resonance frequency and an increase in the unloaded Q-factor of the superconducting Hilbert resonators for increasing side length. The resonator, shown in Figure (18.37), is developed using the YBCO HTS at 77^0K used on a MgO substrate $\varepsilon_r = 9.6$, $h = 0.508$ mm, $\tan\delta = 10^{-6}$. The external side length (S) is related to microstrip width w, and separation g between the strip conductors, and the number of iterations n as follows [J.69]:

$$S = 2^n(w + g) - g. \tag{18.7.8}$$

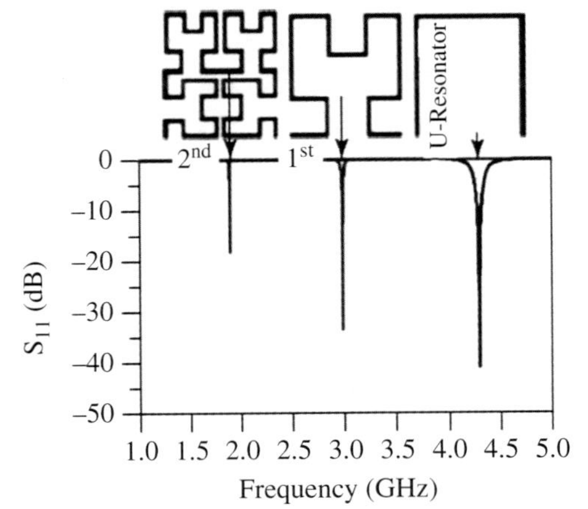

(a) Effect of increasing order of iterations on the resonance frequency of a line resonator[B.25].

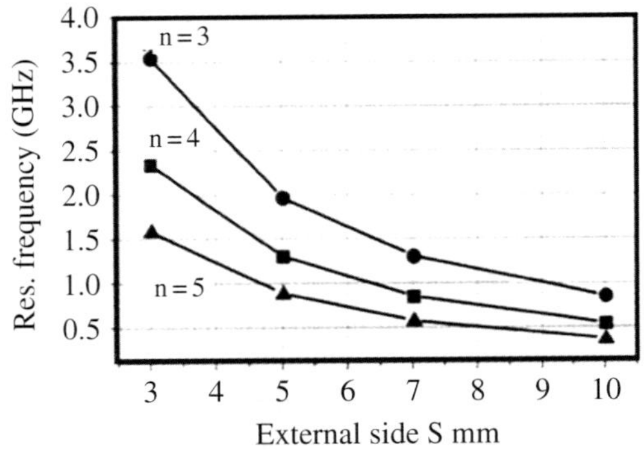

(b) Decrease in resonance freq. of Hilbert resonator with an increased side length. n is the number of iterations[J.69].

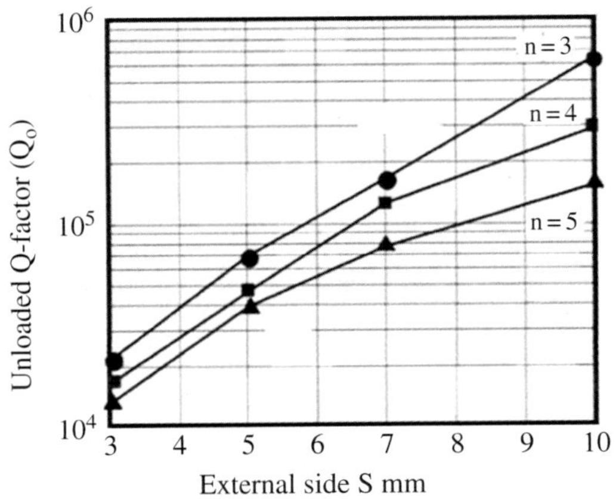

(c) Increase in Q_0 of Hilbert resonator with an increased side length. n is the number of iterations[J.69].

(i) Microstrip and Hilbert resonator on the same substrate.

(ii) Microstrip and Hilbert resonator at two different substrate.

(d) Hilbert resonator and broadside-coupled microstrip and Hilbert resonators[J.78].

Figure 18.42 Behavior of microstrip Hilbert fractal line resonators – simple and superconducting. *Source:* From Jarry etal. [B.25]. © 2009, John Wiley & Sons. From Barra et al. [J.69]. © 2005, IEEE. From Crnojevic-Bengin and Budimir [J.78]. © 2005, IEEE.

Like the previous case, for a fixed side length (S), the resonance frequency decreases with an increase in the number of iterations from n = 3 to 5. However, the Q-factor decreases significantly with an increase in the number of iterations, from n = 3 to 5. It is due to the increase in current density at the edges, causing significant conductor loss. The Q-factor also decreases with a decrease in the w/g ratio. Thus, at 2 GHz for n = 3, and w/g = 1.5, 1.0, 0.5, the corresponding unloaded Q-factors are 67 700, 64 200, 56 100. At n = 5, these are reduced to 10 400, 8 700, 6 800 [J.69]. Figure (18.42b) also shows that the frequency compression, with higher iteration, is more effective at the lower value of S, i.e. at the higher resonating frequency, as compared to its effectiveness at the lower resonating frequency.

Finally, Figure (18.42d) shows two configurations of the capacitively coupled 3rd iteration $\lambda_g/2$-Hilbert line resonator on an alumina substrate $\varepsilon_r = 9.6$, h = 0.625 mm. The resonator width w = 1.96 mm, g = 0.1 mm, and length L consist of several sections (N) of square Hilbert geometry [J.78]. The feeding capacitive gap is 0.2 mm. In the first case, Hilbert line resonator is on the same substrate as that of the feed microstrip. In the second case, Hilbert line resonator is located at another lower substrate, while the upper substrate supports another microstrip resonator and feed lines. Both substrates are alumina, and the separation between the broadside coupled resonators is 0.05 mm. The simulation results for both cases are summarized in Table (18.2). Hilbert resonator has about 84% line length

Table 18.2 Simulation results of two configurations of microstrip Hilbert line resonator.

Case-1: Hilbert resonator				
N	3	4	5	6
L (mm)	4.7	6.3	7.9	9.45
f_r (GHz)	3.403	2.646	2.162	1.827
S_{21} (dB)	−1.65	−2.4	−3.32	−4.41
Q_L	315	441	569	716
Q_0	997	1038	1065	1123
Case-II: Coupled Hilbert and microstrip resonators				
L (mm)	4.7	6.3	7.9	9.45
f_r (GHz)	1.884	1.428	1.148	0.960
S_{21} (dB)	−8.0	−11.0	−13.5	−16.6
Q_L	1047	1231	1400	1433
Q_0	1244	1337	1465	1467

reductions, as compared to a conventional microstrip line resonator. Of course, there is some reduction in Q-factor also for Hilbert resonators.

The above discussions on the fractal resonators demonstrate that the fractal geometry can provide significant compactness to the resonators, even with higher Q-factors for a few numbers of iterations. However, such performances can also be obtained from other non-fractal folded geometries. Moreover, the suitability of particular fractal geometry has to be decided by the designer. There is no unique answer to these important questions. It is difficult to arrive at any definite correlation between the geometrical parameters, such as the dimensionality and self-similarity, of the fractal shapes to the electro-dynamical property of the line resonators, i.e. to the resonance frequency, Q-factor, and bandwidth. However, a large volume of literature is available for the design of compact microwave components using the fractal structures.

18.8 Dual-Mode Resonators

The planar symmetrical resonating resonators, a square and circular patch, and their ring versions, always support two *orthogonal degenerate modes.* The degenerate modes have identical resonance frequencies. However, their field patterns are orthogonal to each other. Two orthogonal resonating modes act as two independent resonators, without any mutual coupling, although both resonators are packed in the same physical body of a patch. Normally, a physical resonating structure is used as a single-mode device for the development of the bandpass filters. Several such coupled single-mode resonators are required for the development of a narrowband high selectivity BPF. If we can control the coupling between two degenerate modes, then the number of physical resonators can be reduced to half for the design of a BPF. This arrangement provides a compact high-performance BPF, with low insertion loss. A single cavity with five modes has been also used to design a BPF [J.79].

This section presents the basic functioning of the dual-mode resonators – both in form of the patch and ring. The concept of the dual mode also accommodates the SIR and the fractal resonators discussed previously.

18.8.1 Dual-Mode Patch Resonators

Figure (18.43a and b) shows the square microstrip capacitively coupled patch resonators, excited from the horizontal side and vertical side, respectively. The square patch supports the TM_{10} and TM_{01} dual degenerate modes at the same resonant frequency; their fields are mutually orthogonal. The arrow shows the current on the patch. Figure (18.43c) shows the excitation of the dual orthogonal resonating modes. Figure (18.43d) shows that each mode could be modeled as an LC resonant circuit, placed orthogonal to each other. The resistance R accounts for the losses incurred in a resonator. The magnetic flux lines from the inductor #1 of one resonator do not couple to the inductor #2 of another resonator. So, the dual orthogonal modes are uncoupled modes. However, to use the dual degenerate modes supporting patch resonator as a two-pole BPF, there must be a coupling between both the modes. Such coupling can be achieved by breaking the orthogonality of the dual modes, using any of the following methods:

- *Deforming the symmetrical shape of resonators.*
- *Introducing a perturbation in a resonator.*
- *Changing the location of the port from its orthogonal position.*

All three methods rotate the equivalent resonator circuit from its orthogonal position. It permits coupling between two resonators, i.e. the coupling between the dual degenerate modes. In the process of the coupling, the single frequency splits into two frequencies, corresponding to the even and odd modes. The even- and odd-mode frequencies have larger separation for strong coupling. The coupling coefficient can be computed from two frequency peaks obtained on an EM-Simulator. The process is discussed in section (18.5). We discuss below the dual-mode patch resonators, in the square, circular, and triangular forms.

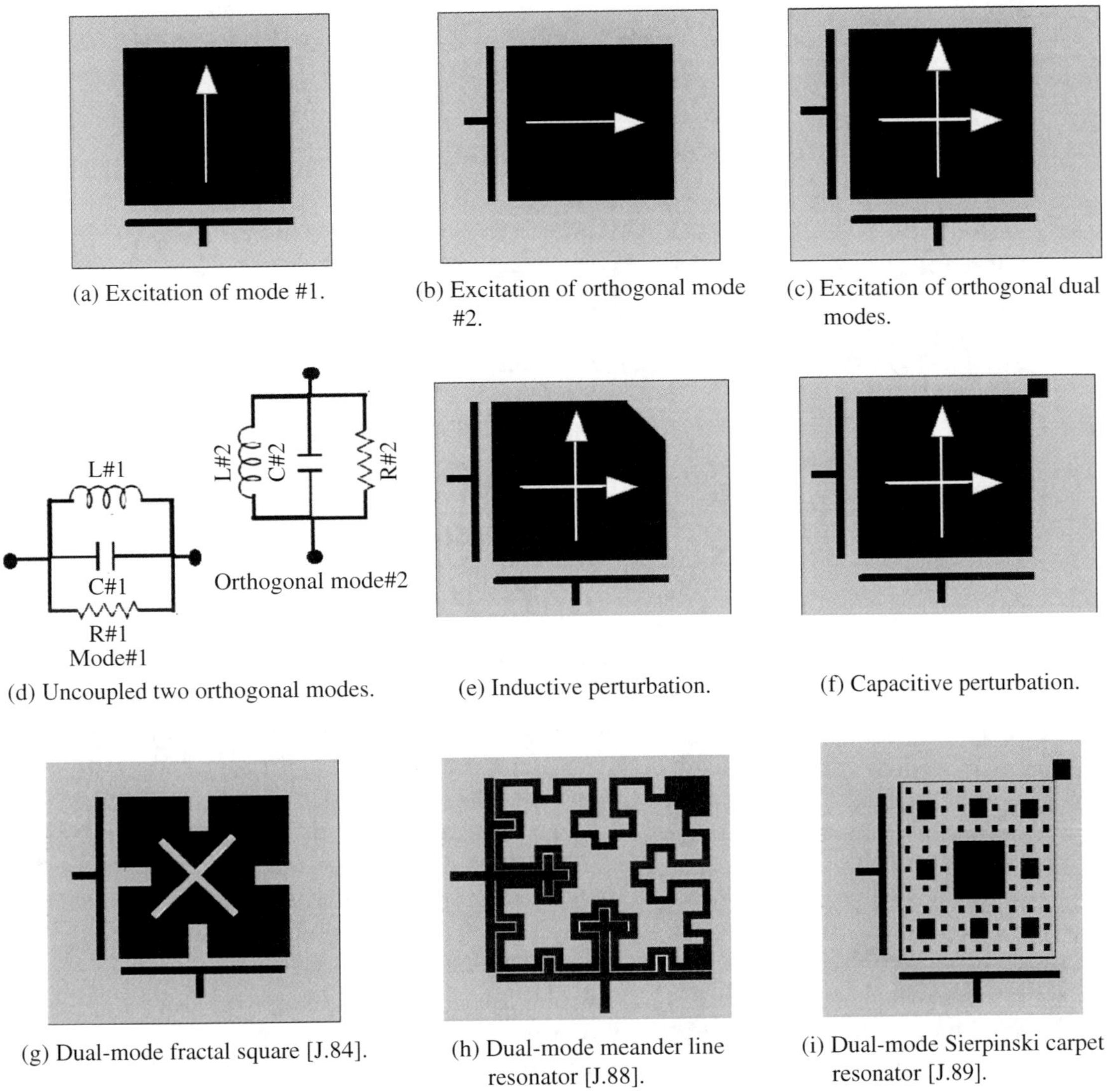

(a) Excitation of mode #1. (b) Excitation of orthogonal mode #2. (c) Excitation of orthogonal dual modes.

(d) Uncoupled two orthogonal modes. (e) Inductive perturbation. (f) Capacitive perturbation.

(g) Dual-mode fractal square [J.84]. (h) Dual-mode meander line resonator [J.88]. (i) Dual-mode Sierpinski carpet resonator [J.89].

Figure 18.43 Excitation of orthogonal dual unperturbed and perturbed modes in a square patch resonator. *Source:* Based Cassinese et al. [J.84]. © 2001, IOP Publishing. Based on Roy [J.88]. Based on Yeh et al. [J.89].

Dual-mode Square Patch Resonator

A slightly deformed square is a rectangle that supports the spitting of dual modes, i.e. the TM_{10} and TM_{01} modes. The impedance locus of a corner-fed square microstrip patch resonator, with both side arms = 11.30 cm on a 1/16″ Rexolite 2200 substrate ($\varepsilon_r = 2.62$, $\tan\delta = 0.001$) is a circle [J.45]. However, the same square patch deformed into a rectangular patch, with arms 11.30 cm and 11.20 cm, provides splitting of the single resonant frequency of the dual modes. Figure (18.44a) shows a cusp in the impedance locus of the deformed square. It gets evolved into a larger loop, and finally into two circular loci. The unloaded Q-factor of the dual-mode resonator is computed from the impedance loop [J.80]. The splitting of the resonance frequency can be also achieved by cutting a corner of the square patch [J.38]. The size of the perturbation, i.e. the cut of the corner, influences the frequency of the dual modes.

Figure (18.43e and f) shows perturbations, at 45° offset from its port, in the square patch by removing a portion of the square, and by adding a small patch at the corner. The first one is the *inductive perturbation,* while the second onc is the *capacitive perturbation*. They produce the inductive and capacitive effects due to the enhanced magnetic and electric field, respectively. So, they determine the inductive and capacitive coupling between the dual degenerate modes [J.74, J.82, J.83].

A patch resonator can handle larger power and has moderate Q-factor (100) without a shield, and 266 with

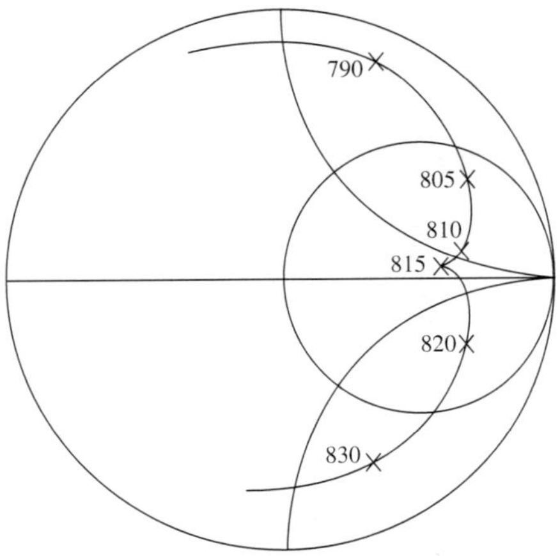

(a) Coupled degenerate modes of rectangular patch (arms: 11.3cm, 11.2cm) [J.45].

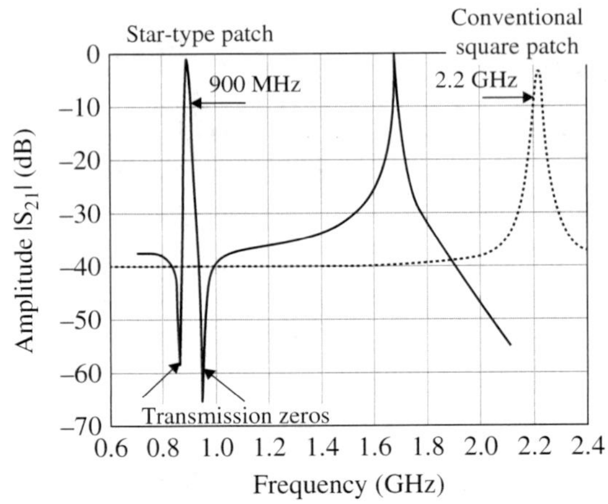

(b) Frequency response of conventional and star-type dual-mode resonators [J.82].

Figure 18.44 Impedance locus of corner-fed deformed square patch and frequency response of star-type dual-mode resonator. *Source:* From Lo et al. [J.45]. © 1979, IEEE. From Banciu et al. [J.82]. © 2004, IEEE.

a shield. However, its size is large [B.6]. The resonating patch can be made compact by using the meander line or fractal geometry. Figure (18.43g) shows a star-type arrangement of the slots with a square patch resonator [J.82, J.84, J.85]. The patch with central orthogonal slots reduces the resonant frequency by more than 30%. If the slot lengths are unequal, it creates a necessary field perturbation to split the resonance frequency of the dual-mode slotted square. An additional 15% frequency reduction is achieved by etching four transverse slots at the center of each arm of the $1/3^{rd}$ of arm-length. It is a Koch, or a Minkowski, type fractal square patch. The center slotted Koch square has unloaded Q-factor between 180 and 310, depending on the crossed slot length – longer the slot length, higher is the unloaded Q-factor. The direct coupling of the external circuits provides 10% fractional BW, whereas the capacitive coupling provides 1% fractional BW. Several other interesting variations are available for the slotted square patch with side slot-cuts. These structures provide the 2-pole BPF having the symmetrical response with transmission zeros on both sides of the passband [J.86, J.87]. Figure (18.44b) compares the typical frequency of a conventional square patch of side arm 20.5 cm on a substrate with $\varepsilon_r = 10.8$, h = 0.635 mm and the star-type dual-mode square patch of the same size. There is a decrease in the resonant frequency from 2.2 GHz to 900 MHz. Two transmission zeros also appear in improving the BPF characteristics [J.82].

Figure (18.43h) further shows a capacitively coupled dual-mode compact 2D square resonator created through the microstrip meandered line. Its size is 35 × 35 mm. It has two field perturbation patches at the corners. The resonator acts like a 2-pole BPF at 605 MHz, with 27 MHz bandwidth, and 1.6 dB insertion loss. Its return-loss is better than 25 dB, and the out of band rejection is also better than 35 dB [J.88]. Finally, Fig (18.43i) shows a compact dual-mode Sierpinski carpet resonator. The Minkowski island dual-mode resonators have also been used to get the wideband BPF [J.89].

Dual-mode Circular Patch Resonator

A circular microstrip patch resonator can also support the orthogonal dual modes – mode #1 and mode #2, traveling in the opposite direction. Figure (18.45a) explains the dual-mode operation [J.81]. A circular patch supports two orthogonal modes #1 and 2 with orthogonal E-field vectors. Again, the circular patch can be loaded with either the inductive or a capacitive perturbation, as shown in Fig (18.45b). The diametrically oppositely placed two perturbations can also be used. An equivalent circuit of the dual resonator is also obtained, where the inductive perturbations are in series with the branch line, and the capacitive perturbations are shunt connected. Four kinds of dual-mode structures are formed out of their combinations [J.81]. Figure (18.45c) shows that in the place of an inductive slit, a circular slot can be used to create the perturbed field [J.90]. Figure (18.45d) shows the frequency response of such a dual-mode resonator at 4.55 GHz designed on a substrate with $\varepsilon_r = 9.6$, h = 0.8 mm.

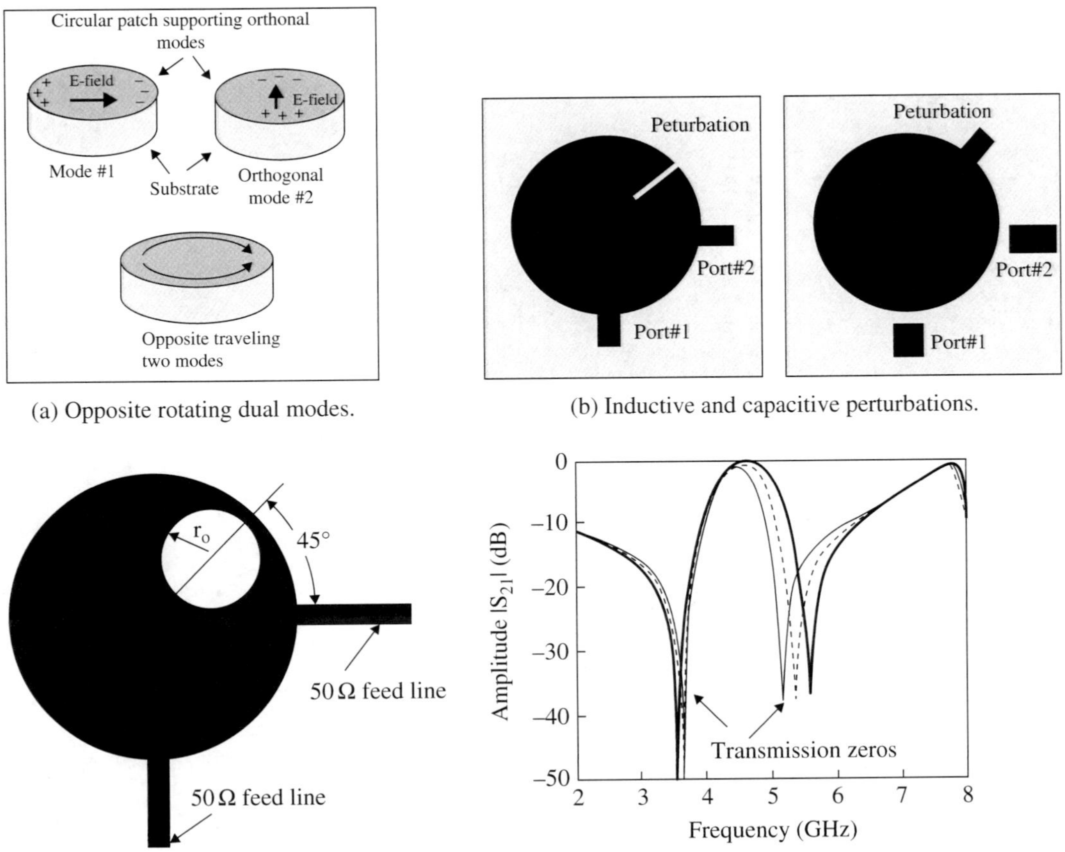

(a) Opposite rotating dual modes.

(b) Inductive and capacitive perturbations.

(c) Circular perturbation [J.90].

(d) Frequency response of Fig (c) [J.90].

Figure 18.45 Dual-mode circular resonators. *Source:* From Li et al. [J.90]. © 2005, John Wiley & Sons.

The radii of the patch and slot are 22.6 mm and 5.75 mm, respectively. Separation of the circular slot gap, from the edge of the patch, influences the upper transmission zero frequency, and the passband bandwidth of the 2-pole BPF. The circular slot can be replaced by several other shapes to get even better responses.

Dual-mode Triangular Patch Resonator

The equilateral triangular patch resonator, shown in Figure (18.46a), is also a symmetrical structure. However, in the XY-coordinate system, it provides only one mode, and its orthogonal dual degenerate mode cannot be obtained [J.91, J.92]. The orthogonal dual mode can be obtained in the rotated X′Y′ coordinate system. Figure (18.46b) shows the current distribution on the patch for both the mode #1 and mode #2. It also shows the orthogonality of the modal currents. Figure (18.46c) shows an equilateral triangular patch of side arm a = 15 cm, with a triangular cut perturbation, on a substrate with $\varepsilon_r = 10.8$, h = 1.27 mm. Figure (18.46d) demonstrates the asymmetrical mode spitting, one resonant mode is almost unchanged with changing the size of the perturbation, whereas the resonance frequency of another mode increases with a larger size of the cut, i.e. for a smaller value of b.

18.8.2 Dual-Mode Ring Resonators

The square ring and circular ring resonators have higher unperturbed Q-factors (161 and 167) as compared to the patch resonators. Figure (18.47a) shows a square ring dual-mode resonator. Like a square patch, the unperturbed square ring also supports the uncoupled degenerate orthogonal dual modes due to its symmetry. Any kind of perturbation, either in form of a cut or in form of a patch, at its diagonal corner located at 135° from the port #1, disturbs the modal field. It results in the mode splitting. The following two aspects of a perturbation are examined: (i) *effect of the size of a perturbation on the modal frequency splitting* and (ii) *effect of nature of coupling due to corner cut, and the addition of a small patch at the corner.*

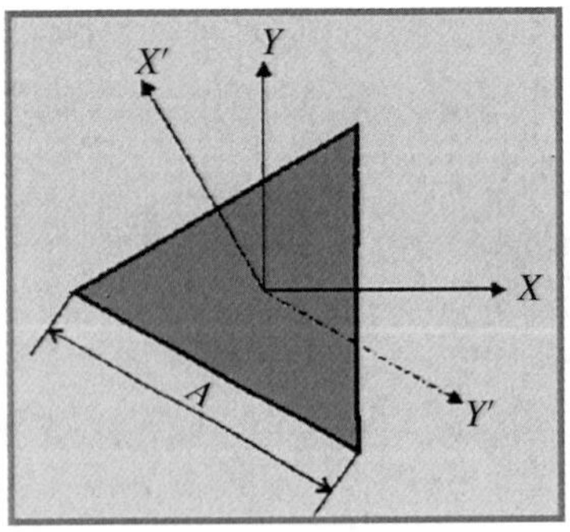

(a) Equilateral triangular patch and rotated coordinate.

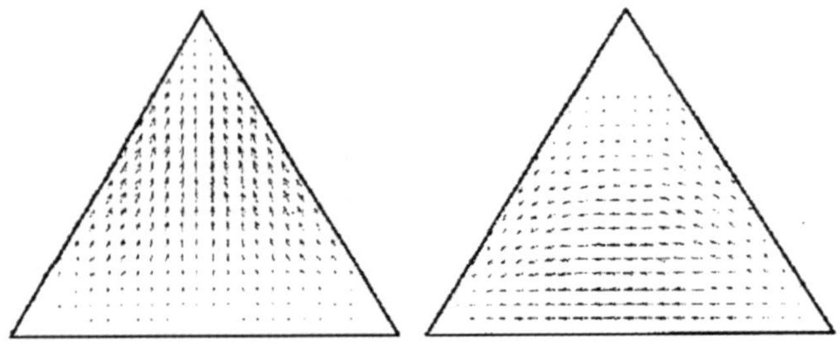

(b) Current distributions for mode #1 and orthogonal mode #2 [J.91].

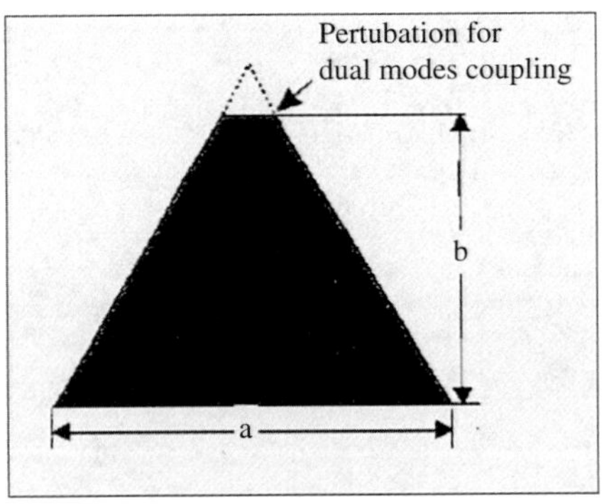

(c) Triangular perturbation cut for dual modes coupling.

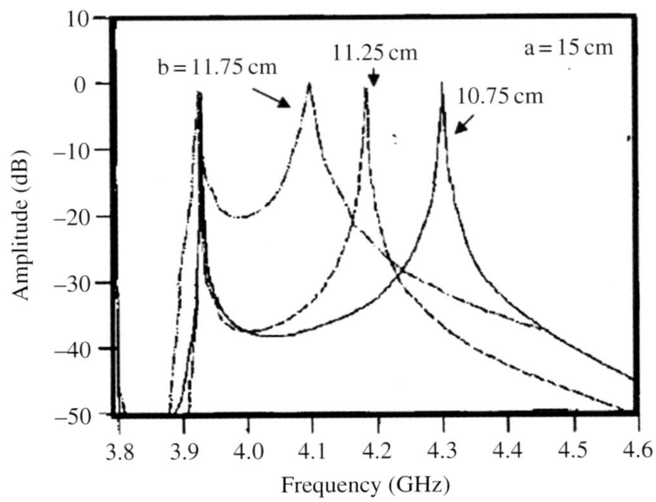

(d) Control of mode splitting by controlling the height of cut [J.91].

Figure 18.46 Dual-mode equilateral triangular resonator. *Source:* From Hong and Li [J.91]. © 2003, IEEE.

Figure (18.47a)) considers a square ring on an RT-Duriod substrate with $\varepsilon_r = 10.8$, h = 1.27 mm. An inner triangular patch of variable size d = 2 mm, 3 mm, 4mm is attached at the corner of the ring. Figure (18.47b) shows its effect on the response at the port #2 [J.93]. Even at d = 2 mm, the degenerate mode frequency breaks into the even- and odd-mode frequencies. Their mutual coupling produces the 2-pole BPF response with a small ripple in the center. Once perturbation size is increased to 3mm, the coupling between dual modes is stronger that gives a larger size ripple, and further separation of two modal frequencies. For 4 mm size perturbation, this effect is enhanced. The data can be generated with the help of an EM-Simulator and the coupling factor is computed. It is needed for the design of a BPF. This has already been discussed in section (18.5).

Figure (18.47c) shows three cases of perturbations: (i) *square shape corner-cut,* (ii) *square shape inner corner attached with the patch,* and (iii) *square shape outer corner attached with a patch* [J.83]. Again, the microstrip square ring is on a substrate with $\varepsilon_r = 10.8$, h = 1.27 mm. The length and width of conducting track of square arm are 20 mm and 2 mm. The perturbation size is d = 2 mm, and the coupling gap is 0.25 mm. The corner cut increases the perimeter forming a longer current path, enhancing the magnetic field. It results in the inductive coupling between the dual modes. Likewise, the addition of a small corner patch, inner or outer, enhances the electric field. It results in the capacitive coupling, i.e. the electric coupling, between the dual modes. Figure (18.47d) of the *case-1* shows that magnetic coupling produces no transmission zero and the BPF response is like a Chebyshev response. However, the

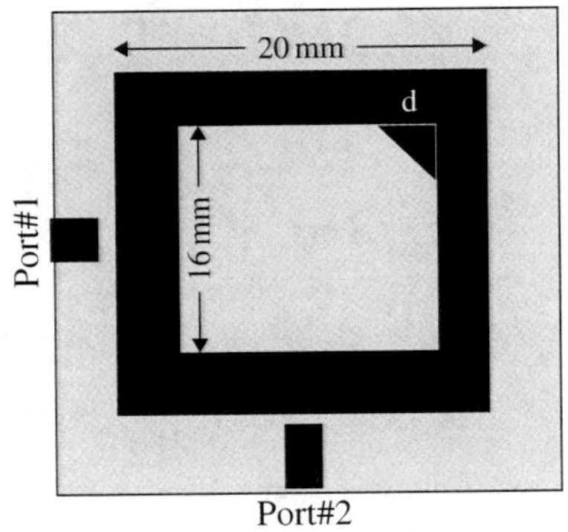

(a) Square ring resonator with triangular patch perturbation.

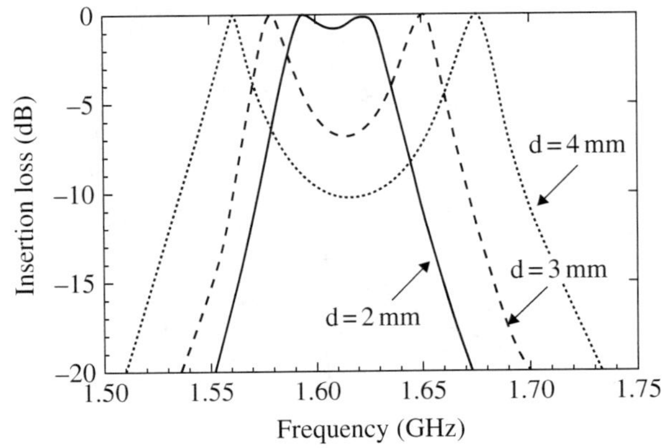

(b) Mode splitting with the size of perturbation [J.93].

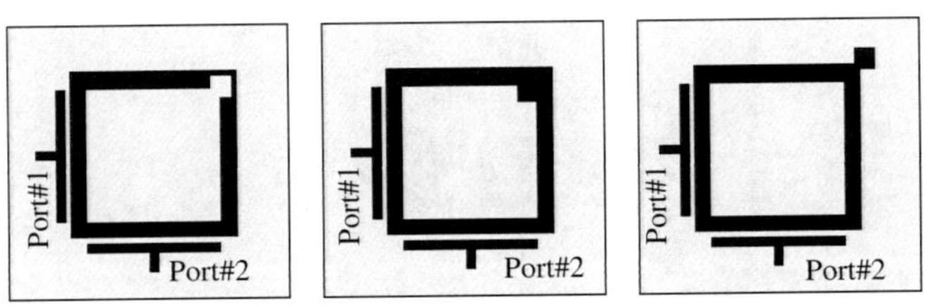

(1) Inductive perturbation. (2) Capacitive perturbation. (3) Capacitive perturbation.

(c) Square ring resonator with square perturbation.

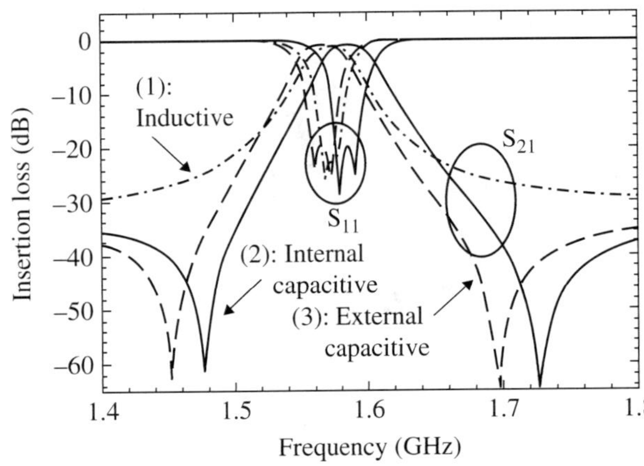

(d) Responses of square ring resonators for a different type of perturbations [J.83].

Figure 18.47 Dual-mode square ring resonators. *Source:* From Görür [J.83]. © 2004, IEEE. From Hong and Lancaster [J.93]. © 1995, IET.

electric coupling produces two symmetrically located transmission zeros. The inner attached patch *of the case-2* moves the response to the right, whereas the outer attached patch of the *case-3* shifts it to the left. Thus, the perturbation's shape and size determine the nature, and also the strength of the coupling. The dual-mode resonator can be analyzed using the transmission line model. An equivalent circuit model of the lumped coupled resonator can also be developed. The dual-mode square ring has also been used to design the CPW-based BPF [J.94].

Figure (18.48a) shows the dual-mode circular ring resonator of diameter 11.6 cm. It is like a square ring resonator on a Polyguide substrate with $\varepsilon_r = 2.33$ [J.95]. However, in this case, the mode splitting takes place due to the non-orthogonal asymmetrical placement of ports. Figure (18.48a) also shows the coupled mode's response at the port #2. It can be controlled by changing the angle of ports placement. Figure (18.48b) further shows that the same ring with a 3 mm wide cut of 4 mm depth. It provides a strong coupling between the dual modes, as two resonance peaks are almost completely separated. The response is controlled by tuning the size of the cut. Finally, Fig (18.48c) shows that the perturbation can also be achieved using the step impedance. The MW and EW can be placed at the plane of symmetry to get even- and odd-mode equivalent circuits, shown in Fig (18.48d). The analysis can be completed, as it has been done for the SIR, discussed in subsection (18.1.4) to get the resonance frequencies [J.25]. Several other variations of the dual mode ring resonator are also available [J.96].

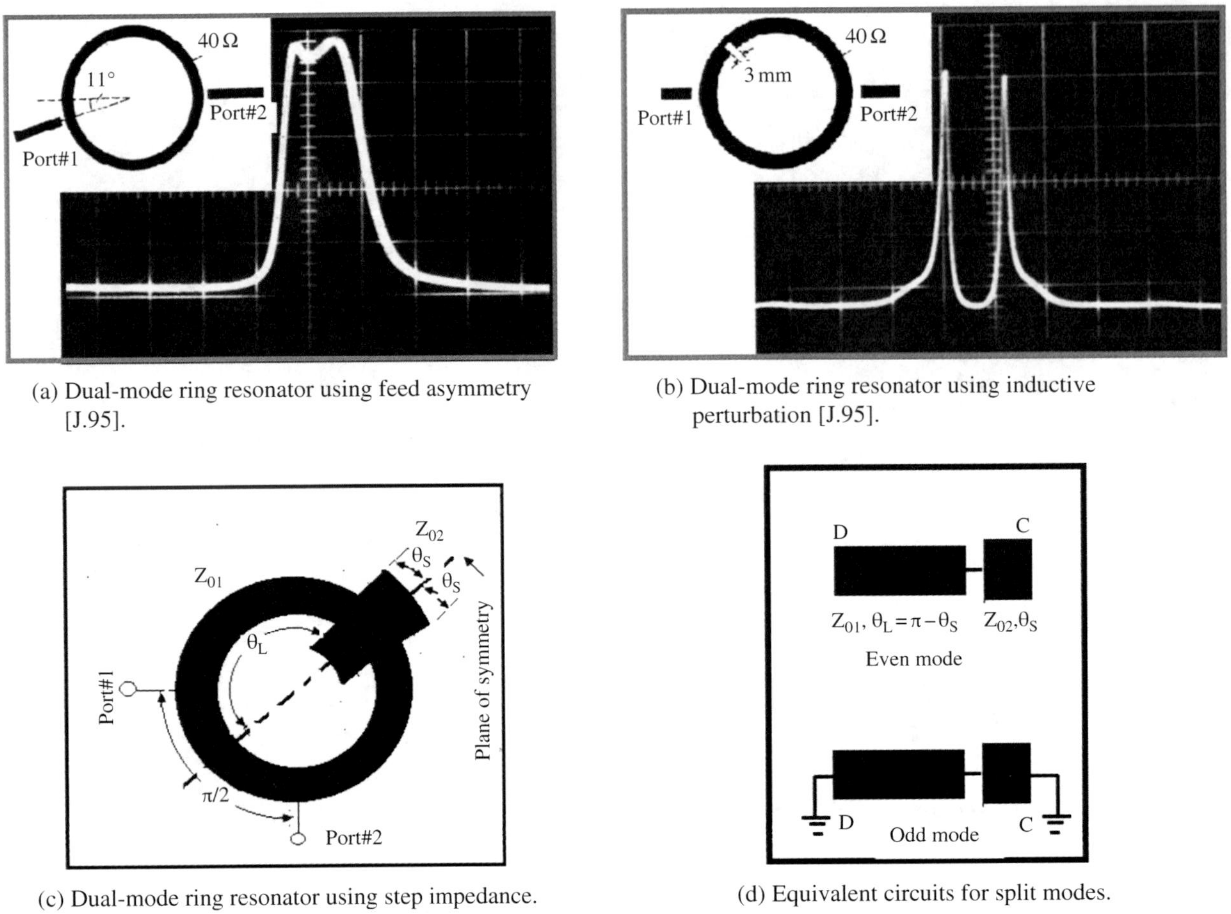

Figure 18.48 Dual-mode ring resonators. *Source:* From Wolff [J.95]. © 1972, IET. By permission of IET.

References

Books

B.1 Chang, K.; Hsieh, L.H.: *Microwave Ring Circuits and Related Structures*, 2nd Edition, John Wiley & Sons, New York, 2004.

B.2 Garg, R.; I. Bahl; P. Bhatia; Ittipiboon, A.: *Microstrip Antenna Design Handbook*, Artech House, Boston, 2000.

B.3 Hoffmann, R.: *Microwave Integrated Circuit Handbook*, Artech House, Boston, 1985.

B.4 Chang, K. (Editor), *Handbook of Microwave and Optical*, Vol. 1, John Wiley & Sons, New York, 1989.

B.5 Lee, K.F.; Dahele, J.S. (Ed. James, J.R.; Hall, P.S.): *Handbook of Microstrip Antennas*, Vol. 1 IEE Publication, 1989.

B.6 Hong, J.S.; Lancaster, M.J.: *Microstrip Filters for RF/Microwave Applications*, John Wiley & Sons, New York, 2001.

B.7 Simon, R.N.: *Coplanar Waveguide Circuits Components and Systems*, John Wiley & Sons, New York 2001.

B.8 Munk, B.A, *Frequency selective surfaces: Theory & Design*, John Wiley & Sons, New York, 2000.

B.9 Rizzi, P.A.: *Microwave Engineering – Passive Circuits*, Prentice-Hall International Edition, Englewood Cliff, NJ, 1988.

B.10 Pozar, D.M.: *Microwave Engineering*, 2nd Edition, John Wiley & Sons, Singapore, 1999.

B.11 Ramo, S.; Whinnery, J.R.; Van Duzer, T.: *Fields, and Waves in Communication Electronics*, 3rd Edition, John Wiley & Sons, Singapore, 1994.

B.12 Collin, R.E.: *Foundations for Microwave Engineering*, 2nd Edition, McGraw-Hill, Inc., New York 1992.

B.13 Ludwig, R.; Bretchko, P.: *RF Circuit Design*, Pearson Education Asia, New Delhi, 2000.

B.14 Rao, N.N.: *Elements of Engineering Electromagnetics*, 3rd Edition, Prentice-Hall, Englewood Cliff, NJ, 1991.

B.15 Staelin, D.H.; Morgenthaler A.W.; Kong J.A.: *Electromagnetic Waves*, Prentice-Hall, Englewood Cliff, NJ, 1994.

B.16 Elliott, R.S.: *An Introduction to Guided-waves and Microwave Circuits*, Prentice-Hall, Englewood Cliff, NJ, 1993.

B.17 Jordan, E.C.; Balmain Keith G. E.: *Electromagnetic Wave and Radiating System*, Prentice-Hall India, New Dehi, 1989.

B.18 Edward, T.C.: *Foundations for Microstrip Circuit Design*, John Wiley & Sons, New York, 1987.

B.19 Sharma, A.; Khanna, A.P.S.; (Ed. Bahl, I.; Bhatia, P.): *Microwave Solid State Circuit Design*, John Wiley & Sons, New York, 1989.

B.20 Schelkunoff, S.A.: *Electromagnetic waves*, D. Van Nostrand Company, 1943.

B.21 Balanis, C.A.: *Antenna Theory Analysis and Design*, 3rd Edition, John Wiley & Sons, 2005.

B.22 Falconer, K, *Fractal Geometry: Mathematical Foundations and Applications*, 2nd Edition, John Wiley & Sons Ltd, 2003.

B.23 Mandelbrot, B.B.: *The Fractal Geometry of Nature*, Freeman, San Francisco, 1982.

B.24 Jaggard, D.L.: *Fractal Electrodynamics: From Super Antenna to Superlattices, Fractal in Engineering*, pp. 204–221, Springer-Verlog, New York, 1997.

B.25 Jarry, P.; Bennett, P.: *Design, and Realizations of Miniaturized Fractal RF and Microwave Filters*, John Wiley & Sons Ltd, 2009.

Journals

J.1 Troughton, P.: *Measurement techniques in microstrip, Electron. Lett.*, Vol. 5, No. 2, pp. 25–26, 1969.

J.2 Gopinath, A.: *Maximum Q-factor of microstrip resonators, IEEE Trans. Microwave Theory Tech.*, Vol. MTT- 29, No. 2, pp. 128–131, 1995.

J.3 Troughton, P.: *High- Q-factor resonators in microstrip, Electron. Lett.*, Vol. 4, No. 24, pp. 520–5216, 1968.

J.4 Hammerstad, E.O.: *Equations for microstrip circuit design*, Eu. Microwave Conf. Proc., pp. 268–272, 1975.

J.5 Krischning, M.; Jansen, R.H.; Koster, N.H.L.: *Accurate model for open-end circuit terminations, Electron. Lett.*, Vol. 17, No. 3, pp. 123–124, 1981.

J.6 Sengupta, D.L.: *Transmission line model analysis of rectangular patch antennas," Electromagnetics*, vol. 4, No. 4, pp. 355–376, 1984.

J.7 Verma, A.K.; Nasimuddin *Accurate determination of input parameters of rectangular microstrip patch antenna on thick substrate and comparison of several models, Electromagnetics, Taylor & Francis, EMG*, Vol. 23(3), pp. 253–276, 2003.

J.8 Pozar, D.M.: *Rigorous closed-form expressions for the surface wave loss of printed antenna, Electron. Lett.*, Vol. 26, pp. 954–956, 1990.

J.9 Hong, J.S.; Lancaster, M.J.: *Capacitively loaded microstrip loop resonator, Electron. Lett.*, Vol. 30, No. 18, pp. 1494–1495, 1994.

J.10 Bengin, V.C; Budimir, D.: *End coupled thick film microstrip resonators with periodical slot loading*, www.telfor.rs/telfor2003/radovi/9-4.pdf.

J.11 Mehmet, K.; McPhun, M.K.; Michie, D.F.: *Simple resonator method for measuring dispersion of microstrip, Electron. Lett.*, Vol. 8, No. 6, pp. 165–166, 1972.

J.12 Wolff, I.; Knoppik, N.: *Microstrip ring resonator and dispersion measurement in microstrip lines, Electron. Lett.*, Vol. 7, No. 26, pp. 779–781, 1971.

J.13 Roberts, R.J.; Easter, B.: *Microstrip resonators having reduced radiation loss, Electron. Lett.*, Vol. 7, No. 8, pp.191–192, 1971.

J.14 El-Tager, A.; Bray, J.; Roy, L.: *High-Q LTCC resonators for millimeter wave applications*, IEEE MTT-S Int. Microwave Symp. Dig., vol. 3, 2003.

J.15 Sarabandi, K; Li Eric S.: *Microstrip ring resonator for soil moisture measurements, IEEE Trans. Geosci. Remote Sensing*, Vol.35, No.5, pp. 1223–1231, September 1997.

J.16 Hsieh, L.H.; Chang, K.: *Equivalent lumped elements G, L, C, and unloaded Q's of closed- and open-loop ring resonators, IEEE Trans. Microwave Theory Tech.*, Vol. 50, No. 2, pp. 454–460, 2002.

J.17 Hopkins, R.; Free, C.: *Equivalent circuit for the microstrip ring resonator suitable for broadband materials characterisation, IET Microw. Antennas Propag.*, Vol.2, No. 1, pp. 66–73, 2008.

J.18 Bray, J.R.; Roy, L.: *Microwave characterization of a microstrip line using a two-part ring resonator with an improved lumped – element model, IEEE Trans. Microwave Theory Tech.*, Vol. 51, No. 5, pp. 1540–1547, 2003.

J.19 Makimoto, M.; Yamashita, S.: *Bandpass filters using parallel-coupled stripline stepped impedance resonators, IEEE Trans. Microwave Theory Tech.*, Vol. MTT- 28, No. 12, pp. 1413–1417, December 1980.

J.20 Banciu, M.G.; Ramer, R.; Joachim, A.: *Compact microstrip resonators for 900 MHz frequency band, IEEE Microwave Wireless Comput. Lett.*, Vol. 13, No. 5, pp. 175–177, 2003.

J.21 Lin, S.C.; Deng, P.H.; Lin, Y.S.; Wang, C.H.; Chen, C. H.: *Wide-stopband microstrip bandpass filters using dissimilar quarter-wavelength stepped impedance* *resonators, IEEE Trans. Microwave Theory Tech.*, Vol. 54, No. 3, pp. 1011–1017, Mar. 2006.

J.22 Sagawa, M.; Takahashi, K.; Makimoto, M.: *Miniaturized hairpin resonator filters and their application to receiver front-end MIC's, IEEE Trans. Microwave Theory Tech.*, Vol. MTT- 37, No. 12. pp. 1991–1997, 1989.

J.23 Lee, S.Y.; C. M. Tsai: *New cross-coupled filter design using improved hairpin resonators, IEEE Trans. Microwave Theory Tech.*, Vol. MTT- 48, No. 12. pp. 2482–2490, Dec. 2000.

J.24 Zhang, H.; Chen, K. J.: *Miniaturized coplanar waveguide bandpass filters using multisection stepped-impedance resonators, IEEE Trans. Microwave Theory Tech.*, Vol. MTT- 54, No. 3. pp. 1090–1095, 2006.

J.25 Sagawa, M.; Makimoto, M.; Yamashita, S.: *Geometrical structures and fundamental characteristics of microwave stepped-impedance resonators, IEEE Trans. Microwave Theory Tech.*, Vol. MTT- 45, No. 7, pp. 1078–1085, July 1997.

J.26 Sagawa, M.; Makimoto, M.; Yamashita, S.: *A design method of bandpass filters using dielectric-filled coaxial resonators, IEEE Trans. Microwave Theory Tech.*, Vol. MTT- 33, No. 2, pp. 152–157, Feb. 1985

J.27 Akgun, C.E.; DelaBarre, L; Yoo, H.; Sohn, S.M; Synder, C.J.; Adrian, G.; Ugurbil, K.; Gopinath, A.; Vaughan: *Stepped-impedance resonators for high-field magnetic resonance imaging, IEEE Trans. Biomedical Eng.*, Vol. 61, No. 2, pp. 327–333, 2014.

J.28 Yen, K.U.; Wollack, E.J.; Doiron, T.A.; Papapolymerou, J.; Lashkar, J.: *A planar bandpass filter design with wide stopband using double split -end stepped impedance resonators, IEEE Trans. Microwave Theory Tech.*, Vol. 54, No. 3, pp. 1237–1243, 2006.

J.29 Ma, K.; Ma, J.; Do, A.V.; Yeo, K.S.; Sun, J.; Miao, J.: *A novel silicon-based CBCPW-fed CBCPS ring resonator*, Proc. 7th Int. Conf. Solid-State and Integrated Circuits Tech., vol.1, pp. 179–182, 2004

J.30 Wu, X.; Awai, I.; Yan, Z.; Wada, K.; Moriyoshi, T.: *Quality factors of coplanar waveguide resonators*, Proc. APMC, pp. 670–673, Singapore, 1999.

J.31 Gao, J.; L. Zhu; W. Menzel; Bögelsack, F.: *Short-circuited CPW multiple-mode resonator for ultra-wideband (UWB) bandpass filter, IEEE Microwave Wireless Components Lett.*, Vol. 16, No. 3, pp. 104–106, Mar. 2006.

J.32 Rahman, A.; Verma, A.K.; Boutejdar, A.; Omar, A.S.: *Control of bandstop response of Hi-Lo microstrip lowpass filter using the slot in the ground plane, IEEE Trans. Microwave Theory Tech.*, Vol. 52, No. 3, pp. 1008–1013, 2004.

J.33 Her, M.L.; Wang, Y.Z.; Chang, C.M.; Lin, K.Y.: *Coplanar waveguide (CPW) defected ground structure (DGS) for bandpass filter application, Microwave Optical Tech. Lett.*; Vol.42, No.4, pp. 331–334, 2004.

J.34 Agrawal, A.K.; Bhat, B.: *Resonant characteristics and end effects of a slot resonator in unilateral fin line, IEEE Proc.*, Vol. 72, No. 10, pp. 1416–1418, 1984.

J.35 Chang, C.C.; Caloz, C.; Itoh, T.: *Analysis of a compact slot resonator in the ground plane for microstrip structures*, Proc. APMC, pp. 1100–1103, Taipei, Taiwan, 2001.

J.36 Kawano, K.; Tomimuro, H.: *Spectral-domain analysis of slot resonator, Trans. IECE (Japan)*, Vol. E65, pp. 480–484, Aug. 1982.

J.37 Hong, J.S.; Lancaster, M.J.: *Couplings of microstrip square open-loop resonators for cross-coupled planar microwave filters, IEEE Trans. Microwave Theory Tech.*, Vol. 44, No. 12, pp. 2099–2109, 1996.

J.38 Hong, J.S.: *Couplings of asynchronously tuned coupled microwave resonators, IEE Proc. Microw. Antennas Propag.*, Vol. 147, No. 5, pp. 354–358, 2000.

J.39 Jensen, R.H.; Koster, N.H.L.: *Accurate results on the end effect of single and coupled microstrip lines for use in microwave integrated circuits, A.E.U.*, Vol. 34, pp. 453–459, 1980.

J.40 Tang, S.C.; Yu, C.H.; Chiou, Y.C.; Kuo, J.T.: Extraction of electric and magnetic coupling for coupled symmetric microstrip resonator bandpass filter with tunable transmission zero, Asia Pacific Microwave Conf, APMC, pp. 2064–2067, 2009.

J.41 Awai, I.; Zhang, Y.: *Separation of coupling coefficient between resonators into magnetic and electric components toward its application to BPF application*, China-Japan Joint Microwave Conf., CJMW, pp. 61–65, 2008.

J.42 Azadegan, R.; Sarabandi, K.: *Miniature high -Q dual -spiral slot-line resonators filters, IEEE Trans. Microwave Theory Tech.*, Vol. 52, No. 5, pp. 1548–1557, May 2004.

J.43 Cheng, F.; Lin, X.Q.; Jiang, Y.; Song, K.J.: *Microstrip bandpass filters based on inductive-coupled stepped-impedance quarter-wavelength resonators, Prog. Electromagn. Res. Lett.*, Vol. 44, pp. 101–106, 2014.

J.44 Cheng, F.; Lin, X.Q.; Jiang, Y.; Song, K.J.; Fan, Y.: *Microstrip filters with adjustable transmission zeros using inductive-coupled open stub-loaded resonators*, Proc. PIERS, pp. 25–28, Guangzhou, China, 2014.

J.45 Lo, Y.T.; Solomon, D.; Richards, W.F.: *Theory, and experiment on microstrip antennas, IEEE Antennas Propag.*, Vol. 27, pp. 137–145, Mar. 1979.

J.46 Watkins, J.: *Circular resonant structures in microstrip, Electron. Lett.*, Vol. 5 No. 21, pp. 524–525, 1969.

J.47 Wolff, I.; Knoppik, N.: *Rectangular and circular microstrip disk capacitors and resonators, IEEE Trans. Microwave Theory Tech.*, Vol. 22, No. 10, pp. 857–864, 1974.

J.48 Cuhai, M.; James, D. S.: *Radiation from triangular and circular resonators in microstrip, G-MTT Int. Symp.*, pp. 438–441, June 1977.

J.49 Helszajn, J.; James, D.S.: *Planar triangular resonators with magnetic walls, IEEE Trans. Microwave Theory Tech.*, Vol. 26, No. 1. pp. 95–100, Feb. 1978.

J.50 Verma, A.K.; Rostamy, Z.: *Modified Wolff Model for resonance frequency of covered rectangular patch antenna; Electron.Lett.*, Vol. 74, No. 20, pp. 1850–1852, 1991.

J.51 Verma, A.K.; Rostamy, Z.: *Modified Wolff Model for determination of resonance, frequency of covered circular microstrip patch antenna, Electron. Lett.*, Vol. 27, No. 24, pp. 2234–2236, 1991.

J.52 Verma, A.K.; Rostamy, Z.: *Resonant frequency of uncovered/ covered rectangular microstrip patch using MWM, IEEE Trans. Microwave Theory Tech.* Vol. 41, No. 1, 109–116, Jan. 1993.

J.53 Verma, A.K.; Bhupal, A.: *Resonance frequency of anisotropic multilayer rectangular microstrip patch using MWM, Electron. Lett.*, Vol. 34, No. 4, pp. 323–325, Feb. 1998.

J.54 Verma, A.K.; Nasimuddin: *Analysis of circular microstrip patch antenna as an equivalent rectangular microstrip patch antenna on Iso/Anisotropic thick substrate, IEE Proc. Microwave, Antenna Propag.*, Vol. 150, No. 4, pp. 223–229, 2003.

J.55 Verma A.K.; Nasimuddin: *Multilayer cavity model for microstrip rectangular and circular patch antenna, Electromagnectics*, Vol. 23, No. 3, pp. 193–217, April 2004.

J.56 Verma, A.K; Nasimuddin; Omar, A.S.: *Microstrip resonator sensors for determination of complex permittivity of materials in sheet, liquid and paste forms, IEE Proc. Microwave Antenna Propag.*, Vol. 151, No. 1, pp. 47–54, Feb. 2005.

J.57 Nasimuddin; Esselle K.; Verma A.K.: *Fast and accurate model for circular microstrip antennas on suspended and composite substrates, IEEE Trans. Antenna Propag.*, Vol. 53, No. 9, pp. 3097–3100, Sept. 2005.

J.58 Nasimuddin; Esselle, K.; Verma A.K.: *Resonance frequency of an equilateral triangular microstrip antenna, Microwave Optical Technol. Lett.*, Vol. 47, No. 5, pp. 485–489, Dec. 2005.

J.59 Verma, A.K.; Rostamy, Z.: *Resonant frequency of covered annular microstrip antenna*, Int EMC Symp. IEMCS-92, pp. 7–9, Dec. 1992, Singapore.

J.60 Verma, A.K.; Rostamy, Z.: *Resonant frequency of equilateral triangular patch resonator with superstrate using Modified Wolff Model*, Int. Symp. Antenna Propag. ISAP, Sept. 1992, Sapporrio, Japan.

J.61 Verma, A.K.; Kumar, R.; Sharma, M.: *Closed-form expressions for accurate determination for capacitance and effective radius of microstrip circular disk, Microwave Opt. Technol. Lett.*, Vol. 14, pp. 217–221, 1997.

J.62 Nasimuddin; Verma, A.K.: *Fast and accurate model for the analysis of equilateral triangular patch antenna, J. Microwave Optoelectron.*, Vol. 3, No. 4, pp. 99–110, April 2004.

J.63 Nachit, A.; Foshi, J.: *Spectral domain integral equation approach of an equilateral triangular microstrip antenna using the moment method, J. Microwave Optoelectron.*, Vol. 2, No. 1, June 2000.

J.64 Gianvittorio, J.P.; Rahmat-Samii, Y.: *Fractal antenna – a novel antenna miniaturization technique, and applications, IEEE Antenna's Propag. Mag.*, Vol. 44, No. 1, pp. 20–36, Feb. 2002.

J.65 Acheron 2.0. Free downloadable, Softpedia, Jan. 2014.

J.66 Krzysztofik, W.J.: *Fractal geometry in electromagnetics applications – from antenna to metamaterials, Microwave Rev.*, Vol. 19, No. 2, pp. 3–14, Dec.2013.

J.67 Werner, D.H.; Haup, R.L.; Werner, P.L.: *Fractal antenna engineering: the theory and design of fractal antenna arrays, IEEE Antennas Propag. Mag.*, Vol. 41, No. 5, pp. 37–59, Oct. I999.

J.68 Crnojevic´-Bengin, V.; Radonic, V.; Jokanovic, B.: *Fractal geometries of complementary split-ring resonators, IEEE Trans. Microwave Theory Tech.* Vol. 56, No. 10. pp. 2312–2321, Oct. 2008.

J.69 Barra, M.; Collado, C.; Mateu, J.; O'Callaghan, J.M.: *Miniaturization of superconducting filters using Hilbert fractal curves, IEEE Trans. App. Superconductivity*, Vol. 15, No. 3, pp. 3841–3846, Sept. 2005.

J.70 Ali, J.K.; Alsaedi, H.; Hasan, M.F.; Hammas, H.A.: *A Peano fractal-based dual-mode microstrip bandpass filters for wireless communication systems*, PIERS Proc., pp. 888–892, Moscow, Russia, Aug. 2012.

J.71 Best, S.R.: *On the performance properties of the Koch fractal and other bent wire monopoles, IEEE Trans. Antenna Propag.*, Vol. 51, No. 6, pp. 1292–1300, June 2003.

J.72 Best, S.R.: *A discussion on the significance of geometry in determining the resonant behavior of fractal and other non-Euclidean wire antennas, IEEE Antenna's Propag. Mag.*, Vol. 45, No. 3, pp. 9–18, June 2003.

J.73 Best, S.R.; Morrow, J.D.: *The effectiveness of space-filling fractal geometry in lowering resonant frequency, IEEE Antenna Propag. Lett.*, Vol. 1, pp.112–115, 2002.

J.74 Curtis, J.A.; Fiedziuszko: *Miniature dual mode microstrip filter, IEEE MTT's Int. Symp. Dig.*, pp. 443–446, 1991.

J.75 Fractus, S.A.: *Fractal miniaturization in RF and microwave networks*, www.fractus.com, Fractus, S.A., 2001.

J.76 Hanna, E.; Jarry, P.; Kerherve, E.; Pham, J.M.: *A novel compact dual-mode bandpass filter usingfractal shaped resonators*, 13th Int. Conf. Electron. Circuits System, ICECS'06, pp. 343–346, Dec.2006.

J.77 Liu, J.C.; Liu, H.H.; Yeh, K.D.; Liu, C.Y.; Zeng, B.H.; Chen, C.C.: *Miniaturized dual-mode resonators with a Minkowski-island-based fractal patch for WLAN dual-band systems, Prog. Electromagn. Res. PIERS Part C*, Vol. 26, pp. 229–243, Aug. 2012.

J.78 Crnojevic-Bengin, V.; Budimir, D.: *Novel compact resonators with multiple 2-d Hilbert fractals curves*, Proc. 35th European Microwave Conference, Paris, pp. 205–207, 2005.

J.79 Lin, W.G.: *Microwave filters employing a single cavity excited in more than one mode, J. Appl. Phys.*, Vol. 22, pp. 989–1001, 1951.

J.80 Kajfez, D.: *Dual resonance, IEE Proc.*, Vol. 135, Pt. H, No. 2, pp. 141–144, April 1988.

J.81 Awai, I.; Yamashita, T.: *Theory on a circular dual-mode resonator and filter with internal coupling scheme*, Asia Pacific Microwave Conf., APMC, pp. 817–820, 1997.

J.82 Banciu, M.G.; Lojewski, G.; Ioachim, A.; Nedelcu, L.; Militaru, N.; Shen, A.; Ramer, R.: *New planar compact dual-mode filters for mobile communications*, IEEE 27th Int. Semiconductor Conf. (CAS 2004), pp. 137–140, Sinaia, Romania, Oct. 2004.

J.83 Görür, A.: *Description of coupling between degenerate modes of a dual-mode microstrip loop resonator using a novel perturbation arrangement and its dual-mode bandpass filter applications, IEEE Trans. Microwave Theory Tech.* Vol. 52, No. 2. pp. 671–677, Feb. 2004.

J.84 Cassinese, A.; Andreone, A.; Di Gennaro, E.; Pica, G.; Vaglio, R.; Malandrino, G.; Perdicaro, L.M.S.; Fragal, I. L.; Granata, C.: *Dual mode cross slotted filter realized with double-sided Tl2Ba2CaCu2O8 films grown by MOCVD, Supercond. Sci. Technol.*, Vol. 14, pp. 406–441, 2001

J.85 Zhu, L.; Wachowski, P.M.; Wu, K.: *New planar dual-mode filter using cross-slotted patch a resonator for simultaneous size and loss reduction, IEEE Trans. Microwave Theory Tech.* Vol.47, No.5. pp.650–654, May 1999.

J.86 Wu, G.-L.; Mu, W.; Dai, X.-W.; Jiao, Y.-C.: *Design of a novel dual-band bandpass filter with microstrip meander-loop resonator and CSRR DGS, Prog. Electromagn. Res.* 78, *pp.*17–24, 2008

J.87 Shen, S.; Ramer, R.: *Design, and realization of star-geometry dual-mode bandpass filter, Opt. Express*, Vol. 13, No. 24, pp. 9753–9757, Nov. 2005.

J.88 Roy, D. *Miniature Filter Screens VHF Band*, Microwave, and RF. July 2009.

J.89 Yeh, K.D.; Liu, H.H.; Liu, J.C.: *An equivalent circuit model for the wide-band band-pass filter with the modified Minkowski- island-based fractal patch, Int. J. RF Microwave Computer-Aided Eng.*, Vol. 24, No. 2, March 2014 pp. 170–176

J.90 Li, J.L.; Chen, J.X, Wang, J.P.; Xue, Q.; Shao, W.; Xia, L.; Xue, L.J.: *Dual-mode microstrip bandpass filter using circular patch resonator with two transmission zeros, Microwave Optical Tech. Lett.*, Vol. 46, No. 1, pp. 28–30, July 2005.

J.91 Hong, J.S.; Li, S.: *Dual-mode microstrip triangular patch resonators and filters, IEEE MTT's Int. Symp. Dig.*, pp. 1901–1904, 2003.

J.92 Hong, J.S.; Li, S.: *Theory, and experiment of dual-mode microstrip triangular patch resonators and filters, IEEE Trans. Microwave Theory Tech.* Vol. 52, No. 4. pp. 1237–143, Apr. 2004

J.93 Hong, J.S.; Lancaster, M.J.: *Bandpass characteristics of new dual-mode microstrip square loop resonators, Electron. Lett.*, Vol. 31, No. 11, pp.391–392, 1995.

J.94 Huang, X. D.; Cheng C. H.: *A novel coplanar waveguide bandpass filter using a dual-mode square-ring resonator, IEEE Microwave Wireless Comput. Lett.*, Vol. 16, No. 1, pp. 13–15, 2006.

J.95 Wolff, I.: *Microstrip bandpass filter using degenerate modes of a microstrip ring resonator, Electron. Lett.*, Vol. 8, No. 12, pp. 302–303, 1972.

J.96 Liu, J.C.; Lu, P.C; Shie, C.H.; Cheng, C.S.; Yao, L.: *Dual-mode double-ring resonators for microstrip band-pass-filter applications, IEE Proc.-Microw. Antennas Propag.*, Vol. 151, No. 5, pp. 430–434, Oct. 2004.

19

Planar Periodic Transmission Lines

Introduction

The unloaded planar lines support the continuous spectrum of wave motion, whereas the reactively loaded 1D periodic lines, also reactively loaded 2D periodic surfaces, exhibit a discrete spectrum of wave motion through alternate passband – stopband phenomena. The stopband frequency region inhibits the wave propagation, thus, creating the electromagnetic frequency bandgap, so the periodic structures are called the *electromagnetic bandgap (EBG)* structures. These are also sometimes called the *photonic bandgap (PBG)* structures, using the terminology from optics. These nomenclatures are due to their parallelism with the electronic bandgap in materials discussed using the quantum mechanics-based Brillouin Zone (BZ) [J.1–J.3]. BZ's description and the language of crystallography are useful for the analysis of the EBG lines and surfaces. The EBG surfaces are discussed in chapter 20. The concept of the periodic lines and periodic surfaces are also associated with the metamaterials lines, i.e. the *metalines* and *metasurfaces* discussed in chapter 22. This chapter is concerned with periodic lines, also known as the *artificial lines*. The periodic lines have the following interesting features:

- Creation of alternate passband and stopband. It is useful for the development of filters and other components.
- Supporting the slow-waves propagation. It is useful to develop a compact, i.e. miniaturized, microwave devices.
- Supporting backward wave propagation. It helps to develop a transmission line-based metamaterials.

Objectives

- *To understand the wave propagation on the 1D and 2D lattice structures.*
- *To understand the Floquet theorem and space harmonics*
- *To discuss the 1D EBG circuit models and realization of the 1D planar EBG in microstrip and CPW.*

19.1 1D and 2D Lattice Structures

The crystal structure of the natural materials has a periodic arrangement of atoms and molecules. It provides a periodic medium for the EM-waves propagation. Below the X-rays wavelengths, the medium is homogeneous and reflection and transmission of waves take place. However, at the X-rays wavelengths, the phenomenon of diffraction takes place. The reactively loaded lines and surfaces are viewed as the scaled-up periodic structures that mimic the one-dimensional (1D) crystal structures. A brief account of the X-rays propagation in the lattice structure is helpful to appreciate the EM-wave propagation on the artificial *electromagnetic bandgap* (EBG)-based lines and surfaces.

19.1.1 Bragg's Law of Diffraction

The periodic array of atoms is located in the stack of parallel planes. The EM-wave propagating through such a periodic structure gets partially reflected from each level, i.e. the plane, of the three-dimensional (3D) crystal. Figure (19.1a) shows the constructive interference of the reflected waves, from three layers of atoms/molecules, resulting in the formation of a diffraction pattern. It is described by the famous *Bragg's law* [B.1–B.4].

Figure (19.1a) shows the rays of incident wave normal forming an angle θ at the location O of the first reflecting plane. The path length A–B–C of the second beam is 2d $\sin\theta$. The path length A′–B′–C′ of the third beam, with respect to the first beam, is $2 \times 2d \sin\theta$. The needed path length is $n\lambda$ to get constructive interference, i.e. the

Introduction to Modern Planar Transmission Lines: Physical, Analytical, and Circuit Models Approach, First Edition. Anand K. Verma.

in-phase interference of beams. So the Bragg's law of diffraction can be written as follows:

$$2d\sin\theta = n\lambda,\quad n:\ \text{an integer} \quad (a)$$
$$\text{wavenumber},\quad k = 2\pi/\lambda \quad (b), \qquad (19.1.1)$$

where d is the separation between two reflecting planes. Figure (19.1b) shows an arrangement of the scatterers, i.e. molecules, at equal distance d to create the 1D periodic structure. Figure (19.1b) also shows the reflecting planes, known as the *Bragg's planes* (BP), satisfying the condition given by equation (19.1.1a). For the normal wave incidence, the angle of incidence is $\theta = 90°$. The following expression is obtained for the distance between the reflecting planes:

$$d = n\frac{\lambda}{2} \quad (a), \qquad \Rightarrow k = \frac{n\pi}{d}, \quad (n = 1, 2, \ldots) \quad (b), \qquad (19.1.2)$$

where k is the propagation constant, i.e. the wavenumber of the incident wave. The reflected wave is discretized by the periodic medium. Thus, the incoming continuous incident wave acquires periodicity due to its propagation through a periodic medium. Figure (19.1b) shows a periodic arrangement of the atoms/molecules in the *real space*, i.e. the *direct space*, with a *periodicity* d. It is also called the *lattice constant* d. A pair of the Bragg's plane defines the *unit cell* of the periodic structure. The wave phenomenon is described in the *wave-vector space*, also called the *k-space*. It is shown in Fig. (19.1c). The wave acquires the periodicity π/d in the k-space. The k-space is further known as the *reciprocal space*, as its periodicity is reciprocal to that of the lattice constant d of the direct space.

The region between the Bragg's planes in the k-space $(-\pi/d < k < \pi/d)$ is known as the *Brillouin zone* (*BZ*). So the EM-wave can propagate for a range of frequency, with the corresponding wavenumber, within the BZ. Only at the edge of the BZ, i.e. at the Bragg's plane (BP), reflection occurs. These concepts are useful to the modern 1D and 2D EBG structures, described in the language of the crystal lattice. Therefore, it is appropriate to discuss briefly the crystal lattice structure.

19.1.2 Crystal Lattice Structures

An artificial crystal structure is a periodic arrangement of *scatterers* (i.e. *inclusions*) in one, two, or three-dimensional *host medium*. The crystal structure is described using the geometrical arrangements called the *lattice*. The identical inclusions, also called *basis*, are attached at each lattice point. Thus, a physical crystal is formed by a combination of the lattice and basis. This arrangement is called the *direct lattice* in the *real space*. The EM-wave propagating in periodic real space is described in the *reciprocal space*, where the wavevector, i.e. the k-vector, forms its own lattice. Both the direct and reciprocal lattices are correlated [B.1].

Direct - Lattice

Figure (19.2a and c) shows the geometrical arrangements of the 2D and 1D direct lattice. The 2D direct lattice is described by two direct *primitive vectors* $\left(\vec{a}_1, \vec{a}_2\right)$. The vectors could be mutually orthogonal or nonorthogonal. They form the following *translational vector* $\vec{T}$ to generate the 2D lattice:

$$\vec{T} = u_1\,\vec{a}_1 + u_2\,\vec{a}_2, \qquad (19.1.3)$$

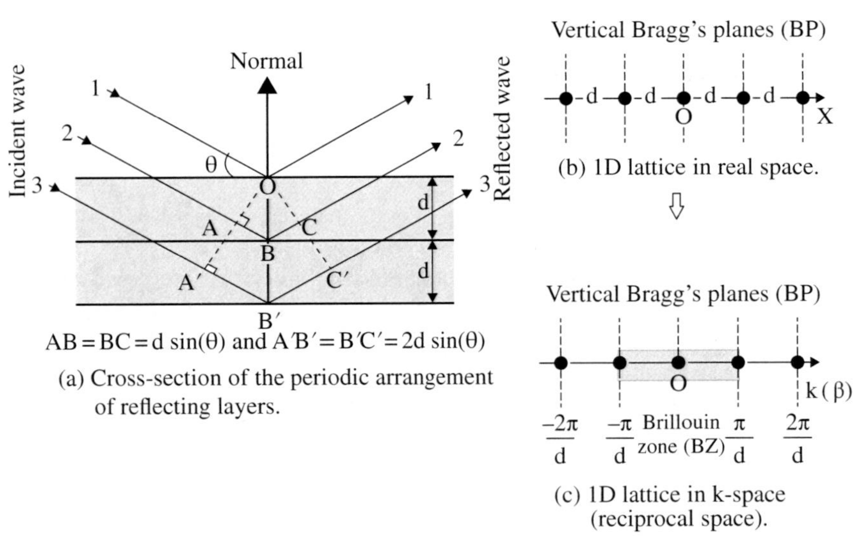

Figure 19.1 Wave propagation in the 1D periodic medium.

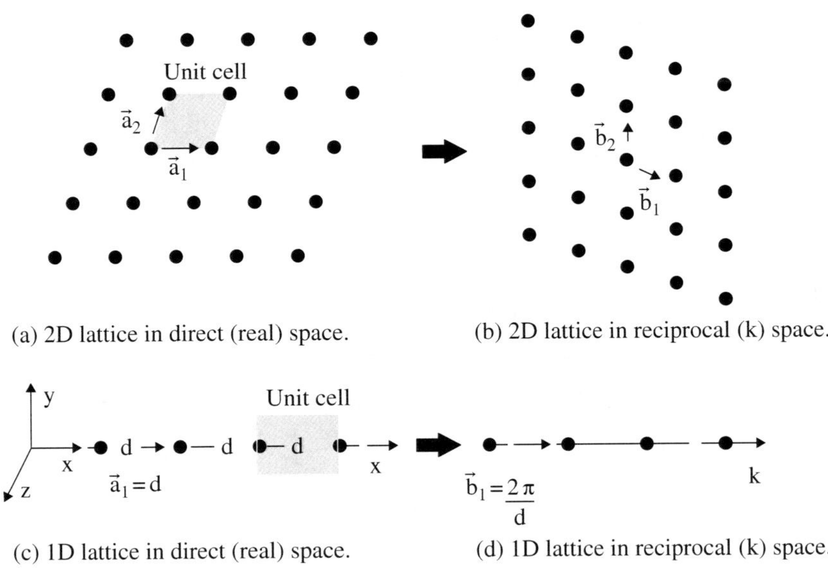

(a) 2D lattice in direct (real) space. (b) 2D lattice in reciprocal (k) space.

(c) 1D lattice in direct (real) space. (d) 1D lattice in reciprocal (k) space.

Figure 19.2 The lattice structures in real and reciprocal space.

where u_1 and u_2 are integers. The translation vector $\vec{T}$ spans the 2D physical space to form the 2D periodic surface. The continuous EM-wave, propagating through the 2D crystal lattice, acquires the periodicity. For instance, the electric field $\vec{E}_0$ located at the distance $\vec{r}$ is expressed as a function of the position vector $\vec{r}$, i.e. $\vec{E}_0(\vec{r})$. The position vector $\vec{r}$ is given by

$$\vec{r} = x\,\vec{a}_1 + y\,\vec{a}_2. \tag{19.1.4}$$

In a 2D lattice, $\vec{E}_0(\vec{r})$ is a periodic function of $\vec{r}$ with periods $\vec{a}_1$ and $\vec{a}_2$ along the direction of two crystal axes. Thus, the electric field $\vec{E}_0(\vec{r})$ acquires its periodicity from the periodicity of the direct lattice:

$$\vec{E}_0\left(\vec{r} + \vec{T}\right) = \vec{E}_0\left(\vec{r}\right). \tag{19.1.5}$$

The electric field at a lattice point of an *infinite periodic structure* is invariant under the translation $\vec{T}$. The 1D direct lattice, Fig. (19.2c), is described by the 1D direct primitive vector $\vec{a}_1$. The direct primitive vectors $\left(\vec{a}_1, \vec{a}_2\right)$ and $\vec{a}_1$, of the 2D and 1D periodic space respectively also define the 2D and 1D *unit cells*. Their infinite repetition creates the infinite extent 2D and 1D periodic lattice. The distances a_1 and a_2 are the *lattice constants*.

Reciprocal Lattice

The EM-wave propagation in a periodic structure has discrete values of the propagation constant. It is described in the 2D and 1D reciprocal lattice space, i.e. in the *k-space*, as shown in Fig. (19.2b and d), respectively. Figure (19.2b) shows the primitive vectors $\left(\vec{b}_1, \vec{b}_2\right)$ of the 2D lattice, defining the reciprocal lattice. Each point of the reciprocal lattice represents a set of planes, identified as the Bragg's planes for the wave reflection. The reciprocal space is mapped by another *translation vector* $\vec{G}$ defined as

$$\vec{G} = v_1\,\vec{b}_1 + v_2\,\vec{b}_2, \tag{19.1.6}$$

where v_1 and v_2 are integers. Figure (19.2d) shows the 1D reciprocal space with the 1D reciprocal lattice. In the case of a 3D crystal lattice, we have $\left(\vec{a}_1, \vec{a}_2, \vec{a}_3\right)$ direct primitive vectors, and $\left(\vec{b}_1, \vec{b}_2, \vec{b}_3\right)$ reciprocal primitive vectors. The reciprocal primitive vectors $\left(\vec{b}_1, \vec{b}_2, \vec{b}_3\right)$ correspond to the vector components $(k_x\hat{x}, k_y\hat{y}, k_z\hat{z})$ of the wavenumber $\left(\vec{k}\right)$ in the k-space. However, the real wave propagates in the physical 3D periodic direct space.

Both sets of primitive vectors, describing the direct space and reciprocal space (i.e. k-space) are related through the following orthogonality expression:

$$\vec{b}_i \cdot \vec{a}_j = 2\pi\,\delta_{ij} = \begin{cases} 2\pi, & i = j \\ 0, & i \neq j \end{cases}. \tag{19.1.7}$$

Expression (19.1.7) shows that the vector $\vec{b}_1$ is normal to both $\left(\vec{a}_2, \vec{a}_3\right)$ vectors, i.e. $\vec{b}_1 . \vec{a}_1 = 2\pi$, $\vec{b}_1 . \vec{a}_2 = 0$, $\vec{b}_1 . \vec{a}_3 = 0$. Similarly, the vector $\vec{b}_2$ is normal to both $\left(\vec{a}_1, \vec{a}_3\right)$ vectors, and $\vec{b}_3$ is normal to both $\left(\vec{a}_1, \vec{a}_2\right)$ vectors. Following this rule, Fig. (19.2b) shows the vectors $\left(\vec{b}_1, \vec{b}_2\right)$ in the 2D reciprocal space. The reciprocal primitive vectors $\left(\vec{b}_1, \vec{b}_2, \vec{b}_3\right)$ of the 3D reciprocal lattice are constructed from the 3D direct primitive vectors $\left(\vec{a}_1, \vec{a}_2, \vec{a}_3\right)$ as follows:

$$\vec{b}_1 = 2\pi \frac{\vec{a}_2 \times \vec{a}_3}{\vec{a}_1 . \left(\vec{a}_2 \times \vec{a}_3\right)} \quad \text{(a)}$$

$$\vec{b}_2 = 2\pi \frac{\vec{a}_3 \times \vec{a}_1}{\vec{a}_1 . \left(\vec{a}_2 \times \vec{a}_3\right)} \quad \text{(b)} \qquad (19.1.8)$$

$$\vec{b}_3 = 2\pi \frac{\vec{a}_1 \times \vec{a}_2}{\vec{a}_1 . \left(\vec{a}_2 \times \vec{a}_3\right)} \quad \text{(c)}.$$

Expressions (19.1.8a–c) are also applicable to the 1D and 2D lattice structures. This chapter is concerned with the 1D and 2D cases only.

1D Reciprocal Lattice

Figure (19.2c) shows the 1D direct space with the lattice constant (i.e. periodicity) d, giving the direct primitive vector, $\vec{a}_1 = d\,\hat{x}$. The reciprocal primitive vector $\vec{b}_1$ is obtained from equation (19.1.8a) as follows:

$$\vec{b}_1 = 2\pi \frac{\hat{y} \times \hat{z}}{d\,\hat{x}.(\hat{y} \times \hat{z})} = \frac{2\pi}{d}\hat{x}. \qquad (19.1.9)$$

As there is no periodicity along the y- and z-directions, so we have taken $\vec{a}_2 = \hat{y}$ and $\vec{a}_3 = \hat{z}$. It is seen that the vector $\vec{b}_1$ is also directed along the *x*-axis, and it is normal to the plane (y–z). It also meets the condition given by equation (19.1.7). As the reciprocal space is also the k-space, so the primitive vector $\vec{b}_1$ could be identified with the wavenumber k_x or propagation constant β_x in the k-space. The x-component of the propagation constant $\vec{\beta}$ is β_x for the wave motion in the *x*-direction.

2D Reciprocal Lattice

The 2D direct periodic lattice could be constructed in five different ways, forming five distinct *Bravais lattices* [B.1, B.2, B.4]. However, the *square*, *rectangular,* and *hexagonal lattices*, shown in Fig. (19.3a and c), are more common formations of the planar EBG structures for the

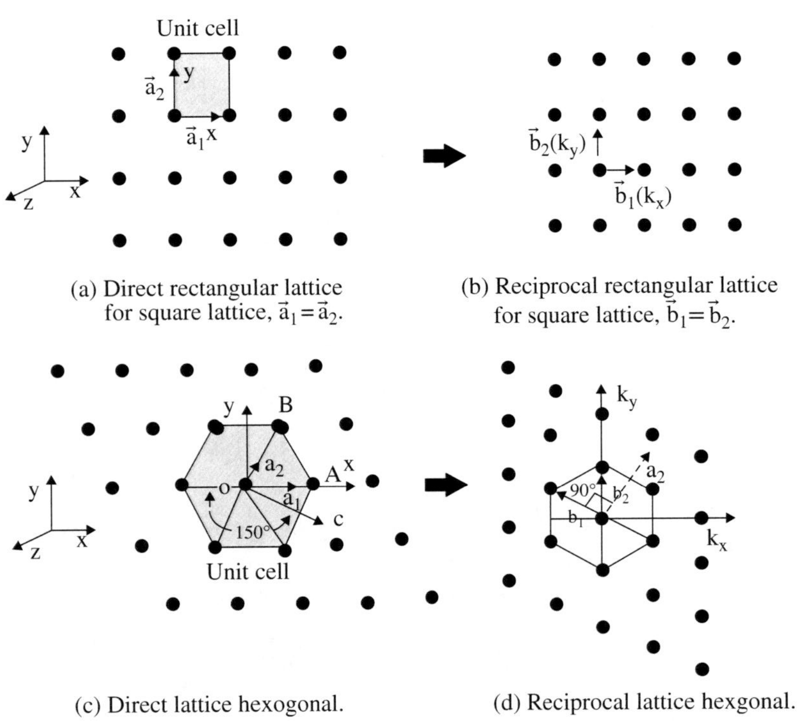

Figure 19.3 Some 2D direct and reciprocal lattices in the k-space.

microwave applications [B.5]. Figure (19.3b and d) also show the corresponding 2D reciprocal lattices. The 2D crystal structures are located in the x–y plane, and the z-axis, with unit vector, $\hat{z}$, and $|k_z| = 1$ is normal to it.

2D Rectangular Lattice

Figure (19.3a) shows the 2D direct rectangular lattice in the (x–y) plane with the primitive vectors $\left(\vec{a}_1, \vec{a}_2\right)$, directed along the x-axis and y-axis. The primitive vectors scan the space to form the 2D periodic structure of an infinite extent. The primitive vectors $\left(\vec{a}_1, \vec{a}_2\right)$ form the rectangular unit cell. The square unit cell is formed for the case $\vec{a}_1 = \vec{a}_2$. The corresponding primitive vectors $\left(\vec{b}_1, \vec{b}_2\right)$ in the reciprocal space are obtained from the equation (19.1.8) as follows:

$$\vec{b}_1 = 2\pi \frac{\vec{a}_2 \times \hat{z}}{\vec{a}_1.\left(\vec{a}_2 \times \hat{z}\right)} = 2\pi \frac{\left|\vec{a}_2\right| \sin 90°\hat{x}}{\left|\vec{a}_1\right|\hat{x}.\left(\left|\vec{a}_2\right| \sin 90°\hat{x}\right)}$$

$$\Rightarrow \vec{b}_1 = \frac{2\pi}{a_1}\hat{x} \quad \text{(a)}, \quad \text{likewise,} \quad \vec{b}_2 = \frac{2\pi}{a_2}\hat{y} \quad \text{(b)}. \tag{19.1.10}$$

The $\vec{a}_3 = \hat{z}$, $|k_z| = 1$ (*unit vector*), a_1, and a_2 are the lattice constants. The reciprocal primitive vector $\vec{b}_1$ is normal to the direct primitive vector $\vec{a}_2$ and parallel to $\vec{a}_1$. Likewise, the reciprocal primitive vector $\vec{b}_2$ is normal to $\vec{a}_1$ and parallel to $\vec{a}_2$. Figure (19.3b) shows the rectangular unit cell in the reciprocal space. *The primitive vectors $\left(\vec{b}_1, \vec{b}_2\right)$ are identified with the components (k_x, k_y) of the wavenumber vector $\vec{k}$ in the k-space for the 2D wave motion in the x–y plane.*

2D Hexagonal Lattice

Figure (19.3c) shows the direct hexagonal unit cell of the 2D oblique crystal structure. It is formed by the equilateral triangular lattice OAB with $a_1 = a_2 = a$. The reciprocal hexagonal unit cell, shown in Fig. (19.3d), is formed by the $\vec{b}_1$ and $\vec{b}_2$ reciprocal primitive vectors. The reciprocal hexagonal unit cell is rotated by 30° with respect to the direct hexagonal unit cell. The primitive direct and reciprocal translation vectors for the 2D case are given by the following expressions:

$$\begin{aligned} \vec{T} &= u_1 \vec{a}_1 + u_2 \vec{a}_2 && \text{(a)} \\ \vec{G} &= v_1 \vec{b}_1 + v_2 \vec{b}_2 && \text{(b)} \\ \vec{a}_3 &= \hat{z} \text{ (unit vector)} && \text{(c).} \end{aligned} \tag{19.1.11}$$

The primitive vectors $\vec{b}_1$ and $\vec{b}_2$ are computed in the reciprocal space, using equation (19.1.8), as follows:

$$\vec{b}_1 = 2\pi \frac{\vec{a}_2 \times \hat{z}}{\vec{a}_1.\left(\vec{a}_2 \times \hat{z}\right)} = 2\pi \frac{a \sin 90°\hat{c}}{(a\hat{x}).(a \sin 90°\hat{c})},$$

where the unit vector $\hat{c}$ is normal to the vectors $\vec{a}_2$ and $\hat{z}$, making an angle 150° with respect to the x-axis. Thus,

$$\Rightarrow \vec{b}_1 = \frac{2\pi}{a \cos\left(\vec{a}_1, \hat{c}\right)}\hat{c} = \frac{2\pi}{a \cos 150°}\hat{c}, \Rightarrow = -\vec{b}_1 \frac{4\pi}{\sqrt{3}a}\hat{c} \quad \text{(a)}$$

$$\text{Likewise, } \vec{b}_2 = 2\pi \frac{\hat{z} \times \vec{a}_1}{\vec{a}_2.\left(\hat{z} \times \vec{a}_1\right)} \quad \left(\vec{a}_1 = a\,\hat{x}\right) \tag{19.1.12}$$

$$\Rightarrow \vec{b}_2 = 2\pi \frac{a\,\hat{y}}{\vec{a}_2.(a\hat{y})} = 2\pi \frac{a\,\hat{y}}{a \cos\left(\vec{a}_2, \hat{y}\right)} = \frac{2\pi\,\hat{y}}{a \cos 30°} = \frac{4\pi}{\sqrt{3}\,a}\hat{y} \quad \text{(b).}$$

So the vector $\vec{b}_2$ is directed along the y-axis. The reciprocal hexagonal unit cell has sides of length $4\pi/(\sqrt{3}\,a)$. It is rotated by 30° from the vector $\vec{a}_2$. It is shown in Fig. (19.3d). The same results are obtained by using equation (19.1.7). The primitive vectors $\left(\vec{b}_1, \vec{b}_2\right)$ are again identified with the components (k_x, k_y) of the wavenumber vector $\vec{k}$ in the k-space.

19.1.3 Concept of Brillouin Zone

BZ is constructed in the reciprocal space, i.e. the k-space. Bragg's diffraction occurs at the boundary of BZ. Therefore, the boundaries of the BZ are a pair of parallel Bragg's planes (BP), acting as the reflecting surfaces. *The reciprocal translation vector $\vec{G}$, with primitive vectors $\left(\vec{b}_1, \vec{b}_2, \vec{b}_3\right)$, is identified as the wave vector $\vec{k}$ with the*

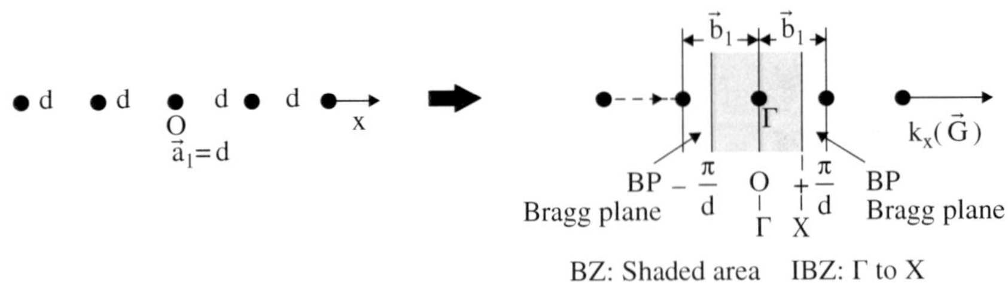

(a) 1D direct lattice with lattice constant d.

(b) 1D reciprocal lattice showing the first Brillouin zone.

Figure 19.4 Formation of 1D first BZ.

components (k_x, k_y, k_z) in the 3D space. We discuss below the 1D and 2D BZ, as applied to the 1D and 2D periodic structures, supporting the EM-wave propagation [B.1, B.4].

1D Brillouin Zone

Figure (19.4a) shows the 1D lattice with period d along the x-axis. The wave propagates in the x-direction with propagation constant k_x. The propagation is described in the 1D reciprocal space, i.e. in the k-space with a primitive vector $\vec{b}_1 = \vec{k}_1 = k_x\hat{x}$. Using equation (19.1.7), the wave vector is $k_x = 2\pi/d$. It is the period of the 1D reciprocal lattice, shown in Fig. (19.4b). Bragg's condition, equation (19.1.2), is satisfied at the Bragg's reflecting planes (BP). Bragg's planes around any lattice point in the k-space are normal to the axis of the lattice and are located at the bisectors of the reciprocal primitive vectors $\pm\vec{b}_1 = \pm\vec{k}_1 = \pm k_x\hat{x} = (2\pi/d)\hat{x}$, so Bragg's planes are located at $\vec{b}_1/2 = \pm(\pi/d)\hat{x}$. The region between two Bragg's planes, $-\pi/d \le k_x \le \pi/d$, is the first *BZ,* and the region $\Gamma \le k_x \le X$, with $\Gamma = 0$ and $X = \pi/d$ is called the *irreducible Brillouin zone (IBZ)* with a minimum distinct dimension that follows the Bragg's law.

2D Brillouin Zone

BZ for the 2D periodic structure, in the reciprocal space, i.e. the k-space, is constructed through the geometrical process in four steps. Figure (19.5a) shows the direct rectangular lattice ABCD and Fig. (19.5b) demonstrates the formation of the first BZ in the reciprocal space.

Step #I: A lattice point in the reciprocal space is taken as the origin of the BZ.

Step #II: The reciprocal primitive lattice vectors $\left(\vec{b}_1, \vec{b}_2\right)$, belonging to the nearest neighboring points A′, B′, C′, D′ to the origin, are connected to the origin Γ(0, 0).

Step #III: The perpendicular bisectors for the vectors $\left(\pm\vec{b}_1, \pm\vec{b}_2\right)$ are drawn. The smallest closed polygon, in the present case, a rectangle, formed is the first BZ. The first BZ, a lightly shaded area, corresponds to the *Wigner– Seitz cell* of the reciprocal lattice [B.1].

Step #IV: Further, by taking into account the rotational symmetry, BZ is reduced to the smallest primitive unit cell called the *IBZ*. For the rectangular lattice, shown in Fig. (19.5b), the IBZ is a darker shaded triangular area.

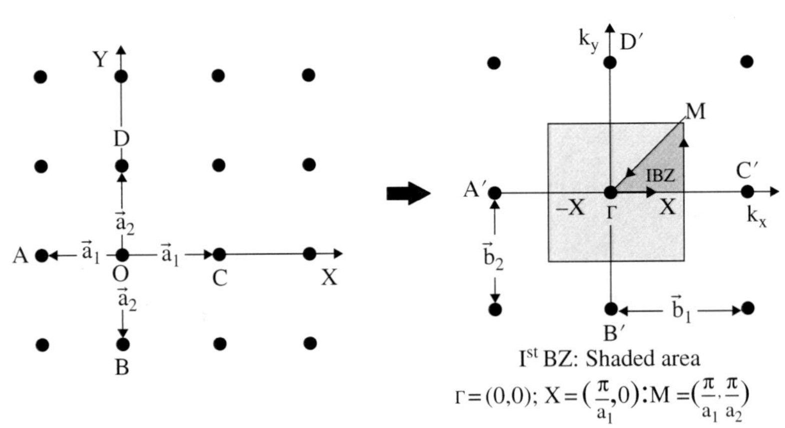

(a) Direct lattice.

(b) Reciprocal lattice.

Figure 19.5 Formation of the BZ and IBZ for a rectangular unit cell.

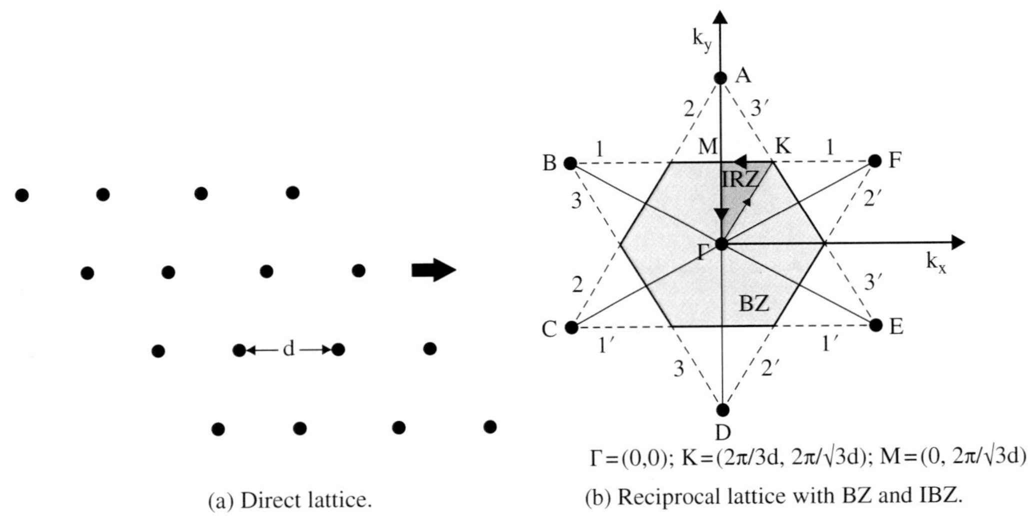

(a) Direct lattice.

(b) Reciprocal lattice with BZ and IBZ.

Figure 19.6 Formation of BZ and IBZ for a hexagonal unit cell.

Figure (19.6a and b) further explains the formation of BZ and triangular IBZ of the hexagonal direct lattice. The coordinates of the IBZ, obtained from equations (19.1.10) and (19.1.12), are shown in Figs. (19.5b and 19.6b). Figure (19.6b) shows the origin Γ(0, 0) of the lattice in the reciprocal space. The locations A, ..., F are the neighboring points of the lattice to the origin Γ(0, 0) with AΓ, ..., FΓ connecting lines and 1–1, 1′–1′, etc. their perpendicular bisectors for creating the BZ shown as a light gray hexagon. The triangle ΓKM of darker gray shade is the IBZ.

The fundamental wave propagation property, i.e. the $(\omega - \beta)$ *or* $(k - \beta)$ *dispersion diagram, is obtained from the IBZ unit cell that is periodically extended to the whole of the k-space.* In a case of the EBG structure, the *eigenmode solver* of EM-simulator is used for this purpose.

19.2 Space Harmonics of Periodic Structures

Floquet–Bloch theorem provides the *periodic boundary condition* to solve Maxwell's wave equation and also the transmission line wave equation, in an infinite extent periodic medium. The solution of the wave equation, in the form of the eigenfunction, provides a discrete solution of a source free periodic medium. The Floquet–Bloch theorem is applied to the 1D, 2D, and 3D unit cell of the direct lattice in the real space. The plane EM-wave propagating through the unit cell acquires its periodicity on the application of the periodic boundary condition, using the Floquet–Bloch theorem. The homogeneous uniform host medium may not have any dispersion. However, a periodic medium, periodically loaded with inclusions, is a dispersive medium and its propagation behavior is described in the reciprocal, i.e. the k-space. It *gives the Brillouin dispersion diagram for the discrete mode of propagation, known as Bloch mode or the Bloch wave* [B.3–B.12]. In this section, the Floquet–Bloch theorem is formulated to get the general nature of the Bloch wave in terms of *space harmonics*. In the next section, it is applied to a cascaded chain of unit cells of the 1D artificial periodic transmission lines. In chapter 20, it is applied to the 2D artificial periodic surface. The EM-simulators also solve the eigenmode problem of the periodic medium under the periodic boundary condition, using the Floquet–Bloch theorem.

19.2.1 Floquet–Bloch Theorem and Space Harmonics

Figure (19.7) shows the periodically arranged *basis (inclusions)* in a 1D *host medium*. It is a cascading of the unit cells of lattice constant (period) d. The unit cell, between two Bragg's planes (BP), contains one inclusion. The Bloch wave propagates along the z-axis with the complex propagation constant $\gamma = \alpha + j\,\beta_0$. The fields inside each unit cell are identical and at the boundary of the unit cell, i.e. at z = 0, d, 2d, ..., they have the same amplitude. They differ only in the phase at the boundary, due to the propagation factor $e^{-\gamma d}$. The periodic physical situation has the following two consequences:

- The field F(x, y, z) is a periodic function of d, i.e.

$$F(x, y, z + nd) = F_p(x, y, z), \tag{19.2.1}$$

where n is an integer. The subscript p is used to emphasize that the field F(x, y, z) is a periodic one.

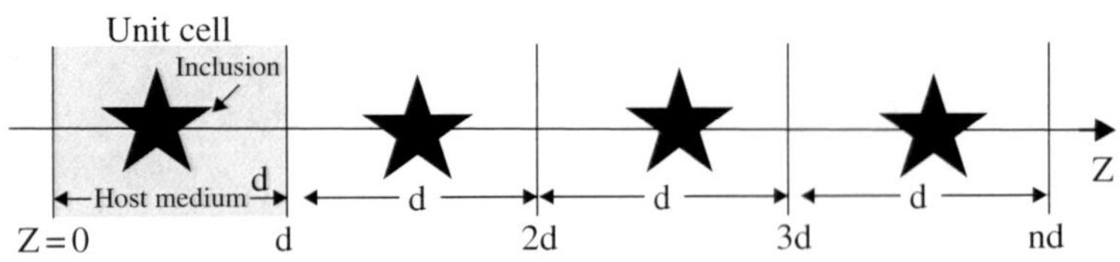

Figure 19.7 Cascaded unit cells of infinitely long 1D periodic line.

- The propagation factor $e^{-\gamma d}$, indicating the phase shift, is associated with the field function at each right-hand side plane of the unit cell, shown in Figure (19.7), with respect to the previous plane.

The above mentioned two properties of the propagating Bloch wave, in the 1D periodic medium, are combined to form the *statement of Floquet–Bloch theorem*:

$$F(x, y, z + d) = F_p(x, y, z)\, e^{-\gamma d}, \quad \text{where,} \quad \gamma = \alpha + j\,\beta_0. \tag{19.2.2}$$

As the field $F_p(x, y, z) = F(x, y, z + d)$ is a periodic function of d, so it can be expanded in form of Fourier series:

$$F_p(x, y, z) = \sum_{n=-\infty}^{+\infty} F_n(x, y)\, e^{-j\frac{2\pi n z}{d}}, \tag{19.2.3}$$

where Fourier coefficient $F_n(x,y)$ is the field amplitude in the x–y plane. It is a solution of the wave equation in the unit cell [B.6–B.8]. The following 1D periodic function is obtained using the above equations:

$$\begin{aligned}
&F(x, y, z + d) = F(x, y, z) = F_p(x, y, z)\, e^{-(\alpha + j\,\beta_0)z} \\
&\Rightarrow F(x, y, z) = e^{-\alpha z} \sum_{n=-\infty}^{+\infty} F_n(x, y)\, e^{-j\left(\beta_0 + \frac{2\pi n}{d}\right) z} \\
&\Rightarrow F(x, y, z) = e^{-\alpha z} \sum_{n=-\infty}^{+\infty} F_n(x, y)\, e^{-j\,\beta_n z} \qquad (a) \\
&\text{where,} \quad \beta_n = \beta_0 + \frac{2\pi n}{d} \qquad (b) \\
&\text{and,} \quad F_n(x, y) = \frac{1}{d}\int_0^d F(x, y, z)\, e^{j\left(\frac{2\pi n}{d}\right) z} dz \qquad (c).
\end{aligned} \tag{19.2.4}$$

If the periodic wave function F(x, y, z) is multiplied by $e^{j\omega t}$, then the propagating Bloch wave, in the 1D periodic medium, is obtained. Bloch wave is a superposition of the ±n *space harmonics,* also called the *spatial harmonics*. Each space harmonics, with positive n, has its phase velocity in the positive z-direction, whereas, with negative n, each space harmonics has separate phase velocity in negative z-direction. The number of space harmonics with the positive and negative phase velocity is the same. However, space harmonics are not distinct individual modes, like the waveguide modes [B.12]. A uniform host medium can support the TEM, TE, TM, or HE mode. If the same medium is periodically loaded, then TEM, TE, TM, or HE modes have their space harmonics. The *space harmonics constitute one group or packet with single group velocity* to transport energy from the source to the load, say in the positive z-direction. On the other hand, *the phase velocity of each space harmonics is different*. They are in the ±n pair, i.e. both in the positive z and negative z directions. A periodic structure supports both the *forward wave* (both group and phase velocities in the same direction) and the *backward wave* (the group and phase velocities in the opposite direction). These waves are discussed in section (3.3) of chapter 3. The phase velocity v_{pn} and group velocity v_g of the nth harmonics are given below:

$$\begin{aligned}
v_{pn} &= \frac{\omega}{\beta_n} = \frac{\omega}{\beta_0 + \frac{2\pi n}{d}} && (a) \\
v_g &= \frac{1}{\partial\beta_n/\partial\omega} = \frac{1}{\partial(\beta_0 + 2\pi n/d)/\partial\omega} = \frac{\partial\omega}{\partial\beta_0} && (b).
\end{aligned} \tag{19.2.5}$$

For a *slow-wave* structure, the +n gives the phase velocity less than the phase velocity in the host medium, say in the free space, i.e. $v_{pn} < c\,(=\omega/\beta_0)$. However, the group velocity is independent of n. For the *fast-wave*, −n gives us $v_{pn} > c\,(\omega/\beta_0)$ causing Cherenkov-type radiation forming the leaky wave. Cherenkov-type radiation is discussed in subsection (9.7.3) of chapter 9. These are obtained from the Brillouin dispersion diagram [J.4–J.7].

19.3 Circuit Models of 1D Periodic Transmission Line

The transmission line is a homogeneous 1D EM-wave supporting *host* medium. It can be periodically loaded with the *inclusions* – the series reactance or shunt susceptance. It can also be loaded with more complex inclusions, such as the L, T, and π-network of lumped reactive elements. Section (3.4) of chapter 3 presents a discussion on the circuit models of the reactance loaded transmission lines, without any consideration of periodicity.

Several 1D periodic media are also modeled, without solving Maxwell's wave equation, as the reactance/susceptance loaded artificial transmission lines [B.6, B.9, B.11]. In some cases, even the 2D periodic medium is modeled as the 1D periodic medium [J.8, J.9]. This section is concerned with the propagation characteristics and the circuit models of the 1D periodic transmission lines, loaded with several kinds of reactive/susceptive inclusions.

19.3.1 Periodically Loaded Artificial Lines

Figure (19.8a and b) shows the shunt susceptance and series reactance loaded periodic transmission lines of an infinite extent. The figures also show the unit cells of these loaded lines with *lattice constant (periodicity)* d. A unit cell is composed of the lumped susceptance (B)/reactance (X) in between two host line sections of length d/2. The characteristic impedance and propagation constant of the unloaded lossless host medium are Z_0 and $k = \omega\sqrt{\mu\varepsilon}$, respectively. The normalized susceptance, $b = B/Y_0 = BZ_0$, and normalized reactance, $x = X/Z_0 = XY_0$ are used as the inclusions. *Bloch wave,* in the form of voltage/current wave, is supported by an *artificial periodic transmission line*, with the complex propagation constant $\gamma = \alpha + j\beta$ and the characteristic impedance, called the *Bloch impedance* Z_B. Due to the periodic loading, the propagation constant β of the Bloch wave has to accommodate the propagation constants of the space harmonics β_n also. As discussed previously, the propagation constant β of a periodic line is located in the reciprocal space, i.e. in the k-space.

The nth unit cell section is described by the [ABCD] matrix of three numbers of the cascaded sections: two sections of length d/2 and one section of the normalized susceptance/reactance. The [A, B, C, D] matrix is discussed in subsection (3.1.3) of chapter 3. The input and output port voltages and currents, shown in Fig. (19.8c), are related as follows:

$$\begin{bmatrix} V_n \\ I_n \end{bmatrix} = \begin{bmatrix} A & B \\ C & D \end{bmatrix} \begin{bmatrix} V_{n+1} \\ I_{n+1} \end{bmatrix}. \tag{19.3.1}$$

The unit cell of a periodically loaded line is modeled as the *uniform artificial transmission line* of length d,

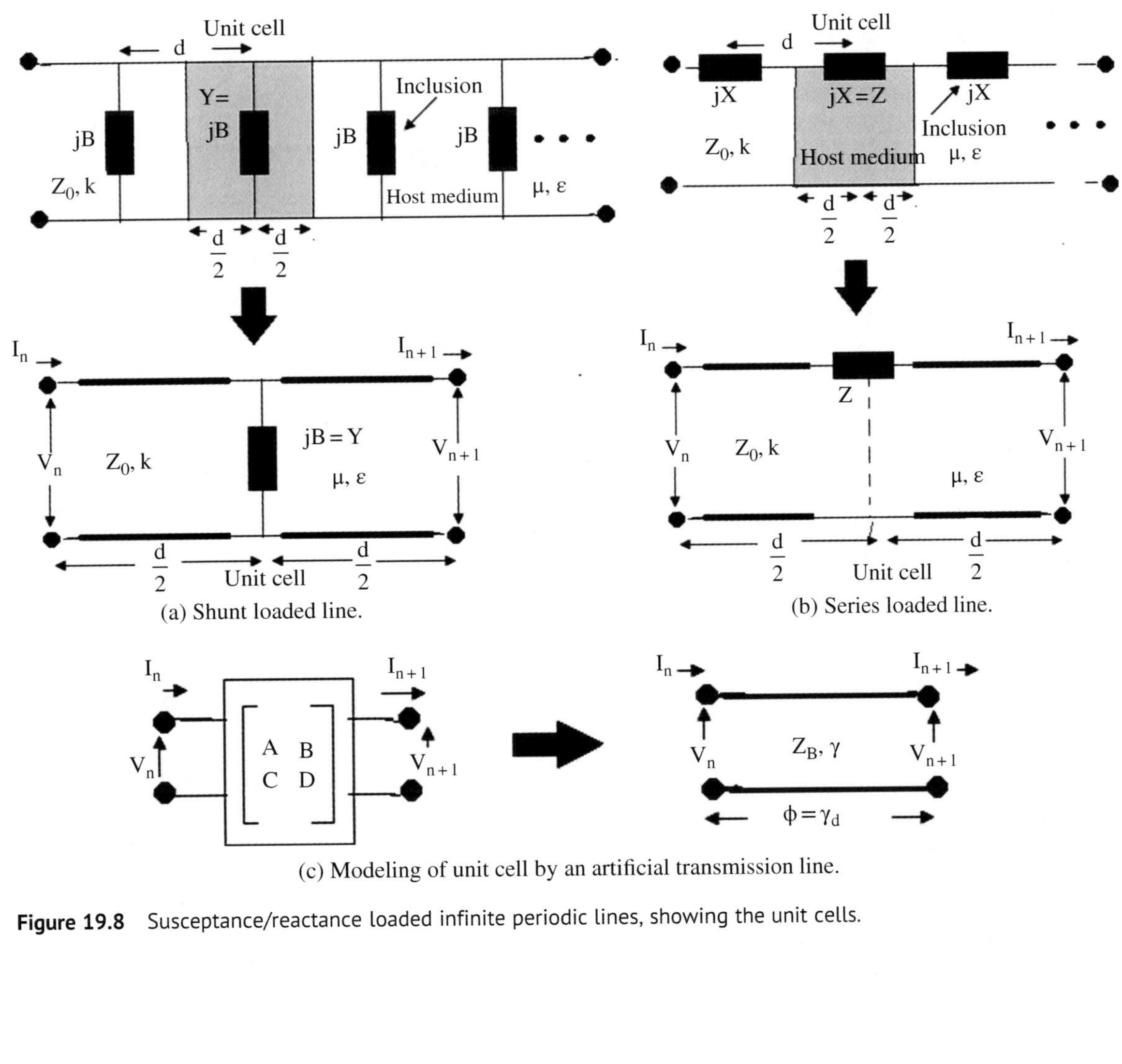

Figure 19.8 Susceptance/reactance loaded infinite periodic lines, showing the unit cells.

($\phi = \gamma$ d). It is shown in Fig. (19.8c). It supports the Bloch waves with propagation constant γ and *Bloch impedance* Z_B. Following Floquet–Bloch theorem, the voltage, and current of the Bloch waves, propagating in the +tive z-direction, at the nth and $(n+1)^{th}$ ports are related as follows:

$$V_{n+1} = V_n e^{-\gamma d} \quad (a), \quad I_{n+1} = I_n e^{-\gamma d} \quad (b)$$
$$\Rightarrow \begin{bmatrix} V_n \\ I_n \end{bmatrix} = \begin{bmatrix} e^{\gamma d} & 0 \\ 0 & e^{\gamma d} \end{bmatrix} \begin{bmatrix} V_{n+1} \\ I_{n+1} \end{bmatrix} \quad (c). \tag{19.3.2}$$

In case of a periodic line, the voltage/current V(z)/I(z) is a periodic function F(z). The Floquet–Bloch theorem is applicable in this case. The following expression, giving the eigenvalue, is obtained from equations (19.3.1) and (19.3.2c):

$$\begin{bmatrix} A - e^{\gamma d} & B \\ C & D - e^{\gamma d} \end{bmatrix} \begin{bmatrix} V_{n+1} \\ I_{n+1} \end{bmatrix} = 0. \tag{19.3.3}$$

The nontrivial solution of equation (19.3.3) is obtained for the determinant $\Delta = 0$:

$$AD - (A + D)e^{\gamma d} + e^{2\gamma d} - BC = 0.$$

The reciprocal network meets the condition, $AD - BC = 1$ [B.13]. Under this condition, the above equation is reduced to

$$e^{2\gamma d} - (A + D)e^{\gamma d} + 1 = 0, \quad \Rightarrow \frac{e^{\gamma d} + e^{-\gamma d}}{2} = \frac{A + D}{2}$$
$$\Rightarrow \cosh(\gamma d) = \frac{A + D}{2} \quad (a)$$
$$\text{also, for } A = D, \Rightarrow \cosh(\gamma d) = A \quad (b). \tag{19.3.4}$$

Equation (19.3.4a) applies to an *asymmetrical network*, and equation (19.3.4b) is for a *symmetrical network*. These expressions help to compute the complex propagation $\gamma = \alpha + j\beta$ of the Bloch-wave modes provided the elements of the [ABCD]-matrix of the unit cell is known. Once the operating frequency is increased from zero onward, a periodic structure exhibits the *passband* over a certain frequency range, followed by the *stopband* over another frequency range. The second passband is followed by the second stopband, and so on. For a circuit arrangement (series capacitor/shunt inductor), the sequence can start with the stopband followed by the passband.

For the *passband*, $\alpha = 0$, $\cosh(\gamma d) = \cosh(j\beta d) = \cos(\beta d)$. In this case, the Bloch wave travels through the periodic line. In the case of the *stopband*, $\beta = 0$, $\cosh(\gamma d) = \cosh(\alpha d)$. Also at $\beta d = \pi$, i.e. at the *Bragg's plane*, we get the stopband. In this case, the Bloch wave is reflected and the periodic line supports only the attenuated evanescent mode. However, the factor $(A + D)/2$ must meet the following conditions to get the passband/stopband:

$$\text{Passband:} \quad \cos(\beta d) = \begin{cases} \dfrac{A + D}{2}, & |A + D| \le 2 \\ \alpha = 0 \end{cases} \quad (a)$$
$$\text{Stopband:} \quad \cosh(\alpha d) = \begin{cases} \dfrac{A + D}{2}, & |A + D| > 2 \\ \beta = 0 \end{cases} \quad (b)$$
$$\text{Stopband:} \quad \cosh(\alpha d) = \begin{cases} \dfrac{A + D}{2}, & |A + D| < -2 \\ \beta d = \pi \end{cases} \quad (c). \tag{19.3.5}$$

Therefore, *the frequency-dependent propagation constant, i.e. the dispersion and loss, of a periodic line is characterized by the unit cell.* The dispersion behavior $\beta(\omega)$ is periodic.

Next, the task is to evaluate the characteristic impedance, i.e. *Bloch impedance,* of an equivalent uniform transmission line, i.e. the artificial transmission line. The normalized Bloch impedance ($Z_{B,n}$) at the nth port of the unit cell is given as follows:

$$Z_{B,n} = \frac{Z_B}{Z_0} = \frac{V_{n+1}}{I_{n+1}}, \quad \Rightarrow Z_B = Z_0 \frac{V_{n+1}}{I_{n+1}}. \tag{19.3.6}$$

Equation (19.3.3) provides the following expression:

$$\left(A - e^{\gamma d}\right) V_{n+1} + B\, I_{n+1} = 0. \tag{19.3.7}$$

Bloch impedance is obtained from equations (19.3.6) and (19.3.7), and for $I_{n+1} \ne 0$:

$$Z_B = \frac{-B Z_0}{A - e^{\gamma d}}. \tag{19.3.8}$$

Equation (19.3.4) is solved for $e^{\gamma d}$:

$$e^{\gamma d} = \frac{(A + D) \pm \sqrt{(A + D)^2 - 4}}{2}. \tag{19.3.9}$$

Bloch impedance for both the asymmetrical ($A \ne D$) and symmetrical ($A = D$) unit cells are obtained from equations (19.3.8) and (19.3.9):

$$\text{case}(A \ne D): \quad Z_B^{asym} = \frac{-2B Z_0}{(A - D) \mp \sqrt{(A + D)^2 - 4}} \quad (a)$$
$$\text{case}(A = D): \quad Z_B^{sym} = \frac{\pm B Z_0}{\sqrt{A^2 - 1}} \quad (b). \tag{19.3.10}$$

For a lossless periodic structure, the parameter B is an imaginary quantity and in the passband $|A| < 1$, leading to a real value of the Bloch impedance Z_B. The positive Z_B is for Bloch wave propagating in the positive z-direction, whereas the negative Z_B shows that it is propagating in the negative z-direction. In the stopband $|A| > 1$, the Bloch impedance is an imaginary quantity. It shows no propagation of the wave in the stopband, due to the reflection. If a finite extent periodic line is terminated in a load Z_L, and the unit cell is symmetric, the reflection coefficient at the input port is computed as

$$\Gamma = \frac{Z_L - Z_B}{Z_L + Z_B}. \quad (19.3.11)$$

To avoid reflection on a periodic line, $Z_L = Z_B$. Further, normally the [ABCD] – parameter of a unit cell is frequency-dependent, giving highly dispersive Bloch impedance. The increase or decrease of Z_B with frequency depends on the inductive or capacitive loading of a line. However, at the band edge, $Z_B = 0$, or ∞, leading to high reflection and creation of the stopband.

In the above discussion, the periodic line is viewed as a normal homogeneous transmission line with the complex propagation constant $\gamma = \alpha + j\beta$ and Bloch impedance Z_B, as its characteristic impedance. However, both lines are not similar in their behavior. For instance, the normal lossless line has $\alpha = 0$ at all operating frequency, without showing any stopband. A periodic line has alternating passband and stopband, even if the host line is lossless. Similarly, the characteristic impedance of a normal line is either frequency-independent or mildly frequency-dependent. However, Bloch impedance, i.e. the characteristic impedance of a periodic line, is highly dispersive giving real value in the passband and an imaginary value in the stopband. Despite these differences, both lines have some similarities. For the matching condition in the passband, $Z_L = Z_B$, and full power is transferred from the source to the load. If the periodic line length is $\ell = m\lambda_g/2$, i. e. $\beta\ell = m\pi$, $m = 1, 2, \ldots$, the input impedance is Z_L, similar to the usual transmission line. The line transformation and design of components in the passband work as usual, giving the compact circuits, as a periodic line can support the slow-wave.

19.3.2 [ABCD] Parameters of Unit Cell

The [ABCD] – parameters of a unit cell is required, to compute the propagation constant and Bloch impedance using equations (19.3.5) and (19.3.10). Figure (19.8a) shows a unit cell comprised of two host line sections of length d/2 (i. e. $\theta/2 = kd/2$) and a shunt admittance Y as an inclusion. Figure (19.8b) shows the unit cell with a series impedance Z as an inclusion. Using equations (3.1.23)–(3.1.25) of chapter 3, the [ABCD] – matrices of both the cases, using the normalized components, are obtained as follows:

Shunt Y:

$$\begin{bmatrix} A & B \\ C & D \end{bmatrix}_{Y-\text{shunt}} = \begin{bmatrix} \cos(\theta/2) & j\sin(\theta/2) \\ j\sin(\theta/2) & \cos(\theta/2) \end{bmatrix} \times \begin{bmatrix} 1 & 0 \\ \bar{y} & 1 \end{bmatrix} \times \begin{bmatrix} \cos(\theta/2) & j\sin(\theta/2) \\ j\sin(\theta/2) & \cos(\theta/2) \end{bmatrix} \quad \text{(a)}$$

Series Z:

$$\begin{bmatrix} A & B \\ C & D \end{bmatrix}_{Z-\text{series}} = \begin{bmatrix} \cos(\theta/2) & j\sin(\theta/2) \\ j\sin(\theta/2) & \cos(\theta/2) \end{bmatrix} \times \begin{bmatrix} 1 & \bar{z} \\ 0 & 1 \end{bmatrix} \times \begin{bmatrix} \cos(\theta/2) & j\sin(\theta/2) \\ j\sin(\theta/2) & \cos(\theta/2) \end{bmatrix} \quad \text{(b).}$$

(19.3.12)

The normalized characteristic impedance and admittance are $Z_0 = 1$ and $Y_0 = 1$. Also, the normalized lumped impedance and admittance are $\bar{z} = Z/Z_0$ and $\bar{y} = Y/Y_0 = YZ_0$, respectively. The propagation constant in the host medium is $k = \omega\sqrt{\mu\varepsilon} = k_0\sqrt{\varepsilon_r}$. In the case of microstrip, or CPW, as the host medium $\varepsilon_r \rightarrow \varepsilon_{reff}$ is taken. The electrical length of the unit cell is $\theta = kd$.

Shunt Y:

$$\begin{bmatrix} A & B \\ C & D \end{bmatrix}_{Y-\text{shunt}} = \begin{bmatrix} \cos(\theta) + j\frac{\bar{y}}{2}\sin(\theta) & j\sin(\theta) + \frac{\bar{y}}{2}\cos\theta - \frac{\bar{y}}{2} \\ j\sin(\theta) + \frac{\bar{y}}{2}\cos\theta + \frac{\bar{y}}{2} & \cos(\theta) + j\frac{\bar{y}}{2}\sin(\theta) \end{bmatrix} \quad \text{(a)}$$

Series Z:

$$\begin{bmatrix} A & B \\ C & D \end{bmatrix}_{Z-\text{series}} = \begin{bmatrix} \cos(\theta) + j\frac{\bar{z}}{2}\sin(\theta) & j\sin(\theta) + \frac{\bar{z}}{2}\cos\theta + \frac{\bar{z}}{2} \\ j\sin(\theta) + \frac{\bar{z}}{2}\cos\theta - \frac{\bar{z}}{2} & \cos(\theta) + j\frac{\bar{z}}{2}\sin(\theta) \end{bmatrix} \quad \text{(b).}$$

(19.3.13)

In order to return to Z_0 of the host line, the B element of the above [ABCD] – matrix is multiplied by Z_0 and C element by $(1/Z_0)$. Equating matrix element A from equations (19.3.4) and (19.3.13), the following dispersion relations of the periodic transmissions are obtained:

$$\text{Shunt Y:} \quad \cosh(\gamma d) = \cos(\theta) + j\frac{\bar{y}}{2}\sin(\theta) \qquad \text{(a)}$$

$$\text{Series Z:} \quad \cosh(\gamma d) = \cos(\theta) + j\frac{\bar{z}}{2}\sin(\theta) \qquad \text{(b)}. \tag{19.3.14}$$

For the capacitive susceptance and inductive reactance as the loading elements, the normalized values are used as, $\bar{y} = jb = j\omega CZ_0$ and $\bar{z} = jx = j\omega L/Z_0$. The equations (19.3.14a, b) are reduced to the following dispersion expressions:

$$\text{Shunt Y:} \quad \cosh(\gamma d) = \cos(\theta) - \frac{b}{2}\sin(\theta) \qquad \text{(a)}$$

$$\text{Series Z:} \quad \cosh(\gamma d) = \cos(\theta) - \frac{x}{2}\sin(\theta) \qquad \text{(b)}. \tag{19.3.15}$$

The dispersion and loss of a *shunt capacitor loaded* line are computed as follows:

$$\cosh(\alpha d + j\beta d) = \cosh(\alpha d)\cos\beta d + j\sinh(\alpha d)\sin\beta d = \cos(\theta) - \frac{b}{2}\sin(\theta).$$

On separating the real and imaginary parts, the following expressions are obtained:

$$\cosh(\alpha d)\cos\beta d = \cos(\theta) - \frac{b}{2}\sin(\theta) \qquad \text{(a)}$$

$$\sinh(\alpha d)\sin\beta d = 0 \qquad \text{(b)}. \tag{19.3.16}$$

Expressions (19.3.16) provide the passband and stopband characteristic impedance of the periodic line.

Passband: For $\sin\beta d \neq 0$, $\sinh(\alpha d) = 0$, therefore, in the passband $\beta \neq 0$ and $\alpha = 0$. The following dispersion expression for the propagating wave in the passband is obtained for $\alpha = 0$:

$$\cos\beta d = \cos(\theta) - \frac{b}{2}\sin(\theta). \tag{19.3.17}$$

Expression (19.3.17) is the matrix element A of the equation (19.3.13). At any frequency $\theta = kd$ is known, as $k = \omega\sqrt{\mu\varepsilon} = (\omega/c)\sqrt{\varepsilon_r}$, where c is the velocity of EM-wave in the free space. The shunt loading element is $b = \omega CZ_0$. Therefore, the above expression gives an infinite number of solutions for βd, when $|\cos(\theta) - (b/2)|\sin\theta \leq 1$, i.e. for $-1 \leq [\cos(\theta) - (b/2)\sin(\theta)] \leq +1$. These β_n solutions correspond to the propagation constants of the space harmonics. The *fundamental propagation constant* is computed for the range $-\pi \leq \beta d \leq +\pi$, i.e. within the first BZ, shown in Figure (19.4b). Other solutions are *periodic extensions.*

Stopband: The stopband frequency range has $\sin\beta d = 0 \Rightarrow \beta_n d = n\pi$ (n = 0, 1, 2, ...) and $\sinh(\alpha d) \neq 0$. Therefore, the propagation constants at the lower and upper edge of the first passband are $\beta = 0$, and $\beta = \pi/d$. The range $0 \leq \beta \leq \pi/d$ defines the IBZ, shown in Figure (19.4b). The $\beta = \pi/d$ also determines the lower edge of the first stopband with $\alpha \neq 0$. The attenuation constant in the stopband is computed from the following expression:

$$\cosh(\alpha d) = |\cos(\theta) - (b/2)\sin(\theta)| \geq 1. \tag{19.3.18}$$

Expression (19.3.18) is the matrix element A of equation (19.3.13). The attenuation constant α has only one solution, as $\cosh(\alpha d)$ is not a periodic function. For the +z direction going wave, the attenuation constant is $\alpha > 0$, and for the negative z-direction going wave, the attenuation constant is $\alpha < 0$. It is noted that at the band edge, $\beta d = \pi, \Rightarrow d = \lambda_g/2$. It shows that the loading elements are $\lambda_g/2$ apart and the load appears to be connected directly to the source with $\beta = 0$.

The above analysis is valid for the *series impedance loaded line* also. The b is replaced by $x = \omega L/Z_0$. The dispersion relation (19.3.17) can be recast in the following compact form, applicable to both cases of the period loading:

$$\cos\beta d = \cos(kd) - p(kd)\sin(kd) = A \qquad \text{(a)}$$

$$\text{for shunt capacitor:} \quad p = \frac{v_{ph}CZ_0}{2d}, \quad v_{ph} = \frac{c}{\sqrt{\varepsilon_r}} \qquad \text{(b)}$$

$$\text{for series inductor:} \quad p = \frac{v_{ph}L}{2Z_0 d}, \quad \text{for microstrip/CPW}, \varepsilon_r \rightarrow \varepsilon_{reff} \qquad \text{(c)}$$

$$Z_0(\varepsilon_r) = \frac{Z_0(\varepsilon_r = 1)}{\sqrt{\varepsilon_r}}, \qquad \text{for microstrip/CPW}, \varepsilon_r \rightarrow \varepsilon_{reff} \qquad \text{(d)}. \tag{19.3.19}$$

In equations (19.3.19a–d), Z_0 is the characteristic impedance of the host transmission line. The matrix element A belongs to the transmission matrix [ABCD]. The parameter p could be treated as a *loading factor* of the reactively loaded line. The loading factor p depends on the type of loading element and its series/parallel connection. For $p = 0$, there is no periodic loading and (19.3.19a) for the loaded line is reduced to the unloaded line with $\beta = k$. *The β is propagation constant of a periodic line and k is the wavenumber (propagation constant) of the host medium.*

Bloch Impedance: Once the [ABCD] – parameters of a unit cell are known, Bloch impedance of both cases of the periodic line can be evaluated. For the *symmetrical and reciprocal* periodic line, ($A = D$, $AD - BC = 1$) and equation (19.3.10b) is reduced to

$$Z_B^{sym} = \pm Z_0\sqrt{\frac{B}{C}}. \tag{19.3.20}$$

Using the parameters B and C from equation (19.3.13) with the above equation, Bloch impedance for both cases of the periodic line are obtained as

$$\text{Shunt Y:}\ Z_B^{sym} = \pm Z_0\left[\frac{j\,\sin\theta + (\bar{y}/2)\cos\theta - \bar{y}/2}{j\,\sin\theta + (\bar{y}/2)\,(\cos\theta + \bar{y}/2)}\right]^{1/2} \quad \text{(a)}$$

$$\text{Series Z:}\ Z_B^{sym} = \pm Z_0\left[\frac{j\,\sin\theta + (\bar{z}/2)\cos\theta + \bar{z}/2}{j\,\sin\theta + (\bar{z}/2)\cos\theta - \bar{z}/2}\right]^{1/2} \quad \text{(b)}. \tag{19.3.21}$$

For the lumped shunt connected capacitor (C) and the series-connected lumped inductor (L), the above equations are reduced to

$$\text{Shunt Y:}\ Z_B^{sym} = \pm Z_0\left[\frac{j\,\sin\theta + p(kd)\cos\theta - p(kd)}{j\,\sin\theta + p(kd)\cos\theta + p(kd)}\right]^{1/2} \quad \text{(a)}$$

$$\text{Series Z:}\ Z_B^{sym} = \pm Z_0\left[\frac{j\,\sin\theta + p(kd)\cos\theta + p(kd)}{j\,\sin\theta + p(kd)\cos\theta - p(kd)}\right]^{1/2} \quad \text{(b)}. \tag{19.3.22}$$

The parameter p for both the cases is different and is given in equation (19.3.19).

19.3.3 Dispersion in Periodic Lines

The dispersion behavior of a periodic transmission line, such as a reactively loaded quasi-TEM mode microstrip or a reactively loaded non-TEM waveguide, is obtained through the ($\omega - \beta$) or ($k - \beta$) dispersion diagram. On a periodic line, apart from the fundamental mode, the space harmonics are also present. However, the dispersion diagram of the space harmonics, i.e. the ($\omega - \beta_n$) diagram, is periodically related to the fundamental mode in the k-space. The range $-\pi \le \beta d \le \pi$ is the domain of the *first* BZ. Figure (19.9a) considers a *hypothetical 1D lattice*, without any inclusion, of the lattice constant (period) d along the x-axis. The period of the propagation vector in the k-space, from the equation (19.1.9), is $b_1 = \beta = 2\pi/d$. Figure (19.9a) further shows a pair of reflecting Bragg's planes (BP) located at $\beta d = \pm\pi$, forming the first BZ. The domains of βd, i.e. $(0, \pi)$ and $(0, -\pi)$, are symmetrically located, so they contain identical information on the dispersion of a periodic medium. In this way, only the domain $(0, \pi)$ of βd is important for the dispersion analysis. This smallest domain is called the IBZ.

Figure (19.9a) shows the dispersion diagram of a host homogeneous medium, both air and dielectric medium with relative permittivity ε_r. The vertical axis of the dispersion diagram is kd, where k is the propagation constant in the nonperiodic homogenous host medium. The parameter kd is also treated as the *normalized frequency*. Thus, in the air medium $k_0 d = \pi$, i.e. $\omega = \pi c/d$ and in the homogeneous dielectric medium, $kd = \pi$, i.e. $\omega = \pi c/d\sqrt{\varepsilon_r}$, where c is the velocity of the EM-wave in free space. The horizontal axis is βd. The propagation constant β is the propagation constant of wave in the periodically loaded medium. The frequency can be treated as a function of β, i.e. *the dispersion diagram is a plot of the $\omega = \omega(\beta)$ function over the IBZ*, $0 \le \beta d \le \pi$. In the case of an air medium, the dispersion diagram is a $k = \pm\beta_0$, ($\omega = c\beta_0$) line, known as the *light lines.* It is shown as a pair of thick black lines in Fig. (19.9a–c). The light lines divide the ($k - \beta$) dispersion diagram into four regions. The first and fourth regions support the forward and backward traveling *slow-waves*, respectively. The second and third regions support the forward and backward traveling *fast-waves*, respectively.

Figure (19.9a) shows the linear dispersion behavior of the *hypothetical unit cell*, without any inclusion or loading. Its gradient, i.e. the gradient of the light line, at the origin gives the phase velocity $v_p(\text{air}) = c = (\omega/\beta)$ in the air medium (*thick black lines*) and $v_p(\text{die}) = c/\sqrt{\varepsilon_r}$ in the dielectric medium (*thin grey lines*). The condition $v_p(\text{die}) < c$ provides the slow-waves propagation in the first region. The slow-waves can also propagate in the negative z-direction, i.e. along the $-\beta d$ direction in the k-space. In the negative z-direction, the phase velocity is negative.

Figure (19.9b) is the dispersion diagrams of the *microstrip type host medium* loaded with the periodic inclusions. It supports the forward traveling slow-wave in

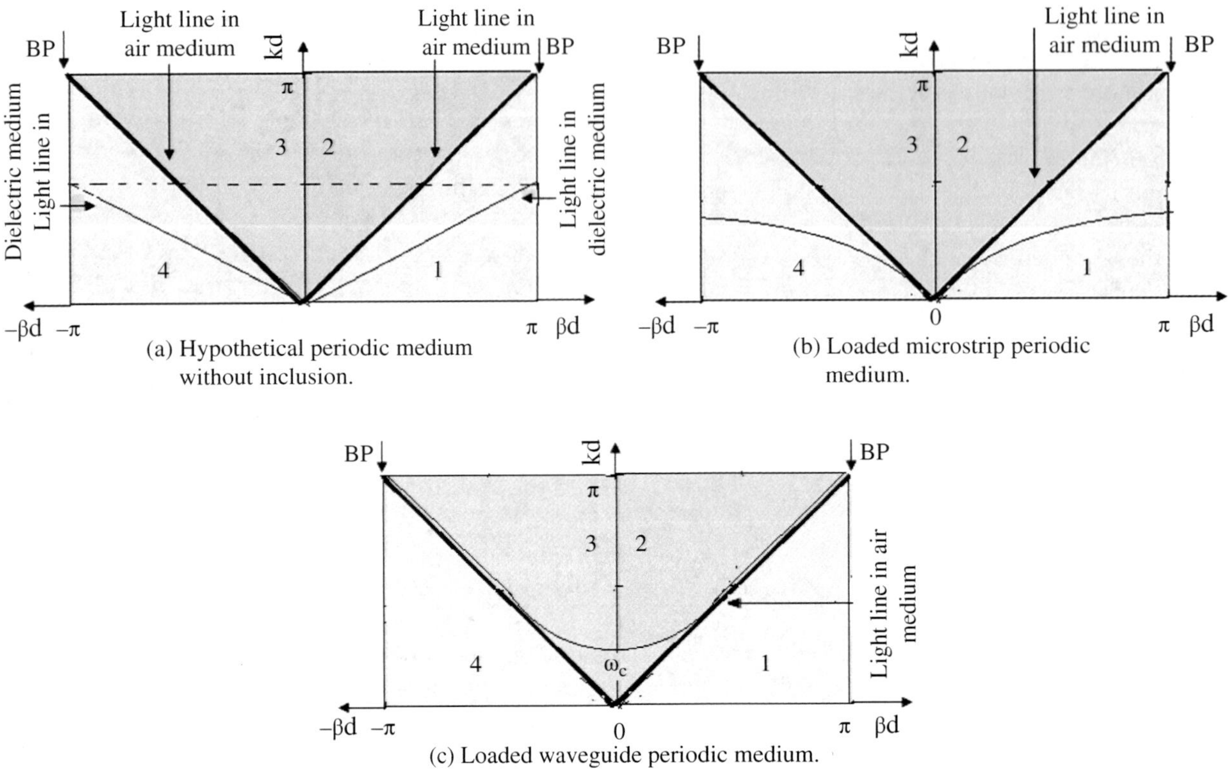

Figure 19.9 Dispersion diagram of the fundamental wave of the periodic media.

the region #1 and backward traveling slow-wave in the region #4. Figure (19.9c) is the dispersion diagrams of the *waveguide type host medium* loaded with the periodic inclusions. It has a cut-off frequency ω_c. The dispersion curve is in the region #2 and region #3, supporting the forward and backward traveling fast-*wave*, respectively, as $v_p > c$. In case, the dispersion diagram of the *open periodic transmission line* is in the region #2 and region #3, the fast wave may result in the leaky wave, i.e. Cherenkov-type radiation, radiating the power to the free space. In the case of a periodic microstrip, the leaky wave can also radiate in the substrate exciting the surface waves [J.4–J.6].

Figure (19.10) demonstrates the space harmonics dispersion diagram of a hypothetical 1D periodic structure, without any inclusion, in the air medium. The horizontal axis shows the period of the propagation constant of the periodic medium, i.e.$\beta d = \beta_n d = \beta_0 + 2\pi n/d$, due to the Floquet–Bloch theorem. Its first period is $0 \le \beta d \le 2\pi$. The Floquet–Bloch boundary condition is satisfied at $\beta d = 0, \pm 2\pi, \pm 4\pi, \ldots$. A pair of Bragg's planes at $\beta d = \pm \pi$, defining the first BZ is shown as the shaded area. It contains the dispersion lines of a pair of the forward and backward moving waves. Figure (19.10) also shows the IBZ.

Figure (19.10) shows the *forward-moving Bloch waves* (→) in the full lines: A...E, from the left to right. It assumed that the source is located on the left-hand side, so the energy is always transported from the source to load, i.e. the group velocity V_g^f is always positive, irrespective of the direction of the phase velocity V_p. It is also assumed that the source is located at the origin 0. The phase – velocity of all space harmonics for (n = 0, +1, +2, ...) are positive. The phase velocities of all space harmonics for (n = 0, −1, −2, ...) are negative.

The waves A and B are the *forward traveling forward waves* with positive v_g^f and positive V_p. However, the waves D and E are the *forward traveling backward waves* with positive v_g^f and negative V_p. The periodic line also supports the *backward traveling Bloch wave* (←), dashed lines: A′...D′, from the right to left. These are the reflected waves from the inclusions located at the 1D direct lattice points. They carry energy with the group velocity v_g^b and treated as a negative quantity; so group velocities v_g^f and v_g^b are opposite to each other. In the case of a matched periodic line, there is no backward traveling wave on the line. However, in Fig. (19.10), the waves B′ and C′ are reflected *backward traveling backward waves* with negative v_g^b and positive V_p. Finally, the waves A′ and D′ are the reflected *backward traveling forward waves* with both v_g^b and V_p negative. An infinitely long periodic line supports an equal number of the forward and backward waves. Even over one period in the k-space, i.e. $\beta d = 2\pi$, one forward traveling forward wave is obtained, and another wave is *the backward traveling backward wave*. Both these waves interact at locations $P_1, P_2, P_3, P_4, \ldots$ for $kd = \pi$. They also interact at locations $Q_1, Q_2, Q_3, \ldots$ for $kd = 2\pi$, at locations $L_1, L_2, \ldots$ for $kd = 3\pi$, at locations $M_1, \ldots$ for $kd = 4\pi$. At these locations, both waves get coupled and produce the standing waves on a periodic line. In the case of a real periodic line with inclusions, the waves split into two traveling Bloch waves separated by the *frequency bandgap,* i.e. by a stopband.

In summary, a periodic line supports four kinds of waves: the *forward traveling forward wave, backward traveling forward wave, forward traveling backward wave, and backward traveling backward wave.* These waves are discussed in chapter 3, section (3.4). Figure (19.11) summarizes the condition of their existence on a line. In the case of the forward wave, the factor $(v_g \times v_p)$ is positive, whereas for the backward wave the factor $(v_g \times v_p)$ is negative. *The positive direction of the movement of waves is always from the source to load.* Thus, for a real finite-length line, with a source located at the left-hand side, the energy always travels from the source to load, i.e. from left to right giving positive group velocity v_g for both the forward wave and backward wave. However, their phase velocities are opposite to each other, positive for the forward wave and negative for the backward wave. Figure (19.9a–c) show that if these waves are outside the light line cone, they are bounded slow-waves. However, if these are inside the *light lines cone*, they are the fast-waves and may turn into the leaky modes.

The backward wave supporting medium is also treated as the metamaterials with both ε_r and μ_r as negative quantities. It has an interesting consequence. Suppose a $\lambda_g/4$ line, supporting the forward wave, is cascaded with the same line length supporting the backward wave, i.e. the metamaterials based line. The total line length forms a $\lambda_g/2$ resonator with the zero phase shift at the output, unlike the π phase shift for a normal line supporting only the forward wave. The first section gives the lagging, i.e. $-\pi/2$, phase shift, while the next section gives the leading, i.e. $+\pi/2$ phase shift, providing overall zero phase-shift at the end of the resonator.

Figure (19.12a) shows, over one period $0 \le \beta d \le 2\pi$ $(-\pi \le \beta d \le \pi)$, the wave splitting of the *shunt capacitors/series inductors* loaded TEM mode line. The dispersion curve starts at $kd = 0$, i.e. at the normalized first lower cut-off frequency $\omega_{c1}^L = 0$ of the first passband. In the

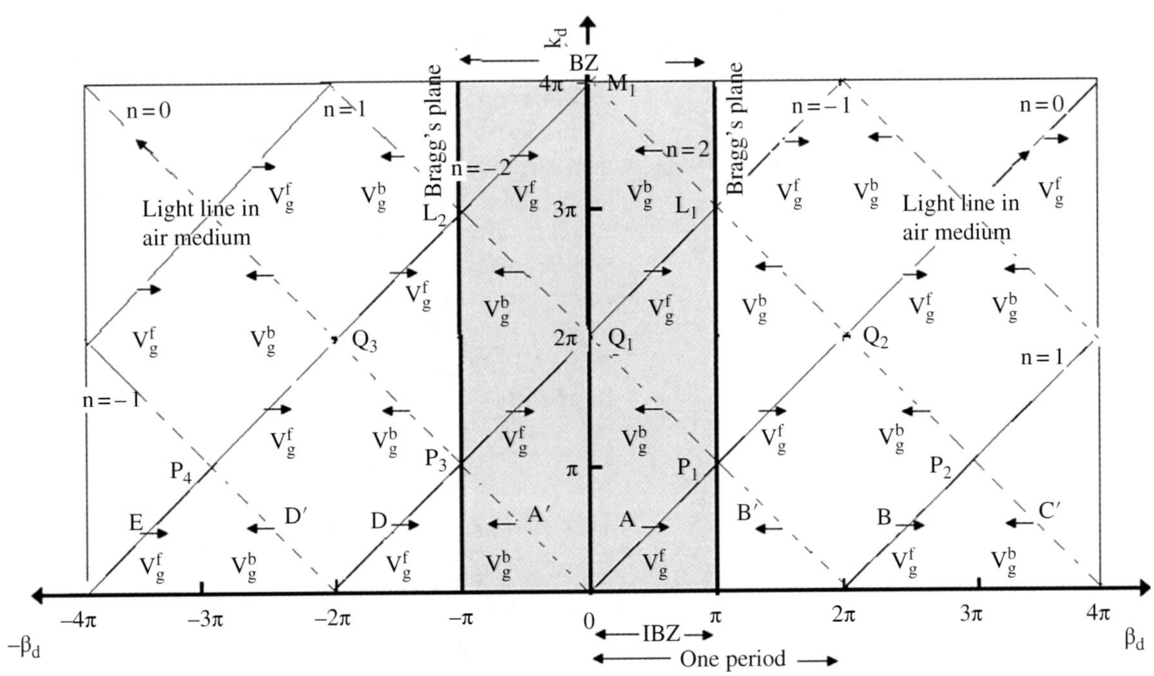

(-) Forward moving wave v_g^f; (--) Backward moving wave v_g^b.

Figure 19.10 Dispersion diagram of the 1D periodic structure.

IBZ region, the dispersion curve increases with frequency and bends to reach the maximum value of $\beta d = \pi$, corresponding to the first upper cut-off frequency ω_{c1}^{U}. The first passband, ω_{c1}^{L} to ω_{c1}^{U}, supports the forward slow-wave propagation in the forward direction. It is *Bloch mode #1*. Figure (19.12a) also shows that the second *passband*, supporting Bloch *mode #2,* starts at the $\beta d = \pi$, corresponding to $kd = \pi$. Bloch mode #2 is the backward wave that is partly fast-wave within the light-line cone and partly slow-wave outside the light-line cone. The normalized lower cut-off frequency at $\beta d = \pi$ is $\omega_{c2}^{L} = \omega(d/c) = \pi$. The upper cut-off frequency of the second passband occurs at ω_{c2}^{U}, corresponding to $\beta d = 0$, or 2π. There is a stopband for Bloch wave over the frequency band from ω_{c1}^{U} to ω_{c2}^{L}. The stopband supports the evanescent mode with the attenuation factor α, and $\beta = 0$ or π/d. It is noted that the propagation constant is symmetrical over the IBZ range $0 \le \beta d \le \pi$ and $-\pi \le \beta d \le 0$. Over $-\pi \le \beta d \le 0$, the replica supports the backward moving slow-waves in the first passband. For a matched terminated line reflected backward moving wave is absent. The dispersion is computed only over the IBZ.

Figure (19.12b) shows the dispersion behavior of the *series capacitors/shunt inductors* loaded TEM type host transmission line. In this case, *Bloch mode #1* inside the passband #1 is a *backward wave*, and mostly the *fast-wave* with the lower cut-off frequency at ω_{c1}^{L} and the upper cut-off frequency at ω_{c1}^{U}, corresponding to $kd = \pi$, i.e. for $\beta d = 2\pi$, i. e. 0 at Bragg's plane. However, the *Bloch mode #2* inside the passband #2 is the *forward fast-wave* in the band ω_{c2}^{L} to ω_{c2}^{U}. Frequency ω_{c2}^{U} corresponds to $kd = 2\pi$, for $\beta d = \pi$ at Bragg's plane. Again, two Bloch modes are separated by a bandgap. The series capacitor/shunt inductor loaded line, supporting the backward wave propagation, is metamaterials-based.

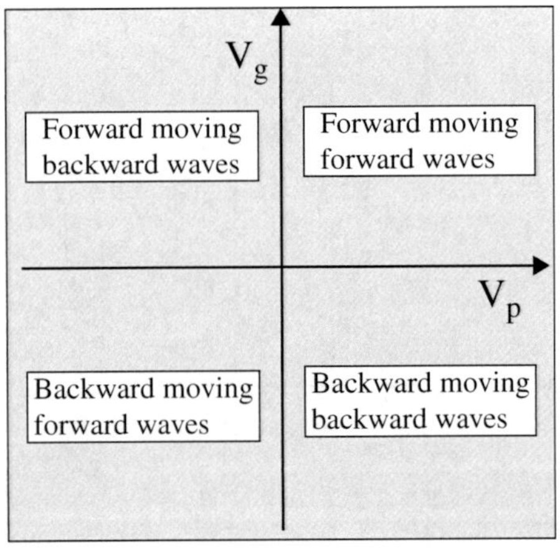

Figure 19.11 Conditions for the existence of four kinds of waves on a periodic structure.

Finally, Fig. (19.9c) shows the dispersion behavior of the waveguide medium with capacitive loading. The host waveguide has a cut-off frequency ω_c at ω_{c1}^{L} for the fundamental TE_{10} mode. The dispersion behavior is obtained for the waveguide periodically loaded with the capacitive irises. In the IBZ, Bloch mode #1 inside the passband #1 is the forward fast-wave between ω_{c1}^{L} and ω_{c1}^{U} corresponding to $\beta d = \pi$. Bloch mode #2 inside the passband #2, between ω_{c2}^{L} to ω_{c2}^{U}, is the backward fast-mode. Other higher-order Bloch modes can also be obtained. The Bloch modes are always separated by a bandgap.

In summary, the propagation constant of the Bloch wave is scanned over the IBZ, $0 \le \beta d \le \pi$, the corresponding frequencies are obtained for the propagating Bloch modes #1, 2, 3, ..., separated by the nonpropagating decaying evanescent modes in the stopband. It also provides the cut-off frequencies, defining the passband and stopband. Using the symmetry about the kd-axis, the dispersion is obtained in the range,$-\pi \le \beta d \le 0$. It is a replica of the dispersion in the IBZ. Combined together, the dispersion in the first BZ, $-\pi \le \beta d \le +\pi$, for the fundamental Bloch wave is obtained. Using this process, EM-simulator in the eigenmode format provides the dispersion behavior of an infinitely long periodic structure. However, the cut-off frequency, propagation constant, attenuation constant, and also Bloch impedance of a periodically loaded planar transmission line are easily computed with the help of the *transmission line model*, discussed in the next section.

The above discussion is centered around the fundamental mode of the space harmonics of the Bloch mode in the IBZ. In the passbands, all the space harmonics of Bloch modes travel at the same group velocity; however, at different phase velocity. At any frequency, the propagation constants β_n of all space harmonics are obtained, if the β of the fundamental mode in the IBZ is known. Figure (19.12c) shows Bloch modes #1, 2, with the space harmonics of the periodically loaded rectangular waveguide with capacitive irises.

19.3.4 Characteristics of 1D Periodic Lines

This subsection discuses both the dispersion and Bloch impedance of the periodic lines. The dispersion characteristics of the fundamental Bloch mode in the passband are computed using equation (19.3.19) for $0 \le \beta d \le \pi$ over the IBZ. Beyond IBZ, for other periods, the dispersion is obtained as the periodic extension. The attenuation factor αd in the stopband is computed using

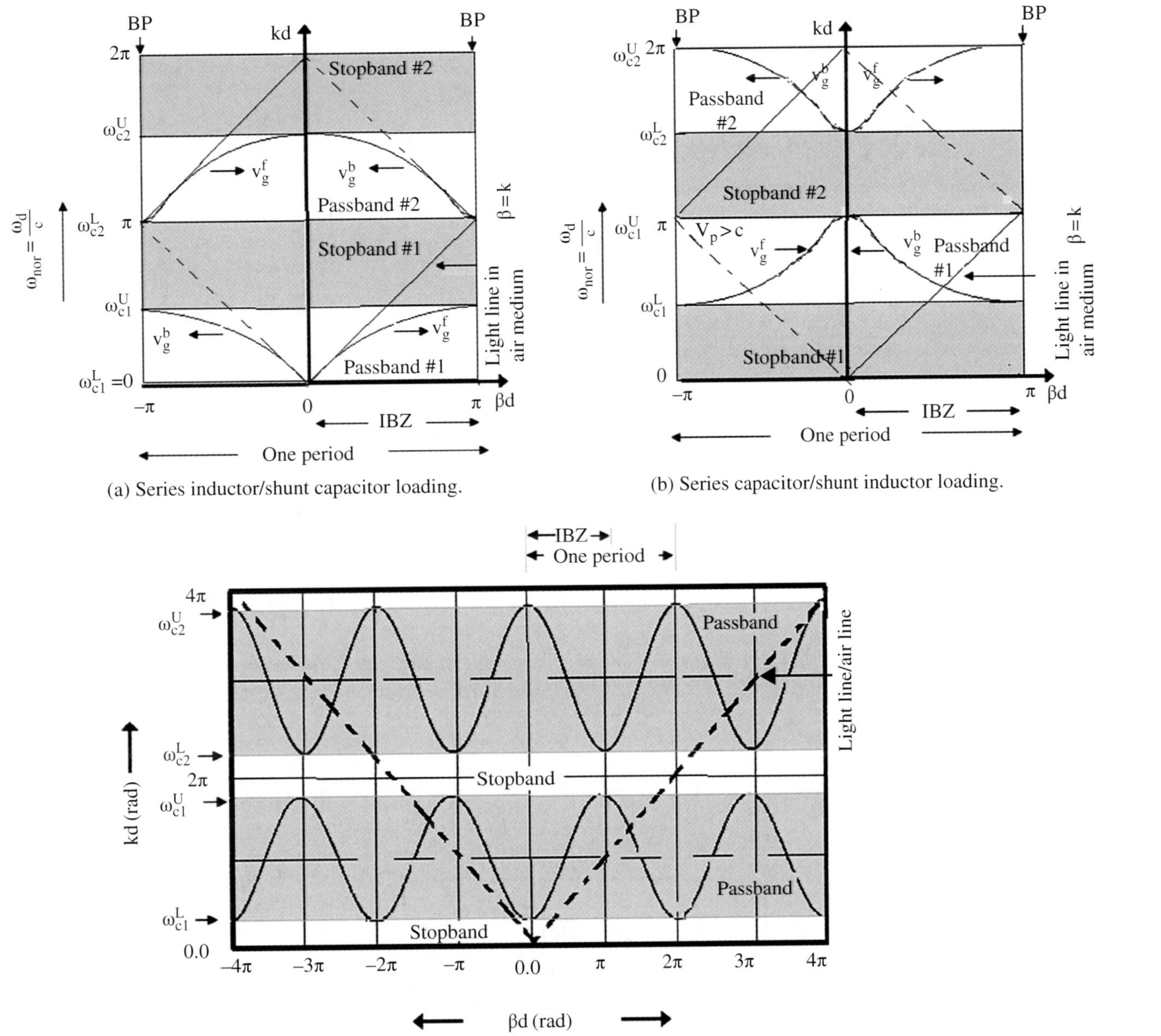

(a) Series inductor/shunt capacitor loading.

(b) Series capacitor/shunt inductor loading.

(c) Capacitive iris-loaded waveguide.

Figure 19.12 Slow-wave/fast-wave supporting periodic TEM-type/waveguide structures.

equation (19.3.18). It is a nonperiodic function. A few loading cases are considered below.

Dispersion Characteristics

Loading Elements: Series Inductor/Shunt Capacitor

Figure (19.8a and b) shows the periodic loading of lines using the shunt and series-connected inclusions. Equations (19.3.19a–c)provides their dispersion relations. The parameter p determined by the reactive loading of the host line is the *loading-factor*. Figure (19.12a) shows similar dispersion behavior for both the loading elements. Figure (19.13a–c) demonstrates computation of the dispersion and attenuation of Bloch modes in the passband and stopband, respectively, using the matrix element A given in equations (19.3.18) and (19.3.19).

The computation of the matrix element A is done with respect to kd, i.e. the *normalized frequency variable* $kd = \omega^{nor}$, $\Rightarrow \omega = \omega^{nor}(v_p/d)$, where, $v_p = c/\sqrt{\varepsilon_r}$. In a case of the unloaded line, i.e. for the uniform host medium, loading-factor, $p = 0$. The passband is obtained for $A = \cos(\beta d), -1 \leq A \leq +1$ and the stopband for $A = \cos(\beta d), |A| > 1$. The shaded region of Fig. (19.13a) shows the passband. Figure (19.13b) shows the corresponding dispersion diagrams of the propagating Bloch modes #1, 2, 3 in three passbands, for p = 0.0, 0.25 and 0.5, i.e. for no loading, the light and strong loading, respectively. The passband is reduced for the increased loading, i.e. for a higher value of the loading-factor. Further, there is a decreasing bandwidth for the higher-order Bloch modes. Figure (19.13c) shows the computed attenuation factor in the stopband. The higher loading (p = 0.5) provides higher attenuation and wider stopband. However, the upper edge of the stopband, i.e. the lower edge of the passband, remains fixed, as it is decided by the lattice constant d. The stopband bandwidth increases for the higher Bloch modes. The dispersion diagram shows the alternating passband/stopband with the lower and upper band-edge cut-off frequencies, $\omega^L_{c,n}$, $\omega^U_{c,n}$; 1, 2, ... of the passband. These edge frequencies also provide the band-edge of the stopband.

The Lower Edge of the Passband It occurs for $A = \pm 1$, i.e. at the normalized frequency $\omega^{nor} = kd = 0, \pi, 2\pi, 3\pi, \ldots$ corresponding to the IBZ edge, $\beta d = 0, \pi, 0, \pi, \ldots$. It is noted that $\beta d = 2\pi$ is identical to $\beta d = 0$ and $\beta d = 3\pi$ is identical to $\beta d = \pi$. The lower band-edges of the n numbers of the passband are $\omega^L_{c,n} = 0, \pi v_p/d, 2\pi v_p/d$, etc., where n = 1, 2, ... is the order of the passband. Therefore, the lattice constant, i.e. the period, d controls the lower edge of the passband, i.e. the upper edge of the stopband.

The Upper Edge of the Passband The upper cut-off frequencies of the n number of passbands $\omega^U_{c,n}$ are obtained at the edge of the IBZ, i.e. for $\beta d = \pi$, 0, etc. The first upper cut-off frequency of the first passband is obtained for A = −1 in equation (19.3.19a). It is rewritten as follows:

$$\cos(\omega^{nor}) - p\,\omega^{nor}\sin(\omega^{nor}) + 1 = 0 \qquad \text{(a)}, \qquad (19.3.23)$$

where $\omega^{nor} = kd$. For a given value of the parameter p, n number of upper cut-off frequencies are obtained corresponding to the odd-numbered passband, i.e. n = 1, 3, ...

Let us illustrate the computation of band structures for p = 2. In this case, by numerically solving equation (19.3.23a), the principal value, i.e. the first normalized cut-off frequency of the passband, is computed $\omega^{nor}_{c,1} = 0.9602$, giving $\omega^U_{c,1} = \omega^{nor}_{c,1}\,c/d\sqrt{\varepsilon_r}$, where c is the velocity of EM-wave in free space. For a line in the air medium, $\varepsilon_r = 1$. However, for microstrip or CPW relative permittivity is replaced by the effective relative permittivity, $\varepsilon_r \to \varepsilon_{reff}$. Normally, the static value of the effective relative permittivity ε_{reff} is used. However, the dispersion in the host microstrip or CPW medium can

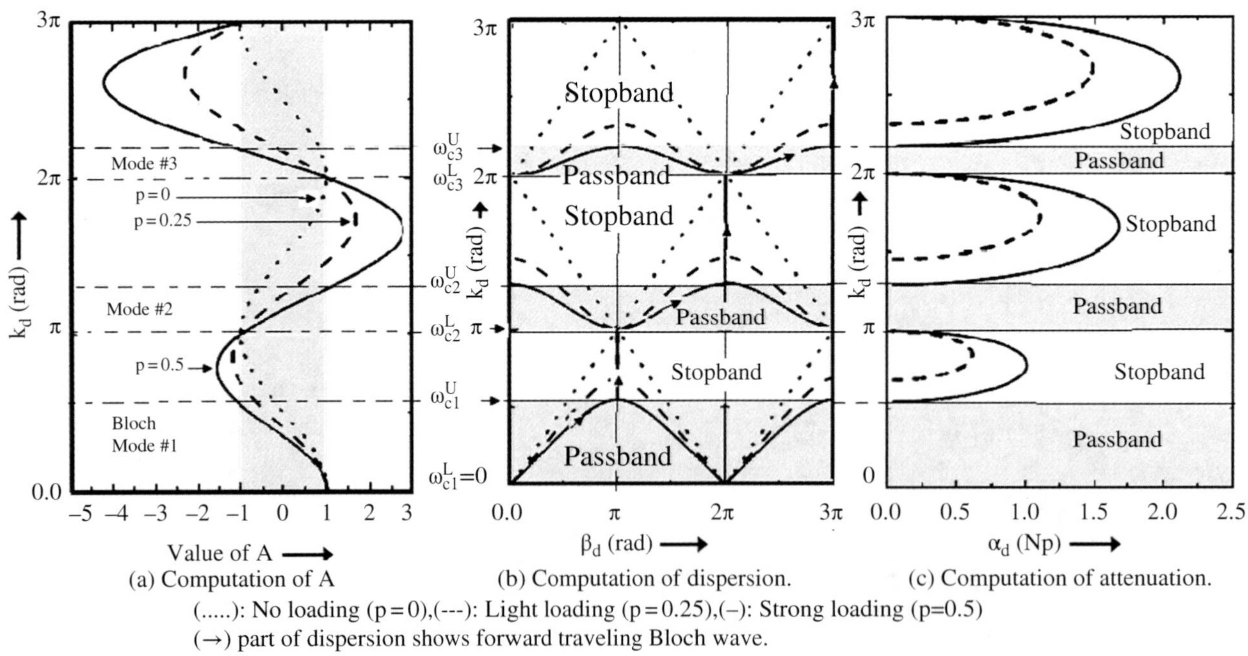

(a) Computation of A (b) Computation of dispersion. (c) Computation of attenuation.
(.....): No loading (p=0),(---): Light loading (p=0.25),(–): Strong loading (p=0.5)
(→) part of dispersion shows forward traveling Bloch wave.

Figure 19.13 Computation of dispersion and attenuation of shunt capacitors/series inductors loaded transmission line.

be easily accounted for. If the period of inclusion is $d = 1$ cm and $\varepsilon_r = 1$, then $f^U_{c,1}$ is 4.5846 GHz. The first lower cut-off frequency of the first passband $\omega^L_{c,1}$ is 0. Thus, the bandwidth of the first passband is $(f^U_{c,1} - f^L_{c,1}) =$ 4.5846 GHz.

Likewise, the lower frequency of the second passband occurs at $kd = \omega^{nor} = \pi$, i.e. for $\omega^L_{c,2} = \pi c/d\sqrt{\varepsilon_r}$. For the present case, $f^L_{c,2}$ is 15 GHz. The upper cut-off frequency $\omega^U_{c,2}$ of the second passband is computed for $A = 1$, i.e. $\beta d = 0$ from the following equation:

$$\cos(\omega^{nor}) - p\,\omega^{nor}\sin(\omega^{nor}) - 1 = 0 \qquad (b). \qquad (19.3.23)$$

The numerical solution of equation (19.3.23) provides the upper cut-off frequencies for the even-numbered passband, i.e. for $n = 2, 4, \ldots$. For the present case, $\omega^{nor}_2 = 3.431$, giving $f^U_{c,2} = 16.382$ GHz. The bandwidth of the second passband is $(f^U_{c,2} - f^L_{c,2}) = 16.382 - 15.000 = 1.832$ GHz.

In a case of the loading factor $p = 0.5$, the first lower cut-off frequency is $\omega^L_{c,1} = 0$. The normalized first upper cut-off frequency $\omega^{nor}_{c,1}$ obtained from equation (19.3.23a) is 1.7207. For $d = 1$ cm and $\varepsilon_r = 1$, $f^U_{c,1}$ is 8.216 GHz, that is the bandwidth of the first passband also. As computed, the $f^L_{c,2} = 15$ GHz and $\omega^{nor}_{c,2} = 4.0457$ for the upper band edge of the second passband. It gives $f^U_{c,2} = 19.373$ GHz and 4.373 GHz wide second passband.

Loading Elements: Series Capacitor/Shunt Inductor

In this case, the host line is periodically embedded with the series capacitors or the shunt inductors. For the series capacitor, the normalized reactance is $\bar{z} = j\bar{x} = -j/\omega CZ_0$, and for the shunt inductor, the normalized susceptance is $\bar{y} = jb = -jZ_0/\omega L$. The dispersion relation, using equation (19.3.19), of these loaded lines is written as follows:

$$\begin{aligned} &\cos(\beta d) = \cos(\omega^{nor}) + (p/\omega^{nor})\sin(\omega^{nor}) && (a)\\ &\text{for the series capacitor: } p = d/2v_pCZ_0(\varepsilon_r) && (b)\\ &\text{for the shunt inductor: } p = d\,Z_0(\varepsilon_r)/2v_pL && (c), \end{aligned} \qquad (19.3.24)$$

where $\omega^{nor} = kd, v_p = c/\sqrt{\varepsilon_r}, Z_0(\varepsilon_r) = Z_0(\varepsilon_r = 1)/\sqrt{\varepsilon_r}$ and in a case of the microstrip/CPW $\varepsilon_r \to \varepsilon_{reff}$.

Figure (19.12b) shows the dispersion diagram, i.e. Brillouin diagram. It starts with the first stopband, at zero lower cut-off frequency. Its upper cut-off frequency is the lower cut-off frequency of the first passband, i.e. $\omega^L_{c,1}$ corresponding to $\beta d = 0$. For $p = 1$, using equation (19.3.24a) $\omega^L_{c,1}$ is 1.3065. The upper cut-off frequency $\omega^U_{c,1}$ of the first passband occurs at $\omega^{nor}_{c,1} = \pi$. Likewise, the lower cut-off frequency $\omega^L_{c,2}$ of the second passband occurs at $\beta d = \pi$. Using the above equation, it is determined as $\omega^L_{c,2} = 3.6773$. Its upper cut-off frequency occurs at $\omega^U_{c,2} = 2\pi = 6.2831$. For the assumed loading factor p, the value of the series capacitor/shunt inductor is computed, if the characteristic impedance $Z_0(\varepsilon_r)$ of the unloaded host line is known. The cut-off frequency in GHz is easily computed, provided the period d, and ε_r or ε_{reff} are also known.

Loading Elements: Shunt Resonant Circuit/Series Resonant Circuit

Sometimes, the reactance/susceptance loading elements are replaced by the LC series resonant circuit or by the LC parallel resonant circuit. Such loadings are involved in the modeling of the 2D planar EBG structures, and even in the modeling of the stub loaded 1D periodic line. Two cases are discussed below:

i) *Shunt-connected series resonant circuit loaded line*

Figure (19.14a) shows the side view of the 2D EBG structure, between a two-dielectric layered parallel plate waveguide, supporting the quasi-TEM mode. It is used to suppress the *simultaneous switching noise* (SSN) in the high-speed PCB technology [J.9]. The structure is modeled through the 1D unit cell, shown in Fig. (19.14a), using a shunt connected series L_1C_1 resonant circuit. The C_1 capacitance is realized by the parallel plate capacitor of thickness h_2 and relative permittivity ε_{r2} between the top metal conductor and the patch. The grounded vias contribute to the series inductance L_1. The parallel plate line has characteristic impedance $Z_0(\varepsilon_r)$ and wavenumber k. The shunt susceptance of the series L_1C_1 resonant circuit is

$$\begin{aligned} &Z = \frac{1}{j\omega C_1} + j\omega L_1,\\ &\Rightarrow Y = \frac{j\omega C_1}{1-(\omega/\omega_0)^2}, \text{ where } \omega_0 = \frac{1}{\sqrt{L_1C_1}}. \end{aligned} \qquad (19.3.25)$$

In equation (19.3.25), ω_0 is the resonance frequency. At $\omega < \omega_0$, Y is capacitive, at $\omega = \omega_0$, $Y = \infty$, and at $\omega > \omega_0$, Y is inductive. At frequency $\omega < \omega_0$, βd follows the shunt capacitance loading. However, at a frequency $\omega > \omega_0$, it follows the shunt inductor loading in the IBZ range of a unit cell. The dispersion relation, from the equation (19.3.19a), is obtained:

$$\cos(\beta d) = |\cos(kd) - p(kd)\ \sin(kd)| \le 1 \qquad \text{(a)}$$

$$\text{where,}\ \ p = \frac{v_{ph} C_1\, Z_0(\varepsilon_r)}{2d\left[1-(\omega/\omega_0)^2\right]},\quad v_{ph} = \frac{c}{\sqrt{\varepsilon_{reff}}} \qquad \text{(b)}$$

$$\cosh(\alpha d) = |\cos(kd) - p(kd)\ \sin(kd)| > 1 \qquad \text{(c)}. \qquad (19.3.26)$$

In expression (19.3.26), the frequency-dependent shunt loading capacitance $C = C_1/(1-(\omega/\omega_0)^2)$ is obtained from equation (19.3.25). Figure (19.14b) shows the dispersion diagram (black line) in the passband for Bloch mode, and also the attenuation (gray line) in the stopband for a periodic line with such loading. Figure (19.14c) again shows the attenuation in stopband for better clarity. The dispersion is shown over IBZ, i.e. in the range $0 \le \beta_x d \le \pi$. For the period d = 5.59 mm, the range is [0, 562] rad/m.

The resonance frequency of the series L_1C_1 resonant circuit is 3.65 GHz. In the first passband, from $f_{c1}^{L} = 0$ to $f_{c1}^{U} = 2$ GHz at $\beta d = \pi$, the loading is capacitive. The lower edge of the first stopband is at 2 GHz and its upper edge is at the lower edge of the second passband at $f_{c2}^{L} = 13.1$ GHz, giving 11.1 GHz wide stopband. However, within the stopband itself, high attenuation occurs due to the resonance at 3.65 GHz. Above the frequency 3.65 GHz, shunt loading is inductive. The second

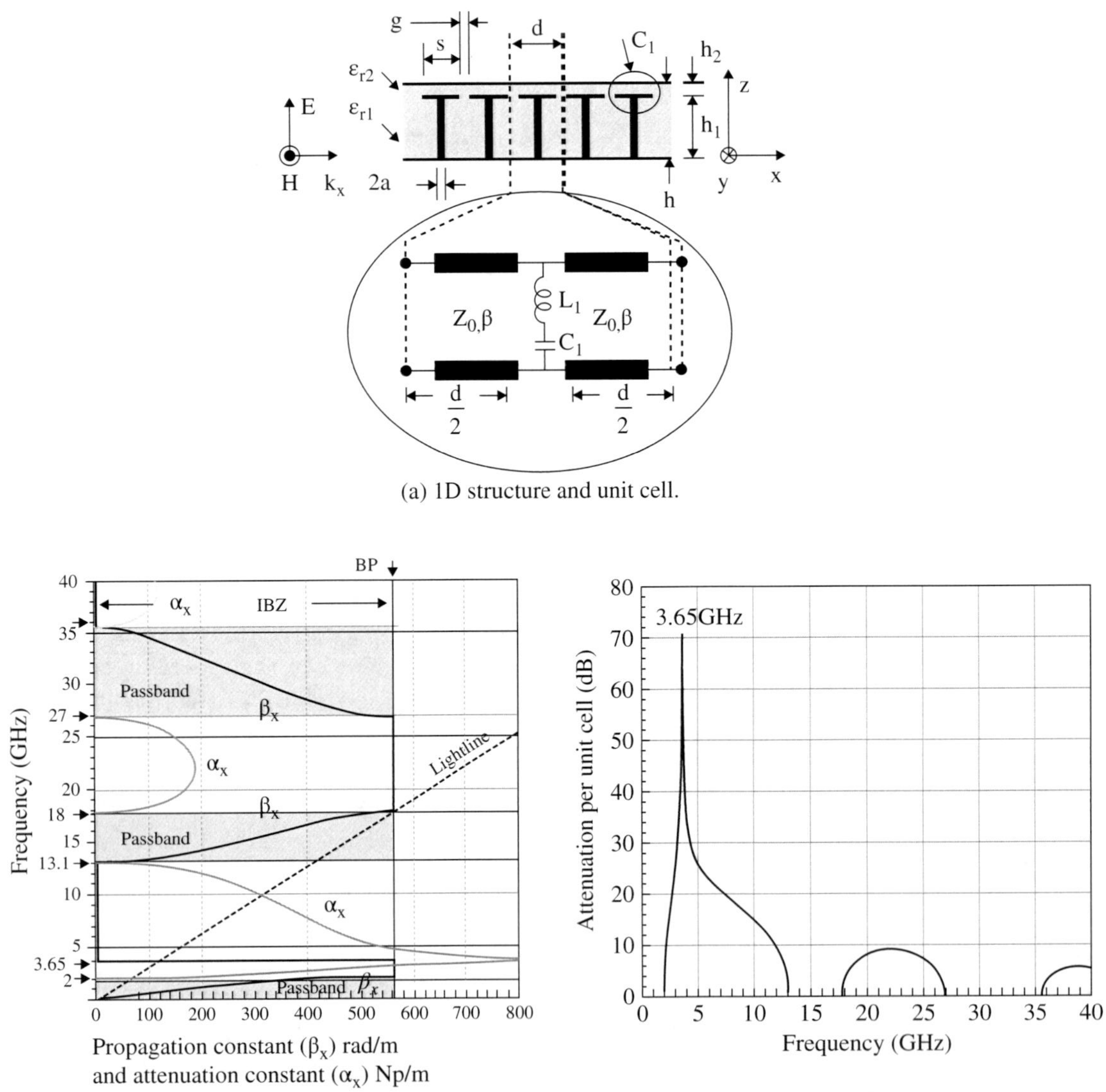

(a) 1D structure and unit cell.

(b) Dispersion diagram over the IBZ [J.9] (Reproduce from [J.9],©2005 IEEE).

(c) Attenuation diagram [J.9].

Figure 19.14 Dispersion and attenuation diagrams of the periodic line with shunt-connected series resonant circuit. *Source*: From Rogers [J.9]. © 2005, IEEE.

passband ends at its upper band edge at $f_{c2}^{U} = 18$ GHz corresponding to $\beta d = \pi$, giving 4.9 GHz bandwidth. The second stopband starts at 18 GHz and ends at 27 GHz, i.e. the third passband starts at 27 GHz and ends with an upper cut-off at $f_{c3}^{U} = 35$ GHz. The edge frequencies are marked in Fig. (19.14).

ii) *Series connected short-circuited stub loaded line*

Figure (19.15a) shows the series-connected short-circuited stubs of length ℓ. The stubs periodically load the transmission line with a lattice constant d. The characteristic impedance and propagation constant of the host line are $Z_0(\varepsilon_r)$ and k, respectively. The characteristic impedance and propagation constant of the stub are $Z_d(\varepsilon_r)$ and k_1. The short-circuited series stub could be modeled as a series-connected parallel resonant circuit. It is discussed in section (17.3.1) of chapter 17. The equivalent lumped circuit element could be obtained. It offers series load Z_1. At a frequency below the resonance frequency, i.e. $\ell < \lambda_g/4$, the stub offers a series inductive loading to the line, where λ_g is the guided wavelength of the stub. At resonance, $\ell = \lambda_g/4$, there is a sudden change in the type of loading, from inductive to capacitive, offering a change in βd by π. The attenuation is also very high due to the open-circuited loading. Above the resonance frequency, i.e. $\lambda_g/4 < \ell < \lambda_g/2$, the line is loaded with a series capacitor. At $\ell = \lambda_g/2$, series resonance occurs. The cycle is repeated at high frequency. In the case of a planar line, normally shunt connected open-circuited stubs are used for the periodic loading [B.13–B.15]. An open-circuited stub is modeled as the shunt connected series resonant circuit offering a capacitive loading below the resonance. It is again discussed in section (17.3.1) of chapter 17. Using equation (19.3.14b), the following dispersion and attenuation expression are obtained for the short-circuited series-connected periodic line shown in Fig. (19.15a):

$$\begin{aligned}
&\text{Series connected load: } Z_1 = j Z_{01} \tan(k_1\ell), \text{ where, } k_1 = 2\pi/\lambda_g && \text{(a)}\\
&\cos(\beta d) = \cos(kd) + j Z_1/(2Z_0)\sin(kd) && \text{(b)} \qquad (19.3.27)\\
&\cosh(\alpha d) = |\cos(kd) + j Z_1/(2Z_0)\sin(kd)| > 1 && \text{(c).}
\end{aligned}$$

In case, the line is in the air medium, we have $\varepsilon_r = 1$, $k = k_1 = k_0$. Figure (19.15b) shows the dispersion behavior of the periodic line, in the air medium, with the characteristic impedance of stub $Z_{01} = 2Z_0$ and $\ell/d = 1.2$ [J.7]. The first passband is governed by the inductive loading between $\omega_{c1}^{L} = 0$, and ω_{c1}^{U} at $\beta d = \pi$. The stopband starts at ω_{c1}^{U} and ends at ω_{c2}^{L} for $\beta d = 2\pi$. At $\beta d = \pi$, the series-connected parallel resonance occurs at a frequency ω_{r1} and βd is changed by π. The electrical length βd is jumped to 2π. The high attenuation at frequency ω_{r1} occurs within the stopband itself. The stopband continues up to ω_{c2}^{L}, i.e. at the edge of the second passband at $\beta d = 2\pi$. In the second passband, $\omega_{c2}^{L} - \omega_{c2}^{U}$, the short-circuited stub offers the series-connected capacitor that decides the nature of dispersion. The series resonance occurs at ω_{r2}.

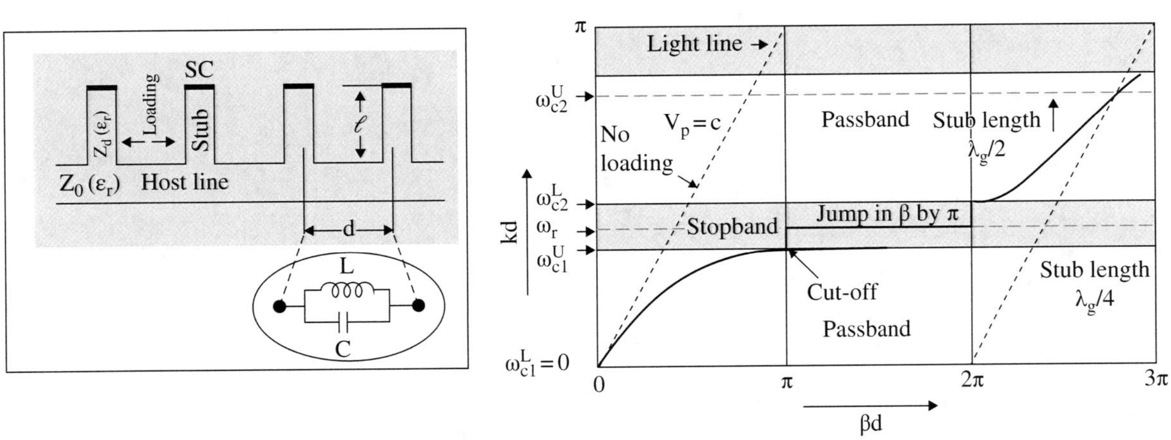

(a) Series connected stub.

(b) Dispersion diagram.

Figure 19.15 Nature of dispersion in the series-connected short-circuited stub-loaded periodic line.

Characteristics of Bloch Impedance

Bloch impedance Z_B, i.e. the characteristic impedance of a periodic line, is highly dispersive. Its nature of dispersion is decided by the capacitive and inductive loading. Equations (19.3.10a) and (19.3.10b) are used to compute Z_B for the asymmetrical and symmetrical unit cells. Using equation (19.3.13), the elements A and B of [ABCD] matrix, of the symmetrical unit cell, are written below:

$$A = \cos(\beta d) = \cos(kd) - p(kd)\sin(kd) \quad (a)$$
$$B = j[\sin(kd) + p(kd)\{\cos(kd) - 1\}] \quad (b), \qquad (19.3.28)$$

where p is the loading-factor is given by equation (19.3.19).

Bloch impedance Z_B^{sym} of the symmetrical unit cell is a real quantity in the passband, for $|A| \le 1$. For the shunt capacitive loading factor p = 0.1, Fig. (19.16a) shows the frequency-dependent Bloch impedance diagram with four passbands (P), corresponding to four propagating Bloch modes [J.10]. Two modes are separated by a stopband (S), supporting the nonpropagating evanescent mode. At the lower frequency, Z_B^{sym} starts with 50 Ω, i.e. the characteristic impedance of the host line. In the first passband, the real Z_B^{sym} decreases with frequency, as the loading element is a shunt capacitor and reaches zero at the upper band edge of the passband #1. In the stopband, Z_B^{sym} is an imaginary quantity (shown in dotted line). In the next passband #2, Z_B^{sym} is a negative real quantity, showing that the second Bloch mode is the backward wave, as suggested in the dispersion diagram Fig. (19.16b) for p = 0.1. Figure (19.16a) also suggests that in the case of the odd-numbered passbands, the Bloch impedance is zero at the band edge and its mid-band value decreases with increasing order of the odd Bloch modes. In the case of the even Bloch mode, Z_B^{sym} is infinitely large at the band edge. The middle frequency band value increases with the increasing order of the even Bloch modes [J.10].

For an asymmetrical unit cell, even in the passband, the Bloch impedance is not a real quantity. It is a complex quantity. However, the propagation constant of the periodic line is not influenced by the asymmetry in the unit cell [B.11, J.11].

19.3.5 Some Loading Elements of 1D Periodic Lines

In the above discussion, the series impedance Z and shunt admittance Y have been taken as the loading elements, i.e. inclusions embedded in the host line. The Z and Y could be a single element reactance/susceptance, or these could be the series/parallel resonant circuit. The L, T, and π networks of the reactive elements, shown in Fig. (19.17a–c), can also be used as inclusions [J.11, J.12]. These networks are also used to model several kinds of loading elements. Figure (19.17d–i) shows a few more loading elements. Some of these elements are useful to realize the metamaterials also, discussed in chapters 21 and 22. We summarize below the [ABCD] parameters of the unit cells of the periodic lines loaded with these inclusions.

The L, T, and π networks of the reactive elements are basic structures. By using the capacitors (C) and inductors (L), in the series/shunt, different kinds of inclusions are created. These have different propagation characteristics. The

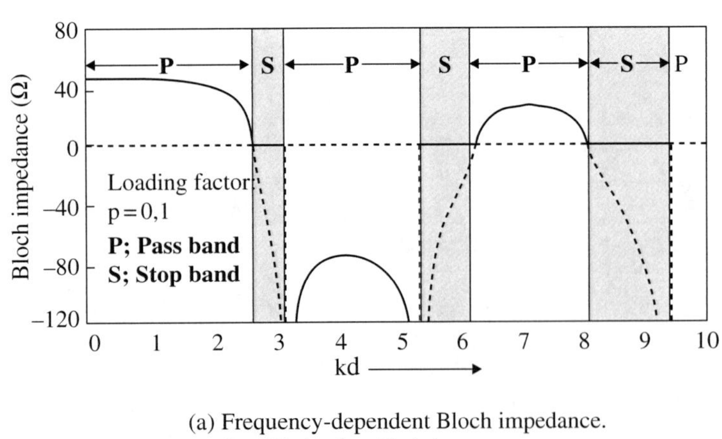

(a) Frequency-dependent Bloch impedance. Real Z_B(-), Img Z_B(...).

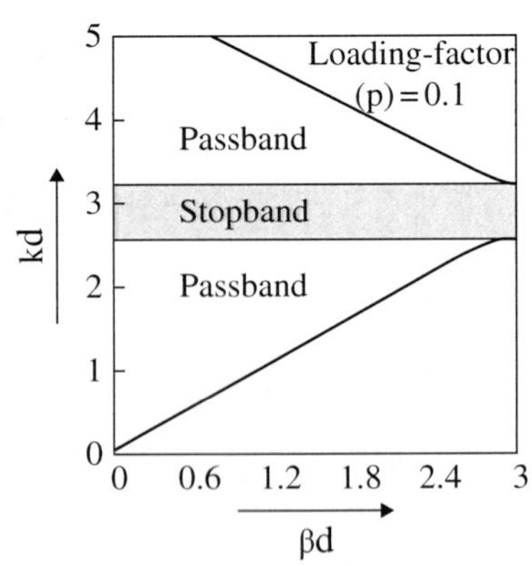

(b) Dispersion for p=0.1.

Figure 19.16 Dispersion diagram and Bloch impedance of shunt capacitance-loaded line. *Source*: From Takagi [J.10]. © 2001, IEEE.

periodic line can also be formed by cascading these loading elements. Figure (19.17d) shows the T-network type inclusion with series L and shunt C between a pair of host line sections of lengths $d/2$ ($\theta/2 = kd/2$). It supports the forward traveling slow-wave, whereas a line with L-network type inclusion with series C and shunt L, shown in Fig. (19.17e), supports the backward wave. The transformer loaded line, shown in Fig. (19.17f), is also used to model the periodic inclusions [J.13]. The step impedance structure, shown in Fig. (19.17g), is also used to get the 1D periodic transmission medium. The L and C of the L, T, and π networks could further be separated by the line sections, shown in Fig. (19.17h and i), supporting the forward and backward waves, respectively.

Using the equations from subsection (3.1.3) of chapter 3, the [ABCD] parameters of the T-network, Fig. (19.17a), are obtained as follows:

$$\begin{bmatrix} A & B \\ C & D \end{bmatrix} = \begin{bmatrix} 1 & Z/2 \\ 0 & 1 \end{bmatrix} \begin{bmatrix} 1 & 0 \\ Y & 1 \end{bmatrix} \begin{bmatrix} 1 & Z/2 \\ 0 & 1 \end{bmatrix}$$
$$= \begin{bmatrix} 1 + ZY/2 & Z(1 + ZY/4) \\ Y & 1 + ZY/2 \end{bmatrix}. \quad (19.3.29)$$

For the L-network shown in Fig. (19.17b), these are

$$\begin{bmatrix} A & B \\ C & D \end{bmatrix} = \begin{bmatrix} 1 & Z \\ 0 & 1 \end{bmatrix} \begin{bmatrix} 1 & 0 \\ Y & 1 \end{bmatrix} = \begin{bmatrix} 1 + ZY & Z \\ Y & 1 \end{bmatrix}. \quad (19.3.30)$$

The [ABCD] parameters of the T-network based unit cell of period d, shown in Fig. (19.17d), are

$$\begin{bmatrix} A & B \\ C & D \end{bmatrix} = \begin{bmatrix} \cos(kd/2) & j Z_0 \sin(kd/2) \\ j Y_0 \sin(kd/2) & \cos(kd/2) \end{bmatrix} \times$$
$$\begin{bmatrix} 1 & j\omega L/2 \\ 0 & 1 \end{bmatrix} \begin{bmatrix} 1 & 0 \\ j\omega C & 1 \end{bmatrix} \begin{bmatrix} 1 & j\omega L/2 \\ 0 & 1 \end{bmatrix} \times$$
$$\begin{bmatrix} \cos(kd/2) & j Z_0 \sin(kd/2) \\ j Y_0 \sin(kd/2) & \cos(kd/2) \end{bmatrix}. \quad (19.3.31)$$

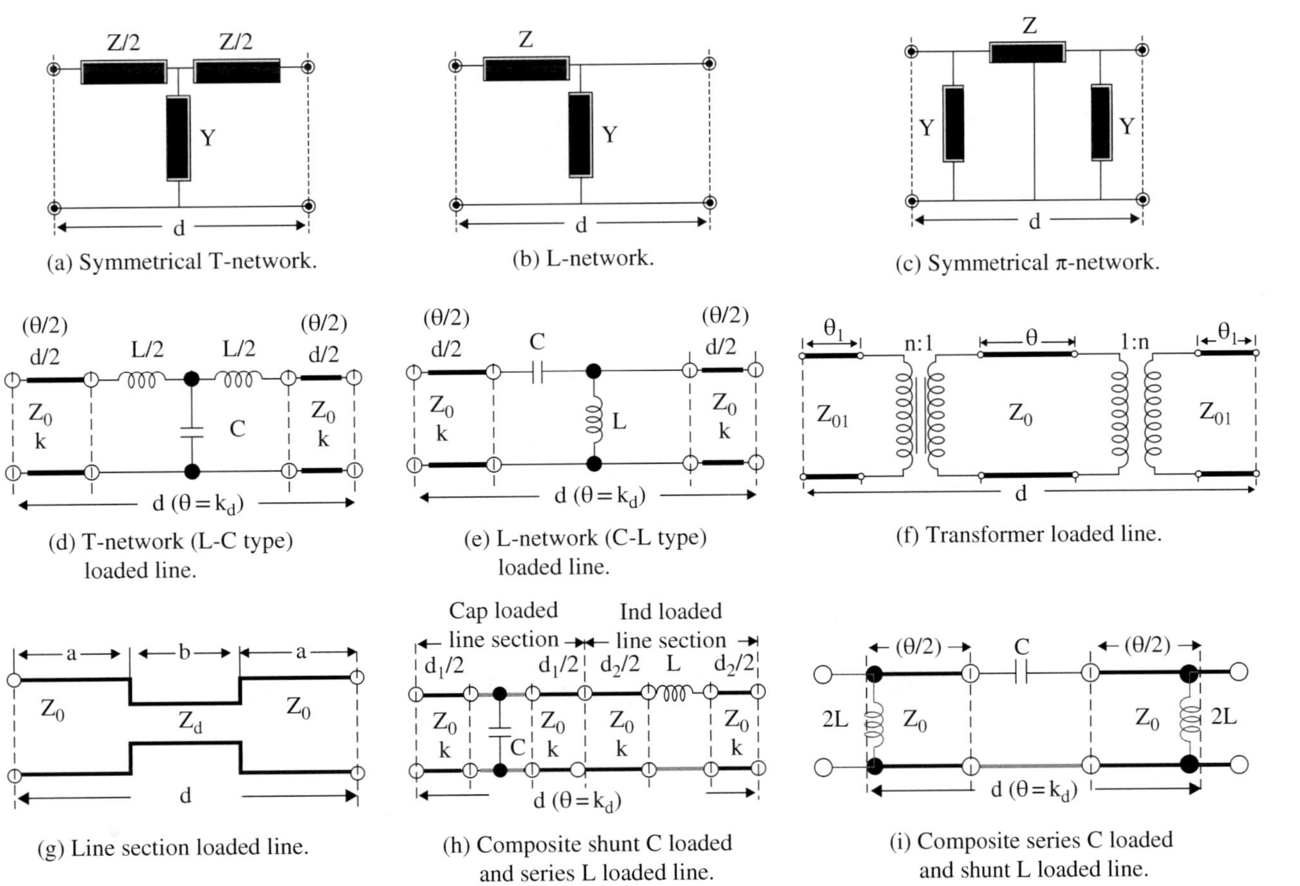

Figure 19.17 Unit cells of some loading elements (inclusions).

The evaluated [ABCD] parameters are summarized below:

$$\begin{aligned}
A = D &= \left(1-\omega^2LC/2\right)\cos(kd) - \left[\omega CZ_0 + \left(2-\omega^2LC/2\right)(\omega L/2Z_0)\right](\sin(kd)/2)\\
B &= j\left[\omega CZ_0^2/2 + \left(2-\omega^2LC/2\right)(\omega L/4)\right]\cos(kd)\\
&\quad + j\left[\left(1-\omega^2LC/2\right)Z_0\sin(kd) - \omega CZ_0^2/2 + \left(2-\omega^2LC/2\right)(\omega L/4)\right]\\
C &= j\left[\omega C/2 + \left(2-\omega^2LC/2\right)\left(\omega L/4Z_0^2\right)\right]\cos(kd)\\
&\quad + j\left[\left(1-\omega^2LC/2\right)(\sin(kd)/Z_0) + \omega C/2 - \left(2-\omega^2LC/2\right)\left(\omega L/4Z_0^2\right)\right].
\end{aligned} \tag{19.3.32}$$

The [ABCD] parameters of the transformer loaded line, shown in Fig. (19.17f), are

$$\begin{aligned}
\begin{bmatrix} A & B \\ C & D \end{bmatrix} &= \begin{bmatrix} \cos(\theta_1) & j Z_{01}\sin(\theta_1) \\ j Y_{01}\sin(\theta_1) & \cos(\theta_1) \end{bmatrix} \begin{bmatrix} n & 0 \\ 0 & 1/n \end{bmatrix} \begin{bmatrix} \cos(\theta) & j Z_0\sin(\theta) \\ j Y_0\sin(\theta) & \cos(\theta) \end{bmatrix}\\
&\quad \times \begin{bmatrix} 1/n & 0 \\ 0 & n \end{bmatrix} \begin{bmatrix} \cos(\theta_1) & j Z_{01}\sin(\theta_1) \\ j Y_{01}\sin(\theta_1) & \cos(\theta_1) \end{bmatrix}.
\end{aligned} \tag{19.3.33}$$

The [ABCD] parameters of the lossless line section (Z_d, k_d) loaded line, shown in Fig. (19.17g), are summarized below:

$$\begin{aligned}
A = D &= \cos(k_d b)\cos(2kd) - \left[\left(Z_d^2 + Z_0^2\right)/2Z_dZ_0\right]\sin(k_d b)\sin(2kd)\\
B &= j Z_0\cos(k_d b)\sin(2kd) + j\sin(k_d b)\left[\left(Z_d^2/Z_0\right)\cos^2(kd) - \left(Z_0^2/Z_d\right)\sin^2(kd)\right]\\
C &= j(1/Z_0)\cos(k_d b)\sin(2kd) + j\sin(k_d b)\left[\left(Z_0^2/Z_d\right)\cos^2(kd) - \left(Z_d/Z_0^2\right)\sin^2(kd)\right].
\end{aligned} \tag{19.3.34}$$

The dispersion relation and Bloch impedance of the unit cell of Fig. (19.17i) are summarized below:

$$A = \cos(\beta d) = [1 + ZY/4]\cos\theta + j(1/4)(ZY_0 + Z_0Y)\sin\theta + ZY/4 \tag{a}$$

$$\text{where, } Z = 1/j\omega C \text{ and } Y/2 = 1/j\omega(2L) \tag{b}$$

$$Z_B = \pm\left[\frac{(Z/2)\cos\theta + jZ_0\sin\theta + Z/2}{\left(Y + ZY^2/4 + ZY_0^2/2\right)\cos\theta + j\left(Y_0 + ZYY_0/2 + Z_0Y^2/4\right)\sin\theta + ZY^2/8 - ZY_0^2/2}\right] \tag{c}$$

(19.3.35)

The elements of the [ABCD] parameters are used to get the dispersion relation and Bloch impedance of a periodic line.

19.3.6 Realization of Planar Loading Elements

Figures (19.18)–(19.20) consider some of the structures to realize the capacitor, inductor, and resonating inclusions, respectively, in the microstrip technology.

The physical size of these elements is within $\lambda_g/10$ to treat them as the lumped elements. The approximate closed-form models are summarized, although more accurate results could be obtained using the EM-simulator and the parameter extraction [B.16]. More elaborate closed-form models of some of these lumped elements are also available [B.17].

Realization of Capacitors in Microstrip

- *Patch capacitor*: It is a usual parallel plate capacitor, shown in Fig. (19.18a). It also appears in the process of modeling more complex inclusions, forming the unit cells. It can acquire different shapes, such as square, rectangle, circle, triangle. An accurate computation of capacitance of a patch requires the computation of its main central capacitance, between the patch and ground conductor, and also the fringe capacitance outside the perimeter of the patch [J.14, J.15]. The total capacitance of a rectangular patch capacitor is computed using the following expressions:

$$
\begin{aligned}
C_{total} &= C_0 + 2C_{f1} + 2C_{f2}; \qquad \text{where, } C_0 = \frac{\varepsilon_0\varepsilon_r A}{h} \\
C_{f1} &= \frac{1}{2}\left[\frac{Z_0(W, \varepsilon_r = 1, h)L}{c\, Z_0^2(W, \varepsilon_r, h)} - \frac{\varepsilon_0\varepsilon_r A}{h}\right] \\
C_{f2} &= \frac{1}{2}\left[\frac{Z_0(L, \varepsilon_r = 1, h)W}{c\, Z_0^2(L, \varepsilon_r, h)} - \frac{\varepsilon_0\varepsilon_r A}{h}\right],
\end{aligned}
\tag{19.3.36}
$$

where c is the velocity of the EM-wave in free space. The central capacitance C_0 and the fringe capacitances C_{f1} and C_{f2} *p.u.l.* of the rectangular patch are computed from the W and L side of the patch. The characteristic impedance of the patch on the air and dielectric substrates are $Z_0(\varepsilon_r = 1)$ and $Z_0(\varepsilon_r)$, respectively.

- *Gap capacitor:* Figure (19.18b) shows a gap S between the microstrip sections. It is modeled as the π-network between the planes of de-embedding T_1 and T_2. The series capacitance C_s is the main capacitance, whereas the shunt capacitance, C_p is the fringe capacitance between the strip ends and the ground conductor. These capacitances are computed using the following set of equations [B.18, B.19]:

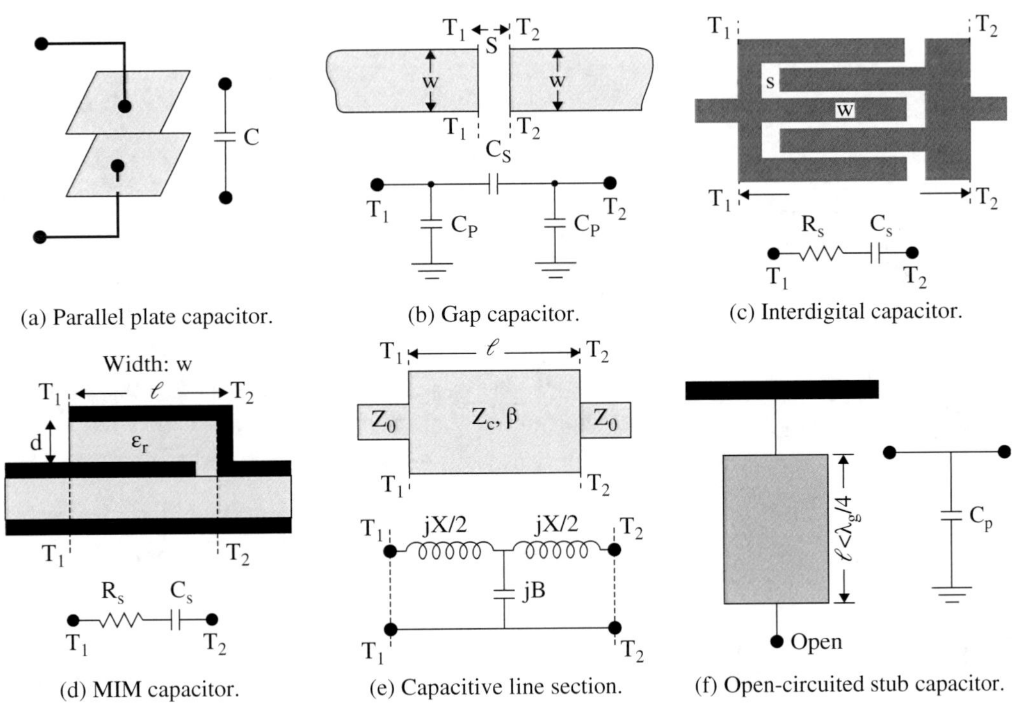

(a) Parallel plate capacitor. (b) Gap capacitor. (c) Interdigital capacitor. (d) MIM capacitor. (e) Capacitive line section. (f) Open-circuited stub capacitor.

Figure 19.18 Some capacitors in microstrip.

$$\begin{aligned}
&C_p = C_e/2,\ C_s = 0.5C_o - 0.25C_e && \text{(a)}\\
&\text{where, } C_o/w = (\varepsilon_r/9.6)^{0.8}(s/w)^{m_o}\exp(k_o)\quad \text{pF/m} && \text{(b)}\\
&C_e/w = 12(\varepsilon_r/9.6)^{0.8}(s/w)^{m_e}\exp(k_e)\quad \text{pF/m} && \text{(c)}\\
&\text{for } 0.1 \le s/w \le 1.0 && \\
&m_o = w/h\,[0.619\ln(w/h) - 0.3853];\ k_o = 4.26 - 1.453\ln(w/h) && \text{(d)}\\
&\text{for } 0.1 \le s/w \le 0.3:\ m_e = 0.8675;\ k_e = 2.043(w/h)^{0.12} && \text{(e)}\\
&\text{for } 0.3 \le s/w \le 1.0:\ m_e = \frac{1.565}{(w/h)^{0.16}} - 1;\ k_e = 1.97 - \frac{0.03}{w/h} && \text{(f).}
\end{aligned} \quad (19.3.37)$$

The expressions for C_s and C_p have accuracy within 7% for $0.5 \le w/h \le 2$, and $2.5 \le \varepsilon_r \le 15$. The capacitances can also be extracted using an EM-simulator [B.16].

- *Interdigital capacitor*: Figure (19.18c) shows the interdigital capacitor, with T_1 and T_2 planes of de-embedding, formed in the microstrip environment. It is an extension of the microstrip gap capacitor with a larger value of the series capacitance C_s, normally $C_s < 10$ pF. The series resistance R_s accounts for the conductor loss. In the case of $w = s$, i.e. the width of the fingers is the same as that of the strip gap, and $h \gg w$, the C_s and R_s are computed using the following approximate equations [J.16, B.19]:

$$\begin{aligned}
&C_s = 3.937 \times 10^{-5}\ell(\varepsilon_r + 1)[0.11(n-3) + 0.252]\quad \text{pF} && \text{(a)}\\
&R_s = 4/3(R_{sur}\ell/n\,w) && \text{(b),}
\end{aligned} \quad (19.3.38)$$

where ℓ in μm is the length of the fingers, n is the number of fingers. ε_r and h are the parameters of a substrate. R_{sur} is the surface resistance of the conductor. More accurate, however more complex, models are also available [J.17].

- *MIM – capacitor:* Figure (19.18d) shows the metal-insulator-metal (MIM) capacitor. It is used to get a higher value of the capacitance in a compact form, say in the range of 30 pF. Its thickness d is about 0.5 μm. It uses a low-loss thin dielectric layer. The MIM is approximately modeled as a patch capacitor [B.19, J.16]:

$$C = \varepsilon_0\varepsilon_r(w \times \ell)/d, \quad (19.3.39)$$

where $(w \times \ell)$ is the patch area. The fringe capacitances can also be estimated, as discussed previously. More accurate expressions for the MIM – capacitor are also available [J.18].

- *Low impedance line section as a capacitor*: Figure (19.18e) shows a wide width microstrip section, of low characteristic impedance between two high impedance host microstrip lines. The low impedance line section acts as a shunt capacitance to the host line. It is normally used to design the low-pass filters [B.13]. It is modeled as the T-network with shunt capacitive susceptance (B) and series fringe reactance (X). These are computed using the following expressions [B.19]:

$$\begin{aligned}
&B = \frac{1}{Z_c}\sin\left(\frac{2\pi\ell}{\lambda_g^c}\right) \approx \frac{1}{Z_c}\left(\frac{2\pi\ell}{\lambda_g^c}\right) && \text{(a)}\\
&\frac{X}{2} = Z_c\tan\left(\frac{\pi\ell}{\lambda_g^c}\right) \approx Z_c\left(\frac{\pi\ell}{\lambda_g^c}\right) && \text{(b),}
\end{aligned} \quad (19.3.40)$$

where Z_c is the characteristic impedance of low impedance capacitive microstrip section, and $\lambda_g^c = 2\pi/\beta_c, \beta_c = \beta_0\sqrt{\varepsilon_{reff}}$ are guided wavelength and propagation constant of the low impedance line section. The above approximation is valid for $\ell < \lambda_g^c/8$.

- *Open-circuited stub as a capacitor*: Figure (19.18f) shows an open-circuited short length stub that is used as a shunt capacitor C_p. Its length is $\ell < \lambda_g^c/4$. The input capacitive admittance is used to compute C_p:

$$\begin{aligned}
&Y_{in} = jY_c\tan\left(\frac{2\pi}{\lambda_g^c}\ell\right) \approx jY_c\frac{2\pi}{\lambda_g^c}\ell \approx j\,\omega C_p\\
&\Rightarrow C_p = \frac{Y_c}{v_p}\ell,\quad \text{where } v_p = \frac{c}{\sqrt{\varepsilon_{reff}}}.
\end{aligned} \quad (19.3.41)$$

The approximate expression (19.3.41) is valid for $\ell < \lambda_g^c/8$, and v_p is the phase velocity of the capacitive stub.

Realization of Inductors in Microstrip

Figure (19.19a–h) show several structures to realize inductors in the microstrips. These structures may also form part of more complex structures. The approximate closed-form expressions, for computation of the inductances, are summarized below. These expressions are useful for the initial design of the periodic lines.

- *Line inductor:* A microstrip of length ℓ, and width w, shown in Fig. (19.19a), offers a series-connected inductance L_s and resistance R_s due to the conductor loss. The microstrip conductor thickness is t. The inductor L_s and resistance R_s are computed using the following expressions [B.19]:

$$\begin{aligned} L_s &= 2.0 \times 10^{-4}\ell\,[\ln\{\ell/(t+w)\} + 1.193 + 0.2235(w+t)/\ell]\,K_g \ \ \text{nH} && \text{(a)} \\ \text{where, } K_g &= 0.57 - 0.145\ \ln(w/h), \ \ \text{for } w/h > 0.05 && \text{(b)} \\ R_s &= (R_{sur}\ell/2(w+t))\,[1.4 + 0.217\ \ln(w/5t)], \ \ \text{for } 5 < w/t < 100 && \text{(c).} \end{aligned} \tag{19.3.42}$$

In equation (19.3.42), R_{sur} is the surface resistance, h is the substrate thickness, and the factor K_g accounts for the decrease in the inductance due to the ground plane of a microstrip. The inductance of a microstrip section increases by cutting an aperture in the ground plane, behind the strip conductor.

- *The inductance of via (pin):* Figure (19.19b) shows a via, i.e. a pin, of the length ℓ and diameter d, connecting the microstrip patch or line to the ground plane. Its inductance in Henry is estimated using the following expression [B.20, J.8, J.19]:

$$\begin{aligned} L &= 2.0 \times 10^{-7}\ell\,[\ln(4\ell/d) + 0.5(d/\ell) - 0.75)] && \text{(a)} \\ \text{or } L &= 2.0 \times 10^{-7}\left[\ell\ \ln\left\{\left(\ell + \sqrt{r^2+\ell^2}\right)/r\right\} + 1.5\left(\ell - \sqrt{r^2+\ell^2}\right)\right] && \text{(b),} \end{aligned} \tag{19.3.43}$$

where r is the radius of via. The substrate thickness is the length of the via. Another expression is also available to compute the wire inductance [B.20].

- *Meander line inductor:* To increase the line inductance, a longer microstrip is bent to form the meandered microstrip line inductor, shown in Fig. (19.19c). The strip width (w) is normally smaller than the substrate thickness (h). The mutual inductance between the adjacent arms, carrying the oppositely flowing currents, is ignored, if the spacing (s) between the adjacent arms is more than three times of the strip width. However, for a compact meandering, the inductance cannot be computed from the total length of an unfolded line, as the mutual inductance cannot be ignored. The corners of the bends also offer additional discontinuities. A more complex model is available for the computation of the inductance of a meander line [J.17, J.20].
- *High-impedance line section as an inductor:* A short length high-impedance (90 Ω – 120 Ω) microstrip section, between two low impedance line sections, acts as a series inductor. The line section, shown in Fig. (19.19d), is modeled as the π-network [B.16, B.19]. The main inductive reactance, giving the series inductor, and the associated fringe capacitive susceptance are obtained from the following expressions

$$\begin{aligned} &X = Z_L \sin\left(\frac{2\pi}{\lambda_g^L}\ell\right) \approx Z_L\left(\frac{2\pi}{\lambda_g^L}\ell\right) = \omega L, \quad \Rightarrow L = \frac{Z_L}{v_p}\ell && \text{(a)} \\ &\text{where, phase velocity: } v_p = c/\sqrt{\varepsilon_{reff}}, \quad \lambda_g^L = \lambda_0/\sqrt{\varepsilon_{reff}} && \text{(b)} \\ &\frac{B}{2} = \frac{1}{Z_L}\tan\left(\frac{\pi}{\lambda_g^L}\ell\right) \approx \frac{1}{Z_L}\left(\frac{\pi}{\lambda_g^L}\ell\right) && \text{(c).} \end{aligned} \tag{19.3.44}$$

In equations (19.3.4a–c), Z_L is the characteristic impedance of the inductive microstrip section and λ_g^L is its propagation constant. The above approximation is valid for $\ell < \lambda_g^L/8$.

- *Short-circuited stub as an inductor:* Figure (19.19e) shows a short-circuited short length stub. It is used as a shunt inductor L_p. For the short-circuited line,

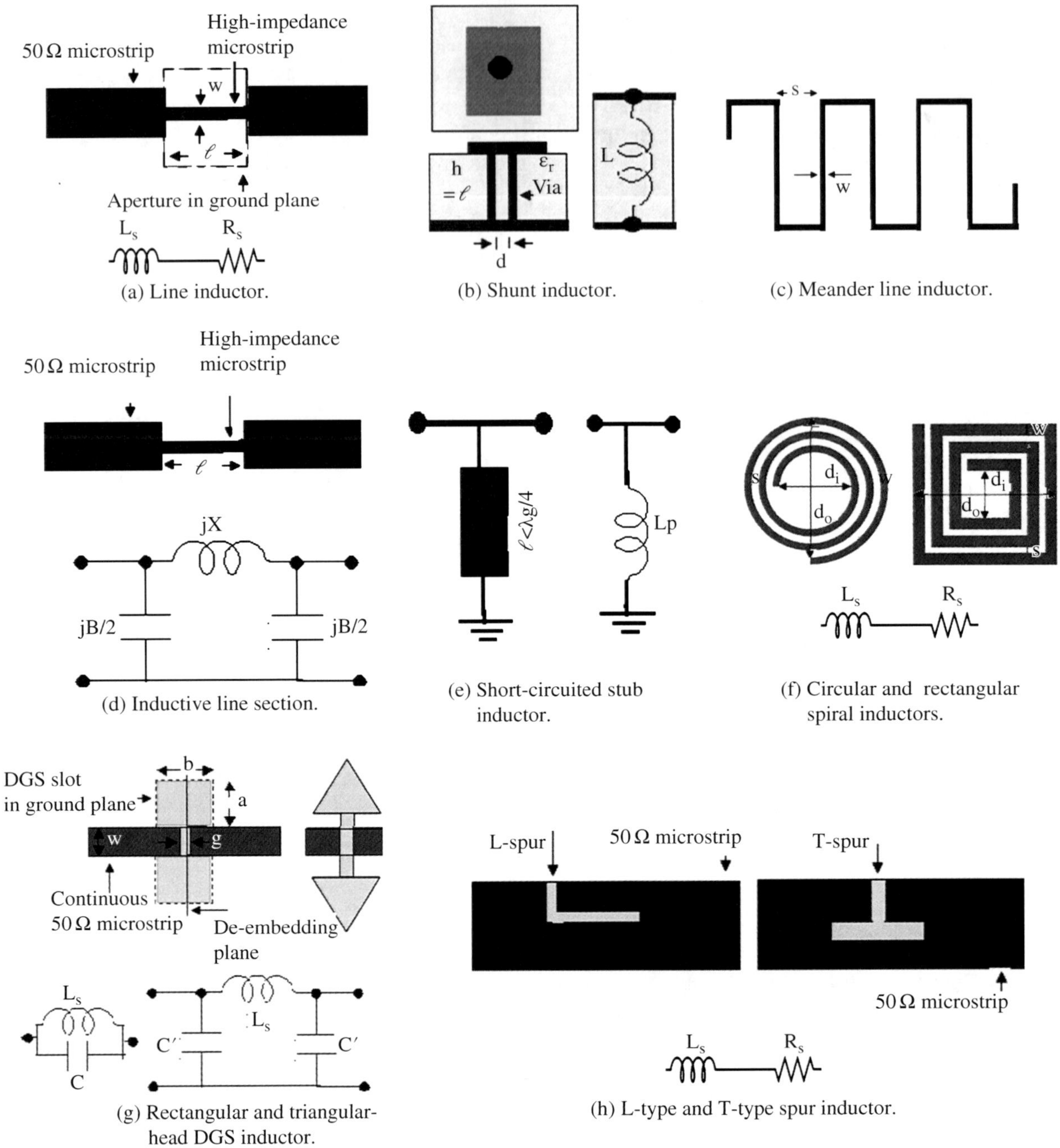

Figure 19.19 Some inductors in microstrip.

with length $\ell < \lambda_g^L/4$, the input impedance is inductive. It is used to compute the L_p:

$$Z_{in} = jZ_L \tan\left(\frac{2\pi}{\lambda_g^L}\ell\right) \approx jZ_L \frac{2\pi}{\lambda_g^L}\ell \approx j\,\omega L_p$$

$$\Rightarrow L_p = \frac{Z_L}{v_p}\ell, \quad \text{where } v_p = \frac{c}{\sqrt{\varepsilon_{reff}}}. \qquad (19.3.45)$$

The approximate expression (19.3.45) is valid for $\ell < \lambda_g^L/8$. The v_p is the phase velocity of the inductive stub.

- *Loop (planar spiral) inductor:* A line in the form of a loop (spiral) offers a higher value of inductance. Normally, it is either in circular or in square form, as shown in Fig. (19.19f). For a circular loop, forming a circular spiral inductor, the inductance, and associated resistance are estimated from the following expression [B.19]:

$$\begin{aligned} L_s &= 0.03937\left[a^2n^2/(8a+11c)\right]\times K_g,\ \ a \text{ in } \mu\text{m} && \text{(a)}\\ a &= (d_o+d_i)/4,\ \ c=(d_o-d_i)/2 && \text{(b)}\\ R_s &= 1.5\,(\pi\, a\, n\, R_{sur}/w) && \text{(c)}. \end{aligned} \qquad (19.3.46)$$

In expressions (19.3.46), d_o and d_i are the outer and inner diameters of a circular loop and n is the number of turns. The factor K_g accounts for the presence of the ground plane, as given in equation (19.3.42). Again, an aperture in the ground plane, behind the spiral inductor, enhances its inductance.

Mohan et al. [J.21, B.17] have given the following modified expression of Wheeler [J.22] to compute the inductance of a square loop:

$$\begin{aligned} L &= 9.375\mu_0\frac{n^2d_{av}^2}{11\,d_o-7\,d_{av}},\ \ d_{av}=\frac{d_o+d_i}{2} && \text{(Wheeler)}\quad \text{(a)}\\ L &= 2.34\mu_0\frac{n^2d_{av}}{1+2.75p}\quad p=\frac{d_o-d_i}{d_o+d_i}, && \text{(Mohan)}\quad \text{(b)}, \end{aligned} \qquad (19.3.47)$$

where p is the filling-factor and d_o and d_i are the outer and inner dimensions of n turns loop. Mohan et al. [J.21] have also provided the following more accurate (8% accuracy) expression to compute the inductance of both the circular and square planar spirals:

$$\begin{aligned} L &= \frac{\mu_0 n^2 d_{av}^2}{2}c_1\left[\ln\left(\frac{c_2}{p}\right)+c_3p+c_4p^2\right]\\ &\text{For square: } c_1,c_2,c_3,c_4 = 1.27, 2.07, 0.18, 0.13\\ &\text{For circle: } c_1,c_2,c_3,c_4 = 1.00, 2.46, 0.00, 0.20. \end{aligned} \qquad (19.3.48)$$

- *DGS inductor:* The defected ground structure (DGS) in the ground plane of a microstrip, behind the strip conductor also acts as an inductor. It is shown in Fig. (19.19g). The DGS loads the host microstrip with a modeled series-connected LC parallel resonant circuit. At the frequency below the resonance frequency, i.e. below the pole frequency f_0; the DGS offers a series inductance L_s to the line. The DGS has been widely used to design improved performance filters and other components. It has also been used in the PBG structures [J.23]. However, the DGS requires a reasonable gap between its patterned ground plane and the shielding enclosure. Moreover, it may suffer from back radiation reducing the Q-factor of the DGS inductor.

Figure (19.19g) further shows an elaborate π-network model of the DGS. The capacitances in the shunt arms model the discontinuity created by the DGS [J.24]. The DGS can acquire several simple and complex geometrical shapes and offer different values of inductance. The complex DGS could be associated with multiple resonances [J.25]. Simple geometrical shapes, such as square, circle, and triangle, are popular. For an equal area, they offer the same inductance; however, their responses have some variations. The elements L_s and C of a simple parallel resonant circuit model of the DGS are evaluated by simulating the structure on an EM-simulator to obtain the 3dB cut-off frequency f_c and the pole frequency f_0. Once f_c and f_0 are known, values of DGS components are computed using the following expressions [J.26]:

$$\begin{aligned} C &= \frac{5\,f_0}{\pi\left(f_0^2-f_c^2\right)}\ \text{pF} && \text{(a)}\\ L &= \frac{250}{C\,(\pi\, f_0)^2}\ \text{nH} && \text{(b)}. \end{aligned} \qquad (19.3.49)$$

Likewise, one can also evaluate L_s and C' of the π-equivalent circuit at low frequency,$f < f_c$, from the de-embedded S-parameters of the DGS, using an EM-simulator. The mid of the DGS neck is taken as the de-embedding plane. Next, the S-parameters are converted to the Y-parameters and L_s and C' at f_c are evaluated as follows:

$$\begin{aligned} Y_1 &= \frac{2\,S_{21}}{(1+S_{11})^2-S_{21}^2} && \text{(a)}\\ Y_2 &= \frac{(1-S_{11})^2+S_{21}^2-2\,S_{21}}{(1+S_{11})^2-S_{21}^2+2\,S_{21}} && \text{(b)}\\ \Rightarrow L_s &= \frac{1}{2\pi f_c}\frac{1}{\text{Im}\,[Y_1]} && \text{(c)}\\ C' &= \frac{\text{Im}\,[Y_2]}{2\pi f_c} && \text{(d)}. \end{aligned} \qquad (19.3.50)$$

However, if the loading inductance L_s is known, the twin slot area of a simple DGS head is computed using the following closed-form expression:

$$S\,(\varepsilon_r, h) = M(\varepsilon_r, h)\,L_s + N(\varepsilon_r, h), \qquad (19.3.51)$$

where the substrate-dependent parameters M and N are determined from the available closed-form expressions [J.24]. The expression (19.3.51) is valid for the 50Ω microstrip line and useful for the slot-head area between 5 mm^2 and 100 mm^2. The substrate parameters are in the range $2.2 \le \varepsilon_r \le 12.9$ and $0.6 \le h \le 2$ mm. For the microstrip with other characteristic impedance, the modified slot-head area can also be computed.

- *Spur inductor*: Figure (19.19h) shows the spur resonators embedded in the conducting strip of a microstrip.

It is a *defected microstrip structure* (DMS), again showing a bandstop behavior. It is modeled as the series-connected parallel RLC resonant circuit. At $f < f_0$, it offers series inductance. The circuit elements are estimated using the EM-simulated results at the pole frequency f_0 and $|S_{21}|$ [J.27]:

$$R = 2Z_0\left[\frac{1}{|S_{21}|} - 1\right] \quad \text{(a)}$$

$$C = \frac{\sqrt{0.5\,(R + 2Z_0)^2 - 4Z_0^2}}{2.8377\pi\, Z_0\, R\, \Delta f} \quad \text{(b)}$$

$$L_s = \frac{1}{4(\pi\, f_0)^2 C} \quad \text{(c)}, \qquad (19.3.52)$$

where Δf is the 3 dB bandwidth of the $|S_{21}|$ – response and Z_0 is the characteristic impedance of the host microstrip line. Expressions (19.3.52a–c) are also applicable to the DGS.

Realization of Resonant Circuits in Microstrip

There are several ways to realize the series and parallel resonant circuits, connected in series/shunt of a host microstrip line. The open/short-circuited stub itself forms the resonant circuit. However, using a combination of DGS, metallic patch, shorting pin (via), etc., more compact and versatile resonating structures could be designed to form the unit cell of a periodic microstrip line. These structures are also used with the CPW and other planar periodic lines. Some of these structures are also useful in suppressing the power line switching noise of the parallel-plate waveguide (PPW) structure, arising in the high-speed PCB [J.8, J.9]. These structures also find application to realize the *high impedance surface*, i.e. the *artificial magnetic conductor (AMC) surface* by suppressing the surface wave supported by a conductor backed dielectric sheet. The high-impedance surface is discussed in chapter 20. Such artificial surfaces are used to develop compact antennas with improved characteristics [J.3]. Figure (19.20a–f) show some resonating structures used with both the 1D and 2D periodic lines and surfaces.

- *Band-accept type DGS:* Figure (19.20a) shows a band-accept type DGS in the microstrip. Figure (19.20a) also shows the equivalent circuit. It is a series combination of the L_pC_p parallel resonant circuit, due to the DGS, and a microstrip gap capacitor C_s. The DGS is placed at a gap g in the microstrip. At $f < f_p$, i.e. at a frequency below the pole frequency, the parallel resonant circuit acts as an inductor L_p. The DGS inductor and gap capacitor provide the series resonance at $f_s = 1/\left(2\pi\sqrt{C_sL_p}\right)$, acting as a single element bandpass filter [J.28]. The circuit could be embedded in a microstrip to form the unit cell of a periodic microstrip line. The gap capacitance can be replaced by the interdigital capacitor, associated with DGS, to get a lower value of the series resonance frequency. The simple DGS can be also replaced by more complex DGS structures to realize versatile microstrip EBG. The circuit arrangement can also be placed in a shunt-connected stub to the main microstrip line.
- *Step-impedance resonator (SIR):* Figure (19.20b) shows that the SIR circuit is also used as the shunt-connected resonating loading element to a host microstrip. The SIR is discussed in subsection (18.1.4) of chapter 18. Its pole frequency is $f_p = 1/\left(2\pi\sqrt{CL}\right)$. The L and C are realized using small sections of high ($100\Omega - 130\Omega$) and low impedance ($10\Omega - 15\Omega$) microstrip sections. This structure is also used to get the microstrip elliptic lowpass filter [B.19].
- *Mushroom type resonators:* Figure (19.20c–e) shows three different structures of a mushroom type resonator embedded in the host microstrip medium, parallel plate waveguide medium, and at the surface of a conductor-backed dielectric sheet to form the unit cells. The *mushroom resonator-I* is used in designing the filters and diplexers [J.29]. It is a two-layered structure. The patch of size $(a \times b)$ is placed at the top of the first layer of the substrate with (ε_{r1}, h_1). The shorting pin of length h_1 connects the patch to the ground conductor. The microstrip line is placed at the top of the second layer of the substrate with (ε_{r2}, h_2). The microstrip and the patch forms a capacitance C and the shorting pin provides the inductance L, so the host microstrip is loaded with the shunt-connected series resonant circuit. The location of the shorting pin has some influence on the resonance frequency [J.29]. The values of the elements of an equivalent circuit can be extracted through the S- and Z-parameters, obtained on an EM-simulator. However, the circuit elements can also be approximately modeled [J.9].

Figure (19.20d) shows the *mushroom resonator-II*. It is placed inside a parallel-plate waveguide (PPW). This structure forms a unit cell, used in the 2D periodic structure, to suppress the *simultaneous switching noise (SSN)* in the modern high-speed PCB [J.9]. The characteristic impedance and propagation constant of the host PPW medium of height h are given by the following expressions:

$$Z_0 = \frac{\eta_0}{\sqrt{\varepsilon_{reff}}}\frac{h}{d} \quad \text{(a)}, \quad k = k_0\sqrt{\varepsilon_{reff}} \quad \text{(b)}$$

$$\text{where, } \varepsilon_{reff} = \frac{h_1 + h_2}{h_1/\varepsilon_{r1} + h_2/\varepsilon_{r2}} \quad \text{(c)}. \tag{19.3.53}$$

The square patch (a × a), on the top of the first layer of the substrate with (ε_{r1}, h_1), forms a capacitor with the upper plate of PPW and the shorting pin provides a series-connected inductance L. The L and C combination forms the shunt-connected series LC resonant circuit, shown in the equivalent circuit of Fig. (19.20d). The C and L [J.13] are computed as follows:

$$C = \frac{\varepsilon_0\varepsilon_{r2}a^2}{h_2}, \quad \text{or more accurately,}$$

$$C = \frac{\varepsilon_0\varepsilon_{r2}(a^2 - \pi r^2)}{h_2} \quad \text{(a)}$$

$$L = \frac{\mu_0 h_1}{4\pi}\left[\ln\left(\frac{1}{\alpha}\right) + \alpha - 1\right] \quad \text{(b)}$$

$$\text{where, } \alpha = \frac{\text{via cross-section}}{\text{cross-section of unit-cell}} = \frac{\pi r^2}{d^2} \quad \text{(c)}. \tag{19.3.54}$$

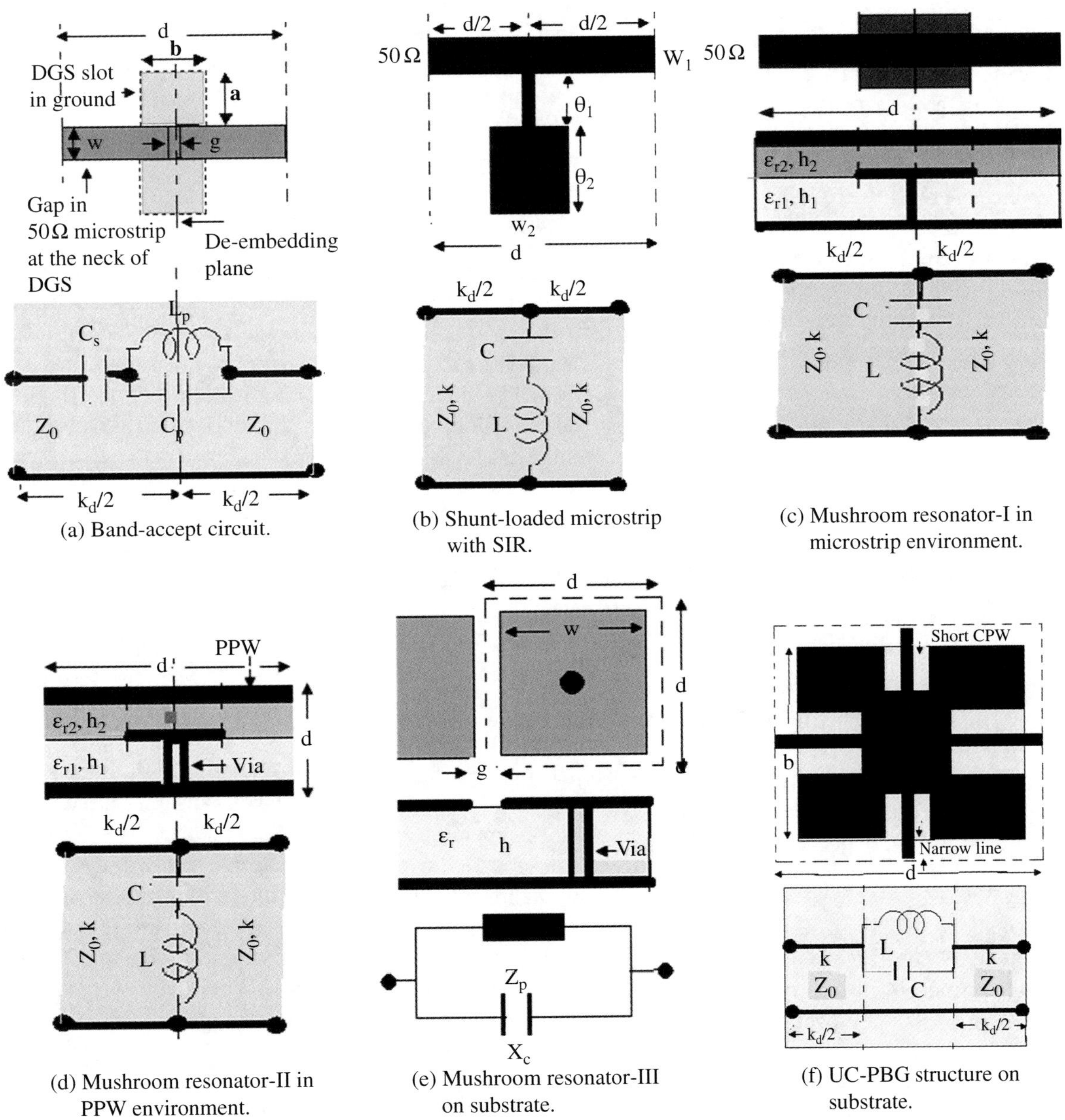

Figure 19.20 Some resonant type loading elements.

In equation (19.3.54), r is the radius of the hollow via pin. In the case of a solid pin r = 0. For the free space, $\mu_0 = 4\pi \times 10^{-7}$ H/m. The PPW line is loaded with the susceptance Y:

$$Y = \frac{j\,\omega C}{1-\omega^2 LC} \quad (a), \quad \text{with } \omega_r = \frac{1}{\sqrt{LC}} \quad (b). \tag{19.3.55}$$

At $f < f_r$, the PPW line is capacitively loaded Fig. (19.20e) shows the single-layer surface mounted mushroom *resonator-III*. It is a popular structure for suppressing the surface -wave on the 2D conductor backed dielectric sheet [J.8, J.30, J.31]. The $(w \times w)$ square patch located at the surface of the dielectric sheet is connected to the ground plane, using a centrally located shorting pin. The pin has inductive reactance X_L. Further, the open-circuited line of width w and length w/2 in the dielectric medium ε_r offers a reactance Z_m, parallel to X_L, $Z_l = X_L Z_m/(X_L + Z_m)$, at the center of the patch. It is transformed as a load Z_p at the edge of the patch. There is an adjacent patch also giving the coupling capacitance C_c and capacitive reactance X_c. A parallel combination of X_c and Z_p forms the parallel resonant circuit distributed over the dielectric surface to suppress the surface wave propagation. The structure is used to form the unit cell of a period d. The X_L and Z_m are computed as follows:

$$\begin{aligned} X_L &= \omega L, \text{where, } L = 2\times 10^{-7} h \times \\ &\quad [\ln(4h/2r) + 0.5\ln(2r/h) - 0.75], && (a)\\ Z_m &= -j\,Z_0 \cot(kw/2) && (b)\\ Z_p &= Z_0 \frac{Z_l + jZ_0 \tan(kw/2)}{Z_0 + jZ_l \tan(kw/2)} && (c), \end{aligned} \tag{19.3.56}$$

where Z_0 and k are the characteristic impedance and propagation constant of the line formed by patch and ground conductor. The line could be treated as a microstrip for $w/h \geq 2$. For the patch $w/h < 2$ and $\varepsilon_r \leq 4$, in presence of the adjacent coupled patches, the line can also be treated as a CPW. The coupling capacitance C_c is obtained from the following expression [J.12, J.31]:

$$C_c = \frac{w\varepsilon_0(\varepsilon_r + 1)}{\pi} \cosh^{-1}\left(\frac{d}{g}\right). \tag{19.3.57}$$

- *Uniplanar compact PBG:* Figure (19.20f) shows another unit cell formed by the uniplanar patch. It is known as the uniplanar compact PBG *(UC-PBG)* structure [J.8, J.32, J.33]. It provides the *patterned ground plane,* useful for the development of the microwave circuits. It also offers parallel resonant circuits distributed over the conductor backed dielectric surface to suppress the surface wave propagation. It further acts as an AMC, useful for the design of the compact high-gain antenna, and the TEM waveguide [J.34, J.35]. It is further discussed in chapter 20.

19.4 1D Planar EBG Structures

The 1D EBG structure is formed by cascading of an infinite number of unit cells. Its $(\omega - \beta)$ or $(kd - \beta d)$ dispersion diagram is obtained with the help of the *eigenmode solver* of an EM-simulator. However, the approximate circuit models are also used for this purpose. In practice, only a finite number of the unit cells form a periodic line. Therefore, there is a difference in bandwidths obtained from the $(\omega - \beta)$ diagram and the S-parameters response. Moreover, from the S-parameters, the Bloch impedance can also be extracted. The stopband rejection is enhanced by increasing the number of unit cells.

19.4.1 1D Microstrip EBG Line

The microstrip line, as a *host* medium, is a composite medium consisting of three uniform transmission media – the strip conductor, dielectric substrate, and ground plane conductor. To get a periodic microstrip line, the *inclusions*, in the form of periodic perturbations, or periodic loading, are added in any one, two, or even in all three uniform segments of a microstrip. The response obtained is significantly influenced by the nature of inclusions, i.e. the loading elements. However, periodicity has a dominant role to play. Normally, only one periodicity is used. However, more than one periodicity, and also cascading of two different periodic microstrips, is used to get wider stopband bandwidth for some applications.

Periodic Loading of Substrate

The host microstrip substrate is periodically loaded with high or low permittivity (even with air holes) inclusions, or the metallic inclusions. For instance, the substrate can be drilled with square or circular holes and arranged in the form of 1D linear periodic array, or the square, triangular, and hexagonal 2D lattice. In the process, we create an *artificial material* with the *photonic crystal*, i.e. the *electromagnetic crystal*. The inclusions can also be magnetodielectric, ferromagnetic, ferroelectric, amplifying, and rectifying elements, etc., to get a dynamically controlled medium.

Figure (19.21a–d) shows examples of periodic loading of the substrate of a microstrip. Figure (19.21a) shows a

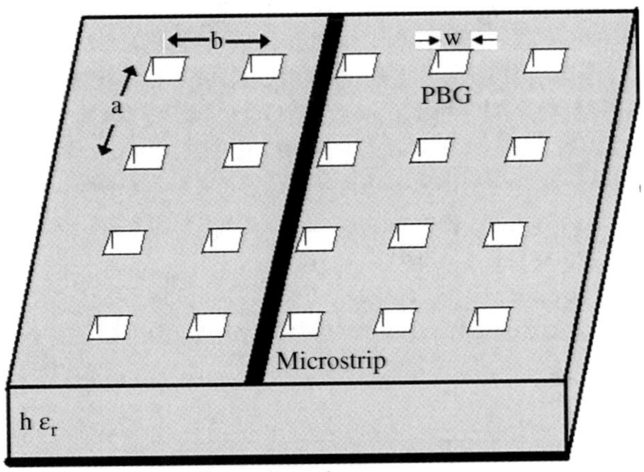

(a) Periodic holes inside the substrate of a microstrip line.

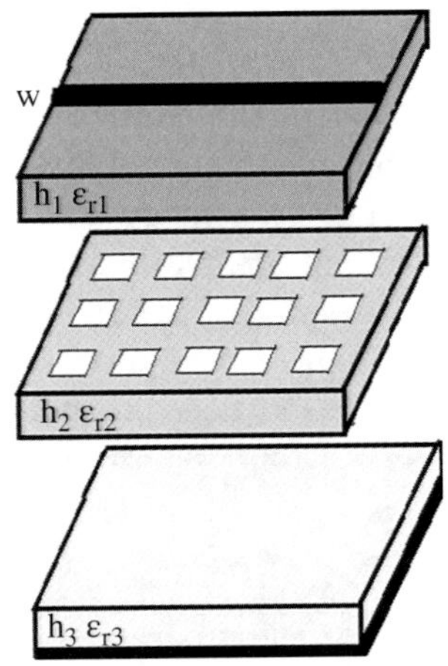

(b) Three-layered microstrip substrate with middle substrate perforation.

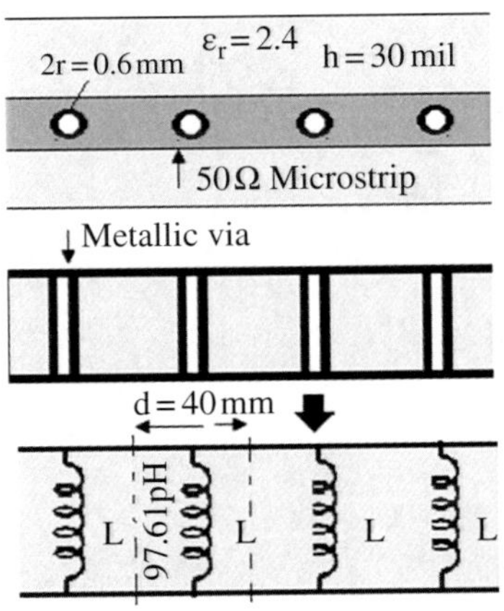

(c) Periodic loading of the substrate of microstrip with metallic vias.

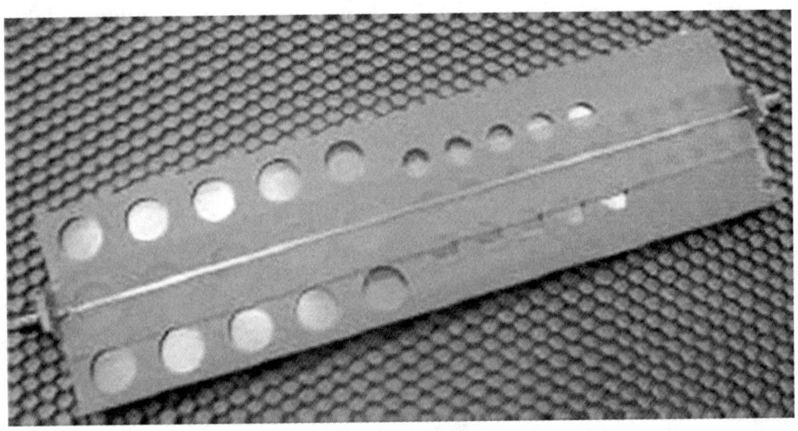

(d) Two-layered EBG with cascaded three periodicities EBG sections [J.36].

Figure 19.21 The periodic loading of the substrate of a microstrip. *Source*: Ian Rumsey et al. [J.36]. © 1998 IEEE.

substrate with embedded rectangular-shaped air-holes [J.37]. The microstrip is placed above one row of the perforation, as the electric field of microstrip is strongest just below the strip conductor. The air-holes, outside of the microstrip width, have an insignificant influence on the quasi-TEM type Bloch wave propagation. The 2D periodic planar surface stops significantly the surface wave propagation. It forms an *AMC* that has several applications, including gain improvement of a compact antenna [J.38, J.39]. However, the embedding of inclusions inside a substrate is a difficult process.

Figure (19.21b) shows a three-layered substrate with the middle-layer periodically perforated [J.37]. The first layer supports the microstrip and the third one supports the ground conductor. These layers can be of different materials. For instance, the first and third layers are alumina, Roger TMM10i substrate ($\varepsilon_{r1} = \varepsilon_{r3} = 9.8$, $h_1 = h_3 = 0.635$mm), and the second layer is a polymer-based GML-1000 substrate ($\varepsilon_{r2} = 3.2$, $h_2 = 0.762$mm). The three-layered medium can be replaced by an equivalent host medium with relative permittivity ε_{rh} and thickness $h = h_1 + h_2 + h_3$. The microstrip (w = 6.5 mm) is placed over two rows of the rectangular air-holes, size 3.0 mm × 6.5 mm, with five air-holes in each row. The period, i.e. the lattice constant of the 2D slots, is $d_x = d_y = 14$ mm. The total length of the microstrip is L = 98 mm. Figure (19.22a) shows the normalized propagation constant β/k_0 and Bloch impedance. The structure has a stopband in the approximate of range 5–6 GHz. The presence of the slow-wave and leaky fast-wave zones in two passbands are also shown. The S-parameters response, shown in Fig. (19.22b), collaborate the stopband range.

Figure (19.21c) shows a periodic line with four numbers of metallic vias (2r = 0.6 mm), embedded inside a substrate with $\varepsilon_r = 2.4, \tan\delta = 0.0015$, h = 30 mil [J.40]. The 50Ω microstrip with width w = 2.286 mm, giving $\varepsilon_{reff} = 2.04$, i. e. $v_p = 2.1 \times 10^8$ m/s is located on the substrate. Each via is modeled as a shunt-connected inductor of value 97.61 pH. The unit cell contains a shunt inductor as an inclusion, at the mid of the period d = 40 mm. The $(kd - \beta d)$ diagram, shown in Fig. (19.22c), is obtained with the help of equation (19.3.15). The four numbers of passbands appear, supporting four propagating Bloch modes of increasing bandwidth, separated by the four stopbands. Due to the shunt inductance, the first band is the stopband followed by the passband. The presence of the first two stopbands and two passbands are further collaborated by the S_{21} and S_{11} responses shown in Fig. (19.22d and e), respectively. The passband occurs at 2.6 GHz and also at 5.2 GHz. These correspond to $kd \approx 3.1$ and 6.2, respectively, obtained from the $(k - \beta)$ diagram of the infinitely long periodic microstrip.

Figure (19.21d) shows the two-layered EBG microstrip [J.36]. The first dielectric layer, $\varepsilon_{r1} = 10$, $h_1 = 62$ mil, accommodates three rows of the circular holes of three sizes and three periodicities. Three sections of the EBG

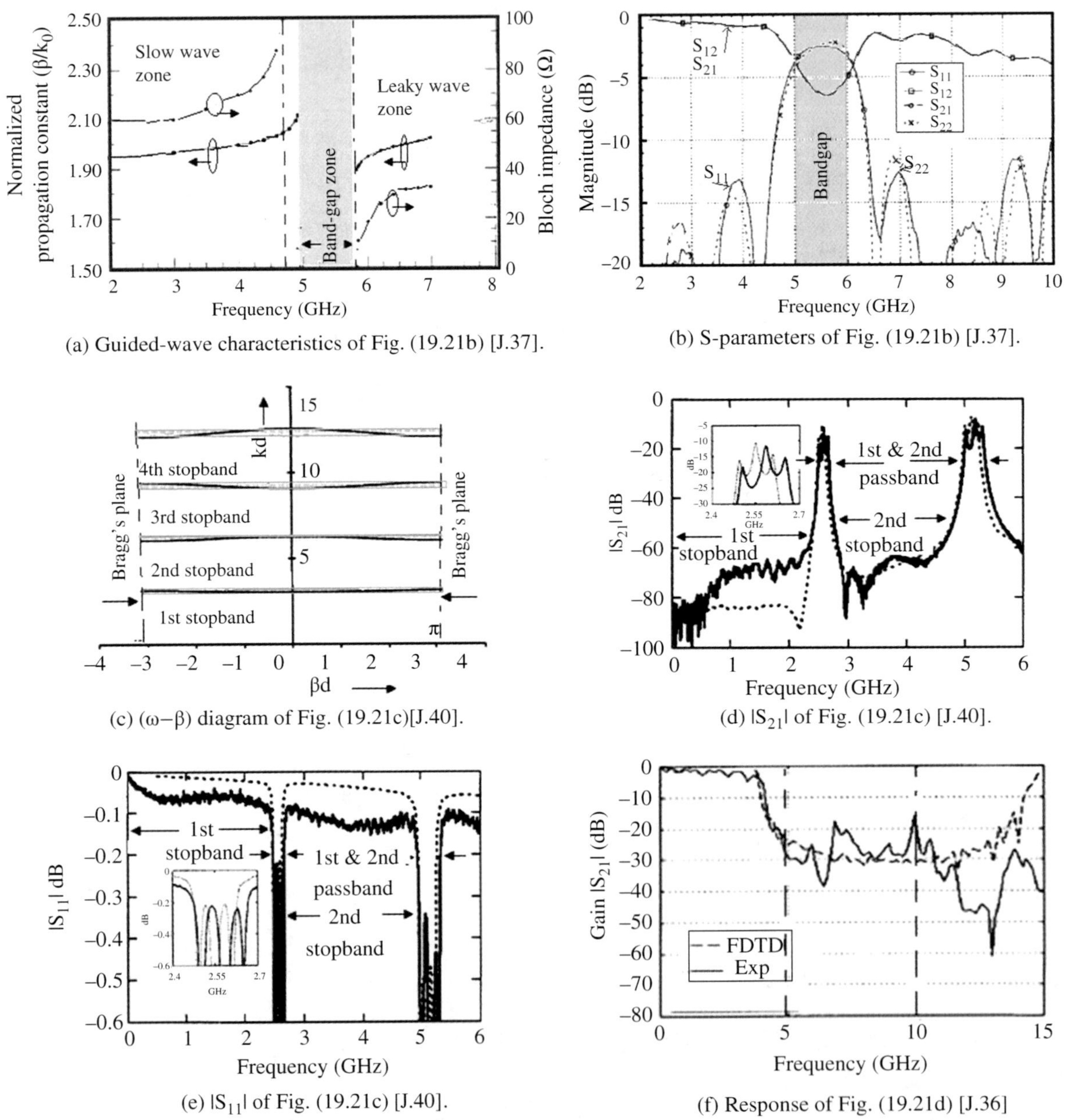

(a) Guided-wave characteristics of Fig. (19.21b) [J.37].

(b) S-parameters of Fig. (19.21b) [J.37].

(c) (ω−β) diagram of Fig. (19.21c)[J.40].

(d) $|S_{21}|$ of Fig. (19.21c) [J.40].

(e) $|S_{11}|$ of Fig. (19.21c) [J.40].

(f) Response of Fig. (19.21d) [J.36]

Figure 19.22 Response of microstrip on a periodic substrate (artificial substrate). *Source*: From Yang [J.37]. © 1999, IEEE. From Laso et al. [J.40]. © 2000, John Wiley & Sons. From Ian Rumsey et al. [J.36]. © 1998, IEEE.

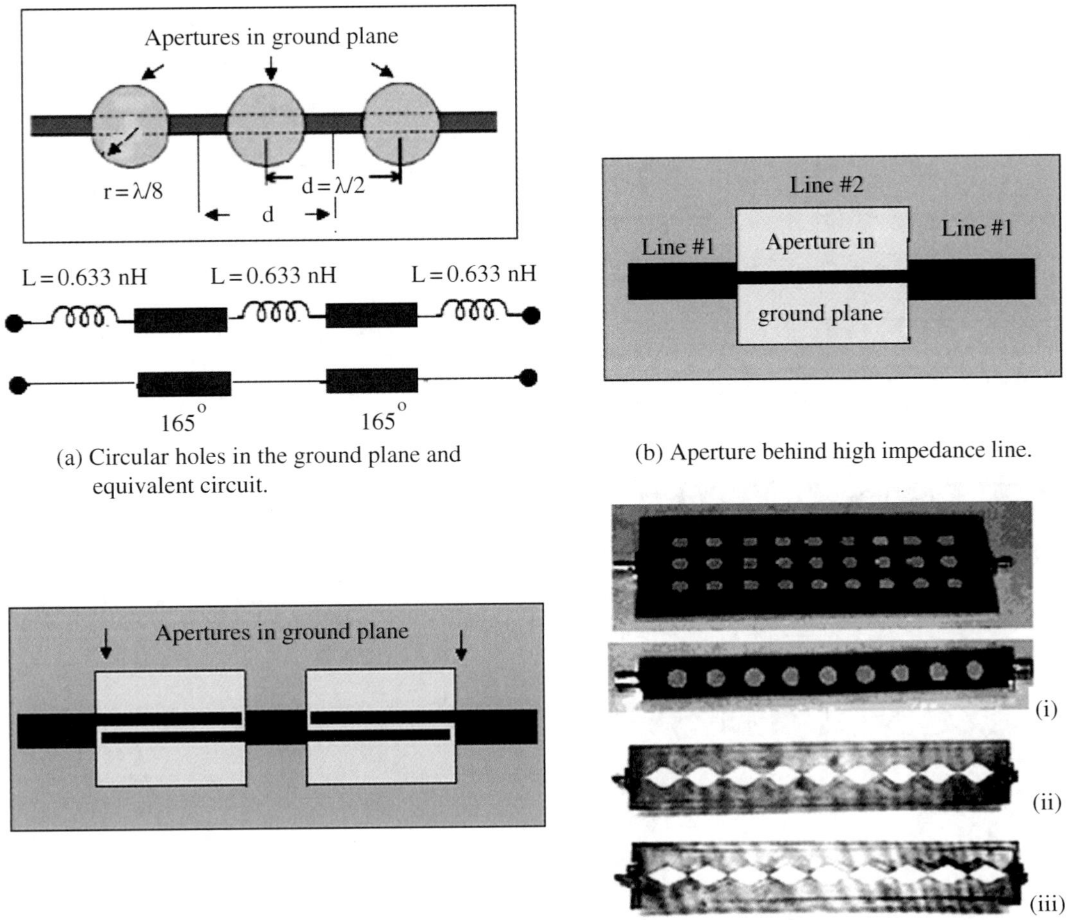

(a) Circular holes in the ground plane and equivalent circuit.

(b) Aperture behind high impedance line.

(c) Aperture backed coupled microstrip.

(d) Discrete and continuously modulated slots [J.41, J.42].

Figure 19.23 The holes/apertures in the ground plane of the microstrip. *Source*: Falcone et al. [J.41] and Lopetegi et al. [J.42], © 2000, Microwave & Optical Tech. Letters

microstrip are cascaded to form one periodic structure. Each EBG section has five circular holes. The first, second, and third sections have period d and hole diameter 2r as (18 mm,12 mm); (12 mm, 8 mm); and (7.4 mm, 5 mm), respectively. These sections are designed at mid-stopband frequency 5 GHz, 7.5 GHz, 12.2 GHz, respectively. The microstrip is fabricated on the second layer of G10 ($\varepsilon_r = 4.8, \tan\delta = 0.017$) thin substrate $h_2 = 5$ mil and placed over the middle row of the EBG. Figure (19.22f) demonstrates the appearance of a very wide 20 dB rejection BW in the $|S_{21}|$ response, from 4.5 GHz to 15 GHz and above. The multiple periodicities cascaded EBG is useful to obtain the wide rejection BW.

Periodic Loading of Ground Conductor

The periodic etching of holes and other patterns in the ground plane of microstrip is a more practical process for obtaining the periodic microstrip line. The periodic loading of the ground planes with defects, i.e. the inclusions, can be put under three broad groups: (i) *Holes and apertures in the ground conductor*; (ii) *Defected ground structures, i.e. the DGS resonators*; and (iii) *Patterned ground conductor.*

i) *Holes and apertures in the ground conductor*

Figure (19.23a) shows the three unit cells periodic microstrip line, using circular apertures in the ground plane [J.43]. The circular aperture of radius r, ($0 < r < d/2$), is placed at the mid of the unit cell of period d. The aperture provides a series inductor loading to the host microstrip. The value of the inductance L is governed by the *filling-factor* F = r/d. Normally, the maximum value of F is about 0.45, and its average value is 0.25. One row of ground plane slots, just behind the microstrip conductor, is adequate to create an effective periodic microstrip line [J.41]. At both sides of the equivalent inductor, the line sections are $\theta/2 = kd/2$ long. The stopband center frequency f_0 is governed by Bragg's resonance condition, $\beta d = \pi$, i.e. $\Rightarrow f_0 = c/2d\sqrt{\varepsilon_{reff}}$, where

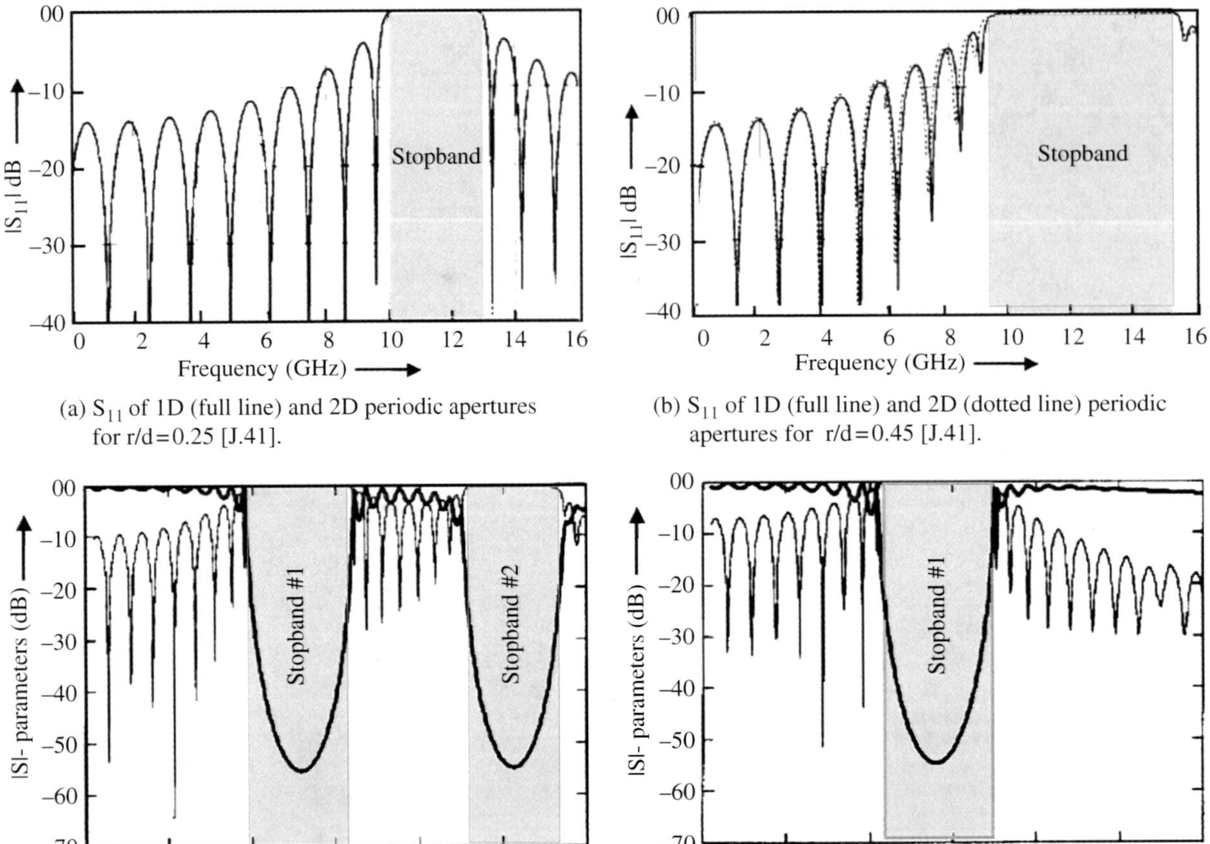

(a) S_{11} of 1D (full line) and 2D periodic apertures for r/d=0.25 [J.41].

(b) S_{11} of 1D (full line) and 2D (dotted line) periodic apertures for r/d=0.45 [J.41].

(c) S_{11} and S_{12} of 1D discrete circular EBG of Fig. (19.23a). d=23.9mm and r/d=0.25 [J.42].

(d) S_{11} and S_{12} of 1D sinusoidal EBG of Fig. (19.23d) d=23.9mm and t/d=0.25 [J.42].

Figure 19.24 Response of the ground plane holes/apertures in the microstrip. *Source*: Falcone et al. [J.41]. © 1999 John Wiley & Sons. Reproduced from Lopetegi et al. [J.42]. © 2000, John Wiley & Sons.

ε_{reff} is the effective relative permittivity of the host microstrip line. The central frequencies of higher-order stopbands are approximately harmonically related. The larger value of the period d results in lower stopband frequency. The width and depth of the stopband are controlled by the number of the unit cells and also by the value of the inductance L.

The circuit model requires the calculation of the value of the inductance L in two steps [J.43]. The unit cell, with a circular aperture in the ground conductor, is simulated on an EM-simulator, to get the $|S_{21}|$. Next, the equivalent circuit of the unit cell, with inductor L, is simulated on a circuit simulator. The inductor L is varied to get the same value of $|S_{21}|$ at the same frequency that was obtained using the EM-simulator. At the matched value of $|S_{21}|$, the value of L is obtained. The known value of L is used to create any number of equivalent unit cells. One unit cell is used to get the $(kd - \beta d)$ dispersion diagram, and the Bloch impedance, as discussed in section (19.3.2). The dispersion diagram can also be obtained using the eigenmode solver of an EM-simulator.

In practice, the stopband characteristic of a finite length periodic structure is obtained from the S-parameter responses on the EM-simulator and validated experimentally. The periodic line, using its equivalent circuit, can also be simulated on the Circuit Simulator. Figure (19.24a and b) shows the appearance of the stopband, showing high reflection, i.e. $|S_{11}| \approx 0$ dB. The 50Ω periodic microstrip, width w = 0.6 mm, period d = 5 mm, is considered on a substrate with $\varepsilon_r = 10.2$, h = 25 mil. Its stopband frequency is $f_0 =$ 11 GHz. Figure (19.24a and b) shows the stopband for filling-factor F = 0.25 and F = 0.45, respectively. A widening of the stopband is seen for the higher filling-factor.

In a case of the identical value of F, say 0.25, a square slot in the ground plane provides about 8 dB higher attenuation in the stopband. For the same linear dimension (2r), a square hole has a larger area ($4r^2$) compared to that of the area (πr^2) of a circular hole, therefore, the inductance of a square hole is more than that of a circular hole. Thus, the EBG based on the square hole provides wider stopband as compared to the EBG of a circular hole of the same linear dimension. The square lattice has also better transmission and reflection as compared to that of the circular lattice [J.44].

The series inductor loading is also obtained using a narrow width microstrip section backed by an aperture on the ground plane. It is shown in Fig. (19.23b). The π-network model of the loading element is extracted using the EM-simulator [J.45]. For a larger size aperture, the extracted inductance value is high. Likewise, Fig. (19.23c) shows a section of coupled microstrip backed by an aperture. It provides a stronger capacitive coupling. In this case, the bandpass filter response is obtained [J.46]. The structure is used as a series capacitor loading element to form the loaded microstrip line.

In place of discrete circular or square slots in a ground plane, the continuously modulated aperture is also used to get an interesting single stopband. Figure (19.23d) shows three types of apertures in the ground plane of microstrip of width w = 1.2 mm on Rogers R1030™ substrate ($\varepsilon_r = 10.2$, h = 1.27mm). The structures have nine cells of period d = 23.9 mm, giving the first stopband centered at 2.5 GHz. The first structure-(i) is for the *perforated ground plane* with discrete circular slots. The second structure-(ii) is for the *sinusoidal modulated continuous* aperture. The third structure-(iii) is for a triangular modulated aperture [J.42]. The first structure-(i) of Fig. (19.23d) also shows three rows of slots [J.41]. Figure (19.24c and d) shows the |S|-parameters, i.e. $|S_{12}|$ and $|S_{11}|$ response of the discrete and *sinusoidal* modulated apertures, respectively. It is interesting to note that the second stopband, around $2f_0$ available for the discrete periodic line, is suppressed for the continuously modulated sinusoidal slot. This structure is useful in the filter design. In the case of a triangular modulated slot, suppression of $2f_0$ stopband is also high, but not complete. There are ripples at the edge of the passband due to the increased value of the Bloch impedance near the band-edge. The ripples are controlled by using the nonuniform tapered slots [J.47, J.48].

In place of the regular discrete or continuous slots in the ground plane, the *fractal* discrete or continuous slots, discussed in section (18.7) of chapter 18, can also be used. The continuous Minkowski curve is useful to get the multifrequency filters [J.49]. Sierpinski carpet slots are used in a two-layered periodic structure, with two periodicities, to obtain a low-pass filter of wide rejection bandwidth.

ii) *Defected ground structure (DGS) as a loading element*

The DGS elements, discussed in section (19.3.6), are also used to load inductively the host microstrip line. Figure (19.25a) shows the DGS loaded meandered microstrip line [J.50]. The meandering of the microstrip line provides a compact periodic structure. The 50Ω microstrip (width w = 1.55 mm) is taken on an alumina substrate with $\varepsilon_r = 10$, h = 1.575 mm. The five numbers of DGS are arranged in the 2D-plane. Figure (19.25b) compares the S-parameters of the DGS-loaded meandered microstrip and DGS loaded straight microstrip lines. The meandered EBG structure has wider stopband, from 4 GHz to 7.18 GHz, at the attenuation level 20 dB, whereas the straight microstrip EBG shows only 1.83 GHz wide stopband. However, the sharpness of cut-off is less for the meandered case, 46.6 dB/GHz, whereas for the straight-line structure, it is 85.11 dB/GHz.

The propagation constant and Bloch impedance of a periodic meandered microstrip line are obtained by converting the de-embedded S-parameters of the structure to its [ABCD] parameters [B.13]:

$$\begin{bmatrix} A & B \\ C & D \end{bmatrix} = \begin{bmatrix} [(1+S_{11})(1-S_{22})+S_{12}S_{21}]/2S_{21} & Z_0\,[(1+S_{11})(1+S_{22})-S_{12}S_{21}]/2S_{21} \\ (1/Z_0)\,[(1-S_{11})(1-S_{22})-S_{12}S_{21}]/2S_{21} & [(1-S_{11})(1+S_{22})+S_{12}S_{21}]/2S_{21} \end{bmatrix}. \quad (19.4.1)$$

The de-embedding of S-parameters is discussed in subsection (3.2.2) of chapter 3. The [ABCD] parameters of the periodic line of length ℓ are written as follows:

$$\begin{bmatrix} A & B \\ C & D \end{bmatrix} = \begin{bmatrix} \cosh(\gamma\ell) & Z_B\sinh(\gamma\ell) \\ \sinh(\gamma\ell)/Z_B & \cosh(\gamma\ell) \end{bmatrix}. \quad (19.4.2)$$

Using equations (19.4.1) and (19.4.2), the complex propagation constant $\gamma = \alpha + j\beta$, and Bloch impedance $Z_B(\omega)$ are computed:

$$\cosh(\gamma\ell) = [(1+S_{11})(1-S_{22})+S_{12}S_{21}]/2S_{21} \quad (a)$$

$$Z_B(\omega) = \sqrt{\frac{B}{C}} = Z_0\left[\frac{(1+S_{11})(1+S_{22})-S_{12}S_{21}}{(1-S_{11})(1-S_{22})-S_{12}S_{21}}\right]^{1/2} \quad (b).$$

(19.4.3)

For the symmetrical and reciprocal periodic line, $S_{11} = S_{22}$, $S_{12} = S_{21}$. The real propagation constant helps to get the slow-wave factor (SWF) = λ_0/λ_g. It can also be obtained from the phase response, $\angle S_{21}$. Figure (19.25c and d) shows the slowing factor, i.e. SWF, and real and imaginary parts of Bloch impedance in the slow-wave region. There is an increase in the Bloch impedance with frequency. It is due to the series inductor loading of the host microstrip. Therefore, the slowing factor, SWF, due to the series inductor loading can be further described in terms of the creation of *artificial permeability*, giving $\text{SWF} = \sqrt{\varepsilon_{reff}\,\mu_{reff}}$ and $Z_B = Z_0(\varepsilon_r = 1)\sqrt{\mu_{reff}/\varepsilon_{reff}}$. The μ_{reff} and ε_{reff} of the artificial line can be recovered from the simulated/measured SWF and Bloch impedance. The ε_{reff} is due to the microstrip line and μ_{reff} is created by the DGS.

iii) *Patterned ground conductor*

Finally, the full surface of a ground plane conductor can also be patterned with the 2D – periodic structures using several geometrical shapes. The patterned ground plane suppresses the propagation of the surface waves. It improves the performances of the microstrip-based circuits and antenna [J.32, J.33, J.51]. The patterned ground plane also acts as an AMC.

Periodic Loading and Variation of Strip Conductor

The periodic microstrip line is also obtained either by periodically loading the strip conductor of microstrip with discrete reactance or by periodically modulating the width of a strip conductor. Several microstrip physical structures are available for both kinds of periodic loading.

i) *Periodic defects in strip conductor (series inductor)*

Figure (19.26a–d) shows some cases of series inductor loading of the host microstrip line due to the defects created in the strip conductor of a microstrip. Figure (19.26a-i to a-iii) show *three kinds of periodic inductive loading* of a microstrip line: (i) inductive loading by the high impedance line section; (ii) inductive loading by the aperture in the ground behind the microstrip; (iii) joint inductive loading of high impedance line and aperture [J.52].

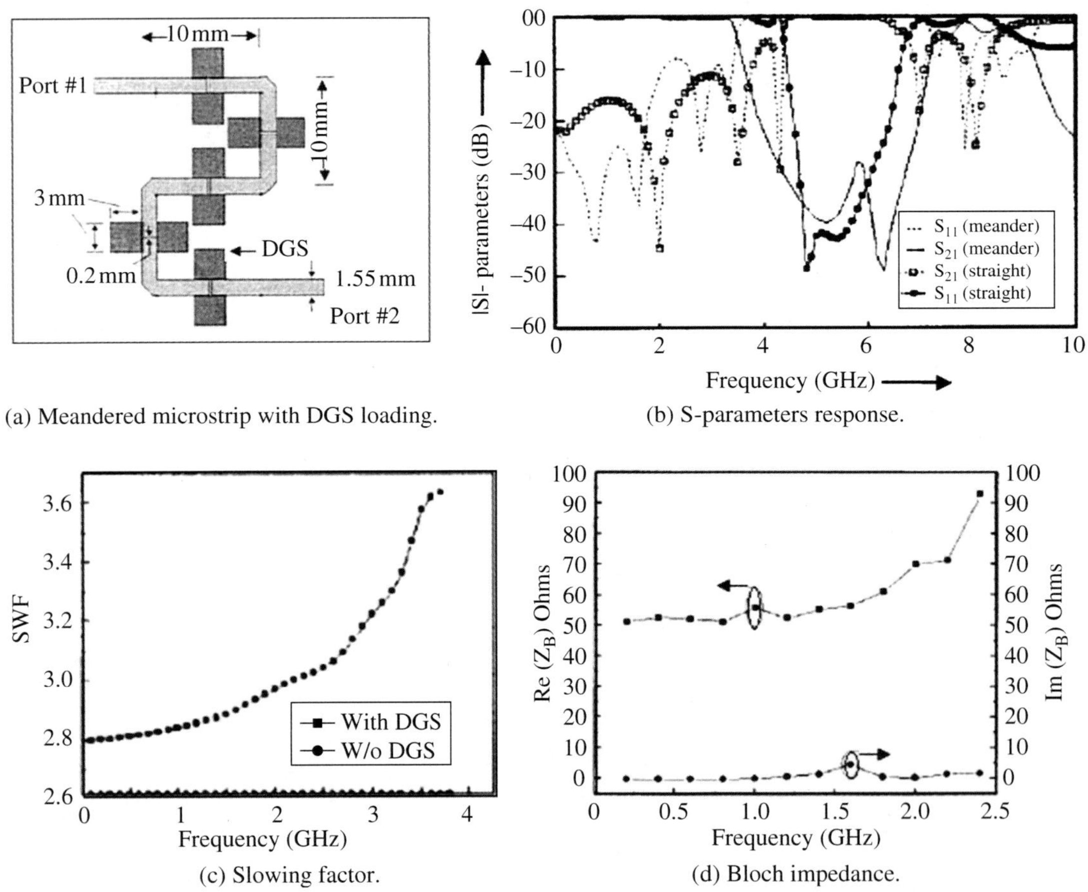

(a) Meandered microstrip with DGS loading. (b) S-parameters response. (c) Slowing factor. (d) Bloch impedance.

Figure 19.25 DGS loaded meandered microstrip periodic line. *Source:* Liu et al. [J.50]. © 2006, John Wiley & Sons.

Figure (19.26a-i) has a 50Ω (width $w_1 = 0.5$ mm) microstrip on a substrate with $\varepsilon_r = 10.2$, $h = 0.635$ mm. A narrow microstrip section ($w_2 = 0.2$ mm) of length 2.8 mm, between 50Ω microstrip sections, offers the series inductance to the host line. A narrow microstrip is a defect in the 50Ω microstrip. The unit cell has a period of $d = 5.6$ mm. The inductor can be modeled through the π-network, discussed in section (19.3.6), with the inductor in the series arm. The shunt arm capacitors can be ignored in the approximate modeling. Figure (19.26a-ii) further shows that the series inductor is realized by etching an aperture behind the strip conductor. Finally, Fig. (19.26a-iii) shows the enhanced series inductance due to aperture backed high-impedance microstrip section. In this case, the unit cell of periodic length $d = 5.6$ mm has two periodic loading elements – high impedance line section in the strip conductor, and the aperture in the ground plane. The periodic line is made of five number of unit cells.

The propagation constant and its characteristic impedance, i.e. Bloch impedance, can be evaluated by obtaining the de-embedded S-parameters at the ports, and next converting these to the [ABCD] parameters using equations (19.4.1) and (19.4.3). Figure (19.27a and b) shows the complex propagation constant $\gamma = \alpha + j\beta$ and Bloch impedance of the Bloch waves. Figure (19.27c) further shows the $|S_{21}|$ response. There is an increase in the normalized propagation constant β of the periodic line with frequency, giving SWF 2.625 at the start of the stopband edge. The first passband supports the *slow-wave Bloch mode* #1 as its SWF is above the SWF of the host microstrip line. The slowing effect increases from structure-i to structure-iii due to increased inductive loading. The increased inductor loading also widens the stopband.

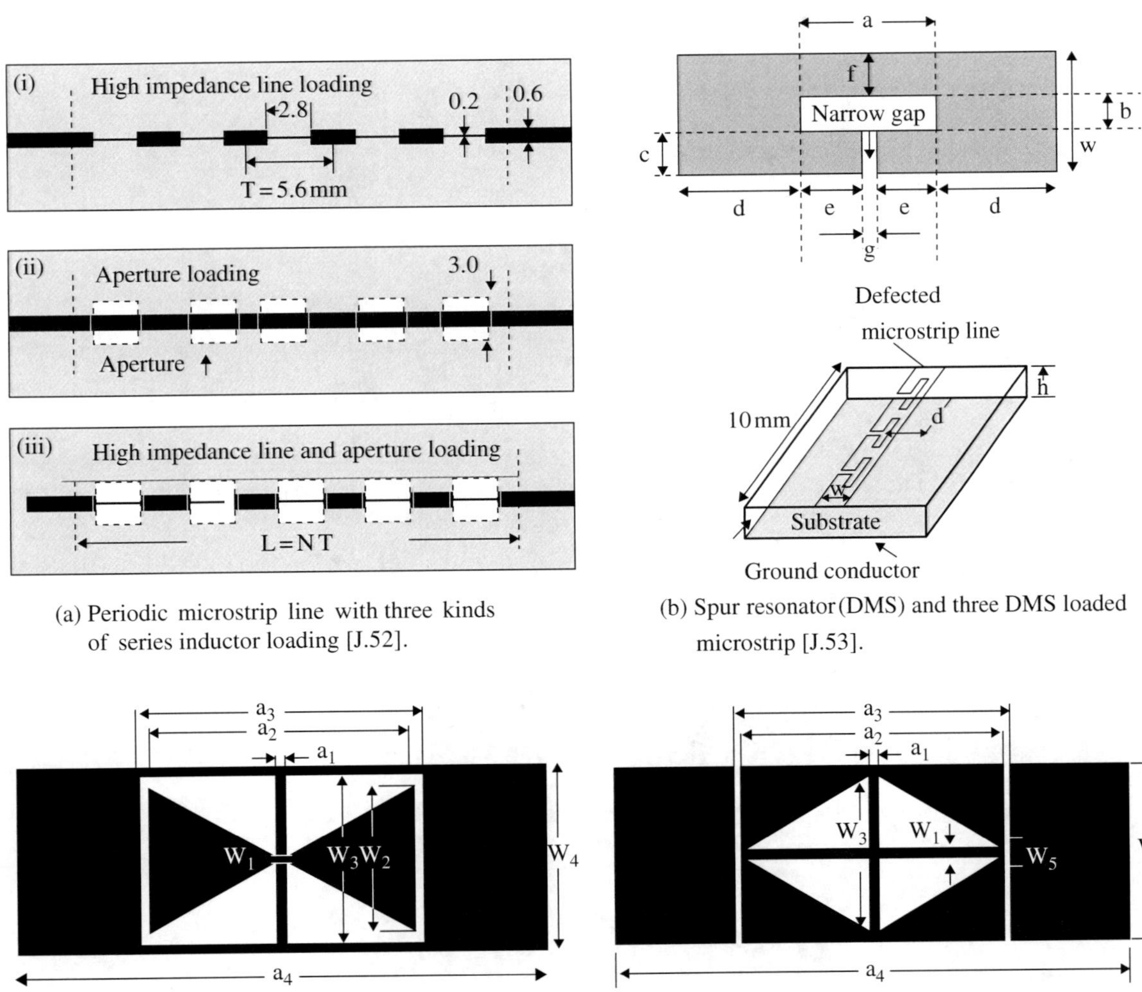

(a) Periodic microstrip line with three kinds of series inductor loading [J.52].

(b) Spur resonator (DMS) and three DMS loaded microstrip [J.53].

(c) Microstrip EBG cell-I.

(d) Microstrip EBG cell-II.

Figure 19.26 Some series inductor loaded microstrip periodic line. *Source*: Reproduced from Zhu [J.52]. © 2004, John Wiley & Sons. Reproduced from Kazerooni et al. [J.53]. © 2010, John Wiley & Sons.

However, the center of the stopband is maintained at 10.2 GHz, while the width of the stopband widens. The period, i.e. the lattice constant, d determines the central stopband frequency. The attenuation in the stopband follows the same order. The second passband supports *Bloch mode #2* that is a *fast-wave*, as its SWF is below the value of SWF of the host microstrip line. It is the leaky wave in the form of the surface wave of the microstrip line.

Figure (19.27b) shows that Re{$Z_B(f)$} is highly dispersive. In the first passband, its value, from 52Ω at 2 GHz, increases to a very high value, 350Ω at the band-edge. The initial 52Ω is due to the inductive loading. Re {$Z_B(f)$} in the second passband is always below 50Ω. In the stopband Re{$Z_B(f)$} = 0, while Img{$Z_B(f)$} starts from a very high value and decreases to zero. Figure (19.27c) shows the increasing order of attenuation in the stopband for the structures of Fig. (19.26a-i to a-iii). There are ripples at the band-edge due to mismatching of the Bloch impedance at the 50Ω ports.

Figure (19.26b) shows another kind of defect in the strip conductor, forming the *DMS* using the T-type *spur resonator*. It also shows a periodic microstrip line of three unit cells, called the cascaded DMS (CDMS) [J.53]. The spur resonator is modeled as a series-connected parallel R, L, C resonator, resonating at pole frequency f_0. Equations (19.3.52a)–(19.3.52c) compute the values of the components R, L, C. At $f < f_0$, the circuit offers the series inductive loading to the host microstrip, supporting the slow-waves. At $f > f_0$, the loading is a series-connected capacitor, and the line supports the fast-waves.

The unit cell is simulated on an EM-simulator to get the de-embedded $|S_{21}|$-response. It provides both the 3 dB cut-off frequency f_c and pole frequency f_0 in GHz. The plane of de-embedding is at the center of the spur slot, similar to the case of a DGS [J.24]. The inductance

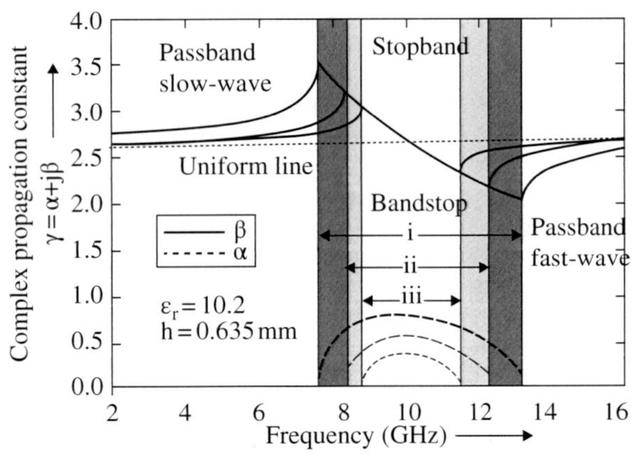

(a) Complex propagation constant of three type EBG. period d = 5.6 mm.

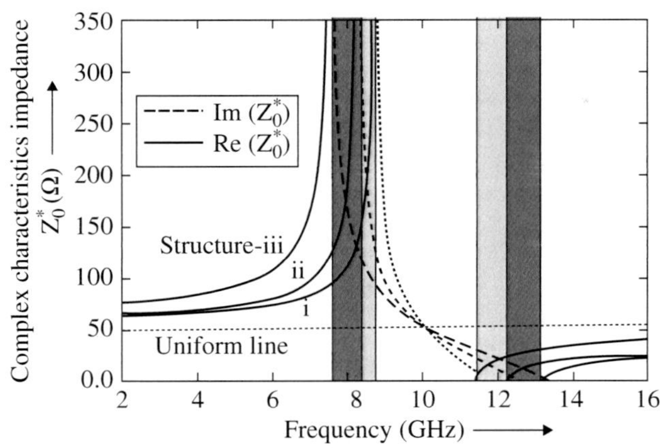

(b) Complex Bloch impedance of three type EBG. period d = 5.6 mm.

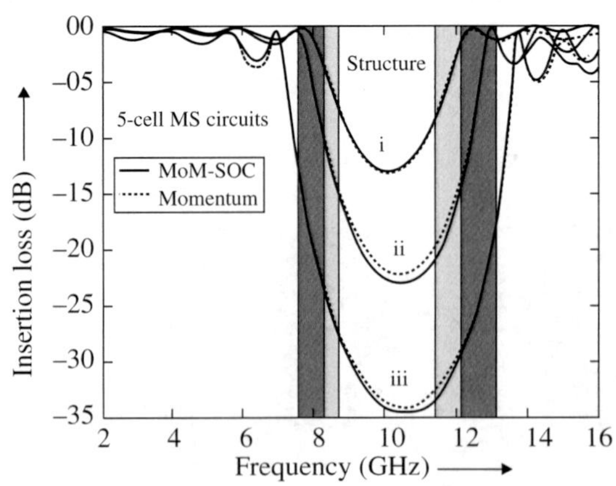

(c) S-parameters of the complex propagation constant of thee type EBG period d = 5.6 mm.

Figure 19.27 Propagation characteristics of three periodic lines shown in Fig. (15.26a). *Source*: Reproduced from Zhu [J.52]. © 2004, John Wiley & Sons.

(L) and capacitance (C) of the RLC circuit model are obtained using equation (19.3.49). At the pole frequency f_0, R of the model is evaluated as follows:

$$S_{21} = \frac{2\,Z_0}{2\,Z_0 + Z}, \quad \text{where at resonance, } Z = R$$
$$\Rightarrow \quad R = 2\,Z_0 \frac{1 - S_{21}(f_0)}{S_{21}(f_0)}. \tag{19.4.4}$$

The series impedance Z of the spur resonator, forming the unit cell, is located between two microstrip sections of length $\theta/2 = kd/2$. The alternative analysis method is also available [J.53]. The unit cell is used to obtain the $(kd - \beta d)$ dispersion diagram. However, the slowing-wave factor (SWF) of the unit cell can also be obtained from its $\angle S_{21}$-phase response with spur and without spur, using an EM-simulator as follows:

Let $\theta_s(f)$ and $\theta_m(f)$ are the phase angles in degree for a unit cell with spur and without spur, i.e. only microstrip line. Due to the loading, $\theta_s(f) > \theta_m(f)$, and the differential phase is obtained:

$$\Delta\theta(f) = \theta_s(f) - \theta_m(f); \quad \text{where } \theta_s(f) = \beta d = \frac{2\,\pi}{\lambda_g} d,$$
$$\text{and} \quad \theta_m(f) = kd = \frac{2\,\pi}{\lambda} d, \quad \lambda = \frac{\lambda_0}{\sqrt{\varepsilon_{reff}}}, \text{SWF} = \frac{\lambda_0}{\lambda_g}. \tag{19.4.5}$$

The λ_g, λ, and λ_0 are the guided wavelength of microstrip with a spur, without spur and in the free space. The periodicity is d. The following expressions are obtained from the above equations:

$$\theta_s(f) = \Delta\theta(f) + \theta_m(f)$$
$$\Rightarrow \frac{2\,\pi}{\lambda_g} d = \frac{\Delta\theta(f)\,\pi}{180} + \frac{2\,\pi}{\lambda} d \qquad \text{(a)}$$
$$\Rightarrow \frac{\lambda_0}{\lambda_g} = \frac{\Delta\theta(f)\,\lambda_0}{360\,d} + \frac{\lambda_0}{\lambda} \qquad \text{(b)} \tag{19.4.6}$$
$$\Rightarrow \text{SWF}(f) = \frac{\Delta\theta(f)\,\lambda_0}{360\,d} + \sqrt{\varepsilon_{reff}} \qquad \text{(c)}.$$

The slowing-factor demonstrates the nature of the Bloch wave propagation, and the S-parameters response helps to get the passband/ stopband characteristics. Let us consider the unit cell for the host 50Ω microstrip on a substrate with $\varepsilon_r = 2.33$, $h = 0.787$mm. Figure (19.28a) compares the performance of the spur loaded microstrip, i.e. the DMS, against the DGS loaded microstrip. The dimensions of the spur resonator are $a = 5.4$ mm, $b = 0.5728$ mm, $g = 0.2$ mm, and the DGS has $a' = b' = 1.4405$ mm and $g' = 0.2$ mm. Figure (19.28a) compares the slowing-factor of both cases at identical pole frequency $f_0 = 17$ GHz. Both the structures have almost identical behavior. At frequency $f < 17$ GHz, due to series-connected inductive loading, the propagating Bloch mode is in the slow-wave zone, as the SWF of the periodic line is above the SWF of the host microstrip. However, at frequency $f > 17$ GHz, the SWF of the loaded line is below that of the host microstrip, giving the fast-wave propagation. Figure (19.28b) shows the S-parameters response of the three spur based unit cells periodic line. It creates a wide stopband, 11.5 GHz–31.75 GHz, and also ripples at the passband edge due to mismatching of Bloch impedance at the band-edge.

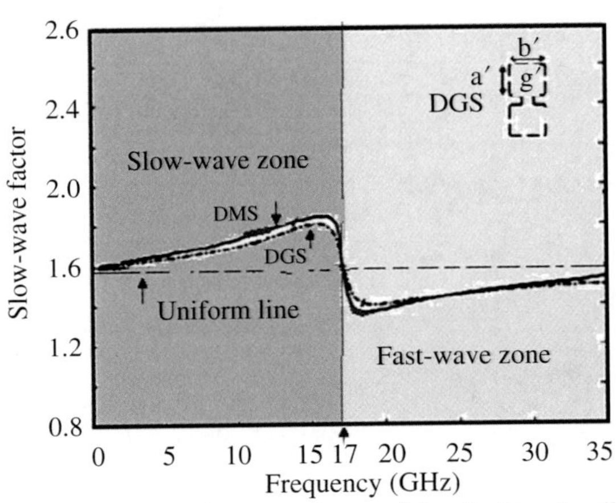

(a) Slow-wave/fast-wave factor of the single unit cell DMS and DGS.

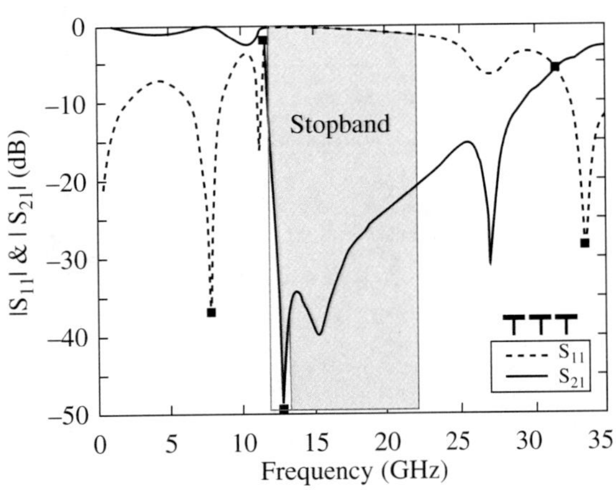

(b) S-parameter response of three unit cells DMS.

Figure 19.28 Propagation characteristics of the periodic line of Fig. (19.26b). *Source*: Reproduced from Kazerooni et al. [J.53]. © 2010, John Wiley & Sons.

Figure (19.26c and d) show two more DMS, giving a high value of the loading series inductor to a 50Ω microstrip line. Figure (19.26d) provides a high value of SWF and wider stopband bandwidth [J.54, J.55]. Several other geometrical shapes, including the fractal shapes, have been reported in the literature [J.56].

ii) *Periodically loaded microstrip (shunt capacitor)*

Figure (19.29a–d) illustrate four cases of the shunt connected capacitively-loaded host microstrip line. The shunt connected capacitors are realized by the straight-line stub, T-stub, or even by the radial stub. A unit cell is comprised of a stub located between two identical length 50Ω microstrip sections [B.13–B.15 J.4, J.6, J.57, J.58]. Bloch wave on these periodic lines, with capacitive inclusions, is a modification of the quasi-TEM mode of the host microstrip line. The propagation behavior is discussed below:

- **Case #I:** It uses the single-sided open stub loaded 50Ω microstrip. Figure (19.29a) shows the loaded line, with a unit cell. The stub characteristic impedance Z_{oc} could be 50Ω or less, such as 20Ω. The loading shunt capacitor, for the stub length $\ell_c < \lambda_{gc}/4$, is computed as follows:

$$B_c = \omega C = \frac{1}{Z_{oc}} \tan\left(\frac{2\pi \ell_c}{\lambda_{gc}}\right). \qquad (19.4.7)$$

However, discontinuity can alter the component value to some extent. It can be accounted for by using a better circuit model [B.19]. The periodic structure supports the *slow-wave* propagation in the *first passband*. The SWF increases with frequency up to the edge of the stopband, while the Bloch impedance Z_B is reduced below 50Ω, and it further decreases with frequency up to the lower edge of the stopband. This frequency-dependent behavior of Z_B for a shunt capacitor-loaded microstrip is opposite to that of the series inductor loaded microstrip. It is discussed earlier that the series inductor line also supports the slow-waves; however, its Bloch impedance is higher than Z_0 (50Ω) that increases with frequency up to the lower band-edge of the stopband. The *second passband* of the shunt capacitor-loaded microstrip supports the *fast-wave*, like a series inductor loaded line, that may cause the leaky wave. It is further discussed for the case-III, shown in Fig. (19.29c). An interesting application of the doubled-sided stub-loaded microstrip is discussed below:

- **Case #II:** In this case, the single-sided capacitive stubs are replaced by the double-sided stubs, shown in

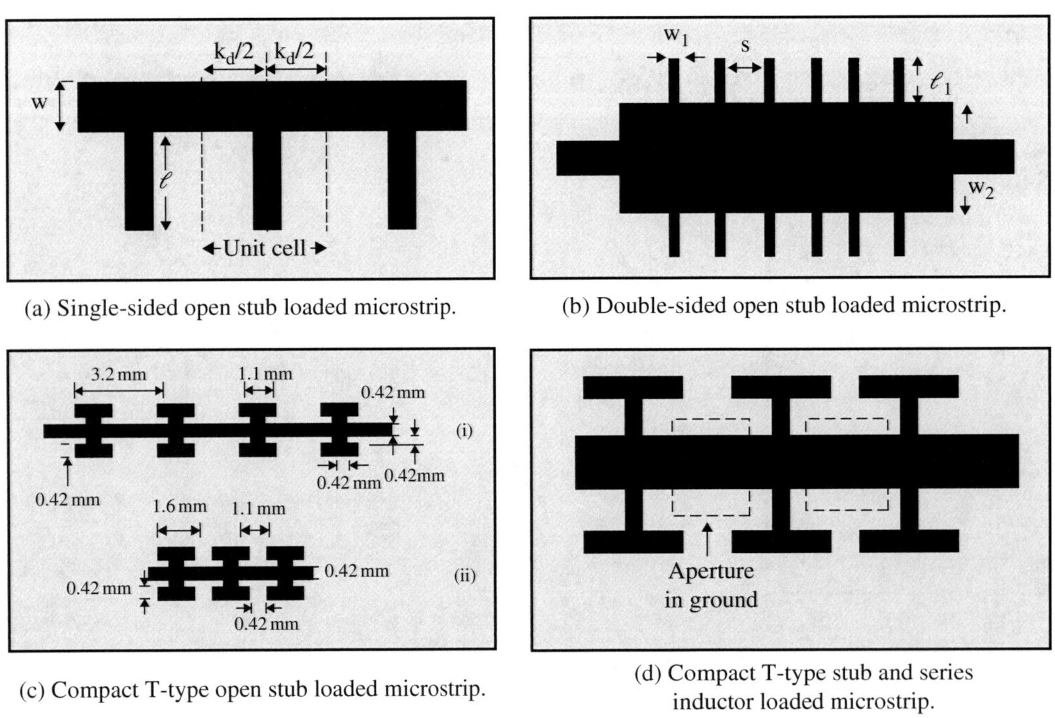

Figure 19.29 Some shunt capacitor-loaded microstrip periodic line.

Fig. (19.29b) [J.58]. The low-frequency behavior of the stub length $\ell_c < \lambda_{gc}/4$ is the same as that of the case-I; however, with increased capacitive loading. At higher frequency, leading to $\ell_c = \lambda_{gc}/4$, the closely placed stubs present the short-circuits, i.e. the electric walls at both sides of the host microstrip. Therefore, over a certain frequency band, around $\ell_c = \lambda_{gc}/4$, the double-sided stub-loaded periodic microstrip acts as a *substrate integrated waveguide (SIW)*, described in section (7.8) of chapter 7. In this version of the SIW, the realization of the electric walls does not require any shorting pins or vias, used in the standard SIW. Even a brittle alumina substrate can be easily used to realize the SIW in its newer version. The structure is called the *corrugated SIW (CSIW)*, and the standard one is the *pin SIW (PSIW)*.

In the case of the CSIW, we use low impedance, i.e. wide width, microstrip. Its width w_2 is computed at the desired center frequency of the CSIW [J.58]. The periodicity of the CSIW, $s \leq \lambda_g/8$ and the stub width w_1, $s/w_1 \geq 2$. The stub length is $\ell_1 = \lambda_g/4 - \Delta\ell$, where $\Delta\ell$ is the open-end discontinuity of the stub. It is computed using the following expression [B.18]:

$$\frac{\Delta\ell}{h} = 0.412\left[\frac{\varepsilon_{reff} + 0.3}{\varepsilon_{reff} - 0.258}\right]\left[\frac{w_1/h + 0.262}{w_1/h + 0.813}\right]. \qquad (19.4.8)$$

λ_g is taken at the center frequency of the CSIW. Figure (19.30a) shows both the EM-simulated and experimentally obtained S-parameters of a CSIW designed on Taconic TXL substrate – $\varepsilon_r = 2.55, \tan\delta = 0.0019$, $h = 1.0$mm. The $w_2 = 10.5$ mm wide CSIW is 62 mm long. The $w_1 = 0.8$ mm wide stub is $\ell_1 = 4.5$ mm long and its period is $s = 2.0$ mm. The 12.2 GHz–17.2 GHz wide 3 dB operating BW is observed for the CSIW. Its insertion loss is better than 1.6 dB, while $|S_{11}|$ is better than 14.0 dB. The lower cut-off region is observed.

- **Case #III:** Figure (19.29c) shows periodic microstrip using the double-sided T-type compact stubs. The $w_2 = 0.42$ mm wide microstrip, on a substrate $\varepsilon_r = 2.95$, $h = 0.508$ mm, is the host medium. For the light capacitive loading, the period is $d = 3.2$ mm. Other dimensions are shown in Fig. (19.29c-i). For a compact microstrip periodic structure of four unit cells, the period is reduced to $d = 1.6$ mm, shown in Fig. (19.29c-ii). In a case of the light loading, the loading capacitors are isolated. However, for the compact case, there is a mutual coupling between the loading stubs that strongly influences the propagation characteristic, discussed below.

Figure (19.30b–d) shows the *new Brillouin diagrams* of lightly and strongly loaded microstrip with T-stubs [J.4–J.6]. In place of one linear light line, it has three pairs of such lines. The line-1 and its replica are dispersion lines of the unloaded host microstrip medium, supporting the quasi-TEM mode with propagation constant, $k = k_0\sqrt{\varepsilon_{reff}}$. The dispersion line-2 and its replica correspond to the TM_0 – surface wave mode, supported by the conductor backed dielectric substrate. The dispersion line-3 and its replica shows free space propagation, i.e. the radiation. The dispersion lines-2 and -3, with $\beta = \pm k_{TM_0}$ and $\beta = k_0$, respectively, are above the $k = k_0\sqrt{\varepsilon_{reff}}$ line. The modal representation of waves on the new dispersion diagram is different from the standard $(\omega - \beta)$ Brillouin diagram. It shows the passband/ stopband regions with the slow-wave and fast-wave presence. Besides, it also explains the origin of the leaky modes, i.e. conversion of any space-harmonics of the Bloch mode to the surface wave modes-TM_0, and other higher-order surface wave modes. It further explains their conversion to the radiation mode in the free space. It is the basis of the leaky-wave antenna that is due to *Cherenkov-type radiation* discussed in subsection (5.5.7) of chapter 5 and subsection (9.7.3) of chapter 9.

The dispersion line-1 divides the Brillouin dispersion diagram into two regions – the *bound region*, below the line-1, supporting the propagating mode, and the *unbound region* above it, supporting the leaky mode wave. The bound region has the first passband, supporting the slow-wave Bloch mode #1 and the closed (*non-radiating*) stopband. In the passband, all components of the space-harmonics have a real propagation constant, $k_{z,n} = \beta_n$, for no dielectric and conductor losses. It is the slow-wave region, as $|\beta_n| > k_0$. In the case of the light loading, shown in Fig. (19.29c-i), for the period $d = 3.2$ mm, Bragg's reflection, corresponding to $\beta d = -\pi$, occurs at $f = 30$ GHz as shown in Fig. (19.30b-i). It is the upper edge of the closed-stopband. Due to the capacitive loading, the slow-wave supporting passband bandwidth is 0 GHz–18 GHz. The EM-simulated stopband bandwidth is 18 GHz–31.5 GHz. The frequency deviates from an estimated 30 GHz, as the T-stub is not purely capacitive. It requires better modeling.

Figure (19.30c) shows the S-parameters of the four unit cells ($d = 3.2$ mm) periodic line of total length

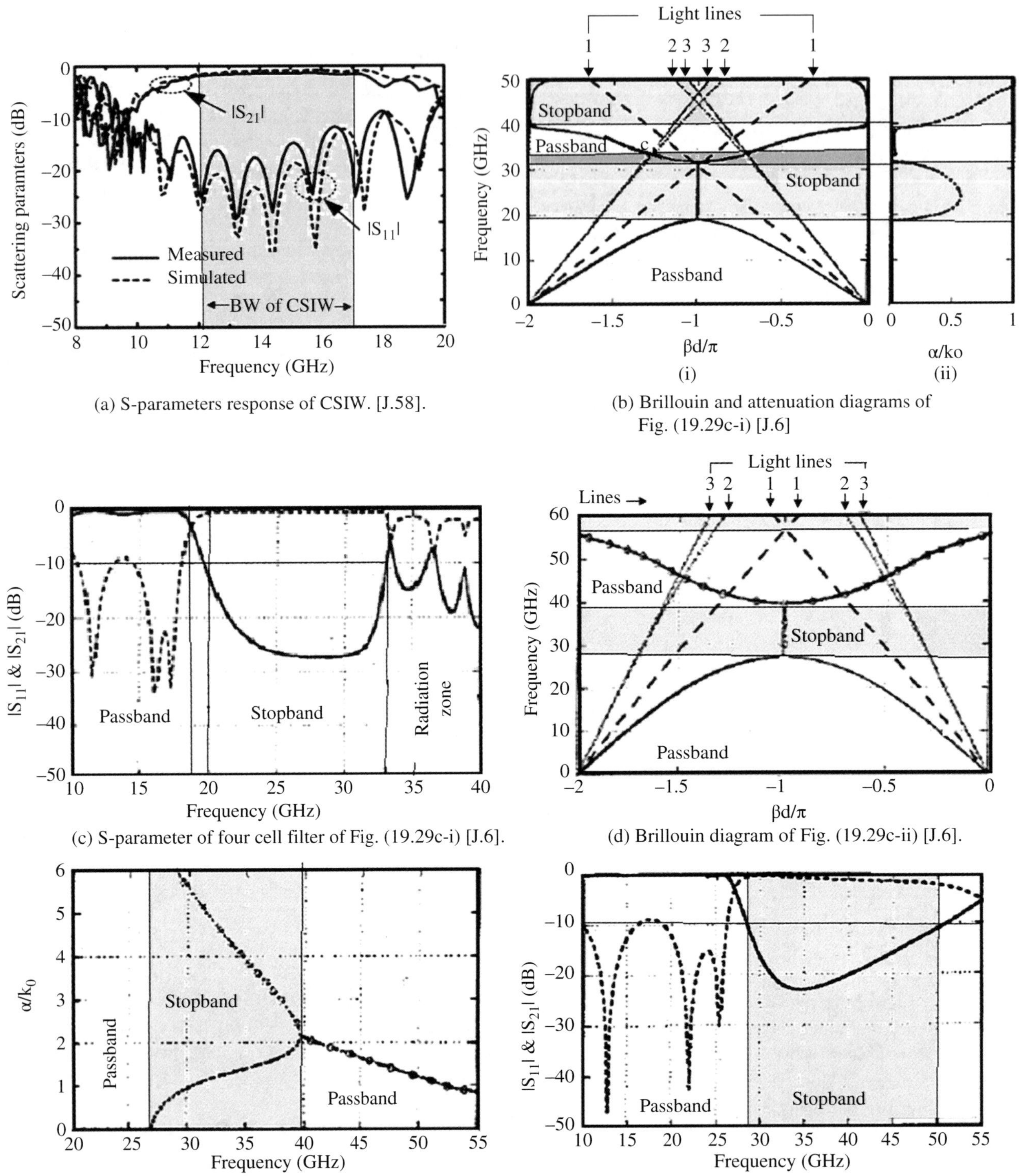

(a) S-parameters response of CSIW. [J.58].

(b) Brillouin and attenuation diagrams of Fig. (19.29c-i) [J.6]

(c) S-parameter of four cell filter of Fig. (19.29c-i) [J.6].

(d) Brillouin diagram of Fig. (19.29c-ii) [J.6].

(e) Attenuation of Fig. (19.29c-ii) [J.6].

(f) S-parameter of four cell filter of Fig. (19.29c-ii) [J.6].

Figure 19.30 Propagation characteristics of the periodic line of Fig. (19.29). *Source*: Reproduced from Liu et al. [J.58]. © 2016,Microwave Optical Tech. Letters. Reproduced from Baccarelli et al. [J.6]. © 2007 ISMOT.

$\ell = 12.8$ mm. It confirms the first passband between 0 and 18 GHz. It shows the stopband, corresponding to $|S_{21}| = -10$dB, from 20 GHz to 32 GHz. The transition region is from 18 GHz to 20 GHz. The $(\omega - \beta)$ dispersion diagram is obtained for an infinitely long periodic line, whereas the S-parameter is for the finite length periodic line. The bandwidth of passband/stopband obtained from both methods have some difference, as ideally the $(\omega - \beta)$-diagram does not show any transition region between the passband and the stopband. Figure (19.30b-ii) shows the $(\omega - \alpha)$ diagram, giving the attenuation in the stopband. In the closed-stopband, space-harmonics propagation constant $k_{z,n}$ is complex. However, the mode of propagation is still the slow-wave, as $|\beta_n| > k_0$. For a lossless microstrip, $\alpha = 0$ in the first passband.

The second passband is from 31.5 GHz to 39.5 GHz. It has two regions. The first one provides the narrow non-radiating passband, shown in Fig. (19.30b), from 31.5 GHz to 33 GHz. It supports the *backward fast-wave* because $|\beta_n| < k_0$. The second one provides the leaky wave region. Above the crossing-point C, one or more space-harmonics meet the condition $|\beta_n| < k_0 < k_{TM_0}$ and leak to the free space in the form of radiation, or leak to the substrate in the form of the surface-wave mode TM_0, etc. The peaks in the response of the S-parameter, shown in Fig. (19.30c), show a presence of the leaky modes. If $\beta_n\alpha > 0$ for any space-harmonics, the leakage increases exponentially in the vertical direction. If $\beta_n\alpha < 0$, space-harmonics leakage decreases. The new Brillouin diagram considers a number of the space-harmonics also [J.4].

Figure (19.29c-ii) shows a very compact three-unit cells periodic line with period $d = 1.6$ mm. It provides stronger loading. Other dimensions are shown in Fig. (19.29c-ii). Figure (19.30d) shows the new $(\omega - \beta)$-dispersion diagram with a widening of both the first and second passband, 0 GHz–26.8 GHz, and 39.5 GHz–55 GHz. The stopband starts around 26.8 GHz, $(\beta d = \pi)$. However, stopband, supporting the complex propagation constant, is the involved one, as shown in the $(\omega - \alpha)$ diagram of Fig. (19.30e). In both stopbands, we have evanescent modes with two different values of attenuation constant α and same β_n. Above 40 GHz, i.e. in the second passband, α is very high, extending the stopband, as shown in Fig. (19.30f). The extended stopband is in the radiating region [J.6].

iii) *Periodically modulated strip conductor of microstrip*

The centers of the multiple stopbands, of a periodic line, are harmonically related. However, on several occasions, for the filter's applications, only one stopband is needed, without disturbing the high-frequency band. The sinusoidally modulated slot in the ground plane provides such a response [J.42]. However, it has a packaging problem and its realization in MMIC is difficult. Normally, the modulated strip conductor of a microstrip also provides multiple passband-stopband characteristics [J.59]. However, if the characteristic impedance of a microstrip follows the sinusoidal variation, say 50Ω to Z_0^{min} back to 50Ω, and to Z_0^{max} and back to 50Ω creating a unit cell, we get only one stopband, suppressing other harmonically related stopbands [J.60–J.62]. The modulated strip profile is obtained from the variation in the characteristic impedance. The modulated microstrips also create the EBG high impedance surface, controlling the surface wave propagation [J.63].

Figure (19.31a) shows characteristic impedance variation of a microstrip over one unit cell of period d. Over the first half period, $0 \le p \le d/2$, the characteristic impedance is below 50Ω, and over the next half period, $d/2 \le p \le d$, it is above 50Ω. The $\left(Z_0^{min}, Z_0^{max}\right)$ pair is shown in Fig. (19.31a). The sinusoidal variation in characteristic impedance is governed by the following expression:

$$Z_0 \le 50\Omega: Z_0 = 50 - \left(50 - Z_0^{min}\right) \sin(2\pi p/d), \quad \text{for } 0 \le p \le d/2 \qquad (a)$$
$$Z_0 \ge 50\Omega: Z_0 = 50 + \left(Z_0^{max} - 50\right) \sin(2\pi p/d), \quad \text{for } d/2 \le p \le d \qquad (b)$$
$$\text{also,} \qquad Z_0 = \sqrt{Z_0^{min} Z_0^{max}} = 50\Omega \qquad (c). \qquad (19.4.9)$$

The variation in the strip width (w/h) at the discrete locations is obtained using the synthesis expression (8.3.38) of a microstrip line given in chapter 8. Figure (19.31b) shows one such structure over one period. The microstrip width W_{50}, W_{Zmin}, and W_{Zmax} corresponding to the characteristic impedance 50Ω, minimum characteristic impedance Z_{min}, and maximum characteristic impedance Z_{max} are shown in Fig. (19.31a and b). Figure (19.31c) shows six numbers of cascaded unit cells on a substrate with $\varepsilon_r = 2.2$, $h = 0.508$ mm, $t = 0.11$ mm. Figure (19.31d–e) shows the S-parameter response for the characteristic impedance pairs (Z_{min}, Z_{max}): (25Ω, 100Ω) and (15.5Ω, 161Ω), period d = 20 mm, giving the bandstop at 5.5 GHz. For both cases, the stopband does not occur at the harmonics 11 GHz and 16.5 GHz. Also, the stopband attenuation and bandwidth are more for the second case, due to the large Z_0^{max}/Z_0^{min} ratio.

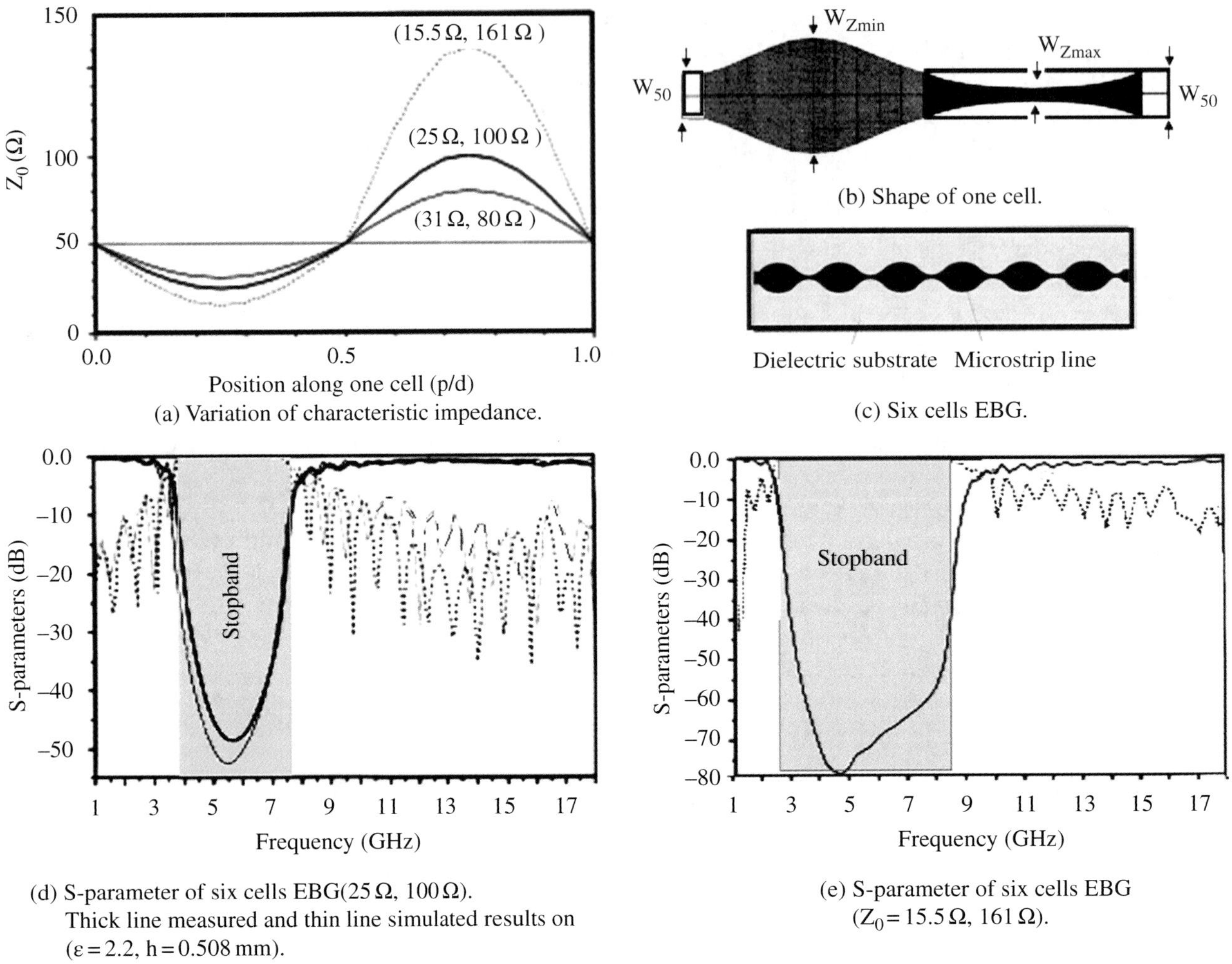

Figure 19.31 Periodic microstrip of modulated strip conductor. *Source:* Reproduced from Nesic and Nesic [J.60]. © 2001 John Wiley & Sons.

19.4.2 1D CPW EBG Line

The CPW, like a microstrip, could be also periodically loaded with series inductor/shunt capacitor or shunt inductor/series capacitor to get the periodic CPW structures. Such loading is also done for the CPS structures. Furthermore, these loaded lines are used to get a high Q-factor resonator [J.64]. The loading is carried out both for the central strip conductor and also for the ground conductors.

Figure (19.32a–j) illustrates several reactive structures to load the central strip conductor of a *host* CPW, without disturbing the ground plane conductors. The geometrical structures of reactive loading elements, i.e. the *inclusion*, have a strong influence on the characteristics of a periodic CPW. The loading elements, in some cases, can be modeled as the lumped elements, using the closed-form expressions [B.21]. Alternatively, the loading elements can be modeled using the T or π-networks. Their component's values are extracted from the Z or Y-parameters obtained using an EM-simulator [B.16].

Figure (19.33a–f) show several structures for the shunt-connected stubs. These stubs are accommodated by creating slots in the ground conductors. Only slots in the ground conductors, without any stub, also provide series inductors loading of a CPW. Similarly, the DGS in the ground plane also loads the CPW with the parallel resonant circuit. The open-circuited shunt stub acts as a shunt capacitor, or even as a shunt connected series resonant circuit. The stubs could be connected to either one side or both sides of the strip conductor. The radial stubs are also used to get a compact periodic CPW. Both the central strip and ground conductors can also be reactively loaded.

There are several variations in the shape of the loading elements forming the unit cells. This section considers one case each for the series/shunt loading of the

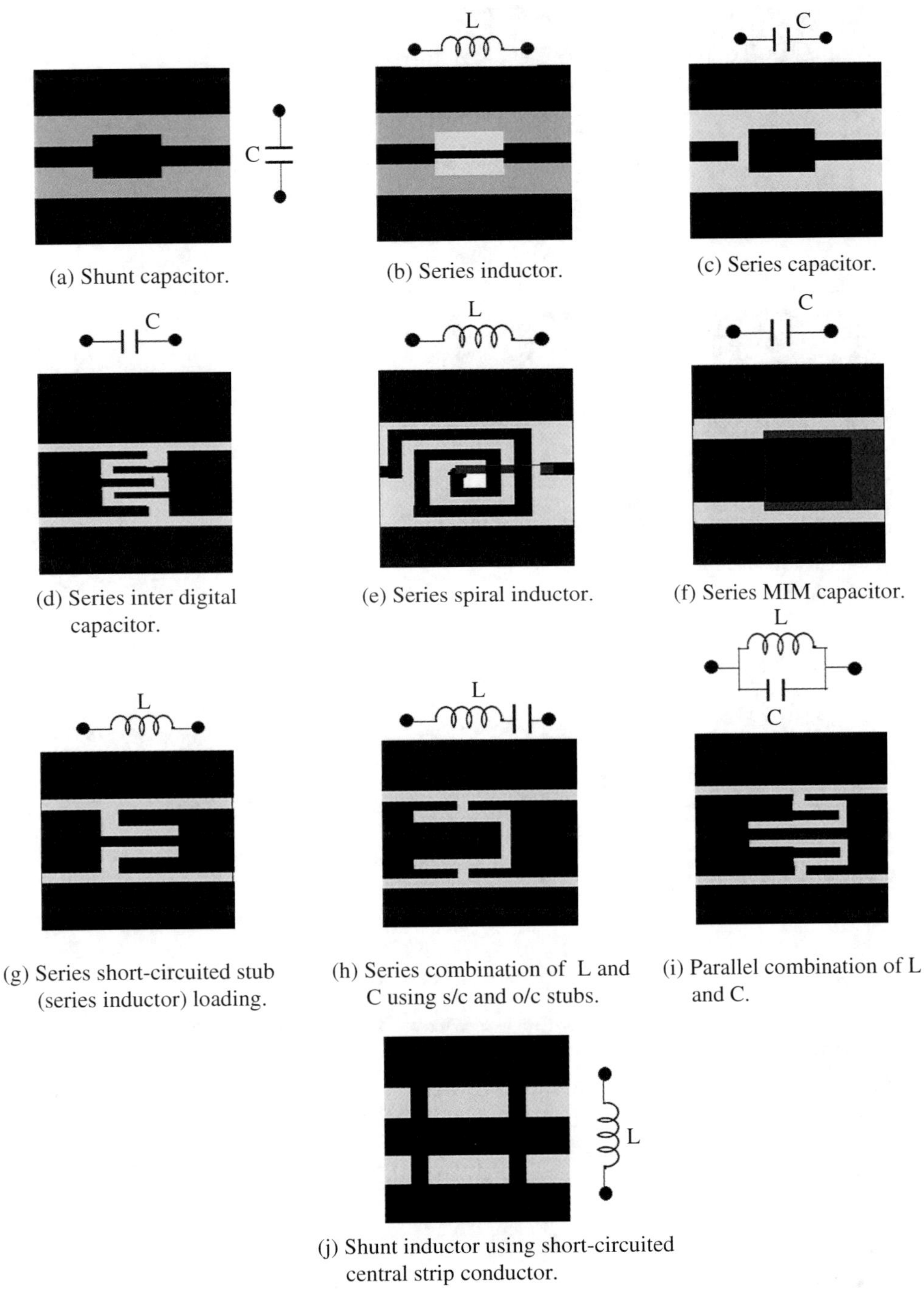

Figure 19.32 Periodic loading of the central strip of the CPW.

inductor/capacitor of the CPW periodic structures, shown in Fig. (19.34a–d).

Series Inductor Loading

Figure (19.34a-i-iii) show three types of series inductor loading of a 50Ω CPW on the substrate with $\varepsilon_r = 10.2$, $h = 0.635$ mm – (i) double-sided inductive T-slots in the ground plane, (ii) double-sided inductive T-slots in the strip conductor, and (iii) double-sided inductive T-slots in both the ground plane and strip conductor. Figure (19.34a) shows three cells of each type of loading with a period $d = 5.08$ mm, slot dimension $\Delta = 2.54$ mm, and slot gap, $g = 0.508$ mm [J.65]. The $(\omega - \beta)$ dispersion diagram and Bloch impedance Z_B are determined by the

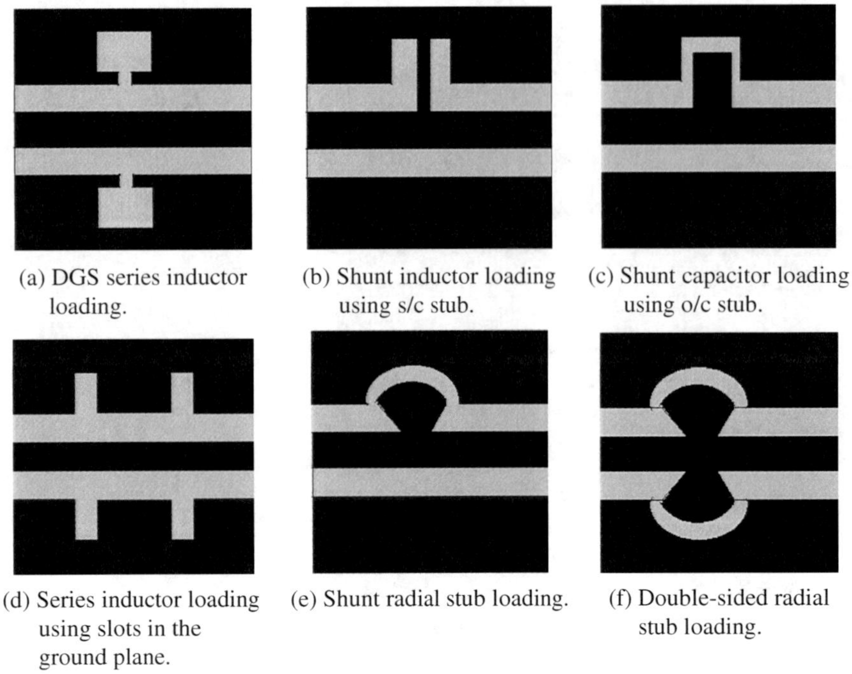

Figure 19.33 Patterned ground planes and CPW stubs to form the unit cells.

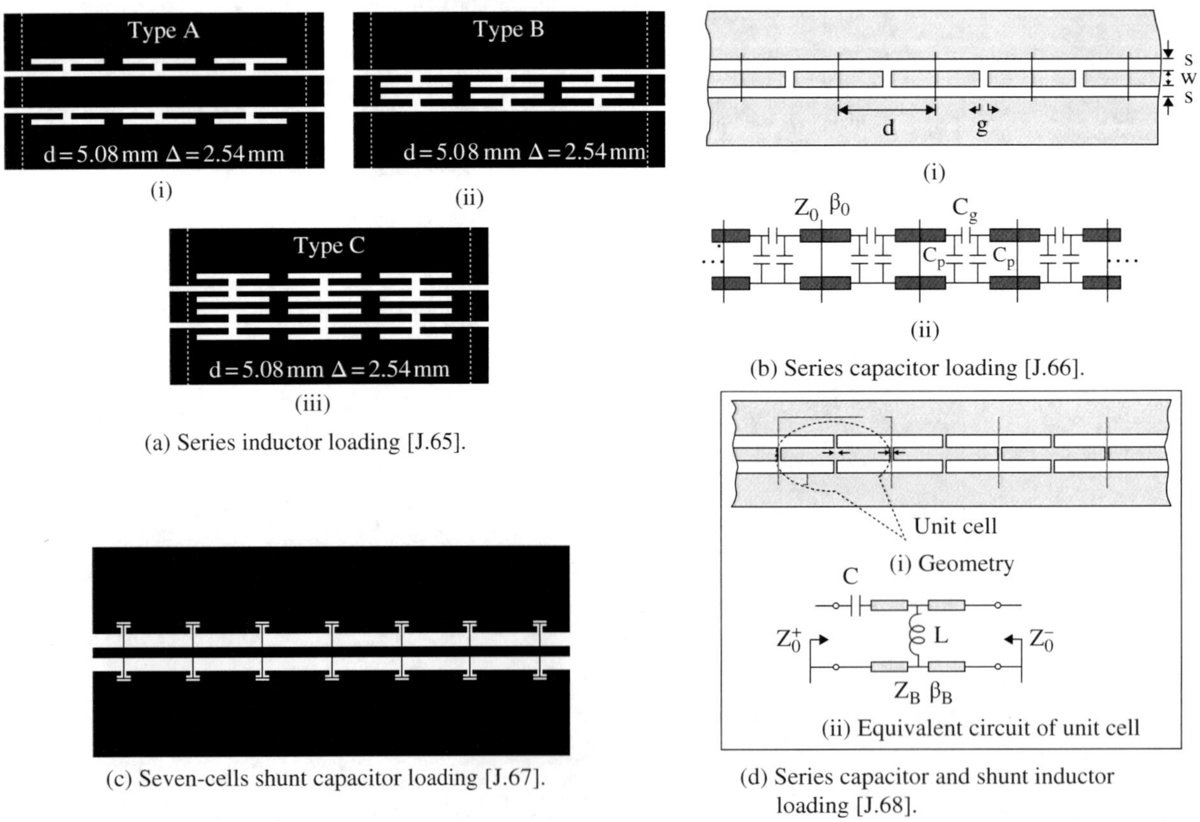

Figure 19.34 Series/shunt-loaded periodic CPW. *Source:* Reproduced from Zhu [J.65]. © 2003, IEEE. Reproduced from Zhu and Shi [J.66]. © 2005 John Wiley & Sons. Reproduced from Martın et al. [J.67]. © 2003, John Wiley & Sons. Reproduced from Gao and Zhu [J.68]. © 2003, IEEE.

de-embedded [ABCD] parameters of the cascaded N unit cells:

$$\begin{bmatrix} A_N & B_N \\ C_N & D_N \end{bmatrix} = \begin{bmatrix} \cosh(\gamma N d) & Z_B \sinh(\gamma N d) \\ Y_B \sinh(\gamma N d) & \cosh(\gamma N d) \end{bmatrix} \quad (a)$$

$$\cosh(\gamma N d) = \frac{A_N + B_N}{2} \quad (b), \quad Z_B = \sqrt{\frac{B_N}{C_N}} \quad (c). \quad (19.4.10)$$

The [ABCD] parameters are obtained from the EM-simulated de-embedded S-parameters. The passband–stopband characteristics are obtained by plotting the extracted value of A_N over a frequency range, 2 GHz–16 GHz, shown in Fig. (19.35a). In the stopband range 8.2 GHz–13.8 GHz, $|A_N| \geq 1$. It is shown in the shaded area. The upper edge of the stopband is nearly fixed at about 13.8 GHz. It is decided only by the lattice constant d. However, the lower edge of the stopband decreases from the type-i to type-iii, creating increasing wider stopband bandwidth. It is due to the increased inductance of the slot in the ground plane (*case-i*), slot in the strip conductor (*case-ii*), and slots in both the strip and ground conductors (*case-iii*). The slot in the strip conductor offers higher inductance, due to the higher current density in the strip, as compared to that of in the ground conductor.

Figure (19.35b) shows the slow-wave region in the passband, as the normalized propagation constant, β/k_0 is above to that of the uniform CPW host medium. The β is propagation constant of the Bloch wave of the periodic line, whereas k_0 is the propagation constant of the host medium. In the stopband range, the propagation constant γ is complex and Bragg's condition is met at $\beta d = \pi$. The first passband of Fig. (19.35c) shows an increase in Bloch impedance with frequency, up to the edge of the stopband. It is always above the 50 Ω impedance of the host CPW. In the second passband, it is much below the 50 Ω impedance. It is the characteristics of a series inductor loaded periodic line. Finally, Fig. (19.35d) compares $|S_{21}|$ of three structures consisting of $N = 6$ cells. The structure-iii provides wide stopband bandwidth and high attenuation level in the stopband due to stronger inductive loading. The ripples at the band-edges are due to the mismatch of Bloch impedance to the 50 Ω impedance of the CPW. The ripples are controlled by tapering the slots [J.69].

The series inductor loading of a CPW is also obtained by using several kinds of the DGS slots and stubs, including the radial stub in the ground plane [J.70, J.71]. More complex loading of both the strip and ground conductors have also been done [J.72].

Series Capacitor Loading

Figure (19.34b) shows four unit cells of the series loaded capacitors to the central strip conductor of a CPW. The loading capacitor is realized by a gap in the strip conductor, modeled by the π-network. Figure (19.36a) shows the extracted value of shunt (C_p) and series (C_g) capacitances of the π-network, using the MOM and de-embedding process [J.66]. These capacitances can also be evaluated, through the de-embedded Y-parameters of the gap discontinuity, using an EM-simulator [B.16]. The value of the main series capacitance C_g is much higher than that of the C_p. The gap capacitance is in the middle of the unit cell of a period d. Two CPW sections have a length $(d - g)/2$, where g is the slot gap.

The [ABCD] parameters of the four unit cells cascaded CPW is obtained using equation (19.4.10). Figure (19.36b) shows the results of the propagation constant and attenuation of the Bloch wave. The periodic structure acts as a bandpass filter between two stopbands. In the first stopband, we get $\beta = 0, \alpha \neq 0$. However, in the second stopband, the complex mode exists with $\beta \neq 0, \alpha \neq 0$, and the passband, occurs between 5.8 GHz and 9.4 GHz. As β/k_0 of the Bloch wave in frequency range 6.5 GHz–9.4 GHz is below that of the host CPW, the passband supports the *fast-wave*. In the frequency 5.8 GHz–6.5 GHz, $\beta/k_0 < 1$, i.e. the normalized propagation constant is below that of the air-medium. In this band, the Bloch wave is a *leaky radiation mode*.

Figure (19.36c) shows that Bloch impedance in the first stopband is capacitive. In the passband, the Re (Z_B) maximum occurs at 20Ω, i.e. much below the 50Ω characteristic impedance of an unloaded host CPW. The value of Re (Z_B) can be controlled by changing the slot-gap g and line gap s of a CPW to some extent. However, it is difficult to obtain 50Ω for a periodically loaded CPW to get proper matching at input and output terminations. Both ends are terminated in 32Ω, $\lambda_g/4$ CPW sections for matching.

The $|S_{21}|$-parameter is computed from [ABCD] parameters of the periodic line, using the following expression [J.66]:

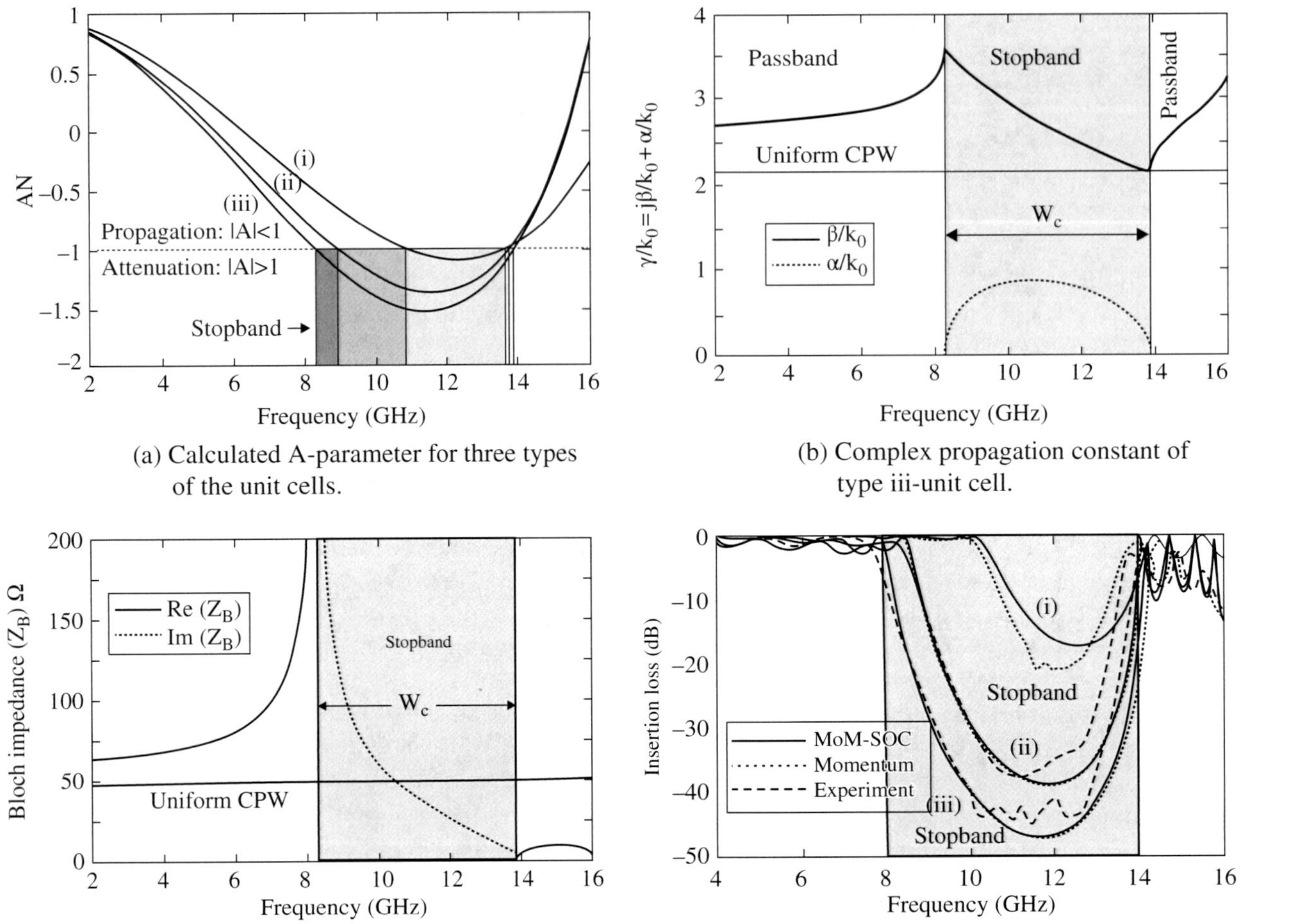

(a) Calculated A-parameter for three types of the unit cells.

(b) Complex propagation constant of type iii-unit cell.

(c) Bloch impedance of type iii unit cell.

(d) Insertion loss of six-cells EBG.

Figure 19.35 Characteristics of series inductor-loaded periodic CPW of Fig. (19.34a). *Source:* Reproduced from Zhu [J.65]. © 2003 IEEE.

$$S_{21} = \frac{2}{2\cosh(\gamma Nd) + (Z_B/Z_0 + Z_0/Z_B)\sinh(\gamma Nd)}$$

$$\Rightarrow |S_{21}|^2 = \left\{\frac{4}{4 + \left[(\mathrm{Im}(Z_B)/Z_0)^2 + (Z_0/\mathrm{Im}(Z_B))^2\right]\sinh^2(\alpha Nd)},\right. \text{ in both stopbands} \quad (19.4.11)$$

$$\text{and } |S_{21}|^2 = \left\{\frac{4}{4 + \left[(\mathrm{Re}\, Z_B/Z_0)^2 + (Z_0/\mathrm{Re}(Z_B))^2 - 2\right]\sinh^2(\beta Nd)},\right. \text{ in passband.}$$

Figure (19.36d) shows the $|S_{21}|$-passband response of the matched terminated periodic line.

In place of the simple gap capacitors, the inter-digital capacitors or the MIM capacitors can be used to form the series capacitors-loaded periodic CPW. This arrangement provides stronger loading.

Shunt Capacitor Loading

Figure (19.34c) shows the seven-sections shunt capacitors-loaded CPW on a substrate with $\varepsilon_r = 10.2$, $h = 1.27$ mm. The 95Ω CPW has a width $w = 11.5$ mm of strip conductor and $g = 3.15$ mm gap. After the shunt capacitor loading, Bloch impedance is about 50Ω. It avoids the $\lambda_g/4$ matching sections at both ends of the periodic CPW.

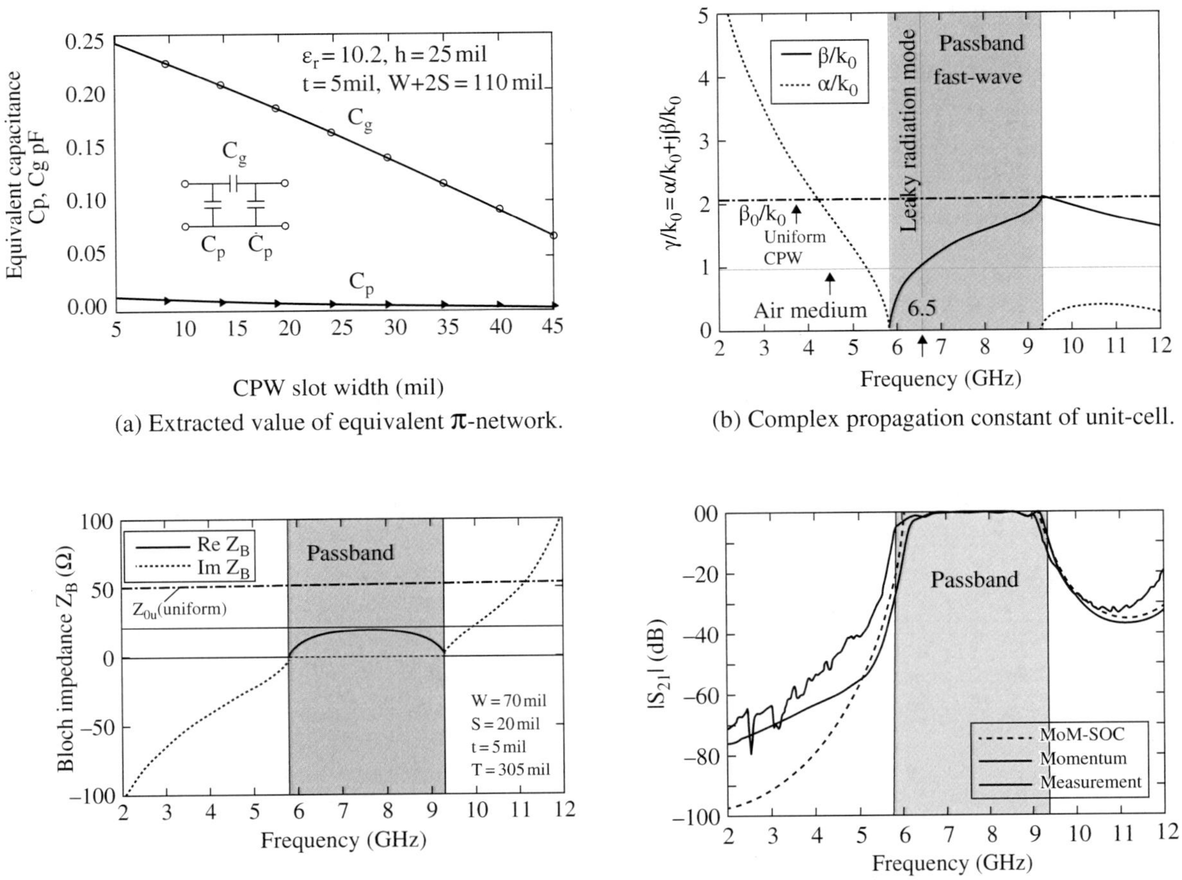

(a) Extracted value of equivalent π-network.

(b) Complex propagation constant of unit-cell.

(c) Bloch impedance of unit cell.

(d) Insertion loss of six cells EBG.

Figure 19.36 Characteristics of series capacitor-loaded periodic CPW shown in Fig. (19.34b). *Source:* Reproduced from Zhu and Shi [J.66]. © 2005 John Wiley & Sons.

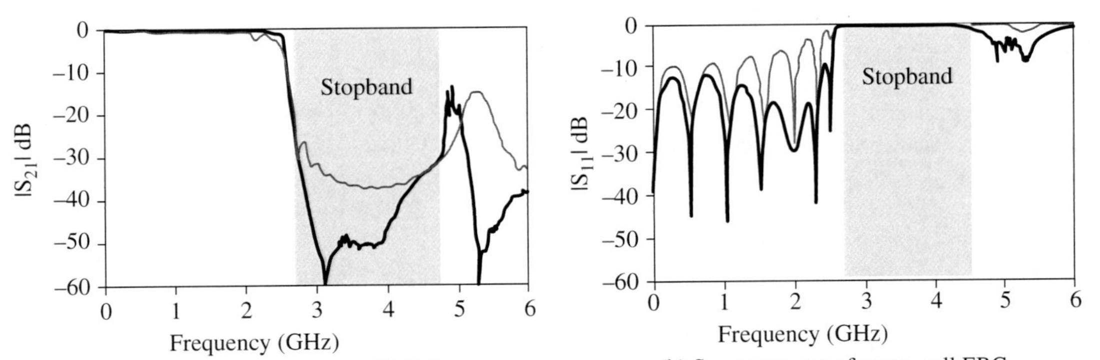

(a) S_{21}-parameter of seven-cell EBG.

(b) S_{11}-parameter of seven-cell EBG.

Figure 19.37 Characteristics of seven-cell shunt capacitor-loaded periodic CPW of Fig. (19.34c). *Source:* Reproduced from Martın et al. [J.67]. © 2003 John Wiley & Sons.

The shunt capacitances are realized by the T-type open-ended stubs, separated by a distance d = 13.3 mm. The total length of the periodic line is 9.2 mm [J.67]. The value of shunt capacitor C_{sh} = 2.3 pF is extracted using an EM-simulator. The $(\omega - \beta)$ diagram of the structure is obtained as discussed previously. In the passband, the periodic line supports the slow-wave with SWF = 2.3. Figure (19.37a,b) shows $|S_{21}|$ and $|S_{11}|$

responses giving the passband and deep attenuation stopband.

The bandwidth of the stopband can be further extended by using the dual periodicity line [J.73]. The use of the shunt connected varactor diode provides a tunable CPW resonator [J.74].

Shunt Inductor Loading

The strip conductor of a CPW can be connected to both the ground conductors, using the short-circuited stubs. These short-circuited stubs provide the shunt inductor loading to a CPW. This structure shows band-pass filter behavior. The series-connected capacitors and the shunt-connected inductors are used to get the 1D metamaterials line structures. Figure (19.34d) shows one such structure [J.68]. The metamaterials lines, i.e. the *metalines*, are discussed in subsection (22.1.5) of chapter 22.

References

Books

B.1 Kittel, C.: *Introduction to Solid State Physics*, John Wiley & Sons, New York, 1986.

B.2 Ashcroft, N.W.; Mermin, N.D.: *Solid State Physics*, Holt. Rinehart and Winston, USA, 1976.

B.3 Yariv, A.; Yeh, P.: *Optical Waves in Crystals: Propagation and Control of Laser Radiation*, Wiley-Interscience, John Wiley & Sons, 1984.

B.4 Brillouin, L.: *Wave Propagation in Periodic Structures – Electric Filters and Crystal Lattice*, 2nd Edition, Dover Pub, USA, 1953.

B.5 Fang, F.; Rahmat-Samii, Y.: *Electromagnetic Band Gap Structures in Antenna Engineering*, Cambridge University Press, UK, 2009.

B.6 Collin, R.E.: *Foundations for Microwave Engineering*, 2nd Edition, McGraw-Hill, Inc., New York, 1992.

B.7 Collin, R.E.: *Field Theory of Guided Waves*, IEEE Press, New York, 1991.

B.8 Hutter, R.G.E.: *Beam and Wave Electronics in Microwave Tubes*, D. Van Nostrand Co., Princeton, NJ, 1960.

B.9 Russer, P.: *Electromagnetics, Microwave Circuit and Antenna Design for Communication Engineering*, 2nd Edition, Artech House, USA, 2006.

B.10 Hwang, R.-B.: *Periodic Structures: Mode-Matching Approach and Applications in Electromagnetic Engineering*, John Wiley & Sons, Hoboken, NJ, 2013.

B.11 Martín, F.: *Artificial Transmission Lines for RF and Microwave Applications*, John Wiley & Sons, Hoboken, NJ, 2015.

B.12 Ramo, S.; Whinnery, J.R.; Van Duzer, T.: *Fields and Waves in Communication Electronics*, 3rd Edition, John Wiley & Sons, Singapore, 1994.

B.13 Pozar, D.M.: *Microwave Engineering*, 2nd Edition, John Wiley & Sons, Singapore, 1999.

B.14 Elliott, R.S.: *An Introduction to Guided-Waves and Microwave Circuits*, Prentice-Hall, Englewood Cliff, NJ, 1993.

B.15 Rizzi, P.A.: *Microwave Engineering – Passive Circuits*, Prentice-Hall International Edition, Englewood Cliff, NJ, 1988.

B.16 Swanson, D.G.; Hoefer, W.J.R.: *Microwave Circuit Modeling Using Electromagnetic Field Simulation*, Artech House, USA, 2003.

B.17 Bahl, I.: *Lumped Elements for RF and Microwave Circuits*, Artech House, USA, 2003.

B.18 Gupta, K.C.; Garg, R.; Bahl, I.; Bhartia, P.: *Microstrip Lines, and Slot Lines*, 2nd Edition, Artech House, Boston, 1996.

B.19 Hong, J.S.: *Microstrip Filters for RF/Microwave Applications*, 2nd Edition, John Wiley & Sons, USA, 2011.

B.20 Bahl, I.J. Bhatia, P.: *Microwave Solid State Circuit Design*, John Wiley & Sons, New York, 1988.

B.21 Simon, R.N.: *Coplanar Waveguide Circuits Components and Systems*, John Wiley & Sons, New York, 2001.

Journals

J.1 Oliner, A.A.: *Periodic structures and photonic-band-gap terminology: historical perspectives*, 29th European Microwave Conf., pp. 295–298, Munich 1999.

J.2 Maagt, P. de; Gonzalo, R.; Vardaxoglou, J.: *Review of electromagnetic bandgap technology and applications*, Electromagnetics Division, European Space Organization.

J.3 Rahmat-Samii, Y.: *The marvels of electromagnetic bandgap (EBG) structures, ACES J.*, Vol. 18, No. 4, pp. 1–10, Nov. 2003

J.4 Baccarelli, P.; Pallotto, S.; Jackson, D.R.; Oliner, A.A.: *A new Brillouin dispersion diagram for 1-D periodic printed structures, IEEE Trans. Microwave Theory Tech.*, Vol. 55, No. 7, pp. 1484–1495, July 2007.

J.5 Baccarelli, P.; Pallotto, S.; Jackson, D.R.; Oliner, A.A.: *Analysis of printed periodic structures on a grounded substrate: a new Brillouin dispersion diagram*, Digest MTT-S Int. Microwave Symp., pp. 12–17, June 2005.

J.6 Baccarelli, P.; Burrghignoli, P.; Frezza, F.; Galli, A.; Lampariello, P.; Pallotto, S.; Valerio, G.: *Dispersive analysis of wide-stopband compact EBG microstrip lines for filter applications*, ISMOT, pp. 57–60, Roma, Itay, Dec. 2007.

J.7 Harvey, A.F.: *Periodic and guiding structures at microwave frequencies, IRE Trans. Microwave Theory Tech.*, Vol. 8, pp. 30–61, Jan. 1960.

J.8 Rahman, M.; Stuchly, M. A.: *Transmission line – periodic circuit representation of planar microwave photonic bandgap structures, Microwave Optical Tech. Letters*, Vol. 30, No.1, pp. 15–19, July 2001.

J.9 Rogers, S.D.: *Electromagnetic-bandgap layers for broadband suppression of TEM modes in power planes, IEEE Trans. Microwave Theory Tech.*, Vol. 53, No. 8, pp. 2495–2505, Aug 2005.

J.10 Takagi, E.: *Frequency dependence of Bloch impedance in a periodic transmission line structure, Digest MTT-S Int. Microwave Symp.*, Vol. 2, pp. 779–782, WE2A-5, 2001.

J.11 Hong, J.S.; Lancaster, M.J.: *Theory and experiment of novel microstrip slow-wave open-loop resonator filters, IEEE Trans. Microwave Theory Tech.*, Vol. 45, No. 12, pp. 2358–2365, Dec. 1997.

J.12 Montagna, M.: *Electromagnetic modeling of mm-wave and optical periodic and quasi-periodic structures*, PhD thesis, Università Degli Studi Di Pavia, Italy.

J.13 Tavallaee, A.A.: *On the stopband characterization of periodic structures*, ME Thesis, Dept. EE, McGill University, Canada, Oct. 2006.

J.14 Wolff, I.; Knoppik, N.: *Rectangular and circular microstrip disk capacitors and resonators, IEEE Trans. Microwave Theory Tech.*, Vol. MTT-22, pp. 857–864, 1974.

J.15 Verma, A.K.; Rostamy, Z.: *Static capacitance of some multilayered microstrip capacitors, IEEE Trans. Microwave Theory Tech.*, Vol. MTT-4, No. 5, pp. 1144–1152, May 1974.

J.16 Alley, G.D.: *Interdigital capacitors and their application to lumped-element microwave integrated circuits, IEEE Trans. Microwave Theory Tech.*, Vol. MTT-18, pp. 1028–1033, 1970.

J.17 Booker, M.R.; Kamyab, M.; Jafargholi, A.; Mousavi, S. M.: *Analytical modeling of the printed dipole antenna loaded with CRLH structures, Prog. in Electromagnetics Research B*, Vol. 20, pp. 167–186, 2010.

J.18 Wang, L.; Xu, R.M.; Yan, B.: *MIM capacitor simple scalable model determination for MMIC application on GaAs, Prog. In Electromagnetics Research*, PIER 66, pp. 173–178, 2006.

J.19 Goldfarb, M.E.; Pucel, R.A.: *Modeling of via hole grounds in microstrip, IEEE Microwave Guided Lett.*, Vol. 3, No. 6, pp. 135–137, June 1991.

J.20 Nravani, B.M.; Ramahi, O.M.: *Wideband circuit model for planar EBG structures, IEEE Trans Adv. Packaging*, Vol. 33, No. 1, pp. 169–179, Feb. 2010.

J.21 Mohan, S.S.; Hershenson, M.D.M.; Boyd, S.P.; Lee, T. H.: *Simple accurate expressions for planar spiral inductances, IEEE J. Solid-State Circuits*, Vol. 34, No. 10, pp. 1419–1424, Oct. 1999.

J.22 Wheeler, H.A.: *Simple inductance formulas for radio coils, Proc. IRE*, Vol. 16, No. 10, pp. 1398–1400, Oct. 1928.

J.23 Karmakar, N.C.; Mollah, M.N.; Padhi, S.K.; Ling, R.L. L.; Roy, S.M.: *Planar electromagnetic bandgap structures, Int. J. RF and Microwave Computer-Aided Eng.* pp. 415–429, July 2006.

J.24 Verma, A.K.; Kumar, A.: *Synthesis of microstrip lowpass filter using defected ground structures, IET Microw. Antennas Propag.*, Vol. 5, No. 12, pp. 1431–1439, 2011

J.25 Verma, A.K.; Kumar, A.: *Design of low pass filters using some defected ground structures, AEU – Int. J. Electronics Communications*, Vol. 65, No. 10, pp. 864–872, 2011.

J.26 Rahman, A.A.; Verma, A.K.; Boutejdar, A.; Omar, A. S.: *Control of bandstop response of Hi-Lo microstrip low-pass filter using slot in ground plane, IEEE Trans. Microwave Theory Tech.*, Vol. MTT-52, No. 3, pp. 1008–1013, Mar. 2004.

J.27 Sumi, M.; Dinesh, R.; Ninjas, C. M.; Mridula, S.; Mohanan, P.: *Frequency coded chipless RFID tag using spurline resonators, Radioengineering*, Vol. 23, pp. 203–208, Apr. 2014.

J.28 Rahman, A.A.; Verma, A.K.; Boutejdar, A.; Omar, A. S.: *Compact stub type microstrip bandpass filter using defected ground plane, IEEE Microwave Wireless Comp. Lett.*, Vol. 14, No. 4, pp. 136–138, April 2004.

J.29 Ning, Y.; Zhining, C.; Yunyi, W.; Chia M.Y.W.: *A novel two-layer compact electromagnetic bandgap (EBG) structure and its applications in microwave circuits, Science in China (Series E)*, Vol. 46, No. 4, pp. 439–447, Aug. 2003.

J.30 Sievenpiper, D.F.; Zhang, L.J.; Broas, R.F.J. Alexopolous, N.G.; Yablonovitch, E.: *High-impedance electromagnetic surfaces with a forbidden frequency band, IEEE Trans. Microwave Theory Tech.*, Vol. 47, pp. 2059–2074, 1999.

J.31 Sievenpiper, D.F.: *High-impedance surfaces*, PhD thesis, University of California, LA, USA, 1999.

J.32 Qian, Y.; Radisic, V.; Itoh, T.: *Simulation and experiment of photonic band-gap structures for microstrip circuits*, Proc. Asia Pacific Microwave Conf., AMPC, pp. 585–588, Hong Kong, 5 Dec. 1997.

J.33 Coccioli, R.; Yang, F.R.; Ma, K.P.; Itoh, T.: *Aperture-coupled patch antenna on UC-PBG substrate, IEEE*

Trans. Microwave Theory Tech., Vol. 47, pp. 2123–2130, 1999.

J.34 Ma, K.P.; Hirose, K.; Yang, F.R.; Qian, Y.; Itoh, T.: *Realization of magnetic conducting surface using novel photonic bandgap structure, Electronics Lett.*, Vol.34, No. 21, pp. 2041–2042, Oct. 1998.

J.35 Yang, F.R.; Ma, K.P.; Qian, Y.; Itoh, T.: *Novel TEM waveguide using uniplanar compact photonic bandgap (UC-PBG) structure, IEEE Trans. Microwave Theory Tech.*, Vol. 47, No. 11, pp. 2092–2098, Nov. 1999.

J.36 Ian Rumsey, I.; Piket-May, M.; Kelly, P.K.: *Photonic bandgap structures used as filters in microstrip circuits, IEEE Microwave Guided Waves Lett.*, Vol. 8, No. 10, pp. 336–338, Oct. 1998.

J.37 Yang, H.Y.D.: *Theory of microstrip lines on artificial periodic structures, IEEE Trans Microwave Theory Tech.* Vol. 47, No. 5, pp. 629–635, May 1999.

J.38 Gauthier, G.P.; Courtney, A.; Rebeiz, G.M.: *Microstrip antenna on synthesized low dielectric constant substrates, IEEE Trans. Microwave Theory Tech.*, Vol. 45, pp. 1310–1314, Aug. 1997.

J.39 Yang, H.Y.D.; Alexopolous, N.G.; Yablonovitch, E.: *Photonic bandgap materials for high gain printed circuit antenna, IEEE Trans. Antenna Propagat.*, Vol. 45, pp. 185–187, Jan. 1997.

J.40 Laso, M.A.G.; Lopetegi, T.; Bacaicoa, T.; Hernández, J.; Gonzalo, R.; Sorolla, M.: *Arrangements of via holes in microstrip lines as metallodielectric periodic structures*, 2000 Asia-Pacific Microwave Conference. Proceedings (Cat. No.00TH8522), Sydney, NSW, Australia, pp. 1257–1260, 2000.

J.41 Falcone, F.; Lopetegi, T.; Sorolla, M.: *1-D and 2-D photonic bandgap microstrip structures, Optical Tech. Letters*, Vol. 22, No. 6, pp. 411–412, Sept. 1999.

J.42 Lopetegi, T.; Laso, M.A.G.; Erro, M.J.; Benito, D.; Garde, M.J.; Falcone, F.; Sorolla, M.: *Novel photonic bandgap microstrip structures using network topology, Optical Tech. Letters*, Vol. 25, No.1, pp. 34–36, Apr. 2000.

J.43 Garde, I.; Yábar, M.J.; del Río, C.: *Simple modelling of DGS to design 1D-PBG low-pass filters, Microwave Optical Tech. Letters*, Vol. 37, No. 3, pp. 228–232, May 2003.

J.44 Karbassian, M.M.; Ghafouri-Shiraz, H.: *Effect of shape of patterns on the performance of microstrip photonic band-gap filters, Microwave & Optical Tech. Letters*, Vol. 48, No.6, pp. 1007–1011, Jun. 2006.

J.45 Zhu, L.; Bu, H.; Wu, K.: *Unified CAD model of microstrip line with backside aperture for multilayer integrated circuit, MTT-S Int. Microwave Symp.*, pp. 981–985, WE4B-4, 2000.

J.46 Zhu, L.; Bu, H.; Wu, K.: *Aperture compensation technique for innovative design of ultra-broadband microstrip bandpass filter, MTT-S Int. Microwave Symp.*, pp. 315–318, TU4C-2, 2000.

J.47 Karmakar, N.C.; Mollah, M.N.: *Investigation into non-uniform photonic bandgap microstrip line low-pass filters, IEEE Trans. Microwave Theory Tech.*, Vol. 51, No.2, pp. 564–572, Feb.2003.

J.48 Huang, S.Y.; Lee, Y.H.: *Tapered dual-plane compact electromagnetic bandgap microstrip filter structures, IEEE Trans. Microwave Theory Tech.*, Vol. 53, No.9, pp. 2656–2664, Sept. 2005.

J.49 Siakavara, K.: *Novel microwave microstrip filters using photonic bandgap ground plane with fractal periodic pattern, Optical Tech. Letters*, Vol. 43, No. 4, pp. 273–276, Nov. 2004.

J.50 Liu, H.-W.; Sun, X.-W.; Li, Z.-F.; Yoshimasu, T.: *Propagation characteristics of microstrip lines with 2D defected ground structures, Int. J. RF Microwave Computer-Aided Engineering*, pp. 280–286, Feb. 2006.

J.51 Qian, Y.; Itoh, T.: *Planar periodic structures for microwave and millimeter wave circuit applications, MTT-S Int. Microwave Symp.*, pp. 1533–1536, TH2C-3, 1999.

J.52 Zhu, L.: *Guided-wave characteristics of periodic microstrip lines with inductive loading: slow-wave and bandstop behaviors, Microwave Optical Tech. Letters*, Vol. 41, No. 2, pp. 77–79, April 2004.

J.53 Kazerooni, M.; Cheldavi, A.; Kamarei, M.: *Analysis, modeling, and design of cascaded defected microstrip structure for planar circuits, Int. J. RF Microwave Computer-Aided Engineering*, pp. 170–182, Jan. 2010.

J.54 Xue, Q.; Shum, K.M.; Chan, C.H.: *Novel 1-D microstrip PBG cells, IEEE Microwave Guided Waves Lett.*, Vol. 10, No. 10, pp. 403–405, Oct. 2000.

J.55 Xue, Q.; Shum, K.M.; Chan, C.H.: *Novel perforated microstrip PBG cell, Microwave Optical Tech. Letters*, Vol. 26, No. 5, pp. 325–327, Sept. 2000.

J.56 Liu, H.W.; Sun, X.W.; Li, Z.F.: *A low-pass filter of wide stopband with a novel multilayer fractal photonic bandgap structure, Optical Tech. Letters*, Vol. 40, No. 5, pp.431–432, Mar. 2004.

J.57 Lee, C.O.; Shin, C.-C.: *Enhanced periodic structures with compact size, Electronics Letters*, Vol. 40, No. 8, pp. 479–481, April 2004.

J.58 Liu, Z.; Xiao, G.; Zhu, L.: *Numerical de-embedding and experimental validation of propagation properties of corrugated substrate integrated waveguide, Microwave Optical Tech. Letters*, Vol. 58, No.10, pp. 2456–2060, Oct. 2016.

J.59 Glandorf, F.J.; Wolff, I.: *A spectral-domain analysis of periodically nonuniform microstrip lines, IEEE Trans.*

Microwave Theory Tech., Vol. 35, No. 3, pp.336–343, Mar.1987.

J.60 Nesic, D.; Nesic, A.: *Bandstop microstrip PBG filter with sinusoidal variation of the characteristic impedance and without etching in the ground plane, Microwave Optical Tech. Letters*, Vol. 29, No.6, pp. 418–420, June 2001.

J.61 Laso, M.A.G.; Lopetegi, T.; Erro, M.J.; Benito, D.; Garde, M.J.; Sorolla Ayza, M.: *Multiple frequency tuned photonic bandgap microstrip structures, IEEE Microwave Guided Wave Lett.*, Vol. 10, pp. 220–222, 2000.

J.62 Lopetegi, T.; Laso, M.A.G.; Erro, M.J.; Benito, D.; Garde, M.J.; Falcone, F.; Sorolla, M.: *Novel photonic bandgap microstrip structures using network topology, Microwave Optical Tech. Letters*, Vol. 25, pp. 33–36, 2000.

J.63 Matekovits, L.; Colemè, G.V.; Orefice, M.: *Controlling the band limits of TE-surface wave propagation along a modulated microstripline-based high impedance surface, IEEE Trans. Antennas Propag.*, Vol. 56, No. 8, pp. 2555–2562, Aug. 2008.

J.64 Yun, T.Y.; Chang, K.: *One-dimensional photonic bandgap resonators and varactor tuned resonators, IEEE Int. Microwave Sym., MTT(s) Digest*, TH3B-4, pp. 1629–1632.

J.65 Zhu, L.: *Guided-wave characteristics of periodic coplanar waveguides with inductive loading – unit-length transmission parameters, IEEE Trans. Microwave Theory Tech.*, Vol. 51, No.10, pp. 2133–2138, Oct. 2003.

J.66 Zhu, L; Shi, H.: *Frequency-dependent guided-wave characteristics of periodically series-capacitive loaded coplanar waveguides, Optical Tech. Letters*, Vol. 46, No. 1, pp. 54–58, Jul. 2005.

J.67 Martın, F.; Falcone, F.; Bonache, J.; Lopetegi, T.; Laso, M.A. G.; Sorolla, M.: *New CPW low-pass filter based on a slow wave structure, Optical Tech. Letters*, Vol. 38, No.3, pp. 190–193, Aug. 2003.

J.68 Gao, J.; Zhu, L.: *Guided wave characteristics of coplanar waveguide metamaterials composed of unsymmetrical unit cells*, Proceeding Asia-Pacific Microwave Conf., APMC, 2006

J.69 Her, M.-L.; Chang, C.-M.; Wang, Y.-Z.; Kung, F.-H.; Chiou, Y.-C. *Improved coplanar waveguide (CPW) bandstop filter with photonic bandgap (PBG) structure, Optical Tech. Letters*, Vol. 38, No.4, pp. 274–277, Aug. 2003.

J.70 Liu, H.; Yoshimasu, T.; Kurachi, S.; Hu, W.; Yamaoka, K.; Sun, L.: *Compact slot resonator-based photonic bandgap for coplanar waveguide and its application to bandstop filter, Optical Tech. Letters*, Vol. 48, No.8, pp. 1602–1606, Aug. 2006.

J.71 Karim, M.F.; Liu, A.-Q.; Alphonse, A.; Zhang, X.J.; Yu, A.B.: *CPW band-stop filter using unloaded and loaded EBG structures, IEE Proc.-Microw. Antennas Propag.*, Vol. 152, No. 6, pp. 434–440, Dec. 2005.

J.72 Mao, S.-G.; Chen, M.-Y.: *Propagation characteristics of finite-width conductor-backed coplanar waveguides with periodic electromagnetic bandgap cells, IEEE Trans. Microwave Theory Tech.*, Vol. 50, No. 11, pp. 2624–2628, Nov. 2002.

J.73 Martín, F.; Falcone, F.; Bonache, J.; Lopetegi, T.; Laso, M.A.G.; Sorolla, M.: *Dual electromagnetic bandgap CPW structures for filter applications, IEEE Microwave Wireless Components Lett.*, Vol.13, No. 9, pp. 393–395, Sept. 2003.

J.74 Gil, I.; Bonache, J.; Garcia, G.J.; Martin, F.: *Application of active electromagnetic bandgaps to the the design of tunable resonators in CPW technology, Optical Tech. Letters*, Vol. 45, No. 3, pp. 229–232, May. 2005.

20

Planar Periodic Surfaces

Introduction

The planar periodic surface is the 2D-extension of the 1D-periodic line discussed in chapter 19. These are the reactively loaded 2D periodic surfaces, exhibiting the discrete spectrum of wave motion in the 2D K-space through alternate passband–stopband phenomena. In this case, Brillouin Zone is two-dimensional that provides complete suppression of the surface waves on a substrate. The *metasurfaces* are also the engineered 2D periodic surface. However, the metasurfaces have no bandgap. It is discussed in chapter 22. So, the planar periodic surface is in general called the electronic bandgap (EBG) surface. It has several forms depending upon the applications. The EBG surface has the following interesting features:

- Creation of alternate 2D passband–stopband over the engineered surface. A natural surface supports the continuous RF spectrum.
- Suppression of surface waves on the planar surface, and creation of the high impedance; or artificial magnetic conductor, i.e. AMC surface, to improve the performances of the planar antennas.
- Supporting the fast-wave propagation leading to Cherenkov-type radiation. It is useful for the design of the leaky-wave antenna.
- Creation of the frequency selective surfaces (FSSs) used in the antenna technology.

Objectives

- *To discuss characteristics of some engineered planar 2D EBG surface.*
- *To discuss the 2D EBG circuit models and realization of the EBG planar surfaces.*

20.1 2D Planar EBG Surfaces

The basic characteristics of the EBG surfaces are obtained through simulations on the EM-Simulators. However, it is a time-consuming process. The initial design and optimization of the EBG surfaces are carried out with the help of the circuit models. The propagating 2D Bloch modes and complete surface waves bandgap are treated through the *2D dispersion diagram* in the k-space. The frequency-dependent phase response, revealing the surface property of the EBG, is explored through the *reflection phase diagram*. This section discusses the basic characteristics of two generic EBG surfaces – the *mushroom-type EBG* and the *uniplanar compact EBG*, i.e. the UC-EBG. The reflection and dispersion diagrams are presented to understand the characteristics of the EBG surface. The models of the EBG surface are further discussed in the next section (20.2).

20.1.1 General Introduction of EBG Surfaces

The 2D-EBG surface is a natural extension of the 1D-periodic transmission lines. The microstrip patch antenna and planar circuits are designed on the conductor-backed dielectric substrates, supporting the undesired surface waves of both TE and TM polarizations. The presence of the surface waves degrades the performance of both the microstrip circuits and antennas. Likewise, horizontally placed dipole near reflecting conducting planes, i.e. near a perfect electric conductor (PEC), suffers a reduction in gain and impedance matching. In general, a special type of artificial surface is needed to suppress the surface waves of both polarizations, in all azimuthal directions. In case such an artificial surface is used as a reflector, it should provide *in-phase reflection* to improve the gain of a wire antenna or other antennas. Such in-phase reflection is possible with the help of a non-existing perfect magnetic conductor (PMC). The 2D-planar EBG creates such artificial surfaces, with both characteristics – *suppression of the surface waves and in-phase reflection*, over a certain frequency band. The EBG surface with unique property is a man-made surface, beyond (*meta*) the natural material

Introduction to Modern Planar Transmission Lines: Physical, Analytical, and Circuit Models Approach, First Edition. Anand K. Verma.

surface. So, it is also called the *metasurface*. However, the periodic surfaces without any bandgap property are treated as the metasurfaces. These are discussed in chapter 22. As a reflecting artificial surface, the EBG or metasurface nearly behaves as the PMC. The EBG surface over a certain frequency band is designated as the *artificial magnetic surface (AMC)* [B.1, J.1–J.4]. The EBG surfaces are the well-known frequency selective surfaces (FSSs), placed on the top of the conductor-backed substrate. The FSS has several applications in antenna technology. It is also used as an ultra-wideband shield [B.2, J.5].

The EBG surface shows alternate 2D passband–stopband in all azimuthal directions on its surface. Each passband corresponds to one Bloch mode, propagating in the form of the TM/TE surface waves. The 2D-dispersion characteristics of the Bloch modes are obtained with the help of the 2D-irreducible Brillouin zone (IBZ) in the k-space, discussed in section (19.1.3) of chapter 19. The unloaded conductor-backed substrate is the host medium. The reactive loading elements, i.e. conducting patches and strips, are inclusions arranged two-dimensionally over the host medium. The square and rectangular lattices are commonly used to get the *isotropic* and *anisotropic* EBG surfaces. However, sometimes hexagonal lattice is also used. The *isotropic EBG surface* has identical reflection coefficients, for the normally incident plane waves, in both the x-direction and y-direction on an EBG plane. The reflection coefficient of the *anisotropic EBG surface* is direction-dependent. The anisotropic surface is used to convert the linearly polarized wire antenna to a circularly polarized antenna [B.1, B.3, J.6–J.10].

Conceptually, the EBG surface is described by the LC parallel resonant circuit distributed over the surface. It works as the planar bandpass and bandstop filters to create alternate passband–stopband for the 2D Bloch mode surface waves. At resonance, the EBG surface has an almost infinite surface impedance, with in-phase reflection. So, at resonance, it acts as a *PMC* surface. However, within its reflection bandwidth, it acts as an AMC surface with high surface impedance. The AMC surface is also called the *high impedance surface (HIS)*. The HIS has a surface impedance equal to, or more than, the intrinsic impedance of free space. The *reflection phase bandwidth* for the AMC surface is defined as the frequency range corresponding to the $\pm 90^\circ$ reflection phase variation [B.1, B.2, J.4, J.11, J.12].

Outside the $\pm 90^\circ$ reflection phase bandwidth, the EBG surface is still reactive and within another BW; over the $(90^\circ \pm 45^\circ)$ reflection phase, the surface is defined as the *reactive impedance surface (RIS)* [J.13]. A compact antenna has better impedance matching over the *RIS bandwidth*. So, $90^\circ \pm 45^\circ$ reflection phase bandwidth is also called the *matching BW*, whereas the $\pm 90^\circ$ reflection phase bandwidth is called the *AMC bandwidth*. The AMC bandwidth may or may not coincide with the stopband bandwidth, i.e. the bandgap, obtained from the complete suppression of the surface waves. The *complete suppression* of the surface waves shows their suppression in all azimuthal directions on the EBG surface. The *partial suppression* of surface waves occurs in a specific direction only. The dispersion diagram provides the *stopband bandwidth*. The $|S_{21}|$-response also provides the information on its stopband, i.e. bandgap [J.6, J.7]. The periodic surface, i.e. the metasurface, in this capacity is called the *electromagnetic bandgap* (EBG) structures. The EBG surface is used as a holographic surface [J.14]. The meaning of various electromagnetic surfaces, derived from the EBG surface or metasurface, and their bandwidths are further explained and computed using the closed-form models.

Figure (20.1a–f) shows two generic EBG surfaces – the *mushroom-type EBG* [J.6, J.7] and the *uniplanar compact EBG, i.e. the UC-EBG* [J.15] and some of their variations. Figure (20.1a–i) shows the structure of a mushroom-type EBG. It is created by connecting an array of square or rectangular patches, located at the substrate, to the ground conductor, using the metallic vias/pins. The hexagonal patches are also used [J.10]. Figure (20.1a–ii) shows the development of a simple parallel LC resonant circuit model. It is distributed over the surface and is used to model the surface impedance of a mushroom EBG. The capacitance of the LC-model is due to the gap-capacitance between adjacent patches. The inductance is caused by two vias and the ground conductor.

Figure (20.1b) shows another *capacitive patch-loaded* EBG surface. In this arrangement, vias are absent. This structure does not provide any stopband characteristics for the surface waves. However, it has a frequency-dependent resonant reflection phase response, as the conductor-backed substrate provides an inductive load at the surface [B.3]. The EBG surface is a RIS, useful for designing the compact antenna with good impedance matching [J.13]. Figure (20.1c) is a further modification of the open mushroom structure. In this case, a top shield is added. The mushroom EBG structure is located within a *parallel-plate waveguide* (PPW). In place of a simple ground conductor, this structure is used as a composite ground plane for the high-speed digital circuits. It suppresses the simultaneous switching noise (SSN) or the ground board noise (GBN) of the high-speed digital circuits [J.16, J.17].

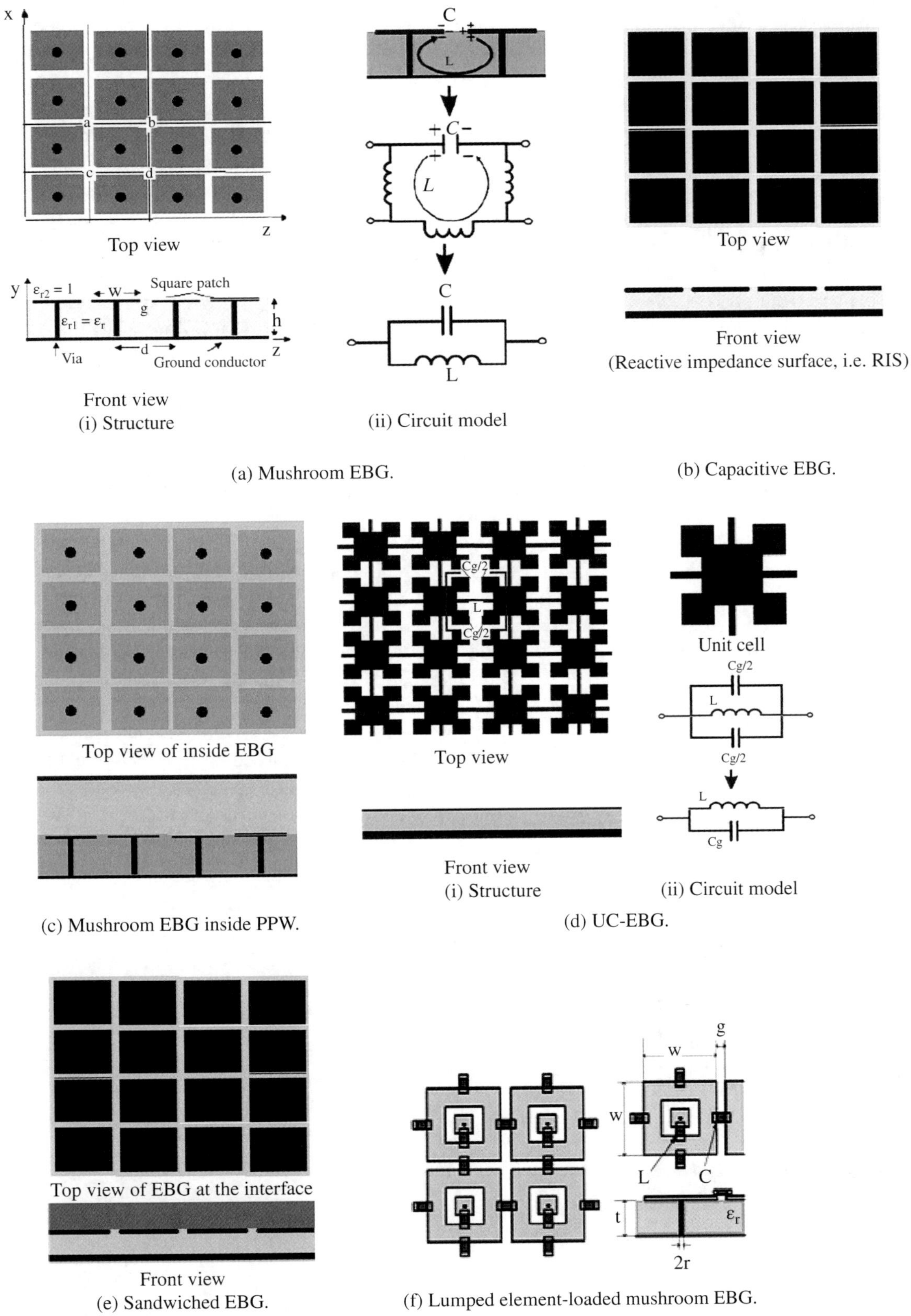

Figure 20.1 Some 2D-planar EBG surfaces.

Figure (20.1d-i) shows the structure of a *uniplanar compact EBG* (UC-EBG), different in concept from that of the mushroom type. There is no via connecting the patch to the ground. A narrow microstrip line connects two unit cells, providing an inductance across the gap capacitance. Figure (20.1d-ii) shows the development of a parallel LC-resonant circuit model for the surface impedance of the structure. Unlike the mushroom-type EBG, the UC-EBG can be easily fabricated even on alumina or on a semiconductor substrate.

Figure (20.1e) shows that the capacitive EBG can also be embedded inside the dielectric media. It is located at the interface of two dielectric layers, forming the *sandwiched EBG*. Over the top of the second layer, an antenna is mounted to improve its performance, say to improve its circular polarization [J.18, J.19]. Figure (20.1f) shows the mounting of the lumped chip capacitors across the gaps between two patches to increase the value of the gap capacitance. It brings down the resonance frequency, so that an EBG surface can operate even in the UHF range. However, this arrangement reduces the BW of the EBG surface. Occasionally, the lumped inductances are also added to the vias, before connecting them to a ground plane. The resonance frequency is further reduced and the BW of the EBG is improved [J.20]. In place of capacitors, varactor diodes are used to get a tunable EBG surface. It is useful for the beam steering of an antenna array [J.21, J.22].

20.1.2 Characteristics of EBG Surface

The artificial periodic surfaces, i.e. the *metasurfaces*, are characterized with the help of the *dispersion diagram* and *reflection phase diagram*. The EM-simulators, such as the CST Microwave Studio, Ansoft HFSS, etc., generate both characteristics diagrams, using their *eigenmode solver*. Figure (20.2a and b) shows the geometrical setups of a rectangular/square unit cell of the EBG surface for simulating its dispersion and reflection phase characteristics, respectively. The periodic boundary conditions (PBCs) are applied to all four faces in the x- and y-directions. An absorbing airbox is located at the top of the setup in the far-field region. Likewise, Floquet port illuminates the EBG surface with polarization-dependent and incident angle-dependent plane waves. In the case of a rectangular lattice, the lattice constant is different in both directions. However, for a square one, it is identical in both the directions. The user's manual of an EM-simulator provides necessary information on such a simulation setup and process.

A periodic metasurface does not need to act simultaneously both as the AMC and EBG, over an identical frequency band. Both frequency bands, i.e. the surface wave bandgap and in-phase reflection band, may not even overlap. It happens only for certain geometrical parameters of a periodic surface. The resonance frequency f_r, giving zero reflection phase, $\phi_{EBG}(f_r) = 0°$, could be within the stopband, or it may be outside the frequency band [J.22, J.23]. The reflection phase of the EBG surface can be polarization-dependent, i.e. an EBG surface can provide different phase variations in the mutually orthogonal directions on its surface. Its resonance frequency can also be polarization-dependent. Such EBG surfaces are *anisotropic*. This kind of surface is used to get the circularly polarized waves, from a

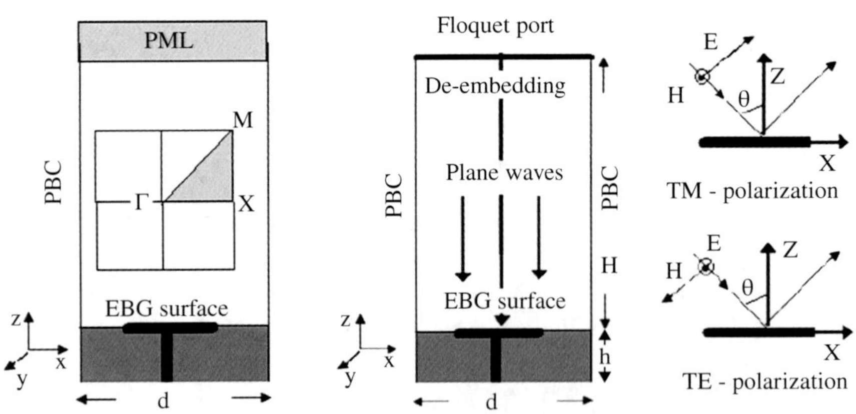

(a) Dispersion simulation setup. (b) Reflection-phase simulation setup.

Figure 20.2 EBG surface setup for EM-simulation.

linearly polarized wire antenna. These are also used to get wideband circular polarization from a narrow band circularly polarized patch antenna [J.8, J.9, J.18, J.19, J.24]. The EBG resonator antenna can provide high gain [J.25]. The EBG surface with the polarization-independent reflection phase is known as the *isotropic EBG* surface. Likewise, the EBG surface can support several kinds of Bloch mode propagations, such as the slow-wave, fast-wave, in both the forward mode and backward mode. For an open structure, it also supports the leaky waves, just like the 1D artificial periodic lines.

The $\omega - \beta$ Dispersion Diagram

The simulation setup shown in Fig (20.2a) is simulated in .the *eigenmode solver* of the EM-simulators to get the $(\omega - \beta)$ diagram of a unit cell of an infinite extent periodic structure. The simulation is carried out over the perimeter of an IBZ, described in the k-space with Γ, X, M as the symmetric points. The x-directed propagation phase $\beta_x d$ of Bloch modes is varied in steps, from Γ (0, 0) to X (π, 0), while taking $\beta_y d = 0$. For each value of $\beta_x d$, the eigenmode solver numerically solves the *eigenfrequencies* of several selected number of Bloch modes of the TM/TE types. It generates the $\omega - \beta$ dispersion diagram for the component of the surface waves propagating in the x-direction. In a case of no real solution, the surface waves do not propagate and the stopband is created in the x-direction. Next, the edge of IBZ, in the XM direction, is scanned by varying the propagation phase $\beta_y d$ from X (π, 0) to M (π, π), while taking $\beta_x d = \pi$. It generates the $\omega - \beta$ dispersion diagram for the component of the surface wave propagating in the y-direction. Finally,

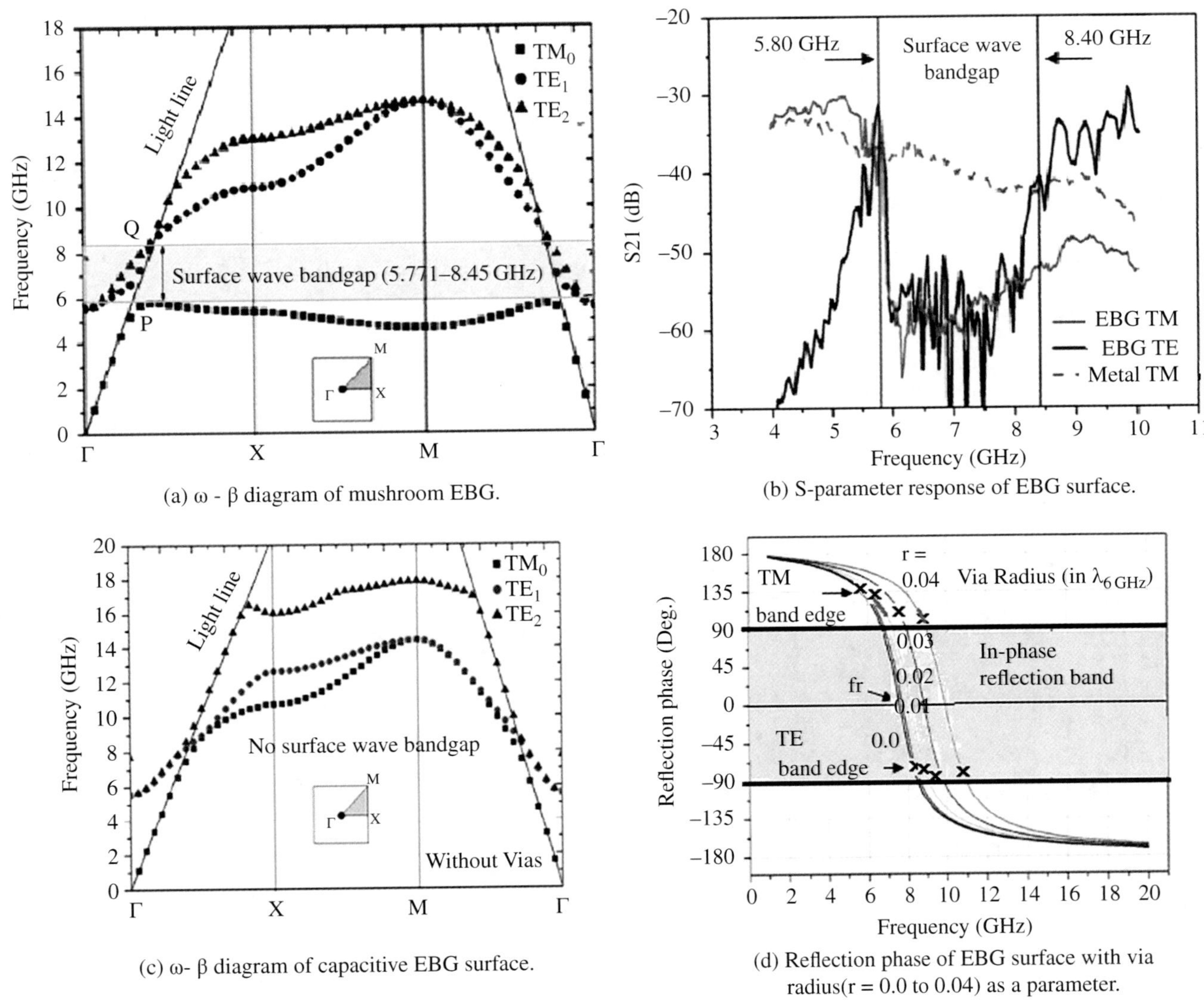

(a) ω - β diagram of mushroom EBG.

(b) S-parameter response of EBG surface.

(c) ω- β diagram of capacitive EBG surface.

(d) Reflection phase of EBG surface with via radius(r = 0.0 to 0.04) as a parameter.

Figure 20.3 Dispersion and reflection phase diagram of the EBG surface. *Source*: Reproduced from Li et al. [J.26]. © 2008 with permission of AIP Publishing.

the edge MΓ is scanned by varying both $\beta_x d$ and $\beta_y d$ simultaneously, from M (π, π) to Γ (0, 0), to get the dispersion in the diagonal direction. The common stopband BW provides the complete stopband BW for the surface wave. It is taken as the bandgap of the 2D EBG surface [B.1, J.26, J.27].

Figure (20.3a–c) illustrates the $\omega - \beta$ dispersion diagrams of the *mushroom-type EBG* and *capacitive surface* (mushroom EBG without vias). These are obtained using the EM-simulator [J.23, J.26]. The parametric study has been done for five structural parameters: patch width (w), patch gap (g), a substrate thickness (h), via radius (r), and relative permittivity (ε_r). Figure (20.1a-i) shows the isotropic EBG surface with a period of d = w + g. The following initial parameters, at 6 GHz, are taken for the mushroom EBG unit cell: w = 0.14 λ_{6GHz}, g = 0.006 λ_{6GHz}, h = 0.03 λ_{6GHz}, r = 0.01λ_{6GHz}, and $\varepsilon_r = 2.65$, where λ_{6GHz} is the free space wavelength at 6 GHz. Figure (20.3a) shows that the first propagating Bloch mode is TM_0 surface wave, with no cutoff frequency. It is below the light line, and closely follows it up to point P at the stopband edge, showing the forward slow-wave propagation, and then it shows a backward slow-wave propagation with the negative group velocity.

The second Bloch mode (TE_1 surface wave mode) commences at the point Q, i.e. at the upper edge of the stopband. The Bloch modes, outside the light cone, are bound modes. The stopband of width 5.771 GHz–8.45 GHz is visible in the dispersion diagram. The third Bloch mode (TE_2 surface wave mode) also appears, without any bandgap. Inside the light cone, TE_1 and TE_2 modes are the fast-waves. They are the leaky modes of the open EBG surface. Figure (20.3b) shows the measured S_{21} response of the 20 × 20 unit cells; size 146 mm × 146 mm. The TM and TE surface modes are excited and received by a pair of very short vertical dipoles and a pair of loops, respectively [J.3, J.28, J.29]. It shows 5.80 GHz–8.4 GHz wide stopband, nearly the same as obtained from the $\omega - \beta$ diagram. The nature of the Bloch modes, as the TM/TE mode, could be decided from their field structures on the EM-simulator [J.23].

Figure (20.3c) shows the dispersion diagram of the capacitive EBG surface of Fig (20.1b). It does not exhibit any bandgap. Therefore, the vias are necessary for the creation of a bandgap. However, Fig (20.3d) shows the reflection phase diagram with an (±90°) in-phase reflection band created for capacitive EBG structure (via radius = 0). Figure (20.3d) also shows the reflection phase of the mushroom-type EBG surface with finite-size via radius. The small cross, marked on the reflection diagram, shows the TM and TE band-edges of the $\omega - \beta$ diagram, corresponding to the lower and upper edge of the bandgap of the surface waves. An increase in the radius of via brings the TM band-edge near to (+90°) reflection phase. Otherwise, the bandgap and in-phase reflection band do not normally overlap. However, in the case of $d/h \leq 2$, there is a simultaneous occurrence of the surface wave stopband and in-phase reflection bandwidth. Figure (20.4a and b) demonstrates it. For Fig (20.4a), the EBG structure has d/h = 4.0, and for Fig (20.4b) d/h = 2.0. For the former case, the TM band edge occurs at reflection phase +135°, and for the latter case, it occurs at +90°. The lower and upper band edges of the bandgap, also the reflection-phase BW, are influenced by the structural parameters of the EBG surface [J.23, J.26]. The simulation process of the reflection diagram is discussed below.

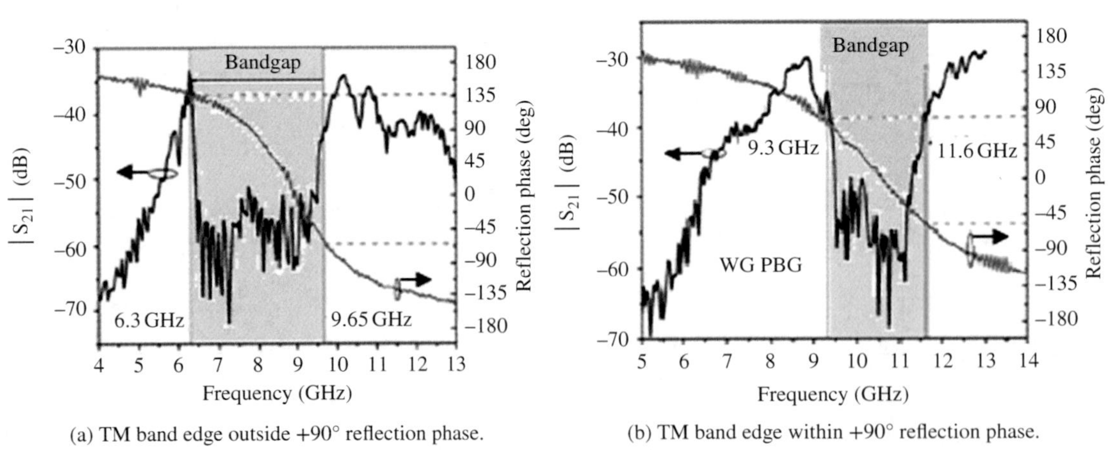

Figure 20.4 Comparison of surface wave bandgap and reflection bandwidth of the mushroom EBG. *Source*: Reproduced from Li et al. [J.26]. © 2008 with permission of AIP Publishing.

The Reflection Phase Diagram

In a case of reflection phase simulation, the Floquet port, located at the height z = H, shown in Fig (20.2b), illuminates the EBG surface with a normal incidence of the TM/TE polarizations. Both the normal incidence and oblique incidence of the plane waves on the EBG surface are used to know the stability of reflection-phase behavior with respect to the angle of incidence. In the case of the TE-polarization, E_y component is normal to the vias. The vias are not excited, and the presence or absence of vias does not influence the reflection-phase response of the TE-polarization. For the normal incidence, the E_x component of the TM-polarization is normal to the vias. Again, the vias are not excited, and both the normal incident TE- and TM-polarized waves have identical reflection phase response.

The reflection phase is needed at the surface of the EBG, as its characteristics. So, the de-embedded phase information at the surface is to be obtained. The EM-simulator provides it. However, understanding of the de-embedding process is useful for the measurement of a reflection phase of an EBG surface [J.3, J.28–J.30]. First, the reflection phase of a conducting surface (PEC) is obtained at height H in the far-field region. Next, at the same height, the reflection phase response of the EBG surface is recorded. At the certain resonance frequency, the reflection phase difference of the PMC is π radian, i.e. 180°, with respect to the reflection phase of the reference PEC. We compute the reflection-phase variation of the EBG surface with frequency from the phase difference of the EBG and PEC surfaces as follows:

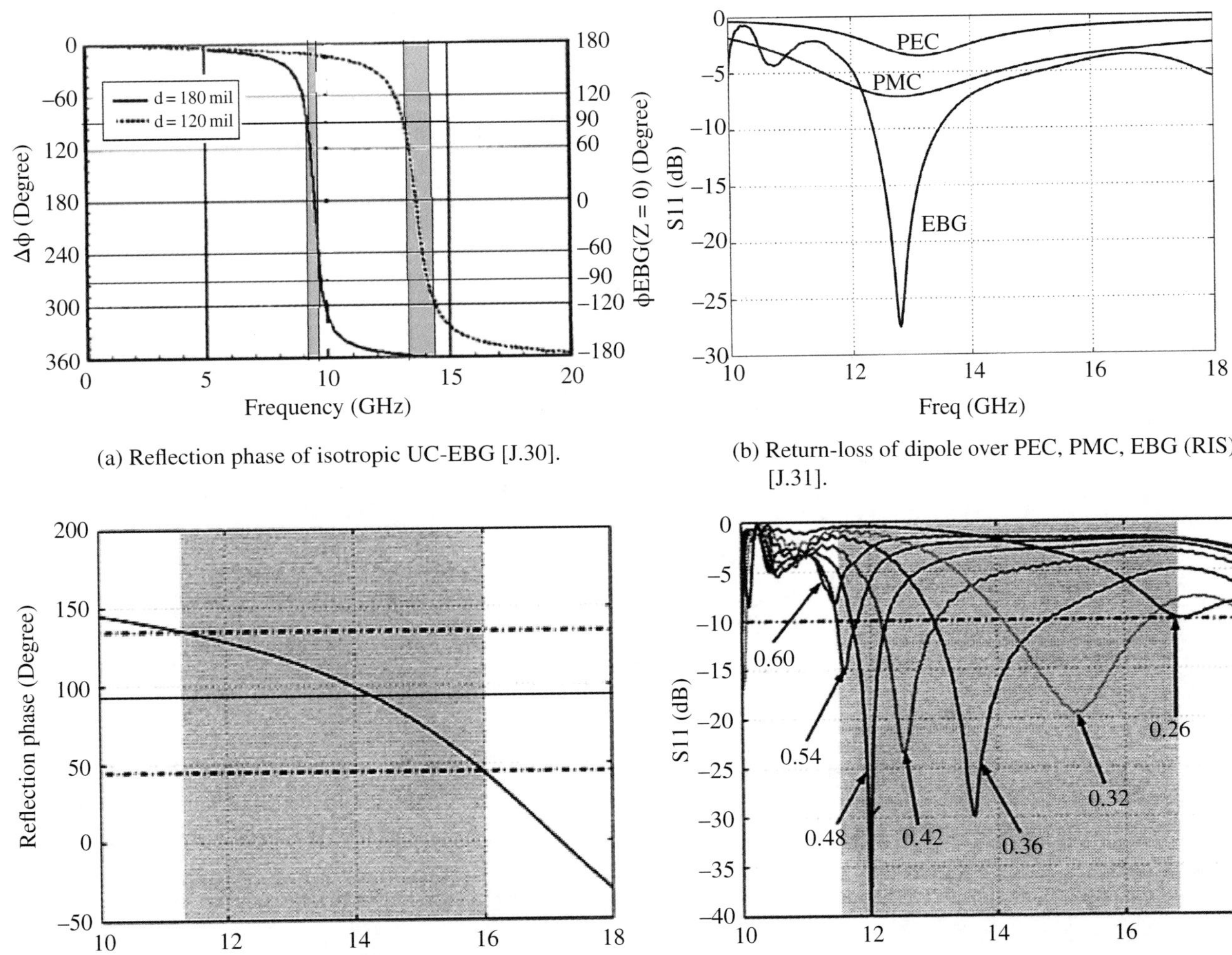

(a) Reflection phase of isotropic UC-EBG [J.30].

(b) Return-loss of dipole over PEC, PMC, EBG (RIS) [J.31].

(c) Reflection phase BW corresponding to 90°± 45° for mushrooms EBG(RIS) [J.31].

(d) RL of dipoles with resonant lengths, shown in the unit of, λ_{12GHz} within the reflection-phase BW of RIS [J.31].

Figure 20.5 Reflection phase of the EBG surface. *Source*: Reproduced from Yang et al. [J.30]. © 1999 IEEE. Reproduced from Yang and Rahmat-Sammi [J.31]. © 2003 IEEE.

$$
\begin{aligned}
&\phi_{PEC}(z = H) = 2kH + \pi && \text{(a)}\\
&\phi_{EBG}(z = H) = 2kH + \phi_{EBG}(z = 0) && \text{(b)}\\
&\Rightarrow \phi_{EBG}(z = 0) = \Delta\phi + \pi = \Delta\phi + 180^\circ && \text{(c)}\\
&\text{where, } \Delta\phi = \phi_{EBG}(z = H) - \phi_{PEC}(z = H) && \text{(d).}
\end{aligned}
\tag{20.1.1}
$$

Figure (20.5a) illustrates the simulated frequency dependence of the reflection phase difference $\Delta\phi$. The EBG reflection phase $\phi_{EBG}(z = 0)$ at the surface of the UC-EBG is also shown. The isotropic UC-EBG is designed on a substrate with $\varepsilon_r = 10.2$, h = 25 mil and period d = 180mil and 120 mils [J.30, J.33]. The EBG surface acts as a PMC at resonance frequencies 13.4 GHz and 9.5 GHz, corresponding to d = 120 mil and 180 mils, respectively. Over the reflection bandwidths, shown in gray strips, the EBG surface acts as an AMC surface [J.4].

20.1.3 Horizontal Wire Dipole Near EBG Surface

A horizontally polarized wire antenna placed near the perfectly conducting surface (PEC) does not radiate properly due to out of phase reflection. A PMC surface provides constructive interference, and such an antenna radiates in a better way. However, Fig (20.5b) shows that the return-loss is not good for an antenna placed over the PMC. The EBG surface, defined by a bandwidth corresponding to the reflection phase $\phi_{EBG}(z = 0) = 90^\circ \pm 45^\circ$, improves the return-loss [J.13, J.31]. Figure (20.5c) shows the reflection-phase diagram with a gray strip defining the reflection-phase bandwidth over $\phi_{EBG}(z = 0) = 90^\circ \pm 45^\circ$. The reflection-phase response is obtained for the square mushroom-type EBG shown in Fig (20.1a). The structure has the following dimensions: $w = 0.12\ \lambda_{12GHz}$, $g = 0.02\ \lambda_{12GHz}$, $h = 0.04\ \lambda_{12GHz}$, $r = 0.005\ \lambda_{12GHz}$, $\varepsilon_r = 2.20$. Over the frequency band, corresponding to $\phi_{EBG}(z = 0) = 90^\circ \pm 45^\circ$, the

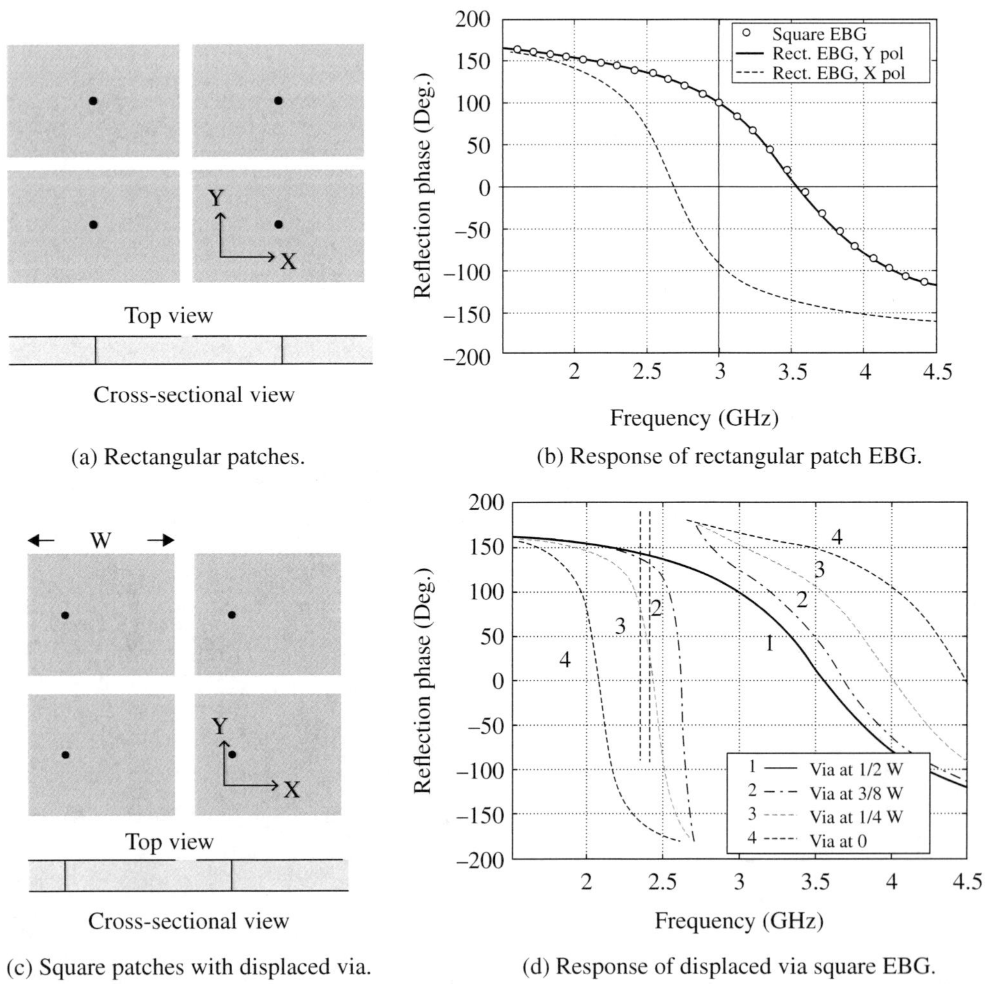

Figure 20.6 Reflection phase of the anisotropic mushroom EBG surface. *Source*: Reproduced from Yang and Rahmat-Sammi [J.32]. © 2004 John Wiley & Sons.

EBG surface is inductive. It is known as the *reactive impedance surface (RIS)*. Usually, the RIS is an inductive surface. Figure (20.5d) shows that any horizontally placed dipole, with a resonant length within the BW of the RIS, has proper return-loss. The frequency band corresponding to $\phi_{EBG}(z=0) = 90^\circ \pm 45^\circ$ is known as the *matching bandwidth* or the *RIS bandwidth*. Figure (20.5b) compares results of a dipole of length 0.40 λ_{12GHz} placed over the PEC, PMC, and EBG (RIS) [J.31].

Anisotropic EBG Surface

The anisotropic EBG surface has polarization-dependent, and also incidence angle-dependent, reflection-phase response. It provides different resonance frequencies and different frequency-dependent phase behavior even for the orthogonal polarizations in the plane of the EBG. The anisotropy is realized by having different lattice constants in the x- and y-directions, or by using the asymmetric loading elements. For instance, a mushroom-type EBG can have anisotropic surface by using a rectangular patch, shown in Fig (20.6a). It is also realized by using rectangular slots-loaded square patch, or by using asymmetrically placed via connected to square patches, shown in Fig (20.6c) [J.9, J.32, J.34]. The rectangular patch has a dimension $0.24\lambda_{3GHz} \times 0.16\lambda_{3GHz}$ on a substrate with $\varepsilon_r = 2.20$, $h = 0.04\lambda_{3GHz}$. The gap between adjacent patches is 0.02 λ_{3GHz} and the radius of vias is 0.0025 λ_{3GHz}. The λ_{3GHz} is the free-space wavelength at 3 GHz. Figure (20.6b) shows the reflection-phase responses of both the rectangular anisotropic and square isotropic EBG surfaces. The isotropic surface has a square patch of dimension $0.16\lambda_{3GHz} \times 0.16\lambda_{3GHz}$. The x-polarized normal incident wave has lower resonance frequency, as compared to that of y-polarized incident waves. It is due to the longer length of the rectangular patch along the x-direction. The identical phase response is obtained for y-polarized waves incident on both the rectangular and square patches. At nearly 2.9 GHz, the $\mp 90^\circ$ reflection phase occurs for the x- and y-polarized waves. Figure (20.6d) shows that the displaced via, along the width w of the patch in the x-direction, has dual resonances with dual reflection BW. The maximum frequency difference between dual resonances is for vias at the edge of the patches.

The anisotropic EBG surface acts both as a *reflecting surface* and as a *ground plane*. The rectangular patch anisotropic mushroom EBG, as a reflecting surface, controls the polarization of the incident right-hand circularly polarized (RHCP) waves [J.32]. The reflected waves can have either RHCP or elliptical polarization. It can also acquire linear polarization. The PEC, PMC, and isotropic EBG surfaces change the polarization of the reflected waves to the LHCP. Likewise, a rectangular patch anisotropic mushroom EBG also acts as the ground plane to an antenna located near to it. In this case, it is used to get the circularly polarized radiated waves from a linearly polarized wire antenna placed at the 45° orientation on the EBG surface, shown in Fig (20.7c) [J.9, J.34]. Both aspects of the anisotropic EBG surface are examined below.

EBG as a Reflecting Surface

Figure (20.7a) shows a reflecting EBG surface with normally incident RHCP waves and reflected waves. The reflecting EBG surface is in the (x-y)-plane. The incident and reflected waves are in the ($\mp$) z-direction. The electric fields of the normally incident RHCP and reflected waves are

$$\vec{E}^{in} = \hat{x}e^{jkz} + \hat{y}je^{jkz} \qquad \text{(a)}$$

$$\vec{E}^{ref} = \hat{x}e^{-jkz+j\theta_x} + \hat{y}je^{-jkz+j\theta_y} \qquad \text{(b)}, \qquad (20.1.2)$$

where θ_x and θ_y are the reflection phases of the EBG surface, for the x-polarized and y-polarized waves, respectively. The reflected waves travel away from the observer, i.e. away from the EBG surface. The time-dependent factor $e^{j\omega t}$ is suppressed in the above equations.

The reflected waves can be rewritten as a linear combination of two circularly polarized waves [B.4]:

$$\vec{E}^{ref} = e^{-jkz}.e^{j\theta_x}\left[\hat{E}_L\left(\frac{1+e^{j(\theta_y-\theta_x)}}{\sqrt{2}}\right) + \hat{E}_R\left(\frac{1-e^{j(\theta_y-\theta_x)}}{\sqrt{2}}\right)\right] \qquad \text{(a)}$$

$$\text{where,}\quad \hat{E}_L = \frac{\hat{x}+j.\hat{y}}{\sqrt{2}} \quad \text{(b)}, \qquad \hat{E}_R = \frac{\hat{x}-j.\hat{y}}{\sqrt{2}} \quad \text{(c)}. \qquad (20.1.3)$$

In the above equations, $\hat{E}_L$ and $\hat{E}_R$ are the unit vectors for the LHCP and RHCP waves, respectively. For the PEC reflecting surface, $\theta_x = \theta_y = \pi$, and for the PMC reflecting surface, $\theta_x = \theta_y = 0$. In both cases, the coefficient of the unit vector $\hat{E}_R$ is zero. Thus, in both cases, the reflected waves are the LHCP waves, because the RHCP is zero. To retain the RHCP, even after reflection, we must have $(\theta_y - \theta_x) = 180^\circ$. It results in zero value for the coefficient of $\hat{E}_L$, resulting in the RHCP for the reflected waves. At 2.9 GHz, Fig (20.6b), the

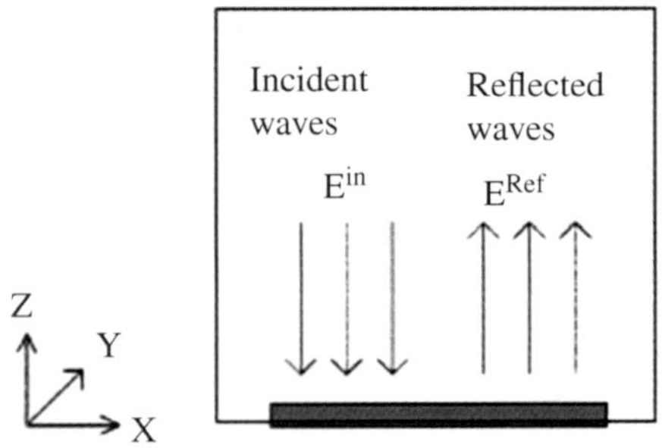

(a) Reflecting EBG surface.

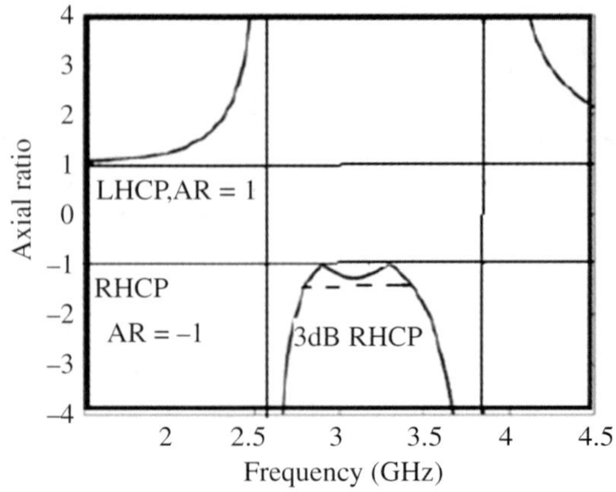

(b) Polarization control [J.32].

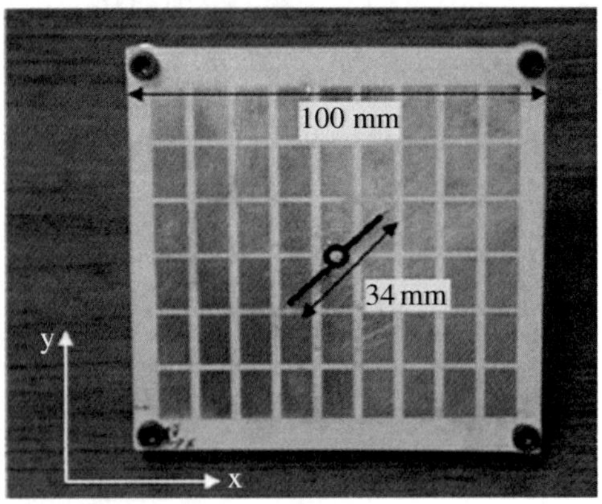

(c) EBG ground plane supporting 45° inclined wire dipole [J.9].

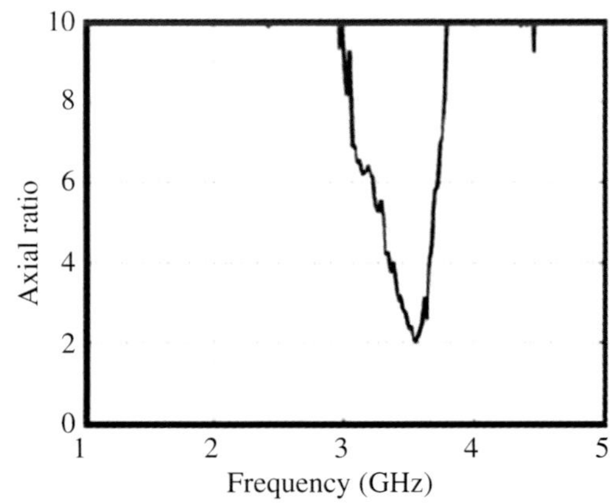

(d) Circularly polarized waves from the wire antenna [J.9].

Figure 20.7 EBG reflection surface and ground plane for polarization control. *Source*: Reproduced from Yang and Rahmat-Samii [J.9]. © 2005, IEEE. Reproduced from Yang and Rahmat-Samii [J.32]. © 2004, John Wiley & Sons.

phase difference 180° is achieved after the reflection. The axial ratio (AR) of the reflected waves is obtained as follows:

$$AR = -\frac{|E_R| + |E_L|}{|E_R| - |E_L|} = -\frac{\left|\left(1 - e^{j(\theta_y - \theta_x)}\right)/\sqrt{2}\right| + \left|\left(1 + e^{j(\theta_y - \theta_x)}\right)/\sqrt{2}\right|}{\left|\left(1 - e^{j(\theta_y - \theta_x)}\right)/\sqrt{2}\right| - \left|\left(1 + e^{j(\theta_y - \theta_x)}\right)/\sqrt{2}\right|}. \tag{20.1.4}$$

Figure (20.7b) shows the variation in the AR with frequency. At 2.90 GHz and 3.30 GHz, AR = −1, and a pure RHCP wave is obtained after reflection because $(\theta_y - \theta_x) = 180°$. It has 3 dB ARBW from 2.81 GHz to 3.43 GHz (19.9%). At 2.58 GHz, the phase difference is 90°. It gives the linear polarization with AR = ∞. Figure (20.7b) also shows the range of LHCP.

EBG as a Ground Plane Surface

Figure (20.7c) shows a 45° inclined wire antenna on a rectangular patch mushroom EBG surface that acts as a virtual anisotropic ground plane [J.9]. The dipole is located at a small distance h = 0.02 λ above the ground plane. The radiated wave $\vec{E}^{rad}$ is a summation of the direct wave $\vec{E}^{dir}$ from the dipole and the reflected wave $\vec{E}^{ref}$ from the conducting ground plane, located at distance h below the EBG surface:

$$\vec{E}^{rad} = \vec{E}^{dir} + \vec{E}^{ref} = \frac{E_0}{2}\left(\hat{x}e^{-jkz} + \hat{y}e^{-jkz}\right) + \frac{E_0}{2}\left(\hat{x}e^{-jkz - 2jk_dh + j\theta_x} + \hat{y}e^{-jkz - 2jk_dh + j\theta_y}\right). \tag{20.1.5}$$

In the present case, $k_dh \approx 0$. If EBG acts as a PEC, then $\theta_x = \theta_y = \pi$. In this case, the above expression provides zero radiated field. If EBG acts as a PMC, then $\theta_x = \theta_y = 0°$. In this case, $\vec{E}^{rad} = E_0 e^{-jkz}(\hat{x} + \hat{y})$, i.e. the radiated wave is linearly polarized. However, if at certain frequency $\theta_x = +90°$, $\theta_y = -90°$ simultaneously, then the above expression for the radiated field is

$$\vec{E}^{rad} = \frac{E_0}{2} e^{-jkz}[(\hat{x} + \hat{y}) + j(\hat{x} - \hat{y})]. \tag{20.1.6}$$

In this case, the radiated field is RHCP waves. The mushroom EBG is designed on the RT-Duriod 6002 substrate, $\varepsilon_r = 2.94 \pm 0.04$, h = 6.1 mm, with rectangular patches of size 8 mm × 13 mm. The gap between patches is 2mm in the x-direction, and 1mm along the y-direction. The size of the surface is 100 mm × 100 mm, with 9 × 6 rectangular patches. The length of the dipole is 34 mm at the height 3 mm and the radius of the wire is 0.34 mm. It is fed by a 50Ω coaxial cable. Figure (20.7d) shows a 2 dB AR in the broadside direction. It occurs at 3.56 GHz with 3 dB AR BW 5.6% (3.45 GHz–3.65GHz). The performance of the antenna as a CP antenna is not satisfactory. However, it demonstrates the concept of the EBG surface as the polarization converter.

20.2 Circuit Models of Mushroom-Type EBG

This section describes lumped circuit models also called the *effective surface models*. They provide the conceptual understanding of the EBG responses, both the reflection and dispersion responses, presented in the previous subsections. More realistic and accurate 2D transmission line-based models are discussed in section (20.4).

20.2.1 Basic Circuit Model

Figure (20.1a-i) shows a mushroom EBG of (w × w) square patch on a substrate with relative permittivity ε_r, and thickness h. Its lattice constant, i.e. period, is d. Its basic circuit model, shown in Fig (20.1a-ii), is the LC parallel resonant circuit distributed over the surface. The gap, between adjacent patches, contributes toward capacitance C, and via provides the inductance. The series resistance R can be further added with an inductor and conductance G across the capacitor to account for the conductor and dielectric losses, respectively. The EBG surface acts as the 2D bandstop filter to suppress the TM surface waves, around the resonance frequency ω_0. The EBG structure creates multiple passband/stopband that suppresses the TE surface waves also. However, the basic LC-circuit model explains the creation of the first bandgap only, and does not capture the higher-order bandgap. The surface impedance (Z_s), without the EBG, is very low, and it supports the propagation of the fundamental TM surface waves. However, the EBG surface shown in Fig (20.8a) provides a HIS suppressing the fundamental TM surface wave mode. At the resonance frequency ω_0 and over a bandwidth $\Delta\omega = \omega_H - \omega_L$ around ω_0, the reactive surface impedance is very high. At the upper and lower band frequencies, (ω_H, ω_L), the surface impedance is equal to the intrinsic impedance of free space, i.e. $|Z_s| = |\pm jX_s| = \eta_0$.

The surface capacitance C F/Sq. and surface inductor L H/Sq. of the basic circuit model are approximately evaluated using the following expressions [J.6, J.7]:

$$C = \frac{w(\varepsilon_{r1} + \varepsilon_{r2})}{\pi} \cosh^{-1}\left(\frac{d}{g}\right) \quad (a), \quad L = \mu_r \mu_0 h \quad (b). \tag{20.2.1}$$

The present structure has $\varepsilon_{r1} = \varepsilon_r$, $\varepsilon_{r2} = 1$, $\mu_r = 1$. The LC-circuit model provides the following expression for evaluating the surface impedance Z_s Ω/sq of the lossless EBG surface:

$$Z_s(\omega) = \frac{j\omega L}{1 - (\omega/\omega_0)^2} \quad (a), \quad \text{where,} \quad \omega_0 = \frac{1}{\sqrt{LC}} \quad (b). \tag{20.2.2}$$

At the resonance frequency, $\omega = \omega_0$, the surface impedance becomes infinite $Z_s \rightarrow \infty$ in the absence of losses. For

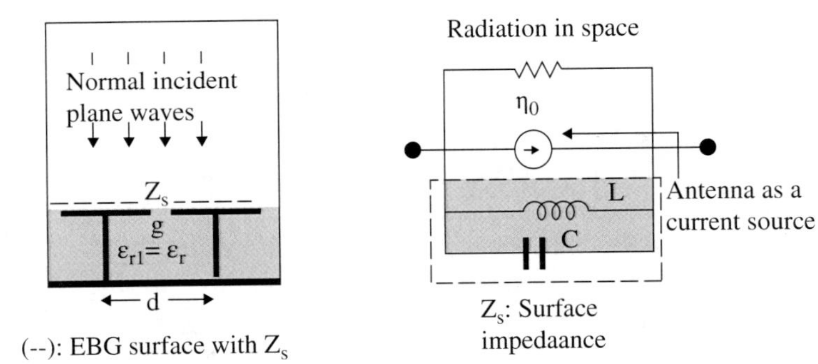

(a) Normal incidence on EBG. (b) Antenna over AMC.

Figure 20.8 Circuit model of the radiating antenna located over the EBG surface.

$\omega < \omega_0$, $Z_s = j\,X_L$, i.e. the EBG surface is inductive, whereas for $\omega > \omega_0$, $Z_s = j\,X_C$, i.e. the EBG surface is capacitive. In the case of the normal incidence of the plane waves, shown in Fig (20.8a), both kinds of reactive surfaces totally reflect the incident waves. However, reflection phases from them are different. The EBG surface contributes to the reflection coefficient. It is computed using the equivalent transmission line concept as follows:

$$\Gamma(\omega) = \frac{Z_s(\omega) - \eta_0}{Z_s(\omega) + \eta_0} \quad \text{(a)}$$

$$\text{at } \omega = \omega_0, \Gamma(\omega_0) = +1\angle 0^\circ \quad \text{(b)}$$

$$\text{for } \omega < \omega_0, \Gamma_L(\omega) = \frac{j\,X_s(\omega) - \eta_0}{j\,X_s(\omega) + \eta_0} = 1\angle\phi_L$$

$$\text{where, } \phi_L = \pi - 2\tan^{-1}(X_s(\omega)/\eta_0) \quad \text{(c)}$$

$$\text{for } \omega > \omega_0, \Gamma_L(\omega) = \frac{-j\,X_s(\omega) - \eta_0}{-j\,X_s(\omega) + \eta_0} = 1\angle\phi_H$$

$$\text{where, } \phi_H = -\pi + 2\tan^{-1}(X_s(\omega)/\eta_0) \quad \text{(d)}. \qquad (20.2.3)$$

In the above equation, free space impedance is $\eta_0 = 120\,\pi$. The frequency-dependent surface impedance $Z_s(\omega)$ of equation (20.2.2a) provides the inductive surface reactance $j\,X_s(\omega)$. The reflection-phase is frequency-dependent. At the resonance frequency, ω_0, the EBG surface acts as a *PMC* with the reflection phase $\phi = 0^\circ$ and the surface impedance $Z_s(\omega = 0) \rightarrow \infty$. The reflection phase ($\phi_L$) is positive forming an inductive surface at the lower frequency below ω_0, i.e. for $\omega \rightarrow 0$, $\phi_L \rightarrow +\pi$. Whereas, at the higher frequency above ω_0, the phase (ϕ_H) is negative forming a capacitive surface, i.e. for $\omega \rightarrow \infty$, $\phi_H \rightarrow -\pi$. Therefore, at very low and at very high frequencies, the EBG surface acts as a *PEC* as the surface impedance $Z_s(\omega \rightarrow 0) = Z_s(\omega \rightarrow \infty) = 0$. Thus, the EBG surface can create an *artificial electric conductor* (AEC) with out-of-phase reflection. At certain lower and higher frequencies, ω_L and ω_H, the relation $|jX_L| = |jX_H| = \eta_0$ is obtained. At ω_L and ω_H, the reflection phases are $\phi_L = \pi/2$ and $\phi_H = -\pi/2$. *Thus, the reflection bandwidth* $\Delta\omega$ *is defined over the reflection phase* $\phi = \pm\pi/2$. In the frequency range (ω_H, ω_L), the EBG surface acts as a *HIS*, as over this frequency band $\eta_0 \leq |Z_s| < \infty$. Almost *in-phase* reflection is maintained within $\Delta\omega$. The EBG surface over (ω_H, ω_L) frequency band acts as an *artificial magnetic conductor (AMC)* [J.4].

The circuit model explains the bandgap for the suppression of surface waves. The bandgap occurs within, or in the neighborhood, of the reflection bandwidth. However, the results of EM-simulations show that the surface wave bandgap and reflection bandwidth may not be identical. In both lower frequency range $0 < \omega \leq \omega_L$ and upper frequency $\omega_H \leq \omega < \infty$ range, $|Z_s| < \eta_0$; the EBG surface is a *low impedance surface (LIS) with* an *out-of-phase* reflection. The impedance matching of an antenna located over the PMC, or even over AMC, is not satisfactory, as surface impedance is high compared to free space impedance. Better impedance matching occurs over the frequency band defined by the reflection phase, $\phi(\omega) = (90^\circ \pm 45^\circ)$ [J.13, J.31]. It is the range of the matched inductive RIS surface. The phase angle $\phi(\omega)$ is measured with respect to the positive $\mathrm{Re}(\Gamma)$ axis. The polar reflection diagram, shown in Fig (20.9), demonstrates the frequency-dependent behavior of the reflection phase, and corresponding changes in the nature of the EBG surface. The outer circle shows the clockwise increasing frequency variation from $\omega = 0$ to $\omega = \infty$. The inner circle shows the corresponding variation in the reflection phase from $\phi = +\pi$ to $\phi = -\pi$. The inner area of the circle shows the corresponding nature of the periodic surface. It is noted that usual matched *inductive* RIS is created around ω_L; however, around ω_H another matched *capacitive* RIS could be developed.

Figure (20.8b) models the radiating antenna as a Norton current source, in the presence of the AMC surface. The current source feeds radiated power to the load η_0, i.e. to free space. The radiated power drops by 3 dB at the reflection bandwidth edge, ω_L and ω_H, for $|Z_s(\omega)| = \eta_0$. It helps to obtain an expression for the *radiation bandwidth* (BW) in terms of circuit elements as follows:

$$\left|\frac{j\omega L}{1-(\omega/\omega_0)^2}\right| = \eta_0, \Rightarrow \omega^2 = \frac{1}{LC} + \frac{1}{2\eta_0^2 C^2} \pm \frac{1}{\eta_0 C}\left[\frac{1}{LC} + \frac{1}{4\eta_0^2 C^2}\right]^{1/2} \qquad (20.2.4)$$

Normally L is about 1nH/Sq., and C is in the range of 0.05–10pF/sq. The term $1/\eta_0^2 C^2$ can be ignored, giving the following expression:

$$\omega^2 \approx \frac{1}{LC} \pm \frac{1}{\eta_0 C\sqrt{LC}}, \Rightarrow \omega = \omega_0\sqrt{1 \pm \frac{Z_0}{\eta_0}}, \omega \approx \omega_0\left[1 \pm \frac{Z_0}{2\eta_0}\right]. \qquad (20.2.5)$$

In the above expression, $Z_0 = \sqrt{L/C}$ is a kind of characteristic impedance $Z_0 << \eta_0$; and $\omega_0 = 1/\sqrt{LC}$. The EBG bandwidth is given as

$$\mathrm{BW} = \frac{\omega_H - \omega_L}{\omega_0} = \left[1 + \frac{1}{2}\frac{Z_0}{\eta_0}\right] - \left[1 - \frac{1}{2}\frac{Z_0}{\eta_0}\right] = \frac{Z_0}{\eta_0}$$

$$\Rightarrow \mathrm{BW} = \frac{1}{\eta_0}\sqrt{\frac{L}{C}}. \qquad (20.2.6)$$

To lower the resonance frequency of an AMC surface, the C and L can be increased together, or individually.

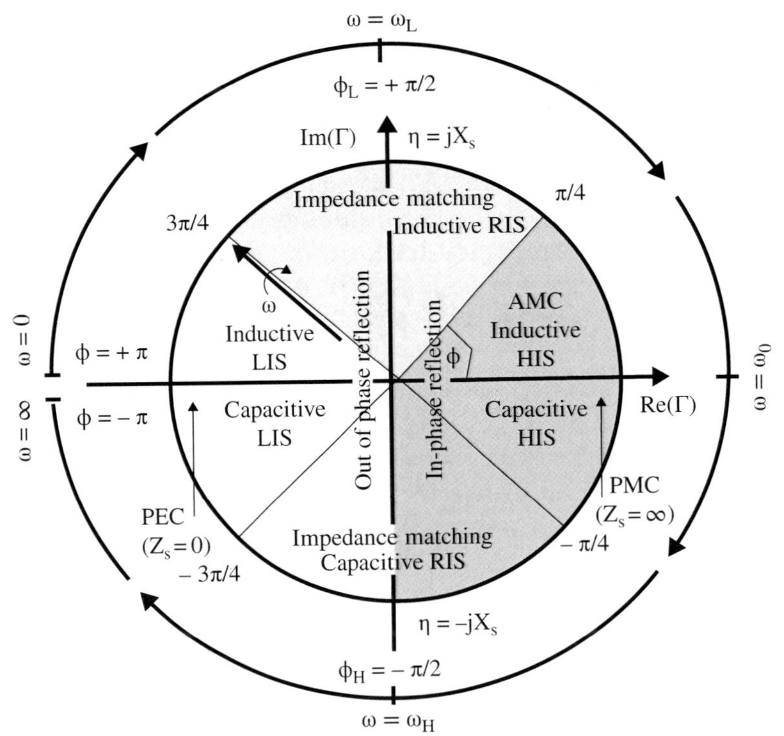

Figure 20.9 Polar reflection diagram of EBG surface showing the nature of the surface impedance and reflection BW. AMC is from $\phi_L = +\pi/2$ to $\phi_H = -\pi/2$.

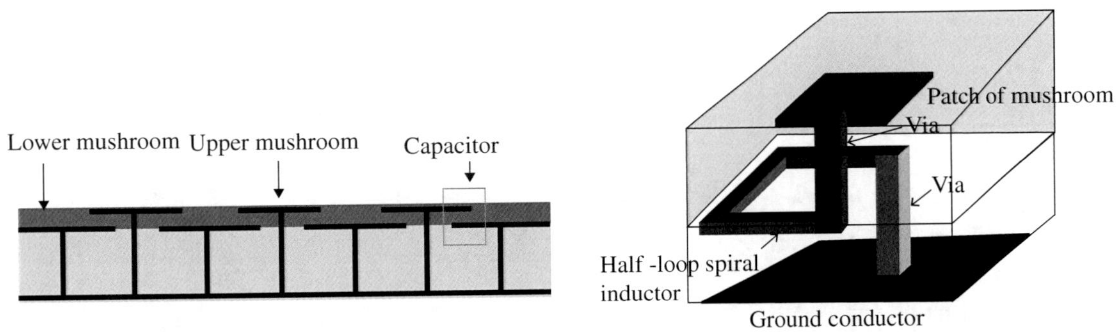

(a) Two-level mushroom EBG.

(b) Half-loop spiral inductor with via.

Figure 20.10 Schemes to increase C and L of a mushroom EBG surface.

However, increasing C reduces the BW, whereas increasing L improves it. Figure (20.10a) shows a multi-layer mushroom scheme to increase the capacitance [J.35]. Figure (20.10b) shows another scheme to increase the inductance of a via adding one-turn spiral inductor. The multi-turn inductor is also used to get a higher value of inductance [J.36, J.37]. These arrangements provide compact EBG at a lower frequency. A simple EBG operates around 10 GHz for a few mm-thick substrates with relative permittivity ε_r in the range 2–10. However, the structure of Fig (20.10a) has a bandgap between 2.2 GHz and 2.5 GHz. The scheme of Fig (20.10b) operates in the frequency range of 1.1 GHz–2.7 GHz. The inductance can also be increased by using the magnetodielectric substrate with the permeability of more than one [J.38]. However, it is difficult to get such substrates.

The simple circuit model also provides the *dispersion relation* for the first two Bloch modes propagating on the EBG surface. However, it is valid only for 1D wave propagation, without any Bragg's reflection. The inductive surface supports the TM-type Bloch mode, whereas the capacitive surface supports the TE-type Bloch mode.

The AMC does not support any surface wave propagation, creating the bandgap. The wave vector k_0 follows the following general dispersion expression:

$$k_o^2 = \mu_o \varepsilon_0 \omega^2 = \beta_x^2 + \beta_y^2 + \beta_z^2. \tag{20.2.7}$$

The EBG surface of Fig (20.1a) is in the (x-z)–plane, where the z-axis is in the direction of propagation of surface waves, and $\beta_x = 0$. In the y-direction, the field is confined in the evanescent mode, i.e. $\beta_y = j\,\alpha$, where α is the attenuation constant. The above general dispersion relation is reduced to

$$\beta_z^2 = \mu_o \varepsilon_0 \omega^2 + \alpha^2. \tag{20.2.8}$$

The surface impedance of the TM and TE-polarized surface waves, given by equation (7.6.1) of chapter 7, is summarized below:

$$Z_s^{TM} = \frac{j\alpha_{TM}}{\omega\varepsilon_0} \quad (a), \qquad Z_s^{TE} = \frac{-j\omega\mu_0}{\alpha_{TE}} \quad (b). \tag{20.2.9}$$

Using equation (20.2.8) with the above equations, the dispersion relations, for both surface waves modes, are obtained below:

$$Z_s^{TM} = \frac{j}{\omega\varepsilon_0}\left[\beta_{z,TM}^2 - \mu_o\varepsilon_0\omega^2\right]^{1/2} \qquad (a)$$

$$Z_s^{TE} = \frac{-j\omega\mu_0}{\left[\beta_{z,TE}^2 - \mu_o\varepsilon_0\omega^2\right]^{1/2}} \qquad (b)$$

$$\Rightarrow \beta_{z,TM} = \frac{\omega}{c}\left[1 - \left(\frac{Z_s^{TM}}{\eta_0}\right)^2\right]^{1/2} = \frac{\omega}{c}\left[1 + \left(\frac{X_s^{TM}}{\eta_0}\right)^2\right]^{1/2} \qquad (c)$$

$$\beta_{z,TE} = \frac{\omega}{c}\left[1 - \left(\frac{\eta_0}{Z_s^{TE}}\right)^2\right]^{1/2} = \frac{\omega}{c}\left[1 + \left(\frac{\eta_0}{X_s^{TE}}\right)^2\right]^{1/2} \qquad (d). \tag{20.2.10}$$

The surface impedance, $Z_s^{TM} = Z_s^{TE} = Z_s$, is taken from equation (20.2.2). The basic circuit model does not provide polarization-sensitive surface impedance. For $\omega < \omega_0$, surface impedance is inductive and an EBG supports the TM-mode with propagation constant $\beta_{z,TM}$. While for $\omega > \omega_0$, surface impedance is capacitive and the EBG supports the TE-mode with propagation constant $\beta_{z,TE}$. Figure (20.11a and b) shows the *reflection phase diagram* and *dispersion diagram* of an EBG surface for the sheet capacitance C = 0.05pF/sq and sheet inductance L = 2nH/sq [J.6, J.7].

In the frequency range,$0 < \omega < \omega_0$, the surface impedance is inductive, i.e. $Z_s^{TM} = j\,X_s^{TM}$. It increases with frequency, resulting in the decreasing reflection phase, as shown in Figs (20.9 and 20.11a), from π to $\pi/2$ (at ω_L), then to 0° at ω_0. Further increase in the frequency results into the capacitive EBG surface and its reflection phase is $\phi(\omega) = -\pi/2$ at ω_H and moves to $-\pi$ at $\omega \to \infty$.

Figure (20.11b) shows the nature of the dispersion of the first two propagating modes supported by the EBG surface. Initially the propagation constant $\beta_{Z,TM}$ of the TM surface wave increases and follows the light line, as $\beta_{Z,TM} = \omega/c$, for $|X_s^{TM}| < < \eta_0$. At ω_L, $|X_s^{TM}| = \eta_0$, corresponding to $\phi_L = \pi/2$, and the propagation constant is $\beta_{Z,TM} = (\omega/c)\sqrt{2}$. For a further increase in frequency, the $\beta_{Z,TM}$ bends horizontally, showing a presence of the TM slow-waves. It is a bounded non-radiating first

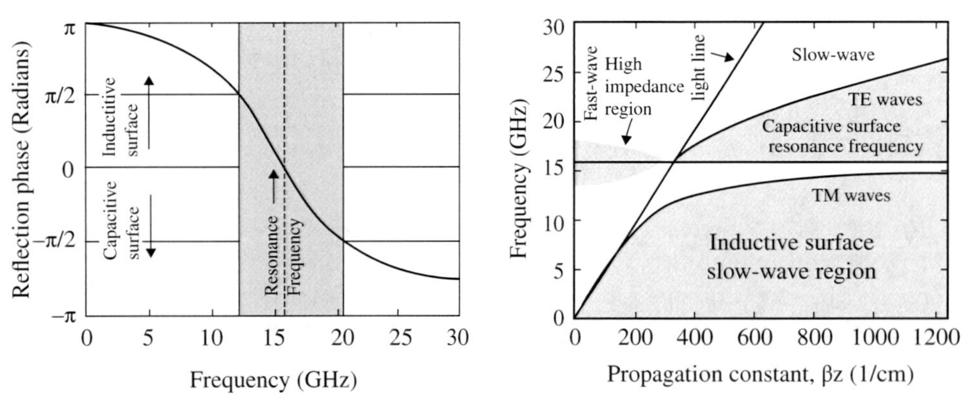

(a) Refection phase diagram (b) Dispersion diagram

Figure 20.11 Reflection and dispersion diagrams of mushroom EBG for C = 0.05pF/Sq and L = 2nH/Sq. *Source*: Reproduced from Sievenpiper et al. [J.7]. © 1999, IEEE.

Bloch mode. At ω_0, $|X_s^{TM}| \rightarrow \infty$; giving $\phi(\omega) = 0°$ and $\beta_{Z,TM} \rightarrow \infty$. The TM mode stops propagation. At ω_0, the propagation constant is $\beta_{Z,TE} = \omega/c$ for the TE surface waves. It starts its propagation at $\omega = \omega_0$, and passes through $|X_s^{TE}| = \eta_0$, corresponding to $\phi_H = -\pi/2$ at ω_H. The reflection phase decreases monotonically with frequency, from 0 to $-\pi/2$, and next to $-\pi$. The TE surface wave mode is the next Bloch wave. It is also a slow-wave.

The basic circuit model, i.e. the *effective surface impedance model*, does not show any bandgap, as the TM-mode ends at ω_0, and the TE-mode starts from there. The basic circuit model does not use the Floquet–Bloch PBC creating the Brillouin zone that is responsible for the bandgap. So, the bandgap is not developed in the basic circuit model. The basic circuit model also does not account for $\beta_{Z,TE}$ above the light line, showing an absence of the fast-wave as the leaky-wave radiation mode. It also does not provide any information on higher-order Bloch modes. However, the reflection phase diagram, Fig (20.11a), shows approximate bandwidth, in the gray strip, of the AMC surface. The reflection phase diagram is valid only for the normal incidence of the plane waves on the EBG surface. It does not provide information on the polarization-sensitive reflection for the oblique incidence. It is taken into account by the improved *dynamic circuit model*. However, complete 2D characterization of the EBG surface is possible only by using the full-wave methods, such as the FEM or the FDTD method [B.1]. The approximate characterization of the EBG surface could be done using the 2D-circuit model discussed in section (20.4).

20.2.2 Dynamic Circuit Model

The basic circuit model of the EBG surface does not consider the interaction between unit cells. The dynamic circuit model is an analytical full-wave model. It is valid for both the TE and TM polarizations of the obliquely and also normal, incident plane waves. The dynamic model is applicable to both the mushroom-type EBG and the capacitive RIS, as shown in Fig (20.1b). The capacitive RIS, without vias, is a special case of the mushroom EBG with vias [B.3, J.39–J.44].

Figure (20.12a) shows a mushroom EBG structure on a substrate with relative permittivity $\varepsilon_{r1} = \varepsilon_r$ and thickness

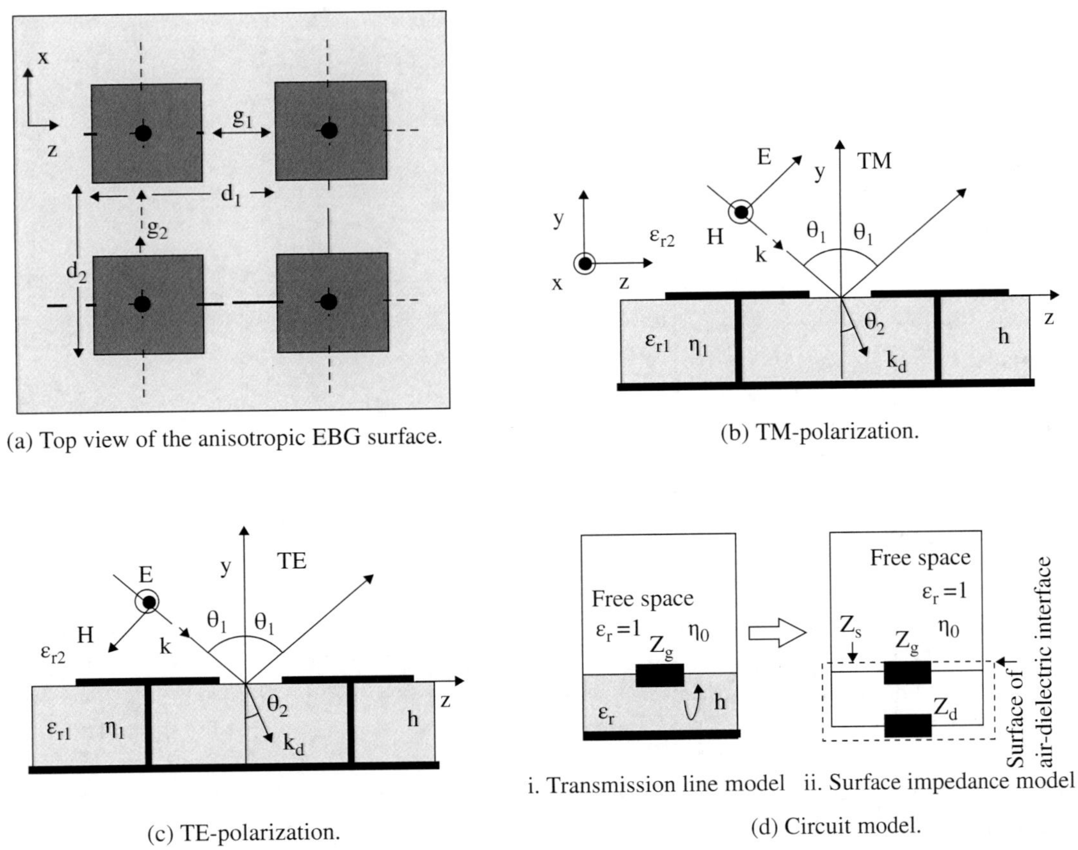

(a) Top view of the anisotropic EBG surface.
(b) TM-polarization.
(c) TE-polarization.
(d) Circuit model.

Figure 20.12 Oblique incidence of plane waves on mushroom EBG.

h. The capacitive patch is located at the interface of the dielectric and air media, $\varepsilon_{r2} = 1$. The structure is in the (x-z)-plane and the EBG has lattice constants d_1 and d_2 in the z- and x-directions. The gaps between adjacent patches are g_1 and g_2 in the z- and x-directions. Thus, the EBG surface is *anisotropic*. In case of the *isotropic* surface, $d_1 = d_2 = d$ and $g_1 = g_2 = g$. The isotropic EBG surface has an identical reflection phase response for both the z- and x-polarized incident plane waves. Whereas, for an anisotropic surface, the reflection phases for both polarizations are different [J.9, J.31, J.34].

Figure (20.12b and c) shows the TM- and TE-polarizations of the incident plane waves. The plane (y-z) is the plane of incidence of the plane waves, striking obliquely to the EBG surface. The angle $\theta_1 = \theta$ is the angle of incidence. Inside the substrate, following Snell's law, waves undergo refraction [B.5]. The plane waves have wavenumber k_o and k_d in the air and dielectric media, respectively. The conductor-backed substrate, $h < \lambda_d/4$, appears as an inductive surface, with Z_d *dielectric surface impedance*. The matrix of patches, located at the interface, is modeled as the capacitive *grid impedance* Z_g. The *surface impedance* Z_s of the patch matrix-loaded conductor-backed substrate is a parallel combination of Z_d and Z_g. The surface impedance is again a parallel resonant circuit, like the simple LC-model, distributed over the surface. Figure (20.12d) demonstrates the modeling process of the surface impedance. First, the transmission line model is obtained; next, it is used to get the surface impedance model.

The substrate relative permittivity is ε_r without a matrix of vias. The presence of vias modifies the substrate relative permittivity. The modified permittivity of the wired dielectric medium is modeled as an anisotropic medium [B.3, J.43]. The vias interact differently with the TM and TE polarized obliquely incident plane waves. For the TM-polarized waves, Fig (20.12b), both E_y and E_z components are present. The E_z component, normal to the via, does not excite it. However, the parallel E_y component induces the current in vias. The induced charges are accumulated at the ends of the vias, creating the induced dipoles. These dipoles are responsible for the creation of anisotropic relative permittivity of a wire medium, that is treated as the plasma medium. The wire medium is further discussed in section (21.1) of chapter 21. However, the densely arranged wire medium, with the period much less than a wavelength, could be treated as the *PPW model*. Two rows of vias arranged in the x-direction form a PPW for the E_z-component of the TM-polarized waves. The PPW is like a substrate integrated waveguide (SIW), discussed in subsection (7.8) of chapter 7. The TE-polarized waves, shown in Fig (20.12c), have only an E_x-field component that is orthogonal to the vias. So, the vias are not excited. The TE-polarized waves have an identical response to the EBG with and without vias. For a normal incidence, the reflection-phase responses of both the TE- and TM-polarized waves are identical; and are not influenced by the presence of vias.

Grid Impedance of a Matrix of Patches

Figure (20.12a) shows a matrix of the rectangular patches on a dielectric substrate, forming an anisotropic EBG surface. The tangential E_z- and E_x-field components create the polarization-dependent gap capacitance, C_g. The tangential field components also induce current density J_z and J_x on the EBG surface. The polarization-dependent grid impedances, $Z_g^{TM} = \hat{E}_z/\hat{J}_z$ and $Z_g^{TE} = \hat{E}_x/\hat{J}_x$, are obtained using the average boundary condition and also the modified Babinet principle. The TM/TE grid impedances for a matrix of the rectangular patches are summarized below [B.3, J.40, J.44]:

$$Z_g^{TM} = \frac{-j\eta_{eff}}{2\alpha_{TM}} \qquad \text{(a)}$$

$$Z_g^{TE} = \frac{-j\eta_{eff}}{2\alpha_{TE}\left[1 - \frac{1}{2}\left(\frac{k_o}{k_{eff}}\right)^2 \frac{\sin^2\theta}{(1 + d_2/d_1)} \frac{d_2}{d_1}\right]} \qquad \text{(b)}, \qquad (20.2.11)$$

where θ is an angle of incidence. The grid parameters, α_{TM} and α_{TE}, are computed from the following expressions:

$$\alpha_{TM} = \frac{k_{eff} d_1}{\pi} \ln\left[\frac{1}{\sin(\pi g_1/2d_1)}\right] \qquad \text{(a)}$$

$$\alpha_{TE} = \frac{k_{eff} d_2}{\pi} \ln\left[\frac{1}{\sin(\pi g_2/2d_2)}\right] \qquad \text{(b)}. \qquad (20.2.12)$$

For $d_1 >> g_1$ and $d_2 >> g_2$, the above expressions, in the simplified forms, are used for the TM- and TE-polarizations:

$$\alpha_{TM} = \frac{k_{eff} d_1}{\pi} \ln\left[\frac{2d_1}{\pi g_1}\right] \quad \text{(a)}, \qquad \alpha_{TE} = \frac{k_{eff} d_2}{\pi} \ln\left[\frac{2d_2}{\pi g_2}\right] \quad \text{(b)}. \qquad (20.2.13)$$

An anisotropic matrix of rectangular patches provides the anisotropic grid impedance, for $d_1 \neq d_2$ and $g_1 \neq g_2$. However, for the isotropic EBG surface, we have $d_1 = d_2 = d$, and $g_1 = g_2 = g$. Also for the normal incidence, $\theta = 0°, Z_g^{TM} = Z_g^{TE} = Z_g$. The effective intrinsic impedance of the inhomogeneous medium and the effective wavenumber k_{eff} are given as

$$\eta_{eff} = \eta_0/\sqrt{\varepsilon_{reff}}, \quad k_{eff} = k_0\sqrt{\varepsilon_{reff}} \quad \text{(a)}$$

$$\text{where,} \quad \varepsilon_{reff} = (\varepsilon_r + 1)/2 \qquad \text{(b)}. \qquad (20.2.14)$$

The characteristic impedance η_0 and wavenumber $k_0 = \omega\sqrt{\mu_0\varepsilon_0} = \omega/c$ are for free space.

Dielectric Surface Impedance

Figure (20.12d) shows that the conductor-backed substrate appears as an inductive surface, $Z_d = j\eta_d \tan(k_d h)$, at the EBG plane. In the case of the oblique incidence, the Z_d is polarization-dependent. The waves entering the dielectric region undergo refraction and follow Snell's law. The TE- and TM-polarized obliquely incident plane waves view the modified intrinsic impedances of the dielectric medium [B.5]. Following the discussion of subsection (5.2.4) of chapter 5, these modified intrinsic impedances are written as follows:

$$\eta_d^{TE} = \frac{\eta_d}{\cos\theta_2} \quad (a), \qquad \eta_d^{TM} = \eta_d \cos\theta_2 \quad (b)$$
$$\text{where,}\quad \eta_d = \eta_0\sqrt{\varepsilon_r}, \quad (c), \qquad \sin\theta_2 = \frac{\sin\theta}{\sqrt{\varepsilon_r}} \quad (d). \tag{20.2.15}$$

Using the above equations and Fig (20.12b and c), the TE-and TM-polarized dielectric surface impedances are obtained as follows:

$$Z_d^{TE} = j\eta_d^{TE}\tan(k_{yd}h) = j\frac{\eta_d}{\cos\theta_2}\tan(k_{yd}h)$$
$$= j\frac{\omega\mu_0}{k_d}\frac{\tan(k_{yd}h)}{\cos\theta_2}, \Rightarrow Z_d^{TE} = j\frac{\omega\mu_0}{(k_{yd}/\cos\theta_2)}\frac{\tan(k_{yd}h)}{\cos\theta_2}$$
$$\Rightarrow Z_d^{TE} = j\omega\mu_0\frac{\tan(k_{yd}h)}{k_{yd}} \quad (a)$$
$$Z_d^{TM} = j\eta_d^{TM}\tan(k_{yd}h) = j\eta_d\cos\theta_2\tan(k_{yd}h)$$
$$= j\frac{\omega\mu_0}{k_d}\cos\theta_2\tan(k_{yd}h)$$
$$\Rightarrow Z_d^{TM} = j\frac{\omega\mu_0}{(k_{yd}/\cos\theta_2)}\cos\theta_2\tan(k_{yd}h)$$
$$\Rightarrow Z_d^{TM} = j\frac{\omega\mu_0}{k_{yd}}\cos^2\theta_2\tan(k_{yd}h) \quad (b). \tag{20.2.16}$$

The TE-polarized dielectric surface impedance is independent of the angle of incidence θ, whereas for the TM polarization, it is dependent on θ. The y-directed normal component of the wave vector, k_{yd}, is obtained from the following dispersion relation:

$$k_{xd}^2 + k_{yd}^2 + k_z^2 = \omega^2\mu_0\varepsilon_0\varepsilon_r. \tag{20.2.17}$$

For the plane waves incident in the (y-z)-plane, the surface waves propagate in the z-direction, i.e. $k_x = 0$ and $k_z = k_0\sin\theta$. The above equation provides k_{yd}:

$$k_{yd}^2 = \omega\sqrt{\mu_0\varepsilon_0}\left[\varepsilon_r - \sin^2\theta\right]^{1/2}. \tag{20.2.18}$$

Surface Impedance of EBG Surface

Figure (20.12d) shows that the polarization-dependent surface impedance is a parallel combination of Z_g and Z_d:

$$Z_s^{TE/TM} = \frac{Z_d^{TE/TM} Z_g^{TE/TM}}{Z_d^{TE/TM} + Z_g^{TE/TM}}. \tag{20.2.19}$$

Using equations (20.2.11) and (20.2.16), with the above equation, the following expressions are obtained for the isotropic surface impedance of the EBG surface for both the polarizations:

$$Z_s^{TE} = \left[\frac{j\omega\mu_0\left(\tan(k_{yd}h)/k_{yd}\right)}{1 - 2k_{eff}\alpha_{TE}\dfrac{\tan(k_{yd}h)}{k_{yd}}F}\right] \quad (a)$$
$$\text{where, } F = \left[1 - \frac{1}{2}\left(\frac{k_o}{k_{eff}}\right)^2 \frac{\sin^2\theta}{(1 + d_2/d_1)}\frac{d_2}{d_1}\right]$$
$$Z_s^{TM} = \frac{j\omega\mu_0\left(\tan(k_{yd}h)\cos^2\theta_2/k_{yd}\right)}{1 - 2k_{eff}\alpha_{TM}\dfrac{\tan(k_{yd}h)}{k_{yd}}\cos^2\theta_2}$$
$$\text{where, } \cos^2\theta_2 = \left[1 - \frac{\sin^2\theta}{\varepsilon_r}\right] \quad (b). \tag{20.2.20}$$

Similarly, the surface impedance of an anisotropic EBG surface can also be obtained.

Another set of approximate expressions for the polarization-dependent isotropic surface impedance of the mushroom EBG is also available. In this case, the polarization-dependent grid impedances Z_g^{TE} and Z_g^{TM} are treated as the impedances of a FSS on an air-substrate, parallel connected to the dielectric surface impedance [B.3]:

$$Z_g^{TE} = \frac{-j\eta_0}{2\alpha\cos^2\theta} \quad (a), \qquad Z_g^{TM} = \frac{-j\eta_0}{2\alpha} \quad (b). \tag{20.2.21}$$

The TE -polarized incident waves are not influenced by vias, and for the TM-polarized waves, it is modeled as a PPW. On using the above expression, with the expressions for Z_d and approximation $\omega\mu_0\tan(k_{yd}h)/k_{yd} \approx \omega\eta_0 h$, the following alternate expressions are obtained for the surface impedance of an isotropic EBG surface [B.3]:

$$Z_s^{TE}(\omega,\theta) = \frac{jk_0\eta_0 h}{1 - 2k_0 h\alpha\cos^2\theta} \quad (a)$$
$$Z_s^{TM}(\omega,\theta) = \frac{jk_0\eta_0 h}{1 - 2k_0 h\alpha} \quad (b). \tag{20.2.22}$$

For the normal incidence (θ = 0°), the surface impedance is polarization-independent. At the *resonance frequency of the EBG surface*, the surface impedance has a pole, i.e. the denominator of the surface impedance is zero. The resonance frequency for the TE-polarization is θ dependent, whereas for the TM-polarization, it almost θ independent.

Reflection Coefficient and Dispersion Relation of EBG Surface

The plane waves, striking the isotropic EBG surface, view the intrinsic impedance of the free space as $\eta_0/\cos\theta$ and $\eta_0\cos\theta$ for the TE- and TM-polarized waves, respectively. The polarization-dependent reflection coefficients of both polarizations, following subsection (5.2.4) of chapter 5, are obtained as follows:

$$\Gamma^{TE}(\omega,\theta) = \frac{Z_s^{TE}(\omega,\theta) - \eta_0/\cos\theta}{Z_s^{TE}(\omega,\theta) + \eta_0/\cos\theta} \quad \text{(a)}$$

$$\Gamma^{TM}(\omega,\theta) = \frac{Z_s^{TM}(\omega,\theta) - \eta_0\cos\theta}{Z_s^{TM}(\omega,\theta) + \eta_0\cos\theta} \quad \text{(b)}. \qquad (20.2.23)$$

In the above equations, surface impedance $Z_s^{TE/TM}(\omega,\theta)$ is taken from either equation (20.2.20) or (20.2.22).

For the normal incidence ($\theta = 0°$) on the isotropic surface, the reflection phase is polarization-independent. However, for the oblique incidence, the reflection phase $\angle\Gamma^{TE}(\omega,\theta)$ depends on the angle θ. Such dependence is weak for the reflection phase $\angle\Gamma^{TM}(\omega,\theta)$. The surface impedance, equation (20.2.22), also shows unstable θ dependent resonance frequency of the mushroom EBG surface [B.3]. However, more accurate expression (20.2.20) provides more stable resonance frequency. Figure (20.13a and b) shows polarization-dependent reflection phases for the obliquely incident plane waves on the isotropic EBG of the patch grid [J.40]. The isotropic EBG has lattice constant d = 2 mm, gap g = 0.2 mm on a substrate with $\varepsilon_r = 10.2$, h = 1 mm. The analytical model closely follows two full-wave results. The

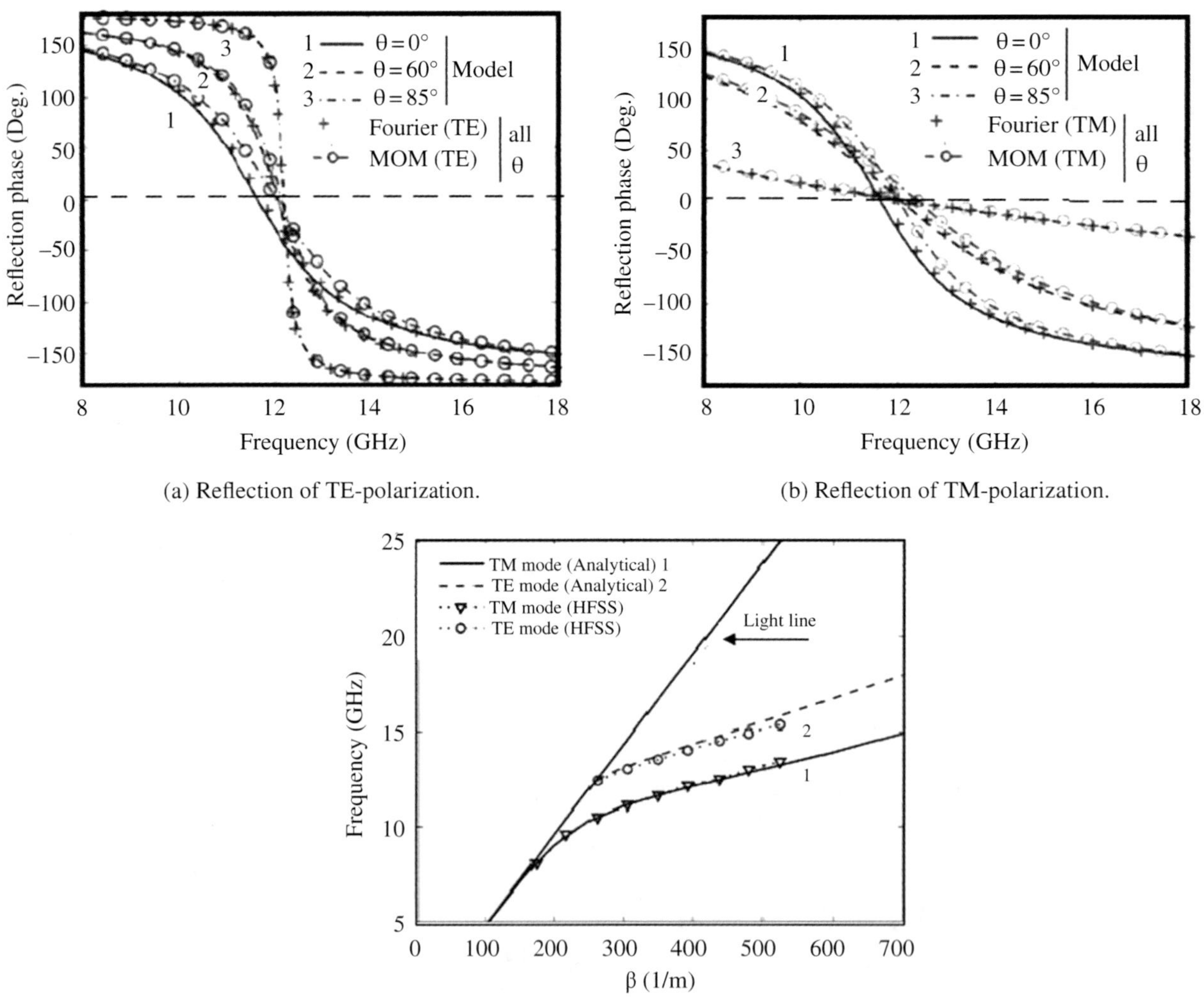

(a) Reflection of TE-polarization.

(b) Reflection of TM-polarization.

(c) Dispersion in EBG.

Figure 20.13 Polarization-dependent reflection coefficient and dispersion of the isotropic EBG surface. *Source*: Reproduced from Luukkonen et al. [J.40]. © 2008, IEEE.

resonance frequency of the TE- polarization is almost stable with respect to the angle of incidence θ, whereas for the TM-polarization, it is more unstable.

Dispersion Model of EBG Surface

The dynamic analytical circuit model can also compute the dispersion of the first two TM and TE Bloch modes, with better accuracy as compared to the basic model. Using the *transverse resonance method*, discussed in section (7.7) of chapter 7, the dispersion relation for the propagation constant k_z is obtained from the following expression:

$$\frac{1}{\eta_0^{TE/TM}} + \frac{1}{Z_s^{TE/TM}} = 0. \quad (20.2.24)$$

For the oblique incidence, frequency-independent $\eta_0^{TE/TM}$ is taken as $\eta_0/\cos\theta$, and $\eta_0 \cos\theta$ for the TE- and TM-polarized waves, respectively. The surface impedance $Z_s^{TE/TM}$ is taken from equation (20.2.20), or also from equation (20.2.22). However, like the basic model, it also provides only the 1D dispersion, without any Bragg's reflection. Again, it is only applicable to the slow-waves outside the light-line cone. Figure (20.13c) compares accurately the results of the dynamic circuit model against the results of HFSS. Again, the dynamic circuit model is not able to account for the fast-waves and leaky waves existing within the light-line cone [J.40].

Polarization-Dependent Patch Capacitance

The dynamic analytical circuit model also provides the polarization-dependent patch capacitances. The polarization-dependent capacitive grid impedances can be written using the polarization-dependent patch capacitances:

$$Z_g^{TE/TM} = \frac{1}{j\omega C_g^{TE/TM}}, \Rightarrow C_g^{TE/TM} = \frac{1}{j\omega Z_g^{TE/TM}}. \quad (20.2.25)$$

Using equations (20.2.11) to (20.2.13) with equation (20.2.22), expressions for $C_g^{TE/TM}$ are obtained as

$$C_g^{TE} = \frac{d_2\varepsilon_0(\varepsilon_r+1)}{\pi}\left[\ln\left(\frac{2d_2}{\pi g}\right)\right] \times \left(1 - \frac{(d_2/d_1)\sin^2\theta}{(\varepsilon_r+1)(1+d_2/d_1)}\right) \quad (a)$$

$$C_g^{TM} = \frac{d_1\varepsilon_0(\varepsilon_r+1)}{\pi}\ln\left(\frac{2d_1}{\pi g}\right) \quad (b). \quad (20.2.26)$$

The isotropic surface has $d_1 = d_2 = d$ (square patch), and for the normal incidence, $C_g^{TE} = C_g^{TM}$. The patch capacitances can be used with the basic LC-circuit model, to compute the resonance frequencies, and also the surface impedances for both the anisotropic and isotropic EBG surfaces.

20.3 Uniplanar EBG Structures

The uniplanar EBG structure shown in Fig (20.1d-i) is popularly called the UC-EBG. It was introduced to create a patterned ground plane, showing EBG behavior, in place of a normal solid ground conductor for the microstrip-based circuits and antennas. By suppressing the surface waves, it improves the stopband characteristics of microstrip-based components and circuits, provided the stopband of the circuits comes within the bandgap of the EBG. It also creates the AMC/RIS surface to improve the radiation performances of an antenna [J.4, J.12]. Figure (20.1d-ii) shows the LC-parallel resonant-type circuit model of the surface impedance of a UC-EBG. It is like the case of a mushroom-type EBG.

Figure (20.14a-i) shows that two adjacent unit cells are connected by a narrow microstrip line, providing the inductance L of the model. The value of L is estimated by equation (19.3.54b) of chapter 19, or simply by equation (20.2.1b). The capacitance C is the gap capacitance between two adjacent patches, given by equation (20.2.1a). Alternatively, equation (20.2.26) can also be used. The surface impedance resonant frequency and associated bandwidths are estimated using equations (20.2.2) and (20.2.6), respectively. Due to the presence of the ground plane, additional shunt capacitances C_p are added to get the π-type equivalent circuit model, shown in Fig (20.14a-ii) of a UC-EBG. The LC resonant circuit is in the series arm of the π-model. The C_p is a parallel plate capacitor between the UC-EBG and the ground conductor. The AMC is created around the resonance frequency, between ±90° reflection phases. The UC-EBG constructed on an RT-Duriod 6010 substrate, with $\varepsilon_r = 10.2$, h = 25 mil has the stopband at resonance frequency 12 GHz. Substrates of very high relative permittivity $\varepsilon_r = 140$ or 200 are used to get lower stopband resonance frequency up to 5 GHz [J.45].

Figure (20.14b and c) suggests a few schemes to inductively load the UC-EBG that help to bring down the resonance frequency, and also to enhance the bandwidth of the EBG surface [J.44–J.47]. Figure (20.14d) demonstrates an application of UC-EBG. The UC-EBG surrounds a patch antenna to improve its directive gain, and also its radiation characteristics, by suppressing the surface waves on the substrate [J.33]. Figure (20.14e) presents an UC-EBG with the fractal unit cell to get the multiband AMC. The smallest patch of the

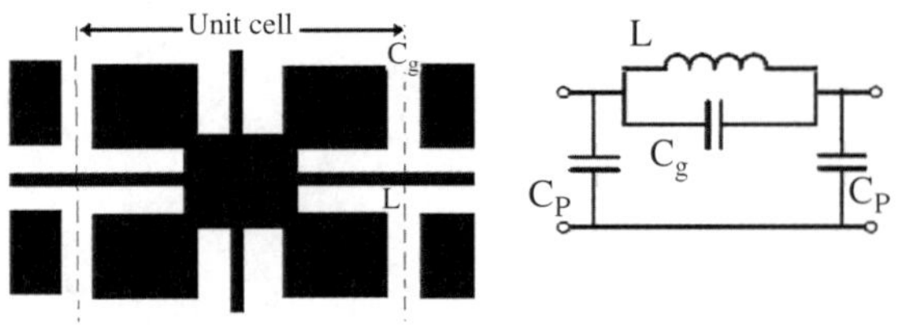

i. Unit cells connected by narrow microstrip

ii. π - circuit model of a unit cell

(a) Standard UC-EBG.

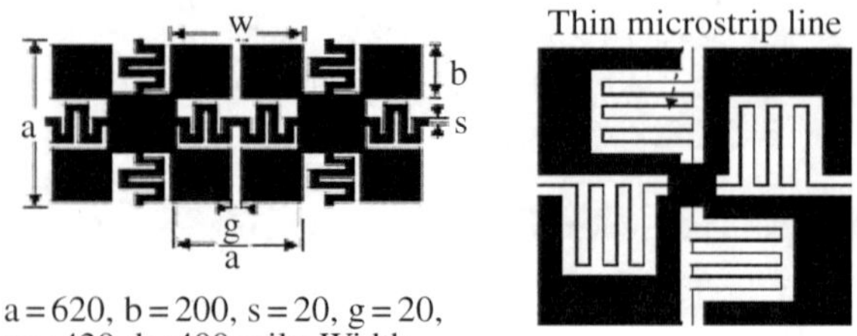

a = 620, b = 200, s = 20, g = 20, w = 420,d = 400 mils. Width 50 Ω microstrip = 48 mils.
i. Two adjacent cells of EBG [J.46].

ii. Very compact EBG [J.47]

(b) Meandered line loaded EBG.

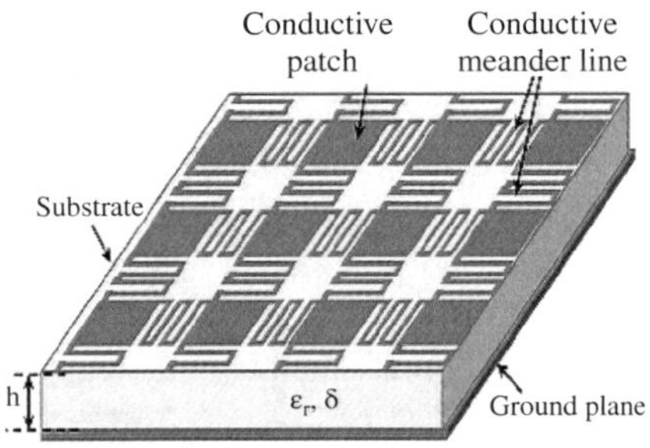

(c) Meander line connected array of patches [J.45].

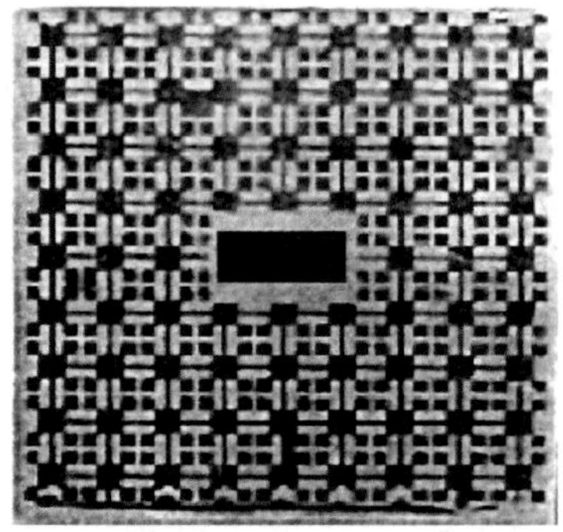

(d) UC-EBG surrounded microstrip patch antenna [J.33].

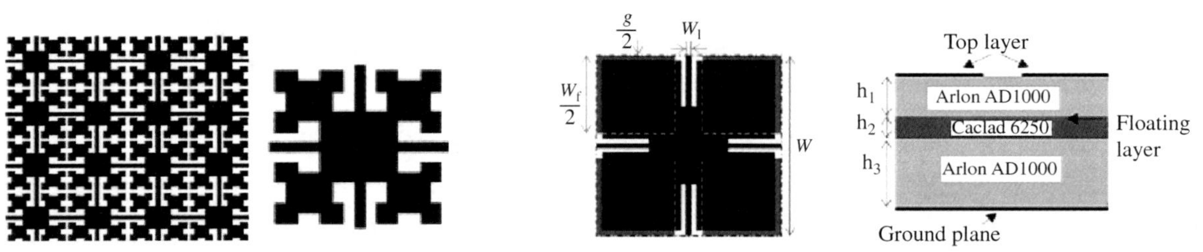

(a) EBG surface.

(b) Unit cell.

(c) Top surface.

(d) Cross-section showing 3 layers.

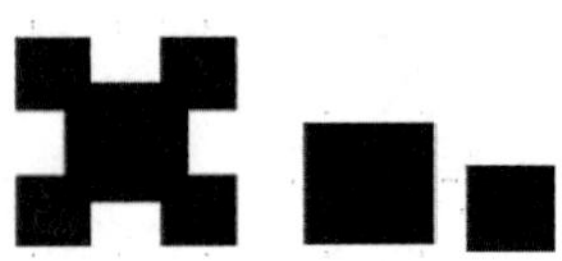

3 patches comprising the unit cell

(e) Fractal EBG for multiband AMC.

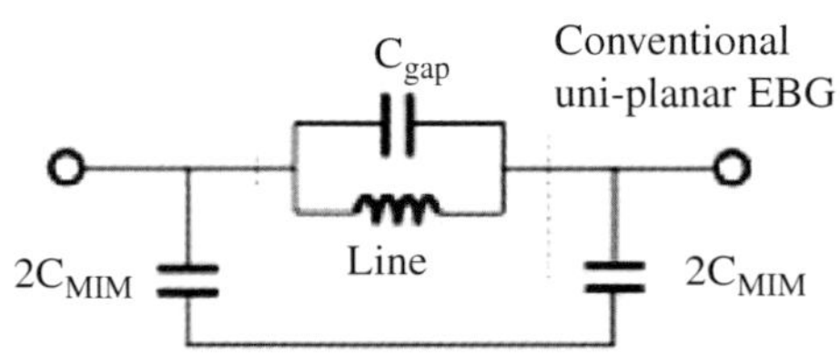

(f) Multilayer EBG structure.

Figure 20.14 Some UC-EBG structures. *Source*: Reproduced from Coccioli et al. [J.33]. © 1999, IEEE. Reproduced from Iravani and Ramahi [J.45]. © 2010, IEEE. Reproduced from Abedin and Ali [J.46]. © 2005, IEEE. Reproduced from Waterhouse and Novak [J.47]. © 2006, IEEE.

unit cell contributes to a higher band of the AMC. The next lower band is due to the larger patch. The complete structure of a fractal patch provides the lowest AMC band [J.48].

Figure (20.14f) shows the multilayer uniplanar EBG structure and its equivalent circuit model. The structure has used the MIM (*metal insulator metal*)-type capacitor loading, to get the EBG resonance in the range of a few

MHz [J.48, J.49]. It is a three-layered UC-EBG structure. The first layer h_1 is Arlon AD 1000 with $\varepsilon_r = 9.1$, $h = 0.127$ mm. The second layer h_2 is a floating metallic layer of Cuclad 250 with $\varepsilon_r = 2.5$, $h = 38$ μm. It rests on the third layer h_3 that is again Arlon AD 1000 with $\varepsilon_r = 10.9$, $h = 3.2$ mm. The UC-EBG is placed at the top of the first layer. Overlap of UC-EBG with the floating conductor forms two MIM capacitors (C_{MIM}), and are in series connected through the floating conductor. The combined C_{MIM} appears across the LC-resonant circuit of the UC-EBG. The equivalent circuit, also given in Fig (20.14f), computes the resonance frequency of the UC-EBG using the expression $\omega_0 = 1/\sqrt{L\left(C_g + C_{MIM}\right)}$. The multilayer EBG has low resonance frequency and its AMC bandwidth is from 170 MHz to 320 MHz.

20.4 2D Circuit Models of EBG Structures

The lumped circuit models of the EBG surface, presented in section (20.2), do not show the existence of the bandgap. However, the 1D-periodic line model of the EBG surface shows the partial bandgap. It is discussed in subsection (19.3.4) of chapter 19. The more realistic and accurate 2D circuit models of the EBG surfaces provide the complete bandgap and correct nature of the first two Bloch modes [J.50–J.54]. The 2D-unit cell, using the 2D Floquet–Bloch theorem is analyzed to get the Bloch modes propagation characteristics of the infinitely large 2D-periodic surface. During the 2D circuit-model-based simulation process, the 2D-IBZ is used to get the dispersion characteristics of Bloch modes. The simulation also computes the attenuation diagram. The 2D models are developed in two forms:

i) *Shunt-connected 2D planar EBG circuit model.*
ii) *Series-connected 2D planar EBG circuit model.*

20.4.1 Shunt-Connected 2D Planar EBG Circuit Model

Figure (20.15a) shows the 2D circuit model of a unit cell of the mushroom-type EBG structure with and without vias. In case the EBG is made of capacitively coupled large patches over the conductor-backed ground plane, the shunt inductor, due to via, is absent. Figure (20.1a-i) shows the boundaries of the 2D unit cell, a-b-c-d marked on the top view of the EBG surface. The magnetic wall-type boundary is located at the center of the coupling gap g. So, the series gap capacitance is taken as 2C in Fig (20.15a). The line section length is taken as $kw/2, k = k_0\sqrt{\varepsilon_r}$. The line section is formed from the PPW between the patch and the ground plane. The *isotropic EBG surface* is realized for identical periodicity, i.e. lattice constant d in both x- and y-directions. For an *anisotropic surface*, the lattice constants are d_1 and d_2 in the x-direction and the y-direction, respectively. The period of the isotropic EBG surface is $d = w + g$, where w is the width of the square patch, and g is the gap between adjacent patches. Figure (20.15b) shows the 2D [ABCD] transmission matrix representation, i.e. the pair of input (T_1, T_2) and pair of output (T_3, T_4) matrices of a unit cell. Figure (20.15c) shows the 4-port matrix network with the *periodic boundaries* (PB) satisfying *Floquet–Bloch periodic conditions* at the ports. At the node #O, the outputs of transmission matrix pair (T_1, T_2), inputs of the transmission matrix pair # (T_3, T_4), and the fifth matrix network (T_5), a model of the grounded via, are connected in parallel.

The transmission matrices, like the 1D-periodic line case, are obtained from the cascaded matrices of the series-connected capacitor (2C) and the transmission line section kd/2 of characteristic impedance Z_0. The transmission matrices are reciprocal matrices resulting in the following expression for their determinants: $\Delta_1 = \Delta_2 = \Delta_3 = \Delta_4 = 1$. The positions of the diagonal elements of output matrices # T_3 and T_4 are interchanged, i.e. $A_3 \rightarrow D_3$ and $A_4 \rightarrow D_4$, due to the order of physical placements of the capacitors and line sections with respect to the node # O. In general, the 2D circuit applies to the anisotropic and nonreciprocal surfaces also. However, the further discussion considers only the isotropic and reciprocal EBG surface, due to the simplicity and popularity.

The input transmission matrix pair (T_1, T_2) and the output transmission matrix pair (T_3, T_4), in the x- and y-directions, are defined as follows:

$$\text{Input matrix along x} - \text{diection}: [T_1]_x = \begin{bmatrix} A_1 & B_1 \\ C_1 & D_1 \end{bmatrix} \quad \text{(a)}$$

$$\text{Input matrix along y} - \text{direction}: [T_2]_y = \begin{bmatrix} A_2 & B_2 \\ C_2 & D_2 \end{bmatrix} \quad \text{(b)}$$

$$\text{Output matrix along x} - \text{diection}: [T_3]_x = \begin{bmatrix} D_3 & B_3 \\ C_3 & A_3 \end{bmatrix} \quad \text{(c)}$$

$$\text{Output matrix along y} - \text{direction}: [T_4]_y = \begin{bmatrix} D_4 & B_4 \\ C_4 & A_4 \end{bmatrix} \quad \text{(d).} \qquad (20.4.1)$$

Figure (20.15c) shows the port voltages and port currents of the 2D unit cell. The voltage and current at the ports #1, #2, #3, and #4; taking care of the 2D

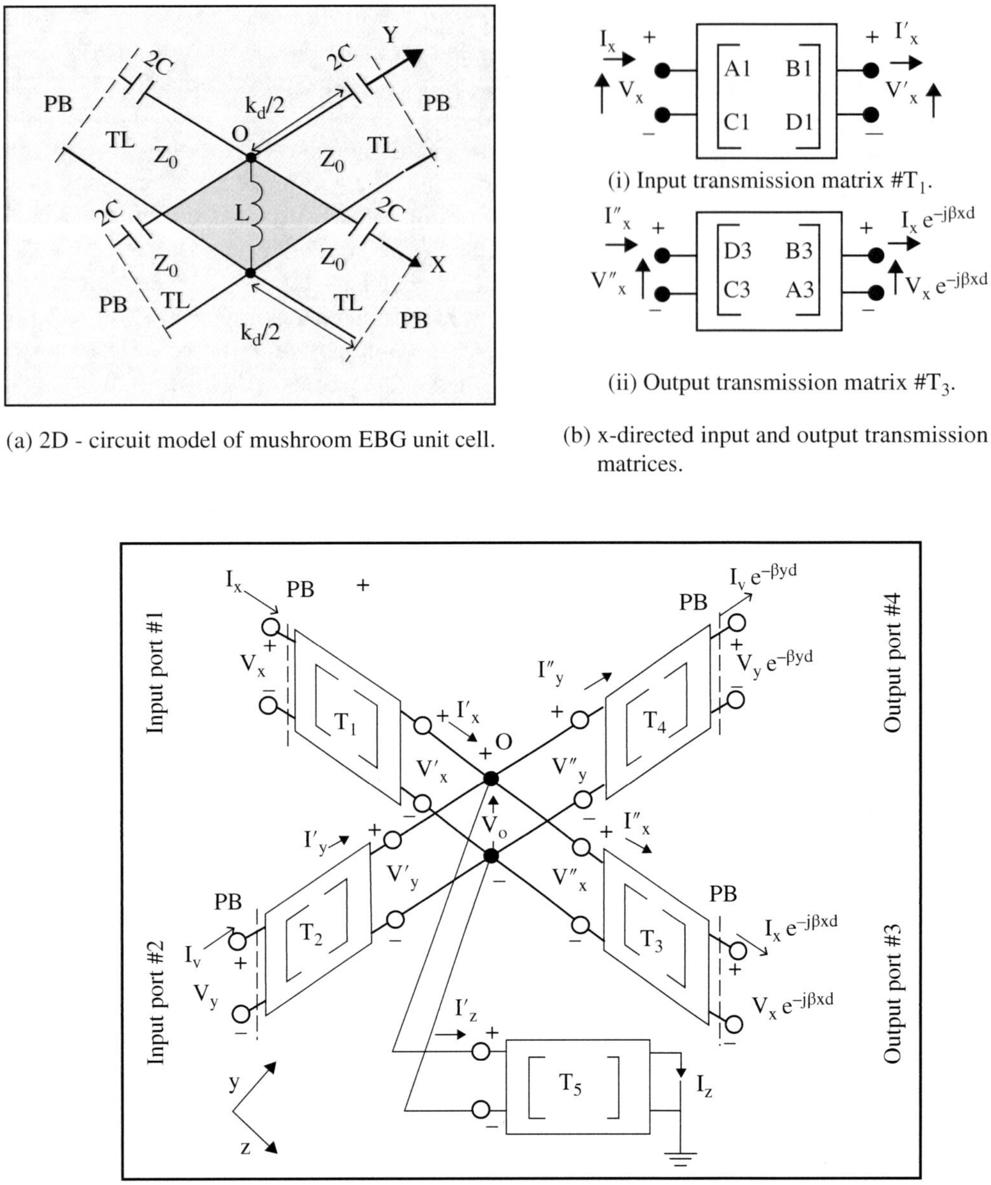

(a) 2D - circuit model of mushroom EBG unit cell.

(i) Input transmission matrix #T_1.

(ii) Output transmission matrix #T_3.

(b) x-directed input and output transmission matrices.

(c) Analytical model of a 2D unit cell.

Figure 20.15 4-Port network of 2D-circuit model of mushroom-type EBG surface in (x-y) plane.

Bloch–Floquet periodic conditions, at the periodic port (PB), are defined as follows:

$$V_1 = V_x, V_3 = V_x e^{-j\beta_x d}; I_1 = I_x, I_3 = I_x e^{-j\beta_x d} \quad \text{(a)}$$
$$V_2 = V_y, V_4 = V_y e^{-j\beta_y d}; I_2 = I_y, I_4 = I_y e^{-j\beta_y d} \quad \text{(b)}. \quad (20.4.2)$$

Figure (20.15b-i) shows x-directed voltage and current pair (V'_x, I'_x) at the output of the input transmission matrix # T_1. Similar relation holds for y-directed voltage and current pair (V'_y, I'_y) at the output of the input transmission matrix # T_2. The voltage and current are computed as follows:

$$\begin{bmatrix} V_x \\ I_x \end{bmatrix} = \begin{bmatrix} A_1 & B_1 \\ C_1 & D_1 \end{bmatrix} \begin{bmatrix} V'_x \\ I'_x \end{bmatrix} \quad \text{(a)}$$

$$\Rightarrow \begin{bmatrix} V'_x \\ I'_x \end{bmatrix} = \frac{1}{\Delta_x} \begin{bmatrix} D_1 & -B_1 \\ -C_1 & A_1 \end{bmatrix} \begin{bmatrix} V_x \\ I_x \end{bmatrix} \quad \text{(b)}$$

$$V'_x = \frac{D_1}{\Delta_x} V_x - \frac{B_1}{\Delta_x} I_x \quad \text{(c)}, \quad I'_x = -\frac{C_1}{\Delta_x} V_x + \frac{A_1}{\Delta_x} I_x \quad \text{(d)}$$

$$\text{Likewise, } V'_y = \frac{D_2}{\Delta_y} V_y - \frac{B_2}{\Delta_y} I_y \quad \text{(e)}$$

$$I'_y = -\frac{C_2}{\Delta_y} V_y + \frac{A_2}{\Delta_y} I_y \quad \text{(f)}$$

where, $\Delta_x = A_1 D_1 - B_1 C_1$ (g), $\Delta_y = A_2 D_2 - B_2 C_2$ (h) . (20.4.3)

Figure (20.15b-ii) shows the output transmission matrix T_3 with the input voltage and current pair (V''_x, I''_x) and the output voltage and current pair $(V_x e^{-j\beta_x d}, I_x e^{-j\beta_x d})$. The voltage and current pairs are correlated as follows:

$$\begin{bmatrix} V''_x \\ I''_x \end{bmatrix} = \begin{bmatrix} D_3 & B_3 \\ C_3 & A_3 \end{bmatrix} \begin{bmatrix} V_x e^{-j\beta_x d} \\ I_x e^{-j\beta_x d} \end{bmatrix}$$

$$\Rightarrow V''_x = (D_3 e^{-j\beta_x d})V_x + (B_3 e^{-j\beta_x d})I_x \qquad \text{(a)}$$

$$I''_x = (C_3 e^{-j\beta_x d})V_x + (A_3 e^{-j\beta_x d})I_x \qquad \text{(b)}$$

Likewise, in the y – direction,

$$V''_y = (D_4 e^{-j\beta_y d})V_y + (B_4 e^{-j\beta_y d})I_y \qquad \text{(c)}$$

$$I''_y = (C_4 e^{-j\beta_y d})V_y + (A_4 e^{-j\beta_y d})I_y \qquad \text{(d)}. \qquad (20.4.4)$$

The voltages at node # O are continuous, i.e. $V'_x = V''_x$, and $V'_y = V''_y$. Under the continuity condition, the following equations are obtained from equations (20.4.3) and (20.4.4):

$$\frac{D_1}{\Delta_x}V_x - \frac{B_1}{\Delta_x}I_x = (D_3 e^{-j\beta_x d})V_x + (B_3 e^{-j\beta_x d})I_x$$

$$\Rightarrow \left[\frac{D_1}{\Delta_x} - D_3 e^{-j\beta_x d}\right]V_x - \left[\frac{B_1}{\Delta_x} + B_3 e^{-j\beta_x d}\right]I_x = 0 \quad \text{(a)}$$

Likewise, $$\left[\frac{D_2}{\Delta_y} - D_4 e^{-j\beta_y d}\right]V_y - \left[\frac{B_2}{\Delta_y} + B_4 e^{-j\beta_y d}\right]I_y = 0 \quad \text{(b)}. \qquad (20.4.5)$$

Across the node #O, the following relation holds for the matrix $[T_5]$:

$$\begin{bmatrix} V_O \\ I'_z \end{bmatrix} = \begin{bmatrix} A_5 & B_5 \\ C_5 & D_5 \end{bmatrix} \begin{bmatrix} 0 \\ I_z \end{bmatrix}, \Rightarrow V_O = B_5 I_z \quad \text{(a)}$$

$$I'_z = \left(\frac{D_5}{B_5}\right)V_O \qquad \text{(b)}. \qquad (20.4.6)$$

At node #O, $V_O = V''_x$, and use of equation (20.4.4a) with the above equation provides current in the shunt arm,

$$I'_z = \frac{D_5}{B_5}\left[(D_3 e^{-j\beta_x d})V_x + (B_3 e^{-j\beta_x d})I_x\right]. \qquad (20.4.7)$$

At node#O, again $V''_x = V''_y = V_O$, and use of equation (20.4.4a) provides

$$(D_3 e^{-j\beta_x d})V_x + (B_3 e^{-j\beta_x d})I_x - (D_4 e^{-j\beta_y d})V_y - (B_4 e^{-j\beta_y d})I_y = 0. \qquad (20.4.8a)$$

Use of the KCL at node #O, i.e. $I'_x + I'_y - I''_x - I''_y - I'_z = 0$, results in the following equation:

$$\left[-\frac{C_1}{\Delta_x} - C_3 e^{-j\beta_x d} - \left(\frac{D_5}{B_5}\right)D_3 e^{-j\beta_x d}\right]V_x + \left[\frac{A_1}{\Delta_x} - A_3 e^{-j\beta_x d} - \left(\frac{D_5}{B_5}\right)B_3 e^{-j\beta_x d}\right]I_x + \left[-\frac{C_2}{\Delta_y} - C_4 e^{-j\beta_y d}\right]V_y + \left[\frac{A_2}{\Delta_y} - A_4 e^{-j\beta_y d}\right]I_y = 0. \qquad (20.4.8b)$$

By combining equations (20.4.5) and (20.4.8), the following matrix equation is obtained:

$$[M]\begin{bmatrix} V_x \\ V_y \\ I_x \\ I_y \end{bmatrix} = 0, \Rightarrow \begin{bmatrix} f_{11} & f_{12} & 0 & 0 \\ 0 & 0 & f_{23} & f_{24} \\ f_{31} & f_{32} & f_{33} & f_{34} \\ f_{41} & f_{42} & f_{43} & f_{44} \end{bmatrix} \begin{bmatrix} V_x \\ I_x \\ V_y \\ I_y \end{bmatrix} = 0, \qquad (20.4.9a)$$

where

$$f_{11} = \frac{D_1}{\Delta_x} - D_3 e^{-j\beta_x d}, \quad f_{12} = -\left(\frac{B_1}{\Delta_x} + B_3 e^{-j\beta_x d}\right)$$

$$f_{23} = \frac{D_2}{\Delta_y} - D_4 e^{-j\beta_y d}, \quad f_{24} = -\left[\frac{B_2}{\Delta_y} + B_4 e^{-j\beta_y d}\right]$$

$$f_{31} = D_3 e^{-j\beta_x d}, \quad f_{32} = B_3 e^{-j\beta_x d}, f_{33} = -D_4 e^{-j\beta_y d}$$

$$f_{34} = -B_4 e^{-j\beta_y d}, f_{41} = -\frac{C_1}{\Delta_x} - C_3 e^{-j\beta_x d} - \left(\frac{D_5}{B_5}\right)D_3 e^{-j\beta_x d}$$

$$f_{42} = \frac{A_1}{\Delta_x} - A_3 e^{-j\beta_x d} - \left(\frac{D_5}{B_5}\right)B_3 e^{-j\beta_x d}$$

$$f_{43} = -\frac{C_2}{\Delta_y} - C_4 e^{-j\beta_y d}, f_{44} = \frac{A_2}{\Delta_y} - A_4 e^{-j\beta_y d}. \qquad (20.4.9b)$$

By eliminating I_x and I_y from equation (20.4.9a), equations are obtained involving only V_x and V_y. From the first and second rows of the above equation, the following expressions are obtained:

$$f_{11}V_x + f_{12}I_x = 0,$$

$$\Rightarrow I_x = \frac{D_1 - D_3\Delta_x e^{-j\beta_x d}}{B_1 + B_3\Delta_x e^{-j\beta_x d}}V_x \qquad \text{(a)}$$

Likewise,

$$I_y = \frac{D_2 - D_4\Delta_x e^{-j\beta_x d}}{B_2 + B_4\Delta_x e^{-j\beta_x d}}V_y \qquad \text{(b)}. \qquad (20.4.10)$$

However, the reciprocal network has $\Delta_x = A_1D_1 - B_1C_1 = 1$. The above equation is reduced to

$$I_x = \frac{D_1 - D_3 e^{-j\beta_x d}}{B_1 + B_3 e^{-j\beta_x d}} \quad \text{(a)}, \quad I_y = \frac{D_2 - D_4 e^{-j\beta_x d}}{B_2 + B_4 e^{-j\beta_x d}} \quad \text{(b)}. \qquad (20.4.11)$$

On substituting I_x and I_y in the third and fourth rows of equation (20.4.9a), the expressions are obtained, involving V_x and V_y. These equations are arranged in the following form:

$$[G]\begin{bmatrix} V_x \\ V_y \end{bmatrix} = \begin{bmatrix} g_{11} & -g_{12} \\ g_{21} & g_{22} \end{bmatrix}\begin{bmatrix} V_x \\ V_y \end{bmatrix} = 0, \quad (20.4.12)$$

where

$$g_{11} = \frac{(B_1D_3 + B_3D_1)e^{-j\beta_x d}}{(B_1 + B_3e^{-j\beta_x d})} \quad (a), \quad g_{12} = \frac{(B_2D_4 + B_4D_2)e^{-j\beta_y d}}{(B_2 + B_4e^{-j\beta_y d})} \quad (b)$$

$$g_{21} = \frac{1 + e^{-j2\beta_x d} - e^{-j\beta_x d}[B_1C_3 + C_1B_3 + A_1D_3 + D_1A_3 + (B_1D_3D_5 + D_1B_3D_5)/B_5]}{(B_1 + B_3e^{-j\beta_x d})} \quad (c)$$

$$g_{22} = \frac{1 + e^{-j2\beta_y d} - e^{-j\beta_y d}[B_2C_4 + C_2B_4 + A_4D_2 + D_4A_2]}{(B_2 + B_4e^{-j\beta_y d})} \quad (d).$$

(20.4.13)

For the nontrivial solution of equation (20.4.12), $\Delta G = 0$. It leads to the following characteristic equation:

$$g_{11}g_{22} + g_{12}g_{21} = 0$$
$$\Rightarrow Me^{-j\beta_x d}(1 + e^{-j2\beta_y d} - Pe^{-j\beta_y d})$$
$$+ Ne^{-j\beta_y d}(1 + e^{-j2\beta_x d} - Qe^{-j\beta_x d}) = 0,$$
$$\Rightarrow M(2\cos(\beta_y d) - P) + N(2\cos(\beta_x d) - Q) = 0, \quad (20.4.14)$$

where

$$M = B_1D_3 + B_3D_1, N = B_2D_4 + B_4D_2, P = B_2C_4 + B_4C_2 + A_4D_2 + A_2D_4$$
$$Q = B_1C_3 + B_3C_1 + A_1D_3 + A_3D_1 + (B_1D_3D_5 + D_1B_3D_5)/B_5.$$

For an isotropic 2D-EBG surface, $[T_1] = [T_2] = [T_3] = [T_4]$, i.e. $A = A_1 = ...A_4$; $B = B_1 = ...\ B_4$; $C = C_1 = ...\ C_4$; and $D = D_1 = ...\ D_4$. However, for an anisotropic surface, the x- and y-directed matrices are different. For isotropic EBG surface, the variables M, N, P, and Q are reduced to

$$M = 2BD, N = 2BD, P = 2(BC + AD)$$
$$Q = 2[BC + AD + (BDD_5)/B_5]. \quad (20.4.15)$$

Equation (20.4.14) is reduced to the following dispersion relations [J.35]:

$$\cos(\beta_x d) + \cos(\beta_y d) - 2(BC + AD) - BDD_5/B_5 = 0 \quad (a)$$

or using AD – BC = 1 for a reciprocal network,

$$\cos(\beta_x d) + \cos(\beta_y d) + 2 - 4AD - BDD_5/B_5 = 0 \quad (b).$$

(20.4.16)

The above expressions show the dispersion relation of the 2D lossless, isotropic, and reciprocal EBG medium. In the case of a lossy EBG surface, the above dispersion relation is modified as

$$\cosh(\gamma_x d) + \cosh(\gamma_y d) - 2(BC + AD) - BDD_5/B_5 = 0. \quad (20.4.17)$$

For the 2D circuit of an isotropic EBG surface, shown in Fig (20.15a), the [ABCD] matrix is obtained from the cascading of the [ABCD] matrices of a series capacitor 2C, followed by the line section kd/2 of characteristic impedance Z_0. For the mushroom EBG, $kd/2 = (w + g)k/2d \approx wk/2d$. The physical line length is d and k is the propagation constant, $k = \omega\sqrt{\mu_0\varepsilon_0}\sqrt{\varepsilon_r}$, of the host line. The results are given below:

$$\begin{bmatrix} A & B \\ C & D \end{bmatrix} = \begin{bmatrix} \cos\left(\frac{kd}{2}\right) + \frac{1}{2Z_0 2\omega C}\sin\left(\frac{kd}{2}\right) & j\left[Z_0\sin\left(\frac{kd}{2}\right) - \frac{1}{2\omega C}\cos\left(\frac{kd}{2}\right)\right] \\ \frac{j}{2Z_0}\sin\left(\frac{kd}{2}\right) & \cos\left(\frac{kd}{2}\right) \end{bmatrix}. \quad (20.4.18)$$

The [ABCD] matrix of the shunt branch of Fig (20.15c) is

$$\begin{bmatrix} A_5 & B_5 \\ C_5 & D_5 \end{bmatrix} = \begin{bmatrix} 1 & j\omega L \\ 0 & 1 \end{bmatrix}. \quad (20.4.19)$$

On substituting the above equations in the dispersion relation (20.4.16), the following dispersion relation is obtained for the 2D circuit shown in Fig (20.15a):

$$\sin^2\left(\frac{\beta_x d}{2}\right) + \sin^2\left(\frac{\beta_y d}{2}\right) = \frac{1}{2}\left[2\sin\left(\frac{kd}{2}\right) - \frac{1}{Z_0\omega C}\cos\left(\frac{kd}{2}\right)\right]\left[2\sin\left(\frac{kd}{2}\right) - \frac{Z_0}{2\omega L}\cos\left(\frac{kd}{2}\right)\right]. \quad (20.4.20)$$

The dispersion diagram, shown in Fig (20.16a), is plotted for the variation in the x-directed propagation

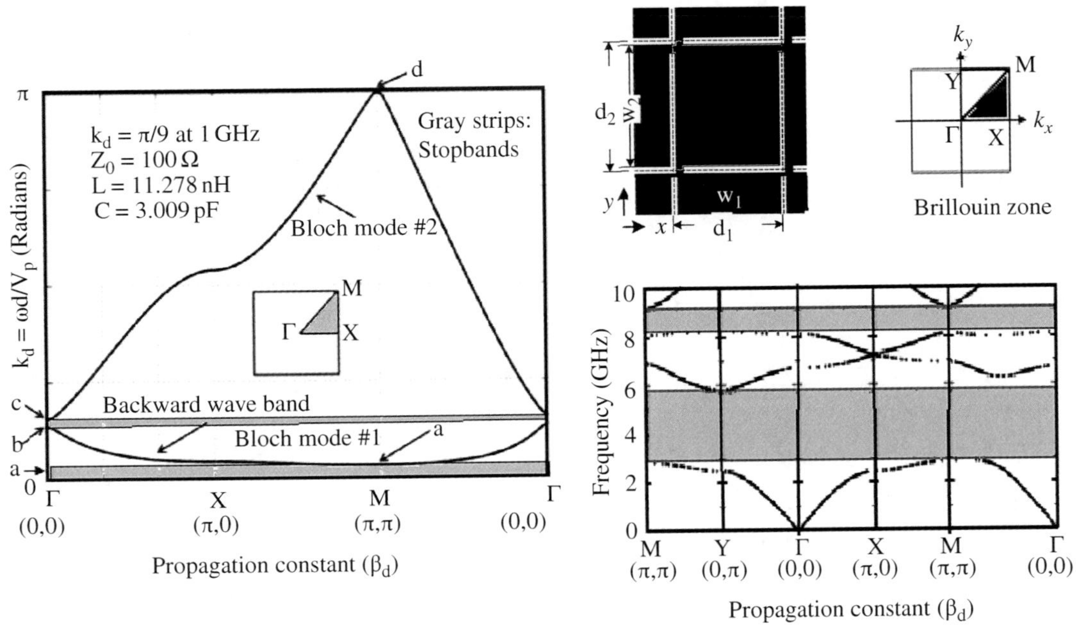

(a) 2D-dispersion diagram of backward wave supporting EBG [J.50].

(b) 2D- dispersion diagram of forward waves supporting EBG. [J.36].

Figure 20.16 Dispersion diagram of the 2D EBG surface. The gray strips show the complete stopband in the azimuthal plane. *Source*: Reproduced from Kamgaing and Ramahi [J.36]. © 2005, IEEE. Reproduced from Gribic and Eleftheriades, [J.50]. © 2003, IEEE.

constant β_x, with respect to frequency, from $\Gamma(0, 0)$ to X $(\pi/d, 0)$; the variation in the β_y from the location $X(\pi/d, 0)$ to $M(\pi/d, \pi/d)$; and finally the variation in the (β_x, β_y) from $M(\pi/d, \pi/d)$ to $\Gamma(0, 0)$. This process scans the edges of the IBZ of the square lattice, shown in the inset of Fig (20.16a). The assumed component values are also shown there [J.50].

In the region $(\Gamma - X)$, β_x is decreasing with increasing frequency, causing propagation of Bloch mode #1 as the backward wave (BW). It is due to the series capacitor and shunt inductor loading. Its passband #1 starts at the frequency location (a), corresponding to the location M, on Fig (20.16a). The bandgap #1 exists from frequency 0 to a. The *partial bandgap* #1, in the range Γ to X, is wider. Bloch mode #1 is followed by a narrow *complete (i.e. the global)* bandgap #2 between frequencies b and c. The Bloch mode #2, from frequency c to π, supports the forward-mode propagation in the passband #2.

The dispersion expression (20.4.17) is applicable to a lossy mushroom EBG also, inside the PPW, as shown in Fig (20.1c). This EBG structure is used to suppress the switching noises of the high-speed digital circuits [J.53]. The dispersion expression can be rewritten for the series-connected impedance Z_1 and the shunt-connected impedance Z_2, shown in the inset of Fig (20.17a), as follows:

$$\cosh(\gamma_x d) + \cosh(\gamma_y d) = \left(2 + \frac{Z_1^{nor}}{2Z_2^{nor}}\right)\cosh(\gamma d) + \left(2Z_1^{nor} + \frac{1}{2Z_2^{nor}}\right)\sinh(\gamma d) + \frac{Z_1^{nor}}{2Z_2^{nor}} \qquad \text{(a)}$$

$$\text{where, } Z_1^{nor} = Z_1/Z_0,\ Z_2^{nor} = Z_2/Z_0,\ Z_0 = \eta_0 h/d\sqrt{\varepsilon_r},$$

$$\gamma = \alpha + jk = j(\omega/c)\sqrt{\varepsilon_r} \qquad \text{(b).}$$

$$(20.4.21)$$

In the above equation, η_0 is the intrinsic impedance of free space, c is the velocity of EM-wave in free space, and the substrate is taken as lossless with $\alpha = 0$.

The structure without the top shield is considered first. The structure has a period of d=10mm. A square patch of width w = 9.6 mm and gap g = 0.2 mm from the adjacent patch is taken on a substrate with $\varepsilon_r = 2.33$, h = 3.08. Figure (20.17a and b) show the dispersion diagram and attenuation diagram, respectively [J.53]. Figure (20.17a) also shows the side-view of the structure and its equivalent circuit. The impedance Z_1 is due to the coupling capacitance C_c, and impedance Z_2 models centrally located via as an inductor. Subsection (19.3.6) of chapter 19 summarizes the model for a via. The computed values of components are shown in Fig (20.17b). Bragg's boundary points, "a" to "e," showing edges of

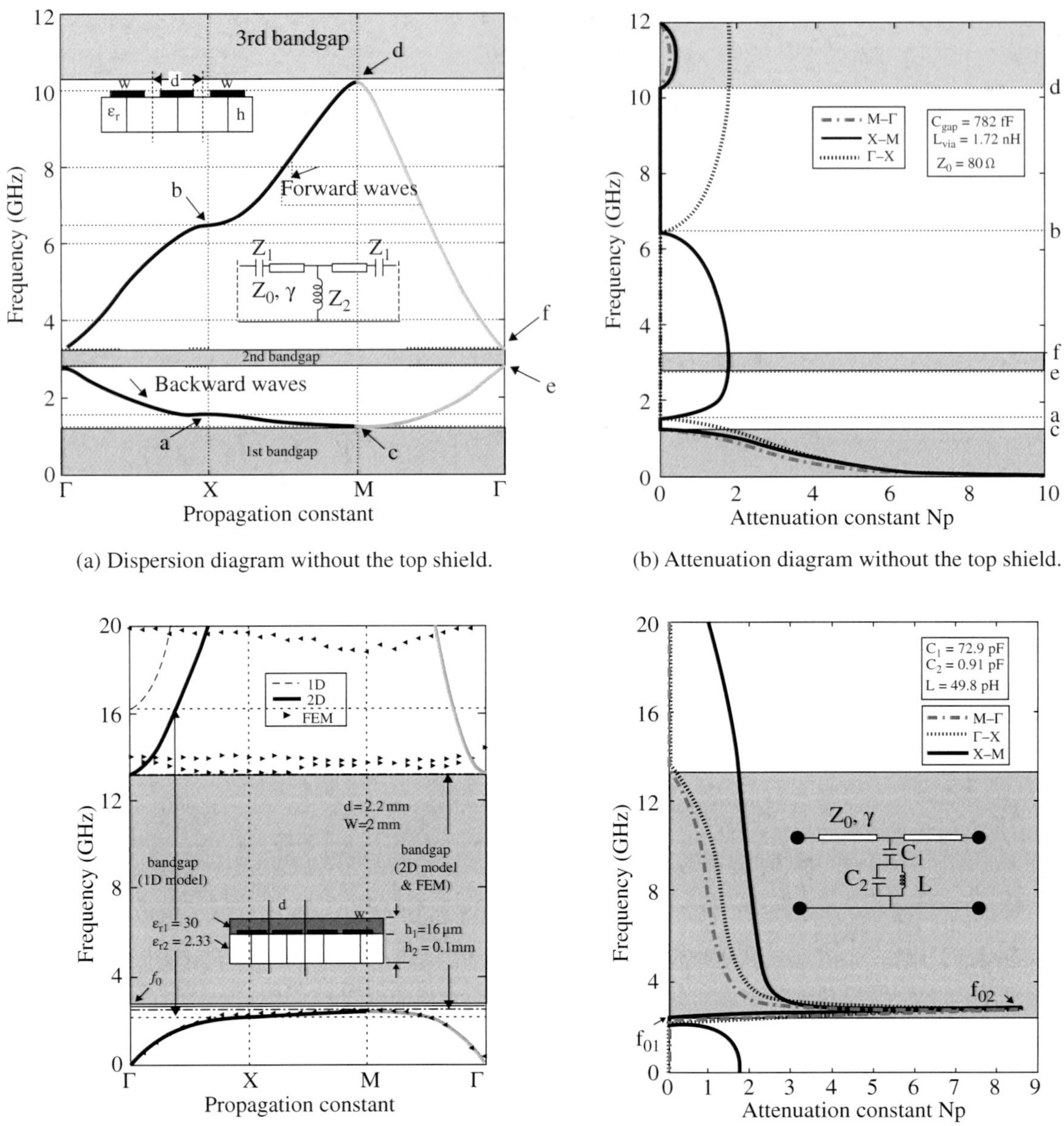

(a) Dispersion diagram without the top shield.

(b) Attenuation diagram without the top shield.

(c) Dispersion diagram with a top shield.

(d) Attenuation diagram with a top shield.

Figure 20.17 The dispersion and attenuation diagrams of the mushroom-type EBG outside/inside a PPW. Source: Reproduced from Tavallaee and Abhari [J.53]. © 2007, IET.

frequency bands, are shown in the diagram. The dispersion diagram follows the previous discussion. In this case, at several frequencies, between 0 and 10 GHz, both the propagation constant and the attenuation constant are computed, using the expression (20.4.21a). Both the complete and partial bandgap are obtained. The global complete bandgap is shown in gray. In bandgap region, attenuation α_x, α_y, α_{xy} are nonzeros, showing the presence of the evanescent mode in all directions. In the range c-a, attenuation is not present and the partial bandgap is enhanced in some directions.

Figure (20.17c and d) shows the dispersion and attenuation diagrams of the EBG inside a PPW. The PPW structure is shown in the inset of Fig (20.17c). The circuit model of the unit cell is modified. It is shown in the inset of Fig (20.17d), along with the values of components. In this case, $Z_1 = 0$, and the Z_2 is a series combination of capacitance C_1 and C_2-L parallel circuit. This model is a little different from that given in subsection (19.3.4) of chapter 19. In the present model, C_1 is the parallel-plate capacitor between the patch and top shield, C_2 is the coupling capacitance C_c between adjacent patches, and L is due to the inductance of via. The PPW is having an inhomogeneous dielectric medium that is replaced with an equivalent homogeneous medium of relative permittivity ε_{req}, and thickness $h = h_1 + h_2$. The PPW is modeled as a transmission line section of length $\gamma d/2$ and characteristic impedance Z_0. The expressions used in the modeling are summarized below:

$$Z_2 = \frac{1}{j\omega C_1} + \frac{1}{j(\omega C_2 - 1/\omega L)} \quad \text{(a)}$$

$$\varepsilon_{req} = \frac{h_1 + h_2}{(h_1/\varepsilon_{r1} + h_2/\varepsilon_{r2})} \quad \text{(b)}. \tag{20.4.22}$$

The propagation relation for this case, for $Z_1 = 0$, is obtained from equation (20.4.21). Figure (20.17d) shows two resonances giving no attenuation (at the parallel resonance, f_{01}) and high attenuation (at the series resonance, f_{02}) in the attenuation diagram. Figure (20.17c) shows that, unlike the previous case, this structure supports the forward Bloch mode, and also a wide bandgap. The dispersions computed from the 1D model and the FEM also are shown in Fig (20.17c) [J.55, J.56]. The 2D circuit model is closer to the results of the FEM method, whereas the 1D circuit model is not so accurate.

Bloch Impedance

Bloch impedances of Bloch waves, both in the x- and y-directions are computed below [J.50]. Bloch impedances are defined as

$$Z_x^{Bloch} = \frac{V_x}{I_x} \quad \text{(a)}, \quad Z_y^{Bloch} = \frac{V_y}{I_y} \quad \text{(b)}. \tag{20.4.23}$$

The following expressions are obtained from the 1st two rows of equation (20.4.9a):

$$f_{11}V_x + f_{12}I_x = 0 \quad \text{(a)}, \quad f_{23}V_y + f_{24}I_y = 0 \quad \text{(b)}$$

$$\Rightarrow Z_x^{Bloch} = -\frac{V_x}{I_x} = \frac{f_{12}}{f_{11}} = \frac{B_1 + B_3 e^{-j\beta_x d}}{D_1 - D_3 e^{-j\beta_x d}} (\text{for } \Delta_x = 1) \quad \text{(c)}. \tag{20.4.24}$$

For an isotropic 2D EBG surface, $B_1 = B_3 = B$ and $D_1 = D_3 = D$. The above expression of Bloch impedance is reduced to

$$Z_x^{Bloch} = -\frac{jB}{D}\frac{1}{\tan(\beta_x d/2)} \quad \text{(a)}$$

$$\text{Likweise, } Z_y^{Bloch} = -\frac{jB}{D}\frac{1}{\tan\left(\beta_y d/2\right)} \quad \text{(b)}. \tag{20.4.25}$$

Bloch impedance provides a relation between the voltage and current at the same port. However, a relation between the cross-ports, i.e. between V_x and V_y, is calculated by using equations (20.4.12), (20.4.13a), and (20.4.13b):

$$V_x = \frac{g_{12}}{g_{11}}V_y = \left[\frac{B_2D_4 + B_4D_2}{B_4 + B_2e^{+j\beta_y d}}\right]\left[\frac{B_3 + B_1e^{+j\beta_x d}}{B_1D_3 + B_3D}\right]V_y \quad \text{(a)}$$

For an isotropic EBG surface,

$$V_x = \left[\frac{1 + e^{+j\beta_x d}}{1 + e^{+j\beta_y d}}\right]V_y \quad \text{(b)}. \tag{20.4.26}$$

Therefore, for the known V_y, the V_x is obtained. Likewise, for the known I_x, the I_y is obtained. The voltages and currents of 4-ports are related through $\beta_x d$ and $\beta_y d$.

Special Case – Uniplanar 2D EBG Structure

The uniplanar EBG surface, shown in Fig (20.16b), is a special case of the mushroom-type EBG. The square capacitive patches are interconnected by the narrow inductive microstrip sections of the patch that form the UC-EBG shown in Fig (20.1d). The shunt arm transmission $[T_5]$ matrix, shown in Fig (20.15c), is absent, i.e. the shunt arm is an open-circuited terminal, $Z_{shunt} \to \infty$,

$$\begin{bmatrix} A_5 & B_5 \\ C_5 & D_5 \end{bmatrix} = \begin{bmatrix} 1 & Z_{shunt} \\ 0 & 1 \end{bmatrix}, \Rightarrow B_5 = \infty. \tag{20.4.27}$$

The parameters of equation (20.4.9) remain unchanged, except f_{41} and f_{42}. These are modified using $B_5 \to \infty$ for the uniplanar EBG, as follows:

$$f_{41} = -\frac{C_1}{\Delta_x} - C_3 e^{-j\beta_x d}, \quad f_{42} = \frac{A_1}{\Delta_x} - A_3 e^{-j\beta_x d}. \tag{20.4.28}$$

Likewise, g_{11}, g_{12}, and g_{22} of equation (20.4.13) remain unchanged. However, g_{21} is modified to

$$g_{21} = \frac{1 + e^{-j2\beta_x d} - e^{-j\beta_x d}[B_1C_3 + C_1B_3 + A_1D_3 + D_1A_3]}{(B_1 + B_3 e^{-j\beta_x d})}. \tag{20.4.29}$$

The reciprocal and isotropic EBG surface has $d_1 = d_2 = d$; $A = A_1 = \dots A_4$; $B = B_1 = \dots B_4$; $C = C_1 = \dots C_4$;

and $D = D_1 = \ldots D_4$, and $\Delta_x = \Delta_y = 1$. The expressions for g_{11}, g_{12}, g_{21}, and g_{22}, given in equations (20.4.13) and (20.4.29), are reduced to

$$g_{11} = \frac{2De^{-j\beta_x d}}{(1+e^{-j\beta_x d})} \quad (a), \quad g_{12} = \frac{2D\,e^{-j\beta_y d}}{(1+e^{-j\beta_y d})} \quad (b)$$

$$g_{21} = \frac{1+e^{-j2\beta_x d} - 2[BC+AD]e^{-j\beta_x d}}{B(1+e^{-j\beta_x d})} \quad (c)$$

$$g_{22} = \frac{1+e^{-j2\beta_y d} - 2[BC+AD]e^{-j\beta_y d}}{B\,(1+e^{-j\beta_y d})} \quad (d). \qquad (20.4.30)$$

The dispersion relations for both the lossless and lossy uniplanar EBG surfaces are obtained by substituting the above parameters in the following expression from equation (20.4.12):

$$g_{11}g_{22} + g_{12}g_{21} = 0$$

$$\Rightarrow \cos(\beta_x d) + \cos\left(\beta_y d\right) = 2\,(AD+BC) \quad (a)$$

Or accounting for loss

$$\cosh(\gamma_x d) + \cos\left(\gamma_y d\right) = 2(AD+BC) \quad (b). \qquad (20.4.31)$$

The expressions of Bloch impedances, equation (20.4.25), remain unchanged. Moreover, this model is also applicable to UC-EBG, as is shown in Fig (20.1d). The inset of Fig (20.1d) shows another form of the UC-EBG, large metallic patches are connected by the narrow metallic strips. The circuit model is also applicable to the planar lumped element EBG, as shown in Fig (20.1f). In all cases, the matrix elements are evaluated using appropriate circuit element models. Figure (20.16b) shows that the dispersion in the uniplanar EBG starts with a passband, supporting the forward Bloch mode #1 [J.33, J.36, J.51, J.53]. This characteristic is different from that of the mushroom-type structure, where Bloch mode #1 is a BW.

20.4.2 Series-Connected 2D Planar EBG Circuit Model

Figure (20.18a) is a series-connected 4-port network, corresponding to the 2D EBG surface with the series-connected loading elements, shown in Fig (20.18b). The EBG constitutes of reactively loaded *coplanar strip lines* to develop the isotropic surface. The *interdigital capacitors* and *meandered inductors* are connected in series with the coplanar strip lines. Figure (20.18c) further shows its equivalent circuit. The equivalent circuit is analyzed below, using the [ABCD] network model [J.54]. Due to the isotropy, the matrices are identical. They further satisfy the reciprocity relation $\Delta = AD - BC = 1$ for the reciprocal EBG surface. Due to the physical placement of components, with respect to the center, the position of matrix elements A and D are interchanged for the output matrices in the x- and y-directions. The present analysis is applied to the isotropic and reciprocal EBG surface. However, the method, in general, applies to the anisotropic and non-reciprocal EBG surface also.

Figure (20.18a) shows the transmission matrices network of the unit cell of the EBG surface of Fig (20.18b). Figure (20.18c) further shows the equivalent 2D circuit diagram of the EBG surface. Applying the KCL and KVL to the network of Fig (20.18a), the following relations are obtained at the junctions of outputs of the input matrices and inputs of the output matrices:

$$\text{KCL: } I'_x = I''_x, I'_y = I''_y, I''_x = -I''_y, \quad (a)$$

$$\text{KVL: } V'_x + V''_y - V''_x - V'_y = 0 \quad (b). \qquad (20.4.32)$$

The set of equations (20.4.3) provide relations between the input and output voltages/currents of an individual matrix. Equations (20.4.4a) to (20.4.4d) relate the input and output of the x- and y-directed voltages/currents, using Floquet–Bloch PBCs.

On using the set of equations (20.4.3) and (20.4.4), with equations (20.4.32a) and (20.4.32b), the following expressions are obtained:

$$C(1+e^{-j\beta_x d})V_x + A(e^{-j\beta_x d} - 1)I_x = 0 \quad (a)$$

$$C(1+e^{-j\beta_y d})V_y + A(e^{-j\beta_y d} - 1)I_y = 0 \quad (b)$$

$$(Ce^{-j\beta_x d})V_x + (Ae^{-j\beta_x d})I_x + (Ce^{-j\beta_y d})V_y$$

$$+ (Ae^{-j\beta_y d})I_y = 0 \quad (c), D(1-e^{-j\beta_x d})V_x - B(1+e^{-j\beta_x d})I_x$$

$$+ D(e^{-j\beta_y d} - 1)V_y + B(1+e^{-j\beta_y d})I_y \quad (d)$$

where, $A_1, \ldots A_4 = A; B_1, \ldots B_4 = B; C_1, \ldots C_4 = C; D_1, \ldots D_4 = D; \Delta = 1.$ (20.4.33)

The above equations are rearranged in the matrix form, given by equation (20.4.9a), where the following expressions compute the matrix elements:

$$f_{11} = C(1+e^{-j\beta_x d}), f_{12} = A(e^{-j\beta_x d} - 1), f_{23} = C(1+e^{-j\beta_y d})$$

$$f_{24} = A(e^{-j\beta_y d} - 1), f_{31} = Ce^{-j\beta_x d}, f_{32} = Ae^{-j\beta_x d}, f_{33} = Ce^{-j\beta_y d}$$

$$f_{34} = Ae^{-j\beta_y d},\ f_{41} = D(1-e^{-j\beta_x d}), f_{42} = -B, (e^{-j\beta_x d} + 1)$$

$$f_{43} = D(e^{-j\beta_y d} - 1), f_{44} = B(e^{-j\beta_y d} + 1). \qquad (20.4.34)$$

Using the above relations with the first two rows of equation (20.4.9a), the following expressions are obtained for the I_x and I_y:

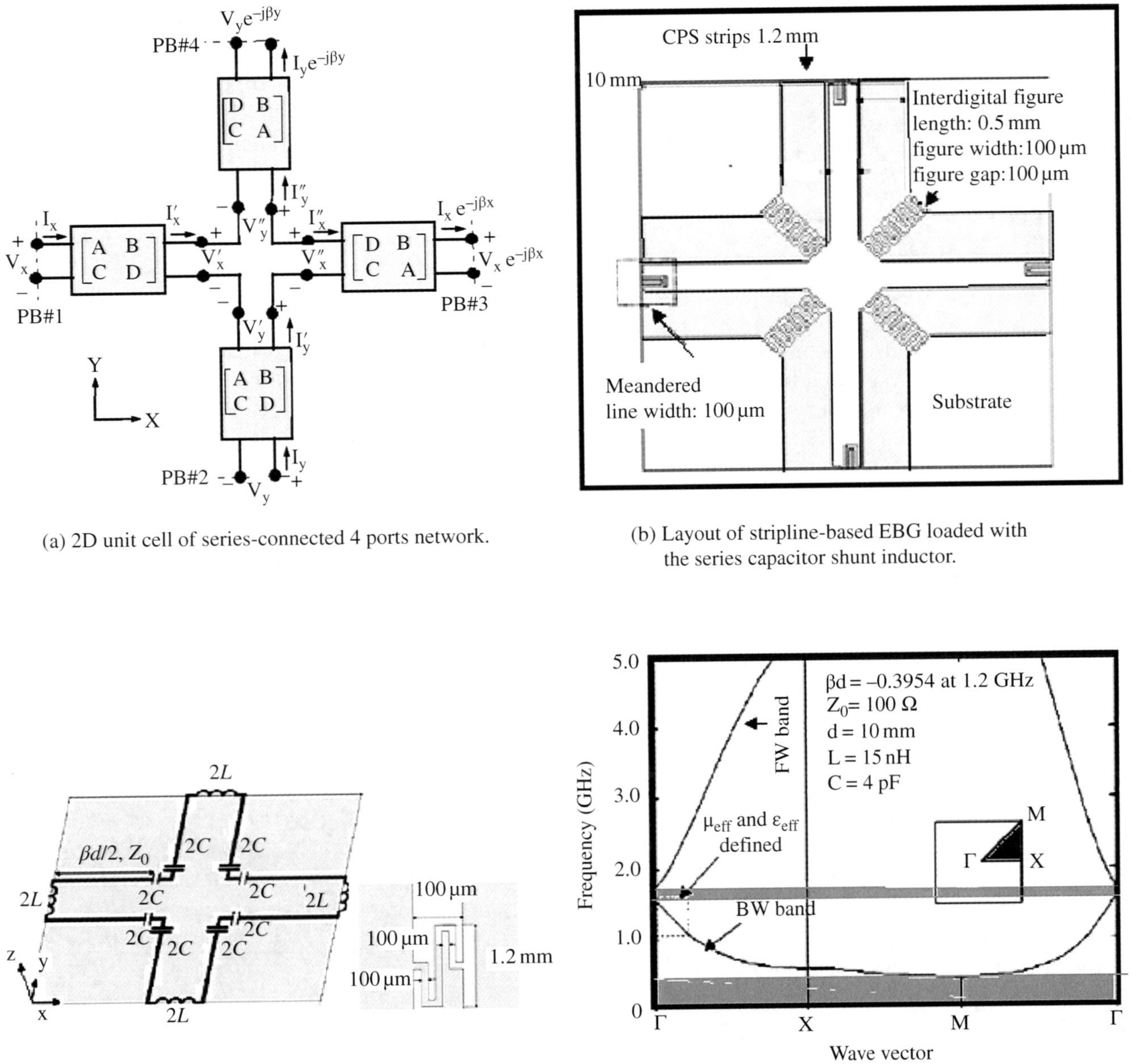

(a) 2D unit cell of series-connected 4 ports network.

(b) Layout of stripline-based EBG loaded with the series capacitor shunt inductor.

(c) 2D equivalent circuit of the unit cell.

(d) Dispersion diagram of structure - (d).

Figure 20.18 Series-connected 2D uniplanar EBG. Source: Reproduced from Elek and Eleftheriades [J.54]. © 2007, IET.

$$f_{11}V_x + f_{12}I_x = 0, \quad f_{23}V_y + f_{24}I_y = 0$$

$$\Rightarrow I_x = \frac{C\left(1 + e^{-j\beta_x d}\right)}{A\left(1 - e^{-j\beta_x d}\right)}V_x \quad (a), \quad I_y = \frac{C\left(1 + e^{-j\beta y d}\right)}{A\left(1 - e^{-j\beta_y d}\right)}V_y \quad (b). \tag{20.4.35}$$

The further relation between V_x and V_y is obtained by substituting I_x and I_y in the expressions from the third and fourth rows of equation (20.4.9a):

$$f_{31}V_x + f_{32}I_x + f_{33}V_y + f_{34}I_y = 0$$

$$f_{41}V_x + f_{42}I_x + f_{43}V_y + f_{44}I_y = 0$$

$$\Rightarrow g_{11}V_x + g_{12}V_y = 0, g_{21}V_x + g_{22}V_y = 0$$

$$\Rightarrow \begin{bmatrix} g_{11} & g_{12} \\ g_{21} & g_{22} \end{bmatrix} \begin{bmatrix} V_x \\ V_y \end{bmatrix} = \begin{bmatrix} 0 \\ 0 \end{bmatrix}, \tag{20.4.36}$$

where g-parameters are given as follows:

$$g_{11} = e^{-j\beta_x d}\left(1 - e^{-j\beta_y d}\right) \ (a), \ g_{12} = e^{-j\beta_y d}\left(1 - e^{-j\beta_x d}\right) \ (b)$$

$$g_{21} = \frac{1 + e^{-j2\beta_x d} - (AD + BC)e^{-j\beta_x d}}{1 - e^{-j\beta_x d}} \quad (c)$$

$$g_{22} = \frac{-\left[1 + e^{-j2\beta_y d} - (AD + BC)e^{-j\beta_y d}\right]}{1 - e^{-j\beta_y d}} \quad (d). \tag{20.4.37}$$

The nontrivial solution of equation (20.4.36) is obtained from $\Delta g = 0$,

$$g_{11}\, g_{22} - g_{12}\, g_{21} = 0$$

$$\Rightarrow \text{for lossless EBG: } \cos(\beta_x d) + \cos\left(\beta_y d\right) = 2(AD + BC) \qquad (a)$$

$$\text{for lossy EBG: } \cosh(\gamma_x d) + \cos\left(\gamma_y d\right) = 2(AD + BC) \qquad (b), \tag{20.4.38}$$

where $\gamma_{x/y} = \alpha_{x/y} + j\,\beta_{x/y}$. The above dispersion relation is the same as that of given in equation (20.4.31), for the uniplanar EBG. However, the Bloch impedances, from equations (20.4.23) and (20.4.35), given below, are different:

$$Z_x^{Bloch} = \frac{A\left(1 - e^{-j\beta_x d}\right)}{C\left(1 + e^{-j\beta_x d}\right)} \; (a), \quad Z_y^{Bloch} = \frac{A\left(1 - e^{-j\beta y d}\right)}{C\left(1 + e^{-j\beta_y d}\right)} \; (b)$$

$$\Rightarrow Z_x^{Bloch} = j\frac{A}{C}\tan\left(\frac{\beta_x d}{2}\right) \; (c), \quad Z_y^{Bloch} = j\frac{A}{C}\tan\left(\frac{\beta_y d}{2}\right) \; (d). \tag{20.4.39}$$

A relation between the port voltages V_x and V_y are also obtained from equations (20.4.36) and (20.4.37):

$$g_{11}V_x + g_{12}V_y = 0, \Rightarrow V_x = -\frac{g_{12}}{g_{11}}V_y$$

$$\Rightarrow V_x = \frac{\left(e^{-j2\beta_x d} - 1\right)}{\left(1 - e^{-j2\beta_y d}\right)}V_y. \tag{20.4.40}$$

In case the wave propagates along the x-direction, $\beta_x d \neq 0$, $\beta_y d = 0$, the above equation provides $V_y = 0$. The above equation is not needed in this, and the 2D EBG surface is treated as a 1D EBG line with transversely loaded short-circuited two stubs, each of length d/2. The layout, shown in Fig (20.18b), provides Bloch mode #1, supporting the BW, in its first passband. It is shown in the dispersion diagram of Fig (20.18d). Next, the Bloch mode #2 is a forward wave (FW). Two numbers of the bandgap are shown as the gray strips.

References

Books

B.1 Fang, F.; Rahmat-Samii, Y.: *Electromagnetic Band Gap Structures in Antenna Engineering*, Cambridge University Press, UK, 2009.

B.2 Munk, B.A.: *Frequency Selective Surfaces -Theory and Design*, John Wiley & Sons, New York,2000.

B.3 Tretyakov, S.: *Analytical Modeling on Applied Electromagnetics*, Artech House, Boston, USA, 2000.

B.4 Balanis, C.: *Advanced Engineering Electromagnetics*, John Wiley & Sons, New York, 1989.

B.5 Simon, R.; Whinnery John R.; Theodore, V.D.: *Fields and Waves in Communication Electronics*, 3rd Edition, John Wiley & Sons, Singapore, 1994.

Journals

J.1 Lim, J.-S.; Kim, H.-S.; Park, J.-S.; Ahn, D.; Nam, S.: *A power amplifier with efficiency improved using defected ground structure, IEEE Microw. Wirel. Compon. Lett.*, Vol. 11, No. 4, pp. 170–172, Apr. 2001.

J.2 Rahmat-Samii, Y.: *The marvels of electromagnetic band gap (EBG) structures. Appl. Comput. Electromagn. Soc. J.*, Vol. 18, No. 4, pp. 1–10, Nov. 2003.

J.3 Sievenpiper, D.; Zhang, L.; Broas, R.; Yablonovich, E.: *High-impedance electromagnetic surfaces with forbidden bands at radio and microwave frequencies*, SPIE Conf. on THz and GHz Photonics, Vol. 3795, pp. 154–161, July 1999.

J.4 Ma, K.P.; Hirose, K.; Yang, F. R., Qian, Y.; Itoh, T.: *Realization of magnetic conducting surface using novel photonic bandgap structure, Electron. Lett.*, Vol. 34, No. 21, pp. 2041–2042, Oct. 1998.

J.5 Syed, I.S.; Ranga, Y.; Matekovits, L.; Esselle, K.P.; Hay, S.G.: *A single-layer frequency-selective surface for ultrawideband electromagnetic shielding, IEEE Trans. Electromag. Compat.*, Vol. 56, No. 6, pp. 1404–1409, Dec. 2014.

J.6 Sievenpiper, D.F.: *High-Impedance Electromagnetic Surfaces*, UCLA, Ph.D. dissertation, Jan. 1999.

J.7 Sievenpiper, V.; Zhang, L.; Broas, R. J.; Alexópolous, N.; Yablonovitch, E.: *High-Impedance electromagnetic surfaces with a forbidden frequency band, IEEE Trans. Microw. Theory Tech.*, Vol. 47, pp. 2059–2074, Nov. 1999.

J.8 Yang, F.; Rahmat-Samii, Y.: *Polarization dependent electromagnetic bandgap characteristics, design, and applications*, IEEE AP – S., Int. Sym. Digest, Vol. 3, pp. 339–342, June 2003.

J.9 Yang, F.; Rahmat-Samii, Y.: *A low profile single dipole antenna radiating circularly polarized waves, IEEE Trans. Antennas Propagat.*, Vol. AP-53, No. 9, pp. 3083–3086, Sept. 2005.

J.10 Lee, R.; Vardaxoglou, Y.C.; Budimir, D.: *Microwave bandgap and bandpass structures using planar metallodielectric periodic arrays,* www.armms.org/media/uploads/1271348629.pdf.

J.11 Ladislau Matekovits, L.; Ranga, Y.: *Controlling the phase of the scattered and/or radiated field from a high impedance surface of quasi-periodic sequences, IEEE Antennas Wirel. Propag. Lett.*, Vol. 12, pp. 321–323, 2013.

J.12 Matekovits, L; Colemè, G.V.; Orefice, M.: *Controlling the band limits of TE-surface wave propagation along a modulated microstrip line- based high impedance surface, IEEE Trans. Antennas Propag.*, Vol. 56, No. 8, pp. 2555–2562, Aug. 2008.

J.13 Mosallaei, H.; Sarabandi, K.: *Antenna miniaturization and bandwidth enhancement using a reactive impedance surface, IEEE Trans. Antennas Propagat.*, Vol. AP-52, No. 9, pp. 2403–2414, Sept. 2004.

J.14 Matekovits, L.: *Analytically expressed dispersion diagram of unit cells for a novel type of holographic surface, IEEE Antennas Wirel. Propag. Lett.*, Vol. 9, pp. 1251–1254, 2010.

J.15 Yang, F.R.; Ma, K.P.; Qian, Y.; Itoh, T.: *A uniplanar compact photonic-bandgap (UC-PBG) structure and its applications for microwave circuits, IEEE Trans. Microw. Theory Tech.*, Vol. 47, No. 8, pp. 1509–1514, Aug. 1999.

J.16 Scogna, A.C.; *Suppression of simultaneous switching noise in power and ground plane pairs, Conformity*, Vol.37, pp. 37–43, June 2007.

J.17 Ki Hyuk Kim, K.H.; Schutt-Ainé, J.E.: *Analysis and modeling of hybrid planar-type electromagnetic-bandgap structures and feasibility study on power distribution network applications, IEEE Trans. Microw. Theory Tech.*, Vol. 56, No. 1, pp. 178–186, Aug. 1999.

J.18 Agarwal, K.; Nasimuddin, Alphones, A.: *RIS-based compact circularly polarized microstrip antennas, IEEE Trans. Antennas Propagat.*, Vol. AP-61, No. 2, pp.547–554, Feb. 2013.

J.19 Xu, H.-X.; Wang, G.-M., Liang, J.-G.; Qi, M.Q.; Gao, X.: *Compact circularly polarized antennas combining meta-surfaces and strong space-filling meta-resonators, IEEE Trans. Antennas Propag.*, Vol. AP-61, No. 7, pp.3442–3450, July 2013.

J.20 Yanagi, T.; Takeshi Oshima, T.; Ohashi, H.; Konishi, Y.; Murakami, S.; Itoh, K.; Sanada, A.: *Lumped-element loaded EBG structure with an enhanced bandgap and homogeneity*, Proc. iWAT2008, P 311, pp. 458–461, Chiba.

J.21 Mias, C.; Yap, J.H.: *A varactor-tunable high impedance surface with a resistive-lumped-element biasing grid, IEEE Trans. Antennas Propagat.*, Vol. AP-55, No. 7, pp. 1955–1962, July 2007.

J.22 Sievenpiper, D.F.; Schaffner, J.H.; Song, H.J.; Lo, R.Y.; Tangonan, G.: *Two-dimensional beam steering using an electrically tunable impedance surface, IEEE Trans. Antennas Propagat.*, Vol. AP-51, No. 10, pp. 2713–2722, Oct.2003.

J.23 Raza, S.: *Characterization of the reflection and dispersion properties of 'mushroom'-related structures and their application to antennas*, M.Sc. Thesis, Department of Electrical And Computer Engineering, University of Toronto, 2012.

J.24 Verma, A.K.; Kumar, A.: *Design of low pass filters using some defected ground structures, Int. J. Electron. Comm. (AEU)*, Vol. 65, No. 10, pp. 864–872, 2011.

J.25 Raheel, M.; Hashmi, R.M.; Zeb, B.A.; Esselle, K.P.: *Wideband high-gain EBG resonator antennas with small footprints and all-dielectric superstructures, IEEE Trans. Antennas Propag.*, Vol. 62, No. 6, pp. 2970–2976, June 2014.

J.26 Li, L.; Chen, Q.; Yuan, Q.; Liang, C.; Sawaya, K.: *Surface wave suppression bandgap and plane-wave reflection phase band of mushroom-like photonic bandgap structures, J. Appl. Phys.*, Vol. 103, No. 023513, pp.103–110, 2008.

J.27 Remski, R.: *Analysis of photonic bandgap surfaces using Ansoft HFSS, Microw. J.*, Vol. 43, pp.190–198, Sept. 2003.

J.28 Sievenpiper, D.F.; Zhang, L.J.; Broas, R.F.J.; Alexopolous, N.G.; Yablonovitch, E.: *High-impedance electromagnetic surfaces with a forbidden frequency band, IEEE Trans. Microw. Theory Tech.*, Vol. 47, pp. 2059–2074, 1999.

J.29 Sievenpiper, D.F.: *High-impedance surfaces, Ph.D. thesis*, University of California, LA, USA, 1999.

J.30 Yang, F.R.; Ma, K.P.; Yongxi Qian, Y.; Itoh, T.: *A novel TEM waveguide using Uniplanar Compact Photonic-Bandgap (UC-PBG) structure, IEEE Trans. Microw. Theory Tech.*, Vol. 47, No. 11, pp. 2092–2098, Nov. 1999.

J.31 Yang, F.; Rahmat-Sammi, Y.: *Reflection phase characterizations of the EBG ground plane for low profile wire antenna application, IEEE Trans. Antennas Propag.*, Vol. 51, No. 10, pp. 2691–2703, Oct. 2003.

J.32 Yang, F.; Rahmat-Sam, Y.: *Polarization-dependent electromagnetic bandgap (PDEBG) structures: Designs and applications, Microw. Opt. Technol. Lett.*, Vol. 41, No. 6, pp. 439–444, June 2004.

J.33 Coccioli, R.; Yang, F.R.; Ma, K.P.; Itoh, T.: *Aperture-coupled patch antenna on UC-PBG substrate, IEEE Trans. Microw. Theory Tech.*, Vol. 47, pp. 2123–2130, Nov. 1999.

J.34 Yang, F.; Rahmat-Samii, Y.: *Polarization dependent electromagnetic bang-gap surfaces: characterization, design, and applications*, IEEE Trans. Antennas Propag. Int. Sym. Digest, Vol. 3, pp. 339–342, 2003.

J.35 Sievenpiper, V.; Zhang, L.; Broas, R.J.; Yablonovitch, E.: *High-Impedance electromagnetic surfaces with forbidden frequency bands at radio and microwave* *frequencies*, SPIE Conf. on THz and GHz Photonics, Vol. 33795, pp. 154–165, July 1999.

J.36 Kamgaing, T.; Ramahi, O.M.: *Design and modeling of High-impedance electromagnetic surfaces for switching noise suppression in power planes, IEEE Trans. Electromagn. Compat.*, Vol. 47, No. 3, pp. 478–488, Aug. 2005.

J.37 Kamgaing, T.; Ramahi, O. M.: *High-impedance electromagnetic surfaces for parallel-plate mode suppression in high-speed digital systems, IEEE 11th Topical Meeting Electr. Perform. Electron. Packag.*, pp. 279–282, 2002.

J.38 Yousefi, L.; Mohajer-Iravani, B.; Ramahi, O.M.: *Enhanced bandwidth artificial magnetic ground plane for low-profile antennas, IEEE Antennas Wirel. Propag. Lett.*, Vol. 6, pp. 289–292, 2007.

J.39 Tretyakov, S.A.; Simovski, C.R.: *Dynamic model of artificial reactive impedance surfaces, J. Electromagn. Waves Appl.*, Vol. 17, No. 1, pp. 131–145, 2003

J.40 Luukkonen, O.; Constantin Simovski, C.; Granet, G.; Goussetis, G.; Dmitri Lioubtchenko, D.; Antti V. Räisänen, A.V.; Sergei A. Tretyakov, S.A.: *Simple and accurate analytical model of planar grids and high-impedance surfaces comprising metal strips or patches, IEEE Trans. Antennas Propag.*, Vol. 56, No. 6, pp. 1624–1632, June 2008.

J.41 Luukkonen, O.; Constantin Simovski, C.; Granet, G.; Goussetis, G.; Dmitri Lioubtchenko, D.; Räisänen, A. V.; Sergei A. Tretyakov, S.A: *Corrections to "Simple and accurate analytical model of planar grids and high-impedance surfaces comprising metal strips or patches", IEEE Trans. Antennas Propag.*, Vol. 58, No. 6, pp. 2162–2162, June 2010.

J.42 Luukkonen, O.; Alitalo, P.; Simovski, C.R.; Tretyakov, S.A.: *Experimental verification of analytical model for high impedance surfaces, Electron. Lett.*, Vol. 45, No. 14, July 2009.

J.43 Luukkonen, O.; Simovski, C.R.; Räisänen, A.V.; Tretyakov, S.A.: *An efficient and simple analytical model for analysis of propagation properties in impedance waveguides, IEEE Trans. Microw. Theory Tech.*, Vol. 56, No. 7, pp.1624–1632, July 2008.

J.44 Luukkonen, O.: *Artificial impedance surfaces, Ph.D. thesis,* Helsinki University of Technology, Dec. 2009.

J.45 Iravani, B.M.; Ramahi, O.M.: *Wideband circuit model for planar EBG structures, IEEE Trans. Adv. Packaging*, Vol. 33, No. 1, pp. 169–179, Feb. 2010.

J.46 Abedin, M.F.; Ali, M.: *Effects of a smaller unit cell planar EBG structure on the mutual coupling of a printed dipole array, IEEE Antennas Wirel. Propag. Lett*, Vol.4, pp. 274–276, 2005.

J.47 Waterhouse, R.B.; Novak, D.: *A small electromagnetic bandgap structure*, Proc. IEEE International Microwave Symposium, pp. 602–605, June 2006. Pharad, Glen Burnie, MD 21061, USA

J.48 Kern, D.J.; Werner, D.H.; Monorchio, A.; Lanuzza, L.; Wilhelm, M.J.: *The design synthesis of multiband artificial magnetic conductors using high impedance frequency selective surfaces, IEEE Trans. Antennas Propag.*, Vol. 53, No. 1, pp. 8–16, Jan. 2005.

J.49 Chen, Z.; Solbach, K.; Erni, D.; Rennings, A.: *A compact electromagnetic bandgap structure based on multi-layer technology for 7-Tesla magnetic resonance imaging applications*, Proc. 9th European Microwave Integrated Circuits Conf., EuMA 6-7, pp. 632–635, Oct. 2014.

J.50 Gribic, A.; Eleftheriades, G. V.: *Periodic analysis of a 2-D negative refractive index transmission line structure, IEEE Trans. Antennas Propag.*, Vol. 51, No. 10, pp. 2604–2611, Oct. 2003.

J.51 Toyota, Y.; Engin, A.E.; Kim, T.H.; Swaminathan, M.: *Stopband analysis using dispersion diagram for two-dimensional electromagnetic bandgap structures in printed circuit boards, IEEE Microw. Wirel. Compon. Lett*, Vol. 16, No. 12, pp. 645–647, Dec. 2006.

J.52 Caloz, C.; Itoh, T.: *Positive/negative refractive index anisotropic 2-D metamaterials, IEEE Microw. Wirel. Compon. Lett.*, Vol. 13, No. 12, pp. 547–550, Dec. 2003.

J.53 Tavallaee, A.; Abhari, R.: *2-D characterization of electromagnetic bandgap structures employed in power distribution networks. IET Microw. Antennas Propag.*, Vol. 1, No. 1, pp. 204–211, Feb. 2007.

J.54 Elek, F.; Eleftheriades, G.V.: *A two-dimensional uniplanar transmission -line metamaterial with a negative index of refraction, New J. Phys.*, Vol. 7, 163 PII, pp. 1–18, 2005.

J.55 Harvey, A.F.: *Periodic and guiding structures at microwave frequencies, IRE Trans. Microw. Theory Tech.*, Vol. 8, No. 1, pp. 30–61, Jan. 1960.

J.56 Rogers, S.D.: *Electromagnetic-bandgap layers for broadband suppression of TEM modes in power planes, IEEE Trans. Microw. Theory Tech.*, Vol. 53, No. 8, pp. 2495–2505, Aug 2005

21

Metamaterials Realization and Circuit Models – I

(Basic Structural Elements and Bulk Metamaterials)

Introduction

The engineered metamaterials [J.1–J.5] are the outcome of the *artificial dielectrics* discussed in subsection (6.2.2) of chapter 6. Their significant properties are presented in section (5.5) of chapter 5. The materials of the EBG group are also artificially engineered materials with unique properties. These are discussed in chapters 19 and 20. However, metamaterials are different from them. In the case of the EBG/PBG bandgap materials, the lattice constant is a half-wavelength long, so these are not treated as homogeneous materials. In the case of the metamaterials, the size of inclusion and lattice constant are less than one-tenth of a wavelength, and Bragg diffraction is avoided. Therefore, the metamaterials are treated as a homogeneous medium, like a natural material. *The homogenized metamaterials are characterized by the effective relative permittivity and effective relative permeability, with unique properties normally not found in nature.*

The metamaterials are wideband materials grouped into two basic categories – *resonant type*, and *nonresonant type*. The resonant type metamaterials are narrow-band materials, whereas the nonresonant types are broadband. This chapter presents the basic resonant metallic inclusion structures for the realization of the 3D bulk epsilon negative (ENG), the mu negative (MNG), and double-negative (DNG) metamaterials. The resonant type inclusions, both metallic and dielectric, embedded in the host dielectric medium is treated as artificial molecules. The nonresonant metamaterials are more commonly used for the 1D metalines, also for the 2D metasurfaces. The metalines and metasurfaces are discussed in chapter 22. Our primary concern in this chapter is to discuss the circuit models and the realization of metamaterials using the planar structures. The homogenization and extraction of the electrical parameters of the metamaterials are also discussed. The detailed theory and applications of the metamaterials are presented in the textbooks [B.1–B.8].

Objectives

- *To discuss the realization of the ENG, MNG, and DNG metamaterials using the wire medium, loop medium, and wire-loop combined medium.*
- *To present a summary of some metallic inclusion structures in the planar technology for the realization of metamaterials.*
- *To discuss the parameter retrieval process of metamaterials.*
- *To discuss the homogenization and the dynamic Maxwell Garnett model of the metamaterials.*

21.1 Artificial Electric Medium

The artificial electric media, i.e. the epsilon negative (ENG) materials, the epsilon near zero (ENZ) materials, and the high permittivity dielectrics, are realized using the *wire medium*. It is also called the *rodded medium*. Figure (21.1a–c) shows the 1D, 2D, and 3D bulk *wire medium*-based *artificial electric medium*. The wires (rods) are the inclusions embedded in the host air medium. The arrangement is important for the realization of the *bulk metamaterials* [J.6, J.7]. Physically, all three arrangements appear to be three-dimensional. However, electrically they are different.

The 1D wire medium of Fig (21.1a), also known as the single-wire medium (SWM), responds to the z-polarized E_z field component, i.e. the *parallel polarized* incident waves, propagating in the *y*-direction with wavenumber *k*. The wave can equally propagate in the *x*-direction with the same wavenumber. The wires of radius r are arranged in the 2D-square lattice with the lattice constant *d*. The E_z field component induces the emf on the conducting wire that causes the *z*-directed electric polarization in the wire. Thus, the permittivity of a wire medium is different along the length of the wires. However, in the (*x*–*y*)-plane, i.e. in the directions normal to

Introduction to Modern Planar Transmission Lines: Physical, Analytical, and Circuit Models Approach, First Edition. Anand K. Verma.

the thin wires, no polarization takes place leaving permittivity of the host air medium unchanged. Therefore, the 1D wire medium is a uniaxial anisotropic medium. The infinitely long, thin wire medium acts as an *electric plasma medium* with a cut-off frequency, i.e. the medium acts as a high-pass filter. The plasma medium, described by the Drude model, is discussed in section (6.5.2) of chapter 6. Below the plasma frequency f_p, the relative permittivity of the wire medium is negative, forming the ENG medium. Above the plasma frequency, the relative permittivity is a positive quantity with the value less than unity. In this frequency range, the wire medium acts as the ENZ medium. The ENZ dielectric medium supports the fast-waves, unlike the natural dielectrics that always support the slow-waves due to its refractive index more than unity. The ENG and ENZ media are discussed in section (5.5) of chapter 5.

The wire medium also suffers from the *spatial dispersion* due to a component of the EM-wave, TEM/TM type, propagating along the wire. Thus, the z-component of the effective permittivity of the wire medium becomes wavenumber-dependent, and the spatial dispersion occurs in the wire medium [B.7, B.9]. In the case of the *temporal dispersion*, the permittivity is frequency-dependent, whereas for *spatial dispersion* the permittivity is wavenumber-dependent. Physically, the finite length wires are supported on dielectric foam. The rectangular waveguide also supports the wire medium with some restrictions [B.10, B.11]. In a planar technology, the wires are replaced by the printed conducting strips of the rectangular cross-section. However, a high permittivity substrate influences the characteristics of the wire medium. The modeling of the 1D wire medium, using in the circuit model, is discussed in this section.

Figure (21.1b) shows the 2D-double wire medium (DWM). It responds to the E_x and E_z field components of the incident waves with polarization in the (x–z)-plane. Likewise, Fig (21.1c) shows the 3D-triple wire medium (TWM). Both the DWM and TWM could be the connected and disconnected wires, i.e. at the crossing points, the wires could be electrically joined together or remain electrically isolated. Both arrangements have different nature of the wave propagation. They form the 2D and 3D plasma media. The 2D wire medium can act as the lightweight reflecting metallic-plate. The TWM supports any arbitrary polarization and could form the isotropic medium [J.7, J.8, B.4]. However, to get a homogeneous medium, the radius r of the wires and also the lattice constant d must be very small, about one-tenth or less than a wavelength.

21.1.1 Polarization in the Wire Medium

Figure (21.2a) shows the cross-section of a part of the arrangement of wires forming the wire medium. The wires, shown as black circles, are arranged periodically using the cubic unit cell of the *lattice constant* d. The infinitely long thin wire element A of radius r, parallel to the z-axis, is located at the center of the arrangement of five wires in the (x–y)-plane. The considered wire length ℓ is within the cubic unit cell ($d \times d \times d$). The four-wire elements around the element A are taken from the neighboring unit cells (UC). The voltage V is induced in the wire due to the time-harmonic external field, $E_z = E_{oz}e^{j\omega t}$. It is assumed that the magnetic field H_ϕ, created by the current I flowing in the wire, extends up to the wires of the neighboring UC without interacting with them. Thus, the wires of the wire medium are isolated and noninteracting. The wire has the self-inductance L per unit length and resistance R *p.u.l.* The resistance R accounts for the losses in the wire medium, giving its relative permittivity by equation $\varepsilon_r^* = 1 - j(\sigma/\varepsilon_0\omega)$.

Figure (21.2b) shows the RL circuit model of the wire in the unit cell. The current $I(t) = I_0e^{j\omega t}$ flowing in a wire of length ℓ in the cubic unit cell is obtained as follows:

$$(L\ell)\frac{dI}{dt} + (R\ell)I = V, \Rightarrow L\frac{dI}{dt} + RI = E_z \qquad (a)$$
$$E_z = (R + j\omega L)I \qquad (b). \qquad (21.1.1)$$

The displaced charges on the wire create the dipole due to the electric polarization P_z. The local field causing the polarization is assumed to be identical to the external field E_z. This simplifying assumption is not valid for the densely packed wire medium. The displacement vector D_z is obtained from equation (6.1.6) of chapter 6. It is also related to the external field through the complex relative permittivity of the wire medium. Both relations are mentioned below, giving an expression for the polarization vector P_z:

$$D_z = \varepsilon_0 E_z + P_z \qquad (a)$$
$$D_z = \varepsilon_0\varepsilon_r^* E_z = \varepsilon_0\left(1 - j\frac{\sigma}{\omega\varepsilon_0}\right)E_z \qquad (b)$$
$$\text{from the above equations}: P_z = \frac{\sigma E_z}{j\omega} = \frac{J_z}{j\omega} = \frac{I}{j\omega d^2} \qquad (c). \qquad (21.1.2)$$

The current density J_z is averaged over the area of the unit cell, i.e. $J_z = I/d^2$. On substituting current from equation (21.1.1b) in equation (21.1.2c), the polarization P_z is obtained:

$$P_z = \frac{1}{j\omega d^2}\frac{E_z}{(R + j\omega L)}. \qquad (21.1.3)$$

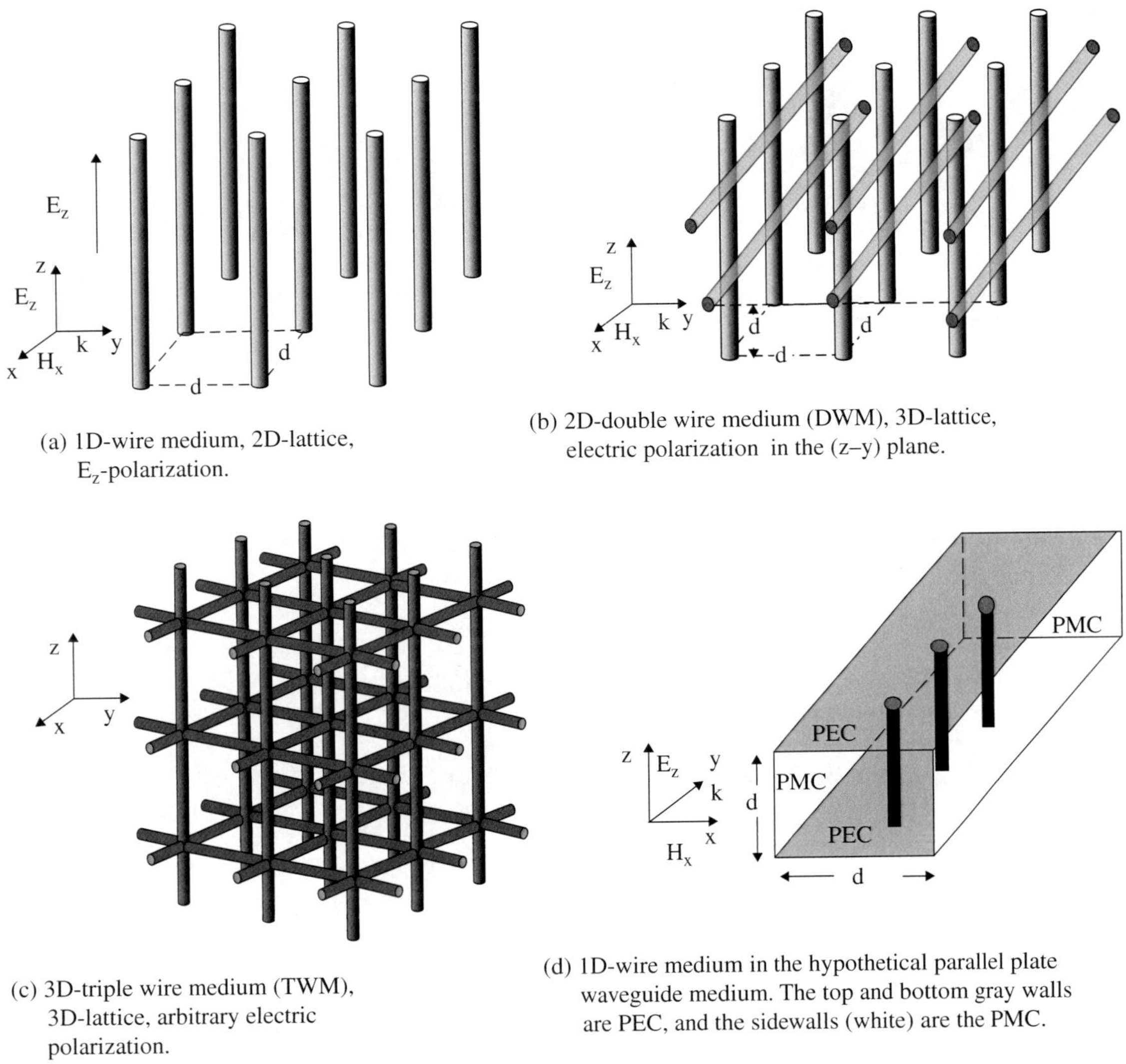

(a) 1D-wire medium, 2D-lattice, E_z-polarization.

(b) 2D-double wire medium (DWM), 3D-lattice, electric polarization in the (z–y) plane.

(c) 3D-triple wire medium (TWM), 3D-lattice, arbitrary electric polarization.

(d) 1D-wire medium in the hypothetical parallel plate waveguide medium. The top and bottom gray walls are PEC, and the sidewalls (white) are the PMC.

Figure 21.1 Geometry of the wire-medium.

The complex relative permittivity of the wire medium is obtained by using equation (21.1.3) with the expression (21.1.2a):

$$\varepsilon_0\varepsilon_r^* E_z = \varepsilon_0 E_z + \frac{1}{j\omega d^2}\frac{E_z}{(R + j\omega L)}$$
$$\Rightarrow \varepsilon_r^* = 1 - \frac{1}{\varepsilon_0 d^2 L[\omega^2 - j(R/L)\omega]}. \qquad (21.1.4)$$

The wire medium could be treated as the electric plasma medium with electric plasma frequency ω_{ep}. Normally, the plasma frequency is expressed as ω_p. However, electric plasma frequency is expressed by the symbol ω_{ep} to distinguish it from magnetic plasma frequency ω_{mp} discussed subsection (21.2.2). On comparing equation (21.1.4) against equation (6.5.16c) of chapter 6, the following expressions are obtained for the electric plasma frequency ω_{ep} and the electric damping coefficient γ_e:

$$\omega_{ep}^2 = \frac{1}{\varepsilon_0 d^2 L} \quad (a), \qquad \gamma_e = \frac{R}{L} \quad (b)$$
$$\text{where, } R = \frac{1}{\sigma\pi r^2} \quad (c). \qquad (21.1.5)$$

The expression for R is obtained from equation (4.2.19) of chapter 4.

The *self-inductance* L of the wire is approximately computed using the model shown in Fig (21.2a). The model could be viewed as a coaxial cable of unit length with inner radius r and outer radius d. The coaxial cable model computes the inductance *L p.u.l.* of the wire as follows:

$$L = \frac{\psi}{I} = \frac{\mu_0}{I}\int_r^d H_\phi d\rho = \frac{\mu_0}{I}\int_r^d \frac{I}{2\pi\rho}d\rho$$
$$\Rightarrow L = \frac{\mu_0}{2\pi}\ln\left(\frac{d}{r}\right). \qquad (21.1.6)$$

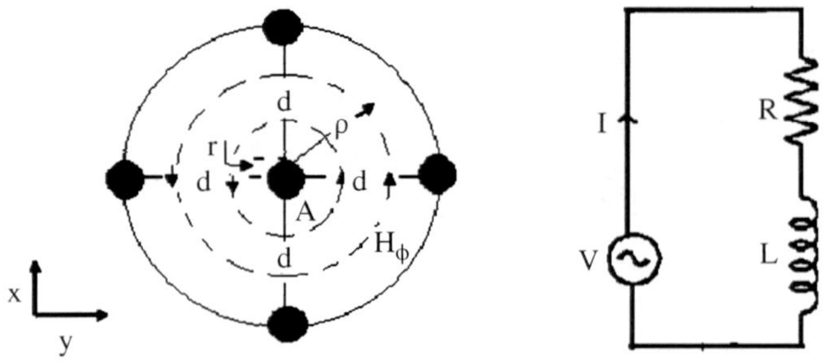

(a) Arrangement of a section of the wire medium in the (x–y)-plane.

(b) Circuit model of an element of wire medium.

Figure 21.2 Circuit model of wire-medium.

We can also compute the inductance of the wire A in presence of neighboring conductors. At distance $\rho = d/2$, the magnetic fields due to the current in the wire A and adjacent wires cancel each other, leading to the creation of the square-shaped *magnetic walls*. So the self-inductance is computed as follows after getting the effective magnetic field H_ϕ in the region $r < \rho < d/2$ due to two current sources separated by a distance d [J.9]:

$$H_\phi = \frac{I}{2\pi}\left(\frac{1}{\rho} - \frac{1}{d-\rho}\right) \qquad \text{(a)}$$

$$L = \frac{\mu_0}{I}\int_r^{d/2} \frac{I}{2\pi}\left(\frac{1}{\rho} - \frac{1}{d-\rho}\right)d\rho, \quad \Rightarrow L = \frac{\mu_0}{2\pi}\ln\left(\frac{d^2}{4r(d-r)}\right) \qquad \text{(b)}$$

$$\Rightarrow L \approx \frac{\mu_0}{2\pi}\left[\ln\left(\frac{d}{r}\right) - 1.387\right], \quad \text{for}\frac{r}{d} < < 1 \qquad \text{(c).}$$

$$(21.1.7)$$

Other expressions are also available to compute approximately the self-inductance L *p.u.l.* of a wire:

$$\text{Reference [J.11]} : L = \frac{\mu_0}{2\pi}\left[\ln\left(\frac{d}{2\pi r}\right) + 0.5275\right] = \frac{\mu_0}{2\pi}\left[\ln\left(\frac{d}{r}\right) - 1.31\right] \qquad \text{(a)}$$

$$\text{Reference [B.12]} : L = \frac{\mu_0}{2\pi}\left[\ln\left(\frac{2d}{r}\right) - \frac{3}{4}\right] = \frac{\mu_0}{2\pi}\left[\ln\left(\frac{d}{r}\right) - 0.057\right] \qquad \text{(b).}$$

$$(21.1.8)$$

Equations (21.1.7) and (21.1.8) provide a reduced value of the wire inductance as compared to that given by equation (21.1.6).

The electric plasma frequency and damping coefficient of the wire medium are obtained by using the inductance from equation (21.1.6) in equation (21.1.5):

$$\omega_{ep}^2 = \frac{2\pi c^2}{d^2 \ln\left(\frac{d}{r}\right)} \qquad \text{(a)}$$

$$\gamma_e = \frac{2}{\mu_0 \sigma r^2 \ln(d/r)} = \frac{\varepsilon_0 d^2}{\pi r^2 \sigma}\omega_{ep}^2 \qquad \text{(b).}$$

$$(21.1.9)$$

Pendry et al. have obtained an identical plasma frequency using the concept of effective mass and effective density of electrons [J.2]. However, their concept has been criticized by other researchers [B.8]. The investigators have also used modified expressions for the self-inductance [J.9, J.10, B.7, B.9]. A higher value of the electric plasma frequency ω_{ep} is obtained by using an alternative expression of inductance mentioned above.

In summary, the relative permittivities of lossy and lossless wire medium are given by the following expressions:

$$\text{lossy medium} : \varepsilon_r^* = 1 - \frac{\omega_{ep}^2}{(\omega^2 - j\omega\gamma_e)} \qquad \text{(a)}$$

$$\text{lossless medium} : \varepsilon_r = 1 - \frac{\omega_{ep}^2}{\omega^2} \qquad \text{(b).}$$

$$(21.1.10)$$

Equations (21.1.10a, b) obtained by using the equivalent circuit model are identical to equation (6.5.16) obtained using the concept of polarizability at the molecular level. Figure (6.9c) of chapter 6 shows the permittivity response, both for the real and imaginary parts, of the plasma medium. The wire medium exhibits the negative value of permittivity below ω_{ep} giving a highly lossy ENG medium. For the frequency above ω_{ep}, permittivity in the range $0 < \varepsilon_r < 1$ provides the ENZ medium. So the wire inclusions, in the host air medium, could be treated as an *artificial molecule*, creating the ENG/ENZ material medium.

The wire medium of Fig (21.1a) is not the isotropic medium, as the polarization occurs only along the z-axis. It is a uniaxial anisotropic medium described by the following complex permittivity tensor:

$$\overline{\overline{\varepsilon}}_r^* = \begin{bmatrix} 1 & 0 & 0 \\ 0 & 1 & 0 \\ 0 & 0 & \varepsilon_r^* \end{bmatrix}. \quad (21.1.11)$$

Like a natural anisotropic medium, discussed in section (4.7) of chapter 4, the wire medium also supports the E_z-polarized *extraordinary waves* propagating along the y-direction, whereas the E_x-polarized wave is the *ordinary wave*. The uniaxial wire medium also supports the TEM/TM kind of wave propagation along the z-directed optic axis, causing the *spatial dispersion* due to the nonlocal field [J.9, J.11, B.7, B.9]. Due to the spatial dispersion, the permittivity of the wire medium becomes wavenumber-dependent. In this case, the z-directed permittivity component gets modified.

21.1.2 Equivalent Parallel Plate Waveguide Model of Wire Medium

The host air medium, with a grid of noninteracting wire inclusions, as shown in Fig (21.1a), could be treated with the help of the *parallel plate waveguide* (PPW) model. The PPW model is created by placing the perfect electric conductors (PEC), separated by the lattice constant d, at the top and bottom of the 1D wire medium. The perfect magnetic conductors (PMC) separated by the lattice constant d, create the side-walls, i.e. the magnetic walls. Figure (21.1d) shows the PPW model of the wire medium of Fig (21.1a). It supports the EM-wave propagation in the y-direction. Using the concept of image formation in the presence of the PEC and PMC, the wire medium-loaded PPW creates the original structure of the wire medium. The periodically loaded PPW could be analyzed by the 1D [ABCD] matrix, i.e. the transmission matrix discussed in section (19.3) of chapter 19. This classical method has been introduced by early investigators [J.6, J.7]. In the above discussion, the wire is replaced by the RL equivalent circuit. The permittivity response of the wire medium could be tailored by loading the z-polarized wire with additional chip reactance, or reactive structures. It is discussed in section (21.1.3).

The empty PPW is modeled by the lumped transmission line model with the series inductance $L = \mu_0$ *p.u.l.*, and shunt capacitance $C = \varepsilon_0$ *p.u.l.* The inductance created by the wire is added across the equivalent transmission line as the shunt inductance L_{sh} *p.u.l.* given by equation (21.1.6), or any other equation presented above. Subsection (3.4.2) of chapter 3 presents the analysis of the wave propagation in the equivalent plasma medium giving the following results:

$$\beta = \frac{\omega}{v_0}\left[1 - \left(\frac{\omega_{ep}}{\omega}\right)^2\right]^{1/2} = \beta_0\sqrt{\varepsilon_r} \quad (a)$$

$$\Rightarrow \varepsilon_r = \left[1 - \left(\frac{\omega_{ep}}{\omega}\right)^2\right] \quad (b)$$

$$\text{where,}\ \omega_c^2 = \omega_{ep}^2 = \frac{1}{L_{sh}C} = \frac{2\pi}{\varepsilon_0\mu_0 \ln(d/r)} \quad (c)$$

$$v_o = \frac{1}{\sqrt{LC}} = \frac{1}{\sqrt{\varepsilon_0\mu_0}} = c \quad (d).$$

$$(21.1.12)$$

The cut-off frequency ω_c of the inductance loaded PPW is the same as the plasma frequency ω_{ep} of the wire medium. However, d^2 could be added in the denominator of equation (21.1.12c) as L_{sh} and C are on *p.u.l* basis. The dispersive frequency response of the medium is discussed above. The wave impedance of the wire medium is obtained, following equation (3.4.15) of chapter 3, as the characteristic impedance of the loaded PPW:

$$Z_{wire} = \frac{\eta_0}{\left[1 - \left(\frac{\omega_{ep}}{\omega}\right)^2\right]^{1/2}}. \quad (21.1.13)$$

The EM-wave cannot propagate below ω_{ep}, as the Z_{wire} becomes reactive. For $\omega > \omega_{ep}$, equations (3.4.10a) and (3.4.10b) of chapter 3 show that the wire medium supports the fast-wave with the group velocity below the velocity of light c.

The predicted Drude model type permittivity response of the wire medium has been experimentally confirmed in the microwave range 7–14 GHz [J.12]. The arrays of *microwires* of radius $r = 10,\ 30\ \mu m$ and lattice constant $d = 4, 5, 6$ mm have been arranged on a stack comprising of 20 numbers of Rohacell HF 51 foam plates with low permittivity $\varepsilon_r = 1.07$ material. The dimension of stack Rohacell HF 51 plates is $200 \times 200 \times 100\ mm^3$. Each plate contains 39 wires. The wire medium is mounted in a window frame surrounded by the absorbing material.

Figure (21.3a) shows the $|S_{21}|$ response of the wire medium for $r = 10\ \mu m$ and $d = 5$ mm. The dashed response is the reference response at −15 dB without the sample of the wire medium. The response of the normal polarized ($E_\perp$) incident waves is identical to the reference response, as there is no polarization in the wire medium. However, below the plasma frequency f_{ep}, the response of the parallel polarized ($E_\parallel$) EM-waves has an attenuation level of less than −60 dB. The plasma frequency f_{ep} is about 9.8 GHz at −35 dB below the

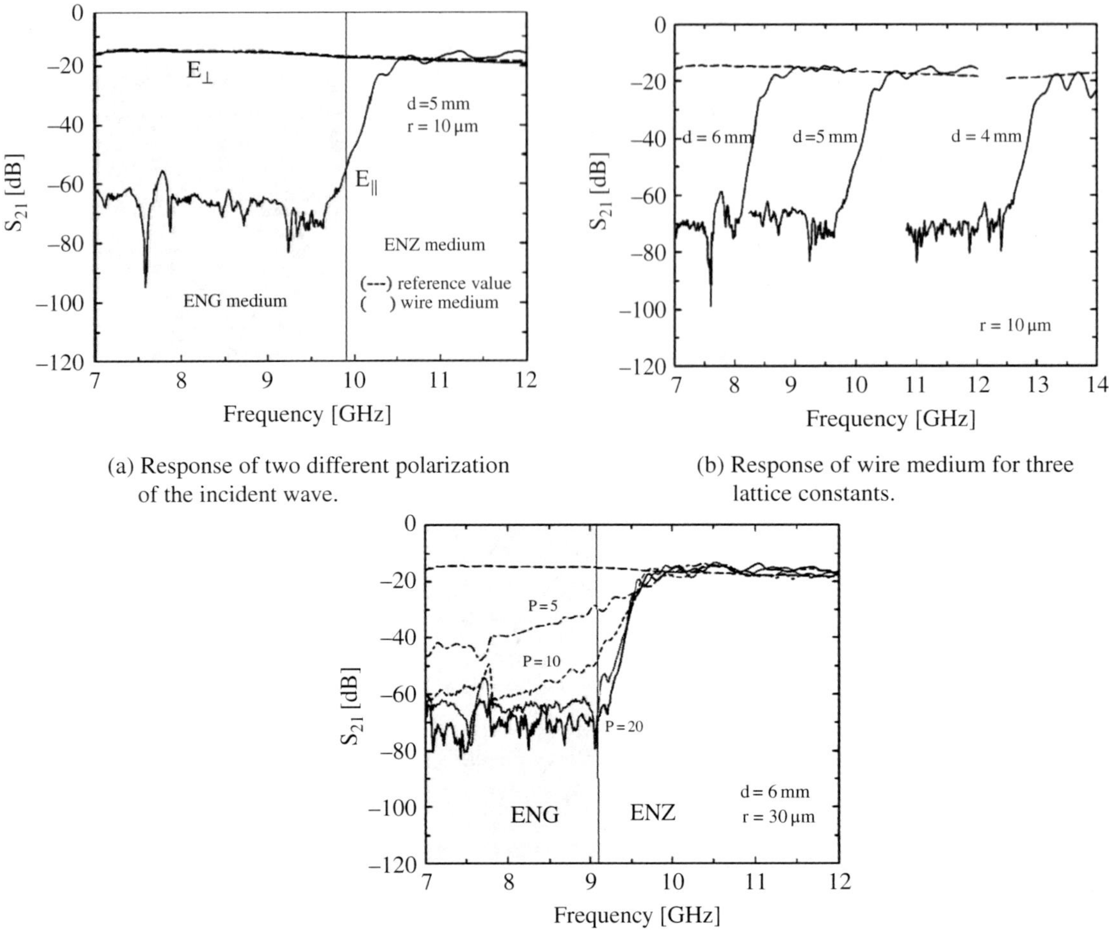

(a) Response of two different polarization of the incident wave.

(b) Response of wire medium for three lattice constants.

(c) Response of wire medium for an increasing number of wire medium plates.

Figure 21.3 Experimental response of wire-medium. *Source*: Reproduced Gay-Balmaz et al. [J.12], with the permission of AIP Publishing. © 2002, AIP Publishing.

reference level. The nonpropagating ENG region is shown in gray. The propagating wave in the ENZ region is above f_{ep}. Figure (21.3b) shows the wire medium response for the lattice constant d = 6, 5, and 4 mm. It is noted that with a decrease in the size of the unit cell, the plasma frequency increases from 8.2 to 9.0 to 12.4 GHz. The plasma frequency also shows an increase in f_{ep} from 8.2 to 9.1 GHz for an increase in the radius of wire from r = 10 to 30 μm. Finally, Fig (21.3c) shows an increase in attenuation level in the ENG medium with an increase in the number of wire carrying plates P, i.e. an increase in the thickness of the wire medium.

The accuracy of the computed plasma frequency, using available expressions for the wire inductance, could be examined against the experimental results. Table 21.1 compares the results of these expressions. Pendry [J.2] provides a 1.5–4.8% deviation from the experimental results. There is an increase in the deviation with an increasing radius of wire, and increasing lattice constant. The expression for inductance given in reference [J.13] provides a little-improved result, 0.8–4.2%. The expression (21.1.8b) results in a high deviation for a small radius of wire, 7.8–11.5%, whereas, for the thicker wire, it is only 4.8%. The expression of Pendry could be used for the most of applications using the wire medium in the air host medium. However, the computed plasma frequency may have more deviation for the printed wire medium on a substrate.

The present discussion illustrates that the engineered artificial ENG and ENZ materials could be realized using the circuit models. The RL circuit model is extended below to the reactively loaded wire medium.

Table 21.1 Computed and exp. f_p.

r in μm	d in mm	Exp. f_p GHz [J.12]	Com. f_p GHz [J.2]	Com. f_p GHz [J.9]	Com. f_p GHz [J.13]
10	4	12.4	12.22	13.82	12.30
10	5	9.80	9.60	10.80	9.65
10	6	8.20	7.88	8.84	7.92
30	6	9.10	8.66	9.09	8.71

21.1.3 Reactance Loaded Wire Medium

Figure (21.4a) shows the impedance loaded wire medium. The loading element could be a lumped inductor, loop or spiral inductor, lumped capacitor, metallic structures of several geometrical forms. It could be even the active elements [J.13–J.17]. Figure (21.4b) shows the inductive loading element L_L in the form of a square, or a circular loop. Figure (21.4c) shows the printed circuit version of the wire medium with the rectangular cross-section. It also shows its RL circuit model. Figure (21.4d) shows the gap capacitor (C_L) loaded printed wire with the RLC_L circuit model. Following the process discussed above, the circuit model is used below to obtain the effective permittivity of the impedance loaded wire medium material.

The equation (21.1.4) is modified below to compute the effective relative permittivity of the loaded wire medium with impedance Z_L:

$$\varepsilon_r^* = 1 + \frac{1}{j\varepsilon_0\omega A(R + j\omega L + Z_L)}. \qquad (21.1.14)$$

In expression (21.1.14), $A = d^2$ is the area of the unit cell in the (x–y)-plane. The Z_L is on a per unit length basis. Expression (21.1.14) does not account for the interaction between the wires of the medium [J.17, J.18]. There are five cases for the load impedance Z_L:

1) Z_L is short-circuited, reducing the loaded wire medium to an unloaded wire medium described by the Drude model as a plasma medium;
2) Inductor loaded wire medium with $Z_L = j\omega L_L$, giving again a plasma medium with lower plasma frequency;
3) Capacitively loaded wire medium with $Z_L = 1/j\omega C_L$, giving the Lorentz type effective permittivity medium;
4) Series resonant circuit loaded wire medium with $Z_L = j(\omega L_L - 1/\omega C_L)$, again giving the Lorentz type medium; and
5) Parallel resonant circuit loaded wire medium, $Z_L = j(\omega C_L - 1/\omega L_L)^{-1}$.

Inductively Loaded Wire Medium

Figure (21.4b) shows the inductively loaded wire medium, using either a square or circular loop. The loops are constructed with the wire of circular cross-section of radius r. The total length of loaded line is ℓ. The loading inductance L_L *p.u.l.* for both cases can be estimated from the following approximate expressions [J.13, J.19, B.12]:

$$\text{Circular loop [J.13]}: L_L = \frac{\mu_o R}{\ell}\left[\ln\left(\frac{8R}{r}\right) - 1.75\right] \quad (a)$$

$$\text{Square loop}: L_L = \frac{\mu_o 4a}{\pi\ell}\sinh^{-1}\left(\frac{4a}{r}\right) \quad (b).$$

$$(21.1.15)$$

In expressions (21.1.15), L_L is the inductance of the loop. The square loop has side length "a", and circular loop radius is R. The expression for the square loop of wire of circular cross-section is obtained from the approximate expression of the square loop made of wire of square cross-section [J.20]. The total inductance in the unit cell is $L_{total} = L_L + L_W$ *p.u.l.*, where wire inductance $L = L_W$ is given equation (21.1.6), or equation (21.1.8a, b). Using total inductance, the plasma frequency of the loop loaded wire medium, from equation (21.1.5), is written as follows:

Circular loop [J.13] :

$$\omega_{ep}^2 = \frac{2\pi(c/d)^2}{(2\pi R/\ell)[\ln(8R/r) - 1.75] + \ln(d/r)} \quad (a)$$

Square loop :

$$\omega_{ep}^2 = \frac{2\pi(c/d)^2}{(8a/\ell)\sinh^{-1}(4a/r) + \ln(d/r)} \quad (b).$$

$$(21.1.16)$$

The inductively loaded wire medium also behaves like a plasma medium, at the reduced frequency. The loading inductor reduces the effective radius of the wire that increases the wire inductance and lowers the plasma frequency [J.13, J.18]. Figure (21.5a) shows the transmission response of the empty test chamber of parallel conducting plates separated by $\ell = 10$ mm. The radius

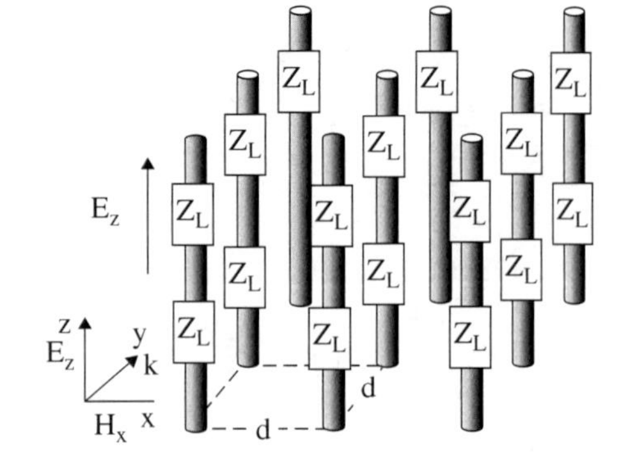

(a) 1D-impedance-loaded wire medium, 2D-lattice, E_z-polarization.

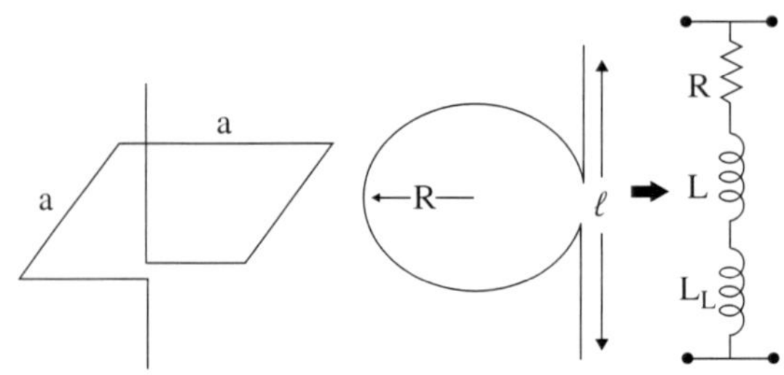

(b) Inductive loop-loaded wire medium.

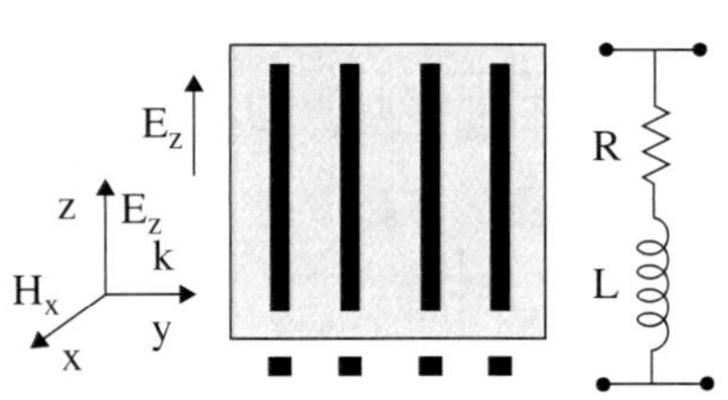

(c) 1D printed wire medium, E_z-polarization.

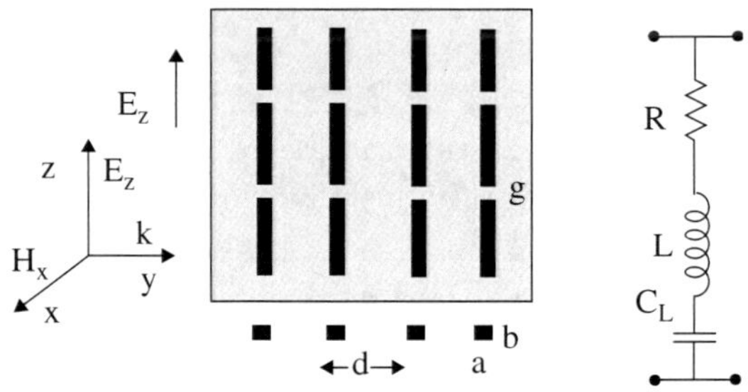

(d) 1D capacitor-loaded printed wire medium, E_z-polarization.

Figure 21.4 Impedance loaded wire medium.

r of unloaded straight wire medium is 0.2 mm, and lattice constant d of the square lattice is 5 mm. The loaded circular loops have radius $R = 1.3$ mm, $r = 0.2$ mm. The array has 5 rows normal to the incident waves, incident on the chamber, and 30 rows in the transverse direction. The unloaded wire medium has an electric plasma frequency f_{ep} above 12 GHz. Therefore, attenuation is more than −50 dB in the lower frequency range, due to the nonpropagating waves in the negative permittivity wire-based plasma medium. However, the

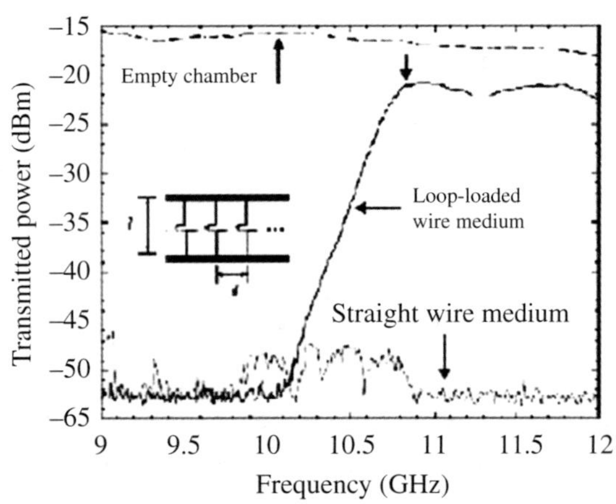

(a) Response of straight wire, and loop-loaded wire medium[J.13].

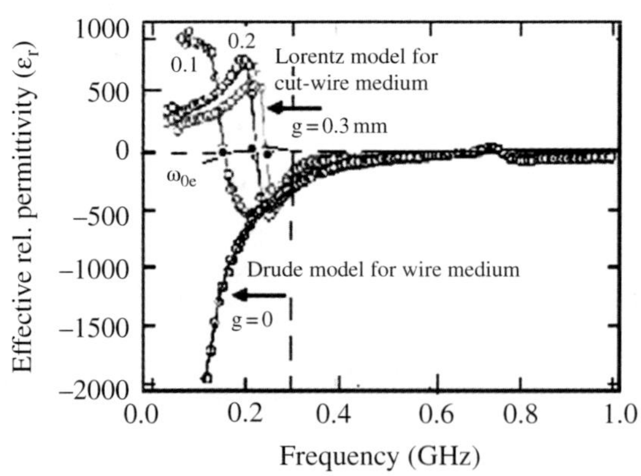

(b) Permittivity response of straight wire, and capacitor-loaded wire medium [J.15].

Figure 21.5 Response of inductor and capacitor loaded wire-medium. *Source:* Reproduced from Smith et al. [J.13], with the permission of AIP Publishing. © 1999, AIP Publishing. Reproduced from Takano et al. [J.15], with the permission of AIP Publishing. © 2010.

computed value of the electric plasma frequency f_{ep} for the loop loaded wire medium is 10.7 GHz. Therefore, above about 10.7 GHz, the response shows the transmission of the waves. For the square loop of side length a = 2 mm, and other parameters remaining unchanged, the computed plasma frequency, using equation (21.1.16b), is 6.24 GHz, whereas the simulated one is 6.8 GHz. The deviation is not high, even if expression (21.1.16b) is approximate.

Table 21.2 Extracted RLC_L parameters.

g (mm)	L (10^{-7} H/m)	C_L (10^{-18} F/m)	R (10^4 Ω/m)
0.0	0.78	>1000	3.66
0.1	1.13	7.67	8.20
0.2	1.92	2.55	6.05
0.3	2.44	1.70	6.38

Source: Reproduced from Takano et al. [J.15]. © 2002, AIP Publishing.

Capacitively Loaded Wire Medium

The capacitive loading of a wire medium is achieved by periodically cutting the wire and inserting a lumped capacitance. Even a cut in the wire provides the loading through the gap capacitance. The capacitor loading creates series resonance in the wire medium below its plasma frequency, and the Drude type wire medium is converted into the Lorentz type wire medium [J.15]. Such a medium, as shown in Fig (21.4d), is known as the *cut-wire medium* [J.15]. The Lorentz type permittivity response is discussed in section (6.5) of chapter 6. The external LC-resonant circuit loading also provides the same type of response with lower resonant and lower plasma frequencies.

For the capacitor-loaded wire medium, the load impedance is $Z_L = -j/\omega C_L$. On substituting it in equation (21.1.14), the complex effective relative permittivity of the capacitor-loaded wire medium is expressed as follows:

$$\varepsilon_r^* = 1 + \frac{\omega_{ep}^2}{\omega_{0e}^2 - \omega^2 + j\omega\gamma_e} \quad \text{(a)}$$

$$\text{where } \omega_{ep}^2 = \frac{1}{\varepsilon_0 AL}, \quad \omega_{0e}^2 = \frac{1}{LC_L}, \quad \gamma_e = \frac{R}{L} \quad \text{(b)}. \tag{21.1.17}$$

In expression (21.1.17a, b), ω_{0e} is the characteristic electric resonance frequency of the Lorentz medium. The positive permittivity below the resonance frequency ω_{0e} is obtained, giving the limiting value as follows: $\omega \rightarrow 0, \varepsilon_r^* \rightarrow (1 + C_L/\varepsilon_0 A)$. Figure (6.9b) of chapter 6 shows the response of the Lorentz medium, so the capacitor-loaded wire medium acts as the *Lorentz artificial dielectric medium.*

The wire medium, and also the capacitor-loaded wire medium, can be realized even in the THz frequency range in the planar technology on a substrate paper. Figure (21.4c) shows the printed wire medium [J.15]. Figure (21.4d) shows the cut-wire grid of cut-width g, creating the loading capacitor. The substrate paper is 0.07 mm thick. It has a refractive index of 1.5 + j0.05. The conductor thickness b of the printed line on substrate paper, using conductive ink, is 2.5 μm. The width of the printed line, the lattice constant, and the capacitive gap are a = 0.2 mm, d = 0.4 mm, and g = 0.1, 0.2, 0.3 mm respectively. Figure (21.5b) shows the experimental results of the *Drude type response* of the unloaded wire medium. It also shows the *Lorentz type response* of the gap-capacitor-loaded wire medium. The resonance frequency is less for a smaller gap, i.e. for a larger value of the loading capacitance. By fitting the parameters in the Lorentz model, the RLC_L parameters are extracted. On using these parameters, Lorentz model responses are shown in Fig (21.5b). Table 21.2 shows the extracted RLC_L parameters of the Lorentz model. Below the resonance frequency, the positive-valued permittivity is dispersive. The Lorentz type artificial dielectric has higher valued permittivity just below the resonance. The permittivity response is similar to the response shown in Fig (6.9b) of chapter 6. In the frequency range $\omega_{0e} < \omega < \omega_{ep}$, the permittivity is negative, i.e. the medium is ENG, and above ω_{ep}, it is the ENZ medium.

21.2 Artificial Magnetic Medium

The artificial magnetic medium with effective relative permeability is engineered by using a periodic array of the current-carrying loops. Such a medium could be called the *loop medium*. A small current-carrying loop, forming the magnetic dipole moment (m) given by equation (6.1.24a) of chapter 6, is the source of diamagnetism. Figure (21.6a) shows the induced current I in a loop excited by the external z-directed axial magnetic field H_z^{ext}. The array of the current-carrying loops creates the mu near-zero (MNZ) material with the effective relative permeability less than unity, i.e. $0 < \mu_r < 1$. In such an arrangement, the paramagnetism showing $\mu_r > 1$ is not achieved. Similarly, the negative permeability is also not realized using the current-carrying continuous loop. However, the capacitor-loaded resonating loop, shown in Fig (21.6b), can provide *Lorentz type magnetic behavior*, similar to the behavior of the ferromagnetic materials [B.13]. The Lorentz model is discussed in section (6.5) of chapter 6. The capacitor-loaded resonating loop can act as an *artificial magnetic molecule,* or *magnetic element,* to create a material with high permeability. It also creates the negative permeability needed by the MNG medium.

Figure (21.6b) shows the capacitor-loaded circular loop that exhibits series resonance. The time-varying external magnetic field excites the resonant current I in the loop. The loop provides the inductance, and the lumped capacitor is externally added to get the series resonance [B.10]. The loop has conductor loss, also a little radiation loss [B.8]. However, the capacitor-loaded loop is usually not a suitable choice for the realization of the MNG and DNG materials, as the lumped capacitors are bulky and lossy, and are not suitable for forming a large array. Figure (21.6c) solves the problem by the use of the concentric *split rings resonator* (SRR) structures. The ring has usually either circular or square shape. The splits in the loops break the flow of the induced conduction currents in the loops. However, the displacement current flows from the outer ring to the inner ring due to a higher valued induced potential at the outer ring. In this arrangement, the narrow gap between two rings acts as the loading capacitance, and the inductance is provided by current flowing in the strip conductors. The RLC equivalent circuit, shown in Fig (21.6d), models the split ring resonator (SRR) [J.21, B.14].

21.2.1 Characteristics of the SRR

Figure (21.7a) shows the double rings circular SRR excited by the external time-varying axial magnetic field H_z^{ext}. The induced voltage and current in both rings are also shown in Fig (21.7a). The SRR is divided into the symmetrical upper and lower parts due to the splits, i.e. due to gap in the strip rings. The splits are placed at the diametrically opposite ends on the outer and inner rings. The induced voltages, V_{out}, and V_{in} at the outer ring and the inner ring, respectively, cause the displacement current I_d to flow from the outer ring to the inner ring, as $V_{out} > V_{in}$. It helps the flow of continuous currents I_{out} and I_{in} in the outer and inner rings. An electric wall (EW) $P_1P_2\ Q_2Q_1$, passing through the middle of the splits, could be placed at the diameter of the SRR. The formation of capacitors shows the accumulation of charges on the rings. Further, the currents of rings have the positional variations along the circumference of the rings. However, the total current at any location on the rings remains unchanged. The voltage and current variations across the rings are shown in Fig (21.7b). The locations of the voltage and current maxima and minima are also indicated in both Fig (21.7a, b). The maxima of

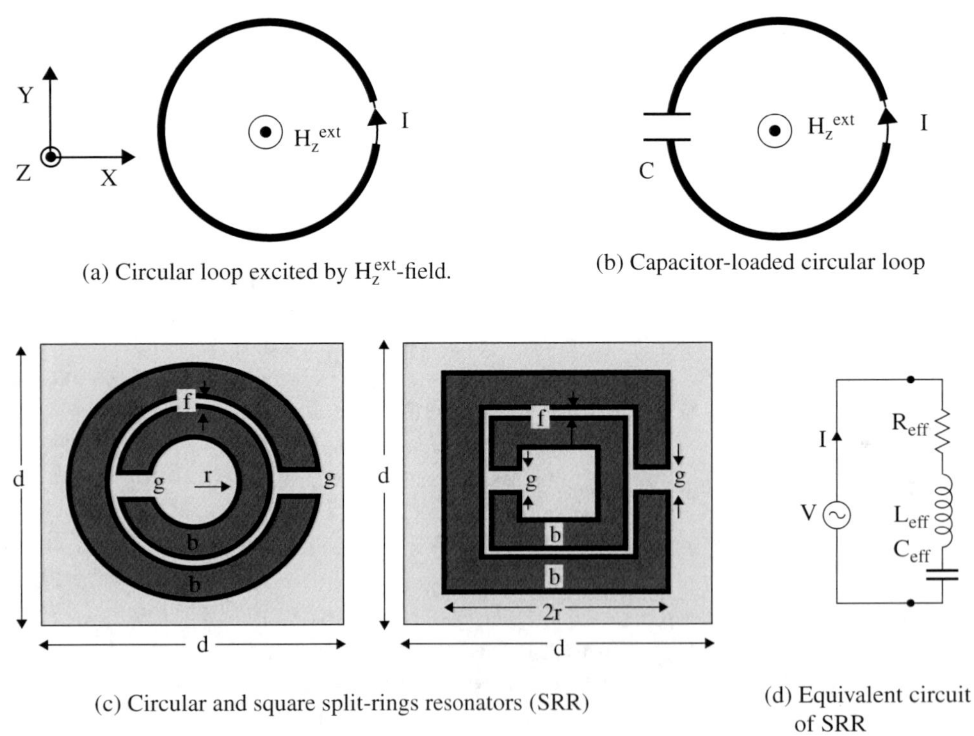

(a) Circular loop excited by H_z^{ext}-field.

(b) Capacitor-loaded circular loop

(c) Circular and square split-rings resonators (SRR)

(d) Equivalent circuit of SRR

Figure 21.6 Artificial magnetic molecules or elements excited by external axial magnetic fields.

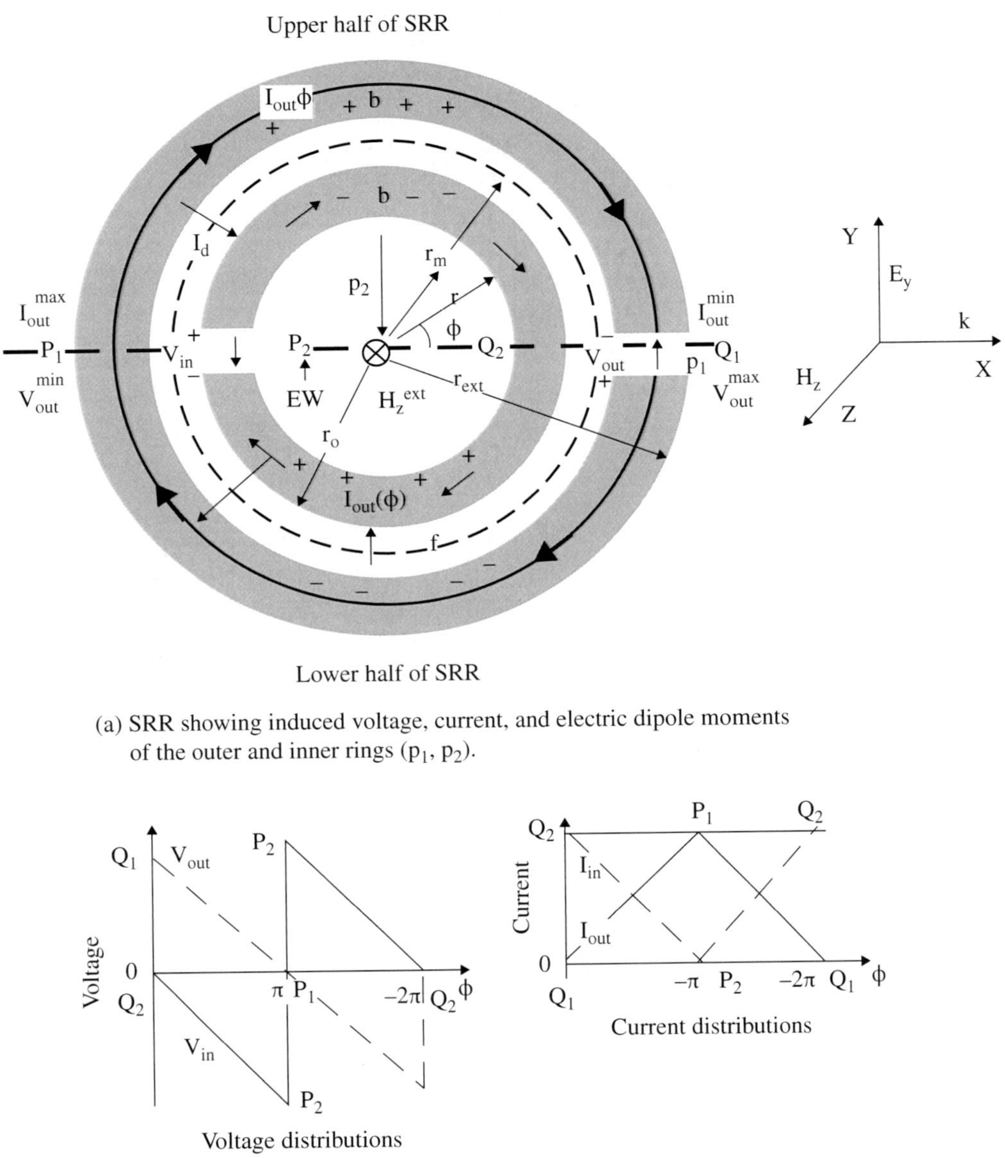

(a) SRR showing induced voltage, current, and electric dipole moments of the outer and inner rings (p_1, p_2).

(b) Voltage and current distributions along the circumference of both rings.

Figure 21.7 The split rings resonator (SRR).

voltages occur at the gaps in the rings and currents are zero at the gaps. The maxima of currents are at the positions opposite to gaps. At these locations, voltages are zero.

The detailed circuit model of the SRR is available [B.5, B.15]. The present section focuses attention on a simplified model to understand its working. The double-ring SRR could be viewed as a single current-carrying loop of *mean radius* $r_m = r + b + f/2$, where b is the width of both the inner and outer strips, and f is the circumferential gap between outer and inner rings [J.22]. The current-carrying loop of the mean radius r_m acts as a magnetic dipole. The arrays of these magnetic dipoles create artificial magnetic materials. The SRR-based magnetic material is primarily a *bianisotropic magnetic medium*. The bianisotropic medium is discussed in section (4.2) of chapter 4. However, ignoring the bianisotropy, we consider below it as an isotropic medium modeled through the RLC – circuit. The section (21.2.4) examines the bianisotropy [J.23].

21.2.2 Circuit Model of the SRR

In our discussion, we ignore capacitance due to the split gaps. Figure (21.6d) shows the series resonance RLC circuit model of the SRR. The circumferential gap f between the strip rings forms the upper and lower capacitances C_U and C_L of equal value, i.e. $C_U = C_L = C$. Both capacitances are connected in series giving the effective gap capacitance as $C_{eff} = C/2$. The current-carrying conducting strips offer the inductance L_{eff}, and the resistance R_{eff} accounts for the losses in the SRR. The axial time-harmonic magnetic field $H_z^{ext} = H_{zo}^{ext}\, e^{j\omega t}$ induces the emf in the *equivalent single loop*, for a narrow gap $f \ll b$, of mean radius r_m. The half capacitance is $C = \pi r_m C_{pul}$, where C_{pul} is the slot gap capacitance *p. u. l.*

Computation of Effective Relative Permeability of Loop Medium

Figure (21.6d) shows the circuit model to obtain the magnetic polarizability α_m of a loop medium due to the axial magnetic field H_z^{ext}. It is needed to compute the effective relative permeability μ_r of the artificial magnetic medium. The loop medium is treated as a homogeneous medium, as the lattice constant d of the periodic loop medium, as shown in Fig (21.8a), the sizes of the SRR $2r_{ext}$, and separation ℓ are much less than the operating wavelength at the resonance frequency of the SRR.

The induced emf V_{ind}, and corresponding loop current I in the SRR, due to the external time-varying magnetic flux, is obtained below with the help of the equivalent circuit shown in Fig (21.6d):

$$V_{ind} = -\frac{d\phi}{dt} = -j\omega\mu_0 S H_z^{ext} \qquad (a)$$

$$I = \frac{V_{ind}}{R_{eff} + j\omega L_{eff} + 1/j\omega C_{eff}}$$

$$\Rightarrow I = \left[\frac{\mu_0\omega^2\omega_{0m}^2 S C_{eff}}{\omega_{0m}^2 - \omega^2 + j\omega R_{eff}/L_{eff}}\right] H_z^{ext}$$

$$\Rightarrow I = \left[\frac{\mu_0\omega^2 S/L_{eff}}{\omega_{0m}^2 - \omega^2 + j\omega\gamma_m}\right] H_z^{ext} \qquad (b)$$

$$\text{damping factor } \gamma_m = \frac{R_{eff}}{L_{eff}} \qquad (c).$$

$$(21.2.1)$$

where S is the area of the single equivalent loop and $\omega_{0m} = 1/\sqrt{L_{eff}C_{eff}} = \sqrt{2/L_{eff}C}$ is the *characteristic magnetic resonance frequency of the SRR*. Using equation (6.1.24a) of chapter 6, the z-oriented magnetic dipole moment m_z, and corresponding magnetic polarizability α_m are obtained as follows:

$$m_z = IS = \left[\frac{\mu_0\omega^2 S^2/L_{eff}}{\omega_{0m}^2 - \omega^2 + j\omega\gamma_m}\right] H_z^{ext} \qquad (a)$$

$$\alpha^m = \frac{m_z}{H_z^{ext}} = \frac{\mu_0\omega^2 S^2/L_{eff}}{\omega_{0m}^2 - \omega^2 + j\omega\gamma_m} \qquad (b).$$

$$(21.2.2)$$

The volume of the cubic unit cell, containing one SRR inclusion, is d^3. Therefore, the magnetic dipole density of the artificial magnetic material is $N = 1/d^3$. Using the modified equation (6.1.27c), i.e. $\mu_r^* = (1 + N\alpha^m)$, the following expression is obtained for the complex effective relative permeability of the artificial magnetic materials:

$$\mu_r^* = 1 + \left[\frac{\mu_0\omega^2 S^2/d^3 L_{eff}}{\omega_{0m}^2 - \omega^2 + j\omega\gamma_m}\right]$$

$$\Rightarrow \mu_r^* = 1 - \frac{F\omega^2}{\omega^2 - \omega_{0m}^2 - j\omega\gamma_m} \qquad (a)$$

$$\mu_r^* = 1 - \frac{F\omega^2(\omega^2 - \omega_{0m}^2)}{(\omega^2 - \omega_{0m}^2)^2 + (\omega\gamma_m)^2} + j\frac{F\omega^3\gamma_m}{(\omega^2 - \omega_{0m}^2)^2 + (\omega\gamma_m)^2} \qquad (b)$$

where, filling – factor :

$$F = \frac{\mu_0\pi^2 r_m^4}{d^3 L_{eff}} = \frac{\mu_0\pi^2 r_m^4\omega_{0m}^2 C_{eff}}{d^3}, S = \pi r_m^2 \qquad (c)$$

$$\text{Also, } \mu_r^* = 1 + \chi_r^* = 1 + \frac{F\omega^2}{\omega_{0m}^2 - \omega^2 + j\omega(\omega_{0m}/Q)}$$

$$\text{where, } Q = \frac{\omega_{om}}{\gamma_m} \qquad (d),$$

$$(21.2.3)$$

where r_m is the mean radius of the equivalent single ring loop of SRR. Expression (21.2.3) of the complex relative permeability μ_r^* applies to a low-density SRR based medium. It does not account for the effect of the local field. Using $\gamma_m = 1/3$ the Clausius–Mossotti expression (6.1.28c) of chapter 6, i.e. the expression $\mu_r^* = [1 + N\alpha_m/(1 - N\alpha_m/3)]$ approximately accounts for the effect of the local field [J.24]. In equations (21.2.3a–d), both μ_r^* and F are dimensionless quantities. The filling-factor F is less than unity. On comparing equation (21.2.3a) against equation (6.5.5b) of chapter 6, it is noted that the SRR-based artificial magnetic materials follow the Lorentz type resonance, except the presence of ω^2 in the numerator. Also, the filling-factor F is not the magnetic plasma frequency ω_{mp}, although it has some relation to it. The permeability response of the SRR element is determined by the parameters F, ω_{0m} and γ_m. To obtain the permeability response analytically, these parameters are determined using the closed-form expressions.

It is difficult to measure the damping factor γ_m. In equation (21.2.3d), it is replaced by the Q-factor. The Q-factor is a measurable quantity for the SRR element. Moreover, the measured Q value of the SRR element can also be used for creating an array to develop the artificial magnetic material. So for a fixed value of the

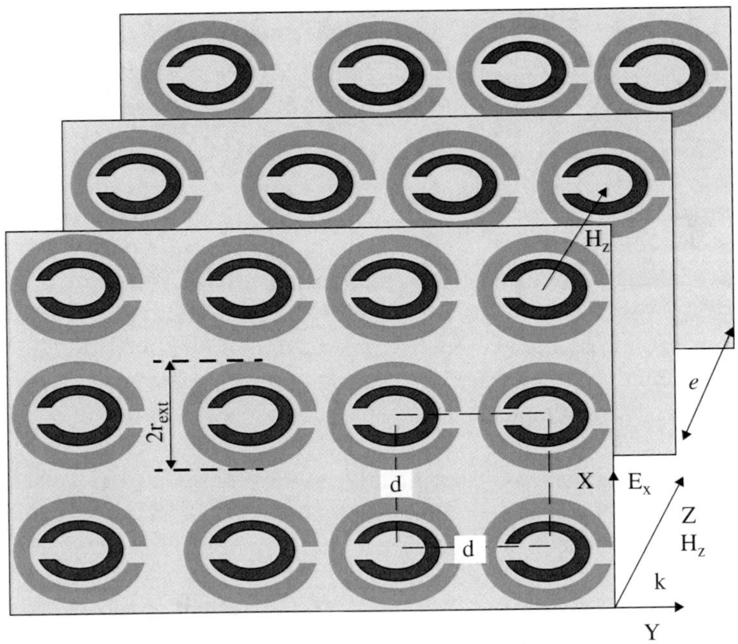

(a) Cubic lattice (ℓ= d) of SRR-based artificial magnetic medium.

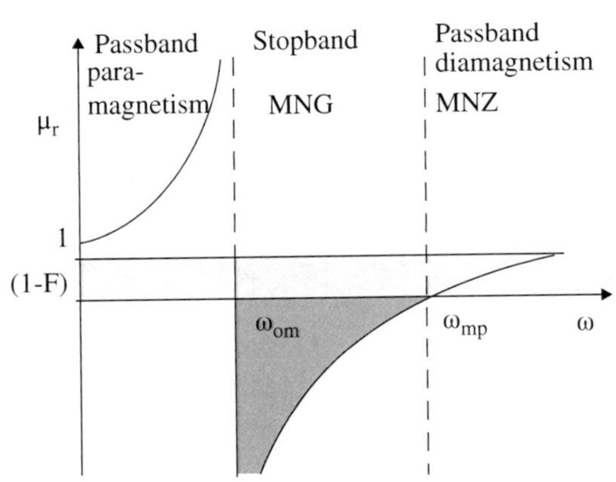

(b) Permeability response the SRR-based artificial magnetic medium.

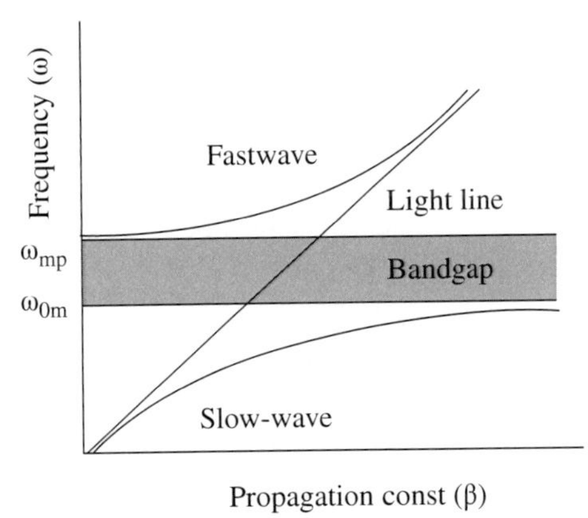

(c) Dispersion diagram of the SRR-based artificial magnetic medium.

Figure 21.8 The SRR lattice and permeability response.

filling-factor F, the measured Q can be used to get the permeability response of the artificial magnetic material accurately [J.25]. However, the maximum value of $\mathrm{Re}\left(\chi_r^*\right) \approx FQ/2$ is associated with a high-loss. Therefore, for a fixed value of F and high Q-factor of the SRR, a high value of μ_r is realized over a narrow bandwidth. The value of F must be above 0.2 to reduce losses in the artificial magnetic materials [J.26, J.27].

The nature of permeability response of a lossless SRR medium, with damping factor $\gamma_m = 0$, is explained from the real part of relative permeability:

$$\mu_r = 1 - \frac{F\omega^2}{\left(\omega^2 - \omega_{0m}^2\right)}, \quad \Rightarrow \mu_r = \frac{\left(\omega^2 - \omega_{mp}^2\right)}{\left(\omega^2 - \omega_{0m}^2\right)}(1 - F) \quad \text{(a)}$$

$$\text{where, magnetic plasma frequency} : \omega_{mp}^2 = \frac{\omega_{0m}^2}{1 - F} \quad \text{(b).} \tag{21.2.4}$$

Figure (21.8b) shows the frequency response of the permeability function with a pole at the *characteristic magnetic resonance frequency* ω_{0m} and zero at the *magnetic plasma frequency* ω_{mp}. The SRR-based medium is the Lorentz type magnetic plasma medium. In the case of a lossy magnetic medium, the infinitely large value of the relative permeability μ_r is replaced by the resonance peak. Figure (21.8b) demonstrates the limiting cases: for $\omega \to 0$, $\mu_r \to 1$, for $\omega \to \omega_{mp}$, $\mu_r \to 0$, and for $\omega \to \infty$, $\mu_r \to (1 - F)$. In the present case, the SRRs are in the host air medium. It is noted that in the frequency range $0 < \omega < \omega_{0m}$, the medium behaves as a *dispersive paramagnetic material* with a positive permeability. Near the characteristic resonance frequency ω_{0m}, the permeability has a high value. In a narrow frequency band $\omega_{0m} < \omega < \omega_{mp}$, the permeability is negative for creating the *MNG material*. In the high-frequency range above the plasma frequency, i.e. for $\omega > \omega_{mp}$ the permeability is again positive. However, its value is less than unity, creating the *MNZ material* medium. The strength of the permeability is controlled by the filling-factor F, i.e. by the size of the SRR inclusion within the unit cell of size $d \times d$. The square-shaped SRR inclusion has the maximum value of F.

The above analysis is approximate, ignoring the effect of depolarization due to the neighboring SRR elements of the array. It could be accounted for through the *interaction constant* [J.28, B.16]. The damping factor γ_m could be treated as the interaction constant used in the Clausius–Mossotti model discussed in subsection (6.1.2) of the chapter 6. The above analysis also ignores the spatial dispersion [J.29]. However, it demonstrates the nature of the low-density resonant artificial magnetic medium. The propagation constant of the EM-waves in the artificial magnetic medium is obtained as

$$\beta = \frac{\omega}{c}\sqrt{\mu_r}, \varepsilon_r = 1, \quad \Rightarrow \beta = \frac{\omega}{c}\left[(1-F)\frac{\left(\omega^2 - \omega_{mp}^2\right)}{\left(\omega^2 - \omega_{0m}^2\right)}\right]^{1/2}. \tag{21.2.5}$$

Figure (21.8c) shows the *slow-wave region* in the range $0 < \omega < \omega_{0m}$, a band-gap in the range $\omega_{0m} < \omega < \omega_{mp}$, due to the negative permeability, and the *fast-wave region* in the range $\omega > \omega_{mp}$. So the artificial magnetic medium behaves as a filter medium with passband–stopband–passband characteristics.

21.2.3 Computation of Equivalent Circuit Parameters of SRR

The five sets of expressions to compute the circuit elements of the equivalent circuit of the SRR element are summarized below:

Set #1: The effective inductance used in the equivalent circuit of a circular SRR, shown in Fig (21.6d), could be approximately computed by using the following expression:

$$\text{Effective inductance [J.30]} : L_{eff} = \mu_0 r_m\left[\ln\left(\frac{32 r_m}{b}\right) - 2\right], \quad \text{where } r_m = r + b + \frac{f}{2}. \tag{21.2.6}$$

Sometimes, the effective inductance is approximated by $L_{eff} \approx \mu_0 r_m$. Alternate approximate expressions for the computation of the inductance can also be used [B.5]. The variational method is also available to compute more accurately the needed inductance [J.22].

The capacitance C_{eff} of the circuit model is computed by treating the concentric strip rings as the parallel conducting strips, ignoring the effect of the curvature, printed on a dielectric sheet. The capacitance of the parallel strips is computed by using equation (9.4.13), or equation (9.4.20) of chapter 9. Other expressions are also available [J.22].

Finally, the resistance R_{eff} of the circuit model is estimated by the following expression:

$$R_{eff} = \begin{cases} \dfrac{2\pi r}{bh\sigma}, & \text{for } \dfrac{h}{2} < \delta \\ \dfrac{\pi r}{b\delta\sigma}, & \text{otherwise.} \end{cases} \tag{21.2.7}$$

In expression (21.2.7) δ is the skin-depth of the strip conductors of conductivity σ, b is the strip width of both rings, and h conductor thickness.

Set #2: Another set of expressions is available to compute the characteristic resonance frequency ω_{0m} and the damping constant $\gamma_m = R/L_{eff}$ of the circular SRR. The resistance R of the ring is replaced by the ring metal impedance $Z(\omega)$ [J.31].

$$C_{eff} = \frac{2\pi r_m}{3}\varepsilon_0\frac{K(k')}{K(k)}, \quad \text{where } k = \frac{f}{2b + f}$$

$$k' = \sqrt{1 - k^2} \tag{a}$$

$$L_{eff} = \frac{\mu_0 \pi r_m^2}{\ell} \tag{b}$$

$$\text{for cubic lattice } \ell = d \quad F = \frac{\pi r_m^2}{d^2} \tag{c}$$

$$R_{eff} = \text{Re}\{Z(\omega)\} = \text{Re}\left\{\frac{2\pi r_m}{b}Z_s(\omega)\right\}$$

$$\text{where, } Z_s(\omega) = \frac{1 - j}{\sigma(\omega)\delta(\omega)} \tag{d}$$

(21.2.8)

where $Z_s(\omega)$ is the surface impedance of the strip conductor. Usually, the skin depth $\delta(\omega)$ is less than the ring conductor thickness, and conductivity $\sigma(\omega)$ is frequency-independent. However, in the optical frequency range, it is frequency-dependent and computed by using the expression $\sigma(\omega) = \omega_{mp}^2\varepsilon_0/(\gamma - j\omega)$ [J.31].

Set #3: More accurate expressions, with accuracy in the range 6–12%, are available to compute the characteristic resonance frequency ω_{0m} of the circular SRR structure [J.32]. The asymmetrical coupled lines-based model is also available to account for the unequal arbitrary strip widths of two SRR [J.33]. The following set of curve-fitted expressions can be used to determine ω_{0m}, between 8 and 12 GHz, with accuracy within 10% [J.34]:

$$L_{eff} = 2.57e^{-w_3/\sqrt{2}}(\pi r_{ext} - 2.2f_1 - \pi/2) \quad \text{(a)}$$

$$C_{eff} = 0.217 + \{[0.059(2r_{ext} + \varepsilon_r - 5)]\ [0.432w_1 - 0.317w_2^2 + 0.07w_2^3]\ [3.3367e^{-3.2d_1} - 0.1955e^{-0.47h}]\} + (0.05\varepsilon_r - 0.218) + \{(0.599h/(0.0248 + h)) - 0.599\} \quad \text{(b)}$$

$$\omega_{0m} = \begin{cases} [1/2\pi\sqrt{L_{eff}C_{eff}}, & \text{for } r_{ext} > 5.2\text{mm} \\ [1/2\pi\sqrt{L_{eff}C_{eff}} + ((5.2 - r_{ext})/2)], & \text{for } r_{ext} < 5.2\text{ mm} \end{cases} \quad \text{(c).}$$

$$(21.2.9)$$

In expression (21.2.9a–c), the parameters are $w_1 = b$, $w_2 = 2b$, $w_3 = b$, $f_1 = f$ for $f < 1$ mm, and $w_1 = b/3$, $w_2 = 1.414b$, $w_3 = b/1.414$, $f_1 = 1.414f$ for $f > 1$ mm.

Set #4: Pendry Expression for Relative Permeability of Loop Medium

Pendry has obtained another set of analytical expressions to compute the relative permeability of the circular SRR-based artificial magnetic medium by averaging the magnetic field within the unit cell. The expressions are summarized below with an illustrative computation [J.3]:

$$\mu_r^* = 1 - \frac{F}{\omega^2 - \omega_{0m}^2 - j\omega\gamma_m} \quad \text{(a)}$$

$$F = \frac{\pi r_m^2}{d^2} \quad \text{(b)}, \quad \gamma_m = \frac{2\ell\rho}{\mu_0 r_m} \quad \text{(c)}$$

$$\omega_{0m}^2 = \frac{3\ell}{\mu_0\pi^2 C_{eff} r_m^3} \quad \text{(d)}$$

$$C_{eff} = \frac{1}{\pi\mu_0 c_{light}^2}\ln\left(\frac{2b}{f}\right) \quad \text{(e).}$$

$$(21.2.10)$$

Figure (21.7a) shows dimensions of an SRR within the unit cell of size $d \times d \times \ell$, ρ is the resistance per unit length of conducting strip, and C_{eff} is the effective capacitance *p.u.l.* of gap f between strip rings, treated as the parallel conducting strips. For the cubic unit cell, shown in Fig (21.8a), the height in the z-direction is $\ell = d$. Figure (21.9a, b) shows the permeability response for the following parameters:

$$d = 1.0 \times 10^{-2}\text{m}, \quad b = 1.0 \times 10^{-3}\text{m}$$
$$f = 1.0 \times 10^{-4}\text{m}, \quad \ell = 2.0 \times 10^{-3}\text{m}$$
$$r_m = 2.0 \times 10^{-3}\text{m},$$
$$\text{for copper} - \text{strip ring}, \quad \rho = 200$$
$$\text{for resistive strip ring}, \quad \rho = 2000.$$

The computed characteristic resonance frequency f_{0m} is 13.41 GHz. Pendry expression computes the resonance frequency f_{0m} within 10% accuracy. Figure (21.9a, b) demonstrate the paramagnetic, MNG, and diamagnetic behavior of the SRR based artificial magnetic materials.

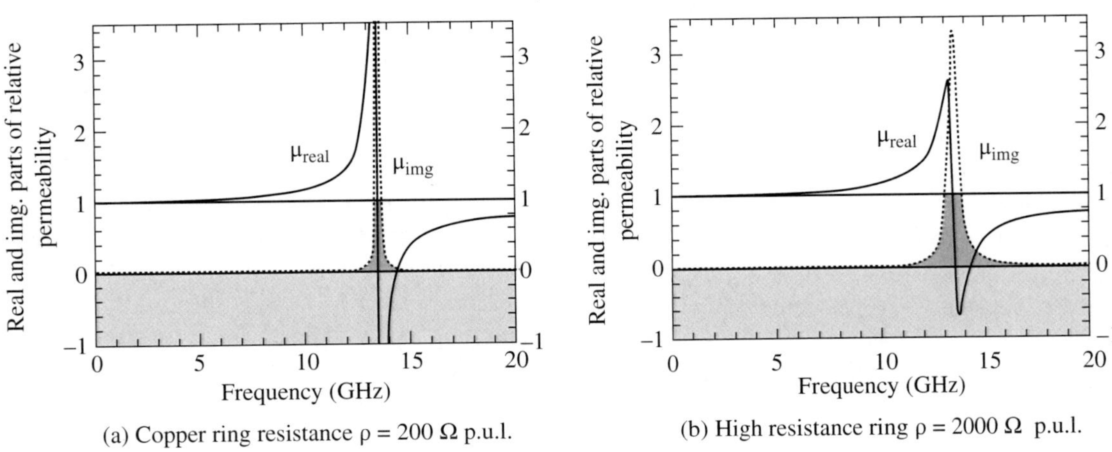

Figure 21.9 The permeability response of the cubic lattice of the SRR based magnetic materials. *Source:* Reproduced from Pendry et al. [J.3]. © 1999, IEEE.

The response of Fig (21.9b) also shows high losses associated with the highly dispersive lossy medium. The magnitude of $(\pm\mu_r)$ is reduced with an increase in loss and finally vanishes for higher losses. Therefore, low-loss is an important condition for the realization of the high-value permeability and also MNG materials.

Set #5: Circuit Parameters of Square Split-ring Resonators

Figure (21.6c) also shows the square split-ring resonator of arm's length 2r within the square unit cell of size "d". The rings of width b are separated by f, and end gap g is the split gap. It is printed on the PCB board. In the vertical direction, the unit cell height is ℓ. For the cubic lattice, it is $\ell = d$. This structure provides the largest value of the filling-factor F. The complex effective relative permeability is computed by using equation (21.2.3a). The parameters are computed using the following expressions [J.35]:

$$\omega_{0m}^2 = \frac{3\ell c_{light}^2}{\pi \ln(2b/f) r^3} \quad (a), \quad F = \frac{4r^2}{d^2} \quad (b)$$
$$\gamma_m = \frac{2\ell\rho}{\mu_0 r} \quad (c). \qquad (21.2.11)$$

21.2.4 Bi-anisotropy in the SRR Medium

The external time-varying axial field, H_z^{ext} applied to the SRR as shown in Fig (21.7a), induces a current in the loop creating the *z-oriented magnetic polarizability* α_{zz}^{mm}. It also induces the charges on the split rings, creating the *y-directed electric polarizability* α_{yz}^{em}. Similarly, the SRR can also respond to the external time-varying electric field E_y^{ext}, creating the y-directed electric polarizability α_{yy}^{ee}, and also the z-directed magnetic polarizability α_{zy}^{me}. Thus, the SRR has *magneto-electric coupling* due to the *cross-polarization* [J.36–J.38]. The cross-polarization is discussed in subsection (6.1.4) of chapter 6. It shows that the external time-varying magnetic field induces both the magnetic and electric polarization. Similarly, the external time-varying electric field generates both of these polarizations. Usually, bianisotropy is disadvantageous. It is stronger in a single split-ring. However, it is reduced in the concentric coplanar double SRR with diametrically opposite splits that partially cancels the electric polarization. However, the z-directed induced magnetic dipoles of both the split-rings are additive, as the direction of the circulating currents in them are in the same direction. It is shown in Fig (21.7a). It further shows oppositely directed induced electric dipole moments, p_1, and p_2 at locations Q_1 and P_2 due to the induced charges on the outer and inner split-rings. The inner ring is smaller than the outer one. Therefore, a complete cancellation of the effect of electric dipoles is not possible for the concentric coplanar rings SRR, as the induced voltages V_{out} and V_{in} at Q_1 and P_2 are not equal. The cancellation of the effect of electric dipoles is achieved with the use of the *broadside coupled SRR*, mentioned in subsection (21.2.5).

Following the discussion in subsection (6.1.4) of chapter 6, the electric and magnetic dipole moments due to the polarization and cross-polarization could be written as follows:

$$p_y = \alpha_{yy}^{ee} E_y^{ext} + \alpha_{yz}^{em} H_z^{ext} \quad (a)$$
$$m_z = \alpha_{zy}^{me} E_y^{ext} + \alpha_{zz}^{mm} H_z^{ext} \quad (b). \qquad (21.2.12)$$

The Onsager symmetry condition provides $\alpha_{zy}^{me} = -\alpha_{yz}^{em}$, and equation (21.2.12) is written as [B.5],

$$\begin{bmatrix} \vec{p}_y \\ \vec{m}_z \end{bmatrix} = \begin{bmatrix} \alpha_{yy}^{ee} & \alpha_{yz}^{em} \\ -\alpha_{yz}^{em} & \alpha_{zz}^{mm} \end{bmatrix} \begin{bmatrix} E_y^{ext} \\ H_z^{ext} \end{bmatrix}. \qquad (21.2.13)$$

The elements of polarization matrix equation (21.2.13) are dyadic as shown in equation (6.1.31) of chapter 6. The edge-coupled SRR creates a bianisotropic medium. The circuit model has been used to obtain the polarization elements [J.30]. Further, the E_x^{ext} field applied to the SRR also creates the x-directed non-resonant electric polarizability $\alpha_{xx}^{ee} = \varepsilon_0(16/3)r_{ext}^3$. The polarizabilities of the SRR, creating the lossy bi-isotropic magnetic medium are summarized below [J.22, J.23, B.5]:

$$\alpha_{zz}^{mm} = \frac{\pi^2 r_m^4}{L_{eff}} \frac{\omega^2}{(\omega_{0m}^2 - \omega^2 + j\omega\gamma_m)} \quad (a), \quad \alpha_{xx}^{ee} = \frac{\varepsilon_0 16}{3} r_{ext}^3 \quad (b)$$
$$\alpha_{yz}^{em} = -\alpha_{zy}^{me} = -j2\pi r_m^3 f_{eff} C_{pul} \left(\frac{\omega_{0m}^2}{\omega}\right) \frac{\omega^2}{(\omega_{0m}^2 - \omega^2 + j\omega\gamma_m)} \quad (c)$$
where $f_{eff} \approx b + f$
$$\alpha_{yy}^{ee} = \frac{\varepsilon_0 16}{3} r_{ext}^3 + 4 f_{eff}^2 r_m^2 C_{pul}^2 L_{eff} \left(\frac{\omega_{0m}^2}{\omega}\right)^2 \left(\frac{1}{(\omega_{0m}^2/\omega^2 - 1)}\right) \quad (d). \qquad (21.2.14)$$

For the lossless SRR-based medium, the damping factor γ_m is ignored, i.e. $\gamma_m = 0$. The polarizabilities α_{zz}^{mm}, α_{yz}^{em}, and α_{yy}^{ee} have resonant nature. The presence of the polarizabilities in the SRR-based medium is observed through the stopband in the transmission response, i.e. in the S_{21} response. The nature of stopband in the S_{21}-response is decided by the polarization of the incident waves, and the orientation of the SRR with respect to the direction of the polarizing incident field. The transmission measurement has been performed in the free space on the SRR based magnetic medium block of the $25 \times 25 \times 25$ UC.

It is also obtained using the numerical method. Figure (21.10a) shows four orientation of the square SRR with respect to the E_y and H_z field components. The wave propagates in the X-direction with the wave-vector k. Figure (21.10b, c) show the S_{21} response of four orientations of the SRR due to different polarization and cross-polarization. Figure (21.10d) shows the current and charge distributions, for both the noncirculating and circulating currents, on a single-ring SRR [J.36].

The ***case #i*** of Figure (21.10a) creates the magnetic dipole, giving the resonant negative permeability, due to the H_z-field component. It appears as the narrow stopband in the simulated and experimental transmission response shown in Fig (21.10b, c). However, for the ***case #ii*** both the magnetic and electric resonant polarizations take place at nearby resonance frequencies. In this case, the E_y-field component induces the charges across the split gap shown in Fig (21.10d) creating the electric dipole. Figure (21.10d-ii) also shows the induced circulating current in the ring creating the magnetic dipole. Thus, the **case *#ii*** creates a wider stopband. The ***case #iii*** does not generate any polarization, so no stopband is seen in the transmission response. Finally, ***case #iv*** generates the electric dipole, creating a stopband in the transmission response. There could be the other two cases for the SRR orientation in the (H–k), i.e. (z–x) plane. These are also the cases of nonpolarization.

A similar study has been conducted by mounting the SRR in the TE_{10} mode rectangular waveguide [J.3, B.17]. The stopbands show the presence of the negative permeability in the Lorentz magnetic materials. The responses of all four orientations have been further experimentally confirmed [J.38].

21.2.5 Variations in SRR Structure

The size of the SRR is about $\lambda_0/10$. It is just the limit of homogenization. To reduce its size further, some variations in the SRR are suggested. The SRR shown in Fig (21.7a) is an *edge-coupled ring* giving a smaller value of the effective capacitance. It also suffers the cross-coupling of the external electric and magnetic fields responsible for the bianisotropy. The effective capacitance has been significantly increased, and the cross-coupling is controlled by the use of the *broadside coupled*

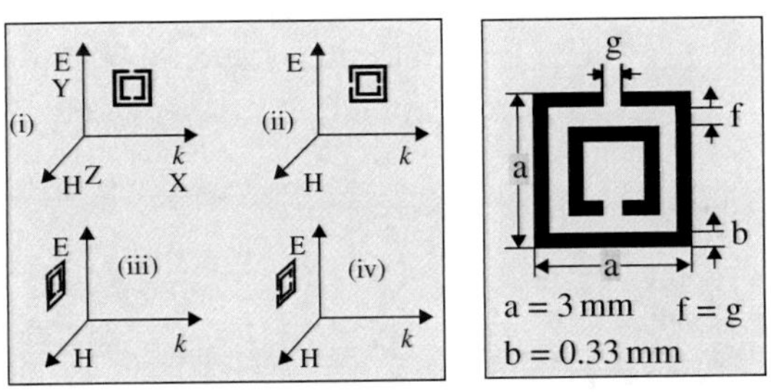

(g: Gap in split rings, f: Gap between two split rings.)

(a) Spatial orientation of the SRR.

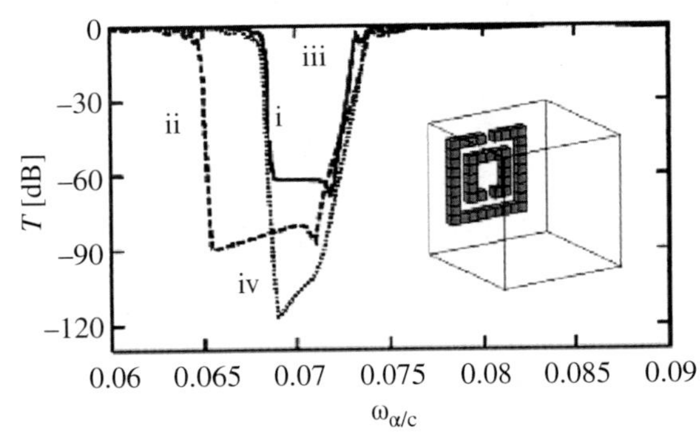

(b) Simulated transmission through the SRR slab [J.36].

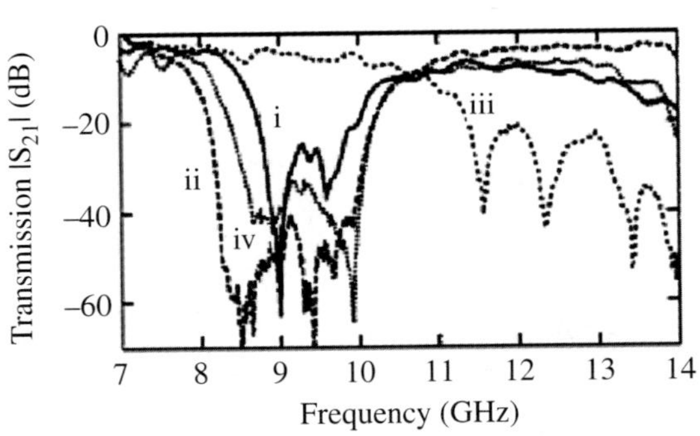

(c) Measured transmission through the SRR slab [J.36].

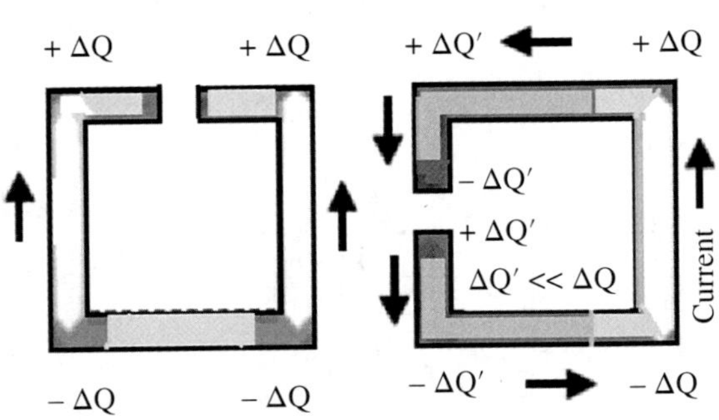

(i) Non-circulating current. (ii) Circulating current.

(d) The induced current on the rings.

Figure 21.10 The permeability response of the SRR of four orientations. *Source:* Reproduced from Katsarakis et al. [J.36], with the permission of AIP Publishing. © 2004, AIP Publishing.

equal size concentric rings, printed on both sides of a thin dielectric sheet. It is shown in Fig (21.11a) [J.22, B.5, B.15]. The use of the ferroelectric substrate and thin-layer technology could be more effective in size reduction [B.15]. Figure (21.11b) shows the chip capacitors C_1, C_2, C_3, loading of the SRR in the gap region of two rings, and also in the split-gaps of both the outer and inner rings. It decreases the resonant frequency without increasing the number of turns. The capacitor loading of the outer ring is a more effective method [J.39]. The complementary SRR, (*CSRR*) is another innovative modification, etched in the conducting plane of the substrate. However, it is an *electric particle* useful in the design of the metasurfaces and metalines in the microstrip and CPW environment. It is discussed in subsection (22.3.2) of chapter 22.

Furthermore, the *multiturns SRR* and the *planar spiral* resonators are used to decrease the resonant frequency of a multiturn SRR particle significantly. It is shown in Fig (21.11c). The resonant frequency decreases rapidly with the increase in the number of turns. These resonators have been modeled using the equivalent RLC circuit [J.22]. Other alternatives to reduce the resonant frequency of the magnetic particles are the use of the *Swiss-rolls* [J.3] and *meta-solenoids*, shown in Fig (21.11d) [J.21]. They significantly increase the effective permeability of the artificial magnetic materials. The Swiss-rolls can operate even in the RF frequency range 21.5–200 MHz useful for the MRI and RF endoscope applications [B.10]. However, these are nonplanar 3D structures.

Figure (21.12a-i, a-ii) show the twin coupled SRR structures, called the *paired split-ring resonators* (PRR). The twin structures form one unit cell [J.40]. The structures have the dimensions as follows: p = 500 nm, b = 80 nm, g = 50 nm, strip conductor thickness t = 50 nm, and lattice constant d = 600 nm. These structures exhibit mixed magnetic and electric responses at two different resonant frequencies for the applied polarization of the incident EM-waves, shown in Fig (21.12a-i). Figure (21.12b, c) shows the simulated *magnetic response (permeability response)* at frequency 83.0 THz, and the *electric response (permittivity response)* at frequency 115.0 THz, respectively. The circuit models of both resonances are shown in the inset. In the first case, the coupled SRR supports the *symmetric mode* of resonance. In the second case, the structure supports the *anti-symmetric* resonating mode. In both cases, strongly coupled modes give the Lorentz type material response, whereas the weak response shows the *anti-Lorentz response*. The orientations of the H-field, E-field, and k-vector of the incident wave on the paired ring resonator are properly taken to get an only magnetic response by the arrangement shown in Fig (21.12a-iii), and only electric response by the

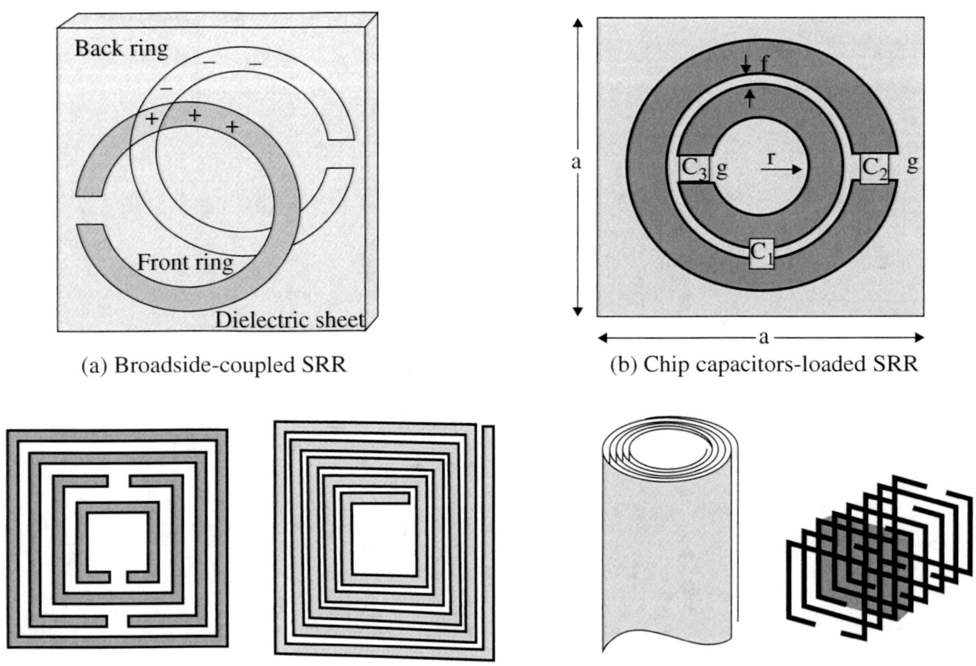

(a) Broadside-coupled SRR

(b) Chip capacitors-loaded SRR

(c) Multi-turns SRR and the planar spiral

(d) Swiss-rolls and meta-solenoids

Figure 21.11 Some more magnetic particles.

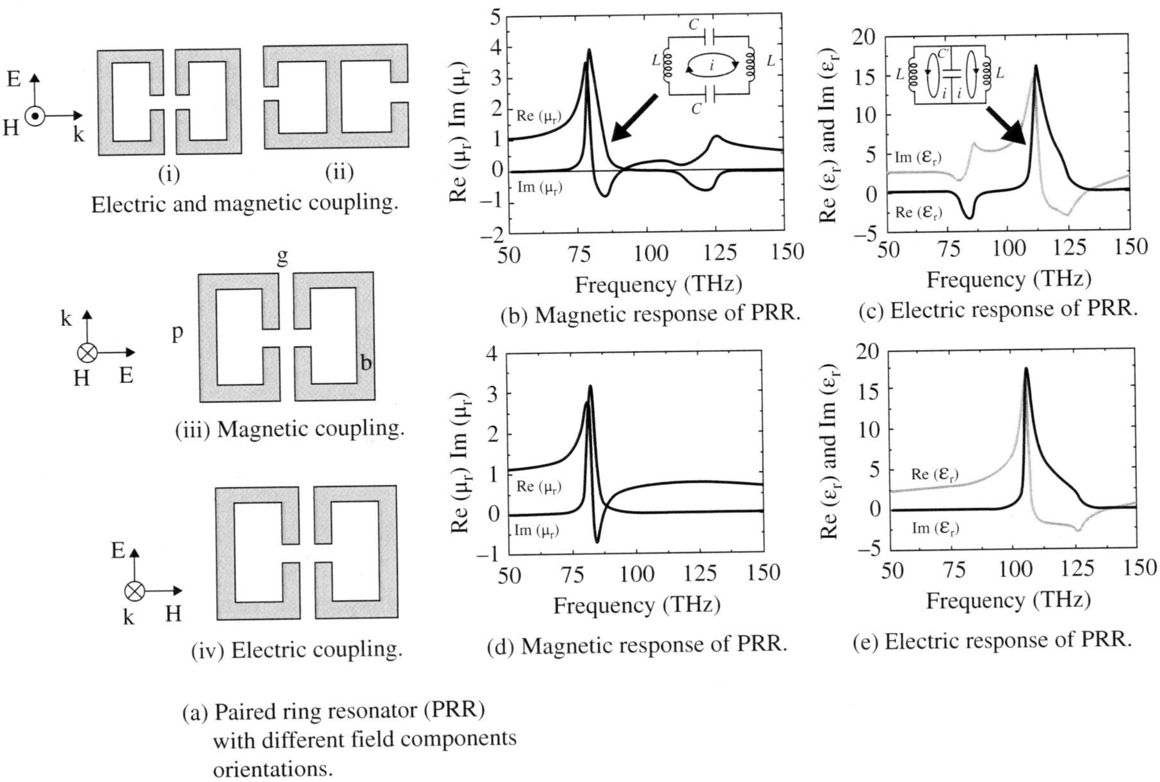

Figure 21.12 The orientation and responses of paired ring resonators (PRR). *Source*: Liu et al. [J.40]. © 2007, Springer-Verlag.

arrangement shown in Fig (21.12a-iv). In the case of Fig (21.12a-iii), the H-field is normal to the loop creating magnetic polarizability. In the case of Fig (21.12a-iv), the E-field is parallel to the strip conductor creating the electric polarizability. Figure (21.12d, e) show that the PRR has been excited only for the magnetic or electric response, respectively, under the proper orientation of the incident fields. The structures have also been implemented, using the printed strips, to realize the DNG medium in the GHz frequency range [B.18].

21.3 Double Negative Metamaterials

The DNG metamaterials have been demonstrated by combining the *negative permittivity electric particle (wire) and negative permeability magnetic particle* (SRR) as entirely separate structures. The combined structure could be called the *composite DNG* (*C-DNG*) metamaterials. The strip-wire (SW) and SRR could be printed either on the same side or on opposite sides of a dielectric sheet. Some innovative structures have also been suggested to combine both the electric and magnetic particles in one geometrical structure. The engineered material based on such *electromagnetic particles* could be called the *single DNG (S-DNG)* metamaterials. The realization of a few C-DNG and S-DNG metamaterials is discussed in this section. Let us first formulate the *composite permittivity-permeability function* of the composite permittivity–permeability medium.

21.3.1 Composite Permittivity–Permeability Functions

There are two kinds of composite permittivity–permeability functions. The first one is the *Drude–Lorentz function*, corresponding to the Drude model type permittivity response and Lorentz model type permeability response. The second one is the *Lorentz–Lorentz function* corresponding to the Lorentz model type permittivity response and also the Lorentz model type permeability response.

Drude–Lorentz Function

The relative permittivity of a rodded or strip wire (SW) medium is described by the Drude model function given by equation (21.1.12b), and the relative permeability of the SRR medium is described by the Lorentz model function given by equation (21.2.4a), respectively. Similarly, the capacitor-loaded wire medium is described by the

Lorentz model function given by equation (21.1.17a). These material functions, for the lossless medium, are summarized below:

$$\varepsilon_r = \frac{\omega^2 - \omega_{ep}^2}{\omega^2} \quad (a), \quad \mu_r \approx \frac{\omega^2 - \omega_{mp}^2}{\omega^2 - \omega_{0m}^2} \quad (b)$$

$$\varepsilon_r = \frac{\left(\omega_{oe}^2 + \omega_{ep}^2\right) - \omega^2}{\left(\omega_{oe}^2 - \omega^2\right)} \approx \frac{\omega^2 - \omega_{ep}^2}{\omega^2 - \omega_{oe}^2} \quad (\text{for } \gamma_m = 0) \quad (c). \qquad (21.3.1)$$

The Drude–Lorentz composite function describes the performance of the composite SW-SRR metamaterial medium. The composite function assumes that the SW and SRR do not interact with each other. However, for a compact unit cell, the assumption is not valid. The Drude–Lorentz composite function, given below, also helps to obtain the propagation constant $\beta_g = \omega/c\sqrt{\varepsilon_r\mu_r}$ of the EM-wave in such medium:

$$\varepsilon_r\mu_r = \left[\frac{\omega^2 - \omega_{ep}^2}{\omega^2}\right]\left[\frac{\omega^2 - \omega_{mp}^2}{\omega^2 - \omega_{0m}^2}\right] \quad (a)$$

$$\beta_g = \frac{1}{c}\left[\frac{\left(\omega^2 - \omega_{mp}^2\right)\left(\omega^2 - \omega_{ep}^2\right)}{\omega^2 - \omega_{0m}^2}\right]^{1/2} \quad (b), \qquad (21.3.2)$$

where $0 < \omega_{0m} < \omega_{mp} < \omega_{ep} < \infty$. The electric and magnetic plasma frequencies could be identical also.

Table 21.3 shows the nature of the frequency response of individual permittivity and permeability function. It also examines the nature of the Drude–Lorentz composite function that is responsible for showing the stopband/passband of EM-wave propagation through the composite medium. The propagation constant β_g determines the nature of the frequency bands. The material responses of the wire medium and SRR medium are examined separately. Assuming unity permeability function, the medium is a dispersive wire medium described by the plasma type frequency-dependent permittivity. It has a *negative permittivity* below the electric plasma frequency ω_{ep} giving the ENG medium. Above ω_{ep}, the medium is a positive ENZ medium with $0 < \varepsilon_r < 1$. Likewise, for the unity permittivity function, the medium is a Lorentz type SRR medium, showing the *negative permeability* in a narrow band $\omega_{0m} < \omega < \omega_{mp}$. The transmission response of the SRR magnetic medium shows a dip in this frequency band, and the medium is the MNG medium. However, above the magnetic plasma frequency, i.e. for $\omega > \omega_{mp}$, the medium is a positive MNZ medium with $0 < \mu_r < 1$. In the case of a composite permittivity–permeability medium, the frequency band $\omega_{0m} < \omega < \omega_{mp}$ creates the DNG medium, and the transmission response shows a passband due to the negative value of both the permittivity and permeability functions. The DNG passband supports the backward wave propagation. For $\omega > \omega_{ep}$, the composite medium is a DPS medium giving another passband, supporting propagation of the standard forward wave.

Figure (21.13a, b) demonstrate the envisaged the material and transmission response of Table 21.3. Figure (21.13a) shows the sketched Drude type permittivity response of the strip-wire medium. It also shows the permeability response of the Lorentz type SRR medium. The medium has negative permittivity and permeability, in the range $\omega_{0m} < \omega < \omega_{mp}$ creating the DNG type medium. In the range, $\omega_{mp} < \omega < \omega_{ep}$ stopband is created due to the negative permittivity. Figure (21.13b) shows the transmission response in the complete frequency range, with a bandgap between two pass-bands. The first passband is due to the DNG supporting the backward wave propagation and the negative refraction, while the second passband supports the forward wave propagation and the positive refraction. In the case of the identical value of the electric and magnetic plasma frequency ($\omega_{mp} = \omega_{ep}$), the stopband disappears and a continuous DNG–DPS medium is obtained. Thus, within a combined wideband passband, the wave propagation is seamlessly varied from the backward wave to forward wave with an increase in frequency. Such a medium is realized by the 1D CRLH metamaterials. It finds application in designing the leaky-wave antenna, discussed in subsection (22.4.1) of chapter 22.

Table 21.3 Frequency response of the composite permittivity–permeability function (Drude–Lorentz function).

	Sign of permeability function			Sign of permittivity function	
Frequency range	$\omega^2 - \omega_{0m}^2$	$\omega^2 - \omega_{mp}^2$	μ_r sign	$\omega^2 - \omega_{ep}^2$	**Nature of band**
$0 < \omega < \omega_{0m}$	(−)	(−)	(+)	(−)	Stop
$\omega_{0m} < \omega < \omega_{mp}$	(+)	(−)	(−)	(−)	DNG pass
$\omega_{mp} < \omega < \omega_{ep}$	(+)	(+)	(+)	(−)	Stop
$\omega > \omega_{ep}$	(+)	(+)	(+)	(+)	DPS pass

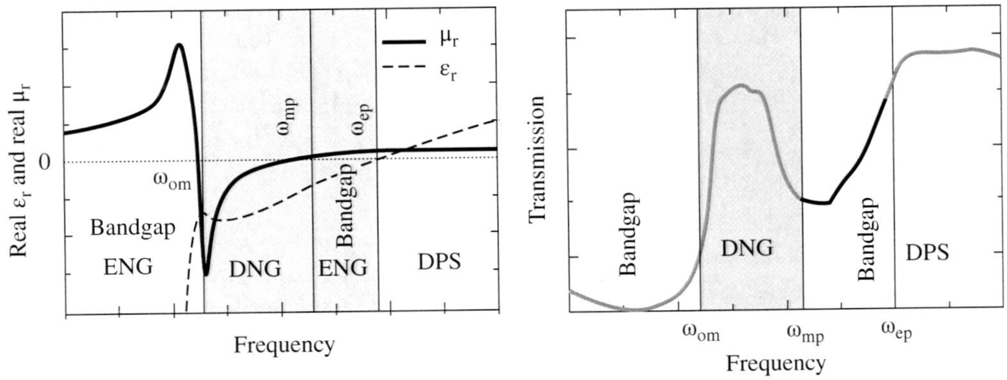

(a) Material response of composite SRR-SW medium.

(b) Transmission response of composite SRR-SW medium.

Figure 21.13 Material and transmission responses of strip wire (SW) and SRR metamaterial.

The above discussion shows that the generation of a passband in the stopband region of the magnetic medium does not ensure the creation of the DNG medium. The DPS medium also supports a passband with the forward wave propagation. *The passband of the DNG medium supports the backward wave.* The nature of the unknown medium – DPS or DNG, is decided by the *differential phase* $\Delta\varphi$ of the wave passing through two distances L_1 and L_2 $(L_2 > L_1)$ in the medium. It is obtained as follows:

$$\phi_i = -\beta_g L_i = -n(\omega) L_i \frac{\omega}{c}, \quad i = 1, 2 \qquad \text{(a)}$$

$$\Delta\phi = -n(\omega)(L_2 - L_1)\frac{\omega}{c} \qquad \text{(b)},$$

$$(21.3.3)$$

where $n(\omega)$ is a dispersive refractive index of the composite medium. For the DPS medium, $n(\omega)$ is a positive quantity giving negative differential phase $\Delta\phi$, whereas for the DNG medium, $n(\omega)$ is negative, giving the positive value of $\Delta\phi$. Using the EM-simulator, the differential phase $\Delta\phi$ is easily obtained from the transmission phase of the wave across the metamaterial medium of two unequal lengths.

Lorentz–Lorentz Function

Following the above discussion, the Lorentz–Lorentz composite function and the propagation constant of the EM-wave in the Lorentz–Lorentz composite medium could be written as follows:

$$\varepsilon_r \mu_r = \left[\frac{\omega^2 - \omega_{ep}^2}{\omega^2 - \omega_{oe}^2}\right]\left[\frac{\omega^2 - \omega_{mp}^2}{\omega^2 - \omega_{0m}^2}\right] \qquad \text{(a)}$$

$$\beta_g = \frac{\omega}{c}\left[\frac{\left(\omega^2 - \omega_{mp}^2\right)\left(\omega^2 - \omega_{ep}^2\right)}{\left(\omega^2 - \omega_{0m}^2\right)\left(\omega^2 - \omega_{oe}^2\right)}\right]^{1/2} \qquad \text{(b)}.$$

$$(21.3.4)$$

where $0 < \omega_{0e} < \omega_{0m} < \omega_{mp} < \omega_{ep} < \infty$. The characteristic electric and magnetic resonance frequencies could be identical also. Another band performance table, similar to Table (21.3), can be prepared for the Lorentz–Lorentz arrangement.

The realization of both the Drude–Lorentz and Lorentz–Lorentz DNG media is discussed in the next subsection. Further examples also are given in subsection (21.3.3).

21.3.2 Realization of Composite DNG Metamaterials

The previous sections have discussed the realization of the negative permittivity and negative permeability material with the help of the Drude electric and Lorentz magnetic type inclusions. The DNG can also be realized by using the Lorentz electric and Lorentz magnetic inclusions. This section discusses a few examples for the realization of the composite DNG (C-DNG) metamaterials by using a unit cell of both arrangements – Drude–Lorentz inclusions, and Lorentz–Lorentz inclusions, with negative permittivity and negative permeability over common frequency band. The DNG metamaterial was first demonstrated by Smith and coworkers, using the rodded (wire) medium with SRR [J.5]. Next, it was realized in the planar technology replacing the *rod-SRR composite* by the composite structure of *strip wire-SRR (SW-SRR)*. It is difficult to solder the strip-wire at the top and bottom conductors in a parallel plate waveguide, so the SW has been replaced by the *capacitively loaded strip (CLS)*. Thus, the DNG medium has also been realized by using the CLS–SRR composite. The complex SRR has been replaced by a simpler *capacitively loaded loop (CLL)*. The CLS–CLL composite provides another structure to realize the DNG medium. The three C-DNG structures are briefly discussed below. The details could be obtained from the literature [J.41–J.44].

Strip Wire and SRR (SW-SRR) Composite Element

The unit cell of the SW-SRR composite element is shown in Fig (21.14a). It is the *Drude–Lorentz* arrangement. The size of unit cell is 2.5 mm × 2.5 mm × 2.5 mm. The electric field component E_z is responsible to create the electric dipole in the strip wire (SW) and the magnetic field component H_x creates the magnetic dipole in the SRR. The z-polarized incident wave propagates in the y-direction with wavevector k_y. The perfect conductor surfaces (PEC) are placed at the top and bottom of the cubic unit cell in the (x–y)-plane. The perfect magnetic surfaces (PMC) are placed at the sidewalls of the cubic unit cell in the (y–z)-planes. The dimensions of SRR are as follows: the size of inner square ring r = 0.506 mm, strip width of SRR b = 0.124 mm, the gap between rings f = 0.15 mm, and split gap g = 0.114 mm. The structure is constructed on a substrate with $\varepsilon_r = 3.02$, h = 0.5 mm. The metallic strip of width w_1 = 0.5 mm is also printed on the backside of the substrate. It is located in the middle of the SRR.

The Z-polarized incident EM-wave on the metamaterial propagates along the y-axis. In the y-direction, 4 UC are considered to get the transmission parameter S_{21}. The incident wave has E_z-field component parallel to the SW and its H_x-field component is normal to the plane of the SRR [J.41]. Figure (21.14b) shows the magnitude of the transmission ($|S_{21}|$) response. The transmission peak in the DNG region, shown in gray, is at 22.6 GHz. The frequency range of the DNG medium is

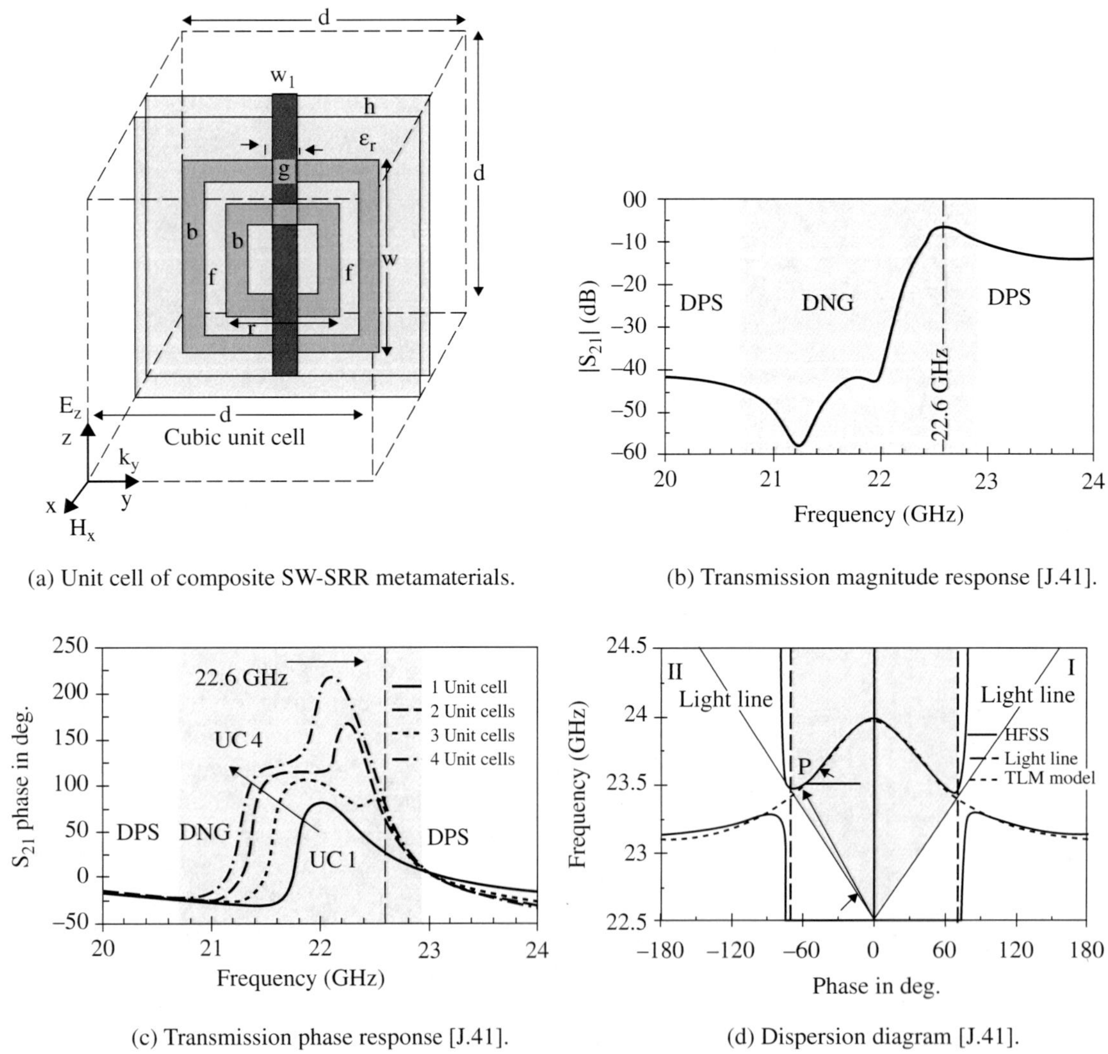

Figure 21.14 Transmission and dispersion responses of strip wire and SRR metamaterial. *Source:* Reprinted (figure) with permission from Woodley et al. [J.41]. © 2005 APS.

obtained from the phase response shown in Fig (21.14c). Outside the DNG, both in the lower and upper-frequency bands, the composite SW-SRR medium behaves like a DPS medium. High attenuation level on the left side of the DNG shows that the DPS of the lower frequency band is the nonperfect ENG due to the negative permittivity of the strip wire medium. Therefore, the electromagnetic nature of the composite cannot be determined from the magnitude of the transmission response alone. Additional confirmation is needed to accept the gray region of Fig (21.14b) as the DNG medium created by the composite SW-SRR. It could be achieved by any of the following processes:

- *By showing that the EM-wave passing through the probable DNG medium has the leading phase, unlike the lagging phase in a normal DPS medium.*
- *The dispersion diagram should exhibit the backward wave propagation in the probable DNG medium.*
- *The probable DNG medium should offer negative refraction for the wave passing through it.*
- *Finally, the real part of the extracted permittivity and permeability parameters must be negative in the probable DNG medium.*

Figure (21.14c) shows the *phase responses* for four lengths of the composite metamaterial medium. The length is expressed in terms of the number of UC. We note that during the frequency interval shown in gray, the differential phase $\Delta\phi$ is positive and the refractive index $n(\omega)$ is negative according to equation (21.3.3b). Thus, the gray interval shows the creation of the DNG material. It is in between two DPS regions. The value of the differential phase $\Delta\phi$ is frequency-dependent, i.e. the refractive index is dispersive. Figure (21.14c) also shows the lowering of the transition frequency in the lower frequency band of DNG with an increase in the number of UC. However, the upper transition frequency of the DNG bandwidth is fixed, i.e. independent of the number of UC.

Figure (21.14d) shows the *dispersion diagram* obtained from the *eigensolver* of the HFSS EM-Simulator and also using the TLM model. The periodic boundary conditions are applied at all faces of the unit-cell and phase is varied only along the y-direction. The gray area is above the light line. The direction of the local slope at the location P in the quadrant-II, and the slope at the origin O shows the anti-parallel phase and group velocities. It indicates the propagation of the *forward traveling backward wave* in the DNG medium. The wave in the quadrant-I is the *backward traveling backward wave*. The HFSS-simulated dispersion diagram, Fig (21.14d), shows a bandgap due to the intersection of the Bloch mode with the light line at the location P. However, the TLM model does not show any bandgap. The DNG frequency range obtained from the phase diagram is different from that of the dispersion diagram. It is noted that the phase diagram of Fig (21.14c) is obtained for a finite structure, whereas the dispersion diagram of Fig (21.14d) is generated for the infinite structure.

CLS–SRR Composite Element

The strip-wire is replaced by the capacitively loaded strip (CLS) to create the DNG medium using the CLS–SRR composite designed on a dielectric substrate [J.42]. It is the *Lorentz–Lorentz arrangement* of inclusions. The composite structure mounted in the parallel plate waveguide (PPW), shown in Fig (21.15a), does not require soldering of the CLS at the top and bottom of the structure. The CLS–SRR composite is mounted in the (y–z)-plane of the PPW along the length (y-axis) of the waveguide. The PPW supports the TEM mode wave propagating in the y-direction with E_z and H_x-field components. The E_z-field component interacts with the CLS structure. *The CLS creates the Lorentz type electric plasma medium with resonance at the frequency* ω_{0e}. It is like a strip cut-wire medium. It could be modeled by the LC resonant circuit. The vertical strip wire of CLS provides the inductance L, and its two horizontal strip wires provide the capacitance C of the equivalent LC resonant circuit. Both the CLS and SRR structure are fabricated on the same substrate. Figure (21.15a) shows the simple DNG medium with an SRR between two CLS. The H_x-field component interacts with the SRR loop. *The SRR creates the Lorentz type magnetic plasma medium.*

The incident wave propagating in the y-direction of the parallel-plate waveguide has E_z and H_x-field components, exciting the electric and magnetic dipoles in the CLS and SRR, respectively. The dimensions of the EM-simulated CLS–SRR structure are available in the reference [J.42]. The complex refractive index response, shown in Fig (21.15b), is extracted from the S-parameter response. It clearly shows the negative refraction over a narrow frequency band around 9.5 GHz, and the medium is DNG type. At this frequency, the extracted value of the permittivity and permeability also exhibits negative value due to the Lorentz type response. The more elaborate arrangement of the structure has also been examined in the reference [J.42].

CLS–CLL Composite Element

Figure (21.15c) shows that the *capacitively loaded loop (CLL)* that has replaced the SRR of the CLS–SRR

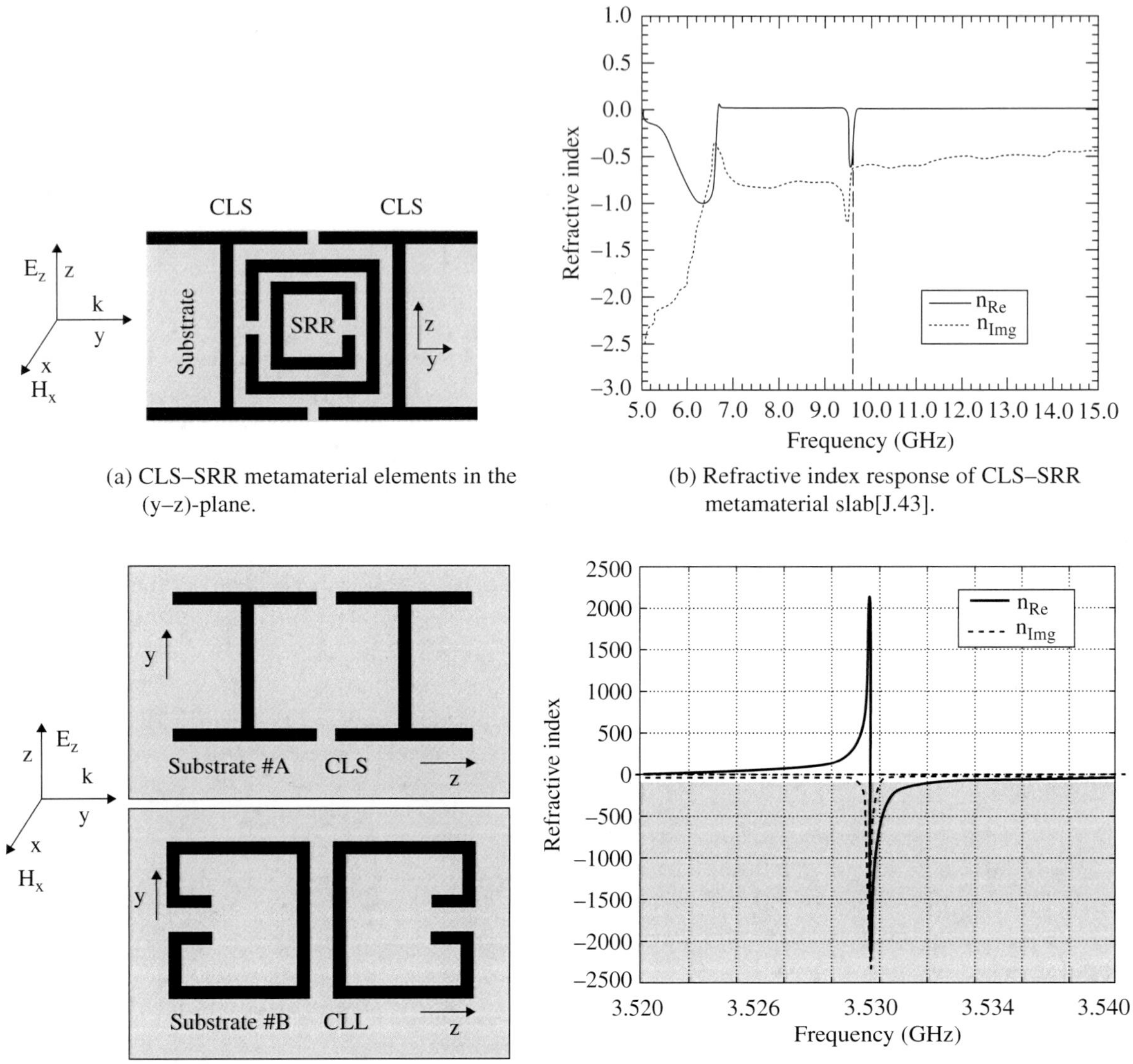

(a) CLS–SRR metamaterial elements in the (y–z)-plane.

(b) Refractive index response of CLS–SRR metamaterial slab[J.43].

(c) CLS–CLL metamaterial elements the (y–z) plane. The CLL is arranged behind the CLS on another substrate.

(d) Permittivity–permeability response of CLS–CLL metamaterial block [J.42].

Figure 21.15 CLS–SRR and CLS–CLL metamaterials responses of the material parameters. *Source:* Reproduced from Cheng and Ziolkowski [J.42], © 2003, IEEE). Reproduced from Ziolkowski [J.43]. © 2003, IEEE.

composite discussed above. The CLS and CLL are constructed on separate substrates #A and B. A block consisting of the alternate layers of the CLS on substrate #A and CLL on substrate #B synthesizes the DNG material. It is the *Lorentz–Lorentz arrangement*. The CLS–CLL composite structure, located in the (y–z)-plane, could be embedded in the substrate region of a microstrip line supporting the quasi-TEM mode wave propagating in the y-direction. The quasi-TEM mode wave has a normal E_z-field component and horizontal H_x-field component. The E_z-field component of the microstrip excites the CLS creating the *Lorentz type permittivity response*. The H_x-field component of the microstrip excites the CLL creating the *Lorentz type permeability response* [J.43]. The resonant refractive index, with its both real and imaginary parts, is shown in Fig (21.15d). The narrow bandwidth negative refractive index material is created at 3.5297 GHz. The details of the CLS–CLL structure are available in the reference [J.43]. The CLL has also been used to create the artificial magnetic conductor (AMC) as a reflector to enhance the gain of an antenna [J.44].

21.3.3 Realization of Single-Structure DNG Metamaterials

The single physical inclusions can also produce both the electric and magnetic response, creating the DNG medium. The composite double array of wire-SRR, over the frequency range $\omega_{0m} < \omega < \omega_{mp}$, forms the narrow bandwidth C-DNG metamaterials. It also suffers from high losses. The DNG and DPS media are separated by a bandgap. However, sometimes a seamless transition, from the DNG state to DPS state with a change in frequency, is needed for practical applications. It is normally obtained in the microstrip line structure using the composite right and left-handed (CRLH) structure, discussed in subsection (22.1.5) of chapter 22. The present subsection discusses the *magneto-electric particles, Ω-particle, and S-ring particles* with both the electric and magnetic responses excited by the incident EM-wave of suitable polarization. Both these particles are used to get the low-loss and wideband S-DNG type metamaterials.

Ω-Particle Metamaterials

Figure (21.16a-i) shows one element of the Ω-particle that is printed on a PCB card without backside metallization. It consists of two parts: the *C-structure*, responding to the magnetic field normal to its plane, and upper and lower connected *strip wires* (SW), responding to the electric field parallel to the SW. It is the *Drude–Lorentz arrangement* for the electric and magnetic responses. However, the metamaterial based on the Ω-particle has bi-anisotropic property, expressed through equation (21.2.12), like the SRR. The closed-form expressions for the bianisotropic polarizabilities have been obtained using the antenna model [J.45, J.46]. However, the electric and magnetic resonances are coupled for the Ω-particles-based metamaterials. The presence of the electric response can significantly influence the strength of the magnetic response in a way that can destroy its negative permeability [J.47]. Even the composite Ω-SW medium may not create the DNG medium. Sometimes, it creates the passband of a DPS medium [J.48].

The unit cell of the *broadside coupled (BC) Ω-particles*, shown in Fig (21.16a-ii), is a composite structure of reversely printed two Ω-particles on both sides of a substrate. This arrangement cancels the inherent bianisotropy of single Ω-particle. In the case of the BC Ω-particles, the electric and magnetic responses are fairly decoupled. Thus, the broadside coupled Ω-particle is a suitable structure for the synthesis of a low-loss and broadband metamaterials [J.49–J.51]. More numbers of BC Ω-particles, connected in series with lattice constant d in the y-direction, form one column of the metamaterial. The periodic repetition of the column in the x-direction, with a lattice constant d, forms the square lattice in the (y–x)-plane. The planar lattice is formed on a PCB substrate with (ε_r, tan δ), and thickness h. The periodic arrangement of the PCB cards in the z-direction forms the volumetric metamaterials, as shown in Fig (21.16b). However, one blank substrate is placed between two printed PCB to form the volumetric metamaterials. The whole assembly is *hot-pressed* to get the 3D solid metamaterials [J.49]. The E_y-polarized incident waves propagate in the x-direction. The Ey-field component

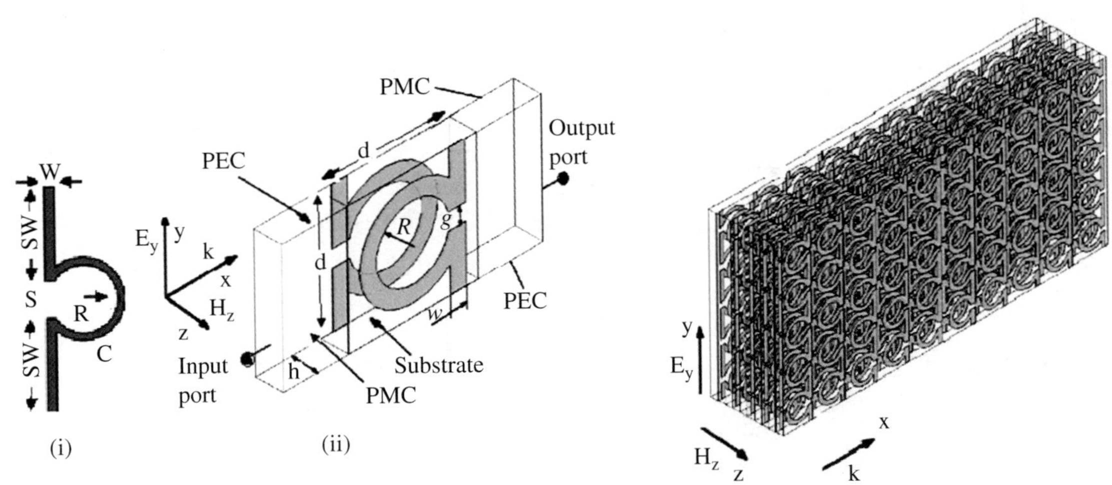

(a) Ω and broadside-coupled Ω-particles. (b) 2D-array of broadside coupled Ω-particles.

Figure 21.16 Broadside coupled Ω-particles-based metamaterial.

excites the *Drude type electric polarization*, and the H_z-field components excite the *Lorentz type magnetic polarization* in the metamaterial of Fig (21.16b).

To get the seamless DNG and DPS medium, without any bandgap, the structure must operate in the *balanced mode,* with identical electric and magnetic plasma frequencies, i.e. $\omega_{ep} = \omega_{mp}$. In the case $\omega_{ep} \neq \omega_{mp}$, the medium operates in the *unbalanced mode* and a bandgap appears between the DNG and DPS [J.51]. The magnetic resonance and magnetic plasma frequencies, ω_{0m} and ω_{mp}, can be estimated by the following expressions:

$$\omega_{0m} = \frac{1}{\sqrt{LC}} \quad (a), \quad \omega_{mp} = \frac{\omega_{0m}}{\sqrt{1-F}} \quad (b)$$

$$F = \frac{\pi R^2}{d^2} \quad (c)$$

$$C = \frac{C_p}{2} = \frac{\varepsilon_0 \varepsilon_r w}{2h}\left[\pi \frac{2R + w}{h} - \frac{g}{2}\right] \quad (d)$$

$$L = \frac{\mu_0 \pi R^2}{h} \quad (e),$$

(21.3.5)

where C_p is the upper and lower C-shape capacitance, treating the structure as the parallel plate capacitor. The dimensions are shown in Fig (21.16a-ii). The electric plasma frequency of the SW medium can be estimated from Pendry formula, or any other expression discussed in subsection (21.1.1). However, dimensions have to be fine-tuned on EM-simulator to equalize both plasma frequencies. It is illustrated below.

The BC Ω-structure-based metamaterial has been engineered on FR-4 substrate of thickness $h = 0.8$ mm and $\varepsilon_r = 4.0$, $\tan\delta = 0.02$. The strip width, slit-gap, radius R of C-section of the Ω-particle, and lattice constant d are as follows: $w = g = 0.5$ mm, $R = 1.1$ mm, $d = 3.9$ mm. The unit cell with front and back PMC, and top and bottom PEC are simulated in the EM-simulator to get the dispersion diagram shown in Fig (21.17a) [J.50]. The metamaterial supports the wideband, approximately from 8.0 to 12.8 GHz, backward wave. At the location P on the dispersion diagram, the local slope and the slope at origin O are oppositely directed supporting the backward wave. The medium also supports the forward wave approximately from 12.8 to 16 GHz, as the slopes at Q and O are in the same direction. So the seamless passband is obtained, between 8.0 and 16.0 GHz, from the DNG to the DPS medium. There is a cut-off frequency of around 8.0 GHz. It is confirmed by the experimental results shown in Fig (21.17b). The experiment has been conducted in the parallel plate waveguide environment [J.49, J.50, B.17, B.19]. It shows the transmission band from 8.3 to 16.0 GHz with minimum insertion loss (IL) of about 7.5 dB. However, the insertion loss of the DNG is high.

The refractive index, permittivity, and permeability of the metamaterial medium have been extracted by treating the unit cell as a two-port network. The extracted material parameters, from the de-embedded S-parameters, help to compute the propagation constant and the characteristic impedance of the medium. Figure (21.17c–f) shows the frequency-dependent extracted parameters. The negative refractive index, permittivity, and permeability are shown in light gray. The hatched area shows the lower cut-off frequency. The process of extraction of the material parameters is discussed in section (21.4).

S-ring Particle Metamaterials

The light gray card of Fig (21.18a-i) shows a column of oppositely printed pairs of S-rings on both sides of a substrate of relative permittivity ε_r and thickness h. It forms the *broadside coupled (BC) S-ring* particle. The cascading of the PCB cards with a blank separating substrate, or air-gap, forms a column. Other PCB cards and blank PCB cards are not shown in Fig (21.18a-i). The BC-S-ring is a self-resonant metallic structure. It is the *Lorentz–Lorentz type arrangement* exhibiting negative permittivity, and also negative permeability in the overlapping frequency band, so the negative refractive index is created over a wideband with small insertion loss. It is unlike the narrowband C-DNG, with higher insertion loss, obtained from the composite SW-SRR structure. The S-ring does not require electrical contact when mounted inside a parallel plate waveguide (PPW). It is a significant advantage for the fabrication and measurement of the metamaterials [J.52–J.55, B.11].

Figure (21.18a-i) shows, in gray, one unit cell of the BCS inclusion on a substrate. The unit cell has dimensions a, b, and d. The dimensions of the S-rings are as follows: conducting strip width c, height p, and S-ring width w. The front and back S-rings are separated by the substrate thickness h. The relative permeability of the structure has been obtained using the circuit model. The results for the identical upper and lower rings are summarized below in terms of the magnetic resonance frequency ω_{om} and the magnetic plasma frequency ω_{mp} [J.52, J.54]:

$$\mu_{reff} = 1 - \frac{2F + jA}{1 - (\omega_{om}/\omega)^2 - B + jG} \quad (a)$$

$$\text{where,} \quad \omega_{om}^2 = \frac{1}{\mu_0 abF(3d/C_s)} \quad (b)$$

$$\omega_{mp}^2 = \omega_{om}^2\left(\frac{1}{1-2F}\right) \quad (c)$$

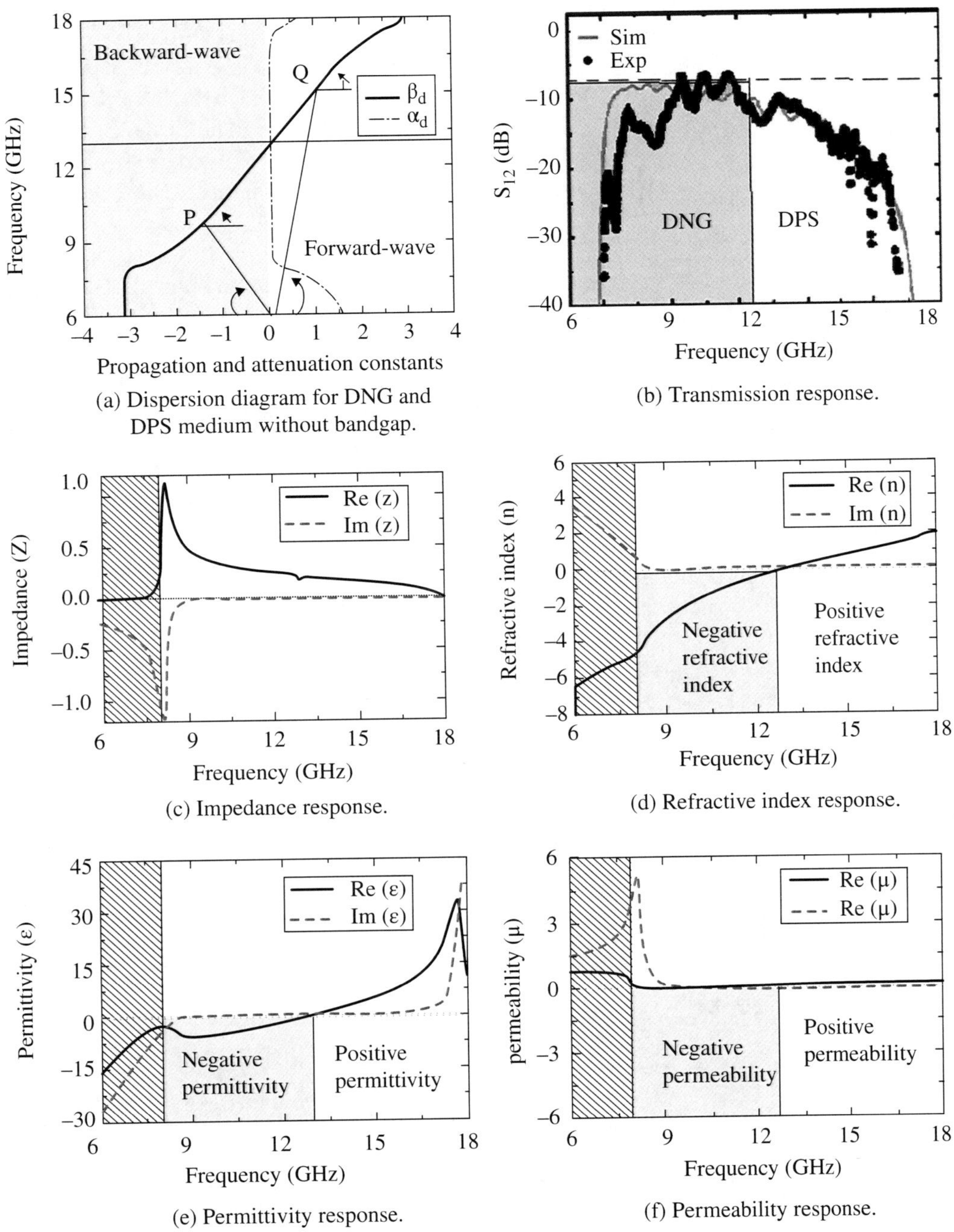

(a) Dispersion diagram for DNG and DPS medium without bandgap.

(b) Transmission response.

(c) Impedance response.

(d) Refractive index response.

(e) Permittivity response.

(f) Permeability response.

Figure 21.17 Responses of broadside coupled Ω-particles-based metamaterial. *Source*: Reproduced from Zhang et al. [J.50], with the permission of AIP Publishing. © 2008, AIP Publishing.

$$\text{Filling - factor } F = \frac{pw}{ab} \qquad \text{(d)}$$

$$C_s = \frac{\varepsilon_r \varepsilon_0 wc}{h} + \frac{\varepsilon_r \varepsilon_0 wc}{p-h} \qquad \text{(e).}$$

(21.3.6)

The remaining parameters, used in equation (21.3.6a), are given in the following expressions:

$$A = \frac{2\omega\mu_0 RF^2 pab}{(\omega\mu_0 Fab)^2[1 - 1/\omega\mu_0 FabC_s]} \qquad \text{(a)}$$

$$B = \frac{(Rd)^2}{(\omega\mu_0 Fab)^2[1 - 1/\omega\mu_0 FabC_s]} \qquad \text{(b)}$$

$$G = \frac{(Rd)\left[2\omega\mu_0 Fab - 2d^2/C_s\omega\right]}{(\omega\mu_0 Fab)^2[1 - 1/\omega\mu_0 FabC_s]} \qquad \text{(c).}$$

(21.3.7)

In equation (21.2.7), R is the resistance of the strip conductor. The magnetic resonance frequency of the S-loop is estimated by minimizing the imaginary part of the permeability function. The accuracy of the model has been tested in the X (8–12.5 GHz), Ku (12.5–18 GHz), and K (18—27 GHz) bands [J.52].

The permittivity of the S-ring rodded medium of Fig (21.18a-ii) follows the Lorentz model:

$$\varepsilon_{reff} = 1 - \frac{\omega_{ep}^2 - \omega_{oe}^2}{\omega^2 - \omega_{oe}^2 + j\gamma_e\omega}. \quad (21.3.8)$$

In the case of the single S-ring structure, the electric resonance frequency ω_{oe} is zero and the medium follows the Drude model. However, ω_{oe} occurs in the S-type cut-wire structure, shown in Fig (21.18a-ii). The S-type cut-wire structure shows the Lorentz electric response, and also the Lorentz magnetic response.

Figure (21.18c) shows the fabricated BC-S ring metamaterial on a low-loss thin R04003C substrate. It also shows the substrate details. For the X-Ku band the structural details are as follows: a = 5.2 mm, b = 4.0 mm, p = 5.0 mm, w = 2.8 mm, c = 0.4 mm, h = 0.508 mm, d = 2.0 mm. The size of the UC and S-ring are less than λ/5. The 10 × 50 UC are fabricated on each PCB card and 120 equally spaced cards are stacked to fabricate the 3D-metamaterial. It is used to construct a prism for the refractive index measurement of the engineered metamaterial. Figure (21.18d) shows the far-field free space transmission measurement in the X-Ku band. It shows wideband, 11.5–13.5 GHz, transmission due to both the DPS–DNG nature without any bandgap. The nature of the transmission band has been examined through the refractive index measurement by using the metamaterial in the prism form. Figure (21.18b) shows the dispersive nature of the positive and negative refractive indices without any bandgap. The resonance frequency is at 12.5 GHz. The negative refraction occurs in the frequency range 12.5–15 GHz, i.e. over the 2.5 GHz bandwidth. It also shows high-valued positive refractive index. The insertion loss of the metamaterial is 1.69 dB/cm (0.422 dB/cell).

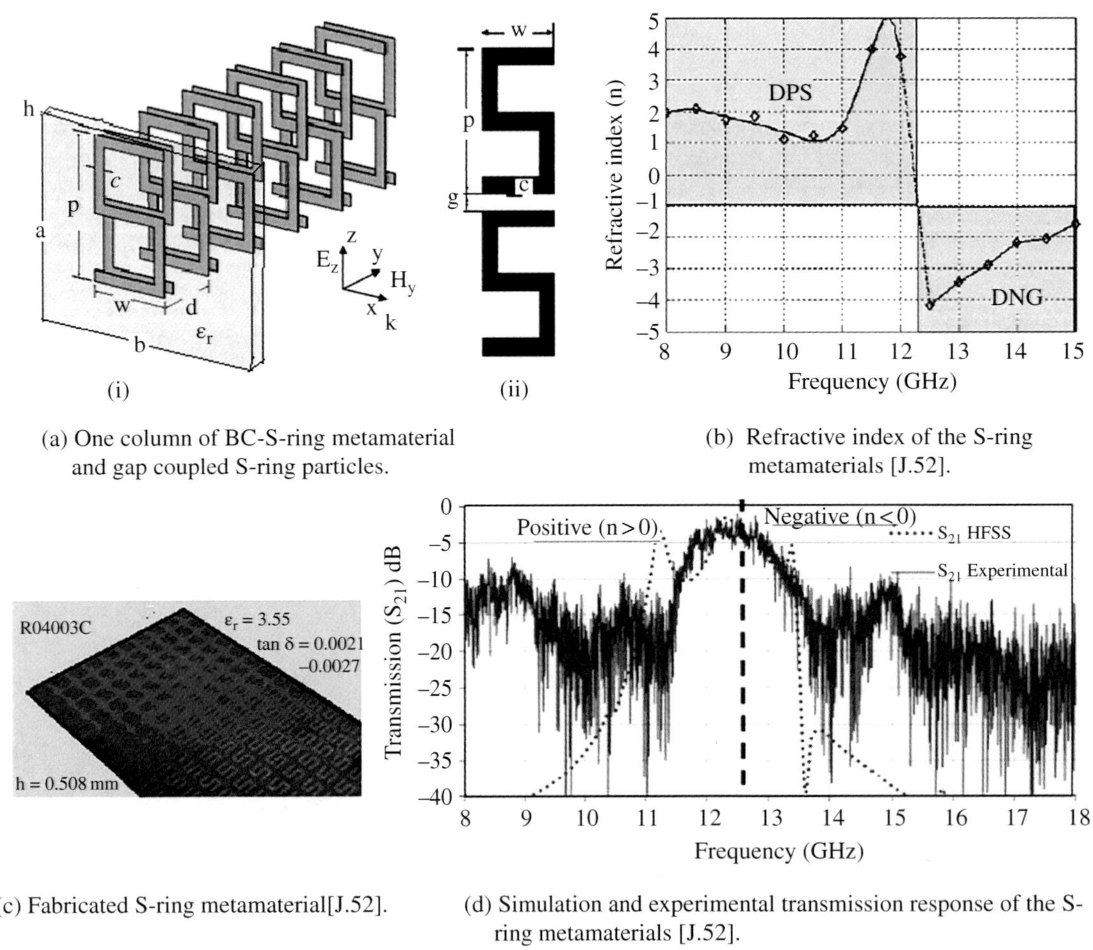

(a) One column of BC-S-ring metamaterial and gap coupled S-ring particles.

(b) Refractive index of the S-ring metamaterials [J.52].

(c) Fabricated S-ring metamaterial[J.52].

(d) Simulation and experimental transmission response of the S-ring metamaterials [J.52].

Figure 21.18 Performance of the broadside coupled S-ring-based metamaterials. *Source:* Reproduced from Lee et al. [J.52], with the permission of AIP Publishing.© 2015, AIP Publishing.

The BC S-rings are also used on the artificially engineered magnetic substrate to enhance the DNG bandwidth. Figure (21.19a) shows the complete metamaterial on the *artificial magnetic substrate, i.e. the meta-substrate,* that is created by the metallic closed-rings on an FR-4 substrate with $\varepsilon_r = 4.0$, $h = 1.0$ mm, $\mu_1 = \mu_0$. The periodic unit cell in the (x–z)-plane is 4.0 mm × 4.0 mm. It has periodicity 2.5 mm in the y-direction. The ring size is $r_{in} = 3.2$ mm and $r_{out} = 3.6$ mm. The fractional filling-factor $F_c = 0.57$ of the closed-rings creates the artificial magnetic substrate. The artificially acquired permeability of the engineered magnetic substrate is $\mu_2 = (1 - F_c)\ \mu_0 = 0.43\mu_0$.

The BC S-ring creates the metamaterial on the magnetic meta-substrate. The relative permeability of the metamaterial is computed by using the following expression [J.55]:

$$\mu_{reff} = \left[(1-F_c)(1-F) - \frac{F}{1-(\omega_{om}/\omega)^2 + jR/\omega L}\right] \quad (a)$$

$$\omega_{mp} = \omega_{om}\left[1 + \frac{1}{(1-F_c)}\frac{F}{(1-F)}\right] \quad (b)$$

$$\omega_{om} = \frac{1}{\sqrt{LC}} \quad (c). \qquad (21.3.9)$$

In equation (21.2.9), F is the filling factor of the BC S-ring of size, p = 10.8 mm, w = 5.6 mm, c = 0.8 mm,

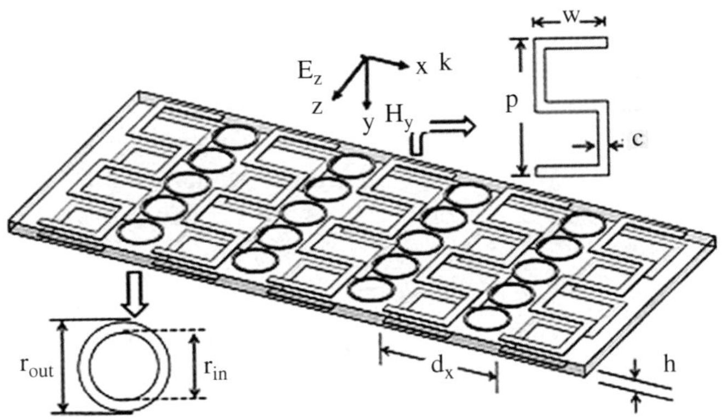

(a) The series-connected BC S-ring particles metamaterials on an artificial magnetic substrate.

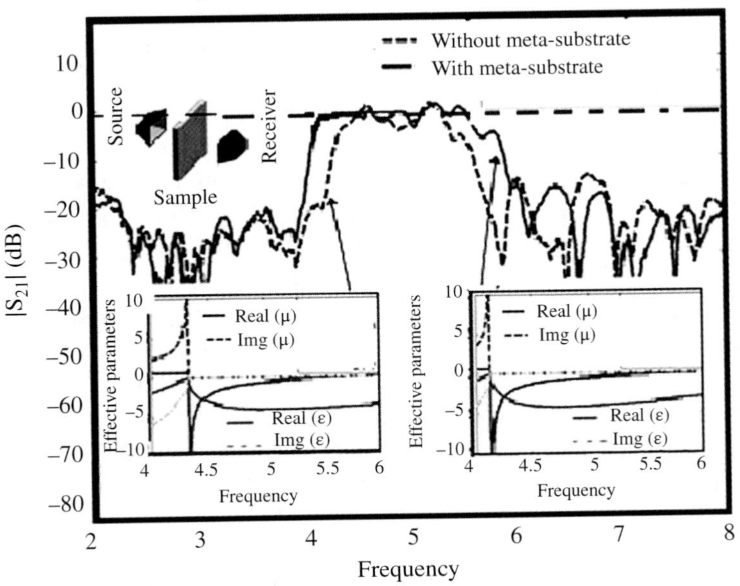

(b) Transmission response of the BC S-ring type metamaterials on the conventional and artificial magnetic substrates.

Figure 21.19 Performance of the broadside coupled S-ring particle-based metamaterials on an artificial magnetic substrate. *Source:* Chen et al. [J.55]. © 2014, Springer Nature.

$d_x = 10.0$ mm. The structure has 10 UC along the direction of propagation. Figure (21.19b) shows the experimental results of the transmission measurement in free space. It also shows the extracted permeability and permittivity responses in insets. The parameters are extracted from the measured S-parameters response of the engineered slab. Both the parameters are negative forming the DNG over the frequency band 4.29–5.61 GHz (30.8%) for the BC S-rings on the conventional FR-4 substrate, whereas it is 44.9% (4.08–5.91 GHz) for the BC S-rings on the meta-substrate. The theoretical bandwidths for the conventional and meta-substrate, using the above equations, are 28 and 44%, respectively. Figure (21.19b) also exhibits a small insertion loss of the S-ring metamaterial.

21.4 Homogenization and Parameters Extraction

The above discussion has treated the periodically arranged discrete metallic structure-based metamaterial medium as a *homogeneous medium* under the condition $kd \ll 1$, where k is wavenumber inside the metamaterial and d is the lattice constant of a unit cell. Thus, the hypothetical electromagnetically continuous medium, forming an *effective medium*, can replace the discrete arrangement of the periodic inclusions in the host medium. The effective medium is described by the primary material constants: complex effective relative permittivity and complex effective relative permeability, or complex refractive index. The medium is also described by the equivalent secondary material parameters: the complex intrinsic impedance and complex propagation constant, i.e. the wavenumber.

At the molecular level, the natural materials are also discrete. Similarly, the metamaterials are based on the discrete metallic and nonmetallic inclusions acting as the artificial atoms or artificial molecules. The permittivity and permeability of the *effective homogeneous medium* created by these discrete inclusions could be numerically computed from the computation of the average field over the unit cell [J.3, J.56, J.57]. However, the numerical computation process is demanding and inefficient. In practice, these parameters are extracted, i.e. retrieved, from the measured S-parameters. The experimental S-parameters of a slab of the metamaterial is obtained from the *free space measurement* [J.43, J.58], by mounting the slab in a *parallel plate waveguide* (PPW) [J.59, B.11], or mounting the slab in the TE_{10} mode supporting *rectangular waveguide* [J.60–J.62]. Such characterization of the metamaterials is useful for their practical applications. The S-parameters of the inhomogeneous engineered metamaterials is also obtained using the EM-simulators [J.43, J.63–J.67]. Therefore, the homogenization of the metamaterials is realized through the S-*parameter method*. The homogenization achieved using the circuit models, discussed above, is approximate, whereas the homogenization obtained from the S-parameter method is more accurate and dynamic. The metamaterial could also be treated as a mixture of inclusions in a host medium. The Maxwell Garnett mixture model of homogenization, discussed in sections (6.3.3) and (6.5.3) of chapter 6 has been extended to metamaterials based on the Mie scattering-based inclusions.

This section presents a brief discussion on the *parameter extraction*, i.e. the *parameter retrieval* from EM-simulated, or experimentally obtained, S-parameters. This section further presents the extension of the Maxwell Garnett mixture model to the metamaterials to get the dynamic frequency-dependent parameters of the homogenized medium. The interested reader should consult the original literature for the detailed discussions [J.60–J.65, J.68–J.74].

21.4.1 Nicolson–Ross–Weir (NRW) Method

The classical NRW method has been used to obtain the complex relative permittivity and complex relative permeability of a homogeneous isotropic material slab [J.60, J.75]. The method has been further extended to get these parameters of the homogenized metamaterials [J.43, J.62–J.66]. Figure (21.20a) shows that the *inhomogeneous metamaterial* slab of thickness d could be replaced by the *homogenized equivalent metamaterial* slab of identical thickness d such that both have almost identical S-parameters response, both in the magnitude and phase. *We can characterize the equivalent homogeneous slab by the effective complex relative permittivity and permeability*. These parameters are extracted from the EM-simulated/measured de-embedded complex S-parameters, i.e. magnitude and phase response of the S-parameters. Inside the material slab of finite thickness *d*, multiple reflections take place. However, these multiple reflections finally provide the reflection and transmission at the input–output interfaces of the slab. The NRW is the three stages process discussed below:

At the *first stage* of the parameter retrieval, the reflection and transmission coefficients are obtained at the interfaces of the slab-air from the measured/simulated de-embedded S-parameters. At the *second stage*, the normalized intrinsic impedance and the refractive index of

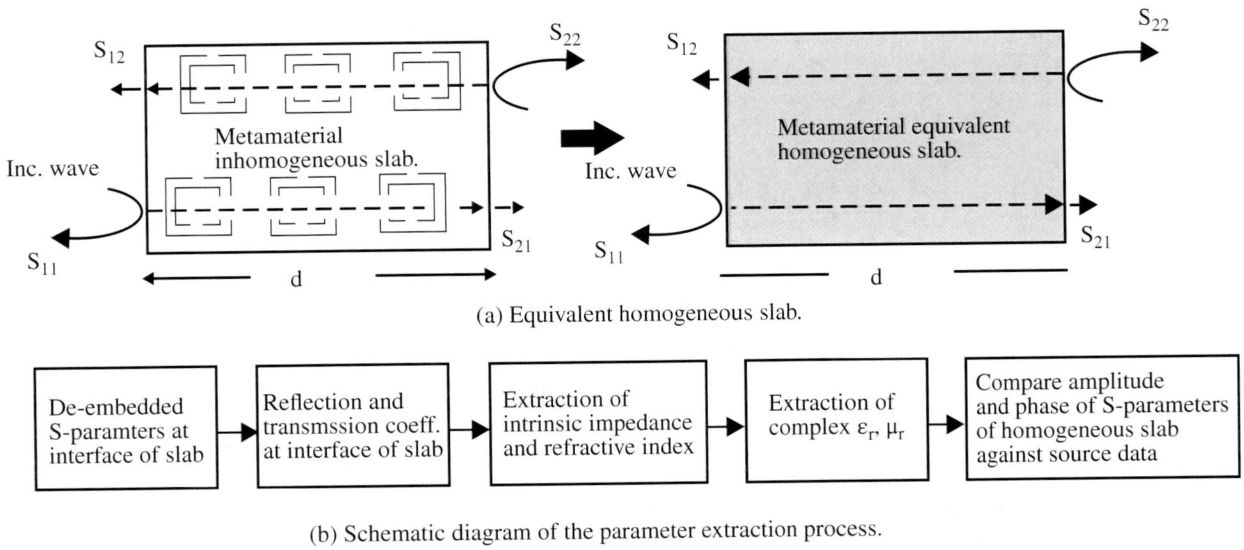

Figure 21.20 Homogenization and extraction process of material parameters.

the slab are obtained. Finally, at the *third stage*, the complex relative permittivity and complex relative permeability of the slab are extracted. Such parameters extraction, at the discrete frequencies, is carried out over a frequency band. The schematic diagram of Fig (21.20b) shows the extraction process and verification of its accuracy. To verify the correctness of the parameter extraction, the extracted metamaterial parameters are used to obtain the S-parameters of the metamaterial slab. The computed S-parameters are compared, both magnitude and phase, against the original S-parameters available at the first stage. The useful expressions are obtained below for the extraction process as applied to both the homogeneous and inhomogeneous slabs in the free space or mounted in the TEM type PPW.

Computation of Relative Permittivity and Permeability of the Slab

Figure (21.21a–c) shows three arrangements of UC of the homogeneous and inhomogeneous material slabs of thickness d. Under the condition $kd \ll 1$, a metamaterial slab of thickness d could be replaced by the equivalent homogeneous isotropic material slab of the same thickness. The wavenumber within the unit cell is k.

Figure (21.21a) shows a slab of the homogeneous, symmetric, and reciprocal structure with $S_{11} = S_{22}$, $S_{21} = S_{12}$. The relative permittivity and permeability of the *equivalent medium* can be extracted from the de-embedded S-parameters at two faces of the slab. To obtain these material parameters, the intrinsic impedance/Bloch characteristic impedance and refractive index/propagation constant of the equivalent homogeneous slab is computed from the S-parameters. Figure (21.21b) shows an inhomogeneous, isotropic, and asymmetric structure with $S_{11} \neq S_{22}$. In this, the Bloch characteristic impedance of the equivalent medium is not uniquely defined, so the relative permittivity and permeability cannot be extracted uniquely. However, Fig (21.21c) shows an inhomogeneous, isotropic, and symmetric slab with $S_{11} = S_{22}$, $S_{21} = S_{12}$. In this case, the Bloch characteristic impedance is uniquely defined and the relative permittivity and permeability of the equivalent homogeneous medium can be extracted uniquely. The slab can also consist of several layers of the material sheets and effective material parameters could be extracted for such a structure also [J.76]. The periodic arrangement of UC can be treated either through the transfer matrix method (TMM) [J.64] or by using the [ABCD] formulation discussed in subsection (19.3.2) of chapter 19.

The needed expressions are presented below for extracting the material parameters of the slab, shown in Fig (21.21a). The expressions correlate the material parameters with the reflection coefficient Γ and the transmission coefficient τ at the interfaces A–A and B–B of the slab of thickness d. The slab material has an intrinsic impedance η_m, refractive index n, relative permittivity ε_r, and relative permeability μ_r. Normally, these are complex quantities due to the losses. The *normalized intrinsic impedance* of the slab is $\eta_m^{nor} = \eta_m/\eta_0 = \sqrt{\mu_r/\varepsilon_r}$. In the case of a semiinfinite slab ($d \to \infty$), the reflection coefficient at the interface A–A and the intrinsic impedance are related as follows:

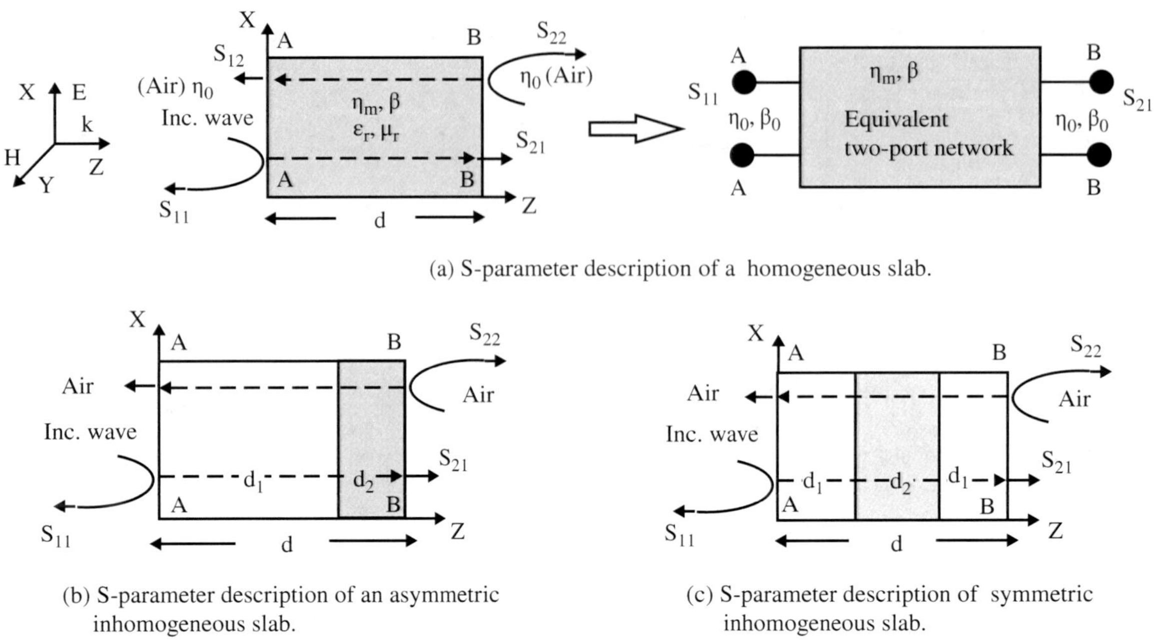

(a) S-parameter description of a homogeneous slab.

(b) S-parameter description of an asymmetric inhomogeneous slab.

(c) S-parameter description of symmetric inhomogeneous slab.

Figure 21.21 S-parameters descriptions of a unit cell of a slab.

$$\Gamma = \frac{\eta_m - \eta_0}{\eta_m + \eta_0} = \frac{\eta_m^{nor} - 1}{\eta_m^{nor} + 1} = \frac{\sqrt{\mu_r/\varepsilon_r} - 1}{\sqrt{\mu_r/\varepsilon_r} + 1} \qquad \text{(a)}$$

$$\eta_m = \eta_0 \frac{\Gamma + 1}{\Gamma - 1} \qquad \text{(b)}. \qquad (21.4.1)$$

In equations (21.4.1a, b), η_0 is the intrinsic impedance of free space. The transmission coefficient at the interface B–B of the slab of thickness d is

$$\begin{aligned} \tau &= e^{-j\beta d} = e^{-j\beta_0\sqrt{\mu_r\varepsilon_r}d} \\ &= e^{-j\beta_0 nd} = e^{-j\beta_0 n' d} e^{-n''\beta_0 d} \\ \Rightarrow \tau &= e^{-j(\beta_0 n' d - 2\pi m)} e^{-n''\beta_0 d} \\ m &= 0, \pm 1, \pm 2, \ldots, \end{aligned} \qquad (21.4.2)$$

where Z-directed propagation constant of the slab of the complex refractive index $n = n' - jn''$ is $\beta = k = \beta_0 n$, and $\beta_0 = k_0$ is the wavenumber in free space. The passive medium has $|n''| > 0$; so that the wave must decay in the direction of propagation. The DPS medium has $|n'| > 0$, giving the *lagging phase* in the direction of propagation. However, the DNG medium has $|n'| < 0$, giving the *leading phase* in the direction of propagation.

NRW Method

Once the reflection and transmission coefficients are known, the *Nicolson–Ross–Weir (NRW)* method [J.60, J.64] helps to extract the complex relative permeability and permittivity of the slab, through the extracted intrinsic impedance and refractive index. The complex intrinsic impedance is obtained from the reflection coefficient Γ using (21.4.1b), whereas the complex refractive index is obtained from the transmission coefficient τ using an equation (21.4.2). However, there is difficulty in extracting a unique value of the refractive index, as the exponential function with j sign is a periodic function giving multiple solutions.

On taking the logarithmic value of the complex transmission coefficient:

$$\begin{aligned} \ln(\tau) &= \mathrm{Re}\{\ln(\tau)\} - j\mathrm{Img}\{\ln(\tau)\} \\ &= -j(\beta_0 n' d - 2\pi m) - n''\beta_0 d. \end{aligned}$$

On separating the real and imaginary parts of the transmission coefficient, the following expressions are obtained:

$$n'' = -\frac{1}{\beta_0 d}\mathrm{Re}\{\ln(\tau)\} \qquad \text{(a)}$$

$$n' = \frac{1}{\beta_0 d}[\mathrm{Img}\{\ln(\tau)\} + 2\pi m]$$

$$\text{where, } m = 0, \pm 1, \pm 2, \ldots. \qquad \text{(b)}. \qquad (21.4.3)$$

It is obvious from equations (21.4.3a, b) that the imaginary part of the refractive index n'' is uniquely determined. However, the value of the real part of the refractive index n' depends on the correct choice of integer m. This is known as the *branching problem* [J.65]. It is not a trivial issue. The choice of m depends on the slab thickness d. However, for the slab thickness, $d < \lambda/2$ the *principal value* $m = 0$ is taken. Even in this case, the wave length λ inside the unit cell is unknown, as the material parameters of the slab are unknown. The branching problem is discussed separately. Assuming that the complex intrinsic impedance and refractive index have been computed correctly, the material parameters are determined as follows:

$$\eta_m = \eta_0\sqrt{\frac{\mu_r}{\varepsilon_r}}, \Rightarrow \frac{\mu_r}{\varepsilon_r} = \left(\frac{\eta_m}{\eta_0}\right)^2 \text{ (a)}, \quad n = \sqrt{\mu_r\varepsilon_r} \quad \text{(b)}$$
$$\mu_r = n\left(\frac{\eta_m}{\eta_0}\right) = n\eta_m^{nor} \quad \text{(c)}, \quad \varepsilon_r = n\left(\frac{\eta_0}{\eta_m}\right) = \frac{n}{\eta_m^{nor}} \quad \text{(d)}. \tag{21.4.4}$$

The real and imaginary parts of the complex intrinsic impedance and the refractive index of a passive medium should satisfy the following *passivity condition* for the complex relative permeability $\mu_r^* = \mu_r' - j\mu_r''$ and complex relative permittivity $\varepsilon_r^* = \varepsilon_r' - j\varepsilon_r''$:

$$\mu_r = (n' - jn'')\left(\eta'^{nor}_m - j\eta''^{nor}_m\right) = \left(n'\eta'^{nor}_m - n''\eta''^{nor}_m\right) - j\left(n''\eta'^{nor}_m + n'\eta''^{nor}_m\right)$$
$$\mathrm{Re}(\mu_r) = |\mu_r'| = \begin{cases} \text{for DPS medium:} \\ \left(n'\eta'^{nor}_m - n''\eta''^{nor}_m\right) > 0 \\ \text{for DNG medium:} \\ \left(n'\eta'^{nor}_m - n''\eta''^{nor}_m\right) < 0 \end{cases} \quad \text{(a)}$$
for both DPS and DNG media:
$$\mathrm{Img}(\mu_r) = |\mu_r''| = \left(n''\eta'^{nor}_m + n'\eta''^{nor}_m\right) > 0 \quad \text{(b)}. \tag{21.4.5}$$

Similarly,

$$\varepsilon_r = \frac{(n' - jn'')}{\left(\eta'^{nor}_m - j\eta''^{nor}_m\right)} = \frac{1}{|\eta_m^{nor}|^2}\left[\left(n'\eta'^{nor}_m + n''\eta''^{nor}_m\right) - j\left(n''\eta'^{nor}_m - n'\eta''^{nor}_m\right)\right]$$
$$\mathrm{Re}(\varepsilon_r) = |\varepsilon_r'| = \begin{cases} \frac{1}{|\eta_m^{nor}|^2}\left(n'\eta'^{nor}_m + n''\eta''^{nor}_m\right) > 0, & \text{for DPS medium} \\ \frac{1}{|\eta_m^{nor}|^2}\left(n'\eta'^{nor}_m + n''\eta''^{nor}_m\right) < 0, & \text{for DNG medium} \end{cases} \quad \text{(a)} \tag{21.4.6}$$
$$\mathrm{Img}(\varepsilon_r) = |\varepsilon_r''| = \frac{1}{|\eta_m^{nor}|^2}\left(n''\eta'^{nor}_m - n'\eta''^{nor}_m\right) > 0, \quad \text{for both DPS and DNG media} \quad \text{(b)}.$$

Therefore, the imaginary parts of the extracted μ_r and ε_r of the DNG should meet the following *passivity condition*:

$$\left|n'\eta''^{nor}_m\right| \le n''\eta'^{nor}_m. \tag{21.4.7}$$

In equation (21.4.7), n'' and η'^{nor}_m are positive quantities for the wave propagation in a lossy medium.

The Relation between S-parameters and Reflection–Transmission Coefficients

The S-parameters of the isotropic symmetric slab, $S_{11} = S_{22}$, $S_{21} = S_{12}$ at the interfaces A–A and B–B are obtained from equation (5.4.15) of chapter 5, equations (21.4.1), and (21.4.2):

$$S_{11} = \left[\frac{\eta_m - \eta_0}{\eta_m + \eta_0}\right]\left[\frac{1 - e^{-j\beta d}}{1 - \left\{(\eta_m - \eta_0)^2/(\eta_m + \eta_0)^2 e^{-j2\beta d}\right\}}\right] = \frac{(1 - \tau^2)\Gamma}{1 - \Gamma^2\tau^2} \quad \text{(a)}$$
$$S_{21} = \frac{4\eta_0\eta_m}{(\eta_m + \eta_0)^2}\left[\frac{e^{-j\beta d}}{1 - \left\{(\eta_m - \eta_0)^2/(\eta_m + \eta_0)^2 e^{-j2\beta d}\right\}}\right] = \frac{(1 - \Gamma^2)\tau}{1 - \Gamma^2\tau^2} \quad \text{(b)}$$
$$\text{where, } 1 - \Gamma^2 = 1 - \left[\frac{\eta_m - \eta_0}{\eta_m + \eta_0}\right]^2 = \frac{4\eta_0\eta_m}{(\eta_m + \eta_0)^2} \quad \text{(c)}. \tag{21.4.8}$$

To use the NRW method for extracting the material parameters, expressions (21.4.8a–c) have to be inverted, i.e. the reflection Γ and transmission τ coefficients have to be expressed in terms of the *known (measured/EM-simulated) de-embedded S-parameters*. The following variables are introduced for the ease of inversion:

$$V_1 = S_{21} + S_{11} \quad \text{(a)}$$
$$V_2 = S_{21} - S_{11} \quad \text{(b)}$$
$$\text{Define another variable, } X = \frac{1 - V_1V_2}{V_1 - V_2} \quad \text{(c)}$$
$$\Rightarrow X = \frac{1 - S_{21}^2 + S_{11}^2}{2S_{11}} \quad \text{(d)}. \tag{21.4.9}$$

By using equations (21.4.8a, b), expressions (21.4.9a–d) for the variables V_1, V_2, and X are rewritten as follows:

$$V_1 = \frac{\tau + \Gamma}{1 + \Gamma\tau} \text{ (a)}, \quad V_2 = \frac{\tau - \Gamma}{1 - \Gamma\tau} \quad \text{(b)}$$
$$X = \frac{1 - \{(\tau + \Gamma)/(1 + \Gamma\tau)\}\{(\tau - \Gamma)/(1 - \Gamma\tau)\}}{\{(\tau + \Gamma)/(1 + \Gamma\tau)\} - \{(\tau - \Gamma)/(1 - \Gamma\tau)\}}$$
$$\Rightarrow X = \frac{1 + \Gamma^2}{2\Gamma} \quad \text{(c)}$$
$$\Rightarrow \Gamma^2 - 2X\Gamma + 1 = 0, \quad \Rightarrow \Gamma = X \pm \sqrt{X^2 - 1} \quad \text{(d)}. \tag{21.4.10}$$

The above expressions (21.4.10b–d) are summarized below to compute the reflection and transmission coefficients in terms of the known S-parameters:

$$\Gamma = X \pm \sqrt{X^2-1} \quad (a), \quad \tau = \frac{V_1-\Gamma}{1-V_1\Gamma} = \frac{S_{11}+S_{21}-\Gamma}{1-(S_{11}+S_{21})\Gamma} \quad (b)$$
$$\text{where,}\ X = \frac{1-S_{21}^2+S_{11}^2}{2S_{11}} \quad (c), \quad V_1 = S_{21}+S_{11} \quad (d). \qquad (21.4.11)$$

Extraction of Refractive Index and Intrinsic Impedance of Equivalent Homogeneous Slab

The expression for the transmission coefficient τ to compute the refractive index of the slab can be rearranged in terms of the measurable S-parameters by using equation (21.4.8) [J.65]:

$$\Gamma S_{11} = \frac{(1-\tau^2)\Gamma^2}{1-\Gamma^2\tau^2} = \frac{\Gamma^2-\Gamma^2\tau+1-1}{1-\Gamma^2\tau^2}$$
$$= 1-\frac{1-\Gamma^2}{1-\Gamma^2\tau^2} = 1-\frac{S_{21}}{\tau} \qquad (21.4.12)$$
$$\Rightarrow \tau = e^{-j\beta_0 nd} = \frac{S_{21}}{1-\Gamma S_{11}}.$$

where Γ is obtained from equation (21.4.11a). Once the transmission coefficient τ is obtained from the S-parameters, the complex refractive index n is extracted from equation (21.4.2).

The next task is to extract the intrinsic impedance of the equivalent homogeneous slab from the known S-parameters. Using equation (21.4.10c), the following expression for the normalized intrinsic impedance η_m^{nor} of the equivalent slab is obtained:

$$X+1 = \frac{(1+\Gamma)^2}{2\Gamma} \quad (a), \quad X-1 = \frac{(1-\Gamma)^2}{2\Gamma} \quad (b)$$
$$\frac{X+1}{X-1} = \left[\frac{1+\Gamma}{1-\Gamma}\right]^2 = \left(\eta_m^{nor}\right)^2 \quad (c)$$
$$\Rightarrow \eta_m^{nor} = \pm\left[\frac{X+1}{X-1}\right]^{1/2} = \pm\left[\frac{(S_{11}+1)^2-S_{21}^2}{(S_{11}-1)^2-S_{21}^2}\right]^{1/2} \quad (d). \qquad (21.4.13)$$

The sign ($\pm$) of the intrinsic impedance is selected such that the passivity requirement for the reflection coefficient $|\Gamma| \le 1$ is met. Moreover, a correct sign of η_m^{nor} also provides the direction of energy flow, i.e. the direction of the group velocity of the EM-wave for both the DPS and DNG media.

Equation (21.4.10d) gives two solutions for the reflection coefficient Γ. It is known as another branching problem. However, once the correct solution for the reflection coefficient is resolved, using the condition $|\Gamma| \le 1$, the intrinsic impedance of the slab η_m is obtained from equation (21.4.1b). The transmission coefficient τ is computed using equation (21.4.11b). Finally, the complex refractive index of the equivalent slab is extracted using equation (21.4.2). However, before discussing it, an alternate set of equations is obtained for the propagation constant and Bloch impedance of the periodic arrangement of the slab.

Alternate Sets of Expressions for Parameters Retrieval

Figure (21.21a) shows the unit cell of a periodic arrangement of slabs constituting the metamaterials. The symmetric unit cell of Figure (21.21c) could be treated as a homogeneous medium that is further treated as a two-port network also, shown in Fig (21.21a). It is analyzed using the normalized [ABCD] parameters. It is similar to the transfer matrix method [J.63, J.64]. The propagation characteristic of the periodic structure, described in terms of the [ABCD] parameters, is discussed in subsection (19.3.2) of chapter 19. For the symmetrical equivalent two-ports networks the results are summarized below:

$$\begin{bmatrix} A & B \\ C & D \end{bmatrix} = \begin{bmatrix} \cos\beta d & j\eta_m \sin\beta d \\ j(1/\eta_m)\sin\beta d & \cos\beta d \end{bmatrix} \quad (a)$$
$$\cos\beta d = A \quad (b)$$
$$\text{Bloch impedamce}\ Z_B = \eta_m = \frac{\pm\eta_{0B}}{\sqrt{A^2-1}} = \pm\eta_0\sqrt{\frac{B}{C}} \quad (c). \qquad (21.4.14)$$

In equations (21.4.14a–c), β is the propagation constant in the equivalent homogeneous medium, and the Bloch impedance is obtained from equation (19.3.10b) or equation (19.3.20) of chapter 19. The S-parameters of the symmetrical 2-port networks are also summarized below [B.13]:

$$S_{21} = \frac{2}{A+B/Z_0+CZ_0+D}, \quad (Z_0=\eta_0, A=D) \quad (a)$$
$$S_{11} = \frac{B/Z_0-CZ_0}{A+B/Z_0+CZ_0+D} \quad (b). \qquad (21.4.15)$$

For the present symmetrical slab, the S-parameters are obtained from equations (21.4.15 a, b) and (21.4.14):

$$S_{21} = \frac{1}{\cos\beta d + j/2\{(\eta_m/\eta_0)+(\eta_0/\eta_m)\}\sin\beta d} \quad (a)$$
$$S_{11} = \frac{j\{(\eta_m/\eta_0)-(\eta_0/\eta_m)\}\sin\beta d}{2[\cos\beta d + j/2\{(\eta_m/\eta_0)+(\eta_0/\eta_m)\}\sin\beta d]} \quad (b). \qquad (21.4.16)$$

Expressions (21.4.16a, b) are also written using the normalized intrinsic impedance of the slab:

$$S_{21} = \frac{1}{\cos\beta d + (j/2)\left[\eta_m^{nor} + 1/\eta_m^{nor}\right]\sin\beta d} \qquad (a)$$

$$S_{11} = \frac{j}{2}\frac{\left[\eta_m^{nor} - 1/\eta_m^{nor}\right]\sin\beta d}{\left[\cos\beta d + (j/2)\left\{\eta_m^{nor} + 1/\eta_m^{nor}\right\}\sin\beta d\right]} \qquad (b)$$

$$\Rightarrow S_{11} = \frac{j}{2}S_{21}\left[\eta_m^{nor} - 1/\eta_m^{nor}\right]\sin\beta d \qquad (c). \qquad (21.4.17)$$

Normally, expressions (21.4.17a–c) are used to compute the S-parameters of a unit cell once the material parameters, leading to η_m^{nor} and β, are known. However, for the extraction of material parameters, expressions (21.4.17a–c) have to be inverted to get the normalized intrinsic impedance of material slab η_m^{nor} and the propagation constant β in terms of the known S_{11} and S_{21}. It is achieved using the following [ABCD] parameters in terms of the *normalized S-parameters* [B.13]:

$$A = \frac{(1 + S_{11})(1 - S_{22}) + S_{21}S_{12}}{2S_{21}}$$
$$B = \frac{(1 + S_{11})(1 + S_{22}) - S_{21}S_{12}}{2S_{21}}$$
$$C = \frac{(1 - S_{11})(1 - S_{22}) - S_{21}S_{12}}{2S_{21}}$$
$$D = \frac{(1 - S_{11})(1 + S_{22}) + S_{21}S_{12}}{2S_{21}}. \qquad (21.4.18)$$

The symmetrical slab unit cell has $S_{11} = S_{22}$, and $S_{21} = S_{12}$. Using expressions (21.4.18) with equations (21.4.14b,c), the following expressions are obtained for the propagation constant and *normalized intrinsic impedance of* the equivalent homogeneous medium:

$$\cos\beta d = \frac{1 - S_{11}^2 + S_{21}^2}{2S_{21}} \qquad (a)$$

$$\eta_m^{nor} = \frac{\eta_m}{\eta_0} = \pm\sqrt{\frac{(1 + S_{11})^2 - S_{21}^2}{(1 - S_{11})^2 - S_{21}^2}} \qquad (b). \qquad (21.4.19)$$

Equation (21.4.19b) is same as equation (21.4.13d). The real and imaginary parts of the refractive index of the slab can be determined from equation (21.4.19a):

$$n'' = \text{Img}(n) = \pm\,\text{Img}\left\{\frac{\cos^{-1}\left[\left(1 - S_{11}^2 + S_{21}^2\right)/2S_{21}\right]}{\beta d}\right\} \qquad (a)$$

$$n' = \text{Re}(n) = \pm\,\text{Re}\left\{\frac{\cos^{-1}\left[\left(1 - S_{11}^2 + S_{21}^2\right)/2S_{21}\right]}{\beta d}\right\} + \frac{2\pi m}{\beta d}, \quad m = 0, \pm 1, \pm 2, \ldots \qquad (b). \qquad (21.4.20)$$

In this case, the imaginary part of the refractive index n'' is determined uniquely. However, the branching problem, due to the ambiguity in the value of *m*, has to be resolved for the correct extraction of the real part of the refractive index n'. The *m* is known as the *branching index.*

Resolving Branching Problem

The correct extraction of the material parameters of a slab of thickness *d* is influenced by two factors: locations of two air-slab interfaces, and resolution of the *branching index m* encountered in equations (21.4.3b) and (21.4.20b). Once correct S-parameters are obtained at the correct interface, the issue of the branching *index* m has to be resolved.

The first air-slab interface influences the correct evaluation of de-embedded S_{11}, whereas the second slab-air interface, giving the effective length of the slab, influences S_{21}. The correct de-embedded values are needed for the correct extraction of the material parameters [J.63–J.65]. In the case of the 1D-symmetric structure of a metamaterial, the first interface is considered to be located at the boundary of the first unit cell, and the second interface is taken at the boundary of the last unit cell. However, these locations of interfaces do not apply to the 1D-asymmetric structure, and 2D structures. For these cases, the locations of interfaces, with respect to physical interfaces of the slab, are numerically optimized such that they provide an identical intrinsic impedance at several frequencies in the considered range of frequencies.

Equations (21.4.3b) and (21.4.20b) show that correct extraction of the real part of the refractive index n' of a slab suffers the branching ambiguity m due to the uncertain value of m. Only for the principal value of n', the branching index m is zero, i.e. $m = 0$. It corresponds to a *thin slab* of the thickness $d < \lambda_m/2$, where λ_m is wavelength inside the material slab whose refractive index is unknown. Therefore, the λ_m is also unknown. However, for a metamaterial slab, it could be approximately estimated using the circuit models discussed above. At the slab thickness $d = \lambda_m/2$, or at the multiple half-wavelengths, the *Fabry–Perot resonances* occur and nonphysical discontinuous material parameters are recovered. However, the materials have a continuous frequency response for their parameters and are independent of the slab thickness. A thick slab, $d > \lambda_m/2$, has nonzero branching index m. It has to be determined correctly for the correct extraction of the material parameters. Several methods have been suggested for this purpose. However, every method has a limitation. A few methods are summarized below.

Using Two Thicknesses of a Slab

Even for a thin slab, several thicknesses should be considered for the parameter extraction and consistent results should be obtained [J.63]. If we consider two thickness d_1 and d_2 such that $(d_2 - d_1) < \lambda_m/2$, then the branching index m could be estimated from the following expression, using the transmission phases $\angle\tau_1$ and $\angle\tau_2$ of slab, corresponding to the thickness d_1 and d_2 [J.66]:

$$m = \frac{d_2\angle\tau_1 - d_1\angle\tau_2}{2\pi(d_1 - d_2)}. \qquad (21.4.21)$$

However, the condition $(d_2 - d_1) < \lambda_m/2$ is uncertain as the $\lambda_m$is unknown inside the slab.

Iterative Method

This method evaluates the real part of the refractive index at two adjacent frequency f_0 and f_1 $(f_1 > f_0)$ using expression (21.4.12) for the transmission coefficient [J.65]. The branching index *m* is estimated by approximating the transmission coefficient at f_1 using the Taylor expansion of the expression (21.4.12):

$$e^{-j\beta_0(f_1)n(f_1)d} = e^{-j\beta_0(f_0)n(f_0)d}\left[e^{j(\beta_0(f_0)n(f_0)d - \beta_0(f_1)n(f_1)d)}\right]$$

$$= e^{-j\beta_0(f_0)n(f_0)d}e^{\Delta}$$

$$\Rightarrow e^{-j\beta_0(f_1)n(f_1)d} \approx e^{-j\beta_0(f_0)n(f_0)d}\left[1 + \Delta + \frac{\Delta^2}{2}\right] \qquad (a)$$

$$\text{where,} \quad \Delta = j\{\beta_0(f_0)n(f_0)d - \beta_1(f_1)n(f_1)d\} \qquad (b). \qquad (21.4.22)$$

To evaluate the parameter, Δ the complex refractive index $n(f_0)$ at the first frequency f_0 has to be computed that also has a branching problem. Using equation (21.4.3a), the Img{$n(f_0)$}, i.e. $n''(f_0)$, is determined uniquely. The computation of Re{$n(f_0)$}, i.e. $n'(f_0)$ involves the determination of the branching index *m*. The value of $n'(f_0)$ is determined for a few branching indexes, say m = 0, 1, 2. The normalized intrinsic impedance is computed using equation (21.4.13d). Each computed value of a complex refractive index corresponding to a distinct value of m must satisfy the passivity condition (21.4.7). If only one value of m satisfies the passivity condition, it is accepted as a correct branching index "*m*." In the case of multiple values for branching index *m* satisfying the passivity condition, the refractive index is computed for each value of m at several other frequencies, and passivity condition is tested. The correctly selected value of m should satisfy the passivity condition at several frequencies. The value of $n'(f_0)$ at frequency f_0 is accepted for the correct branching index. It is used in equation (21.4.22a). The transmission coefficient $e^{-j\beta_0(f_1)n(f_1)d}$ at frequency f_1 appearing in the quadratic equation (21.4.22a), for the variable Δ, is computed from equation (21.4.12).

The approximate quadratic equation (21.4.22) provides two complex roots – $\Delta_1(f_1)$, $\Delta_2(f_1)$ at frequency f_1. Using two roots, two approximate solutions are obtained for two complex refractive indexes $n^1_{appx}(f_1)$ and $n^2_{appx}(f_1)$. Both of complex refractive indexes have their imaginary parts. However, the correct imaginary part $n''(f_1)$ of the refractive index $n(f_1)$ is already known, so the correct choice of the root should provide the corresponding approximate value of the imaginary part, closer to the known value of $n''(f_1)$. The correct choice of root helps to compute the approximate refractive index, i.e. approximate real part $n_{appx}(f_1)$ also. At this stage, a few values of real $n''(f_1)$ are computed, using equation (21.4.3b), for a few values of m = 0, 1, 2... One particular value of a real refractive index $n'(f_1)$, corresponding to the correct value of m, is closer to $n_{appx}(f_1)$ obtained by using the correct root of equation (21.4.22a).

Once correct branching index *m* is determined, the correct value of the refractive index is computed using equation (21.4.3b). As both the intrinsic impedance of the material slab and its refractive index are known, the complex relative permittivity and permeability are extracted using equations (21.4.4c, d). The extracted parameters should have $\mu_r'' \geq 0, \varepsilon_r'' \geq 0$, as the medium is passive. The process has to be repeated at each of the frequency points in the considered range of frequencies. The computer code could be written for this repetitive parameter extraction process. The extracted complex relative permittivity and permeability of the equivalent slab are used to simulate the S-parameters, and it should closely follow the S-parameters of the original inhomogeneous slab. It decides the correctness of the extracted parameters.

Using Kramers–Kronig Relations

This method approximately recovers the real part of the refractive index at a frequency ω' from the known imaginary part of the refractive index at frequency ω [J.68, J.77]:

$$n'^{appx}(\omega') = 1 + \frac{2}{\pi}P\int_0^{\infty}\frac{\omega n''(\omega)}{\omega^2 - \omega'^2}d\omega, \qquad (21.4.23)$$

where P denotes the principal value of improper integral. The imaginary part of the refractive index $n''(\omega)$ is uniquely extracted using equation (21.4.3a). The limits of the integral are replaced by the finite limits of frequency over which the S-parameters are available either experimentally or using EM-simulator. So the process involves the truncation error, and only the approximate value of the refractive index $n'^{appx}(\omega')$ is recovered. However, once it is substituted in equation (21.4.3b), the branching index m is obtained from the following approximation:

$$m = \text{Round}\left[\left\{n'^{appx}(\omega) - n'(\omega, m = 0)\right\}\frac{\beta_0 d}{2\pi}\right]. \quad (21.4.24)$$

In equation (21.4.24), $n'(\omega, m = 0)$ is a real part of the refractive index for the principal value $m = 0$ obtained from equation (21.4.3b). The Round function provides the least integer for the branching index m. The MATLAB code is available for this process [J.68, J.78]. Another code is also available for the extractions process, including the detailed process of the EM-simulation [J.67].

Using the above-computed branching index m, equation (21.4.3b) extracts the correct value of the real part $n'(\omega)$ of the refractive index $n(\omega)$. Again, the complex permittivity and permeability are extracted as discussed above for other methods. The extracted parameters may show a nonphysical discontinuity in the response [J.68]. It should be removed. Any parameter extraction process can have such discontinuity in the response. However, using the Drude–Lorentz model for the permittivity and permeability the discontinuity in the response can be removed [J.69]. A few references could be consulted to examine the extracted parameters using various methods [J.63–J.66, J.68, J.69].

The above discussion is limited to the isotropic slabs with normal incidence. However, the retrieval process has been developed for the oblique incidence also. The parameters of the isotropic homogeneous materials should be independent of the angle of incidence and polarization states of the incident EM-waves. It is difficult to develop such metamaterials. Likewise, the methods of parameter retrieval of the anisotropic and bi-anisotropic slabs have also been investigated by the researchers [J.71, J.79].

21.4.2 Dynamic Maxwell Garnett Model

The *metallic inclusions*, i.e. the strip conductors-based structures have been primarily used to develop both the SNG and DNG engineered metamaterials in the microwave and mm-wave frequency ranges. However, such metamaterials are lossy in the THz to optical frequency ranges. The low-loss SNG and DNG, useful for the microwave to the optical frequency, could be engineered by the all *nonmetallic inclusions* in the dielectric host medium. The spherical dielectric inclusion particles are commonly considered. However, other shapes, such as cylindrical and cubic shaped dielectric resonators, have also been examined. These structures are attractive for the realization of superlens and clocking devices [J.80, J.81].

The dielectric resonators-based metamaterials can be homogenized by using the modified version of the Maxwell Garnett (MG) model discussed subsection (6.3.3) of chapter 6. The *modified MG model* is frequency-dependent that computes both the *effective relative permittivity and effective relative permeability* of a homogenized medium. The spherical dielectric inclusions are modeled using the Mie *scattering* that creates both the *magnetic and electric type Mie resonances* induced by the incident high-frequency waves. The resonating inclusions behavior is also responsible for generating negative permeability and negative permittivity. They follow the Lorentz model response. The negative permittivity is also obtained by using the conducting inclusion, i.e. the plasmonic inclusion modeled by the Drude model, discussed in subsection (6.5.2) of chapter 6. The MG model can accommodate these negative parameters, although the classical MG model has been used only for the positive material parameters [J.82–J.84].

Mie Scattering

The composite material could be developed using the spherical dielectric inclusions of radius "a" with relative permittivity ε_{ri} and relative permeability μ_{ri} embedded in the host medium with relative permittivity ε_{rh} and relative permeability μ_{rh}. The incident EM-wave on the composite material induces both the magnetic and electric dipoles, and also multipoles on the spherical dielectric inclusions. These dipoles and multipoles create the scattered field known as the Mie scattering. The scattered field of the dielectric spherical inclusions is treated as the superposition of the TM and TE normal modes with following Mie coefficients [J.85, B.19]:

$$a_n = \frac{m\psi_n(mx)\psi_n'(x) - \psi_n(x)\psi_n'(mx)}{m\psi_n(mx)\xi_n'(x) - \xi_n(x)\psi_n'(mx)} \quad (a)$$

$$b_n = \frac{\psi_n(mx)\psi_n'(x) - m\psi_n(x)\psi_n'(mx)}{\psi_n(mx)\xi_n'(x) - m\xi_n(x)\psi_n'(mx)} \quad (b)$$

$$\text{where, } m = \frac{n_i}{n_h} = \frac{\sqrt{\varepsilon_{ri}\mu_{ri}}}{\sqrt{\varepsilon_{rh}\mu_{rh}}} \quad (c)$$

$$x = \frac{\omega}{c}\sqrt{\varepsilon_{rh}\mu_{rh}}a = 2\pi n_h\frac{a}{\lambda} \quad (d).$$

(21.4.25)

In equations (21.4.25a–d) m is the *contrast ratio* of the refractive indices, and x is the size parameter of the inclusion. The functions ψ_n and ξ_n are the nth order *Riccati–Bessel* function. The derivatives of ψ_n and ξ_n, with respect to the arguments, are ψ'_n and ξ'_n respectively. The complete solution of the scattered field is presented by Bohren and Huffman in their beautiful book [B.19]. The coefficients a_n and b_n determine magnitudes of fields of the scattered normal modes. The coefficient a_n provides a large value of the E-radial field component of the TM-mode under the resonance condition. The H-radial field is zero. Likewise, the coefficient b_n corresponds to the excited TE-mode with a high value of the H-radial field component. In this case, the E-radial field is zero. To excite these modes by the incident EM-field, the denominators of equation (21.4.25) are nearly zero, giving the following resonance conditions:

$$\text{For TM – mode}: \frac{\psi_n(mx)}{\xi_n(x)} = \frac{1}{m}\frac{\psi'_n(mx)}{\xi'_n(x)} \quad \text{(a)}$$

$$\text{For TE – mode}: \frac{\psi_n(mx)}{\xi_n(x)} = m\frac{\psi'_n(mx)}{\xi'_n(x)} \quad \text{(b).} \tag{21.4.26}$$

The resonance frequency of the TM and TE modes are computed from equations (21.4.26a, b). The TM_{011} mode is the fundamental resonant mode with the lowest frequency, and the TE_{011} mode is the next resonant mode at a higher frequency. Both equations given above are identical except the location of the *contrast-ratio* (m) in the denominator in one equation, and in the numerator for the other. Figure (21.22a, b) show the sketched electric field lines of the TM_{011} and TE_{011} modes on the surface of the dielectric spherical inclusion. The E-field circular loops of the TM_{011} mode suggest the presence of the circular loops of displacement current on the dielectric sphere. The magnetic field is normal to the loops. Thus, the magnetic field excited sphere in the TM_{011} mode behaves as a *magnetic dipole*. The host medium could be treated as loaded with the magnetic inclusions, with the *magnetic polarizability* α^m. Likewise, the E-field lines of the TE_{011} mode of the dielectric sphere behave like an *electric dipole*. In this case, the host medium could be treated as loaded with the electric inclusions, with the *electric polarizability* α^e.

The above description suggests that the composite medium behaves as a magnetic medium at a lower resonant frequency with a narrow bandwidth. The composite medium also behaves like an electric medium at the higher resonance frequency. Using the Lorentz model for the magnetic and electric medium, the negative permeability and negative permittivity can be obtained above their respective resonance frequencies. Thus, at a single frequency, the SNG medium is obtained with inclusions of a uniform fixed value of radius. The DNG medium is not obtained in this way. However, the magnetic and electric polarizabilities of the spherical inclusion could be used to obtain the dynamic Maxwell Garnett model #I. The dynamic Maxwell Garnett model #II could also be obtained using the modified relative permittivity and permeability of the spherical inclusions. A summary of both models is given below.

Dynamic Maxwell Garnett Model # I

The Mie coefficients a_1 and b_1 of the spherical inclusions could be replaced by the following electric and magnetic polarizabilities, corresponding to the electric and

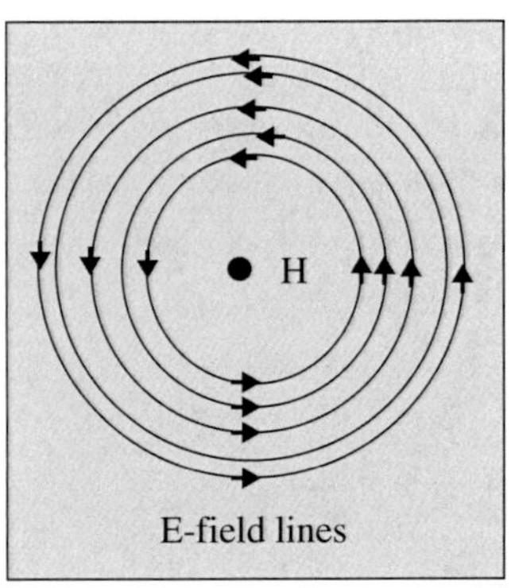

(a) Closed-loop of E-field lines of TM_{011} mode around the sphere. H-field normal to the plane.

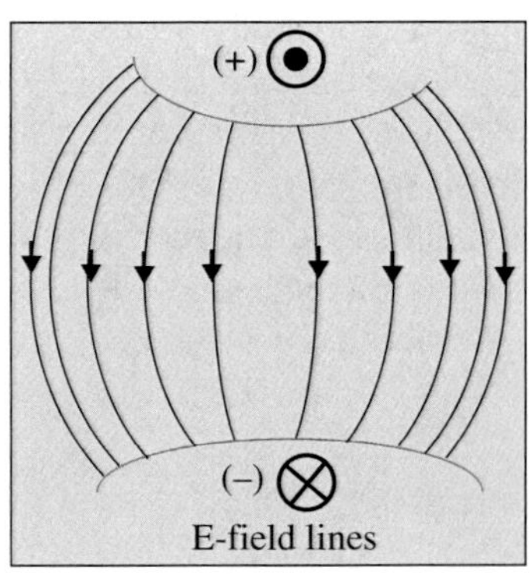

(b) E-field lines of TE_{011} mode on the sphere.

Figure 21.22 The electric field lines of the first two normal modes of Mie resonance.

magnetic fields excited electric and magnetic dipoles [J.86, J.87]:

$$\alpha^e = j4\pi\varepsilon_0\left[\frac{3a^3}{2x^3}a_1\right] \quad (a), \quad \alpha^m = j4\pi\mu_0\left[\frac{3a^3}{2x^3}b_1\right] \quad (b). \tag{21.4.27}$$

In equation (21.4.27), the radius of the spherical inclusion is a, and x is the size-parameter given by equation (21.4.25). For $n = 1$, the Mie coefficients a_1 and b_1 are computed from equations (21.4.25a, b). The FORTRAN code is available to compute the Mie scattering coefficients [B.19]. The *Ricatti–Bessel* functions and their derivatives are expressed in terms of the trigonometric functions [J.88, J.89]:

$$\psi_1 = \frac{\sin\rho}{\rho} - \cos\rho, \qquad \psi_1' = \sin\rho\left(1 - \frac{1}{\rho^2}\right) + \frac{\cos\rho}{\rho}$$
$$\xi_1 = \psi_1 - j\left(\frac{\cos\rho}{\rho} + \sin\rho\right), \quad \xi_1' = \psi_1' + j\left(\psi_1 + \frac{1}{\rho^2}\cos\rho\right). \tag{21.4.28}$$

The spherical inclusions, as the electric dipoles, are arranged in the cubic lattice of volume Δv. The filling-factor f of inclusion is obtained as follows:

$$f = \frac{v_i}{\Delta v} = \frac{4\pi}{3}a^3 N, \quad \Rightarrow N = \frac{3f}{4\pi a^3}, \tag{21.4.29}$$

where N is the inclusion density. The effective relative permittivity ε_{reff}, i.e. the equivalent relative permittivity, of the composite medium, i.e. the mixture medium of host and inclusion, in terms of the electric polarizability of the inclusion, is obtained from the Clausius–Mossotti equation (6.1.13) of chapter 6, and equation (21.4.29):

$$\frac{(\varepsilon_{reff} - \varepsilon_{rh})}{(\varepsilon_{reff} + 2\varepsilon_{rh})} = \frac{N\alpha^e}{3\varepsilon_0} = \left(\frac{f}{4\pi\varepsilon_0 a^3}\right)\alpha^e$$
$$\Rightarrow \varepsilon_{reff} = \frac{\varepsilon_{rh}\left[\dfrac{2f\alpha^e}{4\pi\varepsilon_0 a^3} + 1\right]}{1 - \dfrac{f\alpha^e}{4\pi\varepsilon_0 a^3}}. \tag{21.4.30}$$

On substituting the electric polarizability of the spherical inclusion, resonating in the TE_{011} mode, from equation (21.4.27a) in equation (21.4.30), the following expression of the effective relative permittivity ε_{reff} is obtained for the dynamic MG model #1:

$$\varepsilon_{reff} = \frac{\varepsilon_{rh}\left[j\dfrac{3f}{x^3}a_1 + 1\right]}{1 - j\dfrac{3}{2x^3}a_1}. \tag{21.4.31}$$

Expression (21.4.31) is the high-frequency extension of the classical MG model obtained from the Clausius–Mossotti model.

Similarly, the magnetic composite medium is created by the magnetic dipoles of spherical dielectric inclusions, resonating in the TM_{011} mode. The effective relative permeability of the magnetic composite medium is obtained by recasting equation (6.1.28b) of chapter 6 as follows:

$$\frac{(\mu_{reff} - \mu_{rh})}{(\mu_{reff} + 2\mu_{rh}\varepsilon_{rh})} = \frac{N\alpha^m}{3\mu_0} = \left(\frac{f}{4\pi\mu_0 a^3}\right)\alpha^m$$
$$\Rightarrow \mu_{reff} = \frac{\mu_{rh}\left[\dfrac{2f\alpha^m}{4\pi\mu_0 a^3} + 1\right]}{1 - \dfrac{f\alpha^m}{4\pi\mu_0 a^3}}. \tag{21.4.32}$$

By substituting α^m from equation (21.4.27b) in equation (21.4.32), the dynamic MG model #1 for the effective permeability is obtained:

$$\mu_{reff} = \frac{\mu_{rh}\left[j\dfrac{3f}{x^3}b_1 + 1\right]}{1 - j\dfrac{3f}{2x^3}b_1} = \frac{\mu_{rh}(x^3 + j3fb_1)}{(x^3 - j(3/2)fb_1)}. \tag{21.4.33}$$

Dynamic Maxwell Garnett Model # II

Lewin has suggested a different expression for the dynamic MG model # II by modifying the relative permittivity and permeability of the spherical dielectric inclusion. Thus, relative permittivity ε_{ri} and relative permeability μ_{ri} of the spherical inclusion are replaced by the *frequency-dependent modified relative permittivity* ε_{ri}^{mod} and *permeability* μ_{ri}^{mod} in the static MG model given by equations (6.3.13a) and (6.3.17a) of chapter 6 to get the dynamic Maxwell Garnett model #II [J.90]:

$$\varepsilon_{reff} = \varepsilon_{rh}\left[1 + \frac{3f(\varepsilon_{ri}^{mod} - \varepsilon_{rh})}{\varepsilon_{ri}^{mod} + 2\varepsilon_{rh} - f(\varepsilon_{ri}^{mod} - \varepsilon_{rh})}\right]$$
$$\Rightarrow \varepsilon_{reff} = \varepsilon_{rh}\left[1 + \frac{1}{\{(\varepsilon_{ri}^{mod} + 2\varepsilon_{rh})/3f(\varepsilon_{ri}^{mod} - \varepsilon_{rh}) - 1/3\}}\right] \quad (a)$$
$$\mu_{reff} = \mu_{rh}\left[1 + \frac{3f(\mu_{ri}^{mod} - \mu_{rh})}{\mu_{ri}^{mod} + 2\mu_{rh} - f(\mu_{ri}^{mod} - \mu_{rh})}\right]$$
$$\Rightarrow \mu_{reff} = \mu_{rh}\left[1 + \frac{1}{\{(\mu_{ri}^{mod} + 2\mu_{rh})/3f(\mu_{ri}^{mod} - \mu_{rh}) - 1/3\}}\right] \quad (b)$$
$$\text{where, } \varepsilon_{ri}^{mod} = F(\theta)\varepsilon_{ri}, \mu_{ri}^{mod} = F(\theta)\mu_{ri}, F(\theta) = \frac{2(\sin\theta - \theta\cos\theta)}{(\theta^2 - 1)\sin\theta + \theta\cos\theta} \quad (c). \tag{21.4.34}$$

In equations (21.4.34a–c), the frequency-dependent parameter is $\theta = k_0 a\sqrt{\varepsilon_{ri}\mu_{ri}}$, and the frequency-dependent function $F(\theta)$ is approximately a periodic resonant function that is responsible for the resonant nature of the inclusion. For a spherical dielectric inclusion in the cubic lattice of lattice constant d, the filling-factor f is $4/3\pi(a/d)^3$. The wavenumber of the incident field in free space is $k_0 = 2\pi/\lambda_0$. For a spherical inclusion, the maximum filling-factor in the cubic lattice is $f = v_i/\Delta v = ((4/3)\pi a^3)/8a^3 = \pi/6 \approx 0.5$. The MG model #II works satisfactorily around the filling-factor $f \approx 0.3$.

The TE-mode Mie resonance creates the effective permittivity, while the TM-mode resonance is responsible for the effective permeability of the composite medium with Mie resonance frequency $f_r^{TM} < f_r^{TE}$ for an inclusion particle. Around the Mie resonance, the effective permittivity and permeability of the composite medium are very large and the ε_{reff} and μ_{reff} approach to ε_{rh} and μ_{rh} of the host medium. Thus, around the modal Mie resonance frequencies, the $\varepsilon_{ri}^{mod}(f = f_r^{TE})$ and $\mu_{ri}^{mod}(f = f_r^{TM})$ of a spherical inclusion are obtained from equation (21.4.34) follows:

$$\frac{(\varepsilon_{ri}^{mod} + 2\varepsilon_{rh})}{3f(\varepsilon_{ri}^{mod} - \varepsilon_{rh})} - \frac{1}{3} = 0$$

$$\Rightarrow \varepsilon_{ri}^{mod}(f = f_r^{TE}) = -\varepsilon_{rh}\left[\frac{(2+f)}{(1-f)}\right] \quad \text{(a)}$$

$$\frac{(\mu_{ri}^{mod} + 2\mu_{rh})}{3f(\mu_{ri}^{mod} - \mu_{rh})} - \frac{1}{3} = 0$$

$$\Rightarrow \mu_{ri}^{mod}(f = f_r^{TM}) = -\mu_{rh}\left[\frac{(2+f)}{(1-f)}\right] \quad \text{(b).}$$

(21.4.35)

The above negative values are required for the modified electrical parameters of the inclusion to achieve the resonance for the TM and TE modes of the scattered field. It helps to get an effective homogeneous medium. Using equation (21.4.34), at the given resonance frequency of TE/TM mode, the radius "a" of the spherical inclusion is determined from the relation $\theta = k_0 a\sqrt{\varepsilon_{ri}^{mod}\mu_{ri}^{mod}}$. Once the electrical parameters of the inclusion and host medium are known, the parameter θ is obtained from equations (21.4.34c) and (21.4.35).

Lewin has considered two important limiting cases for the modified relative permittivity and permeability [J.90]:

Case # a: At low frequency of the incident wave, the parameter θ is small with $F(\theta) = 1$. It leads to $\varepsilon_{ri}^{mod} \to \varepsilon_{ri}$, and $\mu_{ri}^{mod} \to \mu_{ri}$. Under this condition, the dynamic MG model #II of equation (21.4.34) is reduced to the static MG model, discussed in chapter 6.

Case # b: The parameter θ could be large and complex. It applies to a metallic, i.e. plasmonic, inclusion with $F(\theta) \approx 2/j\theta$. For this case, the modified parameters of inclusion are given below:

$$\varepsilon_{ri}^{mod}(f) = \frac{2}{j}\left[\sqrt{\frac{\varepsilon_{ri}}{\mu_{ri}}}\right]\frac{\lambda_0}{2\pi a} \quad \text{(a)}$$

$$\mu_{ri}^{mod}(f) = \frac{2}{j}\left[\sqrt{\frac{\mu_{ri}}{\varepsilon_{ri}}}\right]\frac{\lambda_0}{2\pi a} \quad \text{(b).}$$

(21.4.36)

Normally, the frequency-dependent modified material parameters of the inclusion $\varepsilon_{ri}^{mod}(f)$ and $\mu_{ri}^{mod}(f)$ are much smaller than the real material parameters ε_{ri} and μ_{ri} of an inclusion particle. However, around the Mie resonance frequency f_r^{TE} of TE-mode, the modified relative permittivity $\varepsilon_{ri}^{mod}(f)$ of the inclusion is much higher than the relative permittivity ε_{rh} of the host medium, i.e. $\varepsilon_{ri}^{mod}(f) \gg \varepsilon_{rh}$. The resonance frequency f_r^{TE} is the off-resonance frequency for the TM-mode responsible for the creation of the modified relative permeability $\mu_{ri}^{mod}(f)$ of the inclusion. Thus, around resonance frequency f_r^{TE}, $\mu_{ri}^{mod}(f)$ is much less than the relative permeability of the host medium, i.e. $\mu_{ri}^{mod}(f) < < \mu_{rh}$. Under these conditions around the resonance frequency f_r^{TE}, the modified MG model #II of equation (21.4.34) provides the following approximate expression for the effective permittivity and permeability of the mixture:

$$\varepsilon_{ri}^{mod} = \varepsilon_{rh}\left[1 + \frac{3f}{1-f}\right] \quad \text{(a)}$$

$$\mu_{ri}^{mod} = \mu_{rh}\left[1 - \frac{3f}{1-f}\right] \quad \text{(b).}$$

(21.4.37)

The accuracy of both forms of the dynamic MG has been tested against the extracted parameters of the effective homogeneous medium from the simulated S-parameter of the cubic unit cell using the NRW method. The deep sub-wavelength composite of spherical Ag inclusions, radius $a = 3$ nm, arranged in a cubic lattice with lattice constant $d = 15$ nm, in the SiO_2 host medium has been simulated on CST to get the S-parameter of unit-cell. Figure (21.23a) compares the results of both MG models against the extracted results from the S-parameters. All three results show very good agreement for the effective relative permittivity. It shows that the mixture follows the Lorentz model behavior. The effective relative permeability is unity [J.88].

The conditions discussed also prevail around the Mie resonance frequency f_c^{TM} for the resonating TM mode, creating the dominant relative permeability in the inclusion and a much smaller value for the relative permittivity, i.e. $\mu_{ri}^{mod} > > \mu_{rh}$ and $\varepsilon_{ri}^{mod} < < \varepsilon_{rh}$. Again, the approximate expressions for effective permittivity and permeability of the mixture could be obtained.

Several schemes have been suggested and demonstrated to develop the Mie resonance-based DNG

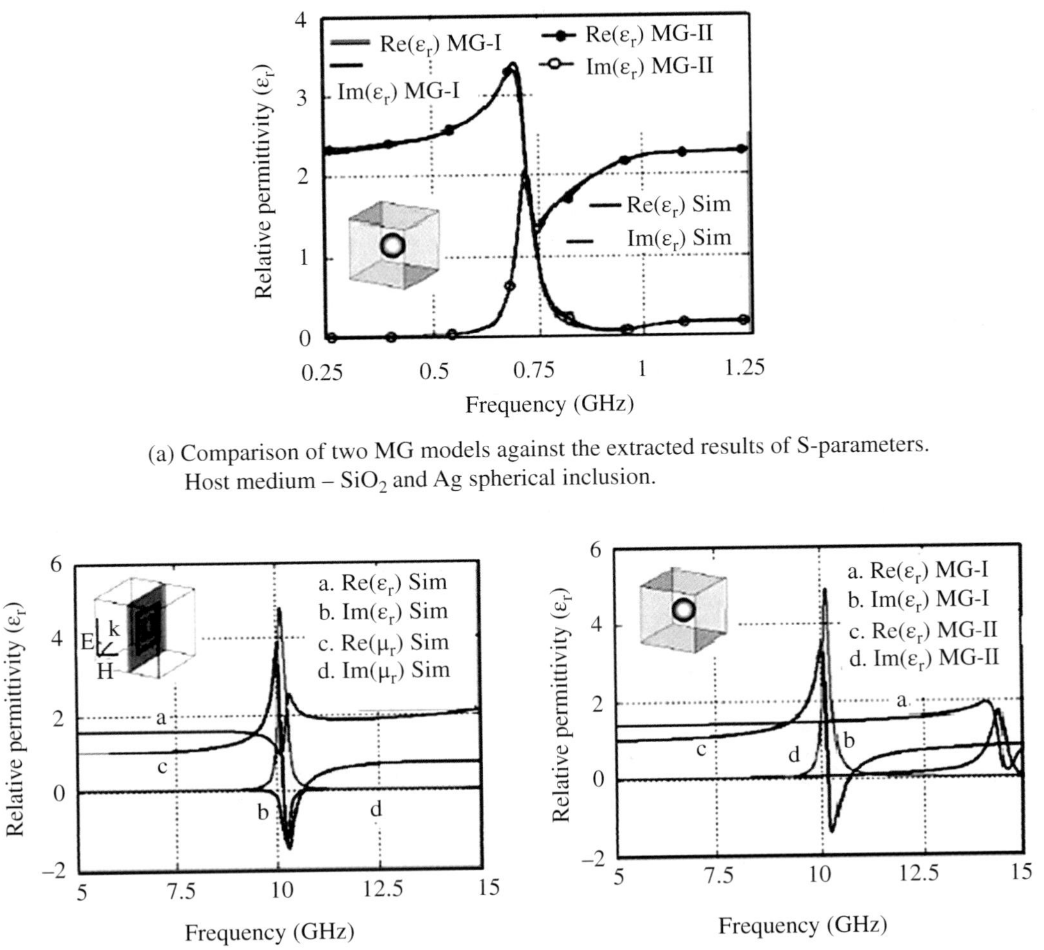

(a) Comparison of two MG models against the extracted results of S-parameters. Host medium – SiO_2 and Ag spherical inclusion.

(b) Extracted parameters of the SRR inclusion.

(c) Parameters of the equivalent spherical dielectric inclusion.

Figure 21.23 Effective medium parameters using the dynamic MG models. *Source*: Reproduced from Szabo and Fuzi [J.88]. © 2014, IEEE.

materials. The Mie resonators could be in several shapes – spherical, cylindrical, cubic, etc. Some of these schemes are nicely summarized in a review paper [J.80]. It is noted above that the magnetic and electric Mie resonances of a dielectric sphere occur at two different frequencies. To obtain the electric resonance and magnetic resonance at the same frequency, two spherical inclusions of the same material but of different sizes could be used within a cubic unit cell [J.91]. The larger size inclusions are needed to bring down the permittivity resonance frequency near to the magnetic resonance of a smaller inclusion. A pair of inclusions of the same size but made of different materials can also achieve identical electric and magnetic resonance frequencies. Normally, the dielectric inclusion resonates at the fundamental frequency giving negative permeability. It is the MNG type of material. However, a metallic pin mounted near or inside the dielectric inclusion provides the negative permittivity [J.92]. The composite inclusion provides the DNG response. The dielectric inclusions can be arranged in a waveguide section also [J.93]. The negative permeability is generated by the Mie resonance and the TE cut-off mode adds the negative permittivity. The combined response is DNG. Similarly, the metal-coated dielectric inclusion provides the DNG response [J.94]. The inner dielectric core responds to the magnetic excitation, while negative permittivity is generated by the outer plasmonic shell. The dynamic Maxwell Garnett model could be helpful to get the homogenized effective medium. Tailoring the effective medium property could be more convenient with the dynamic MG model.

Equivalent Mie Magnetic Resonator of SRR Inclusion

Normally the dynamic MG model, based on the Mie scattering, is not suitable for the metallic SRR type inclusion. However, once the effective permeability of the SRR-based material has been extracted, using the simulated S-parameters, the radius and permittivity of the *equivalent spherical inclusions*, filling-factor, and permittivity of the host medium can be optimized using the dynamic MG model to meet the permeability response of the SRR [J.88]. Figure (21.23b) shows the extracted permeability, and also permittivity of the copper SRR in a cubic unit cell of lattice constant d = 5 mm. The dimensions of the SRR printed on a substrate with $\varepsilon_r = 3.84$, h = 0.25 mm are given: the external size of ring 3 mm, the width of the strip 0.25 mm, the size of the split gap and separation between rings 0.5 mm, and strip thickness 0.02 mm. The *optimized equivalent radius of spherical inclusion* is a = 2.31 mm and filling-factor f = 0.13. The permittivity of the inclusion is 37.67 and the host medium is air. The equivalent inclusion can be implemented practically. Figure (21.23c) shows the response of the *equivalent Mie resonance-based medium*. Its permeability response is almost identical to that of the original SRR response, shown in Fig (21.23b). However, as expected, its permittivity response occurs at a higher frequency, so it is a magnetic medium only. By adding additional wire-medium, the DNG response can be obtained.

Equivalent RLC Circuit Model of Mie Magnetic Resonator

The Mie resonance-based dynamic MG model computes the effective permeability of the composite medium with dielectric spherical inclusions. However, computation of the resonance frequency, the bandwidth of negative permeability response, and computation of magnetic response by varying the geometrical parameters are not done directly. The equivalent RLC circuit model is available for the direct computation of these parameters. The RLC equivalent circuit, as shown in Fig (21.6d), for the SRR resonator is presented in section (21.2.2). The circuit elements of the equivalent RLC, around the first

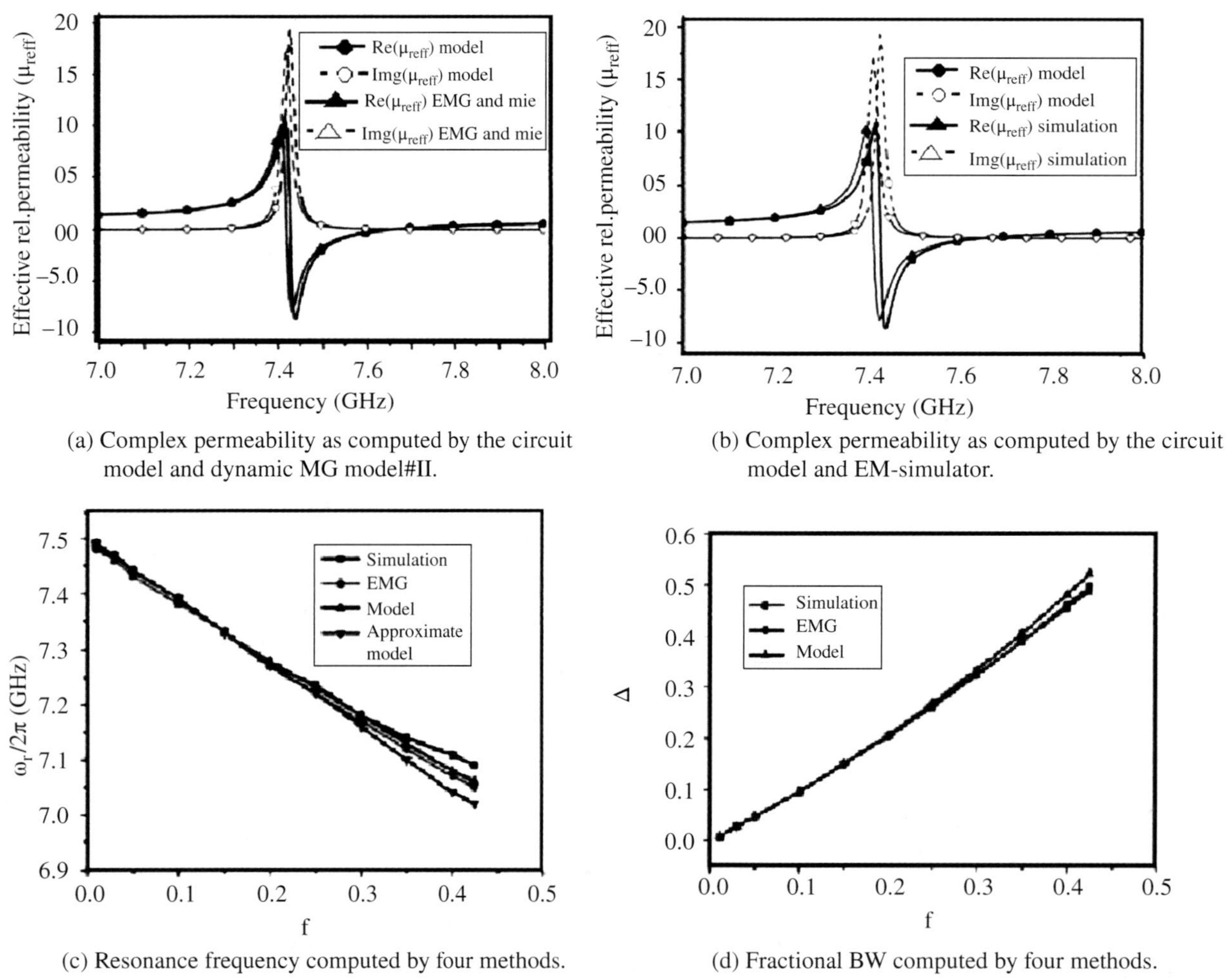

Figure 21.24 Computation magnetic response by three methods. *Source*: From Liu et al. [J.95], reproduced courtesy of The Electromagnetics Academy. © 2011, The Electromagnetics Academy.

resonance, could be computed from the following expressions [J.95]:

$$R = \frac{\mu_0 \pi^4 c n_i''}{6 n_i'} \quad (a), \quad L = \frac{\mu_0 \pi^3 a}{12} \quad (b)$$
$$C = \frac{12 a n_i'^2}{\mu_0 \pi^5 c^2} \quad (c). \tag{21.4.38}$$

In equation (21.4.38), spherical inclusion has a radius a and its low-loss refractive index is $n_i^* = n_i' - j n_i''(n_i'' < < n_i')$. The velocity of EM-wave in free space is c. The resonance frequency of the inclusion is $\omega_0 = 1/\sqrt{LC}$. The effective permeability could be obtained from equation (21.2.3). For the cubic lattice, parameters F, γ_m and frequency ω_{0m} used in equation (21.2.3) are computed from the following expressions [J.95]:

$$F = \frac{9f}{3f + \pi^2} \quad (a), \quad \gamma_m = \frac{2\pi^3 c n_i''}{n_i'^2 a(3f + \pi^2)} \quad (b)$$
$$\omega_{om} = \frac{\pi^2 c}{n_i' a \sqrt{(3f + \pi^2)}} \quad (c). \tag{21.4.39}$$

In expression (21.4.39), f is the volume fraction of inclusion. The relative bandwidth Δ of the negative permeability, and also approximate resonance frequency, can be estimated from the following expressions [J.95]:

$$\Delta = \frac{9f}{\pi^2 - 6f} \quad (a), \quad \omega_r \approx \frac{\pi c}{n_i' a}\left[1 - \frac{3}{2\pi^2} f\right] \quad (b). \tag{21.4.40}$$

The accuracy of the circuit model is tested for the spherical *Barium Strontium Titanate* (BST) material with complex relative permittivity ($\varepsilon_{ri} = 1600 - j4.8$). The sphere of BST with radius a = 0.5 mm is located in the cubic lattice of the lattice constant d = 2 mm. The relative permittivity of the host medium is $\varepsilon_{rh} = 2$. Figure (21.24a) compares the circuit model-based computed results of complex effective permeability against the results of the dynamic MG model #II. Likewise, Fig (21.24b) compares the results of the circuit model against the parameter extracted from the simulated S-parameters. Only a minor variation is seen among the results from three sources. Figure (21.24c, d) further compares the circuit model-based results of resonance frequency and fractional BW of negative permeability of a composite medium against the results obtained from two other sources – dynamic MG model, and EM-simulator. The resonance frequency decreases and fractional BW increases with an increase in the volume fraction.

References

Books

B.1 Engheta, N.; Ziolkowski, R.W. (Editors): *Metamaterials: Physics and Engineering Explorations*, Wiley-Interscience, John Wiley & Sons, Inc., NJ, USA, 2006.

B.2 Christophe Caloz, C.; Itoh, T.: *Electromagnetic Metamaterials: Transmission line Theory and Microwave Applications (The Engineering Approach)*, Wiley-Interscience, John Wiley & Sons, Inc., NJ, USA, 2006.

B.3 Eleftheriades, G.I.; Balmain, K.G. (Editors): *Negative-Refraction Metamaterials: Fundamental Principles and Applications*, Wiley-Interscience, John Wiley & Sons, Inc., NJ, USA, 2005.

B.4 Capolino, F. (Editor): *Theory and Phenomena of Metamaterials*, CRC Press, USA.

B.5 Marques, R.; Martin, F.; Sorolla, M.: *Metamaterials with Negative Parameters. Theory, Design and Microwave Applications*, Wiley, Hoboken, NJ, 2008.

B.6 Sarychev, A.K.; Shalaev, V.M.: *Electrodynamics of Metamaterials*, World Scientific Pub, Singapore, 2007.

B.7 Tretyakov, S.: *Analytical Modeling on Applied Electromagnetics*, Artech House, Boston, USA, 2000.

B.8 Solymar L.; Shamonina, E.: *Waves in metamaterials*, Oxford University Press, NY, USA, 2009.

B.9 Nefedov, I.S.; Viitanen, A.J.: Wire media, Capolino, F. (Editor): *Chapter – 15, Theory and Phenomena of Metamaterials*, CRC Press, USA.

B.10 Wiltshire, M.C.K.; Capolino, F. (Editor): RF metamaterials, *Theory and Phenomena of Metamaterials*, CRC Press, USA

B.11 Grzegorczyk, T.M.; Kong, J.A.; Lixin, R.: Refraction experiments in waveguide environments, Engheta, N.; Ziolkowski, R.W. (Editors): *Chapter – 4 Metamaterials: Physics and Engineering Explorations*, Wiley-Interscience, John Wiley & Sons, Inc., NJ, USA, 2006.

B.12 Grover, F.W.: *Inductance Calculations: Working Formulas and Tables*, Dover Pub. International, New York, 1946.

B.13 Pozar, D.M.: *Microwave Engineering*, 2nd Edition, John Wiley & Sons, Singapore, 1999.

B.14 Schelkunoff, S.A.; Friis, H.T.: *Antennas: Theory and Practice*, John Wiley, New York, 1952.

B.15 Marques, R.; Martin, F.: Split ring resonators and related topologies, Capolino, F. (Editor): Theory and Phenomena of Metamaterials, *Chapter – 16*, CRC Press, USA, 2017.

B.16 Collin, R.E.: *Field Theory of Guided Waves*, IEEE Press, Piscataway, NJ, 1990.

B.17 Hrabar, S.: Waveguide experiments to characterize properties of SNG and DNG metamaterials, Engheta, N.; Ziolkowski, R.W. (Editors): *Metamaterials: Physics and Engineering Explorations*, Wiley-Interscience, John Wiley & Sons, Inc., NJ, USA, 2006.

B.18 Lippens, D.: *Overview of Microwave and Optical Metamaterial Technologies*, Chapter 1 in Metamaterials and Wave Control, Lheurette, E. (Editor): Wiley, 2013.

B.19 Bohren, C.F.; Huffman, D.R.: *Absorption and Scattering of Light by Small Particles*, Wiley-Int. Pub., John Wiley, New York, 1983.

Journals

J.1 Veselago, V.: *The electrodynamics of substances with simultaneously negative values of ε and μ, Sov. Phys. Uspekhi*, Vol. 10, No. 4, pp. 509–514, Jan., Feb. 1968.

J.2 Pendry, J.B.; Holden, A.J.; Stewart, W.J.; Youngs, I.: *Extremely low-frequency plasmons in metallic mesostructures, Phys. Rev. Lett.*, Vol. 76, No. 25, pp. 4773–4776, June 1996.

J.3 Pendry, J.B.; Holden, A.J.; Robbins, D.J.; Stewart, W.J.: *Magnetism from conductors and enhanced non-linear phenomena, IEEE Trans. Microwave Theory Tech.*, Vol. 47, No. 11, pp. 2075–2084, Nov. 1999.

J.4 Shelby, R.A.; Smith, D.R.; Schultz, S.: *Experimental verification of a negative index of refraction, Science*, Vol. 292, No. 5514, pp. 77–79, Apr. 2001.

J.5 Smith, D.R.; Padilla, W.J.; Vier, D.C.; Nemat-Nasser, S. C.; Schultz, S.: *Composite medium with simultaneously negative permeability and permittivity, Phys. Rev. Lett.*, Vol. 84, No. 18, pp. 4184–4187, May 2000.

J.6 Brown, J.: *Artificial dielectrics having refractive indices less than unity, Proc. IEE Monogr. No. 62R*, Vol. 100, pp.51–62, May 1953.

J.7 Rotman, W.: *Plasma simulation by artificial dielectrics and parallel – plate media, IRE Trans. Antenna Propag.*, Vol. 10, No. 1, pp. 82–95, Jan. 1962.

J.8 Forati, E.; Hanson, G.W.: *Scattering from isotropic connected wires medium metamaterials: three, two, and one-dimensional cases, IEEE Trans. Antennas Propag.* Vol. 61, No. 7, pp.3564–3574, July 2013.

J.9 Maslovski, S.I.; Tretyakov, S.A.; Belov, P.A.: *Wire media with negative effective permittivity: a quasistatic model, Microwave Opt. Technol. Lett.*, Vol. 35, pp. 47–51, Oct. 2002.

J.10 Belov, P.A.; Tretyakov, S.A.; Viitanen, A.J.: *Dispersion and reflection properties of artificial media formed by regular lattices of ideally conducting wires, J. Electromagn. Waves Appl.* Vol. 16, pp. 1153–1170, Aug. 2002.

J.11 Hanson, G.W.; Forati, E.; Silveirinha, M.G.: *Modeling of spatially-dispersive wire media: transport representation, comparison with natural materials, and additional boundary conditions, IEEE Trans. Antennas Propag.*, Vol. 60, No. 9, pp. 4219–4232, Sept. 2012.

J.12 Gay-Balmaz, P.; Maccio, C.; Martina, O.J.F.: *Microwire arrays with plasmonic response at microwave frequencies, Appl. Phys. Lett.*, Vol. 81, No. 15, pp. 2896–2898, Oct. 2002.

J.13 Smith, D.R.; Vier, D.C.; Padilla, W.; Nemat-Nasser, S. C.; Schultz, S.: *Loop-wire medium for investigating plasmons at microwave frequencies, Appl. Phys. Lett.*, Vol. 75, No. 10, pp. 1425–1427, Sept. 1999.

J.14 Mousavi Roknabadil, S.M.; Iafargholi, A.; Mirtaheri, S. A.; Kamyab, M.: *Easily implemented miniaturized ENZ metamaterial medium using spiral inductors*, 20th Iranian Conf. on Electrical Engineering, (ICEE2012), pp. 1339–1341, May 2012, Tehran, Iran

J.15 Takano, K; Shibuya, K.; Akiyama, K.; Nagashima, T.; Miyamaru, F.; Hangyol, M.: *A metal-to-insulator transition in cut-wire-grid metamaterials in the terahertz region, J. App. Phys.*, vol.107, pp. 024907-1–024907-6, 2010.

J.16 Weldon, T.P.; Miehle, K.; Adams, R.S.; Daneshvar, K.: *Simulation, measurement, and parameter extraction for a 5.8 GHz negative permittivity metamaterial, SoutheastCon, 2012 Proceedings of IEEE*, pp. 1–4, Mar. 2012.

J.17 Simovski, C.R.; He, S.: *Antennas based on modified metallic photonic bandgap structures consisting of capacitively loaded wires, Microwave Opt. Tech. Lett.*, Vol. 31, No. 3, pp. 214–221, Nov. 2001.

J.18 Belov, P.A.; Simovski, C.R.; Tretyakov, S.A.: *Two-dimensional electromagnetic crystals formed by reactively loaded wires, Phys. Rev. E*, Vol. 66, No. 036610, pp. 1–7, 2002.

J.19 Ikonen, P.; Saenz, E.; Gonzalo, R.; Simovski, C.R.; Tretyakov, S.A.: Mesoscopic effective material parameters for thin layers modeled as single and double grids of interacting loaded wires, *Metamaterials*, Elsevier, Vol.1, pp. 89–105, 2007.

J.20 Thompson, M.T.: *Inductance calculation technique – Part II: approximation and handbook methods, Power Control Intell. Motion*, pp. 1–11, Dec. 1999, http://www.pcitn.coin.

J.21 Maslovski, S.I.; Ikonen, P.; Kolmakov, I.; Tretyakov, S. A.: *Artificial magnetic maerials on the new magnetic particle: metasoleniod, Progr. Electromagn. Res. PIERS*, Vol. 54, pp. 61–81, 2005.

J.22 Marques, R.; Mesa, F.; Martel, J.; Martin, F.: *Comparative analysis of edge and broadside-coupled*

split-ring resonators for metamaterial design. Theory and experiment, IEEE Trans. Antennas Propag., Vol. 5, pp. 2572–2581, 2003.

J.23 Marques, R.; Medina, F.; Rafi-El-Idrissi, R.: *Role of bianisotropy in negative permeability left-handed metamaterials, Phys. Rev. B*, Vol. 65, No. 144440, pp. 1–6, 2002

J.24 Bilotti, F.; Toscano, A.; Vegni, L.; Aydin, K.; Boratay, K.: *Equivalent circuit models for the design of metamaterials based artificial magnetic inclusions, IEEE Trans. Microwave Theory Tech.*, Vol. 55, No. 12, pp. 2865–2872, Dec. 2007.

J.25 Cummer, S.A.; Popa, B.-I.; Hand, T.H.: *Q-based design equations and loss limits for resonant metamaterials and experimental validation, IEEE Trans. Antennas Propag.*, Vol. 56, No. 1, pp. 127–132, Jan. 2008.

J.26 Dong, Y.; Xu, W.; Liu, S.: *Numerical studies for relative bandwidth of left-handed metamaterials with split ring resonators*, IEEE.org, doi:10.1109/mape.2009.5355730, Oct. 2009.

J.27 Kabiri, A.; Yousefi, L.; Ramahi, O.M.: *On the fundamental limitations of artificial magnetic materials, IEEE Trans. Antennas Propag.* Vol. 58, No. 7, pp. 2345–2353, July 2010.

J.28 Belov, P.A.; Simovski, C.A.: *Homogenization of electromagnetic crystals formed by uniaxial resonant scatterers, Phys. Rev. E*, Vol. 72, No. 026615, pp. 1–15, 2005.

J.29 Silveirinha, M.G.; Belov, P.A.: *Spatial dispersion in the lattice of split ring resonators with permeability near zero, Phys. Rev. B*, Vol. 77, No. 233104, pp. 1–4, 2008.

J.30 Sauviac, B.; Simovski, C.A.; Tretyakov, S.A.: *Double split-ring resonators: analytical modeling and numerical simulations, Electromagnetics*, Vol. 24, pp. 317–338, 2004.

J.31 Ishikawa, A.; Tanaka, T.: *Negative magnetic permeability of split-ring resonators in the visible light region, Opt. Commun.*, Vol. 258, pp. 299–305, 2006.

J.32 Shamonin, M.; Shamonin, E.; Kalinin, V.; Solymar, L.: *Resonant frequencies of a double split-ring resonator: analytical solutions and numerical simulations, Microwave Opt. Tech. Lett.*, Vol. 44, No. 2, pp. 133–136, Jan. 2005.

J.33 Zhurbenko, V.; Jensen, T.; Krozer, V.; Meincke, P.: *Analytical model for double split ring resonators with arbitrary ring width, Microwave Opt. Tech. Lett.*, Vol. 50, No. 2, pp. 511–515, Mar. 2008.

J.34 Pradeep, A.; Mridula, S.; Mohanan, P.: *Design of an edge-coupled dual-ring split-ring resonator, IEEE Antenna Propag. Mag.*, Vol. 33, No. 4, pp. 45–54, Aug. 2011.

J.35 Burokur, S.N.; Latrach, M.; Toutain, S.: *Analysis and design of waveguides loaded with split-ring resonators, J. Electromagn. Waves Appl.* Vol. 19, pp. 1407–1421, 2005.

J.36 Katsarakis, N.; Koschny, T.; Kafesaki, M.: *Electric coupling to the magnetic resonance of split-ring resonators, Appl. Phys. Lett.*, Vol. 84, No. 5, pp. 2943–2945, Apr. 2004.

J.37 Gay-Balmaz, P.; Martina, O.J.F.: *Electromagnetic resonances in individual and coupled split-ring resonators, J. Appl. Phys.*, Vol. 92, No. 5, pp. 2929–2936, Sept. 2002.

J.38 Zhang, L.; Tuttle, G.; Soukoulis, C.M.: *GHz magnetic response of split-ring resonators, Photon. Nanostruct.*, Vol. 2, pp. 155–159, 2004.

J.39 Aydin, K.; Ozbay, E.: *Capacitor-loaded split-ring resonators as tunable metamaterials, J. Appl. Phys.*, Vol. 101, pp. 0249111-1–0249111-5, 2007.

J.40 Liu, Y.; Fang, N.; Wu, D.; Sun, C.; Zhang, X.: *Symmetric and antisymmetric modes of electromagnetic resonators, Appl. Phys. A Mater. Sci.*, doi:10.1007/s00339-006-3837-0, 2007.

J.41 Woodley, J.F.; Wheeler, M.S.; Mojahedi, M.: *Left-handed and right-handed metamaterials composed of split-ring resonators and strip wires, Phys. Rev. E*, Vol. 71, No. 066605, pp. 1–6, 2005.

J.42 Cheng, C.Y.; Ziolkowski, R.W.: *Tailoring double-negative metamaterial responses to achieve anomalous propagation effects along microstrip transmission lines, IEEE Trans. Antennas Propag.*, Vol. 51, No. 2, pp. 2306–2314, Dec. 2003.

J.43 Ziolkowski, R.W.: *Design, fabrication, and testing of double negative metamaterials, IEEE Trans. Antennas Propag.*, Vol. 51, No. 7, pp. 1516–1529, July 2003.

J.44 Erentok, A; Luljak, P.L.; Ziolkowski, R.W.: *Characterization of a volumetric metamaterial realization of an artificial magnetic conductor for antenna applications, IEEE Trans. Antennas Propag.*, Vol. 53, No. 1, pp. 160–172, Jan. 2005.

J.45 Simovski, C.R.; Tretyakov, S.A.; Sochava, A.A.: *Antenna model for the conductive omega particles, J. Electromagn. Waves Appl.*, Vol. 11, pp. 1509–1530, 1997.

J.46 Tretyakov, S.A.; Mariotte, F.; Simovski, C.R.; Kharina, T.G.; Heliot, J.-Ph.: *Antenna model for chiral scatterers: comparison with numerical and experimental results, IEEE Trans. Antennas Propag.*, Vol. AP-44, No. 7, pp. 1006–1015, July 1996.

J.47 Aydin, K.; Li, Z.; Bilge, S.; Ozbay, E.: *Experimental and numerical study of omega type bianisotropic metamaterials combined with a negative permittivity medium, Photon. Nanostruct.-Fundam. Appl.*, Vol. 6, pp. 116–121, 2008.

J.48 Li, Z.; Aydin, K.; Bilge, S.; Ozbay, E.: *Retrieval of effective parameters for bianisotropic metamaterials with omega-shaped metallic inclusions, Nanostruct.-Fundam. Appl.*, Vol. 10, No. 9, pp. 329–336 2012.

J.49 Huangfu, J.; Ran, L.; Chen, H.; Zhang, X.; Chen, K.; Grzegorczyk, T.M.; Kong, J.A.: *Experimental confirmation of negative refractive index of a metamaterial composed of Ω-like metallic patterns, Appl. Phys. Lett.*, Vol. 84, pp. 1537–1539, 2004.

J.50 Zhang, F.; Houzet G.; Lheurette, E.; Lippens, D.; Chaubet, M.; Zhao, X.: *Negative-zero-positive metamaterial with omega-type metal inclusions, J. Appl. Phys.*, Vol. 103, pp. 04312-1–04312-8, 2008.

J.51 Zhang, F.; Gaillot, D.P.; Croënne, C.; Lheurette, E.; Mélique, X.; Lippens D.: *Low-loss left-handed metamaterials at millimeter waves, Appl. Phys. Lett.*, Vol. 93, pp. 083104-1–083104-83 2008.

J.52 Lee, D.A.; Vedral, L.J.; Smith, D.A.; Musselman, R.L.; Pinchuk, A.O: *Low-loss negative index metamaterials for X, Ku, and K microwave bands, AIP Adv.*, 5.4, 047119, pp.1–11, 2015.

J.53 Chen, H.; Ran, L.; Huangfu, J.; Zhang, X; Chen, K; Grzegorczyk, T.M.; Kong, J.A.: *Left-handed materials composed of only S-shaped resonators, Phys. Rev. E*, Vol. 70, No. 057605, pp. 1–4, 2004.

J.54 Chen, H.S.; Ran, L.X.; Huangfu, J.T.; Zhang, X.M.; Chen, K.S.: *Magnetic properties of S-shaped split-ring Resonators, Progr. Electromagn. Res. PIERS*, Vol. 51, pp. 231–237, 2005.

J.55 Chen, H.; Wang, Z.; Zhang, R.; Wang, H.; Lin, S.; Yu, F.; Moser, H.O.: *A meta-substrate to enhance the bandwidth of metamaterials, Sci. Rep.*, Vol. 4, No. 5264, pp. 1–5, 2014.

J.56 Smith, D.R.; Pendry, J.B.: *Homogenization of metamaterials by field averaging, J. Opt. Soc. Am. B*, Vol. 23, No. 3, pp. 391–403, Mar 2006.

J.57 Archer, O.; Lerat, J.M.; Mullejac, N.: *Evaluation and illustration of the properties of metamaterials using field summation, Opt. Express*, Vol. 15, No. 3, pp. 1096–1106, Feb. 2007.

J.58 Li, K.; McLean, S.J.; Greegor, R.B.; Parazzoli, C.G.; Tanielian, M.H.: *Free-space focused-beam characterization of left-handed materials, Appl. Phys. Lett.*, Vol. 82, No. 15, pp. 2535–2537, Apr. 2003.

J.59 Shelby, R.; Smith, D.; Schultz, S.: *Experimental verification of a negative index of refraction, Science*, Vol. 292, pp. 77–79, Apr. 2001.

J.60 Weir, W.B.: *Automatic measurement of complex dielectric constant and permeability at microwave frequencies, Proc. IEEE*, Vol. 62, pp. 33–36, Jan. 1974.

J.61 Barroso, J.J.; Paula, A.L. de: *Retrieval of permittivity and permeability of homogeneous materials from scattering parameters, J. Electromagn. Waves Appl.*, Vol. 24, pp. 1563–1574, 2010.

J.62 Chen, H.; Zhang, J.; Bai, Y.; Luo, Y.; Ran, L.; Jiang, Q.; Kong, J.A.: *Experimental retrieval of the effective parameters of metamaterials based on a waveguide method, Opt. Express*, Vol. 14, No. 26, pp. 12944–192947, Dec. 2006.

J.63 Smith, D.R.; Schultz, S.; Markos, P.; Soukoulis, C.M.: *Determination of effective permittivity and permeability of metamaterials from reflection and transmission coefficients, Phys. Rev. B*, Vol. 65, No. 195104, pp. 1–5, 2002.

J.64 Smith, D.R.; Vier, D.C.; Koschny, T.; Soukoulis, C.M.: *Electromagnetic parameter retrieval from inhomogeneous metamaterials, Phys. Rev. E*, Vol. 71, No. 036617, pp. 1–11, 2005.

J.65 Chen, X.; Grzegorczyk, T.M.; Wu, B.I.; Pacheco, J.; Kong, J. A.: *Robust method to retrieve the constitutive effective parameters of metamaterials, Phys. Rev. E*, Vol. 70, No. 016608, pp. 1–7, 2004.

J.66 Arslanagić, S.; Hansenl, T.V.; Mortensen, N.A.; Gregersenl, A.H; Sigmund, O.; Ziolkowski, R.W.; Breinbjerg, O.: *A review of the scattering-parameter extraction method with clarification of ambiguity issues in relation to metamaterial homogenization, IEEE Antennas Propag. Mag.*, Vol. 55, No. 2, pp. 91–100, Apr. 2013.

J.67 Numan, A.B.; Sharawi, M.S.: *Extraction of material parameters for metamaterials using a full-wave simulator, IEEE Antennas Propag. Mag.*, Vol. 55, No. 5, pp. 202–211, Oct. 2013.

J.68 Szabó, Z.; Park, G.H.; Hedge, R, Li, E.P.: *A unique extraction of metamaterial parameters based on Kramers–Kronig relationship, IEEE Trans. Microwave Theory Tech.*, Vol. 5, No. 10, pp. 2646–2633, Oct. 2010.

J.69 Gong, J.Q.; Chu, Q.X.: *Effective electromagnetic parameters extraction method for 1D left-handed material*, Proc. Asia Pacific Conf., APMC, Bangkok, Thailand, 11–14 Dec. 2007.

J.70 Luukkonen, O.; Maslovski, S.I.; Tretyakov, S.A.: *A stepwise Nicolson–Ross–Weir-based material parameter extraction method, IEEE Antenna Wirel. Propag. Lett.*, Vol. 10, pp. 1295–1298, 2011.

J.71 Kriegler, C.E; Rill, M.S.; Linden, S.; Wegener, M.: *Bianisotropic photonic metamaterials, IEEE J. Sel. Top. Quantum Elect.*, Vol. 16, No. 2, pp. 367–375, 2010.

J.72 Chalapat, K.; Sarvala, K.; Li, J.; Paraoanu, G.S.: *Wideband reference-plane invariant method for measuring electromagnetic parameters of materials, IEEE Trans. Microwave Theory Tech.*, Vol. 57, No. 9, pp. 2257–2267, Sept. 2009.

J.73 Galek, T.; Porath, K.; Burkel, E.: Rienen, U. van: *Extraction of effective permittivity and permeability of* *metallic powders in the microwave range, Modell. Simul. Mater. Sci. Eng. IOP Pub.*, Vol. 18, No. 025015, pp. 13, 2010.

J.74 Cohen, D.; Shavit, R.: *Bi-anisotropic metamaterials effective constitutive parameters extraction using oblique incidence S-parameters method, IEEE Trans. Antennas Propag.*, Vol. 63, No. 5, pp. 2071–2078, May. 2005.

J.75 Nicolson, A.M.; Ross, G.F.: *Measurement of the intrinsic properties of materials by time-domain techniques, IEEE Trans. Instrum. Meas.*, Vol. IM-19, pp. 377–382, Nov. 1970.

J.76 Shi, Y.; Li, Z.Y.; Li, L.; Liang, C.H.: *A retrieval method of effective electromagnetic parameters for inhomogeneous metamaterials, IEEE Trans. Microwave Theory Tech.*, Vol. 65, No. 4, pp. 1160–1178, Sept. 2009.

J.77 Peiponen, K.E.; Lucarini, V.; Vartiainen, E.M.; Saarinen, J.J.: *Kramers–Kronig relations and sum rules of negative refractive index media, Eur. Phys. J. B Condens. Matter Complex Syst.*, Vol. 41, No. 1, pp. 61–65, Sept. 2004.

J.78 http://effmetamatparam.sourceforge.net/net

J.79 Jin, J.; Liu, S.; Lin, Z.; Chui, S.T.: *Effective – medium theory for anisotropic magnetic materials, Phys. Rev. B*, Vol. 80, No. 115101, pp. 1–6, 2009.

J.80 Zhao, Q.; Zhoul, J.; Zhang, F.; Lippens, D.: *Mie resonance-based dielectric metamaterials, Mater. Today*, Vol. 12, pp. 60–69, Dec. 2009.

J.81 Holloway, C.L.; Kuester, E.F.; Baker-Jarvis, J.; Kabos, P.: *A double negative (DNG) composite medium composed of magnetodielectric spherical particles embedded in a matrix, IEEE Trans. Antennas Propag.*, Vol. 51, No. 10, pp. 2596–2603, Oct. 2003.

J.82 Vinogradov, A.P.; Dorofeenko, A.V.; Zouhdi, S.: *On the problem of the effective parameters of metamaterials, Phys. Uspechi*, Vol. 51, No. 5, pp. 485–492, 2008.

J.83 Wallén, H.; Kettunen, H.; Qi, J.; Sihvola, A.: *Some Interesting Effects when Homogenizing Plasmonic Composites, XXXII Finnish URSI Convention on Radio Science and Electromagnetics* 2010, pp. 47–50, Oulu, Finland, 2010. http://www.ursi.fi/2010/papers/A33_Wallen.

J.84 Sihvola, A.: *Six-dimensional view of dielectric mixtures as metamaterials, Eur. Phys. J. Appl. Phys.* Vol. 46, No. 32602, pp. 1–5, 2009.

J.85 Mie, G.: *Beitrage zur optic truber medien speziell kolloidaler metallosungen, Ann. Phys.*, Vol. 25, pp. 377–445, 1908.

J.86 Doyle, W.T.: *Optical properties of a suspension of metal particles, Phys. Rev. B*, Vol. 39, No. 14, pp. 9852–9858, 1989.

J.87 Ruppin, R.: *Evaluation of extended Maxwell Garnett theories, Opt. Commun.*, Vol. 182, pp. 273–279, Aug. 2000.

J.88 Szabo, Z.; Fuzi, J.: *Equivalence of magnetic metamaterials and composites in the view of the effective medium theories, IEEE Trans. Magn.*, Vol. 50, No. 4, pp. 2501104-1–2501104-4, Apr. 2014.

J.89 Aspnes, D.E.: *Local field effects and effective – medium theory: a microscopic perspective, Am. J. Phys.*, Vol. 50, No. 8, pp. 704–709, 1982.

J.90 Lewin, L.: *The electrical constants of a material loaded with spherical particles, Proc. Inst. Elec. Eng.*, Vol. 94, pp. 65–68, 1947.

J.91 Vendik, I.; Vendik, O.; Kolmakov, I.; Odit, M.: Modelling of isotropic double negative media for microwave applications, *Opto-Electron. Rev.*, Vol. 14, No. 3, pp. 179–186, 2006.

J.92 Zhao, O.; Kang, L.; Du, B.; Zhao, H.; Xie, Q.; Huang, X.; Li, B.; Zhou, J.; Li, L.: *Experimental demonstration of isotropic negative permeability in a three-dimensional dielectric composite Phys. Rev. Lett.*, Vol. 101, pp. 027402, July 2008.

J.93 Ueda, T.; Lai, A.; Itoh, T.: *Demonstration of negative refraction in a cut-off parallel-plate waveguide loaded with 2-D square lattice of dielectric resonators, IEEE Trans. Microwave Theory Tech.*, Vol. 55, No. 6, pp. 1280–1287, July 2007.

J.94 Andrea Alùa, A.; Enghetab, N.: *Polarizabilities and effective parameters for collections of spherical nanoparticles formed by pairs of concentric double-negative, single-negative, and/or double-positive metamaterial layers, J. Appl. Phys.*, Vol. 97, pp. 094310-1–094310-12, 2005.

J.95 Liu, L.Y.; Sun, J.B.; Fu, X.J.; Zhou, J.; Zhao, Q.: *Artificial magnetic properties of dielectric metamaterials in terms of effective circuit model, Progr. Electromagn. Res. PIERS*, Vol. 116, pp. 159–170, 2011.

22

Metamaterials Realization and Circuit Models – II

(Metalines and Metasurfaces)

Introduction

The electromagnetic properties of the metamaterials are discussed in the section (5.5) of chapter 5. Chapter 21 presents their realization and characterization as the 3D-bulk engineered materials. However, the 1D and 2D forms of the metamaterials, also known as the *metalines* and *metasurfaces,* have their specific properties and applications, from the microwave to optical frequency ranges. In practice, the metalines and metasurfaces are realized by the periodic arrangement of the inclusions in the host medium. However, they are different from the artificially engineered EBG/PBG lines and surfaces, discussed in chapters 19 and 20. The EBG/PBG materials are the *bandgap materials* as their lattice constant is half-wavelength, offering the Bragg diffraction. However, for the metalines and metasurfaces, the size of inclusion and lattice constant both are preferably less than one-tenth of a wavelength, or certainly less than a fourth of a wavelength. Therefore, the Bragg diffraction is avoided in the metalines and metasurfaces. It helps to get the *homogeneous effective medium* characterized by the effective (equivalent) permittivity and permeability. The metalines and metasurfaces are modeled using the circuit models.

The metalines are realized by both the *resonant* and *nonresonant* inclusions providing wide operational bandwidth. The 1D metalines are realized in the microstrip, CPW, and waveguide technology. These are useful for their ability to miniaturize microwave components with controllable dispersion characteristics. The 2D-metasurfaces are also realized in planar technology. This chapter presents the characteristics and realization of the metalines and metasurfaces in planar technology. Several textbooks are available to cover various aspects and applications of the 1D and 2D metamaterials in detail [B.1–B.8].

Objectives

- *To present circuit models to synthesize 1D DNG and other metamaterials.*
- *To discuss the realization of nonresonant metalines in microstrip technology.*
- *To discuss the realization of resonant metalines in microstrip and CPW technologies.*
- *To present a few illustrative application examples of metalines.*
- *To present the characterization and control of 2D-metasurfaces.*
- *To discuss a few applications of 2D-metasurfaces.*

22.1 Circuit Models of 1D-Metamaterials

The 1D-metamaterials, also called the *metalines* [B.7, B.8], are normally designed in the microstrip, CPW, and waveguide technology. The standard 1D transmission line *host* medium supports the forward wave propagation. These DPS media are loaded with the periodic *nonresonant* and *resonant type inclusions* to realize the metalines. Chapter 19 presents the modeling of the 1D-periodic lines using the 1D transmission matrix, i.e. the [ABCD] matrix. However, the periodic medium should be homogenized to get the *effective medium* described by the effective permeability and effective permittivity. The terminology – "*effective*" appears to be confusing in the context of the microstrip-based metalines, as the effective permittivity/permeability is normally used to describe the propagation characteristics of an inhomogeneous microstrip line. To avoid confusion, we prefer to use the terminology *equivalent permeability* and *equivalent permittivity* to characterize the effective medium.

Introduction to Modern Planar Transmission Lines: Physical, Analytical, and Circuit Models Approach, First Edition. Anand K. Verma.

This section considers the basic 1D-circuit models to get the artificial DPS, DNG, ENG, and MNG material media. The electromagnetic properties of these media are discussed in section (5.5) of chapter 5.

22.1.1 Homogenization of the 1D-medium

The usual transmission line and open space are *continuous media*, with or without losses, supporting the forward wave propagation. However, the 1D medium, both the DPS and DNG types, are engineered by the periodic arrangement of the series/shunt lumped inductors and shunt/series capacitor network. Even the series and parallel resonating structures, connected in series and parallel configurations, could be used to get the periodic lines. Thus, an artificial loaded line constitutes the *discrete medium*. To get the homogeneous effective medium, the size of the inclusions and the period must be very small as compared to the operating wavelength. The periodic lines could be modeled by the cascaded symmetric unit cells of the T-network, π-network, or asymmetric L-network. Using equations (19.3.4a) and (19.3.30) of chapter 19, the following dispersion equation is obtained for the lossless 1D medium, formed by the loaded periodic lines, expressed as the L-network [J.1, B.3]:

$$\cos(\beta d) = \frac{A+D}{2} = 1 + \frac{Z_{se}^{pul} Y_{sh}^{pul}}{2}$$
$$\Rightarrow 4\sin^2(\beta d/2) = -Z_{se}^{pul} Y_{sh}^{pul}, \tag{22.1.1}$$

where β is the propagation constant of the loaded transmission line medium, d is the period of the unit cell formed by the L-network with series impedance Z_{se}^{pul} and shunt admittance Y_{sh}^{pul} on the per unit length (*p.u.l.*) basis. Expression (22.1.1) is valid for the symmetrical T-network also. It is obvious that to get the real solution of β, the series and shunt arm impedances must be imaginary quantities with opposite signs. If the impedance (Z_{se}) and admittance (Y_{sh}) are lumped elements, then $Z_{se}^{pul} = Z_{se}/d$, and $Y_{sh}^{pul} = Y_{sh}/d$. Using the *homogeneity condition* $d/\lambda << 1$, equation (22.1.1) is reduced to the following equation for the propagation constant of the effective medium:

$$\beta = \sqrt{-(Z_{se}/d)(Y_{sh}/d)} = \sqrt{-\left(Z_{se}^{pul} Y_{sh}^{pul}\right)}$$
$$\Rightarrow \beta = \sqrt{-\left(j\omega L_{se}^{pul} j\omega C_{sh}^{pul}\right)} = \omega\sqrt{\left(L_{se}^{pul} C_{sh}^{pul}\right)}. \tag{22.1.2}$$

Expression (22.1.2) is a standard expression of the propagation constant of a continuous transmission line discussed in chapter 2. Thus, under the homogeneity condition, the discrete periodic line structure behaves as a continuous effective medium. Of course, under Bragg's condition, the concept of the effective medium is not applicable and the medium exhibits the stopband characteristics [B.2, B.6, B.7]. Equation (22.1.1) is a more general expression. *The series impedance and shunt admittance of a unit cell could be formed by any combination of inductance and capacitance to create several kinds of lossless effective media.* The effective medium is characterized by the *equivalent (i.e. effective) permeability* μ_{eq} and *equivalent (i.e. effective) permittivity* ε_{eq}. These material medium parameters could be computed from series impedance and shunt admittance of the equivalent transmission line model of a DPS medium. However, the equivalence is also applicable to other engineered media.

22.1.2 Circuit Equivalence of Material Medium

Figure (22.1) shows the electromagnetic modeling of a *continuous medium* in terms of the continuous equivalent transmission line with characteristic impedance Z_0 and propagation constant γ, and also in terms of the *discrete circuit elements*: series impedance Z_{se}^{pul} and shunt admittance Y_{sh}^{pul} of the equivalent asymmetrical L-network to model a material medium. However, the symmetrical T-network, and occasionally π-network, is also used for this purpose [B.2, B.3, B.5, B.7]. For the engineered effective medium, the process is in the reverse order, i.e. *the discrete circuit networks are cascaded to synthesize the effective (equivalent) homogeneous*

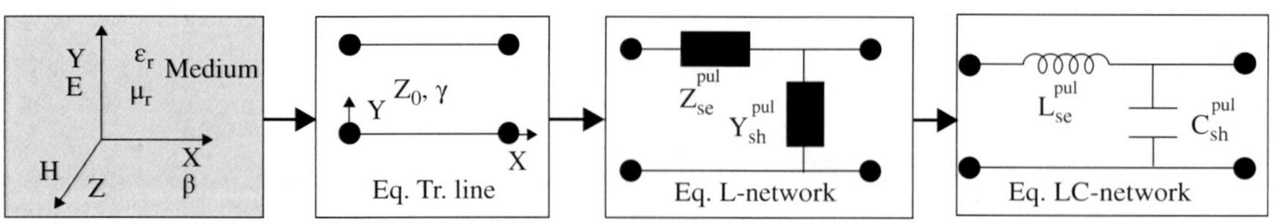

Figure 22.1 Circuit equivalence of a material medium.

medium with μ_{eq}, ε_{eq} *under the homogeneity condition.* The correspondence between the material parameters and the primary line constants is discussed in section (4.3) of chapter 4, and also in subsection (5.5.3) of chapter 5. However, it could also be obtained by using the correspondence between the electric field wave equation of the effective medium and the voltage wave equation of the equivalent transmission line:

$$\frac{d^2E_y}{dx^2} = -\beta^2E_y = -\omega^2\mu_{eq}\varepsilon_{eq}E_y = \left(j\omega\mu_{eq}\right)\left(j\omega\varepsilon_{eq}\right)E_y \quad (a)$$
$$\frac{d^2V}{dx^2} = Z_{se}^{pul}Y_{sh}^{pul}V = \left(jX_{se}^{pul}\right)\left(jB_{sh}^{pul}\right)V = \left(j\omega L_{se}^{pul}\right)\left(j\omega C_{sh}^{pul}\right)V \quad (b), \tag{22.1.3}$$

where E_y is the electric field vector in the medium and V is the voltage across the equivalent line. Figure (22.1) shows the series impedance/reactance *p.u.l* and shunt admittance/susceptance *p.u.l.* of the equivalent L-network, corresponding to electrical parameters of a medium. On comparing the above equations, the equivalent permeability μ_{eq} and equivalent permittivity ε_{eq} of the effective medium are obtained in terms of the line parameters:

$$\mu_{eq} = \frac{Z_{se}^{pul}}{j\omega} \quad (a), \qquad \Rightarrow \mu_{eq} = \frac{X_{se}^{pul}}{\omega} = L_{se}^{pul} \quad (b)$$
$$\varepsilon_{eq} = \frac{Y_{sh}^{pul}}{j\omega} \quad (c), \qquad \Rightarrow \varepsilon_{eq} = \frac{B_{sh}^{pul}}{\omega} = C_{se}^{pul} \quad (d). \tag{22.1.4}$$

In the above equations, the series inductance and shunt capacitance model the transmission line. Therefore, the equivalent line, shown in Fig (22.1), is a usual LC-line that models the DPS medium. However, the above relations (22.1.4a, c) are also valid for other engineered lines, such as the ENG, DNG, and MNG. These engineered media are discussed in chapter 5 using Fig (5.10). The propagation constant β and characteristic impedance Z_c of the equivalent line and effective medium are computed from the following expressions:

$$\text{Transmission line: } \gamma = \sqrt{Z_{se}^{pul}Y_{sh}^{pul}} \quad (a), \quad Z_0 = \sqrt{\frac{Z_{se}^{pul}}{Y_{sh}^{pul}}} \quad (b)$$
$$\text{Effective Medium: } \beta = \omega\sqrt{\mu_{eq}\varepsilon_{eq}} \quad (c), \quad Z_0 = \sqrt{\frac{\mu_{eq}}{\varepsilon_{eq}}} \quad (d). \tag{22.1.5}$$

The effective medium model is also valid for a lossy medium [J.2, B.2]. However, only the lossless medium is considered below. Figure (5.10) of chapter 5 shows that the four kinds of artificial media could be developed by cascading of an infinite number of the unit cells. The following expressions of the effective media are obtained by using the equations (22.1.4a,c) and (22.1.5):

DPS-Medium (L–C) Line (L-Series, C-Shunt)

$$\mu_{eq} = \frac{j\omega L_{se}^{pul}}{j\omega} = L_{se}^{pul} \quad (a), \quad \varepsilon_{eq} = \frac{j\omega C_{sh}^{pul}}{j\omega} = C_{sh}^{pul} \quad (b)$$
$$\beta = \omega\sqrt{L_{se}^{pul}C_{sh}^{pul}}, \quad (c), \quad Z_0 = \sqrt{\frac{L_{se}^{pul}}{C_{sh}^{pul}}} \quad (d). \tag{22.1.6}$$

ENG-Medium (L–L) Line (L-Series, L-Shunt)

$$\mu_{eq} = \frac{j\omega L_{se}^{pul}}{j\omega} = L_{se}^{pul} \quad (a), \quad \varepsilon_{eq} = \frac{1/j\omega L_{sh}^{pul}}{j\omega} = -\left(\frac{1}{\omega^2 L_{sh}^{pul}}\right) \quad (b)$$
$$\beta = j\sqrt{\frac{L_{se}^{pul}}{L_{sh}^{pul}}}, \quad (c), \quad Z_0 = j\omega\sqrt{L_{se}^{pul}L_{sh}^{pul}} \quad (d). \tag{22.1.7}$$

DNG-Medium (C–L) Line (C-Series, L-Shunt)

$$\mu_{eq} = \frac{1/j\omega C_{se}^{pul}}{j\omega} = -\left(\frac{1}{\omega^2 C_{se}^{pul}}\right) \quad (a)$$
$$\varepsilon_{eq} = \frac{1/j\omega L_{sh}^{pul}}{j\omega} = -\left(\frac{1}{\omega^2 L_{sh}^{pul}}\right) \quad (b)$$
$$\beta = -\frac{1}{\omega}\sqrt{\frac{1}{L_{sh}^{pul}C_{se}^{pul}}}, \quad (c), \quad Z_0 = \sqrt{\frac{L_{sh}^{pul}}{C_{se}^{pul}}} \quad (d). \tag{22.1.8}$$

MNG-Medium (C–C) Line (C-Series, C-Shunt)

$$\mu_{eq} = \frac{1/j\omega C_{se}^{pul}}{j\omega} = -\left(\frac{1}{\omega^2 C_{se}^{pul}}\right) \quad (a), \quad \varepsilon_{eq} = \frac{j\omega C_{sh}^{pul}}{j\omega} = C_{sh}^{pul} \quad (b)$$
$$\beta = j\sqrt{\frac{C_{sh}^{pul}}{C_{se}^{pul}}}, \quad (c), \quad Z_0 = \frac{j}{\omega}\sqrt{\frac{1}{C_{se}^{pul}C_{sh}^{pul}}} \quad (d). \tag{22.1.9}$$

In the above equations L_{se}^{pul} and L_{sh}^{pul} are the series inductance and shunt inductance of the equivalent L-network on *p.u.l.* basis. Likewise, C_{se}^{pul} and C_{sh}^{pul} are the series capacitance and shunt capacitance. It is noted from the above equations that for the DPS medium both the equivalent permeability μ_{eq} and equivalent permittivity ε_{eq} are positive, giving positive real values for the propagation constant and characteristic impedance supporting the forward waves with a positive phase and group velocities.

The lossless DPS medium is nondispersive. For the ENG medium ε_{eq} is negative, leading to imaginary values for both β and Z_0. Thus, the medium supports only the nonpropagating evanescent mode. Even during the evanescent mode, the ENG medium is dispersive. For the DNG medium, both equivalent permeability μ_{eq} and equivalent permittivity ε_{eq} are negative quantities, giving the negative value of the frequency-dependent β, and positive value of Z_0. The DNG medium supports the backward wave propagation with phase velocity in the negative direction, and group velocity, indicating power flow, in a positive direction. The direction of group velocity is in the direction of power flow that is from source to load. The lossless DNG medium is dispersive. Finally, for the MNG medium, μ_{eq} is negative, leading to imaginary values for both β and Z_c. Thus, the MNG medium also supports only the nonpropagating evanescent mode. It is a dispersive medium.

22.1.3 Single Reactive Loading of Host Medium

The host transmission line could be treated as a cascaded network of the unit cells of the L-type $L_h^{pul}C_h^{pul}$ – circuit networks forming the 1D DPS medium. Figure (22.2a, c) shows the unit cells of the reactively loaded host line with series arm loading capacitance C_{se}^{pul} in series or parallel to host line inductance L_h^{pul}. Figure (22.2b, d) shows the shunt arm loading inductance L_{sh}^{pul} in parallel or in series to host line capacitance C_h^{pul}. The loading elements are shown in gray boxes. In practice, the host line could be a microstrip, CPW, or waveguide. The reactive loading is achieved by the planar structures, or by the lumped SMT components. Depending upon arrangement and the type of reactive loading, the loaded line creates the MNG, MNZ, ENG, and ENZ type engineered effective medium with either the Drude type or Lorentz type response. Section (3.4) of chapter 3 presents the propagation characteristics of some of the reactively loaded lines [B.9]. However, the 1D-effective medium under the homogeneity condition is not discussed at that stage. The creations of some effective media are discussed below.

Series Capacitance C_{se}^{pul} in the Series Arm

Figure (22.2a) shows the circuit configuration forming an effective medium. The series arm loading modifies the equivalent permeability of the host medium, while its equivalent permittivity remains unchanged. Using

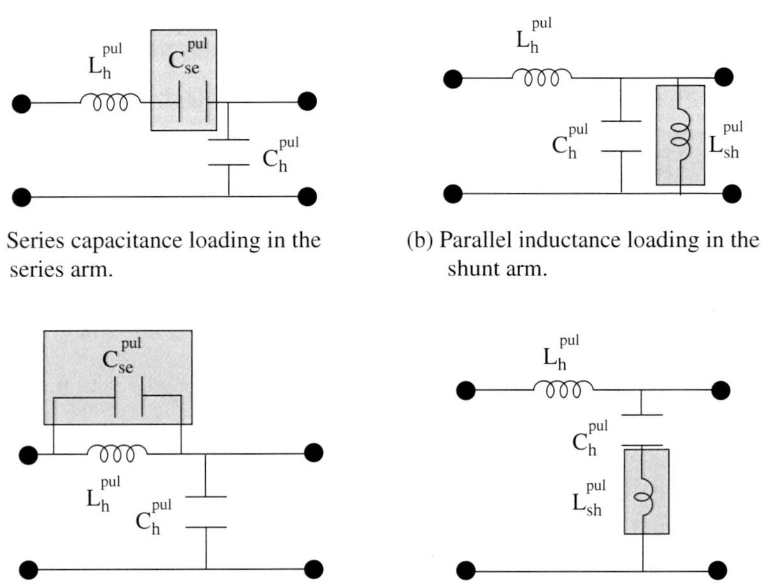

(a) Series capacitance loading in the series arm.

(b) Parallel inductance loading in the shunt arm.

(c) Parallel capacitance loading in the series arm.

(d) Series inductance loading in the shunt arm.

Figure 22.2 Single-arm reactive loading of the host LC-line. Loading components are in gray boxes.

the equations (22.1.4a, c), these are computed as follows:

$$\text{Series impedance: } Z_{se}^{pul} = \left(j\omega L_h^{pul} + \frac{1}{j\omega C_{se}^{pul}}\right) = j\omega L_h^{pul}\left(1 - \frac{1}{\omega^2 L_h^{pul} C_{se}^{pul}}\right) \quad \text{(a)}$$

$$\text{Equivalent Permeability: } \mu_{eq} = \mu_0\mu_{r,eq} = L_h^{pul}\left(1 - \frac{\omega_{mp}^2}{\omega^2}\right) \quad \text{(b)}$$

$$\text{where, } L_h^{pul} = \mu_h = \mu_0\mu_{r,h} \quad \text{(c)}$$

$$\omega_{mp} = \frac{1}{\sqrt{L_h^{pul} C_{se}^{pul}}} \quad \text{(d)}$$

$$\text{Shunt admittance: } Y_{sh} = j\omega C_h^{pul}$$

$$\varepsilon_{eq} = \varepsilon_0\varepsilon_{r,h} = C_h^{pul} \quad \text{(e).} \qquad (22.1.10)$$

In the case of the air host medium, $\mu_h \to \mu_0$ and $\varepsilon_h \to \varepsilon_0$. The line loaded with series capacitance C_{se}^{pul} behaves as an artificial magnetic plasma medium with magnetic plasma frequency ω_{mp}. Therefore, the line loaded with a series capacitance follows the Drude model discussed in subsection (6.5.2) of chapter 6. The propagation constant and characteristic impedance are computed using equations (22.1.5a, b). Next, phase velocity and group velocity can also be computed using equations (3.3.10) and (3.3.11) of chapter 3. Figure (22.3a) shows the Drude model type permeability response of the effective medium. It has negative permeability below the magnetic plasma frequency ω_{mp}.

Thus, for frequency range $\omega < \omega_{mp}$, the effective medium behaves as an MNG medium created by the C–C line. For the host air medium ($\mu_r = 1$) above ω_{mp}, the effective medium acts as an MNZ medium with $0 < \mu_{r,eq} < 1$. The permittivity of the host air medium ($\varepsilon_r = 1$) remains unchanged by the series capacitor loading. The Drude type engineered medium is a wideband nonresonant negative permeability material below ω_{mp}. It is an MNZ medium above ω_{mp}.

Parallel Inductance L_{sh}^{pul} in the Shunt Arm

Figure (22.2b) shows the loading of a host line with parallel inductance L_{sh}^{pul} inclusion in the shunt arm of the L-network. The unloaded L-network corresponds to the

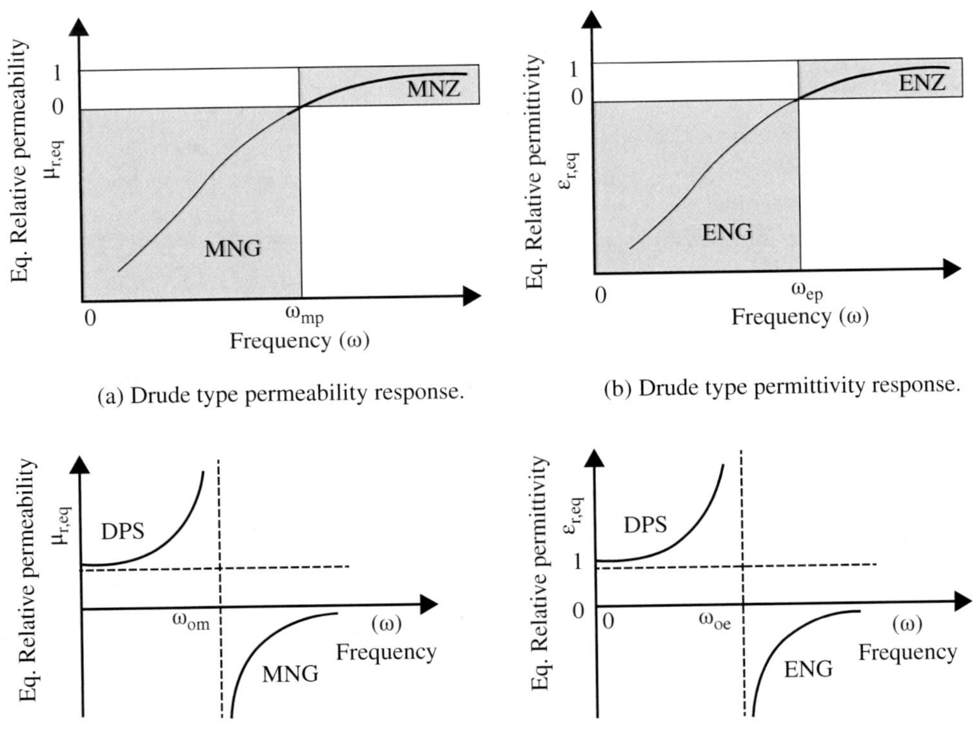

(a) Drude type permeability response.

(b) Drude type permittivity response.

(c) Lorentz type permeability response.

(d) Lorentz type permittivity response.

Figure 22.3 The permeability and permittivity response of the effective medium.

DPS host medium. The permeability of the host medium remains unchanged. However, the equivalent permittivity of the effective medium is modified. The equivalent permittivity is obtained as follows by using equation (22.1.4c):

$$\text{Shunt admittance: } Y_{sh}^{pul} = j\omega C_h^{pul}\left(1 - \frac{1}{\omega^2 L_{sh}^{pul} C_h^{pul}}\right) = j\omega C_h^{pul}\left(1 - \frac{\omega_{ep}^2}{\omega^2}\right) \quad (a)$$

$$\text{Equivalent permittivity: } \varepsilon_{eq} = \varepsilon_0\varepsilon_{r,eq} = C_h^{pul}\left(1 - \frac{\omega_{ep}^2}{\omega^2}\right) \quad (b)$$

$$\text{where, } C_h^{pul} = \varepsilon_h = \varepsilon_0\varepsilon_{r,h}, \qquad \omega_{ep} = \frac{1}{\sqrt{L_{sh}^{pul} C_h^{pul}}} \quad (c). \tag{22.1.11}$$

Again, Fig (22.3b) shows the Drude type equivalent permittivity of the effective medium with electric plasma frequency ω_{ep}. For the frequency range $\omega < \omega_{ep}$, the ENG medium with negative permittivity is created by the L–L line. The air host medium loaded with shunt inductor L_{sh}^{pul} acts as an ENZ medium $(0 < \varepsilon_{r,eq} < 1)$ above the electric plasma frequency ω_{ep}. The propagation constant, characteristic impedance, phase velocity, and the group velocity can be easily computed. Thus, Fig (22.3b) describes the electric plasma-type wire medium discussed in subsection (21.1.2) of chapter 21.

Parallel Capacitance C_{se}^{pul} in the Series Arm

Figure (22.2c) shows the circuit configuration of an effective medium with modified equivalent permeability. It is achieved by loading the host medium line with a capacitance C_{se}^{pul} parallel to L_h^{pul} connected in the series arm. The permittivity of the host medium remains unchanged. Following the previous process, the equivalent permeability is obtained:

$$\text{Series impedance: } Z_{se}^{pul} = \frac{j\omega L_h^{pul}\omega_{0m}^2}{(\omega_{0m}^2 - \omega^2)} \quad (a)$$

$$\text{Equivalent permeability: } \mu_{eq} = \mu_0\mu_{r,eq} = \frac{L_h^{pul}\omega_{0m}^2}{\omega_{0m}^2 - \omega^2} \quad (b)$$

$$\text{where, } L_h^{pul} = \mu_h = \mu_o\mu_{r,h}, \qquad \omega_{0m} = \frac{1}{\sqrt{L_h^{pul} C_{se}^{pul}}} \quad (c). \tag{22.1.12}$$

Figure (22.3c) shows the Lorentz type permeability response of the effective medium. It has no plasma frequency. However, the medium has a *magnetic resonant frequency* ω_{0m}. In the low-pass region $\omega < \omega_{0m}$, the medium has a high value of positive permeability near the resonance. In practice, $\mu_{r,eq}$ is finite due to the losses. In the high-pass frequency range $\omega > \omega_{0m}$, the medium acts as the MNG medium created by the C–C line. Thus, for the frequency $\omega < \omega_{0m}$, medium supports the forward wave propagation, whereas for $\omega > \omega_{0m}$, it is in the evanescent mode. Just below ω_{0m}, *the high value of permeability can be realized in the DPS medium.*

Series Inductance L_{sh}^{pul} in the Shunt Arm

Figure (22.2d) shows the reactively loaded host line with the series-connected inductor L_{sh}^{pul} inclusion in the shunt arm of the L-network corresponding to the host medium. Again, it provides the Lorentz type of equivalent permittivity. In the process of loading, the permeability of the medium remains unchanged. The equivalent permittivity of the effective medium is obtained:

$$\text{Shunt admittance: } Y_{sh}^{pul} = \frac{1}{j\omega L_{sh}^{pul}\left(1 - \frac{\omega_{0e}^2}{\omega^2}\right)} \quad (a)$$

$$\text{Equivalent permittivity: } \varepsilon_{eq} = \varepsilon_0\varepsilon_{r,eq} = \frac{C_h^{pul}\omega_{0e}^2}{\omega_{0e}^2 - \omega^2} \quad (b)$$

$$\text{where, } C_h^{pul} = \varepsilon_h = \varepsilon_0\varepsilon_{r,h}, \qquad \omega_{0e} = \frac{1}{\sqrt{L_{sh}^{pul} C_h^{pul}}} \quad (c). \tag{22.1.13}$$

Figure (22.3d) shows the Lorentz type equivalent permittivity response, with the *electric resonance frequency* ω_{0e}, of the effective medium without any plasma frequency. For $\omega < \omega_{0e}$, the medium is in the low-pass mode and *exhibits high positive permittivity* near the lower side of the resonance. However, for the frequency $\omega > \omega_{0e}$, the ENG effective medium is created by the L–L line.

22.1.4 Single Reactive Loading of Host Medium with Coupling

The effective media discussed above do not account for the coupling between the host line and reactive inclusions. The reactively loaded coupled medium creates the Lorentz type effective medium with the resonance and plasma frequencies. The coupling could be either inductive or capacitive to get the MNG or ENG response respectively below the plasma frequency and above the resonance frequency. The MNG/ENG is created in a narrow frequency band. Above the plasma frequency, the MNZ or ENZ medium is created.

Figure (22.4a) shows a loading capacitor C_{se}^{pul}, in the series arm, parallel to the host medium inductor L_h^{pul}. It also shows that the loaded host line is coupled through

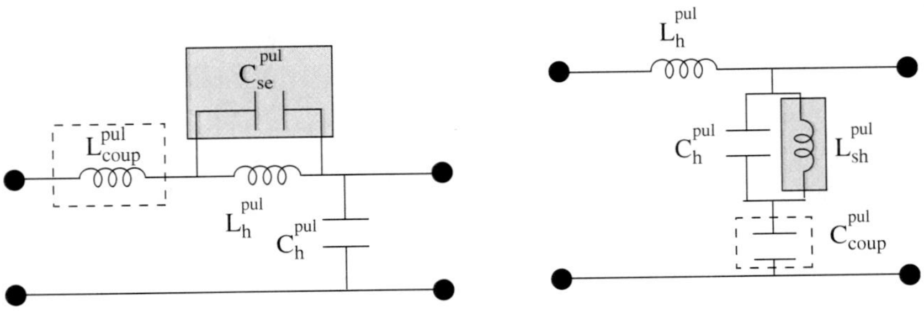

(a) Series inductive coupling. (b) Shunt capacitive coupling.

Figure 22.4 Coupling between host line and reactive loadings. Loading is shown in gray boxes.

the *series inductor* L_{coup}^{pul} *to an external circuit*. The coupling element is shown in a dotted box. The circuit configuration provides the modified equivalent permeability of the effective medium. It could be computed by the series impedance, following the process discussed above. Below the magnetic resonance frequency $\omega_{0m} = 1/\sqrt{L_h^{pul} C_{se}^{pul}}$, the medium is in the low-pass region with higher permeability. However, at the frequency $\omega > \omega_{0m}$, the parallel resonant circuit behaves as a capacitive reactance. It further produces series resonance with the coupling inductance L_{coup}^{pul} to create the magnetic plasma frequency $\omega_{mp} > \omega_{0m}$; so in a narrow frequency band, $\omega_{0m} < \omega < \omega_{mp}$, the MNG medium is created. Above the plasma frequency $\omega > \omega_{mp}$, the MNZ medium is created. It is noted that the presence of the inductive coupling L_{coup}^{pul} modifies the permeability response of Fig (22.3c) in the MNG region. Figure (22.5a) shows the modified Lorentz type permeability response of the effective medium. In Fig (22.4a), *the series capacitive coupling* could also be used, in place to inductive coupling, to realize the bandpass filter type response in the DGS loaded microstrip. The series-connected parallel resonant circuit is realized by a DGS and the coupling capacitor is realized by the gap discontinuity in the microstrip line located at the neck of the DGS [J.3].

Figure (22.4b) shows the parallel inductor loading, i.e. inductive inclusion, in the shunt arm of the host medium. The coupling between the host line and inclusion occurs through the capacitor C_{coup}^{pul} in the shunt arm. Again, the shunt admittance is used to get the Lorentz type equivalent permittivity response shown in Fig (22.5b) with both the electric resonance frequency ω_{0e} and the electric plasma frequency ω_{ep}. In the low-pass frequency range $\omega < \omega_{0e}$, the medium is a DPS medium with high permittivity. In a narrow frequency range $\omega_{0e} < \omega < \omega_{ep}$, the medium is an ENG medium. Finally, above the plasma $\omega > \omega_{ep}$, the effective medium is the ENZ medium. It shows that the presence of the coupling capacitance C_{coup}^{pul} modifies the permittivity response of Fig (22.3b) in the ENG region.

22.1.5 Circuit Models of 1D Metalines

The series or shunt reactive loading of the host transmission line can produce only the SNG, i.e. the ENG/MNG, response. Depending upon the nature of the reactive

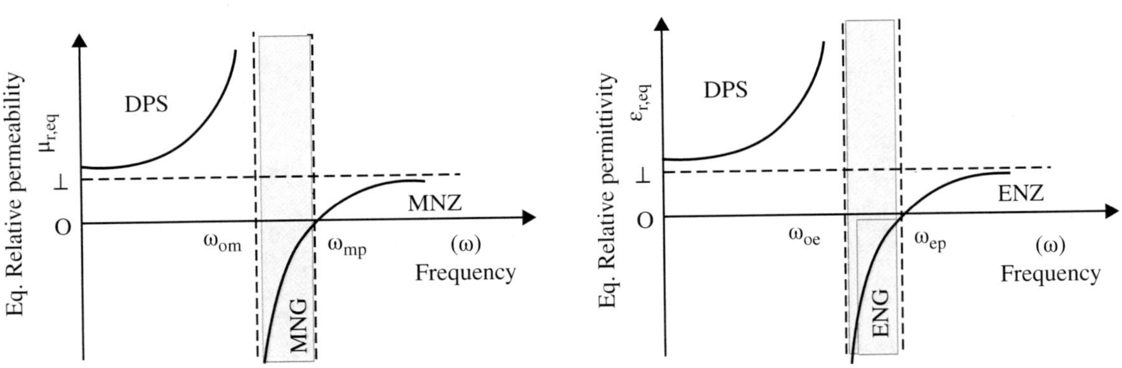

(a) Lorentz type permeability response. (b) Lorentz type permittivity response.

Figure 22.5 Complete Lorentz type response and bandpass response of the effective medium.

loading, the response of the medium could be either the Drude model type broadband response or it could be the resonance type Lorentz narrow band response. However, the cascaded MNG and ENG, with an overlapped frequency band, provide DNG response. The 1D-metalines with the DNG property could be realized by the reactive inclusion loading of both the series and shunt arms of the host L-network unit cell. The doubly loaded host lines can have four configurations: *(i) Series–Parallel configuration, (ii) Parallel–Series configuration, (iii) Parallel–Parallel configuration, and (iv) Series–Series configuration [J.4].*

Figure (22.6a–d) show four circuit configurations for the synthesis of four kinds of metalines. The four circuit configurations create four kinds of material responses. The series–parallel configuration is developed by combining the circuits of Fig (22.2a,b). It provides the *Drude–Drude type response*. This configuration is also known as the *Composite Right/Left Hand (CRLH) configuration*. It has a wideband DNG response [J.5–J.9, B.2, B.3]. Similarly, the *dual-CRLH (D-CRLH) medium, i.e.* the parallel–series configuration, is obtained by combining Fig (22.2c,d). It involves the *Lorentz–Lorentz type response*. It has a limited bandwidth DNG response [J.10]. The parallel–parallel configuration is a combination of Fig (22.2b, c) involving both the Drude and Lorentz responses, so the circuit configuration of Fig. (22.6c) provides the *Lorentz–Drude type response*. It has a narrow band DNG response. Lastly, the series–series configuration is a combination of Fig. (22.2a, d), involving again both the Drude and Lorentz responses. Figure (22.6d) created the *Drude–Lorentz type response*. This configuration also has a narrowband DNG response.

It is concluded that only the CRLH, i.e. the series–parallel circuit configuration can be effectively used to engineer the 1D-DNG medium supporting the backward wave propagation.

Series–Parallel Configuration

Figure (22.6a) shows the unit cell of the series–parallel configuration of the CRLH metalines. The unit cell of the host medium $\left(L_h^{pul}, C_h^{pul}\right)$ is loaded with a series capacitance $\left(C_{se}^{pul}\right)$ in the series arm, and also with a parallel inductance $\left(L_{sh}^{pul}\right)$ in the shunt arm. The loading inclusions are shown in gray boxes. The period d is much smaller than the operating wavelength λ. The condition $d << \lambda$ is called the *homogeneity condition*. The series impedance $\left(Z_{se}^{pul}\right)$ and shunt admittance $\left(Y_{sh}^{pul}\right)$ model

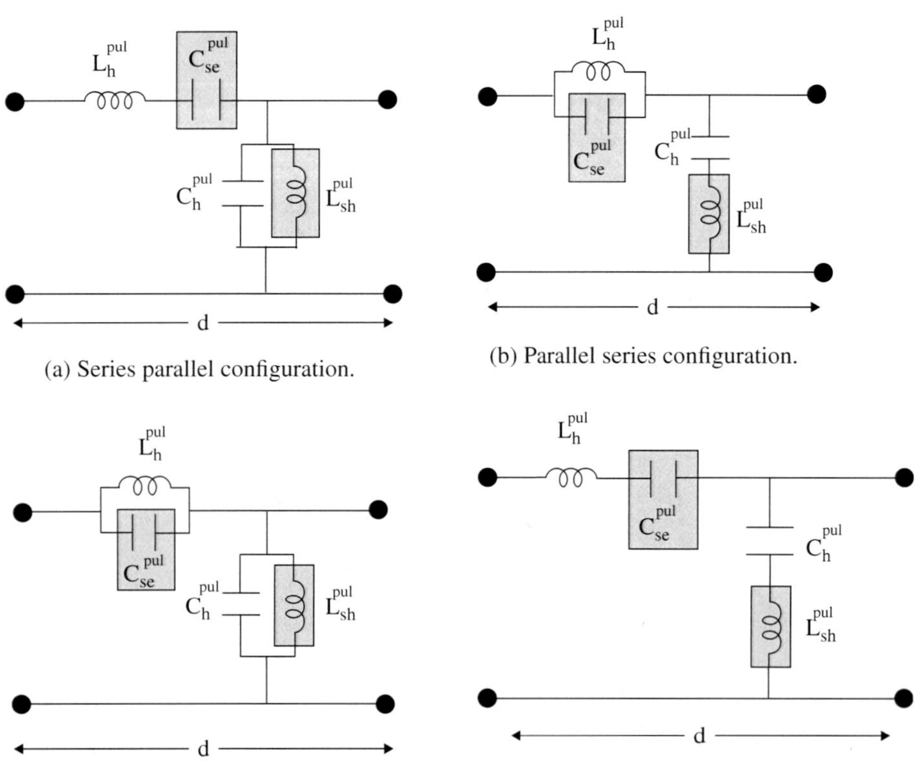

(a) Series parallel configuration.

(b) Parallel series configuration.

(c) Parallel parallel configuration.

(d) Series series configuration.

Figure 22.6 Double arms reactive inclusion loading of the host LC-line forming the unit cell of metalines. Loading elements are shown in gray boxes.

the equivalent permeability and equivalent permittivity of the 1D effective medium. These are given below:

$$Z_{se}^{pul} = j\omega L_h^{pul}\left(1-\frac{\omega_{mp}^2}{\omega^2}\right)\ \text{(a)},\quad Y_{sh}^{pul} = j\omega C_h^{pul}\left(1-\frac{\omega_{ep}^2}{\omega^2}\right)\ \text{(b)}$$

$$\text{where, } \omega_{mp} = \frac{1}{\sqrt{L_h^{pul}C_{se}^{pul}}}\ \text{(c)},\quad \omega_{ep} = \frac{1}{\sqrt{L_{sh}^{pul}C_h^{pul}}}\ \text{(d)}. \tag{22.1.14}$$

In equation (22.1.14), ω_{mp} and ω_{ep} are the magnetic and electric plasma frequencies, responsible for creating the MNG and ENG part of the DNG medium. The equivalent permeability and permittivity of the effective medium are obtained by using equation (22.1.14) with the equations (22.1.4a, c):

$$\mu_{eq} = \mu_0\mu_{r,eq} = L_h^{pul}\left(1-\frac{\omega_{mp}^2}{\omega^2}\right) \quad \text{(a)}$$

$$\Rightarrow \mu_{r,eq} = \mu_{r,h}\left(1-\frac{\omega_{mp}^2}{\omega^2}\right) \quad \text{(b)}$$

$$\varepsilon_{eq} = \varepsilon_0\varepsilon_{r,eq} = C_h^{pul}\left(1-\frac{\omega_{ep}^2}{\omega^2}\right) \quad \text{(c)}$$

$$\Rightarrow \varepsilon_{r,eq} = \varepsilon_{r,h}\left(1-\frac{\omega_{ep}^2}{\omega^2}\right) \quad \text{(d)}$$

$$\text{where, } \mu_{r,h} \equiv L_h^{pul}/\mu_0 \quad \text{(e)}$$

$$\varepsilon_{r,h} \equiv C_h^{pul}/\varepsilon_0 \quad \text{(f)}. \tag{22.1.15}$$

Equations (22.1.15b, d) also compute the equivalent relative permeability $\mu_{r,\,eq}$ and equivalent relative permittivity $\varepsilon_{r,eq}$ of the homogenized effective medium. The relative permeability and relative permittivity of the host medium are $\mu_{r,h}$ and $\varepsilon_{r,h}$ respectively. In the case of the air host medium, these quantities are unity. However, if either microstrip or CPW is the host medium, the geometric parameters-dependent effective permittivity and permeability of these structures could be treated as the permittivity and permeability of the host medium.

The propagation constant β_{sp} and the characteristic impedance Z_{sp} of the *series–parallel (SP) effective medium* are obtained by using the above equations with equations (22.1.5c, d):

$$\beta_{sp} = \omega\sqrt{\mu_{eq}\varepsilon_{eq}} = \omega\sqrt{L_h^{pul}C_h^{pul}} \times \left[\left(1-\frac{\omega_{mp}^2}{\omega^2}\right)^{1/2}\left(1-\frac{\omega_{ep}^2}{\omega^2}\right)^{1/2}\right] \quad \text{(a)}$$

$$Z_{sp} = \sqrt{\frac{\mu_{eq}}{\varepsilon_{eq}}} = \sqrt{\frac{L_h^{pul}}{C_h^{pul}}}\left[\frac{\omega^2-\omega_{mp}^2}{\omega^2-\omega_{ep}^2}\right]^{1/2} \quad \text{(b)}. \tag{22.1.16}$$

It is noted from equation (22.1.16) that both the propagation constant $\omega\sqrt{L_h^{pul}C_h^{pul}}$ and characteristic impedance $\sqrt{L_h^{pul}/C_h^{pul}}$ of a host medium are modified by the plasmonic effect created by double reactive loading. The modifying factors are enclosed in the brackets [] in the above equations. The phase velocity $v_{p,sp}$, and the group velocity $v_{g,sp}$ of the EM-wave on the metaline, i.e. 1D SP-effective medium could also be obtained from β_{sp} [J.4]:

$$v_{p,sp} = \frac{\omega}{\beta_{sp}} = v_{p,h}\Big/\sqrt{\left(1-\frac{\omega_{mp}^2}{\omega^2}\right)\left(1-\frac{\omega_{ep}^2}{\omega^2}\right)} \quad \text{(a)}$$

$$v_{g,sp} = v_{g,h}\left[\left(1-\frac{\omega_{mp}^2}{\omega^2}\right)^{1/2}\left(1-\frac{\omega_{ep}^2}{\omega^2}\right)^{1/2}\right] \times \left[\left(1-\frac{\omega_{mp}^2}{\omega^2}\right)\left(1-\frac{\omega_{ep}^2}{\omega^2}\right) \mp \left(1-\frac{\omega_{ep}^2}{\omega^2}\right) \mp \left(1-\frac{\omega_{mp}^2}{\omega^2}\right)\right]^{-1} \quad \text{(b)}$$

where phase and group velocities in host medium

$$v_{p,h} = v_{g,h} = \frac{1}{\sqrt{L_h^{pul}C_h^{pul}}} \quad \text{(c)}. \tag{22.1.17}$$

The host medium is nondispersion with equal phase and group velocities. However, the reactively loaded SP-metaline medium is highly dispersive. In equation (22.1.17b), (−) is used for $\omega < \omega_{mp}$, $\omega < \omega_{ep}$ and (+) is used for $\omega > \omega_{mp}$, $\omega > \omega_{ep}$.

In equations (22.1.17), the magnetic and electric plasma frequencies are unequal, i.e. $\omega_{mp} \neq \omega_{ep}$ giving the *unbalanced metaline*. In case both plasma frequencies are equal, i.e. $\omega_{mp} = \omega_{ep}$, the *balanced metaline*, also called *matched frequency metaline*, is obtained. The balanced metaline is a more useful configuration [J.5–J.8, J.11, B.2, B.3].

Unbalanced CRLH Metaline

Figure (22.7a) shows the $(\omega - \beta)$ dispersion diagram of the unbalance series–parallel, i.e. the CRLH metaline. It has three distinct frequency regions $\omega < \omega_{mp} < \omega_{ep}$; $\omega_{mp} < \omega < \omega_{ep}$, or $\omega_{ep} < \omega < \omega_{mp}$; $\omega > \omega_{ep} > \omega_{mp}$. These frequency regions have their own characteristics discussed below.

Region $\omega < \omega_{mp} < \omega_{ep}$

Under the above condition, Fig (22.6a) is reduced to the C–L line supporting the backward wave propagation. Equations (22.1.16a) and (22.1.17a) provide negative values for both the propagation constant β_{sp} and the phase velocity $v_{p,sp}$, whereas from equations (22.1.16b) and (22.1.17b) both the characteristic impedance Z_{sp}

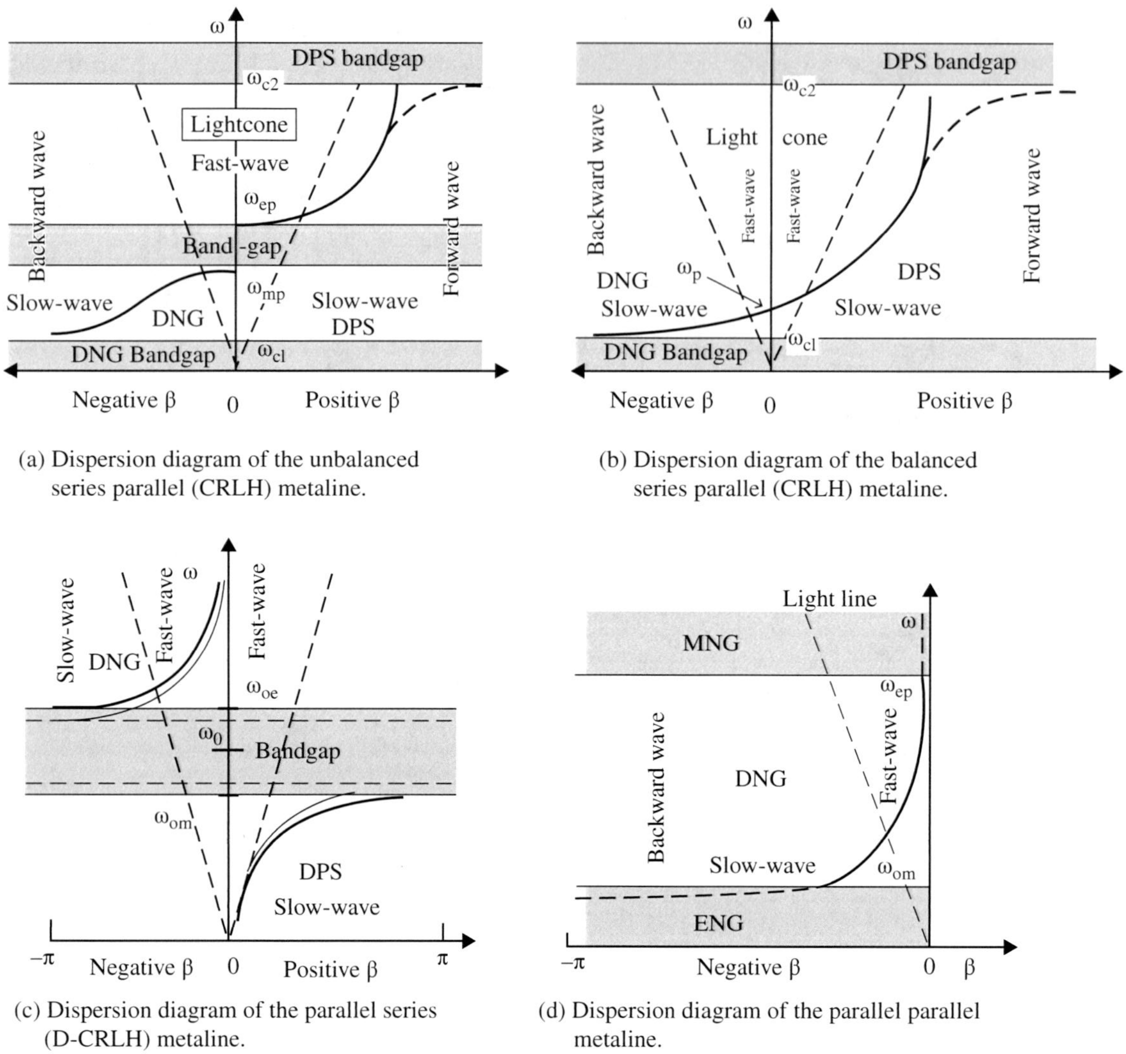

(a) Dispersion diagram of the unbalanced series parallel (CRLH) metaline.

(b) Dispersion diagram of the balanced series parallel (CRLH) metaline.

(c) Dispersion diagram of the parallel series (D-CRLH) metaline.

(d) Dispersion diagram of the parallel parallel metaline.

Figure 22.7 The ($\omega - \beta$) dispersion diagrams of the metalines.

and the group velocity $v_{g,sp}$ are positive. The upper sign of the factor ($\mp$) is taken in this case. It demonstrates the presence of the DNG effective medium in this frequency region, as both the permeability and permittivity are negative from equations (22.1.15a) and (22.1.15c). The DNG medium exists on the left-hand side in Fig (22.7a). It is noted that at $\omega = \omega_{mp}$ and $\omega = \omega_{ep}$, leading to $\beta_{sp} = 0$, showing no propagation. Fig (22.7a) also shows the existence of the first stopband region with a cut-off frequency ω_{c1}. It limits the lower frequency range of the DNG medium. The above-discussed expressions do not show such a cut-off phenomenon.

Region ($\omega_{ep} < \omega < \omega_{mp}$), or ($\omega_{mp} < \omega < \omega_{ep}$)

For the first case ($\omega_{ep} < \omega < \omega_{mp}$), the operating frequency ω is above the shunt resonance and below the series resonance. In this frequency range, the circuit of Fig (22.6a) is reduced to the C–C line. The C–C line corresponds to the MNG medium. However, for the second case ($\omega_{mp} < \omega < \omega_{ep}$), the operating frequency ω is above the series resonance and below the shunt resonance. The circuit of Fig (22.6a) is reduced to the L–L line, creating the ENG medium. Thus, Fig (22.7a) shows a *bandgap* in the frequency band $|\omega_{ep} - \omega_{mp}|$ for the SNG type medium. It supports only the nonpropagating evanescent mode.

Region $\omega > \omega_{ep} > \omega_{mp}$

In this region, the circuit of Fig (22.6a) is reduced to the standard L–C line. The propagation constant β_{sp} and characteristic impedance Z_{sp} are positive quantities. Figure (22.7a) and equations (22.1.17a, b) show that both the phase velocity $v_{p,sp}$, and the group velocity $v_{g,sp}$ are positive. Thus, in this frequency range, the effective medium is DPS type supporting the forward wave propagation, as shown in the right-hand side of Fig (22.7a).

Balanced CRLH metalines

Under the balanced condition, the magnetic and electric plasma frequencies are equal, i.e. $\omega_{mp} = \omega_{ep} = \omega_p$.The effective medium is a plasma medium with the plasmon frequency ω_p. Using equation (22.1.15), the equivalent permeability and permittivity are zero, $\mu_{eq} = \varepsilon_{eq} = 0$ at the plasma frequency ω_p. Likewise, from equation (22.1.17), the phase velocity is infinite at ω_p, whereas the group velocity is finite:

$$v_{g,sp} = \frac{1}{2}\frac{1}{\sqrt{L_h^{pul}C_h^{pul}}} = \frac{1}{2}\frac{1}{\sqrt{L_{sh}^{pul}C_{se}^{pul}}}. \quad (22.1.18)$$

The characteristics impedance of the unbalanced CRLH metaline is frequency-dependent. However, it is frequency-independent for the balanced case. It remains the same for both the DNG and DPS regions. The propagation constant and characteristics impedance of the balanced CRLH metaline are obtained from equations (22.1.16a, b):

$$\beta_{sp}^{bal} = \omega\sqrt{L_h^{pul}C_h^{pul}}\left(1 - \frac{\omega_p^2}{\omega^2}\right) \quad (a)$$

$$Z_{sp}^{bal}(DPS) = \sqrt{\frac{L_h^{pul}}{C_h^{pul}}} \quad (b)$$

$$Z_{sp}^{bal}(DNG) = \sqrt{\frac{L_{sh}^{pul}}{C_{se}^{pul}}} \quad (c)$$

$$Z_{sp}^{bal}(DPS) = Z_{sp}^{bal}(DNG) \quad (d).$$

(22.1.19)

Figure (22.7a,b) shows the $(\omega - \beta)$-dispersion diagram of the balanced CRLH-metaline with a light-cone. In Figure (22.7b), the second bandgap of the unbalanced case disappears and a continuous change in the nature of the medium, from the DNG to DPS, is obtained with an increase in the operating frequency. For $\omega < \omega_p$, the effective medium is DNG, and for the frequency $\omega > \omega_p$, the medium is DPS. It is also evident from expression (22.1.19a). The frequency-independent and equal characteristics impedance, in both the DNG and DPS regions, results in the *broadband impedance matching*.

Figure (22.7b) further shows that a part of the backward wave and forward wave of the DNG and DPS regions are within the light-cone supporting the *fast-wave*. It is the leaky radiation mode, useful for the design of the frequency scanning *leaky-wave antenna*. It scans the upper half of the space from backfire radiation to endfire radiation of the leaky-wave antenna. It is illustrated in Fig (22.20). Outside the light-cone, the medium supports the bounded backward and forward *slow-wave* (SW) propagation. It is useful for the development of several unique and compact microwave devices [J.5, J.7, J.8, J.11, B.2, B.4].

The analysis is valid for the effective medium under the homogeneity condition. However, a more accurate analysis has been done using the *[ABCD] transmission matrix* by treating the metaline as a periodically cascaded structure. The [ABCD] matrix analysis provides the bandgap in the DNG medium called the *DNG bandgap* at the lower cut-off frequency ω_{c1}. It also computes *DPS bandgap* at the upper cut-off frequency ω_{c2}. The DPS bandgap limits the DPS medium upper-frequency range. Figure (22.7a, b) shows both bandgaps with cut-off frequencies and also tapering of the dispersion diagram, shown by the dashed line, at the band-edge. The expressions are available to compute these cut-off frequencies and bandwidth of the DNG and DPS regions [B.2, B.4].

Parallel-Series Configuration

Figure (22.6b) shows a unit-cell of the parallel–series configuration. It is a dual network of the series–parallel network discussed above, so it is also called the *dual-CRLH, i.e. the D-CRLH* metaline [J.10, J.12]. The circuit configuration of Fig. (22.6b) corresponds to the Lorentz-Lorentz type material with the permeability resonance frequency ω_{0m} and the permittivity resonance frequency ω_{0e}. Even the balanced D-CRLH line, under identical resonance frequency in the series and shunt arms, i.e. $\omega_{0m} = \omega_{0e} = \omega_0$, creates a bandgap due to the open-circuit condition in the series arm and the short-circuit condition with the shunt arm. Figure (22.7c) shows that at ω_0, the balanced D-CRLH metaline has a narrow bandgap shown within the dashed-lines. It widens for the unbalanced case. In the low-pass region, the circuit of Fig (22.6b) is reduced to the L–C line showing the DPS effective medium. For the high-pass region, it is reduced to C–L metaline showing the DNG effective medium.

Following the previous process, the equivalent permeability and equivalent permittivity of the effective D-CRLH medium are obtained from the series impedance and shunt admittance *p.u.l.* of the Fig (22.6b):

$$Z_{se}^{pul} = j\omega L_h^{pul}\left(1-\frac{\omega^2}{\omega_{0m}^2}\right)^{-1} \quad (a)$$

$$Y_{sh}^{pul} = j\omega C_h^{pul}\left(1-\frac{\omega^2}{\omega_{0e}^2}\right)^{-1} \quad (b)$$

$$\text{where, } \omega_{0m} = \frac{1}{\sqrt{L_h^{pul}C_{se}^{pul}}} \quad (c)$$

$$\omega_{0e} = \frac{1}{\sqrt{L_{sh}^{pul}C_h^{pul}}} \quad (d). \qquad (22.1.20)$$

$$\mu_{eq} = L_h^{pul}\left(1-\frac{\omega^2}{\omega_{0m}^2}\right)^{-1} \quad (a)$$

$$\varepsilon_{eq} = C_h^{pul}\left(1-\frac{\omega^2}{\omega_{0e}^2}\right)^{-1} \quad (b). \qquad (22.1.21)$$

Equations (22.1.20) and (22.1.21) show that the *parallel–series (ps) medium*, i.e. the D-CRLH medium, is a Lorentz–Lorentz type effective medium, without plasma frequency. Its propagation constant β_{ps} and characteristic impedance Z_{ps}, phase velocity $v_{p,ps}$, and group velocity $v_{g,ps}$ are obtained as [J.4],

$$\beta_{ps} = \omega\sqrt{L_h^{pul}C_h^{pul}}\left[\left(1-\frac{\omega^2}{\omega_{0m}^2}\right)^{-1/2}\left(1-\frac{\omega^2}{\omega_{0e}^2}\right)^{-1/2}\right] \quad (a)$$

$$Z_{ps} = \sqrt{\frac{L_h^{pul}}{C_h^{pul}}\left[\frac{1-\omega^2/\omega_{0e}^2}{1-\omega^2/\omega_{0m}^2}\right]^{1/2}} \quad (b) \qquad (22.1.22)$$

$$v_{p,ps} = \frac{\omega}{\beta_{sp}} = v_{p,h}\sqrt{\left(1-\frac{\omega^2}{\omega_{0m}^2}\right)\left(1-\frac{\omega^2}{\omega_{0e}^2}\right)} \quad (a)$$

$$v_{g,ps} = v_{g,h}\left[\left(\frac{\omega^2}{\omega_{0m}^2}-1\right)^{1/2}\left(\frac{\omega^2}{\omega_{0e}^2}-1\right)^{1/2}\right]\times \left[\left(1-\frac{\omega_{0m}^2}{\omega^2}\right)^{-1}+\left(1-\frac{\omega_{0e}^2}{\omega^2}\right)^{-1}-1\right]^{-1} \quad (b)$$

$$\text{where in host medium } v_{p,ps} = v_{g,h} = \frac{1}{\sqrt{L_h^{pul}C_h^{pul}}} \quad (c). \qquad (22.1.23)$$

Figure (22.7c) shows the dispersion diagram of the D-CRLH metaline with a light-cone. The region within the light-cone supports the fast-wave, and the region outside the light-cone supports the SW. In the frequency region $\omega < \omega_{0m} < \omega_{0e}$, the propagation constant, characteristic impedance, phase velocity, and group velocity are positive. Thus, the effective medium is DPS. At $\omega = \omega_{0m}$, $\beta_{ps} \to \infty$, $Z_{ps} = \infty$, $v_{p,ps} = v_{g,ps} = 0$ and the series arm is open-circuited. Likewise, at ω_{0e}, the same nature is seen as the shunt arm is short-circuited and $Z_{ps} = 0$. For the frequency band $\omega_{0m} < \omega < \omega_{0e}$, the effective medium is MNG, and stopband is created without any wave propagation. For the case $\omega_{0e} < \omega < \omega_{0m}$, stopband is an effective ENG medium. However, for the case $\omega > \omega_{02} > \omega_{01}$, Fig (22.6b) is reduced to the C–L metaline, forming the DNG effective medium. In this frequency range, the propagation constant and phase velocity are negative, whereas the characteristic impedance and group velocity are positive. Unlike the CRLH, the D-CRLH has a narrow stopband even for the balanced case $\omega_{0m} = \omega_{0e} = \omega_0$. It is shown within the dashed-line in Fig (22.7c).

In the high-frequency limiting case with the DNG region, the following results are obtained for the phase velocity and group velocity:

$$\underset{\omega\to\infty}{\mathrm{Lt}}\, v_{p,sp} = -\frac{\omega^2}{\omega_{0m}\omega_{0e}}\frac{1}{\sqrt{L_h^{pul}C_h^{pul}}} \quad (a)$$

$$\underset{\omega\to\infty}{\mathrm{Lt}}\, v_{g,sp} = \frac{\omega^2}{\omega_{0m}\omega_{0e}}\frac{1}{\sqrt{L_h^{pul}C_h^{pul}}} \quad (b). \qquad (22.1.24)$$

Expression (22.1.24) shows that both the phase velocity and group velocity approach to infinity with $\omega \to \infty$. This is unphysical condition. In practice, they have finite values due to the parasitic effect of the components. However, the very high value of group velocity can be realized for the D-CRLH metaline. The D-CRLH metaline has better impedance matching after the steep change in the characteristic impedance [J.10].

Parallel–Parallel Configuration

Figure (22.6c) shows the unit cell of the parallel–parallel (pp) type of artificial metaline. It has the Lorentz–Drude type response. The permeability exhibits Lorentz response with a magnetic resonant frequency ω_{0m}, whereas the permittivity exhibits Drude response with electric plasma frequency ω_{ep}. By following the process applied to the CRLH metaline, the equivalent permeability and permittivity parameters of the *parallel–parallel (pp) effective medium* could also be computed by obtaining the series impedance and shunt admittance *p.u.l.* of the loaded host transmission lines:

$$Z_{se}^{pul} = j\omega L_h^{pul}\left(1 - \frac{\omega^2}{\omega_{0m}^2}\right)^{-1} \quad (a)$$

$$Y_{sh}^{pul} = j\omega C_h^{pul}\left(1 - \frac{\omega_{ep}^2}{\omega^2}\right) \quad (b)$$

$$\text{where, } \omega_{0m} = \frac{1}{\sqrt{L_h^{pul} C_{se}^{pul}}} \quad (c)$$

$$\omega_{ep} = \frac{1}{\sqrt{L_{sh}^{pul} C_h^{pul}}} \quad (d).$$

(22.1.25)

$$\mu_{eq} = L_h^{pul}\left(1 - \frac{\omega^2}{\omega_{0m}^2}\right)^{-1} \quad (a)$$

$$\varepsilon_{eq} = C_h^{pul}\left(1 - \frac{\omega_{ep}^2}{\omega^2}\right) \quad (b).$$

(22.1.26)

The propagation constant β_{pp} and characteristic impedance Z_{pp} of the EM-wave in the parallel–parallel effective medium are obtained as,

$$\beta_{pp} = \omega\sqrt{L_h^{pul} C_h^{pul}}\left[\frac{1 - \omega_{ep}^2/\omega^2}{1 - \omega^2/\omega_{0m}^2}\right]^{1/2} \quad (a)$$

$$Z_{pp} = \sqrt{\frac{L_h^{pul}}{C_h^{pul}}}\left[\frac{1}{\left(1 - \omega^2/\omega_{0m}^2\right)\left(1 - \omega_{ep}^2/\omega^2\right)}\right]^{1/2} \quad (b).$$

(22.1.27)

The phase and group velocities can also be obtained for this medium. However, the characteristic of the medium can be estimated from the circuit shown in Fig (22.6c). The dispersion diagram is shown in Fig (22.7d). In the frequency range $\omega < \omega_{0m} < \omega_{ep}$, the circuit Fig (22.6c) is reduced to the L–L circuit and the effective medium is ENG. Likewise, in the frequency range $\omega > \omega_{ep} > \omega_{0m}$, the parallel–parallel circuit is reduced to the C–C circuit and the effective medium is MNG. Equation (22.1.27) provide the propagation constant and characteristic impedance as the imaginary quantities in both these frequency ranges, creating the stopband.

However, in the middle frequency band $\omega_{0m} < \omega < \omega_{ep}$, the parallel–parallel circuit is reduced to the C–L line creating the DNG medium. For the case $\omega_{ep} < \omega < \omega_{0m}$, the L–C line is obtained, and the DPS medium is created. Thus, the passband could belong to either the DNG or DPS medium, depending on the relative value of the resonance frequencies in two arms of the circuit. The DNG bandwidth can be controlled by controlling both resonance frequencies. The dispersion diagram of Fig (22.7d) shows the real negative value of the propagation constant of the DNG medium in the frequency range $\omega_{0m} < \omega < \omega_{ep}$. Above the light-line, the medium supports the radiative fast-wave. However, below the light-line, the medium supports the backward wave propagating in the slow-wave mode. Outside the bandpass frequency range, the propagation constant is an imaginary quantity shown in dashed-lines. It shows that the wave is in the nonpropagating evanescent mode.

Series–Series Configuration

Figure (22.6d) shows the *series–series (SS) configuration* of the artificial line. It has the Drude–Lorentz response. The permeability exhibits the Drude type response with magnetic plasma frequency ω_{mp}, whereas the permittivity exhibits the Lorentz type response with a resonant frequency ω_{0e}. Like the previous cases, equivalent permeability and equivalent permittivity of the effective medium can be obtained from the circuit components. Its propagation characteristics are identical to the propagation characteristics of the parallel–parallel configuration. It has stopband in the frequency range $\omega < \omega_{mp} < \omega_{0e}$ and $\omega > \omega_{0e} > \omega_{mp}$, and a passband in the middle-frequency range $\omega_{mp} < \omega < \omega_{0e}$ or $\omega_{0e} < \omega < \omega_{mp}$. However, the nature of the frequency bands is different. In the frequency range $\omega < \omega_{mp} < \omega_{0e}$, the circuit shows the C–C line, i.e. the MNG effective medium, whereas in the frequency range $\omega > \omega_{0e} > \omega_{mp}$, the circuit is reduced to L–L, creating the ENG effective medium. Likewise, in the passband, $\omega_{mp} < \omega < \omega_{0e}$ the DPS medium is created, and in the range, $\omega_{0e} < \omega < \omega_{mp}$ the DNG medium is created. However, the dispersion characteristics, as shown in Fig (22.7d), are also applicable to the series–series configuration. The bandwidth of the DNG could also be controlled by controlling the resonance frequencies.

The present section has considered only the lossless *nonresonant metalines* structures by embedding the C and L inclusions in the host transmission lines. The losses could be accounted for by associating resistance and conductance with the inductor and capacitor used in the circuit configuration. They have a noticeable influence on the performance of the engineered media [J.2, J.7]. The practical realizations of the engineered nonresonant media are discussed in section (22.2). The *resonant metalines* are also developed by using the resonant inclusions in the host line [J.13, B.3, B.7]. It is discussed in section (22.3).

22.2 Nonresonant Microstrip Metalines

The capacitor and inductor type inclusion structures are loaded to the host microstrip and CPW to realize the metalines. However, the present discussion is limited to the microstrip. The extended discussion to the CPW-based metaline is available in reference [J.14–J.16]. The microstrip realization of three kinds of the metalines: CRLH, a combination of the ENG-MNG structures, and D-CRLH are discussed in this section.

22.2.1 Series–Parallel (CRLH) Metalines

Figure (22.8a-i) suggests the realization of the CRLH, i.e. the series–parallel configuration, metaline by loading the host microstrip with the series-connected semi-lumped gap capacitor and via. The grounded via forms the shunt connected inductor. Figure (22.6a) shows the loading inclusions C_{se} and L_{sh}. The inclusions could be distributed over the period d of the unit cell to get $C_{se}^{pul} = C_{se}/d$ and $L_{sh}^{pul} = L_{sh}/d$. The host microstrip provides inductance and capacitance L_h^{pul} and C_h^{pul}. However, in this arrangement limited values of loading elements are realized.

Figure (22.8b) shows that the series-connected gap capacitors could be replaced by the chip capacitors, or by the semi-lumped interdigital capacitor as shown in Fig (22.8a)-ii, to get a higher value of series capacitance. However, the interdigital capacitor has self-resonance that limits its operational bandwidth with higher losses. Its physical size is also large to realize the high value of capacitance. Moreover, the interdigital capacitor-based CRLH metalines is an asymmetrical structure with different values of the Bloch impedance at the input and output ports. Thus, the design of the interdigital capacitors at both ends of the line, shown in Fig (22.8c), is different. The end capacitors are different from the other interdigital capacitors.

The low-loss compact chip capacitors solve some of the problems associated with interdigital capacitors. However, these are available only for limited values and normally used up to 3–6 GHz. The metal-insulator-metal (MIM) capacitors are also used to realize higher values for the loading capacitors. The MIM capacitors are helpful to realize the broadband CRLH and

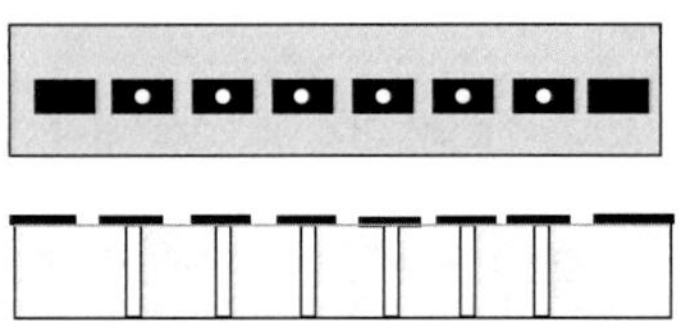

i. series gap capacitance and shunt via inductance

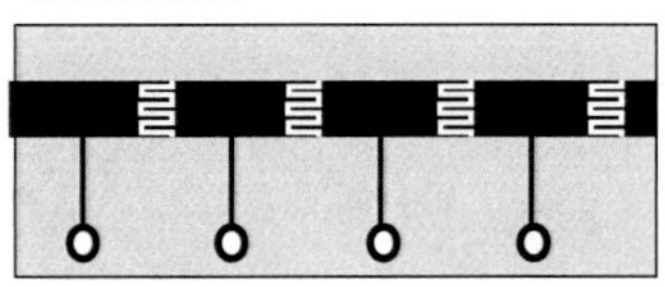

ii. Series interdigital capacitance and grounded shunt stub.

(a) Series capacitance and shunt inductance loaded microstrip.

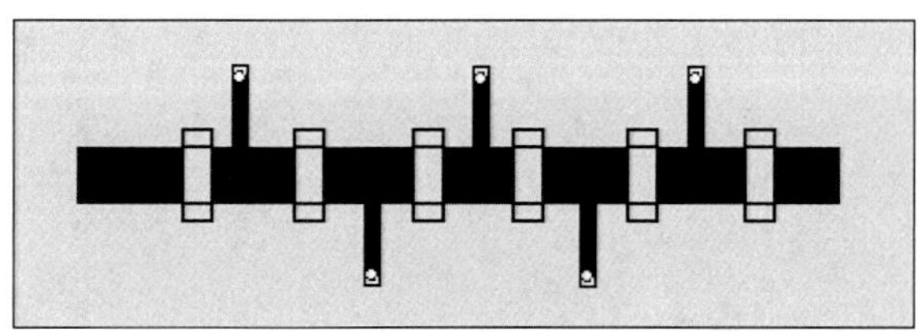

(b) Six series chip capacitors and five grounded shunt stub inductances loaded microstrip.

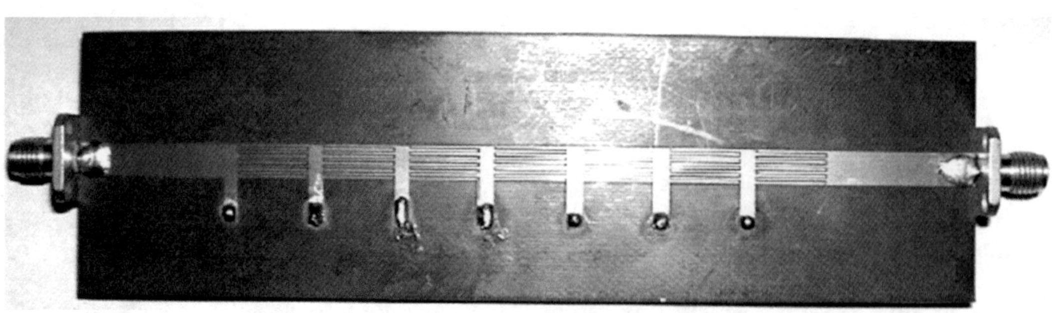

(c) Seven unit cells CL loaded microstrip [J.2].

Figure 22.8 Series capacitance and shunt inductance loaded microstrip metalines. *Source:* Caloz and Itoh [J.2]. © 2002, IEEE.

also D-CRLH metalines. However, the MIM-based metalines require two-level of the substrate [J.10, J.14, J.17].

Figure (22.8a-ii) shows that simple via can be replaced by the high impedance grounded stub to realize higher values for the loading shunt inductors. The microstrip section of the metaline is reduced to a minimum length to support the grounded stub, as shown in Fig (22.8c) [J.2]. In place of the stubs, the meandered-line inductors can also be used to get larger values for the loading inductors. The grounding vias can be replaced by a large patch capacitor to provide the virtual ground to the shunt inductor [J.5].

Figure (22.8b) shows the layout of the chip capacitors-based symmetrical CRLH-metalines. It has six loading series-connected ATC 600S chip capacitors. The end capacitors have a value of 5.6 pF, and other capacitors have a value of 1.8 pF. The host microstrip line sections are 5.14 mm long and 2.78 mm wide and designed on an F4B-1/2 substrate with $\varepsilon_r = 2.65$, h = 1 mm [B.10]. The length and width of the grounded stubs are 9.47 mm and 1 mm, respectively. Figure (22.9a, b) shows its S-parameters magnitude and the phase responses from almost DC to 5.0 GHz. Both the simulated and experimental results are shown. It is the case of the symmetrical CRLH without any bandgap. Its transition frequency is 2.97 GHz. The medium is DNG between 1.0 and 2.97 GHz, and DPS from 2.97 to 5.0 GHz. In the DNG range, the propagation constant β is negative, whereas in the DPS range, it is a positive quantity confirming the backward and forward wave propagation in the DNG and DPS ranges. In the DPS region, almost smooth linear increase of propagation constant β is noted with an increase in frequency. However, the DNG region is dispersive.

Figure (22.8c) shows the 7-cell CRLH metaline developed using eight numbers of interdigital capacitors and seven numbers of the grounded stubs [J.2]. The host microstrip is designed on the RT/Duroid 5880 substrate $\varepsilon_r = 2.2$, h = 1.57 mm. At 1.5 GHz, the interdigital capacitors have a value of 2.06 pF, and the stub has an inductance of value 4.62 nH. The cell length is 12.2 mm. Figure (22.9c,d) shows the experimental magnitude

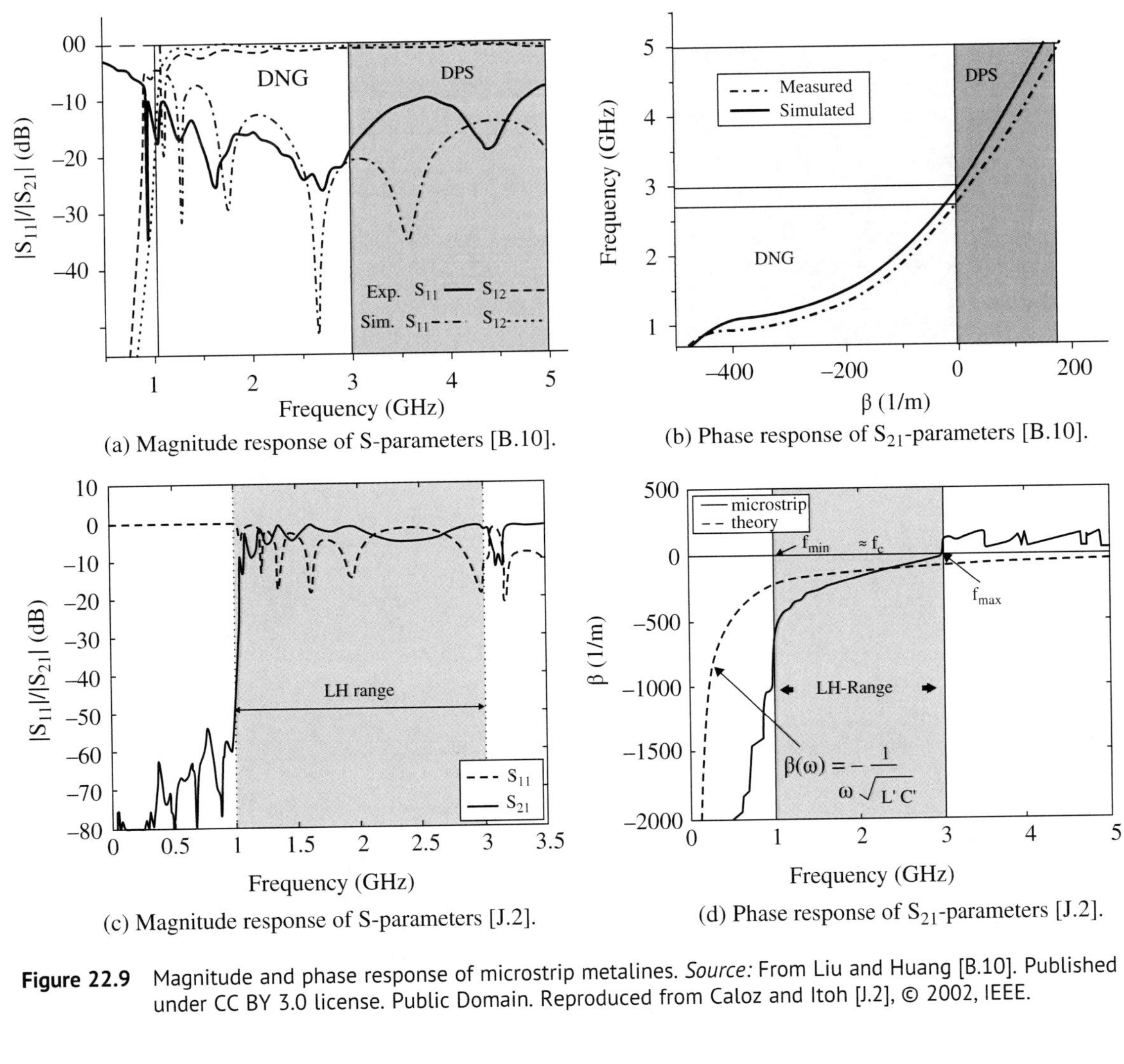

(a) Magnitude response of S-parameters [B.10].

(b) Phase response of S_{21}-parameters [B.10].

(c) Magnitude response of S-parameters [J.2].

(d) Phase response of S_{21}-parameters [J.2].

Figure 22.9 Magnitude and phase response of microstrip metalines. *Source:* From Liu and Huang [B.10]. Published in [short citation] under CC BY 3.0 license. Public Domain. Reproduced from Caloz and Itoh [J.2], © 2002, IEEE.

and phase responses of the S-parameters. The highly dispersive DNG has a bandwidth from 1.0 to 3.0 GHz with negative β. Figure (22.9d) also shows the lumped circuit phase response. In the DPS region, the response is not satisfactory. However, the improved response has been obtained for the 24 unit-cell CRLH structure [J.5, B.2].

22.2.2 Cascaded MNG–ENG (CRLH) Metalines

Individually, the MNG and ENG slabs do not support wave propagation. However, the cascaded MNG–ENG slab or the ENZ–MNZ slab over the common overlapped frequency band supports the wave propagation. This is known as the *wave tunneling* that occurs when tunneling conditions are satisfied. Using the CRLH unit operating in the MNG and ENG modes, the composite MNG–ENG metalines are designed and tunneling is obtained. Its application for the design of a higher-order bandpass filter has also been demonstrated. The CRLH-based circuit model and its realization in microstrip technology are discussed below [J.18–J.21].

Figure (22.10a) considers the cascaded slabs of MNG–ENG [J.20]. Each slab is designed using four unit cells. The unit cells are modeled using the CRLH circuit configuration shown in Fig (22.6a). It reduces the C–C and L–L networks to get the MNG and ENG mediums, respectively. At the interface resonance ω_0, the maximum potential occurs at the interface, and tunneling takes place. Under the tunneling condition, the input signal is transferred to the output. The tunneling

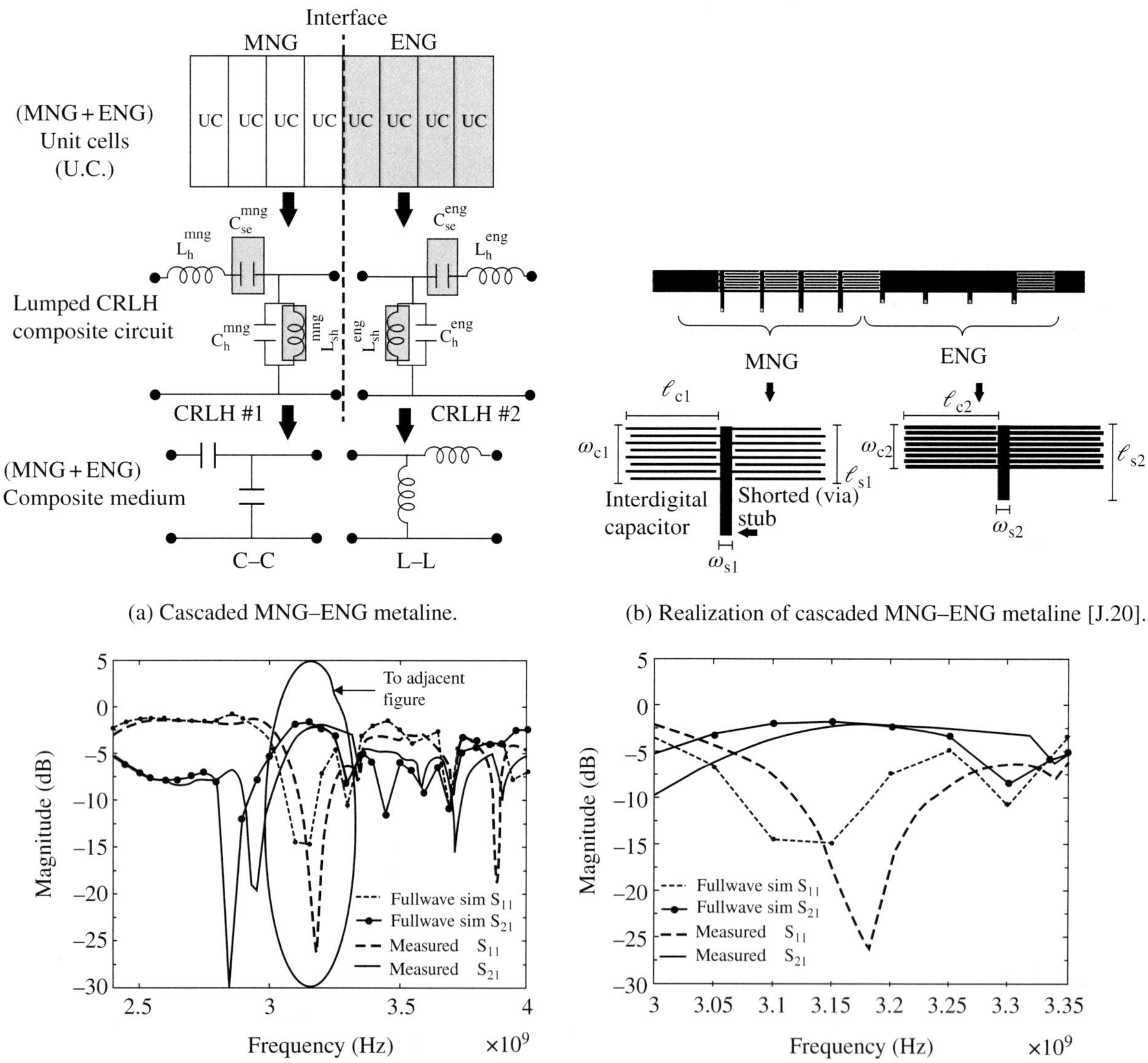

(a) Cascaded MNG–ENG metaline.

(b) Realization of cascaded MNG–ENG metaline [J.20].

(c) Exp. and Sim. S-parameters response [J.20].

(d) Exp. and Sim. S-parameters response of overlapped MNG–ENG region [J.20].

Figure 22.10 (MNG–ENG) Cascaded metalines and tunneling response. *Source:* Fujishiqe et al. [J.20]. © 2005, John Wiley & Sons.

bandwidth is determined by the overlapped frequency band of the MNG and ENG. The homogenized equivalent medium is expressed through the medium parameters μ_{eq}^{mng}, ε_{eq}^{mng}, μ_{eq}^{eng}, and ε_{eq}^{eng}. Using equations (22.1.14) and (22.1.15), these are expressed in terms of the circuit elements of Fig (22.10a):

For the MNG medium: $\mu_{eq}^{mng} < 0, \ \varepsilon_{eq}^{mng} > 0$

$$\mu_{eq}^{mng} = L_h^{mng}\left(1 - \frac{\omega_{mp1}^2}{\omega^2}\right) \quad (a)$$

$$\varepsilon_{eq}^{mng} = C_h^{mng}\left(1 - \frac{\omega_{ep1}^2}{\omega^2}\right) \quad (b)$$

$$\omega_{mp1} = \frac{1}{\sqrt{L_h^{mng} C_{se}^{mng}}} \quad (c)$$

$$\omega_{ep1} = \frac{1}{\sqrt{L_{sh}^{mng} C_h^{mng}}} \quad (d)$$

$$\omega_{ep1} < \omega < \omega_{mp1} \quad (e)$$

(22.1.28)

For the ENG medium: $\mu_{eq}^{eng} > 0, \ \varepsilon_{eq}^{eng} < 0.$

$$\mu_{eq}^{eng} = L_h^{eng}\left(1 - \frac{\omega_{mp2}^2}{\omega^2}\right) \ (a), \quad \varepsilon_{eq}^{eng} = C_h^{pul}\left(1 - \frac{\omega_{ep2}^2}{\omega^2}\right) \ (b)$$

$$\omega_{mp2} = \frac{1}{\sqrt{L_h^{eng} C_{se}^{eng}}} \quad (c)$$

$$\omega_{ep2} = \frac{1}{\sqrt{L_{sh}^{eng} C_h^{eng}}} \quad (d), \qquad \omega_{mp2} < \omega < \omega_{ep2}^{mng} \quad (e).$$

(22.1.29)

The host lines of the unloaded MNG and ENG lines sections are the normal DPS lines with identical propagation constant:

$$\beta_0^{mng} = \beta_0^{eng}, \quad \Rightarrow L_h^{mng} C_h^{mng} = L_h^{eng} C_h^{eng}. \quad (22.1.30)$$

The wave tunneling occurs through the composite MNG–ENG metaline under the *tunneling conditions* that result in the identical operating bandwidth for both the MNG and ENG lines [J.18]:

Tunnelling condition:

$$\mu_{eq}^{mng} = -\mu_{eq}^{eng} \ (a), \quad \varepsilon_{eq}^{mng} = -\varepsilon_{eq}^{eng} \quad (b)$$

$$\text{Upper frequency: } \omega_{mp1} = \omega_{ep2} \quad (c)$$

$$\text{Lower frequency } \omega_{ep1} = \omega_{mp2} \quad (d).$$

(22.1.31)

Satisfying the tunneling conditions, the following relations between the circuit elements of the MNG line are obtained from equations (22.1.29)–(22.1.31) in terms of the known circuit elements of the ENG line:

$$C_h^{mng} = \frac{L_h^{eng} C_h^{eng}}{L_h^{mng}} \quad (a)$$

$$C_{se}^{mng} = \frac{L_{sh}^{eng} C_h^{eng}}{L_h^{mng}} \quad (b)$$

$$L_{sh}^{mng} = \frac{L_h^{mng} C_{se}^{eng}}{C_h^{eng}} \quad (c).$$

(22.1.32)

The inductance L_h^{mng} of the host MNG line is arbitrarily chosen. The complete transparency, i.e. full tunneling occurs at the frequency ω_0. It is the *interface resonance frequency*. It is obtained by substituting equations (22.1.28a) and (22.1.29a) in equation (22.1.31a):

$$\omega_0 = \left[\frac{L_h^{eng}\omega_{mp2}^2 + L_h^{mng}\omega_{mp1}^2}{L_h^{mng} + L_h^{eng}}\right]^{1/2} \quad (a)$$

$$\Rightarrow \omega_0 = \left[\frac{1}{\left(L_h^{mng} + L_h^{eng}\right)}\left(\frac{1}{C_{se}^{eng}} + \frac{1}{C_{se}^{mng}}\right)\right]^{1/2} \quad (b).$$

(22.1.33)

The tunneling bandwidth is $(\omega_{ep1} - \omega_{mp1})$ or $(\omega_{mp2} - \omega_{ep2})$ around the tunneling frequency.

Figure (22.10b) shows the microstrip implementation of the composite MNG and metaline using four unit cells of each MNG and ENG slabs. The series capacitor inclusions are implemented in the interdigital configuration and the shunt inductor inclusions are implemented through the grounded stubs. The component values and dimensions of the unit cell, on the RT/Duroid substrate-5880, $\varepsilon_r = 2.2$, $h = 1.575$ mm, are given the reference [J.20]. Figure (22.10c) shows the simulated and experimental results of the S-parameters. Figure (22.10d) shows the expanded version of the transparency region. The experimental and simulated values for the tunneling frequency f_0 are 3.18 and 3.13 GHz, respectively. At f_0, the insertion loss is −1.9 with −27 dB return loss.

22.2.3 Parallel–Series (D-CRLH) Metalines

Figure (22.11) presents the realization of the D-CRLH metaline in microstrip technology. The *scheme-I*, shown in Fig (22.11a), is based on the MIM capacitor C_{se} in the series arm and C_h in the shunt arm. The inductances L_h in the series arm and L_{sh} in the shunt arm are implemented by the high impedance narrow width lines [J.10]. The *scheme-II*, shown in Fig (22.11b), is based on the series-connected interdigital capacitor C_{se}. The shunt capacitor C_h is obtained from the patch capacitor. The inductances L_h and L_{sh} in the series and shunt are again obtained from the high-impedance microstrip sections [B.10]. The components are tuned to get a wideband

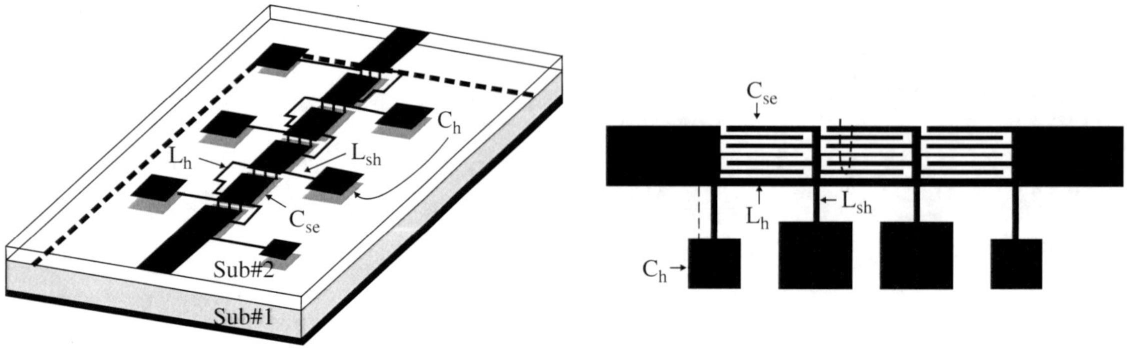

(a) Two-level implementation of D-CRLH using MIM capacitors [J.10].

(b) Single level implementation of D-CRLH using a series interdigital capacitor with parallel inductive strip wire.

Figure 22.11 Microstrip implementation of the D-CRLH metalines. *Source:* Caloz and Nguyen [J.10]. © 2007, Springer Nature.

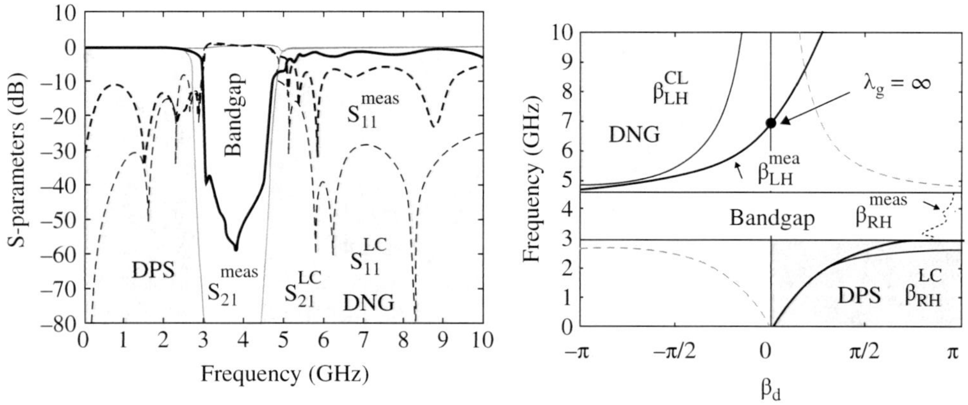

(a) S-parameters response of the D-CRLH.

(b) Phase response of the D-CRLH.

Figure 22.12 S-parameters and phase response of the D-CRLH metalines. *Source:* Caloz and Nguyen [J.10]. © 2007, Springer Nature.

impedance match at both the input and output ports. The complete structure is embedded in the microstrip line.

Figure (22.12a, b) shows the experimental and simulated results based on the LC lumped circuit model for the five-unit cells D-CRLH metaline of the scheme-I [J.10]. A bandgap is present for the D-CRLH metaline. The lower passband is DPS type supporting the forward wave propagation and the upper passband is DNG type supporting the backward wave propagation. There is an insertion loss in the DNG region due to the *leaky-wave radiation* caused by the fast wave. Figure (22.12b) shows that the experimentally obtained propagation constant β_{LH}^{meas} of the DNG has crossed into the DPS region. It is due to the parasitic effect in a practical circuit. However, for the propagation constant β_{LH}^{CL}, computed from the simplified circuit model, the crossing of regions does not occur. The region crossing of propagation constant is a useful phenomenon to design the broadside leaky-wave antenna [J.5, B.2].

22.3 Resonant Metalines

In the previous section, the CRLH-metalines are realized by the series capacitors and shunt inductors embedded in the host lines as inclusions. These are nonresonating inclusions forming wideband metalines. Of course, combined with the host line series inductance and shunt capacitance, the loading elements create series and shunt resonances defining the magnetic and electric plasma frequencies, respectively. The resonance in-line structure is important for the generation of negative permittivity and negative permeability.

The CRLH-type metalines are also realized in the microstrip and CPW host lines by embedding the *resonant inclusions*. Figure (22.13a–d) shows four kinds of the resonating inclusions – *(i)* SRR (*split-ring resonator*) type metallic strips resonator printed on the dielectric substrate, *(ii)* Complementary SRR (CSRR) type resonating apertures cut in the metallic conductor of a metalized

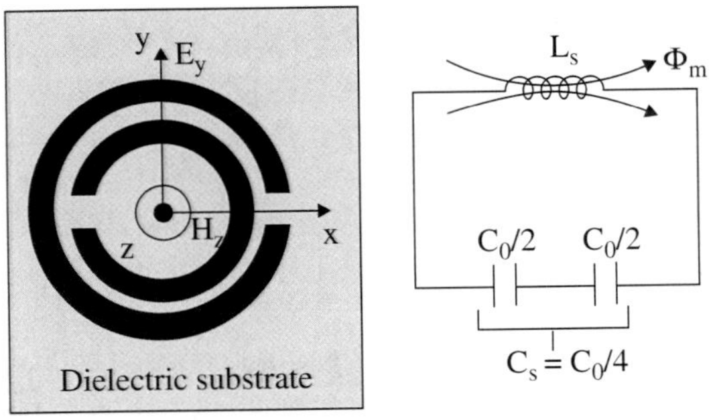

(a) SRR strip on the substrate and its equivalent circuit.

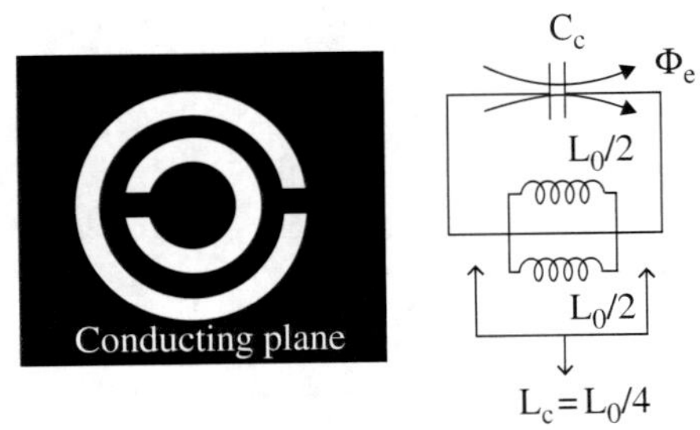

(b) CSRR aperture in the metal conductor and its equivalent circuit. $L_c = L_0/2$.

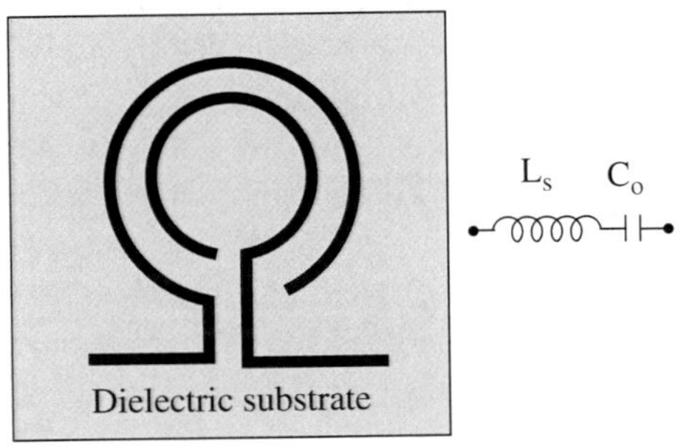

(c) The O-SRR strip on the substrate and its equivalent circuit.

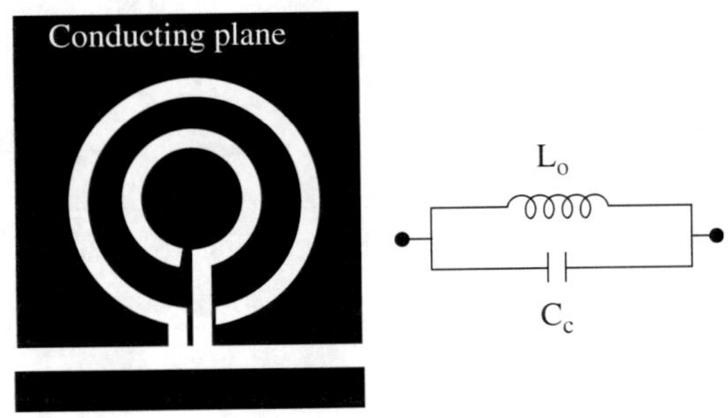

(d) O-CSRR aperture in the metal conductor and its equivalent circuit.

Figure 22.13 Four-kinds of resonating inclusions and their equivalent circuits.

substrate, *(iii)* Open SRR (O-SRR) type metallic strips resonator printed on the dielectric substrate, and *(iv)* Open complementary SRR (O-CSRR) type resonating apertures cut in the metallic conductor of the metalized substrate [J.22–J.27, B.7, B.11, B.12]. *The SRR is a magnetic particle described by the magnetic dipole and the CSRR is an electric particle described by the electric dipole.* Other resonating inclusions, such as the spiral or even the DGS, have also been used to realize the planar metalines [J.28–J31]. This section summarizes the basic property of these four resonating inclusions embedded in the host microstrip and CPW to realize the metalines. Normally, the resonant metalines are narrow bandwidth lines. However, by the suitable design process, the wideband components have been developed [J.23].

22.3.1 Resonant Inclusions

Figure (22.13a) shows the sub-wavelength size resonant SRR inclusion etched on the dielectric substrate. It also shows the equivalent resonant circuit. It is excited by the time-varying H_z axial magnetic field creating the magnetic dipole moment. The MNG type effective medium is created due to negative permeability between the magnetic resonance frequency ω_{0m} and magnetic plasma frequency ω_{mp}. It is shown in Fig (22.5a). The inductance L_s is due to the time-varying magnetic flux Φ_m and capacitance C_s is a series combination of the slot capacitance $C_0/2$ in the upper and lower halves of the SRR. The detailed operation and modeling of the SRR are discussed in subsection 21.2.2 of chapter 21. In the case of a microstrip line, the SRR is located by the side of a strip conductor so that it is magnetically coupled to the microstrip. However, for a CPW structure, the SRRs are located at its backside, underneath the CPW slots to achieve the magnetic coupling. Figure (22.13c) shows a variation of the SRR by opening the gaps of the SRR and connecting them to the strip conductor. It called the *open SRR (O-SRR)* inclusion [J.32]. Figure (22.13c) also shows its equivalent series resonant circuit. Its resonance frequency is half of the resonance frequency of an SRR. It is a more compact resonator.

Likewise, Fig (22.13b) shows the aperture type SRR etched in the conducting ground plane of a microstrip line. It is a complementary structure of the SRR, so it is called the complementary SRR (CSRR). Figure (22.13b) also shows its equivalent circuit. It is excited by the axial time-varying E_z electric field, i.e. electric flux Φ_e and behaves like an electric dipole, creating the ENG type medium with negative permittivity above its resonance frequency ω_{0e} and below the electric plasma frequency ω_{ep} [J.26]. The electric flux is responsible for the capacitance C_c, and inductance L_c is a parallel combination of the upper and lower halves of the CSRR. The CSRR is electrically coupled to microstrip or CPW. Figure (22.13d) shows a variation in the CSRR structure, again by opening the ring aperture. It is called the *open CSRR (O-CSRR)* [J.33]. Its equivalent parallel resonant circuit is also shown in Fig (22.13d). It is also a compact resonator.

22.3.2 Microstrip Resonant Metalines

The microstrip resonant metalines are developed using both the CSRR resonant aperture in the ground plane of microstrip and placing the SRR by the side of the strip conductor of a microstrip.

CSRR-Based Microstrip Metaline

Figure (22.14a) shows the loaded microstrip line with the unit cell of CSRR. The CSRR aperture is etched in the ground plane of a microstrip. As noted above, a periodic arrangement of the CSRR loaded microstrip line creates the narrowband ENG medium. However, the broadband MNG with negative permeability is created by a gap discontinuity of width W_g in the microstrip conductor. To get a larger value of gap capacitance, the gap discontinuity has been replaced by the interdigital capacitor shown in Fig (22.14b). Figure (22.14c) shows the T-type equivalent circuit model with $L_h - C_h$ elements corresponding to the host microstrip. It also shows the π-capacitive network, i.e. the $(C_g - C_f)$ network that models the gap discontinuity or interdigital capacitor. In the $\pi-$ capacitive network, the gap capacitance is C_g and the fringe capacitance is C_f. Figure (22.14c) also shows the parallel resonator tank circuit, i.e. the $(L_c - C_c)$ circuit, coupled to the line. The $(L_c - C_c)$ circuit models the CSRR inclusion.

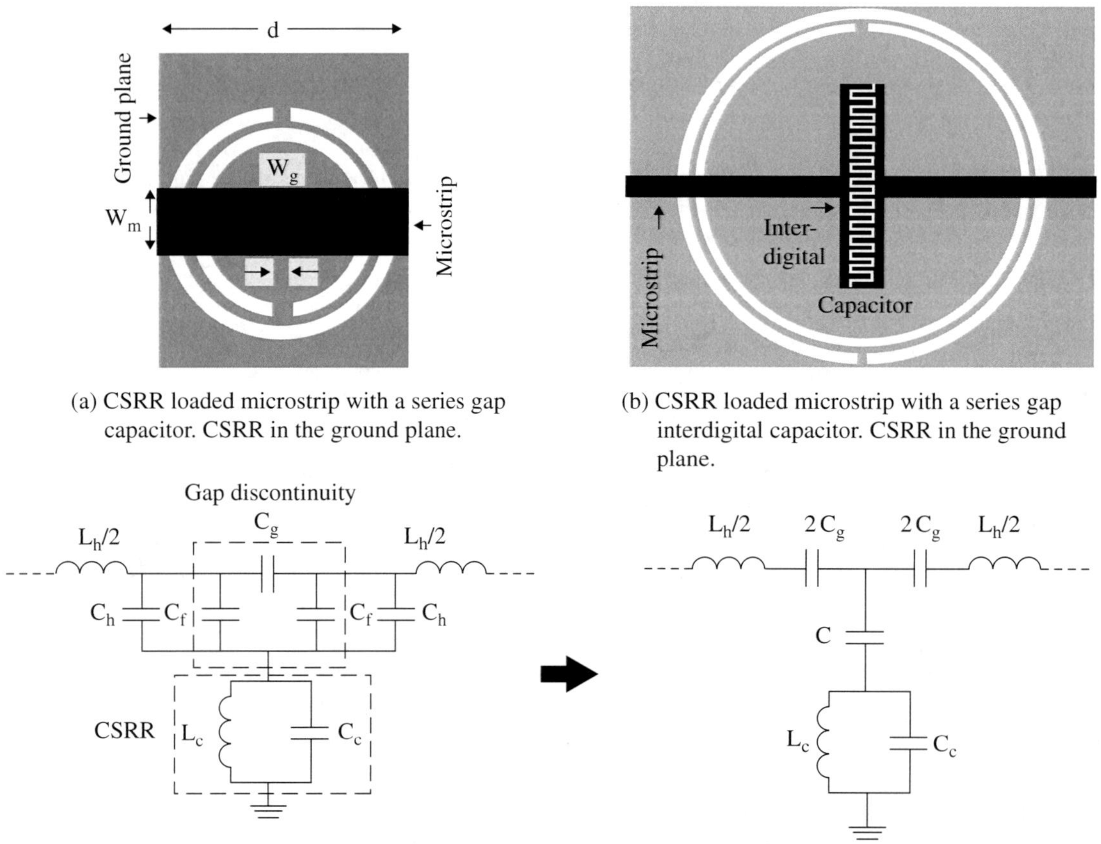

(a) CSRR loaded microstrip with a series gap capacitor. CSRR in the ground plane.

(b) CSRR loaded microstrip with a series gap interdigital capacitor. CSRR in the ground plane.

(c) Equivalent circuit of CSRR and gap embedded in microstrip.

(d) Equivalent circuit of CSRR and gap loaded microstrip forming the CRLH–type line.

Figure 22.14 CSRR and gap capacitor loaded CRLH-metaline.

Figure (22.14d) shows that the circuit of Fig (22.14c) can be transformed into the symmetrical T-type CRLH line, except the presence of a capacitor (C). The circuit model explains the presence of the transmission zero frequency (f_z) at displaced locations in the simulated results of the CRLH line. The frequency f_z is shown both in the absence and also in the presence of gap-capacitor. Figure (22.15a, b) shows the displaced locations of f_z. The values of the five components of the circuit of Fig (22.14d) are extracted from the simulated results [J.23, J.26, J.34, B.11]. The reference [J.23] provides the extracted values of components of the circuit model for the CSRR loaded microstrip, as shown in Fig (22.14a), on Rogers RO3010 substrate with $\varepsilon_r = 10.2$, $h = 1.27$ mm.

Figure (22.15a) shows the EM-simulated and circuit model-based S-parameters response of the CSRR loaded microstrip unit cell without a capacitive gap. It shows the presence of a narrow stopband due to the formation of ENG medium with negative permittivity. However, Fig (22.15b) shows that it is converted into a passband on the addition of the gap capacitor. The gap capacitor is responsible for the wideband plasma-type MNG medium with negative permeability. In a narrow band, the medium acts as a DNG medium. However, using the interdigital capacitor, in place of a gap capacitor, the wideband S_{21} response is obtained. It is shown in Fig (22.15d). Figure (22.15c) shows the dispersion diagram of the CRLH circuit working under a nearly symmetrical condition. A small stopband region appears due to small asymmetry. The wideband passband is due to the combined DNG–DPS mode operation [J.23].

SRR-Based Microstrip Metaline

The microstrip can also be loaded with the SRR inclusions as shown in Fig (22.16a). It forms a bandstop filter due to the negative permeability of the MNG medium. The SRRs placed by the sides of a microstrip are magnetically coupled to it. The equivalent circuit is discussed in

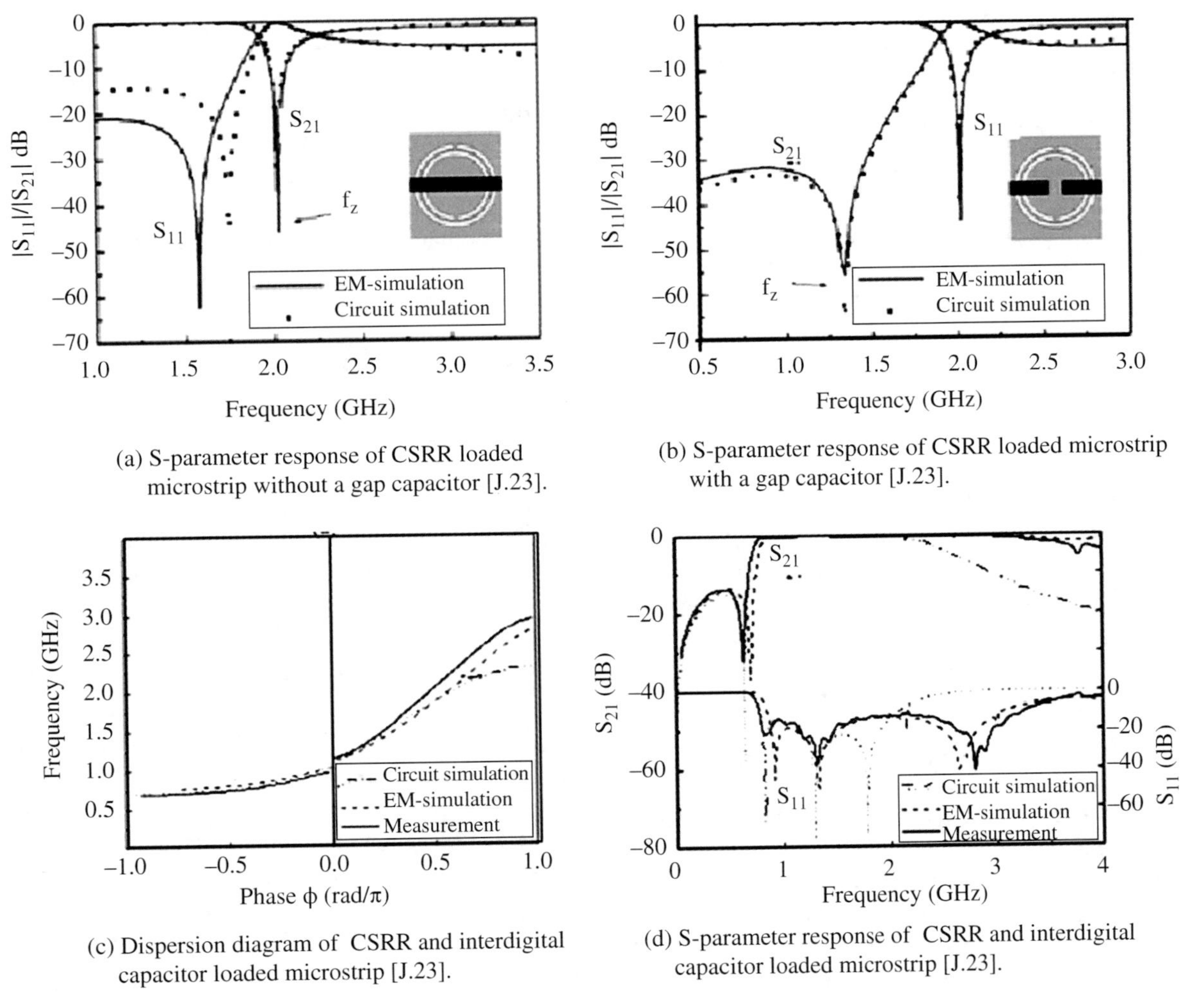

(a) S-parameter response of CSRR loaded microstrip without a gap capacitor [J.23].

(b) S-parameter response of CSRR loaded microstrip with a gap capacitor [J.23].

(c) Dispersion diagram of CSRR and interdigital capacitor loaded microstrip [J.23].

(d) S-parameter response of CSRR and interdigital capacitor loaded microstrip [J.23].

Figure 22.15 Response of CSRR and gap capacitor loaded CRLH-metaline. *Source:* Reproduced from Aznar et al. [J.23]. © 2008, Springer Nature.

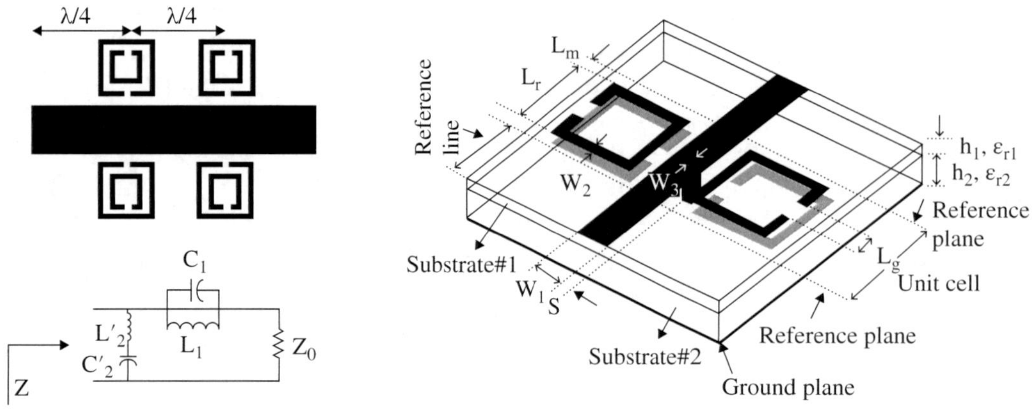

(a) SRR-loaded microstrip bandstop filter [J.36].

(b) BC-SRR-loaded and via loaded microstrip [J.35].

Figure 22.16 Topology of SRR loaded microstrip and SRR-via inductor loaded CRLH-metaline. *Source:* Mao et al. [J.35]. © 2005, IEEE. Lee and Lee [J.36]. © 2006, John Wiley & Sons.

[J.36]. The single-turn SRR can also be used to get the bandstop response [J.37]. The SRR based on tapered width strip conductor improves the rejection bandwidth of a bandstop filter [J.38]. However, by adding via between microstrip and ground plane the DNG passband response could be also obtained. The via creates the plasma medium with negative permittivity. Figure (22.16b) shows such a structure using the broadside-coupled SRR (BC-SRR) that avoids the bi-anisotropy of the SRR. The four unit cells of this structure have been used to design a leaky-wave antenna [J.35]. The SRR loaded microstrip lowpass filter has been used to realize the MNZ medium [J.39].

22.3.3 CPW-Resonant Metalines

Following the development of microstrip metalines, the resonant CPW metalines are also designed using the SRR at the back of the unmetallized face of a CPW, and also by embedding the CSRR aperture in the conducting strips of a CPW.

SRR-Based CPW Metaline

Figure (22.17a, b) show two schemes of periodic loading of the host CPW structure [J.26, J.40, J.41, B.12]. *Scheme #1* is a uniplanar structure, i.e. SRRs and CPW strip conductors are on the same side of the substrate. However, the slot regions of the CPW have been enlarged to accommodate the SRRs. The wide slot-gap CPW structure has very high characteristics impedance, in the range of 190 Ω. Therefore, impedance matching with 50 Ω input/output ports is difficult. *Scheme #2* is a biplanar structure. In this structure, the SRRs are located on the backside of the substrate, just behind the slots of the CPW. The loaded CPW can easily achieve a characteristic impedance of 50 Ω. The SRRs inclusions are magnetically coupled to the CPW.

Figure (22.17c-i) shows the basic equivalent circuit of the unit cell of the magnetically coupled SRR. The SRR is modeled by the parallel LC resonant circuit. The π-network models the host DPS line structure. Figure (22.17c-ii) shows the simplified equivalent circuit. The SRR inclusion appears in the series arm that is responsible for the creation of an MNG medium with negative permeability over a narrow frequency band. It is similar to Fig (22.4a). The SRRs loaded structure can provide only the stopband characteristics.

The topology of Fig (22.17b) shows the strip conductor loading of both slot-gaps of a CPW. The shunt inductor, due to the strip conductors, loaded line acts as an electric plasma medium with negative permittivity. The combined circuit model, shown in Fig (22.17c-iii), is the model of the CRLH configuration, acting as a DNG medium with a passband. The electric plasma frequency should be higher than the resonance frequency of the SRR.

Figure (22.18a) shows the $|S_{21}|$ response of the SRR-loaded CPW in absence of the grounded strip conductors. It has a stopband response near 8.0 GHz due to negative permeability. However, Fig (22.18b) shows the passband response in the same frequency region

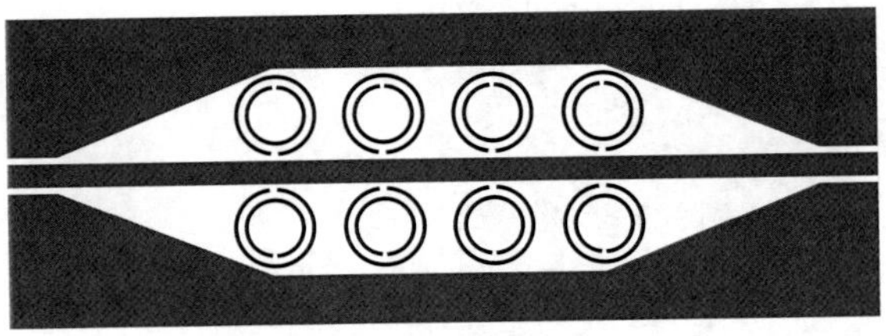

(a) Uniplanar SRRs loaded CPW. SRRs located in wide slot regions [J.41].

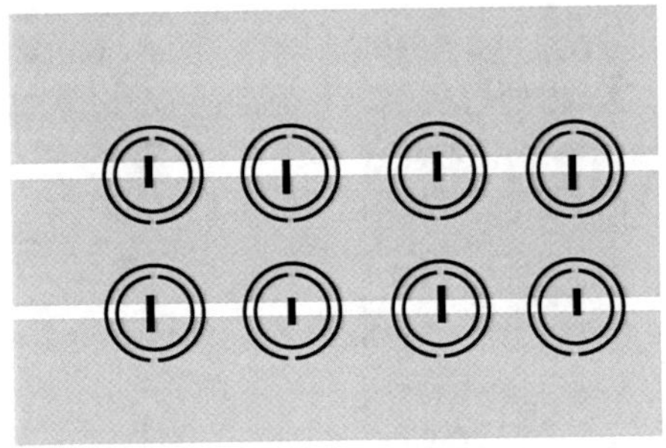

(b) Biplanar SRRs loading at the backside of CPW and strip loading in slot regions [J.26].

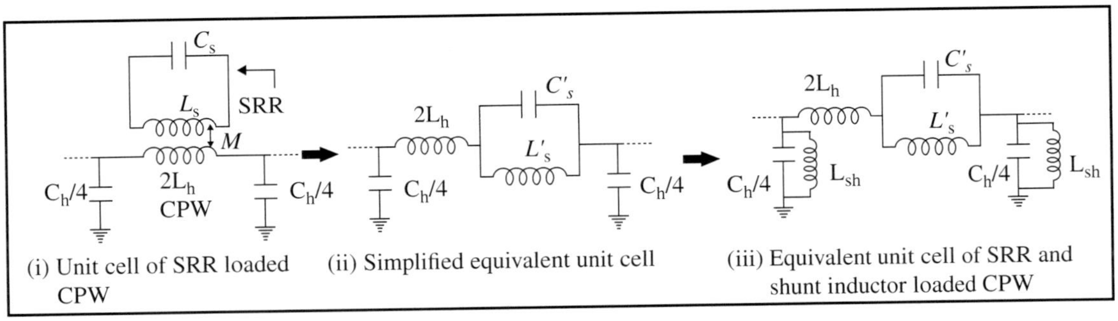

(c) Equivalent circuit of SRR loaded, and also inductor loaded CPW.

Figure 22.17 Topology of SRR loaded CPW, SRR-strip inductor loaded CPW-metaline and circuit model. *Source:* Baena et al. [J.26]. © 2005, IEEE. Falcone et al. [J.41] © 2004, John Wiley & Sons.

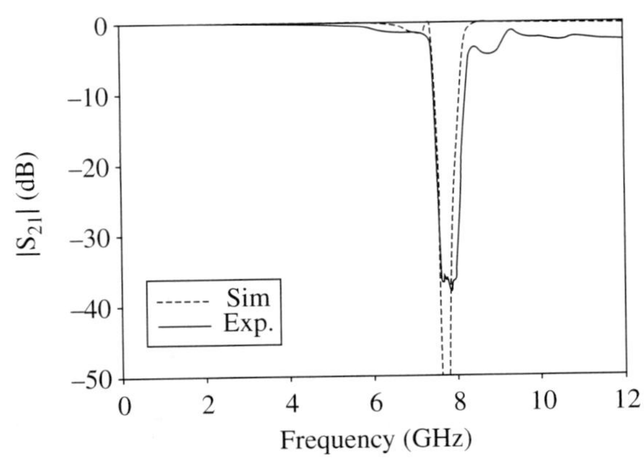

(a) $|S_{21}|$ bandstop response of the SRR loaded CPW.

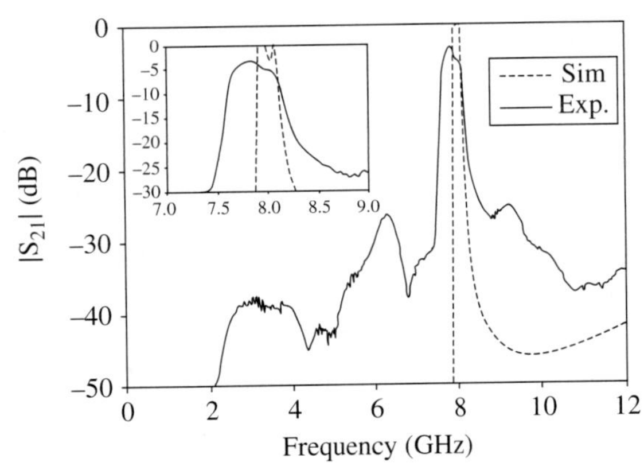

(b) $|S_{21}|$ bandpass response of the SRR and strip loaded CPW.

Figure 22.18 The $|S_{21}|$ response of SRR-strip loaded CPW. *Source:* Baena et al. [J.26]. © 2005, IEEE.

once the metallic strips are added to create negative permittivity. For the case of the SRR and metallic strip-loaded CPW, the medium is DNG. The details of the structures are available in reference [J.26].

CSRR-Based CPW Metalines

The SRR-loaded CPW metaline is not a matched line with uniplanar metallization. In the case of the MMIC and LTCC process, it causes additional complexity.

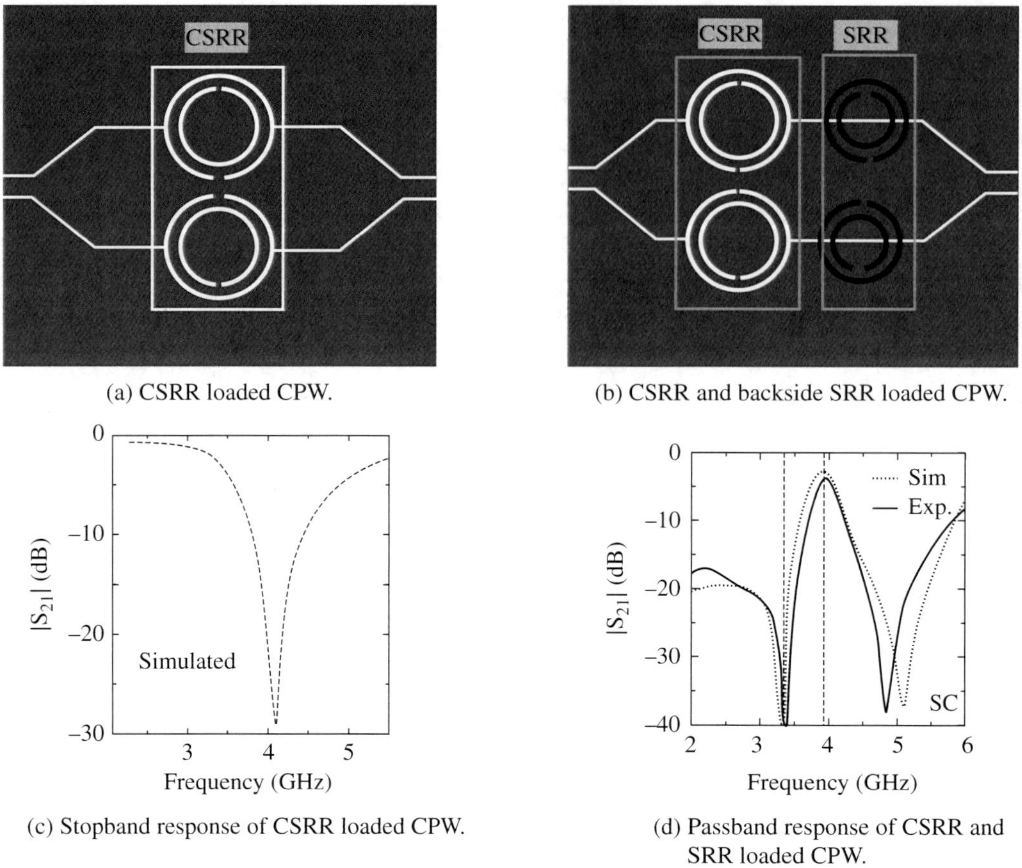

Figure 22.19 Topology of CSRR and CSSR-SRR loaded CPW and S21 response. *Source:* Al-Naib et al. [B.12].© 2012, In Tech.

A pair of CSRR is partly embedded in the ground conductor and partly in the central strip conductor of a CPW. It forms the unit cell of the CSRR loaded CPW, shown in Fig (22.19a). The CSRR inclusion provides a deeper rejection band shown in Fig (22.19c) [J.40, J.41]. It is due to the negative permittivity in the stopband region. The detail of the structure on the FR4 substrate is available in reference [B.12]. Figure (22.19c) shows good impedance matching below the rejection band. However, this structure does not provide a clean passband above the stopband. It has a spurious stopband at about 9.0 GHz, not shown in Fig (22.19c). It can be controlled by using the complementary U-shaped split resonator (CUSR) [B.12]. The structure can be further compacted by using complementary spirals in place of the CSRR. By cascading several unit cells, the rejection bandwidth can be improved.

Figure (22.19b) shows the loading of CPW by the combined CSRR–SRR resonant structures. The SRRs are located at the back of the unmetallized face of the substrate. The CSRR provides negative permittivity and SRR provides negative permeability. The DNG medium is created within the passband, around 4.0 GHz, shown in Fig (22.19d).

22.4 Some Applications of Metalines

The metalines, both nonresonant and resonant types, have been widely used to improve the performance of classical microstrip and CPW-based components and antenna. The metalines have been used to get compact microwave components operating over enhanced bandwidth. These are also used to get the dual-band and quad-band components. The design details of these components are available in numerous open literature and published books [B.1–B.8]. The present section provides just brief accounts of four examples to illustrate novel applications of metalines. The reader can follow the design details from the available resources.

22.4.1 Backfire to Endfire Leaky Wave Antenna

The microstrip based leaky-wave antenna is a traveling wave antenna. It radiates when the antenna operates in

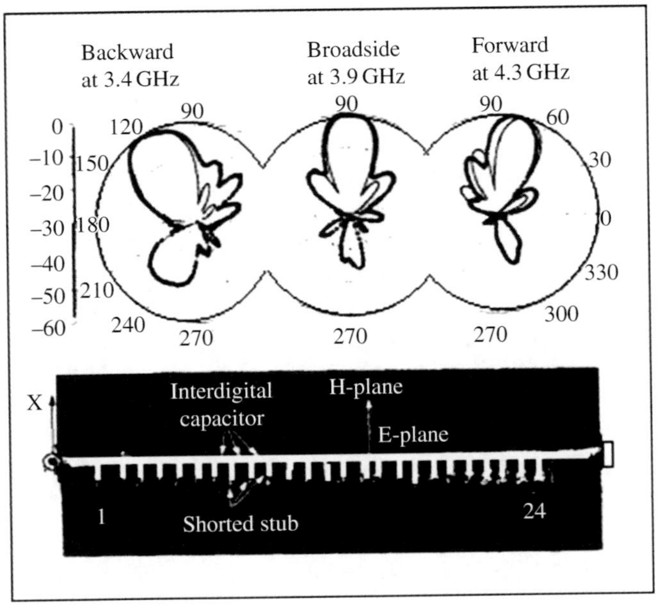

Figure 22.20 24 unit cells CRLH based leaky-wave antenna scanning upper half of sky. *Source:* Liu et al. [J.43]. © 2002, IET.

the higher-order mode. Its propagation constant β is positive, causing radiation in the forward direction with the increase in frequency. Thus, scanning the complete upper half of the sky is not possible. However, the balanced CRLH-based microstrip line, realized by loading the host microstrip line with a series interdigital capacitor and shunt grounded stub, supports both the backward wave propagation with negative β in the DNG region for $f < f_0$ and the forward-wave propagation with positive β in the DPS region for $f > f_0$. For $\beta = 0$, transition in the nature of wave-supporting medium occurs, without any bandgap, at the transition frequency f_0. Thus, the metaline can act as a backfire to endfire leaky-wave antenna scanning complete upper half of the sky [J.42].

Figure (22.20) illustrates the CRLH 24 unit cells-based leaky-wave antenna at $f_0 = 3.9$ GHz. At a frequency below 3.9 GHz, the backward radiation is obtained, at 3.9 GHz the radiation is in the broadside direction, and at a frequency above 3.9 GHz forward radiation occurs. The details of the antenna structure on substrate RT/Duroid 5880 with $\varepsilon_r = 2.2$, $h = 1.57$ mm are available in reference [J.43, B.2].

22.4.2 Metaline-Based Microstrip Directional Coupler

The usual edge coupled microstrip coupler has a broad bandwidth (>25%). However, it has a weak coupling level of −10 dB or less. The metalines-based couplers can provide an arbitrary level of coupling, even nearly up to 0 dB, while retaining the broadband characteristics. There are three kinds of metalines-based directional coupler: (*i*) *CRLH–CRLH coupler,* (*ii*) *CRLH–microstrip coupler and* (*iii*) *(D-CRLH)–CRLH coupler.* Figure (22.21a–c) show the fabricated images of these couplers. Figure (22.21d) shows a schematic diagram of the third case. These

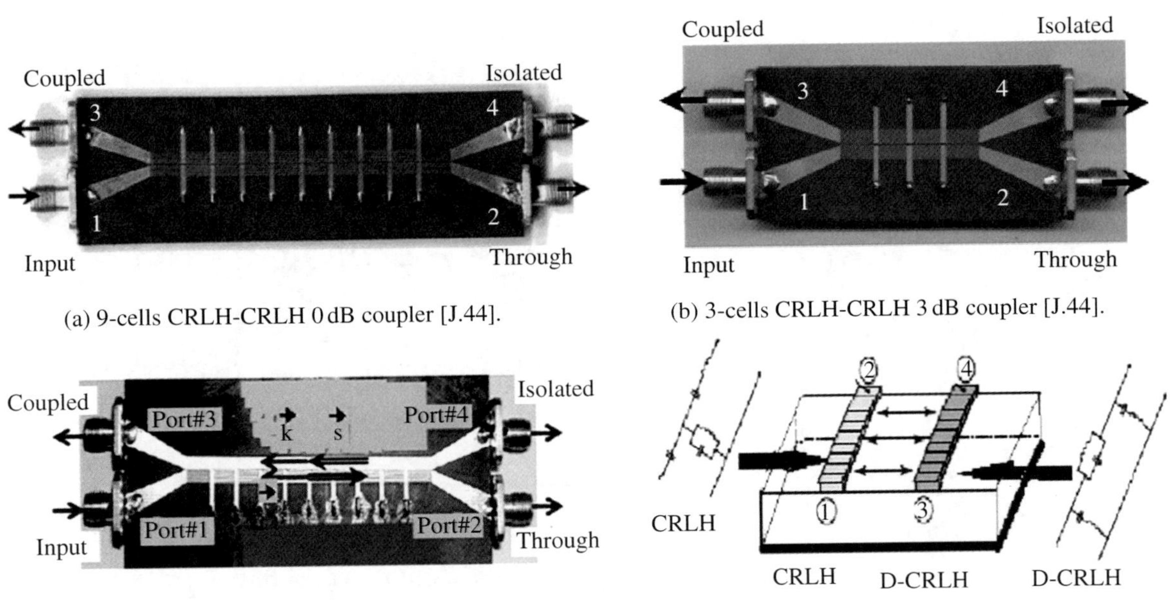

Figure 22.21 Configurations of metalines directional couplers. *Source:* Reproduced from Caloz et al. [J.44], © 2004 IEEE. reproduced from Caloz and Itoh [J.45], © 2004 IEEE.

couplers are analyzed using the *coupled-mode theory*. The design methods, fabrication details, and performance results of these couplers are available in the open literature [J.5, J.10, J.44–J.46, B.2–B.4]. We consider these couplers briefly to illustrate the novelty of design provided by the metalines.

Figure (22.21a) shows the nine unit-cells-long fabricated CRLH–CRLH 0 dB coupler. The mainline with ports 1–2 and the coupled line with port 3–4 are microstrip-based CRLH metalines. Port #1 is the input port and port #3 is the coupled port. The ports #2 and 4 are the through and isolated ports, respectively.

The uncoupled CRLH metaline is described by the equivalent circuit shown in Fig (22.6a). However, the magnetic and electric couplings contribute toward the additional coupling inductor and the coupling capacitor. It is the symmetrical coupler analyzed by the *even–odd mode concepts* [B.1, J.44]. The even and odd mode CRLH equivalent circuits are not identical. It results in the even–odd mode gaps. The coupler operates within these gaps. The couplers are designed such that the gaps overlap each other providing the coupling. The line length, i.e. the number of a unit cell is determined to achieve the desired level of coupling. Figure (22.21a) shows the nine unit cells long coupler with 0 dB coupling, whereas 3 dB coupler using the three unit cells is shown in Fig (22.21b). Both these couplers are developed on Rogers RT/Duroid 5880 substrate with $\varepsilon_r = 2.2$, h = 1.57 mm. The coupling gap between the two lines is 0.3 mm which is a comfortable gap-width from the fabrication point of view. The coupling can be controlled by changing the coupling gap and the number of unit cells.

Figure (22.22a) shows the S-parameters response of the 0 dB coupler. The 0 dB coupling, in practice about 0.5 dB coupling, is achieved over a bandwidth of 36% from 3.2 to 4.6 GHz. Its directivity is about 25 dB. Figure (22.22b) shows the S-parameters response of

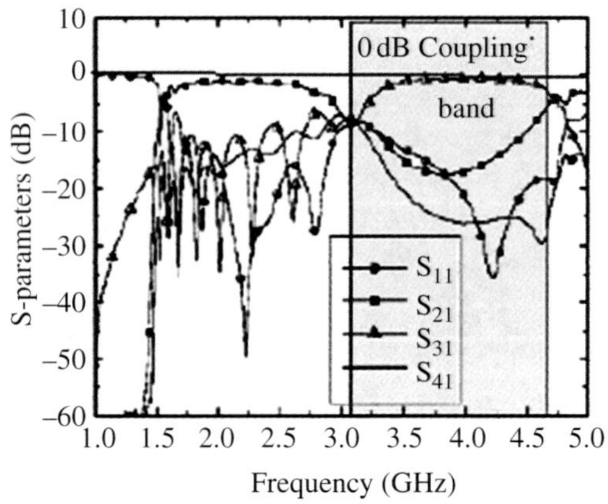

(a) Response of CRLH – CRLH 0 dB coupler [J.44].

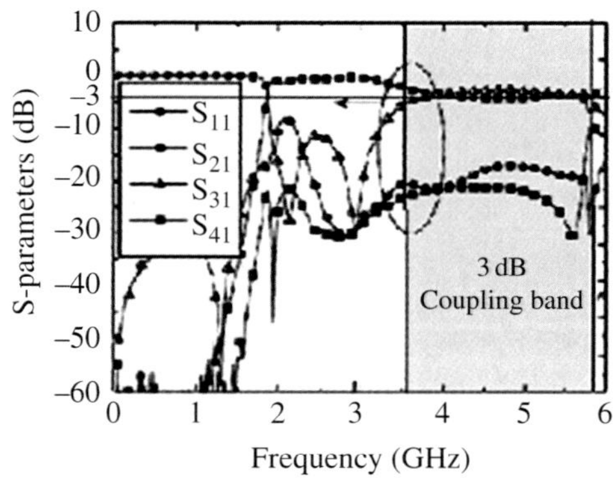

(b) Response of CRLH – CRLH 3 dB coupler [J.44].

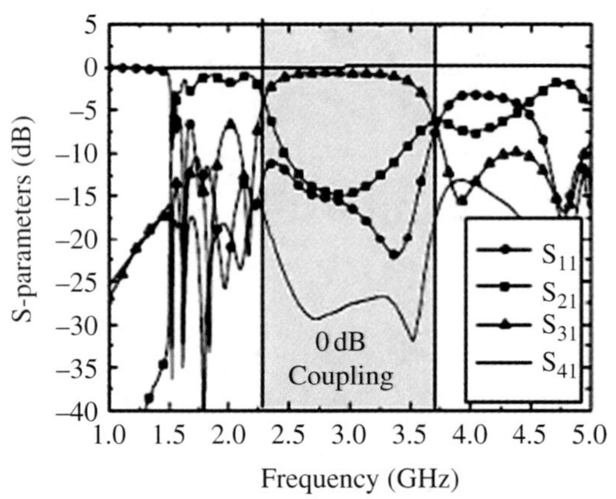

(c) Response of CRLH – microstrip 0 dB coupler [J.45].

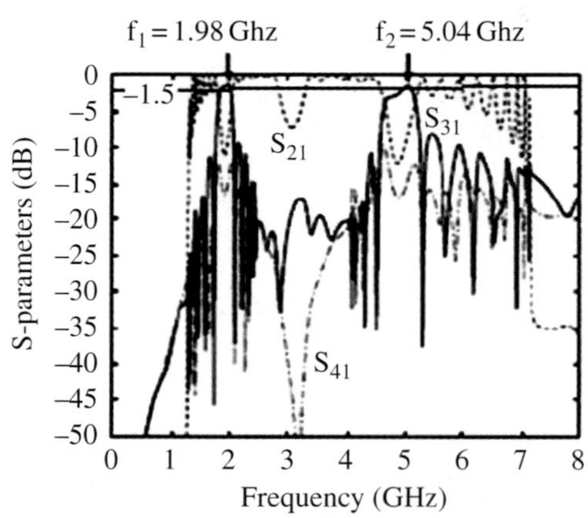

(d) Dual band response of CRLH – microstrip 0 dB coupler [J.10].

Figure 22.22 Frequency response of metalines based directional couplers of Figure (22.21). *Source:* Caloz and Nguyen [J.10]. © 2007, Springer-Verlag. Caloz et al. [J.44]. © 2004, IEEE. Modified from Caloz and Itoh [J.45].

the 3 dB coupler. It has a 50% bandwidth from 3.0–5.8 GHz and 20 dB directivity.

Figure (22.22c) shows the S-parameters response of the nine unit cells long (62 mm) CRLH–microstrip 0dB coupler. The CRLH line supports the backward wave, whereas the microstrip supports the forward wave. Thus, it is an asymmetrical coupler that is analyzed using *the c-mode and π-mode concept* [J.45]. This is also fabricated on Rogers RT/Duroid 5880 substrate with a 0.3 mm coupling gap. The fabricated coupler has a 0.7 dB coupling with 53% bandwidth from 2.2 to 3.8 GHz. It has excellent 30 dB directivity. Again, by reducing the number of unit cells, a compact 3 dB coupler can be designed.

The above-mentioned couplers are single band couplers. However, the 15 unit cells (D-CRLH)–(CRLH) coupler, shown in Fig (22.21d), is a dual-band coupler [J.10]. It also shows the equivalent circuits of both the CRLH and D-CRLH isolated lines. These circuits are modified by the coupling. Figure (22.21d) shows dual coupling bands of (−1.5 dB) coupling, around f_1 = 1.98 GHz and f_2 = 5.04 GHz bands. Its S-parameters are, to some extent, influenced by the choice of the line as the mainline or a coupled line.

22.4.3 Multiband Metaline-Based Components

Usually, microstrip components work in a single specified band. However, for some applications, the microwave components are required to operate in the dual and more frequency bands. These bands are not harmonically related. The dual-band microstrip-based components are designed using the CRLH-metalines [B.2, B.4, B.7]. The unit cell of balanced CRLH-metaline exhibits dual-band response as its dispersion diagram continuously moves, with an increase in frequency, from the DNG region to the DPS region. The dispersion diagram is nonlinear and can be tailored to meet the dual-band requirement. The concept of dispersion diagram tailoring has helped to conceive the concept of *dispersion engineering* [J.7, J.8, J.13, J.47].

Dual-band Components

The microstrip sections are used to realize microwave components. The standard DPS microstrip section has needed ∓90° electrical length at a single frequency. However, the CRLH-metaline can operate at two harmonically unrelated frequencies, as shown in Fig (22.23a), for the dispersion diagrams of a CRLH-metaline. It also shows the working of a DPS microstrip line at harmonically related frequencies. The −90° line section is obtained at frequency f_1 and also −270° (+90°) at frequency f_2 using the metaline, where f_2/f_1 is either integer or noninteger. It is not possible with a DPS line. Figure (22.23b) shows a realization of a metaline section, using SMD (surface-mounted devices) discrete elements, embedded in the microstrip host line [J.48]. It also shows the equivalent circuit of the CRLH line section. Such a line section can be used as an open-circuited λ/4 stub to a microstrip line. Figure (22.23c) shows that the stub creates dual stopband at 890 and 1670 MHz. This concept has been used to design dual-band branch-line coupler and rat-race coupler [J.48]. The resonant inclusions-based metalines are also used to design other dual-band components [J.13, B.7].

Quad-band Components

Figure (22.24a) shows the passband type quad-band metaline. It is realized by embedding the D-CRLH configuration, shown within gray square boxes, in the CRLH type of metaline, shown within light gray elliptical boxes. Let us consider the case when ω_0 is the resonance frequency of all four resonators in the series and shunt arms of the quad-band metaline. In this case, at the operating frequency, $\omega < \omega_0$ the composite network is reduced to the CRLH line #1. At frequency, $\omega > \omega_0$, it is reduced to another CRLH line #2. Each CRLH line supports the dual-band, giving the quad-band for the unit cell of metaline shown in Fig. (22.24a). The dispersion diagram of Fig (22.24b) exhibits an equal phase response at four frequencies. Both CRLH#1 and #2 lines are balanced metalines without bandgap and crossing from the DNG to DPS medium occurs at 1.5 and 2.5 GHz, respectively [J.47].

Figure (22.24c) shows the realization of quad-band metaline in the passband mode. The series resonator in the shunt arm is formed by a combination of the patch and meandered inductor. The parallel resonator in the series arm is formed by a combination of meandered inductor and chip capacitor. Figure (22.24d) shows a wide bandwidth dual passband response, as four passbands have merged into two pass-bands. The dual of the configuration shown in Fig (22.24a) can also be obtained giving quad stopband response [J.47].

22.5 Modeling and Characterization of Metsurfaces

The metasurfaces are considered the 2D equivalents of volumetric 3D metamaterials discussed in chapter 21. Fabrication of the 3D metamaterials is a difficult and expensive process. The fabrication process of the

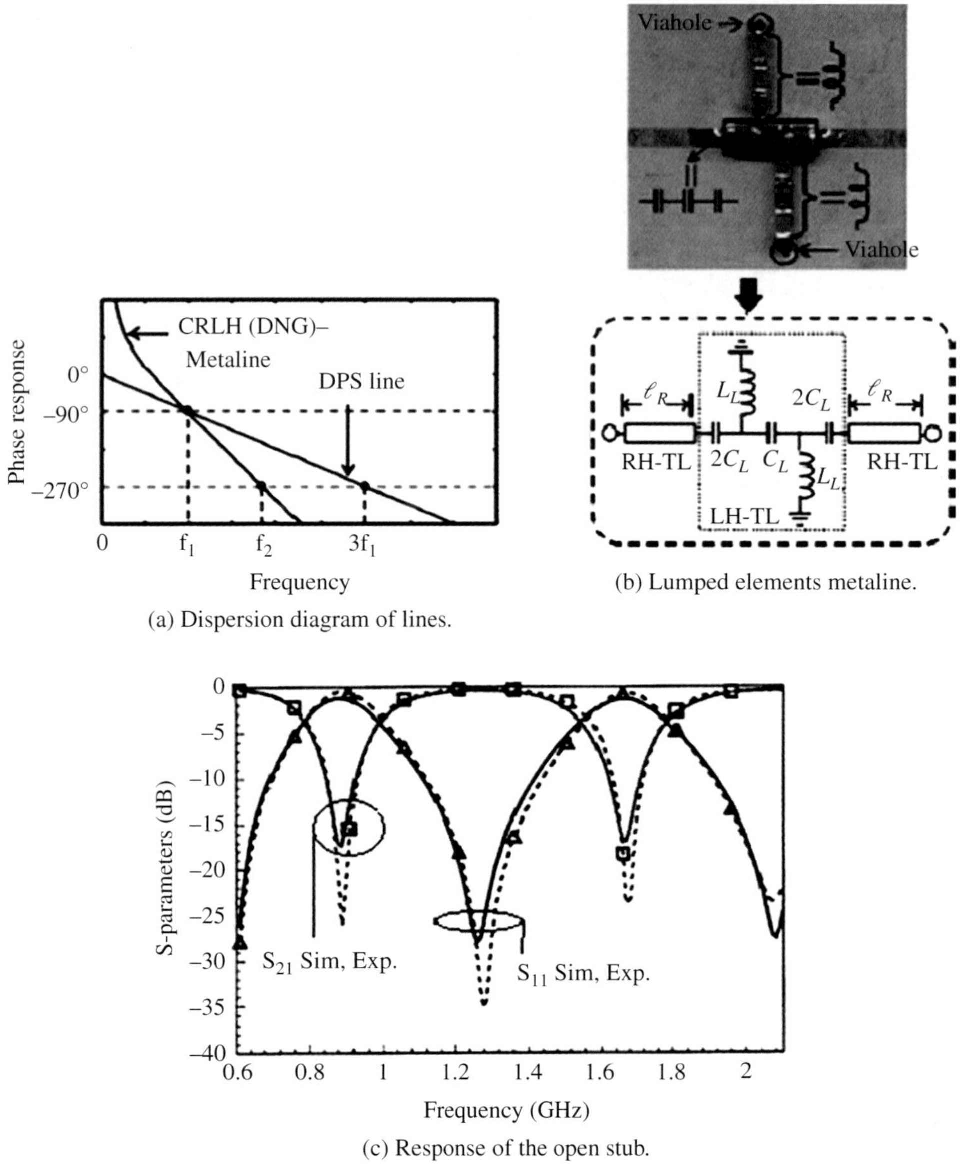

(a) Dispersion diagram of lines.
(b) Lumped elements metaline.
(c) Response of the open stub.

Figure 22.23 Metaline for the dual-band components. *Source:* Hsiang Lin et al. [J.48]. © 2004, IEEE.

monolayer or a few layered metasurfaces is much simpler. Like the metamaterials, the metasurfaces are engineered by periodically arranging the subwavelength metallic/dielectric inclusions on the surface of the host dielectric substrates. These metasurfaces are also engineered using the apertures as inclusions in the metallic screen. Recently, the metasurfaces have been widely explored for their unique applications in the microwave to optical frequency ranges. Reviews of their characteristics, modeling, and applications are available in recent publications [J.49–J.52, B.13].

The periods of inclusions on metasurfaces are at the subwavelength scale. It avoids the Bragg type diffraction. In this respect, the metasurfaces are different from the EBG surfaces, discussed in chapter 20. Under the excitation of the EM-waves, from microwave to optical frequencies, the inclusions get polarized, electrically or magnetically, creating the electric and magnetic dipoles distributed over the metasurfaces. Thus, the *electric metasurfaces* are realized with the electric dipoles and the *magnetic metasurfaces* are engineered using the magnetic dipoles, whereas the *electromagnetic metasurfaces* are engineered using both the electric and magnetic dipoles. The array of polarized inclusions, also called the array of *scatterers* or *meta-atoms*, are treated as the array of antenna causing the reflection and refraction

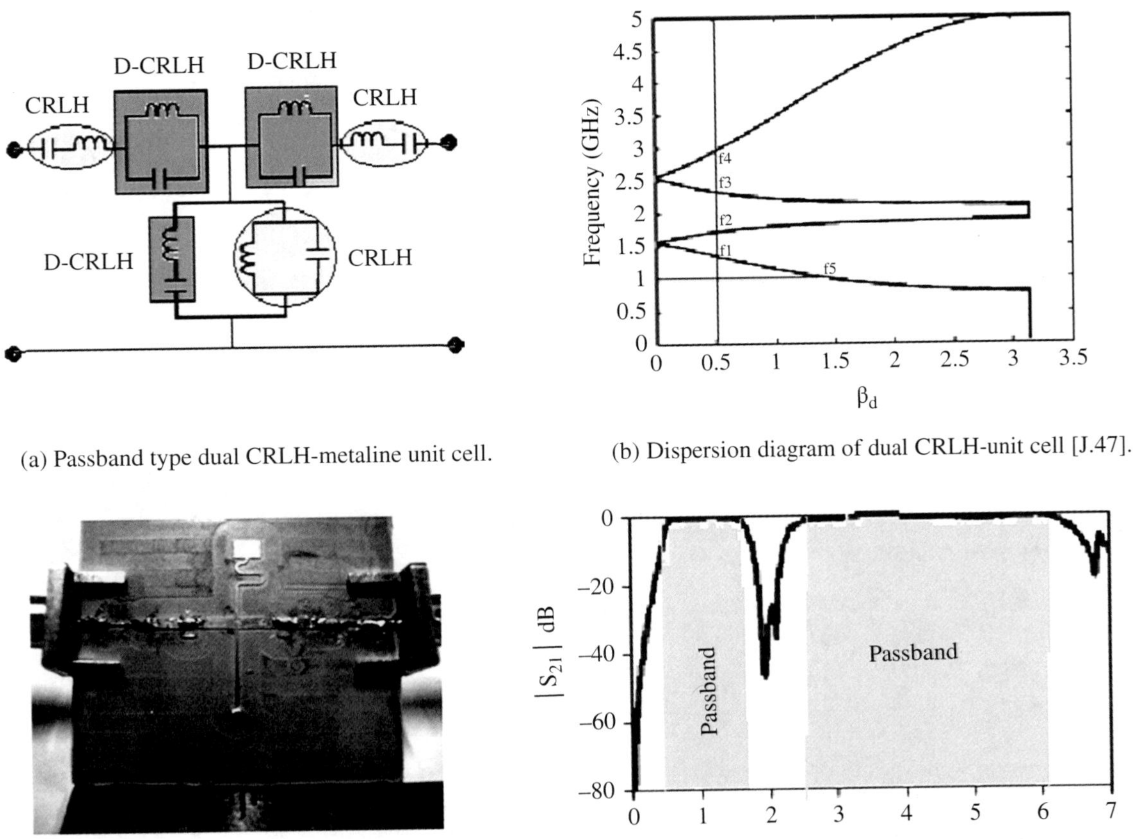

(a) Passband type dual CRLH-metaline unit cell.

(b) Dispersion diagram of dual CRLH-unit cell [J.47].

(c) Realization of dual CRLH-unit cell [J.47].

(d) S_{21}- response of dual CRLH-unit cell [J.47].

Figure 22.24 Quad-band metaline. *Source:* Eleftheriades and Islam [J.47]. © 2007, Elsevier.

of the EM-waves. The inclusions have variable geometric parameters, i.e. variation in their shape, size, and orientation on the metasurfaces. They control the amplitude, phase, and polarization state of the incident waves due to the spatially varying arrangement of inclusions. Thus, the metasurfaces support the *anomalous reflections and refractions* governed by the *generalized Snell's law* resulting in their wavefront controlling ability in the 3D space. The metasurfaces also support the surface-waves propagation. Space-waves can be coupled to the metasurfaces as the surface-waves and further radiated in the desired manner as the leaky-waves. In this way, the metasurfaces are different from similar-looking *frequency selective surfaces (FSS)*.

In principle, the thickness of a metasurface should be almost zero, i.e. very small compared to the operating wavelength. In this case, engineering of the electric dipoles of the surface is simple. However, creating magnetic dipoles on a monolayer metasurface is difficult, as it requires circulatory electric current in a plane normal to the surface of inclusions. It requires at least two-layered structure or presence of a conductor backing to the electric dipoles to generate magnetic dipoles. The electric dipole and its oppositely polarized image dipole create such a current loop acting as the magnetic dipole. The natures of the metasurfaces are determined by the nature of the polarizable inclusions and their periodic arrangements, so the metasurfaces could be *isotropic, anisotropic, bi-isotropic, and bianisotropic*. The materials based on such inclusions are discussed in chapter 6. Under the time-harmonic incident fields, the electric and magnetic dipoles create *discrete electric and magnetic polarization currents*. However, subwavelength size inclusions and periods of their 2D lattices help the homogenization of the metasurfaces. On the homogenized metasurface, both polarization currents are substituted by the equivalent *continuous electric and magnetic surface currents*. The surface currents act as the secondary radiators, like the secondary sources on the Huygens surfaces. The metasurfaces supporting the surface currents could be treated by Schelkunoff's *equivalent surface principle* [B.14].

The 3D bulk metamaterials are characterized by the equivalent (effective) permittivity and permeability. However, usually, it is not a proper description of the 2D metasurface, as it is based on the surface polarization phenomenon. As a practical measure, in some cases, such characterization is used. *The metasurfaces are characterized by surface polarizabilities, surface susceptibility, and surface admittances/impedances*. The surface admittances/impedances are useful to appreciate reflection, refraction, and impedance matching with free space [J.50]. The present selection discusses some of these characterization methods of metasurfaces. The next section (22.6) presents a few illustrative examples for the applications of metasurfaces in the microwave frequency ranges.

22.5.1 Characterization of Metasurface

The wave phenomena of reflection, refraction (transmission), and diffraction from the metasurface are caused by the induced time-harmonic electric and magnetic dipoles periodically arranged on a dielectric sheet. Thus, the incident time-harmonic EM-waves induce the discrete polarization currents and charges distributed over the metasurfaces. The scattering of the fields by metasurfaces is modeled as *Schelkunoff's equivalent surface* to understand the scattering of EM-waves. Further, the metasurface is modeled by several different parameters. These aspects are discussed below.

Schelkunoff's Equivalent Surface

Figure (22.25) shows that a metasurface, supporting the physically realizable discrete electric and magnetic polarization currents could be replaced with Schelkunoff's equivalent surface. It supports the equivalent continuous electric $\vec{J}_s$ and magnetic $\vec{M}_s$ currents. Thus, the equivalent surface is obtained through the process of homogenization of an artificially engineered discrete metasurface. The metasurface is located at the interface of medium #1 and medium #2. The surface currents present discontinuity to the EM-fields $\left(\vec{E}_1, \vec{H}_1\right)$ and $\left(\vec{E}_2, \vec{H}_2\right)$ existing in two media:

$$\left(\vec{E}_2 - \vec{E}_1\right) \times \hat{n} = \vec{M}_s \quad (a) \quad \hat{n} \times \left(\vec{H}_2 - \vec{H}_1\right) = \vec{J}_s \quad (b), \tag{22.5.1}$$

where $\hat{n}$ is the normal unit vector. The field quantities $\left(\vec{E}_1, \vec{H}_1\right)$ account for the total incident and reflected field in the medium #1, and field quantities $\left(\vec{E}_2, \vec{H}_2\right)$ account for the transmitted (refracted) field in the medium #2.

Modeling of Metasurface

Figure (22.26) shows stages of modeling of the engineered metasurface. Each stage of modeling is discussed below:

Stage – 1

It is the physical step. The incident EM-waves induces the time-harmonic electric dipoles in periodically arranged inclusions located at the interface. In a volumetric material, the polarization and magnetization (magnetic polarization) are expressed as the dipoles per unit volume. It is discussed in chapter 6. However, in the case of a metasurface, the *surface polarization* is considered as the dipoles per unit surface area [J.50]. The surface electric and magnetic polarizations are further related to the surface electric and magnetic susceptibilities. Following the equations (6.1.3), (6.1.5), and

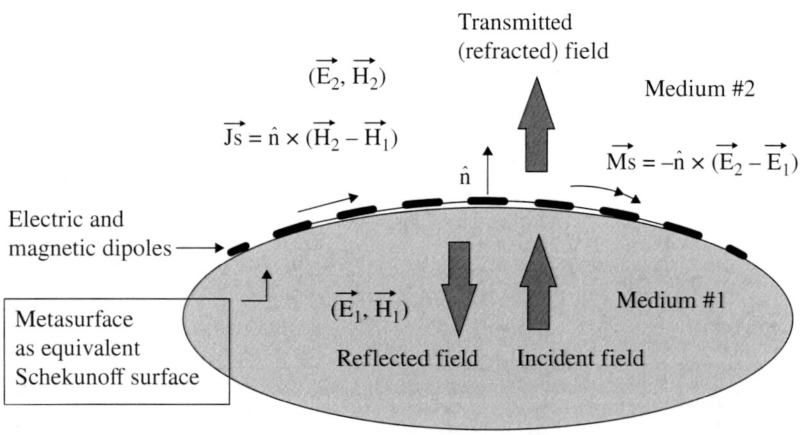

Figure 22.25 Metasurface supporting time-harmonic electric and magnetic dipoles. The corresponding electric and magnetic polarization currents are treated as continuous electric and magnetic currents on Schelkunoff's equivalent surface.

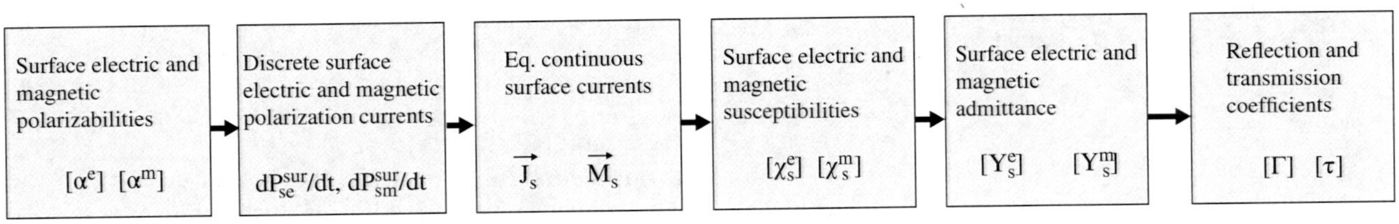

Figure 22.26 Stages of metasurface characterization.

(6.1.24) of chapter 6, the following expressions for the time-harmonic induced dipole moments and susceptibilities could be written:

Electric/magnetic dipole moments:

$$\vec{p} = \alpha^e \vec{E}_0^{loc} e^{j\omega t} \quad (a), \qquad \vec{m} = \alpha^m \vec{H}_0^{loc} e^{j\omega t} \quad (b), \tag{22.5.2}$$

where α^e and α^m are electric and magnetic polarizability respectively.

Electric/magnetic surface polarization (P_{se}, P_{sm}) *and surface susceptibility* (χ_s^e, χ_s^m) are obtained as follows:

$$\vec{P}_{se} = \frac{\vec{p}}{S} = \left(\frac{\alpha^e}{S}\right)\vec{E}_0^{loc} e^{j\omega t} \quad (a)$$

$$\vec{P}_{sm} = \frac{\vec{m}}{S} = \left(\frac{\alpha^m}{S}\right)\vec{H}_0^{loc} e^{j\omega t} \quad (b)$$

$$\text{also,} \quad \vec{P}_{se} = \chi_s^e \vec{E}_0^{loc} e^{j\omega t} \quad (c)$$

$$\vec{P}_{sm} = \chi_s^m \vec{H}_0^{loc} e^{j\omega t} \quad (d)$$

$$\text{and,} \quad \chi_s^e = \frac{\alpha^e}{S} \quad (e)$$

$$\chi_s^m = \frac{\alpha^m}{S} \quad (f), \tag{22.5.3}$$

where S is an area of a unit cell, χ_s^e and χ_s^m are surface electric and magnetic susceptibilities, respectively.

The tangential local fields, causing the polarization, at the interface are treated as the average fields of the fields from the upper (+) and lower (−) sides of the interface:

$$\vec{E}_t^{loc} = \vec{E}_t^{av} = \frac{\vec{E}_+ + \vec{E}_-}{2} \quad (a)$$

$$\vec{H}^{loc} = \vec{H}_t^{av} = \frac{\vec{H}_+ + \vec{H}_-}{2} \quad (b)$$

$$\text{where,} \quad \vec{E}_+ = \vec{E}_2\Big|_{sur}^{U} = \vec{E}^{tra}\Big|_{sur}^{U} \quad (c)$$

$$\vec{E}_- = \vec{E}_1\Big|_{sur}^{L} = \left(\vec{E}^{inc} + \vec{E}^{ref}\right)\Big|_{sur}^{L} \quad (d)$$

$$\vec{H}_+ = \vec{H}_2\Big|_{sur}^{U} = \vec{H}^{tra}\Big|_{sur}^{U} \quad (e)$$

$$\vec{H}_- = \vec{H}_1\Big|_{sur}^{L} = \left(\vec{H}^{inc} + \vec{H}^{ref}\right)\Big|_{sur}^{L} \quad (f). \tag{22.5.4}$$

The total field in the lower-medium #1 is a combination of the incident and reflected fields $\left(\vec{E}^{inc}, \vec{E}^{ref}\right)$, $\left(\vec{H}^{inc}, \vec{H}^{ref}\right)$; and the field in the upper-medium #2 is the transmitted field $(\vec{E}^{tra})$, $(\vec{H}^{tra})$.

Stage - 2

The time-harmonic electric and magnetic dipoles constitute the discrete electric and magnetic polarization currents. Each inclusion is treated as a miniature radiator or antenna radiating in both the lower and upper media resulting in the reflection and transmission, so the metasurfaces behave as *reflectarrays* and *transmit-arrays*. The surface polarization currents, due to the local tangential fields, are obtained as follows from the equations (22.5.3a, b):

Electric/magnetic discrete polarization currents:

$$\frac{d\vec{P}_{se}}{dt} = \frac{j\omega\alpha^e}{S}\vec{E}_t^{loc} = j\omega\alpha_s^e E_t^{loc} \quad (a)$$

$$\frac{d\vec{P}_{sm}}{dt} = \frac{j\omega\alpha^m}{S}\vec{H}_t^{loc} = j\omega\alpha_s^m \vec{H}_t^{loc} \quad (b), \tag{22.5.5}$$

where the *electric and magnetic surface polarizabilities* are α_s^e and α_s^m.

Stage - 3

At this stage, the discrete metasurface is treated as *Schelkunoff's equivalent surface* supporting the continuous electric and magnetic surface currents. These are equivalent to their corresponding physical polarizations currents giving the following relations:

$$\vec{J}_s = j\omega\varepsilon_0\chi_s^e \vec{E}_t^{loc} \quad (a), \qquad \vec{M}_s = j\omega\mu_0\chi_s^m H_t^{loc} \quad (b). \tag{22.5.6}$$

Stage - 4

At this stage, the metasurfaces are characterized by the electric and magnetic surface susceptibilities. These are related to their surface polarizabilities, as the polarization currents are also obtained in terms of the susceptibilities. Therefore, from equations (22.5.5) and (22.5.6), the following relations are obtained:

$$\vec{J}_s = \frac{dP_{se}}{dt} = j\omega\varepsilon_0\chi_s^e \vec{E}_t^{loc} \quad (a), \quad \text{also } j\omega\alpha_s^e = j\omega\varepsilon_0\chi_s^e \quad (b)$$
$$\vec{M}_s = \frac{dP_{sm}}{dt} = j\omega\mu_0\chi_s^m H_t^{loc} \quad (c), \quad \text{also } j\omega\alpha_s^m = j\omega\mu_0\chi_s^m \quad (d). \qquad (22.5.7)$$

Stage – 5

The metasurfaces support both the surface currents and the tangential field components. Thus, the metasurfaces are also characterized by the *electric and magnetic surface admittances* (Y_s^e, Y_s^m) or *surface impedances* (Z_s^e, Z_s^m). The surface impedance/admittances have their origin in the polarizabilities of dipoles and corresponding susceptibilities. Using equation (22.5.7), the following relations are obtained among surface electric and magnetic impedance/admittances, surface electric and magnetic susceptibilities, and surface electric and magnetic polarizabilities of inclusions creating electric and magnetic dipoles:

$$\vec{J}_s = Y_s^e \vec{E}_t^{loc} \quad (a), \quad \text{also } \vec{E}_t^{loc} = Z_s^e \vec{J}_s \quad (b)$$
$$Z_s^e = 1/Y_s^e \quad (c), \quad \vec{M}_s = Z_s^m \vec{H}_t^{loc} \quad (d)$$
$$\text{also } \vec{H}_t^{loc} = Y_s^m \vec{M}_s^m \quad (e), \quad Z_s^m = 1/Y_s^m \quad (f)$$
$$Y_s^e = j\omega\varepsilon_0\chi_s^e = j\omega\alpha_s^e \quad (g), \quad Z_s^m = j\omega\mu_0\chi_s^m = j\omega\alpha_s^m \quad (h). \qquad (22.5.8)$$

Stage – 6

The surface impedances/admittances are responsible for the reflection, transmission, and also impedance matching of the metasurfaces with free space. The impedance-matched metasurfaces avoid the reflection and fully transmit the incident waves. Such metasurfaces are known as the *Huygens metasurfaces*. Thus, the metasurfaces are also characterized by the reflection and transmission coefficients. In the case of the isotropic metasurface, the reflection coefficient (Γ) and transmission coefficient (τ) are scalar quantities. Their computation for the normally incident plane waves is discussed in section (22.5.2).

The above discussion is limited to *isotropic metasurface*. In this case, the surface polarizability, surface susceptibility, and surface admittance/impedance are scalar quantities. However, the metasurfaces could also be anisotropic, bi-isotropic, and bi-anisotropic in nature. In these cases, the metasurface characterizing parameters are described by 2×2 matrices and tensors, shown in Figure (2.26). Following the above discussion, and also section (4.2) of chapter 4, the results related to such metasurfaces are summarized below:

Anisotropic Metasurface

In subsection (4.2.3) of chapter 4, it is discussed that the permittivity and permeability of anisotropic materials are tensor quantities. Likewise, using equations (22.5.2) and (22.5.8) in tensor form, the anisotropic metasurfaces are characterized by the electric and magnetic dipole moments p and m described by the 2×2 tensors (*dyadic*) of the surface electric and magnetic polarizabilities, and the corresponding surface electric and magnetic admittances as follow:

$$\begin{bmatrix} \vec{p} \\ \vec{m} \end{bmatrix} = \begin{bmatrix} [\alpha_s^e] & 0 \\ 0 & [\alpha_s^m] \end{bmatrix} \begin{bmatrix} \vec{E}_t \\ \vec{H}_t \end{bmatrix} \quad (a)$$

where electric polarizability tensor:

$$= \alpha_s^e = [\alpha_s^e] = \begin{bmatrix} \alpha_{xx,s}^e, & \alpha_{xy,s}^e \\ \alpha_{yx,s}^e, & \alpha_{yy,s}^e \end{bmatrix} \quad (b)$$

magnetic polarizability tensor:

$$= \alpha_s^m = [\alpha_s^m] = \begin{bmatrix} \alpha_{xx,s}^m, & \alpha_{xy,s}^m \\ \alpha_{yx,s}^m, & \alpha_{yy,s}^m \end{bmatrix} \quad (c). \qquad (22.5.9)$$

$$\begin{bmatrix} \vec{J}_s \\ \vec{M}_s \end{bmatrix} = \begin{bmatrix} [Y_s^e] & 0 \\ 0 & [Z_s^m] \end{bmatrix} \begin{bmatrix} \vec{E}_t \\ \vec{H}_t \end{bmatrix} \quad (a)$$

$$\overline{\overline{Y}}_s^e = [Y_s^e] = \begin{bmatrix} Y_{xx,s}^e, & Y_{xy,s}^e \\ Y_{yx,s}^e, & Y_{yy,s}^e \end{bmatrix} \quad (b)$$

$$\overline{\overline{Z}}_s^m = [Z_s^m] = \begin{bmatrix} Z_{xx,s}^m, & Z_{xy,s}^m \\ Z_{yx,s}^m, & Z_{yy,s}^m \end{bmatrix} \quad (c). \qquad (22.5.10)$$

In equations (22.5.9) and (22.5.10), the surface admittance and impedance tensors (*dyadic*) $[Y_s^e]$ and $[Z_s^m]$ correspond to the electric and magnetic surface polarizabilities. The electric and magnetic polarizabilities are discussed in section (6.1) of chapter 6. For properly aligned metasurfaces, the cross-coupling terms, i.e. the (xy) and (yx) terms, are zero. The reflection and transmission coefficients of anisotropic surfaces are also tensor quantities:

$$\vec{E}^{ref} = \bar{\bar{\Gamma}}.\vec{E}^{inc} \quad (a), \qquad \vec{E}^{tra} = \bar{\bar{\tau}}.\vec{E}^{inc} \quad (b)$$
$$\bar{\bar{\Gamma}} = \begin{bmatrix} \Gamma_{xx}, & \Gamma_{xy} \\ \Gamma_{yx}, & \Gamma_{yy} \end{bmatrix} \quad (c), \qquad \bar{\bar{\tau}} = \begin{bmatrix} \tau_{xx}, & \tau_{xy} \\ \tau_{yx}, & \tau_{yy} \end{bmatrix} \quad (d). \qquad (22.5.11)$$

The anisotropic metasurface is a useful device for polarization conversion.

Bianisotropic Metasurface

The general linear tensor metasurfaces are characterized with the help of general linear relations of induced surface electric and magnetic dipole moments created by the average local tangential fields:

$$\begin{bmatrix} \vec{p} \\ \vec{m} \end{bmatrix} = \begin{bmatrix} [\alpha_s^{ee}] & [\alpha_s^{em}] \\ [\alpha_s^{me}] & [\alpha_s^{mm}] \end{bmatrix} \begin{bmatrix} \vec{E}_t \\ \vec{H}_t \end{bmatrix} \quad (a)$$
$$[\alpha_s^{em}] = -[\alpha_s^{me}]^T \quad (b), \qquad (22.5.12)$$

where $[\alpha_s^{ee}]$, $[\alpha_s^{mm}]$, $[\alpha_s^{em}]$, and $[\alpha_s^{me}]$ are the electric, magnetic, electromagnetic, and magnetoelectric surface polarizabilities tensors, respectively. In matrix equation (22.5.12), T denotes the transpose matrix. The presence of the magnetoelectric/electromagnetic polarizability is another unique characteristic of bianisotropic metasurfaces as against the anisotropic metasurfaces. It provides additional facility for manipulation of the wavefront by such metasurfaces [J.52, J.53].

Likewise, the general tensor metasurfaces are also characterized by the electric, magnetic, and magnetoelectric/electromagnetic tensor admittances:

$$\begin{bmatrix} \vec{J}_s \\ \vec{M}_s \end{bmatrix} = \begin{bmatrix} [Y_s^{ee}] & [Z_s^{em}] \\ [Y_s^{me}] & [Z_s^{mm}] \end{bmatrix} \begin{bmatrix} \vec{E}_t \\ \vec{H}_t \end{bmatrix} \quad (a)$$

where
$$\begin{bmatrix} \vec{E}_t \\ \vec{H}_t \end{bmatrix} = \begin{bmatrix} \hat{n} \times \vec{E}^{inc} \\ \hat{n} \times \vec{H}^{inc} \end{bmatrix} \quad (b). \qquad (22.5.13)$$

In equation (22.5.13), the surface admittance tensor $[Y_s^{ee}]$, and impedance tensor $[Z_s^{mm}]$ correspond to the surface electric and magnetic susceptibility or polarizability tensor, respectively. The surface impedance tensor $[Z_s^{em}]$ and surface admittance tensor $[Y_s^{me}]$ correspond to the magnetoelectric and electromagnetic polarizability or susceptibility tensor, respectively.

22.5.2 Reflection and Transmission Coefficients of Isotropic Metasurfaces

The reflection and transmission of normally incident plane waves at the interface of two different media are discussed in section (5.1.1) of chapter 5. However, in that discussion, polarizing inclusions are not present at the interface of media. The metasurface as the interface has the specific feature of accommodating polarizable inclusions at the interface. The metasurface could be present even in a homogeneous medium such as air medium. In the present analysis, the metasurface, i.e. *Schelkunoff's equivalent surface* is characterized by the electric and magnetic surface admittances supporting the electric and magnetic surface currents. The boundary condition (22.5.1) is used in terms of admittances given by the equation (22.5.10a).

Figure (22.27) shows the interface of the metasurface located in the (y–z)-plane. The y-polarized EM-wave propagating in the x-direction is normally incident at the metasurface. The total field in the left-hand half-space (medium #1) is a summation of the incident and reflected waves, whereas the transmitted wave is in the right-hand space (medium #2):

$$E^{tot} = E_0\left[e^{-jkx} + \Gamma e^{jkx}\right]\hat{y} \quad (a)$$
$$H^{tot} = \frac{E_0}{\eta_0}\left[e^{-jkx} - \Gamma e^{jkx}\right]\hat{z}, \quad x < 0 \quad (b)$$
$$E^{tra} = \tau E_0 e^{-jkx}\hat{y} \quad (c)$$
$$H^{tra} = \frac{\tau E_0}{\eta_0} e^{-jkx}\hat{z}, \quad x > 0 \quad (d). \qquad (22.5.14)$$

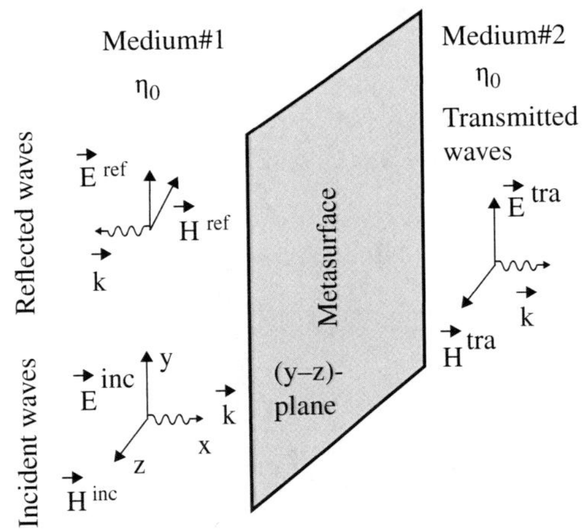

Figure 22.27 Reflection and transmission at the interface of the metasurface.

The time-harmonic variation $e^{j\omega t}$ is suppressed in expressions (22.5.14). The boundary conditions at the interface are given below:

$$\hat{x}\times\left(\vec{H}^{tra}(x)-\vec{H}^{tot}(x)\right)\Big|_{x=0}=\vec{J}_s \quad (a)$$
$$\hat{x}\times\left(\vec{E}^{tra}(x)-\vec{E}^{tot}(x)\right)\Big|_{x=0}=-\vec{M}_s \quad (b). \qquad (22.5.15)$$

On expanding equation (22.5.15a), we get the following expression:

$$\hat{y}\left(H_z^{tot}(x)-H_z^{tra}(x)\right)_{x=0}+\hat{z}\left(H_y^{tra}(x)-H_y^{tot}(x)\right)_{x=0}=\hat{y}J_{sy}+\hat{z}J_{sz}. \qquad (22.5.16)$$

As $H_y^{tra}(x)=H_y^{tot}(x)=0$, so $J_{sz}=0$, and using equation (22.5.14) the following relation is obtained:

$$\left(H_z^{tot}(x)-H_z^{tra}(x)\right)_{x=0}=J_{sy}=j\omega\alpha_s^e E^{av}=\frac{j\omega\alpha_s^e}{2}\left(E^{tra}(x)+E^{tot}(x)\right)_{x=0} \quad (a)$$
$$\Rightarrow \frac{E_0}{\eta_0}[(1-\Gamma)-\tau]=\frac{Y_s^e}{2}[\tau+(1+\Gamma)]E_0$$
$$\Rightarrow \tau+\Gamma=\frac{1-(\eta_0Y_s^e/2)}{1+(\eta_0Y_s^e/2)} \quad (b). \qquad (22.5.17)$$

Likewise, on expanding equation (22.5.15b), we get the following expression:

$$\hat{y}\left(E_z^{tot}(x)-E_z^{tra}(x)\right)_{x=0}+\hat{z}\left(E_y^{tra}(x)-E_y^{tot}(x)\right)_{x=0}=-(\hat{y}M_{sy}+\hat{z}M_{sz})$$
$$\Rightarrow \left(E_y^{tra}(x)-E_y^{tot}(x)\right)_{x=0}=-M_{sz}$$
$$\text{as } E_z^{tot}(x)=E_z^{tra}(x)=M_{sy}=0. \qquad (22.5.18)$$

On using expressions (22.5.7), (22.5.14), and other related expressions the following relation is obtained:

$$E_0[\tau-(1+\Gamma)]=-j\omega\alpha_s^m H^{av}=-\frac{j\omega\alpha_s^m}{2}\left[H_z^{tra}+H_z^{tot}\right]=-\frac{Z_s^m}{2}[\tau+(1-\Gamma)]\frac{E_0}{\eta_0} \quad (a)$$
$$\Rightarrow (\tau-\Gamma)-1=\frac{-Z_s^m}{2\eta_0}[(\tau-\Gamma)+1]$$
$$\Rightarrow \tau-\Gamma=\frac{1-(Z_s^m/2\eta_0)}{1+(Z_s^m/2\eta_0)} \quad (b). \qquad (22.5.19)$$

The transmission and reflections coefficients are obtained from equations (22.5.17b) and (22.5.19b):

$$\tau=\frac{4-Z_s^mY_s^e}{(2+\eta_0Y_s^e)(2+Z_s^m/\eta_0)} \quad (a)$$
$$\Gamma=\frac{2(Z_s^m/\eta_0-\eta_0Y_s^e)}{(2+\eta_0Y_s^e)(2+Z_s^m/\eta_0)} \quad (b)$$
$$\Rightarrow Y_s^e=\frac{2(1-\tau-\Gamma)}{\eta_0(1+\tau+\Gamma)} \quad (c)$$
$$Z_s^m=\frac{2\eta_0(1-\tau+\Gamma)}{(1+\tau-\Gamma)} \quad (d). \qquad (22.5.20)$$

The metasurface constituted of both electric and magnetic surface polarizabilities can provide the *perfect reflection* and *perfect transmission*. The conditions for perfect reflection and perfect transmission are obtained from equation (22.5.20).

Perfect Reflection

The perfect reflection from the metasurface requires zero transmission, i.e. $\tau=0$ in equation (22.5.20a). Thus, using equation (22.5.20a), the condition for the total reflection of normally incident plane wave at the metasurface is

$$Z_s^mY_s^e=4 \quad (a), \qquad \Rightarrow \alpha_s^e\alpha_s^m=-4/\omega^2 \quad (b). \qquad (22.5.21)$$

Equations (22.5.8g, h) are used to get equation (22.5.21b). The complex reflection coefficient of the metasurface under the above condition is obtained from equation (22.5.20d):

$$\Gamma=|\Gamma|\angle\phi^{ref}=\frac{Z_s^m-2\eta_0}{Z_s^m+2\eta_0} \quad (a)$$
$$\Rightarrow \Gamma=-\left[\frac{1-j\omega\alpha_s^m/2\eta_0}{1+j\omega\alpha_s^m/2\eta_0}\right] \quad (b)$$
$$\Rightarrow |\Gamma|=1, \quad \phi^{ref}=2\tan^{-1}(\omega\alpha_s^m/2\eta_0) \quad (c). \qquad (22.5.22)$$

Likewise, the complete reflection can be also achieved using electric polarizability. The reflection coefficient of the metasurface under the perfect reflection acts like the PEC/PMC surface. However, unlike these surfaces, the metasurface facilitates control of the reflection phase from $-\pi$ to π, i.e. full phase control of 2π. *Thus, the magnetic and electric polarizabilities of each inclusion can be tailored at a needed frequency to achieve perfect reflection and the desired reflection phase.* The oblique incidence at the interface of two natural media also provides perfect reflection at a *critical angle*. However, the reflection phase from the interface of natural media cannot be tailored. It is discussed in section (5.3) of chapter 5.

Perfect Transmission

The perfect transmission from the metasurface requires zero reflection, i.e. $\Gamma = 0$ at metasurface. In this case, the metasurface is matched with free space. The metasurface without reflection is known as the *Huygens metasurface* [J.54]. The concept of Huygens metasurface is similar to the concept of Huygens load, discussed in subsection (2.1.4) of chapter 2. Equation (22.5.20b), under the requirement $\Gamma = 0$, provides the following condition for the matching:

$$\eta_0 = \sqrt{\frac{Z_s^m}{Y_s^e}} \quad (a), \qquad \Rightarrow \eta_0 = \sqrt{\frac{\alpha_s^m}{\alpha_s^e}} \quad (b). \tag{22.5.23}$$

The complex transmission coefficient under no reflection condition is obtained from equation (22.5.20c) using the electric polarizability:

$$\tau = |\tau|\angle\varphi^{tra} = \frac{2 - \eta_0 Y_s^e}{(2 + \eta_0 Y_s^e)} \quad (a)$$

$$\Rightarrow \tau = \frac{1 - j(\omega\eta_0\alpha_s^e/2)}{1 + j(\omega\eta_0\alpha_s^e/2)} \quad (b)$$

$$\Rightarrow |\tau| = 1, \quad \phi^{tra} = -2\tan^{-1}(\omega\eta_0\alpha_s^e/2) \quad (c). \tag{22.5.24}$$

A similar result could be obtained for the magnetic polarizability. *The perfectly transmitted waves from the Huygens metasurface can acquire any phase between $-\pi$ to $+\pi$ by tailoring the polarizabilities of each inclusion forming the metasurface.* The perfect transmission also occurs at the Brewster angle for the obliquely incident plane waves at the interface of two natural media. However, the phase of the transmitted waves cannot be tailored. The Brewster angle is discussed in section (5.3.1) of chapter 5.

Metasurface with only Electric Polarization

It is difficult to design an inclusion with the magnetic response, as it requires a current loop normal to the plane of a metasurface. Therefore, only electric type polarization is commonly used to design metasurfaces. In this case, the magnetic polarizability (α_s^m) is zero, leading to $Z_s^m = 0$. The transmission and reflection coefficients of the electric type metasurface are obtained from equation (22.5.20):

$$\tau = \frac{2}{(2 + \eta_0 Y_s^e)} \quad (a), \qquad \Gamma = \frac{-\eta_0 Y_s^e}{(2 + \eta_0 Y_s^e)} \quad (b)$$

$$\Rightarrow \tau = \frac{2}{(2 + j\omega\eta_0\alpha_s^e)} \quad (c), \qquad \Gamma = \frac{-j\omega\eta_0\alpha_s^e}{(2 + j\omega\eta_0\alpha_s^e)} \quad (d). \tag{22.5.25}$$

For the electric type metasurface, both transmission and reflection occur simultaneously. The phase control also has a limited range, i.e. $-\pi/2$ to $\pi/2$. Similar behavior is obtained for the magnetic metasurface designed with inclusions of only magnetic polarizability. However, the electric type metasurfaces backed by conducting sheets act as the *reflectarrays* and provide 0 to 2π reflection phase control [J.51].

22.5.3 Phase Control of Metasurface

The above discussion shows that the reflection and transmission phases can be tailored by controlling the phase of individual inclusions. It helps to control the direction of the reflected and transmitted waves from the metasurface. The realization of such phase control is discussed in the present subsection. In the case of a material slab, the phase of the refracted wave is acquired by the incident wave while traversing the slab. In the case of a metasurface, the inclusions constituting the metasurface act like dipoles receiving the incident waves and reradiating in the forward direction as the refracted waves, and in the backward direction as the reflected waves. However, control of *shape, size, and orientation* of the inclusions at each location on the metasurface in a periodic manner provides the mechanism of the phase control of the scattered incident waves resulting in the *anomalous reflection* and *refraction*. The anomalous reflection and refraction are governed by *generalized Snell's law* that is discussed in next subsection (22.5.4). The working of the metasurface is analogous to the working of the reflect/transmit antenna arrays [J.55]. In the case of the metasurfaces, full 2π i.e. 360° wavefront control of the reflected and refracted waves have been realized. Moreover, metasurfaces are *homogenized surfaces* with subwavelength size periodicity.

The metasurfaces can have a uniform phase distribution of inclusions. The *uniform metasurface* is shown in Fig (22.38a). In this case, there is no wavefront control of the reflected and refracted waves. However, as discussed above, the metasurfaces can have nonuniform phase distribution of inclusions, i.e. the 1D and 2D phases can be the functions of positions along two orthogonal axes over the metasurfaces. Such nonuniform phase metasurfaces are known as the *phase-gradient metasurfaces* (GM). The phase variations of the metasurfaces are obtained by two mechanisms (i) *Controlling the size of inclusions*, (ii) *Controlling rotation (orientation) of identical inclusions* [J.49, J.51]. These mechanisms are applied to electric/magnetic polarization-based inclusions and also to the *Huygens metasurfaces* using both electrically and magnetically polarizable inclusions.

Controlling Size of Inclusions over a Spatial Period

Figure (22.30a) shows the 1D array of variable-sized resonating metallic inclusions of the identical shapes without any rotation. The in-phase current is induced in one inclusion resonating as a half-wavelength electric dipole at the frequency of the incident waves. In other inclusions, with a size smaller or larger than half-wavelength dipole inclusion, the currents are induced with leading/lagging phases. Thus, the total phase of backward radiated (reflected) and forward radiated (transmitted/refracted) fields are controlled by varying the phase of the reradiating induced current. As each inclusion has a different resonating frequency, so it is called the *metallic antenna dispersion method* to control the phase of the metasurface [J.56].

Figure (22.30a) shows that n numbers of inclusions are periodically arranged in square *unit cells* of period d. A *supercell* of length $\Lambda = n\,d$ is formed with a spatial period Λ. The supercell is repeated both horizontally and vertically to form the phase gradient metasurface. In the present arrangement, the induced currents on inclusions acquire the $\pm\pi/2$ lead/lag phase. It results in π, i.e. 180° phase control of reflected/transmitted waves from the metasurface. The full 360° phase control of the wavefront is not achieved by using only electric type inclusions. It is achieved in the presence of a conductor backing. Over a spatial period $\Lambda = n\,d$, 2π phase control is obtained.

Controlling Rotation of Inclusions over a Spatial Period

Figure (22.33c) shows the 1D array of variable rotation of the identical size of resonating metallic inclusions forming one spatial period, i.e. one supercell. The horizontal and vertical repetition of the array forms another type of the phase-gradient metasurface. It is called the *Pancharatnam–Berry (PB) phase-gradient metasurface*. The *Pancharatnam–Berry phase* is associated with polarization conversion. It is realized by the anisotropic subwavelength inclusions of identical geometry (size and shape) parameter; however, with spatially varying orientation (rotation) of inclusions. The full 2π, i.e. 360° phase control of the reradiated wavefront has been achieved by the PB phase-gradient metasurface. The response of a rotated anisotropic inclusion, in terms of the Jones matrix of the anisotropic inclusion without rotation, is presented below. The Jones matrix [J] of the inclusion or metasurface signifies its transmission/ reflection coefficient matrix, i.e. $[\tau]/[\Gamma]$ matrix. It is discussed in subsection (4.6.4) of chapter 4.

The Response of Rotated Anisotropic Inclusion

Figure (22.28a) considers normally incident plane EM-waves incident on an inclusion (scatterer) located in the (y–z) plane. The inclusion is realized by a narrow aperture cut in the metallic sheet. The normally incident wave on the aperture inclusion is moving away from an observer in the negative x-direction. The Jones matrix of inclusion, without rotation, is given by equation (4.6.14) of chapter 4. For simple anisotropic inclusion, the cross-coupling terms are zero, i.e. $J_{yz} = J_{zy} = 0$. Figure (22.28b) considers the incident plane wave of any polarization incident on a rotated inclusion with an angle θ. On using the identities $\cos^2\theta = (1 + \cos 2\theta)/2$ and $\sin^2\theta = (1 - \cos 2\theta)/2$ with equation (4.6.24) of chapter 4, the *rotated Jones matrix* $[J(\theta)]$ of the rotated anisotropic inclusion is given by the following equation:

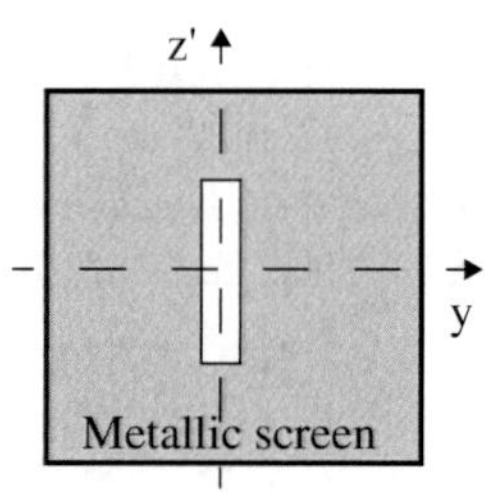

(a) Aperture inclusion (scatterer) in metallic screen.

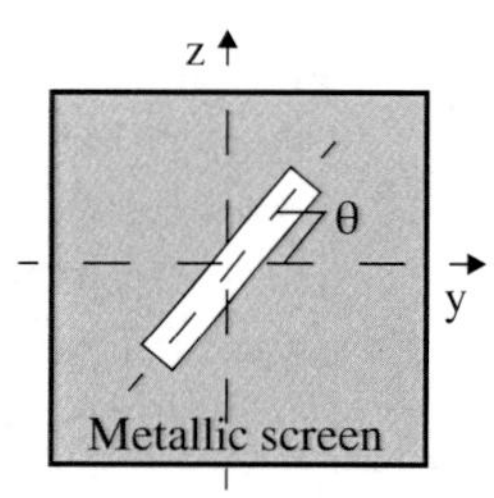

(b) Rotated aperture inclusion at θ.

Figure 22.28 An inclusion scattering the incident waves.

$$[J^{aniso}(\theta)] = \frac{1}{2}\begin{bmatrix} (J_{yy}+J_{zz}) + (J_{yy}-J_{zz})\cos 2\theta & (J_{yy}-J_{zz})\sin 2\theta \\ (J_{yy}-J_{zz})\sin 2\theta & (J_{yy}+J_{zz}) + (J_{zz}-J_{yy})\cos 2\theta \end{bmatrix} \quad \text{(a)}$$

$$\text{For inclusion without rotation,} \quad [J^{aniso}(\theta=0)] = \begin{bmatrix} J_{yy} & 0 \\ 0 & J_{zz} \end{bmatrix} \quad \text{(b).} \qquad (22.5.26)$$

For the RHCP waves' incident at rotated anisotropic inclusion, the above expression of Jones matrix is used to compute the transmitted waves. In the process, the Jones matrix is treated as the transmission matrix $[\tau^{aniso}(\theta)]$ of the metasurface.

$$\left[\vec{E}^{tra}\right] = [\tau^{aniso}(\theta)]\left[\vec{E}^{in}\right], \quad \text{For incident RHCP wave } \vec{E}^{inc} = \begin{bmatrix} 1 \\ j \end{bmatrix}\vec{E}_0^{inc}$$

$$\Rightarrow \begin{bmatrix} E_y^{tra} \\ E_z^{tra} \end{bmatrix} = \frac{1}{2}\begin{bmatrix} (\tau_{yy}+\tau_{zz}) + (\tau_{yy}-\tau_{zz})\cos 2\theta & (\tau_{yy}-\tau_{zz})\sin 2\theta \\ (\tau_{yy}-\tau_{zz})\sin 2\theta & (\tau_{yy}+\tau_{zz}) + (\tau_{yy}-\tau_{zz})\cos 2\theta \end{bmatrix}\begin{bmatrix} 1 \\ j \end{bmatrix}\vec{E}_0^{inc}$$

$$\Rightarrow \begin{bmatrix} E_y^{tra} \\ E_z^{tra} \end{bmatrix} = \frac{1}{2}\begin{bmatrix} (\tau_{yy}+\tau_{zz}) + (\tau_{yy}-\tau_{zz})e^{j2\theta} \\ j(\tau_{yy}+\tau_{zz}) - j(\tau_{yy}-\tau_{zz})e^{j2\theta} \end{bmatrix}\vec{E}_0^{inc}$$

$$\Rightarrow \vec{E}^{tra} = \hat{y}E_y^{tra} + \hat{z}E_z^{tra} = \hat{y}\left[\frac{(\tau_{yy}+\tau_{zz})}{2} + \frac{(\tau_{yy}-\tau_{zz})e^{j2\theta}}{2}\right]\vec{E}_0^{inc} + \hat{z}j\left[\frac{(\tau_{yy}+\tau_{zz})}{2} - \frac{(\tau_{yy}-\tau_{zz})e^{j2\theta}}{2}\right]\vec{E}_0^{inc} \qquad (22.5.27)$$

$$\Rightarrow \vec{E}^{tra} = \frac{(\tau_{yy}+\tau_{zz})}{2}\{\hat{y}+j\hat{z}\}\vec{E}_0^{inc} + \frac{(\tau_{yy}-\tau_{zz})}{2}e^{j2\theta}\{\hat{y}-j\hat{z}\}\vec{E}_0^{inc}$$

$$\Rightarrow \left[\vec{E}^{tra}\right] = \underbrace{\frac{(\tau_{yy}+\tau_{zz})}{2}\begin{bmatrix} 1 \\ j \end{bmatrix}\vec{E}_0^{inc}}_{\text{(In-phase RHCP)}} + \underbrace{\frac{(\tau_{yy}-\tau_{zz})}{2}e^{j2\theta}\begin{bmatrix} 1 \\ -j \end{bmatrix}\vec{E}_0^{inc}}_{\text{(LHCP with leading phase } 2\theta)} \quad \text{(a)}$$

$$\text{For inclusion without rotation}: [\tau^{aniso}(\theta=0)] = \begin{bmatrix} \tau_{yy} & 0 \\ 0 & \tau_{zz} \end{bmatrix} \quad \text{(b).}$$

In the present case, the normally incident wave on the aperture inclusion is moving away from the observer in the (−x)-direction. Therefore, in expression (22.5.27), the column matrices have (+j) for the RHCP and (−j) for the LHCP waves. The present description deviates from the discussion given in subsection (4.6.2) of chapter 6.

Equation (22.5.27) is rearranged using Jones vectors. Expression (22.5.27) shows that the transmitted waves constitute the combined circular polarizations, i.e. it has *in-phase co-polarized* RHCP component and the *cross-polarized* LHCP component with a leading 2θ *Pancharatnam–Berry phase*, i.e. the twice the rotation angle of the inclusion. The transmitted wave is traveling away from an observer. Thus, a supercell of n members of rotated inclusions, with variable rotation angles in the range of $0^\circ \le \theta \le 180^\circ$, can control the wavefront of transmitted waves from 0° to 360°. However, the copolarized component has to be suppressed and power has to be transferred to the cross-polarized wave to enhance the transmission efficiency in the desired direction. Expression (22.5.27) can be rearranged and the information of the LHCP incident waves also added in the following illustrative form:

$$\begin{bmatrix} \vec{E}_+^{tra} \\ \vec{E}_-^{tra} \end{bmatrix} = \begin{bmatrix} \tau_{++} & \tau_{+-} \\ \tau_{-+} & \tau_{--} \end{bmatrix}\begin{bmatrix} \vec{E}_+^{inc} \\ \vec{E}_-^{inc} \end{bmatrix}. \qquad (22.5.28)$$

In expression (22.5.28), the signs (+/−) with the field vectors show the RHCP/LHCP incident waves, respectively. The transmission coefficient components with signs (++/−−) show the copolarized RHCP/LHCP components, whereas the cross-polarized τ with signs (+−/−+) shows the polarization conversion from the LHCP to RHCP, and from the RHCP to LHCP wave after the transmission. The off-diagonal matrix elements are maximized, i.e. the cross-polarized transmitted wave is maximized to achieve the wavefront control of the transmitted wave with high efficiency.

The above discussion is also applicable to the *anisotropic reflective inclusion* or an anisotropic metasurface. The reflected waves from the rotated inclusion could be written as follows:

$$\begin{bmatrix} \vec{E}_+^{ref} \\ \vec{E}_-^{ref} \end{bmatrix} = \begin{bmatrix} \Gamma_{++} & \Gamma_{+-} \\ \Gamma_{-+} & \Gamma_{--} \end{bmatrix} \begin{bmatrix} \vec{E}_+^{inc} \\ \vec{E}_-^{inc} \end{bmatrix} \quad \text{(a)}$$

also $$[\Gamma]_{y-z} = \begin{bmatrix} \Gamma_{yy} & 0 \\ 0 & \Gamma_{zz} \end{bmatrix} \quad \text{(b).}$$

(22.5.29)

Equation (22.5.29b) provides the reflection coefficient matrix of the anisotropic inclusion/metasurface, without rotation, located in the (y–z)-plane. The reflected waves again have both the copolarized and cross-polarized wave components.

The polarization conversion ratio (PCR) shows the efficiency of the metasurface as a polarization converter, for the transmission and reflection mode operation. It is defined as follows [J.57]:

$$\text{Transmission mode: } PCR_+ = \frac{\tau_{++}^2}{\tau_{++}^2 + \tau_{--}^2}$$
$$PCR_- = \frac{\tau_{--}^2}{\tau_{++}^2 + \tau_{--}^2} \quad \text{(a)}$$
$$\text{Reflection mode: } PCR_+ = \frac{\Gamma_{++}^2}{\Gamma_{++}^2 + \Gamma_{--}^2}$$
$$PCR_- = \frac{\Gamma_{--}^2}{\Gamma_{++}^2 + \Gamma_{--}^2} \quad \text{(b).}$$

(22.5.30)

The above analysis has been carried out for the anisotropic metasurface without rotation and without cross-coupling terms. However, it can be applied to the metasurfaces with cross-coupling terms also. Such metasurfaces provide asymmetric transmission, i.e. the wave transmission from the left-hand side to right-hand could be different from the wave transmission from the right-hand side to the left-hand side [J.58].

The unique reflection and transmission (refraction) of the incident waves from the metasurfaces are beautifully explained with the help of *generalized Snell's laws* discussed in next section (22.5.4).

22.5.4 Generalized Snell's Laws of Metasurfaces

The generalized Snell's laws for the metasurface are an extension of the classical Snell's laws applicable to the material surface without any phase discontinuity. The classical Snell's laws explain only in-plane reflection and refraction. It is discussed in section (5.2) of chapter 5. However, the metasurfaces are phase discontinuous interface and follow the generalized Snell's laws. These laws are obtained by conserving the wavevectors across the boundary of the metasurfaces. The laws are obtained more rigorously by using Fermat's principle of least time, formulated as the principle of stationary phase [J.56, J.56, J.60]. These laws explain the *anomalous reflection and refraction* occurring at the metasurfaces. The 2D metasurfaces can provide negative refraction, as well as the negative reflection, whereas the 3D bulk metamaterial slabs provide only negative refraction. It is discussed in subsection (5.5.2) of chapter 5.

Figure (22.29a) shows the 3D reflection and refraction (transmission) of obliquely incident plane wave lying in the (x–z)-plane. The incident wave strikes at the phase-gradient metasurface located in the (x–y)-plane. The wavevector components in three mutually perpendicular planes are as follows:

- *(x–z)-plane*: It is the plane containing the incident wave with the incident wavevector $\vec{k}^{inc} = n_1 \vec{k}_0$, where n_1 is the refractive index of the upper half-space, i.e. the medium #1. The wavevector $\vec{k}^{inc}$ subtends an angle θ^{inc} with respect to the *z*-axis.
- *(x–y)-plane*: It accommodates the phase gradient metasurface with the phase gradients $d\phi/dx$ along the *x*-axis, and $d\phi/dy$ along the *y*-axis. These phase gradients are added to the *x*- and *y*-components of the incident wavevector $\vec{k}^{inc}$ at the surface of the metasurface. Thus, the modified components of the incident wavevector are written as

$$k_{mod,x}^{inc} = k_x^{inc} + \frac{d\phi}{dx}, \quad \text{(a)}$$
$$k_{mod,y}^{inc} = k_y^{inc} + \frac{d\phi}{dy} = \frac{d\phi}{dy} \quad \text{(b).}$$

(22.5.31)

In equation (22.5.31b), $k_y^{inc} = 0$ as the incident wavevector $\vec{k}^{inc}$ is in the (x–z) plane.

- *(y–z)-plane*: It is the orthogonal plane containing the out-of-plane reflected and refracted (transmitted)

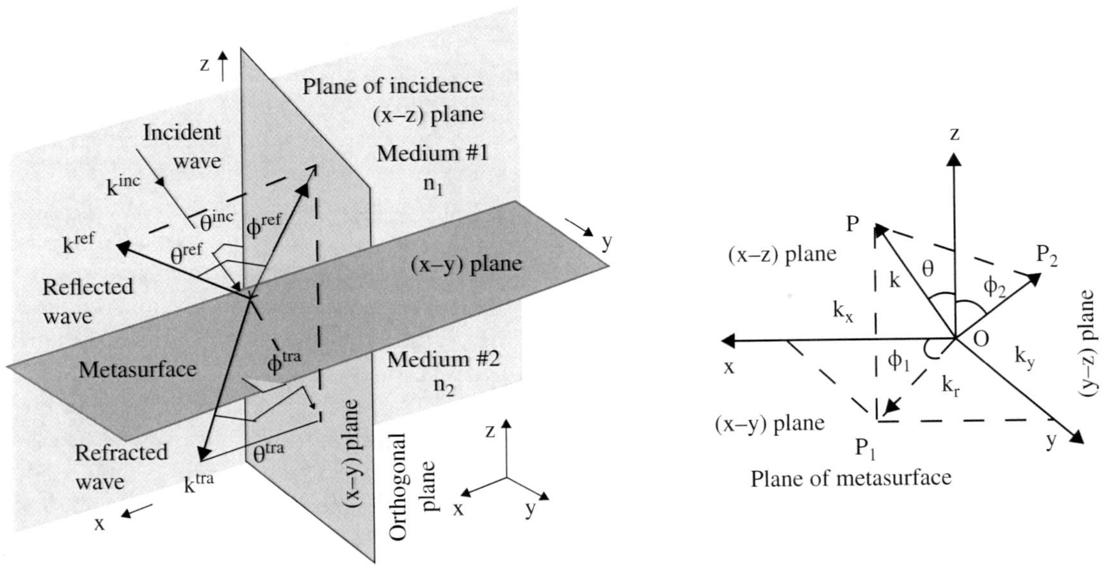

(a) Anomalous reflection and refraction in 3D-space. (b) Wavevector diagram.

Figure 22.29 Out-of-plane anomalous reflection and refraction at the phase-gradient metasurface.

waves in the upper-half and lower-half of the space. The lower half-space is medium #2 with refractive index n_2. The reflected and transmitted wavevectors in media #1 and #2 are $\vec{k}^{ref} = n_1 \vec{k}_o$ and $\vec{k}^{tra} = n_2 \vec{k}_o$, respectively. The reflected wavevector $\vec{k}^{ref}$ subtends an angle of reflections θ^{ref} with the *z*-axis. Its projection in the (*x*–*y*)-plane has an angle $\phi^{ref} = \phi_1$ with respect to the *x*-axis. The reflection angle $\phi^{ref} = \phi_2$ with respect to the *z*-axis in the (*z*–*y*)-plane is also considered. The refracted (transmitted) wavevector $\vec{k}^{tra} = n_2 \vec{k}_o$ subtends an angle of refraction (transmission) θ^{tra} with the *z*-axis and $\phi^{tra} = \phi_1$ with the *x*-axis in the (x–y)-plane. Its projection in the (y–z)-plane subtends $\phi^{tra} = \phi_2$.

The angles of wavevectors could be seen with better clarity on the wavevector diagram. Figure (22.29b) shows the wavevector diagram of arbitrary wavevector $\vec{k}$ with its components in the (x–y)-plane, i.e. in the plane of the metasurface. The wavevector $\vec{k} = \vec{k}^i$ (i = *inc, ref, tra*) could be used for the incident, reflected, and transmitted waves. The angle of the wavevector $OP = \vec{k}$ subtends the angle $\theta = \theta^i$ (i = *inc, ref, tra*) with the *z*-axis. In the planes – (x–y), and (y–z), its projections $O\vec{P}_1$ and $O\vec{P}_2$ subtend angles $\phi_1 = \phi_1^i, \phi_2 = \phi_2^i$ with the *x*-axis and *z*-axis, respectively. The components of the $\vec{k}^i$ in the (x–y) and (y–z) planes could be written from Fig (22.29b):

$$(x-y)-\text{plane}:\ k_x^i = k^i \sin\theta^i \cos\phi_1^i \quad (a)$$
$$k_y^i = k^i \sin\theta^i \sin\phi_1^i \quad (b). \qquad (22.5.32)$$

$$(y-z)-\text{plane}:\ k_y^i = k^i \cos\theta^i \sin\phi_2^i \quad (a)$$
$$k_z^i = k^i \cos\theta^i \cos\phi_2^i \quad (b). \qquad (22.5.33)$$

In this discussion, the incident wavevector $\vec{k}^i$ is in the (x–z) plane. The reflected and transmitted (refracted) plane waves are in the upper-half medium #1 and lower-half medium #2 respectively, as shown in Fig (22.29a). The components of the wavevectors on the metasurface located in the (x–y) plane are obtained as follows:

- *Incident wavevector in the (x–z)-plane,* $\phi_1^{inc} = 0$:

$$k_x^{inc} = n_1 k_0 \sin\theta^{inc} \quad (a), \quad k_y^{inc} = 0 \quad (b). \qquad (22.5.34)$$

- *Reflected wavevector in the (x–y)-plane*:

$$k_x^{ref} = n_1 k_o \sin\theta^{ref} \quad (a)$$
$$k_y^{ref} = n_1 k_o \cos\theta^{ref} \sin\phi^{ref} \quad (b). \qquad (22.5.35)$$

Equation (22.5.35) for k_y^{ref} is obtained from equation (22.5.33a) and the angle $\phi_2^{ref} = \phi^{ref}$ is the out-of-plane angle of reflection in the (y–z)-plane.

- *Refracted wavevector in the (x–y)-plane*:

$$
\begin{aligned}
k_x^{tra} &= n_2 k_o \sin\theta^{tra} && \text{(a)}\\
k_y^{tra} &= n_2 k_o \cos\theta^{tra} \sin\phi^{tra} && \text{(b)}.
\end{aligned}
\qquad (22.5.36)
$$

The angle $\phi_2^{tra} = \phi^{tra}$ is the out-of-plane angle of refraction in the (y–z)-plane

At the plane of the phase-gradient metasurface, the *x*- and *y*-components of the *modified incident wavevector* $\vec{k}_{mod}^{\,inc}$ are obtained from equations (22.5.31) and (22.5.34):

$$
k_{mod,x}^{inc} = n_1 k_0 \sin\theta^{inc} + \frac{d\phi}{dx} \quad \text{(a)}, \quad k_{mod,y}^{inc} = \frac{d\phi}{dy} \quad \text{(b)}. \qquad (22.5.37)
$$

The generalized Snell's laws of anomalous reflection and refraction are obtained from the phase matching of both *x*-and *y*-components of the modified incident wavevector, in the (x–y)-plane of the metasurface, with the components of the wavevectors of the reflected and refracted waves.

Anomalous Reflection Law

Using equations (22.5.35) and (22.5.37), the phase matching of reflected and incident wavevectors provides the anomalous reflection laws:

$$
\begin{aligned}
&k_x^{ref} = k_{mod,x}^{inc}, \quad \text{(a)}, \quad k_y^{ref} = k_{mod,y}^{inc} && \text{(b)}\\
&\Rightarrow \sin\theta^{ref} - \sin\theta^{inc} = \frac{1}{n_1 k_o}\frac{d\phi}{dx} && \text{(c)}\\
&\cos\theta^{ref}\sin\phi^{ref} = \frac{1}{n_1 k_o}\frac{d\phi}{dy} && \text{(d)}.
\end{aligned}
\qquad (22.5.38)
$$

Anomalous Refraction Law

Again, using equations (22.5.36) and (22.5.37), the phase matching of the refracted and incident wavevectors at the plane of the metasurface, the anomalous refraction laws are obtained:

$$
\begin{aligned}
&k_x^{tra} = k_{mod,x}^{inc} \quad \text{(a)}, \quad k_y^{tra} = k_{mod,y}^{inc} && \text{(b)}\\
&\Rightarrow n_2 \sin\theta^{tra} - n_1 \sin\theta^{inc} = \frac{1}{k_o}\frac{d\phi}{dx} && \text{(c)}\\
&\cos\theta^{tra}\sin\phi^{tra} = \frac{1}{n_2 k_o}\frac{d\phi}{dy} && \text{(d)}.
\end{aligned}
\qquad (22.5.39)
$$

Equations (22.5.38c, d) are the *generalized Snell's laws of reflection*. Equations (22.5.39c, d) are the *generalized Snell's laws of refraction*. The phase-gradient $d\phi/dx$ is responsible for the in-plane, i.e. in the plane of incidence, anomalous reflections, and refraction over complete upper and lower halves space respectively. The phase-gradient $d\phi/dy$ is responsible for the out-of-plane, i.e. the 3D anomalous reflections and refraction. In case $d\phi/dy = 0$, the anomalous reflection/refraction laws are limited to the in-plane anomalous reflection/refraction laws. The phase-gradient could be both positive and negative. The negative phase gradients are responsible for the negative reflection and negative refraction. For both $d\phi/dx = d\phi/dy = 0$, the generalized laws are reduced to the classical Snell's laws. The generalized Snell's laws have been confirmed experimentally [J.49, J.50, J.56, J.59, J.60]. These are illustrated in section (22.6.1).

The *linear phase variations*, i.e. $\phi(x) = \xi_x x + \phi_0$ wand $\phi(y) = \xi_y y + \varphi_0$ provide the mechanism to deflect the incident plane waves into the desired direction. The plane waves could be linearly or circularly polarized. Likewise, complete transmission of the incident waves, without reflection, in any direction is possible. It is achieved with Huygens surfaces [J.54]. By using the *quadratic phase profile* $\phi(x, y) = (2\pi/\lambda)\left\{\sqrt{x^2 + y^2 + f^2} - f\right\}$, the flat lens has been realized with focal length f. Several other wavefront controls have been obtained using the metasurfaces [J.49, J.51, J.56].

22.5.5 Surface Waves on Metasurface

The above discussion has explored the ability of the metasurface to control the wavefront of free space waves (FSW). The metasurface also supports the SW propagation that is also controlled by tailoring the inclusions of metasurfaces. The surface-waves are nonradiating bounded modes. It is discussed in section (7.5) of chapter 7. However, under a certain condition, they can become radiating leaky-waves. Thus, there is a coupling mechanism between the FSW to the surface waves, i.e. the FSW–SW coupling, and also SW–FWS coupling. This subsection discusses the surface waves on both the *uniform metasurface* and *nonuniform metasurface* [J.50, J.61].

Surface Waves on Uniform Metasurface

In this case, there is no phase gradient on the uniform metasurface. It is just like a homogenized EBG surface without any bandgap. However, the DNG type metasurface, supporting the backward wave can have a bandgap from zero to cut-off frequency commencing the surface wave. Normally, in the microwave range, the uniform metasurface is a 2D periodic arrangement of the metallic inclusions at the surface of a dielectric sheet. Its 2D circuit model, shown in Fig (20.15c), is discussed in section (20.4) of chapter 20.

Equation (20.4.20) of chapter 20 presents the 2D dispersion relation that gives both the propagating surface waves and nonpropagating Bloch mode. Under the homogeneity condition, i.e. for electrically short line section $kd << 1$ and also for a subwavelength period of the unit cell $\beta_x d << 1$, $\beta_y d << 1$, equation (20.4.20) is reduced to

$$\beta_x^2 + \beta_y^2 = \left(k - \frac{1}{Z_0 \omega C d}\right)\left(2k - \frac{Z_0}{\omega L d}\right). \quad (22.5.40)$$

The homogenized metasurface is described by the equivalent permeability and equivalent permittivity obtained below.

Let us take $\beta_x = \beta \cos\theta$, and $\beta_y = \beta \sin\theta$. The wavevector β of the surface wave, in the (x–y)-plane, subtends an angle θ with the *x*-axis. Using characteristic impedance $Z_0 = \sqrt{\mu/\varepsilon}$, and phase velocity $v = 1/\sqrt{\mu\varepsilon}$ of the host medium, the above dispersion relation of surface-wave can be expressed in the *2D effective medium*:

$$\beta = \omega\sqrt{\mu_{eq}\varepsilon_{eq}} \quad \text{(a)}$$
$$\text{where } \mu_{eq} = \left(\mu - \frac{1}{\omega^2 C d}\right) \quad \text{(b)}, \quad \varepsilon_{eq} = \left(2\varepsilon - \frac{1}{\omega^2 L d}\right) \quad \text{(c)}. \quad (22.5.41)$$

Relation (22.5.41) shows that the homogenized metasurface is isotropic and homogeneous. The loading series capacitor can provide negative magnetic susceptibility and loading shunt inductor can provide negative electric susceptibility. In this case, the metasurface supports the bound backward wave propagation. At a certain higher frequency, both quantities could be positive numbers supporting the usual forward wave propagation. Of course, in some frequency bands, one of the quantities could be a negative number, and a nonpropagating band is created [J.62].

Surface Waves on Nonuniform Metasurfaces

The effective medium concept is only conceptual. The metasurfaces can have spatial variation. The metasurfaces are useful to realize the mode conversion between two bound surface modes, or for the conversion between the bound SW and radiating FSW. Such characteristics are obtained by the nonuniform metasurfaces, tailoring their phase gradient [J.61].

The metasurfaces with unidirectional phase gradient and matching of the wavevectors can couple free-space waves to surface-waves without re-radiation. The propagation constant of the surface-waves is in the bound mode for the condition $\beta > k_0$. However, for $\beta < k_0$, the metasurfaces are in the radiation mode, supporting the Cerenkov type leaky-wave radiation. The Cerenkov radiation is explained in subsection (5.5.7) of chapter 5. These conditions could be realized by the proper design of inclusions. Moreover, the inclusions also control the surface impedance of the metasurfaces. By creating the holographic impedance surface, the holographic leaky-wave antenna has been designed [J.63–J.65].

22.6 Applications of Metasurfaces

The metasurfaces have been used to develop a wide variety of components and devices from microwave to optical frequency ranges. This section covers a few examples of metasurfaces to illustrate some of their unique characteristics, such as the anomalous reflection and refraction exhibiting bending of the EM-waves in the desired direction, complete transmission of waves using the concept of Huygens' surfaces, designing anisotropic metasurfaces, and showing their ability to convert the polarization states.

22.6.1 Demonstration of Anomalous Reflection and Refraction of Metasurfaces

The 1D and 2D linear phase-gradients of *gradient metasurface* (GM) are controlled to get the anomalous reflection and refraction. Both the reflection type and transmission type GM are discussed below.

Reflection Type Metasurfaces

The equal amplitude reflective gradient metasurface has been designed by controlling the phase-gradient of the conductor backed periodically arranged microstrip square inclusions. The metallic patch shows electric polarization. However, due to the conductor backing, it has a magnetic response also and the reflection phase can be controlled over 0–2π in the plane of incidence for the 1D GM. However, 2D phase-gradients provide the out-of-plane reflection also.

1D Phase-gradient Metasurface

The gradient metasurface is realized by n numbers of unit cells with different geometrical parameters of the metallic inclusions without any rotation.

Figure (22.30a) shows one *supercell* of n = 6 numbers of inclusions of the square microstrip patch resonators of varying sizes. Each inclusion is within a unit cell of period $d_x = d_y = d$. The periodicity, i.e. length of the supercell is Λ, i.e. $\Lambda = n \times d$. Figure (22.30b) shows that *isotropic metasurface* is formed by the repetition of the supercells, both horizontally and vertically, on a substrate with conductor backing.

The FR-4 substrate ($\varepsilon_r = 4.3$, $\tan\delta = 0.025$) of thickness h = 1.0 mm is used to design the metasurface. It is located in the (x–y)-plane and the EM-wave with its wave vector k_0 is normally incident on it, i.e. $\theta^{inc} = 0$. The incident wave is reflected at an angle θ^{ref}. The x-directed phase gradient of the metasurface is $\xi_x = d\phi/dx$. Following equation (22.5.38c), the angle of reflection in the air medium ($n_1 = 1$) is

$$\theta^{ref} = \sin^{-1}(\xi_x/k_0), \quad \xi_x < k_0 \qquad (a)$$
$$\Rightarrow \xi_x = k_0 \sin\theta^{ref} \qquad (b). \qquad (22.6.1)$$

The n numbers of unit cells of period d are used to achieve linear discrete 2π phase variation over a supercell of the space periodicity Λ. The phase gradient is obtained below:

$$\text{Phase variation over one unit cell: } \Delta\phi_x = \Delta\phi = 2\pi/n \qquad (a)$$
$$\text{Phase gradient: } \xi_x = \xi = \frac{d\phi}{dx} \approx \frac{\Delta\phi}{d_x} = \frac{2\pi}{nd} \qquad (b). \qquad (22.6.2)$$

The angle of reflection, controlled by the phase-gradient of the metasurface, is obtained from equations (22.6.1) and (22.6.2):

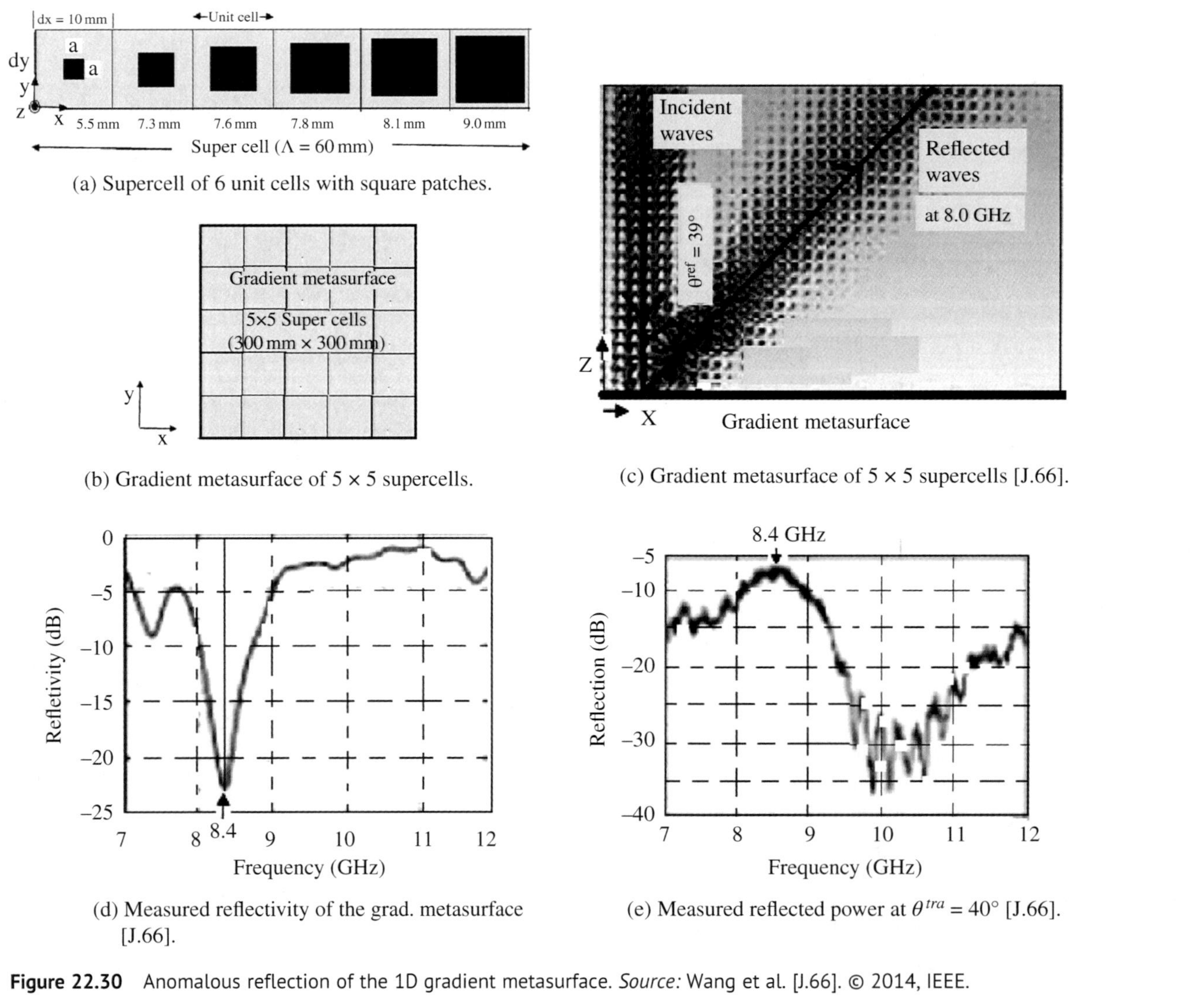

(a) Supercell of 6 unit cells with square patches.

(b) Gradient metasurface of 5 × 5 supercells.

(c) Gradient metasurface of 5 × 5 supercells [J.66].

(d) Measured reflectivity of the grad. metasurface [J.66].

(e) Measured reflected power at $\theta^{tra} = 40°$ [J.66].

Figure 22.30 Anomalous reflection of the 1D gradient metasurface. *Source:* Wang et al. [J.66]. © 2014, IEEE.

$$\theta^{ref} = \sin^{-1}\left(\frac{2\pi}{k_0 nd}\right) = \sin^{-1}\left(\frac{\lambda_0}{nd}\right) = \sin^{-1}\left(\frac{\lambda_0}{\Lambda}\right), \quad (22.6.3)$$

where λ_0 is the wavelength of the incident wave.

A gradient metasurface using the variable size of $n = 6$ numbers of square microstrip patches is examined at 8.0 GHz for the unit cell period of 10 mm [J.66]. The variable patch sizes are marked below patches of Figure (22.30a). The 6 numbers of unit cells constitute one supercell of length (period) $\Lambda = n\,d = 60$ mm. The length of the metasurface in the x-direction is $10\Lambda = 600$ mm. The width of the metasurface in the y-direction is $2\Lambda = 120$ mm. At 8.0 GHz, i.e. $\lambda_0 = 37.5$ mm, the discrete phase change by each patch is $2\pi/n = 360^\circ/6 = 60^\circ$. The theoretical angle of reflection from equation (22.6.3) is $\theta^{ref} = \sin^{-1}(37.5\ \text{mm}/60\ \text{mm}) \approx 38.7^\circ$. The EM-simulation, shown in Fig (22.30c), confirms the theoretical result of the generalized Snell's reflection law with the angle of reflection 39°. Finally, the metasurface is fabricated on the FR-4 substrate. Its size is $5\Lambda \times 5\Lambda = 300\ \text{mm} \times 300$ mm. Figure (22.30d) shows that at 8.4 GHz its reflectivity is −20 dB in the direction of normal. Figure (22.30e) experimentally validates the angle of reflection $\theta^{ref} = 40^\circ$.

2D Phase Gradient Metasurface

The anisotropic gradient metasurface with phase gradients $\xi_x = d\phi/dx$ and $\xi_y = d\phi/dy$ in the x- and y-directions, respectively, facilitates an arbitrary control of reflection direction in the 3D upper space. It has both the in-plane and out-of-plane reflections. Figure (22.29a) shows the in-plane angle of the reflection angle θ^{ref} in the (x–z)-plane. It also shows the out-of-plane reflection angle ϕ^{ref} in the (y–z)-plane. Following the wavevector diagram of Fig (22.29b) and using equation (22.5.37), both angles of reflections are written as

$$k_r = k_0 \sin\theta^{ref} = \left(\xi_x^2 + \xi_y^2\right)^{1/2} \quad (a)$$

$$\tan\phi^{ref} = \frac{k^{inc}_{mod,y}}{k^{inc}_{mod,x}} = \frac{\xi_y}{\xi_x} \quad (b). \quad (22.6.4)$$

Following equation (22.6.2), both phase-gradients are computed. The above expressions are rewritten as follows:

$$\theta^{ref} = \sin^{-1}\left[\lambda_0\left(\frac{1}{(n_x d_x)^2} + \frac{1}{(n_y d_y)^2}\right)^{1/2}\right] \quad (a)$$

$$\phi^{ref} = \tan^{-1}\left(\frac{n_x d_x}{n_y d_y}\right) \quad (b)$$

$$\text{where, } \xi_x = \frac{2\pi}{n_x d_x} \quad (c), \quad \xi_y = \frac{2\pi}{n_y d_y} \quad (d). \quad (22.6.5)$$

In equation (22.6.5), n_x and n_y are the numbers of unit cells in the *x*- and *y*-directions, respectively. Likewise, d_x and d_y are the periods of unit cells in the *x*- and *y*-directions, respectively. In the case of anisotropic metasurfaces, these are unequal quantities. However, for the isotropic metasurface has $n_x d_x = n_y d_y = nd$. The above equations are reduced to

$$\theta^{ref} = \sin^{-1}\left(\frac{\sqrt{2}\lambda_0}{nd}\right) \quad (a), \quad \phi^{ref} = 45^\circ \quad (b). \quad (22.6.6)$$

For the $10\Lambda \times 10\Lambda = 600\ \text{mm} \times 600$ mm gradient metasurface at 8.0 GHz, the computed reflection angles are $\theta^{ref} = 62.12^\circ$ and $\phi^{ref} = 45^\circ$. The experimentally obtained angles $\theta^{ref} = 61^\circ$ and $\phi^{ref} = 45^\circ$ validate the general Snell's laws for the out-of-plane reflection [J.66].

Transmission Type Metasurfaces

Figure (22.31a) shows the multilayered 1D isotropic gradient metasurface to deflect the normally incident plane waves as refracted (transmitted) waves in a predetermined direction. The metasurface structure is constructed with the metallic perforated layer separated by the dielectric layers. It is designed using the PCB of thickness $h = 1.575$ mm, $\varepsilon_r = 2.65$, and the metal layers have thickness $t = 0.035$ mm.

The perforations are electrically polarizable coaxial annular apertures in all three layers and aligned properly [J.70]. The annular apertures are periodically arranged in the (*x*–*y*)-plane with nine columns as shown in Fig (22.31b) with 12 rows. Only three rows are shown in Fig (22.31b). *A thin metallic film perforated with a subwavelength array of apertures exhibits extraordinary transmission due to the localized resonances.* The multilayer arrangement can enhance the transmission over a wider bandwidth [J.67–J.69]. In this case, each aperture is located within a square lattice unit cell of periodicity $d_x = d_y = d = 12$ mm. The $n = 9$ numbers of unit cells constitute one supercell of the period $\Lambda_x = 9 \times 12 = 108$ mm along the *x*-axis. Likewise, in the direction of the *y*-axis, the supercell period is $\Lambda_y = 12 \times 12 = 144$ mm. To achieve, the in-plane full control of the transmitted waves over $0 - 2\pi$ phase along the *x*-axis, $n = 9$ apertures should have a phase difference $\Delta\phi = 2\pi/9 \approx \pi/4$.

The *x*-directed phase gradient obtained from equation (22.6.2) is $\xi_x = d\phi/dx = \Delta\phi/d_x = \pi/4d$. There is no phase gradient along the *y*-axis, i.e. $\xi_y = d\phi/dy = 0$. The gradient metasurface is located in the air medium, i.e. $n_1 = n_2 = 1$. For the normally incident plane waves with $\theta^{inc} = 0$, equation (22.5.39c) provides the angle of in-plane refraction:

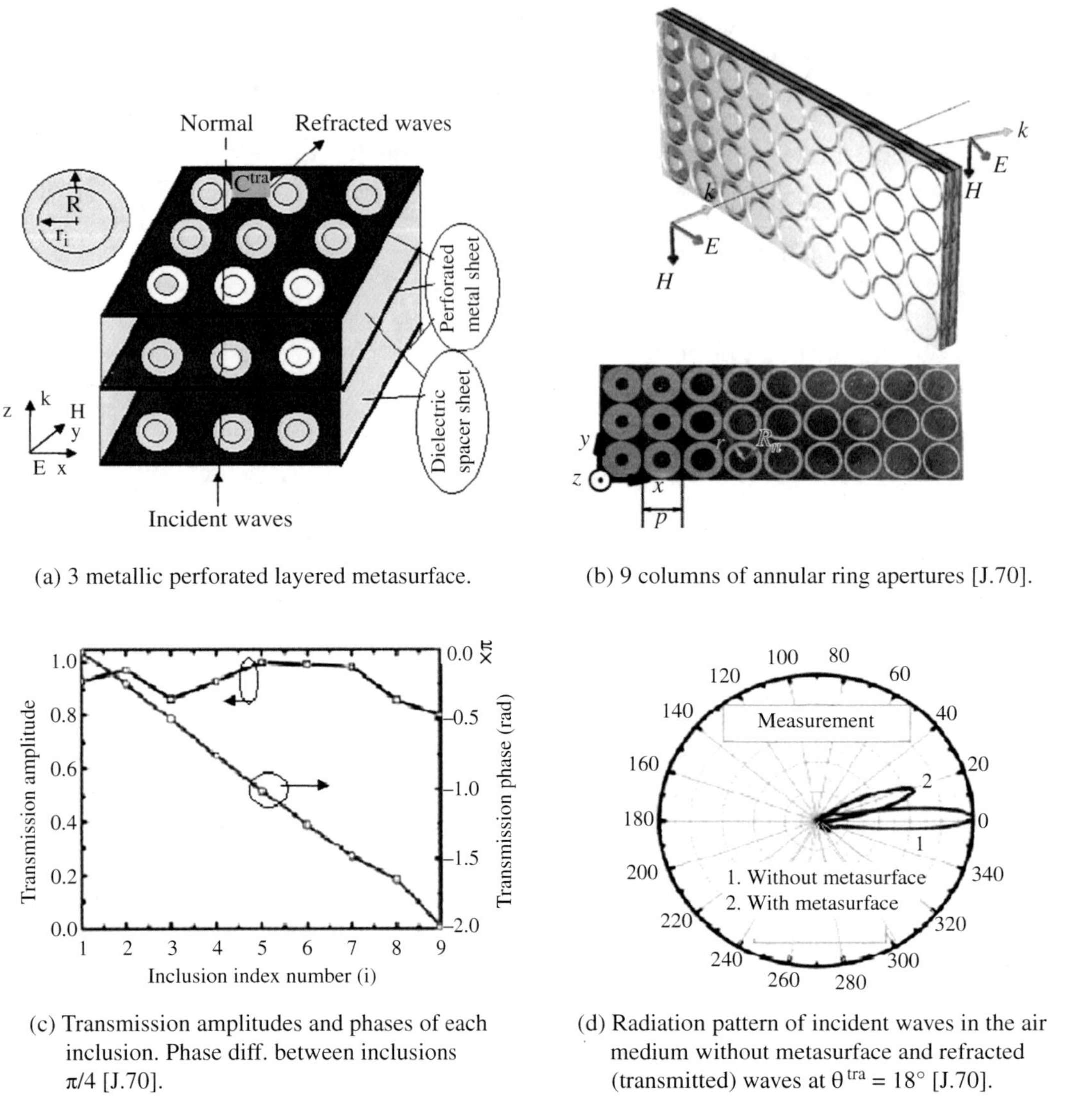

Figure 22.31 Anomalous refraction (transmission) of the 1D multilayer gradient metasurface. *Source:* For the use of figure 1b from Wei et al. [J.70]. © 2013, The Optical Society.

$$\theta^{tra} = \sin^{-1}\left(\frac{\lambda_0}{2\pi}\xi_x\right) \quad (a), \quad \phi^{tra} = 0, \text{ as } \xi_y = 0 \quad (b). \tag{22.6.7}$$

The operating frequency of the incident wave is at 10 GHz, i.e. λ_0 = 30 mm. Therefore, in the present case, the theoretically calculated angle of refraction is $\theta^{tra} = \sin^{-1}[(30/2\pi)(\pi/4\times 12)=18.2^\circ$.

To design the phase-gradient metasurface, 9 numbers of annular apertures are tuned to achieve the phase difference π/4 between the adjacent elements. The outer radius of the aperture is fixed at R = 5.8 mm and the inner radii of nine apertures are tuned at the EM-simulator to achieve the phase difference π/4. The tuned radii in mm are as follows: 1.98, 2.76, 3.66, 4.26, 4.56, 4.76, 4.86, 4.94, and 5.08 [J.70]. In the process, the transmission amplitude of each aperture is maintained above 0.8. Figure (22.31c) shows the amplitude response and the *x*-directed linear phase response of a supercell.

The FDTD simulation of the gradient metasurface with one period along the *y*-axis, using the periodic boundary condition, provides the refraction at 18° with respect to the *z*-axis. It has a high transmission efficiency. The metasurface of size 108 mm × 144 mm has been illuminated with normally incident plane waves. Figure (22.31d) experimentally validates the angle of refraction at 18° of the normally incident waves working at 10 GHz. The experimentally obtained transmission efficiency is 65%.

The polarization-sensitive multilayer metasurfaces have also been developed to get the full control of both the reflected and transmitted waves simultaneously [J.71]. The strategy to improve the transmission efficiency of the metasurface is discussed in section (22.6.2).

22.6.2 Reflectionless Transmission of Metasurfaces

The concepts of Huygens metasurfaces and Fabry–Perot resonance are used to enhance the transmission efficiencies of metasurfaces.

Huygens Metasurfaces

The *electric type metasurfaces*, exclusively based on the electrically polarizable inclusions, do not provide impedance matching with free space. Therefore, reflection occurs at the surface planes of the metasurfaces and their transmission efficiency degrades.

However, the *Huygens' metasurfaces*, based on both the electric and magnetic type polarizable inclusions, provide such impedance matching by controlling the ratio of the imaginary parts of the surface impedance Z_{ms} and surface admittance Y_{es}. Equation (22.5.23) states the matching condition. The polarizabilities of both electric and magnetic inclusions are located at the same location on the metasurfaces. These inclusions are simultaneously controlled to achieve the matching by tuning their dimensions. Moreover, to transmit (refract) the waves in the desired direction, an appropriate phase gradient is also created by varying the dimensions of the inclusion. Figure (22.32a) shows a unit cell consisting of electrically and magnetically polarizable inclusions. The upper face of the substrate, Rogers RO4003 ($\varepsilon_r = 3.55$, $\tan\delta = 0.0027$), supports the capacitively and inductively loaded conducting strips. It is electrically polarized by a *y*-directed electric field, creating Y_{es}. The lower face of the substrate supports the SRR (capacitively-loaded loop). It is magnetically polarized by the *z*-directed magnetic field, creating Z_{ms} [J.54].

Figure (22.32b-i) shows a fabricated (226 mm × 226 mm) metasurface in the (y-z)- plane at 10 GHz. It is composed of a vertical stack of 58 identical PCB strips, with 2.35 mm air gaps between PCB strips. The Huygens' metasurface is developed by the periodic repetition of supercells. One period length Λ_y, in the direction of the *y*-axis, is shown in Fig (22.32b-ii) for both electric and magnetic inclusions. Using equations (22.5.39c) and (22.6.2), the period length Λ_y of the metasurface in the air medium is computed from the specified angles of incidence and transmission:

$$\sin\theta^{tra} - \sin\theta^{inc} = \frac{1}{k_o}\frac{d\phi}{dy} \approx \frac{\lambda_0}{nd} = \frac{\lambda_0}{\Lambda_y}. \quad (22.6.8)$$

The TM polarized EM-wave, at 10 GHz ($\lambda_0 = 30$ mm), is normally incident at the metasurface from its bottom. The intended transmission is required at the refraction angle $\theta^{tra} = 45°$. Equation suggests a period of the supercell $\Lambda_y = 42.36$ mm. The supercell is composed of $n = 12$ numbers of unit cells of $d = 3.53$ mm size. The supercell covers $0–2\pi$ phase variation over its length. Each inclusion provides the phase change $\Delta\phi = \pi/6$. The inclusions of Fig (22.32a) are tuned on the EM-simulator to get around unity transmission and needed phase change. The transmission coefficient τ and reflection coefficient (Γ) of each inclusion pair are also determined from the EM-simulator. Equations (22.5.20c, d) are used to compute complex Y_{es} and complex Z_{ms} of each of 12 numbers of unit cells forming a supercell of the Huygens' metasurface. Figure (22.32c) shows the computed real and imaginary parts of the complex Y_{es} and $Z_{ms.}$ For a typical unit cell, shown in Fig (22.32a), the values are $Y_{es} = (0.02 + j3.14)/\lambda_0)$ and $Z_{ms} = (0.07 + j2.3)\lambda_0)$.

Figure (22.32d) shows the EM-simulated refraction at 45° for the normally incident plane waves at the metasurface. It confirms the design. Finally, Fig (22.32e) shows the transmission efficiency of the fabricated (226 mm × 226 mm) metasurface. Some refraction takes place in the undesired directions also. The peak transmission efficiency is 86% at 10.5 GHz, and the 3 dB bandwidth is 24.2%.

Metasurfaces Using Fabry–Perot Resonance-Based Inclusions

The inclusion, i.e. the scatterer for the design of a metasurface has to be carefully selected to achieve nearly complete transmission of the incident waves. Next, the Panchratanum–Berry rotational arrangement of inclusions, using an identical number of inclusions, is adapted to refract the beam in the desired direction. Both aspects are discussed below.

Choice of Multilayered Inclusion

Nearly total transmission is achieved by a multilayered C-shaped electric type inclusion, i.e. the split ring resonator (SRR) with a gap g, shown in Fig (22.33a-i). The C-shaped inclusions and the metasurface constituted by their periodic arrangement, shown in Fig (22.33c), are located in the (y–z)-plane. The multilayered inclusion is formed by coaxially arranged 4 numbers of C-shaped metallic rings separated by the dielectric spacers of

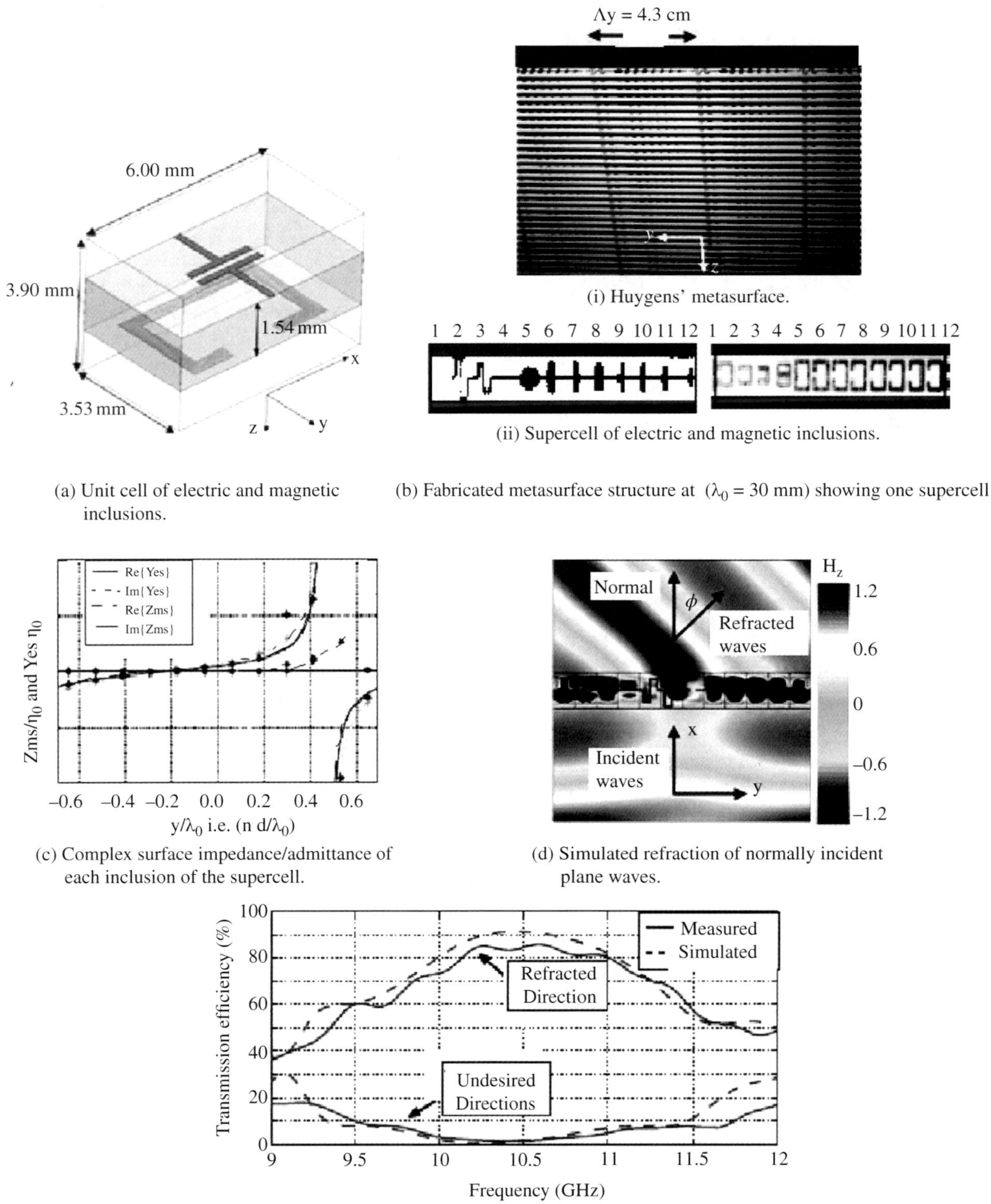

Figure 22.32 Anomalous refraction of Huygens' gradients metasurface. *Source:* Pfeiffer et al. [J.54]. © 2013, American Physical Society.

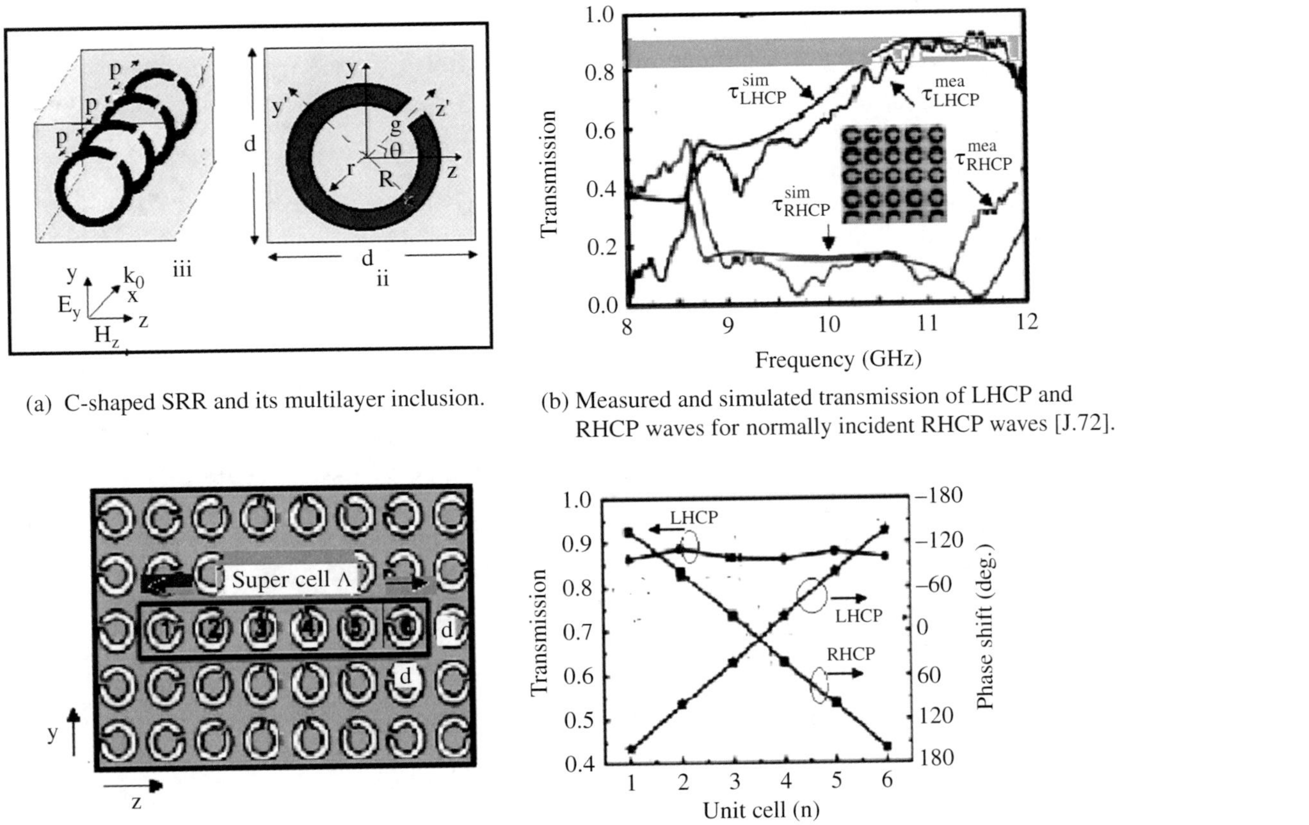

(a) C-shaped SRR and its multilayer inclusion.

(b) Measured and simulated transmission of LHCP and RHCP waves for normally incident RHCP waves [J.72].

(c) Multilayered gradient metasurface [J.72].

(d) Amplitude and phase shift of 6 unit cells (n = 1 to 6) under normally incident RHCP and LHCP waves [J.72].

Figure 22.33 Transmission of normally incident RHCP waves on the multilayered gradient metasurface. *Source:* From Liu et al. [J.72] Springer Nature. Licensed under CC- BY- 4.0. Public Domain.

thickness p. The multilayered inclusion is located within a square lattice of period d. The y-polarized EM-wave, propagating in the *x*-direction, electrically excites the inclusion. The electric current on the metallic strips and displacement current in the dielectric spacers creates a circulating current loop in the (x–y)-plane. The loop is excited by the H_z-magnetic field component and acts as a magnetically polarized inclusion. At local Fabry–Perot type resonance, the impedance of the multilayered inclusion is matched to free space. It provides nearly total transmission of the normally incident waves. Thus, the present arrangement also acts as a Huygens' metasurface.

Figure (22.33a-ii) shows the inner and outer radii r and R of an inclusion. Further, the inclusion could be rotated by an angle θ with respect to the *z*-axis. It provides the Pancharatnam–Berry phase angle 2θ for its transmitted crossed-polarized (CP)-wave component. The copolarized component of the transmitted wave has no phase change. It is discussed in section (22.5.3).

The transmission characteristics of a metasurface developed at 11.0 GHz with a (5 × 5)-array of inclusions, without rotation, have been examined both experimentally and through EM-Simulation [J.72]. The fabrication details of the metasurface on the FR-4 substrate are as follows: r = 1.5 mm, R = 2.5 mm, g = 0.75 mm, p = 2.5, and d = 6.5. The simulated τ^{sim} and experimental transmission τ^{mea} responses, between 8.0 and 12.0 GHz for the right-hand CP (RHCP) wave, are shown in Fig (22.33b). At the resonance frequency, the effective length of the strip forming C-shaped inclusion is $L_{eff} \approx \lambda_0/2$. The maximum transmission of the cross-polarized LHCP transmitted waves at 11.0 GHz is 90% and over broadband (10.3–11.7 GHz) is above 80%. The transmitted level of the copolarized RHCP wave is below 20%. Further improvement in transmission efficiency and bandwidth has been obtained with low-loss Rogers 5880 substrate. The high transmission efficiency is maintained for the oblique angle of incidence between 0° and 60°.

Puncharatnum–Berry Phase Control

Figure (22.33c) shows the gradient metasurface on the FR-5 substrate using the supercell of n = 6 numbers of unit cells. The period of the supercell is $\Lambda_z = n\,d$. Using equation (22.6.7), the angle of refraction of the normally incident RHCP wave at 11.0 GHz is $\theta^{tra} = \sin^{-1}(27.273\ \text{mm}/6 \times 6.5\ \text{mm}) = 44.37^\circ$. In the case of the Puncharatnum–Berry phasing arrangement by the rotation of inclusion, $\theta = 0^\circ - 180^\circ$ rotation covers full 360° phase control of the cross-polarized LHCP refracted waves. Therefore, in Fig (22.33c) each of six scatterers is rotated by $\pi/d = 180^\circ/6 = 30^\circ$ with respect to its neighbor.

The transmission characteristics of the metasurface of Fig (22.33c) are experimentally examined for the normally incident RHCP wave. The results are shown in Fig (22.34a–d). The measured magnetic field H_z in the (x–z)-plane of Fig (22.34b) shows that the LHCP wave is refracted anti-clockwise at (−45°) validating the theoretically calculated angle of refraction. Figure (22.34a) shows that the transmission efficiency is about 80%. Figure (22.34c) shows that the normally incident LHCP is refracted clockwise at +45°. In Fig (22.34d), it is interesting to note that the normally incident linearly polarized (LP) wave is split into two beams transmitted as the LHCP and RHCP waves at angles $\theta^{tra} = \mp 45^\circ$. The LP wave could be treated as a linear combination of the LHCP and RHCP waves.

22.6.3 Polarization Conversion of Incident Plane Wave

The polarization conversions from linear to circular [J.73–J.76], LHCP to RHCP, [J.77] and rotation of linear polarization [J.78] are accomplished with subwavelength thick metasurfaces. The polarization conversions have been achieved with anomalous reflection/refraction in the desired direction. Both the *reflection-type* and *transmission-type* metasurfaces are used to get polarization conversions. Furthermore, the metasurfaces are also used as the substrate and superstrate to convert the linearly polarized microstrip antennas and slot antenna into the circularly polarized antenna [J.79, J.80]. The present subsection discusses briefly illustrative examples

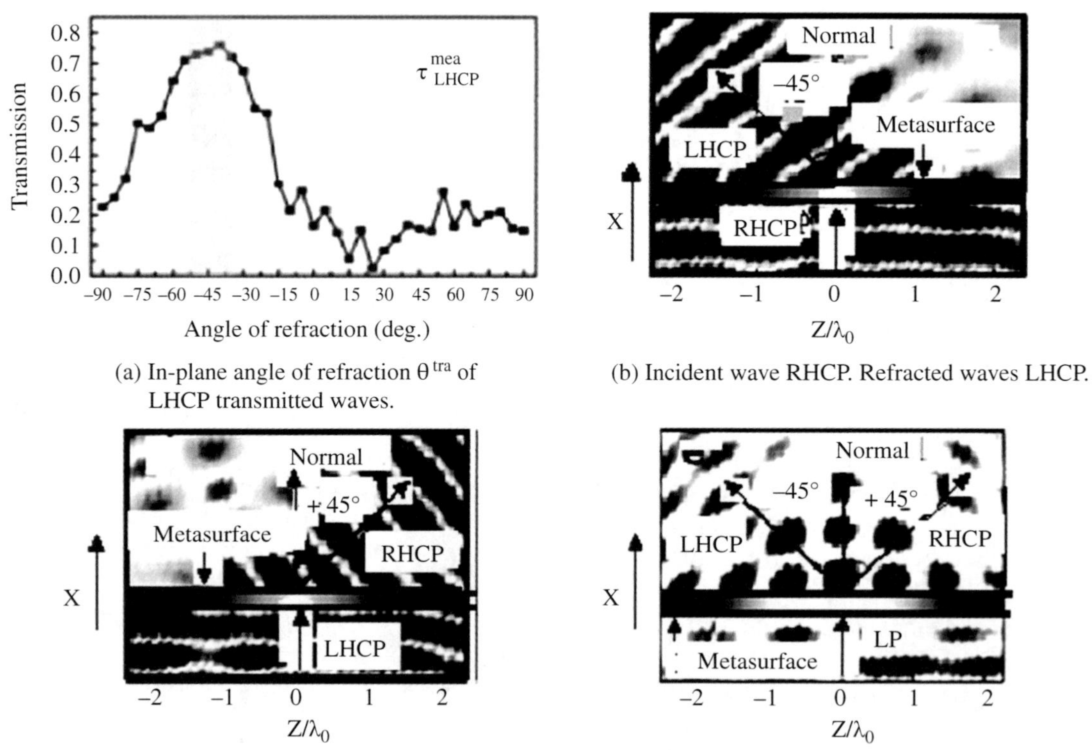

(a) In-plane angle of refraction θ^{tra} of LHCP transmitted waves.

(b) Incident wave RHCP. Refracted waves LHCP.

(c) Incident wave LHCP. Refracted waves RHCP.

(d) Incident wave LP. Refracted waves divided in LPCH and RHCP.

Figure 22.34 Experimental refracted Hz-fields of the normally incident waves on a gradient metasurface. *Source:* From Liu et al. [J.72] Springer Nature. Licensed under CC- BY- 4.0. Public Domain.

of the reflection-type and transmission-type metasurfaces and their applications to the microstrip patch and slot antenna.

Reflection-type Metasurfaces

There are several possible shapes of the microstrip patch type inclusions (scatterers) to develop high-efficiency reflection-type anisotropic metasurface that converts linearly polarized incident waves to the circularly polarized waves. The behavior of the reflection phase of a metasurface is similar to that of the EBG surface, discussed in subsection (20.1.3) of chapter 20. The phase-gradient of metasurface further directs to the CP waves in the desired direction.

Figure (22.35a) shows a unit cell of the orthogonal I-shaped metallic patch inclusions – horizontal I_y and vertical I_z inclusion, printed on a conductor backed F4B substrate with $\varepsilon_r = 2.65, \tan\delta = 0.001$, $h = 2$ mm. The structural details of the unit cell are as follows: $a = 2$ mm, $w = 0.4$ mm, $d = 6$ mm. The arm length p_z of the I_z-inclusion is taken at 3 mm and length p_y is varied from 2 to 5 mm [J.74].

Figure (22.35b) shows at 10 GHz phase response $\phi_z(y)$ and $\phi_y(y)$ of the E_z and E_y polarized normally incident plane waves, propagating in the *x*-direction. Due to conductor backing, the magnitudes of both the horizontal and vertical polarization reflection coefficients are $|\Gamma_{yy}| = |\Gamma_{zz}| \approx 1$. It is noted that the phase response of the reflection phase $\phi_z(y)$ is not influenced by the variation in p_z. However, the reflection phase $\phi_y(y)$ of the I_y-inclusion decreases from 130° to −200° with increasing p_z. At a certain length, the inclusion resonates showing zero reflection phase, like an artificial magnetic conductor (AMC). Thus, the reflection phases of the orthogonal I-shaped inclusion can be controlled independently of each other by varying their lengths.

The anisotropic reflection-type phase-gradient metasurface is designed with 3 × 3 supercells. Each supercell is composed of $n = 12$ unit cells of period 6 mm. The top of Fig (22.35c) shows one supercell $\Lambda_y = n\,d$ with varying size of the inclusions to cover full reflection phase 0°–360° at 10 GHz. The inclusion lengths p_y and p_z are adjusted to achieve the reflection phase difference, i.e. the differential phase $\Delta\phi = |\phi_y - \phi_z|$ for each of 12 inclusions at 10 GHz. The adjusted dimensions are as follows: p_y (mm) = 4.95, 5.1, 5.15, 2.0, 3.7, 4.1, 4.3, 4.45, 4.5, 4.6, 4.75, 4.8; p_z (mm) = 2.0, 3.7, 4.1, 4.3, 4.45, 4.5, 4.6, 4.75, 4.8, 4.95, 5.1, 5.15. Figure (22.35c) shows the EM-simulation of the phase responses of y and z-polarized normally incident EM-waves, for three periods of the supercells at 10 GHz. The differential reflection phase $\Delta\phi = |\phi_y - \phi_x|$ is maintained at 90° to achieve circular polarization.

For a linearly polarized wave incident at 45°, the circularly polarized wave is obtained after reflection, as the magnitudes of the reflected E_y and E_z field components are nearly unity with 90° phase difference. It is the anomalous reflection with an identical angle of reflection for both polarizations. The angle of reflection, for both horizontal (H) and vertical (V) polarization, is computed from equation (22.6.3) $\theta^{ref} = \sin^{-1}(\lambda_0/nd) = \sin^{-1}(30/12 \times 6) = 24.62^0$. The radiation pattern of Fig. (22.35d-i) experimentally confirms the angle of reflection at 24° for both polarizations. Figure (22.35d-ii) further demonstrates the experimentally obtained circular polarization. The bandwidth of the reflection-type linear to the circular converter is about 15% (9.5–11.0 GHz) within 1.0 dB axial ratio [J.73]. This metasurface has also been used to rotate the linearly polarized wave, and further also to split the incident waves after reflection [J.73, J.74]. An ultra-wideband LP to CP converter, (4.7–21.7 GHz) within 3 dB axial ratio, has also been developed using the via connected L-type inclusions [J.75].

Transmission-type Metasurfaces

Figure (22.36a) shows a three-layered unit cell, also called the transmit-arrays, of an anisotropic transmission-type metasurface. It is the symmetrical subwavelength thick structure with identical front and back layers, and different middle layer [J.76]. Each layer (i = 1, 2, 3) is characterized by the diagonalized tensor sheet or surface admittance $\bar{\bar{Y}}_{s,i}$:

$$\bar{\bar{Y}}_{s,i} = \begin{bmatrix} Y^{zz}_{s,i} & 0 \\ 0 & Y^{yy}_{s,i} \end{bmatrix}, \quad i = 1, 2, 3. \tag{22.6.9}$$

The surface admittances are imaginary quantities, as the low-loss structure is ideally treated as lossless. The transmission-type metasurface is a Huygens' surface. Therefore, the final surface admittance $\bar{\bar{Y}}_s$ at the top level of the structure is a combination of the electric and magnetic type surface admittances arising out of the electric and magnetic polarizabilities discussed in subsection (22.5.2). The condition of complete transmission is given by equation (22.5.23). Both diagonal elements of the admittance tensor can be tuned independently to achieve the matching condition of each unit cells. The tuning also provides the phase difference $\Delta\phi = |\phi_y - \phi_z| = 90°$ to get the refracted CP-waves from the incident LP-waves. Moreover, n number of unit cells can form a supercell of Λ periodicity. The unit cells of the supercell can be tuned individually to refract

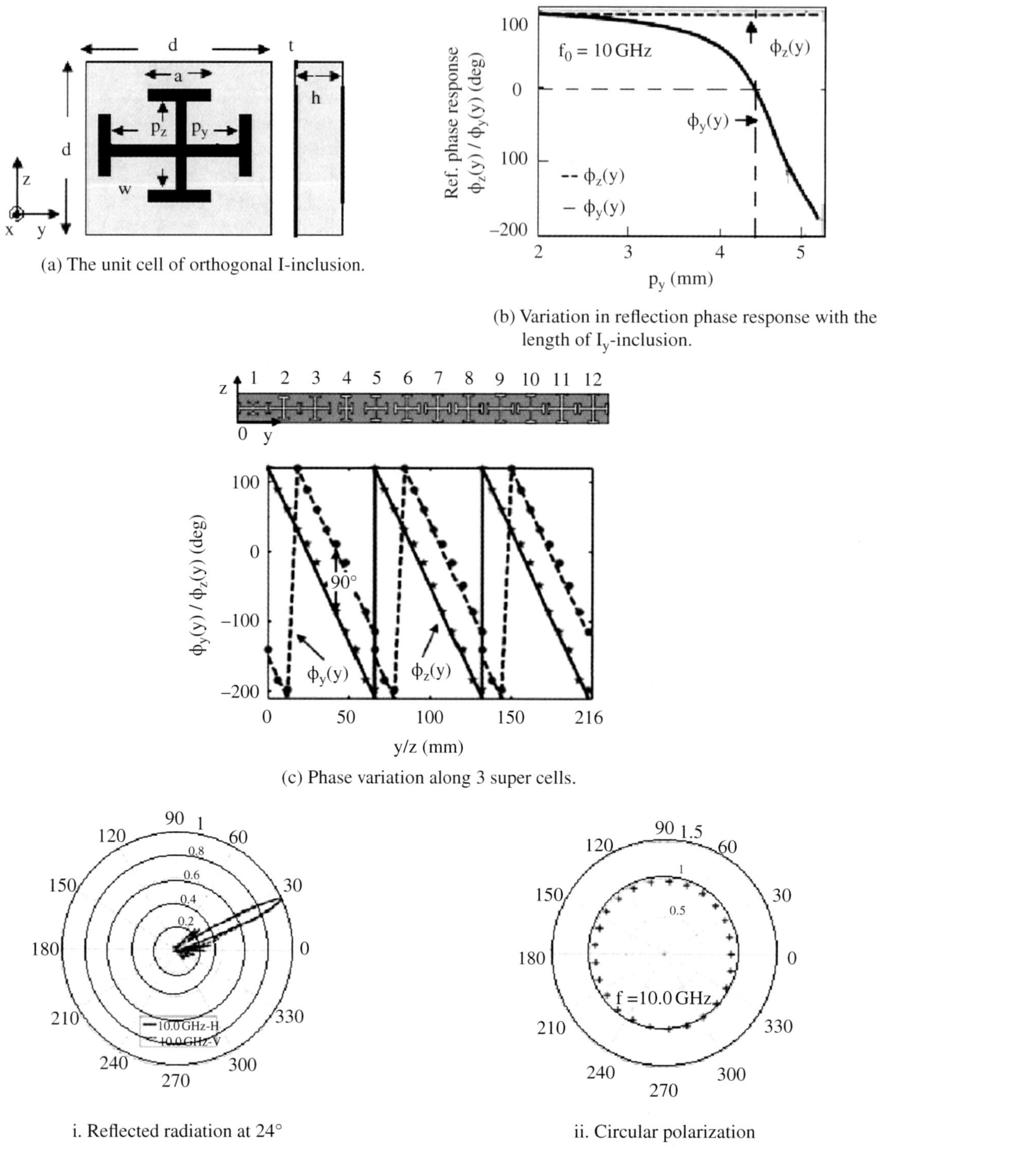

(a) The unit cell of orthogonal I-inclusion.

(b) Variation in reflection phase response with the length of I_y-inclusion.

(c) Phase variation along 3 super cells.

i. Reflected radiation at 24°

ii. Circular polarization

(d) Experimental radiation patterns of reflected waves.

Figure 22.35 Anisotropic reflection – type gradient metasurface to convert incident linearly polarized waves into reflected circularly polarized waves. *Source:* From Ma et al. [J.74], Springer Nature. Licensed under CC- BY- 4.0. Public Domain.

(transmit) the CP-waves in the desired direction following the generalized Snell's Law.

Figure (22.36a) also shows that each layer is a patterned substrate of metallic structures, and separated from other layers by a substrate and bonding layers. Rogers Ultralam 3850 (ε_r = 3.19, $\tan\delta$ = 0.0045, h = 0.05 mm and Rogers 2929 are used as the substrate and bonding film. Figure (22.36b) shows the metallic

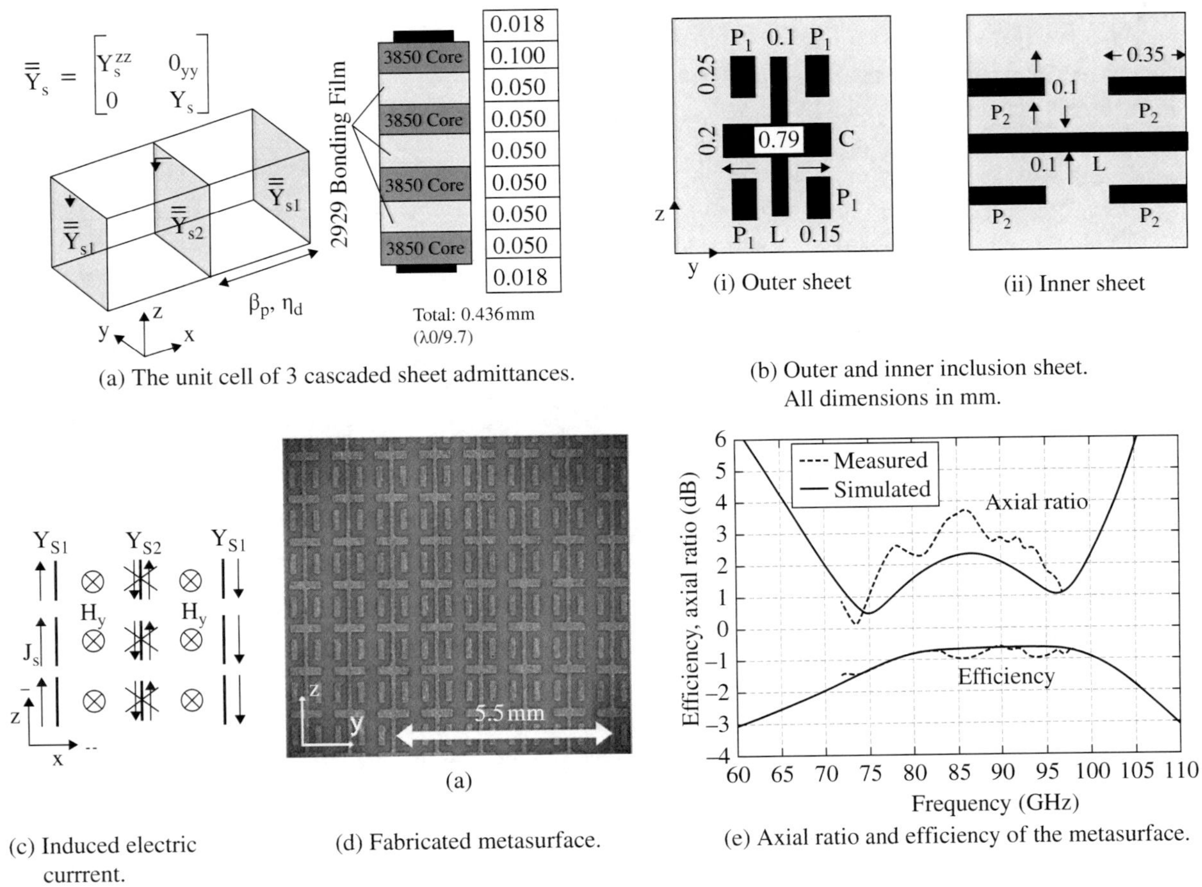

Figure 22.36 Multilayered anisotropic transmission – type metasurface to convert incident linearly polarized waves into refracted circularly polarized waves. *Source:* Pfeiffer and Grbic [J.76]. © 2013, IEEE.

patterns (inclusions/scatterers) of the outer and inner layers. For the outer layer centrally located, thin vertical strip L provides an inductive response that is enhanced by 4 numbers of corner located P_1-patches. The horizontal strip C is a capacitive stub. Likewise, at the inner layer thin horizontal strip is inductive and the capacitive response is obtained by 4 numbers of P_2 patches. The dimensions, shown in Figure (22.36b), provide $\Delta\phi = 90^\circ$ differential phase to convert the incident linearly polarized wave $E^{inc}(x) = E_o^{inc}(\hat{y} \pm \hat{z})\, e^{-k_0 x}$ to get the refracted CP wave.

Figure (22.36c) shows that the incident time-varying electric fields induce the surface current density J_s on the outer layer inclusions. The induced field creates *electric polarizability* and generates an electric response. The induced surface current also forms a circulating loop current through the displacement current between the inner and outer layers. Figure (22.36c) further shows that the effect of surface currents of inner layers is canceled, and a larger current loop is formed between two outer layers. This current loop is responsible for the *magnetic polarizabilities* and magnetic response at the surface of the three-layered metasurface.

The [ABCD] transmission matrix has been used to extract the surface admittances from the simulated S-parameters of the structures on the EM-simulator [J.76].

Quarter – Wave Plate

The above-discussed metasurface is used to develop a quarter-wave plate to get the circular polarization from the incoming linearly polarized wave. It is designed at 77 GHz. Figure (22.36d) shows 81.4 mm × 81.4 mm fabricated metasurface with n = 5 supercells of size (Λ = 5.5 mm). Figure (22.36e) shows the simulated and experimentally obtained axial ratio and efficiency performances of the transmission-type metasurface in the W-band. Its peak axial ratio is 3.7 dB at 85 GHz. The cross-polarization discrimination level is 13.6 dB and the transmission efficiency is better than −1 dB.

Beam Refracting Gradient Metasurface

Figure (22.37a) shows a three-layered unit cell of different geometry to realize anisotropic transmission-type gradient metasurface to convert the incident LP-waves into the CP-wave. It refracts the waves in a predetermined direction. The working of the unit cell is similar to the previous one. The supercell of length $\Lambda = 5.5$ mm is composed of $n = 5$ numbers of unit cells. It spans the complete 0–2π transmission phase. Figure (22.37b) shows the outer and inner layer of a supercell of variable sizes to design the gradient metasurface. The *y*-directed transmission phases $\phi_{y,\,n}$ ($n = 1.5$) of each of five inclusions, located at the outer and inner layers, are tuned to get a π/2.5 phase angle with respect to each other. The tuning also achieves the matching condition for the complete transmission of incident waves. Similar phase tuning is done for the unit cells along the *z*-axis. The dimensions of tuned unit cells of a supercell are available in reference [J.76].

Figure (22.37c) shows a 81.4 mm × 81.4 mm fabricated anisotropic gradient metasurface. It also shows that the linearly polarized wave incident, from the bottom side, on the metasurface at an angle θ^{inc} is refracted at an angle θ^{tra} after conversion to circular polarization. It follows that the generalized Snell's Laws of refraction given by equation (22.5.39c). In the present case, a normally incident 77 GHz LP wave is refracted at an angle $\theta^{tra} = \sin^{-1}(3.896/5.5) = 45.1°$. For the incident LP wave, $E^{inc}(x) = E_o^{inc}(\hat{y} \pm \hat{z})\, e^{-k_0 x}$, the refracted wave is LHCP/RHCP. Figure (22.37d) shows the variations in the axial ratio and high transmission efficiency between 71 and 84.5 GHz. It shows the wideband high-efficiency transmission of the circularly polarized waves. Further, the metasurface can scan the upper space in the range 51°–42° by changing the frequency between 70 and 80 GHz.

Further, the multilayer parabolic phase distributed metasurface has been used to design the flat lens in the infrared range [J.81].

Circularly Polarized Microstrip Patch Antenna on Metasurface Substrate

The reflection-type metasurface can be used as a substrate to design a circularly polarized patch antenna using an LP microstrip patch. Figure (22.38a) shows a sandwiched anisotropic metasurface created by the

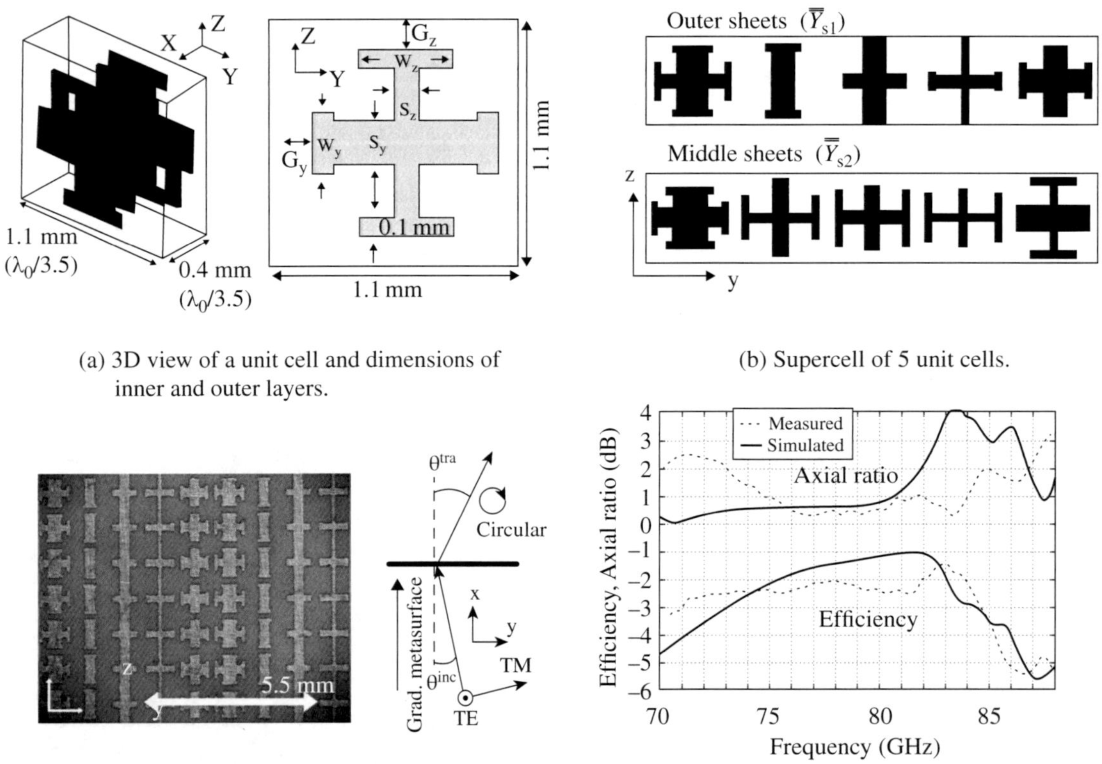

(a) 3D view of a unit cell and dimensions of inner and outer layers.

(b) Supercell of 5 unit cells.

(c) Fabricated metasurface and conversion of incident LP waves to the refracted CP waves.

(d) Axial ratio and efficiency of a grad. metasurface.

Figure 22.37 Multilayered anisotropic transmission – type gradient metasurface to convert incident linearly polarized waves into circularly polarized refracted waves in the desired direction. *Source:* Pfeiffer and Grbic [J.76]. © 2013, IEEE.

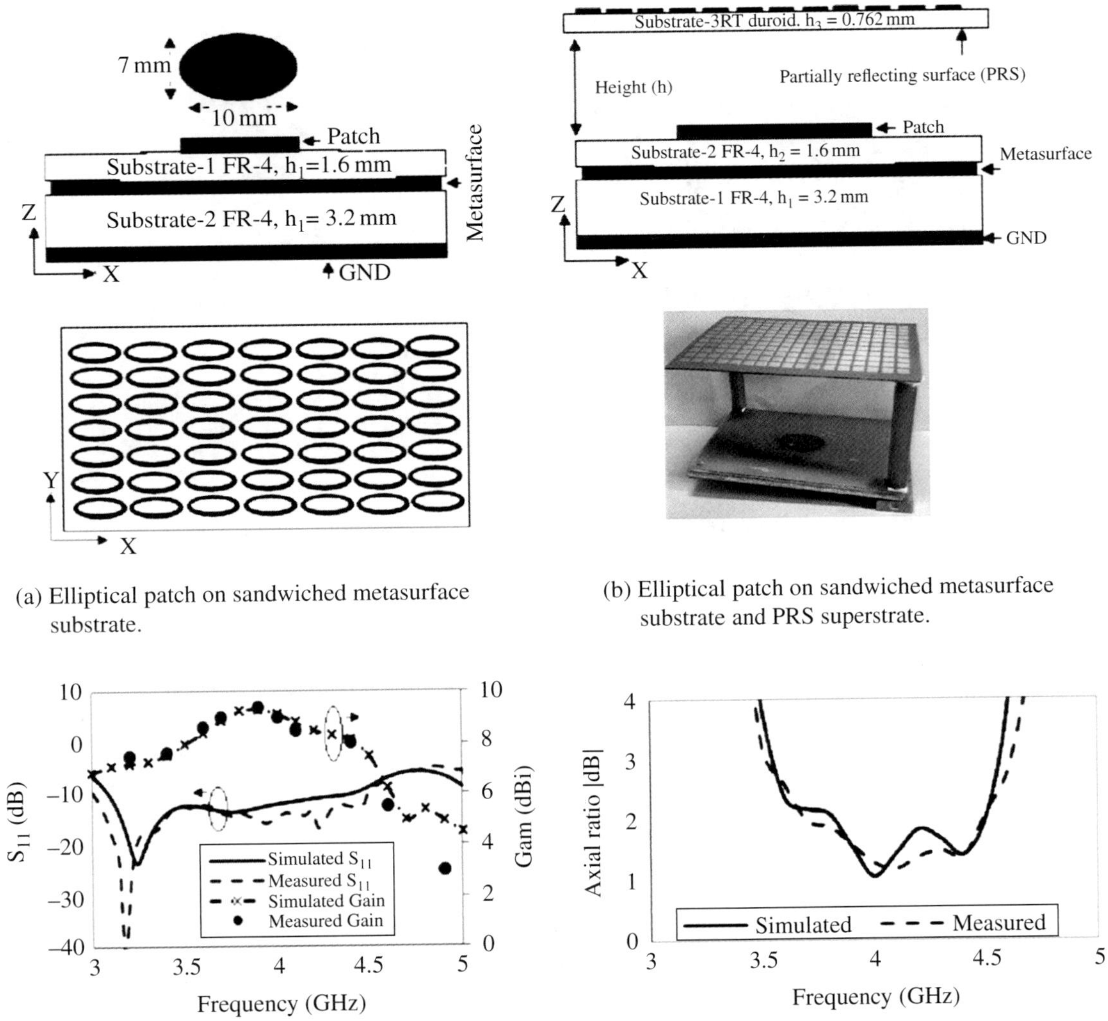

(a) Elliptical patch on sandwiched metasurface substrate.

(b) Elliptical patch on sandwiched metasurface substrate and PRS superstrate.

(c) Return-loss and gain response.

(d) Axial-ratio response.

Figure 22.38 Circularly polarized elliptical patch antenna on sandwiched metasurface substrate and PRS superstrate. *Source:* Srivastava et al. [J.79]. © 2019, IET.

electric type elliptical rings arranged in the rectangular unit cells. FR-4 substrates ($\varepsilon_r = 4.4, \tan\delta = 0.02$), with two thicknesses $h_1 = 3.2$ mm and $h_2 = 1.6$ mm, form a composite layer substrate. The linearly polarized patch is the primary radiating element. The dimensions of the elliptical ring inclusion are tuned on the EM-simulator to achieve the differential reflection phase $\Delta\phi = |\phi_y - \phi_z| = 90^\circ$ for the reflected waves to realize the circularly polarized radiation. The dimensional and other details are available in the reference [J.79].

To enhance the directive gain of the radiator a transmission-type metasurface, designed with a 2D array of square metallic patches, is used as a partially reflecting surface (PRS). The PRS is tuned to achieve both high gain and wider axial ratio BW (ARBW). Figure (22.38b) shows the designed and fabricated structure. Figure (22.38c) shows the $|S_{11}|$, and gain responses, and Figure (22.38d) shows the axial ratio response. The final circularly polarized radiator has 35.3% impedance matching BW, 24.7% 3 dB ARBW with gain 9.3 dBi.

Circularly Polarized Slot Antenna Using Metasurface Superstrate

The transmission-type anisotropic metasurface can be used as a superstrate with a linearly polarized microstrip-fed slot antenna to get the circularly polarized radiation of enhanced gain [J.80]. Figure (22.39a) shows the 3D perspective view of a three-layered structure. The antenna is designed at 5.8 GHz. The ground plane of the first substrate with $\varepsilon_r = 3.6, \tan\delta = 0.003$, $h_1 = 0.762$ mm supports the LP-slot antenna. It is fed

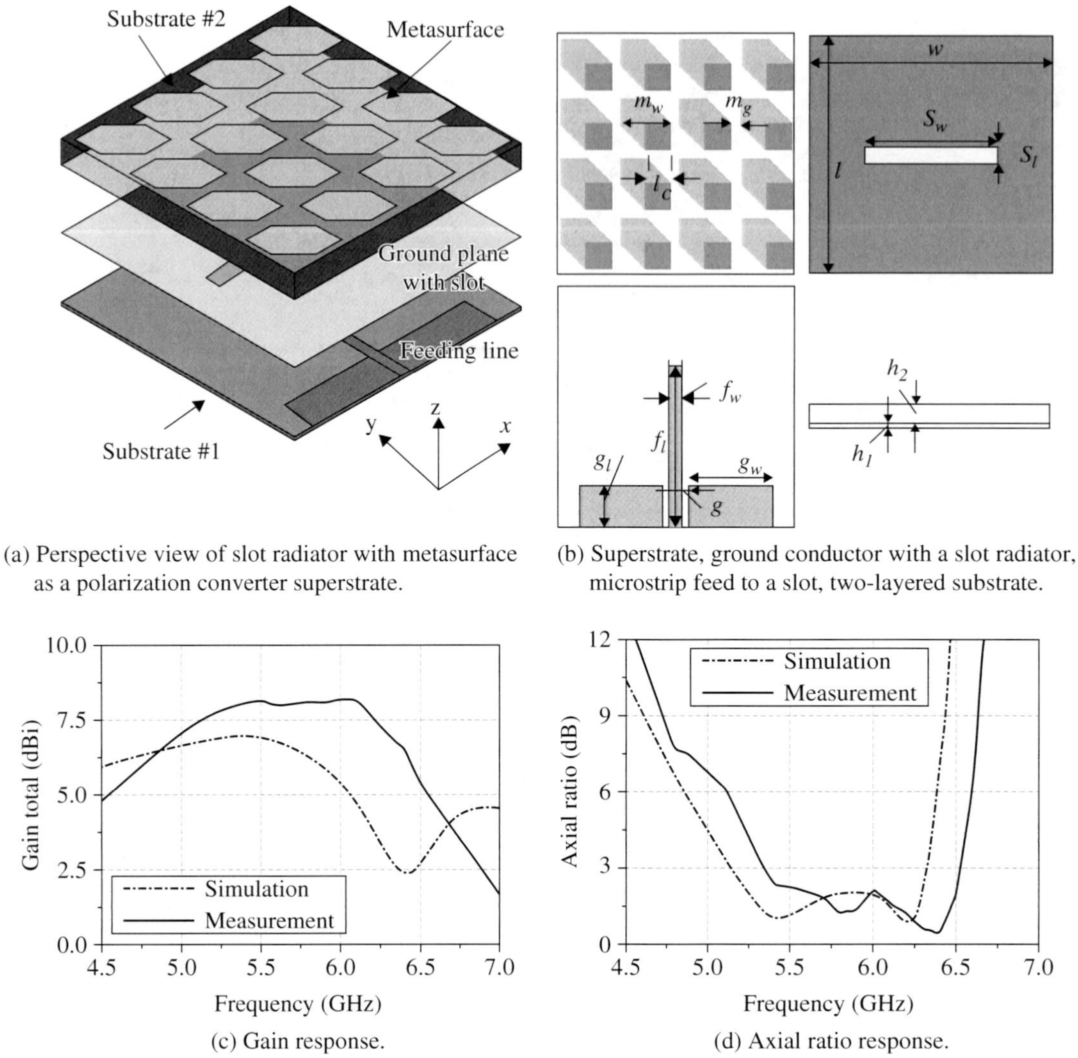

Figure 22.39 Circularly polarized slot radiator using metasurface as a superstrate. *Source:* Huang et al. [J.80]. © 2016, AIP Publishing.

by a microstrip line on the second substrate. The corner-cut square patches are used to design the metasurface-based superstrate on the second substrate. It has a thickness h_2 = 4.0 mm. Figure (22.39b) shows the dimensional details with following optimized values: s_w = 28, s_ℓ = 2.5, f_w = 1.6, f_ℓ = 27, g_w = 6, g_ℓ = 14, g = 0.6, w = 39.2, ℓ = 40, m_w = 8.8, m_g = 1 (all dimensions in mm).

Figure (22.39c) shows the gain response. Figure (22.39d) shows the axial ratio response of the circularly polarized antenna. The experimentally obtained impedance matching BW of the CP antenna is 32% (4.7–6.5 GHz) and its ARBW is 20.6% (5.14–6.32 GHz). The antenna has a directive gain of about 8.0 dBi.

The present subsection has discussed only a few illustrative application examples of metasurfaces. The metasurfaces have been used for many other purposes, for instance, as the polarization-sensitive beam splitters [J.74, J.82], RCS reduction devices [J.83], microwave absorbers [J.84, J.85], holographic antenna [J.63, J.64], and surface-waves-based devices and antenna [J.49, J.50, J.65].

References

Books

B.1 Engheta, N.; Ziolkowski, R.W. (Editors): *Metamaterials: Physics and Engineering Explorations*, Wiley-Interscience, John Wiley & Sons, Inc., NJ, USA, 2006.

B.2 Christophe Caloz, C.; Itoh, T.: *Electromagnetic Metamaterials: Transmission line Theory and Microwave Applications (The Engineering Approach)*, Wiley-Interscience, John Wiley & Sons, Inc., NJ, USA, 2006.

B.3 Eleftheriades, G.I.; Balmain, K.G. (Editors): *Negative-Refraction Metamaterials: Fundamental Principles and Applications*, Wiley-Interscience, John Wiley & Sons, Inc., NJ, USA, 2005.

B.4 Capolino, F. (Editor): *Theory and Phenomena of Metamaterials*, CRC Press, USA.

B.5 Marques, R.; Martin, F.; Sorolla, M.: *Metamaterials with Negative Parameters. Theory, Design and Microwave Applications. Wiley*, Hoboken, NJ, 2008.

B.6 Solymar L.; Shamonina, E.: *Waves in Metamaterials*, Oxford University Press, NY, USA, 2009.

B.7 Martín, F.: *Artificial Transmission Lines for RF and Microwave Applications*, John Wiley & Sons, USA, 2015.

B.8 Lheurette, E.: *Metamaterials and Wave Control*, John Wiley & Sons, USA, 2013.

B.9 Simon, R.; Whinnery J.R.; Van Duzer, T.: *Fields and Waves in Communication Electronics*, 3rd Edition, John Wiley & Sons, Singapore, 1994.

B.10 Liu, C.; Huang, K.: Metamaterial transmission line and its applications: Mukherjee, M. (Ed.), Chapter-13, *Advanced Microwave and Millimeter wave Technologies. Semiconductor Devices Circuits, and Systems*, pp. 249–274, InTech Pub., Online, 2010.

B.11 Gil, M.; Aznar, F.; Velez, A.; Duran-Sindreu, M.; Selga, J.; Siso, G.; Bonache, J.; Martin, F.: Electrically small resonators for metamaterial and microwave circuit design, passive microwave components and antennas: Zhurbenko V. (Ed.), *Passive Microwave Components and Antenna*. ISBN: 978-953-307-083-4, InTech, 2010.

B.12 Al-Naib I.A.I.; Jansen, C.; Koch, M.: Compact CPW metamaterial resonators for high performance filters: Zhurbenko V. (Ed.), Chapter-20, *Passive Microwave Components, and Antenna*. ISBN: 978-953-307-083-4, InTech, 2010.

B.13 Tretyakov, S.: *Analytical Modeling on Applied Electromagnetics*, Artech House, Boston, USA, 2000.

B.14 Roger F.; Harrington, R.F.: *Time-Harmonic Electromagnetic Fields*, Wiley, IEEE Press, 2001.

Journals

J.1 Eleftheriades, G.V., Iyer, A.K.; Kremer, P.C.: *Planar negative refractive index media using periodically L-C loaded transmission lines, IEEE Trans. Microwave Theory. Tech.*, Vol. 50, pp. 2702–2710, 2002.

J.2 Caloz, C.; Itoh, T.: *Left-handed transmission lines and equivalent metamaterials for microwave and millimeter wave application, 32rd Eur. Microwave Conf.*, doi:10.1109/EUMA.2002.339287, Oct. 2002.

J.3 Rahman, A.A.; Verma, A.K.; Boutejdar, A.; Omar, A. S.: *Compact stub type microstrip bandpass filter using a defected ground plane, IEEE Microwave Wirel. Comp. Lett.*, Vol. 14, No. 4, pp. 136–138, Apr. 2004.

J.4 Richard, W.; Ziolkowski, R.W.; Ching-Ying Cheng, C. Y.: *Lumped element models of double negative metamaterial-based transmission lines, Radio Sci. RS*, Vol. 39, No. RS 2017, pp.2–20, 2007.

J.5 Lai, A.; Caloz, C.; Itoh, T.: *Composite right/left – handed transmission line metamaterials, IEEE Microwave Mag.*, pp.35–50, Sept. 2004.

J.6 Caloz, C.; Itoh, T.: *Application of the transmission line theory of left-handed (LH) materials to the realization of a microstrip line, IEEE Int. Sym. Antennas Propag. Dig.*, Vol. 2, pp. 412–415, 2002.

J.7 Christophe Caloz, C.: *Metamaterial dispersion engineering concepts and applications, Proc. IEEE*, Vol. 99, No. 10, pp. 1711–1719, Oct. 2011.

J.8 Caloz, C.; Itoh, T.: *Metamaterials for high-frequency electronic, Proc. IEEE*, Vol. 93, No. 10, pp. 1744–1752, Oct. 2005.

J.9 Cui, T.J.; Ma, H.F.; Liu, R.; Zhao, B.; Cheng, Q.; Chin, J.Y.: *A symmetrical circuit model describing all kinds of circuit metamaterials, Progr. Electromagn. Res. B*, Vol. 5, pp. 63–76, 2008.

J.10 Caloz, C., Nguyen, H.Y.: *Novel broadband conventional and dual-composite right/left-handed (C/D-CRLH) metamaterials: properties, implementation and double-band coupler application, Appl. Phys. A Mater. Sci. Process.*, Vol. 87, pp. 309–316, 2007.

J.11 Caloz, C.; Itoh, T.: *Transmission line approach of left-handed (LH) materials and microstrip implementation of an artificial LH transmission line, IEEE Trans. Antennas Propag.*, Vol. 52, No. 5, pp. 1159–1166, May 2004.

J.12 Caloz, C.: *Dual composite right/left-handed (D-CRLH) transmission line metamaterial, IEEE Wirel. Comp. Lett.*, Vol. 16, No. 11, pp. 585–587, Nov. 2008.

J.13 Siso, G.; Gil, M.; Aznar, F.; Bonache, J.; Ferran Martin, F.: *Dispersion engineering with resonant-type metamaterial transmission lines, Laser Photonics Rev.*, Vol. 3, No. 1–2, pp. 12–29, 2009.

J.14 Gao, J.; Zhu, J.: *Effective complex permittivity and permeability of composite planar left-handed transmission line metamaterial on CPW, Microwave*

Opt. Tech. Lett., Vol. 48, No. 6, pp. 1050–1052, June 2006.

J.15 Borja, A.L.; Belenguer, A.; Cascon, J.; Estenban, H.; Boria, V.E.: *Wideband passband transmission line based on metamaterial – inspired CPW balanced Cells, IEEE Antenna Wirel. Propag. Lett.*, Vol. 10, pp. 1421–1424, Nov. 2011.

J.16 Mao, S.-G.; Wu, M.-S.: *Equivalent circuit modeling of symmetric composite right/left-handed coplanar waveguides, IEEE MTT-S Int. Microwave Symp. Dig.*, Long Beach, CA, USA, pp. 1953–1956, 11–17 June 2005.

J.17 Nguyen, H.V.; Caloz, C.: *Simple – design and compact MIM CRLH microstrip 3-dB coupled line coupler, Proc. IEEE-MTT Int. Symp.*, pp. 1733–1736, San Francisco, CA, 11–16 June June 2006.

J.18 Alù, A.; Engheta, N.: *Pairing an epsilon-negative slab with a mu-negative slab: resonance, tunneling and transparency, IEEE Trans. Antennas Propag.*, Vol. 21, No. 10, pp: 2558–2571, Oct. 2003.

J.19 Zheng, J.; Chen, Y.; Chen, Z.; Wang, X.; Han, P.; Yong, Z.; Wang, Y.; Leung, C.W.; Soukoulis, C.M.: *Investigation of interface states in single-negative metamaterial layered structures based on the phase properties, Opt. Express*, Vol. 21, No. 14, pp. 16742–6752, July 2013.

J.20 Fujishiqe, T.; Caloz, C.; Itoh, T.: *Experimental demonstration of transparency in the ENG-MNG pair in a CRLH transmission line, Microwave Opt. Tech. Lett.*, Vol. 46, No. 6, pp. 476–481, Sept. 2005.

J.21 Liu, C. H.; Behdad, N.: *Theoretical examination of electromagnetic wave tunneling through cascaded ε and μ-negative metamaterial slabs, Prog. Electromagn. Res. B*, Vol. 42, pp. 1–22, 2012.

J.22 Martin, F.: *Metamaterials for wireless communications, radiofrequency identification, and sensors, Int. Scholarly Res. Netw. ISRN Electron.*, doi:https://doi.org/10.5402/2012/780232, Vol. 2012, Article ID 780232, 29 pages, 2012.

J.23 Aznar, F.; Gil, M.; Bonache, J.; Martın, F.: *Modelling metamaterial transmission lines: a review and recent developments, Opto-Electron. Rev.*, Vol. 16, No. 3, pp. 226–236, 2008.

J.24 Duran-Sindreu, M.; Naqui, J.; Bonache, J.; Martin, F.: *Split rings for metamaterials and microwave circuit design: a review of recent development (Invited paper), Int. J. RF Microwave CAD Eng.*, Vol. 22, No. 4, pp. 439–458, July 2012.

J.25 Gil, M.; Bonache, J.; Gil, I.; Garcia-Garcı, J.; Martın, F.: *On the transmission properties of left-handed microstrip lines implemented by complementary split rings resonators, Int. J. Numer. Modell. Electron. Netw. Dev. Fields*, Vol. 19, pp. 87–103, 2006.

J.26 Baena, J.D.; Bonache, J.; Martín, F.; Sillero, R.M.; Falcone, F.; Lopetegi, T.;. Laso, M.A.G.; García-García, J.; Ignacio Gil, J.; Portillo, M.F.; Sorolla, M.: *Equivalent-circuit models for split-ring resonators and complementary split-ring resonators coupled to planar transmission lines, IEEE Trans. Microwave Theory. Tech.*, Vol. 53, No. 4, pp. 1451–1461, Apr.2005.

J.27 Vélez, P.; Naqui, J.; Durán-Sindreu, M.; Bonache, J.; Martin, F.: *Broadband microstrip bandpass filter based on open complementary split ring resonators, Int. J. Antennas Propag.*, doi:10.1155/2012/174023, Vol. 2012, Article ID 174023, 6 pages, 2012.

J.28 Falcone, F.; Martın, F.; Bonache, J.; Laso, M.A.G.; Garcıa-Garcıa J.; Baena, J.D.; Marqués, R.; Sorolla, M.: *Stop-band and band-pass characteristics in coplanar waveguides coupled to spiral resonators, Microwave Opt. Tech. Lett.*, Vol. 42, No. 5, pp. 386–388, Sept. 2004.

J.29 Isik, O.; Esselle, K.P.: *Backward wave microstrip lines with complementary spiral resonators, IEEE Trans. Antennas Propag.*, Vol. 56, No. 10, pp. 3173–3178, Oct. 2008.

J.30 Park, J.H.; Ryu, Y.H.; Lee, J.G.; Jeong-Hae Lee, J.H.: *A novel via-free composite right and left-handed transmission line using defected ground structure, Microwave Opt. Tech. Lett.*, Vol. 49, No. 6, pp. 1989–1993, Aug. 2007.

J.31 Ryu, Y.H.; Park, P.H.; Lee, J.H.; Tae, H.S.: *Generalized negative refractive index transmission line based on defected ground structure, Microwave Opt. Tech. Lett.*, Vol. 52, No. 10, pp. 2223–2225, Oct. 2010.

J.32 Martel, J.; Marqués, R.; Falcone, F.; Baena, J.D.; Medina, F.; Martín F.; Sorolla, M.: *A new LC series element for compact bandpass filter design, IEEE Microwave Wirel. Comp. Lett.*, Vol. 14, pp. 210–212, May 2004.

J.33 Velez, A.; Aznar, F.; Bonache, J.; Velazquez-Ahumada, M.C.; Martel, J.; Martin, F.: *Open complementary split-ring resonators (OCSRRS) and their application wideband CPW bandpass filters, IEEE Microwave Wirel. Comp. Lett.*, Vol. 19, pp. 197–199, Apr. 2009.

J.34 Bonache, J.; Gil, M.; Gil, I.; Garcıa-Garcıa, J.; Martin, F.: *On the electrical characteristics of complementary metamaterial resonators, IEEE Microwave Wirel. Comp. Lett.*, Vol. 16, pp. 543–545, Oct. 2006.

J.35 Mao, S.-G.; Chen, S.-L.; Huang, C.-W.: *Effective electromagnetic parameters of novel distributed left-handed microstrip lines, IEEE Trans. Microwave Theory. Tech.*, Vol. 53, No. 4, pp. 1515–1521, Apr. 2005.

J.36 Lee, J.G.; Lee, J.H.: *Suppression of spurious radiation of patch antennas using split-ring resonators (SRRs), Microwave Opt. Tech. Lett.*, Vol. 48, No. 2, pp. 283–287, Feb. 2006.

J.37 Bojoanic, R.; Milosevic, V.; Jokanovic, B.; Medina-Mena, F.; Mesa, F.: *Enhanced modeling of split-ring resonators coupling in printed circuits, IEEE Trans. Microwave Theory. Tech.*, Vol. 62, No. 8, pp. 1605–1615, Aug. 2014

J.38 Horestani, A.K.; Fumeaux, C.; Al-Rarawi, S.F.; Abbott, D.: *Split ring resonators with tapered strip width for wider bandwidth and enhanced resonance, IEEE Microwave Wirel. Comp. Lett.*, Vol. 22, No. 9, pp. 450–452, Sept. 2012.

J.39 Crnojević-Bengin, V.; Janković, N.; Cselyuszka, N.; Geschke, R.H.: *Mu-near-zero propagation in quasi-TEM microstrip circuits, J. Electromagn. Waves Appl.*, Vol. 27, No. 17, pp. 2198–2212, 2013.

J.40 Martin, F.; Falcone, F.; Bonache, J.; Marqués, R.; Sorolla, M.: *Spilt ring resonator based left-handed the coplanar waveguide, Appl. Phys. Lett.*, Vol. 83, pp. 4652–4654, 2003.

J.41 Falcone, F.; Martin, F.; Bonache, J.; Marques, R.; Sorolla, M.: Coplanar waveguide structures loaded with split-ring resonators, *Microwave Opt. Tech. Lett.*, Vol. 40, No. 1, pp. 3–6, Jan. 2004.

J.42 Lim, S.; Caloz, C.; Itoh, T.: *Metamaterial-based electronically controlled transmission-line structure as a novel leaky-wave antenna with tunable radiation angle and beamwidth, IEEE Trans. Microwave Theory Tech.*, Vol. 52, No. 12, pp. 2678–2690, Dec.2004.

J.43 Liu, L.; Caloz, C.; Itoh, T.: *Dominant mode leaky-wave antenna with backfire to entire scanning capability, IET Electron. Lett.*, Vol. 38, No. 23, pp. 1414–1416, Nov. 2002.

J.44 Caloz, C.; Sanada, A.; Itoh, T.: *A novel composite right-/left-handed coupled-line directional coupler with arbitrary coupling level and broad bandwidth, IEEE Trans. Microwave Theory Tech.*, Vol. 52, No. 3, pp. 980–991, Mar. 2004.

J.45 Caloz, C.; Itoh, T.: *A novel mixed conventional microstrip and composite right/left-handed backward-wave directional coupler with broadband and tight coupling characteristics, IEEE Microwave Wirel. Comp. Lett.*, Vol. 14, No. 1, pp. 31–33, Jan. 2004.

J.46 Nguyen, H.V.; Caloz, C.: *Generalized coupled-mode approach of metamaterial coupled-line couplers: coupling theory, phenomenological explanation and experimental demonstration, IEEE Trans. Microwave Theory Tech.*, Vol. 55, No. 5, pp. 1029–1039, May 2007.

J.47 Eleftheriades, G.V.; Islam, R.: *Miniaturized microwave components and antennas using negative-refractive-index transmission-line (NRI-TL) metamaterials, Metamaterials*, Vol. 1, pp. 53–61, 2007.

J.48 Hsiang Lin, I.-H.; Marc DeVincentis, M.; Caloz, C.; Itoh, T.: *Arbitrary dual-band components using composite right/left-handed transmission lines, IEEE Trans. Microwave Theory Tech.*, Vol. 52, No. 3, pp. 1142–1149, Apr. 2004.

J.49 Chen, H.T.; Taylor, A.T.; Yu, N.: *A review of metasurfaces: physics and applications, Rep. Prog. Phys.*, Vol. 79, No. 076401, pp. 1–40, 2016.

J.50 Holloway, C.L.; Kuester, E.F.; Gordon, J.A.; Hara, J.O.; Booth, J.; Smith D.R.: *An overview of the theory and applications of metasurfaces: the two-dimensional equivalents of metamaterials, IEEE Antennas Propag. Mag.*, Vol. 54, No. 2, pp. 11–35, April 2012.

J.51 Ding, F.; Prs, A.; Bozhevolnyi, S.I.: Gradient metasurfaces: a review of fundamentals and applications, arXiv: 1704.03032v2 [physics.optics], pp. 1–83, 21 Aug. 2017.

J.52 Tretyakov, S.A.: *Metasurfaces for general transformations of electromagnetic fields, Phil. Trans. R. Soc. A*, Vol. 373, No. 20140362, pp. 10, July 2018.

J.53 Viktor, A.; Diaz-Rubio, S.; Tretyakov, S.A.: *Bianisotropic Metasurfaces*, Aalto University, pp. 126, 2018.

J.54 Pfeiffer, C.; Grbic, A.: *Metamaterial Huygens' surfaces: Tailoring wave fronts with reflectionless sheets, Phys. Rev. Lett. PRL.*, Vol. 110, pp. 197401–197405, 2013.

J.55 Xue, F.; Wang, H.; Dong, X.; Liu, Y.; Zhang, X.: *Design of a novel Ku/X-band reflectarray/transmit-array antenna with frequency selective surface, Prog. Electromagn. Res. C*, Vol. 73, pp. 17–25, 2017.

J.56 Nanfang Yu; Genevet, P.; Kats, M.A.; Aieta, F.; Tetienne, J.P.; Capasso, F.; Gaburrol, Z.: *Light propagation with phase discontinuities: generalized laws of reflection and refraction, Science*, Vol. 334, pp. 333–337, Oct. 2011.

J.57 Shi, H.; Li, J.; Zhang, A.; Wang, J.X.Z.; *Broadband cross-polarization converter using plasmon hybridizations in a ring/disk cavity, Opt. Express*, Vol. 22, No. 17, pp. 20973–20981, Aug. 2014.

J.58 Menzel, C.; Rockstuhl, C.; Lederer, F.: *An advanced Jones calculus for the classification of periodic metamaterials*, arXiv: 1008.4117v1 [physics.optics], pp. 1–9, 24 Aug. 2010.

J.59 Yu, N.; Capasso, F.: *Flat optics with designer metasurfaces, Nat. Mater.*, Vol. 13, pp. 139–150, Feb. 2014.

J.60 Aieta, F.; Genevet, P.; Nanfang Yu; Kats, M. A.; Gaburrol, Z.; Capasso, F.: *Out – of- plane reflection and refraction of light by anisotropic optical antenna metasurfaces with phase discontinuities, Nano Lett.*, doi:10.1021/nl300204s, Vol. 12, No. 3, pp. 1702–6, Mar. 2012.

J.61 He, Q.; Xiao, S.; Xu, Q.; Li, X.; Zhou, L.: *Gradient-index metasurfaces as a bridge linking propagating waves and surface waves, Nat. Mater.*, doi:https://doi.org/10.1038/NMAT3292, Apr. 2012.

J.62 Eleftheriades, G.V.; Iyer, A.K.; Keremer, P.C.: *Planar negative refractive index media using periodically L-1. C loaded transmission lines, IEEE Trans. Microwave Theory. Tech.*, Vol. 50, No. 12, pp. 2702–2712, Dec. 2002.

J.63 Fong, B.H.; Colburn, J.S.; Ottusch, J.J.; John L.; Visher, J.L.; Sievenpiper, D.F.: *Scalar and tensor/holographic artificial impedance surface, IEEE Trans. Antennas Propag.*, Vol. 58, No. 10, pp. 3212–3220, Oct. 2010.

J.64 Matekovits, L.; Labate, G.: *Controlling surface waves with metasurfaces: from planar propagation to conformal cloaking, Proc. 46th Eur. Microwave Conf.*, pp. 787–790, Oct. 2016.

J.65 Wang, P.; Shen, Z.; *End-fire surface wave antenna with metasurface coating, IEEE Access*, Vol. 6, pp. 23778–23785, May 2018.

J.66 Wang, J.; Qu, S.; Ma, H.; Chen, H.; Feng, M.; Li, Y.; Zhang, J.: *Manipulating the reflection of electromagnetic waves using reflective metasurfaces, 3rd Asia-Pacific Conference on Antennas and Propagation*, Harbin, China, pp. 657–660, 2014.

J.67 Ebbesen, T.W.; Lezec, H.J.; Ghaemi, H.F.; Thio, T.; Wolff, P.A.: *Extraordinary optical transmission through sub-wavelength hole arrays, Nature*, Vol. 391, No. 6668, pp. 667–66, 1998.

J.68 Ruan, Z.C.; Qiu, M.: *Enhanced transmission through periodic arrays of subwavelength holes: the role of localized waveguide resonances, Phys. Rev. Lett.*, Vol. 96, No. 23, pp. 233901, 2006.

J.69 Wei, Z.Y.; Cao, Y.; Fan, Y.C.; Yu, X.; Li, H.Q.: *Broadband transparency achieved with the stacked metallic multi-layers perforated with coaxial annular apertures, Opt. Express* Vol. 19, No. 22, pp. 21425–21431, 2011.

J.70 Wei, Z.; Cao, Y.; Su, X.; Gong, Z.; Long, Y.; Li, H.: *Highly efficient beam steering with a transparent metasurface, Opt. Express*, doi: https://doi.org/10.1364/OE.21.010739, Vol. 21, No. 9, OSA, May 2013.

J.71 Cai, T.; Wang, G.M.; Tang, S.W.; Xu, H.X.; Duan, J.W.; Guo, H.J.; Guan, F.X.; Sun, S.; He, Q.; Zhou, L.: *High-efficiency and full-space manipulation of electromagnetic wave fronts with metasurfaces, Phys. Rev. Appl.*, Vol. 8, No. 3, pp. 034033-1–034033-11, 2017.

J.72 Liu, C.; Bai, Y.; Zhao, Q.; Yang, Y.; Chen, H.; Zhou, J.; Qiao: *Fully controllable Panchratnam–Berry metasurface array with high conversion efficiency and broad bandwidth, Sci. Rep.*, doi:https://doi.org/10.1038/srep34819, pp. 1–7, Oct. 2016.

J.73 Ma, H.F.; Wang, G.Z.; Kong, G. S.; Cui, T. J.: *Broadband circular and linear polarization conversions realized by thin birefringent reflective metasurfaces, Opt. Mater. Express*, doi:10.1364/OME.4.001717, Vol. 4, No. 8, July 2014.

J.74 Ma, H.F.; Wang, G.Z.; Kong, G.S.; Cui, T.J.: *Independent controls of differently-polarized reflected waves by anisotropic metasurfaces, Sci. Rep.*, doi:10.1038/srep09605, pp. 1–7, April 2015.

J.75 Jia, Y.; Liu, Y.; Zhang, W.; Wang, J.; Wang, Y.; Gong, S.; Liao, G.: *Ultra-wideband metasurface with linear-to-circular polarization conversion of an electromagnetic wave, Opt. Mater. Express*, Vol. 8, No. 3, pp. 597–604, Mar. 1018.

J.76 Pfeiffer, C.; Grbic, G.: *Millimeter-wave transmitarrays for wavefront and polarization control, IEEE Trans. Microwave Theory. Tech.*, Vol. 61, No. 12, pp. 4407–4417, Dec. 2013.

J.77 Guo, Y.; Yan, L.; Pan, W.; Luo, B.: *Achromatic polarization manipulation by dispersion management of anisotropic meta-mirror with dual-metasurface, Opt. Express*, Vol. 23, No. 21, pp. 27566–27575, Oct. 2015.

J.78 Mei, Z.L; Ma, X. M.; Cui Lu, C.; Yao Dong Zhao, Y.D.: *High-efficiency and wide-bandwidth linear polarization converter based on double U-shaped metasurface, AIP Adv.*, Vol. 7, No. 12, Art no. 125323, pp. 1–7, 2017.

J.79 Srivastava, K.; Kumar, A.; Chaudhary, P.; Kanaujia, B. K.; Dwari, S; Verma, A.K.; Esselle, K.P.; Mittra, R.: *Wideband and high-gain circularly polarized microstrip antenna design using sandwiched metasurfaces and partially reflecting surface, IET Microwaves Antennas Propag.*, Vol. 13, No. 3, pp. 305–312, 2019.

J.80 Huang, Y.; Yang, L.; Li, J.; Wang, Y.; Wen, G.: *Polarization conversion of metasurface for the application of wide band low-profile circular polarization slot antenna, Appl. Phys. Lett.*, Vol. 109, pp. 054101-1–054101-5, 2016.

J.81 Pfeiffer, C.; Grbic, G.: *Cascaded metasurfaces for complete phase and polarization control, Appl. Phys. Lett.*; Vol. 102, pp. 231116-1–231116-4, 2013.

J.82 Guo, W.; Li, T.; Li, H.; Hou, H.: *Steer and split electromagnetic waves by employing ultra-thin metamaterials, IOP Conf. Ser. Mater. Sci. Eng.*, doi:10.1088/1757-9X/167/1/012061, Vol. 167, pp. 1–7, 2017.

J.83 Balanis, C.A.; Amiri, M.A.; Modi, A.Y.; Pandi, S.; Birtcher, C.R.: *Applications of AMC-based impedance surfaces, EPJ Appl. Metamater.*, doi:10.1051/epjam/2017010, pp. 1–14, 2018.

J.84 Badloe, T.; Mun, J.; Rho, J.: *Metasurfaces-based absorption and reflection control: perfect absorbers and reflectors, Hindawi J. Nanomater.*, Vol. 2017, Article ID 2361042, pp. 1–18, 2017.

J.85 Zhou, Z.; Chen, K.; Zhu, B.; Zhao, J.; Feng, Y.; Li, Y.: *Ultra-wideband microwave absorption by design and optimization of metasurface Salisbury screen, IEEE Access*, Vol. 6, pp. 26843–26853, Jun. 2018.

Index

f

i

j

k

l

m

n

Printed in the USA/Agawam, MA
September 21, 2026

910661.042